W9-BLC-573

BIOLOGY

OF RELATED INTEREST FROM THE BENJAMIN/CUMMINGS SERIES IN THE LIFE SCIENCES

General Biology

R. J. Kosinski
Fish Farm: Simulation Software (1993)

J. G. Morgan and M. E. B. Carter
Investigating Biology: A Laboratory Manual (1993)

Plant Biology

M. G. Barbour, J. H. Burk, and W. D. Pitts
Terrestrial Plant Ecology, Second Edition (1987)

J. Mauseth
Plant Anatomy (1988)

L. Taiz and E. Zeiger
Plant Physiology (1991)

Biochemistry and Cell Biology

W. M. Becker and D. W. Deamer
The World of the Cell, Second Edition (1991)

C. Mathews and K. H. van Holde
Biochemistry (1990)

G. L. Sackheim
Chemistry for Biology Students, Fourth Edition (1990)

W. B. Wood, J. H. Wilson, R. M. Benbow, and L. E. Hood
Biochemistry: A Problems Approach, Second Edition (1981)

Molecular Biology and Genetics

F. J. Ayala and J. A. Kiger, Jr.
Modern Genetics, Second Edition (1984)

L. E. Hood, I. L. Weissman, W. B. Wood, and J. H. Wilson
Immunology, Second Edition (1984)

R. Schleif
Genetics and Molecular Biology (1986)

J. D. Watson, N. H. Hopkins, J. W. Roberts, J. A. Steitz, and A. M. Weiner
Molecular Biology of the Gene, Fourth Edition (1987)

G. Zubay
Genetics (1987)

Microbiology

I. E. Alcamo
Fundamentals of Microbiology, Third Edition (1991)

R. M. Atlas and R. Bartha
Microbial Ecology: Fundamentals and Applications, Third Edition (1992)

J. Cappuccino and N. Sherman
Microbiology: A Laboratory Manual, Third Edition (1992)

T. R. Johnson and C. L. Case
Laboratory Experiments in Microbiology, Brief Edition, Third Edition (1992)

G. J. Tortora, B. R. Funke, and C. L. Case
Microbiology: An Introduction, Fourth Edition (1992)

Evolution, Ecology, and Behavior

D. D. Chiras
Environmental Science, Third Edition (1991)

M. Lerman
Marine Biology: Environment, Diversity, and Ecology (1986)

R. Trivers
Social Evolution (1985)

Animal Biology

H. E. Evans
Insect Biology: A Textbook of Entomology (1984)

E. N. Marieb
Essentials of Human Anatomy and Physiology, Third Edition (1991)

E. N. Marieb
Human Anatomy and Physiology, Second Edition (1992)

E. N. Marieb and J. Mallatt
Human Anatomy (1992)

L. G. Mitchell, J. A. Mutchmor, and W. D. Dolphin
Zoology (1988)

A. P. Spence
Basic Human Anatomy, Third Edition (1991)

BIOLOGY

THIRD EDITION

NEIL A. CAMPBELL
University of California, Riverside

The Benjamin/Cummings Publishing Company, Inc.
Redwood City, California • Menlo Park, California • Reading, Massachusetts
New York • Don Mills, Ontario • Wokingham, U.K. • Amsterdam • Bonn
Sydney • Singapore • Tokyo • Madrid • San Juan

Sponsoring Editor: Edith Beard Brady

Developmental Editor: Suzanne Olivier

Developmental Manager/Executive Editor: Robin J. Heyden

Consulting Developmental Editors: Pat Burner, Robin Fox, Susan Weisberg

Editorial Assistants: Sissy Lemon, Christine Ruotolo, Kimberly Viano, Thomas Viano

Production and Art Coordinator: Pat Waldo, Publishing Principals, Inc.

Production Assistants: Bradley Burch, Nancy Colman, Amy Head, Joshua King, Carri Mangelli, Publishing Principals, Inc.

Principal Artists for the Third Edition: Carla Simmons, Pamela Drury-Wattenmaker

Contributing Artists to the Third Edition: Barbara Cousins, Nea Bisek, Sandra McMahon

Contributing Artists to Previous Editions: Chris Carothers, Raychel Ciemma, Cecile Duray-Bito, Janet Hayes, Darwen Hennings, Vally Hennings, Georg Klatt, Linda McVay, Kenneth Miller, Fran Milner, Elizabeth Morales-Denney, Jackie Osborn, Carol Verbeeck, John Waller, Judy Waller

Text and Cover Design: Gary Head, Publishing Principals, Inc.

Layout Artists: Michele Mangelli and Gary Head, Publishing Principals, Inc.

Copyeditor: Betsy Dilernia

Proofreaders: Brian Jones, Margot Otway, Judith Hibbard, RoseMary Horner

Indexer: Katherine Pitcoff

Digital Prepress: Publishing Principals, Inc.

Film Preparation: Colotone, Inc.

Managing Editor: Gwen Larson

Composition and Film Buyer: Lillian Hom

Photo Editor: Cecilia Mills

Photo Researcher: Darcy Lanham

Art and Design Manager: Michele Carter

Manufacturing Supervisor: Casimira Kostecki

Executive Marketing Manager: Anne Emerson

About the cover:
"Aspens, Northern New Mexico, 1958."
Photograph by Ansel Adams.

About the frontispiece:
"Aspen grove, Markagunt Plateau, Utah, 1991."
Photograph by Joseph Holmes.

Figure acknowledgments begin on page C-1.

Library of Congress Cataloging-in-Publication Data
Campbell, Neil A., 1946–
Biology/Neil A. Campbell.—3rd ed.
p. cm.—(Benjamin/Cummings series in the life sciences)
Includes bibliographical references and index.
ISBN 0-8053-1880-1
1. Biology. I. Title. II. Series
QH308.2.C34 1993
574—dc20

3 4 5 6 7 8 9 10—DO—97 96 95 94 93

The Benjamin/Cummings Publishing Company, Inc.
390 Bridge Parkway
Redwood City, California 94065

To Rochelle and Allison,
with love

About the Author

BIOLOGY is the product of 24 years of teaching experience and many years of intensive writing and revision by Dr. Neil A. Campbell. This textbook is a natural outgrowth of Dr. Campbell's broad interest in his science. He earned his M.A. in Zoology from UCLA, where he studied the control of protein synthesis during animal development, and went on to the University of California, Riverside, where he earned a Ph.D. in Biology. Dr. Campbell's research efforts on salt transport in plants and the cellular basis of leaf movements have resulted in publications in *Science, The Proceedings of the National Academy of Sciences,* and *Plant Physiology*, among other journals.

In addition to his accomplishments as a research scientist, Dr. Campbell has earned a reputation as an outstanding classroom teacher with a strong commitment to improving undergraduate education. After 10 years of teaching general biology and cell biology at San Bernardino Valley College, he took an academic leave and accepted a faculty position at Cornell University, where he reorganized a two-semester general biology course. After three successful years at Cornell, Dr. Campbell returned to California to reassume his teaching position at San Bernardino Valley College, where in 1986 he received the Outstanding Professor Award for excellence in classroom instruction. He frequently returned to Cornell to teach the summer general biology course to advanced-placement high school students and Cornell undergraduates on a six-week schedule. In 1988, Dr. Campbell accepted an invitation to teach a one-semester general biology course at Pomona College.

During his many years of teaching general biology—most frequently as the sole lecturer—Dr. Campbell has instructed over 13,000 students. His teaching sensibilities have been honed in both large lecture and small classroom environments and with a diverse group of students. He is currently a Visiting Scholar in the Department of Botany and Plant Sciences at the University of California, Riverside.

PREFACE

This third edition of *BIOLOGY* builds on the teaching values that characterized the first two versions. Those teaching values defined the book's two objectives: to explain biological concepts clearly and accurately within a context of unifying themes, and to help students develop more positive and realistic impressions of science as a human activity. Those dual objectives have now shaped the third edition's improvements.

FIVE MAJOR IMPROVEMENTS IN THE THIRD EDITION

1. Evolution, the core theme of *BIOLOGY*, is even more pervasive in this edition. A thematic approach distinguishes this book from an "encyclopedia of biology." The first chapter introduces several themes that resurface throughout the text to help students synthesize connections in their study of life. The overarching theme is evolution, which accounts for the unity and diversity of life and integrates the book's other themes. This edition brings the evolutionary view of life into even sharper focus. For example, several new sections in the first three units will help students link cellular and molecular biology to the core theme of evolution.

2. This edition of *BIOLOGY* features even greater emphasis on the process of science. The power and limitations of science are now presented more thoroughly in Chapter 1, and the hypothetico-deductive method is then applied in several case studies that appear throughout the text. The Methods Boxes, many of them new in this edition, help demystify science by explaining laboratory and field methods in the context of experiments. Eight new interviews with influential biologists (see pp. xiv–xv) personalize science, portraying it as a social activity of creative men and women rather than an impersonal collection of facts.

3. "Science, Technology, and Society" has been added as one of *BIOLOGY*'s themes. Biology and its applications have a profound impact on culture—on our view of the natural world, on our environmental awareness, and on our health and quality of life. It is important for students to understand that ethics has a place in science, even in basic research, and that technology brings with it the need to examine values and to make choices. The third edition of *BIOLOGY* highlights the interrelatedness of science, technology, and society. In particular, complex environmental issues are more prominent in this edition. A new category of Science, Technology, and Society questions at the end of each chapter encourages students to incorporate biology into their world view.

4. Extensive revision of *BIOLOGY's* illustrations helps the third edition teach difficult concepts even more effectively. Most figures have been redesigned, always with an eye toward stronger pedagogy. One important navigational aid is the consistent use of color-coding and icons throughout the book. Proteins, for example, are color-coded purple, and ATP always appears in illustrations as a yellow sunburst. This edition also improves upon the text-figure coordination that distinguished earlier editions. The artists, photo researchers, editors, and I worked together, beginning with the first draft, to embed the figures into the story line of each chapter. In fact, students will find that examining the figures and their self-contained legends is one way to preview or review the content of a chapter.

5. The "Overview–Closer Look–Postview" teaching style has been strengthened in this edition. One reason *BIOLOGY* has so many pages is that it commits much space to methodical presentations of complex topics—carefully paced presentations that add layers of detail judiciously so that the main concepts are reinforced, not obscured. This third edition makes it even easier for students to fit what they are learning into a framework of general concepts. Complicated processes, such as cellular respiration (Chapter 9) and protein synthesis (Chapter 16), are first presented in panoramic view ("overview"), providing the student with a sense of what the whole process accomplishes. Text and figures then dissect the process for "a closer look" at how it works. Orientation diagrams, miniature versions of the overview illustration with appropriate parts highlighted, keep this stepwise development connected to the central concepts. In many chapters, the closer look is followed by a postview, often a comprehensive diagram that helps students reconstruct the overall process from its components. This teaching approach and the other features of *BIOLOGY* continue to evolve from classroom experience with students who share my resolve to survive the information explosion by sorting out each topic's main points.

CONTENT AND ORGANIZATION

BIOLOGY makes no pretense that there is one "correct" way to order the major topics in an introductory biology course. Over the years, I have rearranged my own syllabus in various ways, finding that many different sequences are workable. This book is flexible enough for instructors to adapt its content to a variety of syllabi. The eight units are self-contained, allowing for rearrangement, and most of the chapters within each unit can be assigned in a different sequence without substantial loss of continuity. For example, instructors who integrate plant and animal physiology can merge chapters from Units Six and Seven to fit their courses.

A brief survey of the book's organization will identify the content of each unit. Specific changes in each chapter are too numerous to list here.

Unit One: The Chemistry of Life My colleagues and I have found that many students struggle in their introductory biology courses because of inadequate backgrounds in basic chemistry. Chapters 2–4 are designed to help those students by developing, in carefully paced steps, the concepts of chemistry that are essential for success in biology. This approach makes self-study possible, reducing the need for instructors to spend valuable classroom time on basic chemistry. However, Chapter 5 ("The Structure and Function of Macromolecules") and Chapter 6 ("An Introduction to Metabolism") provide important orientation even to students with solid chemistry backgrounds. This edition builds stronger connections between chemistry and biology, with many new examples of how organisms work at the molecular level. Several new sections relate the chemistry of life to evolution.

Unit Two: The Cell Chapters 7–11 emphasize the correlation of structure and function as a theme for studying cells. For example, the importance of membranes in ordering metabolism is stressed throughout the unit. Substantial refinement of text and illustrations has strengthened the Overview–Closer Look–Postview pedagogy of these chapters. Among the other improvements is a more thorough discussion of how cell division is controlled in Chapter 11.

Unit Three: The Gene Chapters 12–19 take a historical approach to genetics, tracing its development from Gregor Mendel to DNA technology. *BIOLOGY's* extensive coverage of human genetics in Unit Three emerges topically in each chapter, in close proximity to whatever general concept is being applied. The unit has been extensively revised to improve teaching effectiveness and to reflect the exciting progress in molecular genetics. For example, Chapter 19 ("DNA Technology") features many new commercial applications and new techniques, such as antisense-RNA. New sections in Unit Three examine the evolutionary significance of genetics.

Unit Four: Mechanisms of Evolution Evolution is the one theme that surfaces in every part of *BIOLOGY*, but Chapters 20–23 focus on *how* life evolves and how biologists study evolution. Chapter 20 ("Descent with Modification: A Darwinian View of Life") now traces evolutionary theory as a case study in the scientific process. Chapter 20 also features new examples of natural selection in action. Chapter 22 has been updated to compare diverse views of how new species originate. Evolutionary biology is alive with controversy about the tempo and mechanisms of evolution, and students deserve to see that not all issues are settled.

Unit Five: The Evolutionary History of Biological Diversity Chapters 24–30 consider the diversity of life within the context of key evolutionary junctures, such as the origin of prokaryotes, the evolution of the eukaryotic cell, the genesis of multicellular life, and the adaptive radiations of plants, fungi, and animals. The evolutionary theme of this unit contrasts sharply with a "parade of the kingdoms" approach to biodiversity. Among the improvements in the third edition are more extensive coverage of the ecological significance of bacteria and fungi. Chapter 27 includes a new section, "The Value of Plant Diversity," an example of this edition's greater emphasis on Science, Technology, and Society issues.

Unit Six: Plants: Form and Function Chapters 31–35 introduce students to the structure and physiology of plants within the evolutionary context of adaptation to terrestrial environments. A comprehensive revision places new emphasis on molecular and cellular approaches to plant biology. For example, Chapter 34 ("Plant Reproduction and Development") includes new sections on pattern formation, clonal analysis of the shoot apex, and the genetic basis of flower development. Chapter 35 ("Control Systems in Plants") now covers the cellular mechanisms of hormone action and examines signal transduction pathways in plant cells.

Unit Seven: Animals: Form and Function The organism-environment interface is the focus of Chapters 36–45, which take a comparative approach in exploring the diverse adaptations that have evolved in the animal kingdom. Humans fit into this comparative format as an important mammalian example. However, invertebrate and nonmammalian adaptations are even more visible in this edition than in the second edition. Some chapters were reorganized to improve their teaching effectiveness. For example, Chapter 39 now covers the complex topic of immunology more clearly, with the help of many new diagrams. This entire unit was also updated. Chapter 43, for example, explains recent progress in the study of animal development.

Unit Eight: Ecology Chapters 46–50 now represent a more cohesive unit on ecology, with a stronger evolutionary orientation and more connections to the book's other integrating themes. The new version of Chapter 46 is a more effective introduction to Earth's diverse environments, with increased emphasis on marine environments. In keeping with the third edition's new Science, Technology, and Society theme, all ecology chapters now cover environmental issues in greater depth. The chapters also present the different viewpoints in several of ecology's current debates, with the objective of encouraging students to evaluate arguments and evidence critically. Chapter 50 ("Behavior") was expanded and rewritten to stress behavioral ecology, which fits behavior into evolutionary context. This chapter also serves as a capstone for the entire book, relating ecology to other fields of biology, to the other natural sciences, and to the student's general education.

IN-TEXT LEARNING AIDS

Learning aids at the end of each chapter reinforce the chapter's main concepts, vocabulary, and applications. A **Study Outline** is keyed by page number to the major sections of the chapter. A **Self-Quiz** helps students measure their comprehension, but many of these questions also require students to apply knowledge or solve problems. The answers to the Self-Quiz questions are found in Appendix One. **Challenge Questions** encourage students to verbalize their interpretations of concepts, to extrapolate from what they have learned to new situations, to think critically about complex debates in biology, to apply quantitative skills in the context of biological problems, and to generate testable hypotheses of their own. The **Science, Technology, and Society** questions ask students to think about biology's place in culture and about the consequences of applied biology. Short **Further Reading** lists complete the learning aids at the end of each chapter. Students will also find a **Glossary** of key terms at the end of the book. As references, Appendix Two presents a Classification of Life and Appendix Three the Metric System. To assist in still another way, Appendix Four introduces students to the learning tool known as **concept mapping**.

SUPPLEMENTS

Student Study Guide by Martha Taylor, Cornell University.

Investigating Biology: A Laboratory Manual for *BIOLOGY* by Judith Morgan, Emory University, and Eloise Carter, Oxford College of Emory University, with accompanying Annotated Instructor's Edition and Preparation Guide.

***BIOLOGY* ClassNotes** by Nina Caris and Harold Underwood, both of Texas A&M University.

Fish Farm: Simulation Software by Robert J. Kosinski, Clemson University, with accompanying *Student Workbook* and *Instructor's Guide*.

Instructor's Guide by Nina Caris and Harold Underwoood, both of Texas A&M University.

Test Bank edited by William E. Barstow, University of Georgia, with consultants Martha Taylor, Cornell University, Margaret Waterman, Harvard Medical School, Daniel Wivagg, Baylor University, and Betty Ann Wonderley, Richardson Independent School District. This test bank is available on Microtest, a microcomputer test-generation program. (The test bank is available to qualified college and university adopters.)

Laboratory Collection edited by Judith Goodenough, University of Massachusetts.

Overhead Transparencies A set of 300 color acetates of illustrations and micrographs from *BIOLOGY*, Third Edition, is available to qualified college and university adopters.

35 mm Slides The same 300 illustrations available as acetates are available in 35 mm slides to qualified college and university adopters.

Transparency Masters All of the text art from *BIOLOGY*, Third Edition, is available in black-and-white masters to qualified college and university adopters.

BioSHOW: The Videodisc A videodisc of text art, original animations, and motion sequences to accompany *BIOLOGY*, Third Edition, is available to qualified college and university adopters.

* * *

The real test of any textbook is how well it helps instructors teach and students learn. I welcome comments from students and professors who use *BIOLOGY*. Please address your suggestions for improving the next edition directly to me:

Neil A. Campbell
Department of Botany and Plant Sciences
University of California
Riverside, California 92521

ACKNOWLEDGMENTS

Many people have asked me how one person can write an entire general biology textbook. The answer, of course, is that one person *can't,* at least not without a lot of help. Though *BIOLOGY* is in my voice, each chapter is a synthesis of what I have learned from students, teachers, research scientists, contributors, artists, and editors. Their collective influence accounts for the improvements in this third edition.

Almost 75 biology instructors and research specialists reviewed chapters and helped me strengthen the scientific accuracy and teaching effectiveness of the third edition. Many other professors and their students took the time to volunteer their helpful suggestions by writing directly to me. Several scientists became even more involved by actually revising text or submitting early drafts of new material. These contributors are Gary Brusca (Humboldt State University), who helped revise the two chapters on animal diversity (Chapters 29 and 30); Berdell Funke (North Dakota State University), who collaborated on the immunology chapter (Chapter 39); Paul Hertz (Barnard College), who made major improvements in the ecology chapters (Chapters 46–49); Richard Liebaert (Linn Benton College, Oregon), who wrote many of the questions at the ends of chapters; Gary Matthews (State University of New York, Stonybrook), who contributed his expertise to the chapters on nerves, senses, and movement (Chapters 44 and 45); Jane Reece (Benjamin/Cummings) and Lawrence Mitchell (Iowa State University), who both did excellent work on the DNA technology chapter (Chapter 19); Fred Rhoades (Western Washington University), who worked extensively on three chapters in Unit Five (Chapters 25, 26, and 28); Stephen I. Rothstein (University of California, Santa Barbara), who improved the behavior chapter (Chapter 50) in many ways; and Fred Wilt (University of California, Berkeley), who guided revision of the chapters on gene expression in eukaryotes (Chapter 18) and animal development (Chapter 43). Although I am responsible for any errors that remain, they are all the fewer because of the dedication of the reviewers, correspondents, and contributors. They worked hard to help me make this edition more correct, current, and clear, and I thank them for their participation and their commitment to science education.

Numerous UC Riverside colleagues helped shape this revision by sharing insights about their research fields and exchanging ideas about biology education. In particular, I would like to thank Katherine Atkinson, Darleen DeMason, Leah Haimo, Robert Heath, Anthony Huang, Bradley Hyman, Robert Leonard, Elizabeth Lord, Carol Lovatt, John Oross, Kathryn Platt, David Reznick, Rodolfo Ruibal, Clay Sassaman, Irwin Sherman, Vaughan Shoemaker, William Thomson, Giles Waines, Marlene Zuk, and John Moore (whose "Science as a Way of Knowing" essays have influenced the evolution of this book). I am also grateful to Pius Horner, my longtime San Bernardino Valley College colleague, who was such an important mentor during my development as a classroom teacher.

One of the pleasures of revising *BIOLOGY* was the opportunity to conduct the new interviews that open the text's eight units. The interviewees for the third edition are Candace Pert, Michael Bishop and Harold Varmus, David Suzuki, Ernst Mayr, Stephen Jay Gould, Virginia Walbot, Karel Liem, and Ariel Lugo. I thank them for helping *BIOLOGY* communicate the fun of science to students.

The illustration program is such an integral part of *BIOLOGY* that the artists could be considered coauthors. More than two-thirds of the figures in the third edition are new or extensively revised. Carla Simmons and Pamela Drury-Wattenmaker were the principal artists. Carla also served as art consultant, and she is the one artist who has been a major creative force in all three editions of *BIOLOGY.* Nea Bisek, Barbara Cousins, and Sandra McMahon were the other artists who graced this edition with their work. I thank the entire art team for creating illustrations that will enhance *BIOLOGY*'s reputation for art that is as pedagogically innovative as it is visually attractive.

Photo editor Cecilia Mills and photo researcher Darcy Lanham found the many new beautiful, instructive photographs that enrich this third edition. I thank them for their perseverance in locating just the right photos to reinforce key concepts.

Suzanne Olivier, the main developmental editor for this edition, deserves special recognition as my partner throughout the revision process. She worked tirelessly in helping me make every chapter teach more effectively. In particular, Suzanne engineered the ambitious revision of the book's art program, and her stamp is visible on every page. I thank Suzanne for the enormous role she played in improving *BIOLOGY*.

Many other publishing professionals helped me make this edition more useful to students. Pat Burner and Robin Fox did excellent work as developmental editors on several chapters. I also thank Susan Weisberg for editing most of the interviews. Copyeditor Betsy Dilernia was outstanding in improving the clarity and consistency of the text. Proofreaders Brian Jones and Margot Otway were thorough in bringing errors and other problems to my attention. Kathy Pitcoff constructed the much-improved index. Lisa Donohoe and Valerie Kuletz put together the new supplements package. I am also grateful to editorial assistants Sissy Lemon, Christine Ruotolo, Kimberly Viano, and Thomas Viano for their help in making *BIOLOGY* a better book.

Pat Waldo, Gary Head, and Michele Mangelli of Publishing Principals, Inc., coordinated the production of this third edition, transforming manuscript, art, and photographs into a book. Pat is the human funnel through which all of the book's pieces flowed together. I am grateful for her experience, patience, flexibility, and extraordinary effort in what must be one of publishing's most complex jobs. Gary is *BIOLOGY*'s award-winning designer, the person responsible for keeping the book true to our goal of functional beauty. I also thank Gary for another great cover. Michele Mangelli worked with Gary and Pat on page layout, and I am very pleased with the results. I would also like to thank Brad Burch, Nancy Colman, Amy Head, Joshua King, and Carri Mangelli for their production assistance.

Benjamin/Cummings' own production department collaborated with Publishing Principals in designing this third edition and was also responsible for assuring quality in the manufacturing of the book you hold. In particular, I want to thank the production department's art and design manager Michele Carter, manufacturing supervisor Casimira Kostecki, and managing editor Gwen Larson.

The Benjamin/Cummings marketing department keeps *BIOLOGY* in touch with the students and professors it serves. I thank Deborah Phillips-Froese, Karryll Nason, Bob Ting, Rosemarie Forrest, and executive marketing manager Anne Emerson for announcing the third edition with an informative and dignified promotion.

The field staff that represents *BIOLOGY* on campuses is my link to the students and professors who use the text. The field representatives tell me what you like and don't like about the book, and they provide prompt service to biology departments. I thank them for their professionalism in communicating the strengths of our book without slurring other publishers and their competing books.

BIOLOGY originated from a 1979 meeting with Jim Behnke in my Cornell office. Jim was my editor for the first edition, and it took us eight years to craft the new kind of biology textbook we envisioned. Robin Heyden took over as sponsoring editor of the second edition and guided a revision that elevated *BIOLOGY* to the very top of the charts. I thank Jim and Robin for their continuing interest in the book's mission. I am also grateful to Benjamin/Cummings president Sally Elliott and vice president and editorial director Barbara Piercecchi for their sustaining faith in *BIOLOGY* and its author.

Edith Beard Brady, the third edition's sponsoring editor, brought a fresh perspective to the book and inspired me to improve it in many ways. I admire her sincerity, publishing ethics, and genuine interest in the quality of science education. Most of all, I respect Edith for her courage; she is willing to take some risks and try new things. During this adventure of rethinking *BIOLOGY*, Edith has become a trusted friend and valued colleague. I thank her for sharing the goal of continuously improving our book.

Most of all, I thank my family and friends for their encouragement and for continuing to tolerate my obsession with making *BIOLOGY* a better textbook.

PREVIOUS EDITION REVIEWERS

Katherine Anderson, *University of California, Berkeley*, Richard J. Andren, *Montgomery County Community College*, J. David Archibald, *Yale University*, Leigh Auleb, *San Francisco State University*, Katherine Baker, *Millersville University*, William Barklow, *Framingham State College*, Steven Barnhart, *Santa Rosa Junior College*, Tom Beatty, *University of British Columbia*, Wayne Becker, *University of Wisconsin, Madison*, Jane Beiswenger, *University of Wyoming*, Anne Bekoff, *University of Colorado, Boulder*, Marc Bekoff, *University of Colorado, Boulder*, Adrianne Bendich, *Hoffman-La Roche, Inc.*, Barbara Bentley, *State University of New York, Stony Brook*, Darwin Berg, *University of California, San Diego*, Dorothy Berner, *Temple University*, Paulette Bierzychudek, *Pomona College*, Robert Blystone, *Trinity University*, Robert Boley, *University of Texas, Arlington*, Eric Bonde, *University of Colorado, Boulder*, Richard Boohar, *University of Nebraska, Omaha*, James L. Botsford, *New Mexico State University*, J. Michael Bowes, *Humboldt State University*, Barry Bowman, *University of California, Santa Cruz*, Jerry Brand, *University of Texas, Austin*, James Brenneman, *University of Evansville*, Herbert Bruneau, *Oklahoma State University*, Gary Brusca, *Humboldt*

State University, Alan H. Brush, *University of Connecticut, Storrs*, Meg Burke, *University of North Dakota*, Edwin Burling, *De Anza College*, William Busa, *Johns Hopkins University*, John Bushnell, *University of Colorado*, Gregory Capelli, *College of William and Mary*, Nina Caris, *Texas A&M University*, Doug Cheeseman, *De Anza College*, Shepley Chen, *University of Illinois, Chicago*, Henry Claman, *University of Colorado Health Science Center*, William Coffman, *University of Pittsburgh*, J. John Cohen, *University of Colorado Health Science Center*, John Corliss, *University of Maryland*, Stuart Coward, *University of Georgia*, Marianne Dauwalder, *University of Texas, Austin*, Bonnie J. Davis, *San Francisco State University*, Jerry Davis, *University of Wisconsin, La Crosse*, Thomas Davis, *University of New Hampshire*, James Dekloe, *University of California, Santa Cruz*, T. Delevoryas, *University of Texas, Austin*, Jean DeSaix, *University of North Carolina*, Marvin Druger, *Syracuse University*, Betsey Dyer, *Wheaton College*, Robert Eaton, *University of Colorado*, Robert S. Edgar, *University of California, Santa Cruz*, Betty J. Eidemiller, *Lamar University*, David Evans, *University of Florida*, Robert C. Evans, *Rutgers University, Camden*, Sharon Eversman, *Montana State University*, Lincoln Fairchild, *Ohio State University*, Lynn Fancher, *College of DuPage*, Larry Farrell, *Idaho State University*, Jerry F. Feldman, *University of California, Santa Cruz*, Russell Fernald, *University of Oregon*, Milton Fingerman, *Tulane University*, Barbara Finney, *Regis College*, Abraham Flexer, *Manuscript Consultant, Boulder, Colorado*, Norma Fowler, *University of Texas, Austin*, David Fox, *University of Tennessee, Knoxville*, Otto Friesen, *University of Virginia*, Virginia Fry, *Monterey Peninsula College*, Alice Fulton, *University of Iowa*, Sara Fultz, *Stanford University*, Berdell Funke, *North Dakota State University*, Anne Funkhouser, *University of the Pacific*, Arthur W. Galston, *Yale University*, Carl Gans, *University of Michigan*, John Gapter, *University of Northern Colorado*, Reginald Garrett, *University of Virginia*, Patricia Gensel, *University of North Carolina*, Robert George, *University of Wyoming*, Todd Gleeson, *University of Colorado*, William Glider, *University of Nebraska*, Elizabeth A. Godrick, *Boston University*, Lynda Goff, *University of California, Santa Cruz*, Paul Goldstein, *University of Texas, El Paso*, Judith Goodenough, *University of Massachusetts, Amherst*, Ester Goudsmit, *Oakland University*, A. J. F. Griffiths, *University of British Columbia*, William Grimes, *University of Arizona*, Mark Gromko, *Bowling Green State University*, Katherine L. Gross, *Ohio State University*, Gary Gussin, *University of Iowa*, R. Wayne Habermehl, *Montgomery County Community College*, Mac Hadley, *University of Arizona*, Jack P. Hailman, *University of Wisconsin*, Penny Hanchey-Bauer, *Colorado State University*, Laszlo Hanzely, *Northern Illinois University*, Richard Harrison, *Cornell University*, H. D. Heath, *California State University, Hayward*, George Hechtel, *State University of New York at Stony Brook*, Jean Heitz-Johnson, *University of Wisconsin, Madison*, Frank Heppner, *University of Rhode Island*, Ralph Hinegardner, *University of California, Santa Cruz*, William Hines, *Foothill College*, David Ho, *University of Missouri*, Carl Hoagstrom, *Ohio Northern University*, James Holland, *Indiana State University, Bloomington*, Laura Hoopes, *Occidental College*, Nancy Hopkins, *Massachusetts Institute of Technology*, Kathy Hornberger, *Widener University*, Pius F. Horner, *San Bernardino Valley College*, Margaret Houk, *Ripon College*, Ronald R. Hoy, *Cornell University*, Robert J. Huskey, *University of Virginia*, Alice Jacklet, *State University of New York, Albany*, John Jackson, *North Hennepin Community College*, Russell Jones, *University of California, Berkeley*, Alan Journet, *Southeast Missouri State University*, Thomas Kane, *University of Cincinnati*, E. L. Karlstrom, *University of Puget Sound*, George Khoury, *National Cancer Institute*, Robert Kitchen, *University of Wyoming*, Attila O. Klein, *Brandeis University*, Thomas Koppenheffer, *Trinity University*, J. A. Lackey, *State University of New York at Oswego*, Lynn Lamoreux, *Texas A&M University*, Kenneth Lang, *Humboldt State University*, Allan Larson, *Washington University*, Charles Leavell, *Fullerton College*, Robert Leonard, *University of California, Riverside*, Joseph Levine, *Boston College*, Bill Lewis, *Shoreline Community College*, Lorraine Lica, *California State University, Hayward*, Harvey Lillywhite, *University of Florida, Gainesville*, Sam Loker, *University of New Mexico*, Jane Lubchenco, *Oregon State University*, James MacMahon, *Utah State University*, Charles Mallery, *University of Miami*, Lynn Margulis, *Boston University*, Edith Marsh, *Angelo State University*, Karl Mattox, *Miami University of Ohio*, Joyce Maxwell, *California State University, Northridge*, Richard McCracken, *Purdue University*, John Merrill, *University of Washington*, Ralph Meyer, *University of Cincinnati*, Roger Milkman, *University of Iowa*, Helen Miller, *Oklahoma State University*, John Miller, *University of California, Berkeley*, Kenneth R. Miller, *Brown University*, John E. Minnich, *University of Wisconsin, Milwaukee*, Russell Monson, *University of Colorado, Boulder*, Frank Moore, *Oregon State University*, Randy Moore, *Wright State University*, Carl Moos, *Veterans Administration Hospital, Albany, New York*, John Neess, *University of Wisconsin, Madison*, Todd Newbury, *University of California, Santa Cruz*, Harvey Nichols, *University of Colorado, Boulder*, Deborah Nickerson, *University of South Florida*, Bette Nicotri, *University of Washington*, David Norris, *University of Colorado, Boulder*, Cynthia Norton, *University of Maine, Augusta*, Brian O'Conner, *University of Massachusetts, Amherst*, Eugene Odum, *University of Georgia*, Gay Ostarello, *Diablo Valley College*, Peter N. Pappas, *County College of Morris*, Bulah Parker, *North Carolina State University*, Stanton Parmeter, *Chemeketa Community College*, Robert Patterson, *San Francisco State University*, Crellin Pauling, *San Francisco State University*, Kay Pauling, *Foothill Community College*, Patricia Pearson, *Western Kentucky University*, James Platt, *University of Denver*, Jeffrey Pommerville, *Texas A&M University*, Donald Potts, *University of California, Santa Cruz*, David Pratt, *University of California, Davis*, Halina Presley, *University of Illinois, Chicago*, Scott Quackenbush, *Florida International University*, Ralph Quatrano, *Oregon State University*, Charles Ralph, *Colorado State University*, Charles Remington, *Yale University*, Fred Rhoades, *Western Washington State University*, Donna Ritch, *Pennsylvania State University*, Rodney Rogers, *Drake University*, Thomas Rost, *University of California, Davis*, John Ruben, *Oregon State University*, Albert Ruesink, *Indiana University*, Ted Sargent, *University of Massachusetts, Amherst*, Carl Schaefer, *University of Connecticut*, David Schimpf, *University of Minnesota, Duluth*, William H. Schlesinger, *Duke University*, Erik P. Scully, *Towson State University*, Stephen Sheckler, *Virginia Polytechnic Institute and State University,* James Shinkle, *Trinity University,* Barbara Shipes, *Hampton University,* Peter Shugarman, *University of Southern California,* Alice Shuttey, *DeKalb Community College,* James Sidie, *Ursinus College,* Daniel Simberloff, *Florida State University,* John Smarrelli, *Loyola University,* Andrew T. Smith, *Arizona State University,* Andrew J. Snope, *Essex Community College,* Barbara Stewart, *Swarthmore College,* Cecil Still, *Rutgers University, New Brunswick,* John Stolz, *California Institute of Technology,* Stephen Strand, *University of California, Los Angeles,* Daryl Sweeney, *University of Illinois, Urbana-Champaign,* Samuel S. Sweet, *University of California, Santa Barbara,* Samuel Tarsitano, *Southwest Texas State University,* David Tauck, *University of Santa Clara,* James Taylor, *University of New Hampshire,* Roger Thibault, *Bowling Green State University,* John Thornton, *Oklahoma State University,* Robert Thornton, *University of California, Davis,* Robert Tuveson, *University of Illinois, Urbana,* Maura G. Tyrrell, *Stonehill College,* James W. Valentine, *University of California, Santa Barbara,* Joseph Vanable, *Purdue University,* Theodore Van Bruggen, *University of South Dakota,* Frank Visco, *Orange Coast College,* Laurie Vitt, *University of California, Los Angeles,* Susan D. Waaland, *University of Washington,* John Waggoner, *Loyola Marymount University,* Dan Walker, *San Jose State University,* Jeffrey Walters, *North Carolina State University,* Margaret Waterman, *University of Pittsburgh,* Terry Webster, *University of Connecticut, Storrs,* Peter Wejksnora, *University of Wisconsin, Milwaukee,* Kentwood Wells, *University of Connecticut,* Christopher Wills, *University of California, San Diego,* Fred Wilt, *University of California, Berkeley,* Robert T. Woodland, *University of Massachusetts Medical School,* Philip Yant, *University of Michigan,* Hideo Yonenaka, *San Francisco State University,* John Zimmerman, *Kansas State University,* Uko Zylstra, *Calvin College.*

THIRD EDITION REVIEWERS

Focus Group Participants

Ron Basmajian, *Merced College*

Mary C. Nolan, *Irvine Valley College*

Bette H. Nybakken, *Hartnell College*

Christopher Riegle, *Irvine Valley College*

Thomas Rodella, *Merced College*

Manuscript Reviewers

John Alcock, *Arizona State University*

P. Stephen Baenziger, *University of Nebraska*

Mark Bekoff, *University of Colorado*

Gerald Bergstrom, *University of Wisconsin, Milwaukee*

Anna W. Berkovitz, *Purdue University*

Dorothy B. Berner, *Temple University*

Charles Biggers, *Memphis State University*

Carey L. Booth, *Reed College*

Richard Bowker, *Alma College*

Donald P. Briskin, *University of Illinois, Urbana*

Deborah Canington, *University of California, Davis*

William P. Coffman, *University of Pittsburgh*

Stuart J. Coward, *University of Georgia*

Charles Creutz, *University of Toledo*

Richard Cyr, *Pennsylvania State University*

John Dearn, *University of Canberra*

Diane C. DeNagel, *Northwestern University*

Bruce Fall, *University of Minnesota*

Larry Farrell, *Idaho State University*

David Fisher, *University of Hawaii at Manoa*

Berdell Funke, *North Dakota State University*

Lynda Goff, *University of California, Santa Cruz*

Judith Goodenough, *University of Massachusetts, Amherst*

Leah Haimo, *University of California, Riverside*

Caroll Henry, *Chicago State University*

Paul E. Hertz, *Barnard College*

Tuan-hua David Ho, *Washington University*

Carl W. Hoagstrom, *Ohio Northern University*

Laura Hoopes, *Occidental College*

Kenneth C. Jones, *California State University, Northridge*

Janis Kuby, *San Francisco State University*

Carmine A. Lanciani, *University of Florida*

Diane K. Lavett, *State University of New York, Cortland,* and *Emory University*

C. S. Lee, *University of Texas*

John Mutchmor, *Iowa State University*

Charles R. Noback, *College of Physicians and Surgeons, Columbia University*

David O. Norris, *University of Colorado*

Gerard O'Donovan, *University of North Texas*

Wan Ooi, *Houston Community College*

Scott Poethig, *University of Pennsylvania*

Barry Palevitz, *University of Georgia*

Warren Porter, *University of Wisconsin*

C. Gary Reiness, *Pomona College*

David Reznick, *University of California, Riverside*

Wayne Rosing, *Middle Tennessee State University*

Stephen I. Rothstein, *University of California, Santa Barbara*

Mark F. Sanders, *University of California, Davis*

James Schinkle, *Trinity University*

Mark Sheridan, *North Dakota State University*

Susan Sovonick-Dunford, *University of Cincinnati*

Karen Steudel, *University of Wisconsin*

Richard D. Storey, *Colorado College*

Russell Stullken, *Augusta College*

Gerald Summers, *University of Missouri*

Marshall Sundberg, *Louisiana State University*

Lincoln Taiz, *University of California, Santa Cruz*

David Tauck, *Santa Clara University*

James Traniello, *Boston University*

Gordon Uno, *University of Oklahoma*

Robert L. Wallace, *Ripon College*

Stephen Williams, *Glendale Community College*

Fred Wilt, *University of California, Berkeley*

Joseph Woodring, *Louisiana State University*

The Campbell Interviews

UNIT ONE

The Chemistry of Life 20

Candace Pert

Scientific Director, Peptide Design

UNIT TWO

The Cell 112

Michael Bishop and Harold Varmus

University of California, San Francisco

UNIT THREE

The Gene 240

David Suzuki

University of British Columbia

UNIT FOUR

Mechanisms of Evolution 416

Ernst Mayr

Emeritus, Harvard University

UNIT FIVE

THE EVOLUTIONARY HISTORY OF BIOLOGICAL DIVERSITY 500

Stephen Jay Gould

Harvard University

UNIT SIX

PLANTS: FORM AND FUNCTION 670

Virginia Walbot

Stanford University

UNIT SEVEN

ANIMALS: FORM AND FUNCTION 778

Karel Liem

Harvard University

UNIT EIGHT

ECOLOGY 1048

Ariel Lugo

Institute of Tropical Forestry, Caribbean National Forest, USDA Forest Service

Brief Contents

1 Introduction: Themes in the Study of Life 2

UNIT ONE
THE CHEMISTRY OF LIFE 20

2 Atoms, Molecules, and Chemical Bonds 24

3 Water and the Fitness of the Environment 40

4 Carbon and Molecular Diversity 53

5 The Structure and Function of Macromolecules 64

6 An Introduction to Metabolism 91

UNIT TWO
THE CELL 112

7 A Tour of the Cell 116

8 Membrane Structure and Function 151

9 Cellular Respiration: Harvesting Chemical Energy 173

10 Photosynthesis 199

11 The Reproduction of Cells 221

UNIT THREE
THE GENE 240

12 Meiosis and Sexual Life Cycles 244

13 Mendel and the Gene Idea 258

14 The Chromosomal Basis of Inheritance 280

15 The Molecular Basis of Inheritance 300

16 From Gene to Protein 316

17 Microbial Models: The Genetics of Viruses and Bacteria 344

18 Genome Organization and Expression in Eukaryotes 372

19 DNA Technology 390

UNIT FOUR
MECHANISMS OF EVOLUTION 416

20 Descent with Modification: A Darwinian View of Life 420

21 How Populations Evolve 438

22 The Origin of Species 456

23 Tracing Phylogeny: Macroevolution, the Fossil Record, and Systematics 474

UNIT FIVE
THE EVOLUTIONARY HISTORY OF BIOLOGICAL DIVERSITY 500

24 Early Earth and the Origin of Life 504

25 Prokaryotes and the Origins of Metabolic Diversity 515

26 Protists and the Origin of Eukaryotes 533

27 Plants and the Colonization of Land 559

28 Fungi 583

29 Invertebrates and the Origin of Animal Diversity 598

30 The Vertebrate Genealogy 635

UNIT SIX
PLANTS: FORM AND FUNCTION 670

31 Plant Structure and Growth 674

32 Transport in Plants 699

33 Plant Nutrition 718

34 Plant Reproduction and Development 734

35 Control Systems in Plants 756

UNIT SEVEN
ANIMALS: FORM AND FUNCTION 778

36 An Introduction to Animal Structure and Function 782

37 Animal Nutrition 794

38 Circulation and Gas Exchange 818

39 The Body's Defenses 850

40 Controlling the Internal Environment 876

41 Chemical Signals in Animals 907

42 Animal Reproduction 931

43 Animal Development 956

44 Nervous Systems 982

45 Sensory and Motor Mechanisms 1015

UNIT EIGHT
ECOLOGY 1048

46 An Introduction to Ecology: Distribution and Adaptations of Organisms 1052

47 Population Ecology 1083

48 Community Ecology 1106

49 Ecosystems 1132

50 Behavior 1158

Detailed Contents

1 **Introduction: Themes in the Study of Life** 2

A Hierarchy of Organization 3
Emergent Properties 4
The Cellular Basis of Life 7
Heritable Information 8
A Feeling for Organisms 8
The Correlation Between Structure and Function 9
The Interaction of Organisms with Their Environment 10
Unity in Diversity 10
Evolution: The Core Theme of Biology 11
Science as a Process: The Hypothetico-Deductive Method 15
Science, Technology, and Society 19

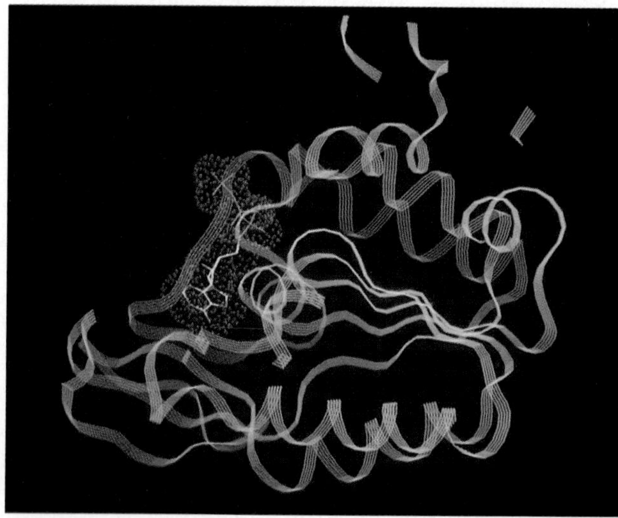

UNIT ONE

The Chemistry of Life 20

Interview: Candace Pert 20

2 **Atoms, Molecules, and Chemical Bonds** 24

Matter: Elements and Compounds 24
  Elements Essential to Life 26
The Structure and Behavior of Atoms 27
  Subatomic Particles 27
  Atomic Number and Atomic Weight 28
  Isotopes 28
  Energy Levels 30
  Electron Orbitals 31
  Electron Configuration and Chemical Properties 31
Chemical Bonds and Molecules 32
  Covalent Bonds 33
  Ionic Bonds 34
  Hydrogen Bonds 36
  The Biological Importance of Weak Bonds 36
Chemical Reactions 36
Chemical Conditions on the Early Earth: Setting the Stage for the Origin and Evolution of Life 37
Methods: The Use of Radioactive Tracers in Biology 29

3 **Water and the Fitness of the Environment** 40

Water Molecules and Hydrogen Bonding 40
Some Extraordinary Properties of Water 41
  Liquid Water Is Cohesive 41
  Water Has a High Specific Heat 42
  Water Has a High Heat of Vaporization 43
  Water Expands When It Freezes 44
  Water Is a Versatile Solvent 45
Aqueous Solutions 46
  Solute Concentration 46
  Acids, Bases, and pH 47
Acid Rain: Upsetting the Fitness of the Environment 49

4 **Carbon and Molecular Diversity** 53

The Foundations of Organic Chemistry 53
The Versatility of Carbon in Molecular Architecture 55
Variation in Carbon Skeletons 56
  Isomers 56
Functional Groups 58
  The Hydroxyl Group 58
  The Carbonyl Group 58
  The Carboxyl Group 59
  The Amino Group 60
  The Sulfhydryl Group 61
  The Phosphate Group 61
The Chemical Elements of Life: A Review 61

5 The Structure and Function of Macromolecules 64

Polymers 64
Polymers and Molecular Diversity 64
Making and Breaking Polymers 65
Carbohydrates 66
Monosaccharides 66
Disaccharides 66
Polysaccharides 67
Lipids 71
Fats 71
Phospholipids 73
Steroids 74
Proteins 74
Amino Acids 74
Polypeptide Chains 77
Protein Conformation 77
Levels of Protein Structure 78
Factors Determining Conformation 81
The Protein-Folding Problem 83
Nucleic Acids 83
The Functions of Nucleic Acids: An Overview 83
Nucleotides 84
Polynucleotides 86
The Double Helix: An Introduction 86
DNA and Proteins as Tape Measures of Evolution 87
Methods: Molecular Models and Computer Graphics 85

6 An Introduction to Metabolism 91

The Metabolic Map: An Overview 91
Energy: Some Basic Principles 92
Forms of Energy 92
Energy Transformations 92
Two Laws of Thermodynamics 93
The Free-Energy Concept: A Criterion for Spontaneous Change 95
Chemical Energy and Life: A Closer Look 96
ATP and Cellular Work 97
The Structure and Hydrolysis of ATP 98
How ATP Performs Work 99
The Regeneration of ATP 99
Metabolic Disequilibrium 99
Enzymes 100
Enzymes and the Lowering of Activation Energy 100
The Specificity of Enzymes 102
The Catalytic Cycle of Enzymes 102
Factors Affecting Enzyme Activity 104
The Control of Metabolism 106
Feedback Inhibition 106
Structural Order and Metabolism 107
Emergent Properties: A Reprise 108

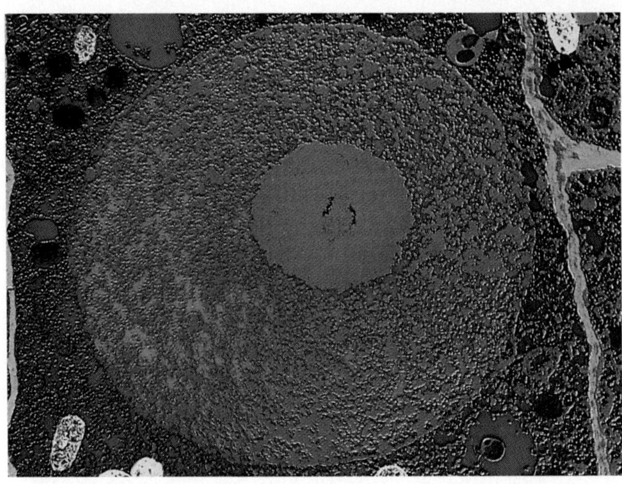

UNIT TWO

The Cell 112

Interview: Harold Varmus and Michael Bishop 112

7 A Tour of the Cell 116

How Cells Are Studied 117
Microscopy 117
Cell Fractionation 120
The Geography of the Cell: A Panoramic View 121
Prokaryotic and Eukaryotic Cells 121
Cell Size 121
The Importance of Compartmental Organization 122
The Nucleus 123
Ribosomes 127
The Endomembrane System 128
Endoplasmic Reticulum 128
The Golgi Apparatus 130
Lysosomes 131
Vacuoles 132
Relationships Among Endomembranes: A Summary 133
Peroxisomes (Microbodies) 133
Energy Transducers: Mitochondria and Chloroplasts 134
Mitochondria 135
Chloroplasts 135
The Cytoskeleton 136
Microtubules 137
Microfilaments and Movement 141
Intermediate Filaments 142
The Cell Surface 143
Cell Walls 143
The Glycocalyx of Animal Cells 144
Intercellular Junctions 144
The Cell: A Living Unit Greater Than the Sum of Its Parts 144

8 Membrane Structure and Function 151

Models of Membrane Structure 151
Two Generations of Membrane Models: A Case Study in the Scientific Process 152
The Fluid Mosaic Model: A Closer Look 155
The Traffic of Small Molecules 158
Selective Permeability 158
Diffusion and Passive Transport 159
Osmosis: A Special Case of Passive Transport 160
Facilitated Diffusion 163
Active Transport 164
The Special Case of Ion Transport 166
Cotransport 166
The Traffic of Large Molecules: Exocytosis and Endocytosis 167
Methods: Freeze-Fracture and Freeze-Etch 154

9 Cellular Respiration: Harvesting Chemical Energy 173

ATP and Cellular Work: A Review 174
Chemical Background: Respiration as an Oxidation-Reduction Process 174
An Introduction to Redox Reactions 174
Respiration: A Stepwise Redox Reaction 176
An Overview of Cellular Respiration 178
Glycolysis: A Closer Look 179
The Krebs Cycle: A Closer Look 182
The Electron Transport Chain and Oxidative Phosphorylation: A Closer Look 184
The Pathway of Electron Transport 184
Chemiosmosis: Coupling Electron Flow to ATP Synthesis 185
The Chemiosmotic Model: Incorporating Many Biological Themes 189
Summarizing Cellular Respiration 189
Fermentation: The Anaerobic Alternative 190
A Comparison of Aerobic and Anaerobic Catabolism 192
The Evolutionary Significance of Glycolysis 193
The Catabolism of Other Molecules 193
Biosynthesis 194
The Control of Respiration 194

10 Photosynthesis 199

Chloroplasts: Sites of Photosynthesis 200
How Plants Make Food: An Overview 201
The Splitting of Water 202
Photosynthesis as a Redox Process 203
The Two Stages of Photosynthesis 203
How the Light Reactions Capture Solar Energy: A Closer Look 204
The Nature of Sunlight 204
Photosynthetic Pigments 205
The Photooxidation of Chlorophyll 207
The Two Photosystems 208
Cyclic Electron Flow 209
Noncyclic Electron Flow 210
Review: A Comparison of Chemiosmosis in Chloroplasts and Mitochondria 211
How the Calvin Cycle Makes Sugar: A Closer Look 212
Photorespiration: An Evolutionary Relic? 214
C_4 Plants 215
CAM Plants 215
Summary: The Fate of Photosynthetic Products 216
Methods: Determining an Absorption Spectrum 206

11 The Reproduction of Cells 221

Bacterial Reproduction 222
An Introduction to Eukaryotic Chromosomes 223
The Cell Cycle: An Overview 224
The Mechanics of Cell Division: A Closer Look 225
The Structure and Function of the Mitotic Spindle 225
The Mechanisms of Cytokinesis 228
The Control of Cell Division 230
Abnormal Cell Division: Cancer Cells 235
Methods: Cell Culture 234

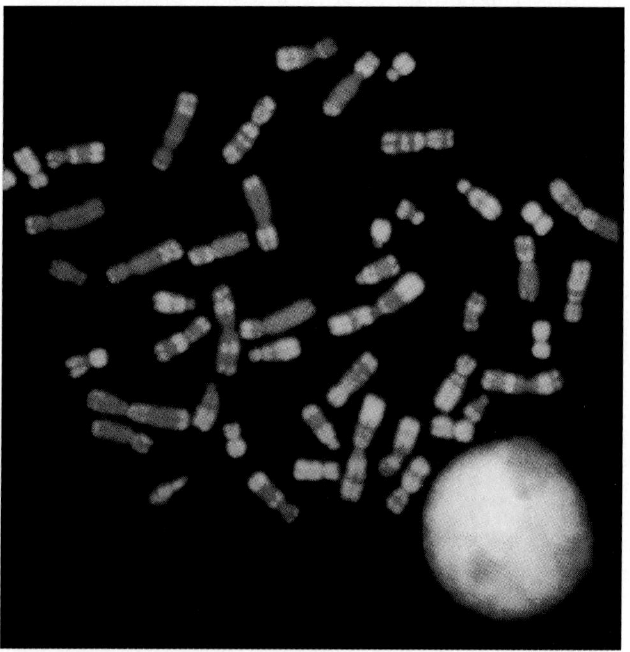

Unit Three

The Gene 240

Interview: David Suzuki 240

12 Meiosis and Sexual Life Cycles 244

Genes, DNA, and Chromosomes: A Brief Orientation 244

Sexual and Asexual Reproduction: A Comparison 245
An Introduction to Sexual Life Cycles: The Human Example 246
The Variety of Sexual Life Cycles 248
Meiosis: A Closer Look 250
A Comparison of Mitosis and Meiosis 253
Sexual Sources of Genetic Variation 253
Independent Assortment of Chromosomes 253
Crossing Over 254
Random Fertilization 255
Genetic Variation and Evolution 255
Methods: Preparation of a Karyotype 247

13 MENDEL AND THE GENE IDEA 258

Mendel's Model: A Case Study in the Scientific Process 258
Mendel's Experimental Approach 259
Mendel's Law of Segregation 260
Inheritance as a Game of Chance 263
Mendel's Law of Independent Assortment 265
Extending Mendelian Genetics 267
Incomplete Dominance 267
What Is a Dominant Allele? 268
Multiple Alleles 269
Pleiotropy 269
Epistasis 269
Polygenic Inheritance 270
Nature Versus Nurture: The Environmental Impact on Phenotype 270
Integrating a Mendelian View of Heredity and Variation 272
Mendelian Inheritance in Humans 272
Human Pedigrees 272
Recessively Inherited Disorders 273
Dominantly Inherited Disorders 274
Genetic Screening and Counseling 275
Multifactorial Disorders 277

14 THE CHROMOSOMAL BASIS OF INHERITANCE 280

The Chromosome Theory of Inheritance 280
Morgan and the *Drosophila* School 280
Linked Genes 283
The Chromosomal Basis of Recombination 284
The Recombination of Unlinked Genes: Independent Assortment 284
The Recombination of Linked Genes: Crossing Over 285
Genetic Maps Based on Crossover Data 285
Sex Chromosomes and Sex-Linked Inheritance: A Closer Look 288
The Chromosomal Basis of Sex in Humans 288
Sex-Linked Disorders in Humans 289
X-Inactivation in Females 290
Chromosomal Alterations 290
Alterations of Chromosome Number 291
Alterations of Chromosome Structure 292
Chromosomal Alterations in Human Disease 293
Parental Imprinting of Genes 295
Extranuclear Inheritance 296

15 THE MOLECULAR BASIS OF INHERITANCE 300

The Search for the Genetic Material: A Case Study in the Scientific Process 300
Evidence That DNA Can Transform Bacteria 301
Evidence That Viral DNA Can Program Cells 302
Additional Evidence That DNA Is the Genetic Material of Cells 302
The Discovery of the Double Helix 303
DNA Replication: The Basic Concept 306
A Closer Look at DNA Replication 308
Getting Started: Origins of Replication 309
Elongating a New DNA Strand 309
Proofreading 312
DNA Repair 313

16 FROM GENE TO PROTEIN 316

Evidence That Genes Specify Proteins 316
How Genes Control Metabolism 317
One Gene–One Polypeptide 317
An Overview of Protein Synthesis 317
The Genetic Code 319
The Evolutionary Significance of a Common Genetic Language 321
A Closer Look at Transcription 322
RNA Polymerase Binding and Initiation of Transcription 323
Elongation of the RNA Strand 324
Termination of Transcription 324
A Closer Look at Translation 325
Transfer RNA 325
Aminoacyl-tRNA Synthetases 326
Ribosomes 327
Building a Polypeptide 327
Polyribosomes 331
From Polypeptide to Functional Protein 331
Protein Targeting 331
Comparing Protein Synthesis in Prokaryotes and Eukaryotes: A Review 332
RNA Processing in Eukaryotes 333
Alteration of the Ends of mRNA 333
RNA Splicing 333
Mutations and Their Effects on Proteins 336
Types of Mutations 336
Mutagenesis 338
What Is a Gene? 338
Methods: The Ames Test 339

17 MICROBIAL MODELS: THE GENETICS OF VIRUSES AND BACTERIA 344

The Discovery of Viruses 344
Viral Structure and Replication: An Overview 345
Viral Genomes 345
Capsids and Envelopes 345
How Viruses Replicate: Viral Infection 346
Bacterial Viruses 348
The Lytic Cycle 348
The Lysogenic Cycle 349

Animal Viruses 350
Diverse Reproductive Cycles of Animal Viruses 350
Viral Diseases in Animals 352
Viruses and Cancer 353
Plant Viruses and Viroids 354
The Evolutionary Origin of Viruses 354
Bacterial Genomes: Replication and Mutation 355
Genetic Recombination and Gene Transfer in Bacteria 356
Transformation 357
Transduction 357
Conjugation and Plasmids 357
Transposons 361
The Control of Gene Expression in Prokaryotes 363
Operons: The Basic Concept 364
Repressible Versus Inducible Enzymes: Two Types of Negative Gene Regulation 366
CAP: An Example of Positive Gene Regulation 367

18 GENOME ORGANIZATION AND EXPRESSION IN EUKARYOTES 372
Genome Organization at the Microscopic Level 373
Nucleosomes, or "Beads on a String" 373
Higher Levels of DNA Packing 373
Genome Organization at the Molecular Level 375
Repetitive Sequences 375
Multigene Families 375
Organization of a Typical Eukaryotic Gene: A Review 376
The Arrangement of Coordinately Controlled Genes 376
Genome Plasticity 378
Gene Amplification and Selective Gene Loss 378
DNA Methylation 378
Rearrangements in the Genome 378
The Control of Gene Expression 379
Transcriptional Control of Gene Expression 380
Posttranscriptional Control of Gene Expression 381
The Role of Small Molecules in Regulating Eukaryotic Gene Expression 383
Chromosome Puffs: Evidence for the Regulatory Role of Steroid Hormones in Insects 383
The Action of Steroid Hormones in Vertebrates 383
Gene Expression and Cancer 384

19 DNA TECHNOLOGY 390
Basic Strategies of Gene Manipulation and Analysis 391
Recombinant DNA and Gene Cloning 391
DNA Synthesis and Sequencing 399
Amplifying DNA by the Polymerase Chain Reaction (PCR) 399
RFLP Analysis 402
Applications of DNA Technology 404
Uses of DNA Technology in Basic Research 404
The Human Genome Project 404
Medical Uses of DNA Technology 405
Forensic Uses of DNA Technology 409
Agricultural Uses of DNA Technology 410
Safety and Ethical Issues 412
Methods: Gel Electrophoresis of Macromolecules 396
Methods: Sequencing of DNA by the Sanger Method 400
Methods: The Polymerase Chain Reaction (PCR) 401
Methods: RFLP Analysis 402
Methods: Chromosome Walking 406

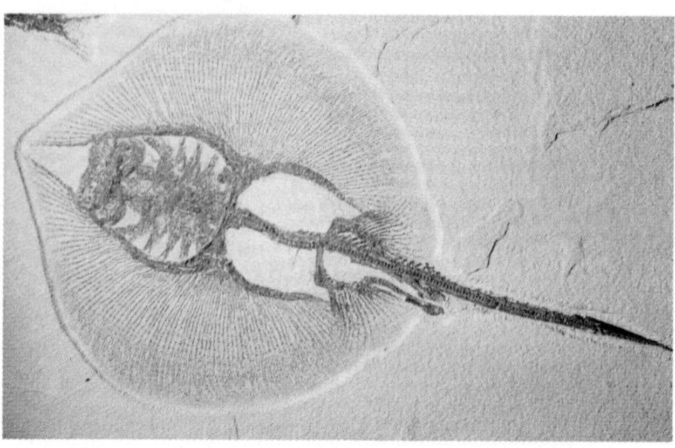

UNIT FOUR
MECHANISMS OF EVOLUTION 416
Interview: Ernst Mayr 416

20 DESCENT WITH MODIFICATION: A DARWINIAN VIEW OF LIFE 420
Pre-Darwinian Views 420
The Scale of Life and Natural Theology 422
Cuvier, Fossils, and Catastrophism 422
Gradualism in Geology 423
Lamarck's Theory of Evolution 424
On the Origin of Darwinism 424
The Voyage of the *Beagle* 424
Darwin Frames His View of Life 426
The Dual Meaning of Darwinism 427
Common Descent 427
Natural Selection and Adaptation 427
The Signs of Evolution 431
Biogeography 431
The Fossil Record 432
Taxonomy 432
Comparative Anatomy 433
Comparative Embryology 434
Molecular Biology 434
Just a Theory? 435

21 HOW POPULATIONS EVOLVE 438
The Modern Evolutionary Synthesis 438
The Genetics of Populations 439
The Gene Pool and Microevolution 439
The Hardy-Weinberg Theorem 440

Causes of Microevolution 442
Genetic Drift 442
Gene Flow 444
Mutation 444
Nonrandom Mating 444
Natural Selection 445
The Genetic Basis of Variation 445
The Nature and Extent of Genetic Variation Within and Between Populations 445
Sources of Genetic Variation 447
How Genetic Variation Is Preserved 448
Is All Genetic Variation Adaptive? 450
Adaptive Evolution 450
Fitness 450
What Does Selection Act On? 451
Modes of Natural Selection 451
Sexual Selection 452
Does Evolution Fashion Perfect Organisms? 453

22 THE ORIGIN OF SPECIES 456

The Species Problem 456
Two Concepts of Species 457
Limitations of the Biological Species Concept 458
Reproductive Barriers 459
Prezygotic Barriers 460
Postzygotic Barriers 461
Introgression 461
The Biogeography of Speciation 462
Allopatric Speciation 462
Sympatric Speciation 465
Genetic Mechanisms of Speciation 467
Speciation by Divergence 467
Speciation by Peak Shifts 468
How Much Genetic Change Is Required for Speciation? 469
Gradual and Punctuated Interpretations of Speciation 469

23 TRACING PHYLOGENY: MACROEVOLUTION, THE FOSSIL RECORD, AND SYSTEMATICS 474

The Record of the Rocks 475
How Fossils Form 475
Limitations of the Fossil Record 475
The Fossil Record and the Geological Time Scale 477
Mechanisms of Macroevolution 478
The Origins of Evolutionary Novelties 480
The Difficulty of Interpreting Evolutionary Trends 482
Continental Drift and the Biogeography of Macroevolution 484
Punctuations in the History of Biological Diversity 486
Systematics: Tracing Phylogeny 489
Taxonomy 489
Sorting Homology from Analogy 491
Molecular Systematics 492
Schools of Taxonomy 495
Is a New Evolutionary Synthesis Necessary? 496
Methods: Radioactive Dating 479

UNIT FIVE

THE EVOLUTIONARY HISTORY OF BIOLOGICAL DIVERSITY 500

Interview: Stephen Jay Gould 500

24 EARLY EARTH AND THE ORIGIN OF LIFE 504

The Antiquity of Life 505
The Origin of Life 506
Abiotic Synthesis of Organic Monomers 507
Abiotic Synthesis of Polymers 508
The Formation of Protobionts 508
The Origin of Genetic Information 509
Some Alternative Views 511
The Kingdoms of Life 512

25 PROKARYOTES AND THE ORIGINS OF METABOLIC DIVERSITY 515

Prokaryotic Form and Function 516
The Morphology of Prokaryotes 516
The Cell Surface 516
The Motility of Prokaryotes 518
Internal Membranous Organization 519
The Prokaryotic Genome 519
Growth, Reproduction, and Gene Exchange 520
Metabolic Diversity 521
The Diversity of Prokaryotes 522
Archaebacteria 523
Eubacteria 524
The Origins of Metabolic Diversity 524

The Origin of Glycolysis 524
The Origin of Electron Transport Chains and Chemiosmosis 524
The Origin of Photosynthesis 525
Cyanobacteria, the Oxygen Revolution, and the Origins of Cellular Respiration 525
The Importance of Prokaryotes 528
Prokaryotes and Chemical Cycles 528
Symbiotic Bacteria 528
Bacteria and Disease 529
Putting Bacteria to Work 530
Methods: The Gram Stain 517

26 PROTISTS AND THE ORIGIN OF EUKARYOTES 533

Characteristics of Protists 534
The Origin of Eukaryotes 534
The Antiquity of Eukaryotes 535
Models of Eukaryotic Origins 535
Boundaries of the Kingdom Protista 536
Protozoa 537
Rhizopoda 537
Actinopoda 538
Foraminifera 539
Apicomplexa 539
Zoomastigophora 539
Ciliophora 541
Algae 542
Dinoflagellata 543
Chrysophyta 544
Bacillariophyta 545
Euglenophyta 545
Chlorophyta 545
Evolutionary Adaptations of Seaweed 548
Phaeophyta 549
Rhodophyta 550
Evolutionary Relationships Among Algae 551
Funguslike Protists 551
Myxomycota 552
Acrasiomycota 553
Oomycota 553
Chytridiomycota 554
The Origins of Multicellularity 554

27 PLANTS AND THE COLONIZATION OF LAND 559

An Introduction to the Plant Kingdom 559
General Characteristics of Plants 559
A Generalized View of Plant Reproduction and Life Cycles 560
Some Highlights of Plant Evolution 560
Classification of Plants 561
The Move onto Land 562
The Case for Green Algae as the Ancestors of Plants 562
Bryophytes: Divisions Bryophyta, Hepatophyta, and Anthocerophyta 562
Terrestrial Adaptations of Vascular Plants 563
The Earliest Vascular Plants 566
Seedless Vascular Plants 566
Division Psilophyta 566
Division Lycophyta 567
Division Sphenophyta 567
Division Pterophyta 568
The Coal Forests 569
Terrestrial Adaptations of Seed Plants 570
Gymnosperms 571
Division Coniferophyta 571
The History of Gymnosperms 572
Angiosperms 574
The Flower 575
The Fruit 575
The Life Cycle of an Angiosperm 576
The Rise of Angiosperms 577
Relationships Between Angiosperms and Animals 577
The Value of Plant Diversity 579

28 FUNGI 583

Characteristics of Fungi 583
Nutrition and Habitats 583
Structure 584
Growth and Reproduction 585
The Diversity of Fungi 586
Division Zygomycota 586
Division Ascomycota 587
Division Basidiomycota 589
Some Unique Lifestyles of Fungi 590
Molds 591
Yeasts 591
Lichens 592
Mycorrhizae 593
The Ecological Importance of Fungi 593
Fungi as Decomposers 593
Fungi as Spoilers 594
Pathogenic Fungi 594
Edible Fungi 594
The Evolution of Fungi 595

29 INVERTEBRATES AND THE ORIGIN OF ANIMAL DIVERSITY 598

Defining "Animals" 598
Clues to Animal Phylogeny 599
Major Branches of the Animal Kingdom 599
Development and Body Plan 600
The Protostome–Deuterostome Dichotomy 601
Parazoa (Phylum Porifera) 603
Eumetazoa: Radiata 604
Phylum Cnidaria 604
Phylum Ctenophora 606
Bilateria: Acoelomates 608
Phylum Platyhelminthes 608
Phylum Nemertea 609

Pseudocoelomates 610
Phylum Rotifera 610
Phylum Nematoda 610
Protostomes (Schizocoelomates) 611
Phylum Mollusca 611
Phylum Annelida 615
Phylum Arthropoda 617
Deuterostomes (Enterocoelomates) 626
The Lophophorate Animals 626
Phylum Echinodermata 627
Phylum Chordata 629
The Origin and Diversification of Animals 629

30 THE VERTEBRATE GENEALOGY 635
Phylum Chordata 635
Chordate Characteristics 635
Chordates Without Backbones 636
Subphylum Cephalochordata 636
Subphylum Urochordata 637
The Origin of Vertebrates 637
Vertebrate Characteristics 638
Class Agnatha 640
Class Placodermi 641
Class Chondrichthyes 641
Class Osteichthyes 643
Class Amphibia 645
Early Amphibians 645
Modern Amphibians 646
Class Reptilia 648
Reptilian Characteristics 648
The Age of Reptiles 648
Reptiles of Today 650
Class Aves 651
Characteristics of Birds 651
The Origin of Birds 653
Modern Birds 653
Class Mammalia 653
Mammalian Characteristics 653
The Evolution of Mammals 655
Monotremes 656
Marsupials 656
Placental Mammals 656
The Human Ancestry 657
Evolutionary Trends in Primates 657
Modern Primates 657
The Emergence of Humankind 660

UNIT SIX
PLANTS: FORM AND FUNCTION 670
Interview: Virginia Walbot 670

31 PLANT STRUCTURE AND GROWTH 674
An Introduction to Plant Biology: The Major Themes 674
Basic Morphology of Flowering Plants: An Evolutionary Perspective 676
The Root System 676
The Shoot System 679
Basic Plant Anatomy: Plant Cells and Tissues 681
Types of Plant Cells 682
The Three Tissue Systems of a Plant 685
An Overview of Plant Growth 686
A Closer Look at Primary Growth 687
Primary Growth of Roots 687
Primary Growth of Shoots 689
A Closer Look at Secondary Growth 693
Secondary Growth of Stems 693
Secondary Growth of Roots 695

32 TRANSPORT IN PLANTS 699
Basic Transport Processes in Plants: An Overview 699
Transport at the Cellular Level 699
Short-Distance (Lateral) Transport at the Level of Tissues and Organs 704
Long-Distance Transport at the Whole-Plant Level 704
The Absorption of Water and Minerals by Roots: A Closer Look 704
The Ascent of Xylem Sap: A Closer Look 706
Pushing Xylem Sap: Root Pressure 706
Pulling Xylem Sap: The Transpiration-Cohesion-Tension Mechanism 706
The Long-Distance Transport of Water in Plants by Solar-Powered Bulk Flow 708

The Control of Transpiration: A Closer Look 708
The Photosynthesis-Transpiration Compromise 708
How Stomata Open and Close 709
Evolutionary Adaptations That Reduce Transpiration 710
Transport in Phloem: A Closer Look 712
Source-to-Sink Transport 712
Phloem Loading and Unloading 713
Pressure Flow (Bulk Flow) of Phloem Sap 714
Methods: Patch Clamp Recording 711

33 PLANT NUTRITION 718

Nutritional Requirements of Plants 718
The Chemical Composition of Plants 718
Essential Nutrients 719
Mineral Deficiencies 720
Soil 722
Texture and Composition of Soils 722
The Availability of Soil Water and Minerals 723
Soil Management 724
Nitrogen Assimilation by Plants 726
Nitrogen Fixation 726
Symbiotic Nitrogen Fixation 727
Improving the Protein Yield of Crops 728
Some Nutritional Adaptations of Plants 729
Parasitic Plants 729
Carnivorous Plants 730
Mycorrhizae 731
Methods: Hydroponic Culture to Identify Essential Nutrients 720

34 PLANT REPRODUCTION AND DEVELOPMENT 734

Sexual Reproduction of Flowering Plants 734
The Angiosperm Life Cycle: An Overview 735
More About Flowers 736
Pollen Development: A Closer Look 738
The Development of Ovules: A Closer Look 738
Pollination and Fertilization 739
The Seed 740
Development of the Fruit 741
Seed Germination 742
Asexual Reproduction 744
Natural Mechanisms of Vegetative Reproduction 744
Vegetative Propagation in Agriculture 746
Comparing Sexual and Asexual Plant Reproduction: An Evolutionary Perspective 748
Some Cellular Aspects of Plant Development 748
An Overview of Plant Development 748
Cellular Differentiation: The Basic Problem 752
Pattern Formation and Positional Information 752
Clonal Analysis of the Shoot Apex 752
Homeotic Mutations in Flower Development: In Search of the Genetic Basis of Pattern Formation 753

35 CONTROL SYSTEMS IN PLANTS 756

The Search for a Plant Hormone: A Case Study in the Scientific Process 757
Functions of Plant Hormones 759
Auxin 759
Cytokinins 761
Gibberellins 762
Abscisic Acid 764
Ethylene 764
Plant Movements 766
Tropisms 766
Turgor Movements 767
Circadian Rhythms and the Biological Clock 769
Photoperiodism 769
Photoperiodic Control of Flowering 769
Phytochrome 771
The Role of the Biological Clock in Photoperiodism 773
Signal-Transduction Pathways in Plant Cells 773
Methods: Bioassay 763

UNIT SEVEN

ANIMALS: FORM AND FUNCTION 778

Interview: Karel Liem 778

36 AN INTRODUCTION TO ANIMAL STRUCTURE AND FUNCTION 782

Levels of Structural Organization 782
Animal Tissues 783
Organs and Organ Systems 787
Size, Shape, and the External Environment 787
The Animal's Internal Environment 789

37 ANIMAL NUTRITION 794

Food Types and Feeding Mechanisms 794
Digestion: A Comparative Introduction 795
Enzymatic Hydrolysis 796

Intracellular Digestion in Food Vacuoles 796
Digestion in Gastrovascular Cavities 797
Digestion in Alimentary Canals 798
The Mammalian Digestive System 799
The Oral Cavity 799
The Pharynx 800
The Esophagus 800
The Stomach 801
The Small Intestine 802
The Large Intestine 806
Evolutionary Adaptations of Vertebrate Digestive Systems 807
Nutritional Requirements 809
Food as Fuel 809
Food for Fabrication 811
Essential Nutrients 811
Methods: Measuring Metabolic Rate 810

38 CIRCULATION AND GAS EXCHANGE 818

Internal Transport in Invertebrates 819
Gastrovascular Cavities 819
Open and Closed Circulatory Systems 819
Circulation in Vertebrates 820
Vertebrate Circulatory Schemes: An Evolutionary Perspective 821
The Heart 822
Blood Flow 825
Capillary Exchange 828
The Lymphatic System 829
Mammalian Blood 830
Plasma 830
Blood Cells 830
Blood Clotting 832
Cardiovascular Disease 832
Gas Exchange in Animals 835
General Problems of Gas Exchange 835
Respiratory Organs: General Structure and Function 835
Gills: Respiratory Adaptations of Aquatic Animals 836
Tracheae: Respiratory Adaptations of Insects 838
Lungs: Respiratory Adaptations of Terrestrial Vertebrates 838
Methods: Measurement of Blood Pressure 827

39 THE BODY'S DEFENSES 850

Nonspecific Defense Mechanisms 850
The Skin and Mucous Membranes 851
Phagocytic White Cells and Natural Killer Cells 851
Antimicrobial Proteins 852
The Inflammatory Response 852
The Immune System and Specific Defense: Some Basic Concepts 854
Basic Features of the Immune System 854
Active Versus Passive Acquired Immunity 855
The Immune System's Dual Defense: An Overview of Humoral and Cell-Mediated Responses 855
The Molecular Basis of Antigen-Antibody Specificity 856
Clonal Selection: The Cellular Basis of Immunological Specificity and Diversity 858
The Cellular Basis of Immunological Memory 859
The Development of Self-Tolerance 860
The Humoral Response: A Closer Look 860
The Activation of B Cells 860
How Antibodies Work 862
Monoclonal Antibodies 863
The Cell-Mediated Response: A Closer Look 863
The Complement System: A Component of Both Nonspecific and Specific Defense 865
Self Versus Nonself: Some Applications 867
Blood Groups 868
Tissue Grafts and Organ Transplants 868
Disorders of the Immune System 869
Autoimmune Diseases 869
Allergy 869
Immunodeficiency 869
Acquired Immunodeficiency Syndrome (AIDS) 870
Immunity in Invertebrates 872
Methods: Production of Monoclonal Antibodies with Hybridomas 864

40 CONTROLLING THE INTERNAL ENVIRONMENT 876

Osmoregulation 877
Osmoconformers and Osmoregulators 877
Problems of Osmoregulation in Different Environments 878
Transport Epithelia and Osmoregulation 880
Excretory Systems of Invertebrates 881
Protonephridia: The Flame-Cell System of Flatworms 881
Metanephridia of Earthworms 882
Malpighian Tubules of Insects 882
The Vertebrate Kidney 883
Anatomy of the Excretory System 883
Structure of the Nephron 885
General Physiology of the Nephron 885
Transport Properties of the Renal Tubule 887
How the Mammalian Kidney Conserves Water: A Closer Look 889
Regulation of the Kidneys 891
Comparative Physiology of the Kidney 893
Nitrogenous Wastes 893
Ammonia 894
Urea 894
Uric Acid 894
The Regulation of Body Temperature 894
Heat Production and Transfer Between Organisms and Their Environment 895
Ectotherms and Endotherms 895
Thermoregulation in Terrestrial Mammals 896
Thermoregulatory Adaptations in Other Animals 897
Temperature Acclimation 902
Torpor 902
The Interaction of Regulatory Systems 903

41 Chemical Signals in Animals 907

An Overview of Chemical Signals 907
- Hormones 908
- Pheromones 909
- Local Regulators 909

Mechanisms of Hormone Action at the Cellular Level 910
- Steroid Hormones and Gene Expression 911
- Peptide Hormones and Second Messengers 912

Invertebrate Hormones 915

The Vertebrate Endocrine System 917
- The Hypothalamus and the Pituitary Gland 917
- The Thyroid Gland 922
- The Parathyroid Glands 922
- The Pancreas 923
- The Adrenal Glands 925
- The Gonads 926
- Other Endocrine Organs 926

Endocrine Glands and the Nervous System 927

42 Animal Reproduction 931

Modes of Reproduction 931
- Asexual Reproduction 931
- Sexual Reproduction 932
- Reproductive Cycles and Patterns 932

Mechanisms of Sexual Reproduction 934
- Patterns of Fertilization and Development 934
- Diversity in Reproductive Systems 935

Mammalian Reproduction 936
- The Anatomy of Reproduction in Humans 937
- Hormonal Control of Mammalian Reproduction 940
- Gamete Formation (Gametogenesis) 945
- Sexual Maturation 947
- Human Sexual Physiology 947
- Conception, Pregnancy, and Birth 947
- Reproductive Technology 953

43 Animal Development 956

An Overview of Developmental Processes 956

Fertilization 957
- The Acrosomal Reaction 957
- The Cortical Reaction 958
- Activation of the Egg 958

Early Stages of Embryonic Development 960
- Cleavage 960
- Gastrulation 961
- Organogenesis 963

The Embryology of Amniotes 965
- Avian Development 965
- Mammalian Development 967

Mechanisms of Development 968
- Polarity of the Embryo 968
- Localized Cytoplasmic Determinants 970
- Mosaic and Regulative Development 971
- Fate Maps and the Analysis of Cell Lineages 971
- Morphogenetic Movements 971
- Induction 974
- Differentiation 974
- Pattern Formation 976

44 Nervous Systems 982

Cells of the Nervous System 983
- Neurons 983
- Supporting Cells 985

Transmission of Electrical Signals Along a Neuron 986
- The Origin of Electrical Membrane Potential 986
- The Action Potential 989
- Propagation of the Action Potential 991
- Speed of Propagation of the Action Potential 992

The Synapse: Transmission Between Cells 993
- Electrical Synapses 993
- Chemical Synapses 994
- Summation: Neural Integration at the Cellular Level 995
- Neurotransmitters and Receptors 997
- Neural Circuits and Clusters 999

Invertebrate Nervous Systems 999

The Vertebrate Nervous System 1001
- The Peripheral Nervous System 1002
- The Central Nervous System 1002
- Integration and Higher Brain Functions 1008

Methods: Measuring Membrane Potentials 987

45 Sensory and Motor Mechanisms 1015

Sensory Receptors 1015
- Sensation and Perception 1015
- The General Function of Sensory Receptors 1016
- Types of Receptors 1017

Vision 1019
- Light Receptors and Vision of Invertebrates 1019
- Vertebrate Vision 1020

Hearing and Balance 1027
- The Mammalian Ear 1027
- Hearing and Equilibrium in Other Vertebrates 1030
- Sensory Organs for Hearing and Equilibrium in Invertebrates 1031

Taste and Smell 1032

An Introduction to Animal Movement 1034

Skeletons and Their Roles in Movement 1035
- Hydrostatic Skeletons 1035
- Exoskeletons 1036
- Endoskeletons 1037

Muscles 1037
- The Structure and Physiology of Vertebrate Skeletal Muscle 1038
- Other Types of Muscle 1043

UNIT EIGHT

ECOLOGY 1048

Interview: Ariel Lugo 1048

46 AN INTRODUCTION TO ECOLOGY: DISTRIBUTION AND ADAPTATIONS OF ORGANISMS 1052

The Scope and Development of Ecology 1053
- The Questions of Ecology 1053
- Ecology as an Experimental Science 1053
- Ecology and Evolution 1054

Terrestrial Biomes 1054
- Tropical Forest 1056
- Savanna 1057
- Desert 1058
- Chaparral 1059
- Temperate Grassland 1059
- Temperate Deciduous Forest 1060
- Taiga 1061
- Tundra 1061

Freshwater Biomes 1062
- Ponds and Lakes 1063
- Streams and Rivers 1064

Marine Biomes 1066
- Estuaries 1066
- The Intertidal Zones 1067
- Coral Reefs 1067
- The Oceanic Pelagic Biome 1068
- Benthos 1068

The Environmental Diversity of the Biosphere 1069
- Important Abiotic Factors 1070
- Climate and the Distribution of Biomes 1070
- Global Climate Patterns 1071
- Local and Seasonal Effects on Climate 1072

Responses of Organisms to Environmental Variation 1075
- Homeostasis and the Principle of Allocation 1076
- Environmental Grain 1077
- Behavioral Responses 1078
- Physiological Responses 1078
- Morphological Responses 1079
- Adaptation Over Evolutionary Time 1079

47 POPULATION ECOLOGY 1083

Density and Dispersion 1084
- Measuring Density 1084
- Patterns of Dispersion 1085

Demography 1087
- Age Structure and Sex Ratio 1087
- Life Tables and Survivorship Curves 1087

The Evolution of Life Histories 1089

Models of Population Growth 1092
- Exponential Population Growth 1092
- Logistic Population Growth 1094

The Regulation of Population Size 1097
- Density-Dependent Factors 1097
- Density-Independent Factors 1099
- The Interaction of Regulating Factors 1100
- Population Cycles 1100

Human Population Growth 1101

Methods: A Mark-Recapture Estimate of Population Size 1085

48 COMMUNITY ECOLOGY 1106

Two General Views of Communities 1106

Population Interactions 1108
- Coevolution 1108
- Predation 1109
- Interspecific Competition 1112
- Symbiotic Interactions 1116

Community Structure: Patterns and Processes 1118
- Community Characteristics 1118
- Determinants of Community Characteristics 1119

Succession 1121
- Causes of Succession 1121
- Human Disturbance 1123
- Community Equilibrium and Species Diversity 1124
- Diversity and Community Stability 1124

Biogeographical Aspects of Diversity 1124
- Limits of Species Ranges 1125
- Global Clines in Species Diversity 1125
- Island Biogeography 1126

Conservation Biology: Lessons from Community Ecology and Biogeography 1128

49 ECOSYSTEMS 1132

Trophic Levels and Food Webs 1132

Energy Flow 1133
- The Global Energy Budget 1134
- Primary Productivity 1134
- Energy Transfers and Ecological Pyramids 1138

Chemical Cycling 1140
- The Carbon Cycle 1142
- The Nitrogen Cycle 1143
- The Phosphorus Cycle 1144
- Variations in Nutrient Cycling Time 1145

Human Intrusions in Ecosystem Dynamics 1145
- Habitat Destruction and the Biodiversity Crisis 1146
- Effects of Deforestation on Chemical Cycling: The Hubbard Brook Forest Study 1147

Effects of Deforestation on Chemical Cycling: The Hubbard Brook Forest Study 1147
Agricultural Effects on Nutrient Cycling 1148
Accelerated Eutrophication of Lakes 1149
Poisons in Food Chains 1149
Intrusions in the Atmosphere 1151
Biotic Effects at the Biosphere Level 1153
Methods: Measuring Gross Primary Productivity in Aquatic Habitats 1138

50 BEHAVIOR 1158
The Evolutionary Logic of Behavioral Ecology 1159
Proximate Versus Ultimate Causation 1160
Innate Components of Behavior 1162
Fixed-Action Patterns 1162
The Nature of Sign Stimuli 1165
Learning and Behavior 1165
Nature Versus Nurture 1165
Learning Versus Maturation 1166
Habituation 1166
Imprinting 1166
Song Development in Birds: A Study of Imprinting 1168
Classical Conditioning 1169
Operant Conditioning 1169
Observational Learning 1169
Play 1169
Insight 1170
The Ultimate and Proximate Bases of Learning 1170
Behavioral Rhythms 1170
Movement 1171
Foraging Behavior 1174
Social Interactions 1176
Agonistic Behavior 1176
Dominance Hierarchies 1177
Territoriality 1177
Mating Behavior 1178
Courtship 1178
Mating Systems 1180
Communication 1180
Selfish Versus Altruistic Behavior 1183
Animal Cognition 1185
Human Sociobiology 1186
Methods: Using Molecular Biology to Answer Behavioral Questions 1181

Appendix One Self-Quiz Answers A-1

Appendix Two Classification of Life A-4

Appendix Three The Metric System A-6

Appendix Four Concept Mapping: A Technique for Learning A-8

Glossary G-1

Photograph Credits C-1

Illustration Credits C-6

Index I-1

BIOLOGY

1 Introduction: Themes in the Study of Life

A Hierarchy of Organization
Emergent Properties
The Cellular Basis of Life
Heritable Information
A Feeling for Organisms
The Correlation Between Structure and Function
The Interaction of Organisms with Their Environment
Unity in Diversity
Evolution: The Core Theme of Biology
Scientific Process: The Hypothetico-Deductive Method
Science, Technology, and Society

Biology, the study of life, is rooted in the human spirit. People keep pets, nurture houseplants, invite avian visitors with backyard birdhouses, and visit zoos and nature parks. This behavior expresses what Harvard biologist E. O. Wilson calls *biophilia,* an innate attraction to life in its diverse forms. Biology is the scientific extension of this human tendency to feel connected to and curious about all forms of life. It is a science for adventurous minds. It takes us, personally or vicariously, into jungles, deserts, seas, and other environments, where a variety of living forms and their physical surroundings are interwoven into complex webs called ecosystems. Studying life leads us into laboratories to examine more closely how living things, which biologists call organisms, work. Biology draws us into the microscopic world of the fundamental units of life known as cells, and into the submicroscopic realm of the molecules that make up those cells. Our intellectual journey also takes us back in time, for biology encompasses not only contemporary life, but also a history of ancestral forms stretching nearly four billion years into the past. The scope of biology is immense. The purpose of this book is to introduce you to this multifaceted science (Figure 1.1). It is unlikely that any biology course could or should cover all 50 chapters, but this textbook is designed to help you succeed in your general biology course *and* to serve as a durable reference in your continuing education.

You are becoming involved with biology during its most exciting era. Using fresh approaches and new research methods, biologists are beginning to unravel some of life's most engaging mysteries. Though stimulating, the information explosion in biology is also intimidating. Most of the biologists who have ever lived are alive today, and they add about 450,000 new research articles to the scientific literature annually. Each of biology's many subfields changes continuously, and it is very difficult for a professional biologist to remain current in more than one narrowly defined specialty. How, then, can beginning biology students hope to keep their heads above water in this deluge of data and discovery? The key is to recognize unifying themes that pervade all of biology—themes that will still apply decades from now, when much of the specific information presented in any textbook will be obsolete. This chapter enunciates some of the broad, enduring themes in the study of life.

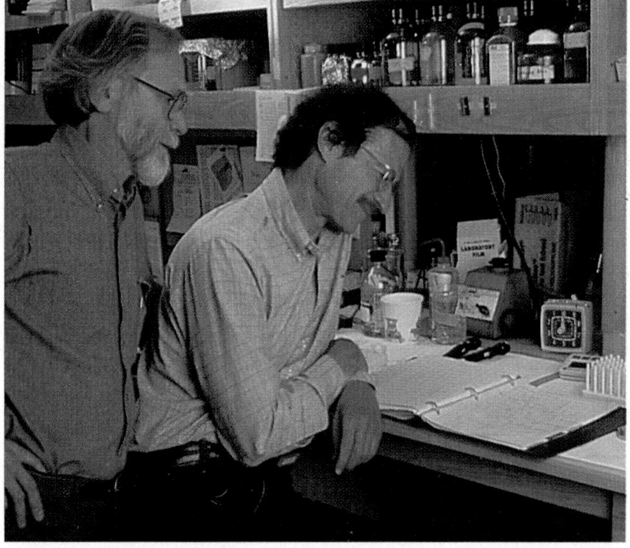
(a)

(b)

(c)

(d)

Figure 1.1
Biologists study life on many different scales of size and time. (**a**) Exploring life at the cellular and molecular levels, Nobel Prize winners Michael Bishop (left) and Harold Varmus have made important discoveries in the biology of cancer. (**b**) Knowledge of whole organisms is fundamental to good biology. Virginia Walbot, who uses corn as an experimental organism to study inheritance, is a keen observer of even the most subtle variations among the plants growing in her research field. (**c**) Biologists also study life at the levels of biological communities and ecosystems. Ariel Lugo (right) and William McDowell are investigating how vegetation in a Puerto Rican rain forest obtains mineral nutrients from clouds. (**d**) Biologists also trace the history of life over the immense scale of geologic time. Paleontologist Stephen Jay Gould is examining animal fossils that are hundreds of millions of years old. You will learn more about the diverse research of these and other scientists in the interviews that precede each unit of chapters in the text.

A HIERARCHY OF ORGANIZATION

A basic characteristic of life is a high degree of order. You can see it in the intricate pattern of veins throughout a leaf or in the colorful pattern of a bird's plumage. If you were to scrutinize the vein of a leaf or the feather of a bird under a microscope, you would discover that biological order also exists at levels below what the unaided eye can see.

Biological organization is based on a hierarchy of structural levels, each level building on the levels below it. Atoms, the chemical building blocks of all matter, are ordered into complex biological molecules such as proteins. The molecules of life are arranged into minute structures called organelles, which are in turn the components of cells. Some organisms consist of single cells, but others, including plants and animals, are aggregates of many specialized types of cells. In such multicellular organisms, similar cells are grouped into tissues, and specific arrangements of different tissues form organs. For example, the nervous impulses that coordinate your movements are transmitted along specialized cells called neurons. The nervous tissue within your brain has billions of neurons organized into a communications network of spectacular complexity. The brain, however, is not pure nervous tissue; it is an organ built of many different tissues, including a type called connective tissue that forms the protective covering of the brain. The brain is itself part of the nervous system, which also includes the spinal cord and the many nerves that transmit messages between the spinal cord and other parts of the body. The nervous system is only one of several organ systems characteristic of humans and other complex animals. Another example of this parts-within-parts anatomy of life is illustrated in Figure 1.2, which shows the levels of structural organization in an aspen grove.

In the hierarchy of biological organization, there are tiers beyond the individual organism. A population is a localized group of organisms belonging to the same species; populations of species living in the same area make up a biological community; and community interactions that include nonliving features of the environment, such as soil and water, form an ecosystem.

Unfolding biological organization at its many levels, from molecular architecture to ecosystem structure, is fundamental to the study of life. This text essentially follows such an organization, beginning by looking at the chemistry of life and ending with the study of ecosystems and the biosphere, the sum of all Earth's ecosystems. However, we will also see that biological processes transcend this hierarchy, with causes and effects at several organizational levels. For example, when a rattlesnake explodes from its coiled posture and strikes a desert mouse, the predator's coordinated movements result from complex interactions at the molecular, cellular, tissue, and organ levels within the snake. But there are also causes and effects of this behavior that operate on the level of the biological community where the snake and its prey live; the feeding response is triggered when the snake senses and locates the nearby mouse. And such episodes of predation have an important cumulative impact on the population sizes of both the mice and the rattlesnakes. Most biologists specialize in the study of life at a particular level, but they gain broader perspective when they connect their discoveries to processes occurring at lower or higher levels. A narrow focus on a single level of biological organization depreciates the fun and power of biology.

EMERGENT PROPERTIES

With each step upward in the hierarchy of biological order, novel properties emerge that were not present at the simpler levels of organization. These emergent properties result from interactions between components. A molecule such as a protein has attributes not exhibited by any of its component atoms, and a cell is certainly much more than a bag of molecules. If the intricate organization of the human brain is disrupted by a head injury, that organ ceases to function properly even though all its parts may still be present. And an organism is a living whole greater than the sum of its parts.

The theme of emergent properties may seem, at first, to support a doctrine known as vitalism, which views life as a supernatural phenomenon beyond the bounds of physical and chemical laws. However, the concept of emergent properties merely accents the importance of structural arrangement and applies to inanimate material as well as to life. Neither the head nor the handle of a hammer alone is very useful for driving nails, but put these parts together in a certain way, and the functional properties of a hammer emerge. Diamonds and graphite have different properties because their carbon atoms are arranged differently. Unique properties of organized matter arise from how parts are arranged and interact, not from supernatural powers. And life is driven not by "vital forces" that defy explanation, but by principles of physics and chemistry extended into a new territory. The emergent properties of life simply reflect a hierarchy of structural organization without counterpart among inanimate objects.

Life resists a simple, one-sentence definition because it is associated with numerous emergent properties. Yet almost any child perceives that a dog or a bug or a tree is alive and a rock is not. We can recognize life without defining it, and we recognize life by what living things do. Figure 1.3 on p. 6 illustrates and describes some of the properties and processes we associate with the state of being alive.

Because the properties of life emerge from complex organization, scientists seeking to understand biological processes confront a dilemma. One horn of the dilemma is that we cannot fully explain a higher level of order by breaking it down into its parts. A dissected animal no longer functions; a cell dismantled to its chemical ingredients is no longer a cell. According to a principle known as holism, disrupting a living system interferes with the meaningful explanation of its processes. The other horn of the dilemma is the futility of trying to analyze something as complex as an organism or a cell without taking it apart. Reductionism—reducing complex systems to simpler components that are more manageable to study—is a powerful strategy in biology. For example, by studying the molecular structure of a substance called DNA that had been extracted from cells, James Watson and Francis Crick deduced, in 1953, how this molecule could serve as the chemical basis of inheritance. The central role of DNA was better understood, however, when it was possible to study its interactions with

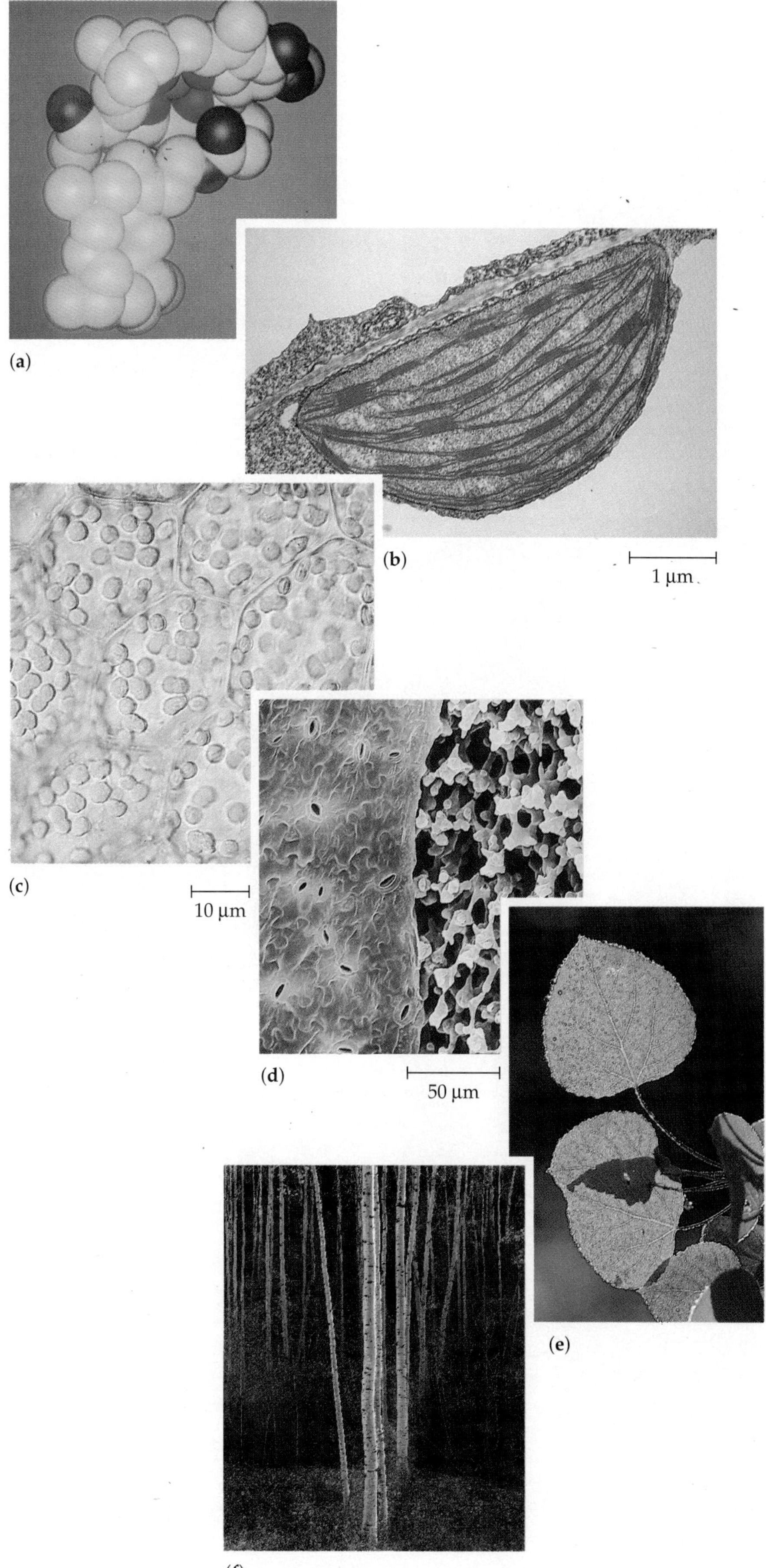

Figure 1.2
The hierarchy of biological organization. This sequence of images takes us all the way from atoms to a biological community of many interacting species. (**a**) Chlorophyll, represented here by a computer graphic, is a molecule built from many atoms. This molecule in the leaves of plants absorbs sunlight as a source of energy for driving photosynthesis, the manufacture of food in the leaf. (**b**) The process of photosynthesis requires participation of many other molecules organized within the cellular organelle called the chloroplast (TEM, transmission electron micrograph). (**c**) Many organelles cooperate in the functioning of the living unit we call a cell. Chloroplasts are evident in these leaf cells (LM, light micrograph). (**d**) In multicellular organisms, cells are usually organized into tissues, groups of similar cells forming a functional unit. The leaf in this micrograph (a photograph taken with a microscope) has been cut obliquely, revealing two different specialized tissues. The honeycomblike tissue consists of photosynthetic cells within the leaf. The tissue with the small pores is the epidermis, the "skin" of the plant. The pores in the epidermis allow carbon dioxide, a raw material that is converted to sugar by photosynthesis, to enter the leaf (SEM, scanning electron micrograph). (**e**) The aspen leaf, a plant organ, has a specific organization of many different tissues, including photosynthetic tissue, epidermis, and the vascular tissue that transports water from the roots to the leaves. (**f**) These aspens are members of a biological community that includes many other species of organisms.

Figure 1.3
Some properties of life. (**a**) **Order:** All other characteristics of life emerge from an organism's complex organization, which is apparent in this skeleton of a sponge called Venus' flower basket. (**b**) **Reproduction:** Organisms reproduce their own kind. Life comes only from life, an axiom known as biogenesis. Here, a Japanese macaque grooms its offspring. (**c**) **Growth and development:** Heritable programs in the form of DNA direct the pattern of growth and development, producing an organism that is characteristic of its species. Shown here are embryos of a Costa Rican species of frog. (**d**) **Energy utilization:** Organisms take in energy and transform it to do many kinds of work, including the maintenance of their ordered state. When the bear eats the fish, it will use the energy stored in the fish's molecules to maintain its own metabolism. (**e**) **Response to the environment:** The growth of this strangler fig's roots and stems adjusts to environmental cues, molding the plant to its substratum—in this case, the ruins of a temple in India. (**f**) **Homeostasis:** Regulatory mechanisms maintain an organism's internal environment within tolerable limits, even though the external environment may fluctuate. This regulation is called homeostasis. For example, regulation of the amount of blood flowing through the blood vessels of this jackrabbit's large ears constantly adjusts heat loss to the surroundings. This contributes to homeostasis of the animal's body temperature. (**g**) **Evolutionary adaptation:** Life evolves as a result of the interaction between organisms and their environments. One consequence of evolution is the adaptation of organisms to their environment. The white fur of the arctic fox makes it nearly invisible against the animal's snowy surroundings.

(a)

(b)

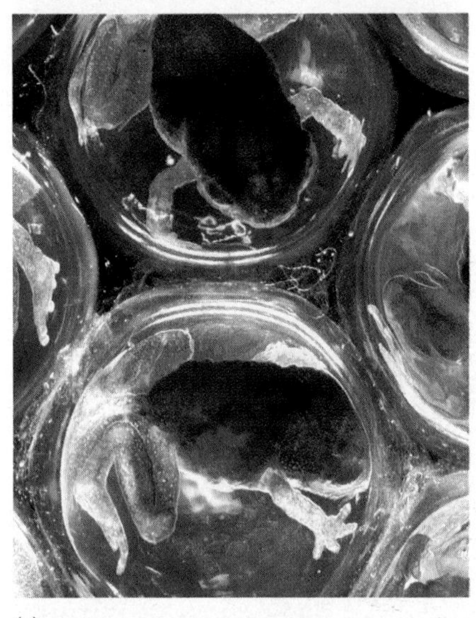
(c)

(d)

(e)

(f)

(g)

other substances in the cell. Biology balances the pragmatic reductionist strategy with the longer-range objective of understanding how the parts of cells and organisms are functionally integrated.

THE CELLULAR BASIS OF LIFE

As the lowest level of structure capable of performing *all* the activities of life, the cell has a very special place in the hierarchy of biological organization. All organisms are composed of cells. They occur singly as a great variety of unicellular organisms, and they occur as the subunits of organs and tissues in plants, animals, and other multicellular organisms. In either case, the cell is an organism's basic unit of structure and function.

Robert Hooke, an English scientist, first described and named cells in 1665, when he observed a slice of cork (bark from an oak tree) with a microscope that magnified 30 times (30×). Apparently believing that the tiny boxes, or "cells," that he saw were unique to cork, Hooke never realized the significance of his discovery. His contemporary, a Dutchman named Anton van Leeuwenhoek, discovered organisms we now know to be single-celled. Using grains of sand that he had polished into magnifying glasses as powerful as 300×, Leeuwenhoek discovered a microbial world in droplets of pond water and also observed the blood cells and sperm cells of animals. In 1839, nearly two centuries after the discoveries of Hooke and Leeuwenhoek, cells were finally acknowledged as the ubiquitous units of life by Matthias Schleiden and Theodor Schwann, two German biologists. In a classic case of inductive reasoning—reaching a generalization based on many concurring observations—Schleiden and Schwann summarized their own microscopic studies and those of others by concluding that all living things consist of cells. This generalization forms the basis of what is known as the cell theory. This theory was later expanded to include the idea that all cells come from other cells. The ability of cells to divide to form new cells is the basis for all reproduction and for the growth and repair of multicellular organisms, such as humans.

Over the past 40 years, a powerful instrument called the electron microscope has revealed the complex structure of cells. All cells are enclosed by a membrane that regulates the passage of materials between the cell and its surroundings. Every cell, at some stage in its life, contains DNA, the heritable material that directs the cell's many activities.

Two major kinds of cells—prokaryotic cells and eukaryotic cells—can be distinguished based on structural organization (Figure 1.4). The cells of the microorganisms known as bacteria are prokaryotic. All other forms of life are composed of eukaryotic cells.

The eukaryotic cell, by far the more complex, is subdivided by internal membranes into many different functional compartments, or organelles. In eukaryotic cells, the DNA is organized along with proteins into structures called chromosomes contained within a nucleus, the largest organelle of most cells. Surrounding the nucleus is the cytoplasm, a thick liquid in which are suspended the various organelles that perform most of the cell's functions. Some eukaryotic cells, including those of plants, have tough walls external to their membranes. Animal cells lack walls.

In the much simpler prokaryotic cell, the DNA is not separated from the rest of the cell into a nucleus. Prokaryotic cells also lack the cytoplasmic organelles typical of eukaryotic cells. Almost all prokaryotic cells (bacteria) have tough external cell walls.

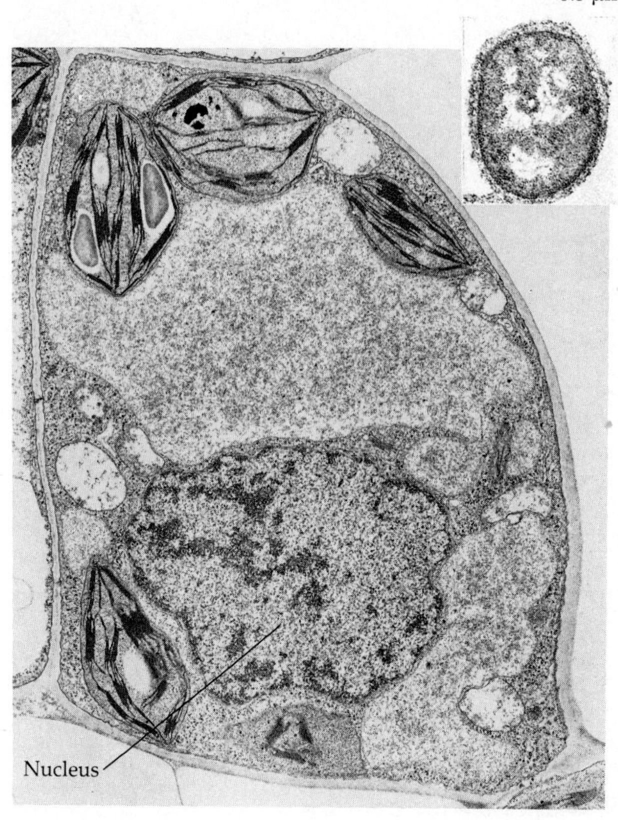

Figure 1.4
Two types of cells, as viewed with the electron microscope. The eukaryotic cell, found in plants, animals, and all other organisms except bacteria, is characterized by an extensive subdivision into many different compartments, or organelles. Only the relatively large organelle known as the nucleus is labeled in this micrograph of a plant cell (TEM). Various organelles can also be seen in the cytoplasm outside the nucleus. The prokaryotic cell (inset), unique to bacteria, is much simpler, lacking most of the organelles found in eukaryotic cells (TEM). Compared to eukaryotic cells, prokaryotic cells are also generally much smaller.

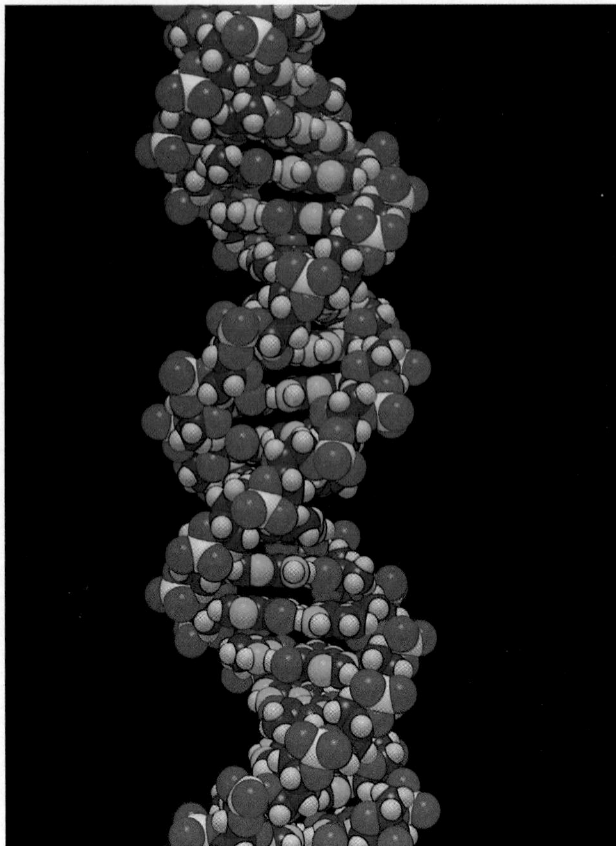

Figure 1.5
The genetic material: DNA. The molecule that conveys biological information from one generation to the next takes the three-dimensional form of a double helix. Along the length of the molecule, information is encoded in specific sequences of nucleotides, the chemical building blocks of DNA.

Although eukaryotic and prokaryotic cells contrast sharply in complexity, we will see that they have some key similarities. Cells vary widely in size, shape, and specific structural features, but all are highly ordered structures that carry out the complicated processes necessary for maintaining life.

HERITABLE INFORMATION

Order implies information; instructions are required to arrange parts or processes in a nonrandom way. Biological instructions are encoded in the molecule known as deoxyribonucleic acid, or DNA. DNA is the substance of genes, the units of inheritance that transmit information from parents to offspring (Figure 1.5).

Each DNA molecule is a long chain made up of four basic chemical building blocks called nucleotides. The way DNA conveys information is analogous to the way we arrange the letters of the alphabet into precise sequences with specific meanings. The word *rat,* for example, conjures up an image of a rodent; but *tar* and *art,* which contain the same letters, mean quite different things. Libraries are filled with books containing information encoded in varying sequences of only 26 letters. We can think of nucleotides as the alphabet of inheritance. Specific sequential arrangements of these four chemical letters encode the precise information in a gene. If the entire library of genes stored within the microscopic nucleus of a single human cell were written in letters the size of those you are now reading, the information would fill more than a hundred books as large as this one. Thus the complex structural organization of an organism is specified by an inherited script conveying an enormous amount of coded information. Inheritance itself is based on a mechanism for copying DNA and passing its sequence of chemical letters on to offspring.

All forms of life employ essentially the same genetic code. A particular sequence of nucleotides says the same thing to one organism as it does to another; differences between organisms reflect genetic programs of different nucleotide sequences. The diverse forms of life are different expressions of a common language for programming biological order.

A FEELING FOR ORGANISMS

In 1983, Barbara McClintock (1902–1992) of the Cold Spring Harbor Laboratory in New York was awarded a Nobel Prize for discoveries she had made three decades earlier. Investigating the inheritance of kernel color in Indian corn, McClintock uncovered a type of genetic change that did not conform to genetic theory of the 1940s and 1950s. In fact, few geneticists understood the importance of McClintock's experimental results until molecular biologists discovered the same phenomenon in bacteria in the 1970s. Other researchers found that similar genetic changes can cause certain kinds of cancer in humans. You will learn about these important discoveries in Chapters 17 and 18. What is important to appreciate now is that Barbara McClintock was only able to see the signs of this unexpected genetic mechanism because she was so familiar with corn, the organism with which she worked for over sixty years. Author Evelyn Keller highlighted this asset in *A Feeling for the Organism,* her biography of McClintock (New York: W. H. Freeman, 1983). Keller tells how McClintock knew the individual corn plants in her experiments and how she enjoyed those attachments. It is a common misconception that scientific objectivity demands that biologists be detached from the organisms they study. In fact, as Keller concluded from her conversations with Barbara McClintock, "Good science cannot proceed without a deep emotional investment on the part of a scientist."

(a)

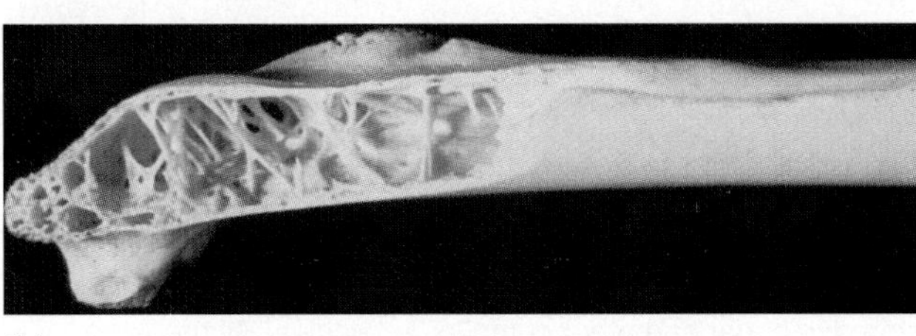

(b)

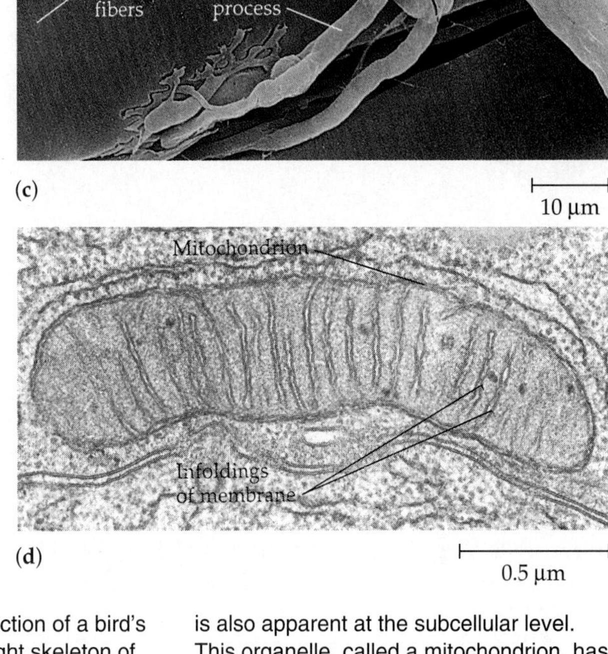

Figure 1.6
Form fits function. (**a**) A bird's build makes flight possible. The correlation between structure and function can apply to the shape of an entire organism, as you can see from this albatross in flight. (**b**) The principle also applies to the structure of organs and tissues. For example, the honeycombed construction of a bird's bones provides a lightweight skeleton of great strength. (**c**) The form of a cell fits its specialized function. Nerve cells, or neurons, have long extensions (processes) that transmit nervous impulses—here, to muscle cells (SEM). (**d**) Functional beauty is also apparent at the subcellular level. This organelle, called a mitochondrion, has an inner membrane that is extensively folded, a structural solution to the problem of packing a relatively large amount of this membrane into a very small container (TEM).

A "feeling for organisms" is what you and the best research specialists in biology have in common. The exploration of cells and their molecular components has revealed much about life. However, most of us do not arrive at biology from an interest in these minute structures, but from our personal experiences with organisms—from observing spiders weaving webs, plant seedlings poking through sidewalk cracks, birds courting mates, and the leaves of trees marking the seasons. Biophilia is manifest in every child. We may go on to study life at different levels of organization, but it is organisms that first attract us. And it is organisms we must keep in mind if our exploration of cells or ecosystems is to have any context.

There are a few questions we should ask about any organism we choose to study. What kind of organism is it? Where does it live? How does it acquire its nutrients and other resources from the environment? What adaptations equip the organism for its way of life? A feeling for the organisms we seek to understand is a theme that enriches our study of life at all levels.

THE CORRELATION BETWEEN STRUCTURE AND FUNCTION

Given a choice of tools, you would not loosen a screw with a hammer or pound a nail with a screwdriver. How a device works is correlated with its structure: Form fits function. Applied to biology, this theme is a guide to the anatomy of life at its many structural levels, from molecules to organisms. Analyzing a biological structure gives us clues about what it does and how it works. Conversely, knowing the function of a structure provides insight about its construction.

An example of this structure-function theme is the aerodynamically efficient shape of a bird's wing (Figure 1.6). Beneath these external contours, the skeleton of the bird also has structural qualities that contribute to flight, with bones that have a strong but light honeycombed structure. The flight muscles of a bird are controlled by neurons. Long extensions of the neurons transmit nervous impulses, making these cells especially well adapted for communication. As an example of functional anatomy at the subcellular level, consider the organelles called mitochondria. They are the sites of cellular respiration, the chemical process that powers the cell by using oxygen to help tap the energy stored in sugar and other food molecules. A mitochondrion is surrounded by an outer membrane, but it also has an inner membrane with many infoldings. Molecules embedded in the inner membrane carry out many of the steps in cellular respiration, and the infoldings pack a large amount of this membrane into a minute container. In exploring life on its different structural levels, we will discover functional beauty at every turn.

THE INTERACTION OF ORGANISMS WITH THEIR ENVIRONMENT

Life does not exist in a vacuum. Each organism interacts continuously with its environment, which includes other organisms as well as nonliving factors. The roots of a tree, for example, absorb water and minerals from the soil, and the leaves take in carbon dioxide from the air. Sunlight absorbed by chlorophyll, the green pigment of leaves, drives photosynthesis, which converts water and carbon dioxide to sugar and oxygen. The tree releases oxygen to the air, and its roots change the soil by breaking up rocks into smaller particles, secreting acid, and absorbing minerals. Both organism and environment are affected by the interaction between them. The tree also interacts with other life, including soil microorganisms associated with its roots and animals that eat its leaves and fruit.

The many interactions between organisms and their environment are interwoven to form the fabric of an ecosystem. The dynamics of any ecosystem include two major processes. One is a cycling of nutrients. For example, minerals acquired by plants will eventually be returned to the soil by microorganisms that decompose leaf litter, dead roots, and other organic debris. The second major process in an ecosystem is a flow of energy from sunlight to photosynthetic life (producers) to organisms that feed on plants (consumers) (Figure 1.7). The theme of organisms interacting with their environment is essential to understanding life on all levels of organization.

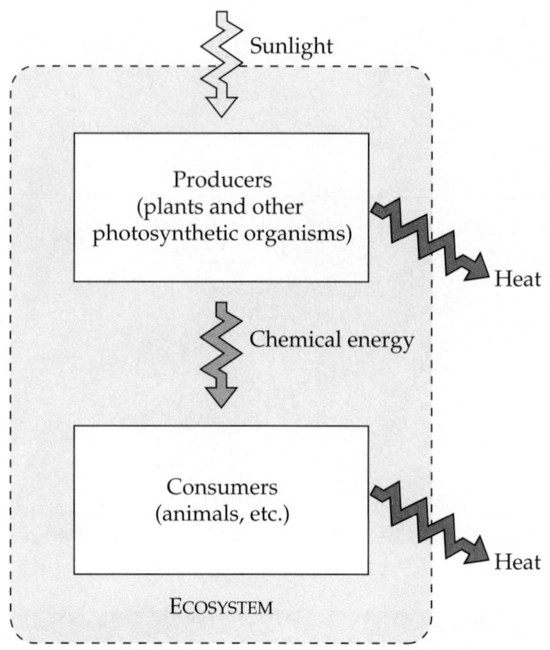

Figure 1.7
An introduction to energy flow in an ecosystem. Living is work, and work requires that organisms obtain and use energy. Most ecosystems are solar-powered. Plants and other photosynthetic organisms convert light energy to the chemical energy stored in sugar and other complex molecules. By breaking these fuel molecules down to simpler waste products, organisms can harvest the stored energy and put it to work. In this scheme of energy flow in an ecosystem, photosynthetic organisms are called producers because the entire ecosystem depends on their photosynthetic products. Animals and other consumers acquire their energy in chemical form by eating plants, by eating animals that ate plants, or by decomposing organic refuse, such as leaf litter and dead animals. The energy that flows into an ecosystem in the form of light flows out in the form of heat, which all organisms dissipate to their surroundings whenever they perform work.

UNITY IN DIVERSITY

Diversity is a hallmark of life. Biologists have identified and named about 1.5 million species, including over 260,000 plants, almost 50,000 vertebrates (animals with backbones), and more than 750,000 insects. Thousands of newly identified species are added to the list each year. Estimates of the total diversity of life range from about 5 million to over 30 million species.

Biological diversity is something to relish and preserve, but it can also be a bit overwhelming. To make the diversity somewhat more comprehensible, people have devised ways of grouping species that are similar. Thus, we may speak of squirrels and pine trees without distinguishing the many different species belonging to these groups. Taxonomy, the branch of biology concerned with naming and classifying species, groups organisms according to a more formal scheme. The scheme consists of different levels of classification, each more comprehensive than those below it (Figure 1.8). The broadest units of classification are the five

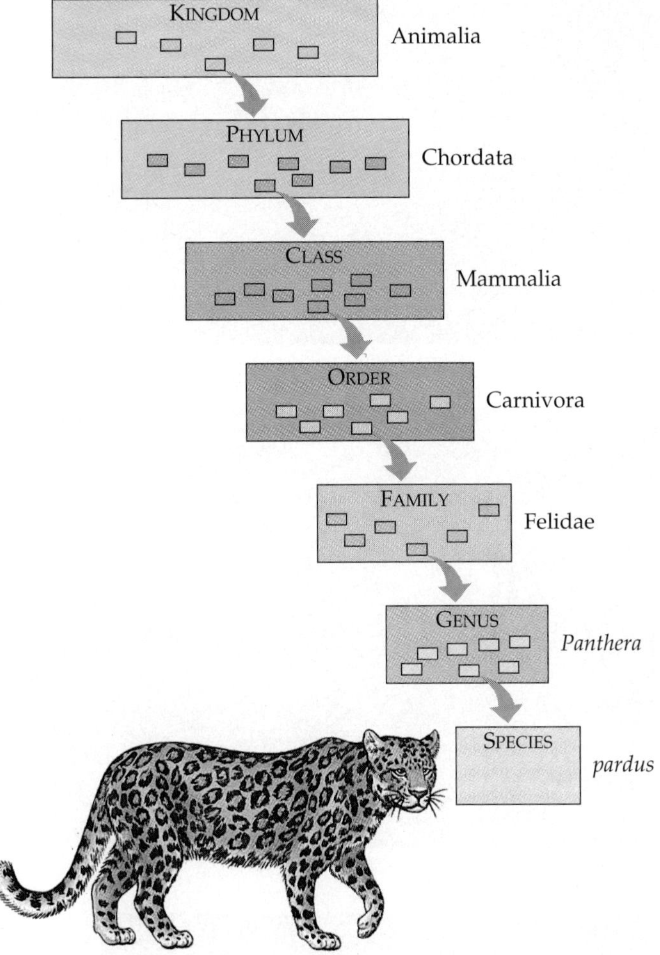

Figure 1.8
Classifying life. The taxonomic scheme classifies species into groups subordinate to more comprehensive groups. Species that are very similar are placed in the same genus, genera are grouped into families, and so on, each level of classification being more comprehensive than those it includes. In this example, a leopard is classified.

kingdoms of life: Monera, Protista, Plantae, Fungi, and Animalia (Figure 1.9).

Kingdom Monera consists of the microscopic organisms known as bacteria. This kingdom is distinguished from the other four kingdoms by cell structure; all members of Monera consist of prokaryotic cells, the simpler of the two major cell types (see Figure 1.4). Organisms with eukaryotic cells are divided among the other four kingdoms.

Kingdom Protista consists mostly of eukaryotic organisms that are unicellular—for example, the microscopic organisms known as protozoa. Also included in this kingdom are certain multicellular forms such as seaweeds that seem to be more closely related to unicellular species than to plants, fungi, or animals.

The remaining three kingdoms—Plantae, Fungi, and Animalia—consist of multicellular eukaryotes. Plants are characterized by photosynthesis. Fungi are mostly decomposers that absorb nutrients obtained by breaking down the complex molecules of dead organisms and waste, such as leaf litter and feces. Animals obtain food by ingestion, eating and digesting other organisms. Thus, plants, fungi, and animals are distinguished partly by their contrasting modes of nutrition. Each kingdom, however, has many other unique features, which will be discussed in Unit Five of this text.

If life is so diverse, how can biology have any unifying themes at all? What, for instance, can a mold, a tree, and a human possibly have in common? As it turns out, a great deal! Underlying the diversity of life is a striking unity, especially at the lower levels of organization. We can see it, for example, in the universal genetic code shared by all organisms. Unity is also evident in certain similarities of cell structure (Figure 1.10). Above the cellular level, however, organisms are so variously adapted to their ways of life that describing biological diversity remains an essential goal of biology. But few biologists view taxonomy as a mere cataloging of seemingly unrelated living objects. In fact, the kinship of all life, though sometimes cryptic, is unmistakable.

EVOLUTION: THE CORE THEME OF BIOLOGY

The history of life is a chronicle of a restless Earth billions of years old, inhabited by a changing cast of living forms (Figure 1.11). Life evolves. Just as an individual has a family history, each species is one tip on a branching tree of life extending back in time through ancestral species more and more remote. Species that are very similar, such as the horse and zebra, share a common ancestor that represents a relatively recent branch point on the tree of life. But through an ancestor that lived much farther back in time, horses and zebras are also related to rabbits, humans, and all other mammals. And mammals, reptiles, birds, and all other vertebrates share a common ancestor even more ancient. Trace evolution back far enough, and there are only the primeval prokaryotes that inhabited Earth more than three billion years ago. All of life is connected. Evolution, the processes that have transformed life on Earth from its earliest beginnings to its apparently unending diversity today, is the one biological theme that ties together all others.

Charles Darwin brought biology into focus in 1859 when he published *On the Origin of Species* (Figure 1.12). Darwin's book had two objectives. First, Darwin argued convincingly from several lines of evidence that contemporary species arose from a succession of

Figure 1.9
The five kingdoms. (**a**) Kingdom Monera includes all prokaryotic organisms. The life forms known as bacteria are placed in this kingdom. This is a mixture of bacteria living in sewage (LM). (**b**) Kingdom Protista consists of unicellular eukaryotes and their relatively simple multicellular relatives. This is an assortment of protists inhabiting pond water (LM). (**c**) Kingdom Plantae, represented by these daffodils, consists of multicellular eukaryotes that carry out photosynthesis. (**d**) Kingdom Fungi is defined, in part, by the nutritional mode of its members, organisms that absorb nutrients after decomposing organic refuse. These are Costa Rican cup fungi. (**e**) Kingdom Animalia consists of multicellular eukaryotes that ingest other organisms. These are rose pelicans in Kenya.

ancestors through a process of "descent with modification," his phrase for evolution. So much additional evidence has accumulated since Darwin's time that nearly all biologists now see evolution as an extensively documented feature of life, much as historians who did not personally witness the U. S. Civil War are convinced, based on an accumulation of evidence, that the war really happened. (The evidence for evolution is discussed in detail in Chapter 20.)

The second objective of Darwin's book was to present his theory for *how* life evolves. This proposed mechanism of evolution is called natural selection. Darwin synthesized the concept of natural selection from observations that by themselves were neither

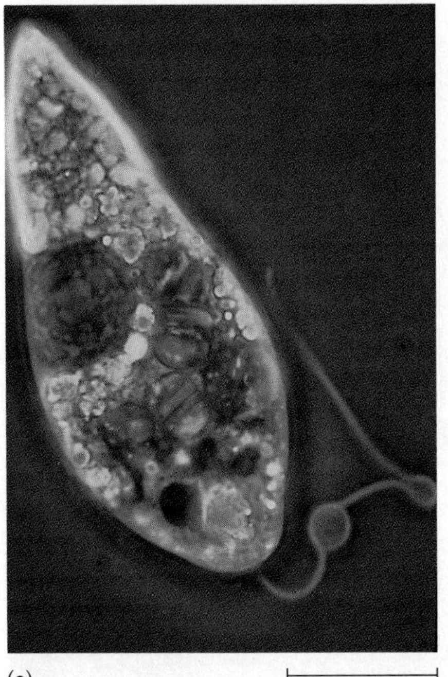

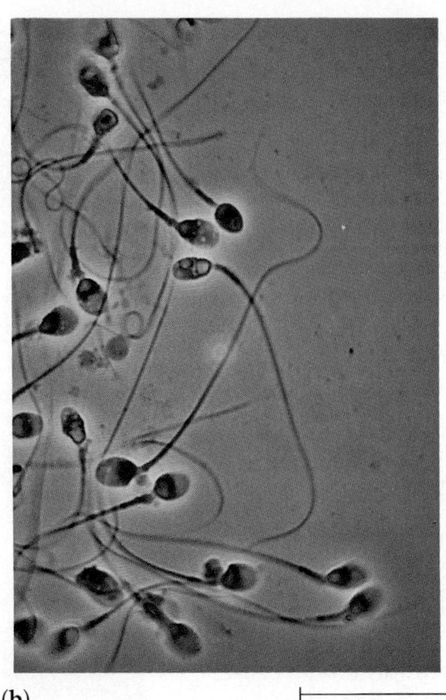

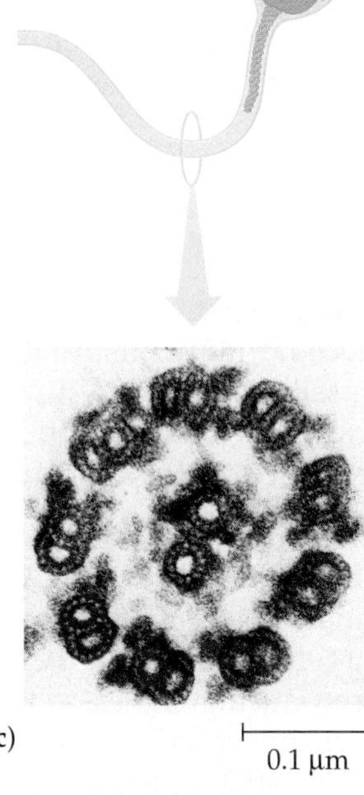

Figure 1.10
An example of unity underlying the diversity of life: the architecture of eukaryotic flagella. Eukaryotic organisms as diverse as protozoa (kingdom Protista) and animals possess flagella, whiplike organelles that propel cells through water. Shown here at approximately the same magnification are (**a**) a flagellated unicellular organism *(Euglena)* and (**b**) some human sperm cells (LMs). (**c**) Using an electron microscope to compare cross sections of flagella from diverse eukaryotes reveals a common structural organization (TEM). Such striking similarity in complex components contributes to the evidence that organisms as different as protozoa and humans are, to some degree, related.

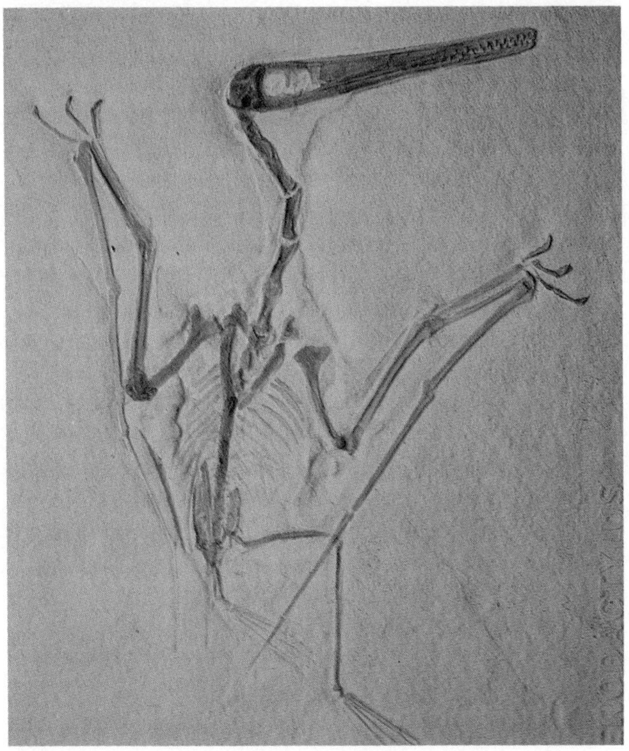

Figure 1.11
The fossil record chronicles evolutionary history. This flying reptile, called a pterosaur, lived over 135 million years ago. The fossil record documents a story of evolving life on an everchanging planet. Many other types of evidence support this evolutionary view of life (see Chapter 20).

Figure 1.12
Charles Robert Darwin (1809–1882). By age 31, when he sat for this portrait, Darwin had already put together the main elements of his theory of evolution by natural selection, though he did not publish *On the Origin of Species* until almost twenty years later. The evolutionary view of life became the theme that unified biology in the twentieth century.

new nor profound. Others had the pieces of the puzzle, but Darwin saw how they fit together. As evolutionary biologist Stephen Jay Gould (see the interview that precedes Unit Five) has put it, Darwin based natural selection on "two undeniable facts and an inescapable conclusion":

- Individuals in a population of any species vary in many heritable traits.
- Any population of a species has the potential to produce far more offspring than the environment can possibly support with food, space, and other resources. This overproduction makes a struggle for existence among the variant members of a population inevitable.
- Those individuals with traits best suited to the local environment generally leave a disproportionately large number of surviving offspring. This selective reproduction increases the representation of certain heritable variations in the next generation. It is this differential reproductive success that Darwin called natural selection, and he envisioned it as the cause of evolution.

We see the products of natural selection in the exquisite adaptations of organisms to the special problems of their environments (Figure 1.13). Notice, however, that natural selection does not create adaptations; rather, it increases the frequencies of certain variations among those that arise randomly in each generation. Adaptation is an editing process, with heritable variations exposed to environmental factors that favor the reproductive success of some individuals over others. The camouflage of the mantids in Figure 1.13 did not result from individuals changing during their lifetimes to look more like their backgrounds and then passing that improvement on to offspring, but by the greater reproductive success in each generation of individuals who were innately better camouflaged than the average mantid.

Darwin proposed that natural selection, by its cumulative effects over vast spans of time, could produce new species from ancestral species. This would occur, for example, when a population fragments into several populations isolated in different environments. In these various arenas of natural selection, what begins as one species may gradually diversify into many, as the geographically isolated populations adapt over many generations to different sets of environmental problems. Descent with modification accounts for both the unity and diversity we observe in life. Features shared by two species are due to their descent from common ancestors, and differences between the species are due to natural selection modifying the ancestral equipment in different environmental contexts. Evolution is the core theme of biology—a theme that will resurface in every unit of this text.

(a)

(b)

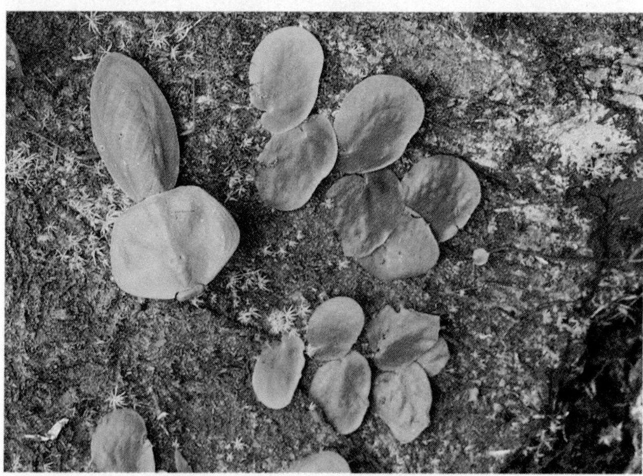

(c)

Figure 1.13
Evolutionary adaptation is a product of natural selection. Diverse forms of camouflage have evolved in various species of insects called mantids. (**a**) A flower mantid in Malaya. (**b**) A Trinidad tree mantid that mimics dead leaves. (**c**) A Central American mantid that resembles a green leaf.

SCIENCE AS A PROCESS: THE HYPOTHETICO-DEDUCTIVE METHOD

Biology is classified as a natural science. Having identified unifying themes that apply specifically to the study of life, we now examine how science works.

Like life, science is better understood by observing it than by trying to create a precise definition. The word *science* is derived from a Latin verb meaning "to know." Science is a way of knowing. It emerges from our curiosity about ourselves, the world, and the universe. Striving to understand seems to be one of our basic drives. At the heart of science are people asking questions about nature and believing that those questions are answerable. Scientists tend to be quite passionate in their quest for discovery. Max Perutz, a Nobel Prize–winning biochemist, puts it this way: "A discovery is like falling in love and reaching the top of a mountain after a hard climb all in one, an ecstasy induced not by drugs but by the revelation of a face of nature that no one has seen before."

A process known as the scientific method outlines a series of steps for answering questions, but few scientists adhere rigidly to this prescription. Science is less structured than most people realize. Like other intellectual activities, the best science is a product of minds that are creative, intuitive, and imaginative. Perhaps science is distinguished by its conviction that natural phenomena, including the processes of life, have natural causes—and by its obsession with evidence. Scientists are generally skeptics. As you begin each unit of study in this text, you will meet such scientists through personal interviews and come to understand a bit more about how they think and why they enjoy their work.

Although it is counterproductive to reduce science to a stereotyped series of steps, we *can* identify the key ingredient of the scientific process: the *hypothetico-deductive method.* The first part of this term refers to *hypothesis,* which is a tentative answer to some question —an explanation on trial. Consider, for example, this imaginary scenario: Scott and Ian, identical twins, are sleepy every day in their 1:00 history class; they often doze and want to know why. Maybe eating a big lunch just before going to history every day makes them sleepy. Or maybe the classroom is too warm. Maybe Scott and Ian are listless because they sit in the back of their history class and are less involved than they are in their other classes, where they always sit in front. Perhaps their history professor is really boring. Or maybe it's just the time of day. These are all hypotheses, possible explanations for this daily behavior of sleeping through history class. It is possible to test these hypotheses.

The *deductive* in hypothetico-deductive method refers to how deductive reasoning is used to test hypotheses. Deduction contrasts with induction. Recall that induction is reasoning from a set of specific observations to reach a general conclusion (as in "All organisms are composed of cells"). In deduction, the reasoning flows in the reverse direction, from the general to the specific. From general premises, we extrapolate to specific results we should expect if the premises are true: If all organisms are made of cells (premise #1) and humans are organisms (premise #2), then humans are composed of cells (predictioı about a specific case). In the scientific process, deduction usually takes the form of predictions about the results of experiments we should expect *if* a particular hypothesis (premise) is correct. We then test the hypothesis by performing the experiments to see whether or not the predicted results occur. This deductive testing takes the form of *If . . . then* logic:

HYPOTHESIS #1: *If* the twins get sleepy because they eat lunch before their 1:00 class,
EXPERIMENT: and Scott postpones lunch until class ends at 2:00 (but Ian still eats at noon),
PREDICTED RESULT #1: *then* Scott should be less sleepy than Ian in history class.
PREDICTED RESULT #2: *then* Scott should get sleepy in his 3:00 class instead.

OR

HYPOTHESIS #2: *If* the twins get sleepy in their 1:00 class because they sit in the back of the room,
EXPERIMENT: and Scott (but not Ian) moves to the front of the class,
PREDICTED RESULT: *then* Scott should be more alert than Ian during the 1:00 class.

Now, suppose the result of the first experiment is that Scott is still as sleepy as Ian (but hungrier) in their 1:00 class. The twins try the second experiment, with the result that now Scott dozes in the front row of class instead of the back row. The twins continue to test and reject various hypotheses. And then they are the subjects of an experiment they did not even plan—daylight savings time begins. Scott and Ian (and most of their classmates) notice that for a day or two, they have more trouble than usual waking up in the morning. However, the twins find that they are more alert than usual in their 1:00 class but hit their low point during the 2:00–3:00 break in their schedule. (Within a few days, patterns are back to normal, and Scott and Ian are listless during their 1:00 class again.) The twins conclude that they are just naturally sleepy this time of day and decide not to schedule 1:00 classes in future semesters.

Although this example is fanciful and there are flaws in the experiments, five important points about hypotheses are evident:

- *Hypotheses are possible causes.* A generalization based on inductive reasoning is not a hypothesis. Based on the twins' daily experience in history class, they may conclude, "We are sleepy every day at about 1:00," but that conclusion merely summarizes a set of observations. A hypothesis, remember, is a tentative *explanation* for what we have observed. "The warm temperature of the room causes Scott and Ian to sleep through class" is a hypothesis.

- *Hypotheses reflect past experience with similar questions.* Sometimes hypotheses are described as *educated* propositions about cause. For example, the hypothesis that a warm classroom causes drowsiness may be based on a general experience of sleepiness under such conditions. Although we should consider any hypothesis that is a possible explanation for what we have observed, the hypotheses we test first should be those that seem the most reasonable, based on what we already know.
- *Multiple hypotheses should be proposed whenever possible.* Proposing alternative explanations that can answer a question is good science. If we operate with a single hypothesis, especially one we favor, we may direct our investigation toward a hunt for evidence in support of this hypothesis.
- *Hypotheses should be testable via the hypothetico-deductive method.* Hypotheses should be phrased in a way that enables us to make predictions that can be tested by experiments or further observation. "Scott falls asleep in history class because the devil makes him do it" is not a testable hypothesis and therefore does not lend itself to the scientific process. The requirement that hypotheses be testable limits the scope of questions that science can answer.
- *Hypotheses can be eliminated but not confirmed with absolute certainty.* If moving to the front of his 1:00 class does not reduce Ian's drowsiness, this casts doubt on one hypothesis. A hypothesis can be falsified by experimental tests, especially if the experiments are repeated with the same results. The onset of daylight savings time supported the "sleepy time of day" hypothesis, and our confidence in this explanation grows if the hypothesis continues to stand up to various kinds of experimental tests. But we can never *prove* that this hypothesis is the true explanation. It is impossible to repeat an experiment enough times to be absolutely certain that the results will *always* be the same. And some false hypotheses make accurate predictions. Consider this hypothesis: "Night and day are caused by the sun orbiting around Earth in an east-west direction." This hypothesis predicts that the sun will rise each morning in the east, move across the sky, and set in the west, which is exactly what we observe. (However, the "geocentric universe" hypothesis makes many other predictions that enable us to falsify the hypothesis.) Of the many hypotheses proposed to answer a particular question, the correct explanation may not even be included. Even the most thoroughly tested hypotheses are accepted only conditionally, pending further investigation.

The "sleepy twins" scenario introduces another important feature of the scientific process: the controlled experiment. In a *controlled experiment,* the subjects (Scott and Ian, in this case) are divided into two groups, an *experimental group* and a *control group.* Ideally, the two groups are treated exactly alike except for the one variable the experiment is designed to test. This provides a basis for comparison, enabling us to draw conclusions about the effects of our experimental manipulation. Scott and Ian attempted to control their experiments. When Scott moved to the front of the 1:00 class as the *experimental,* Ian remained in the back of the room as the *control.*

There are, of course, serious deficiencies in this experiment. The obvious one is that a single individual is an inadequate sample size for a control group or experimental group. But note also that there were uncontrolled variables in this experiment. The twins were specifically trying to test the effect of seating locale on sleepiness in class, but temperature and other unknown factors may have varied between the front and back of the classroom. Some experiments are easier to control than others, but setting up the best possible controls is generally a key element of good experimental design.

Now that we have analyzed the hypothetico-deductive method by using an imaginary example, you should be able to recognize the process in an elegant study actually reported in the scientific literature. For many years, David Reznick of the University of California, Riverside, and John Endler of the University of California, Santa Barbara, have been investigating differences between populations of guppies in Trinidad, a Caribbean island (Figure 1.14). Guppies *(Poecilia reticulata)* are small freshwater fishes you probably recognize as common aquarium pets. In the Aripo River system of Trinidad, guppies live in populations that are relatively isolated from one another. In some cases, two populations inhabiting the same stream live less than 100 m apart, but they are separated by a waterfall that impedes movement of guppies between the two sites.

When Reznick and Endler compared guppy populations, they observed differences in what are called *life history* characteristics. These characteristics included the average age and size of guppies when they reached sexual maturity and began to reproduce, as well as the average number of offspring per brood (a litter of baby guppies). The researchers were able to correlate variations in these life history characteristics with the types

Figure 1.14
David Reznick conducting field experiments on guppy evolution in Trinidad.

of predators present in different locations. In some locales, the main predator is a small fish called a killifish, which preys predominately on small, juvenile guppies. In other locations, a larger predator called a pike-cichlid preys more intensely on guppies and mainly eats relatively large, sexually mature individuals. Guppies in populations exposed to these pike-cichlids have larger broods, reproduce at a younger age, and are smaller at maturity, on average, than guppies that coexist with killifish.

What causes these life history differences between the guppy populations? Correlation with the type of predator present is suggestive, but a correlation does not necessarily imply a cause-and-effect relationship. The type of predator present and the life history characteristics of the guppy populations in a particular location may be independent consequences of some third factor. In fact, Reznick and Endler tested the hypothesis that the life history variations were due to differences in water temperature or other features of the physical environment:

HYPOTHESIS #1: *If* differences in physical environments cause the variations in life histories of guppy populations,

EXPERIMENT: and samples from different wild guppy populations are collected and cultured for several generations in identical environments in predator-free aquaria,

PREDICTED RESULT: *then* the laboratory populations should become more similar in their life history characteristics.

When the researchers performed this experiment, the differences persisted for many generations. This result eliminates hypothesis #1 and also indicates that the life history differences in guppy populations are inherited. Based on the assumption that natural selection can lead to genetic differences in populations, Reznick and Endler tested the following explanation:

HYPOTHESIS #2: *If* the feeding preferences of different predators caused contrasting life histories in different guppy populations to evolve by natural selection,

EXPERIMENT: and guppies are transferred from locations with pike-cichlids (intense predators of mature guppies) to guppy-free sites inhabited by killifish (less intense predators of juvenile guppies),

PREDICTED RESULT: *then* the introduced guppy populations should show a generation-to-generation trend toward later maturation, larger size, and smaller broods—life history characteristics typical of natural guppy populations that coexist with killifish.

In 1976, Reznick and Endler introduced guppies from locations with pike-cichlids to sites that had killifish but no guppies. These transplanted populations were the researchers' experimental groups, and they studied them for 11 years, measuring age and size at maturity, brood size, and other life history characteristics. The scientists compared these measurements to data collected over the same period on control groups, guppies that remained in the original locations inhabited by pike-cichlids. To be certain that only heritable differences were counted, the measurements were made after samples from the experimental and control populations had been reared for two generations in identical aquarium environments (Figure 1.15). Over 11 years, or 30 to 60 generations, the average weight at maturity for guppies in the introduced (experimental) populations increased by about 14% compared to the control populations. Other life history characteristics also changed in the direction predicted by hypothesis #2.

Notice that without a control group for comparison, there would be no way to tell whether it was the killifish or some *other* factor that caused the transplanted guppy population to change. But since control sites and experimental sites were often nearby parts of the same stream, the main variable was probably the presence of different predators. And these careful researchers observed similar results when guppy populations were reared in artificial streams that were identical except for the type of predator.

Of the several hypotheses Reznick and Endler have tested (we examined only two), they are left with natural selection due to differential predation as the most likely explanation for the observed differences between guppy populations. Apparently, when preda-

Figure 1.15
Experimental evidence for natural selection in action. David Reznick and his colleagues transferred guppies to new locations where they were exposed to different predators. (**a**) This photograph compares the sizes of the predators, the larger pike-cichlid (top) and the killifish (middle) to male and female guppies (bottom). (**b**) The controls consisted of guppy populations native to locales where the main predator is the pike-cichlid, which feeds predominately on large guppies. The experimental samples consisted of guppies that were removed from pike-cichlid locales and introduced to stream locations where killifish are found, but where there were no guppies initially. Killifish prey mainly on small, immature guppies. After just 11 years, the introduced guppy populations had evolved noticeably. They reached sexual maturity at an older age and were larger at maturity than controls. The data compared in these histograms (bar graphs) are for guppy samples that were removed from the field sites and reared for two generations in lab aquaria. Thus, the differences had a hereditary basis. Based on the hypothetico-deductive method, the scientists concluded that selective predation can cause key characteristics in the life history of guppies to evolve, an example of natural selection in action.

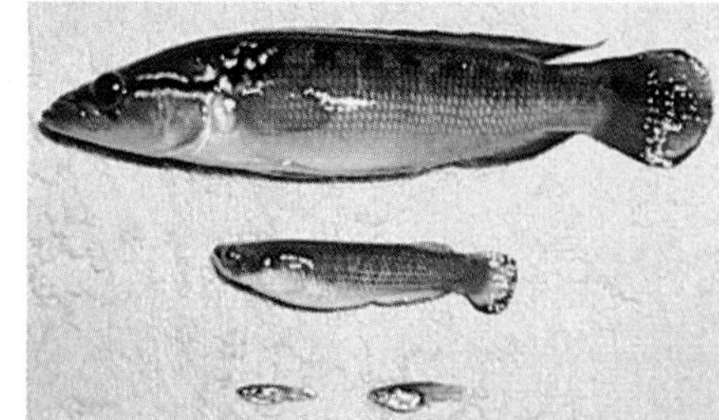

(**a**) Predators and guppies (male, left; female, right) to scale

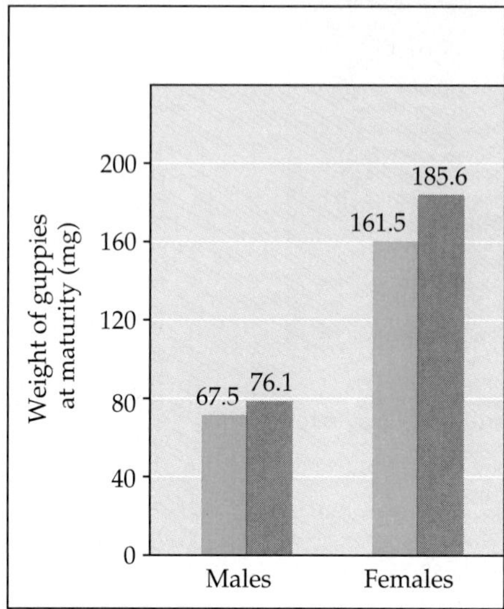

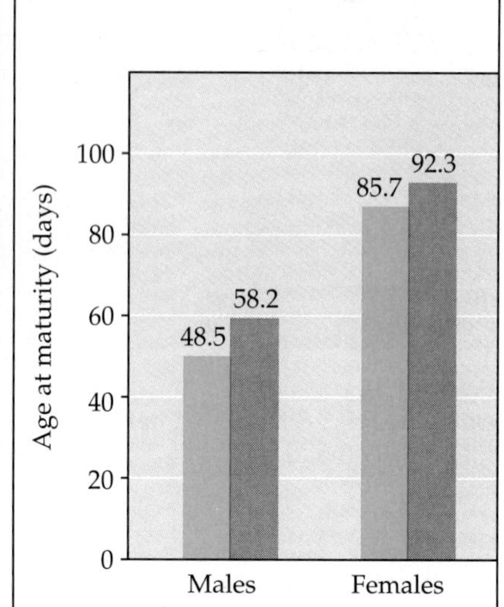

(**b**) Experimental results

tors such as pike-cichlids prey mainly on reproductively mature adults, the chance that a guppy will survive to reproduce several times is relatively low. In this situation, the guppies with greatest reproductive success should be the individuals that mature at a young age and small size and produce at least one brood before growing to a size preferred by the local predator.

The popular press gave Reznick and Endler's research a lot of attention because it documents evolution in a natural setting over a relatively short time period (just 11 years). We have examined the guppy experiments in some detail because this research also provides a fine example of the key role of the hypothetico-deductive method in the scientific process.

Another key feature of science is its progressive, self-correcting quality. A succession of scientists working on the same problem build on what has been learned earlier. It is also common for scientists to check on the conclusions of others by attempting to repeat observations and experiments. Among contemporary scientists working on the same question, there are both cooperation and competition. Scientists share information through publications, seminars, meetings, and personal communication. They also subject one another's work to careful scrutiny. Science thrives on intellectual turbulence and new ideas.

Many people associate the word *discovery* with science. Often, what they have in mind is the discovery of new facts. But accumulating facts is not really what science is about; a telephone book is a catalog of facts, but it has little to do with science. It is true that facts, in the form of observations and experimental results, are the prerequisites of science. What really advances science, however, is a new idea that collectively explains a number of observations that previously seemed to be unrelated. The most exciting ideas in science are those that explain the greatest variety of phenomena. People like Newton, Darwin, and Einstein stand out in the history of science not because they generated a great many facts, but because they synthesized ideas with great explanatory power. Such ideas, much broader in scope than the hypotheses that pose possible causes for one set of observations, are known as theories. Because theories are so comprehensive, they only become widely accepted if they are supported by a large body of evidence. "Natural selection" qualifies as a theory because of its broad application to so many situations. And the theory of natural selection is widely accepted because it has been validated by so many observations and experiments, including those of Reznick and Endler.

This book is only partly about the current state of biological knowledge. It is also important for you to learn, by example and by practice, how the process of science works. The power of the hypothetico-deductive method is one of the themes of this text, and you will read many more examples of how biologists have applied it in their research. However, it is mainly your own experience in the laboratory and field that will teach you how to do science. And your practice with the hypothetico-deductive method will help you think more critically in general.

SCIENCE, TECHNOLOGY, AND SOCIETY

Science and technology are interdependent. You have just learned that science is a process, usually involving the hypothetico-deductive method, that curious people use to answer their questions about nature. Technology, especially in the form of new instruments (electron microscopes, for example), extends our ability to observe and measure and enables scientists to work on questions that were previously unapproachable. In turn, technology depends on science to generate the new information that makes invention possible. In this sense, we can think of technology as scientific discoveries applied to the development of goods and services. Watson and Crick's discovery of the structure of DNA was science. This breakthrough catalyzed a flurry of scientific activity that led to a better and better understanding of DNA chemistry and the genetic code. These discoveries eventually made it possible to manipulate DNA, enabling genetic engineers to transplant foreign genes into microorganisms in order to produce such valuable products as human insulin. The new biotechnology is revolutionizing the pharmaceutical industry. Perhaps Watson and Crick envisioned that their famous discovery would someday have technological applications, but that was probably not what motivated their research, nor could they have predicted exactly what the applications would be. Scientist/writer Lewis Thomas puts it this way in his essay "Making Science Work": "We cannot say to ourselves, we need this or that sort of technology, therefore we should be doing this or that sort of science. . . . Science is useful, indispensable sometimes, but whenever it moves forward it does so by producing a surprise; you cannot specify the surprise you'd like. Technology should be watched closely, monitored, criticized, even voted in or out by the electorate, but science itself must be given its head if we want it to work."*

We have a love-hate relationship with technology. Beginning with ancient agriculture, continuing with the industrial revolution, and presently progressing in the electronic and information age, technology has improved our standard of living in many ways. But it is a double-edged sword. Technology, especially technology that keeps people healthier, has made it possible for the human population to grow more than tenfold in the past three centuries. The environmental consequences are monstrous. Acid rain, deforestation, global warming, nuclear accidents, ozone holes, toxic wastes, and endangered species are just a few of the repercussions from more and more people wielding more and more technology. Science can help us identify these problems and even provide insight about what course of action may prevent further damage. But solutions to these problems have as much to do with politics, economics, culture, and values as with science and technology.

A better understanding of nature must remain the goal of science. Technology, for better or worse, is a by-product of our innate desire to know. Scientists should not distance themselves from technology but instead try to influence how technology applies the discoveries of science. And scientists have a responsibility to help educate politicians, government and corporate leaders, and voters about how science works and about the potential benefits and hazards of specific technologies. The crucial relationship between science, technology, and society is a theme that increases the significance of our study of life.

* * *

In some ways, biology is the most demanding of all sciences, partly because living systems are so complex and partly because biology is a multidisciplinary science that requires a knowledge of chemistry, physics, and mathematics. Modern biology is the decathlon of natural science. If you are a biology major or a preprofessional student, you have an opportunity to become a versatile scientist. If you are a physical science major or an engineering student, you will discover in the study of life many exciting applications for what you have learned in your other science courses. If you are a nonscience student enrolled in biology as part of a liberal arts education, you have selected a course in which you can sample many scientific disciplines.

No matter what brings you to biology, you will find the study of life to be challenging and uplifting. The themes introduced in this chapter are intended to provide you with a conceptual framework for fitting together the many things you will learn about biology and to encourage you to begin asking important questions of your own.

---

*Thomas, L. "Making Science Work." *Late Night Thoughts on Listening to Mahler's Ninth Symphony.* New York: Viking Press, 1983, p. 28.

UNIT ONE | THE CHEMISTRY OF LIFE

AN INTERVIEW WITH CANDACE PERT

Biology and chemistry are allied sciences. Even our thoughts and emotions are based on chemical events in the brain. Candace Pert and Solomon Snyder revolutionized neurochemistry in 1972, when they discovered opiate receptors in the brain. These receptors, molecules on the outer membranes of brain cells, are the sites where morphine, heroin, and other opiate drugs bind and trigger their effects on behavior. The receptors normally function as binding sites for molecules called endorphins, opiate-like substances that the brain itself produces. Opiates evoke euphoria by mimicking the brain's natural pleasure-promoting substances. Pert and Snyder had identified the first step in a chemical pathway that leads to a change in emotional state.

Candace Pert was educated at Bryn Mawr College and Johns Hopkins University, where she earned her Ph.D. in pharmacology in 1974. Dr. Pert continued her neurochemistry research at the National Institutes of Health (NIH) in Bethesda, Maryland. She recently left NIH to devote herself full-time to AIDS research as scientific director of Peptide Design, a biotechnology company. In this interview, Dr. Pert describes how her interest in the behavior of humans led her to study the behavior of molecules.

Dr. Pert, what attracted you to science—and more specifically, to neurochemistry?
The truth is, I grew up thinking of myself as more of an English literature person. The first science that attracted me was psychology. I believe I was interested in psychology because I'm very interested in people—in human behavior. Then in college, when I saw that there was a field called psychology, that was as far into science as I wanted to go. But then I wanted to get to a more scientific basis for behavior. I wanted there to be a field that looked for the biochemical substrate of psychology. That field didn't really exist at the time. But I began to search for it, and it began to exist.

In fact, you helped invent that field, which is sometimes called "molecular psychology." What does the term mean?
It's the idea of finding the chemical basis of human behavior. You can ask what chemicals make up the brain. You can measure those chemicals. You can study how those chemicals change and what chemicals are released in different mood states and different behaviors. And you can ask how different drugs affect behavior by binding to certain receptor molecules in the brain—receptors that normally bind to internal chemical signals in the brain. These are some of the questions of neuroscience and what I've been involved with for twenty years.

You and Solomon Snyder pioneered the field in the early 1970s when you demonstrated the existence of receptor molecules in the brain that bind to opiates such as morphine and heroin. Can you retrace that key discovery?
I was trying to follow from psychology and behavior to an understanding of what's actually going on inside of the black box, the brain. But I had only been exposed to experimental psychology—things like operant learning. So I actually chose Dr. Snyder to work with because he was one of the first to be looking at the biochemistry of the brain, to be grinding up brain and measuring chemicals. And I fell in love with the idea of searching for an opiate receptor as a long-shot project. People have used opiates for thousands of years. The ancient Sumerians had a written symbol for opiates: a combination of two characters, one means "joy," and the other means "juice." The topic of opiates appealed to my interest in emotions. My library research had me really believing that there had to be receptors in the brain for opiates—molecules on certain brain cells that could recognize and bind to the drugs. I was looking for the experimental approach that would be able to demonstrate these receptor molecules. It was my fascination with the topic that kept me working very hard, probably past the point where it was sensible to keep working on the problem. I was really hung up for quite a while and not making any progress. Until there was a breakthrough experiment, then the whole thing opened up.

And what was that experiment?
It was on October 22, 1972, and I have it in my lab notebook. I had been trying to find opiate receptors by adding radioactively labeled morphine to suspensions of membranes collected from brain cells. The drug is tagged with radioactive atoms so that you can detect its presence

if it binds to receptor molecules on the brain cell membranes. At first, the experiments with radioactive morphine were unsuccessful. Then I switched to radioactive naloxone. Naloxone is an opiate antagonist, a drug that reverses the effects of opiates. It was thought that naloxone competes with opiates by binding to the same receptor molecules in the brain. And it was believed that naloxone and other opiate antagonists stay stuck to the receptor, unlike opiates, which bind less tightly. I said: "Oh—if this idea is right, I should be using an antagonist to find the opiate receptor." And so I obtained radioactive naloxone, purified it, and it worked—it stuck to receptors on the brain cell membranes. It was a "Eureka!" moment. And later, by adjusting the salt concentration in our experimental mixture, we showed that morphine binds to the same receptors.

How did it feel to publish such an important discovery?
Absolutely great! The first excitement was really just sharing it with Dr. Snyder, sharing it with my close colleagues in the lab at Johns Hopkins. I think that's probably 90% of the fun. I mean you really do it on some level for yourself. There was finally a method for getting data on the molecular targets of a drug's action. Then we published in the journal *Science*.

And the news media picked up the story?
My theory is that there was not much news at the time. It was a very dull time, right before Watergate had broken and right after Vietnam had died out—you know, '72. There just wasn't much news. So our discovery wound up on the front pages all over the place. It's almost arbitrary what discoveries get promoted. I think I've had other discoveries that are more important, but they didn't get promoted as much as the discovery of opiate receptors in the brain.

Have you been able to determine which regions of the brain are enriched in opiate receptors, which regions are most affected by the drug?
In the beginning, we did crude things; we just divided a brain up into chunks and measured the concentration of opiate receptors. But about seven years later, when I went to NIH, a colleague named Miles Hirkenhand and I developed a method for actually slicing the brain and visualizing the receptor's distribution. It was most concentrated in parts of the brain which are believed to be involved in emotions. So the opiate receptor did lead me to what I'm really interested in, which is emotions. These chemicals can affect mood by binding to receptor molecules in certain parts of the brain.

Why would the brain have receptors that recognize opiates, which, after all, are chemicals produced by poppy plants?
After we discovered the receptor, other researchers found internal opiate-like molecules in the brain that are called endorphins (see the illustration on p. 23). Endorphins are one of several classes of molecules called peptides that function as chemical signals within the brain. Cells have receptors that recognize certain peptide signals, such as endorphins. I guess the idea of receptors for internal opiates led to the discovery that there are natural chemical mechanisms that produce pleasure in some way. There's great survival and hence evolutionary value in rewarding certain behavior with pleasure—so animals will seek to repeat these behaviors. There is now a whole literature showing that endorphins are associated with eating, sexual behavior, and bonding, for instance.

So these "feel good" substances and other brain chemicals that affect emotions not only reinforce survival behavior but play a role in our social development as well?
We learn to control our emotions. We are capable of self-control. For example, there's just the simple thing of training a child not to do certain things, like shouting, disrupting others, or beating up on other children. Maybe this translates as not letting certain chemicals develop to too high a concentration. I don't know what control is on this level, but we do have the potential for altering our brain chemistry and body chemistry.

Is there any validity in the idea that the high some joggers experience from running is associated with elevated levels of endorphins?
The answer's yes. Experiments definitely show that some of the endorphins do increase in the bloodstream under these physically stressful conditions. The increase people have been able to measure is actually quite small, maybe a twofold difference in endorphin levels in the blood. But that's OK, because measuring brain chemicals in the blood is like trying to understand a family by studying the concentrations of wastes in the fluid running in the gutter outside their house. In the case of runner's high, the real action would be at certain receptor sites within the brain. And maybe there, endorphin concentrations have increased much more than what we observe finally coming out in the bloodstream. There's definitely a scientific basis to runner's high. But, as with other emotional states, endorphins are probably only part of a more complex story. There are many other brain peptides; certainly different emotions and physiological states are going to reflect combinations of these things. But people have probably focused a little myopically on the endogenous opiates because it's such a sexy area of research. It may turn out that some brain chemical that hasn't been discovered yet is the main factor in runner's high.

Has research on endorphins and opiate receptors in the brain provided new insights about drug addiction?
This is a tricky area. It's almost like it's political. There is this rationale that if we're funding opiate research, we're going to learn about the biochemical basis of addictive behavior. And I think there are experiments that show that if someone takes external opiates chronically, then the body stops making internal opiates. And then with withdrawal of the drug, the body has to "relearn" how to make endorphins again. The same seems to be true of many other drugs—Valium, for instance, which also mimics a certain brain peptide. But I don't think there's any evidence that certain people are predisposed to drug addiction because they naturally have less endorphins or fewer receptors for these internal opiates than do other people. Social factors and social conditioning are probably more important.

What do you think has been the most important contribution of endorphin research?
It led to the discovery of many other brain peptides. It turned out that almost every biological peptide anybody ever looked at, whether they got it from the skin of a frog or from the gut, is found in the brain also.

The brain is a microcosm of all the chemical signaling that goes on throughout the body?
You've got it! What we experience as emotions depends on the fact that everything going on in the body somehow feeds back into these major parts of the brain. Thirst is an example. You can think of feeling thirsty as a mood. It's a blood molecule called angiotensin that makes us feel thirsty. You drop angiotensin into the brain of a rat, and that rat will start drinking water, no matter how much water it had previously. And if you drop angiotensin on the kidney, the kidney will start to conserve water. So here is one chemical which, in a holistic way, unites the animal into, "Let's get as much water as we can and save as much water as we can."

Are there health implications of the integration of mind and body?
This realization has certainly transformed my own beliefs about myself. The idea that the mind is not confined to the brain has lots of implications. For instance, some of the brain peptides communicate with the immune system. Different kinds of personalities and attitudes can affect both the onset of diseases like cancer and the progression of those diseases. When we start thinking about the mind extending throughout the body, this puts us in the same camp as advocates for a more holistic approach to health care, and that's really outside of the Western tradition of medicine. But the basic ideas are actually ancient in almost every other culture.

Speaking of medicine, your interest in peptides and their receptors has led you into a new area of research, the use of a chemical called peptide-T as a potential treatment for AIDS. What is peptide-T, and why do you think it may help people infected with HIV, the AIDS virus?
Peptide-T is a small piece of the envelope that covers the HIV virus. It's the part of the virus that is probably on the tip of a larger molecule that actually binds to a receptor on the surface of a cell—that's how the virus gets into cells within the human body. We have evidence that purified peptide-T prevents viruses from getting into cells by blocking the receptor sites on the cells. And peptide-T also reduces the toxic effects of the viral envelope. It turns out that many of the symptoms of AIDS are due to the viral envelopes floating in the person's bodily fluids. From the trials just getting underway, it looks as if there's a lot of evidence that peptide-T reverses the symptoms of the disease. It makes people feel a lot better.

How long has this research been going on?
The first work was in 1985. There have been tests all along the way. There was a whole year of lab tests. Then you had to put peptide-T into animals. And finally, the clinical trials with humans. This is like nothing I've ever done. As a scientist, the smartest experiments you try to do are the quick, simple ones that give you results in one day. Nothing like these clinical trials; it takes months just to negotiate who's going to pay for what, and there are all these legal considerations. It's an amazing amount of effort. But I'm doing this research with a group of people. I'm very much into group process and discovery.

On the other hand, there is a competitive side to science.
Yes. I really have mixed feelings about this. In some of my best work, I really

got moving because I was in personal competition with someone else in the lab. So I know how competition can be motivating on some level. But I've really come to see it as a destructive force overall, because I think science at its best is a creative process that prospers when as many people as possible can contribute and come together. It's best if the competitive element can be minimized. I think I was successful at NIH because I could build groups of people working together who hadn't worked together before. Maybe women tend to work better in groups, tend to be less competitive and struggle less than men for recognition. And I think this is good for science. Competitiveness is pretty destructive.

What other encouragement can you offer to women who are considering careers in science?

That I think women are naturally really superb scientists, because they have curiosity and observational powers and good communication skills. And they have the kind of minds that like to consider lots of different things at once and from many angles—analytical minds. So I think women may even be innately well qualified to be scientists. And girls, you're allowed to have babies, and get married, and still be scientists.

Any other advice for students?

Study hard! But don't underestimate how the health of your whole body affects your mood and mental state. If you go away to college, think carefully about how your exercise and diet might change, and how this can really affect the way you feel and how well you do with your studies. Good health should be at the top of your list.

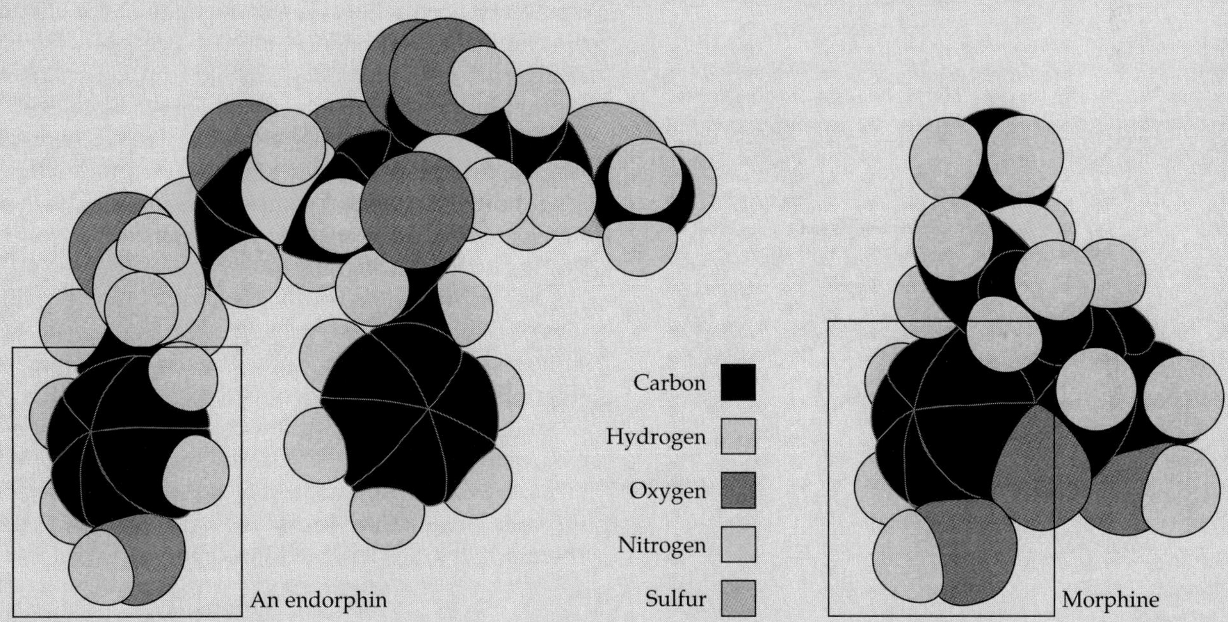

These diagrams compare an endorphin with morphine, an opiate. Receptors on the membranes of brain cells recognize the boxed regions of these molecules. Thus, morphine is a chemical imposter that affects emotional state by mimicking one of the brain's natural chemical signals.

2 ATOMS, MOLECULES, AND CHEMICAL BONDS

MATTER: ELEMENTS AND COMPOUNDS

THE STRUCTURE AND BEHAVIOR OF ATOMS

CHEMICAL BONDS AND MOLECULES

CHEMICAL REACTIONS

CHEMICAL CONDITIONS ON THE EARLY EARTH: SETTING THE STAGE FOR THE ORIGIN AND EVOLUTION OF LIFE

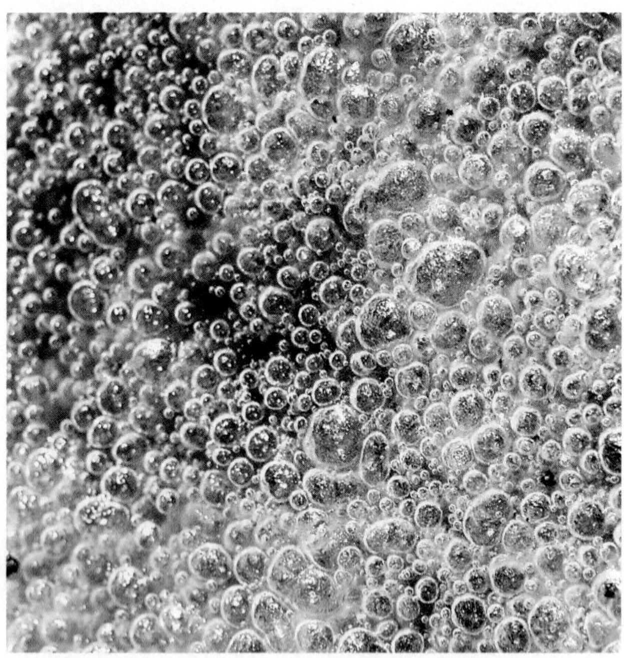

Figure 2.1
Chemistry: fundamental to an understanding of life. These freshwater algae produce sugars and other food by rearranging the atoms of carbon dioxide and water in the chemical process known as photosynthesis. Sunlight powers this chemical transformation. The oxygen bubbles escaping from the algae are by-products of photosynthesis. Humans and other animals ultimately depend on photosynthesis for food and oxygen. In this chapter, you will learn some basic concepts of chemistry that are essential for understanding photosynthesis and many other biological processes.

Life is the cumulative product of interactions among the many kinds of chemical substances that make up the cells of an organism (Figure 2.1). As a preliminary step in understanding life, we might dismantle the cell into its chemical components and then study the structure and behavior of these molecules. From there, we could begin to examine how the molecules aggregate and interact. Somewhere in the transition from molecules to cells, we would cross the blurry boundary between nonlife and life. Life emerges from the integrated organization of the whole organism. To understand a biological process, however, we often must reduce it to simpler steps that can be studied at lower levels of organization. This approach, described in Chapter 1 as reductionism, has been very productive. Our knowledge of the chemical mechanism of inheritance, for instance, grew from the study of DNA molecules extracted from cells. Candace Pert and other molecular psychologists have even extended basic chemistry to the study of the human mind (see the interview that precedes this chapter). After we learn how the chemical components involved in a particular biological process work, the next step is to investigate how those parts interact.

One of the themes of this textbook is the organization of life on a hierarchy of structural levels, with additional properties emerging at each successive level (Figure 2.2). The chapters of Unit One apply the theme of emergent properties to the lowest levels of biological organization—to the ordering of atoms into molecules, and to the interactions of these molecules within the cell. This chapter introduces general principles of chemistry that we will be able to apply to living matter.

MATTER: ELEMENTS AND COMPOUNDS

Chemistry is the study of matter, and **matter** is defined as anything that takes up space and has mass.* Matter

*Sometimes we substitute the term *weight* for *mass,* although the two are not equivalent. We can think of mass as the amount of matter an object contains. The weight of an object measures how strongly that mass is pulled by gravity. An astronaut in a space shuttle is weightless, but the astronaut's mass is the same as it would be on Earth. However, as long as we are earthbound, the weight of an object is a measure of its quantity of matter; so for our purposes, we can use the terms interchangeably.

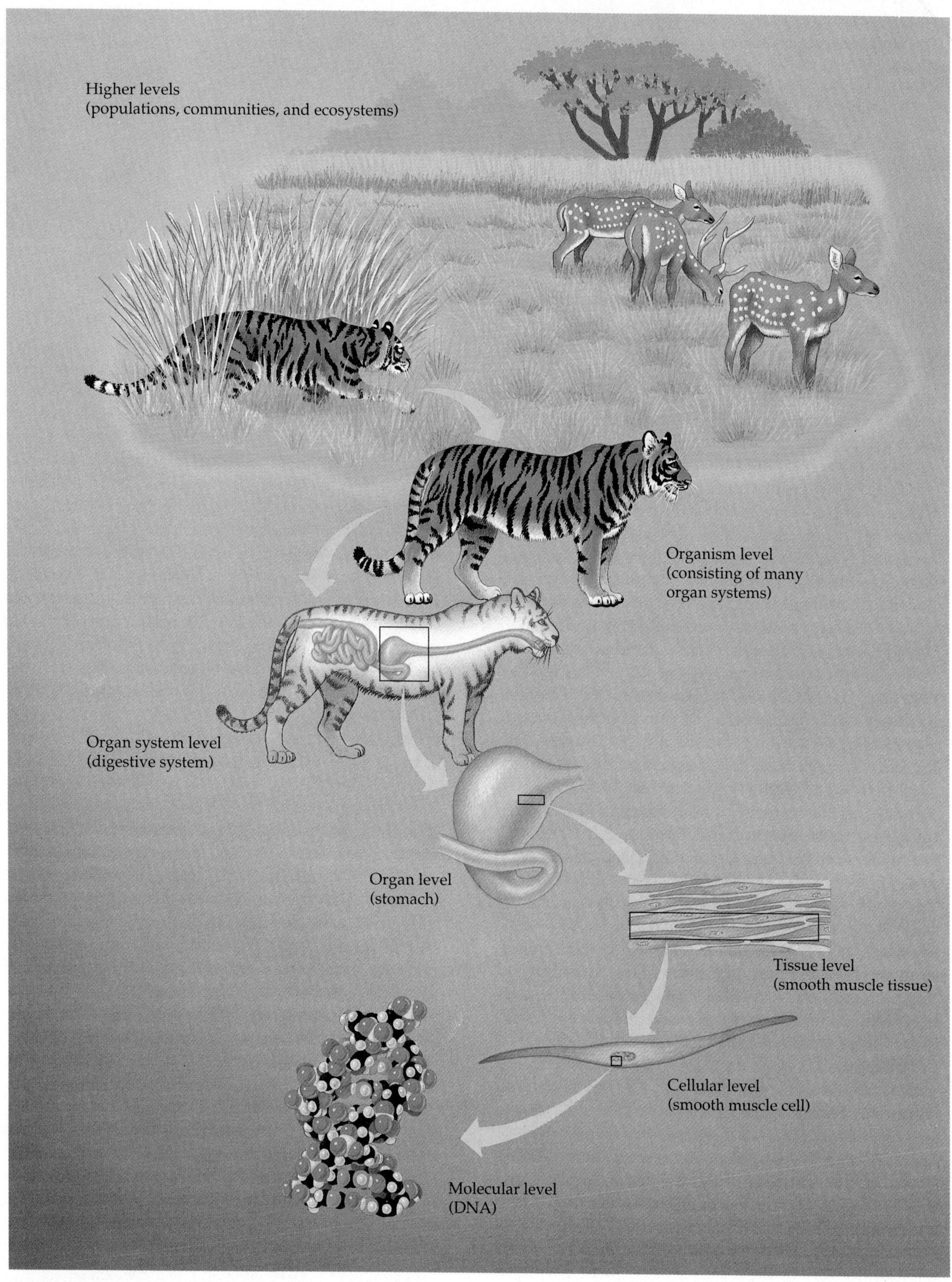

Figure 2.2
Some integrative levels in the hierarchy of biological order.

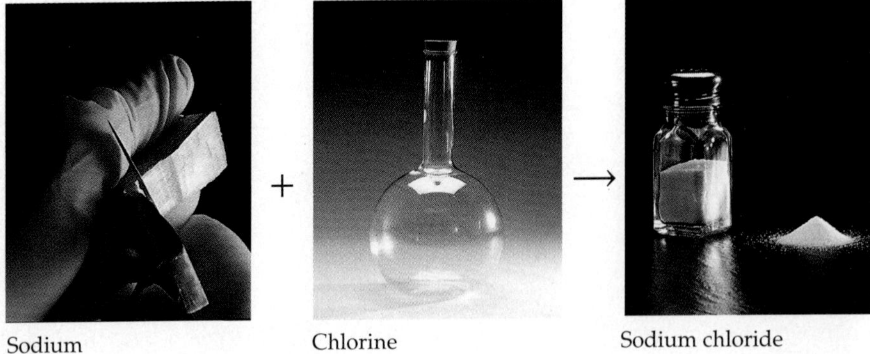

Figure 2.3
The emergent properties of a compound. The metal sodium combines with the poisonous gas chlorine to form the edible compound sodium chloride, or table salt.

exists in many diverse forms, each with its own characteristics. Rocks, metals, wood, glass, and you and I are just a few examples of what seems an endless assortment of matter. Ancient Greek philosophers suggested that the great variety of matter arises from four basic ingredients, or elements.

The Greeks imagined the elements of matter to be air, water, fire, and earth—supposedly pure substances that could not be decomposed to other forms of matter. All other substances were thought to be formed by blending various proportions of two or more of the elements. Even though the classical philosophers proposed the wrong elements, their basic idea was correct.

An **element** is a substance that cannot be broken down to other substances by chemical reactions. Today, chemists recognize 92 elements occurring in nature—for example, gold, copper, carbon, and oxygen. Each element has a symbol, usually the first letter or two of its name. A few of the symbols are derived from Latin or German names; for instance, the symbol for sodium is Na, from the Latin, *natrium*.

Two or more elements may be combined in a fixed ratio to produce a **compound.** Table salt, for example, is actually sodium chloride (NaCl), a compound composed of the elements sodium (Na) and chlorine (Cl). Pure sodium is a metal and pure chlorine is a poisonous gas that was used as a weapon during World War I. Chemically combined, however, sodium and chlorine form an edible compound. This is a simple example of organized matter having emergent properties: A compound has characteristics beyond those of its combined elements (Figure 2.3).

Elements Essential to Life

About 25 of the 92 natural elements are known to be essential to life, but four of these—carbon (C), oxygen (O), hydrogen (H), and nitrogen (N)—make up 96% of living matter. Phosphorus (P), sulfur (S), calcium (Ca), potassium (K), and a few other elements account for most of the remaining 4% of an organism's weight. Table 2.1 lists by percentage the elements found in the human body. Figure 2.4 illustrates a deficiency of an essential element in plants.

Trace elements are those required by an organism in minute quantities. However, because they are mandatory for good health, trace elements are not nutrients of marginal importance to the organism. Some trace elements, such as iron (Fe), are needed by all forms of life; others are required only by certain species. For example, in vertebrates (animals with backbones), the element iodine (I) is a necessary ingredient of a hormone produced by the thyroid gland. A daily intake of only 0.15 milligram (mg) of iodine is adequate for normal activity of the human thyroid, but an iodine deficiency in the diet causes the thyroid gland to grow to abnormal size, producing a deformity called goiter. Where it is available, iodized salt has reduced the incidence of goiter.

Table 2.1 Naturally Occurring Elements in the Human Body

Symbol	Element	Atomic Number	Percent Weight of Human Body
O	Oxygen	8	65.0
C	Carbon	6	18.5
H	Hydrogen	1	9.5
N	Nitrogen	7	3.5
Ca	Calcium	20	1.5
P	Phosphorus	15	1.0
K	Potassium	19	0.4
S	Sulfur	16	0.3
Na	Sodium	11	0.2
Cl	Chlorine	17	0.2
Mg	Magnesium	12	0.1

Trace elements (less than 0.01%): boron (B), chromium (Cr), cobalt (Co), copper (Cu), fluorine (F), iodine (I), iron (Fe), manganese (Mn), molybdenum (Mo), selenium (Se), silicon (Si), tin (Sn), vanadium (V), and zinc (Zn).

Figure 2.4
Nitrogen deficiency in corn. In this controlled experiment, the plants on the left are growing in soil that was fertilized with compounds containing nitrogen, an essential element. The soil on the right is deficient in nitrogen. Even if the poorly nourished crop growing in this soil can be harvested, it will yield less food and its chemical deficiencies will be passed on to livestock or human consumers.

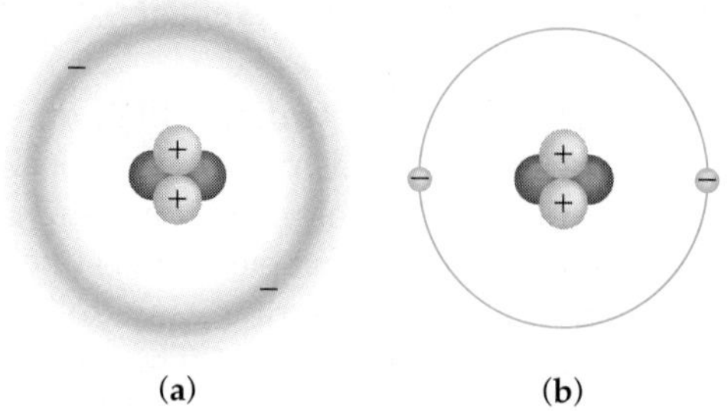

Figure 2.5
Two simplified models of a helium (He) atom. The helium nucleus consists of two neutrons (dark gray) and two protons (light gray). (**a**) The nucleus is surrounded by a cloud of negative charge (blue), owing to the rapid movement of two electrons. (**b**) The circle indicates the average distance of the electrons from the nucleus, although this distance is not drawn to scale compared to the size of the nucleus. (Our model of an atom will be refined as the chapter progresses.)

THE STRUCTURE AND BEHAVIOR OF ATOMS

The units of matter are called **atoms.** They are so small that it would take about a million of them to stretch across the period printed at the end of this sentence. Each element consists of a certain kind of atom, which is different from the atoms of any other element. We symbolize atoms with the same abbreviation used for the element made up of those atoms; thus, C stands for both the element carbon and a single carbon atom. An atom is the smallest possible amount of an element that retains that element's properties.

Subatomic Particles

Although the atom is the smallest unit having the physical and chemical properties of its element, these tiny bits of matter are composed of even smaller parts, called subatomic particles. Physicists have split the atom into more than a hundred types of particles, but only three kinds of particles are stable enough to be of relevance here: **neutrons, protons,** and **electrons.** Neutrons and protons are packed together tightly to form a dense core, or **nucleus,** at the center of the atom. The electrons move about this nucleus at nearly the speed of light (Figure 2.5).

Electrons and protons are electrically charged. Each electron has one unit of negative charge, and each proton has one unit of positive charge. A neutron, as its name implies, is electrically neutral. The atomic nucleus is positive because of the presence of protons, and it is the attraction between opposite charges that keeps the rapidly moving electrons in the vicinity of the nucleus.

The neutron and proton are almost identical in mass, each about 1.7×10^{-24} grams (g). Grams and other conventional units are not very useful for describing the mass of objects so minuscule. Thus, for atoms and subatomic particles, scientists use a unit of measurement called the **dalton,** in honor of John Dalton, the English chemist and physicist who helped develop atomic theory around 1800. Neutrons and protons have a mass of almost exactly 1 dalton apiece (actually 1.007 and 1.009, respectively). Because the mass of an electron is only about $\frac{1}{2000}$ that of a neutron or proton, we can ignore electrons when computing the total mass of an atom (Table 2.2).

Table 2.2 Properties of the Subatomic Particles

Particle	Approximate Weight (Daltons)	Charge
Neutron	1	0
Proton	1	+1
Electron	$\frac{1}{2000}$	−1

c Number and Atomic Weight

f the various elements differ in their number of subatomic particles. All atoms of a particular element have the same number of protons in their nuclei. This number, which is unique to that element, is referred to as the **atomic number** and is written as a subscript to the left of the symbol for the element. The abbreviation $_2$He, for example, tells us that an atom of the element helium has two protons in its nucleus. Unless otherwise indicated, an atom is neutral in electric charge, which means that its protons must be balanced by an equal number of electrons. Therefore, the atomic number tells us the number of protons *and* the number of electrons in a neutral atom.

We can deduce the number of neutrons from a second quantity, the **mass number,** which is the sum of protons plus neutrons in the nucleus of an atom. The mass number is written as a superscript to the left of an element's symbol. For example, we can use this shorthand to write an atom of helium as $^4_2$He. Since the atomic number indicates how many protons there are, we can determine the quantity of neutrons by subtracting the atomic number from the mass number: A $^4_2$He atom has 2 neutrons. An atom of sodium, $^{23}_{11}$Na, has 11 protons, 11 electrons, and 12 neutrons. The simplest atom is hydrogen, $^1_1$H, which has no neutrons—a lone proton with a single electron moving about it constitutes a hydrogen atom.

Essentially all of an atom's mass is concentrated in its nucleus, because the contribution of electrons to mass is negligible. Since neutrons and protons each have a mass that is very close to 1 dalton, the mass number tells us the approximate mass of the whole atom. The term **atomic weight** is often used to refer to what is technically the total atomic mass, or mass number. Thus, the atomic weight of helium ($^4_2$He) is 4 daltons (4.003, to be exact).

Isotopes

All atoms of a given element have the same number of protons, but some atoms have more neutrons than other atoms of the same element, and hence weigh more. These different atomic forms are referred to as **isotopes** of the element. In nature, an element occurs as a mixture of its isotopes. For example, consider the three isotopes of the element carbon, which has an atomic number of 6. The most common isotope is carbon-12, $^{12}_6$C, which accounts for about 99% of the carbon in nature. It has 6 neutrons. Most of the remaining 1% of carbon consists of atoms of the isotope $^{13}_6$C, with 7 neutrons. A third isotope, $^{14}_6$C, has 8 neutrons; it is present in the environment in minute quantities. Notice that all three isotopes of carbon have 6 protons—otherwise, they would not be carbon.

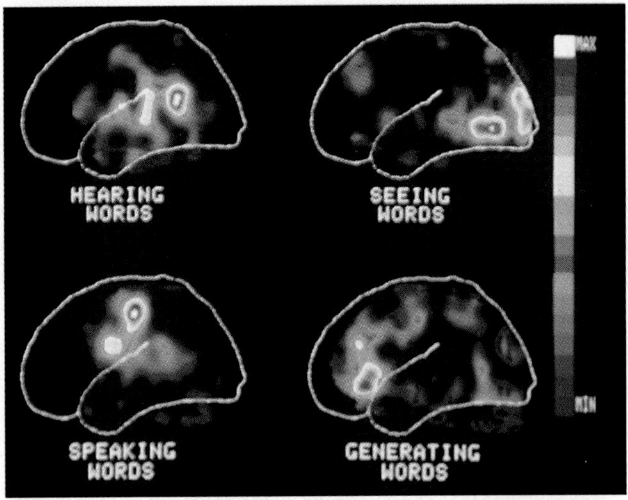

Figure 2.6
PET scans: using radioactivity to study brain function. PET, an acronym for positron-emission tomography, reveals locations of intense chemical activity within the body. Sugar labeled with radioactive isotopes that emit particles called positrons is first injected into a person's bloodstream. These positrons collide with electrons made available by chemical reactions in the body. An instrument called the PET scanner detects the energy released by these collisions and maps metabolic "hot spots"—regions of an organ that are most chemically active at the time. The PET scans shown here illustrate localized brain activity under four different conditions, all related to language. Physicians use the PET scanner to diagnose brain and heart disorders and certain types of cancer.

Both ^{12}C and ^{13}C are stable isotopes, meaning that their nuclei do not have a tendency to lose particles. The isotope ^{14}C, however, is unstable, or radioactive. A **radioactive** isotope is one in which the nucleus decays spontaneously, giving off particles and energy. Loss of nuclear particles transforms the atom to an atom of a different element. For example, radioactive carbon decays to form nitrogen.

Radioactive isotopes have many useful applications in biology. In Chapter 23, you will learn how researchers use the amount of radioactivity in fossils to date those relics of past life. Radioactive isotopes are also useful as tracers to follow atoms through metabolism, the chemical processes of an organism (see the Methods Box). Cells use the radioactive atoms as they would nonradioactive isotopes of the same element, but the radioactive tracers can be detected. Radioactive tracers have thus become important diagnostic tools in medicine. For example, certain kidney disorders can be diagnosed by injecting small doses of substances containing radioactive isotopes into the blood and then measuring the amount of tracer excreted in the urine. The PET scans shown in Figure 2.6 are another example of how radioactivity is applied in medicine.

Although radioactive isotopes are very useful in biological research and medicine, radiation from these

Methods: The Use of Radioactive Tracers in Biology

Radioactive isotopes are among the most important tools in biological research. These isotopes serve as "spies" within an organism: They are used to label certain chemical substances to follow the steps of a metabolic process or to determine the location of the substance within an organism. Organisms do not generally discriminate between radioactive and stable isotopes of the same element; thus, they assimilate and process the labeled substance normally.

The experiments being conducted here were designed to determine how temperature affects the rate at which the genetic material, DNA, replicates in a population of dividing animal cells, and to locate the newly synthesized DNA within the cells. The experiment begins by culturing cells in an artificial medium that contains, among other things, the specific chemical ingredients used by cells to make new DNA. One of those ingredients is labeled with ^3H, a radioactive isotope of hydrogen that will be used to trace the incorporation of the ingredient into new DNA.

After a certain amount of time has elapsed, samples of cells grown at various temperatures in the presence of the radioactive tracer are killed, and their DNA is precipitated onto pieces of filter paper. The papers are then placed in vials containing scintillation fluid, which emits flashes of light whenever certain chemicals in the fluid are excited by radiation from the decay of the radioactive tracer in the DNA. The frequency of flashes, proportional to the amount of radioactive material present, is measured in counts per minute by placing the vials in a scintillation counter (Figure a). The effect of temperature on the rate of DNA synthesis can be determined by plotting the counts per minute for the various DNA samples against the temperatures at which the cells were grown (Figure b).

A technique known as autoradiography can be used to determine the location of the radioactively labeled DNA within the cells. The cells are washed free of any radioactive material that was not incorporated into the DNA. Then they are fixed (preserved), and thin sections of them are placed on glass slides. The sections are covered by a layer of photographic emulsion, and the slides are kept for some time in the dark. Wherever DNA is located in the cells, radiation from the radioactive tracer will expose the photographic emulsion. The emulsion is developed, and the slide is examined with a microscope (Figure c). Black grains of the exposed emulsion are superimposed on the nuclei of the cells, the sites of DNA. The nucleus of the cell on the left has been radioactively labeled.

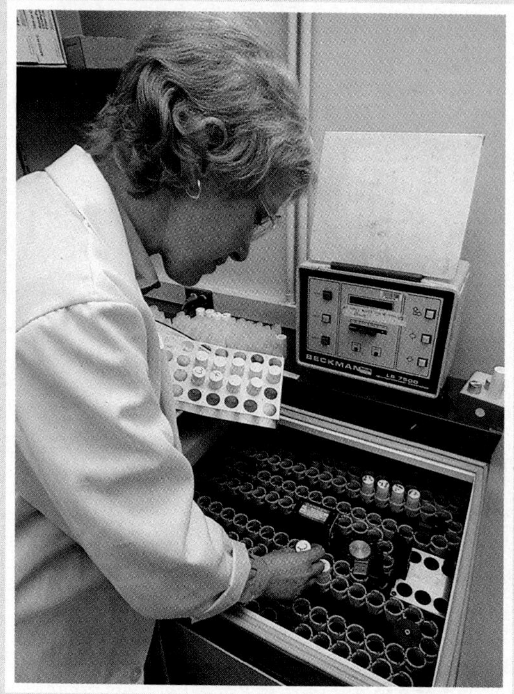

(a)

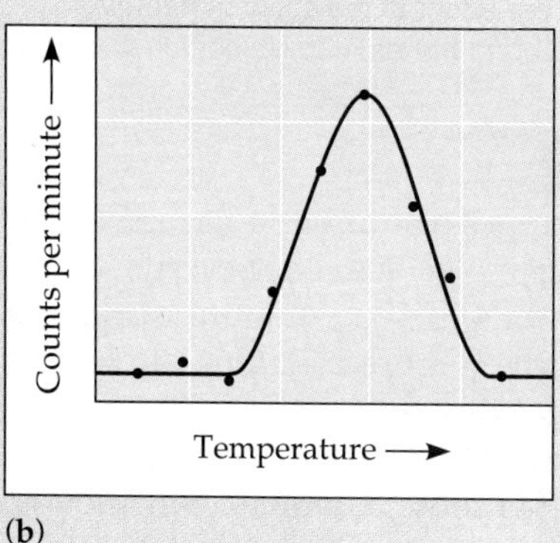

(b)

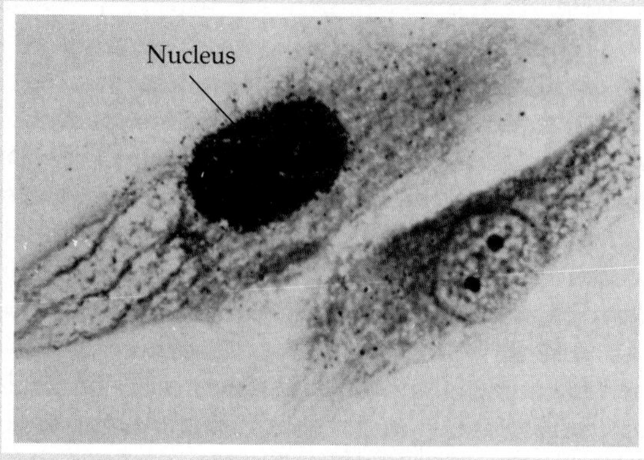

(c)

Figure 2.7
Radiation damage to a forest. In this experiment conducted at the Brookhaven National Laboratory on Long Island, New York, radioactive material was placed in a canister on the top of a pole to assess the effect of radiation on nearby plants. A circle of dead trees around the radioactive source is obvious.

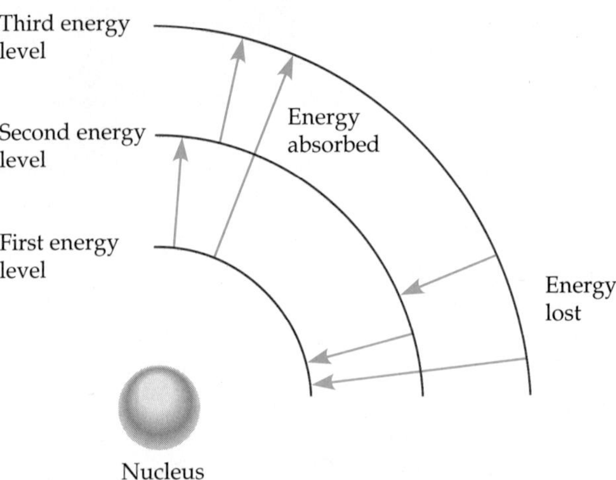

Figure 2.8
Energy levels of electrons. Electrons exist only at fixed levels of potential energy. An electron can move from one level to another only if the energy it gains or loses is exactly equal to the difference in energy between the two levels. Arrows indicate some of the stepwise changes in potential energy that are possible for electrons. Energy levels are also called electron shells.

decaying isotopes also poses a hazard to life by damaging cellular molecules. The severity of this damage depends on the type and amount of radiation an organism absorbs (Figure 2.7). One of the most serious environmental threats is radioactive fallout from nuclear accidents. During the 1986 nuclear power plant disaster in Chernobyl, Ukraine, giant clouds of radioactive materials were belched into the air, producing widespread contamination downwind.

Energy Levels

The simplified models of the atom in Figure 2.5 distort the size of the nucleus relative to the volume of the whole atom. If the nucleus were the size of a golf ball, the electrons would be moving about the nucleus at an average distance of approximately 1 kilometer (km). Atoms are mostly empty space.

When two atoms approach each other, their nuclei do not come close enough to interact. Of the three kinds of subatomic particles we have discussed, only electrons are directly involved in the chemical reactions between atoms.

All electrons vary in the amount of energy they possess. **Energy** is defined as the ability to do work. **Potential energy** is the energy that matter stores because of its position or location. For example, because of its altitude, water in a reservoir on a hill has potential energy. When the gates of the dam are opened and the water runs downhill, the energy is taken out of storage to do work, such as turning generators. Since potential energy has been expended, the water stores less energy at the bottom of the hill than it did in the reservoir. Matter has a natural tendency to move to the lowest possible state of potential energy; in this example, water runs downhill. To restore the potential energy of a reservoir, work must be done to elevate the water against gravity.

The electrons of an atom also have potential energy because of their position in relation to the nucleus. The negatively charged electrons are attracted to the positively charged nucleus; the more distant the electrons are from the nucleus, the greater their potential energy. Unlike the continuous flow of water downhill, changes in the potential energy of electrons can occur only in steps of fixed amounts. An electron having a certain discrete amount of energy is analogous to a ball on a staircase. The ball can have different amounts of potential energy, depending on which step it is on, but it cannot spend much time between the steps (Figure 2.8).

The different states of potential energy for electrons in an atom are called **energy levels,** or **electron shells.** The first shell is closest to the nucleus, and electrons in this shell have the lowest energy. Electrons in the sec-

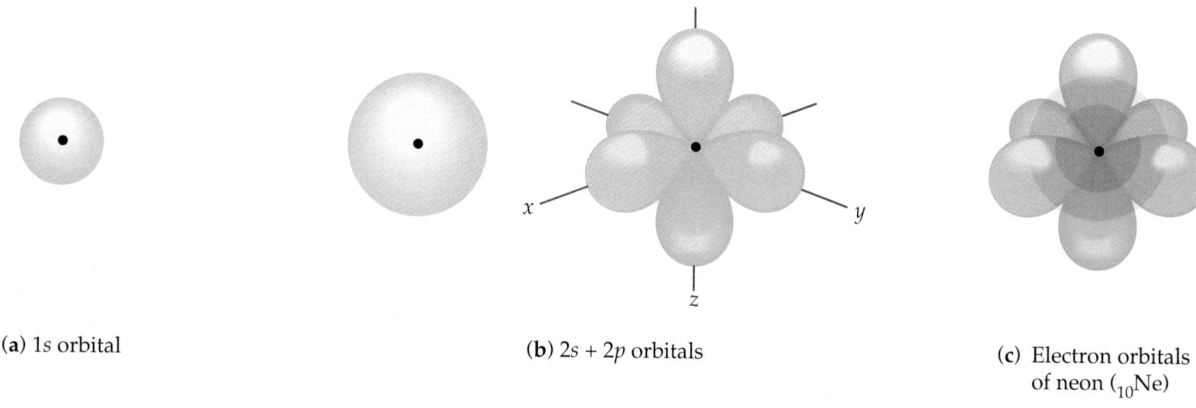

Figure 2.9
Electron orbitals. These three-dimensional shapes represent the volumes of space where the constantly moving electrons are most likely to be found. Each orbital holds a maximum of two electrons. (**a**) The first electron shell has one spherical *(s)* orbital, designated 1*s*. Only this orbital is present in hydrogen, which has one electron, and helium, which has two electrons. (**b**) The second and all higher shells each have one larger *s* orbital (2*s*) plus three dumbbell-shaped orbitals designated *p* orbitals (2*p*). The three 2*p* orbitals are arranged at right angles to one another along imaginary *x, y,* and *z* axes of the atom. The third and higher electron shells can hold additional electrons in orbitals of more complex shapes. (**c**) To symbolize the electron clouds of the element neon, which has a total of ten electrons, we superimpose the 1*s* orbital of the first shell and the 2*s* and three 2*p* orbitals of the second shell.

ond shell have more energy, electrons in the third shell more energy still, and so on. An electron can change its shell, but only by absorbing or losing an amount of energy equal to the difference in potential energy between the old shell and the new shell. To move to a shell farther out from the nucleus, the electron must absorb energy. For example, light can excite an electron to a higher energy level. Indeed, this is the first step when plants harness light energy for photosynthesis. To move to a shell closer in, an electron must lose energy, which is usually released to the environment in the form of heat.

Electron Orbitals

Earlier in this century, the electron shells of an atom were visualized as concentric paths of electrons orbiting the nucleus, something like planets orbiting the sun (see Figure 2.5). The atom is not this simple, however. In fact, we can never know the exact trajectory of an electron. What we can do instead is describe the volume of space in which an electron spends most of its time. The three-dimensional space where an electron is found 90% of the time is called an **orbital.** Unlike a planetary orbit, an orbital is not a defined pathway of movement. The electron orbital is a statistical concept—a volume within which an electron has the greatest probability of being found (Figure 2.9).

No more than two electrons can occupy the same orbital. The first energy shell has a single orbital and can thereby accommodate a maximum of two electrons. This single orbital, which is spherical in shape, is designated the 1*s* orbital. The lone electron of a hydrogen atom occupies the 1*s* orbital, as do the two electrons of a helium atom. Electrons, like all matter, tend to exist in the lowest available state of potential energy, which they have in the first shell. An atom with more than two electrons must use higher shells, because the first shell is full.

The second electron shell can hold eight electrons, two in each of four orbitals. Electrons in the four different orbitals all have virtually the same energy, but they move in different volumes of space. There is a 2*s* orbital, spherical in shape like the 1*s* orbital, but with a slightly greater diameter. The other three orbitals, called 2*p* orbitals, are dumbbell-shaped, each oriented at right angles to the other two.

Electron Configuration and Chemical Properties

The chemical behavior of an atom is determined by its electron configuration—that is, the distribution of electrons in the atom's electron shells. Beginning with hydrogen, the simplest atom, we can imagine building the atoms of other elements by adding one proton and one electron at a time (along with an appropriate number of neutrons). Figure 2.10, an abbreviated version of what is called a periodic table, shows this for the first 18 elements, from hydrogen ($_1H$) to argon ($_{18}Ar$). The elements are arranged in three tiers, or periods, corresponding to the sequential filling of the first three electron shells. The main point is that the chemical properties of an atom depend mostly on the number of

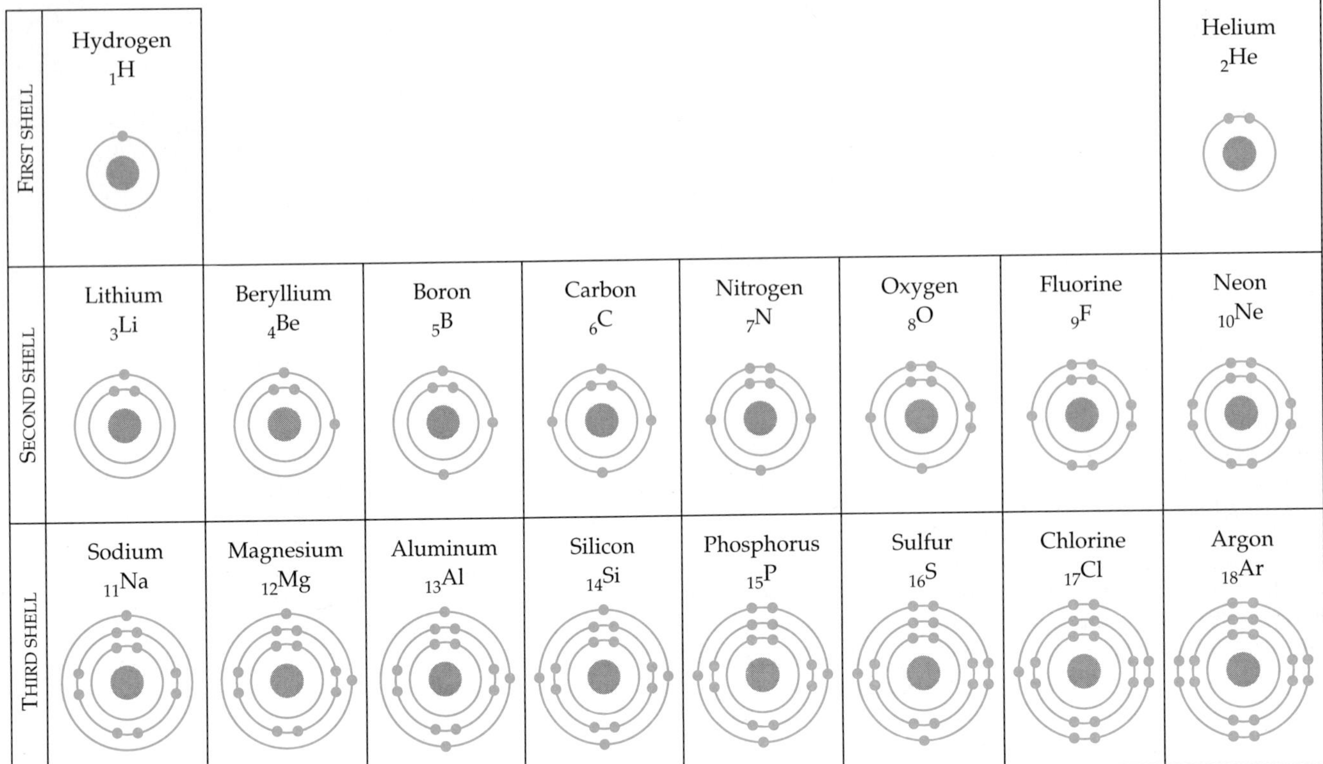

Figure 2.10
Electron configurations of the first 18 elements. The number of electrons in each energy level (shell) is diagrammed as dots on concentric rings. (These symbolic rings do not portray the three-dimensional freedom of electron movement.) The elements are arranged in rows, each representing the filling of an electron shell. As electrons are added, they occupy the lowest available energy level. Hydrogen's one electron and helium's two electrons are located in the first level. The next element, lithium, has three electrons. Two electrons fill the first energy level, while the third electron occupies the second energy level, not some level more distant from the nucleus. This behavior of electrons illustrates the general tendency for matter to exist in its lowest state of potential energy. The outermost energy level occupied by electrons is called the valence shell. Of these first 13 elements, only helium, neon, and argon are made of atoms with full valence shells; such elements are called inert because they are unreactive. All other elements in this figure consist of atoms with incomplete valence shells and are chemically reactive. Those elements with the same number of valence electrons—fluorine and chlorine, for instance—have similar chemical properties.

electrons in its *outermost* shell. We refer to those outer electrons as **valence electrons,** and to the outermost energy shell as the **valence shell.**

An atom with a complete valence shell is unreactive; that is, it will not interact with other atoms it encounters. On the right-hand side of the periodic table are helium, neon, and argon, the only three elements shown that have full valence shells. They are termed inert elements because they are chemically stable (unreactive). All other atoms shown in Figure 2.10 have incomplete valence shells, and all are chemically reactive. Atoms with the same number of electrons in their valence shells exhibit similar chemical behavior. For example, fluorine (F) and chlorine (Cl) both have seven valence electrons, and both combine with the element sodium to form compounds.

CHEMICAL BONDS AND MOLECULES

Now that we have looked at the structure of atoms, we will move up in the hierarchy of organization and see how atoms combine to form molecules. Atoms with incomplete valence shells will interact with certain other atoms in such a way that each partner completes its valence shell. Atoms do this by either sharing or completely transferring valence electrons. These interactions usually result in atoms staying close together, held by attractions called **chemical bonds.** The strongest kinds of chemical bonds are covalent bonds and ionic bonds. We will also examine a third type of bond, the hydrogen bond, which plays an important role in the chemistry of life.

Covalent Bonds

A **covalent bond** is the sharing of a pair of valence electrons by two atoms. For example, let's see what happens when two hydrogen atoms approach each other (Figure 2.11). Recall that hydrogen has one valence electron in the first shell, but the shell's capacity is for two electrons. When the two hydrogen atoms come close enough for their 1*s* orbitals to overlap, they share their electrons. Each hydrogen atom now has two electrons moving through its 1*s* orbital, and the valence shell is complete. In this case, we have formed a hydrogen molecule consisting of two hydrogen atoms held together by a covalent bond. We abbreviate this molecule by writing H—H, where the line represents a covalent bond, that is, a pair of shared electrons. This type of notation, which represents both atoms and bonding, is called a **structural formula.** We can abbreviate even further by writing H_2, a **molecular formula** indicating simply that the molecule consists of two atoms of hydrogen (Figure 2.12a).

With six electrons in its second electron shell, oxygen needs two more electrons to complete this valence shell. Two oxygen atoms form a molecule by sharing two pairs of valence electrons (Figure 2.12b). The atoms are joined by what is called a **double covalent bond.** The structural formula for this molecule is O═O, and its molecular formula is O_2. Nitrogen has five valence electrons, three less than it needs for a complete valence shell. Two nitrogen atoms will join together by a triple covalent bond and share three pairs of valence electrons, forming an N_2, or N≡N, molecule.

Notice that each atom sharing electrons has a bonding capacity: a certain number of covalent bonds that must be formed for the atom to have a full complement of valence electrons. This bonding capacity is called the atom's **valence.** The valence of hydrogen is 1; oxygen, 2; and nitrogen, 3.

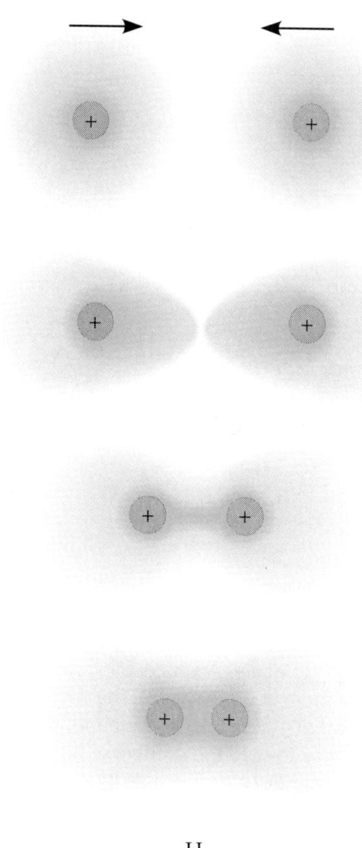

Figure 2.11
Formation of a hydrogen molecule. A hydrogen atom, with its single electron in the first energy level, is chemically reactive. Here, two hydrogen atoms react to form a hydrogen molecule. The atoms' electron clouds first overlap and then unite into a single orbital. The result is a covalent bond—a pair of shared electrons. Sharing completes the valence shells of each partner. The atoms' nuclei are mutually attracted to the dense region of the electron cloud between the nuclei. The length of the bond is determined by a balance between two opposing forces: the attraction of the nuclei for the electrons versus the repulsion of the two positively charged nuclei.

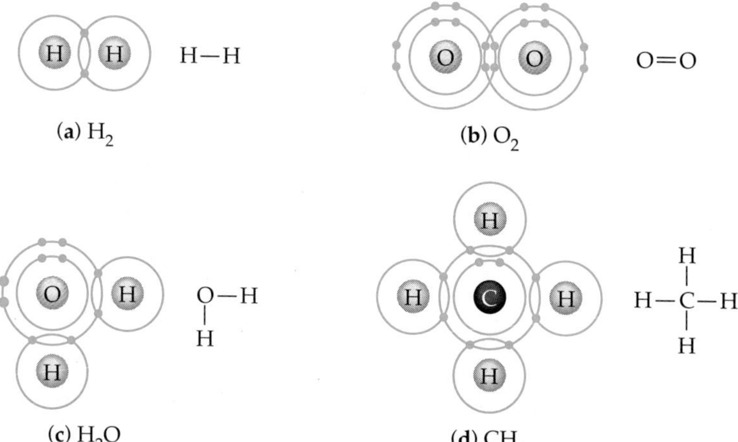

Figure 2.12
Covalent bonds. Each covalent bond consists of a pair of shared electrons. The number of electrons required to complete an atom's valence shell determines how many bonds that atom will form. (**a**) If two unattached hydrogen atoms meet, they will form a single covalent bond. (**b**) Two oxygen atoms form a molecule by sharing two pairs of valence electrons; the atoms are joined by a double covalent bond. (**c**) Two hydrogen atoms can be joined to one oxygen atom by covalent bonds to produce a molecule of water. (**d**) Four hydrogen atoms satisfy the valence of one carbon atom, forming methane.

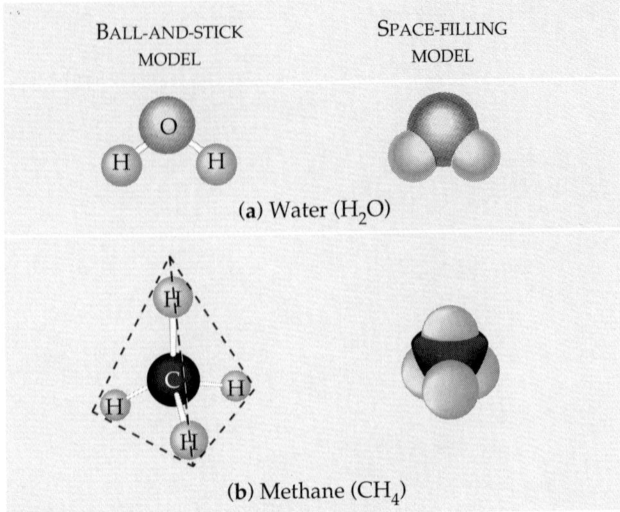

Figure 2.13
Molecular shapes of water and methane. This figure introduces two types of models that represent three-dimensional shapes of molecules. A ball-and-stick model emphasizes the bond angles of the molecule. A space-filling model portrays a molecule's shape more accurately. (**a**) The two covalent bonds of water are angled at 104.5°. (**b**) In methane, carbon's four covalent bonds angle toward the corners of an imaginary tetrahedron, outlined here with a dotted pink line.

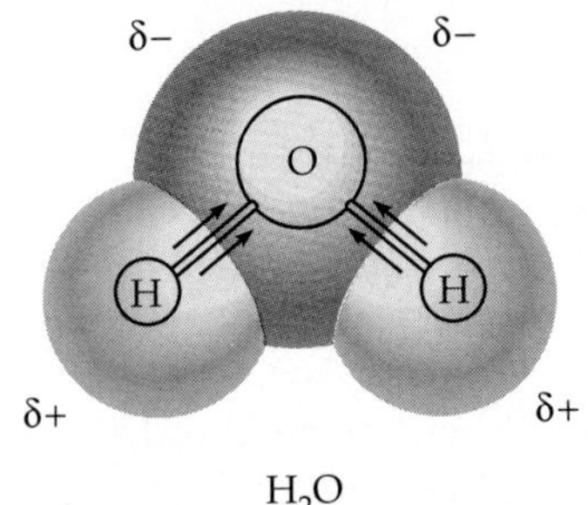

Figure 2.14
Polar covalent bonds in a water molecule. Oxygen, being much more electronegative than hydrogen, pulls the shared electrons of the bond toward itself. This unequal sharing of electrons gives the oxygen a slight negative charge and the hydrogens a small amount of positive charge. The Greek symbol delta (δ) indicates that the charges are less than full units.

The molecules we have looked at so far—H_2, O_2, and N_2—are pure elements, not compounds. (Recall that a compound is a combination of two or more different elements.) Water, for example, is a compound whose molecular formula is H_2O; it takes two atoms of hydrogen to satisfy the valence of one oxygen atom. Figure 2.12c shows the structure of a water molecule. This molecule is so important to life that the next chapter is devoted entirely to its structure and behavior.

Another molecule that is also a compound is methane, a component of natural gas, with the molecular formula CH_4 (Figure 2.12d). Carbon ($_6C$) has four valence electrons, so its bonding capacity, or valence, is 4. It takes four hydrogen atoms, each with a valence of 1, to complement one atom of carbon. We now know the valences of the four most abundant elements in life: Hydrogen forms one bond, oxygen forms two bonds, nitrogen forms three bonds (usually), and carbon forms four bonds.

The Biological Importance of Molecular Shape A molecule has a characteristic size and shape (Figure 2.13). The water molecule, for example, is shaped roughly like a right angle, its two covalent bonds spread apart by 104.5°. The methane molecule is shaped like a tetrahedron, a pyramid with a three-sided base. The nucleus of the carbon atom is at the center, with its four covalent bonds radiating to the hydrogen nuclei at the corners of the tetrahedron. Larger molecules have more complex shapes. Molecular shape is important to biologists because it is the basis for how most molecules of life recognize and respond to one another.

Nonpolar and Polar Covalent Bonds The attraction of an atom for the electrons of a covalent bond is called **electronegativity.** The more electronegative an atom, the more strongly it pulls shared electrons toward itself. In a covalent bond between two atoms of the same element, the outcome of the tug-of-war for common electrons is a standoff; the two atoms are equally electronegative. In a **nonpolar covalent bond,** electrons are shared equally. The covalent bond of H_2 is nonpolar, as is the double bond of O_2. The bonds of methane (CH_4) are also nonpolar; although the partners are different elements, carbon and hydrogen do not differ substantially in electronegativity. This is not always the case in a compound where covalent bonds join atoms of different elements. If one atom is more electronegative than the other, electrons of the bond will not be shared equally. In such cases, the bond is called a **polar covalent bond.** In a water molecule, the bonds between oxygen and hydrogen are polar. Oxygen is one of the most electronegative of the 92 elements, attracting shared electrons much more strongly than hydrogen does. In a covalent bond between oxygen and hydrogen, the electrons spend more time around the oxygen atom than they do around the hydrogen atom. Since electrons have a negative charge, the unequal sharing of electrons in water causes the oxygen atom to have a slight negative charge and each hydrogen atom a slight positive charge (Figure 2.14).

Ionic Bonds

In some cases, two atoms are so unequal in their attraction for valence electrons that the more electronegative atom strips an electron completely away from its

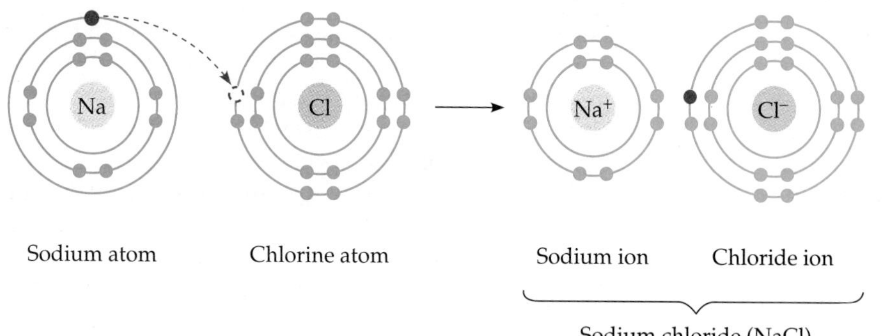

Figure 2.15
Electron transfer and ionic bonding. A valence electron is transferred from sodium (Na) to chlorine (Cl), giving both atoms completed valence shells. The electron transfer leaves the sodium atom with a net charge of +1 and the chlorine atom with a net charge of −1. The attraction between the oppositely charged atoms, or ions, is an ionic bond. Ions can bond not only to the atom they reacted with, but to any other ion of opposite charge.

partner. This is what happens when an atom of sodium ($_{11}Na$) encounters an atom of chlorine ($_{17}Cl$) (Figure 2.15). A sodium atom has a total of 11 electrons, with its single valence electron in the third electron shell. A chlorine atom has a total of 17 electrons, with 7 electrons in its valence shell. When these two atoms meet, the lone valence electron of sodium is transferred to the chlorine atom, and both atoms end up with their valence shells complete. (Since sodium no longer has an electron in the third shell, the second shell is now outermost.)

The electron transfer between the two atoms moves one unit of negative charge from sodium to chlorine. Sodium, now with 11 protons but only 10 electrons, has a net electric charge of +1. A charged atom (or molecule) is called an **ion.** When the charge is positive, the ion is specifically called a **cation.** Conversely, the chlorine atom, having gained an extra electron, now has 17 protons and 18 electrons, giving it a net electric charge of −1. It has become a chloride ion—specifically, an **anion,** or negatively charged ion. Because of their opposite charges, cations and anions attract each other in what is called an **ionic bond.**

Ionic compounds are called salts. We know the compound sodium chloride (NaCl) as table salt. Salts are often found in nature as crystals of various sizes and shapes, each an aggregate of vast numbers of cations and anions bonded by their electrical attraction and arranged in a three-dimensional lattice (Figure 2.16). A salt crystal does not really consist of molecules in the same sense that a covalent compound does, since a covalently bonded molecule has a definite size and number of atoms. The formula for an ionic compound, such as NaCl, indicates only the ratio of elements in a crystal of the salt.

Not all salts have equal numbers of cations and anions. For example, the ionic compound magnesium chloride ($MgCl_2$) has two chloride ions for each magnesium ion. Magnesium ($_{12}Mg$) must lose two outer electrons if the atom is to have a complete valence shell. One magnesium atom can supply valence electrons to two chlorine atoms. After losing two electrons, the magnesium atom is a cation with a new charge of +2 (Mg^{2+}).

The term *ion* also applies to entire covalent molecules that are electrically charged. In ammonium chloride (NH_4Cl), for instance, the anion is a single chloride ion (Cl^-), but the cation is ammonium (NH_4^+), a nitrogen atom with four covalently bonded hydrogen atoms. The whole ammonium ion has an electric charge of +1 because it is one electron short.

There is no distinct line between covalent bonding and ionic bonding. A nonpolar covalent bond and an ionic bond are opposite extremes in a range of situations where atoms share electrons. In the middle zone is the polar covalent bond, in which electrons are shared, but unequally. We might think of an ionic bond as a covalent bond that is so polar that one atom has pulled an electron completely away from its less electronegative partner. Indeed, some compounds spend part of their time in a polar covalent state and the rest of their time as ions.

Figure 2.16
A sodium chloride crystal. The sodium ions (Na^+) and chloride ions (Cl^-) are held together by ionic bonds.

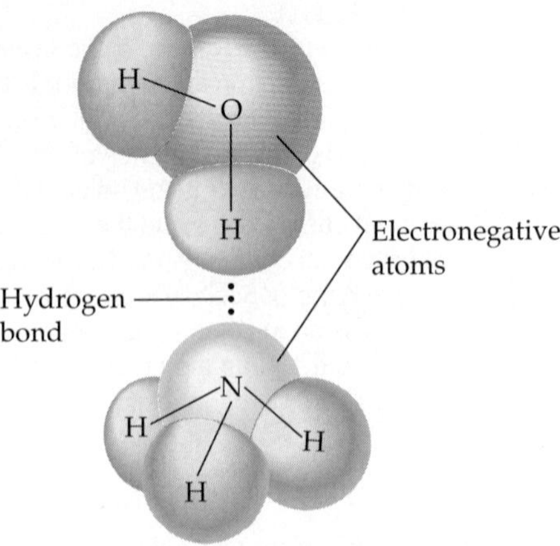

Figure 2.17
A hydrogen bond. A hydrogen atom attached to an electronegative atom is shared with another electronegative atom through a weak electrical attraction. In this figure, a hydrogen bond joins a hydrogen atom of a water molecule (H_2O) with the nitrogen atom of an ammonia molecule (NH_3).

Hydrogen Bonds

A third type of chemical bond important in life is the **hydrogen bond.** A hydrogen bond occurs when a hydrogen atom covalently bonded to one electronegative atom is also attracted to another electronegative atom (Figure 2.17). In living cells, the electronegative partners involved are usually oxygen or nitrogen atoms. You have seen how the polar covalent bonds of water result in the oxygen atom having a slight negative charge and the hydrogen atoms having a slight positive charge. A similar situation arises in the ammonia molecule (NH_3), where an electronegative nitrogen atom has a small amount of negative charge because of its pull on the electrons it shares covalently with hydrogen. If a water molecule and an ammonia molecule are close to each other, there will be a weak attraction between the negatively charged nitrogen atom and a positively charged hydrogen atom of the adjacent water molecule. This attraction is a hydrogen bond.

The Biological Importance of Weak Bonds

Hydrogen bonds are about twenty times weaker (easier to break) than are covalent bonds. Covalent bonds link together the atoms of molecules. But bonding *between* molecules is also important, especially in the chemistry of the cell, where the properties of life emerge from molecular interactions. When two molecules in the cell associate, they may adhere temporarily by weak chemical bonds, including hydrogen bonds. The advantage of weak bonds is that the contact between the molecules can be brief; the molecules come together, respond to one another in some way, and then separate. An example is chemical signaling in the brain, a process that Candace Pert has studied (see the interview preceding this chapter). A nerve cell in the brain stimulates a neighboring nerve cell by releasing molecules that bind to specific receptor molecules on the surface of the receiving cell. The signal molecule uses weak bonds to dock on the receptor just long enough to trigger a momentary response by the receiving cell. If the signal attached by stronger covalent bonds, the receiving cell would continue to respond long after the transmitting cell ceased dispatching the message. (Imagine, for instance, if your brain continued to perceive the ringing sound of a bell for hours after nerve cells transmitted the information from the ears to the brain.)

Hydrogen bonds, ionic bonds, and other weak bonds form not only between molecules; they also may form between different regions of a single large molecule, such as a protein. Although these bonds are individually weak, their cumulative effect is to reinforce the three-dimensional shape of a large molecule. We will learn more about the role of weak bonds in Chapter 5.

CHEMICAL REACTIONS

The making and breaking of chemical bonds, leading to changes in the composition of matter, are called **chemical reactions.** An example is the reaction between hydrogen and oxygen to form water:

$$2\,H_2 + O_2 \longrightarrow 2\,H_2O$$

This reaction breaks the covalent bonds of H_2 and O_2, forming the new bonds of a water molecule. When we write a chemical reaction, we use an arrow to indicate the conversion of the starting materials, called **reactants,** to the **products.** The coefficients indicate the number of molecules involved; for example, the 2 in front of the H_2 means that the reaction starts with two molecules of hydrogen. Notice that all atoms of the reactants must be accounted for in the products. Matter is conserved in a chemical reaction: Reactions cannot create or destroy matter but can only rearrange it in various ways.

Below is the chemical shorthand that summarizes the process of photosynthesis, another example of chemical reactions:

$$6\,CO_2 + 6\,H_2O \longrightarrow\longrightarrow\longrightarrow C_6H_{12}O_6 + 6\,O_2$$

Recall from Figure 1.2 that photosynthesis occurs within the chloroplasts of leaf cells. The raw materials are carbon dioxide (CO_2), which is taken from the air, and water (H_2O), which is absorbed from the soil. Within the chloroplasts, sunlight powers the conversion of these ingredients to a sugar called glucose

($C_6H_{12}O_6$) and oxygen (O_2), which the leaves release into the air. Although photosynthesis is actually a sequence of many chemical reactions, we still end up with the same number and kinds of atoms we had when we started. Matter has been rearranged but conserved.

Some chemical reactions go to completion; that is, all the reactants are converted to products. But most reactions are reversible, the products of the forward reaction becoming the reactants for the reverse reaction. For example, hydrogen and nitrogen molecules combine to form ammonia, but ammonia can also decompose to regenerate hydrogen and nitrogen:

$$3\,H_2 + N_2 \rightleftharpoons 2\,NH_3$$

The double arrows indicate that the reaction is reversible.

One of the factors affecting the rate of reaction is the concentration of reactants. The greater the concentration of reactant molecules, the more frequently they collide with one another and have an opportunity to react to form products. The same holds true for the products. As products accumulate, collisions resulting in the reverse reaction become increasingly frequent. Eventually, the forward and reverse reactions occur at the same rate, and the relative concentrations of products and reactants remain fixed. The point at which the reactions offset one another exactly is called **chemical equilibrium.** This is a dynamic equilibrium; reactions are still going on, but with no net effect on the concentrations of reactants and products. Equilibrium does *not* mean that the reactants and products are equal in concentration, but only that their concentrations have stabilized. For the above reaction involving ammonia, equilibrium is reached when ammonia decomposes as rapidly as it forms. In this case, the forward reaction is much more favorable than the reverse reaction; thus, at equilibrium, there will be far more ammonia than hydrogen and nitrogen.

We will end this chapter by placing the chemical basis of life in an evolutionary context.

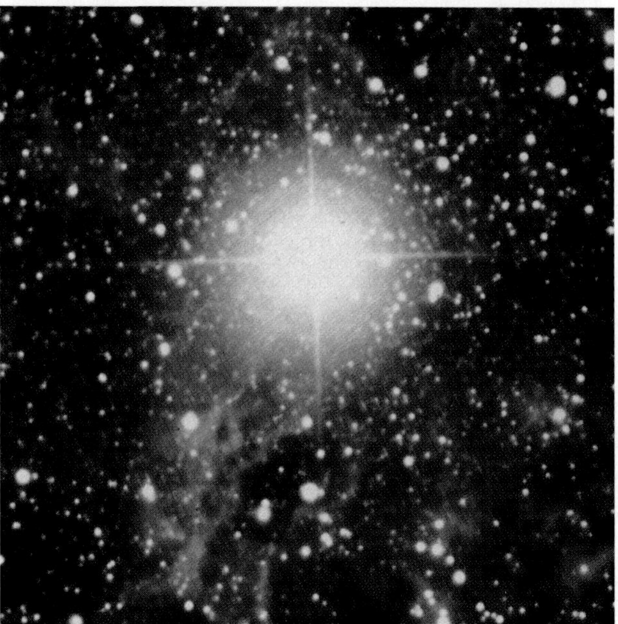

Figure 2.18
Supernova 1987a. The thermonuclear reactions of stars convert hydrogen and helium into heavier chemical elements. When a star consumes all of its hydrogen fuel, it may contract and then explode. Astronomers occasionally witness such explosions as bright objects in the sky, called supernovae. This one was discovered in 1987. The violent death of such stars scatters matter, including heavy elements, in the cosmos. Our solar system formed from such debris.

CHEMICAL CONDITIONS ON THE EARLY EARTH: SETTING THE STAGE FOR THE ORIGIN AND EVOLUTION OF LIFE

Chemical reactions and physical processes on the early Earth created an environment that made life possible. And life, once it began, transformed the planet's chemistry. Biological and geological history are inseparable.

The formation of the planet Earth and its life is a fragment of a much bigger story. Earth is one of nine planets orbiting the sun, one of billions of stars in the Milky Way, which is one of millions of galaxies in the universe. The star closest to our sun, Proxima Centauri, is four light years—40 trillion km—away; we see it by the light it emitted four years ago. Some stars are so distant that even if they burned out millions of years ago, we would still see them in the sky tonight; some new stars are invisible because their light has not yet reached Earth. Gazing at stars, we look back in time.

The universe has not always been so spread out. Based on several lines of evidence, most astronomers now believe that all matter was at one time concentrated in a single mass that blew apart with a "big bang" sometime between 10 and 20 billion years ago, and the universe has been expanding ever since.

Some stars die a violent death when they explode; astronomers call such explosions supernovae (Figure 2.18). Our sun is a second- or third-generation star, born about 5 billion years ago from the fallout of defunct stars. Compared with the overall universe, our solar system is relatively rich in the heavier elements that formed by fusion from smaller atoms in the crucibles of ancestral stars. Most of the swirling matter in the disk-shaped cloud of dust that formed our solar system condensed in the center as the sun. Peripheral material was left spinning around the infant sun in several rings. The planets, including Earth, formed about 4.6 billion years ago from kernels that used gravity to draw together the dust and ice in their zones.

Most geologists believe that Earth began as a cold world that later melted from the heat produced by compaction, radioactive decay, and the impact of meteorites. Molten material sorted into layers of varying density. Most of the nickel and iron sank to the center and formed a core. Less dense material became concentrated in a mantle, and the least dense material solidified into a thin crust. The present continents are attached to plates of crust that float on the flexible mantle (see Chapter 23).

The first atmosphere, which was probably composed mostly of hot hydrogen gas (H_2), escaped because the gravity of Earth was not strong enough to hold such small molecules. Volcanoes belched gases that formed a new atmosphere. Based on analysis of gases vented by modern volcanoes, scientists have speculated that the second early atmosphere consisted mostly of water vapor (H_2O), carbon monoxide (CO), carbon dioxide (CO_2), nitrogen (N_2), methane (CH_4), and ammonia (NH_3). The first seas formed from torrential rains that began when Earth had cooled enough for water in the atmosphere to condense. In addition to an atmosphere very different from the one we know, lightning, volcanic activity, and ultraviolet radiation were much more intense when Earth was young. In such a world, life began.

* * *

In upcoming chapters, you will learn more about how chemical evolution on the early Earth made the origin of life possible. For example, in the next chapter, we will examine the life-promoting properties of water.

STUDY OUTLINE

1. A living organism is the product of the numerous chemical interactions within its cells.

Matter: Elements and Compounds (pp. 24–27)

1. The basic ingredients of matter are the elements, which cannot be broken down to other substances.
2. A compound contains two or more elements in a fixed ratio and has emergent properties very different from those of its constituent elements.
3. Carbon, oxygen, hydrogen, and nitrogen make up 96% of living matter. The remaining 4% includes trace elements, which are required in minute amounts.

The Structure and Behavior of Atoms (pp. 27–32)

1. An atom is the smallest unit of an element.
2. An atom consists of three types of subatomic particles. Uncharged neutrons and positively charged protons are tightly bound in a nucleus; negatively charged electrons move rapidly about the nucleus.
3. The number of protons in an atom is called the atomic number. The number of electrons in an electrically neutral atom is equal to the number of protons.
4. The mass number of an element—also called the atomic weight—indicates the sum of the protons and neutrons and approximates the mass of an atom in daltons.
5. Most elements consist of two or more isotopes, different in neutron number and mass. Some isotopes are unstable and give off particles and energy as radioactivity. Radioactivity has important uses in science and medicine but can also harm organisms.
6. Electron configuration determines the chemical behavior of an atom—that is, the way it reacts with other atoms.
7. Electrons move within orbitals, three-dimensional spaces located within successive electron shells (energy levels) surrounding the nucleus.
8. Chemical properties depend on the number of valence electrons, those in the outermost shell. An atom is chemically unreactive if its valence shell is full; an atom with an incomplete valence shell is reactive (unstable).

Chemical Bonds and Molecules (pp. 32–36)

1. Chemical bonds form when atoms with incomplete valence shells interact to produce complete valence shells.
2. A covalent bond forms when two atoms share a pair of valence electrons.
3. Molecules consist of two or more bonded atoms.
4. A structural formula shows the atoms and bonds in a molecule. A molecular formula indicates only the number and types of atoms.
5. Molecules have characteristic shapes and sizes that affect how they function in biological systems.
6. A nonpolar covalent bond forms when both atoms are equally electronegative. Electrons of a polar covalent bond are pulled closer to the more electronegative atom.
7. An ionic bond is created when two atoms differ so much in electronegativity that one or more electrons are actually transferred from one atom to the other. The recipient atom becomes a negatively charged anion. The donor atom becomes a positively charged cation. Cations and anions attract each other in an ionic bond.
8. Hydrogen bonds are relatively weak bonds between a partially positive hydrogen atom of one polar molecule and a partially negative atom of another polar molecule.
9. Hydrogen bonds and other weak bonds function in adhesion between molecules and help reinforce the shapes of large biological molecules.

Chemical Reactions (pp. 36–37)

1. Chemical reactions break or form chemical bonds to change reactants into products. During a reaction, matter is conserved.
2. Most chemical reactions are reversible. Chemical equilibrium is reached when the forward and backward reaction rates are equal.

Chemical Conditions on the Early Earth: Setting the Stage for the Origin and Evolution of Life (pp. 37–38)

1. Conditions on the early Earth, very different from those of today, made the origin of life possible.

SELF-QUIZ

1. An element is to a (an) ________ as a tissue is to a (an) ________.
   a. atom; organism
   b. compound; organ
   c. molecule; cell
   d. atom; organ
   e. compound; organelle

2. In the term *trace element*, the modifier *trace* means
   a. The element is required in very small amounts.
   b. The element can be employed as a label to trace atoms through an organism's metabolism.
   c. The element is very rare on Earth.
   d. The element enhances health but is not essential for the organism's long-term survival.
   e. The element has a short half-life.

3. Compared to ^{31}P, the radioactive isotope ^{32}P has
   a. a different atomic number
   b. one more neutron
   c. one more proton
   d. one more electron
   e. a different charge

4. What do the four elements most abundant in life—carbon, oxygen, hydrogen, and nitrogen—have in common?
   a. They all have the same number of valence electrons.
   b. Each element exists in only one isotopic form.
   c. They are all relatively light elements, near the top of the periodic table.
   d. They are all about equal in electronegativity.
   e. They are elements produced only by living cells.

5. The atomic number of sulfur is 16. Sulfur combines with hydrogen by covalent bonding to form a compound, hydrogen sulfide. Based on the electron configuration of sulfur, we can predict that the molecular formula of the compound will be
   a. HS b. HS_2 c. H_2S d. H_3S_2 e. H_4S

6. Review the valences of carbon, oxygen, hydrogen, and nitrogen, and then determine which of the following molecules is most likely to exist.

```
a.  O=C—H

b.     H     H
       |     |
    H—C—H—C=O
       |
       H

c.          H   H
            |   |
      H—O—C—C=O
            |
            H

d.      O
        |
     H—N=H
```

7. Which orientation is most likely for two adjacent water molecules? Explain your answer.

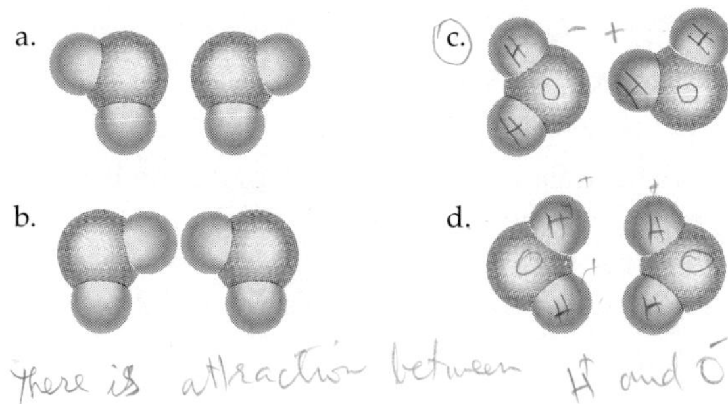

8. Which of these statements is true of *all* anionic atoms?
   a. The atom has more electrons than protons.
   b. The atom has more protons than electrons.
   c. The atom has fewer protons than does a neutral atom of the same element.
   d. The atom has more neutrons than protons.
   e. The net charge is –1.

9. What coefficients must be placed in the blanks to balance this chemical reaction?

$$C_6H_{12}O_6 \longrightarrow \underline{?}\, C_2H_6O + \underline{?}\, CO_2$$

   a. 1; 2
   b. 2; 2
   c. 1; 3
   d. 1; 1
   e. 3; 1

10. Which of the following statements correctly describes *any* chemical reaction that has reached equilibrium?
    a. The concentration of products equals the concentration of reactants.
    b. The rate of the forward reaction equals the rate of the reverse reaction.
    c. Both forward and reverse reactions have halted.
    d. The reaction is now irreversible.
    e. No reactants remain.

CHALLENGE QUESTION

1. Recall from Chapter 1 that vitalism is the belief that life possesses supernatural forces that cannot be explained by physical and chemical principles. Explain why the concept of emergent properties does *not* lend credibility to vitalism.

SCIENCE, TECHNOLOGY, AND SOCIETY

1. The use of radioactive isotopes as tracers in biochemical research is based on cells using the tracers in place of nonradioactive isotopes. Explain why this ability of radioactive isotopes to infiltrate the chemical processes of the cell also compounds the threat posed by radioactive contaminants in air, soil, and water.

2. While waiting at an airport, your author once overheard this claim: "It's paranoid and ignorant to worry about industry or agriculture contaminating the environment with their chemical wastes. After all, this stuff is just made of the same atoms that were already present in our environment." How would you counter this argument?

FURTHER READING

Atkins, P. W. *Molecules.* New York: Scientific American Library, 1987. Beautifully illustrated tour of the world of molecules.

Baker, J. J. W., and G. E. Allen. *Matter, Energy, and Life,* 4th ed. Reading, MA: Addison-Wesley, 1981. A paperback primer covering chemical topics essential to biology.

Flam, F. "COBE Finds the Bumps in the Big Bang." *Science* , May 1, 1992. Visual evidence of how the universe began.

Pennisi, E. "Natureworks." *Science News,* May 16, 1992. How organisms make minerals.

Sackheim, G. *Introduction to Chemistry for Biology Students,* 4th ed. Redwood City, CA: Benjamin/Cummings, 1991. A programmed review of the fundamentals of chemistry.

3 WATER AND THE FITNESS OF THE ENVIRONMENT

WATER MOLECULES AND HYDROGEN BONDING
SOME EXTRAORDINARY PROPERTIES OF WATER
AQUEOUS SOLUTIONS
ACID RAIN: UPSETTING THE FITNESS OF THE ENVIRONMENT

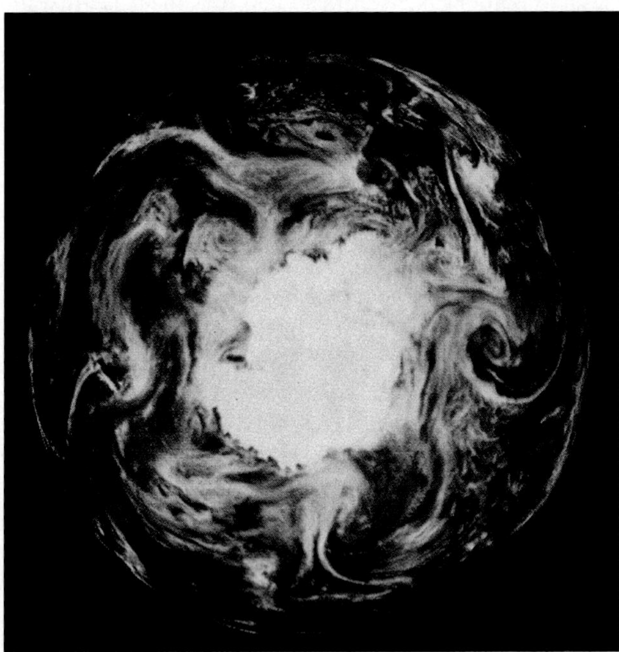

Figure 3.1
Water: an extraordinary compound. The unusual behavior of water is a major factor in the fitness of the environment for life. In this view of Earth's south pole, we see one of water's oddities: its simultaneous presence in three physical states. No other common substance on Earth occurs naturally in the solid, liquid, and gaseous states. The photo is a collage of 21 individual frames shot by the Galileo spacecraft during its December 1990 Earth flyby. In this chapter, you will learn about water's unusual properties and how they contribute to life on Earth.

If we could cruise the universe in a quest for life, we would do well to search for worlds with water. We might not recognize life on dry planets even if it existed. All organisms familiar to us are made mostly of water and live in a world where water dominates climate and many other features of the environment. Here on Earth, water is the biological medium—the substance that makes possible life as we know it.

Life began in water and evolved there for three billion years before spreading onto land. Modern life, even terrestrial (land-dwelling) life, remains inextricably tied to water. Most cells are surrounded by water; in fact, cells contain from about 70% to 95% water. Earth's surface is also wet, with water covering three-quarters of our planet. Although most of this water—enough to cover the United States to a depth of 130 km—is in liquid form, water is also present on Earth as ice and vapor. Water is the only common substance to exist in the natural environment in all three physical states of matter: solid, liquid, and gas (Figure 3.1).

The abundance of water is a major reason Earth is habitable. In a classic book called *The Fitness of the Environment*, Lawrence Henderson highlights the importance of water to life. While acknowledging that life adapts to its environment through natural selection, Henderson emphasizes that for life to exist at all in a particular location, the environment must first be a suitable abode. We will see throughout this chapter how much water contributes to the fitness of Earth for life.

Water is so common that it is easy to overlook the fact that it is an exceptional substance with many extraordinary qualities. Following the theme of emergent properties, we can trace water's unique behavior to the structure and interactions of its molecules.

WATER MOLECULES AND HYDROGEN BONDING

Studied in isolation, the water molecule is deceptively simple. As we saw in Chapter 2, its two hydrogen atoms are joined to an oxygen atom by covalent bonds in a lopsided arrangement (see Figure 2.14). The covalent bonds between oxygen and hydrogen are polar, with the electronegative oxygen nucleus pulling shared electrons toward itself and away from the hydrogen nuclei. The water molecule as a whole is electrically neutral. However, as a result of the unequal electron distribution in the molecule, the regions where the hydrogen atoms are located have a slight positive charge. The opposite side of the molecule has a slight negative charge associated with the oxygen atom. The polarity of its bonds and its asymmetric shape give the water molecule opposite charges on opposite sides; it is a **polar molecule.**

The anomalous properties of water arise from attractions among these polar molecules. The attraction is electrical; a positive hydrogen of one molecule is attracted to the negative oxygen of a nearby molecule. The molecules are thus held together by a hydrogen bond (Figure 3.2). Each water molecule can form hydrogen bonds to a maximum of four neighbors. The extraordinary qualities of water are emergent properties resulting from the hydrogen bonding that orders molecules into a higher level of structural organization.

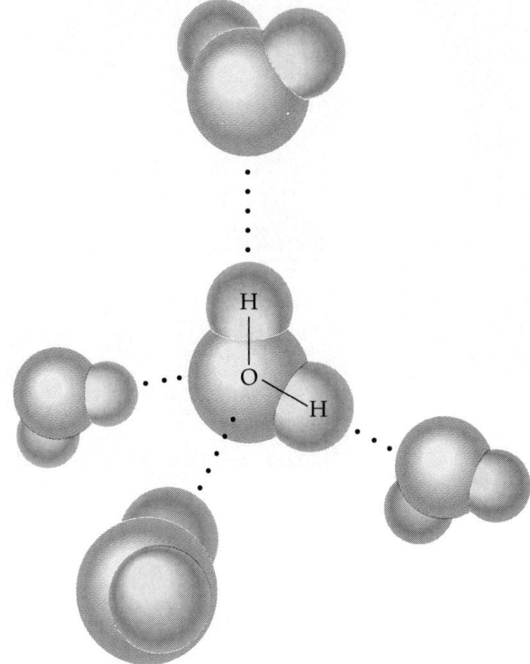

Figure 3.2
Hydrogen bonds between water molecules. The charged regions of a water molecule are attracted to oppositely charged parts of neighboring molecules. (The oxygen has a slight negative charge; the hydrogens have a slight positive charge.) Each molecule can hydrogen-bond to a maximum of four partners. At any instant in liquid water at 37°C (human body temperature), about 15% of the molecules are bonded to four partners in short-lived clusters.

SOME EXTRAORDINARY PROPERTIES OF WATER

In this section, we will concentrate on some of the qualities of water that contribute to the fitness of the environment.

Liquid Water Is Cohesive

Water molecules stick together as a result of hydrogen bonding. When water is in its liquid form, its hydrogen bonds are very fragile, about one-twentieth as strong as covalent bonds. They form, break, and re-form with great frequency. Each hydrogen bond lasts only a few trillionths of a second, but the molecules are constantly forming new bonds with a succession of partners. Thus, at any instant, a substantial percentage of all the water molecules are bonded to their neighbors, giving water more structure than most other liquids. Collectively, the hydrogen bonds hold the substance together, a phenomenon called **cohesion.**

Cohesion due to hydrogen bonding contributes to the transport of water against gravity in plants (Figure 3.3). Water reaches the leaves through microscopic vessels that extend upward from the roots. Water that evaporates from a leaf is replaced by water from the vessels in the veins of the leaf. Hydrogen bonds cause water molecules leaving the veins to tug on molecules farther down in the vessel, and the upward pull is transmitted along the vessel all the way down to the root. **Adhesion,** the clinging of one substance to another, also plays a role. Adhesion of water to the walls of the vessels helps counter the downward pull of gravity.

Related to cohesion is **surface tension,** a measure of how difficult it is to stretch or break the surface of a liquid. At the interface between water and air is an ordered arrangement of water molecules, hydrogen-bonded to one another and to the water below, making the water behave as though it were coated with an invisible film. Surface tension also causes water on a surface to bead into a spherical shape that has the smallest possible ratio of area to volume. Beading maximizes the number of hydrogen bonds that can form (Figure 3.4a).

Water has a greater surface tension than most other liquids. We can observe the surface tension of water by slightly overfilling a drinking glass; the water will stand above the rim. In a more biological example, the insect known as the water strider distributes its weight over enough area for the animal to walk on water without breaking the surface (Figure 3.4b). Water's surface tension also allows us to skip rocks on a pond.

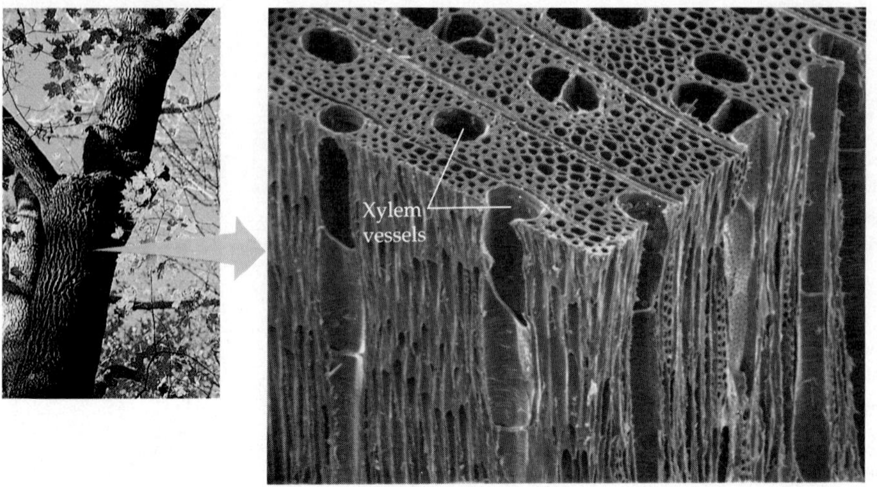

Figure 3.3
Water transport in plants. Evaporation from leaves pulls water upward from the roots through microscopic conduits called xylem vessels, in this case located in the trunk of a maple tree. Cohesion due to hydrogen bonding helps hold together the column of water within a vessel. Adhesion of the water to the vessel wall also helps in resisting the downward pull of gravity (SEM).

Water Has a High Specific Heat

Water stabilizes air temperatures by absorbing heat from air that is warmer and releasing the stored heat to air that is cooler. Water is effective as a heat bank because of its high specific heat: A slight change in its own temperature is accompanied by the absorption or release of a relatively large amount of heat. To understand this quality of water, we must first look briefly at heat and temperature.

Heat and Temperature Anything that moves has **kinetic energy,** the energy of motion. Atoms and molecules have kinetic energy because they are always moving, although in no particular direction. The faster a molecule moves, the greater its kinetic energy. **Heat** is the *total* quantity of kinetic energy due to molecular motion in a body of matter. **Temperature** measures the intensity of heat due to the *average* kinetic energy of the molecules. When the average speed of the molecules increases, a thermometer records this as a rise in temperature. Heat and temperature are related, but they are not the same. A swimmer crossing the English Channel has a higher temperature than the water, but the ocean contains far more heat because of its volume.

Whenever two objects of different temperature are brought together, heat passes from the warmer to the cooler body until the two are the same temperature. Molecules in the cooler object speed up at the expense

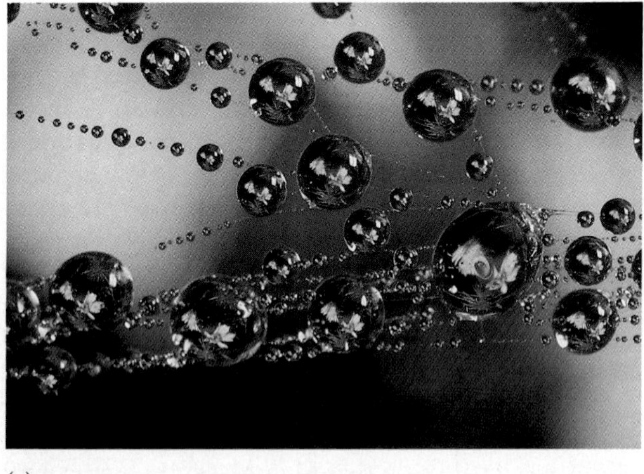
(a)

(b)

Figure 3.4
The high surface tension of water. Water has an unusually high surface tension because of the collective strength of its hydrogen bonds. (**a**) Surface tension causes water to bead on this spider web. (**b**) The water strider, though denser than water, can walk on a pond without breaking the surface.

of the kinetic energy of the warmer object. An ice cube cools a drink not by adding coldness to the liquid, but by absorbing heat as the ice melts.

Throughout this book, we will use the **Celsius scale** to indicate temperature (Celsius degrees are abbreviated as °C). At sea level, water freezes at 0°C and boils at 100°C. The temperature of the human body is 37°C, and comfortable room temperature is about 20°C to 25°C.

The unit of heat used in this book is the **calorie** (cal). A calorie is the amount of heat energy it takes to raise the temperature of 1 g of water by 1°C. Conversely, a calorie is also the amount of heat that 1 g of water releases when it cools down by 1°C. A **kilocalorie** (kcal), 1000 cal, is the quantity of heat required to raise the temperature of 1 kilogram (kg) of water by 1°C. (The "calories" on food packages are actually kilocalories.)

The **specific heat** of a substance is defined as the amount of heat that must be absorbed or lost for 1 g of that substance to change its temperature by 1°C. We already know water's specific heat, because we have *defined* a calorie as the amount of heat that causes water to change its temperature by 1°C. Therefore, the specific heat of water is 1 calorie per gram per degree Celsius, abbreviated as 1 cal/g/°C. Compared with most other substances, water has an unusually high specific heat. For example, ethyl alcohol, the type of alcohol in alcoholic beverages, has a specific heat of 0.6 cal/g/°C.

How Water Stabilizes Temperature Because of the high specific heat of water relative to other materials, water will change its temperature less when it absorbs or loses a given amount of heat. The reason you can burn your fingers by touching the metal handle of a pot on the stove when the water in the pot is still lukewarm is that the specific heat of water is ten times greater than that of iron. In other words, it will take only 0.1 cal to raise the temperature of 1 g of iron 1°C. Specific heat can be thought of as a measure of how well a substance resists changing its temperature when it absorbs or releases heat. Water resists changing its temperature; when it does change its temperature, it absorbs or loses a relatively large quantity of heat for each degree of change.

We can trace water's high specific heat, like many of its other properties, to hydrogen bonding. Heat must be absorbed in order to break hydrogen bonds, and heat is released when hydrogen bonds form. A calorie of heat causes a relatively small change in the temperature of water, because much of the heat energy is used to disrupt hydrogen bonds before the water molecules can begin moving faster. And when the temperature of water drops slightly, many additional hydrogen bonds form, releasing a considerable amount of energy in the form of heat.

What is the relevance of water's high specific heat to life on Earth? By warming up only a few degrees, a large body of water can absorb and store a huge amount of heat from the sun in the daytime and during summer. At night and during winter, the gradually cooling water can warm the air. This is the reason coastal areas generally have milder climates than inland regions. The high specific heat of water also makes ocean temperatures quite stable, creating a favorable environment for marine life. Thus, because of its high specific heat, the water that covers most of planet Earth keeps temperature fluctuations within limits that permit life. Also, because organisms are made primarily of water, they are more able to resist changes in their own temperatures than if they were made of a liquid with a lower specific heat.

Water Has a High Heat of Vaporization

Molecules of any liquid stay close together because they are attracted to one another. Molecules moving fast enough to overcome these attractions can depart from the liquid and enter the air as gas. This transformation from a liquid to a gas is called vaporization, or evaporation. The speed of molecular movement varies. Recall that temperature is the average kinetic energy of molecules. Even at a low temperature, the speediest molecules can escape into the air. Some evaporation occurs at any temperature; a glass of water, for example, will eventually evaporate at room temperature. If a liquid is heated, the average kinetic energy of molecules increases and the liquid evaporates more rapidly.

Heat of vaporization is the quantity of heat a liquid must absorb for 1 g of it to be converted from the liquid to the gaseous state. Compared with most other liquids, water has a high heat of vaporization. To evaporate each gram of water, 540 cal of heat are needed—nearly double the amount needed to vaporize a gram of alcohol or ammonia. Water's high heat of vaporization is another emergent property caused by hydrogen bonds, which restrain the molecules and make their exodus from the liquid state more difficult. A relatively large amount of heat is needed to evaporate water, because hydrogen bonds must be broken first. Related to water's high heat of vaporization is its high boiling point, 100°C. Liquid water is abundant because temperatures that high are rare on Earth.

Water's high heat of vaporization helps moderate Earth's climate. A considerable amount of solar heat absorbed by tropical seas is consumed during the evaporation of surface water. Then, as moist tropical air circulates poleward, it releases heat as it condenses to form rain.

As a substance evaporates, the surface of the liquid that remains behind cools down. This **evaporative**

cooling occurs because the "hottest" molecules, those with the greatest kinetic energy, are the most likely to leave as gas. It is as if the 100 fastest runners at a college transferred to another school; the average speed of the remaining students would decline.

Evaporative cooling of water contributes to the stability of temperature in lakes and ponds and also provides a mechanism that prevents terrestrial organisms from overheating. For example, evaporation of water from the leaves of a plant helps keep the tissues in the leaves from becoming too warm in the sunlight. Evaporation of sweat from human skin cools the surface of the body and helps prevent overheating on a hot day or when excess heat is generated by strenuous activity (Figure 3.5). However, high humidity on a hot day increases discomfort, because the concentration of water vapor in the air inhibits the evaporation of sweat from the body.

Figure 3.5
Evaporative cooling. Because of water's high heat of vaporization, evaporation of sweat cools the surface of the body.

Water Expands When It Freezes

Ice floats. Water is one of the few substances that is less dense as a solid than it is as a liquid. While other materials contract when they solidify, water expands. The cause of this exotic behavior is, once again, hydrogen bonding. At temperatures above 4°C, water behaves like other liquids, expanding as it warms and contracting as it cools. Water begins to freeze when its molecules are no longer moving vigorously enough to break their hydrogen bonds. As the temperature reaches 0°C, the water becomes locked into a crystalline lattice, each water molecule bonded to the maximum of four partners (Figure 3.6). The hydrogen bonds keep the molecules far enough apart from each other to make ice about 10% less dense (10% fewer molecules for the same volume) than liquid water at 4°C. When ice absorbs enough heat for its temperature to increase to above 0°C, hydrogen bonds between molecules are

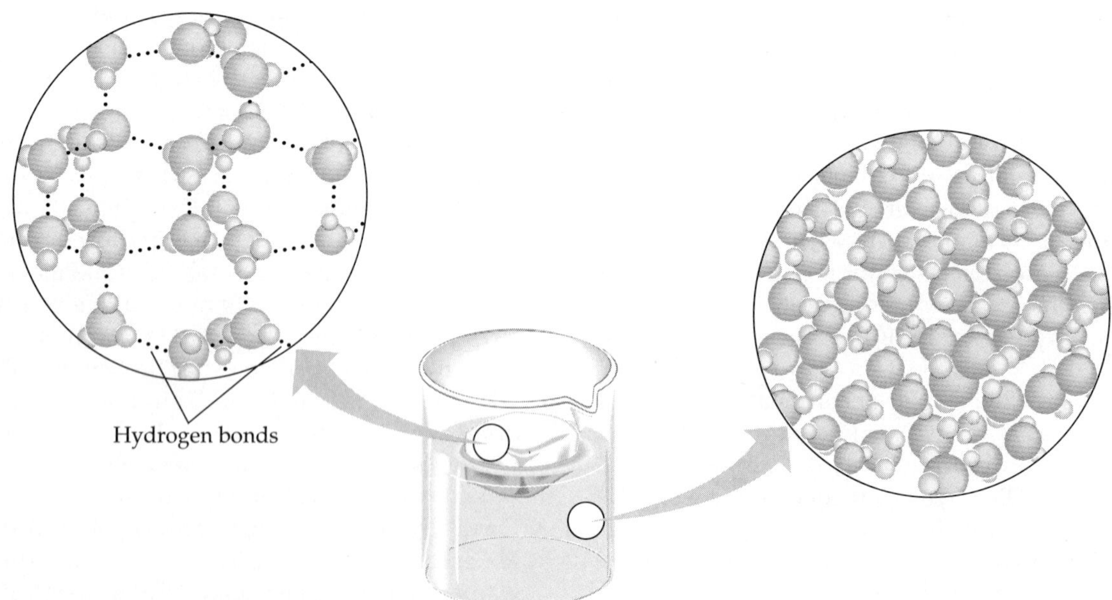

Figure 3.6
The structure of ice. Each molecule is hydrogen-bonded to four neighbors in a three-dimensional crystal with open channels. Because the hydrogen bonds make the crystal spacious, ice has fewer molecules than an equal volume of liquid water. In other words, ice is less dense than liquid water.

Figure 3.7
Floating ice and the fitness of the environment. Floating ice becomes a barrier that protects the liquid water below from the colder air. These are polar cod swimming beneath arctic ice.

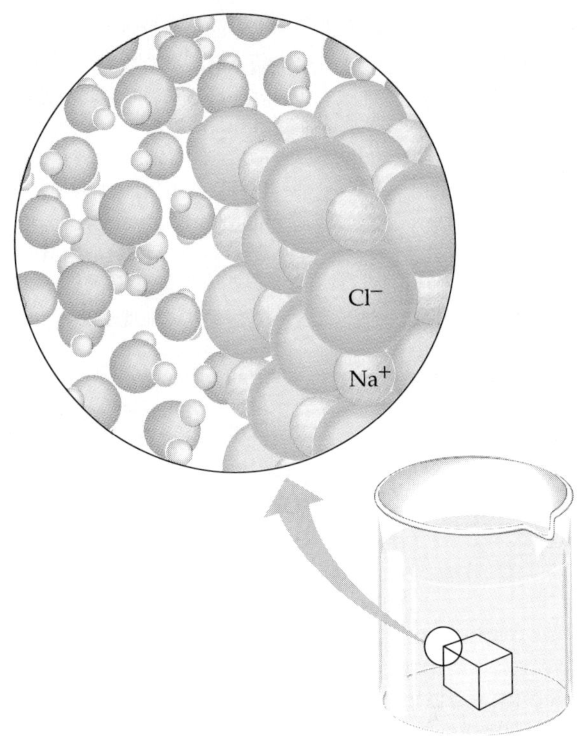

Figure 3.8
A crystal of table salt dissolving in water. The positive hydrogen regions of the polar water molecules are attracted to the chloride anions, whereas the negative oxygen regions cling to the sodium cations.

disrupted. As the crystal collapses, the ice melts, and molecules are free to slip closer together. Water reaches its greatest density at 4°C and then begins to expand, again owing to the increased speed of its molecules. Keep in mind, however, that even liquid water is semi-structured because of transient hydrogen bonds.

The ability of ice to float because of the expansion of water as it solidifies is an important factor in the fitness of the environment (Figure 3.7). If ice could sink, then eventually all ponds, lakes, and even the oceans would freeze solid, making life as we know it impossible on Earth. During summer, only the upper few inches of the ocean would thaw. Instead, when a deep body of water cools, the floating ice insulates the liquid water below, preventing it from freezing and allowing life to exist under the frozen surface.

The freezing of water and the melting of ice also help make seasonal transitions less abrupt, enabling organisms to adjust gradually to the changing climate. Again, water releases heat whenever hydrogen bonds form and absorbs heat whenever hydrogen bonds break. When water solidifies into ice or snow, the heat released warms the surrounding air as hydrogen bonds knit the molecules together into crystals. This process helps to moderate autumn temperatures. During the spring thaw, melting ice absorbs heat as hydrogen bonds are broken, again moderating the change of seasons.

Water Is a Versatile Solvent

A sugar cube placed in a glass of water will dissolve. The glass will then contain a uniform mixture of sugar and water; the concentration of dissolved sugar will be the same everywhere in the mixture. A liquid that is a homogeneous mixture of two or more substances is called a **solution.** The dissolving agent of a solution is the **solvent,** and the substance that is dissolved is the **solute.** In this case, water is the solvent and sugar is the solute. An **aqueous solution** is one in which water is the solvent.

The medieval alchemists tried to find a universal solvent, one that would dissolve anything. They learned that nothing works better than water. However, water is not a universal solvent; if it were, it could not be stored in any container, including our cells. But water is a very versatile solvent, a quality we can trace to the polarity of the water molecule.

Suppose, for example, that a crystal of the ionic compound sodium chloride is placed in water (Figure 3.8). At the surface of the crystal, the sodium and chloride ions are exposed to the solvent. The ions and water molecules have an affinity for one another through electrical attraction. The oxygen regions of the water molecules are negatively charged and cling to sodium cations. The hydrogen regions of the water molecules are positively charged and are attracted to chloride anions. Water surrounds the individual ions, separating the sodium from the chloride and shielding the ions from one another. Working inward from the surface of the salt crystal, water eventually dissolves all the ions, producing a solution of two solutes, sodium and chloride, homogeneously mixed with water, the solvent.

Table 3.1 A Summary of Water's Extraordinary Properties

Property	Explanation	Example of Benefit to Life
Cohesion and high surface tension; adhesion	Hydrogen bonds hold molecules together and adhere them to hydrophilic surfaces.	Leaves pull water upward from roots in microscopic vessels.
High specific heat	Hydrogen bonds absorb heat when they break and release heat when they form, minimizing temperature changes.	Water stabilizes air and sea temperatures, and helps organisms, made mostly of water, resist temperature change.
High heat of vaporization	Hydrogen bonds must be broken for water to evaporate.	Evaporation of water cools the surfaces of plants and animals.
Expansion upon freezing	Water molecules in an ice crystal are spaced relatively far apart because of hydrogen bonding.	Floating ice insulates the water below and prevents seas and lakes from freezing solid.
Versatility as a solvent	Charged regions of polar water molecules are attracted to ions and polar compounds.	The aqueous solutions of life contain a diversity of dissolved substances.

Other ionic compounds also dissolve in water. Seawater, for instance, contains a great variety of dissolved ions, as do living cells.

A compound does not need to be ionic in order to dissolve in water; polar compounds are also water-soluble. An example is ammonia (NH_3), which, like water, has patches of positive and negative charge due to unequal electron sharing (see Figure 2.17). Ammonia dissolves in water because the polar water molecules can arrange themselves around the polar ammonia molecules and form weak hydrogen bonds. Sugars also dissolve in water because they are polar molecules. In fact, many different kinds of polar compounds are dissolved (along with ions) in the water of such biological fluids as blood, the sap of plants, and the liquid within all cells.

Whether ionic or polar, any substance that has an affinity for water is said to be **hydrophilic** (from the Greek *hydro,* "water," and *philios,* "loving"). This term is used even if the substance does not dissolve—because the molecules are too large, for instance. Cotton, a plant product, is an example of a hydrophilic substance that absorbs water without dissolving. Cotton consists of giant molecules of a compound called cellulose. With numerous regions of positive and negative charges associated with polar bonds, the cellulose fibers adhere to water. Thus, a cotton towel does a great job of drying the body, yet does not dissolve in the washing machine. Cellulose is also present in the walls of water-conducting vessels in a plant, and we learned earlier how adhesion of water to these walls functions in water transport.

There are, of course, substances that neither dissolve in water nor have an affinity for water. Such substances, which actually seem to repel water, are termed **hydrophobic** (Gr. *phobos,* "fearing"). They are nonionic, nonpolar substances. For example, vegetable oil and water do not mix. The hydrophobic behavior of the oil molecules results from a prevalence of nonpolar bonds, particularly bonds between carbon and hydrogen, which share electrons almost equally. Hydrophobic molecules related to oils are major ingredients of cell membranes. (Imagine what would happen to a cell if its membrane dissolved.)

The extraordinary properties of water are summarized in Table 3.1. After reviewing these qualities, you will be ready to learn more about aqueous solutions and their importance in biology.

AQUEOUS SOLUTIONS

Biological chemistry is wet chemistry. Most of the chemical reactions that occur in life involve solutes dissolved in water. To understand the chemistry of life, it is important to learn about the properties of aqueous solutions. We will address two quantitative aspects of solutions: solute concentration and pH.

Solute Concentration

Suppose we wanted to prepare an aqueous solution of table sugar having a specified concentration of sugar molecules (a certain number of solute molecules in a certain volume of solution). Because counting or weighing individual molecules is not practical, scientists usually measure substances in units called moles (mol). A **mole** is equal in number to the molecular weight of a substance, but in grams rather than daltons. Suppose we wanted to weigh out 1 mol of table sugar (sucrose), which has the molecular formula $C_{12}H_{22}O_{11}$. A carbon atom weighs 12 daltons, a hydro-

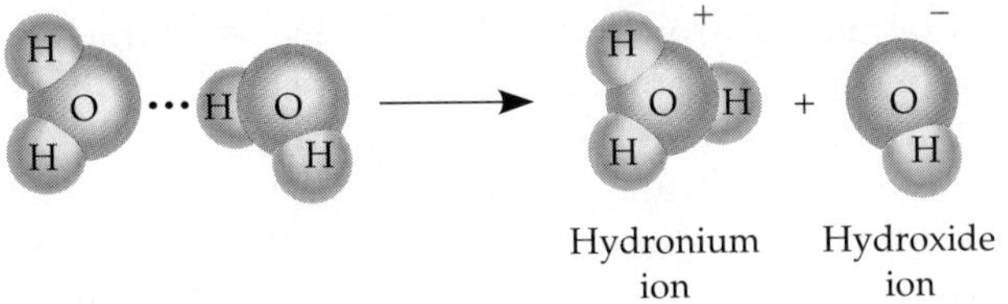

Figure 3.9
The formation of hydronium and hydroxide ions. A proton, the nucleus of a hydrogen atom, shifts from one water molecule to another, forming a hydronium ion and a hydroxide ion.

gen atom weighs 1 dalton, and an oxygen atom weighs 16 daltons. **Molecular weight** is the sum of the weights of all the atoms in a molecule; thus, the molecular weight of sucrose is 342 daltons. To obtain 1 mol of sucrose, we weigh out 342 g, the molecular weight of sucrose expressed in grams.

The practical advantage of measuring a quantity of chemicals in moles is that a mole of one substance has exactly the same number of molecules as a mole of any other substance. If substance A has a molecular weight of 10 daltons and substance B has a molecular weight of 100 daltons, then 10 g of A will have the same number of molecules as 100 g of B. The number of molecules in a mole, called Avogadro's number, is 6.02×10^{23}. A mole of table sugar contains 6.02×10^{23} sucrose molecules and weighs 342 g. A mole of ethyl alcohol (C_2H_6O) also contains 6.02×10^{23} molecules, but it weighs only 46 g because the molecules are smaller than those of sucrose. The mole concept rescales the weighing of molecules from daltons, used for single molecules, to grams, which are more practical units of weight for the laboratory. Measuring in moles also allows scientists to combine substances in fixed ratios of molecules.

Imagine making a liter (L) of solution consisting of 1 mol of sugar dissolved in water. To obtain this concentration, we would weigh out 342 g of sucrose and then gradually add water, while stirring, until the sugar was completely dissolved. We would then add enough water to bring the total volume of the solution up to 1 L. At that point, we would have a one-molar (1 *M*) solution of sucrose. **Molarity**—the number of moles of solute per liter of solution—is the unit of concentration most often used for aqueous solutions.

Acids, Bases, and pH

Dissociation of Water Molecules Occasionally, a hydrogen atom shared between two water molecules in a hydrogen bond shifts from one molecule to the other (Figure 3.9). When this happens, the hydrogen atom leaves its electron behind, and what is actually transferred is a **hydrogen ion,** a single proton with a charge of +1. The water molecule that lost a proton is now a **hydroxide ion** (OH^-), which has a charge of –1. The proton binds to the second water molecule, making that molecule a **hydronium ion** (H_3O^+). We can write the chemical reaction this way:

$$H_2O + H_2O \rightleftharpoons \underset{\text{Hydronium ion}}{H_3O^+} + \underset{\text{Hydroxide ion}}{OH^-}$$

Although this is technically what happens, it is convenient to think of the process as the dissociation (separation) of a water molecule into a hydrogen ion and a hydroxide ion:

$$H_2O \rightleftharpoons \underset{\text{Hydrogen ion}}{H^+} + \underset{\text{Hydroxide ion}}{OH^-}$$

As the double arrows indicate, this is a reversible reaction that will reach a state of dynamic equilibrium when water dissociates at the same rate that it is being reformed from H^+ and OH^-. At this equilibrium point, the concentration of water molecules greatly exceeds the concentrations of H^+ and OH^-. In fact, in pure water, only one water molecule in every 554 million is dissociated. Although the dissociation of water is reversible and statistically rare, it is exceedingly important in the chemistry of life. Hydrogen and hydroxide ions are very reactive. Even slight changes in their concentrations can disrupt a cell's proteins and other complex molecules.

Acids and Bases Since the dissociation of water produces one H^+ for every OH^-, the concentrations of these ions will be equal in pure water. The concentration of each ion is 10^{-7} *M* (at a temperature of 25°C). This means that there is only one ten-millionth of a mole of hydrogen ions per liter of pure water, and an equal number of hydroxide ions.

What would cause an aqueous solution to have an imbalance in its H^+ and OH^- concentrations? When substances called acids dissolve in water, they donate additional hydrogen ions to the solution. An **acid,** according to the definition most biologists use, is a substance that increases the H^+ concentration of a solution. For example, when hydrochloric acid (HCl) is added to water, hydrogen ions dissociate from chloride ions:

$$HCl \longrightarrow H^+ + Cl^-$$

Now there are two sources of H^+ in the solution (dissociation of water is the other), resulting in more H^+ than OH^-. Such a solution is known as an acidic solution.

A substance that reduces the hydrogen ion concentration in a solution is called a **base.** Some bases reduce the H^+ concentration indirectly by dissociating to form hydroxide ions, which then combine with hydrogen ions to form water. An example of a base that acts this way is sodium hydroxide (NaOH), which in water dissociates into its ions:

$$NaOH \longrightarrow Na^+ + OH^-$$

Other bases reduce H^+ concentration directly by accepting hydrogen ions. Ammonia (NH_3), for instance, acts as a base by binding a hydrogen ion from the solution, resulting in an ammonium ion (NH_4^+):

$$NH_3 + H^+ \rightleftharpoons NH_4^+$$

In either case, the base reduces the H^+ concentration, and solutions with a higher concentration of OH^- than H^+ are known as basic solutions. A solution in which the H^+ and OH^- concentrations are equal is said to be neutral.

Notice that single arrows were used in the reactions for HCl and NaOH. These compounds dissociate completely when mixed with water. Hydrochloric acid is called a strong acid and sodium hydroxide a strong base because they dissociate completely. In contrast, ammonia is a weak base. The double arrows in the reaction for ammonia indicate that the binding and release of the hydrogen ion are reversible, and when the reaction reaches equilibrium, there will be a fixed ratio of NH_4^+ to NH_3. There are also weak acids, which dissociate reversibly to release and reaccept hydrogen ions. An example is carbonic acid:

$$\underset{\text{Carbonic acid}}{H_2CO_3} \rightleftharpoons \underset{\text{Bicarbonate ion}}{HCO_3^-} + \underset{\text{Hydrogen ion}}{H^+}$$

However, the equilibrium so favors the reaction in the left direction that when carbonic acid is added to water, only 1% of the molecules are dissociated at any particular time. Still, that is enough to shift the balance of H^+ and OH^- from neutrality.

The pH Scale In any solution, the product of the H^+ and the OH^- concentrations is constant at 10^{-14} *M*. This can be written

$$[H^+][OH^-] = 10^{-14}\ M^2$$

where brackets indicate molar concentration for the substance enclosed within them. In a neutral solution at room temperature (25°C), $[H^+] = 10^{-7}$ and $[OH^-] = 10^{-7}$, so the product is $10^{-14}\ M^2$ ($10^{-7} \times 10^{-7}$). If enough acid is added to a solution to increase $[H^+]$ to 10^{-5} *M*, then $[OH^-]$ will decline by an equivalent amount to 10^{-9} *M* ($10^{-5} \times 10^{-9} = 10^{-14}$). An acid not only adds hydrogen ions to a solution, but also removes hydroxide ions because of the tendency for H^+ to combine with OH^- to form water. A base has the opposite effect, increasing OH^- concentration, but also reducing H^+ concentration by the formation of water. If enough of a base is added to raise the OH^- concentration to 10^{-4} *M*, the H^+ concentration will drop to 10^{-10} *M*. Whenever we know the concentration of either H^+ or OH^- in a solution, we can deduce the concentration of the other ion.

Because the H^+ and OH^- concentrations of solutions can vary by a factor of 100 trillion or more, scientists have developed a way to express this variation more conveniently, by use of the **pH scale,** which ranges from 0 to 14. The pH scale compresses the range of H^+ and OH^- concentrations by employing a common mathematical device: logarithms. The pH of a solution is defined as the negative logarithm (base 10) of the hydrogen ion concentration expressed in moles per liter:

$$pH = -\log [H^+]$$

For a neutral solution, $[H^+]$ is 10^{-7} *M*, giving us

$$-\log 10^{-7} = -(-7) = 7$$

Notice that pH *declines* as H^+ concentration *increases*. Notice, too, that although the pH scale is based on H^+ concentration, it also implies OH^- concentration. A solution of pH 10 has a hydrogen ion concentration of 10^{-10} *M* and a hydroxide ion concentration of 10^{-4} *M* (Figure 3.10).

The pH of a neutral solution is 7, the midpoint of the scale. A pH value less than 7 denotes an acidic solution, and the lower the number, the more acidic the solution. The pH for basic solutions is above 7. Values of pH less than 0 or greater than 14 are rarely encountered, and most biological fluids are within the range pH 6 to pH 8. There are a few exceptions, however, including the strongly acidic digestive juice of the human stomach, which has a pH of about 1.5.

It is important to remember that each pH unit represents a tenfold difference in H^+ and OH^- concentrations. It is this mathematical feature that makes the pH scale so compact. A solution of pH 3 is not twice as acidic as a solution of pH 6, but a thousand times more acidic. When the pH of a solution changes slightly, the actual concentrations of H^+ and OH^- in the solution change substantially.

Buffers The internal pH of most living cells is close to 7. Even a slight change in pH can be harmful, because the chemical processes of the cell are very sensitive to the concentrations of hydrogen and hydroxide ions.

Biological fluids resist changes to their own pH when acids or bases are introduced because of the presence of **buffers,** substances that minimize changes in the concentrations of H^+ and OH^-. Buffers in human blood, for example, normally maintain the blood pH at 7.4. A drop in pH (acidosis) or an increase in pH (alkalosis) is dangerous, and a person cannot survive for more than a few minutes if the blood pH drops to 7 or rises to 7.8. Under normal circumstances, the buffering capacity of the blood prevents such swings in pH.

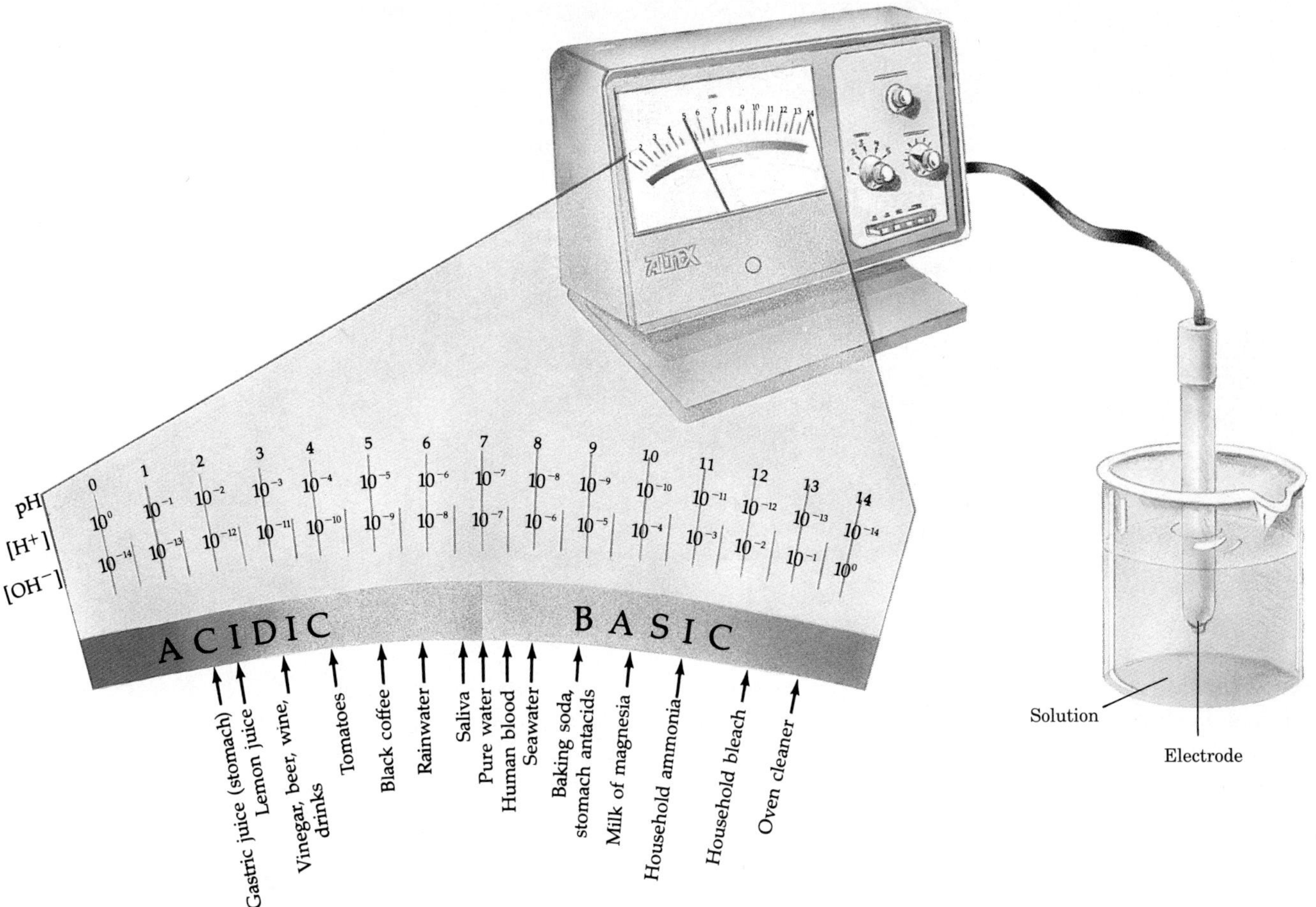

Figure 3.10
The pH of some aqueous solutions. Measurements of pH are usually made with a meter connected to a glass electrode immersed in the solution.

A buffer works by accepting hydrogen ions from the solution when they are in excess and donating hydrogen ions to the solution when they have been depleted. Most buffers are weak acids or weak bases that combine reversibly with hydrogen ions. One of the buffers that contributes to pH stability in human blood and many other biological solutions is carbonic acid (H_2CO_3), which, as already mentioned, dissociates to yield a bicarbonate ion (HCO_3^-) and a hydrogen ion (H^+):

	Response to a rise in pH			
H_2CO_3	$\rightleftharpoons$	HCO_3^-	+	H^+
H^+ donor (acid)	Response to a drop in pH	H^+ acceptor (base)		Hydrogen ion

The chemical equilibrium between carbonic acid and bicarbonate acts as a pH regulator, the reaction shifting left or right as other processes in the solution add or remove hydrogen ions. If the H^+ concentration in blood begins to fall (that is, if pH rises), more carbonic acid dissociates, replenishing hydrogen ions. But when H^+ concentration in blood begins to rise (pH drops), the bicarbonate ion acts as a base and removes the excess hydrogen ions from the solution. Thus, the carbonic acid–bicarbonate buffering system actually consists of an acid and a base in equilibrium with each other. Most other buffers are also acid–base pairs.

ACID RAIN: UPSETTING THE FITNESS OF THE ENVIRONMENT

Acid rain is an environmental problem that has increased public awareness about the sensitivity of life to pH. Uncontaminated rain has a pH of about 5.6, slightly acidic, owing to the formation of carbonic acid from carbon dioxide and water. The term *acid rain* applies to rain more acidic than pH 5.6. The questions we

Figure 3.11
Effects of acid rain on a forest. Rain and snow bearing the acidic products of coal and oil combustion have been blamed for the demise of this forest in Erzgebirge, Czechoslovakia.

must consider are: What causes acid rain? What are its effects on the fitness of the environment? And what can be done to reduce the problem?

Acid rain is caused primarily by the presence in the atmosphere of sulfur oxides and nitrogen oxides, gaseous compounds that react with water in the air to form acids, which fall to Earth with rain or snow. A major source of these oxides is the combustion of fossil fuels by factories and automobiles. Acid rain from air pollution is as old as the Industrial Revolution, but the problem has escalated and become more widespread in the past two decades. By 1985, before tougher restrictions were imposed, the exhaust emitted in the United States alone was adding over 40 million tons of sulfur oxides and nitrogen oxides to the atmosphere annually. One practice that has increased the occurrence of acid rain is the construction of taller smokestacks designed to reduce local pollution by dispersing factory exhaust. Unfortunately, prevailing winds simply move the problem, and acid rain falls hundreds or thousands of miles away from industrial centers, often in once pristine regions. In the Adirondack Mountains of upstate New York, the pH of rainfall averages 4.2, about 25 times more acidic than normal rain. Acid rain falls on many other regions, including the Cascade Mountains of the Pacific Northwest and certain parts of Europe. One West Virginia storm dropped rain having a pH of 1.5.

Experiments and observations have confirmed that acid rain is harming both terrestrial and freshwater ecosystems. Acid rain that falls on land lowers the pH of the soil solution, which affects the solubility of minerals. Some mineral nutrients required by plants are washed out of the topsoil, while other minerals, such as aluminum, reach toxic concentrations when acidification increases their solubility. The effects of acid rain on soil chemistry have contributed to the decline of European forests (Figure 3.11). Acid rain has also lowered the pH of lakes and ponds in some regions, and the accumulation of certain minerals leached from the soil by acid rain further contaminates freshwater habitats. Many species of fish, amphibians, and aquatic invertebrates have been adversely affected (Figure 3.12). More than half the lakes at the higher elevations of the western Adirondacks are now more acidic than pH 5, and fish have completely disappeared from nearly all those lakes. Like the coal mine canary, the fish—or their absence—are a warning that something has gone awry in the environment.

Figure 3.12
The impact of acid precipitation on aquatic life. The curved spine and stunted gills of this salamander are examples of deformities that occur when animals develop in water of pH 5, common for lakes and ponds contaminated by acid precipitation.

Acid rain can be reduced through industrial controls and antipollution devices. In fact, there is evidence that emissions of sulfur oxides have declined by about 30% since 1985. Continued progress can come only from business leaders, voters, consumers, and politicians who are concerned about environmental quality.

* * *

In this chapter, we have seen how the emergent properties of water, a substance so common and yet so extraordinary, are crucial to life. We have also seen how the disturbance of water resources—by a change in the pH of rain, for example—can affect the fitness of the environment. In the next chapter, we will explore another facet of the fitness of the environment: the role of the element carbon in the chemistry of life.

STUDY OUTLINE

1. Water is the medium of life. It is abundant in the environment and makes up a major part of every living cell.
2. Water's structure explains its unusual properties.

Water Molecules and Hydrogen Bonding (p. 40–41)

1. Water is a polar molecule. A hydrogen bond is formed when the oxygen of one water molecule is electrically attracted to the hydrogen of an adjacent molecule.
2. Hydrogen bonding between water molecules is the basis for water's emergent properties.

Some Extraordinary Properties of Water (pp. 41–46)

1. Formation and re-formation of hydrogen bonds makes liquid water cohesive. For example, water is pulled upward in the microscopic vessels of plants.
2. Hydrogen bonding of water molecules on the surface of liquid water is responsible for water's surface tension.
3. Heat is the total kinetic energy of molecules in a body of matter. Temperature measures the average kinetic energy of those molecules. A calorie is the heat energy required to raise the temperature of 1 g of water by 1°C.
4. Hydrogen bonding gives water a high specific heat. Heat is absorbed when hydrogen bonds break and is released when hydrogen bonds form, minimizing temperature fluctuations to within limits that permit life.
5. Evaporative cooling is based on water's high heat of vaporization. Water molecules must have a relatively high kinetic energy to overcome hydrogen bonds. The loss of these energetic water molecules as vapor cools a surface.
6. Ice is less dense than liquid water, owing to more organized hydrogen bonding, which forces water to expand into a characteristic crystal. Floating ice allows life to exist under the frozen surfaces of lakes and polar seas.
7. Water is an unusually versatile solvent because its polarity attracts it to charged and polar substances. When ions or molecules of such substances are surrounded by water molecules, they dissolve and are called solutes. Hydrophilic substances have an affinity for water. Hydrophobic substances repel water.

Aqueous Solutions (pp. 46–49)

1. A mole is the number of grams of a substance equal to its molecular weight in daltons. Molarity is the number of moles per liter of solution.
2. Water can dissociate into H^+ and OH^-. The reaction is reversible, and at equilibrium, $[H^+] = [OH^-] = 10^{-7}$ *M* (for pure water at room temperature).
3. The concentration of H^+ is measured in pH units as follows: $pH = -\log [H^+]$. Each pH unit thus represents a tenfold difference in $[H^+]$.
4. Acids donate additional H^+ in aqueous solutions; bases donate OH^- or accept H^+ in solutions.
5. In a neutral solution, $[H^+] = [OH^-] = 10^{-7}$, and pH = 7. In an acidic solution, $[H^+]$ is greater than $[OH^-]$, and the pH is less than 7. In a basic solution, $[H^+]$ is less than $[OH^-]$, and the pH is greater than 7.
6. $[H^+]$ critically affects the chemistry of life by influencing the structure and function of biological molecules. The internal pH of most living cells must be kept close to 7.
7. Buffers in biological fluids resist changes in pH. A buffer consists of an acid–base pair that combines reversibly with hydrogen ions.

Acid Rain: Upsetting the Fitness of the Environment (pp. 49–51)

1. Acid rain occurs when water in the air reacts with sulfur oxides and nitrogen oxides that result from combustion of fossil fuels. The acids that form give rain and snow a pH less than 5.6, sometimes causing serious environmental consequences.

SELF-QUIZ

1. The main thesis of Lawrence Henderson's *The Fitness of the Environment* is
   a. Earth's environment is constant.
   b. It is the physical environment, not life, that has evolved.
   c. The environment of Earth has adapted to life.
   d. Life as we know it depends on certain environmental qualities on Earth.
   e. Water and other aspects of Earth's environment exist because they make the planet more suitable for life.
2. Air temperature often increases slightly as clouds begin to drop rain or snow. Which behavior of water is *most directly* responsible for this phenomenon?
   a. Water's change in density when it condenses.
   b. Water's reactions with other atmospheric compounds.
   c. Release of heat by formation of hydrogen bonds.
   d. Release of heat by breaking of hydrogen bonds.
   e. Water's high surface tension.

3. For two bodies of matter in contact, heat always flows from
   a. the body with greater heat to the one with less heat
   b. the body of higher temperature to the one of lower temperature
   c. the more dense to the less dense body
   d. the body having more water to the one with less water
   e. the larger to the smaller body
4. A slice of pizza has 500 kilocalories. If we could burn the pizza and use all the heat to warm a 50-L container of cold water, what would be the approximate increase in the temperature of the water? (*Note:* A liter of cold water weighs about a kilogram.)
   a. 50°C
   b. 5°C
   c. 10°C
   d. 100°C
   e. 1°C
5. The bonds that are broken when water vaporizes are
   a. ionic bonds
   b. bonds *between* water molecules
   c. bonds between atoms of individual water molecules
   d. polar covalent bonds
   e. nonpolar covalent bonds
6. We can be sure that a mole of table sugar and a mole of vitamin C are equal in their
   a. weight in daltons
   b. weight in grams
   c. number of molecules
   d. number of atoms
   e. volume
7. How many grams of acetic acid ($C_2H_4O_2$) would you use to make 10 L of a 0.1-*M* aqueous solution of the acetic acid? (*Note:* The atomic weights, in daltons, are approximately 12 for carbon, 1 for hydrogen, and 16 for oxygen.)
   a. 10 g
   b. 0.1 g
   c. 6 g
   d. 60 g
   e. 0.6 g
8. Acid rain has lowered the pH of a particular lake to 4.0. What is the hydrogen ion concentration of the lake?
   a. 4.0 *M*
   b. 10^{-10} *M*
   c. 10^{-4} *M*
   d. 10^{4} *M*
   e. 4%
9. What is the *hydroxide* ion concentration of the lake described in question 8?
   a. 10^{-7} *M*
   b. 10^{-4} *M*
   c. 10^{-10} *M*
   d. 10^{-14} *M*
   e. 10 *M*

10. Which of the following is an example of a hydrophobic material?
    a. paper
    b. table salt
    c. wax
    d. sugar
    e. pasta

CHALLENGE QUESTIONS

1. Explain how panting helps to regulate a dog's body temperature.
2. Adhesion and cohesion cooperate to allow water to rise in thin tubes made of hydrophilic material, such as glass. This phenomenon is called capillary action (see the figure). Water molecules adhering to the glass creep upward, pulling more water along by cohesion. Explain why water rises higher by capillary action in a tube of smaller diameter than in one of wider diameter.

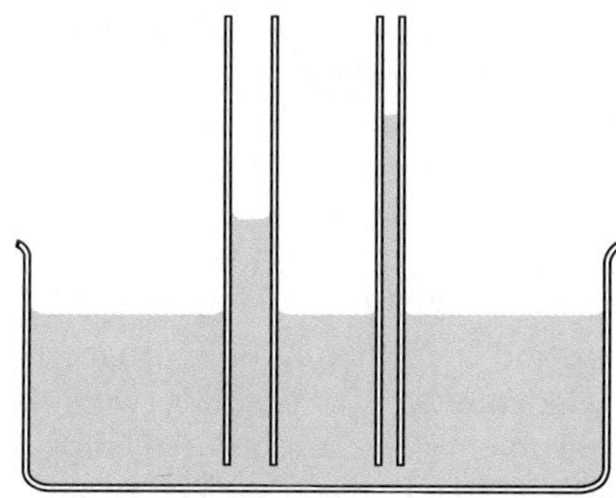

SCIENCE, TECHNOLOGY, AND SOCIETY

1. Discuss the special political obstacles to reducing acid rain (as compared with environmental issues confined to a more localized region).
2. Agriculture, industry, and the growing populations of cities all compete, through political influence, for water. If you were in charge of water resources in an arid region, what would your priorities be for allocating the limited water supply for various uses? How would you defend your position?

FURTHER READING

Henderson, L. J. *The Fitness of the Environment.* New York: Macmillan, 1913. A classic book highlighting the importance of water and carbon to life.

Lehninger, A. L., D. L. Nelson, and M. M. Cox. *Principles of Biochemistry.* New York: Worth, 1992, Chapter 4. A readable biochemistry text with an excellent discussion of water.

Mohner, V. A. "The Challenge of Acid Rain." *Scientific American,* August 1988. An analysis of a complex environmental problem.

4 CARBON AND MOLECULAR DIVERSITY

THE FOUNDATIONS OF ORGANIC CHEMISTRY
THE VERSATILITY OF CARBON IN MOLECULAR ARCHITECTURE
VARIATION IN CARBON SKELETONS
FUNCTIONAL GROUPS
THE CHEMICAL ELEMENTS OF LIFE: A REVIEW

Figure 4.1
Insulin: a relatively small protein. Illustrated here as a computer graphic, insulin belongs to a class of organic compounds known as proteins. Insulin is a regulatory protein called a hormone (see Chapter 41). Compared to the molecules we have studied so far, insulin is huge, consisting of hundreds of atoms joined by covalent bonds. But among proteins, insulin is actually relatively small. It is the ability of carbon atoms (shown in white) to bond to multiple partners, including other carbon atoms, that makes such complex molecules possible. This chapter highlights the importance of carbon in the chemistry of life.

From the origin of the first cells to the current variety of organisms, carbon has played a prominent role in the evolution of life on Earth. Biological diversity reflects molecular diversity; and carbon, of all chemical elements, is unparalleled in its ability to form molecules that are large, complex, and diverse. In this chapter, you will learn the principles of molecular architecture that make carbon so important to life. This approach will further illustrate the theme that emergent properties arise from the organization of living matter.

THE FOUNDATIONS OF ORGANIC CHEMISTRY

Although a single cell is composed of 70% to 95% water, most of the rest consists of carbon-based compounds. Proteins, DNA, carbohydrates, and other molecules that distinguish living matter from inanimate material are all composed of carbon atoms bonded to one another and to atoms of other elements. Hydrogen (H), oxygen (O), nitrogen (N), sulfur (S), and phosphorus (P) are other common ingredients of these compounds, but it is carbon (C) that accounts for the endless diversity of organic molecules.

Compounds containing carbon are said to be organic, and the branch of chemistry that specializes in the study of carbon compounds is called **organic chemistry.** Organic compounds range from simple molecules to colossal ones with thousands of atoms and molecular weights in excess of 100,000 daltons (Figure 4.1). The percentages of the major elements of life—C, O, H, N, S, and P—are quite uniform from individual to individual, and even from species to species. The atoms of organic molecules, however, can be arranged so many different ways that the uniqueness of each organism is ensured. Carbon's versatility allows a limited assortment of atomic building blocks, taken in roughly the same proportions, to be used to build an inexhaustible variety of organic molecules.

For centuries, humans have exploited other organisms as sources of valued substances—everything from wine and food to medicines and fabrics. Organic chemistry originated in various attempts to purify and

Figure 4.2
Applications of organic chemistry in a nineteenth-century laboratory. Working with his students at Tuskegee Institute in Alabama, George Washington Carver (in bow tie) found over a hundred uses for oils and other organic compounds extracted from peanuts.

Figure 4.3
Abiotic synthesis of organic compounds under "early Earth" conditions. Stanley Miller recreates his 1953 experiment, a laboratory simulation demonstrating that environmental conditions on the lifeless, primordial Earth favored the synthesis of some organic molecules. Miller used electric discharges (simulated lightning) to trigger reactions in a primitive "atmosphere" of H_2O, H_2, NH_3 (ammonia), and CH_4 (methane)—some of the gases that are belched into the air by volcanoes. From these ingredients, Miller's apparatus made a variety of organic compounds that play key roles in living cells. Similar chemistry may have set the stage for the origin of life on Earth, a hypothesis we will explore in more detail in Chapter 24.

improve the yield of these products (Figure 4.2). By the early nineteenth century, chemists had learned to make many simple compounds in the laboratory by combining elements under the right conditions, but artificial synthesis of the complex molecules extracted from living matter seemed hopeless. It was at that time that the Swedish chemist Jons Jakob Berzelius first made the distinction between organic compounds, those that seemingly could arise only within living organisms, and inorganic compounds, those that were found in the nonliving world. The new discipline of organic chemistry was first built on a foundation of **vitalism,** the belief in a life force outside the jurisdiction of physical and chemical laws (see Chapter 1).

Chemists began to chip away at the foundation of vitalism when they learned to synthesize organic compounds in their laboratories. In 1828, Friedrich Wöhler, a German chemist who had studied with Berzelius, attempted to make an inorganic salt, ammonium cyanate, by mixing solutions of ammonium (NH_4^+) and cyanate (CNO^-) ions. Wöhler was astonished to find that instead of the expected product, he had made urea, an organic compound present in the urine of animals. Wöhler challenged the vitalists when he wrote, "I must tell you that I can prepare urea without requiring a kidney or an animal, either man or dog." But one of the ingredients used in the synthesis, the cyanate, had been extracted from animal blood, and the vitalists were not swayed by Wöhler's discovery. A few years later, Hermann Kolbe, a student of Wöhler's, made the organic compound acetic acid from inorganic substances that could themselves be prepared directly from pure elements.

The foundation of vitalism was shaking. It finally crumbled after several more decades of laboratory synthesis of increasingly complex organic compounds. In 1953, Stanley Miller, a graduate student at the University of Chicago, helped place this abiotic (nonliving) synthesis of organic compounds into the context of evolution. Miller used a laboratory simulation of chemical conditions on the primitive Earth to demonstrate that spontaneous synthesis of organic compounds may have been an early stage in the origin of life (Figure 4.3).

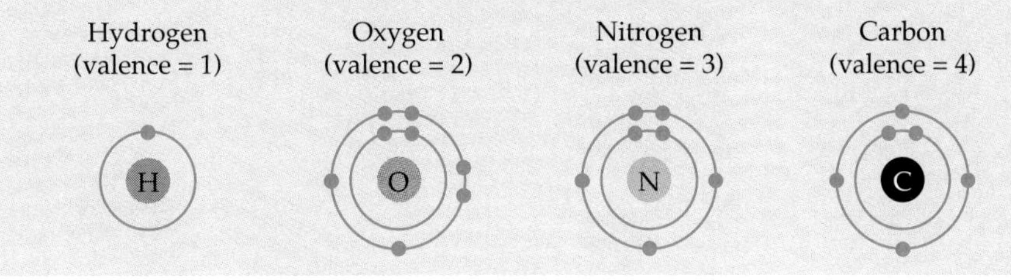

Figure 4.4
Valences for the major elements of organic molecules. Valence is the number of bonds an atom will usually form, equal to the number of electrons required to complete the outermost (valence) electron shell.

The pioneers of organic chemistry helped shift the mainstream of biological thought from vitalism to **mechanism,** the belief that all natural phenomena, including the processes of life, are governed by physical and chemical laws. Organic chemistry was redefined as the study of carbon compounds, regardless of their origin. Most naturally occurring organic compounds are products of organisms, and these molecules present a diversity and range of complexity unrivaled by inorganic compounds. But the same rules of chemistry apply to inorganic and organic molecules alike. The foundation of organic chemistry is not some intangible life force, but the unique chemical versatility of the element carbon.

THE VERSATILITY OF CARBON IN MOLECULAR ARCHITECTURE

The key to the chemical characteristics of an atom, as we learned in Chapter 2, is in its configuration of electrons, which determines the kinds and number of bonds an atom will form with other atoms. Carbon has a total of six electrons, with two in the first electron shell and four in the second shell. Having four valence electrons in a shell that holds eight, carbon has little tendency to gain or lose electrons and form ionic bonds; it would have to donate or accept four electrons to do so. Instead, a carbon atom completes its valence shell by sharing electrons with other atoms in four covalent bonds. Each carbon atom thus acts as an intersection point from which a molecule can branch off in up to four directions. This *tetravalence* is one facet of carbon's versatility that makes large, complex molecules possible.

The electron configuration of carbon also gives it covalent compatibility with many different elements. Figure 4.4 reviews the electron configurations and valences of the four major atomic components of organic molecules: carbon and its most frequent partners, oxygen, hydrogen, and nitrogen. We can think of these valences as the rules of covalent bonding in organic chemistry—the building codes that govern the architecture of organic molecules.

In Chapter 2, we learned that when a carbon atom forms single covalent bonds, the bonds angle toward the corners of an imaginary tetrahedron (see Figure 2.13b). The bond angles in methane (CH_4) are 109°, and they would be approximately the same in any molecule where carbon has four single bonds. For example, ethane (C_2H_6) is shaped like two tetrahedrons joined at their apexes (Figure 4.5). It is convenient to write structural formulas as though molecules were flat, but molecules are three-dimensional, and the shape of an organic molecule can determine its function in a living cell.

A couple of examples will demonstrate the rules of covalent bonding in organic molecules. In the carbon dioxide molecule (CO_2), a single carbon atom is joined to two atoms of oxygen by double covalent bonds. The structural formula for CO_2 is O=C=O. Each line in a structural formula represents a pair of shared electrons. Notice that the carbon atom in CO_2 is involved in a total of four covalent bonds, two with each oxygen atom. The arrangement completes the valence shells of all atoms in the molecule. Carbon dioxide is such a simple molecule that it is generally considered inorganic, even though it contains carbon. Whether we call CO_2 organic or inorganic is an arbitrary distinction, but there is no ambiguity about its importance to the living world. Taken from the air by plants and incorporated into sugar and other foods during photosynthesis, CO_2 is the source of carbon for all the organic molecules found in organisms.

Another relatively simple molecule is urea, $CO(NH_2)_2$, the organic compound from urine that Wöhler learned to synthesize in the early nineteenth century. The structural formula for urea is

```
          O
          ‖
  H       C       H
   \    /   \    /
     N         N
   /             \
  H               H
```

Again, each atom has the correct number of covalent bonds. In this case, one carbon atom is involved in both single and double bonds.

Both urea and carbon dioxide are simple molecules with only one carbon atom. But a carbon atom can also

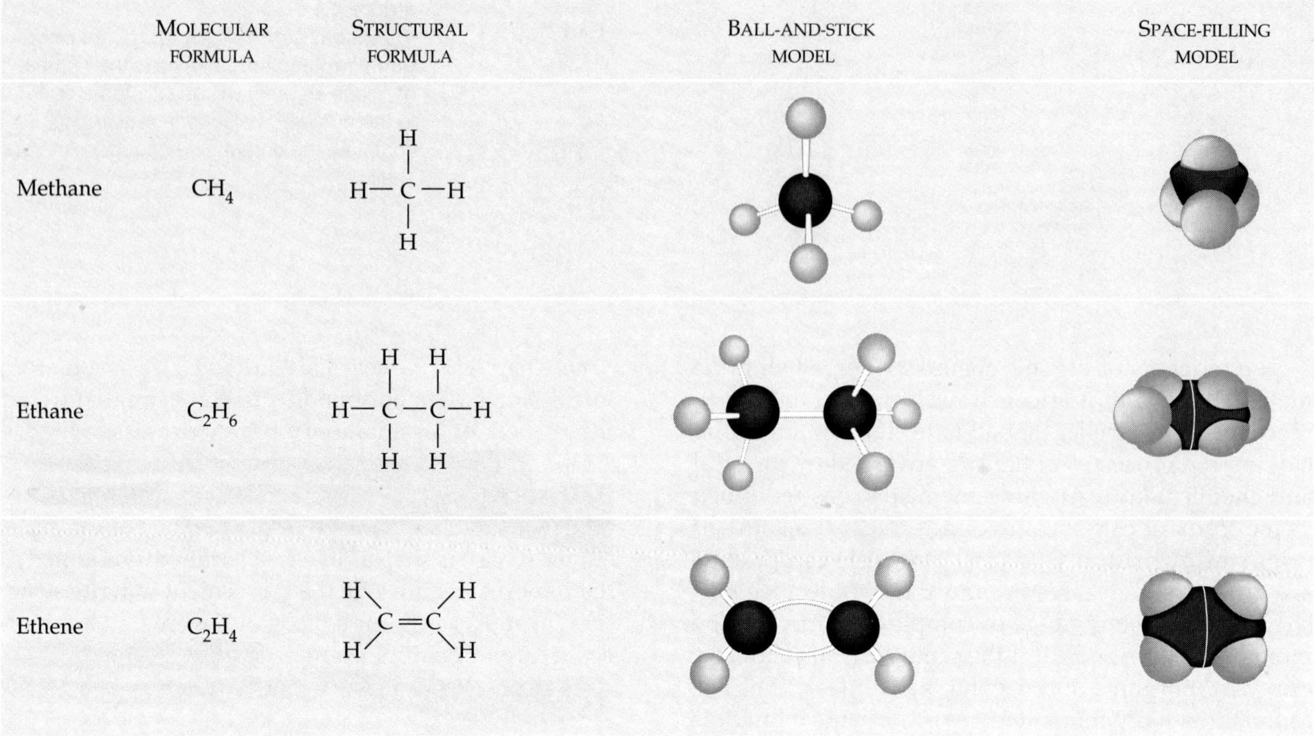

	Molecular Formula	Structural Formula	Ball-and-Stick Model	Space-Filling Model
Methane	CH_4			
Ethane	C_2H_6			
Ethene	C_2H_4			

Figure 4.5
The shapes of three simple organic molecules. Whenever a carbon atom has four single bonds, the bonds angle toward the corners of an imaginary tetrahedron. When two carbons are joined by a double bond, all bonds around those atoms are in the same plane.

use one or more of its valence electrons to form covalent bonds to other carbon atoms, making it possible to link the atoms together into chains of seemingly infinite variety.

VARIATION IN CARBON SKELETONS

Carbon chains form the skeletons of organic molecules. The skeletons vary in length and may be straight, branched, or arranged in closed rings (Figure 4.6). Some carbon skeletons have double bonds, which vary in number and location. Such variation in carbon skeletons is one important source of the molecular complexity and diversity that characterize living matter. In addition, atoms of other elements can be bonded to the skeletons at available sites.

All the molecules shown in Figures 4.5 and 4.6 are **hydrocarbons,** organic molecules consisting only of carbon and hydrogen. Atoms of hydrogen are attached to the carbon skeleton wherever electrons are available for covalent bonding. Hydrocarbons are the major components of petroleum, which is called a fossil fuel because it consists of the partially decomposed remains of organisms that lived millions of years ago. Although hydrocarbons are not prevalent in living organisms, many of the cell's organic molecules have regions consisting only of carbon and hydrogen. For example, the molecules known as fats have long hydrocarbon tails attached to a nonhydrocarbon component (Figure 4.7). Neither petroleum nor fat mixes with water; each is a hydrophobic compound, because the bonds between the carbon and hydrogen atoms are nonpolar (see Chapter 2).

Isomers

Variation in the architecture of organic molecules can be seen in **isomers,** compounds that have the same molecular formula but different structures and hence different properties. Compare, for example, the two butanes in Figure 4.6. Both have the molecular formula C_4H_{10}, but they differ in the covalent arrangement of their carbon skeletons. The skeleton is straight in butane, but branched in isobutane. There are three types of isomers: structural isomers, geometric isomers, and stereoisomers (Figure 4.8).

Structural isomers differ in the covalent arrangements of their atoms. The number of possible isomers increases tremendously as carbon skeletons increase in size. There are only two butanes, but there are 18 variations of C_8H_{18} and 366,319 possible structural isomers

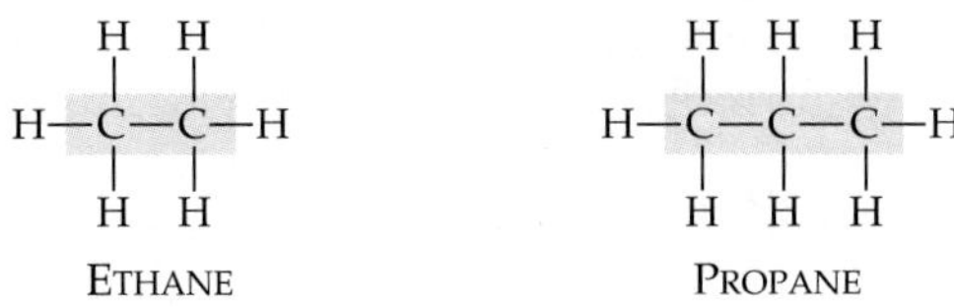

(**a**) Carbon skeletons vary in length.

BUTANE ISOBUTANE

(**b**) Skeletons may be branched or unbranched.

1-BUTENE 2-BUTENE

(**c**) The skeleton may have double bonds, which can vary in location.

CYCLOHEXANE BENZENE

(**d**) Some carbon skeletons are arranged in rings.

Figure 4.6
Variations in carbon skeletons. Hydrocarbons, organic molecules consisting only of carbon and hydrogen, illustrate the diversity of the carbon skeletons of organic molecules.

of $C_{20}H_{42}$. Structural isomers may also differ in the location of double bonds.

Geometric isomers of a molecule have all the same covalent partnerships, but they differ in their spatial arrangements. Geometric isomers arise from the inflexibility of double bonds, which, unlike single bonds, will not allow the atoms they join to rotate freely about the axis of the bonds. The subtle difference in shape between geometric isomers can dramatically affect the biological activities of organic molecules. For example, the complex process of vision is initiated by a light-induced change of rhodopsin, a chemical compound in the eye, from one geometric isomer to another.

Stereoisomers are molecules that are mirror images of each other. In the structural formula shown in Figure 4.8c, the middle carbon is called an *asymmetric carbon* because it is attached to four different atoms or groups of atoms. The four groups can be arranged in space about the asymmetric carbon in two different ways that are mirror images. They are, in a way, left-handed and right-handed versions of the molecule. A cell can distinguish these isomers based on their different shapes. Usually, one isomer is biologically active and the other is inactive. The concept of stereoisomers is important in the pharmaceutical industry, because the two isomers of a drug may not be equally effective; one of the isomers could even produce harmful effects. This may have been the case with thalidomide, the sedative that caused many birth defects in the early 1960s. The drug was a mixture of two stereoisomers,

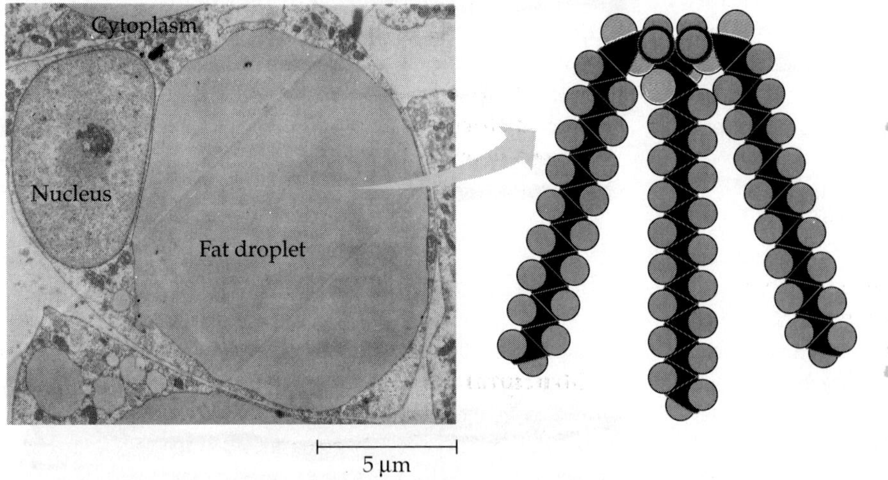

Figure 4.7
The role of hydrocarbons in the characteristics of fats. Humans and other mammals store fats in specialized cells called adipose cells. The cell's fat droplet can stockpile an enormous number of fat molecules (TEM). The drawing is a space-filling model of one fat molecule (black = carbon; gray = hydrogen; pink = oxygen). Three hydrocarbon tails are attached to a headpiece containing oxygen as well as carbon and hydrogen. Hydrocarbon bonds are nonpolar, which accounts for the hydrophobic behavior of fats. Another characteristic of hydrocarbons is that they store a relatively large amount of energy. The gasoline that fuels a car consists of hydrocarbons, and the hydrocarbon tails of fat molecules act as stored fuel for your body.

(a) **Structural isomers:** variation in covalent arrangement.

(b) **Geometric isomers:** variation in arrangement around a double bond.

(c) **Stereoisomers:** variation in spatial arrangement around an asymmetric carbon, resulting in molecules that are mirror images, like left and right hands. Stereoisomers cannot be superimposed on each other.

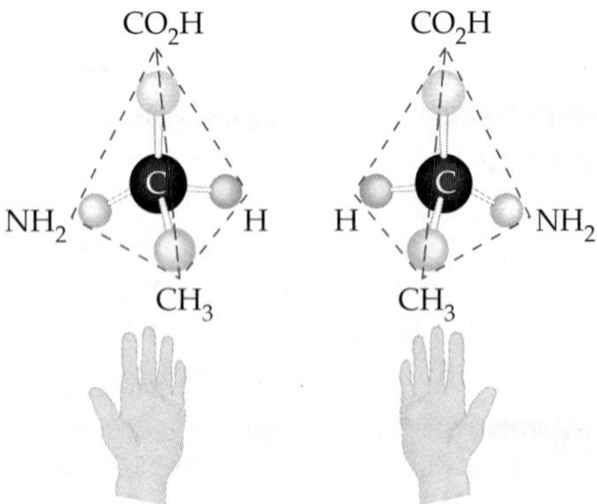

Figure 4.8
Different types of isomers. Compounds with the same molecular formula but different structures, isomers are a source of diversity in organic molecules.

one of which has since been demonstrated to cause birth defects in rats. Organisms are sensitive to even the most subtle variations in molecular construction; once again we see that molecules have emergent properties that depend on the specific arrangement of their atoms.

FUNCTIONAL GROUPS

The distinctive properties of an organic molecule depend not only on the arrangement of its carbon skeleton, but also on the molecular components attached to that skeleton. We will now examine certain groups of atoms that are frequently attached to carbon skeletons of organic molecules. These ensembles of atoms are known as **functional groups** because they are the regions of organic molecules most commonly involved in chemical reactions. The functional groups we will consider are all hydrophilic and thus increase the solubility of organic compounds in water. If we think of hydrocarbons as the simplest organic molecules, we can view functional groups as attachments that replace one or more of the hydrogens bonded to the carbon skeleton of the hydrocarbon.

Each functional group behaves consistently from one organic molecule to another, and the number and arrangement of the groups help give each molecule its unique properties. Consider the differences between estradiol and testosterone, female and male sex hormones, respectively, in humans and other vertebrates (Figure 4.9). Both are steroids, which are organic molecules having a carbon skeleton in the form of four interlocking rings. These sex hormones differ only in the attachment of certain functional groups to the common skeleton. The different actions of these two molecules on many targets throughout the body help produce the contrasting features of females and males. Thus, even our sexuality has its biological basis in the variations of molecular architecture.

The six functional groups most important in the chemistry of life are the hydroxyl, carbonyl, carboxyl, amino, sulfhydryl, and phosphate groups (Table 4.1).

The Hydroxyl Group

In a **hydroxyl group,** a hydrogen atom is bonded to an oxygen atom, which in turn is bonded to the carbon skeleton of the organic molecule. In a structural formula, the hydroxyl group is abbreviated by omission of the covalent bond between the oxygen and hydrogen, and is written as —OH or HO—. (Do not confuse this construction with the hydroxide, OH^-, of bases, such as sodium hydroxide.) The hydroxyl group is polar as a result of the electronegative oxygen atom drawing electrons toward itself. Consequently, water molecules are attracted to the hydroxyl group, and this helps to dissolve organic compounds containing such groups. Sugars, for example, owe their solubility in water to the presence of hydroxyl groups. Organic compounds containing hydroxyl groups are called **alcohols,** and their specific names usually end in *-ol.*

The Carbonyl Group

The **carbonyl group** (—CO) consists of a carbon atom joined to an oxygen atom by a double bond. If this functional group is on the end of a carbon skeleton, the organic compound is called an **aldehyde.** If the carbonyl group is not on the end of a carbon chain, the compound is called a **ketone.** For this to be the case, the

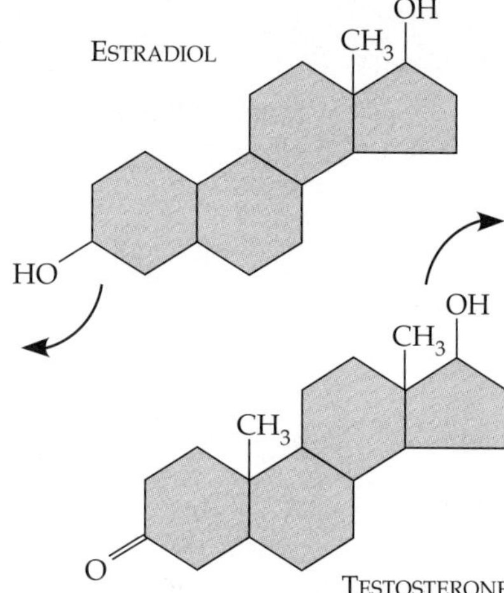

Figure 4.9
Comparison of functional groups of female (estradiol) and male (testosterone) sex hormones. The two molecules differ only in the attachment of functional groups to a common carbon skeleton (which has been simplified here by omitting the double bonds and hydrogens of the four fused rings). This subtle variation in molecular architecture influences the development of the anatomical and physiological differences between female and male vertebrates. Compare, for example, the plumage of the female (left) and male (right) wood ducks.

chain must be at least three carbons long, as it is in acetone, the simplest ketone. Acetone has different properties from propanal, a three-carbon aldehyde (acetone and propanal are isomers). Thus, variation in locations of functional groups along carbon skeletons is another source of molecular diversity in life.

The Carboxyl Group

When an oxygen atom is double-bonded to a carbon atom that is also bonded to a hydroxyl group, the entire assembly of atoms is called a **carboxyl group** (—COOH). Compounds containing carboxyl groups are known as **carboxylic acids,** or organic acids. The simplest is the one-carbon compound called formic acid, the substance some ants inject when they sting (Figure 4.10). Acetic acid, which has two carbons, gives vinegar its sour taste. (In general, acids, including carboxylic acids, taste sour.)

Why does a carboxyl group have acidic properties? A carboxyl group is a source of hydrogen ions. The covalent bond between the oxygen and the hydrogen is so polar that the hydrogen tends to dissociate reversibly from the molecule as an ion (H^+). In the case of acetic acid, we have

$$\underset{\text{Acetic acid}}{CH_3{-}C({=}O){-}OH} \rightleftharpoons \underset{\text{Acetate ion}}{CH_3{-}C({=}O){-}O^-} + \underset{\text{Hydrogen ion}}{H^+}$$

Dissociation occurs as a result of the two electronegative oxygen atoms of the carboxyl group pulling shared electrons away from hydrogen. If the double-bonded oxygen and the hydroxyl group are attached to *separate* carbon atoms, there is less tendency for the —OH group to dissociate because the second oxygen is farther away. Here is another example of how emergent properties result from a specific arrangement of building components.

Figure 4.10
A minute formic acid factory. Formicine ants, which account for about 10% of all the world's ants, manufacture and secrete formic acid:

$$H{-}C({=}O){-}OH$$

They use the organic acid for defense. The 100 trillion formicine ants of the world release about 10 billion kilograms (kg) of formic acid into the atmosphere every year.

Table 4.1 Functional Groups of Organic Compounds

Functional Group	Formula*	Name of Compounds	Example
Hydroxyl	R—OH	Alcohols	CH_3CH_2—OH Ethanol (the drug of alcoholic beverages)
Carbonyl	R—C(=O)—H	Aldehydes	CH_3CH_2—C(=O)—H Propanal
	R—C(=O)—R	Ketones	CH_3—C(=O)—CH_3 Acetone
Carboxyl	R—C(=O)—OH (non-ionized) R—C(=O)—O^- (ionized)	Carboxylic acids	CH_3—C(=O)—OH Acetic acid** (the acid of vinegar)
Amino	R—N(H)—H (non-ionized) R—$^+$N(H)(H)—H (ionized)	Amines	HO—C(=O)—CH_2—N(H)—H Glycine** (an amino acid)
Sulfhydryl	R—SH	Thiols	CH_3CH_2—SH Mercaptoethanol
Phosphate	R—O—P(=O)(O^-)—O^-	Organic phosphates	H—C(OH)(H)—C(OH)(H)—C(H)(H)—O—P(=O)(O^-)—O^- Glycerol phosphate

*The letter R symbolizes the carbon skeleton to which the functional group is attached.
**The ionized forms of the carboxyl and amino groups prevail in cells. However, acetic acid and glycine are represented here in their non-ionized forms.

The Amino Group

The **amino group** (—NH_2) consists of a nitrogen atom bonded to two hydrogen atoms and to the carbon skeleton. Organic compounds with this functional group are called **amines.** An example is glycine, illustrated in Table 4.1. Because glycine *also* has a carboxyl group, it is both an amine and a carboxylic acid. Most of the cell's organic compounds have two or more different functional groups attached to their carbon skeletons. Glycine belongs to a group of organic compounds named amino acids, which are the molecular building blocks of proteins. (In the next chapter, you will learn how the cell links amino acids together to form proteins.)

The amino group acts as a base; the nitrogen atom has a pair of unshared electrons it can use to bond to a hydrogen ion, which is thus removed from the solution in which the reaction takes place.

$$-NH_2 + H^+ \rightleftharpoons -\overset{+}{N}H_3$$

This process gives the amino group a charge of +1, its most common state within the cell.

The Sulfhydryl Group

Sulfur is grouped with oxygen in the periodic table; both have six valence electrons and form two covalent bonds. The organic functional group known as the **sulfhydryl group** (—SH), which consists of a sulfur atom bonded to an atom of hydrogen, resembles a hydroxyl group in shape (see Table 4.1). Organic compounds containing sulfhydryls are called **thiols.** In the next chapter, we will see how sulfhydryl groups help stabilize the intricate structure of many proteins.

The Phosphate Group

Phosphate is an anion formed by dissociation of an inorganic acid called phosphoric acid (H_3PO_4). The loss of hydrogen ions by dissociation leaves the phosphate with a negative charge. Organic compounds containing **phosphate groups** have a phosphate ion covalently attached by one of its oxygen atoms to the carbon skeleton (see Table 4.1). One function of phosphate groups is the transfer of energy between organic molecules. In Chapter 6, you will learn how cells harness the transfer of phosphate groups to perform work, such as the contraction of muscle cells.

THE CHEMICAL ELEMENTS OF LIFE: A REVIEW

Living matter, as you have learned, consists mainly of carbon, oxygen, hydrogen, and nitrogen, with smaller amounts of sulfur and phosphorus. These elements share the characteristic of forming strong covalent bonds, a quality that is essential in the architecture of complex organic molecules. Of all these elements, carbon is the virtuoso of the covalent bond. The chemical behavior of carbon makes it exceptionally versatile as a building block in molecular architecture: It can form four covalent bonds, link together into intricate molecular skeletons, and join with several other elements. The versatility of carbon makes possible the great diversity of organic molecules, each with special properties that emerge from the unique arrangement of its carbon skeleton and the functional groups appended to that skeleton. At the foundation of all biological diversity lies this variation at the molecular level.

Now that we have examined the basic principles of organic chemistry, we can move on to the next chapter, where we will explore the specific structures and functions of the most elegant molecules made by living cells: carbohydrates, lipids, proteins, and nucleic acids.

STUDY OUTLINE

1. Carbon is unparalleled in its ability to form the large, complex, and diverse molecules that characterize living matter.

The Foundations of Organic Chemistry (pp. 53–55)

1. Organic chemistry had its origins in vitalism, a theory that only living organisms could produce organic compounds because their synthesis required a life force outside physical and chemical laws.
2. The belief in vitalism was challenged when chemists were later able to synthesize organic compounds from inorganic ones. Organic chemistry was then redefined as the chemistry of carbon compounds.

The Versatility of Carbon in Molecular Architecture (pp. 55–56)

1. A bonding capacity of four contributes to carbon's ability to form complex and diverse molecules.

Variation in Carbon Skeletons (pp. 56–58)

1. Carbon chains are the skeletons of organic molecules. These skeletons vary in length and shape and possess bonding sites for atoms of other elements.
2. Hydrocarbons consist only of carbon and hydrogen.
3. Carbon's versatile bonding is the basis for isomers, molecules with the same molecular formula but different structures and thus different properties. The three types of isomers are structural isomers, geometric isomers, and stereoisomers.

Functional Groups (pp. 58–61)

1. Functional groups consist of specific groups of atoms that covalently bond to carbon skeletons and give the overall molecule distinctive chemical properties.
2. The hydroxyl group, found in alcohols, has a polar covalent bond, which helps alcohols dissolve in water.
3. The carbonyl group can be either at the end of a carbon skeleton (aldehyde) or within the molecule (ketone).
4. The carboxyl group is found in carboxylic, or organic, acids. The hydrogen of this group can dissociate to some extent, making the molecule a weak acid.
5. The amino group can accept an H^+, thereby acting as a base. Amino acids, the building blocks of proteins, contain both amino and carboxyl groups.
6. The sulfhydryl group helps stabilize the structure of some proteins.
7. The phosphate group can bond to the carbon skeleton by one of its oxygen atoms and has an important role in the transfer of cellular energy.

The Chemical Elements of Life: A Review (p. 61)

1. Living matter is made mostly of carbon, oxygen, hydrogen, and nitrogen, with some sulfur and phosphorus.
2. Biological diversity has its molecular basis in carbon's ability to form an incredible array of molecules with characteristic shapes and chemical properties.

SELF-QUIZ

1. Organic chemistry is currently defined as
   a. the study of compounds that can be made only by living cells
   b. the study of carbon compounds
   c. the study of vital forces
   d. the study of natural (as opposed to synthetic) compounds
   e. the study of hydrocarbons
2. Choose the pair of terms that completes this sentence: Hydroxyl is to ________ as ________ is to aldehyde.
   a. carbonyl; ketone
   b. oxygen; carbon
   c. alcohol; carbonyl
   d. amine; carboxyl
   e. alcohol; ketone
3. Which of these hydrocarbons has a double bond in its carbon skeleton?
   a. C_3H_8
   b. C_2H_6
   c. CH_4
   d. C_2H_4
   e. C_2H_2
4. The gasoline consumed by an automobile is a fossil fuel consisting mostly of
   a. aldehydes
   b. amino acids
   c. alcohols
   d. hydrocarbons
   e. thiols
5. Choose the term that correctly describes the relationship between these two sugar molecules:

   H—C=O
   H—C—OH
   H—C—OH
   H

   H
   H—C—OH
   C=O
   H—C—OH
   H

   a. structural isomers
   b. geometric isomers
   c. stereoisomers
   d. carbon isotopes
6. Identify the asymmetric carbon in this molecule:

   H—C=O (a)
   H—C—OH (b)
   H—C—H (c)
   H—C—H (d)
   H—C—H (e)
   H

7. Which functional group is not present on this molecule?

   HO—C=O
   H—C—OH
   H—C—OH
   N
   H H

   a. carboxyl
   b. carbonyl
   c. hydroxyl
   d. amino
8. An organic chemist would classify the molecule in question 7 as a (an)
   a. ketone
   b. aldehyde
   c. hydrocarbon
   d. amino acid
   e. thiol
9. Which functional group is most responsible for some organic molecules behaving as bases?
   a. hydroxyl
   b. carbonyl
   c. carboxyl
   d. amino
   e. phosphate

10. Which of the following molecules would be the strongest acid? Explain your answer.

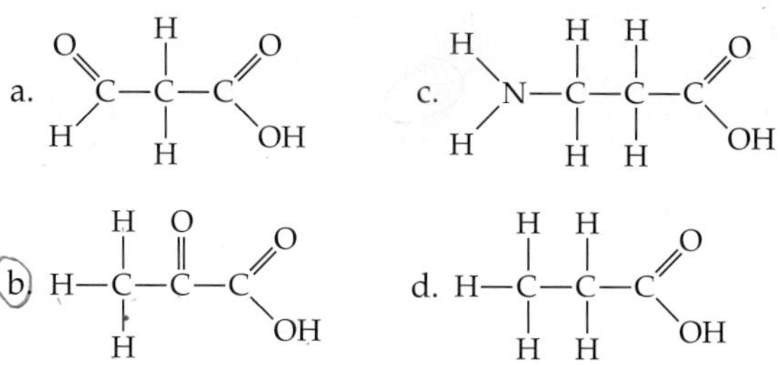

CHALLENGE QUESTIONS

1. Draw an organic molecule having all six functional groups described in this chapter.
2. Alice, in Lewis Carroll's classic *Alice in Wonderland*, poses this question: "Is looking-glass milk good to drink?" Respond, and justify your answer based on what you have learned about the structure of organic molecules, the importance of isomers, and the biological relevance of molecular shape.

SCIENCE, TECHNOLOGY, AND SOCIETY

1. How would you respond to a modern-day vitalist who argues as follows: "Biologists have been unable to discover a physical or chemical basis for human thoughts or ideas because such abstract aspects of being human *have* no basis in natural processes."
2. The role of the Food and Drug Administration (FDA) in monitoring the testing of new drugs continues to be controversial. The thalidomide tragedy (p. 57) is often cited by those who argue that the FDA should not approve any drug until no doubt remains that the drug can be safely prescribed. Others would like to see the FDA requirements relaxed for drugs that may help individuals suffering from terminal diseases, such as AIDS. Where do you stand on this issue? Defend your position.

FURTHER READING

Amato, I. "Looking Glass Chemistry." *Science*, May 15, 1992. The pharmaceutical relevance of isomers.

Asimov, I. *The World of Carbon*. 2nd ed. New York: Macmillan, 1962. A primer on the basics of organic chemistry, as told by one of America's most popular science writers.

Hegstrom, R. A., and D. K. Kondepudi. "The Handedness of the Universe." *Scientific American*, January 1990. The evolutionary significance of "left-handed" and "right-handed" molecules.

Ourisson, G., P. Albrecht, and M. Rohmer. "The Microbial Origin of Fossil Fuels." *Scientific American*, August 1984. The biology behind our important energy resources.

5 THE STRUCTURE AND FUNCTION OF MACROMOLECULES

POLYMERS
CARBOHYDRATES
LIPIDS
PROTEINS
NUCLEIC ACIDS
DNA AND PROTEINS AS TAPE MEASURES OF EVOLUTION

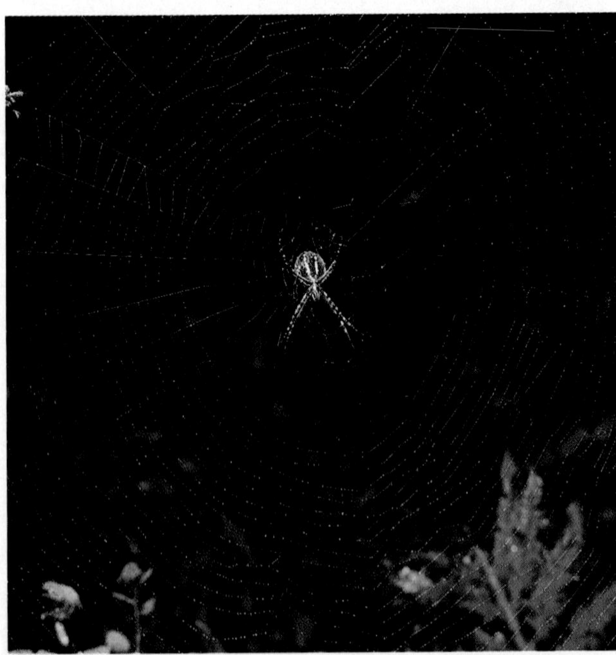

Figure 5.1
The diverse functions of macromolecules. Spiders such as this orb spider weave their webs from silk, a specialized protein. In this chapter, you will learn how the functions of proteins and other biological molecules are determined by their structures.

We have applied the concept of emergent properties to our study of water and relatively simple organic molecules. These substances are central to life, each one having unique behavior arising from the orderly arrangement of its atoms. Another level in the hierarchy of biological organization is attained when cells join together small organic molecules to form larger molecules that belong to four classes: carbohydrates, lipids, proteins, and nucleic acids (Figure 5.1). Many of these cellular molecules are, on the molecular scale, huge. For example, a protein may consist of thousands of covalently connected atoms that form a molecular colossus weighing more than 100,000 daltons. Such a molecule is called a **macromolecule,** the term biologists use for the giant molecules of living matter.

Considering the size and complexity of macromolecules, it is remarkable that the structures of so many of them have been determined in detail by molecular biologists (Figure 5.2). Understanding the architecture of a particular macromolecule helps explain how that molecule works in the living cell. In molecular biology, as in the study of life at all levels, form and function are inseparable.

We begin our investigation of the macromolecules of life with a generalization to help us understand their structure: Most macromolecules are polymers.

POLYMERS

Cells make macromolecules by linking relatively small molecules together, forming chains called polymers (from the Greek *polys,* "many," and *meris,* "part"). A **polymer** is a large molecule consisting of many identical or similar subunits strung together, much as a train consists of a chain of cars. The subunits that serve as the building blocks of a polymer are called **monomers.** Some of the small molecules that function as monomers also have other functions of their own.

Polymers and Molecular Diversity

Variation in the structure of macromolecules is fundamental to the diversity of life. Each cell has thousands

(a)

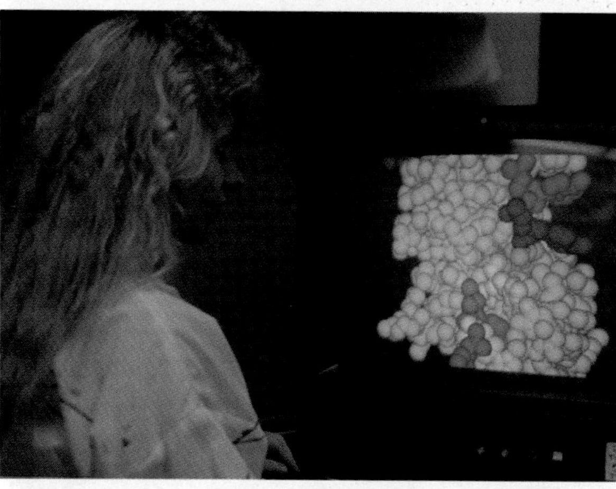

(b)

Figure 5.2
Building models to study the structure and function of macromolecules. (**a**) Linus Pauling poses with a model of part of a protein. In the 1950s, Pauling discovered several of the basic structural principles for proteins. (**b**) Today, scientists like this graduate student use computers to help build molecular models. Though methods have improved, the goal remains the same: to correlate the structure of macromolecules with their functions.

of different kinds of macromolecules, many of which vary from one type of cell to another in the same organism. The inherent differences between human siblings reflect variations in polymers, particularly DNA and proteins. Molecular differences between unrelated individuals are more extensive, and between species greater still. The diversity of macromolecules in the living world is vast—and the potential variety is essentially infinite.

What is the basis for such diversity in life's polymers? All macromolecules are constructed from only 40 to 50 common monomers and some others that occur rarely. Building a limitless variety of polymers from such a limited list of monomers is analogous to constructing hundreds of thousands of words from only 26 letters of the alphabet. The key is arrangement—variation in the linear sequence in which the subunits are strung together. However, this analogy falls far short of describing the great diversity of macromolecules, because most polymers are much longer than the longest word. Proteins, for example, are built from 20 kinds of amino acids arranged in chains that are typically more than 100 amino acids long.

Polymers account for the molecular uniqueness of an organism. The monomers used to make these polymers, however, are universal. Your proteins and those of a carrot or a cow are assembled from the same 20 amino acids, but these common building blocks are arranged in polymers in varying sequence. The molecular logic of life is simple but elegant: Small molecules common to all organisms are ordered into distinctive macromolecules.

Making and Breaking Polymers

The classes of macromolecules differ in the nature of their monomers, but the chemical mechanisms that cells use to make and break polymers are basically the same for all macromolecules. Monomers are linked together by a process known as **condensation synthesis.** The net effect of this process is the removal of a water molecule for each monomer added to the chain. (The actual mechanism, as it occurs in cells, is actually more complicated.) Whenever two monomers are joined, each contributes part of the H_2O molecule that is removed; one monomer loses a hydroxyl group (OH), and the other loses a hydrogen (H). Both monomers, having each lost a covalent partner, now bond covalently with each other (Figure 5.3a). The cell must expend energy to form these new bonds, and the process occurs only with the help of enzymes, specialized proteins that speed up chemical reactions in cells. You will learn how enzymes work in Chapter 6.

Polymers are disassembled to monomers by **hydrolysis,** a process that is essentially the reverse of condensation synthesis (Figure 5.3b). Hydrolysis means to break with water (from the Greek *hydro,* "water," and *lysis,* "break"). Bonds between monomers are broken by the addition of water molecules, a hydrogen from the water attaching to one monomer, and a hydroxyl attaching to the adjacent monomer. An example of hy-

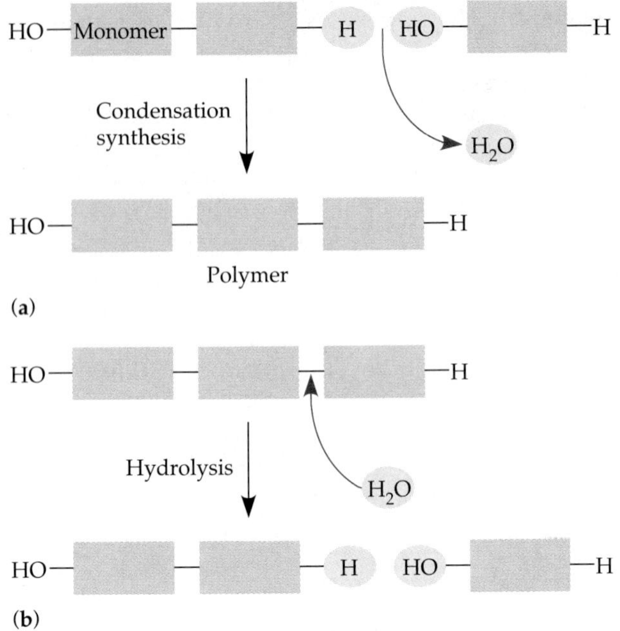

Figure 5.3
The synthesis and breakdown of polymers. (**a**) Monomers are joined by condensation synthesis. The net effect is removal of a water molecule. (**b**) The reverse of this process, hydrolysis, breaks bonds between monomers by adding water molecules.

drolysis working in our bodies is the process of digestion. The bulk of the organic material in our food is in the form of polymers that are much too large to enter our cells. Within the digestive tract, various enzymes attack the polymers, speeding up hydrolysis. The released monomers are then absorbed into the bloodstream for distribution to all body cells.

Having developed an overview of macromolecules as polymers assembled from monomers by condensation synthesis, we are now prepared to investigate the specific structures and functions of the four major classes of organic compounds found in cells: carbohydrates, lipids, proteins, and nucleic acids. For each class, we will see that the macromolecules have emergent properties not found in their individual monomers.

CARBOHYDRATES

Carbohydrates include sugars and their polymers. The simplest carbohydrates are the monosaccharides, single sugars also known as simple sugars. Disaccharides are double sugars, consisting of two monosaccharides joined by condensation synthesis. Carbohydrates also include macromolecules in the form of polysaccharides, polymers of many sugars.

Monosaccharides

Monosaccharides (from the Greek *monos,* "single," and *sacchar,* "sugar") generally have molecular formulas that are some multiple of CH_2O (Figure 5.4). Glucose ($C_6H_{12}O_6$), the most common monosaccharide, is of central importance in the chemistry of life. In the structure of glucose, we can see all the trademarks of a sugar. A hydroxyl group is attached to each carbon except one, which is double-bonded to an oxygen to form a carbonyl group. Depending on the location of the carbonyl group, a sugar is either an aldose (aldehyde sugar) or a ketose (ketone sugar) as well as an alcohol. Glucose, for example, is an aldose; fructose, a structural isomer of glucose, is a ketose. (Notice that most names for carbohydrates end in *-ose.*) Another criterion for classifying sugars is the size of the carbon skeleton, which ranges from three to seven carbons long. Glucose, fructose, and other sugars that have six carbons are called hexoses. Trioses and pentoses are also common.

Still another source of diversity for simple sugars is in the spatial arrangement of their parts around asymmetric carbons. (Recall from Chapter 4 that an asymmetric carbon is one attached to four different kinds of covalent partners.) Glucose and galactose, for example, differ only in the placement of parts around one asymmetric carbon (see Figure 5.4). What may seem at first a small difference is significant enough to give the two sugars distinctive shapes, and shape is a major means by which molecules within cells recognize and interact with one another.

Although it is convenient to draw glucose as though its carbon skeleton were linear, this representation is not accurate. In aqueous solutions, glucose molecules, as well as most other sugars, form rings (Figure 5.5).

Monosaccharides, particularly glucose, are major nutrients for cells. In the process known as cellular respiration, cells extract the energy stored in the glucose molecules. Not only are simple sugar molecules a major fuel for cellular work, but their carbon skeletons serve as raw material for the synthesis of other types of small organic molecules, including amino acids and fatty acids. Sugar molecules that are not immediately used by cells as fuel or as a source of carbon skeletons are generally incorporated as monomers into disaccharides or polysaccharides.

Disaccharides

A **disaccharide,** or double sugar, consists of two monosaccharides joined by a **glycosidic linkage,** a covalent bond formed between two monosaccharides (Figure 5.6). For example, maltose is a disaccharide formed by linking two molecules of glucose. Also known as malt

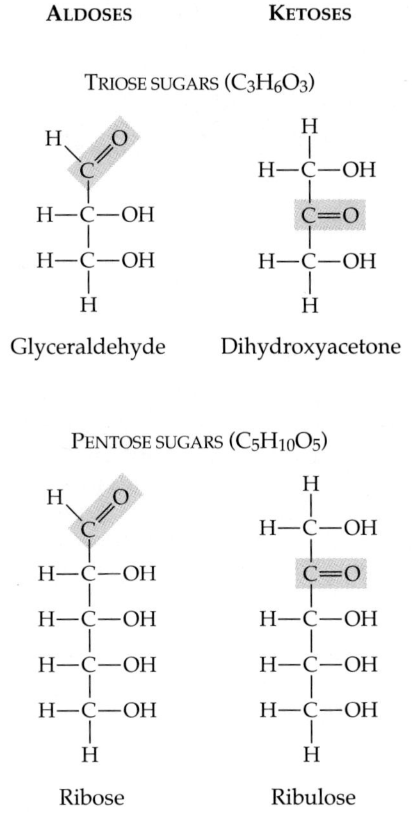

ALDOSES | KETOSES

TRIOSE SUGARS ($C_3H_6O_3$)

Glyceraldehyde | Dihydroxyacetone

PENTOSE SUGARS ($C_5H_{10}O_5$)

Ribose | Ribulose

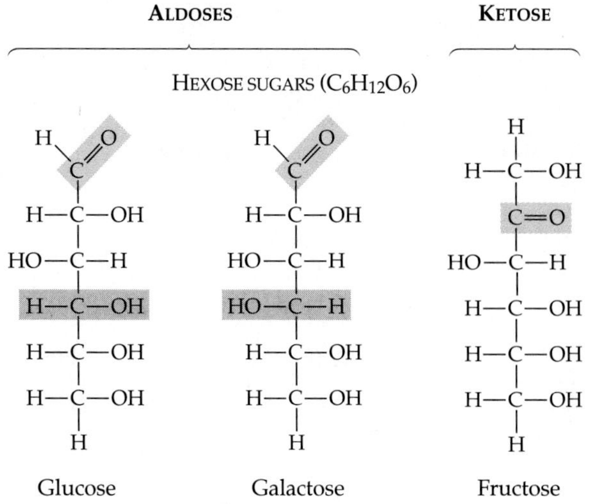

ALDOSES | KETOSE

HEXOSE SUGARS ($C_6H_{12}O_6$)

Glucose | Galactose | Fructose

Figure 5.4
The structure and classification of common simple sugars. Sugars may be aldoses (aldehydes) or ketoses (ketones), depending on the location of the carbonyl group (pink). Sugars are also classified according to the length of their carbon skeletons. A third point of variation is in the spatial arrangement around asymmetric carbons (compare, for example, the gray portions of glucose and galactose).

sugar, maltose is an important ingredient in the brewing of beer. Lactose, the sugar present in milk, is another disaccharide, consisting of a glucose molecule joined to a galactose molecule. The most prevalent disaccharide is sucrose, better known as table sugar. Its two monomers are glucose and fructose. Plants generally transport carbohydrates from one part of the plant to another in the form of sucrose.

Polysaccharides

Polysaccharides are macromolecules. They are polymers in which a few hundred to a few thousand monosaccharides are linked together. Some polysaccharides are storage material, hydrolyzed as needed to provide sugar for cells. Other polysaccharides serve as building

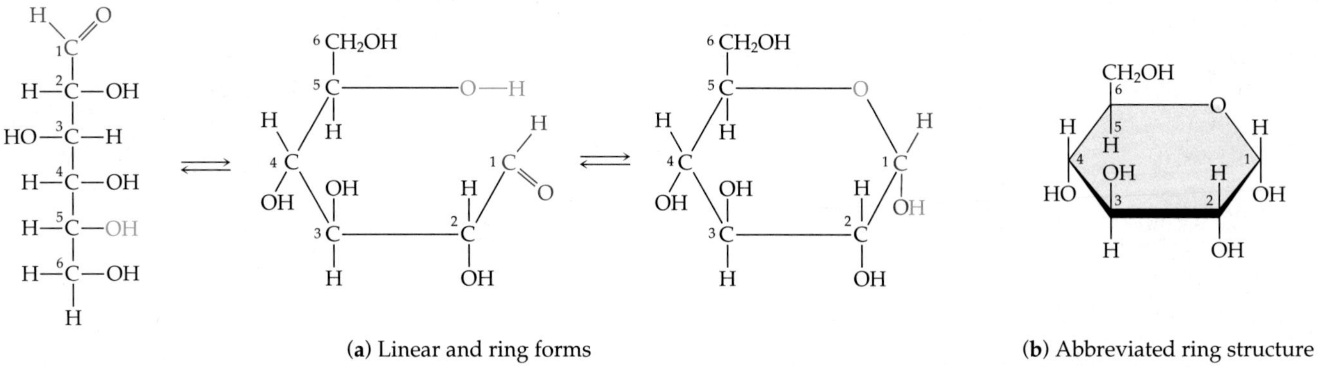

(**a**) Linear and ring forms

(**b**) Abbreviated ring structure

Figure 5.5
Linear and ring forms of glucose. (**a**) Chemical equilibrium between the linear structures and rings greatly favors the formation of rings. To form the glucose ring, carbon number 1 bonds to the oxygen attached to carbon number 5. (**b**) In this abbreviated ring formula, the thicker edge indicates that you are looking at the ring edge-on; that is, the components attached to the ring by the vertical lines lie above or below the plane of the ring.

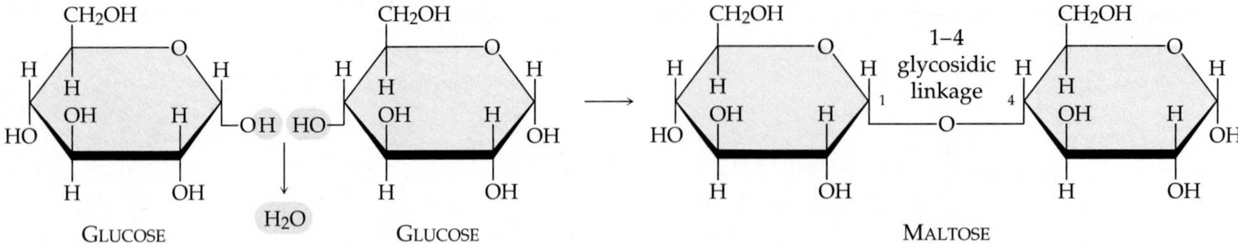

(a) Condensation synthesis of maltose

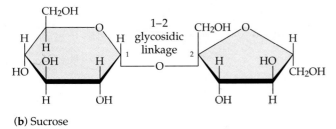

1–2 glycosidic linkage

(b) Sucrose

Figure 5.6
Examples of disaccharides. (**a**) The bonding of two glucose units forms maltose. Notice that the glycosidic link joins the number 1 carbon of one glucose to the number 4 carbon of the second glucose. Joining the glucose monomers at different places would result in different disaccharides. (**b**) Sucrose is a disaccharide formed from glucose and fructose. (Notice that fructose forms a *five*-sided ring rather than a six-sided ring.)

material for structures protecting the cell or the whole organism.

Storage Polysaccharides **Starch,** a storage polysaccharide of plants, is a polymer consisting entirely of glucose (Figure 5.7). The monomers are joined by 1–4 linkages (number 1 carbon to number 4 carbon), like the monomers of maltose (see Figure 5.6a). The angle of these bonds results in the polymer's helical shape. The simplest form of starch, amylose, is unbranched. Amylopectin, a more complex form of starch, is a branched polymer.

Plants store starch as granules within cellular structures called plastids, including chloroplasts (Figure 5.8a). By synthesizing starch, the plant can stockpile surplus sugar. Because glucose is a major cellular fuel, starch represents stored energy. The sugar can later be withdrawn from this carbohydrate bank by hydrolysis,

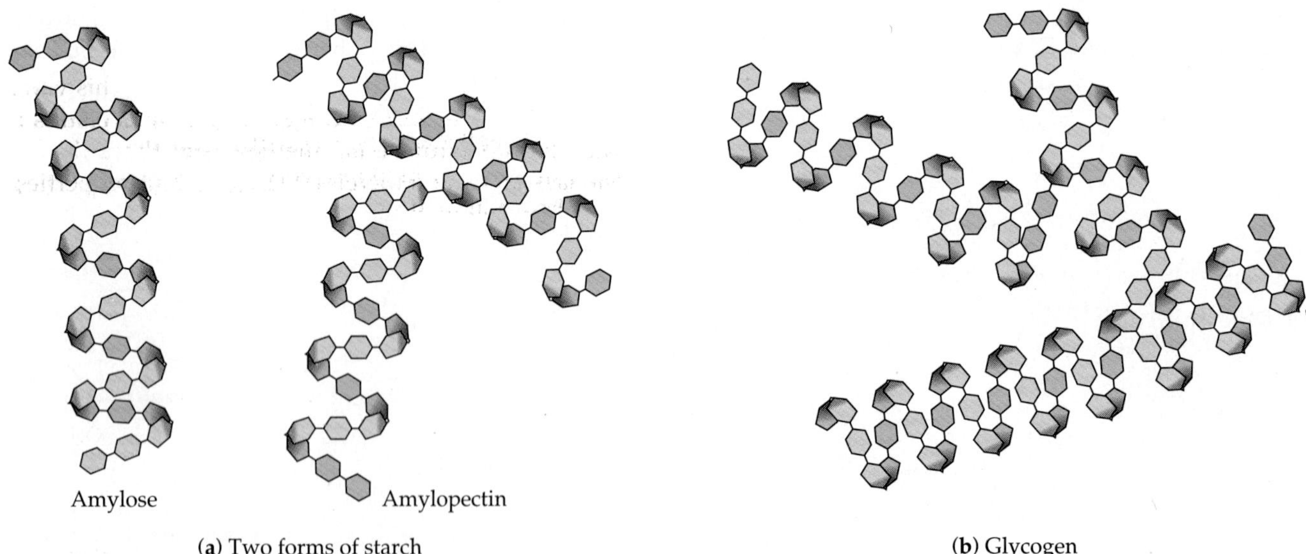

Figure 5.7
Storage polysaccharides. These examples are composed entirely of glucose monomers, which are abbreviated here as hexagons. Starch and glycogen have helical shapes. (**a**) Two forms of starch are amylose (unbranched) and amylopectin (branched). (**b**) Glycogen is more extensively branched than amylopectin.

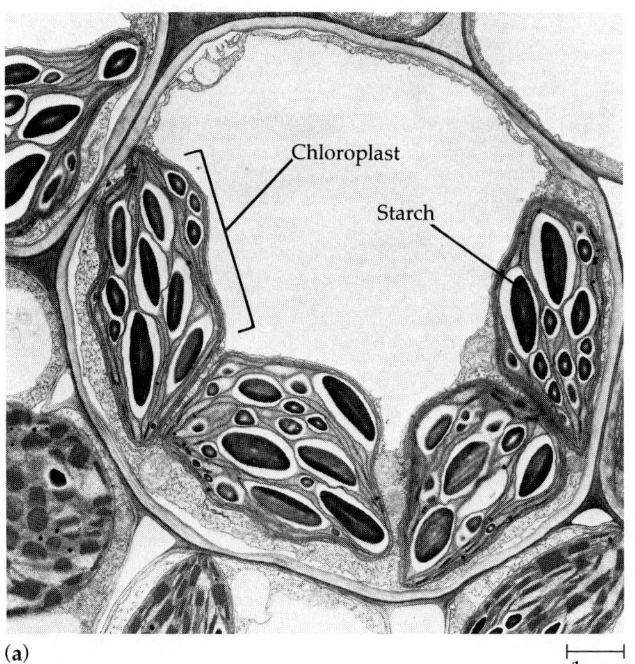

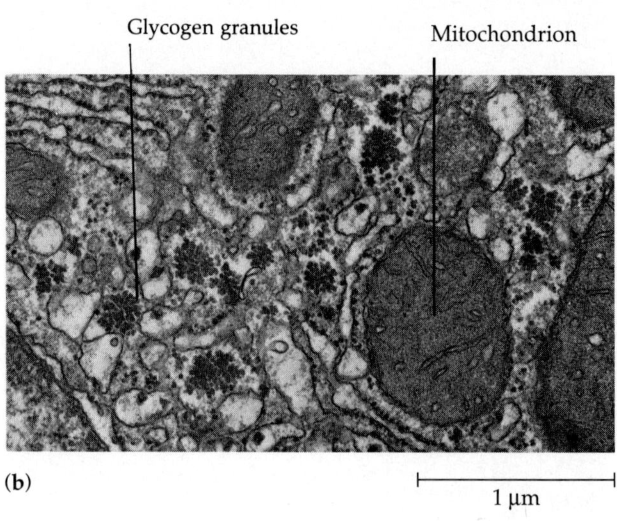

Figure 5.8
Stockpiles of storage polysaccharides. (**a**) The dark ovals are granules of starch within the chloroplasts of a plant cell (TEM). (**b**) Animal cells store the polysaccharide glycogen as dense clusters of granules within liver and muscle cells. Hydrolysis withdraws the glucose from storage (TEM of a portion of a liver cell).

which breaks the bonds between the glucose monomers. Most animals, including humans, also have enzymes that can hydrolyze plant starch, making glucose available as a nutrient for cells. Potato tubers and grains—the fruits of wheat, corn, rice, and other grasses—are the major sources of starch in the human diet.

Animals store a polysaccharide called **glycogen,** a polymer of glucose that is more extensively branched than the amylopectin of plants (see Figure 5.7). Humans and other vertebrates store glycogen mainly in liver and muscle cells, hydrolyzing the glycogen to release glucose when the demand for sugar increases (see Figure 5.8b). This stored fuel cannot sustain an animal for long, however. The glycogen bank of humans, for example, is depleted in about a day unless it is replenished by food.

Structural Polysaccharides Organisms build strong materials from structural polysaccharides. For example, the polysaccharide called **cellulose** is a major component of the tough walls that enclose plant cells. On a global scale, plants produce almost 10^{11} (100 billion) tons of cellulose per year; it is the most abundant organic compound on Earth. Like starch, cellulose is a polymer of glucose, but the glycosidic linkages of these two polymers differ. The difference is based on two possible ring structures for glucose. When the carbon chain of glucose forms a ring, the hydroxyl group attached to the number 1 carbon at the site where the ring closes is locked into one of two alternative positions: it can lie either below or above the plane of the ring. These two ring forms for glucose are called alpha (α) and beta (β), respectively (Figure 5.9a). In starch, all the glucose monomers are in the α configuration (Figure 5.9b). In contrast, the glucose monomers of cellulose are all in the β configuration (Figure 5.9c). This variation in the geometry of the glycosidic links results in starch and cellulose having different three-dimensional shapes, and therefore very different properties.

In the cell wall of a plant, many parallel cellulose molecules, held together by hydrogen bonds between hydroxyl groups of the glucose monomers, are arranged as units called microfibrils (Figure 5.10). Several intertwined microfibrils form a cellulose fibril, and several fibrils may in turn be supercoiled. These strong cables are an excellent building material for plants as well as for humans, who use cellulose-rich wood for lumber.

Enzymes that digest starch by hydrolyzing the α bonds are unable to hydrolyze the β linkages of cellulose. In fact, few organisms possess enzymes that can digest cellulose. Humans do not; the cellulose fibrils in our food pass through the digestive tract and are eliminated with the feces. Along the way, the fibrils abrade the wall of the digestive tract and stimulate the lining

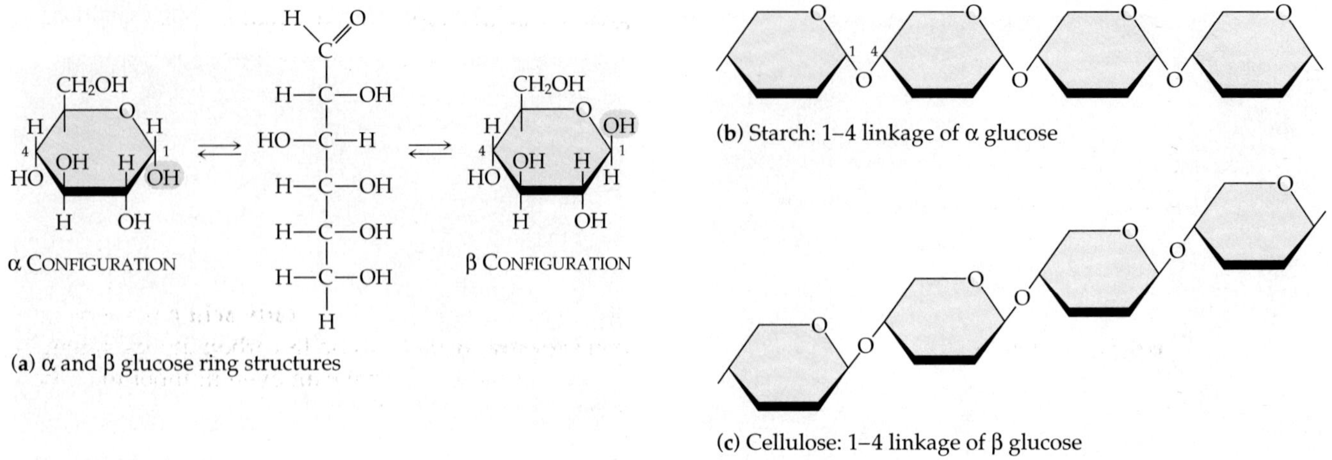

α CONFIGURATION

β CONFIGURATION

(a) α and β glucose ring structures

(b) Starch: 1–4 linkage of α glucose

(c) Cellulose: 1–4 linkage of β glucose

Figure 5.9
A comparison of starch and cellulose structures. (**a**) Glucose forms two interconvertible ring structures, designated α and β, which differ in the placement of the hydroxyl group attached to the number 1 carbon. (**b**) The α ring form is the monomer for starch. (**c**) Cellulose consists of glucose monomers in the β configuration.

to secrete mucus, which aids in the smooth passage of food through the tract. Thus, although cellulose is not a nutrient for humans, it is an important part of a healthful diet. Most fresh fruits, vegetables, and grains are rich in cellulose, or fiber.

Some bacteria and other microorganisms can digest cellulose into glucose. A cow harbors cellulose-digesting bacteria in the rumen, a pouch attached to its stomach. The bacteria hydrolyze the cellulose of hay and grass, converting the glucose molecules to other nutrients that nourish the cow. Similarly, a termite, unable to digest cellulose for itself, has unicellular organisms living in its gut that make a meal of wood. Some molds (fungi) can also digest cellulose, and thus serve as decomposers that are crucial in the web of life on Earth (see Chapter 28).

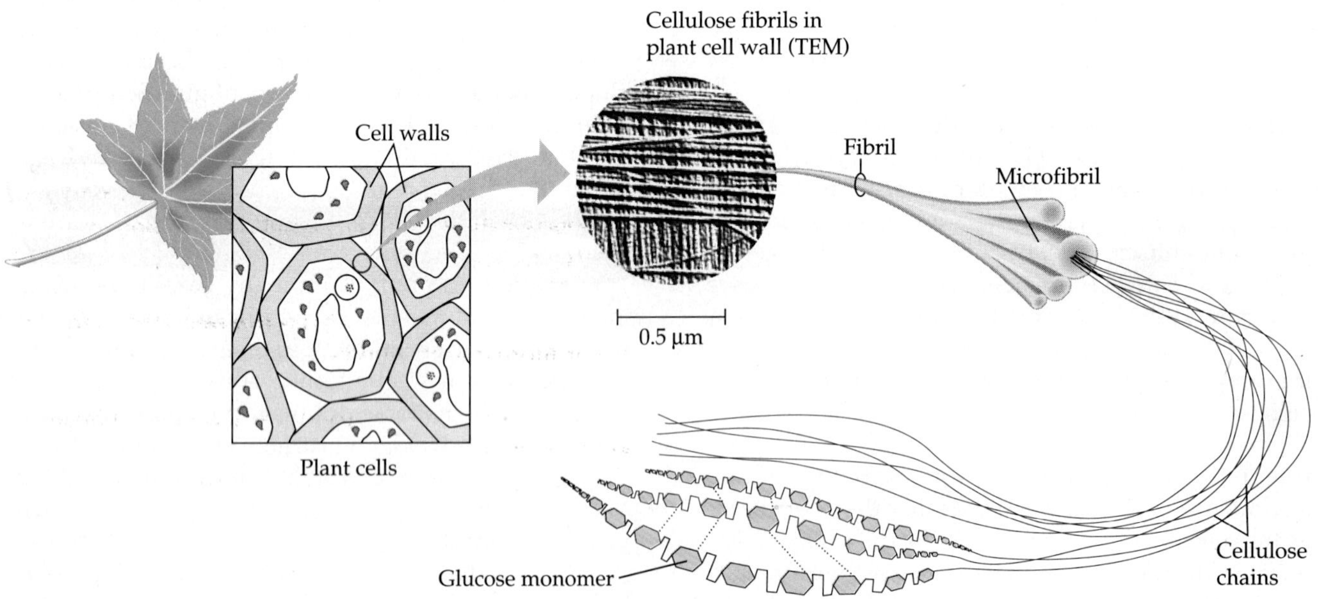

Figure 5.10
Cellulose and plant cell walls. Cellulose is a linear, unbranched molecule. Parallel cellulose molecules are held together by hydrogen bonds (dotted lines) between hydroxyl groups projecting from both sides. About 80 cellulose molecules associate to form a microfibril, and several microfibrils intertwine to form a cellulose fibril (TEM).

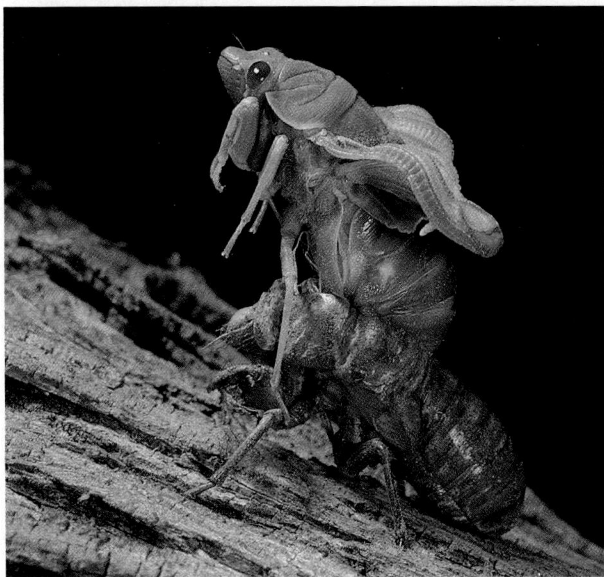

Figure 5.11
Chitin. A structural polysaccharide, chitin forms the exoskeleton of arthropods. This cicada is molting, shedding its old exoskeleton and emerging in adult form. Chitin is similar in structure to cellulose, except the monomer in chitin is an amino sugar (a sugar with nitrogen).

Another important structural polysaccharide is **chitin,** the carbohydrate used by arthropods (insects, spiders, crustaceans, and related animals) to build their exoskeletons (Figure 5.11). An exoskeleton is a hard case that surrounds the soft parts of the animal. Pure chitin is leathery, but it becomes hardened when encrusted with calcium carbonate, a salt. Chitin is also found in many fungi, which use this polysaccharide rather than cellulose as building material for their cell walls. The monomer of chitin, an amino sugar, is a glucose molecule with a nitrogen-containing appendage:

O (ring) — NH_2 — C=O — CH_3

LIPIDS

The chemically diverse compounds called **lipids** are grouped together because they share one important trait: They have little or no affinity for water. Lipids' hydrophobic behavior is based on their molecular structure. Although they may have some polar bonds associated with oxygen, lipids consist mostly of hydrocarbon regions with nonpolar bonds. Three important families of lipids are fats, phospholipids, and steroids.

Fats

Fats are large molecules constructed from two kinds of smaller molecules: glycerol and fatty acids (Figure 5.12a). Glycerol is an alcohol with three carbons, each bearing a hydroxyl group. A **fatty acid** has a long carbon skeleton, usually 16 or 18 carbon atoms in length (nearly all fatty acids have an even number of carbon atoms). At one end of the fatty acid is a "head" consisting of a carboxyl group, the functional group that gives these molecules the name fatty *acids.* Attached to the carboxyl group is a long hydrocarbon "tail." The nonpolar C—H bonds in the tails of fatty acids are the reason fats are insoluble in water (see Chapter 4). Fats separate from water because the water molecules hydrogen-bond to one another and exclude the fats. A common example of this phenomenon is the separation of vegetable oil from the aqueous vinegar solution in a bottle of salad dressing.

A fatty acid can be joined to glycerol by an ester linkage, a bond between a hydroxyl group and a carboxyl group. Glycerol now has two remaining hydroxyls, and each can also bond to a fatty acid. The product is a fat, or **triacylglycerol,** which consists of three fatty acids linked to one glycerol molecule. (Still another name for a fat is triglyceride, a word often found in the list of ingredients on packaged foods.) All the fatty acids in a fat can be the same (Figure 5.12b), or they can be of two or three different kinds.

Fatty acids vary in length and in the number and locations of double bonds. The terms *saturated fats* and *unsaturated fats* are commonly used in the context of nutrition. These terms derive from the structure of the hydrocarbon tails of the fatty acids. If there are no double bonds between the carbon atoms composing the tail, then as many hydrogen atoms as possible are bonded to the carbon skeleton, creating a **saturated fatty acid** (Figure 5.13a). An **unsaturated fatty acid** has one or more double bonds, formed by the removal of hydrogen atoms from the carbon skeleton. The fatty acid will have a kink in its shape wherever a double bond occurs (Figure 5.13b).

Most animal fats are saturated; they have fatty acids that lack double bonds. Saturated animal fats—such as bacon grease, lard, and butter—solidify at room temperature. In contrast, plant fats are generally unsaturated. Usually liquid at room temperature, plant fats are referred to as oils—for instance, corn oil, peanut oil, and olive oil. The kinks where the double bonds are located prevent the molecules from packing together closely enough to solidify at room temperature. The phrase "hydrogenated vegetable oils," often found on

Figure 5.12
The structure of a fat, or triacylglycerol. The molecular building blocks of a fat are one molecule of glycerol and three molecules of fatty acids. (**a**) One water molecule is removed for each fatty acid joined to the glycerol. (**b**) The result is a fat, such as the one shown here.

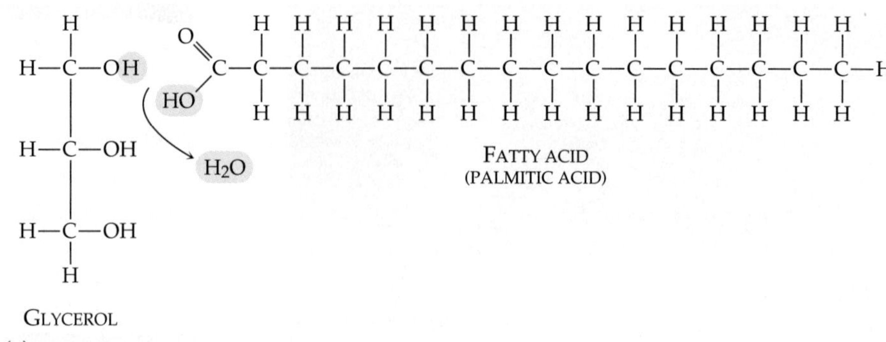

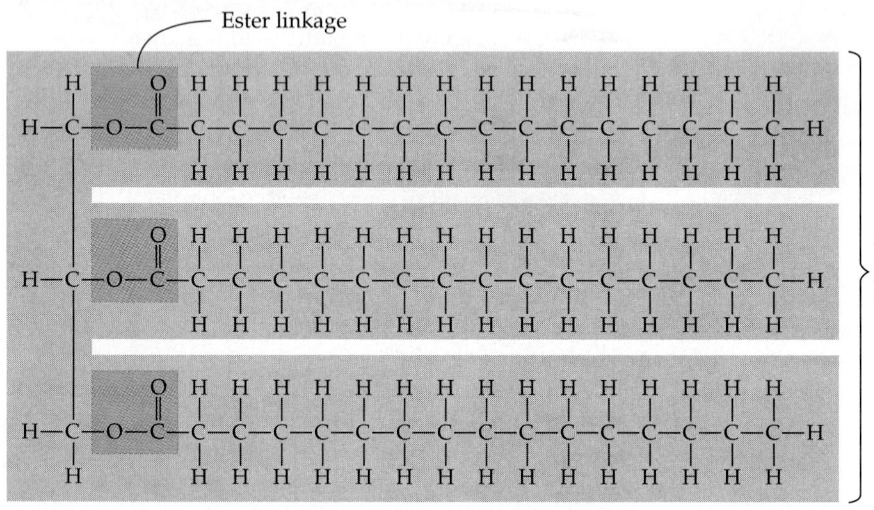

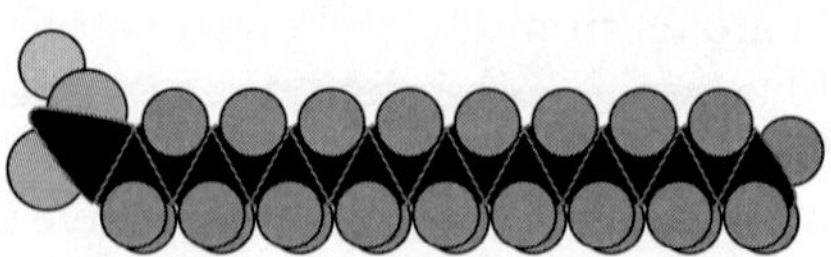

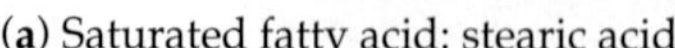

(a) Saturated fatty acid: stearic acid

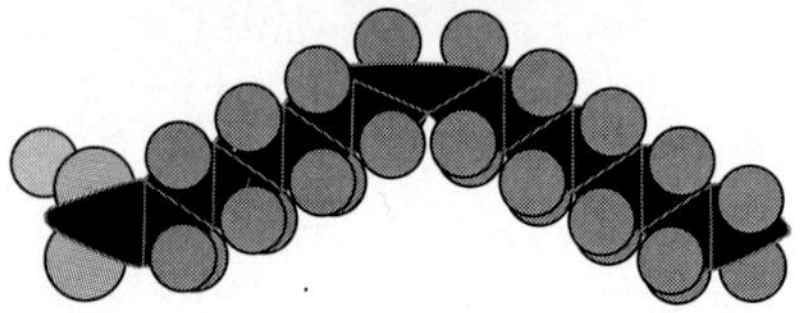

(b) Unsaturated fatty acid: oleic acid

Figure 5.13
A comparison of saturated and unsaturated fats. (**a**) Saturated fatty acids, such as stearic acid, have as many hydrogen atoms as possible bonded to the carbon skeleton. If its fatty acids are saturated (with hydrogen), then a fat is also said to be saturated. Most animal fats, such as those in butter, are saturated. They form solids at room temperature. (**b**) Unsaturated fatty acids, such as oleic acid, have one or more double bonds between carbons. Where a double bond forms, two hydrogen atoms must be removed. The fatty acid bends where these double bonds are located. One such location can be seen in oleic acid. Unsaturated fats have unsaturated fatty acids. Most vegetable fats are unsaturated, and they are called oils because they form liquids at room temperature. The kinks in the fatty acids prevent the fats from packing together closely enough to solidify.

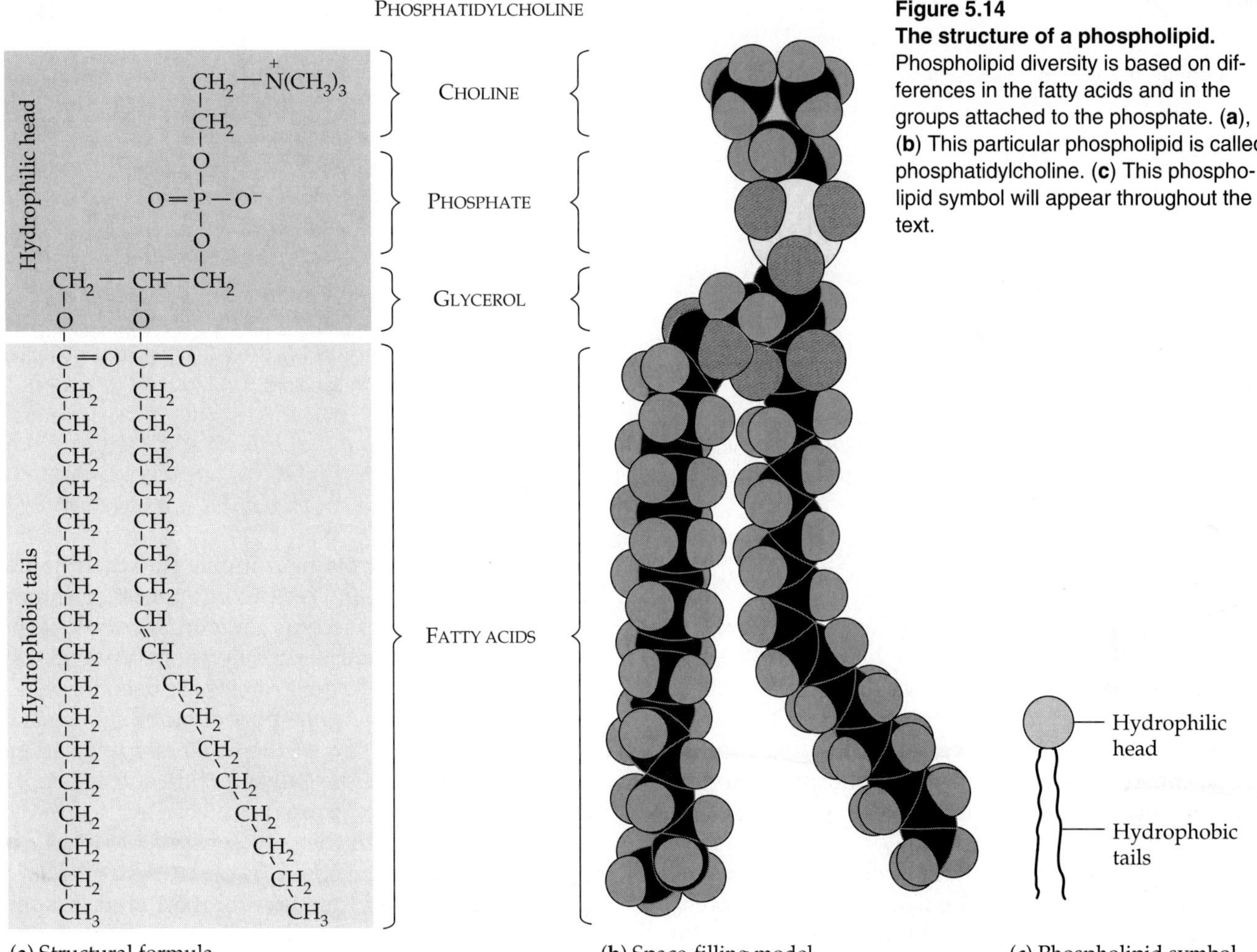

(a) Structural formula (b) Space-filling model (c) Phospholipid symbol

Figure 5.14
The structure of a phospholipid. Phospholipid diversity is based on differences in the fatty acids and in the groups attached to the phosphate. (**a**), (**b**) This particular phospholipid is called phosphatidylcholine. (**c**) This phospholipid symbol will appear throughout the text.

food labels, means that unsaturated fats have been synthetically converted to saturated fats by adding hydrogen. Peanut butter, margarine, and many other products are hydrogenated to prevent lipids from separating out in liquid (oil) form.

A diet rich in saturated fats is one of several factors that may contribute to the cardiovascular disease known as atherosclerosis. In this condition, deposits called plaques develop on the internal lining of blood vessels, impeding blood flow and reducing the resilience of the vessels (see Chapter 38).

Fat has come to have such negative connotations in our society that one might wonder whether fats serve any useful purpose. The major function of fats is energy storage. The hydrocarbons of fats are similar to gasoline molecules and just as rich in energy. A gram of fat stores more than twice as much energy as a gram of a polysaccharide, such as starch. Because plants are relatively immobile, they can function with bulky energy storage in the form of starch. (Vegetable oils are generally obtained from seeds, where more compact storage is an asset to the plant.) Animals, on the other hand, must carry their energy baggage with them, so there is an advantage to a more compact reservoir of fuel—fat. Humans and other mammals stock their long-term food reserves in adipose cells, which swell and shrink as fat is deposited and withdrawn from storage (see Figure 4.7). In addition to storing energy, adipose tissue also cushions such vital organs as the kidneys, and a layer of fat beneath the skin insulates the body. This subcutaneous (below the skin) layer is especially thick in whales, seals, and other marine animals.

Phospholipids

Phospholipids are structurally related to fats, but they have only two fatty acids rather than three (Figure 5.14). The third carbon of glycerol is joined to a phosphate group, which is negative in electrical charge. Additional small molecules, usually charged or polar, can be linked to the phosphate group to form a variety of phospholipids.

Phospholipids show ambivalent behavior toward water. Their tails, which consist of hydrocarbons, are

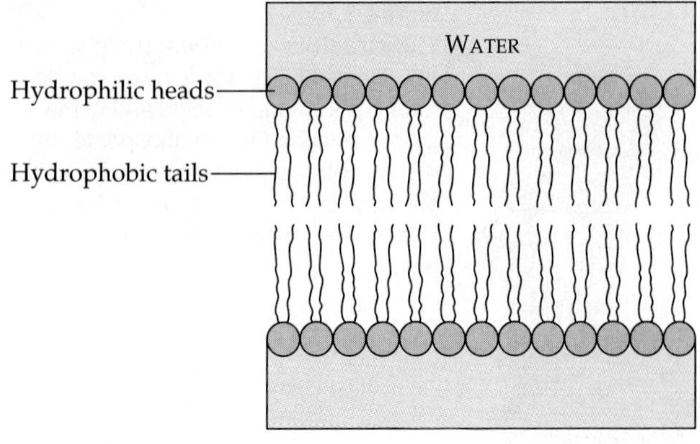

Figure 5.15
The phospholipid bilayer. Such bilayers are the main fabric of biological membranes. The hydrophilic heads (spheres) of the phospholipids are in contact with water. The diagram shows a cross section through a bilayer.

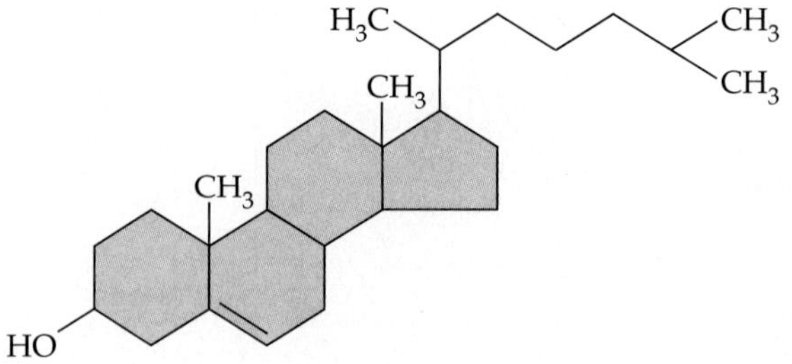

hydrophobic and are excluded from water. However, the phosphate group and its attachments form a hydrophilic head that mixes with water.

Because of their structure, phospholipids are well suited to their function as a major constituent of cell membranes. At the surface of a cell, phospholipids are arranged in a bilayer, or double layer (Figure 5.15). The hydrophilic heads of the molecules are on the outside of the bilayer, in contact with the aqueous solutions inside and outside the cell. The hydrophobic tails point toward the interior of the membrane, away from the water. The phospholipid bilayer forms a boundary between the cell and its external environment. Thus, we can see that phospholipids have emergent properties arising from the arrangement of their atoms; a phospholipid bilayer (membrane) is yet a higher level of organization with its own emergent properties. (We will explore the structure and function of membranes in detail in Chapter 8.)

Steroids

Steroids are lipids characterized by a carbon skeleton consisting of four interconnected rings (Figure 5.16). Different steroids vary in the functional groups attached to this ensemble of rings. An important steroid is **cholesterol,** a common component of the membranes of animal cells. Cholesterol is also the precursor from which most other steroids are synthesized. For example, many hormones, including the sex hormones of vertebrates, are steroids modified from cholesterol (see Figure 4.9). Thus, cholesterol has important functions in animals, although a high concentration of cholesterol in the blood may contribute to atherosclerosis.

In addition to fats, phospholipids, and steroids, other families of lipids include waxes and certain pigments in plants and animals.

Figure 5.16
Cholesterol: a steroid. Cholesterol is the precursor from which other steroids, including the sex hormones, are synthesized. Steroids vary in the functional groups attached to their four interconnected rings (shown in color).

PROTEINS

The importance of proteins is implied by their name, which comes from the Greek word *proteios,* meaning "first place." **Proteins** account for more than 50% of the dry weight of most cells, and they are instrumental in almost everything cells do. Proteins are used for structural support, storage, transport of other substances, signaling from one part of the organism to another, movement, and defense against foreign substances. In addition, as enzymes, proteins selectively accelerate chemical reactions in the cell. A human has tens of thousands of different kinds of proteins, each with a specific structure and function. Table 5.1 on p. 76 summarizes some of these functions.

Proteins are the most structurally sophisticated molecules known. Consistent with their diverse functions, they vary extensively in structure, each type of protein having a unique three-dimensional shape. But as diverse as proteins are individually, they are all polymers constructed from the same set of amino acids, the universal monomers of proteins.

Amino Acids

Amino acids are organic molecules possessing both carboxyl and amino groups (see Chapter 4). Cells build their proteins from 20 kinds of amino acids. There are many other amino acids with important functions in organisms, but they are not incorporated into proteins.

Most amino acids consist of an asymmetric carbon, termed the alpha (α) carbon, bonded to four different covalent partners. Each amino acid has a hydrogen atom, a carboxyl group, and an amino group bonded to the α carbon; the 20 kinds of amino acids that make up proteins differ only in what is attached by the fourth bond to the α carbon (Figure 5.17). This variable part of the amino acid is symbolized by the letter R. The R group, also called **the side chain,** may be as simple as a hydrogen atom, as in the amino acid glycine, or it may be a carbon skeleton with various functional groups at-

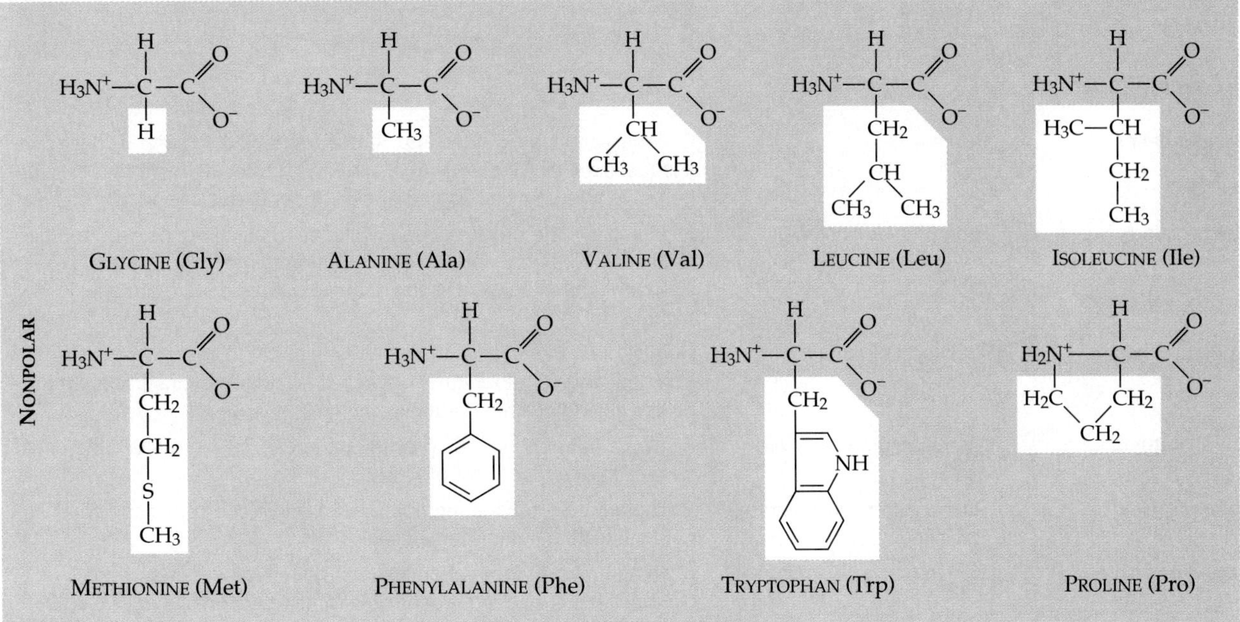

POLAR

SERINE (Ser) THREONINE (Thr) CYSTEINE (Cys) TYROSINE (Tyr) ASPARAGINE (Asn) GLUTAMINE (Gln)

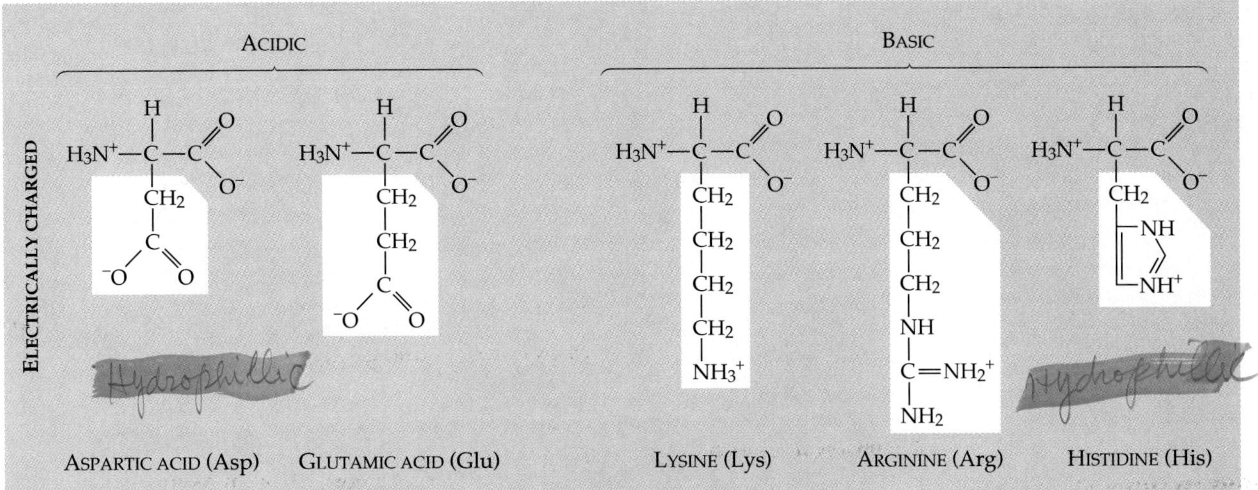

Amino group & Carboxyl group-

Figure 5.17
The 20 amino acids of proteins. The amino acids are grouped here according to the properties of their side chains (R groups), shown in white. The amino acids are shown in their prevailing ionic forms at pH 7, the pH of the cell. In parentheses are the three-letter abbreviations for the amino acids.

Table 5.1 A Summary of Protein Functions

Type of Protein	Function	Examples
Structural proteins	Support	Collagen and elastin provide a fibrous framework in animal connective tissues, such as tendons and ligaments. Keratin is the protein of hair, horns, feathers, quills, and other skin appendages of animals.
Storage proteins	Storage of amino acids	Ovalbumin is the protein of egg white, used as an amino acid source for the developing embryo. Casein, the protein of milk, is the major source of amino acids for baby mammals. Plants store proteins in seeds.
Transport proteins	Transport of other substances	Hemoglobin, the iron-containing protein of blood, transports oxygen from the lungs to other parts of the body. Other proteins transport molecules across cell membranes.
Hormonal proteins	Coordination of bodily activities	Insulin, a hormone secreted by the pancreas, helps regulate the concentration of sugar in the blood.
Receptor proteins	Response of cell to chemical stimuli	Receptors built into the membrane of a nerve cell detect chemical signals released by other nerve cells.
Contractile proteins	Movement	Actin and myosin are responsible for the movement of muscles. Contractile proteins are responsible for the undulations of cilia and flagella, which propel many cells.
Defensive proteins	Protection against disease	Antibodies combat bacteria and viruses.
Enzymatic proteins	Acceleration of chemical reactions	Digestive enzymes hydrolyze the polymers in food.

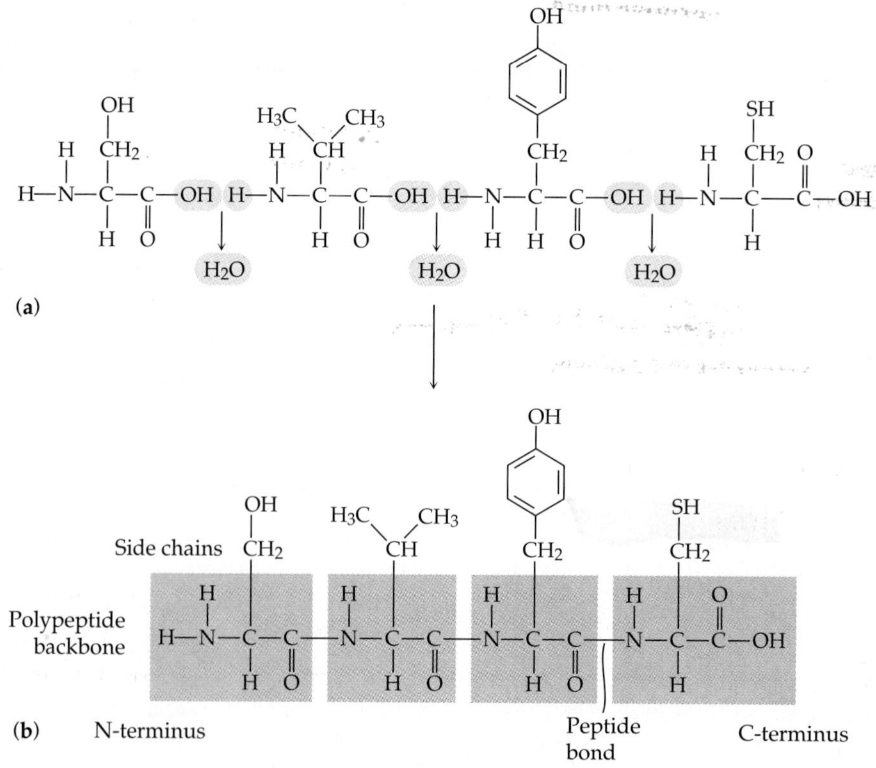

Figure 5.18
Polypeptide chains. (**a**) Peptide bonds formed by condensation synthesis link the carboxyl group of one amino acid to the amino group of the next. (**b**) The polypeptide has a repetitive backbone (purple) with various appendages, the amino acid side chains.

tached, as in glutamine. The physical and chemical properties of the side chain determine the unique characteristics of a particular amino acid.

In Figure 5.17, the amino acids are grouped according to the properties of their side chains. One group consists of amino acids with nonpolar side chains, which are hydrophobic. Another group comprises amino acids with polar side chains, which are hydrophilic. Acidic amino acids are those with side chains that are generally negative in charge, owing to the presence of a carboxyl group, which is usually dissociated at cellular pH. Basic amino acids have amino groups in their side chains that are generally positive in charge. (Notice that *all* amino acids have carboxyl groups and amino groups bonded to the α carbon, and the terms *acidic* and *basic* in this context refer only to the nature of the side chains.) Because they are ionic, acidic and basic side chains are hydrophilic.

Now that we have examined amino acids, let's see how they are linked to form polymers.

Polypeptide Chains

When two amino acids are arranged so that the carboxyl group of one is adjacent to the amino group of the other, an enzyme can join the amino acids by means of condensation synthesis, forming a covalent linkage called a **peptide bond** (Figure 5.18). A polymer of many amino acids linked by peptide bonds is a **polypeptide chain.** At one end of the chain is a free amino group, and at the opposite end is a free carboxyl group. Thus, the chain has a polarity, with an N-terminus (for the nitrogen of the amino group) and a C-terminus (for the carbon of the carboxyl group). All other amino and carboxyl groups of the monomers (except those of side chains) are tied up in the formation of the peptide bonds. The repeating sequence of atoms along the chain (—N—C—C—N—C—C—) is referred to as the polypeptide backbone. Attached to this repetitive backbone are different kinds of appendages, the side chains of the amino acids. Polypeptides range in length from a few monomers to a thousand or more. Each specific polypeptide has a unique linear sequence of amino acids.

Protein Conformation

A protein consists of one or more polypeptide chains twisted, wound, and folded upon themselves to form a macromolecule with a definite three-dimensional shape, or **conformation.** A protein's function depends on its unique conformation, which is a consequence of the specific linear sequence of the amino acids that make up the polypeptide chain. Many proteins are globular (roughly spherical), while others are fibrous in shape.

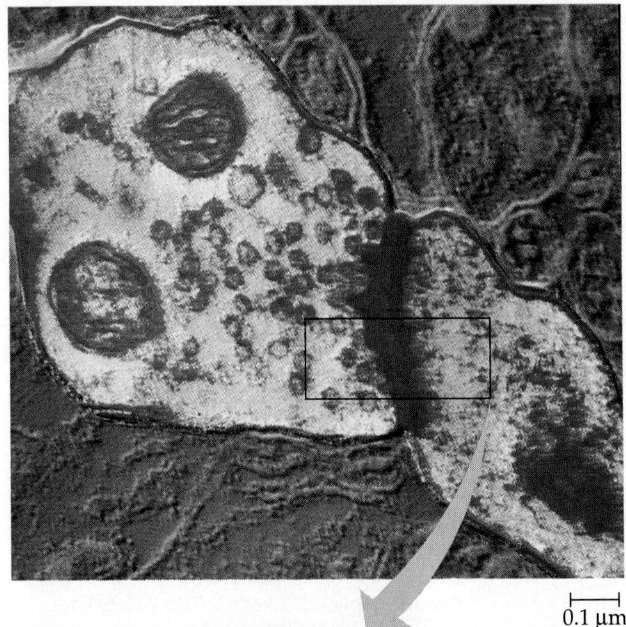

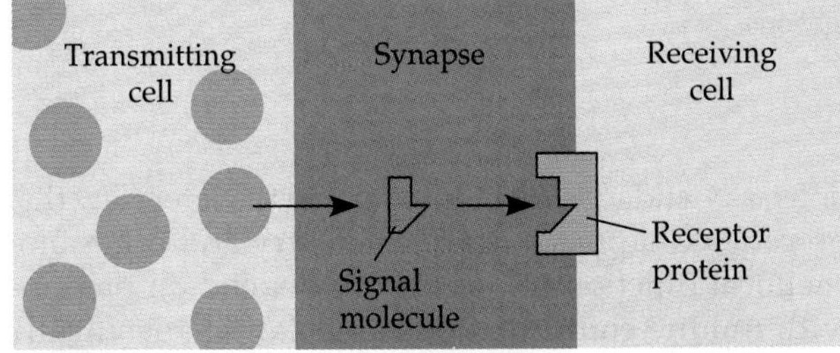

Figure 5.19
Receptor proteins in the brain. One nerve cell in the brain signals another by releasing a chemical messenger that fits a protein receptor molecule on the receiving cell. In the micrograph, the transmitting cell contains numerous vesicles. A signal is transmitted when these vesicles release specific molecules into the synapse, the tiny gap between the nerve cell and its neighbor. (In this artificially colored micrograph, the synapse is dark red.) The signal molecules pass across the synapse and stimulate the receiving cell (TEM). In the diagram, we see a much simplified version of the chemical communication that occurs between two cells. The signal molecule and the receptor protein are shown here as simple blocks of specific shape; the actual molecules have much more complex geometries.

In almost every case, the function of a protein depends on its ability to recognize and bind to some other molecule. For instance, a hormonal protein binds to a cell receptor, an antibody binds to a particular foreign substance that has invaded the body, and an enzyme recognizes and binds to its substrate, the substance the enzyme works on. Information processing in your brain also depends on the ability of proteins to bind selectively to other molecules. One nerve cell in the brain signals another by dispatching specific molecules that have a unique shape. On the surface of the receiving cell are receptor proteins, molecules that fit the signal chemicals something like a lock and key (Figure 5.19). Some of these signals even affect feelings and emo-

Figure 5.20
The native conformation of a protein. The polypeptide chain folds into a specific shape. The protein shown here is lysozyme, an enzyme. Its structure is simplified by showing only the pattern of folding of the polypeptide backbone. The yellow lines symbolize one type of chemical bond that stabilizes the conformation.

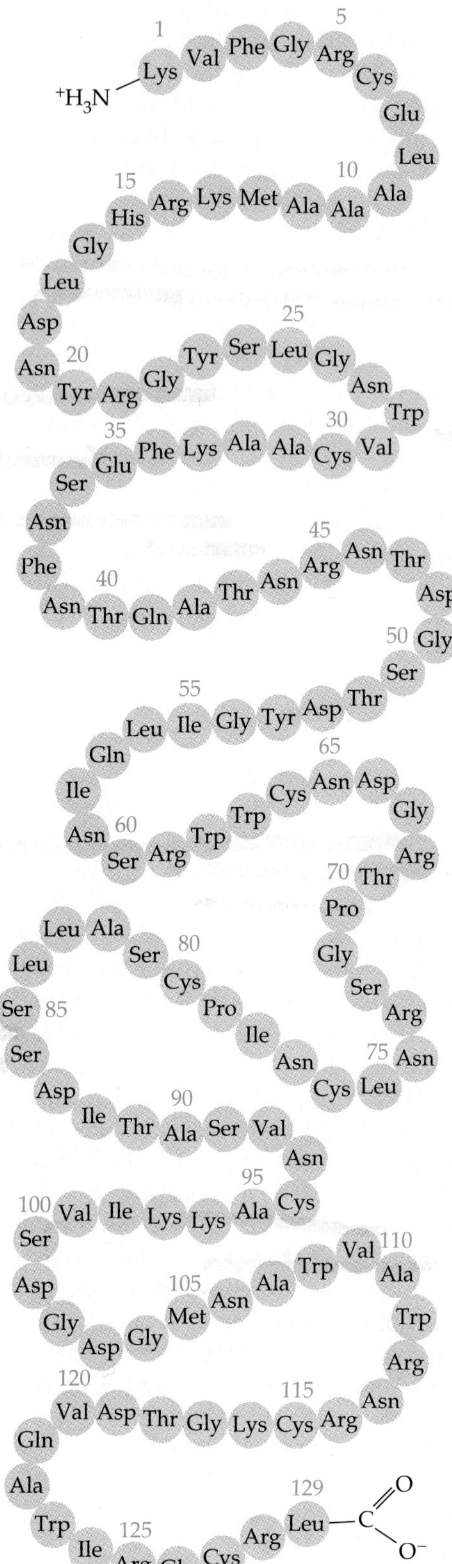

Figure 5.21
The primary structure of a protein. This is the unique amino acid sequence, or primary structure, of the enzyme lysozyme. The names of the amino acids are given as their three-letter abbreviations. (The chain was drawn in this serpentine fashion only so that the entire sequence would be visible on the page. The actual shape of lysozyme is shown in Figure 5.20.)

tions, as Candace Pert pointed out in the interview that precedes Chapter 2. Brain chemistry and other biochemical functions of organisms depend on proteins with unique conformations associating specifically with other molecules.

When a cell synthesizes a polypeptide, the chain folds spontaneously to assume the functional conformation for that protein: its native conformation (Figure 5.20). As this process occurs, the protein's conformation is reinforced by a variety of chemical bonds between parts of the chain.

Levels of Protein Structure

In the complex conformation of a protein, we can recognize three superimposed levels of architecture, known as primary, secondary, and tertiary structure. A fourth level, quaternary structure, occurs when a protein consists of two or more polypeptide chains.

Primary Structure The **primary structure** of a protein is its unique sequence of amino acids. As an example, we will examine lysozyme, an antibacterial enzyme present in our tears, saliva, and other body secretions. Lysozyme is a relatively small protein, only 129 amino acids long (Figure 5.21). A specific one of the 20 amino acids occupies each of the 129 positions along the chain. The primary structure is like the order of letters in a very long word. If left to chance, there would be 20^{129} different ways of arranging amino acids into a polypeptide chain of this length. However, the precise primary structure of a protein is determined not by

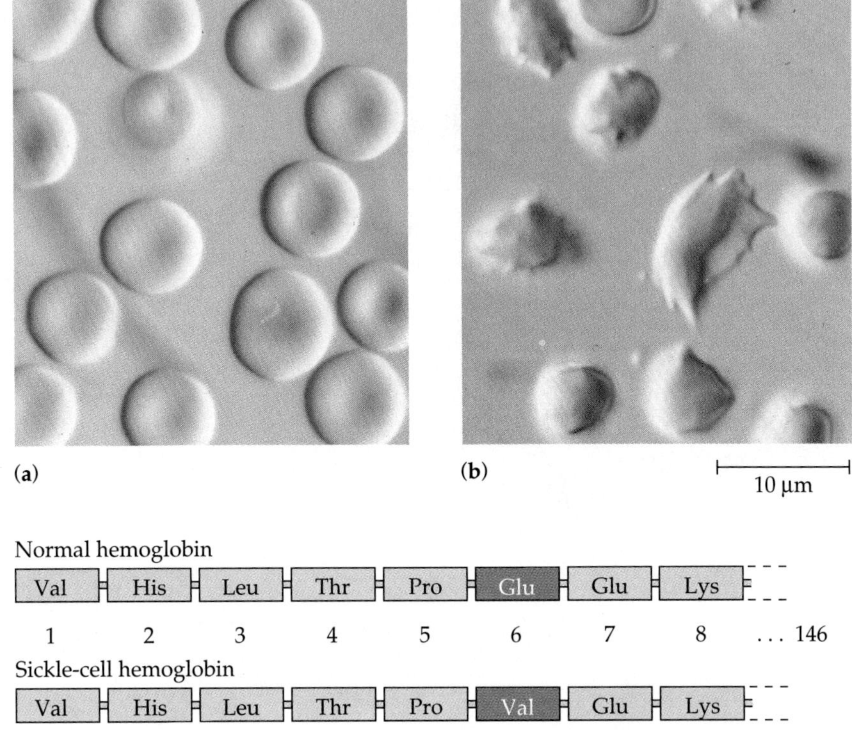

Figure 5.22
Sickle-cell anemia is caused by an amino acid substitution in a single protein. (**a**) The function of normal human red blood cells, which are disk-shaped, is to transport oxygen from the lungs to the other organs of the body. Each cell contains millions of molecules of hemoglobin, the protein that carries the oxygen (LM). (**b**) A slight change in the primary structure of hemoglobin—an inherited substitution of one amino acid—causes sickle-cell anemia. The abnormal hemoglobin molecules tend to link together and crystallize, deforming the cells (LM). The disease is named for the "sickle" shape of some of these deformed cells. The life of someone with the disease is punctuated by "sickle-cell crises," which occur when the angular cells clog tiny blood vessels, impeding blood flow. Blood transfusions can relieve the symptoms, but there is no cure for this disease. Sickle-cell anemia kills about 100,000 people annually in the U.S. (**c**) The inherited amino acid substitution occurs in the number 6 position of a polypeptide chain that is 146 amino acids long.

random linking of amino acids, but by inherited genetic information.

Even a slight change in primary structure can affect a protein's conformation and ability to function. For instance, sickle-cell anemia is an inherited blood disorder in which one amino acid is substituted for another in a single position in the primary structure of hemoglobin, the protein that carries oxygen in red blood cells (Figure 5.22). We will learn more about the inheritance of sickle-cell anemia in Chapter 13 and about its evolutionary significance in Chapter 21.

Molecular biologists have ascertained the primary structures of hundreds of proteins. The pioneer in this work was Frederick Sanger, who, with his colleagues at Cambridge University in England, determined the amino acid sequence of the hormone insulin in the late 1940s and early 1950s. His approach was to use protein-digesting enzymes and other catalysts that break polypeptides at specific places rather than completely hydrolyzing the chain. Treatment with one of these agents would cleave the polypeptide into fragments that could be separated by a technique called chromatography. Hydrolysis with another agent would break the polypeptide at different sites, yielding a second group of fragments. Sanger used chemical methods to determine the sequence of amino acids in these small fragments. Then he searched for overlapping regions among the pieces obtained by hydrolyzing with the different agents. Consider, for instance, two fragments with the following sequences:

Cys-Ser-Leu-Tyr-Gln-Leu
                    Tyr-Gln-Leu-Glu-Asn

We can deduce from the overlapping regions that the intact polypeptide contains in its primary structure the following segment:

Cys-Ser-Leu-Tyr-Gln-Leu-Glu-Asn

Just as we could reconstruct this sentence from a collection of fragments with overlapping sequences of letters, Sanger and his co-workers were able, after years of effort, to reconstruct the complete primary structure of insulin. Since then, most of the steps involved in sequencing a polypeptide have been automated. But it was Sanger's analysis of insulin that first demonstrated what is now a fundamental axiom of molecular biology: Each type of protein has a unique primary structure, a precise sequence of amino acids.

Secondary Structure Most proteins have segments of their polypeptide chain repeatedly coiled or folded in patterns that contribute to the protein's overall conformation. These configurations, collectively referred to as **secondary structure,** are the result of hydrogen bonds at regular intervals along the polypeptide backbone (Figure 5.23). Because they are electronegative, both the oxygen and the nitrogen atoms of the backbone have weak negative charges (see Chapter 2). The weakly positive hydrogen atom attached to the nitrogen atom has an affinity for the oxygen atom of a nearby peptide bond. Individually, these hydrogen

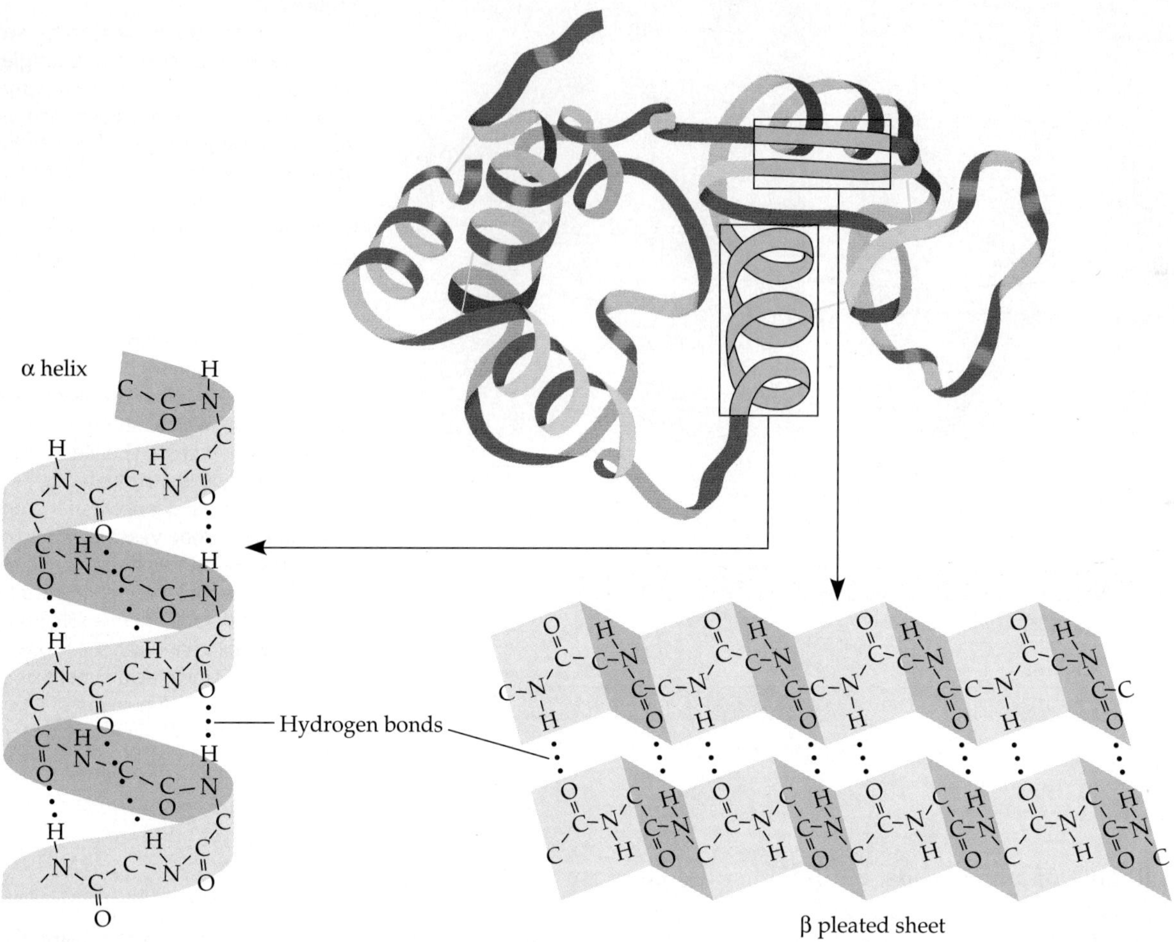

Figure 5.23
The secondary structure of a protein. Two types of secondary structure, the α helix and the β pleated sheet, can both be found in the protein lysozyme. Both patterns depend on hydrogen bonding along the polypeptide chain. The R groups of the amino acids are omitted in these drawings.

bonds are weak, but because they are repeated many times over a relatively long region of the polypeptide chain, they can support a particular shape for that part of the protein. One such secondary structure is the **alpha (α) helix,** a delicate coil held together by hydrogen bonding between every fourth peptide bond. Linus Pauling and Robert Corey first described the α helix in 1951 while working on protein structure at the California Institute of Technology. The regions of α helix in the enzyme lysozyme are evident in Figure 5.23, where one α helix is enlarged to show the hydrogen bonds. Lysozyme is fairly typical of a globular protein in that it has a few stretches of α helix separated by nonhelical regions. In contrast, some fibrous proteins, such as α-keratin, the structural protein of hair, have the α-helix formation over most of their entire length.

Another type of secondary structure is the **beta (β) pleated sheet,** in which the polypeptide chain folds back and forth, or where two regions of the chain lie parallel to each other. Hydrogen bonds between the parallel regions hold the structure together. The β sheets make up the dense core of many globular proteins, and we can recognize one such region in lysozyme (see Figure 5.23). Also, β sheets dominate some fibrous proteins, including fibroin, the structural protein of silk (Figure 5.24).

Tertiary Structure Superimposed on the patterns of secondary structure is a protein's **tertiary structure,** consisting of irregular contortions from bonding between side chains (R groups) of the various amino acids (see Figure 5.17). (In contrast, recall that secondary structure results from hydrogen bonds formed at regular intervals along the polypeptide's backbone.) One of the factors that contributes to tertiary structure is called a **hydrophobic interaction.** As a polypeptide folds into its native conformation, amino acids with hydophobic (nonpolar) side chains usually congregate at the core of the protein, out of contact with water. Their mutual exclusion from water keeps the hydro-

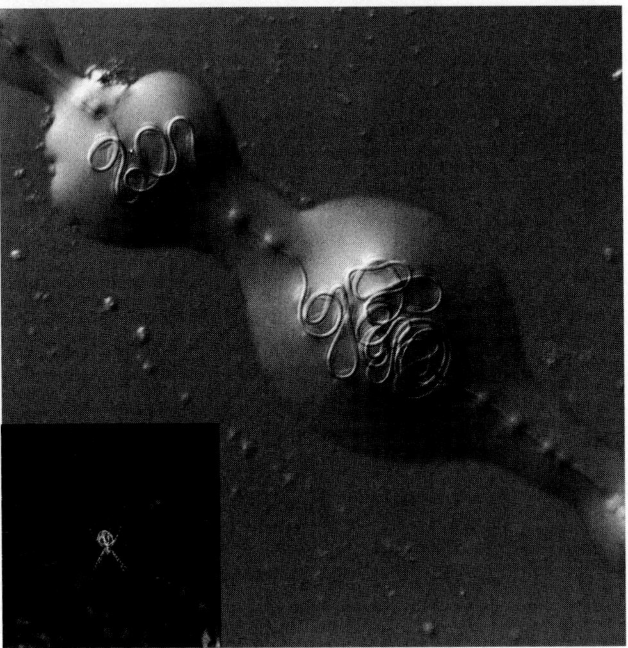

Figure 5.24
Silk: a structural protein. This orb spider (inset) secretes silk protein (fibroin) from abdominal glands. Secreted in liquid form, the protein solidifies upon contact with air. The silk protein owes its strength to its secondary structure. The polypeptides are folded back and forth in the β configuration. The teamwork of so many hydrogen bonds makes each silk fiber stronger than a steel wire of the same thickness. The strands that radiate out from the web's center are made of dry silk protein; their strength maintains the basic shape of the web (see Figure 5.1). The elastic strands forming the spiral—called capture strands—stretch and coil in response to wind, rain, and insects that are unfortunate enough to fly into the web. The micrograph helps us understand this resilience. A capture strand consists of a coiled silk fiber coated by a sticky fluid. Force on the strand unwinds the silk fiber, absorbing the shock before the fluid's surface tension rewinds the fiber.

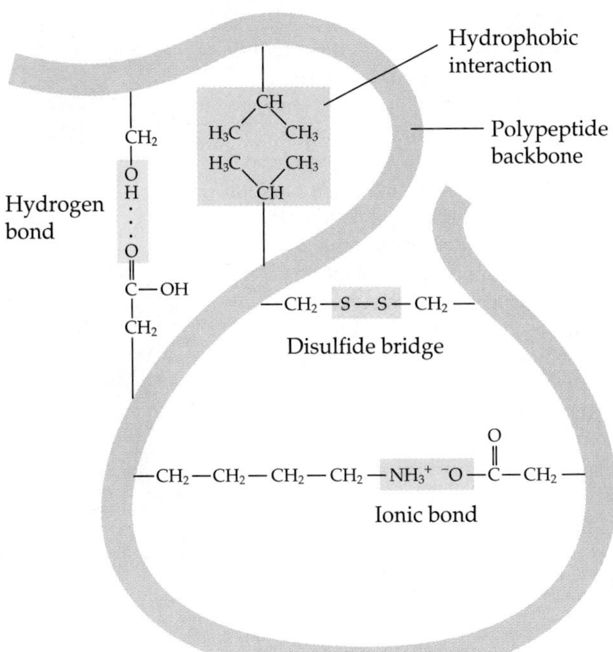

Figure 5.25
Examples of bonds contributing to the tertiary structure of a protein. Hydrogen bonds, ionic bonds, and hydrophobic interactions are weak bonds between side chains that collectively hold the protein in a specific conformation. Much stronger are the disulfide bridges, covalent bonds between the side chains of cysteine pairs.

phobic side chains together in localized clusters. Thus, what we call a hydrophobic interaction is actually caused by the behavior of water molecules, which tend to hydrogen-bond to one another and to hydrophilic molecules. Hydrogen bonds between certain side chains are also important, as are ionic bonds between positively and negatively charged side chains. These are all weak interactions, but their cumulative effect helps give the protein a stable shape.

The conformation of a protein may be reinforced further by strong covalent bonds called **disulfide bridges.** Disulfide bridges form where two cysteine monomers—amino acids with sulfhydryl groups on their side chains—are brought close together by the folding of the protein. The sulfur of one cysteine bonds to the sulfur of a second, and the disulfide bridge rivets parts of the protein together. (The yellow lines in Figures 5.20 and 5.23 represent disulfide bridges.) Note that all of these different kinds of bonds can occur in one protein, as shown hypothetically in Figure 5.25.

Quaternary Structure As mentioned previously, some proteins consist of two or more polypeptide chains aggregated into one functional macromolecule. Each polypeptide chain is called a **subunit** of the protein. **Quaternary structure** is the structure that results from the relationship between the subunits. For example, collagen is a fibrous protein that has helical subunits supercoiled into a larger triple helix (Figure 5.26a). This supercoiled organization of collagen, which is similar to the design of a rope, gives the long fibers great strength, appropriate to their function as the girders of connective tissue, such as tendons and ligaments. Hemoglobin is an example of a globular protein with quaternary structure (Figure 5.26b). It consists of two kinds of polypeptide chains, with two of each kind per hemoglobin molecule.

We have now considered the four levels of the structural organization of proteins. But it is the overall product, a macromolecule with a unique conformation, that works in a cell. The specific function of a protein is an emergent property that arises from the architecture of the molecule (Figure 5.27).

Factors Determining Conformation

Since unique conformation endows each protein with a specific function, what are the key factors determining

Figure 5.26
The quaternary structure of proteins. At this level of structure, two or more polypeptide subunits interact to form a functional protein. (**a**) Collagen is a fibrous protein consisting of three helical polypeptides that are supercoiled to form a ropelike structure of great strength. Accounting for 40% of the protein in the human body, collagen strengthens connective tissue in our skin, bones, ligaments, tendons, and other body parts. (**b**) Hemoglobin is a globular protein with four subunits, two of one kind (α chains) and two of another kind (β chains). (Each subunit has a nonpolypeptide component, called heme, with an iron atom that binds oxygen.)

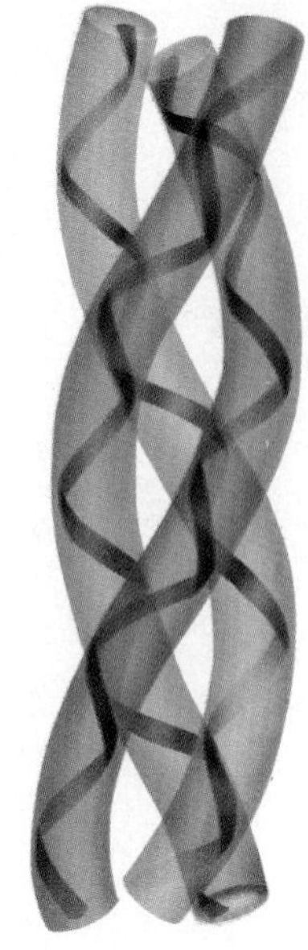

(**a**) Collagen

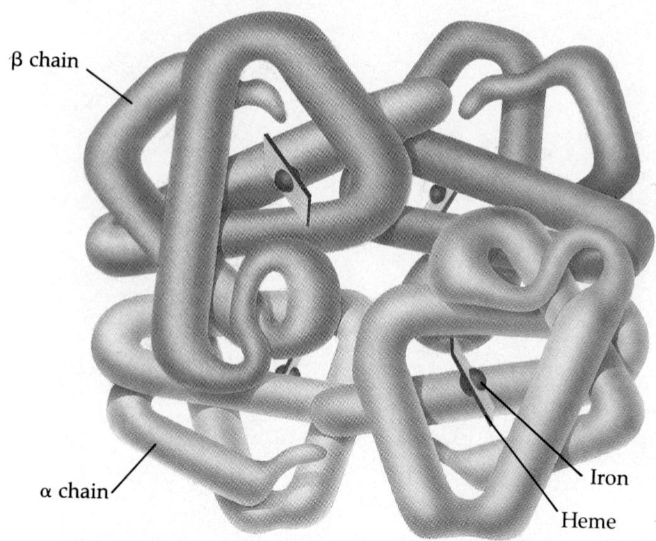

(**b**) Hemoglobin

conformation? As we have seen, a polypeptide chain of a given amino acid sequence will spontaneously arrange itself into a three-dimensional shape maintained by the interactions responsible for secondary and tertiary structure. This event normally occurs when the protein is being synthesized within the cell.

If the pH, salt concentration, temperature, or other aspects of its environment are altered, the protein may unravel and lose its native conformation in a process called **denaturation** (Figure 5.28). Misshapen, the denatured protein becomes biologically inactive. Most proteins become denatured if they are transferred from

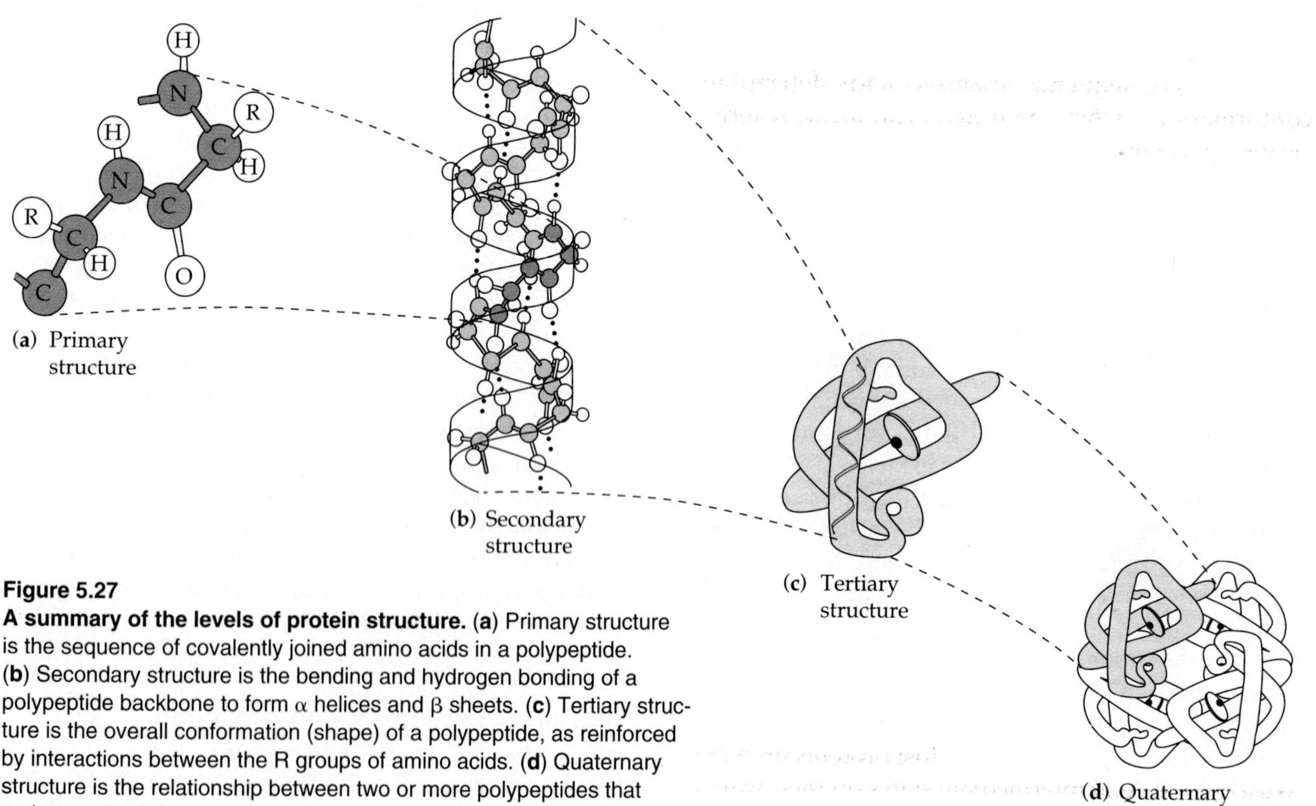

Figure 5.27
A summary of the levels of protein structure. (**a**) Primary structure is the sequence of covalently joined amino acids in a polypeptide. (**b**) Secondary structure is the bending and hydrogen bonding of a polypeptide backbone to form α helices and β sheets. (**c**) Tertiary structure is the overall conformation (shape) of a polypeptide, as reinforced by interactions between the R groups of amino acids. (**d**) Quaternary structure is the relationship between two or more polypeptides that make up a protein.

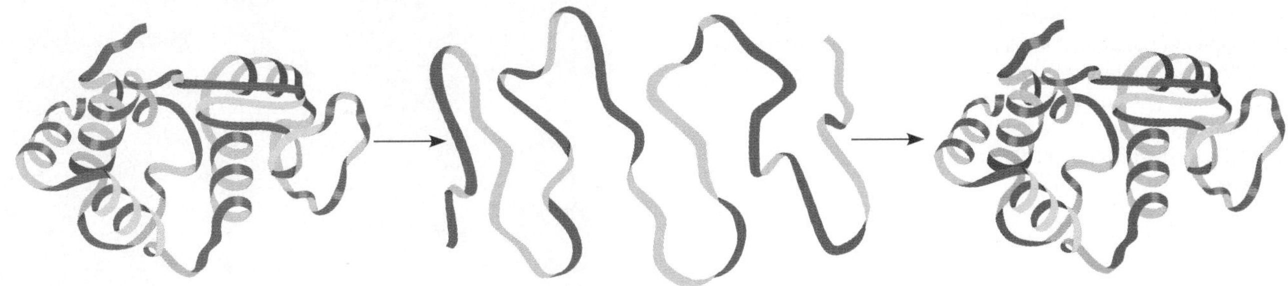

Figure 5.28
Denaturation and renaturation of a protein. High temperatures or various chemical treatments will denature a protein, causing it to lose its conformation and ability to function. If the denatured protein remains dissolved, it may renature when the environment is restored to normal.

an aqueous environment to an organic solvent, such as ether or chloroform—the protein turns inside out, its hydrophobic regions changing places with its hydrophilic portions. Other agents of denaturation include chemicals that disrupt the hydrogen bonds, ionic bonds, and disulfide bridges that shape a protein. Denaturation can also result from excessive heat, which agitates the polypeptide chain enough to overpower the weak interactions that stabilize conformation. The white of an egg becomes opaque during cooking because the denatured proteins are insoluble and solidify.

When a protein is denatured, it may re-form to its original shape when returned to its normal environment. We can conclude that the information for building specific shape is intrinsic in the protein's primary structure. The sequence of amino acids determines conformation—where an α helix can form, where β sheets can occur, where disulfide bridges are located, and so on. In the future, computer programs may be able to predict the conformation of any protein of known primary structure. Before such programs can be written, however, much more must be learned about the multiple factors that contribute to conformation.

The Protein-Folding Problem

Molecular biologists have determined the amino acid sequences of hundreds of proteins, and the three-dimensional shapes of many of those proteins are also known (see the Methods Box, p. 85). One would think that by correlating the primary structures of many proteins with their conformations, it would be possible to discover the rules of protein folding, especially with the help of computers. Unfortunately, the protein-folding problem is not that simple. Most proteins probably go through several intermediate states on their way to a stable conformation, and looking at the "mature" conformation does not reveal the stages of folding required to achieve that form (Figure 5.29). As Thomas Creighton, a British scientist who works on the problem, puts it: "It is like folding the flaps of a cardboard box. To fold them together, you have to put the flaps in a specific order and then distort them."

The protein-folding problem is as important as it is challenging. Once the rules of protein folding are known, it should be possible to design proteins that will carry out specific tasks by making polypeptide chains with appropriate amino acid sequences. Based on what has already been learned, a University of Colorado research team recently made an enzyme that digests proteins.

NUCLEIC ACIDS

If primary structure determines the conformation of a protein, what determines primary structure? As stated earlier, the amino acid sequence is programmed by a unit of inheritance known as a **gene.** Genes consist of DNA, which is a polymer belonging to the class of compounds known as **nucleic acids.**

The Functions of Nucleic Acids: An Overview

There are two types of nucleic acids: **deoxyribonucleic acid (DNA)** and **ribonucleic acid (RNA).** These are the molecules that enable living organisms to reproduce their complex equipment from one generation to the next. Unique among molecules, DNA provides directions for its own replication; this molecular reproduction is the basis for the continuity of life.

DNA is the genetic material that organisms inherit from their parents. A DNA molecule is very long and consists of hundreds of thousands of genes, each occupying a specific position along the single molecule.

Figure 5.29
Protein folding. As it is being synthesized within a cell, a protein begins to assemble into its active conformation (here, we start with a completed polypeptide chain). The intermediate stages of folding illustrated in this figure are hypothetical; molecular biologists are just beginning to learn the protein-folding "rules" that transform a specific primary structure into a tertiary structure.

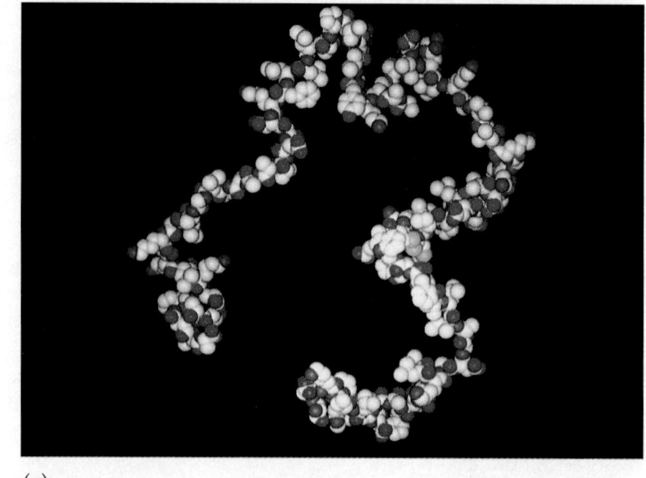
(a)

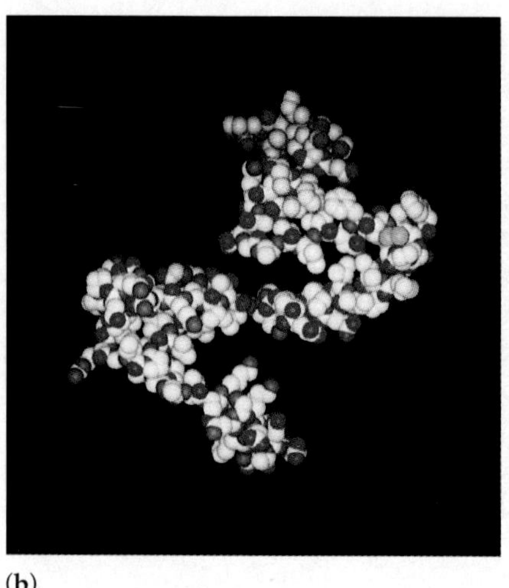
(b)

(c)

(d)

When a cell reproduces itself by dividing, its DNA is copied and passed along from one generation of cells to the next. Encoded in the structure of DNA are the instructions that program all the cell's activities. The DNA, however, is not directly involved in running the operations of the cell, any more than computer software by itself can print a bank statement or read the bar code on a box of cereal. Just as a printer is needed to print out a statement and a scanner is needed to read a bar code, proteins are required for the implementaion of genetic programs; proteins are the molecular hardware of the cell. For example, it is the protein hemoglobin that carries oxygen in the blood, not the DNA that specifies the structure of hemoglobin.

How does RNA, the other type of nucleic acid, fit into the flow of genetic information from DNA to proteins? Although each gene along the length of a DNA molecule stores the coded instructions for the synthesis of a specific protein, it does not actually make the protein. Instead, the gene directs the synthesis of a type of RNA called messenger RNA (mRNA). The mRNA molecule then interacts with the protein-synthesizing machinery to direct the production of a polypeptide. We can summarize this flow of genetic information this way: DNA ⟶ RNA ⟶ protein. The actual sites of protein synthesis are cellular structures called ribosomes. In a eukaryotic cell, ribosomes are located in the cytoplasm, but DNA resides in the nucleus. Messenger RNA conveys the genetic instructions for building proteins from the nucleus to the cytoplasm (Figure 5.30, p. 86). Prokaryotic cells (bacteria) lack nuclei, but they still use RNA to send a message from the DNA to the ribosomes and other equipment that translate the coded information into amino acid sequences.

Nucleotides

Nucleic acids are polymers of monomers called **nucleotides.** Each nucleotide is itself composed of three parts: a nitrogenous base, which is joined to a pentose (five-carbon sugar), which in turn is bonded to a phosphate group (Figure 5.31a, p. 87).

There are two families of nitrogenous bases: pyrimidines and purines. A **pyrimidine** is characterized by a six-membered ring made up of carbon and nitrogen atoms. (The nitrogen tends to take up hydrogen ions from solution, which explains the term *nitrogenous*

Methods: Molecular Models and Computer Graphics

Three-dimensional structures of biological macromolecules provide important insights into molecular function. Determining the structures of macromolecules as complex as proteins, each made up of thousands of atoms, is a formidable task. Pauling, Watson and Crick, and the other pioneers of molecular biology built models from wood, wire, and plastic model sets. Computers have made it possible to build models much more quickly (see Figure 5.2).

In the illustrations in this box (from the Department of Biochemistry at the University of California, Riverside), we follow the development of a computer model for the structure of an enzymatic protein called ribonuclease, whose function involves binding to a nucleic acid molecule. The first step is to crystallize the protein, in this case the protein combined with a short strand of nucleic acid. Then, using a method called X-ray crystallography, an instrument aims an X-ray beam through the crystal, and the regularly spaced atoms of the crystal diffract (deflect) the X-rays into an orderly array (Figure a). The diffracted X-rays can expose photographic film, producing a pattern of spots (Figure b). From such diffraction patterns, computer programs generate electron density maps of successive, cross-sectional slices through the protein (Figure c).

By combining the information from electron density maps with the primary structure of the protein, as determined by chemical methods, the three-dimensional (x, y, z) coordinates of each atom can be plotted. Finally, graphics software generates a picture showing the position of each atom in the molecule (Figure d). The scientist can move the image around on the screen to simulate the molecule's appearance from various angles—even from *inside* the molecule. Thus, computers have expanded the ways we can view a molecule while retaining the advantages of being able to manipulate the model by hand.

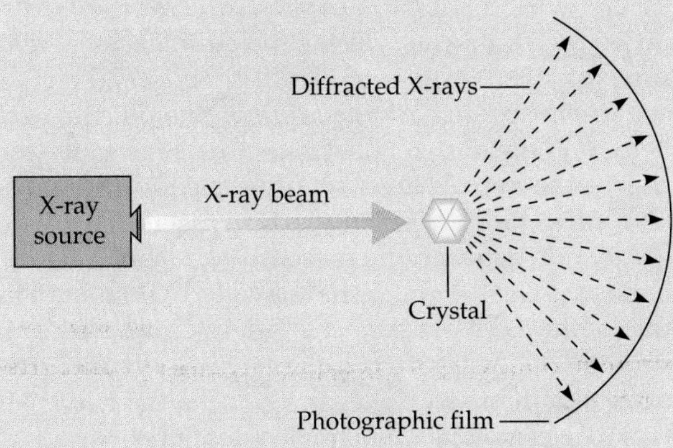

(a) X-ray crystallography.

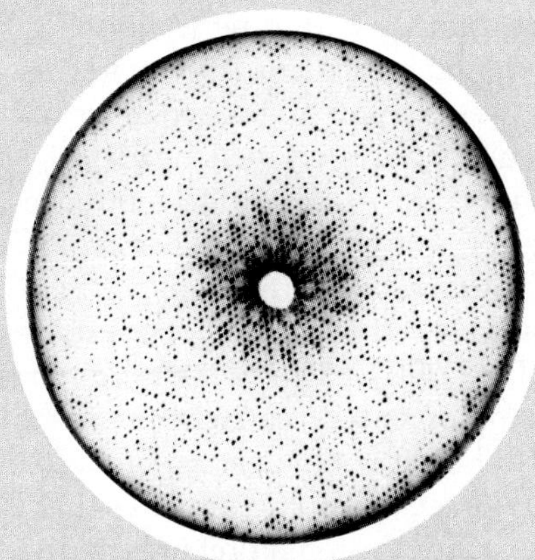

(b) X-ray diffraction pattern from the crystal of a protein.

(c) Electron density map.

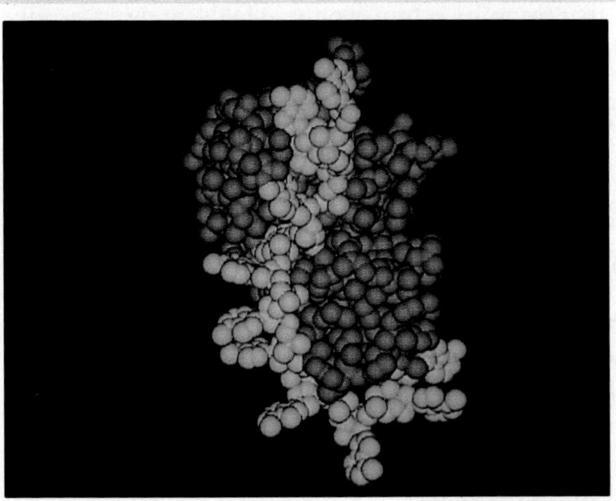

(d) A computer graphic model of the protein ribonuclease (purple) bound to a short strand of nucleic acid (green).

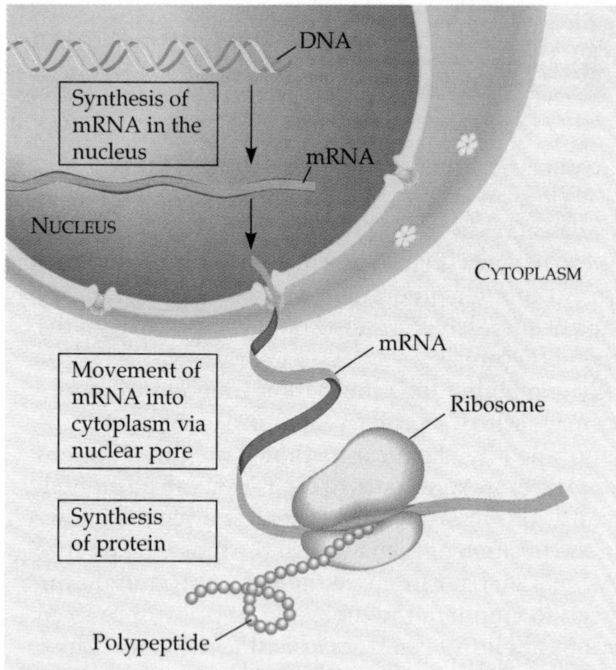

Figure 5.30
The control of protein synthesis by DNA: a diagrammatic view. In a eukaryotic cell, DNA in the nucleus programs protein production in the cytoplasm by producing messenger RNA (mRNA), which travels to the cytoplasm and binds to ribosomes. As a ribosome (greatly enlarged in this drawing) moves along the mRNA, the genetic message conveyed from the nucleus is translated into a polypeptide of specific amino acid sequence.

base.) The members of the pyrimidine family are nitrogenous bases named cytosine (C), thymine (T), and uracil (U). In the second family, the **purines,** a five-membered ring is fused to the pyrimidine type of ring. The purines are adenine (A) and guanine (G). The specific kinds of pyrimidines and purines differ in the functional groups attached to the rings. Note in Figure 5.31 that thymine is found only in DNA and uracil only in RNA.

The pentose connected to the nitrogenous base is **ribose** in the nucleotides of RNA and **deoxyribose** in DNA. The only difference between these two sugars is that deoxyribose lacks one hydroxyl group. So far, we have built a **nucleoside,** a molecule consisting of a nitrogenous base joined to a sugar. To complete the construction of a nucleotide, we attach a phosphate group to the number 5 carbon of the sugar. The molecule is now a nucleoside monophosphate, better known as a nucleotide.

Polynucleotides

A nucleic acid is a **polynucleotide,** a term that emphasizes the way these macromolecules are constructed. In a polynucleotide, the monomers are joined by covalent bonds called **phosphodiester linkages** between the phosphate of one nucleotide and the sugar of the next monomer, resulting in a backbone with a repeating pattern of sugar–phosphate–sugar–phosphate (Figure 5.31b). All along this sugar–phosphate backbone are appendages consisting of the nitrogenous bases. Unlike the regular backbone, the sequence of bases along a DNA polymer is unique for each gene. Because genes are typically hundreds of nucleotides long, the number of possible base sequences is essentially limitless. A gene's meaning to the cell is encoded in its specific sequence of the four bases. For example, the sequence AGGTAACTT means one thing, while the sequence CGCTTTAAC has a different translation (real genes, of course, are much longer). The linear order of bases encoded in a gene specifies the amino acid sequence—the primary structure—of a protein, which then specifies that protein's function in the cell.

The Double Helix: An Introduction

The DNA molecules of cells actually consist of two polynucleotide chains that spiral around an imaginary axis to form a **double helix.** James Watson and Francis Crick, working at Cambridge University, first proposed the double helix as the three-dimensional structure of DNA in 1953 (Figure 5.32). The two sugar–phosphate backbones are on the outside of the helix, and the nitrogenous bases are paired in the interior of the helix. The two polynucleotide chains, or strands, as they are called, are held together by hydrogen bonds between the paired bases. The individual bonds are weak, but they have a collective strength like that of the teeth of a zipper. Most DNA molecules are very long—some several millimeters—with thousands or even millions of base pairs holding the two chains together. One long DNA molecule represents a large number of genes, each one a particular segment of the double helix.

Only certain bases in the double helix are compatible with each other. Adenine (A) always pairs with thymine (T), and guanine (G) always pairs with cytosine (C). If we were to read the sequence of bases along one strand as we traveled the length of the double helix, we would know the sequence of bases along the other strand. If a stretch of one strand has the base sequence AGGTCCG, then the base-pairing rules tell us that the same stretch of the other strand must have the sequence TCCAGGC. The two strands of the double helix are complementary, each the predictable counterpart of the other. It is this feature of DNA that makes possible the precise copying of genes that is responsible for inheritance. As a cell prepares to divide, the two strands of each gene separate. Each existing strand then serves as a template to order nucleotides into a new complementary strand. Once again, we see that the properties of life emerge from structural order.

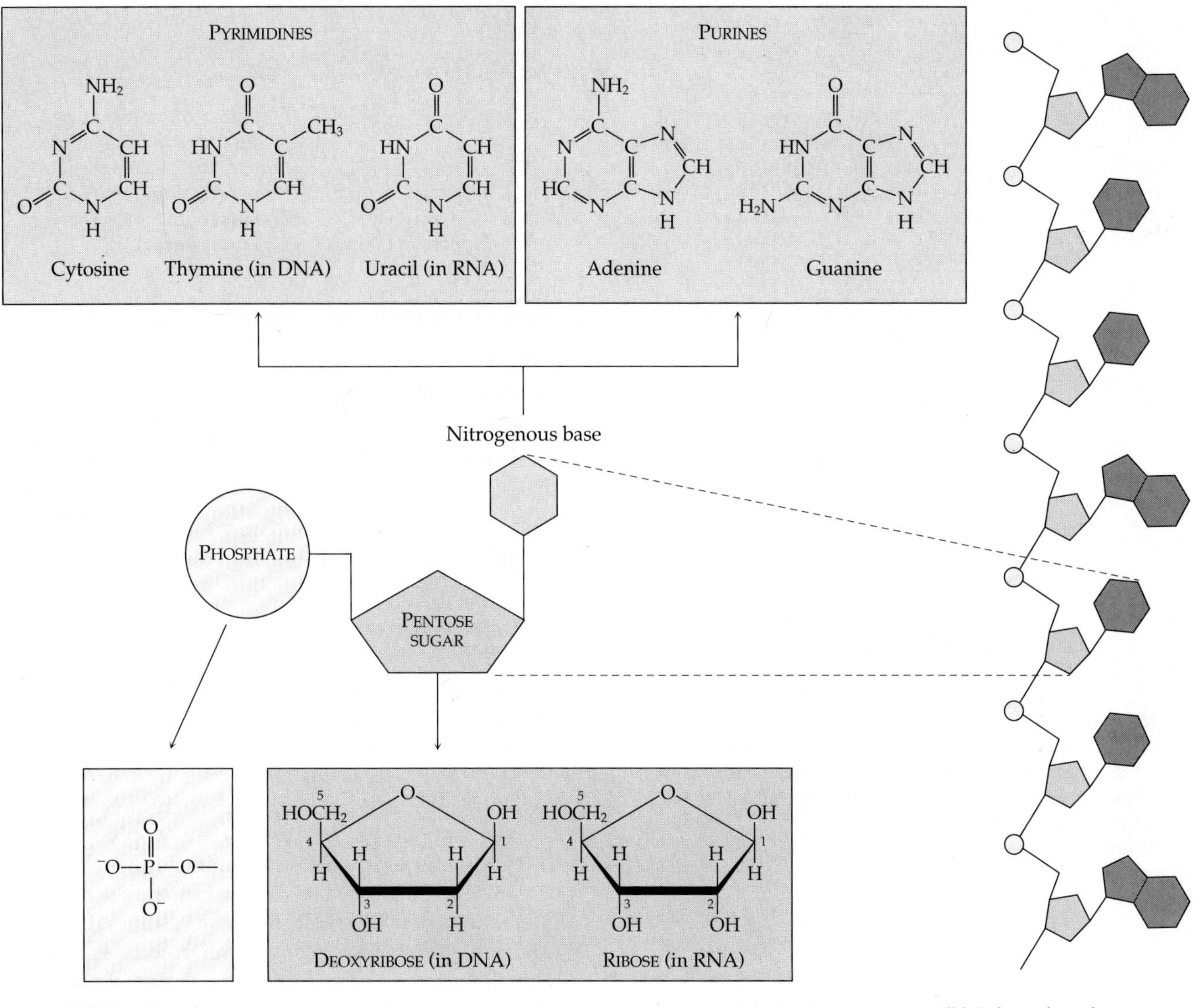

Figure 5.31
The structures of nucleotides and polynucleotides. (**a**) Nucleotides, the monomers of nucleic acids, are themselves composed of three smaller molecular building blocks: a nitrogenous base, either a purine or a pyrimidine; a pentose sugar; and a phosphate group. (**b**) In polynucleotides, each nucleotide monomer has its phosphate group bonded to the sugar of the next nucleotide. The polymer has a regular sugar–phosphate backbone with variable appendages, the four kinds of nitrogenous bases.

The DNA molecule is so central to life that we will devote Chapters 15 and 16 to its structure and function.

DNA AND PROTEINS AS TAPE MEASURES OF EVOLUTION

Genes (DNA) and their products (proteins) document the hereditary background of an organism. The linear sequences of nucleotides in DNA molecules are passed from parents to offspring, and these DNA sequences determine the amino acid sequences of proteins. Siblings have greater similarity in their DNA and proteins than do unrelated individuals of the same species. If the evolutionary view of life is valid, then we should be able to extend this concept of "molecular genealogy" to relationships *between* species: We should expect two species that appear to be closely related based on fossil and anatomical evidence to also share a greater proportion of their DNA and proteins than do more distantly related species. And that is the case. For example, Table 5.2 compares a polypeptide chain of the protein hemoglobin in humans with the hemoglobin in eight other vertebrates. In this chain of 146 amino acids, humans and gorillas differ in just one amino acid. More distantly related species have chains that are less similar. Molecular biology has added a new

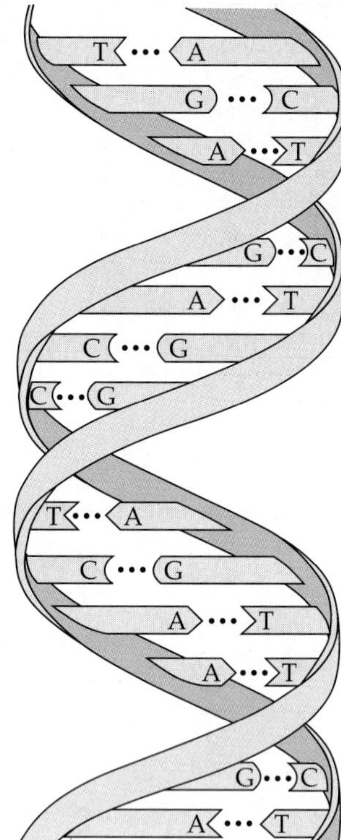

Figure 5.32
The double helix. The DNA molecule is usually double-stranded, with the sugar–phosphate backbone of the polynucleotides (shown here by blue ribbons) on the outside of the helix. In the interior are pairs of nitrogenous bases, holding the two strands together by hydrogen bonds. Hydrogen bonding between the bases is specific. As illustrated here with symbolic shapes for the bases, adenine (A) can pair only with thymine (T), and guanine (G) can pair only with cytosine (C).

Table 5.2 Evolutionary Relationships and Similarities in a Polypeptide Chain of Hemoglobin

Species	Number of Amino Acid Differences in β Chain of Hemoglobin, Compared to Human Hemoglobin (Total Chain Length = 146 Amino Acids)
Human β chain	0
Gorilla	1
Gibbon	2
Rhesus monkey	8
Mouse	27
Gray kangaroo	38
Chicken	45
Frog	67
Lamprey (jawless fish)	125

tape measure to the toolkit researchers use to assess evolutionary kinship.

* * *

We have concluded our survey of macromolecules, but not our study of the chemistry of life. Applying the reductionist strategy, we have examined the architecture of molecules, but we have yet to explore the dynamic interactions between molecules that result in the chemical changes in cells collectively referred to as metabolism. Chapter 6, the concluding chapter of this unit, will take us another step up the hierarchy of biological order by introducing the fundamental principles of metabolism.

STUDY OUTLINE

1. Cells can combine small organic molecules into large macromolecules to form a higher level in the hierarchy of biological order.
2. Carbohydrates, lipids, proteins, and nucleic acids are the four major classes of organic compounds in cells.

Polymers (pp. 64–66)

1. Macromolecules are polymers, chains of identical or similar subunit molecules called monomers. Although there is a limited number of monomers common to organisms, each organism is unique because of the specific arrangement of these monomers into polymers.
2. Monomers of all four classes of macromolecules form larger molecules by condensation synthesis, a chemical reaction in which one monomer donates a hydroxyl group and the other a hydrogen atom, forming a water molecule.
3. Polymers can be disassembled to monomers by the reverse process, called hydrolysis. In this way, large macromolecules in food are digested into monomers small enough to enter our cells.

Carbohydrates (pp. 66–71)

1. Carbohydrates are sugars and their derivatives.
2. Monosaccharides are the simplest carbohydrates. They are used directly for fuel, converted to other types of organic molecules, or used as monomers for polymers.
3. Disaccharides consist of two monosaccharide monomers connected by a glycosidic bond.
4. Polysaccharides may consist of thousands of monosaccharide monomers connected by glycosidic bonds. Starch in plants and glycogen in animals are both storage polymers of glucose. Cellulose is an important structural polysaccharide in the cell walls of plants.

Lipids (pp. 71–74)

1. Lipids make up the most structurally heterogeneous class of macromolecules, but all share the property of being wholly or partly hydrophobic.
2. Fats are high-energy, compact storage molecules also known as triacylglycerols. They are constructed by joining a glycerol molecule to three fatty acids.
3. Saturated fatty acids have the maximum number of hydrogen atoms because of single bonding between all the carbons. Unsaturated fatty acids (present in oils) have one or more double bonds between the carbons.
4. Phospholipids substitute the third fatty acid of triacylglycerols with a negatively charged phosphate group, which may be joined, in turn, to another small molecule. Such bonding introduces polarity and hydrophilic behavior to this part of the molecule. Phospholipids are ideally suited for construction of cell membranes.
5. Steroids, such as cholesterol and the sex hormones, are also classified as lipids.

Proteins (pp. 74–83)

1. Proteins are polymers constructed from 20 different amino acids. They are the most complex and versatile macromolecules.
2. An amino acid is composed of a central carbon singly bonded to a hydrogen atom, a carboxyl group, an amino group, and a variable side chain that confers unique properties on each amino acid.
3. The carboxyl and amino groups of adjacent amino acids link together in a peptide bond, forming long polymers.
4. Protein conformation can be described by three or four hierarchical levels. Primary structure is the first level and describes the unique sequence of amino acids.
5. Secondary structure describes how the primary structure is folded into particular, localized configurations, the α helix and the β pleated sheet, which result from hydrogen bonding between peptide linkages.
6. Tertiary structure describes the additional, less regular contortions of the molecule caused by the involvement of side groups in hydrophobic interactions, hydrogen bonds, ionic bonds, and disulfide bridges.
7. Proteins made of more than one polypeptide chain also show a specific arrangement of their constituent subunits in a quaternary level of structure.
8. The function of a protein is an emergent property of its conformation, which is sensitive to conditions such as pH, salt concentration, and temperature. Changing these conditions can cause denaturation, or alteration of the protein's shape, which renders it unable to perform its biological function.
9. Protein shape is ultimately determined by its primary structure. Molecular biologists are looking for rules that will predict protein folding and final conformation from an amino acid sequence.

Nucleic Acids (pp. 83–87)

1. Nucleic acids are polymers of nucleotides, complex monomers consisting of a pentose (five-carbon sugar) covalently bonded to a phosphate group and to one of five different kinds of nitrogenous bases.
2. In the formation of a polynucleotide, the pentose of one nucleotide joins to the phosphate of another in a phosphodiester linkage, forming a sugar–phosphate backbone from which the nitrogenous bases project.
3. The two kinds of nucleic acids, deoxyribonucleic acid (DNA) and ribonucleic acid (RNA), are named for their characteristic pentoses (deoxyribose or ribose).
4. The five nitrogenous bases are members of two families, the purines and the pyrimidines. The pyrimidines consist of cytosine (C), thymine (T), and uracil (U); the purines consist of adenine (A) and guanine (G). T is unique to DNA, and U is found only in RNA.
5. DNA is a helical, double-stranded polymer with bases projecting into the interior of the molecule. Because A always hydrogen-bonds to T, and C to G, the nucleotide sequence of the two strands is complementary, and one strand can serve as a template for the formation of the other. This unique feature of DNA provides a mechanism for the continuity of life.

DNA and Proteins as Tape Measures of Evolution (pp. 87–88)

1. Molecular comparisons are helping biologists sort out the evolutionary connections among species.

SELF-QUIZ

1. Which of these terms includes all others in the list?
   a. monosaccharide
   b. disaccharide
   c. starch
   d. carbohydrate
   e. polysaccharide

2. The molecular formula for glucose is $C_6H_{12}O_6$. What would be the molecular formula for a polymer made by linking ten glucose molecules together by condensation synthesis?
   a. $C_{60}H_{120}O_{60}$
   b. $C_6H_{12}O_6$
   c. $C_{60}H_{102}O_{51}$
   d. $C_{60}H_{100}O_{50}$
   e. $C_{60}H_{111}O_{51}$

3. There are two ring forms of glucose (alpha and beta) because
   a. The two forms are made from two structural isomers of glucose.
   b. They arise from different line formulas for glucose.
   c. Different carbons of the line structure join to form the rings.
   d. When the ring closes, the hydroxyl group at the point of closure can be trapped in either one of two possible positions.
   e. One is an aldose and the other is a ketose.

4. Choose the pair of terms that completes this sentence: Nucleotides are to ________ as ________ are to proteins.
   a. nucleic acids; amino acids
   b. amino acids; polypeptides
   c. glycosidic linkages; polypeptide linkages
   d. genes; enzymes
   e. polymers; polypeptides

5. Which of the following statements concerning *unsaturated* fats is correct?
   a. They are more common in animals than in plants.
   b. They have double bonds in the carbon chains of their fatty acids.

c. They generally solidify at room temperature.

d. They contain more hydrogen than do saturated fats having the same number of carbon atoms.

e. They have fewer fatty acid molecules per fat molecule.

6. Human sex hormones are classified as:

a. proteins
b. lipids
c. amino acids
d. triacylglycerols
e. carbohydrates

7. For a protein to have a quaternary structure, it *must*

a. have four amino acids
b. consist of two or more polypeptide subunits
c. consist of four polypeptide subunits
d. have at least four disulfide bridges
e. exist in several alternative conformational states

8. What does a protein lose when it denatures?

a. its primary structure
b. its three-dimensional shape
c. its peptide bonds
d. its sequence of amino acids

9. Which of the following is a complication contributing to the difficulty of the protein-folding problem?

a. A specific protein has several alternative amino acid sequences.
b. There are no methods for revealing the three-dimensional shape of a protein.
c. Intermediate stages in folding affect the final shape.
d. It is impossible to determine the precise primary structure of a protein.
e. One must identify the gene that codes for a particular protein before the protein's folding pattern can be determined.

10. Which of these terms includes all others in the list?

a. nucleoside
b. nucleotide
c. nitrogenous base
d. purine
e. pyrimidine

CHALLENGE QUESTIONS

1. Most amino acids can exist in two forms:

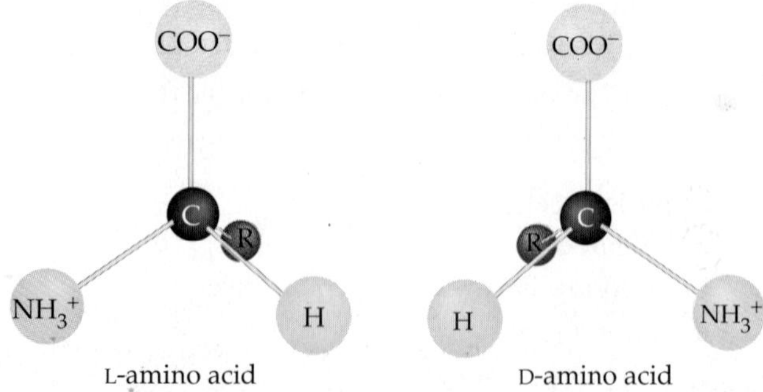

(a) Apply what you learned in Chapter 4 to explain the structural difference between these two molecules.

(b) When an organic chemist synthesizes amino acids, a mixture of these two forms is made; the result was probably the same when the first amino acids formed by abiotic synthesis in the "primordial soup" of the early Earth. However, with very few exceptions, only the L-form of amino acids occurs in proteins. This is one example of molecular "handedness" in life. What are the possible advantages of life's becoming locked into the exclusive use of one of these two amino acid versions? Speculate on the possible evolutionary significance of this "handedness."

2. A particular small polypeptide is nine amino acids long. Using three different enzymes to hydrolyze the polypeptide at various sites, we obtain the following five fragments (N denotes the amino terminus of the chain): Ala-Leu-Asp-Tyr-Val-Leu; Tyr-Val-Leu; N-Gly-Pro-Leu; Asp-Tyr-Val-Leu; N-Gly-Pro-Leu-Ala-Leu. Determine the primary structure of this polypeptide.

SCIENCE, TECHNOLOGY, AND SOCIETY

1. Some amateur and professional athletes take anabolic steroids to help them "bulk up" or build strength. The health risks of this practice are extensively documented. Apart from health considerations, how do you feel about the use of chemicals to enhance athletic performance? Is an athlete who takes anabolic steroids cheating, or is his or her use of such chemicals just part of the preparation required to succeed in a competitive sport? Defend your answer.

2. Molecular biologists have learned how to splice the genes of one species into the DNA of a different species. For example, the gene that codes for the protein insulin can be transferred from a human cell to a bacterial cell, resulting in a culture of bacteria that produce human insulin, which is then used to treat diabetes. Some critics think we should not manipulate Mother Nature and refer to this new biotechnology as "playing god." Given that our ability to modify nature is nothing new, do you feel that genetic engineering is different from other technology in any fundamental way? In your opinion, what determines whether a particular technology is acceptable?

FURTHER READING

Dickerson, R. E., and I. Geis. *The Structure and Action of Proteins.* Menlo Park, CA: Benjamin/Cummings, 1969. A classic, brief text on protein structure, with excellent illustrations.

Dushesne, L. C., and D. W. Larson. "Cellulose and the Evolution of Plant Life." *Bioscience,* April 1989. The chemistry and natural history of the most abundant organic molecule in the biosphere.

Flannery, M. C. "Collagen: Complex and Crucial." *The American Biology Teacher,* November/December 1990. The structure and function of our body's most abundant protein.

Hoffman, M. "Straightening Out the Protein Folding Puzzle." *Science,* September 20, 1991. Recent progress on a key biological problem.

Mathews, C. K., and K. E. van Holde. *Biochemistry.* Redwood City, CA: Benjamin/Cummings, 1990. Excellent explanations and beautiful art.

Richards, F. M. "The Protein-Folding Problem." *Scientific American,* January 1991.

Sharon, N. "Carbohydrates." *Scientific American,* November 1980. The roles of carbohydrates in organisms.

Vollrath, F. "Spider Webs and Silk." *Scientific American,* March 1992. What can materials scientists learn from spiders?

6 An Introduction to Metabolism

The Metabolic Map: An Overview
Energy: Some Basic Principles
Chemical Energy and Life: A Closer Look
ATP and Cellular Work
Enzymes
The Control of Metabolism
Emergent Properties: A Reprise

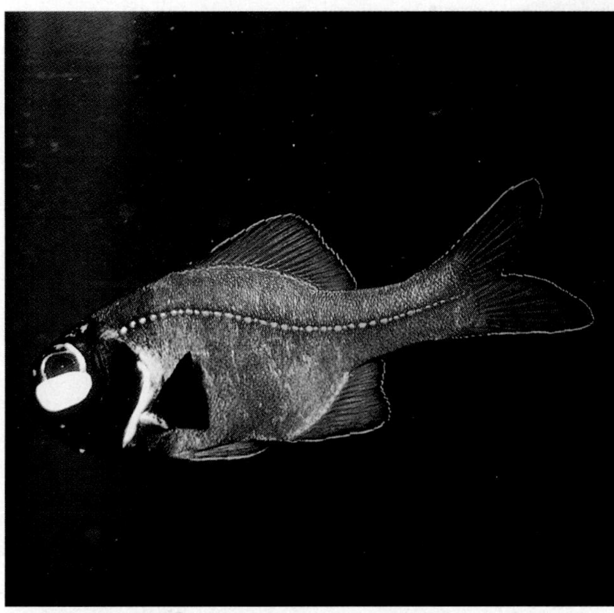

Figure 6.1
"Flashlights" powered by metabolism. Some organisms can use energy extracted from food to generate light—a process called bioluminescence. This flashlight fish *(Photoblepharon palpebratus)* inhabits the dark depths of the sea. Beneath each eye is a large bioluminescent organ, which the fish uses to light its way, to lure light-seeking prey, and to signal potential mates. This light is actually generated by bioluminescent bacteria living in the "flashlight" organs. These bacteria use chemical reactions to convert the energy stored in their food to the energy of light. This phenomenon is just one example of metabolism, which includes all the chemical changes that occur in an organism. The principles of metabolism you learn in this chapter will help you understand how organisms work.

The living cell is a chemical industry in miniature, where thousands of reactions occur within a microscopic space. Metabolism is an emergent property of life that arises from specific interactions between molecules within the orderly environment of the cell. Sugars are converted to amino acids, and vice versa. Small molecules are assembled into polymers, which may later be hydrolyzed as the needs of the cell change. Many cells export chemical products that are used in other parts of the organism. The chemical process known as cellular respiration drives the cellular economy by extracting the energy stored in sugars and other fuels. Cells apply this energy to perform various types of work (Figure 6.1). And always, the myriad reactions going on in the cell are precisely coordinated. In its complexity, its efficiency, its integration, and its responsiveness to subtle changes, the cell is peerless as a chemical institution.

THE METABOLIC MAP: AN OVERVIEW

Metabolism (from the Greek *metabole*, "change") is the totality of an organism's chemical processes. We can think of a cell's metabolism as an elaborate road map of the thousands of reactions that occur in that cell (Figure 6.2). These reactions are arranged in intricately branched metabolic pathways, along which molecules are transformed by a series of steps. The cell routes matter through the metabolic pathways by means of enzymes, which selectively accelerate each of the steps in the labyrinth of reactions. Analogous to the red, green, and yellow lights that control the flow of traffic and prevent snarls, mechanisms that regulate enzymes balance metabolic supply and demand, averting deficiencies and surpluses of chemicals.

As a whole, metabolism is concerned with managing the material and energy resources of the cell. Some metabolic pathways release energy by breaking down complex molecules to simpler compounds. These degradative processes are called **catabolic pathways.** A major thoroughfare of catabolism is cellular respiration, in which the sugar glucose and other organic fuels are broken down to carbon dioxide and water. Energy that was stored in the organic molecules becomes available to do the work of the cell, such as the

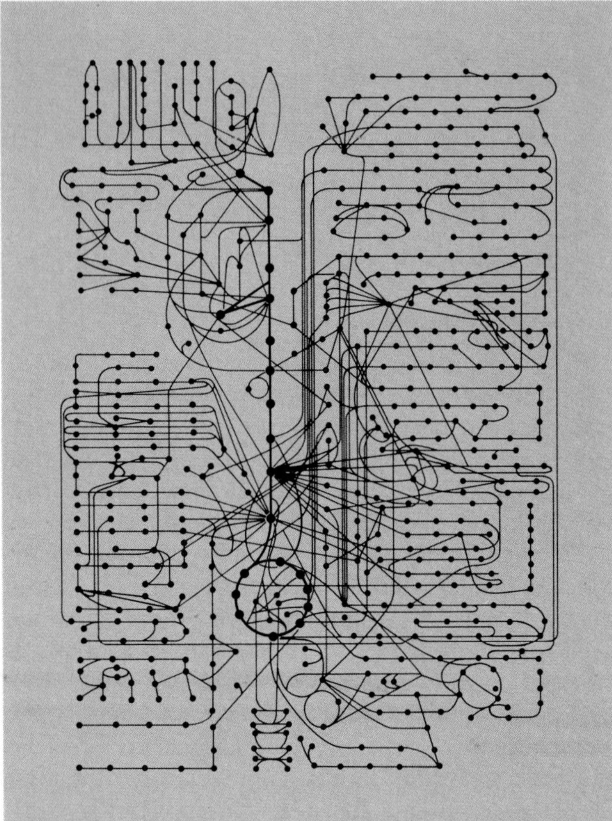

Figure 6.2
The metabolic map: an overview of the complexity of metabolism. This highly schematic diagram traces only a few hundred of the thousands of reactions that occur in a cell. The dots represent molecules, and the lines represent the chemical reactions that transform them. The reactions proceed in stepwise sequences called metabolic pathways, each step catalyzed by a specific enzyme.

bioluminescence illustrated in Figure 6.1. There are also **anabolic pathways,** which consume energy to build complicated molecules from simpler ones. An example of anabolism is the synthesis of a protein from amino acids. Catabolic and anabolic pathways are the downhill and uphill avenues of the metabolic map. The metabolic pathways intersect in such a way that energy released from the "downhill" reactions of catabolism can be used to drive the "uphill" reactions of the anabolic pathways. This interaction between catabolism and anabolism is called *energy coupling.*

In this chapter, rather than attempting to track specific metabolic pathways, we will focus on the mechanisms common to these pathways. Since energy is fundamental to all metabolic processes, a basic knowledge of energy is essential to understanding how the living cell works. We will use many physical and mechanical examples to study energy. But keep in mind that the same principles demonstrated by these examples also apply to **bioenergetics,** the study of how organisms manage their energy resources. An understanding of energy is as important for students of biology as it is for students of physics.

ENERGY: SOME BASIC PRINCIPLES

The two concepts most basic to science are matter and energy. In Chapter 2, we introduced these concepts, defining matter as anything that has mass and takes up space. Energy, however, is more abstract. It can only be described and measured by how it affects matter. You cannot see the energy that enables you to turn the pages of this book, but the resulting movement of matter is clearly visible. Recall that **energy** is the capacity to do work—that is, to move matter against opposing forces, such as gravity and friction. Put another way, energy is the ability to change the way a collection of matter is arranged.

Forms of Energy

Anything that moves possesses a form of energy called **kinetic energy.** As we discussed in Chapter 3, any moving object, including an atom or molecule, has kinetic energy—the energy of motion. Moving objects perform work by imparting motion to other matter: A pool player uses the motion of the cue stick to push the cue ball, which in turn moves the other balls; water gushing through a dam turns turbines; electrons flowing along a wire run household appliances; the contraction of leg muscles pushes bicycle pedals. Heat, or thermal energy, is kinetic energy that results from the random movement of molecules. Light also represents energy, which can be harnessed to perform work, such as powering photosynthesis in green plants.

A resting object not presently at work may also possess energy, which, remember from Chapter 2, is the *capacity* to do work. Stored energy, or **potential energy,** is energy that matter possesses because of its location or arrangement. Water behind a dam, for instance, stores energy because of its altitude. Chemical energy, a form of potential energy especially important to biologists, is stored in molecules because of the arrangement of the atoms that are bonded together.

Energy Transformations

Energy can be converted from one form to another. Consider, for example, the playground scene in Figure 6.3. The girl at the bottom of the slide transformed kinetic energy to potential energy when she climbed the

Figure 6.3
Kinetic and potential energy. The children in this playground scene have more potential energy at the top of the slide (because of the effect of gravity) than they do at the bottom. They convert kinetic energy to potential energy when they climb the ladder, up to the slide, and convert stored energy back to kinetic energy during their descent.

ladder up to the slide. This stored energy was converted back to kinetic energy as she slid down. It was another source of potential energy, chemical energy, that enabled the girl to climb the ladder in the first place.

Chemical energy can be tapped when chemical reactions rearrange the atoms of molecules in such a way that potential energy stored in the molecules is converted to kinetic energy. This transformation occurs, for example, in the engine of an automobile when the hydrocarbons of gasoline react explosively with oxygen, releasing the energy that pushes the pistons. Similarly, chemical energy fuels organisms. Cellular respiration and other anabolic pathways unleash energy stored in sugar and other complex molecules and make that energy available for cellular work. Each child who climbed the ladder in Figure 6.3 transformed some of the chemical energy that was stored in the organic molecules of his or her food to the kinetic energy of movements. The chemical energy stored in these fuel molecules had itself been converted from light energy by plants during photosynthesis.

Whether it is an engine or an organism that converts one form of energy to another, the transformations are all governed by two unbreakable laws.

Two Laws of Thermodynamics

The study of the energy transformations that occur in a collection of matter is called **thermodynamics.** Scientists use the word *system* to denote the collection of matter under study and refer to the rest of the universe—everything outside the system—as the *surroundings.* A closed system, such as that approximated by liquid in a thermos bottle, is isolated from its surroundings. In an open system, energy can be transferred between the system and its surroundings. Organisms are open systems. They absorb light energy or chemical energy in the form of organic molecules and release heat and metabolic waste products, such as carbon dioxide, to the surroundings.

According to the **first law of thermodynamics,** the energy of the universe is constant. *Energy can be transferred and transformed, but it cannot be created or destroyed.* The first law is also known as *conservation of energy.* The electric company does not manufacture energy, but merely converts it to a form that is convenient to use. By converting light to chemical energy, a green plant acts as an energy transformer, not an energy producer. What happens to energy after it has performed work in a machine or an organism? If energy cannot be destroyed, what prevents organisms from behaving like closed systems and recycling their energy? The second law answers these questions.

The **second law of thermodynamics** can be stated many ways. Let's begin with the following concept: Every energy transfer or transformation makes the universe more disordered. There is a quantitative measure of disorder, called **entropy,** whose value increases as disorder increases. We can now state the second law as follows: *Every energy transfer or transformation increases the entropy of the universe.* There is an unstoppable trend toward randomization. In many cases, increased entropy is evident in the physical disintegration of a system's organized structure. Consider, for example, the degradation of an unmaintained building (Figure 6.4). Much of the increasing entropy of the universe is less apparent, however, because it takes the form of an increasing amount of heat, which is the energy of random molecular motion.

In most energy transformations, ordered forms of energy are at least partly converted to heat. Only about 25% of the chemical energy stored in the fuel tank of an automobile is transformed into the motion of the car; the remaining 75% is lost from the engine as heat, which dissipates rapidly through the surroundings. Similarly, the children in Figure 6.3 convert only a frac-

Figure 6.4
Entropy. Entropy measures randomness: The more randomly organized a system, the greater its entropy; the more orderly a system, the lower its entropy. If left alone, any system increases in entropy, losing order. (**a**) Unless energy is expended for maintenance and repairs, buildings fall apart. (**b**) Soluble dyes added to water will spread out spontaneously to become randomly distributed throughout the water. (**c**) Cells of living organisms maintain order only by increasing the randomness of the energy of their surroundings. After a cell, such as this *Paramecium,* dies, its entropy increases (LMs).

(a)

(b)

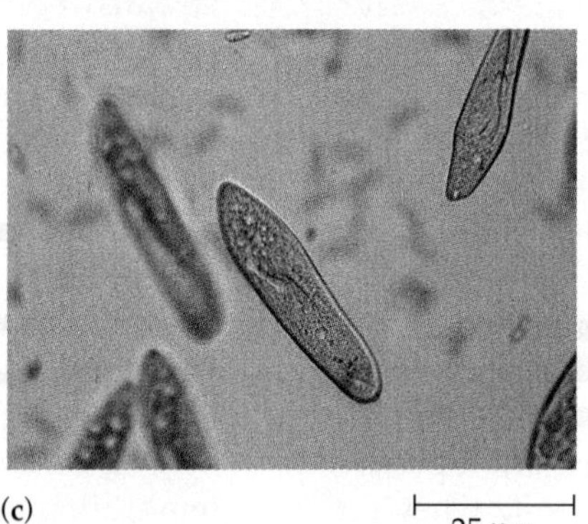

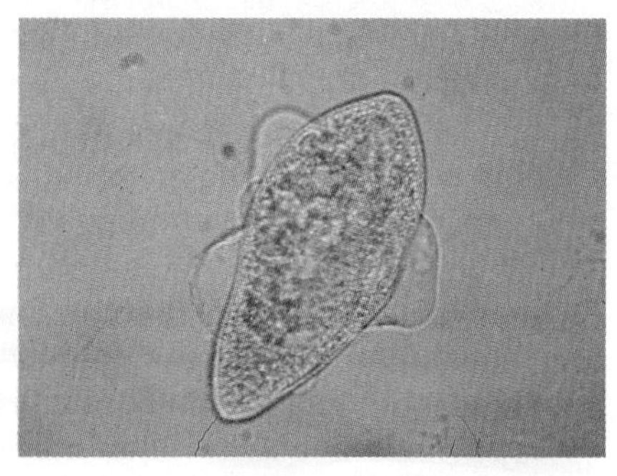

(c)

tion of the energy stored in their food to the kinetic energy of ladder climbing and other play. In performing various kinds of work, living cells unavoidably convert organized forms of energy to heat. (This can make a room crowded with people uncomfortably warm.)

In machines and organisms, even energy that performs useful work is eventually converted to heat. The organized energy of an automobile's forward movement becomes heat when the friction of the brakes and tires stops the car. Conversion to heat is the fate of *all* the chemical energy a child uses to climb a slide: Metabolic breakdown of food generates heat during the climb, and the fraction of energy temporarily stored as gravitational potential energy is converted to heat on the way down, as friction between child and slide warms the surrounding air.

Conversion of other forms of energy to heat does not violate the first law of thermodynamics. Energy has been conserved, because heat is a form of energy, albeit in its most random state. By combining the first and second laws of thermodynamics, we can conclude that the quantity of energy in the universe is constant, but its quality is not. In a sense, heat is the lowest grade of energy, because it is an uncoordinated movement of molecules that many systems cannot harness in order to perform work. A system can only put heat to work when there is a temperature difference that results in the heat flowing from a warmer location to a cooler one (see Chapter 3). If temperature is uniform throughout a system, as it is in a living cell, then the only use for heat energy is to warm a body of matter, such as an organism.

Thus, an organism takes in organized forms of matter and energy from the surroundings and replaces them with less ordered forms. For example, an animal obtains starch, proteins, and other complex molecules from the food it eats, and in turn releases carbon dioxide and water, relatively small and simple molecules that store less chemical energy than the food. The depletion of chemical energy is accounted for by heat generated during metabolism. On a larger scale, energy flows into an ecosystem in the form of light and leaves in the form of heat. Living systems increase the entropy of their surroundings, as predicted by thermodynamic law.

How can we reconcile the second law of thermodynamics—the unstoppable increase in the entropy of the universe—with the orderliness of life, which is one of this book's themes? The key is to remember that organisms are open systems that exchange energy and materials with their surroundings. Cells create ordered structures from less organized starting materials. For example, amino acids are ordered into the specific sequences of polypeptide chains. Complex organisms evolve from simpler ancestors (Figure 6.5). But this high degree of organization in no way violates the second law, because the entropy of a particular system, such as an organism, may actually decrease, as long as the total entropy of the *universe*—the system plus its surroundings—increases. Thus, organisms are islands of low entropy in an increasingly random universe. The evolution of biological order is perfectly consistent with the laws of thermodynamics.

The Free-Energy Concept: A Criterion for Spontaneous Change

How can we predict what can and cannot occur in nature? How can we distinguish the possible from the impossible? We know from experience that certain events occur spontaneously and others do not. For instance, we know that water flows downhill, that objects of opposite charge move toward each other, that an ice cube melts at room temperature, and that a sugar cube dissolves in water. But explaining *why* these processes occur spontaneously is tricky.

Let's begin by defining a spontaneous process as a change that can occur without outside help. A spontaneous change can be harnessed in order to perform

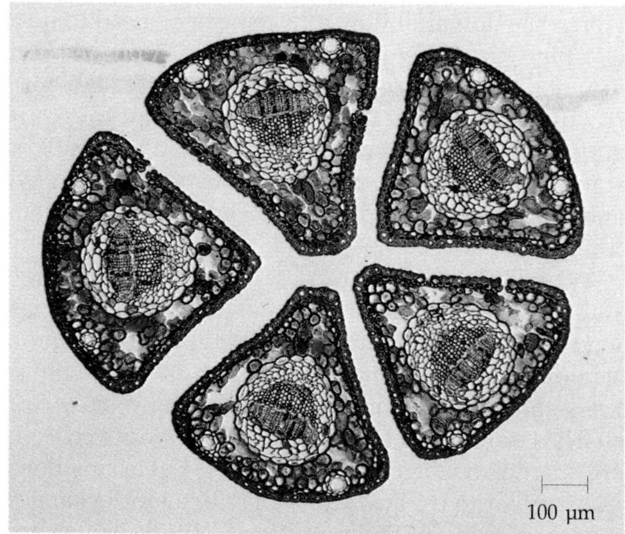

(a)

(b)

Figure 6.5
Does evolution contradict thermodynamic law? (a) Order is a characteristic of life. It is evident, for example, in this cross section of pine needles (LM). Organisms decrease their entropy when they order raw materials, such as organic monomers, into macromolecules and then organize these macromolecules into cellular structures. **(b)** Biological order has, at times, also increased over the grander scale of geological time. This fern fossil represents one group of complex plants that evolved from simpler ancestors. An evolutionary interpretation of the fossil record does not violate thermodynamic law. The second law requires only that processes increase the entropy of the universe. Open systems can increase their order at the expense of the order of their surroundings. Because living systems are open, we can account for the organized structure of an individual organism and for the long evolutionary transformation of relatively simple organisms into more complex ones.

work. The downhill flow of water can be used to turn a turbine in a power plant, for example. A process that cannot occur on its own is said to be nonspontaneous; it will happen only if an external energy source is added. Water moves uphill only when a windmill or some other machine pumps the water against gravity, and a cell must expend energy to synthesize a protein from amino acids.

When a spontaneous process occurs in a system, the stability of that system increases. Unstable systems tend to change in such a way that they become more stable. A body of elevated water, such as a reservoir, is less stable than the same water at sea level. A system of charged particles is less stable when opposite charges are apart than when they are together. But in situations less familiar to us, how can we predict which changes lead to greater stability in a system? Which changes are spontaneous? The concept of entropy teaches us that a process can only occur spontaneously if it increases the disorder of the universe. This principle is helpful in theory, but it does not give us a practical criterion to apply to biological systems, because it requires that we measure changes in the surroundings. We need some standard for spontaneity that is based on the system alone. That criterion is called free energy. The concept of free energy is not easy to grasp, but the effort is worthwhile because the idea can be applied to many biological problems.

***Free energy** is the portion of a system's energy that can perform work when temperature is uniform throughout the system, as in a living cell.* A system's quantity of free energy is symbolized by the letter G. There are two components to G: the system's total energy (symbolized by H) and its entropy (symbolized by S). Free energy is related to these factors in the following way:

$$G = H - TS$$

T stands for absolute temperature (°C + 273). Notice that temperature amplifies the entropy term of the equation. This makes sense, because, as we learned earlier, temperature measures the intensity of random molecular motion (heat), which tends to disrupt order. What does this equation tell us about free energy? Not all the energy stored in a system (H) is available for work. The system's disorder, the entropy factor, is subtracted from total energy in computing the maximum capacity of the system to perform useful work. We are then left with free energy, which is somewhat less than the system's total energy.

How does the concept of free energy help us determine whether a particular process can occur spontaneously? Think of free energy, G in the above equation, as a measure of a system's instability—its tendency to change to a more stable state. Systems that are rich in energy, such as stretched springs or separated charges, are unstable. So are highly ordered systems, such as complex molecules. Thus, those systems that tend to change spontaneously to a more stable state are those that have high energy, low entropy, or both. The free-energy equation weighs these two factors, which are consolidated in the system's G content. Now we can state a versatile criterion for spontaneous change: *In any spontaneous process, the free energy of a system decreases.*

The change in free energy as a system goes from a starting state to a different state is represented by ΔG:

$$\Delta G = G_{final\ state} - G_{starting\ state}$$

Or, put another way:

$$\Delta G = \Delta H - T\Delta S$$

T stands for absolute temperature (in Kelvin units, K, °C + 273—see Appendix Three). For a process to occur spontaneously, the system must either give up energy (a decrease in H), give up order (an increase in S), or both. When these changes in H and S are tallied, ΔG must have a negative value. The greater this decrease in free energy, the greater the maximum amount of work the spontaneous process can perform. This is a formal, mathematical way of stating the obvious: Nature runs "downhill" (downhill in the broad metaphorical sense of a loss in useful energy—the capacity to perform work). The relationships between free energy, stability, spontaneous change, and work are summarized in Figure 6.6. We can now apply what we have learned to metabolism, the chemical changes of life.

CHEMICAL ENERGY AND LIFE: A CLOSER LOOK

Based on their free-energy changes, chemical reactions can be classified as either exergonic (meaning "energy outward") or endergonic (meaning "energy inward"). An **exergonic reaction** proceeds with a net release of free energy. Since the chemical mixture loses free energy, ΔG is negative for an exergonic reaction. In other words, exergonic reactions are those that occur spontaneously. The magnitude of ΔG for an exergonic reaction is the maximum amount of work the reaction can perform. We can use cellular respiration as an example:

$$C_6H_{12}O_6 + 6\ O_2 \rightarrow 6\ CO_2 + 6\ H_2O$$
$$\Delta G = -686\ \text{kcal/mol}$$

For each mole (180 g) of glucose broken down by respiration, 686 kcal of energy are made available for work. (Recall from Chapter 3 that a kilocalorie is the amount of energy required to raise the temperature of a kilogram of water by 1°C.) Since energy must be con-

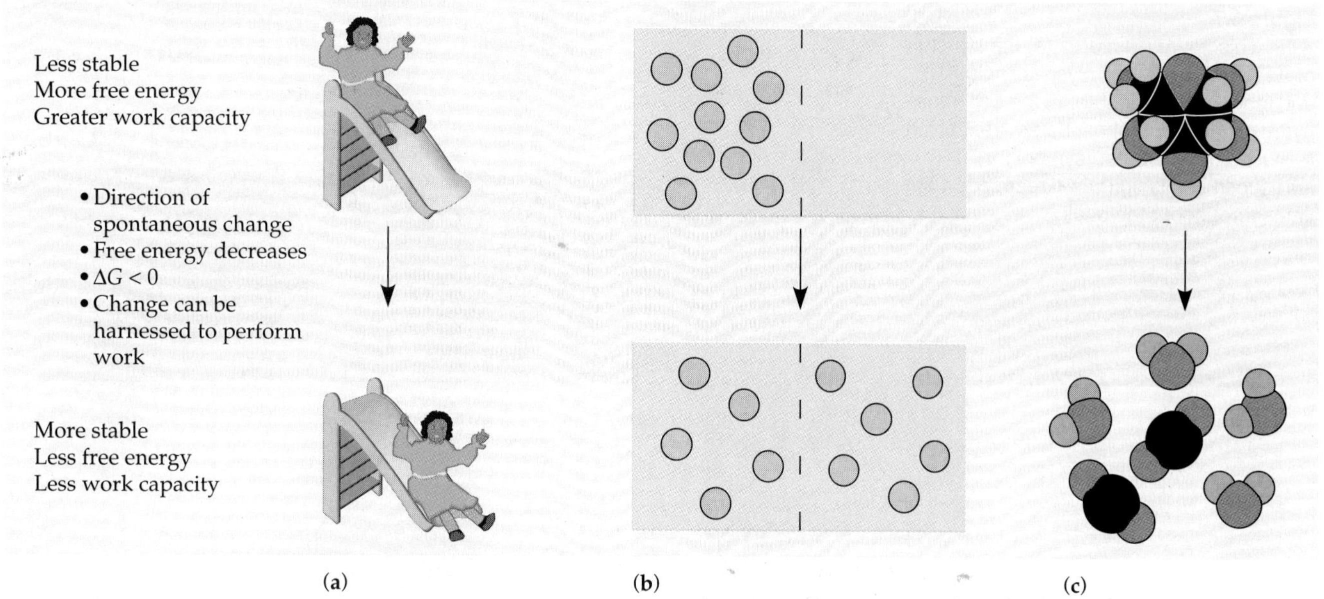

Less stable
More free energy
Greater work capacity

- Direction of spontaneous change
- Free energy decreases
- $\Delta G < 0$
- Change can be harnessed to perform work

More stable
Less free energy
Less work capacity

(a)

(b)

(c)

Figure 6.6
The relationship between stability, free energy, spontaneous change, and work. An unstable system is rich in free energy. It has a tendency to change spontaneously to a more stable state, and it is possible to harness this "downhill" change in order to perform work. (**a**) In this case, free energy is proportional to the girl's altitude. (**b**) The free-energy concept also applies on the molecular scale, in this case to a physical movement of molecules called diffusion. Here, a membrane separates two aqueous compartments. Molecules of a particular solute are distributed unequally across the membrane. This ordered state is unstable—it is rich in free energy. If the solute molecules can cross the membrane, there will be a net movement (diffusion) of the molecules until they are equally concentrated in both compartments. Diffusion is one spontaneous process responsible for the uptake and loss of certain substances by a cell. For example, dissolved oxygen will pass into a cell if it is initially more concentrated outside the cell. (**c**) Chemical reactions also involve free energy. The sugar molecule in the top part is less stable than the simpler molecules in the bottom part. By breaking down complex organic molecules in the process of cellular respiration, a cell can harness the free energy stored in the molecules to perform work.

served, the chemical products of respiration store 686 kcal less free energy than the reactants. The products are, in a sense, the spent exhaust of a process that tapped most of the free energy stored in the sugar molecules (see Figure 6.6c).

An **endergonic reaction** is one that absorbs free energy from its surroundings. Because this kind of reaction stores more free energy in the molecules, ΔG is positive. Such reactions are nonspontaneous, and the magnitude of ΔG is the minimum quantity of energy required to drive the reaction. If a chemical process is exergonic (downhill) in one direction, then the reverse process must be endergonic (uphill). A reversible process cannot travel downhill in both directions. If $\Delta G = -686$ kcal/mol for respiration, then for photosynthesis to produce sugar from carbon dioxide and water, $\Delta G = +686$ kcal/mol. Sugar production in the leaf cells of a plant is steeply endergonic, an uphill process powered by the absorption of light energy from the sun.

There is an important relationship between free energy and chemical equilibrium. Recall from Chapter 2 that most chemical reactions are reversible and proceed until the forward and backward reactions occur at the same rate. The reaction is then said to be at chemical equilibrium, and there is no further change in the concentration of products or reactants. As a reaction proceeds toward equilibrium, the free energy of the mixture of reactants and products decreases. Free energy increases when a reaction is somehow pushed away from equilibrium. For a reaction at equilibrium, $\Delta G = 0$, because there is no net change in the system. We can think of equilibrium as an energy sink, the bottom of a hill. A process at equilibrium performs no work. A reaction is spontaneous and exergonic when sliding toward equilibrium. Movement away from equilibrium is nonspontaneous; it is an endergonic process that can occur only when an outside energy source pushes the reaction uphill. A key strategy in cellular metabolism is the driving of endergonic reactions by coupling them to exergonic reactions through an energy shuttle called ATP.

ATP AND CELLULAR WORK

A cell does three main kinds of work:

1. *Mechanical work*, such as the beating of cilia, the contraction of muscle cells, the flow of cytoplasm

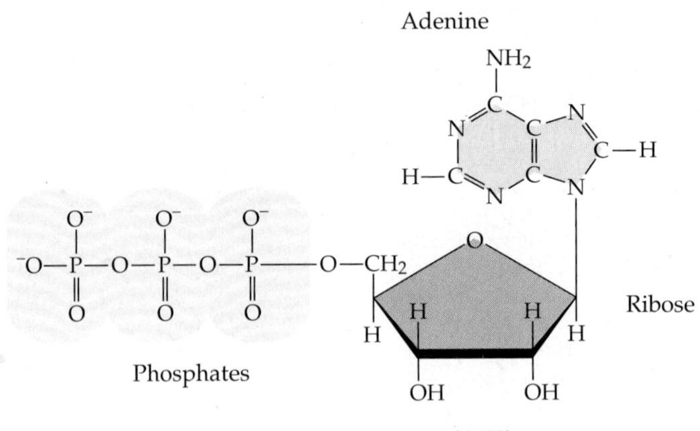

Figure 6.7
ATP. This figure illustrates (**a**) the structure of ATP and (**b**) its hydrolysis to yield ADP and inorganic phosphate. In the cell, most hydroxyl groups of phosphates are ionized (—O^-). The "sunburst" symbol for ATP introduced in this figure will reappear many times throughout the book.

within cells, and the movement of chromosomes during cellular reproduction;

2. *Transport work,* the pumping of substances across membranes against the direction of spontaneous movement; and
3. *Chemical work,* the pushing of endergonic reactions that would not occur spontaneously, such as the synthesis of polymers from monomers.

In most cases, the immediate source of energy that drives cellular work is a molecule called adenosine triphosphate, or ATP.

The Structure and Hydrolysis of ATP

ATP (adenosine triphosphate) is a nucleoside triphosphate consisting of adenine, bonded to the sugar ribose, which in turn is connected to a chain of three phosphate groups (Figure 6.7a). The only difference between ATP and the adenosine monophosphate found as a monomer in the nucleic acid RNA, discussed in Chapter 5, is the two additional phosphate groups on ATP.

The triphosphate tail of ATP is unstable, and the bonds between the phosphate groups can be broken by hydrolysis. When water hydrolyzes the terminal phosphate bond, a molecule of inorganic phosphate (abbreviated Ⓟ$_i$) is removed from ATP, which then becomes adenosine diphosphate, or ADP (Figure 6.7b). The reaction is exergonic, and under laboratory conditions, releases 7.3 kcal of energy per mole of ATP hydrolyzed:

$$\text{ATP} + H_2O \longrightarrow \text{ADP} + Ⓟ_i \qquad \Delta G = -7.3 \text{ kcal/mol}$$

When the reaction occurs in the cellular environment rather than in a test tube, the actual ΔG is estimated to be -10 to -12 kcal/mol.

Because their hydrolysis releases energy, the phosphate bonds of ATP are sometimes referred to as high-energy phosphate bonds, but the term is misleading. The phosphate bonds of ATP are not strong bonds, as the words "high-energy" imply. In fact, these bonds are relatively weak, and it is *because* they are weak, or unstable, that their hydrolysis yields energy. The products of hydrolysis (ADP and Ⓟ$_i$) are more stable than ATP. When a system changes in the direction of greater stability—as when a compressed spring relaxes, for instance—the change is exergonic. Thus, the release of energy during the hydrolysis of ATP comes from the chemical change to a more stable condition, and not from the phosphate bonds themselves. Why are the phosphate bonds so fragile? If we reexamine the ATP molecule in Figure 6.7a, we can see that all three phosphate groups are negatively charged. These like

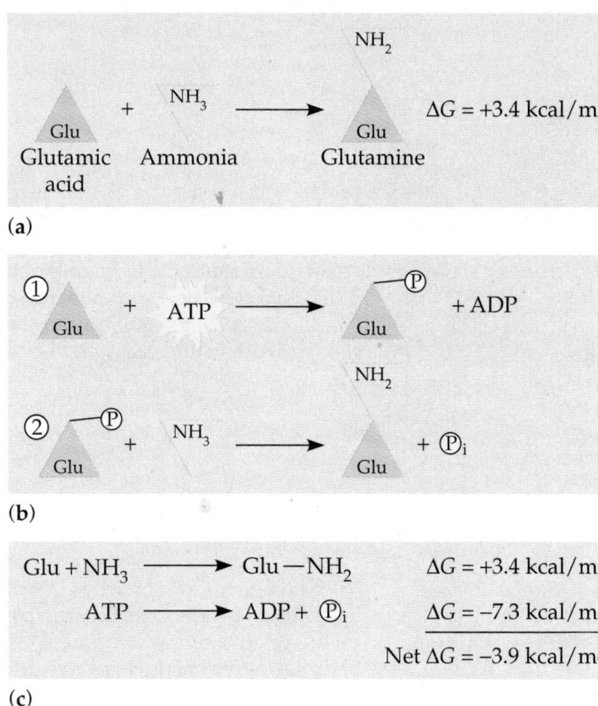

Figure 6.8
Energy coupling by phosphate transfer. In this example, ATP hydrolysis is used to drive an endergonic reaction, the conversion of the amino acid glutamic acid (Glu) to another amino acid, glutamine (Glu—NH_2). (**a**) Without the help of ATP, the conversion is nonspontaneous. (**b**) As it actually occurs in the cell, the synthesis of glutamine is a two-step reaction driven by ATP. The formation of a phosphorylated intermediate couples the two steps. ① In the first step, ATP phosphorylates glutamic acid, transferring chemical instability to the amino acid. ② In the second step, ammonia displaces the phosphate group from the phosphorylated intermediate, forming glutamine. (**c**) We can calculate the free-energy change for the overall reaction by adding together ΔG for each step of the two-step reaction. Because the overall process is exergonic (has a negative ΔG), it occurs spontaneously.

charges are crowded together, and their repulsion contributes to the instability of this region of the ATP molecule. The triphosphate tail of ATP is the chemical equivalent of a loaded spring.

How ATP Performs Work

When ATP is hydrolyzed in a test tube, the release of free energy merely heats the surrounding water. In the cell, that would be an inefficient and wasteful use of valuable chemical energy. With the help of specific enzymes, the cell is able to couple the energy of ATP hydrolysis directly to endergonic processes by transferring a phosphate group from ATP to some other molecule, which is then said to be phosphorylated. The key to the coupling is the formation of a **phosphorylated intermediate,** which is more reactive (less stable) than the original molecule (Figure 6.8). Nearly all cellular work depends on ATP's energizing of other molecules by transferring phosphate groups. For instance, ATP powers the movement of muscles by transferring phosphates to contractile proteins.

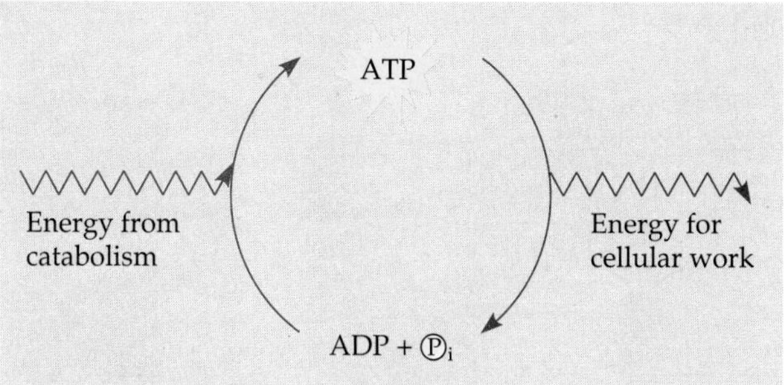

Figure 6.9
The ATP cycle. Energy released by breakdown reactions (catabolism) in the cell is used to phosphorylate ADP, regenerating ATP. Energy stored in ATP drives most cellular work.

The Regeneration of ATP

An organism at work uses ATP continuously, but ATP is a renewable resource that can be regenerated by the addition of phosphate to ADP (Figure 6.9). The ATP cycle moves at an astonishing pace. For example, a working muscle cell recycles its entire pool of ATP about once each minute. That turnover represents 10 million molecules of ATP consumed and regenerated per second per cell. If ATP could not be regenerated by the phosphorylation of ADP, human beings would consume nearly their body weight in ATP each day.

Since a reversible process cannot go downhill both ways, the regeneration of ATP from ADP is necessarily endergonic:

$$\text{ADP} + \text{Ⓟ}_i \longrightarrow \text{ATP} \qquad \Delta G = +7.3 \text{ kcal/mol}$$

Catabolic (exergonic) pathways, especially cellular respiration, provide the energy to make ATP, an endergonic process. Plants can also use light energy to produce ATP.

Cellular respiration is a stepwise pathway by which enzymes decompose glucose and other complex organic molecules. The process is overwhelmingly exergonic, and the energy it releases drives phosphorylation of ADP to regenerate ATP. The ATP cycle is a turnstile through which energy passes to be transferred from catabolic to anabolic pathways.

Metabolic Disequilibrium

The chemical reactions of respiration and other catabolic pathways are reversible and would reach equilib-

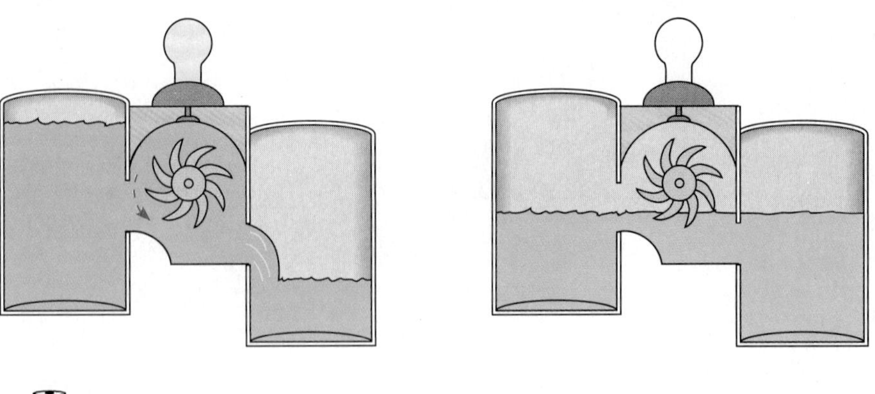

(**a**) The water generates electric energy only while it is falling. Once the levels in the two containers are equal, the turbine ceases to turn and the light goes out. The individual steps of respiration, in isolation, would also come to equilibrium, and cellular work would cease.

(**b**) If there is a series of drops in water level, electric energy can be generated at each drop. In respiration, there is a series of drops in free energy between glucose, the starting material, and the metabolic wastes at the end. The overall process never reaches equilibrium as long as the organism lives: a cell continues to acquire glucose, the product of each reaction becomes the reactant for the next, and the metabolic wastes are expelled from the cell.

Figure 6.10
Keeping metabolism away from equilibrium: a hydraulic analogy.

rium if they occurred in the isolation of a test tube. Because chemical systems at equilibrium have a ΔG of zero and can do no work, in the cell, some of the reversible reactions of respiration must be pulled in one direction and kept out of equilibrium. The key to maintaining this disequilibrium is that the product of one reaction does not accumulate, but instead becomes a reactant in the next step along the metabolic pathway (Figure 6.10). The overall sequence of reactions is pulled by the huge free-energy difference between glucose at the uphill end of respiration and carbon dioxide and water at the downhill end of the pathway. As long as the cell has a steady supply of glucose or other fuels and is able to expel the CO_2 waste to the surroundings, the cell does not reach equilibrium and keeps making ATP. We see once again how important it is to think of organisms as open systems.

ENZYMES

The laws of thermodynamics tell us what can and cannot happen but say nothing about the speed of these processes. A spontaneous chemical reaction may occur so slowly that it is imperceptible. For example, the hydrolysis of sucrose (table sugar) to glucose and fructose is exergonic, occurring spontaneously with a release of free energy ($\Delta G = -7.0$ kcal/mol). Yet a solution of sugar dissolved in sterile water will sit for years at room temperature with no appreciable hydrolysis. However, if we add a small amount of the enzyme sucrase to the solution, then all the sugar may be hydrolyzed within seconds. **Enzymes** are catalysts, chemical agents that change the rate of a reaction without being consumed by the reaction. In the absence of enzymes, chemical traffic through the pathways of the metabolic map would become hopelessly congested. What impedes a spontaneous reaction, and how does an enzyme lower the barrier?

Enzymes and the Lowering of Activation Energy

A chemical reaction involves bond breaking and bond forming. When a reaction rearranges the atoms of molecules, existing bonds in the reactants must be broken

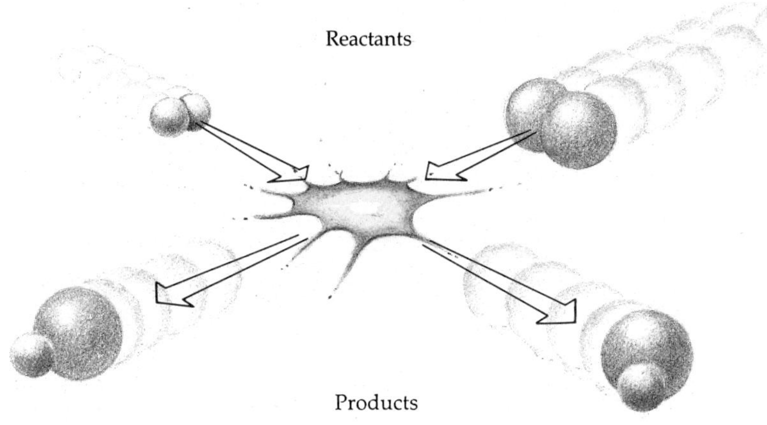

Figure 6.11
A chemical reaction: breaking old bonds and forming new ones. In this illustration, two different molecules react to form two identical molecules of product. For the reaction to occur, the reactants must collide with enough energy to break the bonds holding their atoms together. Formation of new bonds releases energy.

and the new bonds of the products formed. These processes require exchanges of energy between the mixture of molecules and the surrounding environment. The reactant molecules must absorb energy for their bonds to break, and energy is released when the bonds of the product molecules are formed (Figure 6.11).

The initial investment of energy for starting a reaction—the energy required to break bonds in the reactant molecules—is known as the **free energy of activation,** or **activation energy,** abbreviated $\Delta G^{\ddagger}$. It is usually provided in the form of heat absorbed by the reactant molecules from the surroundings. If the reaction is exergonic, $\Delta G^{\ddagger}$ will be repaid with dividends, as the formation of new bonds releases more energy than was invested in the breaking of old bonds. Figure 6.12 graphs these energy changes for a hypothetical reaction that swaps portions of two reactant molecules:

$$A—B + C—D \longrightarrow A—C + B—D$$

The bonds of the reactants break only when the molecules have absorbed enough energy to become unstable (recall that systems rich in free energy are intrinsically unstable, and unstable systems are reactive). The activation energy is represented by the uphill portion of the graph, with the free-energy content of the reactants increasing. The absorption of thermal energy increases the speed of the reactants, so they are colliding more often and more forcefully. Moreover, thermal agitation of the atoms that make up the molecules has made the bonds more fragile and more likely to break. At the summit, the reactants are in an unstable condition known as the *transition state;* they are primed, and the reaction can occur. As the molecules settle into their new bonding arrangements, energy is released to the surroundings. This phase of the reaction corresponds to the downhill portion of the curve, which indicates a loss of free energy by the molecules. The difference in the free energy of the products and reactants is ΔG for the overall reaction, which is negative for an exergonic reaction.

As Figure 6.12 shows, even for an exergonic reaction, which is energetically downhill overall, the barrier of activation energy must be scaled before the reaction can occur. For some reactions, $\Delta G^{\ddagger}$ is modest enough that even at room temperature there is sufficient thermal energy for the reactants to reach the transition state. In most cases, however, the $\Delta G^{\ddagger}$ barrier is loftier, and the reaction will occur at a noticeable rate

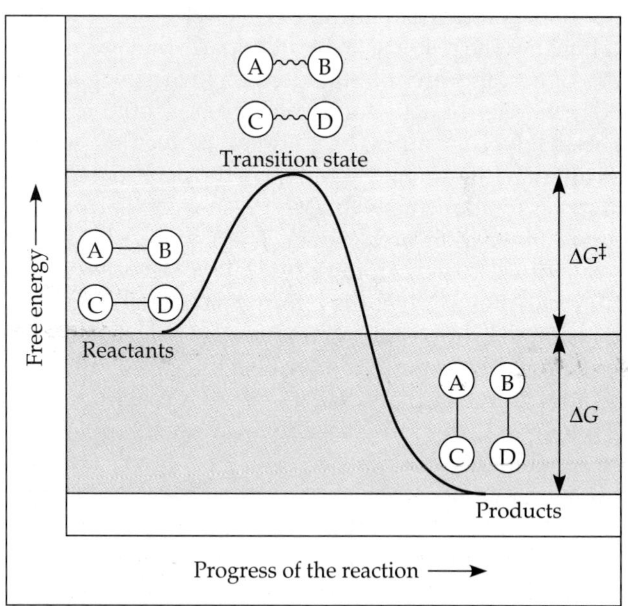

Figure 6.12
An energy profile of a reaction. In this hypothetical reaction, the reactants A—B and C—D must absorb enough energy from the surroundings to surmount the hill of activation energy ($\Delta G^{\ddagger}$) and reach the transition state. The bonds can then break, and as the reaction proceeds, energy is released to the surroundings during the formation of new bonds. This is an exergonic reaction, which has a negative ΔG; the products have less free energy than the reactants.

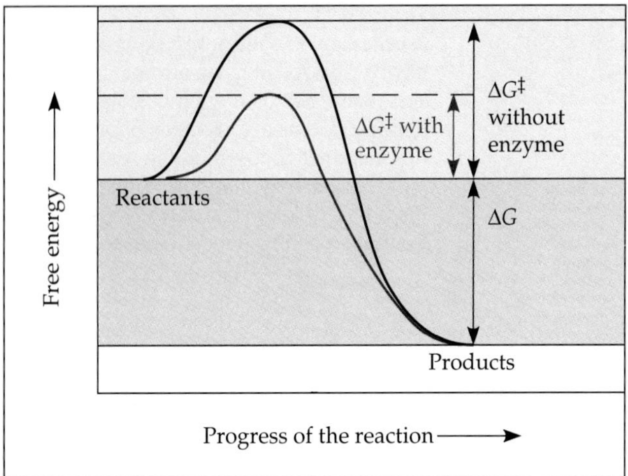

Figure 6.13
Enzymes: lowering the barrier of activation energy. Without affecting the free-energy change (ΔG) for the reaction, an enzyme speeds the reaction by reducing the uphill climb to the transition state. The black curve shows the course of the reaction without an enzyme, and the magenta curve shows the course of the reaction with an enzyme.

only if the material is heated. The spark plugs in an automobile engine heat the gasoline–oxygen mixture so that the molecules reach the transition state and react; only then can there be the explosive release of energy that pushes the pistons. Without a spark, the hydrocarbons of gasoline are too stable to react with oxygen.

The barrier of activation energy is essential to life. Proteins, DNA, and other complex molecules of the cell are rich in free energy and have the potential to decompose spontaneously; that is, the laws of thermodynamics favor their breakdown. These molecules exist only because at temperatures typical for cells, few molecules can make it over the hump of activation energy. Occasionally, however, the barrier for selected reactions must be surmounted, or else the cell would be metabolically stagnant. Heat speeds a reaction, but high temperatures kill cells. Organisms must therefore use an alternative: a catalyst.

Enzymes, most of which are proteins, are biological catalysts. An enzyme speeds a reaction by lowering the barrier of activation energy, so that the precipice of the transition state is within reach even at moderate temperatures (Figure 6.13). An enzyme cannot change the ΔG for a reaction. It cannot make a nonspontaneous reaction spontaneous, or an endergonic reaction exergonic. Enzymes can only hasten reactions that would occur eventually anyway, but this function makes it possible for the cell to have a dynamic metabolism. Further, because enzymes are very selective in the reactions they catalyze, they determine which chemical processes will be going on in the cell at any particular time.

The Specificity of Enzymes

The reactant an enzyme acts on is referred to as the enzyme's **substrate.** The enzyme binds to its substrate (or substrates, when there are two or more reactants), and while the two are joined, the catalytic action of the enzyme converts the substrate to the product (or products) of the reaction. This process can be generalized in this way:

$$\text{Substrate} \xrightarrow{\text{Enzyme}} \text{Product}$$

For example, the enzyme sucrase (many enzyme names end in *-ase*) breaks the double sugar sucrose into its two monosaccharides, glucose and fructose:

$$\text{Sucrose} \xrightarrow{\text{Sucrase}} \text{Glucose} + \text{Fructose}$$

An enzyme can distinguish its substrate from even closely related compounds, such as isomers, so that each type of enzyme catalyzes a particular reaction. For instance, sucrase will act only on sucrose and will reject other disaccharides, such as maltose. What accounts for this molecular recognition? Recall that most enzymes are proteins, and proteins are macromolecules with unique three-dimensional conformations. The specificity of an enzyme is based on its shape.

Only a restricted region of the enzyme molecule actually binds to the substrate. This region, called the **active site,** is typically a pocket or groove on the surface of the protein (Figure 6.14). Usually, the active site is formed by only a few of the enzyme's amino acids, with the rest of the protein molecule providing a framework that reinforces the configuration of the active site.

The specificity of an enzyme is attributed to a compatible fit between the shape of its active site and the shape of the substrate. The active site, however, is not a rigid receptacle for the substrate. As the substrate enters the active site, it induces the enzyme to change its shape slightly so that the active site fits even more snugly around the substrate. This **induced fit** is like a clasping handshake. The embrace of the substrate by the active site not only tailors fit, but also brings chemical groups of the active site into positions that enhance their ability to work on the substrate and catalyze the chemical reaction.

The Catalytic Cycle of Enzymes

In an enzymatic reaction, the substrate binds to the active site to form an enzyme–substrate complex (Figure 6.15). In most cases, the substrate is held in the active site by weak interactions, such as hydrogen bonds and ionic bonds. Side chains (R groups) of a few of the amino acids that make up the active site catalyze the

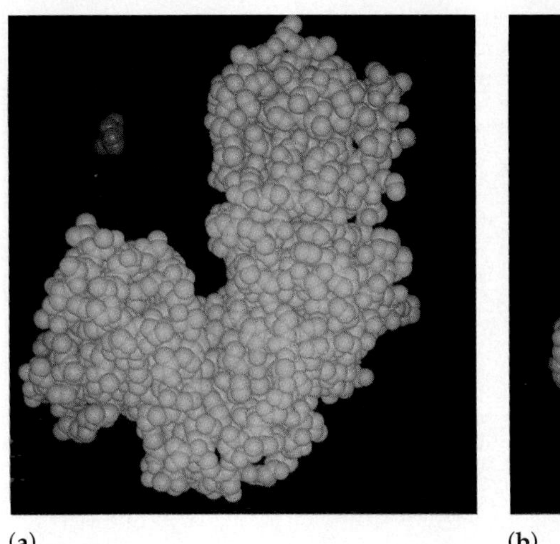

(a)

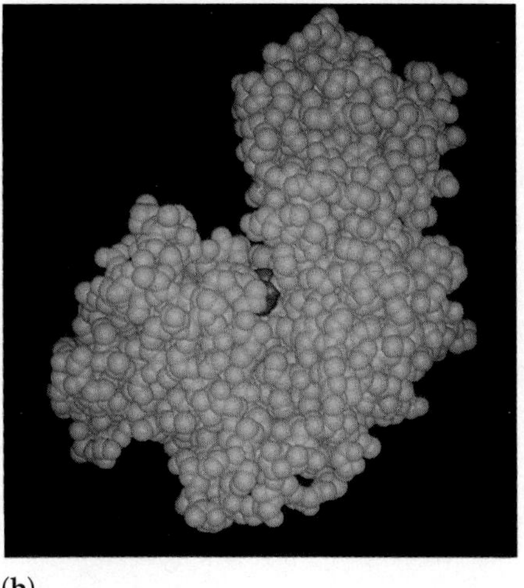

(b)

Figure 6.14
The induced fit between an enzyme and its substrate. (**a**) The active site of this enzyme, called hexokinase, can be seen here as a groove on the surface of the protein. (**b**) On entering the active site, the substrate, which is glucose (red), induces a slight change in the shape of the protein, causing the active site to embrace the substrate.

conversion of substrate to product, and the product then departs from the active site. The enzyme is then free to take another substrate molecule into its active site. The entire cycle happens so fast that a single enzyme molecule typically converts about a thousand substrate molecules per second. Some enzymes are much faster. Enzymes, like all catalysts, emerge from the reaction in their original form. Therefore, very small amounts of enzyme can have a huge metabolic impact by functioning over and over again in catalytic cycles.

Enzymes use a variety of mechanisms for lowering activation energy and speeding up a reaction. In reactions involving two or more reactants, the active site provides a template for the substrates to come together in the proper orientation for a reaction to occur between them. As induced fit causes the active site to clinch the substrates, the enzyme may stress the substrate molecules, stretching and bending critical chemical bonds that must be broken during the reaction. Since $\Delta G^{\ddagger}$ is proportional to the difficulty of breaking bonds, distorting the substrate reduces the amount of

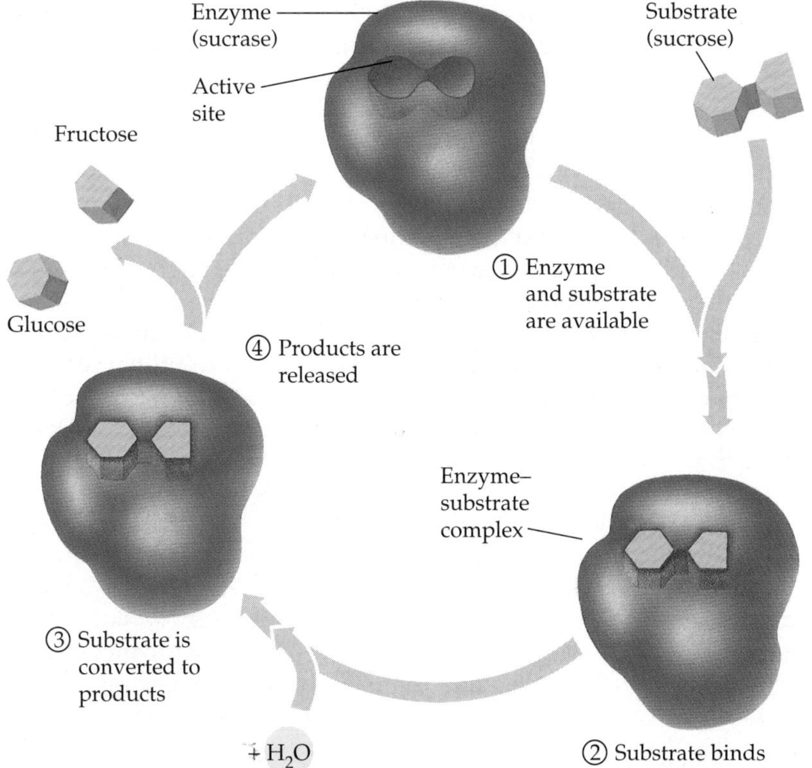

Figure 6.15
The catalytic cycle of an enzyme. In this example, the enzyme sucrase catalyzes the hydrolysis of sucrose to glucose and fructose. ① When the active site of an enzyme is unoccupied and its substrate is available, the cycle begins as ② an enzyme–substrate complex forms when the substrate enters the active site and attaches by weak bonds. The active site changes shape to fit snugly around the substrate (induced fit). ③ The substrate is converted to products while in the active site. ④ The enzyme releases the products, and ① its active site is then available for another molecule of substrate. Most metabolic reactions are reversible, and an enzyme can catalyze both the forward and the reverse reactions. Which reaction prevails depends mainly on the relative concentrations of reactants and products; that is, the enzyme catalyzes the reaction in the direction of equilibrium.

thermal energy that must be absorbed in order to achieve a transition state.

The active site may also provide a microenvironment that is conducive to a particular type of reaction. For example, if the active site has a concentration of amino acids with acidic side chains (R groups), the active site may be a pocket of low pH in an otherwise neutral cell. Still another aspect of catalysis is the direct participation of the active site in the chemical reaction. Sometimes this process even involves brief covalent bonding between the substrate and a side chain of an amino acid of the enzyme. Subsequent steps of the reaction restore the side chains to their original states, so the active site is the same after the reaction as it was before.

The rate at which a given amount of enzyme converts substrate to product is partly a function of the initial concentration of substrate: The more substrate molecules available, the more frequently they wander into the active sites of the enzyme molecules. However, there is a limit to how fast the reaction can be pushed by adding more substrate to a fixed concentration of enzyme. At some point, the concentration of substrate will be high enough that all enzyme molecules have their active sites engaged, and as soon as the product exits an active site, another substrate molecule enters. At this substrate concentration, the enzyme is said to be saturated, and the rate of the reaction is determined by the speed at which the active site can convert substrate to product. When an enzyme population is saturated, the only way to increase productivity is to add more enzymes. Cells sometimes do this by making more enzyme molecules.

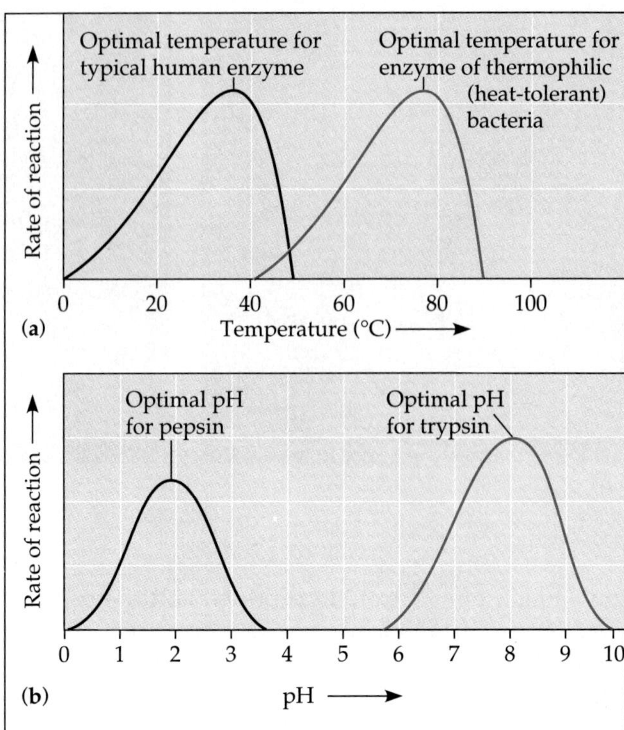

Figure 6.16
Environmental factors affecting enzymes. Each enzyme has an optimal (**a**) temperature and (**b**) pH that favor the active conformation of the protein molecule.

Factors Affecting Enzyme Activity

The activity of an enzyme is affected by general environmental factors, such as temperature and pH, and also by particular chemicals that specifically influence that enzyme.

Environmental Conditions Recall from Chapter 5 that the three-dimensional structures of enzymes and other proteins are sensitive to their environment. Each enzyme has optimal conditions in which it works best, because that environment favors the most active conformation for the enzyme molecule.

Temperature is one environmental factor important in the activity of an enzyme (Figure 6.16a). Up to a point, the velocity of an enzymatic reaction increases with increasing temperature, partly because substrates collide with active sites more frequently when the molecules move rapidly. But at some point on the temperature scale, the speed of the enzymatic reaction drops sharply with additional increase in temperature. The thermal agitation of the enzyme molecule disrupts the hydrogen bonds, ionic bonds, and other weak interactions that stabilize the active conformation, and the protein molecule denatures. Each type of enzyme has an optimal temperature at which its reaction rate is fastest. This temperature allows the greatest number of molecular collisions without denaturing the enzyme. Most human enzymes have optimal temperatures of about 35°C to 40°C (close to human body temperature). Bacteria that live in hot springs contain enzymes with optimal temperatures of 70°C or higher.

Another environmental factor that influences the shape of proteins is pH. Just as each enzyme has an optimal temperature, it also has an optimal pH at which it is most active (Figure 6.16b). The optimal pH values for most enzymes fall in the range of 6 to 8, but there are exceptions. For example, pepsin, a digestive enzyme in the stomach, works best at pH 2. Such an acidic environment denatures most enzymes, but the active conformation of pepsin is adapted to the acidic environment of the stomach. In contrast, trypsin, a digestive enzyme residing in the alkaline environment of the intestine, has an optimal pH of 8.

Enzymes are also sensitive to salt concentration. Most enzymes cannot tolerate extremely saline (salty) solutions because the inorganic ions interfere with ionic bonds within the protein molecule. Again, there are exceptions. Some algae and bacteria inhabit pools

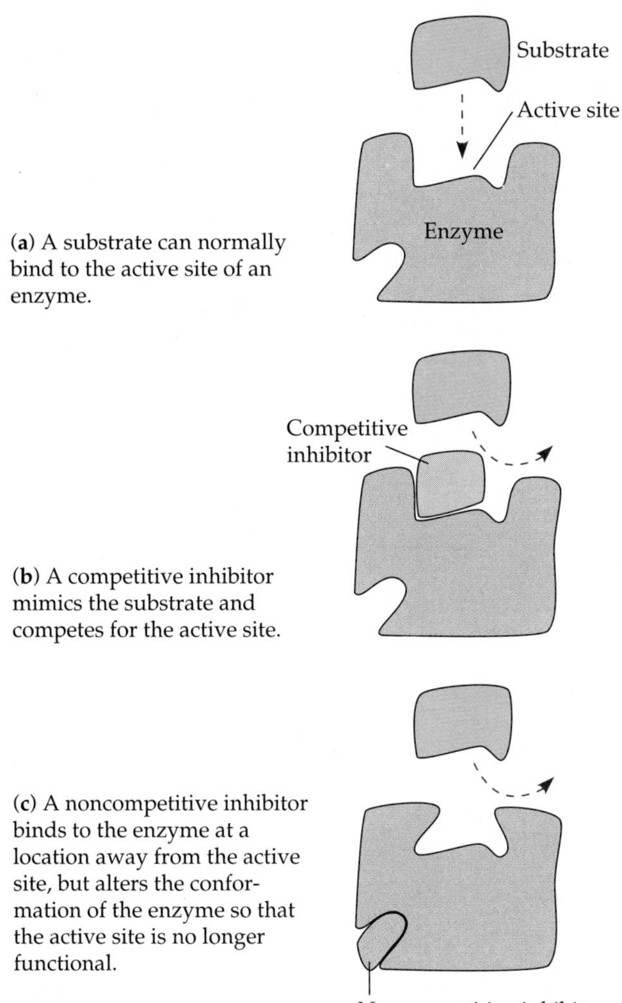

Substrate
Active site
Enzyme
(a) A substrate can normally bind to the active site of an enzyme.
Competitive inhibitor
(b) A competitive inhibitor mimics the substrate and competes for the active site.
(c) A noncompetitive inhibitor binds to the enzyme at a location away from the active site, but alters the conformation of the enzyme so that the active site is no longer functional.
Noncompetitive inhibitor

Figure 6.17
Enzyme inhibition.

where the salt concentration is many times greater than seawater; their enzymes and other proteins are active under conditions that would denature the proteins of other organisms.

Cofactors Many enzymes require nonprotein helpers for catalytic activity. These adjuncts, called **cofactors,** may be bound tightly to the active site as permanent residents, or they may bind loosely and reversibly along with the substrate. The cofactors of some enzymes are inorganic, such as the metal atoms zinc, iron, and copper. If the cofactor is an organic molecule, it is more specifically called a **coenzyme.** Most vitamins are coenzymes or raw materials from which coenzymes are made. Cofactors function in various ways, but in all cases they are necessary for catalysis to take place.

Enzyme Inhibitors Certain chemicals selectively inhibit the action of specific enzymes (Figure 6.17). If the inhibitor attaches to the enzyme by covalent bonds, inhibition is usually irreversible. The inactivation is reversible, however, if the inhibitor binds to the enzyme by weak bonds.

Some inhibitors resemble the normal substrate molecule and compete for admission into the active site. These mimics, called **competitive inhibitors,** reduce the productivity of enzymes by blocking the substrate from entering active sites. If the inhibition is reversible, it can be overcome by increasing the concentration of substrate so that as active sites become available, more substrate molecules than inhibitor molecules are around to gain entry to the sites.

Noncompetitive inhibitors impede enzymatic reactions by binding to a part of the enzyme away from the active site. This interaction causes the enzyme molecule to change its shape, rendering the active site unreceptive to substrate, or leaving the enzyme less effective at catalyzing the conversion of substrate to product.

Enzyme inhibitors may act as metabolic poisons. Some pesticides, including DDT and parathion, are inhibitors of key enzymes in the nervous system. Many antibiotics are inhibitors of specific enzymes in bacteria. For instance, penicillin blocks the active site of an enzyme that many bacteria use to make their cell walls. These examples of enzyme inhibitors as metabolic poisons may give the impression that enzyme inhibition is generally abnormal and harmful. In fact, the selective inhibition and activation of enzymes by molecules naturally present in the cell are essential mechanisms in metabolic control.

Allosteric Regulation In most cases, the molecules that affect enzyme activity bind to an **allosteric site,** a specific receptor site on some part of the enzyme molecule remote from the active site. Most enzymes having allosteric sites are proteins constructed from two or more polypeptide chains, or subunits. Each subunit has its own active site, and allosteric sites are usually located where subunits are joined (Figure 6.18a). The entire complex oscillates between two conformational states, one catalytically active and the other inactive. The binding of an activator to an allosteric site stabilizes the conformation that has a functional active site, while the binding of an allosteric inhibitor stabilizes the inactive form of the enzyme (Figure 6.18b). The joints between the subunits of an allosteric enzyme articulate in such a way that a conformational change in one subunit is transmitted to all others. Through this interaction of subunits, a single activator or inhibitor molecule that binds to one allosteric site will affect the active sites of all subunits.

Because allosteric regulators attach to an enzyme by weak bonds, the activity of the enzyme changes from moment to moment in response to fluctuating concentrations of the regulators. In some cases, an inhibitor

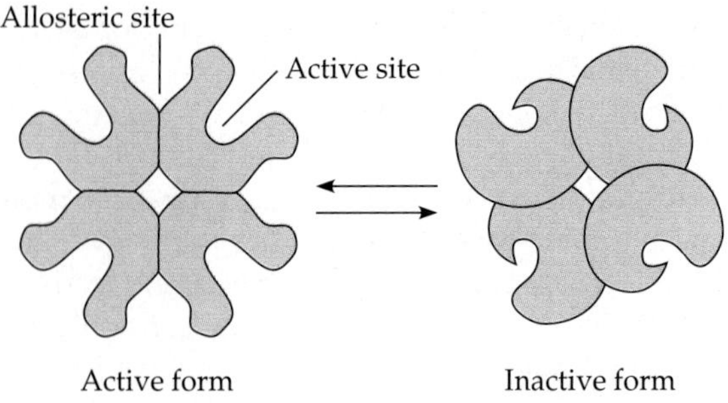

(**a**) Conformational changes in allosteric enzymes

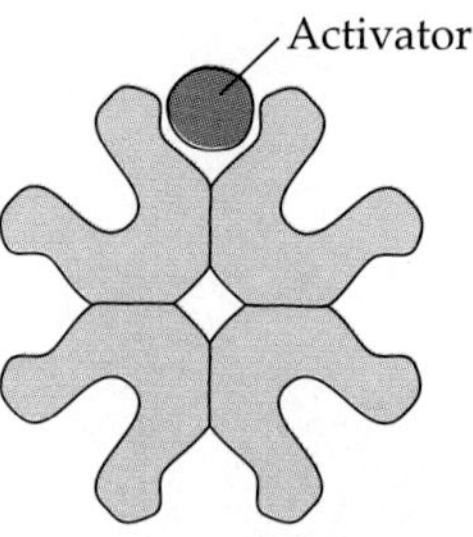

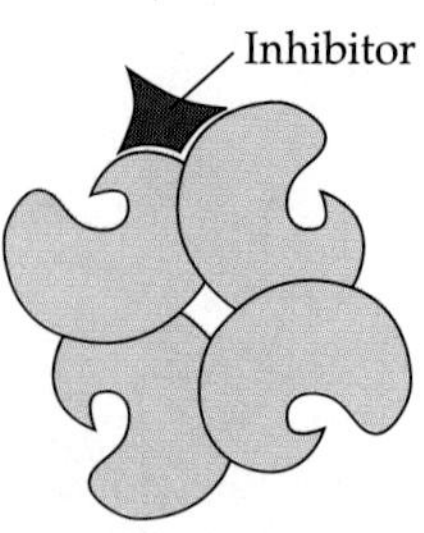

(**b**) Allosteric regulation

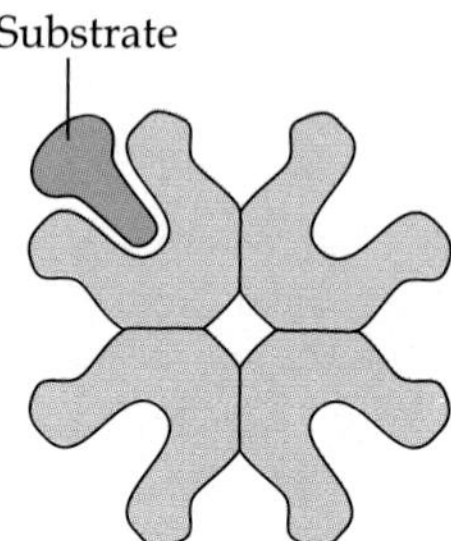

(**c**) Cooperativity

Figure 6.18
Allosteric regulation and cooperativity. (**a**) Most allosteric enzymes are constructed from two or more subunits, each having its own active site. The enzyme oscillates between two conformational states, one active and the other inactive. Remote from the active sites are allosteric sites, specific receptors for regulators of the enzyme, which may be activators or inhibitors. (**b**) Here we see the opposing effects of an allosteric inhibitor and activator on the conformation of all four subunits of an enzyme. (**c**) Similarly, through a phenomenon called cooperativity, one substrate molecule can activate all subunits of the enzyme by the mechanism of induced fit.

and an activator are similar enough in shape to compete for the same allosteric site. For example, an enzyme that catalyzes a step in a catabolic pathway, such as respiration, may have an allosteric site that fits both ATP and ADP. The enzyme is inhibited by ATP and activated by ADP. This control seems logical because a major function of catabolism is to regenerate ATP from ADP. If ATP production lags behind its use, ADP accumulates and activates key enzymes that speed up catabolism. If the supply of ATP exceeds demand, then catabolism slows down as ATP molecules outnumber ADP molecules in competition for allosteric sites. In this way, allosteric enzymes act as valves that control the rates of key reactions in metabolic pathways.

Cooperativity By a mechanism that resembles allosteric activation, substrate molecules themselves may stimulate the catalytic powers of an enzyme (Figure 6.18c). Recall that the binding of a substrate to an enzyme induces a favorable change in the shape of the active site (induced fit). If an enzyme has two or more subunits, this interaction with one substrate molecule triggers the same favorable conformational change in all other units of the enzyme. Called **cooperativity,** this mechanism amplifies the response of enzymes to substrates: One substrate molecule primes an enzyme to accept additional substrate molecules.

Next we will see how the regulation of enzymes fits into the overall metabolic economy of the cell.

THE CONTROL OF METABOLISM

Chemical chaos would result if all of a cell's metabolic pathways were open simultaneously. Imagine, for example, a substance synthesized by one pathway and broken down by another. If the two pathways were to run at the same time, the cell would be spinning its metabolic wheels. Actually, the operation of each metabolic pathway is tightly regulated. Pathways are switched on and off by controlling enzyme activity.

Feedback Inhibition

One of the most common methods of metabolic control is **feedback inhibition.** This occurs when a metabolic pathway is switched off by its end-product, which acts as an inhibitor of an enzyme within the pathway. A specific example of feedback inhibition will reveal the logic of this control mechanism. The cell uses a pathway of five steps to synthesize the amino acid isoleucine from threonine, another amino acid (Figure

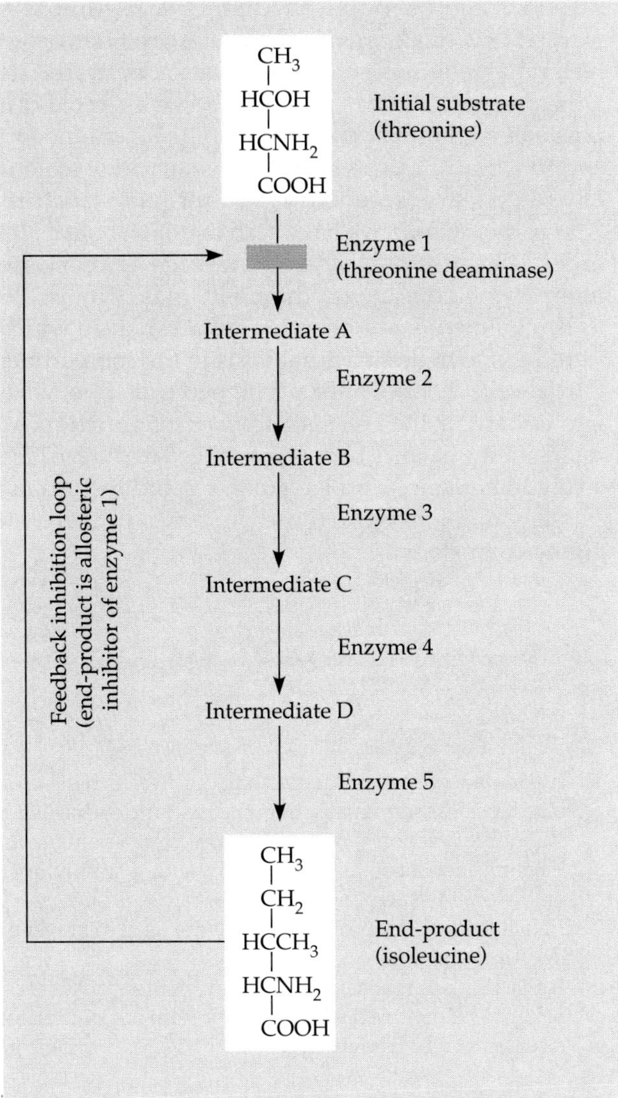

Figure 6.19
Feedback inhibition. The initial reactions of many metabolic pathways are switched off by the end-products of the metabolic sequence, which act as allosteric inhibitors of the first enzyme of the pathway. In this case, the amino acid isoleucine inhibits the enzyme that uses threonine as its substrate.

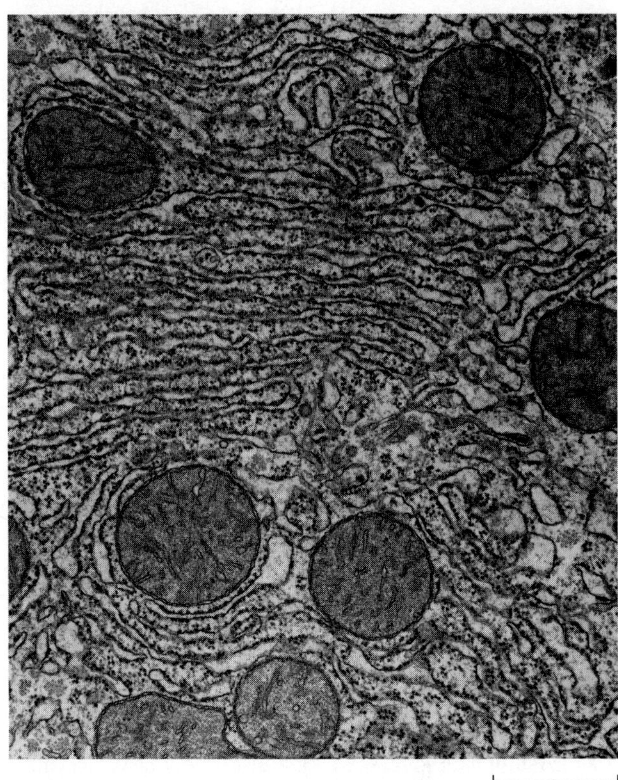

Figure 6.20
Structural order and metabolism. Membranes partition a cell into various metabolic compartments, or organelles, each with a corps of enzymes that carry out specific functions (TEM).

6.19). As isoleucine, the end-product of the pathway, accumulates, it switches off its own synthesis. This happens because isoleucine is an allosteric inhibitor of the enzyme that catalyzes the very first step of the pathway, the enzyme for which threonine is the substrate. Feedback inhibition thereby prevents the cell from wasting chemical resources to synthesize more isoleucine than is necessary.

Structural Order and Metabolism

The cell is not just a bag of chemicals with thousands of different kinds of enzymes and substrates wandering about randomly. The complex structure of the cell orders metabolic pathways in space and time (Figure 6.20). In some cases, a team of enzymes for several steps of a metabolic pathway is assembled together as a multienzyme complex. The arrangement orders the sequence of reactions, as the product from the first enzyme becomes substrate for the adjacent enzyme in the complex, and so on, until the end-product is released. Some enzymes have fixed locations within the cell because they are incorporated into the structure of a specific membrane and thus help to organize metabolism. Even when the enzymes for a metabolic pathway are individually dissolved, they may be highly concentrated along with their substrates within specialized organelles of the cell. Membranes partition the cell into many kinds of compartments, each with its own internal chemical environment and special blend of enzymes. For example, the enzymes for cellular respiration reside within organelles called mitochondria. If the cell had the same number of enzymes for respiration, but they were diluted throughout the entire volume of the cell, respiration would be very inefficient.

By examining the structural basis of metabolic order, we have returned to the theme with which this unit of chapters began.

EMERGENT PROPERTIES: A REPRISE

Recall that life is organized along a hierarchy of structural levels. With each increase in the level of order, new properties emerge in addition to those of the component parts. In these chapters, we have dissected the chemistry of life using the strategy of the reductionist. But we have also begun to develop a more integrated view of life as we have seen how properties emerge with increasing order.

We have seen that the peculiar behavior of water, so essential to life on earth, results from interactions of the water molecules, themselves an ordered arrangement of hydrogen and oxygen atoms. We reduced the great complexity and diversity of organic compounds to the chemical characteristics of carbon, but we also saw that the unique properties of organic compounds are related to the specific structural arrangements of carbon skeletons and their appended functional groups. We learned that small organic molecules are often assembled into giant molecules, but we also discovered that a macromolecule does not behave like a simple composite of its monomers. For example, the unique form and function of a protein are consequences of a hierarchy of primary, secondary, and tertiary structures. And in this chapter we have seen that metabolism, that orderly chemistry that characterizes life, is a concerted interplay of thousands of different kinds of molecules.

By completing our overview of metabolism with an introduction to its structural basis in the compartmentalized cell, we have built a bridge to Unit Two, where we will study the cell's structure and function. We maintain the balance between our need to reduce life to a conglomerate of simpler processes and the ultimate satisfaction of viewing those processes in their integrated context.

STUDY OUTLINE

1. Metabolism is the sum of all the chemical reactions occurring in the cells of an organism.

The Metabolic Map: An Overview (pp. 91–92)

1. Metabolism manages the material and energy resources of the cell. Aided by enzymes, metabolism proceeds by steps along interrelated pathways.
2. Catabolic pathways, such as those of cellular respiration, break complex molecules into simpler compounds, releasing energy in the process. Anabolic pathways build up complex molecules from simpler compounds, requiring the energy input usually provided by catabolism.

Energy: Some Basic Principles (pp. 92–96)

1. Energy is the capacity to do work by moving matter against an opposing force.
2. Kinetic energy, the energy of motion, does its work by transferring motion from one body of matter to another.
3. Potential energy is stored energy based on the specific location or arrangement of matter. Chemical energy is potential energy stored in molecular structure.
4. Energy can be changed from one form to another, governed by the laws of thermodynamics.
5. The first law of thermodynamics, conservation of energy, states that energy cannot be created or destroyed.
6. The second law of thermodynamics states that every time energy changes form, there is an increase in the entropy *(S)*, or disorder, of the universe. Whenever matter becomes more ordered, it does so only by contributing to the disorder of the surroundings.
7. A system's free energy is the amount of energy that can actually be put to work under cellular conditions—that is, in the absence of temperature gradients. Free energy *(G)* is directly related to total energy *(H)* and inversely related to entropy *(S)*: $\Delta G = \Delta H - T\Delta S$.
8. Every spontaneous change in a system proceeds with a decrease in free energy ($-\Delta G$).

Chemical Energy and Life: A Closer Look (pp. 96–97)

1. A spontaneous chemical reaction, one in which the products have less free energy than the reactants, is termed an exergonic reaction ($-\Delta G$). Endergonic (nonspontaneous) reactions are those that occur only with a supply of energy from the surroundings ($+\Delta G$).
2. In metabolism, exergonic reactions are used to power endergonic reactions.
3. A reaction approaches equilibrium spontaneously (ΔG is negative). To move a reaction away from its equilibrium, a cell must add free energy.

ATP and Cellular Work (pp. 97–100)

1. ATP (adenosine triphosphate) serves as the main energy shuttle in cells. Hydrolysis of one of its weak phosphate bonds releases ADP (adenosine diphosphate) and inorganic phosphate. This is an exergonic reaction that releases free energy.
2. ATP drives endergonic reactions in the cell by the enzymatic transfer of the phosphate group to specific reactants. The phosphorylated intermediates formed are more reactive than the original molecules. In this way, cells can carry out work, such as movement, pumping solutions across membranes, and anabolism.
3. The regeneration of ATP from ADP and phosphate is an endergonic reaction, driven primarily by cellular respiration and light-driven reactions in photosynthesis.
4. The removal of metabolic end-products prevents metabolism from reaching equilibrium.

Enzymes (pp. 100–106)

1. Enzymes, most of which are proteins, function as biological catalysts, agents that change the rate of a reaction without being consumed in the reaction.
2. Before a reaction can occur, the reactants must absorb enough energy to break existing bonds. This free energy

of activation ($\Delta G\ddagger$) is usually provided in the form of heat absorbed from the surroundings, which causes the reactants to reach an unstable transition state required for the reaction to proceed. Biological macromolecules would decompose spontaneously if not for high activation energies.

3. Enzymes enable molecules to react during metabolism by lowering activation energies, which allows bonds to break at the moderate body temperatures characteristic of most organisms.
4. Each type of enzyme has a uniquely shaped active site, which gives it specificity in combining with its particular substrate, the reactant molecule on which an enzyme acts.
5. The active site of an enzyme can lower activation energy in a number of ways: by providing a template for substrates to come together in proper orientation; by binding to the substrate in such a way that critical bonds of the substrate are strained; and by providing suitable microenvironments.
6. As proteins, enzymes are very sensitive to environmental conditions that influence the weak chemical bonds responsible for their three-dimensional structure. Each enzyme has optimal conditions of temperature, pH, and salt concentration.
7. Cofactors are nonprotein ions or molecules required for the function of some enzymes. If the cofactor is organic, it is known as a coenzyme.
8. Enzyme inhibitors selectively reduce enzyme function, either reversibly, through the formation of weak bonds, or irreversibly, through covalent bonds.
9. A competitive inhibitor is structurally similar to the substrate and can bind to the active site in its place. A noncompetitive inhibitor binds to a place on the enzyme other than the active site, disrupting the shape and function of the active site.
10. Some enzymes change shape when regulatory molecules, either activators or inhibitors, bind to specific allosteric sites. Allosteric sites are usually located between the subunits of complex enzymes. Induced fit by the binding of substrate activates other attached subunits by the mechanism known as cooperativity.

The Control of Metabolism (pp. 106–107)

1. One of the most common methods of regulating metabolism is by feedback inhibition, in which the end-product of a metabolic pathway inhibits the first enzyme in that pathway. In this way, a cell conserves resources by producing certain molecules only when they are in low concentration.
2. Membranes partition the cell into metabolic compartments (organelles). Enzymes may be built into membranes or dissolved in relatively high concentration within organelles.

Emergent Properties: A Reprise (p. 108)

1. In this unit, we have seen how increasing levels of organization result in the emergence of properties that are different from those of lower levels. Organization is the key to the chemisty of life.

SELF-QUIZ

1. Choose the pair of terms that completes this sentence: Catabolism is to anabolism as ________ is to ________.
   a. exergonic; spontaneous
   b. exergonic; endergonic
   c. free energy; entropy
   d. work; energy
   e. entropy; order
2. Most cells cannot harness heat in order to perform work because
   a. heat is not a form of energy
   b. cells do not have much heat; they are relatively cool
   c. temperature is usually uniform throughout a cell
   d. there are no mechanisms in nature that can use heat to do work
   e. heat denatures enzymes
3. According to the first law of thermodynamics
   a. matter can be neither created nor destroyed
   b. energy is conserved in all processes
   c. all processes increase the entropy of the universe
   d. systems rich in energy are intrinsically unstable
   e. the universe constantly loses energy because of friction
4. Which of the following metabolic processes can occur without a *net* influx of energy from some other process?
   a. ADP + Ⓟ$_i \longrightarrow$ ATP + H_2O
   b. $C_6H_{12}O_6 + 6\ O_2 \longrightarrow 6\ CO_2 + 6\ H_2O$
   c. $6\ CO_2 + 6\ H_2O \longrightarrow C_6H_{12}O_6 + 6\ O_2$
   d. amino acids $\longrightarrow$ protein
   e. glucose + fructose $\longrightarrow$ sucrose
5. Which molecule binds to the active site of an enzyme?
   a. allosteric activator
   b. allosteric inhibitor
   c. noncompetitive inhibitor
   d. competitive inhibitor
6. If an enzyme solution is saturated with substrate, the most effective way to obtain an even faster yield of products would be to
   a. add more of the enzyme
   b. heat the solution to 90°C
   c. add more substrate
   d. add an allosteric inhibitor
   e. add a noncompetitive inhibitor
7. An enzyme accelerates a metabolic reaction by
   a. altering the overall free-energy change for the reaction
   b. making an endergonic reaction occur spontaneously
   c. lowering the free energy of activation
   d. pushing the reaction away from equilibrium
   e. making the substrate molecule more stable
8. Some bacteria are metabolically active in hot springs because
   a. they are able to maintain an internal temperature much cooler than that of the surrounding water

b. the high temperatures facilitate active metabolism without the need of catalysis
c. their enzymes have high optimal temperatures
d. their enzymes are insensitive to temperature
e. they use molecules other than proteins as their main catalysts

9. Which metabolic process in bacteria is directly inhibited by the antibiotic penicillin?
a. cellular respiration
b. ATP hydrolysis
c. synthesis of fats
d. synthesis of chemical components of the cell wall
e. replication of DNA, the genetic material

10. In the following branched metabolic pathway, a dotted arrow with a minus sign symbolizes inhibition of a metabolic step by an end-product:

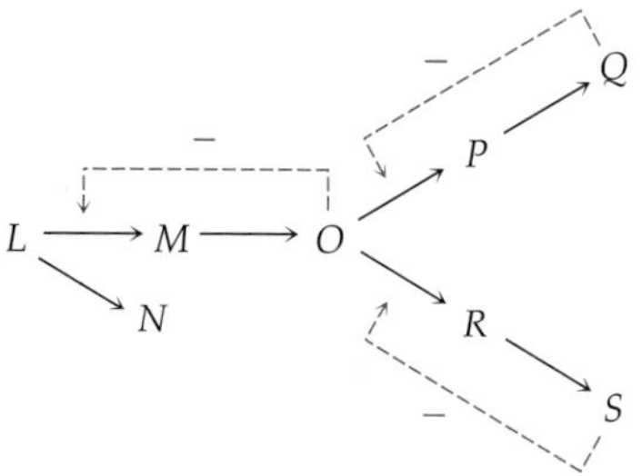

Which reaction would prevail if both *Q* and *S* are present in the cell in high concentrations?
a. $L \longrightarrow M$
b. $M \longrightarrow O$
c. $L \longrightarrow N$
d. $O \longrightarrow P$
e. $R \longrightarrow S$

CHALLENGE QUESTIONS

1. A particular enzyme has an optimal temperature of 37°C and begins to denature at 45°C. During denaturation, entropy increases (the protein loses much of its organization). The protein also increases its energy content (energy must be absorbed from the surroundings to break the numerous weak bonds that reinforce the native conformation). Thus, for denaturation, ΔS and ΔH are both positive. Using the free-energy equation ($\Delta G = \Delta H - T\Delta S$), explain why denaturation becomes spontaneous at a certain temperature.

2. In an experiment, the enzyme sucrase is mixed with various concentrations of its substrate, sucrose. Each test tube begins with a certain sucrose concentration, and all tubes contain the same concentration of the enzyme. The rate of the reaction—conversion of substrate to product—is measured for each of the samples, and the results are plotted on the following graph:

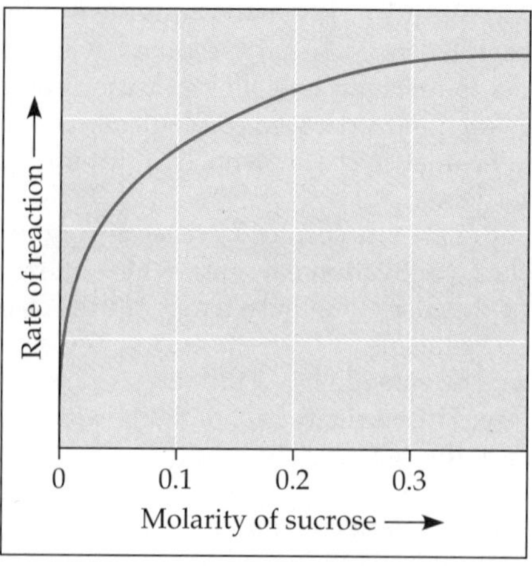

Explain the shape of the curve in the graph.

3. Ice crystals, like this snowflake, are highly ordered structures. When liquid water freezes to form ice, order increases—that is, entropy decreases. For the conversion of water from the liquid state to the solid state:

$$\Delta H = -1440 \text{ cal/mol}$$
$$\Delta S = -5.3 \text{ cal/mol}$$

Using the free-energy equation, prove in terms of ΔG that water will crystallize spontaneously at –10°C (263K) but not at +10°C (283K).

SCIENCE, TECHNOLOGY, AND SOCIETY

1. Biologists often hear the following type of argument: "Evolutionists claim that the complexity of organisms has increased during the history of life. Such evolution of greater biological order contradicts the second law of

thermodynamics, which is known to be unbreakable. Therefore, biological evolution is a scientifically invalid concept." How would you respond to this argument?

2. In the early 1970s, the United States was forced to take its energy consumption more seriously, when a cartel of oil-exporting countries began to regulate their production in order to control prices. "Energy crisis" became part of our everyday vocabulary, and our dependence on imported oil still affects our economy and foreign policies. Your author remembers a faculty colleague (*not* a science professor) who questioned the very premise of an energy crisis by invoking the first law of thermodynamics: "If energy can't be destroyed," he said, "then how can there by an energy crisis? We just need to develop some clever ways of recycling our energy." What are the scientific flaws in this wishful thinking? How would you explain thermodynamic reality to this professor?

FURTHER READING

Becker, W. M., and D. W. Deamer. *The World of the Cell,* 2nd ed. Redwood City, CA: Benjamin/Cummings, 1991. A lucid explanation of cellular energetics and enzymes in Chapters 5 and 6.

Harold, F. M. *The Vital Force: A Study of Bioenergetics.* New York: Freeman, 1986. A challenging but clear introduction to energy and life in Chapters 1–3.

Koshland, D. E. "Protein Shape and Biological Control." *Scientific American,* October 1973. A discussion of how enzymes are regulated.

Mathews, C. K., and K. E. van Holde. *Biochemistry.* Redwood City, CA: Benjamin/Cummings, 1990. An introduction to enzymes and metabolism in Chapters 10 and 12.

Stryer, L. *Biochemistry,* 3rd ed. New York: Freeman, 1989. An introduction to enzymes and metabolism in Chapters 6 and 11.

Waldrop, M. "Spontaneous Order, Evolution, and Life." *Science,* March 30, 1990. A controversial hypothesis about the ability of living systems to self-organize.

THE CELL

AN INTERVIEW WITH HAROLD VARMUS AND MICHAEL BISHOP

In 1989, Harold Varmus (left) and Michael Bishop (right) of the University of California, San Francisco, shared the Nobel Prize in physiology and medicine for their discovery of genetic changes that can cause cancer. This interview reveals a passion for science that characterizes the best researchers and *the best teachers. Professors Bishop and Varmus teach medical students and graduate students at the University of California, San Francisco, in addition to directing research laboratories that study viruses and cancer genes. They are also active in the university's Science Education Project, which provides hands-on laboratory experience for elementary and secondary teachers in the San Francisco area.*

How did each of you first hear the news of your Nobel Prize?

VARMUS: Almost everybody in the United States who hears about the Nobel Prize hears about it in the same way. Someone calls you in the middle of the night, a reporter who's spreading the word and very eager to be the first person to give you a phone call.

BISHOP: It's a shame that the people who award the prize are hardly ever now the first to tell you. They hold a press conference at noon in Stockholm, which is 3 A.M. in California. It goes out on the wires immediately, and within ten minutes the people in New York are on the horn trying to reach the winners. So you almost always hear from the press. It's an interesting reflection of how much more seriously the press and the public take this prize than other prizes in science.

What were your initial feelings when you got those middle-of-the-night calls?

BISHOP: When the phone rang that early in the morning, despite some initial anxieties that it might be trouble in the family, it did cross my mind that, good grief, it is October (the time the prize is traditionally awarded). I can't admit to any immediate exhilaration. My first thought was, "I'm hardly awake and some reporter is going to ask me to explain this work and it's 3:15 in the morning." What a ridiculous circumstance to have to give out your first international sound bites. I didn't start to enjoy it until we had gotten rid of the first round of phone calls, which took an hour or so. It wasn't a total surprise because Harold and I knew that we had been in contention for some years. It's not something we ever discussed in all candor, but we knew.

VARMUS: People always ask how you learned about your prize, but what I find more interesting is how people learn about the Nobel Prize as an institution in the first place and what it means to them. In my own life that was actually fairly significant. I was about thirteen and I was looking around for topics for a public speaking contest and my father, who was a doctor, said, "You ought to learn something about this man John Enders, who recently won the Nobel Prize." I didn't know what the Nobel Prize was, so I was sent off to the library to read a few things about Enders. What I found interesting was that this was a man who didn't cure anything, but developed the technology for growing poliovirus, setting the stage for the polio vaccine. And I learned that the Nobel Prize awards contributions to basic science made without any necessary, explicit practical goal but that generate some fundamental piece of information that allows science to move forward.

How did the two of you come together?

VARMUS: I was shopping for a place to do postdoctoral work. I'd been at NIH working on a problem in bacterial genetics, and I wanted to move into the area of cancer and learn to work with tumor viruses. I learned that there was a group in San Francisco that was starting to work on tumor viruses, so I came by to see what was going on. I found that Mike had the same kind of approach I was hoping to find. The freshness of getting involved with people who were starting up in a field that was obviously ready for some excitement also was an inducement.

Is it difficult to switch from one research field to another?

BISHOP: It takes self-assurance. It takes something bordering on arrogance to do that, because typically the fresh insights seem naive to the experienced. If you read the Watson-Crick story, the people who had been working on the DNA problem were absolutely furious with Watson and Crick at their effrontery in bringing in these simpleminded ideas when everyone knew that this was a complex chemical problem that only a giant like Pauling could solve. Watson and Crick, playing around with little cardboard and tin models, nailed it. So it takes a lot of self-assurance, sometimes in a form that others perceive as arrogance, to make the jump. If you stop and look at what you are doing, you say to yourself, "I'm going to make a fool out of myself. I don't know what I'm doing." And yet, the history of science is full of examples of this.

Scientists take risks?

BISHOP: Risk taking is important. There are a variety of risks. The most fundamental one is that if you divert some of your energy and research resources to explore something unusual, you have to sustain your overall productivity at the same time. Then there is the fear of failure, the fear of making a fool of yourself in your own eyes. No one else would ever know you ran that experiment, but you'd look at yourself and say, "My goodness, how could I have been so stupid as to think that that would have worked!" So there's the practical risk and a psychological risk. But I think if you look at most of the great breakthroughs in the history of science, they involve what I call intellectual courage. You see it over and over again. People who were absolutely so self-assured that they had just bashed ahead when other people had backed off.

At the time you were doing the research that eventually led to the Nobel Prize, did you realize the great significance of the work?

VARMUS: I don't think either of us did. The question that we were working on had been in the air due to the combination of experimental observation and theory described in the late 1960s. The kinds of experimental questions we were asking were not unique. The methods were generally available; the specific applications varied from lab to lab. The way we went about answering the questions turned out to give us clear-cut answers in a way no one else had been able to obtain them. Of course there was a time when our findings were controversial and viewed with skepticism by many. There was a period when it wasn't clear how widespread the phenomenon we were describing actually was, and it took time before the full significance sank in. And it was not only our own work but the work of many others as well that led to the general acceptance of the conclusions we had drawn in our initial papers. So it was a long time before we saw not only that what we said was right but that it was important and prizeworthy. But what Mike and I find more important than the prize is the growth of this field, the basic ideas that come out of it, and the questions that drive us onward.

What was your key discovery that had such an important impact on cell biology?

BISHOP: In conceptual terms, what we did was to reveal that there are normal genes in the cell that have the capability to become cancer genes. Typically, they become cancer genes by being damaged; cancer arises from genetic damage in the cells of our bodies as we mature and age. The same genes are also found in certain viruses that can cause cancer in some animals. By studying one kind of virus with a peculiar growth cycle, and by asking where it got its cancer gene, we came upon the first example of explicit genes in normal cells that have the potential to become cancer genes if damaged. We found it in chickens first; in the long run it didn't matter. We had a glimpse of the idea in our minds; there is a suggestion of it in our first paper. But to be able to make the statement as decisively as I just did requires the intervening ten to fifteen years of data.

VARMUS: We knew from our medical school background that cancers were likely to be caused by mutations. We knew that cells were very complex with respect to their genetic constituents and that it was difficult to imagine what kind of mutations and what affected genes would lead to the development of a cancer in a human. We also knew that there were viruses in nature that could cause cancer and that viruses were a lot simpler than cells; they have very few genes. In the era before gene cloning, in the late sixties and early seventies, the only way you could lay your hand on one or a few genes explicitly involved in cancer might be to study viruses. So despite the likelihood that viruses were not major causes of cancer in humans, there was tremendous interest in developing experimental models for cancer using viruses. That idea, of course, was criticized by many who said "Isn't it ridiculous to try to understand cancer with viruses, when most human cancers probably are not virus-induced? Furthermore, isn't it ridiculous to look

at a chicken cancer virus; after all, chickens aren't even mammals. It's very unlikely that what you are looking at is relevant." But it turned out that the virus we were working with, Rous sarcoma virus from chickens, was the only virus that met certain criteria that would allow us to do a decisive experiment about the origin of the gene involved in inducing cancer. Furthermore, the gene that we ended up working with is extremely widespread in nature, and we could ultimately make statements about mammals, including humans, based on work done with the chicken virus.

How do we reconcile a genetic theory of cancer with the observation that environmental hazards can trigger cancer?

BISHOP: In the simplest formulation, environmental causes of cancer—cigarette smoke, chemicals in water, whatever—act by damaging genes. Our work laid hands on the first example of the kind of genes that might be damaged in the genesis of cancer.

You mentioned earlier that what's really important now is where this work goes from here. What are the important questions now?

BISHOP: We know there are multiple genetic changes involved in causing most, if not all, cancers. And although we can count five or six of these in some tumors (cancer of the human colon is the most well explored so far), in many other instances, we haven't a clue yet what all the changes are, and we don't know whether there is a particular sequence that's essential. And then we need to know how these genes work so we can understand why they give rise to the cancer cell, so that we can think about counteracting their actions in a therapeutic effort. I think those are two of the most important lines of inquiry that will probably flourish over the next decade.

VARMUS: I agree those are the two general areas in which information is lacking. We know from work with viruses that there are perhaps forty or fifty genes that can act during the development of a cancer by overriding the action of a normal copy of the gene. A major concern is the mechanism by which such genes exert their activity. What that means is figuring out exactly what stage in growth or development is influenced by the action of these genes.

BISHOP: It's probably worth emphasizing that what's at stake here is not just cancer. The whole farm is at stake. I mean, these genes run the part of the machinery that directs the everyday lives of ourselves from conception to death, and so we're wrestling with problems here that pertain to growth and development of the normal organism as well as the anomaly of cancer. By studying a disease state, we've been led to insights and whole new avenues of

study on probably the broadest and grandest problem in biology now: How does one cell become an organism (as ugly and glorious as *Homo sapiens* !)?

VARMUS: The genes that were discovered initially as being cancer-causing genes have been used to identify genes that play a role in virtually every step in growth control: Genes that make factors that are secreted and act as intercellular signals; genes that make the receptors for those signals that fly across the membrane and tell the inside of a cell what the outside of the cell is perceiving; genes that make proteins that translate the signals that are coming into the cell through the receptors; genes that we know have a role in cell division; and genes that function in development of the nervous system. Virtually every process that's of interest in cell biology now has been influenced by some of the genes that have been isolated initially as cancer-causing genes, but in their normal guise play explicit and obviously important roles in growth and development.

Can you envision possible applications of your discoveries in the diagnosis and treatment of cancer?

VARMUS: They've already been applied to a very limited extent. There are ways in which knowledge of the small set of genes commonly involved in cancer has affected diagnosis and prognosis. And in those several cases—certain forms of leukemia, certain forms of colon cancer, some childhood cancers—it's possible by asking a molecular question to make a more definitive diagnosis and to stage tumors in a way that's useful for effective prognosis and the design of therapy. I think that will ultimately influence clinical decisions in several areas of oncology.

You've written that progress in cancer research is an unexcelled vindication of fundamental research. Can you expand on this idea?

BISHOP: When I first came here, a colleague of mine was studying the Rous sarcoma virus that causes tumors in chickens. There was at the time not a shred of evidence that viruses caused cancer in humans. But just to look through a microscope and watch the transition the cells undergo when a chicken virus infects them was all it took to get me hooked on the problem. I wanted to understand why it happened. I didn't get out of bed in the morning thinking I was going to cure human cancer by studying this. All I said was, "This must be a profound change in a vertebrate cell that's worth understanding." We started to work on it because it was just so fundamentally engaging. It was fifteen years before I started to relate what we were doing to the problem of human cancer in anything I wrote or said in public.

You've both had medical educations. What influenced you to become research scientists instead of practicing physicians?

VARMUS: I was academically inclined. I went to medical school thinking I was going to end up writing psychoanalytic interpretations of literature. I slid very slowly into research. It was during the Vietnam era and some sort of government service was required. I ended up at NIH, where I was given my first taste of real research. I had dabbled in the past, but not to any serious degree. I sometimes date my addiction to a day in maybe the third or fourth month I was at NIH. I'd been trying to get a simple assay to work so that I could measure levels of a certain RNA in a bacterial cell. I'd been trying a number of things that gave me very soft results, when suddenly I had a new assay. I remember standing at the scintillation counter, with a few counts in the background and thousands of counts in the positive part of the experiment. I still remember the charge I felt. I didn't have any answers at that point. I just had the tool and I knew the next day the experiments were going to work. It was that thrill that made me feel like . . . I was hooked now. I still had responsibilities for patients and I was just praying that nobody was going to call me from the lab and interrupt a day of joy in the immersion into your own experiments that occurs once you have something powerful to answer questions with. It is really pretty exhilarating. It's as addictive as any drug.

BISHOP: I evolved a little more gradually. I went to medical school without a single notion of what I wanted to do, except I was pretty sure that I didn't want to practice medicine. I became interested in a succession of subjects; neurobiology was the earliest. The most important influence was my peers. The faculty could have been on the moon. My peers were a remarkable group of people—articulate, informed, with diverse interests. These people had a respect and knowledge about research that I had not had. I befriended several of them.

I was frustrated and so confused that I took a year off after my second year of medical school with a fellowship to work in a pathology department. I did routine pathology and essentially was fully prepared for the rest of medical school. I dabbled in the research lab. I had read about molecular biology, but I hadn't a prayer of getting into a mainstream molecular biology lab given my background. So in the third year I went looking for a subject that was ready for molecular biology but was untrammeled. And I found animal viruses. I took an elective course on viruses that cause disease in humans and other vertebrates. It was the first time in my life I had ever been required to read the literature and present it, and I was in absolute ecstasy.

Then I hooked up with one of the young instructors, who took me into his lab. I showed up there one Saturday at 9:00 A.M. He was going to teach me to do the assay for the virus that would allow me to do the experiments that I had decided I wanted to do. It was things I'd cooked up myself—just completely off the wall. We walked out of that place at midnight; between 9 A.M. and midnight he and I had worked together doing this assay. And when I got home, I woke my wife up and said, I've found it.

It was never the same after that, even though I had to go through two years of clinical training to work my way into the establishment to find research training. I knew from that time on what I wanted to do. I didn't know whether I would be able to do it, whether I'd make it, but I knew what I wanted to do.

How do you view the current state of medical education?

BISHOP: There are problems. We don't teach science very well to medical students. It's too didactic. The answer seems to lie somewhere in flexibility and in faculty commitment and involvement, and in getting away from didacticism. But, it's not easy to do. Medical school students are very diverse in their preparation. And most are not engaged by science. They see medicine as some other sort of career. We don't usually succeed in engaging them. We'd like to.

VARMUS: A major problem in medical education is the tremendous strain of competing interests. You have an essential core of curriculum; there must be one because students have to come out of medical school knowing about an essential list of diseases, knowing a certain amount of anatomy, fundamental metabolic pathways, and a certain number of drugs. They simply can't be accredited to practice medicine, which is the primary goal of medical education, without exposure to a certain vocabulary and a certain body of knowledge. The demands to teach that body of knowledge are always competing with the specialized interests.

And, of course, a medical school has a huge faculty. Each of us has special interests that we believe are extremely important. I certainly think it would be a sin for a student to leave medical school without at least some exposure to the idea of oncogenes and some notion of how retroviruses grow—after all they're the cause of AIDS. So we all want our hours of time to teach students.

And then you have the basic issue about what are the best ways to learn: from lectures, from books, from individual projects. And you have the students' demand for getting contact with patients early on. I take a somewhat conservative position here, which leads me to believe that you are perfectly well off delaying that all to the third year of medical school. But the students don't like that. They want to lay hands on early. They also want time off. All these factors make it very difficult to teach not only the fundamentals that are required to be a practitioner of medicine, but also some of the excitement in biology that's applicable to medicine now.

What do you find so alluring about science?

VARMUS: People sometimes say to me, "God, you work so damn hard. I respect what you do so much because you're reading all night and you work these incredibly long hours." I feel just the opposite. The kind of attitude that I think has to exist among at least a significant number of scientists is that you are playing—you're asking questions that are interesting and solvable, regardless of what might be their ultimate effect on the marketplace. And there are always new questions. I'm often struck, looking back, at how uninterested I am now in the questions that obsessed me ten or fifteen years ago, because now we have answers to a lot of those questions. What I care about now are the questions that have come out of those answers and the privilege of going out and answering the new questions. There is always the excitement of new questions and people to answer the questions with us.

7 A Tour of the Cell

How Cells Are Studied
The Geography of the Cell: A Panoramic View
The Nucleus
Ribosomes
The Endomembrane System
Peroxisomes (Microbodies)
Energy Transducers: Mitochondria and Chloroplasts
The Cytoskeleton
The Cell Surface
The Cell: A Living Unit Greater Than the Sum of Its Parts

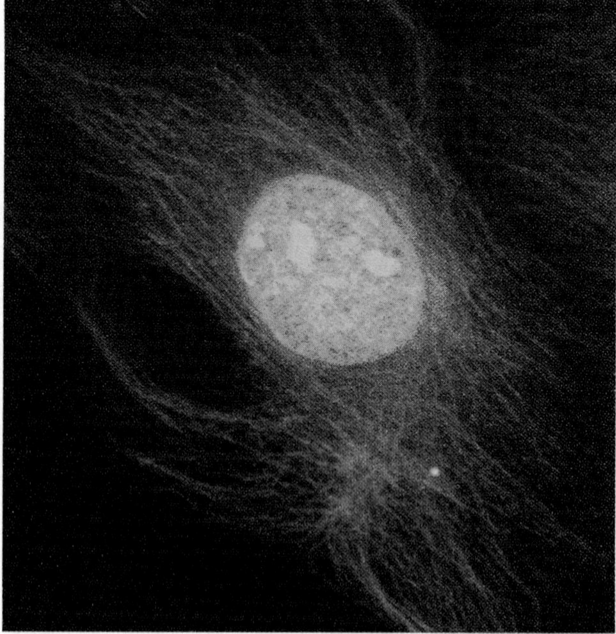

Figure 7.1
Ordered functions are based on ordered structures. This animal cell, called a fibroblast, can change its shape and move along a substratum. Cell motility depends on precise interactions of numerous structures within the cell, including the cellular skeleton surrounding the nucleus in this micrograph (LM). Throughout our study of the cell, we will correlate cellular functions with structural organization.

The cell is as fundamental to biology as the atom is to chemistry: All organisms are made of cells. In the hierarchy of biological organization, the cell is the simplest collection of matter that can live. Indeed, there are diverse forms of life existing as single-celled organisms. More complex organisms, including plants and animals, are multicellular; their bodies are cooperatives of many kinds of specialized cells that could not survive for long on their own. However, even when they are arranged into higher levels of organization, such as tissues and organs, cells can be singled out as the organism's basic units of structure and function. The contraction of muscle cells moves your eyes as you read this sentence; when you decide to turn this page, nerve cells will transmit that decision from your brain to the muscle cells of your hand. Everything an organism does is ultimately happening at the cellular level. This chapter introduces the microscopic world of the cell by surveying its overall geography and describing its various functional components, or organelles.

This text takes a thematic approach to the study of life, and the cell is a microcosmic model of many of the themes introduced in Chapter 1. We will see that life at the cellular level arises from structural order, reinforcing the theme of emergent properties. For example, the movement of an animal cell depends on an intricate interplay of tubules and filaments within the cell (Figure 7.1). A related theme is the correlation of structure and function. Because every ordered process is based on an ordered structure, analyzing the anatomy of the cell rewards us with clues about how the cell works. Another recurring theme in biology is the interaction of organisms with their environment. Cells are excitable units that sense and respond to environmental fluctuations. As open systems, they continuously exchange both materials and energy with their surroundings. And keep in mind the one biological theme that unifies all others: evolution. Though all cells are related to some extent by their descent from earlier cells, they have been modified in various ways during the long evolutionary history of life on Earth. For example, if one unicellular organism lives in fresh water and another inhabits the sea, we can expect these cells to be somewhat differently equipped as a result of their divergent adaptations to disparate environments. Evolution is the basis for the correlations of structure and function we observe in cells.

Perhaps the greatest obstacle to becoming acquainted with the cell is imagining how something too small to be seen by the unaided eye can be so complex. How can cell biologists possibly dissect so small a package to investigate its inner workings? Before we actually tour the cell, it will be helpful to learn how cells are studied.

HOW CELLS ARE STUDIED

The evolution of a science often parallels the invention of instruments that extend human senses to new limits. The discovery and early study of cells progressed with the invention and improvement of microscopes in the seventeenth century. Microscopes of various types are still indispensable tools for the study of cells.

Microscopy

The microscopes first used by Renaissance scientists, as well as the microscopes you are likely to use in the laboratory, are all **light microscopes (LMs).** Visible light is passed through the specimen and then through glass lenses. The lenses refract (bend) the light in such a way that the image of the specimen is magnified as it is projected into the eye.

Two important factors in microscopy are magnification and resolving power, or resolution. **Magnification** is how much larger the object appears compared to its real size. **Resolving power** is a measure of the clarity of the image, or the minimum distance two points can be separated and still be distinguished as two separate points. For example, what appears to the unaided eye as one star in the sky may be resolved as twin stars with the help of a telescope.

Just as the resolving power of the human eye is limited, the resolving power of telescopes and microscopes is limited. Microscopes can be designed to magnify objects as much as desired, but the light microscope can never resolve detail finer than about 0.2 µm, the size of a small bacterium or mitochondrion (Figure 7.2). This resolution cannot be improved upon; it is limited by the wavelength of the visible light used to illuminate the specimen. Light microscopes can magnify effectively to about 1500 times the size of the actual specimen; greater magnifications increase blurriness. Most of the improvements in light microscopy since the beginning of this century have involved new methods for enhancing contrast, a quality that makes the details that *can* be resolved stand out better to the eye (Table 7.1).

Although cells were discovered by Robert Hooke in 1665, the geography of the cell was largely uncharted until the past few decades. Most subcellular structures,

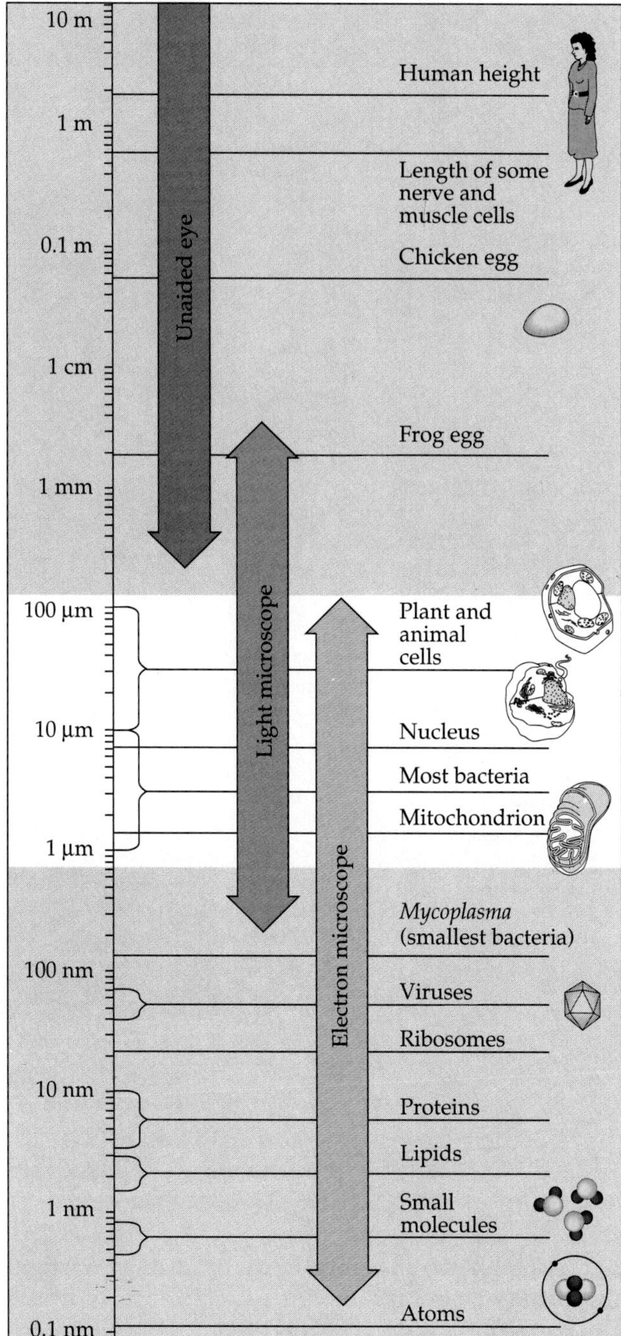

MEASUREMENTS
1 centimeter (cm) = 10^{-2} meter = 0.4 inch
1 millimeter (mm) = 10^{-3} meter
1 micrometer (µm) = 10^{-3} mm
1 nanometer (nm) = 10^{-3} µm

Figure 7.2
The size range of cells. Most cells are between 1 and 100 µm in diameter and are therefore visible only under a microscope. Notice that the scale is logarithmic to accommodate the range of sizes shown. Starting at the top of the scale with 10 meters and going down, each reference measurement along the left side marks a *tenfold* decrease in size.

Table 7.1 Different Types of Light Microscopy: A Comparison

Type of Microscopy	Light Micrographs of Human Cheek Epithelial Cells		Type of Microscopy
Brightfield (unstained specimen): Passes light directly through specimen; unless cell is naturally pigmented or artificially stained, image has little contrast.			**Phase-contrast:** Enhances contrast in unstained cells by amplifying variations in density within specimen; especially useful for examining living, unpigmented cells.
Brightfield (stained specimen): Staining with various dyes enhances contrast, but most staining procedures require that cells be fixed (preserved).			**Differential-interference-contrast (Nomarski):** Also uses optical modifications to exaggerate differences in density.
Darkfield: Passes light through specimen obliquely, and only light scattered by particles can be seen.	50 µm	50 µm	**Confocal:** Uses lasers and special optics for "optical sectioning." Only those regions within a narrow depth of focus are imaged. Regions above and below the selected plane of view appear black rather than blurry.

or **organelles,** are too small to be resolved by the light microscope. Cell biology advanced rapidly in the 1950s, with the introduction of the electron microscope. Instead of using visible light, the **electron microscope (EM)** focuses a beam of electrons through the specimen (Figure 7.3). Resolving power is inversely related to the wavelength of radiation a microscope uses, and electron beams have wavelengths much shorter than the wavelengths of visible light. Modern electron microscopes can achieve a resolution of about 0.2 nanometer (nm), a thousandfold improvement over the light microscope. Biologists use the term **cell ultrastructure** to refer to a cell's anatomy as resolved by an electron microscope.

There are two types of electron microscopes: the **transmission electron microscope (TEM)** and the **scanning electron microscope (SEM).** The TEM aims an electron beam through a thin section of the specimen, similar to the way the light microscope transmits light through a slide. However, instead of using glass lenses, which are opaque to electrons, the TEM uses electromagnets as lenses to focus and magnify the image by bending the trajectories of the charged electrons. The image is ultimately focused onto a screen for viewing or onto photographic film for a permanent record. To enhance contrast in the image, very thin sections of preserved cells are stained with heavy atoms of metals, which attach to certain places in the cells. Cell biologists use the TEM mainly to study the internal ultrastructure of cells (Figure 7.4a).

The SEM is especially useful for detailed study of the surface of the specimen (Figure 7.4b). The electron beam scans the surface of the sample, which is usually coated with a thin film of gold. The beam excites elec-

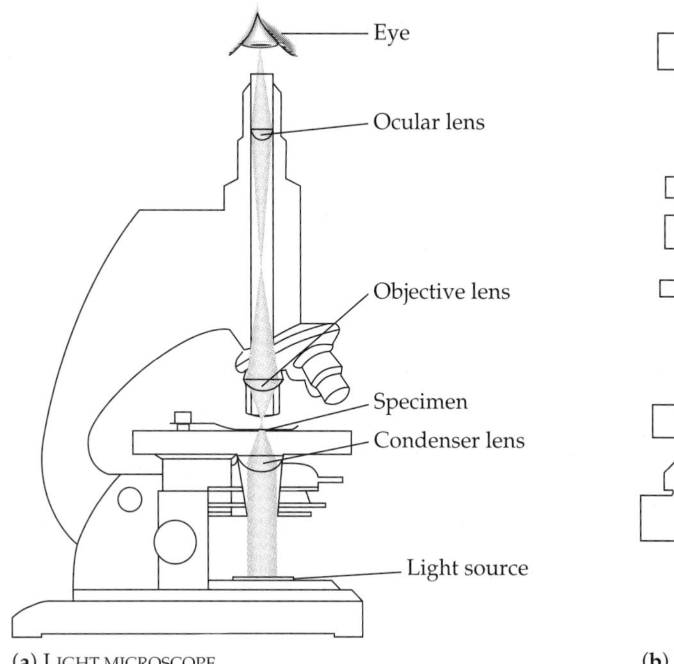

(a) LIGHT MICROSCOPE

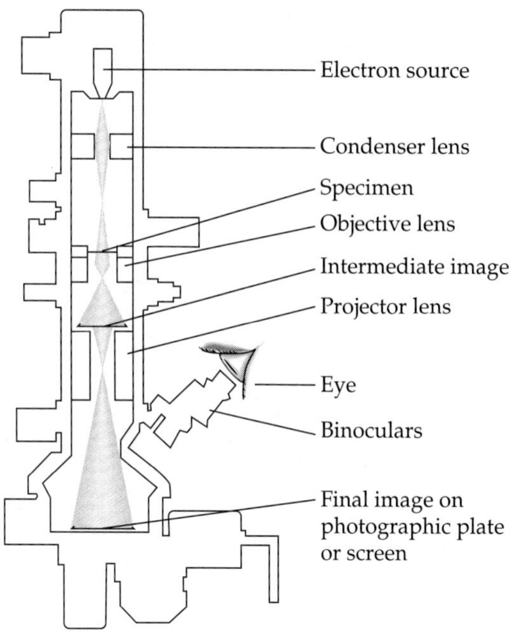

(b) ELECTRON MICROSCOPE

Figure 7.3
The light microscope and transmission electron microscope compared. (**a**) In light microscopy, light is focused on a specimen by a glass condenser lens; the image is then magnified by an objective lens and an ocular lens, for projection on the eye or photographic film. (**b**) In electron microscopy, a beam of electrons (top of the microscope) is used instead of light, and electromagnets instead of glass lenses. The electron beam is focused on the specimen by a condenser lens; the image is magnified by an objective lens and a projector lens, for projection on a screen or photographic film.

trons on the sample surface itself, and these secondary electrons are collected and focused onto a screen, forming an image showing the topography of the specimen. An important attribute of the SEM is its great depth of field, which results in an image that appears three-dimensional.

Electron microscopes reveal many organelles that are impossible to resolve with the light microscope. But the light microscope offers many advantages, especially for the study of live cells. A disadvantage of electron microscopy is that the chemical and physical methods used to prepare the specimen not only kill cells, but also may introduce artifacts, structural features seen in micrographs that do not exist in the living cell.

Microscopes of various kinds are the most important tools of **cytology,** the study of cell structure. But simply describing the diverse organelles within the cell reveals little about their function. Modern cell biology developed from an integration of cytology with biochemistry, the study of metabolism and its products. A biochemical approach called cell fractionation has been particularly important in this multidisciplinary synthesis of cell biology.

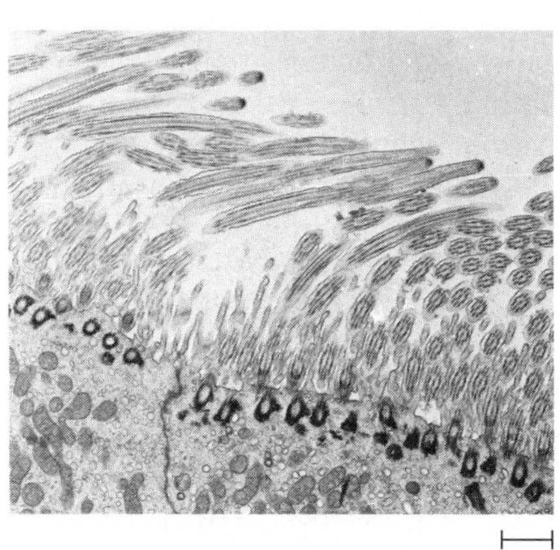

(a) TEM

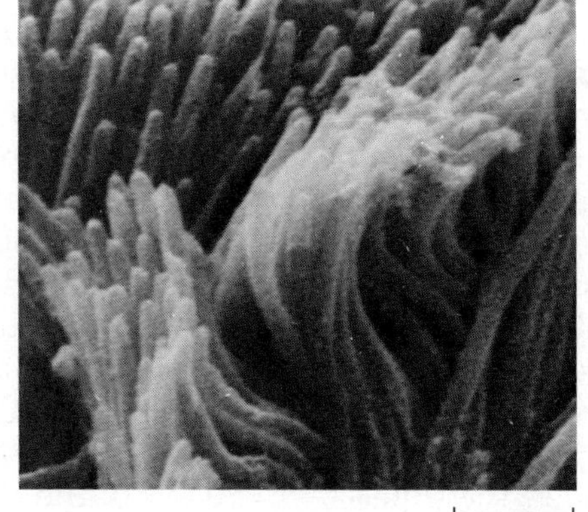

(b) SEM

Figure 7.4
Electron micrographs: photographs taken with electron microscopes.
(**a**) This micrograph, taken with a transmission electron microscope (TEM), profiles a thin section of a cell from a rabbit trachea (windpipe), revealing its ultrastructure.
(**b**) The scanning electron microscope (SEM) produces a three-dimensional image of the surface of the same type of cell. Both micrographs show motile organelles called cilia. Beating of the cilia that line the windpipe helps move inhaled debris upward back toward the pharynx (throat).

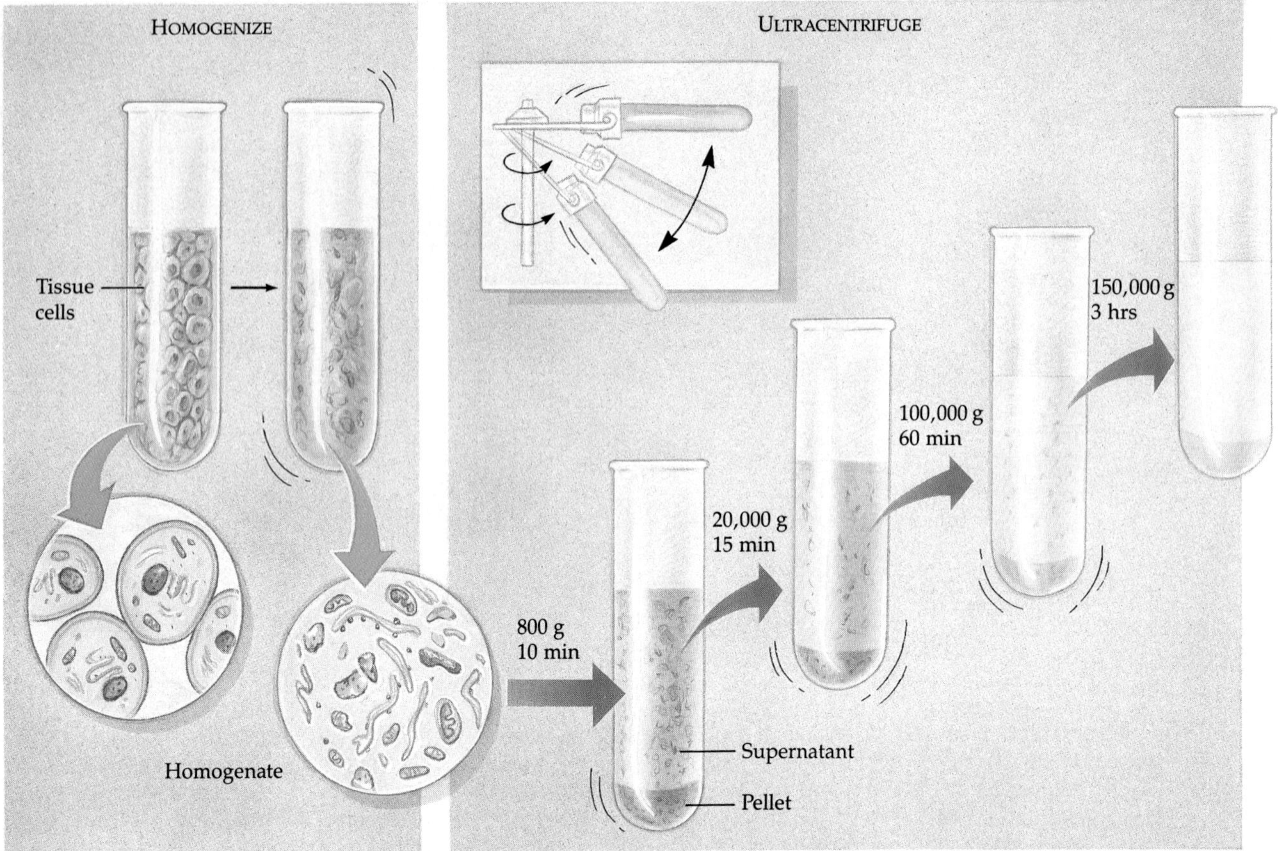

Figure 7.5
Cell fractionation. Disrupted cells are centrifuged at various speeds and durations to isolate components of different sizes, densities, and shapes. The process begins with homogenization, the disruption of a tissue and its cells with the help of such instruments as kitchen blenders or ultrasound devices. The homogenate, a soupy mixture of organelles, bits of membrane, and molecules from the broken cells, is then fractionated by a series of spins in a centrifuge. A slow spin for a short time will cause only the largest components of the homogenate to settle to the bottom of the centrifuge tube as a pellet, which can be resuspended for study. The unpelleted portion, or supernatant, is then decanted and centrifuged again, at a higher speed. The process is repeated, increasing the speed and/or duration of each spin until very small organelles, or even large molecules, are collected in pellets. By determining which cell fractions are associated with particular metabolic processes, those functions can be tied to certain organelles.

Cell Fractionation

The objective of **cell fractionation** is to take cells apart, separating the major organelles so that their individual functions can be studied (Figure 7.5). The instrument used to fractionate cells is the **centrifuge,** a merry-go-round for test tubes capable of spinning at various speeds. The most powerful machines, called **ultracentrifuges,** can spin as fast as 80,000 revolutions per minute (rpm) and apply forces on particles of up to 500,000 times the force of gravity (500,000 g).

Fractionation begins with homogenization, the disruption of cells. The usual objective is to break the cells without severely damaging their organelles. The parts of the soupy homogenate are then separated by spinning in a centrifuge, resulting in a pellet. The supernatant (unsedimented) fraction is decanted into another tube and centrifuged again. The process is repeated, increasing the speed with each step, collecting smaller and smaller components of the homogenized cells (see Figure 7.5).

Cell fractionation enables the researcher to prepare specific components of cells in bulk quantity in order to study their composition and metabolism. By following this approach, biologists have been able to assign various functions of the cell to the different organelles, a task that would be far more difficult if they had to study intact cells. For example, one cellular fraction collected by centrifugation has enzymes that function in the metabolic process known as cellular respiration (see Chapter 6). The prevalent organelles in that fraction match the structures called mitochondria, as visualized in the electron microscope. This discovery helped cell biologists determine that mitochondria are

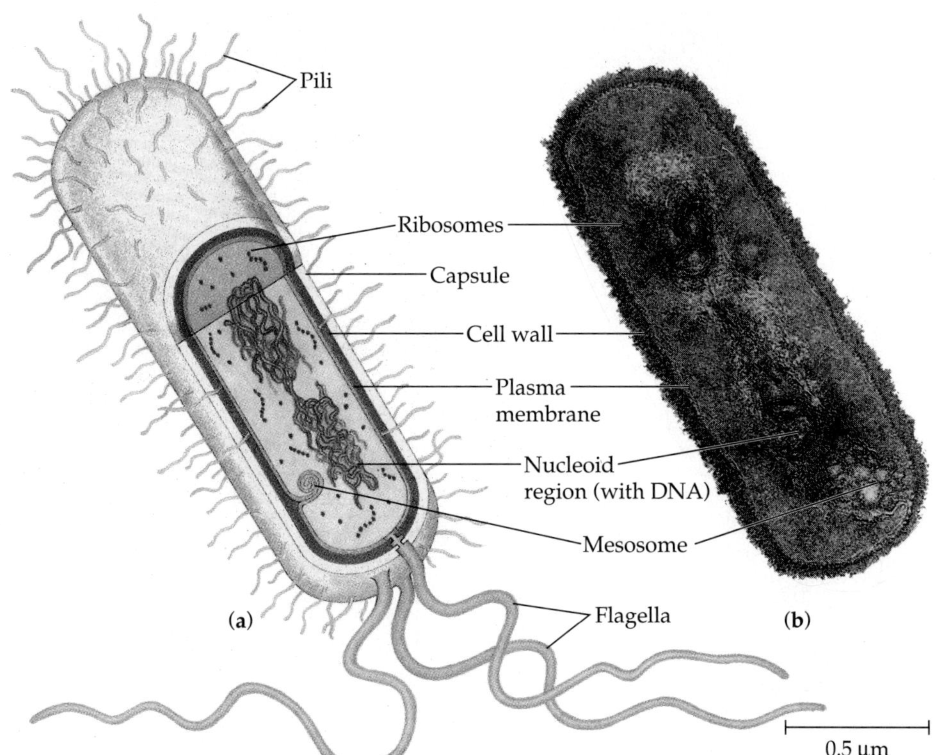

Figure 7.6
The prokaryotic cell. Prokaryotes include the bacteria and the cyanobacteria. (**a**) The diagram depicts a typical rod-shaped bacterium. Lacking the membrane-enclosed organelles of a eukaryote, the prokaryote is much simpler in structure. The DNA is in the nucleoid, and no membrane separates the DNA from the rest of the cell. A prokaryote has a large number of ribosomes, where proteins are synthesized. The border of the cell is the plasma membrane, which in some prokaryotes folds in to form structures called mesosomes. Outside the plasma membrane are a fairly rigid cell wall and often an outer capsule, usually jellylike. Some bacteria have flagella (locomotion organelles), pili (attachment structures), or both projecting from their surface. (**b**) The electron micrograph shows a thin section through the bacterium *Bacillus coagulans* (TEM).

the sites of cellular respiration. And that deduction stimulated further investigation of mitochondrial architecture, which, in turn, helped clarify how cellular respiration works. Cytology and biochemistry complement each other in the correlation of cellular structure and function.

THE GEOGRAPHY OF THE CELL: A PANORAMIC VIEW

Prokaryotic and Eukaryotic Cells

Every contemporary organism is composed of one of two structurally different types of cells: prokaryotic cells or eukaryotic cells. Prokaryotic cells are found only in the kingdom Monera, which consists of bacteria and cyanobacteria (formerly called blue-green algae). Protists, plants, fungi, and animals—four of the five kingdoms of life—are all eukaryotes (see Chapter 1). The line separating prokaryotes from eukaryotes is the sharpest division in the diversity of life.

The two types of cells differ markedly in their internal organization. One difference is denoted by their names. The **prokaryotic cell** has no true nucleus (Figure 7.6). (The word *prokaryote* is from the Greek *pro,* "before," and *karyon,* "kernel," referring here to the nucleus.) Its genetic material (DNA) is concentrated instead in a region called the nucleoid, but no membrane separates this region from the rest of the cell. In contrast, the **eukaryotic cell** (Gr. *eu,* "true," and *karyon*) has a true nucleus enclosed by a membranous nuclear envelope. The entire region between the nucleus and the membrane bounding the cell is called the **cytoplasm.** It consists of a semifluid medium called the **cytosol,** in which are suspended organelles of specialized form and function, most of them absent in prokaryotic cells. Thus, the presence or absence of a true nucleus is just one example of the disparity in structural complexity between the two types of cells. The prokaryotic cell will be described in detail in Chapters 17 and 25, and the possible evolutionary relationships between the two types of cells will be discussed in Chapter 26. Most of the discussion of cell structure that follows in this chapter applies to eukaryotes.

Cell Size

Size is a general feature of cell structure that relates to function. The logistics of carrying out metabolism sets limits on the size range of cells (see Figure 7.2). The smallest cells known are bacteria called mycoplasmas, which have diameters of between 0.1 and 1.0 micrometer (μm). Perhaps these are the smallest packages

Figure 7.7
Why are most cells microscopic? In this comparison diagram, cells are represented as boxes. (**a**) In arbitrary units of length, this small cell measures 1 on each side. We can calculate the cell's surface area (in square units), volume (in cubic units), and surface-area-to-volume ratio. (**b**) As the size of the cell increases to 5 units of length per side, the ratio of surface area to volume decreases compared to the smaller cell. The number of chemical exchanges that could be performed with the extracellular environment in a cell of this size would be inadequate to maintain the cell, because most of its cytoplasm is relatively far from the outer membrane. (**c**) By dividing the large cell into many smaller cells, we can restore a surface-area-to-volume ratio that can serve each cell's need for acquiring nutrients and expelling waste products. These geometric relationships explain why most cells are microscopic, and why larger organisms do not generally have *larger* cells than smaller organisms, but *more* cells.

Surface area increases while total volume remains constant
1
5
1

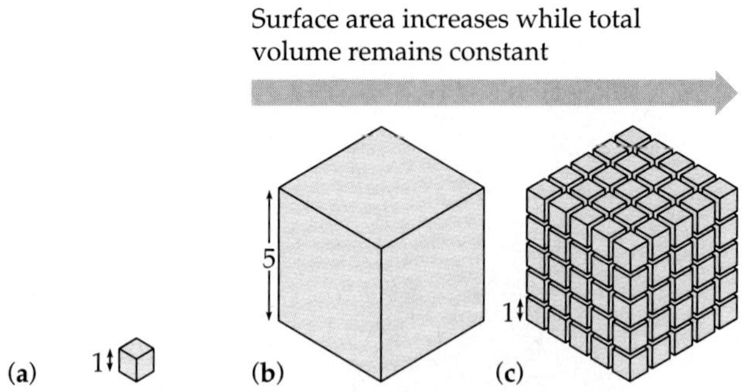

	(a)	(b)	(c)
Total surface area (height × width × number of sides × number of boxes)	6	150	750
Total volume (height × width × length × number of boxes)	1	125	125
Surface-area-to-volume ratio (area ÷ volume)	6	1.2	6

with enough DNA to program metabolism, and enough enzymes and other cellular equipment to carry out the activities necessary for a cell to sustain itself and reproduce. Most bacteria are 1 to 10 μm in diameter, about ten times larger than mycoplasmas. Eukaryotic cells are typically 10 to 100 μm in diameter, ten times larger than bacteria.

Metabolic requirements also impose upper limits on the size that is practical for a single cell. As an object of a particular shape increases in size, its volume grows proportionately more than its surface area. (Area is proportional to a linear dimension squared, whereas volume is proportional to the linear dimension cubed.) For objects of the same shape, the smaller the object, the greater its ratio of surface area to volume (Figure 7.7).

At the boundary of every cell, the **plasma membrane** functions as a selective barrier that regulates the cell's chemical composition. This membrane allows sufficient passage of oxygen, nutrients, and wastes to service the entire volume of the cell. For each square micrometer of membrane, only so much of a particular substance can cross per second. The need for a surface sufficiently large to accommodate its volume helps to explain the microscopic size of most cells. Another factor limiting cell size is the requirement for a single nucleus to control the entire cytoplasmic volume of the cell. The nucleus can better control a smaller cell, just as a single administrator can manage a smaller factory more efficiently than a larger one.

The Importance of Compartmental Organization

The internal complexity of a eukaryotic cell is another general structural feature correlated with function. The average eukaryotic cell, with a diameter about ten times greater than the average prokaryotic cell, has a thousand times the volume but only a hundred times the surface area of the prokaryotic cell. The area of the plasma membrane alone may be inadequate to satisfy the metabolic needs of the eukaryotic cell. It compensates for this small ratio of plasma membrane area to cytoplasmic volume by having internal membranes. These membranes serve as partitions, dividing the cell into compartments, and also participate directly in much of the cell's metabolism; many enzymes are built right into the membranes. Because the compartments of the eukaryotic cell provide different local environments to facilitate specific metabolic functions, processes that are incompatible can go on simultaneously in separate subcellular compartments.

Clearly, membranes of various kinds are fundamental to the complex organization of the cell. In general, biological membranes consist of a double layer of phospholipids and other lipids in which diverse proteins are embedded (Figure 7.8; see also Chapter 5). However, each membrane, whether the plasma membrane or the membrane of a cytoplasmic organelle, has a unique composition of lipids and proteins suited to

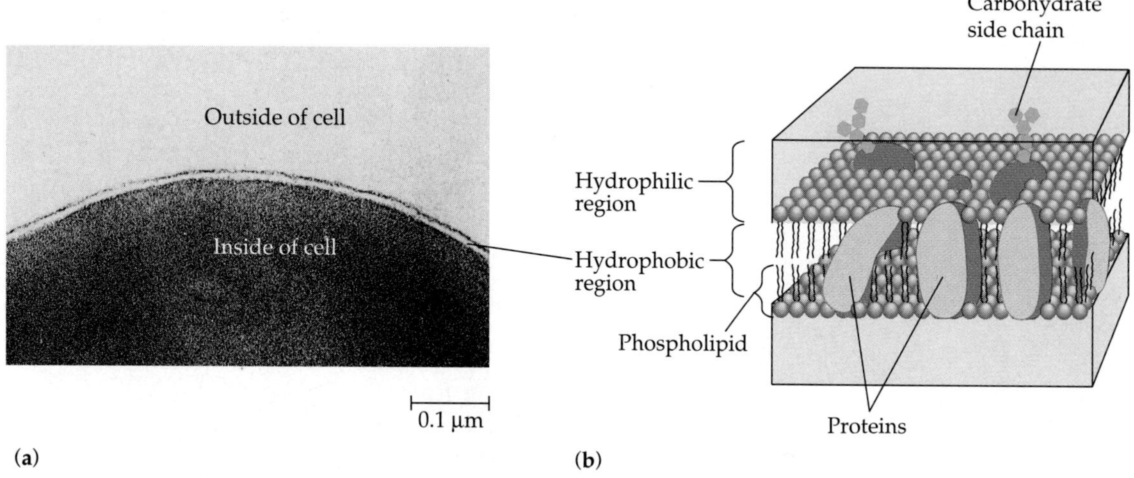

Figure 7.8
The plasma membrane. (**a**) In electron micrographs of sufficient magnification, the plasma membrane appears as a pair of dark bands separated by a light band. This is the membrane of a red blood cell (TEM). (**b**) The plasma membrane and the various internal membranes of cells consist of proteins of diverse functions embedded in a double layer (bilayer) of phospholipids. The specific functions of the plasma membrane and the various types of membranes within the cell depend on the kinds of phospholipids and proteins present. The plasma membrane also has carbohydrates attached to its outer surface.

that membrane's specific functions. For example, many of the enzymes that function in cellular respiration are embedded in the internal membranes of organelles called mitochondria. To a large extent, then, the study of cells is the study of membranes and the functional compartments (organelles) that are sequestered by these diverse membranes. Membranes are so important to how a cell works that they are the subject of Chapter 8.

Before continuing with this chapter, examine the overviews of eukaryotic cells in Figures 7.9 and 7.10. These figures and their legends introduce the various organelles and provide a map of the cell for the detailed tour upon which we will soon embark. Figures 7.9 and 7.10 also contrast animal and plant cells. As eukaryotic cells, they have much more in common with each other than either has with *any* prokaryote. There are, however, important differences between plant and animal cells. For example, a plant cell has a relatively thick cell wall external to its outer membrane, whereas an animal cell lacks walls.

The first stop on our detailed tour of the cell is one of the membrane-enclosed organelles, the nucleus.

THE NUCLEUS

The **nucleus** contains most of the genes that control the cell (some genes are located in mitochondria and chloroplasts). It is generally the most conspicuous organelle in a eukaryotic cell, averaging about 5 μm in diameter (Figure 7.11a and b on p. 126). The **nuclear envelope** encloses the nucleus, separating its contents from the cytoplasm.

The nuclear envelope is a double membrane. The two membranes are separated by a space of about 20 to 40 nm. Closely associated with the nuclear side of the inner membrane is a layer of protein that helps maintain the shape of the nucleus and may also help maintain the organization of the genetic material. The nuclear envelope is perforated by pores that are about 100 nm in diameter (Figure 7.11c and d). At the lip of each pore, the inner and outer membranes of the nuclear envelope are fused. The pore complex somehow regulates the entrance and exit of certain large macromolecules and particles.

Within the nucleus, the DNA is organized along with proteins into **chromosomes.** Unless the cell is in the process of dividing, its chromosomes are too stringy and entangled to be identified as individual structures. Instead, the aggregate of chromosomes appears through both light microscopes and electron microscopes as a mass of stained material collectively referred to as **chromatin.** Only when the nucleus prepares to divide do the chromosomes condense, becoming thick enough to be discerned as separate structures. Each eukaryotic species has a characteristic number of chromosomes. A human cell, for example, has 46 chromosomes in its nucleus; the exceptions are the sex cells—eggs and sperm—which have only 23 chromosomes in humans.

The most visible structure within the nondividing nucleus is the **nucleolus,** which functions in the syn-

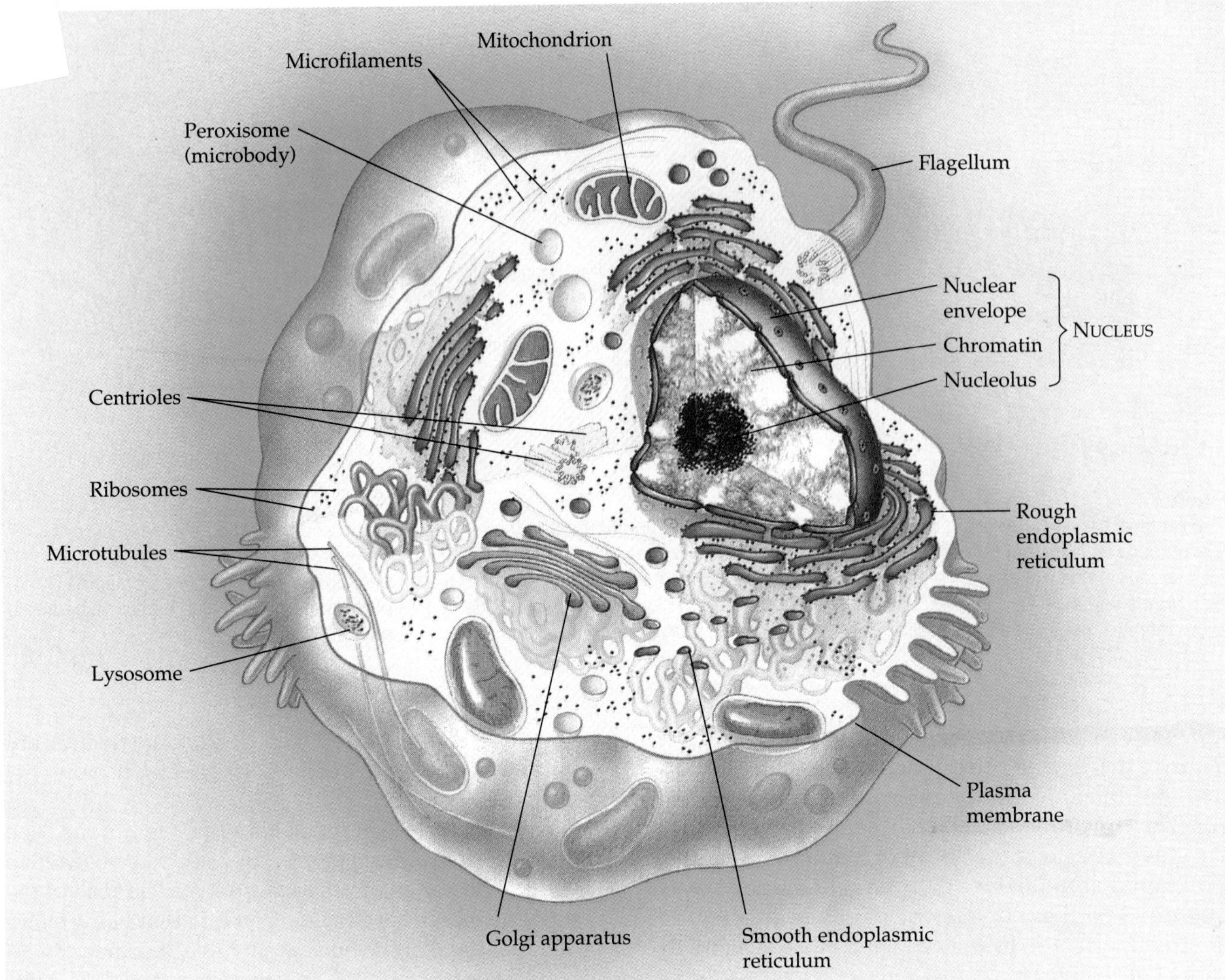

Figure 7.9
An animal cell. This schematic drawing of a generalized animal cell combines the most common structural features found in animal cells. No one cell looks like this. Within the cell are a variety of components collectively called organelles ("little organs"). The most prominent organelle in an animal cell is usually the nucleus, in which inherited genes reside in the form of DNA. The DNA is organized along with proteins into structures called chromosomes; in the nondividing cell, they are not seen as individual structures, but as a diffuse material called chromatin. Also present in the nucleus are one or more nucleoli (singular, nucleolus). Nucleoli are involved in the assembly of particles called ribosomes, which function in the synthesis of proteins. The nucleus is bordered by an envelope consisting of two membranes.

Most of the cell's metabolic activities occur in the cytoplasm, the entire region between the nucleus and the plasma membrane surrounding the cell. The cytoplasm is full of specialized organelles suspended in a semifluid medium called the cytosol. Pervading much of the cytoplasm is the endoplasmic reticulum (ER), a labyrinth of membranes forming flattened sacs and tubes that segregate the contents of the ER from the cytosol. The ER takes two forms: rough (studded with ribosomes) and smooth. Many types of proteins are made by ribosomes attached to ER membranes, and the ER also plays a major role in assembling the other membranes of the cell. The Golgi apparatus, another type of membranous organelle in the cytoplasm, consists of stacks of flattened sacs that play an active role in the synthesis, refinement, storage, sorting, and secretion of chemical products by the cell.

Other classes of organelles enclosed by membranes are: lysosomes, which contain mixtures of digestive enzymes that hydrolyze macromolecules; peroxisomes, a diverse group of organelles containing specialized enzymes for performing specific metabolic processes; and vacuoles, which have a variety of storage and metabolic functions. The mitochondria (singular, mitochondrion) are organelles that generate ATP from such organic fuels as sugar in the process of cellular respiration.

Nonmembranous structures within the cells include microtubules and microfilaments, forming a framework called the cytoskeleton, which reinforces the cell's shape and functions in cell movement. The cell in the drawing has a flagellum, an organelle of locomotion, which is an assembly of microtubules. Also made of microtubules are centrioles, located near the nucleus. These play an important role in cell division and in the formation of flagella.

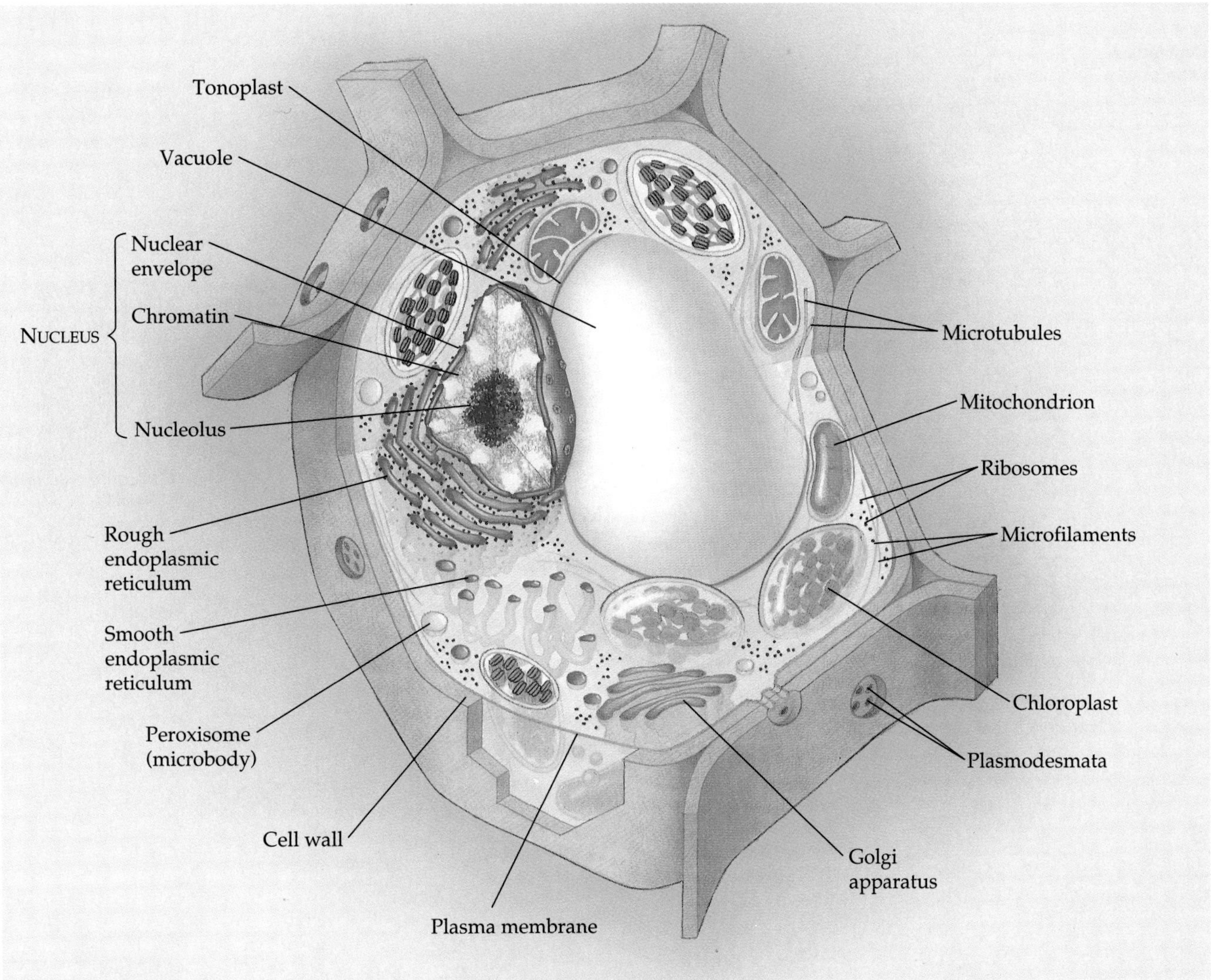

Figure 7.10
A plant cell. This schematic drawing of a generalized plant cell reveals the similarities and differences between an animal and a plant cell. Like the animal cell, the plant cell is surrounded by a plasma membrane and contains a nucleus, ribosomes, ER, Golgi apparatus, mitochondria, peroxisomes, and microfilaments and microtubules. However, a plant cell also contains organelles called plastids. One important type of plastid is the chloroplast, which carries out photosynthesis, converting sunlight to chemical energy stored in sugar and other organic molecules. Another prominent organelle in many plant cells, especially older ones, is a large central vacuole. It stores chemicals, breaks down macromolecules, and, by enlarging, plays a major role in plant growth. The membrane of the vacuole is called the tonoplast. Outside a plant cell's plasma membrane (as well as that in fungi and some protists) is a thick cell wall, which helps maintain the cell's shape and protects the cell from mechanical damage. The cytosol of adjacent cells connects through trans-wall channels called plasmodesmata.

thesis of ribosomes. Sometimes, there are two or more nucleoli; the number depends on the species and the stage in the cell's reproductive cycle. The nucleolus is roughly spherical, and in the electron microscope it appears as a mass of densely stained granules and fibers. It consists of **nucleolar organizers,** specialized regions of some chromosomes with multiple copies of genes for ribosome synthesis, along with a considerable amount of RNA and proteins representing ribosomes in various stages of production. An actively growing cell can produce about 10,000 ribosomes per minute.

The nucleus controls protein synthesis in the cytoplasm by sending molecular messengers in the form of RNA. This **messenger RNA (mRNA),** as it is called, is synthesized in the nucleus according to instructions provided by the DNA, after which the mRNA conveys the genetic messages to the cytoplasm via the nuclear pores. Once in the cytoplasm, the mRNA attaches to ri-

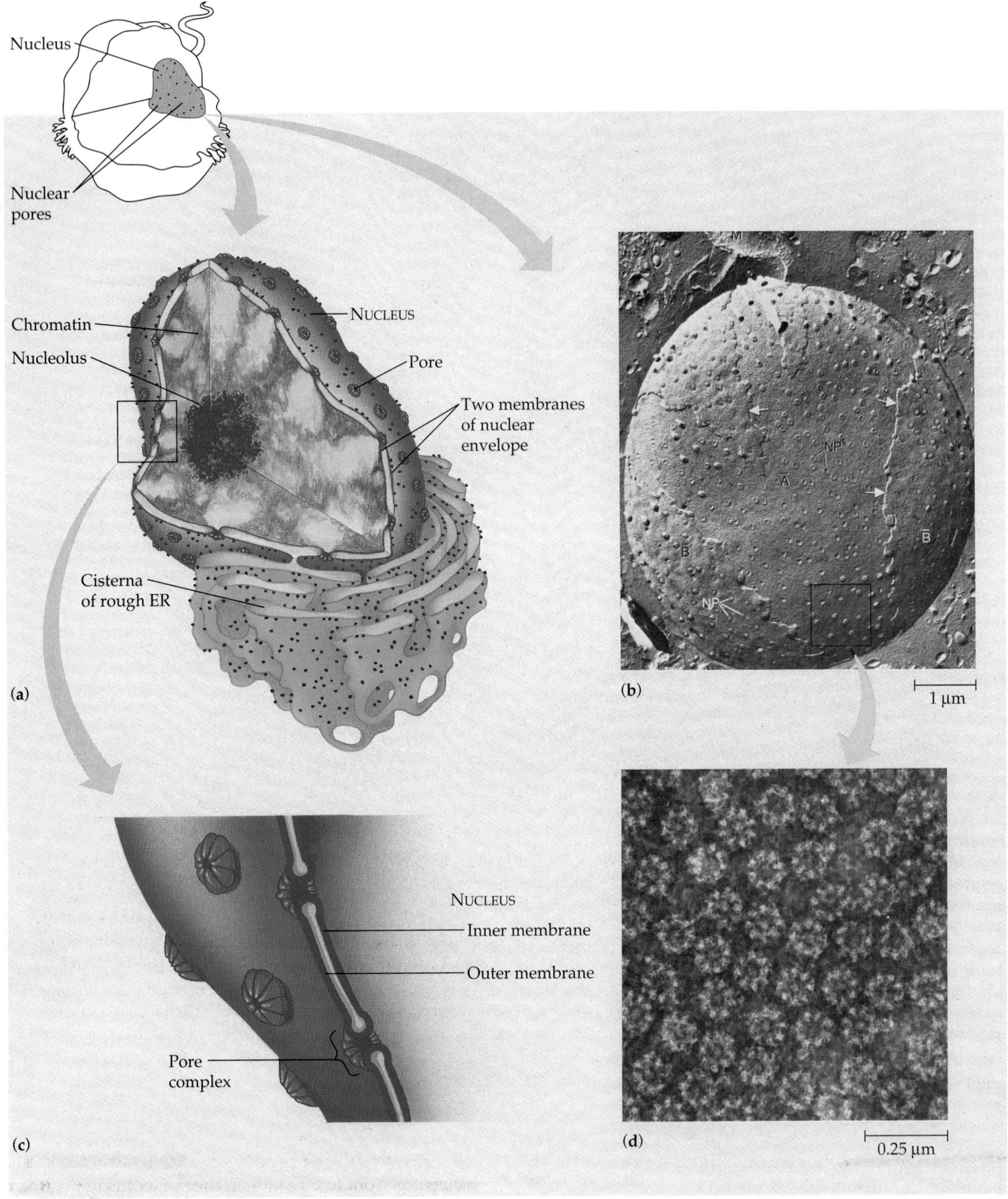

Figure 7.11
The nucleus and its envelope. (**a**) The nucleus. Within the nucleus, the genetic material is in the dispersed form known as chromatin; a darker-staining nucleolus is also visible. The nucleolus is the site of ribosome synthesis. The nuclear envelope, which consists of two membranes separated by a narrow space, is perforated with pores. (**b**) Numerous nuclear pores (NP) through the envelope are evident in this electron micrograph, prepared by a method called freeze-fracture (TEM). (**c**) The nuclear envelope. (**d**) An electron micrograph of the outer surface of the envelope reveals that each pore is bordered by a ring of eight protein particles (TEM).

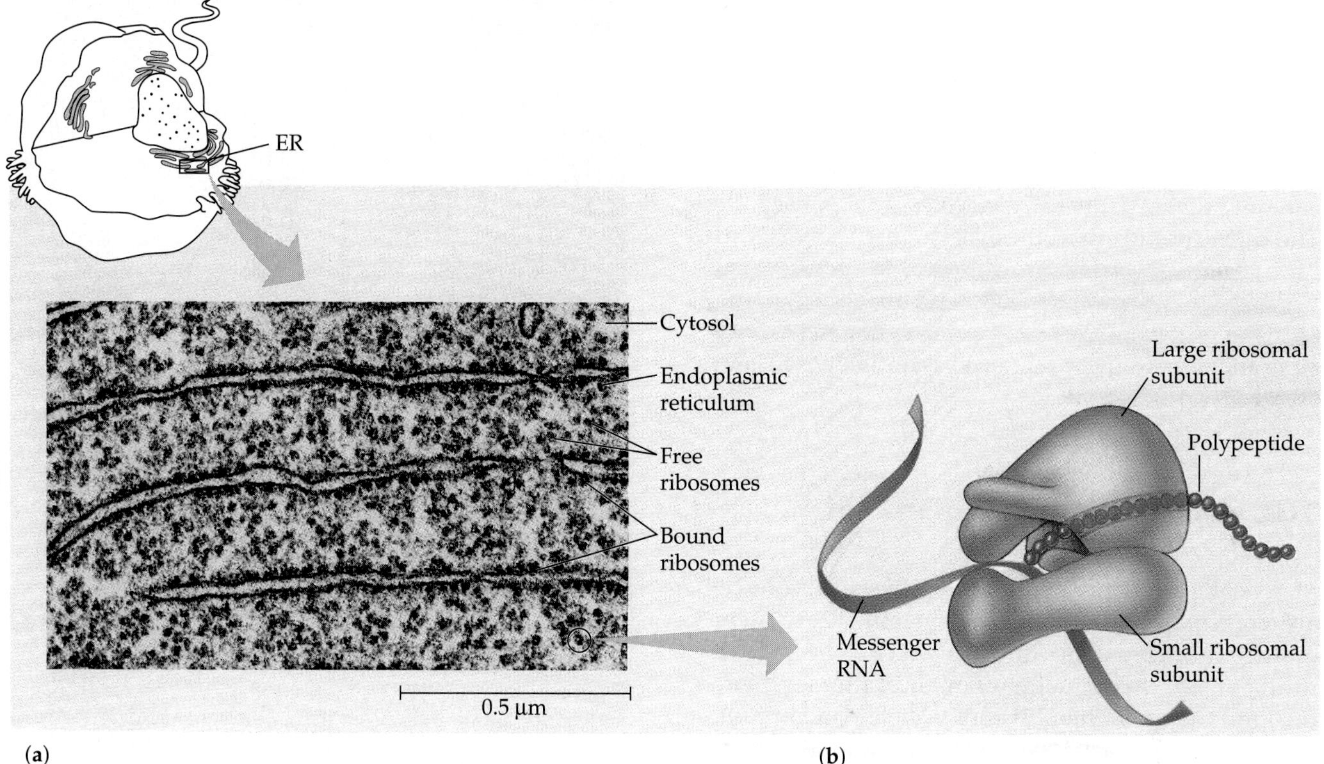

Figure 7.12
Ribosomes. (**a**) Both free and bound ribosomes are abundant in this electron micrograph of a cell from the pancreas (TEM). The pancreas is a gland specialized for the secretion of proteins. It secretes hormones, including the protein insulin, into the bloodstream, and secretes digestive enzymes, which are proteins, into the intestine. The bound ribosomes, those presently attached to the endoplasmic reticulum (ER), produce the secretory proteins. Free ribosomes make proteins that will remain dissolved in the cytosol. The bound and free ribosomes are identical and can alternate between these two roles. (**b**) The structure of a eukaryotic ribosome. The ribosome is generally considered the cell's smallest organelle. Not enclosed by a membrane, a ribosome consists of two subunits, which join together to form a functional ribosome only when they attach to a messenger RNA molecule.

bosomes, and the genetic message is translated into the primary structure of a specific protein (see Figure 5.30). This process is described in detail in Chapter 16.

RIBOSOMES

Ribosomes are the sites where the cell assembles proteins according to genetic instructions. A bacterial cell may have a few thousand ribosomes, whereas a human liver cell has a few million. Cells that have high rates of protein synthesis have a particularly great number of ribosomes, another example of cell structure fitting function. Cells active in protein synthesis also have prominent nucleoli, which make the ribosomes.

Ribosomes function in two cytoplasmic locales (Figure 7.12). *Free* ribosomes are suspended in the cytosol, while *bound* ribosomes are attached to the outside of a membranous network called the endoplasmic reticulum. Most of the proteins made by free ribosomes will function within the cytosol, and free ribosomes are especially abundant in cells growing by the addition of cytoplasm. Bound ribosomes generally make proteins that are destined either for inclusion into membranes or for export from the cell. Cells that specialize in protein secretion—for instance, the cells of the pancreas and other glands that secrete digestive enzymes—frequently have a high proportion of bound ribosomes. Bound and free ribosomes are structurally identical and interchangeable, and the cell can adjust the relative numbers of each as its metabolism changes.

Each ribosome is built from two subunits. In eukaryotes, the ribosomal subunits are constructed in the nucleolus from RNA made in the nucleolus itself, RNA produced elsewhere in the nucleus, and proteins imported from the cytoplasm. The subunits join to form functional ribosomes only when they attach to messenger RNA after export to the cytoplasm. The ribosome plays a key role in translating the genetic message, carried by mRNA from the nucleus to the cytoplasm, into the specific primary structure (amino acid sequence) of a polypeptide chain. In Chapter 16, we will learn more about how the anatomy of a ribosome is correlated with this function.

The ribosomes of prokaryotes are smaller than those of eukaryotes and differ somewhat in their molecular composition. Dissimilarity in a structure as fundamental as the ribosome is an example of the profound dichotomy that separates prokaryotes from eukaryotes. The difference in ribosomes is also medically significant; certain drugs can paralyze prokaryotic ribosomes without inhibiting the ability of eukaryotic ribosomes to make protein. These drugs, which include tetracycline and streptomycin, are used as antibiotics to combat bacterial infections.

THE ENDOMEMBRANE SYSTEM

Many of the different membranes of the eukaryotic cell are part of an **endomembrane system.** These membranes are related either through direct physical continuity or by the transfer of membrane segments through the movement of tiny vesicles (membrane-bounded sacs). These relationships, however, do not mean that the various membranes are alike in structure and function. The thickness, molecular composition, and metabolic behavior of a membrane are not fixed, but may be modified several times during the membrane's history. The endomembrane system includes the nuclear envelope, endoplasmic reticulum, Golgi apparatus, lysosomes, peroxisomes, various kinds of vacuoles, and the plasma membrane (not actually an *endo*membrane in physical location, but nevertheless related to the endoplasmic reticulum and other internal membranes). We have already discussed the nuclear envelope and will now focus on the endoplasmic reticulum and the other endomembranes to which it gives rise.

Endoplasmic Reticulum

The **endoplasmic reticulum (ER)** is a membranous labyrinth so extensive that it accounts for more than half the total membrane in many eukaryotic cells (Figure 7.13). (The word *endoplasmic* means "within" the cyto*plasm,* and *reticulum* is derived from a Latin word meaning "network.") The ER consists of a network of membranous tubules and sacs called **cisternae** (Lat. *cisterna,* "box" or "chest"). The ER membrane separates its internal compartment, the cisternal space, from the cytosol. And because the ER membrane is continuous with the nuclear envelope, the space between the two membranes of the envelope is continuous with the cisternal space of the ER (Figure 7.14).

There are two distinct, though connected, regions of ER that differ in structure and function: **rough ER** and **smooth ER.** Rough ER appears rough through the electron microscope because ribosomes stud the cytoplasmic surface of the membrane. Ribosomes are also attached to the cytoplasmic side of the nuclear envelope's outer membrane, which is confluent with rough ER. Smooth ER is so named because its cytoplasmic surface lacks ribosomes.

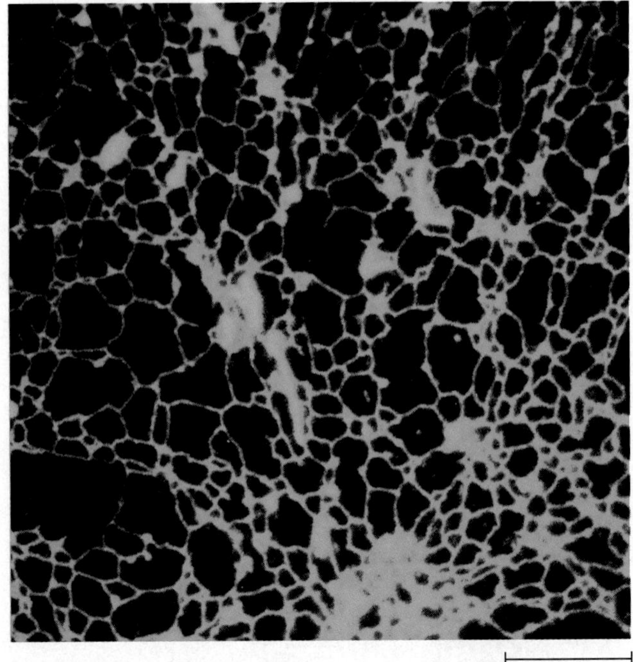

Figure 7.13
The endoplasmic reticulum (ER). This light micrograph shows a portion of a kidney cell, which has been treated with a fluorescent dye that causes the ER to glow green. This treatment gives some idea of the extent of ER, which branches throughout the cytoplasm.

Functions of Smooth ER The smooth ER of various cell types functions in diverse metabolic processes, including the synthesis of lipids, carbohydrate metabolism, and the detoxification of drugs and other poisons.

Enzymes of the smooth ER are important to the synthesis of fats, phospholipids, steroids, and other lipids (see Chapter 5). Among the steroids produced by smooth ER are the sex hormones of vertebrates and the various steroid hormones secreted by the adrenal glands. The cells that actually synthesize and secrete these hormones—in the testes and ovaries, for example—are rich in smooth ER, a structural feature that fits the function of these cells.

Liver cells provide one example of the role of smooth ER in carbohydrate metabolism. Liver cells store carbohydrate in the form of glycogen, a polysaccharide (see Figure 5.8). The hydrolysis of glycogen leads to the release of glucose from the liver cells, which is important in the regulation of sugar concentration in the blood. However, the first product of

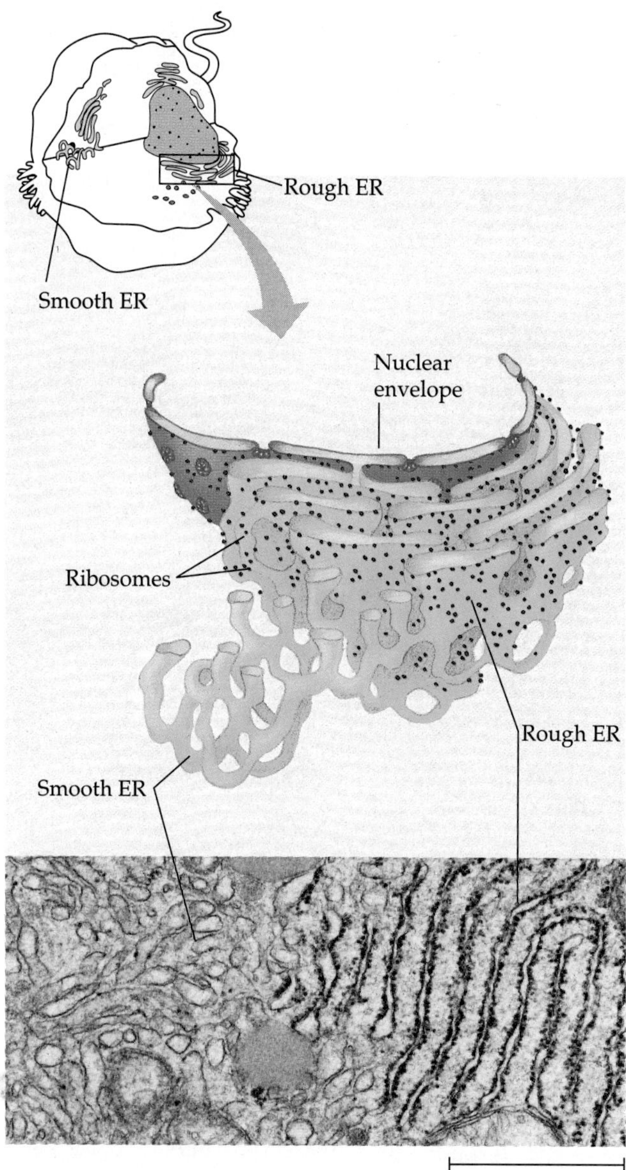

Figure 7.14
The detailed structure of ER. A membranous system of interconnected tubules and flattened sacs called cisternae, the ER is also continuous with the nuclear envelope. The membrane of the ER encloses a compartment called the cisternal space. Rough ER, which is studded on its cytoplasmic surface with ribosomes, can be distinguished from smooth ER in this electron micrograph (TEM).

glycogen hydrolysis is glucose phosphate, an ionic form of the sugar that cannot exit from the cell and enter the blood. An enzyme embedded in the membrane of the liver cell's smooth ER removes the phosphate from the glucose, which can then leave the cell and elevate blood sugar concentration.

Enzymes of the smooth ER help detoxify drugs and other poisons, especially in liver cells. Detoxification usually involves adding hydroxyl groups to drugs, increasing their solubility and making it easier to flush the compounds from the body. The sedative phenobarbital and other barbiturates are examples of drugs metabolized in this manner by smooth ER in liver cells. In fact, barbiturates, alcohol, and many other drugs induce the proliferation of smooth ER and its associated detoxification enzymes. This increases tolerance to the drugs, meaning that higher doses are required to achieve a particular effect, such as sedation. Also, because some of the detoxification enzymes have relatively broad action, the proliferation of smooth ER in response to one drug can also increase tolerance to other drugs. Barbiturate abuse, for example, may decrease the effectiveness of certain antibiotics and other useful drugs.

Muscle cells exhibit still another specialized function of smooth ER. The ER membrane pumps calcium ions from the cytosol into the cisternal space. When a muscle cell is stimulated by a nerve impulse, calcium leaks back across the ER membrane into the cytosol and triggers contraction of the muscle cell.

Rough ER and Protein Synthesis Many types of specialized cells secrete proteins produced by the rough ER (see Figure 7.12). For example, white blood cells in humans and other vertebrates secrete antibodies. Proteins destined for secretion are synthesized by ribosomes attached to the rough ER. As a polypeptide chain grows from a bound ribosome, it is threaded through the ER membrane into the cisternal space, possibly through a pore. As it enters the cisternal space, the protein folds into its native conformation. Most secretory proteins are **glycoproteins,** proteins that are covalently bonded to carbohydrates. In the cisternal space, the carbohydrate is attached to the protein by enzymes built into the ER membrane. The carbohydrate appendage of a glycoprotein is an oligosaccharide, a relatively small polymer of sugar units.

Once the secretory proteins are formed, the ER membrane keeps them separate from the proteins produced by free ribosomes that will remain in the cytosol. Secretory proteins depart from the ER wrapped in the membranes of vesicles budded from a specialized region called **transitional ER.** Such vesicles in transit from one part of the cell to another are called **transport vesicles,** and we will soon learn their fate.

Rough ER and Membrane Production In addition to making secretory proteins, rough ER is a membrane factory that grows in place by adding proteins and phospholipids. As the proteins elongate from the ribosomes, they are inserted into the ER membrane itself and are anchored there by hydrophobic portions of the proteins. The ER, both rough and smooth, also makes its own membrane phospholipids; enzymes built into the ER membrane assemble the phospholipids from precursors obtained from the cytosol. The ER membrane expands and can be transferred in the form of transport vesicles targeted for other components of the endomembrane system.

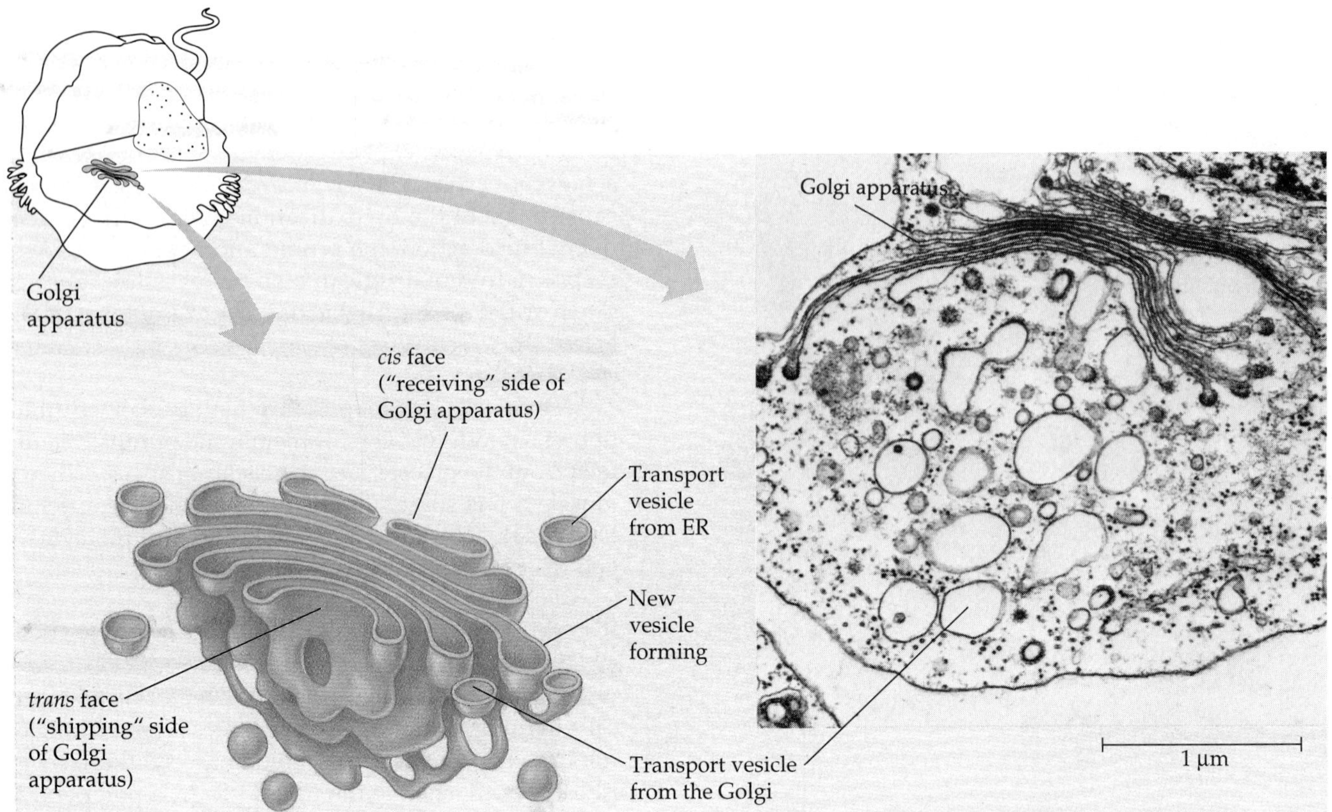

Figure 7.15
The Golgi apparatus. A Golgi apparatus consists of stacks of flattened membranous sacs. The apparatus receives and dispatches vesicles and the products they contain. Materials received from the ER are modified and stored in the Golgi and eventually shipped to the cell surface or other destinations. Note the vesicles forming at the edges of the stack and also the free vesicles that have just arisen. The stacks have a definite structural and functional polarity, with a *cis* face that receives vesicles and a *trans* face that dispatches vesicles (right, TEM).

The Golgi Apparatus

After leaving the ER, many transport vesicles travel to the **Golgi apparatus.** We can think of the Golgi as a center of manufacturing, warehousing, sorting, and shipping. Here, products of the ER are modified and stored, and then sent to other destinations. Not surprisingly, the Golgi apparatus is especially extensive in cells specialized for secretion.

The Golgi apparatus consists of flattened membranous sacs, looking like a stack of pita bread (Figure 7.15). A cell may have several of these stacks, all interconnected. Each cisterna in a stack consists of a membrane that separates its internal space from the cytosol. Vesicles concentrated in the vicinity of the Golgi apparatus may be engaged in the transfer of material between the Golgi and other structures.

The Golgi apparatus generally has a distinct polarity, with the membranes of cisternae at opposite ends of a stack differing in thickness and molecular composition. The two poles of a Golgi stack are referred to as the ***cis* face** (or forming face) and the ***trans* face** (or maturing face); these act, respectively, as the receiving and shipping departments of the Golgi apparatus. The *cis* face is usually located near ER. Transport vesicles move material from the ER to the Golgi. A vesicle that buds from the ER will add its membrane and the contents of its lumen (cavity) to the *cis* face by fusing with a Golgi membrane. The *trans* face gives rise to vesicles, which pinch off from and travel to other sites.

Products of the ER are usually modified during their transit from the *cis* pole to the *trans* pole of the Golgi. Proteins and phospholipids of membranes may be altered. In particular, various Golgi enzymes modify the oligosaccharide portions of glycoproteins. When first added to proteins in the ER, the oligosaccharides of all glycoproteins are identical. The Golgi removes some sugar monomers and substitutes others, producing diverse oligosaccharides. In addition to this finishing work, the Golgi apparatus manufactures certain macromolecules by itself. Many polysaccharides secreted by cells are Golgi products, including hyaluronic acid, a sticky substance that helps glue animal cells together. Golgi products that will be secreted depart from the *trans* faces of Golgi in the lumen of membranous vesicles that eventually fuse with the plasma membrane.

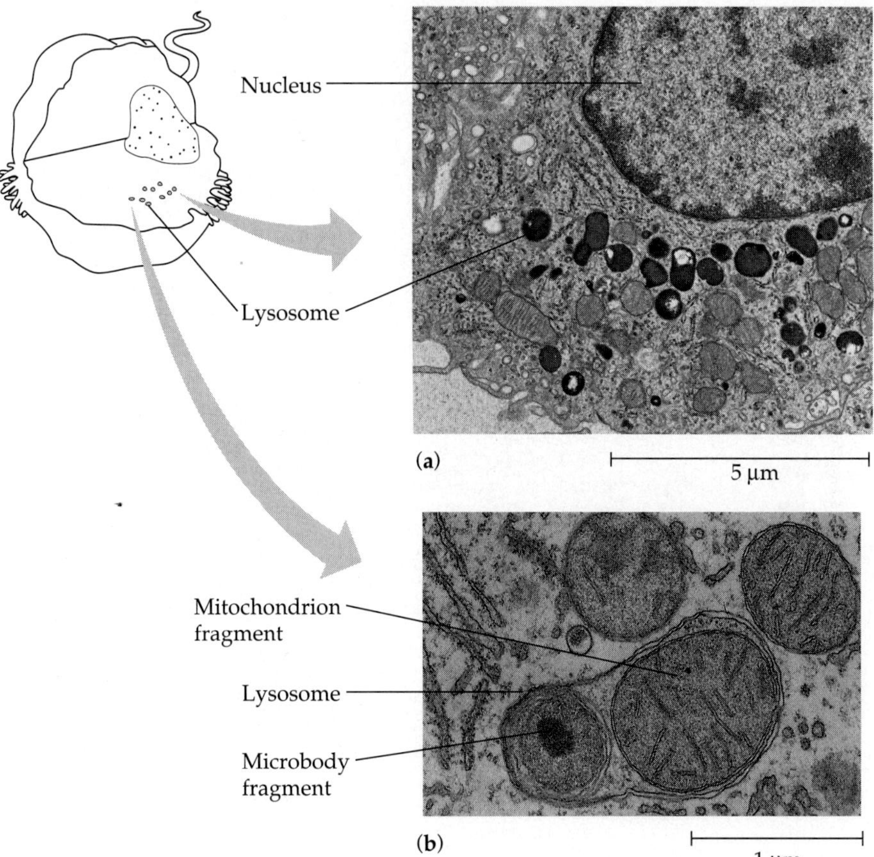

Figure 7.16
Lysosomes. (**a**) In this white blood cell from a rat, the lysosomes are very dark because of a specific stain that reacts with one of the products of digestion within the lysosome (TEM). This type of white blood cell ingests bacteria and viruses and destroys them in the lysosomes. (**b**) In the cytoplasm of this liver cell, we see an autophagic lysosome that has engulfed two disabled organelles, a mitochondrion and a microbody (TEM).

The Golgi manufactures and refines its products in stages, with different cisternae between the *cis* and *trans* ends containing unique teams of enzymes. Products in various stages of processing are transferred from one cisterna to the next by vesicles.

Sorting Functions of the Golgi Before the Golgi apparatus dispatches its products by budding vesicles from the *trans* face, it must sort these products and target them for various parts of the cell. Molecular identification tags, such as phosphate groups and specific oligosaccharides that have been added to the Golgi products, may aid in sorting. And membranous vesicles budded from the Golgi may have external molecules that recognize "docking sites" on the surface of specific organelles.

Lysosomes

A **lysosome** is a membrane-enclosed bag of hydrolytic enzymes that the cell uses to digest macromolecules (Figure 7.16). There are lysosomal enzymes that can hydrolyze proteins, polysaccharides, fats, and nucleic acids—all the major classes of macromolecules. These enzymes work best in an acidic environment, at about pH 5. The lysosomal membrane maintains this low internal pH by pumping hydrogen ions from the cytosol into the lumen of the lysosome. If the lysosome should break open or leak its contents, the enzymes would not be very active in the neutral environment of the cytosol. However, excessive leakage from a large number of lysosomes can destroy a cell by autodigestion. From this example, we can see once again how important compartmental organization is to the functions of the cell: The lysosome provides a space where the cell can digest macromolecules safely, without the general destruction that would occur if active hydrolytic enzymes roamed at large.

The hydrolytic enzymes and lysosomal membrane are made by rough ER and then transferred to a Golgi apparatus for further processing. Lysosomes probably arise by budding from the *trans* face of the Golgi apparatus. The proteins of the inner surface of the lysosomal membrane and the digestive enzymes themselves are probably spared from self-destruction by having three-dimensional conformations that protect vulnerable bonds from enzymatic attack.

Lysosomes function in intracellular digestion in a variety of circumstances. *Amoeba* and many other protozoa eat by engulfing smaller organisms or other food particles, a process called **phagocytosis** (Gr. *phagein*, "to eat," and *kytos*, "vessel," referring here to the cell). The food vacuole formed in this way then fuses with a

Figure 7.17
The formation and functions of lysosomes. Lysosomes digest materials taken into the cell and recycle materials from intracellular refuse. During phagocytosis, the cell encloses food in a vacuole with a membrane that pinches off internally from the plasma membrane. This food vacuole fuses with a lysosome, and hydrolytic enzymes digest the food. After hydrolysis, simple sugars, amino acids, and other monomers pass across the lysosomal membrane into the cytosol as nutrients for the cell. By the process of autophagy, lysosomes recycle the molecular ingredients of organelles. The ER and Golgi may cooperate in the production of lysosomes containing active enzymes, although some lysosomes may bud directly from specialized regions of the ER.

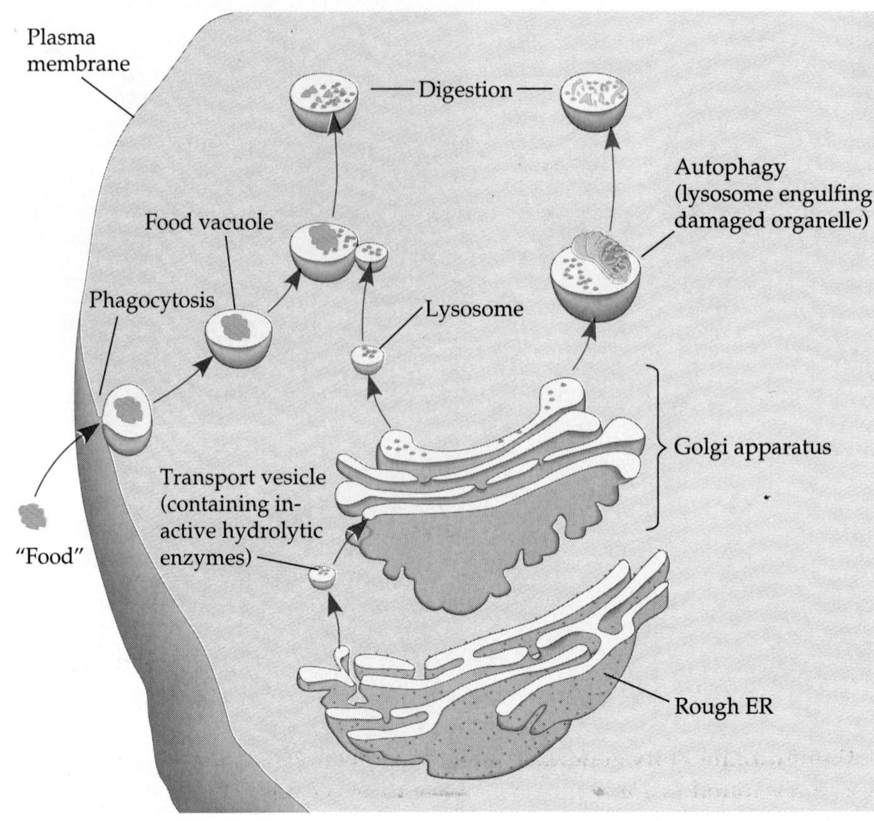

lysosome, whose enzymes digest the food (Figure 7.17). Some human cells also carry out phagocytosis. Among them are macrophages, cells that help defend the body by destroying bacteria and other invaders.

Lysosomes also use their hydrolytic enzymes to recycle the cell's own organic material, a process called autophagy. This occurs when a lysosome engulfs another organelle or a small parcel of cytosol (see Figure 7.16b). The lysosomal enzymes dismantle the ingested material, and the organic monomers are returned to the cytosol for reuse. With the help of lysosomes, the cell continually renews itself. A human liver cell, for example, recycles half of its macromolecules each week.

Programmed destruction of cells by their own lysosomal enzymes is important in the development of many organisms. During the transforming of a tadpole into a frog, for instance, lysosomes destroy the cells of the tail. And the hands of human embryos are webbed until lysosomes digest the tissue between the fingers.

Lysosomes and Human Disease A variety of inherited disorders called storage diseases affect lysosomal metabolism. A person afflicted with a storage disease lacks one of the active hydrolytic enzymes normally present in lysosomes. The lysosomes become engorged with indigestible substrates, which begin to interfere with other cellular functions. In Pompe's disease, for example, the liver is damaged by an accumulation of glycogen due to the absence of a lysosomal enzyme needed to break down the polysaccharide. In Tay-Sachs disease, a lipid-digesting enzyme is missing or inactive, and the brain becomes impaired by an accumulation of lipids in the cells. Fortunately, storage diseases are rare in the general population. Research indicates they may some day be treated by injecting the missing enzymes into the blood along with adaptor molecules that target the enzymes for engulfment by cells and fusion with lysosomes. In the future, it may also be possible to repair a disorder directly by inserting genes (DNA) for the missing enzyme into the appropriate cells (see Chapter 19).

Vacuoles

Vacuoles and vesicles are both membrane-enclosed sacs within the cell, but vacuoles are larger than vesicles. Vacuoles have various functions. **Food vacuoles,** formed by phagocytosis, have already been mentioned (see Figure 7.17). Many freshwater protists have **contractile vacuoles** that pump excess water out of the cell. Mature plant cells generally contain a large **central vacuole** enclosed by a membrane called the **tonoplast,** which is part of their endomembrane system (Figure 7.18).

The plant cell vacuole is a versatile compartment. It is a place to store organic compounds, such as the pro-

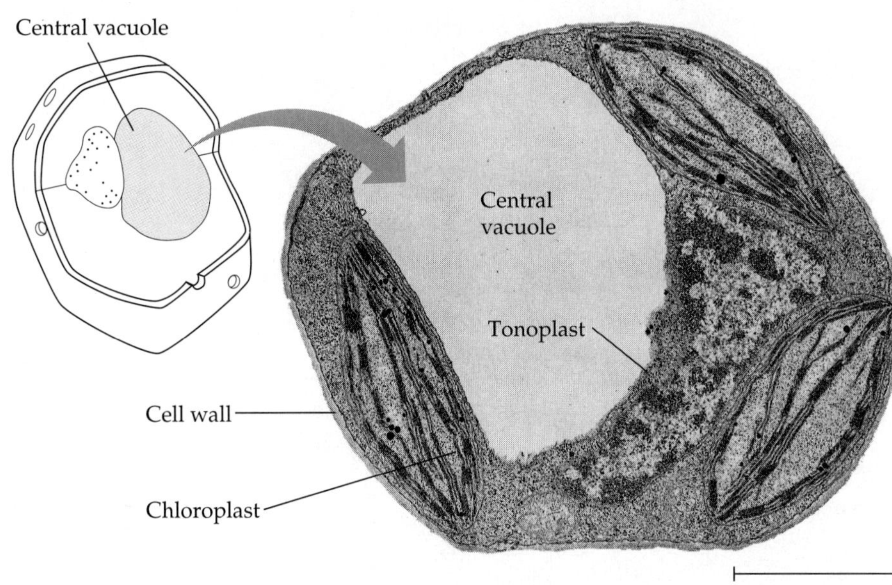

Figure 7.18
The plant cell vacuole. The central vacuole is usually the largest compartment in a plant cell, comprising 80% or more of a mature cell. The cytoplasm is generally confined to a narrow zone between the vacuole and the plasma membrane. The membrane bounding the vacuole, the tonoplast, separates the cytosol from the solution inside the vacuole, which is called cell sap. Like all cellular membranes, the tonoplast is selective in transporting solutes; therefore, cell sap differs in composition from the cytosol. The vacuole functions in storage, waste disposal, hydrolysis, protection, and growth (TEM).

teins that are stockpiled in the vacuoles of storage cells in seeds. The vacuole is also the plant cell's main repository of inorganic ions, such as potassium and chloride. Although plant cells generally lack the specialized lysosomes found in animal cells, the vacuole functions as the plant cell's lysosomal compartment, containing hydrolytic enzymes that digest stored macromolecules and recycle molecular components from organelles. Many plant cells use their vacuoles as disposal sites for metabolic by-products that would be dangerous if they accumulated in the cytoplasm. Some vacuoles are enriched in pigments that color the cells, such as the red and blue pigments of petals that help attract pollinating insects to flowers. Vacuoles may also help protect the plant against predators because they contain compounds that are poisonous or unpalatable to animals. The vacuole has a major role in the growth of plant cells, which elongate as their vacuoles absorb water, enabling the cell to become larger with a minimal investment in new cytoplasm. And because the cytoplasm occupies a thin shell between the plasma membrane and the tonoplast, the ratio of membrane surface to cytoplasmic volume is great, even for a large plant cell.

The large vacuole of a plant cell develops by the coalescence of smaller vacuoles, themselves derived from the endoplasmic reticulum and Golgi apparatus. Through these relationships, the vacuole is an integral part of the endomembrane system.

Relationships Among Endomembranes: A Summary

Through the fusion of transport vesicles, or, in some cases, through direct membrane continuity, the components of the endomembrane system are all related. The nuclear envelope is an extension of the rough ER, which is also confluent with the smooth ER. Membrane produced by the ER flows in the form of transport vesicles to the Golgi, which in turn pinches off vesicles that give rise to lysosomes and the tonoplast of the plant cell vacuole. Even the cell's outermost membrane, the plasma membrane, grows by the fusion of vesicles born in the ER and the Golgi (Figure 7.19). (Coalescence of vesicles with the plasma membrane also releases cellular products, such as secretory proteins, to the outside of the cell.) As membranes of the system flow from the ER to the Golgi and then elsewhere, their molecular compositions and metabolic functions are modified. The endomembrane system is a complex and dynamic player in the cell's compartmental organization.

Three important organelles that are *not* closely related to the endomembrane system are the peroxisome, the mitochondrion, and the chloroplast.

PEROXISOMES (MICROBODIES)

The **peroxisome,** also called a microbody, is a specialized metabolic compartment bounded by a single membrane (Figure 7.20). Peroxisomes contain enzymes that transfer hydrogen from various substrates to oxygen, producing hydrogen peroxide (H_2O_2) as a by-product, from which the organelle derives its name. These reactions may have many different functions. Some peroxisomes use oxygen to break fats down into smaller molecules that can then be transported to mitochondria as fuel for cellular respiration. Peroxisomes in the liver detoxify alcohol and other harmful compounds by transferring hydrogen from the poisons to oxygen. The H_2O_2 formed by peroxisome metabolism

Figure 7.19
Relationships among endomembranes. Although each membrane has a unique molecular composition and function, components of the endomembrane system are related through direct physical contact or through the transfer of vesicles. The latter occurs by the pinching off of membrane and its later fusion with the membrane of another organelle or with the plasma membrane. In this diagram, the lumen (cavity) of the ER and its derivatives are the same color (blue) as the *outside* of the cell to indicate that these areas are all topologically related. A secretory protein or some other product in the cisternal space of the ER can eventually turn up outside the cell without crossing another membrane.

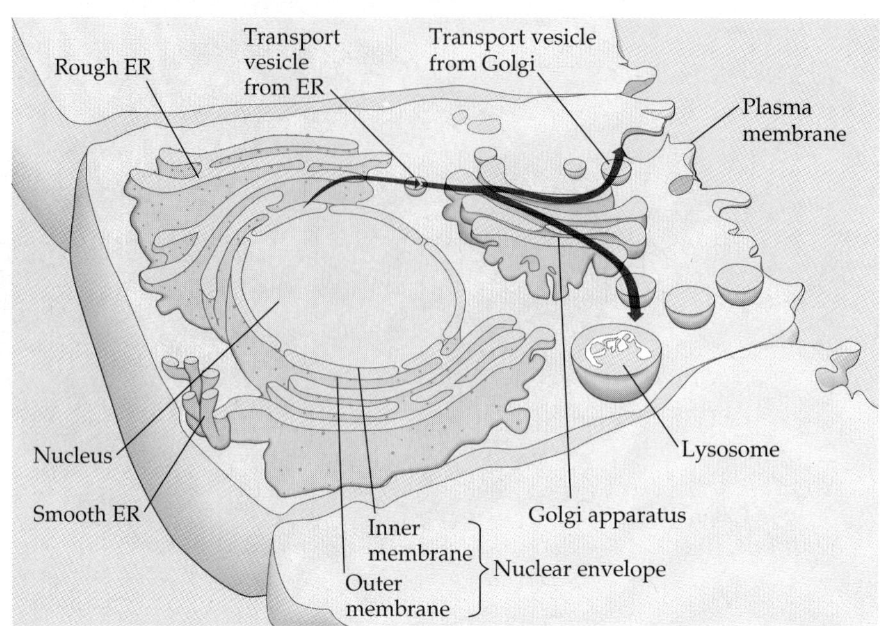

is itself toxic, but the organelle contains an enzyme that converts the H_2O_2 to water. Packaging the enzymes that produce hydrogen peroxide in an enclosure where an enzyme that disposes of this by-product is also concentrated, is another example of how the cell's compartmental structure is crucial to its function.

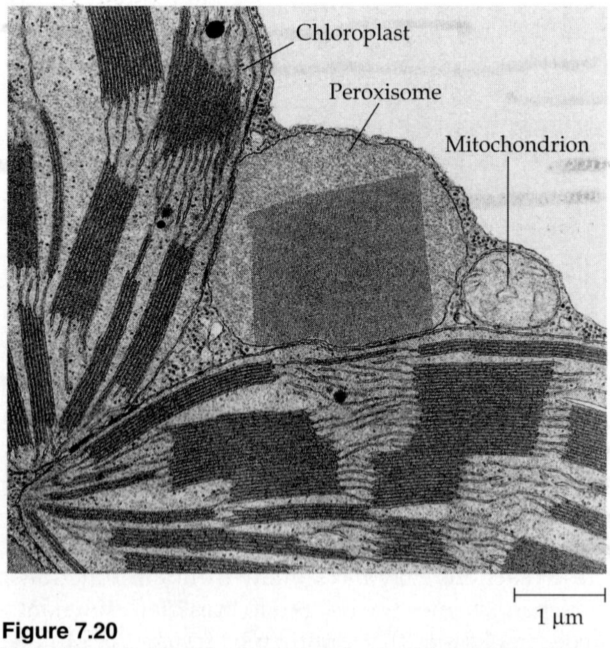

Specialized peroxisomes are found in the fat-storing tissues of the germinating seeds of plants. These organelles contain enzymes that initiate the conversion of fats to sugar, a process that makes the energy stored in the oils of the seed available until the seedling is able to produce its own sugar by photosynthesis.

Unlike lysosomes, peroxisomes are not budded from the endomembrane system. They grow by incorporating proteins and lipids produced in the cytosol, and increase in number by splitting in two when they reach a certain size.

Chloroplast
Peroxisome
Mitochondrion
1 µm

Figure 7.20
Peroxisomes. Peroxisomes range in size from about 0.15 to 1.2 µm. They are usually roughly spherical and often have a granular or crystalline core believed to be a dense collection of enzymes. This peroxisome is in a leaf cell. Notice its close relationship with mitochondria and choloroplasts, which cooperate with peroxisomes in certain metabolic functions (TEM).

ENERGY TRANSDUCERS: MITOCHONDRIA AND CHLOROPLASTS

Mitochondria (singular, mitochondrion) and chloroplasts are organelles that convert energy to forms the cell can use for its various kinds of work. Mitochondria are the sites of cellular respiration, the elaborate catabolic process that generates ATP by extracting energy from sugars, fats, and other fuels with the help of oxygen. Chloroplasts, found only in plants and eukaryotic algae (kingdom Protista), convert solar energy to chemical energy by absorbing sunlight and using it to drive the synthesis of organic compounds from carbon dioxide and water. Recall that the cell is an open system, and both mitochondria and chloroplasts are involved in putting energy from the cell's surroundings to work.

Although mitochondria and chloroplasts are enclosed by membranes, they are not considered part of the endomembrane system. Their membrane proteins are made not by the ER, but by free ribosomes in the cy-

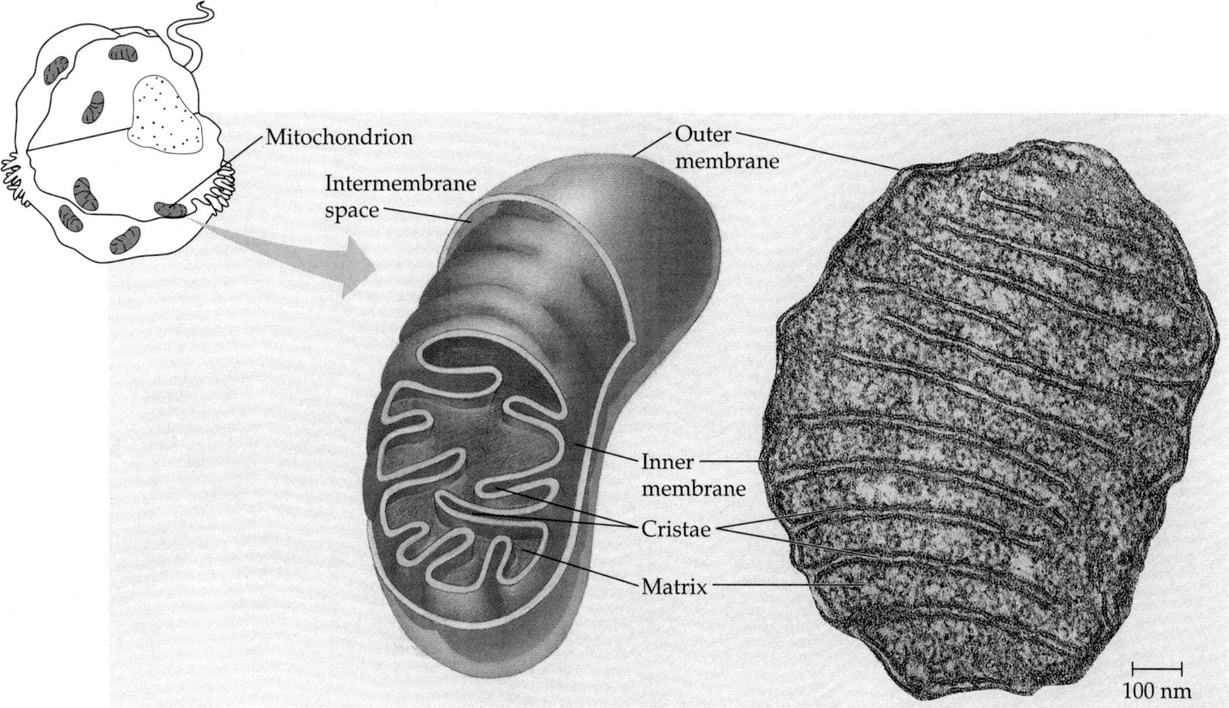

Figure 7.21
The mitochondrion. The double membrane of the mitochondrion is evident in the drawing and the micrograph (TEM). The cristae are infoldings of the inner membrane. The three-dimensional drawing emphasizes the relationships between the two membranes and the compartments they bound: the intermembrane space and the mitochondrial matrix.

tosol and by ribosomes contained within the mitochondria and chloroplasts themselves. Not only do these organelles have ribosomes, but they also contain a small amount of DNA that programs the synthesis of some of their own proteins (although most of these are made in the cytosol, programmed by messenger RNA sent by nuclear genes). Mitochondria and chloroplasts are semiautonomous organelles that grow and reproduce within the cell. In Chapters 9 and 10, we will focus on how mitochondria and chloroplasts work. The evolution of these organelles will be considered in Chapter 26. Here, we are concerned mainly with the structure of these energy transducers.

Mitochondria

Mitochondria are found in nearly all eukaryotic cells. In some cases, there is a single large mitochondrion, but more often, a cell has hundreds or even thousands of mitochondria; the number is generally correlated with the cell's metabolic activity. Mitochondria are about 1 to 10 μm long. Time-lapse films of living cells reveal mitochondria moving around, changing their shapes, and dividing in two, unlike the static cylinders seen in electron micrographs of dead cells.

The mitochondrion is enclosed in an envelope of two membranes, each a phospholipid bilayer with a unique collection of embedded proteins (Figure 7.21). The outer membrane is smooth, but the inner membrane is convoluted, with infoldings called **cristae.** The membranes divide the mitochondrion into two internal compartments. The intermembrane space is the narrow region between the inner and outer membranes. The **mitochondrial matrix** is the compartment enclosed by the inner membrane. Many of the metabolic steps of cellular respiration occur in the matrix, where several different enzymes are concentrated. Other proteins that function in respiration, including the enzyme that makes ATP, are built into the inner membrane. With its cristae, the inner mitochondrial membrane has a large surface area that enhances the productivity of cellular respiration. This feature is another example of the correlation of structure and function.

Chloroplasts

The chloroplast is a specialized member of a family of closely related plant organelles called **plastids.** *Amyloplasts* (also called leucoplasts) are colorless plastids that store starch, particularly in roots and tubers. *Chromoplasts* are enriched in pigments that give fruits and flowers their orange and yellow hues. **Chloroplasts** contain the green pigment chlorophyll along

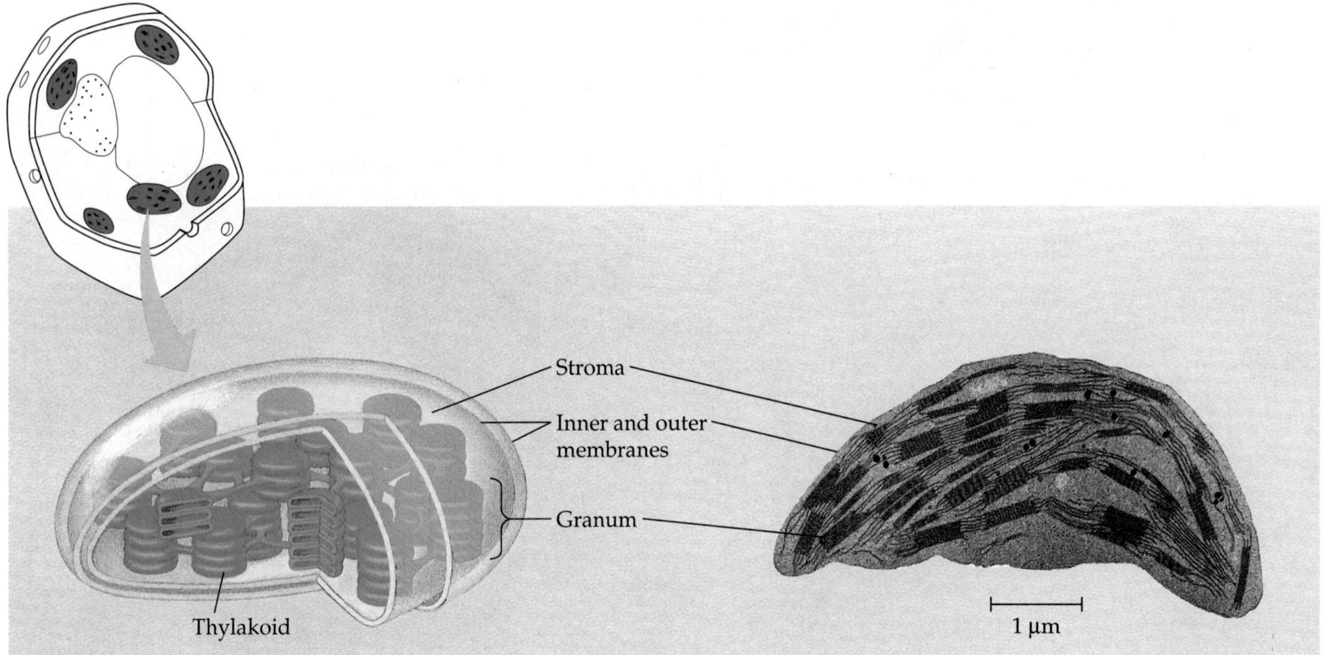

Figure 7.22
The chloroplast. This organelle is the site of photosynthesis. Chloroplasts, like mitochondria, are enclosed by two membranes separated by a narrow intermembrane space. The inner membrane encloses fluid called stroma. The stroma surrounds a third compartment, delineated by its own membrane, the thylakoid membrane. Throughout the chloroplast, thylakoid sacs are stacked to form structures called grana. Individual thylakoids of one granum are continuous with other grana through extensions that traverse the stroma (TEM).

with enzymes and other molecules that function in the photosynthetic production of food. These lens-shaped organelles, measuring about 2 μm by 5 μm, are found in leaves and other green organs of plants and in eukaryotic algae (Figure 7.22).

Chloroplasts, amyloplasts, and chromoplasts can all develop from *proplastids* found in unspecialized cells. As the plant grows, the fate of the proplastids depends on the location of the cells and the environment to which they are exposed. For example, a proplastid will become a chloroplast only if exposed to light. Chloroplasts can also arise from the division of previously existing chloroplasts. Under certain conditions, mature plastids are interconvertible. When a fruit ripens, for instance, chloroplasts in its green tissue are transformed into nonphotosynthetic chromoplasts that contribute to the characteristic color indicating that the fruit is ripe.

The contents of a chloroplast are partitioned from the cytosol by an envelope consisting of two membranes separated by a very narrow intermembrane space. Inside the chloroplast is another membranous system, arranged into flattened sacs called **thylakoids.** In some regions, thylakoids are stacked like poker chips to form structures called **grana** (singular, granum). The fluid outside the thylakoids is called the **stroma.** Thus, the thylakoid membrane segregates the interior of the chloroplast into two compartments: the thylakoid space and the stroma. In Chapter 10, we will learn how this compartmental organization enables the chloroplast to convert light energy to chemical energy during photosynthesis.

As with mitochondria, the static and rigid appearance of chloroplasts in electron micrographs belies their dynamic behavior in the living cell. Their shapes are plastic, and they occasionally pinch in two. They are mobile and move around the cell with mitochondria and other organelles along tracks made up of the tubules and filaments of the cytoskeleton, which will be described next.

THE CYTOSKELETON

In the early days of electron microscopy, the cell seemed to consist of a variety of organelles suspended or floating in a formless, jellylike cytosol. But new techniques in both light microscopy and electron microscopy have revealed a network of fibers throughout the cytoplasm. This mesh is called the **cytoskeleton** (Figure 7.23; see also Figure 7.1).

One function of the cytoskeleton is to give mechanical support to the cell and help maintain its shape. This is especially important for animal cells, which lack walls. Organelles and even cytoplasmic enzymes may be held in place by anchoring to the cytoskeleton. The cytoskeleton also enables a cell to change its shape; like

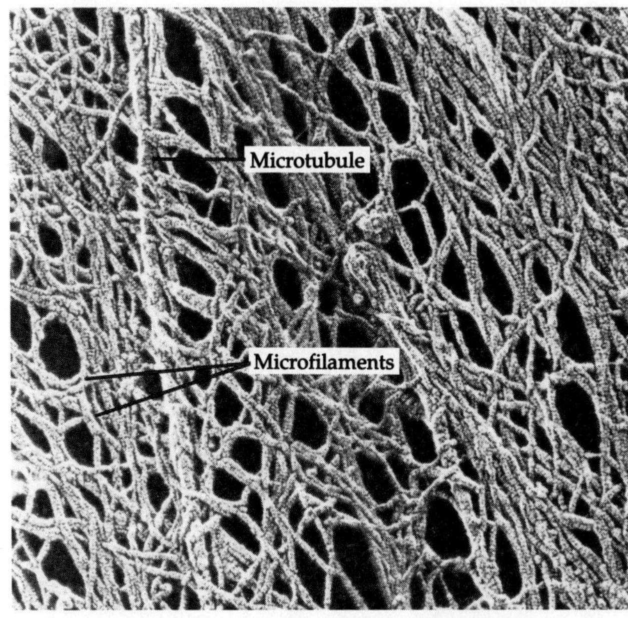

Figure 7.23
The cytoskeleton. The cytoskeleton gives the cell shape, anchors some organelles and directs the movement of others, and in some cases enables the entire cell to change shape or move. In this electron micrograph prepared by a method known as deep-etching, microtubules and microfilaments are visible. A third component of the cytoskeleton, intermediate filaments, is not evident here.

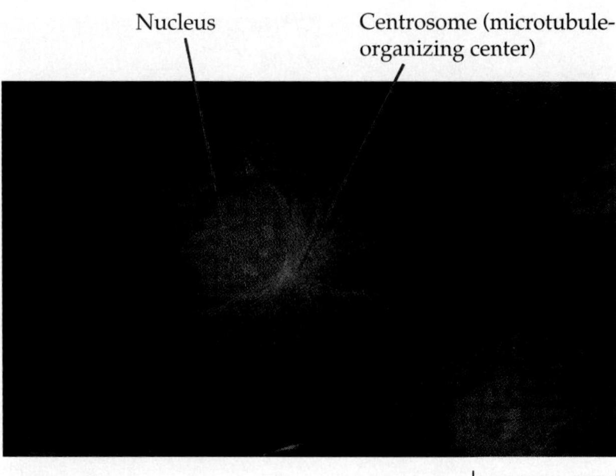

Figure 7.24
The centrosome (microtubule-organizing center). A fluorescent dye reveals microtubules (green) radiating from the microtubule-organizing center, located just outside the nucleus (orange). These are human cells called fibroblasts, which produce the extracellular fibers of cartilage and other connective tissue (LM).

a scaffold, the cytoskeleton can be dismantled in one part of the cell and reassembled in a new location. The cytoskeleton is also associated with motility: movement of the entire cell or movement of organelles within the cell. The fibers of the cytoskeleton are not only the cell's "bones" but also its "muscles." Components of the cytoskeleton wiggle cilia and flagella and enable muscle cells to contract. The cytoskeleton extends the pseudopodia of *Amoeba* and also functions in the streaming of cytoplasm that circulates materials within many large plant cells. Vesicles may travel to their destinations in the cell along "monorails" provided by the cytoskeleton, and contractile components of the cytoskeleton manipulate the plasma membrane to form food vacuoles during phagocytosis.

The cytoskeleton is constructed from at least three types of fibers (Table 7.2). **Microtubules** are the thickest of the three types; **microfilaments** (also called actin filaments) are the thinnest. **Intermediate filaments** are a collection of fibers whose diameters fall in a middle range.

Microtubules

Microtubules are straight, hollow rods measuring about 25 nm in diameter and from 200 nm to 25 µm in length. The wall of the hollow tube is constructed from globular proteins called tubulins, of which there are two closely related kinds, α-tubulin and β-tubulin. A microtubule can elongate by the addition of tubulin proteins to one end of the tubule. Microtubules can be disassembled and their tubulin used to build microtubules elsewhere in the cell.

Microtubules are found in the cytoplasm of all eukaryotic cells. In many cells, they radiate from a **centrosome** (also called the **microtubule-organizing center**), a mass located near the nucleus (Figure 7.24). The microtubules that emanate from this center are strong girders that support the cell. In addition, strategically located bundles of microtubules near the plasma membrane reinforce the cell's shape.

While shaping and supporting the cell, microtubules also serve as tracks along which organelles can move in the cell. For example, microtubules probably help guide secretory vesicles from the Golgi apparatus to the plasma membrane. Specialized proteins function as **motor molecules** to move vesicles and other organelles along the microtubules (Figure 7.25). Figure 7.26 shows pigment granules that move along microtubules. Microtubules are also involved in the separation of chromosomes during cell division, discussed in Chapter 11.

Within the centrosome of an animal cell are a pair of **centrioles.** Each centriole is composed of nine sets of triplet microtubules arranged in a ring (Figure 7.27). When a cell divides, the centrioles replicate. Although centrioles may help organize microtubule assembly, they are not mandatory for this function in all eukaryotes; centrosomes of plant cells lack centrioles altogether.

Table 7.2 The Structure and Function of the Cytoskeleton

Property	Microtubules	Microfilaments	Intermediate Filaments
Structure	Hollow tubes; wall consists of 13 columns of tubulin proteins	Two intertwined strands of actin	Fibrous proteins supercoiled into thicker cables
Diameter	25 nm with 15-nm lumen	7 nm	8–12 nm
Monomers	α-tubulin β-tubulin	Actin	One of several different proteins depending on cell type
Functions	Cell motility (as cilia or flagella) Chromosome movements Movement of organelles Maintenance of cell shape	Muscle contraction Cytoplasmic streaming Cell motility (as pseudopodia) Cell division (cleavage furrow formation) Maintenance of cell shape Changes in cell shape	Structural support Maintenance of cell shape
	5 μm Tubulin dimer 25 nm	10 μm Actin subunit 7 nm	5 μm Fibrous subunits 10 nm

Source: Adapted from W. M. Becker and D. W. Deamer, *The World of the Cell,* 2nd ed. Redwood City, CA: Benjamin/Cummings, 1991, p. 555.

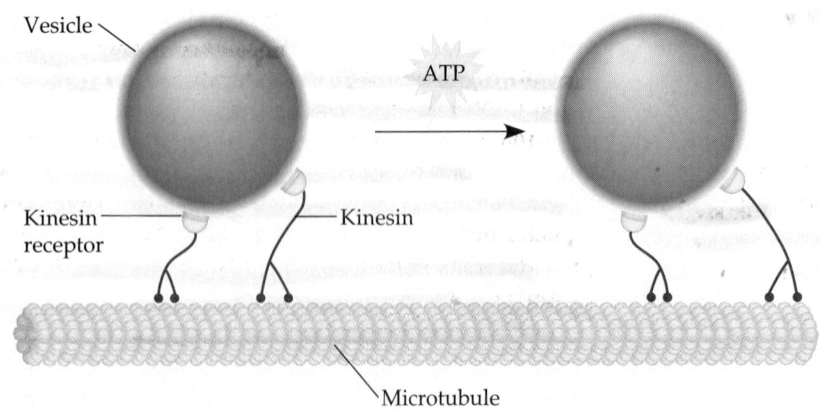

Figure 7.25
Motor molecules and the cytoskeleton. Motor molecules are proteins that convert chemical energy to movement. One example is the protein kinesin, which drives organelles such as vesicles along microtubules. It attaches to specific receptors on the organelles, and its "feet," movable extensions of the protein, "walk" along a microtubule. Kinesin and other motor molecules power movement by hydrolyzing ATP.

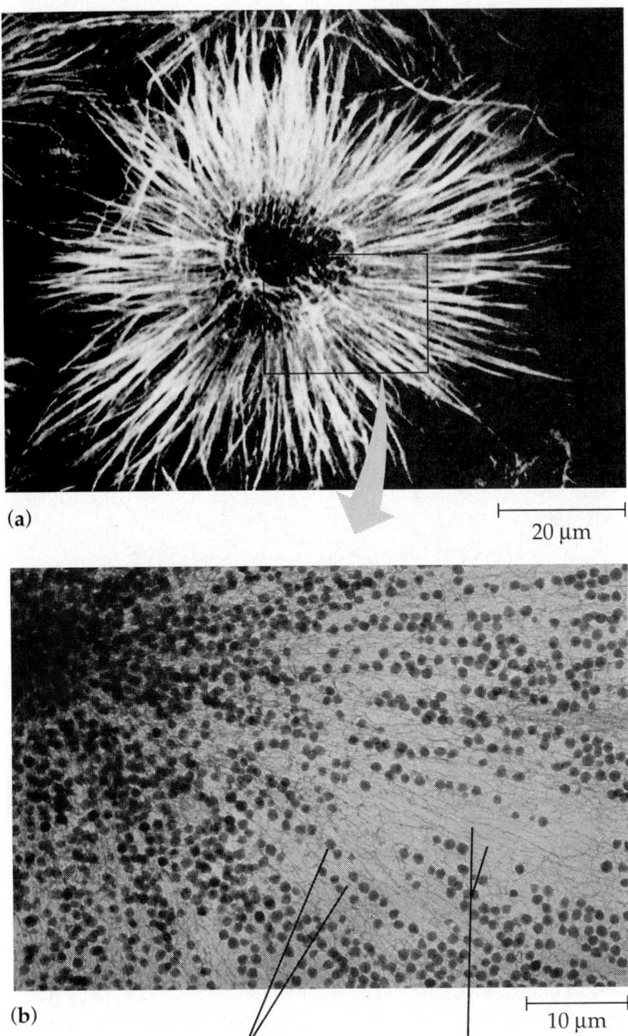

Figure 7.26
Microtubules as tracks. (**a**) Microtubules, stained with a fluorescent dye, radiate from the microtubule organizing center in this skin cell from a fish (confocal LM). (**b**) Electron microscopy reveals pigment granules that migrate back and forth along the microtubules. When the pigment is concentrated near the centers of cells, the fish has a light, speckled coloration. When the granules spread out along the cytoskeleton and fill the cells, the skin of the fish turns dark. This process enables the fish to match its coloration to its background, an adaptation that camouflages the animal and helps protect it from predators (TEM).

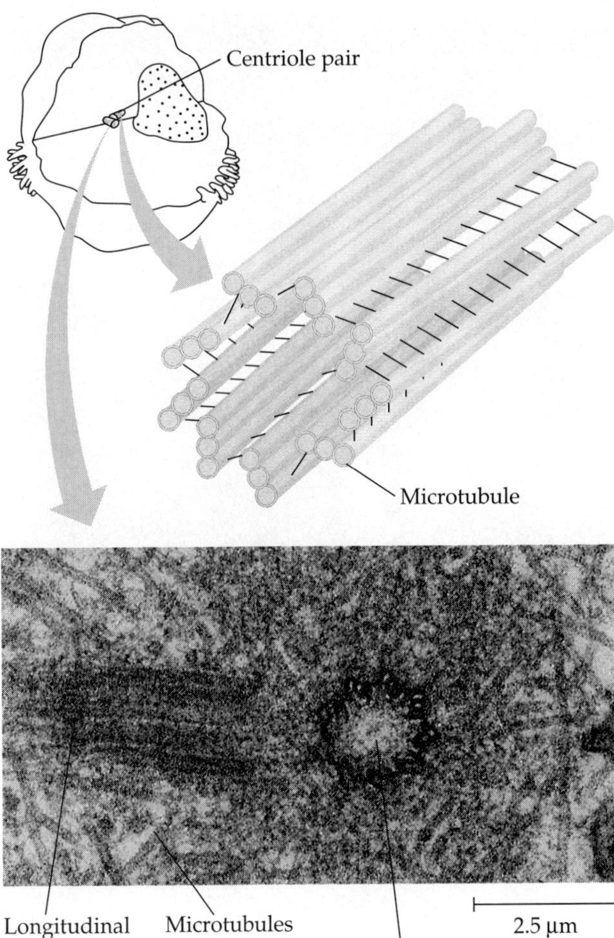

Figure 7.27
Centrioles. An animal cell has a pair of centrioles within its centrosome, the region near the nucleus where the cell's microtubules are initiated. The centrioles, each about 250 nm (0.25 μm) in diameter, are arranged at right angles to each other, and each is made up of nine sets of three microtubules (TEM).

Cilia and Flagella A specialized arrangement of microtubules is responsible for the beating of **flagella** and **cilia,** locomotive appendages that protrude from some cells. Many single-celled organisms (kingdom Protista) are propelled through water by cilia or flagella, and the sperm of animals, algae, and some plants are flagellated. (To a microscopic entity moving through water, the medium is as viscous as liquid asphalt would be to a human swimmer. Flagellated and ciliated organisms actually crawl or climb through the water.) If cilia or flagella extend from cells that are held in place as part of a tissue layer, then they function to draw fluid over the surface of the tissue. For example, the ciliated lining of the windpipe sweeps mucus with trapped debris out of the lungs (see Figure 7.4).

Cilia usually occur in large numbers on the cell surface (Figure 7.28). They are about 0.25 μm in diameter and about 2 to 20 μm in length. Flagella are the same diameter but longer than cilia, measuring 10 to 200 μm in length. Also, flagella are usually limited to just one or a few per cell.

Flagella and cilia also differ in their patterns of beating. A flagellum has an undulating motion that generates force in the same direction as the flagellum's axis.

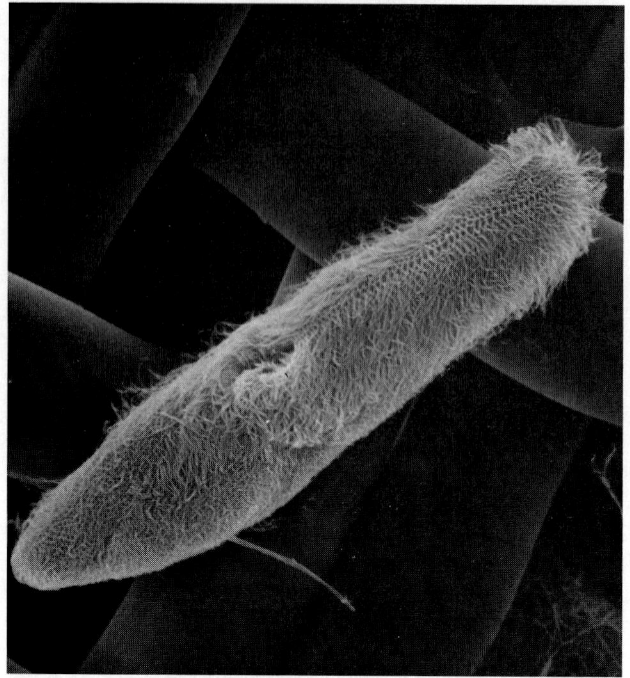

Figure 7.28
Cilia. A dense nap of beating cilia covers this *Paramecium*, a motile protist. The cilia beat at a rate of about 40 to 60 strokes per second (SEM).

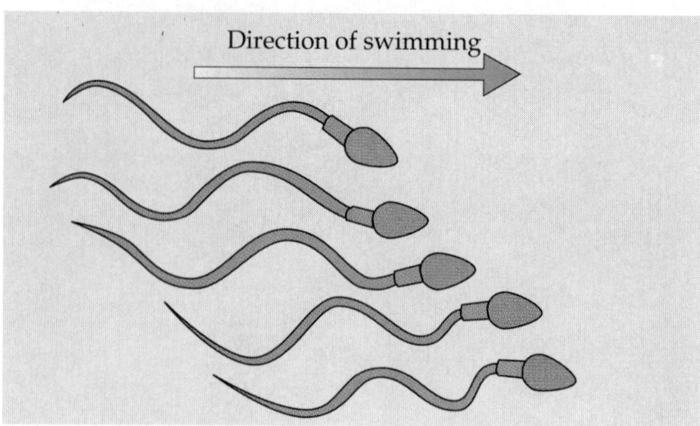

(**a**) Motion of flagella

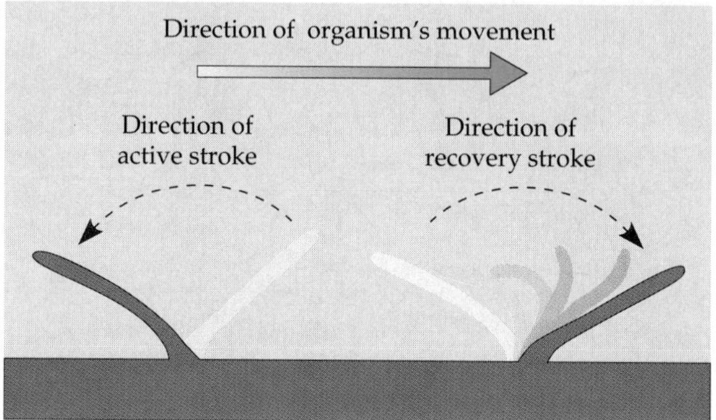

(**b**) Motion of cilia

Figure 7.29
A comparison of the beating of flagella and cilia. (**a**) A flagellum usually undulates, its snakelike motion driving a cell in the same direction as the axis of the flagellum. Propulsion of a sperm cell is an example of flagellate locomotion. (**b**) Cilia usually beat with a back-and-forth motion, alternating active strokes with recovery strokes. This moves the cell, or moves a fluid over the surface of a stationary cell, in a direction perpendicular to the axis of the cilium.

In contrast, cilia work more like oars, with alternating power and recovery strokes generating force in a direction perpendicular to the cilium's axis (Figure 7.29).

Though different in length, number per cell, and beating pattern, cilia and flagella actually share a common ultrastructure. A cilium or flagellum has a core of microtubules ensheathed in an extension of the plasma membrane (Figure 7.30). Nine doublets of microtubules, the members of each pair sharing part of their walls, are arranged in a ring. In the center of the ring are two single microtubules. This arrangement, referred to as the "9 + 2" pattern, is found in nearly all eukaryotic flagella and cilia. (The flagella of motile prokaryotes, which will be discussed in Chapter 25, are entirely different.) The doublets of the outer ring are connected to the center of the cilium or flagellum by radial spokes that terminate near the central pair of microtubules. Each doublet of the outer ring, as seen in cross sections, also has a pair of side-arms reaching toward the neighboring doublet of microtubules. Numerous pairs of these side-arms are evenly spaced along the length of each doublet.

The microtubule assembly of a cilium or flagellum is anchored in the cell by a **basal body,** which is structurally identical to a centriole. (In fact, basal bodies can convert into centrioles, and vice versa.) When a cilium or flagellum first begins to grow, the basal body may act as a template for ordering tubulin into microtubules. Once the building of cilia and flagella begins, however, the tubulin subunits are added to the tips of the microtubules, and not to their bases.

The side-arms extending from each microtubule doublet to the next play a major role in the bending movements of cilia and flagella (see Figure 7.30). The side-arms are made of a very large protein called **dynein,** one of the cell's motor molecules (another example is kinesin, illustrated in Figure 7.25). A dynein side-arm performs a complex cycle of movements caused by changes in the conformation (shape) of the protein, with ATP providing the energy for these changes (see Chapter 6). The mechanics of dynein "walking" are reminiscent of a cat climbing a tree by attaching its claws, moving its legs, releasing its claws, and grabbing again farther up the tree. Similarly, the dynein side-arms of one doublet attach to an adjacent doublet and pull so that the doublets slide past each other in opposite directions. The side-arms then release from the other doublet and reattach a little farther along its length. Without any restraints on this movement, one doublet will continue to "walk" along the

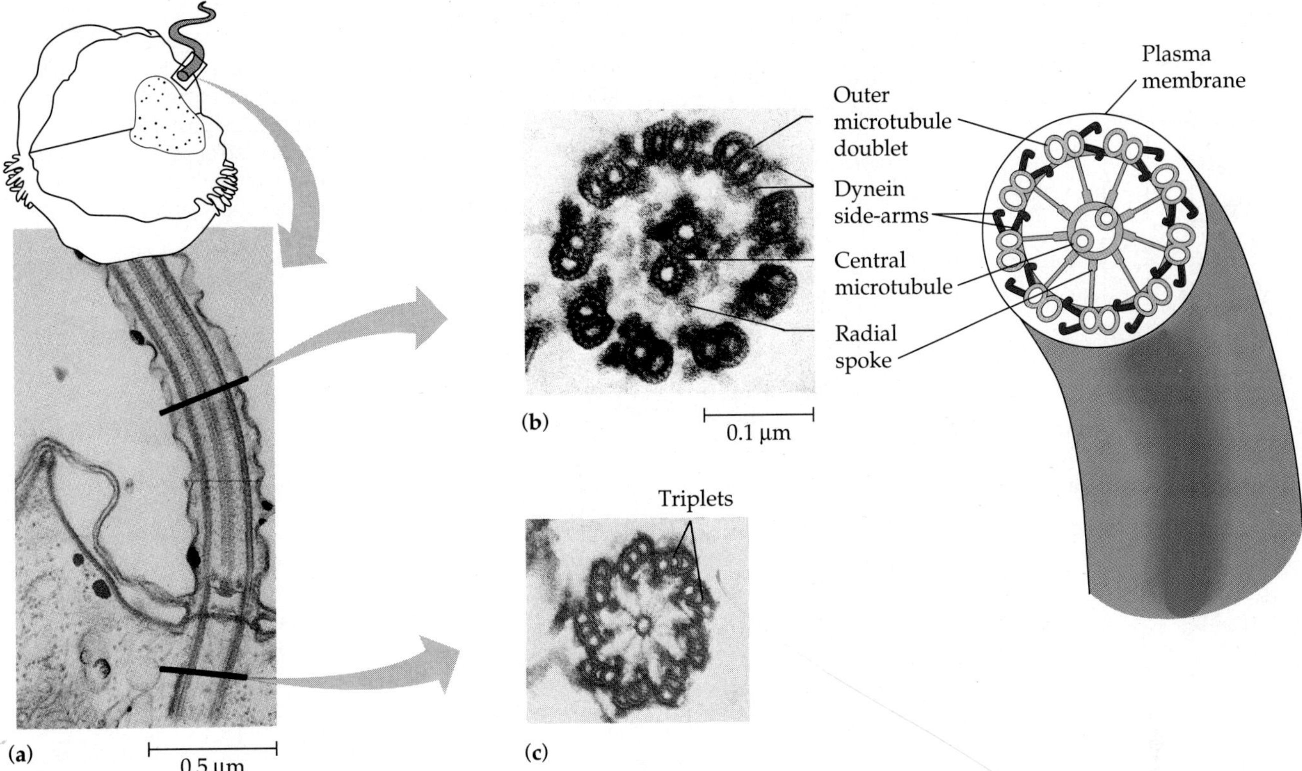

Figure 7.30
Ultrastructure of a eukaryotic flagellum or cilium. (**a**) In this electron micrograph of a longitudinal section of a cilium, microtubules can be seen running the length of the structure (TEM). (**b**) A cross section through the cilium shows the "9 + 2" arrangement of microtubules (TEM). (**c**) The basal body anchoring the cilium or flagellum to the cell has a ring of nine microtubule triplets (structurally identical to a centriole). The nine doublets extend into the basal body, where each doublet joins another microtubule to form the ring of nine triplets. The two central microtubules terminate above the basal body (TEM).

surface of the other, as demonstrated by experiments where portions of microtubule doublets are extracted from cilia and then energized with ATP (Figure 7.31a). However, if this process occurred in an intact cilium, the organelle would simply elongate without moving anywhere. For lateral movement of a cilium, the dynein "walking" must be restrained; that is, it must have something to pull against, as when the muscles in your arm pull against your bones to move your elbow. In cilia and flagella, this restraint is provided by the radial spokes that anchor the microtubule doublets to the center of the organelle. These spokes prevent neighboring doublets from sliding past each other very far. Instead, they cause the microtubules to bend, thus moving the whole organelle (Figure 7.31b and c). In the beating mechanism of cilia and flagella, we see once again that structure fits function.

Microfilaments and Movement

Microfilaments are solid rods about 7 nm in diameter. They are built from molecules of **actin**, a globular protein. The actin molecules are linked into chains; two of these chains twisted about each other in a helix form the microfilament.

Microfilaments are best known for their role in muscle contraction. Thousands of actin microfilaments are arranged parallel to one another along the length of a muscle cell, interdigitated with thicker filaments made of a protein called **myosin** (Figure 7.32a). Contraction of the cell results from the actin and myosin microfilaments sliding past one another, which shortens the cell. The motor molecules are the myosin proteins, which have moving "arms" extending to the actin filaments. ATP powers the movement of these arms. (Muscular contraction will be described more thoroughly in Chapter 45.)

Although microfilaments are especially concentrated and well ordered in muscle cells, they seem to be present to some extent in all eukaryotic cells. Along with the rest of the cytoskeleton, microfilaments function as support. For example, bundles of microfilaments make up the core of microvilli, delicate projections that increase the surface area of cells specialized for the transport of material across the plasma membrane. An example of such specialized cells is the nutrient-absorbing cells that line the human intestine (Figure 7.32b).

In some parts of the cell, microfilaments are associated with myosin in miniature versions of the arrangement found in muscle cells. These actin–myosin aggre-

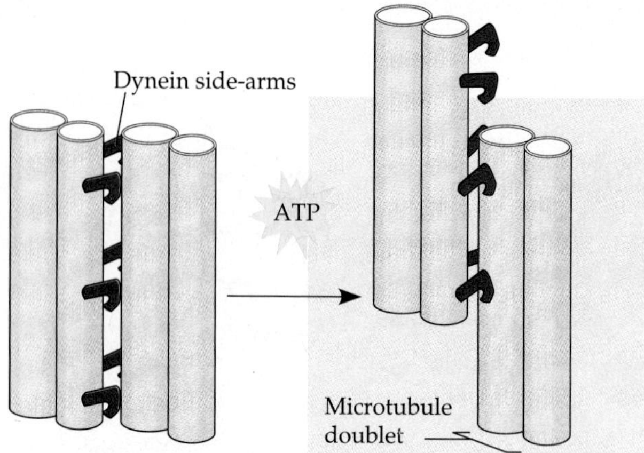

(a) Dynein "walking" in microtubule doublets isolated from cilia

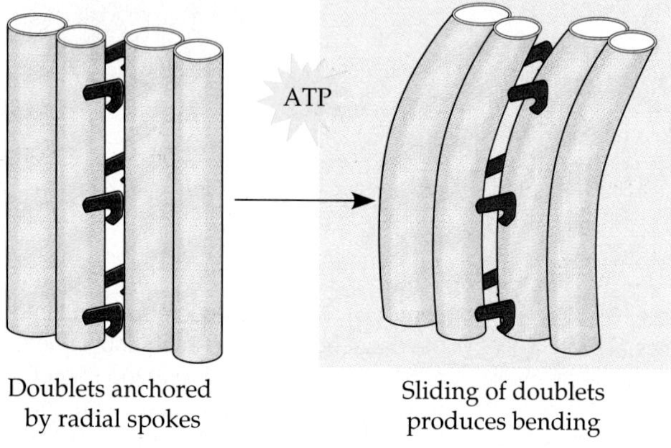

(b) Dynein "walking" in intact cilium or flagellum

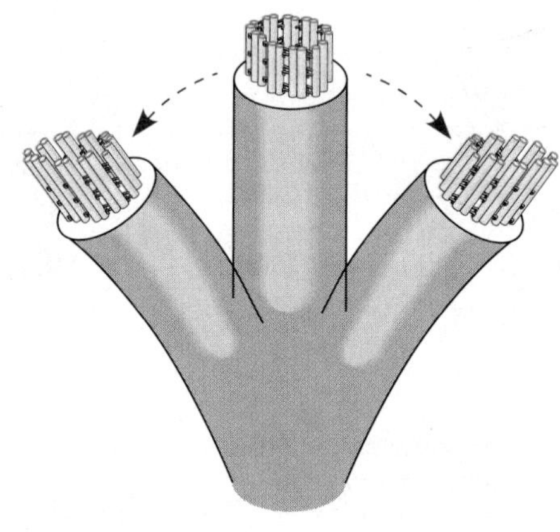

(c) Bending of intact cilium or flagellum

Figure 7.31
How dynein "walking" moves cilia and flagella. (**a**) In a solution containing ATP as an energy source, the grabbing and pulling actions of dynein side-arms cause microtubule doublets that have been removed from cilia to slide past each other. (**b**) In an intact cilium or flagellum, the radial spokes limit linear displacement of microtubules. This restraint causes dynein "walking" to distort the microtubules, causing the whole flagellum or cilium to bend, as shown in (**c**).

gates are responsible for localized contractions of cells. For example, when an animal cell divides, it is pinched in two by a contracting belt of microfilaments. In addition, microfilaments function in the elongation and retraction of cellular extensions called **pseudopodia** (Gr. *pseudes,* "false," and *pod,* "foot") that enable the whole cell to move along a surface. Plant cells also have microfilaments, which are involved in **cytoplasmic streaming,** the phenomenon in which the entire cytoplasm flows around and around the cell in the space between the vacuole and plasma membrane. This movement, which is especially common in large plant cells, speeds the distribution of materials within the cell and from cell to cell.

Intermediate Filaments

Intermediate filaments are named for their diameter, which, at 8 to 12 nm, is larger than the diameter of microfilaments but smaller than that of microtubules. Intermediate filaments actually comprise a diverse class of cytoskeletal elements, differing in protein composition from one type of cell to another. This heterogeneity contrasts with microtubules and microfilaments, which are consistent in diameter and composition in all eukaryotic cells.

Intermediate filaments are also more permanent fixtures of cells than are microfilaments and microtubules, which are often disassembled and reassembled in various parts of a cell. Chemical treatments that remove microfilaments and microtubules from the cytoplasm leave a web of intermediate filaments that retains its original shape. Such experiments suggest that intermediate filaments are especially important in reinforcing the shape of a cell and fixing the position of certain organelles. For example, the nucleus commonly sits within a cage made of intermediate filaments, fixed in location by branches of the filaments that extend into the cytoplasm. In cases where the shape of the entire cell is correlated with function, intermediate filaments support that shape. For instance, the long extensions (axons) of nerve cells that transmit impulses are strengthened by one class of intermediate filament.

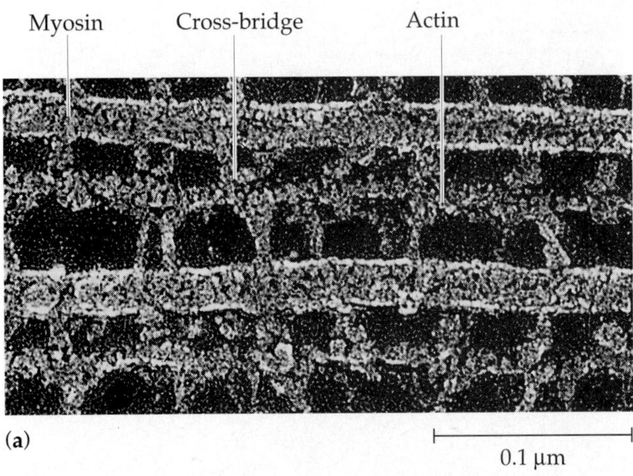

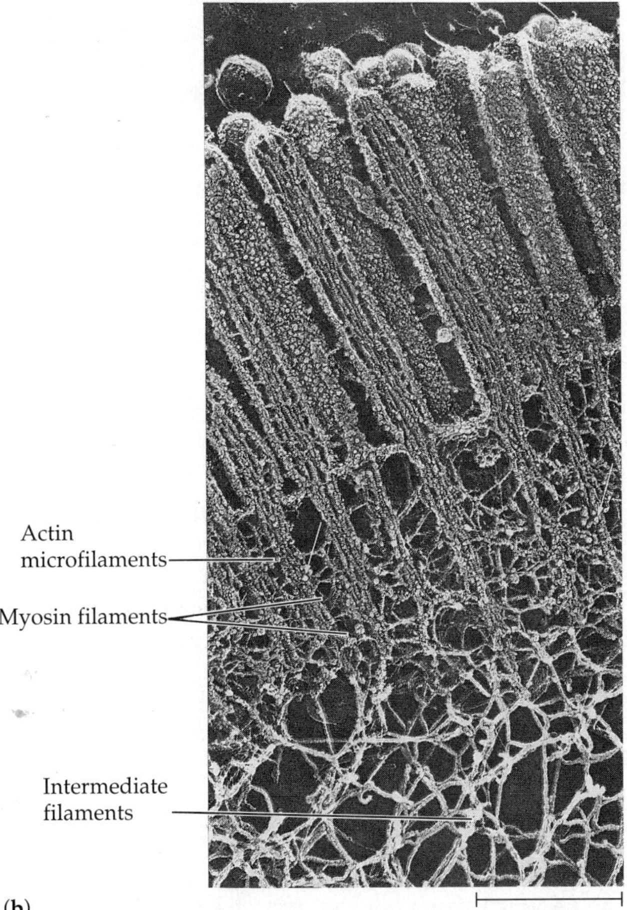

Figure 7.32
Functions of microfilaments. (**a**) In muscle cells, actin filaments interdigitate with thicker filaments made of the protein myosin, a motor molecule. Myosin has cross-bridges that interact with the actin, "walking" the two types of filaments past each other. The teamwork of many such sliding filaments enables the entire muscle cell to shorten (TEM). (**b**) The surface area of this intestinal cell is increased by its many microvilli, cellular extensions reinforced by bundles of microfilaments. The actin is associated with myosin filaments at the bases of the microvilli, and the whole apparatus is anchored to a network of intermediate filaments (TEM).

Specialized for bearing tension, the various kinds of intermediate filaments may function as the framework of the entire cytoskeleton.

THE CELL SURFACE

Having criss-crossed the interior of the cell to explore various organelles, we complete our tour of the cell by returning to the surface of this microscopic world, where there are additional structures with important functions. Although the plasma membrane is usually regarded as the boundary of the living cell, most cells synthesize and secrete coats of one kind or another that are external to the plasma membrane.

Cell Walls

The **cell wall** is one of the features of plant cells that distinguish them from animal cells. The wall protects the plant cell, maintains its shape, and prevents excessive uptake of water (see Chapter 8). On the level of the whole plant, the strong walls of specialized cells hold the plant up, against the force of gravity. Prokaryotes, fungi, and some protists also have cell walls, but we will postpone discussion of them until Unit Five.

Plant cell walls are much thicker than the plasma membrane, ranging from 0.1 to several μm. The exact chemical composition of the wall varies from species to species and from one cell type to another in the same plant, but the basic design of the wall is consistent (see Figure 5.10). Fibers made of the polysaccharide cellulose are embedded in a matrix of other polysaccharides and protein. This arrangement of strong fibers surrounded by an amorphous matrix is the same basic architectural design found in steel-reinforced concrete and in fiberglass.

A young plant cell first secretes a relatively thin and flexible wall called the **primary cell wall** (Figure 7.33). Between primary walls of adjacent cells is the **middle lamella,** a thin layer rich in sticky polysaccharides called pectins. The middle lamella glues the cells together (pectin is used as a thickening agent in jams and jellies). When the cell matures and stops growing, it strengthens its wall. Some cells do this simply by secreting hardening substances into the primary wall. Other plant cells add a **secondary cell wall** between the plasma membrane and the primary wall. The secondary wall, often deposited in several laminated layers, has a strong and durable matrix that affords the cell protection and support. Wood, for example, consists mainly of secondary walls. (The development of cell walls will be discussed in more detail in Chapter 31.)

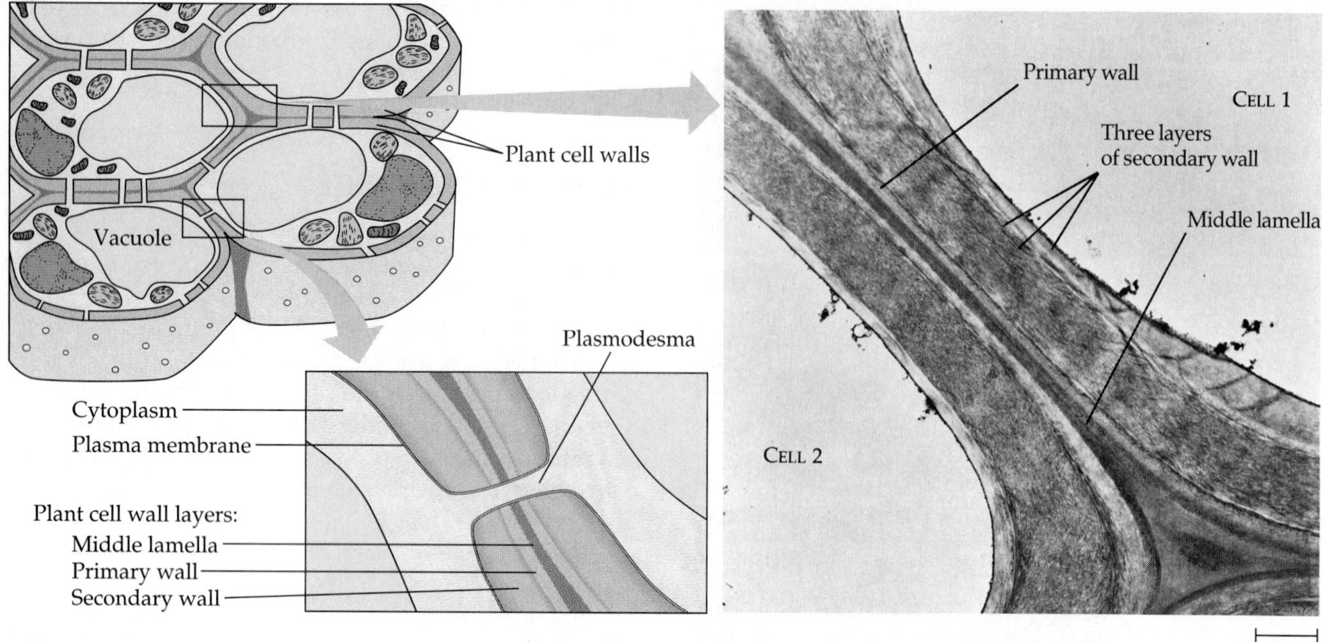

Figure 7.33
Plant cell walls. Young cells first construct thin primary walls, often adding stronger secondary walls to the inside of the primary wall when growth ceases. A sticky middle lamella cements adjacent cells together. Thus, the multilayered partition between these cells consists of adjoining walls individually secreted by the cells. The walls do not isolate the cells: The cytoplasm of one cell is continuous with the cytoplasm of its neighbors via plasmodesmata, channels through the walls (TEM).

The Glycocalyx of Animal Cells

Animal cells lack structured walls, but many have a fuzzy coat called the **glycocalyx,** which is made of sticky oligosaccharides. The glycocalyx strengthens the cell surface and helps glue cells together. The oligosaccharides also contribute to cell–cell recognition by serving as unique identification tags for specific types of cells. The glycocalyx includes carbohydrates attached by covalent bonds to both the proteins and the lipids of the plasma membrane, as well as glycoproteins (carbohydrate–protein complexes) secreted by the cell.

The adhesion of animal cells made possible by their cell coats is often augmented by intercellular junctions between the membranes of bordering cells.

Intercellular Junctions

The many cells of an animal or plant are integrated into one functional organism. Neighboring cells often adhere, interact, and communicate through special patches of direct physical contact.

It might seem that the nonliving cell walls of plants would isolate cells from one another. In fact, the walls are perforated with channels called **plasmodesmata** (singular, plasmodesma; Gr. *desmos,* "to bind"), through which strands of cytoplasm connect the living contents of adjacent cells (see Figure 7.33). This unifies most of the plant into one living continuum. The plasma membranes of adjacent cells are continuous through a plasmodesma; the membrane lines the channel. Water and small solutes can pass freely from cell to cell, a transport that is enhanced by cytoplasmic streaming.

In animals, there are three main types of intercellular junctions: **tight junctions, desmosomes,** and **gap junctions.** These are illustrated and described in detail in Figure 7.34.

THE CELL: A LIVING UNIT GREATER THAN THE SUM OF ITS PARTS

From our panoramic view of the cell's overall compartmental organization to our closeup inspection of each organelle's architecture, this tour of the cell has provided many opportunities to correlate structure with function. Table 7.3, on p. 146, reviews the functions of the cell's parts. But even as we dissect the cell, remember that none of its organelles works alone. As an example of cellular integration, consider the microscopic scene in Figure 7.35. The large cell is a macrophage. It helps defend the body against infec-

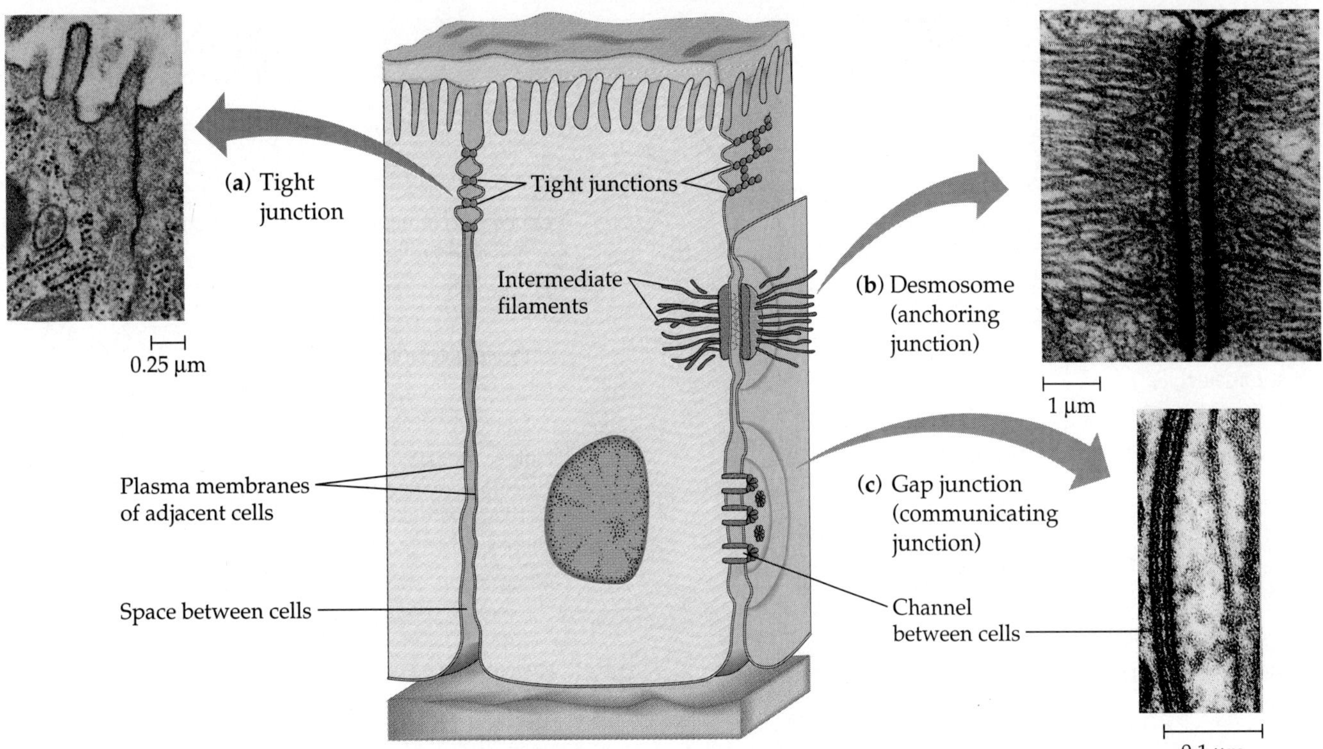

Figure 7.34
Intercellular junctions in animals. These specialized connections are especially common in epithelial tissue, which lines the internal surfaces of the body. Here, we use epithelial cells lining the intestine to describe three kinds of intercellular junctions, each with a structure well adapted for its function. (**a**) Tight junctions. These connections form continuous belts around the cell. The membranes of neighboring cells are actually attached at a tight junction, forming a seal that prevents leakage of extracellular fluid across a layer of epithelial cells. For example, the tight junctions of the intestinal epithelium keep the contents of the intestine separate from the body fluid on the opposite side of the epithelium (TEM). (**b**) Desmosomes. These junctions function like rivets, fastening cells together into strong epithelial sheets. Intermediate filaments made of the sturdy protein keratin, the same protein found in your hair and fingernails, reinforce desmosomes (TEM). (**c**) Gap junctions. These connections provide cytoplasmic channels between adjacent cells. Special membrane proteins surround each pore, which is wide enough for salts, sugars, amino acids, and other small molecules to pass (TEM). In the muscle tissue of the heart, the flow of ions through gap junctions coordinates the contractions of the cells (see Chapter 38). Gap junctions are especially common in animal embryos, where chemical communication between cells is essential for development (see Chapter 43).

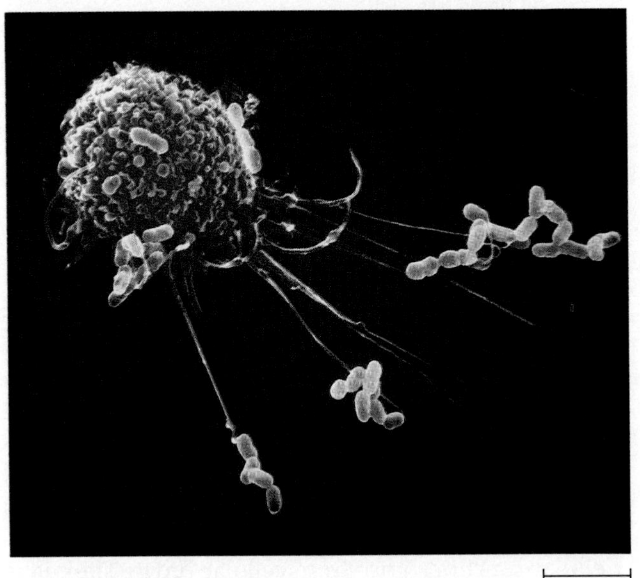

Figure 7.35
The emergence of cellular functions from the cooperation of many organelles. The ability of this macrophage (orange) to recognize, apprehend, and destroy bacteria (green) is a coordinated activity of the whole cell. The cytoskeleton, lysosomes, and plasma membrane are among the components that function in phagocytosis. Other cellular functions are also emergent properties that depend on interactions of the cell's parts (colorized SEM).

Table 7.3 A Summary of Eukaryotic Cell Structure and Function

General Function(s)	Organelle(s)	Specific Function(s)
Genetic control of the cell	Nucleus	
	Chromosomes	Location of inherited DNA, which programs cell's synthesis of proteins
	Nucleolus	Production of ribosomes
Manufacture of macromolecules	Ribosomes	Polypeptide synthesis
	Rough ER	Final steps in synthesis of membane proteins and secretory proteins; membrane production
	Smooth ER	Lipid synthesis Calcium ion storage and release in muscle cells Carbohydrate metabolism in liver cells Detoxification in liver cells
	Golgi apparatus	Modification, storage, sorting, and dispatching of cell's chemical products
Cellular maintenance	Lysosomes	Digestion of food, foreign materials, and damaged organelles
	Peroxisomes	Lipid and carbohydrate metabolism Enzymatic disposal of hydrogen peroxide
	Central vacuole (in plants)	Intracellular digestion, storage, protection, growth
Energy processing	Mitochondria	Cellular respiration, synthesis of ATP
	Chloroplasts (in plants and some protists)	Conversion of light energy to chemical energy
Support, movement, and communication between cells	Cytoskeleton	Maintenance of cell shape Anchorage for organelles Cell movement Movement of organelles within cells
	Cell wall (in plants, fungi, and some protists)	Maintenance of cell shape and skeletal support Surface protection Fusion of cells in tissues
	Glycocalyx (in animals)	Surface protection Binding of cells in tissues Intercellular recognition
	Intercellular junctions	Communication between cells Binding of cells in tissues

tions by ingesting bacteria (the smaller green cells). The macrophage creeps along a surface and reaches out with cellular extensions that help it catch the bacteria. Actin filaments interact with other elements of the cytoskeleton in these movements. After the macrophage engulfs the bacteria, they are destroyed by lysosomes. The elaborate endomembrane system, which includes the ER and the Golgi apparatus, produces the lysosomes. The digestive enzymes of the lysosomes and the proteins of the cytoskeleton are all made on ribosomes. And the synthesis of these proteins is programmed by genetic messages dispatched from the DNA in the nucleus. All of this takes energy, which mitochondria supply in the form of ATP. Cellular functions arise from cellular order: The cell is a living unit greater than the sum of its parts.

STUDY OUTLINE

1. The cell is life's basic unit of structure and function.
2. Cells are open systems, requiring the exchange of materials and energy with their surroundings. They can sense and respond to environmental changes.
3. Evolution is the basis for the correlation of structure and function, a theme encountered repeatedly in our study of the cell.

How Cells Are Studied (pp. 117–121)

1. Most organelles of cells are too small to be viewed with the light microscope. Our current understanding of cell structure and function is due largely to the development of electron microscopy and cell-fractionation techniques.
2. Cytology is the study of cell structure. Modern cell biology integrates cytology and biochemistry.

The Geography of the Cell: A Panoramic View (pp. 121–123)

1. Prokaryotic cells lack true nuclei and membrane-enclosed organelles; bacteria and cyanobacteria (kingdom Monera) are prokaryotes. All other organisms are made up of eukaryotic cells with membrane-enclosed nuclei surrounded by cytoplasm, in which are suspended specialized organelles not found in prokaryotic cells.
2. The requirement for a favorable ratio of membrane surface to cell volume sets upper limits on cell size.
3. Eukaryotic cells are surrounded by a plasma membrane and are partitioned into various compartments by a complex system of internal membranes. These internal membranes provide local environments for specific metabolic processes, some of which are catalyzed by enzymes built into the membranes themselves.
4. All membranes consist of phospholipids and proteins. The diversity of membrane function reflects the variation in a membrane's specific molecular composition.

The Nucleus (pp. 123–127)

1. The trademark of a eukaryotic cell is its distinctive nucleus, enclosed in the nuclear envelope. This envelope is a double membrane that maintains the structure of the nucleus and, through its pores, allows for extensive exchange of macromolecules between the nucleus and the cytoplasm.
2. The nucleus contains the genetic material, DNA, organized in a characteristic number of chromosomes in each eukaryotic species. Specialized regions of the chromosomes, called nucleolar organizers, form one or more nucleoli, which synthesize ribosomal subunits. These ribosomal subunits, as well as mRNA produced at other sites on the chromosomes, pass into the cytoplasm and function together in the synthesis of proteins.

Ribosomes (pp. 127–128)

1. Ribosomes are composed of two subunits, which are constructed in the nucleolus from RNA and proteins.
2. Once in the cytoplasm, the ribosomal subunits join with mRNA to form functional ribosomes, which carry out protein synthesis while suspended in the cytoplasm (free ribosomes) or attached to the outside of the membranous endoplasmic reticulum (bound ribosomes).

The Endomembrane System (pp. 128–133)

1. Many of the membranes of a eukaryotic cell are interrelated directly through physical continuity or indirectly through tiny saclike vesicles, pinched-off portions of membrane in transit from one membrane site to another.
2. This interrelated endomembrane system consists of the nuclear envelope, endoplasmic reticulum, Golgi apparatus, lysosomes, vacuoles, and plasma membrane.
3. The endoplasmic reticulum (ER) is a network of membrane-enclosed compartments called cisternae. Rough ER, that portion of the ER with bound ribosomes, is continuous with the nuclear envelope and functions in producing cell membrane and manufacturing proteins for secretion. Membrane and secretory proteins can be transferred to other locations in the cell by the budding of transport vesicles from transitional ER.
4. Smooth ER lacks ribosomes and can synthesize steroids, metabolize carbohydrates, store calcium in muscle cells, and detoxify poisons in liver cells.
5. The Golgi apparatus consists of stacks of membranous sacs that synthesize various macromolecules and also modify, store, sort, and export products of the ER. One side of a Golgi stack, the *cis* face, receives secretory proteins from the transitional ER through ER transport vesicles. Once inside, these proteins can be chemically modified and sorted before release from the *trans* face of the Golgi stack in vesicles.
6. A lysosome is a membrane-enclosed bag of hydrolytic enzymes originating in the rough ER and processed and released from the Golgi apparatus. Its acidic microenvironment is optimal for the functioning of its enzymes in recycling monomers from cell macromolecules and in digesting substances ingested by phagocytosis.
7. Genetic defects responsible for the absence of one or more lysosomal enzymes cause a variety of specific storage diseases, in which substances that cannot be digested accumulate in the cell and interfere with its functioning.
8. The central vacuole of plant cells is formed by the coalescence of smaller vacuoles from the ER and the Golgi apparatus. It is an extremely versatile organelle, functioning in storage, the breakdown of macromolecules, waste disposal, cell elongation, and protection. The vacuole's membrane is called the tonoplast.
9. The endomembrane system is a dynamic assemblage of related components that compartmentalize the cell and facilitate its diverse functions.

Peroxisomes (Microbodies) (pp. 133–134)

1. Peroxisomes function in a variety of metabolic processes that produce hydrogen peroxide as a waste product. An enzyme in the peroxisome then converts the peroxide to water. Certain peroxisomes also have other functions. For example, plant seeds have peroxisomes that convert fat to sugar.

Energy Transducers: Mitochondria and Chloroplasts (pp. 134–136)

1. As open systems, cells obtain energy from their surroundings. In eukaryotic cells, mitochondria and chloroplasts act as energy transducers, converting energy acquired from the environment to forms the cell can use.
2. Both mitochondria and chloroplasts are enclosed in membranes. These organelles contain a small amount of DNA, which programs the synthesis of some of their proteins.
3. Mitochondria are sites of cellular respiration in eukaryotic cells, where energy is released, with the help of oxygen, from chemical fuels such as sugars and fats and is used to restock the cellular supply of ATP.
4. Mitochondria are compartmentalized by an outer smooth membrane and an inner membrane folded into convolutions called cristae. Some of the metabolic reactions of respiration take place in the space enclosed by the inner membrane, the mitochondrial matrix. Enzymes built into the inner membrane also function in respiration.
5. Chloroplasts, a specialized member of a family of plant organelles called plastids, contain chlorophyll and other pigments, which function in photosynthesis. Chloroplasts are enclosed by two membranes surrounding the fluid called stroma, in which are embedded the thylakoids. These flattened sacs may be stacked to form grana.

The Cytoskeleton (pp. 136–143)

1. The cytoskeleton is an integrated network of fibers in the cytoplasm that dictates pathways of cell circulation, anchors organelles, and gives the cell support, shape, and the ability to move. The cytoskeleton is constructed from microtubules, microfilaments, and intermediate filaments.
2. Microtubules are hollow cylinders made of globular proteins called tubulins. In many cells, microtubules radiate out from the centrosome (microtubule-organizing center), an area near the nucleus that surrounds the centrioles in animal cells. Microtubules shape and support the cell, guide the movement of organelles, and participate in chromosome separation during cell division.
3. Cilia and flagella are motile cellular appendages consisting of a "9 + 2" arrangement of microtubules anchored in the basal body, an organelle structurally identical to a centriole. Movement of cilia and flagella occurs when the side-arms, consisting of the motor protein dynein, move the microtubule doublets past each other.
4. Microfilaments, thinner than microtubules, are solid rods built from the protein actin. Microfilaments in muscle cells interact with myosin to cause contraction. They also function in cell division and amoeboid movement, cytoplasmic streaming in plants, and support for cellular projections, such as microvilli.
5. In addition to microtubules and microfilaments, most cells have a variety of intermediate filaments that are important in supporting cell shape and fixing various organelles in place.

The Cell Surface (pp. 143–144)

1. The cells of plants, prokaryotes, fungi, and some protists are reinforced by cell walls external to the plasma membrane. Plant cell walls are composed of cellulose fibers embedded in other polysaccharides and protein.
2. Animal cells possess a glycocalyx, a sticky carbohydrate coat that strengthens the cell surface.
3. Various kinds of intercellular junctions help integrate many cells of a plant or animal into one functional organism. Plants have plasmodesmata, cytoplasmic channels that pass through adjoining cell walls. Cell-to-cell contact in animals is provided by desmosomes, tight junctions, and gap junctions.

The Cell: A Living Unit Greater Than the Sum of Its Parts (pp. 144–146)

1. Organelles do not function in isolation; they cooperate with other organelles. At the cellular level, life emerges from these complex interactions of a cell's parts.

SELF-QUIZ

1. Cell #1 is a spherical cell with 1 square micrometer (μm^2) of membrane surface for every microliter (μL) of volume. Cell #2 is a spherical cell with a diameter five times smaller than the diameter of cell #1. What is the surface-area-to-volume ratio for cell #2?
   a. 0.2 μm^2 per μL
   b. 1 μm^2 per μL
   c. 1.2 μm^2 per μL
   d. 5 μm^2 per μL
   e. 25 μm^2 per μL
2. From the following, choose the statement that correctly characterizes bound ribosomes.
   a. Bound ribosomes are enclosed in their own membrane.
   b. Bound ribosomes are structurally different from free ribosomes.
   c. Bound ribosomes generally synthesize membrane proteins and secretory proteins.
   d. The most common location for bound ribosomes is the cytoplasmic surface of the plasma membrane.
   e. Bound ribosomes are concentrated in the cisternal space of rough ER.
3. Which of the following organelles is *least* closely associated with the endomembrane system?
   a. nuclear envelope
   b. chloroplast
   c. Golgi apparatus
   d. plasma membrane
   e. ER
4. Cells of the pancreas will incorporate radioactively labeled amino acids into proteins. This "tags" newly synthesized proteins and enables a researcher to track the location of these proteins in a cell. In this case, we are tracking an enzyme that is eventually secreted by pan-

creatic cells. Which of the following is the most likely pathway for movement of this protein in the cell?

a. ER ⟶ Golgi ⟶ nucleus
b. Golgi ⟶ ER ⟶ lysosome
c. nucleus ⟶ ER ⟶ Golgi
d. ER ⟶ Golgi ⟶ vesicles that fuse with plasma membrane
e. ER ⟶ lysosomes ⟶ vesicles that fuse with plasma membrane

5. For each of the following electron micrographs, name the structure and describe its functions.

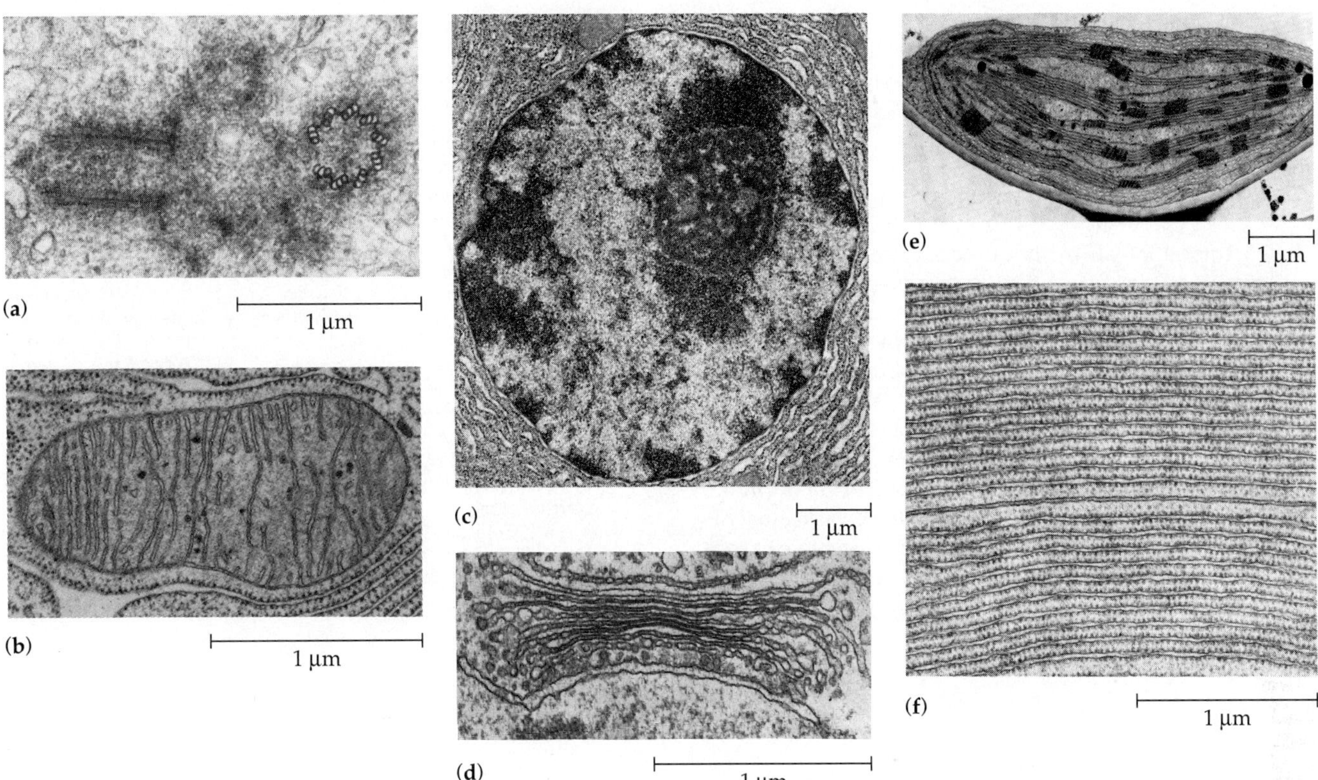

6. Which of the following organelles is common to plant *and* animal cells?
   a. chloroplasts
   b. wall made of cellulose
   c. tonoplast
   d. mitochondria
   e. centrioles

7. Which component is present in a prokaryotic cell?
   a. mitochondria
   b. ribosomes
   c. nuclear envelope
   d. chloroplasts
   e. ER

8. Moving outward from the plasma membrane of a plant cell, which is the correct sequence of wall components?
   a. primary wall—secondary wall—middle lamella
   b. middle lamella—secondary wall—primary wall
   c. primary wall—middle lamella—secondary wall
   d. secondary wall—middle lamella—primary wall
   e. secondary wall—primary wall—middle lamella

9. Which type of cell would probably provide the best opportunity to study lysosomes?
   a. muscle cell
   b. nerve cell
   c. phagocytic white blood cell
   d. leaf cell of a plant
   e. bacterial cell

10. Which of the following structure-function pairs is mismatched?
    a. nucleolus—ribosome production
    b. lysosome—intracellular digestion
    c. ribosome—protein synthesis
    d. Golgi—secretion of cell products
    e. microtubules—muscle contraction

CHALLENGE QUESTIONS

1. An inherited disorder in humans results in the absence of dynein in flagella and cilia. The disease causes respiratory problems and, in males, sterility. What is the ultrastructural connection between these two symptoms?
2. When very small viruses infect a plant cell by crossing its membrane, the viruses often spread rapidly throughout the entire plant without crossing additional membranes. Explain how this occurs.
3. Vinblastine is a drug that interferes with the assembly of microtubules. It is widely used for chemotherapy in treating cancer patients. Suggest a hypothesis to explain how vinblastine slows tumor growth by inhibiting cell division.

SCIENCE, TECHNOLOGY, AND SOCIETY

1. Doctors at a California university removed a man's spleen, standard treatment for a type of leukemia. The disease did not recur. Researchers kept some of the spleen cells alive in a nutrient medium. They found that some of the cells produced a blood protein called GM-CSF, which they are now testing to fight cancer and AIDS. The researchers patented the cells. The patient sued, claiming a share in profits from any products derived from his cells. In 1988, the California Supreme Court ruled against the plaintiff (patient), stating that his suit "threatens to destroy the economic incentive to conduct important medical research." The U.S. Supreme Court agreed. The plaintiff's attorney argued that the ruling left patients "vulnerable to exploitation at the hands of the state." Do you think the plaintiff was treated fairly? Is there anything else you would like to know about this case that might help you make up your mind?

FURTHER READING

Alberts, B., D. Bray, J. Lewis, M. Raff, K. Roberts, and J. D. Watson. *Molecular Biology of the Cell,* 2nd ed. New York: Garland, 1989. A popular text; lucidly written, well illustrated, and comprehensive.

Becker, W. M., and D. W. Deamer. *The World of the Cell,* 2nd ed. Redwood City, CA: Benjamin/Cummings, 1991. A very readable and student-oriented text.

Bretscher, M. S. "How Animal Cells Move." *Scientific American,* December 1987. The role of vesicles in amoeboid movement.

Darnell, J., H. Lodish, and D. Baltimore. *Molecular Cell Biology,* 2nd ed. New York: Scientific American Books, 1990. An authoritative survey of modern cell biology.

DeDuve, C. *A Guided Tour of the Living Cell.* New York: Scientific American Books, 1986. A beautifully illustrated introduction to the cell by the discoverer of lysosomes.

Ezzel, C. "Sticky Situations." *Science News,* June 13, 1992. Structure and function of the glue that holds animal cells together.

Murray, M. "Life on the Move." *Discover,* March 1991. How cellular motors work.

Rothman, J. E. "The Compartmental Organization of the Golgi Apparatus." *Scientific American,* September 1985. A closeup of Golgi structure and function.

Symmons, M., A. Prescott, and R. Warn. "The Shifting Scaffolds of the Cell." *New Scientist,* February 18, 1989. The dynamics of the cytoskeleton.

Weiss, R. "Cellular Transit System Gets Meter Reading." *Science News,* November 24, 1990. Laser beam measurement of the force of organelle movement along microtubules.

8 MEMBRANE STRUCTURE AND FUNCTION

MODELS OF MEMBRANE STRUCTURE
THE TRAFFIC OF SMALL MOLECULES
THE TRAFFIC OF LARGE MOLECULES: EXOCYTOSIS AND ENDOCYTOSIS

The plasma membrane is the edge of life, the boundary that separates the living cell from its nonliving surroundings. A remarkable film only about 8 nm thick—you would have to stack over 8000 such membranes to equal the thickness of this page—the plasma membrane controls traffic into and out of the cell it surrounds. Like all biological membranes, the plasma membrane has **selective permeability;** that is, it allows some substances to cross it more easily than others. One of the earliest episodes in the evolution of life may have been the formation of a membrane that could enclose a solution of different composition from the surrounding solution, while still permitting the selective uptake of nutrients and elimination of waste products (Figure 8.1). This ability of the cell to discriminate in its chemical exchanges with the environment is fundamental to life, and it is the plasma membrane that makes this selectivity possible.

The biological membrane and its control over the passage of substances is the subject of this chapter. We will concentrate on the plasma membrane, the outermost membrane of the cell. However, the general principles of membrane traffic also apply to the many varieties of internal membranes that partition the eukaryotic cell. As we have seen repeatedly in our study so far, cellular structure fits function. To understand how membranes work, therefore, we begin by examining their architecture.

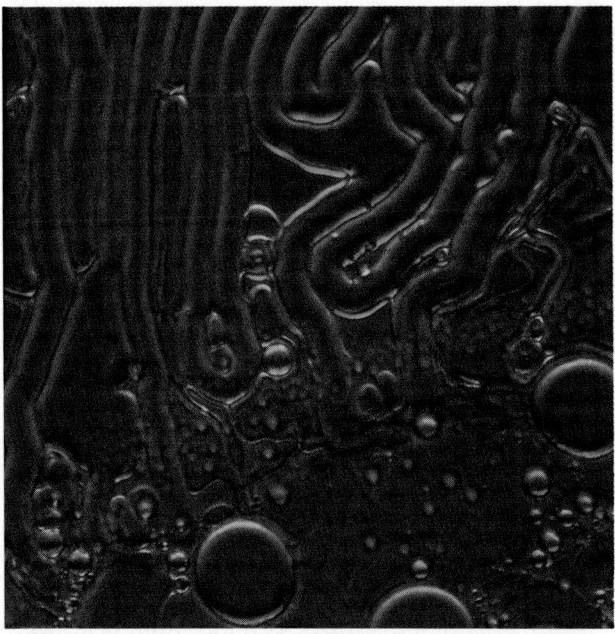

Figure 8.1
The spontaneous formation of membranes: a key step in the origin of life. When phospholipids are mixed with water, they self-assemble to form films. Agitation breaks the films into spheres (LM). Even these primitive membranes have some ability to control the passage of substances between the contents of a sphere and the aqueous environment outside. Phospholipids were probably among the organic molecules that predated life on the primordial Earth, and their spontaneous assembly to form membranes was a step toward protocells that could maintain internal environments differing from the surroundings. In this chapter, you will learn how the structure of cell membranes enables them to control the traffic of molecules.

MODELS OF MEMBRANE STRUCTURE

Lipids and proteins are the staple ingredients of membranes, although carbohydrates are also present. Currently, the most widely accepted model for the arrangement of these molecules in membranes is the fluid mosaic model. We will trace the evolution of this model in some detail as an example of how scientists build on earlier observations and ideas and how they construct models as working hypotheses.

Two Generations of Membrane Models: A Case Study in the Scientific Process

Scientists began building molecular models of the membrane decades before membranes were first resolved by the electron microscope in the 1950s. In 1895, Charles Overton postulated that membranes are made of lipids, based on his observations that substances that dissolve in lipids enter cells much more rapidly than substances that are insoluble in lipids. Twenty years later, membranes isolated from red blood cells were chemically analyzed and found to be composed of lipids and proteins.

Phospholipids are the most abundant lipids in most membranes. The ability of phospholipids to form membranes is built into their molecular structure. A phospholipid is an **amphipathic** molecule, meaning it has both a hydrophilic region and a hydrophobic region (see Figure 5.15). Other types of membrane lipids are also amphipathic.

In 1917, I. Langmuir made artificial membranes by adding phospholipids dissolved in benzene (an organic solvent) to water. After the benzene evaporated, the phospholipids remained as a film covering the surface of the water, with only the hydrophilic heads of the phospholipids immersed in the water (Figure 8.2). In 1925, two Dutch scientists, E. Gorter and F. Grendel, reasoned that cell membranes are actually phospholipid bilayers, two molecules thick. Such a bilayer could exist as a stable boundary between two aqueous compartments, because the molecular arrangement shelters the hydrophobic tails of the phospholipids from water, while exposing the hydrophilic heads to water. Gorter and Grendel measured the phospholipid content of membranes isolated from red blood cells and found just enough of the lipid to cover the cells with two layers. (Ironically, Gorter and Grendel underestimated both the phospholipid content and the surface area of the cells, but the two errors canceled each other. Thus, what turned out to be a correct conclusion was based on flawed measurements.)

If we assume that a phospholipid bilayer is the main fabric of the membrane, where do we place the proteins? Although the heads of phospholipids are hydrophilic, the surface of an artificial membrane consisting of a phospholipid bilayer is less water-absorbent than the surface of an actual biological membrane. This difference could be accounted for if the membrane were coated on both sides with proteins, which generally absorb water. In 1935, H. Davson and J. Danielli incorporated this hypothesis into a molecular model of the membrane. The Davson–Danielli model was a sandwich: a phospholipid bilayer between two layers of globular protein (Figure 8.3a).

Not until the 1950s did biologists finally see membranes with the help of electron microscopes (Figure 8.4). With a thickness of only 7 to 8 nm, the plasma membrane was a little too thin to be the molecular sandwich that Davson and Danielli had predicted. Modifying the model by replacing the globular proteins with continuous layers of protein in the β pleated-sheet configuration fit the model to the observed thickness of the actual membrane (see Chapter 5).

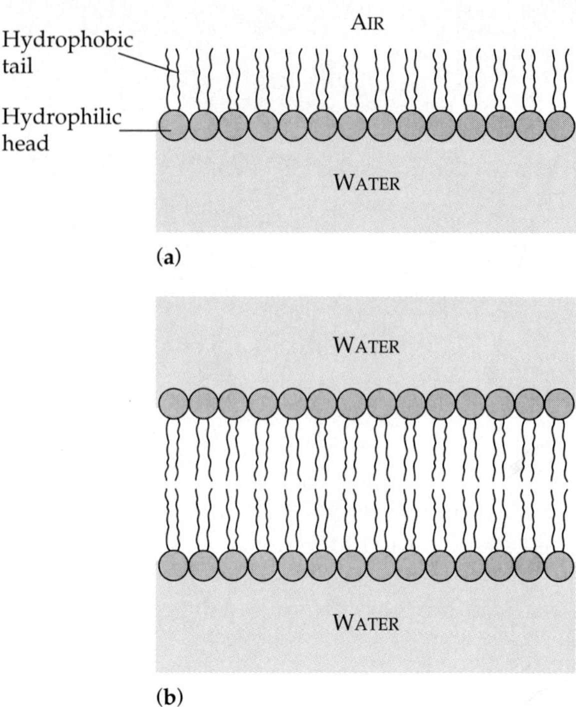

Figure 8.2
Artificial membranes. (**a**) Water can be coated with a single layer of phospholipids. The hydrophilic heads of the phospholipids are immersed in water, and the hydrophobic tails are excluded from water. (**b**) A bilayer of phospholipids forms a stable boundary between two aqueous compartments. The arrangement exposes the hydrophilic parts of the molecules to water and shields the hydrophobic parts from water.

In electron micrographs of cells stained with heavy metal atoms, the plasma membrane is triple-layered, having two electron-dense bands separated by an unstained (electron-transparent) layer. This, too, was interpreted as evidence for the Davson–Danielli model. Most early electron microscopists assumed that the stain adhered to the proteins and hydrophilic heads of the phospholipids, leaving the hydrophobic core of the membrane unstained. In a somewhat circular manner, the electron micrographs and the membrane model became more and more associated as explanations for each other. By the 1960s, the Davson–Danielli sandwich had become widely accepted as the structure of not only the plasma membrane, but of all the internal membranes of the cell. However, by the end of that decade, many cell biologists recognized two problems with the model.

First, the generalization that all membranes of the cell are identical was challenged. Not all membranes

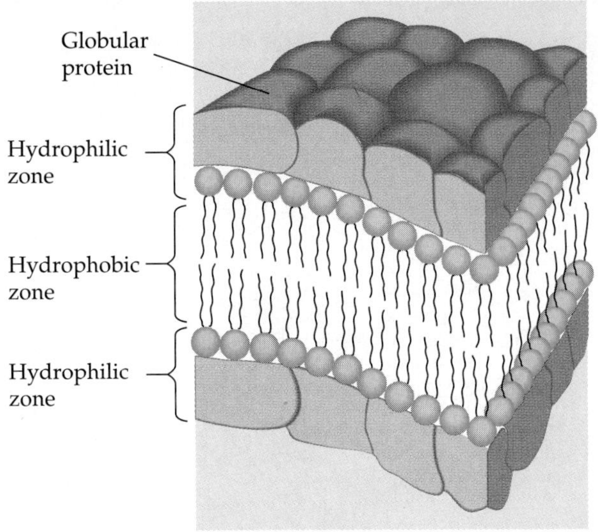

(**a**) Original Davson–Danielli model

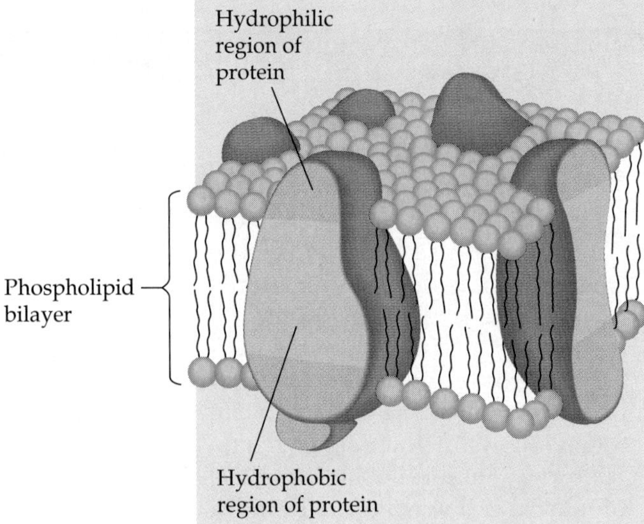

(**b**) Current fluid mosaic model

Figure 8.3
The evolution of membrane models. (**a**) The Davson–Danielli model, proposed in 1935, sandwiched the phospholipid bilayer between two protein layers. With later modifications, this model was widely accepted until about 1970. (**b**) The fluid mosaic model disperses the proteins and immerses them in the phospholipid bilayer, which is in a fluid state. This is our present working model of the membrane.

look alike in the electron microscope. For example, whereas the plasma membrane is 7 to 8 nm thick and has the three-layered structure, the inner membrane of the mitochondrion is only 6 nm thick and in electron micrographs looks like a row of beads. Mitochondrial membranes also have a substantially greater percentage of proteins than plasma membranes, and there are differences in the specific kinds of phospholipids and other lipids. Membranes with different functions differ in chemical composition and structure.

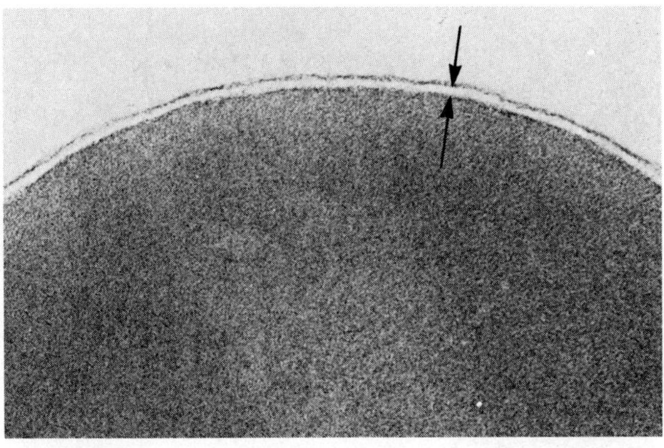

Figure 8.4
The plasma membrane of a red blood cell. In cross section under the transmission electron microscope, the plasma membrane has a three-layered appearance of two dark bands (arrows) separated by a light zone.

The second problem with the sandwich model is in the placement of the proteins. Unlike proteins dissolved in the cytosol, membrane proteins are not very soluble in water. Membrane proteins have hydrophobic and hydrophilic regions; they are amphipathic, as are their phospholipid partners in membranes. If proteins were layered on the surface of the membrane, their hydrophobic parts would be in an aqueous environment, and the proteins would separate the hydrophilic heads of the phospholipids from water.

In 1972, S. J. Singer and G. Nicolson advocated a revised membrane model that placed the proteins in a location compatible with their amphipathic character. Instead of seeing the phospholipid bilayer as coated with solid sheets of protein, Singer and Nicolson proposed that membrane proteins are dispersed and individually inserted into the phospholipid bilayer, with only their hydrophilic portions protruding far enough from the bilayer to be exposed to water. This molecular arrangement would maximize contact of hydrophilic regions of proteins and phospholipids with water, while providing their hydrophobic parts with a nonaqueous environment. According to this model, the membrane is a mosaic of protein molecules bobbing in a fluid bilayer of phospholipids; hence the term **fluid mosaic model** (see Figure 8.3b).

A method of preparing cells for electron microscopy called freeze-fracture has provided the most compelling evidence that proteins are embedded in the phospholipid bilayer of the membrane, rather than being spread upon the surface (see the Methods Box). Freeze-fracture can delaminate a membrane along the

METHODS: FREEZE-FRACTURE AND FREEZE-ETCH

The specimen is frozen at the temperature of liquid nitrogen, then a cold knife is used to fracture the cells. The knife does not cut cleanly through the frozen cells; instead, it cracks the specimen with the fracture plane following the path of least resistance. The fracture plane often follows the hydrophobic interior of a membrane, splitting the lipid bilayer down the middle into a P (protoplasmic) face and an E (exterior) face. The topography of the fractured surface may be enhanced by etching, the removal of water by sublimation (direct evaporation of frozen water to water vapor). The membrane proteins are not split but go with one or the other of the phospholipid layers.

A fine mist of platinum is sprayed from an angle onto the fractured surface of the cell. There will be "shadows" where the platinum is blocked by elevated regions of the fractured cell. A film of carbon is added to strengthen the platinum coat.

The original specimen is digested away with acids and enzymes, leaving the platinum-carbon film as a replica of the fractured surface. It is this replica, not the membrane itself, that is examined in the electron microscope.

The electron micrographs have been superimposed on a drawing of a delaminated membrane (bottom). Note the protein particles.

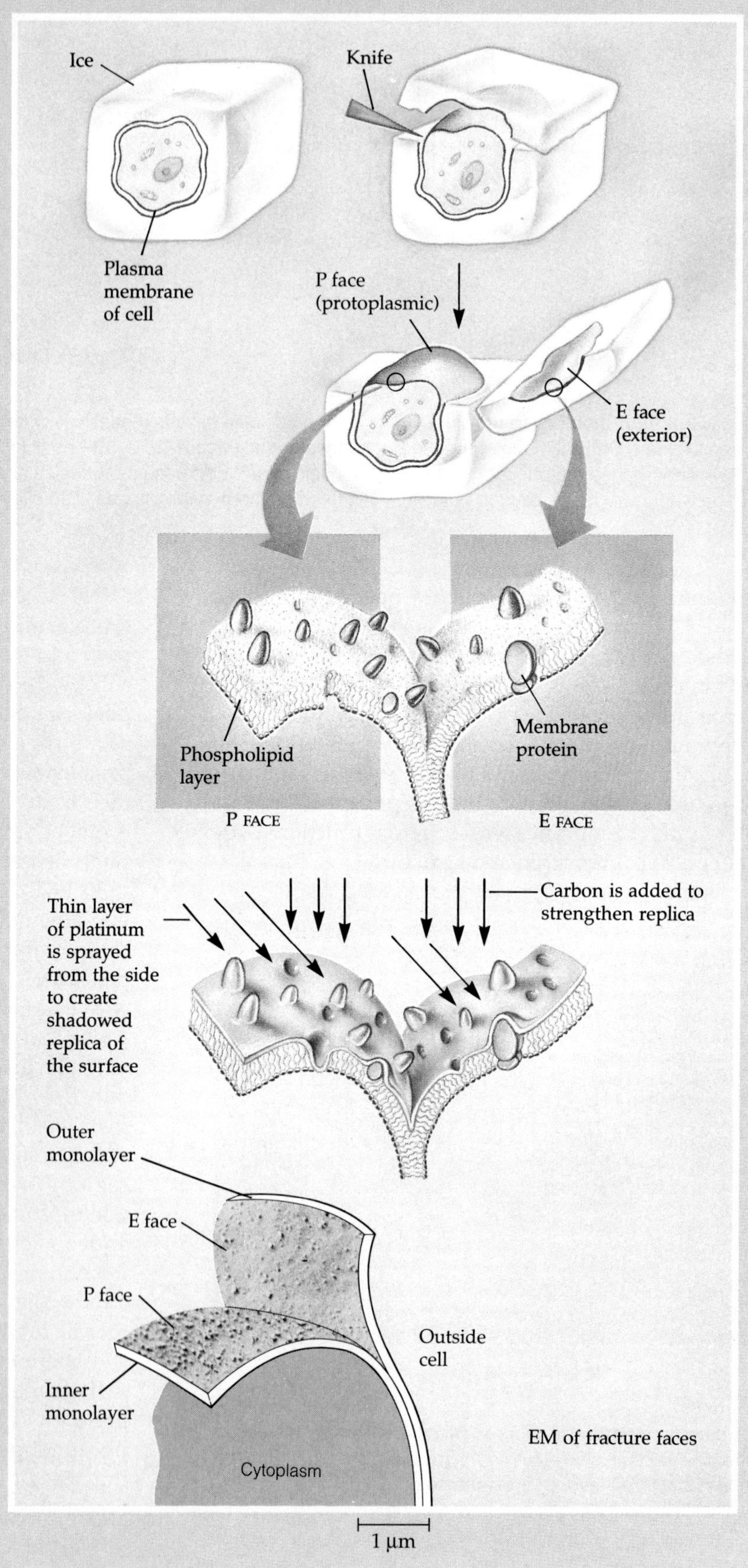

middle of the bilayer, splitting the membrane into outer and inner faces. When the halves of the fractured membrane are viewed in the electron microscope, the interior of the bilayer appears cobblestoned, with protein particles interspersed in a smooth matrix. Proteins penetrate into the hydrophobic interior of the membrane, which would not be the case if the Davson–Danielli sandwich were correct.

We have examined the evolution of our understanding of membrane structure as a case history of how science works. Models are proposed by scientists as ways of organizing and explaining existing information. Replacing one model of membrane structure with another does not imply that the original model was worthless. The acceptance or rejection of a model depends on how well it fits observations and explains experimental results. A good model also makes predictions that shape future research. Models inspire experiments, and few models survive these tests without modification. New findings may make a model obsolete; but even then, it may not be totally scrapped, just revised to incorporate the new observations. Like its predecessor, which endured for 35 years, the fluid mosaic model may eventually be retailored to fit new observations and experiments. For now, it is our most acceptable working model of membrane structure.

The Fluid Mosaic Model: A Closer Look

What exactly does it mean to describe a membrane as a fluid mosaic? Let us begin with the word *fluid*.

The Fluid Quality of Membranes Membranes are not static, solid sheets of molecules locked rigidly in place. A membrane is held together primarily by hydrophobic attractions, which are weaker than covalent bonds (see Chapter 5). Most of the lipids and some of the proteins can drift about laterally in the plane of the membrane (Figure 8.5a). It is rare, however, for a molecule to flip-flop transversely across the membrane, switching from one phospholipid layer to the other; to do so, the hydrophilic part of the molecule would have to cross the hydrophobic core of the membrane.

Phospholipids move along the plane of the membrane rapidly, averaging about 2 μm—the length of a large bacterial cell—per second. Proteins are much larger than lipids and move more slowly, but there is evidence that some membrane proteins do, in fact, drift. This was elegantly demonstrated by experiments in which a human cell was fused to a mouse cell to form a hybrid cell with a continuous plasma membrane, with each animal species contributing part of the membrane. The membrane proteins of the two species were soon mixed, indicating that the proteins must have drifted (Figure 8.6). However, many membrane proteins are unable to move far, because of their attachment to the cytoskeleton.

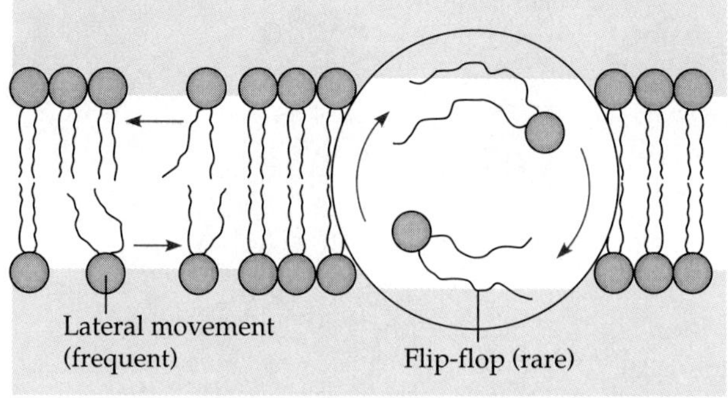

(a)

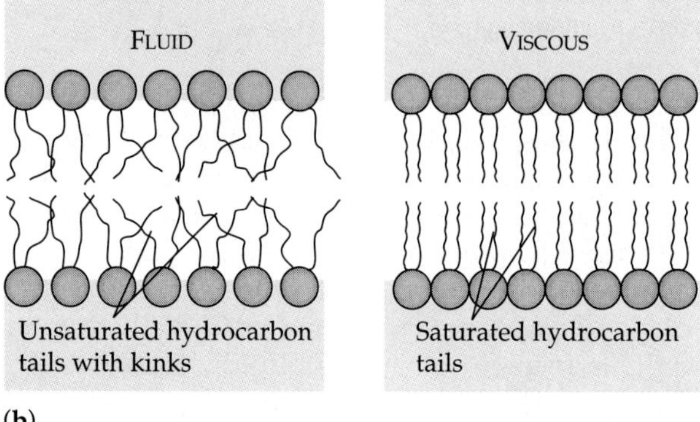

(b)

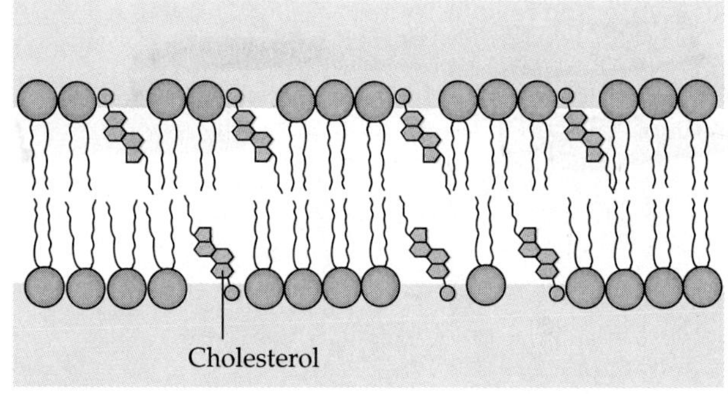

(c)

Figure 8.5
The fluidity of membranes. (**a**) Lipids are free to move laterally (that is, in two dimensions) in a membrane, but flip-flopping across the membrane (in the third dimension) is rare. (**b**) Unsaturated hydrocarbon tails of phospholipids have kinks that keep the molecules from packing together, enhancing membrane fluidity. (**c**) Cholesterol reduces membrane fluidity at moderate temperatures but prevents solidification of membranes at cold temperatures.

A membrane remains fluid as temperature decreases, until finally, at some critical temperature, it solidifies, much as bacon grease forms lard when it cools. The temperature at which a membrane solidifies depends on its lipid composition. The membrane remains fluid to a lower temperature if it is rich in phospholipids with unsaturated hydrocarbon tails (see Chapter 5). Because of kinks where double bonds are located, unsaturated hydrocarbons do not pack together as closely as saturated hydrocarbons (Figure 8.5b).

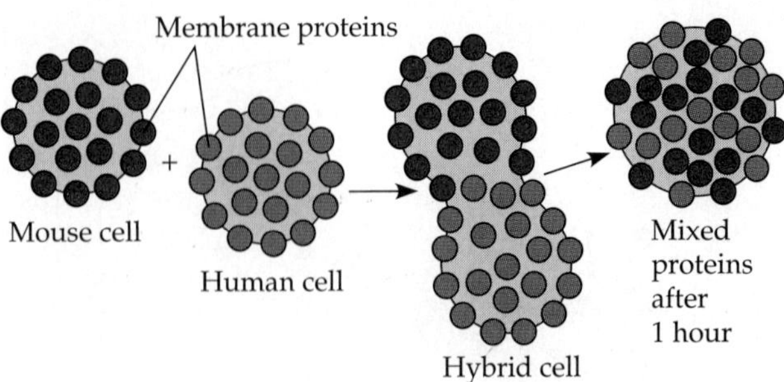

Figure 8.6
Evidence for the drifting of membrane proteins. When researchers fuse a human cell with a mouse cell, it takes less than an hour for the membrane proteins of the two species to completely intermingle.

The steroid cholesterol, which is wedged between the phospholipid molecules in the plasma membranes of eukaryotes, has complex effects on membrane fluidity (Figure 8.5c). At relatively warm temperatures—at 37°C, the body temperature of humans, for example—cholesterol makes the membrane less fluid by restraining the movement of phospholipids. However, because cholesterol also hinders the close packing of phospholipids, it lowers the temperature required for the membrane to solidify.

Membranes must be fluid to work properly. When a membrane solidifies, its permeability changes, and enzymatic proteins in the membrane may become inactive. In renewing its membranes, a cell may alter their lipid composition to some extent as an adjustment to changing temperature. For instance, in many varieties of plants that tolerate extreme cold, such as winter wheat, the percentage of unsaturated phospholipids increases in autumn, an adaptation that keeps the membranes from solidifying during winter.

The functioning membrane, then, is a liquid film consisting of proteins dissolved in a lipid bilayer. It is normally about as fluid as salad oil.

Membranes as Mosaics of Structure and Function Now we come to the word *mosaic.* A membrane is a collage of many different proteins embedded in the fluid matrix of the lipid bilayer (Figure 8.7). The lipid bilayer is the main fabric of the membrane, but proteins determine most of the membrane's specific functions (Figure 8.8). The plasma membrane and the membranes of the various organelles each have their unique collections of proteins. More than 50 kinds of proteins have been found in the plasma membrane of red blood cells, for example, and there are probably many more that have not yet been detected.

Notice in Figure 8.7 that there are two major populations of membrane proteins. **Integral proteins** are inserted into the membrane and penetrate far enough for their hydrophobic regions to be surrounded by the hydrocarbon portions of lipids. Some integral proteins may be unilateral, reaching only partway across the

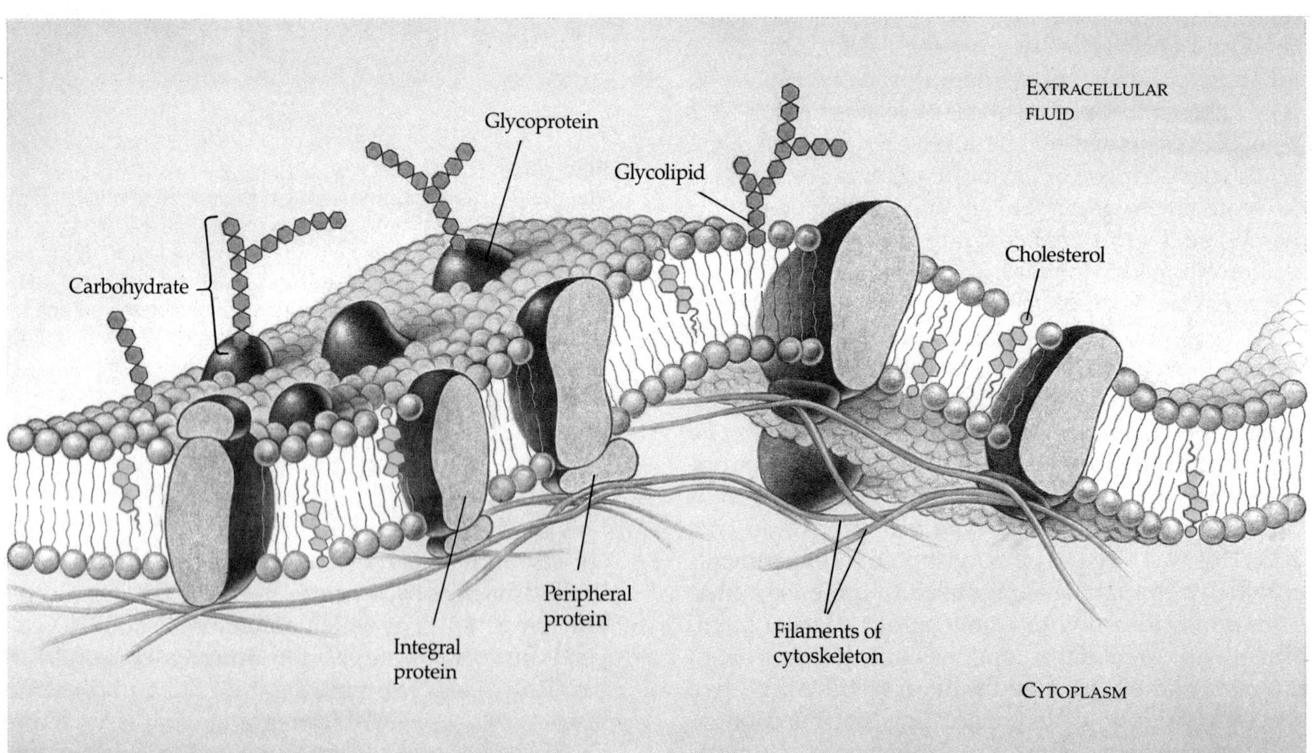

Figure 8.7
The detailed structure of an animal cell's plasma membrane.

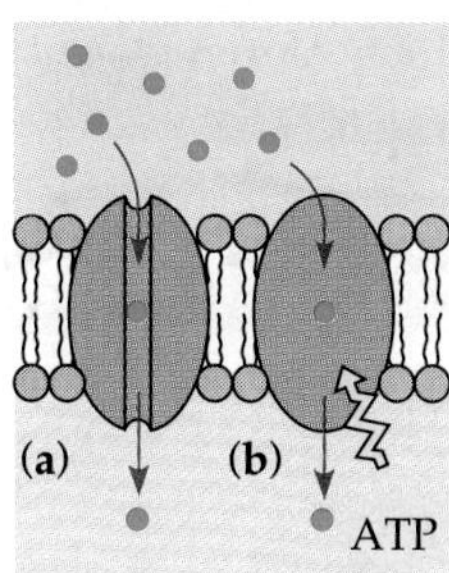

Transport proteins (**a**) A protein that spans the membrane may provide a hydrophilic channel across the membrane that is selective for a particular solute. (**b**) Some transport proteins hydrolyze ATP as an energy source to actively pump substances across the membrane.

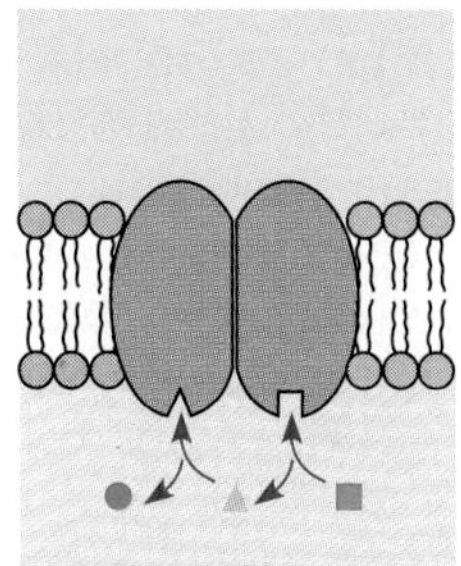

Enzymes A protein built into the membrane may be an enzyme with its active site exposed to substances in the adjacent solution. In some cases, several enzymes are ordered in a membrane as a team that carries out sequential steps of a metabolic pathway.

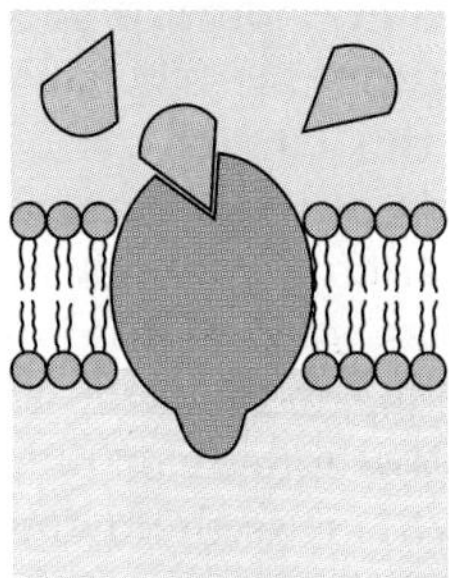

Proteins as receptor sites The portion of a membrane protein exposed to the outside of the cell may have a binding site with a specific shape that complements the shape of a chemical messenger, such as a hormone. If the receptor protein spans the membrane, the external signal may induce a conformational change that activates the portion of the protein facing the cytoplasm to initiate a chain reaction of chemical changes in the cell.

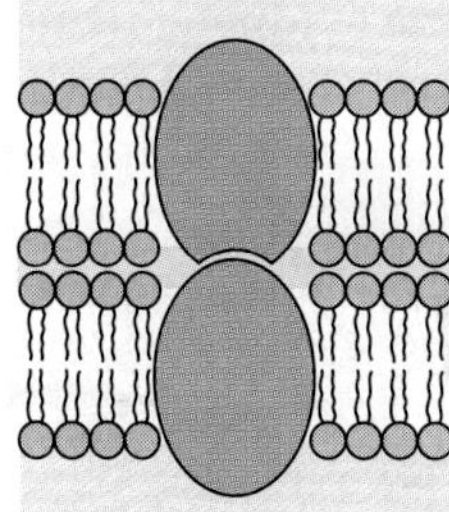

Cell adhesion Membrane proteins of adjacent cells may be hooked together in various kinds of intercellular junctions.

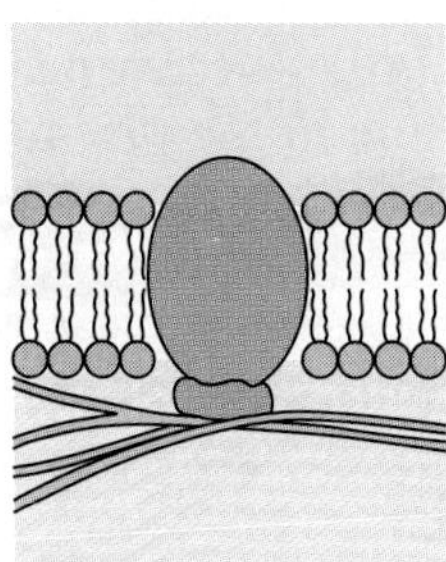

Attachment to the cytoskeleton Actin microfilaments or other elements of the cytoskeleton may be bonded to membrane proteins, a function that is important in maintaining cell shape and in fixing the location of certain membrane proteins.

Figure 8.8
Some functions of membrane proteins. A single protein may perform some combination of these tasks.

membrane. Others—probably most—completely span the membrane. These transmembrane proteins have hydrophobic midsections between hydrophilic ends exposed to the aqueous solutions on both sides of the membrane. **Peripheral proteins** are not embedded in the lipid bilayer at all; they are appendages attached to the surface of the membrane, often to the exposed parts of integral proteins. On the cytoplasmic side of the plasma membrane, some peripheral proteins and their integral protein partners may be held in place by filaments of the cytoskeleton.

Membranes are bifacial, with distinct inside and outside faces. The two lipid layers may differ in specific lipid composition, and each protein has directional orientation in the membrane. The plasma membrane also has carbohydrates, which are restricted to the exterior surface. This asymmetric distribution of proteins, lipids, and carbohydrates is determined as the membrane is being built by the endoplasmic reticulum (Figure 8.9). Thus, the exterior surface of the plasma membrane is topologically equivalent to the interior surfaces of the ER and the other organelles of the endomembrane system.

Membrane Carbohydrates and Cell-Cell Recognition

Cell-cell recognition, the ability of a cell to distinguish one type of neighboring cell from another, is crucial in the functioning of an organism. It is important, for example, in the sorting of cells into tissues and organs in an animal embryo. It is also the basis for rejection of foreign cells (including those of transplanted organs) by the immune system, an important line of defense in vertebrate animals (see Chapter 39). The way cells recognize other cells is by keying on surface molecules, often carbohydrates, on the plasma membrane.

Membrane carbohydrates are usually branched oligosaccharides with fewer than 15 sugar units. An oligosaccharide is a sugar polymer that is shorter than a polysaccharide. Some of these oligosaccharides are covalently bonded to lipids, forming a special class of molecules called glycolipids. (Recall that *glyco* refers to the presence of carbohydrates.) Most, however, are covalently bonded to proteins, which are thereby glycoproteins.

The oligosaccharides on the external side of the plasma membrane vary from species to species, among individuals of the same species, and even from one cell type to another in a single individual. The diversity of the molecules and their location on the cell's outer fringe are features making oligosaccharides likely candidates as markers that distinguish one cell from another. For example, the four human blood groups—designated A, B, AB, and O—reflect variation in the oligosaccharides on the surfaces of red blood cells.

The biological membrane is an exquisite example of a supramolecular structure—many molecules ordered into a higher level of organization—with emergent properties beyond those of the individual molecules.

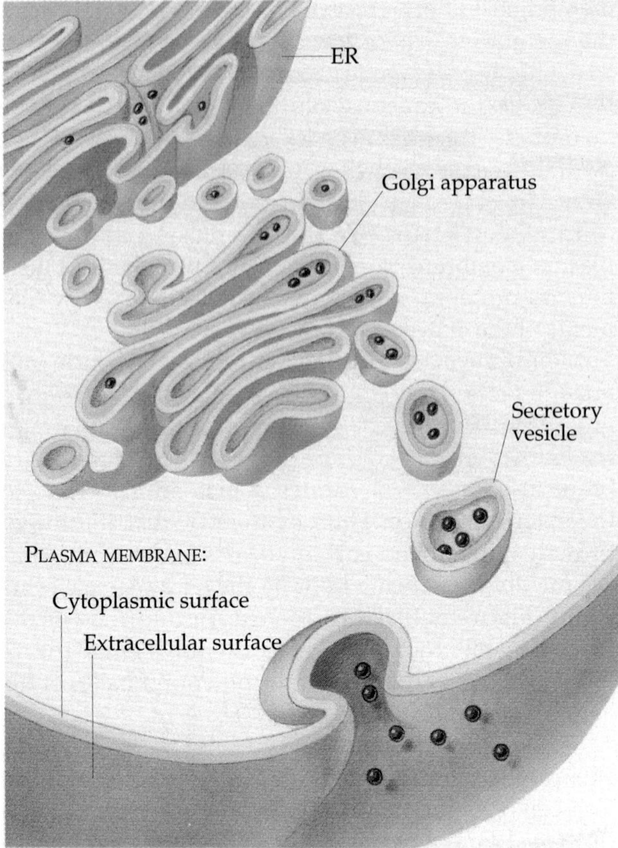

Figure 8.9
Sidedness of the plasma membrane. The membrane is bifacial, with distinct cytoplasmic and extracellular sides. This bifacial quality is determined when the membrane is first synthesized and modified by the ER and Golgi. The diagram color codes the two sides of the membrane, to illustrate that the one facing the lumen (cavity) of the ER, Golgi, and vesicles is topologically equivalent to the extracellular surface of the plasma membrane. The other side faces the cytoplasm, from the time the membrane is produced by the ER to the time it is added to the plasma membrane by fusion of a vesicle.

The remainder of this chapter addresses one of the most important of those properties: the ability to regulate transport across cellular boundaries, a function essential to the cell's existence as an open system. We will see once again that form fits function: The fluid mosaic model of membrane architecture helps to explain how molecules cross membranes.

THE TRAFFIC OF SMALL MOLECULES

There is steady traffic of small molecules across the plasma membrane. Consider the chemical exchanges between a human muscle cell and the extracellular fluid that bathes it. Sugars, amino acids, and other nutrients enter the cell, and waste products of metabolism leave. The cell takes in oxygen for cellular respiration and expels carbon dioxide. It also regulates its concentrations of inorganic ions, such as Na^+, K^+, Ca^{2+}, and Cl^-, by shuttling them one way or the other across the plasma membrane.

Selective Permeability

As mentioned earlier, biological membranes are selectively permeable. This means that although traffic through the membrane is extensive, substances do not cross the barrier indiscriminately. The cell is able to take up many varieties of small molecules and exclude others. Moreover, substances that move through the membrane do so at different rates.

Permeability of the Lipid Bilayer The hydrophobic core of the membrane impedes the transport of ions and polar molecules, which are hydrophilic. The ability of various substances to cross this barrier can be tested by measuring the rate of their transport through an artificial phospolipid bilayer, such as the one in Figure 8.2b. Hydrophobic molecules, such as hydrocarbons and oxygen, can dissolve in the membrane and cross it with ease. If two molecules are equally soluble in lipids, the smaller of the two will cross the membrane faster. Very small molecules that are polar but uncharged can also pass through the synthetic membrane rapidly. Examples are water and carbon dioxide, which are tiny enough to pass between the lipids of the membrane. To larger, uncharged polar molecules, such as glucose and other sugars, the lipid bilayer is not very permeable. The lipid membrane is also relatively impermeable to all ions, even such small ones as H^+ and Na^+. A charged atom or molecule and its shell of water find the hydrophobic layer of the membrane difficult to penetrate (see Chapter 3).

To some extent, we can extrapolate the selective permeability of the artificial membrane to the plasma membrane. Water, carbon dioxide, and small nonpolar molecules all cross the plasma membrane rapidly. However, unlike artificial lipid bilayers, the plasma membrane has proteins, which greatly affect its permeability.

Transport Proteins Biological membranes, unlike artificial ones, are permeable to specific ions and certain polar molecules. These hydrophilic substances avoid contact with the lipid bilayer by passing through **transport proteins** that span the membrane (see Figure 8.8). Transport proteins may be classified into three types: A *uniport* transports a single type of molecule; a *symport* transports two molecules in the same direction; an *antiport* transports two molecules in opposite directions (Figure 8.10). Some transport proteins func-

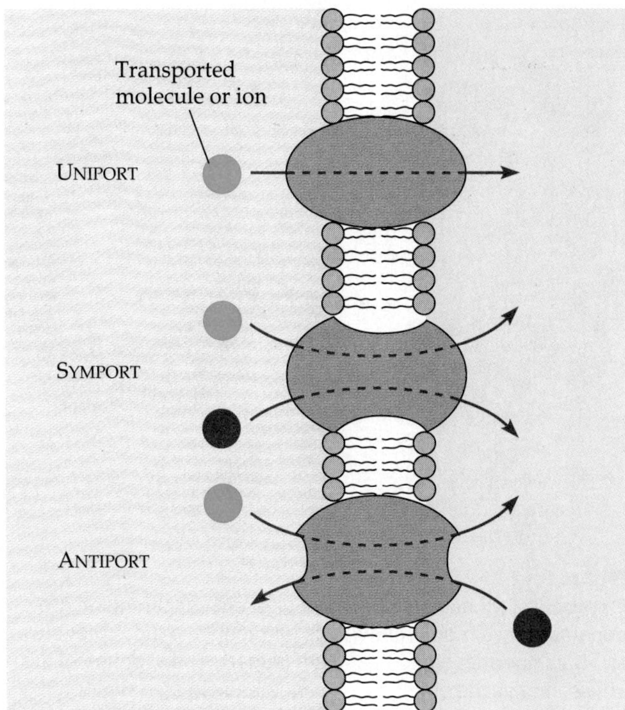

Figure 8.10
Three types of transport proteins. A uniport carries a single solute across the membrane. A symport translocates two different solutes simultaneously in the same direction; transport occurs only if *both* solutes bind to the protein. For example, the plasma membranes of many animal cells have symports that transport a sodium ion and a glucose molecule together from the extracellular solution into the cell. If there is no Na^+ outside the cell, the symport is unable to transport glucose. An antiport exchanges two solutes by transporting one into and the other out of the cell. One antiport, for example, admits sodium ions while expelling calcium ions.

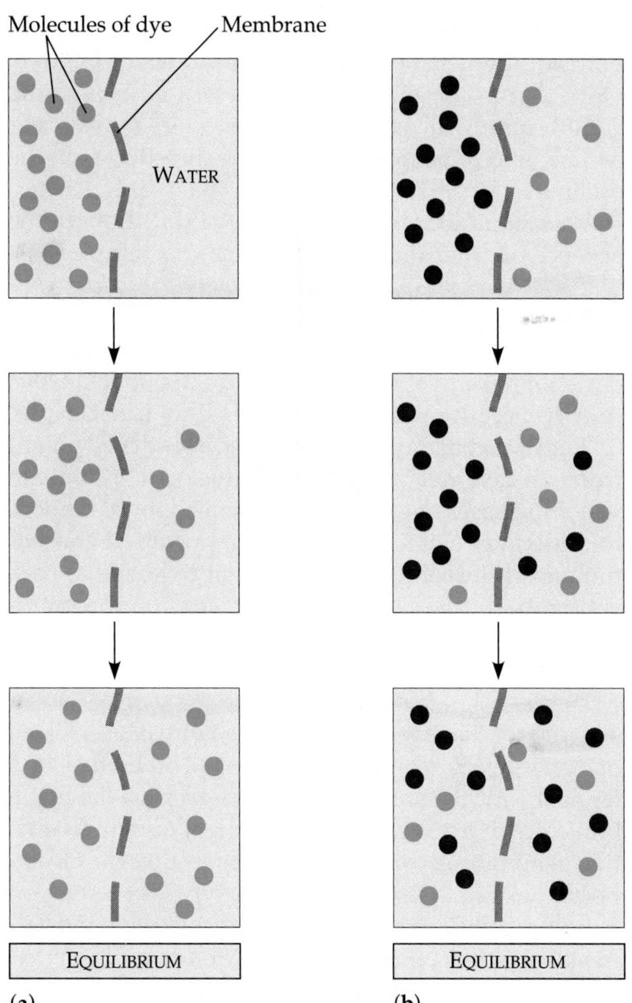

Figure 8.11
The diffusion of a solute. (**a**) A substance will diffuse from where it is more concentrated to where it is less concentrated. Here, molecules of a dye dissolved in water diffuse across a membrane until dynamic equilibrium is reached. (**b**) In this case, solutions of two different dyes are separated by a membrane that is permeable to both dyes. Each dye diffuses down its own concentration gradient. There will be a net diffusion of the green dye toward the left, even though the total solute concentration was initially greater on the left side.

tion by having a channel that certain molecules use as a hydrophilic tunnel through the membrane. Other transport proteins bind to their passengers and physically move them across the membrane. In any case, each transport protein is very specific for the substances it translocates, allowing only a certain molecule or class of closely related molecules to cross the membrane. For example, glucose carried to the human liver in blood enters the liver cells rapidly through specific transport proteins inserted in the plasma membrane. The protein is so selective that it even rejects fructose, a structural isomer of glucose.

The selective permeability of a membrane depends on both the discriminating barrier of the lipid bilayer and the specific transport proteins built into the membrane. But what determines the *direction* of traffic across a membrane? At a given time, will a particular substance enter the cell through the membrane, or leave? Part of the answer can be found in the phenomenon known as diffusion.

Diffusion and Passive Transport

Molecules have intrinsic kinetic energy called thermal motion, or heat. One result of thermal motion is **diffusion,** the tendency for molecules of any substance to spread out into the available space. Each molecule moves randomly, yet diffusion of a population of molecules may be directional. For example, imagine a membrane separating pure water from a solution of a dye dissolved in water. Assume that this membrane is permeable to the dye molecules (Figure 8.11a). Each dye molecule wanders randomly, but there will be a *net* movement of the dye molecules across the membrane to the side that began as pure water. The spreading of the dye across the membrane will continue until both solutions have equal concentrations of the dye. Once that point is reached, there will be a dynamic equilibrium, with as many dye molecules moving per second across the membrane in one direction as in the other.

We can now state a simple rule of diffusion: In the absence of other forces, a substance will diffuse from where it is more concentrated to where it is less concentrated. Put another way, any substance will diffuse down its **concentration gradient.** No work must be done to make this happen; diffusion is a spontaneous process because it decreases free energy (see Chapter 6). Recall that in any system there is a tendency for entropy, or disorder, to increase. Diffusion of a solute in water increases entropy by producing a more random mixture than exists when there are localized concentrations of the solute. It is important to note that each substance diffuses down its *own* concentration gradient, unaffected by concentration differences of other substances (Figure 8.11b).

Much of the traffic across cell membranes occurs by diffusion. Whenever a substance is more concentrated on one side of a membrane than on the other, there is a tendency for the substance to diffuse across the membrane down its concentration gradient (assuming that the membrane is permeable to that substance). One important example is the uptake of oxygen by a cell performing cellular respiration. Dissolved oxygen diffuses into the cell across the plasma membrane. As long as cellular respiration consumes the O_2 as it enters, diffusion into the cell will continue, because the concentration gradient favors movement in that direction.

The diffusion of a substance across a biological membrane is called **passive transport,** because the cell does not have to expend energy to make it happen. (The concentration gradient itself represents potential energy and drives diffusion.) Remember, however, that membranes are selectively permeable and therefore affect the rates of diffusion of various molecules. One molecule that diffuses freely across membranes is water, a fact that has important consequences for cells.

Osmosis: A Special Case of Passive Transport

In comparing two solutions of unequal solute concentration, the solution with a greater concentration of solutes is said to be **hyperosmotic.** The solution with a lesser solute concentration is **hypoosmotic.** These are relative terms that are meaningful only in a comparative sense. For example, tap water is hyperosmotic to distilled water, but hypoosmotic to seawater. In other words, tap water has a higher solute concentration than distilled water, but a lower concentration than seawater. Solutions of equal solute concentration are said to be **isosmotic.**

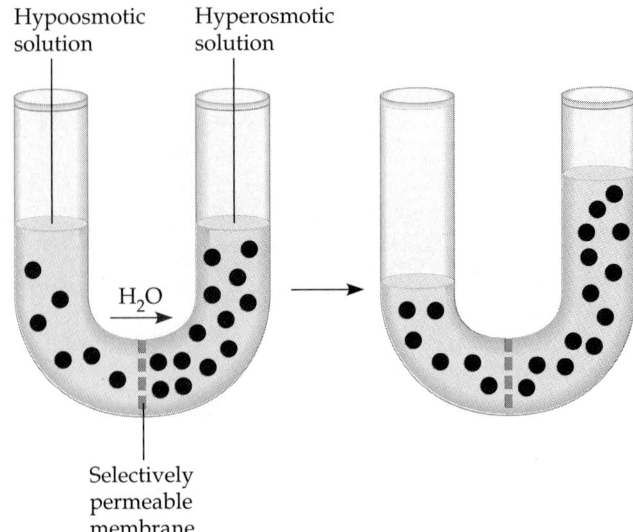

Figure 8.12
Osmosis. Two sugar solutions of different concentration are separated by a membrane that is permeable to the solvent (water) but impermeable to the solute (sugar). The U shape of the vessel makes the changes in volume easy to see. Water will diffuse from the hypoosmotic solution to the hyperosmotic solution. Ideally, osmosis—the diffusion of water across a selectively permeable membrane—would continue until the sugar concentrations on opposite sides of the membrane were equal, and this would probably be the result if the experiment were performed in the weightless conditions of a space shuttle. The two solutions would eventually be isosmotic. Here on Earth, however, the added weight of the rising column of solution into which water moves will eventually force water back across the membrane fast enough to offset movement resulting from the remaining difference in sugar concentrations.

Picture a U-shaped vessel with a selectively permeable membrane separating two sugar solutions of different concentrations (Figure 8.12). The synthetic membrane in this example is permeable to water but impermeable to sugar. Water will diffuse across the membrane from the more dilute (hypoosmotic) sugar solution into the more concentrated (hyperosmotic) solution. (The words *dilute* and *concentrated* here refer to the solute, not to the water.) The concentrations of sugar on opposite sides of the membrane become less different as one solution loses water to the other. Meanwhile, the volume of the more concentrated solution increases and the volume of the more dilute solution decreases, as a result of the water transport.

The diffusion of water across a selectively permeable membrane is a special case of passive transport called **osmosis.** During osmosis, is water diffusing down its own concentration gradient? It seems logical that the solution with the greater concentration of solute (sugar, in this case) has the lesser concentration of solvent (water). But it is not that simple. For a dilute

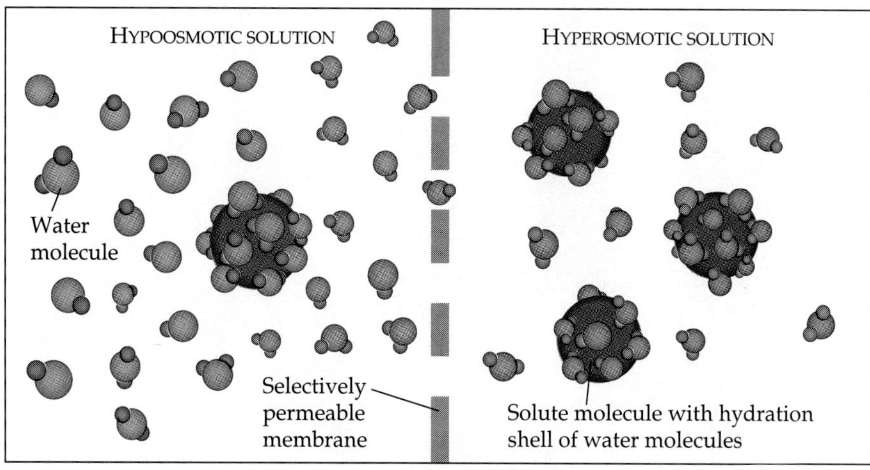

Figure 8.13
The effect of solutes on water m
Without reducing the overall concen of water (the number of water molec per unit volume), solute molecules d duce the proportion of water molecules that can move freely. Water molecules clustered around solute molecules are not free to diffuse across a membrane. Water will move by osmosis from a hypoosmotic to a hyperosmotic solution because the hypoosmotic solution has a greater concentration of unbound water molecules that can cross the membrane.

solution—and most biological fluids are dilute solutions—the presence of solutes does not alter the water concentration significantly. The space taken up by the solute molecules is offset because some of the water becomes more compact as water molecules cluster tightly around molecules of the hydrophilic solutes (see Chapter 3). This bound water is not free to move across the membrane (Figure 8.13). It is not really a difference in total water concentration that causes osmosis, but a difference in the concentration of unbound water that is free to cross the membrane. The effect, however, is the same: Water tends to diffuse across a membrane from a hypoosmotic solution to a hyperosmotic solution.

The direction of osmosis is determined only by a difference in *total* solute concentration, not by the nature of the solutes. Water will move from a hypoosmotic to a hyperosmotic solution even if the hypoosmotic solution has more *kinds* of solutes. Seawater, which has a great variety of solutes, will lose water to a very concentrated sugar solution, because the total solute concentration of the seawater is lower. Water moves across a membrane separating isosmotic solutions at an equal rate in both directions; that is, there is no net osmotic movement of water between isosmotic solutions.

The tendency for a solution to take up water by osmosis can be measured with instruments called osmometers. One type is an apparatus in which pure water is separated from a solution by a membrane that is permeable to the water but not to the solute (Figure 8.14). Ordinarily, osmosis causes the volume of the solution to increase. We can counteract this tendency for the solution to take up water by pushing against the top of the solution with a piston. We can keep the volume of the solution constant by applying just enough pressure with the piston to exactly negate the tendency for water to enter the solution by osmosis. The piston and the solution are thus exerting equal amounts of pressure against each other, called the osmotic pressure of the solution. (Plant biologists use another measurement, called water potential, which will be introduced in Chapter 32.) **Osmotic pressure,** then, is a measure of the tendency for a solution to take up water when separated from pure water by a selectively permeable membrane. The osmotic pressure of pure water is zero. If our osmometer had pure water on both sides of the membrane, no pressure would have to be applied to the piston. A solution's osmotic pressure is proportional to its total solute concentration, or its osmotic concentration. The greater the osmotic concentration, the greater the osmotic pressure and the

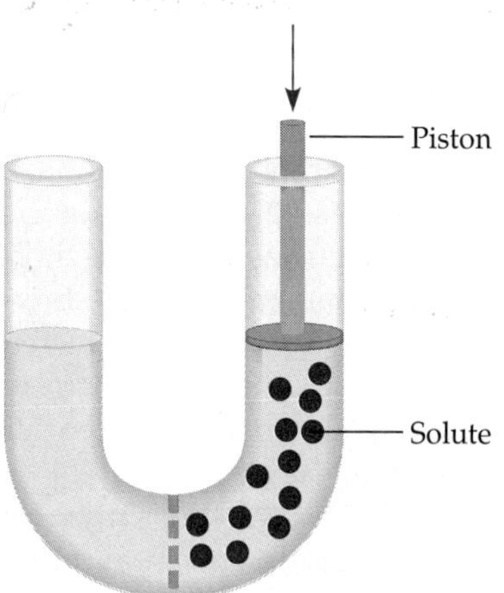

Figure 8.14
Measuring osmotic pressure. In one type of osmometer, a solution is separated from pure water by a synthetic membrane that is permeable to water but not to the solute. The amount of pressure that must be applied to the solution to counteract the tendency for water to enter by osmosis is the solution's osmotic pressure. The greater the osmotic concentration (total solute concentration), the greater the osmotic pressure.

Figure 8.15
The water balance of living cells. How living cells react to changes in the solute concentrations of their environments depends on whether or not they have cell walls. Animal cells do not have cell walls; plant cells do. Unless it has special adaptations to offset the osmotic uptake or loss of water, an animal cell fares best in an isosmotic environment. Plant cells are firm and generally healthiest in a hypotonic environment, where the tendency for continued uptake of water is balanced by the elastic wall pushing back on the cell. (Arrows indicate *net* water movement when the cells are first placed in these solutions.)

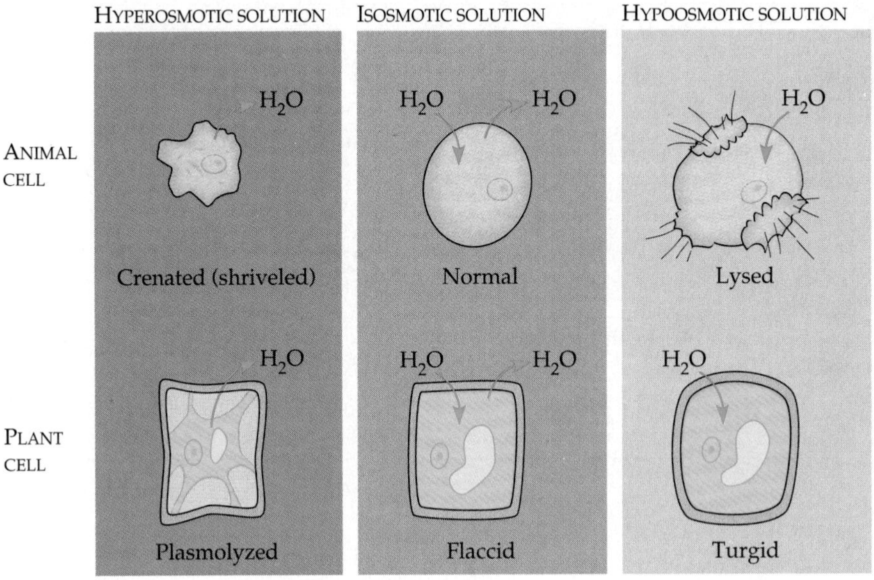

greater the tendency for the solution to take up water from a reservoir of pure water. And if two solutions are separated by a membrane, water will pass from the solution with lower osmotic pressure to the solution with higher osmotic pressure—another way of saying that water moves from a hypoosmotic solution to a hyperosmotic solution.

The movement of water across cell membranes and the balance of water between the cell and its environment are crucial to organisms. Let us now apply to living cells what we have learned about osmosis in artificial systems.

Water Balance of Cells Without Walls If an animal cell is immersed in an environment that is isosmotic to the cell, there will be no net movement of water across the plasma membrane. Water is flowing across the membrane, but at the same rate in both directions. In an isosmotic environment, the volume of an animal cell is stable (Figure 8.15). Now let's transfer the cell to a solution that is hyperosmotic to the cell. The cell will lose water to its environment, crenate (shrivel), and probably die. This is one reason why an increase in the salinity (saltiness) of a lake can kill the animals there. However, taking up too much water can be just as hazardous to an animal cell as losing water. If we place the cell in a solution that is hypoosmotic to the cell, water will enter faster than it leaves, and the cell will swell and lyse (pop) like an overfilled water balloon.

A cell without rigid walls can tolerate neither excessive uptake nor excessive loss of water. This problem of water balance is automatically solved if such a cell lives in isosmotic surroundings. Many marine invertebrates are isosmotic to seawater. The cells of most terrestrial (land-dwelling) animals are bathed in an extracellular fluid that is isosmotic to the cells. Animals and other organisms without rigid cell walls living in hyperosmotic or hypoosmotic environments must have special adaptations for **osmoregulation,** the control of water balance. For example, the protist *Paramecium* lives in pond water, which is hypoosmotic to the cell. Water continually tends to enter the cell, but *Paramecium* has a plasma membrane that is much less permeable to water than the membranes of most other cells. Also, *Paramecium* is equipped with a contractile vacuole, an organelle that functions as a bilge pump to force water out of the cell as fast as it enters by osmosis (Figure 8.16). We will examine other adaptations for osmoregulation in Chapter 40.

Water Balance of Cells With Walls The cells of plants, prokaryotes, fungi, and some protists have walls. Under certain conditions, the wall plays a major role in maintaining water balance between the cell and its external environment. But a wall is of no advantage if the cell is immersed in a hyperosmotic environment. In this case, a plant cell, like an animal cell, will lose water to its surroundings and shrink (see Figure 8.15). As the plant cell shrivels, its plasma membrane pulls away from the wall. This phenomenon, called **plasmolysis,** is usually lethal. The walled cells of bacteria and fungi also plasmolyze in hyperosmotic environments.

When a plant cell is in a hypoosmotic solution—when bathed by rainwater, for example—the wall is a factor in water balance. Again like an animal cell, the plant cell swells as water enters by osmosis. However, the elastic wall will expand only so much before it exerts a back pressure on the cell that offsets the tendency for further water uptake from the hypoosmotic surroundings. When the wall pressure exerts a force equal and opposite to the osmotic pressure of the cell, then water enters and leaves the cell at the same rate, and a dynamic equilibrium is reached. At this point, the cell

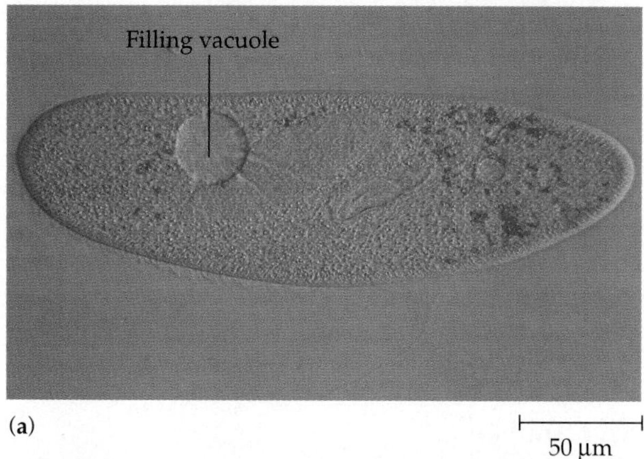

(a)

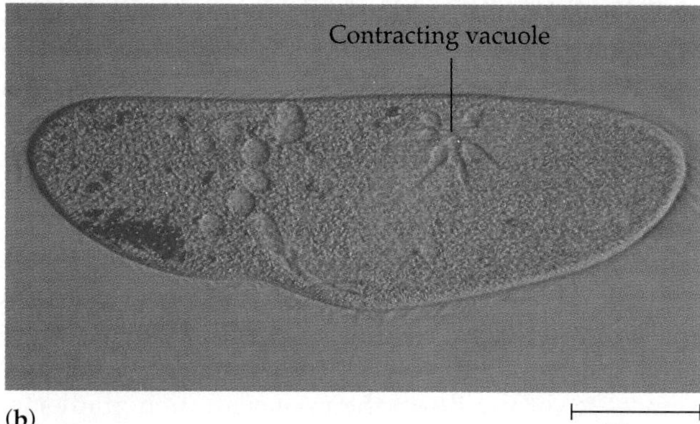

(b)

Figure 8.16
Evolutionary adaptations for osmoregulation in *Paramecium.* *Paramecium* is a genus of freshwater protists. Pond water, which is hypoosmotic, surrounds a *Paramecium* cell, and water tends to enter the cell by osmosis. What keeps the cell, which lacks a cell wall, from bursting? Adaptations enabling the cell to counter the osmotic uptake of water have evolved in *Paramecium.* Compared to the plasma membranes of most other organisms, the *Paramecium* membrane is less permeable to water. This adaptation only slows the uptake of water. The contractile vacuole offsets osmosis by bailing water out of the cell. (**a**) A contractile vacuole fills with fluid that enters from a system of canals radiating throughout the cytoplasm (LM). (**b**) When full, the vacuole and canals contract, expelling fluid from the cell (LM).

is **turgid** (very firm). This is the ideal state for most plant cells. Plants that are not woody, such as most houseplants, depend on turgid cells for mechanical support. This requires that the cells be hyperosmotic to the solution on the outside of their plasma membranes. If a plant cell and its surroundings are isosmotic, then there is no net tendency for water to enter, and the cell is **flaccid** (limp). A plant wilts when its cells are flaccid (Figure 8.17).

Facilitated Diffusion

Let's now turn our attention from the transport of water across a membrane to the traffic of specific solutes dissolved in the water. As mentioned earlier, many polar molecules and ions impeded by the lipid bilayer of the membrane diffuse with the help of transport proteins that span the membrane. This phenomenon is called **facilitated diffusion.**

(a)

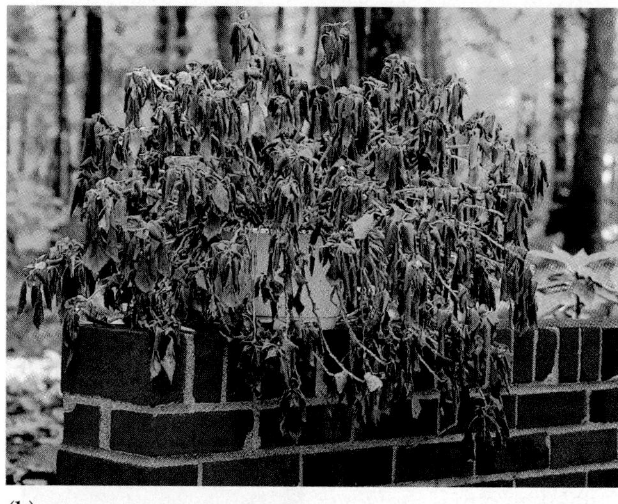

(b)

Figure 8.17
Plants with turgid and flaccid cells. (**a**) The appearance of a whole plant, such as this *Impatiens,* reflects the osmotic status of its individual cells. Nonwoody plants rely mainly on turgid cells to support their stems and leaves against the downward pull of gravity. If a houseplant is adequately watered, its cells are bathed by hypoosmotic solution absorbed from the soil by the roots, which keeps the cells turgid. (**b**) An inadequate water supply causes wilting, the visible evidence that individual cells are flaccid. Unless the cells have actually plasmolyzed, watering will perk up a wilted plant by making its cells turgid again. Overfertilizing the soil can also cause wilting. Fertilizers add solutes to the soil. In excess, these solutes make the soil solution hyperosmotic to the fluids in the plant's roots; the plant may actually lose water to the soil by osmosis.

A transport protein has many of the properties of an enzyme (see Chapter 6). Just as an enzyme is specific for its substrate, a protein is specialized for the solute it transports and has a specific binding site akin to the ac-

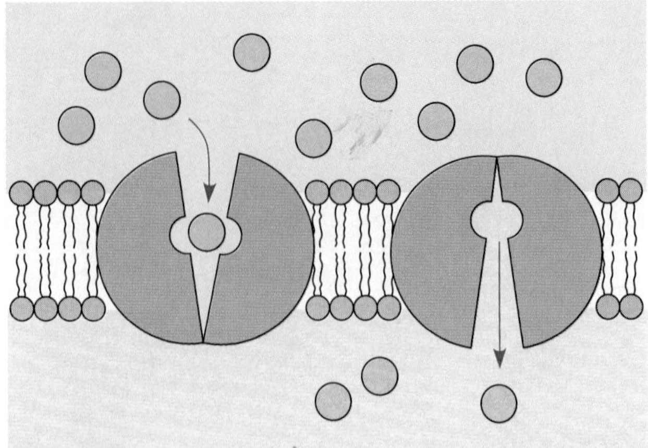

Figure 8.18
One model for facilitated diffusion. The transport protein (violet) alternates between two conformations, moving a solute across the membrane as the shape of the protein changes. The protein can transport the solute in either direction, with the net movement being down the concentration gradient of the solute.

tive site of an enzyme. Like enzymes, transport proteins can be saturated. There are only so many molecules of each type of transport protein built into the plasma membrane, and when these molecules are binding and translocating passengers as fast as they can, transport is occurring at a maximum rate. Also like enzymes, transport proteins can be inhibited by molecules that resemble the normal "substrate." This occurs when the imposter competes with the normally transported solute by binding to the transport protein. Unlike enzymes, however, transport proteins do not usually catalyze chemical reactions. Their function is to catalyze a physical process—the transport of a molecule across a membrane that would otherwise be impermeable to the substance.

Cell biologists are still trying to learn how membrane proteins facilitate diffusion. It is unlikely that the protein acts as a ferry that picks up its passenger on one side of the membrane and then travels across the width of the membrane to deposit the molecule on the opposite side. It is also improbable that the transport protein functions as a revolving door that spins in the membrane, transporting a molecule with each turn. Both of these models are unattractive because the mechanisms would bring the hydrophilic parts of the protein into contact with the hydrophobic interior of the membrane. A model more consistent with what is known about membrane structure is one in which the protein remains in place in the membrane and helps a molecule across by undergoing a subtle change in shape that translocates the binding site from one side of the membrane to the other (Figure 8.18). The changes in shape could be triggered by the binding and release of the transported molecule.

In certain inherited diseases, specific transport systems are either defective or missing altogether. An example is cystinuria, a human disease characterized by the absence of a transport system that carries cystine and other amino acids into kidney cells. Kidney cells normally reabsorb these amino acids from the urine and return them to the blood, but an individual afflicted with cystinuria develops painful stones from amino acids that accumulate and crystallize in the kidneys.

Active Transport

Despite the help of a transport protein, facilitated diffusion is still considered passive transport because the solute is moving down its concentration gradient. Facilitated diffusion speeds the transport of a solute by providing a specific corridor through the membrane, but it does not alter the direction of transport. Some transport proteins, however, can move solutes against their concentration gradients, across the plasma membrane from the side where they are less concentrated to the side where they are more concentrated. This transport is "uphill"; it goes against the tendency for substances to diffuse down their concentration gradients, and therefore requires work. To pump a molecule across a membrane against its gradient, the cell must expend its own metabolic energy; therefore, this type of membrane traffic is called **active transport.**

Active transport is a major factor in the ability of a cell to maintain internal concentrations of small molecules that differ from concentrations in the surrounding environment. For example, compared to its surroundings, an animal cell has a much higher concentration of potassium ions and a much lower concentration of sodium ions. The plasma membrane helps maintain these steep gradients by pumping sodium out of the cell and potassium into the cell.

The work of active transport is performed by specific proteins inserted in membranes. These transport proteins share many of the enzymelike properties of the proteins that function in facilitated diffusion, but membrane proteins engaged in active transport must harness cellular energy to pump molecules against concentration gradients. As in other types of cellular work, ATP supplies the energy for most active transport. One way ATP can power active transport is by transferring its terminal phosphate group directly to the transport protein. This may induce the protein to change its conformation in a manner that translocates a solute bound to the protein across the membrane. One transport system that seems to work this way is the **sodium-potassium pump,** which exchanges sodium (Na^+) for potassium (K^+) across the plasma membrane of animal cells (Figure 8.19). In Chapter 44,

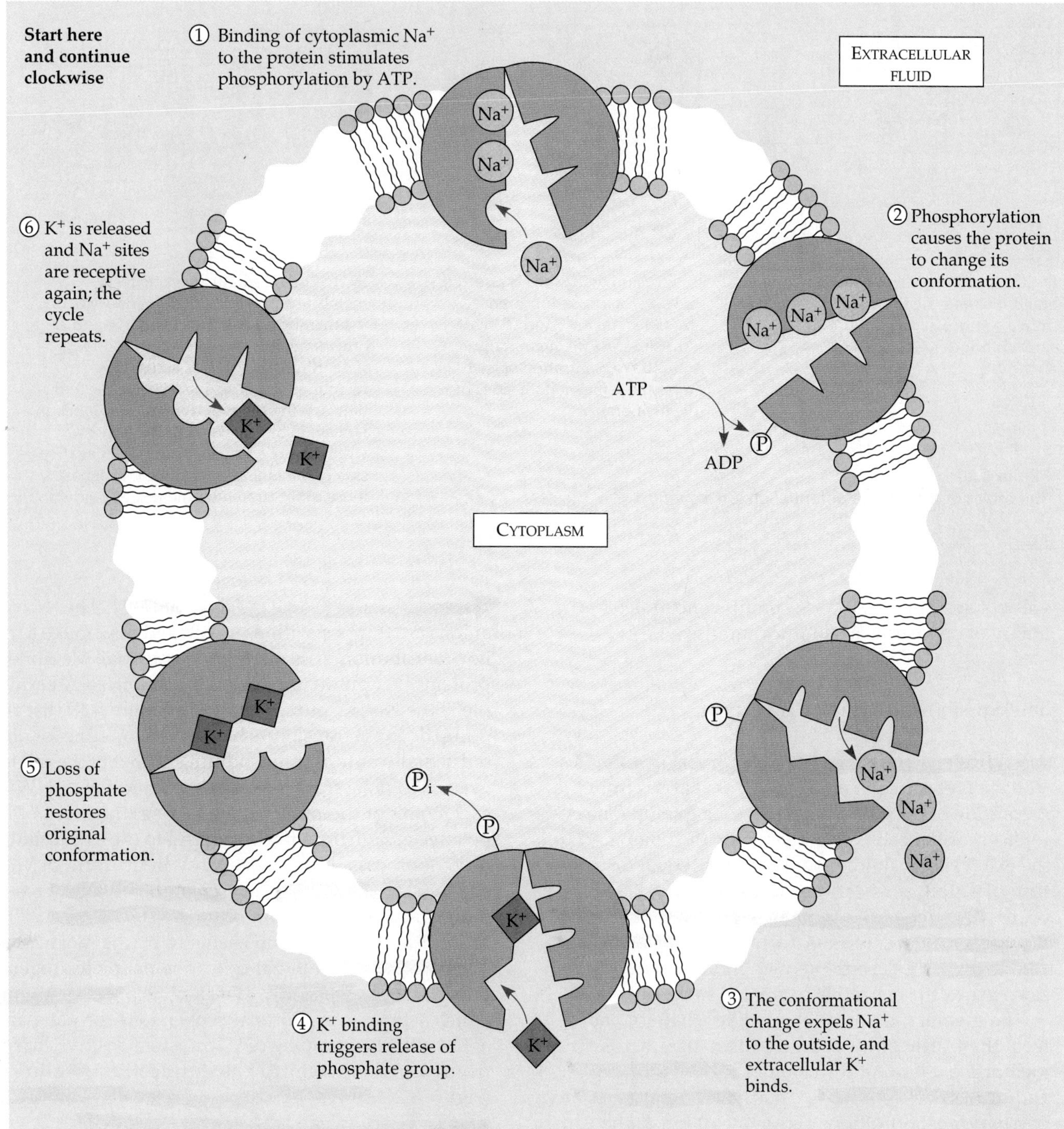
Start here and continue clockwise
① Binding of cytoplasmic Na+ to the protein stimulates phosphorylation by ATP.
EXTRACELLULAR FLUID
Na+
Na+
Na+
② Phosphorylation causes the protein to change its conformation.
Na+
Na+
Na+
ATP
P
ADP
CYTOPLASM
⑥ K+ is released and Na+ sites are receptive again; the cycle repeats.
K+
K+
P
Na+
Na+
Na+
③ The conformational change expels Na+ to the outside, and extracellular K+ binds.
K+
K+
⑤ Loss of phosphate restores original conformation.
Pi
P
K+
K+
④ K+ binding triggers release of phosphate group.

Figure 8.19
The sodium-potassium pump: a specific case of active transport. This transport system pumps ions against steep concentration gradients. The pump oscillates between two conformational states in a pumping cycle that translocates three Na^+ ions out of the cell for every two K^+ ions pumped into the cell. ATP powers the changes in conformation by phosphorylating the transport protein (that is, by transferring a phosphate group to the protein). The two conformational states differ in the affinity for Na^+ and K^+ and in the directional orientation of the ion-binding sites. Prior to phosphorylation, the binding sites face the cytoplasm, and only the Na^+ sites are receptive. Sodium binding induces phosphate transfer from ATP to the pump, triggering the conformational change. In its new conformation, the pump's binding sites face the extracellular side of the membrane, and the protein now has a greater affinity for K^+ than it does for Na^+. Potassium binding causes release of the phosphate, and the pump returns to its original conformation. Because the pump also behaves like an enzyme that removes phosphate from ATP, it is sometimes called the Na^+-K^+ ATPase. The pump is actually an aggregate of four proteins, treated as one hinged protein in this figure.

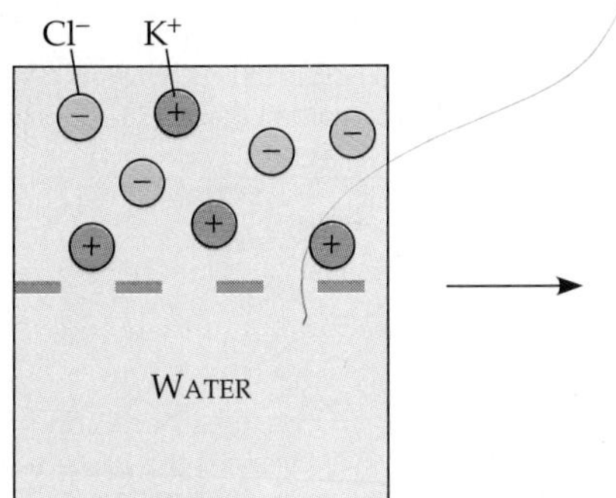

In this artificial system, a membrane separates distilled water from a solution of the salt potassium chloride (KCl).

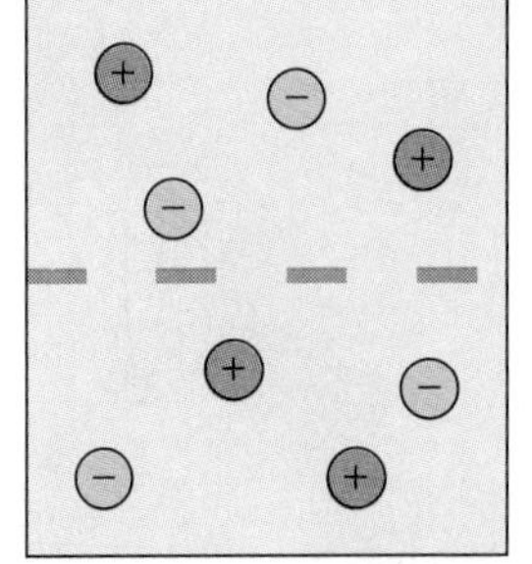

If diffusion of the ions is affected only by their concentration gradients, then K^+ and Cl^- will diffuse across the membrane until the concentration of each ion is the same on both sides of the membrane.

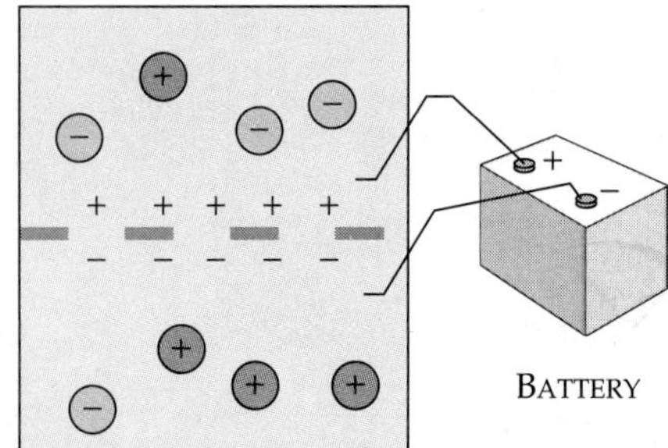

Switching on a battery produces a membrane potential, making one side of the membrane negative in charge and the other side positive. The K^+ and Cl^- will redistribute themselves across the membrane by diffusing in response to this new driving force. A dynamic equilibrium will again be reached, but this time, K^+ will be more concentrated on the negative side of the membrane, and Cl^- will be more concentrated on the positive side.

Figure 8.20
The effect of membrane potential on ion transport.

we will see how the Na^+-K^+ pump contributes to the ability of nerve cells to conduct impulses.

The Special Case of Ion Transport

All cells have voltages across their plasma membranes. Voltage is electric potential energy that arises from the separation of opposite charges (see Chapter 6). The cytoplasm of the cell is negative in charge compared to the extracellular fluid, because of an unequal distribution of anions and cations across the plasma membrane. Voltage across membranes, called the **membrane potential,** ranges from about −50 to −200 millivolts (mV). (The minus sign indicates that the inside of the cell is negative compared to the outside.)

The membrane potential acts like a battery and affects the traffic of all charged substances across the membrane, favoring diffusion of cations into the cell and anions out of the cell. Thus, two forces drive the passive transport of ions across membranes: the concentration gradient of the ion and the effect of the membrane potential on the ion. Figure 8.20 illustrates how these forces interact.

Because of the complication of the membrane potential, it is not correct to say that an ion always diffuses down its concentration gradient. The ion *does* diffuse down its **electrochemical gradient,** a diffusion gradient that combines the influences of the electric force (membrane potential) and the chemical force (concentration gradient). For uncharged solutes, only the concentration gradient is relevant.

Several factors contribute to the membrane potential of a cell. At cellular pH, most proteins and other macromolecules are negatively charged. These large anions are trapped within the cell and may make a minor contribution to its membrane potential. Membrane proteins that pump ions have a bigger effect. An example is the Na^+-K^+ pump. Notice in Figure 8.19 that the pump does not translocate Na^+ and K^+ ions one for one, but actually pumps three sodium ions out of the cell for every two potassium ions it pumps into the cell. With each crank of the pump, there is a net transfer of one positive charge from the cytoplasm to the extracellular fluid, a process that stores energy in the form of voltage. A transport protein that generates voltage across a membrane is called an **electrogenic pump**. The sodium-potassium pump seems to be the major **electrogenic pump** of animal cells. The main electrogenic pump of plants, bacteria, and fungi is a proton pump, which actively transports hydrogen ions (protons) out of the cell. The pumping of H^+ transfers positive charge from the cytoplasm to the extracellular solution (Figure 8.21). Proton pumps are also key features of the membranes of mitochondria and chloroplasts.

By generating voltage across membranes, electrogenic pumps store energy that can be tapped for cellular work, including a type of membrane traffic called cotransport.

Cotransport

A single ATP-powered pump that transports a specific solute can indirectly drive the active transport of several other solutes in a mechanism called **cotransport.** A substance that has been pumped across a membrane can do work as it leaks back by diffusion, analogous to

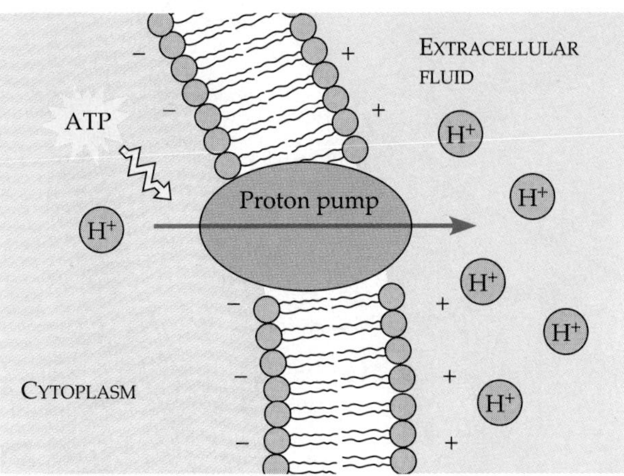

Figure 8.21
An electrogenic pump. Proton pumps are examples of membrane proteins that store energy by generating voltage (charge separation) across membranes. Using ATP for power, a proton pump translocates positive charge in the form of hydrogen ions. The voltage and H^+ gradient represent a dual energy source that can be tapped by the cell to drive other processes, such as the uptake of sugar and other nutrients. Proton pumps are the main electrogenic pumps of plants, fungi, and bacteria, and they also store energy in chloroplasts and mitochondria.

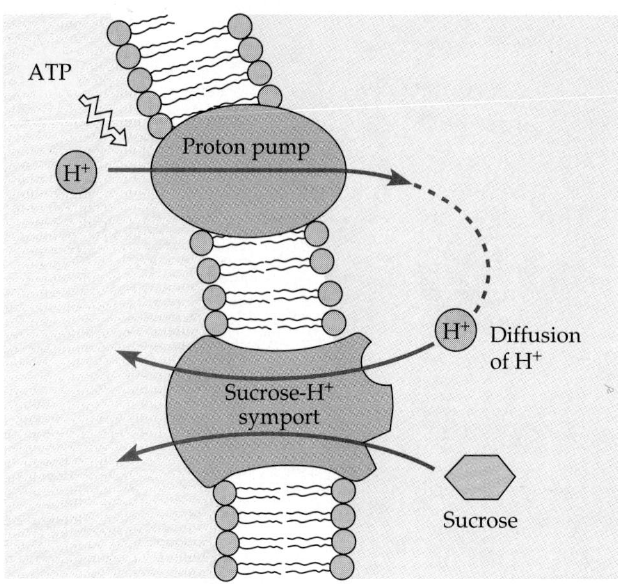

Figure 8.22
Cotransport. An ATP-driven pump stores energy by concentrating a substance (H^+, in this case) on one side of the membrane. As the substance leaks back across the membrane through specific transport proteins, other substances are cotransported, through the same proteins, against their own concentration gradients. In this case, the proton pump of the membrane is indirectly driving sucrose accumulation by a plant cell, with the help of a sucrose-H^+ symport.

water that has been pumped uphill and performs work as it flows back down. Another specialized transport protein, separate from the pump, can couple the "downhill" diffusion of this substance to the "uphill" transport of a second substance against its own concentration gradient. For example, a plant cell uses the gradient of hydrogen ions generated by its proton pumps to drive the active transport of amino acids, sugars, and several other nutrients into the cell (Figure 8.22). One specific transport protein couples the return of hydrogen ions to the transport of sucrose into the cell. The protein can translocate sucrose into the cell against a concentration gradient, but only if the sucrose travels in the company of a hydrogen ion. The sucrose molecule rides on the coattails of the hydrogen ion, which uses the common transport protein as an avenue to diffuse down the concentration gradient maintained by the proton pump. Plants use this mechanism to load sucrose produced by photosynthesis into specialized cells in the veins of leaves. The sugar can then be distributed by the vascular tissue of the plant to nonphotosynthetic organs, such as roots.

THE TRAFFIC OF LARGE MOLECULES: EXOCYTOSIS AND ENDOCYTOSIS

Water and small solutes enter and leave the cell by passing through the lipid bilayer of the membrane, or they are pumped or carried across the membrane by transport proteins. Large molecules, such as proteins and polysaccharides, generally cross the membrane by a different mechanism. In the process called **exocytosis,** the cell secretes macromolecules by the fusion of vesicles with the plasma membrane. In **endocytosis,** the cell takes in macromolecules and particulate matter by forming vesicles derived from the plasma membrane (Figure 8.23). During exocytosis, a vesicle, usually budded from the ER or the Golgi apparatus, is moved by the cytoskeleton to the plasma membrane. When the vesicle membrane and plasma membrane come into contact, the lipid molecules of the two bilayers rearrange themselves. The two membranes then fuse to become continuous, and the contents of the vesicle spill to the outside of the cell. The steps are basically reversed during endocytosis. A localized region of the plasma membrane sinks inward to form a pocket. As the pocket deepens, it pinches into the cytoplasm from the plasma membrane, forming a vesicle containing material that had been outside the cell.

Many secretory cells use exocytosis to export their products (Figure 8.24). For example, certain cells in the pancreas manufacture the hormone insulin and secrete it into the blood by exocytosis. Another example is the neuron, or nerve cell, which uses exocytosis to release chemical signals that stimulate other neurons or muscle cells. When plant cells are making walls, exocytosis

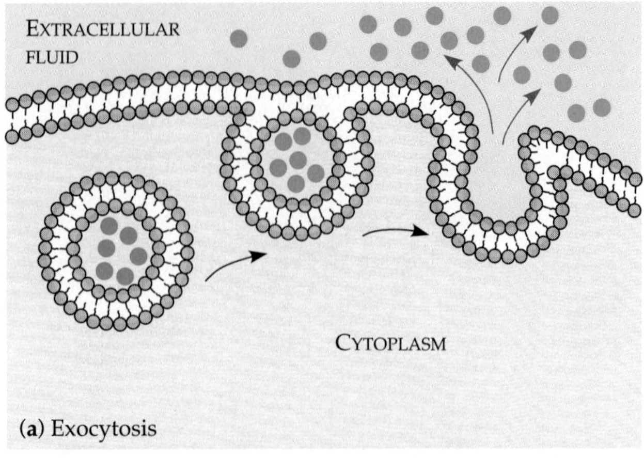

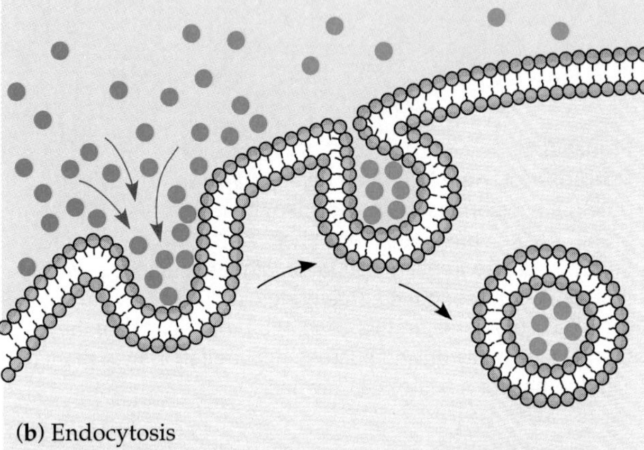

Figure 8.23
Exocytosis and endocytosis. (**a**) During exocytosis, vesicles fuse with the plasma membrane and dump their contents to the outside of the cell. (**b**) During endocytosis, extracellular substances are incorporated into the cell in vesicles formed by an inward budding of the plasma membrane.

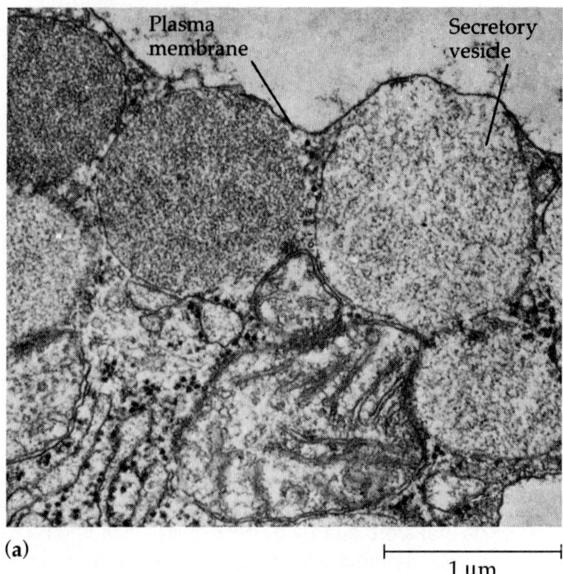

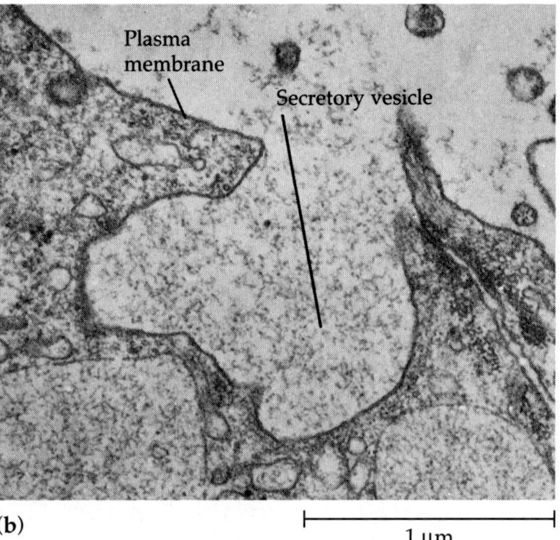

Figure 8.24
Exocytosis in a tear gland. These electron micrographs show two cells at different stages of secretion. (**a**) A transport vesicle containing secretory products fuses with the plasma membrane (TEM) and (**b**) releases its contents outside the cell (TEM).

delivers carbohydrates from Golgi vesicles to the outside of the cell.

There are three types of endocytosis: phagocytosis ("cellular eating"), pinocytosis ("cellular drinking"), and receptor-mediated endocytosis (Figure 8.25). As mentioned in Chapter 7, in **phagocytosis,** a cell engulfs a particle by wrapping pseudopodia around it and packaging it within a membrane-enclosed sac large enough to be classified as a vacuole (Figure 8.25a). The particle is digested after the vacuole fuses with a lysosome containing hydrolytic enzymes. In **pinocytosis,** the cell gulps droplets of extracellular fluid in tiny vesicles (Figure 8.25b). Since any and all solutes dissolved in the droplet are taken into the cell, pinocytosis is unspecific in the substances it transports. In contrast, **receptor-mediated endocytosis** is very specific (Figure 8.25c). Embedded in the membrane are proteins with specific receptor sites exposed to the extracellular fluid. The receptor proteins are usually clustered in regions of the membrane called **coated pits,** which are lined on their cytoplasmic side by a fuzzy layer consisting of the protein **clathrin.** The extracellular substances that bind to the receptors are called **ligands,** a general term for any molecule that binds specifically to a receptor site of another molecule. When appropriate ligands bind to the receptor sites, they are carried into the cell by the inward budding of a coated pit to form a **coated vesicle.**

Receptor-mediated endocytosis enables the cell to acquire bulk quantities of specific substances, even though those substances may not be very concentrated in the extracellular fluid. For example, animal cells use the process to take in cholesterol for use in the synthe-

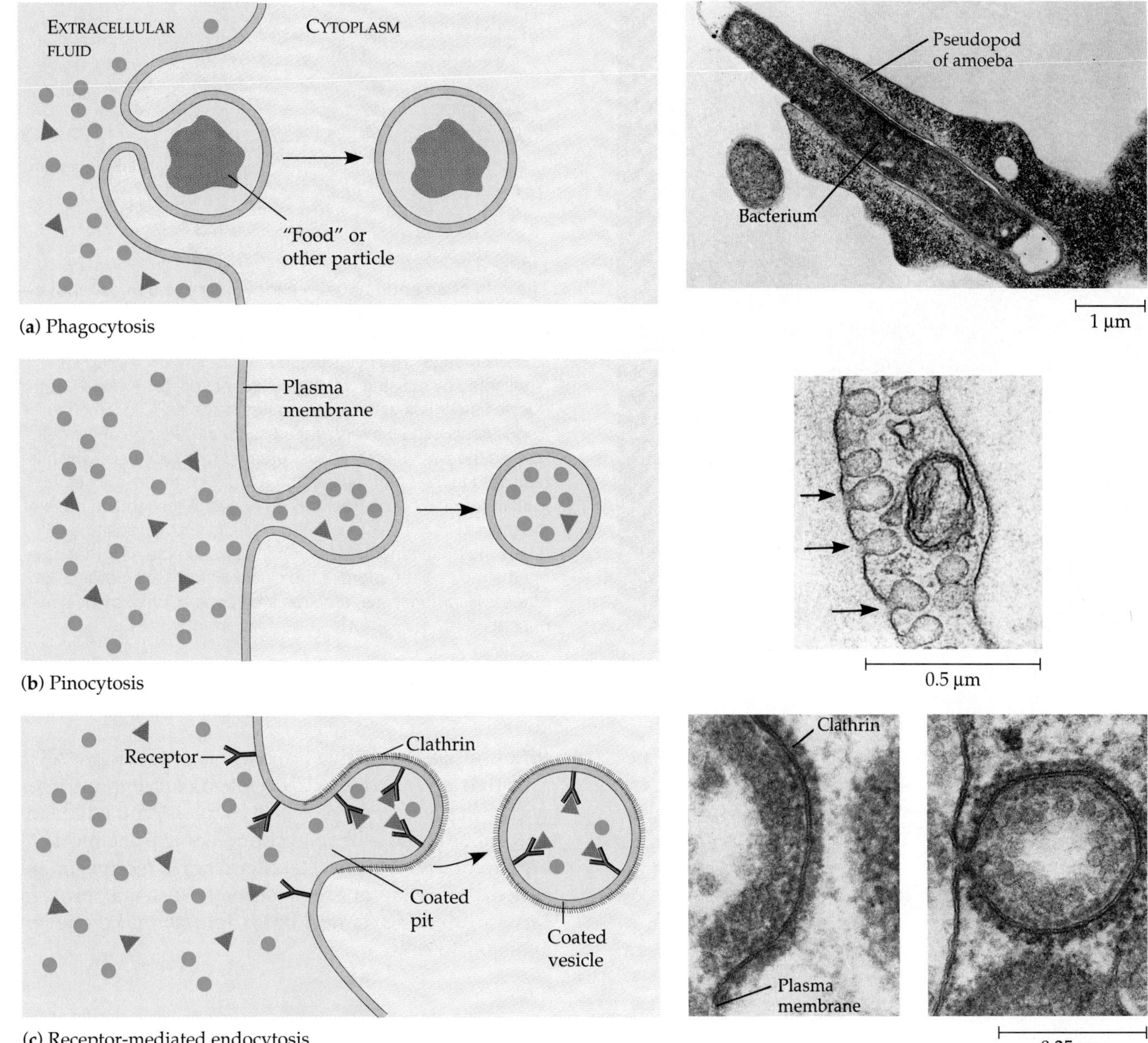

Figure 8.25
The three types of endocytosis in animal cells. (**a**) In phagocytosis, pseudopodia engulf a particle and package it in a vacuole. The electron micrograph shows an amoeba engulfing a bacterium (TEM). (**b**) In pinocytosis, droplets of extracellular fluid are incorporated into the cell in small vesicles. The electron micrograph shows pinocytotic vesicles forming (arrows) in a cell lining a capillary, a small blood vessel (TEM). (**c**) In receptor-mediated endocytosis, coated pits form vesicles when specific molecules bind to receptors on the cell surface. Coated pits are reinforced on their cytoplasmic side by the fibrous protein clathrin. The electron micrographs show two progressive stages of receptor-mediated endocytosis (TEMs). After the ingested material is liberated from the vesicle for metabolism, the receptors are recycled to the plasma membrane.

sis of membranes and as a precursor for the synthesis of other steroids. In humans with familial hypercholesterolemia, an inherited disease characterized by a very high level of cholesterol in the blood, the protein receptor sites are missing. Unable to enter the cells, the cholesterol accumulates in the blood, where it contributes to early atherosclerosis (the development of fat deposits on blood vessel linings).

Vesicles not only transport substances between the cell and its surroundings, but also provide a mechanism for rejuvenating or remodeling the plasma membrane. Endocytosis and exocytosis occur continually to some extent in most eukaryotic cells, yet the amount of plasma membrane in a nongrowing cell remains fairly constant over the long run. Apparently, the addition of membrane by one process offsets the loss of membrane

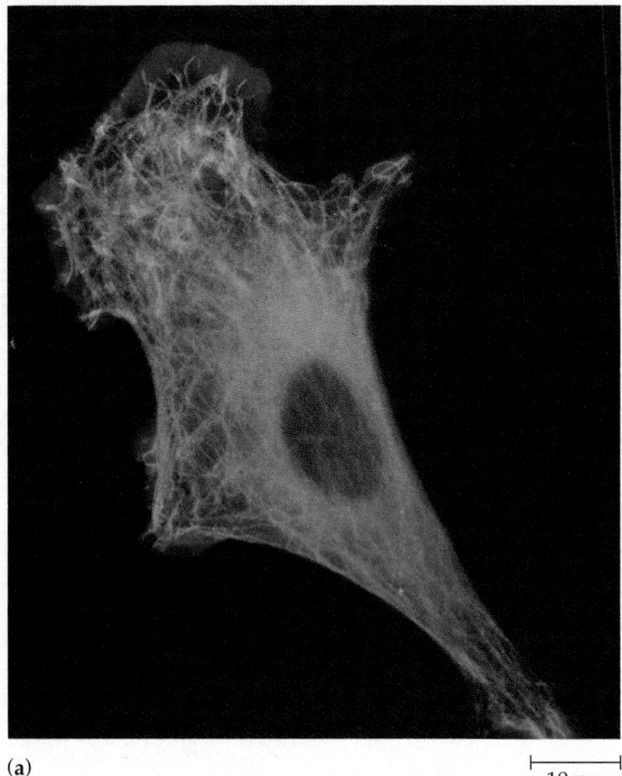

(a)

10 μm

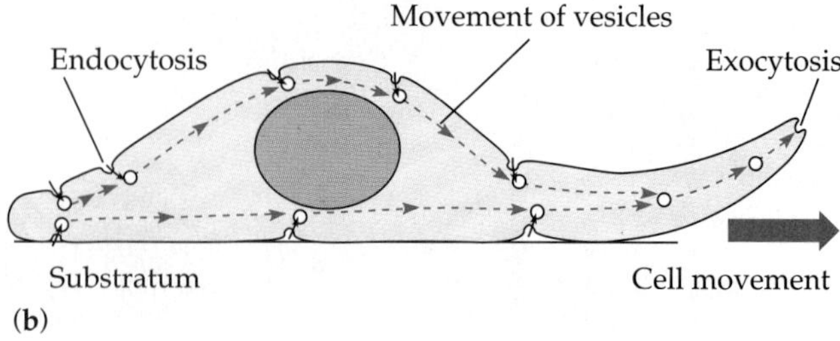

(b)

Figure 8.26
The role of an endocytosis–exocytosis cycle in cell movement. (**a**) This mammalian cell, called a fibroblast, can creep along a substratum by extending its leading edge and retracting its trailing edge. The cytoskeleton in this light micrograph is highlighted by staining with fluorescent dyes. (**b**) A cycle of endocytosis and exocytosis contributes to the cell's locomotion. Endocytosis occurs everywhere except at the edge of the cell leading the way. Membrane is returned by exocytosis, which is restricted to the tip of the leading edge. The result is a net transfer of membrane from trailing edges to the leading edge of the moving cell. The cytoskeleton directs the vesicles and drives their movement. Mitochondria provide the ATP that powers the molecular motors of the cytoskeleton. Cell motility, like other cellular functions, requires the cooperation of many components.

through the other. Researchers have observed a specialized application of these membrane dynamics in cells that move by extending pseudopodia (Figure 8.26).

* * *

Energy and cellular work have figured prominently in our study of membranes and transport. We have seen, for example, that active transport is powered by ATP, and that a working cytoskeleton conveys vesicles within the cell. In the next two chapters, you will learn more about how cells acquire and harvest chemical energy to do the work of life. Membranes receive an encore in these chapters, because they play a major role in how mitochondria and chloroplasts make energy available to cells.

STUDY OUTLINE

1. All cells are separated from their surroundings by the plasma membrane, which controls the traffic of substances into and out of the cell. Both the plasma membrane and the cell's internal membranes, which partition the cell into organelles, are selectively permeable.

Models of Membrane Structure (pp. 151–158)

1. A long-standing model of the membrane, proposed in 1935, placed layers of proteins on either side of the lipid bilayer. This Davson–Danielli sandwich model seemed consistent with later evidence provided by electron microscopy, which revealed a triple-layered ultrastructure for the plasma membrane.
2. In 1972, Singer and Nicolson proposed the fluid mosaic model, in which the membrane is a mosaic of dispersed proteins, individually inserted and floating laterally in a fluid bilayer of phospholipids. The fluid mosaic model is consistent with all known properties of cell membranes and with evidence provided by freeze-fracture electron micrographs of delaminated membranes.
3. The fluid quality and distinctive molecular composition of membranes are essential for their dynamic functioning.
4. Membranes are bifacial, with specific inside and outside faces arising from differences in the lipid composition of the two bilayers and/or the directional orientation of proteins and any attached carbohydrates.
5. Proteins are either embedded in the lipid bilayer (integral proteins) or occur only on the surface (peripheral proteins).
6. Carbohydrates linked to the proteins and lipids are important for cell-cell recognition.

The Traffic of Small Molecules (pp. 158–167)

1. The cell requires an extensive interchange of small nutrient and waste molecules, respiratory gases, and inor-

ganic ions. The plasma membrane regulates the passage of these substances.

2. The selective permeability of the plasma membrane results from its structure. Hydrophobic substances pass through rapidly because of their solubility in the lipid bilayer. Small polar molecules, such as H_2O and CO_2, can also pass through the membrane. Larger polar molecules and ions require specific transport proteins, which provide channels.
3. Diffusion is the spontaneous movement of a substance down its concentration gradient. The diffusion of a substance across a membrane is called passive transport.
4. Osmosis is the diffusion of water across a selectively permeable membrane. Water flows across a membrane from the side with a lesser concentration of solute (hypoosmotic) to the side with the greater solute concentration (hyperosmotic). No net osmosis occurs across membranes separating solutions of equal concentration (isosmotic). The osmotic pressure of a solution is proportional to its solute concentration.
5. Cells lacking walls (as in animals and some protists) are either isosmotic with their environments or else have adaptations for osmoregulation.
6. Plants, prokaryotes, fungi, and some protists have an elastic wall around their cells, which keeps the cells from bursting in a hypoosmotic environment. Under such conditions, these cells are turgid. In isosmotic solutions, such cells are flaccid; in hyperosmotic solutions, they plasmolyze.
7. In facilitated diffusion, transport proteins hasten the movement of certain substances across a membrane down their concentration gradients. The transport protein is specific for its solute.
8. Diffusion, osmosis, and facilitated diffusion are all passive transport processes that do not require the input of energy from the cell.
9. Active transport, on the other hand, requires energy to pump substances across the membrane against their concentration gradients. The energy, usually in the form of ATP, is harnessed to specific transport proteins that mediate the work.
10. The diffusion of uncharged solutes depends only on their concentration gradients. However, ions have both a concentration (chemical) gradient and an electric gradient (voltage). These two forces are combined in an overall force called the electrochemical gradient, which determines the net direction of ionic diffusion.
11. The voltage across the membrane, the membrane potential, depends on an unequal distribution of ions across the plasma membrane and exists in all living cells. Electrogenic pumps, such as the sodium-potassium pump and proton pump, are transport proteins that generate voltage across a membrane.
12. Special membrane proteins can cotransport two solutes, coupling the "downhill" diffusion of one to the "uphill" transport of the other.

The Traffic of Large Molecules: Exocytosis and Endocytosis (pp. 167–170)

1. Large macromolecules leave the cell by exocytosis and enter by endocytosis, two processes involving whole segments of membrane rather than individual membrane molecules.
2. In exocytosis, intracellular vesicles migrate to the plasma membrane, fuse with it, and release their contents.
3. Endocytosis is essentially the reverse of exocytosis. There are three types of endocytosis: Phagocytosis is the ingestion of large particles or whole cells; pinocytosis is the intake of tiny droplets of extracellular fluid with all its contained solutes; and receptor-mediated endocytosis is the ingestion of specific substances that bind to receptor proteins located in coated pits on the membrane.

SELF-QUIZ

1. In what way do the various membranes of a eukaryotic cell differ?
   a. Phospholipids are only found in certain membranes.
   b. Certain proteins are unique to each membrane.
   c. Only certain membranes of the cell are selectively permeable.
   d. Only certain membranes are constructed from amphipathic molecules.
   e. Some membranes have hydrophobic surfaces exposed to the cytoplasm, while others have hydrophilic surfaces facing the cytoplasm.
2. According to the fluid mosaic model of membrane structure, proteins of the membrane are
   a. spread in a continuous layer over the inner and outer surfaces of the membrane
   b. confined to the hydrophobic core of the membrane
   c. embedded in a lipid bilayer
   d. randomly oriented in the membrane, with no fixed inside–outside polarity
   e. free to depart from the fluid membrane and dissolve in the surrounding solution
3. When a plasma membrane is split by freeze-fracture into its outer and inner leaflets, the "bumps" seen in electron micrographs are replicas of
   a. integral proteins
   b. peripheral proteins
   c. phospholipids
   d. cholesterol molecules
   e. clathrin
4. Which of the following factors would tend to increase membrane fluidity?
   a. a greater proportion of unsaturated phospholipids
   b. a lower temperature
   c. a relatively high protein content in the membrane
   d. a greater proportion of relatively large glycolipids compared to lipids having smaller molecular weights
   e. a high membrane potential
5. The sodium-potassium pump is termed electrogenic because
   a. it hydrolyzes ATP
   b. it pumps positive charges out of the cell and negative charges into the cell
   c. it pumps three positive charges out of the cell for every two positive charges it pumps into the cell
   d. it pumps H^+ out of the cell along with Na^+
   e. it pumps electrons into the cell

6. Plant cells are turgid when bathed in a solution that is
   a. hypoosmotic to the cell
   b. hyperosmotic to the cell
   c. isosmotic to the cell
   d. isosmotic to seawater
   e. lower in water concentration than the cell

Questions 7–10

An artificial cell with an aqueous solution enclosed in a selectively permeable membrane has just been immersed in a beaker containing a different solution.

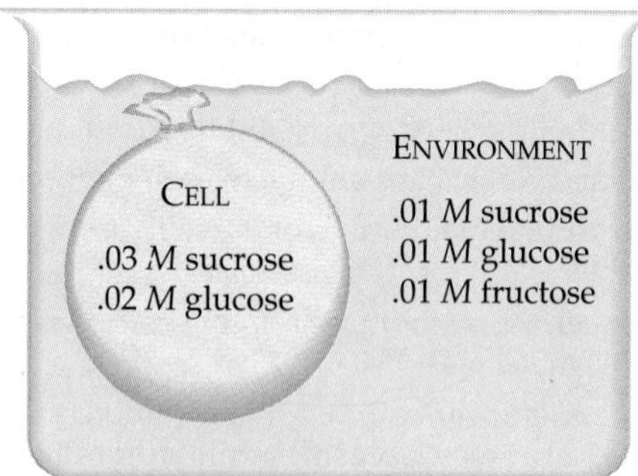

The membrane is permeable to water and to the simple sugars glucose and fructose, but is completely impermeable to the disaccharide sucrose.

7. Which solute(s) will exhibit a net diffusion into the cell?
8. Which solute(s) will exhibit a net diffusion out of the cell?
9. In which direction will there be a net osmotic movement of water?
10. After the cell is placed into the beaker, which of the following changes would occur?
    a. The artificial cell would become more flaccid.
    b. The artificial cell would become more turgid.
    c. The entropy of the system (cell plus surrounding solution) would decrease.
    d. The overall free energy stored in the system would increase.
    e. Both b and d would occur.

CHALLENGE QUESTIONS

1. An experiment is designed to study the mechanism of sucrose uptake by plant cells. Cells are immersed in a sucrose solution, and the pH of the surrounding solution is monitored with a pH meter. The measurements show that sucrose uptake by the plant cells raises the pH of the surrounding solution. The magnitude of the pH change is proportional to the starting concentration of sucrose in the extracellular solution. A metabolic poison that blocks the ability of the cells to regenerate ATP also inhibits the pH change in the surrounding solution. Explain these results.

2. In an adaptation of the preceding experiment, the rates of sucrose uptake from solutions of different sucrose concentrations are compared.

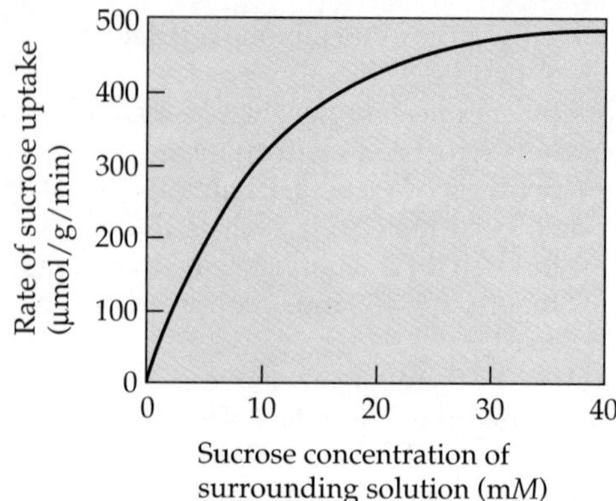

   Explain the shape of the curve in terms of what is happening at the membranes of the plant cells.

3. The cells of plant seeds store oils in the form of droplets enclosed by membranes. Unlike the membranes you studied in this chapter, the oil-droplet membrane probably consists of a single layer of phospholipids rather than a bilayer. Draw a model for a membrane around an oil droplet and explain why this arrangement is more stable than a bilayer.

SCIENCE, TECHNOLOGY, AND SOCIETY

1. A U.S. Government panel has recommended that all people over age 20 should have their blood cholesterol tested, and that those with high cholesterol levels should be put on a special diet or drug therapy. The annual cost of screening and treating everyone in the United States with elevated cholesterol could be between 10 and 50 billion dollars. Research suggests that relatively few people benefit from drug treatment, mainly those suffering from abnormal conditions like familial hypercholesterolemia. Do you think it is worth the effort and expense of testing and medicating everyone with high cholesterol to help only 1% or 2% of the population? Why or why not? Would you have the same opinion about nationwide testing if dietary change, rather than drugs, were the prevalent prescription for those with high cholesterol levels?

FURTHER READING

Alberts, B., D. Bray, J. Lewis, M. Raff, K. Roberts, and J. D. Watson. *Molecular Biology of the Cell*, 2nd ed. New York: Garland, 1989. A description of the structures and functions of the plasma membrane in Chapter 6.

Barinaga, M. "Genetic Disease: Novel Functions Discovered for the Cystic Fibrosis Gene." *Science*, April 24, 1992. How a defective Cl^- pump impairs health.

Bretscher, M. S. "The Molecules of the Cell Membrane." *Scientific American*, October 1985. A superb depiction of the shapes and relationships of membrane molecules.

Kartner, N., and V. Ling. "Multidrug Resistance in Cancer." *Scientific American*, March 1989. An article about tumor cells having membrane proteins that pump out poisons used in chemotherapy.

Rennie, J. "Leaky Channels." *Scientific American*, January 1991. The association between muscular dystrophy and a faulty membrane.

9 CELLULAR RESPIRATION: HARVESTING CHEMICAL ENERGY

ATP AND CELLULAR WORK: A REVIEW

CHEMICAL BACKGROUND: RESPIRATION AS AN OXIDATION-REDUCTION PROCESS

AN OVERVIEW OF CELLULAR RESPIRATION

GLYCOLYSIS: A CLOSER LOOK

THE KREBS CYCLE: A CLOSER LOOK

THE ELECTRON TRANSPORT CHAIN AND OXIDATIVE PHOSPHORYLATION: A CLOSER LOOK

SUMMARIZING CELLULAR RESPIRATION

FERMENTATION: THE ANAEROBIC ALTERNATIVE

A COMPARISON OF AEROBIC AND ANAEROBIC CATABOLISM

THE EVOLUTIONARY SIGNIFICANCE OF GLYCOLYSIS

THE CATABOLISM OF OTHER MOLECULES

BIOSYNTHESIS

THE CONTROL OF RESPIRATION

Figure 9.1
Fuel for cellular work. Organisms are open systems that depend on external energy sources. Animals, such as this desert locust, obtain their energy in chemical form by eating other organisms. After the food is digested and distributed to cells, sugar and other organic molecules are consumed as fuel in the process of cellular respiration. In this chapter, you will learn how cellular respiration harnesses this chemical energy to generate the ATP used for most cellular work, such as the contraction of muscle cells.

Living is work. A cell organizes small organic molecules into polymers, such as proteins and DNA. It pumps substances across membranes. Many cells move or change their shapes. They grow and reproduce. A cell must work just to maintain its complex structure, because order is intrinsically unstable. As we have been emphasizing, the cell is not an autonomous, closed system. To perform their many tasks, cells require transfusions of energy from outside sources (Figure 9.1). Energy enters most ecosystems in the form of sunlight, the energy source for plants and other photosynthetic organisms (Figure 9.2). Animals obtain fuel by eating plants, or by eating other organisms that eat plants. All organisms use the organic molecules in their food not only as energy resources but also as building materials for growth and repair.

How do cells harvest the energy stored in food? With the help of enzymes, the cell systematically degrades complex organic molecules that are rich in potential energy to simpler waste products that have less energy. Some of the energy taken out of chemical storage can be used to do work; the rest is dissipated as heat. As we learned in Chapter 6, metabolic pathways that release stored energy by breaking down complex molecules are called catabolic pathways. One catabolic process, called **fermentation,** is a partial degradation of sugars that occurs without the help of oxygen. However, the most prevalent and efficient catabolic pathway is **cellular respiration,** in which oxygen is consumed as a reactant along with the organic fuel.

In eukaryotic cells, mitochondria house most of the metabolic equipment for cellular respiration. Although very different in mechanism, respiration is in principle similar to the combustion of gasoline in an automobile engine after oxygen is mixed with the fuel (hydrocarbons). Food is the fuel for respiration, and the exhaust is carbon dioxide and water. The overall process can be summarized as follows:

$$\text{Organic compounds} + \text{Oxygen} \longrightarrow \text{Carbon dioxide} + \text{Water} + \text{Energy}$$

Although carbohydrates, fats, and proteins can all be processed and consumed as fuel, it is traditional to learn the steps of cellular respiration by tracking the degradation of the sugar glucose ($C_6H_{12}O_6$):

$$C_6H_{12}O_6 + 6\ O_2 \longrightarrow 6\ CO_2 + 6\ H_2O + \text{Energy (ATP + Heat)}$$

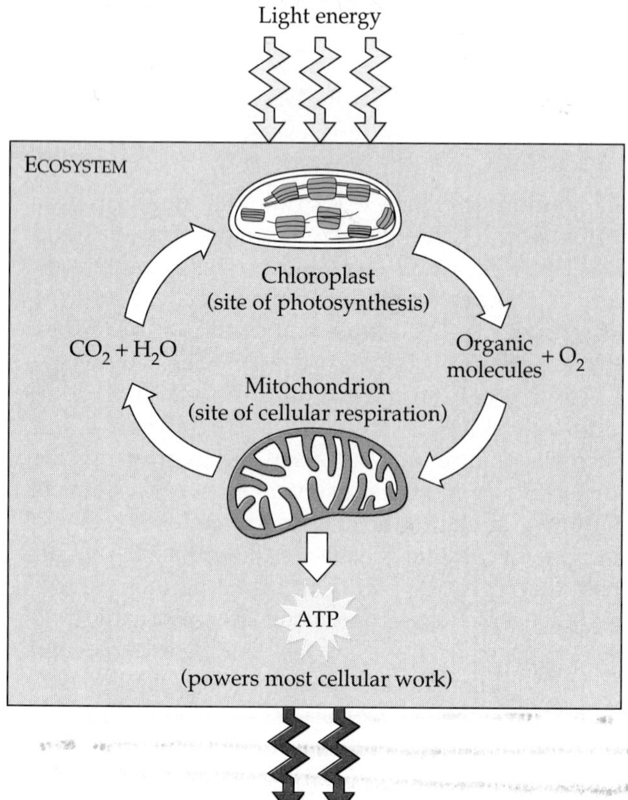

Figure 9.2
Energy flow and chemical recycling in ecosystems. The mitochondria of eukaryotic cells use the organic products of photosynthesis as fuel for cellular respiration, which also consumes the oxygen produced by photosynthesis. Respiration harvests the energy stored in organic molecules to generate ATP, which powers most cellular work. The waste products of respiration, carbon dioxide and water, are the very substances that chloroplasts use as raw materials for photosynthesis. Thus, the chemical elements essential to life are recycled. But energy is not: It flows into an ecosystem as sunlight and back out as heat.

This breakdown of glucose is exergonic, having a free energy change of −686 kilocalories per mole of glucose decomposed ($\Delta G = -686$ kcal/mol; recall that a negative ΔG indicates that the products of the chemical process store less energy than the reactants).

Catabolic pathways like cellular respiration do not directly move flagella, pump solutes, polymerize monomers, or perform other cellular work. Catabolism is linked to work by a chemical drive shaft: ATP. In this chapter, you will learn how catabolism, especially respiration, generates the ATP that is expended by the working cell. The processes are complex and challenging to learn. Throughout the chapter, it is important to keep in mind the objective: discovering how cells use the energy stored in food molecules to make ATP.

ATP AND CELLULAR WORK: A REVIEW

The molecule known as ATP, short for adenosine triphosphate, is the central character in cellular energetics. Recall from Chapter 6 that the triphosphate tail of ATP is the chemical equivalent of a loaded spring; the close packing of the three negatively charged phosphate groups is an unstable, energy-storing arrangement (like charges repel). The chemical "spring" tends to "relax" from the loss of the terminal phosphate (see Figure 6.7). The cell taps this energy source by using enzymes to transfer phosphate groups from ATP to other compounds, which are then said to be phosphorylated. Phosphorylation primes a molecule to undergo some kind of change that performs work, and the molecule loses its phosphate group in the process (Figure 9.3). The price of most cellular work, then, is the conversion of ATP to ADP and inorganic phosphate, products that store less energy than ATP. To keep working, the cell must regenerate its supply of ATP from ADP and inorganic phosphate. To understand how cellular respiration powers this ATP synthesis, we need to learn about the fundamental chemical processes known as oxidation and reduction.

CHEMICAL BACKGROUND: RESPIRATION AS AN OXIDATION-REDUCTION PROCESS

Just what happens when cellular respiration decomposes glucose and other organic fuels? And why does this metabolic pathway yield energy? The answers are based on the transfer of electrons during the chemical reactions known as oxidation and reduction. The relocation of electrons releases the energy stored in food molecules, and this energy is used to synthesize ATP.

An Introduction to Redox Reactions

In many chemical reactions, there is a transfer of one or more electrons (e^-) from one reactant to another. These electron transfers are called oxidation-reduction reactions, or **redox reactions** for short. During a redox reaction, the loss of electrons from one substance is called **oxidation**, and the addition of electrons to another sub-

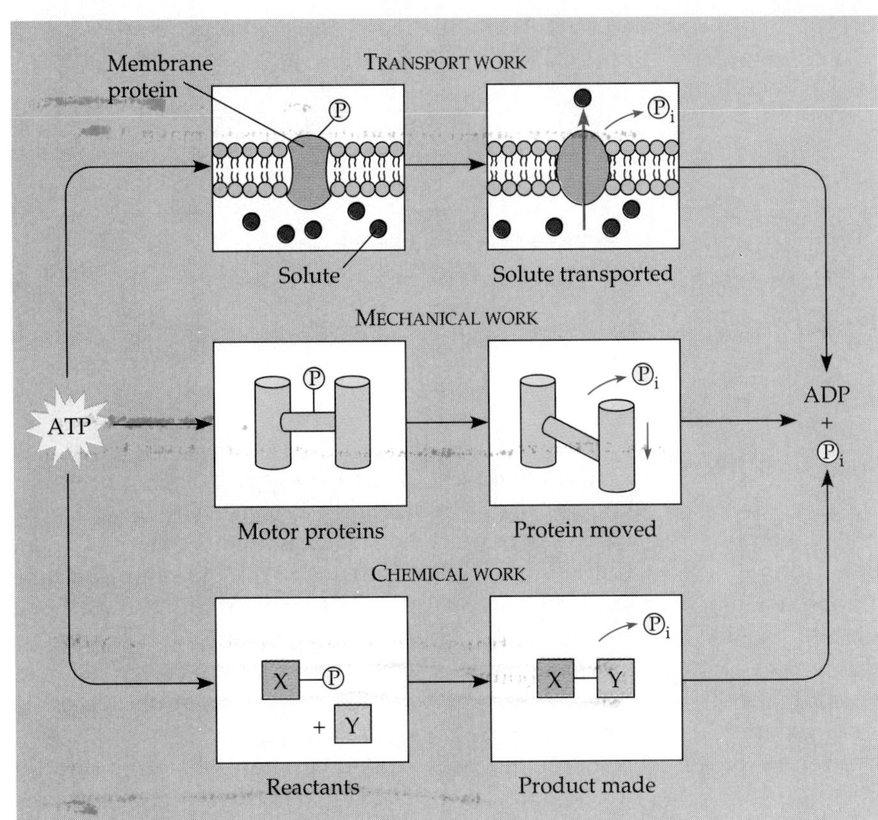

Figure 9.3
A review of how ATP drives cellular work. Phosphate-group transfer is the mechanism responsible for most types of cellular work. Enzymes shift a phosphate group from ATP to some other molecule, and this phosphorylated molecule undergoes a change that performs work. For example, ATP drives active transport by phosphorylating specialized proteins built into membranes; drives mechanical work by phosphorylating motor proteins, such as the dynein responsible for the sliding of microtubules in cilia and flagella; and drives chemical work by phosphorylating key reactants. The phosphorylated molecules lose the phosphate groups as work is performed, leaving ADP and inorganic phosphate (Ⓟi) as products. Cellular respiration replenishes the ATP supply by powering the phosphorylation of ADP.

stance is known as **reduction.*** Consider, for example, the reaction between sodium and chlorine to form table salt:

$$\text{Na} + \text{Cl} \longrightarrow \text{Na}^+ + \text{Cl}^-$$

(Oxidation: Na → Na^+; Reduction: Cl → Cl^-)

Or we could generalize a redox reaction this way:

$$\text{X}e^- + \text{Y} \longrightarrow \text{X} + \text{Y}e^-$$

(Oxidation: Xe^- → X; Reduction: Y → Ye^-)

In the preceding hypothetical reaction, substance X, the electron donor, is called the **reducing agent;** it reduces Y. Substance Y, the electron acceptor, is the **oxidizing agent;** it oxidizes X. Since an electron transfer requires both a donor and an acceptor, oxidation and reduction always go together.

Not all redox reactions involve the complete transfer of electrons from one substance to another; some change the *degree* of electron sharing in covalent bonds. The reaction between methane and oxygen to form carbon dioxide and water, shown in Figure 9.4, is an example. As explained in Chapter 2, the covalent electrons in methane are shared equally between the bonded atoms, because carbon and hydrogen have about the same affinity for valence electrons; they are about equally electronegative. But when methane reacts with oxygen to form carbon dioxide, electrons are shifted away from the carbon atoms to their new covalent partner, oxygen, which is very electronegative. Methane has thus been oxidized. The two atoms of the oxygen molecule also share their electrons equally. But when the oxygen reacts with the hydrogen from methane to form water, the electrons of the covalent bonds are drawn closer to the oxygen; the oxygen molecule has been reduced. Because oxygen is so electronegative, it is one of the most potent of all oxidizing agents.

Energy must be added to pull an electron away from an atom, just as energy must be added to push a large ball uphill. The more electronegative the atom (the stronger its pull on electrons), the more energy is required to keep the electron away from it, just as more energy is required to push a ball up a steeper hill. An electron *loses* potential energy when it shifts from a less electronegative atom *toward* a more electronegative one, just as a ball loses potential energy when it rolls downhill. A redox reaction that relocates electrons closer to oxygen, such as the burning of methane, releases chemical energy that can be put to work.

*This term defies intuition; *adding* electrons is called *reduction*. The term was derived from the electrical effects of adding electrons. When negatively charged electrons are added to a cation, the electrons reduce the amount of positive charge possessed by the cation.

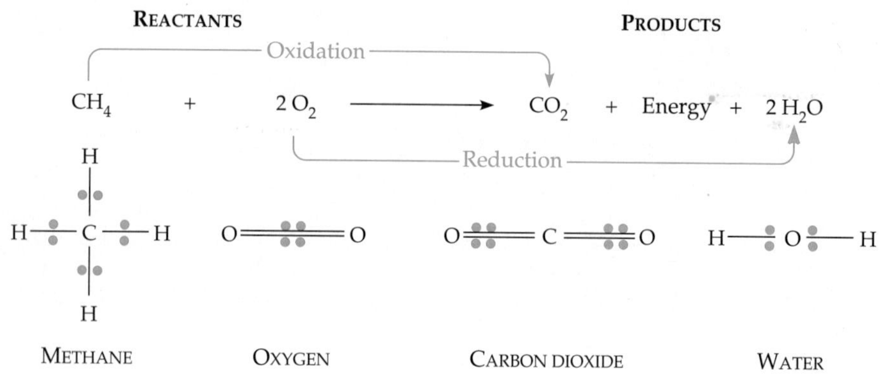

Figure 9.4
Methane combustion as a redox reaction. During the reaction, covalently shared electrons move away from carbon and hydrogen atoms and closer to oxygen, which is very electronegative. The reaction releases energy to the surroundings, because the electrons lose potential energy as they move closer to electronegative atoms.

Respiration: A Stepwise Redox Reaction

The oxidation of methane by oxygen is the main combustion reaction that occurs at the burner of a gas stove. Combustion of gasoline in an automobile engine is also a redox reaction, and the energy released pushes the pistons. But the energy-yielding redox process of greatest interest here is respiration: the oxidation of glucose and other fuel molecules in food. Examine again the summary equation for cellular respiration, but this time think of it as a redox process:

$$\overbrace{C_6H_{12}O_6 + 6\,O_2 \longrightarrow 6\,CO_2}^{\text{Oxidation}} + \underbrace{6\,H_2O}_{\text{Reduction}}$$

As in the combustion of methane or gasoline, the fuel (sugar) is oxidized and oxygen is reduced, and the electrons lose potential energy along the way.

In general, organic molecules that have an abundance of hydrogen are excellent fuels because their bonds are a source of "hilltop" electrons with the potential to "drop" closer to oxygen. The summary equation for respiration indicates that hydrogen is transferred from glucose to oxygen. But the important point, not visible in the summary equation, is that the change in the covalent status of electrons as hydrogen is transferred to oxygen is what liberates energy. By oxidizing glucose, respiration takes energy out of storage and makes it available for ATP synthesis.

The main energy foods, carbohydrates and fats, are reservoirs of electrons associated with hydrogen. Only the barrier of activation energy holds back the flood of electrons to a lower energy state (see Chapter 6). Without this barrier, a food substance like glucose would combine spontaneously with oxygen. When we supply the activation energy by igniting glucose, it burns in air, releasing 686 kcal of heat per mole (about 180 g). Body temperature is not high enough to initiate burning, which is the rapid oxidation of fuel accompanied by an enormous release of energy as heat. But swallow some glucose in the form of a sugar cube, and when the molecules reach your cells, enzymes will lower the barrier of activation energy, allowing the sugar to be oxidized more slowly.

The wholesale release of energy from a fuel is difficult to harness efficiently for constructive work: The explosion of a gasoline tank cannot drive a car very far. Cellular respiration does not oxidize glucose in a single explosive step that would transfer all the hydrogen from the fuel to the oxygen at one time. Rather, glucose and other organic fuels are broken down gradually in a series of steps, each one catalyzed by a coenzyme (see Chapter 6). At key steps, hydrogen atoms are stripped from the glucose, but they are not transferred directly to oxygen. They are usually passed first to a coenzyme (see Chapter 6) called nicotinamide adenine dinucleotide, or **NAD$^+$**, which thus functions as an oxidizing agent.

How does NAD$^+$ trap electrons from glucose and other fuel molecules? Enzymes called dehydrogenases remove a pair of hydrogen atoms from the substrate, a sugar or some other fuel. We can think of this as the removal of two electrons and two protons (the nuclei of hydrogen atoms). The enzyme delivers the *two* electrons along with *one* proton to NAD$^+$ (Figure 9.5). The other proton is released as a hydrogen ion (H$^+$) into the surrounding solution:

$$H{-}\overset{|}{\underset{|}{C}}{-}OH + NAD^+ \xrightarrow{\text{Dehydrogenase}} \overset{|}{\underset{|}{C}}{=}O + NADH + H^+$$

While the oxidized form, NAD$^+$, has a positive charge, the reduced form, NADH, is electrically neutral. By receiving two negatively charged electrons but only one positively charged proton, NAD$^+$ has its charge neutralized. The name NADH for the reduced form shows the hydrogen that has been received in the reaction. Since NAD$^+$ gains electrons, it is an electron acceptor (a synonym for oxidizing agent). The most versatile electron acceptor in cellular respiration, NAD$^+$ functions in many of the redox steps during the breakdown of sugar.

Electrons lose very little of their potential energy when they are transferred by dehydrogenases from food to NAD$^+$. Thus, each NADH molecule formed during respiration represents stored energy that can be tapped to make ATP when the electrons complete their "fall" by flowing from NADH to oxygen.

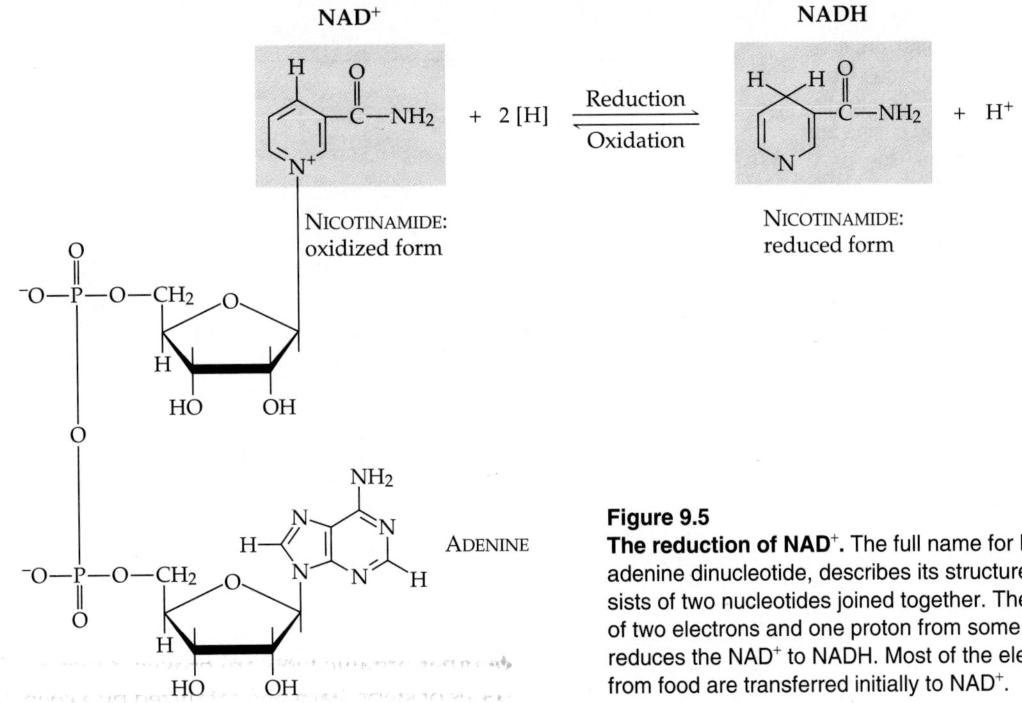

NAD⁺ + 2 [H] $\underset{\text{Oxidation}}{\overset{\text{Reduction}}{\rightleftharpoons}}$ **NADH** + H^+

NICOTINAMIDE: oxidized form

NICOTINAMIDE: reduced form

ADENINE

Figure 9.5
The reduction of NAD^+. The full name for NAD^+, nicotinamide adenine dinucleotide, describes its structure; the molecule consists of two nucleotides joined together. The enzymatic transfer of two electrons and one proton from some substrate to NAD^+ reduces the NAD^+ to NADH. Most of the electrons removed from food are transferred initially to NAD^+.

How do electrons extracted from food and stored by NADH finally reach oxygen? It will help to compare this complex redox chemistry of cellular respiration to a much simpler reaction: the reaction between hydrogen and oxygen to form water (Figure 9.6a). Mix H_2 and O_2, provide a spark for activation energy, and the gases combine explosively. The explosion represents a release of energy as the electrons of hydrogen draw closer to the electronegative oxygen. Cellular respiration also brings hydrogen and oxygen together to form water, but there are two important differences. First, in cellular respiration, the hydrogen that reacts with oxygen is derived from organic molecules. Second, respiration uses an **electron transport chain** to break the "fall" of electrons to oxygen into several energy-releasing steps instead of one explosive reaction (Figure

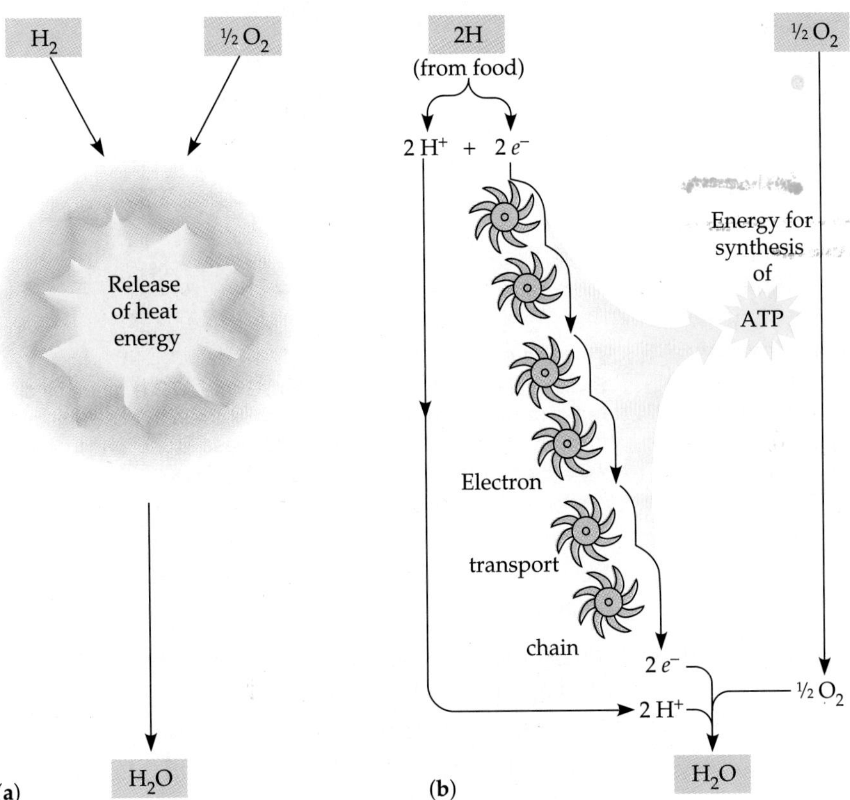

Figure 9.6
An introduction to electron transport chains. (**a**) The exergonic reaction of hydrogen with oxygen to form water releases a large amount of energy in the form of heat and light: an explosion. (**b**) In cellular respiration, an electron transport chain breaks the "fall" of electrons in this reaction into a series of smaller steps and stores some of the released energy in a form that can be used to make ATP (the rest of the energy is released as heat).

Figure 9.7
An overview of cellular respiration. In a eukaryotic cell, glycolysis occurs outside the mitochondria in the cytosol. The Krebs cycle and the electron transport chains are located inside the mitochondria. During glycolysis, each glucose molecule is broken down into two molecules of the compound pyruvate. The pyruvate crosses the double membrane of the mitochondrion to enter the matrix, where the Krebs cycle decomposes it to carbon dioxide. NADH transfers electrons from glycolysis and the Krebs cycle to electron transport chains, which are built into the membrane of the cristae. The electron transport chain transforms the chemical energy into a form that can be used to drive oxidative phosphorylation, which accounts for most of the ATP generated by cellular respiration. A smaller amount of ATP is formed directly during glycolysis and the Krebs cycle by substrate-level phosphorylation.

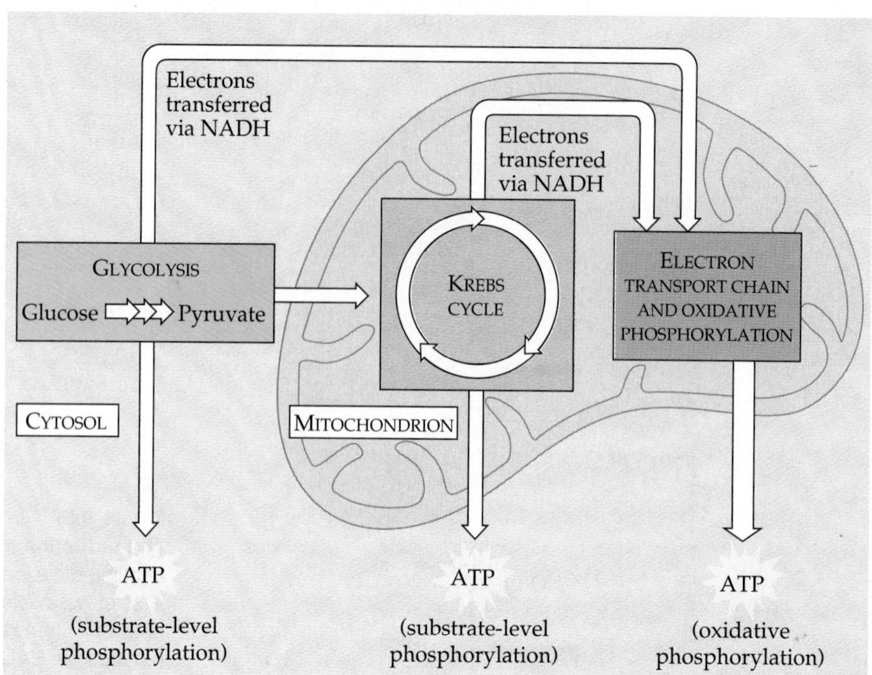

9.6b). The transport chain consists of several molecules, mostly proteins, built into the inner membrane of a mitochondrion. Electrons removed from food are conveyed by NADH to the "top" end of the chain. At the "bottom" end, oxygen captures these electrons along with hydrogen nuclei, forming water.

Electron transfer from NADH to oxygen is exergonic, having a free energy change of –53 kcal/mol. Instead of this energy being released and wasted in a single explosive step, electrons cascade down the chain from one carrier molecule to the next, losing a small amount of energy with each step until they finally reach oxygen, the terminal electron acceptor. What keeps the electrons moving is that each carrier has a greater affinity for electrons than its "uphill" neighbor in the chain. At the bottom of the chain is oxygen, which has a very great affinity for electrons. Thus, electrons removed from food by NAD^+ fall down the electron transport chain to a far more stable location in the electronegative oxygen atom. Put another way, oxygen *pulls* electrons down the chain in an energy-yielding tumble analogous to gravity pulling objects downhill.

In the next four sections, you will learn more about NADH formation and the workings of the electron transport chain. The important point for now is that this chain converts the chemical energy stored in food to a form that can be used to make ATP.

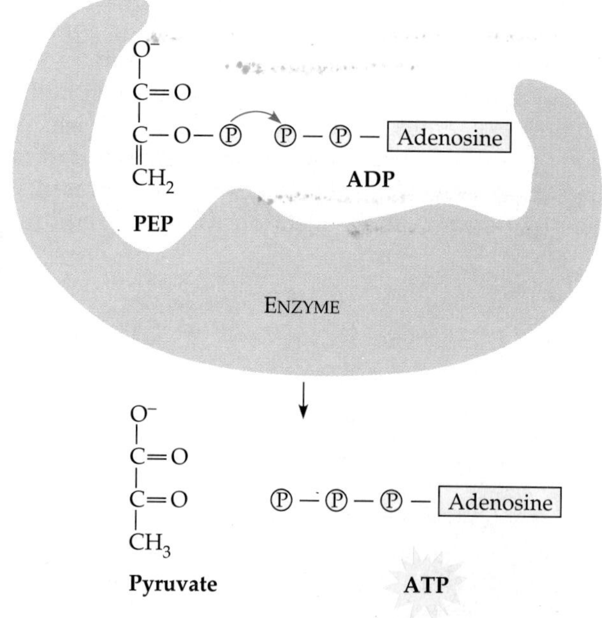

Figure 9.8
Substrate-level phosphorylation. Some ATP is made by direct enzymatic transfer of a phosphate group from a substrate to ADP. The phosphate donor in this case is phosphoenolpyruvate (PEP), which is formed during glycolysis.

AN OVERVIEW OF CELLULAR RESPIRATION

Now that we have covered the basic redox mechanisms of respiration, let's look at the entire process.

Respiration is a cumulative function of three metabolic stages, which are diagrammed in Figure 9.7:

1. Glycolysis (color-coded blue-green throughout the chapter).
2. The Krebs cycle (color-coded salmon).
3. The electron transport chain and oxidative phosphorylation (color-coded violet).

The first two stages, glycolysis and the Krebs cycle, are the catabolic pathways that decompose glucose and other organic fuels. **Glycolysis,** which occurs in the cytosol, begins the degradation by breaking glucose into two molecules of a compound called pyruvate. The **Krebs cycle,** which takes place within the mitochondrial matrix, completes the job by decomposing a derivative of pyruvate to carbon dioxide.

Thus, the carbon dioxide produced by respiration represents fragments of oxidized organic molecules. Some of the steps of glycolysis and the Krebs cycle are redox reactions that transfer electrons from substrates to NAD^+, forming NADH. The third stage of respiration, the electron transport chain, accepts electrons from the breakdown products of the first two stages (via NADH) and passes these electrons from one molecule to another. At the end of the chain, the electrons are combined with hydrogen ions and molecular oxygen to form water. The energy released at each step of the chain is stored in a form the mitochondrion can use to make ATP. This mode of ATP synthesis is called **oxidative phosphorylation,** because it is powered by the exergonic transfer of electrons from food to oxygen.

The site of electron transport and oxidative phosphorylation is the inner membrane of the mitochondrion (see Figure 7.21). Oxidative phosphorylation accounts for almost 90% of the ATP generated by respiration. A smaller amount of ATP is formed directly in some of the reactions of glycolysis and the Krebs cycle by a mechanism called **substrate-level phosphorylation.** This mode of ATP synthesis occurs when an enzyme transfers a phosphate group from a substrate to ADP (Figure 9.8).

Respiration cashes in the large denomination of energy banked in glucose for the small change of ATP, which is more practical for the cell to spend on its work. For each molecule of glucose degraded to carbon dioxide and water by respiration, the cell makes more than 30 molecules of ATP.

Once we understand this overview, we can proceed to a more detailed investigation of glycolysis, the Krebs cycle, and the electron transport chain.

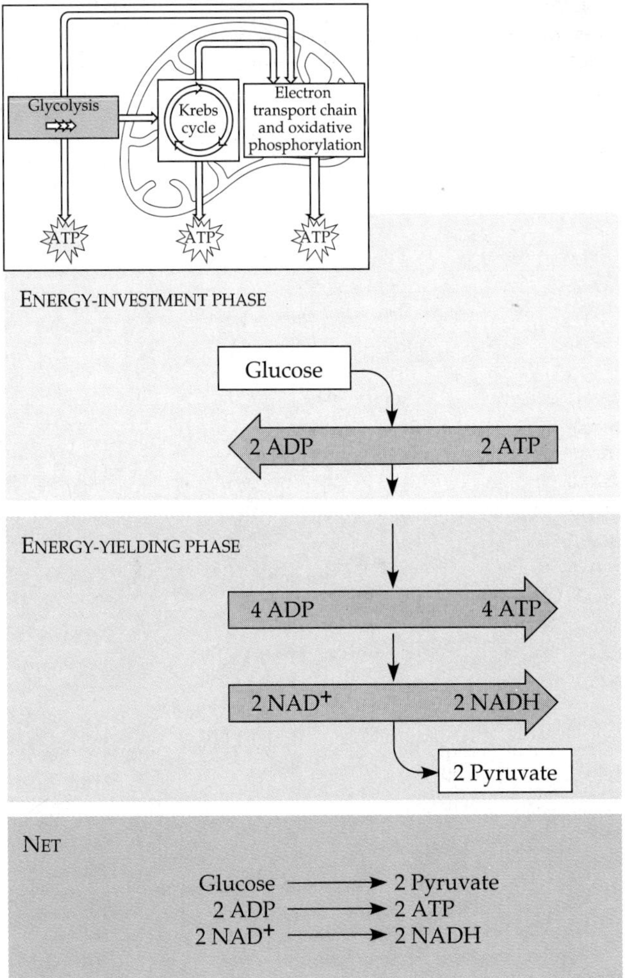

Figure 9.9
A preview of glycolysis. Glycolysis consumes ATP energy during the energy-investment phase, but the energy-yielding phase, which generates ATP and NADH, more than compensates.

GLYCOLYSIS: A CLOSER LOOK

The word *glycolysis* means "splitting of sugar," and that is exactly what happens during this pathway. Glucose, a six-carbon sugar, is split into two three-carbon sugars. These smaller sugars are then oxidized, and their remaining atoms are rearranged to form two molecules of pyruvate.

Glycolysis occurs in the cytosol, outside mitochondria. The catabolic pathway consists of ten steps, each catalyzed by a specific enzyme. We can divide these ten steps of glycolysis into two phases, as shown in Figure 9.9. The energy-investment phase includes the first five steps of glycolysis, and the energy-yielding phase includes the next five steps. All ten steps are shown in Figure 9.10. During the energy-investment phase, the cell actually spends ATP to phosphorylate the fuel molecules. This investment is repaid with dividends during the energy-yielding phase, when ATP is produced by substrate-level phosphorylation and NAD^+ is reduced to NADH by oxidation of the food. The net yield from glycolysis, per glucose molecule, is 2 ATP plus 2 NADH. Notice that all of the carbon originally present in glucose is accounted for in the two molecules of pyruvate; no CO_2 is released during glycolysis. Also notice that glycolysis occurs whether or not oxygen is present. However, if oxygen *is* present, the energy stored in NADH can be converted to ATP energy when the electron transport chain drives oxidative phosphorylation.

Now use Figure 9.10 to follow the steps of glycolysis before proceeding to the Krebs cycle.

Figure 9.10
A detailed look at glycolysis.

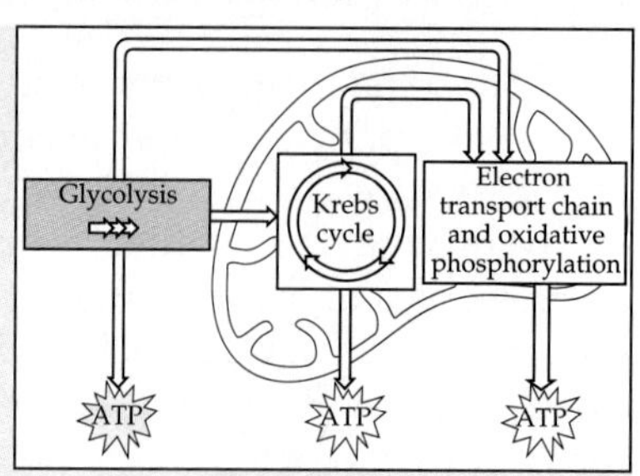

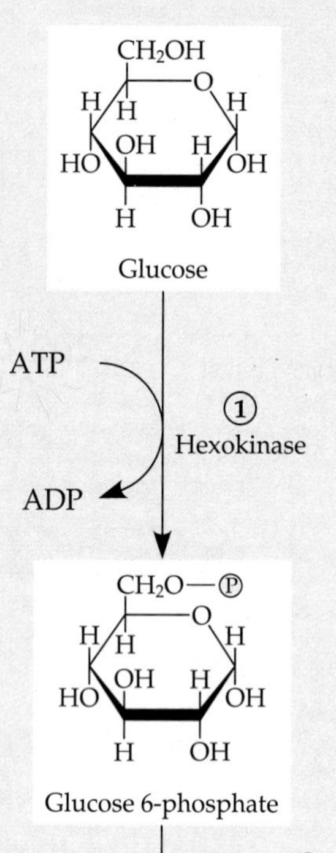

ENERGY-INVESTMENT PHASE

Step 1 Glucose enters the cell and is phosphorylated by the enzyme hexokinase, which transfers a phosphate group from ATP to the sugar. The electrical charge of the phosphate group traps the sugar in the cell because of the impermeability of the plasma membrane to ions. Phosphorylation of glucose also makes the molecule more chemically reactive. In this diagram, coupled arrows (⤺) indicate the transfer of a phosphate group or pair of electrons from one reactant to another.

2
Phosphoglucoisomerase

Step 2 Glucose 6-phosphate is rearranged to convert it to its isomer, fructose 6-phosphate.

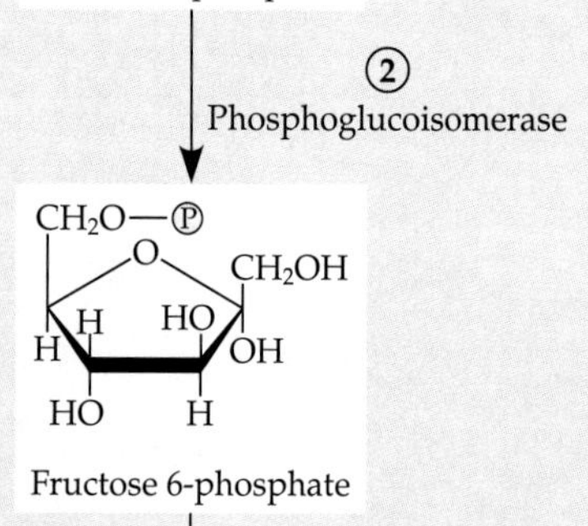

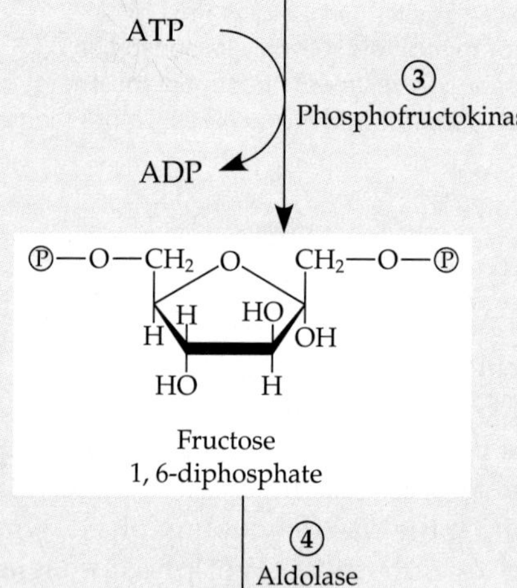

Step 3 In this step, still another molecule of ATP is invested in glycolysis. An enzyme transfers a phosphate group from ATP to the sugar. So far, the ATP ledger shows a debit of 2. With phosphate groups on its opposite ends, the sugar is now ready to be split in half.

Step 4 This is the reaction from which glycolysis gets its name. An enzyme cleaves the sugar molecule into two different three-carbon sugars: glyceraldehyde phosphate and dihydroxyacetone phosphate. These two sugars are isomers of each other.

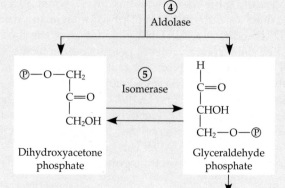

5
Isomerase

Dihydroxyacetone phosphate

Glyceraldehyde phosphate

Step 5 Another enzyme catalyzes the reversible conversion between the two three-carbon sugars, and if left alone in a test tube, the reaction reaches equilibrium. This does not happen in the cell, however, because the next enzyme in glycolysis uses only glyceraldehyde phosphate as its substrate and is unreceptive to dihydroxyacetone phosphate. This pulls the equilibrium between the two three-carbon sugars in the direction of glyceraldehyde phosphate, which is removed as fast as it forms. Thus, the net result of steps 4 and 5 is cleavage of a six-carbon sugar into two molecules of glyceraldehyde phosphate; each will progress through the remaining steps of glycolysis.

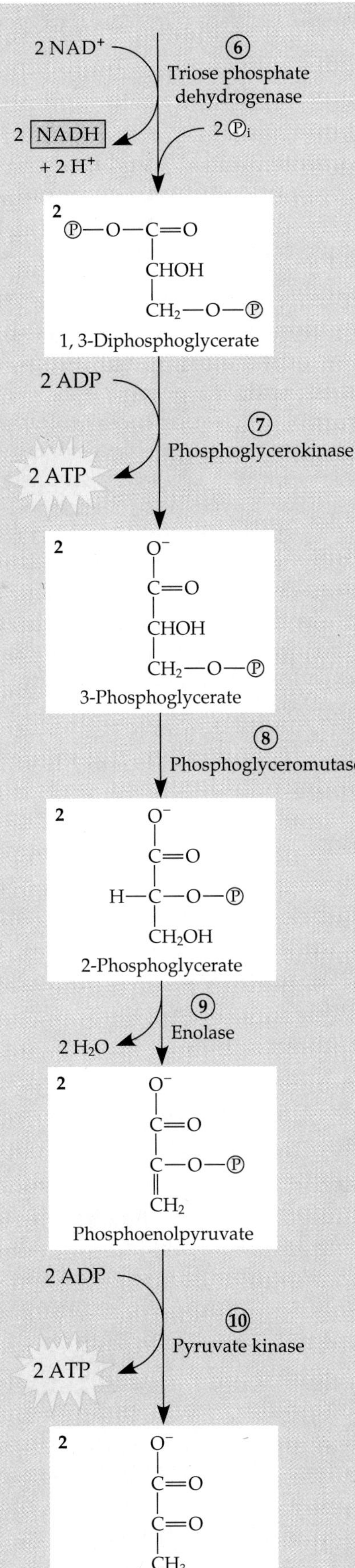

ENERGY-YIELDING PHASE

Step 6 An enzyme now catalyzes two sequential reactions while it holds glyceraldehyde phosphate in its active site. First, the sugar is oxidized by the transfer of electrons and H^+ to NAD^+, forming NADH. Here we see in metabolic context the type of redox reaction described earlier. This reaction is very exergonic ($\Delta G = -10.3$ kcal/mol), and the enzyme capitalizes by attaching a phosphate group to the oxidized substrate. The source of the phosphate is inorganic phosphate, which is always present in the cytosol. Notice that the coefficient 2 precedes all molecules in the energy-yielding phase; these steps occur after glucose is split into two three-carbon sugars.

Step 7 Finally, glycolysis produces some ATP. The phosphate group added in the previous step is transferred to ADP. For each glucose molecule that began glycolysis, step 7 produces two molecules of ATP, since every product after the sugar-splitting step (step 4) is doubled. Of course, two ATPs were invested to get sugar ready for splitting. The ATP ledger now stands at zero. By the end of step 7, glucose has been converted to two molecules of 3-phosphoglycerate. This compound is not a sugar. The carbonyl group that characterizes a sugar has been oxidized to a carboxyl group, the hallmark of an organic acid. The sugar was oxidized in step 6, and now the energy made available by that oxidation has been used to make ATP.

Step 8 Next, an enzyme relocates the remaining phosphate group. This prepares the substrate for the next reaction.

Step 9 An enzyme forms a double bond in the substrate by extracting a water molecule to form phosphoenolpyruvate, or PEP. This results in the electrons of the substrate being rearranged in such a way that the remaining phosphate bond becomes very unstable, preparing the substrate for the next reaction.

Step 10 The last reaction of glycolysis produces more ATP by transferring the phosphate group from PEP to ADP. Since this step occurs twice for each glucose molecule, the ATP ledger now shows a net gain of two ATPs. Steps 7 and 10 each produce two ATPs for a total credit of four, but a debt of two ATPs was incurred from steps 1 and 3. Glycolysis has repaid the ATP investment with 100% interest. Additional energy was stored by step 6 in NADH, which can be used to make ATP by oxidative phosphorylation (if oxygen is present). In the meantime, glucose has been broken down and oxidized to two molecules of pyruvate, the end-product of the glycolytic pathway.

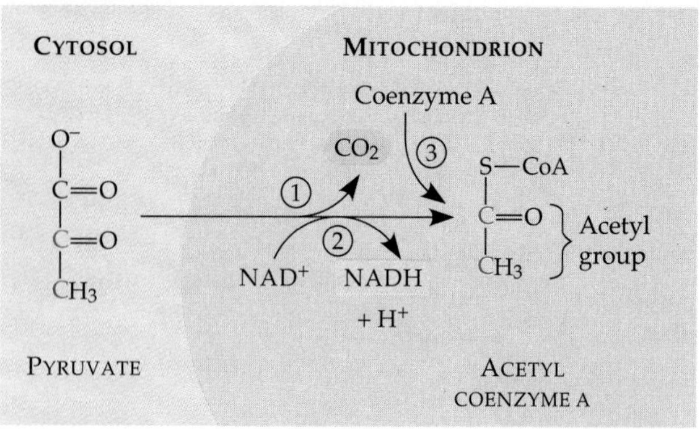

Figure 9.11
The formation of acetyl CoA. A protein built into the mitochondrial membrane translocates pyruvate from the cytosol into the mitochondrion. Then: ① The carboxyl group of pyruvate, already fully oxidized, is removed as a CO_2 molecule, which diffuses out of the cell. ② The remaining two-carbon fragment is oxidized while NAD^+ is reduced to NADH. ③ Finally, the two-carbon acetyl group is attached to coenzyme A (CoA). This coenzyme has a sulfur atom, which bonds to the acetyl fragment by an unstable bond. This activates the acetyl group for the first reaction of the Krebs cycle.

THE KREBS CYCLE: A CLOSER LOOK

Glycolysis releases less than a quarter of the chemical energy stored in glucose; most of the energy remains stocked in the two molecules of pyruvate. If oxygen is present, the pyruvate enters the mitochondrion, where the enzymes of the Krebs cycle complete the oxidation of the organic fuel.

Upon entering the mitochondrion, pyruvate is first converted to a compound called **acetyl CoA** (Figure 9.11). This step, the junction between glycolysis and the Krebs cycle, is catalyzed by a multienzyme complex. First, pyruvate's carboxyl group, which has little chemical energy, is removed and given off as a molecule of CO_2. (This is the first step in respiration where CO_2 is released.) The remaining two-carbon fragment is oxidized to form a compound named acetate (the ionized form of acetic acid). An enzyme transfers the extracted electrons to NAD^+, storing energy in the form of NADH. Finally, coenzyme A, a sulfur-containing compound, is attached to the acetate by an unstable bond that makes the acetyl group (the attached acetate) very reactive. The product of the multienzyme complex, acetyl CoA, is now ready to feed its acetate into the Krebs cycle for further oxidation.

The Krebs cycle is named after Hans Krebs, the German–British scientist who was largely responsible for elucidating the pathway in the 1930s. The cycle has eight steps, each catalyzed by a specific enzyme in the mitochondrial matrix. Figure 9.12 summarizes the overall functioning of the cycle, and Figure 9.13 shows the eight steps in more detail.

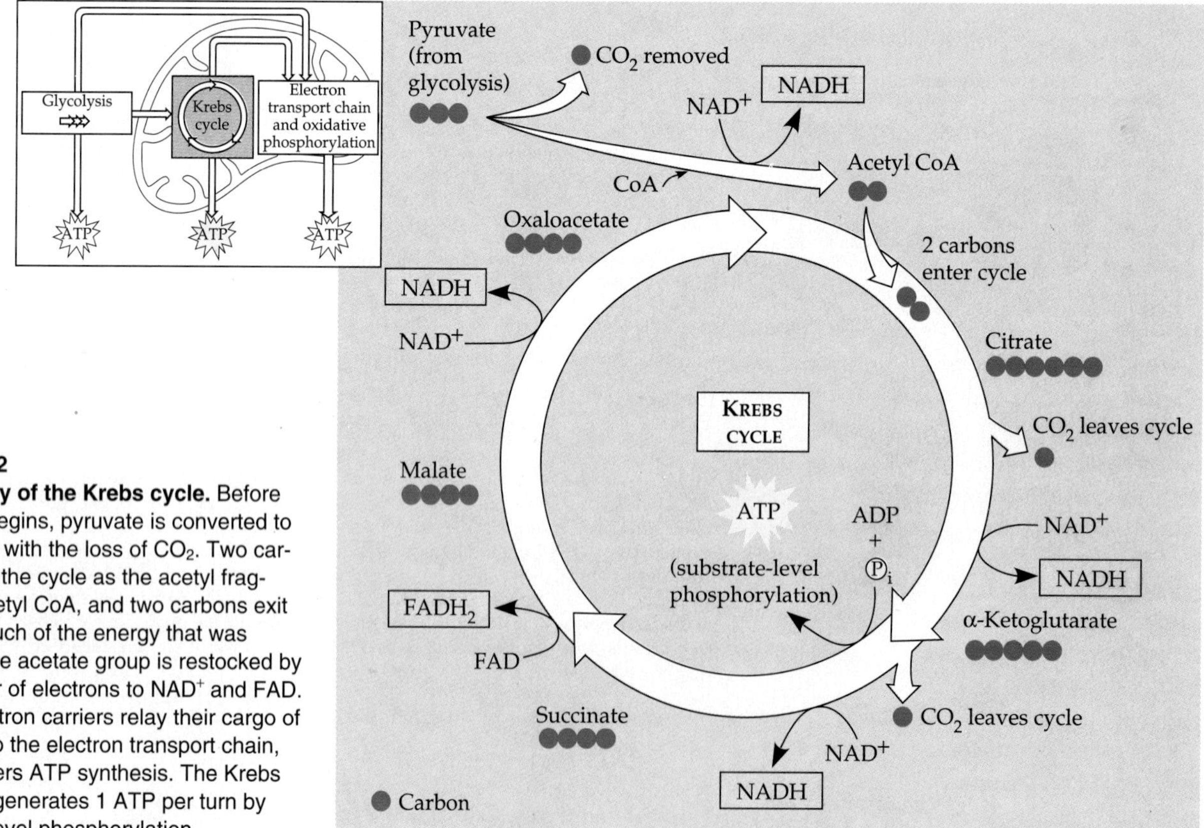

Figure 9.12
A summary of the Krebs cycle. Before the cycle begins, pyruvate is converted to acetyl CoA with the loss of CO_2. Two carbons enter the cycle as the acetyl fragment of acetyl CoA, and two carbons exit as CO_2. Much of the energy that was stored in the acetate group is restocked by the transfer of electrons to NAD^+ and FAD. These electron carriers relay their cargo of electrons to the electron transport chain, which powers ATP synthesis. The Krebs cycle also generates 1 ATP per turn by substrate-level phosphorylation.

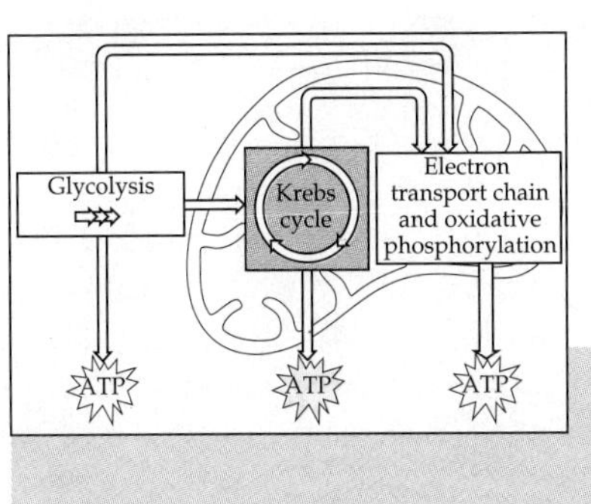

Figure 9.13

The Krebs cycle. Red type is used to trace the fate of the two carbon atoms that enter the cycle via acetyl CoA (step 1), and blue type is used to follow the two carbons that are given off as carbon dioxide in steps 3 and 4. It takes *two* turns of the cycle to complete the oxidation of each glucose molecule broken down by glycolysis. (Notice that here, as in previous figures in this chapter, carboxylic acids are represented in their ionized, $—COO^-$, forms. For example, citrate is the ionized form of citric acid.)

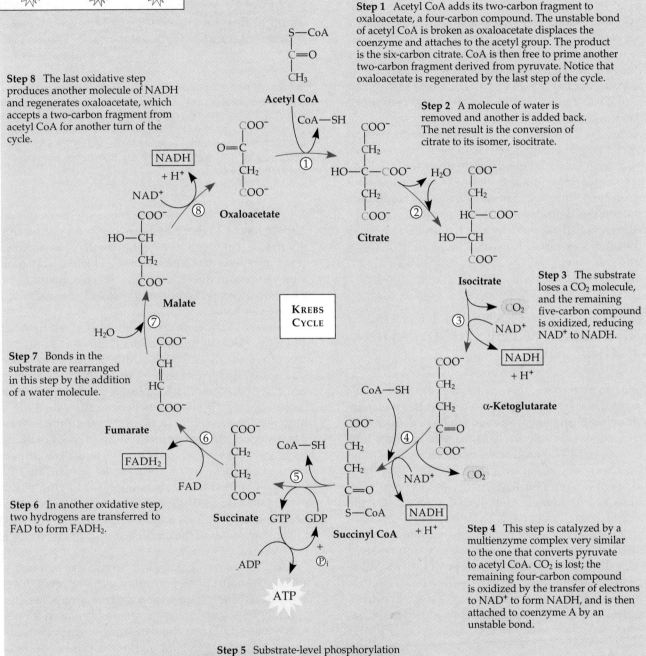

Step 1 Acetyl CoA adds its two-carbon fragment to oxaloacetate, a four-carbon compound. The unstable bond of acetyl CoA is broken as oxaloacetate displaces the coenzyme and attaches to the acetyl group. The product is the six-carbon citrate. CoA is then free to prime another two-carbon fragment derived from pyruvate. Notice that oxaloacetate is regenerated by the last step of the cycle.

Step 2 A molecule of water is removed and another is added back. The net result is the conversion of citrate to its isomer, isocitrate.

Step 3 The substrate loses a CO_2 molecule, and the remaining five-carbon compound is oxidized, reducing NAD^+ to NADH.

Step 4 This step is catalyzed by a multienzyme complex very similar to the one that converts pyruvate to acetyl CoA. CO_2 is lost; the remaining four-carbon compound is oxidized by the transfer of electrons to NAD^+ to form NADH, and is then attached to coenzyme A by an unstable bond.

Step 5 Substrate-level phosphorylation occurs in this step. CoA is displaced by a phosphate group, which is then transferred to GDP to form guanosine triphosphate (GTP). GTP is similar to ATP, which is formed when GTP donates a phosphate group to ADP.

Step 6 In another oxidative step, two hydrogens are transferred to FAD to form $FADH_2$.

Step 7 Bonds in the substrate are rearranged in this step by the addition of a water molecule.

Step 8 The last oxidative step produces another molecule of NADH and regenerates oxaloacetate, which accepts a two-carbon fragment from acetyl CoA for another turn of the cycle.

KREBS CYCLE

Acetyl CoA · Oxaloacetate · Citrate · Isocitrate · α-Ketoglutarate · Succinyl CoA · Succinate · Fumarate · Malate

For each turn of the Krebs cycle, two carbons enter in the relatively reduced form of acetate, and two different carbons leave in the completely oxidized form of CO_2 (see Figure 9.12). The acetate joins the cycle by its enzymatic addition to the compound oxaloacetate, to form citrate. Subsequent steps decompose the citrate back to oxaloacetate, giving off CO_2 as "exhaust." It is this regeneration of oxaloacetate that accounts for the "cycle" in the Krebs cycle.

Most of the energy made available by the oxidative steps of the cycle is conserved in NADH. For each acetate that enters the cycle, three molecules of NAD^+ are reduced to NADH. In one oxidative step, electrons are transferred not to NAD^+, but to a different electron acceptor, FAD (flavin adenine dinucleotide, a compound closely related to NAD^+). The reduced form, $FADH_2$, donates its electrons to the electron transport chain, as does NADH. There is also a step in the Krebs cycle that forms an ATP molecule directly by substrate-level phosphorylation, similar to the ATP-generating steps of glycolysis. But most of the ATP output of respiration results from oxidative phosphorylation, when the NADH and $FADH_2$ produced by the Krebs cycle relay the electrons extracted from food to the electron transport chain.

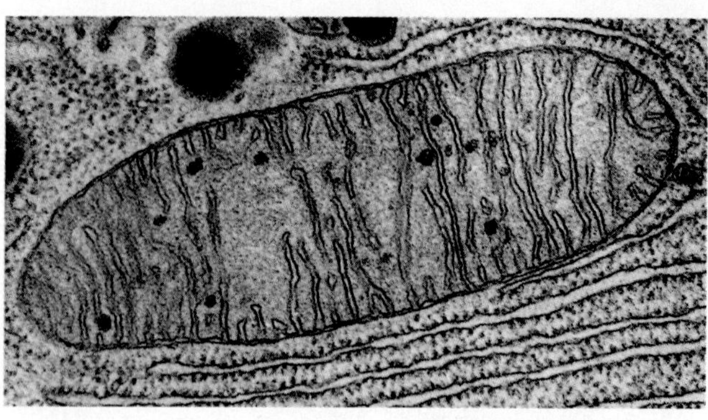

Figure 9.14
Mitochondrion structure: a review. Each of these organelles has thousands of copies of the electron transport chain, which converts energy released by the oxidation of food to a form that can be used to make ATP. The chains are built into the inner mitochondrial membrane, which is expanded by the infoldings called cristae. The cristae are immersed in the matrix, where the Krebs cycle occurs (TEM).

THE ELECTRON TRANSPORT CHAIN AND OXIDATIVE PHOSPHORYLATION: A CLOSER LOOK

Our main objective in this chapter is to learn how cells harvest the energy of food to make ATP. But the metabolic components of respiration we have dissected so far, glycolysis and the Krebs cycle, produce only four molecules of ATP per glucose molecule by substrate-level phosphorylation: two net ATPs from glycolysis and two ATPs from the Krebs cycle. At this point, molecules of NADH (and $FADH_2$) account for most of the energy extracted from food. These electron escorts link glycolysis and the Krebs cycle to the machinery for oxidative phosphorylation, which uses energy released by the electron transport chain to power ATP synthesis. In this section, you will first learn how the electron transport chain works, then how the mitochondrion couples ATP synthesis to electron flow down the chain.

The Pathway of Electron Transport

As explained earlier, the electron transport chain is a collection of molecules embedded in the inner membrane of the mitochondrion. The folding of the inner membrane to form cristae provides space for thousands of copies of the chain in each mitochondrion—once again, we see that structure fits function (Figure 9.14). Most components of the chain are proteins. Tightly bound to these proteins are prosthetic groups, nonprotein components essential for the catalytic functions of certain enzymes. During electron transport along the chain, these prosthetic groups alternate between reduced and oxidized states as they accept and donate electrons.

Figure 9.15 traces the sequence of electron transfers along the electron transport chain. Electrons removed from food during glycolysis and the Krebs cycle are transferred by NADH to the first molecule of the electron transport chain. This molecule is a flavoprotein, so named because it has a prosthetic group called flavin mononucleotide (FMN in Figure 9.15). In the next redox reaction, the flavoprotein returns to its oxidized form as it passes electrons to an iron-sulfur protein (Fe·S in Figure 9.15), one of a family of proteins with both iron and sulfur tightly bound. The iron-sulfur protein then passes the electrons to a compound called ubiquinone (Q in Figure 9.15). This electron carrier is the only member of the electron transport chain that is not a protein.

Most of the remaining electron carriers between Q and oxygen are proteins called **cytochromes** (Cyt). Their prosthetic group, called a **heme group,** has four organic rings surrounding a single iron atom (Figure 9.16). It is similar to the iron-containing prosthetic group found in hemoglobin, the red protein of blood that transports oxygen. But the iron of cytochromes transfers electrons, not oxygen. The electron transport chain has several types of cytochromes, each a different protein with a heme group. The last cytochrome of the chain, cytochrome a_3, passes its electrons to oxygen, which also picks up a pair of hydrogen ions from the aqueous medium to form water. (An oxygen atom is represented in Figure 9.15 as $\frac{1}{2}O_2$ to emphasize that the electron transport chain reduces molecular oxygen,

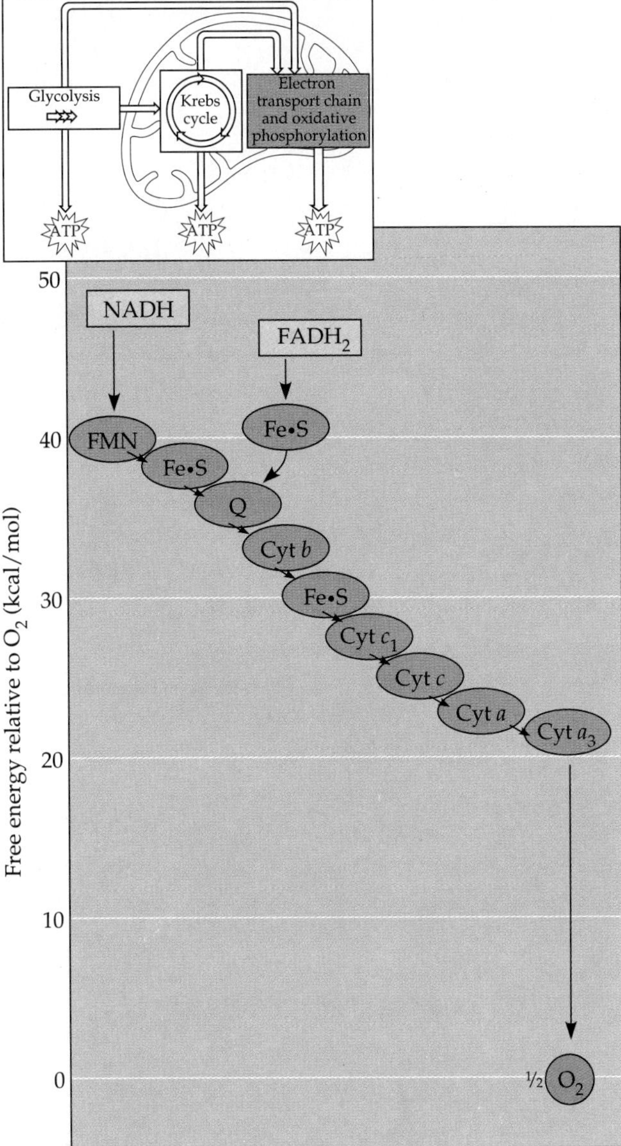

Figure 9.15
The electron transport chain. Each member of the chain oscillates between a reduced state and an oxidized state. A component of the chain becomes reduced when it accepts electrons from its uphill neighbor (which has a lower affinity for the electrons). Each member of the chain returns to its oxidized form as it passes electrons to its downhill neighbor (which has a greater affinity for the electrons). At the bottom of the chain is oxygen, which is *very* electronegative. The overall energy drop for electrons traveling from NADH to oxygen is 53 kcal/mol, but this fall is broken up into a series of smaller steps by the electron transport chain.

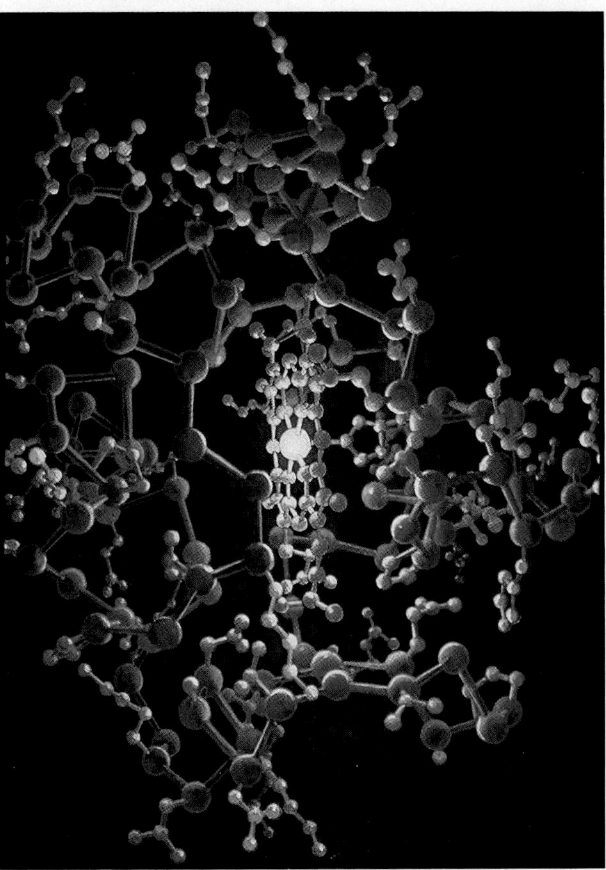

Figure 9.16
The structure of a cytochrome. Cytochromes participate in the energy-yielding transport of electrons from food to oxygen. At the core of each of these proteins is a prosthetic group (nonprotein attachment) called the heme group, colored orange in this illustration. The highlighted atom at the center of the heme group is an iron atom. It alternates between reduced and oxidized states as it gains and loses electrons that are passed along the transport chain. Each type of cytochrome has a heme group bound to a slightly different protein.

O_2, not individual oxygen atoms. For every two NADH molecules, one O_2 molecule is reduced to two molecules of water.)

Another source of electrons for the transport chain is $FADH_2$, the other reduced product of the Krebs cycle. Notice in Figure 9.15 that $FADH_2$ adds its electrons to the electron transport chain at a lower energy level than that of NADH. Consequently, the electron transport chain provides about one-third less energy for ATP synthesis when the electron donor is $FADH_2$ rather than NADH.

The electron transport chain makes no ATP directly. Its function is to ease electrons from food to oxygen, breaking a large free energy drop into a series of smaller steps that release energy in manageable amounts. How does the mitochondrion couple this process to ATP synthesis? The answer is a mechanism called chemiosmosis.

Chemiosmosis: Coupling Electron Flow to ATP Synthesis

Populating the inner membrane of the mitochondrion are many copies of a protein complex called an **ATP synthase,** the enzyme that actually makes ATP. It works like an ion pump running in reverse. Recall from Chapter 8 that ion pumps use ATP as an energy

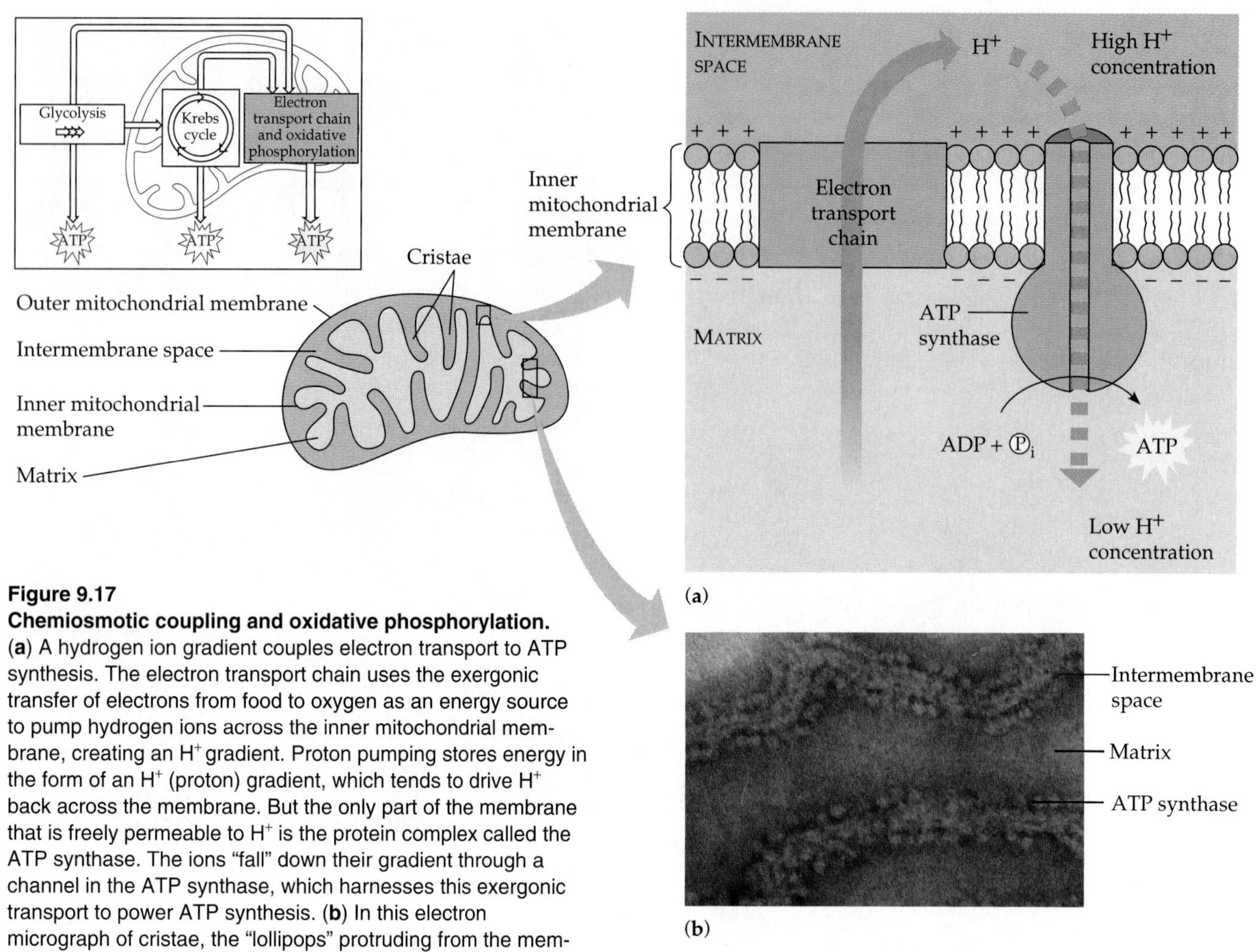

Figure 9.17
Chemiosmotic coupling and oxidative phosphorylation.
(**a**) A hydrogen ion gradient couples electron transport to ATP synthesis. The electron transport chain uses the exergonic transfer of electrons from food to oxygen as an energy source to pump hydrogen ions across the inner mitochondrial membrane, creating an H^+ gradient. Proton pumping stores energy in the form of an H^+ (proton) gradient, which tends to drive H^+ back across the membrane. But the only part of the membrane that is freely permeable to H^+ is the protein complex called the ATP synthase. The ions "fall" down their gradient through a channel in the ATP synthase, which harnesses this exergonic transport to power ATP synthesis. (**b**) In this electron micrograph of cristae, the "lollipops" protruding from the membrane are the ATP synthases (TEM).

source to transport ions against their gradients. In the reverse of that process, an ATP synthase uses the energy of an existing ion gradient to power ATP synthesis. The ion gradient that drives oxidative phosphorylation is a proton (hydrogen ion) gradient; that is, the power source for the ATP synthase is a difference in the concentration of H^+ on opposite sides of the inner mitochondrial membrane. We can also think of this gradient as a difference in pH, since pH is a measure of H^+ concentration (see Chapter 3).

How does the mitochondrial membrane generate and maintain an H^+ gradient? *That* is the function of the electron transport chain. The chain is an energy converter that uses the exergonic flow of electrons to pump H^+ across the membrane, from the matrix into the intermembrane space (Figure 9.17). The H^+ leaks back across the membrane, diffusing down its gradient. But the ATP synthases are the only patches of the membrane that are freely permeable to H^+. The ions pass through a channel in an ATP synthase, and the complex of proteins functions as a mill that harnesses the exergonic "fall" of H^+ to drive the phosphorylation of ADP. Thus, an H^+ gradient couples the redox reactions of the electron transport chain to ATP synthesis.

This coupling mechanism for oxidative phosphorylation is called **chemiosmosis** (Gr. *osmos*, "push"), a term that highlights the relationship between chemical reactions and transport across the membrane. We have previously used the word *osmosis* in discussing water transport, but here the word refers to the pushing of H^+ across a membrane.

If you have followed this complex story of chemiosmosis so far, you should have at least two questions. How does the electron transport chain pump hydrogen ions? And how does the ATP synthase use H^+ backflow to make ATP? Researchers have made some progress on the first problem. Certain members of the electron transport chain must accept and release protons (H^+) along with electrons, while other carriers transport only electrons. Therefore, at certain steps along the chain, electron transfers cause H^+ to be taken up and released back into the surrounding solution. The electron carriers are spatially arranged in the membrane in such a way that H^+ is accepted from the mitochondrial matrix and deposited in the intermembrane space (Figure 9.18). The H^+ gradient that results is referred to as a **proton-motive force,** emphasizing the capacity of the gradient to perform work. The force drives H^+ back

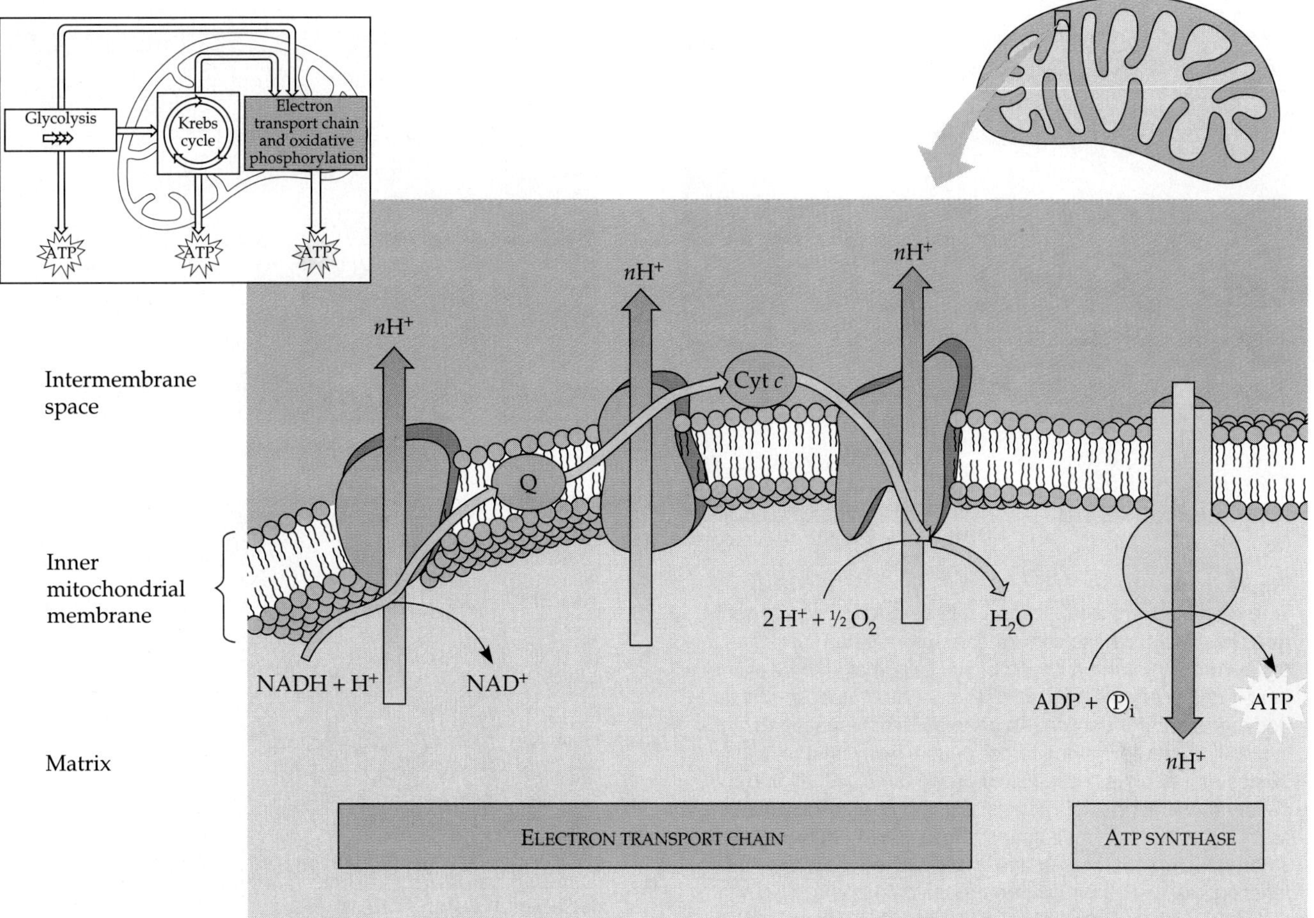

Figure 9.18
A closer look at chemiosmosis in mitochondria. The electron transport chain is an energy converter: it transforms chemical energy into a gradient of H^+. This function depends on the spatial organization of the mitochondrial membrane. Most of the electron carriers of the transport chain are collected into three complexes, each an asymmetric particle with a specific orientation in the membrane. Electrons are relayed between these complexes by two mobile carriers, ubiquinone (Q) and cytochrome *c*, which move rapidly in the plane of the membrane. The electron path from NADH to O_2 is traced through the transport chain in this diagram by the thin black arrow. As each complex of the chain accepts and then donates electrons, hydrogen ions are picked up from the matrix and deposited in the intermembrane space. Functioning collectively as a proton pump, the electron transport chain can achieve a hundred-fold difference in H^+ concentration across the membrane. However, the exact number of hydrogen ions pumped for each NADH that passes electrons to the transport chain is not yet known (the coefficient n for the H^+ reflects this uncertainty). The circuit of H^+ is completed when the ions pass down their gradient through an ATP synthase, which uses the energy to make ATP. This is the mechanism of oxidative phosphorylation, which accounts for about 90% of the ATP yield of cellular respiration.

across the membrane through the specific channels provided by ATP synthase complexes. How the ATP synthase uses the downhill H^+ current to attach inorganic phosphate to ADP is not yet known. The hydrogen ions may participate directly in the reaction, or they may induce a conformation (shape) change of the ATP synthase that facilitates phosphorylation. Research has revealed the general mechanism of energy coupling by chemiosmosis, but many details of the process are still uncertain.

Studying the effects of respiratory poisons has provided important evidence for the chemiosmotic model. A variety of poisons inhibit cellular respiration by disrupting chemiosmosis. Some of these deadly poisons block electron flow along the electron transport chain. For example, cyanide blocks the passage of electrons from cytochrome a_3 to oxygen, plugging the electron transport chain. The result is that protons are not pumped and ATP is not made. Another class of poisons, which includes the antibiotic oligomycin, inhibits the ATP synthase directly. As a result, the electron transport chain generates a proton gradient of greater magnitude than normal because the gradient is not dissipated by ATP synthesis. A third class of poisons, called uncouplers, short-circuits the proton current by making the lipid bilayer of the membrane leaky to hydrogen ions. The electron transport chain works furiously, and oxygen consumption increases, but leakage of H^+ abolishes the proton gradient, and no ATP is made. Instead, the energy released by electron trans-

(a)

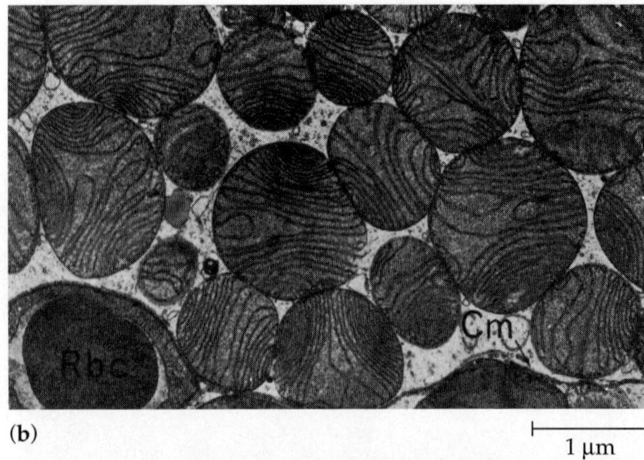

(b)

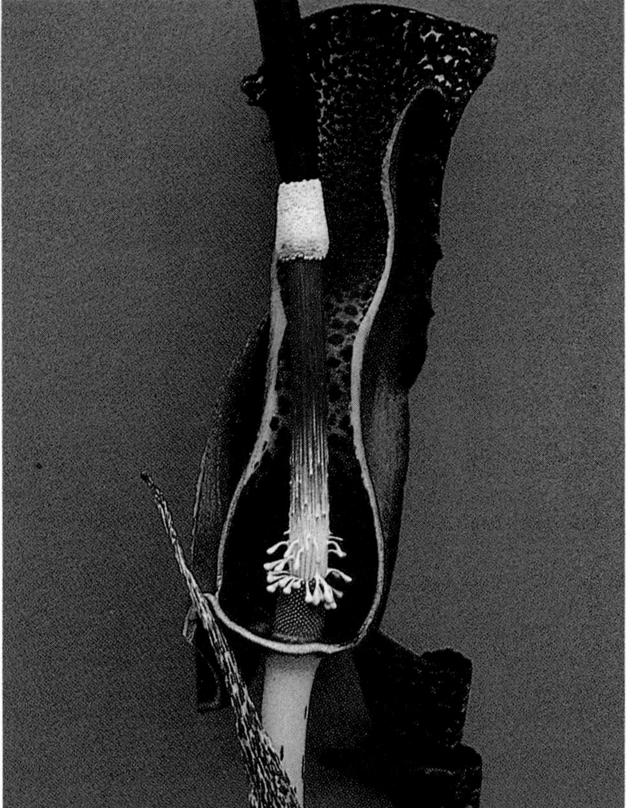

(c)

Figure 9.19
Two evolutionary adaptations for metabolic heat generation. (**a**) Most bats, including these white Costa Rican tent bats, exhibit a daily torpor, a period of inactivity that alternates with a period of active feeding. During torpor, the metabolic rate and body temperature decline. An adaptation of respiration warms the body at the end of the torpor period. Between the shoulder blades of a bat is a tissue called brown fat. (**b**) Of all body tissues, brown fat has the greatest number of mitochondria per cell, as this electron micrograph shows (TEM). In brown fat, the inner mitochondrial membranes have a special protein that functions as an H^+ conduit through the membrane. This short-circuits the usual proton current; instead of diffusing through ATP synthase, the pumped protons simply reenter the mitochondrion through the special protein. Thus, the energy released by oxidation of fat generates heat instead of ATP. (**c**) A different type of mitochondrial adaptation has evolved in heat-generating tissues of certain plants. For example, this plant, the voodoo lily *(Sauromatum guttatum),* warms certain parts of its flowers. The plant is pollinated by flies, which apparently mistake the flowers for dead meat. (This is a cutaway view of a flower.) The flower emits odors that help attract the flies, and heating helps disperse the smelly gases. In the mitochondria that generate the heat, electrons extracted from organic fuel during respiration bypass the normal transport chain and pass along an alternative pathway that does not pump protons across the mitochondrial membrane. This transforms the energy of electron transport into heat.

port is transformed into heat. An example of an uncoupler is the poison dinitrophenol (DNP). Humans and other animals poisoned by DNP may die because they are unable to dissipate the excessive heat generated by the metabolic "wheel spinning" that occurs in mitochondria. On the other hand, some organisms have natural mechanisms that heat the body by occasionally uncoupling electron transport from ATP synthesis (Figure 9.19).

Let's review the key feature of chemiosmosis. It is an energy-coupling mechanism that uses exergonic chemical reactions to store energy in the form of an H^+ gradient, which then drives other kinds of work, including ATP synthesis. Chemiosmosis is not unique to mitochondria. Chloroplasts also use the mechanism to generate ATP during photosynthesis; the main difference is that light drives electrons along an electron transport chain. Bacteria, which lack both mitochondria and chloroplasts, generate H^+ gradients across their plasma membranes. They then tap the proton-motive force to make ATP, to pump nutrients and waste products across the membrane, and even to move by rotating their flagella. In 1961, British biochemist Peter Mitchell first proposed chemiosmosis as an energy-coupling mechanism based on his experiments with bacteria. Nearly two decades later, after many other scientists had confirmed the central importance of chemiosmosis for energy conversions within bacteria, mitochondria, and chloroplasts, Mitchell was awarded the Nobel Prize. His model has helped unify the study of bioenergetics.

The Chemiosmotic Model: Incorporating Many Biological Themes

Many of the themes that integrate this textbook are manifest in our working model of how mitochondria harvest the energy of food. Energy conversion, the main function of mitochondria, is one of our key themes in the study of life, and other themes are apparent in the mechanism of this energy conversion. Oxidative phosphorylation is an emergent property of an intact mitochondrion that uses a precise interaction of molecules to make ATP. The chemiosmotic model is elegant in its correlation of structure and function; it is the spatial arrangement of membrane proteins that enables the mitochondrion to transform chemical energy into an H^+ gradient and then use that proton-motive force to drive ATP synthesis. Viewing oxidative phosphorylation in the context of evolution, life's overarching theme, has helped biologists understand the relationship of the chemiosmotic machinery in mitochondria to very similar equipment in chloroplasts and bacteria. In fact, both mitochondria and chloroplasts, the energy-converting organelles of eukaryotes, probably evolved from bacteria (see Chapter 26). Most of the ideas we learn about in biology turn out to be specific examples of general concepts.

SUMMARIZING CELLULAR RESPIRATION

Now that we have looked more closely at the key processes of cellular respiration, let's return to its overall function: harvesting the energy of food for ATP synthesis. We can do some bookkeeping to calculate the net ATP profit when cellular respiration oxidizes a molecule of glucose to six molecules of carbon dioxide. The three main departments of this metabolic enterprise are glycolysis, the Krebs cycle, and the electron transport chain, which drives oxidative phosphorylation. Figure 9.20, a follow-up to the overview of respiration presented in Figure 9.7, gives a detailed accounting of the ATP yield per glucose molecule oxidized. First let's tally the few molecules of ATP produced directly by substrate-level phosphorylation during glycolysis and the Krebs cycle. (These tallies are

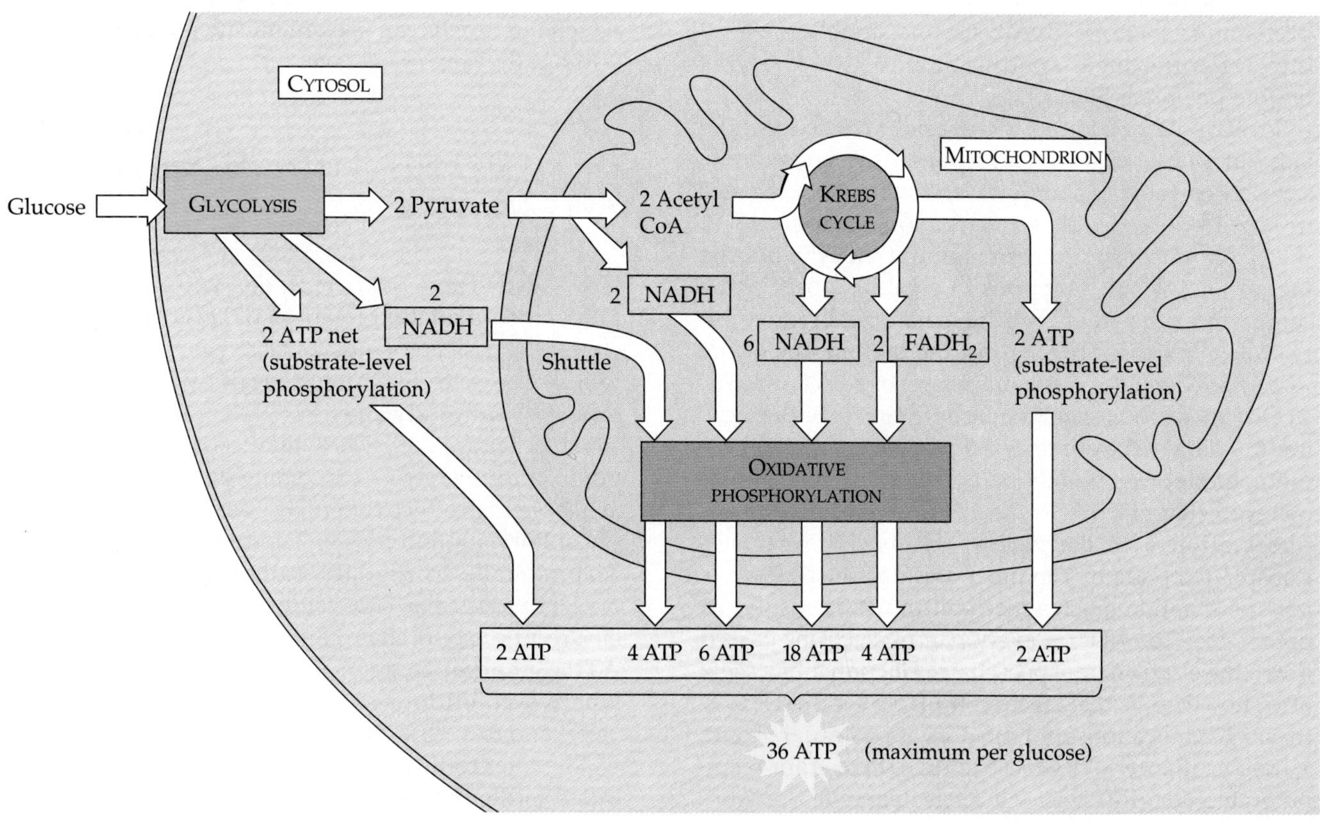

Figure 9.20
The maximum ATP yield for cellular respiration in a eukaryotic cell. For reasons discussed in the accompanying text, the maximum yield of 36 ATP per glucose molecule is only an estimate—and a high one, at that. Our estimate is based on a yield of 3 ATP per NADH, but actual measurements place the ratio at about 2.7. A maximum ATP yield of 32 per glucose is probably more realistic.

shown at the beginning and end of the "bottom line" in Figure 9.20.) To this number we add the many more molecules of ATP generated when chemiosmosis couples electron transport to oxidative phosphorylation. Each NADH that transfers a pair of electrons from food to the electron transport chain contributes enough to the proton-motive force to generate a maximum of about three ATPs. (The average ATP yield per NADH is probably between two and three; we are rounding off to three here to simplify the bookkeeping.) The Krebs cycle also supplies electrons to the electron transport chain via $FADH_2$, but each molecule of this electron carrier is worth a maximum of only two molecules of ATP. (Notice again in Figure 9.15 that $FADH_2$ does not donate its electrons to the top of the transport chain.) In most eukaryotic cells, this lower ATP yield per electron pair also applies to the NADH produced by glycolysis in the cytosol. The mitochondrial membrane is impermeable to NADH, so NADH in the cytosol is segregated from the machinery of oxidative phosphorylation. The electrons of NADH captured by glycolysis must be shuttled across the membrane to electron acceptors within the mitochondrion. In the most common shuttle system, the electrons are received within the mitochondrion not by NAD^+, but by FAD, and this downgrades the "ATP value" of those electrons.

Subtracting the debit of 2 ATP incurred during the preparatory steps of glycolysis, and doubling everything after the sugar-splitting step of glycolysis, the bottom line reads 36 ATP.

Prokaryotes, of course, do not have to discount the value of NADH from glycolysis, because no membrane separates glycolysis from the electron transport chains. In prokaryotes capable of respiration, components of the electron transport chain are built into the plasma membrane, which functions in chemiosmosis in the same way as the inner mitochondrial membrane of eukaryotes. For these organisms, the bottom line of the respiration ledger reads a maximum of 38 ATP.

Our bookkeeping gives only an estimate of the maximum ATP yield from respiration. Before the chemiosmotic model was widely accepted, most biochemists believed that electron transport and oxidative phosphorylation were coupled by the direct transfer of phosphate bonds in a manner akin to substrate-level phosphorylation. Such a mechanism would result in a precise yield of ATP for every pair of electrons passed down the electron transport chain. The prevailing view now, however, is that electron transport and ATP synthesis are more loosely linked by a proton gradient. Variables that would affect the ATP yield from respiration include differences in the leakiness of the mitochondrial membrane to hydrogen ions, and use of the proton-motive force to drive other kinds of work. For example, the proton-motive force powers the mitochondrion's uptake of pyruvate from the cytoplasm. Also, by rounding off the number of ATPs produced per NADH to three, we have inflated the ATP yield of respiration by about 10%.

We can now calculate a rough estimate of the efficiency of respiration—that is, the percentage of chemical energy stored in glucose that has been restocked in ATP. Recall that complete oxidation of a mole of glucose releases 686 kilocalories of energy ($\Delta G = -686$ kcal/mol). Given the chemical conditions of the cell, the phosphorylation of ADP to form ATP stores about 12 kcal/mol of ATP. Therefore, the efficiency of respiration is 12 times 36 (maximum ATP yield per glucose) divided by 686, or about 63%. Even if we have overestimated by 10% to 20%, cellular respiration is still remarkably efficient in its energy conversion. (By comparison, the most efficient automobile converts about 25% of the energy stored in gasoline to movement of the car.) The rest of the stored energy is lost as heat. We use some of this heat to maintain our relatively high body temperature (37°C), and we dissipate the rest through sweating and other cooling mechanisms (to be discussed in detail in Chapter 40).

Because most of the ATP generated by cellular respiration is the work of oxidative phosphorylation, our estimate of ATP yield from respiration is contingent upon an adequate supply of oxygen to the cell. Without the electronegative oxygen to pull electrons down the transport chain, oxidative phosphorylation ceases. However, fermentation provides a mechanism by which many cells can use organic fuel to generate ATP without the help of oxygen.

FERMENTATION: THE ANAEROBIC ALTERNATIVE

How can food be oxidized without oxygen? Remember, oxidation refers to the loss of electrons to *any* electron acceptor, not just to oxygen. Glycolysis oxidizes glucose to two molecules of pyruvate. The oxidizing agent of glycolysis is NAD^+, *not* oxygen (see step 6 in Figure 9.10). The oxidation of glucose is exergonic, and glycolysis uses some of the energy made available to produce two ATPs (net) by substrate-level phosphorylation. If oxygen *is* present, then additional ATP is made by oxidative phosphorylation when NADH passes electrons removed from glucose to the electron transport chain. But glycolysis generates two ATPs whether oxygen is present or not—that is, whether conditions are *aerobic* or *anaerobic* (Gr. *aer*, "air," and *bios*, "life"; the prefix *an* means "without").

The anaerobic catabolism of organic nutrients is called fermentation, as mentioned at the beginning of the chapter. Fermentation can generate ATP by substrate-level phosphorylation, as long as there is a sufficient supply of NAD^+ to accept electrons during the oxidation step of glycolysis. Without some mechanism to recycle NAD^+ from NADH, glycolysis would soon de-

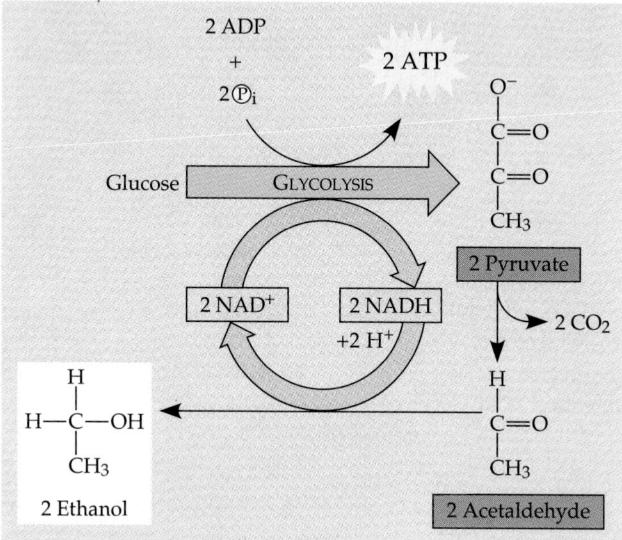

(**a**) Alcohol fermentation

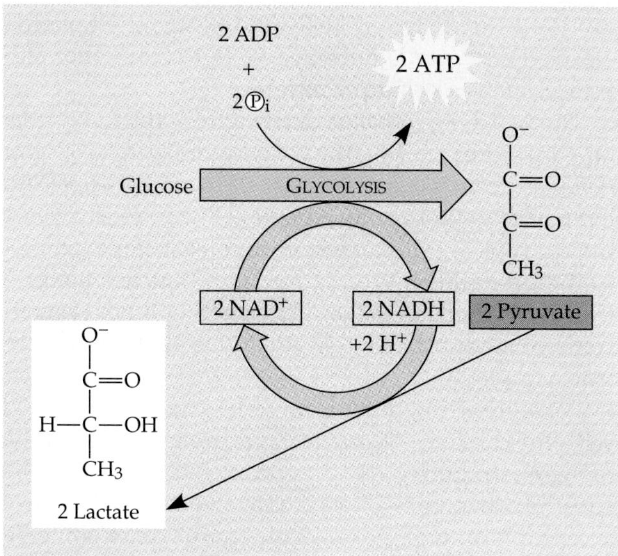

(**b**) Lactic acid fermentation

Figure 9.21
Fermentation. Pyruvate, the end-product of glycolysis, serves as an electron acceptor for oxidizing NADH back to NAD^+. The NAD^+ can then be reused to oxidize sugar during glycolysis, which yields two net molecules of ATP by substrate-level phosphorylation during fermentation. Two of the common waste products formed from fermentation are (**a**) ethanol and (**b**) lactate.

(**a**)

(**b**)

Figure 9.22
Commercial applications of alcohol fermentation. The basic steps used to make wine from grapes were the same in (**a**) ancient times as they are (**b**) today. The process employs yeast to convert some of the sugar in fruit juice to alcohol. Yeast must be cultured in the absence of oxygen for fermentation to occur. One-way gas valves allow carbon dioxide to escape from the ferment without letting air in. To make a sparkling wine, such as champagne, the CO_2 is left dissolved in the wine.

plete the cell's pool of NAD^+ and shut itself down for lack of an oxidizing agent. Under aerobic conditions, NAD^+ is recycled productively from NADH by the transfer of electrons to the electron transport chain. The anaerobic alternative is to transfer electrons from NADH to pyruvate, the end-product of glycolysis.

Fermentation consists of glycolysis plus reactions that regenerate NAD^+ by transferring electrons from NADH to pyruvate or derivatives of pyruvate. There are many types of fermentation, differing in the waste products formed from pyruvate. Two common types are alcohol fermentation and lactic acid fermentation (Figure 9.21).

In **alcohol fermentation,** pyruvate is converted to ethanol, or ethyl alcohol, in two steps. The first step releases carbon dioxide from the pyruvate, which is converted to the two-carbon compound acetaldehyde. In the second step, acetaldehyde is reduced by NADH to ethyl alcohol. This regenerates the supply of NAD^+ needed for glycolysis. Alcohol fermentation by yeast, a fungus, is used in brewing (Figure 9.22). Many bacteria also carry out alcohol fermentation under anaerobic conditions.

During **lactic acid fermentation,** pyruvate is reduced directly by NADH to form lactate as a waste product, with no release of CO_2. (Lactate is the ionized form of lactic acid.) Lactic acid fermentation by certain fungi and bacteria is used in the dairy industry to make

Table 9.1 A Comparison of Catabolic Processes

	Aerobic Respiration	Anaerobic Respiration	Fermentation
Growth conditions	Aerobic	Anaerobic	Anaerobic
Electron transport chain	Yes	Yes	No
Final hydrogen (electron) acceptor	Free oxygen (O_2)	Usually an inorganic substance (such as NO_3^- SO_4^{2-}, or CO_3^{2-}), but not free oxygen (O_2)	Organic molecule, such as lactate or ethanol
Type of phosphorylation used to build ATP	Mostly oxidative; some substrate-level	Mostly oxidative; some substrate-level	Substrate-level

cheese and yogurt. Acetone and methyl alcohol are among the by-products of other types of microbial fermentation that are commercially important.

Human muscle cells make ATP by lactic acid fermentation when oxygen is scarce. This occurs during the early stages of strenuous exercise, when sugar catabolism for ATP production outpaces the muscle's supply of oxygen from the blood. Under these conditions, the cells switch from aerobic respiration to fermentation. The lactate that accumulates as a waste product may cause muscle fatigue and pain, but it is gradually carried away by the blood to the liver. Lactate is converted back to pyruvate by liver cells.

A COMPARISON OF AEROBIC AND ANAEROBIC CATABOLISM

There are actually three major processes for harvesting the chemical energy of food molecules: aerobic respiration, anaerobic respiration, and fermentation. In all three catabolic processes, glucose or other energy-rich substrates are oxidized and their electrons passed to NAD^+, reducing it to NADH. However, the ultimate fate of these electrons differs in each process (Table 9.1).

Organisms that rely on **aerobic respiration** for ATP, such as plants and animals, are called **strict aerobes,** which means that they can survive only in an environment that has oxygen. The final electron acceptor for aerobic respiration is oxygen, which is reduced by electrons that flow down the electron transport chain from NADH.

Anaerobic respiration occurs only in a few groups of bacteria that live deep in the soil, in stagnant ponds, or in other anaerobic environments. Their metabolism does not require oxygen. In fact, these microorganisms are poisoned by oxygen; they are among the bacteria referred to as **strict anaerobes.** During anaerobic respiration, electrons removed from substrates are passed along an electron transport chain to generate a proton gradient for ATP synthesis. However, the terminal electron acceptor is not oxygen, but some other substance, such as sulfate or nitrate.

The word *respiration* is derived from the Latin *respirare,* meaning "to breathe." Therefore, *aerobic respiration* may seem like a redundant term, and *anaerobic respiration* may seem contradictory. But in recent years, the concept of cellular respiration has been expanded to include all types of catabolism that use electron transport chains to make ATP, whether it is oxygen or some other substance that functions as the final electron acceptor.

Fermentation, as you have learned, makes ATP without the help of an electron transport chain. No oxygen is required, ATP is produced exclusively by substrate-level phosphorylation, and the final electron acceptor is an organic molecule (pyruvate or some derivative of pyruvate). Among the bacteria that are strict anaerobes are some that lack electron transport chains and rely entirely on fermentation for ATP synthesis.

Facultative anaerobes, such as yeasts and many bacteria, can make their ATP by either fermentation or aerobic respiration, depending on whether oxygen is available. On the cellular level, our muscle cells behave as facultative anaerobes. In a facultative anaerobe, pyruvate is a fork in the metabolic road that leads to two alternative catabolic routes (Figure 9.23). Under aerobic conditions, pyruvate is converted to acetyl CoA, and oxidation continues in the Krebs cycle. Under anaerobic conditions, pyruvate is diverted from the Krebs cycle, serving instead as an electron acceptor to recycle NAD^+.

Notice that fermentation not only operates without electron transport chains, but without the Krebs cycle as well. Without oxygen, the energy still stored in pyruvate acid is unavailable to the cell. Oxygen enables the cell to harvest much more energy from sugar than the cell can tap by fermentation. In fact, for each

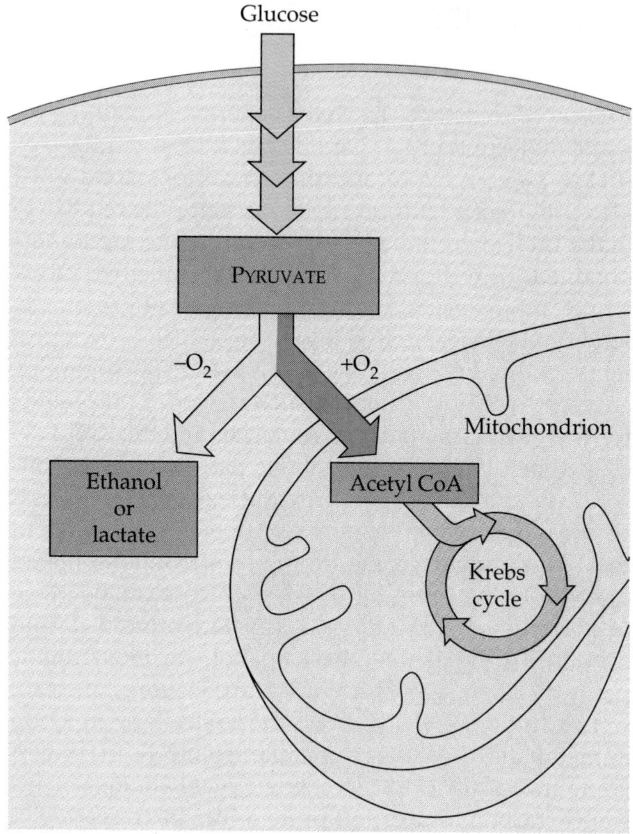

Figure 9.23
Pyruvate as a key juncture in catabolism. Glycolysis is common to fermentation and respiration. The end-product of glycolysis, pyruvate, represents a fork in the catabolic pathways of glucose oxidation. In a cell capable of both respiration and fermentation, pyruvate is committed to one of those two pathways, depending on whether or not oxygen is present.

glucose molecule, aerobic respiration yields 18 times more ATP than does fermentation (36 ATP for respiration, compared to 2 ATP produced by substrate-level phosphorylation in fermentation).

THE EVOLUTIONARY SIGNIFICANCE OF GLYCOLYSIS

Glycolysis is the one catabolic pathway common to fermentation and respiration, a similarity with an evolutionary basis. Ancient prokaryotes probably used glycolysis to make ATP long before oxygen was present in the Earth's atmosphere. The oldest known fossils of bacteria date back over 3.5 billion years, but appreciable quantities of oxygen probably did not begin to accumulate in the atmosphere until about 2.5 billion years ago. (According to fossil evidence, the cyanobacteria that produce O_2 as a by-product of photosynthesis had first evolved by then.) Therefore, the first prokaryotes must have generated ATP exclusively from glycolysis, which does not require oxygen. In addition, glycolysis is the most widespread metabolic pathway, which suggests that it evolved very early in the history of life. The cytoplasmic location of glycolysis also implies great antiquity; the pathway does not require any of the membrane-enclosed organelles of the eukaryotic cell, which evolved nearly 2 billion years after the prokaryotic cell. (The evolution of metabolism is covered more thoroughly in Chapters 24 and 25.) Glycolysis is a metabolic heirloom from the earliest cells that continues to function in the fermentation of modern anaerobes and as the first stage in the breakdown of organic molecules by respiration.

THE CATABOLISM OF OTHER MOLECULES

Throughout this chapter, we have used glucose as the fuel for cellular respiration. But free glucose molecules are not common in the diets of humans and other animals. We obtain most of our calories in the form of fats, proteins, sucrose and other disaccharides, and starch, a polysaccharide. All these food molecules can be used by cellular respiration to make ATP (Figure 9.24).

In the digestive tract, starch is hydrolyzed to glucose, which can then be broken down in the cells by glycolysis and the Krebs cycle. Similarly, glycogen, the polysaccharide that humans and many other animals store in their liver and muscle cells, can be hydrolyzed to glucose between meals as fuel for respiration. The digestion of disaccharides, including sucrose, provides glucose and other monosaccharides that can then be enzymatically converted to glucose as additional fuel for respiration. Thus, glycolysis can accept a wide range of carbohydrates for catabolism.

Proteins can also be used for fuel, but first they must be digested to their constituent amino acids. Many of the amino acids, of course, are used by the organism to build new proteins. Amino acids present in excess are converted by enzymes to intermediates of glycolysis and the Krebs cycle. The entry point into respiration depends on the structure of the amino acid. Common sites are pyruvate, acetyl CoA, and α-ketoglutarate, an intermediate of the Krebs cycle. Before amino acids can feed into glycolysis or the Krebs cycle, their amino groups must be removed, a process called deamination. The nitrogenous refuse is excreted from the animal in the form of ammonia, urea, or other waste products (see Chapter 40).

Catabolism can also harvest energy stored in fats obtained either from food or from storage cells in the body. After fats are digested, the glycerol is converted to glyceraldehyde phosphate, an intermediate of glycolysis. Most of the energy of a fat is stored in the fatty acids. A metabolic sequence called **beta oxidation**

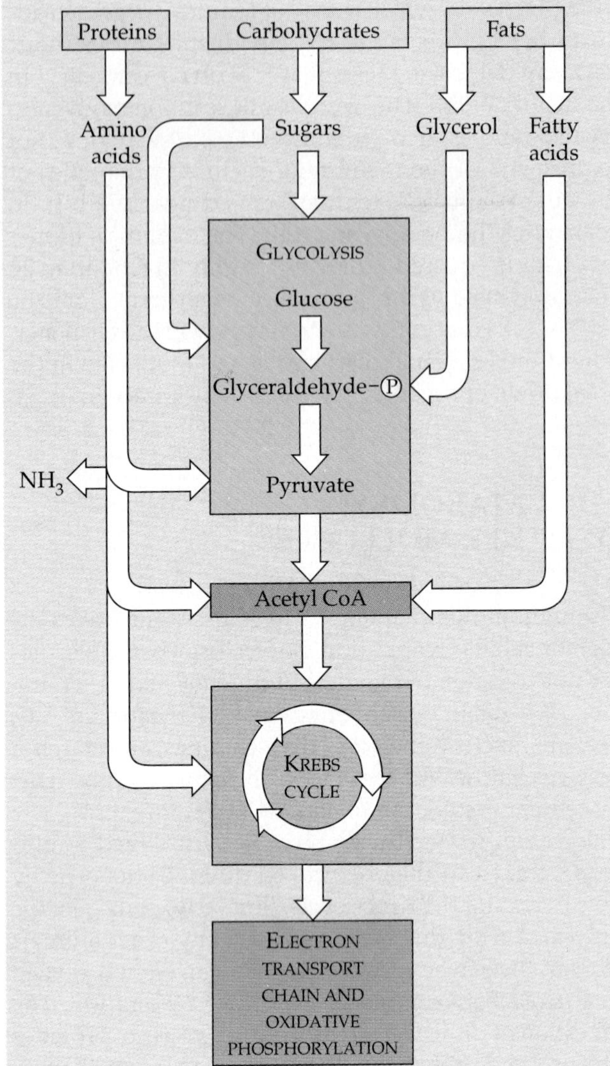

Figure 9.24
The catabolism of various food molecules. Carbohydrates, fats, and proteins can all be used as fuel for respiration. Monomers of these food molecules enter glycolysis or the Krebs cycle at various points. Glycolysis and the Krebs cycle are catabolic funnels through which electrons from all kinds of food molecules flow on their exergonic fall to oxygen.

breaks the fatty acids down to two-carbon fragments, which enter the Krebs cycle as acetyl CoA. Fats make excellent fuel. A gram of fat oxidized by respiration produces more than twice as much ATP as a gram of carbohydrate. Unfortunately, this also means that a dieter must be patient while using fat stored in the body, because so many calories are stockpiled in each gram of fat.

We see, then, that respiration is flexible in the fuels it can oxidize to make ATP. The intermediates of glycolysis and the Krebs cycle also provide carbon skeletons for the cell to use in synthesizing the molecules it needs.

BIOSYNTHESIS

Cells need substance as well as energy. Not all the organic molecules of food are destined to be oxidized as fuel to make ATP. In addition to calories, food must also provide the carbon skeletons that cells require to make their own molecules. Some organic monomers obtained from digestion can be used directly. For example, amino acids from the hydrolysis of proteins in food can be incorporated into the organism's own proteins. Often, however, the body needs specific molecules that are not present as such in food. Compounds formed as intermediates of glycolysis and the Krebs cycle can be diverted into anabolic pathways as precursors from which the cell can synthesize the molecules it requires. For example, humans can make about half of the 20 amino acids by modifying compounds siphoned away from the Krebs cycle. Also, glucose can be made from pyruvate, and fatty acids can be synthesized from acetyl CoA. Of course, these anabolic or biosynthetic pathways do not generate ATP, but consume it instead.

In addition, glycolysis and the Krebs cycle function as metabolic interchanges, enabling our cells to convert some molecules to others as we need them. For instance, carbohydrates and proteins can be converted to fats through intermediates of glycolysis and the Krebs cycle. If we eat more food than we need, we will store fat even if our diet is fat-free. Metabolism is remarkably versatile and adaptable.

THE CONTROL OF RESPIRATION

The basic principles of supply and demand govern the metabolic economy. The cell does not waste energy making more of a particular substance than it needs. If there is a glut of a certain amino acid, for example, the anabolic pathway that synthesizes that amino acid from an intermediate of the Krebs cycle is switched off. The most common mechanism for this control is feedback inhibition: The end-product of the anabolic pathway inhibits the enzyme that catalyzes the first step of the pathway (see Chapter 6). This prevents the needless diversion of key metabolic intermediates from uses that are more urgent.

The cell also controls its catabolism. If the cell is working hard and its ATP concentration begins to drop, respiration speeds up. When there is plenty of ATP to meet demand, respiration slows down, sparing valuable organic molecules for other functions. Again, control is based mainly on regulating the activity of enzymes at strategic points in the catabolic pathway. One important switch is phosphofructokinase, the enzyme that catalyzes step 3 of glycolysis (see Figure 9.10). That is the earliest step that commits substrate irreversibly to the glycolytic pathway. By controlling the rate of this step, the cell can speed up or slow down the entire catabolic process; phosphofructokinase is thus the pacemaker of respiration (Figure 9.25).

An allosteric enzyme with receptor sites for specific inhibitors and activators, phosphofructokinase is inhibited by ATP and stimulated by ADP. As ATP accumulates, inhibition of the enzyme slows down glycolysis. The enzyme becomes active again as cellular work converts ATP to ADP faster than ATP is being regenerated. Phosphofructokinase is also sensitive to citrate, the first product of the Krebs cycle. If citrate accumulates in mitochondria, some of it passes into the cytosol and inhibits phosphofructokinase. This mechanism helps synchronize the rates of glycolysis and the Krebs cycle. As citrate accumulates, glycolysis slows down and the supply of acetate to the Krebs cycle decreases. If citrate consumption increases, either because of a demand for more ATP or because anabolic pathways are draining off intermediates of the Krebs cycle, glycolysis accelerates and meets the demand. Metabolic balance is augmented by control of other enzymes at other key locations in glycolysis and the Krebs cycle. Cells are thrifty, expedient, and responsive in their metabolism.

We have looked at respiration within cells on such a microscopic level that you may have lost the sense of its place in the life of real organisms. But when cells harvest the energy stored in food molecules and use it to make ATP, they are supplying the energy for cellular work that is manifest in all the activities of the whole organism. Without the generation of ATP within our individual cells and, more precisely, within mitochondria, we would not be able to walk, to think, to see—to live.

* * *

The energy that keeps us alive is *released*, but not *produced*, by cellular respiration. We are tapping energy that was stored in food by photosynthesis. In the next chapter, you will learn how photosynthesis captures light and converts it to chemical energy.

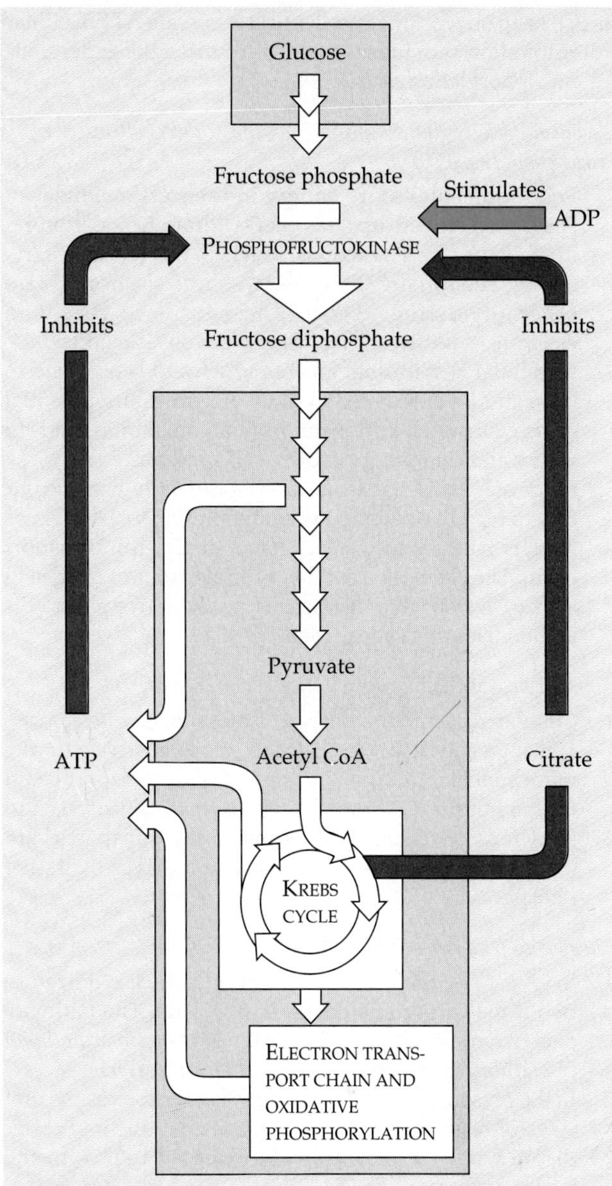

Figure 9.25
The control of cellular respiration. Allosteric enzymes at certain points in the respiratory pathway respond to inhibitors and activators to set the pace of glycolysis and the Krebs cycle. Phosphofructokinase, the enzyme that catalyzes step 3 of glycolysis, is one such key enzyme. It is stimulated by ADP but inhibited by ATP and by citrate.

STUDY OUTLINE

1. The energy that powers most organisms on Earth ultimately comes from the sun. Photosynthetic organisms use solar energy to produce organic molecules from carbon dioxide and water. Animals obtain food by eating plants or by eating animals that eat plants.
2. In eukaryotic cells, cellular respiration occurs in the mitochondria. Starting with glucose or other organic fuel and using oxygen, respiration yields water, carbon dioxide, and energy in the form of ATP and heat.

ATP and Cellular Work: A Review (p. 174)

1. ATP drives cellular work by transferring phosphate groups to various substrates, priming them to undergo change.

2. To keep on working, a cell must regenerate ATP. Cellular respiration provides the energy to drive the endergonic phosphorylation of ADP.

Chemical Background: Respiration as an Oxidation-Reduction Process (pp. 174–178)

1. Food molecules store energy in their arrangement of electrons. The cell taps this energy through oxidation-reduction, or redox, reactions, in which one substance (reducing agent) partially or totally shifts electrons to another (oxidizing agent). The substance receiving electrons is reduced; that losing electrons is oxidized.
2. In cellular respiration, glucose ($C_6H_{12}O_6$) is oxidized to CO_2, and O_2 is reduced to H_2O. Electrons lose potential energy during their transfer from organic compounds to water, and this energy drives ATP synthesis.
3. Electrons extracted from food are usually transferred first to NAD^+, reducing the compound to NADH.
4. NADH passes these electrons to an electron transport chain. The chain then conducts the electrons to oxygen in a series of steps that release energy. Oxidative phosphorylation uses this energy to make ATP.

An Overview of Cellular Respiration (pp. 178–179)

1. Cellular respiration consists of three metabolic components: glycolysis (in the cytosol), the Krebs cycle (in the mitochondrial matrix), and the electron transport chain (in the inner mitochondrial membrane). Glycolysis and the Krebs cycle supply electrons to the transport chain (via NADH), and the transport chain drives oxidative phosphorylation.

Glycolysis: A Closer Look (pp. 179–181)

1. In glycolysis, the six-carbon sugar glucose is oxidized to two three-carbon molecules of pyruvate. The pathway nets two molecules of ATP, produced by substrate-level phosphorylation, and two molecules of NADH.
2. In the presence of oxygen, glycolysis functions as the first stage of respiration. Pyruvate moves into the mitochondrion, where it is completely oxidized to CO_2 in the Krebs cycle.

The Krebs Cycle: A Closer Look (pp. 182–184)

1. The link between glycolysis and the Krebs cycle is the conversion of pyruvate to acetyl CoA in the matrix of the mitochondrion.
2. The acetate of acetyl CoA joins a four-carbon molecule, oxaloacetate, to form the six-carbon citrate molecule, which is subsequently degraded back to oxaloacetate in a series of steps constituting one turn of the cycle. In the process, carbon dioxide is given off, one molecule of ATP is formed by substrate-level phosphorylation, and electrons are passed to three molecules of NAD^+ and one molecule of FAD, another electron acceptor.

The Electron Transport Chain and Oxidative Phosphorylation: A Closer Look (pp. 184–189)

1. Most of the ATP created from the energy stored in glucose is produced by oxidative phosphorylation when NADH and $FADH_2$ donate their electrons to a system of electron carriers embedded in the inner mitochrondrial membrane.
2. The electron transport chain consists of a series of increasingly electronegative components, including a series of cytochrome proteins with iron-containing heme groups. The chain receives electrons from NADH and $FADH_2$. At the end of the chain, electrons are passed to oxygen, reducing it to water.
3. Electron transport is coupled to ATP synthesis by a mechanism called chemiosmosis. The structural arrangement of the electron carriers causes electron transfers at certain steps along the chain to translocate H^+ from the matrix to the intermembrane space, storing energy in an electrochemical gradient known as the proton-motive force. As hydrogen ions diffuse back into the matrix through ATP synthase, the exergonic passage of H^+ drives the endergonic phosphorylation of ADP.

Summarizing Cellular Respiration (pp. 189–190)

1. The complete oxidation of glucose to carbon dioxide during respiration in eukaryotes produces a maximum net yield of about 36 molecules of ATP. The actual yield is usually less than this theoretical maximum.

Fermentation: The Anaerobic Alternative (pp. 190–192)

1. Glycolysis produces two ATP molecules per sugar molecule by substrate-level phosphorylation, whether under aerobic or anaerobic conditions. Fermentation is the anaerobic catabolism of organic nutrients. It yields the two ATPs from glycolysis.
2. The electrons from NADH are transferred to pyruvate or some derivative of that glycolytic end-product. In alcohol fermentation, pyruvate is converted to the two-carbon ethyl alcohol and CO_2 is released. Lactic acid fermentation takes place in our muscles during strenuous exercise.

A Comparison of Aerobic and Anaerobic Catabolism (pp. 192–193)

1. The ultimate electron acceptor for the electrons removed in the oxidation of glucose differs for three major metabolic processes. In aerobic respiration, oxygen is the final electron acceptor. In anaerobic respiration, sulfate or nitrate serves as the electron acceptor. Both types of respiration use electron transport chains and oxidative phosphorylation to make ATP. In fermentation, the electron acceptor is pyruvate.
2. Yeast and certain bacteria are facultative anaerobes, capable of making ATP by either aerobic respiration or fermentation, depending on the availability of oxygen.

The Evolutionary Significance of Glycolysis (p. 193)

1. Glycolysis, the breakdown of glucose to pyruvate, is a catabolic pathway common to fermentation and respiration. It occurs in nearly all organisms and probably evolved in ancient prokaryotes before oxygen was available in the atmosphere.

The Catabolism of Other Molecules (pp. 193–194)

1. Fats, proteins, and carbohydrates can all be consumed by cellular respiration to form ATP. Thus, glycolysis and the Krebs cycle are catabolic pathways that funnel electrons

from all kinds of food molecules into the electron transport chain, which powers ATP synthesis.

Biosynthesis (p. 194)

1. In addition to providing energy, food must also provide the carbon skeletons needed by cells to make molecules for growth and repair.
2. Organic precursors for anabolism come either directly from digestion or from glycolysis and the Krebs cycle, which donate intermediates for use in biosynthesis.

The Control of Respiration (pp. 194–195)

1. Cellular respiration is controlled by allosteric enzymes at key places in glycolysis and the Krebs cycle. This control strikes a moment-to-moment balance between cell catabolism and anabolism.

SELF-QUIZ

1. The *direct* energy source that drives ATP synthesis during oxidative phosphorylation is
   a. the oxidation of glucose and other organic compounds
   b. the endergonic flow of electrons down the electron transport chain
   c. the affinity of oxygen for electrons
   d. a difference of H^+ concentration on opposite sides of the inner mitochondrial membrane
   e. the transfer of phosphate from Krebs cycle intermediates to ADP

2. In the following reaction

   $$\text{phosphoenolpyruvate} + NAD^+ \longrightarrow \text{pyruvate} + NADH + H^+$$

   the oxidizing agent is
   a. oxygen
   b. NAD^+
   c. NADH
   d. phosphoenolpyruvate
   e. pyruvate

3. Which metabolic pathway is common to both fermentation and aerobic respiration?
   a. Krebs cycle
   b. electron transport chain
   c. glycolysis
   d. synthesis of acetyl CoA from pyruvate
   e. reduction of pyruvate to lactate

4. In a eukaryotic cell, most of the enzymes of the Krebs cycle are located in the
   a. plasma membrane
   b. cytosol
   c. inner mitochondrial membrane
   d. mitochondrial matrix
   e. intermembrane space

5. The *final* electron acceptor of the electron transport chain that functions in oxidative phosphorylation is
   a. oxygen
   b. water
   c. NAD^+
   d. pyruvate
   e. ADP

6. When electrons flow along the electron transport chains of mitochondria, which of the following changes occurs?
   a. The pH of the matrix increases.
   b. The ATP synthase pumps protons by active transport.
   c. The electrons gain free energy.
   d. The cytochromes of the chain phosphorylate ADP to form ATP.
   e. NAD^+ is oxidized.

7. In the presence of a metabolic poison that specifically inhibits the mitochondrial ATP synthase, you would expect
   a. a decrease in the pH difference across the mitochondrial membrane
   b. an increase in the pH difference across the mitochondrial membrane
   c. increased synthesis of ATP
   d. oxygen consumption to cease
   e. proton pumping by the electron transport chain to cease

8. Most of the ATP made during cellular respiration is generated by
   a. glycolysis
   b. oxidative phosphorylation
   c. substrate-level phosphorylation
   d. direct synthesis of ATP by the Krebs cycle
   e. transfer of phosphate from glucose-phosphate to ADP

9. Which of the following is a true distinction between fermentation and cellular respiration?
   a. Only respiration oxidizes glucose.
   b. NADH is oxidized by the electron transport chain only in respiration.
   c. Fermentation, but not respiration, is an example of a catabolic pathway.
   d. Substrate-level phosphorylation is unique to fermentation.
   e. NAD^+ functions as an oxidizing agent only in respiration.

10. The rate of glycolysis is
    a. stimulated by ATP
    b. stimulated by ADP
    c. inhibited by ADP
    d. stimulated by citrate
    e. inhibited by O_2

CHALLENGE QUESTIONS

1. A century ago, Louis Pasteur, the great French biochemist, investigated the metabolism of yeast, a facultative anaerobe. He observed that the yeast consumed sugar at a much faster rate under anaerobic conditions than it did under aerobic conditions. Explain this Pasteur effect, as the observation is called.
2. Trematol is a metabolic poison produced by white snakeroot, the plant in this photo:

   Cows that stray from pastures into wooded areas sometimes ingest white snakeroot, a common inhabitant of North American forests. Trematol becomes concentrated in the milk of these cows, and may poison humans who drink the milk. In the early 1800s, milk sickness killed thousands of settlers in the U.S. Midwest, including Abraham Lincoln's mother, Nancy Hanks Lincoln, who died when the future president was only 7 years old. Milk sickness is very rare today because white snakeroot is removed from grazing areas, and commercial dairies mix the milk of many cows, a practice that dilutes any toxins present in one or a few cows. Also, the biochemical basis of the disease is now understood: Trematol inhibits an enzyme that helps convert lactate to other compounds in the liver. Why does physical exertion intensify the symptoms of milk disease? Why does the pH of blood decrease in a person afflicted with milk disease?
3. In the 1940s, some physicians prescribed low doses of dinitrophenol (DNP) to help patients lose weight. This unsafe method was abandoned after a few patients died. Recall that DNP uncouples the chemiosmotic machinery. Explain how this causes weight loss.

SCIENCE, TECHNOLOGY, AND SOCIETY

1. Nearly all human societies use fermentation to produce alcoholic drinks such as beer and wine. The practice dates back to the earliest days of agriculture. How do you suppose this use of fermentation was first discovered? Why did wine prove to be a more useful beverage, especially to a preindustrial culture, than the grape juice from which it was made?

FURTHER READING

Becker, W. M., and D. W. Deamer. *The World of the Cell,* 2nd ed. Redwood City, CA: Benjamin/Cummings, 1991. A particularly clear and comprehensive coverage of how cells harvest energy in Chapters 10 and 11.

Duffy, D. C. "Land of Milk and Poison." *Natural History,* July 1990. How nineteenth-century medical sleuths solved the mystery of milk sickness.

Harold, F. M. *The Vital Force: A Study of Bioenergetics.* New York: Freeman, 1986. An examination of the proton-motive force.

Mathews, C., and K. van Holde. *Biochemistry.* Redwood City, CA: Benjamin/Cummings, 1989. Effective diagrams on catabolism.

Weiss, R. "Blazing Blossoms." *Science News,* June 24, 1989. How skunk cabbage and other plants generate heat.

10 PHOTOSYNTHESIS

CHLOROPLASTS: SITES OF PHOTOSYNTHESIS
HOW PLANTS MAKE FOOD: AN OVERVIEW
HOW THE LIGHT REACTIONS CAPTURE SOLAR ENERGY: A CLOSER LOOK
HOW THE CALVIN CYCLE MAKES SUGAR: A CLOSER LOOK
PHOTORESPIRATION: AN EVOLUTIONARY RELIC?
C_4 PLANTS
CAM PLANTS
SUMMARY: THE FATE OF PHOTOSYNTHETIC PRODUCTS

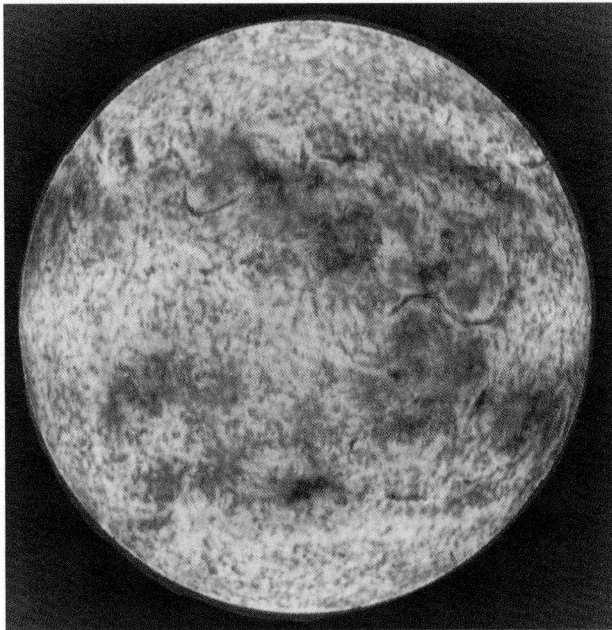

Figure 10.1
The sun: power source for the biosphere. By converting the energy of sunlight to the chemical energy stored in organic molecules, photosynthesis is the metabolic foundation of most ecosystems. This chapter traces the flow of energy from the sun to organic matter.

Life on Earth is solar-powered. The chloroplasts of plants capture light energy that has traveled 160 million kilometers from the sun and convert it to chemical energy stored in sugar and other organic molecules made from carbon dioxide and water (Figure 10.1). The process is called **photosynthesis.**

Photosynthesis nourishes almost all of the living world directly or indirectly. An organism acquires the organic compounds it uses for energy and carbon skeletons by one of two major modes: **autotrophic** or **heterotrophic nutrition.** At first, the term autotrophic (Gr. *autos,* "self," and *trophos,* "feed") may seem to contradict the principle that cells are open systems, taking in resources from their environment. Autotrophs are not totally self-sufficient, however. They are self-feeders only in the sense that they sustain themselves without eating or decomposing other organisms: They make their organic molecules from inorganic raw materials. Plants are autotrophs; the only nutrients they require are carbon dioxide from the air, and water and minerals from the soil. Specifically, plants are **photoautotrophs,** organisms that use light as a source of energy to synthesize carbohydrates, lipids, proteins, and other organic substances. Photosynthesis also occurs in algae, including certain protists (see Chapter 1), and in some prokaryotes (Figure 10.2). In this chapter, our emphasis will be on plants. Variations in photosynthesis that occur in algae and bacteria will be discussed in Unit Five. A much rarer form of self-feeding is unique to those bacteria that are **chemoautotrophs.** These organisms produce their organic compounds without the help of light, obtaining their energy by oxidizing inorganic substances, such as sulfur or ammonia. (We will postpone further discussion of this type of autotrophic nutrition until Chapter 25.)

Heterotrophs obtain their organic material by the second major mode of nutrition. Unable to make their own food, they live on compounds produced by other organisms (the prefix *heteros* means "other, different"). The most obvious form of this "other-feeding" is when an animal eats plants or other animals. But the process may be more subtle. Some heterotrophs do not kill prey, but instead decompose and feed on organic litter—such as carcasses, feces, and fallen leaves—and thus are known as decomposers. Most fungi and many types of bacteria get their nourishment this way. Almost all heterotrophs, including humans, are completely dependent on photoautotrophs for food, and

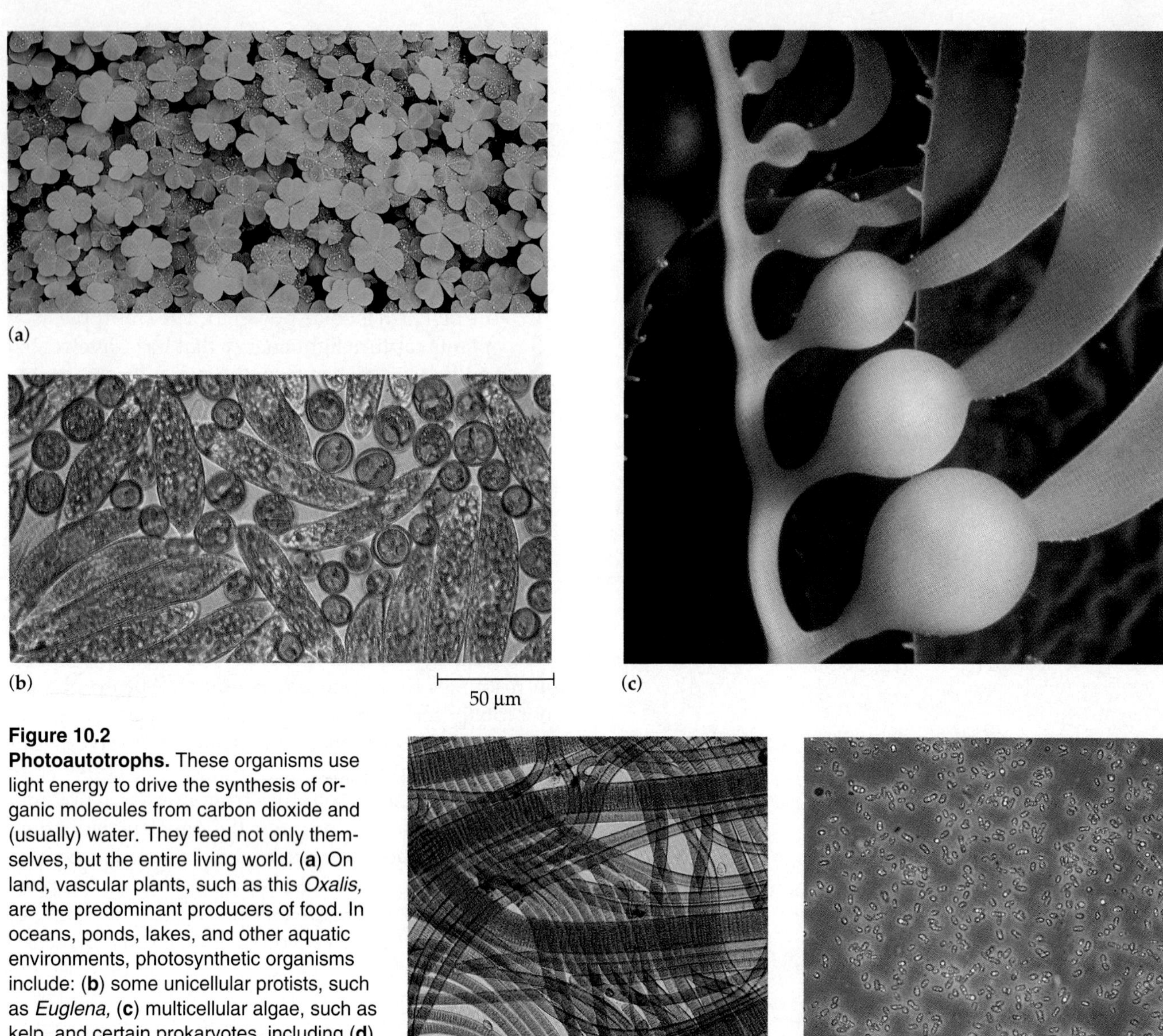

Figure 10.2
Photoautotrophs. These organisms use light energy to drive the synthesis of organic molecules from carbon dioxide and (usually) water. They feed not only themselves, but the entire living world. (**a**) On land, vascular plants, such as this *Oxalis,* are the predominant producers of food. In oceans, ponds, lakes, and other aquatic environments, photosynthetic organisms include: (**b**) some unicellular protists, such as *Euglena,* (**c**) multicellular algae, such as kelp, and certain prokaryotes, including (**d**) cyanobacteria and (**e**) purple sulfur bacteria (b, d, e: LMs).

also for oxygen, a by-product of photosynthesis. Thus, we can trace the food we eat and the oxygen we breathe to the chloroplast.

CHLOROPLASTS: SITES OF PHOTOSYNTHESIS

All green parts of a plant, including green stems and unripened fruit, have chloroplasts, but the leaves are the major sites of photosynthesis in most plants (Figure 10.3). There are about half a million chloroplasts per square millimeter of leaf surface. The color of the leaf is from **chlorophyll,** the green pigment located within the chloroplasts. It is the light energy absorbed by chlorophyll that drives the synthesis of food molecules in the chloroplast. Chloroplasts are found mainly in the cells of the **mesophyll,** the green tissue in the interior of the leaf. Carbon dioxide enters the leaf, and oxygen exits, by way of microscopic pores called **stomata** (singular, stoma, Gr. "mouth"). Water absorbed by the roots is delivered to the leaves in veins. Leaves also use veins to export sugar to nonphotosynthetic parts of the plant.

A typical mesophyll cell has about 30 to 40 chloroplasts, each a lens-shaped organelle measuring about 2–4 μm by 4–7 μm (Figure 10.4a). An envelope of two membranes bounds the stroma, the dense fluid within the chloroplast. An elaborate system of thylakoid membranes segregates the stroma from another compartment, the thylakoid space (see Chapter 7). In some places, thylakoid sacs are layered in dense stacks called grana. Chlorophyll resides in the thylakoid membranes. Thylakoids function in the steps of photosynthesis that initially convert light energy to chemical

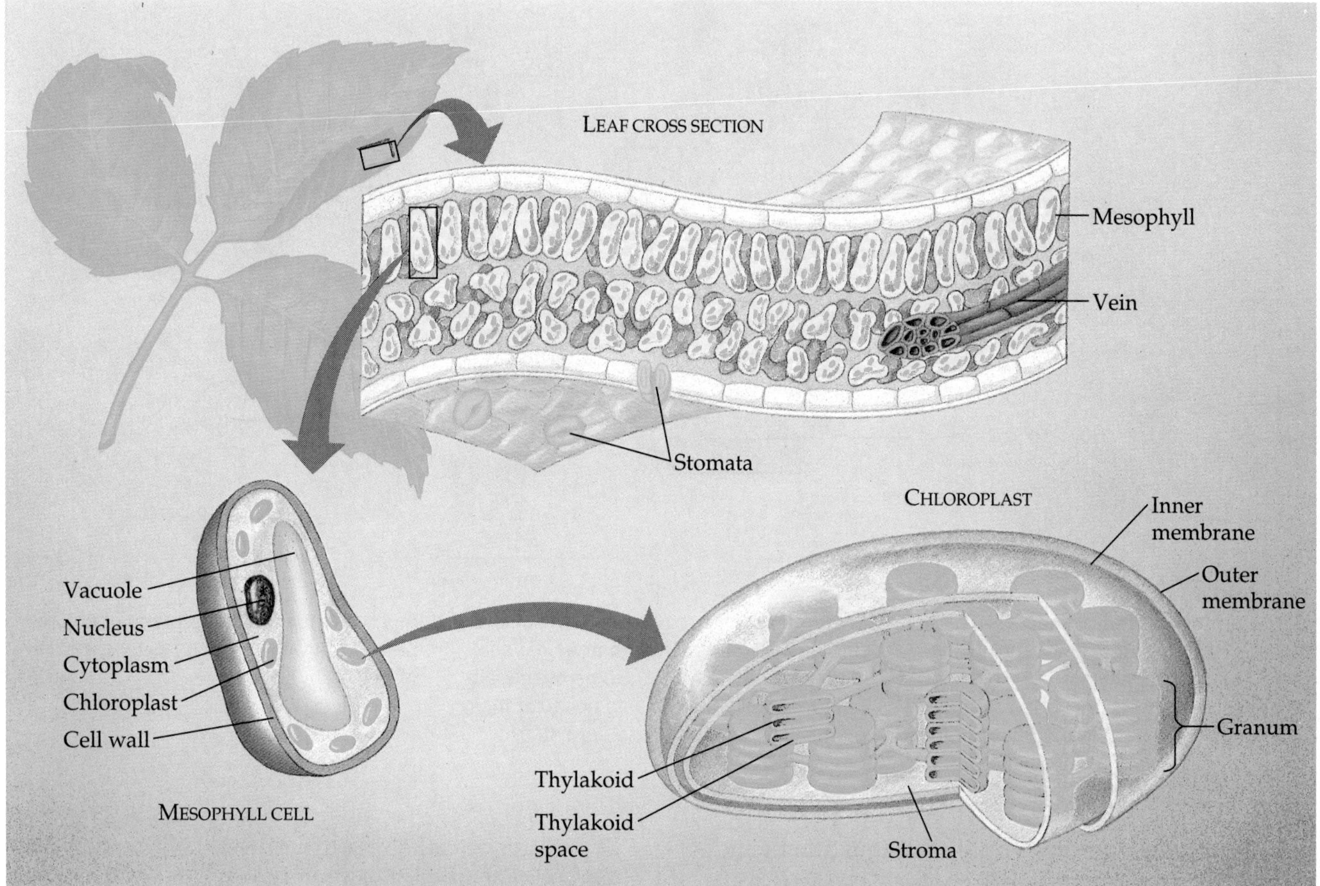

Figure 10.3
The site of photosynthesis in a plant. Leaves are the major organs of photosynthesis in plants. Chloroplasts are found mainly in the mesophyll, the green tissue in the interior of the leaf. Gas exchange between the mesophyll and the atmosphere occurs through microscopic pores called stomata. The chloroplast is bounded by a double membrane that encloses the stroma, the dense fluid content of the chloroplast. Membranes of the thylakoid system separate the stroma from the thylakoid space. Thylakoids are concentrated in stacks called grana.

energy, but the steps that actually use that chemical energy to convert carbon dioxide and water to sugar and oxygen occur in the stroma.

Photosynthetic prokaryotes lack chloroplasts, but they do have membranes that function in a manner similar to the thylakoid membranes of chloroplasts. The chlorophyll of photosynthetic bacteria is built into the plasma membrane or into the membranes of numerous vesicles within the cell (Figure 10.4b). The photosynthetic membranes of cyanobacteria are usually arranged in parallel stacks of flattened sacs, much like the thylakoids of chloroplasts (Figure 10.4c).

How do chloroplasts convert light energy to the chemical energy stored in organic molecules? Scientists have tried for centuries to piece together the process by which plants make food. Although some of the steps are still not understood, the overall photosynthetic equation has been known since the early 1800s: In the presence of light, the green parts of plants produce organic material and oxygen from carbon dioxide and water. This simple statement describes what photosynthesis does, but provides no hints as to how the process works. Only in this century have scientists begun to understand how plants trap sunlight and convert it to chemical energy.

HOW PLANTS MAKE FOOD: AN OVERVIEW

Photosynthesis can be summarized with this chemical equation:

$$6\,CO_2 + 12\,H_2O + \text{Light energy} \longrightarrow C_6H_{12}O_6 + 6\,O_2 + 6\,H_2O$$

The carbohydrate $C_6H_{12}O_6$ is glucose.* Water appears on both sides of the equation, because 12 molecules are consumed and 6 molecules are newly formed during

*The main products of photosynthesis are actually other carbohydrates. Glucose is used here only to simplify the relationship between photosynthesis and respiration.

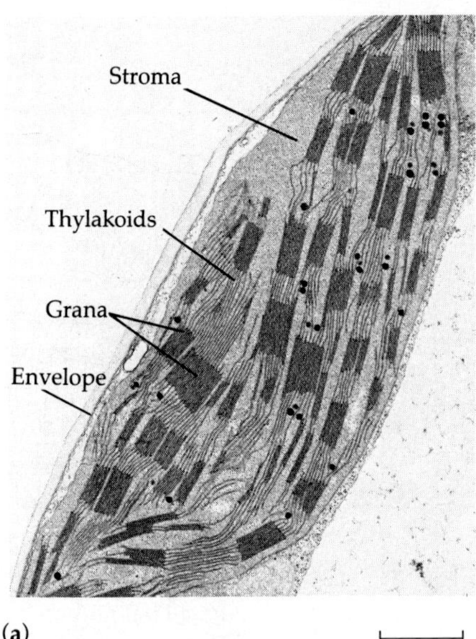

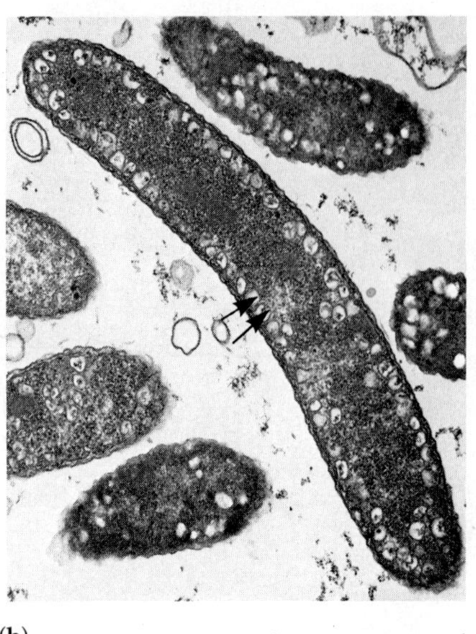

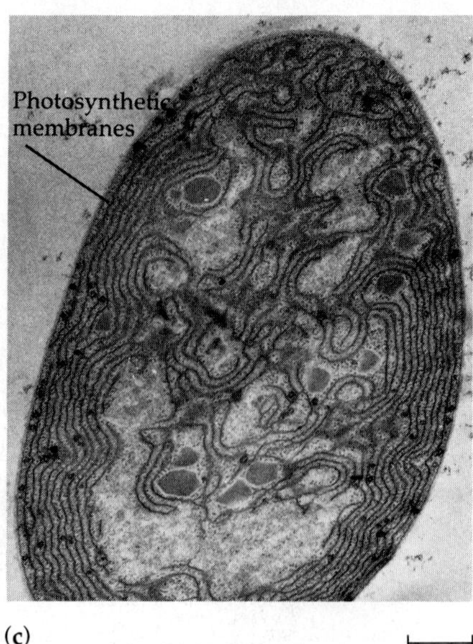

Figure 10.4
Photosynthetic membranes. (**a**) This electron micrograph of a chloroplast displays the grana, stacks of thylakoid sacs where light energy is converted to chemical energy (TEM). (**b**) In the photosynthetic bacterium *Rhodospirillum rubrum*, chlorophyll is built into the membranes of vesicles (arrows) (TEM). (**c**) Some cyanobacteria resemble chloroplasts because their photosynthetic membranes are stacked, much as they are in the grana of chloroplasts. Shown here is *Anabaena azollae* (TEM).

photosynthesis. We can simplify the equation by indicating the net consumption of water:

$$6\ CO_2 + 6\ H_2O + \text{Light energy} \longrightarrow C_6H_{12}O_6 + 6\ O_2$$

Writing the equation in this form, we can see that the chemical change during photosynthesis is the reverse of the one that occurs during cellular respiration. Both of these metabolic processes occur in plant cells. However, it will soon become apparent that plants do not make food by simply reversing the steps of respiration.

Now let's rewrite the photosynthetic equation in its simplest possible form:

$$CO_2 + H_2O \longrightarrow CH_2O + O_2$$

Here, CH_2O is not a true sugar but symbolizes the general formula for a carbohydrate. In other words, we are imagining the synthesis of a sugar molecule one carbon at a time. Six repetitions would produce a glucose molecule. Let's now use this simplified formula to track the chemical elements (C, H, and O) from the reactants of photosynthesis to the products.

The Splitting of Water

One of the first clues to the mechanism of photosynthesis came from the discovery that the oxygen given off by plants is derived from water and not from carbon dioxide. The chloroplast splits water into hydrogen and oxygen. Before this discovery, the prevailing hypothesis was that photosynthesis split carbon dioxide and then added water to the carbon:

$$\text{Step 1: } CO_2 \longrightarrow C + O_2$$
$$\text{Step 2: } C + H_2O \longrightarrow CH_2O$$

This model predicted that the O_2 released during photosynthesis came from CO_2. The idea was challenged in the 1930s by C. B. van Niel, of Stanford University. Van Niel was investigating photosynthesis in bacteria that make their carbohydrate from CO_2 but do not release O_2. Van Niel concluded that, at least in bacteria, CO_2 is not split into carbon and oxygen. One group of bacteria required hydrogen sulfide (H_2S) rather than water for photosynthesis, forming yellow globules of sulfur as a waste product:

$$CO_2 + 2\ H_2S \longrightarrow CH_2O + H_2O + 2\ S$$

Van Niel reasoned that the bacteria split H_2S and used the hydrogen to make sugar. He generalized that all photosynthetic organisms require a hydrogen source, but that the source varies:

$$\text{General: } CO_2 + 2\ H_2X \longrightarrow CH_2O + H_2O + 2\ X$$
$$\text{Sulfur bacteria: } CO_2 + 2\ H_2S \longrightarrow CH_2O + H_2O + 2\ S$$
$$\text{Plants: } CO_2 + 2\ H_2O \longrightarrow CH_2O + H_2O + O_2$$

Thus, van Niel hypothesized that plants split water as a source of hydrogen, releasing oxygen as a by-product.

Nearly 20 years later, scientists confirmed van Niel's hypothesis by using oxygen-18 (^{18}O), a heavy isotope, as a tracer to follow the fate of oxygen atoms during

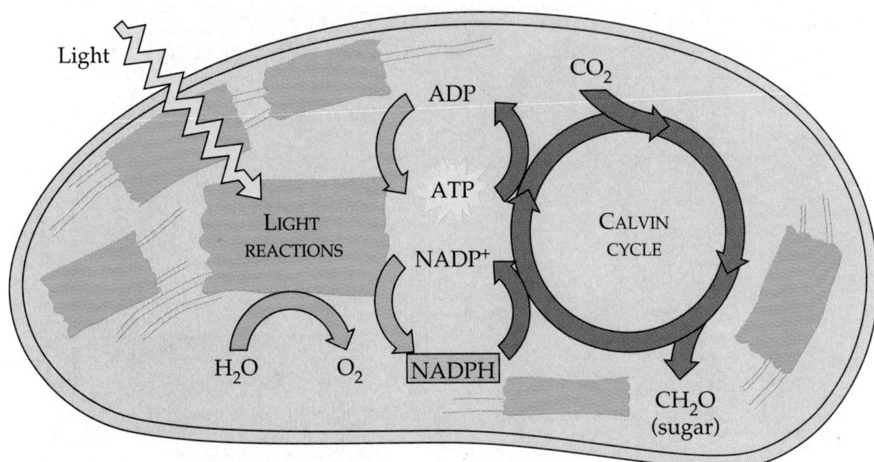

Figure 10.5
An overview of photosynthesis: integration of the light reactions and the Calvin cycle. The light reactions use solar energy to make ATP and NADPH, which function as chemical energy and reducing power, respectively, in the Calvin cycle. CO_2 is incorporated into organic molecules in the Calvin cycle. Thylakoid membranes, especially those of the grana stacks, are the sites of the light reactions, whereas the Calvin cycle occurs in the stroma. A smaller version of this diagram will reappear in several subsequent figures as a reminder of whether the events being described occur in the light reactions or in the Calvin cycle.

photosynthesis (see Chapter 2). The O_2 that came from plants was ^{18}O *only* if water was the source of the tracer. If the ^{18}O was introduced to the plant in the form of CO_2, the label did not turn up in the released O_2. In the following summary of these experiments, red denotes labeled atoms of oxygen:

Experiment 1: $CO_2 + 2\,H_2O \longrightarrow CH_2O + H_2O + O_2$
Experiment 2: $CO_2 + 2\,H_2O \longrightarrow CH_2O + H_2O + O_2$

The most important result of the shuffling of atoms during photosynthesis is the extraction of hydrogen from water and its incorporation into sugar. The waste product of photosynthesis, O_2, restores the atmospheric oxygen consumed during cellular respiration.

Photosynthesis as a Redox Process

Let's briefly contrast photosynthesis with respiration. During respiration, energy is released from sugar when electrons associated with hydrogen are transported by carriers to oxygen, forming water as a byproduct. The electrons lose potential energy as electronegative oxygen pulls them down the electron transport chain, and the mitochondrion uses the energy to synthesize ATP. Photosynthesis, also a redox process, reverses the direction of electron flow. Water is split, and electrons are transferred along with hydrogen ions from the water to carbon dioxide, reducing it to sugar. Since electrons cannot travel downhill in both directions, they must increase their potential energy when moved from water to sugar. The energy boost is provided by light.

The Two Stages of Photosynthesis

The equation for photosynthesis is a deceptively simple summary of a very complex process. Actually, photosynthesis is not a single process, but two, each with multiple steps. These two stages of photosynthesis are known as the **light reactions** and the **Calvin cycle** (Figure 10.5).

The light reactions are the steps of photosynthesis that convert solar energy to chemical energy. Light absorbed by chlorophyll drives a transfer of electrons and hydrogen from water to an acceptor called **$NADP^+$** (nicotinamide adenine dinucleotide phosphate), which temporarily stores the energized electrons. Water is split in the process, and thus it is the light reactions of photosynthesis that give off O_2 as a by-product. The electron acceptor of the light reactions, $NADP^+$, is first cousin to NAD^+, which functions as an electron carrier in cellular respiration; the two molecules differ only by the presence of an extra phosphate group in the $NADP^+$ molecule. The light reactions use solar power to reduce $NADP^+$ to NADPH by adding a pair of electrons along with a hydrogen nucleus, or H^+. The light reactions also generate ATP by powering the addition of a phosphate group to ADP, a process called **photophosphorylation.** Thus, light energy is initially converted to chemical energy in the form of two compounds: NADPH, a source of energized electrons, and ATP, the versatile energy currency of cells. Notice that the light reactions produce no sugar; that happens in the second stage of photosynthesis, the Calvin cycle.

The Calvin cycle is named for Melvin Calvin, who began to elucidate its steps along with his colleagues in the late 1940s. The cycle begins by incorporating CO_2 from the air into organic molecules already present in the chloroplast. This initial incorporation of carbon into organic compounds is known as **carbon fixation.** The Calvin cycle then reduces the fixed carbon to carbohydrate by the addition of electrons. The reducing power is provided by NADPH, which acquired energized electrons in the light reactions. To convert CO_2 to carbohydrate, the Calvin cycle also requires chemical energy in the form of ATP, which is also generated by the light reactions. Thus, it is the Calvin cycle that makes sugar, but it can do so only with the help of the

Figure 10.6
The electromagnetic spectrum. Visible light and other forms of electromagnetic energy radiate through space as waves of various lengths. We perceive different wavelengths of visible light as different colors. Violet and blue have the shortest wavelengths, orange and red the longest. Ultraviolet (UV) light and infrared light are just beyond the range of human vision, though some animals can detect these wavelengths. White light is a mixture of wavelengths. A prism can sort white light into its component colors by bending light of different wavelengths varying degrees. Visible light drives photosynthesis.

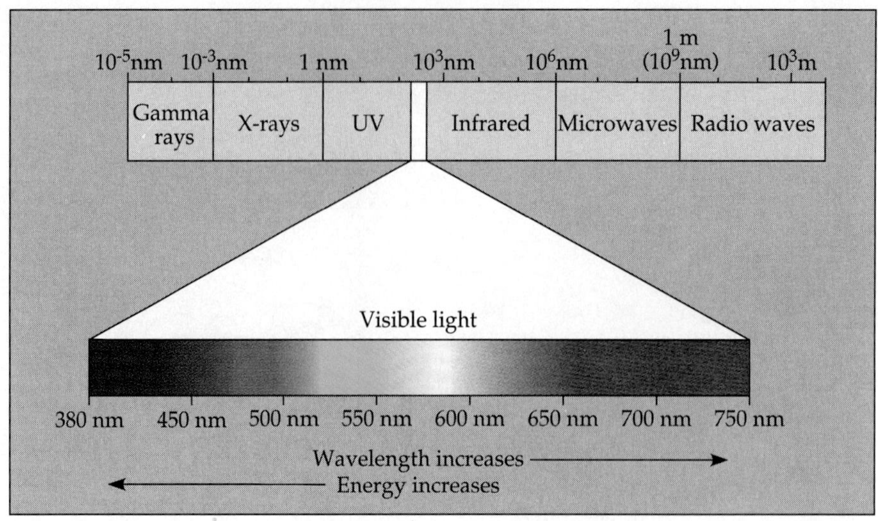

NADPH and ATP produced by the light reactions. The metabolic steps of the Calvin cycle are sometimes referred to as the dark reactions, because none of the steps requires light *directly*. Nevertheless, the Calvin cycle in most plants occurs during daylight, for only then can the light reactions regenerate the NADPH and ATP spent in the reduction of CO_2 to sugar. In essence, the chloroplast uses light energy to make sugar by coordinating the two stages of photosynthesis.

As Figure 10.5 shows, the thylakoids of the chloroplast are the sites of the light reactions, while the Calvin cycle occurs in the stroma. As molecules of $NADP^+$ and ADP bump into the thylakoid membrane, they pick up electrons and phosphate, respectively, and then transfer their high-energy cargo to the Calvin cycle. The two stages of photosynthesis are treated in this figure as metabolic modules that take in ingredients and crank out products. Our next step toward understanding photosynthesis is to look more closely at how the two stages work, beginning with the light reactions.

HOW THE LIGHT REACTIONS CAPTURE SOLAR ENERGY: A CLOSER LOOK

Chloroplasts are chemical factories powered by the sun. Their thylakoids transform light energy into the chemical energy of ATP and NADPH. To understand this conversion better, it is first necessary to learn some important properties of light.

The Nature of Sunlight

The sun is a giant thermonuclear reactor. The energy it emits comes from fusion reactions much like those that occur in a hydrogen bomb. Four hydrogen atoms fuse to form one helium atom, which has a mass slightly less than the total mass of the hydrogen atoms. The lost mass has been converted to energy, as predicted by Albert Einstein in his famous equation $E = mc^2$. Each minute, about 120 million tons of solar matter are converted to a colossal amount of energy that radiates out into space. A minuscule fraction of that energy reaches Earth, taking only a few minutes to make the trip (the speed of light is about 300,000 km/sec). Some of the energy that arrives happens to strike the leaves of plants, where it is put to work by chloroplasts.

Light is a form of energy known as **electromagnetic energy,** also called radiation. Electromagnetic energy travels in rhythmic waves analogous to those created by dropping a pebble into a puddle of water. Electromagnetic waves, however, are disturbances of electric and magnetic fields rather than disturbances of a material medium.

The distance between the crests of electromagnetic waves is called the **wavelength.** Wavelengths range from less than a nanometer (for gamma rays) to more than a kilometer (for radio waves). This entire range of radiation is known as the **electromagnetic spectrum** (Figure 10.6). The segment most important to life is the narrow band that ranges from about 380 to 750 nm in wavelength. This radiation is known as **visible light,** because it is detected as various colors by the human eye.

The model of light as waves explains many of light's properties, but in certain respects, light behaves as though it consists of discrete particles called quanta, or **photons.** Photons are not tangible objects, but they act like objects in that each of them has a fixed quantity of energy. The amount of energy is inversely related to the wavelength of the light; the shorter the wavelength, the greater the energy of each photon of that light. Thus, a photon of violet light packs nearly twice as much energy as a photon of red light.

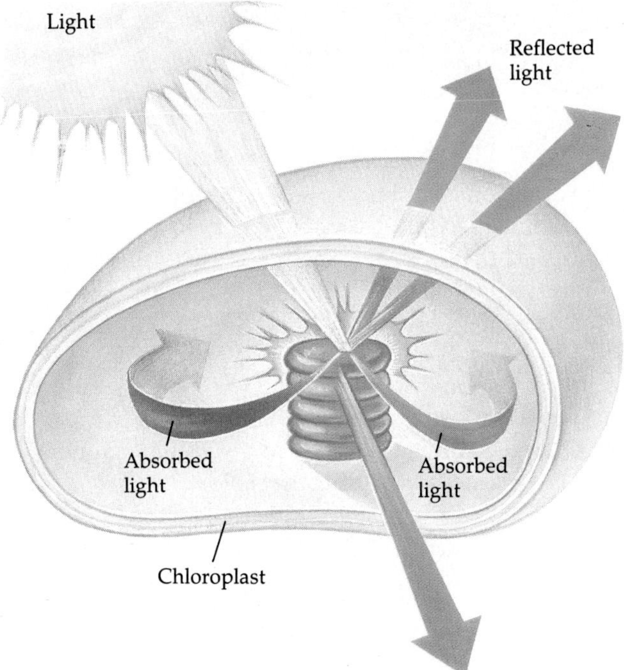

Figure 10.7
Interactions of light with matter. The pigments of chloroplasts absorb blue and red light, the colors most effective in photosynthesis. The pigments reflect or transmit green light, which is why leaves appear green.

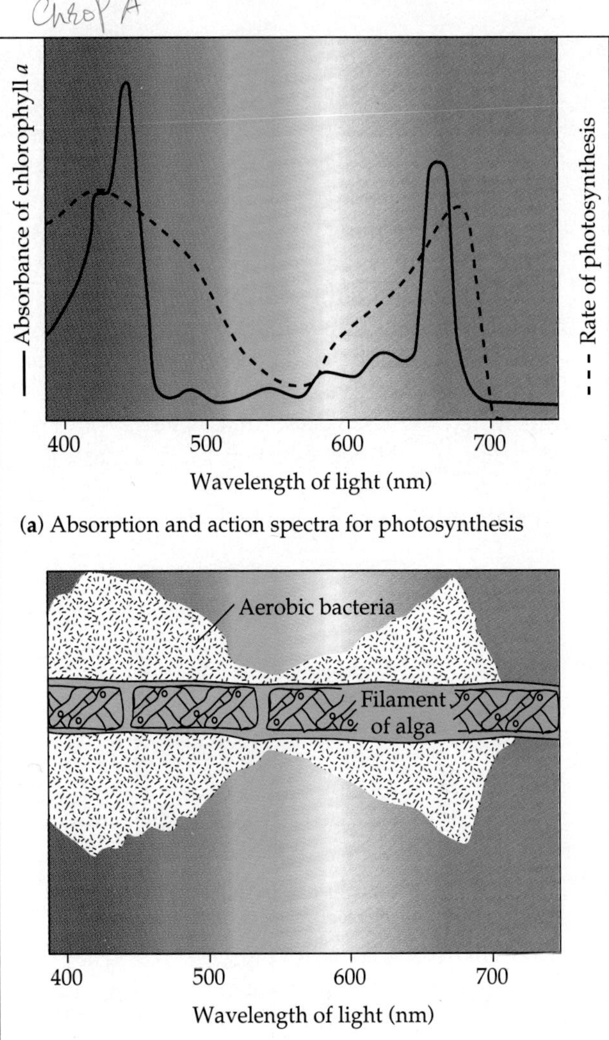

(**a**) Absorption and action spectra for photosynthesis

(**b**) Results of Engelmann's experiment

Figure 10.8
Absorption and action spectra for photosynthesis. (**a**) The dashed line is an action spectrum, which profiles the effectiveness of different wavelengths of light in driving photosynthesis. Compared to the peaks in the absorption spectrum for chlorophyll *a* (solid line), the peaks in the action spectrum are broader, and the valley is narrower and not as deep. This is partly due to absorption of light by chlorophyll *b* and carotenoids, which function as accessory pigments that broaden the spectrum of colors that can be used for photosynthesis. (**b**) An elegant experiment of this type was first performed in 1883 by Thomas Engelmann, a German botanist. He illuminated a filamentous alga with light that had been passed through a prism, thus exposing different segments to different wavelengths of light. Engelmann used aerobic bacteria, which seek oxygen, to determine which segments of the plant were releasing the most O_2. Bacteria congregated in greatest density around the parts of the alga illuminated with red and blue light.

Although the sun radiates the full spectrum of electromagnetic energy, the atmosphere acts like a selective window, allowing visible light to pass through while screening out a substantial fraction of other radiation. The same part of the spectrum we can see—visible light—is also the radiation that drives photosynthesis. Blue and red, the two wavelengths most effectively absorbed by chlorophyll, are the colors most useful as energy for the light reactions.

Photosynthetic Pigments

As light meets matter, it may be reflected, transmitted, or absorbed (Figure 10.7). Substances that absorb visible light are called **pigments.** Different pigments absorb light of different wavelengths, and the wavelengths that are absorbed disappear. If a pigment is illuminated with white light, the color we see is the color most reflected or transmitted by the pigment. (If a pigment absorbs all wavelengths, it appears black.) We see green when we look at a leaf, because chlorophyll absorbs red and blue light while transmitting and reflecting green light. The ability of a pigment to absorb various wavelengths of light can be measured by placing a solution of the pigment in a **spectrophotometer** (see the Methods Box, p. 206). A graph plotting a pigment's light absorption versus wavelength is called an **absorption spectrum.**

The absorption spectrum of the type of chlorophyll called **chlorophyll *a*** is shown in Figure 10.8a. The absorption spectrum provides clues to the relative effectiveness of different wavelengths for driving photosynthesis, since light can only perform work in

METHODS: DETERMINING AN ABSORPTION SPECTRUM

A spectrophotometer measures the proportions of light of different wavelengths absorbed and transmitted by a pigment solution. Inside the spectrophotometer, white light is separated into colors (wavelengths) by a prism. Then, one by one, the different colors of light are passed through the sample. The transmitted light strikes a photoelectric tube, which converts the light energy to electricity, and the electric current is measured by a meter. Each time the wavelength of light is changed, the meter indicates the proportion of light transmitted through the sample or, conversely, the proportion of light absorbed. A graph that profiles absorbance at different wavelengths is called an absorption spectrum. The absorption spectrum for chlorophyll *a*, the form of chlorophyll most important in photosynthesis, has two peaks, corresponding to blue and red light, the colors chlorophyll *a* absorbs best (see Figure 10.8a). The absorption spectrum has a valley in the green region because the pigment transmits that color.

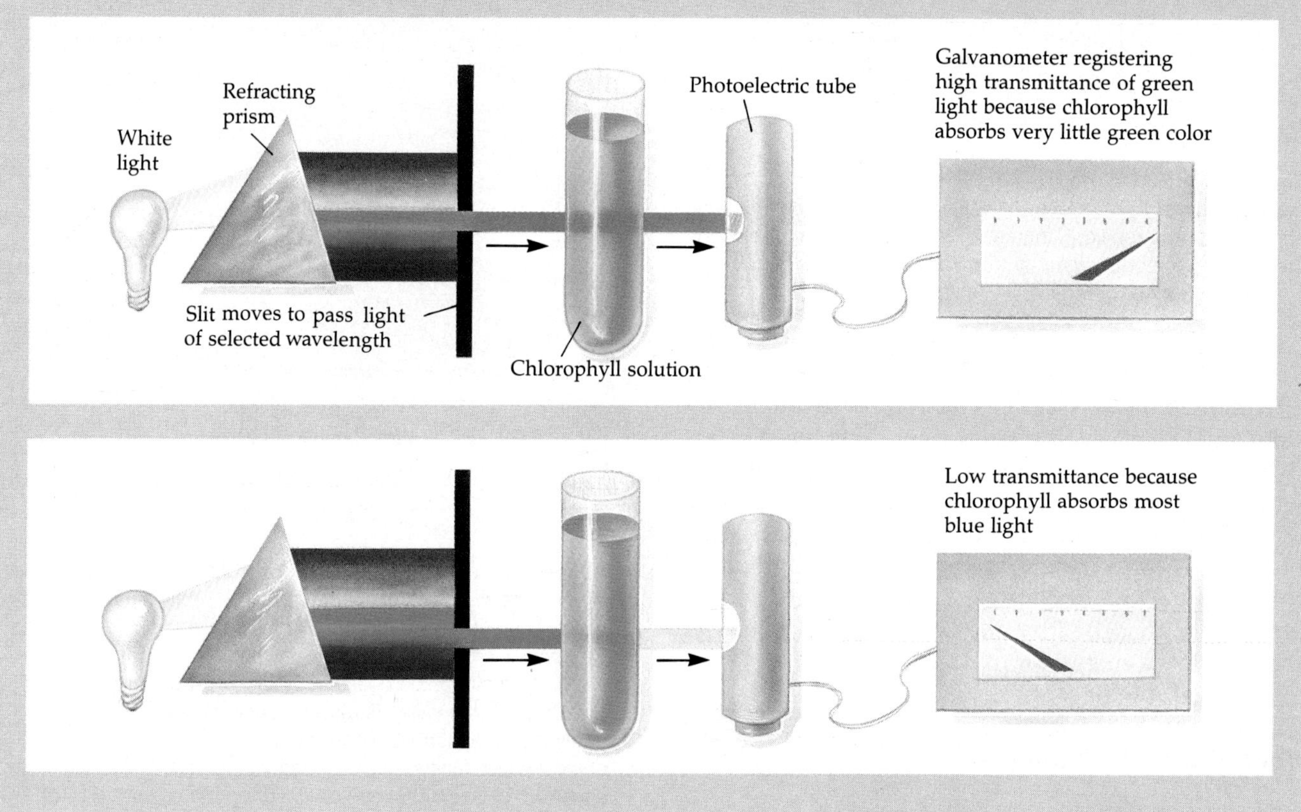

chloroplasts if it is absorbed. Indeed, as previously mentioned, blue and red light work best for photosynthesis, while green is the least effective color. An **action spectrum** (Figure 10.8a) profiles the relative performance of the different wavelengths more accurately than an absorption spectrum. An action spectrum is prepared by illuminating chloroplasts with different colors of light and then plotting wavelength against some measure of photosynthetic rate, such as oxygen release or carbon dioxide consumption (Figure 10.8b).

We can see that the action spectrum for photosynthesis does not exactly match the absorption spectrum of chlorophyll *a*. The absorption spectrum underestimates the effectiveness of certain wavelengths in driving photosynthesis. This is partly because chlorophyll *a* is not the only pigment in chloroplasts important to photosynthesis. Only chlorophyll *a* can participate directly in the light reactions, which convert solar energy to chemical energy. But other pigments can absorb light and transfer the energy to chlorophyll *a*, which then initiates the light reactions. One of these accessory pigments is another form of chlorophyll, **chlorophyll *b*.** Chlorophyll *b* is almost identical to chlorophyll *a* (Figure 10.9), but the slight structural difference be-

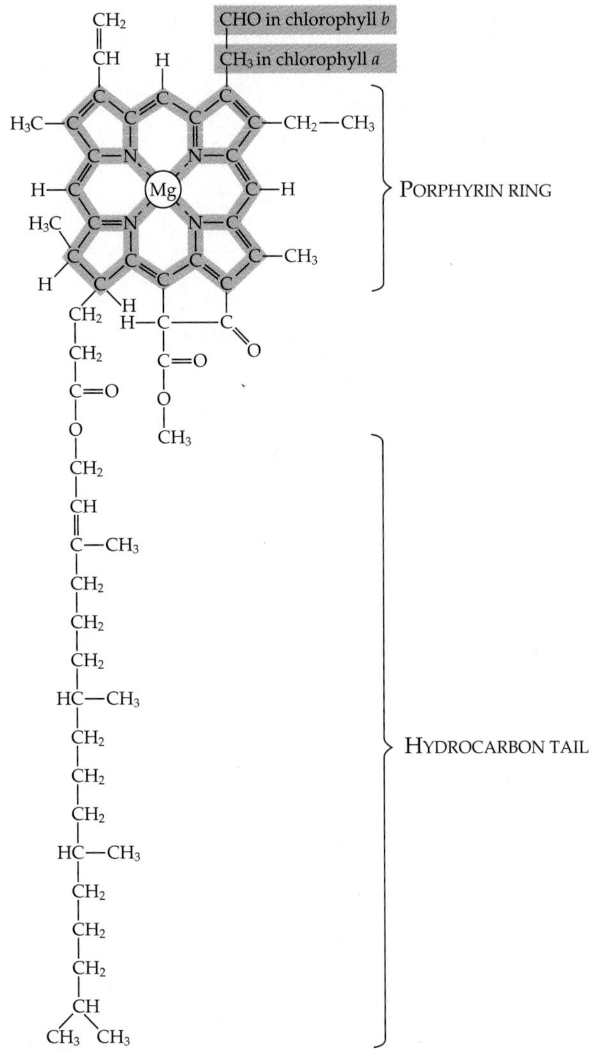

Figure 10.9
The structure of chlorophyll. Chlorophyll *a*, the pigment that participates directly in the light reactions of photosynthesis, has a "head" called a porphyrin ring with a magnesium atom at its center. Attached to the porphyrin is a hydrophobic tail, which anchors the pigment to the thylakoid membrane. Chlorophyll *b* differs from chlorophyll *a* only in one of the functional groups bonded to the porphyrin.

tween them is enough to give the two pigments slightly different absorption spectra and hence different colors. Chlorophyll *a* is blue-green, whereas chlorophyll *b* is yellow-green. The chloroplast also has a family of accessory pigments called **carotenoids,** which are various shades of yellow and orange. These hydrocarbons are built into the thylakoid membrane along with the two kinds of chlorophyll. Carotenoids can absorb wavelengths of light that chlorophyll cannot, thus broadening the spectrum of colors that can drive photosynthesis. If a photon of sunlight strikes an accessory pigment—a carotenoid or chlorophyll *b*—energy is conveyed to chlorophyll *a,* which then behaves just as though it had absorbed the photon.

The Photooxidation of Chlorophyll

What exactly happens when chlorophyll and other pigments absorb photons? The colors corresponding to the absorbed wavelengths disappear from the spectrum of the transmitted and reflected light, but energy cannot disappear. When a molecule absorbs a photon, one of the molecule's electrons is elevated to an orbital where it has more potential energy (see Chapter 2). When the electron is in its normal orbital, the pigment molecule is said to be in its **ground state.** After absorption of a photon boosts an electron to an orbital of higher energy, the pigment molecule is said to be in an **excited state.** The only photons absorbed are those whose energy is exactly equal to the energy difference between the ground state and an excited state, and this energy difference varies from one kind of atom or molecule to another. Thus, a particular compound absorbs only photons corresponding to specific wavelengths, which is why each pigment has a unique absorption spectrum.

The energy of an absorbed photon is converted to the potential energy of an electron raised from the ground state to an excited state. But the electron cannot remain there long; the excited state, like all high-energy states, is unstable. Generally, when pigments absorb light, their excited electrons drop back down to the ground-state orbital in a billionth of a second, releasing their excess energy as heat. This conversion of light energy to heat is what makes the top of an automobile so hot on a sunny day. (White cars are coolest because their paint reflects all wavelengths of visible light, although it may absorb ultraviolet and other invisible radiation.) Some pigments, including chlorophyll, emit light as well as heat after absorbing photons. The electron jumps to a state of greater energy, and as it falls back to ground state, a photon is given off. This afterglow is called **fluorescence.** The fluorescence has a longer wavelength, and hence less energy, than the light that excited the pigment. This follows from the second law of thermodynamics (see Chapter 6). The difference in energy between the incoming photon and the outgoing photon is dissipated as heat.

If a solution of chlorophyll isolated from chloroplasts is illuminated, it will fluoresce in the red part of the spectrum and also give off heat (Figure 10.10a). But the result of illuminating chlorophyll is quite different when it is in its native environment in the thylakoid membrane of the chloroplast. In the membrane, a nearby molecule, referred to as the **primary electron acceptor,** traps a high-energy electron that has absorbed a photon (Figure 10.10b). A redox reaction occurs, with chlorophyll being oxidized by the absorption of light energy (photooxidized) and the electron acceptor being reduced. Isolated chlorophyll fluoresces because there is no electron acceptor to prevent the electrons of the excited chlorophyll molecule from

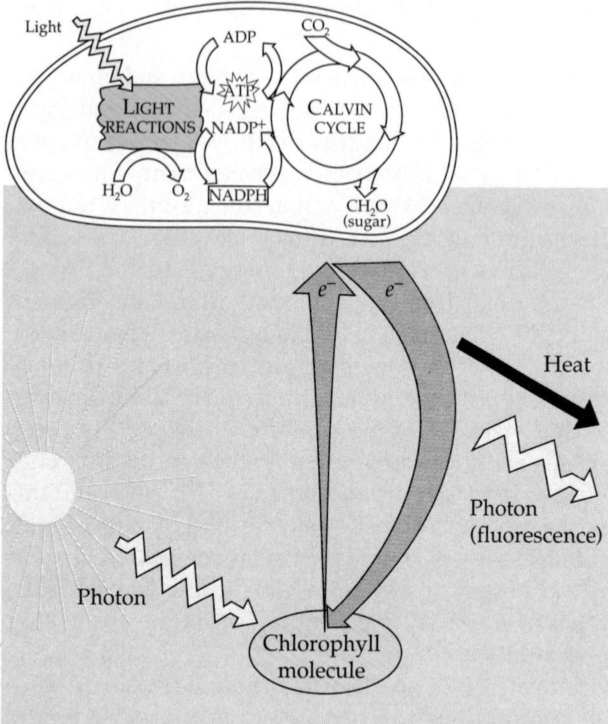

(a)

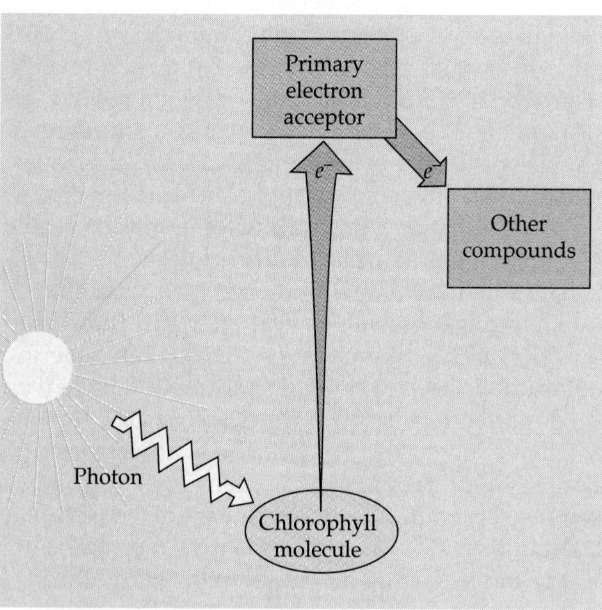

(b)

Figure 10.10
The excitation of chlorophyll. (**a**) Absorption of a photon causes a transition of the chlorophyll molecule from its ground state to its excited state. The photon boosts an electron to an orbital where it has more potential energy. If isolated chlorophyll is illuminated, its excited electron immediately drops back down to the ground-state orbital, giving off its excess energy as heat and fluorescence (light). (**b**) In the intact chloroplast, illuminated chlorophyll loses its excited electron to a neighboring molecule, the primary electron acceptor, which in turn passes the electron on to other molecules.

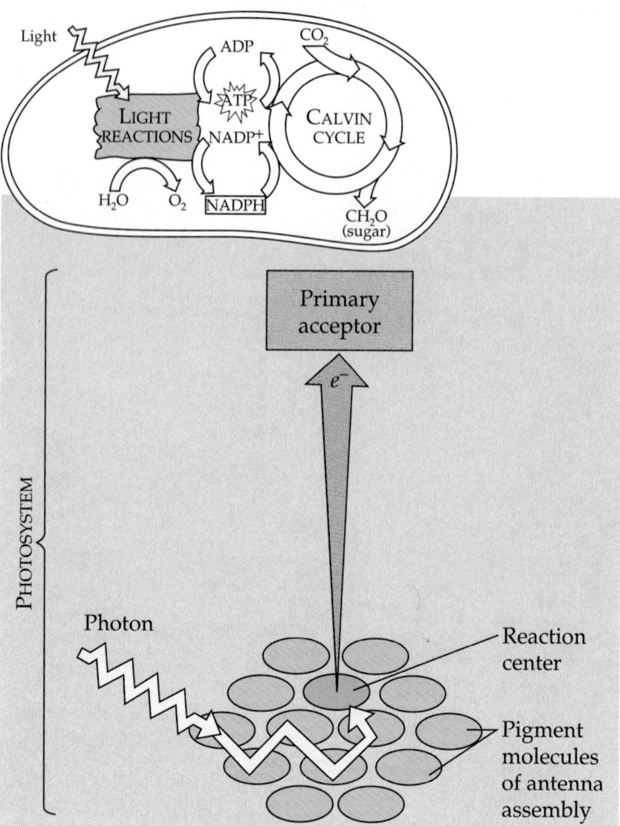

Figure 10.11
How a photosystem harvests light. Photosystems are the light-harvesting units of the thylakoid membrane. The entire apparatus—antenna assembly, reaction center, and primary electron acceptor—makes up the photosystem. Each photosystem has an antenna of a few hundred pigment molecules, including various chlorophyll and carotenoid molecules. When a photon strikes a pigment molecule, the energy is passed from molecule to molecule until it reaches the reaction center, a pair of specialized chlorophyll *a* molecules located next to the primary electron acceptor. (This diagram simplifies the structure of the reaction center and exaggerates the relative distance between the reaction center and the primary acceptor.)

dropping right back to ground state. In a chloroplast, the acceptor molecule functions like a dam that prevents the electrons of excited chlorophyll from plunging immediately back to the ground state. The solar-powered transfer of electrons from chlorophyll to the primary acceptor is the first step of the light reactions. Subsequent steps tap the energy stored in the trapped electrons to power the synthesis of ATP and NADPH.

The Two Photosystems

Chlorophyll *a*, chlorophyll *b*, and the carotenoids are clustered in the thylakoid membrane in assemblies of a few hundred pigment molecules. Of the many chlorophyll *a* molecules in each assembly, only one localized pair of molecules can trigger the light reactions by donating their excited electrons to the primary electron acceptor. The location of these specialized chlorophyll *a* molecules in the pigment assembly is called the **reaction center.** The other chlorophyll *a* molecules and the molecules of chlorophyll *b* and the carotenoids function collectively as a light-gathering antenna that absorbs photons and passes the energy from molecule to molecule until it reaches the reaction center (Figure 10.11). The entire apparatus—the antenna complex along with its reaction-center chlorophyll and the primary electron acceptor—is called a **photosystem.** Photosystems are the light-harvesting units of the thylakoid membrane.

Two types of photosystems have been found in the thylakoid membrane, referred to as **photosystem I** and **photosystem II,** in order of their discovery. At the reaction center of photosystem I is a pair of chlorophyll *a* molecules known as P700, so named because the light that this pigment absorbs best has a wavelength of 700 nm, in the far-red part of the spectrum. At the reaction center of photosystem II is a pair of chlorophyll *a* molecules designated P680; the absorption spectrum of this pigment has a peak at 680 nm, also in the red part of the spectrum. The two pigments, P700 and P680, are actually identical chlorophyll *a* molecules, but they are associated with different proteins, which affects their distribution of electrons and accounts for the slight difference in absorption spectra. In fact, the two reaction-center chlorophylls are probably no different in structure from the many chlorophyll *a* molecules of the antenna complex. What makes P700 and P680 so special are their particular locations in the thylakoid membrane, where they are bound to specific proteins and are in close proximity to their respective primary electron acceptors. Photosynthesis, like any ordered function, is based on ordered structure.

Cyclic Electron Flow

During the light reactions, there are two possible routes for electron flow: cyclic and noncyclic. **Cyclic electron flow** is the simpler pathway (Figure 10.12). It involves only photosystem I and generates only ATP; it does not produce NADPH or O_2. This pathway is called cyclic because excited electrons that depart from chlorophyll at the reaction center eventually return. A series of redox reactions recycles the electrons to chlorophyll along an electron transport chain in the thylakoid membrane that is much like the electron transport chain of the mitochondrial membrane. With each redox reaction along the chain of electron carriers in the thylakoid membrane, electrons lose potential energy, finally returning to their ground-state orbital in P700.

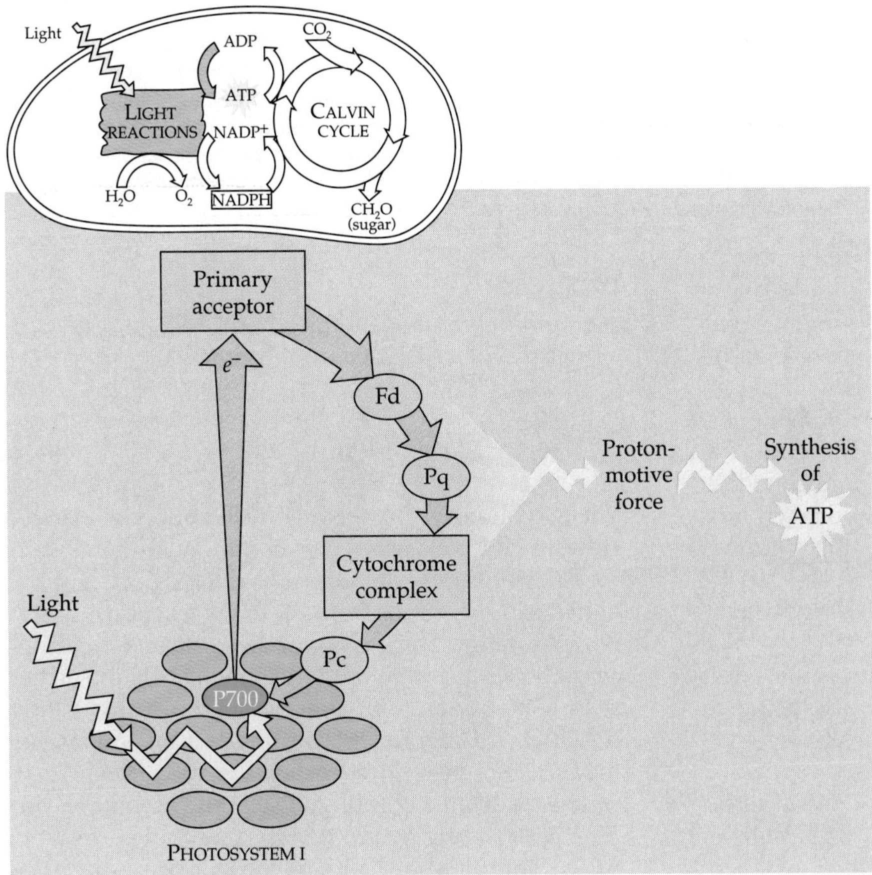

Figure 10.12
Cyclic electron flow. When pigments of the antenna assembly of photosystem I absorb light, the energy reaches P700, the specialized chlorophyll *a* at the reaction center. This boosts electrons of P700 to their excited-state orbital, from which the high-energy electrons are trapped by the primary electron acceptor. The primary acceptor supplies electrons to a chain that includes an iron-containing protein named ferredoxin (Fd), an electron carrier called plastoquinone (Pq), a complex of two cytochromes similar to the cytochromes of mitochondria, and a copper-containing protein named plastocyanin (Pc). The cycle is completed when Pc returns the electrons to P700, the reaction-center chlorophyll. The gold arrows track this cyclic flow of electrons. At each step along the transport chain, electrons lose potential energy. Their trip, energetically downhill, is used by the transport chain to pump H^+ across the thylakoid membrane. The proton-motive force in turn powers an enzyme that phosphorylates ADP.

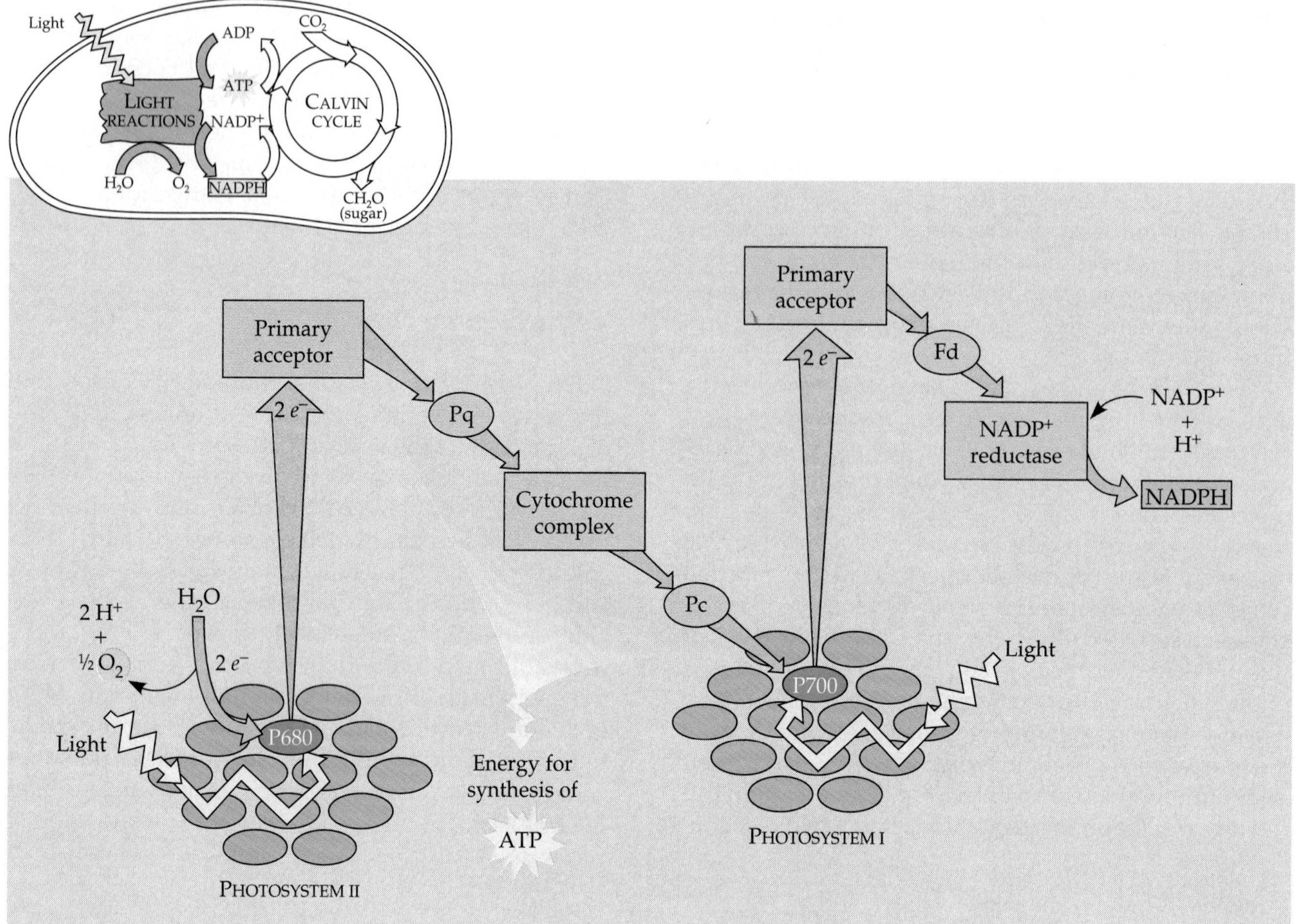

Figure 10.13

Noncyclic electron flow. When both photosystems are illuminated, there is a continuous current of electrons flowing from water to $NADP^+$. (Although each photon excites a single electron, the gold arrows here track two electrons, the number required to reduce $NADP^+$.) Electrons ejected from P680 are replaced by electrons removed from water, a redox process that evolves oxygen. An enzyme complex catalyzes electron transfer from water to P680. The excited electrons from P680 flow down an electron transport chain to P700, providing energy to synthesize ATP. Illumination of photosystem I boosts electrons to a high-energy state again, and these electrons are passed with the help of a reductase enzyme to $NADP^+$, reducing it to NADPH. The net products of noncyclic electron flow are ATP, NADPH, and O_2.

As the excited electrons give up energy on their way to P700 along the transport chain, the thylakoid membrane harnesses the flow of electrons to drive ATP synthesis. The process is very much like oxidative phosphorylation in the mitochondrion. The coupling mechanism is chemiosmosis, as it is in mitochondria. The electron transport chain pumps hydrogen ions across the thylakoid membrane, generating a proton-motive force (see Chapter 9). An ATP synthase enzyme similar to mitochondrial ATP synthase uses the proton-motive force to make ATP. In chloroplasts, ATP synthesis is called photophosphorylation because it is driven by light energy. Specifically, the production of ATP during the cyclic pathway of electron flow is referred to as **cyclic photophosphorylation.**

Notice again that cyclic electron flow uses a single photosystem to make ATP and produces neither NADPH nor O_2. The other pathway of electron flow in the light reactions generates both ATP and NADPH and also evolves oxygen by splitting water.

Noncyclic Electron Flow

The two photosystems cooperate in **noncyclic electron flow,** in which electrons pass continuously from water to $NADP^+$ (Figure 10.13). As in cyclic electron flow, light excites electrons from P700, the reaction center of photosystem I. During noncyclic electron flow, however, the electrons do not return to the reaction center, but are stored as high-energy electrons in NADPH. NADPH will later function as the electron donor when the Calvin cycle reduces carbon dioxide to sugar.

The oxidized chlorophyll itself now becomes a very potent oxidizing agent; its electron "holes" must be filled. This is where photosystem II comes in: It restores electrons to the reaction center of photosystem I. When light is absorbed by the antenna assembly of photosystem II, the energy reaches P680, the specialized chlorophyll *a* of the reaction center. The primary electron acceptor of photosystem II traps electrons ejected from P680 and transfers them to the same transport chain that functions in cyclic electron flow. The electrons cascade down the chain, losing potential energy as they go, until they reach P700 and fill the vacancies left when photosystem I reduced $NADP^+$. As the electrons slide from photosystem II to photosystem I, the transport chain pumps H^+ across the thylakoid membrane. The proton-motive force can then drive ATP synthesis. When noncyclic electron flow generates ATP, the overall process is termed **noncyclic photophosphorylation.** Notice, however, that the actual mechanism of ATP synthesis is the same as in cyclic photophosphorylation.

So far, noncyclic electron flow has generated NADPH and ATP and has restored electrons to the reaction center of photosystem I. But now P680, the chlorophyll at the reaction center of photosystem II, has electron holes to fill, and it obtains the electrons from water. A water-splitting enzyme extracts electrons from water and supplies them to P680. Electron removal splits the water into two hydrogen ions and an oxygen atom, which immediately combines with another oxygen atom to form O_2. This is the water-splitting step of photosynthesis that releases the O_2.

At this point, let's summarize the light reactions. Noncyclic electron flow pushes electrons from water, where they are at a low state of potential energy, to NADPH, where they are stored at a high state of potential energy. The light-driven electron current also generates ATP. Thus, the equipment of the thylakoid membrane converts light energy to the chemical energy stored in NADPH and ATP. Oxygen is a by-product.

What is the function of cyclic photophosphorylation? If we compare Figures 10.12 and 10.13, we can see that the cyclic flow is a short circuit; when electrons ejected from P700 reach ferredoxin, they are shunted back to the chlorophyll molecule rather than being diverted to $NADP^+$. Noncyclic electron flow produces roughly equal quantities of ATP and NADPH, but the Calvin cycle consumes more ATP than NADPH. Cyclic photophosphorylation makes up the difference, by producing ATP but not NADPH.

Review: A Comparison of Chemiosmosis in Chloroplasts and Mitochondria

Chloroplasts and mitochondria generate ATP by the same basic mechanism: chemiosmosis. An electron transport chain assembled in a membrane translocates protons across the membrane as electrons are passed through a series of carriers that are progressively more electronegative. Built into the same membrane is an ATP synthase complex that couples the diffusion of hydrogen ions down their gradient to the phosphorylation of ADP. Some of the electron carriers, including quinones and cytochromes, are very similar in chloroplasts and mitochondria, and the ATP synthase complexes of the two organelles are also very much alike. But there are noteworthy differences between oxidative phosphorylation in mitochondria and photophosphorylation in chloroplasts. In mitochondria, the high-energy electrons dropped down the transport chain are extracted by the oxidation of food molecules. Chloroplasts do not need food to make ATP; their photosystems capture light energy and use it to drive electrons to the top of the transport chain. In other words, mitochondria transfer chemical energy from food molecules to ATP, while chloroplasts transform light energy into chemical energy. It is an important difference.

The spatial organization of chemiosmosis also differs in chloroplasts and mitochondria (Figure 10.14).

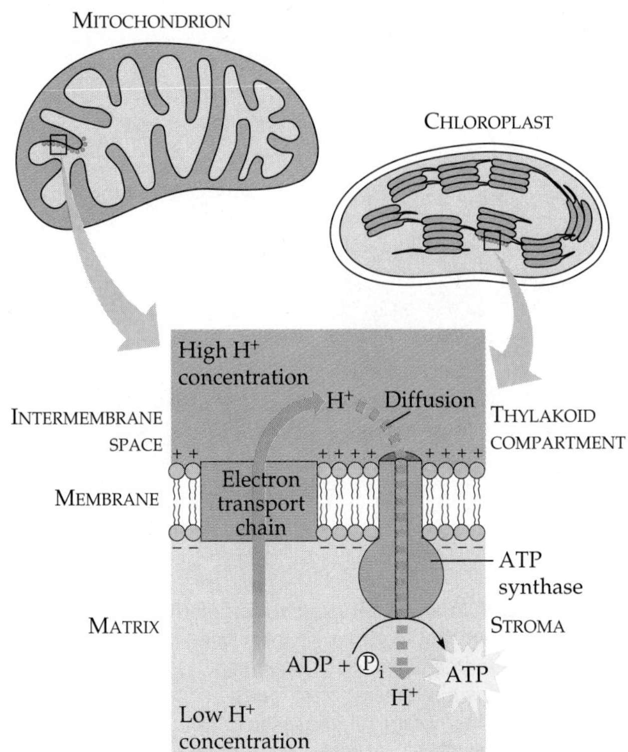

Figure 10.14
The logistics of chemiosmosis in mitochondria and chloroplasts. The inner membrane of the mitochondrion translocates protons (H^+) from the matrix into the intermembrane space (darker brown). ATP is made on the matrix side of the membrane as hydrogen ions diffuse through ATP synthase complexes. In chloroplasts, the thylakoid membrane pumps protons from the stroma into the thylakoid compartment. As the hydrogen ions leak back across the membrane through the ATP synthase, phosphorylation of ADP occurs on the stroma side of the membrane.

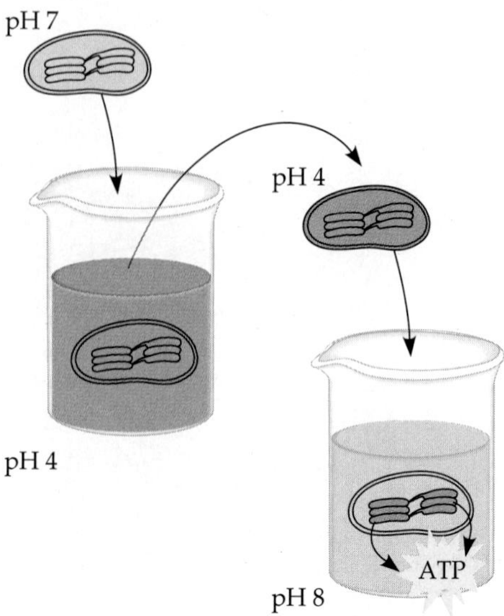

Figure 10.15
An artificially imposed pH gradient drives ATP synthesis. Chloroplasts are first made acidic by soaking them in a solution with pH 4. After the thylakoid compartment has reached pH 4, the chloroplasts are transferred to a basic solution with pH 8. The thylakoid membranes harness the artificially imposed pH gradient between the thylakoid compartment and the stroma to make ATP in the dark.

The inner membrane of the mitochondrion pumps protons from the matrix out to the intermembrane space, which then serves as a reservoir of hydrogen ions that powers the ATP synthase. The thylakoid membrane of the chloroplast pumps protons from the stroma into the thylakoid compartment, which functions as the H^+ reservoir. The membrane makes ATP as the hydrogen ions diffuse from the thylakoid compartment back to the stroma through ATP synthase complexes, whose catalytic heads are on the stroma side of the membrane. Thus, ATP forms in the stroma, where it is used to help drive sugar synthesis during the Calvin cycle.

The proton gradient, or pH gradient, across the thylakoid membrane is substantial. When chloroplasts are illuminated, the pH in the thylakoid compartment drops to about 5, and the pH in the stroma increases to about 8. This gradient of three pH units corresponds to a thousandfold difference in H^+ concentration. When the lights are turned off, the pH gradient is abolished, but it can quickly be restored by turning the lights back on. Such experiments add to the evidence described in Chapter 9 in support of the chemiosmotic model. Perhaps the most compelling evidence for chemiosmosis resulted from experiments performed in the 1960s by André Jagendorf and his colleagues. They induced thylakoids to make ATP in the dark by using artificial means to impose a pH gradient across the membrane (Figure 10.15). Such experiments have demonstrated that the function of the photosystems and the electron transport chain in photophosphorylation is to generate a proton-motive force that drives ATP synthesis.

Considerably more research is required before the precise organization of the thylakoid membrane can be discerned. Figure 10.16 shows a tentative model based on studies in several laboratories. Notice that proton pumping by the thylakoid membrane depends on an asymmetric placement of electron carriers that accept and release H^+. Notice, too, that NADPH, like ATP, is produced on the side of the membrane facing the stroma, where sugar is synthesized by the Calvin cycle. The Calvin cycle uses the ATP and NADPH generated by the light reactions to reduce carbon dioxide to sugar.

HOW THE CALVIN CYCLE MAKES SUGAR: A CLOSER LOOK

The Calvin cycle is similar to the Krebs cycle in that a starting material is regenerated after materials enter and leave the cycle. Carbon enters the Calvin cycle in the form of CO_2 and leaves in the form of sugar. The cycle spends ATP as an energy source and consumes NADPH as reducing power for adding high-energy electrons to make the sugar (Figure 10.17).

The carbohydrate produced directly from the Calvin cycle is actually not glucose, but a three-carbon sugar named **glyceraldehyde 3-phosphate,** referred to in this text as glyceraldehyde phosphate. For the net synthesis of one molecule of this sugar, the cycle must take place three times, fixing three molecules of CO_2 (recall that carbon fixation refers to the initial incorporation of CO_2 into organic material). As we trace the steps of the cycle, keep in mind that we are following three molecules of CO_2 through the reactions.

The Calvin cycle incorporates each CO_2 molecule by attaching it to a five-carbon sugar named ribulose bisphosphate (abbreviated RuBP). The enzyme that catalyzes this first step is RuBP carboxylase, or **rubisco,** the most abundant protein in chloroplasts and probably the most abundant protein on Earth. The product of the reaction is a six-carbon intermediate that is so unstable it immediately splits in half to form two molecules of 3-phosphoglycerate. In the next step of the cycle, each molecule of 3-phosphoglycerate receives an additional phosphate group. An enzyme transfers the phosphate group from ATP, forming 1,3-diphosphoglycerate as a product. Next, a pair of electrons donated from NADPH reduces 1,3-diphosphoglycerate to glyceraldehyde phosphate. Specifically, the electrons from NADPH reduce the carboyxl group of 3-phosphoglycerate to the carbonyl group of glyceraldehyde phosphate, which stores more potential energy. Glyceraldehyde phosphate is a sugar—the same three-

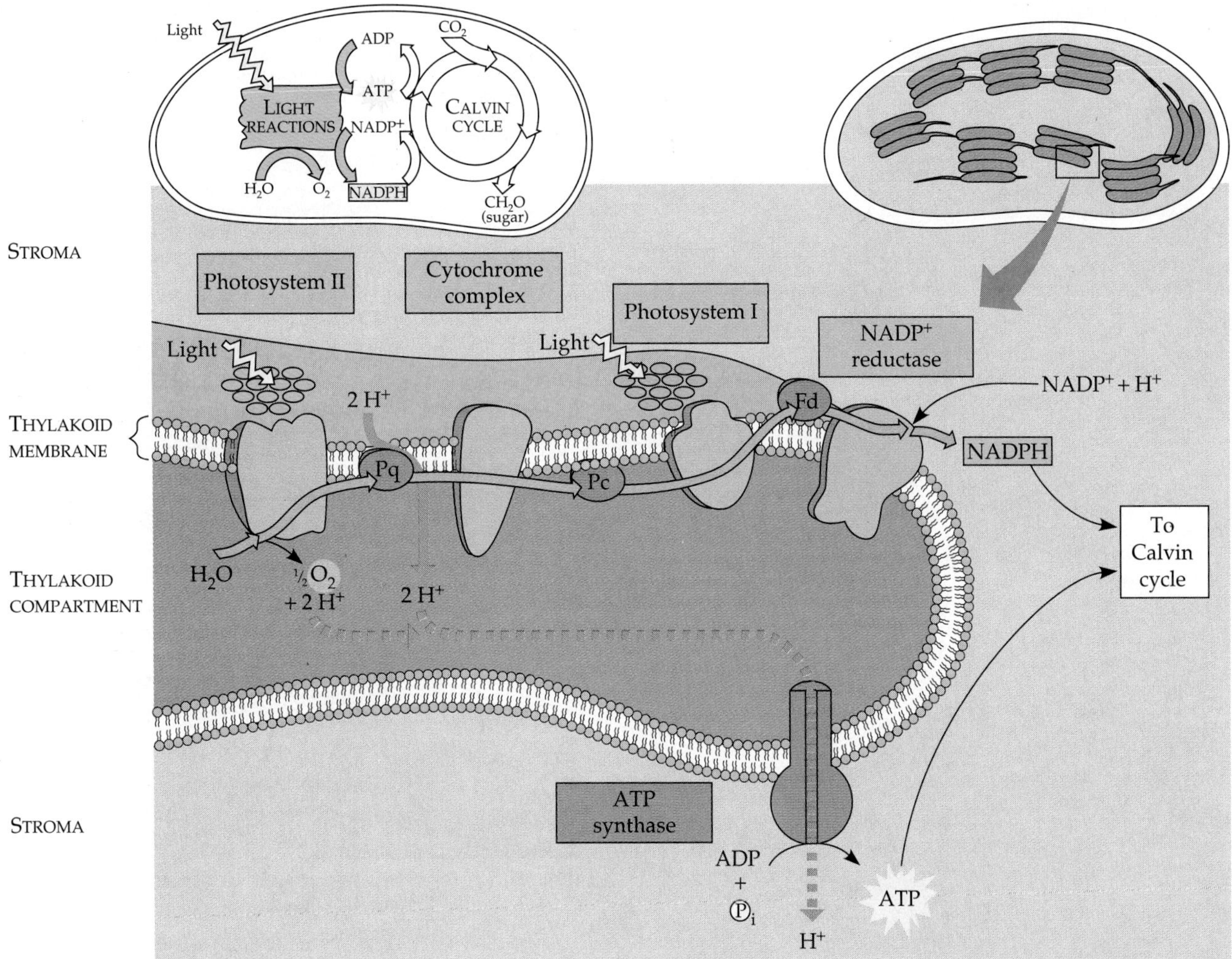

Figure 10.16
A tentative model for the organization of the thylakoid membrane. Electron carriers are arranged in such a way that electrons pass from one side of the membrane to the other. (The gold arrows track electron flow.) As electrons crisscross the membrane, hydrogen ions removed from the stroma are deposited in the thylakoid compartment. There are at least three steps in the light reactions that contribute to the proton gradient: Water is split by photosystem II on the side of the membrane facing the thylakoid compartment; as plastoquinone (Pq), a mobile carrier, transfers electrons to the cytochrome complex, protons are translocated across the membrane; and a hydrogen ion in the stroma is taken up by $NADP^+$ when it is reduced to NADPH. The diffusion of H^+ from the thylakoid compartment to the stroma (along the H^+ concentration gradient) powers the ATP synthase. These light-driven reactions store chemical energy in NADPH and ATP, which shuttle the energy to the sugar-producing Calvin cycle.

carbon sugar formed in glycolysis by the splitting of glucose.

Notice in Figure 10.17 that for every *three* molecules of CO_2, there are *six* molecules of glyceraldehyde phosphate. But only one molecule of this three-carbon sugar can be counted as a net gain of carbohydrate. The cycle began with 15 carbons' worth of carbohydrate in the form of three molecules of the five-carbon sugar RuBP. Now there are 18 carbons' worth of carbohydrate in the form of six molecules of glyceraldehyde phosphate. One molecule exits the cycle to be used by the plant cell, but the other five molecules must be recycled to regenerate the three molecules of RuBP. In a complex series of reactions, the carbon skeletons of five molecules of glyceraldehyde phosphate are rearranged by the last steps of the Calvin cycle into three molecules of RuBP. To accomplish this, the cycle spends three more molecules of ATP. The RuBP is now prepared to receive CO_2 again, and the cycle continues.

For the net synthesis of one glyceraldehyde phosphate molecule by the Calvin cycle, a total of nine molecules of ATP and six molecules of NADPH are consumed. The light reactions regenerate the ATP and NADPH. The glyceraldehyde phosphate spun off from the Calvin cycle becomes the starting material for metabolic pathways that synthesize other organic com-

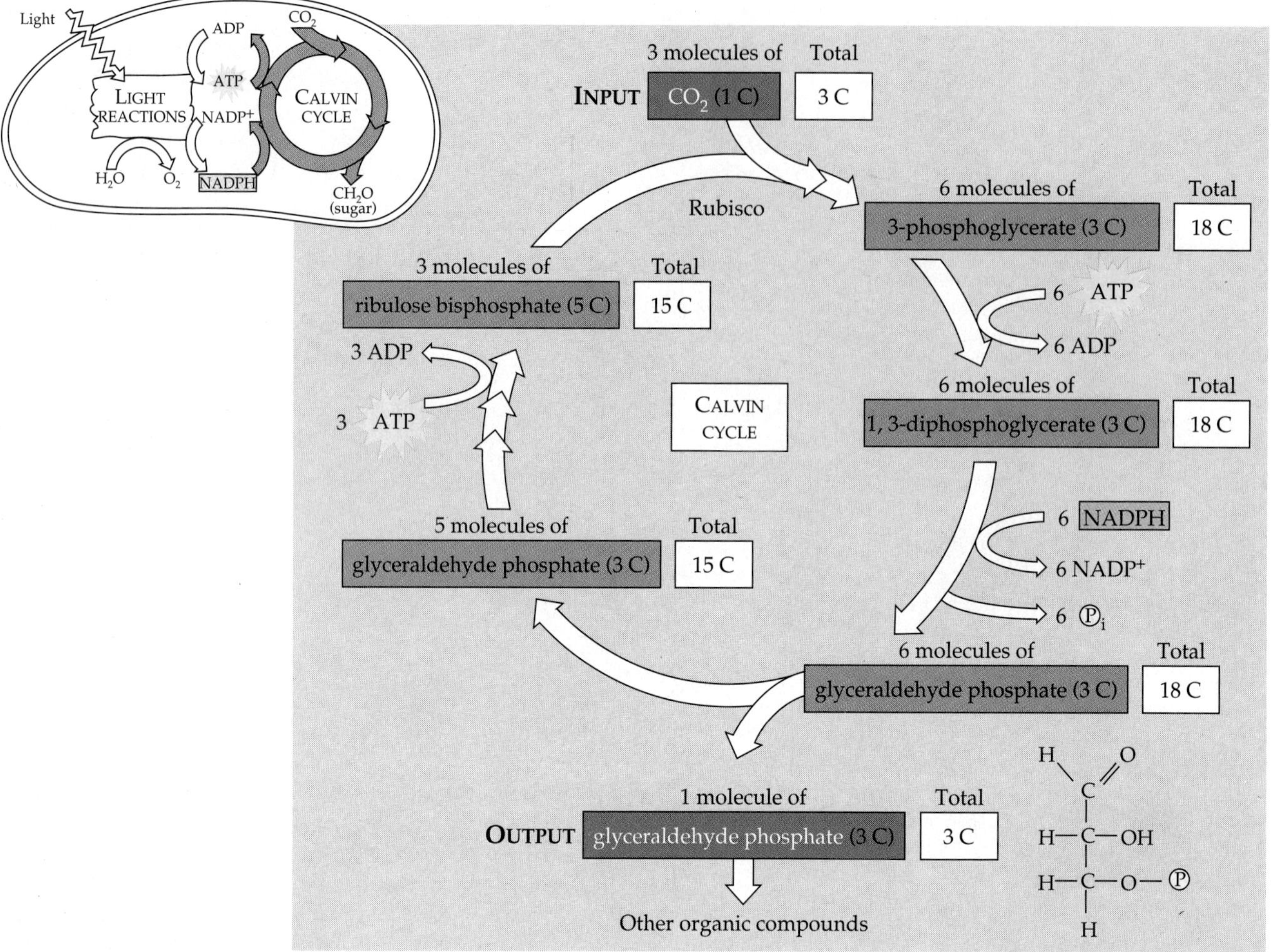

Figure 10.17
The Calvin cycle. For every three molecules of CO_2 that enter the cycle, the net output is one molecule of glyceraldehyde phosphate, a three-carbon sugar. For each glyceraldehyde phosphate synthesized, the cycle spends nine molecules of ATP and six molecules of NADPH. The intermediates involved in the conversion of glyceraldehyde phosphate to ribulose bisphosphate (RuBP) are not shown in this simplified version of the cycle.

pounds, including a variety of carbohydrates. Neither the light reactions nor the Calvin cycle alone can make sugar from CO_2. Photosynthesis is an emergent property of the intact chloroplast, which integrates the two stages of photosynthesis.

PHOTORESPIRATION: AN EVOLUTIONARY RELIC?

A seemingly wasteful process called **photorespiration** reduces the yield of photosynthesis. Photorespiration occurs because the active site of rubisco, the carbon-fixing enzyme of the Calvin cycle, can accept O_2 in place of CO_2. Since these two gases compete for a common enzyme, a decrease in the CO_2 concentration within the air spaces of a leaf favors photorespiration. Under these conditions, rubisco adds O_2, rather than CO_2, to RuBP. The product splits into a three-carbon molecule that remains in the Calvin cycle and a two-carbon compound that exits the cycle. In fact, the two-carbon compound, glycolate, leaves the chloroplast altogether and enters a peroxisome (see Chapter 7). The metabolic pathway continues in the peroxisome and then in a mitochondrion, breaking down glycolate and releasing CO_2. This entire process is photorespiration. The word *respiration* is used because the process consumes oxygen and produces carbon dioxide. *Photo* refers to the fact that photorespiration generally occurs in the light, when photosynthesis reduces the CO_2 concentration in the leaf's air spaces and increases the O_2 concentration. Unlike the cellular respiration you learned about in Chapter 9, photorespiration generates no ATP. Furthermore, it decreases photosynthetic output by siphoning organic material from the Calvin cycle.

How can we explain the existence of a metabolic process that seems to be counterproductive to the plant? According to one hypothesis, photorespiration is evolutionary baggage—a metabolic relic from a much earlier time, when the atmosphere had less O_2 and more CO_2 than it does today. In that ancient atmosphere, when rubisco first evolved, the inability of the enzyme's active site to distinguish CO_2 from O_2 would have made little difference. This hypothesis goes on to speculate that modern rubisco retains some of its ancestral affinity for O_2, which is so concentrated in the present atmosphere that a certain amount of photorespiration is inevitable.

It is not known if photorespiration is beneficial to plants in any way. It *is* known that in many types of plants, including some of agricultural importance, such as soybeans, photorespiration drains away as much as 50% of the carbon fixed by the Calvin cycle. As heterotrophs that depend on carbon fixation in chloroplasts for our food, we naturally view photorespiration as wasteful. Indeed, if photorespiration could be reduced in certain plant species without otherwise affecting photosynthetic productivity, crop yields and food supplies would increase.

The environmental conditions that foster photorespiration are hot, dry, bright days. On such days, plants close their stomata, the pores through the leaf surface (see Figure 10.3). This is an adaptation that prevents dehydration by slowing water loss from the leaf. But photosynthesis soon depletes CO_2 and increases O_2 within the leaf's air spaces, and photorespiration commences as rubisco accepts O_2. In certain species, alternate modes of carbon fixation that minimize photorespiration—even in hot, arid climates— have evolved. The two most important of these photosynthetic adaptations are C_4 photosynthesis and CAM.

C_4 PLANTS

The evolutionary adaptation known as C_4 photosynthesis enhances the ability of certain plants to fix CO_2 under conditions that cause most plants to lose organic material from photorespiration. Recall that most plants use the Calvin cycle for the initial steps that incorporate CO_2 into organic material. They are called **C_3 plants,** because the first stable intermediate formed by carbon fixation is 3-phosphoglycerate, a three-carbon compound. Many plant species, however, preface the Calvin cycle with reactions that incorporate CO_2 first into four-carbon compounds; these species are known as **C_4 plants.** Several thousand species in at least 19 plant families use the C_4 pathway. Among the C_4 plants important to agriculture are sugarcane and corn, members of the grass family.

A unique leaf anatomy is correlated with the mechanism of C_4 photosynthesis (Figure 10.18a). In C_4 plants, there are two distinct types of photosynthetic cells: bundle-sheath cells and mesophyll cells. **Bundle-sheath cells** are arranged into tightly packed sheaths around the veins of the leaf. Between the bundle sheath and the leaf surface are the more loosely arranged **mesophyll cells.** The Calvin cycle is confined to the chloroplasts of the bundle sheath. However, the cycle is preceded by incorporation of CO_2 into organic compounds in the mesophyll (Figure 10.18b). The first step is the addition of CO_2 to phosphoenolpyruvate (PEP) to form the four-carbon product oxaloacetate. The enzyme called **PEP carboxylase** adds CO_2 to PEP. Unlike rubisco, the active site of PEP carboxylase has no affinity for O_2. Therefore, PEP carboxylase can fix CO_2 efficiently when rubisco cannot—that is, when it is hot and dry and stomata are partially closed, causing CO_2 concentration in the leaf to fall and O_2 concentration to rise. After mesophyll in the C_4 plant fixes CO_2, the cells convert oxaloacetate to another four-carbon compound: malate, in some species. The mesophyll cells then export the malate to bundle-sheath cells through plasmodesmata (see Figure 7.33). Within the bundle-sheath cells, malate releases CO_2, which is reassimilated into organic material by rubisco and the Calvin cycle. In effect, the mesophyll cells pump CO_2 into the bundle sheath, keeping the CO_2 concentration in the bundle-sheath cells high enough for rubisco to accept carbon dioxide rather than oxygen. In this way, C_4 photosynthesis minimizes photorespiration and enhances sugar production. This adaptation is especially advantageous in hot regions with intense sunlight, and it is in such environments that C_4 plants evolved and thrive today. The spatial ordering of photosynthesis in a C_4 plant is an adaptation that clearly illustrates the correlation between structure and function.

CAM PLANTS

A second photosynthetic adaptation to arid conditions has evolved in succulent (water-storing) plants (including ice plants), many cacti, pineapples, and representatives of several other plant families. These plants open their stomata during the night and close them during the day, just the reverse of how other plants behave. Closing stomata during the day helps desert plants conserve water, but it also prevents CO_2 from entering the leaves. During the night, when their stomata are open, these plants take up CO_2 and incorporate it into a variety of organic acids. This mode of carbon fixation is called crassulacean acid metabolism, or CAM, after the plant family Crassulaceae, in which the process was first discovered. The mesophyll cells of

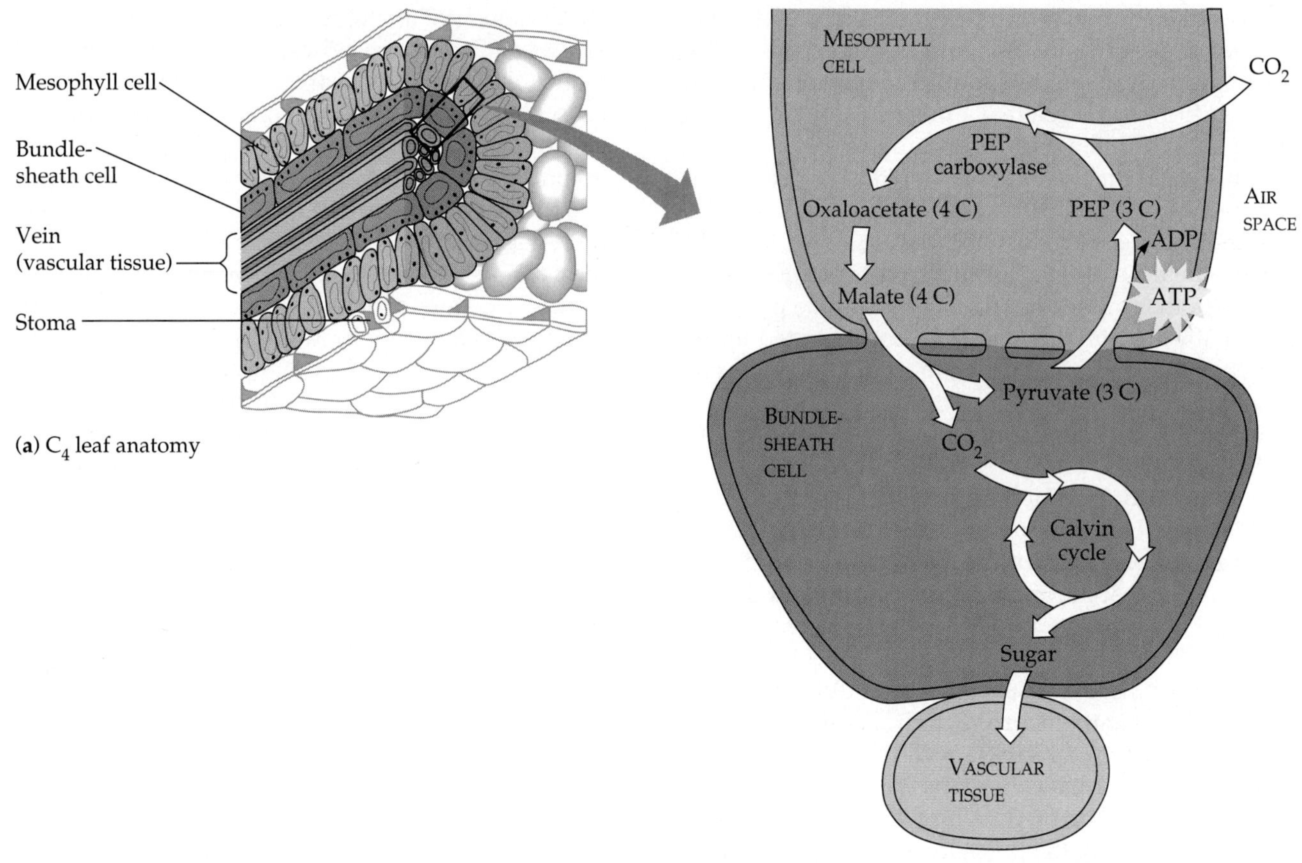

Figure 10.18
The C_4 anatomy and pathway. (**a**) C_4 plant leaves contain two types of photosynthetic cells: a cylinder of bundle-sheath cells surrounding the vein, and mesophyll cells located outside the bundle sheath. (**b**) Carbon dioxide is fixed in mesophyll cells by the enzyme PEP carboxylase. A four-carbon compound—malate, in this case—conveys the CO_2 via plasmodesmata into a bundle-sheath cell, where the enzymes of the Calvin cycle are located. In effect, the mesophyll pumps CO_2 into the bundle sheath. Pumping CO_2 against its concentration gradient requires energy; the C_4 pathway consumes more ATP than does the C_3 type of photosynthesis. In a moderate climate, C_4 plants are actually at a disadvantage because of this greater energy cost; but in a hot, dry environment, C_4 plants excel. Such conditions cause stomata to close, which reduces water loss from the leaf but also reduces diffusion of CO_2 into the leaf. This results in greater photorespiration in a C_3 plant. In contrast, the CO_2 pump (mesophyll) of a C_4 plant can maintain a CO_2 concentration in the bundle sheath that favors photosynthesis over photorespiration.

CAM plants store the organic acids they make during the night in their vacuoles until morning, when the stomata close. During the day, when the light reactions can supply ATP and NADPH for the Calvin cycle, CO_2 is released from the organic acids made the night before to become incorporated into sugar in the chloroplasts. The CAM pathway is similar to the C_4 pathway in that carbon dioxide is first incorporated into organic intermediates before it enters the Calvin cycle. The difference is that in C_4 plants, the initial steps of carbon fixation are separated structurally from the Calvin cycle, whereas in CAM plants the two steps occur at separate times (Figure 10.19). Keep in mind that CAM, C_4, and C_3 plants all eventually use the Calvin cycle to make sugar from carbon dioxide.

SUMMARY: THE FATE OF PHOTOSYNTHETIC PRODUCTS

In this chapter, we have followed photosynthesis from photons to food (Figure 10.20). The light reactions capture solar energy and use it to make ATP and transfer electrons from water to $NADP^+$. The Calvin cycle uses the ATP and NADPH to produce sugar from carbon dioxide. The energy that entered the chloroplasts as sunlight becomes stored as chemical energy in organic compounds.

The sugar made in the chloroplasts supplies the entire plant with chemical energy and carbon skeletons to synthesize all the major organic molecules of cells. About 50% of the organic material made by photosyn-

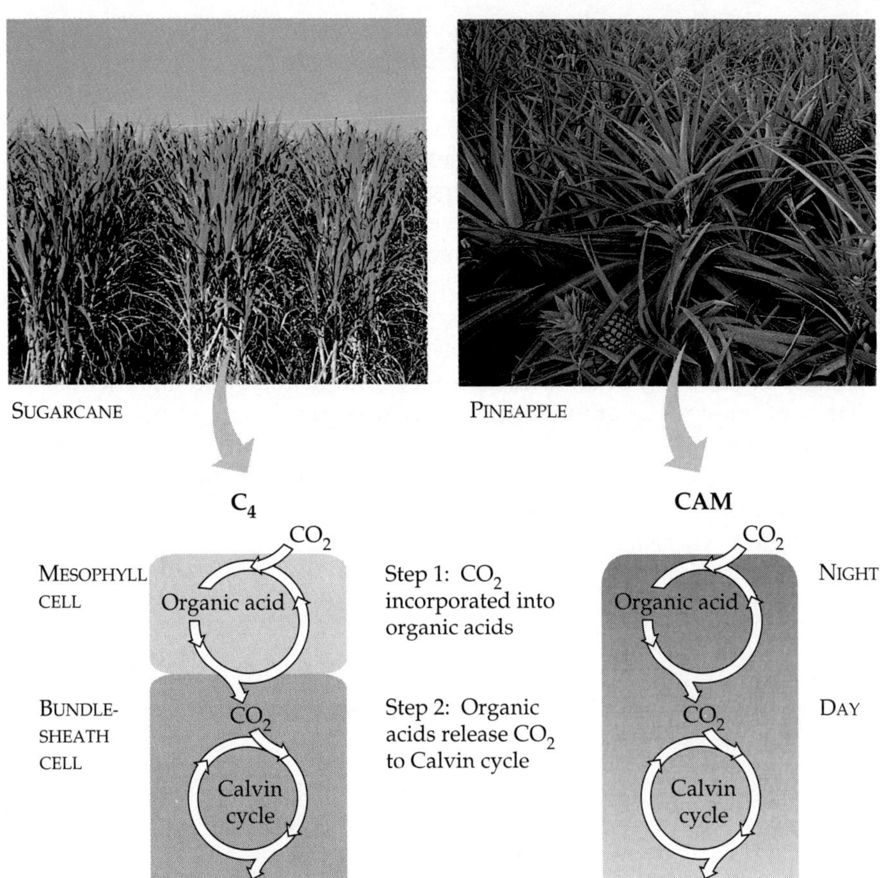

Figure 10.19
A comparison of C_4 and CAM photosynthesis. Both adaptations are characterized by preliminary incorporation of CO_2 into organic acids, followed by transfer of the CO_2 to the Calvin cycle. In C_4 plants, such as sugarcane, these two steps are separated spatially; they are segregated into two cell types. In CAM plants, such as pineapple, the two steps are separated temporally; carbon fixation into organic acids occurs at night, and the Calvin cycle operates during the day. C_4 and CAM are two evolutionary solutions to the problem of maintaining photosynthesis with stomata partially or completely closed on hot, dry days.

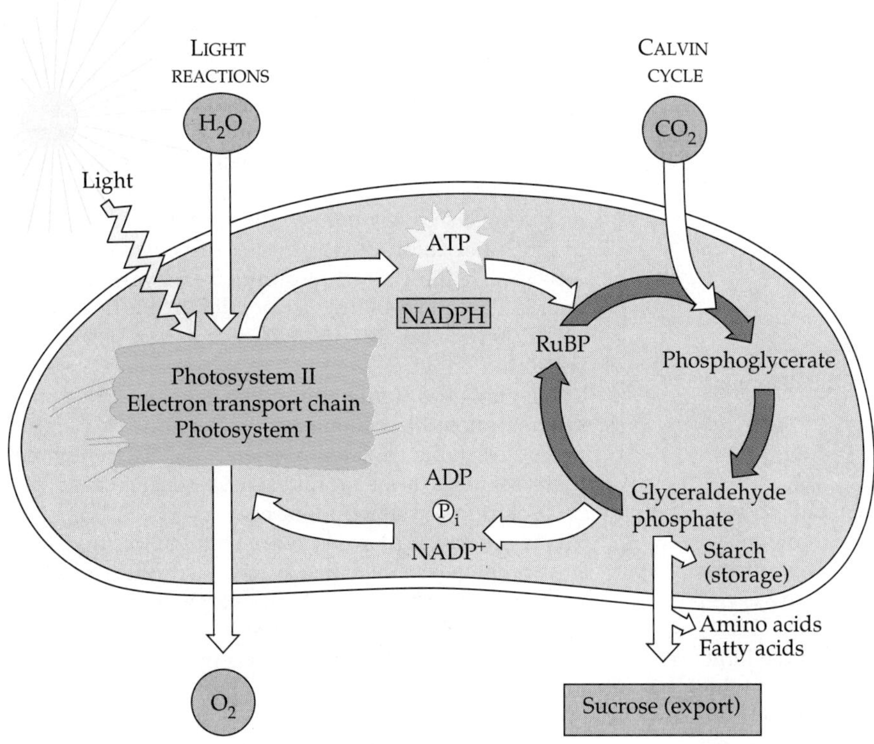

Figure 10.20
A summary of photosynthesis. This diagram outlines the main reactants and products of photosynthesis as it occurs in the chloroplasts of plant cells. The light reactions convert light energy into the chemical energy of ATP and NADPH. The pigment and protein molecules that carry out the light reactions are found in the thylakoid membranes and include the molecules of two photosystems and an electron transport chain. The light reactions split H_2O and release O_2 to Earth's atmosphere. The Calvin cycle, which takes place in the stroma of the chloroplast, uses ATP and NADPH to convert CO_2 to carbohydrate (three key compounds of the cycle are shown). The direct product of the Calvin cycle is the three-carbon sugar glyceraldehyde phosphate. Enzymes in the chloroplast and cytoplasm convert this small sugar to a diversity of other organic compounds. The Calvin cycle returns ADP, inorganic phosphate, and $NADP^+$ to the light reactions. The entire ordered operation depends on the structural integrity of the chloroplast and its membranes.

thesis is consumed as fuel for cellular respiration in the mitochondria of the plant cells. Sometimes there is a loss of photosynthetic products to photorespiration.

Technically, green cells are the only autotrophic parts of the plant. The rest of the plant depends on organic molecules exported from leaves in veins. In most plants, carbohydrate is transported out of the leaves in the form of sucrose, a disaccharide. After arriving at nonphotosynthetic cells, the sucrose provides raw material for cellular respiration and a multitude of anabolic pathways that synthesize proteins, lipids, and other products. A considerable amount of sugar in the form of glucose is linked together to make the polysaccharide cellulose, especially in plant cells that are still growing and maturing. Cellulose is the most abundant organic molecule in the plant—and probably on the surface of the planet.

Over the 24 hours in a day, most plants manage to make more organic material than they need to use as respiratory fuel and precursors for biosynthesis. They stockpile the extra sugar by synthesizing starch, storing some in the chloroplasts themselves and some in storage cells of roots, tubers, and fruits. In accounting for the consumption of the food molecules produced by photosynthesis, let's not forget that most plants lose leaves, roots, stems, fruits, and sometimes their entire bodies to heterotrophs, including humans.

On a global scale, the collective productivity of the minute chloroplasts is prodigious; it is estimated that photosynthesis makes about 160 billion metric tons of carbohydrate per year (a metric ton is 1000 kg, about 1.1 tons). No other chemical process on the planet can match this output. And no process is more important than photosynthesis to the welfare of life on Earth.

STUDY OUTLINE

1. Autotrophs sustain themselves without ingesting organic molecules. Photoautotrophs use the energy of sunlight to synthesize organic molecules from CO_2 and H_2O.
2. Heterotrophs must ingest other organisms or their byproducts to obtain energy and carbon skeletons.

Chloroplasts: Sites of Photosynthesis (pp. 200–201)

1. In autotrophic eukaryotes, photosynthesis occurs inside chloroplasts, organelles containing an elaborate system of thylakoid membranes that are stacked in places as grana. These membranes separate the thylakoid compartment from the chloroplast's stroma.
2. All chloroplasts contain the green pigment chlorophyll, which resides in the thylakoid membranes and absorbs the light energy that drives photosynthesis.
3. The chloroplasts in plants are especially dense in the cells of the mesophyll in the interior of the leaf. CO_2 enters and O_2 exits the mesophyll through stomata.
4. The conversion of CO_2 to sugar occurs in the stroma of the chloroplast.

How Plants Make Food: An Overview (pp. 201–204)

1. The complex process of photosynthesis can be summarized by the following equation:

$$6\,CO_2 + 12\,H_2O + \text{Light energy} \longrightarrow C_6H_{12}O_6 + 6\,O_2 + 6\,H_2O$$

2. The chloroplast splits water into hydrogen and oxygen, incorporating the electrons of hydrogen into the bonds of sugar molecules. Photosynthesis is thus a redox process in which water is oxidized and carbon dioxide is reduced.
3. There are two linked stages of photosynthesis: the light reactions and the Calvin cycle. The light reactions in the grana produce ATP and split water, evolving oxygen and forming NADPH by transferring electrons from water to $NADP^+$.
4. The Calvin cycle occurs in the stroma and uses ATP for energy and NADPH for reducing power to form sugar from CO_2.

How the Light Reactions Capture Solar Energy: A Closer Look (pp. 204–212)

1. Light is a form of electromagnetic energy, which travels in waves. The range of wavelengths of this radiation constitutes the electromagnetic spectrum, part of which we detect as the colors of visible light.
2. Light is emitted in discrete energy packets called photons, each with a fixed quantity of energy inversely proportional to the wavelength of the light.
3. A pigment is a substance that absorbs specific wavelengths of light. A spectrophotometer can be used to produce an absorption spectrum for a specific pigment. The action spectrum of photosynthesis and the absorption spectrum of chlorophyll *a* do not directly correspond. Accessory pigments, chlorophyll *b* and various carotenoids, have molecular structures that enable them to absorb different wavelengths of light and pass their energy on to chlorophyll *a*.
4. A pigment goes from a ground state to an excited state when a photon boosts one of its electrons to a higher energy orbital. In isolated pigments, the electron immediately returns to the ground state, releasing the energy as light (fluorescence) and/or heat.
5. The pigments of chloroplasts are built into the thylakoid membrane near molecules known as primary electron acceptors, which trap the high-energy electrons before they return to the ground state. Energy stored by this light-driven redox reaction powers the synthesis of ATP and NADPH.
6. Accessory pigments in chloroplasts are clustered in an antenna complex of a few hundred molecules surrounding two molecules of chlorophyll *a* at the reaction center.

Photons absorbed anywhere in the antenna can pass their energy along to energize this chlorophyll *a*, which then passes an electron to a nearby primary electron acceptor. The antenna complex, the reaction-center chlorophyll, and the primary electron acceptor make up a photosystem, a light-harvesting unit built into the thylakoid membrane.

7. There are two kinds of photosystems. Photosystem I contains P700, and photosystem II contains P680, chlorophyll *a* molecules at the reaction center that have different absorption characteristics.
8. The flow of energized electrons in the light reactions can be cyclic or noncyclic. In cyclic electron flow, P700 donates electrons to an electron transport chain located on the thylakoid membrane. A series of redox reactions returns the electrons to the ground state in P700, generating a proton-motive force across the thylakoid membrane. The passage of hydrogen ions through an ATP synthase enzyme drives the chemiosmotic formation of ATP from ADP.
9. Noncyclic electron flow involves both photosystems and produces NADPH and oxygen in addition to ATP. Electrons from P700 in photosystem I are trapped by $NADP^+$, which stores them in the form of NADPH. Electrons from P680 in photosystem II restore the electrons lost from P700. These electrons travel between the two photosystems via an electron transport chain that powers ATP synthesis. The electron "holes" in P680 are filled by electrons from water, which is split into hydrogen ions and oxygen. The overall process powers the transfer of electrons from water to $NADP^+$ and the phosphorylation of ADP to form ATP.

How the Calvin Cycle Makes Sugar: A Closer Look (pp. 212–214)

1. The Calvin cycle is organized in a cyclic metabolic pathway in the stroma. An enzyme (rubisco) combines carbon dioxide with ribulose bisphosphate (RuBP), a five-carbon sugar. Then, using electrons from NADPH and energy from the hydrolysis of ATP, the cycle synthesizes the three-carbon sugar glyceraldehyde phosphate in a series of reactions. Most of the glyceraldehyde phosphate is reused in the cycle as an intermediate for reconversion to RuBP, but some exits the cycle and is converted to other essential organic molecules.

Photorespiration: An Evolutionary Relic? (pp. 214–215)

1. On dry, hot days, plants close their stomata, which conserves water. Oxygen from the light reactions builds up. When O_2 substitutes for CO_2 in the active site of rubisco, an intermediate is formed that leaves the cycle and is oxidized to CO_2 and H_2O in the peroxisomes and mitochondria. This process, called photorespiration, consumes organic fuel without producing ATP.

C_4 Plants (p. 215)

1. Most plants are C_3 plants, named after the number of carbons in the first stable intermediate of the Calvin cycle.
2. C_4 plants are adapted to hot, dry conditions. They avert photorespiration by prefacing the Calvin cycle with a series of reactions that incorporate CO_2 into four-carbon compounds in specialized mesophyll cells. This four-carbon compound (malate in some plants) is exported to photosynthetic bundle-sheath cells, where it releases carbon dioxide for use in the Calvin cycle.

CAM Plants (pp. 215–216)

1. Some plants use CAM metabolism for carbon fixation in a hot, dry environment. These plants open their stomata during the night and incorporate the CO_2 that enters into organic acids, which are stored in the vacuoles of mesophyll cells. During the day, the stomata close, conserving water, and the CO_2 is released from the organic acids for use in the Calvin cycle.

Summary: The Fate of Photosynthetic Products (pp. 216–218)

1. Veins export sucrose made in green cells to nonphotosynthetic parts of the plant. Mitochondria degrade about 50% of the carbohydrate made by photosynthesis to form ATP. Much of the remaining carbohydrate is converted to a variety of molecules, including cellulose.
2. Excess organic material is stockpiled as starch, oils, and proteins and stored in leaves, roots, tubers, and fruit. Heterotrophs consume much of this organic material.

SELF-QUIZ

1. The light reactions of photosynthesis supply the Calvin cycle with
   a. light energy
   b. CO_2
   c. H_2O
   d. ATP and NADPH
   e. sugar
2. Which sequence correctly portrays the flow of electrons during photosynthesis?
   a. NADPH $\longrightarrow O_2 \longrightarrow CO_2$
   b. $H_2O \longrightarrow$ NADPH $\longrightarrow$ Calvin cycle
   c. NADPH $\longrightarrow$ chlorophyll $\longrightarrow$ Calvin cycle
   d. $H_2O \longrightarrow$ photosystem I $\longrightarrow$ photosystem II
   e. NADPH $\longrightarrow$ electron transport chain $\longrightarrow O_2$
3. Which of the following conclusions does *not* follow from studying the absorption spectrum for chlorophyll *a* and the action spectrum for photosynthesis?
   a. Not all wavelengths are equally effective for photosynthesis.
   b. There must be accessory pigments that broaden the spectrum of light that contributes energy for photosynthesis.
   c. The red and blue areas of the spectrum are most effective in driving photosynthesis.
   d. Chlorophyll owes its color to the absorption of green light.
   e. Chlorophyll *a* has two absorption peaks.
4. Cooperation of the *two* photosystems of the chloroplast is required for
   a. ATP synthesis
   b. reduction of $NADP^+$
   c. cyclic photophosphorylation
   d. oxidation of the reaction center of photosystem I
   e. generation of a proton-motive force

5. In mechanism, photophosphorylation is most similar to
   a. substrate-level phosphorylation in glycolysis
   b. oxidative phosphorylation in cellular respiration
   c. the Calvin cycle
   d. carbon fixation
   e. reduction of $NADP^+$
6. In what respect are the photosynthetic adaptations of C_4 plants and CAM plants similar?
   a. In both cases, the stomata normally close during the day.
   b. Both types of plants make their sugar without the Calvin cycle.
   c. In both cases, an enzyme other than rubisco carries out the first step in carbon fixation.
   d. Both types of plants make most of their sugar in the dark.
   e. Neither C_4 plants nor CAM plants have grana in their chloroplasts.
7. Light-driven electron transport in the chloroplast pumps H^+ where?
   a. into the stroma
   b. into the intermembrane space between the two layers of the chloroplast envelope
   c. into the thylakoid space
   d. from NADPH to CO_2
   e. out of the chloroplast
8. The stage of photosynthesis that actually produces sugar is
   a. the Calvin cycle
   b. photosystem I
   c. photosystem II
   d. the light reactions
   e. splitting of water
9. For every CO_2 molecule fixed by photosynthesis, how many molecules of O_2 are released?
   a. 1
   b. 2
   c. 3
   d. 6
   e. 12
10. Which of the following statements is a correct distinction between autotrophs and heterotrophs?
    a. Only heterotrophs need to acquire chemical compounds from the environment.
    b. Cellular respiration is unique to heterotrophs.
    c. Only heterotrophs have mitochondria.
    d. Autotrophs, but not heterotrophs, can nourish themselves beginning with nutrients that are entirely inorganic.
    e. Only heterotrophs require oxygen.

CHALLENGE QUESTIONS

1. This graph tracks a bean plant's stomata over 24 hours. The dotted line indicates partial closing of stomata on a hot, dry afternoon. How would this affect CO_2 and O_2 concentrations within the leaf? What would be the impact on photosynthesis and photorespiration? How would these effects differ for a C_4 plant? How would the behavior of a CAM plant's stomata differ from the behavior recorded by this graph?

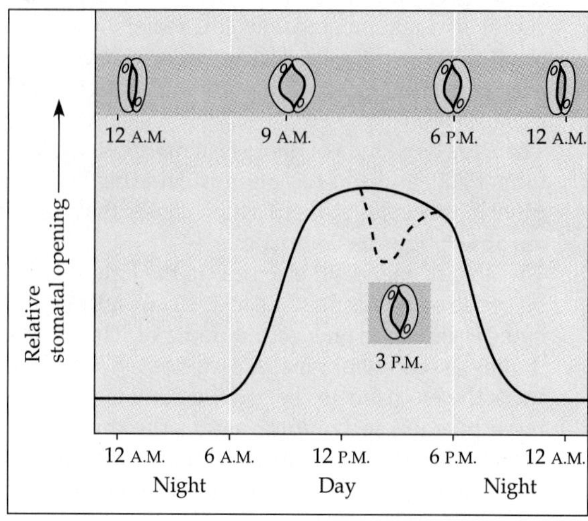

2. Compare and contrast the mechanisms of photosynthesis and cellular respiration. How has the chemiosmotic model helped unify current ideas on the mechanisms of these two processes?
3. The photosynthetic rate of aquatic plants in a test tube can be determined by collecting and measuring the amount of oxygen that gases out of the water. If bicarbonate, the source of CO_2 for aquatic plants, is added to the water, the rate of oxygen evolution increases. If CO_2 is fixed by the Calvin cycle but oxygen is evolved by the light reactions, how can an increase in CO_2 supply increase the rate of oxygen evolution?

SCIENCE, TECHNOLOGY, AND SOCIETY

1. Tropical rain forests cover only about 3% of Earth's surface, but they are estimated to be responsible for more than 20% of global photosynthesis. It seems reasonable to expect that the lush growth of jungle foliage would produce large amounts of oxygen and reduce global warming by consuming carbon dioxide. But in fact, many experts now believe that rain forests make little or no *net* contribution to global oxygen production or reduction of global warming. Using your knowledge of photosynthesis and cellular respiration, explain what the basis of this belief might be. What happens to the food produced by a rain forest tree when it is eaten by animals or the tree dies?

FURTHER READING

Bazzazz, F. A., and E. D. Fajer. "Plant Life in a CO_2-Rich World." *Scientific American*, January 1992. How will increasing atmospheric CO_2 and global warming affect the relative success of C_3 and C_4 plants?

Becker, W. M., and D. W. Deamer. *The World of the Cell*, 2nd ed. Redwood City, CA: Benjamin/Cummings, 1991. Chapter 12.

Cowen, R. "Lab-Made Molecule Taps Light's Energy." *Science News*, April 21, 1990. How technology imitates life to harness solar power.

Govindjee and W. J. Coleman. "How Plants Make Oxygen." *Scientific American*, February 1990.

Hendry, G. "Making, Breaking and Remaking Chlorophyll." *Natural History*, May 1990. The physiology of fall color.

11 The Reproduction of Cells

Bacterial Reproduction
An Introduction to Eukaryotic Chromosomes
The Cell Cycle: An Overview
The Mechanics of Cell Division: A Closer Look
The Control of Cell Division
Abnormal Cell Division: Cancer Cells

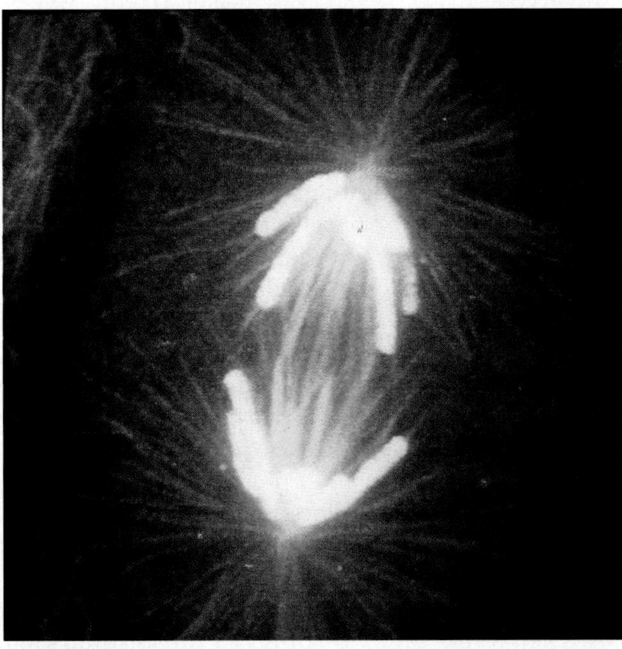

Figure 11.1
All cells from cells. In an unbroken chain extending from the first bacteria inhabiting the primeval seas to the diverse organisms of today, cell division has propagated life on Earth. All cells are derived from the division of previously existing cells. This epithelial cell from a kangaroo rat is in the process of dividing. Its chromosomes, stained here with a yellow fluorescent dye, have duplicated, and the two identical sets are separating toward opposite ends of the cell (LM). The cytoplasm will subsequently divide to form two cells, where before there was one. In this chapter, you will learn about cell division, the basis of all biological reproduction. (Photo courtesy of J. M. Murray, University of Pennsylvania Medical School.)

Life arises only from life. The ability of organisms to reproduce their kind is the one phenomenon that best distinguishes life from nonliving matter. (Though the analogy of making photocopies is frequently used, the reproduction of an organism is actually more analogous to a photocopier making more photocopiers!) This unique capacity to procreate, like all biological functions, has a cellular basis. Rudolf Virchow, a German physician, put it this way in 1855: "Where a cell exists, there must have been a preexisting cell, just as the animal arises only from an animal and the plant only from a plant." He summarized with the axiom, *"Omnis cellula e cellula,"* meaning "All cells from cells." The perpetuation of life is based on the reproduction of cells, or **cell division** (Figure 11.1).

In some cases, division of one cell to form two reproduces an entire organism, as when a unicellular organism, such as *Amoeba,* divides to form duplicate offspring (Figure 11.2a). But cell division also enables a multicellular organism, such as a human, to grow and develop from a single cell—the fertilized egg (Figure 11.2b). Even after the organism is fully grown, cell division continues to function in renewal and repair, replacing cells that die from normal wear and tear or accidents. For example, dividing cells in your bone marrow continuously supply new blood cells (Figure 11.2c).

The reproduction of an ensemble as complex as a cell cannot occur by a mere pinching in half; the cell is not like a soap bubble that simply grows and splits in two. Cell division involves the distribution of identical genetic material—DNA—to the two daughter cells. A cell's total hereditary endowment of DNA is called its **genome.** The genome consists of one or more very long DNA molecules. Arranged along the length of each DNA molecule are hundreds or thousands of genes, the hereditary units that specify an organism's traits (we shall refine our definition of a gene in Chapter 16). In eukaryotes, compact coiling and folding of the DNA molecules form chromosomes, visible with a light microscope just before and during cell division (see

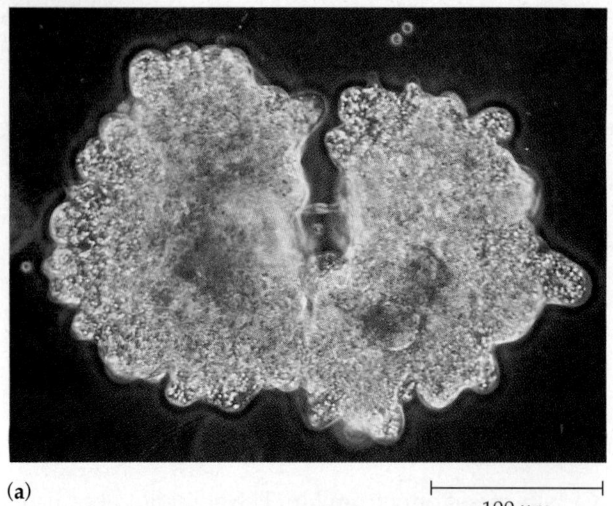

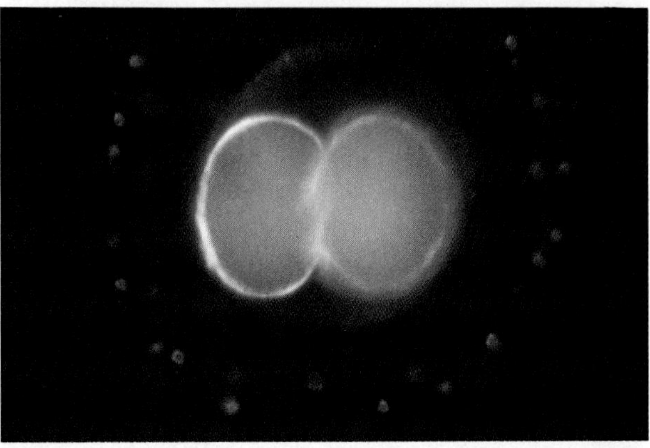

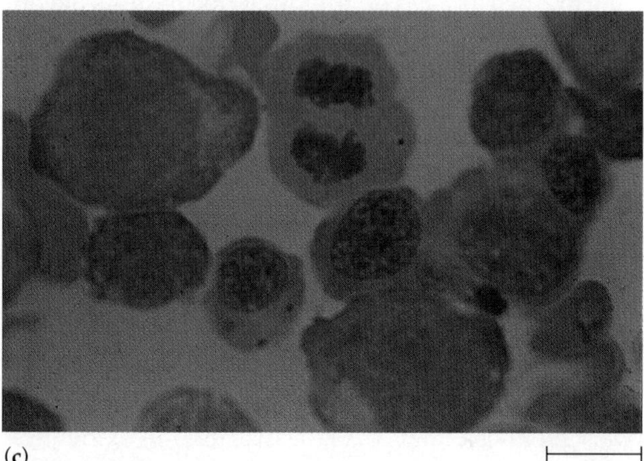

Figure 11.2
The functions of cell division. (**a**) *Amoeba*, a one-celled eukaryote, divides to form two cells, each an individual organism. In this case, cell division functions in reproduction (LM). (**b**) For multicellular organisms, division of embryonic cells is important in growth and development. This darkfield micrograph shows a sand dollar embryo shortly after the fertilized egg divided to form two cells. (**c**) Even in a mature organism, cell division continues to function in the renewal and repair of tissues. These dividing cells in bone marrow give rise to new blood cells (LM).

Figure 11.1). What is most remarkable about cell division is the fidelity with which the genome is passed along without dilution from one generation of cells to the next. A cell preparing to divide first copies all its genes, allocates them equally to opposite ends of the cell, and then separates into two daughter cells.*

In this chapter, you will learn how cells reproduce to form genetically equivalent daughter cells. The focus on *physical* processes that are visible in the light microscope will seem like a departure from the preceding three chapters, which considered submicroscopic, metabolic processes. However, the correlation of structure and function, a recurrent theme throughout our study of cells, will prove useful once again, as we investigate the mechanism of cell division. Although the emphasis in this chapter will be on the division of eukaryotic cells, we will begin with a brief look at prokaryotic cell division.

BACTERIAL REPRODUCTION

Prokaryotes (bacteria and cyanobacteria) reproduce by a type of cell division called **binary fission,** meaning literally "division in half." Most bacterial genes are carried on a single chromosome that consists of a circular DNA molecule and associated proteins. (As we will see, the chromosomes of eukaryotes have a very different organization.) Although bacteria are smaller and simpler than eukaryotic cells, the problem of replicating their genomes in an orderly fashion and distributing the copies equally to two daughter cells is still formidable. Consider, for example, the chromosome of the bacterium *Escherichia coli;* when fully stretched out, this chromosome is about 500 times longer than the length of the cell. Clearly, such a chromosome must be highly folded within the cell.

After a bacterial cell copies its chromosome in preparation for fission, allocation of the duplicate chromosomes to daughter cells depends on the attachment of both chromosomes to the plasma membrane. Growth of the membrane between the two attachment sites separates the two copies of the chromosome (Figure 11.3). When the bacterium has reached about twice its initial size, its plasma membrane pinches inward, and a cell wall forms between the two chromosomes, dividing the parent cell into two daughter cells. Each cell inherits a complete genome. (Bacterial reproduction will be discussed further in Chapters 17 and 25.)

*Although the terms *daughter cells* and *sister chromatids* are traditional and will be used throughout this book, the structures they refer to are actually genderless.

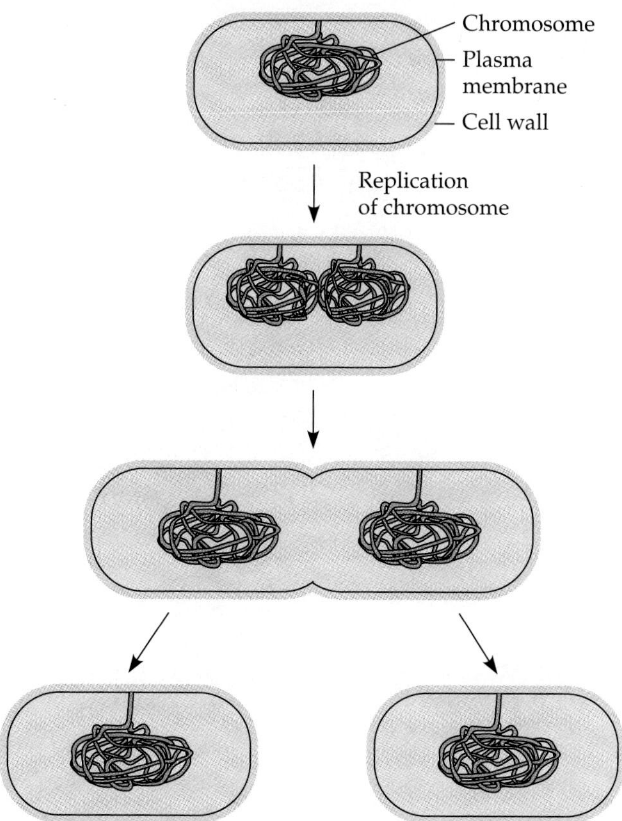

Figure 11.3
Bacterial cell division. In prokaryotic cell division, called binary fission, a membrane attachment mechanism is used to allocate chromosome copies to the two daughter cells. When the bacterial chromosome replicates, it is attached to the plasma membrane; after replication, the duplicate chromosomes attach to the membrane at separate points. Continued growth of the cell gradually separates the chromosomes. Eventually, the plasma membrane pinches inward to divide the cell in two, as new cell wall material is deposited between the daughter cells.

AN INTRODUCTION TO EUKARYOTIC CHROMOSOMES

Eukaryotes have much larger genomes than prokaryotes. Yet each of the tens of thousands of genes in a typical eukaryotic cell must be copied, and then the mechanics of cell division must apportion complete genomes to the two daughter cells. The problem of replicating and distributing so many genes is only manageable because the genes are grouped into multiple chromosomes (Figure 11.4a). Every species of organism has a characteristic number of chromosomes in each cell nucleus. For example, there are 46 chromosomes in human somatic cells (all body cells except the reproductive cells). Reproductive cells—sperm cells and egg cells—have half as many chromosomes as do

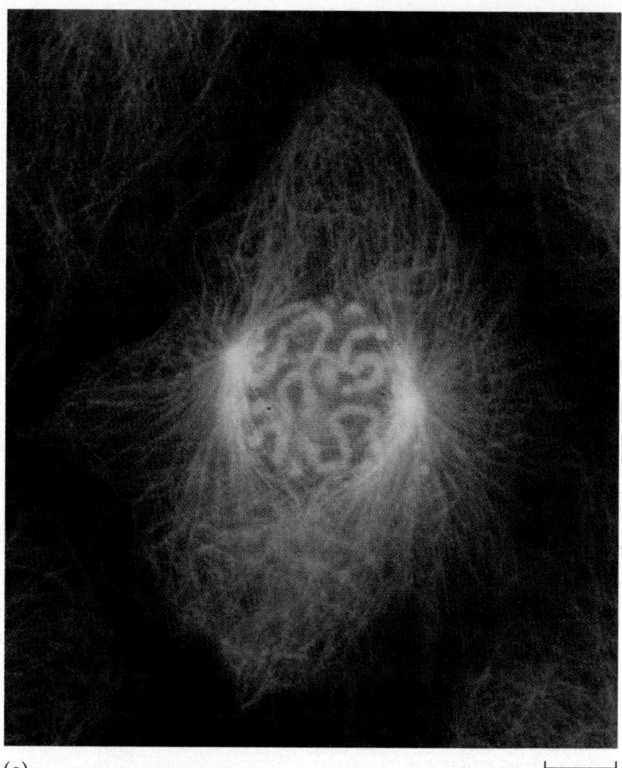

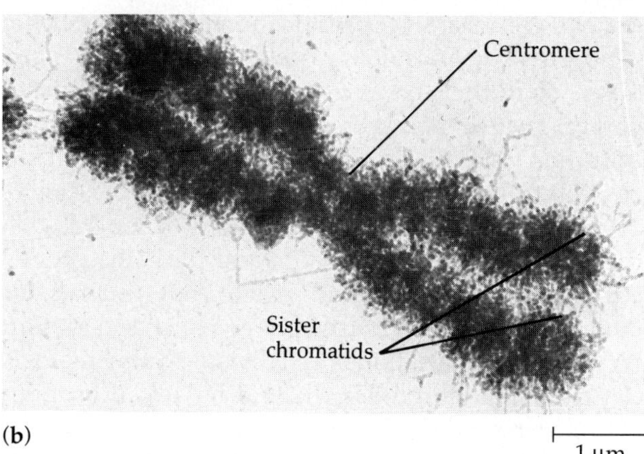

Figure 11.4
Chromosomes in a eukaryotic cell. (**a**) A tangle of threadlike chromosomes (orange) is visible within the nucleus of this kangaroo rat epithelial cell, which is preparing to divide (LM). Chromosomes are so named because they take up certain dyes used in microscopy (*chromo,* "colored," and *somes,* "bodies"). Before dividing, a cell must copy each chromosome and apportion complete sets of the duplicated chromosomes to the daughter cells. (Photo courtesy of J. M. Murray, University of Pennsylvania Medical School.) (**b**) As a cell prepares to divide, each chromosome replicates to make two genetically identical sister chromatids attached at their centromeres. This electron micrograph reveals a human chromosome in the duplicated state. The chromosome has a "hairy" appearance because it consists of a very long chromatin fiber, folded and coiled in a compact arrangement. Nucleic acid and protein make up the chromatin fiber (TEM).

somatic cells, or 23 chromosomes in the case of humans.

Incorporated into each chromosome is one very long DNA molecule representing thousands of genes. The DNA is associated with various proteins, which maintain the structure of the chromosome and help control the activity of the genes. The DNA-protein complex, called **chromatin,** is organized into a long, thin fiber that is compactly folded and coiled to form the chromosome (the details and importance of this compact packing of DNA will be discussed in Chapter 18). As part of its preparation for division, a cell copies its entire genome by duplicating each chromosome. After replication, each chromosome consists of two **sister chromatids,** each a collection of the same genes present in single copy prior to replication (Figure 11.4b). A specialized region of the chromosome called the **centromere** holds the two chromatids together. Then, in the process of **mitosis,** the sister chromatids are pulled apart and repackaged as complete chromosome sets in two nuclei, one at each end of the cell. Mitosis, the division of the nucleus, is usually followed immediately by **cytokinesis,** the division of the cytoplasm. Where there was one cell, there are now two, each the genetic equivalent of the parent cell.

Let's see what happens to chromosome number as we follow the human life cycle through the generations. You inherited 46 chromosomes, 23 from each parent. They were combined in the nucleus of a single cell when a sperm cell from your father united with an egg cell from your mother to form a fertilized egg, or zygote. Mitosis produced the trillions of somatic cells that now make up your body, and the same process continues to generate new cells to replace dead and damaged ones. In contrast, your egg or sperm cells are produced by a variation of cell division called **meiosis,** which yields daughter cells that have half as many chromosomes as the parent cell. Meiosis occurs only in your gonads (your ovaries or testes). In each generation, meiosis reduces the chromosome number from 46 to 23 (in humans). Fertilization doubles the chromosome number to 46 again, and mitosis conserves that number in every somatic cell of the new individual. In Chapter 12, we will examine the role of meiosis in reproduction and inheritance in much more detail. In the rest of this chapter, we will focus on mitosis.

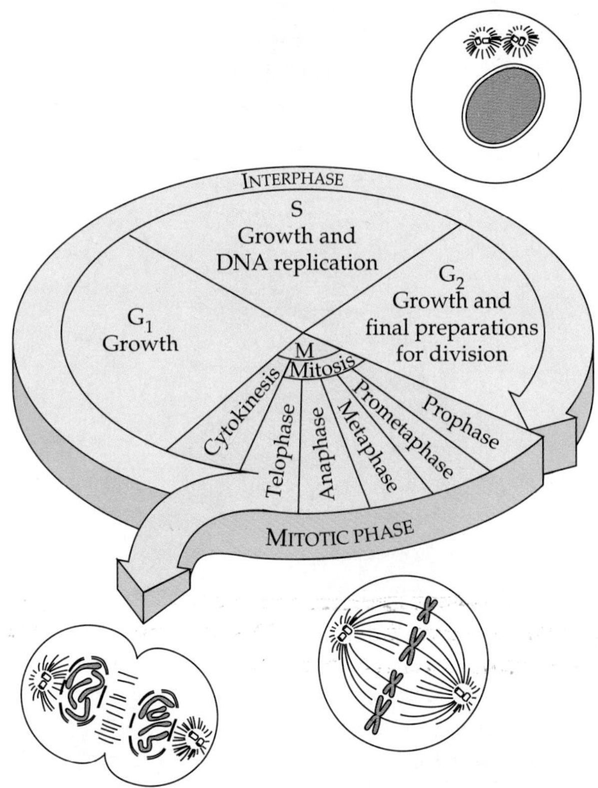

Figure 11.5
The cell cycle. In a dividing cell, the mitotic (M) phase alternates with an interphase, or growth period. The first part of interphase is called G_1, followed by the S phase, during which the chromosomes replicate; the last part of interphase is called G_2. Next, mitosis divides the nucleus and its chromosomes. Finally, cytokinesis divides the cytoplasm, producing two daughter cells. The stages of mitosis are described in Figure 11.6.

THE CELL CYCLE: AN OVERVIEW

Mitosis is just one part of a cell-division cycle, or **cell cycle** (Figure 11.5). In fact, the **M phase,** when mitosis and cytokinesis actually divide the nucleus and cytoplasm, is the shortest part of the cell cycle. Successive mitotic divisions alternate with a much longer **interphase,** which usually accounts for about 90% of the time that elapses during each cell cycle. It is during interphase that the cell grows and copies its chromosomes in preparation for cell division. Interphase consists of three periods of growth, called, in order, the **G_1 phase,** the **S phase,** and the **G_2 phase.** During all three phases, the cell grows by synthesizing proteins and producing cytoplasmic organelles. However, chromosomes are duplicated only during the S phase (S stands for synthesis, as in DNA synthesis). Thus, a cell grows (G_1), continues to grow as it copies its chromosomes (S), grows more as it completes preparations for cell division (G_2), and divides (M). The daughter cells may then repeat the cycle.

Now that we have broken down the cell cycle into its four phases, let's examine the mechanics of cell division more closely, and then investigate how the cell cycle is controlled.

THE MECHANICS OF CELL DIVISION: A CLOSER LOOK

Time-lapse films of living, dividing cells reveal the dynamics of mitosis and cytokinesis as a continuum of changes. For purposes of description, however, mitosis is conventionally described as occurring in five stages: **prophase, prometaphase, metaphase, anaphase,** and **telophase.** The details of these stages for an animal cell are shown in Figure 11.6 (pp. 226–227), which you should study before going on to the next section.

The Structure and Function of the Mitotic Spindle

Many of the events of mitosis depend on a structure called the **mitotic spindle,** which begins to form in the cytoplasm during prophase. This structure consists of fibers made of microtubules and associated proteins (Figure 11.7a). While the mitotic spindle is being assembled, the microtubules of the cytoskeleton are partially disassembled, probably providing the material used to construct the spindle. The spindle microtubules elongate by incorporating more subunits of the protein tubulin (see Table 7.2). Several parallel microtubules form bundles called spindle fibers that are large enough to see with a light microscope.

The assembly of spindle microtubules is initiated in the centrosome (microtuble-organizing center; see Chapter 7). Microtubules are polar, with distinct ends that cell biologists symbolize as the + and – ends. A microtubule changes in length by the addition or removal of tubulin proteins only at the + end. The + end of spindle microtubules is the end *away from* the centrosome. Thus, once initiated in the centrosome, spindle microtubules extend or retract at their tips, not at their bases. In animal cells, a pair of centrioles is located at the center of the centrosome, but these structures are not essential for cell division. In fact, the centrosomes of plant cells lack centrioles. And if the centrioles of animal cells are destroyed with a laser microbeam, spindles form and function during mitosis.

During interphase of the cell cycle, the single centrosome replicates to form two centrosomes located just outside the nucleus (see Figure 11.6). As the two centrosomes move farther apart during prophase and prometaphase, spindle microtubules radiate from them, elongating at their + ends. By the end of prometaphase, the two centrosomes are arranged at opposite poles of the cell, and some of their spindle microtubules attach to chromosomes at specialized regions of the centromeres called **kinetochores** (Figure 11.7b).

Each of the two joined chromatids of a chromosome has its own kinetochore. Microtubules from one pole of the cell may attach to a kinetochore first, and the chromosome begins to move toward that pole. However, this movement is checked as soon as microtubules from the opposite pole attach to the chromosome's other kinetochore. What happens next is like a tug-of-war that ends in a draw. The chromosome moves first in one direction, then the other, to-and-fro, finally settling at the cell's midpoint. Apparently, microtubules can only remain attached to a kinetochore when there is a force exerted on the chromosome from the opposite end of the cell. If microtubules from just one pole attach to a chromosome, they lose their grip. Attachment to one kinetochore is stabilized only when microtubules from the opposite pole hook to the other kinetochore. This check-and-balance system equalizes the number of microtubules attached to the two kinetochores of a chromosome and moves the chromosome to the midline of the cell.

At metaphase, all the duplicated chromosomes are aligned on the cell's midline, or metaphase plate. About 15 to 35 microtubules, called kinetochore microtubules, are attached to each kinetochore. Many more microtubules, called nonkinetochore microtubules, radiate from each centrosome toward the metaphase plate without attaching to chromosomes. Some of these tubules are too short to reach the metaphase plate, but others actually extend across the plate and overlap with nonkinetochore microtubules from the opposite pole of the cell. Let's now see how the structure of this entire spindle apparatus is correlated with its function during mitosis.

Anaphase commences when the chromosome's centromeres divide and the sister chromatids, now separate chromosomes, move toward opposite ends of the cell. How do the kinetochore microtubules function in this poleward movement of chromosomes? It may seem reasonable that the microtubules somehow pull a chromosome toward a pole by depolymerizing (losing protein subunits) at their centrosome end. But remember that it is the + end of a microtubule—the end attached to a kinetochore—where addition or removal of tubulin proteins must occur. There is experimental evidence that kinetochore microtubules do, in fact, shorten during anaphase by depolymerizing at their kinetochore ends (Figure 11.8). A chromosome creeps poleward, as its kinetochore somehow hangs onto the remaining tips of microtubules just ahead of the zone of depolymerization. The exact mechanism of this interaction between kinetochores and microtubules is still unresolved. However, researchers have recently found the motor protein dynein in kinetochores. Recall from Chapter 7 that dynein drives microtubule movement in flagella. The kinetochore dynein may "walk" a chromosome along the shortening microtubules in a similar fashion.

Figure 11.6
The stages of mitotic cell division in an animal cell. The light micrographs show dividing cells from a fish embryo. The highly schematic drawings show details not visible in the micrographs. For the sake of simplicity, only four chromosomes are drawn. In plant cells, centrioles are lacking, and cytokinesis occurs by a different mechanism (see Figures 11.11 and 11.12). A dividing cell does not "click" from one mitotic stage to the next, as it appears in these static micrographs. Cell division is a dynamic continuum of changes.

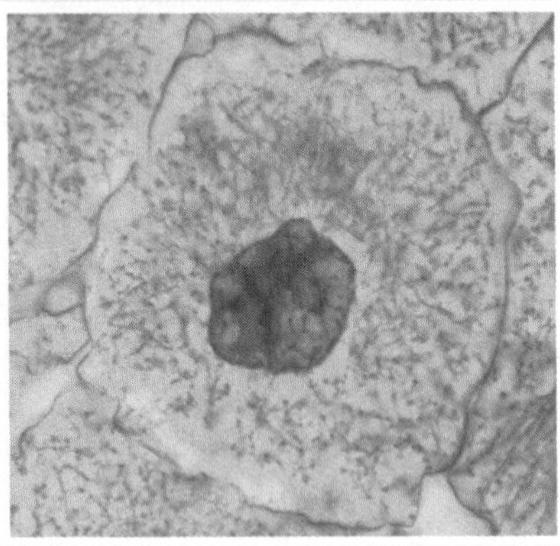

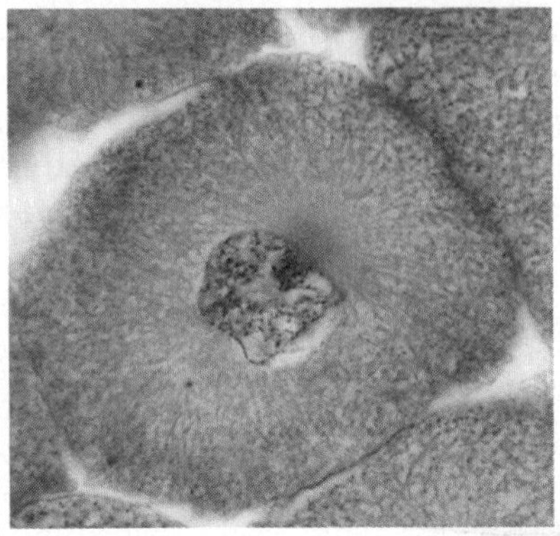

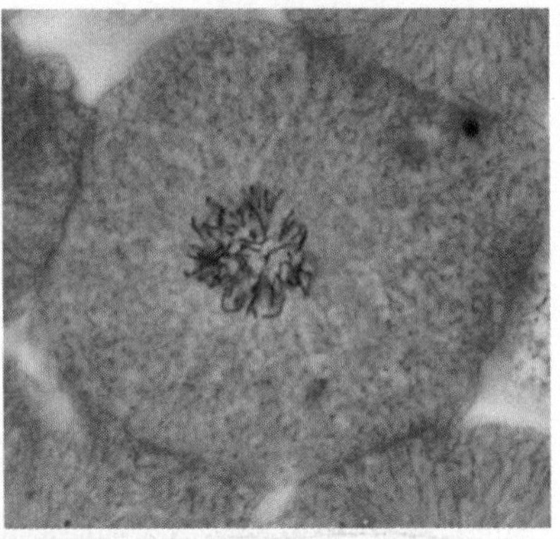

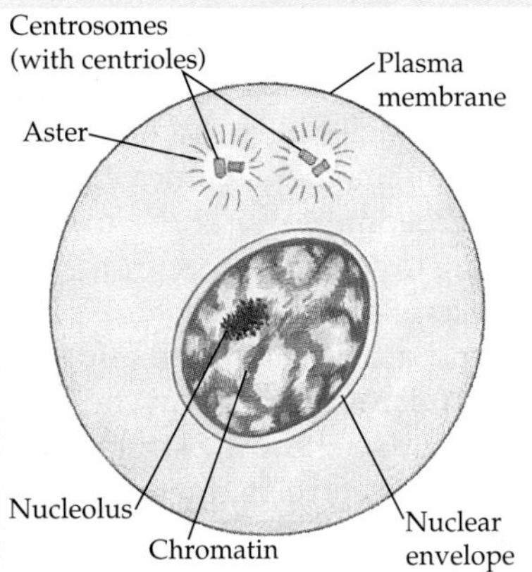

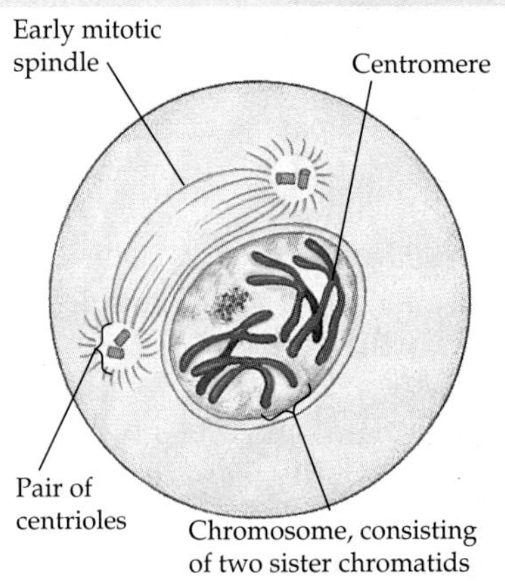

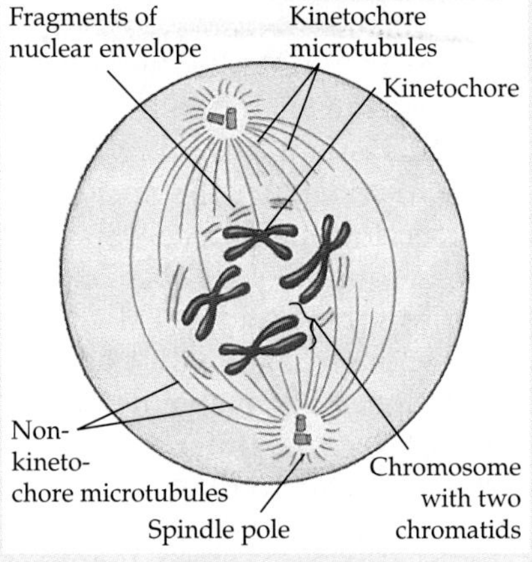

G_2 of INTERPHASE

During late interphase, the nucleus is well defined and bounded by the nuclear envelope. It contains one or more nucleoli. Just outside the nucleus are two centrosomes (microtubule-organizing centers), formed during early interphase by replication of a single centrosome. (In animal cells, each centrosome features a pair of centrioles.) Microtubules extend from the centrosomes in radial arrays called **asters** ("stars"). The chromosomes have already duplicated (during the S phase), but at this stage they cannot be distinguished individually because they are still in the form of loosely packed chromatin fibers.

PROPHASE

During prophase, changes occur in both the nucleus and the cytoplasm. In the nucleus, the nucleoli disappear. The chromatin fibers become more tightly coiled and folded into discrete chromosomes observable with a light microscope. Each duplicated chromosome appears as two identical sister chromatids joined at the centromere. In the cytoplasm, the mitotic spindle forms; it is made of microtubules and associated proteins arranged between the two centrosomes. During prophase, the centrosomes move away from each other, apparently propelled along the surface of the nucleus by the lengthening bundles of microtubules between them.

PROMETAPHASE

During prometaphase, the nuclear envelope fragments. The microtubules of the spindle can now invade the nucleus and interact with the chromosomes, which have become even more condensed. Bundles of microtubules, called **spindle fibers**, extend from each pole toward the equator of the cell. Each of the two chromatids of a chromosome now has a specialized structure called the kinetochore, located at the centromere region. The microtubules that attach to the kinetochores are specifically called kinetochore microtubules. They throw the chromosomes into agitated motion. Many more microtubules, the nonkinetochore microtubules, radiate from the poles toward the cell's equator without attaching to chromosomes.

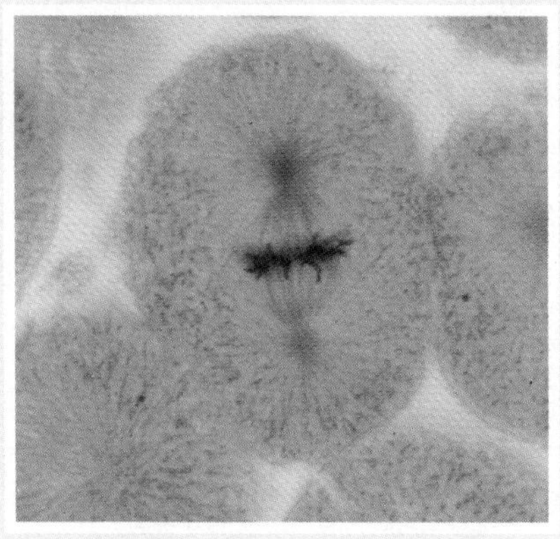

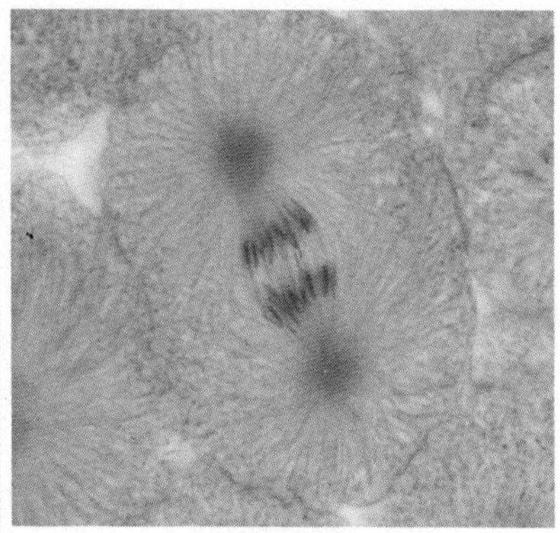

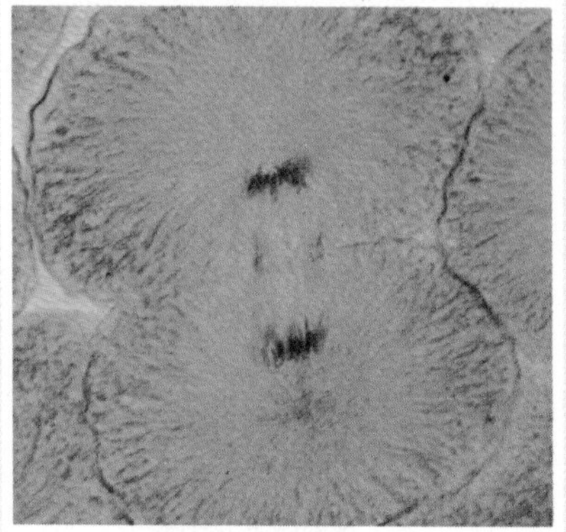

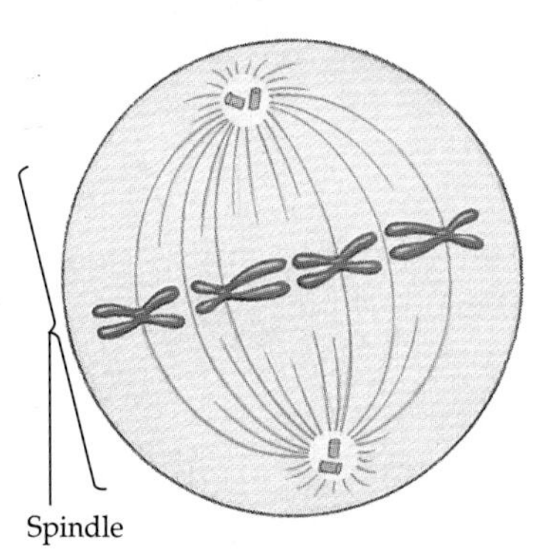

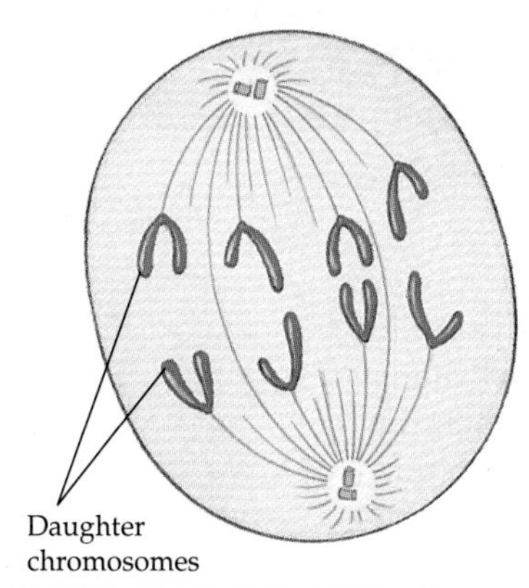

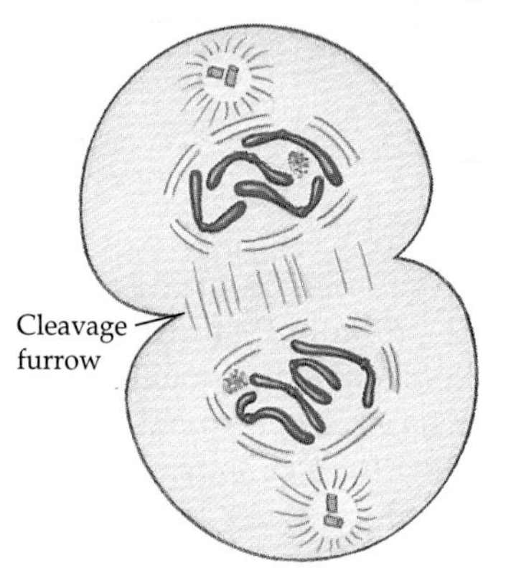

Metaphase

The centrosomes are now at opposite ends, or poles, of the cell. The chromosomes convene on the **metaphase plate**, the plane that is equidistant between the spindle's two poles. The centromeres of all the chromosomes are aligned with one another at the metaphase plate. For each chromosome, the kinetochores of the sister chromatids face opposite poles of the cell. Thus, the identical chromatids of each chromosome are attached to kinetochore microtubules radiating from opposite ends of the parent cell. The entire apparatus of nonkinetochore microtubules plus kinetochore microtubules is called the spindle because of its shape.

Anaphase

Anaphase begins when the paired centromeres of each chromosome divide, liberating the sister chromatids from each other. Each chromatid is now considered a full-fledged chromosome. The spindle apparatus then begins moving the once-joined sisters toward opposite poles of the cell. Because the kinetochore microtubules are attached to the centromere, the chromosomes move centromere first (their pace is about 1 µm/sec). The kinetochore microtubules shorten as the chromo-somes approach the cell poles. At the same time, the poles of the cell also move farther apart. By the end of anaphase, the two poles of the cell have equivalent—and complete—collections of chromosomes.

Telophase and Cytokinesis

At telophase, the nonkinetochore microtubules elongate the cell still more, and the daughter nuclei begin to form at the two poles of the cell where the chromosomes have gathered. Nuclear envelopes are formed from the fragments of the parent cell's nuclear envelope and other portions of the endomembrane system. In a further reversal of prophase and prometaphase events, the nucleoli reappear, and the chromatin fiber of each chromosome becomes less tightly coiled. Mitosis, the equal division of one nucleus into two genetically identical nuclei, is now complete. Cytokinesis, the division of the cytoplasm, is usually well under way by this time, so the appearance of two separate daughter cells follows shortly after the end of mitosis. In animal cells, cytokinesis involves the formation of a cleavage furrow, which pinches the cell in two.

Figure 11.7
The mitotic spindle. (**a**) This electron micrograph illustrates the spindle apparatus as it appears at metaphase (TEM). The chromosomes are arranged on the metaphase plate, attached to spindle microtubules radiating from the poles of the cell. Other microtubules, not attached to chromosomes, overlap at the metaphase plate. (**b**) The kinetochores of a chromosome's two chromatids face opposite poles of the cell. Numerous kinetochore microtubules are attached to each kinetochore (TEM).

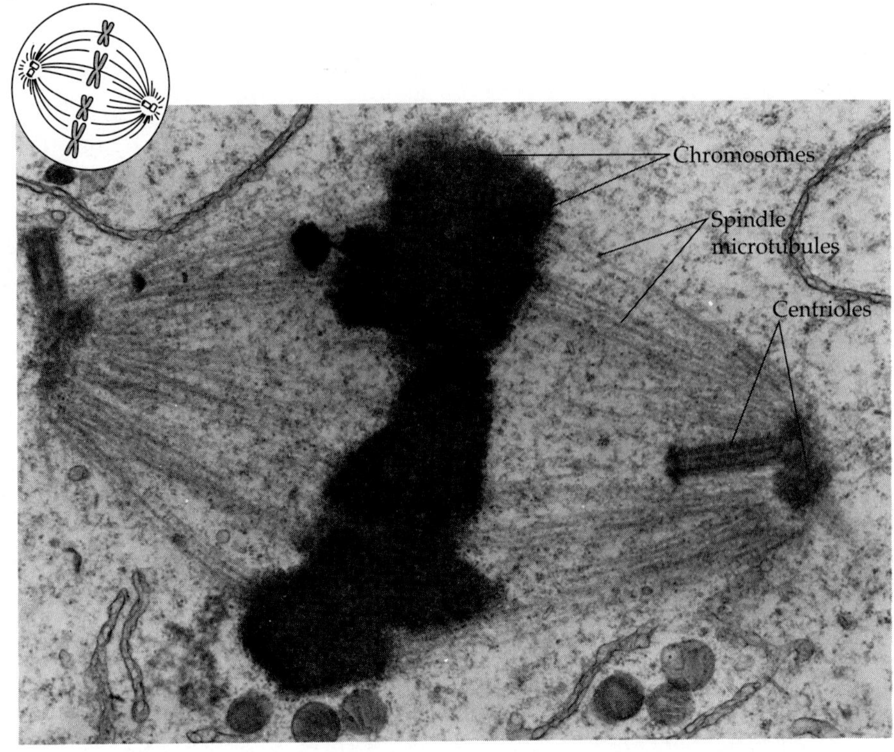

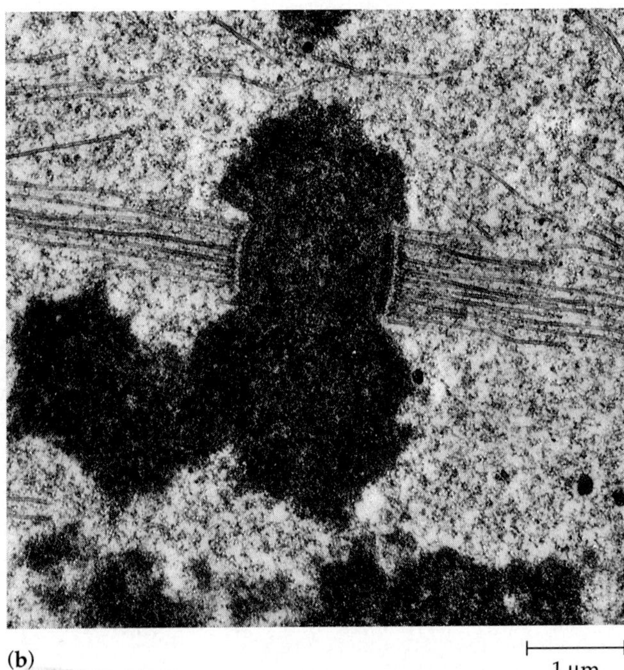

What is the function of the nonkinetochore microtubules? The tubules that overlap at the middle of the cell are responsible for elongating the whole cell along the polar axis during anaphase. This occurs when the interdigitating tubules slide past each other away from the cell's equator (Figure 11.9). The movement may be quite similar to the sliding of neighboring microtubules in a flagellum (see Chapter 7). According to this hypothesis, molecular motors, perhaps like the dynein cross-bridges in flagella, drive nonkinetochore fibers past one another. ATP provides the energy.

At the end of anaphase, then, duplicate sets of chromosomes are clustered at opposite ends of the parent cell, which has elongated in its polar axis. Nuclei reform during telophase. Concurrent with this last stage of mitosis, the process of cytokinesis generally divides the cell's cytoplasm.

The Mechanisms of Cytokinesis

In animal cells, cytokinesis occurs by a process known as **cleavage.** The first sign of cleavage is the appearance of a **cleavage furrow,** which begins as a shallow groove in the cell surface near the old metaphase plate (Figure 11.10). On the cytoplasmic side of the furrow is a **contractile ring** of microfilaments made of the protein actin, the same protein that plays a key role in muscle contraction, as well as many other kinds of cell movement (see Chapter 7). As the dividing cell's ring of microfilaments contracts and its diameter shrinks, the effect is like the pulling of drawstrings. The cleavage furrow deepens until the parent cell is pinched in two. The last bridge between the two daughter cells, containing the remains of the mitotic spindle, finally breaks, leaving two completely separated new cells.

Cytokinesis in plant cells, which have walls, is markedly different. There are no cleavage furrows. Instead, a structure called the **cell plate** forms during telophase across the midline of the parent cell where the old metaphase plate was located (Figure 11.11). Vesicles derived from the Golgi apparatus are driven

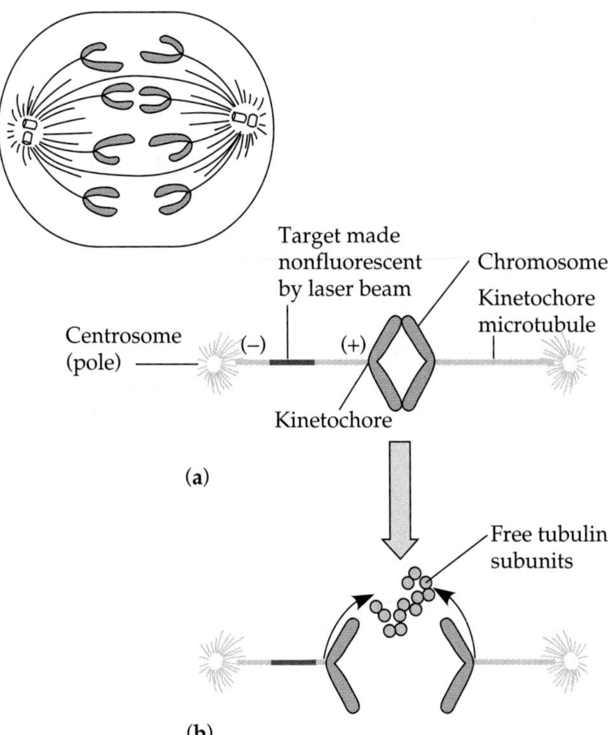

Figure 11.8
Testing a hypothesis for chromosome separation during anaphase. (**a**) In this experiment, the kinetochore microtubules are labeled with a fluorescent dye that glows in the microscope. During early anaphase, a laser microbeam is aimed at microtubules about midway between the pole and the kinetochore. The laser eliminates fluorescence at the target region, but the microtubules still function. This treatment marks a fixed point on a microtubule, making it possible to monitor changes in the length of the microtubule on either side of the mark. (**b**) As anaphase proceeds, the chromosomes move toward the poles, and the segment of microtubule on the kinetochore side of the laser mark shortens. The portion of the tubule on the centrosome side of the mark retains its length. This is one of the experiments supporting the hypothesis that a chromosome tracks along shortening microtubules that depolymerize at their kinetochore ends.

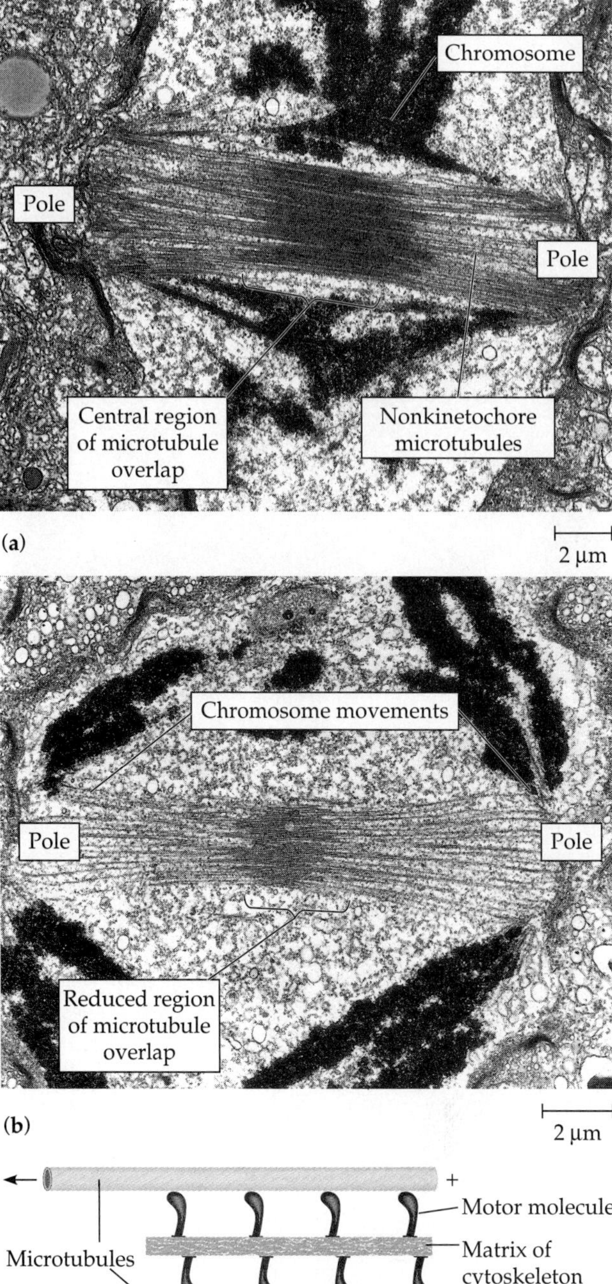

Figure 11.9
The mechanism of cell elongation during anaphase. (**a**) During metaphase and early anaphase, nonkinetochore microtubules from opposite poles of the cell overlap at the metaphase plate (TEM). (**b**) As anaphase proceeds, the two poles of the cell are moved farther apart; the whole cell elongates (TEM). This change in the cell's shape is accompanied by a reduction in the degree of overlap between microtubules. The sliding of microtubules past each other probably pushes the two poles of the cell apart. (**c**) According to one hypothesis, motor molecules that link the microtubules to other components of the cytoskeleton may drive the sliding.

along microtubules to the middle of the cell, where they coalesce to form the cell plate. The fusion of vesicles forms two membranes, which eventually unite laterally with the existing plasma membrane. This results in the formation of two daughter cells, each with its own plasma membrane. A new cell wall forms between the two membranes of the cell plate. Figure 11.12 shows one complete turn of the cell-division cycle for a plant cell. Examining this figure will help you review the process of mitosis.

Researchers still disagree about many of the details of mitotic mechanics, and only additional experiments will resolve these debates. Almost all cell biologists *would* agree, however, that the complex cellular choreography of mitosis represents an evolutionary breakthrough that solved the problem of perpetuating the large genomes of eukaryotes (Figure 11.13).

THE CONTROL OF CELL DIVISION

The timing and rate of cell division in different parts of a plant or animal are critical to normal growth, development, and maintenance. We can observe several patterns of cell division in different types of cells. For example, human skin cells divide frequently throughout life, whereas liver cells maintain the ability to divide, but keep it in reserve until an appropriate situation arises—for example, to repair a wound. Some of the most specialized cells, such as nerve cells and muscle cells, do not divide at all in a mature human.

By growing cells in artificial cultures, researchers have been able to identify many chemical and physical factors that can stimulate or inhibit cell division (see the Methods Box, p. 234). For example, cells fail to divide if an essential nutrient is left out of the culture medium. Poisons that inhibit protein synthesis also arrest cell division. Even if all other conditions are favorable, some types of mammalian cells will only divide in cultures if the growth medium includes specific regu-

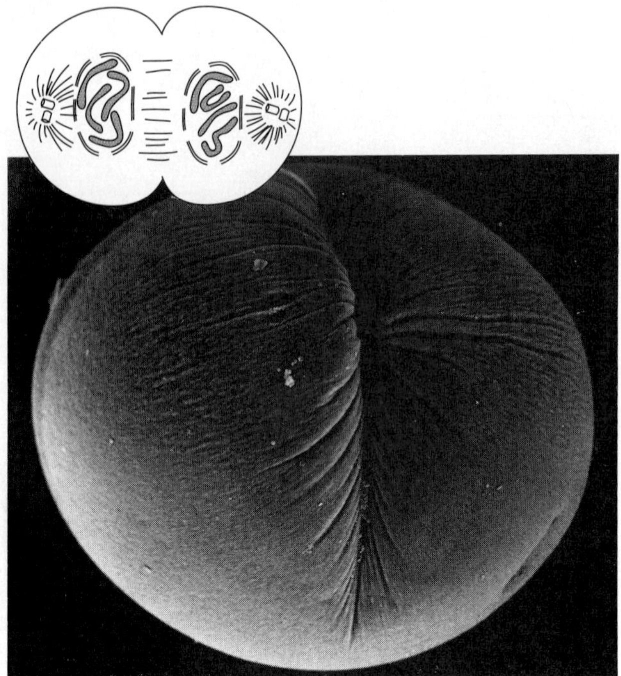

100 µm

Figure 11.10
Cytokinesis in an animal cell. In this electron micrograph, we see the cleavage furrow of a dividing animal cell as it appears on the cell surface (SEM). Microfilaments form a ring just inside the plasma membrane at the location of furrowing. These microfilaments are made of contractile proteins, which cause the cleavage furrow to deepen until the cell is pinched in two.

Membrane vesicles

Double membrane enclosing cell plate

Secreted material, the new cell wall

(a)

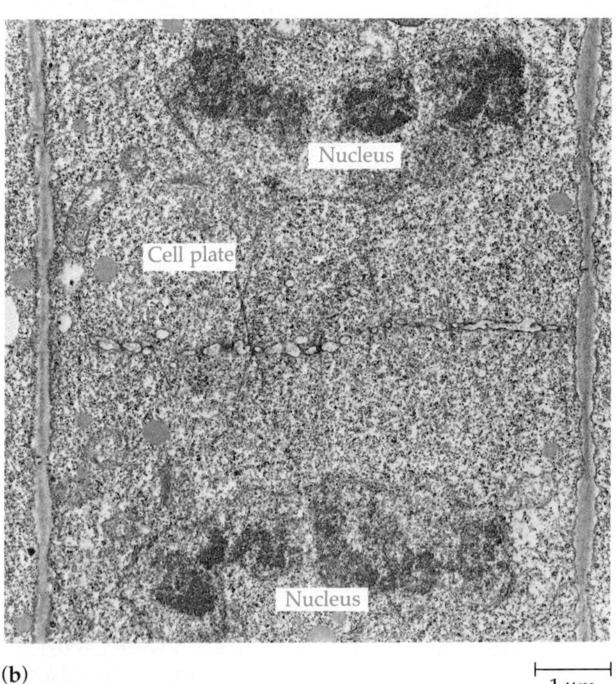

(b)

1 µm

Figure 11.11
Cytokinesis in a plant cell by cell plate formation. The telophase cell has nuclei forming at its two poles. Meanwhile, (**a**) membrane vesicles fuse to form a double membrane, which encloses the cell plate at the equator of the cell, and materials are secreted into the space between the membranes to form the new cell wall. (**b**) This electron micrograph shows a cell plate forming in a cell from a soybean root tip (TEM).

Figure 11.12
Mitosis in a plant cell. These micrographs show cells of *Tradescantia* (wandering Jew) as viewed by differential-interference-contrast microscopy.

Nucleus
Nucleoli

INTERPHASE

Chromosomes

PROPHASE

METAPHASE

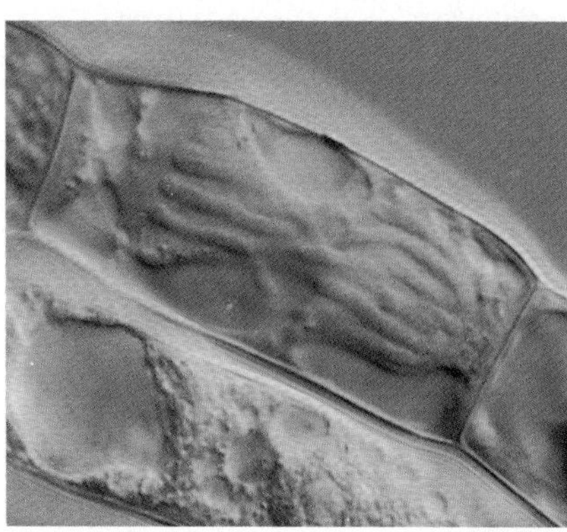

ANAPHASE

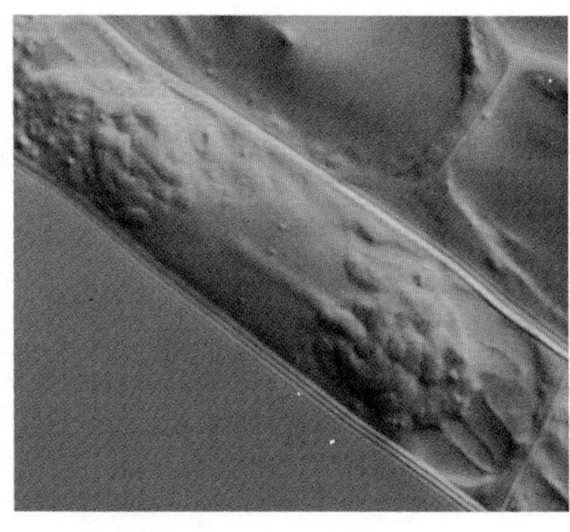

LATE ANAPHASE

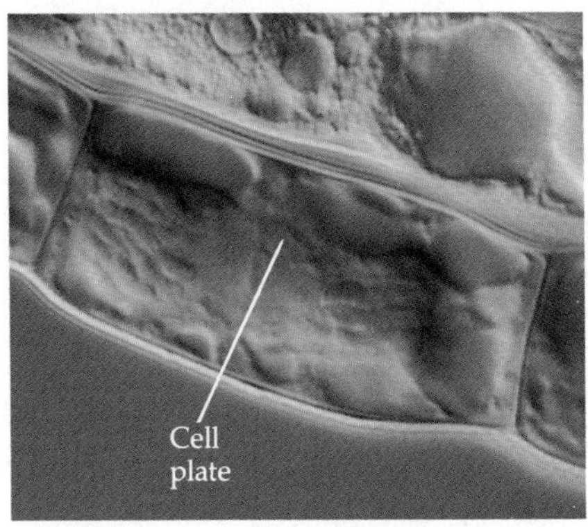

TELOPHASE

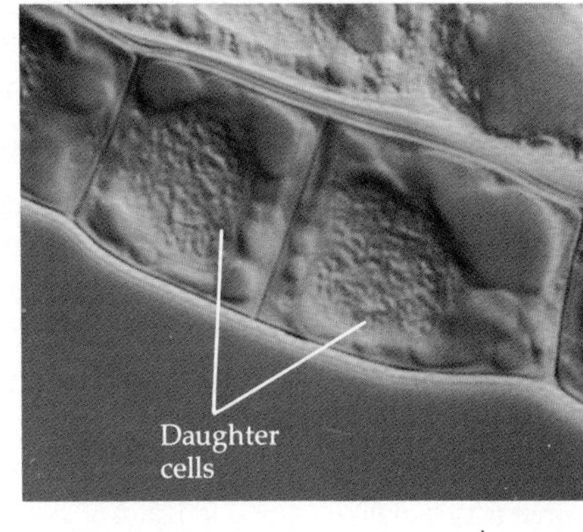

CYTOKINESIS

latory substances called **growth factors.** The fibroblast, for instance, requires platelet-derived growth factor, or PDGF. Fibroblast cells have receptors for PDGF on their plasma membranes, and binding of the growth factor to the cell stimulates cell division. This regulation occurs not only in the artificial conditions of cell culture, but in the body of the animal as well. Blood cells called platelets fragment and release PDGF in the vicinity of an injury. This triggers cell division of fibroblasts in that location, a response that helps heal the wound. Several other growth factors have been discovered. Each cell type probably responds specifically to a certain growth factor or combination of factors.

The density of cells is another important condition

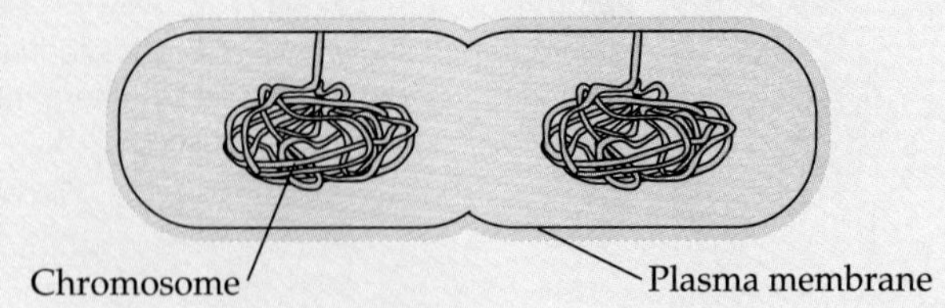

During binary fission of bacteria, the daughter chromosomes are attached to the plasma membrane. Elongation of the cell separates the chromosomes.

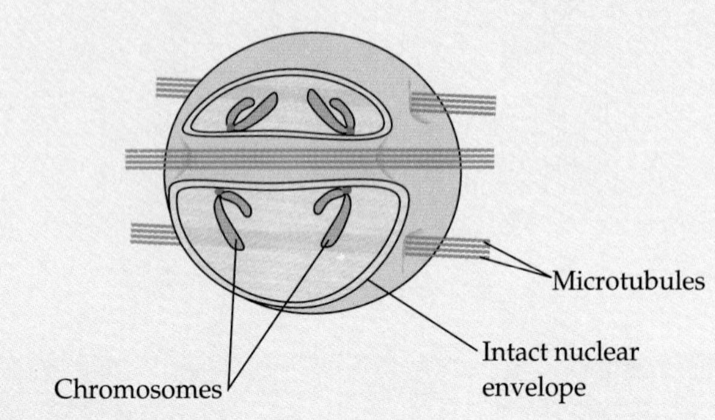

A nucleus contains the multiple chromosomes of a eukaryotic cell. In single-celled algae called dinoflagellates, the nuclear envelope remains intact during cell division. Chromosomes attach to the nuclear envelope, much as the bacterial chromosome attaches to the plasma membrane. Microtubules pass through cytoplasmic tunnels through the nucleus. The microtubules do not attach to chromosomes. They function instead to reinforce the spatial orientation of the nucleus, which then divides in a fission process reminiscent of bacterial reproduction.

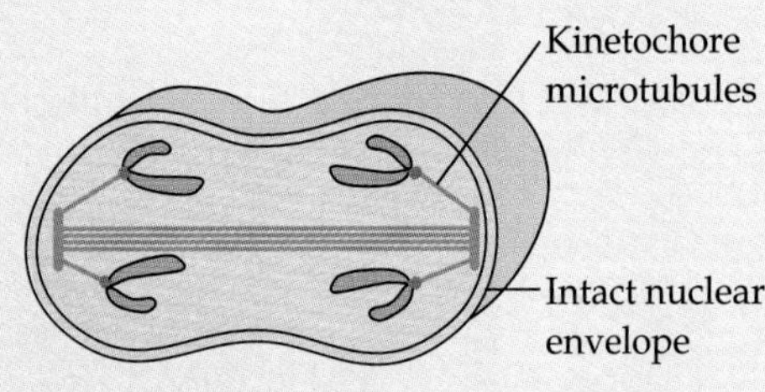

In another group of unicellular algae called diatoms, the nuclear envelope also remains intact during cell division. But in these organisms, the microtubules form a spindle *within* the nucleus. The microtubules separate the chromosomes, and the nucleus splits into two daughter nuclei.

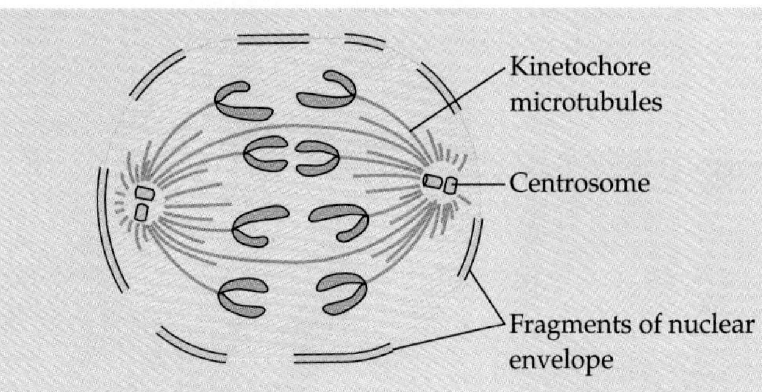

In most other eukaryotes, including plants and animals, the spindle forms outside the nucleus, and the nuclear envelope breaks down during mitosis. Microtubules separate the chromosomes, and the nuclear envelope re-forms.

Figure 11.13
The evolution of mitosis. Mitotic cell division is unique to eukaryotes. Prokaryotes (bacteria) have much smaller genomes and reproduce by the simpler process of binary fission. As eukaryotes, with their larger genomes, evolved, their ancestral process of binary fission somehow gave rise to mitosis. Researchers interested in this evolutionary problem have observed what they believe are mechanisms of cell division intermediate between binary fission and mitosis. Two of these variations in cell division are illustrated here, along with bacterial fission and mitosis as it occurs in plants and animals.

regulating cell division (Figure 11.14). Cultured cells normally divide until they form a single layer of cells on the inner surface of the culture container, at which point the cells stop dividing. If some cells are removed to create a space, those bordering the open space begin dividing again until the vacancy is filled. Crowding inhibits cell division. This phenomenon is called **density-dependent inhibition** of cell division. A population of cells competes for nutrients and for minute quantities of growth regulators. Apparently, when cells reach a certain density, the amount of these required substances per cell is insufficient to allow continued growth of the cell population. This regulatory mechanism probably functions in the body's tissues as well as in cell culture, checking the proliferation of cells at some optimal population density. Related to density-dependent inhibition is a requirement for the adhesion of cells to a substratum: the inside of a culture jar or the extracellular matrix of a tissue. Cells normally stop dividing if they loose their anchorage. Cancer cells, which we will discuss later in the chapter, do not exhibit density-dependent inhibition.

The G_1 phase of the cell cycle is a key period in the control of cell division. A crucial checkpoint occurs late in G_1, just before DNA synthesis (the S phase) would begin. This **restriction point** (or "**start**") is when the "go/no-go decision" to divide is made. If all systems are "go" at this point—if all the internal and external cues are favorable—the cell proceeds to copy its DNA. Then the cell divides. Alternatively, the cell may "exit" from the cell cycle at the restriction point and switch to a nondividing state called the **G_0 phase.** Most cells of the body are actually in the G_0 phase. The most specialized cells, such as nerve and muscle cells, will never divide again. Other cells, such as liver cells, can be "called back" to the cell cycle by certain environmental cues, such as growth factors released during injury.

For cells that *are* actively proliferating in favorable conditions, cell size seems to be the most important criterion for passing the restriction point. The cell must grow to a certain size during G_1 before DNA synthesis begins. How does the cell sense its size? The ratio of cytoplasmic volume to genome size seems to be the most important indicator. As a cell grows by adding cytoplasm, the amount of DNA in the nucleus remains constant. The cell may pass the restriction point and copy its DNA once the cell's volume-to-genome ratio reaches a certain threshold value. Without some regulatory mechanism based on cell size, we can imagine daughter cells getting smaller and smaller with each cell cycle.

The restriction point is a point of no return. The onset of the S phase commits the cell to continue through the G_2 and M phases and divide, regardless of external conditions, such as nutrient supply. This does not mean, however, that the cell is out of control. The precise sequence of events required for successful division seems to depend on the completion of each task before the cell can progress to the next stage. For example, chromosomes do not condense until they have replicated, the nuclear envelope does not break down until the chromosomes are condensed, the spindle does not separate sister chromatids until all the chromosomes have arrived at the metaphase plate, and cytokinesis does not occur until anaphase moves chromosomes toward opposite poles of the cell. Cell biologists are only beginning to work out the "switches" that control this precise sequence.

Researchers *have* identified the "OK signal" required for a cell to progress from late interphase (G_2) to mitosis. It is a complex of proteins called **MPF** (which we can think of as M-phase promoting factor, although the letters originally stood for maturation promoting factor). The amount of MPF in the cell rises and falls in a cycle correlated with the cell cycle: MPF appears in

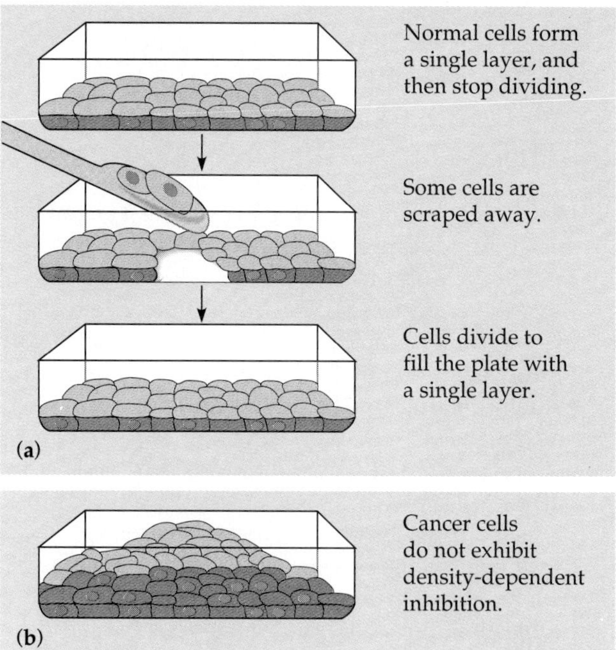

Normal cells form
a single layer, and
then stop dividing.
Some cells are
scraped away.
Cells divide to
fill the plate with
a single layer.
(a)
Cancer cells
do not exhibit
density-dependent
inhibition.
(b)

Figure 11.14
Density-dependent inhibition of cell division. (**a**) When grown in cell culture, normal cells will multiply only until they form a single layer. The availability of nutrients, growth factors, and substratum for attachments limits the density of the cell population. If some cells are removed, those at the border will divide until the gap is filled with a single layer. (**b**) In contrast, cancer cells usually continue to grow after they are crowded, often piling up. (Individual cells are shown disproportionately large in this figure.)

late interphase and reaches its highest concentration during mitosis, only to disappear by the end of mitosis. When MPF reaches a certain threshold concentration in a G_2 cell, prophase commences, and the cell goes on to divide. The high concentration of MPF during mitosis is required for the cell to progress through the mitotic stages, and the decline of MPF signals the end of mitosis and the transition to G_1 of the cell cycle's next interphase. Like most breakthroughs in science, the discovery of MPF raised new questions. How does MPF induce the changes that occur in mitosis? What causes the cyclical change in MPF concentration? There has been considerable progress on both of these problems.

MPF induces mitosis by acting as an enzyme. Specifically, MPF belongs to a family of enzymes called protein kinases. A **protein kinase** is an enzyme that activates other proteins, including other enzymes, by catalyzing the transfer of a phosphate group from ATP to each target protein. Some of these targets are *themselves* protein kinases, which phosphorylate still other proteins in a cascade of activation steps (Figure 11.15). This enables one master switch—MPF, in the case of the cell cycle—to cause many different changes

METHODS: CELL CULTURE

Many types of animal and plant cells can be removed from the organism and cultured in an artificial environment. This makes it possible to study the cell cycle and many other activities of cells in controlled conditions.

In the experiment illustrated here, we test the effect of a specific growth factor on the ability of human fibroblast cells to divide. Fibroblasts are the cells of connective tissue that secrete collagen, the protein making up the extracellular fibers that strengthen our tissues and organs (see Chapter 5 and Figure 36.4). Our source of fibroblasts is a small sample of tissue removed during a biopsy. Two steps are required to isolate these cells. First, small pieces of connective tissue are dissected from the tissue sample. Second, specific enzymes are added to digest the collagen fibers and other components of the extracellular matrix, resulting in a suspension of free fibroblast cells. The isolated fibroblasts are placed in a culture vessel, a flat flask that sits on its side. The cells adhere to the glass on the inside of the vessel, and they are bathed in a liquid culture medium of known composition. It is essential to sterilize the vessels and the culture media before use, in order to kill microorganisms that would contaminate the culture. The culture medium is a complex recipe of glucose, amino acids, salts, and antibiotics (as a further precaution to inhibit bacterial growth). The cultures are incubated at an optimal temperature, 37°C for human fibroblast cultures.

In our experiment, some culture vessels contain only this basic medium, while others contain the basic medium plus a protein called platelet-derived growth factor, or PDGF. In the intact animal, blood cells called platelets release this growth factor at the site of an injury, stimulating nearby fibroblasts to proliferate. This is an important step in the healing of wounds. Our experiment confirms that PDGF stimulates the cell division of cultured fibroblasts.

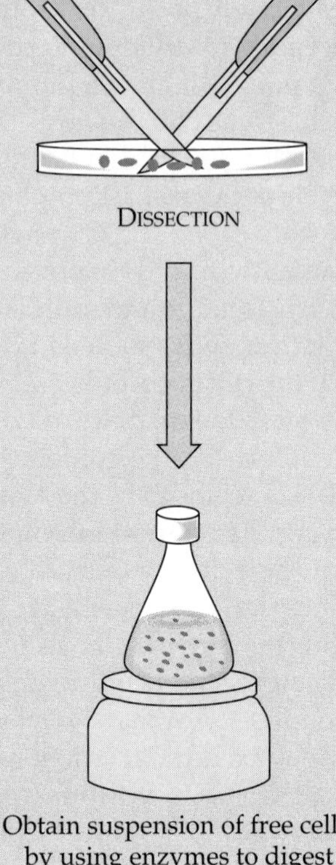
DISSECTION
Obtain suspension of free cells by using enzymes to digest extracellular matrix.

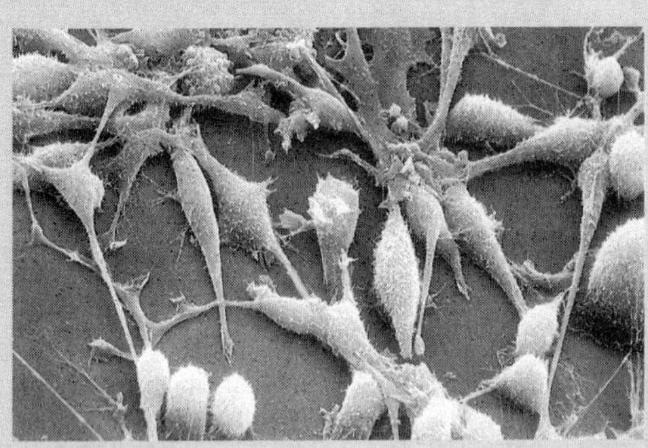

SEM of cultured fibroblasts

10 µm

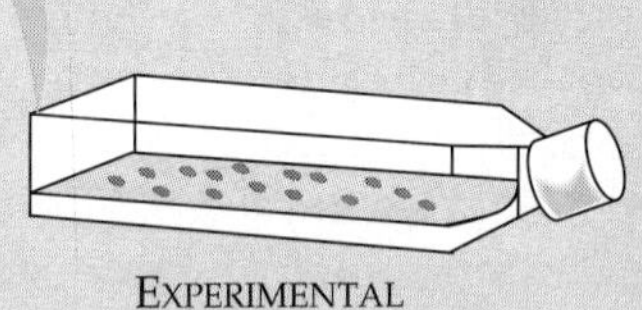

EXPERIMENTAL CELL CULTURE

Basic medium plus PDGF: Cells proliferate.

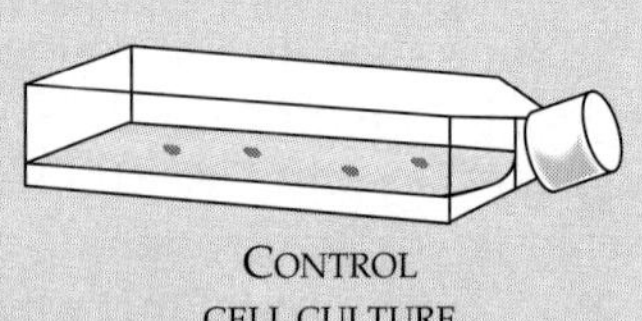

CONTROL CELL CULTURE

Basic medium minus PDGF: Cells fail to divide.

throughout the cell. For example, activation of MPF leads to phosphorylation of certain chromatin proteins, causing chromosomes to condense during prophase. And during prometaphase, the nuclear envelope disperses when some of its membrane proteins are phosphorylated. Active MPF is probably required for the correct performance of the entire mitotic sequence.

We need to learn a little more about the structure of MPF to understand its rhythmic change in concentration during the cell cycle. Active MPF actually consists of two associated proteins. One is called ***cdc2;*** it is one

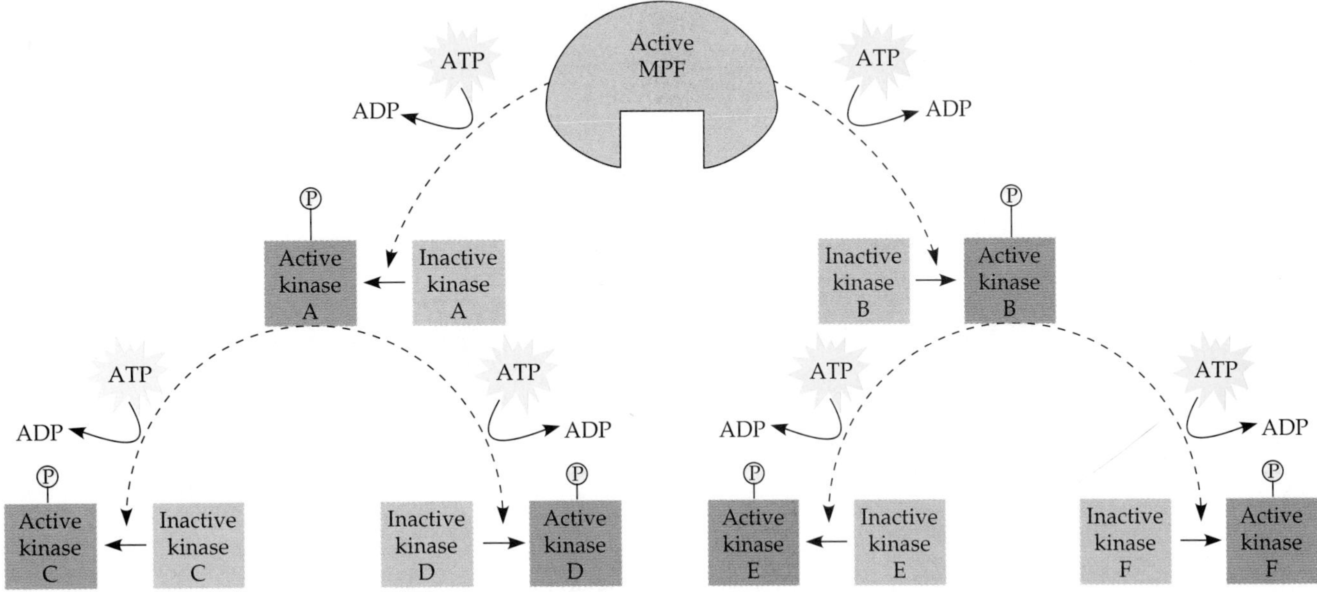

Figure 11.15
MPF: triggering a chain reaction of protein activations. MPF switches on mitosis. It is a protein kinase, an enzyme that activates other proteins by phosphorylating them (ATP provides the phosphate). Some of the proteins activated by MPF are themselves protein kinases, and they activate additional protein kinases. Various kinases also activate proteins that function directly in the mechanics of mitosis. This chain reaction amplifies and diversifies the impact of MPF on the whole cell.

of several cell-division control proteins that have been discovered in cells. The *cdc2* component of MPF is the protein kinase, the enzyme responsible for the transition from interphase to M phase. But *cdc2* is inactive unless attached to the second component of MPF, the protein **cyclin.** Although the *cdc2* concentration is constant throughout the cell cycle, the cyclin concentration fluctuates rhythmically (hence its name). Cyclin is actually produced at a uniform rate throughout the cell cycle, but near the end of mitosis it is destroyed by an enzyme. Continuing cyclin synthesis raises the concentration again during interphase. As the cyclin concentration in the cell rises or falls, the amount of active MPF changes (Figure 11.16). During interphase, newly synthesized cyclin binds to *cdc2* to form active MPF, and mitosis begins. But in an indirect way, this cyclin commits suicide: One of the enzymes activated by MPF is the one that destroys cyclin. The destruction of cyclin, in turn, causes the decline in active MPF at the end of mitosis. These elaborate molecular interactions function as a mitotic clock that keeps the sequential changes in a dividing cell on schedule.

Let's review control of the cell cycle. The G_1 phase of the cycle is the most variable phase, both in duration and in the variety of external and internal controls over cell division. Nutritional status, growth factors, the density of the cell population, and the developmental state of the cell all affect the length of G_1 and whether the cell will pass the restriction point and divide. A cell that does not pass the restriction point will diverge from the cell cycle, temporarily or permanently, as a nondividing G_0 cell. If all other conditions favor cell division, the cell will proceed through the restriction point when it grows enough to achieve a certain volume-to-genome ratio. This cell is now irreversibly committed to divide. It duplicates its chromosomes in the S phase and continues to grow in the G_2 phase, the last period of interphase. The transition from interphase to mitosis (M phase) requires a threshold concentration of MPF, the molecular clock that synchronizes the steps of cell reproduction. Thus, the cell cycle has a "divide/don't-divide" decision point (restriction point) and a control system that choreographs division once the cell is committed to reproduce.

ABNORMAL CELL DIVISION: CANCER CELLS

Cancer cells do not respond normally to the body's control mechanisms. They divide excessively and invade other tissues. If unchecked, they can kill the whole organism.

By studying cancer cells in culture, researchers have learned that they do not heed the normal signals that stop growth. In particular, they ignore density-dependent inhibition when growing in culture (see Figure 11.14b). The cultured cells continue to multiply even after contacting one another, piling up until the nutrients in the growth medium are exhausted.

There are other important differences between nor-

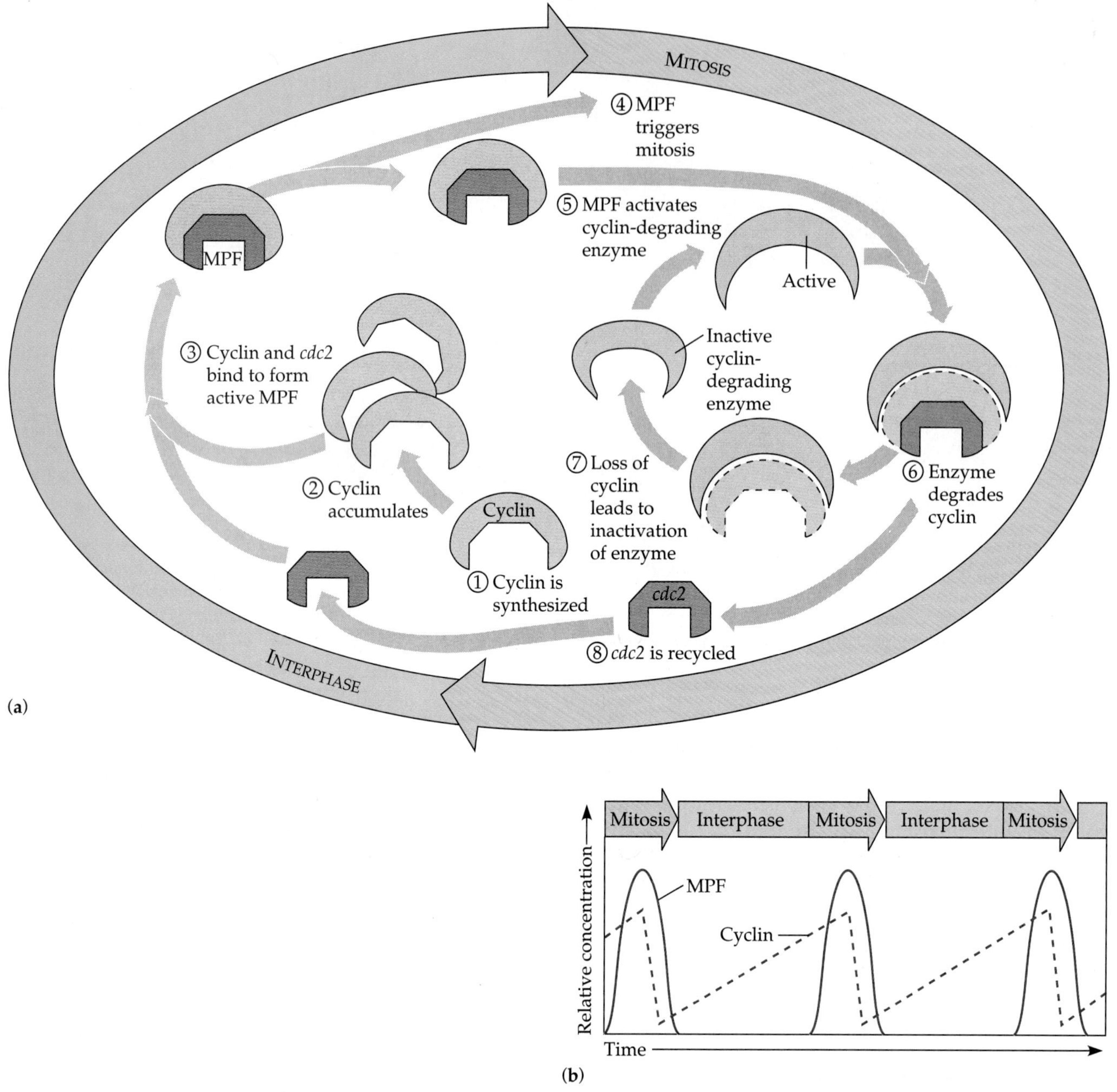

Mitosis
④ MPF triggers mitosis
⑤ MPF activates cyclin-degrading enzyme
Active
MPF
③ Cyclin and cdc2 bind to form active MPF
Inactive cyclin-degrading enzyme
② Cyclin accumulates
Cyclin
⑦ Loss of cyclin leads to inactivation of enzyme
⑥ Enzyme degrades cyclin
① Cyclin is synthesized
cdc2
⑧ cdc2 is recycled
Interphase
(a)
Mitosis
Interphase
Mitosis
Interphase
Mitosis
MPF
Cyclin
Relative concentration
Time
(b)

Figure 11.16
The mitotic clock. (**a**) In a cycle that times key events in cell division, the activity of MPF, which triggers mitosis, fluctuates with a rhythmic change in cyclin concentration. ① Cyclin is synthesized throughout the cell cycle and ② accumulates during early interphase. ③ The cyclin associates with another protein, *cdc2,* to form active MPF. ④ MPF functions as a protein kinase, triggering the activation of numerous proteins that facilitate mitosis. ⑤ One of the proteins activated by MPF is an enzyme that ⑥ brings MPF activity to a halt by degrading cyclin. ⑦ Since the cyclin-degrading enzyme is only active in the presence of MPF, it ceases to function as it destroys cyclin. ⑧ The *cdc2* component of MPF is recycled, and, back to step ①, synthesis of new cyclin leads to another peak of MPF activity. (**b**) The graph relates these peaks of MPF to the cyclin rhythm.

mal cells and cancer cells that reflect derangements of the cell cycle. If and when they stop dividing, cancer cells seem to do so at random points in the cycle, rather than just at the restriction point of G_1. Moreover, in culture, cancer cells can go on dividing indefinitely, if they have a continual supply of nutrients, and thus are said to be "immortal." A striking example is a cell line that has been reproducing in culture since 1951. (Cells of this line are called HeLa cells because their original source was a tumor removed from a woman named Henrietta Lacks.) By contrast, nearly all normal mammalian cells growing in culture divide only about 20 to 50 times before they stop dividing, age, and die.

The abnormal behavior of cancer cells can be catastrophic when it occurs in the body. The potential problem begins when a single cell in a tissue is trans-

Figure 11.17
The growth and metastasis of a malignant breast tumor. The cells of malignant (cancerous) tumors grow in an uncontrolled way and can spread to neighboring tissues and, via the circulatory system, to other parts of the body. The spread of cancer cells beyond their original sites is called metastasis.

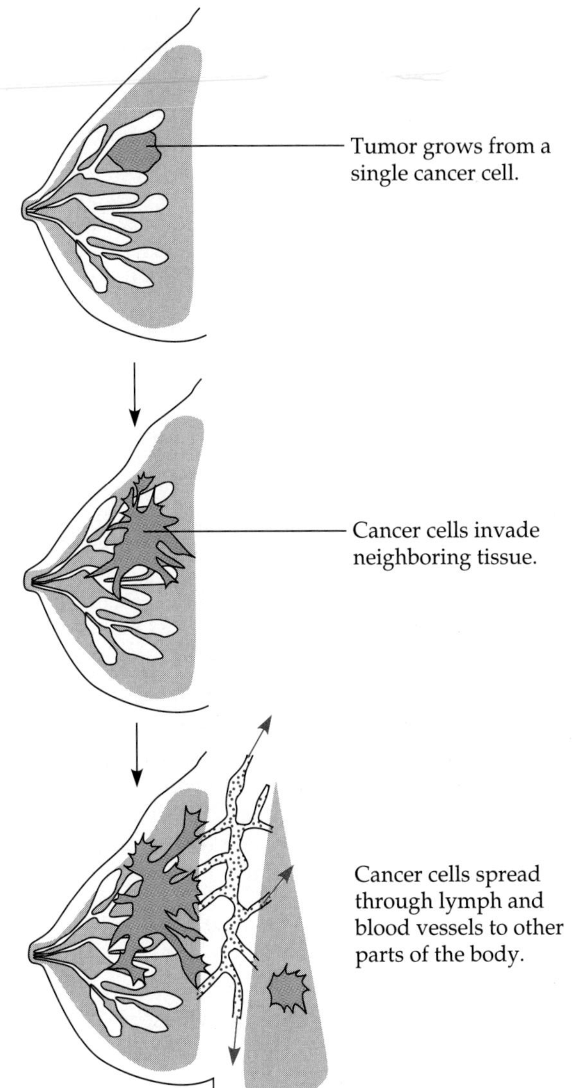

Tumor grows from a single cancer cell.
Cancer cells invade neighboring tissue.
Cancer cells spread through lymph and blood vessels to other parts of the body.

formed, **transformation** being the term for the conversion of a normal cell to a cancer cell. The body's defense system—the immune system—normally destroys these insurgent cells. (The immune system will be discussed in detail in Chapter 39.) However, if the cancer cell somehow evades destruction, it may proliferate to form a **tumor,** a mass of cancer cells within an otherwise normal tissue (Figure 11.17). If the cells remain at this original site, the lump is called a **benign tumor.** Benign tumors may not cause serious problems, and they can be completely removed by surgery. In contrast, a **malignant tumor** spreads to other parts of the body. An individual with a malignant tumor is said to have cancer.

The cells of malignant tumors are abnormal in many ways besides their lack of self-control over cell division. They may have unusual numbers of chromosomes. Their metabolism may be deranged, and they cease to function in any constructive way. Also, due to abnormal changes of the cells' surfaces, they lose their attachments to neighboring cells and the extracellular substratum. This enables the cancer cells to spread into other tissues surrounding the original tumor. Cancer cells may also separate from the tumor and enter the blood and lymph vessels of the circulatory system. These cells can invade other parts of the body and proliferate to form more tumors. This spread of cancer cells beyond their original sites is called **metastasis.** If a tumor metastasizes, it is usually treated with high-energy radiation and poisonous chemicals that are especially harmful to actively dividing cells.

Researchers are beginning to understand how a normal cell is transformed into a cancer cell. Michael Bishop and Harold Varmus, interviewed at the beginning of this unit, won the Nobel Prize for demonstrating that genetic alterations can cause cancer. (We will consider the role of DNA in cancer in detail in Chapter 18.) But knowledge of how changes in the genome lead to the various abnormalities of cancer cells is still rudimentary. Perhaps the reason we understand so little about cancer cells is that there are still many unanswered questions about how normal cells work. The cell, life's basic unit of structure and function, holds enough secrets to engage researchers well into the future.

* * *

With this chapter on the reproduction of cells, we have built a bridge to the next unit, which features genes and their role in inheritance.

STUDY OUTLINE

1. The perpetuation of life has its basis in cell division. Unicellular organisms reproduce by cell division. Multicellular organisms depend on cell division for development, growth, and repair.
2. The ability of DNA to replicate itself is essential to the production of genetically equivalent daughter cells.

Bacterial Reproduction (p. 222)

1. Prokaryotes undergo binary fission, in which the cell splits in two after replication of the single chromosome.

An Introduction to Eukaryotic Chromosomes (pp. 223–224)

1. Eukaryotic cell division consists of mitosis (division of

the nucleus) and cytokinesis (division of the cytoplasm).

2. The grouping of genes in chromosomes makes it possible for a eukaryotic cell to replace and distribute an enormous number of genes.
3. Chromosomes are composed of chromatin, a threadlike complex of DNA and protein that becomes more condensed during mitosis.
4. When chromosomes replicate, they form identical sister chromatids joined by a centromere. These chromatids separate during mitosis, thereby becoming the chromosomes of the new daughter cells.

The Cell Cycle: An Overview (p. 224)

1. Mitosis and cytokinesis make up the M (mitotic) phase of the cell cycle, a sequence of events in the life of dividing cells.
2. Between divisions, cells are in interphase, an active period of growth and metabolism composed of the G_1, S, and G_2 phases. The cell grows throughout interphase, but DNA is replicated only during the S (synthesis) phase.

The Mechanics of Cell Division: A Closer Look (pp. 225–232)

1. The M phase is a dynamic continuum of sequential changes, often described by the five stages of prophase, prometaphase, metaphase, anaphase, and telophase.
2. The mitotic spindle is a complex of microtubules that orchestrates chromosome movement during mitosis. During prophase, the spindle begins to form from the centrosome, a region near the nucleus associated with centrioles in animal cells.
3. The spindle includes kenetochore microtubules, which attach to the kinetochores of chromatids.
4. During prometaphase, the kinetochore microtubules move all the chromosomes to the metaphase plate. When the chromosomes are completely aligned, the cell is in metaphase.
5. During anaphase, sister chromatids separate and move toward opposite poles of the cell. Microtubules disassemble at their kinetochore ends, and the kinetochore somehow moves along the shortening microtubules. At the same time, a sliding of nonkinetochore microtubules elongates the whole cell in the polar axis.
6. During telophase, daughter nuclei form at opposite ends of the dividing cell. In most cases, mitosis is followed by cytokinesis, the formation of cleavage furrows in animals and cell plates in plants.

The Control of Cell Division (pp. 230–235)

1. The rate, timing, and location of cell division in an organism are crucial for normal growth, development, and maintenance.
2. Cell culture is a powerful method for studying the cell cycle. It has enabled researchers to determine the nutrients, growth factors, and other requirements for cell division.
3. The restriction point, near the end of the G_1 phase, is a critical time in the cell cycle. At this point, the cell either becomes a nondividing (G_0) cell or commits to DNA synthesis and cell division. The cell's volume-to-genome ratio must reach a certain threshold value for the cell to pass the restriction point, and other environmental and developmental conditions must also favor cell division.
4. A protein complex named MPF triggers mitosis and activates other proteins that function in cell division. Cyclin, one component of MPF, undergoes rhythmic changes in concentration that help sequence the events of the cell cycle. Cyclin must associate with another protein, *cdc2,* to form active MPF.

Abnormal Cell Division: Cancer Cells (pp. 235–237)

1. Cancer cells elude normal regulation and divide out of control, forming tumors.
2. Malignant tumors are those that spread to surrounding tissues or export cancer cells via the circulatory system to other parts of the body. The latter process is called metastasis.

SELF-QUIZ

1. Which process is *not* associated with the reproduction of bacteria?
   a. replication of DNA
   b. binary fission
   c. mitosis
   d. synthesis of new cell wall material
   e. elongation of the parent cell
2. Through a microscope, you can see a cell plate beginning to develop across the middle of the cell and nuclei reforming at opposite poles of the cell. This cell is most likely a (an)
   a. animal cell in the process of cytokinesis
   b. plant cell in the process of cytokinesis
   c. animal cell in the S phase of the cell cycle
   d. bacterial cell dividing
   e. plant cell in metaphase
3. DNA replicates during
   a. G_1 phase
   b. S phase
   c. G_2 phase
   d. M phase
   e. cytokinesis
4. If a specialized cell no longer divides, it is generally locked in which stage of the cell cycle?
   a. S
   b. G_1
   c. G_2
   d. M
   e. G_0
5. In a typical cell cycle, cytokinesis generally overlaps in time with which stage?
   a. S phase
   b. prophase
   c. telophase
   d. anaphase
   e. metaphase
6. A chromosome is in its most extended (least condensed) form during
   a. interphase
   b. prophase
   c. metaphase
   d. anaphase
   e. telophase

7. A particular cell has half as much DNA as some of the other cells in a mitotically active tissue. The cell in question could be in
   a. G_1
   b. G_2
   c. prophase
   d. metaphase
   e. anaphase

8. The decline of active MPF at the end of mitosis is caused by
   a. destruction of *cdc2*
   b. decreased synthesis of cyclin
   c. destruction of cyclin
   d. synthesis of DNA
   e. an increase in the cell's volume-to-genome ratio

9. One difference between a cancer cell and a normal cell is that
   a. the cancer cell is unable to synthesize DNA
   b. the cell cycle of the cancer cell is arrested at the S phase
   c. cancer cells continue to divide even when they are tightly packed
   d. cancer cells cannot function properly because they suffer from density-dependent inhibition
   e. cancer cells are always in the M phase of the cell cycle

10. In the following light micrograph of dividing cells near the tip of an onion root, identify a cell in interphase, prophase, metaphase, and anaphase. Describe the major events occurring at each stage.

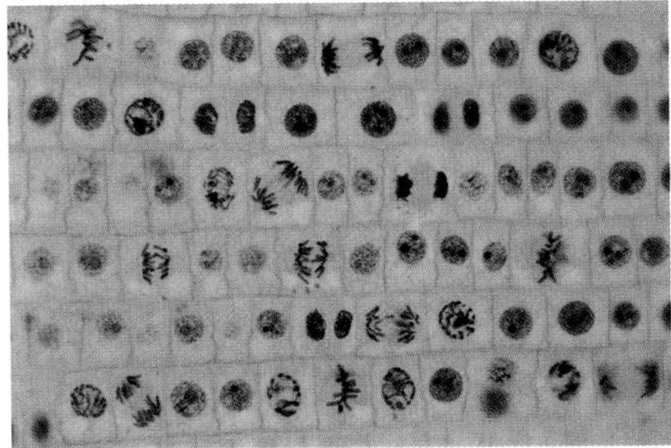

25 µm

CHALLENGE QUESTIONS

1. When a population of cells is examined with a microscope, the percentage of the cells in the M phase is called the mitotic index. The greater the proportion of cells that are dividing, the higher the mitotic index. In a particular study, cells from a cell culture are spread on a slide, preserved and stained, and then inspected in the microscope. A hundred cells are examined: 9 cells are in prophase; 5 cells are in metaphase; 2 cells are in anaphase; 4 cells are in telophase; the remainder, 80 cells, are in interphase. Answer the following questions:
   a. What is the mitotic index for this cell culture?
   b. The average duration for the cell cycle in this culture is known to be 20 hours. What is the duration of interphase? Of metaphase?
   c. Going back to the living culture of these cells, the average quantity of DNA per cell is measured. Of the cells in interphase, 50% contain 10 ng (1 nanogram = 10^{-9} g) of DNA per cell; 20% contain 20 ng DNA per cell; the remaining 30% of the interphase cells have amounts of DNA between 10 and 20 ng. Based on these data, determine the duration of the G_1, S, and G_2 portions of the cell cycle.

2. About a day after a human egg is fertilized by a sperm cell, the zygote (fertilized egg) divides for the first time. The two daughter cells usually stick together, and their repeated cell divisions give rise to a multicellular embryo. On rare occasions, however, the two daughter cells formed by the first division of the zygote separate. Each of these cells can go on to form a normal embryo—not a half-embryo or an otherwise defective embryo. Based on what you have learned in this chapter, explain why these "monozygotic twins" are essentially genetically identical.

SCIENCE, TECHNOLOGY, AND SOCIETY

1. Imagine that the federal government has awarded you a $100,000 research grant from funds budgeted for cancer research. However, you do not work on cancer at all. Your research is very basic: You culture cells from mouse embryos in order to track the concentrations of key proteins through the normal cell cycle. A friend who cannot make the connection between cancer research and your work kids you about squandering tax dollars on dishes of mouse cells. How would you convince your friend that this is money well spent? Does public interest in how research funds are used threaten basic science? How does science benefit from skillful educators who can communicate the excitement of current scientific research to general audiences?

2. A biotechnology company has succeeded in extracting natural vanilla from cells grown in cell culture. Vanilla extracted from vanilla beans costs about $1200 per pound. Vanillin, a synthetic flavor used in many foods, costs about $7 per pound, but lacks the many trace compounds that give natural vanilla its complex flavor. Cell culture vanilla costs less than $200 per pound, which might make it economical for use in more foods. Can you think of other products that might be produced by means of cell culture? What would be the advantages and disadvantages of producing them this way?

FURTHER READING

Alberts, B., D. Bray, T. Lewis, M. Raff, K. Roberts, and J. D. Watson. *Molecular Biology of the Cell*, 2nd ed. New York: Garland, 1987. Chapter 13.

Becker, W. M., and D. W. Deamer. *The World of the Cell*, 2nd ed. Redwood City, CA: Benjamin/Cummings, 1991. Chapter 14.

Benditt, J. "Genetic Skeleton." *Scientific American*, July 1988. A short description of a model integrating the control of the cytoskeleton and the regulation of cell division.

McIntosh, J. R., and K. L. McDonald. "The Mitotic Spindle." *Scientific American*, October 1989.

Murray, A. W., and M. W. Kirschner. "What Controls the Cell Cycle." *Scientific American*, March 1991.

UNIT THREE The Gene

An Interview with David Suzuki

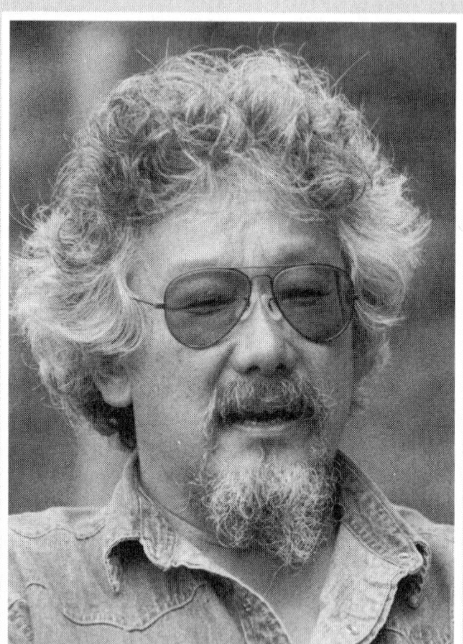

When I passed through Canadian customs at the Vancouver airport on my way to conduct this interview, the officer inquired about the nature of my business. I said that I had an appointment to interview David Suzuki. "Oh," the officer said, "everyone in Canada knows David Suzuki!" Dr. Suzuki, professor of genetics at the University of British Columbia, is the host of "The Nature of Things," *one of the most popular programs on Canadian television. He is also the author or co-author of numerous research articles and books, including the leading genetics textbook,* An Introduction to Genetic Analysis *(Freeman, 1989), and* Genethics: The Ethics of Engineering Life *(Harvard University Press, 1989). In this interview, Professor Suzuki discusses the special responsibilities of research scientists and science educators.*

How did your interest in biology and genetics evolve?

My father was the greatest inspiration and role model for me. It began right from my earliest memories, which were basically of camping and fishing. My father was never interested in material success, and he was regarded as a bit of a dreamer by his father. But he certainly inspired my great love of nature. When I was about nine, my mother made me a butterfly net out of mosquito netting and I started an insect collection, which was a big hobby of mine for years and years. But my great love of nature came from fishing and I wanted to be an ichthyologist. I met a man from the Royal Ontario Museum who was one of the great naturalists on staff there. He told me, "Look, this is a great hobby to have, but for God's sake don't go into it to make a living." I often think about him. Here was a man who had what I thought was one of the greatest jobs in the world, a curator in a museum, and he was discouraging me, advising me instead to go into medicine or something to make a living. That was very sad. I always tell young people to go where their heart leads because they'll be doing it for a long time. I went to Amherst College and basically I was a pre-med, but in my third year I took genetics because all honors students had to. Once I encountered genetics as an incredibly precise, elegant way of looking at the world, I fell madly in love with it. I guess working with fruit flies was an immediate joy because I'd always loved insects as a child.

You've written about how your experience teaching undergraduates influenced your own development. Can you give an example?

With introductory classes, not a year goes by when a student doesn't ask me a question about something I've never thought of before. It really brings you up short. You realize that students come in without all of the prejudices and all of the preconceptions, so they can ask a question that really makes you look at things in a different way.

The story of Martin Greenall is especially instructive. Martin was probably the best student I ever had in genetics at the University of British Columbia. He had organized evening seminars for interested students, and he asked if I would come and talk about the prospects of genetic engineering. It was very easy back in the sixties—you could talk about cloning and all of this science-fiction stuff and the students' eyes would pop open and they were very impressed. But at the end of it, Martin got up and said, "Well, you're so concerned about the misapplication of science, and you know about the history of Nazism and so on. The fact is that genetics hasn't had a great track record. So how do you justify continuing to work in an area that has a high probability of being misused?" I gave him a glib answer: "I'm just working with fruit flies; I have nothing to do with the application of science. I'm just looking for truth." Which is the answer that is generally given by most of my colleagues. And Martin just kept right on and he said, "Look, think of knowledge as a lake, a body of information, and everyone contributes to that lake of knowledge so its level rises. You can't tell whether you, Suzuki, studying genetic recombination in fruit flies, might contribute something to that body of knowledge that someone might draw on to apply. You can't escape responsibility by saying, 'Well I'm just searching for truth.' What you are doing is adding to the *totality* of knowledge." That was a very, very profound insight for me. This undergraduate was telling me that there is a greater responsibility that comes simply by participating in the scientific enterprise.

And that experience led to your interest in science ethics?

It touched off a whole questioning in me. I came through university right after the Soviet spacecraft Sputnik went up in 1957. In the years immediately following that event, a wave of fear swept through the western world when we realized that the Soviet Union was very powerful and very advanced in math and science. One result was that the United States set up NASA and began a massive program of training people in science. I received a big scholarship at the University of Chicago. And as graduate students, we were trained to believe that there was nothing beyond the inquiring mind of science. That the potential benefits of science were enormous. I believed it, and that's what I taught my students. It was only when Martin asked me about the other side of responsibility that I really had to look back at myself. It's one thing to look at cloning human beings and say, "Oh no, that's dangerous." But then to ask yourself what responsibility do I bear as a fruit fly geneticist—that began a long period of introspective questioning.

But my transformation really began when I started seriously doing television programs for a national audience in the mid-sixties. I got started in TV to raise the public awareness and support of science, because science in Canada has traditionally been very poorly funded. I thought that if the public understood why science was important, they would then give more money to the scientific community. But being in a public medium like television, I began to see there was a radically different perspective on science held by nonscientists. They weren't interested in whether or not science was well funded. They were interested in: How is your work going to affect my kids? How will it affect the quality of my life? Is somebody going to make money from what they learn about me? I realized there was a public out there that had a whole set of questions that I as a scientist had to confront. So while Martin raised the issue of my responsibility as a practicing scientist, working in the media opened my mind to a wider set of conditions and questions.

How do your colleagues in science react to your fame as a television celebrity?

When I started my work in television I received a tremendous amount of criticism. You have to remember that this was back in the sixties. I had hair down to my shoulders; I wore a headband. My colleagues were outraged that this "hippie" was speaking as a scientist. And not only speaking as a scientist, but even being critical of the scientific community. Very few of them would ever come and say it to my face; I would hear from my students who would go to parties and end up having to defend me. People have always felt I'm making a great deal of money, but anyone who knows how television is supported in Canada would know that that is not the case. It has always been my concern that I make only as much as I would make at the university. Whatever money I earn over that I use for my causes, which have been civil rights and environmental issues. So, sure, there's a lot of resentment, but since I don't spend much time in the scientific milieu anymore, I don't hear much of it.

Is an idealized view of science common in the general public?

I think there are misconceptions even held by the scientific community. The people practicing science are human beings, and as human beings they show the entire range of human foibles. There are, of course, noble people driven to try to improve the lot for humankind. There are also greedy people, selfish people—the whole range. There's enormous pressure on scientists now, especially in my area of genetics, to reap economic payoffs from biotechnology. All kinds of terrible things can result from economic and political pressure, and the scientific community has to face up to that.

So there's pressure for basic researchers to do research that promises some kind of practical application?

Yes. In Canada, because we have been so starved for funding, the government has enormous power to influence scientific priorities. Say you have a guy studying the basic molecular biology of cell division, and suddenly there is a high priority on doing cancer research. Of course he immediately says, "Oh well, I'm doing work on cancer because I'm looking at the mechanism of how cells divide," and everybody begins to rewrite grant applications this way. The tragedy is that few scientists have gotten up and said, "The very nature of science is that we don't know where our results will lead. If we knew where our research was going to lead, for God's sake, we would have all the solutions by now."

Look at the history of science. I think of Barbara McClintock. When I was at graduate school in the sixties, Barbara McClintock was a very eminent geneticist, a past president of the Genetics Society of America, a member of the National Academy [of Sciences]. We studied her work on jumping genes in corn. We always said, "This is something that is unique to corn and has nothing to do with mainstream genetics. But because she is so eminent we must study this as an example of genetics." Now, thirty years later, her ideas are at the base of much genetic manipulation,

and she belatedly won a Nobel Prize. Thirty years ago, nobody had any notion that her work would ever be mainstream. If she had tried to get a grant back then, I doubt that it would have had a high priority. So it's a sell-out of science to say, "Yes, we will do research that is mission-oriented or goal-oriented. Give us the money and we'll do it." You've got to be led by where your best minds find an interesting problem on the faith—a faith corroborated by history—that good scientists will make discoveries that are ultimately very important and useful.

You mentioned the potential for perversion of science, particularly genetics. Can you give some examples?
All human beings, including scientists, have prejudices that are determined very early by our culture and socioeconomic background. And we're very quick to leap on notions that seem to corroborate our own prejudices. In *The Mismeasure of Man,* Stephen Jay Gould gives a good example. It was taken for granted earlier in this century that females had an inability to do certain kinds of calculations. It was assumed that the ability to do those calculations resides in certain parts of the brain, and when they looked at that part of the brain, sure enough, the researchers concluded that those parts were smaller in females than males. What Gould did was to go back and re-examine the data, often the very same skulls. He showed that, in fact, there is no statistical difference between parts of the brains of women and men. The prejudices of the earlier researchers were reflected in the gathering and interpretation of the data. Every student should take that very seriously as a lesson.

One of the dangers today is that our students don't get any history of science; that's regarded as a frill. We assume that today's research hot-shots are smarter than anybody in the past. We always assume that someone who's dead couldn't possibly have been as smart as we are today. When I was a graduate student in 1961, we thought that DNA was arranged in short pieces about gene size, hooked together by protein linkers in a linear sequence. When I tell my students that, they fall off their chairs laughing because they know that DNA is a single molecule in a chromosome. And I tell them, "You're right; it does seem silly today. Well, you're not going to believe this, but what you think is a hot idea now will make your students fall off their chairs laughing in twenty years." They don't believe me. We don't remember any of the history, and we think what we are getting is truth, which means that we are going to make the same old mistakes. Look back at the great claims that were made early in the century that fueled the whole eugenics movement.

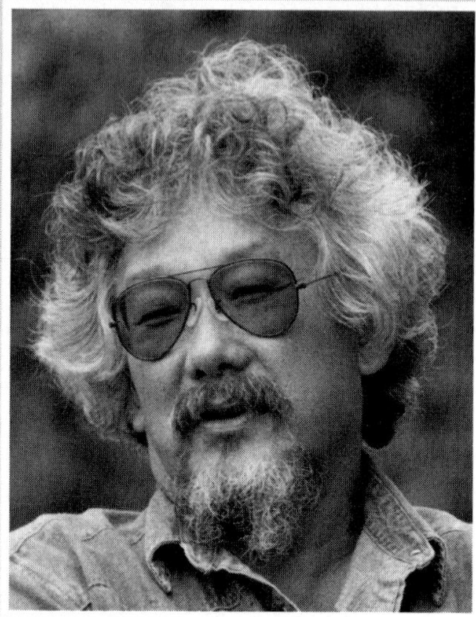

With the rediscovery of Mendel's laws and the study of genetic recombination and mutation, people thought that they really had their hands on the levers of life, that through this powerful science we could come to understand and be able to control human heredity. It was assumed at that time that drunkenness, syphilis and all those things were hereditary and they could be eliminated.

I hear the same kinds of claims being made again. And with the Human Genome Project, once again there is a sense that virtually every aspect of the human makeup is biologically determined. I find this horrifying because people are finding all kinds of correlations and assuming that correlations show causation. Just last year, a discovery was announced that there were two alleles of a gene for alcohol dehydrogenase. They showed that 80% of alcoholics have one form of this gene and 80% of nonalcoholics have the other form. That's a *correlation.* It was immediately assumed by the scientists themselves as well as the popular press that they had found the gene that causes alcoholism. I think that that's a danger we're going to run into again and again. Think about this. Suppose I perform an autopsy on 100 people who died of lung cancer and discover that 90% of them have brown-stained teeth and yellow fingers. That's a correlation. But it would be absolutely wrong to conclude that brown-stained teeth and yellow fingers *cause* lung cancer. Yet that's exactly what they are doing with this alcohol dehydrogenase allele.

With the Human Genome Project, as we begin to be able to sequence larger and larger amounts of DNA and have a complete profile of DNA fragments, of course we will be able to say, "Let's look at heart disease, let's look at stroke, let's look at various kinds of cancers and show whether we can correlate them with certain patterns of DNA fragments or sequences." I have no doubt that those correlations will be found. Of course the leap from finding a correlation to a cure is a long jump. But I think because we have so much computer power now, we'll start looking for other correlations—with alcoholism, with people on welfare, with criminals. I have no doubt that correlations will be found, and you can be pretty sure that it will be immediately assumed this is a causal relationship. Claims of biological determinism will continue to fulfill all of our preconceptions.

Do you see other ethical issues associated with the Human Genome Project?
There is a great deal of interest in what are called ethnic diseases. It arose from the study of sickle cell anemia, Tay-Sachs disease, and cystic fibrosis, which tend to be associated with certain ethnic backgrounds. In principle, it sounds good: If you can show that Tay-Sachs disease is found primarily in Ashkenazi Jews and you can focus prenatal diagnosis on that group, you can avoid terrible hereditary conditions. The problem, of course, is that the information can be used in many other ways. Employers can use it as a reason not to employ people who carry certain genes, for instance. I was particularly struck by an article titled "Ethnic Weapons" in the December 1990 issue of *Military Review.* The author, Carl Larson, a Swedish human geneticist, pointed out that there are a number of different allelic makeups of ethnic groups. For example, Asians don't make the enzyme that digests milk sugar. So if we consume large quantities of milk, Asians tend to get very sick with diarrhea. That's an ethnic complaint. What Larson pointed out was that if you could catalog a number of ethnic differences, you should be able to construct a weapon with total ethnic

specificity. An ethnic weapon is a very powerful way of selecting your target so you don't kill off your own troops along with the enemy. I don't think that it's just paranoid to think that ethnic weapons are of great interest to the military. If you look at military history, there is nothing that lies beyond the capacity of the military mind to consider using for war, and I think biologists better think very, very seriously about that. Again, it's something that most scientists don't like to face.

Does the potential for gene therapy also raise ethical issues?

There are always ethical issues. My own feeling when we wrote *Genethics* was that we really ought to have a moratorium on all the experimentation in humans until there has been a very broad public airing of the issues. But the fact is, we're already into gene therapy. It's not hypothetical; it's already being used. They're using it for people with SCIDS, severe combined immunodeficiency syndrome, and I don't doubt that it's going to be used in a number of other hereditary defects. In *Genethics*, what we said is that we see the coming of somatic cell manipulation, but that there should be an absolute prohibition against manipulating germ cells—gametes and gamete-producing cells. In this way, you confine the effects of manipulation to one generation. When you start manipulating the germ line, then you're talking about passing the consequences on to future generations. And I think we ought to be very careful about that. But today this germ-cell/somatic-cell distinction is being clouded as well.

Do you think there's enough emphasis on such ethical issues in science courses?

Definitely not. It's always regarded as a frill, or if we're forced to, maybe we say a little bit in a couple of lectures. But scientists do it very reluctantly, because there's never enough time in a course to cover even the basic material. I was shocked to discover when I became a professor that I had been taught a history of science that worshipped the hero in science, the bold intellect that makes great discoveries. I had been taught a version of science that simply didn't correlate with the reality of what I saw around me. I think we do a great disservice to our students by perpetuating a mythical idea of what science is like and in keeping them ignorant about the enormous abuses that have resulted from it. Of course, we do society a tremendous disservice as well.

You have argued for what you call a "new mythology" to help resolve the inevitable conflicts between science and human values. What "new mythology" do you envision?

Albert Einstein was asked by a good friend of his, "Do you think that absolutely everything in the universe can be explained through science?" And Einstein's answer was that it could be done, but it would be a meaningless explanation. A physicist could describe a Beethoven symphony accurately as simply variations in wave pressure, but that would tell you nothing about what Beethoven's work really is. What Einstein was saying is that a straight description misses the deeper meaning. If people involved in the Human Genome Project think that by logging the entire three billion letters in the genome they are suddenly going to have an understanding of what it is to be human, they are missing the boat. They'll have a set of linear instructions, but that blueprint has to operate in a four-dimensional world, and we know very little about that.

In the late eighties, a group of prominent scientists signed a declaration that says, in effect, that we as scientists have experienced reverence and awe in looking out at the universe and that we must now come to regard the Earth with reverence, as a sacred place to be treated with greater care. To me that was an astonishing document. What these scientists were saying is that without a sense of reverence and awe, science is missing something and invariably becomes destructive. The traditional scientific paradigm is that a scientist looks at objects from a distance. We try to be objective and detached because if we feel an emotional involvement in what we are examining, it colors the way we look at it. The problem with that idea is that by distancing yourself from nature, you no longer care.

This came through to me very strikingly when I was involved in a fight to stop logging on the Queen Charlotte Islands of British Columbia. When we were up there filming, a professor from the Botany Department at the University of British Columbia who had made his world reputation studying a group of plants that are found in the Charlottes and nowhere else ran into the crew. When he found out what they were doing, he said I had lost all credibility because I was involved in the logging controversy, which is a social issue. He didn't believe I deserve to be called a professor. When I heard that, of course I was mad as hell, but much more strongly, I felt very, very sad. Here was a man who undoubtedly loved those plants—that's why he went into botany—who had made his reputation on these plants, yet he wouldn't lift a finger to save them because he didn't want to sully his credibility. I think that's one of the great problems that we face in science: By virtue of looking at the world from a distance, we no longer have any sense of involvement.

If you look at aboriginal people around the world, they would never think of distancing themselves from nature. They are intimately immersed in their environment, and by being embedded in it, they have come to understand that they are related to all living things. They speak of the frogs and the whales as their brothers and sisters, and they really mean it. And now E. O. Wilson is saying that we must come to know our *kin*, the other animals and plants with whom we share the world, who are related to us, who share our DNA. I find it interesting that through a sense of total involvement and immersion in the world, aboriginal people have a sense of wonder and awe and kinship with the rest of life that some of the leading ecologists and other scientists are only now beginning to see. We must begin to have greater respect for aboriginal perspectives and realize that these two very different ways of knowing have things to tell each other.

12 MEIOSIS AND SEXUAL LIFE CYCLES

GENES, DNA, AND CHROMOSOMES: A BRIEF ORIENTATION

SEXUAL AND ASEXUAL REPRODUCTION: A COMPARISON

AN INTRODUCTION TO SEXUAL LIFE CYCLES: THE HUMAN EXAMPLE

THE VARIETY OF SEXUAL LIFE CYCLES

MEIOSIS: A CLOSER LOOK

A COMPARISON OF MITOSIS AND MEIOSIS

SEXUAL SOURCES OF GENETIC VARIATION

GENETIC VARIATION AND EVOLUTION

Figure 12.1
Cattle breeders of ancient Africa. About 5000 years ago, the artist of this rock painting portrayed North African people with their various breeds of cattle. Thousands of such paintings document a keen awareness of variations in the physical characteristics of domesticated animals. Early farmers propagated certain variations by selectively breeding livestock and crops. In this unit, you will learn about the science of genetics, the modern manifestation of human curiosity and pragmatic interest in heredity and variation.

Life's most exclusive distinction is the ability of organisms to reproduce their kind. Like begets like. Only oak trees produce oaks, and only condors can make more condors. Furthermore, offspring resemble their parents more than they do less closely related individuals of the same species. This continuity of traits from one generation to the next is called **heredity** (L. *heres,* "heir"). Along with inherited similarity, there is also **variation:** Offspring exhibit individuality, differing somewhat in appearance from parents and siblings. These observations have been exploited for the thousands of years that people have bred plants and animals (Figure 12.1). Curiosity about human similarities and differences is just as ancient. But the mechanisms of heredity and variation eluded biologists until the development of genetics in this century. **Genetics,** the scientific study of heredity and variation, is the subject of this unit. You will learn how biologists are answering questions about life that have endured for centuries; how discoveries in genetics are catalyzing progress in every other biological field, including physiology, evolutionary biology, ecology, and the behavioral sciences; how modern genetics is revolutionizing the pharmaceutical industry and other technologies; and how all these achievements are affecting society and raising new philosophical and ethical questions.

GENES, DNA, AND CHROMOSOMES: A BRIEF ORIENTATION

Friends sometimes tell my daughter that she has her father's freckles, but *I* still *have* mine. Parents do not, in any literal sense, give their children freckles, eyes, hair, or any other traits. What, then, actually *is* inherited? Parents endow their offspring with coded information in the form of hereditary units called **genes.** Our genomes consist of the tens of thousands of genes we inherit from our mothers and fathers. Our genetic link to our parents accounts for family resemblance. Among my daughter's complement of genes is the one for freckles, which she inherited from her father. Our genes program the emergence of specific traits as we develop from fertilized eggs into adults.

Genes are made of DNA. You learned in Chapters 1 and 5 that DNA is a polymer of four different kinds of monomers called nucleotides. Inherited information is passed on in the form of each gene's specific sequence of nucleotides, much as printed information is communicated in the form of meaningful sequences of letters. Language is abstract. The brain translates words and sentences into mental images and ideas; for example, the object you imagine when you read "apple" looks nothing like the word itself. Analogously, cells translate genetic "sentences" into freckles and other features that bear no resemblance to genes. Most genes program cells to synthesize specific enzymes and other proteins, and it is the action of these proteins that produce an organism's inherited traits.

Inheritance has its chemical basis in the precise replication of DNA, which produces copies of genes that can be passed along from parents to offspring. The cellular vehicles for these genes are sperm and ova (unfertilized eggs). When a sperm cell unites with an ovum (a single egg), the genes from the two parents are combined in the nucleus of the fertilized egg. Biological order based on heritable programs in the form of DNA is one of the unifying themes of biology.

The DNA of a eukaryotic cell is subdivided into chromosomes located within the nucleus. Each species of life has a characteristic number of chromosomes. For example, humans have 46 chromosomes (except in their reproductive cells). Each chromosome consists of a single DNA molecule that is much longer than the chromosome itself. Along with various kinds of proteins, the DNA is elaborately folded and coiled, making up the structure of the chromosome. One chromosome represents hundreds or thousands of genes, each of which is a specific region of the DNA molecule. A gene's specific location along the length of a chromosome is called the gene's **locus.**

The physical mechanism of inheritance—the actual transmission of genes from parents to offspring—depends on the behavior of chromosomes. Our genetic endowment consists of whatever genes happened to be carried by the chromosomes we acquired from our parents. In this chapter, we begin our study of genetics by learning how sexual reproduction passes chromosomes from parents to offspring.

SEXUAL AND ASEXUAL REPRODUCTION: A COMPARISON

Strictly speaking, "Like begets like" really applies only to organisms that reproduce asexually. In **asexual reproduction,** a single individual is the sole parent and passes on all its genes to its offspring. For example, one-celled organisms can reproduce asexually by cell division, in which DNA is copied and allocated equally to two daughter cells (see Chapter 11). The offspring are exact copies of the parent. Some multicellular organisms are also capable of reproducing asexually. *Hydra,* a relative of the jellyfish, can reproduce by budding (Figure 12.2). A new individual begins as a mass of dividing cells growing on the side of the parent. The mass develops into a small *Hydra,* the bud, which eventually detaches from the parent to take up life on its own. Since the cells of the bud were derived by mitosis in the parent, the "chip off the old block" is almost genetically identical to its parent. Any genetic differences are due to relatively rare changes in the DNA called **mutations,** which will be discussed in Chapter 16. An individual that reproduces asexually gives rise to a **clone,** a group of genetically identical individuals.

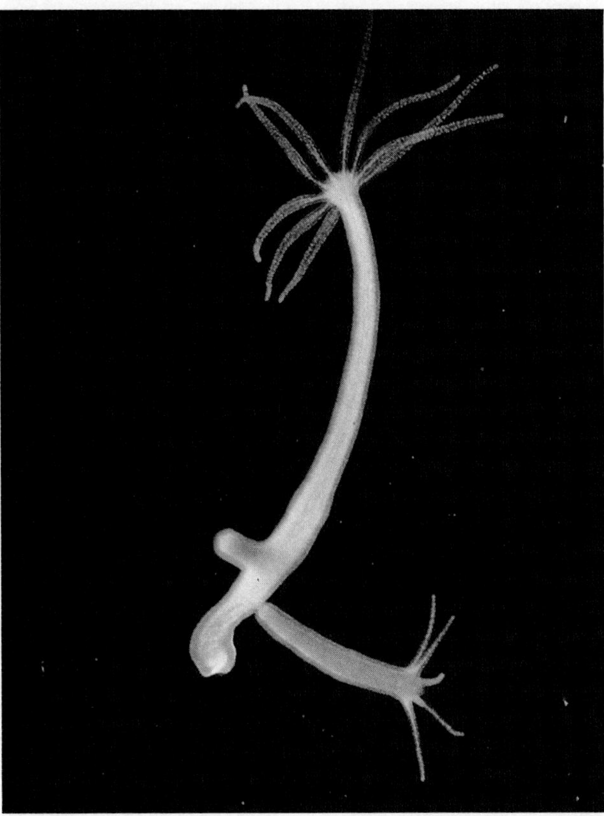

1 mm

Figure 12.2
The asexual reproduction of *Hydra.* This relatively simple multicellular animal reproduces by budding. The bud, a localized mass of mitotically dividing cells, develops into a small *Hydra,* which detaches from the parent (LM).

Compared to asexual reproduction, **sexual reproduction** usually results in greater variation; two parents give rise to offspring that have unique combinations of genes inherited from both parents. In contrast to a clone, offspring of sexual reproduction vary genetically from their siblings and both parents (Figure 12.3). What mechanisms generate this genetic variation? The key is the activity of chromosomes during the sexual life cycle.

Figure 12.3
Two families. Two sets of parents make up the top row, but the photos are randomly arranged. Each couple has two children represented among the four photos in the bottom row, also randomly arranged. (All individuals were photographed at about the same age; these are senior pictures from high school annuals.) Can you match the offspring with their parents? (See the bottom of the page for the answers.*) "Like begets like" in the general sense of family resemblance, but notice that each offspring is unique in her or his appearance, differing from parents and siblings. This genetic variation is an important consequence of sexual reproduction.

AN INTRODUCTION TO SEXUAL LIFE CYCLES: THE HUMAN EXAMPLE

The term **life cycle** refers to the generation-to-generation sequence of stages in the reproductive history of an organism, from conception to production of its own offspring. In this section, we track the behavior of chromosomes through the human life cycle. In humans, each **somatic cell**—any cell other than a sperm or egg cell—has 46 chromosomes (see Chapter 11). With a light microscope, chromosomes can be distinguished from one another by their appearance. The size and position of their centromeres differ. Each chromosome also has a distinctive pattern of bands, which is visible after staining with certain dyes.

Careful examination of a micrograph of the 46 human chromosomes reveals that there are two of each type. This becomes clear when the chromosomes are arranged in pairs, starting with the longest chromosomes. The resulting display is called a **karyotype** (see the Methods Box). The chromosomes that make up a pair—that have the same length, centromere position, and staining pattern—are called **homologous chromosomes,** or homologues. The two chromosomes of each pair carry genes controlling the same inherited traits. For example, if a gene for eye color is situated at a particular locus on a certain chromosome, then the homologue of that chromosome will also have a gene specifying eye color at the equivalent locus.

There is an important exception to the rule of homologous chromosomes for human somatic cells: the two distinct chromosomes referred to as *X* and *Y*.

*Answer: The boy and girl on either end of the bottom row are children of the parents on the right. The boy and girl in the center of the bottom row are children of the other parents.

Methods: Preparation of a Karyotype

Karyotypes, ordered displays of an individual's chromosomes, are useful in identifying certain abnormalities in the chromosomes. Medical technicians often prepare karyotypes by using lymphocytes, a type of white blood cell.

The cells are treated with a drug to stimulate mitosis and grown in culture for several days. They are then treated with another drug to arrest the cell cycle at metaphase, when the chromosomes, each consisting of two joined sister chromatids, are very condensed. This is the stage when chromosomes are easiest to identify in the microscope. The drawings here outline the further steps in the preparation of a karyotype from lymphocytes. The micrograph in step 6 shows the karyotype of a normal human male, with one *X* and one *Y* chromosome. (Human females have two *X* chromosomes.) The chromosomes are stained to reveal band patterns, which help identify specific chromosomes and parts of chromosomes. Karyotyping can be used to screen for abnormal numbers of chromosomes or defective chromosomes associated with congenital disorders, such as Down syndrome. The causes and effects of chromosomal disorders are discussed in Chapter 14.

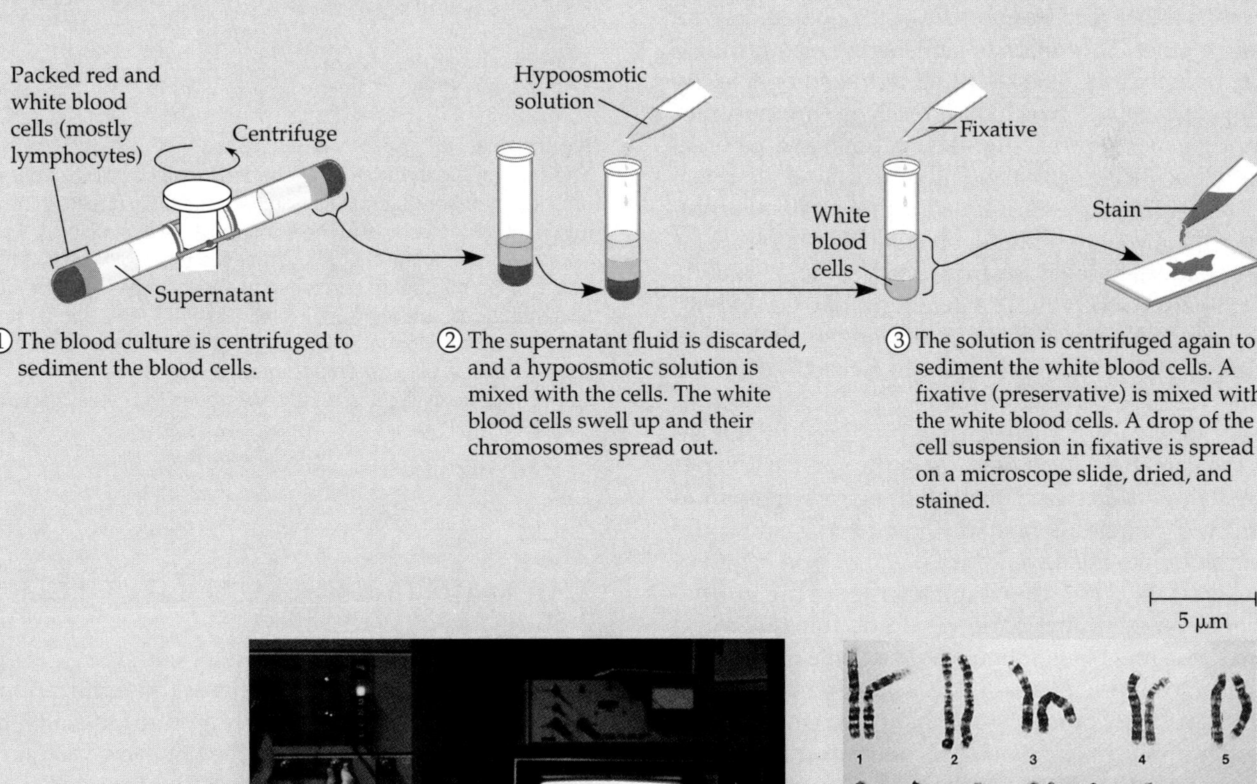

① The blood culture is centrifuged to sediment the blood cells.

② The supernatant fluid is discarded, and a hypoosmotic solution is mixed with the cells. The white blood cells swell up and their chromosomes spread out.

③ The solution is centrifuged again to sediment the white blood cells. A fixative (preservative) is mixed with the white blood cells. A drop of the cell suspension in fixative is spread on a microscope slide, dried, and stained.

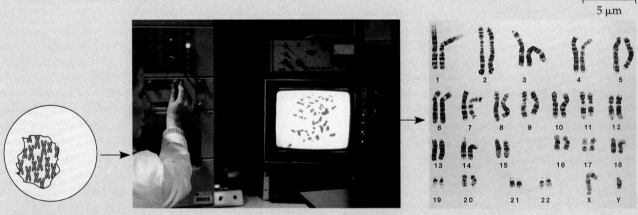

④ The slide is viewed with a microscope, and the chromosomes are photographed.

⑤ The photograph is entered into a computer, and the chromosomes are electronically rearranged into pairs according to size and shape.

⑥ The resulting display is the karyotype.

Human females have a homologous pair of X chromosomes (XX), but males have one X and one Y chromosome (XY). Because they determine an individual's sex, the X and Y chromosomes are called **sex chromosomes.** The other chromosomes are called **autosomes.**

The occurrence of pairs of chromosomes in our karyotype is a consequence of our sexual origins. We inherit one member of each chromosome pair from each parent. So the 46 chromosomes in our somatic cells are actually two sets of 23 chromosomes—a maternal set (from the mother) and a paternal set (from the father).

Sperm cells and ova are distinct from somatic cells in their chromosome count. Each of these reproductive cells, or **gametes,** has a single set of the 22 autosomes plus a single sex chromosome, either X or Y. A cell with a single chromosome set is called a **haploid cell.** For humans, the haploid number (abbreviated n) is 23.

By means of sexual intercourse, a haploid sperm cell from the father reaches and fuses with a haploid ovum of the mother. This union of gametes is called **fertilization,** or **syngamy.** The resulting fertilized egg, or **zygote,** contains the two haploid sets of chromosomes bearing genes representing the maternal and paternal family lines. The zygote and all other cells having two sets of chromosomes are called **diploid cells.** For humans, the diploid number (abbreviated $2n$) is 46.

As a human develops from a zygote to a sexually mature adult, the zygote's genes are passed on with precision to all somatic cells of the body by the process of mitosis. Thus, somatic cells, like the zygote from which they are derived, are diploid.

The only cells of the body *not* produced by mitosis are the gametes, which develop in the gonads (ovaries in females and testes in males). Imagine what would happen if gametes *were* made by mitosis: They would be diploid like the somatic cells. At the next round of fertilization, when two gametes fused, the normal chromosome number of 46 would double to 92, and each subsequent generation would double the number of chromosomes yet again. But sexually reproducing organisms carry out a process that halves the chromosome number in the gametes, compensating for the doubling that occurs at fertilization. This process is a form of cell division called **meiosis,** and it occurs only in the ovaries or testes. While mitosis conserves chromosome number, meiosis reduces the chromosome number by half. As a result, human sperm and egg cells have haploid sets of 23 chromosomes. Fertilization restores the diploid condition, and the human life cycle goes on generation after generation (Figure 12.4).

The processes of meiosis and fertilization are the unique trademarks of sexual reproduction. The life cycles of *all* sexually reproducing organisms follow a basic pattern of alternation between the diploid and haploid conditions, even though the details of the life cycles differ.

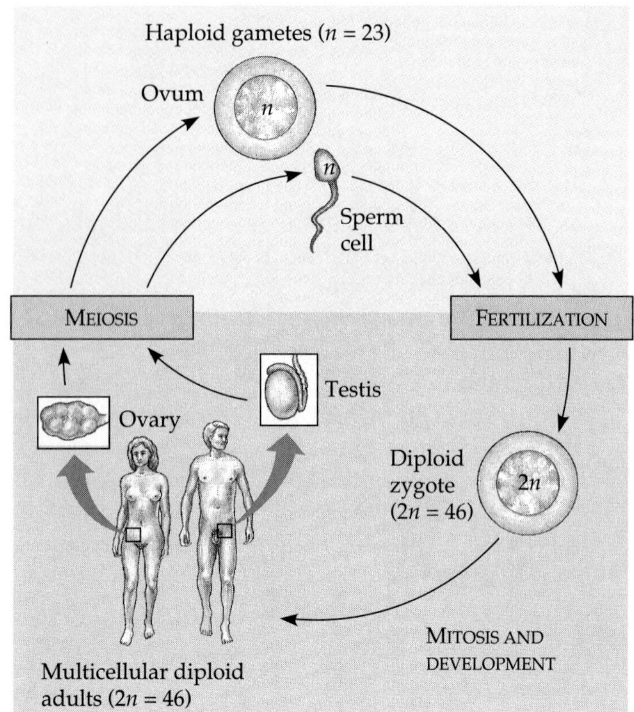

Figure 12.4
The human life cycle. Each generation, the doubling of chromosome number that results from fertilization is offset by the halving of chromosome number that results from meiosis. For humans, the number of chromosomes in a haploid cell is 23 ($n = 23$); the number of chromosomes in the diploid zygote and all somatic cells arising from it is 46 ($2n = 46$). This figure introduces a color-code that will be used for all life cycles throughout the book. A blue-green background represents haploid stages of a life cycle, and a tan background indicates diploid stages.

THE VARIETY OF SEXUAL LIFE CYCLES

Although the alternation of meiosis and fertilization is common to all organisms that reproduce sexually, the timing of these two events in the life cycle varies, depending on the species (Figure 12.5). These variations can be grouped into three main types of life cycles. The human life cycle is an example of one type, characteristic of most animals. Gametes are the only haploid cells. Meiosis occurs during the production of gametes, which undergo no further cell division prior to fertilization. The diploid zygote divides by mitosis, producing a multicellular organism that is diploid.

A second type of life cycle occurs in many fungi and some protists (including some algae). Meiosis occurs immediately after the gametes fuse, and mitosis then creates a multicellular adult organism that is haploid. Subsequently, gametes are produced from the haploid organism by mitosis rather than by meiosis. The only diploid stage is the zygote. (Notice that *either* haploid

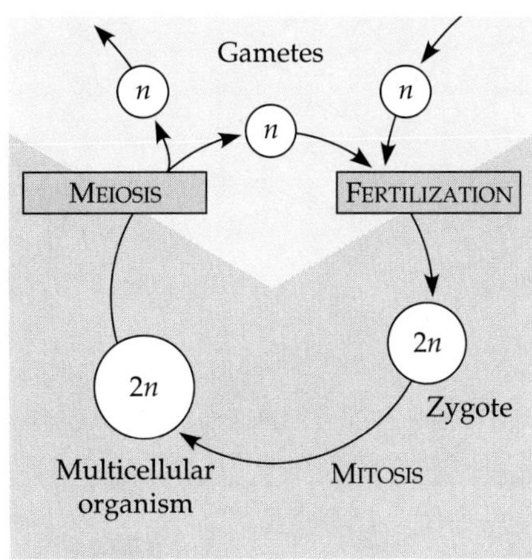

(a) Animals

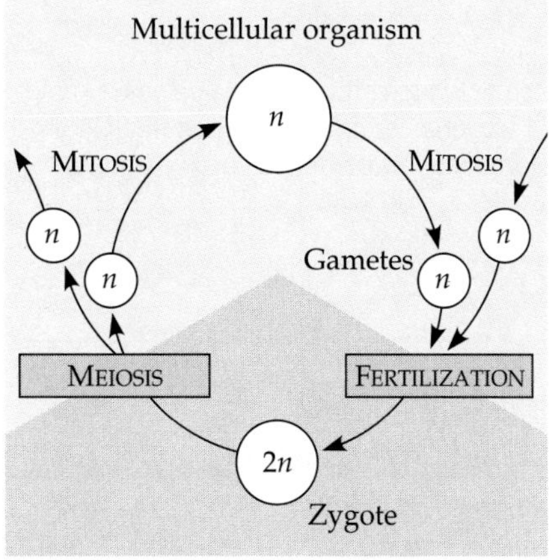

(b) Some fungi and some algae

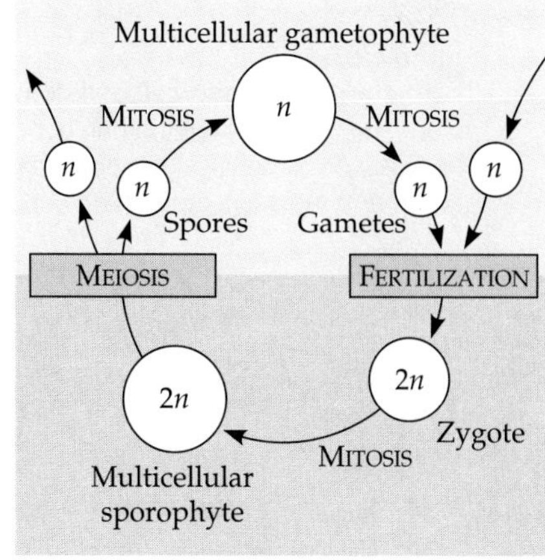

(c) Plants and some algae

□ Haploid
□ Diploid

Figure 12.5
Three sexual life cycles differing in the timing of meiosis and fertilization (syngamy). The common feature of all three cycles is the alternation of these two key events, which contribute to genetic variation among offspring.

or diploid cells can divide by mitosis depending on the type of life cycle. Only diploid cells, however, can undergo meiosis.)

Plants and some species of algae go through a third type of life cycle called **alternation of generations.** In this type of life cycle, there are both diploid and haploid multicellular stages. The multicellular diploid stage is called the **sporophyte.** Meiosis in the sporophyte produces haploid cells called **spores.** Unlike a gamete, a spore gives rise to a multicellular individual without fusing with another cell. A spore divides mitotically to generate a multicellular haploid stage called the **gametophyte.** The haploid gametophyte makes gametes by mitosis. Fertilization results in a diploid zygote, which develops into the next sporophyte generation. In this type of life cycle, therefore, the sporophyte and gametophyte generations take turns reproducing each other (Figure 12.6).

Though the three types of sexual life cycles differ in the timing of meiosis and fertilization, they share a fun-

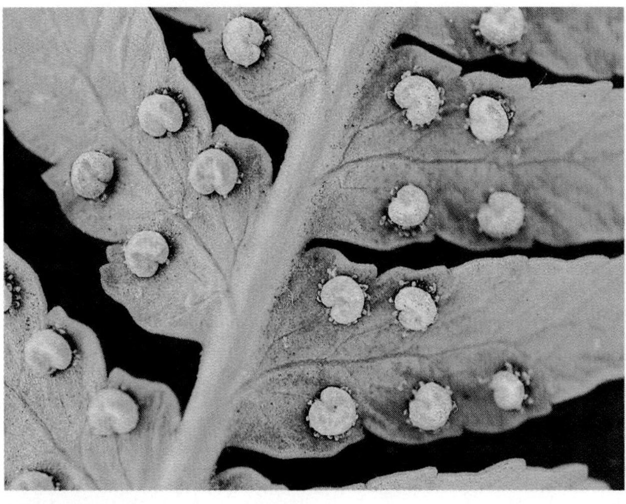

(a) Sporophyte

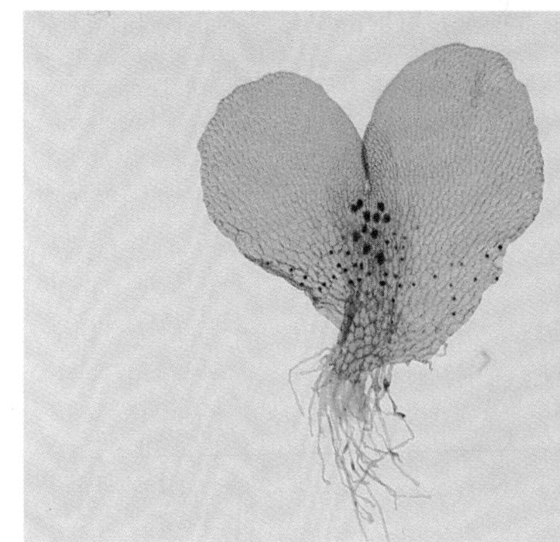

(b) Gametophyte

Figure 12.6
Alternation of generations in a fern.
(**a**) The fern plants familiar to us are sporophytes, the multicellular diploid stage in the life cycle. The globular organs on the underside of this woodfern frond (leaf) contain chambers where meiosis forms haploid reproductive cells called spores. The sporophyte is named for this production of spores. (**b**) If a spore released from the plant lands in a hospitable environment, it developes into a tiny heart-shaped gametophyte, the multicellular haploid stage in the life cycle. The gametophyte is named for its production of haploid gametes, which unite to form diploid zygotes. A zygote gives rise to a new sporophyte, and the alternation of haploid and diploid plant generations continues.

Figure 12.7
The stages of meiotic cell division. These drawings show meiotic cell division for an animal cell with a diploid number of 4. The behavior of the chromosomes is emphasized. For a discussion about spindle formation and other features common to mitosis and meiosis, see Figure 11.6.

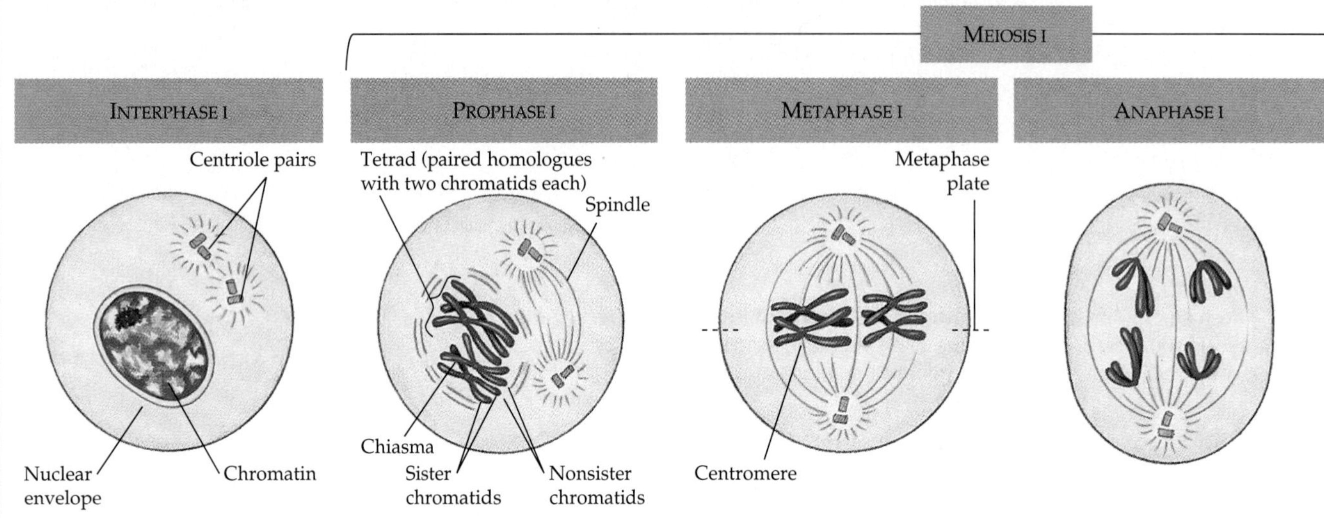

INTERPHASE I

Meiosis is preceded by an interphase, during which each of the chromosomes replicates. This process is similar to the chromosome replication preceding mitosis. For each chromosome, the result is two genetically identical sister chromatids attached at their centromeres. The centriole pairs (in an animal cell) also replicate to form the two pairs represented in this drawing.

PROPHASE I

Some important differences between meiosis and mitosis occur in prophase I. Meiotic prophase I lasts longer and is more complex than prophase in mitosis. The chromosomes begin to condense and attach at their ends to the nuclear envelope. In the process of synapsis, homologous chromosomes, each made up of two chromatids, come together as pairs. Each chromosome pair is now visible in the microscope as a tetrad, a complex of four chromatids. At numerous places along their length, nonsister chromatids (chromatids belonging to homologous chromosomes, in contrast to sister chromatids belonging to the same chromosome) are criss-crossed. These crossings, which help hold homologous chromosomes together, are called chiasmata (singular, chiasma). (We will examine the genetic significance of these crossings later in the chapter.) The chromosomes now thicken further and detach from the nuclear envelope.

As prophase I continues, the cell prepares for the division of the nucleus in a manner similar to that observed during mitosis. The centriole pairs move away from each other, and spindle microtubules form between them. The nuclear envelope and nucleoli disperse. Finally, the chromosomes begin their migration to the metaphase plate, midway between the two poles of the spindle apparatus. Prophase I, which can last for days or even longer, typically occupies more than 90% of the time required for meiosis.

METAPHASE I

Chromosomes are now arranged on the metaphase plate, still in homologous pairs. Spindle fibers from one pole of the cell attach to one chromosome of each pair, while spindle fibers from the opposite pole attach to the homologue.

damental result: Each cycle of chromosome halving and doubling contributes to genetic variation among offspring. A closer look at meiosis will reveal the source of this variation.

MEIOSIS: A CLOSER LOOK

Many of the steps of meiosis closely resemble corresponding steps in mitosis. Meiosis, like mitosis, is pre-

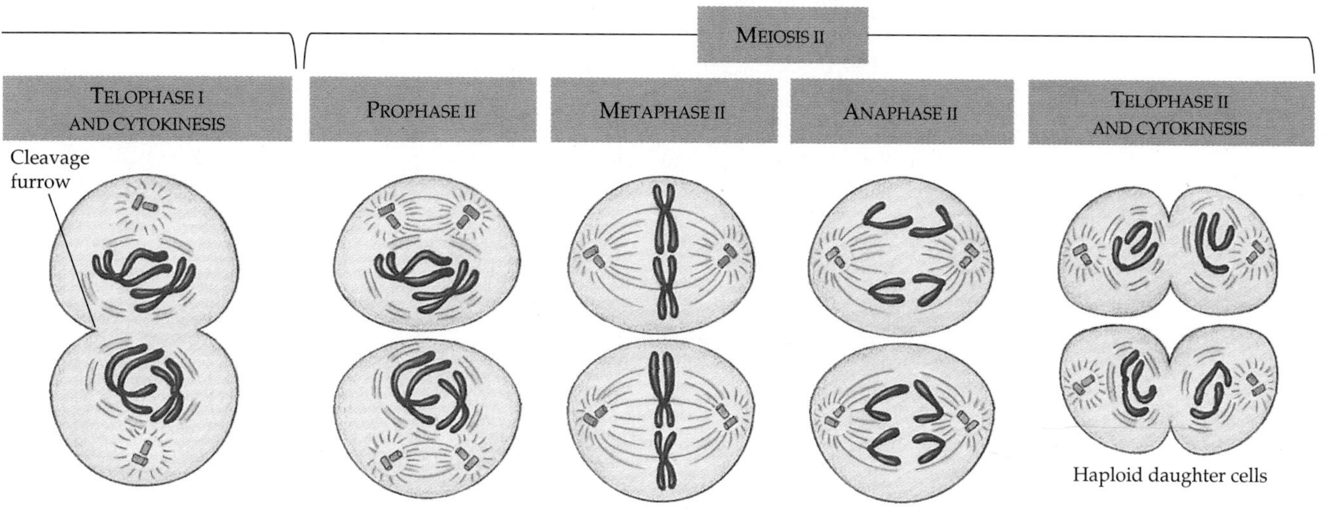

ANAPHASE I

As in mitosis, the spindle apparatus moves the chromosomes toward the poles. However, sister chromatids remain attached at their centromeres and move as a single unit toward the same pole. The homologous chromosome moves toward the opposite pole. This contrasts with the behavior of chromosomes during mitosis. In mitosis, chromosomes align individually on the metaphase plate rather than in pairs, and the spindle separates sister chromatids of each chromosome.

TELOPHASE I AND CYTOKINESIS

The spindle apparatus continues to separate the homologous chromosome pairs until the chromosomes reach the poles of the cell. Each pole now has a haploid chromosome set, but each chromosome still has two chromatids. Usually cytokinesis (division of the cytoplasm) occurs simultaneously with telophase I, forming two daughter cells. Cleavage furrows form in animal cells, and cell plates appear in plant cells. In some species, nuclear membranes and nucleoli re-form, and there is a period of time, called interkinesis (or interphase II), before meiosis II. In other species, daughter cells of telophase I immediately begin preparation for the second meiotic division. With or without an interkinesis, there is no further replication of the genetic material prior to the second division of meiosis II.

PROPHASE II

A spindle apparatus appears, and the chromosomes progress toward the metaphase II plate.

METAPHASE II

The chromosomes align on the metaphase plate in mitosislike fashion, with the kinetochores of sister chromatids of each chromosome pointing toward opposite poles.

ANAPHASE II

The centromeres of sister chromatids finally separate, and the sister chromatids of each pair, now individual chromosomes, move toward opposite poles of the cell.

TELOPHASE II AND CYTOKINESIS

Nuclei begin to form at opposite poles of the cell, and cytokinesis occurs. There are now four daughter cells, each with the haploid number of chromosomes.

ceded by the replication of chromosomes. However, this single replication is followed by *two* consecutive cell divisions, called meiosis I and meiosis II. These divisions result in four daughter cells (rather than the two daughter cells of mitosis), each with only half as many chromosomes as the parent. The drawings and text in Figure 12.7 describe in some detail the two divisions of meiosis for an animal cell whose diploid num-

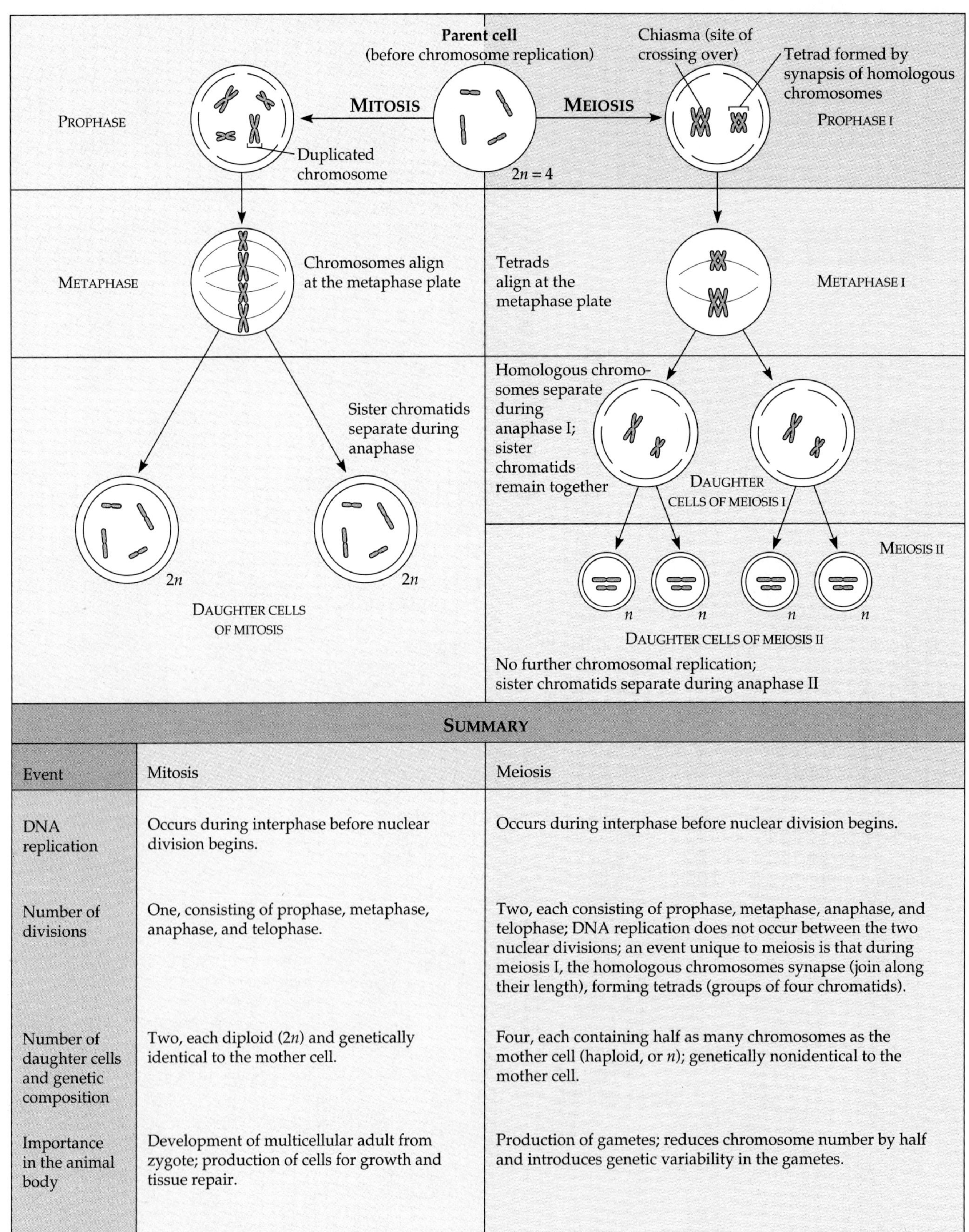

SUMMARY

Event	Mitosis	Meiosis
DNA replication	Occurs during interphase before nuclear division begins.	Occurs during interphase before nuclear division begins.
Number of divisions	One, consisting of prophase, metaphase, anaphase, and telophase.	Two, each consisting of prophase, metaphase, anaphase, and telophase; DNA replication does not occur between the two nuclear divisions; an event unique to meiosis is that during meiosis I, the homologous chromosomes synapse (join along their length), forming tetrads (groups of four chromatids).
Number of daughter cells and genetic composition	Two, each diploid ($2n$) and genetically identical to the mother cell.	Four, each containing half as many chromosomes as the mother cell (haploid, or n); genetically nonidentical to the mother cell.
Importance in the animal body	Development of multicellular adult from zygote; production of cells for growth and tissue repair.	Production of gametes; reduces chromosome number by half and introduces genetic variability in the gametes.

Figure 12.8
A comparison of mitosis and meiosis in amimals.

ber is 4. Examine Figure 12.7 thoroughly before going on to the next section of the text.

A COMPARISON OF MITOSIS AND MEIOSIS

Now that we have followed chromosomes through meiosis in Figure 12.7, let's summarize the key differences between meiosis and mitosis. The chromosome number is reduced by half in meiosis but not in mitosis. The genetic consequences of this difference are important. Whereas mitosis produces daughter cells genetically identical to their parent cell and to each other, meiosis produces cells that differ genetically from their parent cell and from each other.

Figure 12.8 compares the key steps in the processes of mitosis and meiosis. Although meiosis involves two nuclear divisions, the three events that are unique to meiosis all occur during the first division, meiosis I:

1. During prophase I of meiosis, the duplicated chromosomes pair with their homologues, a process called **synapsis.** The four closely associated chromatids are visible in the light microscope as tetrads. Also visible in the light microscope are X-shaped regions called **chiasmata** (singular, **chiasma**). They represent a crossing of nonsister chromatids, which are two chromatids belonging to separate but homologous chromosomes. Chiasmata are the physical manifestations of a genetic rearrangement called crossing over, discussed in the next section. Neither synapsis nor chiasmata occur during mitosis.

2. At metaphase I of meiosis, homologous pairs of chromosomes, rather than individual chromosomes, align on the metaphase plate.

3. At anaphase I of meiosis, centromeres do not divide and sister chromatids do not separate, as they do in mitosis. Rather, the sister chromatids of each chromosome go to the same pole of the cell. *Meiosis I separates homologous pairs of chromosomes, not sister chromatids of individual chromosomes.*

The second meiotic division, meiosis II, separates sister chromatids and is virtually identical in mechanism to mitosis. However, since the chromosomes do not replicate between meiosis I and meiosis II, the final outcome of meiosis is a halving of the number of chromosomes per cell.

SEXUAL SOURCES OF GENETIC VARIATION

How do we account for the genetic variation apparent in Figure 12.3? In species that reproduce sexually, the activity of chromosomes during meiosis and fertilization is responsible for most of the variation that arises each generation. Let's examine three sexual mechanisms that contribute to genetic variation: independent assortment of chromosomes, crossing over, and random fertilization.

Independent Assortment of Chromosomes

One way sexual reproduction generates genetic variation is shown in Figure 12.9, which color-codes the chromosomes so that we can track them as they are transmitted to gametes. The two colors are used to distinguish chromosomes in a diploid cell inherited from the mother from those inherited from the father. At

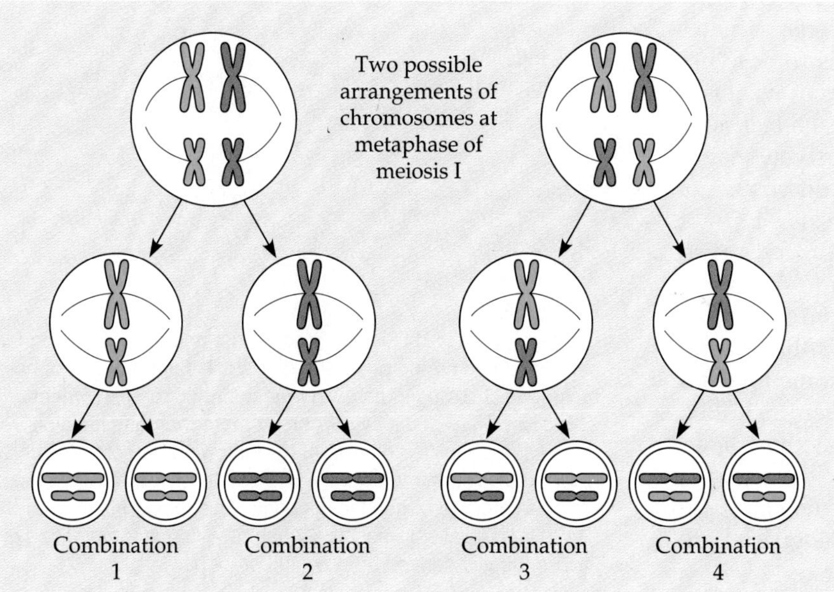

Figure 12.9
Two possible arrangements of chromosomes on the metaphase plate in meiosis I. In this figure, we consider the consequences of meiosis in a hypothetical organism with a diploid chromosome number of 4 ($2n = 4$). The origins of the chromosomes are symbolized with color-coding, blue representing chromosomes inherited from one parent, salmon for chromosomes from the other parent. The positioning of each homologous pair of chromosomes at metaphase of meiosis I is a matter of chance, like the flip of a coin. The arrangement of chromosomes at metaphase I determines which chromosomes will be packaged together in the haploid daughter cells.

metaphase of meiosis I, each homologous pair of chromosomes, consisting of one maternal and one paternal chromosome, is situated on the metaphase plate. The orientation of the homologous pair relative to the two poles of the cell is random; there are two possibilities. Thus, there is a fifty-fifty chance that a particular daughter cell of meiosis I will get the maternal chromosome of a certain homologous pair, and a fifty-fifty chance of receiving the paternal chromosome.

Because each homologous pair of chromosomes is oriented independently of the other pairs at metaphase I—it is as random as the flip of a coin—the first meiotic division results in independent assortment of maternal and paternal chromosomes into daughter cells. Each gamete represents one outcome of all possible combinations of maternal and paternal chromosomes. The number of combinations possible for gametes formed by meiosis starting with two homologous pairs of chromosomes ($2n = 4, n = 2$) is four, as shown in Figure 12.9. In the case of $n = 3$, there are eight combinations of chromosomes possible for gametes. More generally, the number of combinations possible when meiosis packages chromosomes into gametes by independent assortment is 2^n, where n is the haploid number.

In the case of humans, the haploid number (n) in the formula is 23. Thus, the number of possible combinations of maternal and paternal chromosomes in the resulting gametes is 2^{23}, or about 8 million. The variations are analogous to the 8 million combinations of heads and tails possible for the simultaneous tossing of 23 coins. Thus, each gamete that a human produces contains one of 8 million possible assortments of chromosomes inherited from that individual's mother and father.

Crossing Over

Because of the independent assortment of chromosomes during meiosis, each of us produces gametes containing diverse combinations of the chromosomes we inherited from our two parents. But from what you have learned so far, it would seem that each *individual* chromosome in a gamete would be exclusively maternal or paternal in origin; that is, it would consist of DNA derived from the mother or the father, but not from both. This, in fact, is *not* the case. A process called **crossing over** produces individual chromosomes that combine genes inherited from the two parents. It occurs during prophase of meiosis I. Recall that during prophase I, homologous chromosomes come together as pairs (see Figure 12.7). A protein apparatus called the **synaptonemal complex** functions something like a zipper to bring the chromosomes into close association, but the exact mechanism of synapsis is not yet known. The pairing is precise, the homologues aligning with each other gene by gene.

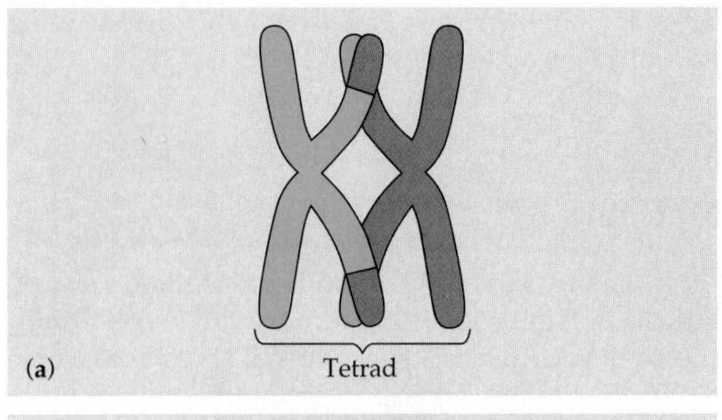

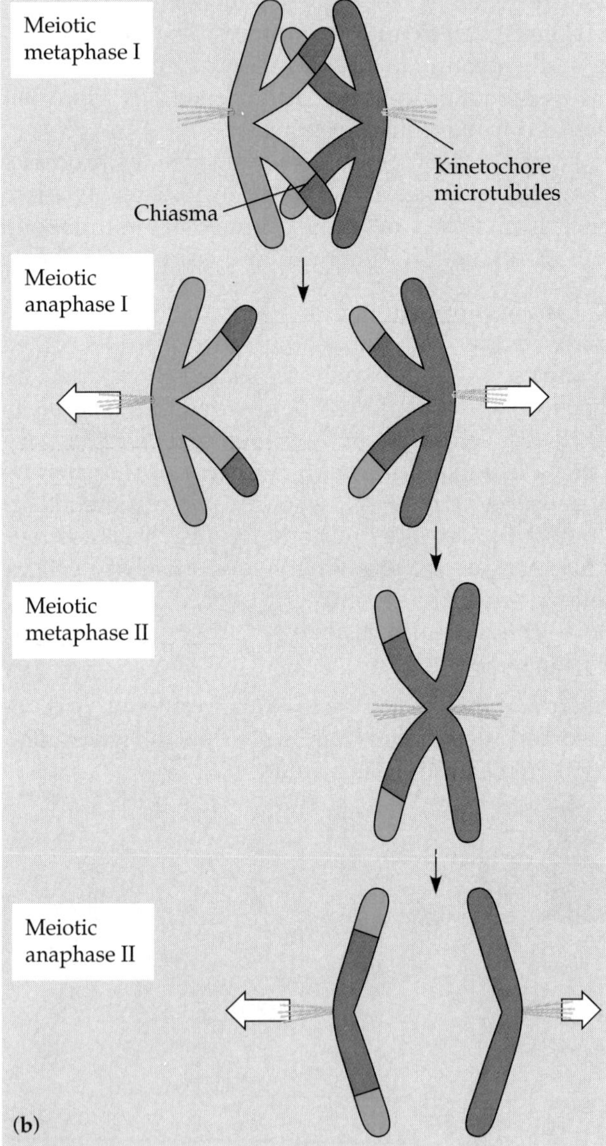

Figure 12.10
Crossing over. (**a**) During synapsis, the pairing of homologous chromosomes during prophase of meiosis I, chromatids of homologous chromosomes exchange corresponding segments. This crossing over usually occurs at more than one location (two in this simplified diagram). (**b**) Following one of these chromosomes through meiosis, we can see that crossing over is an important source of genetic variation. This mechanism gives rise to individual chromosomes that have some combination of DNA originally derived from two different parents.

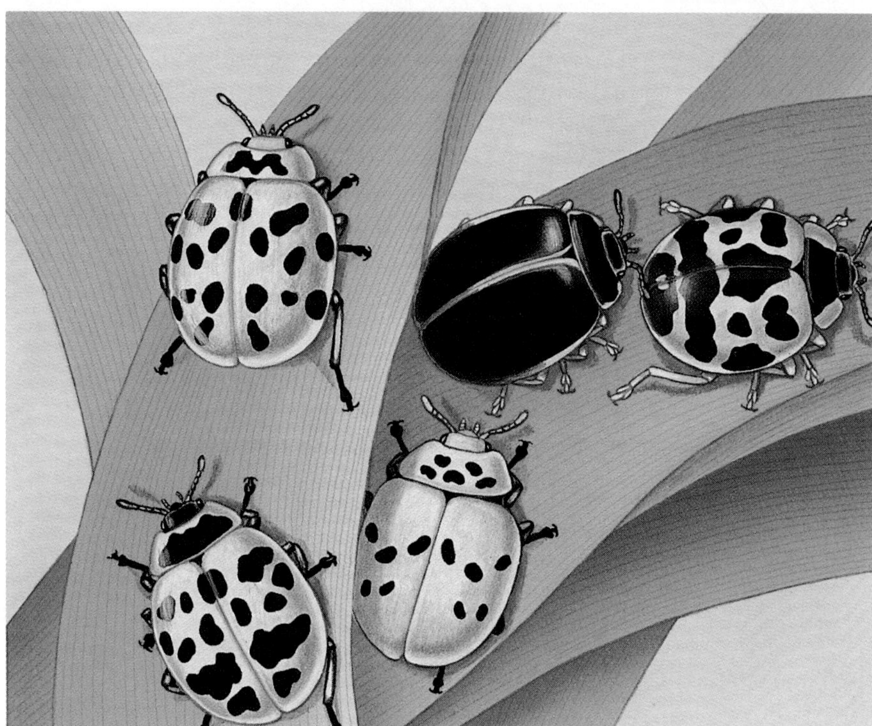

Figure 12.11
Heritable variations: the raw material of natural selection. These are a few of the color variations in a population of Asian lady beetles.

Crossing over occurs when homologous portions of two nonsister chromatids trade places (Figure 12.10). In the case of humans, an average of two or three such crossover events occur per chromosome pair. The locations of these genetic exchanges are visible in light micrographs as chiasmata. The mechanism of crossing over will be discussed in more detail in Chapter 14. The important point for now is that crossing over, by combining DNA inherited from two parents into a single chromosome, is an important source of genetic variation in sexual life cycles.

Random Fertilization

The random nature of fertilization adds to the genetic variation arising from meiosis. A human ovum, representing one of approximately 8 million possible chromosome combinations, will be fertilized by a single sperm cell, which represents one of 8 million *different* possibilities. Thus, even without considering crossing over, any two parents will produce a zygote with any of 64 trillion (8 million × 8 million) diploid combinations (the exact number is 70,368,744,177,664). No wonder brothers and sisters can be so different. Each one of us is unique.

So far, we have seen that there are three sources of genetic variability in a sexually reproducing population of organisms:

- Independent assortment of homologous chromosome pairs during meiosis I.
- Crossing over between homologous chromosomes during prophase of meiosis I.
- Random fertilization of an ovum by a sperm.

All three mechanisms reshuffle the various genes carried by the individual members of a population. However, as you will learn in later chapters, mutations are what ultimately create a population's diversity of genes.

GENETIC VARIATION AND EVOLUTION

Having considered how sexual reproduction contributes to genetic variation in a population, we should bridge these concepts to evolution, biology's core theme. Darwin recognized the importance of heritable variation in the evolutionary mechanism he called natural selection (Figure 12.11). Recall from Chapter 1 that a population evolves through the differential reproductive success of its variant members. On average, those individuals best suited to the local environment leave the most offspring, transmitting their genes in the process. This natural selection results in adaptation, the accumulation of those heritable variations that are favored by the environment. As the environment changes or a population moves, a population may survive if in each generation, at least some of its members can cope effectively with the new conditions. Different heritable variations may work better than those that prevailed in the old time or place. Sex and mutations are the two sources of this variation, and we have considered the sexual contribution in this chapter.

Although Darwin realized that heritable variation is what makes evolution possible, he had no satisfactory explanation for the precise mechanism that makes offspring resemble—but not be identical to—their parents. Ironically, Gregor Mendel, a contemporary of Darwin, published a theory of inheritance that helps explain genetic variation, but his discoveries had no impact on biologists until 1900, more than 15 years after Darwin (1809–1882) and Mendel (1822–1884) died. In the next chapter, you will learn how Mendel discovered the basic rules governing the inheritance of specific traits.

STUDY OUTLINE

1. Genetics is the study of heredity and individual variation in a population.

Genes, DNA, and Chromosomes: A Brief Orientation (pp. 244–245)

1. Genetic material consists of DNA organized into genes, each with a specific locus on a certain chromosome.

Sexual and Asexual Reproduction: A Comparison (p. 245)

1. In asexual reproduction, a single parent gives rise to genetically identical offspring by mechanisms involving mitosis.
2. Sexual reproduction combines two different sets of genes, carried by gametes from two different parents, to form offspring that are genetically diverse.

An Introduction to Sexual Life Cycles: The Human Example (pp. 246–248)

1. Normal human somatic cells contain 46 chromosomes, half inherited from the father and half from the mother.
2. Each of 22 autosomes in the maternal set has a corresponding homologous chromosome in the paternal set. The twenty-third pair, the sex chromosomes, determines whether the person is a female (*XX*) or a male (*XY*).
3. Single, haploid (n) sets of chromosomes in maternal and paternal gametes unite during fertilization to produce a diploid ($2n$), single-celled zygote. The zygote develops into a multicellular individual by mitosis.
4. At sexual maturity, ovaries and testes (the gonads) produce haploid gametes by meiosis.
5. All sexually reproducing organisms alternate diploid and haploid states through fertilization and meiosis.

The Variety of Sexual Life Cycles (pp. 248–250)

1. Differences in the timing of meiosis with respect to fertilization characterize a variety of sexual life cycles. Multicellular organisms may be diploid (as in animals), haploid (as in some fungi), or may alternate generations between haploid and diploid (as in plants).

Meiosis: A Closer Look (pp. 250–253)

1. Meiosis consists of two cell divisions, meiosis I and meiosis II, resulting in four daughter cells, each with half the chromosome number of the original cell. Thus, meiosis reduces the chromosome count from diploid to haploid.

A Comparison of Mitosis and Meiosis (p. 253)

1. Meiosis is distinguished from mitosis by a series of distinctive events that occur during meiosis I.
2. In prophase I of meiosis, replicated homologous chromosomes, each with two chromatids, undergo synapsis. This association allows the exchange of genetic material by the crossing over of homologous segments of nonsister chromatids (chromatids of different but homologous chromosomes). The crossing-over sites are visible as chiasmata.
3. The paired chromosomes align on the metaphase plate, and at anaphase I, the two chromosomes of each homologous pair (rather than sister chromatids) are pulled toward separate poles. This halves the number of chromosomes in the daughter cells.
4. Meiosis II separates the sister chromatids to form four haploid daughter cells.

Sexual Sources of Genetic Variation (pp. 253–255)

1. The sexual processes that contribute to genetic variation in a population are independent assortment of chromosomes, crossing over, and random fertilization.

Genetic Variation and Evolution (pp. 255–256)

1. Heritable variation among a population's members is the raw material for evolution. Sex and mutations are the two processes that generate this variation.

SELF-QUIZ

1. A human cell containing 22 autosomes and a *Y* chromosome is
   a. a somatic cell of a male
   b. a zygote
   c. a somatic cell of a female
   d. a sperm cell
   e. an ovum
2. Homologous chromosomes segregate toward opposite poles of a dividing cell during
   a. mitosis
   b. meiosis I
   c. meiosis II
   d. fertilization

3. Meiosis II is similar to mitosis in that
   a. homologous chromosomes synapse
   b. DNA replicates before the division
   c. the daughter cells are diploid
   d. sister chromatids separate during anaphase
   e. the chromosome number is reduced
4. The DNA content of a diploid cell in the G_1 phase of the cell cycle is measured (see Chapter 11). If this DNA content is X, then the DNA content of the same cell at metaphase of meiosis I would be
   a. $0.25X$
   b. $0.5X$
   c. X
   d. $2X$
   e. $4X$
5. If we continued to follow the cell lineage from question 4, then the DNA content at metaphase of meiosis II would be
   a. $0.25X$
   b. $0.5X$
   c. X
   d. $2X$
   e. $4X$
6. Crossing over most commonly occurs during
   a. prophase I
   b. anaphase I
   c. interphase
   d. prophase II
   e. telophase II
7. How many different combinations of maternal and paternal chromosomes can be packaged in gametes made by an organism with a diploid number of 8 ($2n = 8$)?
   a. 2
   b. 4
   c. 8
   d. 16
   e. 32
8. The most direct product of meiosis in a plant is a
   a. spore
   b. gamete
   c. zygote
   d. sporophyte
   e. gametophyte
9. Somatic cells of the adult body are haploid in many
   a. vertebrates
   b. invertebrates
   c. fungi
   d. sporophytes
10. The following choices indicate chromosome number before and after a process. Which choice corresponds to the process of fertilization?
   a. $2n \longrightarrow n$
   b. $n \longrightarrow 2n$
   c. $2n \longrightarrow 2n$
   d. $n \longrightarrow n$

CHALLENGE QUESTIONS

1. In domestic turkeys, viable offspring are sometimes produced by the development of an unfertilized egg cell, in a process called parthenogenesis. Such offspring, like the mother, are diploid. What variation in meiosis could produce a diploid organism without fertilization?
2. Many species can reproduce *either* asexually or sexually. In a favorable, stable environment, these species generally reproduce asexually. It is usually when the environment changes in some way that is unfavorable to an existing population that the organisms begin to reproduce sexually. Based on your knowledge of natural selection, speculate about the evolutionary significance of this switch from asexual to sexual reproduction.

SCIENCE, TECHNOLOGY, AND SOCIETY

1. Imagine that you are a biologist called as an expert witness in the trial of an alleged serial killer. Genetic analysis of blood and tissue samples link the defendant to the crimes. The defense attorney moves that *all* genetic evidence be excluded, because it is possible that the crimes were committed by another individual with the same combination of genes and chromosomes as his client. How would you respond to this argument? How do standards of evidence and certainty in the laboratory and the courtroom differ?
2. Studies of human characteristics indicate that most variation in the human population is due to the differences within racial groups. Only a small amount of variation is due to differences *between* races. If these findings are correct, is the concept of "race" valid? How might a better understanding of biology help us reduce racial conflict and better appreciate human diversity?

FURTHER READING

Anderson, A. "The Evolution of Sexes." *Science,* July 17, 1992. Why do humans and many other organisms exist in only two sexes?

Becker, W. M., and D. W. Deamer. *The World of the Cell,* 2nd ed. Redwood City, CA: Benjamin/Cummings, 1991. Chapter 15.

Cunningham, P. "The Genetics of Thoroughbred Horses." *Scientific American,* May 1991. The history of how humans have exploited heritable variation in one species.

Kinoshita, J. "Swap Meet." *Discover,* April 1991. How did sex evolve?

Suzuki, D., A. Griffiths, J. Miller, and R. Lewontin. *An Introduction to Genetic Analysis,* 4th ed. New York: Freeman, 1989. A good basic undergraduate genetics text.

13 MENDEL AND THE GENE IDEA

MENDEL'S MODEL: A CASE STUDY IN THE SCIENTIFIC PROCESS
EXTENDING MENDELIAN GENETICS
MENDELIAN INHERITANCE IN HUMANS

Figure 13.1
Gregor Mendel (1822–1884). Based on his experiments with garden peas, Mendel built a model of inheritance that became the foundation of modern genetics. His published records of these experiments provide a window to a great scientific mind. This chapter examines Mendel's experiments and conclusions and applies Mendelian concepts to human heredity.

A person's eyes can be blue, brown, green, gray, or hazel; a person's hair can be different shades of blond, brown, red, or black; a parakeet's feathers can be green, blue, or yellow, with black or gray markings. What causes these biological spectra of colors? We can frame the question in more general terms: What is the genetic basis of variation among a population's individuals? Or, what principles account for the transmission of these variations from parents to offspring?

One possible explanation of heredity is a "blending model," the idea that genetic material contributed by the two parents mixes in a manner analogous to the way blue and yellow paints blend to make green. This hypothesis predicts that mating a blue parakeet with a yellow one would result in green offspring, and once blended, the hereditary material of the two parents would be as inseparable as the colors of mixed paint. If the blending model were accurate, over many generations a freely mating population of blue and yellow parakeets would give rise to a uniform population of green birds. The actual results of parakeet breeding, however, contradict such a prediction. The blending theory also fails to explain other phenomena of inheritance, such as traits skipping a generation.

An alternative to the blending model is a "particulate model" of inheritance: the gene idea. According to this model, parents pass on discrete heritable units—genes—that retain their separate identities in offspring. An organism's collection of genes is more like a bucket of marbles than a pail of paint. Like marbles, genes can be sorted and passed along, generation after generation, in undiluted form.

Modern genetics had its genesis in an abbey garden when a monk named Gregor Mendel documented a particulate mechanism of inheritance (Figure 13.1). In this chapter, you will learn how Mendel developed his theory and how the Mendelian model applies to the inheritance of human variations.

MENDEL'S MODEL: A CASE STUDY IN THE SCIENTIFIC PROCESS

Gregor Mendel discovered the basic principles of heredity by breeding garden peas in carefully planned experiments. As we retrace Mendel's work, we will be able to recognize the key elements of the scientific process that were introduced in Chapter 1.

Johann Mendel (he took the name Gregor when he entered the Augustinian brotherhood) grew up on his parents' small farm in a region of Austria that is now part of the Czech Republic. In this agricultural area, crops and orchards were of great local interest, and at school Mendel and the other children received agricultural training along with basic education. Later Mendel overcame financial hardships and a series of illnesses to excel in high school and at the Olmutz Philosophical Institute.

Mendel entered the Augustinian monastery in 1843. After three years of theological studies, he was assigned to a school as a temporary teacher but failed the teacher's examination. An administrator sent Mendel to the University of Vienna, where he studied from 1851 to 1853. These were very important years for Mendel the scientist. Two professors were major influences. One was the physicist Doppler, who encouraged his students to learn science through experimentation and trained Mendel to apply mathematics to help explain natural phenomena. The second was a botanist named Unger, who aroused Mendel's interest in the causes of variation in plants. These influences came together in Mendel's subsequent experiments with garden peas.

After attending the university, Mendel was assigned to teach at the Brünn Modern School, where several teachers shared his enthusiasm for scientific research. At the monastery where Mendel lived, there were also stimulating colleagues, many of them university professors and active researchers. There had also been a long tradition of interest in the breeding of plants, including peas, at the monastery. Thus, it was probably not extraordinary when, around 1857, Mendel began breeding garden peas in the abbey garden to study inheritance. What *was* extraordinary was Mendel's fresh approach to very old questions about heredity.

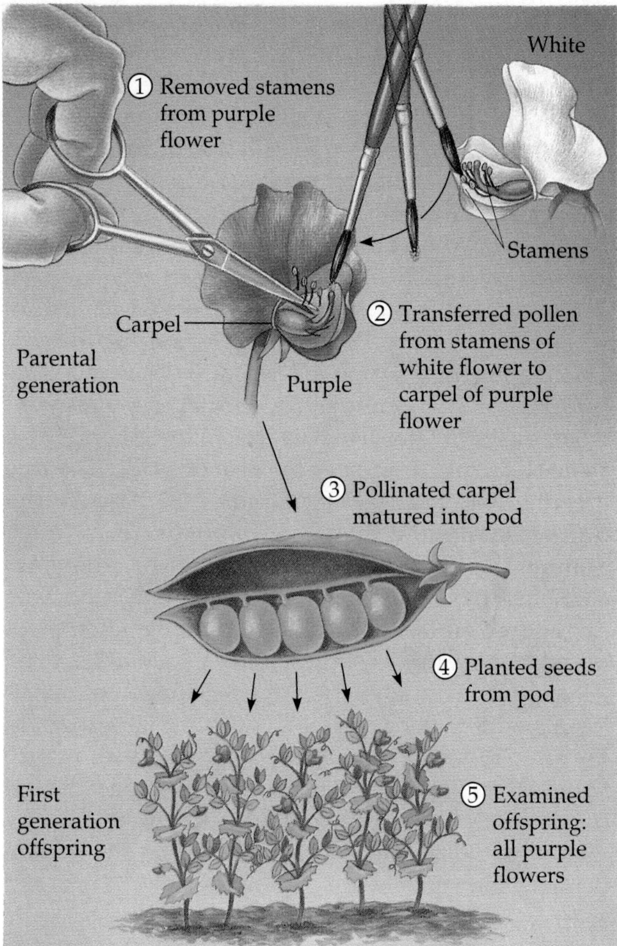

Figure 13.2
A genetic cross. To hybridize between pea varieties, Mendel used an artist's brush to transfer pollen. In this case, the character of interest is flower color, and the two varieties are purple versus white flowers. Seeds develop within the female organ, or carpel, which develops into the fruit (pod). Germination of the seeds produces the first generation hybrids, which all have purple flowers. The result is the same for the reciprocal cross, the transfer of pollen from purple flowers to white flowers.

Mendel's Experimental Approach

Mendel probably chose to work with peas because they are available in many varieties. For example, one variety has purple flowers, while a contrasting variety has white flowers. Geneticists use the term **character** for a heritable feature, such as flower color, that varies among individuals. Each variant for a character, such as purple or white flowers, is called a **trait.**

The use of peas also gave Mendel strict control over which plants mated with which. The petals of the pea flower almost completely enclose the female and male parts (carpel and stamens); normally, the plants self-fertilize after pollen grains released from the stamens land on the carpel. When Mendel wanted cross-pollination (fertilization between different plants), he would remove the immature stamens of a plant before they produced pollen and then dust pollen from another plant onto the emasculated flowers (Figure 13.2). Whether ensuring self-pollination or executing artificial cross-pollination, Mendel could always be sure of the parentage of new seeds.

Mendel was careful to track the inheritance of only categorical variations, that is, inherited characters that varied in an "either-or" rather than a "more-or-less" manner. For example, his plants had either purple or white flowers; there was nothing intermediate between these two varieties. Had Mendel focused instead on characters that vary in a continuum among individuals—seed weight, for example—he would not have discovered the particulate nature of inheritance.

Mendel also made sure that he started his experiments with varieties that were **true-breeding,** which means that when the plants self-pollinate, all their offspring are of the same variety. For example, a plant with purple flowers is true-breeding if its seeds produced by self-pollination all give rise to plants that also have purple flowers.

In a typical breeding experiment, Mendel would cross-pollinate between two contrasting, true-breeding pea varieties—between purple-flowered plants and white-flowered plants, for example (see Figure 13.2). This mating, or crossing, of two varieties is called **hybridization.** Our example is specifically a **monohybrid cross,** the term for a cross that tracks the inheritance of a single character—flower color, in this case. The true-breeding parents are referred to as the **P generation** (for parental), and their hybrid offspring are the **F_1 generation** (for first filial, referring to offspring). Allowing these F_1 hybrids to self-pollinate produces an **F_2 generation** (second filial). Mendel generally followed traits for at least these three generations: the P, F_1, and F_2 generations. Had Mendel stopped his experiments with the F_1 generation, the basic patterns of inheritance would have eluded him. It was mainly Mendel's analysis of F_2 plants that revealed the two fundamental principles of heredity that are now known as the law of segregation and the law of independent assortment.

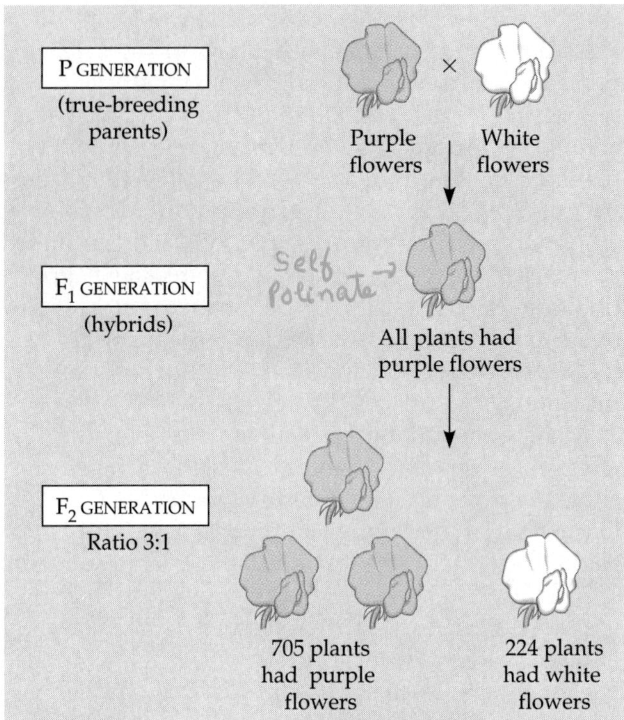

Figure 13.3
Mendel tracked heritable characters for three generations. When F_1 hybrids were allowed to self-pollinate, or when they were cross-pollinated with other F_1 hybrids, a 3:1 ratio of the two varieties occurred in the F_2 generation. An "×" sign symbolizes a genetic cross, or mating.

Mendel's Law of Segregation

If the blending model of inheritance were correct, the F_1 hybrids from a cross between purple-flowered and white-flowered pea plants would have pale purple flowers, intermediate between the two varieties of the P generation. Notice in Figure 13.2 that the experiment yielded a very different result: The F_1 offspring all had flowers just as purple as the purple-flowered parents. What happened to the white-flowered plants' genetic contribution to the hybrids? If it were lost, then the F_1 plants could produce only purple-flowered offspring in the F_2 generation. But when Mendel allowed the F_1 plants to self-pollinate and planted their seeds, the white-flower trait reappeared in the F_2 generation. Mendel used very large sample sizes and kept accurate records of his results: 705 of the F_2 plants had purple flowers, and 244 had white flowers. These data fit a ratio of 3 purple to 1 white (Figure 13.3). Mendel reasoned that the heritable factor for white flowers did not disappear in the F_1 plants, but only the purple-flower factor was affecting flower color in these hybrids. In Mendel's terminology, purple flowers is a dominant trait and white flowers is a recessive trait. The occurrence of white-flowered plants in the F_2 generation was evidence that the heritable factor causing that recessive trait had not been diluted in any way by coexisting with the purple-flower factor in the F_1 hybrids.

Mendel observed the same pattern of inheritance in six other characters, each represented by two contrasting varieties (Table 13.1). For example, the parental pea seeds either had a smooth, round shape, or they were wrinkled. In a monohybrid cross for this character, all the F_1 hybrids produced round seeds; this is the dominant trait. In the F_2 generation, 75% of the seeds were round and 25% were wrinkled—the typical 3:1 ratio. How did Mendel explain this pattern, which he consistently observed in his monohybrid crosses? He developed a hypothesis that we can break down into four related ideas. (We will replace some of Mendel's original terms with modern words; for example, "gene" will be used in place of Mendel's "heritable factors.")

1. *Alternative versions of genes account for variations in inherited characters.* The gene for flower color, for example, exists in two versions, one for purple flowers and the other for white. These alternative versions of a gene are now called **alleles.** Today, we can relate this concept to chromosomes and DNA. As we saw in Chapter 12, each gene resides at a specific locus on a specific chromosome. The DNA at that locus, however, can vary somewhat in its sequence of nucleotides, and

Table 13.1 The Results of Mendel's F_1 Crosses for Seven Characters in Pea Plants

Character	Dominant Trait	×	Recessive Trait	F_2 Generation Dominant:Recessive	Ratio
Flower color	Purple	×	White	705:224	3.15:1
Flower position	Axial	×	Terminal	651:207	3.14:1
Seed color	Yellow	×	Green	6022:2001	3.01:1
Seed shape	Round	×	Wrinkled	5474:1850	2.96:1
Pod shape	Inflated	×	Constricted	882:299	2.95:1
Pod color	Green	×	Yellow	428:152	2.82:1
Stem length	Tall	×	Dwarf	787:277	2.84:1

hence in its information content. The purple-flower allele and the white-flower allele are two DNA variations possible at the flower-color locus on one of a pea plant's chromosomes.

2. *For each character, an organism inherits two genes, one from each parent.* Mendel made this deduction without knowing about the role of chromosomes, but what we learned about chromosomes in Chapter 12 can help us understand Mendel's idea. Recall that a diploid organism has homologous pairs of chromosomes, one chromosome of each pair inherited from each parent. Thus, a genetic locus is actually represented twice in a diploid cell. These homologous loci may have matching alleles, as in the true-breeding plants of Mendel's P generation. Or, the two alleles may differ, as in the F_1 hybrids. In the flower-color example, the hybrids inherited a purple-flower allele from one parent and a white-flower allele from the other parent. This brings us to the third aspect of Mendel's hypothesis.

3. *If the two alleles differ, then one, the* ***dominant allele,*** *is fully expressed in the organism's appearance; the other, the* ***recessive allele,*** *has no noticeable effect on the organism's appearance.* According to this part of the hypothesis, Mendel's F_1 plants had purple flowers because the allele for that variation is dominant and the allele for white flowers is recessive.

4. *The two genes for each character segregate during gamete production.* Thus, an ovum and a sperm each receive only one of the genes that are present in two copies in the somatic cells of the organism. (In the case of peas, "sperm" refers to a nucleus in a pollen grain.) In terms of chromosomes, this segregation corresponds to meiotic reduction of chromosome count from the diploid to the haploid number. Notice that if an organism has matching alleles for a particular character—that is, the organism is true-breeding for that character—then that allele exists in a single copy in all gametes. But if contrasting alleles are present, as in the F_1 hybrids, then 50% of the gametes receive the dominant allele, while 50% receive the recessive allele. It is this last part of the hypothesis, the *sorting of alleles into separate gametes*, for which Mendel's **law of segregation** is named.

One test of Mendel's segregation hypothesis is whether or not it can account for the 3:1 ratio he observed in the F_2 generation of his numerous monohybrid crosses. The hypothesis predicts that the F_1 hybrids will produce two classes of gametes. When alleles separate, half the gametes receive a purple-flower allele, while the other half get a white-flower allele. During self-pollination, these two classes of gametes unite randomly. An ovum with a purple-flower allele has an equal chance of being fertilized by a sperm with a purple-flower allele or one with a white-flower allele. Since the same is true for an ovum with a white-flower allele, there are a total of four equally likely combinations of sperm and ovum. Figure 13.4 illustrates these combinations using a type of diagram called a Punnett square, a handy device for predicting the results of a genetic cross. Notice that a capital letter symbolizes a dominant allele. In our example, *P* is the allele for purple flowers.

What will be the physical appearance of these F_2 plants? One-fourth of the plants have two alleles specifying purple flowers; clearly, these plants will have purple flowers. But one-half of the F_2 progeny have inherited one allele for purple flowers and one allele for white flowers; like the F_1 plants, these plants will also have purple flowers, the dominant trait. Finally, one-fourth of the F_2 plants have inherited two alleles specifying white flowers and will, in fact, express the recessive trait. Thus, Mendel's model accurately explains the 3:1 ratio that he observed in the F_2 generation.

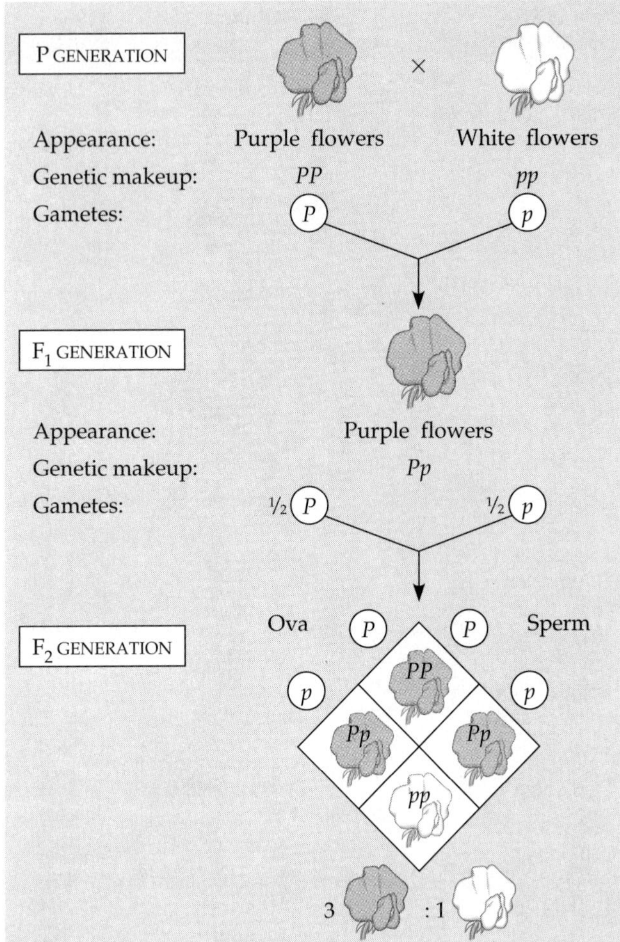

Figure 13.4
Mendel's law of segregation. A more detailed version of Figure 13.3, this diagram illustrates Mendel's model for monohybrid inheritance. The gene for flower color exists in two forms, or alleles. The purple-flower allele *(P)* is dominant, and the white-flower allele *(p)* is recessive. Each plant has two genes for flower color, one inherited from each parent. Each true-breeding plant of the parental generation has matching alleles, either *PP* (purple-flower parentals) or *pp* (white-flower parentals). Gametes, symbolized with circles, contain only one gene for flower color. Union of the parental gametes produces F_1 hybrids having nonmatching alleles, a *Pp* combination. Because the purple-flower allele is dominant, all these hybrids have purple flowers. When the hybrids produce gametes, the two alleles segregate, half the gametes receiving the *P* allele and the other half receiving the *p* allele. Random combination of these gametes results in the 3:1 ratio that Mendel observed in the F_2 generation. The box at the bottom of the figure is a Punnett square, a useful tool for showing all possible combinations of alleles in offspring.

Some Useful Genetic Vocabulary An organism having a pair of identical alleles for a character is said to be **homozygous** for that character. A pea plant that is true-breeding for purple flowers *(PP)* is an example. Pea plants with white flowers are homozygous for the recessive allele *(pp)*. If we cross dominant homozygotes with recessive homozygotes, as in the parental cross (P generation) of Figure 13.4, all the offspring will have a nonmatching combination of alleles—*Pp* in the case of the F_1 hybrids of our flower-color experiment. Organisms having two different alleles for a character are said to be **heterozygous** for that character. Unlike homozygotes, heterozygotes are not true-breeding, because they produce gametes having one *or* the other of the different alleles. We have seen that a *Pp* plant of the F_1 generation will produce both purple-flowered and white-flowered offspring when it self-pollinates.

Because of dominance and recessiveness, an organism's appearance does not always reflect its genetic

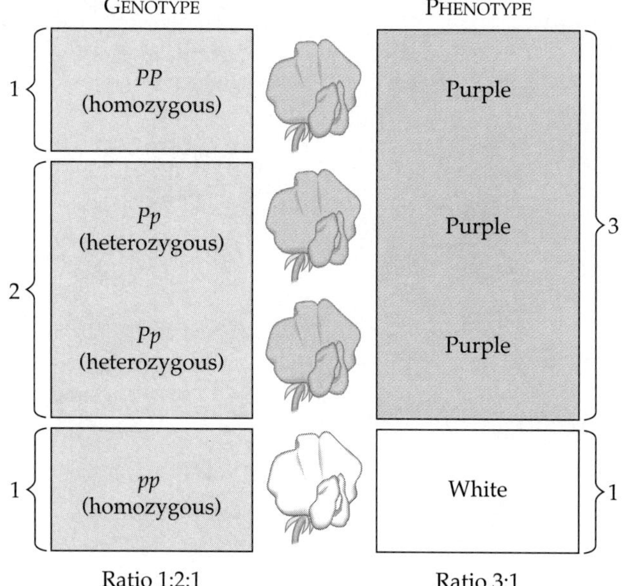

Figure 13.5
Genotype versus phenotype. Grouping F_2 offspring from a monohybrid cross for flower color according to phenotype—the physical appearance of the plants—results in the typical 3:1 ratio. In terms of genotype, however, there are actually two categories of purple-flowered plants: *PP* (homozygous) and *Pp* (heterozygous). Notice that there are two possible ways of generating a *Pp* genotype, depending on whether the ovum or sperm supplies the dominant allele. Also notice that the ratio of genotypes is 1*PP*:2*Pp*:1*pp*.

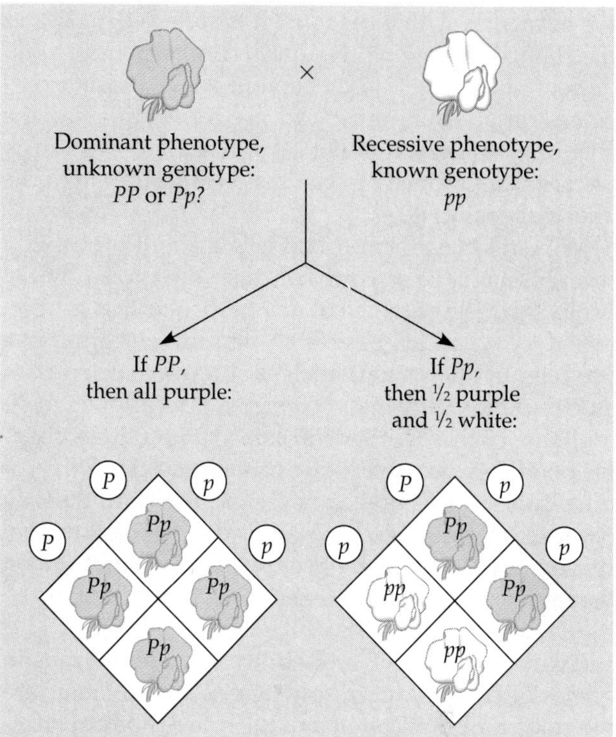

Figure 13.6
A testcross. A testcross is designed to reveal the genotype of an organism that exhibits a dominant trait, such as purple flowers in pea plants. Such an organism could be either homozygous or heterozygous for the dominant allele. The most efficient way to resolve the genotype is to cross the organism with an individual expressing the recessive trait. Since we know the genotype of the white-flowered parent (it must be homozygous for this trait to be expressed), we can deduce the genotype of the purple-flowered parent by observing the phenotypes of the offspring.

composition. Therefore, we have to distinguish between an organism's appearance, called its **phenotype,** and its genetic makeup, its **genotype.** In the case of flower color in peas, *PP* and *Pp* plants have the same phenotype (purple), but different genotypes. Figure 13.5 reviews these concepts. (Phenotype refers to physiological traits as well as physical traits. For example, one pea variety is incapable of the normal trait of self-pollination.)

The Testcross Suppose we have a pea plant that has purple flowers. How can we determine whether it is homozygous or heterozygous, since the genotypes *PP* and *Pp* result in the same phenotype? If we cross this pea plant with one having white flowers, the appearance of the offspring will reveal the genotype of the purple-flowered parent. The genotype of the plant with white flowers is known: Because this is the recessive trait, the plant must be homozygous. If all the offspring of the cross have purple flowers, then the other parent is homozygous also, since a *PP* × *pp* cross produces nothing but *Pp* offspring. But if both the purple and white phenotypes appear among the offspring, then the purple-flowered parent must be heterozygous. The offspring of a *Pp* × *pp* cross will have a 1:1 phenotypic ratio of *Pp* and *pp* (Figure 13.6). This breeding of a recessive homozygote with an organism of dominant phenotype, but unknown genotype, is called a **testcross.** It was devised by Mendel and continues to be an important tool of geneticists.

Inheritance as a Game of Chance

The law of segregation is a specific case of the same general rules of probability that apply to tossing coins, rolling dice, or drawing cards. A basic understanding of these rules of chance is essential for genetic analysis.

The probability scale ranges from 0 to 1. An event that is certain to occur has a probability of 1, while an event that is certain *not* to occur has a probability of 0. With a two-headed coin, the probability of tossing heads is 1, and the probability of tossing tails is 0. With a normal coin, the chance of tossing heads is ½ and the chance of tossing tails is ½. The probability of rolling the number 3 with a die, which is six-sided, is $\frac{1}{6}$, and the chance of drawing the queen of spades from a full deck of cards is $\frac{1}{52}$. The probabilities of all possible outcomes for an event must add up to 1. With a die, the chance of rolling a number other than 3 is $\frac{5}{6}$. In a deck of cards, the chance of drawing a card other than the queen of spades is $\frac{51}{52}$.

We can learn an important lesson about probability from tossing a coin. For every toss, the probability of heads is $\frac{1}{2}$. The outcome of any particular toss is unaffected by what has happened on previous attempts. We refer to phenomena such as successive coin tosses as independent events. (The term also applies to simultaneous tosses of several coins.) It is entirely possible that five successive tosses of a normal coin will produce five successive heads. Before the sixth toss, an observer might predict, "A tail is due to come up, because there have already been so many heads." But on the sixth toss, the chance that the outcome will again be heads is still $\frac{1}{2}$.

Two basic laws of probability that can help us in games of chance and in solving genetic problems are the rule of multiplication and the rule of addition.

The Rule of Multiplication If two coins are tossed simultaneously, the outcome for each coin is independent of what happens with the other coin. What is the chance that both coins will land heads up? How do we determine the chance that two or more independent events will occur together in some specific combination? The solution is in computing the probability for each independent event, then multiplying these individual probabilities to obtain the overall probability of these events occurring in combination. According to this rule of multiplication, the probability that both coins will land heads up is $\frac{1}{2} \times \frac{1}{2} = \frac{1}{4}$. A Mendelian F_1 cross is analogous to this game of chance. With flower color as the heritable character, the genotype of an F_1 plant is *Pp*. What is the probability that a particular F_2 plant will have white flowers? For this to happen, both the ovum *and* the sperm must carry the *p* allele, so we invoke the rule of multiplication. Segregation in the heterozygous plant is like flipping a coin. The probability that an ovum will have the *p* allele is $\frac{1}{2}$. The chance that a sperm will have the *p* allele is also $\frac{1}{2}$. Thus, the overall probability that two *p* alleles will come together at fertilization is $\frac{1}{2} \times \frac{1}{2} = \frac{1}{4}$, equivalent to the probability that two independently tossed coins will land heads up (Figure 13.7).

The Rule of Addition What is the probability that an F_2 plant will be heterozygous? Notice in Figure 13.7 that there are two ways F_1 gametes can combine to produce a heterozygous result. The dominant allele can come from the ovum and the recessive allele from the sperm, or vice versa. According to the rule of addition, the probability of an event that can occur in two or more different ways is the *sum* of the separate probabilities of those ways. Using this rule, we can calculate the probability of an F_2 heterozygote as $\frac{1}{4} + \frac{1}{4} = \frac{1}{2}$.

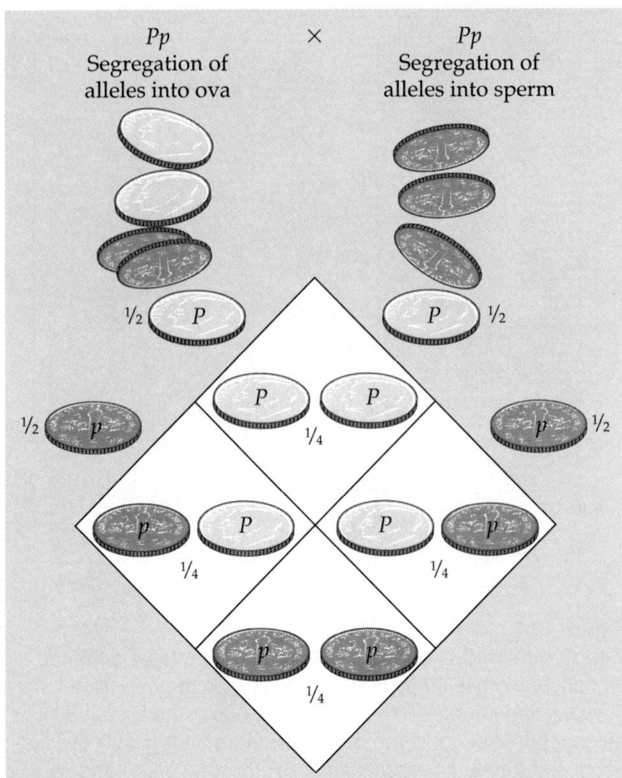

Figure 13.7
Segregation and fertilization as chance events. When a heterozygote *(Pp)* forms gametes, segregation of alleles is like the toss of a coin. An ovum has a 50% chance of receiving the dominant allele and a 50% chance of receiving the recessive allele. The same odds apply to a sperm cell. Like two separately tossed coins, segregation during sperm and ovum formation occurs as two independent events. To determine the probability that both the ovum and the sperm will carry the dominant allele, we multiply the probabilities of each required event: $\frac{1}{2} \times \frac{1}{2} = \frac{1}{4}$. Similarly, we can predict the probability for any genotype among offspring, as long as we know the genotypes of the parents.

The Statistical Nature of Inheritance If we plant a seed from the F_2 generation of Figure 13.4, we cannot predict with absolute certainty that the plant will yield white flowers, any more than we can predict with certainty that two tossed coins will both come up heads. What we *can* say is that there is exactly a $\frac{1}{4}$ chance that the plant will have white flowers. Stated another way, among a large sample of F_2 plants, one-fourth (25%) will have white flowers. Usually, the larger the sample size, the closer the results will conform to our predictions. The fact that Mendel counted so many offspring from his crosses suggests that he understood this statistical feature of inheritance. We will see more evidence of Mendel's keen sense of the rules of chance as

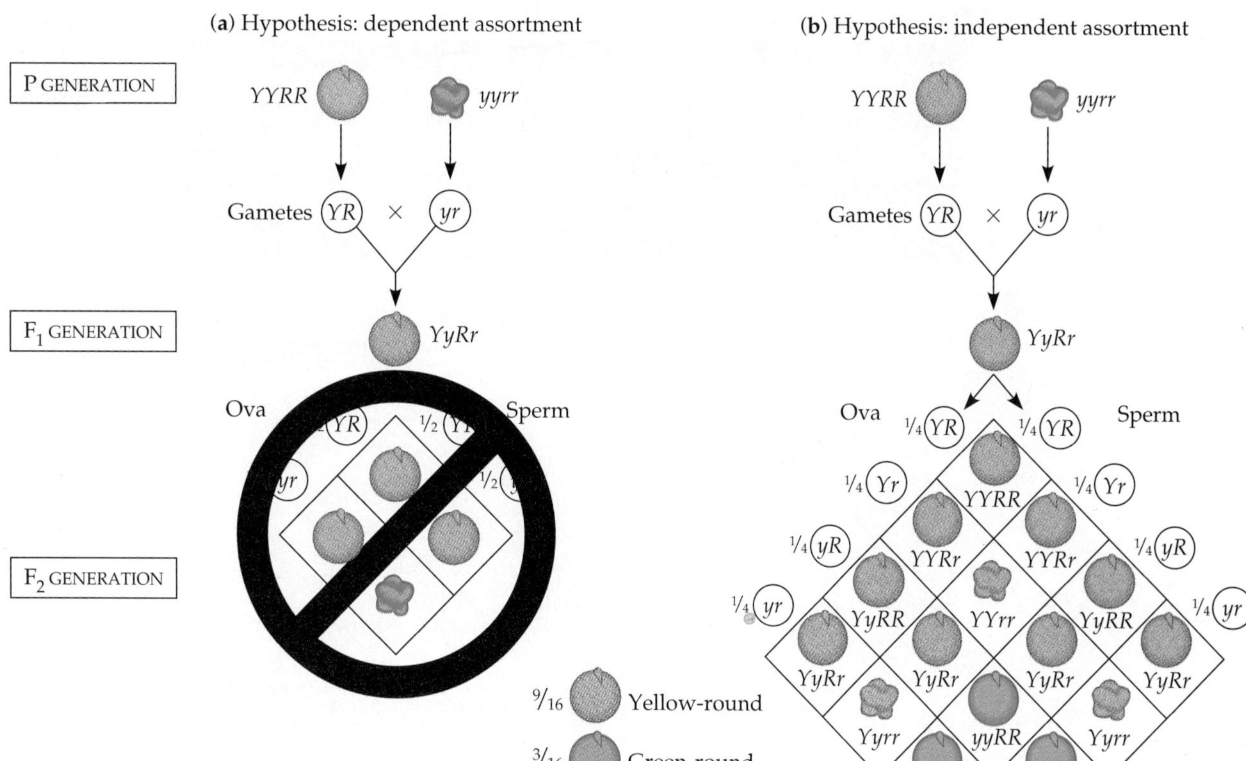

Figure 13.8
A comparison of two hypotheses for segregation in a dihybrid cross. A cross between true-breeding parent plants that differ in two characters produces F_1 hybrids that are heterozygous for both characters. In this example, the two characters are seed color and seed shape. Yellow color *(Y)* and round shape *(R)* are dominant. (**a**) If the two characters segregate dependently (together), the F_1 hybrids can only produce the same two classes of gametes that they received from the parents, and the F_2 progeny will show a 3:1 phenotypic ratio. (**b**) If the two characters segregate independently, four classes of gametes will be produced by the F_1 generation, and there will be a 9:3:3:1 phenotypic ratio in the F_2 generation. Mendel's results supported this latter hypothesis, called independent assortment.

we follow the discovery of his second law of inheritance.

Mendel's Law of Independent Assortment

Mendel derived the law of segregation by performing monohybrid crosses, breeding experiments using parental varieties that differ in a single character, such as flower color. What would happen in a mating of parental varieties differing in *two* characters—a **dihybrid cross?** For instance, two of the seven characters Mendel studied were seed color and seed shape. Seeds may be either yellow or green. They also may be either round or wrinkled. From monohybrid crosses, Mendel knew that the allele for yellow seeds is dominant, and the allele for green seeds is recessive. For the seed-shape character, round is dominant and wrinkled is recessive. In monohybrid crosses, such as between a pea plant with yellow-round seeds and a plant with yellow-wrinkled seeds, each character follows a simple Mendelian pattern of inheritance, producing a 3:1 ratio in the F_2 generation. But what happens if we breed two pea varieties differing in *both* of these characters—a parental cross between a plant with yellow-round seeds *(YYRR)* and a plant with green-wrinkled seeds *(yyrr)*? Are these two characters, seed color and seed shape, transmitted from parents to offspring as a package? Put another way, will the *Y* and *R* alleles always stay together, generation after generation? Or, are seed color and seed shape inherited independently of each other? Figure 13.8 illustrates how a dihybrid cross can determine which of these two hypotheses is correct.

In the F_1 generation of this dihybrid cross, the genotype is *YyRr*, and the plants exhibit both dominant phenotypes, yellow seeds with round shapes. The key step in the experiment is to see what happens when F_1 plants self-pollinate to produce F_2 offspring. If the hy-

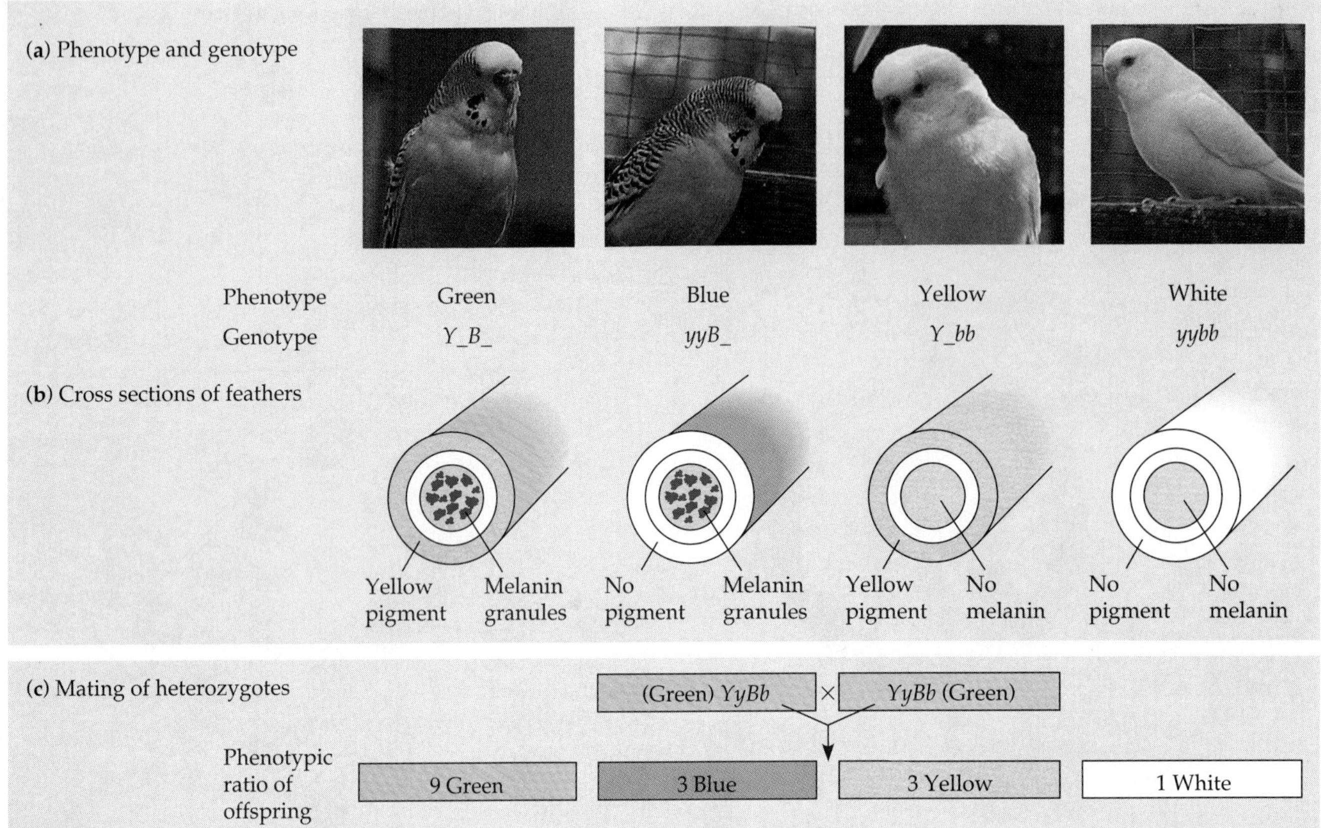

Figure 13.9
Independent assortment and variations in the feather color of budgies. (**a**) Two different and independently inherited genes affect the color of plumage of these small parrots called budgies. One gene determines pigmentation in the outer region of a feather, and the second gene determines pigmentation of the feather's core. Two alleles exist for each gene. For the outer zone, the allele for yellow color is dominant *(Y)* and the allele for a colorless condition is recessive *(y)*. For the core of the feather, an allele for the presence of a pigment called melanin is dominant *(B)* and the allele for absence of melanin is recessive *(b)*. Various combinations of these alleles result in the four phenotypes illustrated by the photos. (The blanks in the genotypes indicate that homozygous dominant and heterozygous conditions result in the same phenotype.) (**b**) Cross sections of feathers reveal the anatomical and biochemical bases for the four phenotypes. A dihybrid cross between true-breeding green budgies *(YYBB)* and white birds *(yybb)* results in green offspring that are heterozygous for both genes *(YyBb)*. (**c**) When these dihybrids mate, a Mendelian F_1 cross, the four possible phenotypes occur in the F_2 generation in a 9:3:3:1 ratio. If we construct a Punnett square for the results of the F_1 cross, we can derive this ratio by grouping genotypes that produce the same phenotype.

brids must transmit their alleles in the same combinations in which they were inherited from the P generation, then there will only be two classes of gametes: *YR* and *yr*. This hypothesis predicts that the phenotypic ratio of the F_2 generation will be 3:1, just as in a monohybrid cross (Figure 13.8a).

The alternative hypothesis is that the two pairs of genes segregate independently of each other. In other words, genes are packaged into gametes in all possible allelic combinations, as long as each gamete has one gene for each character. In our example, four classes of gametes would be produced in equal quantities: *YR*, *Yr*, *yR*, and *yr*. If four classes of sperm are mixed with four classes of ova, there will be 16 (4 × 4) equally probable ways in which the alleles can combine in the F_2 generation, as shown in Figure 13.8b. These combinations make up four phenotypic categories with a ratio of 9:3:3:1 (nine yellow-round to three green-round to three yellow-wrinkled to one green-wrinkled). When Mendel did the experiment and categorized ("scored") the F_2 progeny, he obtained a ratio of 315:108:101:32, which is approximately 9:3:3:1.

The experimental results supported the hypothesis that each character is independently inherited; that in the dihybrids *(YyRr)*, the two alleles for seed color segregate independently of the two alleles for seed shape. Mendel tried his seven pea characters in various dihybrid combinations and always observed a 9:3:3:1 phenotypic ratio in the F_2 generation. Notice in Figure 13.8b, however, that there remains a 3:1 phenotypic ratio for each character: three yellow to one green; three round to one wrinkled. As far as an individual character is concerned, the segregation behavior is the same as if this were a monohybrid cross. This behavior of alleles during gamete formation is now called Mendel's **law of independent assortment.** Figure 13.9 shows how this law applies to a mating of budgies that are heterozygous for two characters.

We can use the rules of probability, as applied to segregation and independent assortment, to solve some complex genetics problems. For instance, Mendel crossed pea varieties that differed in three characters, and the results of these *trihybrid crosses* could be explained by an independent assortment of alleles. Consider a trihybrid cross between two organisms with these genotypes: $AaBbCc \times AaBbCc$. What is the probability that an offspring from this cross will be a recessive homozygote for all three characters *(aabbcc)*? Since the two alleles for each character assort into gametes independently, we can treat this as three separate monohybrid crosses:

$Aa \times Aa$: Probability for *aa* offspring = $\frac{1}{4}$
$Bb \times Bb$: Probability for *bb* offspring = $\frac{1}{4}$
$Cc \times Cc$: Probability for *cc* offspring = $\frac{1}{4}$

We use the rule of multiplication to calculate the probability that the offspring will be *aabbcc:*

$$\frac{1}{4}aa \times \frac{1}{4}bb \times \frac{1}{4}cc = \frac{1}{64}$$

Mendel's two laws, segregation and independent assortment, explain heritable variations in terms of discrete genes that are passed along, generation after generation, according to simple rules of chance. This particulate theory of inheritance, first discovered in garden peas, is equally valid for figs, flies, fish, birds, and human beings. Mendel's impact endures, not only on genetics, but on all of science, as testimony to the power of the hypothetico-deductive method (see Chapter 1).

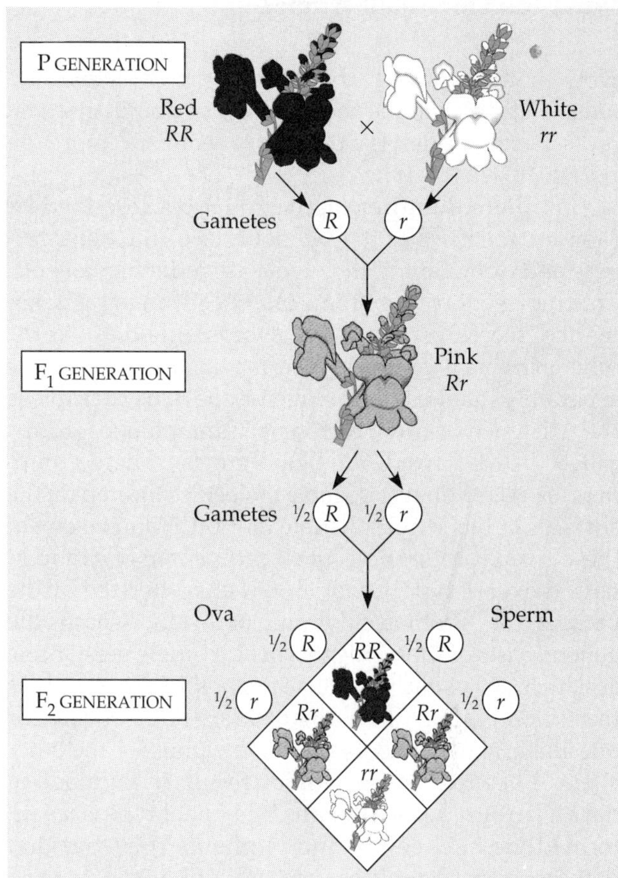

Figure 13.10
Incomplete dominance in snapdragon color. When red snapdragons are crossed with white ones, the F_1 hybrids have pink flowers. Segregation of alleles into the gametes of the F_1 plants results in an F_2 generation with a 1:2:1 ratio for both genotype and phenotype.

EXTENDING MENDELIAN GENETICS

In this century, geneticists have extended Mendelian principles to patterns of inheritance more complex than Mendel actually described. It was brilliant (or lucky) that Mendel chose pea plant characters that turned out to have a relatively simple genetic basis: Each character is determined by one gene, for which there are only two alleles, one completely dominant to the other. But these conditions are not met by all heritable characters, not even in garden peas. This does not diminish the utility of Mendelian genetics, for the basic principles of segregation and independent assortment apply even to more complex cases. In this section, we will extend Mendelian genetics to patterns of inheritance that were not reported by Mendel.

Incomplete Dominance

The F_1 offspring of Mendel's classic pea crosses always looked like one of the two parental varieties because of the complete dominance of one allele over another. But for some traits, there is **incomplete dominance,** where the F_1 hybrids have an appearance somewhere in between the phenotypes of the two parental varieties. For instance, when red snapdragons are crossed with white snapdragons, all the F_1 hybrids have pink flowers (Figure 13.10). This third phenotype results from the heterozygote flowers having less red pigment than the red homozygotes. We should not, however, regard incomplete dominance as evidence of the blending theory, which would predict that the red or white traits could never be retrieved from the pink hybrids. In fact, breeding the F_1 hybrids produces F_2 offspring with a phenotypic ratio of 1 red to 2 pink to 1 white. (Notice that when dominance is incomplete, we can distinguish the heterozygotes from the two homozygous varieties, and the genotypic and phenotypic ratios for the F_2 generation are the same, 1:2:1.) The segregation of the red and white alleles in the gametes produced by the pink-flowered plants confirms that the genes for flower color are heritable factors that maintain their identity in the hybrids; that is, inheritance is particulate.

What Is a Dominant Allele?

Now that you have learned about incomplete dominance, let's reexamine the meaning of dominance and recessiveness. What is a dominant allele? Or, more importantly, what is it *not?*

Complete dominance, the situation described by Mendel, means that the phenotypes of the heterozygote and dominant homozygote are indistinguishable. This represents one extreme of a spectrum in the dominance/recessiveness relationships of alleles. At the other extreme is **codominance,** in which *both* alleles are separately manifest in the phenotype. One example is the existence of three different human blood groups called the M, N, and MN blood groups. These groupings are based on two specific molecules located on the surfaces of blood cells. People of group M have one of these two types of molecules, and people of group N have the other type. Group MN is characterized by the presence of *both* molecules on blood cells. What is the genetic basis of these phenotypes? A single gene locus, at which two allelic variations are possible, determines these blood groups. M individuals are homozygous for one allele; N individuals are homozygous for the other allele. A heterozygous condition results in the blood of the MN group. Notice that the MN phenotype is *not* intermediate between M and N phenotypes, but that both of these phenotypes are individually expressed by the presence of the two types of molecules on blood cells. In contrast, incomplete dominance is characterized by an intermediate phenotype, as in the pink flowers of snapdragon hybrids. Thus, the range of relationships between alleles includes complete dominance, codominance, and different degrees of incomplete dominance. These variations are reflected in the phenotypes of heterozygotes.

For any character, the dominance/recessiveness relationship we observe depends on the level at which we examine phenotype. For example, consider Tay-Sachs disease, an inherited disorder in humans. The brain cells of a baby with Tay-Sachs disease are unable to metabolize gangliosides, a type of lipid, because a crucial enzyme does not work properly. As the lipids accumulate in the brain, the brain cells gradually cease to function normally, leading to death (you will learn more about this disease later in the chapter). Only children who inherit two copies of the Tay-Sachs allele (homozygotes) have the disease. Thus on the *organismal* level of normal versus Tay-Sachs phenotype, the Tay-Sachs allele qualifies as a recessive. At the *biochemical* level, however, we observe an intermediate phenotype characteristic of incomplete dominance: The enzyme deficiency that causes Tay-Sachs disease can be detected in heterozygotes, who have an activity level of the lipid-metabolizing enzyme that is intermediate between individuals homozygous for the normal allele and individuals with Tay-Sachs disease. Heterozygotes lack symptoms of the disease, apparently because half the normal amount of functional enzyme is sufficient to prevent lipid accumulation in the brain. In fact, heterozygous individuals produce equal numbers of normal and dysfunctional enzymes. Thus, at the *molecular* level, the normal allele and the Tay-Sachs allele are codominant. As you can see, dominance/recessiveness relationships are rarely as straightforward as Mendel reported.

It is also important to understand that an allele is not termed *dominant* because it somehow subdues and mutes a recessive allele. Recall that alleles are simply variations in a gene's nucleotide sequence. When a dominant allele coexists with a recessive allele in a heterozygous genotype, they do not actually interact at all. It is in the pathway from genotype to phenotype that dominance and recessiveness come into play. We can use one of Mendel's characters—round versus wrinkled pea seeds—as an example. The dominant allele codes for the synthesis of an enzyme that helps convert sugar to starch in the seed. The recessive allele codes for a defective form of this enzyme. Thus, in a recessive homozygote, sugar accumulates in the seed because it is not converted to starch. As the seed develops, the high sugar concentration stimulates the osmotic uptake of water, and the seed swells (see Chapter 8). When the mature seeds dry, they have wrinkles, analogous to the stretch marks of a person who has lost a lot of weight. In contrast, if a dominant allele is present, sugar is converted to starch, and the seeds do not wrinkle when they dry. One dominant allele results in enough of the enzyme to convert sugar to starch, and thus dominant homozygotes and heterozygotes have the same phenotype: round seeds. By exploring the mechanisms responsible for phenotype, we can demystify the concepts of dominance and recessiveness.

There is another important lesson about the meaning of the term *dominance.* Because an allele for a particular character is dominant does not necessarily mean that it is more common in a population than the recessive allele for that character. For example, about one baby out of 400 in the United States is born with extra fingers or toes, a condition known as polydactyly. The allele for polydactyly is *dominant* to the allele for five digits per appendage. In other words, 399 out of every 400 people are recessive homozygotes for this character; the recessive allele is far more prevalent than the dominant allele in the population. In Chapter 21, you will learn that the relative frequencies of alleles in a population result mainly from natural selection.

Let's summarize three important points about dominance/recessiveness relationships. (1) They range from complete dominance, through various degrees of incomplete dominance, to codominance. (2) They reflect the mechanisms by which specific alleles are expressed in phenotype and do not involve the ability of

Blood Group Phenotype	Genotypes	Antibodies Present in Blood Serum	Reacts (Clumps) When Red Blood Cells from Groups Below Are Added to Serum from Groups at Left?			
			O	A	B	AB
O	ii	Anti-A Anti-B	No	Yes	Yes	Yes
A	$I^A I^A$ or $I^A i$	Anti-B	No	No	Yes	Yes
B	$I^B I^B$ or $I^B i$	Anti-A	No	Yes	No	Yes
AB	$I^A I^B$	—	No	No	No	No

Figure 13.11
Multiple alleles for the ABO blood groups. There are three alleles. Beca each person carries two alleles, six genotypes are possible. Whenever the I^A or the I^B allele is present, the corresponding factor (A or B, respectively) is present on the surface of red blood cells. Both of these alleles, which are codominant, are dominant to the i allele, which does not code for any surface factor. A person produces antibodies against foreign blood factors, causing clumping of red blood cells if a transfusion is performed with incompatible blood.

one allele to subdue another at the level of the DNA. (3) They do not determine the relative abundance of alleles in a population.

Multiple Alleles

Most genes actually exist in more than two allelic forms. The ABO blood groups in humans are one example of multiple alleles. There are four phenotypes for this character. A person's blood group may be either A, B, AB, or O. These letters refer to two carbohydrates, the A substance and the B substance, which may be found on the surface of red blood cells. (These groups are based on different molecular markers on blood cells than those used for the MN classification.) A person's blood cells may be coated with one substance or the other (type A or B), with both (type AB), or with neither (type O). Matching compatible blood groups is critical for blood transfusions. If the donor's blood has a factor (A or B) that is foreign to the recipient, specific proteins called antibodies produced by the recipient bind to the foreign molecules and cause the donor's blood cells to agglutinate (clump together). This agglutination can cause the recipient to die.

The four blood groups result from various combinations of three different alleles, symbolized as I^A (for the A carbohydrate), I^B (for B), and i (giving rise to neither A nor B). Six genotypes are possible (Figure 13.11). Both the I^A and I^B alleles are dominant to the i allele. Thus, I^AI^A and I^Ai individuals have type A blood, and I^BI^B and I^Bi individuals have type B. Recessive homozygotes, ii, have type O blood, because neither the A nor the B substance is produced. The I^A and I^B alleles are codominant; both are expressed in the phenotype of the I^AI^B heterozygote, who has type AB blood.

Pleiotropy

So far, we have treated Mendelian inheritance as though each gene affects one phenotypic character. Most genes, however, have multiple phenotypic effects. This situation is called **pleiotropy.** For example, the same allele causes both abnormal pigmentation and a cross-eye condition in tigers. And alleles that are responsible for certain hereditary diseases in humans, including sickle-cell anemia, usually cause multiple symptoms. Considering the intricate molecular and cellular interactions responsible for an organism's development, it is not surprising that a gene can affect many of the organism's characteristics.

Epistasis

In some cases, a gene at one locus alters the phenotypic expression of a second gene, a condition known as **epistasis,** (Gr. for "standing still on"). An example will help clarify this concept. In mice and many other mammals, black coat color is dominant to brown. The two alleles for this character are designated B and b. For a

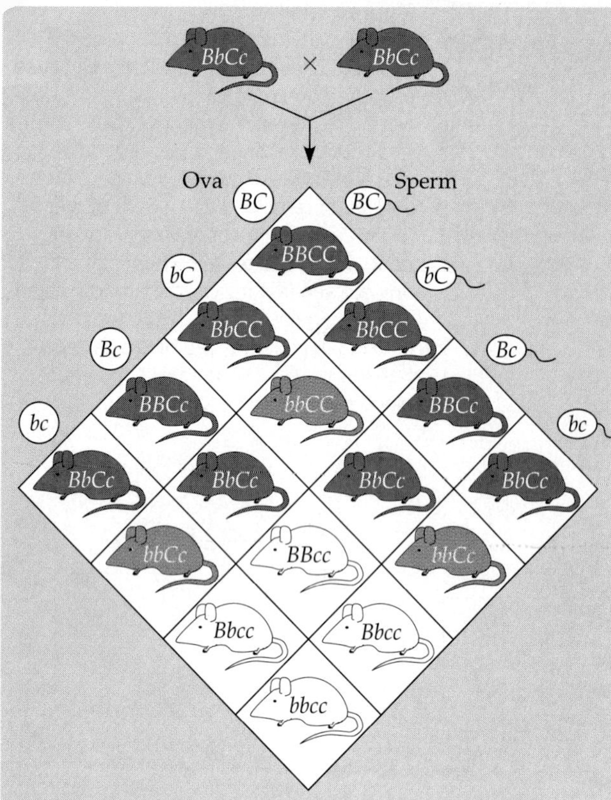

Figure 13.12
Epistasis. This Punnett square illustrates the genotypes and phenotypes for offspring of matings between two black mice. The parents are heterozygous for two genes, which assort independently of each other. Thus, our breeding experiment corresponds to a Mendelian F_1 cross between dihybrids. One gene determines whether the coat will be black (dominant, *B*) or brown (recessive, *b*). The second gene controls whether or not pigment of *any* color will be deposited in the hair, with the allele for the presence of color *(C)* dominant to the allele for the absence of color *(c)*. All offspring of the *cc* genotype are white (albino), regardless of the genotype for the black/brown genetic locus. This epistatic relationship of the color gene to the black/brown gene results in an F_2 phenotypic ratio of 9 black to 3 brown to 4 white.

mouse to have brown fur, its genotype must be *bb*. But there is more to the story. A second gene locus determines whether or not pigment will be deposited in the hair. For this second gene, the dominant allele, symbolized by *C*, results in the deposition of pigment. This allows either black or brown color, depending on the genotype at the first locus. But if the mouse is homozygous recessive for the second locus *(cc)*, then the coat is white (albino), regardless of the genotype at the black/brown locus.

What happens if we mate black mice that are heterozygous for both genes *(BbCc)*? Although the two genes affect the same phenotypic character (coat color), they follow the law of independent assortment (the two genes are inherited separately). Thus, our breeding experiment represents an F_1 dihybrid cross, which produced a 9:3:3:1 ratio in Mendel's experiments. In the case of coat color, however, the ratio of phenotypes among F_2 offspring is 9 black to 3 brown to 4 white. Figure 13.12 uses a Punnett square to clarify this ratio in terms of epistasis. Other types of epistatic interactions produce different ratios.

Polygenic Inheritance

Mendel studied characters that could be classified on an either-or basis, such as purple versus white flowers. There are many characters, however, including human skin color and height, for which an either-or classification is impossible, because the characters vary in the population along a continuum (in gradations). These are called **quantitative characters.** Quantitative variation usually indicates **polygenic inheritance,** an additive effect of two or more genes on a single phenotypic character (the converse of pleiotropy, where a single gene affects many phenotypic characters).

There is evidence, for instance, that skin pigmentation in humans is controlled by at least three separately inherited genes (probably more, but we will simplify). Let's consider three genes, with a dark-skin allele for each gene *(A, B, C)* contributing one "unit" of darkness to the phenotype and being incompletely dominant over the other alleles *(a, b, c)*. An *AABBCC* person would be very dark, while an *aabbcc* individual would be very light. An *AaBbCc* person would have skin of an intermediate shade. Because the alleles have a cumulative effect, the genotypes *AaBbCc* and *AABbcc* would make the same genetic contribution (three units) to skin darkness. Figure 13.13 shows how this system could result in a bell-shaped curve, called a normal distribution, for skin darkness among the members of a hypothetical population. (You are probably familiar with the concept of a normal distribution for class curves of test scores.) Environmental factors, such as exposure to the sun, would modulate the skin-color phenotype and help make the graph a smooth curve rather than a stairlike histogram.

Nature Versus Nurture: The Environmental Impact on Phenotype

Phenotype depends on environment as well as on genes. A single tree, locked into its inherited genotype, has leaves that vary in size, shape, and greenness, depending on exposure to wind and sun. For humans, nutrition influences height, exercise alters build, suntanning darkens the skin, and experience increases IQ.

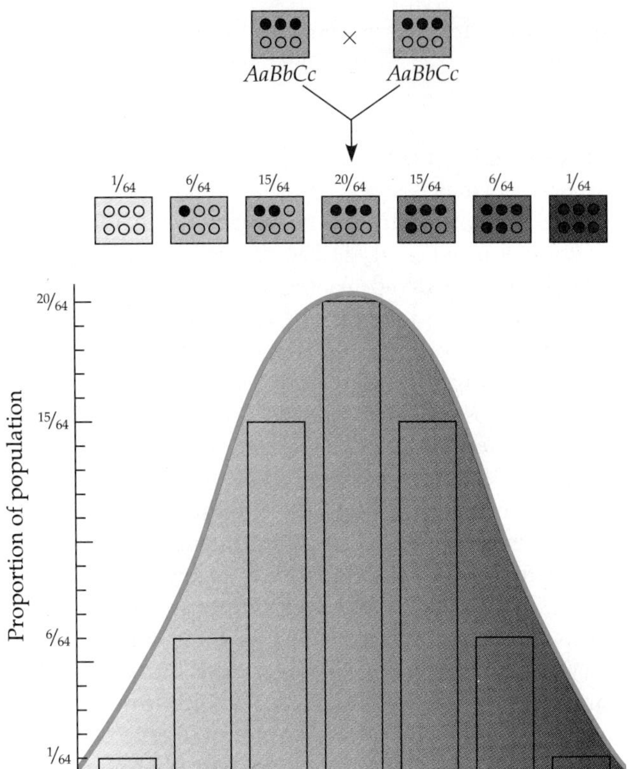

Figure 13.13
A simplified model for polygenic inheritance of skin color. According to this model, three separately inherited genes affect the darkness of skin. For each gene, an allele for dark skin is incompletely dominant over an allele for light skin. Thus, an individual who is heterozygous for all three genes *(AaBbCc)* has inherited three "units" of darkness, indicated in this figure by the three dots in the small square. An individual with the genotype *AABbcc* would also have a total of three units for dark skin. Imagine a large number of matings between individuals who are heterozygous for all three genes. Along the top of the graph are the variations that can occur among offspring. The *y*-axis represents the proportions of these variations among offspring of such trihybrid matings. (You can calculate the proportions with the help of a Punnett square or by using the rules of multiplication and addition to determine the probabilities of different genotypes and phenotypes.) The resulting histogram is smoothed into a bell-shaped curve by environmental factors, such as exposure to the sun, that affect skin color. This model is an oversimplification, but it should help clarify how polygenic inheritance can contribute to phenotypic characters that vary continuously among the members of a population.

Even identical twins, who are genetic equals, accumulate phenotypic differences as a result of their unique experiences.

Whether it is genes or the environment—nature or nurture—that most influences human characteristics is a very old and hotly contested debate that we will not attempt to settle here. We can say, however, that the product of a genotype is generally not a rigidly defined phenotype, but a range of phenotypic possibilities over which there may be variation due to environmental influence. This phenotypic range is called the **norm of reaction** for a genotype (Figure 13.14). There are cases where the norm of reaction has no breadth whatsoever; that is, a given genotype mandates a very specific phenotype. An example is the gene locus that determines a person's ABO blood group. In contrast, a person's blood count of red and white cells varies, depending on such factors as the altitude of one's home, the person's customary level of physical activity, and the presence of infectious agents.

Generally, norms of reaction are broadest for polygenic characters, including behavioral traits. Environment contributes to the quantitative nature of these characters, as we have seen in the continuous variation of skin color. Geneticists refer to such characters as **multifactorial,** meaning that many factors, both genetic and environmental, collectively influence phenotype.

Figure 13.14
The effect of environment on phenotype. The effect of a genotype is defined by its norm of reaction, a phenotypic range that depends on the environment in which the genotype is expressed. Hydrangea flowers of the same genetic variety range in color from blue-violet to pink, depending on the acidity of the soil.

Integrating a Mendelian View of Heredity and Variation

Over the past several pages, we have broadened our view of Mendelian inheritance by exploring incomplete dominance and other variations in dominance/recessiveness relationships, multiple alleles, pleiotropy, polygenic inheritance, and the phenotypic impact of the environment. How can we integrate these refinements into a comprehensive theory of Mendelian genetics? The key is to make the transition from the reductionist emphasis on single genes and phenotypic characters to the idea of the organism as a whole. In fact, the term *phenotype* does double duty. We have been using the word in the context of specific characters, such as flower color and blood group. But phenotype is also used to describe the organism in its entirety—*all* aspects of its physical appearance, internal anatomy, physiology, and behavior. Similarly, the term *genotype* can also refer to an organism's entire genetic makeup, not just its alleles for a single genetic locus. In most cases, a gene's impact on phenotype is affected by other genes and by the environment. In this integrated view of heredity and variation, an organism's phenotype reflects its overall genotype and unique environmental history.

Considering all that can occur in the pathway from genotype to phenotype, it is indeed impressive that Mendel could simplify the complexities to reveal the fundamental principles governing the transmission of individual genes from parents to offspring. By extending the principles of segregation and independent assortment to help explain such hereditary patterns as epistasis and quantitative characters, we begin to see how broadly Mendelism applies. From the abbey garden came a theory of particulate inheritance that anchors modern genetics. In the last section of this chapter, we will apply Mendelian genetics to human inheritance, especially the transmission of hereditary diseases.

MENDELIAN INHERITANCE IN HUMANS

Whereas peas are convenient subjects for genetic research, humans are not. The human generation span is about 20 years, and human parents produce relatively few offspring (compared to peas and most other species). Furthermore, well-planned breeding experiments like the ones Mendel performed are impossible (or at least socially unacceptable) with humans. In spite of these difficulties, the study of human genetics continues to advance, powered by the incentive to understand our own inheritance. New techniques in molecular biology have led to many breakthrough discoveries, as we will see in Chapter 19, but basic Mendelism endures as the foundation of human genetics.

Human Pedigrees

Unable to manipulate the mating patterns of people, geneticists must analyze the results of matings that have already occurred. As much information as possible is collected about a family's history for a particular trait, and this information is assembled into a family tree describing the interrelationships of parents and children across the generations—the family **pedigree.** The pedigree in Figure 13.15a traces the occurrence of a trait called wooly hair for three generations of one family. Caucasian individuals with this trait have curly hair that is fuzzy in texture, superficially similar to the hair of many Blacks. Because it is very brittle and breaks at the tips, wooly hair cannot grow long. The trait is due to a dominant allele that we will symbolize by *W*. Wooly hair is rare in the overall human population; thus, most people are recessive homozygotes *(ww)*. However, the family in the pedigree of Figure 13.15a has a heritage of the wooly hair phenotype, and we can apply Mendelian principles to analyze the pedigree. First, we can deduce the genotypes of the couple at the top of the pedigree (first generation in this pedigree). We know that the man is *ww*, because he lacks wooly hair. The woman, who has wooly hair, must be heterozygous *(Ww)*; we deduce this because three of her six children had nonwooly hair. Had she been homozygous for wooly hair *(WW)*, all children born to the couple would have had wooly hair. This pedigree fits (more perfectly than usual) the 1:1 phenotypic ratio expected from a Mendelian testcross *(Ww × ww)*. Figure 13.15b illustrates pedigree analysis for a recessive trait.

A pedigree not only helps us understand the past; it also helps us predict the future. Suppose that one of the grandsons with wooly hair marries a woman with nonwooly hair (see Figure 13.15a, bottom right). They plan to have three children. What is the probability that all three children will have wooly hair? The pedigree tells us that the grandson must have the genotype *Ww*. Since each of his children is an independent event, genetically speaking, each has a $\frac{1}{2}$ chance of inheriting the wooly allele. Applying the rule of multiplication, the overall probability that all three of the anticipated children will have wooly hair is $\frac{1}{2} \times \frac{1}{2} \times \frac{1}{2} = \frac{1}{8}$. The overall probability that *not* all three children will have wooly hair is therefore $\frac{7}{8}$ (since there is a total probability of 1 that the children will have one or the other type of hair).

When the genetic trait to be analyzed can lead to a disabling or lethal disorder, the stakes are much higher than when examining a harmless trait like hair type. But for disorders that are inherited as simple Mendelian traits, these same basic techniques for pedi-

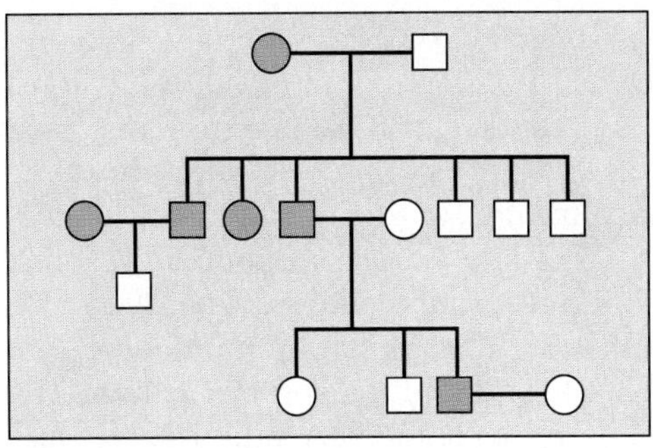

(a) Pedigree of a dominant trait

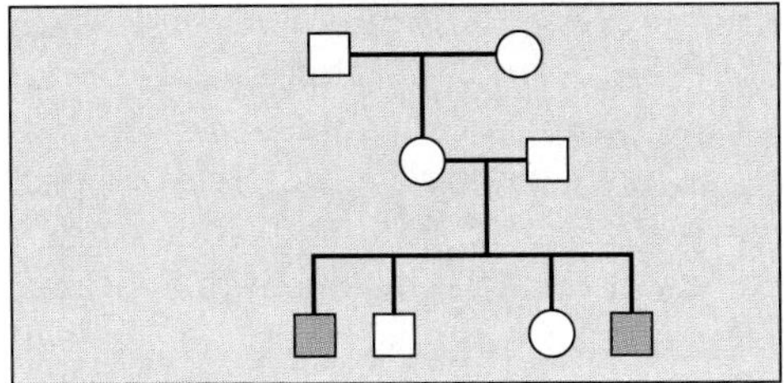

(b) Pedigree of a recessive trait

Figure 13.15
Pedigree analysis. In these family trees, squares symbolize males and circles represent females. Horizontal lines indicate matings, with offspring listed below in their order of birth, from left to right. Colored symbols stand for individuals with the trait we are tracing through the pedigree. (**a**) This pattern fits the case of a dominant trait—wooly hair, in our example. You should be able to fill in the genotypes for most individuals in this pedigree. One clue is the mating in the second generation (left) between two people with wooly hair. If the trait were recessive, then the child of this couple would have wooly hair. (**b**) This pedigree traces the recessive trait albinism (lack of skin pigmentation) through three generations. Both parents in the second generation have pigmented skin, and yet two of their four children have the albino trait. We can deduce that both of these parents are heterozygous and have pigmented skin because the allele for that trait is dominant. Why do the proportions of the phenotypes in the third generation not conform exactly to a Mendelian ratio of 3:1? (Remember the importance of sample size.)

gree analysis are applied by geneticists, physicians, and genetic counselors.

Recessively Inherited Disorders

Several thousand genetic disorders are known to be inherited as simple recessive traits. These disorders range in severity from traits that are usually nonlethal, such as albinism (lack of skin pigmentation), to deadly diseases, such as cystic fibrosis.

How can we account for the recessive behavior of the alleles causing these disorders? Recall that genes code for proteins of specific function. An allele that causes a genetic disorder codes either for a malfunctional protein or for no protein at all. In the case of disorders classified as recessive, heterozygotes are normal in phenotype, because one copy of the "normal" allele produces a sufficient amount of the specific protein. Thus, a recessively inherited disorder shows up only in the homozygous individuals who inherit one recessive allele from each parent. We can symbolize the genotype of such persons as *aa*, with normal individuals being either *AA* or *Aa*. The heterozygotes (*Aa*), who are phenotypically normal, are called carriers of the disorder, and they may transmit the recessive allele to their offspring.

The vast majority of people afflicted with recessive disorders are born to normal parents who are both carriers. A mating between two carriers corresponds to a Mendelian F_1 cross ($Aa \times Aa$), with the zygote having a $\frac{1}{4}$ chance of inheriting a double dosage of the recessive allele. A normal child from such a cross has a $\frac{2}{3}$ chance of being a carrier. (Genotypic ratio for the offspring is 1*AA*:2*Aa*:1*aa*. Thus, two out of three individuals of *normal* phenotype—*AA* or *Aa*—are heterozygous carriers.) Recessive homozygotes could also result from $Aa \times aa$ and $aa \times aa$ matings, but if the disorder is lethal before reproductive age or results in sterility, no *aa* individuals will reproduce. Even if recessive homozygotes are able to reproduce, such individuals will still account for a much smaller percentage of the population than heterozygous carriers, for reasons we will discuss in Chapter 21.

In general, a genetic disorder is not evenly distributed among all racial or cultural groups. These disparities result from the different genetic histories of the world's peoples during less technological times, when populations were more geographically, hence genetically, isolated (see Chapter 21). We will now examine three examples of such recessively inherited disorders.

The most common lethal genetic disease in the United States is cystic fibrosis, which strikes one out of every 2500 Caucasians but is much rarer in other races. One out of 25 Caucasians (4%) is a carrier (heterozygote). The dominant allele for this gene codes for a membrane protein that pumps chloride ions out of cells. These chloride pumps are defective or absent in children who inherit two copies of the recessive allele that causes cystic fibrosis. Chloride accumulates in cells, which take up water by osmosis from the mucus that coats the cells. The mucus thickens and builds up in the pancreas, lungs, digestive tract, and other organs, a condition that favors pneumonia and other infections. Untreated, most children with cystic fibrosis die by the time they are 4 or 5 years old. A special diet, daily doses of antibiotics to prevent infection, and other preventive treatments can prolong life; in the United States, the average longevity of individuals with cystic fibrosis is 27 years.

Another lethal disorder inherited as a recessive allele is Tay-Sachs disease, which was described earlier in the chapter. Recall that the disease is caused by a dysfunctional enzyme that fails to break down a class of brain lipids. The symptoms of Tay-Sachs disease usually become manifest a few months after birth. The infant begins to suffer seizures, blindness, and degeneration of motor and mental performance. Inevitably, the child dies within a few years. There is a disproportionately high incidence of Tay-Sachs disease among Ashkenazic Jews, Jewish people whose ancestors lived in central Europe. In that population, the frequency of the disease is one case out of 3600 births, about 100 times greater than the incidence among non-Jews or Mediterranean (Sephardic) Jews.

By far, the most common inherited disease among blacks is sickle-cell anemia, which affects one out of 400 African-Americans. This disease is caused by substitution of a single amino acid in the hemoglobin protein of red blood cells (see Chapter 5). This causes the cells to deform to a sickle shape, which in turn triggers blood clotting and many other symptoms. The multiple effects of the sickle-cell allele exemplify pleiotropy.

Individuals who are heterozygous for the sickle-cell allele are said to have sickle-cell trait. These carriers are usually healthy, although a fraction of the heterozygotes suffer some symptoms of sickle-cell anemia when there is an extended reduction of blood oxygen—at very high altitude, for instance. (The two alleles are codominant at the molecular level; both normal and abnormal hemoglobins are made.) About one out of ten African-Americans has sickle-cell trait, an unusually high frequency of heterozygotes for an allele with severe detrimental effects in homozygotes. The reason for the prevalence of this allele appears to be that while only individuals who are homozygous for the sickle-cell allele suffer from the disease, in certain environments, heterozygotes have an advantage over people who carry no copies of the sickle-cell allele. A single copy of the sickle-cell allele increases resistance to malaria. Thus, in tropical Africa, where malaria is common, the sickle-cell allele is both boon and bane. The relatively high frequency of African-Americans with sickle-cell trait is a vestige of their African roots.

Consanguinity Although it is relatively unlikely that two carriers of the same rare harmful allele will meet and mate, the probability increases greatly if the man and woman are close relatives (for example, siblings or first cousins). These are called **consanguineous** ("same blood") matings, and they are indicated in pedigrees with double lines. Because people with recent common ancestors are more likely to carry the same recessive alleles than are unrelated people, it is more likely that a mating of close blood relatives will produce offspring homozygous for a harmful recessive trait. Such effects can be observed in many types of domesticated and zoo animals that have become inbred.

There is debate among geneticists about the extent to which human consanguinity increases the risk of inherited diseases. Many deleterious alleles have such severe effects that a homozygous embryo spontaneously aborts long before birth. Most societies and cultures have laws or taboos forbidding marriages between close relatives. These rules may have evolved out of empirical observation that in most populations, stillbirths and birth defects are more common when parents are closely related. But social and economic factors have also influenced the development of customs and laws against consanguineous marriages, which could concentrate wealth in a few families.

Some geneticists argue that consanguinity is just as likely to concentrate favorable alleles as harmful ones. It must be admitted that some very important people have resulted from matings between close relatives. Charles Darwin and his wife, Emma Wedgewood, were first cousins. Among their ten children were very successful scientists and professionals. Darwin was concerned enough, however, to advocate (without success) that the British census include questions about the effects of cousin marriages. In some human populations, such as the Tamils of India, marriage between close relatives (such as first cousins) has long been the rule. In these cases, there is no observable ill effect of continued inbreeding. And in some zoo populations of endangered species, inbreeding, unavoidable because there are so few individuals, is saving small groups of animals from extinction. Such breeding programs are carefully planned. For example, a new technique called DNA fingerprinting (explained in Chapter 19) is used to match condors in mating pairs that are as genetically dissimilar as possible.

Dominantly Inherited Disorders

Although most harmful alleles are recessive, several human disorders are due to dominant alleles. One example is achondroplasia, a form of dwarfism with an incidence of one case among every 10,000 people. Heterozygous individuals have the dwarf phenotype (in fact, homozygosity for this dominant allele is usually lethal, causing spontaneous abortion). Therefore, all people who are not achondroplastic dwarfs—99.99% of the population—are homozygous for the recessive allele.

Lethal dominant alleles are much less common than lethal recessives. One reason for this difference is that the effects of lethal dominant alleles are not masked in heterozygotes. Many lethal dominant alleles are the result of new changes (mutations) in a gene of the sperm or egg that subsequently kill the developing organism (mutations will be discussed in Chapter 16). An individual who does not survive to reproductive maturity will not pass on the new form of the gene. This is in contrast to lethal recessive mutations, which are perpetuated from generation to generation by the reproduction of heterozygous carriers who have normal phenotypes.

However, a lethal dominant allele can escape elimination if it is late-acting, causing death at a relatively advanced age. By the time the symptoms become evident, the afflicted individual may have already transmitted the lethal gene to his or her children. For example, Huntington's disease, a degenerative disease of the nervous system, is caused by a lethal dominant allele that has no obvious phenotypic effect until the individual is 35 to 45 years old. Once the deterioration of the nervous system begins, it is irreversible and inevitably fatal. Any child born to a parent who has the allele for Huntington's disease has a 50% chance of inheriting the allele and the disorder. (The mating can be symbolized as *Aa* × *aa*, with *A* being the dominant allele that causes Huntington's disease.) Until recently, it was impossible to tell before the onset of symptoms if a person at risk for Huntington's disease had actually inherited the allele. But in the late 1980s, molecular geneticists developed a test for the Huntington's allele by analyzing DNA extracted from members of a large family with a high incidence of the disease (Figure 13.16). (The methods that make such tests possible will be discussed in Chapter 19.) For those with a family history of Huntington's disease, the availability of this new test poses an agonizing dilemma: Under what circumstances is it beneficial for a presently healthy person to find out whether or not he or she has inherited a fatal and not yet curable disease?

Figure 13.16
Large families: living laboratories of human genetics. Here, Nancy Wexler, of Columbia University and the Hereditary Disease Foundation, stands in front of a huge pedigree that traces Huntington's disease through several generations of one large family. Classical Mendelian analysis of this family, coupled with new techniques of molecular biology, enabled scientists to develop a test for the presence of the dominant allele that causes Huntington's disease—before symptoms are evident. Dr. Wexler, who has studied the afflicted family, is herself at risk for developing Huntington's disease. Her mother died of the disorder, and there is a 50% chance that Dr. Wexler inherited the dominant allele that causes the disease.

Genetic Screening and Counseling

A preventive approach to genetic disorders is sometimes possible, because in some cases the risk that a particular genetic disorder will occur can be assessed before a child is conceived or in the early stages of the pregnancy. Many hospitals have genetic counselors who can provide information to prospective parents concerned about a family history for a specific disease.

Let's consider the example of an imaginary couple, John and Carol, who are planning to have their first child and are seeking genetic counseling because of family histories of a lethal disease known to be recessively inherited. John and Carol each had a brother who died of the disorder, so they want to determine the risk of their having a child with the disease. From the information about their brothers, we know that both parents of John and both parents of Carol must have been carriers of the recessive allele. Thus, John and Carol are both products of *Aa* and *Aa* crosses, where *a* symbolizes the allele that causes this particular disease. We know that John and Carol are not homozygous recessive *(aa)*, because they do not have the disease. Therefore, their genotypes are either *AA* or *Aa*. Given a genotypic ratio of 1*AA*:2*Aa*:1*aa* for offspring of an *Aa* × *Aa* cross, John and Carol each have a ⅔ chance of being carriers *(Aa)*. Using the rule of multiplication, we can determine that the probability that *both* John *and* Carol are carriers is ⅔ × ⅔ = 4/9. Taking these chance factors into consideration, the overall probability of their firstborn having the disorder is ⅔ (the chance that John is a carrier) multiplied by ⅔ (the chance that Carol is a carrier) multiplied by ¼ (the chance of two carriers having a child with the disease), which equals 1/9. Suppose that Carol and John decide to take the risk and have a child; after all, there is an 8/9 chance that their baby will be normal. But their child is born with the disease. We no longer have to guess about John's and Carol's genotypes. We now know that both John and Carol are, in fact, carriers. If the couple decides to have another child, they now know there is a ¼ chance that the second child will have the disease.

When we use Mendel's laws to predict possible outcomes of matings, it is important to recall that chance has no memory: Each child represents an independent event in the sense that its genotype is unaffected by the genotypes of older siblings. Suppose that John and Carol have three more children, and *all three* have the hypothetical hereditary disease. This is an unfortunate family, for there is only one chance in 64 (¼ × ¼ × ¼) that such an outcome will occur. But this run of misfortune will in no way affect the result if John and Carol decide to have still another child. There is still a ¼ chance that the additional child will have the disease and a ¾ chance that it will not. Mendel's laws, remember, are simply rules of probability applied to heredity.

Carrier Recognition Because most children with recessive disorders are born to parents with normal phenotypes, the key to assessing the genetic risk for a par-

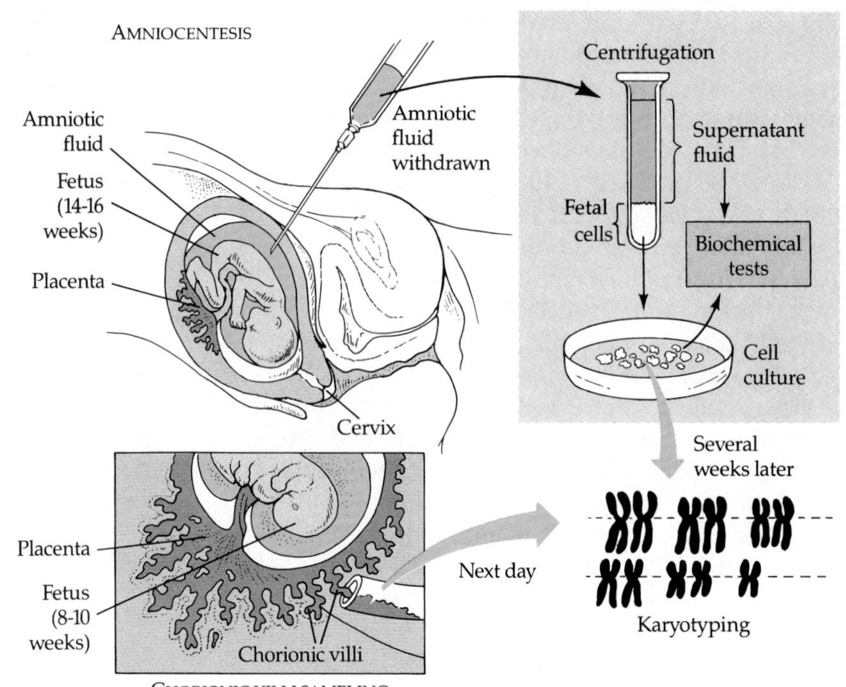

Figure 13.17
Fetal diagnosis. In amniocentesis, ultrasound is used to locate the fetus, and a small amount of amniotic fluid is extracted for testing. Physicians can diagnose some disorders from chemicals in the fluid itself, while other disorders may show up in tests performed on cells cultured from fetal cells present in the fluid. These tests include biochemical tests, to detect the presence of certain enzymes, and karyotyping, to determine whether the chromosomes of the fetal cells are normal in number and microscopic appearance. In chorionic villi sampling, a physician inserts a narrow tube through the cervix and suctions out a tiny sample of fetal tissue (chorionic villi) from the placenta, the organ that transmits nourishment and wastes between the fetus and the mother. The fetal tissue sample is used for immediate karyotyping.

ticular disease is determining whether the prospective parents are heterozygous carriers of the recessive trait. For some heritable disorders, there are tests that can distinguish between normal individuals who are dominant homozygotes and normal individuals who are heterozygous, and the number of such tests is increasing. Examples are tests that can identify carriers of the alleles for Tay-Sachs disease, sickle-cell anemia, and the most common form of cystic fibrosis. On one hand, these tests enable people with family histories of genetic disorders to make informed decisions about having children. On the other hand, these new methods for genetic screening could be abused. If confidentiality is breached, will carriers be stigmatized? Will they be denied health or life insurance, even though they are themselves healthy? Will misinformed employers equate "carrier" with disease? And will sufficient genetic counseling be available to help a large number of individuals understand their test results? New biotechnology offers possibilities of reducing human suffering, but not before key ethical issues are resolved. The dilemmas posed by human genetics reinforce one of this book's themes: the immense social implications of biology.

Fetal Testing Suppose a couple learns that they are both Tay-Sachs carriers, but they decide to have a child anyway. Tests performed in conjunction with a technique known as **amniocentesis** (Figure 13.17) can determine, beginning at the fourteenth to sixteenth week of pregnancy, whether the developing fetus has Tay-Sachs disease. To perform this procedure, a physician inserts a needle into the uterus and extracts about 10 milliliters of amniotic fluid, the liquid that bathes the fetus. Some genetic disorders can be detected from the presence of certain chemicals in the amniotic fluid itself. Tests for other disorders, including Tay-Sachs disease, are performed on cells grown in the laboratory from the fetal cells that had been sloughed off into the amniotic fluid. These cultured cells can also be used for karyotyping to identify certain chromosomal defects (see the Methods Box in Chapter 12, p. 247).

In an alternative technique called **chorionic villi sampling,** the physician suctions off a small amount of fetal tissue from the projections (villi) of the embryonic membrane, or chorion, which forms part of the placenta. Because the cells of the chorionic villi are proliferating rapidly, enough cells are undergoing mitosis to allow karyotyping to be carried out immediately, giving results within 24 hours. The rapidity of this method is an advantage over amniocentesis, in which the cells must be cultured and karyotyping results typically take several weeks. Another advantage of chorionic villi sampling is that it can be performed as early as the eighth to tenth week of pregnancy. However, the risks of the procedure are still being determined.

Other techniques allow a physician to examine a fetus directly for major abnormalities. One such technique is ultrasound, which uses sound waves to produce an image of the fetus by a simple noninvasive procedure. This procedure has no known risk to either fetus or mother. With another technique, fetoscopy, a needle-thin tube containing a viewing scope and fiber optics (to transmit light) is inserted into the uterus. Fetoscopy allows the physician to examine the fetus for certain anatomical deformities (see Chapter 42).

In about 1% of the cases, amniocentesis or fetoscopy causes complications, such as maternal bleeding or fetal death. Thus, these techniques are usually reserved for cases in which the risk of a genetic disorder or other type of birth defect is relatively great. (For example, amniocentesis is now performed routinely for pregnant women over age 35.) If the fetal tests reveal a serious disorder, the parents must choose between terminating the pregnancy or preparing to care for a child with a genetic disorder.

Newborn Screening Some genetic disorders can be detected at birth by simple tests that are now routinely performed in most hospitals in the United States. One of the most significant screening programs is for the recessively inherited disorder called phenylketonuria (PKU), which occurs in about one out of every 15,000 births in the United States. Children with this disease cannot properly break down the amino acid phenylalanine. This compound and its by-product, phenylpyruvate, can accumulate to toxic levels in the blood, causing mental retardation. But if the deficiency is detected in the newborn, retardation can be prevented with a special diet, low in phenylalanine, that allows normal development. Thus, screening newborns for PKU and other treatable disorders is vitally important. Unfortunately, very few genetic disorders are treatable at the present time.

Multifactorial Disorders

The hereditary diseases we have studied so far are sometimes described as "simple Mendelian disorders," because they result from certain alleles at a single genetic locus. Many more people are susceptible to diseases that have a multifactorial basis—a genetic component plus a significant environmental influence. The long list of multifactorial diseases includes heart disease, diabetes, cancer, alcoholism, and some forms of mental illness, such as schizophrenia and manic/depressive psychosis. In many cases, the hereditary component is polygenic. For example, many genes affect our cardiovascular health, making some of us more prone than others to heart attacks and strokes. But our lifestyle intervenes tremendously between genotype and phenotype for cardiovascular health and other multifactorial characters. Exercise, a healthful diet, lack of smoking, and an ability to put stressful situations in perspective all reduce our risk of heart disease and some types of cancer.

At present, so little is understood about the genetic contributions to most multifactorial diseases that the best public-health strategy is to educate people about the importance of environmental factors and to promote healthful behavior.

* * *

In this chapter, you have learned about the Mendelian model of inheritance and its application to human genetics. We owe the "gene idea," the concept of heritable factors that are transmitted according to simple rules of chance, to the elegant experiments of Gregor Mendel. Mendel's quantitative approach was foreign to the biology of his era, and even the few biologists who read his papers apparently missed the importance of his discoveries. It wasn't until the beginning of the twentieth century that Mendelian genetics was rediscovered by biologists who studied the role of chromosomes in inheritance. In the next chapter, you will learn how Mendel's laws have their physical basis in the behavior of chromosomes during sexual life cycles, and how the synthesis of Mendelism and a chromosome theory of inheritance catalyzed progress in genetics.

STUDY OUTLINE

1. In the 1860s, Gregor Mendel developed a particulate theory of inheritance based on his experiments with garden peas. Today, we call the hereditary units genes.

Mendel's Model: A Case Study in the Scientific Process (pp. 258–267)

1. By breeding garden peas, Mendel demonstrated that parents pass on to their offspring discrete genes that retain their identity generation after generation.
2. Mendel's success is attributed to his quantitative approach and his choice of organism. He worked with seven pea characters, each of which occurred in two distinct, alternative forms. His peas were true-breeding, and he was able to control mating.
3. By producing hybrid offspring and allowing them to self-pollinate, Mendel arrived at the law of segregation. The hybrids (F_1) exhibited the dominant trait. In the next generation (F_2), 75% of offspring had the dominant trait and 25% had the recessive trait, for a 3:1 ratio.
4. To account for these results, Mendel postulated that genes have alternative forms (now called alleles) and that each organism inherits one allele for each gene from each parent. These separate (segregate) from each other during gamete formation, so that a sperm or an egg carries only one allele. After fertilization, if the two alleles of the pair are different, one (the dominant allele) is fully expressed in the offspring and the other (the recessive allele) is completely masked.
5. Homozygous individuals have two identical alleles for a given character and are true-breeding. Heterozygous individuals have two different alleles for a given character.
6. The genotype of an organism showing a dominant trait can be determined by breeding it to a recessive homozygote in a testcross.

7. The law of segregation operates according to the rules of probability. According to the rule of multiplication, the probability of a compound event is equal to the product of the separate probabilities of the independent single events. The rule of addition states that the probability of an event that can occur in two or more independent ways is the sum of the separate probabilities.
8. Mendel proposed the law of independent assortment based on dihybrid crosses between plants contrasting in two or more characters, such as flower color and seed shape. Alleles for each character segregate into gametes independently of alleles for other characters.

Extending Mendelian Genetics (pp. 267–272)

1. Some heterozygous genotypes result in incomplete dominance. Such individuals have an appearance that is intermediate between the phenotypes of the two parents.
2. There are other variations in dominance/recessiveness relationships. In codominance, a heterozygous organism expresses *both* phenotypes of its two alleles. The type of dominance we observe often depends on the size scale used for examination.
3. Many genes have *multiple* (more than two) alleles, like the gene for the human ABO blood groups.
4. Pleiotropy is the ability of a single gene to affect multiple phenotypic traits.
5. In epistasis, one gene interferes with the expression of another gene.
6. Certain characters, such as human skin color, are quantitative characters that vary in a continuous fashion, indicating polygenic inheritance, an additive effect of two or more genes on a single phenotypic character.
7. Environment also influences quantitative characters. Such characters are said to be multifactorial.

Mendelian Inheritance in Humans (pp. 272–277)

1. Family pedigrees can be used to deduce the *possible* genotypes of individuals and make predictions about future offspring. Any predictions are usually statistical probabilities rather than absolute statements.
2. Certain genetic disorders—such as sickle-cell anemia, Tay-Sachs disease, and cystic fibrosis—are inherited as simple recessive traits from phenotypically normal, heterozygous carriers.
3. Consanguineous matings between close relatives may increase the chance that the offspring will be homozygous for a rare deleterious allele.
4. Although they are far less common than the recessive type, some human disorders are due to dominant alleles. Dominant alleles that are lethal may kill the organism as an embryo or act later, as in Huntington's disease.
5. Using family histories, genetic counselors aid couples in determining the odds that their children will have genetic disorders. For certain diseases, tests that can identify carriers can more accurately define those odds.
6. Once a child is conceived, the techniques of amniocentesis and chorionic villi sampling can help determine whether a suspected genetic disorder is present.
7. Medical researchers are just beginning to sort out the genetic and environmental components of multifactorial disorders, such as heart disease and cancer.

GENETICS PROBLEMS

1. Flower position, stem length, and seed shape were three characters that Mendel chose to study. Each is controlled by an independently assorting gene and has dominant and recessive expression as follows:

Trait	*Dominant*	*Recessive*
Flower position	Axial *(A)*	Terminal *(a)*
Stem length	Tall *(L)*	Dwarf *(l)*
Seed shape	Round *(R)*	Wrinkled *(r)*

   If a plant that is heterozygous for all three traits were allowed to self-fertilize, what proportion of the offspring would be expected to be as follows? (*Note:* Use the rules of probability instead of a huge Punnett square.)
   a. homozygous for the three dominant traits
   b. homozygous for the three recessive traits
   c. heterozygous for the three traits
   d. homozygous for axial and tall, heterozygous for round
2. A black guinea pig crossed with an albino one gave 12 black offspring. When the albino was crossed with a second black one, 7 blacks and 5 albinos were obtained. What is the best explanation for this genetic situation? Write genotypes for the parents, gametes, and offspring.
3. A rooster with blue (actually gray) feathers is mated with a hen of the same phenotype. Among their offspring, 15 chicks are blue, 6 are black, and 8 are white. What is the simplest explanation for the inheritance of these colors in chickens? What offspring would you predict from the mating of a blue rooster and a black hen?
4. In some flowers, a true-breeding, red-flowered strain gives all pink flowers when crossed with a white-flowered strain: *RR* (red) × *rr* (white) ⟶ *Rr* (pink). If flower position is inherited as it is in peas (see problem 1), what will be the ratios of genotypes and phenotypes of the generation resulting from the following cross: Axial-red (true-breeding) × terminal-white? What will be the ratios in the F_2 generation?
5. In sesame plants, the one-pod condition *(P)* is dominant to the three-pod condition *(p)*, and normal leaf *(L)* is dominant to wrinkled leaf *(l)*. These traits are inherited independently. Determine the genotypes for the two parents for all possible matings producing the following offspring:
   a. 318 one-pod normal, 98 one-pod wrinkled
   b. 323 three-pod normal, 106 three-pod wrinkled
   c. 401 one-pod normal
   d. 150 one-pod normal, 147 one-pod wrinkled, 51 three-pod normal, 48 three-pod wrinkled
   e. 223 one-pod normal, 72 one-pod wrinkled, 76 three-pod normal, 27 three-pod wrinkled
6. A man with group A blood marries a woman with group B blood. Their child has group O blood. What are the genotypes of these individuals? What other genotypes, and in what frequencies, would you expect in offspring from this marriage?
7. Color pattern in a species of duck is determined by a single pair of genes with three alleles. Alleles *H* and *I* are codominant, and allele *i* is recessive to both. How many phenotypes are possible in a flock of ducks that contains all the possible combinations of these three alleles?

8. Phenylketonuria (PKU) is an inherited disease determined by a recessive allele. If a woman and her husband are both carriers, what is the probability of each of the following?
   a. All three of their children will be normal.
   b. One *or* more of the three children will have the disease.
   c. All three children will be afflicted with the disease.
   d. *At least* one child will be normal.

   (*Note:* Remember that the probabilities of all possible outcomes always add up to 1.)
9. The genotype of F_1 individuals in a tetrahybrid cross is *AaBbCcDd*. Assuming independent assortment of these four genes, what are the probabilities that F_2 offspring would have the following genotypes?
   a. *aabbccdd*
   b. *AaBbCcDd*
   c. *AABBCCDD*
   d. *AaBBccDd*
   e. *AaBBCCdd*
10. In 1981, a stray black cat with unusual rounded, curled-back ears was adopted by a family in California. Hundreds of descendants of the cat have since been born, and cat fanciers hope to develop the "curl" cat into a show breed. Suppose you owned the first curl cat and wanted to develop a true-breeding variety. How would you determine whether the curl allele is dominant or recessive? How would you breed true-breeding cats? How would you know they are true-breeding?

11. What is the probability that each of the following pairs of parents will produce the indicated offspring (assume independent assortment of all gene pairs)?
   a. *AABBCC* × *aabbcc* ⟶ *AaBbCc*
   b. *AABbCc* × *AaBbCc* ⟶ *AAbbCC*
   c. *AaBbCc* × *AaBbCc* ⟶ *AaBbCc*
   d. *aaBbCC* × *AABbcc* ⟶ *AaBbCc*
12. Karen and Steve each have a sibling with sickle-cell anemia. Neither Karen, Steve, nor any of their parents has the disease, and none of them has been tested to reveal sickle-cell trait. Based on this incomplete information, calculate the probability that if this couple has a child, the child will have sickle-cell anemia.
13. Imagine that a newly discovered, recessively inherited disease is only expressed in individuals with group O blood, although the disease and blood group are independently inherited. A normal man with A blood and a normal woman with B blood have already had one child with the disease. The woman is now pregnant for a second time. What is the probability that the second child will also have the disease? Assume both parents are heterozygous for the "disease" gene.
14. This pedigree traces the inheritance of alkaptonuria, a biochemical disorder. Affected individuals, indicated here by the filled-in circles and squares, are unable to break down a substance called alkapton, which colors the urine and stains body tissues. Does alkaptonuria appear to be caused by a dominant or recessive allele? Fill in the genotypes of the individuals whose genotypes you know. What genotypes are possible for each of the other individuals?

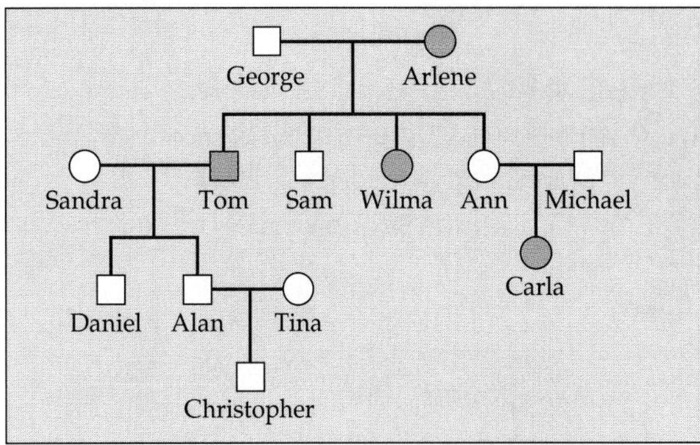

15. Eric and Kristin are cousins who would like to marry and have children. Suppose Eric has inherited a very rare allele for a disorder from their common grandparent, who is a carrier. What is the probability that Kristin has inherited the same harmful allele?

SCIENCE, TECHNOLOGY, AND SOCIETY

1. Imagine that one of your parents suffered from Huntington's disease. What would be the probability that you, too, would someday manifest the disease? There is no cure for Huntington's. Would you want to be tested for the Huntington's allele? Why or why not? Suppose you tested positive. In what ways would this information change your life?
2. Proponents of the discredited eugenics movement believe that the human population would be improved if individuals suffering from genetic defects, such as cystic fibrosis or Tay-Sachs disease, could be sterilized or persuaded not to have children. Would implementation of this strategy eliminate these defects? Why or why not? How might new tests that identify carriers of harmful alleles bolster eugenics? Are such tests an opportunity or a danger for society? Why?

FURTHER READING

Bower, B. "Gene in the Bottle." *Science News*, September 21, 1991. To what extent is alcoholism inherited?

Diamond, J. "Curse and Blessing of the Ghetto." *Discover*, March 1991. The biology and sociology of Tay-Sachs disease.

Ezzell, C. "Fetal Tissue: A Hope for Huntington's?" *Science News*, November 16, 1991. Ethical issues complicate progress on treating some genetic disorders.

Roberts, L. "To Test or Not to Test." *Science*, January 5, 1990. How genetic screening is a double-edged sword.

Stone, J. "The Marfan Family." *Discover*, November 1990. A genetic disorder that may have affected Abraham Lincoln.

Suzuki, D., A. Griffith, J. Miller, and R. Lewontin. *An Introduction to Genetic Analysis*, 4th ed. New York: Freeman, 1989. A good basic undergraduate genetics text.

14 The Chromosomal Basis of Inheritance

The Chromosome Theory of Inheritance

The Chromosomal Basis of Recombination

Genetic Maps Based on Crossover Data

Sex Chromosomes and Sex-Linked Inheritance: A Closer Look

Chromosomal Alterations

Parental Imprinting of Genes

Extranuclear Inheritance

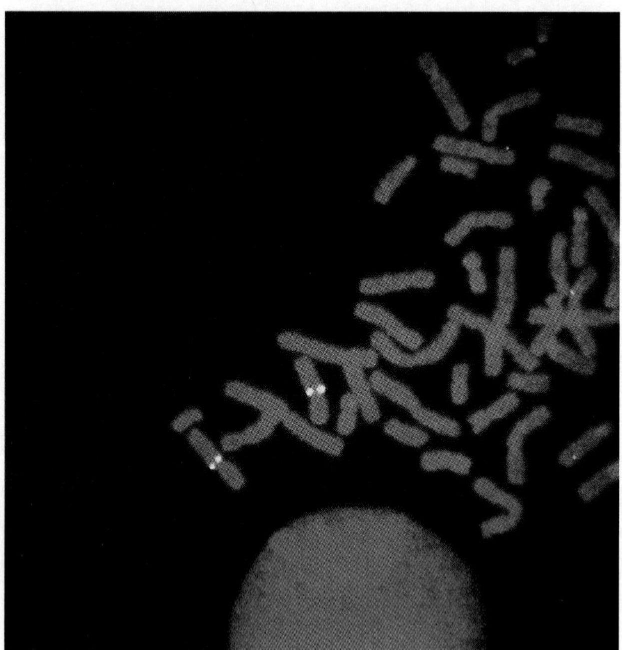

Figure 14.1
Genes, Mendel's "hereditary factors," have specific loci on chromosomes. In this light micrograph of human chromosomes, a fluorescent dye (yellow) marks the locus of the gene that codes for synthesis of an enzyme called glycogen phosphorylase. Notice that in this diploid cell, the gene is represented on two chromosomes. (The large sphere is the cell's nucleolus.) In this chapter, you will learn how the behavior of chromosomes during sexual life cycles provides the physical basis for Mendel's laws of segregation and independent assortment.

It was not until the year 1900 that biology finally caught up with Gregor Mendel. At that time, three botanists, working independently on plant breeding experiments, reproduced Mendel's results. By searching the literature, the German Karl Correns, the Austrian Erich von Tschermak, and the Dutchman Hugo de Vries all found that Mendel had explained the same results 35 years before. During the intervening years, biology had grown more experimental and quantitative and thus more receptive to Mendelism. Nevertheless, many biologists remained incredulous about Mendel's laws of segregation and independent assortment until evidence had mounted that these principles of heredity had a physical basis in the behavior of chromosomes (Figure 14.1). This chapter integrates and extends what you have learned in the previous two chapters by describing the chromosomal basis for the transmission of genes from parents to offspring.

THE CHROMOSOME THEORY OF INHERITANCE

Cytologists worked out the process of mitosis in 1875 and the process of meiosis in the 1890s. Then, around the turn of the century, cytology and genetics converged as biologists began to see parallels between the behavior of chromosomes and the behavior of Mendel's factors (Figure 14.2). For example, chromosomes and genes are both present in pairs in diploid cells; homologous chromosomes separate and alleles segregate during meiosis; and fertilization restores the paired condition for both chromosomes and genes. Around 1902, Walter S. Sutton, Theodor Boveri, and others independently noted these parallels, and a **chromosome theory of inheritance** began to take form. According to this theory, Mendelian genes are located on chromosomes, and it is the chromosomes that undergo segregation and independent assortment.

Morgan and the *Drosophila* School

Thomas Hunt Morgan, an embryologist at Columbia University, first associated a specific gene with a specific chromosome early in this century. Although

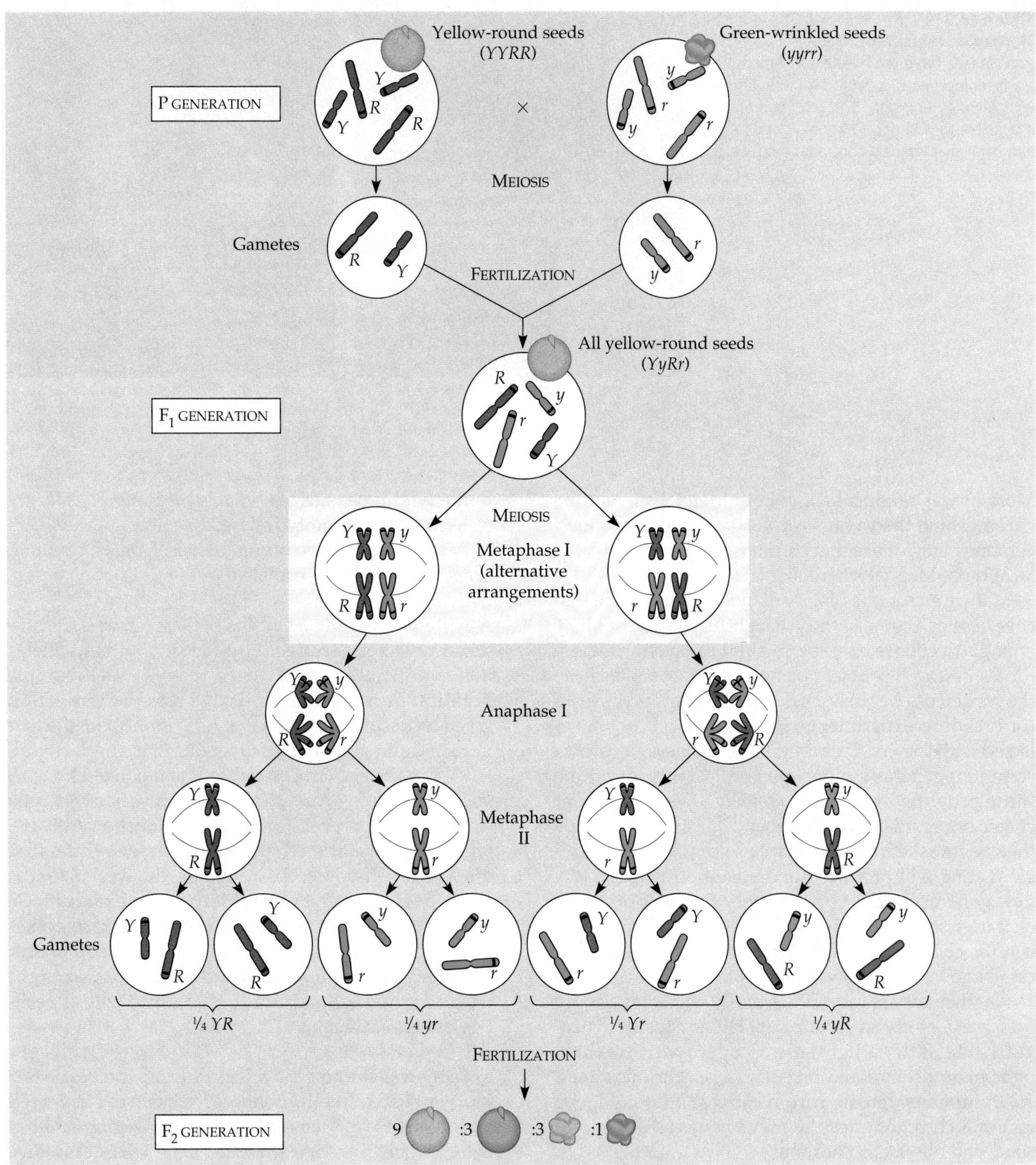

Figure 14.2
The chromosomal basis of Mendel's laws. Here we correlate the results of one of Mendel's dihybrid crosses with the behavior of chromosomes. The two characters we are tracking are the color and shape of pea seeds. The two genes are on different chromosomes, and black bands symbolize the loci in these drawings. (Peas actually have a total of seven chromosome pairs, but only two are illustrated here.) The movements of chromosomes during metaphase and anaphase of meiosis I account for the segregation and independent assortment of the alleles for seed color and shape. The two alleles for each character segregate when homologous chromosomes move toward opposite poles of the cell and become packaged in different cells during the first meiotic division. Mendel's law of independent assortment has its physical basis in the random arrangement of chromosomes on the metaphase plate during meiosis I. For each homologous pair of chromosomes, the polar orientation of the maternal and paternal versions (color-coded salmon and blue, respectively) of the chromosome is unrelated to the orientation of the other chromosome pair. This results in the alternative metaphase I arrangements illustrated in the yellow box. If we continue to trace the results of these alternative alignments of chromosomes to the F_2 generation, we see the physical explanation for Mendel's 9:3:3:1 ratio of phenotypes.

Figure 14.3
Morgan's first mutant. Wild-type *Drosophila* flies have red eyes (left). Among his flies, Morgan discovered a mutant male with white eyes (right). This variation made it possible for Morgan to trace a gene for eye color to a specific chromosome (LMs).

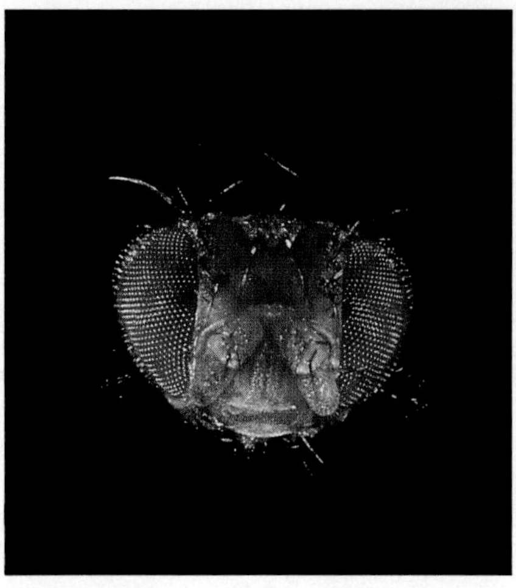

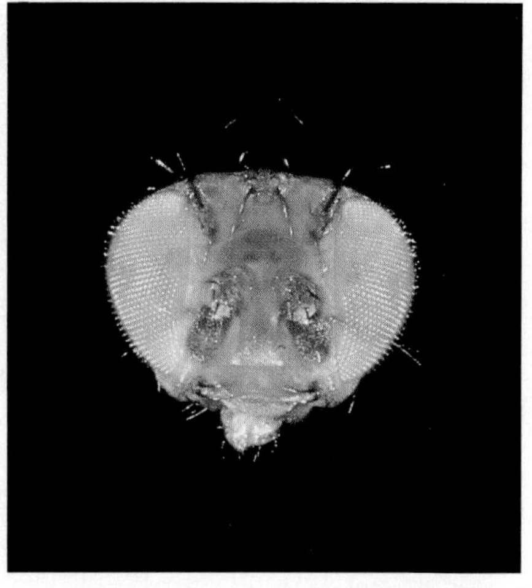

Morgan was skeptical about both Mendelism and the chromosome theory, his early experiments provided convincing evidence that chromosomes are indeed the location of Mendel's heritable factors.

Many times in the history of biology, important discoveries have come to those insightful enough or lucky enough to choose an experimental organism suitable for the research problem being tackled. Mendel chose the garden pea because it offered some key advantages for breeding experiments. For his work, Morgan selected a species of fruit fly, *Drosophila melanogaster*, a common, generally innocuous pest that feeds on the fungi growing on fruit. *Drosophila* was an excellent choice of organism for the genetic questions that were then being asked. Fruit flies are prolific breeders; a single mating will produce hundreds of offspring, and a new generation can be bred every two weeks. These characteristics make the fruit fly a convenient organism for genetic studies. Morgan's laboratory soon became known as "the fly room."

Another advantage of the fruit fly is that it has only four pairs of chromosomes, which are easily distinguishable under a light microscope. There are three pairs of autosomes and one pair of sex chromosomes. As in humans, female fruit flies have a homologous pair of *X* chromosomes and males have one *X* chromosome and one *Y* chromosome.

While Mendel could readily obtain different pea varieties, there were no convenient suppliers of fruit fly varieties for Morgan to employ. Indeed, he may have been the first person to want different varieties of this common insect. After a year of breeding flies and looking for variant individuals, Morgan was rewarded with the discovery of a single male fly with white eyes instead of the usual red. The normal phenotype for a character (the phenotype most common in natural populations), such as red eyes in *Drosophila*, is called the **wild type** (Figure 14.3). Traits that are alternatives to the wild type, such as white eyes in *Drosophila*, are called **mutant phenotypes,** because they are due to alleles assumed to have originated as changes, or mutations, in the wild-type allele.

A Note on Genetic Symbols Morgan and his students invented a convention for symbolizing alleles that is now more widely used than Mendel's simple notation of uppercase and lowercase letters. For a given character, the gene takes its symbol from the first mutant (nonwild type) discovered. For example, the allele for white eyes in *Drosophila* is symbolized by w. A superscript + represents the allele for the wild-type trait—w^+ for red eyes, for example. Notice that the letter is lowercase if the mutant is recessive. If the first mutant trait discovered is dominant, the symbol for the allele begins with a capital letter. For example, *Cy* stands for a mutant allele called "curly," which causes the wings of a fly to form an abnormal, curled shape. The allele is dominant. Flies with normal, straight wings are homozygous for the recessive, wild-type allele, symbolized as Cy^+.

Tracing a Gene to a Specific Chromosome After Morgan discovered his white-eyed male fly, he mated it with a red-eyed female. All the F_1 offspring had red eyes, suggesting that the wild type was dominant. When Morgan bred the F_1 flies to each other, he observed the classical 3:1 phenotypic ratio among the F_2 offspring. However, there was a surprising result: The white-eye trait showed up only in males. All the F_2 females had red eyes, while half the males had red eyes and half had white eyes. Somehow, a fly's eye color was linked to its sex.

From this and other evidence, Morgan deduced that the gene for eye color is located exclusively on the *X* chromosome; there is no corresponding eye-color site, or locus, on the *Y* chromosome (Figure 14.4). Thus, fe-

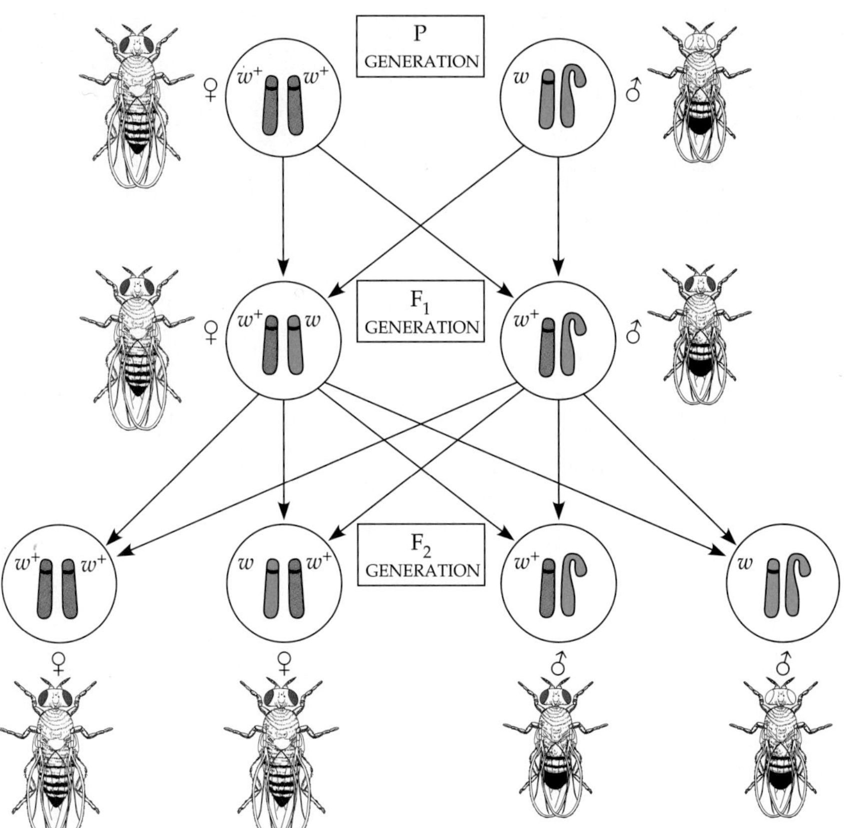

Figure 14.4
Sex-linked inheritance. When Morgan bred his white-eyed male to a wild-type female, all F_1 offspring had red eyes. The F_2 generation showed a typical Mendelian 3:1 ratio of traits, but the recessive trait—white eyes—was linked to sex. All females had red eyes, but half the males had white eyes. Morgan hypothesized that the gene responsible was located on the *X* chromosome (symbolized as a straight chromosome here) and that there was no corresponding locus on the *Y* chromosome (hooked, in this diagram). In this figure, the dominant allele (for red eyes) is symbolized by w^+, and the recessive allele (for white eyes) is symbolized by *w*. The symbols ♀ and ♂ stand for female and male, respectively. (Derived from Roman mythology, these symbols represent the hand-mirror and comb of Venus and the shield and spear of Mars.) Notice that female *Drosophila* flies are larger than males.

males *(XX)* carry two copies of the gene for this character, while males *(XY)* carry only one. Because the mutant allele is recessive, a female will have white eyes only if she receives that allele on both *X* chromosomes—an impossibility for the F_2 females in Morgan's experiment. For a male, on the other hand, a single copy of the mutant allele confers white eyes. Since a male has only one X chromosome, there can be no wild-type allele present to offset the recessive allele.

Genes located on a sex chromosome are called **sex-linked genes,** and this term is commonly applied, in fact, only to genes on the *X* chromosome. Morgan's evidence that a specific gene is carried on the *X* chromosome added credibility to the chromosome theory of inheritance. Recognizing the importance of this work, many bright students were attracted to Morgan's fly room, and his laboratory dominated genetics research for the next three decades. We will see the influence of Morgan and his colleagues as we consider some other important aspects of the chromosomal basis of inheritance.

Linked Genes

The number of genes in a cell is far greater than the number of chromosomes; in fact, each chromosome has hundreds or thousands of genes. Genes located on the same chromosome tend to be inherited together in genetic crosses, because they are part of a single chromosome that is passed along as a unit. Such genes are said to be **linked genes.** (Notice that the use of the word *linked* in this way is different from its use in the term *sex-linked.*) When geneticists follow linked genes in breeding experiments, the results deviate from those expected according to the Mendelian principle of independent assortment.

Let's examine another of Morgan's *Drosophila* experiments to see how linkage affects the inheritance of two different characters. In this case, the two characters are body color and wing size. Wild-type flies have gray bodies and normal wings. Mutant phenotypes for these characters are black body and vestigial wings, which are much smaller than normal wings (Figure 14.5). The alleles for these traits are represented by the following symbols: b^+ = gray, b = black; vg^+ = normal wings, vg = vestigial wings. (Neither gene is sex-linked; the loci are on autosomes.) Morgan crossed female dihybrids (b^+bvg^+vg) with males that had both mutant phenotypes, black bodies and vestigial wings $(bbvgvg)$. Notice that this corresponds to a Mendelian testcross rather than to an F_1 cross between two dihybrids (see Chapter 13). According to Mendel's law of independent assortment, Morgan's *Drosophila* testcross would produce four phenotypic classes of offspring, approximately equal in number: 1 gray-normal (wings) : 1

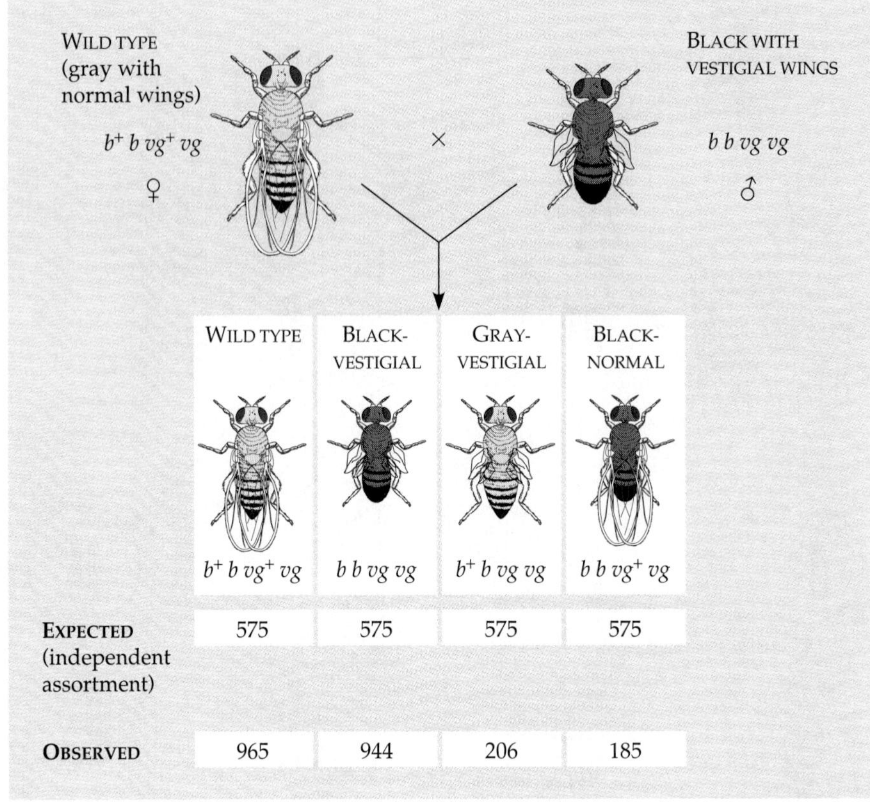

Figure 14.5
Evidence for linked genes in *Drosophila.* This is a testcross between flies differing in two characters, body color and wing size. The females are heterozygous for both genes and their phenotypes are wild type (gray bodies and normal wings). The males are homozygous recessive and express the mutant phenotypes for both characters (black bodies and vestigial wings). Morgan "scored" (classified according to phenotype) 2300 offspring from such matings. The top row of numbers (Expected) represents the 1:1:1:1 ratio of phenotypes we would expect if the two genes assort independently of each other. The bottom row shows the actual (Observed) results. The parental phenotypes are disproportionately represented among offspring. Morgan concluded that genes for body color and wing size are usually transmitted together from parents to offspring because they are located on the same chromosome—that is, they are linked genes. Crossing over accounts for the offspring that have combinations of traits different from either parent.

black-vestigial : 1 gray-vestigial : 1 black-normal (see Figure 14.5). The actual results were very different. There were disproportionate numbers of wild-type (gray-normal) and double mutant (black-vestigial) flies among the offspring. Notice that these two phenotypes corresponded to the phenotypes of the two parents. Morgan reasoned that body color and wing shape are usually inherited together in a specific combination because the genes for these two characters are located on the same chromosome:

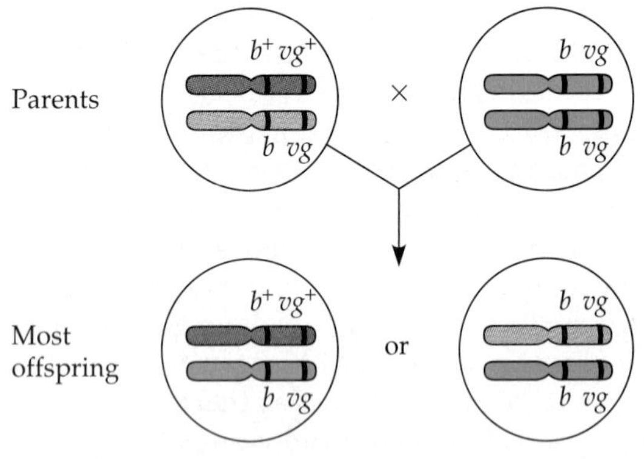

Although the other two phenotypes (gray-vestigial and black-normal) numbered fewer than expected based on independent assortment, these phenotypes *were* represented among the offspring of Morgan's cross. These new combinations of the two characters resulted from the crossing over, one of the topics discussed in the next section.

THE CHROMOSOMAL BASIS OF RECOMBINATION

In Chapter 12, we saw that meiosis and random fertilization generate genetic variation among offspring of sexually reproducing organisms. The general term for the production of offspring with new combinations of traits inherited from two parents is **genetic recombination.** Here, we will examine the basis of recombination in more detail.

The Recombination of Unlinked Genes: Independent Assortment

Mendel learned from his dihybrid crosses that some offspring have combinations of traits that do not match either parent. For example, in a testcross between a pea plant with yellow-round seeds that is heterozygous for both traits and a plant with green-wrinkled seeds (homozygous for both recessive traits), half the offspring are unlike either parent (Figure 14.6). The gene loci for these two traits are on separate chromosomes: Seed shape and seed color are unlinked. Notice that one-

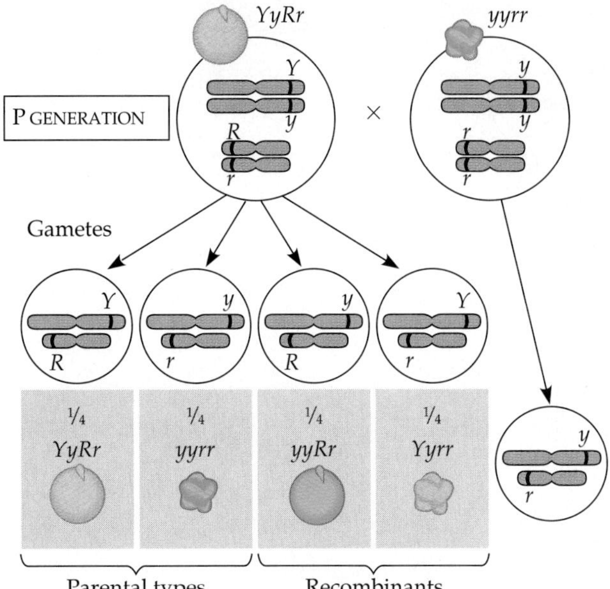

Figure 14.6
The recombination of unlinked genes. The inheritance of two genes carried on different pea chromosomes is analyzed in this testcross. A heterozygous plant with yellow-round seeds is mated with a homozygous plant with green-wrinkled seeds. The heterozygous parent produces four classes of gametes. One-half of the offspring have parental phenotypes; one-half have different trait combinations, or recombinant phenotypes. The recombination is due to the independent assortment of the two characters, which is possible because the genes for the two characters are on different chromosomes.

fourth of the offspring have yellow-round seeds, and one-fourth have green-wrinkled seeds. Hence, one-half of the offspring have the same phenotype as one or the other of the parents; these offspring are called **parental types.** But other phenotypes are also represented: one-fourth of the offspring have green-round seeds, and one-fourth have yellow-wrinkled seeds. Because these offspring have different combinations of seed shape and color than either parent, they are called **recombinants.** When half of all offspring are recombinants, geneticists say that there is a 50% frequency of recombination.

A 50% frequency of recombination is observed for any two genes that are located on different chromosomes. The physical basis of recombination between unlinked genes is the random alignment of homologous chromosomes during metaphase I of meiosis, which leads to the independent assortment of alleles (see Figure 14.2).

The Recombination of Linked Genes: Crossing Over

Linked genes do not assort independently, because they are located on the same chromosomes and tend to move together through meiosis and fertilization. We would not expect linked genes to recombine into assortments of alleles not found in the parents. But, in fact, recombination between linked genes *does* occur. To see how, let's return to Morgan's fly room.

How can we explain the results of the *Drosophila* cross illustrated in Figure 14.5? The offspring of the testcross for body color and wing shape did not conform to the 1:1:1:1 phenotypic ratio we would expect if the genes for these two characters were on different chromosomes and assorted independently. But if the two genes were *completely* linked because their loci are on the same chromosome, then we should observe a 1:1 ratio, with only the parental phenotypes represented among offspring. The actual results fit neither of these expectations. Most of the offspring had parental phenotypes, suggesting linkage between the two genes, but about 17% of the flies were recombinants (Figure 14.7). Although there was linkage, it appeared incomplete. Morgan proposed that some mechanism, such as an exchange of segments between homologous chromosomes, must occasionally break the linkage between the two genes. Subsequent experiments have demonstrated that such an exchange—crossing over—accounts for the recombination of linked genes (see Chapter 12). While homologous chromosomes are paired in synapsis during prophase of meiosis I, nonsister chromatids may break at corresponding points and switch fragments. A crossover between chromatids of homologous chromosomes breaks linkages in the parental chromosomes to form recombinants that may bring together alleles in new combinations. The subsequent events of meiosis distribute the recombinant chromosomes to gametes.

GENETIC MAPS BASED ON CROSSOVER DATA

The discovery of linked genes and recombination due to crossing over led Morgan's research team to a method for constructing genetic maps. A genetic map lists the sequence of genetic loci along a particular chromosome. In this section, you will learn how to use crossover data to construct a type of genetic map called a linkage map. Newer techniques for mapping chromosomes, based on methods in molecular biology, will be covered in Chapter 19.

As discussed earlier, a gene for *Drosophila* body color—gray (b^+) versus black (b)—is linked to a gene for wing size—normal (vg^+) versus vestigial (vg). The recombination frequency between these two genetic loci is about 17%; that is, crossing over results in 17% of the offspring having a combination of body color and wing size unlike either parent. Carried on the same chromosome as these two genes is a third gene called

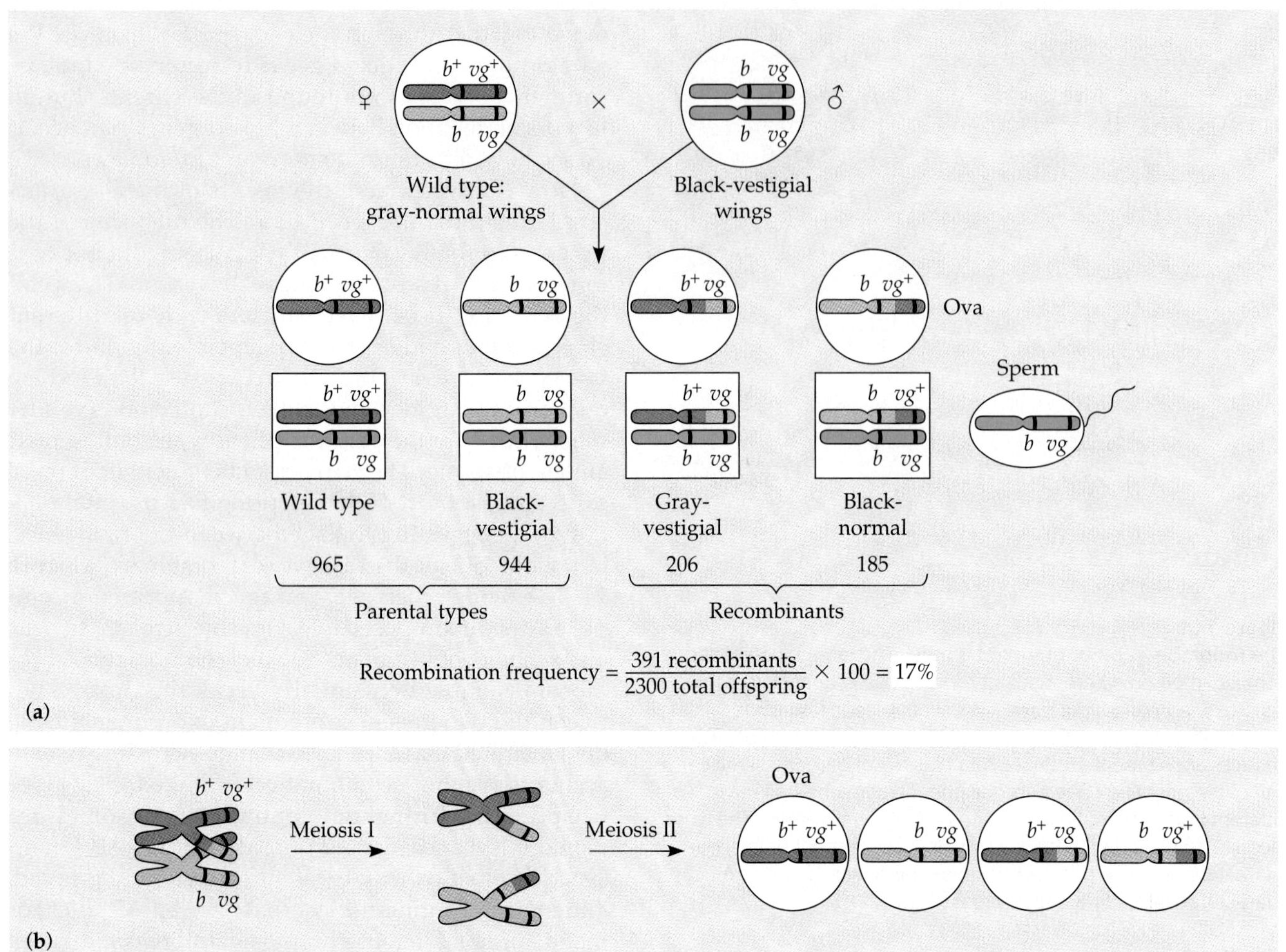

Figure 14.7
Recombination due to crossing over. (**a**) This diagram recreates the testcross in Figure 14.5, but this time we track chromosomes as well as genes. The maternal chromosomes are color-coded with two shades of salmon so that we can distinguish one homologue from the other. Linkage of the genes for body color and wing size explains the prevalence of parental phenotypes among offspring. Recombinant offspring, those with genotypes and phenotypes different from either parent, are due to crossing over. This results in the transmission of a chromosome bearing the allele for gray color *(b^+)* along with the allele for vestigial wings *(vg)*, a genetic combination not found in the chromosomes of either parent. The other product of this crossover is a chromosome combining the *b* and *vg^+* alleles. We can calculate the recombination frequency from the proportion of recombinant flies out of the total pool of offspring. (**b**) Crossing over occurs when homologous chromosomes pair during prophase of meiosis I. Nonsister chromatids break, and the fragments join the homologous chromosome. In this case, the crossover occurs in the region between the *b* and *vg* loci. If we follow the chromosomes through meiosis, we see that crossing over results in the ova that, when fertilized, give rise to the recombinant offspring.

cinnabar *(cn)*, one of the many *Drosophila* genes affecting eye color. Cinnabar eyes, a mutant phenotype, are brighter red than the wild-type color. The recombination frequency between the cinnabar gene *(cn)* and the *b* locus is 9%. Thus, crossovers between the *b* and *vg* loci are about twice as frequent as crossovers between *b* and *cn* (17% versus 9%). In 1917, Alfred H. Sturtevant, one of Morgan's students, reasoned that different recombination frequencies reflect different distances between genes on a chromosome; that is, if two genes are far apart on a chromosome, there is a higher probability that a crossover event will separate them than if the two genes are close together. If we assume that the probability of crossing over between two genes is directly proportional to the distance between them, then the distance along the *Drosophila* chromosome between *b* and *vg* must be about twice as great as the distance between *b* and *cn*.

Sturtevant began using recombination data to assign genes positions on a map. He defined one "map unit" as the equivalent of a 1% recombination frequency. (Today, the word *centimorgan* is often used in place of map unit, in honor of Morgan.) Thus, the *b* and *cn* loci are separated by 9 map units, while the *b* and *vg* genes are 17 map units apart. But what is the sequence of the three genes? We can eliminate the sequence *b-vg-*

cn, since we know that *cn* is closer to *b* than is *vg*. This leaves two possible sequences: *cn-b-vg* or *b-cn-vg*. The frequency of recombination between *cn* and *vg* should reveal the correct sequence of the three genes (Figure 14.8). The first possible sequence predicts that *cn* and *vg* are about 26 map units (9 + 17) apart, while the second sequence predicts a separation of about 8 map units (17 – 9). Sturtevant found that the frequency of recombination between *vg* and *cn* was 9.5%; he therefore proposed that the genes were arranged along a chromosome in the sequence *b-cn-vg*. This method was soon extended to map the other identified *Drosophila* genes in linear arrays.

Some genes on a chromosome are so far apart from each other that crossovers between them occur very often. The frequency of recombination measured between such genes can have a maximum value of 50%, a result indistinguishable from that for genes on different chromosomes. In fact, the seven characters that Mendel studied in his peas are not all on separate chromosomes, although the pea coincidentally has seven chromosome pairs. Seed color and flower color, for instance, are now known to be on chromosome 1. But they are so far apart on that chromosome that linkage is not observed in genetic crosses. Only for one pair of the genes Mendel studied, the genes for plant height and pod shape, do modern biologists observe linkage. Although Mendel observed segregation of alleles for each of these characters, he did not report the results of dihybrid crosses for this particular combination of characters. Genes located far apart on a chromosome are mapped by adding the recombination frequencies from crosses involving each of the distant genes and an intermediate gene.

Using crossover data, Sturtevant and his colleagues were able to cluster the known mutations (and hence the wild-type alleles) of *Drosophila* into four groups of linked genes. Because microscopists had found four sets of chromosomes in *Drosophila* cells, this clustering of genes was additional evidence that genes are located on chromosomes. Each chromosome has a linear array of specific gene loci (Figure 14.9).

A linkage map—a genetic map based on recombination frequencies—is *not* a picture of an actual chromosome. The frequency of crossing over is not uniform over the length of the chromosome, and thus map units do not have absolute size (in nanometers, for instance). A genetic map portrays the sequence of genes along a chromosome, but it does not pinpoint the precise locations of genes. Additional methods enable geneticists to construct **cytological maps** of chromosomes, which actually pinpoint genes. One of these methods locates a gene by associating a mutant phenotype with the position of a chromosomal defect or other feature that can be seen in the microscope. In a comparison of a linkage map with a cytological map of the same chromosome, the sequence of genetic loci matches, but the spacing

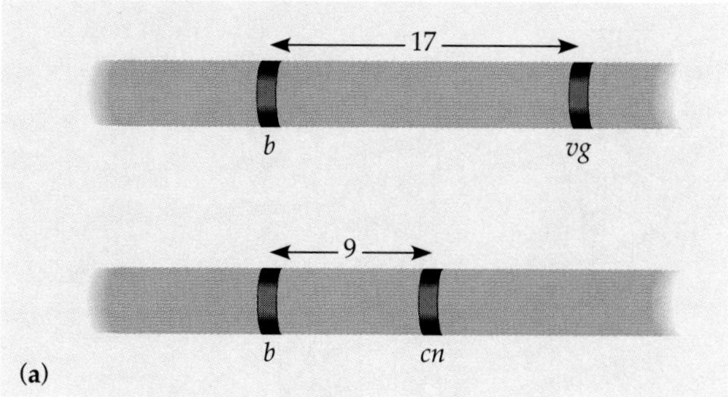

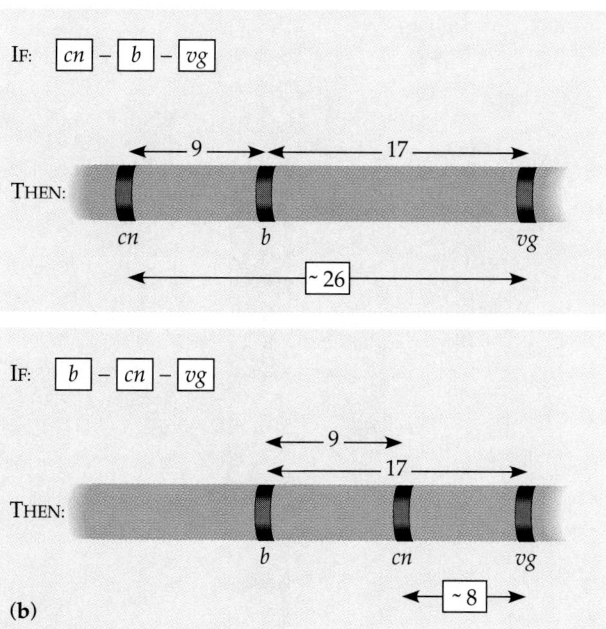

Figure 14.8
Using crossover data to construct a genetic map. A genetic map represents the linear sequence of genes along a chromosome. One method for constructing a genetic map is to determine how frequently crossovers occur in the region between two genes. This method is based on the assumption that the probability of a crossover between two genetic loci is proportional to the distance separating the loci. In this example, we use crossover data (recombination frequencies) to sequence three *Drosophila* genes: *b*, *vg*, and *cn*. (**a**) The *b* and *vg* loci are separated by 17 map units, each map unit being equal to a 1% recombination frequency. The *b* and *cn* loci are separated by 9 map units. (**b**) In order to decide which of two possible sequences of genes is correct, we must determine the frequency of crossing over between the *vg* and *cn* genes. Experiments reveal this recombination frequency to be 9.5%, which best fits the sequence *b-cn-vg*. Notice that if we add the *b-cn* and *cn-vg* distances, we get a *b-vg* distance of 18.5 map units (9 + 9.5). This exceeds the actual *b-vg* recombination frequency of 17%. The discrepancy is explained by multiple crossovers that cancel some of the recombination between these two loci. For example, if crossovers occur at two locations between *b* and *vg*, then the second crossover cancels the effect of the first, reuniting alleles on their original chromosomes. Thus, for loci that are relatively far apart on a chromosome, recombination frequency underestimates actual map distance.

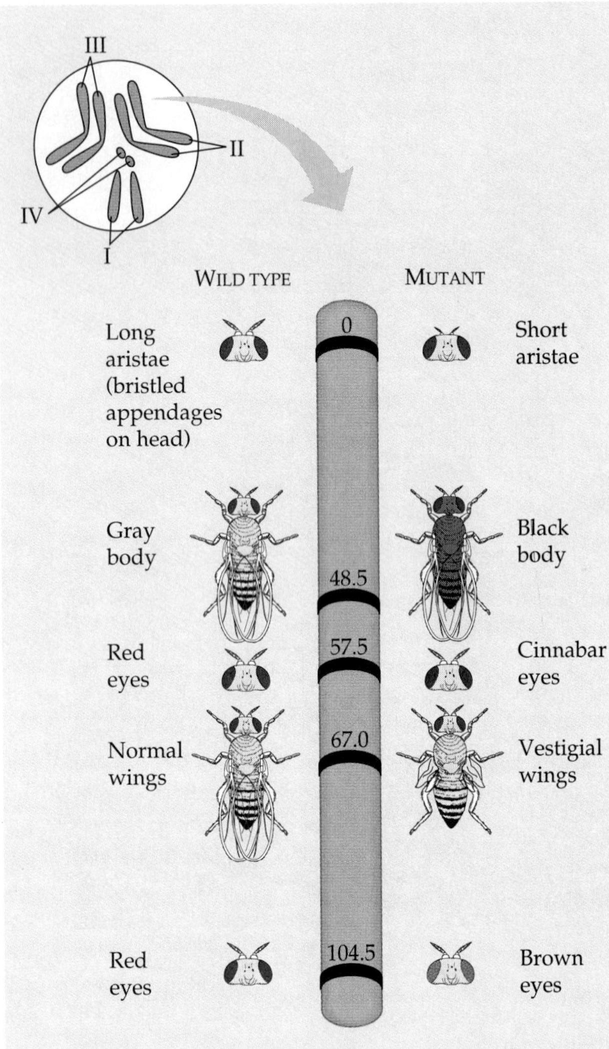

Figure 14.9
A partial genetic map of a *Drosophila* chromosome. *Drosophila* has four pairs of chromosomes, and genetic cartographers have been able to map many loci along each chromosome. The simplified map shown here represents just a few of the genes that have been mapped on chromosome II. Included are the *b* (black body), *cn* (cinnabar eyes), and *vg* (vestigial wings) loci, which we mapped in Figure 14.8. (Notice that more than one gene can affect a given character, such as eye color.) The map is based on recombination frequencies, which cannot exceed 50%, even for genes on opposite ends of a chromosome. How, then, can we account for distances of more than 50 units on this map? Remember that for genes that are far apart, recombination frequency underestimates map distance, because multiple crossovers of even numbers restore alleles to their original chromosomes. Geneticists derive map distances from the recombination frequencies of genes that are close together; it is rare for more than a single crossover to occur within such short intervals. The map is then constructed by adding the map distances between these short intervals. This "stretches out" the map by discounting multiple crossovers that cancel each other. Thus, although the loci for short aristae (antenna bristles) and brown eyes are separated by 104.5 map units, the actual frequency of recombination between these genes is 50%. A genetic map based on linkage studies is a collection of crossover data, not a "picture" of a chromosome.

between loci does not. One reason a 1% recombination frequency does not correspond exactly to a fixed length of chromosome is that crossovers are more common for some chromosomal regions than for others.

SEX CHROMOSOMES AND SEX-LINKED INHERITANCE: A CLOSER LOOK

You learned earlier that Morgan's discovery of a sex-linked allele (white eyes) was a key episode in the development of a chromosome theory of inheritance. In this section, we consider the role of sex chromosomes in inheritance in more detail. We begin by reviewing the genetics of sex in humans.

The Chromosomal Basis of Sex in Humans

Our sex is one of our more obvious phenotypic characters. Although the anatomical and physiological differences between women and men are numerous, the chromosomal basis of sex is rather simple. In humans and other mammals, there are two varieties of sex chromosomes, designated the *X* and *Y* chromosomes (see Chapter 12). When sperm and ovum unite to form a zygote, each individual inherits one of two possible combinations of sex chromosomes. A person who inherits two *X* chromosomes, one from each parent, usually develops as a female. A male usually develops from a zygote containing one *X* chromosome and one *Y* chromosome (Figure 14.10).

When meiosis occurs in gonads (male testes and female ovaries), the two sex chromosomes segregate, and each gamete receives one. Each ovum contains one *X* chromosome. In contrast, a male produces sperm representing two categories, based on sex chromosomes: half the sperm cells contain an *X* chromosome, and half contain a *Y* chromosome. We can trace the sex of each offspring to the moment of conception: If a sperm cell bearing an *X* chromosome happens to fertilize an ovum, the zygote is *XX;* if a sperm cell containing a *Y* chromosome fertilizes an ovum, the zygote is *XY*. Sex is a matter of chance.

The anatomical evidence of sex begins to emerge when the human embryo is about ten weeks old. Before then, the rudiments of the gonads are generic—that is, they can develop into either ovaries or testes, depending on hormonal conditions within the embryo. (Sexual development will be discussed in more detail in Chapter 42.) Which of these two possibilities occurs depends on whether or not a *Y* chromosome is present, and it is probably just a small region of the *Y* chromosome that confers masculinity. In 1990, a British re-

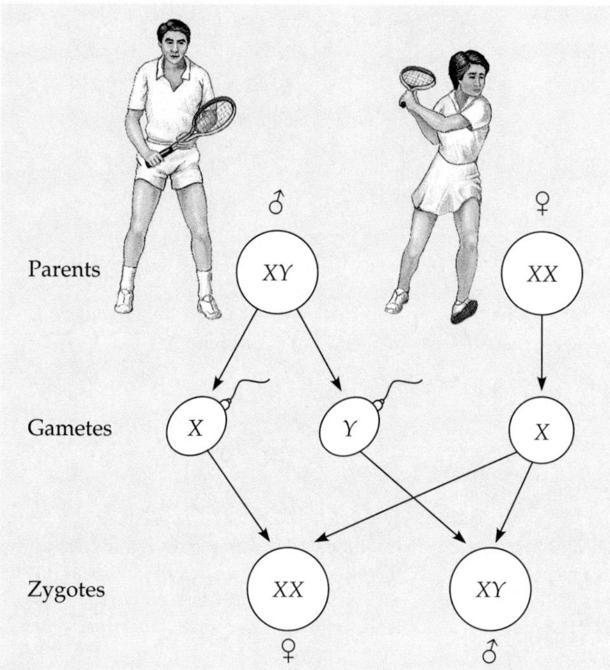

Figure 14.10
The chromosomal basis of sex in humans. An individual's sex is a consequence of a chance event: The combination of sex chromosomes in a zygote depends on which type of sperm cell, *X*-bearing or *Y*-bearing, happens to fertilize the ovum.

search team identified a single gene that seems to be required for the development of testes. They named the gene *Sry*, for sex-determining region *Y* gene. In the absence of *Sry*, the gonads develop into ovaries. The researchers emphasized that the presence (or absence) of *Sry* is just a trigger. The biochemical, physiological, and anatomical features of sex are complex, and many genes are involved in their development. It is likely that *Sry* codes for a protein that regulates other genes, but these control mechanisms are not yet known.

Although chromosomes determine sex in most species, the *X*-*Y* mechanism is only one of several mechanisms. For example, bees and ants lack special sex chromosomes, and yet sex has a chromosomal basis. Fertilized eggs develop into females, which are diploid. Males, which are haploid, develop from unfertilized eggs.

Sex-Linked Disorders in Humans

In addition to their role in determining sex, the sex chromosomes, especially *X* chromosomes, have genes for many characters unrelated to sex. In humans, the term sex-linked usually refers to *X*-linked characters. These traits all follow the same pattern of inheritance that Morgan observed for the white-eye locus in *Drosophila*. Fathers pass *X*-linked alleles to all their daughters but to none of their sons (Figure 14.11). In contrast, mothers can pass sex-linked alleles to both sons and daughters.

If a sex-linked trait is due to a recessive allele, a female will express the phenotype only if she is a homozygote. Because males have only one locus, the terms *homozygous* and *heterozygous* lack meaning for describing their sex-linked genes (the term *hemizygous* is used in such cases). Any male receiving the recessive allele from his mother will express the trait. For this reason, far more males than females have disorders that are inherited as sex-linked recessives. However, even though the chance of a female's inheriting a double dose of the mutant gene is much less than the probability of a male's inheriting a single dose, there *are* females with sex-linked disorders. For instance, color blindness is a mild disorder inherited as a sex-linked trait. A color-blind daughter may be born to a color-blind father whose mate is a carrier (see Figure 14.11c). However, because the sex-linked allele for color blindness is rare, the probability that such a man and woman will come together is very low. An example of a sex-linked disorder much more serious than color blindness is **Duchenne's muscular dystrophy,** which affects about one out of every 3500 males born in the United States. People with Duchenne's muscular dystrophy rarely live past their early twenties. The disease is characterized by a progressive weakening of the muscles and loss of coordination. Researchers have recently traced the disorder to the absence of a key muscle protein and have even tracked the gene to a specific locus on the *X* chromosome. Perhaps in the future this new information will lead to treatments that will prevent the disease from progressing.

Hemophilia is a sex-linked recessive trait with an interesting history. Hemophiliacs bleed excessively when injured, because they are missing a certain protein required for blood clotting. The most seriously afflicted individuals may bleed to death after relatively minor skin abrasions, bruises, or cuts. The ancient Hebrews must have had some understanding of the hereditary pattern of hemophilia, because sons born to women having a family history of hemophilia were exempted from circumcision.

A high frequency of sex-linked hemophilia has plagued the royal families of Europe. The first hemophiliac in the royal line seems to have been Leopold, son of Queen Victoria (1819–1901) of England. The recessive allele for hemophilia was probably introduced to the royal family through a mutation in one of the sex cells of Victoria's mother or father, making Victoria a heterozygote, or carrier, of the deadly allele. Leopold survived to father a daughter who was also a carrier, transmitting hemophilia to one of her sons. Hemophilia was eventually introduced to the royal families of Prussia, Russia, and Spain through the marriages of

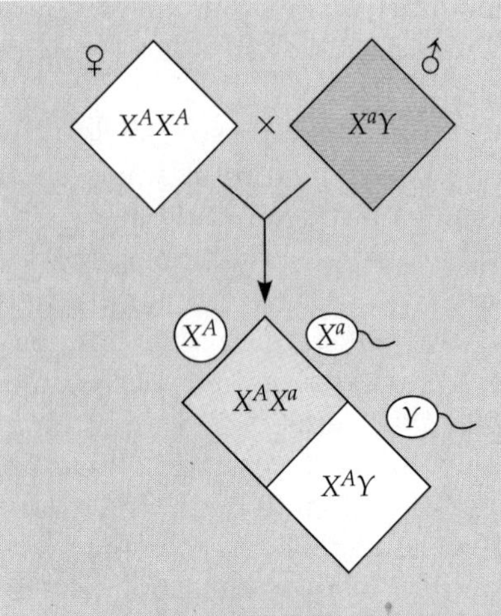

(**a**) A father with the trait will transmit the mutant allele to all daughters, but to no sons. When the mother is a dominant homozygote, the daughters will have the normal phenotype but will be carriers of the mutation.

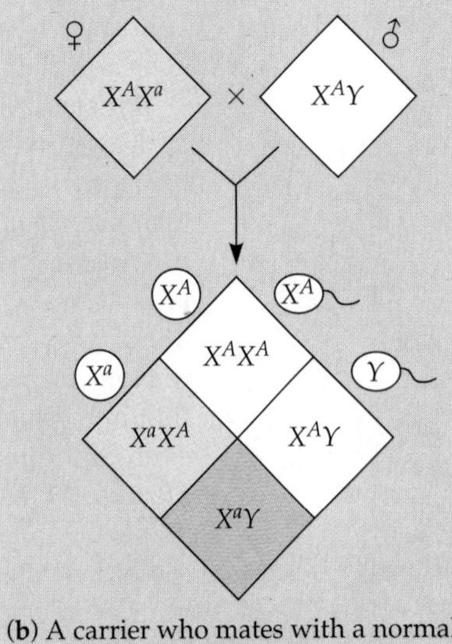

(**b**) A carrier who mates with a normal male will pass the mutation to half her sons and half her daughters. The sons with the mutation will have the disease. The daughters who have inherited the mutation in single dosage will have the normal phenotype but will be carriers like their mother.

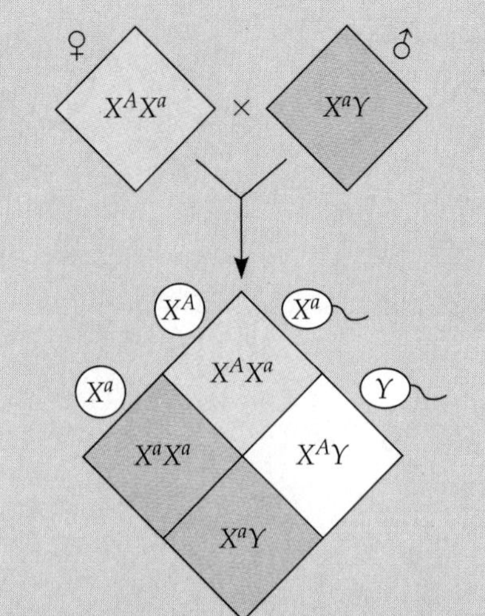

(**c**) If a carrier mates with a male who has the trait, there is a 50% chance that each child born to them will have the trait, regardless of sex. Daughters who do not have the trait will be carriers, whereas males without the trait will be completely free of the deleterious recessive allele.

Figure 14.11
The transmission of sex-linked recessive traits. In this diagram, *X* and *Y* symbolize the sex chromosomes. The superscript *A* represents a dominant allele carried on the *X* chromosome, and the superscript *a* represents a recessive allele. Imagine that this recessive allele is a mutation that causes a sex-linked disease. White boxes indicate unaffected individuals, light-colored boxes indicate carriers, and dark-colored boxes indicate afflicted individuals.

two of Victoria's daughters, Alice and Beatrice, both carriers. The age-old practice of strengthening international alliances by having royalty marry royalty effectively spread hemophilia through the royal families of several European kingdoms.

X-Inactivation in Females

Although female mammals, including humans, inherit two *X* chromosomes, one *X* chromosome in each cell becomes almost completely inactivated during embryonic development. The inactive *X* condenses into a compact object, called a **Barr body,** which lies along the inside of the nuclear envelope in the cells of females. Most of the genes of the *X* chromosome that forms the Barr body are not expressed, although small regions of that chromosome remain active. (Barr bodies are "reactivated" in the cells of gonads that undergo meiosis to form gametes.)

British geneticist Mary Lyon has demonstrated that the selection of which of the two Xs will be inactivated occurs randomly and independently in each of the embryonic cells present at the time of *X*-inactivation. As a consequence, females consist of a mosaic of two types of cells: those with the active *X* derived from the father and those with the active *X* derived from the mother. After an *X* chromosome is inactivated in a particular cell, all mitotic descendants of that cell have the same inactive X. Therefore, if the female is heterozygous for a sex-linked trait, approximately half her cells will express one allele, while the others will express the alternate allele. This mosaicism can be seen graphically in the coloration of a calico cat (Figure 14.12). In humans, there is a recessive *X*-linked mutation that prevents development of sweat glands. A woman who is heterozygous for this trait will have patches of normal skin and patches of skin lacking sweat glands.

CHROMOSOMAL ALTERATIONS

Physical and chemical disturbances, as well as errors during meiosis, can damage chromosomes or alter their number in a cell. In this section, we survey these chromosomal alterations and apply this information to some important disorders in humans.

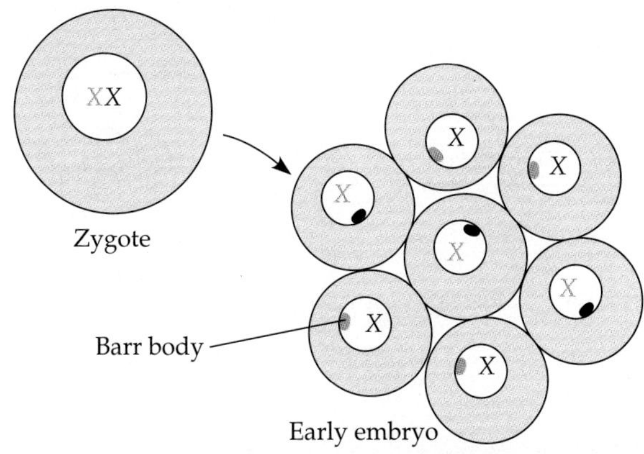

Figure 14.12
***X*-inactivation and the calico cat.** On the *X* chromosome is a gene controlling fur color, with one allele causing black fur and another causing orange fur. (A separate gene is responsible for a patchlike pattern of colored and white fur.) A male *(XY)* can inherit one of these alleles, but not both. Thus, calicos, which have both black and orange patches, are almost always females. A female calico is heterozygous for the patch color locus, inheriting an allele for black on one *X* chromosome and an allele for orange on the other *X*. This is symbolized in the diagram by color-coding the *X* chromosomes according to the allele they carry. During the cat's early embryonic development, one or the other *X* chromosome is randomly inactivated in each cell. The inactivated *X* condenses as a Barr body located just inside the nuclear envelope. As the embryonic cells divide by mitosis, each gives rise to a population of cells having a specific *X* chromosome active and the other inactive. The patchwork coat of the calico results from these dual cell populations.

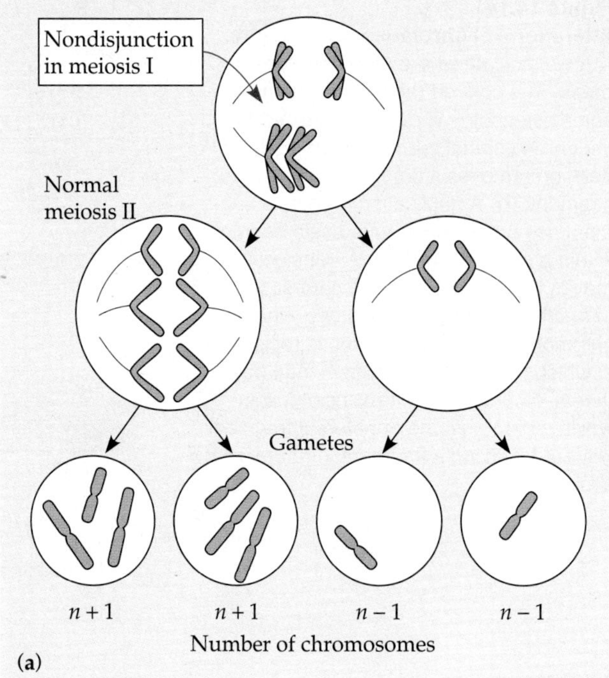

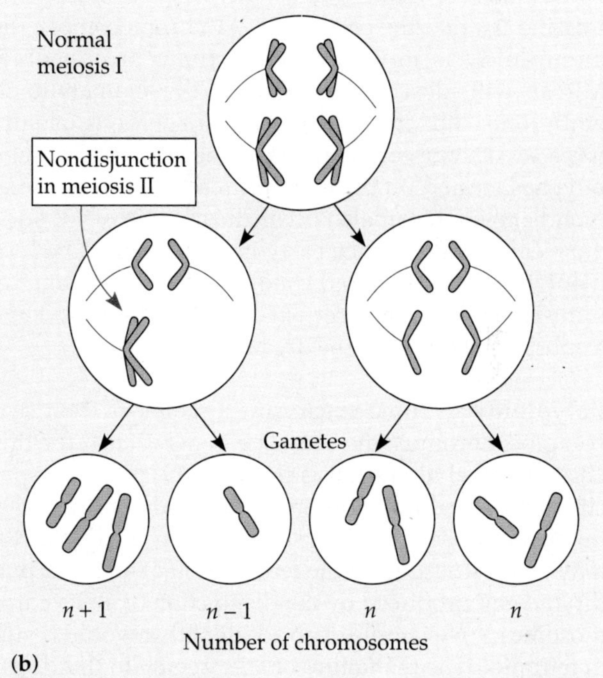

Figure 14.13
Meiotic nondisjunction. (**a**) Homologues may fail to separate during anaphase of meiosis I, or (**b**) chromatids may fail to separate during anaphase of meiosis II. Either type of accident will produce gametes with an anomalous chromosome number.

Alterations of Chromosome Number

Aneuploidy Ideally, the meiotic spindle distributes chromosomes to daughter cells without error. But there is an occasional accident, called a **nondisjunction,** in which the members of a pair of homologous chromosomes do not move apart properly during meiosis I, or in which sister chromatids fail to separate during meiosis II. In these cases, one gamete receives two of the same type of chromosome and another gamete receives no copy (Figure 14.13). The other chromosomes are usually distributed normally. If either of these aberrant gametes unites with a normal one, the offspring will have an abnormal chromosome number, known as **aneuploidy.** If the chromosome is present in triplicate in the fertilized egg (so that the cell has a total of $2n + 1$ chromosomes), the aneuploid cell is said to

Figure 14.14
Alterations of chromosome structure. Arrows indicate where chromosomes break. The colored regions of chromosomes symbolize the genes affected by the chromosomal rearrangements. (**a**) A deletion removes a chromosomal segment. (**b**) A duplication repeats a segment. (**c**) An inversion reverses a segment within a chromosome. (**d**) A translocation moves a segment from one chromosome to another, nonhomologous one. The most common type of translocation is reciprocal, in which nonhomologues exchange fragments. Nonreciprocal translocations, in which a chromosome transfers a fragment without receiving a fragment in return, also occur.

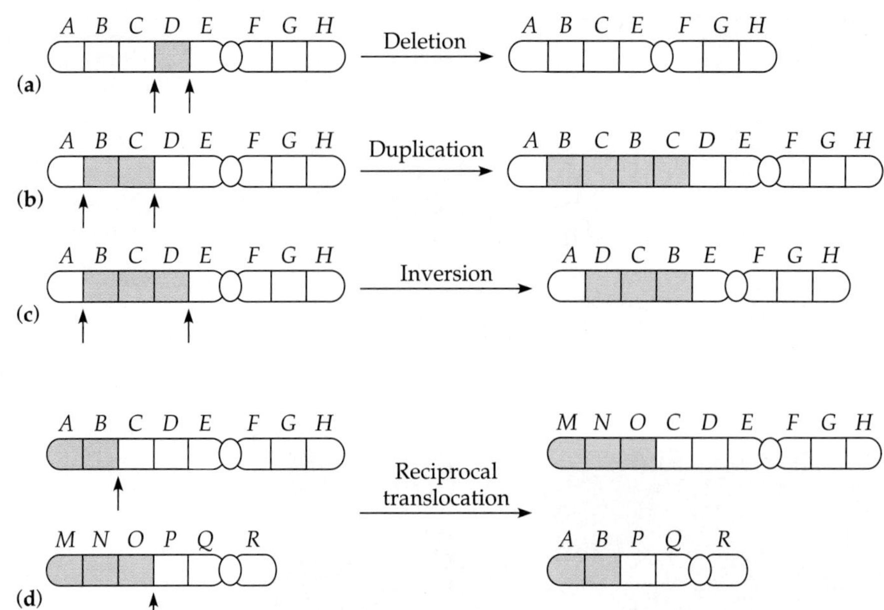

be **trisomic** for that chromosome. If a chromosome is missing (so that the cell has $2n - 1$ chromosomes), the aneuploidy is **monosomic** for that chromosome. Mitosis will subsequently transmit the anomaly to all embryonic cells. If the organism survives, it usually has a set of symptoms caused by the abnormal dosage of genes located on the extra or missing chromosome. Nondisjunction can also occur during mitosis. If such an accident takes place early in embryonic development, then the aneuploid condition is passed along by mitosis to a large number of cells and is likely to have a substantial effect on the organism.

Polyploidy Some organisms have more than two complete chromosome sets. The general term for this chromosomal alteration is **polyploidy,** with the specific terms **triploidy** ($3n$) and **tetraploidy** ($4n$) indicating the number of haploid sets. One way a triploid cell may be produced is by the fertilization of an abnormal diploid egg produced by nondisjunction of all its chromosomes. An example of an accident that would result in tetraploidy is the failure of a $2n$ zygote to divide after replicating its chromosomes. Subsequent mitosis would then produce a $4n$ embryo.

Polyploidy is relatively common in the plant kingdom, and we will see in Chapter 22 that the spontaneous origin of polyploid individuals plays an important role in the evolution of plants. In the animal kingdom, the natural occurrence of polyploids seems to be extremely rare, although polyploidy can be induced experimentally in certain animals, such as frogs and rabbits. In general, polyploids are more normal in appearance than aneuploids. One extra (or missing) chromosome apparently disrupts genetic balance more than having an entire extra set of chromosomes does. More common than complete polyploid animals are mosaic polyploids, animals with patches of polyploid cells. If the sister chromatids for all the chromosomes fail to separate during a *mitotic* division, so that one daughter cell gets all the replicated chromosomes, a tetraploid cell results, which can subsequently produce a localized clone of tetraploid cells.

Alterations of Chromosome Structure

Breakage of a chromosome can lead to a variety of rearrangements affecting the genes of that chromosome. Fragments without centromeres are usually lost when the cell divides. The chromosome from which the fragment originated will then be missing certain genes, a chromosomal alteration known as **deletion.** In some cases, however, the fragment may join to the homologous chromosome, producing a **duplication** there. It also may reattach to the original chromosome but in the reverse orientation, a change called an **inversion,** or it may join a nonhomologous chromosome, an event called a **translocation.** Figure 14.14 illustrates these different types of structural alterations of chromosomes.

Another source of deletions and duplications is error during crossing over. Nonsister chromatids sometimes break at different places, and one partner consequently gives up more genes than it receives. The products of such a nonreciprocal crossover are one chromosome with a deletion and one chromosome with a duplication.

An organism that inherits a homozygous deletion (or a single *X* chromosome with a deletion, in a male) has a genetic imbalance that is usually lethal. This observation is evidence that most genes are vital to an organism's existence. Duplications and translocations also tend to have deleterious effects. In reciprocal

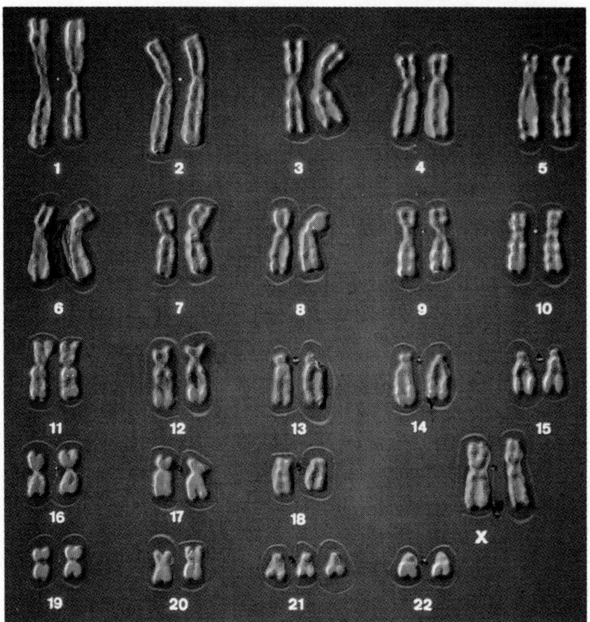

(a)

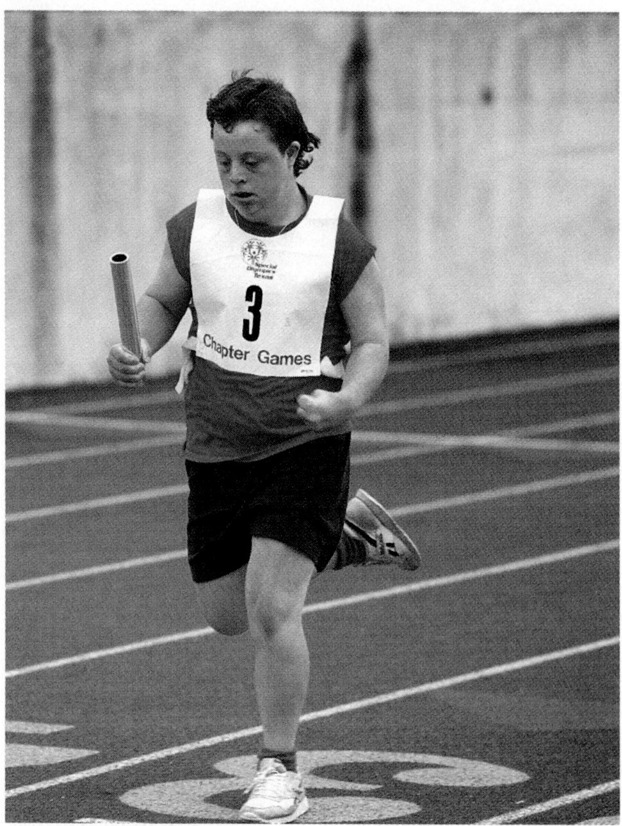

(b)

Figure 14.15
Down syndrome. (**a**) The karyotype shows trisomy 21. (**b**) Thousands of individuals with Down syndrome are among those who participate nationwide in the Special Olympics.

translocations, in which segments are exchanged between chromosomes, and in inversions, the balance of genes is not abnormal—all genes are present in their normal doses. Nevertheless, inversions and translocations can alter phenotype, because of more subtle **position effects:** A gene's expression can be influenced by its location among neighboring genes.

Chromosomal Alterations in Human Disease

Alterations of chromosome number and structure are associated with a number of serious human disorders. When nondisjunction occurs in meiosis, the result is aneuploidy, an abnormal number of chromosomes in the gamete produced and, later, in the zygote. Although the frequency of aneuploid zygotes may be quite high in humans, most of these chromosomal alterations are so disastrous to development that the embryos are spontaneously (naturally) aborted long before birth. However, some types of aneuploidy appear to upset the genetic balance less than others, so that individuals are occasionally born with a characteristic condition, or syndrome. Genetic diseases caused by aneuploidy can be diagnosed before birth by amniocentesis (see Chapter 13).

Down syndrome is the most common serious birth defect in the United States, affecting approximately one out of every 700 children born. Down syndrome is usually the result of aneuploidy: There is an extra chromosome 21, so that each body cell has a total of 47 chromosomes (Figure 14.15). Although chromosome 21 is the smallest human chromosome, its trisomy severely alters the individual's phenotype. Down syndrome includes characteristic facial features, short stature, heart defects, susceptibility to respiratory infection, and mental retardation. Down syndrome is by far the most common form of severe mental retardation. Furthermore, individuals with Down syndrome are prone to developing leukemia and Alzheimer's disease. (It is probably not a coincidence that genes associated with the latter two diseases have been found to be on chromosome 21.)

Although people with Down syndrome, on average, have a lifespan much shorter than normal, some live to middle age or beyond. Most are sexually underdeveloped and sterile, but a few women with Down syndrome have had children. Since half of these women's eggs have the extra chromosome 21, there is a 50% chance that a woman with Down syndrome will transmit it to her child.

Among parents with normal karyotypes, the frequency of Down syndrome in offspring correlates with the age of the mother (Figure 14.16). Down syndrome strikes 0.04% of children born to women under age 30. The risk climbs to 1.25% for mothers in their early thirties and is even higher for older mothers. Because of

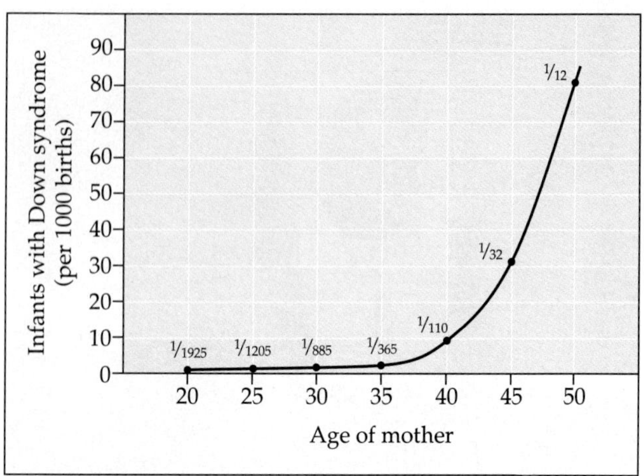

Figure 14.16
Maternal age and Down syndrome. The incidence of Down syndrome increases with maternal age. No other chromosomal disorder is known to follow this pattern.

this relatively high risk, pregnant women who are over 35 are candidates for amniocentesis in order to check for trisomy 21. The correlation of Down syndrome with maternal age has not yet been explained. But recent evidence supports the hypothesis that older women are more likely than younger women to carry Down babies to term rather than spontaneously aborting the trisomic embryos. No other chromosomal disorder is known to follow this pattern of increased incidence with maternal age.

Far rarer than Down syndrome are several other human diseases caused by autosomal aneuploidy. Patau syndrome, caused by trisomy for chromosome 13, is characterized by serious eye, brain, and circulatory defects, as well as harelip and cleft palate. Patau syndrome occurs once in every 5000 live births. Trisomy of chromosome 18 causes the condition known as Edwards syndrome, which affects almost every organ system in the body. It occurs about once in every 10,000 live births. In both these syndromes, most victims survive less than a year.

Nondisjunction of sex chromosomes produces a variety of aneuploid conditions in humans (Table 14.1). Most of these conditions appear to upset the genetic balance less than aneuploid conditions involving autosomes. This may be because the *Y* chromosome carries very few genes and because extra copies of the *X* chromosome become inactivated as Barr bodies in the somatic cells.

An extra *X* chromosome in a male, producing *XXY*, occurs approximately once in every 2000 live births. People with this disorder, called **Klinefelter syndrome,** have male sex organs, but the testes are abnormally small and the man is always sterile. The syndrome often leads to breast enlargement and other feminine body characteristics. The affected individual is usually of normal intelligence. Klinefelter syndrome is also associated with males having more than one additional sex chromosome *(XXYY, XXXY, XXXXY,* and *XXXXXY).* Such individuals are more likely to be mentally retarded than *XXY* individuals.

Human males with a single extra *Y* chromosome *(XYY)* are not characterized by any well-defined syndrome, although they tend to be somewhat taller than the average male. Females with trisomy *X (XXX)* occur once in approximately 1000 live births. These **metafemales** have limited fertility. Monosomy *X*, called **Turner syndrome,** occurs about once in every 5,000 births and is the only known viable human monosomy. Although these *XO* individuals are phenotypically female, their sex organs do not mature at adolescence, and secondary sex characteristics fail to develop. Such individuals are sterile and of short stature. Turner syndrome patients usually have no mental deficiency.

In addition to aneuploid conditions, there are also structural alterations of chromosomes associated with specific disorders in humans. Many deletions in human chromosomes, even in a heterozygous state, cause severe physical and mental defects. One such chromosomal deficiency is known as *cri du chat* ("cry of the cat") syndrome. A human child born with this specific deletion in chromosome 5 is mentally retarded and has a small head with unusual facial features and a cry that sounds like the mewing of a distressed cat. Such individuals usually die in infancy or early childhood.

Another type of structural alteration of chromosomes associated with human disorders is the chromosomal translocation, the attachment of a fragment from one chromosome to another, nonhomologous chromosome. Chromosomal translocations have been implicated in certain cancers. One example is chronic myelogenous leukemia (CML). Leukemia is a cancer affecting the cells that give rise to white blood cells, and in the cancerous cells of CML patients, a reciprocal

Table 14.1 Abnormalities of Sex Chromosome Number in Humans

Genotype	Phenotype	Origin of Nondisjunction	Frequency in Population
XO	Turner syndrome (female)	Meiosis in egg or sperm formation	1/5000
XXX	Metafemale	Meiosis in egg formation	1/1000
XXY	Klinefelter syndrome (male)	Meiosis in egg or sperm formation	1/2000
XYY	Normal male	Meiosis in sperm formation	1/2000

translocation has occurred. A portion of chromosome 22 has switched places with a small fragment from a tip of chromosome 9. (How such a switch might cause cancer will be discussed in Chapter 18.)

A small fraction of individuals with Down syndrome have a chromosomal translocation of a different sort. All the cells of such persons have the normal number of chromosomes, 46. Close inspection of the karyotype, however, shows the presence of part or all of a third chromosome 21 attached to another chromosome by translocation.

PARENTAL IMPRINTING OF GENES

Throughout our discussions of Mendelian genetics and the chromosomal basis of inheritance, we have been assuming that a specific allele will have the same effect regardless of whether it was inherited from the mother or the father. This is probably a safe assumption most of the time. For example, when Mendel crossed purple-flowered peas with white-flowered peas, he observed the same results regardless of whether the purple-flowered parent supplied the ova or pollen. Recently, however, geneticists have identified some traits, including some inherited disorders in humans, that seem to depend on which parent passed along the alleles for those traits.

Consider, for example, two different disorders called Prader-Willi syndrome and Angelman syndrome. The symptoms are different. Prader-Willi is characterized by mental retardation, obesity, short stature, and unusually small hands and feet. People with Angelman syndrome exhibit spontaneous (uncontrollable) laughter, jerky movements, and other motor and mental symptoms. For both diseases, the genetic cause seems to be the same: deletion of a particular segment of chromosome 15. If a child inherits the defective chromosome from the father, the result is Prader-Willi syndrome. If the defective chromosome is inherited from the mother, the result is Angelman syndrome. Put another way, an individual with Prader-Willi lacks those genes that have been deleted from the paternal version of chromosome 15; an individual with Angleman lacks those genes that have been deleted from the maternal version of the chromosome. The implication is that the genes of the deleted region normally behave differently in offspring, depending on whether they belong to the maternal or the paternal chromosome.

A process called **genomic imprinting** may explain the Prader-Willi/Angelman enigma and some similar phenomena. According to this hypothesis, certain genes are imprinted in some way each generation, and the imprint is different depending on whether the genes reside in females or in males (Figure 14.17). In

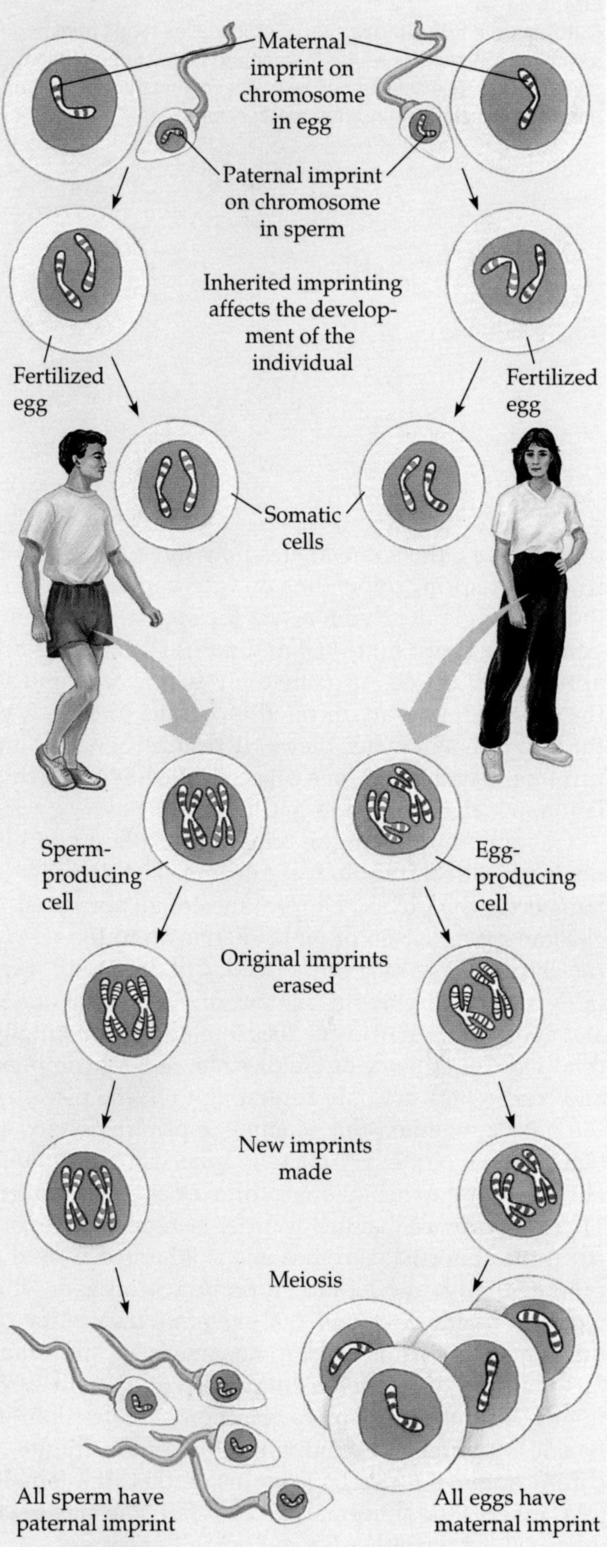

Figure 14.17
Genomic imprinting. Sperm and ova may convey chromosomes that are differently imprinted. According to this hypothesis, the same allele may have a different phenotypic effect, depending on whether it has the maternal or the paternal imprint. Each generation, the old imprints are "erased" when sperm or ova are produced, and all the chromosomes are newly imprinted according to the sex of the individual.

Figure 14.18
Cytoplasmic inheritance. Variegated leaves result from genes located in the plastids rather than on the nuclear chromosomes of plant cells. The cells of yellow areas contain plastids that are unable to develop into normal green chloroplasts.

other words, the same alleles may have different effects on offspring depending on whether they arrive in the zygote via the ovum or via the sperm. In the new generation, both maternal and paternal imprints are apparently "erased" in gamete-producing cells, and all the chromosomes are recoded according to the sex of the individual in which they now reside. (Genomic imprinting may depend on a process called DNA methylation, which is discussed in Chapter 18.)

Genomic imprinting may help to explain another disorder called **fragile-*X* syndrome.** The disorder is named for the physical appearance of an abnormal *X* chromosome, the tip of which hangs on to the rest of the chromosome by a thin thread. Children with fragile-*X* syndrome—about one out of every 1000 males and about one out of every 2000 females—are mentally retarded. Inheritance of the disorder is very complex, and researchers are only beginning to fit the pieces of the puzzle together. But genomic imprinting seems to be involved. Not everyone who inherits the fragile-*X* chromosome exhibits symptoms of the syndrome. There is evidence that the disorder is more likely to occur if the abnormal chromosome is inherited from the mother than from the father. Some researchers who work on fragile-*X* believe that maternal imprinting of the damaged part of the chromosome is an important factor in the disease. This may help to explain why fragile-*X* syndrome is more common in males than in females. If a male (*XY* individual) inherits a fragile-*X* chromosome, it *has* to be from his mother. If a female (*XX*) inherits the abnormal chromosome, she may have inherited it from either her mother *or* her father.

A chromosomal alteration called uniparental disomy provides further evidence for genomic imprinting. Let's return to the Prader-Willi/Angelman problem to see how this works. Only about 60% of individuals with Prader-Willi syndrome have karyotypes exhibiting the deletion in the paternal copy of chromosome 15. What accounts for the disorder in the other 40% of cases? These children have inherited *both* copies of normal chromosome 15 from their mother. This uniparental disomy probably results from the union of a sperm cell lacking chromosome 15 with an ovum having two copies of the chromosome (that is, a nondisjunction affecting the same chromosome during formation of both gametes). Why would such an individual have Prader-Willi syndrome? Apparently, certain genes on the portion of chromosome 15 that is critical to the Prader-Willi and Angelman syndromes are imprinted differently by one's father than by one's mother. And normal development requires that both versions of the imprinted chromosomes are inherited. If this hypothesis is correct, then uniparental disomy affecting chromosomes other than chromosome 15 should also have dramatic effects on phenotype. Several geneticists are now searching for genetic disorders that can be explained in this way.

EXTRANUCLEAR INHERITANCE

Although our focus in this chapter has been on the chromosomal basis of inheritance, we end with an important amendment: Not all of a eukaryotic cell's genes are located on nuclear chromosomes, or even in the nucleus. Most of the extranuclear genes are found in cytoplasmic organelles, such as mitochondria and, in plants, plastids, the self-replicating organelles that include chloroplasts. Both mitochondria and plastids replicate and transmit their genes to daughter organelles. These cytoplasmic genes do not display Mendelian inheritance, because they are not distributed to offspring according to the same rules that direct the distribution of nuclear chromosomes during meiosis.

Cytoplasmic genes were first observed in plants. In 1909, Karl Correns studied the inheritance of yellow or white patches on the leaves of an otherwise green ornamental plant (Figure 14.18). He found that the coloration of the offspring was determined only by plants with flowers bearing the seeds and not by plants with flowers producing the pollen. Subsequent research has shown that such variegated (striped or spotted) colorations of leaves are due to differences in pigment production by the plastids. Such differences may be controlled by genes located in the plastids. In most plants, a zygote receives all its plastids from the cytoplasm of the ovum, and none from pollen. Thus, as the zygote of these plants develops, its pattern of leaf coloration depends only on maternal cytoplasmic genes.

Maternal inheritance is also the rule for the mitochondrial genes in mammals. The mitochondria are situated in the cytoplasm of a cell, and the ovum always contributes much more cytoplasm to the zygote than does the sperm. Thus, mitochondria come from the mother. For example, when laboratory rats carrying one genetic type of mitochondrial DNA are mated to rats carrying another type, all the offspring contain only mitochondria of the maternal type. Mitochondrial genes can be used to trace evolutionary relationships, as will be discussed in Chapter 23.

Wherever DNA is located in a cell, its fundamental structure and function are the same. In the next chapter, we will begin to examine the behavior of DNA on a molecular level.

STUDY OUTLINE

1. In the early 1900s, geneticists demonstrated that Mendel's laws of segregation and independent assortment have their physical basis in the behavior of chromosomes.

The Chromosome Theory of Inheritance (pp. 280–284)

1. Thomas Hunt Morgan was the first to associate a specific gene with a specific chromosome. Like Mendel, he used an organism that had intrinsic advantages for genetic research. The prolific and fast-breeding fruit fly, *Drosophila melanogaster,* has only four pairs of chromosomes.
2. After a year of breeding flies, Morgan obtained a spontaneously mutant white-eyed male that led him to discover the first known sex-linked gene. This gene for eye color, carried on the *X* chromosome, gave powerful support to the chromosome theory of inheritance.
3. Each chromosome has hundreds or thousands of genes. Genes near each other on a chromosome are said to be linked and do not assort independently.

The Chromosomal Basis of Recombination (pp. 284–285)

1. The events of meiosis and random fertilization are responsible for genetic recombination, the production of offspring with new combinations of traits inherited from two parents.
2. Offspring that have the same phenotype as one or the other parent are called parental types. The recombinant offspring, however, have combinations of traits that do not match either of the parents, owing largely to independent assortment of alleles during the first meiotic division. A 50% recombination frequency for two genes usually indicates that the genes are located on separate chromosomes and are thus unlinked.
3. A recombination frequency of less than 50% indicates that the genes are linked but that crossing over has occurred. In this process, homologous chromosomes in synapsis during prophase of meiosis I break at corresponding points and switch fragments, thereby creating new combinations of alleles that are subsequently passed on to the gametes.

Genetic Maps Based on Crossover Data (pp. 285–288)

1. Methods have been devised to assign, or map, genes to specific regions on particular chromosomes.
2. Morgan's group was the first to map genes by deducing relative distances between them based on crossover data. Genes that are far apart on a chromosome are more likely to be separated during crossover than are genes that are close together. One map unit is defined as the equivalent of a 1% recombination frequency.
3. Cytological mapping is a technique that pinpoints the physical locus of a gene by associating a mutant phenotype with a chromosomal defect seen in the microscope.
4. Genetic loci determined from crossover data and loci determined by cytological mapping techniques are not completely consistent, because the frequency of crossing over varies among different regions of the chromosome.

Sex Chromosomes and Sex-Linked Inheritance: A Closer Look (pp. 288–290)

1. Sex is an inherited phenotypic character usually determined by the presence or absence of special chromosomes, but the exact mechanism for sex determination varies among different species.
2. In humans and other mammals, an *X-Y* system is operative. The *XY* males apportion either an *X* or a *Y* chromosome to a given sperm, which combine with ova containing only *X* chromosomes from the *XX* females. Thus, the sex of the offspring is determined at conception by whether the sperm cell bears an *X* or a *Y* chromosome.
3. Researchers have recently tracked the hereditary basis of maleness to a specific locus on the *Y* chromosome.
4. Certain genes for traits that are unrelated to maleness or femaleness are located on the sex chromosomes. Hemophilia is one of many sex-linked recessive traits whose gene is carried on the *X* chromosome.
5. In mammalian females, only one *X* chromosome per diploid cell is active. The other is condensed into a Barr body located on the inside of the nuclear envelope. One of the two *X* chromosomes in each cell is randomly inactivated during early embryonic development. Small regions of the inactive chromosome may remain functional.

Chromosomal Alterations (pp. 290–295)

1. Environmental disturbance or errors in meiosis can change the number of chromosomes per cell or the structure of individual chromosomes. Such alterations can

significantly affect the phenotype, especially if several linked genes or an entire chromosome is affected.

2. Aneuploidy is an abnormal chromosome number. It can arise when a normal parental gamete unites with its counterpart that contains, for example, either two copies or no copies of a particular chromosome as a result of nondisjunction during meiosis. The resulting zygote will be trisomic or monosomic for the given chromosome.
3. Polyploidy, a condition in which there are more than two entire sets of chromosomes, can occur as a result of complete nondisjunction during gamete formation. Polyploidy is more common in plants, but it can also occur rarely in animals. Mitotic nondisjunction can produce patches of polyploid cells in an otherwise diploid individual.
4. A variety of rearrangements can result from breakage of a chromosome. A lost fragment leaves the original chromosome with a deletion, but may produce a duplication, translocation, or inversion by reattaching to a chromosome.
5. Chromosomal alterations cause a variety of human disorders in individuals whose malfunction does not prevent them from surviving gestation. For example, partial or total trisomy of chromosome 21 is responsible for Down syndrome, a relatively common, serious birth defect that shows increasing incidence with maternal age.

Parental Imprinting of Genes (pp. 295–296)

1. Genomic imprinting is helping to explain some baffling hereditary disorders, including fragile-X syndrome.
2. Individuals apparently imprint chromosomes in their gamete-producing cells with either a male or a female "stamp." This affects the way some genes are expressed in offspring. Offspring "erase" the imprints in gametes and recode according to their own sex.
3. Nondisjunction affecting chromosome count in both sperm and ovum can produce an individual with a uniparental disomy, in which *both* copies of a particular chromosome are inherited from the same parent. The absence of imprinted genes from the other parent can have an adverse effect on phenotype.

Extranuclear Inheritance (pp. 296–297)

1. Mitochondria and chloroplasts contain some of their own genes. Such cytoplasmic genes do not display Mendelian inheritance.
2. In both plants and animals, the zygote receives almost all its cytoplasm from the ovum. Such inheritance causes certain aspects of the offspring's phenotype to be dependent solely on maternal cytoplasmic genes.

SELF QUIZ

1. A key discovery in the early development of the chromosome theory of inheritance was
   a. Mendel's realization that the behavior of chromosomes paralleled the behavior of his "heritable factors" (genes) in peas
   b. the association of specific traits with chromosomes that determine sex in fruit flies
   c. the discovery of the chromosomal basis of Down syndrome
   d. the discovery that DNA is found in chromosomes
   e. the discovery that chromosomes are located in the nucleus
2. Two genes will probably assort independently if
   a. they are very close together on the same chromosome
   b. they are very far apart on the same chromosome
   c. they are on homologous chromosomes
   d. they are both located on the *X* chromosome
   e. one is dominant and the other is recessive
3. An aneuploid person is obviously female, but her cells have two Barr bodies. Which aneuploid condition probably accounts for these observations?
   a. *XXX*
   b. *XYY*
   c. *XXY*
   d. *XO*
   e. *YYY*
4. The genetic event that results in Down syndrome can best be described as
   a. nonreciprocal crossingover
   b. nondisjunction
   c. a chromosomal duplication
   d. a deletion
   e. uniparental disomy
5. Determine the order of genes along a chromosome based on the following recombination frequencies: *A* - *B*, 8 map units; *A* - *C*, 28 map units; *A* - *D*, 25 map units; *B* - *C*, 20 map units; *B* - *D*, 33 map units.
   a. *A* - *B* - *C* - *D*
   b. *A* - *C* - *D* - *B*
   c. *B* - *A* - *C* - *D*
   d. *D* - *A* - *B* - *C*
   e. *C* - *D* - *B* - *A*
6. One consequence of *X*-inactivation is that
   a. females are a genetic mosaic due to random nonseparation of chromatids during mitosis
   b. both males and females have equal dosages of most *X*-linked genes because of the formation of Barr bodies
   c. males inherit their sex-linked traits from their mothers
   d. the *Y* chromosome can exert its genetic influence in the egg cytoplasm
   e. the calico condition is lethal in male cats because the *X* chromosome is inactive in males
7. In the chromosomal rearrangement known as a duplication,
   a. a fragment of a chromosome may join to its homologous chromosome
   b. homologous chromosomes do not separate in meiosis I
   c. sister chromatids do separate in meiosis II
   d. a fragment of a chromosome joins to a nonhomologous chromosome
   e. an aneuploid condition results from an extra chromosome

GENETICS PROBLEMS

1. The normal daughter of a man with hemophilia (a recessive, sex-linked condition) marries a man who is normal for the trait. What is the probability that a daughter will be a hemophiliac? A son? If the couple has four sons, what is the probability that all four will be born with hemophilia?

2. Pseudohypertrophic muscular dystrophy is a disorder that causes gradual deterioration of the muscles. It is seen only in boys born to apparently normal parents and usually results in death in the early teens. Is pseudohypertrophic muscular dystrophy caused by a dominant or recessive allele? Is its inheritance sex-linked or autosomal? How do you know? Explain why this disorder is always seen in boys and never in girls.

3. Red-green color blindness is caused by a sex-linked recessive allele. A color-blind man marries a woman with normal vision whose father was color-blind. What is the probability that they will have a color-blind daughter? What is the probability that their first son will be color-blind? (*Note:* The two questions are worded a bit differently.)

4. A wild-type fruit fly (heterozygous for gray body color and normal wings) was mated with a black fly with vestigial wings. The offspring gave the following distribution: wild type, 778; black-vestigial, 785; black-normal, 158; gray-vestigial, 162. What is the recombination frequency between these genes for body color and wing type?

5. In another cross, a wild-type fruit fly (heterozygous for gray body color and red eyes) was mated with a black fruit fly with purple eyes. The offspring were as follows: wild type, 721; black-purple, 751; gray-purple, 49; black-red, 45. What is the recombination frequency between these genes for body color and eye color? Following up on this question and question 4, what fruit flies (genotypes and phenotypes) would you mate to determine the order of the body color, wing shape, and eye color genes on the chromosome?

6. A space probe discovers a planet inhabited by creatures who reproduce according to the same genetic laws as humans. Three phenotypic characters are height (*T* = tall, *t* = dwarf), head appendages (*A* = antennae, *a* = no antennae), and nose morphology (*S* = upturned snout, *s* = downturned snout). Since the life was not "intelligent," Earth scientists were able to do some controlled breeding experiments, using various heterozygotes in testcrosses. For a tall heterozygote with antennae, the offspring were tall-antennae, 46; dwarf-antennae, 7; dwarf-no antennae, 42; tall-no antennae, 5. For a heterozygote with antennae and an upturned snout, the offspring were antennae-upturned snout, 47; antennae-downturned snout, 2; no antennae-downturned snout, 48; no antennae-upturned snout, 3. Calculate the recombination frequencies for both experiments.

7. Using the information from problem 6, a further testcross was done using a heterozygote for height and nose morphology. The offspring were tall-upturned nose, 40; dwarf-upturned nose, 9; dwarf-downturned nose, 42; tall-downturned nose, 9. Calculate the recombination frequency from these data; then use your answer from problem 6 to determine the correct sequence of the three linked genes.

8. Imagine that a geneticist has identified two disorders that appear to be caused by the same chromosomal defect and affected by genomic imprinting: blindness and numbness of the hands and feet. A blind woman (whose mother suffered from numbness) has four children, two of whom—a son and daughter—have inherited the abnormality. If this defect works like Prader-Willi and Angelman syndromes, what disorders do they display? What disorders would be seen in *their* sons and daughters?

9. What pattern of inheritance would lead a geneticist to suspect that an inherited disorder of cell metabolism is due to a defective mitochondrial gene?

SCIENCE, TECHNOLOGY, AND SOCIETY

1. A chromosome test is used to determine whether women Olympic competitors are actually female. In the past, the test checked for the presence of a Barr body. The method used in the 1992 Olympics checked for genes found on the *Y* chromosome. Athletes who fail the test—about one in 500—are barred from competition. Some *XY* individuals are anatomically female, although they lack ovaries and a uterus; their *Y* chromosome fails to cause development of testes, or their cells are insensitive to the effects of the male hormone testosterone. Athletes and physicians argue that the chromosome test unfairly bars *XY* females from competition. What is the purpose of the test? Is it fair? Can you suggest an alternative?

2. Gregor Mendel never saw a gene, yet he concluded that these "heritable factors" were responsible for inheritance in peas. Similarly, Morgan and Sturtevant never actually saw linked genes on chromosomes; their maps were deduced from patterns of inheritance. Is it legitimate science for biologists to claim the existence of things and processes they cannot actually see? Why or why not?

FURTHER READING

Charlesworth, B. "The Evolution of Sex Chromosomes." *Science*, March 1, 1991. How did the *X-Y* mode of sex determination evolve?

Chefas, J. "Sex and the Single Gene." *Science*, May 10, 1991. The genetic cause of maleness has been traced to a specific region of the *Y* chromosome.

Hoffman, M. "Unraveling the Genetics of Fragile-X Syndrome." *Science*, May 24, 1991. Geneticists are beginning to understand the inheritance of a major cause of mental retardation.

Patterson, D. "The Causes of Down Syndrome." *Scientific American*, August 1987. How genes thought responsible for the symptoms of Down syndrome are being identified and mapped on chromosome 21.

Sapienza, C. "Parental Imprinting of Genes." *Scientific American*, October 1990. A new explanation for some genetic disorders.

Suzuki, D., A. Griffith, J. Miller, and R. Lewontin. *An Introduction to Genetic Analysis*, 4th ed. New York: Freeman, 1989.

15 The Molecular Basis of Inheritance

The Search for the Genetic Material: A Case Study in the Scientific Process

The Discovery of the Double Helix

DNA Replication: The Basic Concept

A Closer Look at DNA Replication

DNA Repair

Deoxyribonucleic acid, the substance of genes, is the most celebrated molecule of our time (Figure 15.1). Better known as DNA, this substance was largely ignored by biologists for nearly a century after its discovery, because it seemed far too simple and uniform in structure to serve a very significant purpose. Today, we know that this macromolecule is the genetic material—that Mendel's heritable factors and Morgan's genes on chromosomes are, in fact, composed of DNA. Chemically speaking, your genetic endowment consists of the DNA inherited from your mother and father.

Of all nature's molecules, nucleic acids are unique in their ability to direct their own replication. Indeed, the resemblance of offspring to their parents has its molecular basis in the precise replication and transmission of DNA from one generation to the next. In other words, DNA is the substance behind the adage "Like begets like." Hereditary information is encoded in the chemical language of DNA and reproduced in all cells of your body. It is this DNA program that directs the development of your biochemical, anatomical, physiological, and, to some extent, behavioral traits.

Today, molecular biologists can make or alter DNA in the laboratory and insert it into a cell, changing the cell's heritable characteristics (see Chapter 19). Earlier in the century, however, no one realized the relationship between DNA and heredity, and identification of the molecules of inheritance then loomed as a major challenge to biologists. As with the work of Mendel and Morgan, a key factor in meeting this challenge was the choice of appropriate experimental organisms. Because microscopic organisms—bacteria and the viruses that infect them—are far simpler than peas, fruit flies, or humans, the role of DNA in heredity was first worked out by studying such microbes.

Figure 15.1
James Watson and Francis Crick with their model of DNA (the double helix). In 1953, Watson (left) and Crick won a very competitive race to discover the molecular structure of DNA. In this chapter, you will learn how biologists deduced that DNA is the genetic material, how Watson and Crick discovered the double helix, and how cells replicate their DNA—the molecular basis of inheritance.

THE SEARCH FOR THE GENETIC MATERIAL: A CASE STUDY IN THE SCIENTIFIC PROCESS

By the 1940s, scientists realized that chromosomes, which were known to carry hereditary information,

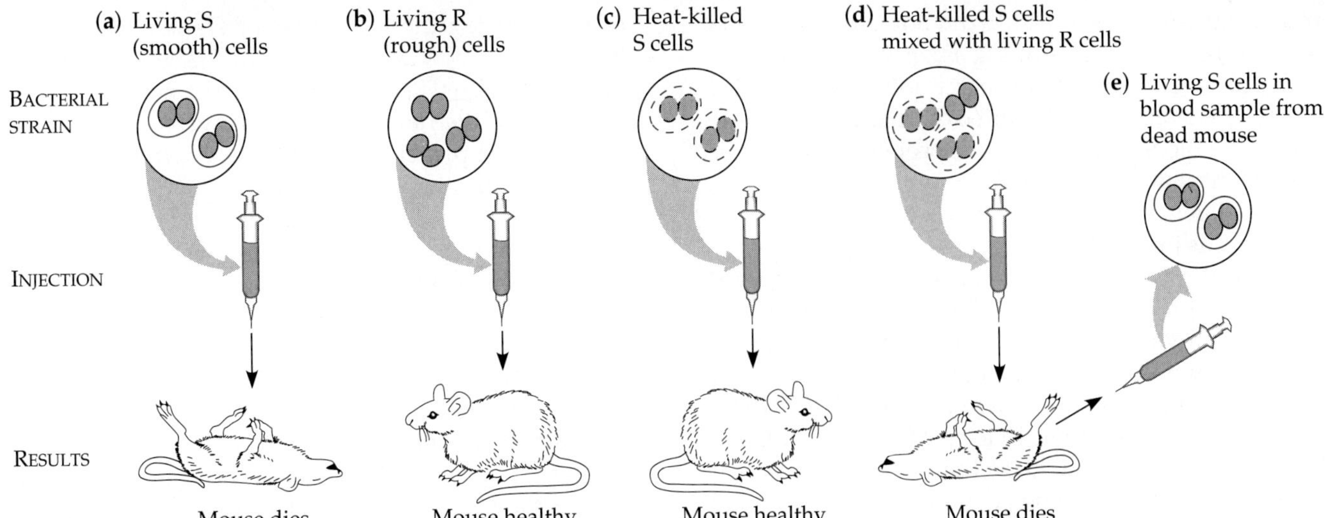

Figure 15.2
Transformation of bacteria. Griffith discovered that (**a**) the S strain of the bacterium *Streptococcus pneumoniae*, which was protected from a mouse's defensive system by a capsule, was pathogenic; (**b**) the R strain, a mutant lacking the coat, was nonpathogenic; (**c**) heat-killed S cells were harmless; but (**d**) a mixture of heat-killed S cells and live R cells caused pneumonia and death. (**e**) Live S bacteria could be retrieved from the dead mice injected with the mixture. Griffith concluded that some molecule from the dead S cells had genetically transformed some of the living R bacteria into S bacteria.

consisted of two substances, DNA and protein; but most researchers thought it was the protein that was the material of genes. The case for proteins seemed strong, especially since biochemists had identified them as a class of macromolecules with great heterogeneity and specificity of function, essential requirements for the elusive hereditary material. Moreover, little was known about nucleic acids, whose physical and chemical properties seemed far too uniform to account for the multitude of specific inherited traits expressed by every organism. This view gradually changed, as experiments with microorganisms yielded unexpected results. In this section, we will trace the search for the genetic material in some detail, as a case study in "science as a way of knowing."

Evidence That DNA Can Transform Bacteria

The first evidence that genes are specific molecules was found in 1928. Frederick Griffith, a British medical officer, was studying *Streptococcus pneumoniae*, a bacterium that causes pneumonia in mammals. When Griffith grew colonies of the bacteria in Petri dishes, he could distinguish between two genetic varieties, or strains. One strain produced colonies that appeared smooth, while colonies of the other strain appeared rough. Cells of the smooth strain (abbreviated S) synthesized a polysaccharide that surrounded the cells with a mucous coat, or capsule, that was not formed by the rough cells (R). These alternative phenotypes were inherited: Each strain reproduced its own kind.

When Griffith injected the bacteria into mice, he found that only the S strain was pathogenic (disease-causing). Mice injected with S cells died of pneumonia, while those injected with R cells survived (Figure 15.2). But it was not the polysaccharide of the coat that caused pneumonia, for Griffith found that S cells that had been killed by heat were harmless to the mice.

Then Griffith tried something remarkable. He mixed heat-killed S cells with live R cells and injected the mixture into mice. Although neither the dead S cells nor the live R cells alone were pathogenic, mice injected with the mixture developed pneumonia and died. More startling, Griffith found live S cells in blood samples taken from the dead mice, although only dead S cells had been injected. Somehow, some of the R cells had acquired from the dead S cells the ability to make polysaccharide coats. Furthermore, this new-found ability was heritable: When Griffith cultured S cells taken from the dead mice, the dividing bacteria produced daughter cells with coats. The phenomenon that Griffith discovered is now called **transformation,** the assimilation of external genetic material by a cell.* (Many bacteria live in a "broth" enriched with molecules derived from decomposing organic material. In Chapter 17, you will learn how bacteria take up macromolecules, including genetic material, from such sur-

*This usage of *transformation* should not be confused with the conversion of a normal animal cell to a cancerous one; see Chapter 11.

roundings.) Although Griffith did not know the chemical nature of the transforming agent, his observations spurred other scientists to search intensively for the elusive genetic material. Moreover, his use of heat to inactivate the S cells hinted that protein might not be the genetic material. Heat denatures most proteins, yet the S-cell genetic material retained its ability to transform R cells into S cells.

For a decade, American bacteriologist Oswald Avery tried to identify Griffith's transforming agent. He purified various chemicals from the heat-killed S cells, then tested each substance with live R cells to see if it could transform the bacteria. Finally, in 1944, Avery and his colleagues Maclyn McCarty and Colin MacLeod announced that the transforming agent had to be DNA. Their discovery, however, was greeted with considerable skepticism, in part because of the lingering belief that proteins were better candidates for the genetic material. Moreover, many biologists were not convinced that the genes of bacteria would be similar in composition and function to those of more complex organisms. But the major reason for the continued doubt was that so little was known about DNA. No one could imagine how DNA could carry genetic information.

Evidence That Viral DNA Can Program Cells

Additional evidence involving DNA as the genetic material came from studies of a virus that infects bacteria. Viruses are much simpler than cells. A virus is little more than DNA, or sometimes RNA, enclosed by a protective coat of protein. To reproduce, a virus must infect a cell and take over the cell's metabolic machinery.

Viruses that infect bacteria have been widely used in laboratory research. These viruses are called bacteriophages (meaning "bacteria eaters"), or just **phages,** for short. In 1952, Alfred Hershey and Martha Chase discovered that DNA was the genetic material of a phage known as T2. This is one of many phages that infect the bacterium *Escherichia coli (E. coli),* which normally lives in the intestines of mammals. At that time, biologists already knew that T2, like other viruses, was composed almost entirely of DNA and protein. They also knew that the phage could quickly turn an *E. coli* cell into a T2-producing factory that released phages when the cell ruptured. Somehow, T2 could reprogram its host cell to produce viruses, but which viral component—protein or DNA—was responsible?

Hershey and Chase answered this question by devising an experiment to determine which substance was transferred from the phage to the *E. coli* during infection (Figure 15.3). They used different radioactive isotopes to tag the molecules of DNA and protein. First, they grew T2 with *E. coli* in the presence of radioactive sulfur. Because protein, but not DNA, contains sulfur, the radioactive atoms were incorporated only into the protein of the phage. Next, in a similar way, the DNA of a separate batch of phage was labeled with atoms of radioactive phosphorus; because nearly all the phage's phosphorus is in its DNA, this procedure left the phage protein unlabeled. Then the protein-labeled and DNA-labeled batches of T2 were each allowed to infect separate samples of nonradioactive *E. coli* cells. Shortly after the onset of infection, the cultures were agitated in a kitchen blender to shake loose any parts of the phages that remained outside the bacterial cells. The mixtures were then spun in a centrifuge, forcing the heavier bacterial cells to form a pellet at the bottom of the centrifuge tubes, but allowing the lighter viral parts to remain suspended in the liquid, or supernatant. Radioactivity in the pellet and supernatant was then measured and compared.

Hershey and Chase found that when the bacteria had been infected with the T2 containing labeled proteins, most of the radioactivity was found in the supernatant, which contained virus particles (but not bacteria). This suggested that the phage protein did not enter the host cells. But when the bacteria had been infected with T2 phage whose DNA was tagged with radioactive phosphorus, then the pellet of mainly bacterial material contained most of the radioactivity. Moreover, when these bacteria were returned to culture medium, the infection ran its course; the *E. coli* released phages containing radioactive phosphorus.

Hershey and Chase concluded that the DNA of the virus is injected into the host cell, whereas most of the proteins remain outside. The injected DNA molecules cause the cells to produce additional viral DNA and proteins—indeed, additional intact viruses—providing powerful evidence that nucleic acids, rather than proteins, are the hereditary material, at least in viruses.

Additional Evidence That DNA Is the Genetic Material of Cells

Additional circumstantial evidence pointed to DNA as the genetic material in eukaryotes. Prior to mitosis, a eukaryotic cell doubles its DNA content, and during mitosis, this DNA is distributed equally to the two daughter cells. Also, diploid sets of chromosomes have twice as much DNA as the haploid sets found in the gametes of the same organism.

Still more compelling evidence came from the laboratory of biochemist Erwin Chargaff. Recall from Chapter 5 that DNA is a polymer of monomers called nucleotides, each consisting of three components: a nitrogenous base, a pentose sugar called deoxyribose,

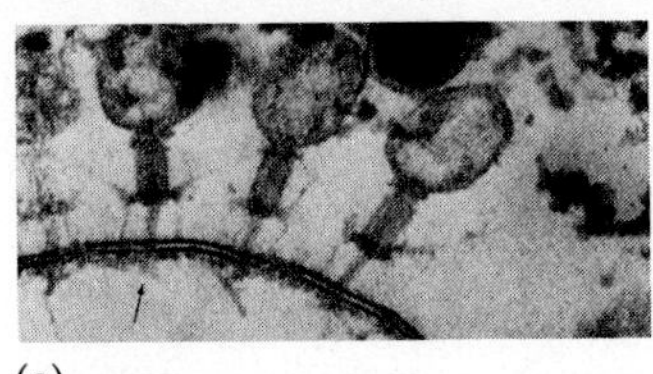
(a)

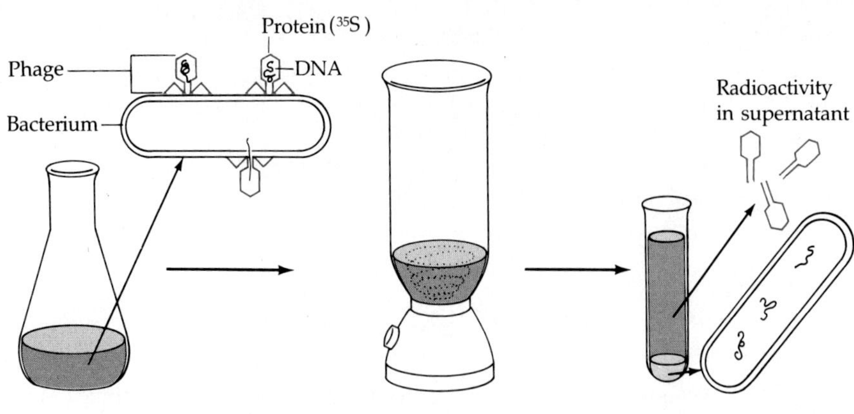

(b)

Mix radioactively labeled phage with bacteria. The phage infects the bacterial cells.

Agitate in a blender to separate phage outside the bacteria from the cells and their contents.

Centrifuge and measure the radioactivity in the pellet and supernatant.

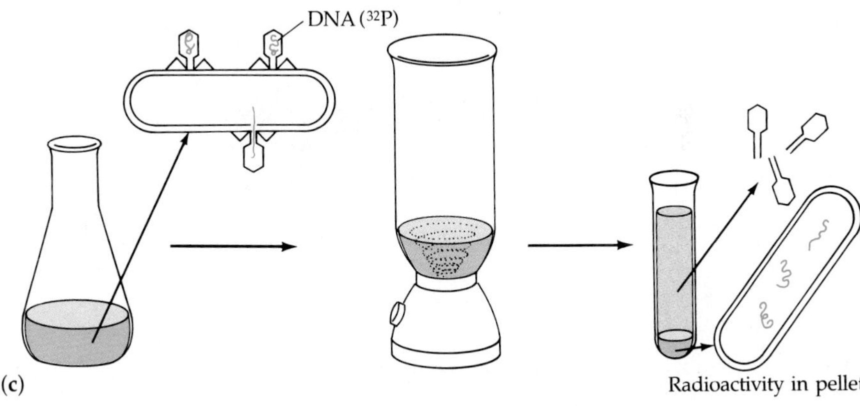

(c)

Figure 15.3
The Hershey-Chase experiment.
(**a**) Phages are viruses that infect bacteria (TEM). They use their tail pieces to attach to the host cell and inject genetic material (arrow). (**b**) In their famous 1952 experiment, Hershey and Chase demonstrated that it was DNA, not protein, that functioned as the phages' genetic material. Viral proteins, labeled with radioactive sulfur, remained outside the host cell during infection. (**c**) In contrast, viral DNA, labeled with radioactive phosphorus, entered the bacterial cell.

and a phosphate group (Figure 15.4). The base of each nucleotide can be any one of four different bases: adenine (A), thymine (T), guanine (G), or cytosine (C). Chargaff analyzed the base composition of DNA from a number of different organisms. In 1947, he reported that DNA composition is species-specific: In the DNA of any one organism, the amounts of the four nitrogenous bases are not all equal, and the ratios of nitrogenous bases vary from one species to another. Such evidence of molecular diversity, which had been presumed absent from DNA, made DNA a more credible candidate for the genetic material. Chargaff also found a peculiar regularity in the base ratios. In the DNA of each species he studied, the number of adenine residues approximately equaled the number of thymines, and the number of guanines approximately equaled the number of cytosines. In human DNA, for example, the four bases are present in these percentages: A = 30.9% and T = 29.4%; G = 19.9% and C = 19.8%. The A = T and G = C equalities, later known as Chargaff's rules, remained unexplained until the discovery of the double helix.

THE DISCOVERY OF THE DOUBLE HELIX

Once most biologists were convinced that DNA was the genetic material, the race was under way to determine how the structure of DNA could account for its role in inheritance. By the beginning of the 1950s, the arrangement of covalent bonds in a nucleic acid polymer was well established (see Figure 15.4), and the competition focused on discovering the three-dimensional structure of DNA. Among the scientists working on the problem were Linus Pauling, in California, and Maurice Wilkins and Rosalind Franklin, in London. First to the finish line, however, were two scientists who were relatively unknown at the time—the

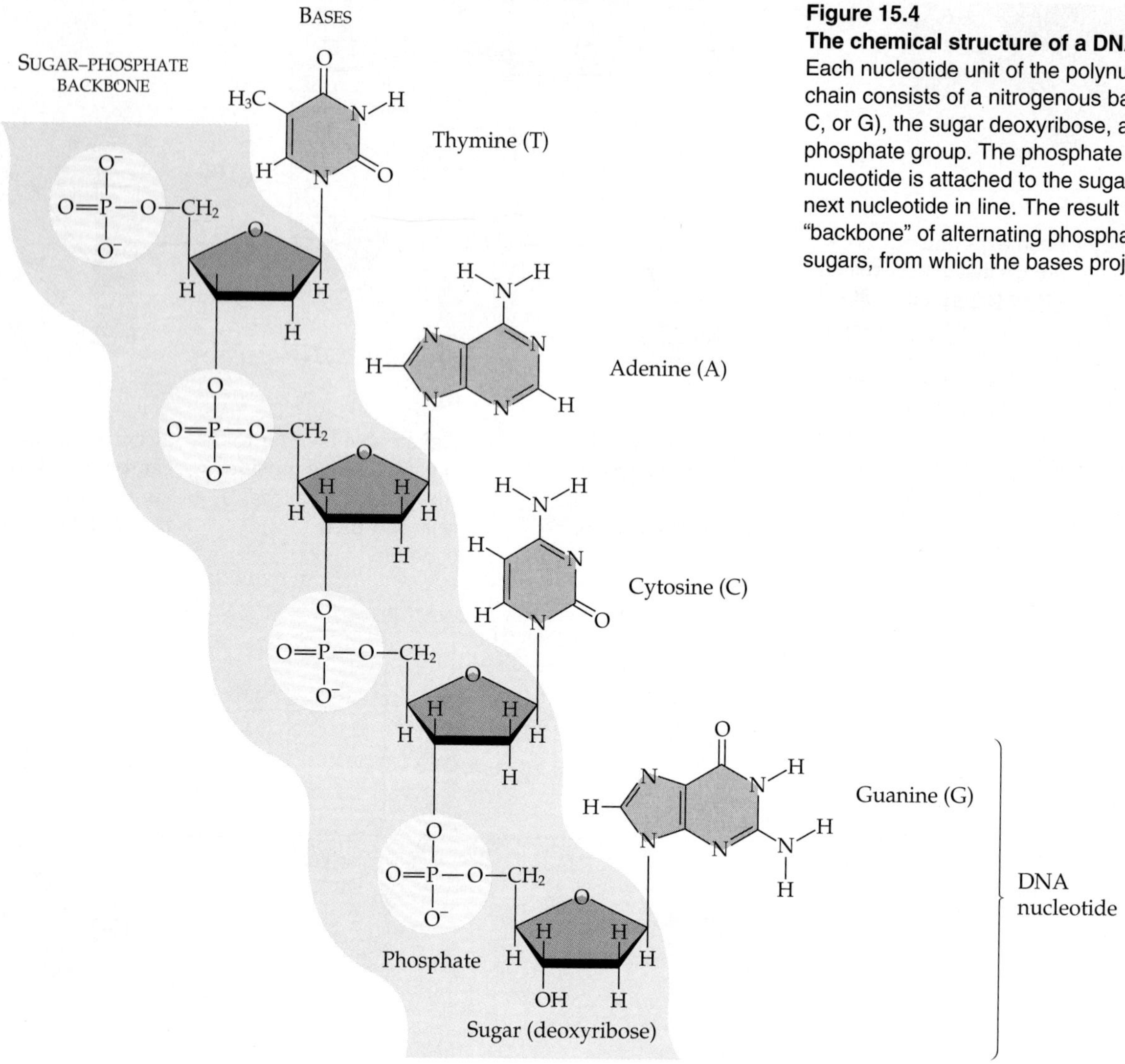

Figure 15.4
The chemical structure of a DNA strand. Each nucleotide unit of the polynucleotide chain consists of a nitrogenous base (T, A, C, or G), the sugar deoxyribose, and a phosphate group. The phosphate of one nucleotide is attached to the sugar of the next nucleotide in line. The result is a "backbone" of alternating phosphates and sugars, from which the bases project.

American James Watson and the Englishman Francis Crick (see Figure 15.1).

The brief but celebrated partnership that solved the DNA puzzle began soon after the young Watson journeyed to Cambridge University, where Crick was studying protein structure with a technique called X-ray crystallography (see the Methods Box in Chapter 5, p. 85). While visiting the laboratory of Maurice Wilkins at King's College in London, Watson saw an X-ray photograph of DNA, produced by Wilkins' colleague, Rosalind Franklin (Figure 15.5a). Photographs produced by the X-ray crystallography method are not actually "pictures" of molecules. The spots and smudges in Figure 15.5b were produced by X-rays that were diffracted (deflected) as they passed through crystallized DNA. Crystallographers combine experience and mathematical equations to translate such patterns of spots into information about the three-dimensional shapes of molecules. Watson and Crick based their model of DNA on data they were able to extract from Franklin's X-ray diffraction photo. They interpreted the pattern of spots on the X-ray photograph to mean that DNA was helical in shape. Based on Watson's recollection of the photograph, he and Crick deduced that the helix had a uniform width of 2 nanometers (nm), with its nitrogenous bases stacked 0.34 nm apart. The width of the helix suggested that it was made up of two strands—contrary to a three-stranded model that Linus Pauling had recently proposed. The presence of two strands accounts for the now-familiar term **double helix.**

Using molecular models made of wire, Watson and Crick began building scale models of a double helix that would conform to the X-ray measurements and what was then known about the chemistry of DNA. After failing to make a satisfactory model that placed the sugar–phosphate chains on the inside of the molecule, Watson tried putting them on the outside and forcing the nitrogenous bases to swivel to the interior of the double helix. Imagine this double helix as a rope ladder having rigid rungs, with the ladder twisted into a spiral. The side ropes are the equivalent of the sugar–phosphate backbones, and the rungs represent pairs of nitrogenous bases. Franklin's X-ray data indi-

(a)

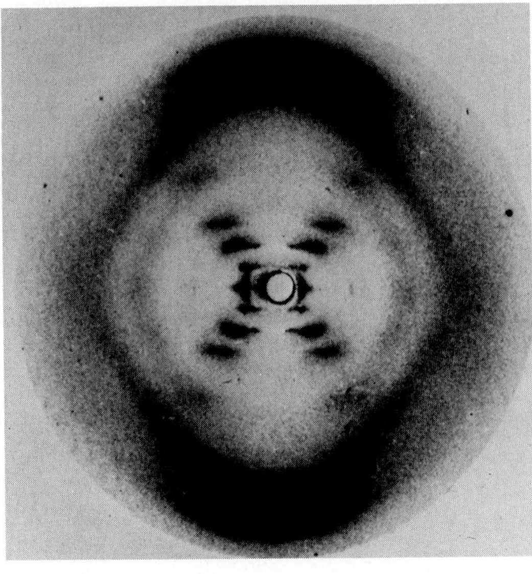

(b)

Figure 15.5
Rosalind Franklin and her X-ray diffraction photo of DNA.
(**a**) Franklin was the X-ray crystallographer who took the photo that enabled Watson and Crick to deduce the double-helical structure of DNA. Franklin died of cancer when she was only 38, in 1958. Her colleague, Maurice Wilkins, received the Nobel Prize in 1962 along with Watson and Crick. Because the Nobel Prize is not awarded posthumously, science historians can only speculate about whether the committee would have recognized Franklin's contribution to the discovery of the double helix. (**b**) Franklin's X-ray diffraction photograph of DNA.

cated that the helix makes one full turn every 3.4 nm along its length. Because the bases are stacked just 0.34 nm apart, there are ten layers of base pairs, or rungs on the ladder, in each turn of the helix (Figure 15.6). This arrangement was appealing, because it put the more hydrophobic nitrogenous bases in the molecule's interior and thus away from the surrounding aqueous medium.

Notice in Figure 15.6 that the nitrogenous bases of the double helix are paired in specific combinations, adenine (A) with thymine (T), and guanine (G) with cytosine (C). It was mainly by trial and error that Watson and Crick arrived at this key feature of DNA. At first, Watson imagined that the bases paired like-with-like—for example, A with A and C with C. But this model did not fit with the X-ray data, which suggested that the double helix had a uniform diameter. Why is this requirement inconsistent with like-with-like pairing of bases? Adenine and guanine are purines, nitrogenous bases with two organic rings. In contrast, cytosine and thymine belong to the family of nitrogenous bases known as pyrimidines, which have a single ring. Thus, purines (A and G) are about twice as wide as pyrimidines (C and T). A purine-purine pair is too wide and a pyrimidine-pyrimidine pair too narrow to account for the 2-nm diameter of the double helix. The solution is to always pair a purine with a pyrimidine:

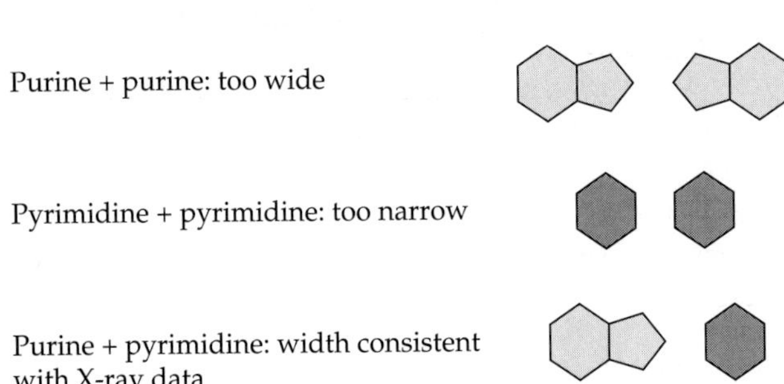

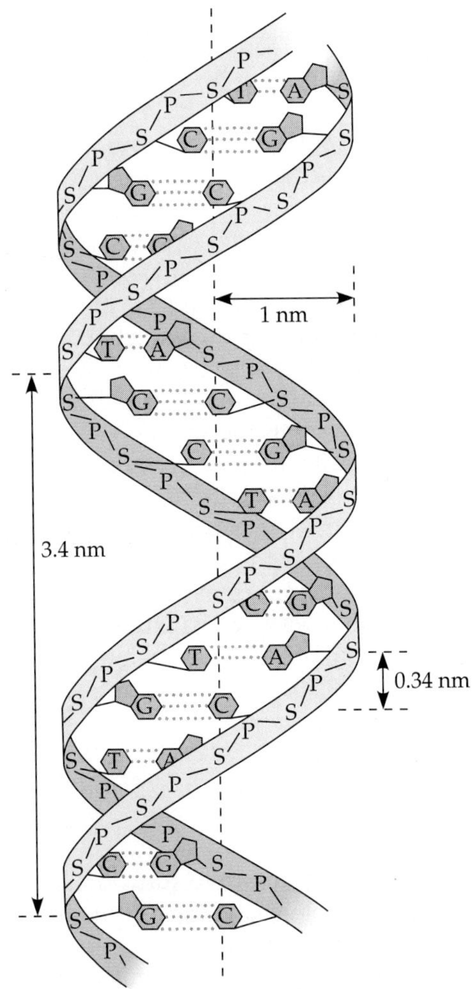

Figure 15.6
The double helix. The "ribbons" in this diagram represent the sugar–phosphate backbones of the two DNA strands. The two strands are held together by hydrogen bonds (dotted lines) between the nitrogenous bases, which are paired in the interior of the double helix. The base pairs are 0.34 nm apart; there are ten pairs per turn of the helix.

Moreover, Watson and Crick realized that there must be additional specificity of pairing dictated by the structure of the bases. Each base has chemical side groups that can form hydrogen bonds with its appropriate partner: Adenine can form hydrogen bonds only with thymine, and guanine only with cytosine. In biologists' shorthand, A pairs with T, and G pairs with C:

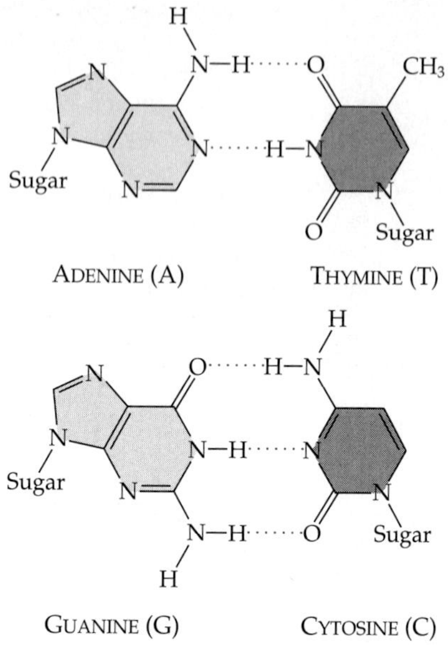

ADENINE (A) THYMINE (T)

GUANINE (G) CYTOSINE (C)

The Watson–Crick model explained Chargaff's rules. Wherever one strand of a DNA molecule has an A, the partner strand has a T. And a G in one strand is always paired with a C in the complementary strand. Therefore, in the DNA of any organism, the amount of adenine equals the amount of thymine, and the amount of guanine equals the amount of cytosine.

In April 1953, Watson and Crick surprised the scientific world with a succinct, one-page paper in the British journal *Nature*. The paper reported a new molecular model for DNA: the double helix, which has since become the symbol of molecular biology. The beauty of the model was that its structure suggested the basic mechanism of DNA replication.

DNA REPLICATION: THE BASIC CONCEPT

The relationship of structure to function, one of the themes that drives biological science, is manifest in the double helix. The molecular architecture of DNA provides the mechanism for the replication of genes. The idea that there is specific pairing of nitrogenous bases in DNA was the flash of inspiration that led Watson and Crick to the double helix. At the same time, they understood the functional significance of the base-pairing rules. They ended their classic paper by saying: "It has not escaped our notice that the specific pairing we have postulated immediately suggests a possible copying mechanism for the genetic material."* In this section, you will learn about this basic mechanism of DNA replication. Some important details of the process will be presented in the next section.

Although the base-pairing rules dictate the combinations of nitrogenous bases that form the "rungs" of the double helix, they do not restrict the sequence of nucleotides along a DNA strand. Thus, the linear sequence of the four bases can be varied in countless ways, and each gene has a unique base sequence. Figure 15.7 untwists a short section of double helix so that it is easier to follow replication of the DNA. Our objective here is to visualize how a cell copies this genetic information. Cover one of the two DNA strands of Figure 15.7a with a piece of paper, and you can still determine its linear sequence of bases by referring to the unmasked strand and applying the base-pairing rules. The two strands are complementary; each stores the information necessary to reconstruct the other. When a cell copies a DNA molecule, the two strands separate, and each strand then serves as a template (mold) for ordering nucleotides into a new complementary strand. One at a time, nucleotides line up along the template strand according to the base-pairing rules. Enzymes link the nucleotides to form the new strands. Where there was one double-stranded DNA molecule at the beginning of the process, there are now two, each an exact replica of the "parent" molecule. The copying mechanism is analogous to using a photographic negative to make a positive image, which can in turn be used to make another negative, and so on.

This model of gene replication remained untested for several years following publication of the DNA structure. The requisite experiments were simple in concept but difficult to perform. Watson and Crick's model predicts that when a double helix reproduces, each of the two daughter molecules will have one old strand derived from the parent molecule and one newly made strand. This **semiconservative model** can be distinguished from a conservative model of replication, in which the parent molecule remains intact (is conserved) and the new molecule is formed entirely from scratch. In a third model, called the dispersive model, all four strands of DNA, after the double helix is replicated, have a mixture of "old" and "new" DNA (Figure 15.8). In the late 1950s, Matthew Meselson and Franklin Stahl demonstrated that DNA replication is indeed semiconservative, as predicted by the Watson–Crick model (Figure 15.9).

---

*Watson, J. D., and F. H. C. Crick. "Molecular Structure of Nucleic Acids: A Structure for Deoxynucleic Acids." *Nature* 171 (1953), p. 738.

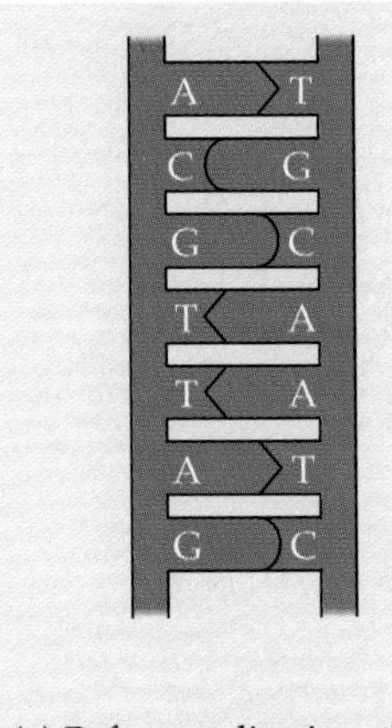

(a) Before replication, the parent molecule has two complementary strands of DNA. Each base is paired with its specific partner, A with T and G with C.

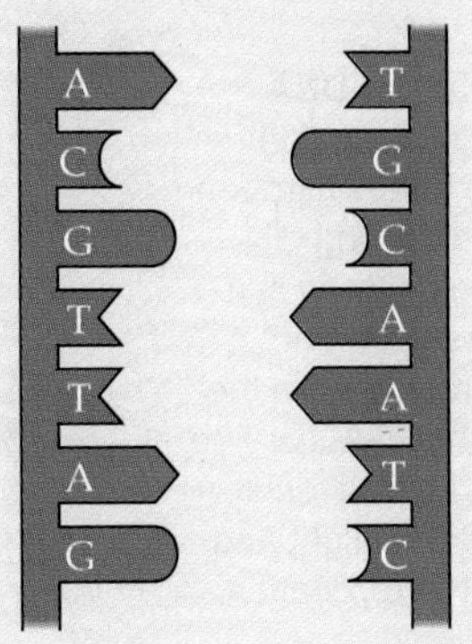

(b) The first step in replication is separation of the two DNA strands.

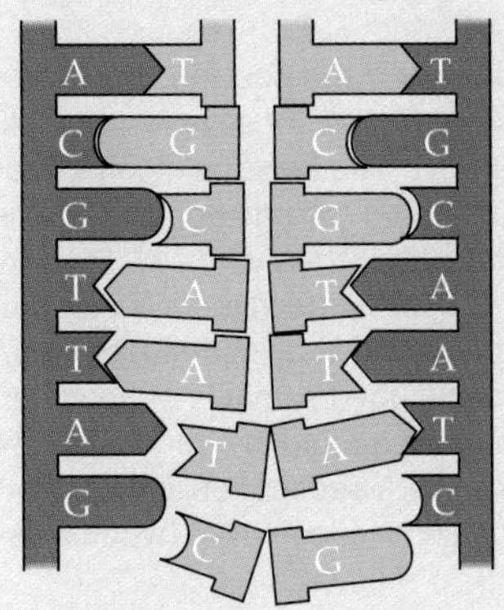

(c) Each "old" strand now serves as a template that directs synthesis of "new" complementary strands. Nucleotides plug into specific sites along the template surface according to the base-pairing rules.

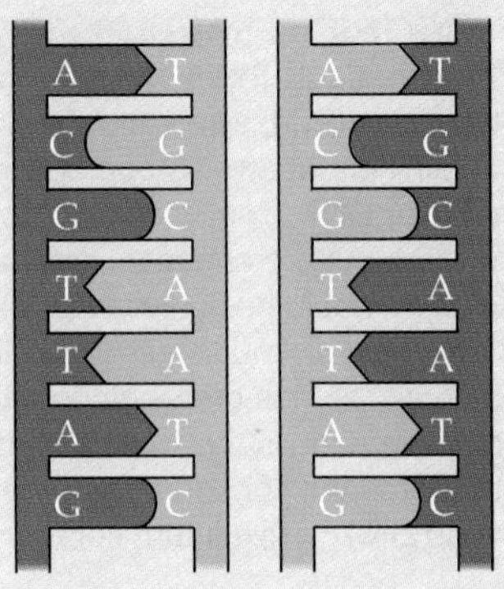

(d) The nucleotides are connected to form the sugar–phosphate backbones of the new strands. Each DNA molecule now consists of one "old" strand and one "new" strand, resulting in two copies identical to the one DNA molecule with which we started.

Figure 15.7
DNA replication: the basic concept. In this simplification, a short segment of DNA has been untwisted to convert the double helix to a two-dimensional version of the molecule that resembles a ladder. The rails of the ladder are the sugar–phosphate backbones of the two DNA strands. The rungs are the pairs of nitrogenous bases. Simple shapes are used to symbolize the four kinds of bases. Dark blue represents DNA strands originally present in the parent cell. Newly synthesized DNA is represented by light blue.

PARENT CELL

FIRST REPLICATION

SECOND REPLICATION

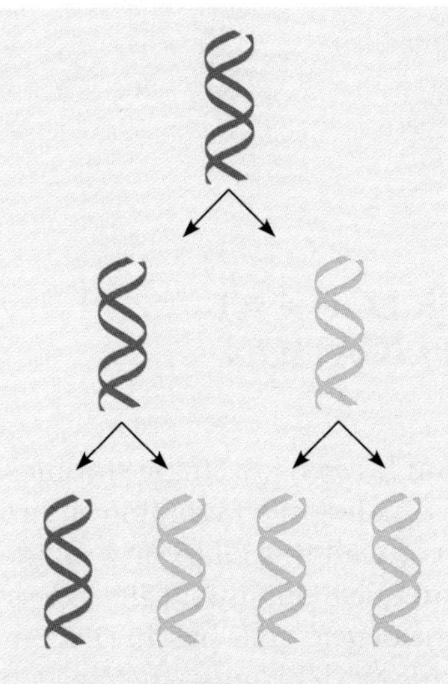

(a) **Conservative model:** The parental double helix remains intact and a second, all-new copy is made.

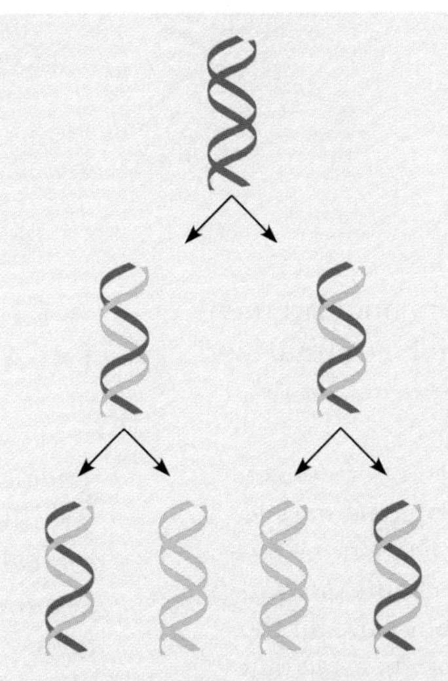

(b) **Semiconservative model:** The two strands of the parental molecule separate, and each functions as a template for synthesis of a new complementary strand.

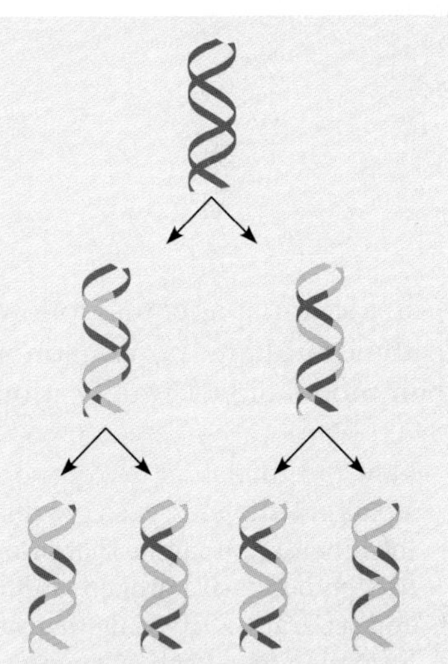

(c) **Dispersive model:** Each strand of *both* daughter molecules would contain a mixture of old and newly synthesized parts.

Figure 15.8
Three models of DNA replication. The short segments of double helix illustrated here symbolize the genetic material within a cell. Beginning with a parent cell, we follow the DNA for two generations of cells—two replications of the genetic material.

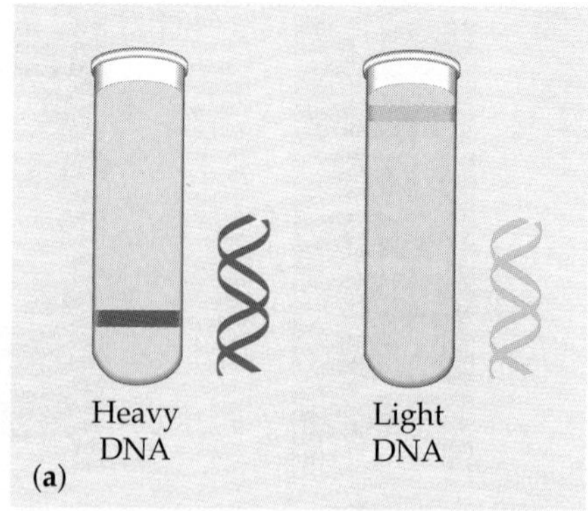

Figure 15.9
The Meselson-Stahl experiment confirmed the semiconservative nature of DNA replication. Meselson and Stahl cultured *E. coli* for several generations on a medium containing a heavy isotope of nitrogen, ^{15}N. The bacteria incorporated the heavy nitrogen into their DNA. The scientists then transferred the bacteria to a medium containing ^{14}N, the lighter, more common isotope of nitrogen. Thus, any new DNA that the bacteria synthesized would be lighter than the "old" DNA made in the ^{15}N medium. Meselson and Stahl could distinguish DNA of different densities by centrifuging DNA extracted from the bacteria. (**a**) This drawing compares centrifuged samples of heavy and light DNA. (**b**) The first replication in the ^{14}N medium produced a band of hybrid (^{15}N-^{14}N) DNA. This result eliminated the conservative hypothesis. (**c**) A second replication produced both light and hybrid DNA, a result that confirmed the semiconservative hypothesis of DNA replication.

In a second paper that followed their announcement of the double helix, Watson and Crick summarized their model of DNA replication:

> Now our model for deoxyribonucleic acid is, in effect, a pair of templates, each of which is complementary to the other. We imagine that prior to duplication the hydrogen bonds are broken, and the two chains unwind and separate. Each chain then acts as a template for the formation onto itself of a new companion chain, so that eventually we shall have two pairs of chains, where we only had one before. Moreover, the sequence of the pairs of bases will have been duplicated exactly.*

The copying of the genetic material is, in basic concept, elegantly simple. However, the actual process involves complex biochemical gymnastics, as we will now see.

*Crick, F. H. C., and J. D. Watson."The Complementary Structure of Deoxyribonucleic Acid." *Proc. Roy. Soc.* (A) 223 (1954), p. 80.

A CLOSER LOOK AT DNA REPLICATION

The bacterium *E. coli* has a single chromosome of about 5 million base pairs. In a favorable environment, an *E. coli* cell can copy all this DNA and divide to form two genetically identical daughter cells in less than 30 minutes. Each one of *your* cells has 46 DNA molecules, one giant molecule per chromosome (see Chapter 11). In all, that represents about 6 billion base pairs, or over a thousand times more DNA than is found in a bacterial cell. If we were to print the one-letter symbols for these bases (A, G, C, and T) the size of the letters you are reading, the 6 billion bases of a single human cell would fill about 900 books as thick as this text. Yet it takes a cell just a few hours to copy all this DNA. This replication of an enormous amount of genetic information is achieved with very few errors—only about one

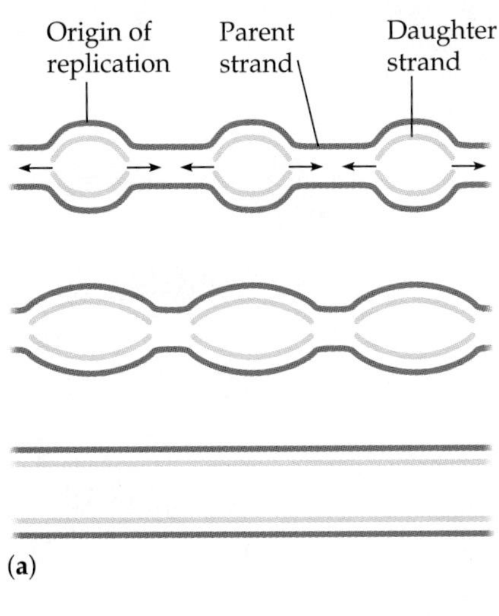

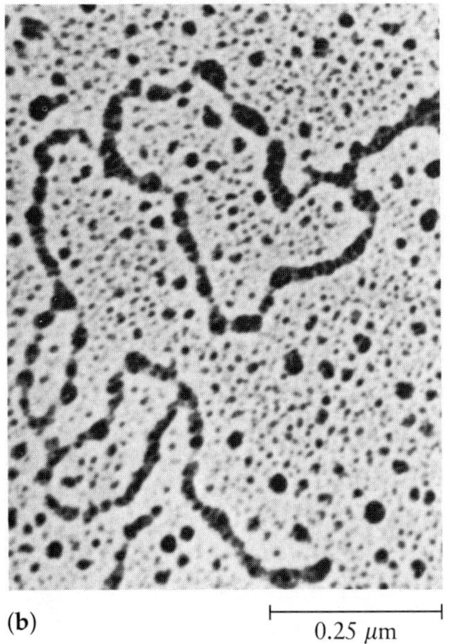

Figure 15.10
Origins of replication. (**a**) DNA replication begins at specific sites where the two parental strands of DNA separate to form replication "bubbles" (top). In eukaryotes, there are hundreds or thousands of origin sites along the giant DNA molecule of each chromosome. A replication bubble expands laterally, as DNA replication proceeds in both directions. Eventually, the replication bubbles fuse (center), and synthesis of the "daughter" strands of DNA is complete (bottom). (**b**) In this micrograph, three replication bubbles are visible along the DNA of cultured Chinese hamster cells (TEM).

per billion nucleotides. The copying of DNA is remarkable in its speed and accuracy.

More than a dozen enzymes and other proteins participate in DNA replication. Much more is known about how this "replication machine" works in bacteria than in eukaryotes. However, most of the basic steps of replication seem to be similar for prokaryotes and eukaryotes. In this section, we examine some of these steps.

Getting Started: Origins of Replication

The replication of a DNA molecule begins at special sites called **origins of replication.** The bacterial chromosome has a single origin marked by a stretch of DNA having a specific sequence of nucleotides. Proteins that initiate DNA replication recognize this sequence and attach to the DNA, separating the two strands and opening up a replication "bubble." Replication of DNA then proceeds in both directions, until the entire molecule is copied. Specific nucleotide sequences probably mark origins of replication in eukaryotes as well, but molecular biologists have not yet identified such sequences. It *is* known that, in contrast to the bacterial chromosome, each eukaryotic chromosome has hundreds or thousands of replication origins. Multiple replication bubbles form and eventually fuse, thus speeding up the copying of a very large DNA molecule (Figure 15.10). As in bacteria, DNA replication proceeds in both directions from each point of origin. At each end of a replication bubble is a **replication fork,** a Y-shaped region where the new strands of DNA are elongating.

Elongating a New DNA Strand

Elongation of new DNA at a replication fork is catalyzed by enzymes called **DNA polymerases.** As nucleotides align with complementary bases along the "old" template strand of DNA, they are added by polymerase, one by one, to the growing end of the new DNA strand. The rate of elongation is about 500 nucleotides per second in bacteria and 50 per second in human cells.

What is the source of energy that drives polymerization of nucleotides to form new DNA strands? The substrates for DNA polymerase are not actually nucleotides, but related compounds called nucleoside triphosphates (Figure 15.11). Instead of the one phosphate group characteristic of nucleotides, nucleoside triphosphates have three, just like ATP. (In fact, ATP is a nucleoside triphosphate. The only difference between ATP and the nucleoside triphosphate that supplies adenine to DNA is the sugar component, which is ribose for ATP and deoxyribose for the building block of DNA; see Chapter 6.) Like ATP, the monomers for DNA synthesis are chemically reactive, because their triphosphate tails have an unstable cluster of negative charge. As each monomer joins the growing end of a DNA strand, it loses two phosphate groups. Hydrolysis of the phosphate is the exergonic reaction that drives polymerization of nucleotides to form DNA.

The Problem of Antiparallel DNA Strands There is more to the scenario of DNA synthesis at the replication fork. Until now, we have ignored an important feature of the double helix: The two DNA strands are

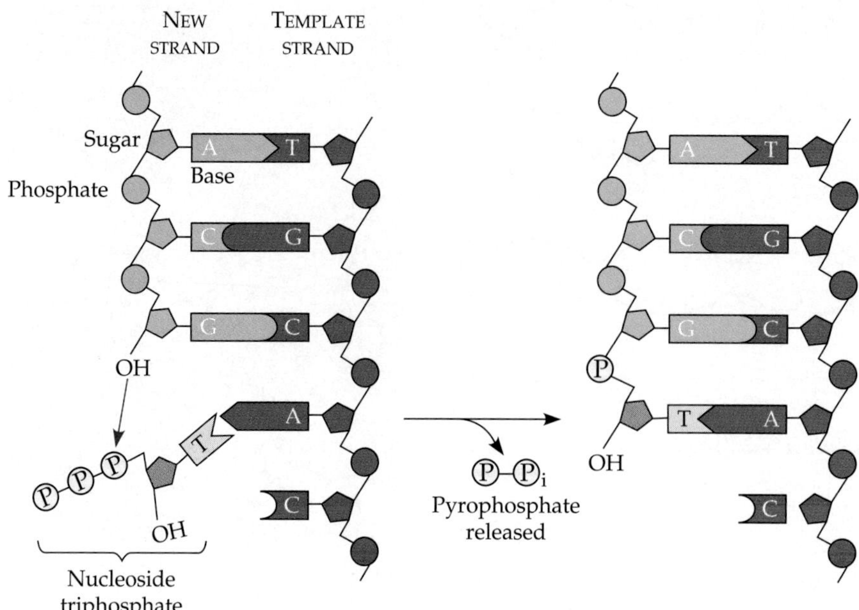

Figure 15.11
Incorporation of a nucleotide into a DNA strand. When a nucleoside triphosphate links to the sugar–phosphate backbone of a growing DNA strand, it loses two of its phosphates as a pyrophosphate molecule. The enzyme DNA polymerase catalyzes the reaction, and hydrolysis of the bonds between the phosphate groups provides the energy.

antiparallel; that is, their sugar–phosphate backbones run in opposite directions. Notice in Figure 15.12 that a nucleotide's phosphate group is attached to the 5′ carbon of deoxyribose. (The prime sign is used to distinguish the numbered carbons of the sugar from the numbers representing the atoms of the nitrogenous bases.) Notice also that the phosphate group of one nucleotide is joined to the 3′ carbon of the adjacent nucleotide. The result is a DNA strand of distinct polarity. The terminal carbon at one end of the sugar–phosphate backbone is the 3′ carbon, which is not attached to a phosphate group. This terminus is called the 3′ end of

5′ END
3′ END
3′ END
5′ END

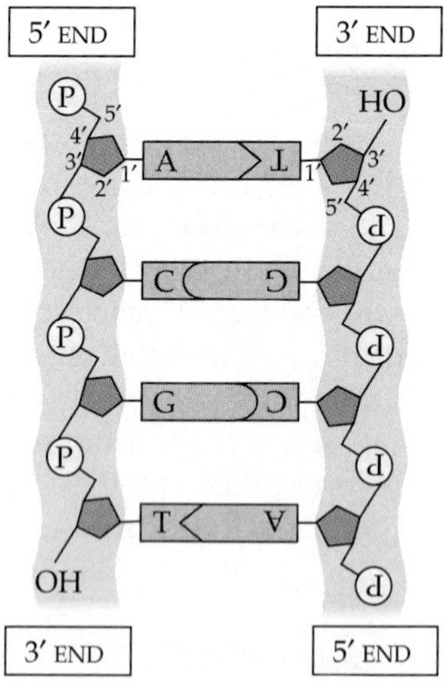

Figure 15.12
The two strands of DNA are antiparallel. The 5′ ⟶ 3′ direction of one strand runs counter to the other strand. The carbon atoms of the deoxyribose sugars at the top are numbered for orientation.

molecule. At the opposite end, the sugar–phosphate backbone terminates with the phosphate group attached to the 5′ carbon of the last nucleotide. This is the 5′ end of the DNA strand. In the double helix, the two sugar–phosphate backbones are essentially upside-down relative to each other. Because the strands are antiparallel, if one strand is said to have a 5′ ⟶ 3′ orientation (polarity), then the complementary strand has a 3′ ⟶ 5′orientation.

How does the antiparallel structure of the double helix affect replication? The enzyme DNA polymerase can only add nucleotides to the free 3′ end of a growing DNA strand, never to the 5′ end. Thus, a new DNA strand can only elongate in the 5′ ⟶ 3′ direction. With this in mind, let's reexamine a replication fork (Figure 15.13). Along one template strand, DNA polymerase can synthesize a continuous complementary strand by elongating the new DNA in the mandatory 5′ ⟶ 3′ direction. The polymerase simply nestles in the replication fork and moves along the template strand as the fork progresses. The DNA strand made by this mechanism is called the **leading strand.** To elongate the other new strand of DNA, polymerase must work along the template *away from* the replication fork. The DNA synthesized in this direction is called the **lagging strand.** The process is analogous to back-stitching. As a replication bubble opens, polymerase can work its way away from a replication fork and synthesize a short segment of DNA. As the bubble widens, another short segment of the lagging strand can be made by a polymerase working away from the fork. In contrast to the leading strand, which can be elongated continuously, the lagging strand is first synthesized as a series of segments. These pieces are called **Okazaki fragments,** after the Japanese scientist who discovered them. The fragments are about 1000 to 2000 nucleotides long in

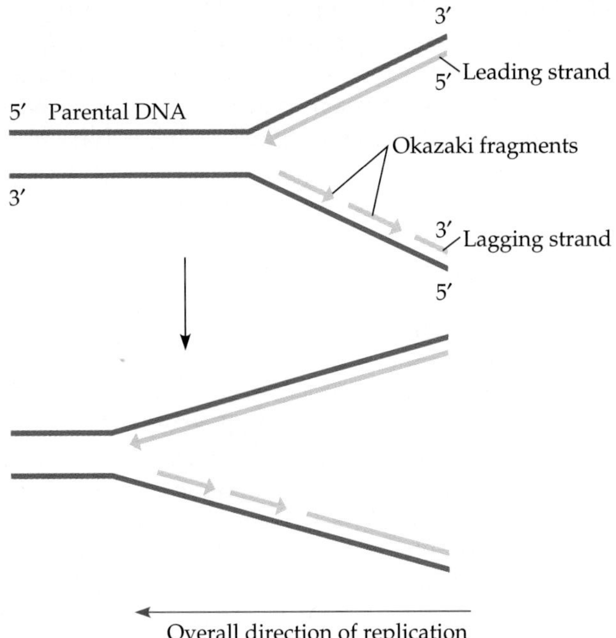

Figure 15.13
Discontinuous synthesis of the lagging strand. DNA polymerase elongates strands in the 5′ ⟶ 3′ direction. One new strand, called the leading strand, can therefore elongate continuously in the 5′ ⟶ 3′ direction as the replication fork progresses. But the other new strand, the lagging strand, must grow in an overall 3′ ⟶ 5′ direction by the addition of short segments, Okazaki fragments, that individually grow 5′ ⟶ 3′. An enzyme called ligase connects the fragments.

bacteria and about 100 to 200 nucleotides long in eukaryotes. Another enzyme, **DNA ligase,** joins the Okazaki fragments into a single DNA strand. Figure 15.14 illustrates a mechanism by which a team of two DNA polymerases may coordinate synthesis of the leading and lagging strands of new DNA.

Priming There is another important restriction for DNA polymerase. It can only add a nucleotide to a polynucleotide that is already correctly paired with the complementary strand. (This requirement is evident in Figure 15.11.) This means that DNA polymerase cannot actually *initiate* synthesis of a DNA strand by joining the first nucleotides. Nucleotides must be added to the end of an already existing chain, called a **primer.** The primer is not DNA, but a short stretch of RNA. Still another enzyme, **primase,** makes the primer, which is about 10 nucleotides long in eukaryotes (Figure 15.15). Only one primer is required for polymerase to begin synthesizing the leading strand of new DNA. For the lagging strand, each fragment must be primed. An enzyme then replaces the RNA nucleotides of the primers with DNA versions, and ligase joins all the DNA fragments into a strand.

Other Proteins Assisting DNA Replication You have learned about three of the proteins that function in DNA synthesis: DNA polymerase, ligase, and primase. Many other proteins also participate, and here we examine two of them: helicase and single-strand binding proteins. **Helicase** is an enzyme that works at the crotch of the replication fork, untwisting the double

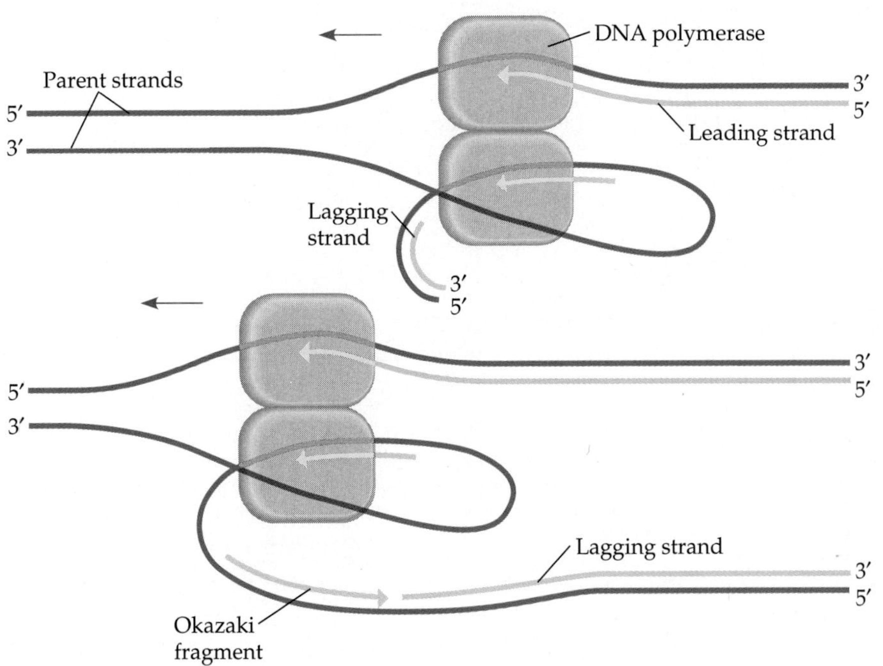

Figure 15.14
A model for coordinated synthesis of the leading and lagging strands of DNA. The two parent strands of DNA thread through an enzyme dimer, a unit of two DNA polymerase molecules. The parent strand that serves as the template for synthesis of the lagging strand forms a loop before passing through the active site of polymerase. The red arrows indicate the direction in which the new DNA strands are elongated.

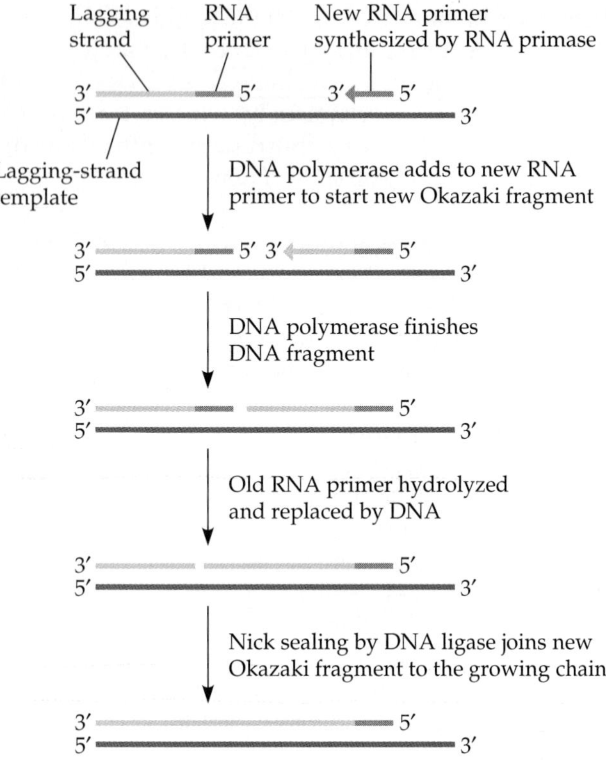

Figure 15.15
Priming DNA synthesis. DNA polymerase cannot initiate a polynucleotide strand, but can only add to the 3′ end of an already-started strand. The primer is a short segment of RNA synthesized by the enzyme primase.

helix and separating the two "old" strands. **Single-strand binding proteins** then attach in chains along the unpaired DNA strands, holding these templates straight until new complementary strands can be synthesized. Figure 15.16 summarizes the activity at a replication fork.

Proofreading

We cannot attribute the accuracy of DNA replication solely to the specificity of base pairing. Although errors in the completed DNA molecule only amount to one in 1 billion nucleotides, initial pairing errors between incoming nucleotides and those in the template strand are 100,000 times more common—an error rate of one in 10,000 base pairs. In bacteria, these pairing errors are almost always corrected by the DNA polymerase itself, which checks each nucleotide against its template as soon as it is added to the strand. Upon finding an incorrect nucleotide, the polymerase backs up, removes the incorrect nucleotide, and replaces it before continuing with synthesis. (This action resembles a command in a word-processing program, in which you backspace to an error, then delete and replace the error before continuing.) In eukaryotes, it is not yet known whether DNA polymerase or other enzymes proofread the new strand and remove errors.

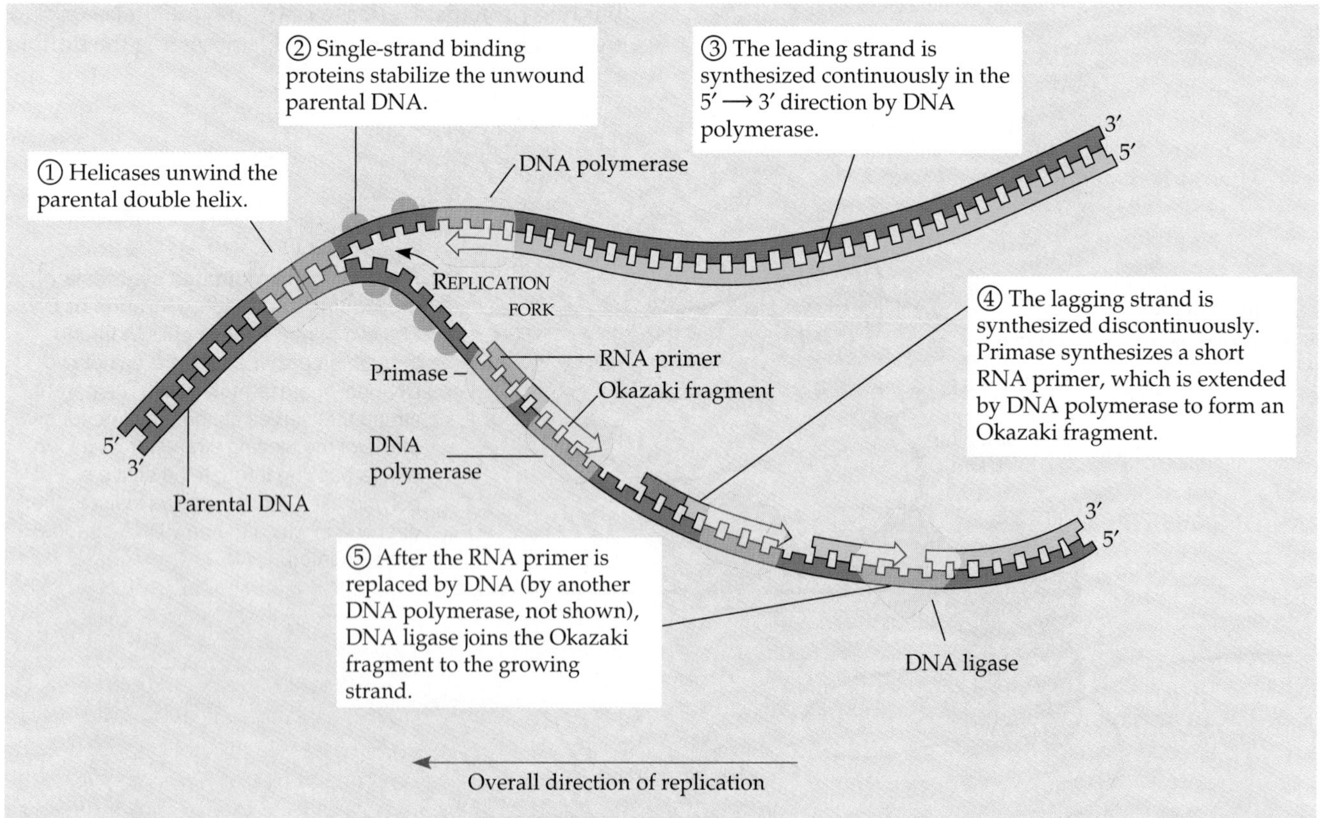

Figure 15.16
A summary of activities at a replication fork.

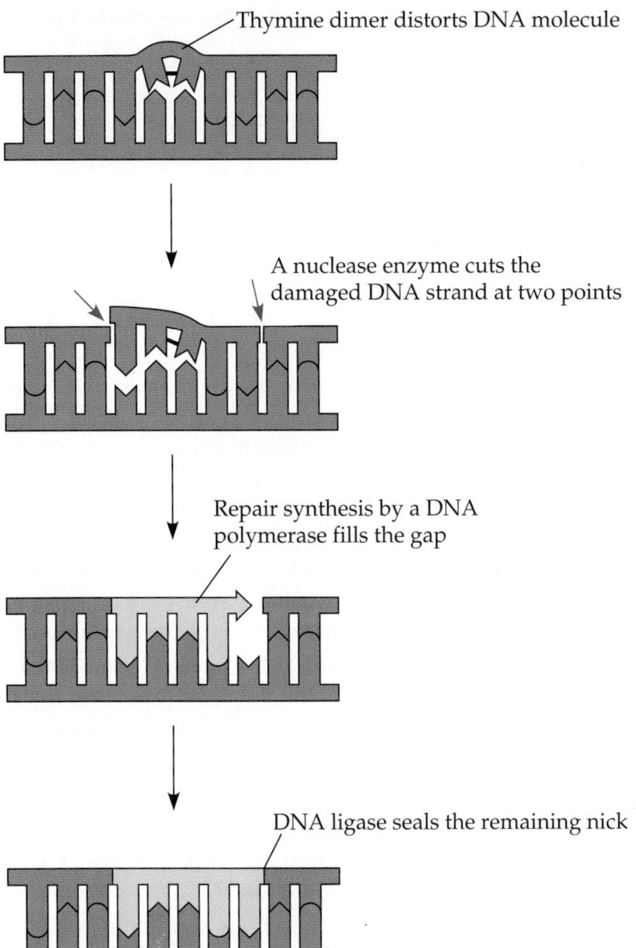
Thymine dimer distorts DNA molecule
A nuclease enzyme cuts the damaged DNA strand at two points
Repair synthesis by a DNA polymerase fills the gap
DNA ligase seals the remaining nick

Figure 15.17
Excision repair of DNA. A team of enzymes detects and repairs damaged DNA. One type of damage, shown here, is the covalent linking of thymine bases that are adjacent on a DNA strand. Such thymine dimers, induced by ultraviolet radiation, cause the DNA to buckle and will lead to errors during DNA replication. Repair enzymes can excise the damaged region from the DNA and replace it with a normal DNA segment.

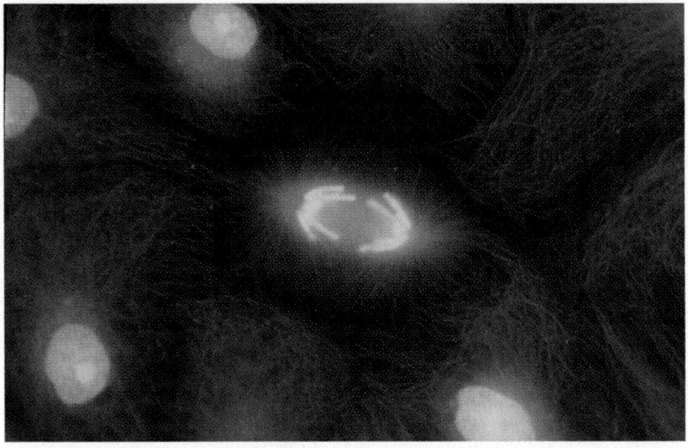
20 µm

Figure 15.18
DNA replication and the cell cycle. Before we leave the topic of DNA replication, let's connect this process to other concepts you have learned about cells. Before a cell divides, its DNA is replicated during the S phase of the cell cycle (see Chapter 11). The duplication of chromosomes is the microscopic evidence that the genetic material has been copied. Mitosis then distributes duplicated chromosomes to daughter cells. The cell in the center of this micrograph is in late anaphase, with the spindle apparatus moving chromosomes toward the two poles of the parent cell (LM, stained with fluorescent dye). Thus, DNA replication provides copies of genes, and the mechanics of cell division transmit these genes from one cellular generation to the next—and from one organismal generation to the next when the genes' cellular vehicles are gametes. (Courtesy of J. M. Murray, University of Pennsylvania Medical School.)

DNA REPAIR

In addition to proofreading, precise maintenance of the genetic information encoded in DNA also requires a way of repairing accidental changes that occur in existing DNA. DNA molecules are constantly subjected to many potentially damaging physical and chemical agents. Reactive chemicals, radioactive emissions, X-rays, and ultraviolet light can change nucleotides in ways that can affect encoded genetic information, usually adversely. Fortunately, these changes, or mutations, are usually corrected. Each cell continuously monitors and repairs its genetic material. Biochemists have identified more than 50 different types of DNA repair enzymes. Sometimes, the damage can be directly reversed by an appropriate enzyme. Most often, however, the repair process, like the replication process, takes advantage of the base-paired structure of DNA. For example, in **excision repair,** a segment of the strand containing the damage is cut out by one repair enzyme, and the resulting gap is filled in with nucleotides properly paired with the nucleotides in the undamaged strand. The enzymes involved in filling the gap are a DNA polymerase and DNA ligase (Figure 15.17).

One function of the DNA repair enzymes in our skin cells is to repair the genetic damage caused by the ultraviolet rays of sunlight. The importance of this function to healthy people is underscored by the skin disease xeroderma pigmentosum, which is caused by an inherited defect in an excision repair enzyme. Sunlight readily kills skin cells in people with this disease and invariably causes skin cancers.

* * *

In this chapter, we have concentrated on the structure of DNA and how this genetic material is copied. Mitosis distributes these copies to daughter cells every time one of our cells divides (Figure 15.18). And DNA replication provides the copies of genes that parents pass to offspring via gametes. However, it is not enough that genes be copied and transmitted; they must also be expressed. How can genes manifest themselves in such phenotypic characters as eye color, for instance? In the next chapter, we will examine the molecular basis of gene expression—how the cell translates genetic information encoded in DNA.

STUDY OUTLINE

1. DNA is the genetic material, and its replication is the chemical basis of inheritance.

The Search for the Genetic Material: A Case Study in the Scientific Process (pp. 300–303)

1. The ability of DNA from a pathogenic strain of bacteria to transform harmless bacteria into pathogens provided the first evidence that the genetic material was DNA.
2. The ability of phages to take over bacterial cells by injecting DNA was further evidence that DNA is the genetic material.
3. Correlation between a cell's DNA content and replication of its chromosomes before cell division added more evidence for DNA as the genetic material.

The Discovery of the Double Helix (pp. 303–306)

1. Scientists from several laboratories quickly realized that unraveling the three-dimensional structure of DNA was the key to understanding its function.
2. Basing their model on data from Franklin's X-ray diffraction photo of DNA, Watson and Crick discovered that DNA is a double helix. Two antiparallel sugar–phosphate chains wind around the outside of the molecule; the nitrogenous bases project into the interior, where they hydrogen-bond in specific pairs, A with T and G with C. Base pairing explained Chargaff's earlier discovery that A and T are present in equal amounts in DNA, as are G and C.

DNA Replication: The Basic Concept (pp. 306–308)

1. Meselson and Stahl demonstrated that DNA replication is semiconservative, confirming Watson and Crick's hypothesis that the parent molecule unwinds and each strand then serves as a template for the synthesis of a new half-molecule according to base-pairing rules.

A Closer Look at DNA Replication (pp. 308–312)

1. More than a dozen enzymes and other proteins cooperate in the rapid and accurate copying of DNA.
2. Replication begins at special sites called origins of replication. Y-shaped replication forks form at opposite ends of a replication bubble, where the two DNA strands separate.
3. Using energy from the hydrolysis of nucleoside triphosphate bonds, DNA polymerases catalyze the synthesis of the new DNA strand working in the $5' \longrightarrow 3'$ direction.
4. Simultaneous $5' \longrightarrow 3'$ synthesis of antiparallel strands at a replication fork yields a continuous leading strand and short, discontinuous segments of lagging strand called Okazaki fragments. The fragments are later joined together with the help of DNA ligase.
5. DNA synthesis must start on the end of a primer, a short segment of RNA synthesized by the enzyme primase.
6. As a replication fork advances, helicases separate the two parent strands of DNA. Single-strand binding proteins hold the template strands straight.
7. Bacterial DNA polymerases also proofread replication errors, correcting any improperly paired nucleotides. A similar proofreading mechanism exists in eukaryotes.

DNA Repair (p. 313)

1. DNA molecules require continuous monitoring and repair because of ongoing assault by physical and chemical agents.
2. DNA repair enzymes restore the integrity of the molecule by such processes as excision repair, which replaces a damaged DNA segment with a normal one.

SELF-QUIZ

1. In his work with pneumonia-causing bacteria and mice, Griffith found that
   a. the protein coat from smooth (S) cells was able to transform rough (R) cells
   b. heat-killed S cells were able to cause pneumonia only when they were transformed by the DNA of R cells
   c. some chemical from S cells was transferred to R cells to transform them into S cells
   d. the polysaccharide coat of R cells caused pneumonia
   e. bacteriophages injected DNA from S cells into R cells
2. Which of the following experimental techniques were used by Hershey and Chase to establish DNA as the genetic material of phage?
   a. X-ray crystallography
   b. electron microscopy
   c. using radioactive isotopes of sulfur and phosphorus to distinguish DNA from protein
   d. analysis of the chemical makeup of chromosomes
   e. incorporation of ^{15}N into DNA
3. *E. coli* cells grown on ^{15}N medium are transferred to ^{14}N medium and allowed to grow for two generations (two cell replications). DNA extracted from these cells is centrifuged. What density distribution of DNA would you expect in this experiment? Explain your answer.
   a. one high-density and one low-density band
   b. one intermediate-density band
   c. one high-density and one intermediate-density band
   d. one low-density and one intermediate-density band
   e. one low-density band
4. The elongation of the *leading* strand during DNA synthesis
   a. progresses away from the replication fork
   b. occurs in the $3' \longrightarrow 5'$ direction
   c. produces Okazaki fragments
   d. depends on the action of DNA polymerase
   e. does not require a template strand
5. A biochemist isolated and purified components needed for DNA replication. When she added some DNA, replication occurred, but the DNA molecules formed were defective. Each consisted of a normal DNA strand paired with numerous segments of DNA a few hundred nucleotides long. What had she probably left out of the mixture? Explain your answer.
   a. DNA polymerase
   b. ligase
   c. nucleotides
   d. Okazaki fragments
   e. primers

6. What is the basis for the difference in the synthesis of the leading and lagging strands of DNA molecules?
   a. The origins of replication occur only at the 5′ end of the molecule.
   b. Helicases and single-strand binding proteins work at the 5′ end.
   c. DNA polymerase can join new nucleotides only to the 3′ end of the growing strand.
   d. DNA ligase works only in the 3′ ⟶ 5′ direction.
   e. Polymerase can only work on one strand at a time.
7. In analyzing the number of different bases in a DNA sample, which result would be consistent with the base-pairing rules?
   a. A = G
   b. A + G = C + T
   c. A + T = G + T
   d. A = C
8. Why does cytosine pair with guanine and not adenine?
   a. A C-A pair would be too wide to fit in the double helix.
   b. C and A are both polar.
   c. A C-A pair would not reach across the double helix.
   d. The functional groups that form hydrogen bonds are not complementary between C and A.
   e. C and A are both purines.
9. The "primer" required to initiate synthesis of a new DNA strand consists of
   a. RNA
   b. DNA
   c. an Okazaki fragment
   d. a structural protein
   e. a thymine dimer
10. A particular gene measures about 1 μm in length along a double-stranded DNA molecule. What is the approximate number of base pairs in this gene?
   a. 3
   b. 10
   c. 1000
   d. 3000
   e. 30,000

CHALLENGE QUESTIONS

1. Mature nerve cells no longer divide, so they do not replicate their DNA. A cell biologist found that there was *X* amount of DNA in a human nerve cell. The biologist then measured the amount of DNA in four other types of human cells; the results are recorded in the chart below. Complete the chart by filling in the type of cell from these choices: (a) sperm cell; (b) bone marrow cell just beginning interphase of the cell cycle; (c) skin cell in the S phase of the cell cycle; (d) intestinal cell beginning mitosis.

Cell	Amount of DNA	Type of Cell
A	2 *X*	
B	1.6 *X*	
C	0.5 *X*	
D	*X*	

2. Cells proofread and repair errors in the genetic information encoded in DNA. But lasting mistakes, or mutations, can happen. The most error-prone stage in DNA replication occurs after parent DNA strands have separated but before new complementary strands are in place. Why might permanent errors be more likely then?

SCIENCE, TECHNOLOGY, AND SOCIETY

1. Using the right ingredients, DNA replication can be carried out in a test tube. This process is routine now, but when it was first accomplished, it was hailed as "the creation of life in a test tube." In fact, replicating DNA (using molecules from living cells) is a long way from building a living organism from scratch. Do you think the creation of test tube life is a worthwhile goal? How would it affect your perspective if a living thing, however simple, were produced in a laboratory someday?
2. DNA molecules with any desired base sequence can be synthesized and replicated in large quantities in the laboratory. It is also possible to determine the base sequence of a DNA molecule. For these reasons, it has been suggested that oil shipments could be tagged with DNA of known sequences. If an oil slick were found, the DNA could be sequenced and the offending ship tracked down. Suggest some other potential applications for "DNA labeling."

FURTHER READING

Becker, W. M., and D. W. Deamer. *The World of the Cell,* 2nd ed. Redwood City, CA: Benjamin/Cummings, 1991. Chapter 14 integrates DNA replication, the cell cycle, and mitosis.

Freedman, D. H. "Life's Off-Switch." *Discover,* July 1991. Arthur Kornberg, who won the Nobel Prize for discovering DNA polymerase, is now trying to learn what turns *off* DNA synthesis.

Judson, H. F. *The Eighth Day of Creation: Makers of the Revolution in Biology.* New York: Simon & Schuster, 1979. An engaging history of molecular biology.

McCarty, M. *The Transforming Principle: The Discovery That Genes Are Made of DNA.* New York: Norton, 1985. A first-person account of the process of science.

Radman, M., and R. Wagner. "The High Fidelity of DNA Duplication." *Scientific American,* August 1988. A well-illustrated article explaining why so few errors occur when DNA replicates.

Watson, J. D. *The Double Helix.* New York: Atheneum, 1968. The brash, controversial best-seller by the codiscoverer of the double helix.

Watson, J. D., and F. H. C. Crick. "Molecular Structure of Nucleic Acids: A Structure for Deoxynucleic Acids." *Nature* 171 (1953): 737–738. Watson and Crick's classic paper.

16 From Gene to Protein

Evidence That Genes Specify Proteins

An Overview of Protein Synthesis

The Genetic Code

The Evolutionary Significance of a Common Genetic Language

A Closer Look at Transcription

A Closer Look at Translation

Protein Targeting

Comparing Protein Synthesis in Prokaryotes and Eukaryotes: A Review

RNA Processing in Eukaryotes

Mutations and Their Effects on Proteins

What Is a Gene?

Figure 16.1
Connecting molecular biology to Mendelism. The tall and dwarf varieties of peas studied by Mendel differ in a single gene. Dwarfs fail to produce an enzyme required for synthesis of a growth hormone. More generally, genes control metabolism by instructing cells to produce specific enzymes and other proteins. This chapter explores the steps in the flow of information from genes to proteins.

Written in the DNA of a human zygote's nucleus are the basic instructions for how to build a person—not a generic human, but one with unique traits. Each individual's phenotype is the cumulative product of a one-of-a-kind genetic makeup combined with environmental influences. The information content of DNA, the genetic material, is in the form of specific sequences of nucleotides along the DNA strands. But how is this information related to an organism's inherited traits? Put another way, what does a gene actually *say*, and how is its message translated by cells into a specific character, such as hair color or blood group?

Consider, once again, Mendel's peas. One of the characters Mendel studied was stem length. Variation in a single gene accounts for the difference between tall and dwarf varieties of peas (Figure 16.1). Mendel did not know the physiological basis of this phenotypic difference, but plant scientists have since worked out the explanation: Dwarf peas lack growth hormones called gibberellins, which stimulate the normal elongation of stems (see Chapter 35). A dwarf plant treated with gibberellins grows to normal height. Why do dwarf peas fail to make their *own* gibberellins? They are missing a key protein, an enzyme required for gibberellin synthesis. This example illustrates the main point of this chapter: The DNA inherited by an organism leads to specific traits by dictating the synthesis of certain proteins. *Proteins are the links between genotype and phenotype.*

EVIDENCE THAT GENES SPECIFY PROTEINS

The relationship between genes and proteins was first proposed in 1909, when British physician Archibald Garrod suggested that genes dictate phenotypes through enzymes that catalyze specific chemical processes in the cell. Garrod postulated that inherited diseases reflect a person's inability to make a particular enzyme, and he referred to such diseases as "inborn errors of metabolism." He gave as one example the hereditary condition called alkaptonuria, in which the urine appears black because it contains the chemical alkapton, which darkens upon exposure to air. Garrod reasoned that normal individuals have an enzyme that breaks down alkapton, whereas alkaptonuric individuals have inherited an inability to make the enzyme that metabolizes alkapton.

How Genes Control Metabolism

Garrod's hypothesis was ahead of its time, but research conducted several decades later supported the idea that the function of a gene is to dictate production of a specific enzyme. Biochemists accumulated much evidence that cells synthesize and degrade most organic molecules via metabolic pathways, in which each chemical reaction in a sequence is catalyzed by a specific enzyme. Such metabolic pathways lead, for instance, to the synthesis of the pigments that give fruit flies their eye color (see Figure 14.3). In the 1930s, George Beadle and Boris Ephrussi speculated that each of the various mutations affecting eye color in *Drosophila* blocks pigment synthesis at a specific step by preventing production of the enzyme that catalyzes that step. However, neither the chemical reactions nor the enzymes that catalyze them were known at the time.

A breakthrough in demonstrating the relationship between genes and enzymes came a few years later, after Beadle and Edward Tatum began to search for mutants of a bread mold, *Neurospora crassa,* that differed from the wild type in their nutritional needs (Figure 16.2). Wild-type *Neurospora* has modest food requirements. It can survive in the laboratory on agar (a moist support medium) that is mixed only with inorganic salts, sucrose, and the vitamin biotin. From this **minimal medium,** the mold uses its metabolic pathways to produce all the other molecules it needs. Beadle and Tatum identified mutants that apparently could not synthesize certain essential molecules, because they could not survive on minimal medium. Such nutritional mutants are called **auxotrophs** (Gr. *auxely,* "to increase"; *trophikos,* "nourishment"), because most of them *can* survive on a **complete growth medium,** minimal medium supplemented with all 20 amino acids and some other nutrients. To pinpoint an auxotroph's metabolic defect, Beadle and Tatum took samples from the mutant growing on complete medium and distributed them to several different vials, each containing minimal medium plus a single additional nutrient. The particular supplement that allowed growth indicated the metabolic defect. For example, if the only supplemented vial that supported growth of the mutant was the one fortified with the amino acid arginine, one could conclude that the mutant was defective in the pathway that normally synthesizes arginine.

With further experimentation, a defect could be described even more specifically. For instance, consider three of the steps in the synthesis of arginine: A precursor nutrient is converted to ornithine, which is converted to citrulline, which is converted to arginine (see Figure 16.2). Beadle and Tatum could distinguish among the various arginine auxotrophs that they found. Some required arginine, others required either arginine or citrulline, and still others could grow when any of the three compounds—arginine, citrulline, or ornithine—was provided. These three classes of mutants, Beadle and Tatum reasoned, must be blocked at different steps in the pathway that synthesizes arginine. Beadle and Tatum concluded that each mutant lacked a different enzyme. Assuming that each mutant was defective in a single gene, they formulated what came to be known as the one gene–one enzyme hypothesis, which states that the function of a gene is to dictate the production of a specific enzyme.

One Gene–One Polypeptide

As more was learned about proteins, it became necessary to make minor revisions in the one gene–one enzyme hypothesis. With a few exceptions, all enzymes are proteins, but not all proteins are enzymes. Keratin, the structural protein of animal hair, and the hormone insulin are two examples of nonenzyme proteins. Because proteins that are not enzymes are nevertheless gene products, molecular biologists began to think in terms of one gene–one protein. However, many proteins are constructed from two or more different polypeptide chains, and each subunit is specified by its own gene. Thus, Beadle and Tatum's axiom has come to be restated as *one gene–one polypeptide.* However, because most proteins consist of single polypeptides, in this chapter we will sometimes simplify by referring to proteins as the gene products.

AN OVERVIEW OF PROTEIN SYNTHESIS

Genes are the instructions for making specific proteins. But a gene does not build a protein directly. The bridge between genetic information and protein synthesis is ribonucleic acid, or RNA. You learned in Chapter 5 that RNA and DNA are both nucleic acids, polymers of nucleotides. Recall that there are two important structural differences between the two nucleic acids: The sugar component of RNA nucleotides is ribose rather than deoxyribose, and the nitrogenous base uracil (U) is found instead of thymine in RNA (see Figure 5.31).

It is customary to describe the flow of information from gene to protein in linguistic terms, because both nucleic acids and proteins have specific sequences of monomers that confer information, much as specific sequences of letters communicate information in written English. In nucleic acids, the monomers are the four types of nucleotides, which differ in their nitrogenous bases. Genes are typically hundreds or thousands of

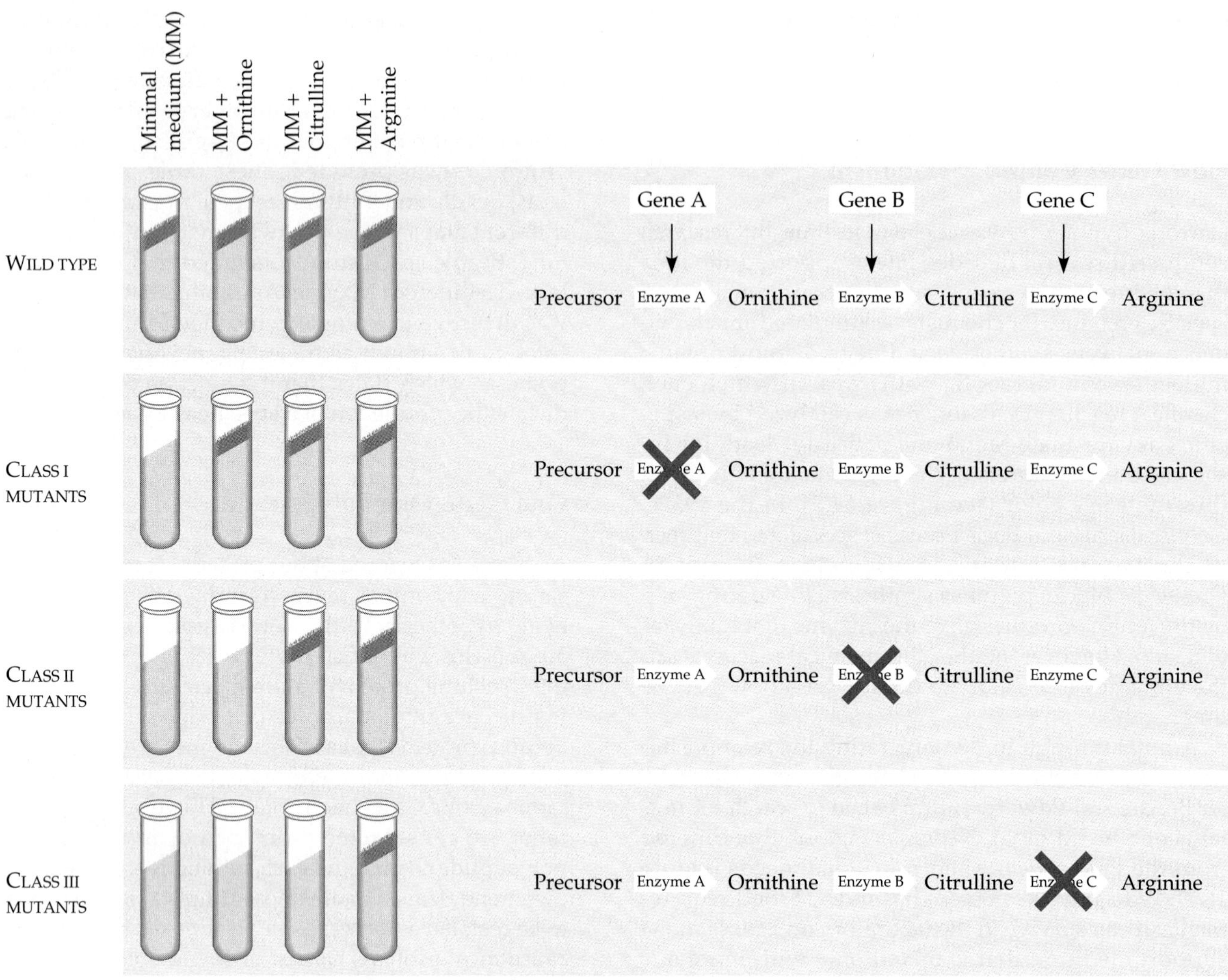

Figure 16.2
The one gene–one enzyme hypothesis. Beadle and Tatum based this idea on their study of nutritional mutants of the red bread mold, *Neurospora crassa.* The wild-type strain requires only a minimal nutritional medium containing sucrose, essential minerals (inorganic salts), and one vitamin. The mold uses a multistep pathway to synthesize the amino acid arginine from a precursor. Beadle and Tatum identified three classes of mutants unable to synthesize arginine for three different reasons. Each mutant had a metabolic block (X in this diagram) at a different step in the pathway. For example, Class II mutants failed to grow on minimal medium or minimal medium supplemented with ornithine. However, adding either citrulline or arginine to the nutritional medium enabled these mutants to grow. Beadle and Tatum deduced that Class II mutants lacked the enzyme that converts ornithine to citrulline. Citrulline bypasses the metabolic block and allows the mold to survive. The other classes of mutants lacked different enzymes. Beadle and Tatum concluded that various mutations were abnormal variations of different genes, each gene dictating the production of one enzyme; hence, the one gene–one enzyme hypothesis.

nucleotides long, each gene having a specific sequence of bases. A protein also has monomers arranged in a particular linear order (the protein's primary structure; see Chapter 5), but its monomers are the 20 amino acids. Thus, nucleic acids and proteins contain information written in two different chemical languages. To get from one to the other requires two major steps, called transcription and translation (Figure 16.3).

Transcription is the synthesis of RNA under the direction of DNA. Both nucleic acids use the same language, and the information is simply transcribed, or copied, from one molecule to the other. A gene's unique sequence of DNA nucleotides provides a template for assembling a unique sequence of RNA nucleotides, and the resulting RNA molecule is a transcript of the gene's protein-building instructions. This type of RNA molecule is called **messenger RNA (mRNA)** because it functions as a genetic message from DNA to the protein-synthesizing machinery of the cell. (There are also RNA molecules with different functions, which we will discuss later in the chapter.) **Translation** is the actual synthesis of a polypeptide, which occurs under the direction of mRNA. In this step there is a change in language: The cell must translate the base sequence of an mRNA molecule into the amino acid sequence of a polypeptide. The sites of

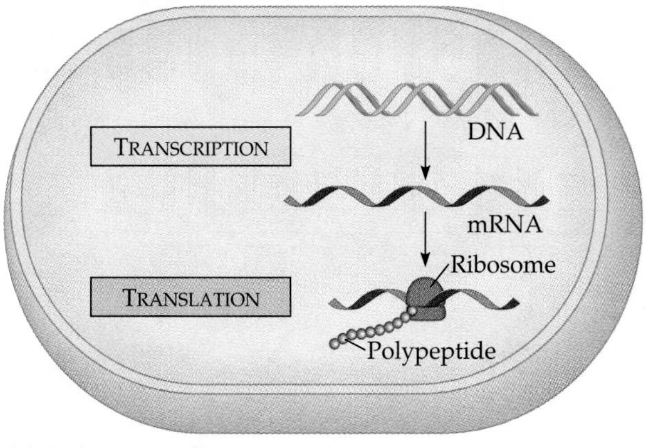

(**a**) Prokaryotic cell

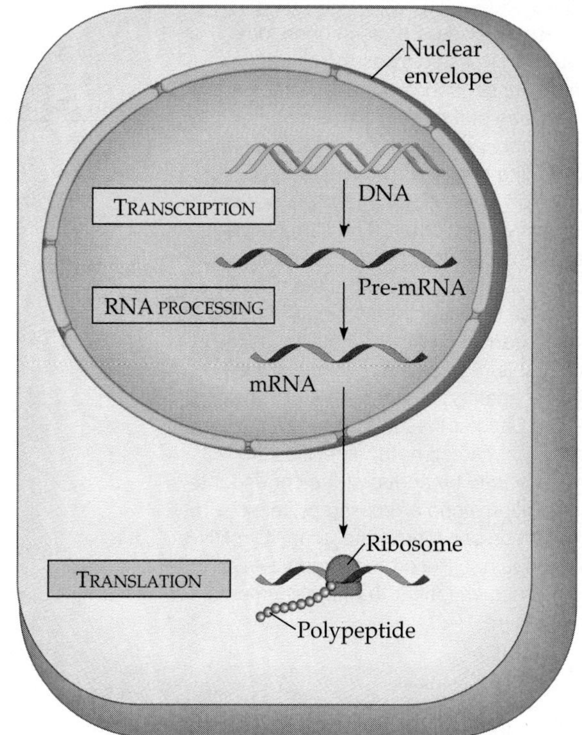

(**b**) Eukaryotic cell

Figure 16.3
The flow of genetic information. In a cell's chain of command, inherited information flows from DNA to RNA to protein. The two major steps in this passing of commands are called transcription and translation. During transcription, a gene provides the instructions for synthesizing a messenger RNA (mRNA) molecule. During translation, the genetic message encoded in mRNA orders amino acids into a protein of specific amino acid sequence. Ribosomes are the sites of translation. (**a**) In a prokaryotic cell, which lacks a nucleus, mRNA produced by transcription is immediately translated without additional processing. (**b**) In a eukaryotic cell, the two main steps of protein synthesis occur in separate compartments: transcription in the nucleus and translation in the cytoplasm. Thus, mRNA must be translocated from nucleus to cytoplasm via pores in the nuclear envelope. The RNA is first synthesized as pre-mRNA, which is processed by enzymes before leaving the nucleus as mRNA.

translation are ribosomes, complex particles with many enzymes and other agents that facilitate the orderly linking of amino acids into polypeptide chains.

Although the basic mechanics of transcription and translation are similar for prokaryotes and eukaryotes, there is an important difference in how the overall process of protein synthesis is organized within the cells. Because bacteria lack nuclei, their DNA is not segregated from ribosomes and the other protein-synthesizing equipment. As a result, transcription and translation occur in rapid succession (see Figure 16.3a). In contrast, the nuclear envelope of a eukaryotic cell separates transcription from translation in space and time. Transcription occurs in the nucleus, and mRNA is dispatched to the cytoplasm, where translation occurs (see Figure 16.3b). This compartmentalization of transcription and translation in eukaryotes provides an opportunity to modify mRNA in various ways before it leaves the nucleus. This modification, called **RNA processing,** occurs only in eukaryotes. You will learn more about the mechanisms and functions of RNA processing later in the chapter.

Let's summarize the main point of our overview of protein synthesis: Genes program protein synthesis via genetic messages in the form of RNA. Put another way, cells are governed by a chain of command: DNA $\longrightarrow$ RNA $\longrightarrow$ protein. The next section discusses in more detail how the instructions for assembling amino acids into a specific order are encoded in the language of DNA, and we'll see how biologists "cracked" this genetic code.

THE GENETIC CODE

When biologists began to suspect that the instructions for protein synthesis are encoded in DNA, they recognized a problem: There are only 4 nucleotides to specify 20 amino acids. Thus, the genetic code cannot be a language like Chinese, where each written symbol corresponds to a single word. If each nucleotide base were translated into an amino acid, only 4 of the 20 amino acids could be specified. Would a language of two-let-

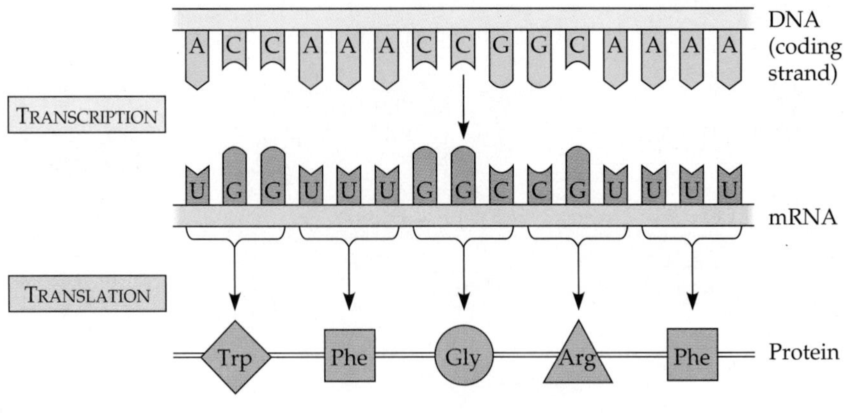

Figure 16.4
The triplet code. For each gene, one strand of DNA serves as the coding strand, its sequence of bases specifying production of a protein with a certain amino acid sequence. During transcription, the coding strand provides a template for synthesis of an mRNA molecule of complementary sequence. The same base-pairing rules that apply to DNA synthesis also guide transcription, but the base uracil (U) takes the place of thymine (T) in RNA. During translation, the genetic message (mRNA) is read as a sequence of base triplets, analogous to three-letter code words. Each of these triplets, called a codon (bracketed here), specifies the amino acid to be added at the corresponding position along a growing protein chain. All of these polymers—the gene, its mRNA transcript, and the protein product—are typically much longer than the segments shown here.

ter code words suffice? The base sequence, AG, for example, could specify one amino acid, and GT could specify another. Since there are four bases, this would give us 16 (4^2) possible arrangements—still not enough to code for all 20 amino acids.

Triplets of bases are the smallest units of uniform length that can code for all the amino acids. If each arrangement of 3 consecutive bases specifies an amino acid, there can be 64 (4^3) possible code words—more than enough to specify all the amino acids. Experiments have verified that the flow of information from gene to protein is based on a **triplet code:** The genetic instructions for a polypeptide chain are written in the DNA as a series of three-nucleotide words, called **codons.**

A cell cannot directly translate a gene's codons into amino acids. The intermediate step is transcription, during which the gene determines the codon sequence of an mRNA molecule. For each gene, only one of the two DNA strands is transcribed. This is called the coding strand of that gene. The noncoding strand serves as a template for making a new coding strand when the DNA replicates. A given DNA strand can be the coding strand in some regions of a DNA molecule and the noncoding strand in others; that is, the same strand that serves as a template for transcription from one gene may be the noncoding strand of another gene that is part of the same long double helix.

An mRNA molecule is complementary rather than identical to its DNA template, because RNA bases are assembled on the template according to the base-pairing rules (Figure 16.4). For example, where the coding strand of a gene has a codon CCG, the codon at the corresponding position along the mRNA molecule will be GGC, an mRNA code word for glycine. Note that U, the RNA substitute for T, pairs with A. Thus, the DNA codon GCA is transcribed as an mRNA codon CGU, the code word for arginine.

During translation, the sequence of codons along a genetic message (mRNA) is decoded, or translated, into a sequence of amino acids making up a polypeptide chain. Each codon along the mRNA molecule specifies which one of the 20 amino acids will be incorporated at the corresponding position along a polypeptide. Because codons are base triplets, the number of nucleotides making up a genetic message must be three times the number of amino acids making up the protein product. For example, it takes an RNA strand 300 nucleotides to code for a protein that is 100 amino acids long.

Molecular biologists cracked the code of life in the early 1960s, when a series of elegant experiments disclosed the amino acid translations of each of the codons of nucleic acids. The first codon was deciphered in 1961 by Marshall Nirenberg, of the National Institutes of Health. Nirenberg had synthesized an artificial mRNA by linking identical RNA nucleotides containing uracil as their base. No matter where this message started or stopped, it could contain only one triplet codon: UUU. Nirenberg added this "poly U" to a test-tube mixture containing amino acids, ribosomes, and the other components required for protein synthesis. His artificial system translated the poly U into a polypeptide containing a single amino acid, phenylalanine (Phe), showing that the mRNA codon UUU specifies this amino acid. Soon, the amino acids specified by the codons AAA, GGG, and CCC were also determined.

First base	Second base: U	C	A	G	Third base
U	UUU Phe	UCU Ser	UAU Tyr	UGU Cys	U
	UUC	UCC	UAC	UGC	C
	UUA Leu	UCA	UAA Stop	UGA Stop	A
	UUG	UCG	UAG Stop	UGG Trp	G
C	CUU Leu	CCU Pro	CAU His	CGU Arg	U
	CUC	CCC	CAC	CGC	C
	CUA	CCA	CAA Gln	CGA	A
	CUG	CCG	CAG	CGG	G
A	AUU Ile	ACU Thr	AAU Asn	AGU Ser	U
	AUC	ACC	AAC	AGC	C
	AUA	ACA	AAA Lys	AGA Arg	A
	AUG Met or start	ACG	AAG	AGG	G
G	GUU Val	GCU Ala	GAU Asp	GGU Gly	U
	GUC	GCC	GAC	GGC	C
	GUA	GCA	GAA Glu	GGA	A
	GUG	GCG	GAG	GGG	G

Figure 16.5
The dictionary of the genetic code. The three bases of an mRNA codon are designated here as the first, second, and third bases. Practice using this dictionary by finding the codon UGG, the first codon in Figure 16.4. Locate U along the left (first base) reference bar. Move across this U-row to the G column. Now that you have the first two bases, find G along the right reference bar (third base bar). Note that the codon UGG is translated as the amino acid tryptophan, as in Figure 16.4. This is the only codon for tryptophan, but most amino acids are specified by two or more codons. For example, both UUU and UUC stand for the amino acid phenylalanine (Phe). When either of these codons is read along an mRNA molecule, phenylalanine will be incorporated into the growing polypeptide chain. We can think of UUU and UUC as synonyms in the genetic code. Note that the codon AUG not only stands for the amino acid methonine (Met), but also functions as a "start" signal. Ribosomes, the particles that actually assemble proteins, recognize AUG as the start of a genetic message and begin translating the mRNA at that location. Three of the 64 codons function as "stop" signals. Any one of these termination codons marks the end of a genetic message, and the completed polypeptide chain is released from the ribosome.

Although more elaborate techniques were required to decode mixed triplets such as AUA and CGA, all 64 codons were deciphered by the mid-1960s. As Figure 16.5 shows, 61 of the 64 triplets code for amino acids. The codons are presented in their mRNA versions rather than their (complementary) DNA versions. Notice that the triplet AUG has a dual function: It not only codes for the amino acid methionine (Met), but also functions as a "start" signal. Genetic messages begin with the mRNA codon AUG, which signals the protein-synthesizing machinery to begin translating the mRNA at that location. (Since AUG also stands for methionine, all polypeptide chains begin with methionine when they are synthesized. However, an enzyme may subsequently remove this "starter" amino acid from a chain.) The remaining three codons do not designate amino acids. Instead, they are "stop" signals marking the end of a genetic message.

Notice in Figure 16.5 that there is redundancy in the genetic code, but no ambiguity. For example, although codons GAA and GAG both specify glutamic acid (redundancy), neither of them ever specifies any other amino acid (no ambiguity). The redundancy in the code, often called degeneracy, is not altogether random. In many cases, codons that are "synonyms" for a particular amino acid differ only in the third base of the triplet. We will consider a possible explanation for this redundancy later in the chapter.

Our ability to extract the intended message from written language depends on reading the symbols in the correct sequence and groupings. This ordering is called the **reading frame.** Consider this statement: "The red dog ate the cat." Read the words out of order, and you may get the unintended message, "The red cat ate the dog." Group the letters incorrectly, as in "Thereddogatethecat," and you may think that the first word of the message is "There." The reading frame is also important in the molecular language of cells. The amino acids shown in Figure 16.4, for instance, can only be assembled in the correct order if the mRNA codons UGGUUUGGCCGUUUU are read from start to finish in the correct sequence and groups. Although a genetic message is written with no spaces between the codons, the cell's protein-synthesizing machinery reads the message in the correct frame as a series of nonoverlapping three-letter words: UGG-UUU and so on. The message is *not* read as a series of overlapping words—UGG-GGU-GUU, and so on—which would convey a very different message. (Some viruses that *do* employ an overlapping code are exceptions.)

Let's summarize what we have just covered. In nearly all organisms, genetic information is encoded as a sequence of nonoverlapping base triplets, or codons, each of which is translated into a specific amino acid during protein synthesis.

THE EVOLUTIONARY SIGNIFICANCE OF A COMMON GENETIC LANGUAGE

The genetic code is nearly universal, shared by organisms as diverse as bacteria and humans. The RNA

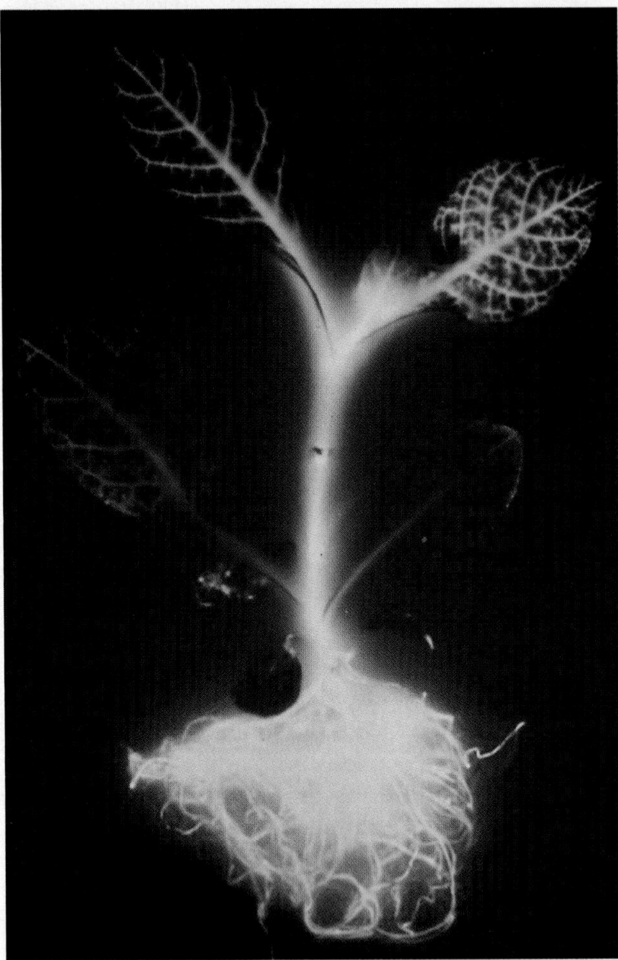

Figure 16.6
A tobacco plant expressing a firefly gene. Because diverse forms of life share a common genetic code, it is possible to program one species to produce proteins characteristic of another species by transplanting DNA. In this experiment, researchers were able to incorporate a gene from a firefly into the DNA of a tobacco plant. The gene codes for the enzyme that catalyzes the chemical reaction that releases energy in the form of light in fireflies.

codon CCG, for instance, is translated as the amino acid proline in all organisms whose genetic code has been examined. In laboratory experiments, genes can be transcribed and translated after they are transplanted from one species to another (Figure 16.6). For example, bacteria can be programmed by the insertion of a human gene to synthesize the protein insulin, a product that can be used to treat diabetes. Such applications have produced many exciting developments in biotechnology, which you will learn about in Chapter 19.

There are some interesting exceptions to the universality of the genetic code. In several single-celled eukaryotes called ciliates (including *Paramecium*, an organism you may know from the lab), biologists have found a variation from the standard code. In these organisms, the RNA codons UAA and UAG are not the usual stop signals (see Figure 16.5); instead they code for the amino acid glutamine. Researchers have discovered other exceptions to the standard genetic code in mitochondria and chloroplasts, organelles that contain DNA coding for some of their own proteins (see Chapter 7).

Although biologists do not yet understand how these variations of the genetic code evolved, they generally agree on the evolutionary significance of the code's *near* universality. A language shared across all of life must have been operating very early in the history of life—early enough to be present in the organisms that were the common ancestors of all modern organisms, from the simplest bacteria to humans. A shared genetic vocabulary is a reminder of the kinship that bonds all life on Earth.

Now that we have considered the linguistic logic and evolutionary significance of the genetic code, we can reexamine, in more detail, the key processes of transcription and translation.

A CLOSER LOOK AT TRANSCRIPTION

Messenger RNA, the carrier of information from DNA to protein, is transcribed from the coding strand of a gene. Enzymes called **RNA polymerases** pry the two strands of DNA apart and hook together the RNA nucleotides as they base pair along the DNA template. Like the DNA polymerases that function in DNA replication, RNA polymerases can only add nucleotides to the 3′ end of the growing polymer (see Figure 15.15). Thus, an RNA molecule elongates in its 5′ ⟶ 3′ direction. Specific sequences of nucleotides along the DNA mark the initiation and termination sites, where transcription of a gene begins and ends. Including these "bookends" and the hundreds or thousands of nucleotides in between, the entire stretch of DNA that is transcribed into a single RNA molecule is called a **transcription unit** (Figure 16.7). In eukaryotes, a transcription unit represents a single gene; the mRNA codes for synthesis of one protein.

In prokaryotes, one transcription unit may include a few genes that code for proteins of related function—enzymes catalyzing sequential steps of a metabolic pathway, for example. In these cases, the mRNA is punctuated with start and stop codons that signal the ends of each segment of mRNA coding for one protein.

Bacteria have a single type of RNA polymerase that synthesizes not only mRNA but also other types of RNA that function in protein synthesis (discussed in the next section). In contrast, eukaryotes are equipped

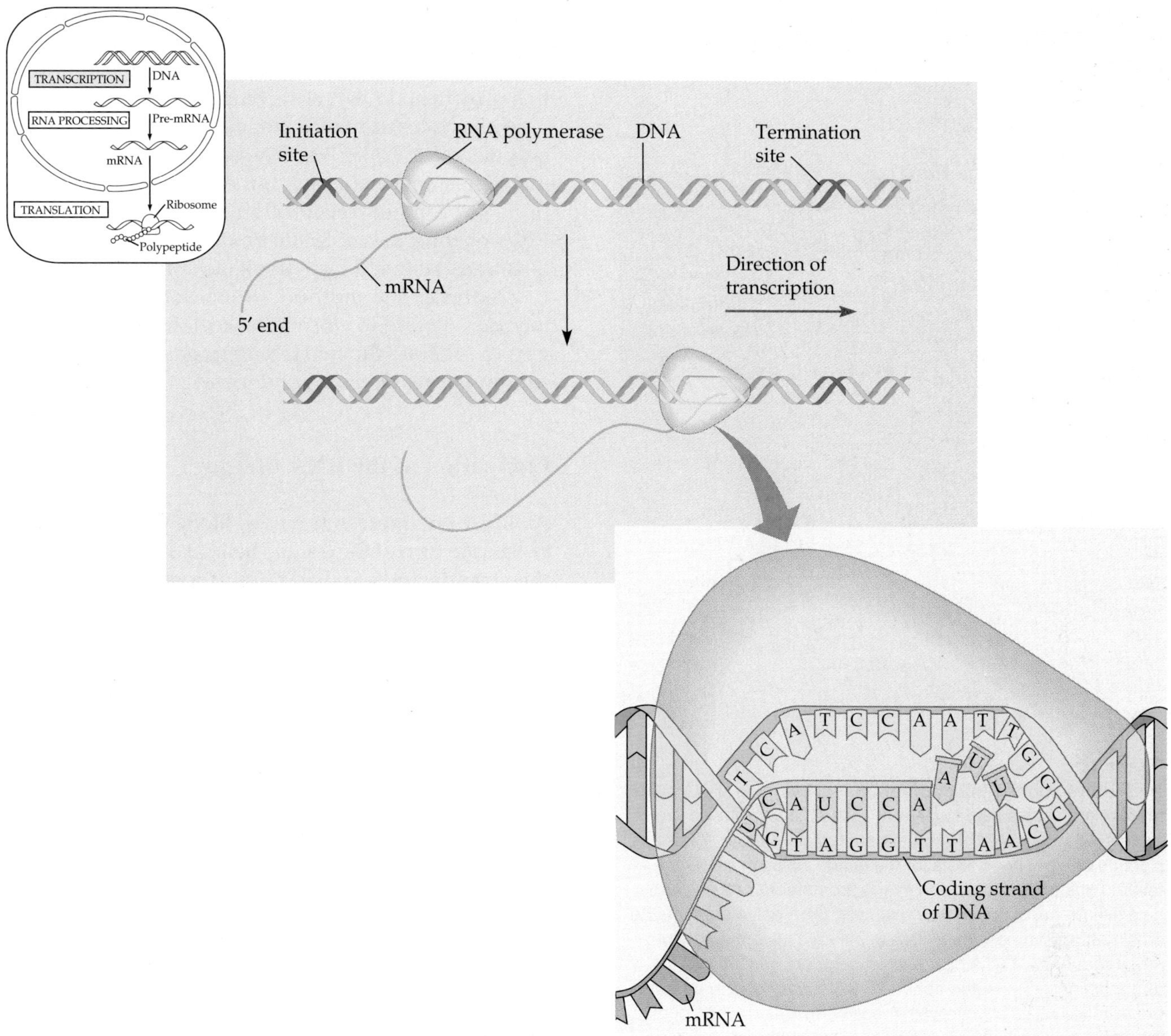

Figure 16.7
Transcription. As an RNA polymerase molecule moves along a gene from the initiation site to the termination site, it synthesizes an RNA molecule that consists of the nucleotide sequence determined by the coding strand of the gene. The entire stretch of DNA that is transcribed is called a transcription unit. Transcription begins at the initiation site when the polymerase separates the two DNA strands and exposes the coding strand for base pairing with RNA nucleotides. The RNA polymerase works its way "downstream" from the initiation site, prying apart the two strands of DNA and elongating the mRNA. In the wake of transcription, the two DNA strands re-form the double helix. The RNA polymerase continues to elongate the RNA molecule until it reaches the termination site, a specific sequence of nucleotides along the DNA that signals the end of the transcription unit. The mRNA, a transcript of the gene, is released, and the polymerase subsequently dissociates from the DNA.

with three types of RNA polymerase. The one specialized for mRNA synthesis is called RNA polymerase II. About 40,000 RNA polymerase II molecules are found in the nucleus of a human cell.

The three key steps in transcription are polymerase binding and initiation, elongation, and termination. Let's examine each of these processes in greater detail.

RNA Polymerase Binding and Initiation of Transcription

RNA polymerases bind to regions of DNA known as **promoters** (Figure 16.8). A promoter includes the initiation site, where transcription actually begins, and

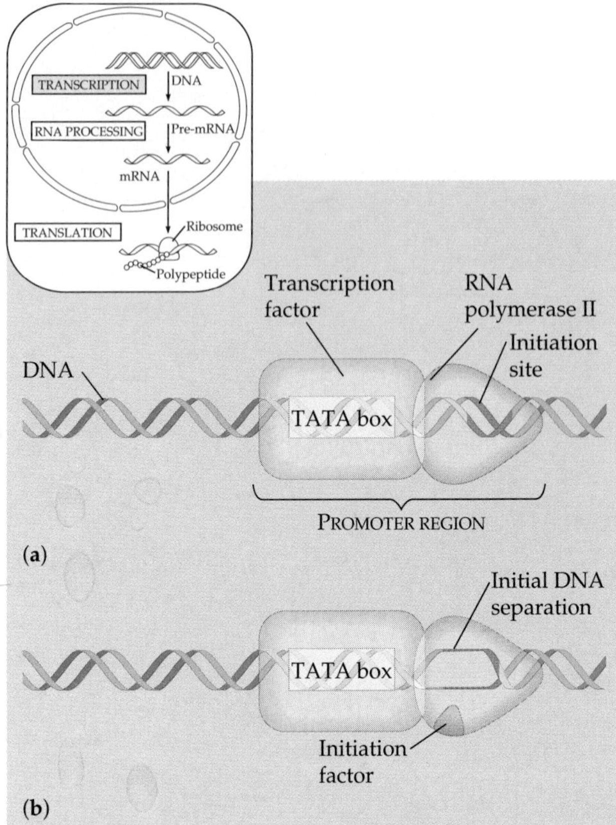

Figure 16.8
Promoters and the initiation of transcription. (**a**) RNA polymerase molecules bind to DNA at special regions called promoters. In eukaryotes, promoters are typically about 100 nucleotides long. They consist of the actual initiation site and a few regions of DNA that identify the promoter to proteins that help initiate transcription. For example, RNA polymerase II, the enzyme that synthesizes mRNA in eukaryotes, binds to a promoter that has a TATA box, a short segment of thymines (T) and adenines (A) located about 25 nucleotides upstream from the initiation site. However, RNA polymerase cannot recognize the TATA box and other landmarks of the promoter region on its own. Another protein, a transcription factor that recognizes the TATA box, binds to the DNA before the RNA polymerase can do so. (**b**) Once RNA polymerase attaches to the promoter region, it probably associates with other transcription factors before RNA synthesis begins (only one of these accessory proteins, an initiation factor, is illustrated here).

dozens of nucleotides "upstream" from the initiation site. Certain regions within the promoter are especially important for recognition by RNA polymerases. For example, the RNA polymerase II of eukaryotes keys on a region called the **TATA box,** so named because it is enriched with the nitrogenous bases thymine (T) and adenine (A). TATA boxes are centered at locations that are 25 nucleotides upstream from initiation sites.

RNA polymerase II evidently cannot recognize and bind to a promoter on its own. Other proteins, called **transcription factors,** aid the polymerases in their search for promoter regions along DNA molecules. One type of transcription factor, shown in Figure 16.8, is a protein that must attach to a promoter *before* an RNA polymerase II molecule can bind. Apparently, RNA polymerase II recognizes the complex between this protein and DNA as its binding site.

Prokaryotes use a different mechanism to bind RNA polymerase to DNA. The RNA polymerase of bacteria includes a detachable subunit called sigma that functions in promoter recognition and binding. After an RNA polymerase molecule binds to the DNA and begins transcribing, it loses its sigma subunit.

Whatever the method, when active RNA polymerase is bound to a promoter region, the enzyme begins to separate the two DNA strands at the initiation site, and transcription is under way.

Elongation of the RNA Strand

As RNA polymerase II moves along the DNA, it untwists one turn of the double helix at a time, separating the strands and exposing about ten DNA bases for pairing with RNA nucleotides (see Figure 16.7). The enzyme adds nucleotides to the 3′ end of the growing RNA molecule as it continues along the double helix. In the wake of this advancing wave of RNA synthesis, the mRNA molecule peels away from its DNA template, as the noncoding strand of DNA re-forms a double helix by pairing with the coding strand. Transcription progresses at a rate of about 60 nucleotides per second.

A single gene can be transcribed simultaneously by several molecules of RNA polymerase II, following each other like trucks in a convoy. The growing strands of RNA trail off from each polymerase, with the length of each new strand reflecting how far along the template the enzyme has traveled from the initiation site. The congregation of many polymerase molecules simultaneously transcribing a single gene allows a cell to produce a particular protein in large amounts.

Termination of Transcription

Transcription proceeds until the RNA polymerase reaches a termination site on the DNA. The sequence of nitrogenous bases that marks this site signals RNA polymerase to stop adding nucleotides to the RNA strand and release the RNA molecule. In eukaryotes, the most common termination sequence is AATAAA. Additional proteins probably cooperate with RNA polymerase in the termination step.

In bacteria, mRNAs are ready for translation as soon as they peel away from their DNA templates. In contrast, the RNA products of transcription in eukaryotes are processed before they leave the nucleus as mRNA molecules. We will postpone discussing this RNA processing until after we have studied the process of translation more closely.

A CLOSER LOOK AT TRANSLATION

In the process of translation, a cell interprets a genetic message and builds a protein accordingly. The message is a series of codons along an mRNA molecule, and the interpreter is another type of RNA called **transfer RNA (tRNA).** The function of tRNA is to transfer amino acids from the cytoplasm's amino acid pool to a ribosome. A cell keeps its cytoplasm stocked with all 20 amino acids, either by synthesizing them from other compounds or by taking them up from the surrounding solution. The ribosome adds each amino acid brought to it by tRNA to the growing end of a polypeptide chain (Figure 16.9).

Molecules of tRNA are not all identical. The key to translating a genetic message into a specific amino acid sequence is that each type of tRNA molecule associates a particular mRNA codon with a particular amino acid. As a tRNA molecule arrives at a ribosome, it bears a specific amino acid at one of its ends. At the other end is a base triplet called the **anticodon,** which binds, according to the base-pairing rules, to a complementary codon on mRNA. For example, the mRNA codon UUU is translated as the amino acid phenylalanine (see Figure 16.5). The tRNA that plugs into this codon by hydrogen bonding has AAA as its anticodon and always carries phenylalanine at its other end. Thus, as an mRNA molecule slides through a ribosome, phenylalanine will be added to the polypeptide chain whenever the codon UUU is presented for translation. Codon by codon, the genetic message is translated, as tRNAs deposit amino acids in the order prescribed, and ribosomal enzymes join the amino acids into a chain. The tRNA molecule is like a flashcard with a "nucleic acid word" (anticodon) on one side and a "protein word" (amino acid) on the other.

Translation is simple in principle, but complex in its actual biochemistry and mechanics. To dissect the process, let's take a closer look at some of the major players in this cellular drama of protein synthesis, then see how they work together to make a polypeptide.

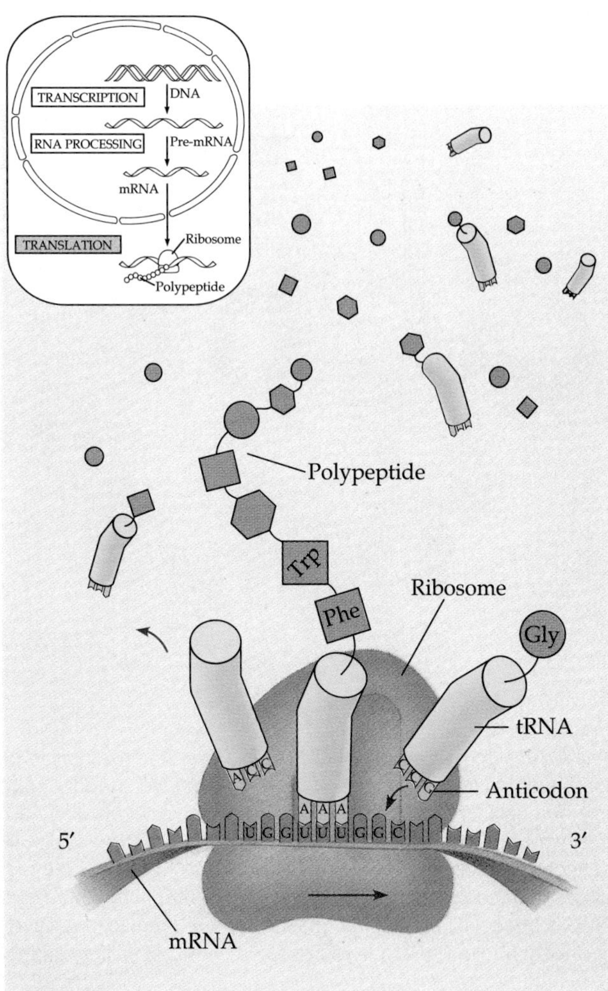

Figure 16.9
Translation: the basic concept. As a strand of mRNA slides through a ribosome, codons are translated into amino acids, one by one. The interpreters are tRNA molecules, each type with a specific anticodon at one end and a certain amino acid at the other end. A tRNA adds its amino acid cargo to a growing polypeptide chain when the anticodon binds to a complementary codon on the mRNA.

Transfer RNA

Transfer RNA molecules, like mRNA and other types of RNA, are transcribed from DNA templates within the nucleus of a eukaryotic cell. Like mRNA, tRNA must travel from the nucleus to the cytoplasm, where translation occurs. Each tRNA molecule can then be used repeatedly, picking up its designated amino acid in the cytoplasm, depositing this cargo at the ribosome, and leaving the ribosome to pick up another load. The structure of a tRNA molecule fits its function as a shuttle for a specific amino acid (Figure 16.10).

A tRNA molecule consists of a single RNA strand that is only about 80 nucleotides long (compared to hundreds of nucleotides for most mRNA molecules). This single RNA strand folds back upon itself to form a molecule with a secondary structure, a three-dimensional structure reinforced by interactions between different parts of the nucleotide chain. Certain regions of the tRNA strand form hydrogen bonds with complementary bases of other regions. Flattened into one plane to reveal these regions of hydrogen bonding, a tRNA molecule has a cloverleaf shape. The two-dimensional cloverleaf of the typical tRNA molecule twists and folds into a fairly compact three-dimensional structure that is roughly L-shaped. The loop protruding from one end of the L includes the anticodon, the specialized base triplet that binds to a specific mRNA codon. From the other end of the L-shaped tRNA molecule protrudes its 3′ end, which is the attachment site for an amino acid.

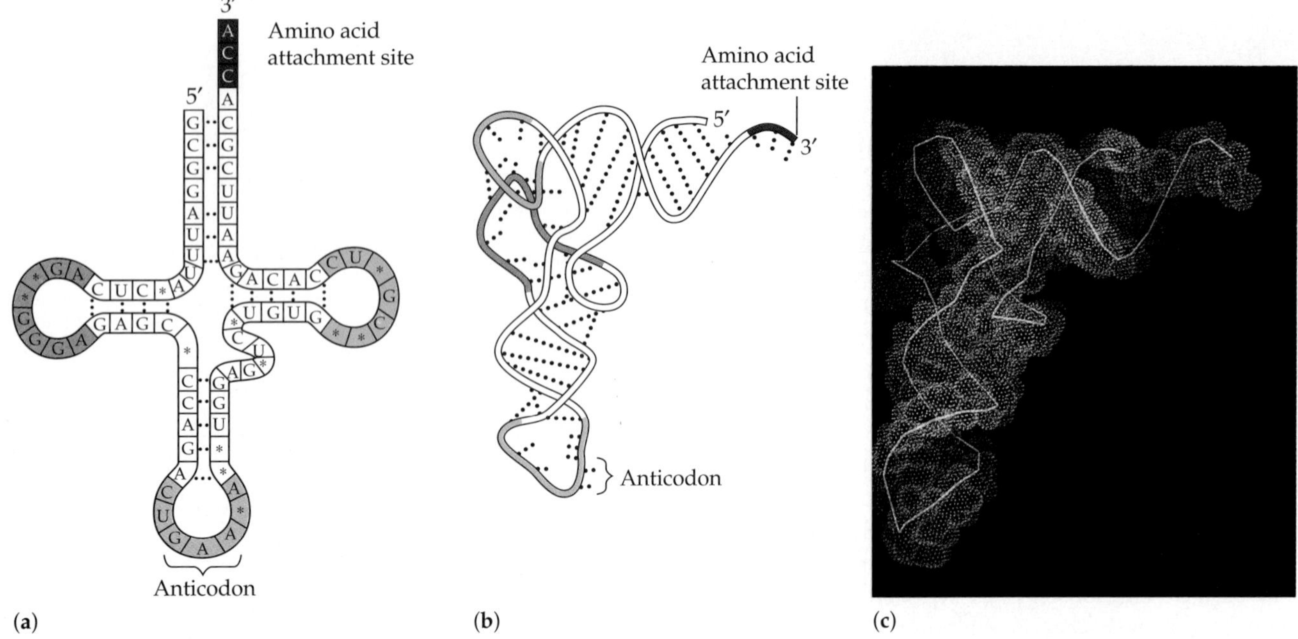

Figure 16.10
The structure of transfer RNA. (**a**) The two-dimensional structure of a tRNA specific for the amino acid leucine. Notice the four double-stranded regions and three loops characteristic of all tRNAs. At one end of the molecule is the amino acid attachment site, which has the same base sequence for all tRNAs; within the middle loop is the anticodon triplet, which is unique to each tRNA type. (The asterisks mark unusual bases that are unique to tRNAs.) (**b**) A diagram of the three-dimensional structure of an L-shaped tRNA molecule. (**c**) A computer graphic of a tRNA molecule. (**d**) In the figures that follow, tRNA is represented by this simplified shape.

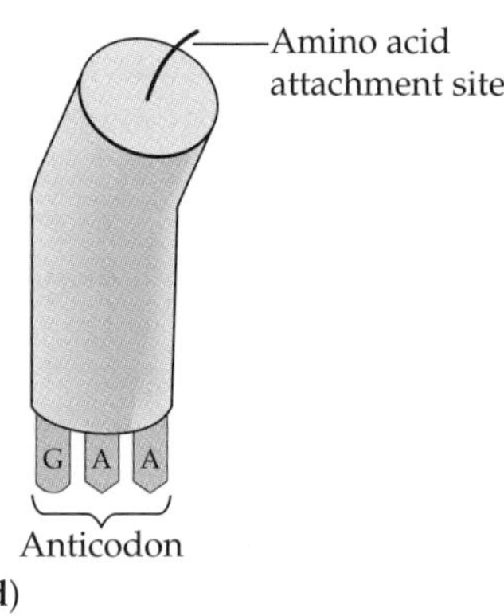

If one tRNA variety existed for each of the mRNA codons that specifies an amino acid, there would be 61 tRNAs (see Figure 16.5). The actual number is smaller: about 45. This number is sufficient, because some tRNAs have anticodons that can recognize two or more different codons. Such versatility is possible because the rules for base pairing between the third base of a codon and the corresponding base of a tRNA anticodon are not as strict as those for DNA and mRNA codons. For example, the base U of a tRNA anticodon can pair with either A or G in the third position of an mRNA codon. This relaxation of the base-pairing rules is called **wobble.** The most versatile tRNAs are those with inosine (I), a modified base, in the wobble position of the anticodon. Inosine is formed by enzymatic alteration of adenine after tRNA is synthesized. When anticodons associate with codons, the base I can hydrogen-bond with any one of three bases: U, C, or A. Thus, the tRNA molecule that has CCI as its anticodon can bind to the codons GGU, GGC, and GGA, all of which code for the amino acid glycine. Wobble explains why the synonymous codons for a given amino acid can differ in their third base, but usually not in their other bases.

Aminoacyl-tRNA Synthetases

Codon-anticodon bonding is actually the second of two recognition steps required for the accurate translation of a genetic message. It must be preceded by a correct match between tRNA and an amino acid. A tRNA that binds to an mRNA codon specifying a particular amino acid must carry *only* that amino acid to the ribosome. Each amino acid is matched with the correct tRNA by a specific enzyme called an **aminoacyl-tRNA synthetase.** There is a whole family of these enzymes, one enzyme for each amino acid. The active site of each type of aminoacyl-tRNA synthetase fits only a specific combination of amino acid and tRNA. The synthetase catalyzes the attachment of the amino acid to its tRNA in a two-step process driven by the hydrolysis of ATP (Figure 16.11). The resulting amino acid–tRNA complex is released from the enzyme and delivers its

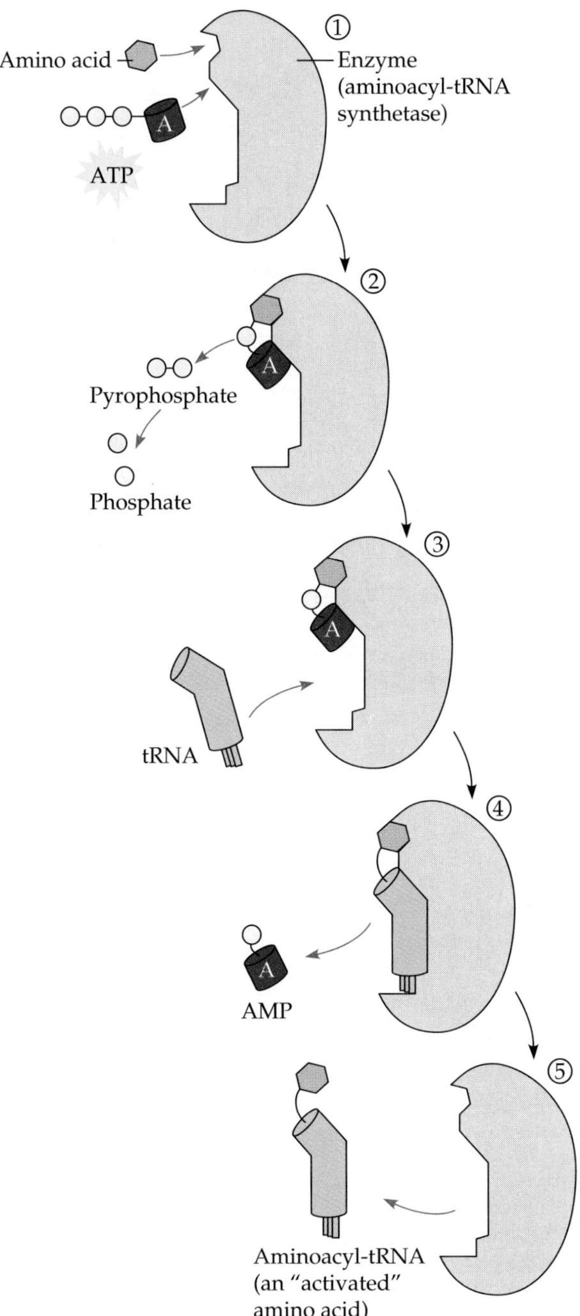

Figure 16.11
The function of an aminoacyl-tRNA synthetase. Each enzyme attaches a specific amino acid to its appropriate tRNA. ① The active site of the enzyme binds the amino acid and an ATP molecule. ② The ATP loses two phosphate groups and joins to the amino acid as AMP (adenosine monophosphate). ③ The appropriate tRNA covalently bonds to the amino acid, ④ displacing the AMP from the enzyme's active site. ⑤ The enzyme releases the amino acid–tRNA complex, also known as aminoacyl-tRNA.

amino acid to a growing polypeptide chain on a ribosome.

Ribosomes

Ribosomes facilitate the specific coupling of tRNA anticodons with mRNA codons during protein synthesis. A ribosome, which can be seen with the electron microscope, is made up of two subunits (Figure 16.12a). (When they are not involved in protein synthesis, the two subunits are not bound together.) Each ribosomal subunit is an aggregate of numerous proteins and yet another form of specialized RNA called **ribosomal RNA (rRNA).** About 60% of the weight of each ribosome is rRNA. Because most cells contain thousands of ribosomes, rRNA is the most abundant type of RNA. (For further discussion about the molecular composition of ribosomes, see Chapter 7.) In addition to a binding site for mRNA, each ribosome has two binding sites for tRNA (Figure 16.12b). The **P site** (peptidyl-tRNA site) holds the tRNA carrying the growing polypeptide chain, while the **A site** (aminoacyl-tRNA site) holds the tRNA carrying the next amino acid to be added to the chain. Acting like a vise, the ribosome holds the tRNA and mRNA molecules close together, while one of its several proteins catalyzes the transfer of an amino acid to the carboxyl end of the growing polypeptide chain.

Building a Polypeptide

We can divide translation, the synthesis of a polypeptide, into three stages: chain initiation, chain elongation, and chain termination. All three stages require protein factors (mostly enzymes) that aid mRNA, tRNA, and ribosomes in the translation process. For chain initiation and elongation, energy is also required. It is provided by GTP (guanosine triphosphate), a cellular energy currency closely related to ATP.

Initiation The initiation stage of translation brings together mRNA, a tRNA bearing the first amino acid of the polypeptide, and the two subunits of a ribosome. First, a small ribosomal subunit binds to both mRNA and a special initiator tRNA (Figure 16.13). The small ribosomal subunit attaches to a specific sequence of nucleotides at the 5′ (upstream) end of the mRNA. Just downstream from this loading site is the initiation codon, AUG, where translation actually begins. The initiator tRNA, which carries the amino acid methionine, attaches to the initiation codon.

The union of mRNA, initiator tRNA, and small ribosomal subunit is followed by the attachment of a large ribosomal subunit to form a functional ribosome. Proteins called **initiation factors** are required to bring all these components together. The cell also spends en-

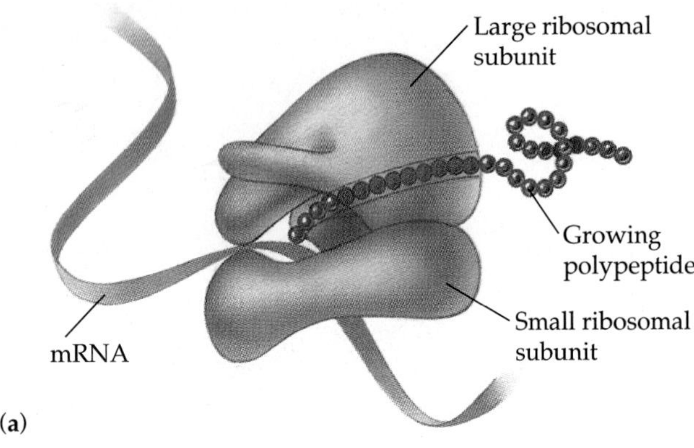

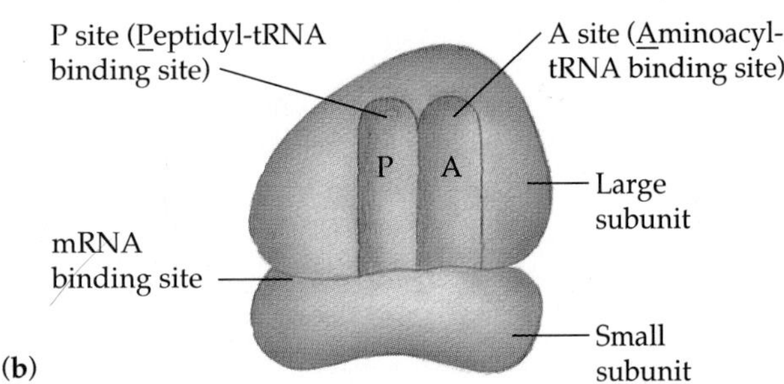

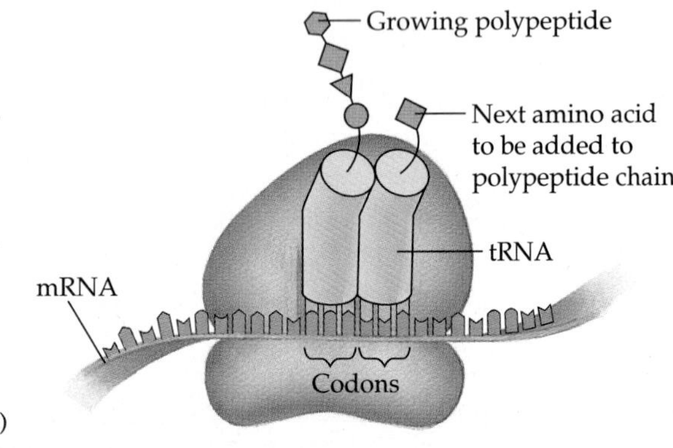

Figure 16.12
The anatomy of a ribosome. (**a**) A functional ribosome consists of two subunits, each an aggregate of ribosomal RNA and several proteins. This is a model of a bacterial ribosome. The eukaryotic ribosome is similar in shape, but larger, with more proteins and rRNA molecules. (**b**) A ribosome has an mRNA binding site and two tRNA binding sites, known as the P and A sites. This is a simplified version of the ribosomal shape that will appear in the next several figures. (**c**) A tRNA fits into a binding site when its anticodon base pairs with an mRNA codon. The P site holds the tRNA attached to the growing polypeptide. The A site holds the tRNA carrying the next amino acid to be added to the polypeptide chain.

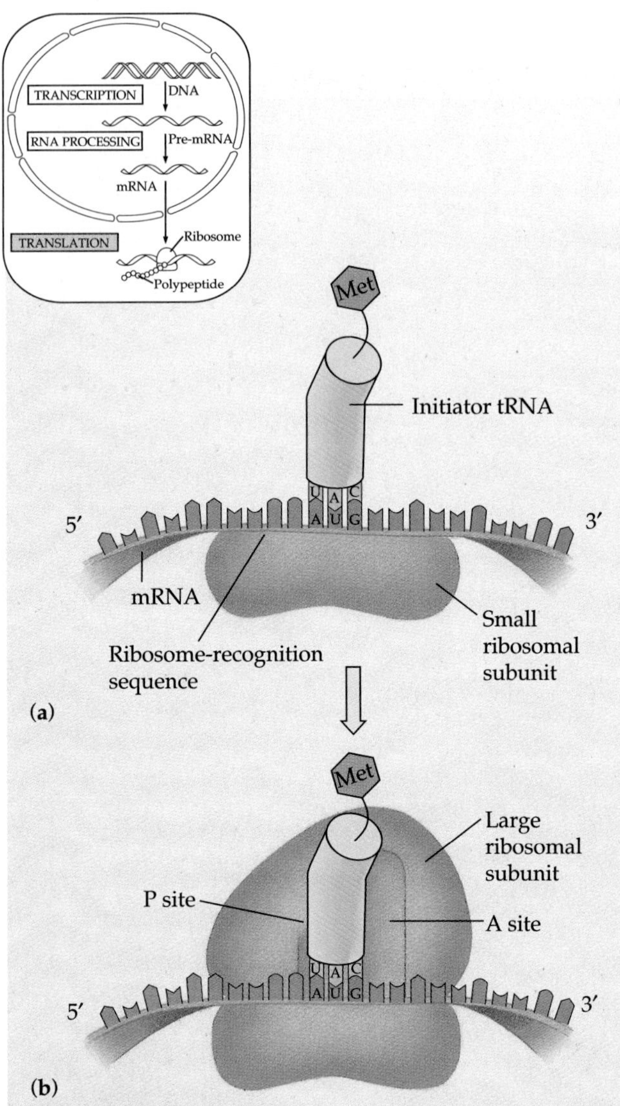

Figure 16.13
Initiation of translation. (**a**) A small ribosomal subunit binds to a molecule of mRNA. The positioning of the mRNA is signaled by a ribosome-recognition sequence on the mRNA that base pairs with rRNA. At the same time, the initiator tRNA, with the anticodon UAC, base pairs with the initiation codon AUG. This tRNA carries the amino acid methionine (Met). (**b**)The large ribosomal subunit joins the initiation complex. The initiator tRNA is in the P site. The A site is available to the tRNA bearing the next amino acid. Proteins called initiation factors are also required to bring these translation components together. GTP provides the energy for the initiation process.

ergy in the form of one GTP to form the initiation complex. At the completion of the initiation process, the initiator tRNA sits in the P site of the ribosome, and the vacant A site is ready for the next tRNA.

Elongation In the elongation stage of translation, amino acids are added one by one to the initial amino acid. Each addition, which involves the participation of several proteins called **elongation factors,** occurs in a three-step cycle (Figure 16.14):

1. *Codon recognition.* In the first step of elongation, the mRNA codon in the A site of the ribosome forms hydrogen bonds with the anticodon of an incoming mol-

TRANSCRIPTION
DNA
RNA PROCESSING
Pre-mRNA
mRNA
TRANSLATION
Ribosome
Polypeptide

Amino end
5′
3′
mRNA
P site
A site

Figure 16.14
The elongation cycle of translation.

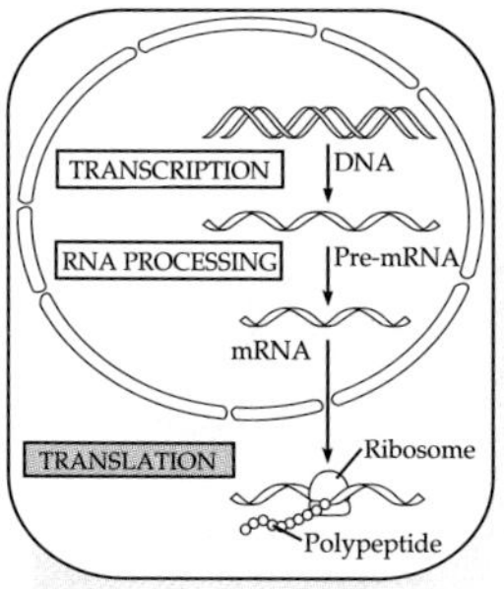

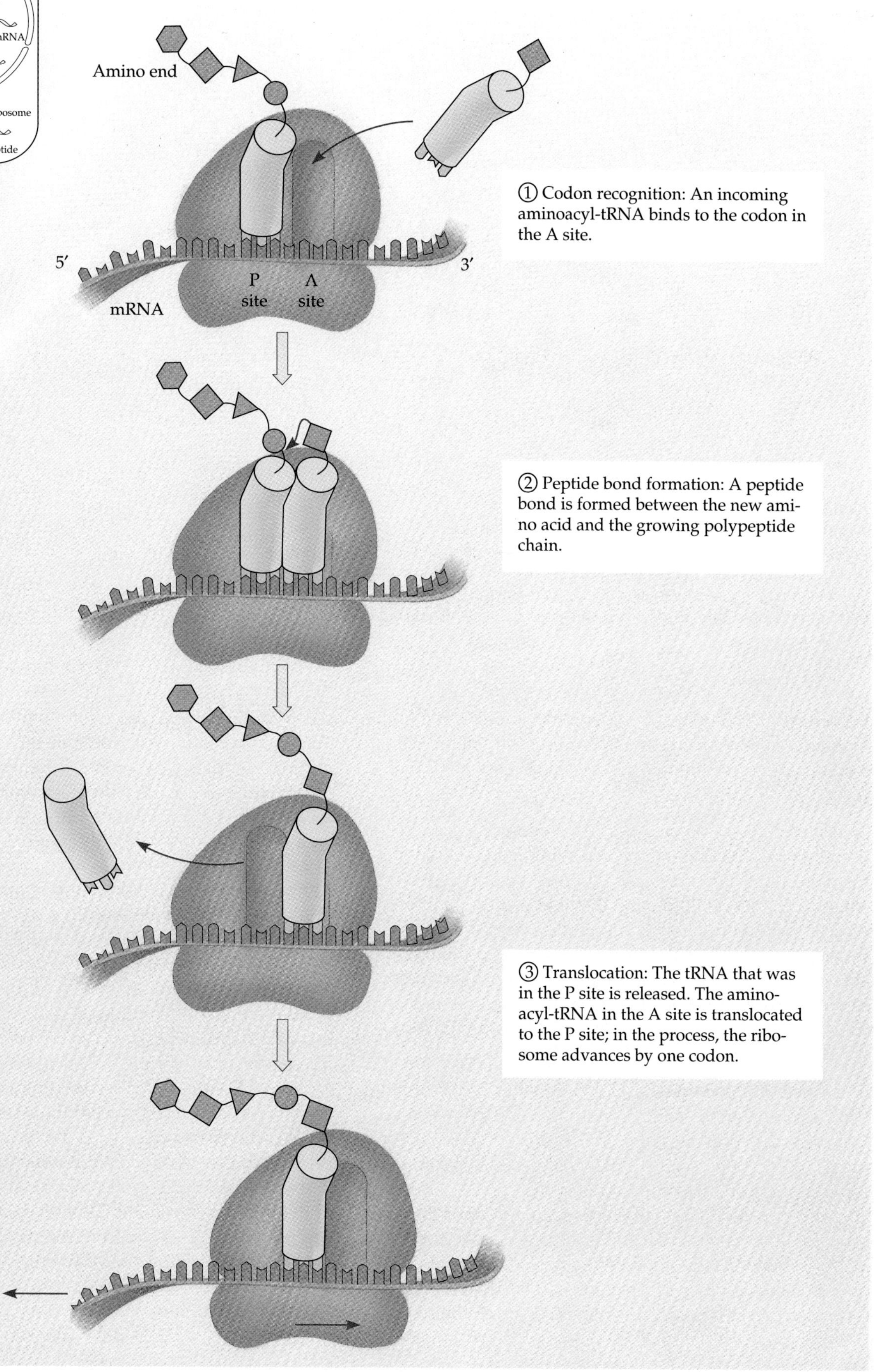

① Codon recognition: An incoming aminoacyl-tRNA binds to the codon in the A site.

② Peptide bond formation: A peptide bond is formed between the new amino acid and the growing polypeptide chain.

③ Translocation: The tRNA that was in the P site is released. The aminoacyl-tRNA in the A site is translocated to the P site; in the process, the ribosome advances by one codon.

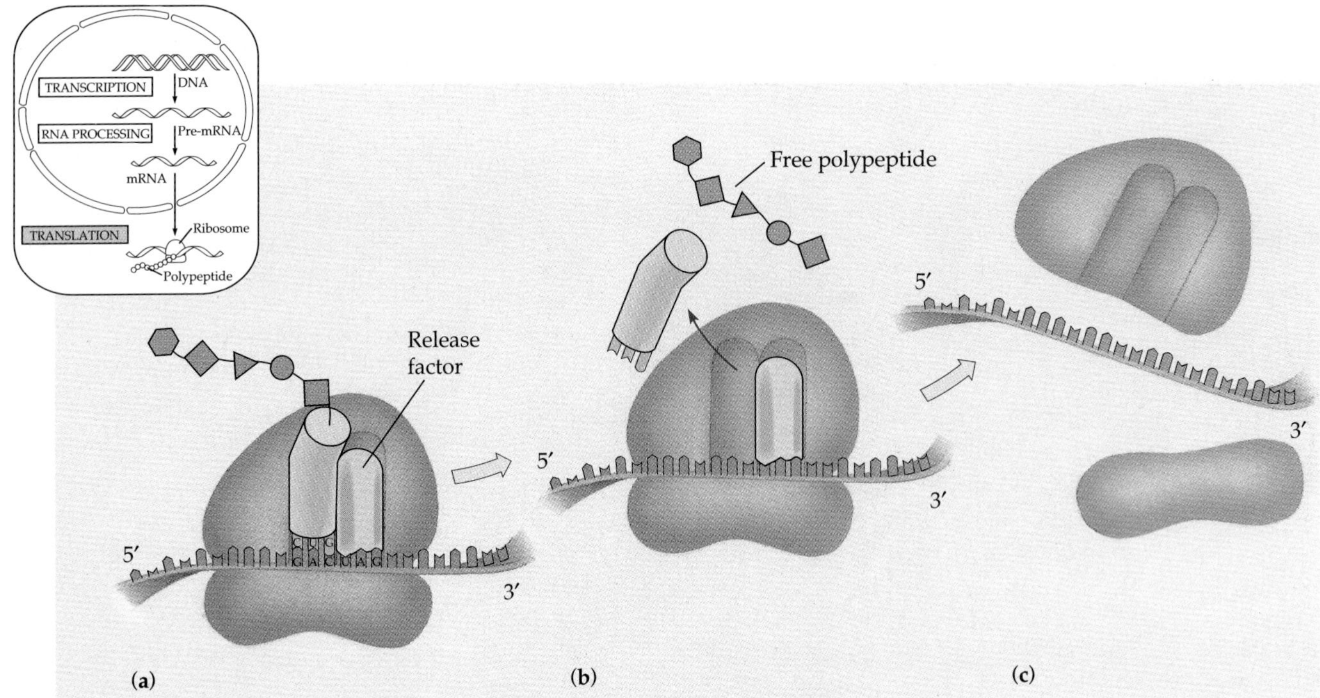

Figure 16.15
Termination of translation. (**a**) When a ribosome reaches a stop codon on a strand of mRNA, the A site of the ribosome accepts a protein called a release factor instead of tRNA. (**b**) The release factor hydrolyzes the bond between the tRNA in the P site and the last amino acid of the polypeptide chain. Both polypeptide and tRNA are then free to depart from the ribosome. (**c**) The two ribosomal subunits dissociate from the mRNA.

ecule of tRNA carrying its appropriate amino acid. An elongation factor ushers the tRNA into the A site. This step also requires the hydrolysis of a phosphate bond from GTP.

2. *Peptide bond formation.* Next, an enzyme that is an integral part of the large ribosomal subunit catalyzes the formation of a peptide bond between the polypeptide extending from the P site and the newly arrived amino acid in the A site. The enzyme, called **peptidyl transferase,** consists of a collection of ribosomal proteins in association with rRNA. In this step, the polypeptide separates from the tRNA to which it was bound and is transferred to the amino acid carried by the tRNA in the A site.

3. *Translocation.* The tRNA in the P site dissociates from the ribosome. The tRNA in the A site, now attached to the growing polypeptide, moves (is translocated) to the P site. As the tRNA changes sites, its anticodon remains hydrogen-bonded to the mRNA codon, allowing the mRNA and tRNA molecules to move as a unit. This movement, in turn, brings the next codon to be translated into the A site. The translocation step requires energy, which is provided by hydrolysis of a GTP molecule. The mRNA is moved through the ribosome in the 5′ ⟶ 3′ direction only, much as a ratchet allows a mechanical device to be turned in only one direction; or perhaps it is the *ribosome* that moves. The important point is that the ribosome and the mRNA move relative to each other, unidirectionally, codon by codon.

The elongation cycle takes only about 60 milliseconds and is repeated as each amino acid is added to the chain until the polypeptide is completed.

Termination The final stage of translation is termination (Figure 16.15). Elongation continues until a **termination codon** reaches the A site of the ribosome. These special base triplets—UAA, UAG, and UGA (see Figure 16.5)—do not code for amino acids but instead act as signals to stop translation. A protein called a **release factor** binds directly to the termination codon in the A site. The release factor causes the peptidyl transferase to add a water molecule instead of an amino acid to the polypeptide chain. This reaction hydrolyzes the completed polypeptide from the tRNA that is in the P site, thereby freeing the polypeptide from the ribosome. The ribosome then separates into its small and large subunits (Figure 16.15c).

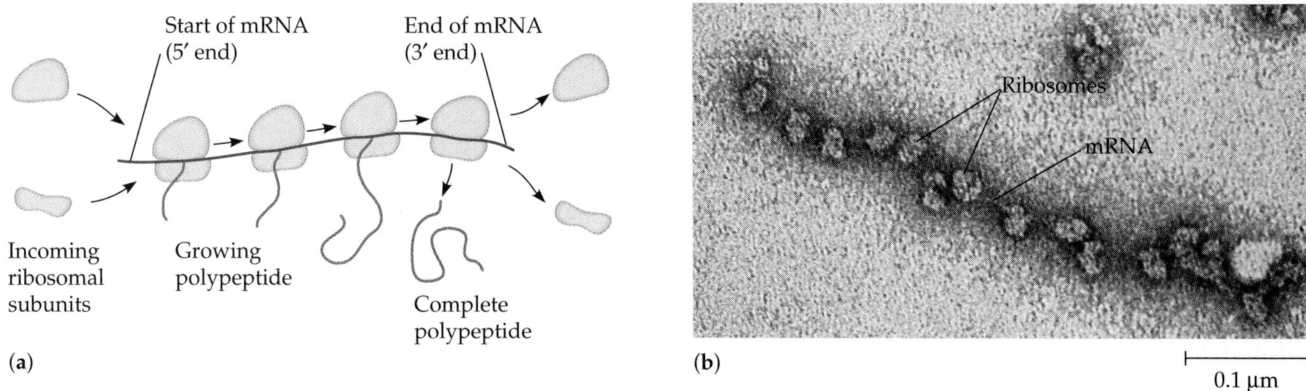

Figure 16.16
Polyribosomes. (**a**) An mRNA molecule is generally translated simultaneously by several ribosomes in clusters called polyribosomes. (**b**) This micrograph shows a large polyribosome in a prokaryotic cell (TEM).

Polyribosomes

A single ribosome can make an average-sized polypeptide in less than a minute. Typically, however, a single mRNA is used to make many copies of a polypeptide simultaneously, because several ribosomes work on translating the message at the same time. Once a ribosome moves past the initiation codon, a second ribosome can attach to it, and thus several ribosomes may trail along the same mRNA. Such clusters, called **polyribosomes,** can be seen with the electron microscope (Figure 16.16).

From Polypeptide to Functional Protein

During and after its synthesis, a polypeptide chain begins to coil and fold spontaneously to form a functional protein of specific conformation: a three-dimensional molecule with secondary and tertiary structures. A gene determines primary structure, and primary structure determines conformation (see Chapter 5).

Additional steps may be required before the protein can begin doing its particular job in the cell. Certain amino acids may be chemically modified by the attachment of sugars, lipids, phosphate groups, or other additives. Enzymes may remove one or more amino acids from the leading (amino) end of the polypeptide chain. In some cases, a single polypeptide chain may be enzymatically cleaved into two or more pieces. For example, the protein insulin is first synthesized as a polypeptide chain but becomes active only after an enzyme excises a central part of the chain, leaving a protein made up of two polypeptide chains connected by disulfide bridges. In other cases, two or more polypeptides that are synthesized separately may join to become the subunits of a protein that has quaternary structure. (See Chapter 5 to review the levels of protein structure.)

PROTEIN TARGETING

In electron micrographs of eukaryotic cells active in protein synthesis, two populations of ribosomes (and polyribosomes) are evident: free and bound (see Figure 7.12). Free ribosomes are suspended in the cytosol and mostly synthesize proteins that dissolve in the cytosol and function there. In contrast, bound ribosomes are attached to the cytoplasmic side of the endoplasmic reticulum. They make membrane proteins and proteins that are secreted from the cell. Insulin is an example of a secretory protein. The ribosomes themselves are identical and can switch their status from free to bound.

What determines whether a ribosome will be free in the cytosol or bound to rough ER at any particular time? The synthesis of all proteins begins in the cytosol when a ribosome starts to translate a messenger RNA molecule. The growing polypeptide chain itself cues the ribosome to either remain in the cytosol or attach to the ER. Secretory proteins are marked by a **signal sequence** of about 20 amino acids (Figure 16.17). This signal sequence, usually the first part of the polypeptide made, enables the ribosome to attach to a receptor site on the ER membrane. Synthesis of the protein continues there, and as the growing polypeptide snakes across the membrane into the cisternal space, the signal sequence is removed by an enzyme. In contrast, if the mRNA molecule lacks a segment that programs synthesis of the ER signal sequence, the ribosome translating that RNA remains free in the cytosol, where the fin-

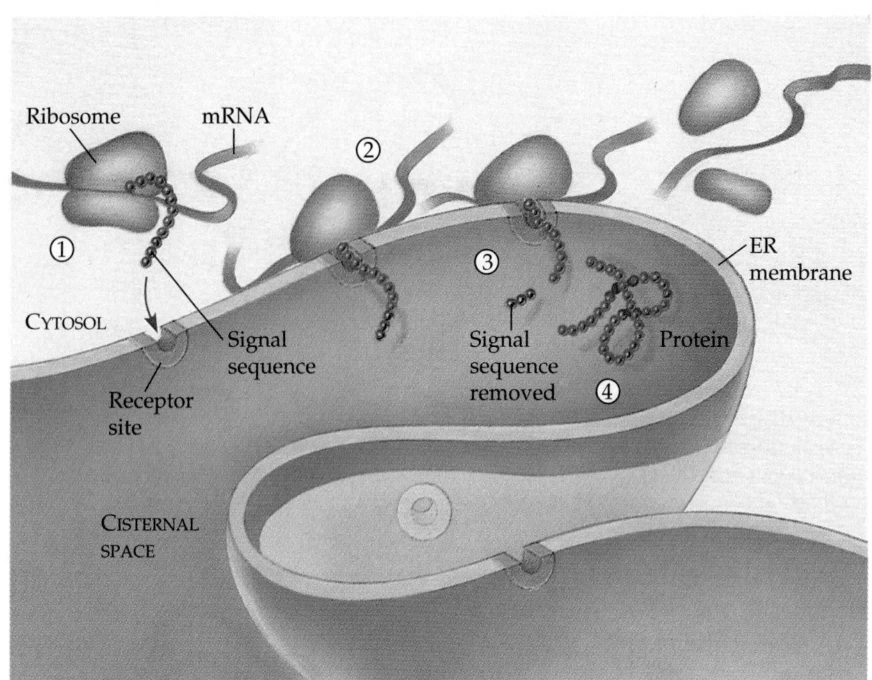

Figure 16.17
The signal mechanism for targeting proteins. Many polypeptide chains begin with a signal sequence, a stretch of amino acids that targets the protein for some organelle in the cell. ① In this example, a secretory protein, one that will eventually be secreted from the cell, is targeted for the endoplasmic reticulum (ER). The signal sequence causes the ribosome producing the polypeptide to bind to the cytoplasmic surface of rough ER. The signal is abbreviated here; it is actually about 20 amino acids long. ② The polypeptide threads through the ER membrane as synthesis continues. ③ The signal sequence is removed by an enzyme. ④ Released from the ribosome, the completed polypeptide folds into the conformation of a specific protein. Within the ER compartment, enzymes may modify the protein before it is exported to other destinations.

ished protein will be released. Thus, the function of the signal sequence is to dispatch proteins to their target: in this case, the ER.

The use of signal sequences to target secretory proteins to the ER is only one example of a general mechanism for dispatching proteins to specific sites. Other signal sequences target proteins for mitochondria or chloroplasts after the proteins are released from ribosomes. Signal sequences function like ZIP codes, addressing proteins to certain locations in the cell.

COMPARING PROTEIN SYNTHESIS IN PROKARYOTES AND EUKARYOTES: A REVIEW

Bacteria and eukaryotes carry out transcription and translation in very similar ways, although some details of each step in the overall process of protein synthesis differ. There are also differences in some of the basic equipment, such as RNA polymerases and ribosomes. But these distinctions are not essential for a basic understanding of the pathway from gene to protein. What is important to learn is the significance of a eukaryotic cell's compartmental organization of protein synthesis. This point was made earlier in the chapter (see Figure 16.3), but it is worth reinforcing now that we know more about transcription and translation. These two processes are coupled in prokaryotes. In fact, bacterial ribosomes may attach to a growing mRNA molecule, and translation may begin before transcription has even been completed (Figure 16.18).

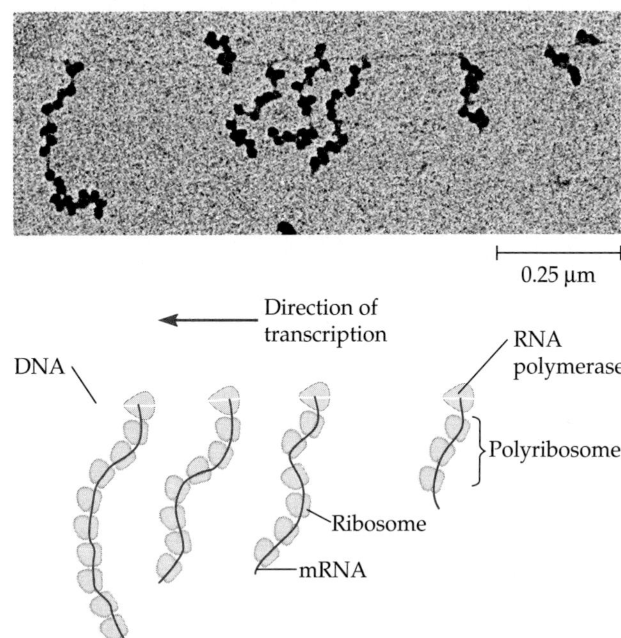

Figure 16.18
Coupled transcription and translation in bacteria. In bacterial cells, where transcription and translation are not segregated by a nuclear envelope, the translation of mRNA can begin as soon as the leading end (5′) of the mRNA molecule peels away from the DNA template. The micrograph shows a strand of *E. coli* DNA being transcribed by RNA polymerase molecules (TEM). Attached to each RNA polymerase molecule is a growing strand of mRNA, which is already being translated by ribosomes. The newly synthesized polypeptides are not visible here.

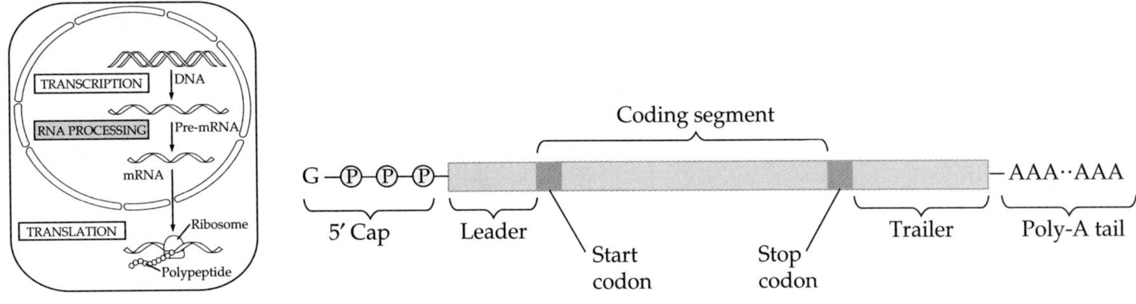

Figure 16.19
RNA processing: addition of the 5′ cap and poly-A tail. The two ends of an mRNA molecule are altered by enzymes after transcription. A cap consisting of a modified guanosine triphosphate is added to the 5′ end. A poly-A tail consisting of 150 or more adenine nucleotides is attached to the 3′ end of the mRNA. These modified ends probably help protect the mRNA from degradation. Also, the 5′ cap, along with a leader segment of RNA that is not translated into amino acids, functions as an "attach here" sign for ribosomes. The poly-A tail may be required for the mRNA to leave the nucleus. The tail is not atached directly to the stop codon, but to a trailer segment of RNA that is not translated.

In contrast, the nuclear envelope of a eukaryotic cell separates transcription from translation. This provides the time for RNA processing, an extra step between transcription and translation that does not occur in prokaryotes. Let's examine the mechanisms and functions of RNA processing in eukaryotic cells.

RNA PROCESSING IN EUKARYOTES

Enzymes in the eukaryotic nucleus modify mRNA in various ways before the genetic messages are dispatched to the cytoplasm. First, both ends of the mRNA are altered, then the molecule is cut apart and spliced together again.

Alteration of the Ends of mRNA

Figure 16.19 shows how the ends of an mRNA molecule are altered during RNA processing. The 5′ end, the end formed first during transcription, is capped off with a modified form of a guanine (G) nucleotide. This **5′ cap** has at least two important functions. First, it helps protect the mRNA from hydrolytic enzymes. Second, after the mRNA reaches the cytoplasm, the 5′ cap functions as an "attach here" sign for small ribosomal subunits. The other end of an mRNA molecule, the 3′ end, is also modified before the message exits the nucleus. To the 3′ end, which is synthesized last during transcription, an enzyme adds a **poly-A tail** consisting of 150 to 200 adenine nucleotides. Like the 5′ cap, the poly-A tail helps prevent degradation of the RNA. Addition of the tail may also play a regulatory role in protein synthesis by somehow facilitating the export of mRNA from the nucleus to the cytoplasm.

RNA Splicing

The most remarkable stage of RNA processing in the eukaryotic nucleus is the removal of a large portion of the molecule that is initially synthesized during transcription: a cut-and-paste job called **RNA splicing** (Figure 16.20). The average length of a transcription unit along a DNA molecule is about 8000 nucleotides, so the RNA product of transcription is also about that long. But it only takes about 1200 nucleotides to code for an average-sized protein of 400 amino acids. (Remember, each amino acid is coded for by a *triplet* of nucleotide bases.) This means that most eukaryotic genes and their RNA transcripts have long noncoding stretches of nucleotides—regions that are not translated. Even more surprising is that these noncoding sequences are interspersed between coding segments of the gene, and thus between coding segments of the mRNA transcript. In other words, the sequence of nucleotides that codes for a protein does not occur as an unbroken continuum. Interrupting the coding sequence of a gene are noncoding segments of DNA called intervening sequences, or **introns** for short. The coding regions are called **exons,** because they are eventually expressed (translated into protein).

Both introns and exons are transcribed to form an oversized RNA molecule. This pre-mRNA is more technically called **heterogeneous nuclear RNA (hnRNA),** a reference to the wide size range of these molecules. The hnRNA never leaves the nucleus; the mRNA molecule that enters the cytoplasm is an edited version of the original transcript. Enzymes excise the introns from the molecule and join the exons to form an mRNA molecule with a continuous coding sequence. This RNA splicing also occurs during the posttranscriptional processing of transfer RNA and ribosomal RNA.

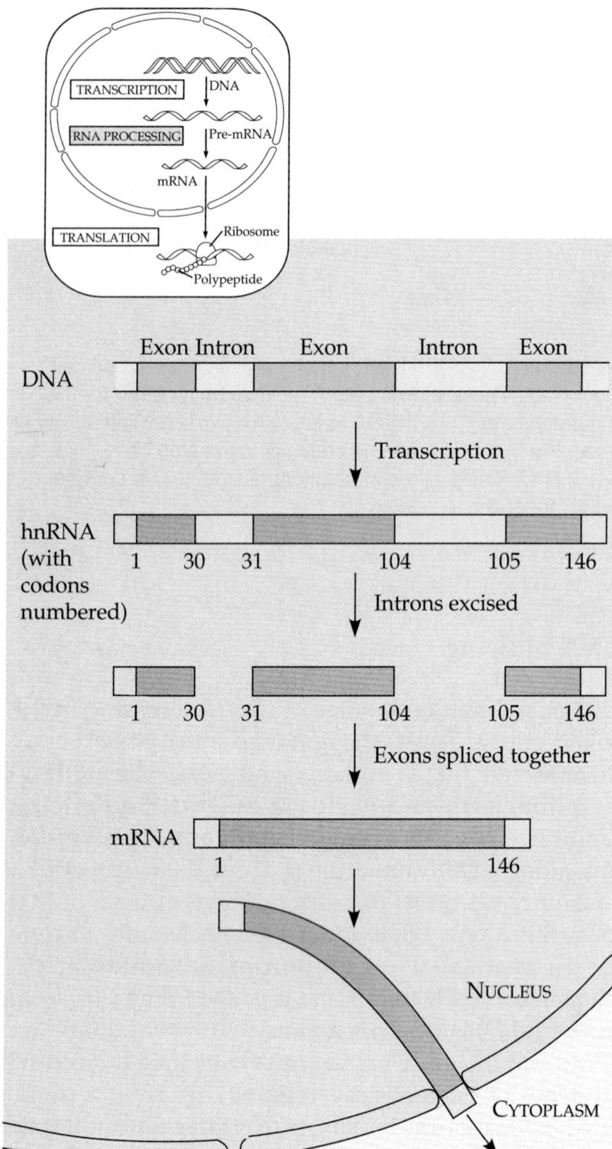

Figure 16.20
RNA processing: mRNA splicing. The gene depicted here codes for β-globin, one of the polypeptides of hemoglobin. β-globin is 146 amino acids long. Its gene has three segments containing coding regions, called exons, that are separated by noncoding introns. The entire gene is transcribed to form an hnRNA (heterogeneous nuclear RNA) molecule. However, before the molecule leaves the nucleus as mRNA, the introns are excised and the exons are spliced together. The mRNA also contains noncoding sequences at either end (see Figure 16.19).

The Mechanisms of RNA Splicing The details of RNA splicing are still being worked out. Biologists have learned that the signals for RNA splicing are sets of a few nucleotides located at either end of each intron. Particles called **small nuclear ribonucleoproteins,** or **snRNPs** (pronounced "snurps"), play a key

Table 16.1 The Major Types of Cellular RNA

Type of RNA	Prokaryotic or Eukaryotic Cells	Function
Messenger RNA (mRNA)	Both	Carries information specifying amino acid sequences of proteins from DNA to ribosomes.
Transfer RNA (tRNA)	Both	Serves as adaptor molecule in protein synthesis; translates mRNA nucleotide sequences into protein amino acid sequences.
Ribosomal RNA (rRNA)	Both	Plays structural and probably enzymatic roles in ribosomes, the sites of protein synthesis.
Small nuclear RNA (snRNA)	Eukaryotic	Plays structural and enzymatic roles in snRNP particles, which help carry out mRNA splicing within spliceosomes.

role in RNA splicing (Figure 16.21). As the name implies, these small particles are located in the cell nucleus and are composed of RNA and protein molecules. The RNA in a snRNP particle is called **small nuclear RNA (snRNA),** and it is typically a single molecule about 150 nucleotides long (about twice as long as a tRNA molecule but much shorter than mRNA). Each snRNP particle also has seven or more proteins. There appear to be a variety of snRNPs, and their functions have been only partly determined. The several kinds already known to be involved in RNA splicing carry out their roles as components of a larger, more complex assembly called a **spliceosome.** The spliceosome interacts with the ends of an RNA intron. It cuts at specific points to release the intron, then immediately joins the two exons that were adjacent to the intron.

The splicing of other kinds of RNA transcripts, such as those for tRNA and rRNA, occurs in several different ways. However, as with mRNA splicing, RNA is often involved in catalyzing the reactions. In some cases—for example, in the splicing of rRNA in the ciliated protozoan *Tetrahymena*—the splicing occurs completely without proteins or even extra RNA molecules: The intron RNA itself catalyzes the process! In these cases, RNA is acting as an enzyme.

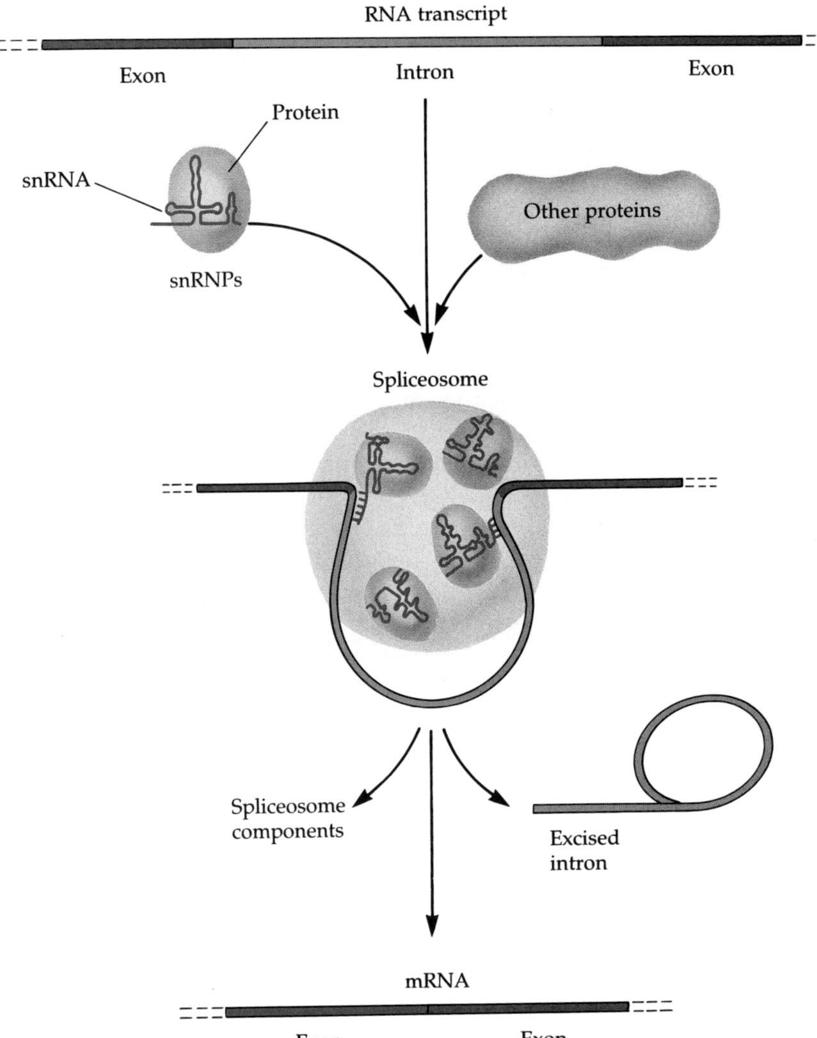

Figure 16.21
The roles of snRNPs and spliceosomes in mRNA splicing. After a eukaryotic gene containing exons and introns is transcribed, the RNA transcript combines with small nuclear ribonucleoproteins (snRNPs, or "snurps") and other proteins to form a molecular complex called a spliceosome. Within the spliceosome, the RNA of certain snRNPs base pairs with the ends of each intron, the RNA transcript is cut to release the intron, and the exons are spliced together. The spliceosome then comes apart, releasing mRNA, which now contains only exons.

The discovery of **ribozymes,** RNA molecules that function as enzymes, rendered the statement "All enzymes are proteins" obsolete. We have just seen that ribozymes catalyze reactions during RNA splicing. In addition, molecular biologists have recently learned that ribosomal RNA has enzymatic functions during the translation process. Table 16.1 summarizes the diverse roles of RNA molecules during protein synthesis. The ability of RNA to perform so many different functions is based on variations in the three-dimensional shapes of different types of RNA. DNA may be the genetic material of cells, but RNA is much more versatile. As you will learn in Chapter 17, many viruses even use RNA as their genetic material. We can add "All genes consist of DNA" and "All enzymes are proteins" to the list of obsolete ideas. Molecular biology has produced many reminders that the phrase "scientific dogma" is an oxymoron.

The Functional and Evolutionary Importance of Introns What are the biological functions of introns and gene splicing? One hypothesis is that introns play a regulatory role in the cell. Perhaps intron DNA includes sequences that control gene activity in some way, or perhaps the splicing process itself is part of a mechanism that regulates passage of mRNA from the nucleus to the cytoplasm. Introns may also enable different kinds of cells in the same organism to make different proteins from a common gene. This can occur if all introns are removed from a particular transcript in one cell type, but one or more of the introns are left in place in the same transcript in another cell type. The maintained introns are then translated along with the mRNA's exons to produce a protein different from the one formed by a cell type where all introns were removed.

Introns also play an important role in the evolution

of protein diversity. Many proteins have a modular architecture, consisting of structural and functional components called **domains**. One domain of an enzymatic protein, for instance, might include the active site, while another might attach the protein to some cellular membrane. In many cases, the exons of a "split gene" code for the different domains of a protein. It is possible for genetic recombination to modify the function of a protein by changing just one of its domains without altering the other domains. Because coding regions for a particular protein can be separated by considerable distances along the DNA, the frequency of recombination *within* a "split gene" can be higher than for a continuous coding region lacking introns. Thus, introns facilitate the recombination of exons between different alleles of a gene by increasing the probability that a crossover will switch one variation of an exon for another variation found on the homologous chromosome. This can lead to a novel protein with a single one of its multiple domains altered. The existence of genetic elements called transposons provides another mechanism for generating new combinations of exons. (Transposons are discussed further in Chapter 17.) Whether by recombination or by the action of transposons, the modular arrangement of many genes, with their axons separated by introns, makes it possible to shuffle preexisting DNA segments to produce new proteins.

Mutations in one or more of a gene's exons can also contribute to protein evolution. Mutation, the ultimate source of genetic diversity, is the topic of the next section.

MUTATIONS AND THEIR EFFECTS ON PROTEINS

Mutations are changes in the genetic makeup of a cell. In Chapter 14, we considered mutations that alter the structure of chromosomes. Now that you have learned about the genetic code and its translation, we can examine **point mutations,** which are chemical changes in just one nucleotide or a few nucleotides in a single gene.

If a point mutation occurs in a gamete, or in a cell that gives rise to gametes, it may be transmitted to offspring and to a succession of future generations. If the mutation has a noticeably adverse effect on the phenotype, then the mutant condition is referred to as a genetic disorder, or hereditary disease. For example, we can trace the genetic basis of sickle-cell anemia to a mutation affecting a single nucleotide in the gene that codes for one of the polypeptides of hemoglobin (see Chapter 5). Let's see how different types of point mutations translate into altered proteins.

Types of Mutations

Point mutations within a gene can be divided into two general categories: base-pair substitutions and base-pair insertions or deletions. While reading about how these mutations affect proteins, refer to the appropriate parts of Figure 16.22.

Substitutions A **base-pair substitution** is the replacement of one nucleotide and its partner from the complementary DNA strand with another pair of nucleotides. Some substitution mutations have no effect on the protein coded for, because of the redundancy of the genetic code. In other words, a change in a base pair may transform one codon into another that is translated into the same amino acid. For example, if CCG mutated to CCA, the mRNA codon that used to be GGC would become GGU, and a glycine would still be inserted at the proper location in the protein (see Figure 16.5). Other changes of a single nucleotide pair may alter an amino acid but have little effect on the encoded protein. The new amino acid may have properties similar to those of the amino acid it replaces, or may be in a portion of the protein where the exact sequence of amino acids is not essential to the protein's activity.

However, the base-pair substitutions of greatest interest are those that cause a readily detectable change in a protein. The alteration of a single amino acid in a crucial area of a protein—in the active site of an enzyme, for example—will significantly alter protein activity. Occasionally, such a mutation leads to an improved protein or one with novel capabilities that enhance the success of the mutant organism and its descendants. But much more often, such mutations are detrimental, creating a useless or less active protein that impairs cellular function.

Substitution mutations are usually **missense mutations;** that is, the altered codons still code for amino acids and thus make sense, although not necessarily the right sense. But if a point mutation changes a codon for an amino acid into a codon that signals termination, translation will be terminated prematurely, and the resulting polypeptide will be shorter than the polypeptide encoded by the normal gene. Alterations that change an amino acid codon to a stop signal are called **nonsense mutations,** and nearly all nonsense mutations lead to nonfunctional proteins.

Insertions and Deletions **Insertions** and **deletions** are additions or losses of one or more nucleotide pairs in a gene. These mutations usually have a more disastrous effect than substitutions on the resulting protein. Because mRNA is read as a series of nucleotide triplets during translation, the insertion or deletion of nucleotides may alter the reading frame (triplet group-

WILD TYPE

mRNA

Protein

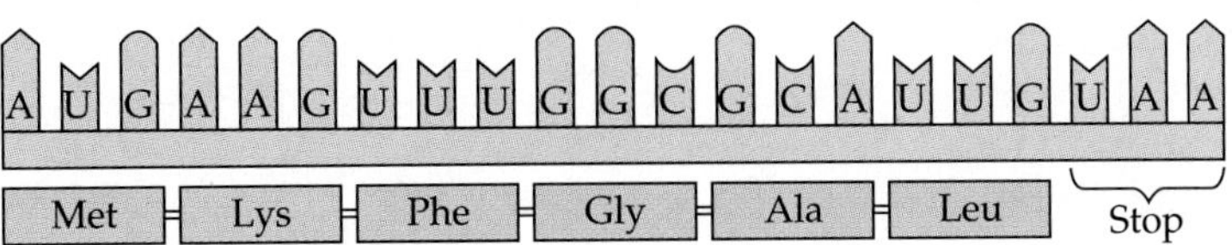

BASE-PAIR SUBSTITUTION

No effect on amino acid sequence

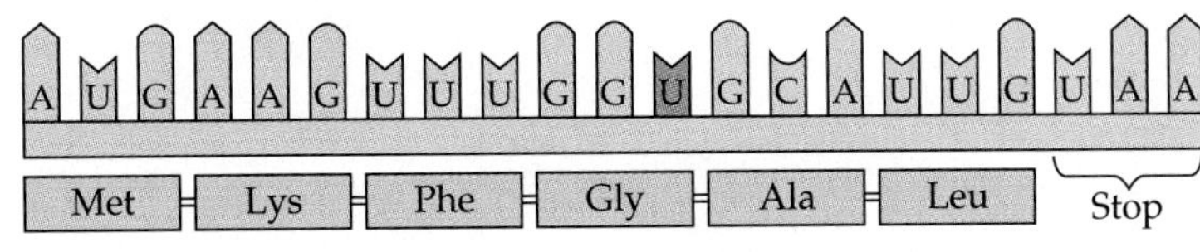

Missense

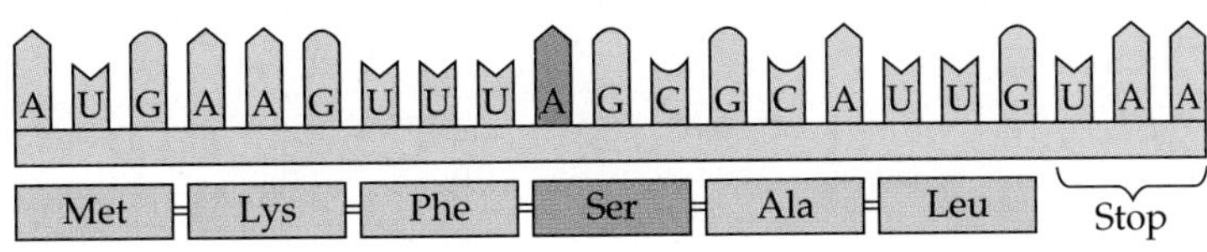

Nonsense

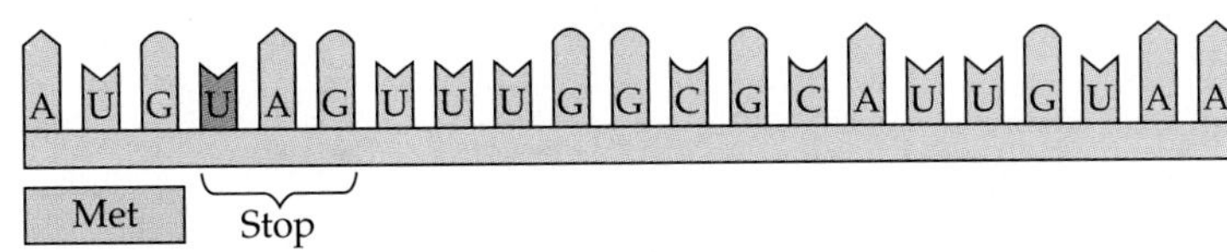

BASE-PAIR INSERTION OR DELETION

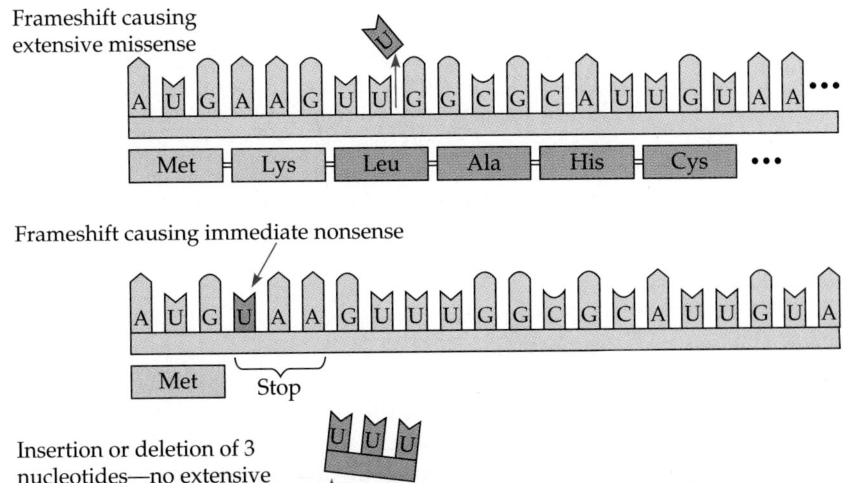

Figure 16.22
Categories and consequences of base-pair mutations. The two major categories of mutations that involve changes in only one or a few base pairs are base substitutions and base insertions or deletions. They are represented here as they are reflected in mRNA and its protein product. Base substitutions have a range of effects on the resulting protein, from no effect whatsoever to premature termination of the protein. Insertions or deletions usually result in frameshift mutations (shifting of the translation reading frame) and in extensive missense in the protein, eventually resulting in incorrect termination. The only situation where an insertion or deletion might not be extremely deleterious is if the number of nucleotides added or deleted is a multiple of 3, which would restore the correct reading frame at some point.

ing) of the genetic message. Such a mutation, called a **frameshift mutation,** will occur whenever the number of nucleotides inserted or deleted is not a multiple of 3. All the nucleotides that are downstream of the deletion or insertion will be improperly grouped into codons, and the result will be extensive missense ending sooner or later in nonsense—premature termination. Unless the frameshift is very near the end of the gene, it will produce a protein that is almost certain to be nonfunctional.

Mutagenesis

In the 1920s, Hermann Muller discovered that if he subjected fruit flies to X-rays, genetic changes increased in frequency. Using this method, Muller was able to obtain mutant *Drosophila* that he could use in his genetic studies. But he also recognized an alarming implication of his discovery: X-rays and other forms of radiation pose hereditary hazards to people as well as to laboratory animals. Since then, many other causes of mutations have been discovered.

Mutagenesis, the creation of mutations, can occur in a number of ways. Errors during DNA replication, repair, or recombination can lead to base-pair substitutions, insertions, or deletions. Mutations resulting from such errors are called **spontaneous mutations.**

A number of physical and chemical agents, called **mutagens,** interact with DNA to cause mutations. X-rays and ultraviolet (UV) light are examples of physical mutagens. In Chapter 15, we saw how the UV of sunlight can produce thymine dimers in DNA. Chemical mutagens fall into several categories, including base analogues, chemicals that are similar to normal DNA bases but that pair incorrectly (Figure 16.23). Researchers have developed various methods to test the mutagenic activity of different chemicals. One of the simplest and most popular methods is the Ames test (see the Methods Box).

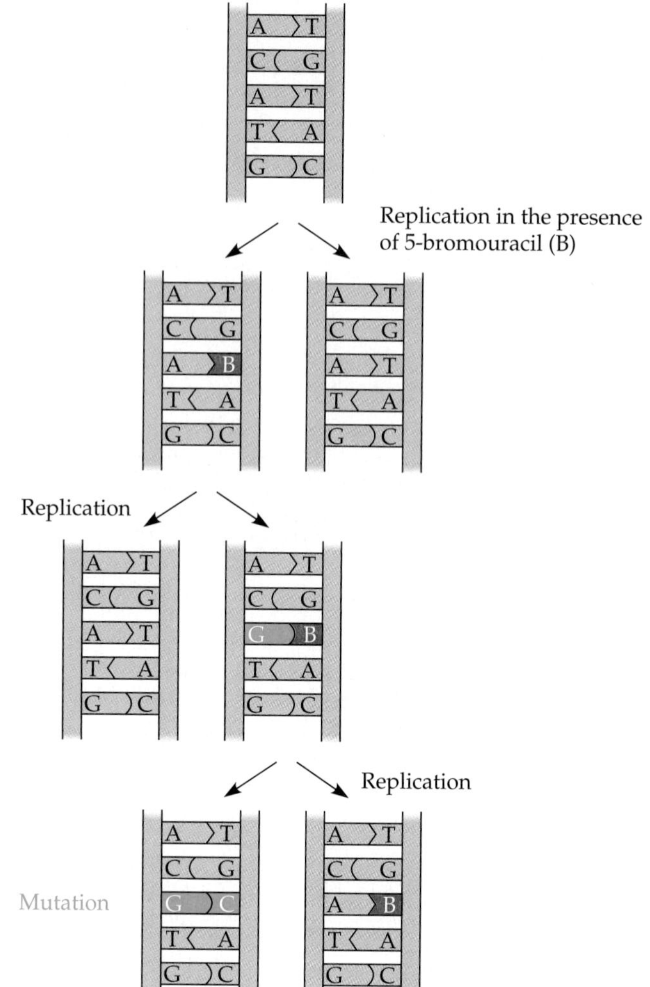

Figure 16.23
Base analogues as chemical mutagens. Bromouracil (B in this diagram) is an example of a modified nitrogenous base that acts as a mutagen. Bromouracil resembles the DNA base thymine, but a bromine atom replaces a $—CH_3$ group attached to thymine's pyrimidine ring (see Figure 15.4). By imitating thymine, bromouracil pairs with adenine during DNA replication. But after bromouracil is incorporated in DNA, it sometimes changes to an isomer that pairs better with G than with A. During the next DNA replication, a G will be substituted for an A in the strand that is complementary to the one containing the bromouracil. The mutagen may revert to the more common isomer, which pairs with A, but the damage has already been done. The mutation, which substituted a G-C pair for an A-T pair, will persist, generation after generation.

WHAT IS A GENE?

Our definition of a gene has evolved during the past few chapters. We began with the Mendelian concept of a gene as a discrete unit of inheritance that affects a phenotypic character (Chapter 13). We saw that Morgan and his colleagues assigned such genes to specific loci on chromosomes and that geneticists sometimes use the term *locus* as a synonym for gene (Chapter 14). We went on to view a gene as a region of specific nucleotide sequence along the length of a DNA molecule that includes thousands of genes (Chapter 15). Finally, in this chapter, we have moved toward a functional definition of a gene as a DNA sequence coding for a specific polypeptide chain. All these definitions are useful, depending on the context in which genes are being studied.

Even our one gene–one polypeptide definition must be refined and applied selectively. Most eukaryotic genes contain noncoding segments (introns), so large portions of these genes have no corresponding segments in polypeptides. Most molecular biologists also include promoters and other regulatory regions of DNA within the boundaries of a gene. These DNA sequences are not transcribed, but they can be considered part of the functional gene because they must be pre-

Methods: The Ames Test

The Ames test, named for its developer, microbiologist Bruce Ames, measures the mutagenic strength of various chemicals. The suspected mutagen is mixed with a culture of bacteria. It is also necessary to add a rat liver extract, which contains enzymes that convert certain chemicals from nonmutagenic to mutagenic forms. (These liver enzymes normally function to metabolize toxic substances, but unfortunately the chemical modification sometimes makes a bad situation worse by making the substances more mutagenic.) The bacteria, *Salmonella,* represent a mutant strain unable to produce the amino acid histidine. These bacteria fail to survive when plated on a growth medium lacking histidine. Some of the bacteria, however, undergo a back-mutation that restores the ability to make histidine. Such revertant bacteria give rise to colonies on the histidine-free medium. A mutagen will increase the frequency of these back-mutations and therefore result in more colonies on the histidine-free medium. The mutagenic activity of a chemical can be tested by comparing the number of colonies for samples treated with the chemical to control samples, which were not treated with the suspected mutagen. One application of the Ames test is the screening of chemicals to identify those that can cause cancer. This works because most carcinogens (cancer-causing chemicals) are mutagenic and, conversely, most mutagens are carcinogenic.

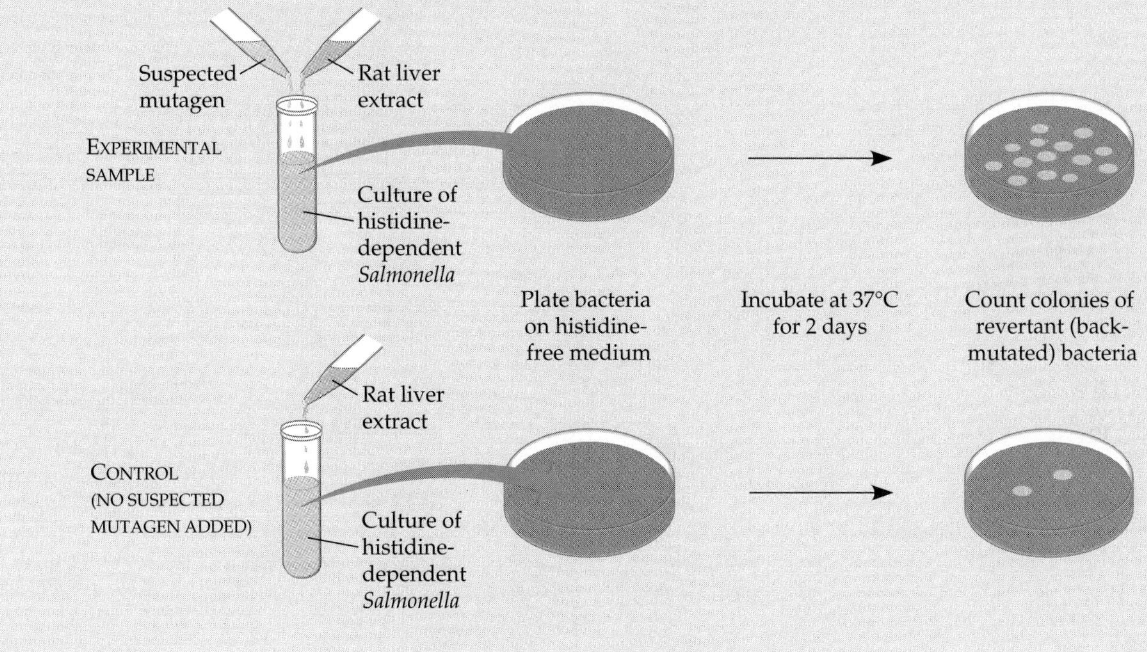

sent for transcription to occur. Our molecular definition of a gene must also be broad enough to include the DNA that codes for rRNA, tRNA, and snRNA. These genes have no polypeptide products. The following definition of a gene applies more generally at the molecular level than does the one gene–one polypeptide idea: A gene is a region of DNA that is required for the production of an RNA molecule.

* * *

We have seen in this chapter how genetic information is expressed by its translation into proteins of specific structure and function, which in turn bring about an organism's phenotype. Figure 16.24 summarizes this path from gene to protein.

Genes are subject to regulation. The control of gene expression enables a bacterium, for example, to vary the amounts of particular enzymes as the metabolic needs of the cell change. In eukaryotes, the control of gene expression makes it possible for cells with the same DNA to diverge during their development into different cells, such as muscle and nerve cells. The regulation of gene expression in eukaryotes is one of the topics of Chapter 18. In the next chapter, Chapter 17, we begin our discussion of gene regulation by focusing on the simpler biology of bacteria and viruses.

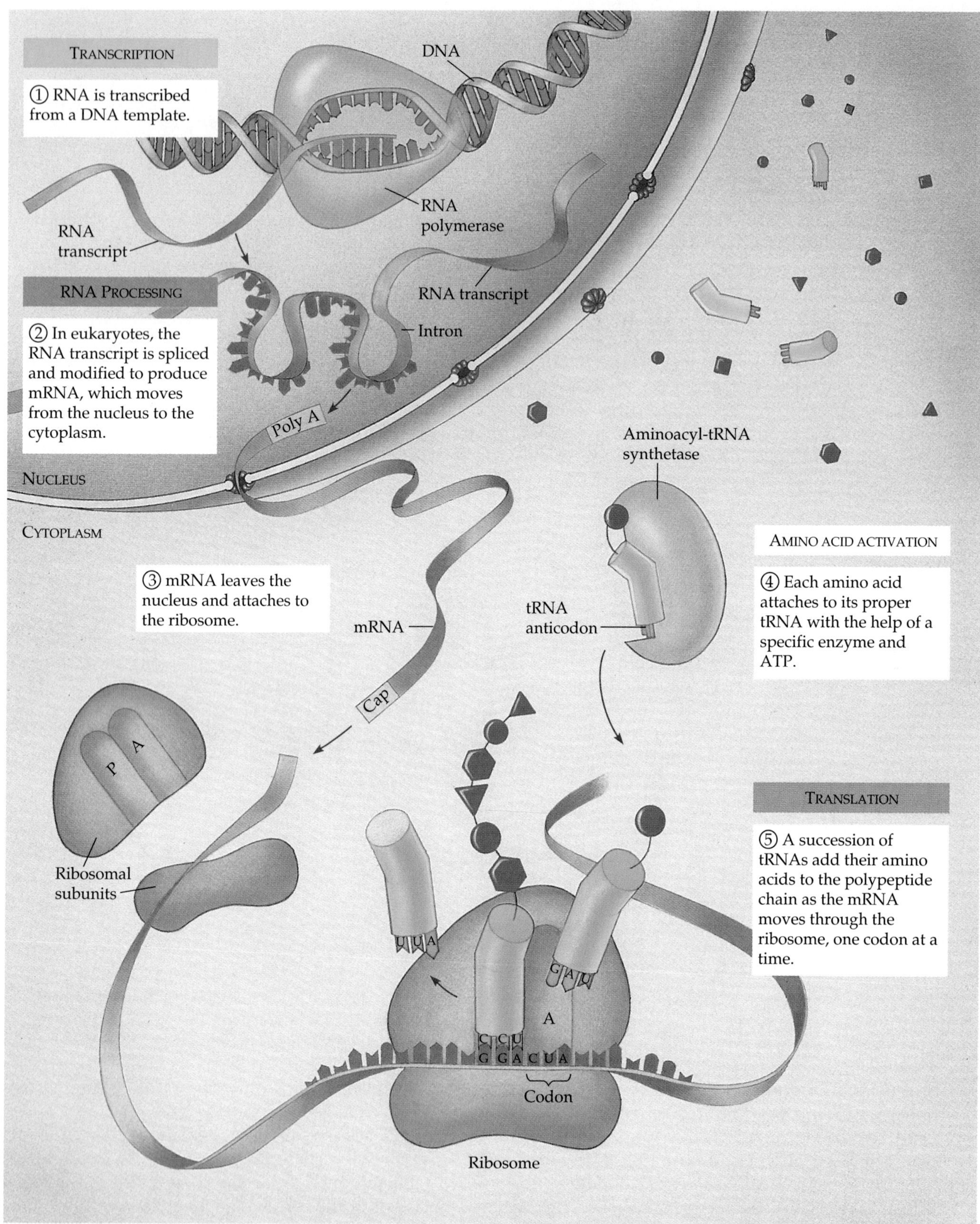

TRANSCRIPTION
① RNA is transcribed from a DNA template.
DNA
RNA polymerase
RNA transcript
RNA PROCESSING
② In eukaryotes, the RNA transcript is spliced and modified to produce mRNA, which moves from the nucleus to the cytoplasm.
RNA transcript
Intron
Poly A
NUCLEUS
CYTOPLASM
Aminoacyl-tRNA synthetase
AMINO ACID ACTIVATION
④ Each amino acid attaches to its proper tRNA with the help of a specific enzyme and ATP.
③ mRNA leaves the nucleus and attaches to the ribosome.
mRNA
tRNA anticodon
Cap
P
A
TRANSLATION
⑤ A succession of tRNAs add their amino acids to the polypeptide chain as the mRNA moves through the ribosome, one codon at a time.
Ribosomal subunits
U U A
G A U
A
C C U
G G A C U A
Codon
Ribosome

Figure 16.24
A summary of transcription and translation in eukaryotes. In general, the processes are similar in prokaryotic and eukaryotic cells. The major difference is the occurrence of RNA processing in the eukaryotic nucleus.

STUDY OUTLINE

1. DNA controls metabolism by commanding cells to make specific enzymes and other proteins.

Evidence That Genes Specify Proteins (pp. 316–317)

1. The early idea that inherited diseases were a result of "inborn errors of metabolism" was supported by data obtained by Beadle and Tatum on mutant strains of *Neurospora* bread mold. These classic experiments gave rise to the one gene–one enzyme hypothesis.
2. Beadle and Tatum's hypothesis was later modified to one gene–one polypeptide. The amino acid sequence of each polypeptide chain is determined by a gene.

An Overview of Protein Synthesis (pp 317–319)

1. Both nucleic acids and proteins are informational polymers assembled from linear sequences of nucleotides and amino acids, respectively.
2. Messenger RNA (mRNA) is the intermediate in the flow of information from DNA to proteins.
3. The nucleotide-to-nucleotide transfer of information from DNA to RNA is called transcription, whereas the informational transfer from RNA nucleotides to polypeptide amino acids is called translation. In eukaryotes, those two steps of protein synthesis are separated by the nuclear envelope.
4. Transfer RNA (tRNA) interprets the genetic code during translation. Each kind of tRNA brings a specific amino acid to ribosomes.

The Genetic Code (pp. 319–321)

1. Experiments have verified that genetic instructions from DNA are written in three-nucleotide units called codons, which indirectly specify particular amino acids.
2. Of the 64 codons, 61 code for amino acids. In most cases, 2 or more codons are synonyms for the same amino acid. A few codons function as start and stop signals that mark the ends of a genetic message.

The Evolutionary Significance of a Common Genetic Language (pp. 321–322)

1. Although there are interesting exceptions, nearly all organisms share a common genetic code. This suggests that the code had already evolved in ancestors common to all five kingdoms of life.

A Closer Look at Transcription (pp. 322–324)

1. RNA synthesis on a DNA template is catalyzed by RNA polymerase. It follows the base-pairing rules governing DNA replication, except that uracil substitutes for thymine.
2. Promoters, specific nucleotide sequences flanking the start of a gene, signal the initiation of mRNA synthesis. Transcription factors (proteins) help RNA polymerase recognize promoter sequences and bind to the DNA. Transcription continues until the RNA polymerase reaches the terminator sequence of nucleotides on the DNA template. As the mRNA peels away, the DNA double helix re-forms.

A Closer Look at Translation (pp. 325–331)

1. Transfer RNA (tRNA) molecules pick up specific amino acids and line up by means of their anticodon triplets at complementary codon sites on the mRNA molecule. Wobble is the ability of a single tRNA anticodon to bond mRNA codons that differ in their third base.
2. The binding of a specific amino acid to its particular tRNA is a precise, ATP-driven process catalyzed by a family of aminoacyl-tRNA synthetase enzymes.
3. Ribosomes coordinate the coupling of tRNAs to mRNA codons. They provide a site for the binding of mRNA, as well as P and A sites for holding adjacent tRNAs, as amino acids are linked in the growing polypeptide chain. Each ribosome is composed of two subunits made of aggregates of protein and ribosomal RNA (rRNA).
4. Initiation, the first of three stages of translation, is a complex process that requires energy from GTP and protein initiation factors. Initiation brings together the mRNA, the first amino acid attached to its tRNA, and the two subunits of the ribosome.
5. In the second stage, called the elongation cycle, amino acids are added one by one to the initial amino acid until the polypeptide chain is completed. This GTP-requiring stage includes the binding of the incoming tRNA to the A site, peptide bond formation, and translocation of the tRNAs and mRNA along the ribosome.
6. Termination, the final stage, occurs when one of three special termination codons reaches the A site of the ribosome, triggering the action of a protein release factor. This release factor causes the freeing of the polypeptide chain and the dissociation of the ribosomal subunits.
7. Several ribosomes often read a single mRNA, forming polyribosome clusters.
8. A protein often undergoes one or more alterations during and after translation that affect its three-dimensional structure and hence its final activity in the cell.

Protein Targeting (pp. 331–332)

1. In eukaryotic cells, proteins destined for membranes or for export from the cell are synthesized on ribosomes bound to the endoplasmic reticulum. A signal sequence on the leading end of the growing polypeptide enables the ribosome to bind to the ER. Proteins that will remain in the cytosol lack this signal and are manufactured on free ribosomes.

Comparing Protein Synthesis in Prokaryotes and Eukaryotes: A Review (pp. 332–333)

1. Prokaryotic cells and eukaryotic cells use the same basic steps for protein synthesis, but the ribosomes and many of the enzymes vary between these two major cell types.
2. In comparing protein synthesis in prokaryotes and eukaryotes, the most important difference is in the spatial and temporal relationships of transcription and translation. In a bacterial cell, which lacks a nuclear envelope, translation of an mRNA can begin while transcription is still in progress. In a eukaryotic cell, the nuclear envelope

separates transcription from translation, making RNA processing possible.

RNA Processing in Eukaryotes (pp. 333–336)

1. Eukaryotic mRNA molecules are processed before leaving the nucleus by modification of their ends and by RNA splicing.
2. The mRNA molecule receives a modified guanosine triphosphate cap at the 5′ end and a poly-A tail of nucleotides at the 3′ end. The cap and tail probably protect the molecule from degradation and enhance translation.
3. Most eukaryotic genes are interrupted by long noncoding regions, called introns, interspersed among coding regions, known as exons. RNA splicing involves removing the introns and joining the exons.
4. RNA splicing is catalyzed by small nuclear ribonucleoproteins (snRNPs), consisting of small nuclear RNA (snRNA) and proteins; they operate within larger assemblies called spliceosomes.
5. In some cases, only RNA is needed to catalyze RNA splicing. Enzymatic RNA molecules are called ribozymes.
6. Introns may play some regulatory role in the expression of genes. The shuffling of exons due to recombination contributes to the evolution of protein diversity.

Mutations and Their Effects on Proteins (pp. 336–338)

1. Point mutations involve changes in a single nucleotide pair.
2. Base-pair substitutions within a gene have a variable effect, depending on whether or not an amino acid is actually altered, and if so, whether the alteration has any effect on the function of the protein. Many substitutions are detrimental, causing missense or nonsense mutations.
3. Base-pair insertions or deletions are almost always disastrous, often resulting in frameshift mutations that disrupt the codon messages downstream of the insertion or deletion.
4. Spontaneous mutations can occur during DNA replication or repair. In addition, various chemical and physical mutagens can affect the gene.

What Is a Gene? (pp. 338–340)

1. Different definitions of a gene suit different situations. At the molecular level, a gene can be defined as a region of DNA required for the production of one RNA molecule.

SELF-QUIZ

1. The creation of an RNA molecule from a section of DNA is known as
   a. transcription
   b. translation
   c. RNA splicing
   d. replication
   e. recombination
2. Which of the following is *not* true of a codon?
   a. It consists of three nucleotides.
   b. It may code for the same amino acid as another codon does.
   c. It never codes for more than one amino acid.
   d. It extends from one end of a tRNA molecule.
   e. It is the basic unit of the genetic code.
3. Which of the following statements about RNA polymerase is correct?
   a. It functions in translation.
   b. It transcribes both introns and exons.
   c. It creates hydrogen bonds between nucleotides on the DNA strand and their complementary RNA nucleotides.
   d. It starts transcribing at an AUG triplet on one DNA strand.
   e. It can produce several polypeptide chains at one time through the creation of polyribosomes.
4. Beadle and Tatum discovered several classes of *Neurospora* mutants that were able to grow on minimal medium with arginine added. Class I mutants were also able to grow on medium supplemented with either ornithine or citrulline, whereas class II mutants could grow on citrulline medium but not on ornithine medium. The metabolic pathway of arginine synthesis is as follows:

   $$\text{Precursor} \xrightarrow[A]{} \text{Ornithine} \xrightarrow[B]{} \text{Citrulline} \xrightarrow[C]{} \text{Arginine}$$

   From these growth results, they could conclude that
   a. one gene codes for the entire metabolic pathway
   b. the genetic code of DNA is a triplet code
   c. class I mutants have their mutations later in the nucleotide chain than do class II mutants, and thus have more functional enzymes
   d. class I mutants have a nonfunctional enzyme at step A, and class II mutants have a nonfunctional enzyme at step B
   e. class I mutants have a nonfunctional enzyme at step B, and class II mutants have a nonfunctional enzyme at step C
5. The bonds between the anticodon of a tRNA molecule and the complementary codon of mRNA are
   a. catalyzed by peptidyl transferase
   b. formed by the input of energy from ATP
   c. hydrogen bonds that form while the codon is in the A site
   d. catalyzed by aminoacyl-tRNA synthetase
   e. covalent bonds formed with energy from GTP
6. The phenomenon known as wobble refers to
   a. the movement of a tRNA from the A site to the P site
   b. the ability of DNA to make more than one type of RNA
   c. the ability of a tRNA to pair with codons that may differ in the third base
   d. the shifting of the reading frame in a deletion or insertion mutation
   e. the movement of multiple ribosomes along the same mRNA
7. Which of the following is *not* true of RNA processing?
   a. Exons are excised and hydrolyzed before mRNA moves out of the nucleus.
   b. The existence of exons and introns may facilitate

crossing over between regions of a gene that code for polypeptide domains.

c. Ribozymes function in RNA splicing.

d. RNA splicing may be catalyzed by spliceosomes.

e. An initial RNA transcript is much longer than the final RNA molecule that may leave the nucleus.

8. Using the genetic code in Figure 16.5, identify a possible sequence of nucleotides in the DNA that would code for the polypeptide sequence Phe-Pro-Lys.

a. AAA-GGG-UUU

b. TTC-CCC-AAG

c. TTT-CCA-AAA

d. AAG-GGC-TTC

e. UUU-CCC-AAA

9. Which of the following mutations would be *most* likely to have a harmful effect on an organism?

a. a base-pair substitution

b. a deletion of three consecutive bases near the middle of the gene

c. a single base deletion near the middle of an intron

d. a single base deletion close to the end of the coding sequence

e. a single base insertion near the start of the coding sequence

10. Which component is *not directly* involved in the process known as translation?

a. mRNA

b. DNA

c. tRNA

d. ribosomes

e. GTP

CHALLENGE QUESTIONS

1. A biologist inserted a gene from a human liver cell into the chromosome of a bacterium. The bacterium then transcribed this gene into mRNA and translated the mRNA into protein. The protein produced was useless; it contained many more amino acids than the protein made by the eukaryotic cell, and the amino acids were in a different sequence. Explain why.

2. The base sequence of the gene coding for a short polypeptide is CTACGCTAGGCGATTATC. What would be the base sequence of the mRNA transcribed from this gene? Using the genetic code chart (Figure 16.5), give the amino acid sequence of the polypeptide translated from this mRNA.

SCIENCE, TECHNOLOGY, AND SOCIETY

1. As part of the Human Genome Project (discussed in Chapter 19), researchers are determining the nucleotide sequences of human genes and identifying the proteins coded by the genes. Labs of the U.S. National Institutes of Health (NIH), for example, have worked out thousands of sequences, and similar analysis is being carried out by many private companies. Knowing the nucleotide sequence of a gene and identifying its product can be useful; this information might be used to treat genetic defects or produce life-saving medicines. U.S. law allows the first person to isolate a pure protein or a gene to patent it, whether or not a practical use for the discovery has been demonstrated. The NIH and biotechnology companies have applied for patents on their discoveries. What are the purposes of a patent? How might the discoverer of a gene benefit from a patent? How might the public benefit? What kinds of negative impacts might result from patenting genes? Do you think individuals and companies should be able to patent genes and gene products? Why or why not? Under what conditions should such patenting be permitted?

FURTHER READING

Darnell, J. E., Jr. "RNA." *Scientific American*, October 1985. A clear presentation of the role of RNA in protein synthesis and its relationship to DNA.

Hoffman, M. "RNA Editing: What's in a Mechanism?" *Science*, July 12, 1991. Ribozymes can correct errors in genetic messages.

Moore, P. B. "The Ribosome Returns." *Nature* 331 (1988):223–227. A refreshingly written overview of ribosome biochemistry and the recent reawakening of interest in these organelles.

Radetsky, P. "Genetic Heretic." *Discover*, November 1990. The story of how the discovery of ribozymes led to a Nobel Prize.

Steitz, J. A. "'Snurps.'" *Scientific American*, June 1988. The story of the discovery of snRNPs and the splicing of eukaryotic mRNA.

Waldrop, M. "The Structure of the Second Genetic Code." *Science*, December 1, 1989. How enzymes match tRNAs and amino acids.

Watson, J. D., N. H. Hopkins, J. W. Roberts, J. A. Steitz, and A. M. Weiner. *Molecular Biology of the Gene*, 4th ed. Menlo Park, CA: Benjamin/Cummings, 1987. Chapters 13–16 and 20.

17 Microbial Models: The Genetics of Viruses and Bacteria

The Discovery of Viruses
Viral Structure and Replication: An Overview
How Viruses Replicate: Viral Infection
Bacterial Viruses
Animal Viruses
Plant Viruses and Viroids
The Evolutionary Origin of Viruses
Bacterial Genomes: Replication and Mutation
Genetic Recombination and Gene Transfer in Bacteria
The Control of Gene Expression in Prokaryotes

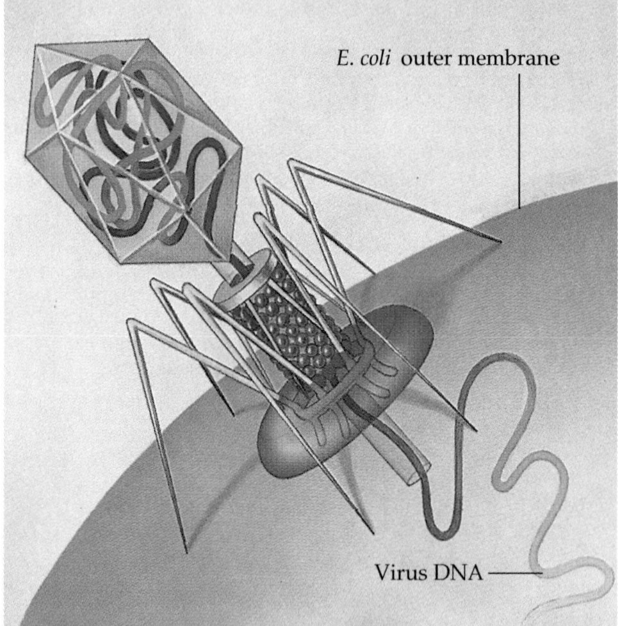

Figure 17.1
A virus shooting its DNA into a bacterium. This drawing shows the first step of one of the most remarkable events in biology: the genetic takeover of a cell by a virus. Infection begins when the virus injects its DNA into the host cell. In this case, the cell is the bacterium *E. coli* and the virus is T4. By studying viruses and bacteria, the simplest biological systems, biologists caught their first glimpses of the elegant molecular mechanisms of heredity. In this chapter, you will learn more about the genetics of these microbes.

Molecular biology was born in the laboratories of microbiologists studying viruses and bacteria. Microbiologists were the ones who provided most of the evidence that DNA is the genetic material (see Chapter 15). And microbiologists were the ones who outlined the major steps of replication, transcription, and translation, the three key processes in the flow of genetic information. Viruses and bacteria are the simplest biological systems—microbial models where scientists find life's fundamental molecular mechanisms in their most basic, accessible forms. Great breakthroughs in biology have often involved scientists who were clever enough or lucky enough to match their questions with appropriate experimental organisms (Figure 17.1).

The value of viruses and bacteria as model systems in biological research is just one reason for beginning students to learn about these microbes. While microbial models have helped biologists understand the molecular genetics of more complex organisms, viruses and bacteria also have unique features that make microbial genetics interesting in its own right. These specialized mechanisms have important applications for understanding how viruses and bacteria cause disease. In addition, new techniques enabling scientists to manipulate genes and transfer them from one organism to another have emerged from the study of the genetic mechanisms of microorganisms. These techniques are having an important impact on both basic research and biotechnology (see Chapter 19).

In this chapter, we will explore the genetics of viruses and bacteria. Recall that bacteria are prokaryotic organisms, characterized by cells much more simply organized than those of eukaryotes, such as plants and animals. Viruses are simpler still, lacking the structures and most of the metabolic machinery found in cells. In fact, most viruses are little more than aggregates of nucleic acids and proteins—genes packaged in protein coats. It is with these simplest of all genetic systems that we begin.

THE DISCOVERY OF VIRUSES

Microbiologists were able to observe viruses indirectly long before they were actually able to see them. The

Figure 17.2
Tobacco mosaic disease. The disease stunts the growth of tobacco plants and produces a mottled (mosaic) discoloration of the leaves. The infectious agent that causes the disease, tobacco mosaic virus (TMV), was the first virus to be discovered.

story of how viruses were discovered begins in 1883 with A. Mayer, a German scientist who was seeking the cause of tobacco mosaic disease. This disease stunts the growth of tobacco plants and gives their leaves a mottled, or mosaic, coloration (Figure 17.2). Mayer discovered that the disease was contagious when he found he could transmit it from plant to plant by spraying sap extracted from diseased leaves onto healthy plants. He searched for a microbe in the infectious sap, but found none. Mayer concluded that the disease was caused by unusually small bacteria that could not be seen with the microscope. This hypothesis was tested a decade later by D. Ivanowsky, a Russian, who passed sap from infected tobacco leaves through a filter designed to remove bacteria. But even after filtering, the sap produced mosaic disease.

Ivanowsky clung to the hypothesis that bacteria caused tobacco mosaic disease. Perhaps, he reasoned, the pathogenic bacteria were so small they could pass through the filter. Or perhaps the bacteria made a filterable toxin that caused the disease. This latter possibility was ruled out in 1897 when Dutch microbiologist M. Beijerinck discovered that the infectious agent in the filtered sap could reproduce. Beijerinck sprayed plants with the filtered sap, and after these plants developed mosaic disease, he used their sap to infect more plants, continuing this process through a series of infections. The pathogen must have been reproducing, for its ability to cause disease was undiluted after several transfers from plant to plant.

In fact, the pathogen could reproduce only within the host it infected. Unlike bacteria, the mysterious agent of mosaic disease could not be cultivated on nutrient media in test tubes or Petri dishes. Also, the pathogen was not inactivated by alcohol, which is generally lethal to bacteria. Beijerinck imagined a reproducing particle much smaller and simpler than bacteria. His suspicions were confirmed in 1935, when American scientist Wendell Stanley crystallized the infectious particle, now known as tobacco mosaic virus (TMV). Subsequently, TMV and many other viruses were actually seen with the help of the electron microscope.

VIRAL STRUCTURE AND REPLICATION: AN OVERVIEW

The tiniest viruses are only 20 nm in diameter—smaller than a ribosome. Millions could easily fit on a pinhead, and even the largest viruses can barely be resolved with the light microscope. Stanley's discovery that viruses could be crystallized was exciting and puzzling news. Not even the simplest of cells can aggregate into regular crystals. But if viruses are not cells, then what are they? They are infectious particles consisting only of the viral genes enclosed in a shell made of proteins. Let's examine these two components of viruses more closely, then take a preliminary look at how they function in viral replication.

Viral Genomes

We usually think of genes as being made of double-stranded DNA—the conventional double helix—but viruses often defy this convention. Their genomes (sets of genes) may consist of double-stranded DNA, single-stranded DNA, double-stranded RNA, or single-stranded RNA, depending on the specific type of virus. A virus is called a DNA virus or an RNA virus, according to the type of nucleic acid that makes up its genome. In either case, the genome is usually organized as a single linear or circular molecule of nucleic acid. The smallest viruses have only four genes, while the largest have several hundred.

Capsids and Envelopes

The protein shell that encloses the viral genome is called a **capsid** and may be rod-shaped (more precisely, helical), polyhedral, or more complex in shape. Capsids are built from a large number of protein subunits called capsomeres, but the number of different *kinds* of proteins is usually small. Tobacco mosaic virus, for example, has a rigid, rod-shaped capsid made from over a thousand molecules of a single type of protein (Figure 17.3a). Adenoviruses, which infect the respiratory tracts of animals, have 252 identical protein molecules arranged into a polyhedral capsid

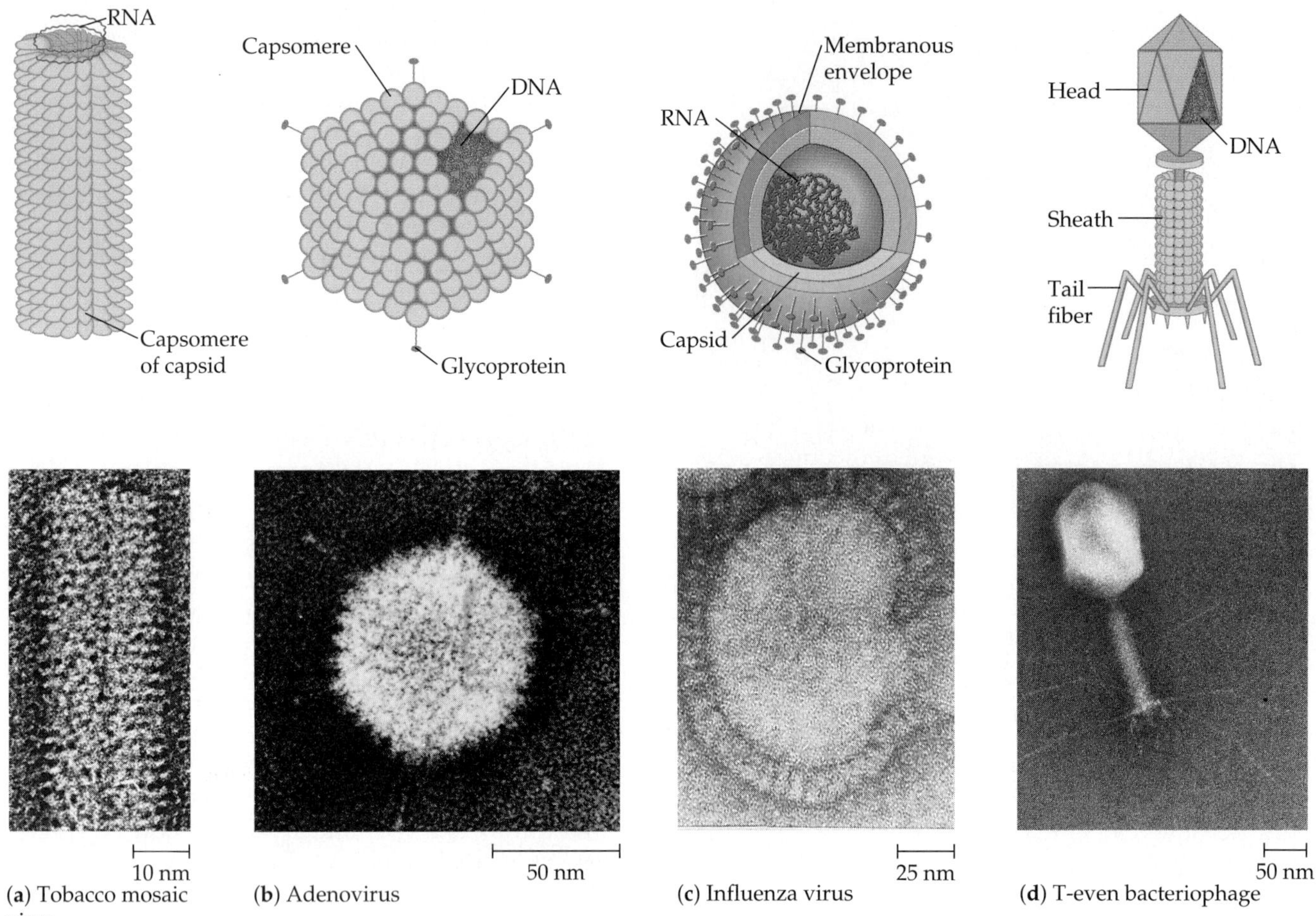

(**a**) Tobacco mosaic virus (**b**) Adenovirus (**c**) Influenza virus (**d**) T-even bacteriophage

Figure 17.3
Viral structure. Viruses are made up of nucleic acid (DNA or RNA) enclosed in a protein coat (the capsid) and sometimes further wrapped in a membranous envelope. The individual protein subunits making up the capsid are called capsomeres. Although viruses are diverse in size and shape, there are common structural motifs, most of which appear in the four examples shown here (TEMs). (**a**) Tobacco mosaic virus has a helical capsid with the overall shape of a rigid rod. (**b**) Adenovirus has a polyhedral capsid with a protein spike at each vertex. Some adenoviruses cause upper respiratory infections in humans. (**c**) Influenza virus has a flexible helical capsid and an outer membranous envelope studded with glycoprotein spikes. (**d**) Phages are viruses that infect bacteria. A T-even phage, such as T4, has a complex capsid consisting of a polyhedral head and a tail apparatus. DNA is stored in the head, and the tail apparatus functions in the injection of this DNA into a bacterium (see Figure 17.1).

with 20 triangular facets—an icosahedron (Figure 17.3b).

Some viruses have accessory structures that help them infect their hosts. Influenza viruses, and many other viruses found in animals, have membranous **envelopes** cloaking their capsids (Figure 17.3c). These envelopes are derived from membrane of the host cell, but in addition to host cell phospholipids and proteins, they also contain proteins and glycoproteins (proteins with carbohydrate covalently attached) of viral origin.

The most complex capsids are found among viruses that infect bacteria. Bacterial viruses are called bacteriophages, or simply **phages** (see Chapter 15). The first phages studied included seven that infect the bacterium *Escherichia coli*. These seven phages were named type 1 (T1), type 2 (T2), and so forth, in the order of their discovery. By coincidence, the three T-even phages—T2, T4, and T6—turned out to be very similar in structure. Their capsids have icosahedral (20-sided) heads that enclose the genetic material. Attached to the head is a protein tailpiece with tail fibers that the phage uses to attach to a bacterium (Figure 17.3d; also see Figure 17.1).

HOW VIRUSES REPLICATE: VIRAL INFECTION

Viruses are obligate intracellular parasites; they can only reproduce within a host cell. An isolated virus is unable to replicate itself—or do anything else, for that matter, except infect an appropriate host cell. Viruses lack the enzymes for metabolism and have no ribosomes or other equipment for making their own proteins. Thus, isolated viruses are merely protein-coated sets of genes in transit from one host cell to another.

Each type of virus can infect and parasitize only a limited range of host cells, called its **host range.** This host specificity depends on the evolution of recognition systems by the virus. Viruses identify their host cells by a "lock-and-key" fit between proteins on the outside of the virus and specific receptor molecules on the surface of the cell. Some viruses have host ranges broad enough to include several species. Swine flu virus, for example, can infect both hogs and humans, and the rabies virus can infect a number of mammalian species, including rodents, dogs, and humans. In other cases, viruses have host ranges so narrow that they infect a single species or a single type of tissue within a species. For instance, there are several phages that can parasitize only the bacterium *E. coli*. Human cold viruses usually infect only the cells lining the human upper respiratory tract, ignoring other tissues. And the virus that causes AIDS binds to a specific receptor on certain types of white blood cells.

A viral infection begins when the genome of a virus makes its way into a cell. The mechanism by which this nucleic acid enters the host varies, depending on the type of virus. For example, the T-even phages use their elaborate tail apparatus to inject DNA into a bacterium (see Figures 17.1 and 17.3d). Once inside, the viral genome can commandeer its host, reprogramming the cell to copy the viral genes and manufacture capsid proteins (Figure 17.4). Most DNA viruses use the DNA polymerases of the host cell to synthesize new genomes along the templates provided by the viral DNA. In contrast, RNA viruses usually contain enzymes of their own to initiate replication of their genomes within the host. A cell has no native enzymes for copying RNA; it never produces its own RNA by transcribing one RNA molecule from another. We will describe the replication of DNA and RNA viruses in more detail later in the chapter, when we discuss specific viral infections.

Regardless of the type of viral genome, the parasite diverts its host's resources for viral production. The host provides the nucleotides for nucleic acid synthesis. It also uses its enzymes, ribosomes, tRNAs, amino acids, ATP, and other machinery and ingredients to produce the viral proteins dictated by mRNA transcribed from the parasite's genes.

After the viral nucleic acid molecules and capsomeres are produced, their assembly into new viruses is often a spontaneous process—a process of **self-assembly.** In fact, the RNA and capsomeres of TMV can be separated in the laboratory and then reassembled to form perfect viruses simply by mixing the components together again.

The simplest type of viral life cycle is completed when hundreds or thousands of viruses emerge from the infected host cell. The cell is often destroyed in the process. In fact, some of the symptoms of human viral infections, such as colds and influenza, result from cellular damage and death and from the body's responses to this destruction. The viral offspring that exit from the cell that produced them have the potential to infect additional cells, spreading the viral infection.

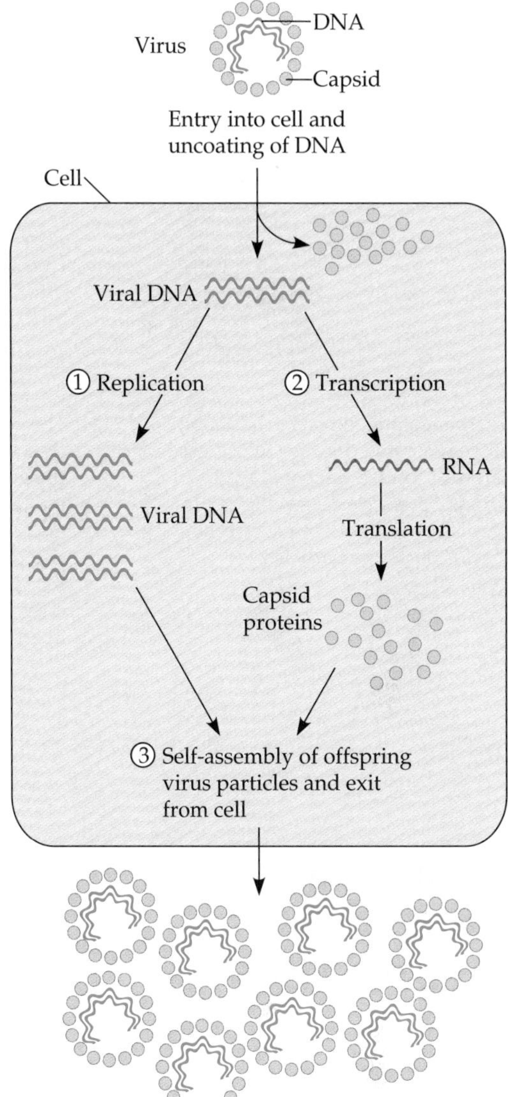

Figure 17.4
A simplified viral life cycle. A virus is an obligate intracellular parasite that uses the equipment of its host cell to reproduce. In this simplest of all possible viral life cycles, the parasite is a DNA virus with a capsid consisting of a single type of protein. ① After entering the cell, the viral DNA uses host nucleotides and enzymes to replicate itself. ② It uses other host materials and machinery to produce its own capsid proteins. ③ Viral DNA and capsid proteins then assemble into new virus particles.

There are many variations on the simplified life cycle we have traced in this overview, a few of which we will see as we take closer looks at some bacterial viruses (phages), plant viruses, and animal viruses.

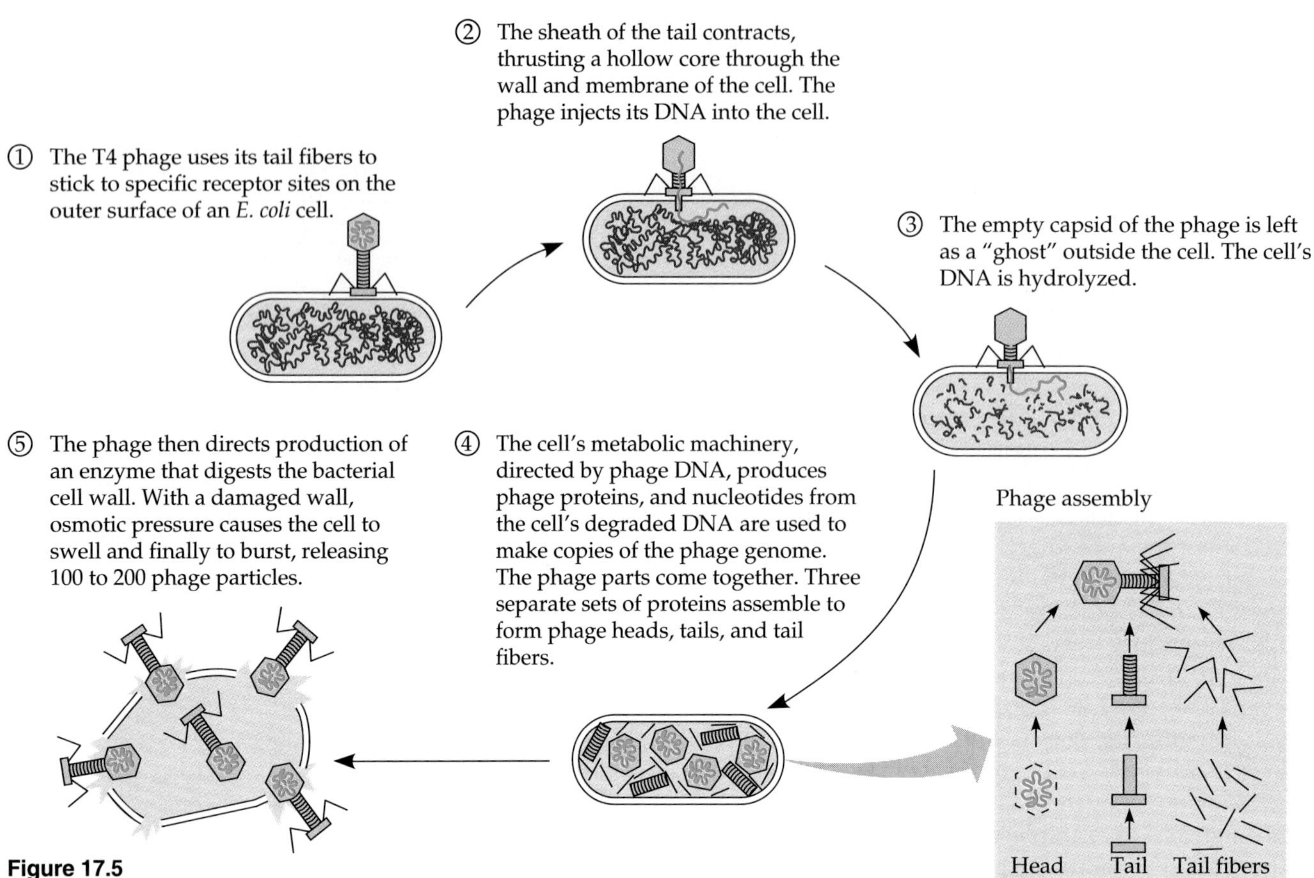

② The sheath of the tail contracts, thrusting a hollow core through the wall and membrane of the cell. The phage injects its DNA into the cell.
① The T4 phage uses its tail fibers to stick to specific receptor sites on the outer surface of an E. coli cell.
③ The empty capsid of the phage is left as a "ghost" outside the cell. The cell's DNA is hydrolyzed.
⑤ The phage then directs production of an enzyme that digests the bacterial cell wall. With a damaged wall, osmotic pressure causes the cell to swell and finally to burst, releasing 100 to 200 phage particles.
④ The cell's metabolic machinery, directed by phage DNA, produces phage proteins, and nucleotides from the cell's degraded DNA are used to make copies of the phage genome. The phage parts come together. Three separate sets of proteins assemble to form phage heads, tails, and tail fibers.
Phage assembly
Head
Tail
Tail fibers

Figure 17.5
The lytic cycle of phage T4.

BACTERIAL VIRUSES

The phages are the best understood of all viruses, although some of them are also among the most complex. Research on phages led to the discovery that double-stranded DNA viruses can reproduce by two alternative mechanisms, the lytic cycle and the lysogenic cycle.

The Lytic Cycle

A reproductive cycle of a virus that culminates in death of the host cell is known as a **lytic cycle.** The term refers to the last stage of infection, during which the bacterium lyses (breaks open) and releases the phages that were produced within the cell. Each of these phages can then infect a healthy cell, and a few successive lytic cycles can destroy an entire bacterial colony in just hours. Viruses that depend on lytic cycles to reproduce are called **virulent viruses.** We will use the virulent phage T4 to illustrate the steps of a lytic cycle (Figure 17.5).

The lytic cycle begins when the tail fibers of a T4 virus stick to specific receptor sites on the outer surface of an *E. coli* cell. The sheath of the tail then contracts, thrusting a hollow core through the wall and membrane of the cell. Molecules of ATP stored in the T4 tailpiece power this penetration. Functioning like a miniature syringe, the phage injects its DNA into the cell, leaving an empty capsid as a "ghost" outside the cell. We saw in Chapter 15 that early experiments with phages provided evidence for DNA as the genetic material, by demonstrating that it was the phages' DNA, not their protein, that enters bacteria during infection.

Once infected by the phage DNA, the *E. coli* cell quickly begins to transcribe and translate the viral genes. Phage T4 has about 100 genes, and most of their functions are known. One of the first phage genes translated by the *E. coli* cell codes for an enzyme that chops up the host cell's own DNA. The phage DNA itself is protected because it contains a modified form of cytosine that is not recognized by the enzyme. The cell now submits completely to the genetic commands of its invader.

Once the phage genome gains control of the cell, it commands the host's metabolic machinery to produce phage components. Nucleotides salvaged from the cell's degraded DNA are recycled to make many copies of the phage genome. Three separate sets of capsid proteins are made and assembled into phage tails, tail fibers, and polyhedral heads. The phage completes its subversion of the cell when one of its genes directs production of an enzyme (lysozyme) that digests the bacterial cell wall. With the cell wall damaged, osmotic

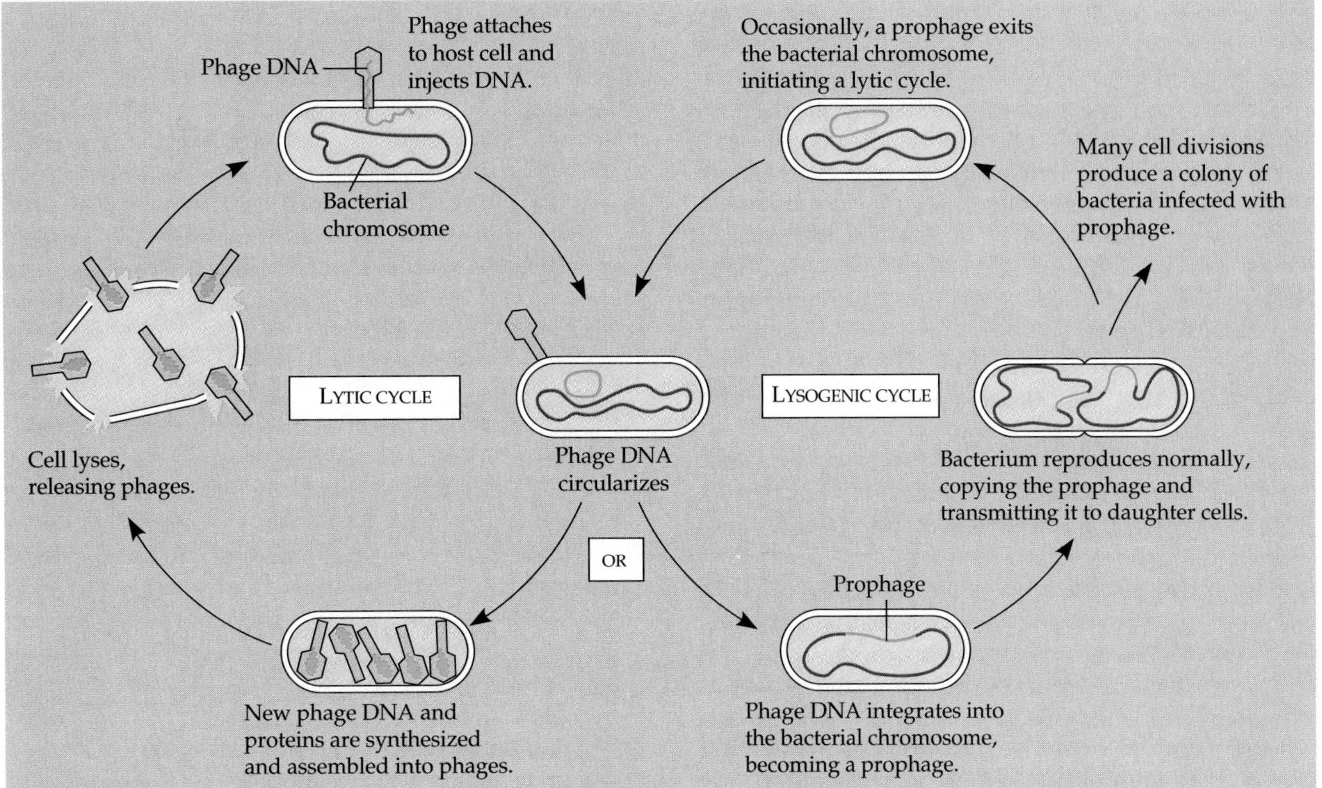

Figure 17.6
The lysogenic and lytic reproductive cycles of phage λ. After entering the bacterial cell, the λ DNA can either integrate into the bacterial chromosome (lysogenic cycle) or immediately initiate the production of a large number of offspring (lytic cycle). In most cases, the lytic pathway is followed, but once the lysogenic cycle is started, the prophage may be carried in the host cell's chromosome for many generations.

pressure causes the cell to swell and finally to burst. The lysed bacterium releases 100 to 200 phage particles, which can then infect other cells nearby.

The entire lytic cycle, from the phage's contact with the cell surface to lysis, takes only 20 to 30 minutes at 37°C. In that time, a T4 population can increase in size more than one-hundredfold, whereas even a fast-growing *E. coli* population can only double within that period. Thus, phage reproduction greatly outpaces the growth of a bacterial colony. If we add a single T4 virus to a susceptible culture of *E. coli* growing in a thin layer on a solid medium in a Petri dish, an expanding clear zone will soon appear in the bacterial "lawn." The clear spot results from the lysis of cells by successive generations of phage.

After reading about the lytic cycle, you may wonder why phages haven't exterminated all bacteria. Actually, bacteria are not defenseless. Natural selection favors bacterial mutants with receptor sites that are no longer recognized by a particular type of phage. And when phage DNA successfully enters a bacterium, various types of degradative enzymes may break it down. Enzymes called **restriction enzymes,** for example, recognize and cut up DNA that is foreign to the cell, including certain phage DNA. The bacterial cell's own DNA is chemically modified in a way that prevents attack by restriction enzymes. But just as natural selection favors bacteria with effective restriction enzymes, natural selection favors phage mutants that are resistant to these enzymes. Thus, the parasite-host relationship is in a constant evolutionary flux.

There is still another important reason bacteria have been spared from extinction as a result of phage activity. Many phages can check their own destructive tendencies and, instead of lysing their host cells, coexist with them in what is called the lysogenic cycle.

The Lysogenic Cycle

Viruses that are capable of the two different modes of reproducing within a bacterium are called **temperate viruses.** In contrast to the lytic cycle, which kills the host cell, the **lysogenic cycle** reproduces the viral genome without destroying the host. To compare these alternative life cycles, we will examine a temperate virus called lambda, abbreviated with the Greek letter λ. Phage λ resembles T4 but lacks tail fibers.

Infection of an *E. coli* cell by λ begins when the phage binds to the surface of the cell and injects its DNA (Figure 17.6). Within the host, the λ DNA molecule forms a circle. What happens next depends on the type of reproductive mode: lytic cycle or lysogenic cycle. During a lytic cycle, the viral genes immediately turn the host cell into a λ-producing factory, and the cell soon lyses and releases its viral products. The viral genome behaves differently during a lysogenic cycle. The λ DNA molecule is incorporated into a specific site of the host cell's chromosome, and it is then known as a **prophage.** One prophage gene codes for a repressor protein, so named because it represses most of the other prophage genes. Thus, the phage genome is mostly silent within the bacterium. How, then, does the phage reproduce? Every time the *E. coli* cell prepares to divide, it replicates the phage DNA along with its own and passes the copies on to daughter cells. A single infected cell can soon give rise to a large population of bacteria carrying the virus in prophage form. This mechanism enables viruses to propagate without eliminating the host cells upon which they depend.

The term *lysogenic* implies that prophages can, at some point, give rise to active phages that lyse their host cells. This occurs when the λ genome exits from the bacterial chromosome. At this time, the λ genome commands the host cell to manufacture phages and then self-destruct, releasing the infectious phages. It is usually an environmental trigger, such as radiation or the presence of certain chemicals, that switches the virus from the lysogenic to the lytic mode.

In addition to the gene for the repressor protein, a few other prophage genes may also be expressed during lysogenic cycles, and these genes may alter the phenotype of the host bacteria. This can have important medical significance. For example, the bacteria that cause the human diseases diphtheria, botulism, and scarlet fever would be harmless if it were not for certain prophage genes that induce bacterial production of toxins.

Table 17.1 Classes of Animal Viruses, Grouped by Type of Nucleic Acid

Class*	Examples/Diseases
I. dsDNA**	
Papovavirus	Papilloma (human warts, cervical cancer); polyoma (tumors in certain animals)
Adenovirus	Respiratory disease; some that cause tumors in certain animals
Herpesvirus	Herpes simplex I (cold sores); herpes simplex II (genital); varicella zoster (chicken pox, shingles); Epstein-Barr virus (mononucleosis, Burkitt's lymphoma)
Poxvirus	Smallpox; vaccinia; cowpox
II. ssDNA (parvovirus)	
	Most depend on coinfection with adenoviruses for growth
III. dsRNA (reovirus)	
	Diarrhea viruses
IV. ssRNA that can serve as mRNA (+ strand RNA)	
Picornavirus	Poliovirus; rhinovirus (common cold); enteric viruses
Togavirus	Rubella virus; yellow fever virus; encephalitis viruses
V. ssRNA that is a template for mRNA (– strand RNA)	
Rhabdovirus	Rabies
Paramyxovirus	Measles, mumps
Orthomyxovirus	Influenza viruses
VI. ssRNA that is a template for DNA synthesis (retrovirus)	
	RNA tumor viruses (e.g., leukemia); AIDS

*The subclasses within each class differ mainly in capsid structure and in the presence or absence of a membrane envelope.
**ds = double-stranded; ss = single-stranded.

ANIMAL VIRUSES

Everyone has suffered from viral infections, whether chicken pox, influenza, or the common cold. Table 17.1 lists some important classes of animal viruses. Like all viruses, those that cause illness in humans and other animals are obligate intracellular parasites that can reproduce only after infecting host cells.

Diverse Reproductive Cycles of Animal Viruses

Many variations on the basic scheme of viral infection are represented among the animal viruses. The two variations we will examine are viruses with envelopes and viruses with RNA genomes. (These adaptations are not mutually exclusive—that is, some viruses have both envelopes and RNA genomes—but we will discuss them separately here for convenience.)

Viruses with Envelopes Some animal viruses are equipped with an outer membrane, or envelope, outside the capsid that helps the parasite enter the host cell (Figure 17.7). This membrane is generally a lipid bilayer, like cell membranes, with glycoproteins protruding from the outer surface. The glycoprotein spikes bind to specific receptor molecules on the surface of the host cell and the viral envelope fuses with the host's plasma membrane, transporting the capsid and viral genome into the cell. After cellular enzymes remove the capsid, the viral genome can replicate and

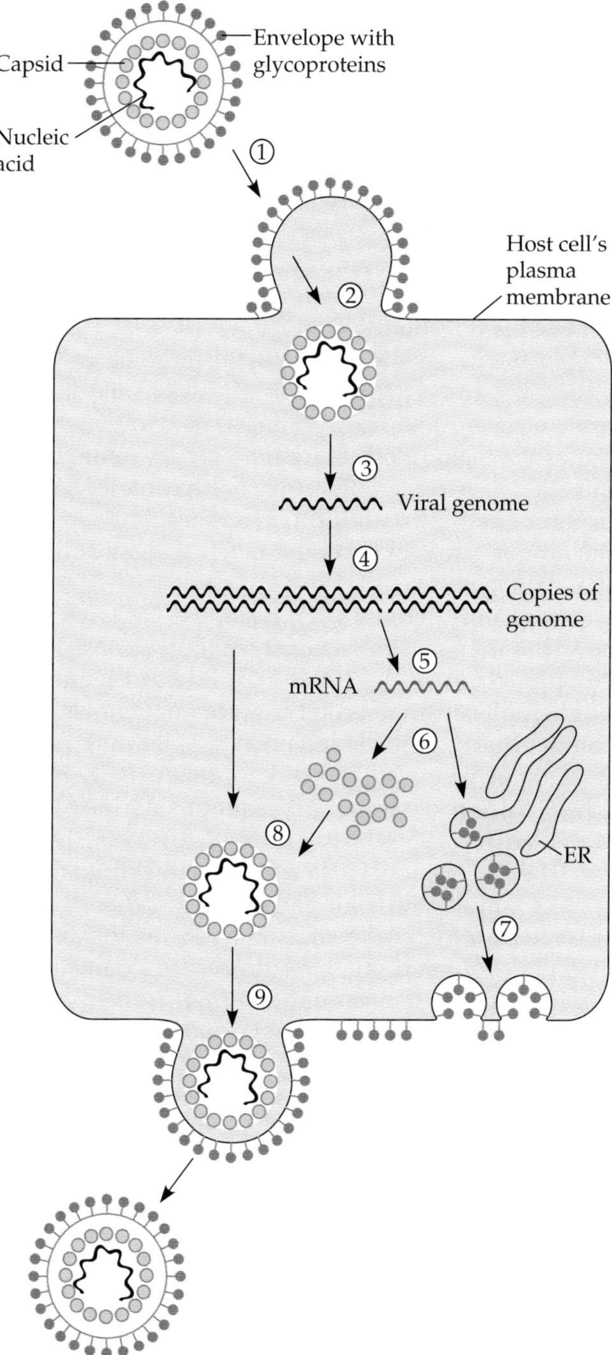

Figure 17.7
The reproductive cycle of an enveloped virus. ① Glycoproteins projecting from the viral envelope recognize and bind to specific receptor molecules on the surface of the host cell. The viral envelope fuses with the cell's plasma membrane. ② The capsid and viral genome enter the cell. ③ Cellular enzymes remove the capsid. ④ The viral genome is replicated, and ⑤ the copies are transcribed into messenger RNA. ⑥ The mRNA is translated into both capsid proteins and glycoproteins characteristic of the viral envelope. The host cell's endoplasmic reticulum (ER) synthesizes the glycoproteins. ⑦ Vesicles transport the glycoproteins to the cell's plasma membrane. ⑧ Capsids assemble around the viral nucleic acid molecules. ⑨ The virus buds from the cell. Its envelope, studded with glycoproteins, is derived from the cell's plasma membrane.

direct the synthesis of viral proteins, including glycoproteins for new viral envelopes. The endoplasmic reticulum of the host makes these membrane proteins, which are transported to the plasma membrane, where they are clustered in patches that serve as exit points for the viral offspring. The viruses bud from the cell surface at these points, wrapping themselves in membrane as they go. In other words, the viral envelope is derived from the host cell's membrane, although some of the molecules of this membrane are specified by viral genes. The enveloped viruses are now free to spread the infection to other cells. Notice that this reproductive cycle does not necessarily kill the host cell, in contrast to the lytic cycles of phages.

Other viruses have envelopes that are not derived from plasma membrane. Herpesviruses, for example, have envelopes derived from the nuclear membrane of the host. The genomes of herpesviruses are double-stranded DNA, and these viruses reproduce within the cell nucleus, using a combination of viral and cellular enzymes to replicate and transcribe their DNA. While within the nucleus, herpesvirus DNA may become integrated into the cell's genome as a **provirus,** similar to a bacterial prophage. Once acquired, herpes infections (including cold sores and genital sores) tend to recur throughout a person's life. Between these episodes, the virus apparently remains latent within the body. From time to time, physical or emotional stress may cause the herpes proviruses to be excised from the host's genome and reproduce, resulting in active infections.

RNA Viruses The full range of viral genomes is represented among animal viruses. Particularly interesting are the RNA viruses, and we will discuss their molecular biology here, even though some phages and most plant viruses are also RNA viruses. Both DNA and RNA viruses are classified according to the relationship of their messenger RNA to the genome, that is, to the DNA or RNA molecule of the infectious virus. Figure 17.8 correlates the molecular genetics of these viral classes to the examples in Table 17.1.

The RNA viruses with the most complicated reproductive cycles are the **retroviruses.** *Retro,* meaning "backward," refers to the reverse direction in which genetic information flows in these viruses. Retroviruses are equipped with a unique enzyme called **reverse transcriptase,** which can transcribe DNA from an RNA template, providing an RNA ⟶ DNA information flow. The newly formed DNA then integrates as a provirus into a chromosome within the nucleus of the animal cell. The host's RNA polymerase transcribes the viral DNA into RNA molecules, which can function both as mRNA for synthesis of viral proteins and as new genomes for viral offspring released from the cell. A retrovirus of particular importance is **HIV (human immunodeficiency virus),** the virus that causes **AIDS**

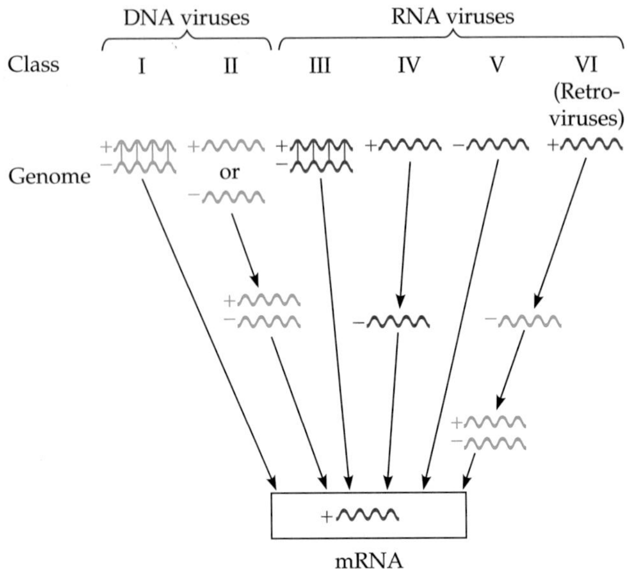

Figure 17.8
Classes of animal viruses. All possible types of viral genomes are represented among viruses that infect animal cells. Here, the viruses are classified according to the relationship of the genome to the messenger RNA that is translated into viral proteins. The plus (+) strand of a genome indicates that the nucleotide sequence codes for proteins. The minus (–) strand does not encode proteins, but serves as a template for synthesis of a plus strand. For viruses with DNA genomes, whether double-stranded (Class I) or single-stranded (Class II), mRNA is transcribed from the minus DNA strand. For viruses having genomes consisting of double-stranded RNA (Class III), the minus strand is the template for mRNA synthesis. A Class IV virus has an RNA genome consisting of a single plus strand. The genome can function directly as mRNA, but also acts as a template for synthesis of minus RNA. This minus RNA, in turn, provides the template for synthesis of additional plus strands. Viral enzymes are required for this synthesis of RNA from RNA templates. There are also single-stranded RNA viruses in which the genome is a minus strand (Class V), and mRNA is transcribed directly from this genomic RNA. The single plus RNA strand of a retrovirus (Class VI) serves as a template for synthesis of a complementary DNA strand. Reverse transcriptase, an enzyme unique to retroviruses, catalyzes this RNA ⟶ DNA information transfer. Transcription then forms mRNA from a DNA template.

(acquired immunodeficiency syndrome). Figure 17.9 traces the reproductive cycle of HIV, but we will postpone a detailed discussion of AIDS until Chapter 39.

Viral Diseases in Animals

The link between a viral infection and the symptoms it produces is often obscure. Some viruses damage or kill cells by causing the release of hydrolytic enzymes from lysosomes. Some viruses cause the infected cells to produce toxins that lead to disease symptoms, and some have toxic components themselves, such as envelope proteins. How much damage a virus causes depends partly on the ability of the infected tissue to regenerate by cell division. We usually recover completely from colds because the epithelium of the respiratory tract, which the viruses infect, can efficiently repair itself. In contrast, the poliovirus attacks nerve cells, which do not divide and cannot be replaced. Polio's damage to such cells, unfortunately, is permanent. Many of the temporary symptoms associated with viral infections, such as fever, aches, and inflammation, may actually result from the body's own efforts at defending itself against the infection.

As we will see in Chapter 39, the immune system is a complex and critical part of the body's natural defense mechanisms. The immune system is also the basis for the major medical weapon for preventing viral infections—vaccines. **Vaccines** are harmless variants or derivatives of pathogenic microbes that stimulate the immune system to mount defenses against the actual pathogen.

The term is derived from *vacca,* the Latin word for cow; the first vaccine, against smallpox, consisted of cowpox virus. In the late 1700s, Edward Jenner, an English physician, learned from his patients in farm country that milkmaids and other people who had contracted cowpox (a milder disease that usually infects cows) were resistant to subsequent smallpox infections. In his famous experiment of 1796, Jenner scratched a farmboy with a needle bearing fluid from a sore of a milkmaid who had cowpox. When the boy was later exposed to smallpox, he resisted the disease.

The cowpox and smallpox viruses are so similar that the immune system cannot distinguish them. Vaccination with the mimic, the cowpox virus, sensitizes the immune system to react vigorously in defending the body, if it is ever exposed to the actual smallpox virus. This strategy has eradicated smallpox, which was once a devastating scourge in many parts of the world. There are also effective vaccines against many other viral diseases, including polio, rubella, measles, and mumps. Modern biotechnology has opened up new approaches to the development of antiviral vaccines and to the pharmaceutical production of certain other natural antiviral agents, such as interferons (see Chapter 39).

Although vaccines can prevent illnesses caused by certain viruses, medical technology can do little, at present, to cure most viral infections once they occur. The antibiotics that help us recover from bacterial infections are powerless against viruses. Antibiotics kill bacteria by inhibiting enzymes or biosynthetic processes specific to the pathogens, but viruses have few or no enzymes of their own; they use their hosts'. However, a few antiviral drugs have been developed in recent years. Several of these drugs are chemical analogues of purine nucleosides, which interfere with viral nucleic acid synthesis. One such drug is adenine arabinoside (also called Ara-A or vidarabine), which acts on several human viruses at concentrations well

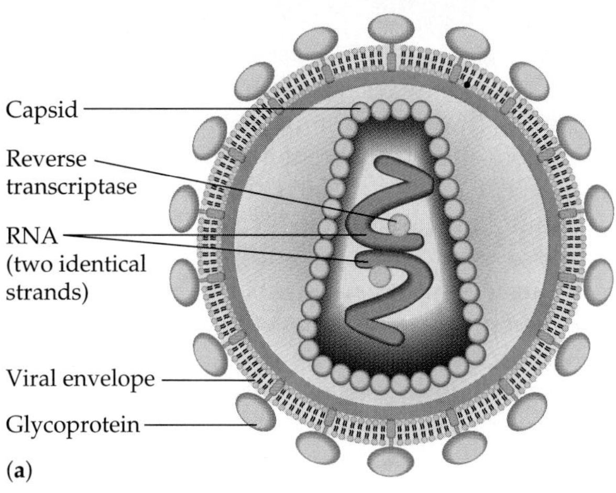

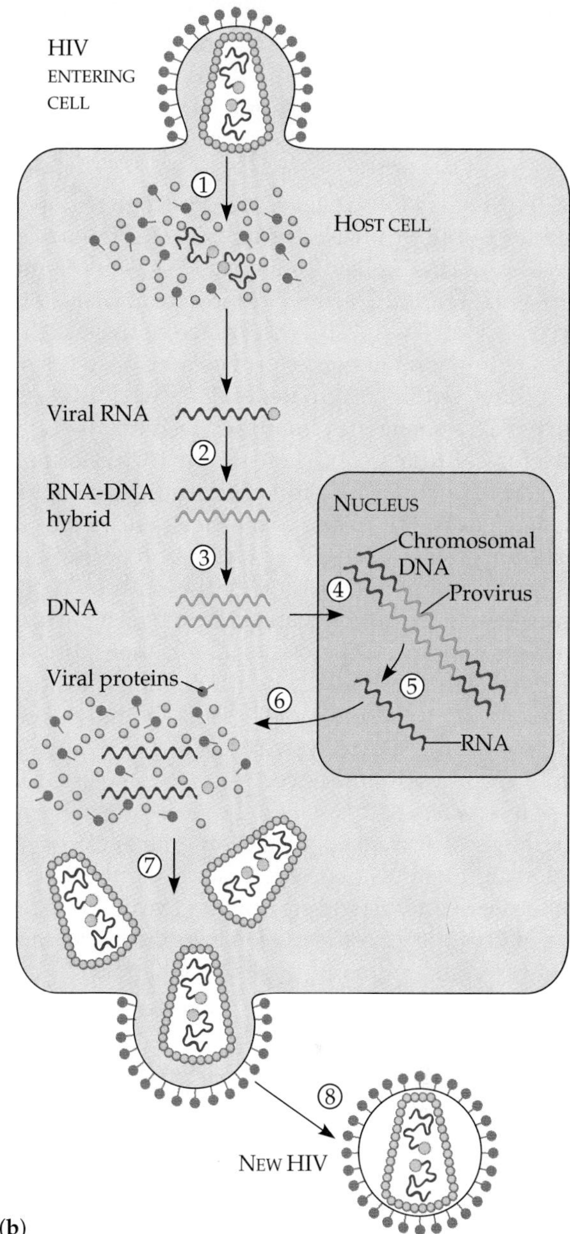

Figure 17.9
The life cycle of HIV, a retrovirus. (**a**) HIV structure. The human immunodeficiency virus is the infectious agent that causes AIDS. The glycoproteins of the envelope enable the virus to bind to specific receptors on the surface of certain white blood cells. The genome of HIV is single-stranded RNA. (Although there are two RNA molecules, they are both plus strands.) The virus also contains reverse transcriptase. (**b**) The reproductive cycle of HIV. ① The genome enters a host cell when the virus fuses with the plasma membrane and the proteins of the capsid are enzymatically removed. ② Reverse transcriptase then catalyzes synthesis of DNA that is complementary to the RNA template provided by the viral genome. ③ The new DNA strand then serves as a template for synthesis of a complementary DNA strand, and ④ the double-stranded DNA is incorporated as a provirus into the host cell's genome. ⑤ Proviral genes are transcribed into mRNA molecules that are ⑥ translated in the cytoplasm into HIV proteins. The RNA transcribed from the provirus also provides the genomes for the next generation of viruses. ⑦ The assembly of capsids around genomes is followed by ⑧ budding of the new viruses from the host cell.

below those that inhibit synthesis of host nucleic acid. Another is acyclovir, which seems to inhibit herpesvirus DNA synthesis. Two drugs of a different type, amantadine and rimantadine, are effective in the prevention of influenza. They seem to inhibit the influenza virus after it enters a cell, but their mechanism of action remains unknown.

Viruses and Cancer

For many years, scientists have recognized that some viruses can cause cancer in animals. Research on these **tumor viruses,** which include members of the retrovirus, papovavirus, adenovirus, and herpesvirus groups, has been facilitated by techniques for growing them in cell cultures. When certain tumor viruses infect animal cells growing in laboratory culture, the cells undergo transformation to a cancerous state. They assume the rounded shapes characteristic of cancer cells and abandon their orderly growth (see Figure 11.17).

In a few cases, there is strong evidence that viruses cause certain types of human cancer. The virus that causes hepatitis B also seems to cause liver cancer in individuals with chronic infections. And the Epstein-Barr virus, the herpesvirus that causes infectious mononucleosis, has been linked to several types of cancer prevalent in parts of Africa, notably Burkitt's lymphoma. Papilloma viruses (of the papovavirus group) have been associated with cancer of the cervix. Among the retroviruses, one called HTLV-1 causes a type of adult leukemia. All tumor viruses transform cells through the integration of viral nucleic acid into host cell DNA. This viral insertion is permanent; the provirus is never excised, as prophages are.

Scientists have identified a number of viral genes directly involved in triggering cancerous characteristics in cells. Many of these genes, called **oncogenes,** are not peculiar to tumor viruses or tumor cells, but are also found in normal cells of many species (see the interview with Harold Varmus and Michael Bishop that precedes Unit Two). The occurrence of these cellular genes within viral genomes provides a clue to the puzzle of how viruses originated, which we will discuss shortly. The oncogenes identified to date all seem to code for cellular growth factors or for proteins involved in growth factor action (for example, receptors). In some cases, the tumor virus lacks oncogenes and transforms the cell simply by turning on or increasing the expression of one or more of the cell's own oncogenes. Whatever the mechanism by which a particular virus causes cancer, there is evidence that more than one oncogene must usually be activated to transform a cell to a fully cancerous state. It is likely that infection with most cancer-causing viruses is effective only in combination with other events—and vice versa. This leads us to suspect that *non*viral cancer-causing agents (carcinogens) also act by turning on cellular oncogenes. Understanding how gene expression is controlled may be the key to understanding cancer. (Oncogene expression, eukaryotic gene expression in general, and cancer will be covered more thoroughly in Chapter 18.)

PLANT VIRUSES AND VIROIDS

Plant viruses are serious agricultural pests that can stunt plant growth and diminish crop yields. Most plant viruses discovered so far are RNA viruses. Many of them, including the tobacco mosaic virus, have rod-shaped capsids with protein arranged in a spiral.

There are two major routes by which a plant viral disease may spread. By the first route, called horizontal transmission, a plant is infected from an external source of the virus. Since the invading virus must get past the plant's outer protective layer of cells (the epidermis), the plant becomes more susceptible to viral infections if it has been damaged by wind, chilling, injury, or insects. Insects are a double threat, because they often also act as carriers, or vectors, of viruses, transmitting disease from plant to plant. Farmers and gardeners themselves may transmit plant viruses inadvertently on pruning shears and other tools. The other route of viral infection is vertical transmission, in which a plant inherits a viral infection from a parent. Vertical transmission can occur in asexual propagation (for example, by taking cuttings) or in sexual reproduction via infected seeds.

Once a virus enters a plant cell and begins reproducing, virus particles can spread throughout the plant by passing through plasmodesmata, the cytoplasmic connections that penetrate the walls between adjacent plant cells (see Figure 7.33). Agricultural scientists have found no cure for most viral diseases of plants. Therefore, their efforts have largely focused on reducing the incidence and propagation of such diseases and on breeding genetic varieties of crop plants that are resistant to certain viruses.

As small and simple as viruses are, they dwarf another class of plant pathogens called **viroids.** These are tiny molecules of naked RNA, only several hundred nucleotides long. Somehow, these RNA molecules can foul up the metabolism of a plant cell and stunt the growth of the whole plant. One viroid disease has killed over ten million coconut palms in the Philippines. Another viroid nearly wiped out the chrysanthemum industry in the United States before growers began keeping stock plants in sterile greenhouses. Viroids also threaten potato and tomato crops.

Clues to what viroids do within cells may emerge from an intriguing discovery that relates viroids to certain normal eukaryotic genes, including rRNA genes. The nucleotide sequences of viroid RNA turn out to be similar to the sequences of introns found within those genes—introns that can excise themselves from their RNA transcript without help from enzymatic proteins (see Chapter 16). Perhaps viroids originated as "escaped introns." An alternative hypothesis would be that viroids and the self-splicing introns both evolved from a common ancestral molecule.

In any case, it seems likely that viroids somehow cause errors in the regulatory systems that control the genes of the cell. Indeed, the symptoms that are typically associated with viroid diseases are abnormal development and stunted growth.

THE EVOLUTIONARY ORIGIN OF VIRUSES

Viruses are in the semantic fog between life and nonlife. Do we think of them as nature's most complex molecules or as the simplest forms of life? Either way, we must bend our usual definitions. An isolated virus is biologically inert, unable to replicate its genes or regenerate its own supply of ATP. Yet it has a genetic program written in the universal language of life. Although viruses are obligate intracellular parasites that cannot reproduce independently, it is hard to deny their evolutionary connection to the living world.

How did viruses originate? Since they depend on cells for their own propagation, it is reasonable to assume that viruses are not the descendants of precellular prototypes of life, but that they evolved *after* the first cells. Most molecular biologists favor the hypothesis that viruses originated from fragments of cellular nucleic acids that could move from one cell to another. Consistent with this idea is the observation that a viral genome usually has more in common with the host cell's genome than with the genomes of viruses infecting other hosts. Indeed, some viral genes are essentially identical to genes of the host, as in the case of oncogenes, for example. Perhaps the earliest viruses were naked bits of nucleic acid, similar to plant viroids, that made it from one cell to another via injured cell surfaces. The evolution of genes coding for capsid proteins may have facilitated the infection of undamaged cells.

The most likely candidates for potential sources of viral genomes are two genetic elements of cells called plasmids and transposons. Plasmids are small, circular DNA molecules that are separate from chromosomes. They are found in bacteria and in yeasts, which are unicellular eukaryotes of the kingdom Fungi. Plasmids, like most viruses, can replicate independently of the rest of the cell's genome and are occasionally transferred between cells. Transposons are DNA segments that can move from one location to another in a cell's genome. Thus plasmids, transposons, and viruses all share an important feature: They are mobile genetic elements. (We will discuss plasmids and transposons in more detail later in the chapter.)

It is the evolutionary relationship between viruses and the genomes of their host cells that makes viruses such useful model systems in molecular biology. By studying how the replication of viruses is controlled, researchers are learning more about the mechanisms that regulate DNA replication and gene expression (transcription and translation) in cells. Bacteria are equally valuable as microbial models in genetics research, but for different reasons. Unlike viruses, bacteria are true cells. But as prokaryotic cells, bacteria provide researchers with the opportunity to investigate molecular genetics in the simplest organisms. In fact, *E. coli*, the bacterium sometimes called "the laboratory rat of molecular biology," is the most completely understood of all organisms at the molecular level. Let's leave the topic of viruses and learn more about the genetics of bacteria.

BACTERIAL GENOMES: REPLICATION AND MUTATION

Bacteria are very adaptable, both in the evolutionary sense of adaptation via natural selection and in the physiological sense of adjustment to changes in the environment by individual bacteria. These sections on bacterial genetics will help clarify how these microbes can be so malleable.

The major component of the bacterial genome is a single double-stranded DNA molecule arranged in a circle. Although we will refer to this structure as the bacterial chromosome, it is very different from eukaryotic chromosomes, which have linear DNA molecules associated with a considerable amount of protein. In the case of the common intestinal bacterium *E. coli*, the chromosome consists of approximately 4 million base pairs representing about 3000 genes. This is 100 times more DNA than is found in a typical virus, but only about one-thousandth as much DNA as in an average eukaryotic cell. Still, this is a lot of DNA to be packaged in such a small container. Stretched out, the DNA of an *E. coli* cell would measure about a millimeter in length, 500 times longer than the cell. Within a bacterium, however, the chromosome is so tightly packed that it does not even fill the whole cell, but forms a structure something like a long loop of yarn tangled into a ball. This dense region of DNA, called the **nucleoid,** is not bounded by a membrane like the true nucleus of a eukaryotic cell. In addition to the chromosome, many bacteria also have plasmids, much smaller circles of DNA. Each plasmid has only a small number of genes, from just a few to about two dozen. You will learn about the structure and function of plasmids in the next section.

Bacterial cells divide by binary fission, which occurs without the complex spindle apparatus characteristic of mitosis in eukaryotic cells (see Chapter 11). Fission is preceded by replication of the bacterial chromosome. From a single origin of replication, copying of DNA progresses in both directions around the circular chromosome (Figure 17.10).

Bacteria can proliferate very rapidly in a favorable environment, whether in a natural habitat or in a laboratory culture. For example, *E. coli*, growing under optimal conditions, can divide every 20 minutes. A laboratory culture started with a single cell can produce a colony of 10^7 to 10^8 bacteria overnight (12 hours). Reproductive rates in the organism's natural habitat, the large intestines (colons) of mammals, are just as impressive. For example, in the human colon, *E. coli* reproduces rapidly enough to replace the 2×10^{10} bacteria lost each day in feces.

Since fission is an asexual process—the production of offspring from a single parent—most of the bacteria in a colony are genetically identical to the parent cell. Due to mutation, however, some of the offspring *do* differ slightly in genetic makeup. For a given *E. coli* gene, the probability of a mutation averages only about 1×10^{-7} per cell division. But since about 2×10^{10} new *E. coli* cells are produced daily in a human colon, mutations in each gene will give rise to approximately 2000 ($2 \times 10^{10}/1 \times 10^{-7}$) *E. coli* mutants daily in just one human host. The total number of mutations when all 3000 *E. coli* genes are considered is about 6×10^6 (3000 $\times$ 2000) per day. The important point is that new muta-

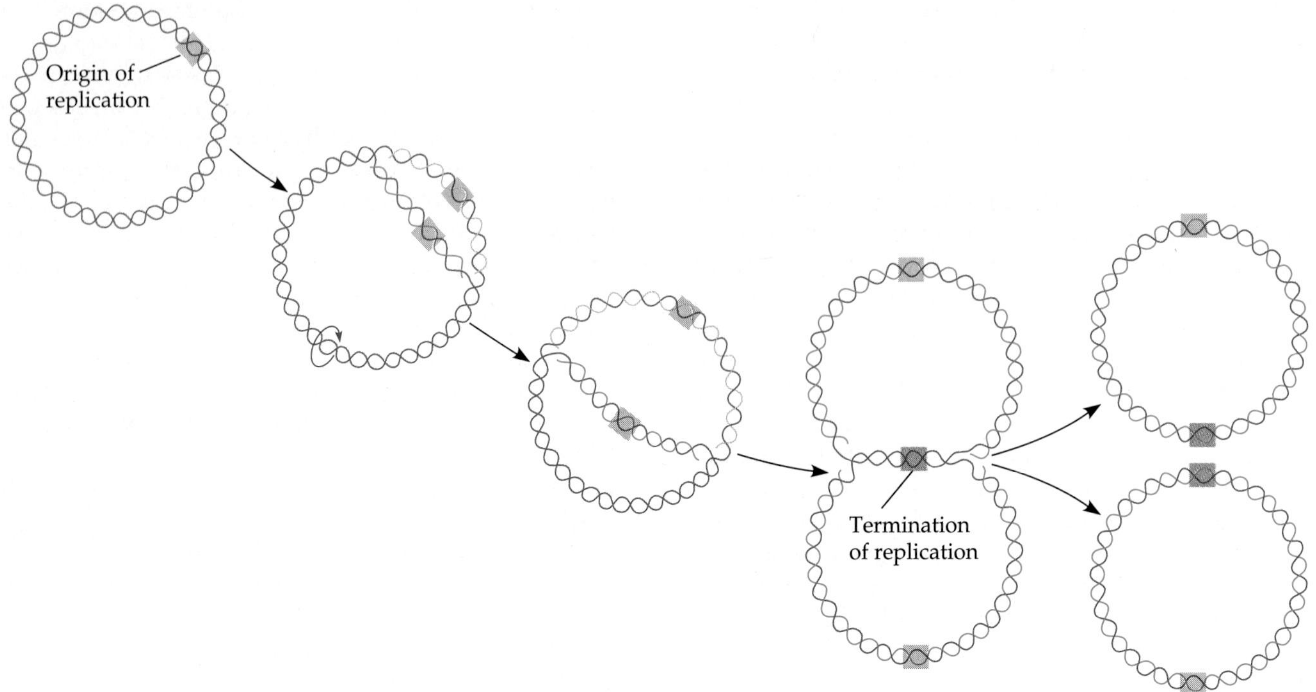

Figure 17.10
Replication of the bacterial chromosome. From one origin, DNA replication progresses in both directions around the circular chromosome until the entire chromosome has been reproduced.

tions, though individually rare, can have a significant impact on genetic diversity when reproductive rates are very high due to short generation spans. This diversity, in turn, affects the evolution of bacterial populations: Individual bacteria that are genetically well-equipped for the local environment clone themselves more prolifically than do less fit individuals.

In contrast, new mutations make a relatively small contribution to genetic variation among individuals of a population of slowly reproducing organisms, such as humans. Most of the heritable variation we observe in a human population is not due to the creation of novel alleles by *new* mutations, but to the sexual recombination of existing alleles (see Chapter 14). Even in bacteria, where new mutations *are* a major source of individual variation, genetic recombination adds more diversity to a population, as we will now see.

GENETIC RECOMBINATION AND GENE TRANSFER IN BACTERIA

Natural selection depends on heritable variation among the individuals of a population (see Chapter 1). In addition to mutations, genetic recombination generates diversity within bacterial populations. We will define recombination here as the combining of genetic material from two individuals into the genome of a single individual.

How can we detect genetic recombination in bacteria? Consider two *E. coli* strains (genetic varieties), each unable to synthesize one of its required amino acids. Wild-type *E. coli* can grow on a minimal medium containing only glucose as a source of organic carbon. The bacteria use the carbon skeletons of glucose molecules to synthesize all their other organic compounds, including amino acids. Our two mutant strains cannot grow on this culture medium of minimal nutrients, because one of them cannot synthesize tryptophan and the other cannot synthesize arginine.

Let's suppose we have grown the two strains together on a medium containing both tryptophan and arginine. After a few hours, we transfer a small sample of this culture to a Petri dish containing minimal medium and incubate this dish overnight. The next morning we observe numerous colonies of bacteria on the minimal medium. Each of these colonies must have started with a cell capable of making *both* tryptophan *and* arginine, but their number far exceeds what can be accounted for by mutation. Most of the cells with genes for making both amino acids must have acquired these genes from two different cells, one from each strain. Genetic recombination has occurred.

In mechanism, genetic recombination in bacteria is different from recombination in eukaryotes. The sexual processes of meiosis and fertilization combine genetic material from two individuals in a single zygote in eukaryotes (see Chapter 12). Sex, in this eukaryotic manner, is absent in prokaryotes: Meiosis and fertilization do not occur. The bacterial mechanisms of genetic recombination are three processes called transformation, transduction, and conjugation.

Transformation

Transformation is the alteration of a bacterial cell's genotype by the uptake of naked, foreign DNA from the surrounding environment. For example, we saw in Chapter 15 that harmless *Streptococcus pneumoniae* bacteria lacking coats could be transformed to pneumonia-causing individuals having protective coats by the uptake of naked DNA from a medium containing dead cells of the coated strain. This transformation occurs when a live, coatless cell absorbs a piece of DNA that happens to include the gene for producing the protective coat. The foreign allele is then incorporated into the bacterial chromosome by replacing the native allele (for the "coatless" condition, in this case) in a DNA exchange similar to crossing over in eukaryotic meiosis. The transformed cell now has a chromosome containing DNA derived from two different cells, which fits our definition of genetic recombination.

For many years after transformation was discovered in laboratory cultures, most biologists believed the process to be too rare and haphazard to play an important role in natural bacterial populations. But researchers have since learned that many bacterial species possess proteins on their surfaces that are specialized for the uptake of naked DNA from the surrounding solution. These proteins specifically recognize and transport only DNA from closely related species of bacteria. Not all bacteria have such membrane proteins. For instance, *E. coli* does not seem to have any specialized mechanism for the uptake of foreign DNA. However, placing *E. coli* in a culture medium containing a relatively high concentration of calcium ions will artificially stimulate the cells to take up small pieces of DNA. In biotechnology, this technique is applied to introduce foreign genes into bacteria—genes coding for valuable proteins, such as human insulin and growth hormone. It is interesting that bacterial transformation has occupied such a central place in the history of molecular genetics. Recall that the discovery of transformation provided early evidence that DNA is the genetic material (see Chapter 15). More recently, transformation has made it possible to engineer the genomes of bacteria so that the cells produce proteins characteristic of other species.

Transduction

In the recombination mechanism known as **transduction,** phages (the viruses that infect bacteria) transfer bacterial genes from one host cell to another. There are two forms of transduction, generalized transduction and specialized transduction (Figure 17.11). Both result from aberrations in the infection cycles of phages.

First let's consider **generalized transduction,** shown in Figure 17.11a. Recall that at the end of a phage's lytic cycle, the viral nucleic acid molecules are packaged within capsids, and the completed phages are released when the host cell lyses. Occasionally, a small piece of the host cell's degraded DNA is packaged within a phage capsid in place of the phage genome. Such a virus is defective because it lacks its own genetic material. However, after its release from the lysed host, the defective phage can attach to another bacterium and inject the piece of bacterial DNA acquired from the first cell. This piece of DNA can subsequently replace the homologous region of the second cell's chromosome by a molecular rearrangement similar to crossing over. The cell's chromosome now has a combination of genetic material derived from two cells; recombination has occurred. This mode of recombination is called generalized transduction because it does not selectively transfer certain genes. Whatever piece of bacterial DNA happens to get packaged within a phage is the genetic material that is moved between cells.

Let's contrast this result with the result of **specialized transduction,** shown in Figure 17.11b. This form of transduction requires infection by a temperate phage. Recall that in the lysogenic cycle, the genome of a temperate phage integrates as a prophage into the host bacterium's chromosome, usually at a specific site. Later, when the phage genome is excised from the chromosome, it sometimes takes with it small regions of the bacterial DNA that were adjacent to the prophage. When such a virus carrying bacterial DNA infects another host cell, the bacterial genes are injected along with the phage's genome. This mode of recombination is called specialized transduction because it specifically transfers genes that are near the chromosomal location where the prophage is incorporated.

Conjugation and Plasmids

Conjugation is the direct transfer of genetic material between two bacterial cells that are temporarily joined. This mechanism of genetic recombination, the bacterial version of sex, has been studied most extensively in *E. coli.* The DNA transfer is one-way, one cell donating DNA and its "mate" receiving the genes. The DNA donor, referred to as the "male," uses appendages

Figure 17.11
Transduction. Phages occasionally carry bacterial genes from one cell to another. (**a**) In generalized transduction, random pieces of the host chromosome are packaged within a phage capsid. (**b**) In specialized transduction, a prophage exits incorrectly from the host chromosome in such a way that it carries adjacent bacterial genes along with it. In both types of transduction, some of the transferred DNA may recombine with the genome of the new host cell.

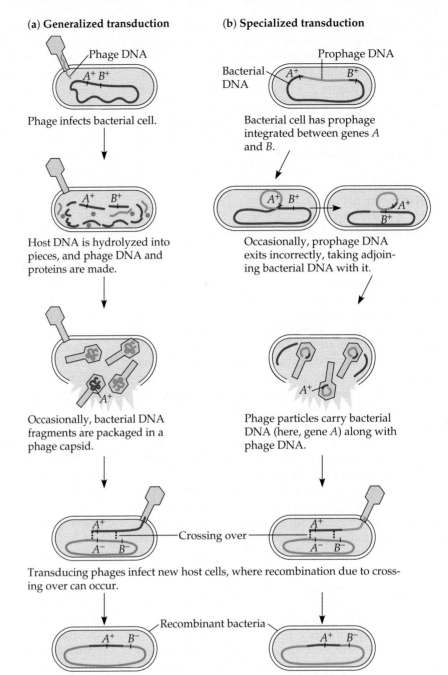

(a) Generalized transduction

Phage DNA

A^+ B^+

Phage infects bacterial cell.

A^+ B^+

Host DNA is hydrolyzed into pieces, and phage DNA and proteins are made.

A^+

Occasionally, bacterial DNA fragments are packaged in a phage capsid.

(b) Specialized transduction

Prophage DNA

Bacterial DNA

A^+ B^+

Bacterial cell has prophage integrated between genes *A* and *B*.

A^+ B^+

A^+ B^+

Occasionally, prophage DNA exits incorrectly, taking adjoining bacterial DNA with it.

A^+

Phage particles carry bacterial DNA (here, gene *A*) along with phage DNA.

A^+ A^- B^-

Crossing over

A^+ A^- B^-

Transducing phages infect new host cells, where recombination due to crossing over can occur.

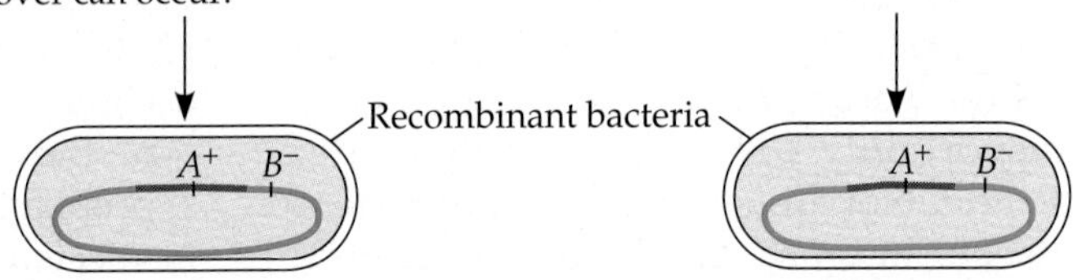

The recombinants have genotypes ($A^+ B^-$) different from either the donor ($A^+ B^+$) or recipient ($A^- B^-$).

called sex pili to attach to the DNA recipient, the "female" (Figure 17.12). Then a temporary cytoplasmic bridge forms between the two cells, providing an avenue for DNA transfer. "Maleness," the ability to form sex pili and donate DNA during conjugation, requires the presence of a special plasmid called the F plasmid. This is one of several types of plasmids that have been discovered in bacteria. Before learning about the specialized role of the F plasmid in conjugation, we should examine plasmids more generally.

Plasmids: General Characteristics A **plasmid** is a small, circular DNA molecule separate from the bacterial chromosome. Plasmids replicate independently, but some do so in synchrony with the chromosome. Other plasmids replicate on their own schedule, and this can alter the number of these plasmids present in the cell at a particular time.

Certain plasmids have another interesting behavior: They can undergo a reversible incorporation into the cell's chromosome. Such a genetic element, which can replicate either as a free molecule in the cytoplasm or as part of the main bacterial chromosome, is called an **episome.** In addition to some plasmids, temperate viruses, such as lambda phage, also qualify as episomes. Recall that the genomes of these phages repli-

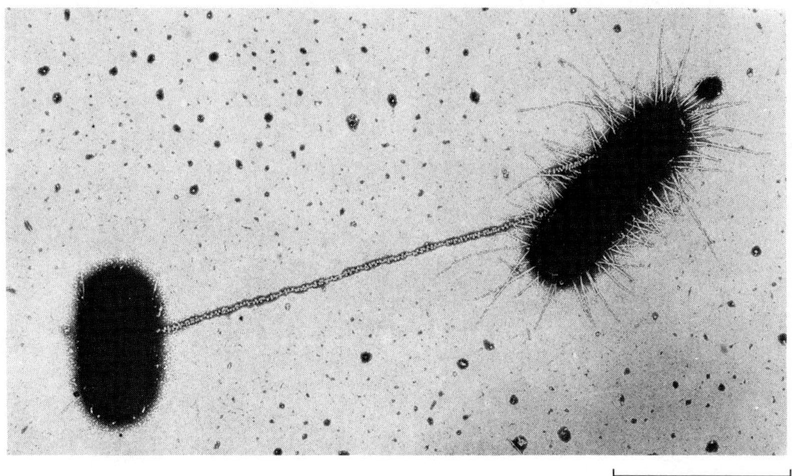

Figure 17.12
Bacterial mating. The *E. coli* "male" extends sex pili, one of which is attached to a "female" cell (TEM). Later, the mates will briefly join by a cytoplasmic bridge through which the "male" transfers DNA to the "female." This mechanism of DNA transfer is called conjugation.

cate separately in the cytoplasm during a lytic cycle, and as an integral part of the host's chromosome during a lysogenic cycle. The hypothesis that some viruses evolved from plasmids was introduced earlier. Of course, there are important differences between plasmids and viruses. Plasmids, unlike viruses, lack an extracellular stage. And plasmids are generally beneficial to the cell, while viruses are parasites that usually harm their hosts.

Plasmids have only a few genes, and these genes are not required for the survival and reproduction of the bacterium under normal conditions. However, the genes of plasmids can confer advantages to bacteria in stressful environments. For example, the F plasmid facilitates genetic recombination, which may be advantageous in a changing environment that no longer favors existing strains in a bacterial population.

The F Plasmid and Conjugation The **F plasmid** (F for fertility) consists of about 25 genes, most required for the production of sex pili. Geneticists use the symbol F^+ to denote a cell that contains the F plasmid (a "male" cell). The F^+ condition is heritabie: The F plasmid replicates in synchrony with the chromosomal DNA, and division of an F^+ cell usually gives rise to two offspring that are both F^+. Cells lacking the F plasmid are designated F^-, and they function as DNA recipients ("females") during conjugation. The F^+ condition is "contagious" in the sense that an F^+ cell can convert an F^- cell to F^+ when the two cells conjugate. The F plasmid replicates within the "male" cell, and a copy is transferred to the "female" through the conjugation tube joining the cells (Figure 17.13a). In such an $F^+ \times F^-$ mating, only an F plasmid is transferred.

Under what circumstances is DNA of the main bacterial chromosome transferred during conjugation? The F plasmid is an episome that occasionally becomes integrated into the main chromosome (Figure 17.13b). A cell with the F genes built into its chromosome is called an Hfr cell (for high frequency of recombination). An Hfr cell continues to function as a male during conjugation, transferring the F genes to its F^- partner. But now, the F episome takes chromosomal DNA along with it (Figure 17.13c). The Hfr cell's chromosome is replicating as DNA is transferred, so this donor cell retains all of its own genes. Random movements of the bacteria usually disrupt conjugation before an entire copy of the Hfr chromosome can be passed to the F^- cell. Temporarily, the recipient cell is a partial diploid, containing its own chromosome plus DNA copied from part of the donor's chromosome. Recombination occurs when the newly acquired DNA aligns with the homologous region of the cell's own chromosome and a crossover exchanges DNA (Figure 17.13d). Binary fission of this cell gives rise to a colony of recombinant bacteria with genes derived from two different cells.

Interrupting Conjugation to Map Bacterial Chromosomes Microbial geneticists have used Hfr $\times F^-$ matings to map the sequence of genes around the *E. coli* chromosome. When the F plasmid joins a chromosome to form an Hfr cell, it always inserts into the chromosome at a specific site (though this site differs from one genetic variety of *E. coli* to another). As a copy of the chromosome snakes through the conjugation tube, the F episome leads the way. Thus, during conjugation, the chromosome's other genes are always transferred from donor to recipient in a specific sequence that is determined by the chromosomal position of the F episome. In Figure 17.13c, for example, gene *A*, which is closest to the episome, precedes gene *B* into the F^- cell. If conjugation lasted longer, then gene *C* would follow, then gene *D*, and so on. Researchers can track this sequence, and thereby map the linear order of genes, by interrupting conjugation at different time intervals. In such an experiment, an Hfr strain is mated with an F^- strain having different alleles for the genes being mapped. The two cultures of bacteria are mixed, and small samples are taken at different times: 5 minutes after mixing, 10 minutes after mixing, and so on. A kitchen blender is used to agitate

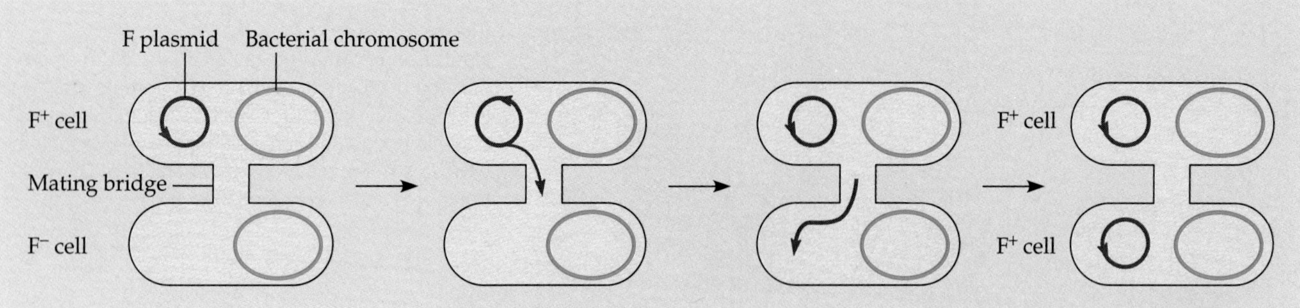

(a) Conjugation between an F⁺ (male) and an F⁻ (female) bacterium. Cells that carry an extrachromosomal fertility factor, the F plasmid, are called F⁺ cells. They are "male" in that they can transfer an F plasmid to a "female" F⁻ cell during conjugation. In this way, an F⁻ cell can become F⁺. The F factor replicates as it is transferred, so that the donor cell remains F⁺. The arrowhead marks the point where replication begins.

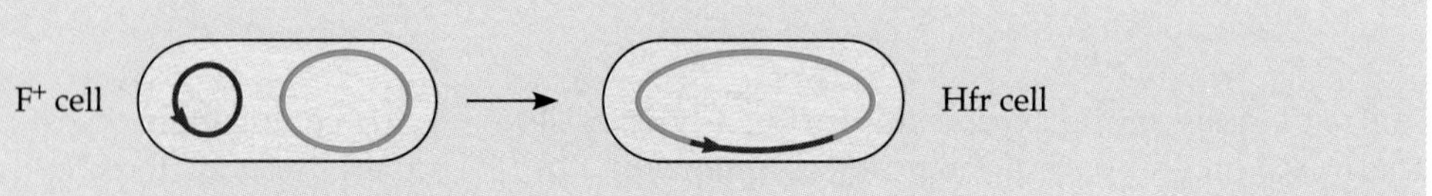

(b) Conversion of an F⁺ male into an Hfr male by integration of the F plasmid (an episome) into the chromosome.

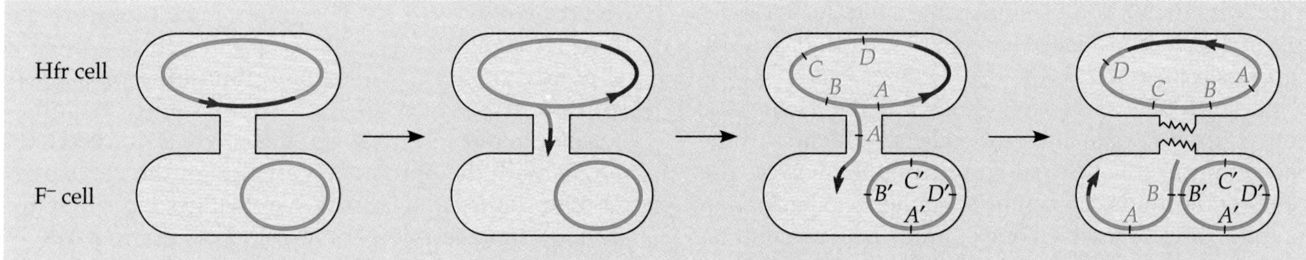

(c) Conjugation between an Hfr and an F⁻ bacterium. Replication of the "male's" chromosome begins at a fixed point (arrowhead) within the F episome (the location where the episome inserts into the chromosome varies from one genetic strain of *E. coli* to another). The location of this point determines the sequence in which genes are transferred to the "female" during conjugation. For example, in this *E. coli* strain, the transfer sequence for four genes is symbolized as *A-B-C-D*. The conjugation bridge usually breaks before the entire chromosome and the tail end of the F episome are transferred.

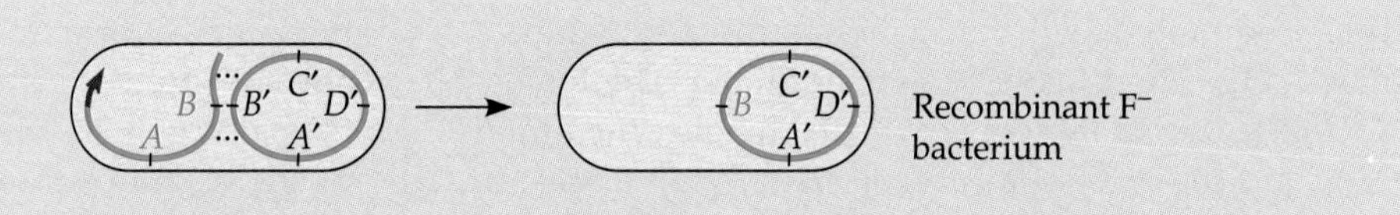

(d) Recombination between the Hfr chromosome fragment and the F⁻ chromosome. Crossing over can occur between genes on the fragment of bacterial chromosome transferred from the Hfr cell and the same (homologous) genes on the recipient (F⁻) cell's chromosome. A recombinant F⁻ cell will result. Pieces of DNA ending up outside the bacterial chromosome will eventually be degraded by the cell's enzymes or lost in cell division.

Figure 17.13
Conjugation and recombination in *E. coli*.

each sample, a treatment that separates the two members of any mating pair. The bacteria in each sample are then cultured, and each culture will include offspring of F⁻ cells that received genes during conjugation. Genetic analysis of these recombinants reveals which genes were transferred during each time interval. Using this method and other approaches, geneticists have mapped the locations of more than half of *E. coli's* genes, a remarkable achievement contributing to the claim that *E. coli* is biology's best known organism.

R Plasmids and Antibiotic Resistance In the 1950s, Japanese physicians began to notice that some hospital patients suffering from bacterial dysentery, which causes severe diarrhea, did not respond to antibiotics that had generally been effective in treating this type of infection. Apparently, resistance to these antibiotics had evolved in certain strains of *Shigella,* the pathogen that causes bacterial dysentery. Many years later, researchers began to identify the specific genes that confer antibiotic resistance, not only in *Shigella,* but in many other pathogenic bacteria. Some of these genes, for example, code for enzymes that specifically destroy certain antibiotics, such as tetracycline or ampicillin. The genes conferring resistance are not chromosomal in location, but are carried by a special class of plasmids now known as **R plasmids** (R for resistance).

Exposure of a bacterial population to a specific antibiotic—whether in a laboratory culture or within a host organism, such as a human—will kill antibiotic-sensitive bacteria but not those that happen to have R plasmids that counter that antibiotic. The theory of natural selection predicts that, under these circumstances, an increasing number of bacteria will inherit genes for antibiotic resistance, and that is exactly what happens. The medical consequences are also predictable: Resistant strains of pathogens are becoming more common, making the treatment of certain bacterial infections more difficult. The problem is compounded by the fact that R plasmids, like F plasmids, can be transferred from one bacterial cell to another during conjugation. Making the problem still worse, some plasmids carry as many as ten genes for resistance to that many antibiotics. How do so many antibiotic-resistant genes become part of a single plasmid? The answer involves another type of mobile genetic element called a transposon, which we will investigate next.

Transposons

Transposons, also called transposable genetic elements, were introduced earlier in the chapter. They are pieces of DNA that can move from one location to another in a cell's genome. In bacteria, a transposon may move from one locus to another within the chromosome; from a plasmid to the chromosome, or vice versa; or from one plasmid to another. For example, transposons are responsible for combining several genes for antibiotic resistance into a single plasmid by moving the genes to that location from different plasmids.

Transposons are sometimes called "jumping genes," but the phrase is misleading. Genes *do* jump from one genomic location to another in what is called a conservative transposition. In this case, the transposon's genes are not replicated before the move, so the number of copies of these genes is conserved. However, in another type of transposition, called a replicative transposition, the transposon replicates at its original site and a *copy* inserts at some other location in the genome; that is, the transposon's genes are added at some new site without being lost from the old site.

A transposon does not seem to have a single specific target in the genome (although some regions of DNA are more likely than others to receive transposons). This ability to scatter certain genes throughout the genome makes transposition fundamentally different from all other mechanisms of genetic shuffling. Crossing over in eukaryotic meiosis and the three mechanisms of recombination in prokaryotes—transformation, transduction, and conjugation—all depend on exchange of alleles between homologous regions of DNA. While the integration of an episomic plasmid into a chromosome does not require that a target site have an extensive stretch of DNA homologous to the plasmid, the insertion *is* site-specific; plasmids are generally incorporated into specific positions on the chromosome. In contrast, a transposon may move genes to a site where no such genes have ever before existed.

Insertion Sequences The simplest transposons are called **insertion sequences.** They consist of only the DNA necessary for the act of transposition itself—no other genes are present. The one gene found in an insertion sequence codes for transposase, an enzyme that catalyzes transposition. The transposase gene is bracketed by a pair of DNA sequences called inverted repeats, noncoding sequences about 20 to 40 nucleotides long (Figure 17.14a). They are called inverted repeats because these two regions of DNA on the ends of an insertion sequence are upside-down, backward versions of each other. In the following abbreviated example, note that each base sequence is repeated, in reverse, along the opposite DNA strand of the inverted repeat at the other end of the transposon:

DNA strand #1	...ATCCGGT...	...ACCGGAT...
DNA strand #2	...TAGGCCA...	...TGGCCTA ...

Transposase recognizes these inverted repeats as the boundaries of the transposon. The enzyme binds to these two regions, bringing them close together and catalyzing the DNA cutting and resealing required for transposition (Figure 17.14b). Other enzymes are also required for transposition. For example, DNA polymerase helps form identical regions of DNA, called direct repeats, which flank a transposon in its new target site (Figure 17.15).

When an insertion sequence invades a gene, it usually fouls the function of that gene by interrupting the

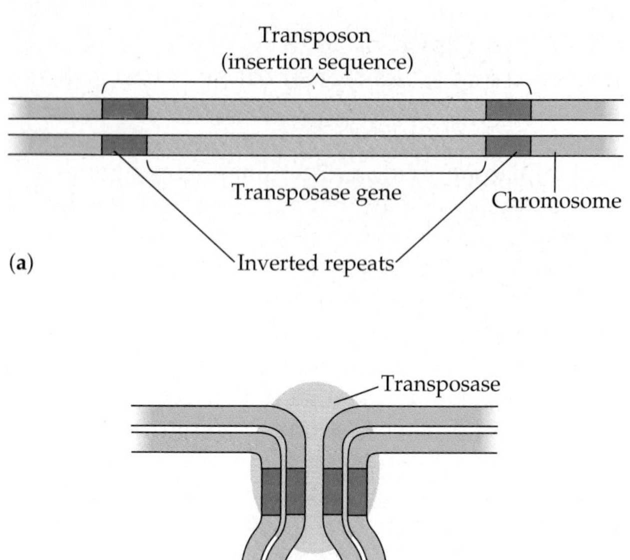

Figure 17.14
Insertion sequences, the simplest transposons. (**a**) The one and only gene of an insertion sequence codes for transposase, the enzyme that catalyzes movement of the transposon from one location to another in the genome. On either end of the transposase gene are inverted repeats, nucleotide sequences that are backward, upside-down versions of each other. (**b**) During transposition, transposase holds the inverted repeats close together and catalyzes the cutting and resealing of DNA required for insertion of the transposon at some target site.

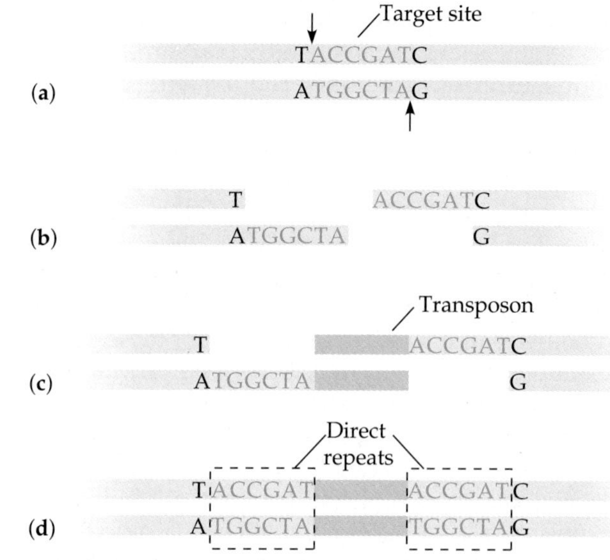

Figure 17.15
Insertion of a transposon. (**a**) An enzyme, probably transposase, cuts the two DNA strands at the target site in a staggered fashion (arrows). (**b**) This leaves short segments of unpaired DNA at each end of the cut chromosome or plasmid. (**c**) Transposase inserts the transposon into the opening of the target DNA. (**d**) The gaps in the two DNA strands are filled in when nucleotides pair with the single-stranded regions and DNA polymerase joins the nucleotides. This results in direct repeats, identical segments of DNA on either side of the transposon. Notice that the direct repeats did not exist at the target site until they were created by the act of transposition.

coding sequence that specifies a protein. In other words, insertion sequences cause mutations when they move. But notice that this mechanism of mutation is intrinsic to the cell, in contrast to mutagenesis caused by extrinsic factors, such as radiation and foreign chemicals. In addition to inserting within the coding regions of genes, insertion sequences may also increase or decrease the production of a protein by inserting within regulatory regions of DNA that control transcription rates. Wherever they land in the genome, insertion sequences are likely to alter the phenotype of the cell in some way. Insertion sequences account for about 1.5% of the *E. coli* genome, but they transpose and cause mutations rarely—only about once in every ten million generations. However, this is about the same as the mutation rate due to extrinsic causes. Given the rapid proliferation of bacteria, transposition of insertion sequences probably plays a significant role in bacterial evolution as a source of genetic variation.

Because biologists have not yet documented any positive role for insertion sequences, these mobile genetic elements are sometimes categorized as "selfish DNA"—DNA that is perpetuated by replication of the genome without providing any known benefit to the cell. Indeed, insertion sequences may be related to certain viruses, such as a bacteriophage known as Mu. Phage Mu, for "mutator," is so named because it causes a very high mutation rate in the host cells it infects. It is a temperate phage that uses the lysogenic cycle to reproduce. But unlike λ and most other temperate phages, Mu inserts at random in the bacterial genome rather than invading a specific site in the chromosome. As copies of the Mu DNA transpose to additional sites, the mutation rate soars to ten thousand times the normal rate. In comparison, insertion sequences behave as parasitic DNA that transposes much less frequently than Mu and lacks the genes for the capsid proteins that make an extracellular stage possible.

Complex Transposons and R Plasmids Transposons larger and more complex than insertion sequences also move about in the bacterial genome. In addition to the DNA required for transposition, complex transposons include other genes that go along for the ride, such as genes for antibiotic resistance. These genes are sandwiched between two insertion sequences (Figure 17.16). It is as though two insertion sequences happened to land relatively close together in the genome and now travel together, along with all the DNA be-

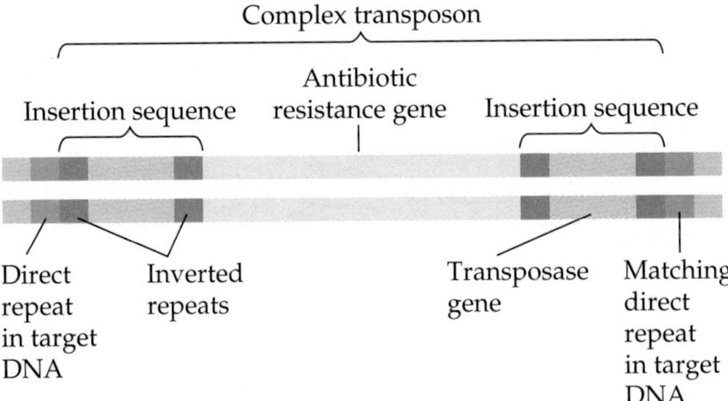

Figure 17.16
A complex transposon. A complex transposon consists of one or more genes located between twin insertion sequences. In this case, a gene for resistance to a specific antibiotic will be carried along as part of the complex transposon when this DNA inserts at some new site in the genome. For example, such transposition can add a gene for antibiotic resistance to a plasmid that already carries genes for resistance to other antibiotics. When such composite plasmids are transmitted to bacteria by either cell division or conjugation, the microbes can resist a variety of antibiotics.

Figure 17.17
Barbara McClintock, discoverer of transposable genetic elements. In 1947, when this photograph was taken at the Cold Spring Harbor laboratory, in New York, McClintock was already regarded as a leading geneticist. It was decades, however, before colleagues recognized the general importance of her greatest discovery, the identification of transposable genes in corn. In 1983, at age 81, Barbara McClintock was awarded the Nobel Prize. She continued her experiments at Cold Spring Harbor until her death in 1992.

tween them, as a single transposon. In contrast to simple insertion sequences, which are not known to benefit bacteria in any specific way, complex transposons may help bacteria adapt to new environments. We have already mentioned the example of transposition packaging several genes for resistance to different antibiotics into a single plasmid. Apparently, natural selection favors bacterial clones that have built up these composite R plasmids through a series of transpositions.

Transposable genetic elements are not unique to bacteria, but are important components of eukaryotic genomes as well. In fact, the first evidence for such wandering genes came from American geneticist Barbara McClintock's breeding experiments with Indian corn in the 1940s and 1950s (Figure 17.17). McClintock identified changes in the color of corn kernels that only made sense if she postulated the existence of mobile genetic elements capable of moving from other locations in the genome to the genes for kernel color. She called these mobile elements "controlling elements" because they seemed to insert next to the genes responsible for kernel color, either activating or inactivating those genes. McClintock's discovery received little attention until transposons were discovered in bacteria many years later, and microbial geneticists learned more about the molecular basis of transposition. In 1983, more than 30 years after she discovered transposable genetic elements, McClintock was awarded a Nobel Prize.

You will learn more about transposable elements in eukaryotes in Chapter 18. We end this chapter by examining how bacterial genes are switched on and off in different environments.

THE CONTROL OF GENE EXPRESSION IN PROKARYOTES

Mutations and the various mechanisms of genetic recombination we have been studying generate the genetic variation that makes natural selection possible. And natural selection, acting over many generations, can increase the proportion of individuals in a bacterial population that are adapted to some new environmental condition, such as the introduction of a specific antibiotic. But how can an individual bacterium, locked into the genome it has inherited, cope with environmental fluctuation?

Think, for instance, of an *E. coli* cell living in the erratic environment of a human gut, dependent for its nutrients on the whimsical eating habits of its host. If the bacterium is deprived of the amino acid tryptophan, which it needs in order to survive, it responds by activating a metabolic pathway to make its own tryptophan from another compound. Later, if the human host eats a tryptophan-rich meal, the cell stops produc-

Figure 17.18
Regulation of a metabolic pathway. Cells can adjust the rates of specific metabolic pathways by regulating gene expression (synthesis of new enzyme molecules) or by regulating the catalytic activity of existing enzymes. In the pathway for tryptophan synthesis, the end-product of the pathway (tryptophan) can inhibit the activity of the first enzyme in the pathway (feedback inhibition) and repress the expression of the genes for all the enzymes needed for the pathway.

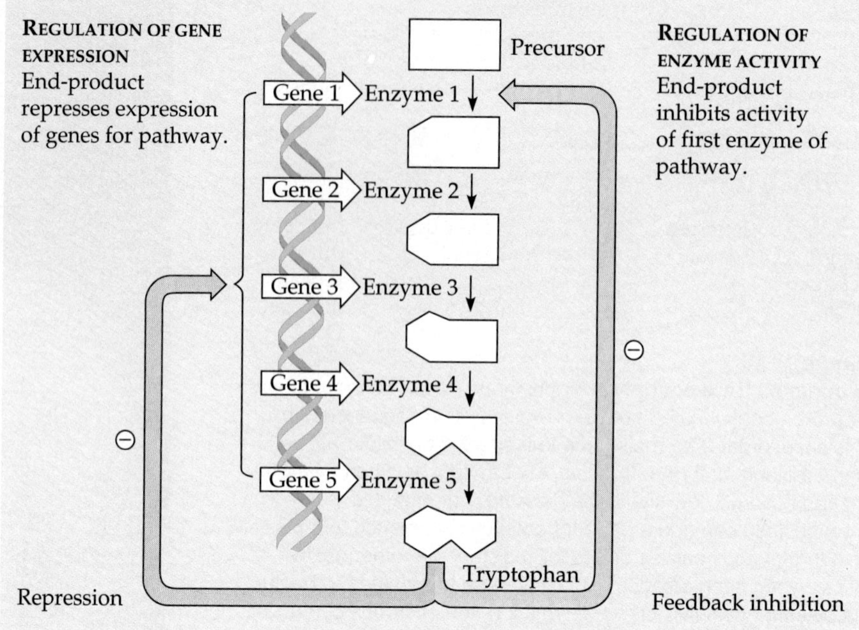

ing tryptophan for itself, thus saving the cell from squandering its resources to produce a substance that is available from the surrounding solution in prefabricated form. This is just one example of how bacteria tune their metabolism to changing environments.

Metabolic control occurs on two levels (Figure 17.18). First, cells can vary the numbers of specific enzyme molecules; that is, they can regulate the expression of a gene. Second, cells can vary the activities of enzymes already present. The latter mode of control, which is more immediate, depends on the sensitivity of many enzymes to chemical cues that increase or decrease their catalytic activity (see Chapter 6). For example, activity of the first enzyme of the tryptophan-synthesis pathway is inhibited by the pathway's end-product. Thus, if tryptophan accumulates in a cell, it shuts down its own synthesis. Such feedback inhibition, typical of anabolic (biosynthetic) pathways, allows a cell to adapt to short-term fluctuations in levels of a substance it needs. If, in our example, the environment continues to provide all the tryptophan the cell needs, then regulation of gene expression also comes into play: The cell stops making enzymes of the tryptophan pathway. This control of enzyme number occurs at the level of transcription, the synthesis of messenger RNA coding for these enzymes. More generally, many genes of the bacterial genome are switched on or off by changes in the metabolic status of the cell. The basic mechanism for this control of gene expression, described as the operon model, was discovered in 1961 by François Jacob and Jacques Monod at the Pasteur Institute in Paris. Let's see what an operon is and how it works, using control of tryptophan synthesis as an example.

Operons: The Basic Concept

E. coli synthesizes tryptophan from a precursor molecule in a series of steps, each reaction catalyzed by a specific enzyme (see Figure 17.18). The five genes coding for the polypeptide chains that make up these enzymes are clustered together on the chromosome. A single promoter serves all five genes, which constitute a transcription unit. (Recall from Chapter 16 that a promoter is a site where RNA polymerase can bind to DNA and begin transcribing genes.) Thus, transcription gives rise to one long mRNA molecule representing all five genes for the tryptophan pathway. The cell can translate this transcript into separate polypeptides because the mRNA is punctuated with start and stop codons signaling where the coding sequence for each polypeptide begins and ends.

Genes that code for polypeptides are called **structural genes.** A key advantage of grouping structural genes of related function into a single transcription unit is that a single "on-off switch" can control the whole cluster of functionally related genes. When an *E. coli* cell must make tryptophan for itself because the nutrient medium lacks this amino acid, all the enzymes for the metabolic pathway are synthesized at one time. The switch is a segment of DNA called an **operator.** Its location and name both suit its function: Positioned within the promoter or between the promoter and the

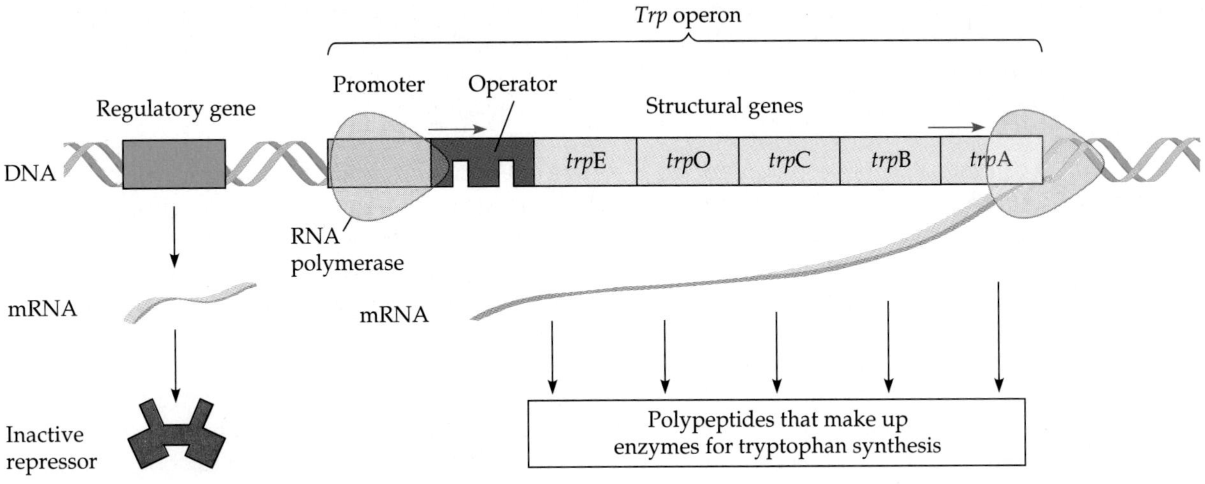

(a) Tryptophan absent, repressor inactive, operon on

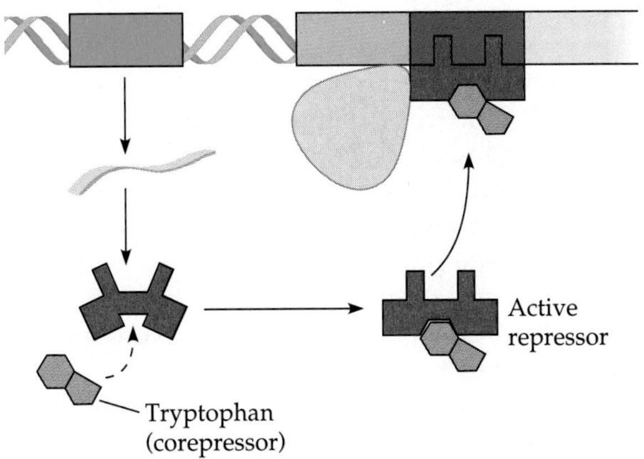

(b) Tryptophan present, repressor active, operon off

Figure 17.19
The *trp* operon: regulated synthesis of repressible enzymes. (**a**)Tryptophan is an amino acid produced by an anabolic pathway catalyzed by repressible enzymes. Accumulation of tryptophan, the end-product of the pathway, represses synthesis of the enzymes. The mechanism for this regulation in an *E. coli* cell is shown here. Five structural genes, which code for the polypeptides that make up the enzymes of the pathway, are grouped into an operon. The operon also includes a promoter and an operator. (The operator region is actually located within the promoter, but it is illustrated as a separate region in this simplified diagram of the *trp* operon.) When the operon is in the "on" mode, RNA polymerase molecules attach to the DNA at the promoter region and transcribe the structural genes. A regulatory gene, located outside the operon, codes for a repressor protein. The repressor can switch the *trp* operon off by binding to the operator and blocking access of RNA polymerase to the promoter. The repressor protein is synthesized in an inactive form, and remains inactive in the absence of tryptophan. With no repressor bound to the operator, the operon is on, producing mRNA for the enzymes that synthesize tryptophan. (**b**) As tryptophan accumulates in the cell, it inhibits its own production by activating the repressor protein. The repressor can now bind to the operator and switch the operon off.

structural genes, the operator controls access of RNA polymerase to the structural genes. All together, the structural genes, the operator, and the promoter—the entire stretch of DNA required for production of enzymes for the tryptophan pathway—is called an **operon** (Figure 17.19). Here we are dissecting one of many operons that have been discovered in *E. coli:* the *trp* operon (*trp* for tryptophan).

If the operator is the control point for transcription, what determines whether the operator is in the on or off mode? By itself, the operator is on; RNA polymerase can bind to the promoter and transcribe the structural genes. The operon must be switched off by a protein called the **repressor.** The repressor binds to the operator and blocks attachment of RNA polymerase to the promoter, stopping transcription of the structural genes. Repressor proteins are specific; that is, they recognize and bind only to the operator of a certain operon. The repressor that switches off the *trp* operon has no effect on other operons in the *E. coli* genome.

The repressor is the product of a gene called a **regulatory gene.** The regulatory gene for the *trp* repressor is located some distance away from the operon it controls (see Figure 17.19). Transcription of the regulatory gene produces an mRNA molecule that is translated into the repressor protein, which can then reach the operator of the *trp* operon by diffusion. Regulatory genes are transcribed continuously, although at a slow rate, and a few repressor molecules are always present in the cell. Why, then, is the *trp* operon not switched off permanently? First of all, the binding of repressors to operators is reversible. An operator vacillates between the on and off modes, with the relative duration of each state depending on the number of active repressor molecules around. Secondly, the *trp* repressor is first synthesized in an inactive form, which has little affinity for the *trp* operator. It assumes its active conformation and attaches to the operator only if it first binds to a molecule of tryptophan. Tryptophan functions in this regulatory system as a **corepressor,** a metabolite that cooperates with a repressor protein to switch an operon off. (A metabolite is a small organic molecule that is a precursor, intermediate, or end-product of a metabolic pathway.) As tryptophan levels rise, more tryptophan molecules bind to *trp* repressors, which can then bind to operators, shutting down tryptophan production. If the cell's tryptophan concentration drops, transcription of the operon's structural genes resumes. This is one example of how gene expression is sensitive to changes in the cell's internal and external environment.

Repressible Versus Inducible Enzymes: Two Types of Negative Gene Regulation

The enzymes of the tryptophan pathway are said to be **repressible enzymes** because their synthesis is inhibited by a metabolite (tryptophan, in this case). In contrast, the synthesis of **inducible enzymes** is stimulated, rather than inhibited, by specific metabolites. Let's investigate an example.

The disaccharide lactose (milk sugar) is available to *E. coli* if the host human drinks milk. The bacteria can absorb the lactose and break it down for energy or use it as a source of organic carbon for synthesizing other compounds. Metabolism of lactose begins with hydrolysis of the disaccharide into its two component monosaccharides, glucose and galactose. The enzyme that catalyzes this reaction is called β-galactosidase. Only a few molecules of this enzyme are present in an *E. coli* cell that has been growing in the absence of lactose—in the gut of a person who does not drink milk, for example. But if lactose is added to the bacterium's nutrient medium, it takes only about fifteen minutes for the number of β-galactosidase molecules in the cell to increase one thousandfold.

The gene for β-galactosidase is part of an operon, the *lac* operon (*lac* for lactose metabolism), that includes two other structural genes coding for proteins that function in lactose metabolism (Figure 17.20). This entire transcription unit is under the command of a single operator and promoter. The regulatory gene, located outside the operon, codes for a repressor protein that can switch off the *lac* operon by binding to the operator. So far, this sounds just like regulation of the *trp* operon, but there is one important difference. Recall that the *trp* repressor was innately inactive and required tryptophan as a corepressor in order to bind to the operator. The *lac* repressor, in contrast, is active all by itself, binding to the operator and switching the *lac* operon off. In this case, a specific metabolite, called an **inducer,** *inactivates* the repressor. For the *lac* operon, the inducer is allolactose, an isomer of lactose formed in small amounts from lactose that enters the cell. In the absence of lactose (and hence allolactose), the *lac* repressor is in its active configuration, and the structural genes of the *lac* operon are silent. If lactose is added to the cell's nutrient medium, allolactose binds to the *lac* repressor and alters its conformation, nullifying the repressor's ability to attach to the operator. Now, on demand, the *lac* operon produces mRNA for the enzymes of the lactose pathway. In the context of gene regulation, these enzymes are referred to as inducible enzymes because their synthesis is induced by a metabolite (allolactose, in this case).

Let's contrast repressible enzymes and inducible enzymes in terms of the metabolic economy of the *E. coli* cell. Repressible enzymes generally function in anabolic pathways, which synthesize essential end-products from raw materials (precursors). By suspending production of this end-product when it is already present in sufficient quantity, the cell can allocate its organic precursors and energy for other uses. On the other hand, inducible enzymes usually function in catabolic pathways, which break a nutrient down to simpler molecules. By only producing the appropriate enzymes when the nutrient is available, the cell avoids making proteins that have nothing to do. Why, for example, bother to make the enzymes that break down milk sugar when no milk is present?

Although repressible and inducible enzymes are solutions to two very different types of metabolic problems, at the level of gene regulation, they are variations on a common theme. In both cases, a cluster of structural genes for a metabolic pathway is under the control of a single operator, and binding of a repressor protein to the operator makes the structural genes inaccessible to RNA polymerase. In both cases, the ability of the repressor to bind to the operator is affected by a chemical cue from metabolism: the presence or absence of a key metabolite. The two systems differ only in how the metabolite affects the repressor. In a repressible system, such as the *trp* operon, the repressor

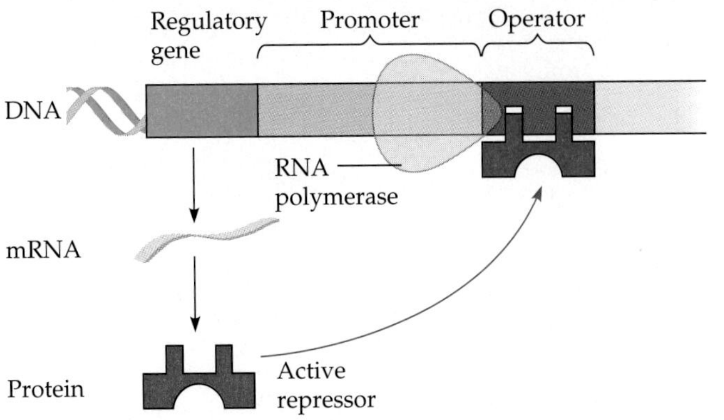

(**a**) Lactose absent, repressor active, operon off

Figure 17.20
The *lac* operon: regulated synthesis of inducible enzymes. *E. coli* uses three enzymes to take up and metabolize lactose. The structural genes for these three enzymes are clustered in a single operon, the *lac* operon. One gene, *lacZ*, codes for β-galactosidase, which hydrolyzes lactose to glucose and galactose. Another gene, *lacY*, codes for a permease, the membrane protein that transports lactose into the cell. The third gene, *lacA*, codes for an enzyme called transacetylase, whose function in lactose metabolism is still uncertain. The gene for the *lac* repressor happens to be adjacent to the *lac* operon, an unusual situation. (**a**) The *lac* repressor is innately active, and in the absence of lactose, switches off the operon by binding to the operator region. (**b**) Allolactose, an isomer formed from lactose, derepresses the operon by inactivating the repressor. Thus, the enzymes for lactose metabolism are induced.

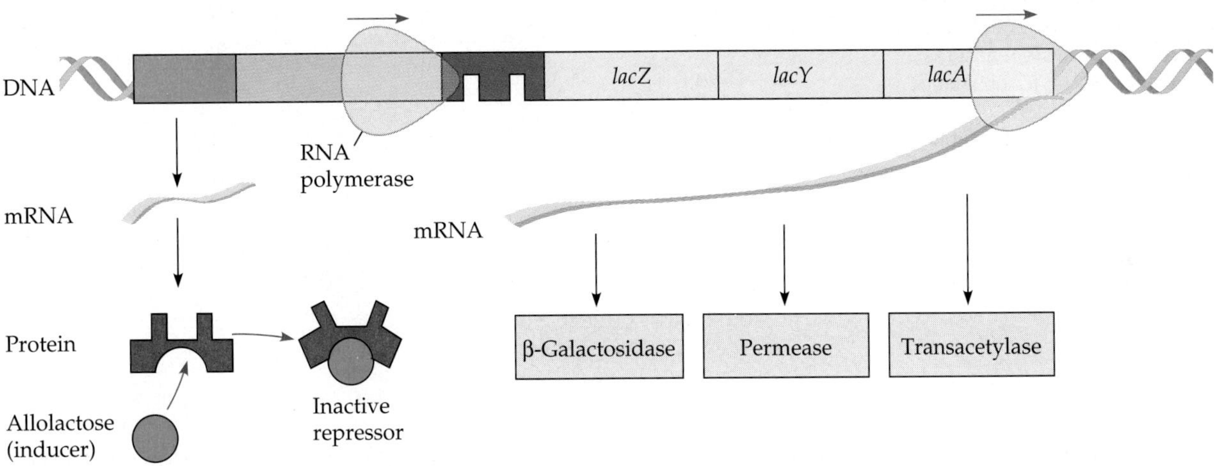

(**b**) Lactose present, repressor inactive, operon on

is synthesized in inactive form and must be activated by a metabolite behaving as a corepressor. The operon is in the on mode until repressed by accumulation of the end-product of a specific anabolic pathway. In an inducible system, such as the *lac* operon, the repressor is synthesized in an active form that will bind the operator unless inactivated by a metabolite behaving as an inducer. The operon is in the off mode until induced by the substrate of a specific catabolic pathway.

In comparing repressible and inducible enzymes, there is one more important point: Both systems are examples of *negative* control of genes, because the operons are switched off by the active form of the repressor protein. It may be easier to see this in the case of the *trp* operon, but it is true for the *lac* operon as well. Allolactose does not induce enzyme synthesis by acting directly on the genome, but by freeing the *lac* operon from the negative effect of the repressor. Technically, allolactose is more of a *derepressor* than an inducer of genes. Gene regulation is termed positive only when an activator molecule interacts directly with the genome to switch transcription on. Let's look at an example, again involving the *lac* operon.

CAP: An Example of Positive Gene Regulation

For the enzymes that break down lactose to be synthesized in appreciable quantity, it is not enough that lactose be present in the bacterial cell. The other requirement is that the simple sugar glucose be absent. Given a choice of substrates for glycolysis and other catabolic pathways, *E. coli* preferentially uses glucose, the sugar most reliably present in the nutrient medium. Starting with glucose as its only source of organic carbon, *E. coli* cells have both fuel and carbon skeletons for synthesis of their other organic molecules. Most of the genes required for catabolism of glucose are **constitutive genes,** meaning that the genes are continuously transcribed (though not necessarily at a fast rate). This contrasts with the inducible genes for catabolism of lactose and other catabolites, which are active only if the appropriate catabolite is present. Even then, these genes are quiet if glucose is in abundant supply.

How does the *E. coli* cell sense the glucose concentration, and how is this information relayed to the genome? The key is a protein called **catabolite activa-**

Figure 17.21
Positive control: catabolite activator protein. RNA polymerase has a low affinity for the promoter of the *lac* operon unless helped by catabolite activator protein (CAP), which binds to the DNA. The CAP molecule can attach to the DNA only when associated with cyclic AMP (cAMP), whose concentration in the cell is inversely proportional to the concentration of glucose. (**a**) If glucose is scarce, cAMP activates CAP, and the *lac* operon produces abundant mRNA for the lactose pathway. (**b**) But when glucose is present, cAMP is scarce, and CAP is unable to stimulate transcription. Thus, even if lactose is available, the cell will preferentially catabolize glucose, using enzymes that are always present. This regulatory system assures that *E. coli* will only gear up for consumption of lactose and other secondary catabolites when glucose is unavailable.

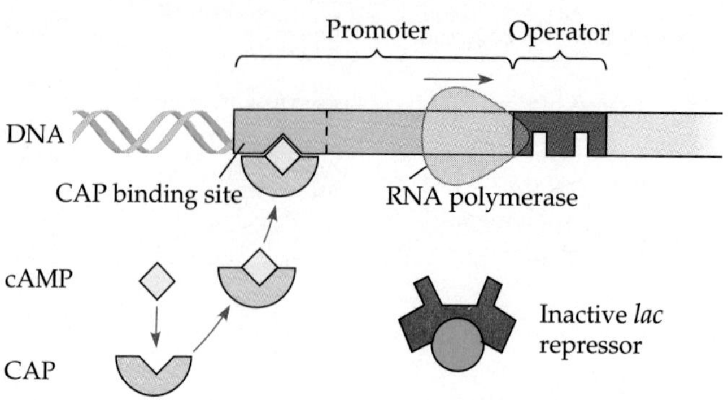

(**a**) Lactose present, glucose absent (cAMP level high): abundant *lac* mRNA synthesized

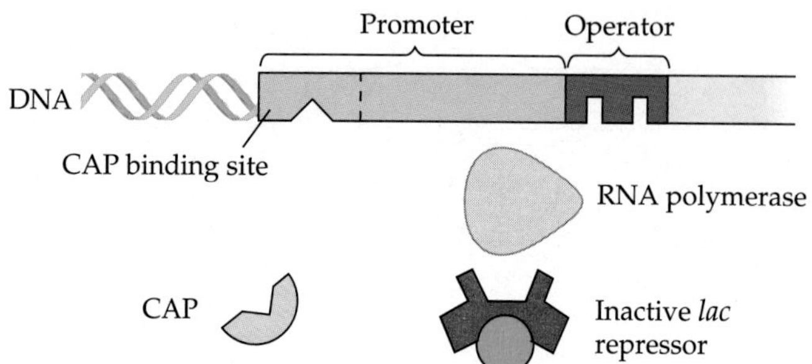

(**b**) Lactose present, glucose present (cAMP level low): little *lac* mRNA synthesized

tor protein (CAP). CAP accelerates transcription of an operon, such as the *lac* operon, by adhering to the promoter and facilitating the binding of RNA polymerase (Figure 17.21). Since CAP associates directly with the DNA to stimulate gene expression, this mechanism qualifies as positive regulation.

The activity of CAP is sensitive to the cell's glucose concentration. The absence of glucose results in the accumulation of a molecule called **cyclic AMP (cAMP),** which is derived from ATP. The CAP molecule has a binding site for cAMP, and it is actually the CAP-cAMP complex that attaches to the *lac* promoter and stimulates transcription of the genes for lactose catabolism. If glucose is added, the cAMP concentration falls, and CAP molecules disengage from the *lac* promoters. Thus, the *lac* operon is under dual control: negative control by the *lac* repressor, described earlier, and positive control by CAP. The state of the *lac* repressor (active or inactive) determines whether or not transcription of the *lac* operon's structural genes can occur; the state of CAP (plus or minus cAMP) controls the rate of transcription if the operon is repressor-free. It is as though the operon has both an on-off switch and a volume control.

Although we have used the *lac* operon as an example, CAP, unlike specific repressor proteins, works on several different operons. When glucose is present and CAP is inactive, there is a general slowdown in the synthesis of enzymes required for utilization of all catabolites except glucose. The ability of the cell to utilize alternative catabolites, such as lactose, provides backup systems enabling a cell deprived of glucose to survive. The specific catabolites present then determine which operons are switched on. These elaborate contingency mechanisms suit an organism that cannot control what its human host eats. Bacteria are remarkable in their ability to adapt—over the longer term by evolutionary changes in their genetic makeup, and over the shorter term by the control of gene expression in individual cells. Of course, the individual control mechanisms are also evolutionary products that exist because they have been favored by natural selection.

* * *

Molecular genetics was founded on the study of viruses and bacteria, the microbial models that have been the subjects of this chapter. Eukaryotic organisms are much more complex, and researchers are only beginning to learn how the control of gene expression can bring about this complexity. How, for example, can the genome present in a human zygote program the development of so many different kinds of cells in the adult organism? This is one of the problems we will investigate next, in Chapter 18.

STUDY OUTLINE

1. Many researchers study viruses and bacteria as microbial models to learn more about the molecular basis of heredity.
2. Microbial genetics has also made possible a greater understanding of many diseases and the emergence of biotechnology.

The Discovery of Viruses (pp. 344–345)

1. In the late 1800s, researchers discovered that tobacco mosaic disease is caused by an infectious agent much smaller than bacteria.

Viral Structure and Replication: An Overview (pp. 345–346)

1. Viruses are not cells, but in their simplest form consist only of nucleic acid enclosed in a protein shell called a capsid.
2. The viral genome may be single- or double-stranded DNA, or single- or double-stranded RNA, depending on the specific virus.
3. Some animal viruses have a membranous envelope outside the capsid made of both host and viral material.

How Viruses Replicate: Viral Infection (pp. 346–347)

1. Viruses are obligate intracellular parasites whose genome uses the enzymes, ribosomes, and small molecules of host cells to synthesize multiple copies of itself and the viral capsid.
2. Each type of virus has a characteristic host range, determined by specific receptor sites on host cells.

Bacterial Viruses (pp. 348–350)

1. Viruses that infect bacteria are called phages.
2. In the lytic cycle of phage replication, injection of a phage genome into a bacterium programs destruction of host DNA, production of new viruses, and digestion of the bacterial cell wall, which bursts and releases the new virus particles.
3. In a lysogenic cycle, temperate viruses coexist with their hosts by inserting their genome into the bacterial chromosome as a prophage. In this innocuous form, the virus can be passed on indefinitely to host daughter cells until at some point it is stimulated to leave the bacterial chromosome and initiate a lytic cycle.

Animal Viruses (pp. 350–354)

1. Animal viruses are often equipped with an envelope acquired from host cell membrane. The envelope allows entry and exit through the plasma membrane of host cells.
2. All types of viral genomes are represented among animal viruses. RNA viruses called retroviruses exhibit the most complex reproductive cycles. They use an enzyme called reverse transcriptase to synthesize DNA from their RNA template. The DNA can then integrate into the host genome as a provirus. HIV, the virus that causes AIDS, is among the important retroviruses.
3. Viral infections in animals create a spectrum of effects, depending on the precise mechanism of viral damage, the host's defensive responses to the pathogens, and the ability of the infected tissue to regenerate.
4. Vaccines against specific viruses stimulate the immune system to defend the host against an infection.
5. Viruses can cause or contribute to the onset of certain kinds of cancer. Tumor viruses insert viral DNA into host cell DNA, triggering subsequent cancerous changes through their own or host cell oncogenes. Oncogenes code for cellular growth factors or proteins involved in growth factor action.

Plant Viruses and Viroids (p. 354)

1. Most plant viruses are RNA viruses that seriously compromise plant growth and development.
2. Plant diseases can also be caused by viroids, tiny molecules of naked RNA that are believed to interfere with plant growth and development.

The Evolutionary Origin of Viruses (pp. 354–355)

1. Various lines of evidence point to the origin of viruses as fragments of cellular nucleic acid that came to acquire specialized packaging.

Bacterial Genomes: Replication and Mutation (pp. 355–356)

1. The bacterial chromosome is a circular DNA molecule with few associated proteins. Accessory genes are carried on smaller rings of DNA called plasmids.
2. During binary fission of a bacterial cell, chromosomal replication proceeds bidirectionally from a single origin of replication.
3. Because bacteria proliferate rapidly and have a short generation span, new mutations can affect a population's genetic variation in a relatively short period of time.

Genetic Recombination and Gene Transfer in Bacteria (pp. 356–363)

1. Bacteria have three mechanisms of transferring genes between cells: transformation, transduction, and conjugation. Along with mutations, these recombination mechanisms generate the genetic variation that makes natural selection possible.
2. In transformation, naked DNA enters the cell from the surroundings.
3. In transduction, bacterial DNA is carried from one cell to another by phages.
4. In conjugation, a primitive kind of mating, an F^+ or Hfr cell transfers DNA to an F^- cell. The transfer is brought about by a plasmid called the F (fertility) plasmid, which carries genes for the sex pili and other functions needed for mating. In an Hfr cell, the F episome is integrated into the bacterial chromosome, and the Hfr cell will transfer chromosomal DNA along with the F episome DNA in conjugation.
5. Because genes are always transferred in the same sequence during Hfr conjugation, the chromosome can be mapped by disrupting the process after various durations.
6. Genetic elements such as the F plasmid and some viral genomes can replicate either separately or as part of a chromosome.
7. R plasmids confer resistance to various antibiotics. Their

transfer between bacterial cells poses serious medical problems.

8. Transposons, DNA segments that can insert at multiple sites in the genome or move from one site to another, also contribute to genetic shuffling in bacteria.
9. Insertion sequences, the simplest transposons, may affect gene function as they move about. They consist of inserted repeats of DNA flanking a gene for transposase, the enzyme that catalyzes transposition.
10. Complex transposons include additional genes, such as genes for antibiotic resistance.

The Control of Gene Expression in Prokaryotes (pp. 363–368)

1. Cells control metabolism by regulating enzyme activity or by regulating enzyme synthesis through the activation or inactivation of selected genes.
2. In bacteria, regulated genes are often clustered into units called operons, consisting of a single promoter serving adjacent structural genes. A region called the operator serves as the on-off switch controlling the operon. Binding of a specific repressor protein to the operator shuts off transcription by blocking the attachment of RNA polymerase.
3. A repressible operon is switched off in the presence of a key metabolite, usually an end-product of a biochemical pathway; the metabolite acts as a corepressor by binding to the normally inactive repressor protein and enhancing its ability to bind to the operator.
4. Inducible enzymes are quiescent until activated by a key metabolite or inducer. In contrast with the repressible system, binding of the metabolite to the innately active repressor prevents its attachment to the operator, thereby turning on structural genes only when necessary. Inducible enzymes usually function in catabolic pathways.
5. Operons can also involve positive control via a stimulatory activator protein. For example, catabolite activator protein (CAP) stimulates transcription by binding to the promoter and enhancing its ability to associate with RNA polymerase. The promoter-binding ability of CAP is, in turn, dependent on the presence of cyclic AMP, which accumulates when glucose is scarce.

SELF-QUIZ

1. What aspect of viruses contributed most to the difficulty of their discovery?
   a. their very small size
   b. their inability to reproduce
   c. the unusual nature of some of their RNA genomes
   d. their complex protein coats
   e. their ambiguous state between living and nonliving material
2. Scientists have discovered how to put together a bacteriophage with the protein coat of phage T2 and the DNA of phage T4. If this composite phage were allowed to infect a bacterium, the phages produced in the host cell would have
   a. the protein of T2 and the DNA of T4
   b. the protein of T4 and the DNA of T2
   c. a mixture of the DNA and proteins of both phages
   d. the protein and DNA of T2
   e. the protein and DNA of T4
3. Horizontal transmission of a plant viral disease may involve
   a. the movement of viral particles through plasmodesmata
   b. the inheritance of an infection from a parent plant
   c. the spread of an infection by vegetative (asexual) propagation
   d. insects as vectors carrying viral particles between plants
   e. the transmission of proviruses via cell division
4. RNA viruses require their own supply of certain enzymes because
   a. the viruses are rapidly destroyed by host cell defenses
   b. host cells do not have RNA ⟶ RNA or RNA ⟶ DNA enzymes
   c. the enzymes translate viral mRNA into proteins
   d. the viruses use these enzymes to penetrate host cell membranes
   e. these enzymes cannot be made in host cells
5. A microbiologist found that some bacteria infected by phages had developed the ability to make a particular amino acid that they could not make before. This new ability was probably a result of
   a. transformation
   b. induction
   c. conjugation
   d. transduction
   e. transposition
6. Specialized transduction occurs when
   a. restriction enzymes cut up viral DNA
   b. reverse transcriptase creates a DNA copy from an RNA template
   c. a prophage includes some bacterial genes when it is excised from the bacterial chromosome
   d. naked pieces of double-stranded DNA are picked up by bacterial cells
   e. the genetic transfer between a Hfr cell and an F^- cell is disrupted, and only part of the donor chromosome moves into the recipient
7. Transposition differs from other mechanisms of genetic recombination because it
   a. occurs only in bacteria
   b. moves genes between homologous regions of the DNA
   c. plays little or no role in evolution
   d. occurs only in eukaryotes
   e. scatters genes to new loci in the genome
8. A particular operon produces enzymes that manufacture an important amino acid. If this process works as it does in other operons,
   a. the amino acid inactivates the repressor
   b. the enzymes produced are called inducible enzymes
   c. the repressor binds to the operator in the absence of the amino acid

d. the amino acid acts as a corepressor
e. the amino acid "turns on" enzyme synthesis

9. A mutation that renders the regulatory gene of a repressible operon nonfunctional would result in
   a. continuous transcription of the structural genes
   b. continuous synthesis of an inducer
   c. accumulation of large quantities of a substrate for the catabolic pathway controlled by the operon
   d. irreversible binding of the repressor to the promoter
   e. excessive synthesis of a catabolic activator protein
10. Which of the following information transfers is catalyzed by reverse transcriptase?
   a. RNA ⟶ RNA
   b. DNA ⟶ RNA
   c. RNA ⟶ DNA
   d. DNA ⟶ DNA
   e. RNA ⟶ protein

CHALLENGE QUESTIONS

1. When bacteria infect an animal, the number of bacteria in the body increases gradually. A graph of the growth of the bacterial population is a smoothly increasing curve, as shown in graph (a). A virus infection shows a different pattern. For a while, there is no evidence of infection, and then there is a sudden rise in the number of viruses. The number stays constant for a time, then there is another sudden increase. The graph of viral population growth looks like a series of steps, as shown in graph (b). Explain the difference in these growth curves.

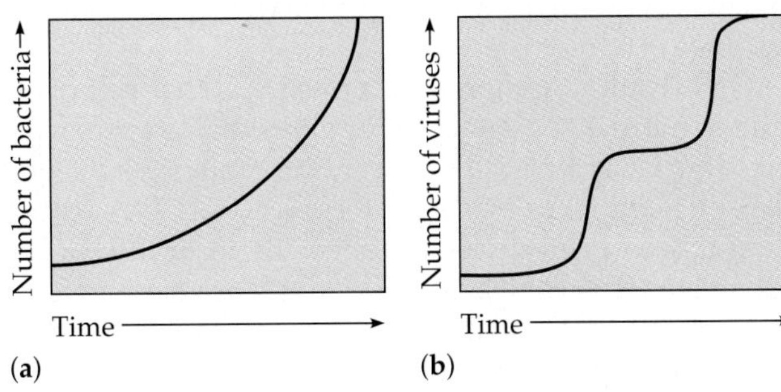

SCIENCE, TECHNOLOGY, AND SOCIETY

1. AIDS is not a new disease, just new to the Western world. It probably originated in central Africa and may have been infecting monkeys for thousands of years. There are many other serious viral diseases that infect humans and animals in remote areas. In what ways might modern technology contribute to the spread of these viruses and make new epidemics like AIDS possible?
2. Smallpox is one of humankind's oldest and most devastating viral diseases, but vaccines have wiped it out. The last known outbreak occurred in Africa in 1977. Since then, the virus has existed only in laboratory freezers. (The last smallpox victim, a British lab technician, died in 1978.) Health organizations worldwide have agreed that the last remaining smallpox viruses will be destroyed in 1993—the first time humans will have deliberately caused the extinction of a life form. Although many believe the eradication of all smallpox viruses is cause for celebration, some researchers are reluctant to destroy the remaining viruses. What are some reasons for destroying the viruses? What might be some reasons for preserving them? What do you think should be done and why?

FURTHER READING

Diamond, J. "The Arrow of Disease." *Discover*, October 1992. Columbus and other explorers imported European pathogenic viruses and bacteria to the New World.
Langone, J. "Emerging Viruses." *Discover*, December 1990.
Prusiner, S. B. "Molecular Biology of Prion Diseases." *Science*, June 14, 1991. The case for certain infectious agents (prions) that are *proteins* rather than nucleic acids.
Ptashne, M. "How Gene Activators Work." *Scientific American*, January 1989. How regulatory proteins bind to DNA.
Tiollais, P. and M.-A. Buendia. "Hepatitis B Virus." *Scientific American*, April 1991. Genetic engineering may lead to new vaccines against this liver disease.
Varmus, H. "Reverse Transcription." *Scientific American*, September 1987. Discusses the evidence for the surprisingly widespread occurrence of reverse transcription in eukaryotic cells.
Watson, J. D., N. H. Hopkins, J. W. Roberts, J. A. Steitz, and A. M. Weiner. *Molecular Biology of the Gene*, 4th ed. Menlo Park, CA: Benjamin/Cummings, 1987. Chapters 7, 8, 16, 17, and 24.
"What Science Knows About AIDS." *Scientific American*, October 1988. An entire issue devoted to AIDS, with ten articles on various aspects of the disease and the virus that causes it.

18 Genome Organization and Expression in Eukaryotes

Genome Organization at the Microscopic Level

Genome Organization at the Molecular Level

Genome Plasticity

The Control of Gene Expression

The Role of Small Molecules in Regulating Eukaryotic Gene Expression

Gene Expression and Cancer

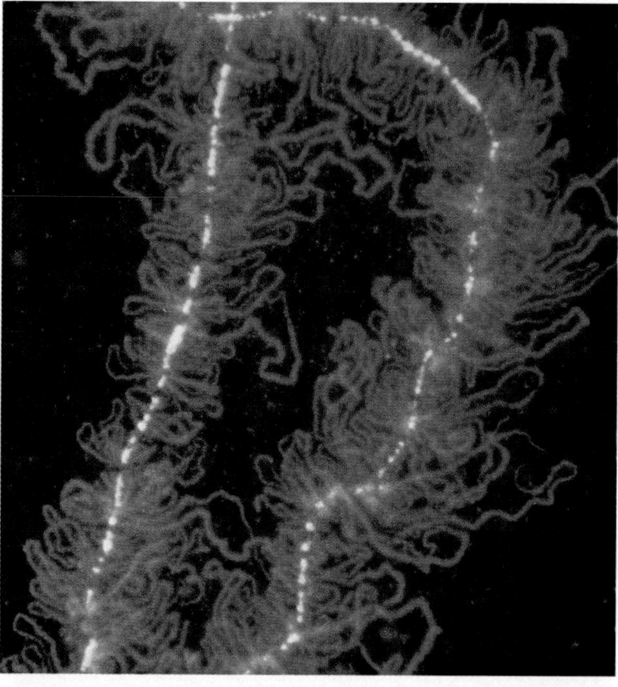

25 μm

Figure 18.1
A eukaryotic chromosome in action. This micrograph gives a sense of how much DNA is packed into a eukaryotic chromosome, in this case a "lampbrush" chromosome from a salamander oocyte (developing ovum, or unfertilized egg) (LM). The red-stained loops of chromatin are actively synthesizing RNA (transcription). A much larger amount of chromatin, stained white, is compactly packed into the main axis of each of this chomosome's two chromatids. In this chapter, you will learn more about how eukaryotic DNA is organized and how its expression is regulated.

Eukaryotic cells face the same challenges in expressing their genes as prokaryotic cells, with two main differences: the vastly greater size of the typical eukaryotic genome and the importance of cell specialization in eukaryotes. These two complications present a formidable information-processing task for the eukaryotic cell.

Consider the genome of a human cell, which has 50,000 to 100,000 genes—about fifty times more than a typical bacterium. The genomes of humans and other eukaryotes also include a large amount of nongene DNA, DNA that does not program for synthesis of RNA or protein. The entire mass of DNA must be precisely replicated with each turn of the cell cycle, as we saw in Chapter 11. Managing such a large amount of DNA requires that the eukaryotic genome be more complexly organized than prokaryotic DNA. In both prokaryotes and eukaryotes, DNA is associated with proteins, but in eukaryotes, the DNA-protein complex, called chromatin, is ordered into higher structural levels (Figure 18.1).

One result of genomic complexity is that cell division is more elaborate in eukaryotes than in prokaryotes (see Chapter 11). However, the challenges the eukaryotic cell faces during interphase may be greater than those of cell division, especially in multicellular organisms. Interphase is when genes are expressed—transcribed and translated into the products that enable each cell to carry out its role in the organism. Like unicellular organisms, the cells of multicellular organisms must continually turn certain genes on and off in response to signals from their external and internal environments. In addition, however, gene expression must be controlled on a long-term basis for **cellular differentiation**—the divergence in structure and function of different types of cells as they become specialized during an organism's development. Highly specialized cells, such as those of muscle or nervous tissue, express only a tiny fraction of their genes. In fact, a typical human cell expresses only 3% to 5% of its genes at any given time. The enzymes that transcribe DNA must locate the right genes at the right time, which is like finding—and threading—a needle in a haystack. When gene action goes awry, serious imbalances and diseases, including cancer, can arise. Thus, the question of how eukaryotic genes are regulated is paramount for medical research as well as for basic biology.

Only twenty years ago, the mechanisms that control gene expression in eukaryotes seemed almost hopelessly out of reach. Since then, new research methods have empowered molecular biologists to begin solving some of these once impenetrable mysteries. Equipped with recombinant DNA techniques for cloning genes and increasingly fast methods of sequencing DNA (discussed in Chapter 19), biologists are uncovering many of the details about how genes are organized within the eukaryotic genome and how their expression is regulated.

Despite the extra "nuts and bolts" it requires, the control of gene activity in eukaryotes involves the same basic principles as prokaryotic gene regulation. In all organisms, gene activity is regulated by DNA-binding proteins that also interact with other proteins and with environmental factors. And in most cases, it is the transcription of the DNA that is specifically controlled. However, the greater complexity of chromosome structure and gene organization in eukaryotes and the greater complexity of eukaryotic cell structure do offer additional opportunities for controlling gene expression. So before examining eukaryotic gene expression and its control, we will look at how eukaryotic DNA is physically packaged into chromatin and chromosomes, and how genes and other DNA sequences are organized within the genome. We will also consider how the physical and chemical plasticity (changeability) of the eukaryotic genome affects the availability of its genes for expression.

GENOME ORGANIZATION AT THE MICROSCOPIC LEVEL

Although both prokaryotic and eukaryotic cells contain hereditary material in the form of double-stranded DNA, their genomes are organized differently. Prokaryotic DNA is usually circular, and the nucleoid it comprises is so small that it can only be seen with an electron microscope. The "chromosome" of *E. coli*, for example, is composed of only 4.3×10^6 nucleotide pairs, and while it is associated with various proteins to form a sort of chromatin, the protein molecules are relatively few. (The bacterial chromosome does have some additional structure: The DNA-protein fiber forms a number of loops, with the bases of the loops apparently anchored to the plasma membrane.) In contrast, eukaryotic chromatin consists of DNA precisely complexed with a large amount of protein. During interphase, the chromatin fibers are usually highly extended and tangled. When interphase cells are stained with basic dyes, the chromatin (which is acidic because of its DNA) appears as a diffuse colored mass. You learned in Chapter 11, however, that as a cell prepares for mitosis, its chromatin coils and folds up ("condenses") to form a number of short, thick, discrete chromosomes that, when stained, are clearly visible with a light microscope.

Eukaryotic chromosomes contain an enormous amount of DNA relative to their size. Each chromosome contains a single uninterrupted DNA double helix that, in humans, typically contains about 2×10^8 nucleotide pairs. If extended, such a DNA molecule would be about 6 cm long, thousands of times longer than the diameter of a cell nucleus. All this DNA—and the DNA of the other 45 human chromosomes, as well—can fit into the nucleus because of an elaborate, multilevel system of packing.

Nucleosomes, or "Beads on a String"

Small proteins called **histones** are responsible for the first level of DNA packing in eukaryotic chromatin. In fact, the amount of histone in chromatin is approximately equal to the amount of DNA. Histones have a high proportion of positively charged amino acids (lysine and arginine), and they bind tightly to the negatively charged DNA, forming chromatin. There are five types of histone found in most eukaryotic cells. Histones are very similar from one eukaryote to another, suggesting that evolution has been relatively conservative with the histone genes.

In electron micrographs, unfolded chromatin has the appearance of beads on a string (Figure 18.2a). Each "bead" is a **nucleosome,** the basic unit of DNA packing. The nucleosome consists of DNA wound around a protein core composed of two molecules each of four types of histone. A molecule of the fifth histone, called H1, may be attached to the outside of the "bead." Chromatin organization in nucleosomes may influence gene expression by limiting the access of transcription proteins to DNA.

Higher Levels of DNA Packing

The beaded string undergoes higher-order packing. This is most strikingly evident when the extended interphase chromatin coils and folds further to produce the thickened, compact chromosomes we see during mitosis. In the laboratory, mitotic chromosomes can be isolated and unraveled to reveal several orders of chromatin coiling. Figure 18.2b–d illustrates the various structures in order of increasing compaction. With the aid of histone H1, the beaded string can coil tightly to make a cylinder 30 nm in diameter, dubbed the 30-nm chromatin fiber (Figure 18.2b; some cell biologists call it the 30-nm solenoid, using the term for a coil of insulated electrical wire). The 30-nm fiber, in turn, forms loops called *looped domains* (Figure 18.2c).

In a mitotic chromosome, the looped domains themselves coil and fold, further compacting all the chromatin to produce the characteristic metaphase chro-

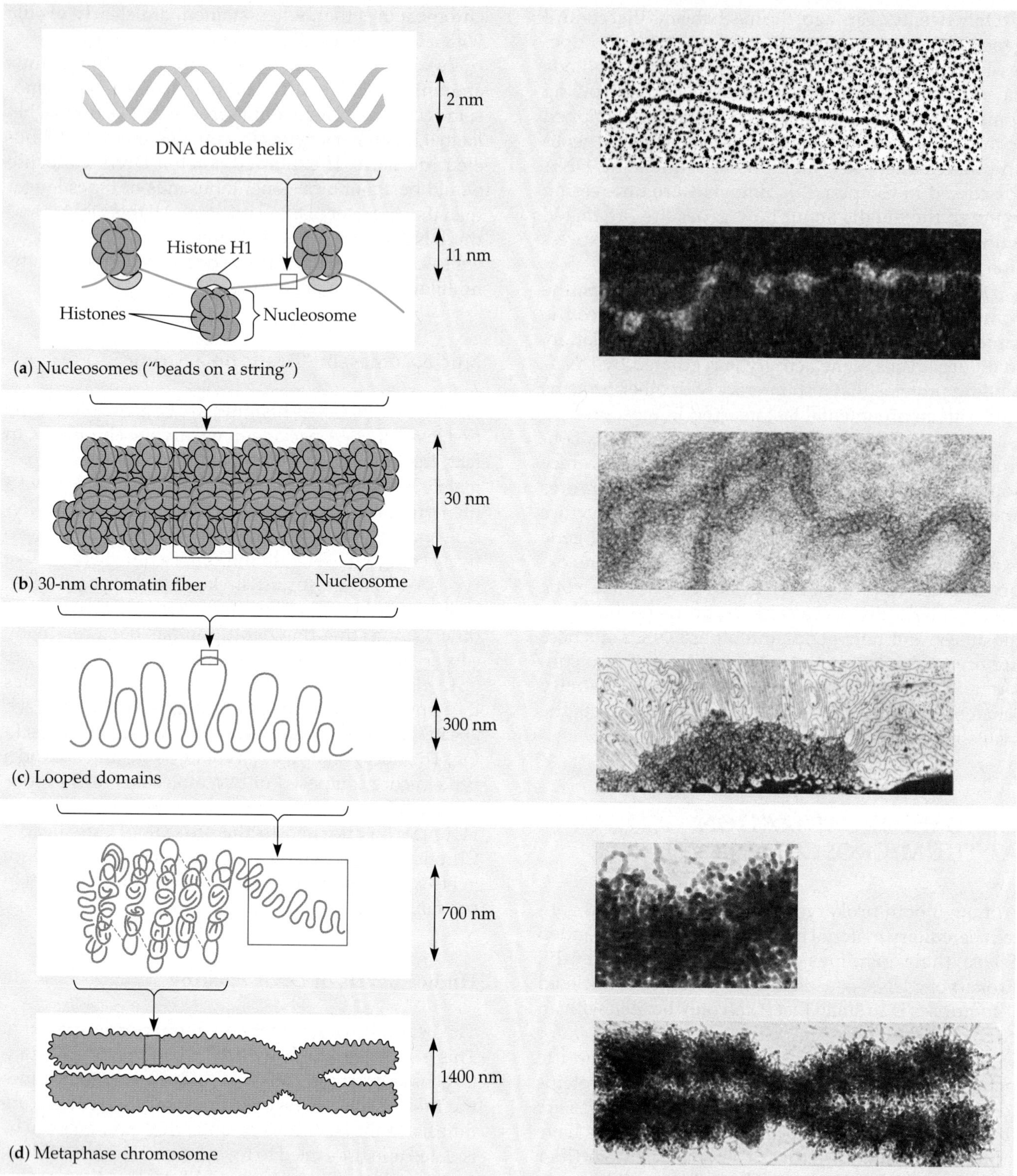

Figure 18.2
Levels of chromatin packing. This series of diagrams and transmission electron micrographs shows a current model for the progressive stages of DNA coiling and folding, which culminate in the highly condensed metaphase chromosome. (**a**) DNA in association with histone to form "beads on a string," consisting of nucleosomes in an extended configuration. Each nucleosome has two molecules each of four types of histone. The fifth histone (called H1) may be present on DNA adjacent to the "bead." (**b**) The 30-nm chromatin fiber, thought to be a tightly wound coil with six nucleosomes per turn. (**c**) Looped domains of 30-nm fibers, visible here because a compact chromosome has been experimentally unraveled. (**d**) A metaphase chromosome, with all its DNA highly condensed into a compact structure.

mosome (Figure 18.2d). The loops seem to be anchored onto a framework of nonhistone proteins (visible at the bottom of the micrograph in Figure 18.2c). Viewed as a whole, Figure 18.2 will give you a sense of how the various levels of folding can pack a huge amount of DNA into one chromosome.

Though interphase chromatin is generally much less condensed than the chromatin of mitotic chromosomes, it shows several of these same levels of higher-order packing. Its "beaded string" is usually coiled into a 30-nm fiber, and the 30-nm fiber is folded into looped domains. There is evidence that these looped domains are attached to an interphase scaffolding on the inside of the nuclear envelope. Recent experiments suggest that the chromatin of each chromosome occupies a restricted area within the nucleus, and that the chromatin fibers of different chromosomes do not become entangled with each other.

Even during interphase, portions of certain chromosomes in some cells exist in the highly condensed state represented in Figure 18.2d. Such interphase chromatin, which is visible with a light microscope, is called **heterochromatin,** to distinguish it from the less compacted **euchromatin** ("true chromatin"). What is the function of this selective condensation in interphase cells? Active transcription takes place on euchromatin. The formation of heterochromatin may be a sort of coarse adjustment in the control of gene expression, for it is known that the DNA of heterochromatin is not transcribed. The most striking example in mammalian cells is the Barr body, an interphase *X* chromosome that is almost entirely heterochromatin. One of the two *X* chromosomes in each female somatic cell is a Barr body (see Figure 14.12), and only the genes of the other *X* chromosome are expressed. It is not known whether the chromatin condensation of the Barr body is what prevents transcription, but it certainly minimizes the space this nontranscribed DNA occupies in the interphase nucleus.

GENOME ORGANIZATION AT THE MOLECULAR LEVEL

In prokaryotes, most of the DNA in a genome codes for protein (or tRNA and rRNA), with the small amount of noncoding DNA consisting mainly of control sequences, such as promoters. Moreover, the coding sequence of nucleotides along a prokaryotic gene proceeds from start to finish without interruption. In eukaryotic genomes, in contrast, most of the DNA does *not* encode protein or RNA. Also, certain DNA sequences may be present in multiple copies in a eukaryotic genome, and coding sequences may be interrupted by long stretches of noncoding DNA (introns; see Chapter 16). In this section, we will examine the functional significance of how DNA is organized in a eukaryotic genome.

Repetitive Sequences

Approximately 10% to 25% of the total DNA of multicellular eukaryotes is made up of short sequences (typically five to ten nucleotides) repeated in series thousands or even millions of times. The copies may be identical or merely similar. In either case, the nucleotide compositions of highly repetitive sequences are often different enough from the rest of the cell's DNA to have a different density, so that researchers can isolate the repetitive DNA by ultracentrifugation. DNA that can be isolated in this way is called **satellite DNA,** because it appears as a "satellite" band separate from the rest of the DNA in the centrifuge tube. In chromosomes, most of the satellite DNA is located at the tips and the centromeres. Apparently, this DNA plays a *structural* rather than a genetic role in the cell, functioning in the replication of the chromosome and the separation of chromatids in mitosis and meiosis.

Multigene Families

As in prokaryotes, the eukaryotic DNA sequences that code for proteins or RNA—the genes—are usually present in single copies in the genome (called unique sequences). However, some genes are represented by more than one copy, and others resemble each other in nucleotide sequence. A collection of identical or similar genes is called a multigene family, and each family probably evolved from a single ancestor gene.

The members of a multigene family may be clustered or dispersed in the genome, with identical genes usually clustered. The genes for the histone proteins exist as families of multiple identical genes, but almost all other families of identical genes consist of genes whose final product is RNA, rather than protein. An example is the family of identical genes for the major ribosomal RNA (rRNA) molecules (Figure 18.3). These genes are repeated in series (tandemly) hundreds to thousands of times in the genomes of multicellular eukaryotes, forming huge tandem arrays of genes that enable the cells to make the millions of ribosomes needed for active protein synthesis.

The classic examples of multigene families of *nonidentical* genes are the two related families of genes that encode globins, the α and β polypeptide subunits of hemoglobin (see Figure 5.26b). One family, located on chromosome 16 in humans, encodes various versions of α-globin; the other, on chromosome 11, encodes versions of β-globin (Figure 18.4). The different versions

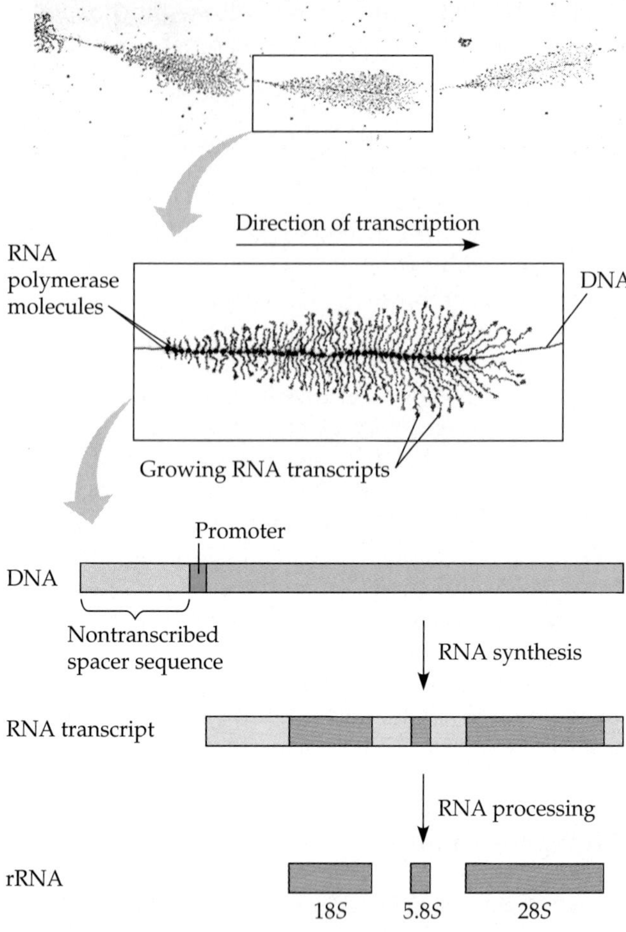

Figure 18.3
Part of a family of identical genes for ribosomal RNA. The micrograph at the top shows three of the hundreds of copies of rRNA genes in a salamander genome (TEM). Each "feather" corresponds to an rRNA gene being transcribed, from left to right, by about 100 molecules of RNA polymerase (the dark dots along the DNA). The growing RNA transcripts extend out from the DNA. The genes are arranged in tandem, each one separated from the next by a spacer sequence of nontranscribed DNA. The RNA transcripts are processed to yield three kinds of rRNA molecules: 18*S*, 5.8*S*, and 28*S*. (The *S* designations refer to their sedimentation rates in the ultracentrifuge.)

of each globin subunit are expressed at different times in development, allowing the hemoglobin to change and function effectively in the changing environment of the developing animal. Similarities in the sequences of the various globin genes indicate that the α-like globins and β-like globins all evolved from a common ancestral globin.

How do families of genes arise from a single gene? The most likely explanation for families of identical genes is that they arise by repeated gene duplication. This occurrence, called tandem gene duplication, results from mistakes made in DNA replication and recombination (see Chapter 16). Families of nonidentical genes probably arise from mutations that accumulate in duplicated genes over a long time. The existence of DNA segments called pseudogenes is evidence for this process of gene duplication and mutation. **Pseudogenes** have sequences very similar to real (functional) genes but lack the sites (for example, promoters) necessary for gene expression. The globin gene families include several pseudogenes within the stretches of noncoding DNA between the functional genes.

The globin pseudogenes also provide evidence that duplicated genes may move about in the genome by a transposition process involving reverse transcription (synthesis of DNA from an RNA template, as in the molecular biology of retroviruses; see Chapter 17). Two characteristics of the pseudogenes point to a messenger-RNA intermediate: They lack introns, and they have poly-A tails, just like mRNA.

Pseudogenes and repetitive DNA do not account for all of the enormous amount of noncoding DNA in the genomes of multicellular eukaryotes. Significant amounts of additional noncoding DNA are found *within* genes, as introns (see Chapter 16). Let's now take a closer look at how introns and other features are organized in a typical protein-encoding gene.

Organization of a Typical Eukaryotic Gene: A Review

The DNA that makes up a gene and its control regions is typically organized as shown in Figure 18.5, which reviews and extends what you learned about eukaryotic genes in Chapter 16. The most striking difference between this gene and a prokaryotic gene is the presence of introns, noncoding sequences that are interspersed within the coding sequence. Recall from Chapter 16 that RNA polymerase attaches to a promoter sequence at the 5′ ("upstream") end of the gene and proceeds to transcribe the introns along with the coding sequences, called exons. The introns are removed in later RNA processing, so that they do not appear in the mature mRNA. The processing of the initial RNA transcript also includes the addition of a modified guanosine triphosphate cap at the 5′ end and a poly-A tail at the 3′ end.

Another distinctive feature of the eukaryotic gene illustrated in Figure 18.5 is the presence of other noncoding control sequences that may be located thousands of bases away from the promoter. These sequences, called **enhancers,** can have a powerful influence on transcription of the associated gene, an important control mechanism we will examine later in the chapter.

The Arrangement of Coordinately Controlled Genes

Coordinately controlled genes—genes of related function that are switched on and off as a unit—are arranged differently on a eukaryotic chromosome than

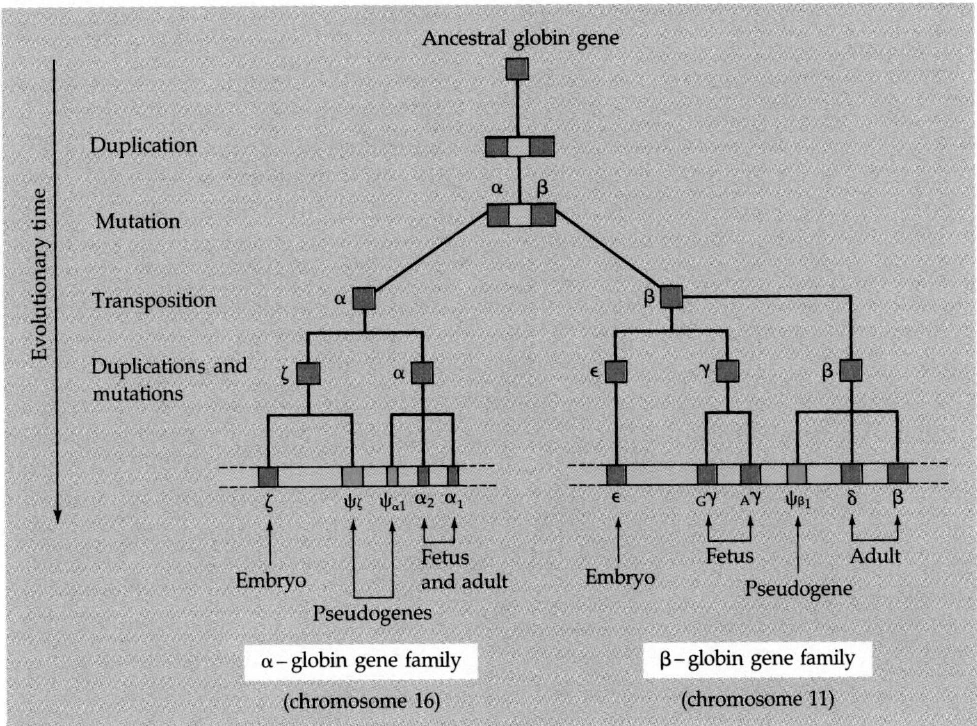

Figure 18.4
The evolution of α-globin and β-globin. Each gene family consists of a group of similar, but not identical, genes clustered together on a chromosome. The various genes are represented by different Greek letters. In each gene family, the genes are arranged in order of their expression during development; expression of individual genes is turned on or off in response to the organism's changing environment as it develops from an embryo into a fetus, and then into an adult. At all times in development, functional hemoglobin consists of two α-like and two β-like polypeptides. Separating the functional genes within each family cluster are long stretches of noncoding DNA, which include pseudogenes, nonfunctional nucleotide sequences very similar to the functional genes. Presumably, the different genes and pseudogenes in each family arose from gene duplication of an original α or β gene followed by mutation. In fact, the original α and β genes themselves undoubtedly arose in much the same way from a common ancestral globin gene. Transposition (see Chapter 17) put the α-globin and β-globin families on different chromosomes, probably early in their evolution.

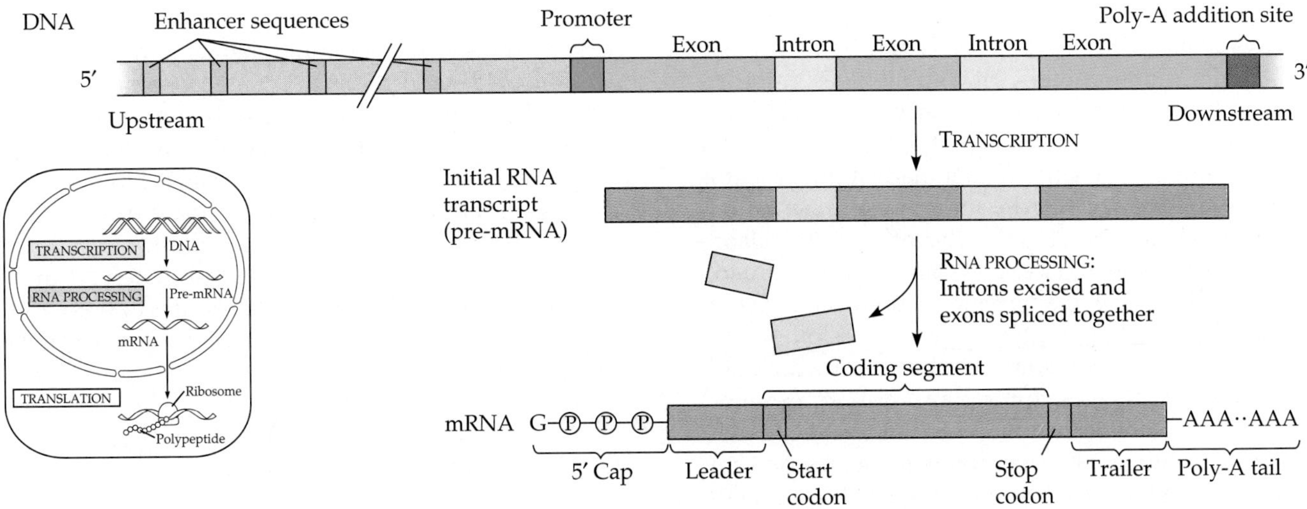

Figure 18.5
A review of the molecular anatomy of a eukaryotic gene and its transcript. The promoter and associated neighboring sequences function in the initiation of transcription, which proceeds "downstream" in the 5′⟶3′ direction (see Chapter 16). Located as many as thousands of nucleotides upstream or downstream from the promoter are enhancer sequences, sites where proteins that regulate transcription of the gene can bind. After the initial RNA transcript is made, processing enzymes excise introns and add the methylated guanosine 5′ cap and poly-A tail. The mRNA is now ready for export to the cytoplasm.

in the genomes of prokaryotes. In Chapter 17, you learned that in prokaryotes, genes that are turned on and off together are often clustered into an operon; they are adjacent to each other in the DNA molecule and share the regulatory sites located at one end of the cluster. All the genes of the operon are transcribed into a single mRNA molecule and are translated together. After prokaryotic operons were described in the 1960s, many molecular biologists expected that operonlike systems would also be the basis for gene regulation in eukaryotes. However, such operons have not been found in eukaryotic cells. Genes coding for the enzymes of a metabolic pathway, for example, are often scattered over different chromosomes in the eukaryotic genome. Even when genes for related functions are located near one another on the same chromosome, each gene has its own promoter and is individually transcribed. Nevertheless, scattered collections of eukaryotic genes are often coordinately expressed.

A plausible mechanism for coordinate gene expression in eukaryotes involves the presence of a specific nucleotide sequence in regulatory elements associated with every gene of a scattered group. This sequence would be recognized by a single regulatory protein (just as a prokaryotic operator sequence is recognized by a repressor or activator). Examples of this sort of arrangement are known in bacteria, and there is evidence for such shared sequences in eukaryotic cells. The sequences may serve as signals to specific regulatory molecules, saying, in effect, "Transcribe or repress all these genes in synchrony."

GENOME PLASTICITY

A gene that is present in an organism's genome is not always available to be expressed. It may be available in some cells and not others, or at some times in the organism's development and not others. Molecular biologists have also discovered that a gene may, under some conditions, be amplified or made more available than usual. In other words, an organism's genome is plastic, or changeable, in ways that affect the availability of specific genes for expression. We have already observed that DNA changes in its *physical arrangement* and that these changes affect gene expression. (Genes in highly folded chromatin, including heterochromatin and mitotic chromosomes, are not expressed.) However, we are accustomed to the idea that, except for rare mutations, the chemical composition and nucleotide sequence of an organism's DNA is constant during its lifetime. So it may come as a surprise that there are important exceptions, enough of them to allow us to say that the *structural organization* of an organism's genome is also somewhat plastic. Movement of DNA within the genome and chemical modification of DNA have a major influence on gene expression.

Gene Amplification and Selective Gene Loss

Sometimes the number of copies of a gene or gene family may temporarily increase in some tissues during a particular stage of development. For instance, consider the genes for ribosomal RNA (rRNA) in amphibians. As in most eukaryotes, multiple copies of these genes are built into the genome of every cell. A developing ovum, however, synthesizes a million or more additional copies of the rRNA genes, which exist as extrachromosomal circles of DNA. This selective replication of certain genes, or **gene amplification,** is a potent way of increasing expression of the rRNA genes, enabling the developing egg cell to make enormous numbers of ribosomes. These ribosomes make possible a burst of protein synthesis once the egg is fertilized. The extra copies of the rRNA genes are hydrolyzed during early embryonic development.

In other instances, especially in certain insects, genes are selectively lost in certain tissues (although not in the cells that give rise to gametes, of course). In fact, as we will see in Chapter 43, whole chromosomes or parts of chromosomes may be eliminated from certain cells early in embryonic development.

DNA Methylation

In another type of chemical change of the genome, the nitrogenous bases of DNA are sometimes modified by a process called methylation. **DNA methylation** is the addition of methyl groups ($—CH_3$) to bases of DNA after DNA synthesis. The DNA of most plants and animals has methylated bases, usually cytosine. About 5% of the cytosine bases in eukaryotic DNA are methylated. Inactive DNA, such as that of Barr bodies, is generally highly methylated compared to DNA that is actively transcribed, although there are important exceptions. Comparison of the same genes in different types of cells (say, from different tissues) shows that the genes are usually more heavily methylated in the cells where they are not expressed. In addition, drugs that inhibit methylation can induce gene reactivation, even in Barr bodies. Is DNA methylation a cellular mechanism for the long-term control of gene expression? Molecular biologists are still trying to figure out whether methylation is an initial cause of reduced gene expression or a result of it.

Rearrangements in the Genome

A surprisingly common change in the genome is the shuffling of substantial stretches of DNA. Here we are not talking about the genetic recombination that goes on in meiosis but rather about rearrangements that occur in somatic cells of an organism. Such rearrangements may have powerful effects on gene expression.

Transposons All organisms seem to have transposons, stretches of DNA that are particularly prone to moving from one location to another within the genome. Transposons were discussed in detail in Chapter 17. Recall that if a transposon "jumps" into the middle of a coding sequence of another gene, it prevents the normal functioning of the interrupted gene. If the transposon inserts within a sequence that is involved in regulating transcription, the transposition may increase or decrease the production of one or more proteins. In some cases, the transposon itself carries a gene that is activated when it is inserted just downstream from an active promoter.

Yeast Mating-Type Genes Rearrangements of DNA, which can activate or inactivate certain genes, have known functions in certain kinds of cells. An example is the ability of yeast cells to change from one mating type to another as a result of genetic shuffling. Yeasts, which are unicellular fungi, exist in two mating types, α (alpha) and *a,* and can mate only if they are of opposite mating type. The mating type is determined by whether the α allele or the *a* allele is present at a site in the genome called the mating-type locus *(MAT).* However, yeast cells can change mating type by switching the allele at the mating-type site (Figure 18.6). "Silent" (inactive) copies of both alleles are located elsewhere in the same chromosome, and the allele in the mating-type locus is sometimes excised and replaced with a copy of the other allele from another site. Silent copies of both alleles remain available, enabling a cell's descendants to change mating type again and again. This process is called the **cassette mechanism,** because the mating-type locus is like a tape-player slot into which either the α or *a* "cassette" (allele) can be inserted and "played" (transcribed).

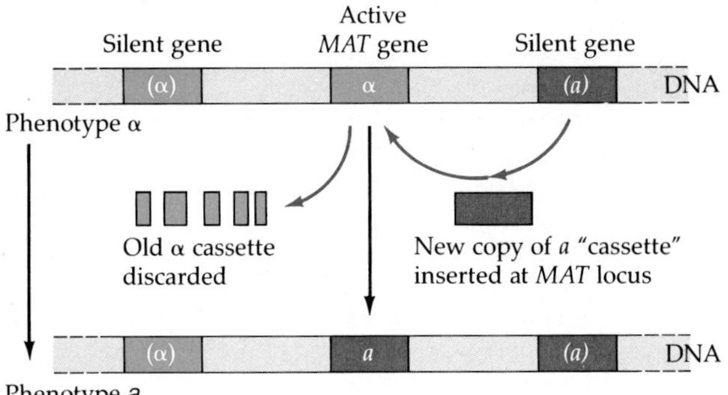

Figure 18.6
Cassette mechanism for mating-type switches in yeast. The mating type of a yeast cell, α or *a*, is determined by which of two alleles is present at a genetic locus called the mating-type locus *(MAT).* Yeast cells carry silent (unexpressed) versions of both α and *a* alleles at loci distant from *MAT* and can move copies of these "cassettes" to *MAT.* Such switching can occur as frequently as once in every cell cycle.

Immunoglobulin Genes In multicellular eukaryotes, at least one set of genes undergoes permanent rearrangements of DNA segments during cellular differentiation, and the function of these rearrangements is well known. These changes occur in the mammalian genes that encode antibodies, or **immunoglobulins,** proteins that specifically recognize and help combat viruses, bacteria, and other invaders of the body.

Immunoglobulins are made by cells of the immune system called B lymphocytes, which are a type of white blood cell. B lymphocytes are highly specialized, with each differentiated cell and its descendants producing one specific type of antibody that attacks a specific invader. This extreme cell specialization is made possible by the unique organization of the antibody gene in each differentiated cell and its descendants. As an unspecialized cell of the immune system differentiates into a B lymphocyte, its antibody gene is pieced together randomly from several DNA segments that are physically separated in the genome of an embryonic cell. The human immune system, with its millions of subpopulations of B lymphocytes, can make millions of different kinds of antibody molecules (see Chapter 39).

The basic immunoglobulin (antibody) molecule is shown at the bottom right of Figure 18.7. It consists of four polypeptide chains held together by disulfide bridges. Each chain has two major parts: a constant region, which is the same for all antibodies of a particular class, and a variable region, which gives a particular antibody its unique function—the ability to recognize and bind to a specific foreign molecule. In the genome of an embryonic cell, the DNA region coding for the constant part of each type of antibody polypeptide is separated by a long stretch of DNA from a location containing hundreds of variable-region–coding segments (top of Figure 18.7). As a B lymphocyte differentiates, a specific variable segment of the DNA is connected to a constant segment by the deletion of intervening DNA. The joined segments form the continuous sequence of nucleotides that functions as the gene for one of the immunoglobulin polypeptides. Much of the variation in antibodies thus arises from the different combinations of variable and constant regions in the immunoglobulin polypeptides, as well as from the different combinations of polypeptides that form the complete antibody molecules. The formation of functional antibody genes shows how a chemical rearrangement of DNA can produce individual somatic cells with distinctive genomes.

THE CONTROL OF GENE EXPRESSION

The presence or absence of a particular DNA sequence in a cell, its physical state (as heterochromatin or eu-

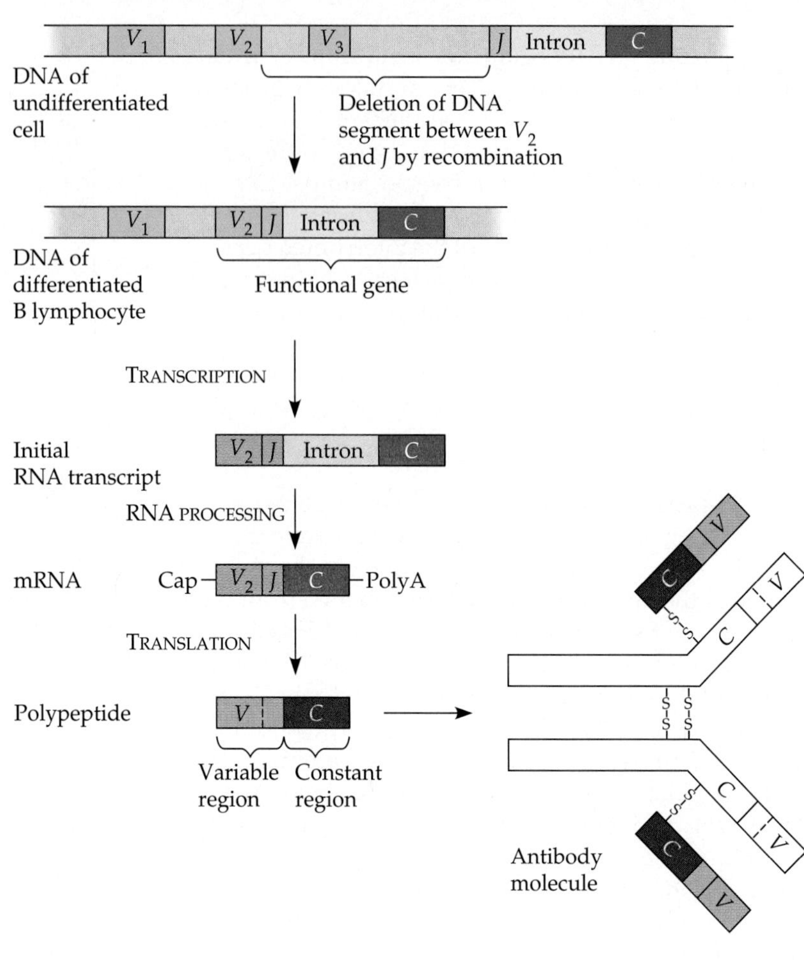

Figure 18.7
DNA rearrangement in the maturation of an antibody gene. The DNA of antibody genes in undifferentiated cells carries coding segments for hundreds of different antibody variable *(V)* regions (only three are shown here), for several different junction *(J)* regions, and for one or more different constant *(C)* regions. During differentiation of white blood cells called B lymphocytes, a long segment of DNA, from the end of one of the *V* segments to the beginning of one of the *J* segments, is deleted. This deletion brings a *V* segment (in this case V_2) adjacent to a *J* segment and produces a gene that can be transcribed. The RNA transcript is processed in the usual way to remove introns (and any extra *J* segments), and the resulting mRNA is translated into one of the polypeptide chains for an antibody molecule. The amino acids coded by the *J* segment are considered part of the variable region of the polypeptide. By bringing together *V*, *J*, and *C* regions of DNA in random combinations (many more than would be possible from the simplified version in this diagram), this genome plasticity helps arm the immune system with diverse antibody-producing lymphocytes, each keyed to a particular foreign invader (see Chapter 39). Different combinations of the polypeptide chains that make up each antibody are another source of diversity in these proteins.

chromatin), and its location within the genome all determine whether its genes are available to be expressed. We now turn our attention to mechanisms that regulate expression of the *available* genes. Figure 18.8 gives an overview of the steps in gene expression in eukaryotes, highlighting the stages at which expression of a given gene can be regulated. In this section, we will consider each of these levels of regulation, starting with the control of transcription.

Transcriptional Control of Gene Expression

In both prokaryotes and eukaryotes, RNA synthesis depends on RNA polymerase enzymes acting in concert with numerous proteins called transcription factors. The polymerase and transcription factors bind to specific sequences within the promoter region just upstream from the coding sequence of a gene, and the polymerase moves along the DNA template, producing the complementary strand of RNA. In eukaryotes, additional transcription factors bind selectively to enhancer regions of DNA that may be thousands of nucleotides away from the promoter and coding sequence (see Figure 18.5).

The specific associations between transcription factors and enhancer sites in the genome play an important role in the control of gene expression in eukaryotes. However, molecular biologists do not yet understand just *how* enhancers stimulate transcription of specific genes. According to one hypothesis, transcription is activated when a hairpin loop in the DNA brings the transcription factor attached to the enhancer into contact with the transcription factors and polymerase at the promoter (Figure 18.9). Diverse transcription factors keyed to different enhancer sequences in the genome may selectively activate expression of specific genes at appropriate stages in the development of a cell.

Over 100 transcription factors, including those that bind to enhancers, have been discovered so far in eukaryotes. As diverse as these proteins are, they can be grouped into three main classes. Each class is defined by the structure of the functional region, or domain, by which the protein binds to DNA. These three types of DNA-binding domains are the helix-turn-helix do-

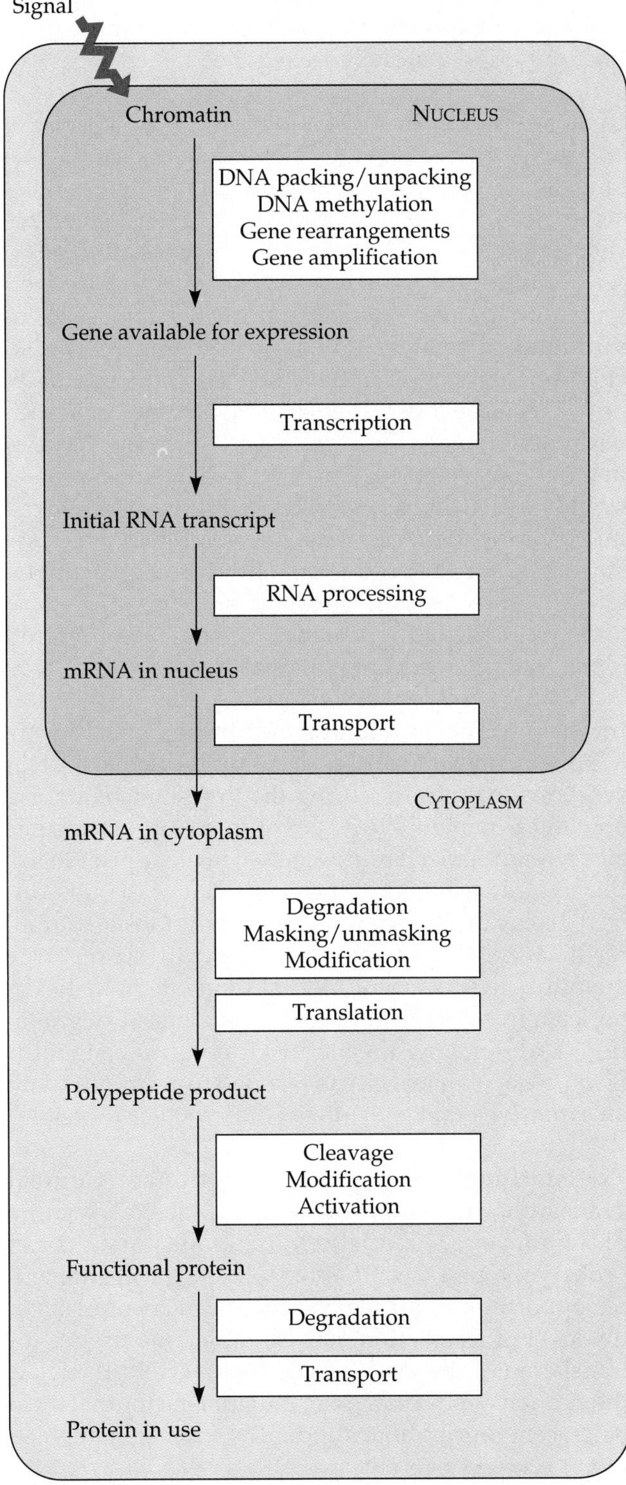

Figure 18.8
Opportunities for the control of gene expression in eukaryotic cells: an overview. Unlike a prokaryotic cell, a eukaryotic cell has a nuclear envelope that separates transcription and translation in both space and time. This feature offers greater opportunity for posttranscriptional control in the form of RNA processing. In addition, eukaryotes have a greater variety of control opportunities at the gene level (before transcription) and at the protein level (after translation). In this diagram, the processes that offer opportunities for regulation are highlighted by white boxes.

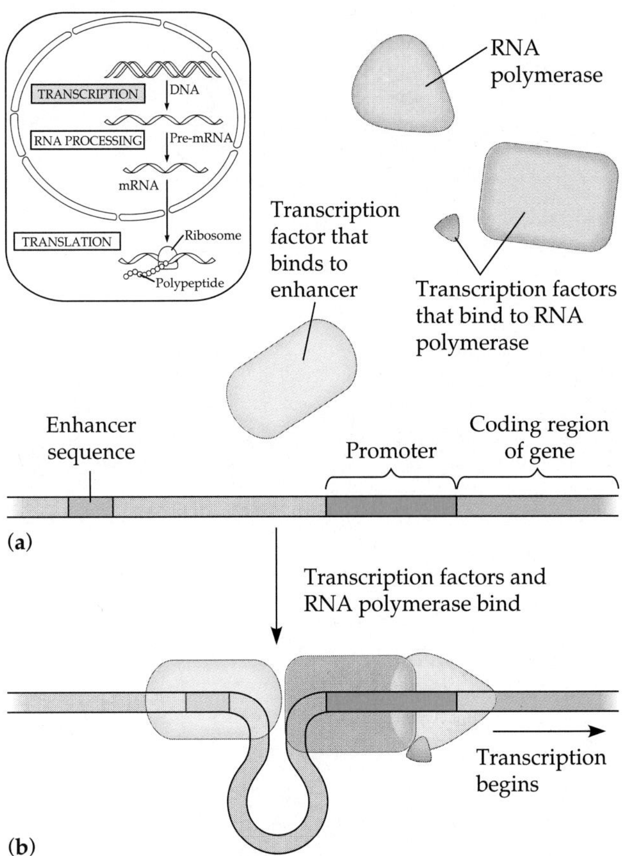

Figure 18.9
A hypothesis for the role of enhancers in the control of eukaryotic gene expression. (**a**) RNA polymerase requires additional proteins called transcription factors in order to recognize and bind to the promoter region at the upstream end of a gene. (**b**) Remote from the promoter is an enhancer region. Binding of a specific transcription factor to the enhancer stimulates the entire complex of polymerase and transcription factors.

main, the zinc-finger domain, and the leucine zipper domain. Figure 18.10 illustrates hypotheses for how these domains bind to DNA and affect gene expression at the transcriptional level.

Posttranscriptional Control of Gene Expression

Transcription alone does not constitute expression. The expression of protein-coding genes is measured in terms of the types and amounts of functional proteins that a cell makes, and much happens between synthesis of the RNA transcript and the creation of a functional protein. Gene expression may be blocked or stimulated at any posttranscriptional step (see Figure 18.8).

RNA Processing and Export The segregation of translation from transcription is a feature of eukaryotic cells that opens new possibilities for controlling gene expression. As described in Chapter 16, the eukaryotic

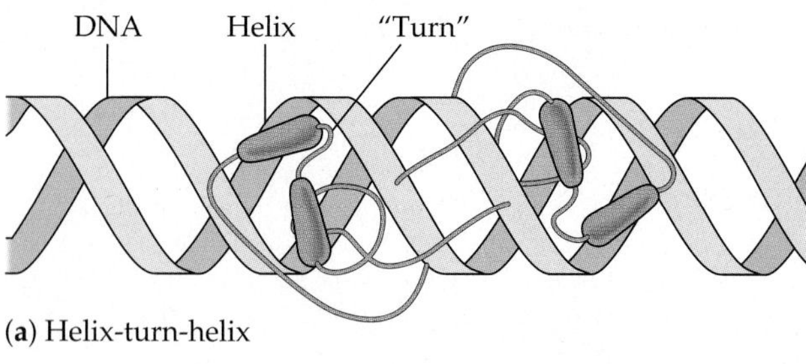

(a) Helix-turn-helix

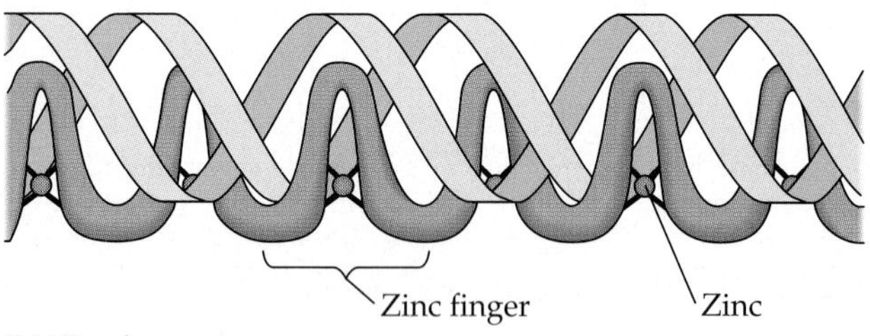

(b) Zinc fingers

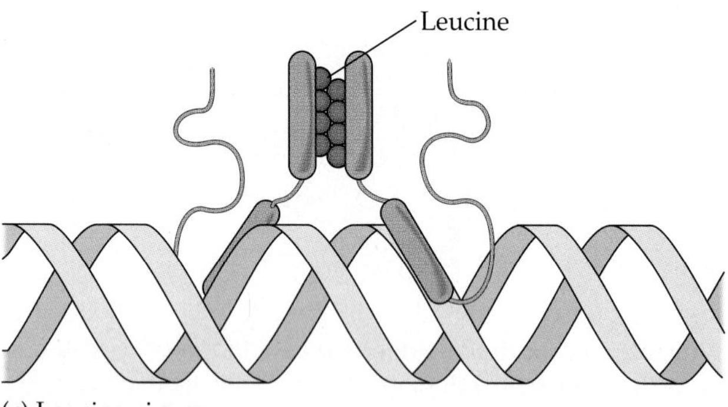

(c) Leucine zipper

Figure 18.10
Three structural motifs of DNA-binding proteins that regulate gene expression. The illustrated interactions with DNA are hypothetical. (**a**) The "helix-turn-helix" domain employs two regions of α helix to bind to specific DNA sequences. The binding protein is actually a dimer, with each part having a helix-turn-helix structure. (**b**) Other DNA-binding proteins are equipped with "zinc fingers," protein loops stabilized by atoms of the metal zinc. (**c**) The third motif of DNA-binding proteins is the "leucine zipper" domain. Two polypeptide chains, each with a region rich in the amino acid leucine, zip together to form the active DNA-binding protein.

cell must process its initial transcripts before they can act as mRNA, tRNA, or rRNA. An mRNA transcript must receive a 5′ cap and a poly-A tail, and the RNA segments representing the introns of the genes must be removed and the exons (coding segments) spliced together (see Figure 18.5). By means that are not well understood, the processed transcript then passes from the nucleus to the cytoplasm, probably via a nuclear pore. In the cytoplasm, the mRNA interacts with a number of specific proteins and may associate with ribosomes to undergo translation. Each step in RNA processing and in the mechanisms responsible for the export of mRNA from the nucleus and translation in the cytoplasm represents an opportunity for controlling gene expression. However, little is known about how these posttranscriptional steps are regulated.

Regulation of mRNA Degradation The lifespan of an mRNA molecule in the cytoplasm is also an important factor in controlling the pattern of protein synthesis in a cell. Prokaryotic mRNA molecules have very short lives; they are degraded by enzymes after only a few minutes. This is one reason bacteria can vary their patterns of protein synthesis quickly in response to environmental changes. In contrast, the mRNA molecules of eukaryotes can have lifetimes of hours, days, or even weeks. If two species of mRNA molecules differ in how rapidly they are broken down by enzymes in the cytoplasm, they may differ in how much protein synthesis each directs. A striking example of long-lived mRNA is found in vertebrate red blood cells, which are "factories" for the production of the protein hemoglobin. The mRNAs for hemoglobin are unusually stable and are translated repeatedly in the developing red blood cells of most vertebrate species.

In other cases, eukaryotic mRNA accumulates, but translation is delayed until some control signal triggers it. For example, the mRNA used for the active protein synthesis that occurs during the first stage of embryonic development (cleavage) of many organisms has all been synthesized by the egg cell nucleus prior to fertilization. It is stored in the cytoplasm of the unfertilized egg as inactive mRNA and is not translated until fertilization (see Chapter 43). By synthesizing large quantities of specific mRNAs, stocking them in the cytoplasm in some form safe from enzymatic degradation, and delaying their translation until a signal is given, a developing cell can respond to a stimulus with an explosive burst of synthesis of particular proteins.

Translational and Posttranslational Control Translation in eukaryotic cells involves many more protein factors, especially initiation factors, than in prokaryotic cells (see Chapter 16). Thus, there are ample opportunities for the control of gene expression at the level of translation. For example, one trigger for translation of the inactive mRNA in fertilized eggs is the sudden appearance of an initiation factor that must be present for ribosomes and mRNA to begin interacting.

The last opportunities for controlling gene expression occur after translation. Often, eukaryotic polypeptides must be cut up to yield the active final products. The posttranslational processing of the hormone insulin is an example (see Chapter 16). Many polypeptides must be modified by the addition of chemical groups, such as chains of sugars, in order to be active. And for many proteins to reach their final destinations in the cell, the signal mechanism must target these proteins (see Figure 16.17). In theory at least, regulation might occur at any of these steps involved in modifying or transporting a protein. Finally, the extent of gene expression could be controlled by the cell's selective degradation of particular proteins.

THE ROLE OF SMALL MOLECULES IN REGULATING EUKARYOTIC GENE EXPRESSION

Having traced the stages of gene expression from transcription to translation and beyond, we now backtrack to ask what initially triggers a cell's transcription machinery to start operating on a specific gene. How does the cell's external or internal environment influence gene expression? In bacteria (see Chapter 17), small organic metabolites such as a lactose derivative and tryptophan affect transcription by combining with regulatory proteins, such as the repressor protein of the *lac* operon. The regulatory protein's shape is changed in such a way that its ability to bind to DNA is altered. In eukaryotes, too, certain small organic molecules influence transcription by combining with regulatory proteins. Among the molecules that alter gene expression by this mechanism are the steroid hormones of animals. Carried in the blood, these molecules serve as signals from cells in one part of the animal to cells elsewhere in the body. The effect of steroid hormones on gene transcription in target cells was first clearly recognized in insects.

Chromosome Puffs: Evidence for the Regulatory Role of Steroid Hormones in Insects

Studies of giant (polytene) chromosomes in the salivary gland and other tissues of certain insect larvae have provided evidence for both the control of gene expression at the transcriptional level and the role of steroids in the process. As described in Chapter 14, polytene chromosomes consist of hundreds of parallel chromatids. At characteristic stages in the larva's development, chromosome puffs appear at specific sites on the polytene chromosomes (Figure 18.11). A puff forms when DNA loops out from the chromosome axis, perhaps making the DNA in that region more accessible to RNA polymerase. Analysis using autoradiography (see Chapter 2) shows that the puffs indeed correspond to regions of intense RNA synthesis (that is, transcription).

The locations of chromosome puffs along a chromosome change as the larva develops. When the larva prepares to molt, some puffs disappear and others form at new sites. The shifting puffs are visual indicators of the selective switching on and off of specific genes during development. These changes in puffing patterns can be induced by ecdysone, the steroid hormone that initiates molting. This effect demonstrates that gene regulation in the insect is responsive to specific chemical signals, and in particular to a steroid.

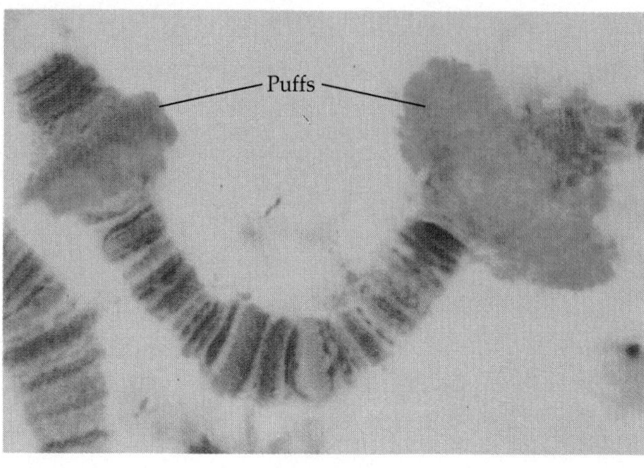

Figure 18.11
Chromosome puffs. This micrograph shows two puffs in a polytene chromosome of an insect, *Trichosia pubescens* (LM). The locations of puffs along the chromosomes change as a cell develops. Puffs are a visible indication of gene activity—sites of active transcription.

The Action of Steroid Hormones in Vertebrates

Researchers have worked out the key steps by which sex hormones and other steroids alter gene expression in target cells of vertebrates. Steroids are soluble in lipids. When a cell is exposed to a steroid, the hormone diffuses across the plasma membrane and across the cytoplasm. It then enters the nucleus, where it encounters a soluble receptor protein. In the absence of the steroid, the receptor protein is associated with an inhibitory protein that prevents the receptor from binding to DNA (Figure 18.12a). Binding of the steroid hormone to the receptor causes the release of the inhibitory protein, and the activated receptor protein can now attach to specific sites on the DNA (Figure 18.12b). These sites are within enhancer regions that control steroid-responsive genes. Binding of the receptor protein to the enhancer activates transcription, and in this way, a steroid acts as a chemical signal to switch on specific genes in certain cells. You will learn more about the pathways by which hormones affect cells in Chapter 41.

This is a good time to summarize five key points about the control of gene expression in eukaryotes:

1. The various cell types of a multicellular organism express different genes.
2. Physical and chemical rearrangements of the genome make certain genes available for expression and other genes unavailable.

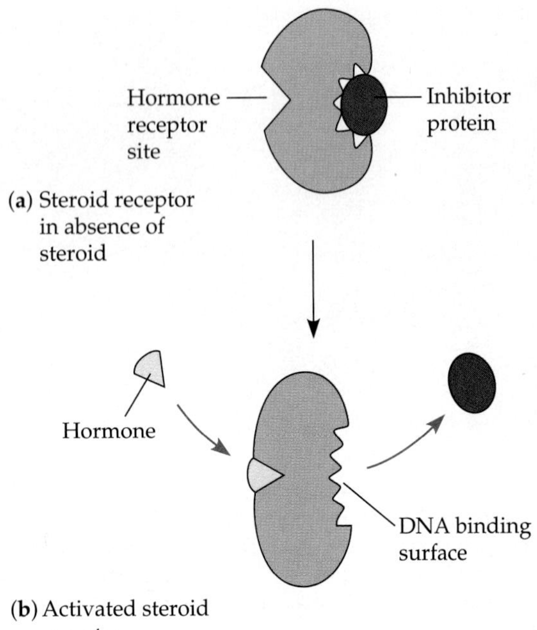

Figure 18.12
Activation of a steroid receptor. (**a**) The steroid receptor is a DNA-binding protein that stimulates transcription of specific genes. In the absence of the steroid, an inhibitor protein keeps the steroid receptor from binding to DNA. (**b**) Arrival of the steroid causes the receptor protein to release the inhibitor, and the receptor can now bind to specific regulatory sites on DNA. By this mechanism, steroid hormones activate expression of certain genes in target cells.

3. For genes that are available for expression, regulatory opportunities exist at each step in the pathway from gene to functional protein.
4. Control of transcription is especially important in determining which genes are expressed; in eukaryotes, selective binding of regulatory proteins to enhancer sequences in DNA stimulates transcription of specific genes.
5. The regulatory activity of some of these DNA-binding proteins is sensitive to certain hormones and other chemical cues.

You will learn more about the role of selective gene expression in cellular differentiation and the development of plants and animals in Chapters 34 and 43. We conclude this chapter with a look at how changes in gene expression can cause cancer.

GENE EXPRESSION AND CANCER

In Chapter 11, we considered cancer as a variety of diseases in which cells escape from the control mechanisms that normally limit their growth and division. Researchers are beginning to understand the genetic basis of a cancer cell's aberrant behavior.

Cancer can be caused by physical agents such as X-rays and by chemicals called **carcinogens** (see Chapter 16). All these agents cause mutations that alter the expression of certain genes. But what kinds of genes are involved? An important clue came with the discovery of "cancer-causing" genes, or **oncogenes** (*onco-* comes from the Greek word for tumor). These genes were found in certain RNA viruses (retroviruses) that cause uncontrolled growth of infected cells in culture. Then Michael Bishop, Harold Varmus, and other researchers found close counterparts of these viral genes in the genomes of humans and other animals. (See the interview that precedes Unit Two.) These normal cellular genes, called **proto-oncogenes,** code for protein products that normally regulate cell growth, cell division, and cell adhesion.

How might a gene that has an essential function in normal cells become a cancer-causing gene? In general, the answer seems to be that the gene is not under its normal controls. There are four mechanisms that can convert a proto-oncogene to an oncogene. They are (1) gene amplification, (2) chromosome translocation, (3) gene transposition, and (4) point mutation (a change in a single nucleotide). In the case of amplification, oncogenes are present in more copies per cell than is normal. Frequently, malignant cells contain chromosome translocations, chromosomes that have broken and rejoined, juxtaposing pieces of different chromosomes. In these cases, oncogenes may be found in the new joint region, suggesting that such translocations separate the oncogene from its normal control regions. In gene transposition, the oncogene itself or a regulatory gene may have been transposed to an unusual locus, leading to abnormal expression of the oncogene. In still other cases, the oncogene has a mutated nucleotide sequence, slightly different from that of the corresponding proto-oncogene. Such a change might create a growth-stimulating protein that is more active or more resistant to degradation than the normal protein.

In addition to mutations affecting growth-stimulating proteins, changes in genes whose products *inhibit* cell division can also be involved in cancer. Such genes are called **tumor-suppressor genes,** because the proteins they encode normally help prevent uncontrolled cell growth. Any mutation that prevents the normal tumor-repressor protein from being made may contribute to the onset of cancer. The first tumor-suppressor to be recognized was the retinoblastoma allele, which in homozygous dosage allows formation of a malignant tumor in the eyes of young children. When a normal retinoblastoma allele is inserted into a cell with mutated versions of the gene, the aberrant growth of the cancerous cells in culture is abated.

There is evidence that more than one somatic mutation is needed to produce a full-fledged cancer cell. One of the best understood types of human cancer,

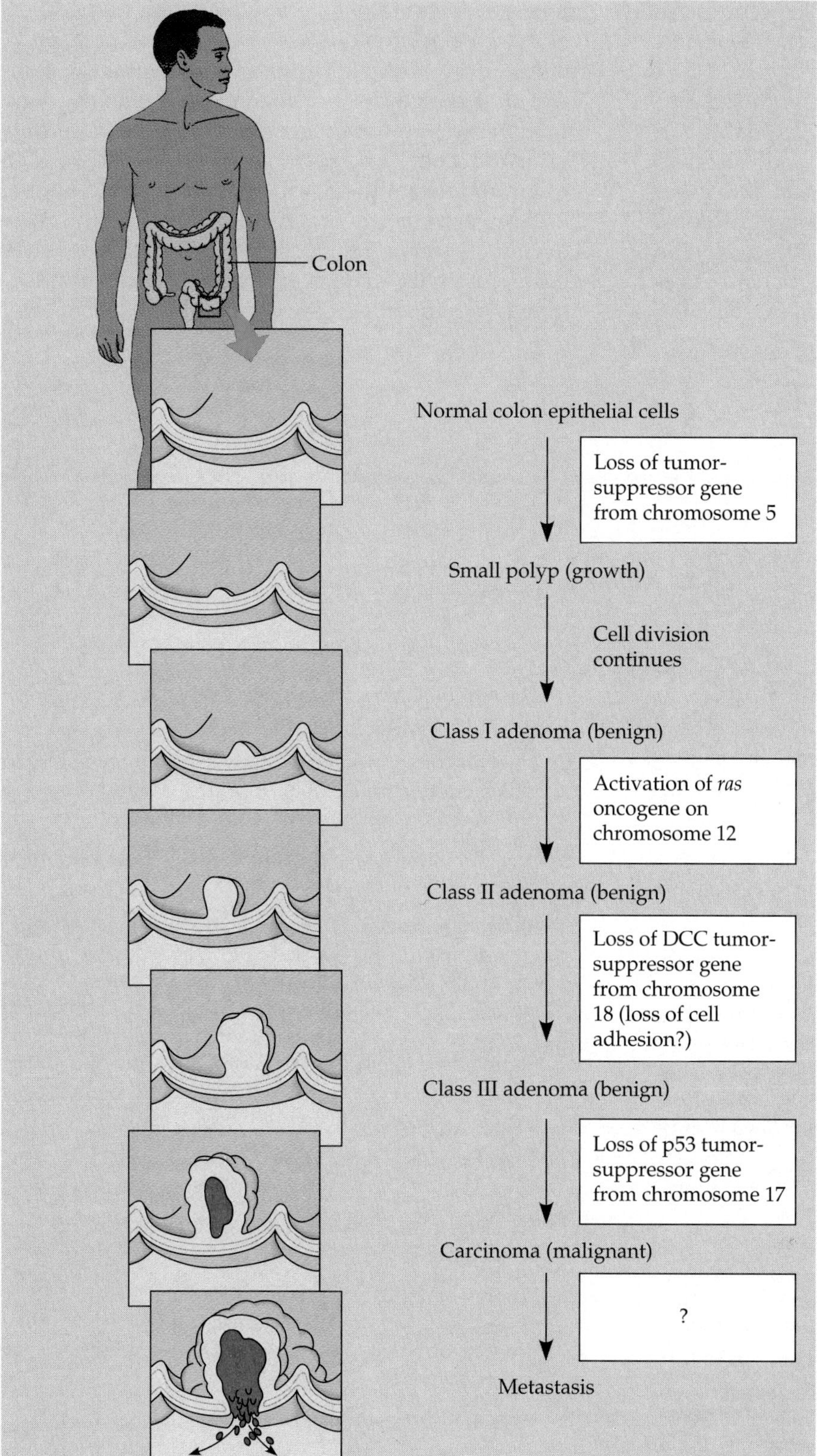

Figure 18.13
A model for the development of colorectal cancer. Changes in the tumor parallel a series of genetic alterations, including changes in tumor-suppressor genes on chromosomes 5, 18, and 17, as well as a mutation of an oncogene called *ras.* Other mutation sequences can also cause cancer. In each case, the mutations change genes from forms that normally regulate cellular processes into forms that cause a breakdown in regulation.

colorectal cancer, illustrates this principle (Figure 18.13). Like many cancers, the development of a metastasizing colorectal cancer is gradual. The first sign is unusually frequent division of apparently normal cells in the colon lining; later a benign tumor (polyp) appears, and eventually a malignant tumor may develop. These cellular changes are paralleled by a gradual accumulation of oncogenes and tumor-suppressor gene mutations, and only after a number of genes have changed is a full-blown malignant tumor present. In most cases studied, at least four changes occur at the DNA level: the activation of a cellular oncogene and the inactivation of three tumor-suppressor genes.

How are oncogenes involved in virus-induced cancers, which are thought to account for 15% of human cancer cases worldwide? The virus might add an oncogene to the cell or disrupt the cell's DNA in a way that affects a proto-oncogene or a tumor-suppressor gene. However, the change in the cell's DNA induced by the virus is probably only one in a series of changes required for cancer. This requirement could explain why virus-associated cancers often take a long time to develop. The additional DNA changes needed must be triggered by other environmental agents, the same ones that trigger nonviral cancers. Thus, the term "virus-associated cancers" is more accurate than "virus-induced."

Our understanding of cancer will become clearer as basic research on the organization and control of eukaryotic genomes progresses. Much of this research employs new techniques for experimentally manipulating DNA, as will be discussed in the next chapter.

STUDY OUTLINE

1. Eukaryotic genomes are more complexly organized than prokaryotic genomes. The control of gene expression is also more elaborate in eukaryotes than in prokaryotes.

Genome Organization at the Microscopic Level (pp. 373–375)

1. The enormous amount of DNA in a eukaryotic chromosome is compacted by a multilevel system of folding.
2. Chromatin is composed of DNA and five types of histone proteins that bind to the DNA and dictate its first level of folding into nucleosomes, basic units of DNA packing.
3. The next order of DNA packing is the 30-nm chromatin fiber, a coil with six nucleosomes per turn.
4. The chromatin fibers fold to form looped domains, the next higher level of DNA organization. Further folding coils the DNA into its most highly compacted form, the metaphase chromosome. In certain cases interphase chromatin also exists in a highly condensed form, called heterochromatin. Active transcription occurs on euchromatin, the more open, unfolded form of the chromosome.

Genome Organization at the Molecular Level (pp. 375–378)

1. Much of the DNA in a eukaryotic genome does not encode protein.
2. Some DNA sequences may be present in hundreds or thousands of copies.
3. A multigene family is a collection of genes that are similar or identical in nucleotide sequence. They may be clustered or dispersed in the genome.
4. Along with pseudogenes and introns, highly repetitive sequences of five to ten nucleotides tandemly repeated thousands of times contribute to the enormous reservoir of noncoding DNA in the eukaryotic genome.
5. Unlike prokaryotic operons, eukaryotic genes that need to be transcribed at the same time are often scattered throughout the genome. Coordinated control is believed to be mediated by specific nucleotide sequences common to all the genes of the group.

Genome Plasticity (pp. 378–379)

1. The amphibian oocyte can selectively amplify (make extra copies of) the genes for ribosomal RNA.
2. The cells of some species show selective gene loss, in which entire chromosomes or parts of chromosomes of certain cells are eliminated.
3. Attachment of methyl groups to DNA may diminish transcription of that DNA.
4. Rearrangements of DNA can activate or inactivate specific genes. By moving from one location to another within the genome, transposons can affect gene expression. In yeast, the alternation of mating types results from a cassette mechanism that switches genes at the mating-type locus. In vertebrates, a rearrangement and selective deletion of DNA segments in differentiating B lymphocytes accounts for antibody diversity.

The Control of Gene Expression (pp. 379–383)

1. Eukaryotic gene expression can be controlled at any of the steps in the flow of information from DNA to functional protein.
2. Binding of regulatory proteins to enhancer regions in the DNA selectively stimulates transcription of genes. These proteins are among the diverse DNA-binding proteins that affect gene expression.
3. Before mRNA can leave the nucleus, it must undergo processing, which presents opportunities for controlling gene expression.
4. mRNA degradation is precisely regulated, and this influences how much protein may be made from a given molecule of mRNA.
5. Translation itself is regulated; some mRNAs, especially in unfertilized eggs, may be sequestered so that they are not translated until after fertilization.

6. After translation, proteins may be extensively modified and targeted for specific sites in the cell.

The Role of Small Molecules in Regulating Eukaryotic Gene Expression (pp. 383–384)

1. Some small molecules in the cell affect gene expression by associating with DNA-binding proteins and altering the conformations of these regulatory proteins.
2. Ecdysone, a steroid hormone in insects, selectively activates genes during development, a regulation evident in the changing pattern of chromosome puffs.
3. Steroid hormones in vertebrates interact with receptor proteins, which are thereby activated to bind to enhancers and influence transcription.

Gene Expression and Cancer (pp. 384–386)

1. Oncogenes (called proto-oncogenes when the cell is in the noncancerous state) are thought to be key genes in controlling cellular growth and differentiation. When such genes escape normal control mechanisms, they allow the formation of tumors. Physical agents, chemical carcinogens, or viruses may activate oncogenes in various ways.
2. Tumor-supressor genes encode proteins that keep cell division in a normal state. Loss or mutation of these genes seems to be a factor in the appearance of cancers.
3. Tumors show progressive accumulation of defects in proto-oncogenes and tumor-supressor genes.

SELF-QUIZ

1. In a nucleosome, the DNA is wrapped around
   a. polymerase molecules
   b. ribosomes
   c. histones
   d. the nucleolus
   e. satellite DNA
2. Apparently, our muscle cells are different from our nerve cells mainly because
   a. they express different genes
   b. they contain different genes
   c. they use different genetic codes
   d. they have unique ribosomes
   e. they have different chromosomes
3. Chromosome puffs on the giant chromosomes of *Drosophila* probably represent regions where
   a. genes are inactivated by repressor proteins
   b. hormones are produced
   c. genes have been damaged
   d. genes are especially active in transcription
   e. ribosomes are being synthesized
4. The function of enhancers is an example of
   a. a transcriptional control that affects gene expression
   b. a posttranscriptional mechanism for editing mRNA
   c. initiation factors that stimulate translation
   d. posttranslational control activating proteins
   e. a eukaryotic equivalent of the prokaryotic promoter
5. Multigene families are
   a. groups of enhancers that control key developmental events
   b. pseudogenes that appear to have arisen by reverse transcription of mRNA
   c. equivalent to the operons of prokaryotes
   d. collections of genes whose expression is controlled by the same regulatory proteins
   e. often collections of identical genes, arranged in tandem
6. Which of the following statements about the DNA in one of your brain cells is true?
   a. Some DNA sequences may be present in multiple copies.
   b. Most of the DNA codes for protein.
   c. The majority of genes are likely to be transcribed.
   d. Each gene lies next to an enhancer that helps control transcription.
   e. Many genes are grouped into operonlike clusters.
7. Permanent rearrangement of DNA segments is known to be involved in genes coding for
   a. ribosomal RNA
   b. most proteins in eukaryotes
   c. hemoglobin
   d. histone proteins
   e. antibodies
8. Which of the following is an example of a possible step for posttranscriptional control of gene expression?
   a. the addition of methyl groups to cytosine bases of DNA
   b. the binding of transcription factors to a promoter region
   c. the removal of introns and splicing of exons
   d. gene amplification during a particular stage in development
   e. the folding of DNA to form heterochromatin
9. The amount of protein made from a given mRNA molecule depends partly on
   a. the degree of DNA methylation
   b. the rate at which the mRNA is degraded
   c. the presence of certain transcription factors and enhancers
   d. the number of introns present in the mRNA
   e. the types of ribosomes present in the cytoplasm
10. All our cells contain proto-oncogenes, which can change into oncogenes that cause cancer. Which of the following is the best explanation for the presence of these potential time bombs in our cells?
    a. Proto-oncogenes first arose from viral infections.
    b. Proto-oncogenes normally help regulate cell growth and division.
    c. Proto-oncogenes are genetic "junk" with no known function.

d. Proto-oncogenes are mutant versions of normal genes.

e. Cells produce proto-oncogenes as a by-product of the aging process.

CHALLENGE QUESTIONS

1. The amino acid sequences encoded by the genes for certain DNA-binding proteins have been found to be remarkably similar between vertebrates and invertebrates, despite the evolutionary divergence of these animals over 500 million years ago. Speculate on why evolution has been so conservative with genes that control development.

2. As a multicellular eukaryote develops from a zygote, cells differentiate (specialize) in structure and function. These light micrographs illustrate four examples of specialized cells in a mammal: (**a**) three muscle cells, each with numerous nuclei (dark discs); (**b**) a nerve cell (the large cell in the center with the fibrous extensions); (**c**) sperm cells; and (**d**) a white blood cell surrounded by smaller red blood cells.

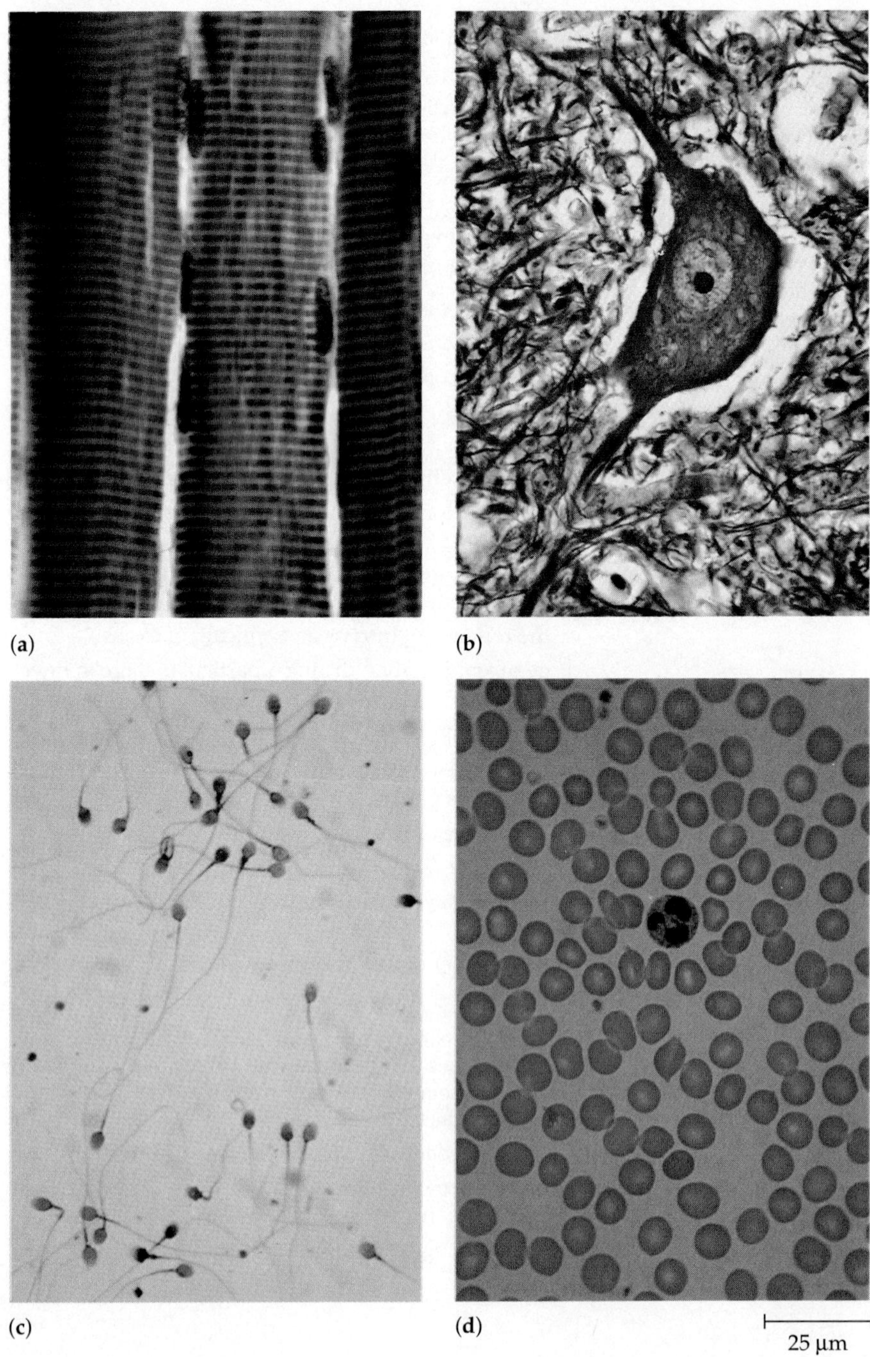

How could you demonstrate that these specialized cells differ from one another in some of the genes they express? Based on what you learned in this chapter, describe at least three regulatory mechanisms that could contribute to these differences in gene expression.

SCIENCE, TECHNOLOGY, AND SOCIETY

1. Some researchers believe that environmental tobacco smoke (passive smoking) poses a cancer risk to nonsmokers. If this is true, what might be done to protect the public from environmental tobacco smoke?

2. A chemical called dioxin, or TCDD, is produced as a contaminant of some chemical manufacturing processes. Trace amounts of this substance were present in Agent Orange, a defoliant sprayed on vegetation during the Vietnam War. There has been a continuing controversy over its effects on soldiers exposed to it during the war. Animal tests have shown that dioxin can be lethal and can cause birth defects, cancer, liver and thymus damage, and immune system suppression. But its effects on humans are unclear, and even animal tests are equivocal; a hamster is not affected by a dose that can kill a guinea pig. Researchers have discovered that dioxin exerts its effects like some hormones. It enters a cell and binds to a receptor protein, which in turn attaches to the cell's DNA. How might this mechanism help explain the variety of dioxin's effects on different body systems and in different animals? How might you determine whether a type of illness is related to exposure to dioxin or whether a particular individual became ill as a result of exposure to dioxin? Which would be more difficult to demonstrate? Why?

FURTHER READING

Barinaga, M. "Signals into Unknown Territory." *Science,* March 27, 1992. How cancer researchers and developmental biologists are helping one another.

Gilbert, S. F. *Developmental Biology,* 3rd ed. Sunderland, MA: Sinauer Associates, 1991. An excellent textbook used in upper-division developmental biology courses.

Grunstein, M. "Histones as Regulators of Genes." *Scientific American,* October 1992. How one type of chromosomal protein helps control gene expression.

Holliday, R. "A Different Kind of Inheritance." *Scientific American,* June 1989. A possible role of DNA methylation in development.

Lowenstein, J. "Genetic Surprises." *Discover,* December 1992. The more molecular biologists learn about the organization of eukaryotic genomes, the more interesting the strory becomes.

Ross, J. "The Turnover of Messenger RNA." *Scientific American,* April 1989. What factors control rates of RNA synthesis and degradation?

Weinberg, R. "Tumor-Supressor Genes." *Science,* November 22, 1991. The molecular biology of cancer.

19 DNA Technology

Basic Strategies of Gene Manipulation and Analysis
Applications of DNA Technology
Safety and Ethical Issues

Figure 19.1
The manufacture of a human protein by genetically engineered bacteria. Bacteria that carry a human gene and synthesize its protein product can be grown in large fermentation tanks. The tank shown here holds 1500 L. In this chapter, you will learn about the methods of DNA manipulation that are driving revolutions in biotechnology and basic research.

In the photograph on this page, a microbiologist employed by a pharmaceutical company monitors growth of a huge culture of bacteria (Figure 19.1). The tank behind this maze of pipes and gauges holds 1500 L of liquid culture medium supporting trillions of *Escherichia coli.* The bacteria are a clone derived from a genetically engineered *E. coli* cell. Spliced into their own DNA are human genes that encode a valuable protein. The human genes make the bacteria synthesize the protein, which the bacteria secrete into the growth medium. The protein will be separated from the medium, purified, and marketed.

Today, literally hundreds of useful products are produced by **genetic engineering,** the manipulation of genetic material for practical purposes. In the last decade, it has become routine to combine genes from different sources—often different species—in test tubes, and then transfer this **recombinant DNA** into living cells, where it can be replicated and expressed. *E. coli* is often used because it is easy to grow and its biochemistry is well understood. Some of the early successes of genetic engineering were the creation in the early 1980s of strains of *E. coli* that carried genes for making and secreting insulin and human growth hormone.

Although the use of genetic engineering to create valuable new products is impressive, the most important achievements resulting from recombinant DNA technology to date have been advances in our basic understanding of eukaryotic molecular biology. For example, only through the use of gene-splicing techniques have the details of eukaryotic gene arrangement and regulation (see Chapter 18) been opened to experimental analysis. Basic biological research has been further aided by techniques for manipulating and analyzing DNA that have developed along with recombinant DNA technology. For instance, it is now possible to rapidly determine the exact sequence of nucleotides in a DNA molecule. Today, biologists have a DNA technology toolkit more powerful than they could have imagined even a decade ago. There is hardly an area of biology that is not being influenced by the new methods.

DNA technology has launched an industrial revolution in biotechnology. In broad terms, **biotechnology** is the use of living organisms or their components to perform practical tasks. Practices that go back centuries, such as the use of microorganisms to make wine and cheese and the selective breeding of livestock and field crops, are aspects of biotechnology. So are the production of antibiotics from microorganisms and the synthesis of monoclonal antibodies using modern techniques of immunology (see Chapter 39). Biotechnology based on manipulation of DNA in vitro (outside of living cells) is distinct from earlier phases in that it is more precise. It is also much more powerful, allowing genes to be moved between organisms as distinct as bacteria, plants, and animals.

In this chapter, we will examine the main techniques and applications of genetic manipulation. We will see that DNA technology is in the process of revolutionizing biological research, human medicine, criminal law, and agriculture. Finally, following the lead of David Suzuki (see the interview that precedes this unit), we will consider some of the social and ethical issues that we face as our ability to manipulate genes becomes increasingly powerful.

BASIC STRATEGIES OF GENE MANIPULATION AND ANALYSIS

Before 1975, the technology for altering the genes of organisms for study or practical use was severely constrained. The primary strategy was to find desirable mutants and then grow (or breed) them. Sometimes, geneticists relied on nature to produce the mutations. For example, microbiologists looking for new antibiotics combed through vast collections of microbes from soil, sewage, and seawater. In other cases, beginning with Morgan's work with fruit flies, scientists used mutagenic radiation or chemicals to induce mutations. The screening process—checking the phenotype of each organism for the desired mutant characteristics—was often very laborious. Geneticists worked to develop tricks for the selection of organisms with desired traits—that is, methods that would allow only the desired organisms to survive and grow. A simple example of such a method is the inclusion of an antibiotic in the growth medium for bacteria normally sensitive to it; only cells with mutations that confer antibiotic resistance will grow to form colonies on the medium.

The selection of mutants does not involve the transfer of a desired gene from one organism to another. Prior to 1975, gene transfer was not generally possible except by cumbersome and relatively nonspecific breeding procedures. Only the bacteria and their viruses (phages) were exceptions. As you learned in Chapter 17, genes can be transferred from one bacterial strain to another strain with a different genotype by the natural biological processes of transformation, conjugation, or transduction. Putting these processes to work in the laboratory, geneticists were able to investigate in molecular detail the arrangement, structure, and functioning of prokaryotic and phage genes. The ease with which bacteria and phages can be propagated, and the relatively small size and simplicity of their genomes, contributed to large advances in prokaryotic molecular biology in the 1950s and 1960s. Eukaryotic organisms were another story. Even though both animal and plant cells could be grown in culture (see Chapters 11 and 34), the detailed workings of their genes remained an apparently impenetrable mystery until the advent of recombinant DNA techniques.

Recombinant DNA and Gene Cloning

The first recombinant DNA constructed in vitro used bacterial plasmids, which can replicate autonomously within bacterial cells (see Chapter 17). Biotechnologists and basic researchers continue to use plasmids in genetic engineering. Foreign genes are inserted into isolated plasmids, which are then returned to bacterial cells. The cells with this recombinant DNA reproduce, and the bacterial clone produces the protein encoded by the foreign gene (Figure 19.2). Scientists often use this same technique to obtain multiple copies of genes for basic research. (As we saw in Chapter 18, most genes exist in only one copy per genome—something on the order of one part in a million of DNA.)

In the next few pages, we will examine in some detail how the steps mentioned in Figure 19.2 are carried out. We begin with a discussion of the special enzymes that allow biologists to make recombinant DNA in test tubes.

Restriction Enzymes The major tools of recombinant DNA technology are bacterial enzymes called **restriction enzymes,** which were first discovered in the late 1960s. In nature, these enzymes protect bacteria against intruding DNA from other organisms. They work by cutting up the foreign DNA, a process called *restriction.* Most restriction enzymes are very specific, recognizing short, specific nucleotide sequences in DNA molecules and cutting at specific points within these sequences. The bacterial cell protects its own DNA from restriction by adding methyl groups (—CH_3) to adenines or cytosines within the sequences that would otherwise be recognized by the restriction enzyme. (This methylation of DNA is catalyzed by separate enzymes that recognize the same sequences.) There are hundreds of restriction enzymes and more than 150 different recognition sequences.

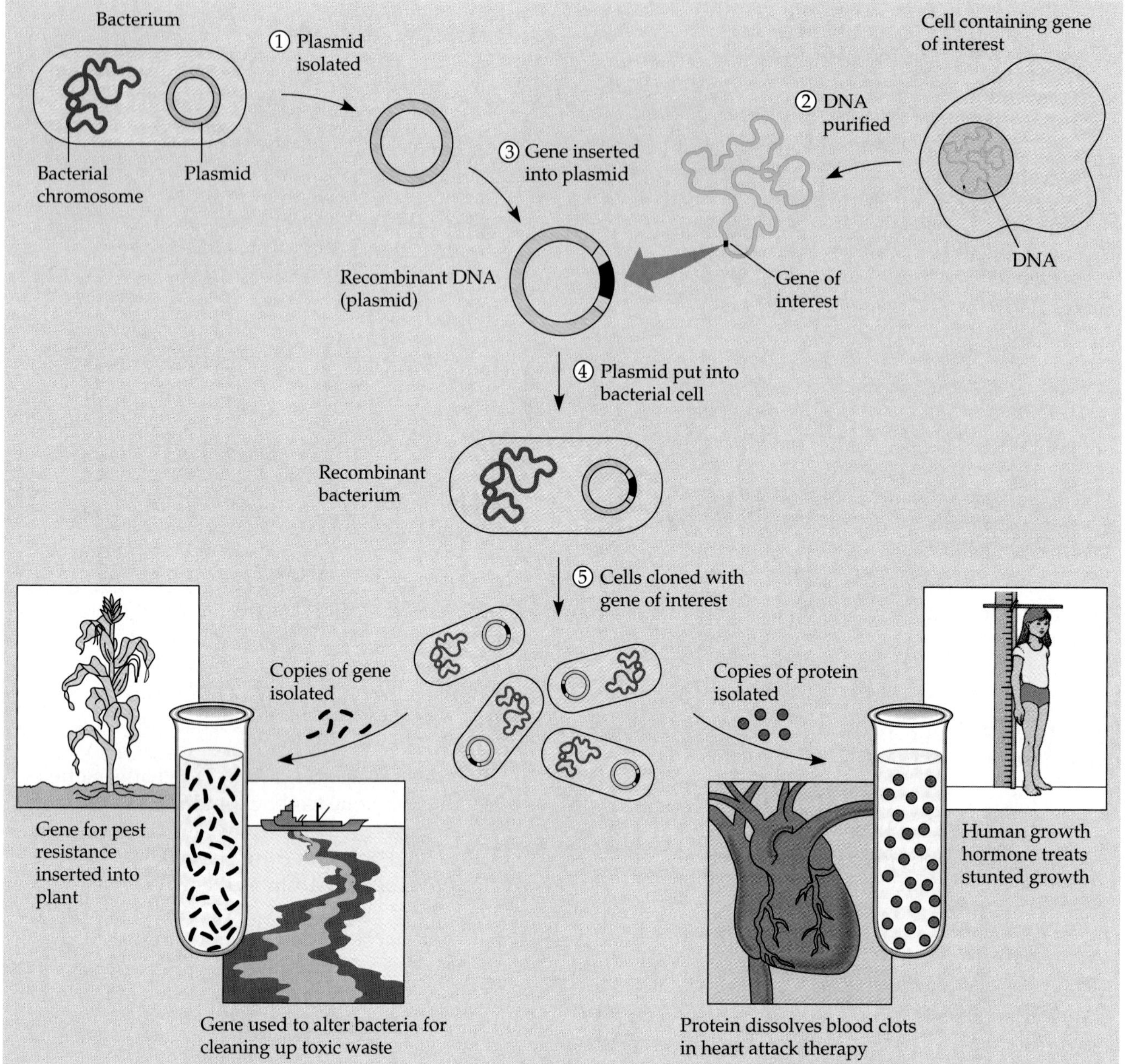

Figure 19.2

An overview of how plasmids are used in genetic engineering. ① The genetic engineer isolates plasmid DNA from bacteria and ② purifies DNA containing a gene of interest from another cell. This cell might be a bacterium or a plant or animal cell, and the gene of interest might be, for instance, a plant gene that confers resistance to pest insects or a human gene that encodes a hormone. ③ A piece of DNA containing the gene is inserted into the plasmid, producing recombinant DNA, and ④ the plasmid is put back into a bacterial cell. ⑤ This genetically engineered bacterium is then cloned (grown in culture). Since the foreign DNA spliced into the plasmid does not impair the plasmid's ability to replicate within a bacterial cell, the result is the cloning of the gene of interest. The bacterial culture now contains many copies of the gene (one per cell in this example). The illustrations at the bottom represent some current applications of genetically engineered bacteria. In the examples on the left, copies of the gene itself are the desired product. In the cases on the right, useful protein products are harvested in large quantities from the bacterial cultures.

The top of Figure 19.3 is a diagram of a molecule of DNA containing two recognition sequences for a particular restriction enzyme. As shown in this example, a restriction-enzyme recognition sequence (darker blue) is symmetric: The same sequence of four to eight nucleotides (here, six) is found on both strands but running in opposite directions. Restriction enzymes cut covalent phosphodiester bonds of both strands, usu-

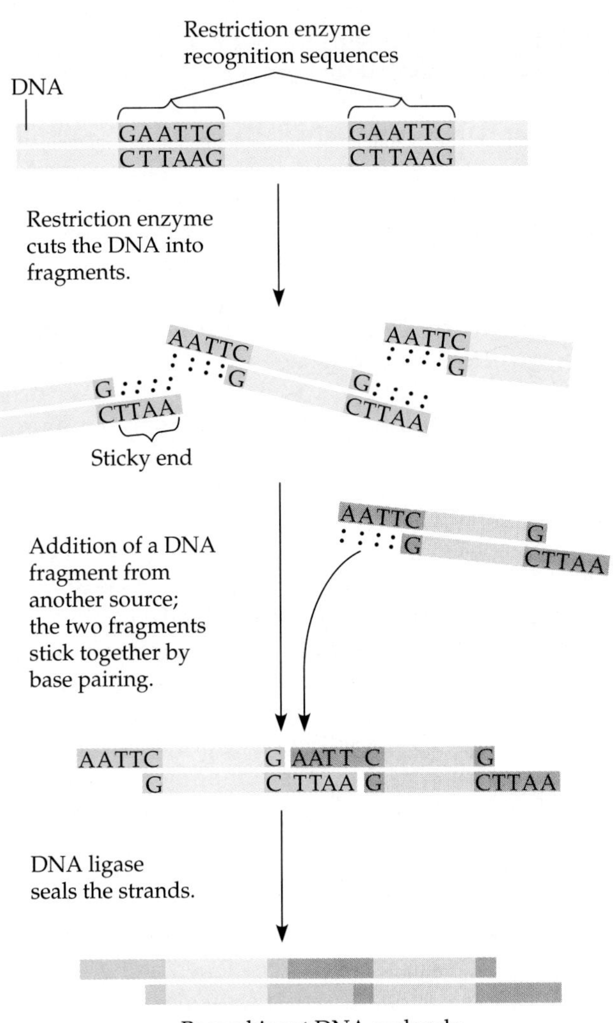

Figure 19.3
Using a restriction enzyme and DNA ligase to make recombinant DNA. The restriction enzyme in this example (called *Eco*RI) recognizes a six-base-pair sequence and makes staggered cuts in the sugar–phosphate backbone within this sequence. Notice that the recognition sequence along one DNA strand is the exact reverse of the sequence along the complementary strand. Because of this symmetry of sequences, the result of enzyme action is fragments of DNA with single-stranded "sticky" ends. Complementary ends will stick to each other by hydrogen bonding, transiently rejoining fragments in their original combinations or in new, recombinant arrangements. When restriction fragments come together by base pairing, the enzyme DNA ligase can catalyze the formation of covalent bonds joining their ends. If the fragments are from two different sources, the result is recombinant DNA.

ally in a staggered way, as indicated in the figure. The result is a set of double-stranded DNA fragments with single-stranded ends, called sticky ends. These short extensions will form hydrogen-bonded base pairs with complementary single-stranded stretches on other DNA molecules cut with the same enzyme.

The sticky ends of restriction fragments can be used in the laboratory to join DNA pieces originating from different sources. Such unions are only temporary, because only a few hydrogen bonds hold them together. They can be made permanent, however, by the enzyme DNA ligase, which seals the strands by catalyzing the formation of phosphodiester bonds. (Recall from Chapter 15 that DNA ligase is a key enzyme in DNA replication and repair.) The biochemical process used by the molecular biologist differs from the natural genetic recombination that goes on within living cells, which does not involve restriction enzymes. But the outcome is similar: the production of recombinant DNA, a DNA molecule carrying a new combination of genes.

After a large molecule of DNA is treated with a restriction enzyme, the different **restriction fragments** that result can be separated by **gel electrophoresis** (see the Methods Box, p. 396). This method separates macromolecules—either nucleic acids or proteins—on the basis of size, electric charge, and other physical properties. A molecule's properties determine how rapidly an electric field can move the molecule through a gelatinous medium. Gel electrophoresis is one of the staple tools in molecular biology and is of critical value in many aspects of genetic manipulation and study. One use is the identification of particular DNA molecules by the band patterns they yield in gel electrophoresis after being cut with various restriction enzymes. Viral DNA, plasmid DNA, and particular segments of chromosomal DNA can all be identified in this way. Another use is the isolation and purification of individual fragments containing interesting genes, which can be recovered from the gel with full biological activity. We will see still other uses for gel electrophoresis later in the chapter.

Gene Cloning in a Plasmid: A Closer Look Let's now examine more closely the gene-cloning procedure that was outlined in Figure 19.2. A typical version of this procedure is shown in Figure 19.4. Step ① is the isolation of two kinds of DNA: the bacterial plasmid and eukaryotic DNA containing the gene of interest. In this hypothetical example, the DNA containing the gene of interest comes from human tissue cells that have been growing in laboratory culture. The plasmid comes from the bacterium *E. coli* and carries two genes that confer antibiotic resistance on its *E. coli* host cell: amp^R (for ampicillin resistance) and tet^R (for tetracycline resistance). Furthermore, the plasmid has a single restriction-enzyme recognition sequence for the enzyme used, and the sequence lies within the tet^R gene. (The antibiotic-resistance genes will be useful later, as we shall see.)

In step ②, both the plasmid and the human DNA are treated with the same restriction enzyme. The enzyme cuts the plasmid DNA at its single recognition sequence, or **restriction site,** disrupting the tet^R gene. It also cuts the human DNA, generating many thousands of fragments; one of these fragments carries the gene of interest. In making the cuts, the restriction enzyme creates sticky ends on both the human DNA fragments and the plasmid. For simplicity, the figure here shows the step-by-step processing of one human DNA fragment and one plasmid, but actually millions of plasmids and a heterogeneous mixture of millions of human DNA fragments are treated simultaneously.

In step ③, the human DNA is mixed with the clipped plasmid. The sticky ends of the plasmid base pair with the complementary sticky ends of the human DNA fragment. In step ④, the enzyme DNA ligase joins the two DNA molecules by covalent bonds. The result is a new plasmid consisting of recombinant DNA; it is a single molecule combining DNA from two sources. In step ⑤, the recombinant plasmid is introduced into a bacterial cell by adding the naked DNA to a bacterial culture. Under the right conditions, some of the bacteria will take up the plasmid DNA from solution by the process of transformation (see Chapter 17).

Step ⑥ is the actual **gene cloning,** the production of multiple copies of the gene of interest. The bacterium, with its recombinant plasmid, is allowed to reproduce. As the bacterium forms a cell clone, any genes carried by the recombinant plasmid are also cloned. It is now that we take advantage of the plasmid's antibiotic resistance genes. We can identify colonies of bacteria that carry recombinant plasmids by the fact that their cells are ampicillin-resistant (since they carry the amp^R gene) but tetracycline-sensitive (since the tet^R gene has been inactivated by the insertion of foreign DNA). Thus, the cells we want will grow on medium containing ampicillin but not on medium with tetracycline. (Most often today, another gene plays the role of the tet^R gene here, but the principle remains the same: The inserted DNA disrupts a gene whose activity is easily determined.)

Inserting DNA into Cells Because isolated plasmids can be introduced into bacterial cells by transformation, they can serve as carriers—vectors—for moving recombinant DNA from test tubes back into cells. And because a plasmid can replicate in its new home, cloning any genes it carries, it is called a **cloning vector.** However, plasmids are not the only kinds of cloning vectors. Bacteriophages can play a similar role. Special strains of phage λ (see Chapter 17) have been widely used. In these strains, part of the phage's genome (containing nonessential genes) has been deleted, with the cuts made at restriction sites. In place of the deleted DNA, restriction fragments of foreign DNA can be inserted. The recombinant phage DNA is then introduced into an *E. coli* cell in the same way that plasmid DNA would be. Once inside the cell, the phage DNA replicates itself and produces new phage particles, each of which carries the foreign DNA "passenger." These phages, in turn, can carry the foreign DNA into other bacterial cells by normal infection.

As we will see later, it is sometimes desirable to clone DNA in eukaryotic cells rather than in bacteria. Both yeast cells and animal cells growing in culture can also take up foreign DNA from the surrounding medium under the right conditions, and if the new DNA becomes incorporated into chromosomal DNA or can replicate itself, it will be cloned along with the cell. When yeast cells are used for gene cloning, scientists can take advantage of the fact that yeast cells have plasmids; genetic engineers have even constructed recombinant plasmids that combine yeast and bacterial DNA and can replicate in either type of cell. Viruses are another type of vector that can be effectively used with eukaryotic cells.

A variety of more aggressive techniques for getting DNA into eukaryotic cells have also been developed. In **electroporation,** a brief electric pulse applied to a solution containing cells causes temporary holes to open in the plasma membrane, through which DNA can enter. Alternatively, DNA can be injected directly into single eukaryotic cells using microscopically thin needles. And, in a technique used primarily for plant cells, DNA can be attached to microscopic particles of metal and fired into cells with a gun.

Sources of Genes for Cloning Where do biologists get genes for cloning? There are two main sources: DNA isolated directly from an organism and "complementary DNA" made in the laboratory from mRNA templates.

Scientists isolate DNA directly by starting with all the DNA from cells of an organism with the gene they want. They use a restriction enzyme to snip the DNA

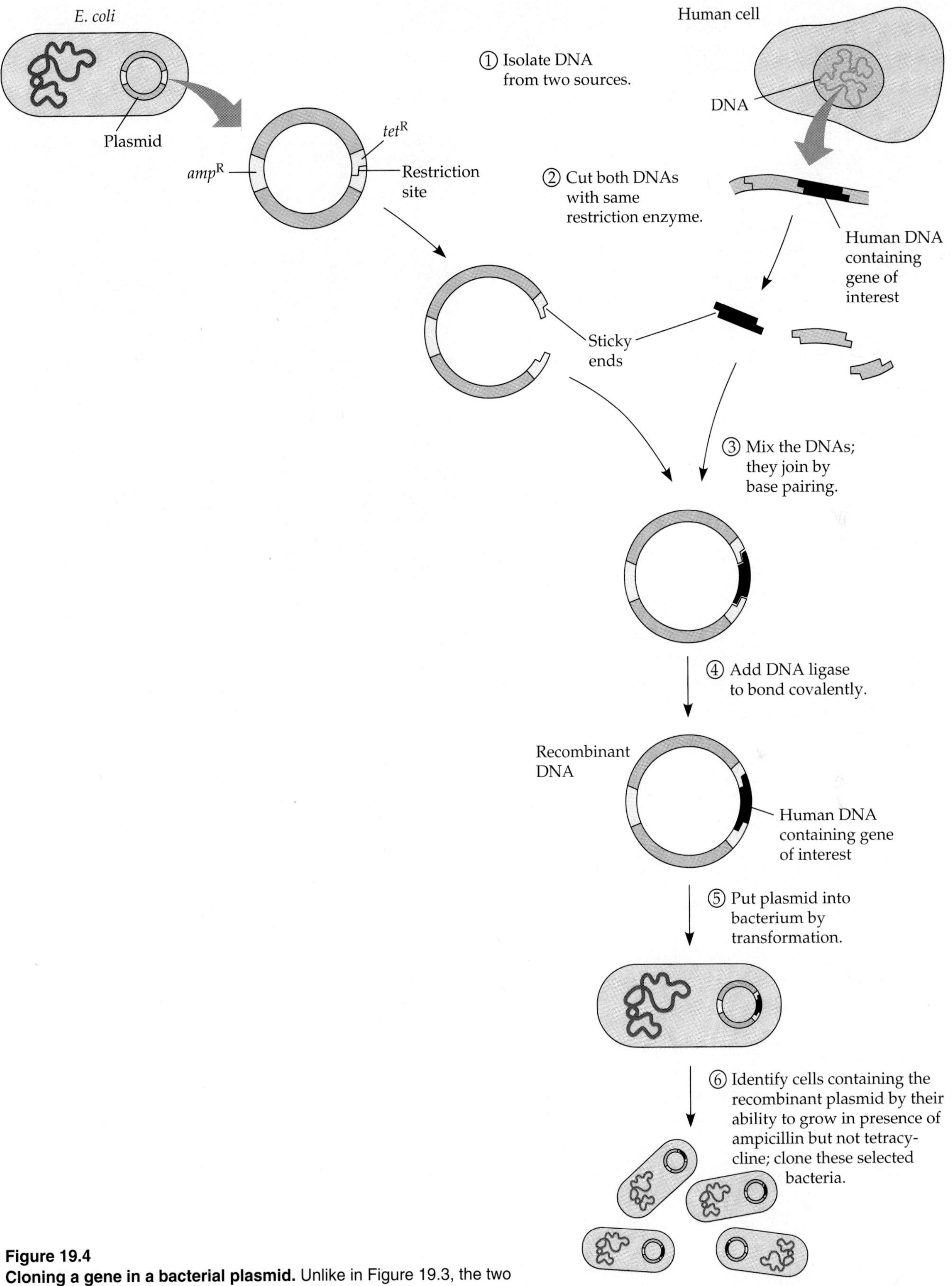
E. coli
Human cell
① Isolate DNA from two sources.
DNA
Plasmid
tetR
ampR
Restriction site
② Cut both DNAs with same restriction enzyme.
Human DNA containing gene of interest
Sticky ends
③ Mix the DNAs; they join by base pairing.
④ Add DNA ligase to bond covalently.
Recombinant DNA
Human DNA containing gene of interest
⑤ Put plasmid into bacterium by transformation.
⑥ Identify cells containing the recombinant plasmid by their ability to grow in presence of ampicillin but not tetracy-cline; clone these selected bacteria.
Bacterial clone carrying many copies of the human gene

Figure 19.4
Cloning a gene in a bacterial plasmid. Unlike in Figure 19.3, the two strands of DNA are not depicted separately in this figure. The numbered steps in the text correspond to the circled numbers in this diagram.

Methods: Gel Electrophoresis of Macromolecules

Gel electrophoresis separates macromolecules on the basis of their rate of movement through a gel under the influence of an electric field. As Figure a indicates, mixtures of nucleic acids or proteins are placed in wells near one end of a thin slab of a polymeric gel. (The gel is supported by glass plates and bathed in an aqueous solution.) Electrodes are attached to both ends, and the current is turned on. Each macromolecule then migrates toward the electrode of opposite charge at a rate determined mostly by the molecule's charge and size. Usually several different samples, each a mixture of molecules, are run simultaneously on a slab gel.

For nucleic acids, the rate of migration—how far a molecule travels while the current is on—is inversely proportional to molecular size. Nucleic acids carry negative charges (phosphate groups) proportionate to their lengths, but the gel impedes longer fragments more than it does shorter ones. The gel in Figure b has been treated with a DNA-binding dye that fluoresces pink in ultraviolet light. In each "lane," you can see a number of pink bands, which correspond to DNA molecules of different sizes. The bands contain DNA restriction fragments. Six samples were run, each a mixture of fragments from a DNA sample digested with a restriction enzyme.

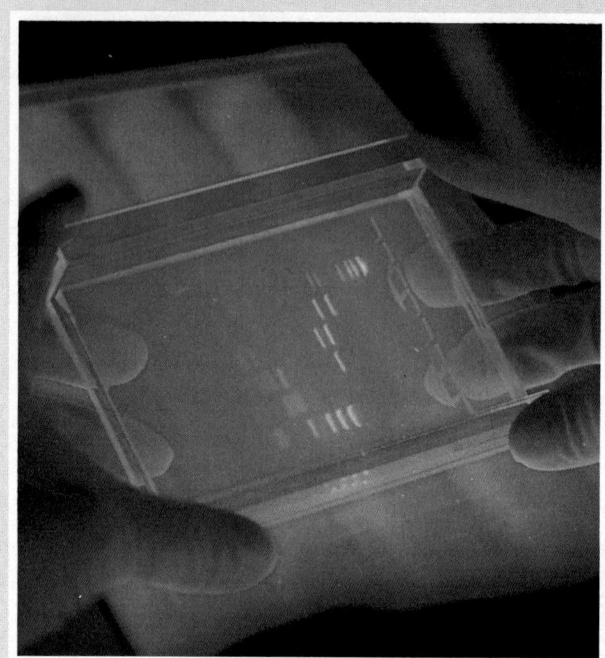

(b)

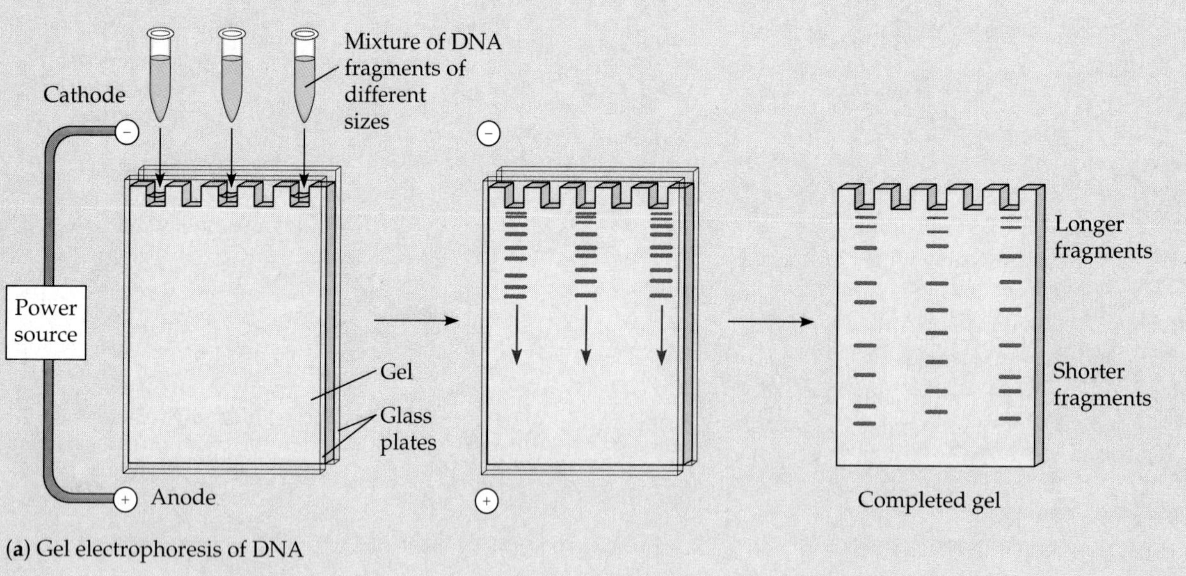

(a) Gel electrophoresis of DNA

into many thousands of fragments; the fragments vary in size, but a typical one is long enough to carry a few genes. Next, all the fragments are mixed with copies of cloning vector DNA that have been cut with the same restriction enzyme, and DNA ligase is added to seal the recombinant molecules that form. Finally, the vectors, now carrying foreign DNA, are introduced into bacteria. If the vector is a plasmid, each bacterium multiplies to form a bacterial clone, in which all cells carry the same DNA segment from the foreign genome. If

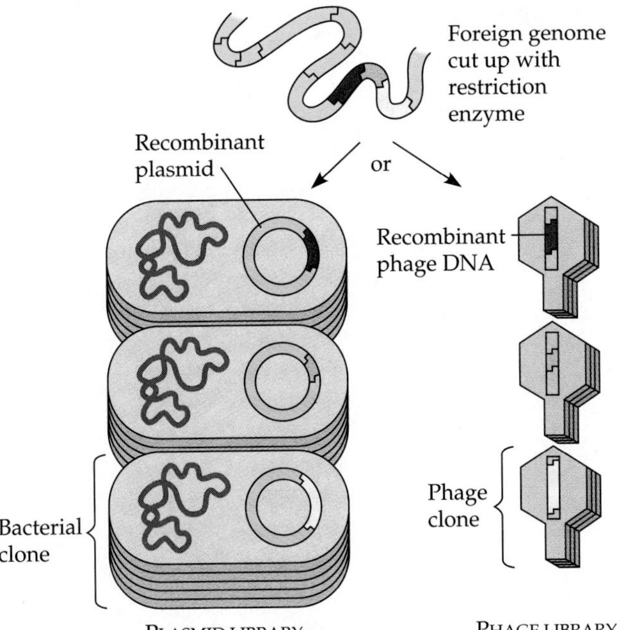

Figure 19.5
Genomic libraries. A genomic library is a collection of a large number of bacterial or phage clones, each containing copies of a DNA segment from a foreign genome. In a complete genomic library, the foreign DNA segments represented cover the entire genome of an organism. This diagram shows parts of two genomic libraries. On the left are three of the thousands of "books" in a plasmid (bacterial) library. Each "book" is a bacterial clone containing one particular variety of foreign genome fragment—red, orange, or yellow here—in its recombinant plasmids. On the right, the same three foreign genome fragments are shown in three "books" of a phage library. Notice that the recombinant vectors in each library vary somewhat in size from clone to clone, reflecting variations in the lengths of the restriction fragments from the foreign genome.

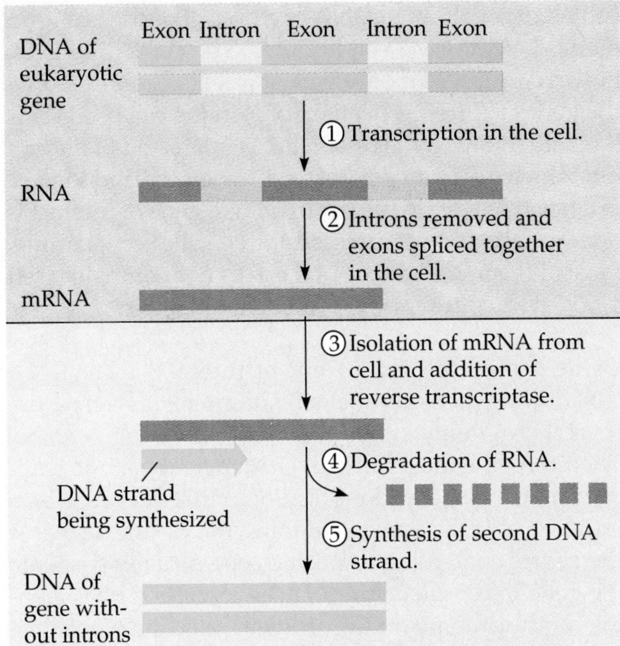

Figure 19.6
Making complementary DNA (cDNA) for a eukaryotic gene. Complementary DNA is DNA made in the laboratory using mRNA as a template and the enzyme reverse transcriptase obtained from retroviruses. Complementary DNA lacks introns and is therefore smaller than the original gene and easier to clone. It is also more likely to be functional in bacterial cells, which lack the machinery for removing introns from RNA transcripts. To be transcribed, the cDNA will have to be joined to an appropriate bacterial promoter.

the vector is phage DNA, each one replicates within its host bacterium and forms a phage clone containing multiple copies of the same foreign DNA segment. In either case, the resulting set of thousands of clones, each carrying copies of a particular segment from the foreign genome, is referred to as a **genomic library.** Because this gene-cloning procedure uses a mixture of fragments from the entire genome of an organism, it is often called a "shotgun" approach—no single gene is targeted for cloning. Figure 19.5 illustrates both a plasmid library and a phage library.

One problem with cloning DNA directly from a eukaryotic genome is that because eukaryotic genes often contain long noncoding regions (introns), they can span enormous lengths of DNA (see Chapter 16). Such a gene can destabilize a plasmid vector or prevent a phage DNA vector from fitting into a phage coat. To avoid this problem, scientists can sometimes make an artificial gene that lacks introns.

The starting material for the procedure, mRNA, is prepared by the eukaryotic cell, as shown in the first two steps in Figure 19.6. ① Transcription of an intron-containing gene in the cell nucleus yields an RNA molecule. ② Spliceosomes then remove the intron RNA and splice the exon RNA together to produce mRNA. ③ The biologist then isolates the mRNA molecules from the cell and uses them as templates to synthesize a complementary DNA strand. This synthesis of DNA on an RNA template is the reverse of transcription. It is catalyzed by the enzyme reverse transcriptase, which is obtained from retroviruses (see Chapter 17). ④ After a single strand of DNA is synthesized, the RNA is degraded. ⑤ The second DNA strand is made using the first as a template. The result is a double-stranded molecule of DNA carrying a gene made of **complementary DNA (cDNA)**—an artificial gene that has no introns.

The cDNA gene created by this method is more manageable in size than the original gene. It also has the potential for being transcribed and translated by bacterial cells, which do not have RNA-splicing machinery. However, in order for the cDNA gene to be transcribed in bacteria, it must be attached to DNA containing a bacterial promoter and other essential transcription signals (see Chapter 16).

Because the mRNA molecules for a particular gene cannot usually be isolated from the other mRNA molecules of a cell, the cDNA method, like the shotgun method, produces gene libraries. The cDNA libraries, however, represent only part of a cell's genome—only the genes expressed (transcribed) in the cell used. This is an advantage if the researcher is interested in finding out which genes are expressed by a specialized kind of cell (such as a brain cell) or if a single gene of interest is one of only a few expressed in the cell. For example, most of the mRNA in the precursors of mammalian red blood cells codes for the protein hemoglobin.

Using Probes to Find a Gene of Interest The DNA constituting a eukaryotic gene of interest is very rare, representing only about one millionth of the genome. Therefore, the most difficult challenge in genetic engineering is often the identification of the very rare bacterial or phage clones containing the desired gene. If the clones containing a specific gene actually translate the gene into protein, they can be identified by screening for the presence of the protein. Detection of the protein can be based on either its activity (as with an enzyme) or its structure, using antibodies that combine specifically with it (see Chapter 39). More often, screening techniques rely on detecting the gene itself.

Methods for detecting a gene directly all depend on base pairing between the gene and a complementary sequence on another nucleic acid molecule, either RNA or DNA. When at least part of the nucleotide sequence of the gene is known, or can be guessed from knowledge of the amino acid sequence of the protein, short nucleic acid molecules complementary to it can be chemically synthesized. Such a nucleic acid molecule, which will hydrogen bond specifically to the desired gene, is called a **probe,** and its location is traced by labeling it with a radioactive isotope. Figure 19.7 shows how a number of bacterial clones, growing as colonies on a plate of solid medium, can be simultaneously screened for the presence of DNA complementary to a specific RNA probe. The radioactive probe tags the correct clone—finds the needle in the haystack—by hydrogen bonding to its DNA.

Once a clone carrying the desired gene is identified, the clone can be grown in a large culture, and the gene of interest can easily be isolated in large amounts and used for further study. Also, the cloned gene itself can be used as a probe to identify similar or identical genes.

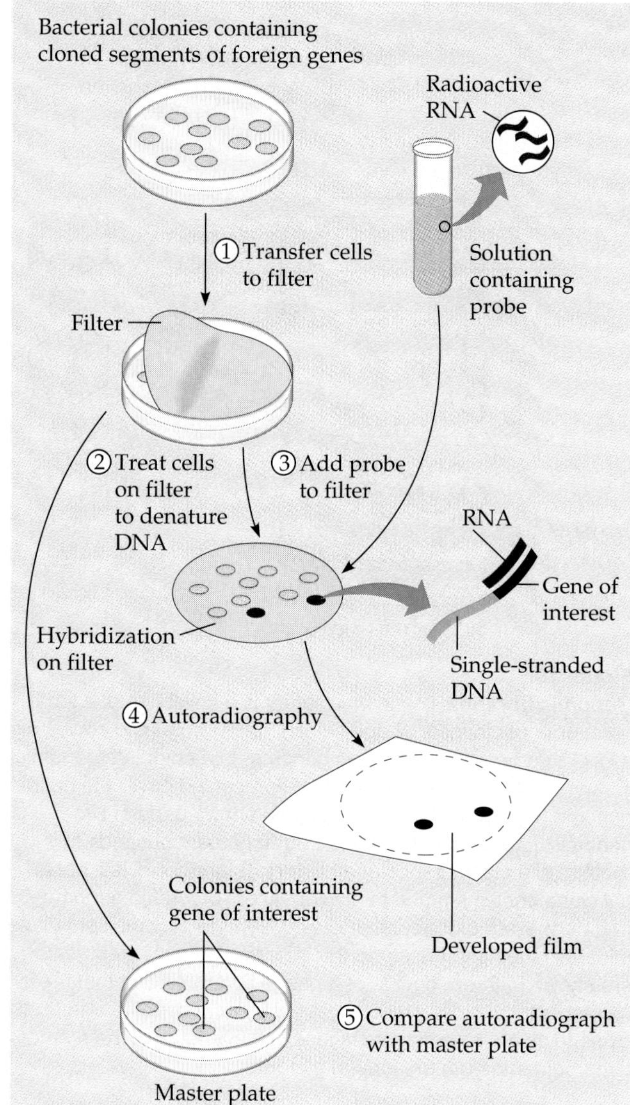

Figure 19.7
Using a nucleic acid probe to identify a cloned gene. This technique depends on the fact that complementary nucleotide sequences will base pair—bind together by hydrogen bonds. Here the cloned gene of interest is carried on a bacterial plasmid, and the probe is a short length of synthetic RNA complementary to part of the gene. ① Bacterial colonies on agar are pressed against special filter paper, transferring cells to the filter. ② The filter is treated to break open the cells and denature their DNA; the resulting single DNA strands stick to the filter. ③ A solution of probe molecules is incubated with the filter. The RNA hybridizes (base pairs) with any complementary DNA on the filter; excess RNA is rinsed off. ④ The filter is laid on photographic film, allowing any radioactive areas to expose the film (autoradiography). ⑤ The developed film, an autoradiograph, is compared with the master culture plate to determine which colonies carry the desired gene.

Making Gene Products Using Genetic Engineering Transferring a gene between two different kinds of cells is one thing; getting it to function in a new setting is quite another. Despite the universality of the genetic code, the problem of getting bacteria to express eukaryotic genes, even cDNA genes, was expected to be very difficult. Many details of transcription and translation are different in prokaryotes and eukaryotes. Molecular biologists were pleasantly surprised to find that, with a little genetic trickery, they could get bacteria to produce many eukaryotic proteins.

When the goal is to make as much of a gene product as possible, such as for commercial purposes, bacteria are usually the organisms of choice. They can be grown rapidly and cheaply in large fermenters, like the one in Figure 19.1, and their genomes are relatively simple and easy to manipulate. A number of genetic and biochemical methods can help maximize the expression of a eukaryotic gene in a bacterial cell. They include using a plasmid vector that will produce many copies of itself per cell, changing the promoter controlling the gene's expression to a highly active one, and attaching the eukaryotic gene to a bacterial gene encoding a protein that is produced in large quantities. (Enzymes are later used to clip off the unwanted, bacterial portion of the resulting protein.) Bacterial cells can be engineered to secrete a protein as it is made, thereby simplifying the task of purifying it.

Yeast cells offer some of the advantages of bacteria, such as ease of growth, and they are the organisms of choice for some commercial applications of genetic engineering. As eukaryotes, yeast cells are also useful in basic research. To study eukaryotic mechanisms of gene function and regulation, eukaryotic cells as well as eukaryotic genes must be used. Eukaryotic genes of interest are initially cloned in bacteria, but they are later returned to eukaryotic cells to allow gene expression and its study. Reinserting genes into eukaryotic cells is relatively easy with yeast, because yeast cells can be made to take up DNA by transformation, and they even have plasmids that facilitate the process. Biological information derived from yeast studies has exploded in the past few years.

Even the combination of bacteria and yeast, however, does not suit every project involving DNA technology. The cells of more complex eukaryotes carry out some biochemical processes that are absent in yeast. Thus, cultured animal cells or plant cells are used for certain kinds of genetic research and for certain commercial applications. For example, animal cells produce specialized glycoproteins called antibodies in response to foreign molecules called antigens. So the cells that are engineered to produce antibodies in commercial quantities are usually animal cells.

We have now covered the basic components of gene cloning using recombinant DNA methods. Next, we turn to some other methods that are important in DNA technology.

DNA Synthesis and Sequencing

Probes are not the only kind of nucleic acid that can be synthesized in a test tube. In some cases, entire genes for cloning have been chemically synthesized in the laboratory without using any sort of nucleic acid template; an example is the artificial genes for the two polypeptide chains of the hormone insulin. The laboratory synthesis of genes was once a tedious process, but machines are now available that can rapidly produce genes several hundred nucleotides long. Still, this approach is currently only practical for short genes whose exact nucleotide sequence is known.

Thanks to methods developed during the late 1970s in the United States and England, the sequences of even some large genes can now be determined in little more than a day. Sequencing (determining the nucleotide sequences) of DNA molecules is made possible by restriction enzymes, which can be used to cut the very long DNA molecules of cells and viruses into unique fragments of manageable length. The two main methods of DNA sequencing use gel electrophoresis to separate strands of DNA differing in length by only a single nucleotide. For a description of the most commonly used method, see the Methods Box on p. 400.

As they are being determined, thousands of DNA sequences are being collected in computer data banks. These storehouses of genetic information are of great value for understanding genes and genetic control elements, as well as for biotechnology. With the help of computers, long sequences can be scanned for shorter sequences known to be regulatory-protein recognition sites, such as enhancers (see Chapter 18), or control sequences, such as promoters. It is also possible to scan a DNA sample for similarities to known sequences in other genes or other organisms. Furthermore, the amino acid sequence of a polypeptide can quickly be determined once the nucleotide sequence of its gene is known; in fact, this is often the fastest method for determining amino acid sequence.

Amplifying DNA by the Polymerase Chain Reaction (PCR)

The **polymerase chain reaction (PCR)** is a technique by which any piece of DNA can be quickly amplified (copied many times) in vitro (in test tubes). The DNA is simply incubated under appropriate conditions with the enzyme DNA polymerase and special short pieces of nucleic acid called primers. Billions of copies of a segment of DNA can be made in a few hours, whereas it usually takes weeks to clone a piece of DNA by attaching it to a plasmid or viral genome. The Methods Box on p. 401 describes the PCR procedure. Devised in 1985, PCR has revolutionized research in molecular biology and is being used in many practical applications, as we will see later in the chapter.

Almost as impressive as the speed of PCR is its specificity. Because the primers determine the DNA sequence that is amplified, there is no need to isolate the desired segment of DNA from the starting material before carrying out the reaction. In fact, there is no need to prepare a purified sample of DNA. Only minute amounts of DNA need be present in the starting material, and this DNA can be in a partially degraded state. Furthermore, because DNA is an unusually stable biological molecule, it can often be amplified by PCR from sources thousands and even millions of years old.

PCR has been used to amplify DNA from a wide variety of sources: fragments of ancient DNA from a 40,000-year-old, frozen woolly mammoth; DNA from tiny amounts of tissue or semen found at the scenes of

Methods: Sequencing of DNA by the Sanger Method

Developed by Frederick Sanger, this method and its variations are the most commonly used techniques for determining the nucleotide sequence of DNA molecules. The Sanger method involves synthesizing in vitro DNA strands complementary to one of the strands of the DNA being sequenced. The method is based on the incorporation of a modified nucleotide (a *di*deoxyribonucleotide, which lacks *two* oxygen atoms) that blocks further DNA synthesis. Before beginning the synthesis procedure, the DNA to be sequenced is cut up into restriction fragments. Then the procedure illustrated below is carried out with each fragment.

Most of the work of DNA sequencing is now automated. As an alternative to labeling the fragments with radioactive primers, some of the newest sequencing machines use fluorescent dyes to tag the dideoxyribonucleotides, one color for each of the four types of nucleotide. This approach allows the four reactions to be performed in a single tube; the dideoxy ends of the reaction-product DNA strands can be distinguished by the color of their fluorescence.

A preparation of one of the strands of the DNA fragment is divided into four portions, and each portion is incubated with all the ingredients needed for the synthesis of complementary strands: a primer (radioactively labeled), DNA polymerase, and the four deoxyribonucleoside triphosphates. In addition, each reaction mixture contains a different *one* of the four nucleotides in the modified, dideoxy (dd) form.

Synthesis of the new strands starts with the primer and continues until a dideoxyribonucleotide is incorporated, which prevents further synthesis. Since the reaction mixture contains both deoxy and dideoxy forms of one nucleotide, the two forms "compete" for incorporation into the strand. Eventually, a set of radioactive strands of various lengths will be generated. This is shown here only for the reaction mixture with ddGTP.

The new DNA strands in each reaction mixture are separated by electrophoresis on a polyacrylamide gel, which can separate strands differing by as little as one nucleotide in length. The sequence of the newly synthesized strands can be read directly from the bands produced in the gel, and from that, the sequence of the original template strand is deduced. In this example, since the longest fragment terminates with ddG, this means that G is the last base in the new DNA strand. Notice that the second longest fragment terminates with ddA, meaning that A is the second to last base, and so on.

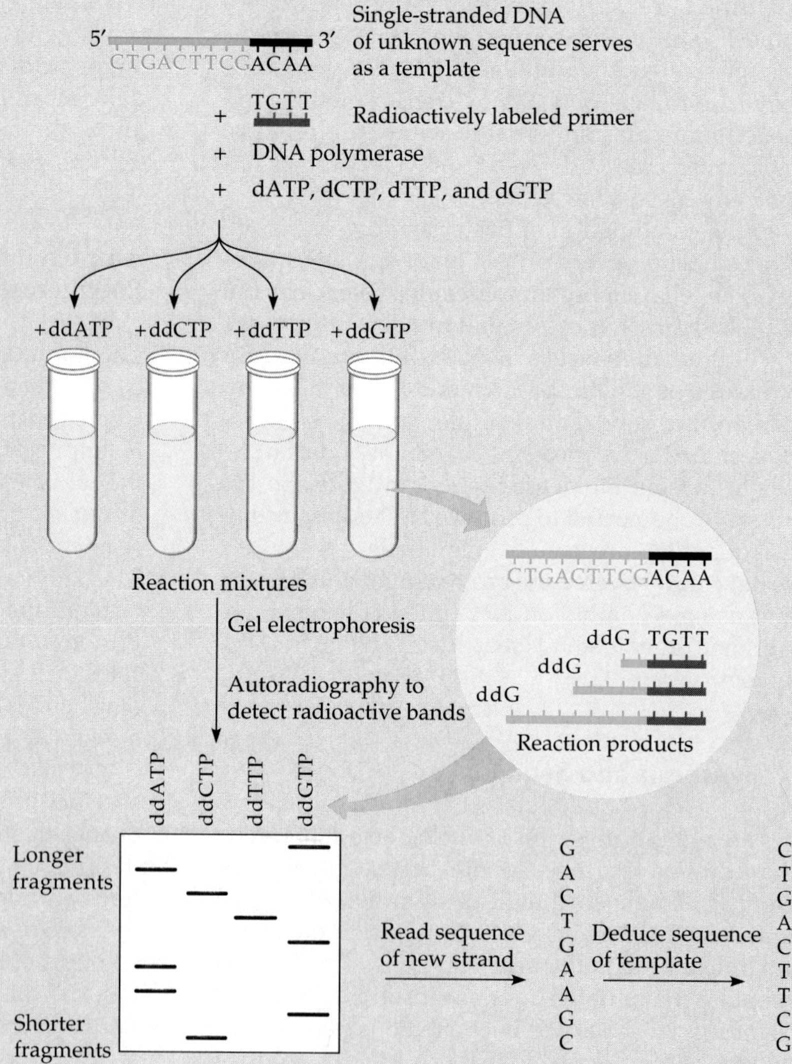

METHODS: THE POLYMERASE CHAIN REACTION (PCR)

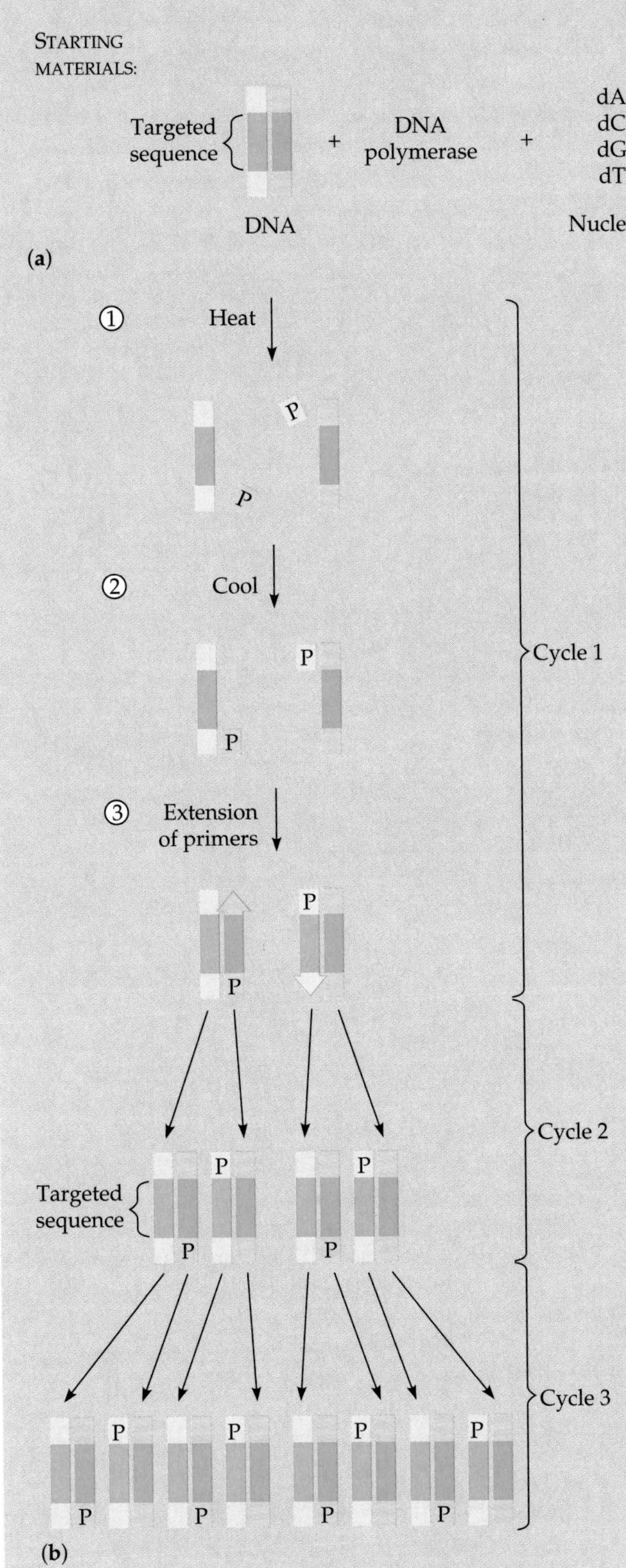

PCR is a method for making many copies of a specific segment of DNA. It is much faster than gene cloning with plasmid or phage DNA and is performed completely in vitro. The starting material for PCR (Figure a) is a solution of double-stranded DNA containing the nucleotide sequence that is "targeted" for copying (it need not be a gene). The scientist then adds DNA polymerase (which catalyzes the reaction), a supply of all four nucleotides (for assembly into new DNA), and primers. The primers are required for the DNA polymerase enzyme to initiate DNA synthesis (see Chapter 16). The particular primers used in PCR are chemically synthesized with sequences that are complementary to the ends of the targeted segment of DNA.

Figure b outlines the PCR procedure. ① The DNA is briefly heated to separate its strands and then ② cooled to allow the primers to bind by hydrogen bonds to the ends of the target sequence, one primer on each strand. Then ③ the DNA polymerase extends the primers by adding nucleotides, using the longer DNA strands as templates. Within a short time, the amount of the target DNA sequence has been doubled. The solution is then heated again, starting another cycle of strand separation, primer binding, and DNA synthesis. Again and again the cycle is repeated, at about 5 minutes per cycle, until the targeted sequence has been duplicated enough times.

Methods: RFLP Analysis

Mapping the human genome, diagnosing genetic diseases, and solving violent crimes—these are only some of the applications of the powerful method called RFLP analysis. This method is based on naturally occurring minor differences (polymorphisms) in DNA sequences, which can be detected because they result in restriction fragment length polymorphisms, or RFLPs. For practical applications in human medicine or law, the problem is usually to determine which variants of a particular RFLP marker (or set of markers) are carried by the individuals being tested.

As indicated in these diagrams, RFLP analysis uses four laboratory techniques that have already been discussed: restriction enzyme treatment of DNA, gel electrophoresis, the use of DNA probes, and autoradiography. RFLP analysis also requires another technique, called *Southern blotting* (after its

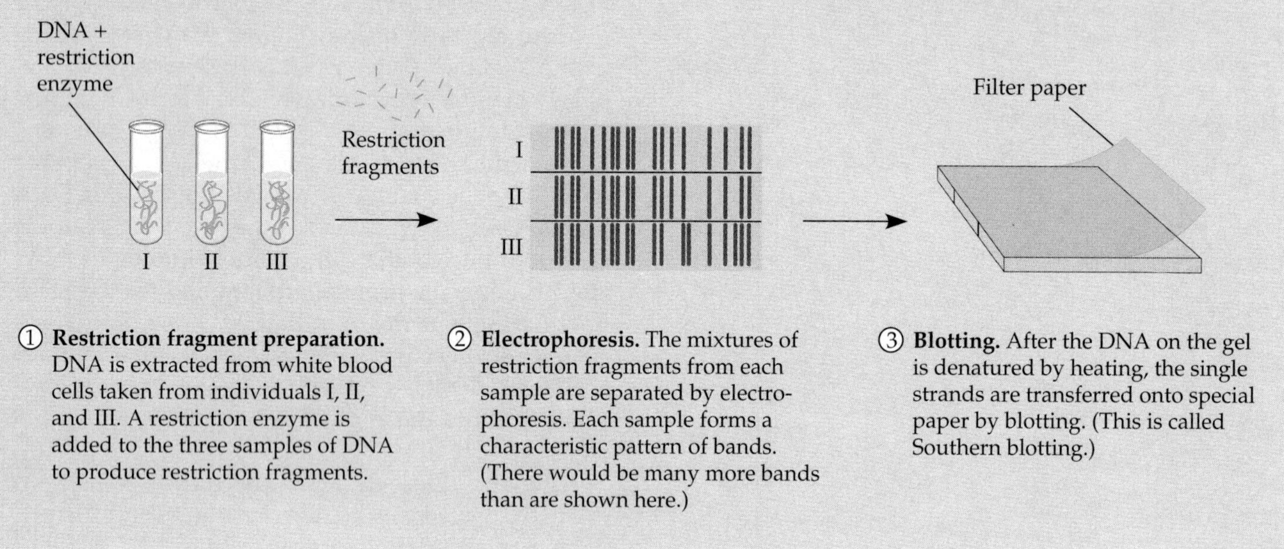

① **Restriction fragment preparation.** DNA is extracted from white blood cells taken from individuals I, II, and III. A restriction enzyme is added to the three samples of DNA to produce restriction fragments.

② **Electrophoresis.** The mixtures of restriction fragments from each sample are separated by electrophoresis. Each sample forms a characteristic pattern of bands. (There would be many more bands than are shown here.)

③ **Blotting.** After the DNA on the gel is denatured by heating, the single strands are transferred onto special paper by blotting. (This is called Southern blotting.)

violent crimes; DNA from single embryonic cells for rapid prenatal diagnosis; and DNA of viral genes from cells infected with such difficult-to-detect viruses as HIV (the AIDS virus).

RFLP Analysis

Earlier in the chapter, we saw that the DNA fragments resulting from cutting a particular piece of DNA with a specific restriction enzyme give a characteristic pattern of bands upon gel electrophoresis. Each band corresponds to a DNA restriction fragment of a certain length. Using this technique to examine homologous segments of DNA known to carry different alleles of a gene, researchers found that DNA of different alleles would show different band patterns after treatment with certain restriction enzymes. This result was not surprising, since differences in nucleotide sequence would be expected to result in differences in the number and locations of restriction sites. Figure 19.8 shows why this is so. The diagram illustrates the DNA of two different alleles of the same gene that differ in sequence by a single base pair (outlined in black) (Figure 19.8a). As a consequence of this difference, allele 1 has one more restriction site than is found in allele 2. Therefore, when the DNA segments representing these two alleles are treated with a restriction enzyme, they are cut into fragments that differ in both number (three versus two) and length. Upon gel electrophoresis (Figure 19.8b), the DNA from the two alleles shows different patterns of bands.

Biologists were excited to discover similar results when homologous but *noncoding* segments of DNA were used as starting material. In fact, differences in band patterns were observed much more often than they had expected. Differences in DNA sequence on homologous chromosomes that result in different restriction fragment patterns turn out to be scattered abundantly throughout genomes, including the human genome. Such differences have been named **restriction fragment length polymorphisms (RFLPs)** (pronounced "riflips"). A given RFLP marker frequently occurs in numerous variants in a population (the word *polymorphism* comes from the Greek for "many forms"). The Methods Box on this page describes in more detail how RFLPs are detected and analyzed. As you can see there, radioactively labeled nu-

inventor). For starting material, DNA from an entire genome can be used, typically DNA from white blood cells in the case of humans. This huge length of DNA will produce so many restriction fragments that, if all are made visible (with a dye, for example), they will appear as a smear in gel electrophoresis, rather than as discrete bands. However, the DNA bands of interest can be selectively visualized by using a radioactive probe. The probe consists of multiple copies of a radioactively labeled piece of DNA that will base pair with DNA of the segment being tested (imperfections in base pairing resulting from the RFLP variations are not extensive enough to interfere with the hybridization). The example illustrated here is RFLP analysis on three relatives. The results indicate that individuals I and II carry the same version(s) of the RFLP marker but that individual III carries a different version.

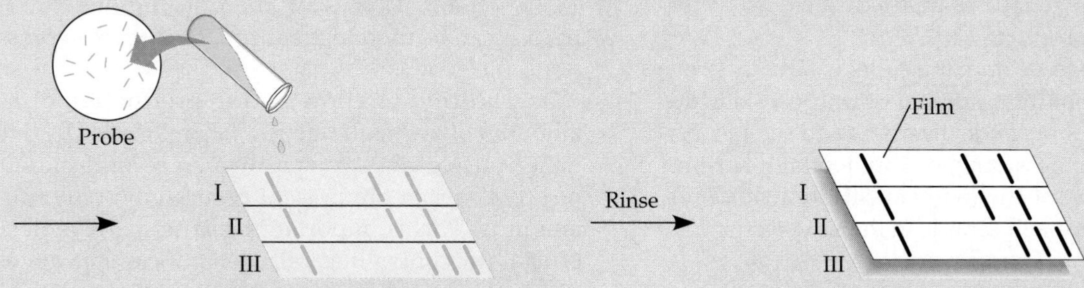

④ **Radioactive probe.** A radioactive probe is added to the DNA bands. The probe is a single-stranded DNA molecule that is complementary to the DNA making up the restriction fragments of the genetic marker. The probe attaches only to those fragments by base pairing.

⑤ **Autoradiography.** After the probe is added and the excess probe is rinsed off, a sheet of photographic film is laid over the gel. The radioactivity in the bound probe exposes the film to form an image corresponding to specific DNA bands—the bands containing DNA that base pairs with the probe.

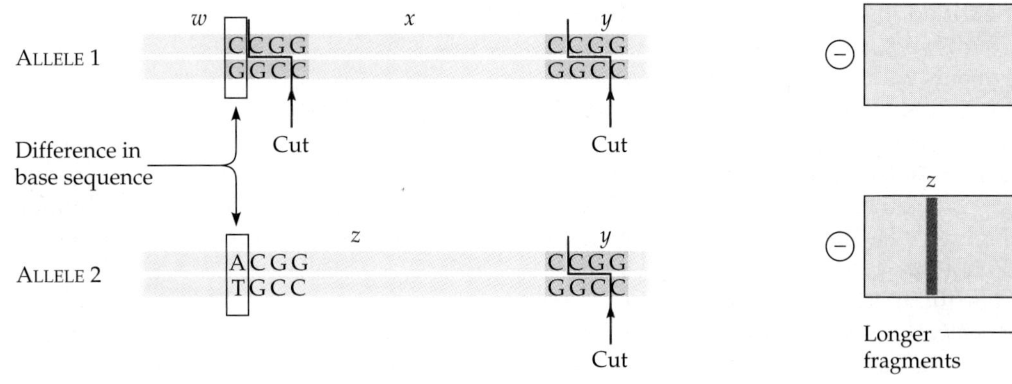

(**a**) DNA from two alleles

(**b**) Electrophoresis of restriction fragments

Figure 19.8
Using restriction fragment patterns to distinguish DNA from different alleles. (**a**) Restriction site differences in the DNA of two alleles (only the relevant bases are shown). The particular restriction enzyme for which the recognition sequences are shown here cuts the DNA from allele 1 into three pieces (*w*, *x*, and *y*) but cuts the DNA from allele 2 into only two pieces (*z* and *y*). (**b**) The results of gel electrophoresis of the two restriction fragment mixtures. A clear difference between the two alleles is revealed by their band patterns on the gel. Allele 1 has three bands, corresponding to fragments *w*, *x*, and *y;* allele 2 has two bands, corresponding to *z* and *y*.

cleic acid probes are key to the procedure. A photograph of the end-product of RFLP analysis, an autoradiograph, is shown in Figure 19.11 (p. 409).

Whether or not RFLP sites are within coding DNA, they are inherited in a Mendelian fashion. They can serve as genetic markers for making linkage maps; the geneticist uses the same reasoning you learned about in Chapter 14. Because of the abundance of RFLP sites and because RFLPs can be detected whether or not they lead to a difference in the organism's phenotype, they are extremely useful for chromosome mapping. The discovery of RFLPs has greatly increased the number of markers available for mapping the human genome. No longer are human geneticists limited to genetic variations that lead to obvious phenotypic differences (such as genetic diseases) or even to differences in protein products.

The current map of human RFLPs is already proving a powerful tool for a number of purposes, including the diagnosis of genetic disorders and the identification of symptomless carriers of potentially harmful genes. This is one of the many applications of DNA technology that we will sample in the next section.

APPLICATIONS OF DNA TECHNOLOGY

Uses of DNA Technology in Basic Research

DNA technology has triggered research advances in almost all fields of biology by allowing biologists to tackle more specific questions with finer tools. As mentioned earlier, the new techniques have opened up the study of the molecular details of eukaryotic gene structure and function. Once a gene has been cloned, and a clone containing the gene identified, the gene can be readily produced in large amounts and used as a probe to search out similar DNA segments within the same or other genomes. By virtue of its ability to hydrogen bond to complementary nucleotide sequences, the DNA probe acts like a magnet used to locate needles in a haystack. Radioactive isotopes or other kinds of labels tag the probe itself (see Figure 19.7). The beauty of these methods is that they do not depend on whether or not genes are expressed. For the first time, geneticists can easily study genes directly, without having to infer genotype from phenotype as in classical genetics. However, they are now often faced with the opposite problem: determining what a cloned gene does—that is, inferring phenotype from genotype.

The DNA segments fished out by the probe can help the biologist answer important questions. They can provide information about the evolutionary relationships between the gene of interest and other genes of the same organism or of other organisms. They can allow the biologist to learn about the natural form of the gene (the probe may have been cDNA), including regulatory sequences and other noncoding sequences adjacent to the gene or within it. Such information, combined with the results from gene expression experiments, is greatly increasing our understanding of how genes are organized and controlled.

A DNA probe can even be used to map a gene on a eukaryotic chromosome. In the technique called **in situ hybridization,** a radioactive DNA probe is allowed to base pair (hybridize) with complementary sequences on intact chromosomes on a microscopic slide (*in situ* means "in place"). Autoradiography and chromosome staining are then used to reveal which band on which chromosome the probe has attached to (see Figure 14.1). (In a similar way, the binding of a probe to mRNA can be used to identify cells that are expressing a specific gene.)

In addition to allowing the production of large amounts of particular genes, recombinant DNA technology also enables the production of large quantities of proteins that are present naturally in only minute amounts. This is important because many molecules crucial for controlling cell metabolism and development are scarce and therefore cannot be purified and characterized by traditional biochemical methods.

Some researchers, emboldened by the power of these new techniques, are attempting to catalog all the proteins made by particular organs, such as the brain. The task has become feasible now that biologists are no longer limited to studying proteins produced in large enough quantities to detect with assays for enzymatic activity or antibody binding. Instead, complementary DNA from all the messenger RNA molecules present in the organ can be prepared, cloned in bacteria, and used to prepare enough of each protein for study. Early estimates are that the brain catalog includes tens of thousands of different proteins.

The Human Genome Project

The most ambitious research project made possible by DNA technology is the effort to map the entire human genome. Four complementary approaches are being used:

1. *Genetic (linkage) mapping of the human genome.* The initial goal is to locate at least 3000 genetic markers (genes or other identifiable loci on the DNA) spaced evenly throughout the chromosomes. This approach is made feasible by the abundance of RFLPs in the human genome, and many of the markers used will undoubtedly be RFLP markers. The resulting map will make it easy for researchers to find the loci of other markers—including genes—by testing for genetic linkage to known markers.

2. *Physical mapping of the human genome.* This is done by cutting each chromosome into a number of identifiable fragments, and then determining their actual order in the chromosome.

3. *Sequencing the human genome.* This is the process of determining the exact order of the nucleotide pairs of each chromosome. Since a haploid set of human chromosomes contains approximately 3 billion nucleotide pairs, this is potentially the most time-consuming part of the project.

4. *Analyzing the genomes of other species.* The project also includes similar analysis of the genomes of other species important in genetic research, such as *E. coli*, yeast, the plant *Arabadopsis* (see Chapter 31), and mouse.

These four approaches are yielding data that together will give a complete map of the human genome and an understanding of how the human genome compares to those of other organisms. Work is well under way to map the genome of the yeast *Saccharomyces cerevisiae.* In 1992, an international team sequenced one entire chromosome (chromosome 3, consisting of 315,357 nucleotide pairs), and by the end of the decade, the sequencing of most of the other 15 yeast chromosomes will probably be completed. Interestingly, more than half the genes on chromosome 3 encode proteins whose functions are unknown. Researchers predict that many of these genes will also turn up in other eukaryotes.

The potential benefits of genomic information are huge. In the area of health care, the identification and mapping of the genes responsible for genetic diseases will surely aid in the diagnosis, treatment, and prevention of those conditions. In the area of basic science, detailed knowledge of the genomes of humans and other species will give insight into fundamental questions of genome organization, control of gene expression, cellular growth and differentiation, and evolutionary biology. Analysis of the human genome data and comparison with data from other species will proceed throughout all phases of the project, which involves researchers in many countries.

The methods used to build the human genetic map combine classical pedigree analysis of large families (see Chapter 13), the use of many of the molecular techniques described earlier in this chapter, and a variety of other techniques. A method called **chromosome walking,** for instance, makes use of several of the techniques already described to put in order the large number of DNA fragments that result when the very long DNA molecules of eukaryotic chromosomes are treated with restriction enzymes. In chromosome walking, the researcher isolates *overlapping* DNA segments, which have been obtained by cutting two samples of the original DNA with different restriction enzymes. The procedure is outlined in the Methods Box on p. 406. The overlaps between DNA segments reveal the order in which the segments occur in the original DNA and hence the rough order of any genes or other markers found on the segments.

Another technique that should prove valuable to the Human Genome Project is PCR amplification. For example, PCR can amplify specific portions of DNA from individual sperm cells. In this way, researchers can obtain the immediate products of meiotic recombination in amounts sufficient for study, and they can analyze as large a sample as they need, even thousands of sperm. Thus, based on the frequencies of crossovers between genes, researchers can perform human linkage mapping without having to find large families for pedigree analysis.

Finally, a significant part of the technological power to achieve the goals of the Human Genome Project will come from advances in automation and utilization of the latest electronic technology, including computer software.

Medical Uses of DNA Technology

Modern biotechnology has already made enormous contributions to the field of medicine. Major advances continue in the diagnosis of human genetic disorders and other diseases, in the first applications of gene therapy, and in the development of new vaccines and other pharmaceutical products.

Diagnosis of Diseases A new chapter in the diagnosis of infectious diseases is being opened by DNA technology, in particular the use of PCR and labeled DNA probes. With these sensitive methods, elusive pathogens may be tracked down. In one dramatic case, PCR was used to amplify, and thus detect, HIV DNA in tissue samples of a British sailor who died in 1959.

The use of DNA technology to diagnose genetic diseases is proceeding even faster. Medical scientists can now diagnose more than 200 human genetic disorders using DNA technology. In an increasing number of cases, it is possible to identify individuals with genetic diseases before the onset of symptoms, even before birth. It is also possible to identify symptomless carriers of potentially harmful recessive alleles. Genes have been cloned for a number of human diseases, including hemophilia, phenylketonuria (PKU), cystic fibrosis, and the deadly degenerative disorder Duchenne's muscular dystrophy (see Chapter 14).

DNA technology makes it possible to detect which allelic forms of genes are present in DNA samples directly. Once cloned, the gene (either a normal or a mutant allele) can be used as a probe to find the corresponding alleles in the cells being tested. The alleles are then compared with normal and mutant standards,

Methods: Chromosome Walking

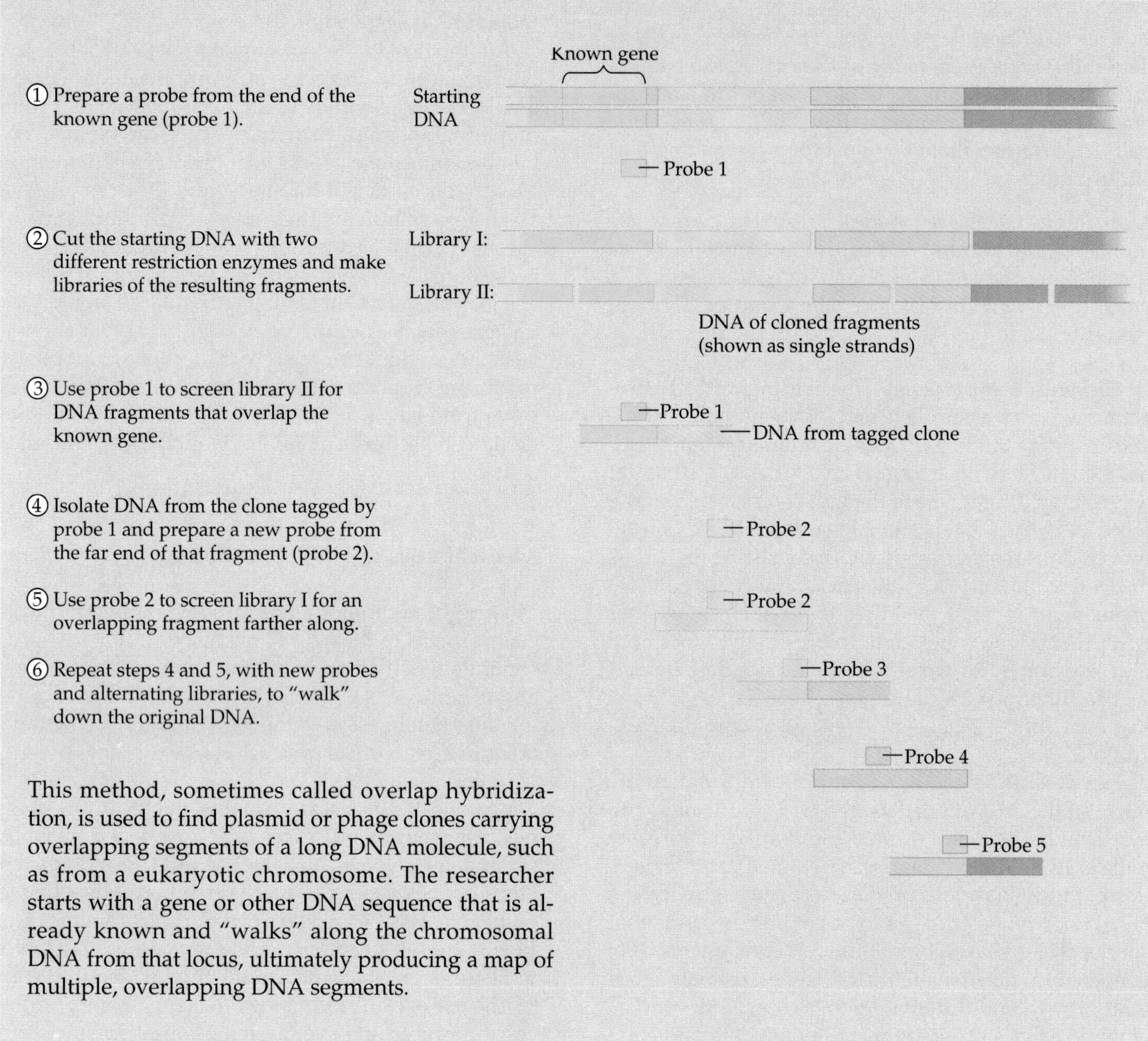

Known gene
① Prepare a probe from the end of the known gene (probe 1).
Starting DNA
Probe 1
② Cut the starting DNA with two different restriction enzymes and make libraries of the resulting fragments.
Library I:
Library II:
DNA of cloned fragments (shown as single strands)
③ Use probe 1 to screen library II for DNA fragments that overlap the known gene.
Probe 1
DNA from tagged clone
④ Isolate DNA from the clone tagged by probe 1 and prepare a new probe from the far end of that fragment (probe 2).
Probe 2
⑤ Use probe 2 to screen library I for an overlapping fragment farther along.
Probe 2
⑥ Repeat steps 4 and 5, with new probes and alternating libraries, to "walk" down the original DNA.
Probe 3
Probe 4
Probe 5

This method, sometimes called overlap hybridization, is used to find plasmid or phage clones carrying overlapping segments of a long DNA molecule, such as from a eukaryotic chromosome. The researcher starts with a gene or other DNA sequence that is already known and "walks" along the chromosomal DNA from that locus, ultimately producing a map of multiple, overlapping DNA segments.

usually by RFLP analysis: The segments of DNA containing the known and unknown alleles are cut up with various restriction enzymes, and the electrophoresis patterns formed by the sets of restriction fragments are compared. Because a mutation changes the nucleotide sequence, it may also change a restriction enzyme cutting site and, hence, the band patterns on gels.

Even in cases where the gene has not yet been cloned, the presence of an abnormal allele can be diagnosed with reasonable accuracy if a closely linked RFLP marker has been found. The idea is that if a mapped RFLP marker is co-inherited at high frequency with a disease, it is probable that the defective gene is located close to the RFLP marker on the chromosomal DNA. Blood samples from relatives of the person at risk must be studied in order to determine which variant of the RFLP marker is linked to the abnormal allele in that family, and this variant must be different from the RFLP variant(s) linked to that family's normal allele (Figure 19.9). If these conditions are met, the RFLP marker variants found in the genome of the person at risk reveal whether the abnormal allele is also likely to be present. Alleles for Huntington's disease and a

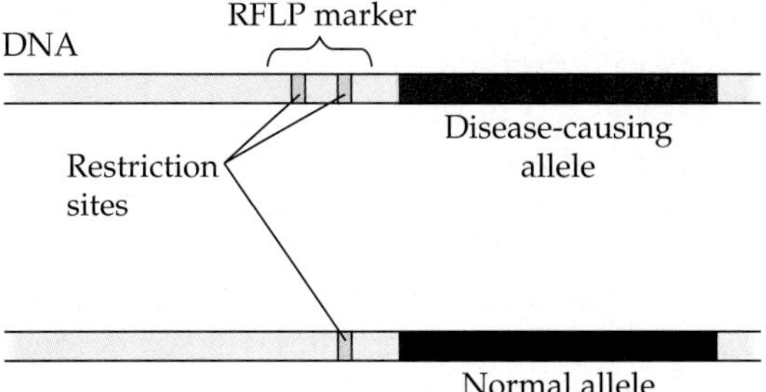

Figure 19.9
RFLP markers close to a gene. Even if a disease-causing gene has not been cloned and its precise locus is unknown, its presence can sometimes be detected with high (though not perfect) accuracy by testing for the status of RFLP markers that are very close to the gene in question. This diagram shows homologous segments of DNA from a family in which some members have a genetic disease. In this family, different versions of a RFLP marker are associated with the different forms of the allele—allowing the test to be applied. If a family member has inherited the version of the RFLP marker with two restriction sites (rather than one), there is a high probability that the individual has also inherited the disease-causing allele.

number of other genetic diseases are currently detected in this indirect way. Once a gene is mapped precisely, it can easily be cloned for study and for use as the ultimate probe for identical or similar DNA.

Human Gene Therapy Genetic engineering has the potential to actually correct some genetic disorders in individuals. For any genetic disorder traceable to a single defective gene, it should theoretically be possible to replace or supplement the defective gene with a functional, normal gene using recombinant DNA techniques. The new gene could be inserted into some of the somatic cells of a child or adult, or into germ (gamete-producing) cells or embryonic cells.

Gene therapy has begun with efforts to correct somatic cells of individuals with certain life-threatening genetic disorders. A number of human trials are under way to test the effectiveness of gene therapy against certain forms of cancer and a number of enzyme deficiency diseases.

Gene therapy is especially suited for single-enzyme deficiency diseases. Normal genes are introduced into a patient's own somatic cells. The cells must be ones that actively reproduce in the body (so that the normal gene will be replicated in the individual) and whose normal protein products can correct the disorder. Several children with an immunodeficiency disease (see Chapter 39) that results from a lack of the enzyme adenosine deaminase (ADA) are currently being treated successfully by gene therapy. They are receiving intravenous injections of their own T lymphocytes (a type of white blood cell) that have been engineered to carry a normal ADA allele. Prior to injection, T lymphocytes are removed from the patients and cultured in vitro with harmless retroviral vectors carrying the normal ADA allele. The recombinant retroviruses infect the cells and insert the normal ADA allele into the cells' genomes along with a DNA version of the viral genome. The engineered cells are then returned to the patient's body. Also in progress are efforts to put the normal ADA allele into patients' bone marrow cells (Figure 19.10). This approach could effect a broader, more long-lasting treatment, since the bone marrow cells multiply and give rise to all the cells of the immune system.

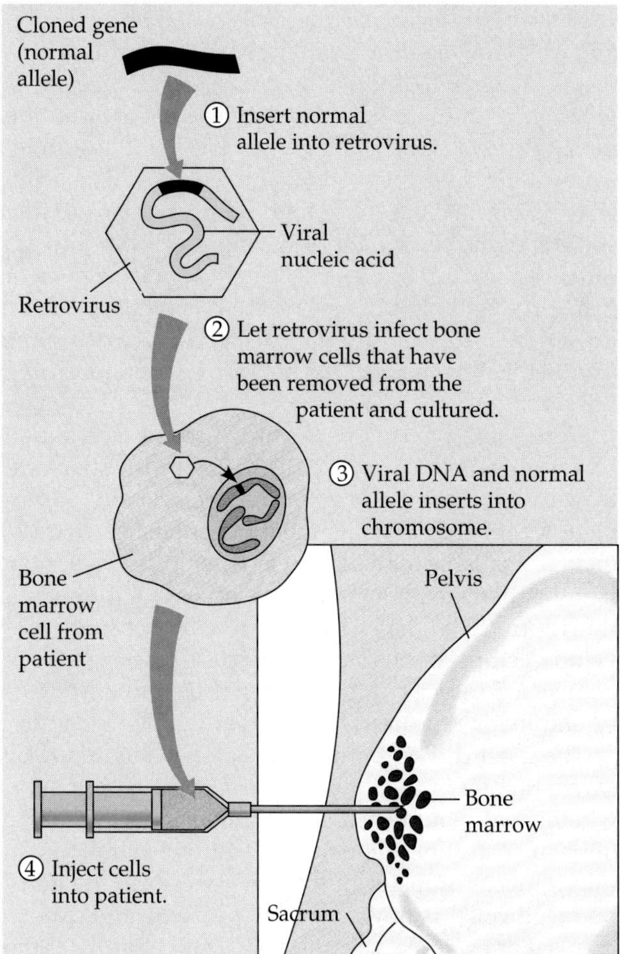

Figure 19.10
One type of gene therapy procedure. In this procedure, a harmless retrovirus is used as a vector to introduce a normal allele of a gene into the cells of a patient who lacks it. The method takes advantage of the fact that a retrovirus inserts a copy of its nucleic acid (actually a DNA transcript of its RNA genome; see Chapter 17) into the chromosomal DNA of its host cell. If its nucleic acid includes a foreign gene and that gene is expressed, the cell—and its mitotic descendants—will be cured. Cells that continue to reproduce throughout life, such as bone marrow cells, are ideal for gene therapy (the insertion of such cells is shown here in a side view of a pelvis).

Some of the technical questions posed by gene therapy are: How can the proper genetic control mechanisms be made to operate on the transferred gene? At what stage of biological development is intervention most effective, and when is it too late? Is it best to introduce the gene into cells that have not finished differentiating, such as those in the bone marrow, or into specific tissue cells, such as T lymphocytes in the case of ADA therapy? How can we get the correct gene or gene product to the tissues where it is required? How do we treat developmental diseases, where a gene product may be needed for only a brief period during development?

Gene therapy also raises some difficult ethical and social questions. At present, the procedures are extremely expensive and require expertise and equipment found only in major medical centers. Once treatment is available for a particular genetic disorder, how many patients will actually have access to it? One authority predicts that researchers will eventually develop vectors—perhaps engineered viruses—that can be injected directly into patients, thus reducing costs by eliminating the need for engineering cells in culture.

The most difficult ethical question—one raised by David Suzuki in the interview preceding this unit—is whether we should try to treat human germ cells in the hope of correcting the defect in future generations. In mice at least, transfer of foreign genes into the germ line (egg cells) is now a routine procedure in experiments. For example, scientists have successfully transplanted a human gene for one of the polypeptides of hemoglobin into the germ line of mice defective in the corresponding mouse gene. In many of the recipient mice and their descendants, the human gene is active in producing the polypeptide, not only in the correct location (red blood cells) but also at the correct time in development (at the appropriate fetal stage). So far, getting good expression of foreign genes in live animals has been much less successful in other experimental systems, but it is clear that the technical problems will eventually be overcome. Thus, we may soon have to face the question of whether to intervene in human germ lines.

Some critics have said flatly that tampering with human genes in any way, even to treat individuals who have life-threatening diseases, is wrong. They argue that it will inevitably lead to the practice of eugenics, a deliberate effort to control the genetic makeup of human populations. Other observers see no fundamental difference between genetic engineering of somatic cells and other conventional medical interventions to save lives.

Vaccines Recombinant DNA techniques are helping medical scientists develop vaccines for the prevention of infectious diseases. Especially for the many viral diseases for which there are no effective drug treatments, prevention by vaccination is virtually the only way to fight the disease. Traditional vaccines for viral diseases are of two types: particles of a virulent virus that have been inactivated by chemical or physical means, and active virus particles of an attenuated (nonpathogenic) viral strain. In both cases, the virus particles are similar enough to the active pathogen to trigger an immune response that will protect against it. In the immune response, the animal produces antibodies that react very specifically with invading pathogens (see Chapter 39).

There are several ways in which the new biotechnology is being used to modify current vaccines or provide new vaccines against previously recalcitrant diseases. First, recombinant DNA techniques can generate large amounts of a specific protein molecule from the protein coat of a particular disease-causing virus, bacterium, or other microbe. If the protein, referred to as a subunit, is one that triggers an immune response against the intact pathogen, it can be used as a vaccine. Second, genetic engineering methods can be used to modify the genome of the pathogen in order to attenuate it. Vaccination with a live but attenuated organism is often more effective than a subunit vaccine, because a small amount of material triggers a greater response by the immune system, and pathogens attenuated by gene-splicing techniques may be safer than the natural mutants traditionally used.

Another scheme that uses a live vaccine employs vaccinia, the virus that is the basis of the smallpox vaccine. Researchers benefit from a wealth of medical experience with smallpox vaccine. With recombinant DNA techniques, the viral genes that induce immunity to smallpox can be replaced with genes that induce immunity to other diseases. In fact, the vaccinia virus can be made to carry the genes required to vaccinate against several diseases simultaneously. Therefore, in the future, a single live vaccinia inoculation could conceivably protect people from as many as a dozen diseases, perhaps including AIDS.

Other Pharmaceutical Products One of the first practical applications of gene splicing was for producing mammalian hormones and other mammalian proteins in bacteria. Insulin, growth hormone, and several proteins of the immune system, such as molecules called interferons, were among the early examples. Insulin was the first, and human growth hormone the second, polypeptide hormone made by recombinant DNA procedures to be approved for use in treating human patients in the United States.

About two million diabetics in the United States depend on insulin treatment to help control their disease. Before 1982, the principal sources of therapeutic insulin were pig and cattle pancreatic tissues obtained from slaughterhouses. Although the insulin extracted from pigs and cattle closely resembles the human version, it is not identical, and causes adverse reactions in some people. Now there are several ways of producing insulin in genetically engineered bacteria, and the insulin produced is chemically identical to that made in the human pancreas.

Human growth hormone (HGH) has almost 200 amino acids, making it far larger than insulin. Moreover, growth hormones are more species-specific than insulin, meaning that growth hormone from other animals is generally not an effective growth stimulator if administered to humans. Before genetically engineered HGH became available in 1985, children born with hypopituitarism—a form of dwarfism caused by inadequate amounts of HGH—had to rely on scarce supplies of the hormone obtained from human cadavers. Gene splicing made much more HGH available for treating children. It also gave researchers a chance to learn about other potential uses for the hormone. In the future, HGH may be widely used to help heal wounds and broken bones, to treat severe burns, and to retard, if not reverse, the loss of muscle mass that often accompanies the aging process.

In 1989, a biotechnology firm used recombinant DNA methods to produce still another medically important human hormone, erythropoietin (EPO). This protein is normally produced by the kidney as a hormone that stimulates the production of red blood cells in the bone marrow. Biotechnology now provides a source of EPO to treat anemia due to a variety of causes, including kidney damage.

The most recent developments in pharmaceutical products involve some truly novel ways to fight disease-causing viruses, which are much less susceptible to drug treatment than are bacterial or fungal pathogens. Current research may soon lead to drugs that prevent viral messenger RNAs from being translated into viral proteins in an infected cell. For example, one approach is to synthesize **antisense nucleic acid,** single-stranded molecules of DNA or RNA that base pair with crucial viral mRNA molecules and block their translation. Another approach is to use recombinant DNA techniques to make drugs that either block or mimic surface receptors on cell membranes. One such experimental drug substitutes for a receptor protein that HIV binds to when it attacks white blood cells. The idea is to decoy the AIDS virus and keep it from infecting white blood cells. Both the antisense approach and surface-receptor drugs are also being tried for selectively destroying cancer cells.

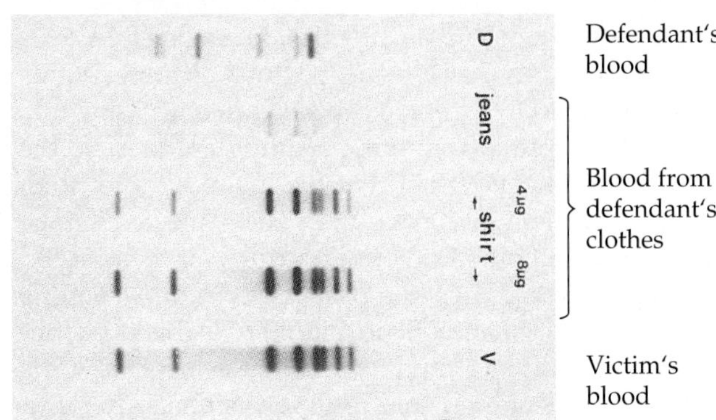

Figure 19.11
RFLP autoradiograph from a murder case. As revealed by RFLP analysis, DNA from bloodstains on the defendant's clothes matches the DNA fingerprint of the victim but differs from the DNA fingerprint of the defendant. This is evidence that the blood on the defendant's clothes came from the victim, not the defendant.

Forensic Uses of DNA Technology

In violent crimes, blood or small fragments of other tissue may be left at the scene or on the clothes or other possessions of the victim or assailant. If rape is involved, small amounts of semen may be recovered from the victim's body. If enough tissue or semen is available, forensic laboratories can perform tests to determine the blood type or tissue type. However, such tests have limitations. First, they require fairly fresh tissue in a sufficient amount for testing. Second, because there are many people in the population with the same blood type or tissue type, this approach can only exclude a suspect; it is not evidence of guilt.

DNA testing, on the other hand, can theoretically identify the guilty individual with certainty, since the DNA base sequence of every individual is unique (except for identical twins). For this forensic application, the most common DNA technology is a RFLP autoradiograph. The autoradiograph displays selected DNA restriction fragments as bands separated by electrophoresis. The method is used to compare DNA samples from the suspect (a murder suspect, for example), the victim, and a small amount of semen, blood, or other tissue found at the scene of the crime. Radioactive probes mark the bands that contain certain RFLP markers. Even a small set of RFLP markers from an individual can provide a **DNA fingerprint** that is of forensic use, since the probability that two people (who are not identical twins) would have the exact same set of RFLP markers is very small. The autoradiograph in Figure 19.11 is the type of evidence presented (with explanation) to juries. It shows a match between a victim's DNA fingerprint and that of blood found on the defendant's clothes. In support of such evidence, the prosecution presents statistical calculations of the chance that more than one person's RFLP markers would match those in the blood sample.

Just how reliable is DNA fingerprinting? When we say that the DNA fingerprint from each individual is absolutely unique, we are speaking of the theoretical situation where restriction fragment analysis is done on the entire genome. In practice, forensic DNA tests focus on only five or ten tiny regions of the genome. However, the DNA regions chosen are ones known to be highly variable from one person to another. So the probability is very small that two people (other than identical twins) will have exactly the same sequences, and hence matching DNA fingerprints, for these tested regions. In most cases, the probability is between one chance in 100,000 and one in a million. Is this probability small enough for a jury to convict a suspect? Some legal experts contend that juries are so heavily influenced by this kind of scientific data that the evidence should be virtually flawless before it is admissible. Some courts have even thrown out DNA evidence because of uncertainty about its probability of being correct. As with most new technology, forensic applications of DNA fingerprinting raise important ethical questions.

There is also the question of what happens to DNA data once they are available. Should DNA fingerprints be filed or destroyed? Some states are now saving DNA data from convicted criminals. We will examine more of the ethical issues associated with DNA technology later.

Agricultural Uses of DNA Technology

Scientists concerned with feeding Earth's human population are working to learn more about the genetics of the plants and animals important to agriculture, and they have begun using genetic engineering to improve agricultural productivity.

Animal Husbandry Farm animals are already being treated with products that have been made by recombinant DNA methods. These products include new or redesigned vaccines, antibodies, and growth hormones. For example, some milk cows are being injected with bovine growth hormone (BGH), made by *E. coli,* in order to raise milk production (it usually increases by about 10%). BGH has passed all safety tests so far, and it is likely that it will be used extensively in dairy herds in the near future. BGH also improves weight gain in beef cattle. Another protein made by engineered *E. coli* that is useful for agriculture is the enzyme cellulase, which hydrolyzes cellulose and makes it possible for virtually all parts of a plant to be used for animal feeds.

A number of **transgenic organisms,** ones that contain genes from another species, have been developed for potential agricultural use. Transgenic animals, including beef and dairy cattle, hogs, sheep, and several species of commercially raised fishes, are being produced by injecting foreign DNA into the nuclei of egg cells or early embryos. For instance, rainbow trout, widely raised for human food, are being genetically engineered with a cloned growth hormone gene. Use of gene-splicing techniques to engineer fishes and other animals for human food is still in the experimental stages, but many scientists and farmers are optimistic about future applications.

Manipulating Plant Genes More important than an increase in livestock productivity will be the effects of genetic engineering on crop plants. In one striking way, plant cells have so far proved easier to engineer than the cells of many animals. For many plant species, an adult plant can be regenerated from a single cell grown in tissue culture (see Figure 34.14). This is an important advantage, because many genetic manipulations, such as the introduction of genes from other species, are easier to perform and assess on single cells than on whole organisms. Commercially important plants that readily grow from single somatic cells include asparagus, cabbage, citrus fruits, sunflowers, carrots, alfalfa, millet, tomatoes, potatoes, and tobacco.

Like researchers working with animal cells, molecular biologists working with plants commonly use DNA vectors to move genes from one organism to another. The best-developed DNA vector is a plasmid of the bacterium *Agrobacterium tumefaciens.* In nature, *A. tumefaciens* infects plants and causes tumors called crown galls. The tumors are induced by the plasmid, called the **Ti plasmid** (Ti for tumor-inducing). The Ti plasmid integrates a segment of its DNA, known as T DNA, into the chromosomal DNA of its host plant cells. Researchers have developed ways to eliminate the plasmid's disease-causing properties while maintaining its ability to move genetic material into plant cells.

Foreign genes can be inserted into the Ti plasmid using recombinant DNA techniques. The recombinant plasmid is either put back into *Agrobacterium,* which can then be used to infect plant cells growing in culture, or introduced directly into plant cells. Then, taking advantage of the capacity of those cells to regenerate whole plants, it is possible to produce plants that contain and express the foreign gene and pass it on to their offspring (Figure 19.12).

A major drawback to using the Ti plasmid as a vector is that only dicots (plants with two seed leaves) are susceptible to infection by *Agrobacterium.* Monocots, including agriculturally important grasses such as corn and wheat, cannot be infected by *Agrobacterium.* Fortunately, newer techniques, such as electroporation and the DNA particle gun, are helping plant scientists overcome this limitation. The DNA gun is being used for a wide variety of plants, especially monocots. It explodes .22-caliber plastic bullets tipped with tiny metal pellets coated with DNA. The bullet shell stays in the gun, but the pellets are driven through the cell walls into the cytoplasm. The cells quickly repair the tiny punctures in the cell wall and plasma membrane, the

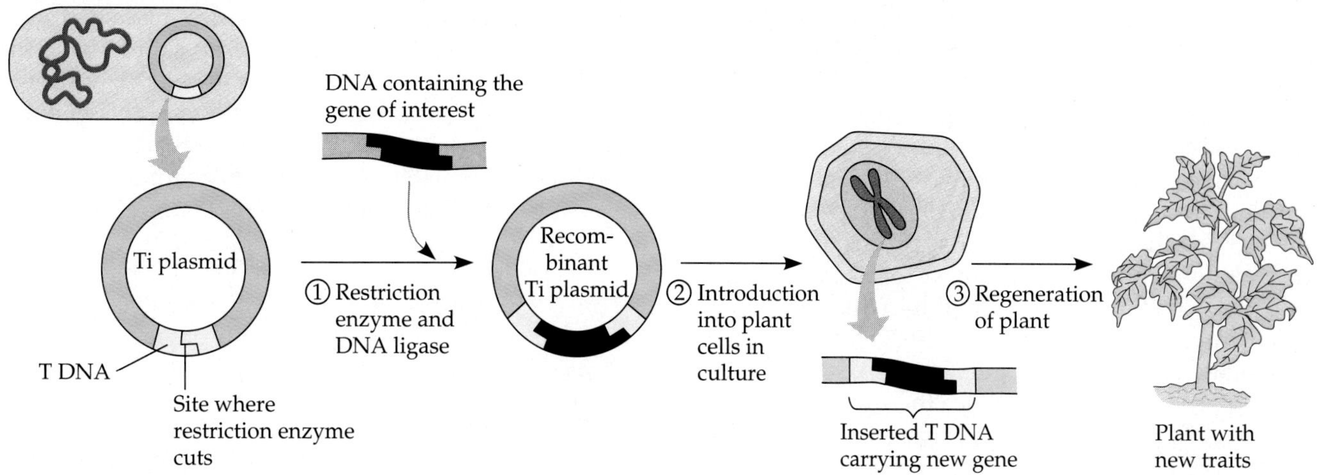

Figure 19.12
Using the Ti plasmid as a vector for genetic engineering in plants. ① The Ti plasmid is isolated from the bacterium *Agrobacterium tumefaciens*, and a fragment of foreign DNA is inserted into its T region by standard recombinant DNA techniques. ② When the recombinant plasmid is introduced into cultured plant cells, the T DNA integrates into the plant's chromosomal DNA. ③ As the plant cell divides, each of its descendants receives a copy of the T DNA and any foreign genes it carries. If an entire plant is regenerated, all its cells will carry—and may express—the new genes.

metal pellets remain in the cytoplasm but cause no harm, and the DNA becomes integrated into the DNA of the host cell.

Even with these new techniques, engineering plant genomes is a formidable challenge. The cloning of plant DNA is straightforward, but identifying genes of interest may be very difficult. Furthermore, many agriculturally desirable plant traits, such as crop yield, are polygenic, involving many genes.

Some Early Successes of Genetic Engineering in Plants Despite its difficulty, genetic engineering of plants is already yielding some positive results, mostly in cases where useful traits are determined by one or only a few genes. For example, several chemical companies have developed strains of wheat, cotton, and soybeans carrying a bacterial gene that makes the plants resistant to the herbicides used by many farmers to control weeds. This gene would make it easier to grow crops while still ensuring that weeds are destroyed. The first gene-spliced fruits approved by the FDA for human consumption were tomatoes engineered with antisense genes that retard spoilage (see Figure 35.8). After cloning the tomato gene that codes for the enzyme mainly responsible for ripening, researchers cloned a complementary (antisense) gene—one with the complementary sequence of bases. When spliced into the DNA of a tomato plant, the antisense gene transcribes messenger RNA that is complementary to the ripening gene's mRNA. Then, when the ripening gene transcribes the normal mRNA, the complementary (antisense) mRNA binds to the normal RNA, blocking the synthesis of the ripening enzyme. The engineered tomatoes produce only about 1% of the normal amount of ripening enzyme.

A number of crop plants are being engineered to resist infectious pathogens and pest insects. Tomato and tobacco plants, for example, have been engineered to carry and express certain genes of viruses that normally can infect and damage plants. With these versions of the viral genes, however, the plants are resistant to—in effect, vaccinated against—attack by the viruses. Other crop plants have been engineered to resist insect attack. In one approach, the plants receive genes of the bacterium *Bacillus thuringiensis*, which produces insecticidal proteins. Field tests of strains of corn, cotton, and potatoes containing bacterial genes that encode these proteins show that the plants resist attack by several of the most damaging insects. Growing insect-resistant plants will reduce the need to apply chemical insecticides to crops.

These early successes in plant engineering point to an agricultural revolution just around the corner. It is likely that many crop plants will soon be made more productive by enlarging their agriculturally valuable parts—whether they be roots, leaves, flowers, or stems. Researchers are also making strides toward improving the food value of plants, as in engineering corn or wheat to produce a mix of amino acids more suitable for the human diet. At present, more than 30 different crop plants developed with recombinant DNA techniques are being tested in field trials.

The Nitrogen-Fixation Challenge Perhaps the most exciting potential use of DNA technology in agriculture involves nitrogen fixation, a bacterial process that benefits plants and their consumers. Nitrogen fixation is the conversion of atmospheric, gaseous nitrogen (N_2), which is useless to plants, into nitrogen-containing compounds that plants can take up from the soil and use for making essential organic molecules, such as amino acids and nucleotides (see Chapter 33). The bacteria that fix nitrogen live either in the soil or within the roots of certain plants, such as beans and alfalfa. Since the availability of appropriate nitrogen compounds is often the limiting factor in plant growth and crop yield, modern agriculture makes extensive use of chemically synthesized nitrogen fertilizers to supplement bacterial nitrogen fixation. Gene-splicing techniques offer ways to reduce the use of expensive fertilizers by increasing the nitrogen-fixing capability of bacteria. In the future, it may be possible to design nitrogen-fixing bacteria that can live in the tissues of nitrogen-demanding plants, such as corn and wheat. Ultimately, DNA technology may produce crop plants that fix nitrogen themselves.

SAFETY AND ETHICAL ISSUES

As soon as scientists realized the potential power of DNA technology, they also began to worry that it could have dangerous consequences. The earliest concerns were that genetic manipulations of microorganisms could create hazardous new pathogens, which might escape from the laboratory. In response to their own concerns, scientists developed a self-monitoring approach—an honor system whereby researchers would adhere to a set of voluntary and self-imposed guidelines to ensure safety. This early approach soon led to formal regulatory programs administered by federal agencies. Today, governments and regulatory agencies throughout the world are grappling with how to promote the potential industrial, medical, and agricultural revolution engendered by biotechnology while ensuring that new products are safe. In the United States, the FDA, the National Institutes of Health (NIH) Recombinant DNA Advisory Committee, the Department of Agriculture (USDA), and the Environmental Protection Agency (EPA) share responsibility for setting policies and regulating new developments in genetic engineering.

In the past decade, hundreds of genetically designed products and new strains of organisms have been developed. Many have been exhaustively tested, and these, at least, seem to pose little or no threat to humans or the environment. The tests also confirm that genetic engineering holds enormous potential for improving human health and increasing agricultural productivity.

Virtually all new developments in DNA technology have ethical overtones. Obtaining a complete map of the human genome will open the door for significant advances in gene therapy. It will also raise significant ethical questions. For example, who should have the right to examine someone else's genes? How should that information be used? Should a person's genome be a factor in their suitability for a job? Should insurance companies have the right to examine an applicant's genes? The benefits of gene therapy itself must be weighed against the need to assure that gene vectors are safe. In the case of environmental problems—oil spills, for example, and chemical wastes that threaten our soil, water, and air—genetically engineered organisms may be part of the solution, but their impact on native species must be considered before they are widely used.

With new medical products, the main cause for concern is the potential for harmful side effects, both short-term and long-term. Hundreds of new genetically engineered diagnostic products, vaccines, and drugs, including some designed to treat AIDS and certain forms of cancer, currently await federal approval. Before the FDA considers a new medical product for general marketing, the substance must pass exhaustive tests in laboratory animals and humans.

There is heated debate about genetically engineered agricultural products because of the potential dangers of introducing new organisms into the environment. Some scientists argue that producing transgenic organisms by gene splicing is only an extension of traditional crossbreeding, or hybridization—the procedure that has given us the tangelo (a tangerine-grapefruit hybrid) and beefalo (a cow-buffalo hybrid). Since no hybrid crops or animals have ever been tested for safety before they were marketed, it is argued that genetically engineered organisms need not be treated differently. In general, the FDA has held that if the result of genetic engineering is not significantly different from a product already on the market, testing is not required.

On the other side of the argument are scientists who believe that creating transgenic organisms by splicing genes from bacteria or animals into plants and vice versa is radically different from hybridizing closely related species of plants or animals. There is a concern that foods produced by gene splicing will contain new proteins that are toxic or that cause severe allergies in some people. There is also concern that genetically engineered crop plants could become "superweeds." For example, plants that have genetically engineered resistance to herbicides or microbial diseases and pest insects could escape into the wild and overgrow native species. They could also colonize croplands where other plant species are grown and become difficult to control there.

Another consideration is that engineered crop plants may hybridize and pass their new genes to close relatives in neighboring wild areas. Lawn and crop grasses, for example, commonly exchange genes with wild relatives, and there is little doubt that if some species received new genes, they would pass them to wild plants. In the state of California alone, a number of important crop plants—including carrots, asparagus, celery, onions, and potatoes—either grow in the wild or can hybridize with wild species. If domestic plants pollinate wild ones, the offspring could become superweeds, exceedingly difficult to control because they have genes that confer resistance to herbicides, natural diseases, and insect pests. Researchers are looking for ways to prevent the escape of engineered plant genes. Among the possibilities are crop isolation and the engineering of plants so that they cannot hybridize.

Concerns about potential environmental and health hazards will slow the application of the products of the new biotechnology. There is always a danger that too much regulation will stifle basic research and its potential benefits. However, the power of genetic engineering—our ability to profoundly and rapidly alter species that have been evolving for millennia—demands that we proceed with humility and caution.

STUDY OUTLINE

1. DNA technology is a powerful set of techniques that allows biologists to manipulate and analyze genetic material. Recombinant DNA is new genetic material constructed in vitro from the DNA of different organisms (usually different species). Genetic engineering is the creation of useful new products and organisms using techniques of gene manipulation.

Basic Strategies of Gene Manipulation and Analysis (pp. 391–404)

1. The first recombinant DNA was made in 1975 with bacterial plasmids, and this technique is still used to make multiple copies of genes and many useful gene products, including growth hormones and pharmaceutical products.
2. A variety of bacterial restriction enzymes recognize short, specific nucleotide sequences in DNA and cut the sequences at specific points on both strands to yield a set of double-stranded DNA fragments with single-stranded sticky ends. The sticky ends readily form base pairs with complementary single-stranded segments on other DNA molecules. Gel electrophoresis is used to identify, isolate, and purify these individual fragments.
3. Either plasmids or bacteriophages can serve as vectors (carriers) to introduce recombinant DNA molecules into host cells. Recombinant DNA is made by inserting a piece of DNA containing a gene of interest into the plasmid or phage DNA that has been clipped by restriction enzymes. In either case, gene cloning results when the foreign genes replicate inside the host bacterium.
4. In addition to vectors for bacterial cells, DNA can be introduced directly into eukaryotic cells by electroporation (use of an electric pulse to open tiny holes in the plasma membrane), microinjection, and by a "gene gun" that fires DNA attached to fine metal particles across the cell walls of plant cells.
5. The two major sources of desired genes for insertion into vectors are (a) genomic libraries that contain all the plasmid- or phage-carried DNA segments isolated directly from an organism, and (b) complementary DNA (cDNA) synthesized from mRNA templates. The latter source is especially useful for manipulating and expressing intron-rich eukaryotic DNA.
6. The presence of a particular gene can be determined directly using radioactively labeled nucleic acid segments of complementary sequence called probes. A gene can also be detected indirectly by identifying its protein product.
7. Genes are cloned in yeasts and other eukaryotic cells, as well as bacteria. Only plant and animal cells have the requisite complex biosynthetic machinery to produce certain specialized proteins.
8. Machines are now used to synthesize short genes of known nucleotide sequence and to sequence long stretches of DNA, using restriction enzymes and gel electrophoresis as key tools. DNA sequences collected in computer banks allow rapid analysis and automatic translation into amino acid sequences.
9. The polymerase chain reaction (PCR) is a technique for quickly making many copies of DNA in vitro.
10. Restriction fragment length polymorphisms (RFLPs) are differences in DNA sequence on homologous chromosomes that result in different patterns of restriction fragment lengths, which are visualized as bands on gel electrophoresis. RFLP analysis has many applications, including genetic mapping for basic research and diagnosing genetic disorders.

Applications of DNA Technology (pp. 404–412)

1. Recombinant DNA technology has allowed investigators to answer questions about molecular evolution, probe details of gene organization and control, and produce and catalog proteins of interest. An international research effort, called the Human Genome Project, involves linkage mapping, physical mapping, and sequencing of the entire human genome. A technique called chromosome walking is used to determine the correct order of all the pieces of human DNA in genomic libraries.
2. Medical applications of recombinant DNA technology include the development of diagnostic tests for detecting mutations that cause genetic disease; the design of safer, more effective vaccines; the large-scale production of many new, and some previously scarce, pharmaceutical products; and the ultimate prospect of curing and preventing genetic disorders caused by single defective genes.

3. DNA "fingerprints" obtained from RFLP analysis of organic materials found at the scenes of violent crimes are often introduced as evidence in trials.
4. In agriculture, transgenic plants and animals (those containing genes from other species) are being designed to improve productivity and food quality. Transgenic animals are often produced by microinjection of fertilized eggs; transgenic plants are engineered by using a plasmid called Ti from *Agrobacterium* and by the gene gun.

Safety and Ethical Issues (pp. 412–413)

1. Several U.S. government agencies are responsible for setting policies and regulating the use of recombinant DNA technology.
2. Genetic engineering raises safety and ethical questions that have triggered scientific and public debate. The potential benefits of genetic engineering must be carefully weighed against the potential hazards of creating products that are harmful to humans or the environment.

SELF-QUIZ

1. Which of the following tools of recombinant DNA technology is *incorrectly* paired with its use?
   a. restriction enzyme—production of RFLPs
   b. DNA ligase—enzyme that cuts DNA, creating the sticky ends of restriction fragments
   c. DNA polymerase—used in a polymerase chain reaction to amplify sections of DNA
   d. reverse transcriptase—production of cDNA from mRNA
   e. electrophoresis—DNA sequencing

2. Which of the following is *not* true of complementary DNA?
   a. It can be amplified by a polymerase chain reaction.
   b. It can be used to create a complete genomic library.
   c. It is produced from mRNA using reverse transcriptase.
   d. It can be used as a probe to locate a gene of interest.
   e. It eliminates the introns of eukaryotic genes and thus is more easily introduced into and cloned by bacterial cells.

3. Plants are more readily manipulated by genetic engineering than are animals because
   a. plant genes do not contain introns
   b. more vectors are available for transferring recombinant DNA into plant cells
   c. a somatic plant cell can grow into a complete plant
   d. recombinant genes can be inserted into plant cells by microinjection
   e. plant cells have larger nuclei

4. A paleontologist has recovered a bit of tissue from the 400-year-old preserved skin of an extinct dodo. He would like to compare DNA from the sample with DNA from living birds. Which of the following would be most useful for increasing the amount of DNA available for testing?
   a. RFLP analysis
   b. polymerase chain reaction (PCR)
   c. electroporation
   d. gel electrophoresis
   e. in situ hybridization

5. The Human Genome Project involves all of the following goals *except*
   a. the location of RFLP markers
   b. the sequencing of the entire nucleotide sequence of the human genome
   c. the physical mapping of the chromosomes
   d. the analysis of the genomes of other species
   e. altering the human genome

6. Recombinant DNA technology has many medical applications. Which of the following has *not yet* been attempted or achieved?
   a. production of hormones for treating diabetes and dwarfism
   b. production of subunits of viruses that may serve as vaccines
   c. introduction of genetically engineered genes into human germ cells
   d. prenatal identification of genetic disease genes
   e. genetic testing for carriers of harmful alleles

7. RFLP analysis is being used as evidence to link suspects with blood and tissues found at crime scenes. DNA fingerprints look something like supermarket bar codes. The pattern of bars in a DNA fingerprint shows
   a. the order of bases in a particular gene
   b. the individual's genotype
   c. the order of genes along particular chromosomes
   d. the presence of dominant or recessive alleles for particular traits
   e. the presence of certain DNA fragments

8. Which of the following sequences along a double-stranded DNA molecule may be recognized as a cutting site for a particular restriction enzyme?
   a. A A G G
T T C C
   b. A G T C
T C A G
   c. G G C C
C C G G
   d. A C C A
T G G T
   e. A A A A
T T T T

9. In recombinant DNA methods, the term *vector* refers to
   a. the enzyme that cuts DNA into restriction fragments
   b. the sticky end of a DNA fragment
   c. a RFLP marker
   d. a plasmid or other agent used to transfer DNA into a living cell
   e. a DNA probe used to identify a particular gene

10. The template used to make cDNA is
    a. DNA
    b. mRNA
    c. a plasmid
    d. a DNA probe
    e. a restriction fragment

CHALLENGE QUESTIONS

1. A neurophysiologist hopes to study a gene that codes for a neurotransmitter protein in human brain cells. She knows the amino acid sequence of the protein. Briefly explain how she might (a) isolate only the genes that are expressed in a specific type of brain cell, (b) identify the gene that codes for the neurotransmitter, (c) produce multiple copies of the gene for study, and (d) produce a quantity of the neurotransmitter for evaluation as a potential medication.
2. Hemophilia A is a hereditary disease characterized by a shortage of clotting factor VIII, a blood protein normally synthesized by liver cells. Without the clotting factor, blood fails to clot normally, and a hemophiliac can bleed to death from a minor injury. Tay-Sachs disease results from a lack of the enzyme hexosaminidase A, which normally breaks down a fatty substance called ganglioside GM2 inside nerve cells. Accumulation of ganglioside GM2 in brain cells causes blindness, seizures, the breakdown of sensory and motor function, and eventually death. Which of these diseases do you think would be easiest to treat with future gene therapy? Why? Describe how this might be done.

SCIENCE, TECHNOLOGY, AND SOCIETY

1. Which uses of DNA technology do you think will prove to be of most value in the coming decades? Why do you think so? Which safety and ethical issues do you think are most important? Why? How do you think we should deal with these concerns?
2. Do you think there is a potential in our society for genetic discrimination based on testing for "harmful" genes? What policies can you suggest that would prevent abuses of genetic testing?
3. Under pressure from the biotechnology industry, the U.S. government recently loosened up some regulations affecting DNA technology. What trade-offs do you see between government regulation of biotechnology and the desire of U.S. industry to remain competitive with other nations in this field?

FURTHER READING

Beardsley, T. "Diagnosis by DNA." *Scientific American,* October 1992. Clinical applications of biotechnology.

Gilbert, W. "Toward a Paradigm Shift in Biology." *Nature,* January 10, 1991. How the ability to dissect genomes is changing the kinds of questions biologists ask.

Haffie, T. "Mendel to Monctezuma: The End-of-Term Review in Genetics." *BioScience,* April 1989. Sample questions that will help you review many of the topics covered in this unit.

Kahn, P. "Germany's Gene Law Begins to Bite." *Science,* January 31, 1992. A case study in attempts to regulate DNA technology.

Moffat, A. "Improving Plant Disease Resistance." *Science,* July 24, 1992. DNA technology on the farm.

Moody, M. D. "DNA Analysis in Forensic Science." *BiosSience,* January 1989. DNA fingerprinting to help help solve violent crimes.

Roberts, L. "Two Chromosomes Down, 22 to Go." *Science,* October 2, 1992. Progress of the Human Genome Project.

Watson, J. D., J. Tooze, and D. T. Kurtz. *Recombinant DNA: A Short Course,* 2nd ed. New York: Scientific American Books, 1989. An overview of the principles and applications of recombinant DNA technology.

Weiss, R. "Predisposition and Prejudice." *Science News,* January 21, 1989. As molecular biologists develop more tests to detect "bad" genes, will genetic discrimination emerge?

UNIT FOUR | MECHANISMS OF EVOLUTION

AN INTERVIEW WITH ERNST MAYR

Ernst Mayr is one of the greatest influences on evolutionary biology since Darwin. Mayr was one of the architects of the evolutionary synthesis of the 1930s and 1940s, which unified biology by integrating Darwin's theory of natural selection with new discoveries in genetics, paleontology, and taxonomy. Mayr based his views on evolution mainly on relationships among bird species that he studied on Pacific islands. Now 89 years old, Mayr, Professor Emeritus at Harvard, is still going strong and generating exciting new ideas. His latest book, One Long Argument *(Harvard University Press, 1991), analyzes Darwin's theories. I interviewed Professor Mayr at his summer cottage in New Hampshire.*

Dr. Mayr, how did you become a naturalist?
I have been a naturalist, I would say, ever since I could walk. My parents were very much interested in nature and took me and my two brothers out to watch birds, to collect fossils, to find spring flowers and everything else. So I ardently followed all of these things, but particularly birds. Since I came from a medical family—there were four generations of medical doctors preceding me—I was supposed to be the medical doctor of my generation. So I went to medical school, but I also became a volunteer at the Berlin Natural History Museum, and that led to my giving up my medical studies. After I had my Ph.D., which I got at the age of 21, I was introduced to Lord Rothschild in England, who sent me on an expedition to New Guinea. While I was there, the American Museum of Natural History in New York needed an ornithologist on their expedition, which was cruising in the seas near New Guinea, and so I joined them. I spent the next nine months on a schooner in the Solomon Islands. It was 2½ years before I finally came back to Berlin.

Later I was invited by the American Museum of Natural History to come to New York for a year, but my temporary visiting job became a permanent job when the museum bought the Rothschild bird collection and I stayed with the American Museum from 1931 to 1953. Then I became a professor at the Museum of Comparative Zoology at Harvard.

So you, like Darwin, were a naturalist on a major expedition in your early twenties. Tell us a little more about your journey and how that early fieldwork influenced your thinking about evolution.
My task in New Guinea, given to me by Lord Rothschild, was to find the home of some rare birds of paradise that had been collected by the plume hunters for hat decorations, but had not since been encountered. Each mountain range in New Guinea has its endemic species of birds; the mountains are like islands in the sky, as they have sometimes been called. Geographic speciation takes place on these mountaintops just as it takes place on islands in the ocean.

After New Guinea I came to the Solomon Islands, where we went on our schooner from island to island. There is no place in the world in which geographic speciation is as well demonstrated as in the changes from island to island in the Solomon Islands. And that, more than almost anything else, was the basis of my 1942 book *Systematics and the Origin of Species,* which was one of the books that contributed to the evolutionary synthesis of the 1930s and '40s. It is indeed true, just as Wallace had found in the Indonesian Islands, and Darwin in the Galapagos, that islands are the best place in which to demonstrate speciation. One of the things that I followed up later in life was to see if geographic separation also caused speciation on continents. Indeed it does. Wherever there are barriers, due to water, mountains, or vegetational changes, speciation can occur.

Can you provide some historical perspective about the evolutionary synthesis and your role in its development?
When I was a student in Germany, I was a Lamarckian, which shocks some people. But at that time it made more sense than the view of the Mendelians that all new species originate from macromutations and that a new species is formed at once by one individual becoming the first representative of the new species. The gradual speciation that I saw fitted in far better with Lamarckism.

In the 1920s, there was still total disagreement among biologists as to what was the truth about evolution and what were the mechanisms of evolution. The theory of natural selection was clearly a minority viewpoint. Then, rather rapidly, in the 1930s and '40s, all the difficulties seemed to evaporate and a modern picture of evolution originated.

The three major contributions of the synthesis were the following. First, it refuted all the opposing theories, such as Lamarckism, that believed there was some intrinsic force in organisms that would lead to improvement or evolutionary change and a more perfect adaptation. The second contribution was to supplement the evolutionary concepts of the geneticists of that period, who concentrated entirely on adaptive changes within a population. You won't find anything in their writings about the origin of evolutionary diversity. Adaptation and the origin of diversity are the two major components of evolution. My contribution to the evolutionary synthesis was to introduce the study of diversity, the explanation of species, of how they originate, and how one gets from there to the higher taxonomic categories. The third major contribution was that there were many disparate disciplines at that time and people in different fields were hardly on speaking terms with each other, like paleontologists and geneticists; the botanists were way outside. The evolutionary synthesis brought them together, and that's why the word synthesis is so often used for this unification of biology through the acceptance of Darwinian evolution.

Did you and the other architects of the Evolutionary Synthesis actually collaborate?

Curiously enough, very little. [Theodore] Dobzhansky and I saw each other a couple of times before I wrote my book, and I showed him my island birds. But otherwise we had very little contact. [George Gaylord] Simpson did his thing entirely by himself, and so did [G. Ledyard] Stebbins in botany. The major evolutionary theorists never had a joint symposium until after the synthesis was completed and they wrote their books independently.

Dr. Mayr, you mentioned the Darwinian stamp on the modern evolutionary synthesis. When did Darwin really begin to influence your view of the world?

Well, I wasn't nearly as much aware at that time that it was a Darwinian influence. I just gradually realized that natural selection can do far more than I thought when I was younger. I didn't read Darwin's *Origin of Species* until after I had published my *Systematics and the Origin of the Species*. But then, particularly in the 1960s, I became very much aware of the really great importance of Darwin.

And your latest book is about Darwin's ideas.

The reason I wrote this book is that most of the books on Darwin speak about Darwinism as if it was a single, monolithic theory. In reality, Darwin had a great many theories. What I've tried to do is to separate these major theories and discuss them individually—where Darwin got his ideas, how valid they were, when they were accepted by other people, and so forth. I've found in the literature that there are seven or eight different meanings of the word Darwinism for different people. In Darwin's own time, for instance, Darwinism meant evolution without supernatural causation—nothing more, nothing less. Today, Darwinism means the theory of natural selection. But natural selection was almost totally ignored in Darwin's time.

Speaking of Darwin's time, if you could go back to that time and actually visit with Darwin, what would you ask him?

You will be very much surprised when you hear my answer. I would ask him about his relation to religion, to a belief in a personal God. Did his loss of belief precede or follow his development of the concept of natural selection? What went on in his mind, particularly in relation to general things like belief in religion, is something that he always carefully concealed in his writings, particularly since his wife Emma was a deeply devout person. But it is rather obvious from his handwritten notes that he didn't believe in a personal God. Furthermore, to me it seems obvious that he lost his belief in God at least a year, if not two, before he developed his theory of natural selection. So the claim that biology and a belief in natural selection is dangerous because it may make

you lose your faith in God is not substantiated. That's a very important thing to know.

What could you tell Darwin that would help him with his ideas about evolution?
I would just convey to him what we know about genetic variation in populations, because that's the area where he was most puzzled and where he was most anxious to learn more and never succeeded.

Many of your books and articles highlight the importance of variation among individuals and contrast what you call typological thinking with population thinking. What is the issue here, and why is it so important to evolutionary theory?
Typological thinking, which is also sometimes called essentialism, has a very long history. It goes back to the ideas of Plato, who spoke of the underlying *eidos,* the unchanging, ideal forms that underlie all the variable phenomena of this world. The variations observed are simply incomplete and imperfect manifestations or reflections of the *eidos.* Population thinking is exactly the opposite. It says that there is no constant value to any variable population. Every individual is unique and is different from all other individuals. Even in the human species, with now five billion individuals, every single one of them is uniquely different. In the human body and in all animal bodies, most likely every cell is somehow or other individually different from every other cell. This emphasis on the variability of populations—on the uniqueness of the individual—is characteristic of the so-called population thinking.

Darwin hadn't completed his conversion to population thinking, though he had it essentially. You can never accept evolution by natural selection unless you have adopted population thinking. For a typologist, natural selection is ineffective, because the variation does not affect the underlying ideal.

You view physiology as being largely typological in its traditional approach. This must have important implications for medicine.
It certainly does. I remember that I once talked about population thinking to an audience in which there were several doctors, and one of the doctors got up and said, "I can't accept this! Unless I treat every patient as if he were 'normal' just like all other patients, I wouldn't know what to do." This is one of the big difficulties in medicine. For instance, no two people react necessarily exactly the same way to particular drugs. So this variability of the human species is something that a modern doctor should be very much aware of.

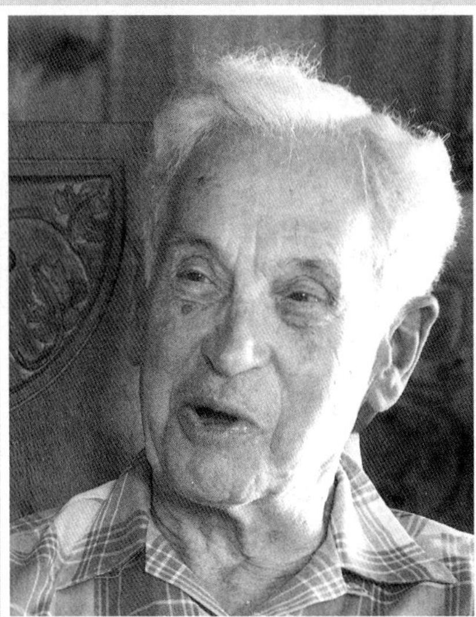

Another point your books make is that it is the whole organism—the whole phenotype—that is the object of natural selection.
This is absolutely so. One of the greatest weaknesses of the mathematical geneticists was that they acted as if the gene was the target of selection. The gene is not the target of selection. A given gene can be highly favorable in one genotype and lethal in another context. Those who believed that the gene is the target of selection said that the larger a population—the more individuals a species has—the faster it would evolve, because it has a bigger reservoir of genetic differences. The naturalists said, we see just the opposite. We see that species with very large populations don't do anything in evolution. They may stay the same for five, ten, or twenty-five million years. The most rapid evolutionary changes take place in small, isolated populations. The disagreement was strictly due to the fact that the two groups of people accepted different targets of natural selection.

How do you think new species most commonly arise?
We might as well start with a few words about what a species is. A population which does not interbreed with another population, even though the two occur at the same time and the same place, is a different species from the other population. Now, the question is how two populations acquire the characteristic that they don't fuse with each other; that they are, as we say, reproductively isolated. That is the problem of speciation.

Up to the 1930s and '40s, there was a widespread view that much of speciation is due to a group of individuals acquiring a new ecological niche at the same place and then becoming a different species. Darwin himself believed that this happens all the time. Analyzing the situation, I came to the conclusion that this must be very rare, and all the cases quoted in the literature could be much better explained as examples of geographical speciation, or as it is sometimes technically called, allopatric speciation.

There are two ways in which geographic speciation can take place. Either there is a continuous area somewhere in the range of the species in which a new barrier arises—for example, a new mountain range, or an arm of the ocean, or a new vegetational belt that prevents a meeting of the individuals on either side of this new barrier. That is the classical geographic speciation that has always been described in the literature. But in 1954, I published a paper, which I've always considered to be the most important paper I ever published in my life, in which I showed that a far more important process of speciation is when some founders, a few individuals, establish a new founder colony beyond the existing species border. This new founder population rapidly changes genetically and acquires the reproductive characteristics that are needed for becoming a different species. Then it can exist as a separate species even if comes again in contact with the mother species.

In these founder populations, more important genetic changes can take place than when you just separate a species into two large populations. In these rather large parts of the species, even when temporarily separated, not very much happens. The really important evolutionary changes probably happened in the populations that came out of founder populations. And that is why the paleontologists find so little evidence for the gradual origin of new species. They find a lineage that changes somewhat, but not very drastically, and then suddenly a new species originates somewhere else, next to it. [Niles] Eldredge and [Stephen Jay] Gould based their theory of punctuated equilibrium on my 1954 paper. The major point which they added to my analysis is that once a species has become widespread and successful, it no longer changes very much; it enters a stage of stasis. It may continue in this unchanging condition for many millions of years until finally becoming extinct.

The theory of punctuated equilibrium also maintains that new species usually arise relatively rapidly. Does this contest the traditional Darwinian view of gradual evolution?
A mistake people make is thinking that if something evolves very rapidly, it is no longer Darwinian gradual evolution. But as long as the evolution occurs at the populational level and not at the level of individuals, then it is gradual evolution, occurring over many generations. Some paleontologists call the origin of a species relatively sudden if it takes place during 1% of the total lifespan of the species. But that 1% is 50,000 years or 100,000 years in a species living 5 or 10 million years. Hardly anyone else would call that sudden.

Can you speculate about how your ideas on geographic speciation apply to the origin of humans?
If you look at the fossil record of the hominids, you find that every new type did exactly what the theory of speciational evolution postulates. They turned up suddenly and are not a direct continuation of the previous lineage. There's nothing intermediate between *Australopithecus africanus* and *Homo habilis.* There's nothing really in between *Homo habilis* and *Homo erectus.* My interpretation, which the anthropologists are beginning to accept more and more, is that hominids at any time were geographically variable and had isolated groups of populations. Invariably, one of these was the next step in human evolution.

What's our current evolutionary status?
All populations of the human species are heavily exchanging genes with each

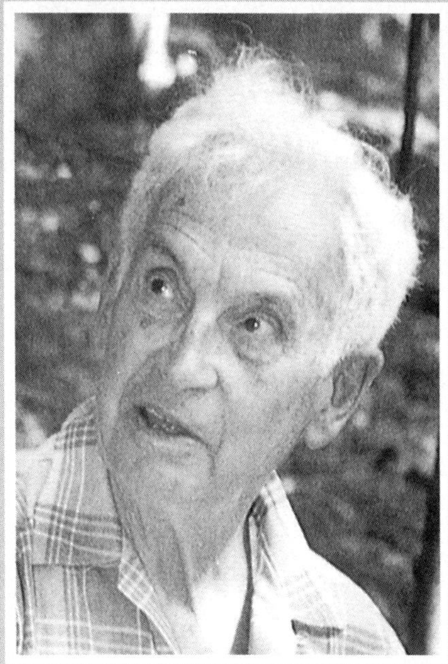

other, and there is no isolated population that could become a different species. Any very widespread species is usually evolutionarily inert, more or less.

You once wrote, "It's a miracle that humankind ever happened."
Well, that was broadly speaking. On earth, there have probably been more than a billion species of organisms since the origin of life. How many of these billion species have become humans? One. A chance of one in a billion is almost like a miracle. And even that one human species might never have happened. We might have had several billion species on Earth without even one of them acquiring the characteristics of mankind. The presumption that wherever there is life and evolution, then there must eventually be intelligence is teleological thinking.

I argue with the people who want to receive radio signals from intelligent life from all over the universe. I said that the chance of our ever getting such signals is so small that I would consider it indistinguishable from zero. Let's say we had a billion species on earth. Only one developed the intelligence of mankind. And then we have had at least fifteen civilizations on Earth, but only one of them produced the electronic revolution that can send radio signals into space. Furthermore, how many planets are there that can produce life? The physical constraints are incredible. The fact that one of the nine planets in the solar system produced conditions suitable for life doesn't mean that of the next thousand planets, even one will have the conditions suitable for life.

I've made up a little story to illustrate this point. Imagine when the Earth originated there was an intelligent civilization somewhere in the universe, and they said, "Well, let's send signals to this planet, it looks very promising." So for the 3.8 billion years that the Earth had life they sent signals to the earth and no response ever came. Finally, in the year 900 they said, "We'll send signals for another thousand years, and if we don't have a response by then we'll give up." Along came the year 1900 and no signal had been received. So they said, "No, there's obviously no intelligent life on Earth," and they quit just half a century before we developed the capability to respond. If you multiply all these high improbabilities, you come to a figure which, as I said, is indistinguishable from zero.

You've also written that we humans have extraordinary responsibility because of our uniqueness as a species.
Yes, humans are basically responsible for all the bad things that at the present time happen to our planet, and we are the only ones who can see all these things and do something about them. If we would stop the human population explosion, we would have already won two-thirds of the battle. That we live here just as exploiters of this planet is an ethic that does not appeal to me. Having become the dominant species on our planet, we have the responsibility to preserve the well-being of this planet. I feel that it should be a part of our ethical system that we should preserve and maintain and protect this planet that gave origin to us.

20 DESCENT WITH MODIFICATION: A DARWINIAN VIEW OF LIFE

PRE-DARWINIAN VIEWS

ON THE ORIGIN OF DARWINISM

THE DUAL MEANING OF DARWINISM

THE SIGNS OF EVOLUTION

JUST A THEORY?

Figure 20.1
Charles Darwin (1809–1882). Darwin and his son William posed for this photograph in 1842. The author of numerous books and monographs on topics as diverse as barnacles, plant movements, and island geology, Darwin would be remembered as one of the greatest naturalists of the nineteenth century even if he had never published on the topic of evolution. But it was *The Origin of Species* that established Darwin's place as the most influential person in the development of modern biology. He is buried next to Isaac Newton in London's Westminster Abbey. This chapter introduces the unit on mechanisms of evolution by placing the Darwinian revolution into historical context.

Biology came of age on November 24, 1859, the day Charles Darwin published *On the Origin of Species by Means of Natural Selection* (Figure 20.1). His book, which presented a convincing case for evolution, connected what had previously seemed a bewildering array of unrelated facts into a cohesive view of life. In biology, **evolution** refers to the processes that have transformed life on Earth from its earliest forms to the vast diversity that characterizes it today (Figure 20.2). Darwin addressed the sweeping issues of biology: the great diversity of organisms, their origins and relationships, their similarities and differences, their geographical distribution, and their adaptations to the surrounding environment. None of these aspects of life makes sense except in the context of evolution. Thus, evolution is the most pervasive principle in biology, and a thematic thread woven throughout this book (see Chapter 1). This unit focuses specifically on the *mechanisms* by which life evolves.

Darwin made two points in *The Origin of Species.* First, he argued from evidence presented that species were not specially created in their present forms but had evolved from ancestral species. Second, he proposed a mechanism for evolution, which he termed **natural selection.** According to Darwin's concept of natural selection, a population of organisms can change over time as a result of individuals with certain heritable traits leaving more offspring than other individuals. Darwin's first point—that evolution occurs—can stand on its own, whether or not natural selection is the cause.

There is a recurrent theme throughout the chapters of this unit: Evolutionary change is based mainly on the interactions between populations of organisms and their environments. This first chapter defines the Darwinian view of life and traces its historical development.

PRE-DARWINIAN VIEWS

To put the Darwinian view in perspective, we must compare it to earlier ideas about Earth and its life. The impact of an intellectual revolution such as Darwinism depends as much on timing as on logic. The timeline in

Figure 20.2
A small sample of biological diversity. Shown here are just some of the thousands of species of moths and butterflies in the Lepidoptera collection of the National Museum of Natural History. As diverse as Lepidoptera are, all share a common body plan. A major goal of evolutionary biology is to explain how such diversity arose, while also accounting for characteristics common to different species. Darwin described evolution as descent with modification and attributed similarities of related species to common ancestry.

Figure 20.3 places Darwin's ideas in a historical context.

The Origin of Species was truly radical, for not only did it challenge prevailing scientific views, but it also shook the deepest roots of Western culture. Darwin's view of life contrasted sharply with the conventional paradigm of an Earth only a few thousand years old, populated by immutable (unchanging) forms of life that had been individually made by the Creator during the single week in which he formed the entire universe. Darwin's book subverted a world view that had been taught for centuries.

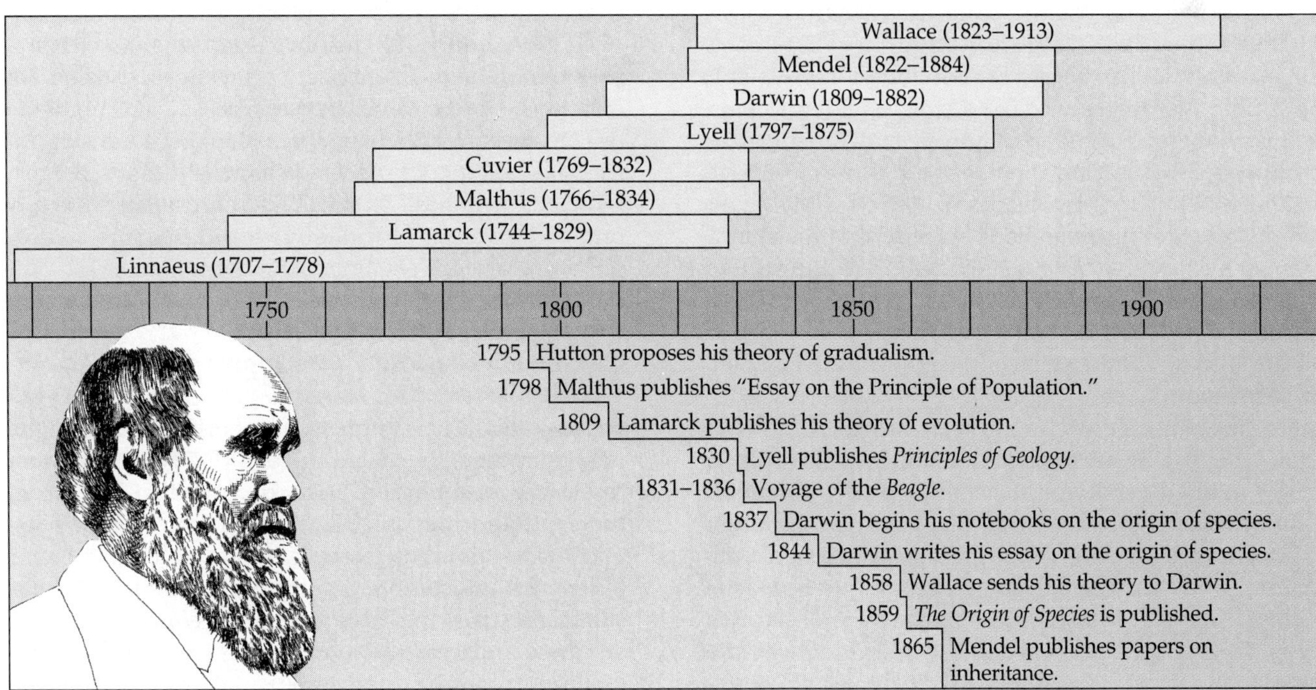

Figure 20.3
Darwinism in historical context.

The Scale of Life and Natural Theology

A number of classical Greek philosophers believed in the gradual evolution of life. But the philosophers who influenced Western culture most, Plato (427–347 B.C.) and his student Aristotle (384–322 B.C.), held opinions that were inimical to any concept of evolution. Plato believed in two worlds: a real world, ideal and eternal, and an illusory world of imperfection that we perceive through our senses. The variations we see in plant and animal populations were to Plato merely imperfect representatives of ideal forms, or essences, and only these perfect forms were real. This Platonic view of life is at the heart of the "typological thinking" criticized by Ernst Mayr in the interview preceding this chapter. Plato's philosophy, known as idealism, or essentialism, ruled out evolution, which would be counterproductive in a world where ideal organisms were already perfectly adapted to their environments.

Although Aristotle questioned the Platonic philosophy of dual worlds, his own beliefs also precluded evolution. A careful student of nature, Aristotle recognized that organisms ranged from relatively simple to very complex. He believed that all living forms could be arranged on a scale of increasing complexity, later called the *scala naturae* ("scale of nature"). Each form along this ladder of life had its allotted rung, and every rung was taken. In this view of life, which prevailed for over 2000 years, species are fixed, or permanent, and do not evolve.

Prejudice against evolution was fortified in Judeo-Christian culture by the Old Testament account of creation. The creationist-essentialist dogma that species were individually designed and permanent became firmly embedded in Western thought. There were many evolutionists before Darwin, but none was able to topple the doctrine of fixed species. Even as Darwinism emerged, biology in Europe and America was dominated by natural theology, a philosophy dedicated to discovering the Creator's plan by studying his works. Natural theologians saw the adaptations of organisms as evidence that the Creator had designed each and every species for a particular purpose. A major objective of natural theology was to classify species in order to reveal the steps of the scale of life that God had created.

In the eighteenth century, Carolus Linnaeus (1707–1778), a Swedish physician and botanist, sought order in the diversity of life *ad majorem Dei gloriam* — "for the greater glory of God." Linnaeus was the father of **taxonomy,** the branch of biology concerned with naming and classifying the diverse forms of life. He developed the two-part, or binomial, system of naming organisms according to genus and species that is still used today. In addition, Linnaeus adopted a filing system for grouping species into a hierarchy of increasingly general categories. For example, similar species are grouped in the same genus, similar genera (plural of genus) are grouped in the same family, and so on. (Taxonomy will be discussed in more detail in Chapter 23.)

Figure 20.4
Fossils in sedimentary rock. These marine invertebrates called trilobites were found in sedimentary rock that is over 500 million years old. The fossil record provides unequivocal evidence for the transformation of life on Earth.

Linnaeus sought and found order in the diversity of life with his hierarchy of taxonomic categories. But clustering certain species under taxonomic banners implied no evolutionary kinship to Linnaeus. As a natural theologian, he believed that species were permanent creations, and he developed his classification scheme only to reveal God's plan. Or, as Linnaeus himself put it, *Deus creavit, Linnaeus disposuit*—"God creates, Linnaeus arranges." Ironically, a century later, the taxonomic system of Linnaeus would become a focal point in Darwin's arguments for evolution.

Cuvier, Fossils, and Catastrophism

Fossils are relics or impressions of organisms from the past, hermetically sealed in rock (Figure 20.4). Most fossils are found in **sedimentary rocks** that form from the sand and mud that settle to the bottom of seas, lakes, and marshes. New layers of sediment cover older ones and compress them into rock, such as sandstone and shale. In places where shorelines repeatedly advance and retreat, sedimentary rock will be deposited in many superimposed layers called strata. Later, erosion may scrape or carve through upper

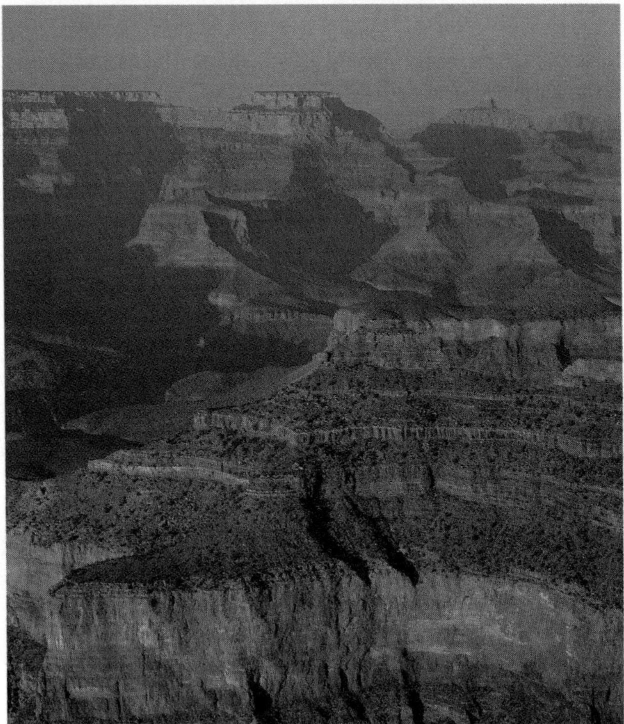

Figure 20.5
Stratification of sedimentary rock at the Grand Canyon. The Colorado River cut through 2000 m of sedimentary rock and unearthed many strata of varying colors and thicknesses. Each stratum, or layer, represents a particular period in Earth's history and is characterized by a collection of fossils of organisms that lived at that time.

(younger) strata and reveal more ancient strata that had been buried (Figure 20.5). The fossil record thus displays graphic and incontrovertible evidence that the Earth has had a succession of floras (plant life) and faunas (animal life).

Paleontology, the study of fossils, was largely developed by Georges Cuvier (1769–1832), the great French anatomist. Realizing that the history of life is recorded in strata containing fossils, he documented the succession of fossil species in the Paris Basin. He noted that each stratum is characterized by a unique suite of fossil species, and the deeper (older) the stratum, the more dissimilar the flora and fauna are from modern life. Cuvier even understood that extinction had been a common occurrence in the history of life. From stratum to stratum, new species appear and others disappear. Yet Cuvier was a staunch and effective opponent to the evolutionists of his day. How did he reconcile the dynamic story told by the fossil record with the concept that species are immutable? Cuvier speculated that the boundaries between the fossil strata corresponded in time to catastrophic events, such as floods or drought, that had destroyed many of the species that had lived at that location at that time. Where there were multiple strata, there had been many catastrophes. This view of Earth's history is known as **catastrophism.**

If Cuvier believed that species were fixed, how did he account for the appearance of species in younger strata that were not present in older rocks? He proposed that the periodic catastrophes that caused mass extinctions were usually confined to local geographical regions. After the extinction of much of the native flora and fauna, the ravaged region would be repopulated by foreign species immigrating from other areas.

Some of Cuvier's followers had more extreme theories of catastrophism. One theory held that the catastrophes were global, and after each holocaust, God created life anew. Although Cuvier himself left religion out of his writing, his aversion to evolution came through loud and clear. But even as Cuvier was winning debates against advocates of evolution, a theory of Earth's history that would help pave the way for Darwin was gaining popularity among geologists.

Gradualism in Geology

Competing with Cuvier's theory of catastrophism was a very different idea of how geological processes had shaped the Earth's crust. In 1795, Scottish geologist James Hutton proposed that it was possible to explain the various land forms by looking at mechanisms currently operating in the world. For example, canyons were cut by rivers running down their lengths, and sedimentary rocks with marine fossils were built of particles that had eroded from the land and been carried by rivers to the sea. Hutton explained the state of the Earth by applying the principle of **gradualism,** which holds that profound change is the cumulative product of slow but continuous processes.

The leading geologist of Darwin's era, Charles Lyell (1797–1875), embellished Hutton's gradualism into a theory known as **uniformitarianism.** The term refers to Lyell's idea that geological processes are so uniform that their rates and effects must balance out through time. For example, processes that build mountains are eventually balanced by the erosion of mountains. Darwin rejected this extreme version of uniformity in geological processes, but he was strongly influenced by two conclusions that followed directly from the observations of Hutton and Lyell: First, if geological change results from slow, continuous actions rather than sudden events, then Earth must be very old, certainly much older than the 6000 years assigned by many theologians on the basis of biblical inference. Second, very slow and subtle processes persisting over a long period of time can cause substantial change. Darwin was not the first to apply the principle of gradualism to biological evolution, however.

Lamarck's Theory of Evolution

Toward the end of the eighteenth century, several naturalists suggested that life had evolved along with the evolution of Earth. But only one of Darwin's predecessors developed a comprehensive model that attempted to explain how life evolves: Jean Baptiste Lamarck (1744–1829).

Lamarck published his theory of evolution in 1809, the year Darwin was born. Lamarck was in charge of the invertebrate collection at the Natural History Museum in Paris. By comparing current species to fossil forms, Lamarck could see what appeared to be several lines of descent, each a chronological series of older to younger fossils leading to a modern species.

Where Aristotle saw one ladder of life, Lamarck saw many, and they were more analogous to escalators. On the ground floor were the microscopic organisms, which Lamarck believed were continually generated spontaneously from inanimate material. At the top of the evolutionary escalators were the most complex plants and animals. Evolution was driven by an innate tendency toward greater and greater complexity, which Lamarck seemed to equate with perfection. As organisms attained perfection, they became better and better adapted to their environments. Thus, Lamarck believed that evolution responded to organisms' *sentiments interieurs*, or "felt needs."

Lamarck is remembered most for the mechanism he adopted to explain how specific adaptations evolve. It incorporates two ideas that were popular during Lamarck's era. The first was use and disuse, the idea that those parts of the body used extensively to cope with the environment become larger and stronger, while those that are not used deteriorate. Among the examples Lamarck cited were the blacksmith developing a bigger bicep in the arm that works the hammer and a giraffe stretching its neck to new lengths in pursuit of leaves to eat. The second idea Lamarck borrowed was called the inheritance of acquired characteristics. In this concept of heredity, the modifications an organism acquires during its lifetime can be passed along to its offspring. The long neck of the giraffe, Lamarck reasoned, evolved gradually as the cumulative product of a great many generations of ancestors stretching higher and higher. There is, however, no evidence that acquired characteristics can be inherited. Blacksmiths may increase strength and stamina by a lifetime of pounding with a heavy hammer, but these acquired traits do not change genes transmitted by gametes to offspring.

The Lamarckian theory of evolution is ridiculed by some today because of its erroneous assumption that acquired characteristics are inherited; but in Lamarck's times, that concept of inheritance was generally accepted (and, indeed, Darwin could offer no acceptable alternative). To most of Lamarck's contemporaries, however, the mechanism of evolution was an irrelevant issue. In the creationist-essentialist view that still prevailed, species were fixed, and *no* theory of evolution could be taken seriously. Lamarck was vilified, especially by Cuvier, who would have no part of evolution. In retrospect, Lamarck deserves credit for his unorthodox theory, which was visionary in many respects: in its claim that evolution is the best explanation for both the fossil record and the current diversity of life; in its emphasis on the great age of Earth; and in its stress on adaptation to the environment as a primary product of evolution.

ON THE ORIGIN OF DARWINISM

We have set the scene for the Darwinian revolution. Natural theology, with its view of an ordered world where each living form fit its environment perfectly because it had been specially created, still dominated the intellectual climate as the nineteenth century dawned. A few clouds of doubt about the permanence of species were beginning to gather, but no one could have forecast the thundering storm just over the horizon.

Charles Darwin was born in Shrewsbury, in western England, in 1809. Even as a boy, Darwin's consuming interest in nature was evident. When he was not reading nature books, he was in the fields and forests fishing, hunting, and collecting insects. His father, an eminent physician, could see no future for a naturalist and sent Charles to the University of Edinburgh to study medicine. Only 16 years old at the time, Charles found medical school boring and distasteful, although he managed decent grades. He left Edinburgh without a degree and shortly thereafter enrolled at Christ College at Cambridge University, with the intent of becoming a clergyman. At that time in England, most naturalists and other scientists belonged to the clergy, and nearly all saw the world in the context of natural theology. Darwin became the protégé of the Reverend John Henslow, professor of botany at Cambridge. Soon after Darwin received his B.A. degree in 1831, Professor Henslow recommended the young graduate to Captain Robert FitzRoy, who was preparing the survey ship *Beagle* for a voyage around the world.

The Voyage of the *Beagle*

Darwin was 22 years old when he sailed from England with the *Beagle* in December 1831. The primary mission of the voyage was to chart poorly known stretches of the South American coastline (Figure 20.6). While the

Figure 20.6
The Galapagos Islands. (**a**) These islands, located about 900 km off the west coast of Ecuador, are inhabited by many plant and animal species found nowhere else in the world. (**b**) These marine iguanas, on Fernandina Island, are among the relatively young species that evolved from mainland ancestors that colonized the volcanic islands.

South America
Galapagos Islands
Rio de Janeiro
South Pacific Ocean
Tierra del Fuego
92°W
0°
Pinta
Marchena
Genovesa
Equator
San Salvador
Baltra
Santa Cruz
Fernandina
Isabela
Pinzón
Santa Fe
San Cristobál
Santa María
Española
Galapagos Islands (Ecuador)
0 Kilometers 75

(a)

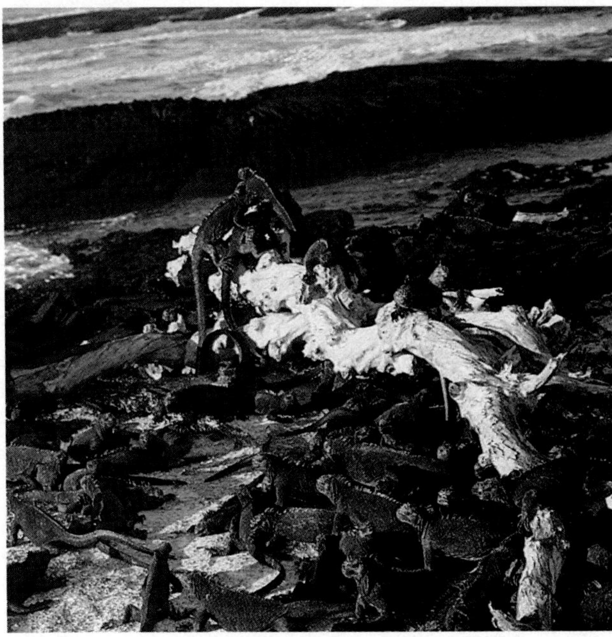

(b)

crew of the ship surveyed the coast, Darwin spent most of his time on shore, collecting thousands of specimens of the exotic and exceedingly diverse faunas and floras of South America. As the ship worked its way around the continent, Darwin was able to observe the various adaptations of plants and animals that inhabited such diverse environments as the Brazilian jungles, the expansive grasslands of the Argentine pampas, the desolate lands of Tierra del Fuego near Antarctica, and the towering heights of the Andes Mountains.

Their unique adaptations notwithstanding, the fauna and flora of the different regions of the continent all had a definite South American stamp, very distinct from the life forms of Europe. That in itself may not have been surprising. But the plants and animals living in temperate regions of South America were taxonomically closer to species living in tropical regions of that continent than to species in temperate regions of Europe. Furthermore, the South American fossils that Darwin found, though clearly different from modern species, were distinctly South American in their resemblance to the living plants and animals of that continent. Darwin was perplexed by the peculiarities of the geographical distribution of species.

A particularly puzzling case of geographical distribution was the fauna of the Galapagos, islands of relatively recent volcanic origin that lie on the equator about 900 km west of the South American coast (Figure 20.6). Most of the animal species on the Galapagos live nowhere else in the world, although they resemble species living on the South American mainland. It was as though the islands were colonized by plants and animals that strayed from the South American mainland and then diversified on the different islands. Among the birds Darwin collected on the Galapagos were 13 types of finches that, although quite similar, seemed to be different species. Some were unique to individual islands, while other species were distributed on two or more islands that were close together. However, Darwin did not keep careful records of which birds came from which islands, apparently because he did not yet appreciate the full significance of the Galapagos fauna and flora.

By the time the *Beagle* sailed from the Galapagos, Darwin had read Lyell's *Principles of Geology*. Lyell's ideas, together with his experiences on the Galapagos, had Darwin doubting the church's position that the Earth was static and had been created only a few thousand years ago. By acknowledging that the Earth was very old and constantly changing, Darwin had taken an important step toward recognizing that life on Earth had also evolved.

(a)

(b)

Figure 20.7
The Galapagos finches. The Galapagos Islands have a total of 13 species of closely related finches, some found on only a single island. The most striking difference among species is in their beaks, which are adapted for specific diets. (**a**) A medium-billed (left) and a large-billed (right) ground finch, both of which have beaks adapted for cracking seeds. (**b**) A tree-dwelling finch using its beak to hold a stick and probe for food.

Darwin Frames His View of Life

At the time Darwin collected the Galapagos finches, he was not sure whether they were actually different species or merely varieties of a single species. Soon after returning to England in 1836, he learned from ornithologists (bird specialists) that the finches were indeed separate species. He began to reassess all that he had observed during the voyage of the *Beagle,* and in 1837 began the first of several notebooks on the origin of species.

Darwin began to perceive the origin of new species and adaptation as closely related processes. A new species would arise from an ancestral form by the gradual accumulation of adaptations to a different environment. For example, if one species became fragmented into several localized populations isolated in different environments by geographical barriers, the populations would diverge more and more in appearance as each adapted to local conditions, gradually, over many generations, becoming dissimilar enough to be designated separate species. This is apparently what happened to the Galapagos finches. Among the differences between the birds are their beaks, which are adapted to the specific foods available on their home islands (Figure 20.7). Darwin anticipated that explaining how such adaptations arise was essential to understanding evolution.

By the early 1840s, Darwin had worked out the major features of his theory of natural selection as the mechanism of evolution. However, he had not yet published his ideas. He was in poor health, and he rarely left home. Despite his reclusiveness, Darwin was not isolated from the scientific community. Already famous as a naturalist because of the letters and specimens he sent to England during the voyage of the *Beagle,* Darwin had frequent correspondence and visits from Lyell, Henslow, and other scientists.

In 1844, Darwin wrote a long essay on the origin of species and natural selection. Realizing the significance of this work, he asked his wife to publish the essay should he die before writing a more thorough dissertation on evolution. Evolutionary thinking was emerging in many areas by this time, but Darwin was reluctant to introduce his theory publicly. Apparently, he understood its subversive quality and quite correctly anticipated the stir it would cause. While he procrastinated, he continued to compile evidence in support of his theory. Lyell, not yet convinced of evolution himself, nevertheless advised Darwin to publish on the subject before someone else came to the same conclusions and published first.

In June 1858, Lyell's prediction came true. Darwin received a letter from Alfred Wallace, a young naturalist working in the East Indies. The letter was accompanied by a manuscript in which Wallace developed a theory of natural selection essentially identical to Darwin's. Wallace asked Darwin to evaluate the paper and forward it to Lyell if it merited publication. Darwin complied, writing to Lyell: "Your words have come true with a vengeance. . . . I never saw a more striking coincidence . . . so all my originality, whatever it may amount to, will be smashed." That was not to be Darwin's fate, however. Lyell and a colleague presented Wallace's paper, along with extracts from Darwin's unpublished 1844 essay, to the Linnaean Society of London on July 1, 1858. Darwin quickly finished *The Origin of Species* and published it the next year. Although Wallace wrote up his ideas for publication first, Darwin developed and supported the theory of natural selection so much more extensively than Wallace that he is known as its main author. Darwin's notebooks also prove that he formulated his theory of natural selection 15 years before reading Wallace's manuscript. Even Wallace felt that Darwin deserved most of the credit.

Within a decade, Darwin's book and its proponents had convinced the majority of biologists that biological diversity was the product of evolution. Darwin succeeded where previous evolutionists had failed, partly because science was beginning to shift away from natural theology, but mainly because he convinced his readers with immaculate logic and an avalanche of evidence in support of evolution.

(a)

(b)

Figure 20.8
Adaptations that camouflage. (**a**) The ptarmigan's winter plumage hides it against the snow. (**b**) This sea horse looks so much like its kelp (seaweed) environment that it lures prey into seeming safety and then eats them.

THE DUAL MEANING OF DARWINISM

The Darwinian view of life has a dual character. One facet is the acknowledgment of evolution as the basis of life's unity and diversity. The second facet is the Darwinian concept of natural selection as the cause of evolution. This section highlights these two main points of Darwin's book.

Common Descent

In the first edition of *The Origin of Species,* Darwin did not use the word evolution, referring instead to **descent with modification,** a term that condensed his view of life. Darwin perceived unity in life, with all organisms related through descent from some unknown prototype that lived in the remote past. As the descendants of that inaugural organism spilled into various habitats over millions of years, they accumulated diverse modifications, or adaptations, that fit them to specific ways of life. In the Darwinian view, the history of life is like a tree, with multiple branching and rebranching from a common trunk all the way to the tips of the living twigs, symbolic of the current diversity of organisms. At each fork of the evolutionary tree is an ancestor common to all lines of evolution branching from that fork. Species that are closely related, such as the domestic cat and the lion, share many characteristics because their lineage of common descent extends to the smallest branches of the tree of life. Most branches of evolution, even some major ones, are dead ends; about 99% of all species that have ever lived are extinct.

Natural Selection and Adaptation

Despite the title of his book, Darwin actually devoted little space to the origin of species, concentrating instead on how populations of individual species become better adapted to their local environments through natural selection (Figure 20.8).

Ernst Mayr has dissected the logic of Darwin's theory of natural selection into three inferences based on five facts*:

Fact 1: All species have such great potential fertility that their population size would increase exponentially if all individuals that are born reproduced successfully.

Fact 2: Most populations are normally stable in size, except for seasonal fluctuations.

*Adapted from Mayr, E. *The Growth of Biological Thought: Diversity, Evolution and Inheritance.* Cambridge, MA: Harvard University Press, 1982.

Fact 3: Natural resources are limited.

INFERENCE 1: Production of more individuals than the environment can support leads to a struggle for existence among individuals of a population, with only a fraction of offspring surviving each generation.

Fact 4: Individuals of a population vary extensively in their characteristics; no two individuals are exactly alike.

Fact 5: Much of this variation is heritable.

INFERENCE 2: Survival in the struggle for existence is not random, but depends in part on the hereditary constitution of the surviving individuals. Those individuals whose inherited characteristics fit them best to their environment are likely to leave more offspring than less fit individuals.

INFERENCE 3: This unequal ability of individuals to survive and reproduce will lead to a gradual change in a population, with favorable characteristics accumulating over the generations.

Figure 20.9
Artificial selection. The broccoli, cauliflower, and cabbages shown here, as well as kale, kohlrabi, and Brussels sprouts, have a common ancestor in one species of wild mustard. By selecting different parts of the plant to accentuate, breeders have obtained these divergent results.

Natural selection is this differential success in reproduction, and its product is adaptation of organisms to their environment. Even if the advantages of some variations over others are slight, the favorable variations will accumulate in the population after many generations of being disproportionately perpetuated by natural selection.

Thus, natural selection occurs through an interaction between the environment and the variability inherent in any population. Variations arise by the chance mechanisms of mutation and genetic recombination (see Chapter 14), but natural selection is not a chance phenomenon. Environmental factors set definite criteria for reproductive success.

A struggle for life is ensured by the excessive production of new individuals. Darwin was already aware of the struggle for existence when he read an influential essay on human population written by the Reverend Thomas Malthus in 1798. Malthus contended that much of human suffering—disease, famine, homelessness, and war—was the inescapable consequence of the potential for the human population to increase faster than food supplies and other resources. The capacity to overproduce seems to be characteristic of all species. Of the many eggs laid, young born, and seeds spread, only a tiny fraction complete their development and leave offspring of their own. The rest are eaten, frozen, starved, diseased, unmated, or unable to reproduce for some other reason.

Each generation, environmental factors screen heritable variations, favoring some over others. Differential reproduction results in the favored traits being disproportionately represented in the next generation. But can selection actually cause substantial change in a population? Darwin found evidence in **artificial selection,** the breeding of domesticated plants and animals. Humans have modified useful species over many generations by selecting individuals with the desired traits as breeding stock. The plants and animals we grow for food bear little resemblance to their wild ancestors (Figure 20.9). The power of selective breeding is especially apparent in our pets, which have been bred more for fancy than utility.

If so much change can be achieved by artificial selection in a relatively short period of time, Darwin reasoned, then natural selection should be capable of considerable modification of species over hundreds or thousands of generations. He postulated that natural selection operating in varying contexts over vast spans of time could account for the entire diversity of life. Darwin did not see life evolving abruptly by quantum leaps, but envisioned instead a gradual accumulation of minute changes. Gradualism is fundamental to the Darwinian view of evolution.

We can now summarize the two features of Darwin's view of life: The diverse forms of life have arisen by descent with modification from ancestral species, and the mechanism of modification has been natural selection working continuously over enormous tracts of time.

Some Subtleties of Natural Selection There are some subtleties of natural selection that require clarification. One is the importance of populations in evolution. For now, we will define a population as a group of interbreeding individuals belonging to a particular species and sharing a common geographic area. A population is the smallest unit that can evolve. Natural se-

(a)

(b)

Figure 20.10
Natural selection in action: industrial melanism. The English peppered moth, *Biston betularia,* occurs in a light gray variety and a dark variety. In regions where the landscape was darkened when industrial pollution killed lichens, dark moths increased in relative number and light moths nearly disappeared. The two varieties are shown on (**a**) a tree trunk covered with lichens and (**b**) a dark tree trunk lacking lichens due to pollution. In either case, birds find and eat a greater proportion of conspicuous moths than camouflaged individuals (although other factors probably contribute to the comparative success of the two varieties of moths).

lection involves interactions between individual organisms and their environment, but individuals do not evolve. Evolution can be measured only as change in relative proportions of variations in a population over a succession of generations. Furthermore, natural selection can amplify or diminish only those variations that are heritable. As we have seen, an organism may become modified through its own experiences during its lifetime, and such acquired characteristics may even adapt the organism to its environment; but there is no evidence that acquired characteristics can be genetically inherited. We must distinguish adaptations an organism acquires by its own actions from innate adaptations that evolve in a population over many generations as a result of natural selection.

It must also be emphasized that the specifics of natural selection are regional and timely; environmental factors vary from place to place and from time to time. An adaptation in one situation may be useless or even detrimental in different circumstances. A few examples will reinforce this situational quality of natural selection.

Natural Selection in Action: Three Examples The most cited and extensively documented example of natural selection in action involves the English peppered moth, *Biston betularia.* It is found throughout the English midlands, occurring in two varieties that differ in coloration. The form for which the peppered moth is named is light, with splotches of pigment. The other variety is uniformly dark. Peppered moths feed at night and rest during the daytime, sometimes on trees and rocks encrusted with light-colored lichens. Against this background, light individuals are camouflaged, but the dark moths, being very conspicuous, are easy prey for birds (Figure 20.10). Before the Industrial Revolution, dark peppered moths were very rare, presumably becoming bird food before they could reproduce and pass the genes for darkness on to the next generation. But industrial pollution darkened the landscape of much of the countryside in the late 1800s, mainly by killing lichens that covered rocks and the dark bark of trees. Against this darkened background, light moths are easier to see by humans, and probably by birds. The frequency of dark individuals in populations of *Biston* began to increase. By the turn of the century, the population in the Manchester region consisted almost entirely of dark moths. This phenomenon, known as industrial melanism, occurred in hundreds of other species of moths in polluted areas.

It is important to realize that industrial melanism is not a case of the inheritance of acquired characteristics. The environment did not *create* favorable characteristics, as in Lamarck's theory, but only acted upon the inherited variations manifest in any population, favoring the survival and reproduction of some individuals over others. Natural selection edits populations. The dark moths were reproductively favored because the newly visible light moths were more commonly eaten by birds and consequently left fewer offspring. Experiments support the hypothesis that natural selection in the form of predation by birds contributed to the shift in the composition of *Biston* populations in industrial regions (although recent research suggests that other factors may also be involved in the relative success of dark and light moths).

The case of *Biston* reinforces the point that natural selection operates in the here and now, tending to adapt organisms to their local environment. Natural selection is utilitarian, picking traits that work best for the present situation. In recent years, the case of the peppered moth has taken a satisfying turn, for much of the pollution has been curbed, enabling some parts of the countryside in industrial areas to return to natural hues. In those places, the light form of *Biston* has made a strong comeback. (However, researchers are still uncertain if birds and coloration of trees and rocks are the most important selection factors responsible for this evolutionary reversal of industrial melanism.)

Many other examples document the evolution of protective coloration by natural selection. One is particularly quaint. In April 1962, mutant mice with pale yellow fur were discovered among a population of house mice *(Mus musculus)* living on the farm of Burl and Alfred McCrosky in Little Sac River, Missouri. This color, which is much lighter than the more common dark brown color, is inherited as a simple Mendelian recessive trait. Larry Brown, a University of Wyoming mammalogist, was interested in how this color difference affected survival of the mice on the McCrosky farm and studied the proportions of yellow and brown mice over a two-year period. He sampled the mouse population at four-month intervals by placing live traps in a granary made of dark wood. Brown recorded the number of yellow and brown individuals and then released the mice. During the early months of the study, Brown convinced the McCroskys to keep the granary tightly closed so that none of the dozen domestic cats living on the farm could enter. In December 1962, out of 58 mice trapped in the granary, 27 of them (46%) had light-yellow fur (the number of mice trapped is only a small sample of the actual population). By early January 1963, the population had become so dense that the mice were severely damaging the stored grain, and the McCroskys made an opening in one wall of the granary so the cats could enter. By April 1963, the mouse population had dropped by 62%, and none of the 22 mice trapped had yellow fur (Figure 20.11). Of the 22 mice that Brown trapped, 38% were brown juveniles, the offspring of mice that managed to survive and reproduce. Brown concluded that the addition of cats as predators in the environment selected against the light-yellow mice, probably because these mice were more visible than brown mice against the dark background within the granary.

(a)

Trapping Date	No. of Mice Trapped	No. of Light-Yellow Mutants
December 1962	58	27
April 1963	22	0

(b)

Figure 20.11
Natural selection in action: protective coloration in a population of house mice. (**a**) In this "cat's-eye view," a brown mouse is less visible than a mutant with light-yellow fur against a dark background. (**b**) These data are the results of an experimental test of the prediction that coloration of mice affects their survival (and hence reproductive success) in an environment that includes domestic cats. The mice were trapped in a granary made of dark wood. The December 1962 data were collected before cats were allowed access to the granary. In January 1963, the farmers who owned the granary made an opening in a wall so that cats could control the population of mice. By April 1963, the number of both brown and light-yellow mice had decreased, but the cats apparently killed a disproportionately large number of the yellow mice. (From Brown, L. N. *Journal of Mammalogy, 46,*1965.)

Another investigation of natural selection in action tests Darwin's hypothesis that the beaks of Galapagos finches are evolutionary adaptations to different food sources. For twenty years, Peter and R. Grant of Princeton University have been studying the population of medium ground finches *(Geospiza fortis)* on Daphne Major, a tiny islet of the Galapagos. These birds use their strong beaks to crush seeds. They feed preferentially on small seeds, which are produced in abundance by certain plant species during wet years. The finches resort to larger seeds, which are harder to crush, only when small seeds are in short supply. Such shortages occur during dry years, when the plants produce fewer seeds, both small and large. The Grants have discovered that the average depth of beaks in this finch population has oscillated over the years (Figure 20.12). During droughts, beak depth (the dimension from the top to bottom of the beak) has increased, only to decrease again during wet periods. This is a heritable trait. The Grants attribute the change to the availability of small seeds in different years. Birds with stronger beaks may have an advantage during dry periods when survival and reproduction depend on cracking large seeds in greater proportion than is necessary during wet periods. Again, we see that natural selection is situational: What works best in one environmental context may be inferior equipment in some other context.

Among the other examples of natural selection in action we have already studied are the evolution of an-

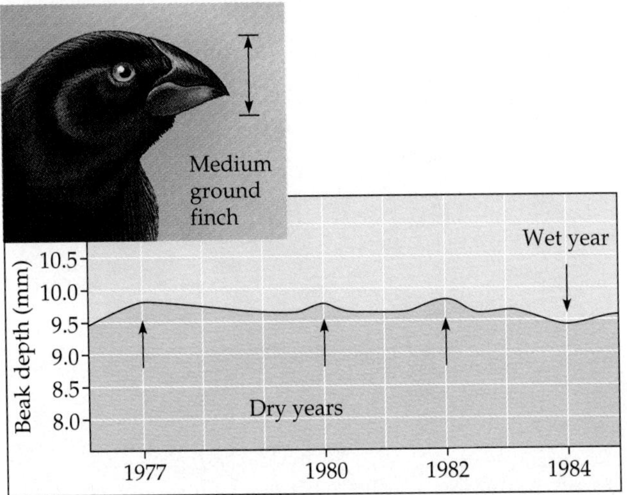

Figure 20.12
Natural selection in action: beak evolution in one of Darwin's finches. The medium ground finch, one of the birds Darwin found on the Galapagos, uses its strong beak to crush seeds (see inset). Given a choice of small seeds or large seeds, the birds eat mostly small ones, which are easier to crush. During wet years, small seeds are produced in such abundance that ground finches consume relatively few large seeds. This changes during dry years, when both small and large seeds are in short supply and the birds resort to eating a greater than usual proportion of large seeds. The change in diet is correlated with a change in the average depth (top to bottom dimension) of the birds' beaks. Field studies comparing offspring to parents confirm that this trait is inherited rather than acquired (by exercising the beak on large seeds, for example). The most likely explanation is that those birds that happen to have stronger beaks have a feeding advantage during droughts and pass the genes for this trait on to their offspring.

tibiotic resistance in bacteria (Chapter 17) and change in the body size of guppies exposed to different predators (Chapter 1). Researchers have published more than 100 other accounts of natural selection in the wild. And hundreds of other studies have documented natural selection in laboratory populations of such organisms as *Drosophila.* Scientists do not accept natural selection solely on the face of its logic, but because it has been confirmed repeatedly by the hypothetico-deductive method, in which predictions based on hypotheses are tested by observation and experimentation (see Chapter 1). Ironically, Darwin himself thought that natural selection always operated too slowly to be observed. He was also unable to satisfactorily account for the genetics of variation (a problem we will address in Chapter 21). For these and other reasons, the theory of natural selection as the mechanism of evolution won relatively few advocates during Darwin's time. In contrast, within just a few years after Darwin published *The Origin of Species,* most biologists were convinced that evolution does, in fact, occur, whatever the mechanism.

THE SIGNS OF EVOLUTION

Evolution leaves observable signs. Such data are the clues to the past, essential to any historical science. For example, historians of human civilization can study written records from earlier times, and they can also piece together the evolution of societies by recognizing vestiges of the past in modern cultures. Even if we did not know from written documents and archaeological evidence that Spaniards colonized the Americas, we would deduce this from the Hispanic stamp on much of Latin America. Similarly, biological evolution has left marks—in the fossil record and in the historical vestiges evident in modern life. In this section, we briefly survey some of these signs of evolution (which will be discussed more thoroughly in Chapter 23). Darwin documented common descent mainly with evidence from the geographic distribution of species and from the fossil record, but we will not limit ourselves to these two categories of evidence. As biology has continued to progress, new discoveries, including the revelations of molecular biology, continue to validate the evolutionary view of life.

Biogeography

It was the geographical distribution of species—**biogeography**—that first suggested common descent to Darwin. Islands have many species of plants and animals that are both endemic (native, found nowhere else) and closely related to species of the nearest mainland or neighboring island. Some revealing questions arise: Why are two islands with similar environments in different parts of the world *not* populated by closely related species, but instead inhabited by species taxonomically affiliated with the plants and animals of the nearest mainland, where the environment is often quite different? Why are the tropical animals of South America more closely related to species of South American deserts than to species of the African tropics? Why is Australia home to a great diversity of pouched mammals (marsupials) but almost no placental mammals (those in which embryonic development is completed in the uterus)? It is not because Australia is inhospitable to placental mammals; in recent years, humans have introduced rabbits to Australia, and the rabbit population has exploded. Apparently, placental animals are not part of the original fauna of Australia because that continent has been isolated from places where the ancestors of placental mammals lived. The geographic distribution of species makes sense only in the historical context of evolution.

Although such biogeographic patterns are incongruous if one imagines that species were individually placed in suitable environments, they make sense in

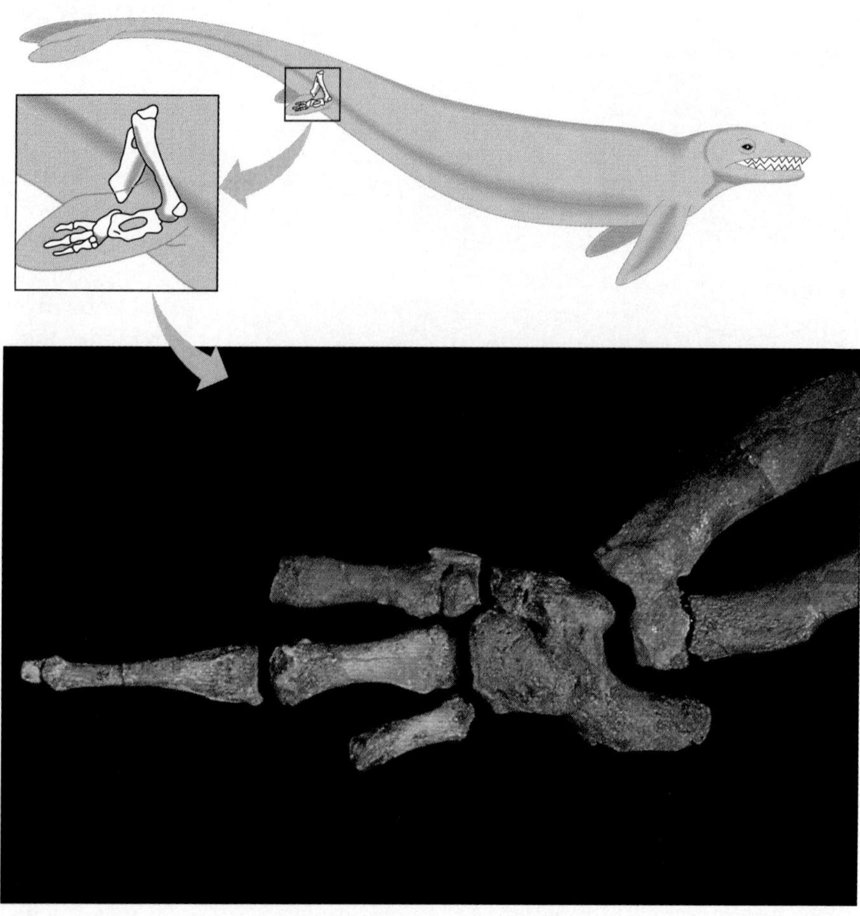

Figure 20.13
Transitional fossils linking past and present. Whales evolved from terrestrial ancestors, an evolutionary transition that left many signs, including fossil evidence. Paleontologists digging in Egypt have recently identified extinct whales that had hind limbs. (Modern whales have forelimbs in the form of flippers but lack hind limbs, although there are tiny skeletal remnants of hind limbs that do not extend from the body.) Shown here are the fossilized leg bones of *Basilosaurus*, one of those ancient whales. The legs were half a meter long and were well muscled, but these whales were already aquatic animals that no longer used their legs to support their weight and walk.

the historical context of evolution. The Darwinian interpretation is that we find modern species where they are because they evolved from ancestors that inhabited those regions. Consider armadillos, the armored mammals that live only in the Americas. The evolutionary view of biogeography predicts that contemporary armadillos are modified descendants of earlier species that occupied these continents, and the fossil record confirms that such ancestors existed. This example brings us to the more general importance of the fossil record as a chronicle of evolution.

The Fossil Record

The succession of fossil forms is compatible with what is known from other types of evidence about the major branches of descent in the tree of life. For instance, evidence from biochemistry, molecular biology, and cell biology places prokaryotes as the ancestors of all life, and indeed the oldest known fossils are prokaryotes. Another example is the chronological appearance of the different classes of vertebrate animals in the fossil record. Fossil fishes predate all other vertebrates, with amphibians next, followed by reptiles, then mammals and birds. This sequence is consistent with the history of vertebrate descent as revealed by many other types of evidence.

The Darwinian view of life predicts that evolutionary transitions should leave signs in the fossil record. Indeed, paleontologists have discovered many transitional forms that link even older fossils to modern species. For example, a series of fossils documents the changes in skull shape and size that occurred as mammals evolved from reptiles. Every year, paleontologists turn up other important links between contemporary forms and their ancestors. In the past few years, for instance, researchers have found fossilized whales that link these aquatic mammals to their terrestrial predecessors (Figure 20.13).

Taxonomy

Ironically, Linnaeus, who apparently believed that species are fixed, provided Darwin with some of the most potent evidence for common descent by recognizing that the great diversity of organisms could be

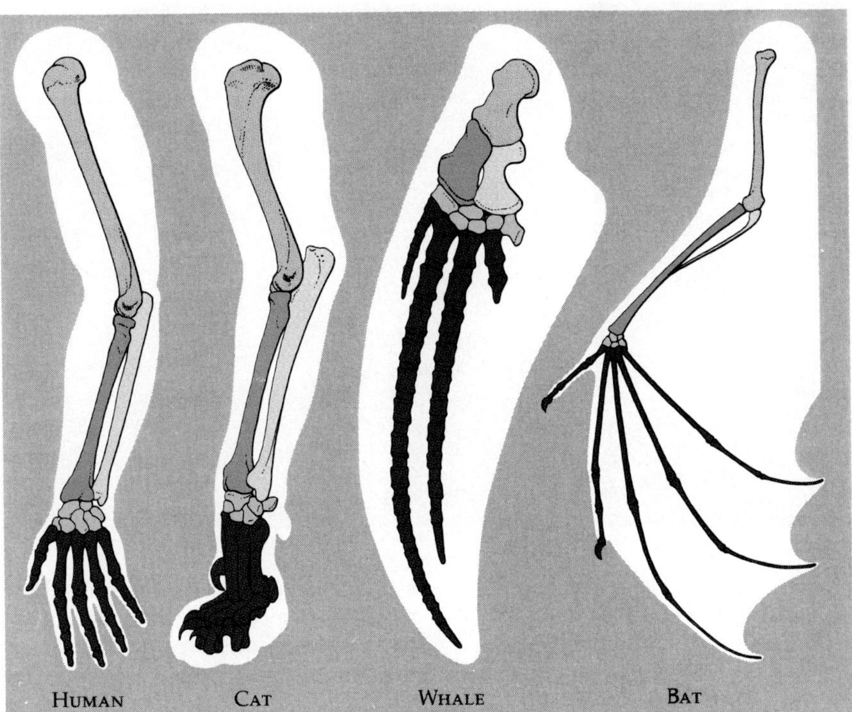

Figure 20.14
Homologous structures: anatomical signs of evolution. The forelimbs of all mammals are constructed from the same skeletal elements, suggesting that a common ancestral forelimb has been modified for many different functions.

ordered into "groups subordinate to groups" (Darwin's phrase). The major taxonomic categories were introduced in Chapter 1, but let's review them here:

Kingdom
Phylum
Class
Order
Family
Genus
Species

To Darwin, the natural hierarchy of the Linnaean scheme reflected the branching genealogy of the tree of life, with organisms at the different taxonomic levels related through descent from common ancestors. If we acknowledge that lions and tigers are more closely related than are lions and horses, then we have recognized that evolution has left signs in the form of different degrees of kinship among modern species. Because taxonomy is a human invention, it cannot, by itself, confirm common descent. But taken along with the many other types of evidence, the evolutionary significance of taxonomy is unmistakable. For example, genetic analysis reveals that such species as lions and tigers, thought to be closely related on the basis of anatomical features and other criteria, are indeed "blood" relatives with a common hereditary background.

Comparative Anatomy

Common descent is evident in anatomical similarities between species grouped in the same taxonomic category. For example, the same skeletal elements make up the forelimbs of humans, cats, whales, bats, and all other mammals, although these appendages have very different functions (Figure 20.14). Surely, the best way to construct the infrastructure of a bat's wing is not also the best way to build a whale's flipper. Such anatomical peculiarities make no sense if the structures are uniquely engineered and unrelated. It is more logical, especially in view of corroborative evidence, that the basic similarity of these forelimbs is the consequence of the descent of all mammals from a common ancestor. The forelegs, wings, flippers, and arms of different mammals are variations on a common anatomical theme that has been modified for divergent functions. Similarity in characteristics resulting from common ancestry is known as **homology,** and such anatomical signs of evolution are called **homologous structures.** Comparative anatomy is consistent with all other evidence in testifying that evolution is a remodeling process in which ancestral structures that functioned in one capacity become modified as they take on new functions.

The oddest homologous structures are **vestigial organs,** rudimentary structures of marginal, if any, use to the organism. Vestigial organs are historical remnants

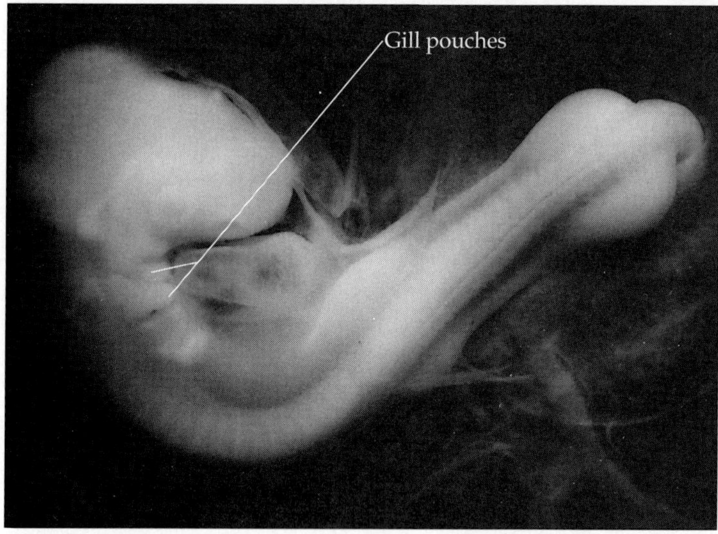

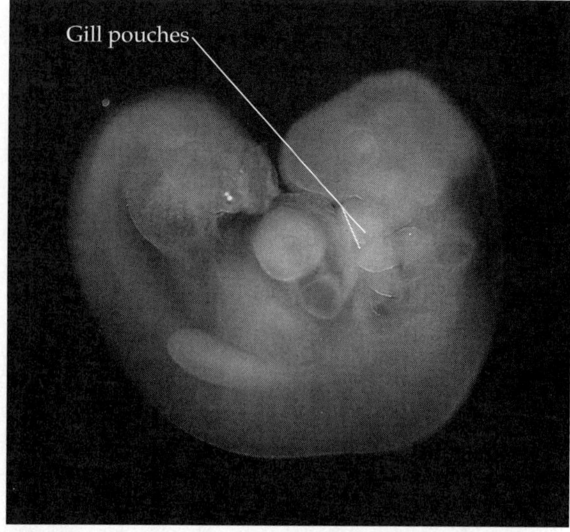

Figure 20.15
Evolutionary signs from comparative embryology. At this early stage of development, the kinship of vertebrates is unmistakable. Notice, for example, the gill pouches in both the bird embryo (left) and the human embryo. Comparative embryology helps biologists identify anatomical homology that is less apparent in adults, because the structures are extensively modified in different ways during later development of the organisms. Although the gill pouches develop into gills and their supporting structures in fishes, they give rise to other anatomical structures in terrestrial vertebrates such as birds and mammals.

of structures that had important functions in ancestors but are no longer essential. For instance, the skeletons of some snakes retain vestiges of the pelvis and leg bones of walking ancestors.*

Comparative Embryology

Closely related organisms go through similar stages in their embryonic development. For example, all vertebrate embryos go through a stage in which they have gill pouches on the sides of their throats. Indeed, at this stage of development, similarities between fishes, frogs, snakes, birds, humans, and all other vertebrates are much more apparent than differences (Figure 20.15). As development progresses, the various vertebrates diverge more and more, taking on the distinctive characteristics of their classes. In fish, for example, the gill slits develop into gills; but in terrestrial vertebrates, these embryonic structures become modified for other functions, such as the Eustachian tubes that connect the middle ear with the throat in humans.

Inspired by the Darwinian principle of descent with modification, many embryologists in the late nineteenth century proposed the extreme view that "ontogeny recapitulates phylogeny." This notion holds that the embryonic development of an individual organism (**ontogeny**) is a replay of the evolutionary history of the species (**phylogeny**). The theory of recapitulation is an overstatement. What recapitulation does occur is a replay of embryonic stages, not a sequence of adultlike stages of ever more advanced vertebrates. Although vertebrates share many features of embryonic development, it is not as though a mammal first goes through a "fish stage," then an "amphibian stage," and so on. Also, because embryonic processes ultimately affect the fitness of the adult organism, they are subject to natural selection. Thus, even relatively early stages of development may become modified in the course of evolution. Nevertheless, ontogeny does provide clues to phylogeny. In particular, comparative embryology can often establish homology among structures that become so altered in later development that their common origin would not be seen by comparing their fully developed forms.

Molecular Biology

Molecular signs of evolution were discussed in Chapter 5. The main point was that evolutionary relationships among species are reflected in their DNA and proteins—in their genes and gene products (Figure 20.16). If two species have libraries of genes and proteins with sequences of monomers that match closely, the sequences must have been copied from a

*Vestigial organs may seem to support the Lamarckian concept of use and disuse, but they can be explained by natural selection. It would be wasteful to continue providing blood, nutrients, and space to an organ that no longer has a major function. Individuals with reduced versions of those organs would be favored by the environment, and natural selection operating over thousands of generations would tend to phase out obsolete structures.

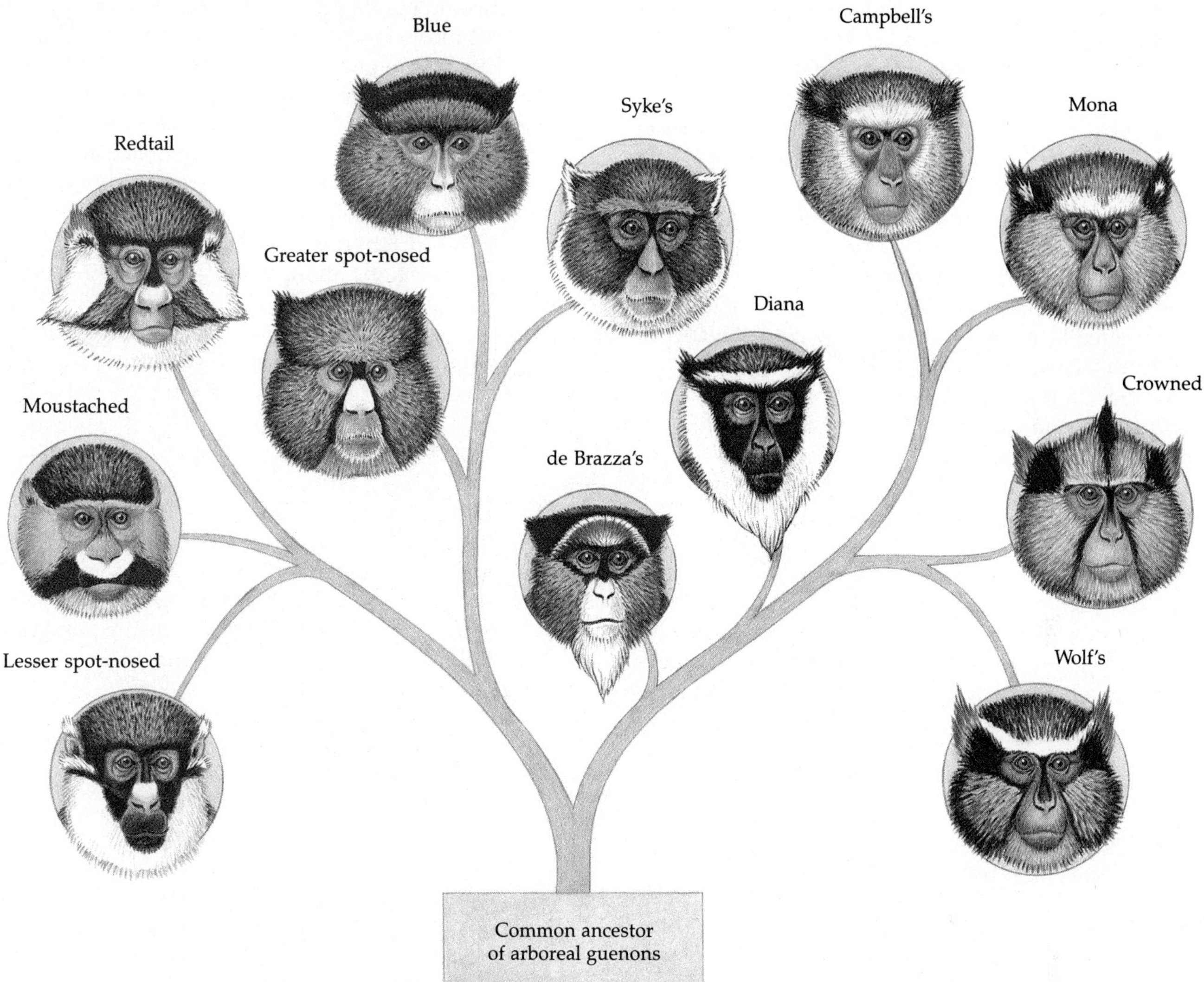

Figure 20.16
An evolutionary tree based on molecular data. This tree traces the evolution of 12 of the 25 known species of guenon monkeys (genus *Cercopithecus*), which inhabit eastern Africa. The relationships among the monkeys were assessed by comparing blood proteins, which are gene products. The molecular data are compatible with evidence based on facial markings, vocal calls, and the biogeographic distribution of the monkeys.

common ancestor. If two long paragraphs were identical except for the substitution of a letter here and there, we would surely attribute them both to a single source.

Darwin's boldest speculation—that *all* forms of life are related to some extent through branching descent from the earliest organisms—has also been substantiated by molecular biology. Even taxonomically remote organisms, such as humans and bacteria, have some proteins in common. An example is cytochrome *c*, the respiratory protein found in all aerobic species (see Chapter 9). Mutations have substituted amino acids at some places in the protein during the long course of evolution, but the cytochrome *c* molecules of all species are clearly akin in structure and function.

A common genetic code is further evidence that all life is related. Evidently, the language of the genetic code has been passed along through all branches of evolution ever since its inception in an early form of life. Molecular biology has thus added the latest chapter to the evidence confirming evolution as the basis for the unity and diversity of life.

JUST A THEORY?

Some people dismiss Darwinism as "just a theory." This tactic for nullifying the evolutionary view of life has two flaws. First, it fails to separate Darwin's two claims: that modern species evolved from ancestral forms, and that natural selection is the main mechanism for this evolution. The conclusion that life has evolved is based on historical facts—the signs of evolution discussed in the previous section. For example, to ignore all the evidence that mammals evolved from reptiles just because none of us was there to observe the transition would be like denying that the American Revolution happened because none of us saw it.

What, then, is theoretical about evolution? Theories are our attempts to explain facts and integrate them with overarching concepts. To biologists, "Darwin's Theory of Evolution" is natural selection—the mechanism Darwin proposed to explain the facts of evolution documented by fossils, biogeography, and other types of historical evidence.

So the "just a theory" argument concerns only Darwin's second claim, his theory of natural selection. This brings us to the second flaw in the "just a theory" case. The term *theory* has a very different meaning in science compared to our general use of the word. The colloquial use of "theory" comes close to what scientists mean by a "hypothesis." In science, a theory is more comprehensive than a hypothesis. A theory, such as Newton's theory of gravity or Darwin's theory of natural selection, accounts for many facts and attempts to explain a great variety of phenomena. Such a unifying concept does not become a widely accepted theory in science unless its predictions stand up to thorough and continuous testing by experiments and observations. Even then, good scientists do not allow theories to become dogma. For example, many evolutionary biologists now question whether natural selection alone is sufficient to account for the evolutionary history observed in the fossil record. The study of evolution is more robust and lively than ever, and we will evaluate some of the current debates in the next three chapters. But these questions about *how* life evolves in no way imply that the observation *that* life evolves is "just a theory." Arguing about evolutionary theory is like arguing about competing theories of gravity; we know that objects keep right on falling while we debate the cause.

* * *

Darwin gave biology a sound scientific basis by attributing the diversity of life to natural causes rather than supernatural creation. Nevertheless, the products of evolution are elegant and inspiring in their variety and harmony. As Darwin said in the closing paragraph of *The Origin of Species*, "There is grandeur in this view of life."

STUDY OUTLINE

1. The first convincing case for evolution, *The Origin of Species,* was published by Charles Darwin in 1859.
2. Evolution encompasses all the changes that have transformed life on Earth throughout its history.

Pre-Darwinian Views (pp. 420–424)

1. The philosophies of Plato and Aristotle ruled out evolution. In particular, Aristotle envisioned a *scala naturae,* in which fixed species occupied allotted rungs on an increasingly complex ladder of life.
2. Prejudice against evolution was fortified by natural theologians, who interpreted the Old Testament account of creation literally. For example, Linnaeus devised a hierarchy for naming and classifying organisms to reveal the specific steps in the divinely created scale of life.
3. Cuvier believed that catastrophic extinctions explained the unique sets of fossil species between successive strata.
4. Geologists James Hutton and Charles Lyell proposed that profound changes in the Earth's surface can result from slow, continuous actions.
5. Before Darwin, Jean Baptiste Lamarck proposed a theory of evolution in which increasing complexity and more perfect adaptations result from inheritance of characteristics acquired by organisms interacting with the environment. There is, however, no evidence for inheritance of acquired characteristics.

On the Origin of Darwinism (pp. 424–426)

1. Darwin's view of life apparently began to change when he served as naturalist on HMS *Beagle.* Darwin was impressed by the peculiar geographical distribution and distinctive interrelationships of species, including those of the Galapagos Islands. This experience eventually led him to the idea that new species originate from ancestral forms by the gradual accumulation of adaptations.
2. The long-delayed publication of Darwin's *The Origin of Species* was catalyzed by Alfred Wallace, who independently arrived at the theory of natural selection.

The Dual Meaning of Darwinism (pp. 427–431)

1. The Darwinian view of life can be dissected into two separate claims: the evolution of new species by descent with modification from ancestral species, and natural selection as the mechanism of this evolution.
2. Natural selection is based on differential success in reproduction, made possible because of variation among the individuals of any population and the tendency for a population to produce more offspring than the environment can support. The individuals best adapted to the local environment leave the most offspring and thereby pass on their adaptive characteristics.

The Signs of Evolution (pp. 431–435)

1. The biogeography of species first suggested common descent to Darwin. He noticed that island species were more closely related to those on the mainland than to those on distant islands with similar environments.
2. The chronological fossil record is compatible with other lines of evidence in support of evolution.
3. The taxonomic hierarchy reflects common descent.
4. Homologous structures testify to an evolutionary remodeling process.

5. The study of embryonic development reveals homologies not apparent in adult species.
6. Closely related species show unmistakable similarities in their DNA and proteins.

Just a Theory? (pp. 435–436)

1. Evolution is documented by historical facts.
2. As applied to evolution, the term *theory* refers to models for *how* evolution occurs—as in Darwin's theory of natural selection.

SELF-QUIZ

1. The ideas of Hutton and Lyell that Darwin incorporated into his theory concerned
   a. the age of Earth and gradual geological processes producing profound change
   b. extinctions evident in the fossil record
   c. adaptation of species to the environment
   d. a hierarchical classification of organisms
   e. the inheritance of acquired characteristics
2. Which of the following is *not* a fact or inference of natural selection?
   a. There is heritable variation among individuals.
   b. Poorly adapted individuals never leave offspring.
   c. Since only a fraction of offspring survive, there is a struggle for limited resources.
   d. Individuals whose inherited characteristics best fit them to the environment will leave more offspring.
   e. Unequal reproductive success leads to adaptations.
3. Which of the following statements best explains changes observed in the English peppered moth populaton?
   a. Bird predation was probably an important factor in natural selection.
   b. Soot incorporated into the moths resulted in industrial melanism.
   c. Pollution caused sterility in light moths.
   d. Natural selection produced new genes adapted to the darkened landscape.
4. The gill pouches of reptile and bird embryos are
   a. vestigial structures
   b. support for "ontogeny recapitulates phylogeny"
   c. homologous structures
   d. used by the embryos to breathe
   e. evidence for the degeneration of unused body parts
5. The best evidence for a common origin of *all* life is
   a. comparative anatomy
   b. comparative embryology
   c. biogeography
   d. molecular biology
   e. the fossil record
6. Perhaps unfairly, Lamarckism is now most associated with
   a. catastrophism
   b. essentialism
   c. creationism
   d. inheritance of acquired characteristics
   e. uniformitarianism
7. Darwin's theory, as presented in *The Origin of Species*, mainly concerned
   a. how new species arise
   b. the origin of life
   c. how adaptations evolve
   d. how extinctions happen
   e. the genetics of evolution
8. The Galapagos Islands are located nearest
   a. the west coast of Africa
   b. the west coast of South America
   c. the west coast of North America
   d. the east coast of Australia
   e. the west coast of Greenland
9. Which person is *incorrectly* matched with a term or idea?
   a. Plato—essentialism
   b. Linnaeus—use and disuse
   c. Malthus—overpopulation
   d. Lyell—uniformitarianism
   e. Aristotle—*scala naturae*

CHALLENGE QUESTIONS

1. Some opponents of evolution have exclaimed, "I just can't believe we came from a chimpanzee!" What is their misconception about evolution?
2. Spadefoot toads from Diamond Craters, Oregon, an area of black lava rock, are much darker than toads of the same species from the surrounding desert. Give a Darwinian explanation that might account for this difference.

SCIENCE, TECHNOLOGY, AND SOCIETY

1. To what extent are humans in a technological society exempt from natural selection? Explain your answer.
2. Is the concept of natural selection relevant in a political or economic context? In other words, if a particular nation or corporation achieves success or dominance, does this mean that it is more fit than its competitors and that oppression of those competitors is valid? Why or why not?

FURTHER READING

Darwin, C. *On the Origin of Species by Means of Natural Selection, or The Preservation of Favored Races in the Struggle for Life.* New York: New American Library, 1963. A modern printing of the historical book that revolutionized biology.

Desmond, A., and J. Moore. *Darwin.* New York: Warner, 1992. A new biography.

Futuyma, E. J. *Evolutionary Biology,* 2nd ed. Sunderland, MA: Sinauer, 1986. An excellent undergraduate text.

Grant, P. R. "Natural Selection and Darwin's Finches." *Scientific American,* October 1991. A single drought can change a population.

Milner, R. *The Encyclopedia of Evolution.* New York: Facts on File, 1990. Many interesting anecdotes.

Symonds, N. "A Fitter Theory of Evolution?" *New Scientist,* September 21, 1991. A controversy about how bacteria adapt to changing environments.

21 How Populations Evolve

The Modern Evolutionary Synthesis

The Genetics of Populations

Causes of Microevolution

The Genetic Basis of Variation

Adaptive Evolution

Figure 21.1
Individuals are selected, but populations evolve. The bent grass *(Agrostis tenuis)* in the foreground is growing on the tailings of an abandoned mine in Wales. These plants tolerate a concentration of heavy metals that is toxic to other plants of the same species growing just meters away, in the pasture on the other side of the fence. Each year, many seeds land on the mine tailings, but most are unable to grow successfully there. The only plants that germinate, grow, and reproduce are those that inherited genes making it possible to tolerate metallic soil. Thus, this adaptation does not evolve by individual plants becoming more metal-tolerant during their lifetimes. We can only observe the evolution of the adaptation as an increasing proportion of tolerant individuals from one generation to the next. Natural selection works by favoring the reproductive success of certain individuals over others among the varying members of a population. But the impact of this individual selection is a generation-to-generation change in the prevalence of certain traits on the level of the whole population. In this chapter, you will learn more about how populations evolve.

Natural selection acts on individual organisms, affecting their survival and reproduction. But it is a population, not its individuals, that actually evolves. The evolution of industrial melanism among peppered moth populations in the British midlands reinforces this important point. (See Chapter 20 for a review of this example.) Individual moths did not change color. But by favoring the reproductive success of dark moths over light ones for many generations, natural selection changed the proportions of these two phenotypes in a population. Figure 21.1 illustrates another example. Populations, not individual organisms, are the smallest units that can evolve. In this chapter, you will learn more about how natural selection and other mechanisms cause populations to evolve. We begin by tracing how biology finally began to accept Darwin's theory of natural selection during the first half of this century.

THE MODERN EVOLUTIONARY SYNTHESIS

The Origin of Species convinced most biologists that species are products of evolution, but Darwin was not nearly as successful in gaining acceptance for natural selection as the mechanism of evolution. A major obstacle was the lack of any theory of genetics that could explain how chance variations arise, while also accounting for the hereditary precision that perpetuates parents' traits in their offspring. Natural selection was based on what seemed to be a paradox: Like begets like—but not exactly. Darwin could observe this quirk of inheritance, but he could not explain it. Although Gregor Mendel and Charles Darwin were contemporaries, Mendel's discoveries were unappreciated at the time, and apparently no one noticed that he had elucidated the very principles of inheritance that could have resolved Darwin's paradox and given credibility to natural selection.

Ironically, when Mendel's research article was rediscovered and reassessed at the beginning of this century, many geneticists believed that the laws of inheritance were at odds with Darwin's theory of natural selection. As the raw material for natural selection, Darwin emphasized characters that vary in a continuum in a population, such as the fur length of mammals or the speed with which an animal can flee from a predator. We know today that such quantitative characters are influenced by multiple genetic loci. (See Chapter 13 to review polygenic inheritance and quantitative characters.) But Mendel, and later the geneticists of the early nineteenth century, recognized only discrete "either-or" characters, such as purple versus white flowers in Mendel's peas, as heritable. Thus, there seemed to be no genetic basis for natural selection to work on the more subtle variations within a population that were central to Darwin's theory.

During the 1920s, genetic research focused on mutations, and, as an alternative to Darwin's theory of natural selection, a widely accepted hypothesis held that evolution occurred in rapid leaps as a result of radical changes in phenotype caused by mutations. This idea contrasted sharply with Darwin's view of gradual evolution due to environmental selection acting on continuous (quantitative) variations among individuals of a population. There was also considerable sentiment for *orthogenesis,* the idea that evolution has been a predictable progression to more and more elite forms of life. This notion of goal-oriented evolution, a throwback to Lamarck's "felt needs," opposed Darwin's mechanistic view that evolution simply reflected differential reproductive success extrapolated over many generations.

An important turning point for evolutionary theory was the birth of **population genetics,** which emphasizes the extensive genetic variation within populations and recognizes the importance of quantitative characters. With progress in population genetics in the 1930s, Mendelism and Darwinism were reconciled, and the genetic basis of variation and natural selection was worked out.

A comprehensive theory of evolution that became known as the **modern synthesis,** or neo-Darwinism, was forged in the early 1940s. It is called a synthesis because it integrated discoveries and ideas from many different fields, including paleontology, taxonomy, biogeography, and, of course, population genetics. Among the architects of the modern synthesis were Theodosius Dobzhansky, Ernst Mayr (see the interview that precedes Chapter 20), George Gaylord Simpson, and G. Ledyard Stebbins. The synthesis emphasizes the importance of populations as the units of evolution, the central role of natural selection as the most important mechanism of evolution, and the use of gradualism to explain how large changes can evolve as an accumulation of small changes occurring over long periods of time. No scientific paradigm is likely to endure without modification for half a century. In the next two chapters, you will learn that many evolutionary biologists are now challenging some of the claims and assumptions of the modern synthesis, and we will evaluate the need for a new evolutionary synthesis at the end of this unit. Still, twentieth-century biology has been profoundly affected by the modern synthesis, which shaped most of the ideas about how populations evolve that are introduced in this chapter.

THE GENETICS OF POPULATIONS

A **population** is a localized group of individuals belonging to the same species. A **species,** simply defined, is a group of populations that have the potential to interbreed in nature (this definition will be examined more critically in Chapter 22). Each species has a geographical range within which individuals are not spread out evenly, but are usually concentrated in several localized populations. One population may be isolated from others of the same species, exchanging genetic material only rarely. Such isolation is particularly common for populations confined to widely separated islands, unconnected lakes, or mountain ranges separated by lowlands. However, populations are not always isolated, nor do they necessarily have sharp boundaries. One dense population center may blur into another in an intermediate region where members of the species occur but are less numerous. Although the populations are not isolated, individuals are still concentrated in centers and are more likely to interbreed with members of the same population than with members of other populations. Therefore, individuals near a population center are, on average, more closely related to one another than to members of other populations (Figure 21.2).

The Gene Pool and Microevolution

The total aggregate of genes in a population at any one time is called the population's **gene pool.** It consists of all alleles at all gene loci in all individuals of the population. For a diploid species, each locus is represented twice in the genome of an individual, who may be either homozygous or heterozygous for those homologous loci (see Chapter 13). If all members of a population are homozygous for the same allele, that allele is said to be fixed in the gene pool. More often, there are two or more alleles for a gene, each having a relative frequency (proportion) in the gene pool. For example, in a peppered moth population living in an unpolluted region of England, the allele for light color is much more frequent in the population than the allele for dark color. But during the industrialization of England, the allele for dark color increased at the expense of the allele for light color. Evolution is occurring on the smallest scale when the relative frequencies of alleles in a population change over a succession of generations; such change in the gene pool is called **microevolution.**

(a)

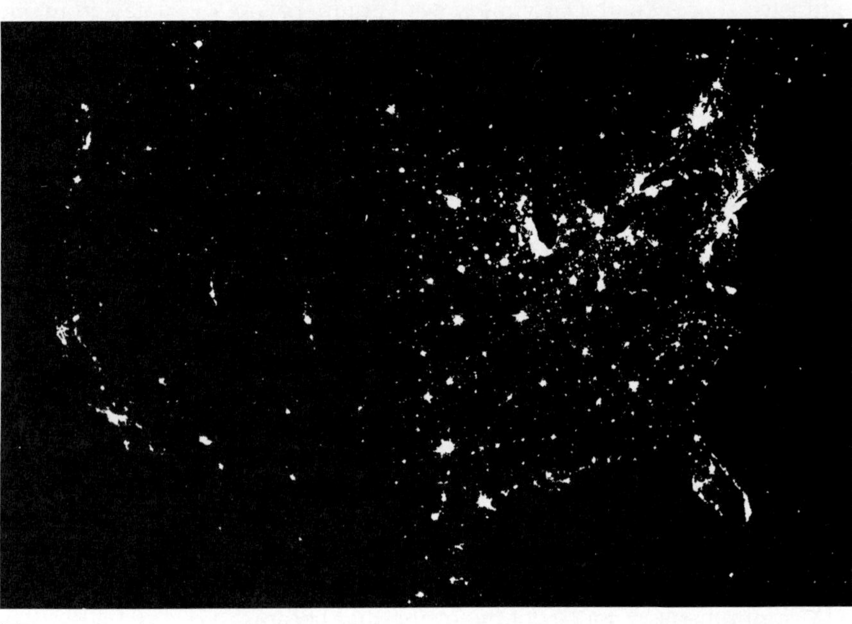

(b)

Figure 21.2
Population distribution. Populations are localized groups of individuals belonging to the same species. (**a**) Here, two dense populations of Douglas fir *(Pseudotsuga memzieii)* are separated by a river bottom where firs are uncommon. The two populations are not totally isolated; interbreeding occurs when wind blows pollen between the populations. Nevertheless, trees are more likely to interbreed with members of the same population than with trees on the other side of the river. (**b**) Humans also tend to concentrate in localized populations. In this nighttime satellite view of the United States, the lights of major population centers, or cities, are visible. These populations are, of course, not isolated; people move around, and there are low-density suburban and rural communities between cities. But city dwellers are most likely to choose mates who live in the same city, often in the same neighborhood.

The Hardy-Weinberg Theorem

Before we learn about the mechanisms of microevolution, it will be helpful to examine, for comparison, the genetics of a nonevolving population. Such a gene pool in evolutionary stasis is described by the **Hardy-Weinberg theorem,** named for the two scientists who derived the principle independently in 1908. It states that the frequencies of alleles in a population's gene pool remain constant over the generations unless acted upon by agents other than sexual recombination. Put another way, the sexual shuffling of alleles due to meiosis and random fertilization has no effect on the overall genetic makeup of a population.

The Hardy-Weinberg theorem will seem less abstract if we test it with an example (Figure 21.3). Imagine a wildflower population with two varieties contrasting in flower color. An allele for pink flowers, which we will symbolize by *A*, is completely dominant to an allele for white flowers, symbolized by *a*. For our simplified problem, these are the only two alleles for this locus in the population. Our imaginary population has 500 plants. Twenty have white flowers because they are homozygous for the recessive allele; their genotype is *aa*. Of the 480 plants with pink flowers, 320 are homozygous *(AA)* and 160 are heterozygous *(Aa)*. Since these are diploid organisms, there are a total of 1000 genes for flower color in the population. The dominant allele accounts for 800 of these genes ($320 \times 2 = 640$ for *AA* plants, plus $160 \times 1 = 160$ for *Aa* individuals). Thus, the frequency of the *A* allele in the gene pool of this population is 80%, or 0.8. And since there are only two allelic forms of the gene, we know that the *a* allele has a frequency of 20%, or 0.2.

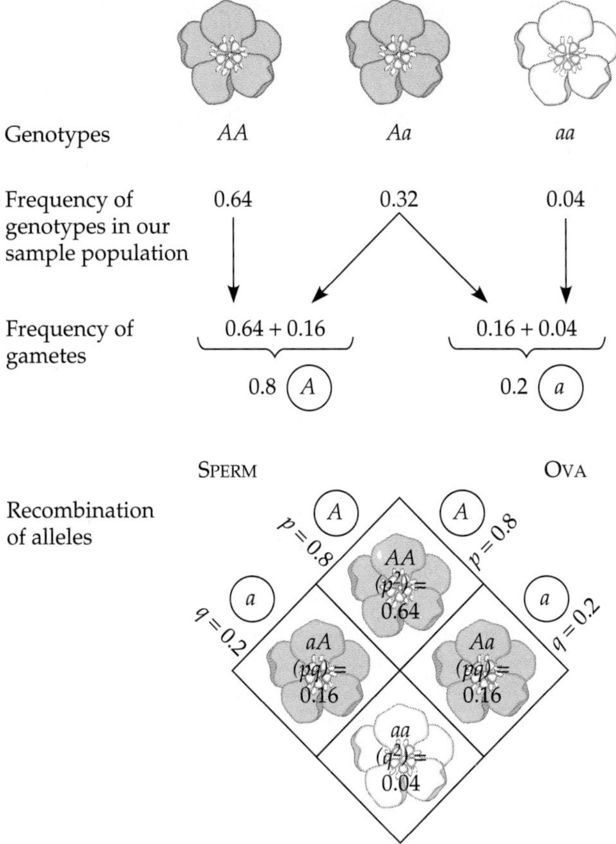

Figure 21.3
The Hardy-Weinberg theorem. The gene pool of a nonevolving population remains constant over the generations. Sexual recombination alone will not alter the relative frequencies of alleles. (p = frequency of *A*; q = frequency of *a*.)

How will genetic recombination during sexual reproduction affect the frequencies of the *A* and *a* alleles in the next generation of our wildflower population? We will assume that the union of sperm and ova in the population is completely random; that is, all male-female mating combinations are equally likely. The situation is analogous to mixing all gametes in a sack and then drawing them randomly, two at a time, to determine the genotype for each zygote. Each gamete has one gene for flower color, and the *A* and *a* alleles will occur in the same frequencies in which they had occurred in the population that made the gametes. Every time a gamete is drawn from the pool at random, the chance that the gamete will bear an *A* allele is 0.8, and the chance that an *a* allele will be present is 0.2.

Using the rule of multiplication (see Chapter 13), we can calculate the frequencies of the three possible genotypes in the next generation of the population. The probability of picking two *A* alleles from the pool of gametes is 0.64 (0.8×0.8). Thus, about 64% of the plants in the next generation will have the genotype *AA*. The frequency of *aa* individuals will be about 4%, or 0.04 (0.2×0.2). And 32%, or 0.32, of the plants will be heterozygous—that is, *Aa* or *aA*, depending on whether it is the sperm or ovum that supplies the dominant allele ($0.8 \times 0.2 = 0.16$, $\times$ 2 ways = 0.32).

Regarding the locus for flower color, the genetic makeup in the second generation of the wildflower population will be 0.64 *AA*, 0.32 *Aa*, and 0.04 *aa*. The *AA* individuals account for 64% of all genes for flower color in the population, and heterozygotes account for 32% of the genes, half of which are represented by the *A* allele. The overall frequency of the *A* allele in the population is therefore $0.64 + (0.32 \div 2)$, or 0.8. The frequency of the *a* allele is $0.04 + (0.32 \div 2)$, or 0.2. Notice that the alleles are present in the gene pool of the current population in the same frequencies as they were in the previous generation. If we were to repeat the process, segregating alleles to make gametes, then picking the gametes two at a time to produce the genotypes of still another generation of plants, the frequencies of the alleles and genotypes would remain the same, generation after generation. Thus, the gene pool of the population would be in a state of equilibrium—referred to as **Hardy-Weinberg equilibrium.** (In this example, the wildflower population was at equilibrium initially. If we had started with the population not yet at equilibrium, only a single generation would be required for equilibrium to be attained. You will have a chance to prove this in Challenge Question #2 at the end of the chapter.)

From the specific case of the wildflower population, we can derive a general formula, called the Hardy-Weinberg equation, for calculating the frequencies of alleles and genotypes in populations. We will restrict our analysis to the simplest case of only two alleles, one dominant over the other. However, the Hardy-Weinberg equation can be adapted to situations in which there are three or more alleles for a particular locus, and there is no clear-cut dominance.

For a gene locus where only two alleles occur in a population, the letter p is used to represent the frequency of one allele and the letter q to symbolize the frequency of the other allele (see Figure 21.3). In the imaginary wildflower population, $p = 0.8$ and $q = 0.2$. Note that $p + q = 1$; the combined frequencies of all possible alleles must account for 100% of the genes for that locus in the population. If there are only two alleles and we know the frequency of one, the frequency of the other can be calculated:

$$1 - p = q, \text{ or } 1 - q = p$$

When gametes combine their alleles to form zygotes, the probability of generating an *AA* genotype is p^2. In the wildflower population, $p = 0.8$, and $p^2 = 0.64$, the probability of an *A* sperm fertilizing an *A* ovum to produce an *AA* zygote. The frequency of individuals homozygous for the other allele (*aa*) is q^2, or $0.2 \times 0.2 = 0.04$ for the wildflower population. Because there are two ways in which an *Aa* genotype can arise, depending on which parent contributes the dominant allele, the frequency of heterozygous individuals in the population is $2pq$ ($2 \times 0.8 \times 0.2 = 0.32$, in our example). If we have calculated the frequencies of all possible genotypes correctly, they should add up to 1:

$$\underset{\text{Frequency of } AA}{p^2} + \underset{\text{Frequency of } Aa \text{ and } aA}{2pq} + \underset{\text{Frequency of } aa}{q^2} = 1$$

For our wildflowers, this is $0.64 + 0.32 + 0.4 = 1$.

The Hardy-Weinberg equation enables us to calculate frequencies of alleles in a gene pool if we know frequencies of genotypes, and vice versa. One use is to calculate the percentage of the human population that carries the allele for a particular inherited disease. For instance, one out of approximately 10,000 babies in the United States is born with phenylketonuria (PKU), a metabolic disorder that, untreated, results in mental retardation and other problems (see Chapter 13). The disease is caused by a recessive allele, and thus the frequency of individuals in the U.S. population born with PKU corresponds to q^2 in the Hardy-Weinberg equation. Given one PKU occurrence per 10,000 births, $q^2 = 0.0001$. Therefore, frequency of the recessive allele for phenylketonuria in the population is $q = \sqrt{0.0001}$, or 0.01. And the frequency of the dominant allele is $p = 1 - q$, or 0.99. The frequency of carriers, heterozygous people who are normal but may pass the PKU allele on to offspring, is

$$2pq = 2 \times 0.99 \times 0.01 = 0.0198$$

About 2% of the U.S. population carries the PKU allele.

How is Hardy-Weinberg equilibrium relevant to our study of microevolution? It tells us what to expect for a nonevolving population, providing a baseline for comparing actual populations where the gene pools may, in fact, be changing. Hardy-Weinberg equilibrium is maintained only if the population meets all five of the following conditions:

1. *Very large population size:* Chance can affect the gene pool of a small population.
2. *Isolation from other populations:* Migration of individuals into or out of a population can change the gene pool.
3. *No net mutations:* By changing one allele into another, mutations alter the gene pool.
4. *Random mating:* If individuals select mates having certain heritable traits, then the random mixing of gametes required for equilibrium does not occur.
5. *No natural selection:* Differential reproductive success alters frequencies of alleles in the gene pool.

It is important to realize that Hardy-Weinberg equilibrium describes the genetics of ideal populations that never exist in nature. Let's look now at how real populations evolve.

Table 21.1 Causes of Microevolution

Mechanism	Action on Gene Pool	Usually Adaptive?*
Genetic drift	Random change in small gene pool due to sampling errors in propagation of alleles	No
Gene flow	Change in gene pools due to immigration or emigration of individuals between populations	No
Mutation	Change in allele frequencies due to net mutation	No
Nonrandom mating	Inbreeding or selection of mates for specific phenotypes (assortative mating) reduces frequency of heterozygous individuals	Unknown
Natural selection	Differential reproductive success increases frequencies of some alleles and diminishes others	Yes

CAUSES OF MICROEVOLUTION

Five potential agents of microevolution are genetic drift, gene flow, mutation, nonrandom mating, and natural selection (Table 21.1). Each is a deviation from one of the five conditions for Hardy-Weinberg equilibrium. Of all the causes of microevolution, only natural selection generally leads to an accumulation of favorable adaptions in a population. The other agents of microevolution are sometimes called non-Darwinian because of their usually nonadaptive nature.*

Genetic Drift

Flip a coin a thousand times, and a result of 700 heads and 300 tails would make you very suspicious about that coin. Flip a coin ten times, and an outcome of seven heads and three tails is within reason. The smaller a sample, the greater the chance deviations from an idealized result—an equal number of heads and tails, in the case of a sample of coin tosses. This disproportion of results in a small sample is known as sampling error, and it is an important factor in the genetics of small populations of organisms. If a new gen-

*"Nonadaptive" does not mean "maladaptive." Microevolution due to non-Darwinian causes may affect populations in positive, negative, or neutral ways. In contrast, the effects of natural selection are almost always positive because selection favors the disproportionate propagation of favorable traits.

eration draws its alleles at random, then the larger the sample size, the better it will represent the gene pool of the previous generation. If a population of organisms is small, its existing gene pool may not be accurately represented in the next generation because of sampling error. Chance events can cause the frequencies of alleles in a small population to drift randomly from generation to generation. For example, consider what could happen if the wildflower population discussed earlier consisted of only 25 plants. Assume that 16 of the plants have the genotype *AA* for flower color, 8 are *Aa*, and only 1 is *aa*. Now imagine that three of the plants are accidentally destroyed by a rock slide before they have a chance to reproduce. By chance, all three plants lost from the population could be *AA* individuals. The event would alter the relative frequencies of the two alleles for flower color in subsequent generations. This is a case of microevolution caused by **genetic drift,** changes in the gene pool of a small population due to chance. Only luck could result in random drift improving adaptation.

Ideally, a population must be infinitely large for genetic drift to be ruled out completely as an agent of evolution. Although that is impossible, many populations are so large that drift may be negligible. However, some populations are small enough for significant genetic drift to occur; chance certainly plays a major role in the microevolution of populations having fewer than 100 or so individuals. The two situations that most often lead to populations small enough for genetic drift to occur are known as the bottleneck effect and the founder effect.

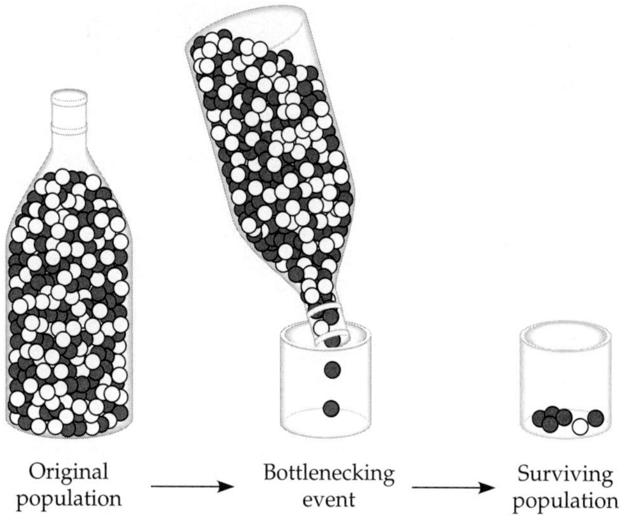

Figure 21.4
The bottleneck effect. A gene pool can drift by chance when the population is drastically reduced by a disaster that kills victims unselectively. In this analogy, a population of magenta and white marbles in a bottle is reduced in the bottleneck. Notice that the composition of the surviving population in the bottleneck is not representative of the makeup of the larger, original population. By chance, the magenta marbles are overrepresented.

The Bottleneck Effect Disasters such as earthquakes, floods, or fires may reduce the size of a population drastically, killing victims rather unselectively. The result is that the small surviving population is unlikely to be representative of the original population in its genetic makeup—a situation known as the **bottleneck effect.** By chance, certain alleles will be overrepresented among survivors, other alleles will be underrepresented, and some alleles may be eliminated completely (Figure 21.4). The genetic drift that has occurred may continue to affect the population for many generations, until the population is again large enough for random drift to be insignificant.

In some cases, bottlenecking reduces the overall genetic variability in a population, since some alleles are lost from the gene pool. One extreme example concerns the population of northern elephant seals, which passed through a bottleneck in the 1890s when hunters reduced the population to about 20 individuals. Since then, the animal has become a protected species, and the population has grown to over 30,000 members. Researchers have examined 24 gene loci in many individuals of the northern elephant seal population, and no genetic variation has been found; a single allele has been fixed at each of the 24 loci, probably due in large part to genetic drift. By comparison, genetic variation abounds in populations of the southern elephant seal, which have not been bottlenecked. Bottlenecking may also explain why the South African population of cheetahs displays greater genetic uniformity than do inbred strains of laboratory mice. The cheetah population was probably severely reduced during the last ice age about 10,000 years ago, and a second time when the animals were hunted to near extinction at the beginning of this century. In contrast to these examples, the extensive diversity of an endangered rhinoceros species indicates that low variation is not an inevitable consequence of bottlenecking (Figure 21.5).

The Founder Effect Genetic drift is also likely whenever a few individuals colonize an isolated island, lake, or some other habitat new to that species. The smaller the sample size, the less the genetic makeup of the colonists will represent the gene pool of the larger population they left. The most extreme case would be the founding of a new population by one pregnant animal or a single plant seed. If the colony is successful, random drift will continue to affect the frequency of alleles in the gene pool until the population is large enough for sampling errors from generation to generation to be minimal. Genetic drift in a new colony is known as the **founder effect.** The effect undoubtedly contributed to the evolutionary divergence of Darwin's finches after strays from the South American mainland reached the remote Galapagos Islands.

The founder effect is probably responsible for the relatively high frequency of certain inherited disorders

Figure 21.5
Genetic diversity in the bottlenecked Indian rhino population. A recent study of the Indian rhino suggests that bottlenecking does not always drastically reduce genetic diversity in a population. By 1962, poaching and farming had decreased the Indian rhino population to less than 40 individuals. Since then, the rhinos have been protected in a national park and their population has rebounded to about 400. When G. McCracken and E. Dinerstein, of the University of Tennessee, compared genetic variation of the Indian rhino population to average variation in other mammals, they discovered that the rhinos were unusually diverse. The researchers speculate that the prebottlenecked population had a wealth of genetic variation, since it was widely distributed and had a relatively low density of individuals. This situation resulted in rhinos mating with partners that were not closely related, and the resulting genetic diversity has been retained in the bottlenecked population. McCracken and Dinerstein do not disclaim the concept that bottlenecking reduces genetic diversity, but they conclude we cannot assume that all species endangered by bottlenecking are genetically impoverished.

among human populations established by a small number of colonists. In 1814, 15 people founded a British colony on Tristan da Cunha, a group of small islands in the Atlantic Ocean midway between Africa and South America. Apparently, one of the colonists carried a recessive allele for retinitis pigmentosa, a progressive form of blindness that afflicts individuals who are homozygous for the allele. Of the 240 descendants who still lived on the island in the late 1960s, 4 had retinitis pigmentosa and at least 9 others were known, based on pedigree analysis, to be carriers. The frequency of this allele is much higher on Tristan da Cunha than in the populations from which the founders came. Although inherited diseases provide striking examples of the founder effect, this form of genetic drift also alters the frequencies of many alleles in the gene pool that affect more subtle characteristics.

Gene Flow

Hardy-Weinberg equilibrium requires the gene pool to be a closed system, but most populations are not completely isolated. A population may gain or lose alleles by **gene flow,** the migration of fertile individuals, or the transfer of gametes, between populations.

For example, if a wind storm blows pollen to our hypothetical wildflower population from another population of the same species, allele frequencies may change; perhaps, for example, the outlying population consists entirely of white-flowered individuals. Gene flow tends to reduce differences between populations that have accumulated because of natural selection or genetic drift. If it is extensive enough, gene flow can eventually amalgamate neighboring populations into a single population. As humans began to move about the world more freely, gene flow undoubtedly became an important agent of microevolutionary change in populations that were previously quite isolated.

Mutation

A new mutation that is transmitted in gametes immediately changes the gene pool of a population by substituting one allele for another. For example, a mutation that causes a white-flowered plant in our hypothetical wildflower population to produce gametes bearing the dominant allele for red flowers would decrease the frequency of the a allele in the population and increase the frequency of the A allele. However, mutation by itself does not have much quantitative effect on a large population in a single generation. This is because a mutation at any given gene locus is a very rare event; although mutation rates vary, depending on the species and the gene locus, rates of one mutation per locus per 10^5 to 10^6 gametes are typical. If an allele has a frequency of 0.50 in the gene pool and mutates to another allele at a rate of 10^{-5} mutations per generation, it would take 2000 generations to reduce the frequency of the original allele from 0.50 to 0.49. The gene pool would be affected even less if the mutation were reversible, as most are. If some new allele produced by mutation increases its frequency in a population, it is not because mutation is generating the allele in abundance, but because individuals carrying the mutant allele are producing a disproportionate number of offspring as a result of natural selection or genetic drift. Over the long run, however, mutation is, in itself, very important to evolution because it is the original source of the genetic variation that serves as raw material for natural selection.

Nonrandom Mating

For Hardy-Weinberg equilibrium to hold, an individual of any genotype must choose its mates at random from the population. But in actuality, individuals usually mate more often with close neighbors than with more distant members of the population, especially in species that do not disperse far. Other individuals in the same "neighborhood" within a larger population tend to be closely related. This promotes **inbreeding,** mating between closely related partners. The most extreme case of inbreeding is self-fertilization ("selfing"), particularly common in plants.

Inbreeding causes the relative frequencies of genotypes to deviate from that which is expected from Hardy-Weinberg equilibrium. For example, in our imaginary wildflower population, self-pollination would tend to increase the frequencies of homozygous genotypes at the expense of heterozygotes. If *AA* individuals and *aa* individuals "self," then their offspring must also be homozygous. If *Aa* plants "self," however, only half their offspring will be heterozygous. With each generation, the proportion of heterozygotes decreases and the proportions of dominant and recessive homozygotes increase. Even in less extreme cases of inbreeding without selfing, the decline of heterozygosity occurs, though more slowly. One visible effect of this change in genotypic frequencies is a greater proportion of individuals expressing recessive phenotypes; the frequency of white-flowered individuals in our wildflower population would be greater than the Hardy-Weinberg equation predicts. Regardless of the impact of inbreeding on the ratio of genotypes and phenotypes in the population, however, the values of p and q, the frequencies of the two alleles, remain the same. It is just that a smaller proportion of recessive alleles are "masked" in heterozygous individuals.

Another type of nonrandom mating is **assortative mating,** in which individuals select partners that are like themselves in certain phenotypic characters. For example, blister beetles *(Lytta magister)* that live in the Sonoran Desert of Arizona most commonly mate with individuals of the same size. To some extent, humans also use size as a criterion for assortative mating; for example, tall women commonly (but not always) pair with tall men.

Notice again that nonrandom mating—inbreeding or assortative mating—increases the number of gene loci in the population that are homozygous, but nonrandom mating does not in itself alter the overall frequencies of alleles in a population's gene pool.

Natural Selection

Hardy-Weinberg equilibrium requires that all individuals in a population be equal in their ability to produce viable, fertile offspring. This condition is probably never completely met. Populations of sexually reproducing organisms consist of varied individuals, and on average, some variants leave more offspring than others. This differential success in reproduction is, of course, natural selection. Selection results in alleles being passed along to the next generation in numbers disproportionate to their relative frequencies in the present generation. For example, in our imaginary wildflower population, plants with red flowers (*AA* or *Aa* genotypes) may for some reason produce more offspring on average than plants having white flowers *(aa)*; perhaps white flowers are more visible to herbivorous insects that eat the flowers. This would disturb Hardy-Weinberg equilibrium; the frequency of the *A* allele would increase and the frequency of the *a* allele would decline in the gene pool.

Of all agents of microevolution that change the gene pool, only selection is likely to be adaptive. Natural selection accumulates and maintains favorable genotypes in a population. If the environment should change, selection responds by favoring genotypes adapted to the new conditions. But the degree of adaptation can be extended only within the realm of the variability present in the population. Before we examine the process of adaptation by natural selection more closely, let's look at the genetic basis of the variation that makes it possible for populations to evolve.

THE GENETIC BASIS OF VARIATION

Heritable variation is at the heart of Darwin's theory of evolution, for variation provides the raw material for natural selection. And an emphasis on variation—what Ernst Mayr calls "population thinking" in the interview preceding this unit—was a major element in the modern synthesis. In what ways do members of a population vary? How extensive is variation? What mechanisms generate and maintain variations in a population? Do all variations function as raw material for selection? These are the questions we will try to answer as we look at the genetic variations so crucial to the process of natural selection.

The Nature and Extent of Genetic Variation Within and Between Populations

You have no trouble recognizing your friends in a crowd. Each person has a unique genome, and this is reflected in individual variations of appearance and temperament. Individual variation occurs in populations of all species of sexually reproducing organisms. We are very conscious of human diversity; we are less sensitive to individuality in populations of other animals and plants, and the diversity may escape our notice because the variations are subtle. But these slight differences between individuals in a population are the variations Darwin wrote most about as the raw material for natural selection.

Not all the variation we observe in a population is heritable. Phenotype is the cumulative product of an inherited genotype and a multitude of environmental influences. For example, a campus population would look quite a bit different after spring break if many people got suntans during their vacation. It is important to remember that only the genetic component of variation can have adaptive impact as a result of natural selection, because it is the only component that transcends generations.

Both discrete and quantitative characters contribute to variation within a population. Most heritable variation consists of polygenic characters that vary quantitatively within a population. For example, plant height may vary continuously in our hypothetical wildflower population, from very short individuals to very tall individuals and everything in between. Discrete characters, such as red versus white flowers, vary categorically, probably because they are determined by a single gene locus with different alleles that produce distinct phenotypes. In such cases, when two or more forms of a Mendelian character are represented in a population, the contrasting forms are called **morphs**—as in the red-flowered and white-flowered morphs of our wildflower population, for example. A population is said to be **polymorphic** for a character if two or more morphs are each represented in high enough frequencies to be readily noticeable. (Obviously, this definition is arbitrary, but a population is not termed polymorphic if it consists almost exclusively of a single morph, with other morphs extremely rare.) Figure 21.6 illustrates a striking example of **polymorphism** (the existence of polymorphic characters) in a population of Oregon garter snakes. Polymorphism is extensive in human populations, both in physical characters, such as the presence or absence of freckles, and in biochemical characters, such as ABO blood groups (for which there are four morphs: Type A, Type B, Type AB, and Type O; see Chapter 13).

Figure 21.6
Polymorphism. Some populations consist of two or more discrete varieties of individuals. These four garter snakes *(Thamnophis ordinoides)*, which differ markedly in their patterns of coloration, were captured in the same Oregon field. Edmund Brodie, of the University of Chicago, has discovered that the behavior of each morph is keyed to its coloration. Spotted snakes generally blend into their background better than striped snakes, but stripes make it harder to judge the speed of the snake in motion. When approached, spotted garter snakes usually freeze, while snakes of the striped morph flee.

The reservoir of genetic variation in a population is much more extensive than Darwin realized. Although much of the variation is invisible, it is manifest in molecular differences that can be detected by biochemical methods. Several laboratories have used electrophoresis, a technique that can separate proteins differing in electric charge (see the Methods Box in Chapter 19, p. 396), to study variations in the protein products of specific gene loci among individuals in a population. Scores of loci have been studied in many different animal species. For example, in populations of the fruit fly *Drosophila,* the gene pool typically has two or more alleles for about 30% of the loci examined, and each fly is heterozygous at about 12% of its loci; that amounts to 700–1200 heterozygous loci per fly. Expressed another way, any two flies in a *Drosophila* population differ in genotype at about 25% of their loci. The extent of genetic variation in human populations, as revealed by electrophoresis, is comparable. And electrophoresis underestimates genetic variation, because proteins produced by different alleles may vary in amino acid composition even though they have the same overall electric charge. Furthermore, variation in DNA that is not expressed as protein is not detected by this method.

Most species exhibit **geographical variation,** differences between populations in their frequencies of alleles. Because at least some environmental factors are likely to be different from one place to another, natural selection can contribute to geographical variation. For example, one population of our now-familiar wildflower species may have a higher frequency of recessive alleles at the flower-color locus than other populations, perhaps because of a local prevalence of pollinators that key on white flowers (recessive homozygotes). Genetic drift can also cause chance variations among different populations. On a more local scale, geographical variation can also occur within a population, either because the environment has patchlike variation or because the population is differentiated into subpopulations due to localized inbreeding.

One particular type of geographical variation, called a **cline,** is a graded change in some trait along a geographic axis. In some cases, a cline may represent a graded region of overlap where individuals of neighboring populations are interbreeding. In other cases, a gradation in some environmental variable may produce a cline. For example, the average body size of many North American species of mammals increases

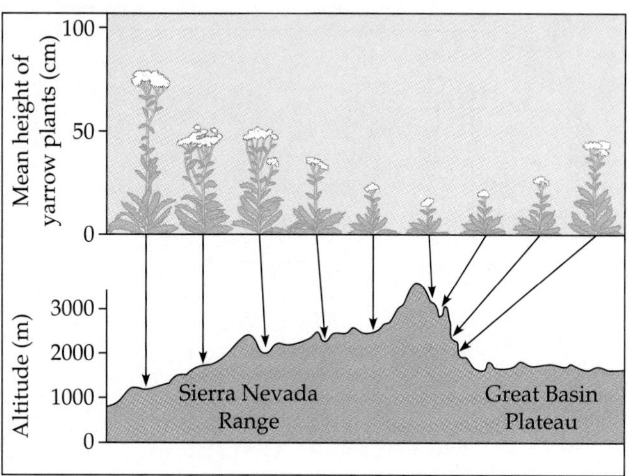

Figure 21.7
A cline. Yarrow plants on the slopes of California's Sierra Nevada Mountains gradually decrease in average size at higher and higher elevations. Although the environment affects growth rates directly to some extent, some of the variation has a genetic basis. Researchers collected seeds at different elevations and grew plants in a uniform garden; the average sizes of the plants were correlated with the altitude at which the seeds were collected.

gradually with increasing latitude. Presumably, the reduced ratio of surface area to volume that accompanies larger size is an adaptation that helps animals living in cold environments conserve body heat. Experimental studies of some clines, such as geographical variation in the height of yarrow plants that grow on the slopes of mountains, confirm the role of genetic variation in the spatial differences of phenotype (Figure 21.7).

Sources of Genetic Variation

Mutation and sexual recombination (see Chapter 14) are the two processes that generate genetic variation.

Mutation New alleles originate by mutation (see Chapter 16). A mutation affecting any gene locus is an accident that is rare and random. Most mutations occur in somatic cells and die with the individual. Geneticists estimate that in humans an average of only one or two mutations occur in each cell line that produces a gamete, and these are the only mutations that can be passed along to children. A mutation is a shot in the dark. Chance determines where it will strike and how it will alter a gene.

Most point mutations, those affecting a single base in DNA, are probably relatively harmless. Much of the DNA in the eukaryotic genome does not code for protein products, and it is uncertain how a change of a single nucleotide base in this silent DNA will affect the well-being of the organism (see Chapters 16 and 18). Even mutations of structural genes, which do code for proteins, may occur with little or no effect on the organism, partly because of redundancy in the genetic code (see Chapter 16). Of course, a single point mutation can have a significant impact on phenotype, as in sickle-cell anemia, for example.

A mutation that alters a protein enough to affect its function is more often harmful than beneficial. Organisms are the refined products of thousands of generations of past selection, and a random change is not likely to improve the genome any more than firing a gunshot blindly through the hood of a car is likely to improve engine performance. On rare occasions, however, a mutant allele may actually fit its bearer to the environment better and enhance the reproductive success of the individual. This is not especially likely in a stable environment, but becomes more probable when the environment is changing and mutations that were once selected against are now favorable under the new conditions. As a result of random mutations, dark peppered moths occurred in populations dominated by the light moths before pollution darkened the English landscape. The mutation became an advantage instead of a liability when the environment changed. Similarly, some mutations that happen to endow house flies with resistance to DDT also reduce growth rate and were deleterious before the pesticide was introduced. A new environmental factor, DDT, tilted the balance in favor of the mutant alleles, and they spread through fly populations by natural selection.

Because chromosomal mutations usually affect many gene loci, they are almost certain to disrupt the development of the organism. But even rearrangements of chromosomes may in rare instances bring benefits. For example, the translocation of a chromosomal piece from one chromosome to another could link alleles that affect the organism in some positive way when they are inherited together as a package.

Duplications of chromosome segments, like other chromosomal mutations, are nearly always harmful. But if the repeated segment does not disrupt genetic balance severely, it can persist over the generations and provide an expanded genome with superfluous loci that may eventually take on new functions by mutation, while the original genes continue to function at their old locations in the genome. New genes may also arise from existing DNA sequences by the shuffling of exons within the genome, either within a single locus or between loci (see Chapter 18).

For bacteria and other microorganisms that have very short generation spans, mutation can be an adequate source of genetic variation. Bacteria reproduce asexually by dividing as often as once every 20 minutes; a single cell can potentially give rise to a billion descendants in just 10 hours. A new mutation that happens to be beneficial can increase its frequency in a bacterial population very rapidly. Imagine, for example, exposing a bacterial population to an antibiotic. If a single individual in the population happens to harbor a mutation that renders it resistant to the poison, in just a few hours there may be millions of resistant bacteria, while bacteria sensitive to the antibiotic may have been almost completely eliminated. Bacterial populations can evolve, one mutation at a time, by the explosive asexual expansion of clones favored by the local environment. But even most bacteria increase genetic variation by occasionally exchanging and recombining genes through processes that resemble sex (see Chapter 17). Animals and plants depend almost entirely on sexual recombination for the genetic variation that makes adaptation possible.

Recombination Although mutations are the source of new genes, they are so infrequent at any one locus that generation to generation, their contribution to genetic variation in a large population is negligible. Members of a population owe nearly all their differences to the unique recombinations of existing alleles each individual draws from the gene pool.

Sex shuffles genes and deals them at random to determine individual genotypes. During meiosis, homologous chromosomes, one inherited from each parent, trade some of their genes by crossing over, and then the homologous chromosomes and the alleles they carry segregate randomly into separate gametes (see Chapter 12). Gametes from one individual vary extensively in their genetic makeup, and each zygote made by a mating pair has a unique assortment of genes resulting from the random union of a sperm and an ovum. A population, of course, contains a vast number of possible mating combinations, each bringing together the gametes of individuals that are likely to have different genetic backgrounds. Sexual reproduction recombines old alleles into fresh assortments every generation.

How Genetic Variation Is Preserved

What prevents natural selection from extinguishing a population's variation by culling unfavorable genotypes? The tendency for natural selection to reduce variation is countered by several mechanisms that preserve or restore variation.

Diploidy The diploid character of most eukaryotes hides a considerable amount of genetic variation from selection in the form of recessive alleles in heterozygotes. Recessive alleles that are less favorable than their dominant counterparts, or even harmful in the present environment, can persist in a population through their propagation by heterozygous individuals. This latent variation is exposed to selection only when both parents carry the same recessive allele and combine two copies in one zygote. This happens only rarely if the frequency of the recessive allele is very low. For example, if the frequency of the recessive allele is 0.01 and the frequency of the dominant allele is 0.99, then 99% of the copies of that recessive allele are protected from selection in heterozygotes, and only 1% of the alleles are present in homozygotes. The rarer the recessive allele, the greater the degree of protection afforded by heterozygosity (Figure 21.8). Heterozygote protection maintains a huge pool of alleles that may not be suitable for present conditions but that could bring new benefits when the environment changes.

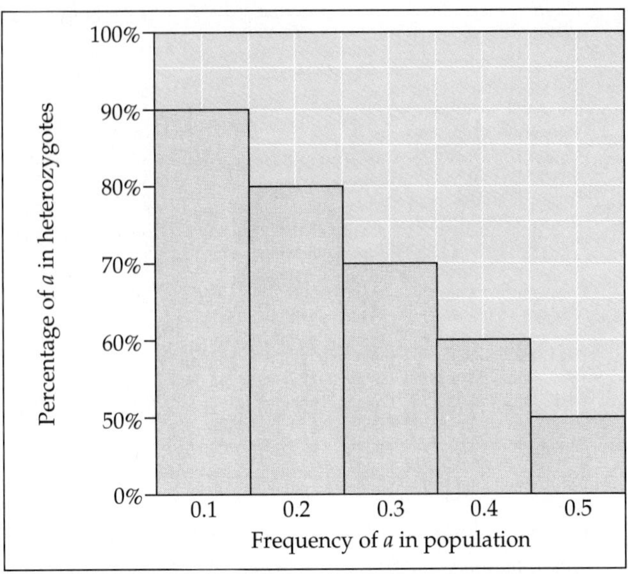

Figure 21.8
Diploidy: preserving genetic variation. Recessive alleles that are selected against in homozygotes persist in a population's heterozygotes. As recessive alleles become rarer, a greater proportion of them are hidden in heterozygotes. The height of each bar in this histogram indicates the degree of heterozygote protection for different frequencies of the recessive allele.

Balanced Polymorphism Selection itself may preserve variation at some gene loci. This ability of natural selection to maintain diversity in a population is called **balanced polymorphism.** One of the mechanisms for this preservation of variation is **heterozygote advantage.** If individuals who are heterozygous for a particular locus have greater reproductive success than any type of homozygote, then two or more alleles will be maintained at that locus by natural selection. An interesting case of heterozygote advantage involves the locus in humans for one chain of hemoglobin, the protein of red blood cells that transports oxygen. A specific recessive allele at that locus causes sickle-cell anemia in homozygous individuals. Heterozygotes, however, are resistant to malaria, an important advantage in tropical regions where that disease is a major cause of death. The environment in these regions favors the heterozygotes over both homozygous dominant individuals, who are susceptible to malaria, and homozygous re-

Figure 21.9
A balanced polymorphism in a finch population. Two different bill sizes occur in a single population of black-bellied seedcrackers, a species of finch that lives in Cameroon, West Africa. There are no individuals with bills of intermediate size. The small-billed individuals (left) feed mainly on soft seeds, while the large-billed birds specialize in cracking hard seeds. Each morph shows a higher feeding efficiency on each respective seed. Thomas Smith, of San Francisco State University, has been studying this population in the field. His hypothesis is that natural selection maintains the polymorphism by selecting against intermediate bill sizes, which crack both classes of seeds relatively inefficiently. (This is an example of diversifying selection; see p. 452.)

Figure 21.10
Frequency-dependent selection. The females of *Papilio dardanus*, an African swallowtail butterfly, occur in several morphs (left column), each a mimic of a different butterfly species that is noxious to predators (right). This mimicry would be less advantageous if any one *Papilio* morph became so common that predators encountered them as often as the distasteful model.

cessive individuals, who are disabled by sickle-cell anemia. The frequency of the sickle-cell allele in Africa is generally highest in areas where the malaria parasite is most common. In some tribes, the recessive allele accounts for 20% of the hemoglobin loci in the gene pool, a very high frequency for a gene that is disastrous in homozygotes. But at this frequency ($q = 0.2$), 32% of the population consists of heterozygotes resistant to malaria ($2pq$), and only 4% of the population suffers from sickle-cell anemia (q^2).

Another example of heterozygote advantage is found in the crossbreeding of crop plants. When corn, for instance, is highly inbred, the number of homozygous gene loci increases, and the corn may gradually become stunted in growth and increasingly sensitive to a variety of diseases. Crossbreeding between two different inbred varieties often produces hybrids that are much more vigorous than either parent stock. This **hybrid vigor** is probably due to two factors: the segregation of deleterious recessives that were homozygous in the inbred varieties, and heterozygote advantage at many loci in the hybrids.

A patchy environment, where natural selection favors different phenotypes in different subregions within a population's geographic boundaries, can also result in balanced polymorphism. For example, in many populations of British land snails *(Cepaea nemoralis)*, there are several morphs, each with a coloration that camouflages its shell in a particular patch of the area inhabited by the population. Protective coloration suited to different backgrounds may also help explain the morphs of garter snakes in Figure 21.6. Figure 21.9 illustrates another example of a polymorphism that may be due to environmental patchiness—in this case, specialization for slightly different foods.

Still another cause of balanced polymorphism is **frequency-dependent selection,** in which the reproductive success of any one morph declines if that phenotypic form becomes too common in the population. A particularly intricate example is a balanced polymorphism that has been observed in populations of *Papilio dardanus*, an African swallowtail butterfly. The males all have similar coloration, but the females occur in several different morphs, each resembling another butterfly species that is noxious to predators (Figure 21.10). *Papilio* females are not noxious, but birds learn to avoid them because they look so much like the distasteful butterflies. This mimicry would be less effective if all *Papilio* females copied the same noxious species, because birds would be slow to associate a particular pattern of coloration with bad taste if they encountered good-tasting mimics as often as the noxious models.

Is All Genetic Variation Adaptive?

Some of the genetic variations observed in populations are probably trivial in their impact on reproductive success. The diversity of human fingerprints is an example of what is called **neutral variation,** which seems to confer no selective advantage for some individuals over others. Much of the protein variation detectable by electrophoresis may represent chemical "fingerprints" that are neutral in their adaptive qualities. For instance, 99 known mutations affect 71 of the 146 amino acids in the β chain of human hemoglobin, one of two kinds of polypeptide chains that make up that protein. Some of those mutations, including the allele for sickle-cell anemia, certainly affect the reproductive potential of the individual. However, according to a **neutral theory** of molecular evolution, many of the variant alleles at this locus and others may convey no selective advantage or disadvantage. The relative frequencies of neutral variations will not be affected by natural selection; some neutral alleles will increase in the gene pool, and others will decrease by the chance effects of genetic drift.

There is no consensus among evolutionary biologists on how much genetic variation is neutral, or even if any variation can be considered truly neutral. Variations appearing to be neutral may, in fact, influence reproductive success in ways that are difficult to measure. It is possible to show that a particular allele is detrimental, but it is impossible to demonstrate that an allele brings no benefits at all to an organism. Furthermore, a variation may be neutral in one environment but not in another. We can never know the degree to which genetic variation is neutral. But we can be certain that even if only a fraction of the extensive variation in a gene pool significantly affects the organisms, that is still an enormous resource of raw material for natural selection and the adaptive evolution it causes.

ADAPTIVE EVOLUTION

Adaptive evolution is a blend of chance and sorting—chance in the origin of new genetic variations by mutation and sexual recombination, and sorting in the workings of selection as it favors the propagation of some chance variations over others. From the pool of variations available to it, natural selection orders combinations of genes and fits organisms to their environments.

Fitness

The phrases *struggle for existence* and *survival of the fittest* are loaded with misleading connotations. There are, of course, species in which individuals, usually the males, lock horns or otherwise do combat that determines mating privilege. But direct and violent confrontations between members of a population are rare; success is generally subtle and passive. A barnacle may produce more eggs than its neighbors because it is more efficient at collecting food from the water. In a population of moths, certain variants may average more progeny than others because their coloration hides them from predators better. Plants in a wildflower population may differ in reproductive success because some are better able to attract pollinators, owing to slight variations in flower color, shape, or fragrance. *Darwinian fitness is measured only by the relative contribution an individual makes to the gene pool of the next generation.*

Survival alone does not guarantee reproductive success. Darwinian fitness is zero for a sterile plant or animal, even if it is robust and outlives other members of the population. But, of course, survival is a prerequisite for reproducing, and longevity increases fitness if it results in certain individuals leaving disproportionately high numbers of descendants. Then again, an individual that matures quickly and becomes fertile at an early age may have a greater reproductive potential than individuals that live longer but mature late. Thus, the components of selection are the many factors that affect both survival and fertility.

In a more quantitative approach to natural selection, population geneticists define **relative fitness** as the contribution of a genotype to the next generation compared to the contributions of alternative genotypes for the same locus. For example, consider our wildflower population, in which *AA* and *Aa* plants have red flowers and *aa* plants have white flowers. Let's assume that, on average, individuals with red flowers produce more offspring than those with white flowers. The relative fitness of the most fertile variants is set at 1 as a basis for comparison; so in this case, the relative fitness of an *AA* or *Aa* plant is 1. If plants with white flowers average only 80% as many offspring, their relative fitness is 0.8. The difference between the relative fitness of the most fit genotype and a less fit genotype is called the **selection coefficient.** In our imaginary wildflower population, the selection coefficient is 0.2 (1 – 0.8). We can think of the selection coefficient as a relative measure of selection *against* the inferior genotype; the more disadvantageous the genotype, the greater the selection coefficient, ranging to a coefficient of 1 for a lethal genotype. Selection coefficients are only statistical estimates, something like a handicapper predicting the order of finish for a horse race that has not yet begun.

The rate at which the frequency of a deleterious allele declines in a population depends on the magnitude of the selection coefficient working against it and on whether the allele is dominant or recessive to the more successful allele. Harmful recessives are rarely eliminated completely because of heterozygote protec-

Table 21.2 Decrease in the Frequency of a Harmful Recessive Allele Due to Natural Selection

Change in Frequency	Selection Coefficient	
	1 (Lethal)	0.10
From q =	Number of Generations	
0.10 to 0.01	90	924
0.01 to 0.001	900	9023

tion. Table 21.2 reinforces these points. Notice that to reduce the frequency of a harmful recessive allele by the same amount, natural selection requires ten times as many generations if the selection coefficient is only 0.1 instead of 1.0. If the coefficient is 0.1, individuals who are homozygous recessive *(aa)* leave 90% as many offspring each generation as *AA* or *Aa* individuals. But even for a lethal recessive (selection coefficient = 1.0), the rate at which the frequency of the allele declines lessens as it becomes more and more rare. This is because the percentage of remaining recessive alleles "hiding" from natural selection in heterozygous individuals increases with the rarity of the allele (see Figure 21.8).

Selection acts faster against dominant alleles that are harmful because they are expressed even in heterozygotes. The rate of increase of a beneficial allele is also affected by whether it is a dominant or recessive. A new recessive mutation spreads very slowly in a population, even if it is very beneficial, because selection cannot act in its favor until the mutation is common enough for two copies to occasionally come together in the same zygote. A new dominant mutation that confers greater advantage than existing alleles can increase in frequency more rapidly, because each individual that inherits even a single copy benefits from the allele. The mutant allele responsible for the rapid replacement of light peppered moths by the dark morph in England was dominant. However, most new mutations, whether dominant or recessive, and whether beneficial or harmful, probably disappear from the gene pool early in their history by genetic drift.

What Does Selection Act On?

An organism exposes its phenotype—its physical traits, metabolism, physiology, and behavior—not its genotype, to the environment. Acting on phenotypes, selection indirectly adapts a population to its environment by increasing or maintaining favorable genotypes in the gene pool.

Throughout this chapter, we have been looking at a wildflower population with alleles for flower color that affect phenotype unambiguously, but the connection between genotype and phenotype is rarely so simple or definite. A genotype may have multiple effects, especially if it influences the development or growth of the organism. This ability of genes to influence many phenotypic characters is called pleiotropy (see Chapter 13). The overall fitness of a genotype depends on whether its positive effects outweigh any harmful effects it may have on the reproductive success of the organism. Another complication in the translation of genotype into phenotype arises when many gene loci influence the same characteristic. These are polygenic, or quantitative, characters (see Chapter 13); human height is an example. Individuals in a population do not usually fit into exclusive categories for a polygenic character, but instead vary continuously over a phenotypic range. This chapter has been simplified by using discrete characters for most of the examples of microevolution, but quantitative characters actually account for most of the variation exposed to natural selection.

The finished organism subjected to natural selection is an integrated composite of its many phenotypic features, not a collage of individual parts. The fitness of a genotype at any one locus depends on the entire genetic context in which it works. For example, alleles that enhance the growth rate of the trunk and limbs of a tree may be useless or even detrimental in the absence of alleles at other loci that enhance the growth rate of roots required to support the tree. On the other hand, alleles that contribute nothing to an organism's success, or may even be slightly maladaptive, may be perpetuated because they are present in individuals whose overall fitness is high. A whole baseball team wins the league pennant, even the player with the worst batting average and most errors.

Modes of Natural Selection

Natural selection can affect the frequency of a heritable trait in a population in three different ways, depending on which phenotypes in a varying population are favored. These three modes of selection are called stabilizing selection, directional selection, and diversifying selection. They can be depicted with graphs that show how the frequencies of different phenotypes change with time (Figure 21.11). This is most meaningful for quantitative traits that depend on many gene loci.

Stabilizing selection acts against extreme phenotypes and favors the more intermediate variants. This occurs when individuals who are "average" for a particular quantitative character are the best adapted individuals. Stabilizing selection reduces phenotypic variation. For example, stabilizing selection keeps the majority of human birth weights in the 3–4 kg range. For babies much smaller or larger than this, infant mortality is greater.

Directional selection is most common during periods of environmental change or when members of a

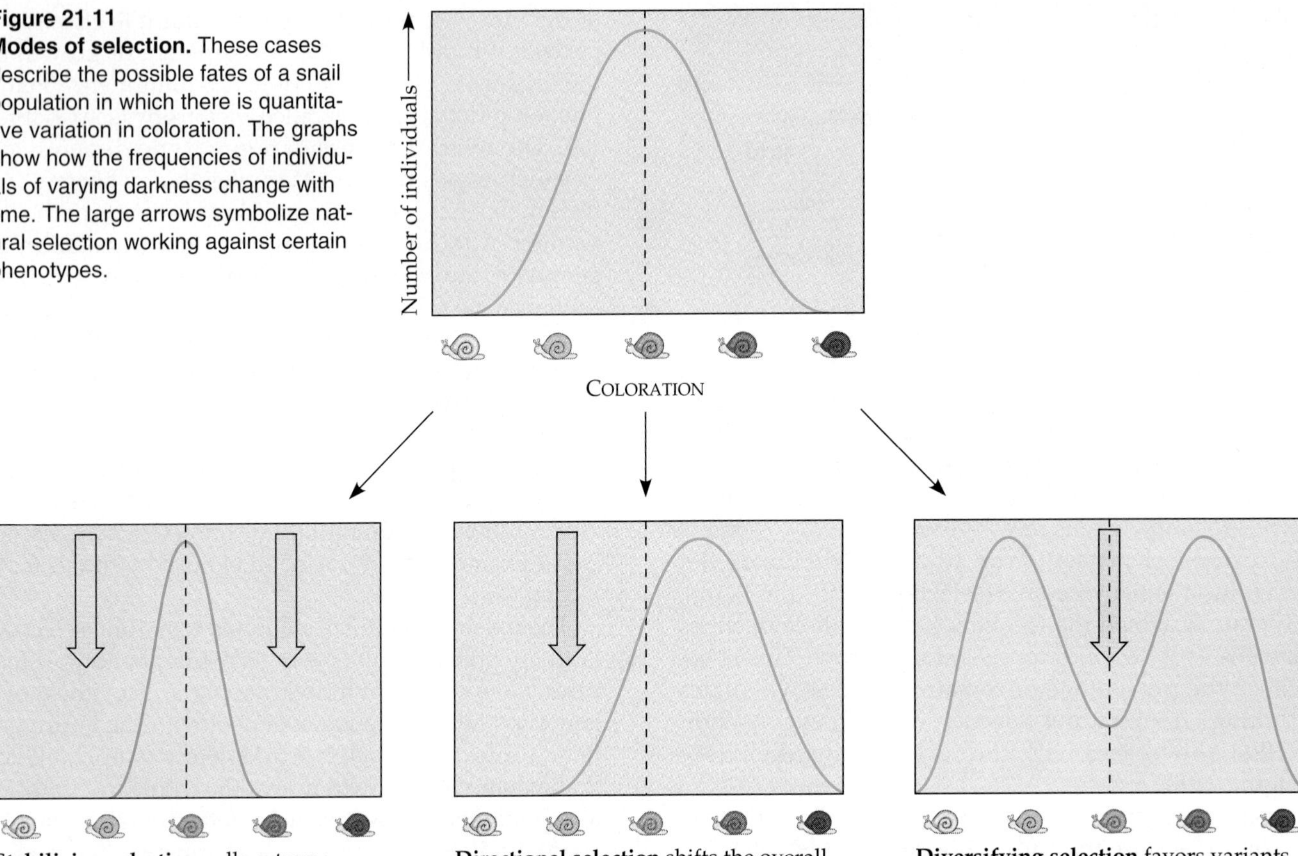

Figure 21.11
Modes of selection. These cases describe the possible fates of a snail population in which there is quantitative variation in coloration. The graphs show how the frequencies of individuals of varying darkness change with time. The large arrows symbolize natural selection working against certain phenotypes.

Stabilizing selection culls extreme variants from the population, in this case eliminating individuals that are unusually light or dark. The trend is toward reduced phenotypic variation and maintenance of the status quo.

Directional selection shifts the overall makeup of the population by favoring variants of one extreme. In this case, the trend is toward darker color, perhaps because the landscape has been blackened by lava.

Diversifying selection favors variants of opposite extremes over intermediate individuals, leading to balanced polymorphism. Here, very light and very dark snails have increased their frequencies relative to intermediate variants. Perhaps the snails have recently colonized a patchy habitat where a background of white sand is studded with lava rocks.

population migrate to some new habitat with different environmental conditions. Directional selection shifts the frequency curve for variations in some phenotypic trait in one direction or the other by favoring relatively rare individuals that deviate from the average for that trait. For instance, there is fossil evidence that the average size of black bears in Europe increased with each glacial period of the ice ages, only to decrease again during the warmer interglacial periods.

Diversifying selection (also called disruptive selection) occurs when environmental conditions are varied in a way that favors individuals on both extremes of a phenotypic range over intermediate phenotypes. We can see an example in populations of *Papilio,* the African butterfly discussed earlier. Butterflies with characteristics intermediate between two different noxious models would resemble neither model enough to gain advantage from mimicry. Thus, diversifying selection can result in balanced polymorphism, such as the multiple mimics that have evolved in *Papilio.*

Although we refer to these three selection trends as "modes of selection," the basic mechanism of natural selection is the same for each case. Selection favors certain heritable traits via differential reproductive success. What determines the mode is which individuals are favored.

Sexual Selection

Males and females of many animal species exhibit marked differences in addition to the differences in the reproductive organs that define the sexes. This distinction between the secondary sex characteristics of males and females is known as **sexual dimorphism.** It is often expressed as a difference in size, with the male usually larger, but it also involves such features as colorful plumage in male birds, manes on male lions, antlers on male deer, and other adornments. In most cases of sexual dimorphism, at least for vertebrates, the male is the showier sex. In some cases, the males with the most impressive masculine features may be the most attractive to females. There are also species in which the secondary sexual structures may be used in direct competition with other males; this is particularly common in species where a single male garners a harem of females. These males may succeed because they defeat smaller, weaker, or less fierce males in combat; more often, however, they are effective in ritual displays that discourage would-be competitors (see Chapter 50).

Darwin was intrigued by **sexual selection**, which he saw as a separate selection process leading to sexual dimorphism. Many secondary sexual features do not seem to be adaptive in the general sense; showy plumage probably does not help male birds cope with their environment and may even attract predators. If such accoutrements give the individual an edge in gaining a mate, however, they will be favored for the most Darwinian of reasons: because they enhance reproductive success. Thus, in many cases, the ultimate evolutionary outcome is a compromise between the two selection forces. In some species, the line between sexual selection and ordinary natural selection blurs because the sexual feature does double duty as an adaptation to the environment. For example, a stag may use his antlers to defend himself against a predator.

Many researchers who study sexual selection and dimorphism have recently shifted their attention to the powerful roles of females in the evolution of these characters. Every time a female chooses a mate based on particular phenotypic traits, she perpetuates the genes that caused her to make that choice and allows a male of particular phenotype to perpetuate his genes (Figure 21.12).

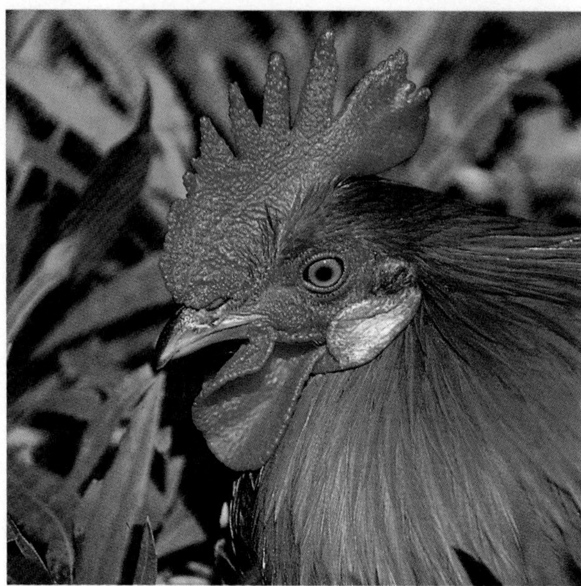

Figure 21.12
Sexual selection and the evolution of male appearance in jungle fowl. Marlene Zuk, of the University of California, Riverside, has been studying mate selection in Asian jungle fowl, the ancestors of domesticated chickens. Hens generally choose roosters, such as the one in this photograph, with bright eyes and large, red combs and wattles (the flap of skin draped from the throat). These traits are correlated with good health and resistance to pathogens. Thus, according to Dr. Zuk's hypothesis, a hen keys on rooster features that advertise vitality. Over the generations, such choices by females shape the appearance of males and play an active role in favoring genes that contribute to good health.

Does Evolution Fashion Perfect Organisms?

In a word, no. There are at least four reasons why natural selection cannot breed perfection.

1. *Organisms are locked into historical constraints.* As we saw in Chapter 20, each species has a legacy of descent with modification from a long line of ancestral forms. Evolution does not scrap ancestral anatomy and build each new complex structure from scratch, but co-opts existing structures and adapts them to new situations. For example, the excruciating back problems some humans endure result in part because the skeleton and musculature modified from the anatomy of four-legged ancestors are not fully compatible with upright posture.

2. *Adaptations are often compromises.* Each organism must do many different things. A seal spends part of its time on rocks; it could probably walk better if it had legs instead of flippers, but it would not swim nearly as well. We owe much of our versatility and athleticism to our prehensile hands and flexible limbs, which also make us prone to sprains, torn ligaments, and dislocations; structural reinforcement has been compromised for agility.

3. *Not all evolution is adaptive.* Chance probably affects the genetic makeup of populations to a greater extent than was once believed. For instance, when a storm blows insects hundreds of miles over an ocean to an island, the wind does not necessarily pick up the specimens that are best suited to the new environment. And not all alleles fixed by genetic drift in the gene pool of the small founding population are better suited to the environment than alleles that are lost. Similarly, the bottleneck effect can cause nonadaptive or even maladaptive evolution.

4. *Selection can only edit variations that exist.* Natural selection favors only the most fit variations from what is available, which may not be the ideal characteristics. New alleles do not arise on demand.

With all these constraints, we cannot expect evolution to craft perfect organisms. Natural selection operates on a "better than" basis. Perhaps the best evidence for evolution is in the subtle imperfections of the organisms it produces.

STUDY OUTLINE

1. Natural selection acts on individuals, but only populations *evolve,* because evolution is change in the prevalence of inherited traits over a succession of generations.

The Modern Evolutionary Synthesis (pp. 438–439)

1. The development of population genetics, with its emphases on quantitative inheritance and variation, brought Darwinism and Mendelism together.
2. In the 1940s, the modern synthesis provided a comprehensive theory of evolution that focused on populations as units of evolution.

The Genetics of Populations (pp. 439–442)

1. A species is a group of populations that have the potential to interbreed in nature.
2. A population is united by its gene pool, the aggregate of all genes in the population. A change in the relative frequencies of alleles in the gene pool over a succession of generations is called microevolution.
3. According to the Hardy-Weinberg theorem, the frequencies of alleles in a population will remain constant if sexual reproduction is the only process that affects the gene pool. The mathematical expression for such a nonevolving population is the Hardy-Weinberg equation, which states that for a two-allele locus, $p^2 + 2pq + q^2 = 1$, where p and q represent the relative frequencies of the dominant and recessive alleles, respectively, p^2 and q^2 the frequencies of the homozygous genotypes, and $2pq$ the frequency of the heterozygous genotype.
4. For Hardy-Weinberg equilibrium to apply, the population must be very large, be totally isolated, have no net mutations, show random mating, and have equal reproductive success for all individuals. Hardy-Weinberg equilibrium is never met in real populations because conditions in nature violate one or more of these prerequisites.

Causes of Microevolution (pp. 442–445)

1. Genetic drift is the change in gene frequencies observed in small populations due to sampling error or chance events. When large segments of a population are destroyed by disasters (the bottleneck effect) or when a small sample of a population colonizes a new habitat (the founder effect), the new, small population is unlikely to be representative of the parent population, and genetic drift will continue until the population grows larger.
2. Gene flow is the exchange of alleles between two populations due to migration.
3. Mutation can theoretically affect allele frequencies in a gene pool, but is usually insignificant over the short term for large populations. Mutation is important in evolution, however, because it generates new variations.
4. Nonrandom mating, such as inbreeding or assortative mating, does not ordinarily affect allele frequencies, but does affect the ratio of genotypes in populations.
5. Natural selection, differential success in reproduction, is the only agent of microevolution that tends to cause adaptive change in the gene pool.

The Genetic Basis of Variation (pp. 445–450)

1. Genetic variation in a population includes individual variation in quantitative characters, geographical variation, and polymorphism.
2. Genetic variation is extensive, but much of it can be detected only at the molecular level.
3. Mutation and sexual recombination produce genetic variation.
4. Populations remain variable despite natural selection. Diploidy maintains a reservoir of latent variation in heterozygotes. Balanced polymorphism may maintain variation at some gene loci as a result of heterozygote advantage or frequency-dependent selection.
5. Biologists debate what fraction of genetic variation is actually acted on by natural selection and what fraction is neutral in its impact on reproductive success.

Adaptive Evolution (pp. 450–453)

1. Adaptive evolution results from the workings of natural selection on the chance variations.
2. Darwinian fitness is measured only by reproductive success. The selection coefficient is the difference between relative fitness values representing the contributions of specific genotypes and is a relative measure of selection against an inferior genotype.
3. Selection maintains favorable genotypes in a population by acting on the phenotype of individual organisms. The whole organism is the object of selection.
4. Natural selection can affect the frequency of a phenotype in three different ways: stabilizing selection, which discriminates against extreme phenotypes; directional selection, which favors relatively rare individuals on one end of the phenotypic range; and diversifying selection, which favors individuals at both extremes of a range over intermediate phenotypes.
5. Sexual selection leads to the evolution of secondary sex characteristics, which give the individual an advantage in mating, but which may or may not help the individual cope with the environment.
6. Evolution by natural selection does not fashion perfect organisms for several reasons: Structures result from modified ancestral anatomy; adaptations are often compromises; the gene pool can be affected by genetic drift; and natural selection can act only on available variation.

SELF-QUIZ

1. A gene pool consists of
   a. all the genes exposed to natural selection
   b. the total of all alleles present in a population
   c. the entire genome of a reproducing individual
   d. the frequencies of the alleles for a gene locus within a population
   e. all the gametes in a population

2. In a population with two alleles for a particular locus, *B* and *b*, the allele frequency of *B* is 0.7. What would be the frequency of heterozygotes if the population is in Hardy-Weinberg equilibrium?
   a. 0.7 d. 0.42
   b. 0.49 e. 0.09
   c. 0.21

3. In a population that is in Hardy-Weinberg equilibrium, 16% of the individuals show the recessive trait. What is the frequency of the dominant allele in the population?
   a. 0.84 c. 0.6 e. 0.48
   b. 0.36 d. 0.4

4. The average length of jackrabbit ears decreases the farther north the rabbits live. This variation is an example of
   a. a cline d. genetic drift
   b. discrete variation e. diversifying selection
   c. polymorphism

5. If a particular genotype has a selection coefficient of 0.4,
   a. its relative fitness would be 0.6
   b. individuals with this genotype would leave 40% as many offspring as individuals with the most common genotype leave
   c. the genotype would increase by 40% in each generation
   d. the genotype would completely disappear from the population
   e. the frequency of the genotype would be 0.4

6. Selection acts *directly* on
   a. phenotype d. each allele
   b. genotype e. the entire gene pool
   c. the entire genome

7. As a mechanism of microevolution, natural selection can be most closely equated with
   a. assortative mating
   b. genetic drift
   c. differential reproductive success
   d. bottlenecking of a population
   e. gene flow

8. Most of the variation we see in coat coloration and pattern in a population of wild mustangs in any generation is probably due to
   a. new mutations that occurred in the preceding generation
   b. sexual recombination of alleles
   c. genetic drift due to the small size of the population
   d. geographic variation within the population
   e. environmental effects

9. In terms of the algebraic symbols used in the Hardy-Weinberg equation (p and q), the most likely effect of assortative mating on the frequencies of alleles and genotypes for a locus would be
   a. a decrease in p^2 compared to q^2
   b. a trend toward zero for q^2
   c. convergence of p^2 and q^2 toward equal values
   d. a change in p and q, the relative frequencies of the two alleles in the gene pool
   e. a decrease in $2pq$ below the value expected by the Hardy-Weinberg theorem

10. A founder event favors microevolution in the founding population mainly because
   a. mutations are more common in a new environment
   b. a small founding population is subject to extensive sampling error in the composition of its gene pool
   c. the new environment is likely to be patchy, favoring diversifying selection
   d. gene flow increases
   e. members of a small population tend to migrate

CHALLENGE QUESTIONS

1. Some species have been rescued from near extinction by conservationists. What evolutionary problems do such species face as their populations rebound?

2. Let's return to the wildflowers with which we derived the Hardy-Weinberg theorem. The frequency of *A*, the dominant allele for red flowers, is 0.8, and the frequency of *a*, the recessive allele for white flowers, is 0.2. In our starting population, the frequencies of genotypes do not conform to Hardy-Weinberg equilibrium: 60% of the plants are *AA* and 40% of the plants are *Aa* (at this point, the population has no plants with white flowers). Assuming that all conditions for the Hardy-Weinberg theorem are met, prove that genotypes will reach equilibrium in the next generation.

SCIENCE, TECHNOLOGY, AND SOCIETY

1. Some scientists have suggested that the increased appreciation of the importance of females in sexual selection is one consequence of increasing numbers of women making contributions to animal behavior and evolutionary biology. Do you think gender makes a difference in the kinds of questions scientists ask and information they uncover? Why or why not?

2. How might modern technologies and lifestyles be altering human gene pools and reshaping human populations and the human species as a whole? Consider the five agents of microevolution discussed in this chapter.

FURTHER READING

Dawkins, R. *The Blind Watchmaker.* New York: Norton, 1986. A master of metaphor explains how complexity can arise in the absence of design.

Diamond, J. "Founding Fathers and Mothers." *Natural History,* June 1988. The importance of genetic drift in human evolution.

Ryan, M. J. "Signals, Species, and Sexual Selection." *American Scientist,* January/February 1990. How sexual selection has composed the songs of frogs.

Smith, J. M. *Evolutionary Genetics.* Oxford: Oxford University Press, 1989. Each chapter includes computer projects.

Smith, T. B. "A Double-Billed Dilemma." *Natural History,* January 1991. Field experiments on a balanced polymorphism.

Symonds, N. "A Fitter Theory of Evolution." *New Scientist,* September 21, 1991. Can environmental changes direct specific mutations in bacteria?

22 The Origin of Species

The Species Problem

Reproductive Barriers

The Biogeography of Speciation

Genetic Mechanisms of Speciation

Gradual and Punctuated Interpretations of Speciation

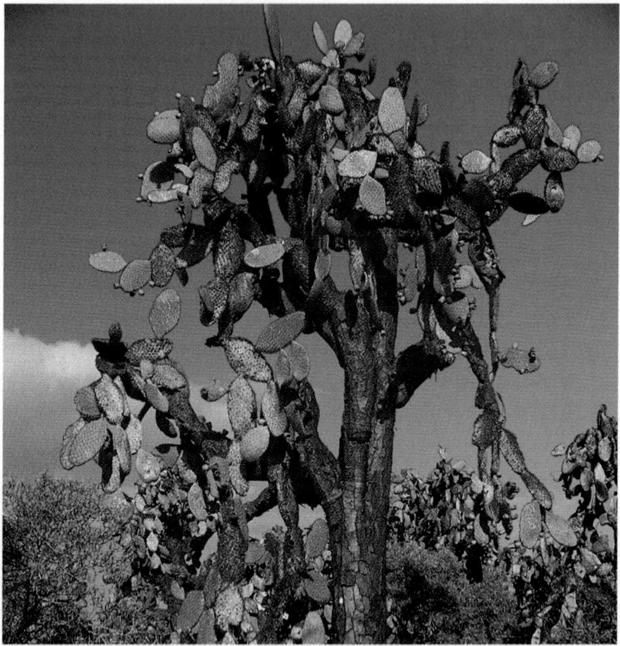

Figure 22.1
The Galapagos, a compact model of speciation. This giant prickly pear cactus *(Opuntia echios gigantea)* on Santa Fe Island is just one of hundreds of plant and animal species inhabiting the Galapagos Islands that are found nowhere else in the world. Darwin, who visited the Galapagos for five weeks in 1835, eventually recognized that speciation on these islands provides a model for how the multiplication of species has populated the diverse environments of Earth. In this chapter, you will learn about the mechanisms of speciation.

When Darwin saw that the geologically young Galapagos Islands had already become populated with many plants and animals known nowhere else in the world, he realized that he was visiting a place of genesis (Figure 22.1). Darwin wrote in his diary: "Both in space and time, we seem to be brought somewhat near to that great fact—that mystery of mysteries—the first appearance of new beings on this Earth." The beginning of new forms of life—the origin of species—is at the focal point of evolutionary theory, for it is in new species that biological diversity arises. It is not enough to explain how adaptations evolve in populations, a topic covered in Chapter 21. Peppered moths *(Biston betularia)* are still the same species even as the relative proportions of dark and light individuals change over the generations. Evolutionary theory must also explain the multiplication of species, the radiation of an existing species that gives rise to two or more new species.

The fossil record chronicles two patterns of **speciation** (origin of new species): anagenesis and cladogenesis (Figure 22.2). **Anagenesis** (Gr. *ana*, "up," and *genesis*, "origin"), also known as **phyletic evolution,** is the transformation of an unbranched lineage of organisms, sometimes to a state different enough from the ancestral population to justify renaming it as a new species. **Cladogenesis** (Gr. *clados*, "branch"), also called **branching evolution,** is the budding of one or more new species from a parent species that continues to exist. Cladogenesis is more important than anagenesis in the history of life, not only because it seems to be the more common pattern, but also because only cladogenesis can promote biological diversity by increasing the number of species.

Our objective in this chapter is to examine and evaluate possible mechanisms of speciation. The first step is to appraise the assumption that species actually exist in nature as discrete biological units distinct from all others.

THE SPECIES PROBLEM

In 1927, a young biologist named Ernst Mayr led an expedition into the remote Arafak Mountains of New Guinea to study the wildlife and collect specimens (see

Figure 22.2
Two patterns of speciation. (**a**) In anagenesis (phyletic evolution), a single population is transformed enough to be designated a new species. (**b**) Cladogenesis is branching evolution, in which a new species arises from a small population that buds from a parent species. Most new species probably evolve by cladogenesis, the branching evolution that is the basis for biological diversity.

Figure 22.3
Ernst Mayr in New Guinea, 1927. During his expedition, the naturalist (on the right, photographed with his guide) was struck by the almost exact match in how he and the native Papuans divided the birds of the Arafak Mountains into separate species. It was one of many experiences that led to Mayr's biological species concept, which emphasizes interbreeding within species and reproductive isolation between species.

the interview preceding Chapter 20). He found a great diversity of birds, identifying 138 separate species on the basis of differences in their appearance (Figure 22.3). Mayr was surprised to learn that the local tribe of Papuan natives, who hunted the animals for food and feathers, had given names of their own to 137 birds (two birds assigned to separate species by Mayr are extremely similar, and the Papuans did not distinguish between them). Although their motives and training could not have been more different, Mayr and the indigenous people agreed almost exactly in their inventory of the local birdlife. There are many such cases of university-trained taxonomists giving specific names to organisms in a particular locale, only to find that the discrete forms they identified correspond to the folk taxonomy of the region. From this we might conclude that species exist in nature as discrete units, demarcated from other species. Yet devising a formal definition of a species poses a formidable challenge.

Two Concepts of Species

Species is a Latin word meaning "kind" or "appearance." Indeed, we learn to distinguish between the kinds of plants or animals—between dogs and cats, for instance—from differences in their appearance. Linnaeus, the founder of modern taxonomy, described individual species in terms of their physical form, or morphology, and this is still the method most often used to characterize species. Species defined by their anatomical features are referred to as **morphospecies.**

There are certainly pitfalls in applying the morphospecies concept in some situations. It is sometimes difficult, for example, to determine whether a set of organisms represents multiple species or a single species with extensive phenotypic variation. Conversely, two populations that are almost indistinguishable by morphological criteria may turn out to be different species based on other criteria. Despite these difficulties, the morphospecies concept, based as it is on observable and measurable anatomy, is usually practical to apply in the field, even on fossils. Most of the species recognized by taxonomists have been designated as separate species based on morphological criteria. From the standpoint of evolutionary theory, however, the morphospecies concept does not help us understand how each species remains distinct from other species. Put another way, why is biological diversity not a continuum of variation, but instead broken up into separate forms that we can identify as species? This is the question addressed by a second species definition known as

(a)

(b)

Figure 22.4
The biological species concept. This view of species emphasizes interfertility rather than physical similarity. (**a**) Members of a single biological species may differ extensively in appearance from one another. (**b**) Conversely, members of two different biological species, such as the eastern meadowlark (top) and the western meadowlark (bottom), may appear very similar.

the biological species concept, first enunciated by Mayr in 1942.

A **biological species** is a population or group of populations whose members have the potential to interbreed with one another in nature to produce fertile offspring, but who cannot successfully interbreed with members of other species (Figure 22.4). In other words, a biological species is the largest unit of population in which gene flow is possible, and that is genetically isolated from other such populations. Put still another way, each species is circumscribed by reproductive barriers that preserve its integrity as a species by blocking genetic mixing with other species. Members of a species, said to be conspecific, are united by being reproductively compatible, at least potentially. A businesswoman in Manhattan has little probability of producing offspring with a dairyman in Outer Mongolia, but if the two should get together, they could have viable babies that develop into fertile adults. All humans belong to the same biological species. In contrast, humans and chimpanzees remain distinct species even where they share territory, because the two species do not successfully interbreed and produce hybrid offspring.

Remember that biological species are defined by their reproductive isolation from other species in *natural* environments. In the laboratory, hybrids can often be produced between two species that do not interbreed in nature.

Limitations of the Biological Species Concept

The biological species concept does not work in all situations. The criterion of interbreeding is useless for organisms that are completely asexual in their reproduction, as are all prokaryotes, some protists, some fungi, and even some plants (such as the commercial banana) and some animals (including certain lizards and other vertebrates). Many bacteria do transfer genes on a limited scale by conjugation and other processes, but there is nothing akin to the equal contribution of genetic material from two parents that occurs in sexual reproduction (see Chapter 17). Different lineages of descent give rise to clones, which, genetically speaking, represent single individuals. Asexual organisms can be assigned to species only by the grouping of clones that have the same morphology and biochemical characteristics.

The biological species concept is also inadequate as a criterion for grouping extinct forms of life, the fossils of which must be classified according to the morphospecies concept.

There is still another situation where the biological species concept does not work. If two populations are geographically segregated from each other, we do not know if they have the potential to interbreed in nature, even if they are so much alike that they are placed in the same species on morphological grounds.

Even for populations that are sexual, contemporaneous, and geographically contiguous, there are cases

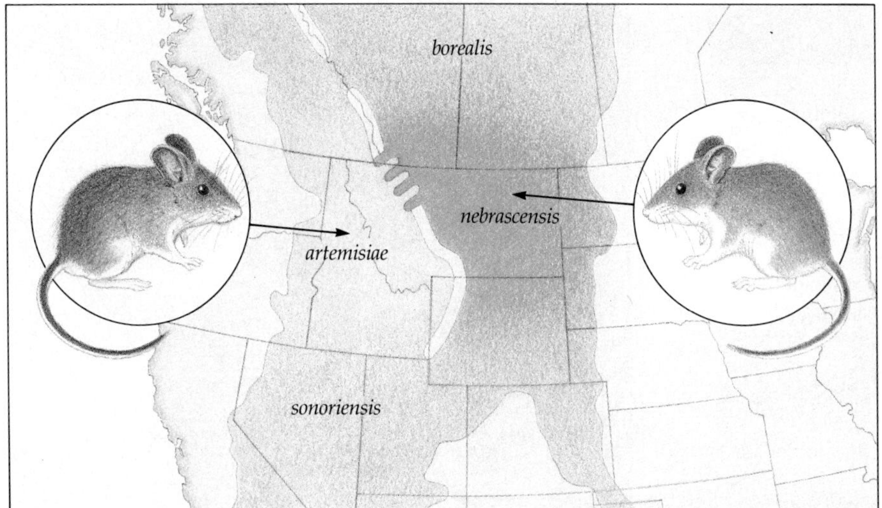

Figure 22.5
Speciation in progress? These four populations of the deer mouse *(Peromyscus maniculatus)* display geographic variation and are designated by some experts as subspecies. Interbreeding between subspecies occurs where their ranges overlap, except for the subspecies *artemisiae* and *nebrascensis.* These two subspecies will not interbreed, yet gene flow between them is possible via the other neighboring populations. Such cases may represent speciation in progress; for example, *artemisiae* and *nebrascensis* could legitimately be called different species if, in the future, the populations connecting them become extinct.

where the biological species concept cannot be applied unambiguously. Consider, for example, four populations of the deer mouse *(Peromyscus maniculatus)* found in the Rocky Mountains (Figure 22.5). Such phenotypically distinct populations that are separated geographically are sometimes called subspecies. The deer mouse populations overlap at certain locations, and some interbreeding occurs in these zones of cohabitation; we would therefore consider the populations to belong to the same species according to the biological species concept. The exception is the subspecies *Peromyscus maniculatus artemisiae* and the subspecies *P. m. nebrascensis.* Though their ranges overlap, these two populations do not interbreed. Nevertheless, their gene pools are not completely isolated, because each of these populations interbreeds with its other neighboring populations, and genes could conceivably flow eventually from the *artemisiae* to the *nebrascensis* population. The extent of gene flow by this circuitous route is probably so slight that even the experts disagree on whether or not these two populations should be called subspecies or be assigned to separate species. If the corridors for gene flow were eliminated by extinction of the other two populations, then *artemisiae* and *nebrascensis* could be named separate species without reservation.

Population biologists are discovering more and more cases where the distinction between populations with limited gene flow and full biological species with segregated gene pools blurs. It is as though we are catching populations at different stages in their evolutionary descent from common ancestors. This is to be expected, if new species usually arise by the gradual divergence of populations.

Many evolutionary biologists now question the usefulness of the biological species concept, and additional species concepts have been proposed in recent years (we will examine one of them later in the chapter). But the species problem may have no completely satisfactory resolution; it is unlikely that any single definition of a species can be stretched to cover all cases. These theoretical problems aside, the concept of morphospecies, still the most practical definition of species for taxonomic purposes, and the conceptual definition of biological species usually result in recognizing the same units as species. Ernst Mayr and the Papuan tribe arrived at the same set of species of Arafak birds based on differences in appearance, and it is reproductive isolation that preserves species boundaries.

REPRODUCTIVE BARRIERS

Any factor that impedes two species from producing fertile hybrids contributes to reproductive isolation. No single barrier may be completely impenetrable to gene flow, but most species are genetically sequestered by more than one type of barrier. Here, we are considering only biological barriers to reproduction, which are intrinsic to the organisms. Of course, if two species are geographically segregated, they cannot possibly interbreed, but a geographical barrier is not considered equivalent to reproductive isolation because it is not intrinsic to the organisms themselves. Reproductive isolation prevents populations belonging to different species from interbreeding, even if their ranges overlap.

Clearly, a fly will not mate with a frog or a fern, but what prevents species that are very similar—that is, closely related—from interbreeding? The various reproductive barriers that isolate the gene pools of species can be categorized as prezygotic or postzygotic, depending on whether they function before or after the formation of zygotes, or fertilized eggs (Table 22.1).

Table 22.1 Reproductive Barriers Between Species

I. *Prezygotic:* Prevent mating or fertilization.
   A. *Habitat isolation:* Populations live in different habitats and do not meet.
   B. *Temporal isolation:* Mating or flowering occurs at different seasons or times of day.
   C. *Behavioral isolation:* There is little or no sexual attraction between females and males.
   D. *Mechanical isolation:* Structural differences in genitalia or flowers prevent copulation or pollen transfer.
   E. *Gametic isolation:* Female and male gametes fail to attract each other or are inviable.

II. *Postzygotic:* Prevent the development of viable, fertile adults.
   A. *Hybrid inviability:* Hybrid zygotes fail to develop or fail to reach sexual maturity.
   B. *Hybrid sterility:* Hybrids fail to produce functional gametes.
   C. *Hybrid breakdown:* The offspring of hybrids have reduced viability or fertility.

Figure 22.6
Courtship ritual as a behavioral barrier between species. These blue-footed boobies, inhabitants of the Galapagos, will mate only after a specific ritual of courtship displays. Part of the "script" calls for the male to high-step, a behavior that advertises the bright blue feet characteristic of the species.

Prezygotic Barriers

Prezygotic barriers impede mating between species or hinder the fertilization of ova if members of different species attempt to mate.

Habitat Isolation Two species that live in different habitats within the same area may encounter each other rarely, if at all, even though they are not technically geographically isolated. For example, two species of garter snakes belonging to the genus *Thamnophis* occur in the same areas, but one lives mainly in water and the other is primarily terrestrial. Habitat isolation also affects parasites, which are generally confined to certain plant or animal host species. Two species of parasites living on different hosts will not have a chance to mate.

Temporal Isolation Two species that breed during different times of the day, different seasons, or different years cannot mix their gametes. The geographic ranges of the western spotted skunk *(Spilogale gracilis)* and the eastern spotted skunk *(Spilogale putorius)* overlap, but these two very similar species do not interbreed because *S. gracilis* mates in late summer and *S. putorius* mates in late winter. Three species of the orchid genus *Dendrobium* living in the same rain forest do not hybridize because they flower on different days. Pollination of each species is limited to a single day, because the flowers open in the morning and wither that evening. A sudden storm induces all three species to flower, but the number of days that lapse between the stimulus and flowering is eight in one species, nine in another, and ten in the third species, so reproductive isolation is still maintained.

Behavioral Isolation Special signals that attract mates, as well as elaborate behavior unique to a species, are probably the most important reproductive barriers among closely related animals. Male fireflies of different species signal to females of their kind by blinking their lights in particular patterns. The females respond only to signals characteristic of their own species, flashing back and attracting the males.

The eastern and western meadowlarks are almost identical in morphology and habitat, and their ranges overlap in the central United States (see Figure 22.4). Yet they remain two separate species, partly because of the difference in their songs, which enables them to recognize individuals of their own kind. Still another form of behavioral isolation is courtship ritual specific to a species (Figure 22.6).

Mechanical Isolation Closely related species may attempt to mate, but fail to consummate the act because they are anatomically incompatible. For example, mechanical barriers contribute to reproductive isolation of flowering plants that are pollinated by insects or other animals. Floral anatomy is often adapted to a specific pollinator that faithfully transfers pollen only among plants of the same species (Figure 22.7).

Gametic Isolation Even if the gametes of different species meet, they rarely fuse to form a zygote. For animals whose eggs are fertilized within the female reproductive tract (internal fertilization), the sperm of one species may not be able to survive in the environment of the female reproductive tract of another

Figure 22.7
Mechanical isolation of flowering plants. Many plants have exclusive relationships with the animals that carry their pollen. Hummingbirds use their needlelike beaks to reach sugary nectar secreted by glands at the bottom of long floral tubes. As a bird feeds, its head is dusted with pollen, which it then transfers to the next flower it visits. In some cases, the beak of a particular species of hummingbird is just the right length for the floral tube of the plant species it pollinates. The unique floral architectures of plants pollinated by animals impose barriers to the transfer of pollen between species.

species. Many aquatic animals release their gametes into the surrounding water, where the eggs are fertilized (external fertilization). Even when two closely related species release their gametes at the same time in the same place, cross-specific fertilization is uncommon. Gamete recognition may be based on the presence of specific molecules on the coats around the egg, which adhere only to complementary molecules on sperm cells of the same species. A similar mechanism of molecular recognition enables a flower to discriminate between pollen of the same species and pollen of different species.

Postzygotic Barriers

If a sperm cell from one species does fertilize an ovum of another species, then **postzygotic barriers** prevent the hybrid zygote from developing into a viable, fertile adult.

Hybrid Inviability When prezygotic barriers are crossed and hybrid zygotes are formed, genetic incompatibility between the two species may abort development of the hybrid at some embryonic stage. Of the several species of frogs belonging to the genus *Rana,* some live in the same regions and habitats, where they may occasionally hybridize. But the hybrids generally do not complete development, and those that do are frail.

Hybrid Sterility Even if two species mate and produce hybrid offspring that are vigorous, reproductive isolation is intact if the hybrids are sterile because genes cannot flow from one species' gene pool to the other. One cause of this barrier is a failure of meiosis to produce normal gametes in the hybrid if chromosomes of the two parent species differ in number or structure. The most familiar case of a sterile hybrid is the mule, a robust cross between a horse and a donkey; horses and donkeys remain distinct species because, except very rarely, mules cannot backbreed with either species.

Hybrid Breakdown In some cases when species cross-mate, the first-generation hybrids are viable and fertile, but when these hybrids mate with one another or with either parent species, offspring of the next generation are feeble or sterile. For example, different cotton species can produce fertile hybrids, but breakdown occurs in the next generation when offspring of the hybrids die in their seeds or grow into weak and defective plants.

Introgression

Alleles may occasionally seep through all reproductive barriers and pass between the gene pools of closely related species when fertile hybrids mate successfully with one of the parent species. This transplantation of alleles between species is called **introgression.** For example, corn *(Zea mays)* has some alleles that can be traced to a closely related wild grass called teosinte *(Zea mexicana).* Introgression occurs when the two species hybridize and a fraction of the hybrids manage to cross with corn plants. The transplant of alleles increases the reservoir of genetic variation that can be exploited by breeders trying to produce new corn varieties by artificial selection. But occasional hybridization does not erase the boundary between corn and teosinte. As long as reproductive barriers hold introgression to a trickle, the isolation of the two gene pools is not seriously breached, and the two species remain distinct.

(a)

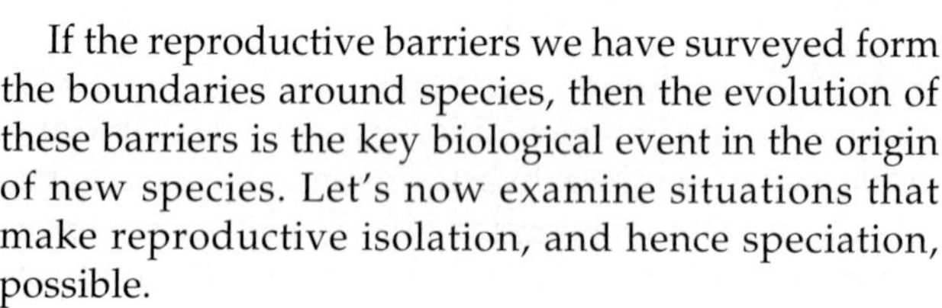

(b)

Figure 22.8
Pupfish speciation in Death Valley. What is now a desert valley was once a great inland lake that began to dry up about 10,000 years ago. Small isolated springs are all that remain. Some of the springs are home to small animals called pupfishes *(Cyprinodon)*, usually with each spring inhabited by a single species found only in that pool. (**a**) The ecologists in this photo are counting pupfish of the species (*C. diabolis*, shown in **b)** that inhabits Devil's hole. The different pupfish species descended from a common ancestral population that became fragmented when the lake began to dry up. The isolated populations of pupfishes diverged through genetic drift and natural selection, and have differentiated to the level of true species that cannot interbreed even when brought together in the laboratory.

If the reproductive barriers we have surveyed form the boundaries around species, then the evolution of these barriers is the key biological event in the origin of new species. Let's now examine situations that make reproductive isolation, and hence speciation, possible.

THE BIOGEOGRAPHY OF SPECIATION

A crucial episode in the origin of a species occurs when the gene pool of a population is severed from other populations of the parent species. With its gene pool isolated, the splinter population can follow its own evolutionary course as changes in allele frequencies caused by selection, genetic drift, and mutations occur undiluted by gene flow from other populations. Speciation episodes can be classified into two modes based on the geographical relationship of a new species to its ancestral species. The initial block to gene flow may be a geographical barrier that physically isolates the population. This speciation mode is termed **allopatric speciation** (Gr. *allos*, "other," and L. *patria*, "homeland"), and populations segregated by a geographical barrier are known as allopatric populations. In the second speciation mode, a subpopulation becomes reproductively isolated in the midst of its parent population; this is **sympatric speciation** (Gr. *sym*, "together"). Populations are said to be sympatric if their ranges overlap.

Allopatric Speciation

Geological processes can fragment a population into two or more isolated populations. A mountain range may emerge and gradually split a population of organisms that can inhabit only lowlands; a creeping glacier may gradually divide a population; a land bridge, such as the Isthmus of Panama, may form and separate the marine life on either side; or a large lake may subside until there are several smaller lakes with their populations now isolated. Alternatively, a small population may become geographically isolated when individuals from the parent population travel to a new location.

Just how formidable a geographical barrier must be to keep allopatric populations apart depends on the ability of the organisms to disperse due to the mobility of animals or the dispersibility of spores, pollen, and seeds of plants. The Grand Canyon is easily crossed by hawks and many other birds, but it is an impassable barrier to populations of small rodents confined to either the north or south rim of the canyon.

Let's consider an example of how geographic isolation can lead to allopatric speciation. About 50,000 years ago, during an ice age, what is now the Death Valley region of California and Nevada had a very rainy climate and a system of interconnected lakes and rivers. A drying trend began about 10,000 years ago, and by 4000 years ago the region had become a desert. Today, all that is left of the network of lakes and rivers is isolated springs scattered in the desert, mostly in deep clefts between rocky walls. The springs vary extensively in water temperature and salinity. Living in many of the springs are tiny fishes called pupfishes, belonging to the genus *Cyprinodon.* Each inhabited spring, often no more than a few meters in diameter, is home to its own species of pupfish adapted to that pool and found nowhere else in the world. The various pupfishes probably descended from a single ancestral species whose range was broken up when the region became arid, cloistering several small populations that diverged in their evolution as they adapted to their home springs (Figure 22.8).

Conditions Favoring Allopatric Speciation Whenever populations become allopatric, it is possible for speciation to occur as the isolated gene pools accumulate differences by microevolution that may cause the populations to diverge in phenotype. But an isolated population that is small is more likely than a large population to change substantially enough to become a new species.

The geographical isolation of a small population usually occurs at the fringe of the parent population's range. The splinter population, or peripheral isolate, is a good candidate for speciation for three reasons:

1. The gene pool of the peripheral isolate probably differs from that of the parent population from the outset. Living near the border of the range, the peripheral isolate represents the extremes of any genotypic and phenotypic clines that existed in the original population. If the peripheral isolate is small, there will be a founder effect, with chance alone resulting in a gene pool that is not representative of the gene pool of the parent population (see Chapter 21).

2. Until the peripheral isolate becomes a large population, genetic drift will continue to change its gene pool at random. New mutations or combinations of existing alleles that are neutral in adaptive value may become fixed in the population by chance alone, causing phenotypic divergence from the parent population. (See Chapter 21 to review genetic drift.)

3. Evolution caused by selection may take a different direction in the peripheral isolate than in the parent population. Because the peripheral isolate inhabits a frontier, where the environment is somewhat different, the peripheral isolate will probably encounter selection factors that are different from, and generally more severe than, those operating on the parent population.

These factors may cause peripheral isolates to follow an evolutionary course that diverges from that of the parent population as long as the gene pools remain isolated. This does not mean that all peripheral isolates persist long enough or change enough to become new species. Life on the frontier is usually harsh, and most pioneer populations probably become extinct. As evolutionary biologist Stephen Jay Gould puts it: "Status as a peripheral isolate merely gives a lottery ticket to a small population. A population can't win (speciate) without a ticket, but there are very few winning tickets."

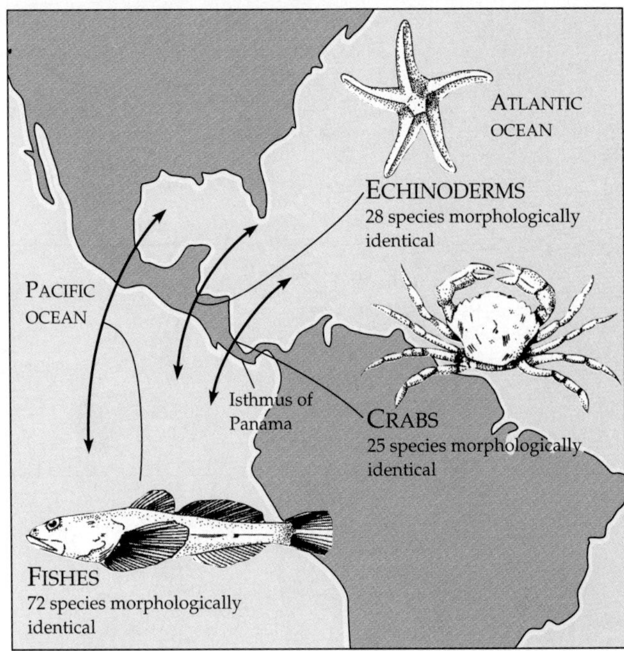

Figure 22.9
Evolutionary stasis of large successful populations. The Panamanian land bridge formed about 3 million years ago, segregating marine life in the Atlantic Ocean from species in the Pacific. The majority of species on opposite sides of the isthmus have not diverged during their 3 million years of allopatry, apparently because the two environments are similar and because most large populations evolve very slowly. New alleles and genetic combinations that arise in a large gene pool are swamped by the huge number of existing alleles.

There is ample evidence that allopatric speciation is much faster in small populations than in very large ones. The North American sycamore tree and the European sycamore represent large populations that have been allopatric for at least 30 million years, but specimens that are brought together still produce fertile hybrids. Another example can be seen in the similarity of species on either side of the Panamanian land bridge (Figure 22.9). In comparison, the relatively small number of organisms that managed to reach the Galapagos Islands has given rise to faunas and floras that have become almost entirely endemic in less than two million years. In fact, evolutionary biologists generally agree that a small population can accumulate enough genetic change to become a new morphospecies in only hundreds to thousands of generations, equivalent to less than a thousand years to a few tens of thousands of years, depending on generation span.

Adaptive Radiation on Island Chains Flurries of allopatric speciation have occurred on island chains where organisms that have strayed or become passively dispersed from their parent populations have

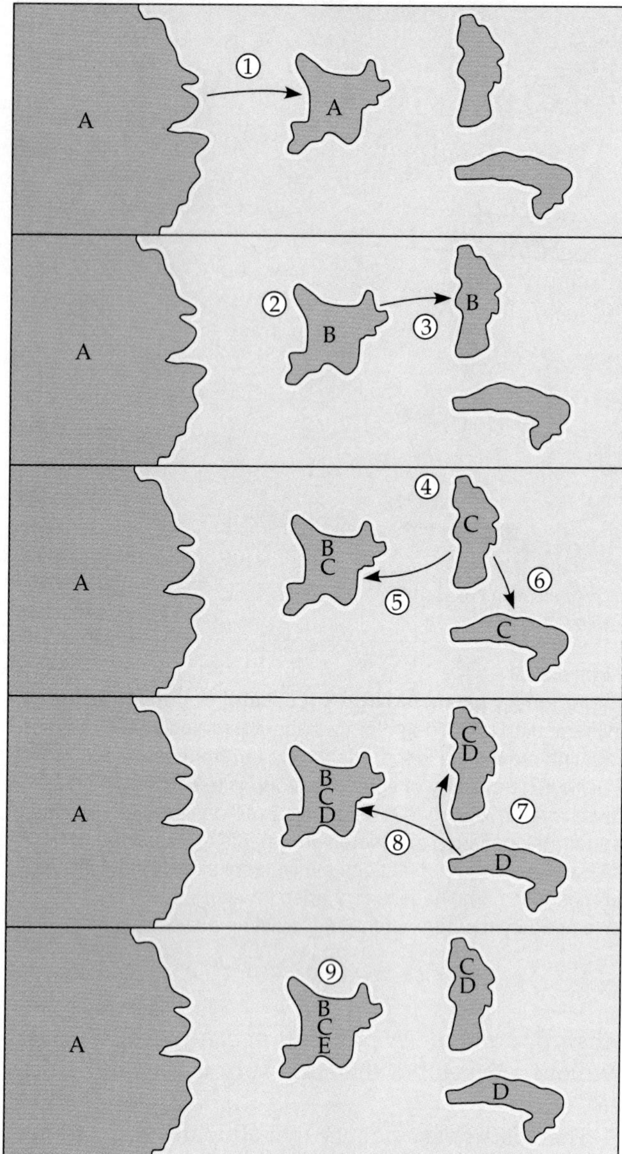

Figure 22.10
A model for adaptive radiation on island chains. ① One island in this cluster of three is seeded by a small colony founded by individuals of species A blown over from a mainland population. ② Its gene pool isolated from the parent species, the island population evolves into species B as it adapts to its new environment. ③ Storms or other agents of dispersion spread species B to a second island, ④ where the isolated colony evolves into species C. ⑤ Later, individuals from species C recolonize the first island and cohabit with species B, but reproductive barriers keep the species distinct, each adapted to certain foods and other requirements. ⑥ A colony of species C may also populate a third island, ⑦ where it adapts and forms species D. ⑧ Species D is dispersed to the two islands of its ancestors, finding an unoccupied niche on one island, ⑨ but forming a new species, E, on the other island. The story could go on and on, with a series of allopatric episodes made possible by the combination of isolation and occasional dispersal.

founded new populations that evolved in isolation. The many endemic species of the Galapagos descended from stragglers that floated, flew, or were blown over the sea from the South American mainland. For example, consider Darwin's finches. A single dispersal event may have seeded one island with a small population of the ancestral finch, and the peripheral isolate formed a new species. Later, a few individuals of this island species may have reached neighboring islands, where geographical isolation permitted more speciation episodes (Figure 22.10). After diverging on the island it invaded, a young species could recolonize the island from which its founding population emigrated and coexist there with its parent species, or form still another species. Multiple invasions of islands by peripheral isolates of species from neighboring islands would eventually lead to the coexistence of several species on each island. The islands are far enough apart to permit populations to evolve in isolation, but close enough together for occasional dispersion events to occur.

Thirteen species of finches have evolved on the Galapagos Archipelago, almost certainly from a single ancestral species. Each island now has multiple species, with as many as ten on some islands. In contrast, Cocos Island, about 700 km north of the Galapagos, has only one finch species, derived from an ancestral species that made it to that remote island. Cocos is so isolated that there has been no opportunity for the kind of island hopping that resulted in the many episodes of allopatric speciation on the Galapagos.

The emergence of numerous species from a common ancestor that spreads to several new environments is called **adaptive radiation.** Adaptive radiation of Darwin's finches is evident in the many types of beaks specialized for different foods (see Figure 20.12).

The Hawaiian Archipelago is perhaps the world's greatest showcase of evolution. The volcanic islands are about 3500 km from the nearest continent. They become progressively younger to the southeast, terminating with the youngest and largest island, Hawaii, which is less than a million years old and still has active volcanoes. Each island was born naked and was gradually clothed by fauna and flora derived from strays that rode the ocean currents and winds from distant islands and continents, or from older islands of the archipelago itself (Figure 22.11). The physical diversity of each island, including a range of altitudes and extensive differences in rainfall, provides many environmental opportunities for evolutionary divergence by natural selection. Multiple invasions and allopatric speciations have ignited an explosion of adaptive radiation; most of the thousands of species of plants and animals that now inhabit the islands are found nowhere else in the world. In contrast, there are no en-

Figure 22.11
Long-distance dispersal. Plant seeds (the black spots) cling to this sea bird, which is capable of long flights. This is one mechanism that can disperse terrestrial organisms to isolated islands.

demic species on the Florida Keys. Apparently, those islands are so close to the mainland that founding populations are not sequestered long enough for the origin of intrinsic reproductive barriers that block their gene pools from the steady stream of immigrants from the parent populations on the mainland.

Sympatric Speciation

In sympatric speciation, new species arise within the range of parent populations; reproductive isolation evolves without geographical isolation. This can occur in a single generation if some genetic change results in a reproductive barrier between the mutants and the parent population.

Many plant species have their origins in accidents during cell division that result in extra sets of chromosomes. Examine the causes of these chromosomal changes in Figure 22.12. An **autopolyploid** results from a single species that doubles its chromosome number to the tetraploid state. The tetraploids can then fertilize themselves or mate with other tetraploids. However, the mutants cannot interbreed successfully with diploid plants of the original population. The hybrids, which would be triploid ($3n$), would be sterile because unpaired chromosomes result in abnormal meiosis. In just one generation, a postzygotic barrier has caused reproductive isolation and interrupted gene flow between a tiny population of tetraploids (maybe only a single plant initially) and the parent diploid population that surrounds it. Sympatric speciation by autopolyploidy was first discovered early in this century by geneticist Hugo de Vries while he was studying the genetics of the evening primrose, *Oenothera lamarckiana,* a diploid species with 14 chromosomes. One day, de Vries noticed an unusual variant that had appeared among his plants, and microscopic inspection revealed that it was a tetraploid with 28 chromosomes. He found that the plant was unable to breed with the diploid primrose, and he named the new species *Oenothera gigas.*

Another type of polyploid species, much more common than autopolyploids, is called an **allopolyploid,** referring to the contribution of two different species to a polyploid hybrid. The potential evolution of an allopolyploid begins when two different species interbreed and combine their chromosomes. Interspecific hybrids are usually sterile because the haploid set of chromosomes from one species cannot pair during meiosis with the haploid set from the other species. Though infertile, a hybrid may actually be more vigorous than its parents and propagate itself asexually (which many plants can do). At least two mechanisms, described in Figure 22.12b, can transform the sterile hybrids into fertile polyploids.

Some allopolyploids are especially vigorous, apparently because they combine the best qualities of their two parent species. Speciation of polyploids, especially allopolyploids, accounts for 25% to 50% of plant species. Some of these new polyploid species have originated and spread with historic times (Figure 22.13). Many of the plants we grow for food are polyploids. The wheat used for bread, *Triticum aestivum,* is an allopolyploid with 42 chromosomes that probably originated about 8000 years ago as a spontaneous hybrid of a cultivated wheat having 28 chromosomes and a wild grass having 14 chromosomes. Oats, cotton, potatoes, and tobacco are among the other polyploid species important to agriculture. Plant geneticists are now hybridizing plants and using chemicals that induce nondisjunction to help create new polyploids with special qualities. For example, there are artificial hybrids that combine the high yield of wheat with the

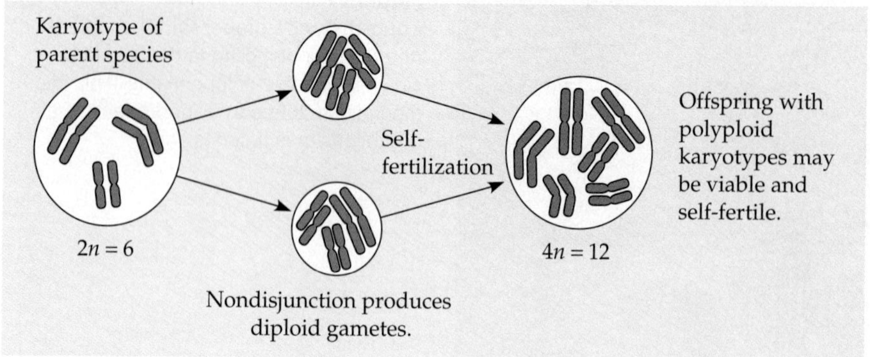

(**a**) Autopolyploidy

Gametes $n = 2$

Hybrid

$2n = 10$

Gametes $n = 5$

Viable fertile hybrid

Species *A* $2n = 4$

Mitosis

Meiosis

Self-fertilization

Nondisjunction produces polyploid condition.

$n = 3$

$n = 5$

$2n = 10$

Species *B* $2n = 6$

Unreduced gamete $2n = 4$

Triploid hybrid $3n = 7$

Unreduced gamete $3n = 7$

Viable fertile hybrid

Species *A* $2n = 4$

$2n = 10$

Normal gamete $n = 3$

Normal gamete $n = 3$

Species *B* $2n = 6$

(**b**) Allopolyploidy

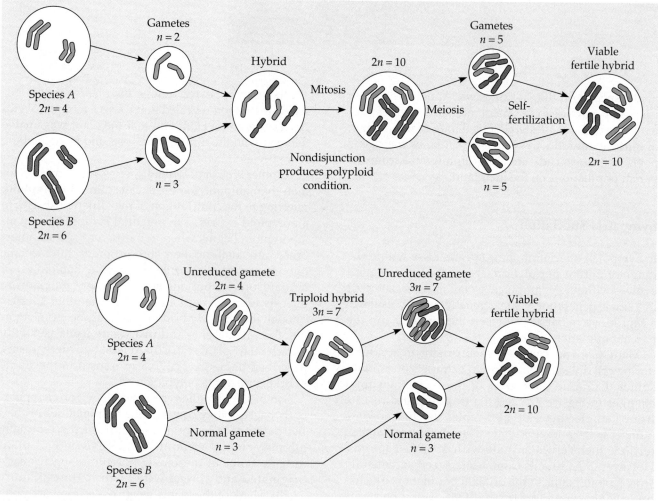

Figure 22.12
Sympatric speciation by polyploidy in plants. (a) Autopolyploidy. Nondisjunction in the germ (reproductive) cell line during either mitosis or meiosis results in diploid gametes. Self-fertilization produces a tetraploid zygote. (**b**) **Allopolyploidy.** A hybrid between two species is normally sterile because chromosomes are not homologous and cannot pair during meiosis. However, the hybrids may be able to reproduce asexually. Two different mechanisms can produce allopolyploid species from such hybrids. (Top) At some instant in the history of the hybrid clone, mitotic nondisjunction affecting the reproductive tissue of an individual may double chromosome number. The hybrid will then be able to make gametes because each chromosome has a homologue with which to synapse during meiosis. Union of gametes from this hybrid may give rise to a new species of interbreeding plants reproductively isolated from both parent species. (Bottom) This figure illustrates a more complex, but probably more common, origin of allopolyploid species. Both mechanisms produce new species having a chromosome number equal to the sum of the chromosomes in the two parent species. These various kinds of accidents during cell division make it possible for a new species to arise without geographic isolation from parent species.

Figure 22.13
***Spartina anglica*, a species with a rapid origin during historic times.** This species of salt-marsh grass originated along the coast of southern England in the 1870s. It is an allopolyploid derived from a European species *(Spartina maritima)* and an American species *(Spartina alternaflora)*. Seeds of the American species, stowaways in the ballast of ships, were accidently introduced to England in the early nineteenth century. The invaders hybridized with the local species, and eventually a third species *(Spartina anglica)*, morphologically distinct and reproductively isolated from its two parent species, evolved as an allopolyploid. (Chromosome numbers are consistent with this mechanism of speciation: For *S. maritima*, $2n = 60$; for *S. alternaflora*, $2n = 62$; and for the new species, *S. anglica*, $2n = 122$). Since its inception, the new marsh grass has spread around the coast of Great Britain, clogging estuaries and becoming something of a pest (weed).

ability of rye to resist disease. (The first attempt at artificial hybridization was not very successful. In 1924, the cabbage was crossed with the radish, resulting in an allopolyploid with the roots of a cabbage and the leaves of a radish.)

Sympatric speciation may also occur in animal evolution, though the mechanisms are different from the chromosome doublings of plants. Animals may become isolated within the geographic range of a parent population, if genetic factors cause them to become fixed on resources not used by the parent population. Consider, for example, the wasps that pollinate figs. Each fig species is pollinated by a particular species of wasp, which mates and lays its eggs in the figs. A genetic change that caused wasps to select a different fig species would segregate mating individuals of this new phenotype from the parent population. This would set the stage for further evolutionary divergence. A balanced polymorphism coupled with assortative mating could also result in sympatric speciation (see Chapter 21). For example, if birds in a population that is dimorphic for beak size began to mate selectively with birds of the same morph, then eventual speciation is possible. Lake Victoria in Africa, which is less than a million years old, is home to almost 200 species of closely related fishes belonging to the cichlid family. Subdivision of populations into groups specialized for exploiting different food sources and other resources in the lake is one process that probably contributed to the explosive radiation of these fishes.

GENETIC MECHANISMS OF SPECIATION

Classifying speciation modes as allopatric or sympatric accentuates the biogeographical factors of speciation but does not emphasize the actual genetic mechanisms that lead to reproductive barriers between populations. Many population geneticists, notably Alan Templeton of Washington University in St. Louis, have advocated an alternative classification that groups speciation mechanisms according to genetic criteria rather than geographical scale. The two general categories can be called speciation by divergence and speciation by peak shifts.

Speciation by Divergence

When two populations adapt to disparate environments, they accumulate differences in the frequencies of genotypes and phenotypes. In the course of this gradual adaptive divergence of two gene pools, reproductive barriers between the two populations may evolve coincidentally, differentiating the populations into two species.

A key point in evolution by divergence is that reproductive barriers can arise without being favored directly by natural selection—there is no drive toward speciation for its own sake. Reproductive isolation is usually a secondary consequence of divergence of the two populations as they adapt to their separate environments. Postzygotic barriers may be pleiotropic effects of interspecific differences in genes controlling development (see Chapter 13 to review pleiotropy). For example, in laboratory hybrids between two very similar species of *Drosophila*, *D. melanogaster* and *D. simulans*, only one of the two sets of genes for synthesis of ribosomal RNA is active. This results in a very low viability for the hybrids. Prezygotic barriers can also evolve as by-products of the gradual genetic divergence of two populations. For instance, if one population of an insect species adapts to a different host plant than other populations do, then a habitat barrier to interbreeding with the other populations is a side effect.

There *are* cases where reproductive isolation evolves more directly by sexual selection in isolated populations (see Chapter 21 to review sexual selection). Such selection may have helped differentiate *Drosophila* on the Hawaiian Islands into hundreds of

Figure 22.14
The role of sexual selection in the radiation of Hawaiian *Drosophila* species. The Hawaiian Islands are home to about 800 *Drosophila* species, one-third of all known species of these fruit flies. Colonization and island hopping made this adaptive radiation possible (see Figure 22.10). Sexual selection probably played a major role in differentiating populations into separate species. Unique courtship rituals, usually dependent on specific morphological features of the males, contribute to prezygotic barriers between closely related species. In *D. heteroneura*, shown here as an example, males have "hammerheads," with eyes set far apart. This trademark probably functions in species recognition as the male approaches the female during courtship. The "pictures" on the wings, species-specific patterns of dots, may also contribute to mate recognition.

Figure 22.15
The adaptive landscape and peak shifts. For any population in a stable environment, there are many alternative states of genetic adaptation symbolized as peaks on an adaptive landscape. The peaks are separated by valleys representing genetic combinations of relatively low average fitness. In a new environment, the adaptive landscape is redefined, and survival of the population may depend on natural selection moving the gene pool to a new adaptive peak. However, population geneticists reserve the term *peak shift* for speciation episodes triggered by nonadaptive changes in the genetic system, such as genetic drift or the origin of a polyploid population. Once a small population is genetically destabilized and dislodged from its original adaptive peak, natural selection may cause a generation-by-generation climb to some new adaptive peak.

species. For example, the wide head of the *D. heteroneura* male enhances reproductive success with females of the same species, but makes successful courtship with females of other species very unlikely (Figure 22.14). However, even when sexual selection results in reproductive barriers, they evolve as adaptations that enhance reproductive success within a single population, not as safeguards against interbreeding with other populations. After all, reproductive barriers usually evolve when populations are allopatric, so they cannot possibly be functioning directly to isolate the gene pools of populations. For this reason, one criticism of the biological species concept is its emphasis on reproductive isolating mechanisms. An alternative that is gaining favor is the **recognition concept of species,** which assumes that the reproductive adaptations of a species consist of a set of characteristics that maximize successful mating with members of the same population. Reproductive isolation from other populations would be a spin-off. Whether a species definition stresses "reproductive isolating mechanisms" or "mate-recognition mechanisms" may seem like stressing opposite sides of the same coin. But the recognition concept may help focus attention on the characteristics that are actually subject to natural selection in an isolated population in the process of speciation.

Speciation by Peak Shifts

In the 1930s, Sewell Wright crafted an evolutionary metaphor known as the adaptive landscape (Figure 22.15). This symbolic landscape has many **adaptive peaks** separated by valleys. An adaptive peak represents an equilibrium state where the gene pool has allele frequencies that maximize the average fitness of a population's members. Even in a stable environment, several adaptive peaks for a given population are possible, but natural selection will tend to maintain the population at a single peak. To reach an alternative adaptive peak by some change in the overall gene pool, a population must go through a period corresponding to a valley on the adaptive landscape, where the average fitness of individuals is low. Thus, if some slight change in the frequency of alleles at one or more loci drives a population off an adaptive peak, natural selection will usually push the population back to its original peak. However, if the environment should change, then the adaptive landscape is redefined. A population that survives in this new environment must climb another adaptive peak through microevolution of its gene pool. But this is just another way of looking at speciation by adaptive divergence. What population geneticists call a *peak shift* is triggered *not* by a new physical environment, but by nonadaptive changes in the genetic system.

Peak shifts can be caused by a founder effect or a bottleneck (see Chapter 21). By randomly changing allele frequencies in the gene pool, genetic drift can knock a small population off its original adaptive peak. If the gene pool is sufficiently destabilized, new adaptive peaks may be within reach. If the population survives, then it will be natural selection that pushes it to a new adaptive peak as the generations pass. Thus, adaptive evolution plays a major role in a peak shift, but it is genetic drift that makes the shift possible. A peak shift can occur in a bottlenecked population even if its environment is stable, because many adaptive peaks are possible under the same environmental conditions. In the case of a founder effect, the small population is not only subject to genetic drift moving the gene pool randomly over the adaptive landscape, but a new set of adaptive peaks now exists. This combination of genetic drift followed by natural selection in a new environment may be responsible for the relatively rapid radiation of island species.

How Much Genetic Change Is Required for Speciation?

It is not possible to generalize about the "genetic distance" between closely related species. In some cases, reproductive isolation may result from the cumulative divergence of populations at many gene loci. In other cases, changes at only a few loci produce reproductive barriers. For example, two species of Hawaiian *Drosophila, D. silvestris* and *D. heteroneura,* differ in alleles at a gene locus determining head shape, a character important in mate recognition by these flies (see Figure 22.14). However, the phenotypic effect of different alleles at this locus is amplified by at least ten other gene loci that interact in an epistatic system. (For a discussion on the genetic interactions known as epistasis, see Chapter 13.) Thus, it took more than one mutation to differentiate these two species of *Drosophila.* However, it is also clear from such examples that massive genetic change involving hundreds of loci is not mandatory for speciation. This conclusion is relevant to a debate among evolutionary biologists about the tempo of speciation, an issue we will now examine.

GRADUAL AND PUNCTUATED INTERPRETATIONS OF SPECIATION

The traditional evolutionary tree that diagrams the descent of species from ancestral forms sprouts branches that diverge gradually, each new species evolving continuously over long spans of time (Figure 22.16a). The theory behind the tree is the extrapolation of the processes of microevolution—changes in the frequencies of alleles in gene pools—to the divergence of species; that is, big changes occur by the accumulation of many small ones. However, paleontologists rarely find gradual transitions of fossil forms. Instead, they often observe species appearing as new forms rather suddenly (in geological terms) in a layer of rocks, persisting essentially unchanged for their tenure on Earth and then disappearing from the record of the rocks as suddenly as they appeared. To Darwin, the origin of species was an extension of adaptation by natural selection, with isolated populations from common ancestral stock evolving differences gradually as they adapted to their local environments (see Chapter 20). But Darwin himself was bewildered by the dearth of connecting fossils and wrote: "Although each species must have passed through numerous transitional stages, it is probable that the periods during which each underwent modification, though many and long as measured by years, have been short in comparison with the periods during which each remained in an unchanged condition."

Advocates of a theory known as **punctuated equilibrium** have redrawn the evolutionary tree to represent the fossil evidence for evolution occurring in spurts of relatively rapid change, instead of a gradual divergence of species (Figure 22.16b). This model, first proposed in 1972 by Niles Eldredge and Stephen Jay Gould, depicts species undergoing most of their morphological modification as they first bud from parent species, and then changing little, even as they produce additional species. The name of the theory is derived from the idea of long periods of stasis punctuated by episodes of speciation. A change in the genome, such as occurs in the origin of new polyploid plants, would be one mechanism of sudden speciation. Punctuationalists, as the proponents of this model of jerky evolution are called, point out that allopatric speciation of a splinter population isolated from its parent population by a geographical barrier can also be quite rapid. Recall that genetic drift and selection can cause significant change in just a few hundred to a few thousand generations in the gene pool of a population cloistered in a challenging new environment.

How can speciation in a few thousand generations, which may require several thousand years, be called an abrupt episode? The fossil record indicates that successful species last for a few million years, on average. Suppose that a particular species survives for 5 million years, but most of its morphological changes occurred during the first 50,000 years of its existence. In this case, the speciation episode was compressed into just 1% of the lifetime of the species. On the time scale that can generally be resolved in fossil strata, the species will appear suddenly in rocks of a certain age and then linger with little or no change before becoming extinct.

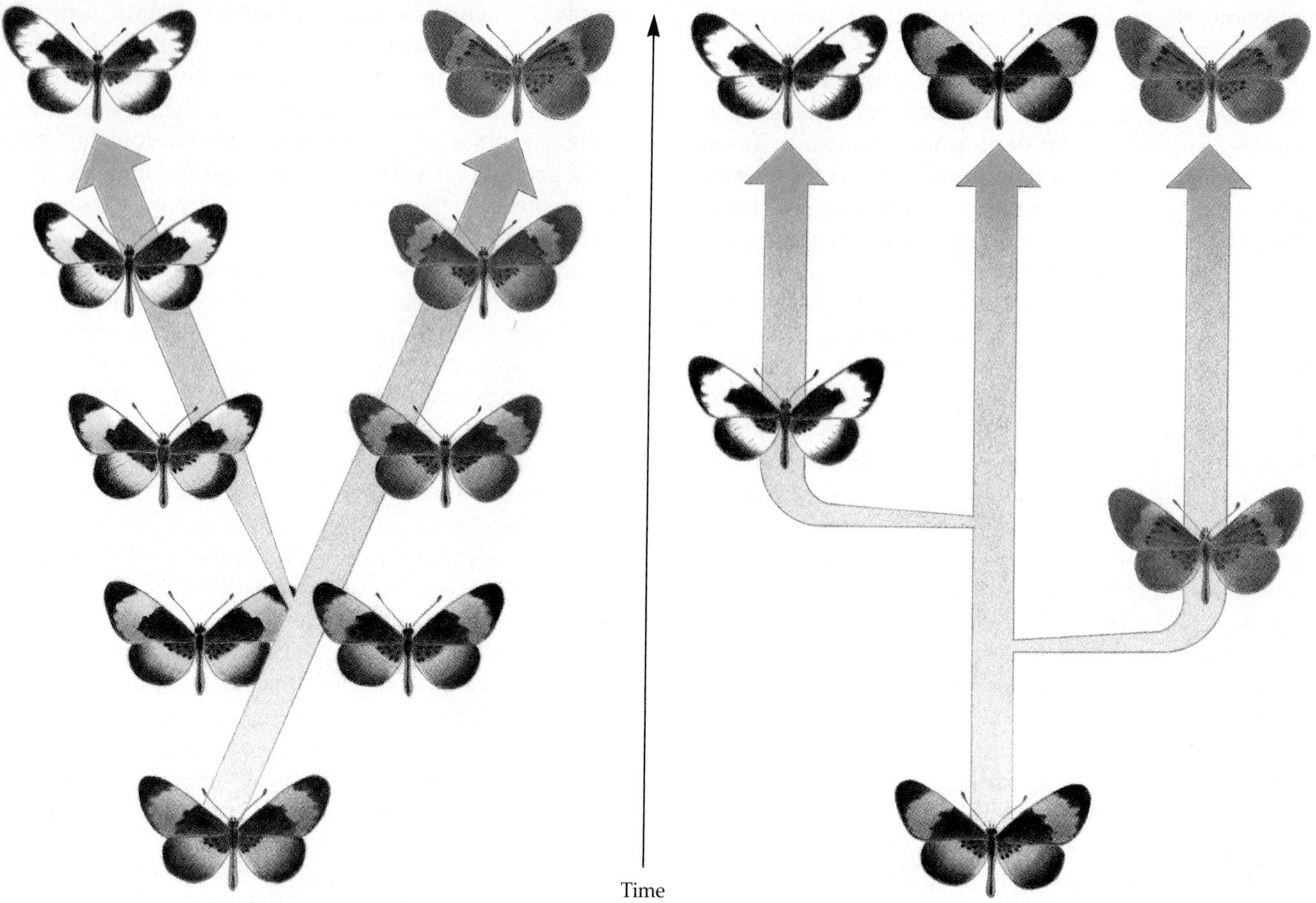

Figure 22.16
Two models for the tempo of speciation. A new species buds from its parent species as a small isolated population, indicated in this figure by the narrow base of the branching arrow. The horizontal axis symbolizes morphological change. (**a**) In the gradualist model, species descended from a common ancestor diverge more and more in morphology as they acquire unique adaptations. (**b**) According to the model known as punctuated equilibrium, a new species changes most as it buds from a parent species, and then changes little for the rest of its existence.

During its formative millennia, the species may have accumulated its modifications gradually, but relative to the overall history of the species, its inception was abrupt. This scenario of an evolutionary spurt preceding a much longer period of morphological stasis would explain why paleontologists find so few transitional fossils.

Once it is acknowledged that "sudden" may be many thousands of years on the vast scale of geological time, the debate between punctuationalists and gradualists over the rate of speciation is muted somewhat. The degree to which a species changes after its origin is another issue. If the species is adapted to an environment that stays the same, then natural selection would counter changes in the gene pool (see Chapter 21). In this view, the tendency for stabilizing selection to hold a population at one adaptive peak results in long periods of stasis.

Some gradualists retort that stasis is an illusion. Many species may continue to change after they come into existence, but in ways that cannot be detected from fossils. By necessity, paleontologists base their theories of descent almost entirely on external anatomy and skeletons. Changes in internal anatomy may go unnoticed, as would modifications in physiology or behavior. Population geneticists point out that many of the effects of microevolution occur at the molecular level without overtly affecting morphology.

Even the claim of long periods of morphological stasis in the history of species is contested. Over the past 13 years, Peter Sheldon of Trinity College in Dublin has analyzed about 15,000 fossils of trilobites from shale deposits in Wales (trilobites are extinct arthropods; see Chapter 29). At this study site, the fossil record of trilobites is unusually complete. Paleontologists have arranged these fossils into several evolutionary lineages, and the youngest and oldest fossils of each lineage have been classified as different species because of morphological differences, including a different number of ridges on the tail sections of their shells. However, after Sheldon's extensive study of the fossils, he finds it impossible to draw such species boundaries. In each evolutionary lineage of the trilobites, the average number of tail ridges in the fossil populations changes gradually in the succession of rock layers. Sheldon argues that the requirement for paleontologists to give their fossils species names can artificially lead to a punctuated interpretation of a fossil series that actually changes gradually. What is needed, Sheldon suggests, are many additional exhaustive studies of fossil morphology where specific lineages are well preserved. Only then will we be able to assess the relative importance of gradual and punctuated tempos in the origin of new species. Whatever the outcome, there is no question that the theory of punctuated equilibrium has stimulated research and catalyzed a new interest in paleontology.

In the next chapter, we will see that the debate between gradualists and punctuationalists extends beyond the issue of speciation to major patterns in the history of life.

STUDY OUTLINE

1. New species can arise by transformation of an entire population (anagenesis) or, more commonly, by budding of small isolated populations from an ancestral species (cladogenesis).
2. Cladogenesis, or branching evolution, increases biological diversity.

The Species Problem (pp. 456–459)

1. There are two main concepts for describing species. Morphospecies are defined in terms of their unique anatomy. Biological species are groups of populations that share a distinctive gene pool by having the potential to interbreed.
2. The biological species concept cannot apply to species that are asexual or extinct.

Reproductive Barriers (pp. 459–462)

1. Any intrinsic factor that prevents two species from producing fertile hybrids is a reproductive barrier that maintains the genetic integrity of a species.
2. Prezygotic barriers prevent cross-specific mating or fertilization. In particular, species that occupy the same geographical area often live in separate habitats (habitat isolation); breed at different times (temporal isolation); possess unique, exclusive mating signals and courtship behaviors (behavioral isolation); and/or have anatomically incompatible reproductive organs (mechanical isolation) or incompatible sex cells (gametic isolation).
3. Even if two different species manage to mate, postzygotic barriers usually prevent the interspecific hybrids from developing into adults, breeding with either parent species, or producing viable, fertile offspring.
4. Occasionally, however, fertile hybrids successfully mate with one of the parent species, causing genes to seep through the reproductive barriers. As long as this introgression is limited, the two species remain distinct.

The Biogeography of Speciation (pp. 462–467)

1. The gene pool of a species can diverge in evolution through allopatric or sympatric speciation.
2. Allopatric speciation occurs when a splinter population diverges in evolution from its parent population after becoming geographically isolated.
3. Small splinter populations are better candidates than large ones for allopatric speciation because genetic drift and selection can change a small gene pool faster than a large one.
4. Adaptive radiation is the generation of numerous species from a common ancestor introduced to diverse environments. Island chains are showcases of radiation. Beginning with an ancestral species that managed to reach an archipelago, cloistering of populations on separate islands and occasional island hopping by founders result in multiple episodes of allopatric speciation.
5. Sympatric speciation occurs without geographical separation when a segment of the population experiences a genetic change that results in reproductive isolation. Sympatric speciation is most common in plants, in which the mutation is usually the doubling of the chromosome number. Autopolyploids are species derived this way from one ancestral species. Allopolyploids are species with multiple sets of chromosomes derived from two different species. Sympatric speciation is also possible in animals if a few individuals in a population switch habitats within the geographic range of the parent population.

Genetic Mechanisms of Speciation (pp. 467–469)

1. Reproductive barriers may arise coincidentally as two populations genetically diverge while adapting to different environments.
2. Sexual selection may lead directly to reproductive barriers. According to the recognition concept of species, natural selection would amplify adaptations that enhance reproductive success with members of the same species.
3. According to Wright's evolutionary metaphor, a population's gene pool is perched on an adaptive peak, one of several possible peaks on an adaptive landscape. Natural selection will tend to maintain a population at an adaptive peak that maximizes fitness. Peak shifts, associated with nonadaptive changes in the gene pool, may be initiated by a founder effect or bottleneck and driven by genetic drift. Adaptive evolution can push a destabilized gene pool to a new adaptive peak.
4. Reproductive isolation may result from a change at just a few loci or by the cumulative divergence at many gene loci.

Gradual and Punctuated Interpretations of Speciation (pp. 469–471)

1. The conventional view is that speciation usually occurs gradually by an accumulation of microevolutionary changes in gene pools.
2. According to the theory of punctuated equilibrium, the relatively sudden appearance of species in the fossil record is a real phenomenon rather than a reflection of an incomplete fossil record. In this view, a species changes most when it buds from an ancestral species and then remains fairly static in morphology for the rest of its duration. Biological history is thus envisioned as long periods of stasis punctuated by episodes of speciation.
3. Additional exhaustive studies of fossil lineages will be necessary to determine the relative importance of gradual and punctuated tempos in the origin of species.

SELF-QUIZ

1. Most of biological diversity has probably arisen by
   a. anagenesis
   b. cladogenesis
   c. phyletic evolution
   d. hybridization
   e. sympatric speciation

2. Bird guides used to list the myrtle warbler and Audubon's warbler as distinct species, but recent books show them as eastern (left) and western forms of a single species, the yellow-rumped warbler. Experts must have found that the two kinds of warblers
   a. live in the same areas
   b. successfully interbreed
   c. look enough alike to be considered one species
   d. are reproductively isolated from each other
   e. are allopatric
3. The largest unit in which gene flow is possible is a
   a. population
   b. species
   c. genus
   d. subspecies
   e. phylum
4. Some species of *Anopheles* mosquito live in brackish water, some in running fresh water, and others in stagnant water. What type of reproductive barrier is most obviously separating these different species?
   a. habitat isolation
   b. temporal isolation
   c. behavioral isolation
   d. gametic isolation
   e. postzygotic barriers
5. The reproductive barrier that maintains the species boundary between horses and donkeys is
   a. mechanical isolation
   b. gametic isolation
   c. hybrid inviability
   d. hybrid sterility
   e. hybrid breakdown
6. According to advocates of the punctuated equilibrium theory,
   a. natural selection is unimportant as a mechanism of evolution
   b. given enough time, most existing species will branch gradually into new species
   c. a new species accumulates most of its unique features as it comes into existence and changes little for the rest of its duration as a species
   d. most evolution is anagenic
   e. speciation is usually due to single mutation
7. The biological species concept cannot be applied to two putative species that are
   a. sympatric
   b. nearly indistinguishable in morphology
   c. reproductively isolated
   d. capable of forming viable hybrids
   e. exclusively asexual
8. Future cladogenesis of human populations to form new hominid (human) species is probably unlikely because
   a. the environment has stabilized
   b. humans are already perfectly adapted organisms
   c. only one adaptive peak is possible for hominids
   d. most human populations are very large and are incompletely isolated from surrounding populations
   e. human variation is not very extensive
9. Plant species A has a diploid number of 12. Plant species B has a diploid number of 16. A new species, C, arises as an allopolyploid from hybridization of A and B. The diploid number of C would probably be
   a. 12
   b. 14
   c. 16
   d. 28
   e. 56
10. The speciation episode described in Question #7 is most likely a case of
    a. allopatric speciation
    b. sympatric speciation
    c. speciation based on sexual selection
    d. adaptive radiation
    e. anagenic speciation

CHALLENGE QUESTIONS

1. According to one hypothesis, two closely related species should be most distinct from each other where their ranges overlap—that is, in regions of sympatry. This phenomenon is called character displacement. The hypothesis goes on to propose that character displacement reduces interspecific competition for food, habitat, and other resources, and also minimizes hybridization between the species in their region of sympatry. Character displacement has been difficult to demonstrate in nature. An experiment has nevertheless addressed the problem by comparing beak sizes of two species of Darwin's ground finches *(Geospiza fortis* and *G. fuliginosa)* in sympatry and allopatry. The experiment controlled for any possible effect on morphology by variation in food supply among locations. Why do you think this control was necessary? After taking this control into account, how would you expect the relative beak sizes of *G. fortis* and *G. fuliginosa* on Santa Cruz Island, the only island they cohabit, to compare with the relative beak sizes of *G. fortis* on the islands of Daphne Major and *G. fuliginosa* on Los Hermanos if character displacement were operative?

2. In 1990, J. Jackson and A. Cheetham, of the Smithsonian Institute, published a study on the reliability of morphology as a taxonomic tool for classifying invertebrates called bryozoans into species. Skeleton morphology alone was sufficient to distinguish modern species. Relate this research to the criticism that the theory of punctuated equilibrium is an inaccurate interpretation of the fossil record because it is based only on measuring morphological change.

SCIENCE, TECHNOLOGY, AND SOCIETY

1. Thousands of Snake River sockeye salmon used to spawn in Redfish Lake, Idaho, but in 1990 not a single sockeye returned to the spawning grounds. The salmon are impeded on their journeys up and down the Columbia River and Snake River by eight hydroelectric dams. In 1991, the U. S. Government added the Snake River sockeye salmon to the list of endangered species. The Snake River sockeye salmon is one of several sockeye populations in the river system. The Endangered Species Act grants protection not just to a whole biological species, but also to "distinct population segments." How would you define "distinct" in this context, and how would you try to find out whether a salmon population meets your definition? What measures would you be willing to implement to protect the Snake River sockeye salmon population? What would be the advantages and disadvantages of these protective measures?

FURTHER READING

Culotta, E. "How Many Genes Had to Change to Produce Corn?" *Science,* June 28, 1991. The evolution of an important food source.

Eldredge, N., and S. J. Gould. "Punctuated Equilibria: An Alternative to Phyletic Gradualism." In *Models in Paleobiology,* ed. T. J. M. Schopf. San Francisco: Freeman, Cooper, 1972. The original proposal for punctuated equilibrium.

Gould, S. J. "Opus 200." *Natural History,* August 1991. Gould's two-hundredth essay for this magazine traces the origin and impact of punctuated equilibrium.

Kaneshiro, K. Y. "Speciation in the Hawaiian Drosophila." *BioScience,* April 1988. The importance of sexual selection in a classic case of adaptive radiation.

Kluger, T. "Go Fish." *Discover,* March 1992. Rapid speciation in Lake Victoria.

Rennie, J. "Are Species Specious?" *Scientific American,* November 1991. Problems with the biological species concept.

Ryan, M. J. "Signals, Species, and Sexual Selection." *American Scientist,* January/February 1990. Frog songs in the context of the mate-recognition concept of species.

Sheldon, P. "Making the Most of Evolutionary Diaries." *New Scientist,* January 21, 1988. An exhaustive study of Welsh trilobites supports a gradualistic interpretation of their evolution.

23 Tracing Phylogeny: Macroevolution, the Fossil Record, and Systematics

The Record of the Rocks

Mechanisms of Macroevolution

Systematics: Tracing Phylogeny

Is a New Evolutionary Synthesis Necessary?

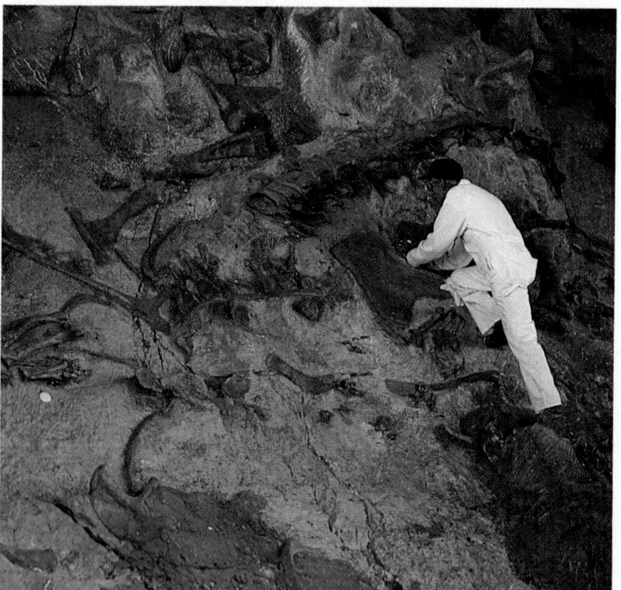

Figure 23.1
The fossil record: a chronicle of macroevolution. This paleontologist is recovering a fossilized dinosaur skeleton at Utah's Dinosaur National Monument. The adaptive radiation of dinosaurs and their subsequent disappearance during a period of mass extinctions are examples of major transitions in the history of life. In this chapter, you will learn more about the fossil record as a chronicle of macroevolution and examine mechanisms of macroevolution.

The term *macro*evolution implies substantial change in organisms. Here is a more specific definition: **Macroevolution** is the origin of taxonomic groups higher than the species level. In other words, macroevolution causes changes substantial enough that we view its products as new genera, new families, or even new phyla. The origin of mammals from their reptilian ancestors is an example (reptiles and mammals are related, but different enough that zoologists place them in separate vertebrate classes). In broader terms, macroevolution is the story of the major events in the history of life that are revealed by the fossil record (Figure 23.1). Evolution on this grand scale encompasses the origin of novel designs, such as the feathers and wings of birds and the upright posture of humans; evolutionary trends, such as increasing brain size in mammals; the explosive diversification of certain groups of organisms following some evolutionary breakthrough, such as the adaptive radiation of flowering plants; and mass extinctions, which cleared the way for new episodes of adaptive radiation, such as the radiation of mammals following the disappearance of the dinosaurs.

This chapter describes how biologists study macroevolution, discusses theories on the origin of new biological designs, and examines a few important episodes in the history of life. The chapter also relates these topics to systematics, the study of biological diversity. The last section, the capstone of our unit on mechanisms of evolution, considers one of the fundamental questions of evolutionary biology: Is macroevolution mainly the cumulative product of microevolution working over vast spans of time? Put another way, is the gradual modification of populations due to natural selection the main cause of the larger changes we observe in the fossil record as macroevolution? Throughout our study of macroevolution, the theme that evolution is a consequence of interactions between organisms and their environments will be extended to environmental and biological change of global proportions.

THE RECORD OF THE ROCKS

Fossils are the historical documents of biology. They are collected and interpreted by specialists called paleontologists (Gr. *palaios*, "old," a reference to the fossils, not the collectors). In this section, you will learn more about the contributions and limitations of the fossil record in the study of macroevolution.

How Fossils Form

A fossil (L. *fossilis*, "dug up") is any preserved remnant or impression left by an organism that lived in the past. Sedimentary rocks are the richest sources of fossils (Figure 23.2a). Sand and silt weathered and eroded from the land are carried by rivers to seas and swamps, where the particles settle to the bottom. Deposits pile up and compress the older sediments below into rock—sand into sandstone, and mud into shale. Aquatic organisms, and terrestrial ones swept into the seas and swamps, settle when they die, along with the sediments. A tiny fraction of them is then preserved as fossils.

The organic substances of a dead organism buried in sediments usually decay rapidly, but hard parts that are rich in minerals, such as the bones and teeth of vertebrates and the shells of many invertebrates and protists, may remain as fossils (Figure 23.1; Figure 23.2a and b). Paleontologists have unearthed nearly complete skeletons of dinosaurs and other forms, but more often, the finds consist of parts of skulls, bone fragments, or teeth. Many of these relics are hardened even more and preserved by a process called petrification. Under the right conditions, minerals dissolved in groundwater seep into the tissues of a dead organism and replace organic material; the plant or animal turns to stone. Bizarre forests of petrified trees can be explored in parts of the southwestern United States that are now desert (Figure 23.2c).

Rarer than mineralized fossils are those that retain organic material. They are sometimes discovered as thin films pressed between layers of sandstone or shale. For example, paleontologists have discovered plant leaves millions of years old that are still green with chlorophyll and well enough preserved for their organic composition to be analyzed and the ultrastructure of their cells to be explored with the electron microscope (Figure 23.2d). In 1990, one research team even managed to clone a minuscule sample of DNA extracted from an ancient magnolia leaf (you will learn more about this feat later in the chapter). The most common fossilized plant material is pollen, which has a hard organic case that resists degradation.

The fossils that paleontologists find in many of their digs are not the actual remnants of organisms at all, but rocks that form as replicas of the organisms. These fossils result when a dead organism captured in sediments decays and leaves an empty mold that becomes filled with minerals dissolved in water. The minerals may subsequently crystallize, forming a cast in the shape of the organism (Figure 23.2e).

Casts called trace fossils form in footprints, animal burrows, or other impressions left in sediments by the activities of animals. These rocks are fossilized behavior; they tell paleontologists something about how the animals that left trace fossils lived. For example, dinosaur tracks provide clues about the animal's locomotion—its gait (pattern of leg movements), stride length, and speed.

Although relics of past life locked in sedimentary rocks are the most common fossils, they are not the only kind. If an organism happens to die in a place where bacteria and fungi cannot decompose the corpse, the entire body, including soft parts, may be preserved as a fossil. For example, the insect in Figure 23.2f got stuck in a drop of resin from a tree about 40 million years ago. The resin eventually hardened into amber, entombing the insect much like a biological specimen encased in a plastic block. There are other mechanisms that preserve whole organisms. Explorers have discovered mammoths, bison, and other extinct mammals frozen in Arctic ice. You have probably read about Bronze Age humans preserved in acid bogs where conditions retard decomposition. Such discoveries make the news, but biologists rely mainly on sedimentary fossils to reconstruct the history of life.

Limitations of the Fossil Record

The discovery of a fossil is the culmination of a sequence of improbable coincidences. First, the organism had to die in the right place at the right time for burial conditions to favor fossilization. Then the rock layer containing the fossil had to escape geological processes that destroy or severely distort rocks, such as erosion, pressure from superimposed strata, or the melting of rocks that occurs at some locations. If the fossil was preserved, there is only a slight chance that a river carving a canyon or some other process will expose the rock containing the fossil. There is only a remote chance that someone will find the fossil, although discovery is more probable for people who are purposefully looking for fossils. No wonder the fossil record is incomplete. A substantial fraction of species that have lived probably left no fossils, most fossils that formed have been destroyed, and only a fraction of the existing fossils has been discovered. The fossil record, far from being a complete sampling of organisms of the past, is slanted in favor of species that existed for a long time, were abundant and widespread, and had shells or hard skeletons. Paleontologists, like all historians,

Figure 23.2
A survey of fossils. (**a**) Sedimentary rocks are the richest hunting grounds for paleontologists. Numerous invertebrate shells and other fossils about 15 million years old are visible in the sediments exposed on the face of this cliff. (**b**) The hard parts of organisms, such as this skull of *Australopithecus africanus,* an ancestor of humans that lived about 2.5 million years ago, are the most common fossils. The fossils may become even harder by the process of petrification, which replaces organic material with minerals. (**c**) These petrified trees stood about 190 million years ago in what is now an Arizona desert. (**d**) Some sedimentary fossils, such as this 40-million-year-old leaf, retain organic material. (**e**) Buried organisms, such as these invertebrates called brachiopods, which lived about 375 million years ago, decay and leave molds that may be filled by minerals dissolved in water. The casts that form when the minerals harden are replicas of the organisms. (**f**) This 40-million-year-old insect embedded in amber (hardened resin from a tree) is almost perfectly preserved.

must reconstruct the past from incomplete records. Even with its limitations, however, the fossil record is a remarkably detailed document of macroevolution over the vast scale of geological time.

The Fossil Record and the Geological Time Scale

Fossils are reliable historical data only if we can determine their ages. Here, we examine the methods for resolving where fossils fit on the geological time scale.

Relative Dating The trapping of dead organisms in sediments freezes fossils in time. At any particular location, sedimentation is not continuous, but occurs in intervals when the sea level changes or lakes and swamps dry up and refill. Even when a region is submerged, the rate of sedimentation and the types of sedimentary particles vary with time. As a result of these different periods of sedimentation, the rock forms in layers, or strata (see Figure 20.5). The fossils in each layer are a local sampling of the organisms that existed at the time that sediment was deposited. Younger sediments are superimposed upon older ones. Thus, this book of sediments tells the relative ages of fossils.

The strata at one location can often be correlated with strata at another location by the presence of similar fossils, known as **index fossils.** The best index fossils for correlating strata that are far apart are the shells of sea animals that were widespread. At any one location where a road cut or canyon wall reveals layered rocks, there are likely to be gaps in the sequence. That area may have been above sea level during different periods, and thus no sedimentation occurred. Some of the sedimentary layers that were deposited when the area was submerged may have been scraped away by subsequent periods of erosion.

By studying many different sites, geologists have established a time scale with a consistent sequence of geological periods (Table 23.1). These periods are grouped into four eras: the Precambrian, Paleozoic, Mesozoic, and Cenozoic eras. Each era represents a distinct age in the history of Earth and its life; the boundaries are marked in the fossil record by explosive radiations of many new forms of life following mass extinctions. For example, the beginning of the Paleozoic era is recorded by a great diversity of fossilized invertebrates that are absent in rocks of the late Precambrian era. And most of the animals that lived during the late Precambrian became extinct at the end of that era. The boundaries between periods within each era also mark major transitions in the forms of life fossilized in rocks, though these changes are less radical than the geological and biological revolutions that divide the eras. The periods within each era are further subdivided into finer intervals called epochs (only the epochs of the current era, the Cenozoic, appear in Table 23.1). The timeline to the right of Table 23.1 will help you understand that the geological eras were not equal in duration. Also, notice that the Jurassic period lasted almost twice as long as the Triassic period. Remember that scientists have not divided geological time in an arbitrary manner, but have located boundaries in the record of the rocks that correspond to times of great change.

The record of the rocks is a serial that chronicles the *relative* ages of fossils; it tells us the order in which groups of species present in a sequence of strata evolved. However, the series of sedimentary rocks does not tell the *absolute* ages of the embedded fossils. The difference is analogous to peeling the layers of wallpaper from the walls of a very old house that has been inhabited by many owners. You could determine the sequence in which the papers had been applied, but not the year that each layer was added.

Absolute Dating "Absolute" dating does not mean errorless dating, but only that age is given in years instead of relative terms such as *before* and *after, early* and *late.* **Radioactive dating** is the method most often used to determine the ages of rocks and fossils on a scale of absolute time. Fossils contain isotopes of elements that accumulated in the organisms when they were alive. Because each radioactive isotope has a fixed rate of decay, it can be used to date a specimen (see Chapter 2). An isotope's **half-life,** the number of years it takes for 50% of the original sample to decay, is unaffected by temperature, pressure, and other environmental variables. For example, carbon-14 has a half-life of 5600 years, a reliable rate of decay that can be used to date relatively young fossils (see the Methods Box, p. 479). Paleontologists use radioactive isotopes with longer half-lives to date older fossils.

Methods other than radioactive dating can be used on some fossils. Amino acids exist in two isomers with either left-handed or right-handed symmetry, designated the L- and D-forms, respectively. Organisms synthesize only L-amino acids, which are incorporated into proteins. After an organism dies, however, its population of L-amino acids is slowly converted, resulting in a mixture of L- and D-amino acids. In a fossil, the ratio of L- and D-amino acids can be measured. Knowing the rate at which this chemical conversion, called racemization, takes place, we can determine how long the organism has been dead. Archeologists have recently used this method to date ostrich eggshells found along with fossils of early humans. The humans probably ate the eggs and used the shells for water bowls. Racemization, unlike radioactive decay, is temperature-sensitive. But for fossils found in locations where climate apparently has not changed significantly since the fossils formed, the two dating methods agree closely on the age of the fossils.

The order of sedimentary strata records the sequence of biological change, and dating methods tell

Table 23.1 The Geological Time Scale

Era	Period	Epoch	Age (millions of years)	Some Important Events in the History of Life
CENOZOIC	Quaternary	Recent		Historic time
			0.01	
		Pleistocene		Ice ages; humans appear
			1.8	
	Tertiary	Pliocene		Apelike ancestors of humans appear
			5	
		Miocene		Continued radiation of mammals and angiosperms
			24	
		Oligocene		Origins of most modern mammalian orders, including apes
			38	
		Eocene		Angiosperm dominance increases; further increase in mammalian diversity
			54	
		Paleocene		Major radiation of mammals, birds, and pollinating insects
			65	
MESOZOIC	Cretaceous			Flowering plants (angiosperms) appear; dinosaurs become extinct at end of period
			144	
	Jurassic			Gymnosperms continue as dominant plants; dinosaurs dominant
			213	
	Triassic			Gymnosperms dominate landscape; first dinosaurs, mammals, and birds
			248	
PALEOZOIC	Permian			Radiation of reptiles; origins of mammal-like reptiles and most modern orders of insects; extinction of many marine invertebrates
			286	
	Carboniferous			Extensive forests of vascular plants; first seed plants; origin of reptiles; amphibians dominant
			360	
	Devonian			Diversification of bony fishes; first amphibians and insects
			408	
	Silurian			Diversity of jawless vertebrates; colonization of land by vascular plants and arthropods
			438	
	Ordovician			First vertebrates (jawless fishes); marine algae abundant
			505	
	Cambrian			Origin of most invertebrate phyla; diverse algae
			590	
PRECAMBRIAN			700	Origin of first animals
			1500	Oldest eukaryotic fossils
			2500	Oxygen begins accumulating in atmosphere
			3500	Oldest definite fossils known (prokaryotes)
			4600	Approximate origin of Earth

us how long ago these episodes of macroevolution occurred. We now shift our attention from the record of macroevolution to the mechanisms that cause these transformations of life.

MECHANISMS OF MACROEVOLUTION

What processes actually cause the large-scale evolutionary changes that we can trace through the fossil record? How, for instance, do the novel features that define taxonomic groups above the species level, such as the flight adaptations of birds, arise? What accounts for evolutionary trends that appear from the fossil record to be progressive, such as the general increase in the size of certain families of reptiles during the age of dinosaurs, or the increase in brain size during human evolution? How have global geological changes affected macroevolution? And how can we explain the major fluctuations in biological diversity evident in the fossil record, such as a proliferation of animal diversity

Methods: Radioactive Dating

Radioactive isotopes can be detected and their amounts measured by the radiation they emit as they decompose to more stable atoms (see Chapter 2). Paleontologists use the clocklike decay of radioactive isotopes to date fossils and rocks. Carbon-14, for example, has a half-life of 5600 years, meaning that half the carbon-14 in a specimen will be gone in 5600 years, half the remainder will be gone in another 5600 years, and so on. The graph to the right traces this exponential decline in radioactivity. In the example illustrated below, we use carbon-14 dating to determine the vintage of a fossilized clam shell. Because the half-life of carbon-14 is relatively short, this isotope is only reliable for dating fossils less than about 50,000 years old. To date older fossils, paleontologists use radioactive isotopes with longer half-lives. For instance, potassium-40, a radioactive isotope with a half-life of 1.3 billion years, can be used to date rocks hundreds of millions of years old and infer the age of fossils embedded in those rocks. Radioactive dating has an error factor of less than 10%.

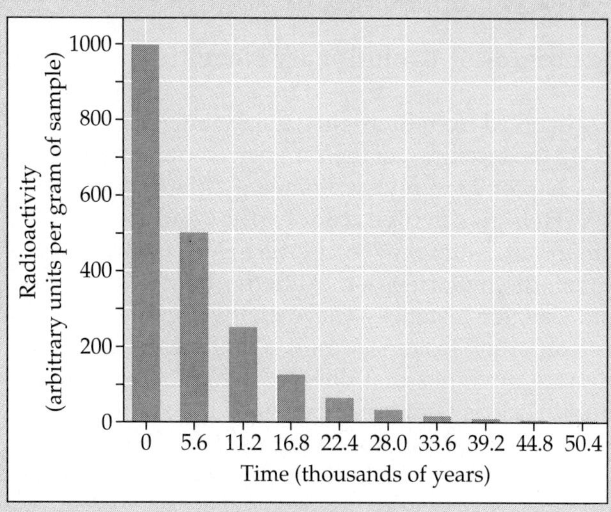

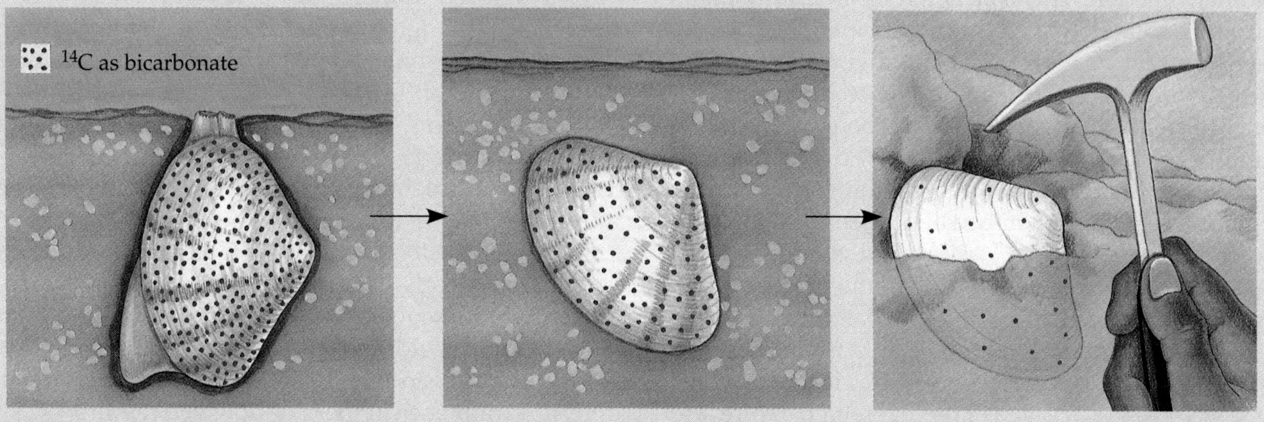

While an organism, in this case a clam, is alive, it assimilates the different isotopes of each element in proportions determined by their relative abundances in the environment. Carbon-14 is taken up in trace quantities, along with much larger quantities of the more common carbon-12.

After the clam dies, it is covered with sediment, and its shell eventually becomes consolidated into a layer of rock as the sediment is compressed. From the time the clam dies and ceases to assimilate carbon, the amount of carbon-14 relative to carbon-12 in its remains declines due to radioactive decay.

After the clam fossil is found, its age can be determined by measuring the ratio of the two isotopes to learn how many half-life reductions have occurred since it died. For example, if the ratio of carbon-14 to carbon-12 in this fossil clam was found to be one-fourth that of a living organism, this fossil would be about 11,200 years old.

at the beginning of the Paleozoic era or the mass extinctions that have occurred in the past? These are some of the questions that interest biologists who study macroevolution.

The Origins of Evolutionary Novelties

Birds evolved from dinosaurs; their wings are homologous to the forelimbs of their modern reptilian cousins. (See Chapter 20 to review homology.) How could flying vertebrates evolve from earthbound ancestors? Humans and chimpanzees are very close relatives, and yet their dissimilarities—the differences in posture and brain size, for instance—place the two animals in entirely different ecological roles. What creates the evolutionary novelties that define the higher taxonomic groups, such as families and classes? Put another way, how do new designs for living evolve? One mechanism may be the gradual refinement of existing structures for new functions.

Preadaptation Most biological structures have an evolutionary plasticity that makes alternative functions possible. From a retrospective vantage point, evolutionists use the term **preadaptation** for a structure that evolved in one context and became co-opted for another function. This concept does not imply that a structure somehow evolves in anticipation of future use. Natural selection cannot predict the future and can only improve a structure in the context of its current utility. The light, honeycombed bones of birds (see Figure 1.6b, p. 9) could not have evolved in earthbound reptilian ancestors as an adaptation for upcoming flights. If these honeycombed bones predated flight, as clearly indicated by the fossil record, then they must have had some function on the ground. The probable ancestors of birds were agile, bipedal dinosaurs that also would have benefited from a light frame. It is possible that winglike forelimbs, as well as feathers, which increased the surface area of these forelimbs, were also co-opted for flight after functioning in some other capacity, such as "netting" insects and other small prey chased by small, fleet-footed dinosaurs (or perhaps feathers functioned mainly in social displays—in courtship, for example). The first flight may have been only a glide down from a tree or an extended hop in pursuit of prey or escape from predator. Once flight itself became an advantage, natural selection would have remodeled feathers and wings to better fit their additional function.

We cannot prove that this scenario for the evolution of birds from reptiles is correct, because the known fossil record provides so little history of this transition. However, preadaptation offers one explanation for how novel designs can arise gradually through a series of intermediate stages, each of which has some function in the organism's current context. Harvard zoologist Karel Liem, who is interviewed preceding Unit Seven, puts it this way: "Evolution is like modifying a machine while it's running." The concept that evolutionary novelties can evolve by the remodeling of old structures for new functions is in the Darwinian tradition of large changes being an accumulation of many small changes crafted by natural selection.

Development and Macroevolution The evolution of complex structures, such as wings and feathers, from their antecedents requires so much remodeling that changes are probably involved at a large number of gene loci. In other cases, relatively few changes in the genome can apparently cause major modifications of morphology, such as some of the differences between humans and chimpanzees. How can slight genetic divergence become magnified into major differences between organisms? Scientists working at the exciting interface of developmental biology and evolutionary biology are trying to answer this question.

Genes that program development control the rate, timing, and spatial pattern of changes in an organism's form as it is transfigured from a zygote into an adult. For example, **allometric growth,** a difference in the relative rates of growth of various parts of the body, helps shape an organism. Figure 23.3a tracks how allometric growth alters human body proportions during development. Change these relative rates of growth even slightly, and you change the adult form substantially. For example, different allometric patterns contribute to the contrasting shapes of human and chimpanzee skulls (Figure 23.3b). Allometry is one mechanism by which a subtle alteration of development becomes compounded in its effects on the adult.

In addition to affecting developmental rates, genetic changes can also alter the timing of developmental events—the sequence in which different organs start and stop developing. For example, the variation in stripes between two zebra species probably results from a difference in the time at which the stripes begin to take form in the embryos (Figure 23.4). Of more general importance are changes in developmental timing that result in **paedomorphosis** (Gr. *paid,* "child," and *morphos,* "form"), in which a sexually mature adult retains features that were juvenile structures in its evolutionary ancestors. For example, most salamander species have a larval stage that undergoes metamorphosis to become an adult. But some species grow to adult size and become sexually mature while retaining gills and certain other larval features (Figure 23.5). Such an evolutionary alteration of developmental timing can produce animals that appear very different from their ancestors.

Changes in developmental chronology have also been important in human evolution. Humans and chimpanzees are very closely related through descent

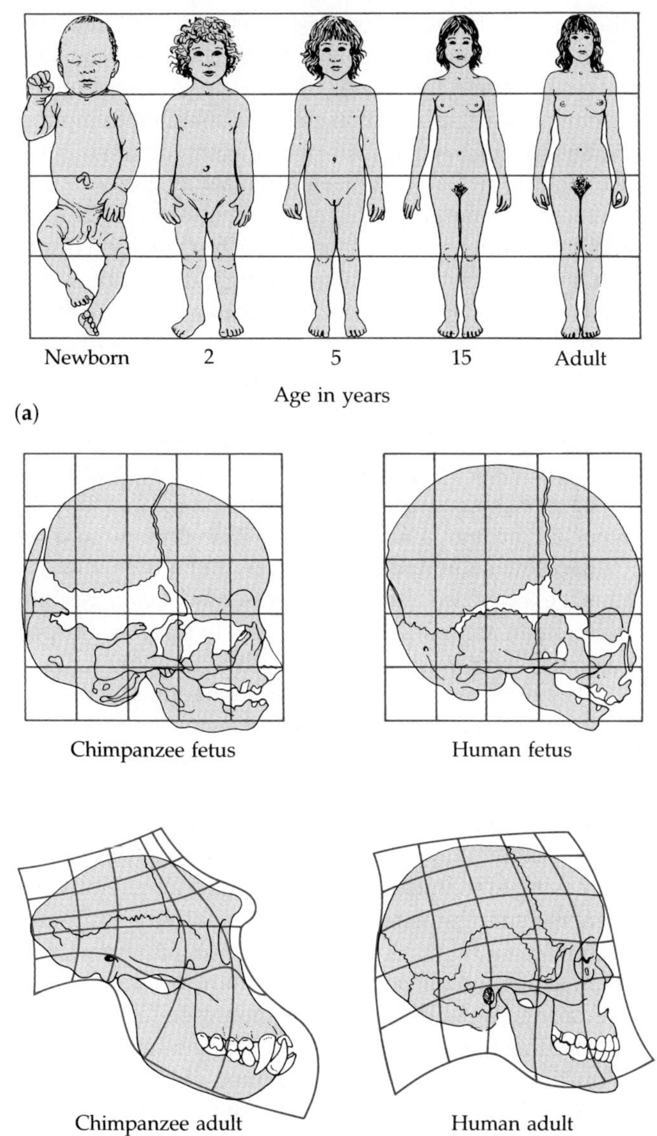

Figure 23.3
Allometric growth. Differences in the growth rates for some parts of the body compared to others determines body proportions. (**a**) During growth of a human, the arms and legs grow faster than the head and trunk, as can be seen in this conceptualization of different-aged individuals all drawn at the same height. (**b**) The fetal skulls of humans and chimpanzees are similar in shape. Allometric growth of the bones transforms the rounded skull of a newborn chimpanzee to the sloping skull characteristic of adult apes. The same allometric pattern occurs in humans, but it is attenuated, and the adult human skull has departed less than the chimpanzee skull from the fetal shape common to primates.

Plains zebra

Grevy's zebra

Figure 23.4
The effect of developmental timing on zebra stripes. The stripes of the plains zebra, *Equus burchelli* (top), are wider and fewer in number than those of Grevy's zebra, *Equus grevyi.* This variation probably reflects the different times at which stripes start to develop in the zebra embryos, three weeks after fertilization in the plains zebra and five weeks after fertilization in Grevy's zebra. To picture this process, imagine painting stripes of equivalent width and spacing around two balloons, one less inflated than the other. If we continue inflating the two balloons to much larger but equal sizes, the one that was smaller when we applied the stripes will have stripes that are fewer and wider than those of the other balloon.

from a common ancestor. Of the numerous anatomical differences between these two modern primates that can be attributed to different allometric properties and variations in developmental timing, the most significant contrast is brain size. Human brains are proportionately larger than chimpanzee brains because growth of the organ is switched off much later in human development. Compared to the brains of chimpanzees, our brains continue to grow for several more years, which can be interpreted as the prolonging of a juvenile process. We owe our culture to this evolutionary novelty, coupled with an extended childhood during which parents and teachers can influence what young, growing brains store.

All these examples of temporal changes in development that create evolutionary novelties fit into the category of **heterochrony,** the general term for evolutionary changes in the timing or rate of development. Equally important in evolution are **homeotic** changes, which alter the placement of different body parts—the arrangement of different kinds of appendages on an

Figure 23.5
Paedomorphosis. Some species retain features as adults that were juvenile in ancestors. This salamander is an axolotl, which grows to full size, becomes sexually mature, and reproduces while retaining certain larval (tadpole) characteristics, including gills. This figure and Figures 23.3 and 23.4 illustrate examples of heterochrony, an evolutionary change in the timing or rate of development.

animal, for example, or the placement of flower parts on a plant. Homeotic alterations of organisms will be discussed in more detail in Chapters 34 and 43. For now, the important point is that change in developmental dynamics, both temporal (heterochrony) and spatial (homeosis), has played an important role in macroevolution.

The Difficulty of Interpreting Evolutionary Trends

Extracting a single evolutionary progression from a fossil record that is likely to be incomplete is misleading; it is like describing a bush as growing toward a single point by tracing the system of branches that leads from the base of the bush to one particular twig. A case in point is the evolution of the modern horse, which is believed to be a descendant of a much smaller ancestor named *Hyracotherium,* which browsed in the woods of the Eocene epoch about 40 million years ago. In comparison to this ancestor, not only are modern horses (genus *Equus*) larger, but the number of toes has been reduced from four on each foot to one, and the teeth have become modified for grazing rather than browsing. By selecting certain species from the available fossils, it is possible to arrange a succession of animals intermediate between *Hyracotherium* and modern horses that shows trends toward increased size, reduced number of toes, and grazing teeth (Figure 23.6a). We might interpret this series of fossils as an unbranched lineage leading directly from *Hyracotherium* to modern horses through a continuum of intermediate stages. If we include all fossil horses known today, however, the illusion of coherent, progressive evolution leading directly to modern horses vanishes. The transition occurs in steps rather than in a smooth gradation of forms; each species appears and disappears in the fossil record without changing noticeably in the interim. *Equus* just happens to be the only surviving twig of a phylogenetic tree that is so branched it is more like a bush (Figure 23.6b). *Hyracotherium* did not become the modern horse by changing gradually, any more than your great grandparents became you. *Equus* descended through a series of speciation episodes that included several adaptive radiations, not all of which led to large, one-toed, grazing horses. Had *Equus* become extinct and some other horse persisted, we might perceive different evolutionary trends.

Evolution *has* produced many trends that seem from the fossil record to be genuine. For example, a family of now-extinct giant mammals called titanotheres, as large as elephants, evolved from a mouse-sized ancestor that lived during the early Cenozoic era (Figure 23.7). There is no evidence of a smooth continuum of increasing size in the various lineages of titanotheres; there is a succession of progressively larger species, but each species remains the same size for its duration in the fossil record. Punctuated equilibrium seems to be the mode of evolution in this case (see Chapter 22). According to this interpretation of the fossil record, evolutionary trends in most groups of organisms are produced not by a phyletic slurring of forms, but by staccato changes occurring in increments during the time that new species branch from ancestral ones.

Branching evolution can produce a trend even if some new species counter the trend. During the Mesozoic era, there was an overall trend in reptilian evolution toward largeness, a progression that produced the dinosaurs, which became the dominant animals of that era. The trend was sustained even though some new species were smaller than their parent species. In fact, even if the descendant species were smaller as often as they were larger than the species from which they budded, an overall trend would still develop if larger reptiles speciated more often than smaller ones or lasted longer than smaller species before becoming extinct (Figure 23.8). In this view of macroevolution, enunciated by Steven Stanley of Johns Hopkins University, species are analogous to individuals. Speciation is their birth and extinction is their death. New species are their offspring. An evolutionary trend, according to the Stanley model, is produced by **species selection,** which is analogous to the pro-

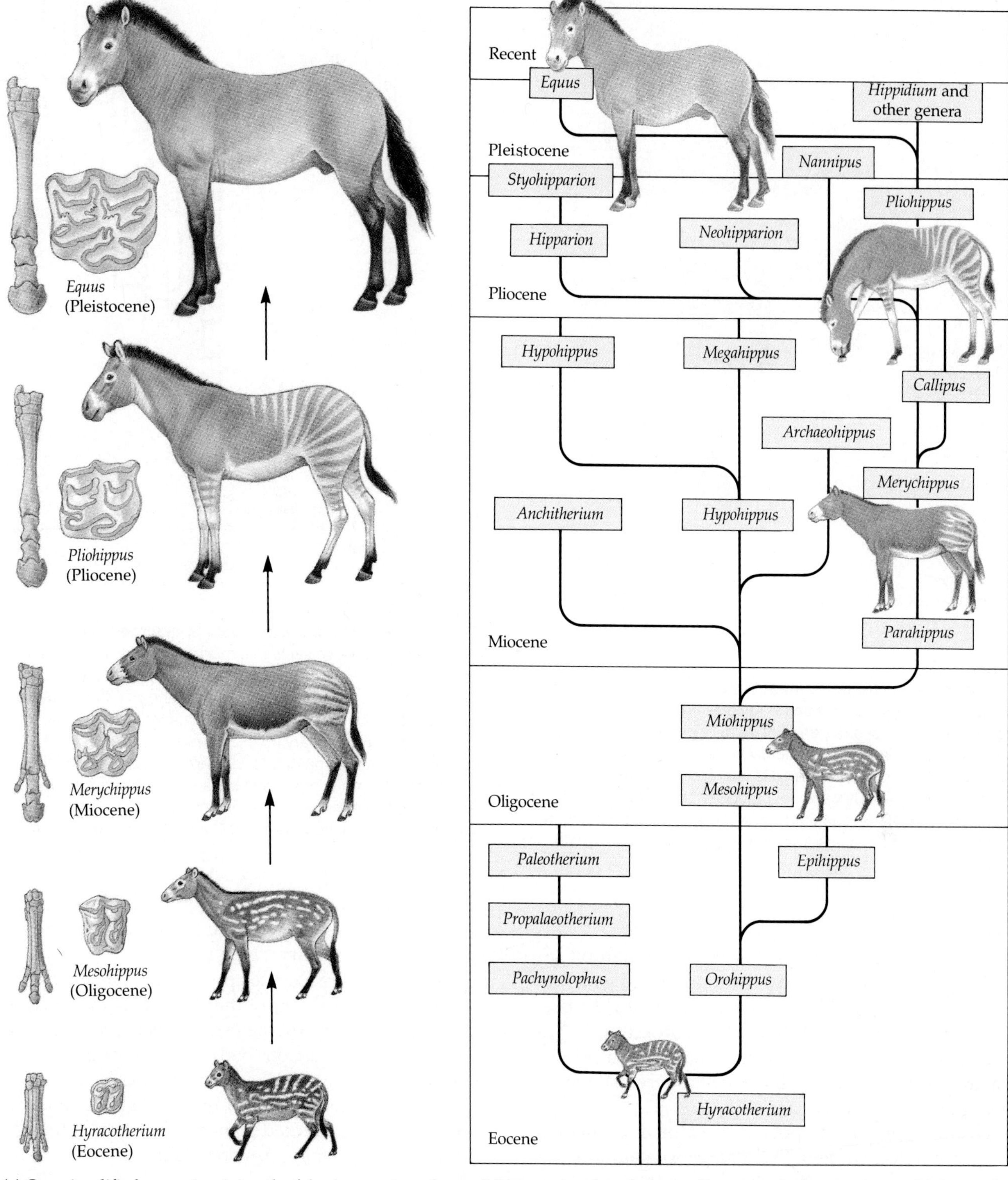

(a) Oversimplified scenario, giving the false impression of a direct evolutionary progression.

(b) More complete analysis of horse evolution, tracing multiple trends.

Figure 23.6
The branched evolution of horses. (**a**) By choosing one sequence of fossil horses that are intermediate in form between the modern horse and its Eocene ancestor, *Hyracotherium*, we create the illusion of phyletic progression with trends toward larger size, reduced number of toes, and teeth modified for grazing. (**b**) A more complete phylogeny reveals that the modern horse is the only surviving twig of an evolutionary bush with many divergent trends.

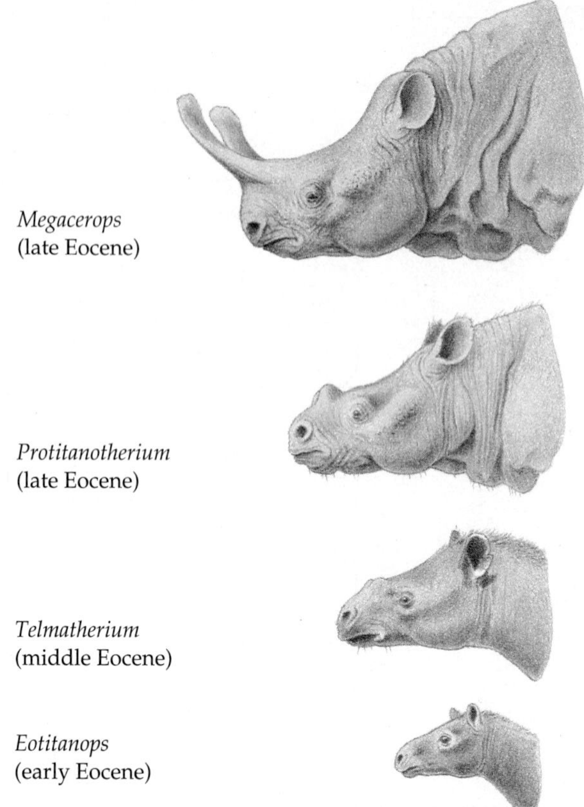

Figure 23.7
Trends in the evolution of titanotheres. In each of the genera of this extinct family of mammals, species became larger and their horns bigger over a period of 45 million years. But no individual species changes significantly in size for its duration in the fossil record. Some paleontologists interpret this to mean that the evolutionary trends were not produced by gradual phyletic evolution of individual populations, but by a series of speciation episodes, with the average size of the species increasing with time.

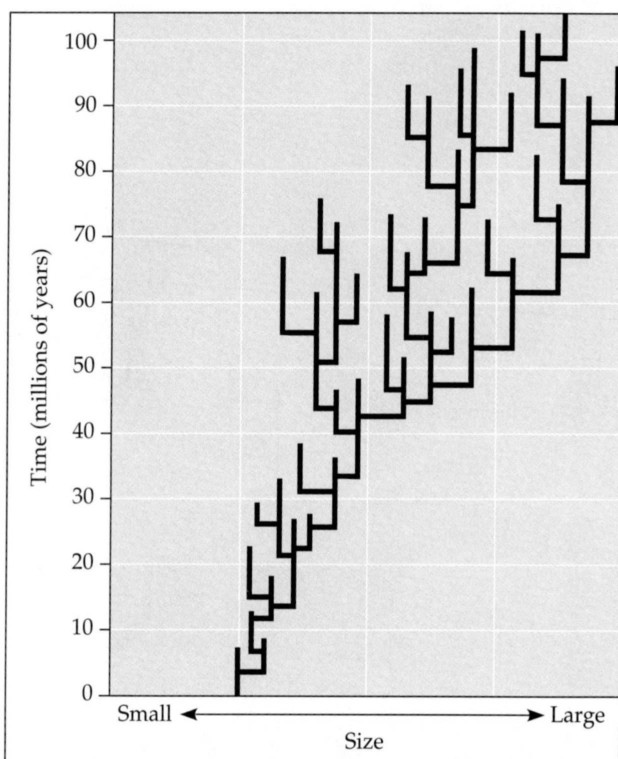

Figure 23.8
The species selection hypothesis of macroevolution. As an alternative to phyletic evolution, this model explains evolutionary trends as the products of the differential longevity and budding of daughter species. This example shows a trend toward species of larger body size. Species smaller than their parent bud from the evolutionary bush as frequently as species that are larger than their parent. Nevertheless, according to this model, there is an overall trend because large species live longer and leave the most descendant species, because of either a faster speciation rate, a slower extinction rate, or both.

duction of a trend within a population by natural selection. The species that live the longest and generate the greatest number of species determine the direction of major evolutionary trends. According to this hypothesis, differential speciation may play a role in macroevolution similar to the role of differential reproduction in microevolution.

To the extent that speciation rates and species longevity reflect success, the analogy to natural selection is even stronger. But qualities unrelated to the overall success of organisms in specific environments may be equally important in species selection. For example, the ability of a species to disperse to new locations may contribute to its giving rise to a large number of "daughter species." The species selection model has many critics who argue that evolutionary trends more commonly result from the gradual modification of populations in response to environmental change. The value of such debates is that they stimulate research, and many paleontologists and other evolutionary biologists are now focusing on the question of what produces evolutionary trends.

Whatever its cause, the appearance of an evolutionary trend does not imply that there is some intrinsic drive toward a preordained state of being. Evolution is a response to interactions between organisms and their current environments. If conditions change, an evolutionary trend may cease or even reverse itself. The Mesozoic world favored giant reptiles, but by the end of that era, smaller species prevailed.

Continental Drift and the Biogeography of Macroevolution

Macroevolution has dimension in space as well as in time. Indeed it was biogeography even more than fossils that first nudged Darwin and Wallace toward an evolutionary view of life. The history of Earth helps explain the current geographical distribution of species.

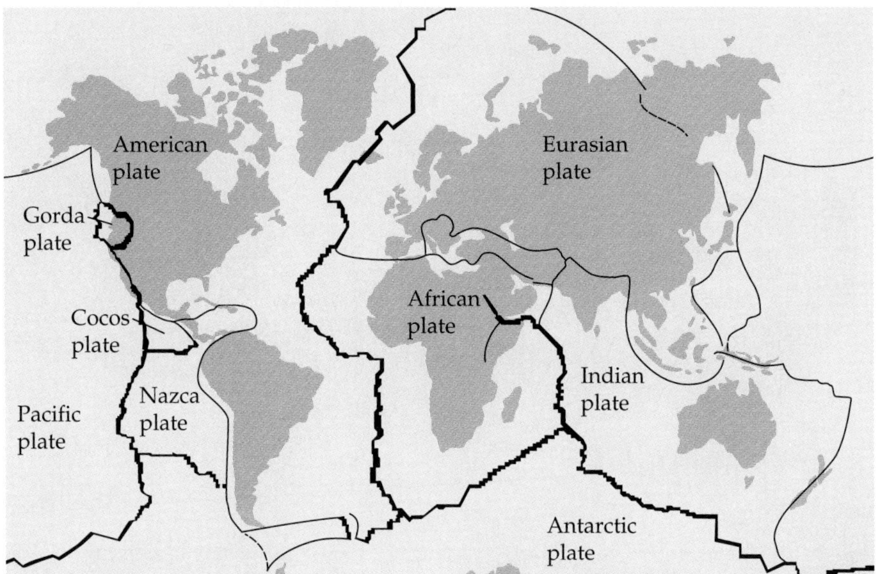

Figure 23.9
Earth's crustal plates. The modern continents are passengers on crustal plates that are swept across Earth's surface by convection currents of the molten mantle below. This map identifies only the major plates.

For example, the emergence of volcanic islands such as the Galapagos opens new environments for founders that reach the outposts, and adaptive radiation fills many of the available niches with new species. On a global scale, the drifting of continents is the major geographical factor correlated with the spatial distribution of life and with such macroevolutionary episodes as mass extinctions and explosive increases of biological diversity.

The continents are not fixed, but drift about Earth's surface like passengers on great plates of crust floating on the molten mantle (Figure 23.9). Unless two land masses are embedded in the same plate, their positions relative to each other change. For example, North America and Europe are presently drifting apart at a rate of about 2 cm per year. Many important geological phenomena, including mountain building, volcanism, and earthquakes, happen at plate boundaries (Figure 23.10). California's infamous San Andreas Fault is part of a border where two plates slide past each other. The Philippines, where Mt. Pinatubo erupted on June 12, 1991, sits right on another plate boundary.

Plate movements rearrange geography incessantly, but two chapters in the continuing saga of continental drift must have been especially significant in their influence on life. About 250 million years ago, near the end of the Paleozoic era, plate movements brought all the land masses together into a supercontinent that has been named **Pangaea,** meaning "all land" (Figure 23.11). Imagine some of the possible effects on life. Species that had been evolving in isolation came together and competed. When the land masses coalesced, the total amount of shoreline was reduced, and there is evidence that the ocean basins increased in depth, which lowered sea level and drained much of the shallow coastal seas that remained. Then, as now, most marine species inhabited shallow waters, and the formation of Pangaea destroyed a considerable amount of that habitat. It was probably a long, traumatic period for terrestrial life as well. The continental interior, which has a harsher climate than coastal regions, increased in area substantially when the land came together. Changing ocean currents also would have affected land life as well as sea life. The formation of Pangaea surely had a tremendous environmental impact that reshaped biological diversity by causing extinctions and providing new opportunities for taxonomic groups that survived the crisis.

Another dramatic chapter in the history of continental drift was written about 180 million years ago, during the early Mesozoic era. Pangaea began to break up, and this caused geographical isolation of colossal proportions. As the continents drifted apart, each became a separate evolutionary arena, and the faunas and floras of the different biogeographical realms diverged (see Chapter 48).

The pattern of continental separations is the solution to many biogeographical puzzles. For example, paleontologists have discovered matching fossils of Triassic reptiles in Ghana (West Africa) and Brazil (Figure 23.12). These two parts of the world, now separated by 3000 km of ocean, were contiguous during the early Mesozoic era. Continental drift also explains why the Australian fauna and flora contrast so sharply with the rest of the world. The great diversity of marsupials (pouched mammals), which fill ecological roles in Australia analogous to those filled by placental mammals on other continents, is just one example of Australia's unique collection of species. Marsupials probably evolved first in what is now North America

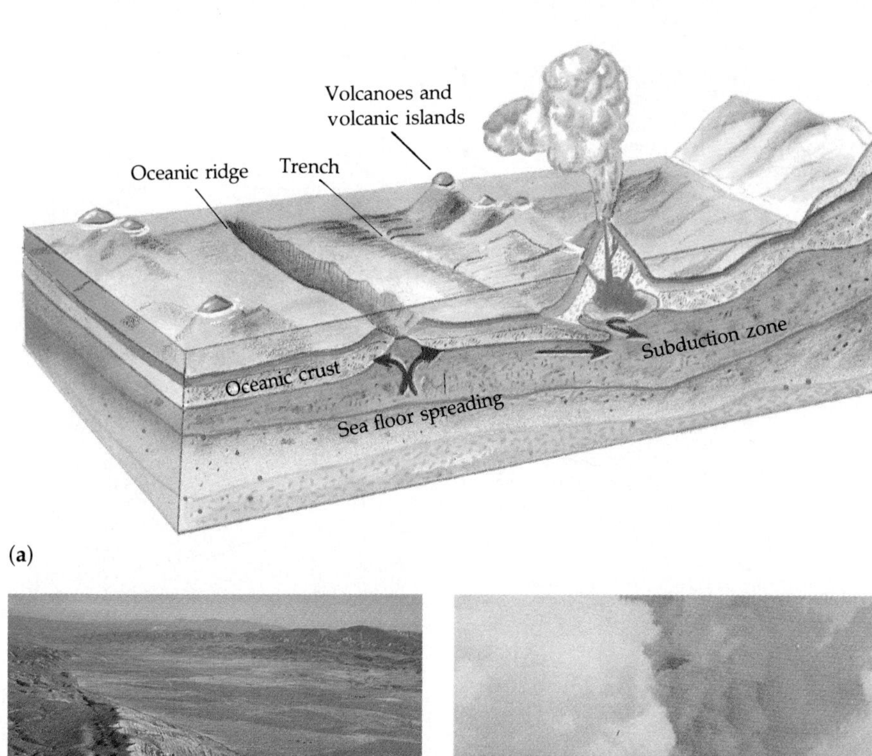

(a)

(b)

(c)

Figure 23.10
Plate tectonics (geological processes resulting from plate movements). (**a**) At some plate boundaries, such as oceanic ridges, the plates separate, and molten rock wells up in the gap. The rock solidifies and adds crust symmetrically to both plates, a phenomenon called sea-floor spreading. There are also areas known as subduction zones, where plates move toward each other, with the denser plate diving below the less dense one and creating a trench. The Marianas Trench of the South Pacific is a subduction zone over 11,000 m deep. The abrasion at subduction zones causes earthquakes and volcanic eruptions. When continents riding on different plates collide, they pile up and build mountains. (**b**) The San Andreas Fault, north of Los Angeles in this photo, is the seismically active boundary between the Pacific and American plates. (**c**) The 1991 eruption of Mt. Pinatubo was a devastating display of plate tectonics. The Philippines, site of this volcano, are located where the Eurasian plate meets a smaller plate (see Figure 23.9).

and reached Australia via South America and Antarctica while the continents were still joined. The subsequent breakup of the southern continents set Australia "afloat" like a great ark of marsupials, while placental mammals evolved and diversified on other continents. Australia has been completely isolated for 50 million years, and bats, rats, mice and humans (and their domesticated animals) are the only placental mammals that have managed to populate the island continent. Had Darwin known about continental drift, Australian life would have made sense to him.

Punctuations in the History of Biological Diversity

The evolutionary byways from ancient to modern life have not been smooth. The fossil record reveals an episodic history, with long, relatively quiescent periods punctuated by briefer intervals when the turnover in species composition was much more extensive. The episodes include explosive adaptive radiations of major taxonomic groups as well as mass extinctions.

Examples of Major Adaptive Radiations Many taxonomic groups have diversified prolifically early in their history after the evolution of some novel characteristic that opened a new **adaptive zone,** a term for a new way of life that presents many previously unexploited opportunities. For example, the development of wings enabled insects to enter an adaptive zone with many new food sources, and adaptive radiation produced hundreds of thousands of variations on the basic insect body plan.

The boundary between the Precambrian era and the Paleozoic era is marked by a large increase in the di-

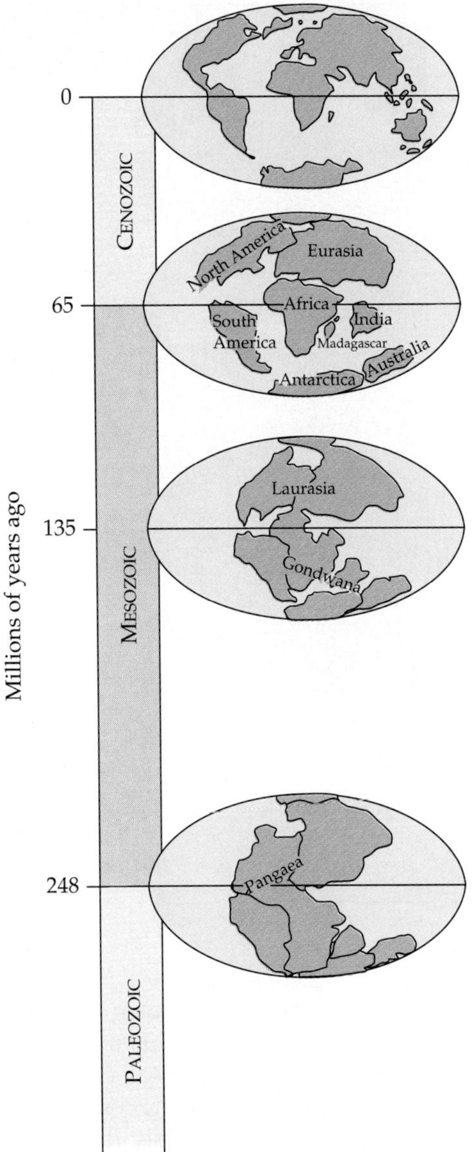

Figure 23.11
Continental drift. About 200 to 250 million years ago, all Earth's land masses were locked together in a supercontinent named Pangaea. About 180 million years ago, Pangaea began to split into northern and southern land masses, which later separated into the modern continents. The continents continue to drift. India collided with Eurasia just 10 million years ago, forming the Himalayas, the tallest and youngest of Earth's mountain ranges.

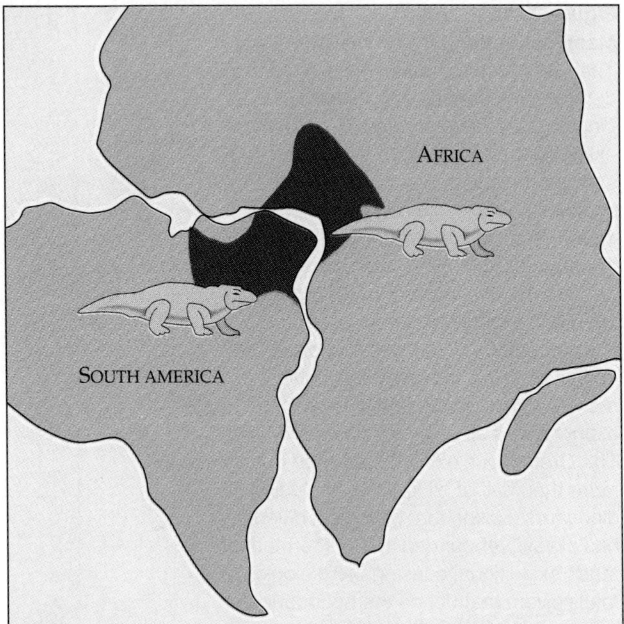

Figure 23.12
Continental drift and biogeography: an example. Although South America and Africa are separated today by the Atlantic Ocean, these continents were joined during the early Mesozoic era. This explains matching fossils of Triassic reptiles and other organisms in the two regions that are shaded in this drawing. The two regions also have matching rock formations.

versity of sea animals, a radiation that was mentioned earlier. The oldest animals are creatures found in late Precambrian rocks about 700 million years old (see Table 23.1). These animals, known from fossilized imprints that have been discovered at several sites around the world, were shell-less invertebrates with body plans quite different from those of their Paleozoic successors. Within the first 10 to 20 million years of the Cambrian, the first period of the Paleozoic era, all the animal phyla that exist today evolved, along with many phyla now extinct. One key evolutionary novelty behind this remarkable diversification may have been the origin of shells and skeletons in a few phyla, an innovation that opened a new adaptive zone by making many new complex designs possible and by rewriting the rules for predator-prey relations.

There are probably empty adaptive zones even today. An empty adaptive zone can be exploited only if the appropriate evolutionary novelties arise. For example, flying insects existed at least 100 million years before the flying reptiles and birds that ate the insects evolved. Conversely, an evolutionary novelty cannot enable organisms to take advantage of adaptive zones that do not exist or are already occupied. Mammals, with the many unique features characteristic of their class, existed at least 75 million years before their first major adaptive radiation. The rise in mammalian diversity during the early Cenozoic era may have been associated with the ecological void left by the extinction of the dinosaurs. New adaptive radiations have often followed mass extinctions that swept away old tenants of adaptive zones.

Examples of Mass Extinctions A species may become extinct because its habitat has been destroyed or because the environment has changed in a direction unfavorable to the species. If ocean temperatures fall by a few degrees, many species that are otherwise

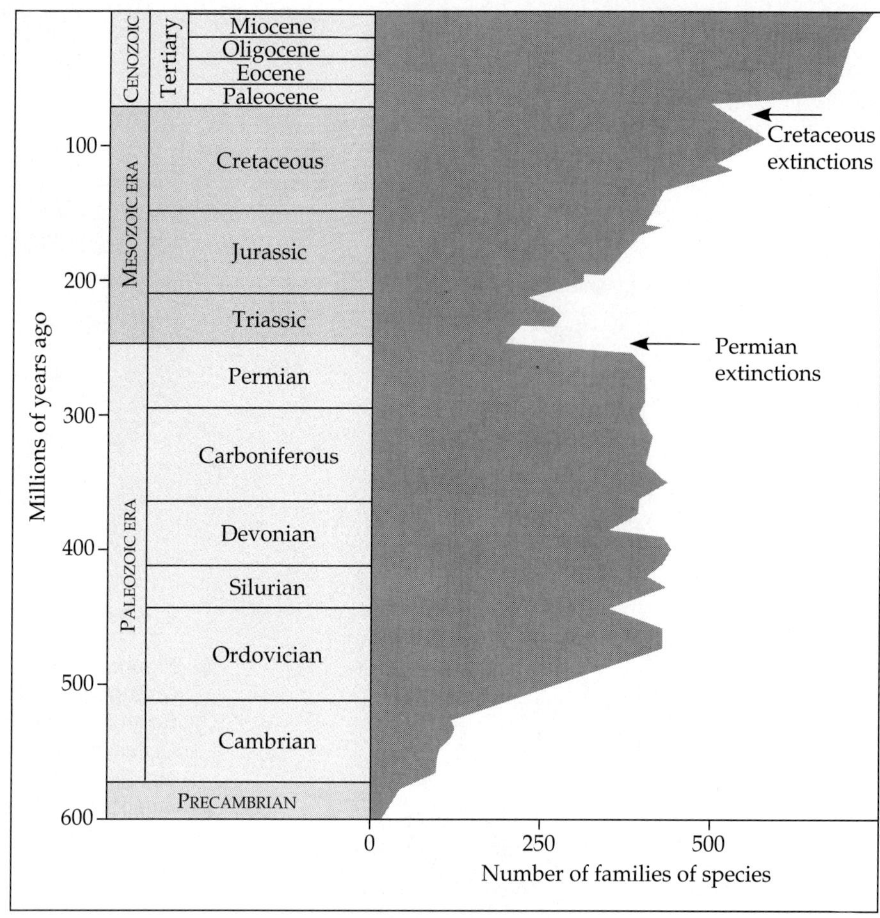

Figure 23.13
Mass extinctions in the history of life. The fossil record profiles mass extinctions during many periods of geological time. Only two are labeled here. The Permian extinctions claimed more than 90% of species on land and in the seas. (The extinctions seem less catastrophic in this figure because we are tracking changes in the number of families, not species. The percentage of species that disappear during mass extinctions is much larger than the percentage of families that disappear. This is because most families consist of many species, and a family is counted as extinct only if all of its species are extinct.) The Cretaceous extinctions wiped out more than half of all species, including all dinosaurs. Notice that biological diversity has always rebounded in the aftermath of mass extinctions, causing the biological make-overs that define the boundaries of the geological periods and eras.

beautifully adapted will perish. Even if physical factors in the environment are stable, the biological factors may change; the environment in which a species lives includes the other organisms that live there, and evolutionary change in one species is likely to have some impact on other species in the community. For example, the evolution of shells by some Cambrian animals may have contributed to the extinction of some shell-less forms.

Thus, extinction is inevitable in a changing world. The average rate of extinction has been between 2.0 and 4.6 families per million years (each family may include many species). However, there have been crises in the history of life when global environmental changes have been so rapid and disruptive that a majority of species were swept away, perhaps rather indiscriminately. During periods of mass extinctions, the rate of destruction escalates to as high as 19.3 families per million years (Figure 23.13).

Of the dozen or so mass extinctions chronicled in the fossil record, two have received the most attention. They are known primarily from the decimation of hard-bodied animals of shallow seas, the organisms for which the fossil record is most complete. The Permian extinctions, which define the boundary between the Paleozoic and Mesozoic eras, claimed over 90% of the species of marine animals about 250 million years ago and probably took a tremendous toll on terrestrial life as well. That was about the time the continents merged to form Pangaea, which probably disturbed many habitats and altered climate. Keep in mind, however, that a temporal correlation of two events does not demonstrate a cause-and-effect relationship.

Another emphatic punctuation, the Cretaceous extinctions of 65 million years ago, delineates the boundary between the Mesozoic and Cenozoic eras. That debacle doomed more than half the marine species and exterminated many families of terrestrial plants and animals, including the dinosaurs. The climate became cooler at that time, and shallow seas receded from continental lowlands. It is possible that increased volcanic activity contributed to the cooling by releasing materials into the atmosphere that blocked sunlight. There is also evidence that an asteroid or comet struck Earth at about the time the Cretaceous extinctions occurred. Separating Mesozoic from Cenozoic sediments is a thin layer of clay enriched in iridium, an element very rare on Earth, but common in meteorites and other extraterrestrial debris that occasionally falls to Earth. Walter and Luis Alvarez and their colleagues at the

University of California, Berkeley, studied the anomalous clay and proposed that it is fallout from a huge cloud of dust that billowed into the atmosphere when an asteroid hit Earth. The great cloud would have blocked light and disturbed climate severely for several months. (The Alvarez scenario is similar to speculations of what a nuclear winter would be like.)

Earth is packed with enough craters to tell us that many large objects have fallen to the planet in the past. Recent research has focused on a 180-km-wide crater located along the Yucatan coast of Mexico. It is about the right size and age to have been caused by an asteroid with a diameter of about 10 km crashing to Earth at the end of the Cretaceous. But the coincidence of an impact there or elsewhere within a period of mass extinctions does not link the two events as cause and effect. Many paleontologists and geologists believe that changes in climate due to continental drift and other processes on Earth are sufficient to account for mass extinctions without looking heavenward for extraterrestrial causes.

Whatever the causes, mass extinctions affect biological diversity profoundly. But there is a creative side to the destruction. The species that manage to survive these crises, by their adaptive qualities or by sheer luck, become the stock for new radiations that fill many of the adaptive zones vacated by the extinctions. The world might be a very different place today if a few families of dinosaurs had escaped the Cretaceous extinctions or if *Purgatorius,* the one known primate that lived in the Cretaceous period, had not survived.

In this section, we have examined some of the geological and biological mechanisms that help us understand the history of macroevolution as told by the fossil record. Paleontology is closely related to the field of systematics, which is concerned with biological diversity, present and past. The fossil record helps systematists determine evolutionary relationships among species, but, as we will see in the next section, evidence from comparing modern organisms is also important.

SYSTEMATICS: TRACING PHYLOGENY

The evolutionary history of a species or group of related species is called **phylogeny** (Gr. *phylon,* "tribe," and *genesis,* "origin"). These genealogies are traditionally diagrammed as phylogenetic trees that trace putative evolutionary relationships (Figure 23.14).

Reconstructing phylogenetic history is part of the scope of **systematics,** the study of biological diversity. The diversity of contemporary life reflects past episodes of speciation and macroevolution. In its search for evolutionary relationships among diverse organisms, systematics encompasses taxonomy, the identification and classification of species.

Taxonomy

The system of taxonomy developed by Linnaeus in the eighteenth century had two main features (see Chapter 20). First, it assigned to each species a two-part Latin name, or **binomial.** The first word of the name is the **genus** (plural, **genera**) to which the species belongs. The second word is the **specific epithet** (name) of the species. For example, the scientific name for the domestic cat is *Felis silvestris.* A given genus may include several similar species, each with its own specific name. For instance, the lynx, *Felis lynx,* belongs to the same genus as the domestic cat. Common names—such as cat, black bear, and mountain lilac—work well in casual communication, but when biologists publish their research, they define the organisms they have studied with scientific names to avoid ambiguity. Many of the scientific names still in use date back to Linnaeus, who assigned binomials to over 11,000 species of plants and animals.

The second major contribution Linnaeus made to taxonomy was adopting a filing system for grouping species into a hierarchy of increasingly general categories. The first step in grouping species is built into binomial nomenclature. Species that are very similar, such as the bobcat and house cat, are placed in the same genus. Grouping species is natural for us, at least in concept. We lump together several trees we know as oaks and distinguish them from several other species of trees we call maples. Indeed, oaks and maples belong to separate genera. The Linnaean system formalizes this grouping of species into genera and extends the scheme to progressively broader categories of classification, some of which have been added since the time of Linnaeus.

Taxonomists place similar genera in the same **family,** group families into **orders,** orders into **classes,** classes into **phyla** (singular, **phylum**), and phyla into **kingdoms.** For example, the genus *Felis* is lumped with various species of the genus *Panthera* (lion, tiger, leopard, and jaguar) in the family Felidae, the cat family. This family belongs to the order Carnivora, which also includes the Canidae (dog family), Ursidae (bear family), and a few other related families. The order Carnivora is grouped with many other orders in the class Mammalia, the mammals. And the class Mammalia is one of several belonging to the phylum Chordata in the kingdom Animalia. Each taxonomic level is more comprehensive than the one below. All members of the family Felidae also belong to the order Carnivora and class Mammalia, but not all mammals are cats. Classifying a species by phylum, class, and so

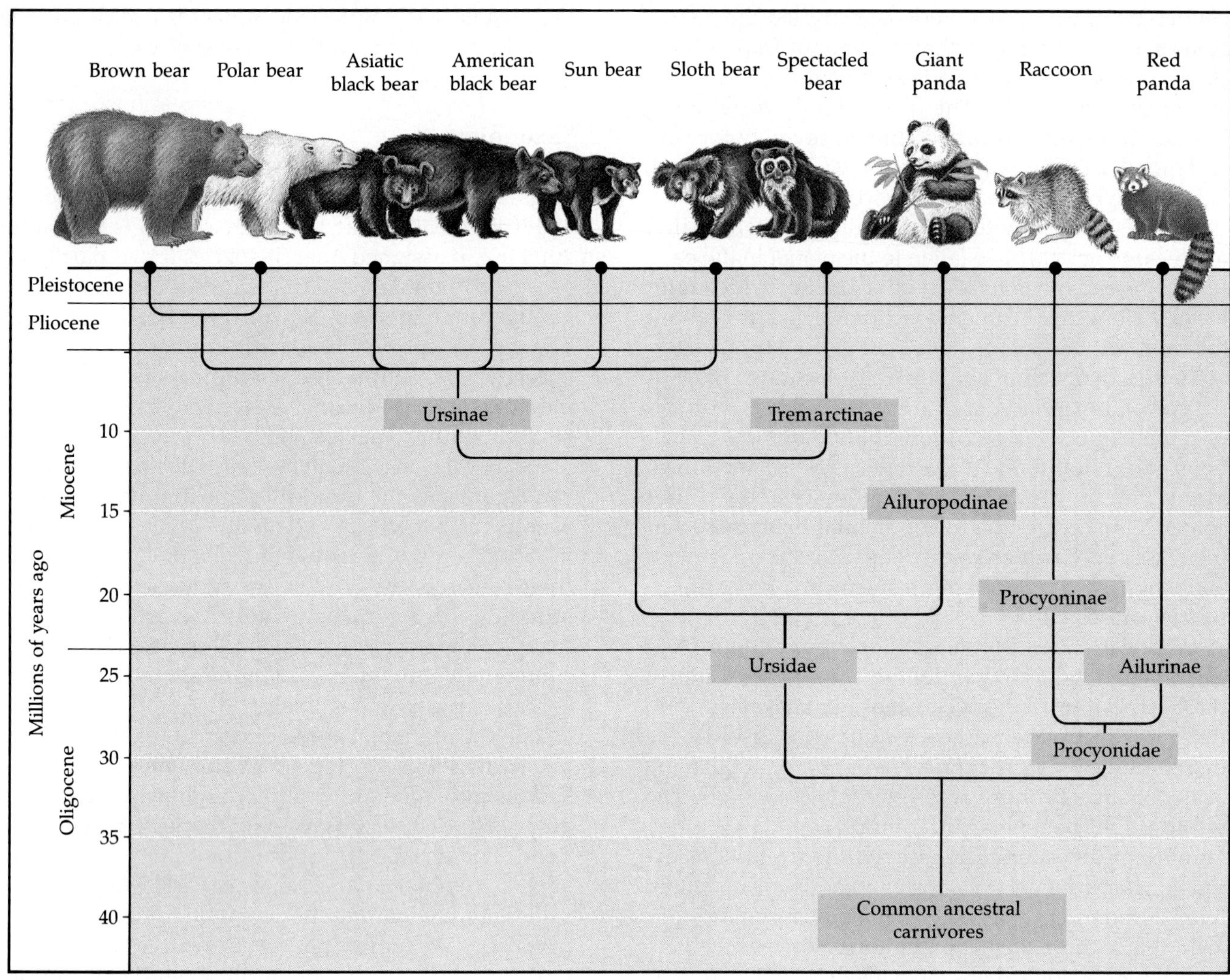

Figure 23.14
A phylogenetic tree. Phylogenetic trees are genealogies of probable evolutionary relationships among species and higher taxonomic groups. In this tree of species of bears and raccoons, the two main branches are labeled with the names of the families (e.g., Ursidae), and the subbranches represent subfamilies (e.g., Ursinae). This tree also shows how long ago each evolutionary divergence occurred. Whenever possible, systematists use the fossil record to help construct phylogenetic trees, but they also apply other methods. For example, this tree is based mainly on comparison of the species' DNA and proteins. A hypothesis of past history based on available data, a phylogenetic tree, like all hypotheses, makes predictions that can be tested by further study. The tree is refined if new information does not fit the existing hypothesis.

on is analogous to a postal worker sorting mail, first by ZIP codes and then by streets and house numbers. Appendix Two gives taxonomic classification of the major groups of organisms discussed in this text down to the level of class.

Taxonomy has two main objectives. The first is to sort out closely related organisms and assign them to separate species, describing the diagnostic characteristics that distinguish the species from one another. Related to this function is the naming of newly discovered species. In the Linnaean tradition, the name is a binomial, with the name of the genus to which the species belongs followed by the specific name, or epithet. The second major objective of taxonomy is to order species into the broader taxonomic categories, from genera to kingdoms. In some cases, there are intermediate categories, such as superfamilies (a category between families and orders) or subclasses (between orders and classes). The named taxonomic unit at any level is called a **taxon** (plural, **taxa**). For example, *Pinus* is a taxon at the genus level, the generic name for the various species of pine trees. Mammalia, a taxon at the class level, includes all the many orders of mammals. Only the genus name and specific epithet are italicized, and all taxa at the genus level or higher (broader) are capitalized. International committees establish rules of nomenclature, which are somewhat different for animals, plants, and bacteria. In Table 23.2,

Table 23.2 Classification of the Domestic Cat and Common Buttercup

Category	Domestic Cat	Common Buttercup
Kingdom	Animalia (animals)	Plantae (plants)
Phylum (animals) or division (plants)	Chordata (chordates)	Anthophyta (flowering plants)
Subphylum	Vertebrata (vertebrates)	—
Class	Mammalia (mammals)	Dicotyledones (dicots—see Chapter 27)
Order	Carnivora (carnivores)	Ranunculales
Family	Felidae (cats)	Ranunculaceae (crowfoot family)
Genus	*Felis*	*Ranunculus*
Specific epithet	*silvestris*	*acris*

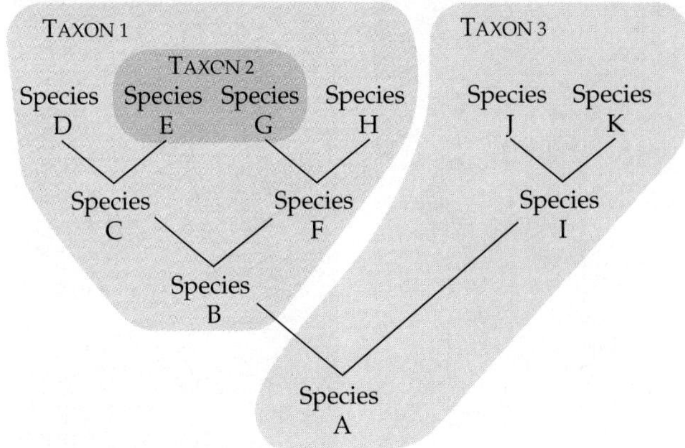

Figure 23.15
Grouping species into higher taxa. Taxon 1, with seven species (B–H), qualifies as a monophyletic grouping, the taxonomic ideal. The taxon includes all descendant species along with their immediate common ancestor (species B). However, taxon 2, a subgrouping within taxon 1, is polyphyletic. Species E and species G are derived from two different immediate ancestors (species C versus species F). Taxon 3 is paraphyletic. It includes species A without incorporating all other descendants of that ancestor. Although taxonomists generally value monophyletic classifications over polyphyletic or paraphyletic versions, there are exceptions.

the house cat and the common buttercup are placed in their appropriate taxa as examples of classification.

Of all taxa, only the species actually exists in nature as a biologically cohesive unit, bonded by intrabreeding and bounded by reproductive isolation from all other species. In most cases, distinguishing between species can be done objectively if enough is known about their characteristics (recall the story in Chapter 22 about Ernst Mayr and the New Guinea birds), but combining species into higher taxa often involves judgment calls. One taxonomist may value fine distinctions and favor a relatively large number of taxa for each category above species, whereas another may stress unification and propose a minimal number of taxa. For instance, some taxonomists who specialize in the cat family lump all cats except the cheetah into a single genus, *Felis*. Other taxonomists (the majority, in this case) split the same group of species into a genus of small cats (*Felis*, which includes the house cat), a genus of large cats (*Panthera*, which includes the lion), a genus of bobtailed cats (*Lynx*), and other genera.

Ever since Darwin, systematics has had a goal beyond simple organization: to have classification reflect the evolutionary affinities of species. Each successive group in the taxonomic hierarchy should represent finer and finer branching of phylogenetic trees. A taxon is said to be **monophyletic** if a single ancestor gave rise to all species in that taxon and to no species placed in any other taxon (Figure 23.15). A taxon is **polyphyletic** if its members are derived from two or more ancestral forms not common to all members. A **paraphyletic** taxon excludes species that share a common ancestor that gave rise to the species included in the taxon. Ideally, each taxon should be monophyletic, but, for reasons to be explained soon, taxonomists sometimes depart from this ideal.

Sorting Homology from Analogy

A taxonomist classifies species into higher taxa based on the extent of similarities in morphology and other characteristics. Recall from Chapter 20 that likeness attributed to shared ancestry is called **homology.** The forelimbs of mammals are homologous; that is, the similarity in the intricate skeleton that supports the limbs has a genealogical basis (see Figure 20.14).

There is a joker in this game of making evolutionary connections by evaluating similarity: Not all likeness is inherited from a common ancestor. Species from different evolutionary branches may come to resemble one another if they have similar ecological roles and natural selection has shaped analogous adaptations. This is called **convergent evolution,** and similarity due to convergence is termed **analogy,** not homology (Figure 23.16). The wings of insects and those of birds, for example, are analogous flight equipment that evolved independently and are built from entirely different structures. Convergent evolution has produced the analogous resemblances between certain Australian marsupials and placental look-alikes that have evolved independently on other continents.

Figure 23.16
Convergent evolution and analogous structures. The ocotillo of southwestern North America (left) looks remarkably similar to the allauidia (right) found in Madagascar. The plants are not closely related and owe their resemblance to analogous adaptations that evolved independently in response to similar environmental pressures.

To reconstruct evolutionary history, we must sort homology from analogy and build phylogenetic trees on the basis of homologous similarities alone. As a general rule, the greater the number of homologous parts between two species, the more closely the species are related, and this should be reflected in their classification. This guideline is simpler in principle than it is in practice. Adaptation can obscure homologies, and convergence can create misleading analogies. As we saw in Chapter 20, comparing the embryonic development of the features in question can often expose homology that is not apparent in the mature structures.

There is another clue to identifying homology and sorting it from analogy: The more complex two similar structures are, the less likely it is they have evolved independently. Consider the skulls of a human and a chimpanzee, for example. The skulls are not single bones, but a fusion of many, and the chimp skull and human skull match almost perfectly, bone for bone. It is highly improbable that such complex structures matching in so many details could have separate origins. The multitude of genes required to build these skulls must have been inherited from a common ancestor.

Molecular Systematics

The comparison of information-rich macromolecules—proteins and DNA—has become a powerful taxonomic tool. For example, molecular data were used to build the phylogenetic tree of bears in Figure 23.14. Sequences of nucleotides in DNA are inherited, and they program corresponding sequences of amino acids in proteins. Molecular comparisons go right to the heart of evolutionary relationships.

Protein Comparison Because the primary structures of proteins are genetically determined (see Chapter 5), a close match in the amino acid sequences of two proteins from different species indicates that the genes for those proteins evolved from a common gene present in a shared ancestor. The degree of similarity is evidence of the extent of common genealogy. One advantage of this taxonomic tool is that it is objective and quantitative.

A second advantage is that it can be used to assess relationships between groups of organisms that are so phylogenetically distant that they share very few morphological similarities. For example, the amino acid sequence of cytochrome *c*, an ancient protein common to all aerobic organisms, has been determined for a wide variety of species ranging from bacteria to complex plants and animals. The sequences for humans and chimpanzees match perfectly for all 104 positions along the polypeptide chain, and the cytochromes of both species differ from the version found in the rhesus monkey by just one amino acid. All three species belong to the same mammalian order, Primates. Comparing these proteins to the forms found in nonprimates, we find that the differences increase as the species become more taxonomically distant. For instance, human cytochrome *c* differs from that of a dog by 13 amino acids, from that of a rattlesnake by 20 amino acids, and from that of a tuna by 31 amino acids. Phylogenetic trees based on cytochrome *c* are consistent with evidence from comparative anatomy and the fossil record.

DNA Comparison Comparing the genes or genomes of two species is the most direct measure of common inheritance from shared ancestors. Comparisons can be made by three methods: DNA-DNA hybridization, restriction mapping, and DNA sequencing.

Whole genomes can be compared by **DNA-DNA hybridization,** which measures the extent of hydrogen bonding between single-stranded DNA obtained from two sources. How tightly the DNA of one species can bind to the DNA of the other depends on the degree of similarity, as base pairing between complementary sequences holds the two strands together. After the DNA is extracted, it is heated to separate the complementary strands. Single-stranded DNA from two species is then mixed and cooled to re-form double-stranded DNA. The hybrid DNA is again heated to separate the paired strands. The temperature required to do this is correlated with the similarity of the DNA from the two species; the more extensive the pairing, the greater the

Coyote

Red wolf

Gray wolf

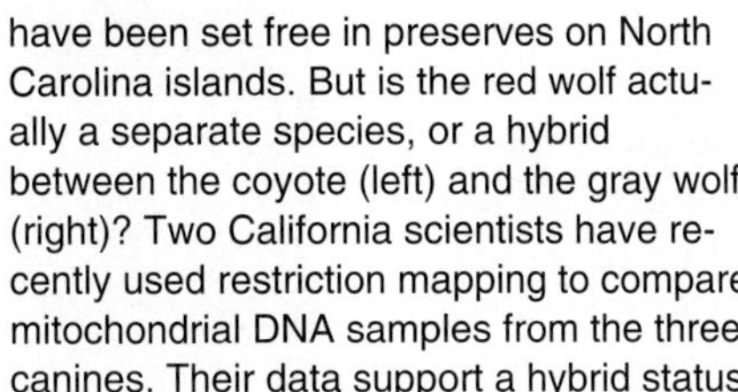

Figure 23.17
Using DNA technology to evaluate the taxonomy of the red wolf. The North American red wolf (center), an inhabitant of the southeastern United States, was on the brink of extinction in 1975 due to hunting and human encroachment on habitat. A few of the remaining mating pairs were bred in captivity, and 25 of their offspring have been set free in preserves on North Carolina islands. But is the red wolf actually a separate species, or a hybrid between the coyote (left) and the gray wolf (right)? Two California scientists have recently used restriction mapping to compare mitochondrial DNA samples from the three canines. Their data support a hybrid status for the red wolf. However, there is fossil evidence that red wolves predate coyotes and gray wolves, which supports status as a separate species. The debate may be resolved when the genomes are compared more completely via restriction mapping of the three animals' nuclear DNA.

heat energy required to pull the strands apart. Using the temperature needed to pry apart double-stranded DNA from a single species as a standard, the temperature at which hybrid DNA separates measures phylogenetic distance.

Evolutionary trees constructed by this technique generally agree with the phylogeny deduced by other methods such as comparative morphology, but DNA-DNA hybridization has the potential to settle some old taxonomic debates. For example, ornithologists (bird specialists) have long disagreed as to whether flamingos are more closely related to storks or geese. Comparison of DNA places the flamingo with storks. There has also been disagreement about whether the giant panda is a true bear or a member of the raccoon family; the evidence from DNA-DNA hybridization groups the giant panda with the bears, but places the red panda in the raccoon family (see Figure 23.14).

Although DNA-DNA hybridization can estimate the overall similarity of two genomes, it does not give precise information about the matchup in specific nucleotide sequences of the DNA. An alternative approach is **restriction mapping** of DNA. This method employs the same restriction enzymes used in recombinant DNA technology (see Chapter 19). Each type of restriction enzyme recognizes a specific sequence of a few nucleotides and cleaves DNA wherever such sequences are found in the genome. The DNA fragments obtained after treatment with a restriction enzyme can be separated by electrophoresis (see the Methods Box in Chapter 19, p. 396) and compared to restriction fragments derived from the DNA of another species. Two samples of DNA with similar maps for the locations of restriction sites will produce similar collections of fragments. In contrast, two genomes that have diverged extensively since their last common ancestor will have a very different distribution of restriction sites, and the DNA will not match closely in the sizes of restriction fragments. Because so many fragments are obtained from the nuclear genome, restriction mapping is more practical for comparing smaller segments of DNA, usually a few thousand nucleotides long. Several laboratories are using restriction maps to compare mitochondrial DNA (mtDNA), which is relatively small. There is the added benefit that mtDNA changes by mutation about ten times faster than does the nuclear genome, which makes it possible to sort out phylogenetic relationships between very closely related species or even between different populations of the same species. For example, molecular systematists are applying this method to help settle a debate about the taxonomy of the North American red wolf (Figure 23.17). And comparison of mtDNA from people of several different ethnicities has corroborated the fossil evidence that our species originated in Africa (see Chapter 30).

The most precise and powerful method for comparing DNA from two species is **DNA sequencing**—that is, actually determining the nucleotide sequences of entire DNA segments that have been cloned by recombinant DNA techniques (see Chapter 19). Such a comparison tells us exactly how much divergence there has been in the evolution of two genes derived from the rived from the same ancestral gene. As the technology for the rapid sequencing of DNA continues to improve, DNA sequencing will become an increasingly powerful taxonomic approach. A related approach is the sequencing of ribosomal RNA (rRNA), gene products found in all organisms. Because genes for rRNA change slowly relative to most other DNA, differences in rRNA sequences can be used to trace some of the earliest branching in the tree of life. Comparison of rRNA sequences has been especially useful in sorting out the phylogenetic relationships among the bacteria (see Chapter 25).

In the past few years, new methods have extended molecular systematics to the study of DNA traces recovered from fossils. In 1990, a team of researchers used polymerase chain reaction, or PCR (see Chapter 19), to amplify DNA extracted from magnolia leaves found within Idaho sediments that are 18 million years old (Figure 23.18a). The scientists were able to compare a short piece of this ancient DNA to homologous DNA from modern magnolias. Only a tiny fraction of fossils retain any DNA at all, but that may be enough for systematists to solve some phylogenetic problems that seemed unapproachable before PCR and other advances in DNA technology.

Anthropology will also benefit from these techniques. Already, scientists have been able to analyze DNA samples taken from a 2400-year-old mummy (Figure 23.18b) and from a human bone over 5500 years old. Such studies will help anthropologists determine the historical relationships of early human cultures. Molecular biology, which has already revolutionized so many fields, is beginning to make its mark in paleontology and anthropology.

(a)

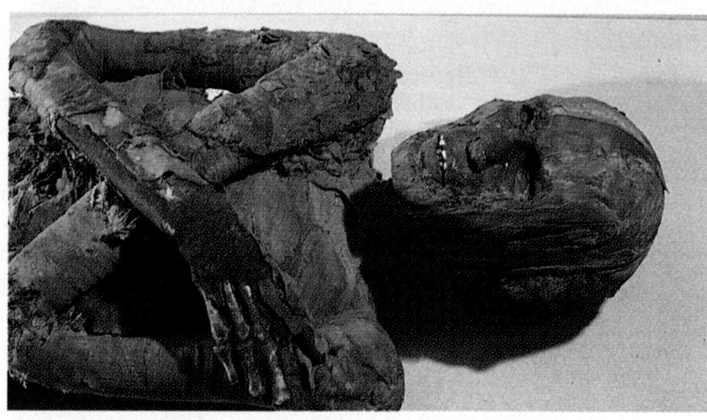

(b)

Figure 23.18
Analysis of ancient DNA. Molecular systematists can now use polymerase chain reaction (PCR) to amplify DNA traces recovered from preserved specimens into quantities of DNA large enough to compare to DNA from other species, both fossil and contemporary. This approach may enable systematists to resolve some phylogenetic questions that were unanswerable before the invention of PCR. (**a**) The oldest DNA to be analyzed so far were small pieces extracted from magnolia fossils that date back 18 million years. Discovered in Idaho, the fossils were so well preserved that they were still green when scientists first cracked open the shale that contained the leaves. (**b**) New DNA technology has also enabled scientists to analyze DNA samples obtained from this 2400-year-old Egyptian mummy.

Molecular Clocks Proteins evolve at different rates, but for a given type of protein—cytochrome *c*, for instance—the rate of evolution seems to be quite constant with time. If homologous proteins are compared for taxa that are known from the fossil record to have diverged from common ancestors during certain periods in the past, the number of amino acid substitutions is proportional to the time that has elapsed since the lineages branched apart. The homologous proteins of bats and dolphins are much more alike than those of sharks and tuna, which is consistent with the fossil evidence that sharks and tuna have been on separate evolutionary lines much longer than bats and dolphins. In this case, molecular divergence has kept better track of the time than superficial changes in body form.

As a clock to date branch points in phylogenetic trees, DNA comparisons are even more promising than protein comparisons. As with protein clocks, the DNA clock may have a reliable beat; dating of phylogenetic branchings based on nucleotide substitutions in DNA generally approximate the dates determined from the fossil record. In many cases, the difference in DNA between two taxa is more closely correlated with how long they have been on separate evolutionary branches than is the degree of morphological difference between the taxa.

Molecular clocks are calibrated by graphing the number of amino acid or nucleotide differences against the times for a series of evolutionary branch points known from the fossil record. The graph can then be used to estimate the time of divergence for species when there is no clear fossil evidence for their time of origin from other forms. For example, this was the method used to determine the antiquity of branch points in the phylogenetic tree of bears illustrated in Figure 23.14.

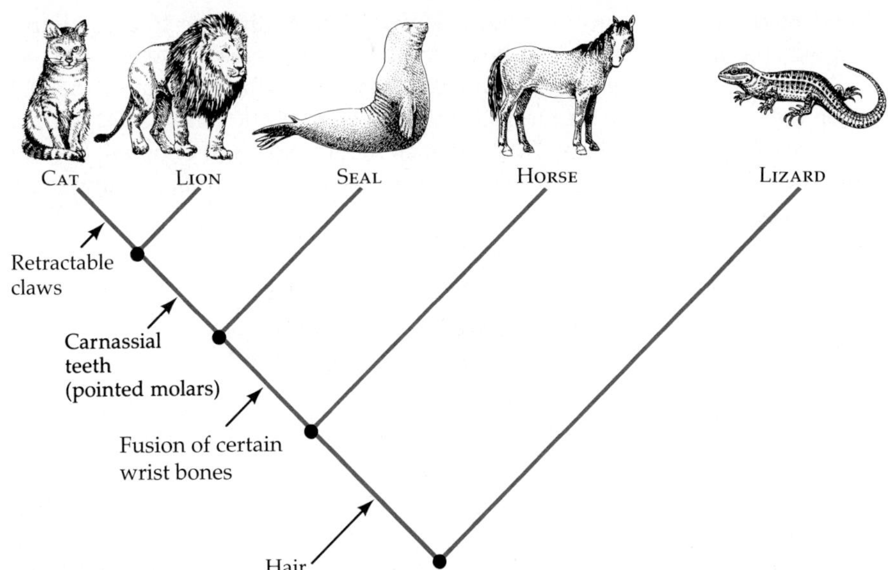

Figure 23.19
A cladogram. Each branch point is defined by apomorphic characters, derived homologies unique to the lineage that arises at that point. The defining characteristics noted here are not the only ones at each branch. Only the sequence of branching is represented in a cladogram, with no consideration of the degree of divergence between branches. The branch points (dots) represent the most recent ancestor common to all species beyond that point. For example, the lion and cat share a common ancestor that lived more recently than did the ancestor that also gave rise to the evolutionary lineage leading to the seal. However, this does not mean that the seal itself evolved before lions.

Both the consistent rate of protein change and the rate of DNA divergence imply that there is a significant background of neutral mutations that gradually changes the genome as a whole more than the specific genetic changes associated with adaptation. Evolutionary biologists disagree about the extent of neutral variations (see Chapter 21). Many evolutionists doubt that neutral evolution is prevalent, so they also question the credibility of molecular clocks as tools for absolute dating of the origin of taxa. There is less skepticism about the value of molecular systematics for determining the relative sequence of branch points in phylogeny. The modern systematist evaluates any available molecular data along with all other taxonomic evidence in order to reconstruct phylogeny.

Schools of Taxonomy

Phylogenetic trees have two significant aspects: the location of branch points along the tree, symbolizing the relative time of origin of different taxa, and the degree of divergence between branches, representing how different two taxa have become since branching from a common ancestor. If taxonomy is to be based on evolutionary history, which property of phylogenetic trees should be given the greatest weight when grouping species into taxa? This question has divided taxonomy into three schools of thought: phenetics, cladistics, and classical evolutionary taxonomy.

Phenetics Endeavoring to make classification less subjective, **phenetics** (Gr. *phainein*, "to appear"; the term *phenotype* is derived from this same root) makes no phylogenetic assumptions and decides taxonomic affinities entirely on the basis of measurable similarities and differences. As many anatomical characteristics (known as characters) as possible are compared, with no attempt to sort homology from analogy. Pheneticists contend that if enough phenotypic characters are examined, then the contribution of analogy to overall similarity will be swamped by the degree of homology. Critics of phenetics argue that overall phenotypic similarity is not a reliable index of phylogenetic proximity. Although a strictly phenetic approach has few proponents today, the methods used in phenetics, especially the emphasis on multiple quantitative comparisons with the help of computers, has had an important impact on taxonomy.

Cladistics Clades (Gr. *clados*, "branch") are evolutionary branches. **Cladistics** classifies organisms according to the order in time that branches arise along a phylogenetic tree, excluding the degree of divergence from consideration. The tree takes the form of a cladogram, a series of dichotomous forks. Each branch point is defined by novel homologies unique to the various species on that branch. Let's apply the strategy to five vertebrates: a lizard, a horse, a seal, a lion, and a cat (Figure 23.19). Each species has a mixture of primitive characters that already existed in the common ancestor, along with characters that have evolved more recently. The sharing of primitive characters tells us nothing about the pattern of evolutionary branching from a common ancestor. For example, we cannot use the presence of five separate toes to divide these vertebrate species among evolutionary branches. According to fossil evidence, the remote ancestor common to all species in our list was five-toed, and hence this homology is termed a primitive character, or **plesiomorphic character.** Seals and horses apparently lost the trait independently, whereas the other species retained the trait. We must seek derived characters, or **apomorphic characters,** which are homologies that evolved after a branch diverged from the phylogenetic tree. Hair and mammary glands are two of the apomorphic characters that define a branching point with the lion, cat, seal, and horse on one limb and the lizard on the other. Now we must determine the branching sequence along the mammalian limb. The lion, cat, and seal share many skeletal and dental modifications not present in the horse, and these are among the apomorphic characters that define the next branch point in our cladogram. The lion and cat branch from the lineage leading to seals at a later point defined by a number of modifications of the skull and teeth.

The cladistic approach produces some taxonomic surprises. For example, the branch point between birds and crocodiles is more recent than the branch point between crocodiles and the other reptiles. That is, birds and crocodiles share apomorphic characters not present in snakes and lizards, and indeed the fossil record supports the conclusion that birds are closer relatives of crocodiles than are lizards and snakes. In the strictly cladistic view, there are no such taxa as the class Aves (birds) and the class Reptilia, as we conventionally know them, because birds intrude in the cladogram of the collective of animals we call reptiles. (Some taxonomists propose retaining the class Reptilia and including birds as a subclass or order.) Birds seem so superficially different because of the extensive morphological remodeling associated with flight that has occurred since birds branched from their reptilian ancestors. Cladistics ignores the extent of morphological divergence between evolutionary branches, information that critics of a strictly cladistic approach argue should be included in a classification scheme.

Classical Evolutionary Taxonomy The classification used in this book and most others is based on **classical evolutionary taxonomy,** an approach that predates phenetics and cladistics and that now attempts to balance the criteria of phenetics and cladistics by considering overall homology along with branching sequence. In cases where this leads to a taxonomic conflict, a subjective judgment is made about which type of information should be given higher priority. For example, classical taxonomists acknowledge that crocodiles share a more recent common ancestor with birds than with lizards, but opt to combine lizards and crocodiles in a taxon that excludes birds because the ability to fly was an evolutionary breakthrough that placed birds in a major new "adaptive zone." This resulted in adaptive divergence of birds so extensive that classical taxonomists assign birds to their own class (Aves).

By using evolutionary criteria for classification—the sequence of branching, the extent of divergence, or some combination of these features of evolution—taxonomists are helping to trace the history of life. It is a goal that Darwin set in *The Origin of Species,* where he wrote: "Our classifications will come to be, as far as they can be so made, genealogies." Darwin defined modern taxonomy, as the Darwinian theme of descent with modification has shaped the entire science of biology. Evolutionary theory has itself evolved, and it is by reviewing some of these developments that we conclude Unit Four.

IS A NEW EVOLUTIONARY SYNTHESIS NECESSARY?

Evolutionary biology has not had a quiet moment since Darwin published *The Origin of Species* more than 130 years ago. The closest thing to a consensus has been the modern synthesis, the paradigm conceived by Ernst Mayr and others that has dominated evolutionary theory for the past 50 years (see Chapter 21). Called a synthesis because the ideas were fashioned from several disciplines, including paleontology, biogeography, systematics, and population genetics, it has continued to absorb the discoveries of such new fields as molecular biology. The modern synthesis reaffirmed the Darwinian view of life and updated it by applying the principles of genetics. The paradigm is distinctly gradualist in its view that large-scale evolutionary changes are the accumulations of many minute changes occurring over vast spans of time. Microevolution, the changes in gene frequencies in populations, is extrapolated to explain most macroevolution.

In the view of the modern synthesis, natural selection is the major cause of evolution at all levels. Populations adapt by natural selection, new species arise when isolated populations diverge as different adaptations evolve, and continued divergence due to natural selection differentiates the higher taxa. The modern synthesis recognizes, and in fact first described, how genetic drift can cause rapid, nonadaptive evolution. But the major emphases of the synthesis are gradualism and natural selection.

A number of evolutionists dissent from the view that the evolution recorded in the fossil record can be explained by extrapolating the processes of microevolution. The debate is partly about the pace of evolution. Many transitions in the fossil record are punctuational, not gradual. Gradualists argue that apparent abruptness partly derives from the imperfection of the fossil record and partly is a semantic issue clouded by the vastness of geological time. Do we call an evolutionary episode that required 10,000 years "sudden" or "gradual"? Punctuationalists counter that the imperfection of the fossil record is not enough to account for the rarity of transitional forms if speciation and the origin of higher taxa were primarily gradual extensions of microevolution. The debate is not just about the tempo of evolution, but also about the degree to which microevolution compounded over time is sufficient to explain macroevolution.

Some evolutionists favor a hierarchical theory, with different mechanisms being most important at different levels of evolution. In this view, natural selection is the key to adaptive evolution of a population but is not usually the most important factor in speciation; it plays even less of a role at the level of macroevolution. Most new species begin as small populations isolated from their parent populations by either geographical barriers or genetic accidents, such as chromosomal mutations. The small, isolated population can evolve relatively rapidly, its divergence from the parent population due at least as much to genetic drift as to selection. Chance may produce a new species even before selection has fashioned new adaptations. Chance also figures prominently in macroevolution. Continental drift and mass extinctions have probably had at least as much effect on the history of biological diversity as gradual adaptation caused by selection operating on gene pools at the population level. Among those who see evolution in this dicey context, contingency, the occurrence of unforeseen events, has become a popular word.

The importance of natural selection is not under fire; the various factions agree that natural selection is the mechanism of adaptation and should therefore be the centerpiece of evolutionary biology. Selection fine-tunes a population to its environment with generation-to-generation changes in the gene pool that are adaptive. When a new species or higher taxon comes into being, it is natural selection that refines unique adaptations. Although the events that lead to speciation and episodes of macroevolution may have more to do with contingency than with adaptation, new species only persist long enough to be entered into the fossil record if they have adapted to their environment through natural selection.

Perhaps those who have challenged the orthodoxy of the modern synthesis are quibbling. The synthesis has never claimed that evolution is always smooth and gradual, or that processes other than changes in gene pools due to selection are unimportant. The questions are not as much about the nature of evolutionary mechanisms as about their relative importance. The modern synthesis may not need radical surgery, only a face-lift.

Vigorous debate about how life evolves is a healthy sign; evolutionary biology is a robust science that will neither stagnate in complacency nor wallow in dogma. The debates will continue as long as we are curious about our origins and our relationship to the rest of the living world.

STUDY OUTLINE

1. Macroevolution is the origin and history of life above the species level. It involves the origin of evolutionary novelties, the study of evolutionary trends, and global episodes of major adaptive radiations and mass extinctions.

The Record of the Rocks (pp. 475–478)

1. The fossil record provides the historical archives biologists use to study macroevolution. Sedimentary strata reveal the relative ages of fossils by vertically chronicling specimens from successive geological periods.
2. The absolute ages of fossils in years can be determined by radioactive dating and other methods. Absolute dating of sedimentary strata has defined the ages for the different geological periods, each of which corresponds to a major transition in the composition of fossil species. The chronology of geological periods and eras makes up the geological time scale.
3. The fossil record is incomplete, owing to the rarity in the formation, endurance, exposure, and discovery of fossil species.

Mechanisms of Macroevolution (pp. 478–489)

1. An important event in the formation of higher taxonomic groups is the appearance of an evolutionary novelty. One mechanism that makes evolutionary novelty possible is preadaptation, the gradual modification of an existing structure for a new function. Changes in the temporal or spatial details of development are also important in the evolution of novel characteristics.
2. Species selection may cause evolutionary trends. According to this hypothesis, species with certain characteristics survive longer and speciate more frequently than species with other characteristics.
3. Continental drift has had a significant impact on the history of life by causing major geographical rearrangements affecting biogeography and evolution. The formation of the supercontinent Pangaea during the late Paleozoic era and its subsequent breakup during the early Mesozoic era explain many puzzling cases of geographical distribution observed today.
4. Evolutionary history has not been a series of smooth gradations. Long, relatively stable periods have been interrupted by brief intervals of extensive species turnover—mass extinctions followed by grand episodes of adaptive radiation.

Systematics: Tracing Phylogeny (pp. 489–496)

1. Systematics is the study of biological diversity. It includes taxonomy, the identification and classification of species.
2. Taxonomic affinity and phylogenetic relationships are decided on the basis of homology, structural similarity due to common ancestry. Sometimes, however, two un-

related species will possess similar, or analogous, structures as a result of convergent evolution.

3. Molecular systematics is the ultimate level for determining homology. Evolutionary relationships can be revealed by comparing amino acid sequences of proteins and nucleotide sequences of DNA. Molecular evolution may occur at a rate consistent enough to function as a crude clock for determining the relative sequence of branch points in phylogeny.
4. Phylogeny has two aspects: the relative time of origin for different taxa and their degree of divergence from a common ancestor at each branch point. These two aspects of phylogeny have generated different schools of taxonomy.
5. Phenetics ignores the timing of branch points and classifies organisms solely on the basis of similar characteristics.
6. Cladistics ignores overall similarity and bases taxonomy on the timing of branch points as decided by the sequence in the origin of apomorphic (derived) characters.
7. Classical evolutionary taxonomy considers overall similarity and branching sequence.

Is a New Evolutionary Synthesis Necessary? (pp. 496–497)

1. The modern synthesis combines a variety of disciplines to explain evolution. It emphasizes the Darwinian concepts of gradualism and natural selection.
2. Some researchers question the degree to which macroevolution is the cumulative product of microevolution due to natural selection. In this view, chance episodes play an important role in speciation and may be the most important factors in macroevolution.
3. In the final analysis, the debate among evolutionists is more about the relative importance of evolutionary mechanisms than about their nature. Natural selection, as the mechanism of adaptation, remains central to evolutionary biology.

SELF-QUIZ

1. A paleontologist estimates that when a particular rock formed, it contained 12 mg of the radioactive isotope potassium-40. The rock now contains 3 mg of potassium-40. The half-life of potassium-40 is 1.3 billion years. About how old is the rock?
   a. 0.4 billion years
   b. 0.3 billion years
   c. 1.3 billion years
   d. 2.6 billion years
   e. 5.2 billion years
2. If humans and pandas belong to the same class, then they must also belong to the same
   a. order
   b. phylum
   c. family
   d. genus
   e. species
3. In the case of comparing birds to other vertebrates, having four appendages is
   a. a plesiomorphic character
   b. an apomorphic character
   c. a character useful for distinguishing the birds from other vertebrates
   d. an example of analogy rather than homology
   e. a character useful for sorting the avian (bird) class into orders
4. Advocates of the model known as species selection propose that most evolutionary trends result from
   a. the tendency for natural selection to perfect adaptations
   b. stepwise progression of an unbranched lineage, with each step furthering the evolutionary trend
   c. phyletic transformation of a single species
   d. preadaptation of species for possible changes in the environment
   e. differences between species in their longevity and/or rates of speciation
5. The greatest adaptive radiation of the animal kingdom occurred during the
   a. early Precambrian era
   b. late Precambrian era
   c. early Paleozoic era
   d. early Mesozoic era
   e. early Cenozoic era
6. The DNA from two species is compared by the method of restriction mapping. Extensive similarity between the species in the collection of DNA fragments from treatment with a restriction enzyme indicates that
   a. the genes being compared have the same functions
   b. most sites recognized by the restriction enzyme have equivalent locations in the DNA samples from the two species
   c. the two species normally possess the same restriction enzyme
   d. the DNA fragments that match in size between the two species have identical base sequences
   e. the genomes of the same species are about the same size in their total amount of DNA
7. Extensive adaptive radiations have usually followed in the wake of mass extinctions mainly because
   a. many adaptive zones are vacated
   b. conditions of the physical environment are usually at their most favorable after some crisis has passed
   c. the survivors have superior adaptations that enable them to spill into many environments when conditions improve after the extinction episode
   d. preadaptation assures that survivors will radiate to give rise to many new species
   e. given a stable environment, biological diversity tends to increase
8. Which of the following would be most useful for constructing a cladogram showing the taxonomic relationships among several fish species?
   a. several analogous characteristics shared by all the fishes
   b. a single homologous characteristic shared by all the fishes
   c. the total degree of morphological similarity among various fish species

d. several characteristics thought to have evolved after different fish diverged from one another

e. a single characteristic that is different in all the fishes

9. The evolutionary transformation of a fish's primitive lung into a swim bladder (float) is an example of
   a. convergent evolution
   b. divergent evolution
   c. preadaptation
   d. adaptive radiation
   e. paedomorphosis

10. The differences between the modern synthesis and a more hierarchical view of evolution include all the following *except*
   a. gradualism versus punctuated equilibrium
   b. natural selection versus chance as central to adaptation
   c. the relative importance of microevolution in macroevolution
   d. phyletic transitions versus species selection
   e. the relative role of chance in speciation and biological diversity

CHALLENGE QUESTIONS

1. In the "DNA clock," some nucleotide changes cause amino acid substitutions in the encoded protein (nonsynonymous changes), and others do not (synonymous changes). In a comparison of rodent and human genes, rodents were found to accumulate synonymous changes 2.0 times faster than humans and nonsynonymous substitutions 1.3 times as fast. How do such data complicate the use of molecular clocks in absolute dating?

2. Imagine that Pangaea were re-formed today. What macroevolutionary changes would you expect?

SCIENCE, TECHNOLOGY, AND SOCIETY

1. Experts estimate that human activities cause the extinction of hundreds of species every year. The natural "background" rate of extinction is thought to be a few species per year. As we continue to alter the global environment, especially by cutting down tropical rain forests, the resulting extinction will probably rival that at the end of the Cretaceous period. Many scientists and environmentalists are alarmed at this prospect. What are some reasons for their concern? Others are not as worried, since life has endured numerous mass extinctions and has always bounced back. Is the coming mass extinction different from previous extinctions? Why? What might be the consequences for the surviving species? For human beings?

2. As described in Figure 23.17, it is not clear whether the red wolf is a hybrid between the coyote and the gray wolf or a separate species. The solution to this taxonomic puzzle has become the center of a debate about whether or not the red wolf qualifies for protection under the Endangered Species Act. If more detailed DNA studies convincingly demonstrate that the red wolf is indeed a hybrid between the coyote and gray wolf, should efforts to protect and breed the red wolf be terminated? Why or why not?

FURTHER READING

Alvarez, W., F. Asaro, and V. Courtillot. "What Caused the Mass Extinction?" *Scientific American,* October 1990. Two articles, a point-counterpoint on asteroids versus volcanoes as the main cause of Cretaceous extinctions.

Bogin, B. "The Evolution of Human Childhood." *BioScience,* January 1990. The importance of heterochrony.

Cherfas, J. "Ancient DNA: Still Busy After Death." *Science,* September 20, 1991. Molecular biologists go to work on fossils.

Ezzell, C. "Conserving a Coyote in Wolf's Clothing? Molecular Systematics and Endangered Species." *Science News,* June 15, 1991.

Gould, S. J."We Are All Monkey's Uncles." *Natural History,* June 1992. Cladists and human classification.

Kerr, R.A. "Huge Impact Tied to Mass Extinction." *Science,* August 14, 1992. Is a Yucatan crater the scar of an asteroid collision?

Raup, D. "Extinction: Bad Genes or Bad Luck?" *New Scientist,* September 14, 1991.

Ravven, W. "They Went That-a-Way." *Discover,* February 1992. What can we learn about dinosaurs from studying their footprints?

Ross, P. "Eloquent Remains." *Scientific American,* May 1992. Molecular archaeology.

Waters, T. "Greetings from Pangaea." *Discover,* February 1992. Environments were extreme on the supercontinent.

UNIT FIVE The Evolutionary History of Biological Diversity

An Interview with Stephen J. Gould

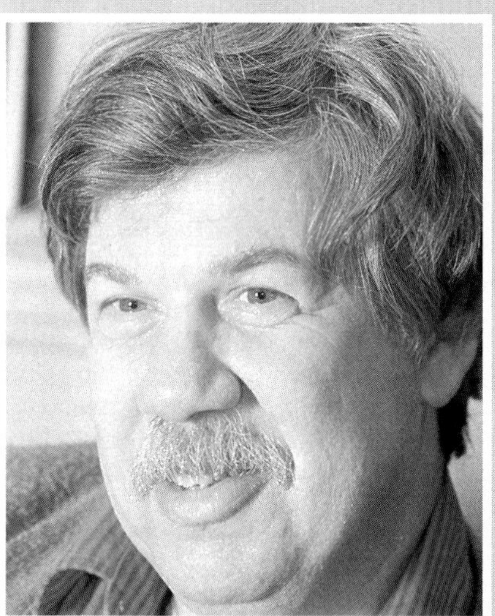

Stephen Jay Gould is one of those rare scientists who become widely known to the general public through popular writing. His monthly columns in Natural History and his many best-selling books, including Wonderful Life *(New York: Norton, 1989) and* Bully for Brontosaurus *(New York: Norton, 1991), have turned hundreds of thousands of readers into biology fans. Many of Dr. Gould's essays highlight the role of chance in the history of life. He is also well known for the theory of punctuated equilibrium, which he developed with fellow paleontologist Niles Eldredge of the American Museum of Natural History. Stephen Jay Gould teaches biology, geology, and the history of science at Harvard*

How did your interest in fossils develop?
There are two paths for kids who get into it young, which I did. One is you live rurally and you collect fossils, and the other is you live in an urban setting and you go to museums. I grew up in New York, so for me it was the Museum of Natural History.

At what age?
Five or six. But this situation is common. Maybe what's rare is that within the profession you find so many people who have been committed to it ever since they were kids, because that's when many people first get interested in fossils. Among paleontologists you find a lot who have been at it since they were little.

If so many kids are interested in fossils, why aren't there more paleontologists?
Well, first of all, they don't ever really get into the intellectually interesting part of it; for them it's just a transient stage. From that they go onto the policeman or the fireman stage, or the Ninja turtle stage, or whatever is next. I don't mean to say that condescendingly; that's what kids do. There are a lot of interests that are meant to be sampled, and then you pass onto something else. You'll be very fascinated with something and learn a little bit about it, and then you move on. And so you lose a lot there, and that's fine, that's how a kid has to grow up. I'm not being at all critical. The ones I'm sad about are the ones you lose because their parents don't encourage it, or because it isn't cool on the playground, or because they're poor and don't get a chance for an education, or because their parents persuade them it's not remunerative and that's too bad.

What motivated you to start writing for general audiences?
My first *Natural History* piece was a regular article, and then I started the series in January 1974. It was never an intention—it's like most things in the evolution of life, to use an analogy—it just happened. They'd always had columns. They asked me because I had done my graduate work at the Museum of Natural History and I already had a reputation for writing technical scientific papers in a style that was very unconventional and fairly literary. When they asked me I thought, "There's a funny idea," and I thought I'd do two or three and see what happened. That was 210 essays ago. It just happened. I tried it and I liked it, as the saying goes. And I still do it.

Several of your essays address the fallacy of searching for moral messages in nature. Why doesn't this work?
The very worst example ever was, of course, Adolph Hitler building a theory of racial purity on what he thought was the way in which nature operated. The fact that his ideas were nonsensical doesn't mean the notion is valid that you can find *good* morals in nature. The main point is that life is 3½ billion years old and we got here yesterday, geologically speaking. Why should a process that's been going on that long in ignorance of us contain moral messages for our lives? It can't. Science as an enterprise is a discipline that deals with the factual state of the world, and you don't derive ethical beliefs from factual statements. The most science can do is to supply information that may be relevant to ethical decisions, but it's never going to tell you what the proper behavior is. It just can't.

Another common theme in your essays is how cultural and social biases affect our view of life. Can you give an example?
Anything is an example because science is inevitably embedded in social context. Because I'm an evolutionary biologist, from my point of view the most interest-

ing cases are biases that we have that distort our interpretation of evolution. Some are very general, some very deep biases like determinism, which don't really hold in the world. But in respect to the interpretation of evolution, surely the most pervasive bias is the various ways in which people still want to see themselves as the top of some evolutionary ladder rather than as one branch on a very elaborate bush of life. The old notion of progress, of evolution as an advancing, "complexifying" force that leads to human beings, is probably still the strongest bias affecting the way that people look at evolution.

You have also pointed out that our view of nature is biased by a provincial sense of scale—our use of human body size and human lifespan as measuring rods. How do scaling problems contribute to misconceptions about life?
The range of time is so vast and the human life is about 70 years, if you're lucky. To a geologist, 70 years is so short you couldn't even measure it. We think of things as long if they last most of a 70-year lifetime, but that's an immeasurable geological instant. People get messed up on concepts of time. When I developed the punctuated equilibrium theory with Niles Eldredge, it was misinterpreted because we kept saying that new species arise with geological abruptness, and people thought we meant overnight in a single macromutation. In fact, we meant 5000 to 10,000 years, which would be glacially slow on the scale of a human life, but is instantaneous geologically. People just don't understand the scale differences.

Size might not be so big a problem except that the forces of nature are different on organisms of very different sizes. Because we're fascinated by the few animals that are larger than we are, most people think that humans are kind of in-between in the range of organic size. It's not true. We're at the very upper end—we must be in the highest half a percent of body size for animals. We're big. There aren't many animals that are bigger than us—the vast majority, about 80%, are arthropods, and they're all tiny compared to us. Because in the world of small creatures surface forces predominate, and in our world volumetric and gravitational forces predominate, we just don't realize that small animals are experiencing a different world. The most amusing examples are the science fiction films that completely mess that up. The giant insect films where they have 20-foot ants flying—well, a 20-foot ant couldn't fly because flight is surface dependent and much of lift-up is volumetric—it just couldn't do it.

Besides the time-scale issue, are there other reasons that punctuated equilibrium is still controversial among biologists?
Part of it is misunderstanding based on scale. But a lot of it is the deep preference that people have for gradualistic explanations. There is a reason why gradualism is preferred in Darwinian theory, which is that if you're going to ascribe change entirely to natural selection building adaptation, then the adaptation has to be built slowly and gradually, step by step. Otherwise how can natural selection be behind it. For selection to be a creative force in evolutionary change, it has to select from a random field of variation and build a direction itself, and it has to do that through intermediary stages. Hence the strong preference for gradualism.

Your 1989 book, *Wonderful Life*, is about the role of contingency in the history of life. What's a good example?
The best example, the one that strikes home closest, is all of us. What are we doing here anyway? A cardinal fact in the history of life is that mammals arose at the same time as dinosaurs. Most people think that mammals came late and pushed the dinosaurs out by superiority, but it's not true. The mammals and the dinosaurs arose pretty much at the same time and for a hundred million years dinosaurs were big creatures that usurped all the space for large organisms. The largest mammals before the extinction of dinosaurs were about rat size. The smallest dinosaur, which was about ostrich size, was much larger than the largest mammal. Mammals were doing fine in their little world, but you never would have gotten intelligent

creatures at this size. It wasn't until dinosaurs died out that mammals got a chance. And the dinosaurs dying was part of a mass extinction that was probably at least triggered by the impact of some extraterrestrial body. Now if that isn't the ultimately contingent, unpredictable event, I don't know what is. It's a bolt from the blue in the literal sense of the term. And it's only because that happened that mammals got a chance to differentiate.

Dinosaurs had ruled for a hundred million years; there's only been 65 million years since they died out. If it hadn't been for that event that knocked them out, I suspect they'd still be the big ones. They weren't moving towards intelligence, probably couldn't, given the nature of their construction, so we wouldn't be here. So in a very direct way, we're here because of that unanticipated extinction of dinosaurs. But even that doesn't make human beings, it just makes mammals. Why have primates at all? Primates are a small group, only 200 species, not a conspicuously successful group. We're lucky to be here. If it weren't for a marked change in African environments 2½ million years ago that produced savannas and reduction of forests, we'd probably still be in the trees and not sentient in the sense we are now. Contingency doesn't just apply to the big changes; it is equally strong for detail of life's history, and we're a detail.

Is there *anything* predictable about the course of evolution?

Yes, but the predictions are very broad ones. I'd certainly be willing to predict that life would arise given earth conditions. That's probably a basic result of chemistry and the physics of self-organizing systems. I would grant you that there are always going to be more prey than predators because there's just no other way to run an ecosystem. I might even give you bilateral symmetry because that's an effective form. I will allow you that if flying things are going to evolve at all, they're likely to have wings because there's no other way to move through air. So there are certain very broad features of life that have predictability. But what most people want to know about when they ask about the history of life are not such broad-scale, abstract questions. What we're usually interested in is why people are here. Why did the dinosaurs rule? Why did mammals take over? Questions like that are all in the domain of contingency.

By whatever contingency, we *are* here. Many students ask if we're still evolving.

I've never given a talk where that question is not asked. *My* question is why everyone anticipates that it should be going on. For a highly successful species

like *Homo sapiens,* we should expect stability. So why is there always the assumption that humans will keep on changing? Why do people think it's peculiar to human beings that we haven't changed for 40 or 50 thousand years (which we haven't)? The assumption is always made that this is anomalous. It isn't. It's the standard, it's what evolution is about—large, successful species don't change for long periods of time. Pigeons, rats, whatever it is—they just hang in there.

The question is asked because of the idea of progress. Most people think that evolution means progress. They think people are the best and they hope that evolution will mean we are going to get better. But if people understood what evolution really meant—which is adaptation to change in local environments—they would see the high level of worldwide success of human beings.

Can you envision any circumstances that would lead to the origin of new hominid species?

Oh yes, but they're science fiction things like sending off little colonies into space somewhere. That's where you get speciation, where you get isolation of small groups. But that's truly science fiction. I like science fiction as a genre, it's just not what I do.

You mentioned the mass extinction that wiped out dinosaurs and many other forms 65 million years ago. How does that rank compared to other mass extinctions?

All paleontologists are probably agreed that there are five really large extinctions that stand out. I don't know the exact ranking, but by far the biggest is the one at the end of the Permian, 225 million years ago. The estimates for that mass extinction range as high as 96% of all marine invertebrate species. The other four came at the end of the Ordovician, of the Triassic, and of the Cretaceous, and the one that came near the top of the Devonian. The one that most people care about because it wiped out dinosaurs and gave us a chance, the Cretaceous extinction, is not the biggest; it's third or fourth.

Extinction is a subject of much interest these days. Have you encountered this argument: If extinction is the inevitable fate of every species, then why should we worry so much about current extinctions caused by human activity?

Yes, that's a standard form of developers' arguments, and a specious one. Saying that is as foolish as saying that you shouldn't cure an easily treatable childhood infection because that child will ultimately die as an old person. Yes, species die eventually, but if it's millions of years down the road, why should we dispense with them now when the reason for their death is our rapacious activities and nothing due to their natural or inevitable state?

Other than mass extinctions, what were some of the most important episodes in the history of life?

There is so much else that happened that's very interesting but isn't actually an episode because it sort of oozes in and out. There are episodes of relatively rapid origination, but they are not quite so sudden as mass extinctions. My main interest and that of almost any paleontologist is the very rapid introduction of essentially all the modern designs of multicellular life at the base of the Cambrian period some 550 million years ago—the so-called Cambrian explosion. That's really almost a facetious term, however, because geological explosions have very long fuses.

What other questions about the history of life and evolution of form are of greatest interest to you now?
Several things. I'd like to know more about the interaction of contingency and predictability and the basic pattern of life's history. I'd like to know the extent to which Darwin was right in thinking that selection working on organisms is dominant if not exclusive. I rather suspect that the hierarchical notion that selection works on species and works on genes and works on cell lineages and works on all those levels at the same time is much closer to what happens, and is actually conceptually a very different evolutionary theory from Darwinism. I'm interested in the role of random processes and mass extinctions in life's history. I'm basically interested in the way in which the very strict version of Darwinism, which has been orthodox in the last twenty or thirty years, is inadequate. I don't think it's wrong; I think it's a correct argument and that it's powerful, but I don't think it's everything,

and I'm interested in those patterns of life's history that need other kinds of explanations.

If you had 30 minutes with Charles Darwin, what would you ask him?
That's a good question. First of all, I'd prepare very carefully. There would be two alternatives, depending on how open he wanted to be. First, since I've studied so much about him personally, there are some personal questions I'd like to ask that would really help us understand him—what his real reactions were about his mother's death when he was eight, and what were his personal religious feelings, which he was very cryptic about. But knowing Mr. Darwin, I don't think he'd be willing to talk about those things, and that's fine, because I wouldn't, either. I'd respect his privacy. In that case, I'd take a second tack. I would very quickly, because we'd only have half an hour, tell him what's happened to evolutionary theory since his time, and I'd love to know what he thought of it—in the same sense that if Bach came back I'd love to show him a modern performance of his work, because I bet it would be much better than anything he heard in his own day. And I'd love to know what Darwin would think about recent discoveries in mass extinction theories, and about discoveries in heredity and molecular genetics. But probably he'd say, "You tell me, because it's so interesting; I want to know," and then my whole purpose would be defeated.

24 Early Earth and the Origin of Life

The Antiquity of Life
The Origin of Life
The Kingdoms of Life

Life is a continuum extending from the earliest organisms through the various phylogenetic branches to the great variety of forms alive today. In Unit Five, we will survey the diversity of contemporary life and trace the evolution of this diversity over 3.5 billion years of history (see Table 23.1).

One recurrent theme in these chapters is the association between biological and geological history. Geological events that alter environments change the course of biological evolution. The formation and subsequent breakup of the supercontinent Pangaea, for instance, affected the diversity of life tremendously (see Chapter 23). Conversely, life has changed the planet it inhabits (Figure 24.1). For example, the evolution of photosynthetic organisms that released oxygen into the air completely altered Earth's atmosphere. Much more recently, the emergence of *Homo sapiens* has changed the land, water, and air on a scale and at a rate unprecedented for a single species. The histories of Earth and its life are inseparable.

These chapters also emphasize key junctures in evolution that have punctuated the history of biological diversity. Earth history and biological history have been episodic, marked by what were in essence revolutions that opened many new ways of life. And, as Stephen Jay Gould pointed out in the interview that precedes this unit, the history of life is a story of contingency, not predictable outcomes.

Historical study of any sort is an inexact discipline, dependent on the preservation, reliability, and interpretation of past records. The fossil record of past life is generally less and less complete the farther into the past we delve. Fortunately, each organism carries traces of its evolutionary history in its molecules, metabolism, and anatomy. As we saw in Unit Four, such traces are clues to the past that augment the fossil record, much as similarities and differences between extant cultures help social scientists understand historical relationships between the cultures. Still, the evolutionary episodes of greatest antiquity are generally the most obscure. This chapter is the most speculative of the unit, for its main subject is the origin of life on a young Earth, and no fossil record of that seminal episode exists. We begin with the oldest known fossils, in order to certify the antiquity of life on Earth. Then we will examine theories about how natural processes

Figure 24.1
The changing Earth and its life. In a scene reminiscent of conditions on the early Earth, violent discharges of lightning and volcanic activity were associated with the birth in 1963 of the island of Surtsey near Iceland in the North Atlantic. Terrestrial organisms began colonizing Surtsey almost immediately after its birth. Since then, the island and its inhabitants have been evolving together, demonstrating on a small scale the inseparable history of Earth and its life. In this chapter, we will evaluate some ideas about how life began on a relatively young Earth.

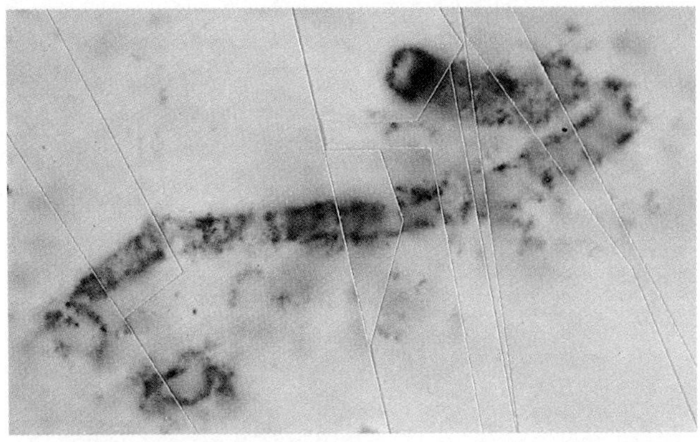

Figure 24.2
An early prokaryote. Precambrian fossils escaped notice until about 30 years ago, partly because they are microscopic. This filamentous prokaryote, about 3.5 billion years old, was collected in Western Australia (LM).

on the youthful planet could have created life. The last section of the chapter introduces the various kingdoms of life, as a prelude to the survey of biological diversity covered in Chapters 25–30.

THE ANTIQUITY OF LIFE

The metaphor of an evolutionary tree implies that the history of life chronicles an increasing diversity of organisms all descended from the primitive creatures that were the first living things. The apparent absence of fossils of ancestral organisms in Precambrian rocks was incongruous with Darwin's view that complex life evolved from simpler forms, and he wrote in *The Origin of Species:* "To the question why we do not find rich fossiliferous deposits belonging to these assumed earliest periods prior to the Cambrian system, I can give no satisfactory answer. . . . The case at present must remain inexplicable, and may be truly urged as a valid argument against the views here entertained." Some of Darwin's adversaries seized the cue and declared the beginning of the Cambrian period as the time of Genesis and all creation. Only in the past few decades has the discovery of older fossils filled in the Precambrian blank. The Cambrian fauna was preceded by a less diverse collection of animals dating back 700 million years (see Chapter 23). Before then there was a succession of microorganisms spanning nearly 3 billion years (Figure 24.2). For most of that time, only prokaryotes inhabited Earth. One would guess from the relatively simple structure of the prokaryotic cell (compared with the eukaryotic cell) that early organisms were primitive bacteria, and the fossil record now supports that presumption.

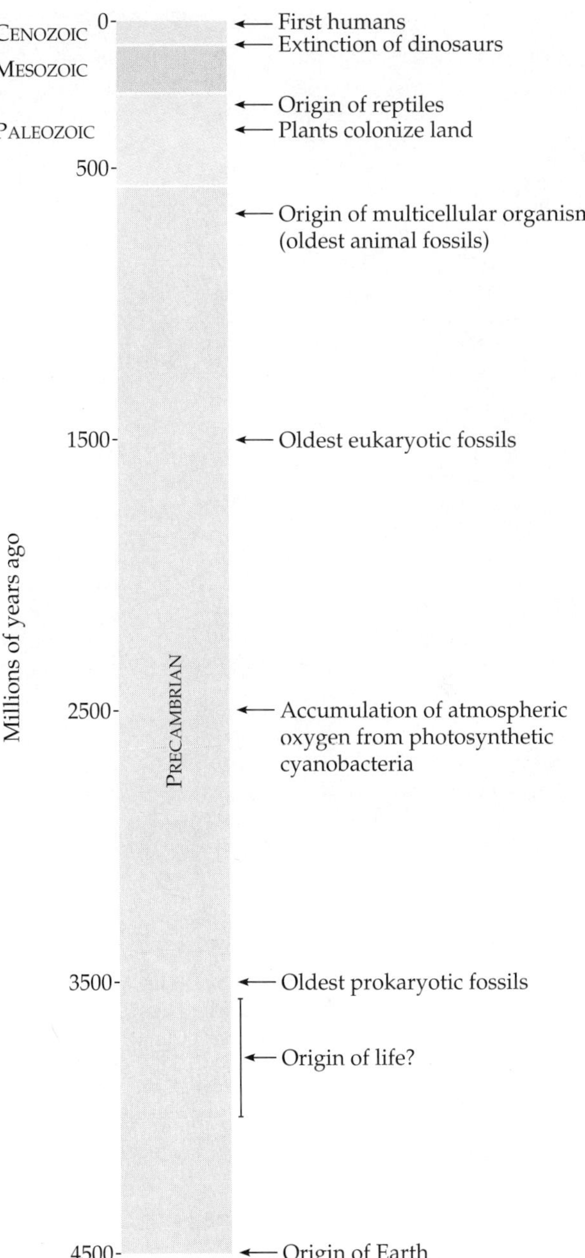

Figure 24.3
Some major episodes in the history of life.

Prokaryotes originated within a few hundred million years after Earth's crust cooled and solidified (Figure 24.3). Though geologists have discovered minerals that crystallized about 4.1 billion years ago and sedimentary rocks that date back 3.8 billion years, no fossils that old have yet been found. However, fossils that appear to be prokaryotes about the size of bacteria *have* been discovered in southern Africa in a rock formation called the Fig Tree Chert, which is 3.4 billion years old. Evidence of even more ancient prokaryotic life has been found in rocks called stromatolites (Gr. *stroma,* "bed," and *lithos,* "rock"). **Stromatolites** are banded domes of sediment strikingly similar to the

(a)

(b)

(c)

Figure 24.4
Bacterial mats and stromatolites. The mats are sedimentary structures produced by colonies of bacteria and cyanobacteria that live, uncropped by predators, in environments inhospitable to most other life. (**a**) Lynn Margulis and Kenneth Nealson, who study the history of life, are collecting bacterial mats in a Baja California lagoon. (**b**) The bands, seen in this section of a mat, are layers of sediment that adhere to the sticky prokaryotes, which produce the succession of layers by migrating. (**c**) Fossilized mats known as stromatolites resemble the layered structures formed by contemporary bacterial colonies. This stromatolite is a western Australian specimen about 3.5 billion years old. Microfossils, such as the one in Figure 24.2, are present in many stromatolites.

layered mats constructed by colonies of bacteria and cyanobacteria living today in very salty marshes. The layers are sediments that stick to the jellylike coats of the motile microbes, which continually migrate out of one layer of sediment and form a new one above, producing the banded pattern (Figure 24.4). Fossils resembling spherical and filamentous prokaryotes have been found in stromatolites that are 3.5 billion years old in western Australia and southern Africa. For now, these are the oldest evidence of life. However, the western Australia fossils appear to be those of photosynthetic organisms, perhaps oxygen producers. If so, it is likely that life had been evolving long before these organisms lived, possibly as early as 4 billion years ago. To put this in perspective, recall from Chapter 2 that Earth is about 4.6 billion years old. On the vast scale of geological time, life originated relatively early.

THE ORIGIN OF LIFE

The question of how life began is more specifically about the genesis of prokaryotes. Sometime between about 4.1 billion years ago, when Earth's crust began to solidify, and 3.5 billion years ago, when the planet was inhabited by bacteria advanced enough to build stromatolites, the first organisms came into being. What was their origin? The majority of biologists subscribe to the hypothesis that life developed on Earth from nonliving materials that became ordered into molecular aggregates that were eventually capable of self-replication and metabolism. Life cannot arise by spontaneous generation from inanimate material today, as far as we know, but conditions were very different when Earth was only a billion years old. The atmosphere was different (there was little oxygen, for instance), and lightning, volcanic activity, meteorite bombardment, and ultraviolet radiation were all more intense than what we experience today (see Chapter 2). In that ancient environment, the origin of life was evidently possible, and it is likely that at least the early stages of biological inception were inevitable. However, debate abounds about what occurred during these early stages.

According to one hypothesis, the first organisms were products of a chemical evolution in four stages: (1) the abiotic (nonliving) synthesis and accumulation of small organic molecules, or monomers, such as amino acids and nucleotides; (2) the joining of these monomers into polymers, including proteins and nucleic acids; (3) the aggregation of abiotically produced molecules into droplets, called protobionts, that had chemical characteristics different from their surroundings; and (4) the origin of heredity (which may have been under way even before the "droplet" stage).

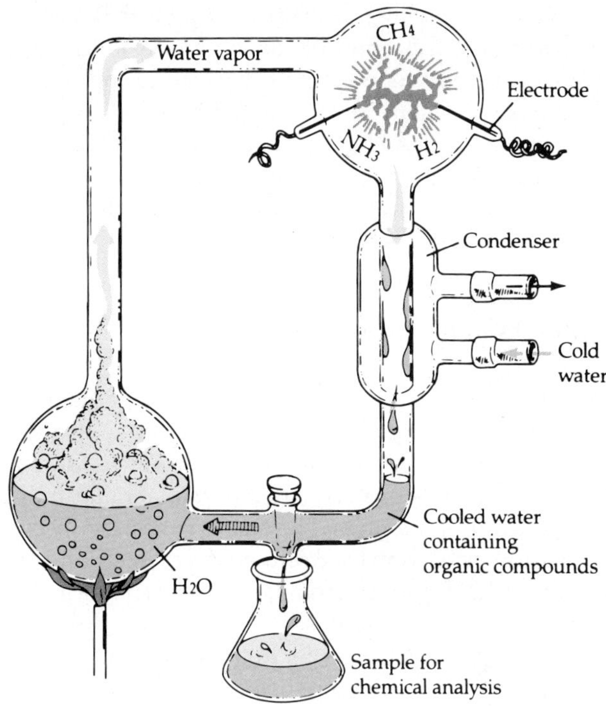

Figure 24.5
Abiotic synthesis of organic molecules in a model system. Stanley Miller and Harold Urey used an apparatus similar to this one to simulate chemical dynamics on the primitive Earth. A warmed flask of water simulated the primeval sea. The "atmosphere" consisted of H_2O, H_2, CH_4, and NH_3. Sparks were discharged in the synthetic atmosphere to mimic lightning. A condenser cooled the atmosphere, raining water and any dissolved compounds back to the miniature sea. As material circulated through the apparatus, the solution in the flask changed from clear to murky brown. After one week, Miller and Urey analyzed the contents of the solution and found a variety of organic compounds, including some of the amino acids that make up the proteins of organisms.

Abiotic Synthesis of Organic Monomers

In the 1920s, A. I. Oparin of Russia and J. B. S. Haldane of England independently postulated that conditions on the primitive Earth favored chemical reactions that synthesized what are now called organic compounds from inorganic precursors present in the early atmosphere and seas. This cannot happen in the modern world, Oparin and Haldane reasoned, because the present atmosphere is rich in oxygen produced by photosynthetic life. The oxidizing atmosphere of today is not conducive to the spontaneous synthesis of complex molecules because the oxygen attacks chemical bonds, extracting electrons. Before oxygen-producing photosynthesis, Earth had a much less oxidizing atmosphere, derived mainly from volcanic vapors. Such a reducing (electron-adding) atmosphere would have enhanced the joining together of simple molecules to form more complex ones. Even with a reducing atmosphere, making organic molecules would require considerable energy, which was probably provided by lightning and the intense ultraviolet radiation that penetrated the primitive atmosphere. The modern atmosphere has a layer of ozone produced from oxygen, and this ozone shield screens out most ultraviolet radiation. Evidence also exists that young suns emit more ultraviolet radiation than older suns. Oparin and Haldane envisioned an ancient world with the necessary chemical conditions and energy resources for the abiotic synthesis of organic molecules.

In 1953, Stanley Miller and Harold Urey tested the Oparin-Haldane hypothesis by creating, in the laboratory, conditions comparable to those of the early Earth (see Figure 4.3). Their apparatus produced a variety of amino acids and other organic compounds found in living organisms today (Figure 24.5).

The atmosphere in the Miller-Urey model was made up of H_2O, H_2, CH_4 (methane), and NH_3 (ammonia), the gases the researchers in the 1950s believed prevailed in the ancient world. This atmosphere was probably more strongly reducing than the actual atmosphere of the early Earth. The vapors of modern volcanoes include CO, CO_2, and N_2, and it is likely that these gases were present in the ancient atmosphere evolved from volcanoes. Traces of O_2 may even have been present, formed from reactions among other gases as they baked under powerful ultraviolet radiation. Many laboratories have repeated the Miller-Urey experiment using a variety of recipes for the atmosphere, including a mixture having a very low concentration of O_2. Abiotic synthesis of organic compounds occurred in these modified models, although yields were generally less than in the original experiment. The most relevant characteristic of the early atmosphere seems to be the rarity of the strong oxidizing agent O_2.

Laboratory analogs of the primeval Earth have been used to make all 20 amino acids commonly found in organisms, several sugars, lipids, the purine and pyrimidine bases present in the nucleotides of DNA and RNA, and even ATP (if phosphate is added to the flask). Before there was life, its chemical building blocks may have been accumulating as a natural stage in the chemical evolution of the planet.

Abiotic Synthesis of Polymers

The abiotic synthesis of more complex organic molecules by the joining together of smaller ones also may have been inevitable on the primitive Earth. Organic polymers such as proteins are chains of similar building blocks, or monomers. They are synthesized by condensation reactions that remove hydrogen and hydroxyl (—OH) groups from the monomers, forming a water molecule as a by-product of each new linkage in the polymer (see Chapter 5). In the living cell, specific enzymes catalyze the dehydration reactions. Abiotic synthesis of polymers would have had to occur without the help of these efficient enzymes, and the dilute concentrations of the monomers dissolved in an excess of water would not favor spontaneous dehydration reactions that form more water. (See Chapter 2 to review chemical equilibrium.) Polymerization does occur in laboratory experiments when dilute solutions of organic monomers are dripped onto hot sand, clay, or rock, a process that vaporizes water and concentrates the monomers on the substratum. Using this method, Sidney Fox of the University of Miami has made what he calls proteinoids, which are polypeptides produced by abiotic means. It is possible to imagine waves or rain splashing dilute solutions of organic monomers onto fresh lava or other hot rocks on the early Earth and then rinsing proteinoids and other polymers back into the water.

Clay, even cool clay, may have been especially important as a substratum for the polymerization reactions prerequisite to life. Clay concentrates amino acids and other organic monomers from dilute solutions because the monomers bind to charged sites on the clay particles. At some of the binding sites, metal atoms, such as iron and zinc, function as catalysts facilitating the dehydration reactions that link monomers together. Clay, having many of these binding sites, could have functioned as a lattice that brought monomers close together and then assisted in joining them into polymers. As an alternative to clay, iron pyrite (fool's gold), which consists of iron and sulfur, has been proposed as the substratum of organic synthesis. This hypothesis is advocated by Günter Wachtershäuser of Germany, who points out properties of pyrite that could have catalyzed abiotic synthesis of organic polymers. Pyrite provides a charged surface, and formation of this mineral from iron and sulfur yields electrons that could support bonding between organic molecules to form more complex products.

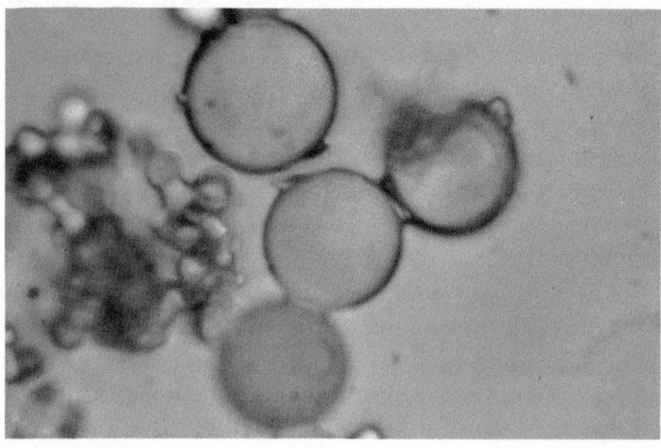

Figure 24.6
Laboratory versions of protobionts, aggregates of organic molecules with some biological properties. These microspheres are made by cooling solutions of proteinoids, polypeptides created abiotically from amino acids polymerized on hot surfaces (LM). Microspheres grow by absorbing free proteinoids until they reach an unstable size, when they split to form daughter microspheres. Of course, this division lacks the precision of cellular reproduction.

The Formation of Protobionts

The properties of life emerge from an interaction of molecules organized into higher levels of order (see Chapter 1). Living cells may have been preceded by **protobionts**—aggregates of abiotically produced molecules not yet capable of precise reproduction but able to maintain an internal chemical environment different from the surroundings and exhibiting some of the properties associated with life, including metabolism and excitability.

Laboratory experiments suggest that protobionts could have formed spontaneously from abiotically produced organic compounds. When mixed with cool water, proteinoids self-assemble into tiny droplets called microspheres (Figure 24.6). Coated by a membrane that is selectively permeable, the microspheres undergo osmotic swelling or shrinking when placed in solutions of different salt concentrations. Some microspheres also store energy in the form of a membrane potential, a voltage across the surface (see Chapter 8). The protobionts can discharge the voltage in nervelike fashion; such excitability is characteristic of all life (which is not to say that microspheres are alive, only that they display *some* of the properties of life). Droplets of another type, called liposomes, form spontaneously when the organic ingredients include certain

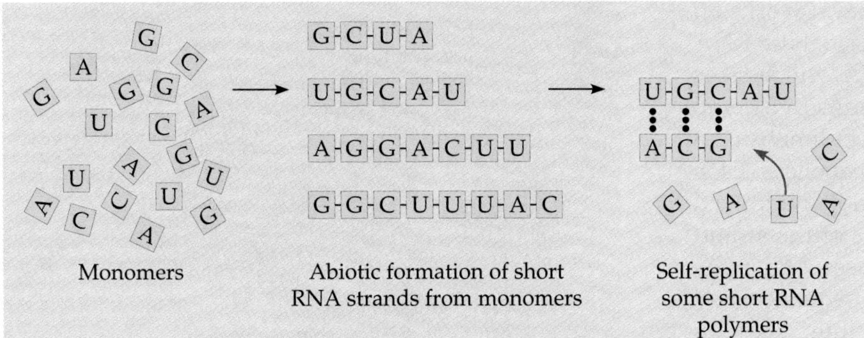

Figure 24.7
Abiotic replication of RNA. According to this hypothesis, the first genes were RNA molecules that polymerized abiotically and replicated themselves autocatalytically while bound to clay surfaces. The letters A, G, C, and U symbolize the four RNA bases, which pair specifically in A-U and G-C combinations (see Chapter 16).

lipids, which organize into a molecular bilayer at the surface of the droplet, much like the lipid bilayer of cell membranes (see Figure 8.1). Oparin has made still a different type of protobiont that he calls coacervates, which are colloidal droplets that form when a solution of polypeptides, nucleic acids, and polysaccharides is shaken. If enzymes are included among the ingredients, they are incorporated into the coacervates, and the protobionts function as miniature chemical factories that absorb substrates from their surroundings and release the products of the reactions catalyzed by the enzymes.

Unlike some laboratory models, protobionts that formed in the ancient seas would not have possessed refined enzymes, which are made in cells according to inherited instructions. Some molecules produced abiotically, however, do have weak catalytic capacities, and there could well have been protobionts that modified the substances they took in across their membranes by a rudimentary metabolism.

The Origin of Genetic Information

Imagine a tidepool, pond, or moist clay on the primeval Earth with a suspension of protobionts varying in chemical composition, permeability, and catalytic capabilities. Those droplets most stable and best able to accumulate organic molecules from the environment would grow and split, distributing their chemical components to the "baby" droplets. Other droplets would fall apart or fail to grow and divide. In this way, the environment may have selected in favor of some molecular aggregates and against others. But competition among the various protobionts could not lead to long-range improvement because there was no way to perpetuate success. As prolific droplets grew, split, grew and split again, their unique catalysts and other functional molecules would become increasingly diluted. The chemical aggregates that were the forerunners of cells could not build on the past and evolve until the development of some mechanism for replicating their characteristics—some mechanism of heredity that would pass along not just samples of key molecules but also instructions for making more of those molecules.

A cell stores its genetic information as DNA, transcribes the information into RNA, and then translates the messages into specific enzymes and other proteins (see Chapters 15 and 16). Instructions are transmitted by the replication of DNA when a cell divides. The DNA ⟶ RNA ⟶ protein axis of cellular control employs intricate machinery that could not have evolved all at once but must have emerged bit by bit as improvements to much simpler processes. Even before DNA, some primitive mechanism may have existed for aligning amino acids along strands of RNA that could replicate themselves. According to this hypothesis, the first genes were not DNA molecules but short strands of RNA that began self-replicating in the prebiotic world.

Some scientists studying the origin of life envision an "RNA world" and are testing the hypothesis of RNA self-replication. Short polymers of ribonucleotides have been produced abiotically in test-tube experiments. If RNA is added to a test-tube solution containing monomers for making more RNA, sequences about five to ten nucleotides long are copied from the template according to the base-pairing rules (see Chapter 16). If zinc is added as a catalyst, sequences up to 40 nucleotides long are copied with less than 1% error. In the 1980s, Thomas Cech and his coworkers at the University of Colorado, Boulder, revolutionized thinking about the evolution of life when they discovered that RNA molecules are important catalysts in modern cells. This disproved the long-held view that only protein enzymes serve as biological catalysts. Cech and others found that modern cells use RNA catalysts, called **ribozymes,** to do such things as remove introns from RNA (see Chapter 16). Ribozymes also catalyze the synthesis of new RNA, notably ribosomal RNA, tRNA, and mRNA. Thus, RNA is autocatalytic, and in the prebiotic world, long before there were protein enzymes or DNA, RNA molecules may have been fully capable of self-replication (Figure 24.7).

Natural selection on the molecular level works on diverse populations of replicating autocatalytic RNA molecules. Unlike double-stranded DNA, which takes the form of a uniform helix, single-stranded RNA molecules assume a variety of specific three-dimensional shapes mandated by their nucleotide sequences. Each sequence folds into a unique conformation, reinforced by hydrogen bonding between regions of the strand having complementary sequences of bases. The molecule thus has a genotype (its nucleotide sequence) and a phenotype (its conformation, which interacts with surrounding molecules in specific ways). In a particular environment, RNA molecules of certain base sequences are more stable and replicate faster with fewer errors than other sequences. Beginning with a diversity of RNA molecules that must compete for monomers to replicate, the sequence best suited to the temperature, salt concentration, and other features of the surrounding solution and having the greatest autocatalytic activity will prevail. Its descendants will be not a single RNA species but a family of closely related sequences (because of copying errors). Selection screens mutations in the original sequence, and occasionally a copying error results in a molecule that folds into a shape even more stable or more adept at self-replication than the ancestral sequence. Natural selection has been observed operating on RNA populations in test tubes, and perhaps it happened in prebiotic times as well.

The rudiments of RNA-directed protein synthesis may have been in the weak binding of specific amino acids to bases along RNA molecules, which functioned as simple templates holding a few amino acids together long enough for them to be linked, perhaps with zinc or some other metal acting as an adjunctive catalyst. If RNA happened to synthesize a short polypeptide that in turn behaved as an enzyme helping the RNA molecule to replicate, then the early chemical dynamics included molecular cooperation as well as competition (Figure 24.8a).

Thus, the first steps toward the replication and translation of genetic information may have been taken by molecular evolution even before RNA and polypeptides became packaged within membranes. Once primitive genes and their products became confined to membrane-enclosed compartments, the protobionts could evolve as units (Figure 24.8b). Molecular cooperation could be refined because components that interacted in ways favorable to the success of the protobiont as a whole were concentrated together in a microscopic volume rather than being spread throughout a puddle or film on the surface of clay. Suppose, for example, that an RNA molecule ordered amino acids into a primitive enzyme that extracted energy from an organic fuel taken up from the surroundings and made the energy available for other reactions within the protobiont, including replication of RNA. Natural selection could favor such a gene only if its diffusible product were kept close by, rather than being shared with competing RNA sequences in its environment.

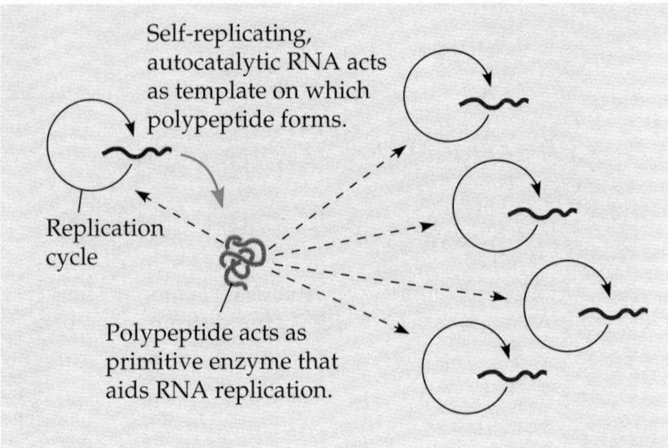

(**a**) Without membranes

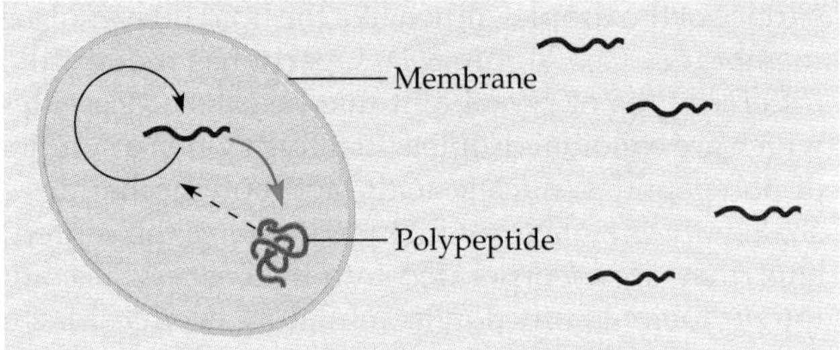

(**b**) Within membrane-enclosed compartment

Figure 24.8
One hypothesis for the beginnings of molecular cooperation. (**a**) An RNA strand enhances its own replication if it orders amino acids into a polypeptide that in turn functions as an enzyme helping the RNA strand replicate. In such cooperation may have been the rudiments for the translation of genetic information into protein structure. The benefits of this protein, however, would have been shared by competing RNA molecules not yet segregated within protobionts. (**b**) Once genes were in membrane-enclosed compartments, they would have benefited exclusively from their protein products.

In this scenario, we have built a hypothetical antecedent of the cell by incorporating genetic information into an aggregate of molecules that selectively accumulates monomers from its surroundings and uses enzymes programmed by genes to make polymers and carry out other chemical reactions. The protobiont grows and splits, distributing copies of its genes to offspring. Even if only one such protobiont arose initially by the abiotic processes that have been described, its descendants would vary because of mutations, errors in the copying of RNA. Evolution in the true Darwinian sense—differential reproductive success of varying individuals—presumably accumulated many refinements to primitive metabolism and inheritance. One trend apparently led to DNA becoming the repository of genetic information. RNA could have provided the template on which DNA nucleotides were assembled. DNA is a much more stable repository for genetic in-

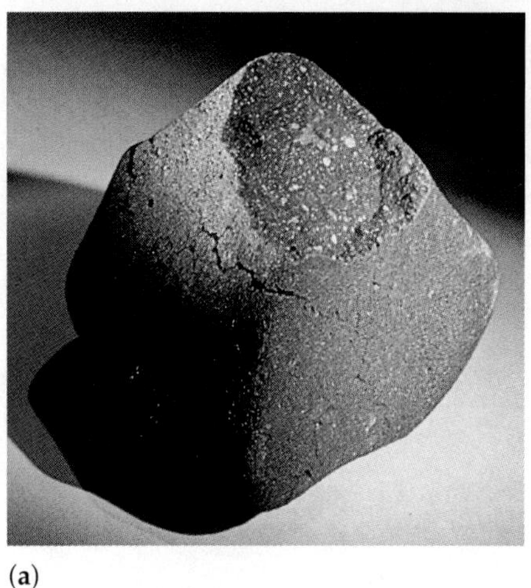

(a)

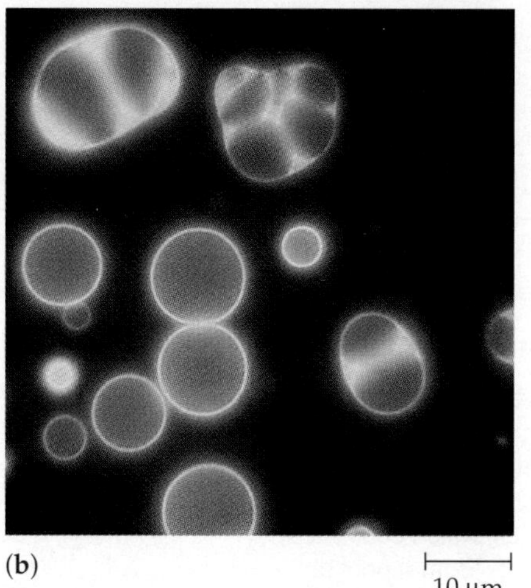

(b)

Figure 24.9
Some meteorites export organic compounds to Earth. (**a**) This golf ball-sized rock (90 g) is a piece of a meteorite that landed near Murchison, Australia, in 1969. It is an example of a carbon-rich meteorite, from which hydrocarbons, alcohols, fatlike molecules, amino acids, and other organic compounds have been extracted. (**b**) Organic molecules obtained from the Murchison meteorite form vesicles such as these. Notice how similar these vesicles are to the synthetic protobionts shown in Figure 24.6.

formation than RNA, and once DNA appeared, RNA molecules would have begun to take on their modern roles as intermediates in the translation of genetic programs.

Some Alternative Views

Laboratory simulations cannot establish that the kind of chemical evolution that has been described here actually created life on the primitive Earth, but only that some of the key steps *could* have happened. The origin of life remains a matter of scientific speculation, and there are alternative views of how several key processes occurred.

Some researchers question whether abiotic synthesis of organic monomers was necessary as a first step in the origin of life. It is possible that at least some organic compounds reached the early Earth from space. This idea, called **panspermia,** holds that hundreds of thousands of meteorites and comets hitting the early Earth brought with them organic molecules formed by abiotic reactions in outer space. Extraterrestrial organic compounds, including amino acids, have been found in modern meteorites, and it seems likely that these bodies could have seeded the early Earth with organic compounds. And biochemists recently demonstrated that organic molecules extracted from a meteorite produce small vesicles when mixed with water (Figure 24.9). Both panspermia and chemical evolution could have contributed to the pool of organic molecules that formed the earliest life.

Where life began is another issue. Until recently, most researchers favored shallow water or moist sediments as the most likely sites for life's origin. Some scientists now question this view, arguing that Earth's surface was very inhospitable during the period when life probably began. Asteroids and comets, debris left over from the formation of the solar system, pounded Earth and the other young planets. (The pocked face of the moon records this violent period, but plate tectonics destroyed most of the evidence on Earth.) An asteroid collision that may have contributed to the mass extinction of dinosaurs and other Mesozoic life (see Chapter 23) was a minor disturbance compared to the bombardment of the young Earth a few billion years earlier. Some scientists speculate that incipient life could not survive this cosmic assault—unless, that is, it began on the less exposed sea floor.

In the late 1970s, marine explorers discovered deep-sea vents where hot water and minerals gush through gaps in Earth's crust. Perhaps earlier vents supplied the energy and chemical precursors for the origin of protobionts.

Some biologists interested in the origin of life also challenge the idea of an "RNA world." They point out that even under optimal test-tube conditions, it is difficult to make RNA without enzymes. Even short RNA strands may be too complicated to be good candidates as the first self-replicating molecules. In 1991, Julius Rebek, Jr., and his colleagues at the Massachusetts Institute of Technology synthesized a simple organic molecule that acts as a template to produce copies of itself (Figure 24.10). This breakthrough strengthens an alternative hypothesis that nucleic acid genes were preceded by simpler heredity systems.

Debate about the origin of life abounds, and we have sampled only a few of the issues. Whatever way prebiotic chemicals accumulated, polymerized, and eventually reproduced, the leap from an aggregate of molecules that reproduces to even the simplest prokaryotic cell is immense and must have been taken in many smaller evolutionary steps. The point at which we stop calling membrane-enclosed compartments that metabolize and replicate their genetic programs protobionts and begin calling them living cells is as fuzzy as our definitions of life. We do know that prokaryotes were already flourishing at least 3.5 billion years ago, and that all kingdoms of life descended from those ancient prokaryotes.

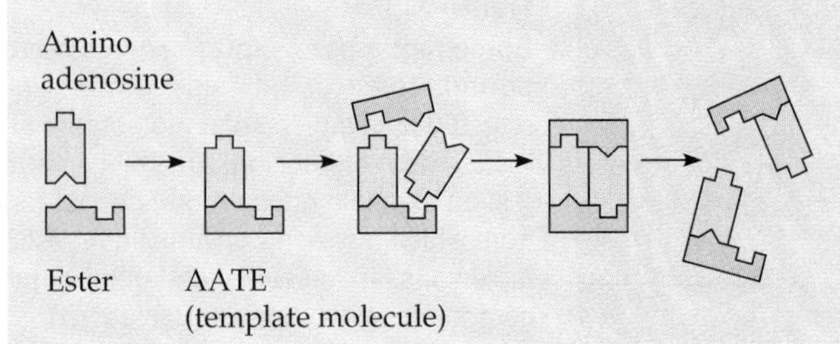

Figure 24.10
A synthetic self-replicating molecule. Were RNA or DNA genes preceded by simpler heredity systems? The 1991 laboratory creation of a relatively simple organic molecule that makes copies of itself is influencing ideas on the origin of life. The molecule is aminoadenosine triacid ester (AATE), which consists of two components, amino adenosine and ester. From a pool of these two building blocks, AATE can catalyze synthesis of additional AATE by acting as a template. No one is arguing that the first genetic material resembled AATE, but this research does demonstrate that molecules much smaller than nucleic acids can replicate.

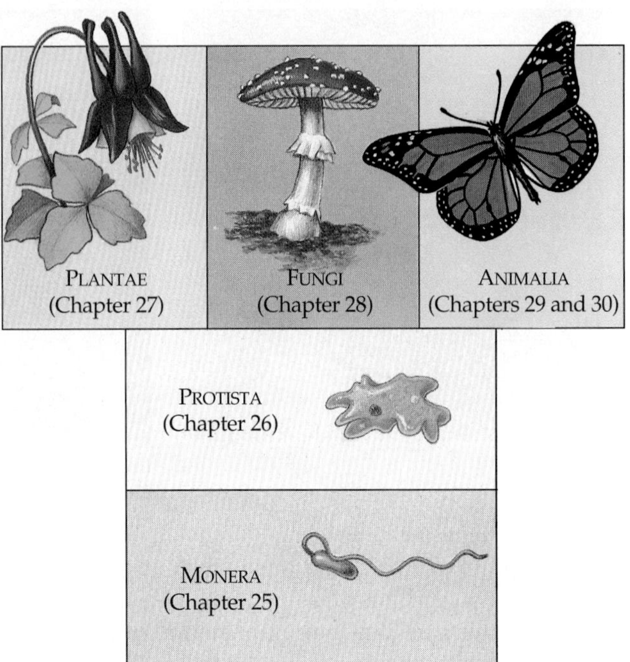

Figure 24.11
A five kingdom system. This classification system used in this text divides the diversity of life into a prokaryotic kingdom (Monera, the bacteria) and four eukaryotic kingdoms. Taxonomists continue to reevaluate the number and boundaries of kingdoms, and several alternative schemes have been proposed.

THE KINGDOMS OF LIFE

In Chapter 23, we looked at taxonomy as a tool scientists use in tracing the evolution of organisms. Now that we have gone backward to the very origin of life on Earth, taxonomy is once again relevant as we attempt to reconstruct evolutionary relationships among the immense diversity of forms that arose from those organisms.

The kingdom is the highest—the most inclusive—taxonomic category. We grow up with the bias that there are only two kingdoms of life—plants and animals—because we live in a macroscopic, terrestrial realm where we rarely encounter organisms that do not fit neatly into a plant–animal dichotomy. The two-kingdom scheme also had a long tradition in formal taxonomy; Linnaeus divided all known forms of life between plant and animal kingdoms. Even with the discovery of the diverse microbial world, the two-kingdom system persisted. Bacteria were placed in the plant kingdom, their rigid cell walls used as justification. Eukaryotic unicellular organisms with chloroplasts were also called plants. Fungi, too, fell under the plant banner, partly because they are sedentary, even though no fungi are photosynthetic and they have little in common structurally with green plants. In the two-kingdom system, unicellular creatures that move and ingest food—protozoa—were called animals. Microbes such as *Euglena* that move but are photosynthetic were claimed by both botanists and zoologists and showed up in the taxonomies of plant *and* animal kingdoms. Schemes with additional kingdoms were proposed, but none became popular with the majority of biologists until Robert H. Whittaker of Cornell University argued effectively for a five-kingdom system in 1969. These five kingdoms are Monera, Protista, Plantae, Fungi, and Animalia (Figure 24.11). Lynn Margulis, now at the University of Massachusetts, helped to promote the five-kingdom scheme and has proposed important modifications, some of which this text uses in its classification system.

The five-kingdom system recognizes the two fundamentally different types of cell—prokaryotic and eukaryotic—and sets the prokaryotes apart from all eukaryotes by placing them in their own kingdom, Monera. The prokaryotes are bacteria, including cyanobacteria, formerly called blue-green algae.

Organisms of the other four kingdoms all consist of cells organized on the eukaryotic plan (see Chapter 7). The kingdoms Plantae, Fungi, and Animalia are multicellular eukaryotes, each kingdom defined by characteristics of structure and life cycle discussed in upcoming chapters. Plants, fungi, and animals also generally differ in their modes of nutrition (the criterion originally used by Whittaker). Plants are autotrophic in nutrition, making their food by photosynthesis. Fungi are heterotrophic organisms that are absorptive in nutrition. Most fungi are decomposers that live embedded in their food source, secreting digestive enzymes and absorbing the small organic molecules that are the products of digestion. Animals live mostly by ingesting food and digesting it within specialized cavities.

We are left with kingdom Protista, a grab bag containing all eukaryotes that do not fit the definitions of plants, fungi, or animals. Most protists are unicellular forms, but in the version of the five-kingdom scheme used in this text, Protista also includes relatively simple multicellular organisms that are believed to be direct descendants of unicellular protists.

The most natural and unambiguous kingdom is Monera, its boundary defined by prokaryotic organization. Prokaryotes were the first forms of life and the *only* ones for at least 2 billion years. They remain tremendously important on modern Earth. In the next chapter, we will address the diversity and history of prokaryotic life.

STUDY OUTLINE

1. Biological and geological history are interwoven. Geological events alter environments that affect biological evolution, and organisms, in turn, change the planet they inhabit.

The Antiquity of Life (pp. 505–506)

1. For the first few billion years after Earth's crust cooled and solidified, only prokaryotes inhabited Earth.
2. The oldest available evidence for life appears in stromatolites containing fossils resembling bacteria dating back 3.5 billion years.

The Origin of Life (pp. 506–512)

1. One hypothesis about the origin of life is based on chemical evolution of protobionts, abiotically produced molecular droplets with distinctive chemical characteristics.
2. Laboratory experiments performed under conditions simulating those of the primitive Earth have produced diverse organic molecules from inorganic precursors.
3. Small organic molecules polymerize when they are concentrated on hot sand, rock, or clay.
4. Organic molecules synthesized in the laboratory have spontaneously assembled into a variety of droplets—microspheres, liposomes, and coacervates—with some of the properties associated with life.
5. The first genes may have been abiotically produced RNA, whose base sequence served as a template for both alignment of amino acids in polypeptide synthesis and alignment of complementary nucleotide bases in a primitive form of self-replication.
6. Once genetic information became incorporated inside membrane-enclosed compartments, protobionts would have acquired heritability and the ability to evolve as units.
7. Debate abounds about how the stepwise origin of life actually occurred. Abiotic synthesis of organic molecules versus import via meteorites, origins in shallow water versus deep-sea vents, and RNA genes versus simpler self-replicating molecules are just three of the issues.

The Kingdoms of Life (pp. 512–513)

1. The five-kingdom system classifies organisms as Monera, Protista, Plantae, Fungi, and Animalia.
2. All prokaryotes belong to kingdom Monera. The remaining four kingdoms, all eukaryotic, are defined by structure, life cycle, and mode of nutrition.

SELF-QUIZ

1. The *main* explanation for the lack of a continuing abiotic origin of life on Earth today is that
   a. there is not sufficient lightning to provide an energy source
   b. our oxidizing atmosphere is not conducive to the spontaneous formation of complex molecules
   c. there is much less visible light reaching Earth to serve as an energy source
   d. there are no molten surfaces on which weak solutions of organic molecules would polymerize
   e. all habitable places are already filled
2. Stromatolites are
   a. aggregates of abiotically produced organic molecules
   b. meteorites that contain amino acids and may have seeded Earth with organic molecules
   c. layers of clay that may have facilitated the polymerization of abiotically produced monomers
   d. a group of ancient prokaryotes
   e. banded domes of sediment that contain the oldest known fossils
3. Which of the following scientists are *incorrectly* paired with their theories or experiments?
   a. Fox—created polypeptides called proteinoids by dripping organic molecules onto hot sand
   b. Cech—discovered RNA catalysts, called ribozymes, that remove introns and catalyze RNA synthesis
   c. Miller—demonstrated that pyrite surfaces can catalyze organic synthesis
   d. Oparin and Haldane—first proposed that conditions on the early Earth favored abiotic synthesis of organic compounds
   e. Whittaker—developed the five-kingdom system
4. Minerals in clay or iron pyrite may have had an important role in the origin of life as
   a. sources of energy for synthesis of organic monomers
   b. oxidizing agents
   c. components in the first selectively permeable membranes
   d. catalysts in the formation of organic polymers
   e. components of early self-replicating RNA molecules
5. Formation of microspheres, liposomes, and coacervates all require

a. RNA
b. a membrane potential
c. self-assembly
d. phospholipids
e. primitive genes

6. Competition among various protobionts would have led to evolutionary improvement only when
   a. they were able to catalyze chemical reactions
   b. some kind of heredity mechanism developed
   c. they were able to grow and reproduce
   d. the protobionts acquired selectively permeable membranes
   e. DNA first appeared

7. In the two-kingdom system of classification,
   a. prokaryotes were placed in the plant kingdom
   b. only multicellular organisms were considered animals
   c. the animal kingdom included all heterotrophs
   d. unicellular eukaryotes were placed only in the animal kingdom
   e. prokaryotes and eukaryotes were separated

8. In his laboratory apparatus, Stanley Miller synthesized
   a. proteins
   b. DNA
   c. amino acids
   d. protobionts
   e. proteinoids

9. Which gas was probably *least* abundant in the early atmosphere?
   a. H_2O
   b. O_2
   c. NH_3
   d. CO
   e. CO_2

10. Which of the following steps has *not* yet been accomplished by scientists studying the origin of life?
   a. abiotic synthesis of small RNA polymers
   b. abiotic synthesis of polypeptides
   c. the formation of molecular aggregates with selectively permeable membranes
   d. the formation of protobionts that use DNA to direct polymerization of amino acids
   e. abiotic synthesis of organic monomers

CHALLENGE QUESTIONS

1. Describe the minimum structural, metabolic, and genetic equipment of a postprotobiont that you would consider to be a true primitive cell.
2. Carl Woese and his colleagues at the University of Illinois compared the ribosomal RNAs of primitive prokaryotes called archaebacteria with rRNAs from other organisms. They found that archaebacteria are all very similar to each other, but different from both other bacteria and eukaryotes. Other characteristics of archaebacteria are actually more like eukaryotes than other bacteria. What problems does this pose for the current version of the five-kingdom classification scheme?

SCIENCE, TECHNOLOGY, AND SOCIETY

1. There is no reason to believe that the processes that led to the origin of life are unique to Earth. Many scientists think that life may be present on many other planets throughout the universe. During the 1990s, astronomers first detected planets circling stars beyond our solar system. In what ways do you think our perspective would be changed by convincing evidence of extraterrestrial life?

FURTHER READING

Amato, I. "Capturing Chemical Evolution in a Jar." *Science,* February 14, 1992. Synthetic self-replicating molecules, complete with mutations.

Freedman, D. H. "The Handmade Cell." *Discover,* August 1992. Will researchers be able to build a cell from scratch?

Gonick, L. "Soup or Sandwich?" *Discover,* January 1991. A cartoon version of a new hypothesis of life's origins.

Horgan, J. "In the Beginning. . . " *Scientific American,* February 1991. An examination of new controversies about the origin of life.

Horgan, J. "Life in a Test Tube?" *Scientific American,* May 1992. Simple self-replicating molecules.

Margulis, L., and Schwartz, K. V. *Five Kingdoms: An Illustrated Guide to the Phyla of Life on Earth,* 2nd ed. New York: Freeman, 1987. A catalog of the world's living diversity.

Radetsky, P. "How Did Life Start?" *Discover,* November 1992. Was clay the life-giving substratum?

Waldrop, M. "Goodbye to the Warm Little Pond?" *Science,* November 23, 1990. Did bombardment of the early Earth by cosmic debris prevent life from originating near the planet's surface?

25 PROKARYOTES AND THE ORIGINS OF METABOLIC DIVERSITY

PROKARYOTIC FORM AND FUNCTION

THE DIVERSITY OF PROKARYOTES

THE ORIGINS OF METABOLIC DIVERSITY

THE IMPORTANCE OF PROKARYOTES

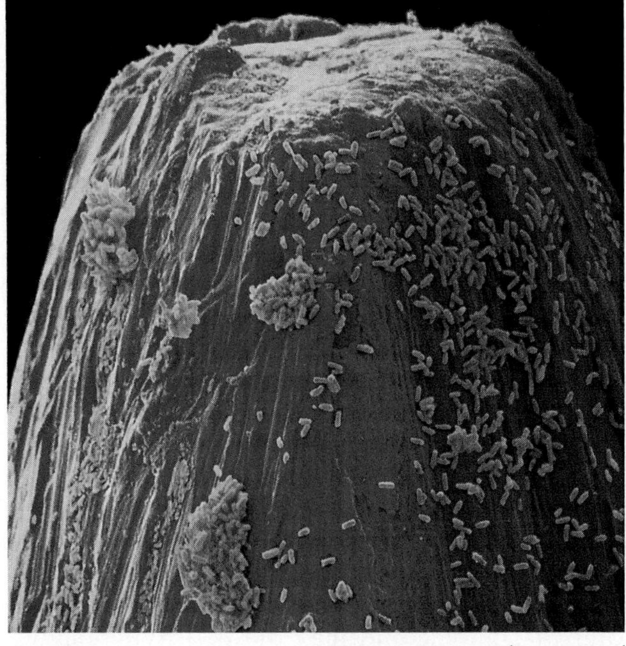

Figure 25.1
Bacteria: the smallest, most numerous, most widespread organisms. This artificially colored scanning electron micrograph demonstrates the size of bacteria in relation to a pinpoint. This chapter features bacterial structure and function and surveys the diversity, evolution, and ecological importance of prokaryotic life.

The history of prokaryotic life, the kingdom Monera, is a success story spanning at least 3.5 billion years. Prokaryotes are commonly called **bacteria** (singular, **bacterium**), a general term that includes organisms that are diverse both structurally and metabolically. In fact, as we will see, some groups of bacteria are as distinct from each other as they are from other kingdoms of life. Prokaryotes were the earliest organisms, and they lived and evolved all alone on Earth for 2 billion years. They have continued to adapt and flourish on a changing Earth, and in turn they have helped to change the Earth.

Unique cellular and molecular features define the kingdom Monera and unify its otherwise diverse members. Prokaryotic cells are relatively small (Figure 25.1) and lack membrane-enclosed organelles. Although most prokaryotes have cell walls, they differ in molecular composition and construction from the cell walls of plants, certain protists, and fungi. Compared to eukaryotes, bacteria also have smaller, simpler genomes, and they differ from eukaryotes in some of the details of genetic replication, expression (protein synthesis), and recombination.

In terms of metabolic impact and numbers, prokaryotes still dominate the biosphere, outnumbering all eukaryotes combined. More bacteria inhabit a handful of dirt or the human mouth or skin than the total number of people who have ever lived. Prokaryotes are not only the most numerous organisms by far but also the most pervasive. Wherever we find life of any kind, prokaryotes are among the organisms present. Prokaryotic species thrive in habitats too hot, too cold, too salty, too acidic, or too alkaline for any eukaryote. Incomparably bountiful and omnipresent, the kingdom Monera is a dynasty that has endured and expanded through its billions of years of descent from the first cells that were the beginnings of all life.

Prokaryotes are individually microscopic, but their collective impact on Earth and all of life is gigantic. We rarely notice these ubiquitous microbes because they are usually invisible to the unaided eye. Illness caused by bacterial infections occasionally reminds us that these tiny organisms exist, but the kingdom Monera is no rogues' gallery. Only a minority of prokaryotes cause disease in humans or any other organisms. The

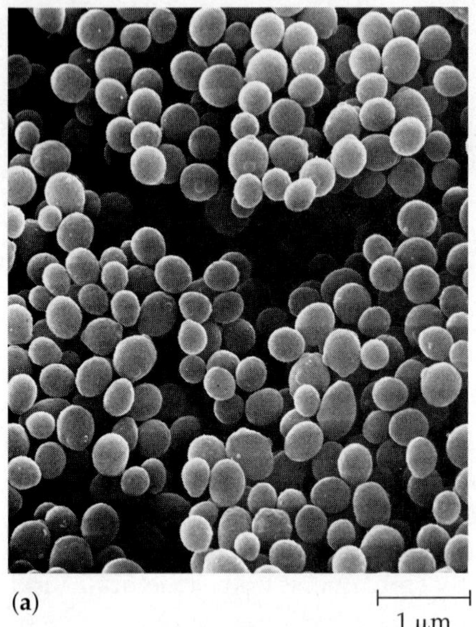

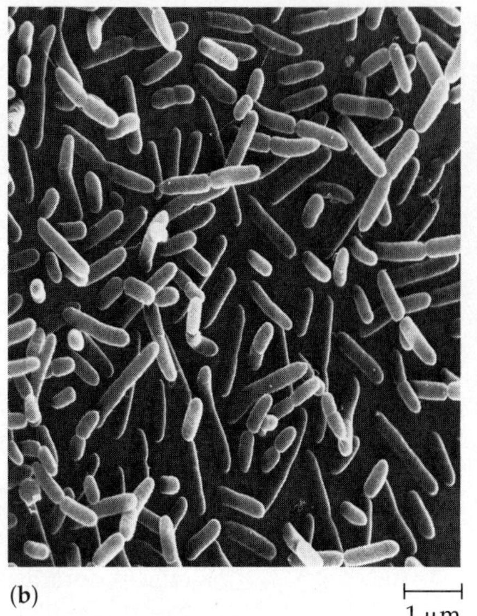

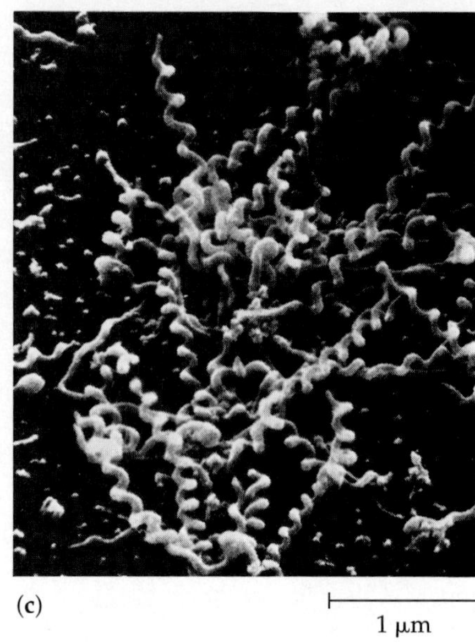

Figure 25.2
Diversity of form in the kingdom Monera. (**a**) Cocci (singular, coccus), or spherical bacteria, occur singly or in pairs (diplococci), in chains of many cells (streptococci), and in clusters resembling bunches of grapes (staphylococci). (**b**) Rod-shaped bacilli (singular, bacillus) are most commonly solitary, but there are also forms with the rods arranged in chains. (**c**) Spirilla (singular, spirillum) are spiral-shaped bacteria reminiscent of corkscrews. (All SEMs.)

great majority of prokaryotic species are essential to all life on Earth. For example, bacteria decompose matter from dead organisms and return vital chemical elements to the environment in the form of inorganic compounds required by plants, which in turn are fed on by animals. If for some reason the entire kingdom Monera were to suddenly perish, the chemical cycles that sustain life would halt and the other four kingdoms would also be doomed. In contrast, prokaryotic life would undoubtedly persist in the absence of eukaryotes, as it once did for a long time.

The prokaryotes often live in close associations among themselves and with organisms in other kingdoms in what are called symbiotic relationships. One theory of the origin of mitochondria and chloroplasts even suggests that all animals, plants, fungi, and protists evolved from symbiotic associations of ancestral prokaryotes. (We will examine this theory in Chapter 26.)

In this chapter, you will become more familiar with prokaryotes by studying their form and physiology, their diversity, their origins and evolution, and their ecological significance.

PROKARYOTIC FORM AND FUNCTION

The Morphology of Prokaryotes

The word *Monera* is derived from the Greek *moneres,* meaning "single," referring here to the single-celled nature of the great majority of prokaryotes. However, some species tend to aggregate reversibly in two-celled to several-celled groups. Others have the form of true **colonies,** which are permanent aggregates of identical cells. And some bacterial species even exhibit a simple multicellular organization in which there is a division of labor between two or more specialized types of cells. Most prokaryotic cells have diameters in the range of 1–5 μm, compared to 10–100 μm for the majority of eukaryotic cells.

There is a diversity of cell shapes among prokaryotes, the three most common being spheres (cocci), rods (bacilli), and spirals (spirilla) (Figure 25.2). An important step in identifying bacteria is to determine their shapes by microscopic examination.

The Cell Surface

Nearly all prokaryotes have cell walls external to their plasma membranes. The wall maintains the shape of the cell, affords physical protection, and prevents the cell from bursting in a hypoosmotic environment (see Chapter 8). Like other walled cells, however, prokaryotes plasmolyze and may die in a hyperosmotic medium, which is why heavily salted meat can be kept so long without being spoiled by bacteria.

The presence of a cell wall is one reason bacteria were grouped with plants in the old two-kingdom system. But the walls of prokaryotes and plants are analogous rather than homologous; they have a completely different molecular composition. Instead of cellulose,

Methods: The Gram Stain

This method, named for Hans Christian Gram, a Danish physician who developed the technique in the late 1800s, distinguishes between two different kinds of bacterial cell walls. Bacteria are stained with a violet dye and iodine, rinsed in alcohol, and then stained again with a red dye. The structure of the cell wall determines the staining response.

Gram-positive bacteria (top), which have cell walls with a large amount of peptidoglycan, retain the violet dye (LM).

Gram-negative bacteria (bottom) have less peptidoglycan, which is located in a periplasmic gel between the plasma membrane and an outer membrane. The violet dye used in the Gram stain is easily rinsed from gram-negative bacteria, but the cells retain the red dye (LM). (The colors in the diagrams do not represent the stains.)

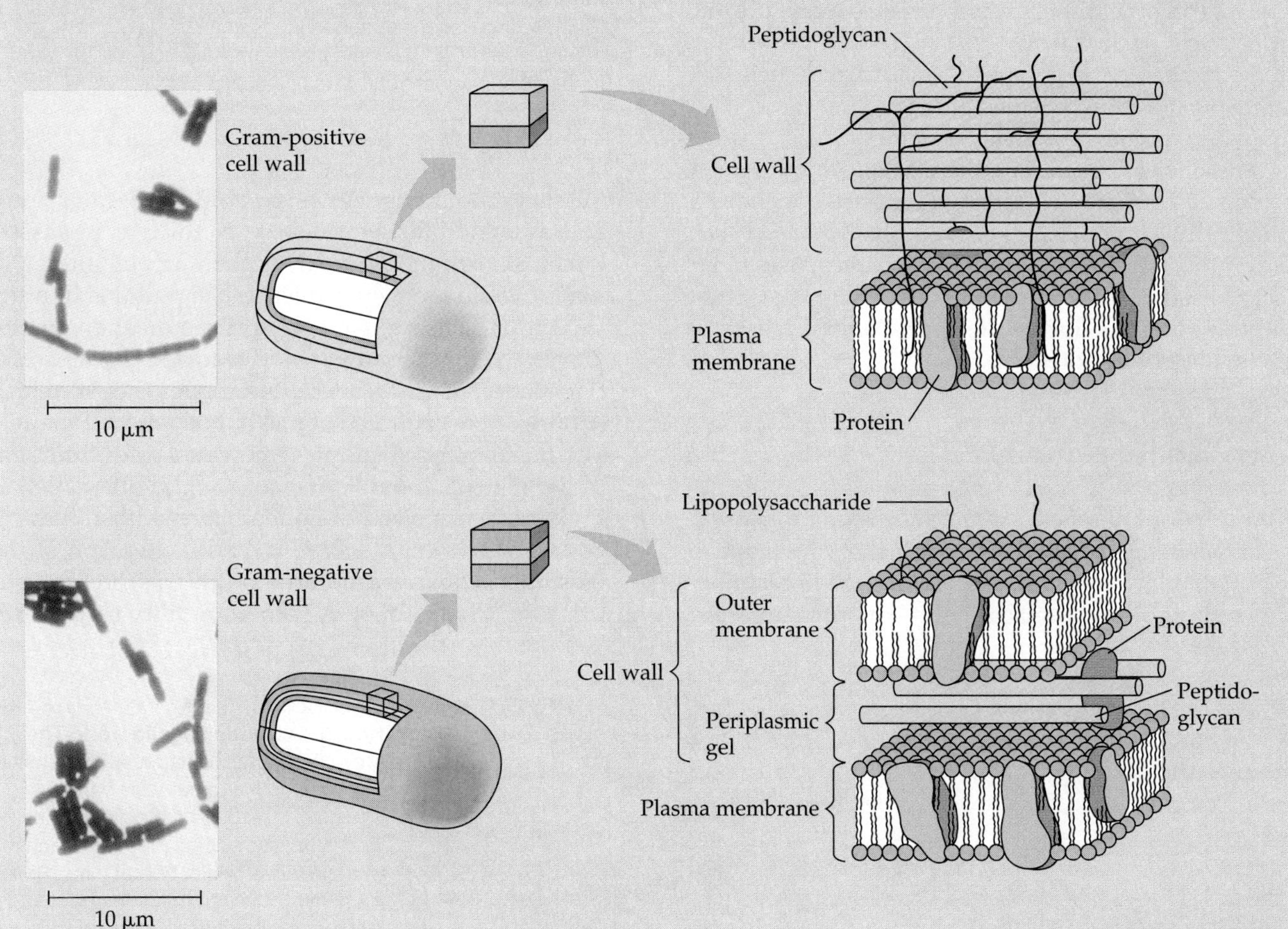

the staple of plant walls, most bacterial walls contain a unique material called **peptidoglycan,** which consists of polymers of modified sugars cross-linked by short polypeptides that vary from species to species. The effect is a single, giant, molecular network enclosing and protecting the cell. External to this fabric are other substances that also differ from species to species.

One of the most valuable tools for identifying bacteria is the **Gram stain,** which can be used to separate many bacteria into two groups based on a difference in their cell walls. **Gram-positive** bacteria have simpler walls, with a relatively large amount of peptidoglycan. The walls of **gram-negative** bacteria have less peptidoglycan and are more complex in structure. An outer membrane on the gram-negative cell wall contains lipopolysaccharides, carbohydrates bonded to lipids (see the Methods Box).

Among pathogenic (Gr. *pathos,* "suffering," and *gignomai,* "cause"), or disease-causing, bacteria, gram-negative species are generally more threatening than gram-positive species. The lipopolysaccharides on the walls of gram-negative bacteria are often toxic, and the outer membrane helps protect the pathogens against the defenses of their hosts. Furthermore, gram-negative bacteria are commonly more resistant than gram-positive species to antibiotics because the outer membrane impedes entry of the drugs.

Many antibiotics, including penicillins, inhibit the synthesis of cross-links in peptidoglycan and prevent the formation of a functional wall, particularly in gram-positive species. These drugs are like selective bullets that cripple many species of infectious bacteria without adversely affecting humans and other eukaryotes, which do not make peptidoglycan.

Many prokaryotes secrete sticky substances that form still another protective layer called a **capsule** outside the cell wall. Capsules provide additional protection and enable the organisms to adhere to their substrate. Gelatinous capsules glue together the cells of many prokaryotes that live as colonies.

Another way bacteria adhere to one another or to some substratum is by means of surface appendages called **pili** (Figure 25.3). For example, *Neisseria gonorrhoeae,* the pathogen that causes gonorrhea, uses pili to fasten itself to mucous membranes of the host. Some pili are specialized for transferring DNA when bacteria conjugate (see Chapter 17).

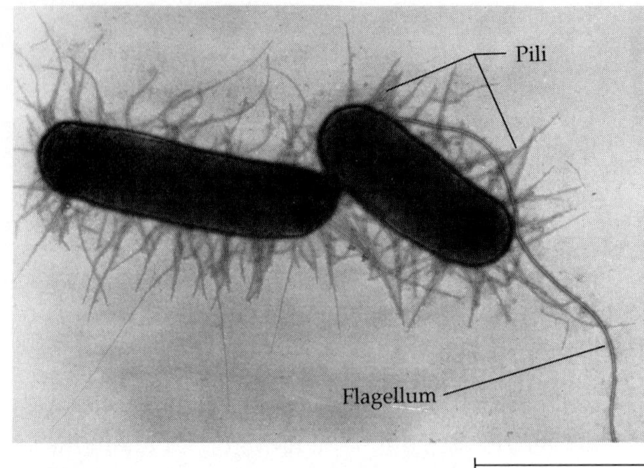

Figure 25.3
Pili. Bacteria use these appendages to attach to surfaces or other bacteria (TEM). Some pili are specialized for DNA transfer between partners during conjugation.

The Motility of Prokaryotes

About half of all bacterial species are capable of directed movement. Before examining the ways that bacteria move, let's consider what it would be like to move through water at the scale of a few micrometers. Relative to our experience, a solution of water, ions, and other solutes would seem as thick as molasses. Other environments, such as the mucous linings of lungs, would seem even thicker, like pudding or mud.

Motile bacteria use one of three mechanisms to move. The most common motive force is by means of flagella, which may be scattered over the entire cell surface or concentrated at one or both ends of the cell. The flagella of prokaryotes and eukaryotes differ entirely. (See Chapter 7 to review eukaryotic flagella.) Prokaryotic flagella are one-tenth the width of those of eukaryotes, are not covered by an extension of the plasma membrane, and are unique in structure and function (Figure 25.4). A second motility mechanism

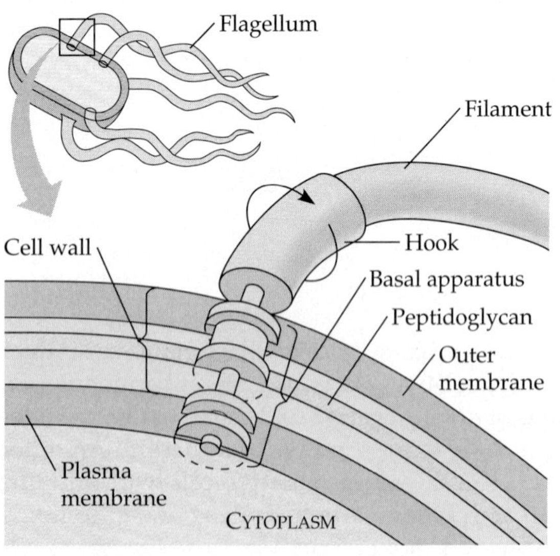

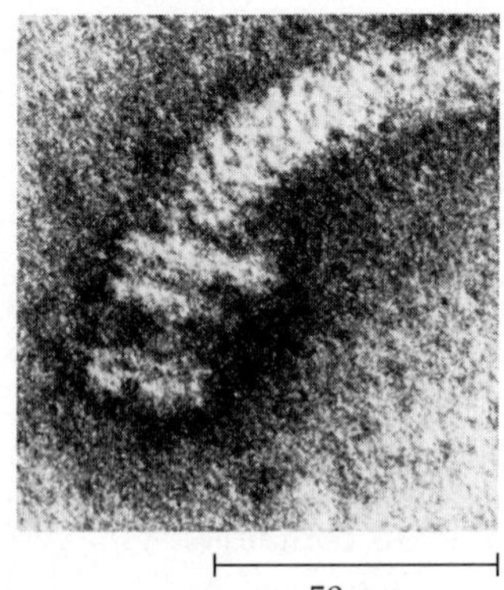

Figure 25.4
How prokaryotic flagella work. The entire flagellum is composed of protein and arranged in three basic parts. Chains of a globular protein, flagellin, wound in a tight spiral produce a filament with a relatively rigid, helical form. This filament is attached to another protein that forms a curved hook, which is, in turn, inserted into a basal apparatus composed of about 35 different proteins. This apparatus consists of a system of rings that sit in the various cell wall layers. (The electron micrograph and drawing are characteristic of gram-negative bacteria; other arrangements are found in motile gram-positive bacteria.) The filament rotates like a corkscrew, with the basal apparatus functioning as a motor. This motor is powered by the diffusion of protons into the cell after these protons have been pumped outward across the plasma membrane at the expense of ATP.

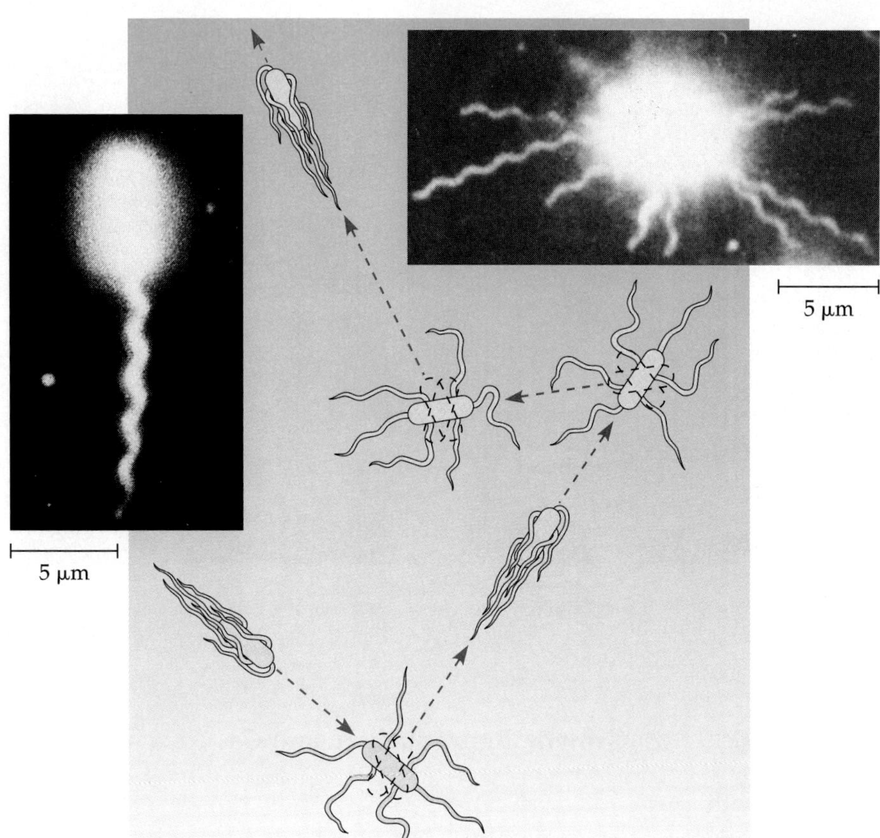

Figure 25.5
Directed movement (taxis) by alternating running and tumbling. The way bacteria use their flagella has been studied in *Salmonella typhimurium*, a motile bacillus. When the many flagella that cover the cells of these bacteria spin counterclockwise, they spiral around one another, propelling the cell forward along a straight path, a movement called running (see left inset). When the flagella reverse and spin clockwise, the flagella separate and cause the cell to move in an uncoordinated movement called tumbling (see right inset) that randomly reorients the direction of the cell. Bacteria have specific receptor molecules on their membranes that detect the presence of certain chemicals. To sense a concentration that varies over space, bacteria seem to use some form of memory to compare present and past concentrations over short time spans, while moving in an environment with a chemical gradient. If the cell senses that it is moving toward an attractant or away from a repellent, its runs between tumbles are relatively long. Movement in the wrong direction causes the organism to tumble more often, shortening the duration of each run. The net effect is movement in a specific direction.

characterizes a group of spiral-shaped bacteria called spirochetes. Several filaments spiral around the cell under the outer sheath of the cell wall. These filaments are very much like flagella in structure. Their basal motors are attached at either end of the cell and filaments attached at opposite ends move relative to each other, behaving somewhat like the microtubules within eukaryotic flagella. When they do so, the flexible cell moves like a corkscrew. This mechanism is particularly effective in moving spirochetes through the highly viscous environments in which they sometimes live. In the third mechanism of motility, some bacteria secrete slimy chemicals and move by a gliding motion that may result from the presence of flagellar motors that lack flagellar filaments.

In an environment that is fairly uniform, flagellated bacteria wander randomly. In a heterogeneous environment, however, many bacteria are capable of **taxis** (Figure 25.5), movement oriented toward or away from a stimulus (Gr. *taxis*, "to arrange or put in order"). With chemotaxis, for example, bacteria respond to chemical stimuli, perhaps moving toward food or oxygen (a positive chemotaxis) or away from some toxic substance (a negative chemotaxis). Several kinds of receptor molecules that detect specific substances are located on the surfaces of chemotactic bacteria. Motile prokaryotes that are photosynthetic generally display a positive phototaxis, a behavior that keeps them in the light. There are even bacteria with tiny magnets that help the microbes distinguish up from down.

Internal Membranous Organization

The prokaryotic cell lacks the extensive compartmentalization by internal membranes characteristic of eukaryotes. Various prokaryotes do have a variety of specialized membranes that perform many of their metabolic functions. These membranes are usually invaginated regions of the plasma membrane (Figure 25.6).

The Prokaryotic Genome

On average, prokaryotes have only about 1/1000 as much DNA as a eukaryotic cell. Recall that prokaryotes are named for their lack of true nuclei enclosed by membranes (see Figure 7.6). In most prokaryotic cells, the DNA is concentrated as a snarl of fibers in a **nucleoid region** that stains less dense than the surrounding cytoplasm in electron micrographs. The snarl of fibers is actually the bacterial chromosome, one double-stranded DNA molecule in the form of a ring. The DNA has very little protein associated with it. The term genophore is often used for the bacterial chromosome

Figure 25.6
Specialized membranes of prokaryotes. (**a**) These infoldings of the plasma membrane, reminiscent of the cristae of mitochondria, function in cellular respiration of aerobic bacteria (TEM). (**b**) Prokaryotes called cyanobacteria have thylakoid membranes, much like those in chloroplasts, that function in photosynthesis (TEM).

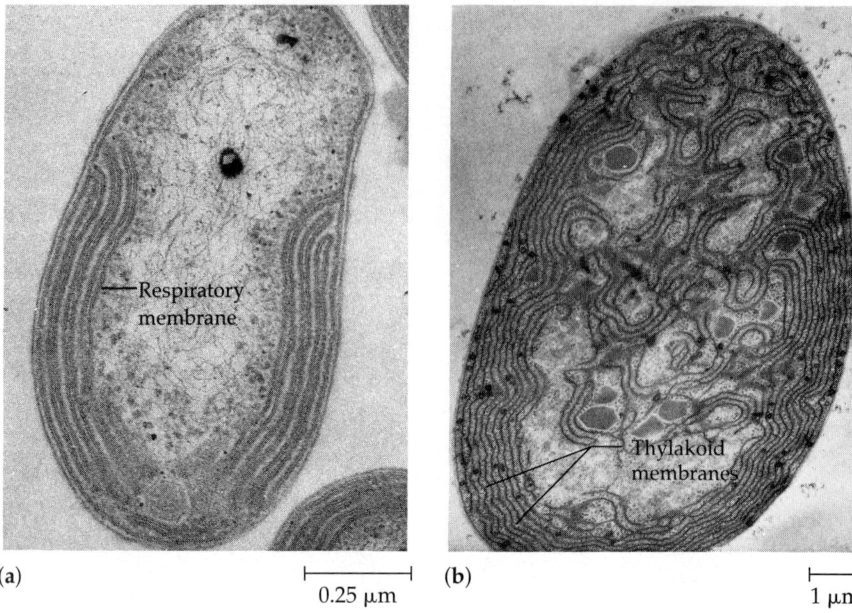

to distinguish it from eukaryotic chromosomes, which have a very different structure. The eukaryotic genome consists of linear DNA molecules packaged along with proteins into a number of chromosomes characteristic of the species.

In addition to its one major chromosome, the prokaryotic cell may also have much smaller rings of DNA called plasmids, each consisting of only a few genes. In most environments, bacteria can survive without their plasmids because all essential functions are programmed by the chromosome. However, plasmids endow the cell with genes for resistance to antibiotics, for metabolism of unusual nutrients not present in the normal environment, and for other special contingencies. Plasmids replicate independently of the main chromosome, and many can be readily transferred between partners when bacteria conjugate (see Chapter 17).

Although the broad outlines for DNA replication and the translation of genetic messages into proteins are alike for eukaryotes and prokaryotes, some of the details differ. For example, the prokaryotic ribosome is slightly smaller than the eukaryotic version and differs in its protein and RNA content. The disparity is great enough that selective antibiotics, including tetracycline and chloramphenicol, bind to the ribosomes of prokaryotes and block protein synthesis, while not inhibiting eukaryotic ribosomes.

Because of the relatively simple organization of prokaryotic genomes and the ability of the cells to take up foreign DNA from the surrounding solution (see Chapter 17), bacteria were the organisms of first choice for recombinant DNA research and genetic engineering (see Chapter 19).

Growth, Reproduction, and Gene Exchange

Neither mitosis nor meiosis occurs in the kingdom Monera; this is another fundamental difference between prokaryotes and eukaryotes. Prokaryotes reproduce only asexually by the mode of cell division called **binary fission,** synthesizing DNA almost continuously. (Binary fission is described in Chapter 11.) A single bacterium in a favorable environment will give rise by repeated divisions to a colony of offspring (Figure 25.7). The word *growth* as applied to bacteria actually refers more to the multiplication of cells and population growth than to the enlargement of individual cells. The conditions for optimal growth—temperature, pH, salt concentrations, nutrient sources, and so on—vary according to species. Refrigeration retards food spoilage because most bacteria and other microorganisms grow only very slowly at such low temperatures.

The resistance of some bacterial cells to environmental destruction is impressive. Some bacteria form resistant cells called **endospores.** The original cell replicates its chromosome, and one copy becomes surrounded by a durable wall. The outer cell disintegrates, but the endospore it contained survives all sorts of trauma, including lack of nutrients and water, extreme heat or cold, and most poisons. Unfortunately, boiling water is not hot enough to kill most endospores in a reasonable length of time. Home canners and the food-canning industry must take precautions to kill endospores of dangerous bacteria. To sterilize media, glassware, and utensils in the laboratory, microbiologists use an appliance called an autoclave, a pressure cooker that kills even endospores by heating to tem-

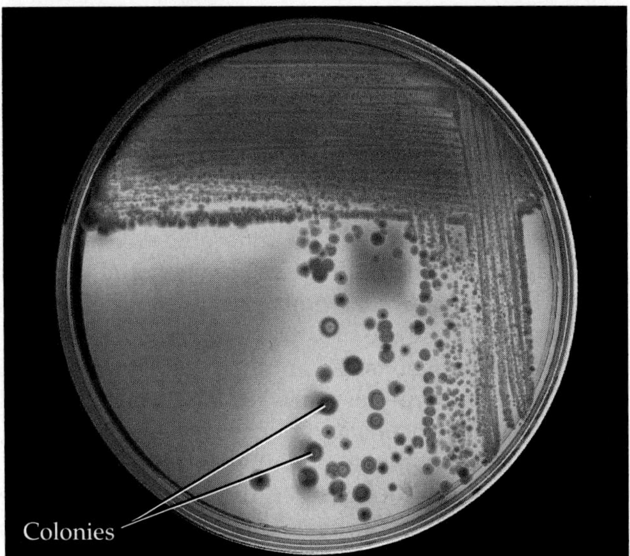

Figure 25.7
Cultured bacterial colonies. Bacteria are grown in the laboratory by culturing them in Petri dishes or test tubes containing liquid or solid media of known composition. The media are sterilized to ensure that no unwanted microbes will grow, then a sample of bacteria, sometimes just a single cell, is introduced. The plates or tubes are incubated at an appropriate temperature. When the bacteria are grown on solid media, colonies are usually large enough to be visible to the unaided eye after a day or two. The size, shape, texture, and color of a colony provide clues to identification of the bacteria, as do the nutrients and physical conditions required for growth. Shown here are the colonies of several species of bacteria. Microscopic examination of bacteria from the colony is another step in identification.

peratures higher than 120°C. In less hostile environments, endospores may remain dormant for centuries. They hydrate and revive to the vegetative (colony-producing) state only in hospitable environments. Even without the ability to form endospores, some bacteria can endure harsh conditions. A recent investigation has uncovered 11,000-year-old bacteria of the genus *Enterobacter* in the intestinal contents of the partially preserved remains of a mastodon. This genus does not form endospores, so the cells must have been slowly reproducing over the years to remain viable. Some prokaryotes even thrive and grow rapidly in environments that are deadly to other organisms. For example, bacteria called thermoacidophiles are adapted to environments as hot as 100°C and as acidic as pH 2.

In an environment without limiting resources, bacterial growth is effectively geometric: One cell divides to form 2, which divide again to produce a total of 4 cells, then 8, 16, and so on, the numbers in an aggregation doubling with each generation. Most bacteria have generation times in the range of 1 to 3 hours, but some species can double every 20 minutes in an optimal environment. If that growth rate were sustained, a single cell would give rise to a colony weighing 1 million kg in just 24 hours. However, growth of bacteria both in the laboratory and in nature is usually checked at some point when the cells exhaust some nutrient or the colony poisons itself with an accumulation of metabolic wastes.

In most natural environments, bacteria must compete for space and nutrients. A general feature of all microorganisms (including monerans, protists, and fungi) is the release of **antibiotics,** chemicals that inhibit the growth of other microorganisms. Humans have discovered some of these compounds and use them to combat pathogenic bacteria. Prokaryotes produce many antibiotics (although you may be most familiar with penicillin, produced by molds of the kingdom Fungi).

The sexual cycle of meiosis and syngamy (the union of haploid nuclei; see Chapter 12) so important as a source of genetic variation in eukaryotes does not occur in the reproduction of prokaryotes. However, there are three mechanisms of genetic recombination in bacteria: (1) **transformation,** in which genes are taken up from the surrounding environment, allowing for considerable genetic transfer in prokaryotes; (2) **conjugation,** in which genes are transferred directly from one bacterium to another; and (3) **transduction,** in which genes are transferred between bacteria by means of viruses (see Chapter 17). These processes, however, involve the unilateral passage of a variable amount of DNA—nothing like the meiotic sex of eukaryotes, in which two parents each contribute homologous genomes to a zygote. Mutation is the major source of genetic variation in prokaryotes. Because generation times are measured in minutes or hours, a favorable mutation can be rapidly propagated to a large number of offspring.

Metabolic Diversity

Metabolic diversity is greater in the kingdom Monera than among all eukaryotes combined. Every type of nutrition observed in eukaryotes is represented among prokaryotes, plus some nutritional modes unique to monerans. Nutrition refers here to how an organism obtains two resources for synthesizing organic compounds: energy and a source of carbon. Species that use light energy are termed *phototrophs. Chemotrophs* obtain their energy from chemicals taken from the environment. If an organism needs only the inorganic compound CO_2 as a carbon source, it is called an *autotroph. Heterotrophs* require at least one organic nutrient—glucose, for instance—as a source of carbon for making other organic compounds. Depending on how they obtain energy and carbon, bacteria can be divided into four major categories:

1. **Photoautotrophs** harness light energy to drive the synthesis of organic compounds from carbon dioxide. The specialized metabolic machinery of these organisms includes internal membranes with light-harvesting pigment systems (see Chapter 10). Among the diverse groups of photosynthetic bacteria are the **cyanobacteria** (formerly known as blue-green algae). All photosynthetic eukaryotes—plants and certain protists—also fit this nutritional category.

2. **Photoheterotrophs** can use light to generate ATP but must obtain their carbon in organic form. This type of nutrition is restricted to certain prokaryotes.

3. **Chemoautotrophs** need only CO_2 as a carbon source, but instead of using light for energy, these bacteria obtain energy by oxidizing inorganic substances, such as hydrogen sulfide (H_2S), ammonia (NH_3), ferrous ions (Fe^{2+}), or some other chemical, depending on the species. This mode of nutrition is unique to certain prokaryotes. For instance, bacteria of the genus *Sulfobolus* oxidize sulfur.

4. **Chemoheterotrophs** must consume organic molecules for both energy and carbon. This type of nutrition is found widely among monerans, protists, fungi, animals, and even some plants.

The majority of bacteria are chemoheterotrophs. This category includes **saprobes,** decomposers that absorb their nutrients from dead organic matter, and **parasites,** which absorb their nutrients from the body fluids of living hosts. There are no known contemporary prokaryotes that are **phagotrophs,** a third subcategory of chemoheterotrophic organisms that ingest their food and enzymatically digest it within cells or multicellular bodies (like animals).

The specific organic nutrients needed for growth vary extensively among chemoheterotrophic bacteria. Some species are very exacting in their requirements; for example, bacteria of the genus *Lactobacillus* will grow well only in a medium containing all 20 amino acids, several vitamins, and other organic compounds. Among species less fastidious in their nutritional needs, *E. coli* can grow on a medium containing glucose as the only organic ingredient, and the metabolism of the organism is so versatile that many other compounds can substitute for glucose as the sole organic nutrient. There is such a diversity of chemoheterotrophs that almost any organic molecule can serve as food for at least some species. For example, bacteria capable of metabolizing petroleum are used to clean up oil spills. Those few classes of synthetic organic compounds, including some kinds of plastics, that cannot be broken down by any chemoheterotrophs are said to be nonbiodegradable.

Another metabolic variation in the kingdom Monera is in the effect that oxygen has on growth (see Chapter 9). **Obligate aerobes** use oxygen for cellular respiration and cannot grow without it. **Facultative anaerobes** will use oxygen if it is present but can also grow by fermentation in an anaerobic environment. **Obligate anaerobes** cannot use oxygen and are poisoned by it. These variations are among the criteria used in classifying the prokaryotes.

Nitrogen metabolism is another facet of nutritional diversity among prokaryotes. Nitrogen is an essential component of proteins and nucleic acids (see Chapter 5). While animals, plants, and other eukaryotes are limited in the forms of nitrogen they can use, diverse prokaryotes are able to metabolize most nitrogenous compounds. Key steps in the cycling of nitrogen through ecosystems are performed only by bacteria. (See Chapter 49 for an overview of the nitrogen cycle.) Chemoautotrophic bacteria such as *Nitrosomonas* convert NH_3 into NO_2^-. Facultatively anaerobic bacteria, such as a few species of *Pseudomonas,* "denitrify" NO_2^- or NO_3^- to atmospheric N_2 gas. And diverse species of prokaryotes, including some cyanobacteria, are able to use atmospheric nitrogen directly as a source of nitrogen. In this process, called **nitrogen fixation,** bacteria convert atmospheric N_2 to NH_3 (ammonia). Nitrogen fixation, unique to certain prokaryotes, is the only biological mechanism that makes atmospheric nitrogen available to organisms for incorporation into organic compounds. In terms of nutrition, the nitrogen-fixing cyanobacteria are the most self-sufficient of all organisms. They require only light energy, CO_2, N_2, water, and some minerals to grow.

THE DIVERSITY OF PROKARYOTES

Moneran taxonomy used by practicing microbiologists, especially in medical labs, remains more a classification of convenience to aid in identifying bacteria than an attempt to group according to evolutionary relationships. The prokaryotic cell is profoundly different from the eukaryotic cell, making the Monera the most clearly defined of all the kingdoms. But grouping the more than 10,000 known species of prokaryotes into subordinate taxa within the kingdom Monera on the basis of phylogenetic affinities has, until recently, been an exercise in futility. There are several reasons for difficulty in devising a realistic phylogeny of the Monera. Prokaryotes diversified extensively so long ago that genealogical ties between groups are hazy. The potential for unrelated genera to exchange genes via transformation and transduction (see Chapter 17) may have mixed up the phylogenetic branches in nonlinear ways. Help from the fossil record is limited because most microfossils found in Precambrian rocks are little more than shadowy outlines of cells.

Molecular systematics offers the best hope for developing a moneran classification that reflects phylogeny (see Chapter 23). Comparing amino acid sequences of homologous proteins and base sequences of DNA and RNA measures genetic similarity and helps establish the timing of phylogenetic branching along the moneran tree. Comparisons of ribosomal RNAs have been particularly illuminating. Based on these studies, the kingdom Monera can be divided into about 20 major groups.

The most significant discovery from molecular systematics is that prokaryotes split into at least two divergent lineages very early in the history of life. One branch produced the **archaebacteria,** with the only survivors being a few genera of prokaryotes confined to extreme environments that may resemble habitats on the early Earth. A second branch, the **eubacteria,** includes nearly all modern prokaryotes.

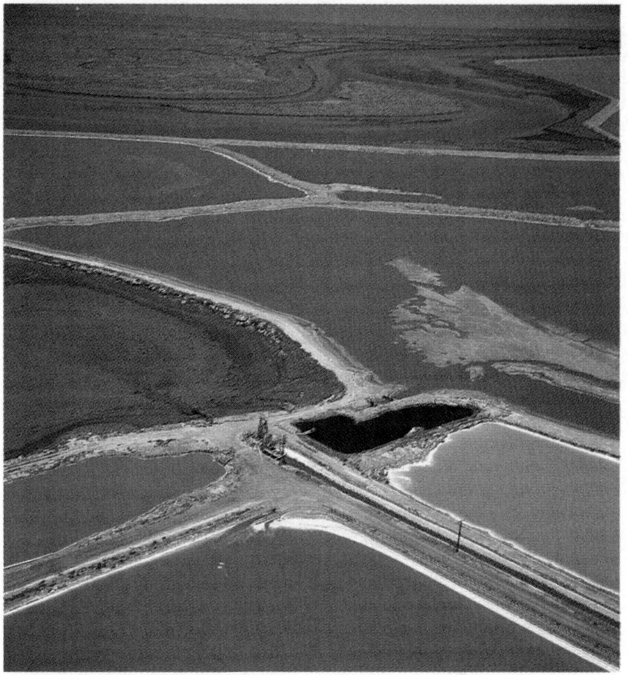

Figure 25.8
Extreme halophiles. Members of the ancient prokaryotic group known as the archaebacteria, these organisms live in extremely saline waters. The colors of these seawater evaporating ponds at the edge of San Francisco Bay result from a dense growth of extreme halophiles that thrive in the ponds when the water reaches a salinity of 15% to 20% (before evaporation, the salinity of seawater is about 3%). The ponds are used for commercial salt production; the halophilic bacteria are harmless.

Archaebacteria

The name of this group refers to the antiquity of its origin from the earliest prokaryotes (Gr. *archaio,* "ancient"). The archaebacteria have many distinctive traits, as should be expected of a taxon that has followed a separate evolutionary path for so long. Their cell walls lack peptidoglycan, a component of all walled eubacteria. The plasma membranes of archaebacteria have lipid compositions unlike any other organisms. The RNA polymerase and a ribosomal protein of archaebacteria are eukaryote-like and very different from those of eubacteria.

Most archaebacteria live in extreme environments too harsh for other forms of modern life but that may have been prevalent habitats during the early evolution of prokaryotes. For example, archaebacteria have been found living in the hot water at the openings of deep-sea vents. A lineage nearly as old as life itself, archaebacteria survive as three subgroups: the methanogens, the extreme halophiles, and the thermoacidophiles.

The **methanogens** are named for their unique form of energy metabolism, in which H_2 is used to reduce CO_2 to methane (CH_4). Methanogens are among the strictest of anaerobes, poisoned by oxygen. They live in swamps and marshes where other microbes have consumed all the oxygen; the methane that bubbles out at these sites is known as marsh gas. Methanogens are also important decomposers employed for sewage treatment. Some farmers have experimented with use of these microbes to convert garbage and dung to methane, a valuable fuel. Other species of methanogens inhabit the anaerobic environment within the guts of animals, playing an important role in the nutrition of cattle, termites, and other herbivores that subsist mainly on a diet of cellulose.

The **extreme halophiles** (Gr. *halo,* "salt," and *philos,* "lover") live in such saline places as the Great Salt Lake and the Dead Sea. Some species merely tolerate salinity, whereas others actually require an environment ten times saltier than seawater to grow (Figure 25.8). Colonies of halophiles form a pink scum that owes its color to a photosynthetic pigment called bacteriorhodopsin.

As their name implies, the **thermoacidophiles** thrive where it is both hot and acidic, a double whammy for nearly all organisms. The optimal conditions for these archaebacteria are temperatures of 60°C to 80°C and pH between 2 and 4. *Sulfolobus* inhabits hot sulfur springs in Yellowstone National Park, obtaining its energy by oxidizing sulfur.

Carl Woese of the University of Illinois has recently proposed that the early evolutionary split between archaebacteria and eubacteria be recognized with a new

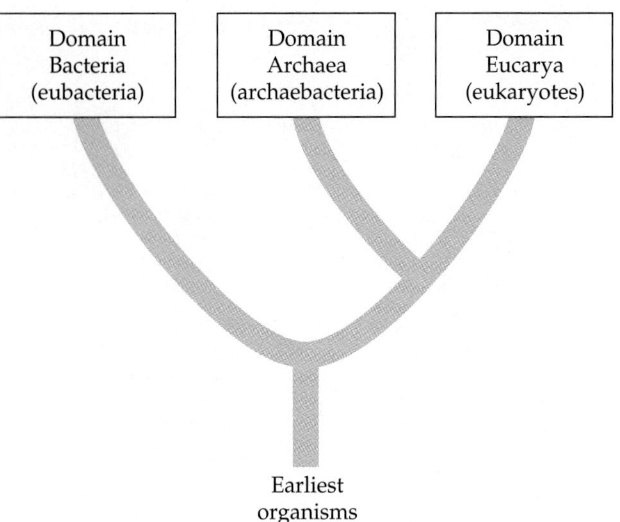

Figure 25.9
The domains of life. Ordering life into these three domains emphasizes the antiquity of the archaebacterial lineage. This simple tree also reflects the current view, based on evidence from molecular systematics, that archaebacteria and eukaryotes share an ancestor more recent than the ancestor common to archaebacteria and eubacteria.

taxonomic category called the **domain,** a classification unit above the kingdom level (Figure 25.9). In this view, the three domains are Bacteria (eubacteria), Archaea (archaebacteria), and Eucarya (eukaryotes). Notice in Figure 25.9, which is based on molecular comparisons, that archaebacteria may actually be related more closely to eukaryotes than to eubacteria. However, the classification scheme adopted by this text still recognizes Monera, a kingdom of life defined by the prokaryotic organization common to archaebacteria and their distant relatives, the eubacteria.

Eubacteria

An exhaustive survey of the eubacteria would be excessive for a general biology textbook. Table 25.1 (pp. 526–527) highlights some eubacterial groups to demonstrate their diversity, especially in their modes of nutrition. You should study Table 25.1 before going on to the next section of the text.

THE ORIGINS OF METABOLIC DIVERSITY

Now that we have surveyed the major prokaryotic groups, let's trace the evolutionary roots of their metabolic diversity. All forms of nutrition and nearly all metabolic pathways evolved in the kingdom Monera before eukaryotes arose. As early prokaryotes evolved, they were met with constantly changing physical and biological environments. In response to these changes, new metabolic capabilities evolved that, in turn, changed the environment faced by the next community of prokaryotes. All the major metabolic capabilities seen among contemporary prokaryotes probably evolved in the first billion years of the history of life. Reasonable speculations about the early history of prokaryotes and the origins of metabolic diversity are now possible. The scenario described here is just one hypothetical sequence based on inferences from molecular systematics, a comparison of energy metabolism among extant prokaryotes, and evidence from the geological record about conditions on the early Earth.

The Origin of Glycolysis

The first prokaryotes, which originated at least 3.5 billion years ago, were probably chemoheterotrophs that absorbed free organic compounds generated in the primordial seas by abiotic synthesis (see Chapter 24). ATP was probably among those nutrients. The universal role of ATP as an energy currency in all modern organisms implies that prokaryotes became fixed on its use very early. As the bacteria began to deplete the supply of free ATP, natural selection would have favored cells with enzymes that could regenerate ATP from ADP using energy extracted from other organic nutrients that were still available. The result may have been the step-by-step evolution of glycolysis, a metabolic pathway that breaks organic molecules down to simpler waste products and uses the energy to generate ATP by substrate phosphorylation (see Chapter 9). Glycolysis is the only metabolic pathway common to nearly all modern organisms, suggesting great antiquity.

Further, glycolysis does not require O_2, and indeed the ancient atmosphere had very little of that gas. Fermentation, in which electrons extracted from nutrients during glycolysis are transferred to organic recipients, became a way of life on the anaerobic Earth. The archaebacteria and other obligate anaerobes that live today by fermentation deep in the soil or in stagnant swamps are believed to have forms of nutrition most like that of the original prokaryotes.

The Origin of Electron Transport Chains and Chemiosmosis

The chemiosmotic mechanism of ATP synthesis is common to all five kingdoms of life, implying a relatively early origin. Recall from Chapter 9 that chemiosmosis uses electron transfers along a chain of membrane proteins to pump hydrogen ions, and then taps

Stage 1. Proton pumps, driven by ATP, are used to regulate cellular pH by extruding hydrogen ions.

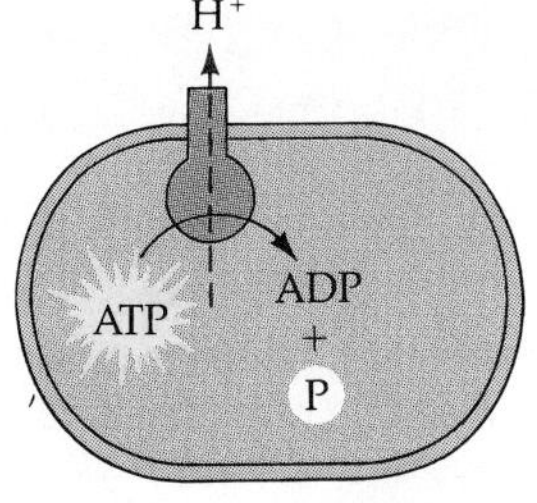

Stage 2. Electron transport chains take over the function of pH regulation by using the oxidation of organic acids to drive proton pumps.

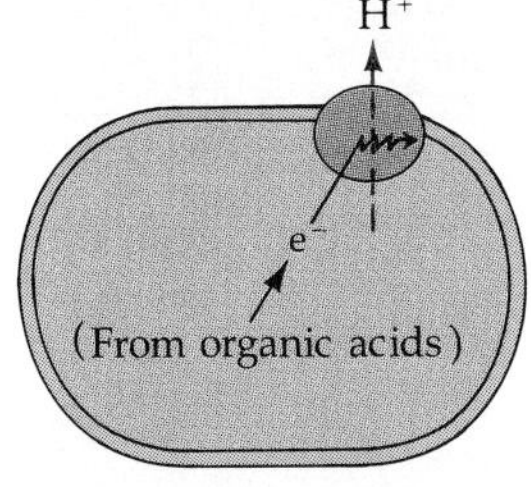

Stage 3. Electron transport chains become efficient enough to generate a gradient of hydrogen ions that can be used to drive ATP synthesis.

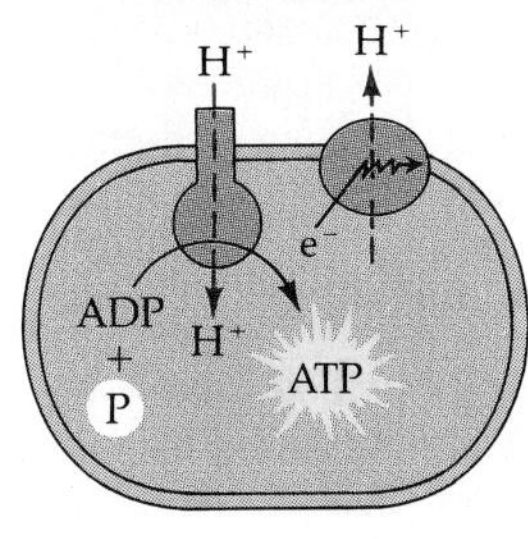

Figure 25.10
A model for the evolution of electron transport chains.

the H^+ gradient to power ATP synthesis. Transmembrane proton pumps may have functioned originally to expel hydrogen ions and help early prokaryotes regulate their internal pH (Figure 25.10, Stage 1). But the cell would have to spend a large portion of its ATP to drive the proton pumps. The first electron transport chains may have saved ATP by coupling oxidation of organic acids to the transport of H^+ out of the cell (Figure 25.10, Stage 2). Finally, in some bacteria, electron transport systems efficient enough to extrude more H^+ than necessary for regulating pH evolved. These cells could then use the inward gradient of the ions to reverse the proton pump, which now generated ATP rather than consuming it (Figure 25.10, Stage 3). This type of energy metabolism, called anaerobic respiration, persists in some modern bacteria. For instance, in waterlogged, anaerobic soils, some species of *Pseudomonas* (see Table 25.1) can pass electrons down transport chains from organic substrates to NO_3^- (instead of O_2, the electron acceptor in aerobic respiration).

The Origin of Photosynthesis

Life confronted its first energy crisis when the supply of free ATP dwindled. It faced its second when the fermenting prokaryotes consumed organic nutrients faster than the compounds could be replaced by abiotic synthesis. An organism that could make its own organic molecules from inorganic precursors would have had a tremendous advantage.

In the earliest prokaryotes, light-absorbing pigments may have been used to absorb excess light energy (particularly ultraviolet) that was harmful to cells growing in surface communities. Later, these energized pigments were coupled with electron transport systems to drive ATP synthesis. In the extreme halophiles, the light-energy-capturing pigment bacteriorhodopsin is built into the plasma membrane. Bacteriorhodopsin absorbs light and uses the energy to pump hydrogen ions out of the cell. The gradient of hydrogen ions then drives the synthesis of ATP (see Figure 8.21). This is the simplest mechanism of photophosphorylation known, and the halophiles are being studied as model systems of solar energy conversion.

Other prokaryotes had pigments and photosystems that used light to drive electrons from hydrogen sulfide (H_2S) to $NADP^+$, generating reducing power that could be used to fix CO_2 (see Chapter 10). They probably co-opted components of electron transport chains that had previously functioned in anaerobic respiration and continued to use the chains to power ATP synthesis as well as to provide reducing power (NADPH). The modern bacteria with nutrition most like the early photosynthetic prokaryotes are believed to be the green sulfur bacteria and purple sulfur bacteria (phototrophic anaerobes; see Table 25.1). They owe their color to bacteriochlorophyll, which functions instead of chlorophyll *a* as the main photosynthetic pigment. And because these bacteria split H_2S instead of H_2O as a source of electrons, they produce no O_2.

Cyanobacteria, the Oxygen Revolution, and the Origins of Cellular Respiration

Some of the photosynthetic bacteria eventually had the metabolic machinery to use H_2O instead of H_2S or other compounds as a source of electrons and hydrogen for fixing CO_2. These were the first cyanobacteria (see Table 25.1). Able to make organic compounds from water and CO_2, they flourished and changed the world by releasing O_2 as a by-product of their photosynthesis.

Cyanobacteria evolved at least 2.5 billion years ago, living along with other bacteria in colonies that built the stromatolites that have been found all over the world (see Figure 24.4). In marine sediments of about that same age are the banded iron formations, red layers rich in iron oxide and valuable as a source of iron ore today. Perhaps the sediments formed over a period when early cyanobacteria released oxygen that reacted with dissolved iron ions, which precipitated as iron ox-

Table 25.1 Some Major Groups of Eubacteria

Group (Examples)	Characteristics and Approximate Number of Genera	Illustration
Actinomycetes *(Mycobacterium, Streptomyces)*	Form colonies of branching hyphae (tubular filaments resembling those of fungi, EM). Reproduction by fragmentation of the end of hyphae into spores (another fungal character). Most actinomycetes live in the organic litter of soil. In spite of their name (*mycetes* means fungi), these organisms are true prokaryotes, not fungi. (40 genera)	*Streptomyces* 1 μm
Chemoautotrophic bacteria *(Nitrobacter, Nitrosomonas)*	Use the energy obtained from oxidizing inorganic substances (such as NH_3, NO_3^-, H_2S, S, Fe^{3+}) plus carbon from CO_2 to build organic molecules, Since O_2 is the terminal electron acceptor, these bacteria are obligate aerobes. Common in aerated soil. (25 genera)	
Cyanobacteria *(Chroococcus, Anabaena, Nostoc, Oscillatoria, Spirulina)*	Photoautotrophs with plantlike photosynthesis; their process uses chlorophyll *a* and two photosystems, splitting water and releasing O_2 as a by-product (see Chapter 10). Chlorophylls and electron-transporting systems are embedded in thylakoid membranes (see Figure 25.6b). Other accessory pigments called phycobilins, found in complexes on the surfaces of the thylakoids, impart the species' typical blue-to-grayish-brown color. Phylum includes solitary, multicellular, and colonial (usually filamentous) species. Cell walls often thick and gelatinous (top LM). Flagella absent, motile forms glide. Most inhabit fresh water, but species also live in seas, damp soils, and lichens. The nitrogen-fixing enzyme complex is often packaged into specialized cells called **heterocysts** (boxed area) in some filamentous genera such as *Anabaena* (bottom LM). (32 genera)	*Chroococcus* 50 μm *Anabaena* 50 μm
Endospore-forming bacteria *(Bacillus, Clostridium)*	Gram-positive, flagellated rods survive periods of harsh conditions by producing endospores (boxed area), internal, dehydrated cells with thick walls (TEM). Includes both obligate anaerobes and aerobes. (6 genera)	*Bacillus* 1 μm
Enteric bacteria *(Escherichia, Salmonella, Vibrio)*	Gram-negative, facultative anaerobes with a wide variety of metabolic capabilities, including anaerobic respiration using NO_3^- as the electron donor. Inhabit the intestinal tracts of animals. Some enteric bacteria are benign, permanent residents (such as *E. coli*), while others are pathogenic. (34 genera)	

Table 25.1 *(continued)*

Group (Examples)	Characteristics and Approximate Number of Genera	Illustration
Mycoplasmas *(Mycoplasma)*	Believed to be the smallest of all cells, with diameters of only 0.10–0.25 μm. The only prokaryotes that lack cell walls (SEM). Unlike the rickettsias and chlamydias, mycoplasmas grow extracellularly. Saprobes and animal pathogens. (6 genera)	*Mycoplasma* 2.5 μm
Myxobacteria *(Myxococcus)*	Soil-dwelling chemoheterotrophs. Individual cells move by gliding. When the soil dries out or food becomes scarce, the cells congregate into a fruiting body (spore mass on a bulbous stalk, SEM) that may be brightly colored and up to 1 mm in diameter. Spores are released and grow into new colonies when conditions are favorable. (8 genera)	*Myxococcus* 10 μm
Nitrogen-fixing aerobic bacteria *(Azotobacter, Rhizobium)*	Includes common free-living and mutualistic species. The genus *Rhizobium,* which lives in root nodules on legumes, plays an important role in the nutrition of those plants (see Figures 33.10 and 33.11). (5 genera)	
Phototrophic anaerobic bacteria *(Chromatium, Rhodospirillum)*	Photoautotrophs that are much less plantlike than cyanobacteria in their photosynthetic equipment (EM; see Figure 10.4b). Reduce $NADP^+$ with electrons extracted from molecules other than H_2O, such as H_2S, therefore do not release O_2; most species are strict anaerobes. Found in pond, lake, and ocean sediments. Includes several groups that may be distantly related, such as the purple bacteria and the green sulfur bacteria. (27 genera)	*Rhodospirillum* 1 μm
Pseudomonads *(Pseudomonas)*	Includes the diverse genus *Pseudomonas,* represented by species in nearly all aquatic and soil habitats. Cells are typically rod-shaped, gram-negative, and flagellated at one end. Chemoheterotrophs capable of metabolizing unusual nutrients. (5 genera)	
Rickettsias and chlamydias *(Rickettsia, Chlamydia)*	Obligate intracellular parasites of animals. Reduced gram-negative cell wall and metabolism, depending on host cells. Rickettsias alternate between arthropod and mammal hosts. Diseases caused by chlamydias are contracted from other humans or birds. (15 genera)	
Spirochetes *(Borrelia, Leptospiru, Treponema)*	Shallowly helical cells, sometimes very long (up to 0.25 mm, but too thin to see without a microscope), with a corkscrewlike movement effected by their unique internal flagellar filaments (EM). Includes both free-living saprobes and parasites. (7 genera)	*Leptospiru* 0.5 μm

ide. This reaction would have prevented any accumulation of free O_2 for perhaps a few hundred million years until precipitation exhausted the dissolved iron. Only then would the seas become saturated with O_2, which began gassing out and accumulating in the atmosphere. Beginning about 2 billion years ago, terrestrial rocks rich in iron were rusted red by oxidation with atmospheric O_2.

The gradual change to a more oxidizing atmosphere created a crisis for Precambrian prokaryotes, because oxygen attacks the bonds of organic molecules. The corrosive atmosphere probably caused the extinction of many bacteria unable to cope. Other species survived in habitats that remained anaerobic, where we find their descendants living today as obligate anaerobes, such as the methanogens, the anaerobic phototrophs, and many fermenting bacteria. The evolution of antioxidant mechanisms enabled other bacteria to tolerate the rising oxygen levels. Among photosynthetic prokaryotes, some species went a step further than mere oxygen tolerance to actually using the oxidizing power of O_2 to pull electrons from organic molecules down existing transport chains. Thus, aerobic respiration may have originated as modifications of electron transport chains co-opted from photosynthesis. Photoheterotrophs, the purple nonsulfur bacteria, still use an electron transport system that is a hybrid of photosynthetic and respiratory equipment. Several other bacterial lineages gave up photosynthesis and reverted to chemoheterotrophic nutrition, their electron transport chains adapted to function exclusively in aerobic respiration.

THE IMPORTANCE OF PROKARYOTES

One of the themes of this unit is changing life on a changing planet—the interactions of geological history and evolving biological forms. Organisms as pervasive, abundant, and diverse as the prokaryotes have a tremendous impact on the Earth and all its inhabitants. Here we consider some of these relationships.

Prokaryotes and Chemical Cycles

Not too long ago, in geological terms, the atoms of the organic molecules in our bodies were parts of the inorganic compounds of soil, air, and water, as they will be again. Ongoing life depends on the recycling of chemical elements between the biological and physical components of ecosystems. Prokaryotes are indispensable links in these chemical cycles (to be discussed in detail in Chapter 49). Along with fungi, bacteria decompose the organic matter of dead organisms and the wastes of live ones, returning the elements to the environment in inorganic forms available for reassimilation by other organisms. If it were not for such **decomposers,** carbon, nitrogen, and other elements essential to life would become locked in the organic molecules of corpses and feces.

Prokaryotes also mediate the return of elements from the nonbiological sources in the environment (air, inorganic soil, and water). Autotrophic bacteria fix CO_2, supporting food chains through which organic nutrients pass from the bacteria to bacteria-eaters, then on to secondary consumers. Because of their many unique metabolic capabilities, bacteria are the only organisms able to transform nonbiological molecules containing elements such as iron, sulfur, nitrogen, and hydrogen. Cyanobacteria not only synthesize food and restore oxygen to the atmosphere but fix nitrogen, stocking the soil and water with nitrogenous compounds that other organisms can use to make proteins. And when plants and the animals that eat them die, it is soil prokaryotes that return the nitrogen to the atmosphere to keep the cycle running. All life on Earth depends on prokaryotes and their unparalleled metabolic diversity.

Symbiotic Bacteria

Prokaryote species rarely function singly in the environment. More often they interact in groups, often consisting of different bacterial species with complementary metabolism. They also often depend on close relationships with organisms from other kingdoms. **Symbiosis,** meaning "living together," is the term used to describe ecological relationships between organisms of different species that are in direct contact. The organisms involved are known as **symbionts.** If one of the symbionts is much larger than the other, the larger is also termed the **host.** There are three categories of symbiotic relationships: mutualism, commensalism, and parasitism. In **mutualism,** both symbionts benefit. In **commensalism,** one symbiont receives benefits while neither harming nor helping the other in any significant way. In **parasitism,** one symbiont, called a parasite in this case, benefits at the expense of the host.

Symbiosis among prokaryotes is undoubtedly common but poorly understood because until recently, study has been restricted to pure cultures of single species. Symbiosis likely played a major role in the evolution of the prokaryotes and also in the evolution of the early members of the kingdom Protista (see Chapter 26). Regarding symbiosis with species in other kingdoms, the kingdom Monera is represented extensively in all three types of symbiosis. For example, plants of the legume family (peas, beans, alfalfa, and others) have lumps on their roots called nodules, which are home to mutualistic bacteria that fix nitrogen used by the host, while the plant reciprocates with a steady supply of sugar and other organic nutrients (see Chapter 33). The bacteria inhabiting the inner and outer surfaces of the human body consist mostly of commensal species, but some species are mutual symbionts. For instance, fermenting bacteria living in the vagina produce acids that maintain a pH between 4.0 and 4.5, suppressing the growth of yeast and other potentially harmful microorganisms. Humans also benefit from the metabolic products of *E. coli,* an inhabitant of the intestine. Among parasitic bacteria are those classified as pathogens because they cause disease.

Bacteria and Disease

Bacteria are everywhere, and exposure to pathogenic ones is a certainty. Most of us are well most of the time because our defenses check the growth of harmful bacteria and other pathogens to which we are exposed. Occasionally, the balance shifts in favor of a pathogen, and we become ill. To be pathogenic, a parasite must invade the host, resist internal defenses well enough to begin growing, then harm the host in some way. Approximately half of all human disease is caused by bacteria.

Bacteria need not be exotic intruders to be pathogenic. Some pathogens are **opportunistic,** meaning they are normal residents of the human body that inflict illness only when defenses have been weakened by such factors as poor nutrition or a recent bout with flu. For example, *Streptococcus pneumoniae* lives in the throats of most healthy people, but this opportunist can multiply and cause pneumonia when the host's defenses are down.

Louis Pasteur, Joseph Lister, and other scientists began linking disease to pathogenic microbes in the late 1800s. The first to actually connect certain diseases to specific bacteria was Robert Koch, a German physician who determined the bacteria responsible for anthrax and tuberculosis. His methods established four criteria, now called **Koch's postulates,** which are still the guidelines for medical microbiology. To substantiate a specific pathogen as the cause of a disease, the researcher must (1) find the same pathogen in each diseased individual investigated, (2) isolate the pathogen from a diseased subject and grow the microbe in a pure culture, (3) induce the disease in experimental animals by transferring the pathogen from the culture, and (4) isolate the same pathogen from the experimental animals after the disease develops. The postulates can be applied for most pathogens, but prudent exceptions must be made for some cases. For example, no one has yet been able to culture the bacterium that causes syphilis *(Treponema pallidum)* on artificial media, but the volume of circumstantial evidence associating organism with disease leaves no doubt in this case. Lyme disease is caused by another unculturable spirochete, *Borrelia burgdorferi.* In the case of this disease, which was first studied in patients from around Lyme, Connecticut, in 1975–1976, careful examination of epidemiological evidence (where and when people were suffering from a peculiar set of symptoms) enabled researchers to trace the pathogen, which is transmitted by ticks.

Some bacteria disrupt the physiology of the host by their actual growth and invasion of tissues. Examples include Rocky Mountain spotted fever and typhus, two of the diseases caused by rickettsias (see Table 25.1). One of the chlamydias is the agent of nongonococcal urethritis (NGU), now suspected to be the most common sexually transmitted disease in the United States. Two actinomycetes that induce diseases by growing into tissues are the species that cause tuberculosis and leprosy.

Pathogenic bacteria more commonly cause illness by producing toxins. These poisons are of two types: exotoxins and endotoxins. **Exotoxins** are proteins secreted by the bacterial cell. The exotoxin can produce symptoms even without the bacteria actually being present. For example, when *Clostridium botulinum* grows anaerobically in poorly canned foods, one of the by-products of its fermentation is an exotoxin that causes the potentially fatal disease botulism. Exotoxins are among the most potent poisons known; one gram of botulism toxin would be sufficient to kill a million humans. Another exotoxin-producing species is the enteric bacterium *Vibrio cholerae,* which can infect the lower intestine of humans and cause cholera, a dangerous disease characterized by severe diarrhea. Cholera has recently become a serious problem in Peru, where consumption of raw fish from polluted waters has impeded efforts to control the disease. Even *E. coli* can be an exotoxin-releasing culprit. Traveler's diarrhea results from toxins released by alien strains of this gut inhabitant.

In contrast to exotoxins, **endotoxins** are not secreted by the pathogens but are instead components of the outer membranes of certain gram-negative bacteria. Endotoxins induce the same general symptoms—fever and aches—regardless of the species of bacteria, whereas exotoxins elicit specific symptoms. Examples of endotoxin-producing bacteria include other enteric bacteria, such as nearly all members of the genus *Salmonella,* which are not normally present in healthy animals. *Salmonella typhi* causes typhoid fever, and several other species of *Salmonella,* some of which are commonly found in poultry, cause food poisoning.

With the discovery in the nineteenth century that "germs" cause disease, public health officials took steps to upgrade hygiene. Sanitation measures played a significant role in reducing infant mortality and extended life expectancy dramatically in developed countries. In the past few decades, medical technology has increased its success at combating bacterial disease with a variety of antibiotics. More than half of our antibiotics (including streptomycin, neomycin, erythromycin, aureomycin, tetracycline, and others) come from soil bacteria of the genus *Streptomyces* (an actinomycete; see Table 25.1). In the wild, these compounds prevent encroachment by other microbes. Pharmaceutical companies culture various species of *Streptomyces* to produce these antibiotics in commercial amounts. Bacterial disease has certainly not been conquered, but its decline over the past century, probably due more to public health policies than to "wonder drugs," has so far been the greatest achievement of biomedical research and its application.

Putting Bacteria to Work

Humans have learned many ways of exploiting the diverse metabolic capabilities of prokaryotes, both for scientific research and for practical purposes. Much of what we know about metabolism and molecular biology has been learned in laboratories using bacteria as relatively simple model systems. In fact, *Escherichia coli,* the "white rat" of so many research labs, is the best understood of all organisms. Methanogens are important decomposers employed for sewage treatment (Figure 25.11). Some soil species of pseudomonads (see Table 25.1) decompose pesticides and other synthetic compounds. (They can also be pests. Their ability to feed on unusual carbon sources enables them to invade hot tubs, drug solutions, and even antiseptic solutions meant to prevent bacterial growth.) The chemical industry grows immense cultures of bacteria that produce acetone, butanol, and several other products. Pharmaceutical companies culture bacteria that make vitamins and antibiotics. The food industry uses bacteria to convert milk to yogurt and various kinds of cheese. Recombinant DNA techniques promise a new era in the commercial importance of prokaryotes (see Chapter 19).

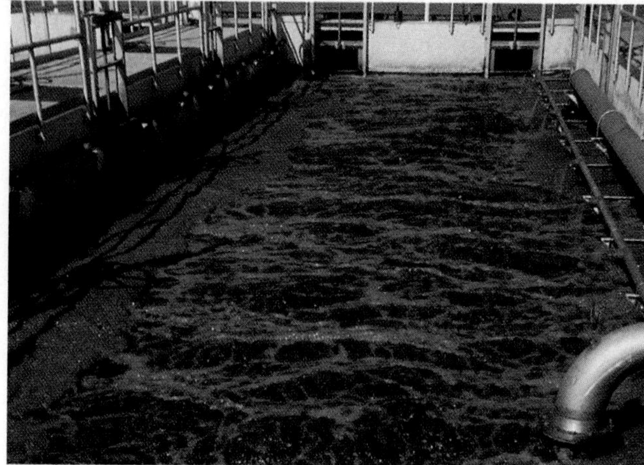

Figure 25.11
Using bacteria in secondary sewage treatment. This facility, called an activated sludge system, aerates sewage in the presence of aerobic bacteria, protists, and fungi. The microbes metabolize large organic molecules into smaller ones and carbon dioxide.

* * *

In this chapter, we have surveyed the prokaryotes and traced their history. On an ancient Earth inhabited only by bacteria, all the diverse forms of nutrition and metabolism evolved. Most subsequent evolutionary breakthroughs were structural rather than metabolic. The most significant was the origin of eukaryotic cells from prokaryotic ancestors, a juncture in the history of life we will explore in the next chapter.

STUDY OUTLINE

1. All prokaryotes (bacteria) are assigned to the kingdom Monera.
2. Prokaryotes were the first organisms, predating eukaryotes by 2 billion years and persisting today as the most numerous and pervasive of all living things.

Prokaryotic Form and Function (pp. 516–522)

1. Prokaryotes are generally single-celled organisms, although some occur as aggregates, colonies, or simple multicellular forms.
2. The three most common prokaryotic shapes are spherical (cocci), rod-shaped (bacilli), and spiral (spirilla) forms.
3. Nearly all prokaryotes have external cell walls, which protect and shape the cell and prevent osmotic bursting. Cell walls typically contain peptidoglycan, a unique polymer. Gram-positive and gram-negative bacteria differ in the structure of their walls and other surface layers.
4. Many species secrete sticky substances that form capsules. Some have surface appendages called pili outside the cell wall. Both structures help the cells adhere to one another, and some pili are specialized for conjugation.
5. Motile bacteria glide on slime secretions, propel themselves by flagella, or use flagellalike filaments positioned inside the cell wall (spirochetes). Directional movement (taxis) occurs in some species in response to chemicals, light, and/or magnetism.
6. Prokaryotic cells are not compartmentalized by endomembranes. However, the plasma membrane may be invaginated to provide internal membrane surface for specialized functions.
7. The prokaryotic genome consists of a single circular DNA molecule in a nucleoid region unbounded by a membrane. Many species also possess smaller separate rings of DNA called plasmids, which code for special metabolic pathways and resistance to antibiotics.

8. Bacteria reproduce asexually by binary fission. The rapid growth of a bacterial colony is usually checked by nutrient exhaustion or accumulation of toxic metabolic wastes.
9. Reproduction and production of genetic variation are not combined in prokaryotes as they often are in eukaryotes. In the absence of meiosis, genetic variation occurs through mutation and gene transfer by bacterial transformation, conjugation, or viral transduction.
10. Prokaryotes are the most metabolically diverse organisms on Earth. Photoautotrophs use light energy and chemoautotrophs use inorganic substances to synthesize their organic compounds from carbon dioxide. Photoheterotrophs require organic molecules for metabolic processes and synthesize ATP using light energy. Most bacteria are chemoheterotrophs, which require organic molecules as a source of both energy and organic carbon.
11. The ability or inability to survive in the presence of oxygen also reflects variation in metabolism. Obligate aerobes require oxygen, obligate anaerobes are poisoned by it, and facultative anaerobes can survive with or without oxygen.
12. Several groups of bacteria metabolize nitrogen compounds unavailable to other organisms. By doing so, these prokaryotes play critical roles in the cycling of nitrogen in the environment.

The Diversity of Prokaryotes (pp. 522–524)

1. Comparisons of selected macromolecules suggest an early split of the prokaryotes into the archaebacteria, which live in harsh environments reminiscent of conditions on the primordial Earth, and all other bacteria, the eubacteria.
2. Methanogens, extreme halophiles, and thermoacidophiles are the three subgroups of archaebacteria.
3. See Table 25.1 to review the eubacteria.

The Origins of Metabolic Diversity (pp. 524–528)

1. All forms of nutrition and virtually every kind of metabolic pathway evolved in prokaryotes.
2. The first prokaryotes were likely chemoheterotrophs that absorbed free organic compounds. Glycolysis evolved early as a mechanism of regenerating ATP.
3. Electron transport chains could have evolved from transmembrane pumps that originally served to regulate internal pH.
4. Early photosynthetic prokaryotes used pigments and light-powered photosystems to fix carbon dioxide. The first cyanobacteria began making organic compounds from water and carbon dioxide, releasing free oxygen as a by-product. This drastically changed the ancient atmosphere and affected subsequent biological evolution.

The Importance of Prokaryotes (pp. 528–530)

1. Prokaryotes, along with fungi, are decomposers (saprobes) that recycle elements between the biological and physical components of ecosystems.
2. Some prokaryotes live with other species in symbiotic relationships of mutualism, commensalism, and parasitism.
3. Some parasitic prokaryotes are pathogenic, causing disease in the host by invading tissues or poisoning with endotoxins or exotoxins.
4. Bacteria have been put to work in laboratories, sewage treatment plants, and the food and drug industry. One especially exciting development has been the use of prokaryotes in recombinant DNA technology.

SELF-QUIZ

1. A prokaryotic genome is different from a eukaryotic genome in that
   a. it has only one-half as much DNA as does a typical eukaryotic genome
   b. it consists of a single-stranded DNA molecule
   c. it has less protein associated with its DNA and is not enclosed in a nuclear envelope
   d. it is made of ribosomes that are smaller and chemically distinct
   e. it consists of RNA rather than DNA
2. Photoautotrophs use
   a. light as an energy source and can use water or hydrogen sulfide as a source of electrons for producing organic compounds
   b. light as an energy source and oxygen as an electron source
   c. inorganic substances for energy and CO_2 as a carbon source
   d. light to generate ATP but need organic molecules for a carbon source
   e. light as an energy source and CO_2 to reduce organic nutrients
3. Which of the following statements about bacterial groups is *not* true?
   a. The lipid composition of the plasma membrane found in archaebacteria is different from that of eubacteria.
   b. The archaebacteria and eubacteria probably diverged very early in evolutionary history.
   c. Both archaebacteria and eubacteria have cell walls, but those of archaebacteria lack peptidoglycan.
   d. The three groups of archaebacteria that are living today are anaerobic and pathogenic.
   e. Eubacteria include the cyanobacteria.
4. Home canners pressure-cook low-acid vegetables as a precaution primarily against the
   a. mycoplasmas
   b. endospore-forming bacteria
   c. enteric bacteria
   d. pseudomonads
   e. actinomycetes
5. The first prokaryotes were probably
   a. cyanobacteria
   b. chemoheterotrophs that used abiotically made organic compounds
   c. anaerobic photosynthetic organisms
   d. mycoplasmas
   e. parasitic bacteria

6. Banded iron formations in marine sediments indicate that
   a. early cyanobacteria were probably producing oxygen during that time period
   b. the early atmosphere was very reducing
   c. purple sulfur bacteria were the dominant life form in the seas at that time
   d. the pH of the early seas was quite low due to early prokaryotes' excretion of organic acids from fermentation
   e. mats of bacterial colonies were forming stromatolites
7. Which of the following statements about prokaryotic flagella is true?
   a. The flagella are composed of several microtubules of protein surrounded by the plasma membrane.
   b. All bacteria have flagella.
   c. There is always only one flagellum per cell.
   d. Motion is produced by lashing back and forth, like a whip.
   e. Motion is produced by rotation of a semirigid, helical filament.
8. Penicillins function as antibiotics mainly by inhibiting the ability of bacteria to
   a. form spores
   b. replicate DNA
   c. synthesize normal cell walls
   d. produce functional ribosomes
   e. synthesize ATP
9. Plantlike photosynthesis that releases oxygen occurs in the
   a. cyanobacteria
   b. purple sulfur bacteria
   c. archaebacteria
   d. actinomycetes
   e. chemoautotrophic bacteria
10. In motile bacteria, the movement of the organism toward a chemical source results from
   a. continuous movement in the direction of increasing concentration of the chemical
   b. longer runs when moving toward the chemical source
   c. more frequent tumbles when moving toward the chemical source
   d. coming to a complete stop whenever moving in a direction of decreasing concentration of the chemical
   e. controlling the direction of the power stroke taken by the flagella

CHALLENGE QUESTIONS

1. Nitrogen-fixing bacteria work either alone or in close conjunction with plants. If you were a scientist investigating the biochemistry of nitrogen fixation, would you choose a solitary or symbiotic species for study? Explain your answer.
2. Lynn Margulis of the University of Massachusetts has suggested that we could probe for extraterrestrial life by simply determining the mixture of gases in atmospheres of planets. If you were to conduct such research, what would you look for? Why?

SCIENCE, TECHNOLOGY, AND SOCIETY

1. Despite improvements in sanitation and antibiotics, people continue to be plagued by bacterial diseases, especially in the developing countries. In the United States, health officials have recently become concerned by a resurgence of tuberculosis (TB), a bacterial lung disease spread by airborne droplets. In particular, they are worrying about the current TB epidemic caused by bacteria resistant to standard drugs. Drugs can alleviate TB symptoms in a few weeks, but it takes much longer to halt the infection; a patient is likely to discontinue treatment while bacteria are still present. Why can bacteria quickly reinfect a patient if they are not wiped out? How might this result in the evolution of drug-resistant bacteria? How might urban poverty, homelessness, AIDS, and drug abuse contribute to the rise in the incidence of TB?

FURTHER READING

Chollar, S. "The Poison Eaters." *Discover,* April 1990. Cultured bacteria are used to destroy hazardous pollutants.

McEvedy, C. "The Bubonic Plague." *Scientific American,* February 1988. A bacterial disease that afflicted humans for over a thousand years.

Myers, F. and A. Anderson. "Microbes from 20,000 Feet Under the Sea." *Science,* January 3, 1992. Deep-sea bacteria are providing clues about the origin of life.

Postgate, J. "Microbial Happy Families." *New Scientist,* January 21, 1989. Using molecular systematics to unravel bacterial phylogeny.

Rietschel, E., and H. Brade. "Bacterial Endotoxins." *Scientific American,* August 1992. Many symptoms of bacterial disease are caused by these compounds.

Tortora, G. J., B. R. Funke, and C. L. Case. *Microbiology: An Introduction,* 4th ed. Redwood City, CA: Benjamin/Cummings, 1992. A general text.

26 Protists and the Origin of Eukaryotes

Characteristics of Protists
The Origin of Eukaryotes
Boundaries of the Kingdom Protista
Protozoa
Algae
Funguslike Protists
The Origins of Multicellularity

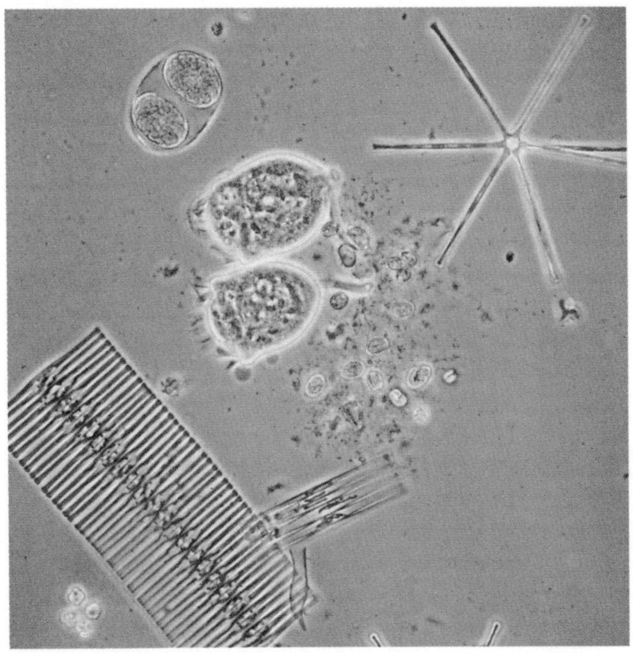

Figure 26.1
A pondwater droplet, home to a variety of protists. In this chapter, you will learn about the diversity and evolution of the kingdom Protista.

No more pleasant sight has met my eye than this of so many thousands of living creatures in one small drop of water," wrote Anton von Leeuwenhoek after his discovery of the microbial world more than three centuries ago. It is a world every biology student rediscovers by peering through a microscope into a droplet of pond water filled with diverse creatures of the kingdom Protista (Figure 26.1). Most protists are unicellular, but we also find colonial forms and even some multicellular organisms with tissues arranged in relatively simple body plans. For example, in the taxonomic scheme this book uses, the giant algae known as seaweeds are classified as protists. These and some other multicellular organisms are included in the kingdom because they are believed to be more closely related to certain unicellular protists than to plants, fungi, or animals.

Protists are eukaryotic, and thus even the simplest are much more complex than the prokaryotes of the kingdom Monera. The first eukaryotes to evolve from prokaryotic ancestors were probably unicellular and would therefore be classified as protists. The very word implies great antiquity (Gr. *protos,* "first"). It was during the genesis of protists that a true nucleus, mitochondria, chloroplasts, endoplasmic reticulum, Golgi bodies, 9 + 2 flagella and cilia (see Chapter 7), mitosis and meiosis, and the other structures and processes unique to eukaryotic organization arose. The primal eukaryotes were not only the predecessors of the great variety of modern protists but were also ancestral to plants, fungi, and animals, the other three kingdoms of eukaryotes. Two of the most significant chapters in the history of life—the origin of the eukaryotic cell and the subsequent emergence of multicellular eukaryotes—unfolded during the evolution of protists.

This chapter characterizes protists, discusses theories on the origins of eukaryotic cells, surveys the diversity and importance of protists, and finally considers the phylogeny within the kingdom and the evolutionary origins of multicellular organization.

CHARACTERISTICS OF PROTISTS

Protists vary so extensively in cellular anatomy, ecological roles, and life cycles that few general characteristics can be cited without exceptions.

Protists are found almost anywhere there is water. They are important constituents of **plankton** (Gr. *planktos*, "wandering"), the communities of organisms, mostly microscopic, that drift passively or swim weakly near the surface of oceans, ponds, and lakes. Among both seawater and freshwater protists are also bottom-dwellers that attach themselves to rocks and other anchorages or creep through the sand and silt. A particularly rich environment is the still-water surface at the edges of lakes and ponds. Here, floating mats of filamentous, photosynthetic protists provide home and food for a diversity of other protists. Protists are also common inhabitants of damp soil, leaf litter, and other terrestrial habitats that are sufficiently moist. In addition to free-living protists are the many symbionts that inhabit the body fluids, tissues, or cells of hosts. These symbiotic relationships span the continuum from mutualism to parasitism. Some parasitic protists are important pathogens of animals, including many that cause potentially fatal diseases in humans.

Nearly all protists are aerobic in their metabolism, using mitochondria for cellular respiration (a few lack mitochondria and either live in anaerobic environments or contain mutualistic respiring bacteria). Nutritional diversity is greater among the protists than in any other kingdom except Monera. Some forms are photoautotrophs with chloroplasts, some are heterotrophs that absorb organic molecules or ingest larger food particles, and still others, called **mixotrophs,** combine photosynthesis *and* eating for their nutrition. Mutualism between two protists, one a photoautotroph and one a heterotroph, is common. It is convenient (though phylogenetically inaccurate) to categorize this nutritional diversity into three groups that show similarities to the three multicellular kingdoms: the photosynthetic, plantlike protists (**algae,** singular **alga**), the ingestive, animal-like protists (**protozoa,** singular **protozoan**), and the absorptive, funguslike protists (these have no other name).

Most members of the kingdom Protista have flagella or cilia at some time in their life cycles. It is important to remember that prokaryotic and eukaryotic versions of flagella are not homologous structures. Bacterial flagella, made of the protein flagellin, are attached to the cell surface (see Figure 25.4). In contrast, eukaryotic flagella and cilia are extensions of the cytoplasm, with bundles of microtubules covered by the plasma membrane (see Figure 7.30). Eukaryotic cilia and flagella have the same basic ultrastructure, but cilia are shorter and more numerous. They move a cell with their rhythmic power strokes, analogous to the oars of a boat (see Figure 7.29).

Although all protists are eukaryotic, nuclear organization and cell division are perplexingly varied in the kingdom. Mitosis occurs in most phyla of protists, but there are many variations in the process unknown in any other kingdom. Reproduction and life histories also vary extensively among the protists. All protists can reproduce asexually. Some forms are exclusively asexual; others can also reproduce sexually or at least use the sexual processes of meiosis and **syngamy** (union of two gametes) to shuffle genes between two individuals that then go on to reproduce asexually. In Chapter 12 you learned about three basic types of sexual life cycles that differ in the timing of meiosis and syngamy (see Figure 12.5). All three types are represented in the kingdom Protista, along with some variations that do not quite fit any of the three basic life cycle patterns. At some point in the life history of many protists, resistant cells called **cysts** that can survive harsh conditions are formed.

Because most protists are unicellular, they are justifiably considered to be the simplest eukaryotic organisms. But at the *cellular* level, many protists are exceedingly complex; indeed, the kingdom Protista can claim among its members the most elaborate of all cells. We should expect this of organisms that must carry out within the bounds of single cells all the basic functions performed by the collective of specialized cells that makes up the bodies of plants and animals. Each unicellular protist is not at all analogous to a single cell from a multicellular organism, but is itself an organism as complete as any whole plant or animal.

THE ORIGIN OF EUKARYOTES

The first protists were also the first eukaryotes, and their origin was one of the most important chapters in the history of life. The many differences between prokaryotic and eukaryotic cells represent a distinction even greater than that between the cells of plants and animals. Among the most fundamental issues in biology are the questions of when and how the complex eukaryotic cell, with its true nucleus, cytoplasmic organelles, and endomembrane system, evolved from much simpler prokaryotic cells.

The small size and relatively simple construction of a prokaryotic cell offer many advantages (see Chapter 25) but also impose limits on the number of different metabolic activities that can be handled at one time. The relatively small size of the prokaryotic genome limits the number of genes coding for the enzymes that control these activities. This is not to say that prokaryotes are less successful than eukaryotes. Bacteria have been evolving and adapting since the dawn of life, and they are the most widespread organisms even today. But in at least some prokaryotic groups, natural selection favored increasing complexity—higher levels of organization with emergent properties. One trend was the evolution of multicellular prokaryotes, such as the filaments of some cyanobacteria, where different cell types are specialized for different functions. A second solution was the evolution of complex bacterial communities, where each species benefited from the metabolic specialties of other species. A third solution was to compartmentalize different functions within single cells, an evolutionary trend that produced the first eukaryotes.

Modern protists represent end-products of at least 2 billion years of evolution compared to the mere 700 million years for animals and 500 million years for terrestrial plants. In these groups we can see all sorts of evolutionary experiments that have been tried and been successful, some as the ancestors of the other eukaryotic kingdoms, others as organisms successful in their own right.

The Antiquity of Eukaryotes

The fossil record of prokaryotic life extends back 3.5 billion years (see Chapter 25). Two billion years of prokaryotic history elapsed before the debut of anything resembling eukaryotic cells. The oldest putative fossils of eukaryotes are among Precambrian structures known as **acritarchs** (Gr., meaning "of uncertain origin"). Some acritarchs are about the right size and appearance to be the ruptured coats of cysts similar to those made by certain algal protists today. The oldest are in rocks that have been estimated by radioisotopic dating to be about 1.5 billion years old. Metabolic evidence supports the fossil evidence of a long prokaryotic history before eukaryotes evolved. For instance, almost all eukaryotes require oxygen, implying that eukaryotes evolved after cyanobacteria had existed long enough for the oxygen they produced to accumulate in the atmosphere. The oxygen revolution occurred about 2.5 billion years ago (see Chapter 25), and it is possible that eukaryotic life dates back almost that far.

Models of Eukaryotic Origins

Currently, there are two models for the origin of the eukaryotic cell. According to the **autogenous model,** eukaryotic cells evolved by the specialization of internal membranes derived originally from the plasma membrane of a prokaryote (Figure 26.2a). The endomembrane system, consisting of nuclear envelope, endoplasmic reticulum, Golgi, and organelles enclosed by single membranes, such as lysosomes, are viewed as the differentiated products of invaginated membranes. Mitochondria and chloroplasts may have acquired their double-membrane status by secondary invagination or more elaborate folding of membranes.

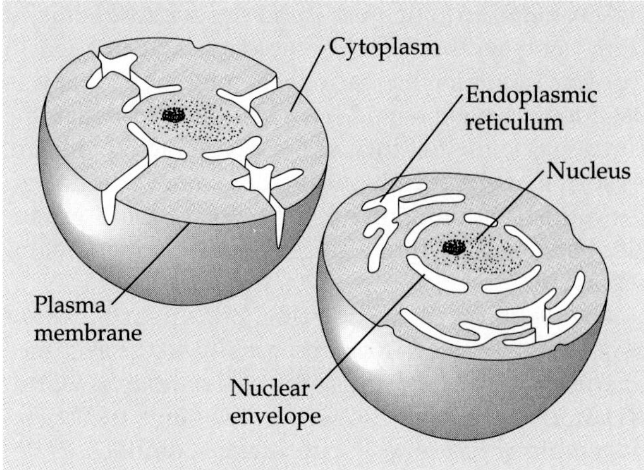

(**a**) The autogenous model

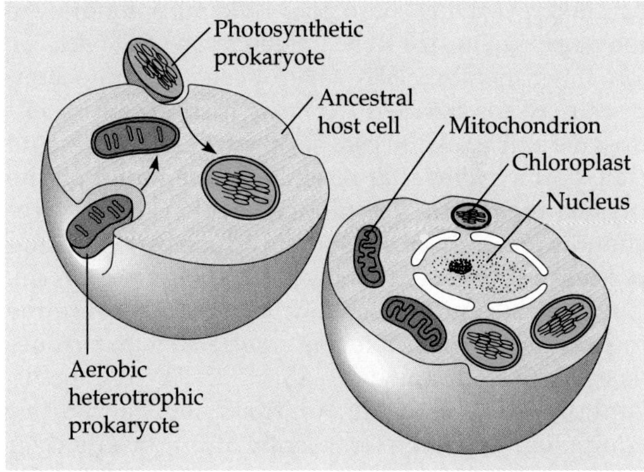

(**b**) The endosymbiotic model

Figure 26.2
Models of eukaryotic origins. (**a**) According to the autogenous model, the complexity of the eukaryotic cell arose by invagination and specialization of the plasma membrane. (**b**) According to the endosymbiotic model, the eukaryotic cell evolved as a consortium of prokaryotes that established symbiotic relationships. The inner membranes of mitochondria and chloroplasts may be remnants of the plasma membranes of early prokaryotes that infected or were engulfed by a larger host cell.

According to the **endosymbiotic model,** the forerunners of eukaryotic cells were symbiotic consortiums of prokaryotic cells, with certain species, termed endosymbionts, living within larger prokaryotes (Figure 26.2b). Developed most extensively by Lynn Margulis, the endosymbiotic model focuses on the origins of chloroplasts and mitochondria. Chloroplasts are probably the descendants of photosynthetic prokaryotes that became endosymbionts within larger cells. The proposed ancestors of mitochondria were endosymbiotic bacteria that were aerobic heterotrophs. Perhaps they first gained entry to the larger cell as undigested prey or internal parasites. By whatever means the relationships began, it is not hard to imagine the symbiosis eventually becoming mutually beneficial. A heterotrophic host could derive nourishment from photosynthetic endosymbionts. And in a world that was becoming increasingly aerobic, a cell that was itself an anaerobe would have benefited from aerobic endosymbionts that turned the oxygen to advantage. As host and endosymbionts became more interdependent, the conglomerate of prokaryotes would gradually be integrated into a single organism, its parts inseparable.

The feasibility of an endosymbiotic origin of chloroplasts and mitochondria rests partially on the existence of endosymbiotic relationships in the modern world. As another line of evidence, proponents of the endosymbiotic hypothesis cite various similarities between eubacteria and the chloroplasts and mitochondria of eukaryotes. Comparisons of structure and function reveal that chloroplasts and mitochondria are the appropriate size to be descendants of eubacteria. The inner membranes of chloroplasts and mitochondria, perhaps derived from the membranes of endosymbiotic prokaryotes, have several enzymes and transport systems that resemble those found on the plasma membranes of modern prokaryotes. Mitochondria and chloroplasts reproduce by a splitting process reminiscent of binary fission in bacteria. Chloroplasts and mitochondria contain DNA in the form of circular molecules not associated with histones or other proteins, as in prokaryotes. The organelles contain the transfer RNAs, ribosomes, and other equipment needed to transcribe and translate their DNA into proteins. In fact, some of the subunits of the cytochromes and ATPases that function in chloroplasts and mitochondria are known to be made in the organelles themselves. The ribosomes of chloroplasts are more similar in size and biochemical characteristics to prokaryotic ribosomes than to the ribosomes outside the chloroplast in the cytoplasm of the eukaryotic cell. Mitochondrial ribosomes vary extensively from one group of eukaryotes to another, but they are generally more similar to prokaryotic ribosomes than to their counterparts in the eukaryotic cytoplasm.

The limited evidence available so far from molecular systematics also suggests eubacterial origins for chloroplasts and mitochondria. Comparisons of base sequences show that the ribosomal RNA of chloroplasts, which is transcribed from genes within the organelles, is more similar to the RNA of certain photosynthetic eubacteria than it is to the ribosomal RNA in eukaryotic cytoplasm, which is transcribed from nuclear DNA. Base-sequence comparisons also suggest a eubacterial origin for ribosomal RNA of mitochondria, although the similarity is not as close as the one between chloroplasts and eubacteria.

Those skeptical about the endosymbiotic model point out that chloroplasts and mitochondria are not even close to being genetically autonomous. The great majority of proteins in the organelles are made by cytoplasmic ribosomes translating messenger RNA transcribed from nuclear genes. Advocates of the endosymbiotic hypothesis retort that a billion years of coevolution has been sufficient time for the host cell to develop extensive nuclear control over its symbionts, either by the accumulation of mutations or, more likely, by the direct transfer of DNA from the symbionts. In fact, the discovery of transposons (see Chapter 17) has revealed that DNA is surprisingly mobile within the nuclear genome, and there is recent evidence that genes have also jumped between the genomes of organelles and the nucleus.

In considering the origin of eukaryotes, we must understand that the autogenous and endosymbiotic models are not mutually exclusive. Possibly, the nuclear envelope and the rest of the endomembrane system arose by modification of a single cell, whereas chloroplasts and mitochondria probably originated as endosymbionts. In addition, neither model requires that an event leading to greater cellular complexity happen just once in the course of evolution. Comparison of algal pigments and chloroplast construction suggests that, by whatever mechanism, photosynthetic protists likely evolved at least three times from separate prokaryotic ancestors.

A comprehensive theory of the eukaryotic cell must also account for the evolution of 9 + 2 flagella and cilia. (The more ardent supporters of the endosymbiotic model have proposed that eukaryotic flagella began as spirochetes or other motile bacteria attached to a host cell, but this particular idea does not seem to be widely accepted.) Related to the evolution of the eukaryotic flagellum is the origin of mitosis and meiosis, processes unique to eukaryotes that also employ microtubules. Mitosis made it possible to reproduce the large genomes of the eukaryotic nucleus, and the closely related mechanics of meiosis became an essential process in eukaryotic sex. In comparing the four eukaryotic kingdoms, sexual life histories are the most varied among the protists. You will learn about some of these variations and their evolutionary significance as we now survey the major phyla of protists.

BOUNDARIES OF THE KINGDOM PROTISTA

A consensus does not yet exist on which groups of organisms should be included in the kingdom Protista. The problem is that multicellularity apparently evolved several times during the history of protists, giving rise not only to plants, fungi, and animals but also to multicellular organisms such as seaweeds that lack the distinctive traits that define those three kingdoms. When Robert H. Whittaker popularized the five-kingdom system in 1969, he assigned unicellular eukaryotes to their own kingdom, Protista. The trend in the past two decades has been to expand the boundaries of the kingdom Protista to include some phyla of multicellular organisms classified in earlier renditions of the five-kingdom format as plants or fungi but that seem to have closer relatives among the unicellular organisms. Because the term *protist* had come to connote unicellular life, some advocates of the new classification also recommended changing the name of the kingdom to Protoctista, but the name Protista is still more widely used.

One modification of the five-kingdom scheme moves the multicellular algae, including kelp and other seaweeds, from the plant kingdom to the kingdom Protista. In its expanded form, the kingdom Protista also encompasses phyla of funguslike organisms, such as the forms known as slime molds, which lack important characteristics that define true fungi. This text incorporates these changes, while conceding that all taxonomic boundaries, especially above the species level, are tentative lines drawn by biologists seeking order in the diversity of life. The goal is to draw those lines so that classification is not only convenient but also as phylogenetically relevant as the evidence allows. What may seem to be hair-splitting taxonomic debates are actually dialogues about the history of life.

In morphology and lifestyles, protists are the most diverse of all organisms. The kingdom Protista, as its limits are defined in this chapter, includes organisms as different as amoebas and giant kelp. In all, more than 60,000 species of extant protists have been described, and about the same number are known from the fossil record. The taxonomy of protists is in a state of flux, and taxonomists do not yet agree on the number or names of phyla. For example, all unicellular organisms with pseudopodia were once grouped in the phylum called Sarcodina. Current classification puts these organisms in at least three phyla—Rhizopoda, Actinopoda, and Foraminifera. We will examine a selection of 17 protistan phyla (some taxonomists define as many as 45 phyla) grouped in these three informal categories: protozoa, algae, and funguslike protists. This informal scheme will make our survey of protistan diversity more convenient, but the categories do *not* reflect evolutionary relationships. A slime mold, for instance, is funguslike only in the sense that a whale is fishlike; the resemblance is due to convergent evolution. Also, many phyla have both photosynthetic and heterotrophic members; there is in the real world no protozoa–algae dichotomy. With these reservations, let's sample the major phyla of protists.

Table 26.1 A Partial List of Protozoa (Animal-like Protists; Nutrition Mainly by Ingestion)

Phylum	Brief Description
Rhizopoda	Naked and shelled amoebas, with broad pseudopodia for motility and feeding
Actinopoda	Occupy planktonic habitats; usually spherically symmetrical; feed with axopodia, slender, radiating pseudopodia supported by internal microtubules; radiolarians posses internal siliceous skeletons, while heliozoans do not
Foraminifera	"Forams": feed and move with slender, interconnected pseudopodia exuding from spirally arranged calcareous compartments
Apicomplexa	Formerly called sporozoans; parasitic with complex life cycles in animal hosts
Zoomastigophora	Zooflagellates: use flagella for motility and feeding; mostly unicellular, but some colonial
Ciliophora	Ciliates: cilia used for motility and feeding; mostly unicellular, with a few sessile, colonial species

PROTOZOA

Protozoa means "first animals," a misnomer in the context of the five-kingdom system. The term persists to refer informally to protists that live primarily by ingesting food, an animal-like mode of nutrition. These heterotrophs actively seek and consume bacteria, other protists, and **detritus** (dead organic matter). There are also symbiotic protozoa, including some parasites that cause human diseases. Note in Table 26.1 that the protozoa are mainly subdivided into phyla according to the means by which they feed or move.

Rhizopoda

Members of the phylum Rhizopoda, the **amoebas** and their relatives, are all unicellular. With or without shells, they are among the simplest of protists (Figure 26.3). No stages in their life histories are flagellated. Instead, amoebas use cellular extensions called **pseudopodia** to move and to feed. You have probably observed this mode of motility in *Amoeba proteus* in the laboratory. Watching a living amoeba, you see one of the most flexible of all cells. Pseudopodia may bulge from virtually anywhere on the cell surface. When an amoeba moves, it extends a pseudopodium and anchors its tip, and then more cytoplasm streams into the pseudopodium (*Rhizopoda* means "rootlike feet"). The cytoskeleton, consisting of microtubules and microfilaments, functions in amoeboid movement. Pseudopo-

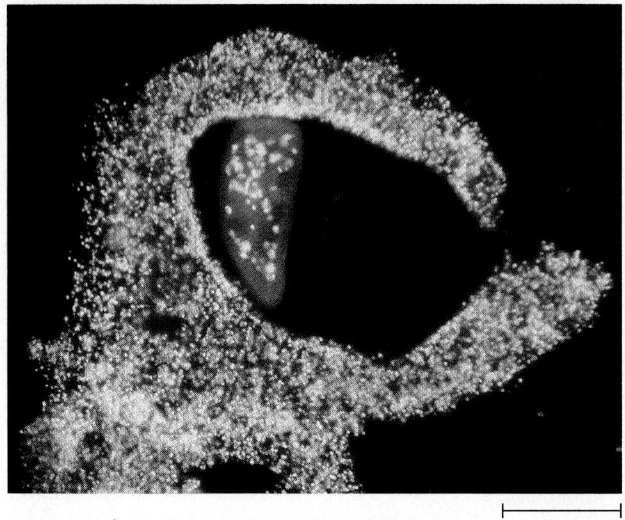

Figure 26.3
Rhizopoda. An amoeba ingests prey (a ciliophoran) by phagocytosis, engulfing the food with pseudopodia (LM).

dial activity may appear chaotic, but in fact amoebas show taxis as they creep slowly toward a food source.

Meiosis and sex do not occur in this phylum. The organisms reproduce asexually by various mechanisms of cell division. Spindle fibers form, but the typical stages of mitosis are not apparent in most amoebas. In many genera, for instance, the nuclear envelope persists during cell division.

Amoebas inhabit both freshwater and marine environments and are also abundant in soils. The majority of amoebas are free-living, but some are important parasites, including *Entamoeba histolytica,* which causes amoebic dysentery in humans. These organisms spread via contaminated drinking water, food, or eating utensils.

Actinopoda

Actinopoda means "ray feet," a reference to the slender pseudopodia called axopodia that radiate from the beautiful protists that compose the phylum (Figure 26.4). Each axopodium is reinforced by a bundle of microtubules, which is covered by a thin layer of cytoplasm. The projections place an extensive area of cellular surface in contact with the surrounding water, help the organisms float, and function in feeding. Smaller protists and other microorganisms stick to the axopodia and are phagocytized by the thin layer of cytoplasm. Cytoplasmic streaming then carries the engulfed prey down to the main part of the cell.

Most actinopoda are components of plankton. Most **heliozoans** ("sun animals") live in fresh water, whereas **radiolarians** are primarily marine. The term *radiolarian* actually applies to several groups of organisms that may not be very closely related, all of which have delicate shells, most commonly made of silica, the material of glass. After radiolarians die, their shells settle to the sea floor, where they have accumulated as an ooze that is hundreds of meters thick in some locations.

Figure 26.4
Actinopoda. (**a**) Heliozoans are mainly freshwater protists with stiff axopodia used for feeding (LM). (**b**) Radiolarians are mostly marine forms with glassy shells, different in shape for each species (LM).

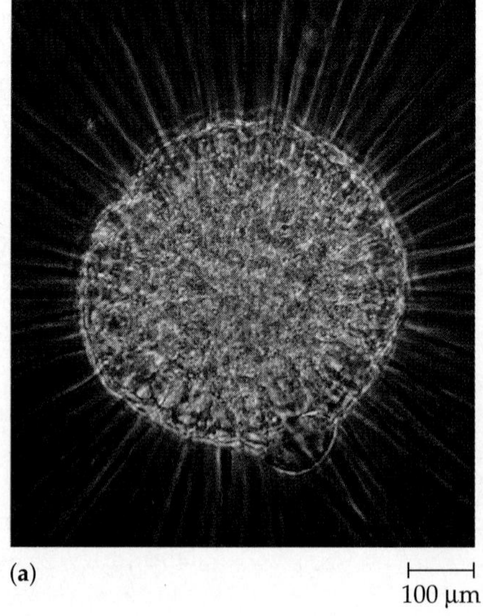

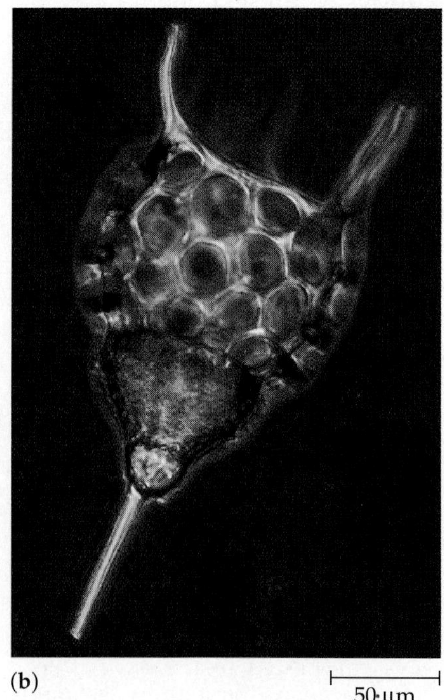

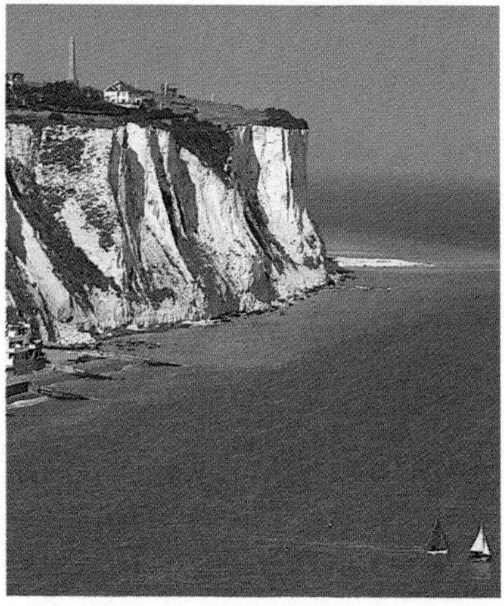

(a) (b)

Figure 26.5
Foraminifera. (**a**) The calcium carbonate shells of these protists have left an excellent fossil record. This living foram has a snail-like shell. The largest forams grow to diameters of several centimeters (LM). (**b**) The white cliffs of Dover, England, are sedimentary deposits of fossil shells of the foraminiferan genus *Globigerina.*

Foraminifera

Exclusively marine, the majority of members of Foraminifera, or **forams,** live in the sand or attach themselves to rocks and algae, but some families are also abundant in plankton. The phylum is named for the porous shells of its members (L. *foramen,* "little hole," and *ferre,* "to bear"). The shells are generally multichambered and consist of organic material hardened with calcium carbonate. Strands of cytoplasm extend through the pores, functioning in swimming, shell formation, and feeding. Many forams also derive nourishment from the photosynthesis of symbiotic algae that live beneath the shells.

Ninety percent of all described foraminifera are fossils. The shells of forams are important components of marine sediments, including sedimentary rocks that are now land formations, such as the chalky white cliffs of Dover (Figure 26.5). Foram fossils are excellent markers for correlating the vintages of sedimentary rocks in different parts of the world (see Chapter 23).

Apicomplexa

All members of the phylum Apicomplexa, which were formerly called sporozoans, are parasites of animals, some causing dangerous human diseases. The parasites disseminate as tiny infectious cells called **sporozoites.** As seen with the electron microscope, one end (the apex) of the sporozoite cell contains a complex of organelles specialized for penetrating host cells and tissues, thus the phylum name. Most apicomplexans have intricate life cycles with both sexual and asexual stages, and these cycles often require two or more different host species for completion. An example is *Plasmodium,* the parasite that causes malaria (Figure 26.6). The incidence of malaria was greatly reduced in the 1960s by the use of insecticides that reduced populations of *Anopheles* mosquitoes, which spread the disease, and by drugs that killed the parasites in humans. However, the multiplication of resistant varieties of both the mosquitoes and *Plasmodium* species have caused a resurgence of the disease. Each year, more than 200 million people are infected in the tropics, and at least a million die from the disease in Africa alone.

Considerable research on possible malarial vaccines has been carried out, with little success. *Plasmodium* is an extremely evasive parasite. It spends most of its time inside human liver and blood cells, thus hiding from the immune system. The problem is compounded by changes in the surface proteins of *Plasmodium*—the parasite changes the "face" that it shows to the infected person's immune system.

Zoomastigophora

Phylum Zoomastigophora is named for the whiplike flagella these protozoa use to propel themselves (Gr. *mastix,* "whip"). Also called **zooflagellates,** these heterotrophs absorb organic molecules from the surrounding medium or engulf prey by phagocytosis. Most live as solitary cells but some form colonies of cells. There are both free-living and symbiotic zoomastigotes. Living within the gut of a termite, for instance, are symbiotic flagellates that digest cellulose in the wood eaten by the host. At the opposite end of the spectrum of symbiotic relationships are parasitic zoomastigotes, some of which are pathogenic to humans. Species of *Trypanosoma* cause African sleeping sickness, which is spread by the bite of the tsetse fly (Figure 26.7).

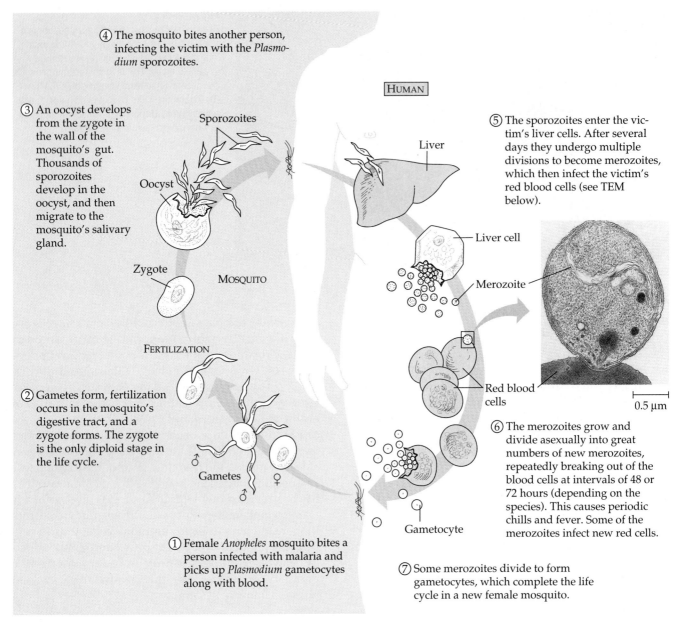

④ The mosquito bites another person, infecting the victim with the *Plasmodium* sporozoites.

HUMAN

③ An oocyst develops from the zygote in the wall of the mosquito's gut. Thousands of sporozoites develop in the oocyst, and then migrate to the mosquito's salivary gland.

Sporozoites

Oocyst

Zygote

MOSQUITO

Liver

⑤ The sporozoites enter the victim's liver cells. After several days they undergo multiple divisions to become merozoites, which then infect the victim's red blood cells (see TEM below).

Liver cell

Merozoite

Red blood cells

0.5 μm

FERTILIZATION

② Gametes form, fertilization occurs in the mosquito's digestive tract, and a zygote forms. The zygote is the only diploid stage in the life cycle.

Gametes

♂ ♀ ♂

⑥ The merozoites grow and divide asexually into great numbers of new merozoites, repeatedly breaking out of the blood cells at intervals of 48 or 72 hours (depending on the species). This causes periodic chills and fever. Some of the merozoites infect new red cells.

Gametocyte

① Female *Anopheles* mosquito bites a person infected with malaria and picks up *Plasmodium* gametocytes along with blood.

⑦ Some merozoites divide to form gametocytes, which complete the life cycle in a new female mosquito.

Figure 26.6
The life history of *Plasmodium*, the apicomplexan that causes malaria.

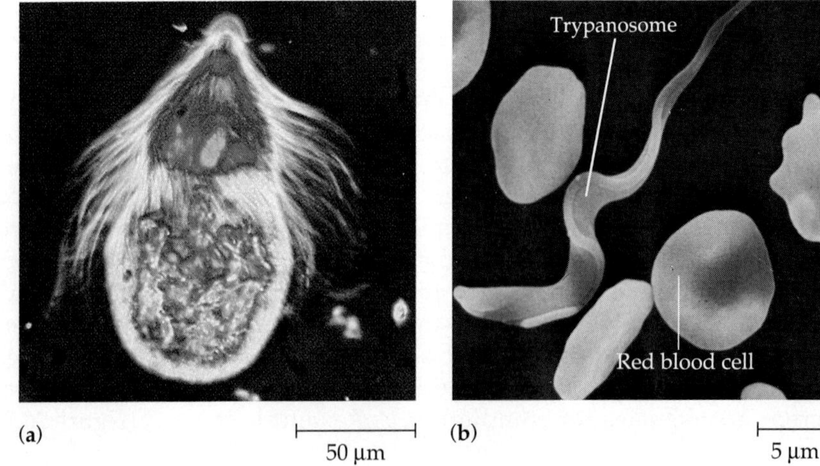

Figure 26.7
Zoomastigophora (Zooflagellata).
(**a**) *Trichonympha*, one of several symbiotic flagellates inhabiting the gut of termites (LM). The posterior region of the cell contains wood particles that are being digested.
(**b**) *Trypanosoma*, seen here in human blood, is the flagellate that causes African sleeping sickness (SEM). The molecular composition of these pathogens' coats changes frequently, preventing immunity from developing in hosts.

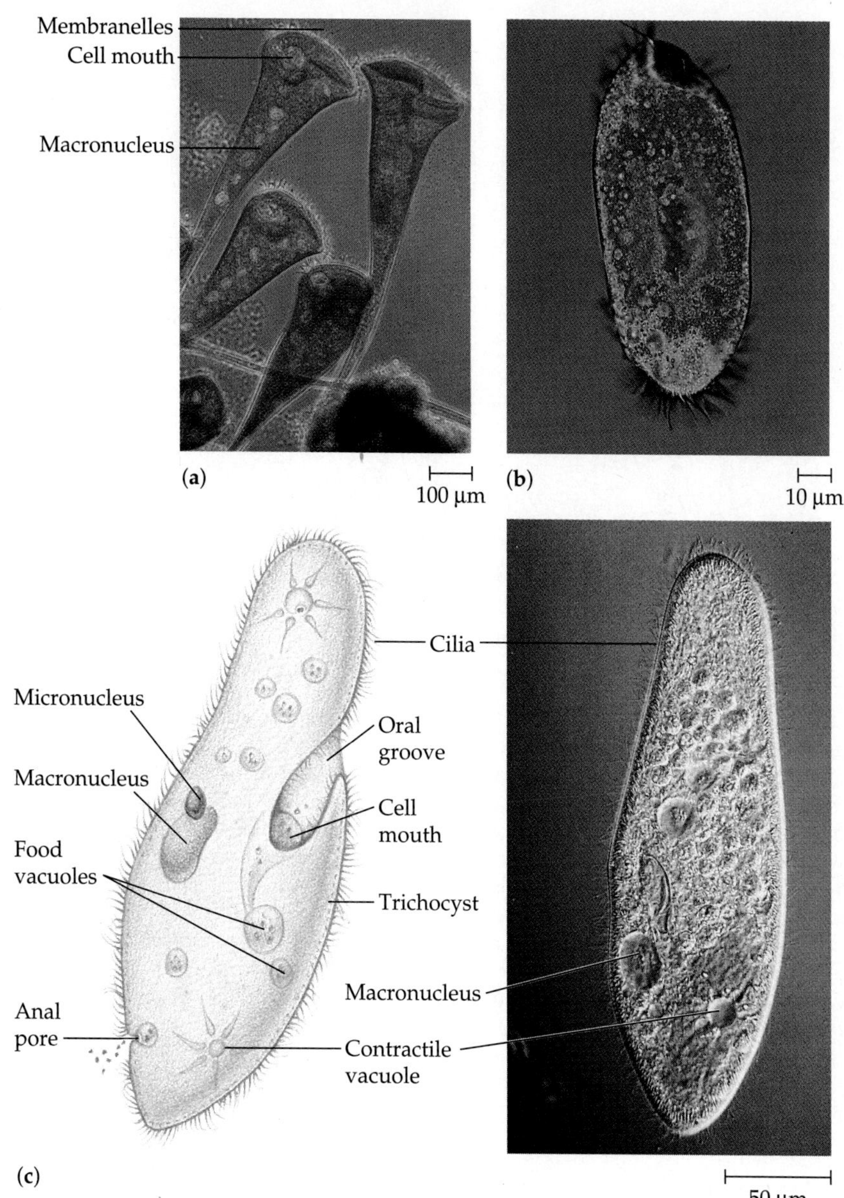

Figure 26.8
Ciliata. (**a**) The beautiful freshwater ciliate *Stentor* moves by individual cilia along the cell sides and by rows of finlike membranelles that spiral around the broad, anterior end of the protozoan (LM). *Stentor* often attaches its narrow (posterior) end to debris and, by beating the anterior membranelles, causes a whirlpool-like current to move food into the cell mouth. The macronucleus can be seen as light beaded strands running the length of these cells. (**b**) *Stylonychia* belongs to a group of ciliates that often have no individual cilia at all (LM). Cilia on the cell mouth side of the cell are united into leglike cirri. These ciliates scurry around, feeding among bits of organic debris. A diatom can be seen inside this cell. (**c**) *Paramecium*, an example of ciliate complexity, is covered by thousands of individual cilia (LM). Associated with each cilium are trichocysts, bulblike organelles that discharge sticky, proteinaceous threads. Although these threads are released in the presence of predators, they have little effect and are thought to function mainly in stabilizing the cell during feeding. Other genera have toxic trichocysts. *Paramecium* feeds mainly on bacteria. Rows of cilia along a funnel-shaped oral groove move food down into the cell mouth, where the food is engulfed by phagocytosis. The food vacuoles combine with lysosomes and, as the food is digested, the vacuoles follow a looping path, first to the anterior end and finally to the posterior end. The undigested contents are released when the vacuoles fuse with a specialized region of the plasma membrane that functions as an anal pore. *Paramecium*, like other freshwater protists, constantly takes in water by osmosis from the hypoosmotic environment. Bladderlike contractile vacuoles accumulate the excess water from radial canals and periodically expel it through the plasma membrane by contractions of the surrounding cytoplasm (see Chapter 8).

Ciliophora

The diverse protists of the phylum Ciliophora are characterized by their use of cilia to move and feed. Most members of Ciliophora, or **ciliates,** live as solitary cells in fresh water. Unlike most flagella, cilia are relatively short and beat synchronously. They are associated with a submembrane system of microtubules that may coordinate the movement of the thousands of cilia. Some ciliates are completely covered by rows of cilia, whereas others have their cilia clustered into fewer rows or tufts. The specific arrangements adapt the ciliates for their diverse lifestyles. Some species, for instance, scurry about on leglike cirri constructed from many cilia bonded together. Other forms, such as *Stentor,* have rows of tightly packed cilia that function collectively as locomotor membranelles. Ciliates are probably among the most complex of all cells (Figure 26.8).

A unique feature of ciliate genetics is the presence of two types of nuclei, a large **macronucleus** and usually several tiny **micronuclei.** The macronucleus has fifty or more copies of the genome. The genes are not distributed in typical chromosomes but are instead packaged into a much larger number of small units, each with hundreds of copies of just a few genes. The macronucleus controls the everyday functions of the cell by synthesizing RNA and is also necessary for asexual reproduction. Ciliates generally reproduce by binary fission, during which the macronucleus elongates and splits, rather than undergoing mitotic division. The micronuclei, of which species of *Paramecium* have from one to as many as 80, do not function in growth, maintenance, and asexual reproduction of the

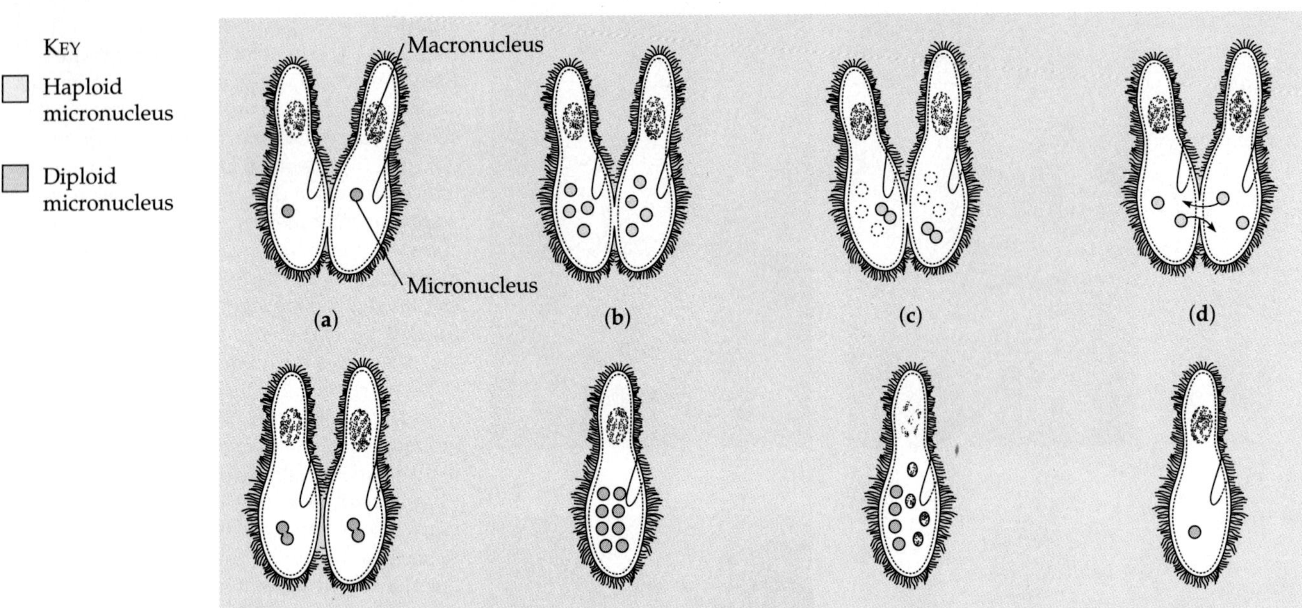

Figure 26.9
Conjugation by *Paramecium caudatum.* (**a**) Two individuals of compatible mating strains align side by side and partially fuse. In each cell, all but one diploid micronucleus disintegrate. (**b**) The micronucleus that remains undergoes meiosis to produce four haploid micronuclei. (**c**) One of these divides by mitosis, and the other three disintegrate. (**d**) Mates then swap one micronucleus. (**e**) Syngamy occurs when the micronucleus a cell acquired from its partner fuses with its remaining micronucleus, forming a fresh diploid nucleus with a mixture of chromosomes derived from the two individuals. The partners separate. (**f**) From now on, only one partner is shown. In each individual, the newly constituted micronucleus divides repeatedly by mitosis, resulting in eight identical micronuclei. (**g**) Subsequently, in each cell the original macronucleus disintegrates. Four micronuclei develop into new macronuclei by repeated replication of the DNA without nuclear division, and four remain as micronuclei. (**h**) After two cell divisions (without nuclear divisions), the new macronuclei and micronuclei are parceled out into four new individuals for each original conjugating cell. (Only one individual is shown here.) Note that all of these eight individuals ultimately resulting from one conjugation are identical genetically. However, they represent a genetic makeup different from either of the two original cells that conjugated.

cell but are required for sexual processes that generate genetic variation. The sexual shuffling of genes occurs during the process known as **conjugation,** diagrammed in Figure 26.9. In ciliates, sexual mechanisms of meiosis and syngamy are processes separate from reproduction.

ALGAE

The phyla grouped here consist mainly of photosynthetic organisms, although some phyla include heterotrophic or mixotrophic members. Recall that the term *alga* refers to relatively simple aquatic organisms that are plantlike in the sense that they are photoautotrophs. Except for the prokaryotic cyanobacteria (formerly called blue-green algae), all organisms generically called algae belong to the kingdom Protista according to the taxonomic scheme used in this text. Some biologists still prefer to place certain multicellular phyla in the plant kingdom (particularly Chlorophyta, Rhodophyta, and Phaeophyta—the green algae, red algae, and brown algae, respectively).

Algae are extremely important ecologically. On a global scale, about half of all photosynthetic production of organic material is achieved by the organisms classified here as algal protists. As freshwater and marine phytoplankton and intertidal seaweeds, algae are the bases of aquatic food webs, supporting countless suspension-feeding (also known as filter-feeding) and predatory animals.

All algae have chlorophyll *a,* the same "primary" pigment found in cyanobacteria and plants. But algae differ considerably in their accessory pigments, pigments that trap wavelengths of light to which chlorophyll *a* is not as sensitive (Figure 10.8). These pigments include other forms of chlorophyll (greenish), carotenoids (yellow-orange), xanthophylls (brownish), and phycobilins (red and blue varieties). The mixture of pigments in the chloroplasts lends characteristic colors to related algae. Many of the scientific and common

Table 26.2 Characteristics of Algae (Plantlike Protists; Nutrition Mainly by Photosynthesis)

Phylum	Approx. Number of Species	Predominant Color (Photosynthetic Pigments)	Carbohydrate Food Reserve	Number and Position of Flagella	Cell Wall Components	Habitat
Dinoflagellata (dinoflagellates)	1,100	Brown (chlorophyll *a*, chlorophyll *c*, carotenoids, xanthophylls)	Starch (an α1–4, branched glucose polymer)	1 lateral, 1 posterior	Submembrane cellulose	Marine and fresh water
Chrysophyta (golden algae)	850	Golden olive (chlorophyll *a*, often chlorophyll *c*, carotenoids, xanthophylls)	Laminarin (a β1–3 glucose polymer)	1 or 2, apical	Pectic compounds with siliceous material	Mostly fresh water
Bacillariophyta (diatoms)	10,000	Olive brown (chlorophyll *a*, chlorophyll *c*, carotenoids, xanthophylls)	Leucosin (a β1–3 glucose polymer)	1 in sperm only	Hydrated silica in organic matrix	Fresh water and marine
Euglenophyta (*Euglena* and its relatives)	800	Green (chlorophyll *a*, chlorophyll *b*, carotenoids, xanthophylls)	Paramylon (a β1–3 glucose polymer)	1 to 3, apical	No cell wall; submembrane protein	Mostly fresh water
Chlorophyta (green algae)	7,000	Green (chlorophyll *a*, chlorophyll *b*, carotenoids)	Plant starch	2 or more, apical or subapical	Cellulose	Mostly fresh water, but some marine
Phaeophyta (brown algae)	1,500	Olive brown (chlorophyll *a*, chlorophyll *c*, carotenoids, xanthophylls)	Laminarin (a β1–3 glucose polymer)	2 lateral in sperm only	Cellulose matrix with other polysaccharides	Almost all marine; flourish in cold ocean waters
Rhodophyta (red algae)	4,000	Red to black (chlorophyll *a*, carotenoids, phycobilins, chlorophyll *d* in some)	Floridean starch (a glycogen-like α1–4 branched glucose polymer)	None	Cellulose matrix with other polysaccharides	Mostly marine, but some fresh water; many species tropical

names of algal phyla are based on these colors (the Chlorophyta, or green algae, for example). Study of these pigment mixes has helped establish taxonomic affinities among the algae. Additional clues have come from chloroplast structure; the chemistry of cell walls (a few algal phyla lack walls); number, type, and position of flagella; and the form of food stored by the cells (Table 26.2). The following survey of selected algal phyla will give us a sense of the diversity of forms that has evolved. In particular, we will learn about the diverse life cycles of algae. (See Chapter 12 to review sexual life cycles.)

Dinoflagellata

Dinoflagellates are abundant components of the vast aquatic pastures of phytoplankton, microscopic algae floating near the surface of the sea that provide the foundation of most marine food chains (Figure 26.10).

Dinoflagellate blooms, episodes of explosive population growth, cause the red tides that occur occasionally in warm coastal waters. These blooms are brownish-red because of predominant xanthophylls in the chloroplasts. When suspension-feeding shellfish such as oysters feed on these blooms, they concentrate the

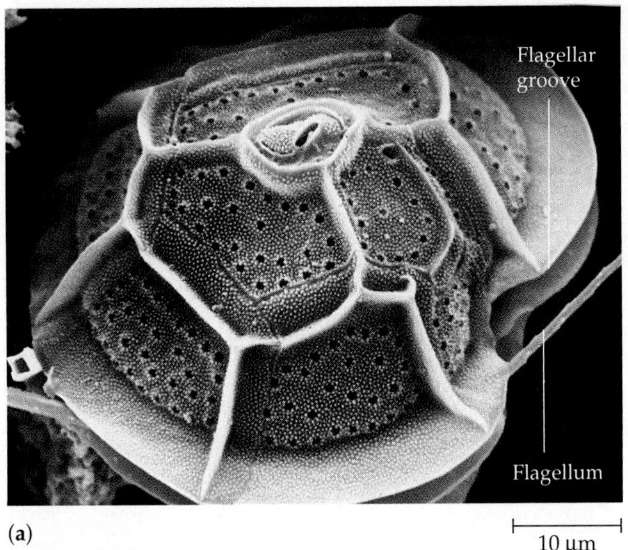

(a) 10 µm

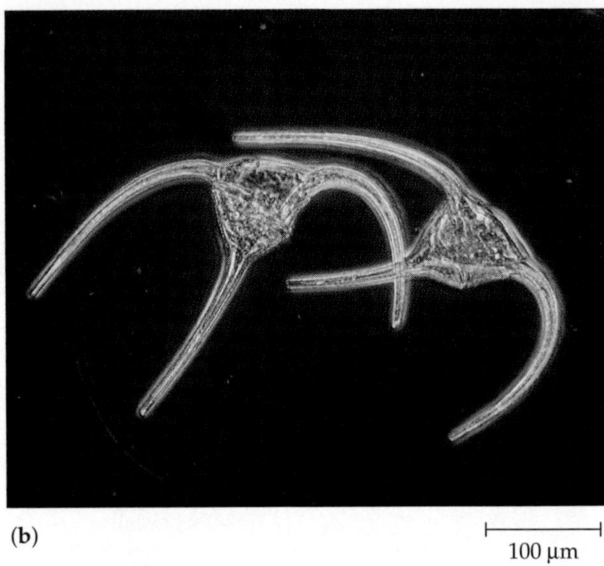

(b) 100 µm

Figure 26.10
Dinoflagellata. These unicellular algae are characterized by a pair of flagella in perpendicular grooves. Beating of the flagella causes the cell to spin as it swims. Each species has a distinctively shaped internal wall. (**a**) *Gonyaulax tamarensis* (SEM). (**b**) *Ceratium* (LM).

algae along with toxic compounds released by the dinoflagellate cells. These toxins are extremely dangerous to humans, who collect and eat the invertebrates. As a consequence, collecting shellfish during red tides is often regulated to prevent the widespread occurrence of "paralytic shellfish poisoning."

Some dinoflagellates live as mutualistic symbionts of animals called cnidarians that build coral reefs; the photosynthetic output of these dinoflagellates is the main food source for reef communities. Other dinoflagellates lack chloroplasts and live as parasites within marine animals. There are even carnivorous species. The existence of both photosynthetic and heterotrophic forms closely related enough to be grouped in the same phylum reinforces the point made earlier that the terms *protozoa* and *algae,* although traditional and somewhat useful, have no basis in phylogeny.

Of the several thousand known species, most dinoflagellates are unicellular, but there are some colonial forms. Each dinoflagellate species has a characteristic shape reinforced by internal plates of cellulose. The beating of two flagella in perpendicular grooves in this "armor" produces a spinning movement for which these organisms are named (Gr. *dinos,* "whirling").

The structure of the dinoflagellate nucleus and its division during asexual reproduction are unusual (see Figure 11.13).

Chrysophyta

Members of this phylum (Gr. *chrysos,* "golden") are named for their color, which results from yellow and brown carotenoid and xanthophyll accessory pigments. Their cells are typically biflagellated, with both flagella attached near one end of the cell. **Golden algae** live among freshwater plankton. Most species are colonial (Figure 26.11). In ponds and lakes that freeze in winter or dry up in summer, golden algae survive by forming resistant cysts, from which active cells emerge when conditions are favorable. Microfossils resembling the ruptured cysts of chrysophytes and other algae have been found in Precambrian rocks.

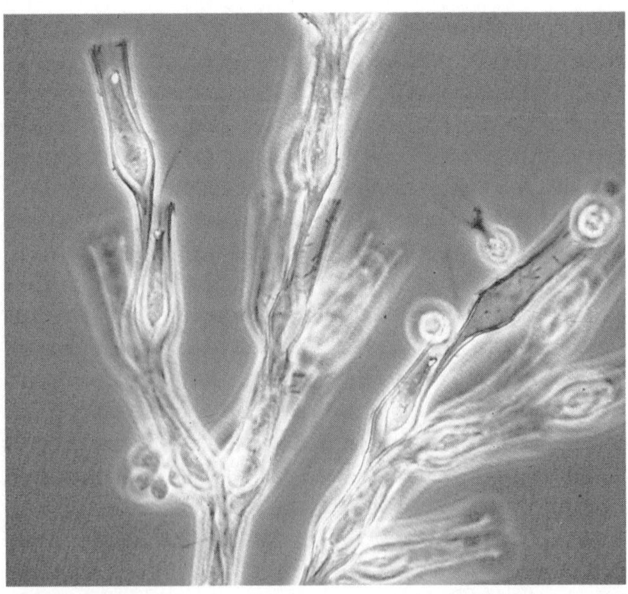

50 µm

Figure 26.11
A golden alga. *Dinobryon,* a freshwater organism, is one of many colonial forms of this algal phylum.

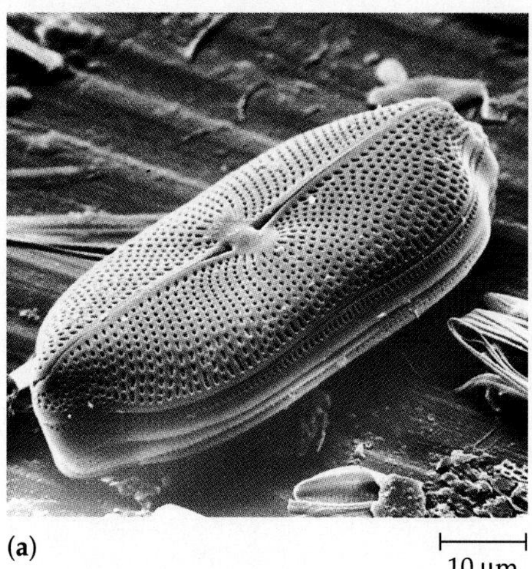

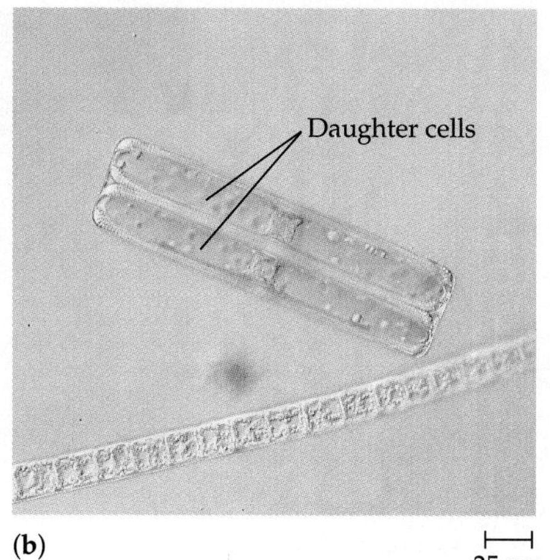

Figure 26.12
Diatoms. (**a**) The glasslike shells consist of two halves that fit like the bottom and lid of a shoe box. Tiny pores in the ornate shells allow for the exchange of gases and other substances between the cell and its environment. The shape of the shell and its pattern of pores are used for classifying diatoms. This species is *Navicula monilifera* (SEM). (**b**) In this side view of a species of *Pinnularia*, the cell has just divided by mitosis (LM). Each daughter cell keeps half of the parent cell's wall and builds a new complementary half.

Bacillariophyta

The members of the phylum Bacillariophyta, or **diatoms,** are yellow or brown in color. They have the same set of photosynthetic pigments as golden algae (Chrysophyta), and the two groups were once assigned to the same phylum. Diatoms, however, have such distinctive cell structure and life cycles that they are now considered to be a separate phylum by most phycologists (specialists in the study of algae). Diatom cells have unique glasslike walls consisting of hydrated silica embedded in an organic matrix (Figure 26.12). Each wall is in two parts that overlap like a shoe box and lid. Many diatoms are capable of a gliding movement caused by chemicals secreted out of slits in their cell walls.

Most of the year, diatoms reproduce asexually by mitotic cell divisions, with each daughter cell receiving half of the cell wall of its parent and regenerating a new second half. Cysts are formed by some species as resistant stages. Sexual stages are not common and involve the formation of eggs or sperm (each with a single flagellum) by meiosis. The sperm are the only flagellated cells in this phylum. Both freshwater and marine plankton are rich in diatoms; a bucket of water scooped from the surface of the sea may have millions of these microscopic algae. Planktonic species store food reserves in the form of oils, which also provide buoyancy to counteract the relatively heavy weight of their walls and keep the diatoms near the water surface.

Massive accumulations of fossilized diatom walls are major constituents of the sediments known as diatomaceous earth, which is mined for its quality as a filtering medium and for many other uses.

Euglenophyta

The best-known members of the phylum Euglenophyta belong to the genus *Euglena,* tiny green flagellates among the most common inhabitants of murky pond water (Figure 26.13). The pigment complex of euglenophytes is identical to that of the Chlorophyta (green algae), but these algae differ in all other respects. For example, the cells of euglenophytes are bounded by flexible internal protein plates rather than cellulose walls. Euglenophyte flagella are "tinseled" and pull the cells through the water. In close quarters, these algae cease flagellar activity and squirm among the debris; this is called euglenoid movement.

Euglenophytes are versatile in their nutrition. In the light, *Euglena* uses its chloroplasts to make a living by photosynthesis, although the organism is not completely autotrophic because it requires minute quantities of vitamin B_{12}. If *Euglena* is placed in the dark, it can live as a heterotroph by ingesting particles of food by phagocytosis. Indeed, some members of the phylum lack chloroplasts and depend entirely on heterotrophism. The euglenophytes remind us once again that no phylogenetically reliable distinction exists between protozoa and unicellular algae; all these organisms are protists, distinct in many ways from the other kingdoms of eukaryotes.

Chlorophyta

This phylum is named for its members' grass-green chloroplasts (Gr. *chloros,* "green"), which are much like those of plants in ultrastructure and pigment composition. In fact, most botanists believe that the ancestors of the plant kingdom were green algae. We will examine the evidence for this relationship in Chapter 27; for now, let's survey the diversity of extant chlorophytes.

More than 7000 species of **green algae** have been identified. Most live in fresh water, but there are also many marine species. Various species of unicellular green algae live as plankton, inhabit damp soil or snow, or occupy the cells or body cavities of protozoa and invertebrates as photosynthetic symbionts that

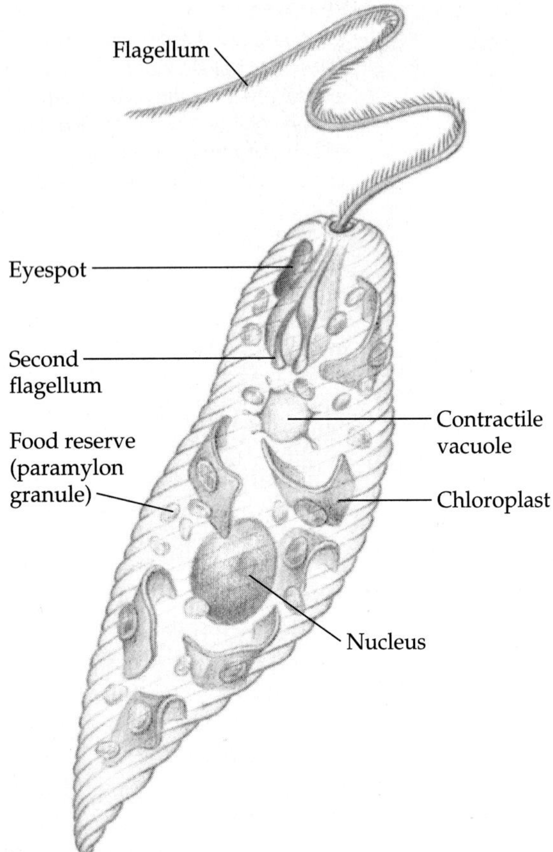

Figure 26.13
Euglena. A motile protist with chloroplasts, this unicellular alga uses its long flagellum for propulsion. Near the base of the flagellum is an eyespot that functions as a pigment shield. Depending on the position of the organism, it allows light from only a certain direction to strike a light detector, the swelling at the base of the long flagellum. These structures seem to function in phototaxis, which is important for these photosynthetic algae. *Euglena* lacks a cell wall but has a strong, flexible covering made of protein beneath its plasma membrane.

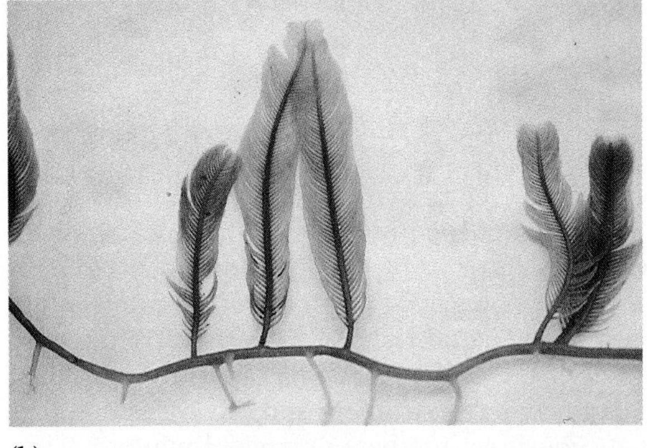

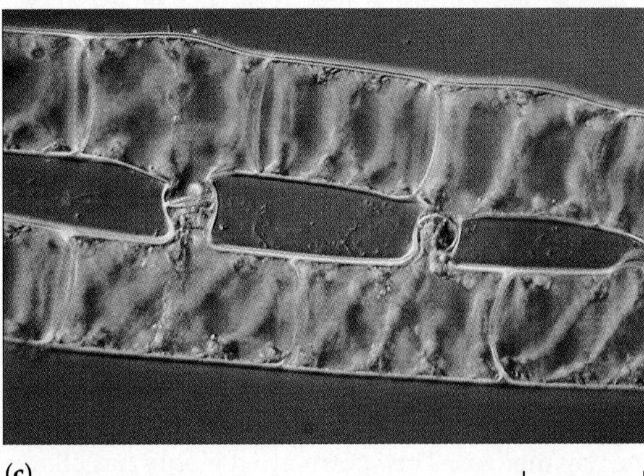

Figure 26.14
Colonial and multicellular green algae. (**a**) Species of *Volvox* are colonial chlorophytes that inhabit fresh water (LM). The colony is a hollow ball, with its wall composed of hundreds or thousands of biflagellated cells embedded in a gelatinous matrix. The cells are usually connected by strands of cytoplasm—if isolated, they cannot reproduce. The large colonies seen here will eventually release the small green and red "daughter" colonies within them. (**b**) Species of *Bryopsis* are found in the marine intertidal zone. They are formed of highly branched filaments lacking cross-walls and thus are multinucleate. (**c**) *Spirogyra* is composed of uninucleate cells arranged in long, unbranched filaments (LM). Each cell contains one or more large, spiral-shaped chloroplasts. The filaments grow in length and disperse by fragmentation as the cells divide by mitosis. During sexual reproduction, as seen here, cells of adjacent mating strands join by conjugation tubes.

contribute to the food supply of the hosts (see Figure 29.11d). Chlorophytes are also among the algae that live symbiotically with fungi in the mutualistic collectives known as **lichens** (see Chapter 28). The simplest chlorophytes are biflagellated unicells such as *Chlamydomonas,* which resemble the gametes of more complex green algae.

In addition to unicellular chlorophytes, there are colonial species, many of them filamentous forms that contribute to the stringy masses known as pond scum. There are even some truly multicellular chlorophytes such as *Ulva,* with bodies large and complex enough that many texts still consider them to be true plants. The similarities to plants are analogous, not homologous; even the most complex algae are more closely related to unicellular members of their phyla than to any true plant.

Three separate evolutionary trends have probably produced the diverse forms of colonial and multicellu-

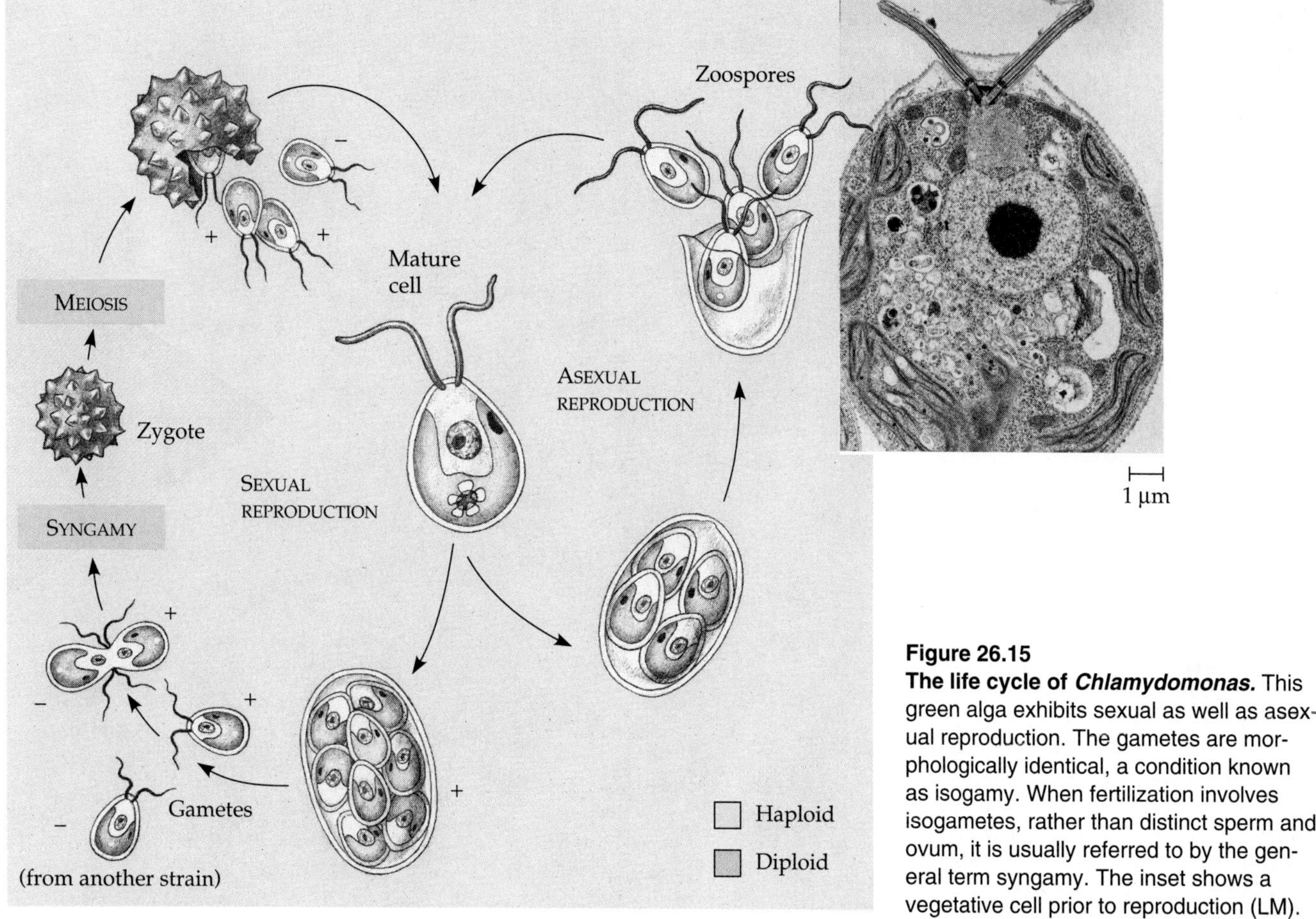

Figure 26.15
The life cycle of *Chlamydomonas.* This green alga exhibits sexual as well as asexual reproduction. The gametes are morphologically identical, a condition known as isogamy. When fertilization involves isogametes, rather than distinct sperm and ovum, it is usually referred to by the general term syngamy. The inset shows a vegetative cell prior to reproduction (LM).

lar chlorophytes from unicellular, flagellated ancestors. Larger size and greater complexity have evolved by (1) the formation of colonies of individual cells, as seen in species of *Volvox* (Figure 26.14a); (2) the repeated division of nuclei with no cytoplasmic division, as seen in multinucleate filaments of *Bryopsis* (Figure 26.14b); and (3) the formation of definite multicellular forms, as in *Ulva* (see Figure 26.16).

Most green algae have complex life histories, with both sexual and asexual reproductive stages. Nearly all reproduce sexually by way of biflagellated gametes having cup-shaped chloroplasts. The exceptions are the **conjugating algae,** such as *Spirogyra* (see Figure 26.14c), which produce amoeboid gametes (the difference is significant enough that many phycologists assign the conjugating algae to a separate phylum).

Let's examine the life cycle of *Chlamydomonas* (Figure 26.15). The mature organism is a single haploid cell. When it reproduces asexually, the cell resorbs its flagella and then divides twice by mitosis to form four cells (more in some species). These daughter cells develop flagella and cell walls and then emerge as swimming zoospores from the wall of the parent cell, which had enclosed them. The zoospores grow into mature haploid cells, completing the asexual life cycle. A shortage of nutrients, drying of the pond, or some other stress triggers sexual reproduction. Within the wall of the parent cell, mitosis produces many haploid gametes. After their release, gametes from opposite mating strains (designated + and –) pair off and cling together by the tips of their flagella. The gametes are morphologically indistinguishable, and their fusion is known as **isogamy,** which literally means a "marriage of equals." The gametes fuse slowly, forming a diploid zygote, which secretes a durable coat that protects the cell against harsh conditions. When the zygote breaks dormancy, meiosis produces four haploid individuals (two of each mating type) that emerge from the coat and grow into mature cells, completing the sexual life cycle.

Though many of the features of *Chlamydomonas* sex are believed to be primitive, refinements of the sexual process have evolved among chlorophytes. Some green algae, for instance, produce gametes that differ morphologically from vegetative cells, and in some species the male and female gametes differ in size (**anisogamy**) or morphology. Many species exhibit **oogamy,** a flagellated sperm fertilizing a nonmotile egg.

In the life cycles of some multicellular green algae, haploid and diploid individuals alternate, each releasing reproductive cells that grow into the other. An example of this phenomenon, called **alternation of generations,** is the life cycle of *Ulva,* the sea lettuce (Figure

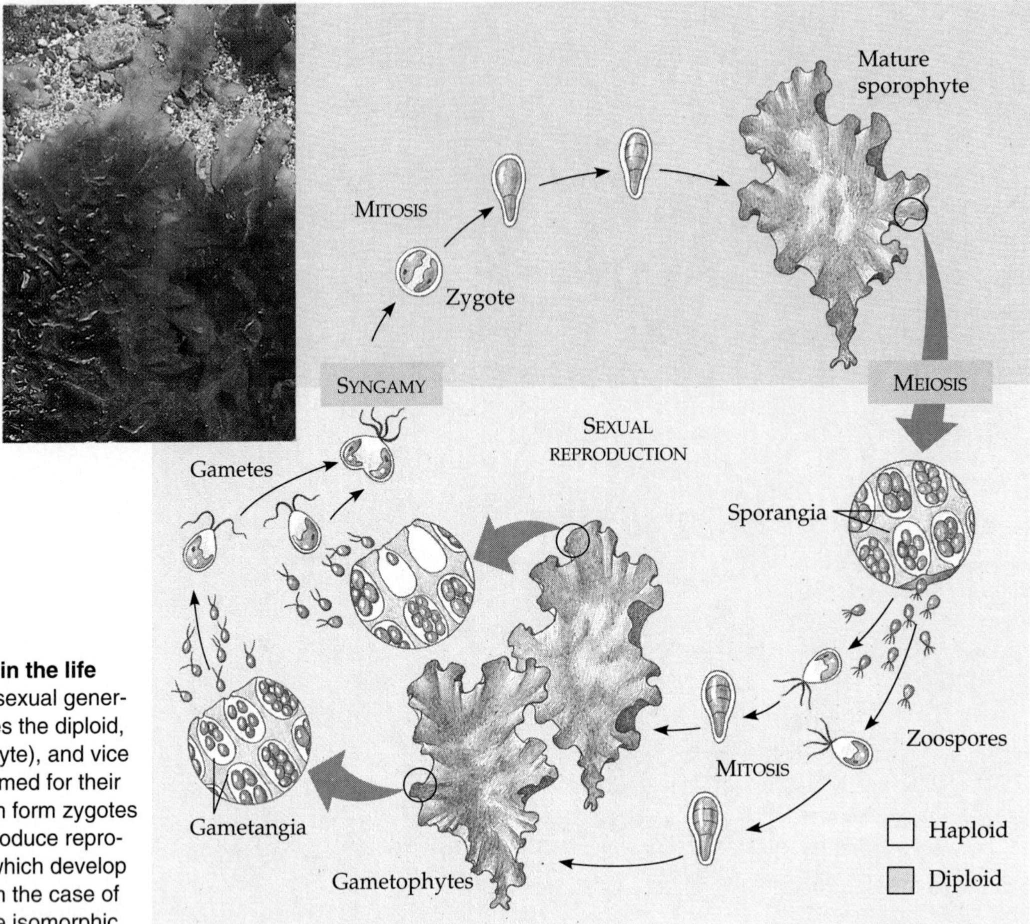

Figure 26.16
Alternation of generations in the life cycle of *Ulva*. The haploid, sexual generation (gametophyte) produces the diploid, asexual generation (sporophyte), and vice versa. Gametophytes are named for their production of gametes, which form zygotes by syngamy. Sporophytes produce reproductive cells called spores, which develop directly into gametophytes. In the case of *Ulva*, the two generations are isomorphic, or identical in appearance.

26.16). This alga is composed of wrinkled sheets of cells, two cells thick, which are attached to rocks in the intertidal zone. Each species of *Ulva* occurs in two forms that are virtually indistinguishable macroscopically but that differ in their chromosome number. Haploid individuals are called **gametophytes** because they produce gametes (but they do so by mitosis, not meiosis). Diploid individuals are called **sporophytes** because they produce reproductive cells called spores (by meiosis). At times of unusually high and low tides, cells along the margins of each form become reproductive. Gametes released by the gametophyte (similar in appearance to those produced by *Chlamydomonas;* see Figure 26.15) swim about and, attracted to each other, fuse in syngamy. The resulting diploid zygotes swim for a while, settle on rock surfaces, absorb their flagella, and divide and grow slowly into multicellular sporophytes. Zoospores released by the sporophyte bear four flagella each and swim about, finally settling on rocks to grow into gametophytes. Alternation of generations is widespread among green, brown, and red algae, and was an important feature in the evolution of plants (see Chapter 27).

Evolutionary Adaptations of Seaweed

The green alga *Ulva* is an example of a *seaweed,* a term that applies generally to certain large marine algae belonging to the phyla Chlorophyta (green algae), Phaeophyta (brown algae), and Rhodophyta (red algae). Seaweeds, along with many animals and other heterotrophs that feed on the algae, inhabit the intertidal and subtidal zones of coastal waters (see Figure 46.18).

The intertidal zone presents unique challenges to life. At times it is violently active, churned by waves and wind. Two times each day, at low tide, intertidal seaweeds are exposed to the drying atmosphere and unfiltered rays of the sun. Twice each day, at high tide, the same seaweeds are covered by up to 5 meters of water. Seaweeds have unique anatomical and biochemical adaptations that enable them to survive and thrive in this rough-and-tumble environment.

Seaweeds have the most complex multicellular anatomy of all protists. Some even have differentiated tissues and organs that resemble those we find in plants. The similarities, however, are analogous, not

Figure 26.17
Seaweeds: adapted to life at the ocean's margins. The sea palm, *Postelsia,* lives on rocks in the crashing surf along the northwest coast of the United States and Canada. Its thallus is well adapted to maintaining a firm foothold in this extreme environment. *Postelsia* is a brown alga (phylum Phaeophyta).

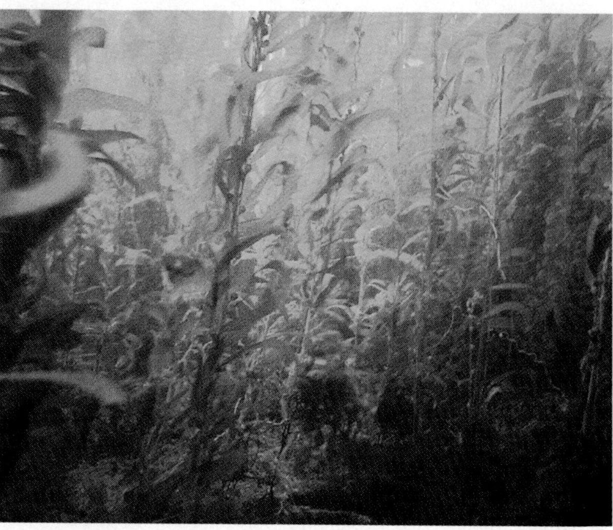

Figure 26.18
Brown algae. The great kelp beds of temperate coastal waters provide habitat and food for a variety of organisms, including many fish caught by humans. The kelps are prodigiously productive. This alga, *Macrocystis,* common along the U.S. Pacific coast, grows to a length of more than 60 m in a single season, the fastest linear growth of any organism. Kelp is a renewable resource reaped by special boats that cut and collect the tops of the algae.

homologous (see Chapter 23). Seaweeds are more closely related to unicellular members of their phyla than to any true plant, and their anatomical complexity evolved independently. The term **thallus** (plural, **thalli;** Gr. *thallos,* "sprout") is used for a seaweed body that is plantlike but lacks true roots, stems, and leaves. A typical seaweed thallus consists of a rootlike **holdfast,** which anchors the alga, and a stemlike **stipe,** which supports leaflike **blades** (Figure 26.17). The blades provide most of the surface for photosynthesis. Some brown algae are equipped with floats, which keep the blades near the water surface.

In addition to these structural adaptations of thalli, some red and brown algae are also endowed with biochemical adaptations to intertidal and subtidal conditions. For example, their cell walls are composed of cellulose and additional gel-forming polysaccharides, accounting for the slimy and rubbery feel of these seaweeds. These substances help cushion the thalli against the agitation of the waves and also help prevent the thalli from drying during low tides. Some red algae in the lower intertidal and subtidal zones incorporate large amounts of calcium carbonate in the cell walls, making them virtually unpalatable to marine invertebrate grazers.

Coastal people, particularly in Asia, harvest seaweeds for food. For example, in Japan and Korea, the brown alga *Laminaria* is used in soups (Japanese "kombu") and the red alga *Porphyra* is used to wrap sushi (Japanese "nori"). Marine algae are rich in iodine and other essential minerals, but much of their organic material consists of unusual polysaccharides that humans cannot digest, which prevents seaweeds from becoming staple foods. They are used mostly for their rich tastes and unusual textures. The gel-forming substances in their cell walls (algin in brown algae, agar and carageenan in red algae) are extracted in commercial operations. These substances are widely used in the manufacture of thickeners for such processed foods as puddings and salad dressing, and as lubricants in oil drilling. Agar is also used as the gel-forming base for microbiological culture media.

We complete our survey of algae with a closer look at Phaeophyta and Rhodophyta, the two phyla that include most seaweeds.

Phaeophyta

The largest and most complex protists are members of the phylum Phaeophyta (Gr. *phaios,* "dusky," "brown"), or **brown algae.** These algae owe their characteristic brown or olive color to pigments in the chloroplasts. All brown algae are multicellular and most are marine, including the most complex seaweeds. The sea palm *(Postelsia)* in Figure 26.17 is an example. Brown algae are especially common along temperate coasts, where the water is cool.

Beyond the intertidal zone in deeper waters live the giant seaweeds known as kelps (Figure 26.18). Their

Figure 26.19
Alternation of generations in the life cycle of *Laminaria*. ①The sporophytes of this seaweed are usually found in water just below the line of lowest tides, attached to rocks with branching holdfasts. ② In early spring, at the end of the main growing season, cells on the surface of the blade develop into sporangia, ③ which produce two types of zoospores by meiosis. ④ One type grows into male gametophytes and the other into female gametophytes. The gametophytes look nothing like the sporophytes, being short, branched filaments that grow on the surface of subtidal rocks, often entangled in one another. ⑤ Male gametophytes release sperm and female gametophytes produce eggs, which remain attached to the gametophyte. ⑥ Sperm fertilize the eggs and ⑦ the zygotes grow into new sporophytes, starting life attached to the remains of the old female gametophyte. The *Laminaria* life cycle is an example of a heteromorphic alternation of generations, in which sporophyte and gametophyte forms are noticeably different in appearance. Contrast this to the isomorphic condition of *Ulva* (Figure 26.16).

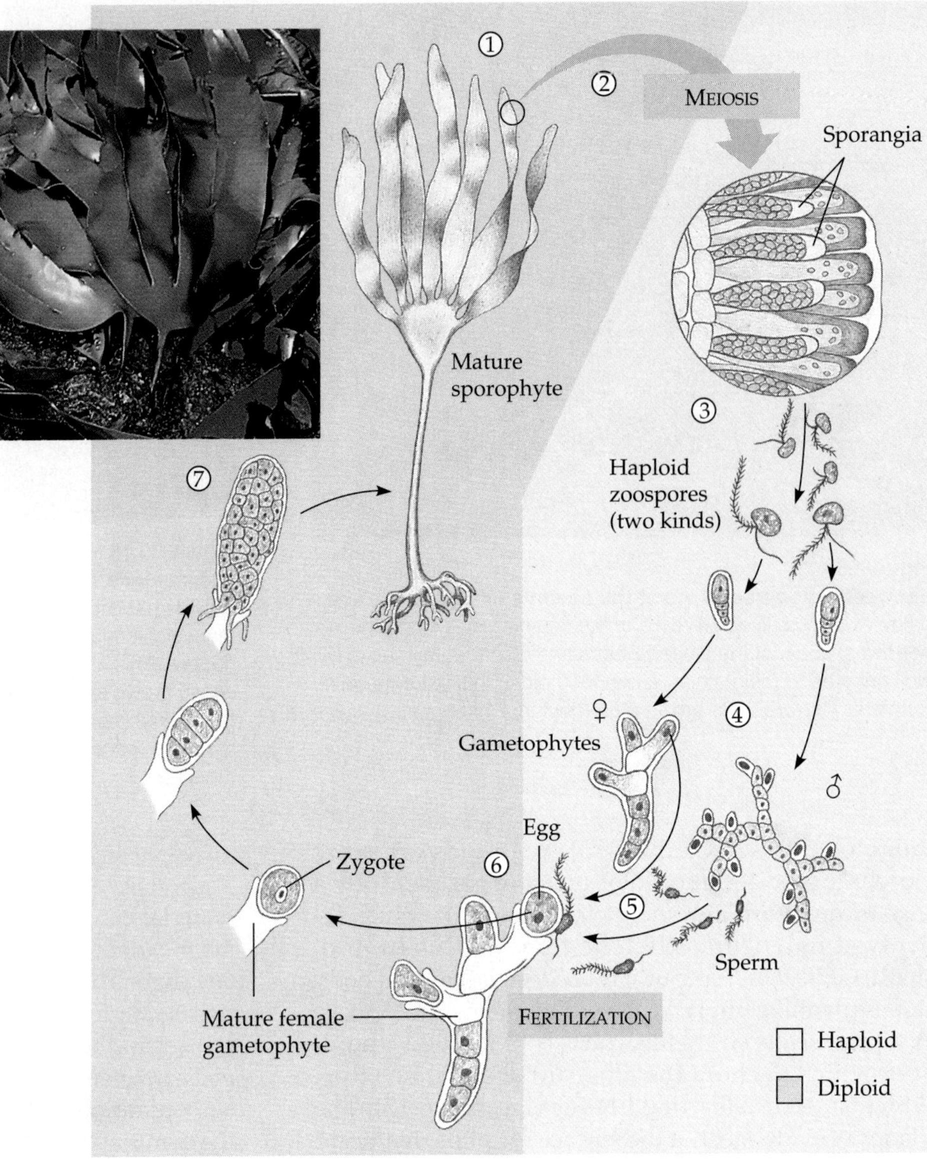

stipes may be as long as 100 m. Sugar produced by photosynthesis in the blades is transported down the stipe in tubular cells similar to those in the vascular tissue of plants. But remember, the anatomical complexity of these giant algae and the complex bodies of true plants arose independently from unicellular ancestors; the resemblance is analogy due to convergent evolution.

A variety of life cycles has evolved in the phylum Phaeophyta. Most species alternate generations, with a haploid gametophyte stage and a diploid sporophyte stage. In some cases, the two generations are **isomorphic,** meaning that the gametophyte and the sporophyte have the same appearance (as in the case of the green alga *Ulva*). In other cases, such as the life cycle of *Laminaria,* the gametophyte and sporophyte are **heteromorphic,** or distinct in appearance (Figure 26.19). In fact, in some species of brown algae, the two generations are so different that it takes considerable detective work to reveal the two types of thalli as members of the same species.

Rhodophyta

The majority of members of Rhodophyta, or **red algae,** live in the ocean, but there are also some freshwater and soil species. Rhodophytes (Gr. *rhodos,* "red") are commonly reddish due to an accessory pigment called phycoerythrin. It belongs to a family of pigments known as phycobilins, found only in red algae and cyanobacteria. Red algae are most abundant in the warm coastal waters of the tropics. Diverse accessory pigments allow some species to absorb filtered wave-

(a)

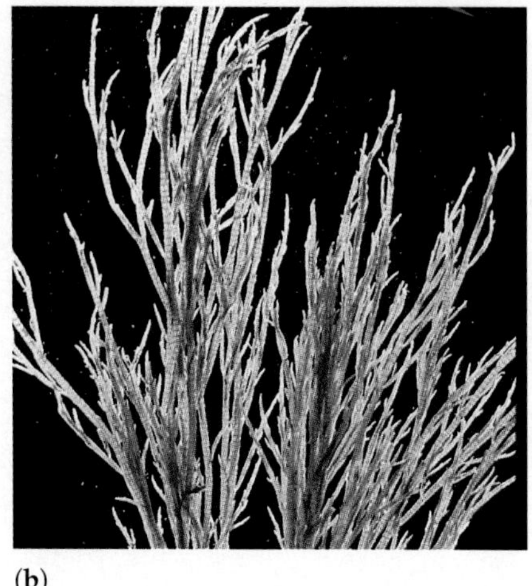

(b)

(c)

Figure 26.20
Red algae. (**a**) Dulce *(Palmaria),* an edible species with a "leafy" form. (**b**) *Polysiphonia,* a filamentous red alga. (**c**) This species is among the coralline algae that contribute to the great coral reefs.

lengths (blues and greens) in deep water. A species of red algae has recently been discovered living near the Bahamas at a depth of more than 260 m.

Despite the name of the phylum, not all rhodophytes are red, and individuals of the same species can change their pigmentation and optimize photosynthesis at different water depths. Individuals of the same species may be almost black in deep water, bright red at more moderate depths, and greenish when living in very shallow water, owing to less phycoerythrin masking the green of the chlorophyll. Some tropical species lack pigmentation altogether and function heterotrophically as parasites on other red algae.

Most red algae are multicellular, and the largest share the name seaweeds with the brown algae, although none of the reds are as big as the giant browns. The thalli of many red algae are filamentous, often branched and interwoven in delicate lacy designs (Figure 26.20). The base of the thallus is usually differentiated as a simple holdfast.

Life cycles are especially diverse among the red algae. Unlike other algal protists, red algae have no flagellated stages in their life cycles. Gametes rely on water currents to get together. Alternation of generations is common in red algae, but details about reproduction and life cycles are known for a very few species.

Evolutionary Relationships Among Algae

The endosymbiotic hypothesis posits that chloroplasts probably had at least three separate origins, perhaps descending from different prokaryotic endosymbionts. In terms of photosynthetic pigments and chloroplast structure, the phyla of algal protists can be divided into three lineages: a red line (Rhodophyta), a green line (Chlorophyta), and a brown line (Phaeophyta and Chrysophyta). Chloroplasts of red algae bear a striking resemblance in thylakoid arrangement and pigment composition to cyanobacteria. According to many advocates of the endosymbiotic hypothesis, red algae descended from a prokaryotic syndicate in which the photosynthetic endosymbionts were cyanobacteria. The chloroplasts of green algae are more similar to a photosynthetic prokaryote named *Prochlorothrix,* a grass-green eubacterium with chlorophyll *b* as well as chlorophyll *a. Prochlorothrix* is one of the few known prokaryotes with chlorophyll *b,* an accessory pigment also found in green algae (and plants). Perhaps an ancient prokaryote similar to *Prochlorothrix* was the endosymbiotic ancestor of the chloroplasts of chlorophytes. The thylakoids of the brown line generally occur in stacks of three, and chlorophyll *c* and characteristic xanthophylls are present as accessory pigments. Biologists are actively searching for modern prokaryotes to fit the bill as relatives of the possible prokaryotic progenitors of brown chloroplasts.

FUNGUSLIKE PROTISTS

Even with five kingdoms instead of only two, the slime molds are a taxonomic enigma. They resemble fungi in appearance and lifestyle, but the similarities are the result of convergence. In their cellular organization, reproduction, and life cycles, slime molds depart from the true fungi and probably have their closest relatives among the amoeboid protists. The two phyla of slime molds described here are now widely regarded as protists, although they are still studied by mycologists (biologists who study fungi; Gr. *mykes,* "fungus"). The

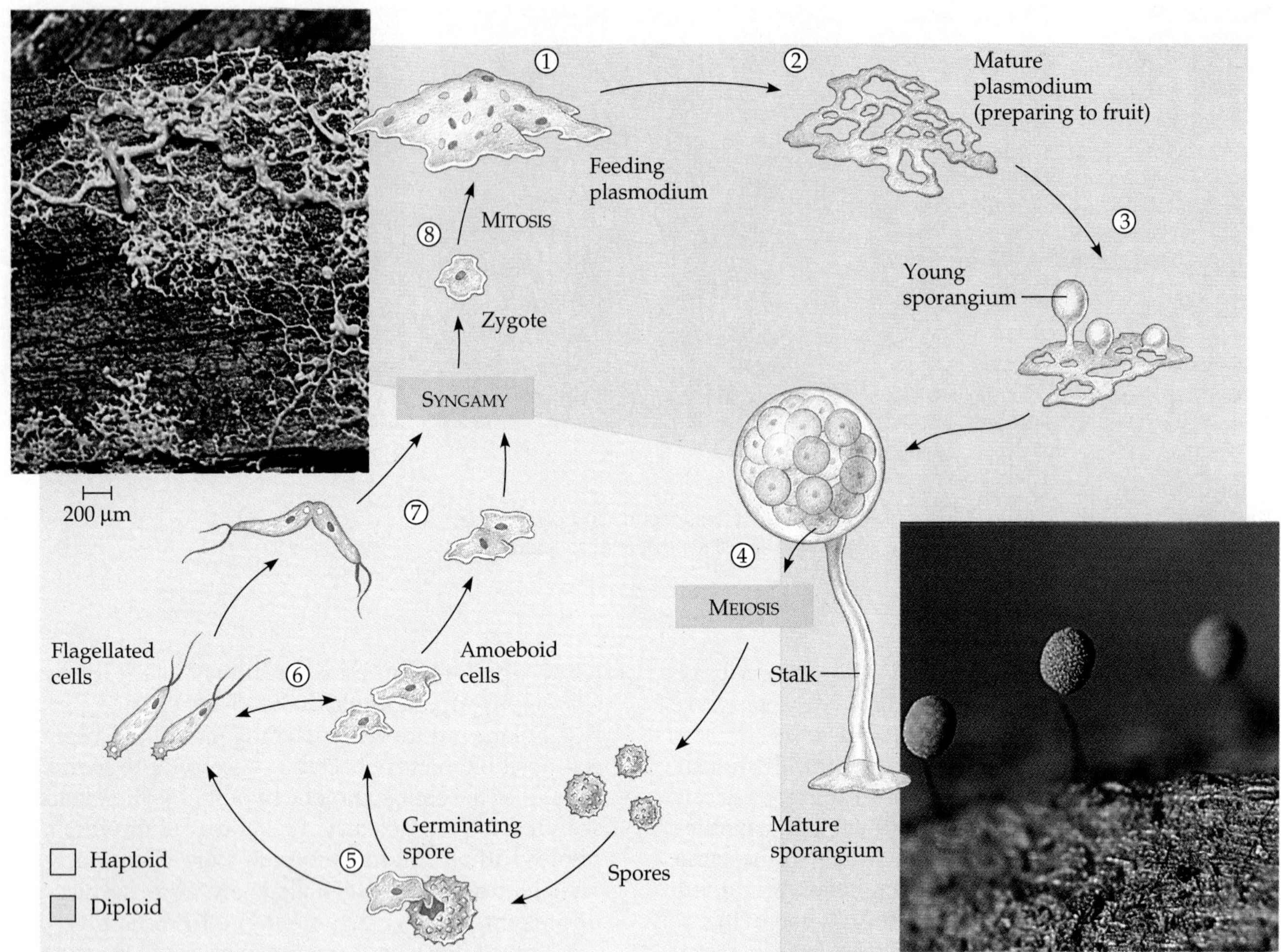

Figure 26.21
The life cycle of a plasmodial slime mold. ① The feeding stage is a multinucleate (coenocytic) plasmodium that lives on organic refuse (see left inset, LM). ② The plasmodium often takes a weblike form, an adaptation that increases the surface area contacting food, water, and oxygen. ③ The plasmodium rounds into a mound and erects stalked fruiting bodies called sporangia ④ when conditions become harsh (see right inset, LM). Within the bulbous tips of the sporangia, meiosis produces haploid spores. ⑤ The resistant spores germinate to become active haploid cells when conditions are again favorable. ⑥ These cells are either amoeboid or flagellated, the two forms readily reverting from one to the other. ⑦ These cells unite in like pairs (flagellated with flagellated and ameboid with ameboid) to form diploid zygotes. ⑧ Repeated division of the nucleus of the zygote by mitosis, without cytoplasmic division, forms a feeding plasmodium and completes the life cycle.

classification used in this text also regards the water molds and chytrids as protists rather than true fungi (Table 26.3, p. 554).

Myxomycota

The phylum Myxomycota consists of **plasmodial slime molds,** which are more attractive than their name implies. Many species are brightly pigmented, usually yellow or orange, but slime molds are not photosynthetic; all are heterotrophs. The feeding stage of the life cycle is an amoeboid mass called a **plasmodium,** which may grow to a diameter of several centimeters (Figure 26.21). Large as it is, the plasmodium is not multicellular; it is a coenocytic mass, a multinucleated continuum of cytoplasm undivided by membranes or walls. In most species, the nuclei of the plasmodium are diploid and divisions are synchronous, with each of thousands of nuclei going through each phase of mitosis at the same time. Because of this characteristic, plasmodial slime molds have been used to study the molecular details of mitosis (see Chapter 11). Within the fine channels of the plasmodium, cytoplasm streams first one way, then the other, in pulsing flows that are beautiful to watch through a microscope. The cytoplasmic streaming apparently helps distribute nutrients and oxygen. The plasmodium engulfs food particles by phagocytosis as it grows by extending pseudopodia through moist soil, leaf mulch, or rotting

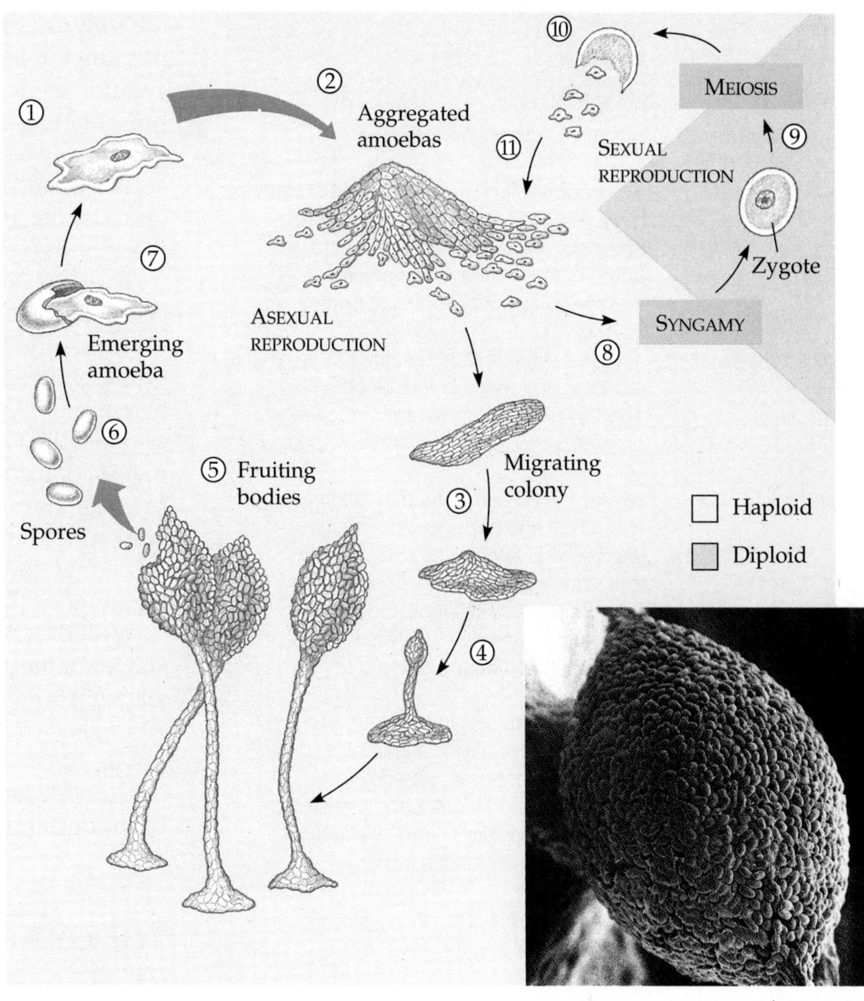

Figure 26.22
The life cycle of a cellular slime mold. ① The feeding stage of the life cycle consists of solitary cells that engulf bacteria while creeping by amoeboid movement through damp compost. ② When food is depleted, the amoeboid cells migrate toward an aggregation center, where hundreds of the cells congregate in response to a chemical attractant they secrete. ③ The sluglike colony of amoeboid cells may migrate as a unit for a while before ④ settling down and developing stalked fruiting bodies that function in asexual reproduction. ⑤ As each fruiting body forms, some cells dry up to form a supportive stalk, while others continue to crawl up over their dried brethren, amass, and then turn into spores. Thus, a cluster of somewhat resistant spores forms at the tip of each fruiting body (see inset, SEM). ⑥ After the spores are released and exposed to a favorable environment, ⑦ amoeboid cells emerge from their protective coats and begin feeding, completing the asexual portion of the life cycle. ⑧ In the sexual phase in *Dictyostelium*, a pair of haploid amoebas fuse to form a zygote, the only diploid stage in the life cycle. ⑨ A zygote becomes a giant cell by consuming surrounding haploid amoebas. The giant cell then becomes surrounded by a resistant wall. ⑩ The encysted giant cell undergoes meiosis, followed by several mitotic divisions. ⑪ New haploid amoebas are released when the cyst ruptures.

logs. If the habitat of a slime mold begins to dry up or there is no food left, the plasmodium ceases growth and differentiates into a stage of the life cycle that functions in sexual reproduction.

Acrasiomycota

Members of Acrasiomycota, the **cellular slime molds,** pose a semantic question about what it means to be an individual organism. The feeding stage of the life cycle consists of solitary cells that function individually. When there is no more food, the cells form an aggregate that functions as a unit (Figure 26.22). Although the mass of cells resembles a plasmodial slime mold, the important distinction is that the cells of a cellular slime mold maintain their identity and remain separated by their membranes.

In addition to not being coenocytes, cellular slime molds differ from plasmodial slime molds in other ways. Cellular slime molds are haploid organisms, whereas the diploid condition predominates in the life cycles of most plasmodial slime molds (compare Figures 26.21 and 26.22). Cellular slime molds have fruiting bodies that function in asexual reproduction. Also, cellular slime molds have no flagellated stages (except in one recently discovered genus).

Oomycota

This phylum includes water molds, white rusts, and downy mildews. These organisms resemble fungi in appearance, consisting of coenocytic hyphae. Oomycota and fungi also have similar nutritional modes. Closer inspection, however, suggests that oomycota are not closely related to the true fungi, and the similarities are analogous, not homologous. These funguslike protists have cell walls most commonly made of cellulose, while the walls of true fungi are made of another polysaccharide, chitin. The diploid condition prevails in the life cycles of most members of Oomycota but is reduced in true fungi. Biflagellated cells occur in the life cycles of oomycotes while true fungi lack flagella.

Oomycota means "egg fungi," a reference to the

Table 26.3 A Partial List of Funguslike Protists (Nutrition Usually by Absorption; Produce Funguslike Spores)

Phylum	Brief Description
Myxomycota	Plasmodial slime molds: feeding stage an amoeboid coenocyte that obtains food by absorption and phagocytosis; mature individuals reproduce by turning into spore-producing bodies
Acrasiomycota	Cellular slime molds: feeding stage solitary, amoeboid cells that obtain food by phagocytosis; mature colonies aggregate to release spores
Oomycota	Water molds and terrestrial relatives: saprobes and parasites; coenocytic, filamentous bodies resembling fungi, but with walls of cellulose; asexual reproduction by biflagellated spores; sexual reproduction involves nonflagellated oospores
Chytridiomycota	Saprobes and parasites on a variety of substrates, usually in aquatic or moist habitats; coenocytic; one-celled chytrids to simple branched filaments; cell walls of chitin; spores, gametes, and zygotes are uniflagellate

mode of sexual reproduction in water molds. A relatively large egg cell is fertilized by a smaller "sperm nucleus" to form a resistant zygote (Figure 26.23).

Most water molds are saprobes that grow as cottony masses on dead algae and animals, mainly in fresh water. They are important decomposers in aquatic ecosystems. There are also parasitic water molds, such as those that grow on the skin and gills of fish in ponds or aquariums, but they usually attack only injured tissue. White rusts and downy mildews are close relatives of water molds but generally live on land as parasites of plants. They are dispersed primarily by windblown spores, but they also form flagellated zoospores at some point during their life cycles. Some of the most devastating plant pathogens are members of the Oomycota, including the downy mildew that threatened the French vineyards in the 1870s and the species that causes late potato blight, which contributed to the Irish famine in the nineteenth century.

Chytridiomycota

Most representatives of this phylum are microscopic. The unicellular, coenocytic chytrids are the simplest of the funguslike protists. More complex members of the phylum Chytridiomycota consist of branched, coenocytic filaments called hyphae (Figure 26.24). The phylum includes both free-living saprobes and parasites that invade algae, pollen grains of plants, and some insects. (One species of Chytridiomycota infects mosquito larvae and has been investigated as a control for these pests, without much success.)

The Chytridiomycota and true fungi share some key characteristics, including the branched (hyphal) structure of some members and cell walls made of chitin (see Chapter 5). Some researchers speculate that the true fungi evolved from Chytridiomycota, a hypothesis we will evaluate in Chapter 27. Chytrids and other members of this phylum are classified in this text as protists mainly because they produce motile spores and gametes having flagella, a trait uncharacteristic of true fungi. Many experts, however, classify chytrids in the kingdom Fungi.

With the Chytridiomycota, you have completed our survey of the diverse protists. We have learned that not all protists are unicellular; in fact, multicellular organization has evolved independently in several protistan phyla. In the last section of this chapter, we examine one model for how multicellular eukaryotes evolved from unicellular ancestors.

THE ORIGINS OF MULTICELLULARITY

The origin of eukaryotic organization could have ignited an explosion of biological diversification. More variations are possible for complex structures than for simpler ones. Unicellular protists, which are organized on the complex eukaryotic plan, are much more diverse in morphology than the simpler prokaryotes. Protists in which multicellular bodies evolved broke through another threshold in structural organization and became the stock for new waves of adaptive radiations. Among the products were the ancestors of plants, fungi, and animals. The most widely held view is that links between multicellular organisms and their unicellular ancestors were colonies, or loose aggregates of interconnected cells. Multicellular algae, plants, fungi, and animals probably descended from several lineages of colonial protists that formed by amalgamations of individual cells. The evolution of multicellularity from colonial aggregates involved increasing cellular specialization and division of labor. Initially, in colonial aggregates ancestral to multicellular algae, plants, and animals, all cells may have been motile with flagella. As cells in the colony tended to become increasingly intimate and interdependent, some of the cells may have lost their flagella and become more proficient in performing functions other than locomotion.

Germ tube
①
②
Cyst
⑤
⑥
Antheridial hyphae with sperm nuclei
Oogonium
Egg nucleus
MEIOSIS
ASEXUAL REPRODUCTION
Zoospore
④
③
Zoosporangium
SEXUAL REPRODUCTION
FERTILIZATION
⑦
Release of zoospores
⑨
Zygotes (oospores)
⑧
Zygote germination
Haploid
Diploid

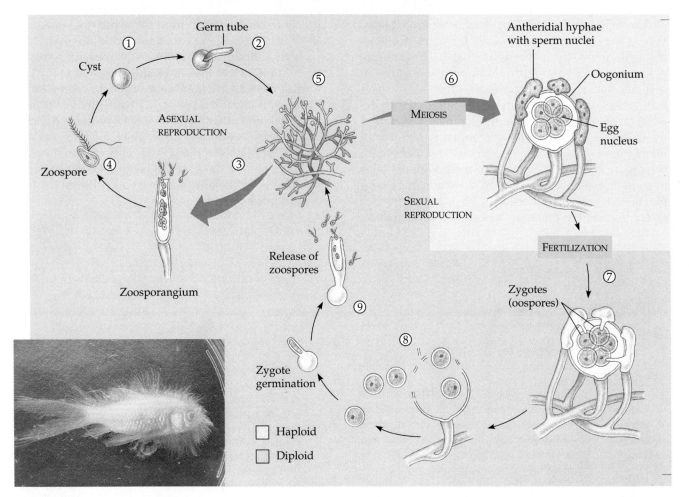

Figure 26.23
The life cycle of a water mold. Water molds commonly help decay dead insects, fish, and other animals. ① Encysted zoospores land on a new substrate and ② germinate to grow into a tufted body of coenocytic hyphae. ③ Within several days, the ends of the hyphae form tubular zoosporangia, ④ each of which produces about thirty biflagellated zoospores asexually (the hyphal mass on the goldfish is in the early zoosporangial stage). ⑤ Several days later, the organism begins to form sexual structures. ⑥ Meiosis produces eggs within structures called oogonia. On separate branches of the same or different individuals, meiosis produces several haploid "sperm nuclei" contained within compartments called antheridial hyphae. ⑦ These hyphae grow like hooks around the oogonium and deposit their nuclei through fertilization tubes that lead to the eggs. The resulting zygotes (oospores) may develop resistant walls but are also protected within the walls of the old oogonia. ⑧ After a period of dormancy, during which the oogonium wall usually disintegrates, ⑨ the oospores germinate to form short hyphae tipped by zoosporangia, and the cycle is completed.

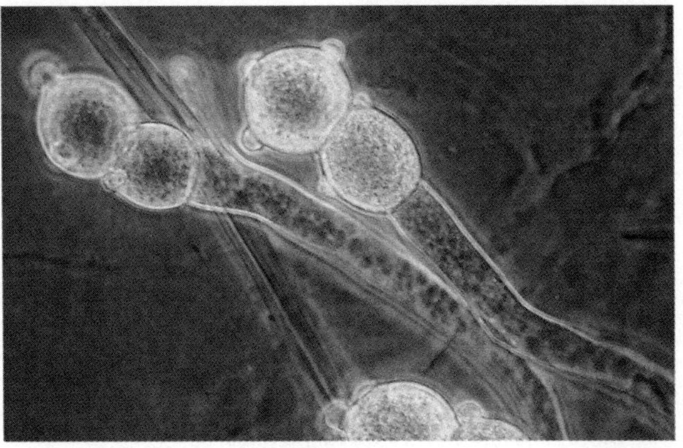

Figure 26.24
A hyphal member of the Chytridiomycota. The hyphae, branched filaments, expose a large surface to the surrounding medium, from which the organism absorbs nutrients. This genus, *Allomyces,* is often grown in culture as a research model to study development in relatively simple organisms (LM).

Another early form of division of labor probably involved the separation of sex cells (gametes) from somatic (nonreproductive) cells. We see this type of specialization and intercellular cooperation today in several colonial protists, such as the green alga *Volvox*. Gametes are specialized for reproduction, and they depend on somatic cells while developing. Evolution of the extensive division of labor required to perform all the nonreproductive functions in multicellular organisms as we know them involved many additional steps in somatic cell specialization. In seaweeds, for example, there is extensive division of labor among the different tissues that make up their thalli. Multicellular life more complex than these algae did not appear until about 700 million years ago during the twilight of the Precambrian era. A variety of animal fossils has been found in late Precambrian strata, and many new forms evolved after the Paleozoic era dawned with the Cambrian period, about 570 million years ago. Seaweeds and other complex algae were also abundant in Cambrian oceans and lakes. The land, however, was barren. About 400 million years ago, certain green algae living along the edges of lakes gave rise to primitive plants. In the next chapter, we will trace the long evolutionary trek of plants onto land.

STUDY OUTLINE

1. Protists are a diverse group of eukaryotic organisms. Most are unicellular, but colonial and simple multicellular forms also exist.
2. As they evolved from prokaryotes, the earliest protists acquired the structures and processes unique to eukaryotic life. Thus, they are the predecessors of not only their modern protistan descendants but also the other three eukaryotic kingdoms.

Characteristics of Protists (p. 534)

1. Protists are found wherever there is water, living as plankton, submerged bottom-dwellers, or inhabitants of moist soil or body fluids of other organisms.
2. Virtually all engage in aerobic metabolism. They are photoautotrophs, heterotrophs, or mixotrophs.
3. Many have cilia or flagella, of the eukaryotic type, during some part of the life cycle.
4. All protists reproduce asexually; some also can reproduce sexually. Many can survive harsh conditions by forming resistant cysts. Asexual reproduction generally occurs early as an organism colonizes a new environment. Sexual reproduction often occurs when environments deteriorate.

The Origin of Eukaryotes (pp. 534–536)

1. The eukaryotic cell made its debut with the advent of the first protist.
2. Acritarch fossils, over 1.5 billion years old, are believed to be the first evidence of eukaryotic life.
3. The autogenous hypothesis of the origin of eukaryotes maintains that eukaryotic cells evolved from invaginations of the prokaryotic plasma membrane that subsequently differentiated into the endomembrane system.
4. Another hypothesis, the endosymbiotic model, envisions eukaryotic cells arising as a result of prokaryotes taking up residence inside other prokaryotes. Chloroplasts and mitochondria are thus purported descendants of photosynthetic symbionts and aerobic, heterotrophic symbionts, respectively.

Boundaries of the Kingdom Protista (pp. 536–537)

1. Ancient protists gave rise to more than one multicellular line, making their classification complicated and controversial. The kingdom Protista includes multicellular descendants that do not clearly belong to one of the other three eukaryotic kingdoms.
2. Protists exhibit the most diverse spectrum of structure and life cycles of all known organisms.

Protozoa (pp. 537–542)

1. The simplest protozoa are the rhizopods, unicellular amoebas and their relatives, all of which move by cellular extensions called pseudopodia.
2. Actinopods are protozoa with slender, raylike axopodia that help them float and feed. Heliozoan and radiolarian species are components of plankton in freshwater and marine environments, respectively.
3. The marine forams are famous for their lovely porous shells, through which strands of cytoplasm extend for swimming, shell formation, and feeding.
4. Apicomplexans are parasitic protozoa with complex life cycles characterized by both sexual and asexual stages that often require two or more host species.
5. Zoomastigotes are a varied group of flagellated heterotrophs.
6. Ciliates use cilia to move and feed. Ciliates are among the most complex cells.

Algae (pp. 542–551)

1. Algae are aquatic, plantlike eukaryotes that possess chlorophyll. A few representatives are heterotrophic. Classification characteristics include variations in chloroplast structure, accessory pigments, cell walls, flagella, and forms of food storage.
2. Dinoflagellates, abundant in marine plankton, are either photosynthetic or heterotrophic. Most are unicellular, but some forms are colonial. They move in a spinning motion by the beating of flagella.

3. Chrysophytes are the biflagellated freshwater golden algae, named for the color of their carotenoid pigments.
4. Bacillariophytes are the diatoms, primarily unicellular organisms with unique glasslike walls of silica.
5. Euglenophytes include *Euglena* and its relatives, a group that blurs the distinction between plantlike and animal-like characteristics. Most species are photosynthetic, but others are either mixotrophic or exclusively heterotrophic.
6. Chlorophytes, the green algae, are the likely ancestors of the plant kingdom. Diverging evolutionary pathways have generated an assortment of unicellular, colonial, multinucleate, and multicellular species that live in a variety of environments. Complicated life histories include both sexual and asexual reproductive stages, including examples of alternation of generations.
7. Seaweeds include thallus-forming, marine species in the green, brown, and red algae. They are well adapted to life along the turbulent margins of the oceans.
8. Phaeophytes are multicellular, primarily marine, brown algae. They are the largest and most morphologically complex protists. Most species show some type of alternation of generations.
9. Rhodophytes, the red algae, possess the red accessory pigment phycoerythrin, which may or may not be masked by pigments of other colors. Red algae are multicellular, often lacy protists that reproduce sexually, usually as part of an alternation of generations.

Funguslike Protists (pp. 551–554)

1. These protists show basic differences from true fungi in cellular organization and life cycles.
2. The myxomycotes are the plasmodial slime molds. These diploid organisms feed by means of a coenocytic amoeboid plasmodium capable of differentiating into sexually reproducing sporangia when moisture or food is scarce.
3. Acrasiomycotes, the cellular slime molds, include a group of haploid organisms that lead unicellular lives until food is depleted. They then aggregate into a multicellular amoeboid mass that erects asexual fruiting bodies.
4. Oomycotes, the water molds, have cell walls of cellulose and biflagellated stages in their life cycles, traits that don't fit the definition of fungi used in this text. The phylum also includes white rusts and downy mildews, some of which are serious plant pathogens.
5. Chytridiomycotes are usually small coenocytic unicells that are saprobic or parasitic on a variety of aquatic substrates. Their walls are composed of chitin, and they have uniflagellated reproductive cells.

The Origins of Multicellularity (pp. 554–556)

1. Multicellularity evolved many times in the kingdom Protista apparently by the aggregation of individual cells into colonial forms, in which cellular specialization and division of labor developed.

SELF-QUIZ

1. Which of the following is the most accurate general description of the kingdom Protista?
   a. eukaryotic, unicellular organisms that may be photosynthetic or heterotrophic
   b. eukaryotic, heterotrophic and/or photosynthetic, unicellular or simple multicellular organisms that are different enough from multicellular plants, fungi, or animals to be placed in this diverse group
   c. eukaryotic plankton, which may be flagellated at some point in their life cycle and which reproduce asexually
   d. eukaryotic, photosynthetic or heterotrophic organisms that inhabit moist environments, form resistant cysts, and reproduce with flagellated gametes
   e. relatively simple versions of plants, animals, and fungi
2. Which of the following protozoa are *incorrectly* paired with their description?
   a. rhizopods—naked and shelled amoebas
   b. actinopods—planktonic with slender, raylike axopodia
   c. forams—flagellated heterotrophs, free-living or symbiotic
   d. apicomplexans—parasites with complex life cycles
   e. ciliates—complex, unicellular organisms with macronucleus and micronuclei
3. Which of the following algae are *incorrectly* paired with their description?
   a. dinoflagellates—marine plankton, whirling, spinning movement, characteristic shell
   b. chrysophytes—golden algae, xanthophylls predominant, flagellated, freshwater plankton
   c. bacillariophytes—diatoms, two-piece shells of silica
   d. phaeophytes—multicellular brown algae, seaweeds
   e. rhodophytes—cause red tides, xanthophylls predominant
4. Plants are believed to have evolved from the
   a. euglenophytes, because they have chlorophyll *a* and *b*
   b. dinoflagellates, because they have similar storage products and cellulose cell walls
   c. chlorophytes, because of similarities in chloroplasts and accessory pigments
   d. phaeophytes, because they have specialized body parts and an alternation of generations
   e. rhodophytes, because they have carotenoids and use a glycogenlike starch as a storage product
5. Unlike plasmodial slime molds, cellular slime molds
   a. exhibit phagocytosis
   b. form fruiting bodies
   c. have more than one nucleus per cell
   d. are haploid organisms except for the giant-celled zygote
   e. can move as an amoeboid mass
6. Acritarchs are
   a. symbiotic associations between algae and fungi
   b. eukaryotes that evolved by the autogenous model
   c. the oldest putative eukaryotic fossils
   d. fossils that resemble the ruptured coats of cyanobacteria
   e. diatom shells

7. Which of the following is an *incorrect* statement about the possible endosymbiotic origins of chloroplasts and mitochondria?
   a. They are the appropriate size to be descendants of bacteria.
   b. They contain their own genome and produce all their own proteins.
   c. They contain circular DNA molecules not associated with histones.
   d. Their membranes have enzymes and transport systems that resemble those found in the plasma membranes of prokaryotes.
   e. Their ribosomes are more similar to those of eubacteria than to those of eukaryotes.
8. The endosymbiotic hypothesis maintains that chloroplasts may have had at least three different origins because
   a. the chloroplasts of green plants show three distinct morphologies and pigment compositions
   b. the pigment composition and thylakoid arrangement of cyanobacteria are most like the chloroplasts of green plants
   c. molecular systematics link the red, green, and brown line to three different present-day eubacteria
   d. the phyla of algal protists can be divided into three lineages based on photosynthetic pigments and chloroplast structure
   e. there are three distinct groups of photosynthetic prokaryotes
9. The organism that caused the Irish potato famine is
   a. an actinopod
   b. an apicoplexan
   c. an oomycotan
   d. a plasmodial slime mold
   e. a cellular slime mold
10. A student collected a flask of pond water and placed it near a window. A brownish-green scum collected on the side of the flask facing the light. When the flask was turned around, the scum moved to the side facing the window. Which of the following phyla are most likely represented in this pond scum?
   a. Zoomastigophora and Ciliophora
   b. Chrysophyta and Euglenophyta
   c. Chlorophyta and Phaeophyta
   d. Euglenophyta and Acrasiomycota
   e. Phaeophyta and Bacillariophyta

CHALLENGE QUESTIONS

1. Different genes are expressed at different stages of the life cycle of the malaria-causing apicomplexan *Plasmodium,* which causes different proteins to appear on the outer coat of the infecting cells. Sporozoites are injected by mosquitoes and travel in the blood to liver cells, where they continue their life cycle. It was discovered that the sporozoites produce protein coats that are sloughed off and continuously replaced. Host antibodies can attack these proteins by specific complementary binding. How does the continual sloughing and replacing of the protein coat work as an adaptive mechanism to prevent immune destruction of the sporozoite before it gets inside the liver cell, where it is protected from bloodborne antibodies?
2. What is the taxonomic justification for placing algae, protozoa, and funguslike protists in the kingdom Protista? Can you suggest an alternative classification scheme? What are the advantages and disadvantages of your scheme?

SCIENCE, TECHNOLOGY, AND SOCIETY

1. The burning of fossil fuels is increasing the amount of carbon dioxide in the atmosphere, and many experts think this will intensify the greenhouse effect and lead to a warming of the global climate. The photosynthesis of diatoms and other microscopic algae in the oceans uses enormous quantities of carbon dioxide. These algae need minute quantities of iron, and some researchers suspect that a shortage of iron may limit photosynthesis, especially in the Antarctic Ocean. Oceanographer John Martin suggests that one way to slow global warming might be to fertilize the ocean with iron. This would stimulate growth of diatom populations, which would remove carbon dioxide from the air. Martin estimates that a single supertanker of iron dust, spread over a wide enough area, could reduce CO_2 levels significantly. Do you think this would be worth a try? Why or why not? Do you see any reasons to be cautious? Why?
2. The malarial parasite is able to evade the immune system, so it has been difficult to develop a malaria vaccine (see Challenge Question 1). An additional problem is that fewer scientists are engaged in malaria research, and less money is spent on it, than on diseases such as cystic fibrosis, which affects far fewer people than does malaria. What are possible reasons for this imbalance?

FURTHER READING

Knoll, A. "The Early Evolution of Eukaryotes: A Geological Perspective. " *Science,* May 1, 1992. What can Precambrian rocks tell us about the origin of eukaryotic cells?

Kunzig, R. "Invisible Garden." *Discover,* April 1990. The global importance of microscopic algae.

Mann, C. "Lynn Margulis: Science's Unruly Earth Mother." *Science,* April 19, 1991. A profile of a remarkable scientist who is the chief advocate of the endosymbiotic theory.

Margulis, L., and R. Gurrero. "Kingdoms in Turmoil." *New Scientist,* March 23, 1991. Current debates about the phylogeny of eukaryotes.

Penny, D., and O'Kelly, C. "Seeds of a Universal Tree." *Nature,* March 14, 1991. An endosymbiotic history of chloroplasts.

Rennie, J. "Proteins 2, Malaria 0." *Scientific American,* July 1991. Encouraging news about a malaria vaccine.

Saffo, M. B. "New Light on Seaweeds." *BioScience,* October 1987. The scientific method reasserts itself, as a century-old hypothesis explaining vertical distribution of marine algae turns out to be too simple.

Vidal, G. "The Oldest Eukaryotic Cells." *Scientific American,* February 1984. Evidence that eukaryotes originally evolved in the form of unicellular plankton.

27 Plants and the Colonization of Land

An Introduction to the Plant Kingdom
The Move onto Land
Seedless Vascular Plants
Terrestrial Adaptations of Seed Plants
Gymnosperms
Angiosperms
The Value of Plant Diversity

It is difficult to picture the continents barren, totally uninhabited by life of any form. But that is how we must imagine Earth for almost the first 90% of the time life has existed. Life was cradled in the seas and ponds, and there it evolved in confinement for three billion years. The long evolutionary pilgrimage onto land finally began about 425 million years ago. Plants led the way, followed by herbivorous animals and their predators. The terrestrial communities founded by green plants transformed the landscape (Figure 27.1).

The evolutionary history of the plant kingdom is a story of increasing adaptation to changing terrestrial conditions. That is the historical context in which this chapter surveys the current diversity of plants and traces their origins.

Figure 27.1
A terrestrial community. This landscape, on the shore of Trapper's Lake, Colorado, is colored by flowering plants (angiosperms) and conifers (gymnosperms). In this chapter, we will trace the evolution of plant diversity from the algal ancestors that first moved onto land.

AN INTRODUCTION TO THE PLANT KINGDOM

General Characteristics of Plants

All plants, as defined in this text, are multicellular eukaryotes that are photosynthetic autotrophs. However, not all organisms with these characteristics are plants; such characteristics also apply to some algae, which we have classified as protists (see Chapter 26).

Plants as we are defining them are nearly all terrestrial organisms, although some plants have returned secondarily to water during their evolution. Living on land poses very different problems from living in the water. As plants have adapted to the terrestrial environment, complex bodies with extensive specialization of cells for different functions have evolved. Aerial parts of most plants, such as stems and leaves, are coated with a waxy **cuticle** that helps prevent desiccation, a major problem on land. Gas exchange cannot occur across the waxy surfaces, but carbon dioxide and oxygen diffuse between the interior of leaves and the surrounding air through microscopic pores on the leaf's surface called **stomata** (singular, **stoma**). Besides their special adaptations for terrestrial life, land plants share many features with their progenitors, the green algae. For example, their photosynthetic cells contain chloroplasts having the pigments chlorophyll *a*, chlorophyll *b*, and a variety of yellow and orange carotenoids. Plant cells also have walls, and the staple

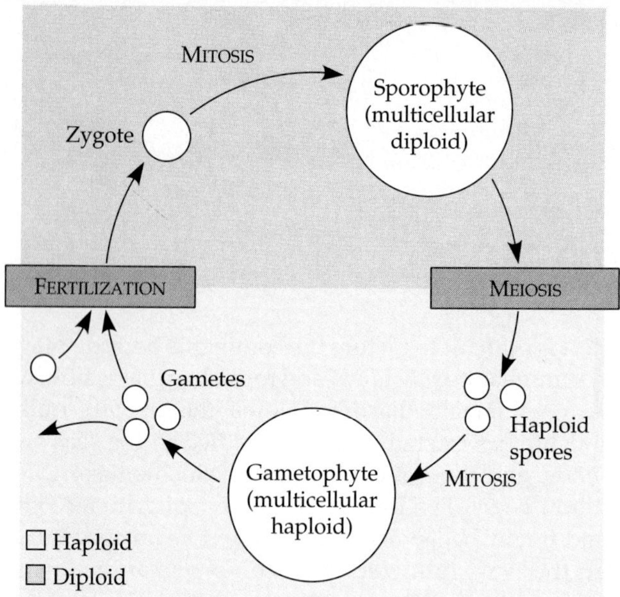

Figure 27.2
Alternation of generations: a generalized scheme. The life cycles of all plants include a gametophyte (haploid generation) and a sporophyte (diploid generation). The two generations alternate, each producing the other. The two plant forms are named for the type of reproductive cells they produce: Gametophytes form gametes by mitosis; sporophytes produce spores by meiosis. Spores develop directly into organisms. Gametes *cannot* develop into organisms but unite, sperm and egg, to form a zygote. The zygote, in turn, gives rise to an organism.

material of their walls is cellulose. Carbohydrate is stored in the form of starch, generally in chloroplasts and other plastids.

A Generalized View of Plant Reproduction and Life Cycles

The move onto land paralleled a new mode of reproduction. In contrast to the reproductive style of seaweeds and other algae, gametes now had to be dispersed in a nonaquatic environment, and embryos, like mature body structures, had to be protected against desiccation.

Nearly all plants reproduce sexually, and most are also capable of asexual propagation. Plants produce their gametes within **gametangia,** organs having protective jackets of sterile (nonreproductive) cells that prevent the delicate gametes from drying out during their development. The egg is fertilized within the female organ, where the zygote develops into an embryo that is retained for some time within the jacket of protective cells.

In the life cycles of all plants, an alternation of generations occurs, in which haploid gametophytes and diploid sporophytes take turns producing one another (Figure 27.2; also see Chapters 12 and 26). In the life cycles of all extant plants, the sporophyte and gametophyte generations differ in morphology; that is, they are heteromorphic. In all plants but the mosses and their relatives, the diploid sporophyte is the larger and more noticeable individual. The physiology of plant reproduction is covered in Chapter 34; in this chapter, we look in some detail at representative plant life cycles. It is valuable to understand these life cycles for two reasons. First, they clarify one of the main trends in plant evolution, toward reduction of the haploid generation and dominance of the diploid. Second, in many cases, features of the life cycle, such as the replacement of flagellated sperm by pollen, can be interpreted as evolutionary adaptations to a terrestrial environment.

Some Highlights of Plant Evolution

The fossil record chronicles four major periods of plant evolution, which are also evident in the diversity of contemporary plants (Figure 27.3). Each period followed the evolution of structures that opened new adaptive zones on the land (see Chapter 23).

The first period of evolution was associated with the origin of plants from aquatic ancestors, probably green algae, during the mid-Silurian period, about 425 million years ago. The first terrestrial adaptations included a cuticle and jacketed gametangia that protected gametes and embryos. **Vascular tissue,** consisting of cells joined into tubes that transport water and nutrients throughout the plant body, also evolved relatively early in plant history. Most mosses lack vascular tissue, and hence they are sometimes categorized as nonvascular plants. However, water-conducting tubes *are* present in some mosses. Researchers have not yet resolved whether this vascular tissue in mosses is analogous (separate evolutionary development) or homologous (common evolutionary development) to the water-conducting tissue of other plants. (See Chapter 23 to review the concepts of analogy and homology.) Evidence from molecular systematics (see Chapter 23) does suggest an early split between mosses and the lineages leading to ancient ferns and other vascular plants.

The second major period of plant evolution was the diversification of vascular plants during the early Devonian period, about 400 million years ago. The earliest vascular plants lacked seeds, a condition still represented by ferns and a few other groups of vascular plants.

The third major period of plant evolution began with the origin of the seed, a structure that advanced

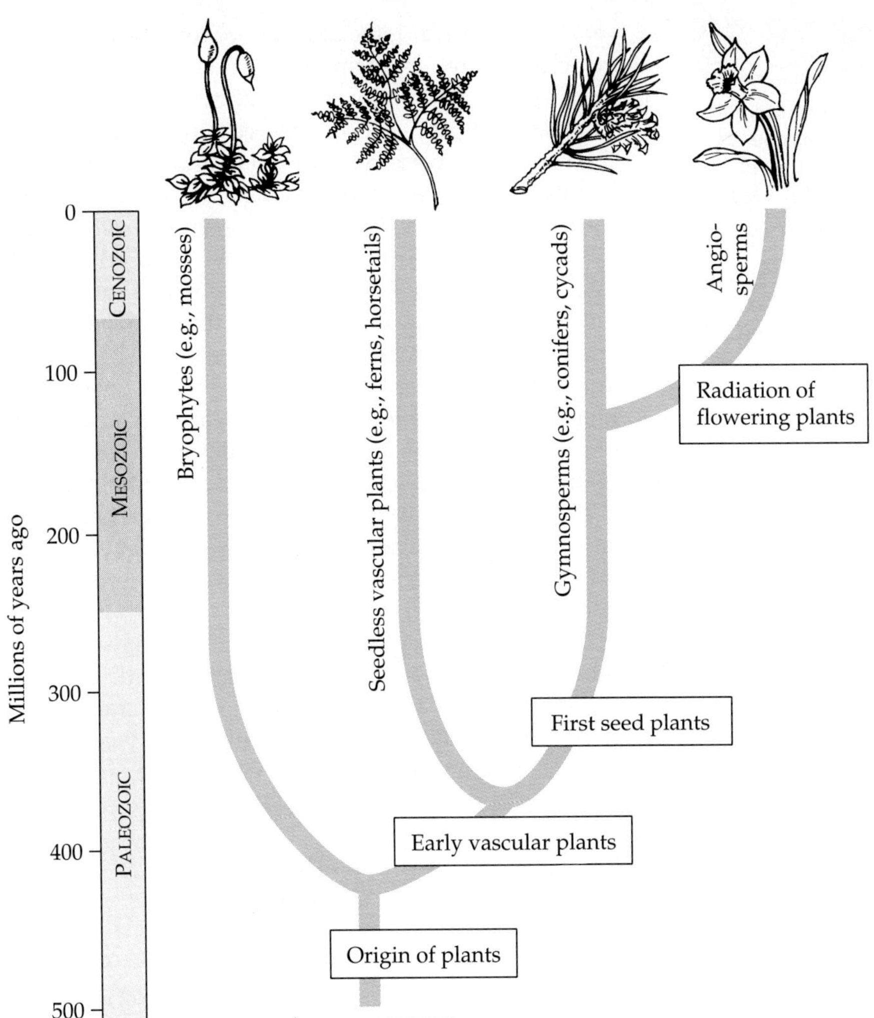

Figure 27.3
Some highlights of plant evolution. The origins of plants from green algae, the adaptive radiation of early vascular plants, the emergence of seed plants, and the origin and diversification of flowering plants (angiosperms) are four important chapters in the history of the plant kingdom. Modern representatives of major plant lineages are illustrated at the top of this evolutionary tree.

the colonization of land by further protecting plant embryos from desiccation and other hazards. A **seed** consists of an embryo packaged along with a store of food within a protective covering. The first vascular plants with seeds arose about 360 million years ago, near the end of the Devonian period. They bore their seeds as naked structures not enclosed in any specialized chambers. Early seed plants gave rise to many types of **gymnosperms** (Gr. *gymnos,* "naked," and *sperma,* "seed"), including the conifers, which are the pines and other plants with cones. Gymnosperms coexisted with ferns and other seedless plants in great forests that dominated the landscape for more than 200 million years.

The fourth major episode in the evolutionary history of plants was the emergence of flowering plants during the early Cretaceous period, about 130 million years ago. The flower is a complex reproductive structure that bears seeds within protective chambers called ovaries, which contrasts with the bearing of naked seeds by gymnosperms. The great majority of contemporary plants are flowering plants, or **angiosperms** (Gr. *angion,* "container," referring to the ovary, and *sperma,* "seed").

Classification of Plants

Plant biologists use the term **division** for the major plant groups within the plant kingdom. This taxonomic category corresponds to phylum, the highest unit of classification within the animal kingdom. Divisions, like phyla, are further subdivided into classes, orders, families, and genera.

The classification scheme used in this text recognizes twelve divisions within the kingdom Plantae. In addition to listing the twelve divisions, Table 27.1 associates them with the stages of evolution discussed in the preceding section. Although these broader groupings do not have the status of formal taxonomic categories, they help fit the current diversity of plants into the historical context of a long evolutionary journey onto land.

Table 27.1 A Classification of Plants

	Common Name	Approximate Number of Extant Species
Nonvascular Plants*		
Division Bryophyta	Mosses	10,000
Division Hepatophyta	Liverworts	6,500
Division Anthocerophyta	Hornworts	100
Vascular Plants		
Seedless Plants		
Division Psilophyta	Whiskferns	10–13
Division Lycophyta	Club mosses	1,000
Division Sphenophyta	Horsetails	15
Division Pterophyta	Ferns	12,000
Seed Plants		
■ Gymnosperms		
Division Coniferophyta	Conifers	550
Division Cycadophyta	Cycads	100
Division Ginkgophyta	Ginkgo	1
Division Gnetophyta	Gnetae	70
■ Angiosperms		
Division Anthophyta	Flowering plants	235,000

*Use of the term *nonvascular* for mosses must be qualified: Water-conducting tissue is present in some species.

THE MOVE ONTO LAND

The Case for Green Algae as the Ancestors of Plants

Plants probably evolved from green algae (chlorophytes). These protists share the following set of key characteristics with plants:

1. Green algae and plants have, in addition to chlorophyll *a*, the same accessory photosynthetic pigments, including chlorophyll *b* and beta-carotene.
2. The chloroplasts of many green algae have their thylakoid membranes stacked locally into grana like those of plants.
3. The cell walls of most green algae, as of plants, are made of cellulose.
4. Green algae, like plants, store their carbohydrate reserves in the form of starch.
5. In some green algae, as in plants, vesicles derived from Golgi bodies form a cell plate that divides the cytoplasm during cytokinesis.

The specific green algae that first colonized land are unknown; the soft thalli of algae left few fossils. The most likely candidates are filamentous chlorophytes, which carpeted the fringes of lakes or salt marshes. During the Silurian period, when the ancestors of plants first established themselves on the beachheads, the continents were relatively flat, and they may have been subject to periodic flooding and draining. As water levels changed during the seasons or over longer cycles, natural selection would have favored those algae that could survive through periods when they were not submerged. Some evolutionary lineages accumulated adaptations that made it possible to live permanently above the water line, including waxy cuticles and jacketed reproductive organs, two hallmarks of plants. The evolutionary novelties of the first plants opened an adaptive zone that had never before been occupied. The new frontier was spacious, the bright sunlight was unfiltered by water and algae, the soil was rich in minerals, and, at least at first, there were no herbivores on land.

In the context of this move onto land, let's examine mosses, liverworts, and hornworts, three divisions of plants that probably diverged along separate evolutionary pathways before the origin of the vascular tissue characteristic of the other plant divisions.

Bryophytes: Divisions Bryophyta, Hepatophyta, and Anthocerophyta

Until recently, the nonvascular plants—mosses, liverworts, and hornworts—were grouped together in a single division, Bryophyta (Gr. *bryon*, "moss"). This text adopts the current view that mosses, liverworts, and hornworts deserve separate divisions because they are probably not closely related. In this taxonomy, only mosses belong to the formal division Bryophyta. However, most plant biologists still use the common name bryophyte to refer to liverworts and hornworts as well as mosses. This usage is appropriate, because these three plant groups share some key characteristics.

Bryophytes display the two adaptations that first made the move onto land possible. They are covered by a waxy cuticle that helps the body retain water, and their gametes develop within gametangia. The male gametangium, known as an **antheridium,** produces flagellated sperm. In each female gametangium, or **archegonium,** one egg is produced. The egg is fertilized within the archegonium, and the zygote develops into an embryo within the protective jacket of the female organ.

Even with their cuticles and protected embryos, bryophytes are not totally liberated from their ancestral aquatic habitat. First of all, these plants need water to reproduce, for their sperm, like those of most green algae, are flagellated and must swim from the an-

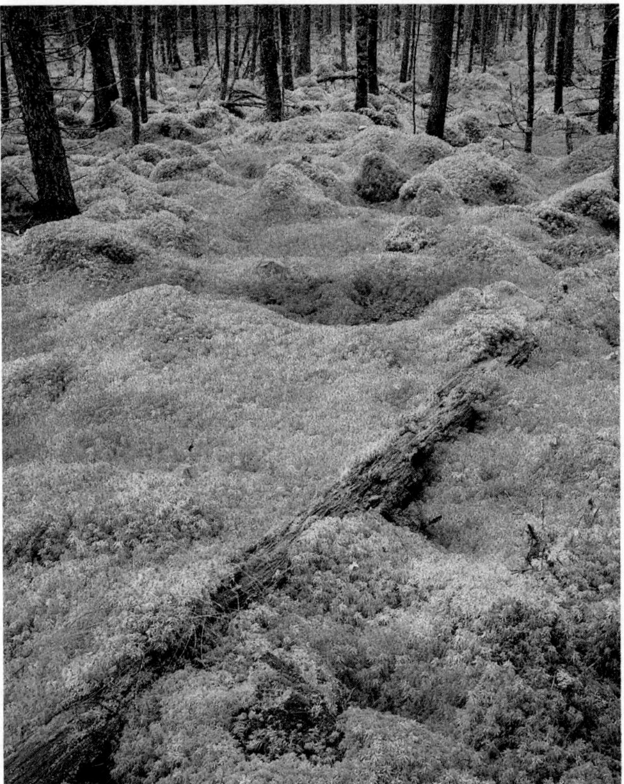

Figure 27.4
A moss bog. Lacking rigid supporting tissue, bryophytes are low-profile plants most common in damp habitats. The matlike plants of this green carpet are gametophytes, the dominant generation in the life cycle of bryophytes.

theridium to the archegonium to fertilize the egg. For many bryophyte species, a film of rainwater or dew is sufficient for fertilization to occur. In addition, most bryophytes have no vascular tissue to carry water from the soil to the aerial parts of the plant (the exceptions, as mentioned earlier, are certain mosses with elongated water-conducting cells). As water moves over the surface of most bryophytes, they must imbibe it like sponges and distribute it throughout the plant by the relatively slow processes of diffusion, capillary action, and cytoplasmic streaming. This mode of hydration helps explain why damp, shady places are the most common habitats of bryophytes.

Bryophytes lack the woody tissue required to support tall plants on land. Although they may sprawl horizontally as mats over a large surface, bryophytes always have a low profile (Figure 27.4). Most are only 1–2 cm in height, and even the largest are usually less than 20 cm tall.

Mosses (Division Bryophyta) The most familiar bryophytes are **mosses.** A mat of moss actually consists of many plants growing in a tight pack, helping to hold one another up. The mat has a spongy quality that enables it to absorb and retain water. Each plant of the mat grips the substratum with elongate cells or cellular filaments called rhizoids. Most photosynthesis occurs in the upper part of the plant, which has many small stemlike and leaflike appendages. The "stems" and "leaves" of a moss, however, are not homologous with these structures in vascular plants.

In the life cycle of a moss, we see a specific example of an alternation of haploid and diploid generations (Figure 27.5). The diploid sporophyte produces haploid spores via meiosis in a structure called a **sporangium;** the spores go on to form new gametophytes. The haploid gametophyte is the dominant generation in mosses and other bryophytes. The sporophyte is generally smaller, shorter lived, and depends on the gametophyte for water and nutrients. This contrasts with the life cycles of vascular plants, where the diploid sporophyte is the dominant generation.

Liverworts (Division Hepatophyta) **Liverworts** are even less conspicuous plants than mosses. The bodies of some are divided into lobes, giving an appearance that must have reminded someone of the lobed liver of an animal (the root *wort* means "herb").

The life cycle of a liverwort is much like that of a moss. Within the sporangia of some liverworts are coil-shaped cells, called elaters, that spring out of the capsule when it opens, helping to disperse the spores. Liverworts can also reproduce asexually from little bundles of cells called gemmae, which are bounced out of cups on the surface of the gametophyte by raindrops (Figure 27.6).

Hornworts (Division Anthocerophyta) **Hornworts** resemble liverworts but are distinguished by their sporophytes, which are elongated capsules that grow like horns from the matlike gametophyte. The photosynthetic cells of hornworts each have a single large chloroplast rather than the many smaller ones more typical of most plants.

The three divisions of bryophytes—mosses, liverworts, and hornworts—have had a long success dating back at least 400 millions years, and there are more than 16,000 species today. However, bryophytes probably never dominated much of the landscape. They are elegantly adapted to a limited range of terrestrial habitats. The vascular plants have additional terrestrial adaptations that enabled them to claim much more territory.

Terrestrial Adaptations of Vascular Plants

During their long evolution from aquatic ancestors, vascular plants accumulated many terrestrial adaptations in addition to cuticles and jacketed sex organs.

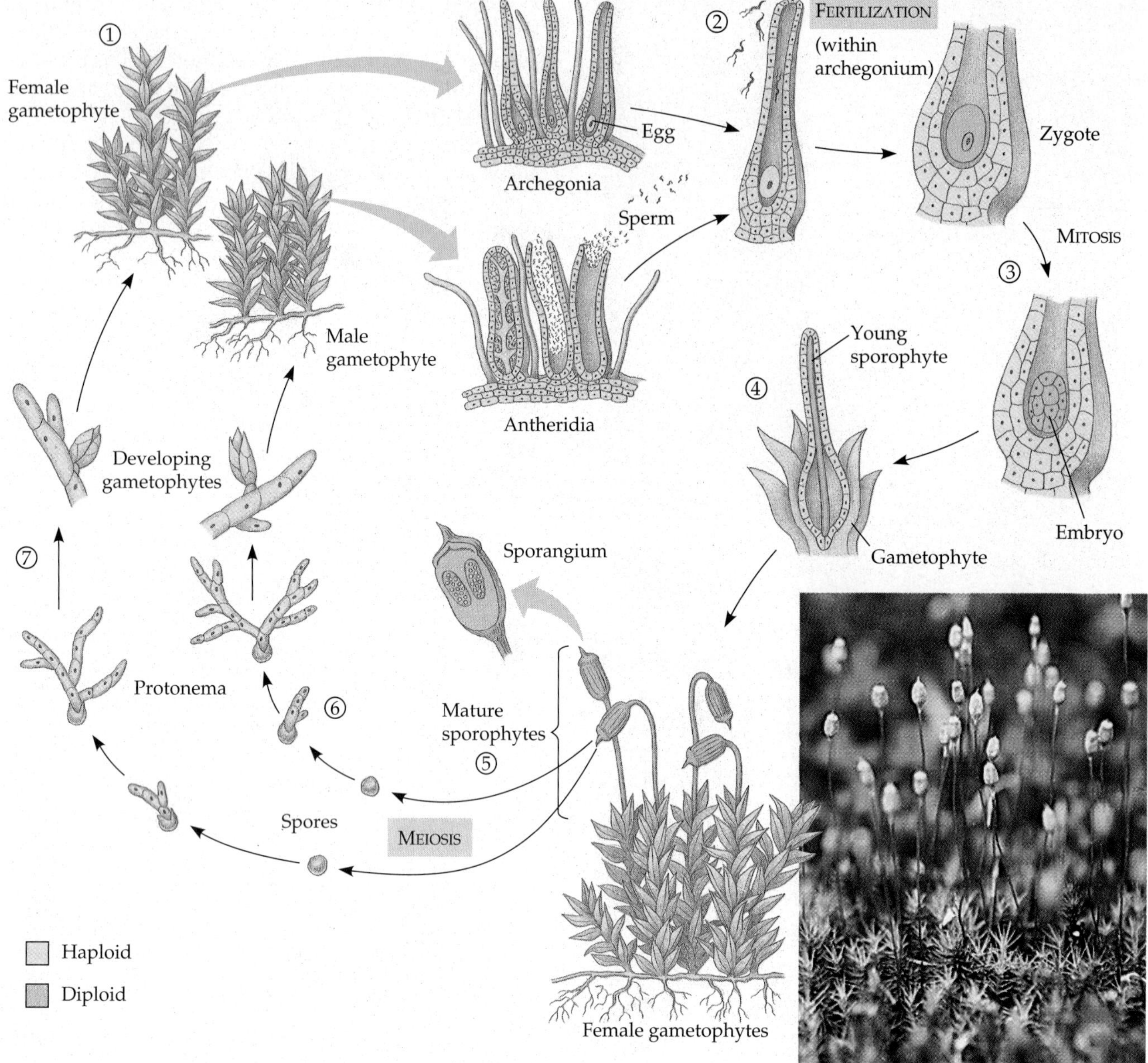

Figure 27.5
The life cycle of a moss. ① Most species of moss have separate male and female gametophytes, which have antheridia and archegonia, respectively. ② After a sperm swims through a film of moisture to an archegonium and fertilizes the egg, ③ the diploid zygote divides by mitosis and develops into an embryonic sporophyte within the archegonium. ④ During the next stage of its development, the sporophyte grows a long stalk that emerges from the archegonium, but the base of the sporophyte remains attached to the female gametophyte. ⑤ At the tip of the stalk is a sporangium, a capsule in which meiosis occurs and haploid spores develop. When the sporangium bursts, the spores scatter. ⑥ A spore germinates by mitotic division to form a small, green, threadlike protonema resembling a green alga. ⑦ The haploid protonema continues to grow and differentiates into a new gametophyte, completing the life cycle. The gametophyte is the more conspicuous generation in bryophytes.

The colonization of land entailed solutions to a new set of problems that aquatic algae did not face (Figure 27.7).

The resources a land plant needs to live are spatially segregated. The soil provides water and minerals, but there is no light underground for photosynthesis. The body of a vascular plant is differentiated into a subterranean root system that absorbs water and minerals and an aerial shoot system of stems and leaves that makes food. Roots, which must absorb water across their surface, generally lack the waxy cuticles that limit evaporation of water from stems and leaves.

Regional specialization of the plant body solved one problem but presented new ones. Roots anchor the

Figure 27.6
Liverworts. The gemmae cups function in asexual reproduction. The tiny plantlets (gemmae) within the cups are dispersed by the impact of raindrops.

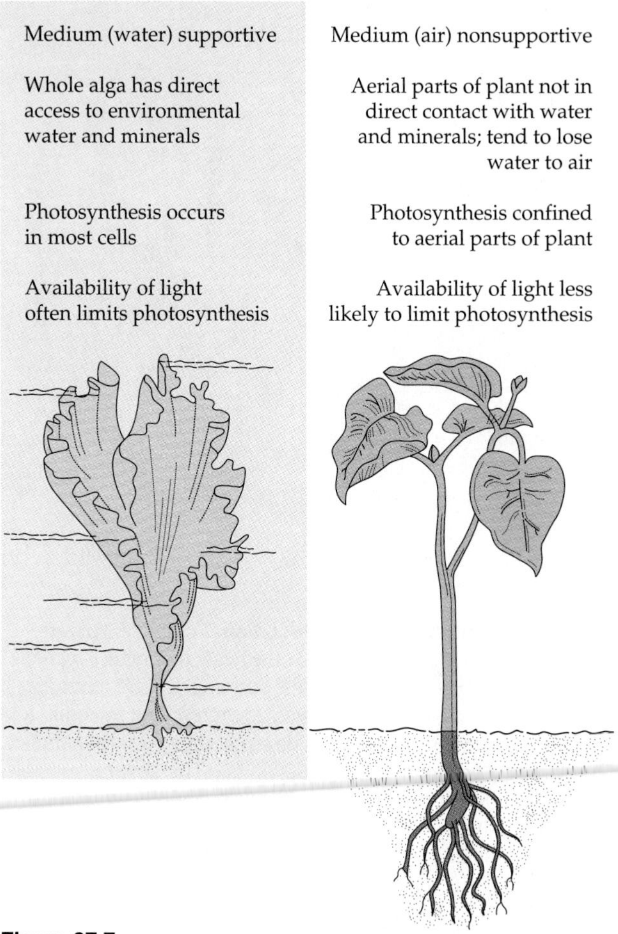

Medium (water) supportive	Medium (air) nonsupportive
Whole alga has direct access to environmental water and minerals	Aerial parts of plant not in direct contact with water and minerals; tend to lose water to air
Photosynthesis occurs in most cells	Photosynthesis confined to aerial parts of plant
Availability of light often limits photosynthesis	Availability of light less likely to limit photosynthesis

Figure 27.7
A comparison of conditions faced by algae and plants.

plant, but for the shoot system to stand up straight in the air, it must have support. This is not a problem in the water: Huge seaweeds need no skeletons because they are essentially weightless in water. An important terrestrial adaptation of vascular plants is **lignin,** a hard material embedded in the cellulose matrix of the walls of cells that function in support. Turgor pressure (see Chapter 8) contributes to the support of small plants, but it is the skeleton of lignified walls that holds up a tree or any other large vascular plant.

With increasing specialization of the root system and shoot system came the new problem of transporting vital materials between the distant organs. Water and minerals must be conducted upward from the roots to the leaves. Sugar and other organic products of photosynthesis must be distributed from leaves to the roots. These problems are solved by an efficient vascular system that is continuous throughout the plant (Figure 27.8). The two conducting tissues of the vascular system are **xylem** and **phloem.** Tube-shaped cells in the xylem carry water and minerals up from the roots. These water-conducting cells are actually dead; only their walls remain to provide a system of microscopic water pipes. The walls are generally lignified, and thus xylem functions in support as well as water transport. Phloem is a living tissue with food-conducting cells arranged into tubes that distribute sugar, amino acids, and other organic nutrients throughout the plant. (Transport in plants is discussed further in Chapter 32.)

Figure 27.8
The vascular system of a plant. An extensive network of veins services all parts of this aspen leaf. The vascular tissue includes xylem, specialized for conducting water and dissolved minerals from the roots up to the shoot system, and phloem, specialized for exporting sugar and other organic nutrients from leaves to roots and other nonphotosynthetic parts of the plant.

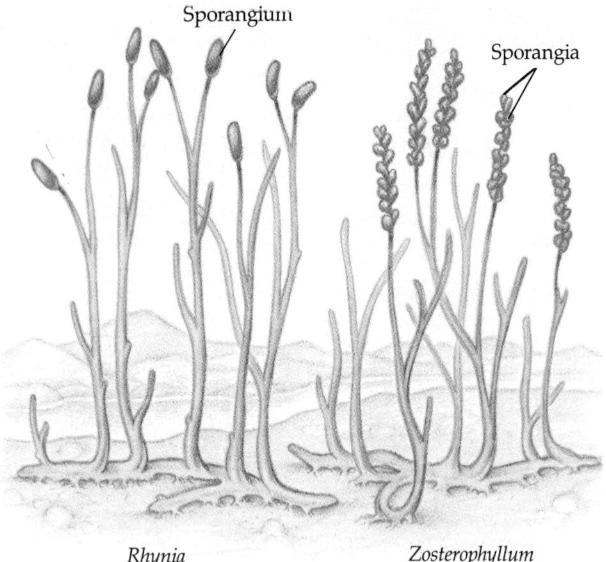

Figure 27.9
Vascular plants of the early Devonian. In form, *Rhynia* resembled *Cooksonia* (the oldest known vascular plant), with dichotomous branching and terminal sporangia. True roots and leaves were absent. The plant was anchored by a rhizome, a horizontal stem. *Rhynia* grew in dense stands around marshes. The largest species was about 50 cm tall. The several species of *Zosterophyllum* differed from *Rhynia* by bearing clusters of lateral sporangia near the tips of stems, rather than single terminal sporangia.

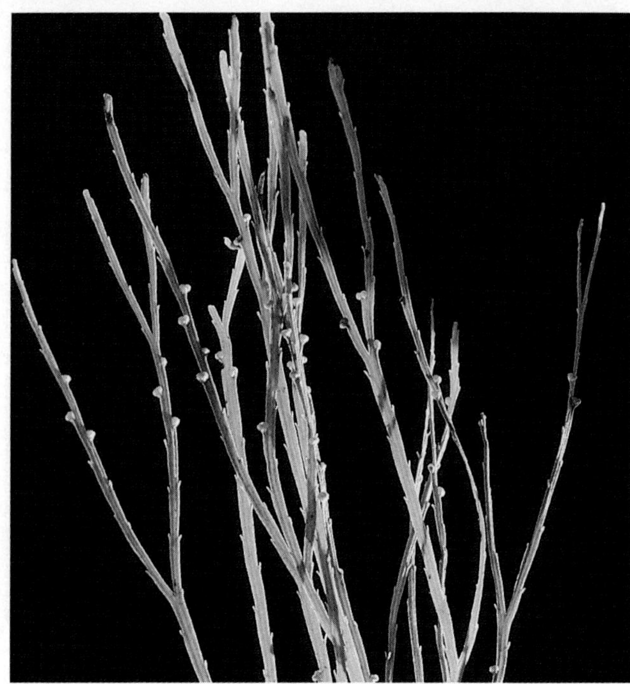

Figure 27.10
The sporophyte of *Psilotum*. The scales on these dichotomously branched stems are not true leaves; they lack vascular tissues. The knobs along the stems are sporangia, which release haploid spores that germinate in the soil. Tiny subterranean gametophytes lack chlorophyll and depend for their food on symbiotic soil fungi that decompose organic litter. Flagellated sperm swim through the moist soil from antheridia to archegonia of the gametophytes. The zygote begins its development within the archegonium, and soon a young sporophyte emerges from the gametophyte, which then dies.

In some groups of vascular plants, additional adaptations to living on land evolved, including the seed, the replacement of flagellated sperm with pollen as a means for delivering gametes outside water, and the increasing dominance of the diploid sporophyte in the alternation of generations.

The Earliest Vascular Plants

Encased in the sedimentary strata of the late Silurian and early Devonian periods are fossils of a variety of vascular plants, among the oldest terrestrial organisms known. Many of these petrified plants are beautifully preserved, right down to the microscopic organization of their tissues. The oldest is *Cooksonia,* which has been discovered in Silurian rocks in both Europe and North America (the two continents were probably joined during the Silurian period). It was a simple plant with dichotomous (repeated "Y") branching. Some of the stems terminated in bulbous sporangia. *Cooksonia* was followed by a diversity of early Devonian species. Two characteristic genera are *Rhynia* and *Zosterophyllum* (Figure 27.9).

Rhynia and *Zosterophyllum* were geographically widespread during the early Devonian. The *Zosterophyllum* type of early Devonian plants is the most likely ancestor of the division Lycophyta, one of the nine divisions of vascular plants. Plants resembling *Rhynia* were probably on the evolutionary limb leading to all other divisions of vascular plants. A group of early Devonian plants called trimerophytes were possibly links between plants of the *Rhynia* type and later groups of seedless plants that began to appear during the middle and late Devonian.

SEEDLESS VASCULAR PLANTS

The earliest vascular plants were seedless. Four extant divisions of plants have retained this early condition.

Division Psilophyta

This division of relatively simple plants has only two genera. The better known is *Psilotum,* which is widespread in the tropics and subtropics. It is known in the United States by the common name whiskfern (but it is not a true fern). In the diploid sporophyte generation, *Psilotum* has dichotomous branching reminiscent of some of the early vascular plants (Figure 27.10). True roots and leaves are absent. The subterranean part of the plant consists of a rhizome (horizontal stem) covered with tiny hairs called rhizoids. The upright stems bear emergences, which, unlike true leaves, lack vascular tissue.

Division Lycophyta

The extant **lycopods,** of the division Lycophyta, are relicts of a far more eminent past. Lycopods first evolved during the Devonian period and became a major part of the landscape during the Carboniferous period, which began about 340 million years ago and lasted until 280 million years ago. By that time, the division Lycophyta split into two evolutionary lines. One group evolved into woody trees that had diameters as large as 2 m and heights of more than 40 m. A second line of lycopods remained small and herbaceous (nonwoody). The giant lycopods thrived in the Carboniferous swamps for millions of years, but became extinct when the swamps began to dry up at the end of that geological period, about 280 million years ago. The small lycopods survived, and they are represented today by about a thousand species, most belonging to the genera *Lycopodium* and *Selaginella.* Common names for these plants are club mosses or ground pines, though they are neither mosses nor pines.

Many species of *Lycopodium* are tropical plants that grow on trees as **epiphytes**—plants that use another organism as a substratum but are not parasites. Other species of *Lycopodium* grow close to the ground on forest floors in temperate regions, including the northeastern United States.

The club moss in Figure 27.11 is the sporophyte, the diploid generation. The sporangia of *Lycopodium* are borne on **sporophylls,** leaves specialized for reproduction. After their discharge, the spores develop into inconspicuous gametophytes that may live underground for ten years or longer. These tiny haploid plants are, like the gametophytes of *Psilotum,* nonphotosynthetic, and are nurtured by symbiotic fungi. Each gametophyte develops archegonia with eggs and antheridia that make flagellated sperm. After a swimming sperm fertilizes an egg, the diploid zygote gives rise to a new sporophyte.

Lycopodium makes a single type of spore, which develops into a bisexual gametophyte having both female and male sex organs (archegonia and antheridia); it is thus said to be **homosporous.** In **heterosporous** plants, such as the lycopod genus *Selaginella,* the sporophyte makes two kinds of spores. **Megaspores** develop into female gametophytes bearing archegonia, and **microspores** become male gametophytes with antheridia. The gametophytes of heterosporous plants are unisexual, either female or male. We will encounter the homosporous and heterosporous conditions again as we continue our survey of vascular plants.

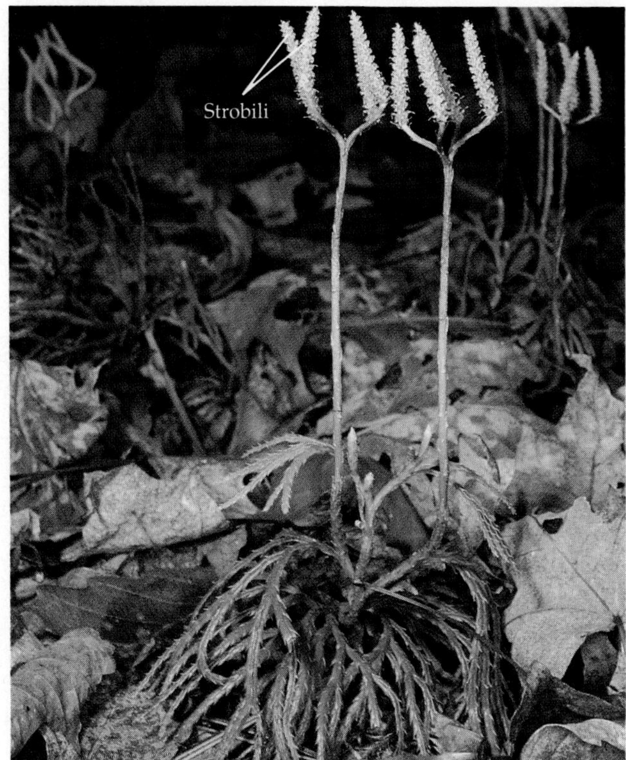

Figure 27.11
***Lycopodium*, a club moss.** Club mosses are common inhabitants of forest floors in the northeastern United States. The small plant has a horizontal rhizome that gives rise to roots and vertical branches and has true leaves containing strands of vascular tissue. The sporangia of *Lycopodium* are borne by specialized leaves called sporophylls. In some species, such as the one shown here, the sporophylls are clustered at the tips of branches into club-shaped structures called strobili (hence the common name club mosses).

Division Sphenophyta

Sphenophyta, whose members are commonly called horsetails, is another ancient lineage of seedless plants dating back to the Devonian radiation of early vascular plants. The group reached its zenith during the Carboniferous period, when many species grew up to 15 m tall. All that survives of this division of plants are about 15 species of a single genus, *Equisetum. Equisetum* is widely distributed but is most common in the Northern Hemisphere, generally in damp locations such as stream banks (Figure 27.12).

The conspicuous horsetail plant is the sporophyte generation. Meiosis occurs in the sporangia, and haploid spores are released. The gametophytes that develop from these spores are only a few millimeters long, but they are photosynthetic and free-living (not dependent on the sporophyte for food). Horsetails are homosporous; the single type of spore gives rise to a bisexual gametophyte with both antheridia and archegonia. Flagellated sperm fertilize eggs in the archegonia, and young sporophytes later emerge.

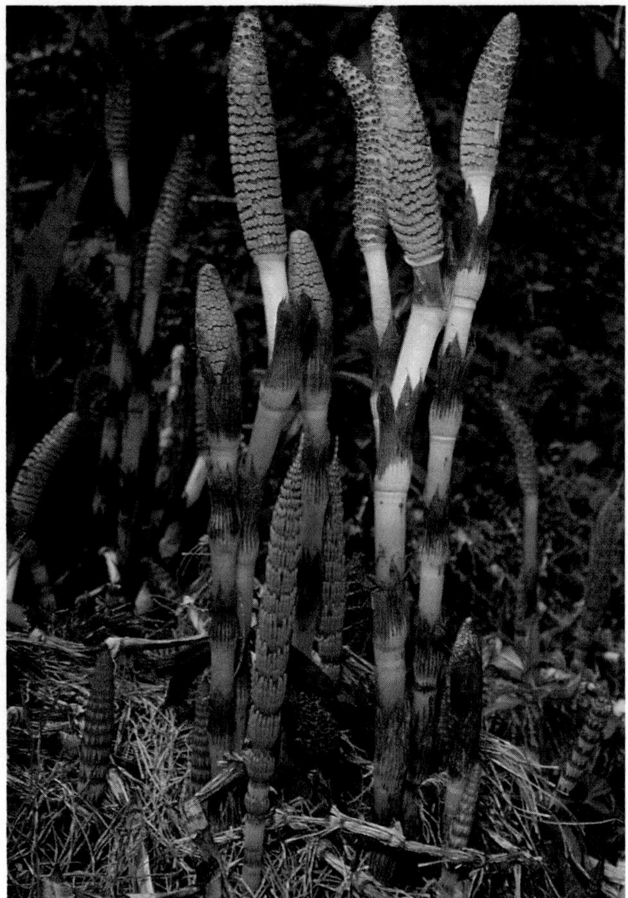

Figure 27.12
***Equisetum* (horsetail).** *Equisetum* has an underground rhizome from which vertical stems arise. The straight, hollow stems are jointed, and whorls of small leaves or branches emerge at the joints. The epidermis, the outer layer of cells, is embedded with silica, which gives the plants an abrasive texture. At the tips of some stems of *Equisetum* are conelike structures bearing sporangia. Horsetails are also called scouring rushes, because their abrasive stems were often used for scrubbing pots and pans before the development of modern methods.

Division Pterophyta

From their Devonian beginning, ferns (division Pterophyta) radiated into the many species that stood alongside tree lycopods and horsetails in the great forests of the Carboniferous period. Of all seedless plants, ferns are by far the most extensively represented in the modern floras (Figure 27.13). More than 12,000 species of ferns live today. They are most diverse in the tropics, but a variety of species is also found in temperate forests.

The leaves of ferns are generally much larger than those of lycopods and probably evolved in a different way. The origin of leaves is currently a subject of much study. The small leaves of lycopods probably evolved as emergences from the stem that contained a single strand of vascular tissue. Leaves with this origin are called **microphylls.** Each leaf of a fern, termed a **megaphyll,** has a branched system of veins. Megaphylls probably evolved by the formation of webbing between many separate branches growing close together.

Most ferns have leaves, commonly called fronds, that are compound, meaning each leaf is divided into several leaflets. The frond grows as its coiled tip, the fiddlehead, unfurls. The leaves may sprout directly from a prostrate stem, as they do in brackens and sword ferns. Large tropical tree ferns, by contrast, have upright stems many meters tall.

The leafy fern plant familiar to us is the sporophyte generation. Some of the leaves are specialized sporophylls with sporangia on their undersides. The sporangia of many ferns are arranged in clusters called sori and are equipped with springlike devices that catapult spores several meters. Once airborne, spores can be blown by the wind far from their origin. Figure 27.14 illustrates the life cycle of a fern. With their swimming sperm and fragile gametophytes, the majority of ferns are restricted to relatively damp habitats.

Figure 27.13
Ferns. Ostrich fern *(Matteuccia)* growing on a forest floor in New York. The fiddleheads in the inset are young fronds ready to unfurl.

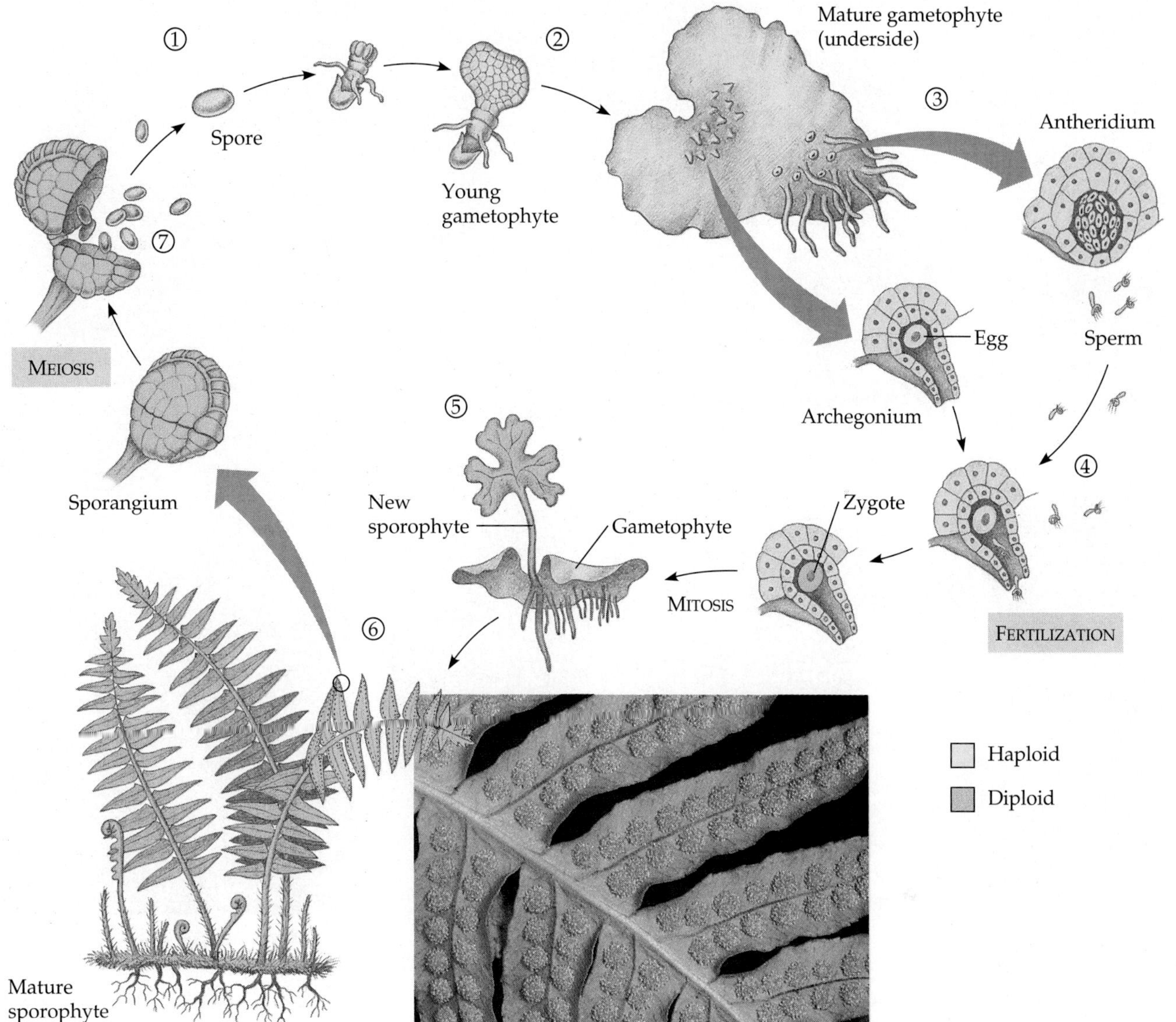

Figure 27.14
The life cycle of a fern. ① After a fern spore settles in a favorable place, ② it develops into a small, heart-shaped gametophyte that sustains itself by photosynthesis. ③ Most ferns are homosporous; each gametophyte has both male and female sex organs, but the archegonia and antheridia usually mature at different times, assuring cross-fertilization between gametophytes. ④ Fern sperm, like those of club mosses and horsetails, use flagella to swim through moisture from antheridia to eggs in the archegonia and then fertilize the egg. A sex attractant secreted by archegonia helps direct the sperm. ⑤ A fertilized egg develops into a new sporophyte, and the young plant grows out from an archegonium of its parent, the gametophyte. ⑥ The spots on the underside of reproductive leaves (sporophylls) are called sori (photograph). Each is a cluster of sporangia, ⑦ which release the spores that give rise to gametophytes.

The Coal Forests

The four divisions of plants that we have just surveyed represent the extant lineages of seedless vascular plants that formed vast forests during the Carboniferous period (about 300 to 350 million years ago). Seedless plants of the Carboniferous forests left not only living relicts but also fossilized fuel in the form of coal. Coal powered the Industrial Revolution, and a resurgence in its use is inevitable as we continue to deplete oil and gas reserves.

Coal formed during several geological periods, but the most extensive beds of coal are found in strata deposited during the Carboniferous, a time when most of the continents were flooded by shallow seas and swamps. Europe and North America, near the equator at that time, were covered by tropical swamp forests. Dead plants did not completely decay in the stagnant waters, and great depths of organic rubble called peat accumulated. The swamps were later covered by the sea, and marine sediments piled on top of the peat. Heat and pressure gradually converted the peat to coal.

(a)

(b)

(c)

(d)

Figure 27.15
The four divisions of gymnosperms. (**a**) Cycadophyta. Cycads, such as this *Cycas revoluta,* resemble palms and in fact are sometimes called sago palms. But cycads are not true palms, which are flowering plants. The cycad is a gymnosperm that bears naked seeds on the scales of cones. (**b**) Ginkgophyta. The ginkgo, also known as the maidenhair tree, has fanlike leaves that turn gold and are deciduous in autumn, an unusual trait for a gymnosperm. The ginkgo is a popular ornamental tree in cities because it can survive air pollution and other environmental insults. (**c**) Gnetophyta. Gnetae is a potpourri class of three gymnosperm genera that are probably not closely related. *Gnetum* grows in the tropics as a tree or vine. *Ephedra* is a shrub found in the American deserts. *Welwitschia,* shown here, is a bizarre plant with straplike leaves—the largest known leaves. It lives only in the deserts of southwestern Africa. (**d**) Coniferophyta. Pines, firs, and redwoods are among the cone-bearing plants called conifers. This giant sequoia *(Sequoiadendron giganteum),* the General Grant tree in California's King's Canyon National Park, is over 80 m tall.

Growing along with the seedless plants in the Carboniferous swamps were primitive seed plants. These gymnosperms were not the dominant plants at that time, but they rose to prominence after the swamps began to dry up at the end of the Carboniferous period.

TERRESTRIAL ADAPTATIONS OF SEED PLANTS

Three life cycle modifications contributed to the success of seed plants as terrestrial organisms:

1. The gametophytes of seed plants became even more reduced than in ferns and other seedless plants. Rather than developing in the soil as an independent generation, the minute gametophytes of seed plants are protected from desiccation by being retained within the moist reproductive tissue of the sporophyte generation. Some plant biologists speculate that the shift toward diploidy in land plants was also related to the harmful impact of the sun's ionizing radiation, which causes mutations. This damaging radiation is more intense on land than in aquatic habitats because of the light-filtering properties of water. Of the two generations of land plants—gametophyte and sporophyte—

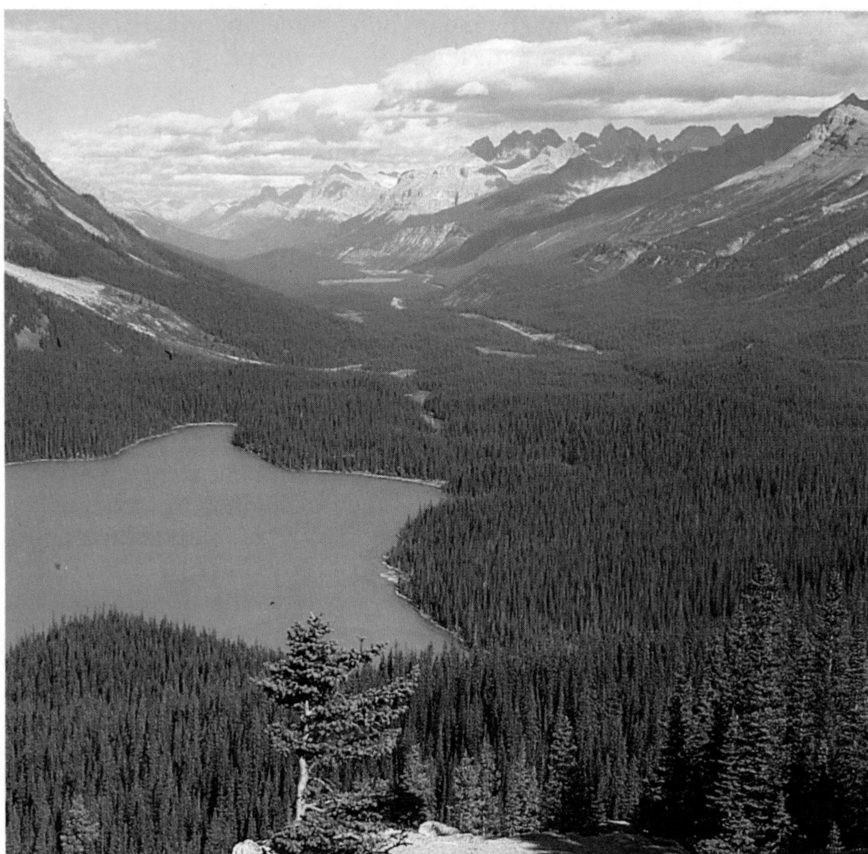

Figure 27.16
A coniferous forest. As shown in this view of the Peyto Lake region of Banff National Park in Alberta, Canada, conifers form an almost solid band across the northern continents at latitudes where the growing season is short.

the diploid form (sporophyte) may cope better with mutagenic radiation. A diploid organism homozygous for a particular essential allele has a "spare tire" in the sense that one copy of the allele may be sufficient for survival if the other is damaged. According to this hypothesis, the increasing prevalence of sporophytes during the evolution of vascular plants can be interpreted as another adaptation to terrestrial conditions.

2. Pollination replaced swimming as the mechanism for delivering sperm to eggs.

3. The seed evolved. Instead of the zygote developing directly into a young sporophyte that fends for itself, the zygote of a seed plant develops into an embryo that is packaged along with a food supply within a seed coat. This protects the dormant embryo from drought, cold, and other harsh conditions. Seeds also function in overland dispersal; they may be carried far from their parents by wind, water, or animals. In seed plants, the seed has replaced the spore as the stage in the life cycle that disperses the species.

GYMNOSPERMS

Of the two groups of seed plants, gymnosperms appear much earlier in the fossil record, and they lack the enclosed chambers in which angiosperm seeds develop. There are four divisions of gymnosperms (Figure 27.15). Three are relatively small: Cycadophyta, Ginkgophyta, and Gnetophyta. By far the largest division of gymnosperms is Coniferophyta.

Division Coniferophyta

The term **conifer** (Gr. *konos,* "cone," and *phero,* "carry") comes from the reproductive structure of these plants, the cone. Pines, firs, spruce, larches, yews, junipers, cedars, cypresses, and redwoods all belong to this division of gymnosperms. Most are large trees. Although there are only about 550 species, conifers dominate vast regions of the Northern Hemisphere, where the growing season is relatively short because of latitude or altitude (Figure 27.16).

Nearly all conifers are evergreens, meaning they retain leaves throughout the year. Even during winter, a limited amount of photosynthesis occurs on sunny days. And when spring comes, conifers already have fully developed leaves that can take advantage of the warmer days.

The needle-shaped leaves of pines and firs are adapted to dry conditions. A thick cuticle covers the leaf, and the stomata are located in pits, further reducing water loss. The conifer needle, despite its shape, is a megaphyll, as are the leaves of all seed plants.

We get most of our lumber and paper pulp from the wood of conifers. What we call wood is actually an accumulation of lignified xylem tissue, which gives the tree structural support.

Coniferous trees are among the tallest, largest, and oldest living organisms on Earth. Redwoods, found only in a narrow coastal strip of Northern California, grow to heights of more than 110 m; only certain eucalyptus trees in Australia are taller. The largest (most massive) organisms alive are the giant sequoias, relatives of redwoods that grow in the Sierra Nevada of California (see Figure 27.15d). One, known as the General Sherman tree, has a trunk with a circumference of 26 m and weighs more than a dozen space shuttles. Bristlecone pines, another species of California conifer, are among the oldest organisms alive. One bristlecone, named Methuselah, is more than 4600 years old; it was a young tree when humans invented writing.

The Life History of a Pine The pine tree, a representative conifer, is a sporophyte, with its sporangia located on cones. The gametophyte generation develops from haploid spores that are retained within the sporangia. Conifers are heterosporous; male and female gametophytes develop from different types of spores produced by separate cones. Each tree usually has both types of cones. Small pollen cones produce small spores that develop into the male gametophytes. Larger, more complex ovulate cones usually develop on separate branches of the tree and make larger spores that develop into female gametophytes (Figure 27.17). From the time young cones appear on the tree, it takes nearly three years for a complicated series of events to produce mature seeds. The scales of the ovulate cone then separate, and the winged seeds travel on the wind. A seed that lands in a habitable place germinates, its embryo emerging as a pine seedling. Table 27.2 puts the complicated life cycle of a conifer into perspective by highlighting some important ways it differs from the life cycles of other plants.

The History of Gymnosperms

Gymnosperms probably descended from a group of Devonian plants called progymnosperms. They were originally seedless plants, but by the end of the Devonian period, seeds had evolved. Adaptive radiation during the Carboniferous and early Permian produced the various divisions of gymnosperms.

In the history of life, the Permian period was one of great crises. Formation of the supercontinent Pangaea (see Chapter 23) may have been one reason that continental interiors became warmer and drier as the Permian progressed. The flora and fauna of Earth changed dramatically, as many groups of organisms disappeared and others emerged as their successors. The changeover was most pronounced in the seas, but terrestrial life was affected as well. In the animal kingdom, amphibians decreased in diversity and were replaced by reptiles, which were better adapted to the arid conditions. Similarly, the lycopods, horsetails, and ferns that dominated the Carboniferous swamps were largely replaced by conifers and cycads, which were more suited to the drier climate. The world and its life had changed so markedly that geologists use the end of the Permian period as the boundary between the Paleozoic and Mesozoic eras (this boundary was originally defined by the changeover in marine fossils). The Mesozoic is sometimes referred to as the "age of dinosaurs," when giant reptiles were supported by a vegetation consisting mostly of conifers and great palmlike cycads. When the climate changed again at the end of the Mesozoic, becoming cooler, the dinosaurs became extinct. Some of the gymnosperms, particularly conifers, persisted, however, and are still an important part of Earth's flora.

Table 27.2 A Comparison of Reproduction for Some Major Plant Groups

Group	Dominant Stage of Life Cycle	Homosporous or Heterosporous	Mechanism for Combining Gametes
Mosses	Gametophyte	Homosporous	Flagellated sperm swims through film of water to egg
Ferns	Sporophyte	Homosporous	Flagellated sperm
Conifers	Sporophyte	Heterosporous	Sperm nuclei transported in windblown pollen
Flowering plants	Sporophyte	Heterosporous	Pollen transferred by wind or animals

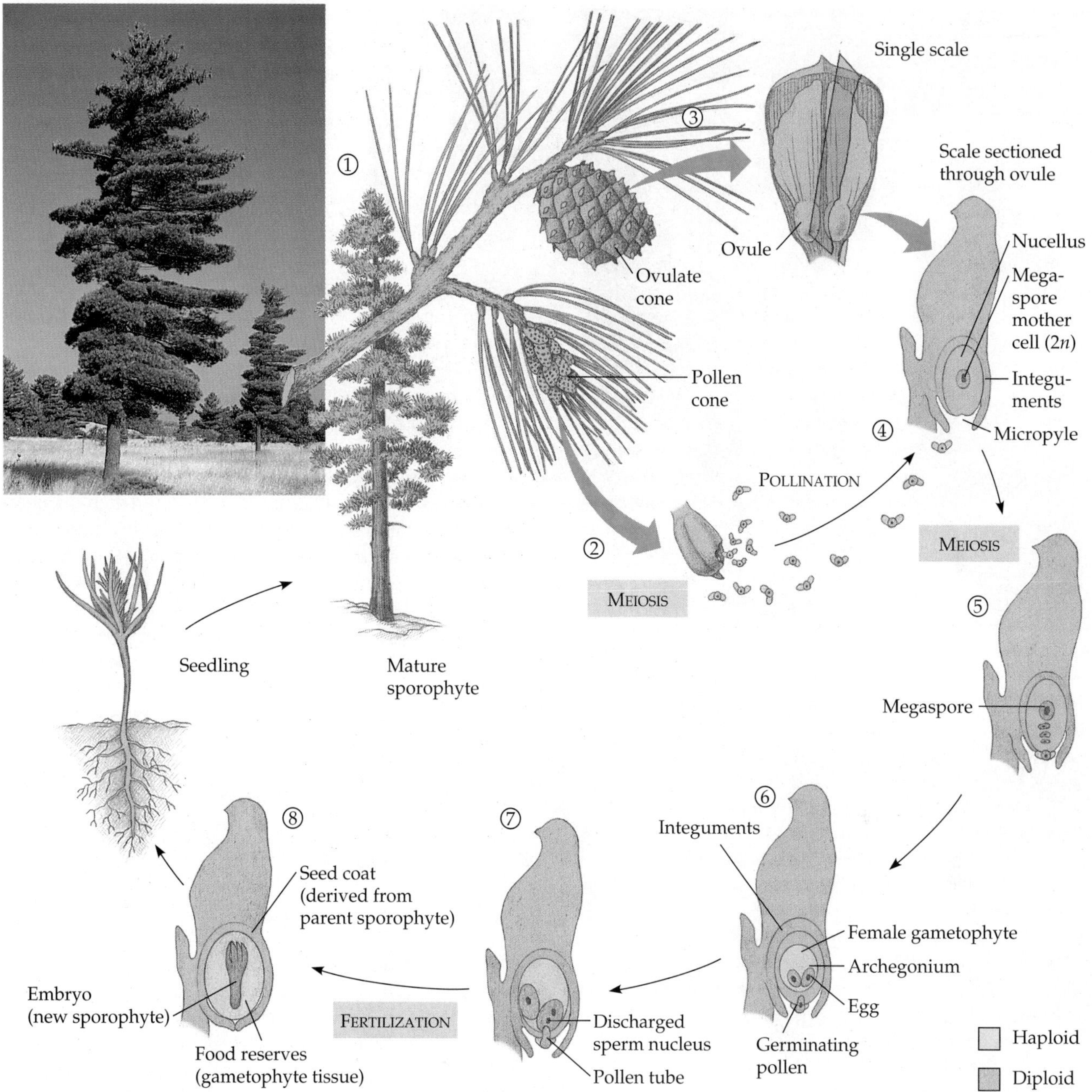

Figure 27.17
The life cycle of a pine. ① Trees (sporophytes) of most species bear both pollen cones and ovulate cones. ② A pollen cone contains hundreds of sporangia held in tiny sporophylls (reproductive leaves). Cells in the sporangia undergo meiosis, giving rise to haploid microspores that develop into pollen grains (immature male gametophytes). ③ An ovulate cone consists of many scales, each with two ovules. Each ovule includes a sporangium, called the nucellus, enclosed in protective integuments with a single opening, the micropyle. ④ During pollination, windblown pollen falls on the ovulate cone and is drawn into the ovule through the micropyle. The pollen grain germinates in the ovule, forming a pollen tube that begins to digest its way through the nucellus. Fertilization usually occurs more than a year after pollination. During that year, ⑤ a megaspore mother cell in the nucellus undergoes meiosis to produce four haploid cells. One of these cells survives as a megaspore, which divides repeatedly, giving rise to the immature female gametophyte. ⑥ Two or three archegonia, each with an egg, then develop within the gametophyte. ⑦ By the time eggs are ready to be fertilized, two sperm cells have developed in the male gametophyte and the pollen tube has grown through the nucellus to the female gametophyte. Fertilization occurs when one of the sperm nuclei, injected into an egg cell by the pollen tube, unites with the egg nucleus. All the eggs in an ovule may be fertilized, but usually only one zygote develops into an embryo. ⑧ The pine embryo, or the new sporophyte, has a rudimentary root and several embryonic leaves called cotyledons. A food supply, consisting of the female gametophyte, surrounds and nourishes the embryo until it is capable of photosynthesis. The ovule has developed into a pine seed, which consists of an embryo, its food supply, and a surrounding seed coat derived from the parent tree.

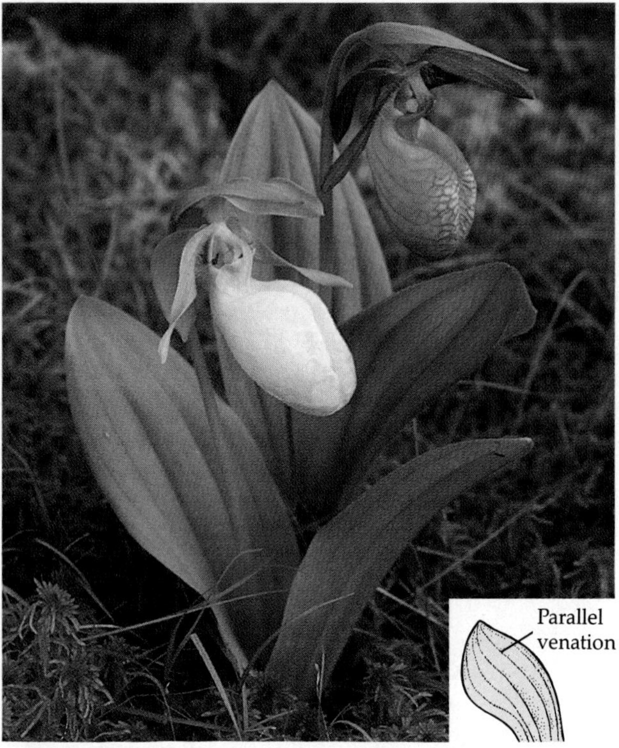

(a)

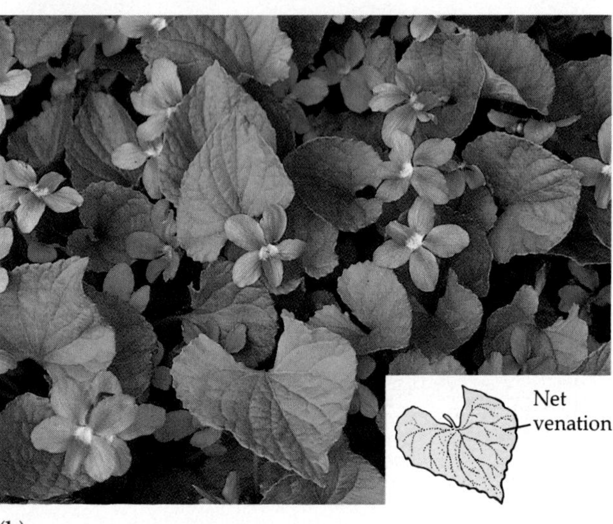

(b)

Figure 27.18
Two classes of angiosperms. (**a**) Monocots, such as these pink lady's slipper orchids, generally have leaves with parallel veins. (**b**) The leaves of dicots, such as this blue violet, usually have netlike veins.

ANGIOSPERMS

Today, angiosperms, or flowering plants, are by far the most diverse and geographically widespread of all plants. About 235,000 species are known, compared with 721 gymnosperm species. All angiosperms are placed in a single division, Anthophyta (Gr. *antho*, "flower"). The division is split into two classes: Monocotyledones (monocots) and Dicotyledones (dicots), which differ in several ways that will be described in Chapter 31. Examples of monocots are lilies, orchids, yuccas, palms, and grasses, including lawn grasses, sugar cane, and grain crops (corn, wheat, rice, and others). Among the many dicot families are roses, peas, buttercups, sunflowers, oaks, and maples (Figure 27.18).

Most angiosperms employ insects and other animals as couriers for transferring pollen to female sex organs, which makes pollination less random than the wind-dependent pollination of gymnosperms. Some flowering plants are wind-pollinated, but we do not know whether this condition is primitive or whether it evolved secondarily from ancestors that were pollinated by animals.

Vascular tissue also became more refined during angiosperm evolution. The cells that conduct water in conifers are **tracheids,** believed to be a relatively early type of xylem cell (Figure 27.19). The tracheid is an elongated, tapered cell that functions in both mechanical support and movement of water up the plant. In most angiosperms, shorter, wider cells called **vessel elements** evolved from tracheids. Vessel elements are arranged end to end to form continuous tubes that are more specialized than tracheids for transporting water but less specialized for support. The xylem of angiosperms is reinforced by a second cell type, the **fiber,** which also evolved from the tracheid. With their thick lignified walls, the xylem fibers are specialized for support. Fibers evolved in conifers, but vessel elements did not.

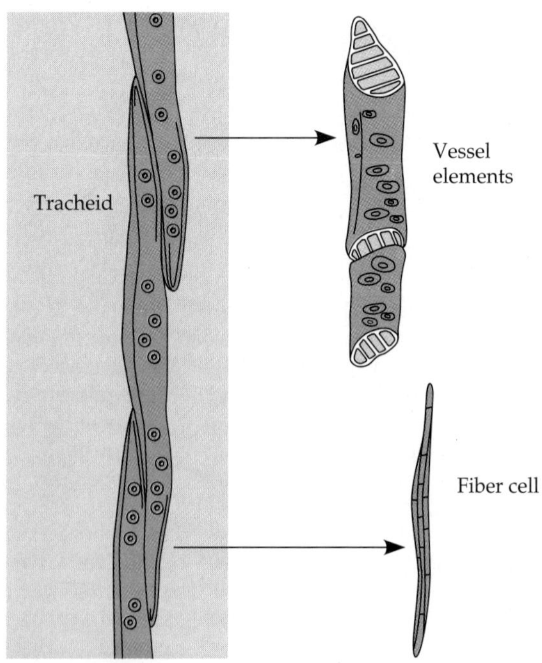

Figure 27.19
The evolution of xylem cells. In angiosperm evolution, tracheids gave rise to vessel elements specialized for conducting water and fiber cells specialized for support.

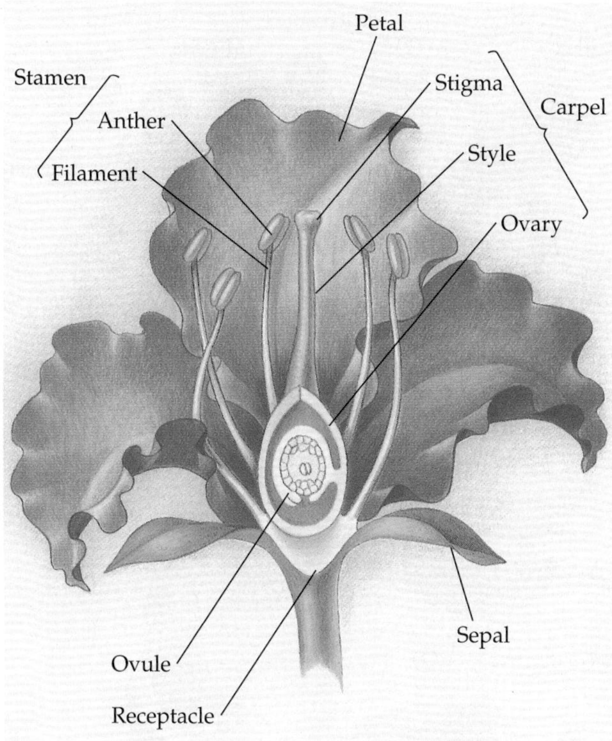

Figure 27.20
The structure of a flower.

Figure 27.21
A developing fruit. Zucchini squash is a ripened ovary containing seeds that develop from fertilized ovules.

The refinements in vascular tissue and other structural advances surely contributed to the success of angiosperms, but the greatest factor in the rise of angiosperms was probably the evolution of the flower, a remarkable apparatus that enhances the efficiency of reproduction by attracting and rewarding pollen-carting animals.

The Flower

The **flower** is the reproductive structure of an angiosperm. A flower is a compressed shoot with four whorls of modified leaves (Figure 27.20). Starting at the bottom of the flower are the **sepals,** which are usually green. They enclose the flower before it opens (think of a rosebud). Above the sepals are the **petals,** brightly colored in most flowers. They aid in attracting insects and other pollinators. Flowers that are wind-pollinated, such as those of many grasses, are generally drab in color. The sepals and petals are sterile floral parts not directly involved in reproduction. Within the ring of petals are the reproductive organs, **stamens** and **carpels.** A stamen consists of a stalk called the **filament** and a terminal sac, the **anther,** where pollen is produced. At the tip of the carpel is a sticky **stigma** that receives pollen. A **style** leads to the **ovary** at the base of the carpel. Protected within the ovary are the ovules, which develop into seeds after fertilization. Recall that the enclosure of seeds within the ovary is one of the features that distinguishes angiosperms from gymnosperms. The carpel probably evolved from a seed-bearing leaf that became rolled into a tube.

Botanists recognize four evolutionary trends in various angiosperm lineages:

1. The number of floral parts has become reduced.
2. Floral parts have become fused. For example, some flowers have compound carpels formed by the fusion of several carpels. The general term **pistil** is sometimes used for a single carpel or several fused carpels.
3. Symmetry has changed from radial, in which any cut down its central axis will divide the flower into two equal halves, to bilateral, in which the flower has distinct left and right halves.
4. The ovary has dropped to a position below the petals and sepals, where the ovules are better protected.

With modification in floral structure, many angiosperms are specialists at using specific animals for pollination, as we will see later in this chapter.

The Fruit

Fruits protect dormant seeds and aid in their dispersal. A **fruit** is a mature ovary (Figure 27.21). As seeds develop after fertilization, the wall of the ovary thickens. A pea pod is an example of a fruit, with seeds (mature ovules) encased in the ripened ovary. Some fruits, such as apples, incorporate other floral parts along with the ovary. Peas and apples are both simple fruits, meaning they develop from a single ovary. Aggregate fruits,

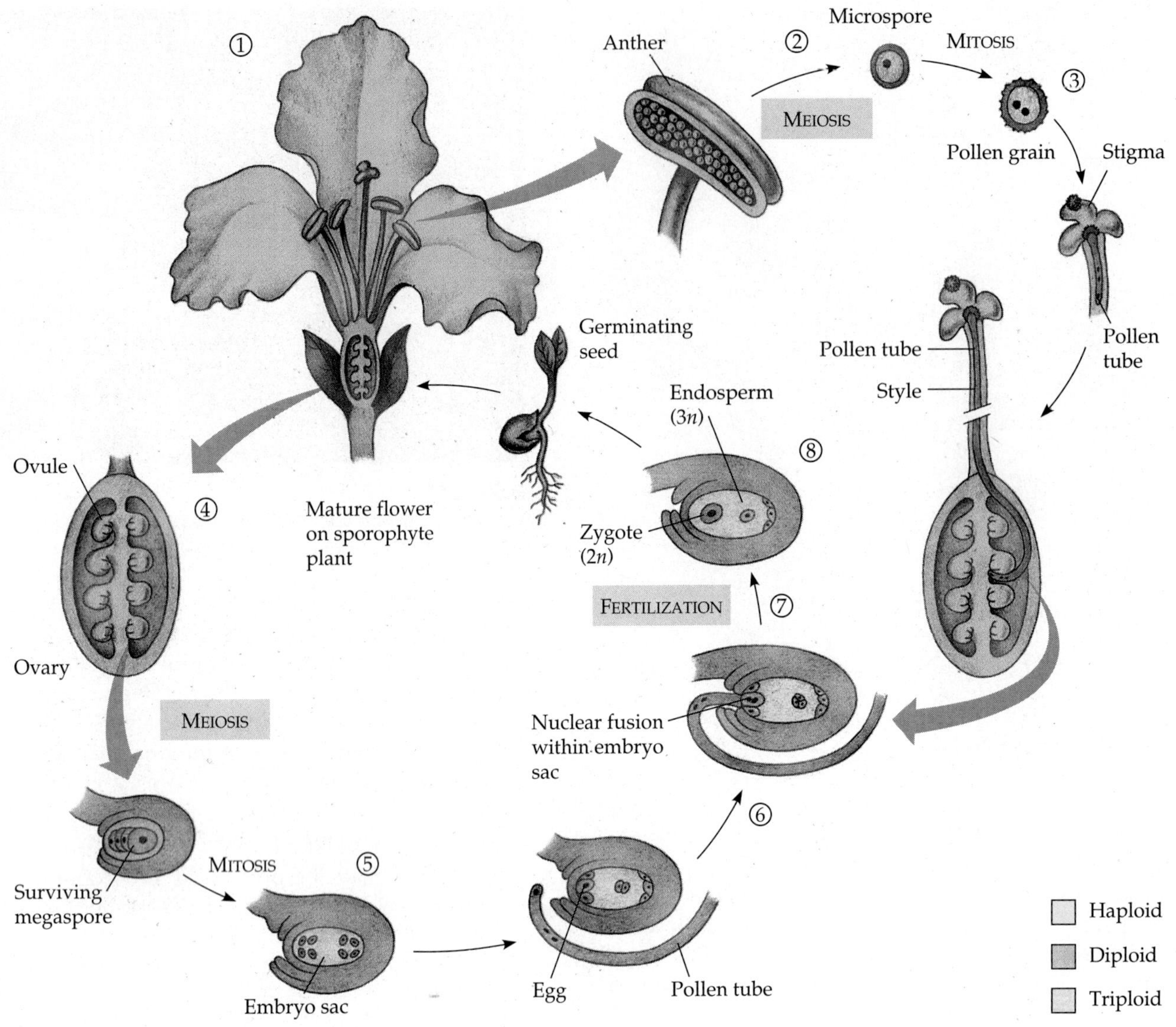

Figure 27.22
The life cycle of an angiosperm. ① The flower of the sporophyte produces ② microspores that form ③ male gametophytes (pollen) and ④ megaspores that produce ⑤ female gametophytes (embryo sacs) within ovules. ⑥ Pollination brings the gametophytes together in the ovary. ⑦ Fertilization occurs, and ⑧ zygotes develop into sporophyte embryos that are packaged along with food into seeds.

such as raspberries, come from several ovaries that were part of the same flower. A pineapple is an example of a multiple fruit, one that develops from several separate flowers (see Chapter 34).

Fruits are modified in various ways that help disperse seeds. Some flowering plants, such as dandelions and maples, have seeds within fruits acting as kites or propellers that aid in dispersal by wind. But most angiosperms use animals to carry seeds. Some of these plants have fruits modified as burrs that cling to animal fur (or the clothes of humans). Other angiosperms produce edible fruits. When it eats the fruit, the animal digests the fleshy part, but the tough seeds usually pass unharmed through the digestive tract. Mammals and birds may deposit seeds, along with a fertilizer supply, miles from where the fruit was eaten. The use of animals to tote seeds and pollen has helped angiosperms become the most successful plants on Earth.

The Life Cycle of an Angiosperm

Angiosperms are heterosporous. The flower of the sporophyte produces microspores that form male gametophytes and megaspores that produce female gametophytes (Figure 27.22). The immature male gametophytes are **pollen grains,** which develop within the anthers of stamens. Each pollen grain has two haploid cells. **Ovules,** which develop in the ovary, include the female gametophyte, an **embryo sac** with eight haploid nuclei in seven cells (a large central cell has two haploid nuclei). One of the cells is the egg. (The development of pollen and the embryo sac will be described in more detail in Chapter 34.)

After its release from the anther, the pollen is carried to the sticky stigma at the tip of a carpel. Although some flowers self-pollinate, most have mechanisms that ensure **cross-pollination,** the transfer of pollen from flowers of one plant to flowers of another plant of the same species. For example, stamens and carpels of a single flower may mature at different times, or the organs may be so arranged within the flower that self-pollination is unlikely (see Chapter 34).

The pollen grain, an immature male gametophyte, germinates after it adheres to the stigma of a carpel. The pollen grain extends a tube that grows down the style of the carpel. After it reaches the ovary, the pollen tube penetrates through a pore in the integuments of the ovule and discharges two sperm cells into the embryo sac. One sperm nucleus unites with the egg to form a diploid zygote. The other sperm nucleus fuses with two nuclei in the center cell of the embryo sac. This central cell now has a triploid ($3N$) nucleus. The pollen of conifers, remember, also releases two sperm nuclei, but one disintegrates. In contrast, both sperm nuclei of angiosperm pollen fertilize cells in the embryo sac. This phenomenon, known as **double fertilization,** is characteristic of angiosperms. (Researchers have recently discovered that double fertilization also occurs in *Ephedra,* a gymnosperm of the division Gnetophyta.)

After double fertilization, the ovule matures into a seed. The zygote develops into a sporophyte embryo with a rudimentary root and one or two seed leaves, the **cotyledons** (monocots have one seed leaf and dicots have two). The triploid nucleus in the center of the embryo sac divides repeatedly to give rise to a triploid tissue called **endosperm,** rich in starch and other food reserves. Monocot seeds such as corn store most of their food in the endosperm. Beans and many other dicots restock most of the nutrients into the developing cotyledons.

The seed is a mature ovule, consisting of the embryo, endosperm, and a seed coat derived from the integuments (outer layers of the ovule). In a suitable environment, the seed germinates. The coat ruptures and the embryo emerges as a seedling, using the food stored in the endosperm and cotyledons.

The Rise of Angiosperms

Charles Darwin called the origin of the angiosperms an "abominable mystery." The mystery endures. The problem is the relatively sudden appearance of angiosperms in the fossil record, with no known transitional links to ancestors. The oldest fossils that are widely accepted as angiosperms are found in rocks of the early Cretaceous period, about 120 million years old. They are sparsely represented among a much greater abundance of ferns and gymnosperms. By the end of the Cretaceous, 65 million years ago, the angiosperms had radiated and become the dominant plants on Earth, as they are today.

Some paleobotanists believe the relative suddenness of the angiosperm appearance in the early Cretaceous is an artifact of an imperfect fossil record. Angiosperms may have originated somewhat earlier in the highlands or some other location where fossilization was unlikely, and their sudden appearance may be the result of their spread to locations where fossils form more frequently. Another view, in the spirit of the evolutionary theory known as punctuated equilibrium (see Chapter 23), holds that angiosperms actually did evolve and radiate rather abruptly (on the scale of geological time).

The lack of transitional forms in the fossil record obscures the ancestry of flowering plants. Nevertheless, nearly all paleobotanists agree that angiosperms evolved from some group of gymnosperms, perhaps seed ferns (not really ferns, because they had seeds). This ancient group of unspecialized gymnosperms, now extinct, lived in Carboniferous forests and persisted into the Mesozoic era.

Whatever and whenever their origin, the rise to prominence of angiosperms during the Cretaceous is amply documented in the fossil record. The end of the Cretaceous was another crisis period when many old groups of organisms were replaced by new ones. Cooler climates may have contributed to the changeover. Again, the frequency of extinctions was greatest in the seas, but significant changes in terrestrial fauna and flora also occurred. The dinosaurs disappeared, as did many of the cycads and conifers that had thrived during the Mesozoic era. They were replaced by mammals and flowering plants. The change in fossils during the late Cretaceous is so extreme that geologists use the end of that period as the boundary between the Mesozoic and Cenozoic eras.

Relationships Between Angiosperms and Animals

Ever since they followed plants onto the land, animals have influenced the evolution of terrestrial plants, and vice versa. The fact that animals must eat affects the natural selection of both animals and plants. For instance, with animals crawling and foraging for food on the forest floor, there must have been selection pressure in favor of plants that kept their spores and gametophytes up in the treetops, rather than dropping these

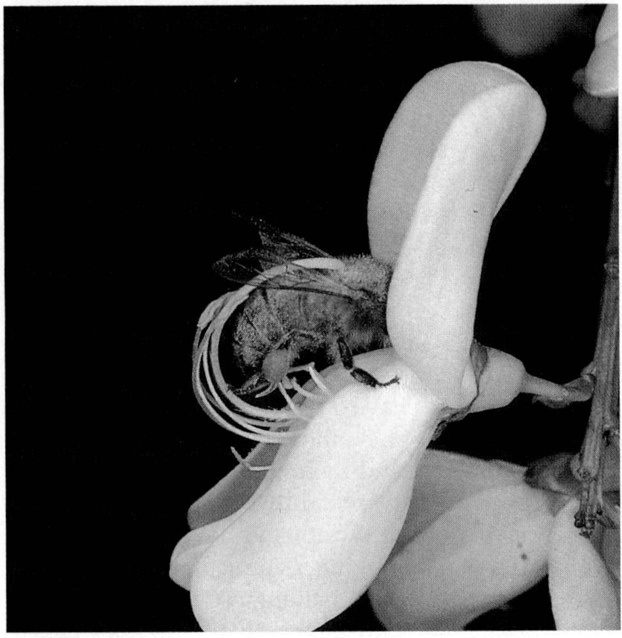

(a)

(b)

(c)

Figure 27.23
Relationships between angiosperms and their pollinators. (**a**) This scotch broom has a tripping mechanism that dusts pollen onto the back of a visiting bee. (**b**) Some flowers have nectaries at the bottom of long tubes. This butterfly *(Heliconius erato)* has a slender proboscis (coiled in this photo) that reaches the nectary. Such mechanisms often result in exclusive relationships between angiosperms and their pollinators. (**c**) Wahlberg's epauleted bat *(Epomophorus wahlbergi)* feeds on the baobab flower. Its body collects and distributes the plant's pollen. The baobab, like many plants that depend on bats, has blossoms that bloom at night. They are light colored, large, and scented, and are therefore easily found by nocturnal feeders.

critical structures to hungry animals on the ground. This, in turn, may have been a selection factor in the evolution of flying insects. On the other hand, as plants with flowers and fruits evolved, some herbivores became beneficial to the plants by carrying the pollen and seeds of plants they used as food. Certain animals became specialists at these tasks, feeding on specific plants. Natural selection reinforced these interactions, for they improved the reproductive success of both partners. The plant got pollinated and the animal got fed. The mutual evolutionary influence between two species is termed **coevolution.** (This definition will be refined in Chapter 48.)

Coevolution of angiosperms and their pollinators is largely responsible for the diversity of flowers. Many flowers are pollinated by a specific animal—say, a particular type of bee, beetle, bird, or bat. These exclusive relationships ensure that the plant's pollen will not be wasted by being carried to the flower of a different species. At the same time, the pollinator has a monopoly on a food source. The color and fragrance of a flower are usually keyed to its pollinator's senses of sight and smell. Flowers pollinated by bees, for instance, often have nectar guides, markings that help direct the bee as it taxies to the nectaries, glands that secrete the sugary solution bees eat. On its way to or from the nectar, the bee becomes dusted with pollen (Figure 27.23). The markings on some bee-pollinated flowers are invisible to humans but are apparently vivid to bees, whose eyes are sensitive to ultraviolet light. Flowers pollinated by birds are usually red, to which bird eyes are especially sensitive. The shape of the flower may also specify a particular pollinator. Flowers pollinated by hummingbirds, for instance, have their nectaries located deep in a floral tube where only the long, thin tongue of the hummingbird is likely to reach.

Relationships between angiosperms and animals are also evident in the edible fruits of angiosperms. Fruits that are not yet ripe are usually green, hard, and distasteful (at least to humans). This helps the plant retain its fruit until the seeds are mature and ready for dispersal. As it ripens, the fruit becomes softer and its sugar content increases. Many fruits also become fragrant and brightly colored, advertising their ripeness to animals. One of the most common colors for ripe fruit is red, which insects cannot see very well. Thus, most of the fruit is saved for birds and mammals, animals large enough to disperse the seeds. Again, we see that one of the keys to angiosperm success has been interaction with animals.

Angiosperms and Agriculture Flowering plants provide nearly all our food. All of our fruit and vegetable crops are angiosperms. Corn, rice, wheat, and the other grains are grass fruits. The endosperm of the grain seeds is the main food source for most of the people of the world and their domesticated animals. We also grow angiosperms for fiber, medications, perfumes, and decoration.

Like other animals, early humans probably collected wild seeds and fruits. Agriculture was gradually invented as humans began sowing seeds and cultivating plants to have a more dependable food source. As they domesticated certain plants, humans began to intervene in plant evolution by selective breeding designed to improve the quantity and quality of the foods the crops produced. We have developed a very special relationship with the plants we cultivate. We water them and fertilize them, try to protect them from insects, and plant their seeds. Many of these plants are so genetically removed from their origins that they probably could not survive in the wild. Agriculture is a unique case of an evolutionary relationship between plants and animals. And it is a precarious relationship for the billions of people now living on Earth, because cultivated crops are very vulnerable to natural and human-caused disasters. In an attempt to protect the world's food supply, many countries try to maintain seed banks of agriculturally important plants. The U.S. National Seed Storage Laboratory at Fort Collins, Colorado, maintains nearly 250,000 varieties representing about 1300 crop species.

(a)

(b)

Figure 27.24
Forest mismanagement, a worldwide problem. (**a**) Clear-cutting to harvest lumber from an old-growth forest in Oregon. (**b**) Slash-and-burn clearing of a tropical rain forest in the Amazon Basin to provide temporary farmland.

THE VALUE OF PLANT DIVERSITY

The exploding human population and its demand for space and natural resources are extinguishing plant species at an unprecedented rate. The problem is especially critical in the tropics, where more than half the human population lives and population growth is fastest. Tropical rain forests are being destroyed at a frightening pace. The most common cause of this destruction is slash-and-burn clearing of the forest for agricultural use. Fifty million acres, an area about the size of the state of Washington, are cleared each year, a rate that would completely eliminate Earth's tropical forests within 25 years. As the forest disappears, so do thousands of plant species. Insects and other rainforest animals that depend on these plants are also vanishing. In all, researchers estimate that destruction of habitat in the rain forest and other ecosystems is claiming hundreds of species per year. The toll is greatest in the tropics, because that is where most species live; but the environmental assault is a generically human tendency. Europeans eliminated most of their forests centuries ago, and in North America, destruction of habitat is endangering many species (Figure 27.24).

Many people are disturbed about contributing to the extinction of living forms. But there are also selfish reasons to be concerned about the loss of plant diversity. We depend on plants for thousands of products, including food, building materials, and medicines. So far, we have explored the potential uses of only a tiny fraction of the 250,000 known plant species. For example, almost all our food is based on the cultivation of

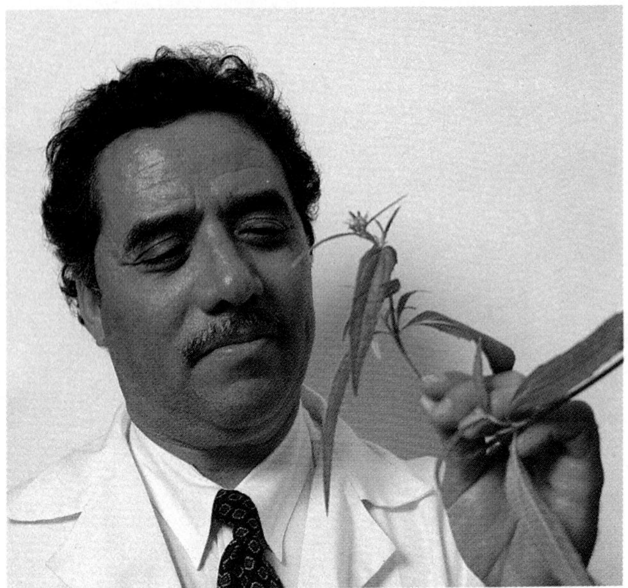

Figure 27.25
***Aspilia:* a medicinal plant.** This scientist, Eloy Rodriguez of the University of California, Irvine, is investigating the medicinal value of thiarubrine-A, a compound present in young leaves of *Aspilia.* Field researchers in Tanzania observed chimpanzees occasionally ingesting these young *Aspilia* leaves, often grimacing as they swallowed. There is evidence that thiarubrine-A kills intestinal parasites that may be making the chimpanzees sick. Researchers are discovering other medicines by observing animals in natural environments. The welfare of humans and other animals is closely tied to the diversity of plants that have evolved on Earth.

only about two dozen species. More than 120 prescription drugs, including digitalis for treating heart patients and taxol, a promising drug for cancer therapy, are extracted from plants. However, researchers have investigated fewer than 5000 plant species as potential sources of medicine. Pharmaceutical companies were led to most of these species by local people who use the plants in preparing their traditional medicines. Another approach is to observe animals, which seem to seek out and eat certain plants when they are sick (Figure 27.25).

The tropical rain forest is a medicine chest of healing plants that may be extinct before we even know they exist. This is only one reason to value what is left of plant diversity and to search for ways to slow the loss. The solutions we propose must be economically realistic. If it is more profitable to clear forests than to preserve them, then we will slash and burn. If, however, we begin to see rain forests and other ecosystems as living treasures that can regenerate only slowly, we may learn to harvest their products at sustainable rates. This will only work if the local people who are the custodians of plant diversity in their region receive a fair share of profits from the development of medicine and other plant products. What else can we do to preserve plant diversity? Few questions are as important.

* * *

In this chapter, we have tracked 425 million years of plant evolution, from the descendants of green algae that moved onto shore to the flowering plants that now dominate most landscapes. We have seen once again that the organisms of today are best understood by their history. This chapter has also reinforced the theme of the connectedness of biology and geology. Transitions such as changes in climate certainly influenced plant evolution, but plants have also changed the course of geology by altering soil and transforming the land in many other ways. The next chapter considers the history and environmental impact of fungi, another kingdom of organisms that spread onto land with the plants.

STUDY OUTLINE

1. Ever since the colonization of land by plants 425 million years ago, plant evolution has been interwoven with major changes in terrestrial conditions.

An Introduction to the Plant Kingdom (pp. 559–561)

1. All plants are photosynthetic multicellular eukaryotes. Plants have cellulose in their cell walls and store their carbohydrates as starch. Stomata and the cuticle of stems and leaves are two important terrestrial adaptations.
2. As plants colonized the land, jacketed sex organs, called gametangia, evolved and protected gametes and embryos from desiccation.
3. All modern plants show a heteromorphic alternation of generations, with distinctive haploid gametophyte and diploid sporophyte forms.
4. Important innovations in plant structure catalyzed four major periods of plant evolution. First, terrestrial adaptations allowed plants to make the transition from water to land about 425 million years ago. The second major period occurred with the emergence of vascular tissue in certain plants. Third, the origin of the seed about 360 million years ago allowed embryos to leave the parent plant. The fourth major episode occurred 130 million years ago with the evolution of the flower, a specialized reproductive structure that produces seeds enclosed within an ovary.

The Move onto Land (pp. 562–566)

1. Plants probably evolved from green algae, which have photosynthetic pigments and other features in common with plants.
2. The mostly nonvascular bryophytes include three plant divisions: Bryophyta (mosses), Hepatophyta (liverworts), and Anthocerophyta (hornworts). Bryophytes have a waxy cuticle and gametangia that protect the gametes and the embryo on land, but they still require a moist habitat for fertilization and for imbibing water in the absence of vascular tissue. Lack of woody tissue dic-

tates their short stature. The haploid gametophyte is the dominant stage in the life cycle of bryophytes.
3. In vascular plants, distant organs are interconnected by the vascular tissue, with water and minerals transported by xylem and organic nutrients by phloem.
4. Over evolutionary time, there was a trend toward increasing dominance of the diploid sporophyte in the life cycles of vascular plants.

Seedless Vascular Plants (pp. 566–570)

1. The division Psilophyta is a small group with simple morphology to which the widespread genus *Psilotum* (whiskferns) belongs.
2. The division Lycophyta consists of the lycopods, or club mosses—small, herbaceous survivors of a dominant ancient division that once included large treelike forms. Modern species use specialized leaves called sporophylls to produce spores, which develop into gametophytes.
3. The horsetails belong to a single surviving genus of the division Sphenophyta.
4. The division Pterophyta consists of the ferns, the most species-rich group of living seedless plants. The fronds of the sporophyte generation form sporangia that produce spores germinating into small gametophytes.

Terrestrial Adaptations of Seed Plants (pp. 570–571)

1. The success of seed plants on land may be attributed to three developments: (a) further reduction of the gametophyte and its retention within the sporophyte; (b) replacement of swimming sperm with pollination; and (c) development of the seed, which functions in protection and dispersal.

Gymnosperms (pp. 571–573)

1. Coniferophyta is the largest of the four gymnosperm divisions. Almost all conifers are evergreens with needle-shaped leaves.
2. The cone is the distinguishing characteristic of conifers. In the pine, two different types of cones produce male and female gametophytes on the sporophyte tree. Pollen grains released as immature male gametophytes land on female cones housing immature female gametophytes inside complex ovules. After a period of gametophyte maturation, fertilization occurs. The zygote develops into an embryo, which is packaged into a winged seed that disperses by the wind.
3. Three smaller divisions of gymnosperms are the Cyadophyta (cycads), Ginkgophyta (ginkgo), and Gnetophyta (for example, *Welwitschia*).

Angiosperms (pp. 574–579)

1. Angiosperms, or flowering plants, belong to the division Anthophyta, which contains the most diverse and widespread members of the plant kingdom. Angiosperms are divided into monocots and dicots.
2. The origin of refined xylem tissue for transport and support contributed to the rise of the angiosperms. But the greatest contribution to their success was probably the evolution of the flower, which greatly improved reproductive efficiency.
3. The flower is a reproductive structure housing stamens and carpels within sterile sepals and petals.
4. Fruits form from the ripened ovaries of one or more flowers. They protect dormant seeds and are modified in various ways that aid in seed dispersal.
5. Pollen grains are immature male gametophytes that form in the anthers of stamens and germinate on the sticky stigma of the carpel. A pollen tube grows down to the ovary, where one sperm cell fertilizes the egg and another cell combines with two female haploid cells to make a triploid endosperm that functions in food storage. This double fertilization produces a seed containing the embryonic sporophyte surrounded by the endosperm and seed coat.
6. Coevolution of plants and pollinating animals has contributed to the reproductive success of the angiosperms and has influenced the color, fragrance, and shape of flowers and the production of nectar and edible fruits.
7. A special case of an evolutionary relationship between plants and animals is the human invention of agriculture.

The Value of Plant Diversity (pp. 579–580)

1. Destruction of rain forests and other ecosystems is eliminating thousands of potentially useful plants.

SELF-QUIZ

1. Which of the following is *not* evidence that plants evolved from chlorophytes?
   a. The chloroplasts of both groups are similar in structure, with thylakoid membranes stacked into grana.
   b. Both groups use chlorophyll *a* and *b* and the same accessory pigments.
   c. Organisms of both groups have jacketed reproductive organs called gametangia.
   d. Cellulose is the major component of their cell walls, and starch is used as a storage product.
   e. Some members of Chlorophyta and all green plants have a similar formation of cell plates during cytokinesis.
2. All bryophytes (mosses, liverworts, and hornworts) share certain characteristics. These are
   a. reproductive cells in protective chambers and a waxy cuticle
   b. a waxy cuticle, true leaves, and reproductive cells in protective chambers
   c. vascular tissues, true leaves, and a waxy cuticle
   d. reproductive cells in protective chambers and vascular tissues
   e. vascular tissues and a waxy cuticle
3. Which of the following is *not* characteristic of all divisions of vascular plants?
   a. the development of seeds
   b. an alternation of generations
   c. differentiation into roots, stems, and leaves
   d. xylem and phloem for transporting materials between roots and leaves
   e. addition of lignin to cell walls to provide aerial support
4. A heterosporous plant is one that
   a. produces a gametophyte that bears both sex organs

b. produces microspores and megaspores in separate sporangia, giving rise to separate male and female gametophytes
c. is a seedless vascular plant
d. produces two kinds of spores, one asexually by mitosis and one type by meiosis
e. reproduces only sexually

5. During the Carboniferous period, the dominant plants, which later formed the great coal beds, were mainly
a. the giant lycopods, horsetails, and ferns
b. the conifers
c. the angiosperms
d. the treelike early vascular seed plants
e. the bryophytes that dominated early swamps

6. The male gametophyte of an angiosperm consists of
a. an anther
b. a sac containing eight haploid nuclei
c. a microspore
d. a germinated pollen grain
e. an ovule

7. A fruit is most commonly
a. a mature ovary
b. a thickened style
c. an enlarged ovule
d. an enlarged aggregate of several flowers
e. a mature female gametophyte

8. Important terrestrial adaptations that evolved *exclusively* in seed plants include all of the following *except*
a. pollination by wind or animal instead of fertilization by swimming sperm
b. transport of water through vascular tissue
c. retention of the gametophyte plant within the sporophyte
d. dispersal of new plants by seeds
e. protection and nourishment of the embryo within the seed

9. A land plant produces flagellated sperm and the dominant generation is diploid. The plant is most likely a
a. fern
b. moss
c. conifer
d. chlorophyte
e. dicot

10. In the evolution of flowers, there have been major trends toward
a. bilateral symmetry, reduction and fusion of parts, and ovary above petals
b. radial symmetry, proliferation and separation of parts, and ovary above petals
c. bilateral symmetry, reduction and fusion of parts, and ovary below petals
d. radial symmetry, reduction and fusion of parts, and ovary below petals
e. bilateral symmetry, proliferation and separation of parts, and ovary below petals

CHALLENGE QUESTIONS

1. The history of life has been punctuated by several mass extinctions. The impact of a meteorite may have wiped out the dinosaurs and many forms of marine life at the end of the Cretaceous period. Fossils indicate that plants were much less severely affected by this and other mass extinctions. What adaptations may have enabled plants to weather these disasters better than animals?

2. Plant evolution has been characterized by compromise. For example, branching patterns that allow growth mainly in the vertical direction are more efficient at gathering light because they can reach beyond the shadows of other plants. However, tall plants must bear the mechanical stresses associated with their height and have woody stems that increase in girth as the plant becomes taller. What trade-off might limit the bushiness of plants, considering the advantage that more extensively branched plants have in outcompeting neighboring plants by shading them?

SCIENCE, TECHNOLOGY, AND SOCIETY

1. Why are tropical rain forests being destroyed at such an alarming rate? What kinds of social, technological, and economic factors are responsible? Most forests in the northern developed countries have already been cut. (In the United States, less than 2% of the forest outside Alaska remains undisturbed.) Do the developed nations have a right to ask the southern developing nations to slow or stop the destruction of their forests? Defend your answer. What kinds of benefits, incentives, or programs might slow the assault on the rain forest?

FURTHER READING

Fearnside, P. M. "Extractive Reserves in Brazilian Amazonia." *BioScience,* June 1989. The economics of the tropical rain forest.

Heyler, D., and C. M. Poplin. "The Fossils of Montceau-les Mines." *Scientific American,* September 1988. Going back in time to a Carboniferous swamp.

Lewington, A. *Plants for People.* New York: Oxford University Press, 1990. The many uses of plant products.

Mauseth, J. *Botany.* Philadelphia: Saunders, 1991. A beautifully illustrated introduction to plants.

Nicholson, R. "Death and Taxus." *Natural History,* September 1992. A promising cancer drug from yew.

Stewart, D. "Green Giants." *Discover,* April 1990. Why redwoods and sequoias, the largest of all organisms, face extinction.

Wickelgren, I. "Plants Poised at Extinction's Edge." *Science News,* December 1988. A brief article describing the impending extinction of nearly 700 plant species in the United States.

Woolf, N. B. "Biotechnologies Sow Seeds for the Future." *BioScience,* May 1990. New methods for preserving samples of rare plants.

28 Fungi

Characteristics of Fungi
The Diversity of Fungi
Some Unique Lifestyles of Fungi
The Ecological Importance of Fungi
The Evolution of Fungi

Figure 28.1
Fungi at work. The mushrooms of this "fairy ring" are reproductive structures that develop at the edge of the main body of the fungus, which consists of an underground mass of tiny filaments within the ring. The filamentous network, a mycelium, secretes enzymes that decompose the dead grass thatch of the lawn. This chapter introduces the diversity, ecological significance, and evolution of fungi.

The words *fungus* and *mold* may evoke some unpleasant images. Fungi rot timbers, attack plants, spoil food, and afflict humans with athlete's foot and worse maladies. However, ecosystems would collapse without fungi to decompose dead organisms, fallen leaves, feces, and other organic materials, thus recycling vital chemical elements back to the environment in forms other organisms can assimilate (Figure 28.1). Virtually all plants depend on mutualistic fungi that help their roots absorb minerals and water from the soil. In addition to these ecological roles, fungi have been used by humans in various ways for centuries. We eat some fungi (mushrooms, for instance), culture fungi to produce antibiotics and other drugs, add them to dough to make bread rise, and use them to ferment beer and wine. Whatever our initial subjective perceptions may be, as objects of study, fungi are fascinating. They are a form of life so distinctive that in the five-kingdom system of taxonomy they have been accorded their own kingdom.

This chapter is an introduction to fungi. We will characterize the members of the kingdom Fungi, survey their diversity, discuss their ecological and commercial impact, and consider speculations about their phylogeny. As we did with the plant kingdom, we will look in some detail at life cycles, mainly for what they tell us about the phylogeny and evolutionary adaptations of fungi.

CHARACTERISTICS OF FUNGI

Fungi are eukaryotes, and nearly all are multicellular. First studied by botanists, fungi were grouped with plants in the two-kingdom system. Now it is clear that fungi are not primitive or degenerate plants lacking chlorophyll, but are unique organisms that generally differ from other eukaryotes in style of nutrition, structural organization, and growth and reproduction.

Nutrition and Habitats

All fungi are heterotrophs. In contrast to animals—heterotrophs that, for the most part, ingest food—fungi

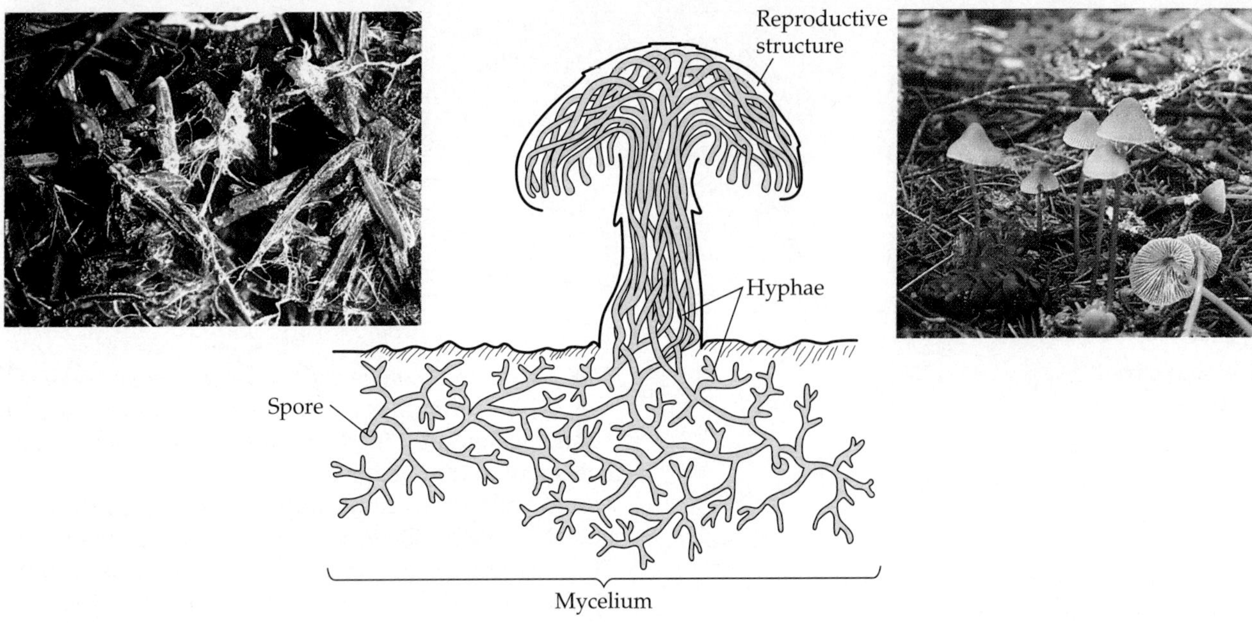

Figure 28.2
The fungal mycelium. The center diagram illustrates the relationship of the microscopic hyphae that make up a mycelium to the visible structure we recognize as a mushroom. The mushroom functions in reproduction by producing tiny cells called spores. (Other kinds of fungi produce spores on plain hyphae or in specialized reproductive structures that differ from mushrooms.) At left, the vegetative part of a mycelium is shown decomposing conifer needles, while at right, the mushrooms *(Mycena)* that grow up from this mycelium each fall are shown. A fungal mycelium begins with the germination of a fungal spore in a suitable habitat.

acquire their nutrients by **absorption.** In this mode of nutrition, small organic molecules are absorbed from the surrounding medium. A fungus digests food outside its body by secreting powerful hydrolytic enzymes into the food. The enzymes decompose complex molecules to the simpler compounds that the fungus can absorb and use.

Their absorptive mode of nutrition specializes fungi as saprobes, parasites, or mutualistic symbionts. Saprobic fungi absorb nutrients from nonliving organic material, such as fallen logs, animal corpses, or the wastes of live organisms, and in the process, the fungi decompose this material. Parasitic fungi absorb nutrients from the cells of living hosts. Many of these fungi, such as certain species infecting human lungs, are pathogenic. Mutualistic fungi also absorb nutrients from another organism, but they reciprocate with functions beneficial to their partners in some way, such as aiding a plant in the uptake of minerals from the soil.

Fungi inhabit diverse environments and are associated symbiotically with many organisms. Although most common in terrestrial habitats, many fungi inhabit aquatic environments where they are associated with both marine and freshwater organisms and their remains. The symbiotic fungi in lichens are found in some of the most inhospitable habitats on Earth: cold, dry deserts on Antarctica and alpine and arctic tundras. Other symbiotic fungi live inside the healthy tissues of plants, and still other species form cellulose-consuming mutualisms with insects, including ants and termites.

Structure

The vegetative (nutritionally active) bodies of most fungi are usually hidden, being diffusely organized around and within the tissues of their food sources. Except in yeasts, these bodies are constructed of basic building units called **hyphae** (singular, **hypha**) arranged in a "feeding" network known as a **mycelium** (Figures 28.1 and 28.2). Hyphae are minute threads composed of tubular walls surrounding their plasma membranes and cytoplasm. The cytoplasm contains the usual eukaryotic organelles.

Most fungal hyphae are divided into cells by cross-walls, or **septa** (singular, **septum**). The septa generally have pores large enough to allow ribosomes, mitochondria, and even nuclei to flow from cell to cell (Figure 28.3a). The cell walls of fungi differ from the cellulose walls of plants. Most fungi build their walls mainly of chitin, a strong but flexible nitrogen-containing polysaccharide similar to the chitin found in the external skeletons of insects and other arthropods. Some fungi are **aseptate;** that is, their hyphae are not divided into cells by cross-walls. These so-called **coenocytic**

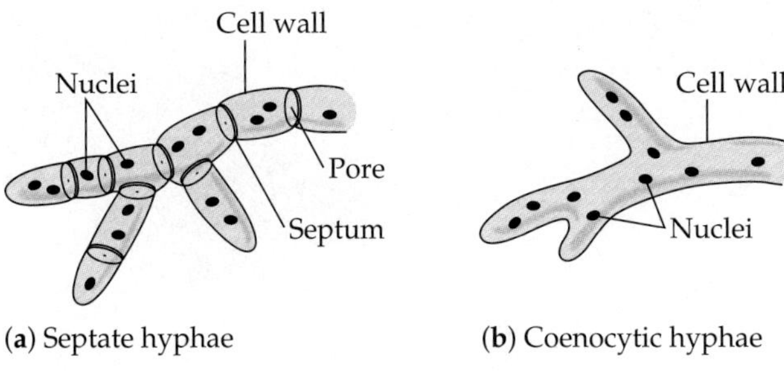

(a) Septate hyphae (b) Coenocytic hyphae

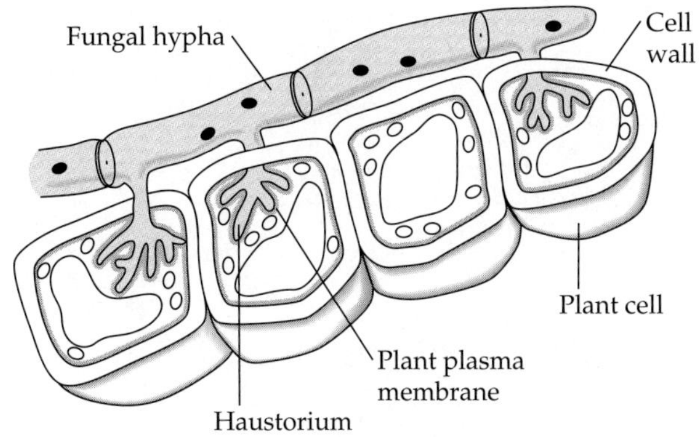

(c) Haustoria growing into host plant cells

Figure 28.3
Characteristics of fungal hyphae.

fungi consist of a continuous cytoplasmic mass with hundreds or thousands of nuclei (Figure 28.3b). The coenocytic condition results from the repeated division of nuclei without cytoplasmic division.

The filamentous structure of the mycelium provides an extensive surface area that suits the absorptive nutritional style of fungi: 10 cm^3 of rich organic soil may have as much as 1 km of hyphae. If these hyphae were 10 μm in diameter, about 314 cm^2 of fungal surface area would interface with the soil. Parasitic fungi usually have some of their hyphae modified as **haustoria,** nutrient-absorbing hyphal tips that penetrate the tissues of the host but remain outside the host cell membranes (Figure 28.3c).

According to the classification scheme used in this text, none of the fungi has flagellated stages in its life cycle. (This feature excludes the phyla Chytridiomycota and Oomycota from the kingdom Fungi; this text classifies these phyla as protists.)

Growth and Reproduction

The fungal mycelium grows rapidly, adding as much as a kilometer of hyphae each day as it ramifies (branches) within a food source. Such fast growth is possible because proteins and other materials synthesized by the entire mycelium are channeled by cytoplasmic streaming to the tips of the extending hyphae. The fungus concentrates its energy and resources on adding hyphal length rather than girth, another growth pattern adapted to the absorptive lifestyle. All fungi are nonmotile organisms; they cannot run, swim, or fly in search of food or mates. But the mycelium makes up for the lack of mobility by swiftly extending the tips of its hyphae into new territory.

The chromosomes and nuclei of fungi are relatively small, and the nuclei divide in a manner different from that of most other eukaryotes. From prophase to anaphase of mitosis, the nuclear envelope remains intact around an internal spindle. Following anaphase, the nuclear envelope pinches in two and the spindle disappears.

Fungi reproduce by releasing spores that are produced either sexually or asexually (following mitosis). Fungal spores come in all shapes and sizes. Usually they are unicellular, but multicellular spores also occur. They are produced in, or from, specialized hyphal compartments. For many fungi, as for protists, sex is a contingency mode of reproduction that occurs (following meiosis) when there has been some change in the environment. When conditions are habitable and stable, fungi generally clone themselves by producing enormous numbers of spores asexually. Carried by wind or water, the spores germinate if they land in a moist place where there is an appropriate substratum, or surface. Spores thus function in dispersal and account for the wide geographic distribution of many species of fungi. The airborne spores of fungi have been found more than 160 km (100 mi) above Earth.

The nuclei of fungal hyphae and spores are haploid, except for transient diploid stages that form during sexual life cycles. However, some mycelia may become genetically heterogeneous, resulting from the fusion of hyphae with different nuclei. In some cases, the different nuclei stay in separate parts of the same mycelium, which is then a mosaic in terms of phenotype. In other cases, the different nuclei mingle and may even exchange chromosomes and genes in a process similar to crossing over. A special case of genetic heterogeneity occurs during the sexual cycle of fungi. Syngamy, the sexual union of cells from two individuals, occurs in two stages that are separated in time. These two stages of syngamy are called **plasmogamy** (the fusion of cytoplasm) and **karyogamy** (the fusion of nuclei). After plasmogamy, the nuclei from each parent pair up but do not fuse, forming a **dikaryon.** The pairs of nuclei may exist and divide in tandem in a dikaryotic cell or mycelium for months or years. This condition has some of the advantages of diploidy; one haploid genome may be able to compensate for harmful mutations in the other nucleus, and vice versa. Finally, the nuclei fuse to form a diploid cell that undergoes immediate meiosis.

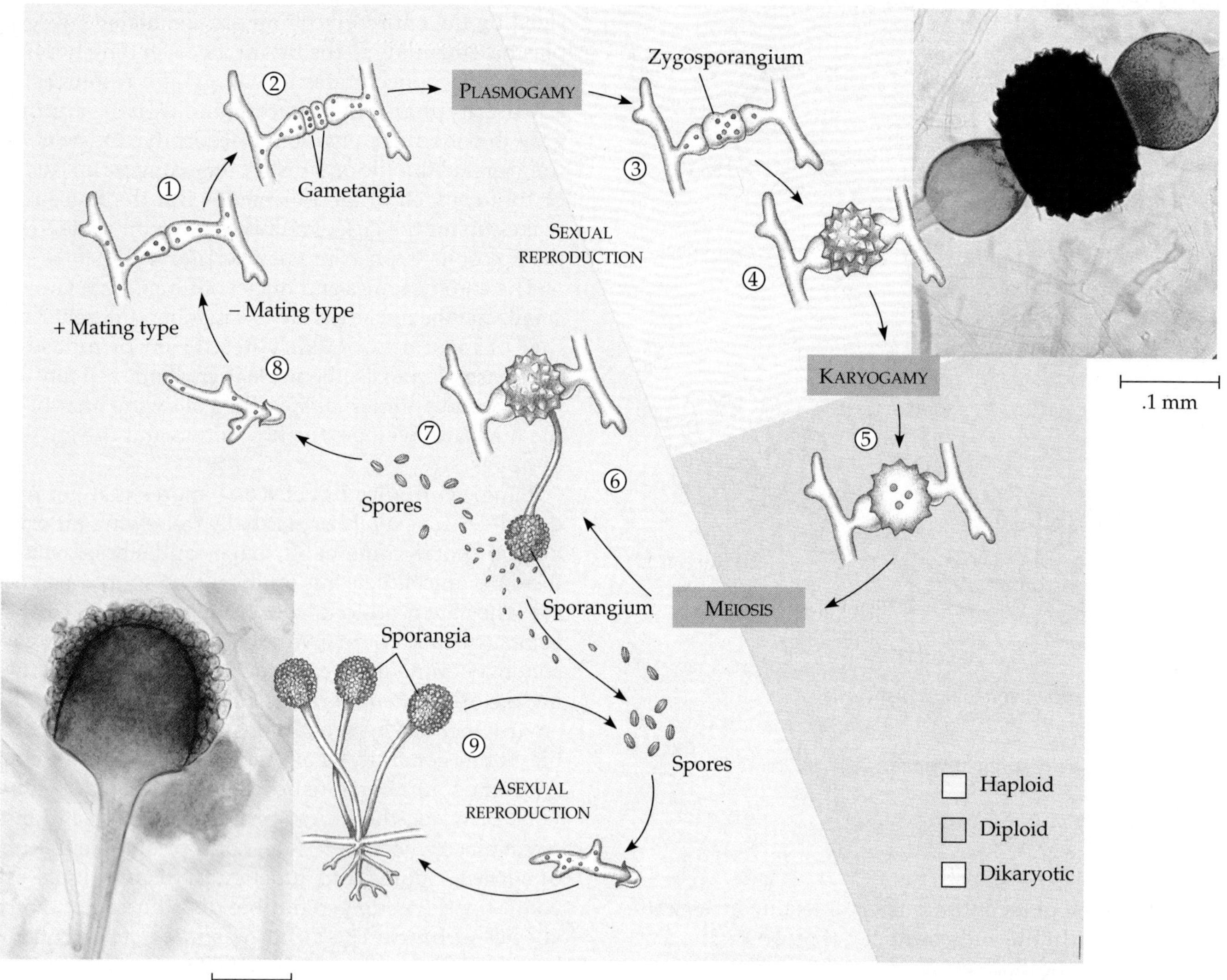

Figure 28.4
The life cycle of the zygomycete *Rhizopus*. ① Neighboring mycelia of opposite mating types (designated + and –) ② form hyphal extensions called gametangia, each walled off around several haploid nuclei by a septum. ③ The gametangia undergo plasmogamy and the haploid nuclei pair off, forming a dikaryotic zygosporangium. ④ This cell develops a rough, thick-walled coating (right LM) that can resist dry conditions and other harsh environments for months. ⑤ When conditions are favorable again, karyogamy between paired nuclei occurs, followed rapidly by meiosis. ⑥ The zygosporangium then breaks dormancy, germinating into a short sporangium that ⑦ disperses the genetically diverse, haploid spores. ⑧ These spores germinate and grow into new mycelia. ⑨ *Rhizopus* can also reproduce asexually by forming sporangia (left LM).

THE DIVERSITY OF FUNGI

More than 100,000 species of fungi are now known, and mycologists describe another thousand or so each year. The taxonomic scheme used in this chapter classifies the species into three divisions. Use of the botanical term *division* instead of *phylum* is a vestige of the two-kingdom system, when fungi were grouped with plants. The three divisions differ in the structures involved in plasmogamy, the length of time spent as a dikaryon, and the location of karyogamy. In fact, the three divisions are named after the sexual cells in which karyogamy occurs.

Division Zygomycota

Mycologists have described about 600 zygomycetes, or zygote fungi. These fungi are mostly terrestrial and live in soil or on decaying plant and animal material. One group of major importance forms **mycorrhizae,** mutualistic associations with the roots of plants (see Figure 28.13). Zygomycete hyphae are coenocytic, with septa found only where reproductive cells are formed. The name of this division comes from **zygosporangia,** resistant structures formed during sexual reproduction.

A common zygomycete is black bread mold, *Rhizopus stolonifer,* still an occasional household pest

(a)

(b)

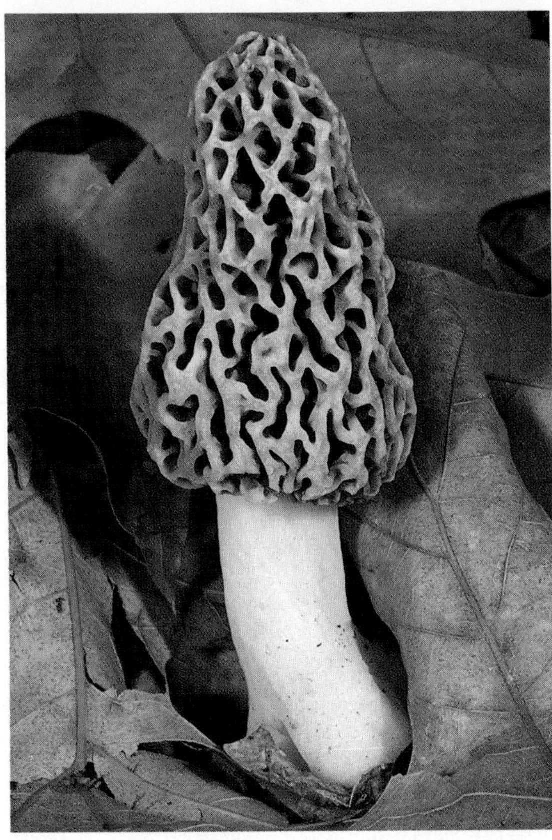

(c)

Figure 28.5
Ascomycota. Ascomycetes range from yeasts to such gastronomical delicacies as truffles and morels. (**a**) Many ascomycetes form ascocarps shaped like tiny flasks. In the "carbon fungus," *Hypoxylon multiforme*, these ascocarps appear in clusters on the wood the mycelium decomposes. (**b**) Truffles are ascocarps that fruit underground and emit strong odors, which attract animals that eat the fungi and disperse the ascospores. *Tuber melanosporum* is highly prized for its flavor by gourmet cooks, who pay over $600 a pound for this truffle. (**c**) *Morchella esculenta*, the succulent morel, is often found under trees in orchards. Many, if not all, morels produce mycorrhizae.

despite the addition of preservatives to most processed foods. Horizontal hyphae spread out over the food, penetrate it, and absorb nutrients into a rapidly expanding mycelium. In the asexual phase, bulbous black sporangia develop at the tips of upright hyphae (Figure 28.4, left). Within each sporangium, hundreds of haploid spores develop and are dispersed through the air. Spores that happen to land on moist food germinate, growing into new mycelia. If environmental conditions deteriorate (for instance, if all the food is used up), and mycelia of opposite mating types are present, this species of *Rhizopus* reproduces sexually (Figure 28.4, right). The zygosporangia formed are resistant to freezing and desiccation and are metabolically inactive. When conditions improve, the zygosporangia release haploid spores to recolonize the new substrate.

Air currents are not a very precise way to disperse spores, but *Rhizopus* releases the tiny cells in great numbers. Though they drift aimlessly, enough land in hospitable places. Some zygomycetes, however, are actually able to aim their spores. One is *Pilobolus*, a fungus that decomposes animal dung. *Pilobolus* bends its sporangium-bearing hyphae toward light, a direction where grass is likely to be growing. The whole sporangium is then shot off on an explosive squirt of cytoplasm out the end of the hypha, sometimes carrying the sporangium as far as 2 m. This adaptation disperses the spores away from the mass of dung and onto surrounding grass, which will be eaten by a herbivore such as a cow. The asexual life cycle is completed when the animal scatters the spores in feces.

Division Ascomycota

Over 60,000 species of ascomycetes, or sac fungi, have been described from a wide variety of habitats. They range in size and complexity from unicellular yeasts to minute leaf-spot fungi to elaborate cup fungi and truffles (Figure 28.5). Ascomycetes include the most devastating plant pathogens (see Figure 28.14), but a great many are important saprobes, particularly of plant material, and about half the species in this division live

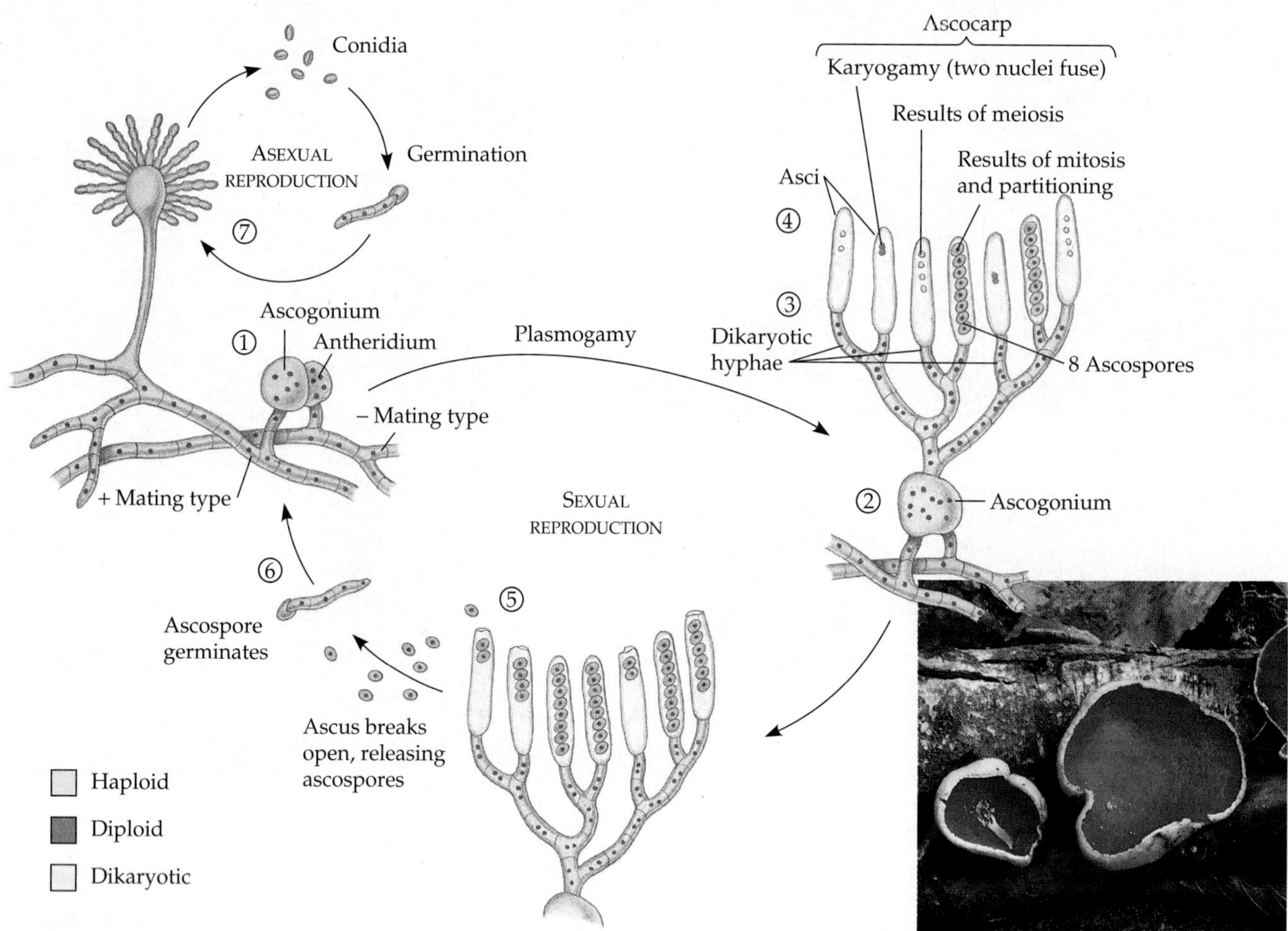

Figure 28.6
The life cycle of an ascomycete.
① Haploid mycelia of opposite mating strains become intertwined. One acts as a "female," producing a structure called an ascogonium, which receives many haploid nuclei from the antheridium of the "male." ② The ascogonium then has a pool of nuclei from both parents but karyogamy does not occur at this time. ③ The ascogonium gives rise to dikaryotic hyphae that are incorporated into an ascocarp—the cup of a cup fungus. The photograph features ascocarps of the scarlet cup *(Sarcoscypha coccinea)* growing on dead twigs in early spring. ④ The tips of the ascocarps' dikaryotic hyphae are partitioned into asci. Karyogamy occurs within these asci, and the diploid nucleus divides by meiosis to yield four haploid nuclei. Each of these haploid nuclei divides once by mitosis, and the ascus now contains eight nuclei. Cell walls develop around these nuclei to form ascospores. ⑤ When mature, all ascospores in an ascus are dispersed at once out the end of the ascus. A collapsing ascus jars neighboring asci and causes them to fire their spores. The chain reaction releases a visible cloud of spores with an audible hiss. ⑥ Germinating ascospores give rise to new haploid mycelia. ⑦ Ascomycetes can also reproduce asexually by producing airborne spores called conidia.

with algae in the symbiotic associations called lichens. Some ascomycetes, including truffles and morels, form mycorrhizae with plants. Others live in leaves on the surface of mesophyll cells, where the fungi apparently help protect these plant tissues from insects by releasing toxic compounds. Marine ascomycetes are the major nonbacterial saprobes on plant and seaweed material in saltwater habitats.

The defining feature of the Ascomycota is the production of sexual spores in saclike **asci** (singular, **ascus**) (Figure 28.6). Unlike the zygote fungi, most sac fungi bear their sexual stages in macroscopic fruiting bodies, or **ascocarps** (see Figure 28.5a). Before developing ascocarps, ascomycetes reproduce by producing enormous numbers of asexual spores, which are often dispersed by wind. The asexual spores are produced at the ends of hyphae, often in long chains or clusters. Because these spores are not formed inside sporangia, as in the Zygomycota, they are distinguished as **conidia** (Gr., "dust").

Compared to zygomycetes, ascomycetes are characterized by a more extensive dikaryotic stage, which is associated with formation of ascocarps (see Figure 28.6). Plasmogamy gives rise to the dikaryotic hyphae, and the cells at the tips of these hyphae become the asci. Within the asci, karyogamy combines the two

(a)

(b)

(c)

Figure 28.7
Basidiomycota. These photos showcase diverse basidiocarps, the fruiting bodies that produce sexual spores. (**a**) A miniature waxy cap *(Hygrophorus)* is mycorrhizal with oaks. (**b**) A shelf fungus growing on the trunk of a tree. Shelf fungi are important decomposers of wood. (**c**) The common puffball *(Lycoperdon gemmatum)* is a saprobe on buried plant litter. When raindrops hit the basidiocarp, basidiospores are expelled.

parental genomes, and then meiosis forms genetically varied **ascospores.** In many asci, eight ascospores are lined up in a row in the order in which they formed from a single zygote. This arrangement provides geneticists with a unique opportunity to study genetic recombination. Genetic differences between mycelia grown from ascospores taken from the same ascus reflect crossing over and independent assortment of chromosomes during meiosis.

Division Basidiomycota

Approximately 25,000 fungi, including mushrooms, shelf fungi, puffballs, and rusts, are classified in the division Basidiomycota (Figure 28.7). The name derives from the **basidium** (L., "little pedestal"), a transient diploid stage in the organism's life cycle. The clublike shape of the basidium also gives rise to the common name club fungus.

Basidiomycetes are important decomposers of wood and other plant material. The division also includes mycorrhiza-forming mutualists and plant parasites. Of all fungi, the saprobic basidiomycetes are best at decomposing the complex polymer lignin, an abundant component of wood. Many shelf fungi (Figure 28.7b) live as parasites on wood of weak or damaged trees and function later as saprobes after the trees die. Half of the mushroom-forming club fungi are saprobic and the other half form mycorrhizae. There are very few strictly parasitic mushrooms. However, two groups of basidiomycetes, the rusts and smuts, include particularly obnoxious plant parasites.

The life cycle of a club fungus usually includes a long-lived dikaryotic mycelium (Figure 28.8). Periodically, in response to environmental stimuli, this mycelium reproduces sexually by producing elaborate fruiting bodies called **basidiocarps.** The numerous basidia of a basidiocarp are the sources of sexual spores. Asexual reproduction in basidiomycetes is much less common than in ascomycetes, but also occurs as conidia.

A mushroom is an example of a basidiocarp. The cap of the mushroom supports and protects a large surface area of basidia on gills; each common, store-bought mushroom has a gill surface area of about 200 cm^2. Such a mushroom may release a billion spores, which drop beneath the cap and are blown away. The giant puffball may be over 1 m in diameter and produces trillions of spores, which are released when the basidiocarp breaks apart due to animal activity, wind, or rain. Basidiomycetes called stinkhorns produce basidiospores in a slimy, stinky mass on upper surfaces of the basidiocarp and attract carrion-feeding insects, such as flies, that disperse the sticky spores.

A ring of mushrooms, popularly called a fairy ring, may appear on a lawn overnight (see Figure 28.1). These mushrooms are saprobes on dead plant material, not parasites on living grass. Although the grass in the center of the ring is normal, after a few days the grass near the ring is stunted and the grass just outside the garland of mushrooms is especially lush. As the underground mycelium grows outward, its center portion and the mushrooms above it die because the mycelium has consumed all the available nutrients. The grass beneath the mushrooms is stunted because it

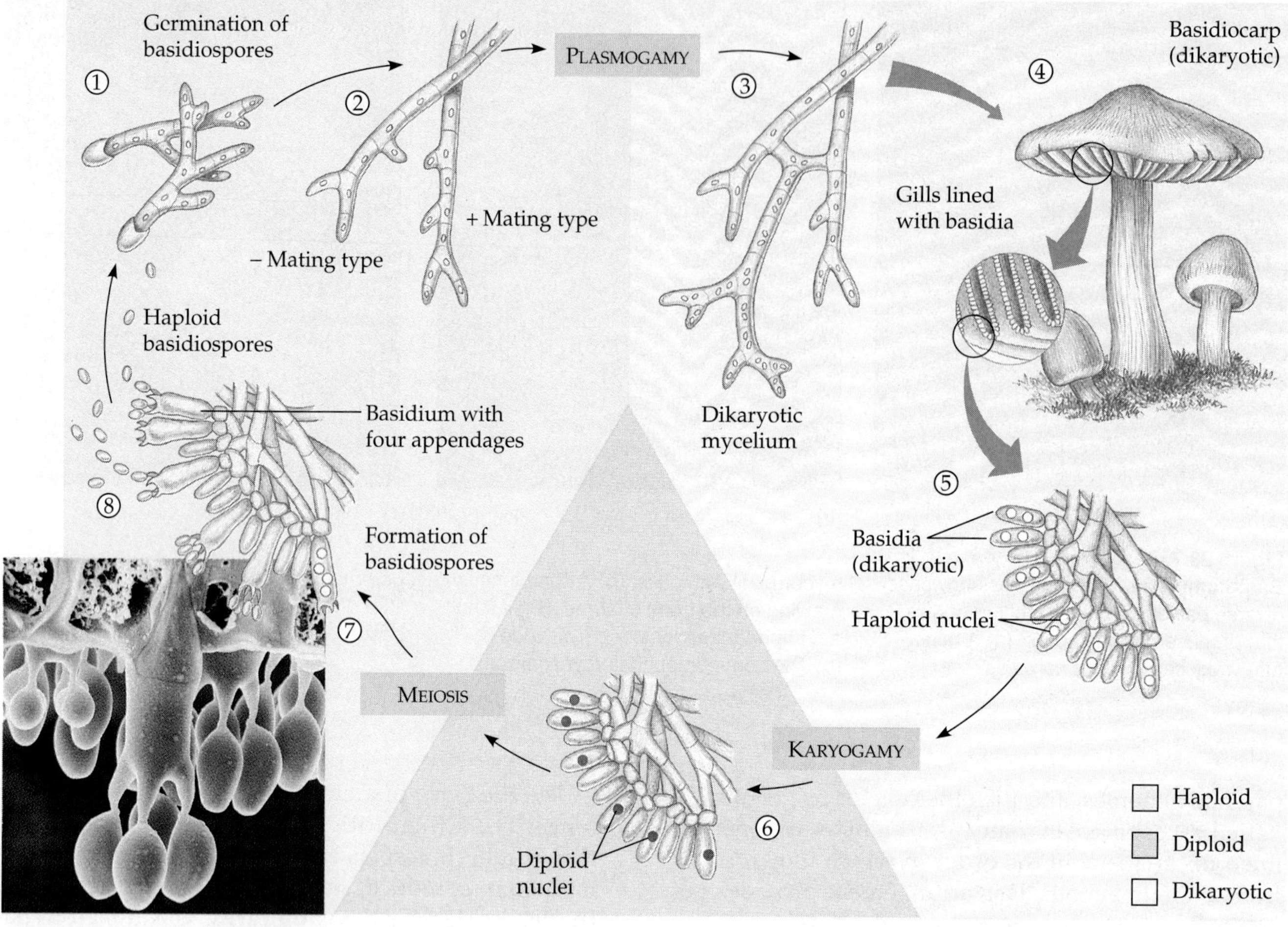

Figure 28.8
The life cycle of a mushroom-forming basidiomycete. ① Haploid basidiospores germinate in a suitable environment and grow into short-lived haploid mycelia. ② Undifferentiated hyphae from two haploid mycelia of opposite mating type undergo plasmogamy, ③ creating a dikaryotic mycelium that grows faster than, and ultimately crowds out, the parent haploid mycelium. ④ The mycelium of the mushroom illustrated here *(Cortinarius)* forms mycorrhizae with trees. Environmental cues such as rain, temperature changes, and, for mycorrhizal species, seasonal changes in the plant host, induce the dikaryotic mycelium to form compact masses that develop into mushrooms. Cytoplasm streaming in from the mycelium and from the attached mycorrhizae swells the hyphae of mushrooms, causing them to "pop up" overnight. The dikaryons of basidiomycetes are long-lived, generally producing a new crop of basidiocarps (mushrooms, in this case) each year. ⑤ Karyogamy occurs in the terminal dikaryotic cells that line the surfaces of the gills (see SEM). ⑥ Each cell swells to form a diploid basidium, which rapidly undergoes meiosis and yields four haploid nuclei. ⑦ The basidium then grows four appendages, and one haploid nucleus enters each appendage and develops into a basidiospore. ⑧ When mature, the basidiospores are propelled slightly (by electrostatic forces) into the spaces between the gills. After the spores drop below the cap, they are dispersed by the wind.

cannot compete for minerals with the active mycelium. But fungal enzymes diffuse ahead of the advancing mycelium, and the grass there is fertilized by the minerals that become available. The fairy ring slowly increases in diameter as the mycelium advances at a rate of about 30 cm per year. Some giant fairy rings may be centuries old. This perennial nature of basidiomycete mycelia was recently highlighted by the discovery, in Michigan, of a genetically uniform mycelium of the honey mushroom, *Armillaria bulbosa,* which occupies about 37 acres, weighs more than 100 tons, and may be more than 1500 years old.

SOME UNIQUE LIFESTYLES OF FUNGI

Certain ways of living that involve both morphological and ecological specialization have evolved independently among the three divisions of the kingdom Fungi. These lifestyles allow the fungi to take advantage of unusual habitats. Humans have learned to exploit the abilities of these fungi for a variety of commercial purposes. This section discusses four fungal forms with unique ways of life: molds, yeasts, lichens, and mycorrhizae.

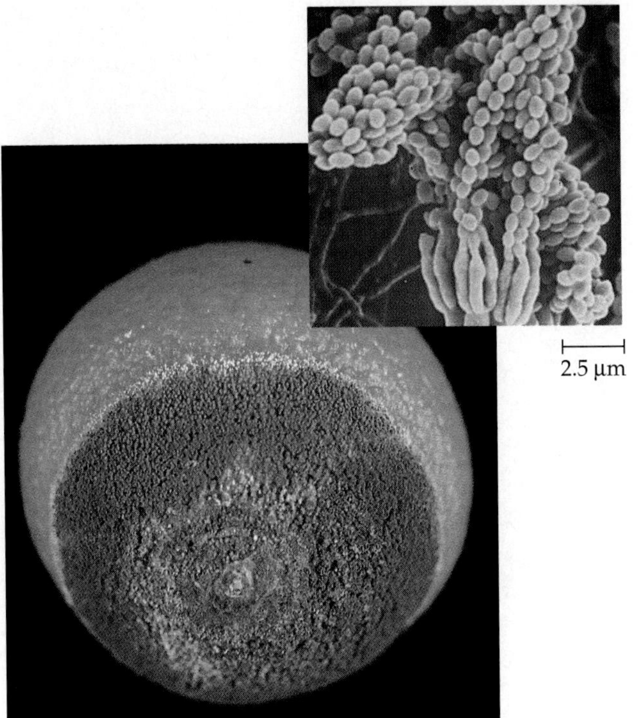

Figure 28.9
A moldy orange. "Blue mold" is usually caused by saprobic species of *Penicillium*, a deuteromycete. This mold reproduces asexually by producing chains of conidia on hyphae called conidiophores (SEM).

Figure 28.10
***Arthrobotrys*, a predatory deuteromycete.** Portions of the hyphae are modified as hoops that constrict around nematodes (roundworms) in less than a second, when an unsuspecting worm rubs the inside of the hoop. The fungus then penetrates its prey with hyphae and digests the inner tissues.

Molds

Mention of fungi may bring the ubiquitous molds to mind. A **mold** is a rapidly growing, asexually reproducing fungus. The mycelia of these fungi grow as saprobes or parasites on a great variety of substrates. You are already familiar with one example, bread mold *(Rhizopus)*. Molds may go through a series of different reproductive stages. Early in life, a mold produces asexual spores. The term *mold* applies only to these asexual stages. Later, the same fungus *may* reproduce sexually, producing zygosporangia, ascocarps, or basidiocarps. There are also molds that cannot be classified as zygomycetes, ascomycetes, or basidiomycetes, because they have no known sexual stages. Those molds are classified as Deuteromycota, or **imperfect fungi** (from the botanical use of the term *perfect* to refer to the sexual stages of life cycles). Imperfect fungi reproduce asexually by producing conidia on specialized hyphae called conidiophores (Figure 28.9).

Humans have found many commercial uses for molds. Imperfect fungi are sources of antibiotics; pharmaceutical companies grow these molds in large liquid cultures, then extract the antibiotics. Penicillin is produced by some species of *Penicillium*, an imperfect fungus. Other *Penicillium* species are important fermenters on the surfaces of blue cheese, Brie, and Camembert and produce the unusual colors and flavors we associate with each type of cheese. Roquefort is goat milk cheese that has been incubated in certain caves in the Roquefort region of France, where wild *Penicillium roquefortii* is allowed to "infect" the cheese naturally. A species of the zygomycete mold *Rhizopus* is used to ferment cakes of soybeans into a tasty product called tempeh. The imperfect mold *Tolypocladium inflatum* is the commercial source of cyclosporine, used as a drug to help prevent rejection of foreign organs in transplant patients (see Chapter 39).

Among the more unusual deuteromycetes are predatory fungi in soil that trap, kill, and consume small animals, especially roundworms, the nematodes (Figure 28.10). This provides additional nitrogen-containing compounds, which are in short supply in the wood these molds decompose.

Yeasts

Yeasts are unicellular fungi that inhabit liquid or moist habitats, including plant sap and animal tissues. Yeasts reproduce asexually, by simple cell division or by the pinching of small "bud cells" off a parent cell. Some yeasts reproduce sexually, by forming asci or basidia, and are then classified as Ascomycota or Basidiomycota. Others are placed in the division Deuteromycota because no sexual stages are known. Some fungi can grow as either a mycelium or a yeast, depending on the amount of liquid in the environment.

Humans have used yeasts to raise bread and ferment alcoholic beverages for thousands of years. Only relatively recently have the yeasts involved been separated into pure culture for more controlled human use. The yeast *Saccharomyces cerevisiae*, an ascomycete, is the most important of all domesticated fungi. The tiny yeast cells, available as many strains of baker's yeast and brewer's yeast, are very active metabolically. When oxygen is present, baker's yeast respires, releasing small bubbles of CO_2 that leaven the dough. Cultured anaerobically in breweries and wineries, *Saccharomyces* ferments sugars to alcohol. Researchers use *Saccharomyces* to study the molecular genetics of eukaryotes because these microbes are easy to culture and manipulate (see Chapter 19).

Some yeasts cause problems for humans. A pink yeast, *Rhodotorula*, grows on shower curtains and other moist surfaces in our homes. Another well-known yeast is *Candida*, one of the normal inhabitants of moist human tissues, such as the vaginal lining. Certain circumstances can cause *Candida* to become pathogenic by growing too rapidly and releasing harmful substances. This can occur, for example, with an environmental change, such as a pH change, or when the immune system of the human host is compromised (by AIDS, for instance).

Figure 28.11
Lichens. Lichens representing three growth forms inhabit the bark of this maple branch. *Parmelia* (lower left) is foliose (having a flattened, leaflike appearance with differentiated upper and lower surfaces). *Ramalina* (upper right) is fruticose (shrub-like). Elsewhere, several crustose (paint smear-like) species in the genera *Lecanora* and *Bacidia* are forming disk-shaped ascocarps. The minute orange lichen is *Xanthoria*, a foliose form.

Lichens

At a distance, lichens are often mistaken for mosses or other simple plants growing on rocks, rotting logs, trees, and roofs (Figure 28.11). In fact, lichens are *not* mosses or any other kind of plant, nor are they even individual organisms. A **lichen** is a symbiotic association of millions of photosynthetic microorganisms tangled in a lattice of fungal hyphae. The fungal component is most commonly an ascomycete, but several basidiomycete lichens are known. The photosynthetic partners are usually unicellular or filamentous green algae or cyanobacteria (blue-green algae). The merger of fungus and alga is so complete that lichens are actually given genus and specific names, as though they were single organisms. The scientific name of the lichen is the name of the fungus. Over 25,000 species have been described.

Lichens vary considerably in the details of their architecture and physiology, but there are some key similarities. The fungus usually gives the lichen its overall shape and structure, and tissues formed of hyphae account for most of the lichen's mass. The algal component usually occurs in an inner layer below the lichen surface. In most cases that have been examined, each partner provides things the other could not obtain on its own. The alga always provides the fungus with food. Cyanobacteria in lichens fix nitrogen (see Chapter 25) and provide organic nitrogen. The fungus provides the alga with an ideal physical environment for growth. Lichens absorb most of the minerals they need either from air in the form of dust or from rain. The physical arrangement of hyphae retains water and minerals, allows for gas exchange, and protects the algae (Figure 28.12). The fungi produce unique organic compounds with several functions. Fungal pigments help shade the algae from intense sunlight. Some fungal compounds are toxic and prevent lichens from being eaten by consumers. The fungi also secrete acids, which aid in the uptake of minerals.

The fungi of many lichens reproduce sexually by forming ascocarps or, rarely, basidiocarps (see Figure 28.12). Lichen algae reproduce independently of the fungus by asexual cell division. As might be expected of "dual organisms," asexual reproduction as symbiotic units also occurs commonly, either by fragmentation of the parental lichen or by formation of specialized structures called **soredia** (see Figure 28.12). Soredia are small clusters of hyphae with embedded algae.

Whether the lichen symbiosis is mutualistic or parasitic is still debated. Lichens are able to live in environments where neither fungi nor algae could live alone. While the fungal components do not grow alone in the wild, some lichen algae also occur as free-living organ-

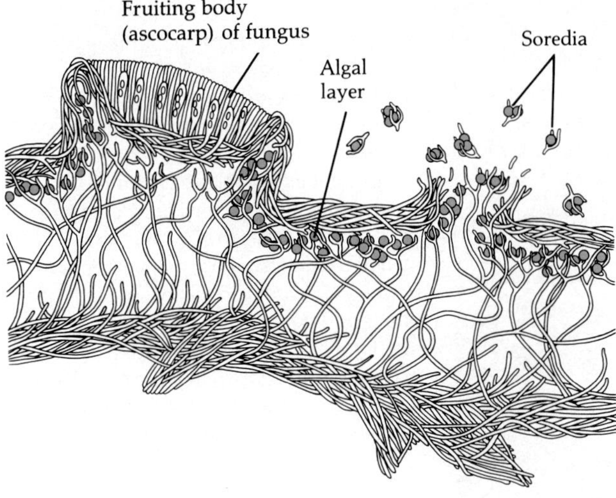

Figure 28.12
Anatomy of a lichen. The upper and lower surfaces are protective layers of tightly packed fungal hyphae. Just beneath the upper surface are the algae, enmeshed in a net of hyphae. The middle of a lichen generally consists of loosely woven hyphae of the fungus. Reproductive structures generally form on the upper surface. A sexual ascocarp of the fungus and several asexual soredia that disperse both fungal and algal components are shown here.

isms. The fungi and algae of some lichens have been experimentally separated into culture. Such cultures look like free-living molds and algae and lose certain physiological behaviors typical of lichens. For instance, the cultured fungi do not produce lichen compounds. Cultured lichen algae do not "leak" carbohydrate from their cells as they do in lichens. Most evidence points to a mutualistic type of symbiosis. In some lichens, however, the fungus invades the algal cells with haustoria and may kill some of them, though not as fast as the alga replenishes its numbers by reproduction. Citing this apparent harm to the algae, some lichenologists refer to lichen symbiosis as "controlled parasitism" rather than mutualism.

Lichens are important pioneers on newly cleared rock and soil surfaces, such as burned forests and volcanic flows. Physical penetration of the outer crystals of rocks and chemical attack of rock by lichen acids help break down the rock and establish soil-trapping lichens. This process makes it possible for a succession of plants to grow. Nitrogen-fixing lichens also add organic nitrogen to some ecosystems.

Some lichens tolerate severe cold. In the arctic tundra, great herds of caribou and reindeer graze on carpets of reindeer lichen at times of the year when other foods are unavailable. Lichens can also survive desiccation. When it is foggy or rainy, lichens may absorb more than ten times their weight in water. Photosynthesis begins almost immediately when the water content reaches 65% to 75%. In dry air, lichens rapidly dehydrate and photosynthesis stops. Thus in arid climates, lichens grow very slowly, often in spurts of less than a millimeter per year. Some lichens are thousands of years old, rivaling the oldest plants as the elder organisms on Earth.

As tough as lichens are, many do not stand up very well to air pollution. Their passive mode of mineral uptake from rain and moist air makes lichens particularly sensitive to sulfur dioxide and other aerial poisons. The death of sensitive lichens and an increase in hardier species in an area is an early warning that air quality is deteriorating.

Mycorrhizae

Physiological cousins of lichens are mycorrhizae, mutualistic associations of plant roots and fungi. The word *mycorrhizae* means "fungus roots," referring to the structures formed by both root cells and hyphae from the associated fungus. The anatomy of this symbiosis varies, depending on the type of fungus (Figure 28.13). The extensions of the fungal mycelium from the hyphae forming the mycorrhizae greatly increase the absorptive surface of the plant roots, and the partners exchange minerals accumulated from the soil by the fungus for organic nutrients synthesized by the plant.

Mycorrhizae are enormously important in natural ecosystems and agriculture. Over 95% of all vascular plants have mycorrhizae. The fungi involved are permanent associates with their hosts and periodically form fruiting bodies. Basidiomycota, Ascomycota, and Zygomycota all have members that form mycorrhizae. In fact, half of all species of mushroom-forming basidiomycetes form mycorrhizae with oak, birch, and pine trees. The mushrooms that sprout around the bases of these trees are the surface evidence of the underground symbiosis. (The structure and physiology of mycorrhizae will be described in more detail in Chapter 33.)

THE ECOLOGICAL IMPORTANCE OF FUNGI

Fungi as Decomposers

Fungi and bacteria are the principal decomposers that keep ecosystems stocked with the inorganic nutrients essential for plant growth. Without decomposers, carbon, nitrogen, and other elements would accumulate in corpses and organic wastes, no longer available as raw material for new generations of life. Imagine what would become of a forest if its decomposers rested for even a few years. Leaves, logs, feces, and dead animals

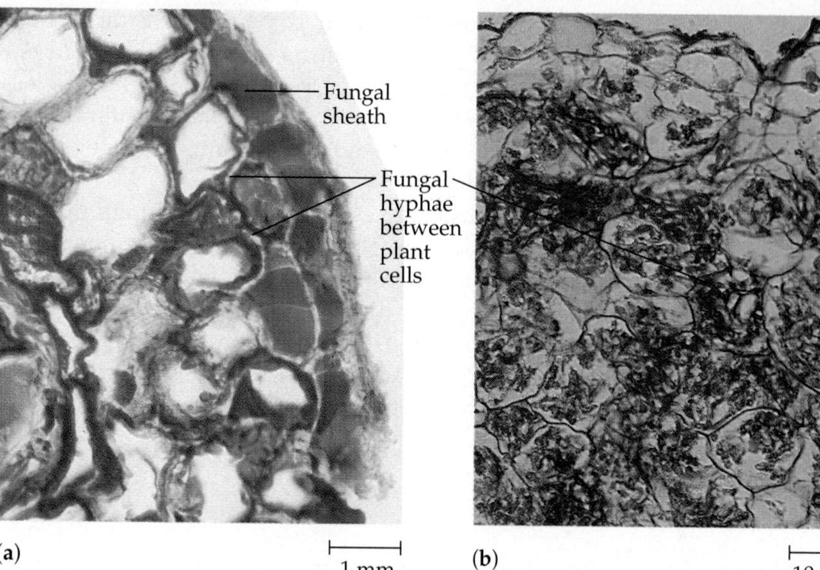

Figure 28.13
Mycorrhizae. (**a**) Ectomycorrhizae (sheathing mycorrhizae). The fungal mycelium forms a sheath around this pine root, with some hyphae extending between the plant cells into the root interior. The fungi of ectomycorrhizae are usually mushroom-forming basidiomycetes (LM). (**b**) Endomycorrhizae. Certain zygomycetes establish these mutualistic associations with roots, in this case, the roots of a fern. Fungal hyphae that penetrate the root form branched haustoria that exchange materials with the plant. These endomycorrhizae do not form dense sheaths around the root (LM).

would pile up on the forest floor. Plants and the animals they feed would starve because elements taken from the soil would not be returned. Gradually, the forest would die. This hypothetical forest is only a symbol of the fate that would befall all ecosystems without decomposers. In reality, an endless recycling of chemical elements occurs between organisms and their nonliving surroundings (see Chapter 49).

Fungi are well adapted as decomposers of plant material. Their invasive hyphae enter the tissues and cells of dead organic matter and hydrolyze polymers, including the cellulose and lignin of plant cell walls. A succession of fungi, in concert with bacteria and, in some environments, invertebrate animals, are responsible for the complete breakdown of plant litter. The air is so loaded with fungal spores that as soon as a leaf falls or an insect dies, it is covered with spores and is soon infiltrated by saprobic hyphae.

Fungi as Spoilers

We may applaud fungi that decompose forest litter or dung, but it is a different story when molds attack our fruit or our shower curtains. A wood-digesting saprobe does not distinguish between a fallen oak limb and the oak planks of a boat. During the Revolutionary War, the British lost more ships to dry rot than to enemy attack. Soldiers stationed in the tropics during World War II watched their tents, clothing, boots, and binoculars be destroyed by molds. Some fungi can even decompose certain plastics. The best way to protect materials from mold is to keep them as dry as possible.

In addition to their rotting action, some molds also taint food by producing poisons. For example, some species of the mold *Aspergillus* contaminate improperly stored grain by secreting aflatoxins, which are carcinogenic.

Pathogenic Fungi

Many fungi are pathogens. Among the diseases that fungi cause in humans are athlete's foot, ringworm, vaginal yeast infections, and lung infections that may be fatal.

Plants are particularly susceptible to fungal diseases. For example, the ascomycete that causes Dutch elm disease has drastically changed the landscape of the northeastern United States. The fungus was accidently introduced to the United States on logs that were sent from Europe to help pay World War I debts. Carried from tree to tree by bark beetles, the fungus is on its way to completely eliminating the American elm. Another ascomycete has almost eliminated the native American chestnut.

Some of the fungi that attack food crops are toxic to humans. One ascomycete forms purple structures called ergots on rye (Figure 28.14). If diseased rye is inadvertently milled into flour and consumed, poisons from the ergots cause gangrene, nervous spasms, burning sensations, hallucinations, and temporary insanity. One epidemic in 944 A.D. killed more than 40,000 people. During the Middle Ages, the disease (ergotism) became known as St. Anthony's fire because many of its victims were cared for by a Catholic nursing order dedicated to Saint Anthony. One of the hallucinogens that has been isolated from ergots is lysergic acid, the raw material from which LSD is made. Toxins extracted from fungi often have medicinal uses when administered in weak doses. For example, an ergot compound is helpful in treating high blood pressure and stopping maternal bleeding after childbirth.

Edible Fungi

In the food webs of most ecosystems, fungi are consumed as food by a variety of animals, including hu-

Figure 28.14
Ergots on rye. The ascomycete *Claviceps purpurea* forms dark purple growths (ergots) on the seed heads of rye and other grasses. Ergots, which are the overwintering bodies of the fungus, contain toxins poisonous to humans and domestic animals.

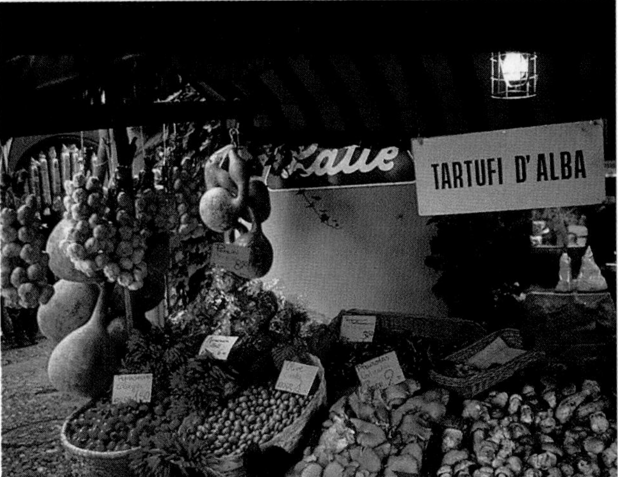

Figure 28.15
Edible mushrooms in an Italian open-air market. On sale along with other delicacies are a variety of fungi: king bolete, *Boletus edulis,* both dried ("Porcini," back left) and fresh (front right); fresh oyster mushroom, *Pleurotus ostreatus* (front center); and chanterelles, *Cantharellus cibarius* (back right in basket). White truffles ("Tartufi d'alba," *Tuber magnatum*) are advertised for sale also.

mans. Most of us have eaten mushrooms (basidiomycetes), but we usually restrict our consumption to one species of *Agaricus,* which is cultivated commercially on compost in the dark. In many countries, however, people eat a variety of cultivated and wild mushrooms (Figure 28.15). The fungi most prized by gourmets are truffles, the underground ascocarps of mycelia that are mycorrhizal on the roots of trees. Their complex flavor is variously described as nutty, musky, cheesy, or all three. The fruiting bodies (ascocarps) release strong odors that attract mammal and insect consumers that excavate the truffles and disperse their spores. In some cases, the odors mimic sex attractants of certain mammals. Truffle hunters traditionally used pigs to locate their prizes, although dogs are now more commonly used, because they have the nose for the scent without the fondness for the flavor. Mycologists make no distinction between poisonous and edible mushrooms, and there are no simple rules to help the novice distinguish them. Only experts in mushroom identification are qualified to collect wild mushrooms for eating.

THE EVOLUTION OF FUNGI

The evolutionary roots of fungi remain unclear. Most researchers agree that fungi arose directly from protistan ancestors; they are not directly related either to plants or animals or to the immediate protistan ancestors of plants or animals. As a kingdom, fungi apparently share a common ancestor, but phylogeny within the kingdom is uncertain.

Among the possible protistan ancestors of fungi are red algae, conjugating green algae, and chytrids (see Chapter 26). Molecular systematics in the form of rRNA sequencing is under way, but so far there are not enough data to say which protists are most closely related to fungi.

The oldest undisputed fossils of fungi are in Silurian strata about 400 to 440 million years old. Plants and fungi likely moved from water to land together. Fossils of the first vascular plants that colonized land during the late Silurian period have petrified mycorrhizae. From their inception, terrestrial communities were dependent on fungi as decomposers and as symbiotic associates that helped other organisms adapt to living on land.

STUDY OUTLINE

1. The fungi constitute a kingdom of organisms that have tremendous impact, both positive and negative, on other organisms.

Characteristics of Fungi (pp. 583–585)

1. The fungi are a eukaryotic, primarily multicellular group that lack flagella. Most have walls made of chitin.
2. All fungi are heterotrophs, acquiring their nutrients by absorption. There are saprobic decomposers, parasitic species, and mutualistic forms.
3. The fungal vegetative body consists of mycelia, netlike collections of branched hyphae ideally suited for absorption. Parasitic fungi penetrate their hosts with specialized hyphae called haustoria.
4. Although aseptate (coenocytic) forms occur, most fungi have their hyphae partitioned into cells by septa, with pores allowing cell-to-cell continuity.
5. Fungi reproduce by dispersing astounding numbers of airborne and aquatic (nonflagellated) spores. Spores germinate to produce short hyphae, which soon grow and branch.
6. Fungal life cycles vary by division but generally start with haploid mycelia. Early reproduction is usually asexual and later reproduction is sexual.
7. The sexual cycle involves cell fusion (plasmogamy) and nuclear fusion (karyogamy), with an intervening dikaryotic stage (with two haploid nuclei) of varying length, depending on the fungal division. The diploid phase is short-lived and rapidly undergoes meiosis to produce haploid spores.

The Diversity of Fungi (pp. 586–590)

1. The zygote fungi of the division Zygomycota, fungi that live in soil or decaying organic matter, include the familiar black bread mold. Asexual spores develop in aerial sporangia. The division is named for its sexually produced zygosporangia, which are tough, dormant, dikaryotic structures capable of persisting through unfavorable conditions.
2. The sac fungi of the division Ascomycota are most important as plant parasites, fungal components of lichens, and saprobes. Their mycelia are predominantly haploid with a short dikaryotic phase associated with sexual reproduction. Asexual reproduction by conidia is abundant. Sexual reproduction involves formation of spores in sacs, or asci, at the ends of dikaryotic hyphae, usually in ascocarps.
3. The division Basidiomycota, or club fungi, contains the mushrooms, shelf fungi, puffballs, and rusts, which are important as saprobes of wood and plant litter, fungal components of mycorrhizae, and plant parasites (rusts and smuts). Mycelia of the club fungi can last years as dikaryons. Sexual reproduction involves formation of spores on club-shaped basidia at the ends of dikaryotic hyphae in fruiting bodies, such as mushrooms.

Some Unique Lifestyles of Fungi (pp. 590–593)

1. Molds are either asexual stages of fungi classified as zygomycetes, ascomycetes, or basidiomycetes, or conidia-producing fungi not known to reproduce sexually (deuteromycetes, or imperfect fungi). Molds are important in the commercial production of antibiotics, such as penicillin.
2. Yeasts are unicellular fungi adapted to life in liquids such as plant saps. They may be classified as ascomycetes, basidiomycetes, or deuteromycetes. The yeast *Saccharomyces* is known as "bread and wine yeast."
3. Lichens are such highly integrated symbiotic associations of algae and fungi that they are classified as single organisms. The fungal component is usually an ascomycete and the algal component is generally a green alga or cyanobacterium. Lichens are rugged organisms that set the stage for plant growth by slowly breaking down bare rocks.
4. Mycorrhizae are mutualistic associations of fungi with the roots of vascular plants. The plant is more efficiently able to absorb minerals and water with aid of the fungus, which is provided with its carbohydrate needs in return.

The Ecological Importance of Fungi (pp. 593–595)

1. Without fungi as decomposers, we would be surrounded by the organic debris of dead organisms and deprived of the essential recycling of chemical elements between the biological and nonbiological world. Fungi are particularly important decomposers of wood.
2. Fungi also decompose food and other useful objects.
3. Some fungi cause disease, harming humans with a variety of ills from athlete's foot to fatal lung infections. Plants are especially vulnerable to fungal infections.
4. Many fungi are food for humans and other animals.

The Evolution of Fungi (p. 595)

1. The origin of fungi is unknown, but taxonomists generally agree that this kingdom arose directly from ancestral protists rather than from plants or animals. Fungi accompanied plants in the move from water to land.

SELF-QUIZ

1. The pores in the septa of a fungal mycelium function as
   a. regions of high rates of absorption of dissolved organic molecules through the cell walls
   b. the openings through which spores are expelled
   c. a means for proteins and other materials to move to the rapidly growing tips of hyphae
   d. devices for the unusual mitosis found in many fungi
   e. the dividers between the two kinds of nuclei of a dikaryon

2. In the sexual life cycles of fungi, _______ usually occurs shortly after _______.
   a. syngamy; karyogamy
   b. meiosis; fusion of haploid nuclei
   c. asexual reproduction; sexual reproduction
   d. spore production; formation of a dikaryotic mycelium
   e. syngamy; meiosis

3. Which of the following cells or structures are associated with *asexual* reproduction in fungi?

a. ascospores
b. basidiospores
c. conidia
d. zygosporangia
e. ascogonia

4. Which of the following lists three lifestyles that have evolved independently in the three divisions of fungi?
   a. sac fungi, club fungi, molds
   b. zygote fungi, yeasts, mycorrhizae
   c. yeasts, molds, sac fungi
   d. lichens, molds, yeasts
   e. sac fungi, zygote fungi, club fungi
5. Sporangia on erect hyphae that produce asexual spores are characteristic of
   a. Ascomycota
   b. Basidiomycota
   c. Deuteromycota
   d. Zygomycota
   e. lichens
6. Which of the following statements is true about Basidiomycota?
   a. Hyphae fuse to give rise to a dikaryotic mycelium.
   b. Spores line up in a sac after they are formed by meiosis.
   c. The vast majority of spores are formed asexually.
   d. Basidiomycetes are the most important commercial sources of antibiotics.
   e. Basidiomycetes have no known sexual stages.
7. Mycorrhizae are
   a. asexual reproductive structures formed by lichens
   b. thin hyphae that grow directly into host tissues
   c. the mycelium that forms fairy rings
   d. compact, dikaryotic hyphae that form a basidiocarp
   e. mutualistic associations between plant roots and fungi
8. Parasitic fungi have specialized hyphae called
   a. haustoria
   b. ascogonia
   c. conidia
   d. soredia
   e. ergots
9. The fungus responsible for Dutch elm disease is a(n)
   a. zygomycete
   b. ascomycete
   c. basidiomycete
   d. deuteromycete
   e. oomycete
10. The photosynthetic symbiont of a lichen is most commonly a(n)
   a. moss
   b. green alga
   c. red alga
   d. ascomycete
   e. small vascular plant

CHALLENGE QUESTIONS

1. In what way might use of a wide-spectrum fungicide that kills all fungal species have a harmful effect on vegetation, such as a forest?
2. Zygomycetes fuse dikaryotic nuclei to form zygotes, only to restore the haploid state again by meiosis before growth of new mycelia. What does this formation of a transient diploid stage accomplish?
3. Many fungi produce antibiotics, such as penicillin, that are valuable in medicine. Of what value are antibiotics to the fungi that produce them? Similarly, fungi often produce compounds with unpleasant tastes and odors as they digest their food. What might be the value of these chemicals to the fungi? To other organisms that might eat the decomposing food? How might the production of antibiotics and odors have evolved?

SCIENCE, TECHNOLOGY, AND SOCIETY

1. Many fungi and lichens in the United States are threatened with extinction due to destruction of habitats. Should they come under legal protection by their inclusion on lists of endangered species? Defend your answer.
2. American chestnut trees once made up more than 25% of the hardwood forests of the eastern United States. These trees were wiped out by a fungus accidentally introduced on imported Asian chestnuts, which are not affected. More recently, a fungus has killed large numbers of dogwood trees from New York to Georgia; some experts suspect the parasite was accidentally introduced from elsewhere. Why are plants particularly vulnerable to fungi imported from other regions? What kinds of human activities might contribute to the spread of plant diseases? Do you think the introduction of plant pathogens like chestnut blight are more or less likely to occur in the future? Why?

FURTHER READING

Abelson, P. H. "Plant-Fungal Symbiosis." *Science* 229:617, 1985. A general article presenting hopeful prospects for efficient reforestation by infecting trees with symbiotic fungi.

Alexopoulos, C. J., and C. W. Mims. *Introduction to Mycology,* 3rd ed. New York: Wiley, 1979. A general text.

Barron, G. "Jekel-Hyde Mushrooms." *Natural History,* March 1992. How some wood-decomposing fungi supplement their diets by trapping nematodes.

Brune, J. "Porky's Defeat." *Discover,* January 1991. A new electronic instrument is putting pigs and dogs out of business as truffle rooters.

Cherfas, J. "Disappearing Mushrooms: Another Mass Extinction?" *Science,* December 6, 1991. Fungi are vanishing from Europe, causing concern about forest ecology.

Dusheck, J. "Fungus Degrades Toxic Chemicals." *Science News,* June 22, 1985. The versatile metabolism of fungi appears to have exciting potential for environmental cleanup.

Gould, S. J. "A Humungous Fungus Among Us." *Natural History,* July 1992. Is a 30-acre fungus the world's largest organism?

Kiester, E. "Prophets of Gloom." *Discover,* November 1991. Lichen death means more bad news about air quality.

Newhouse, J. R. "Chestnut Blight." *Scientific American,* July 1990. A new parasite may help control the blight.

29 Invertebrates and the Origin of Animal Diversity

Defining "Animals"
Clues to Animal Phylogeny
Parazoa (Phylum Porifera)
Eumetazoa: Radiata
Bilateria: Acoelomates
Pseudocoelomates
Protostomes (Schizocoelomates)
Deuterostomes (Enterocoelomates)
The Origin and Diversification of Animals

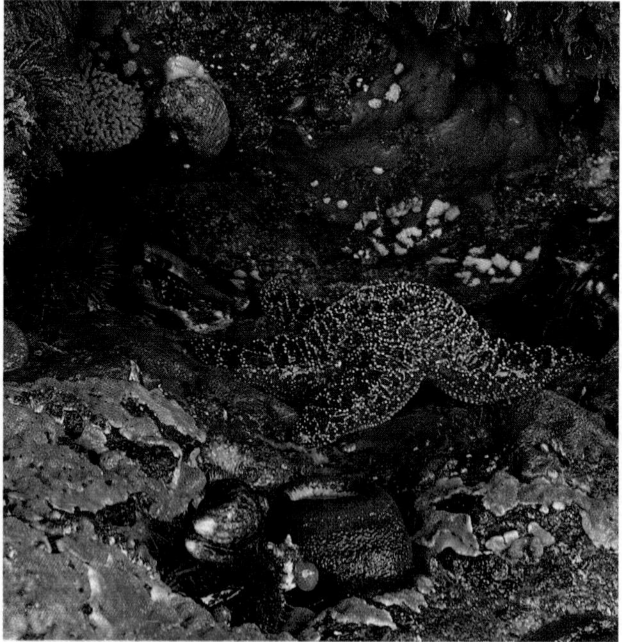

Figure 29.1
The fauna of a tidepool. Living as we do on land, where vertebrates are dominant animals, we have a biased sense of animal diversity. Tidepools and coral reefs reveal a multitude of invertebrate phyla that live in the sea. In this chapter, you will learn about some of these invertebrates and their evolutionary heritage.

Animal life began in Precambrian seas with the evolution of multicellular forms that lived by eating other organisms. This new way of life allowed the exploitation of previously untapped resources and adaptive zones and led to an evolutionary radiation of diverse forms. Early animals populated the seas, fresh water, and eventually the land. The dazzling diversity of animal life on Earth today is the result of over half a billion years of evolution from those ancestral forms.

More than a million extant species of animals are known, and at least as many more will probably be identified by future generations of zoologists. Based largely on anatomical and embryological criteria, animals are grouped into about 35 phyla, the exact number depending on the views of different taxonomists. We are most familiar with the vertebrates, a single subphylum of backboned animals within the phylum Chordata. All other animals, over 95% of the species, lack backbones and are collectively called **invertebrates** (Figure 29.1), the main subject of this chapter. The chapter also describes general characteristics of animals, addresses theories on the origin and early radiation of animals, and discusses possible relationships among the phyla.

DEFINING "ANIMALS"

Constructing a good definition of animals is not as easy as it might first appear. There are exceptions to nearly every criterion for distinguishing an animal from other life forms. However, when taken together, the following characteristics of animals will serve our purposes.

1. Animals are multicellular, heterotrophic eukaryotes. In contrast to the autotrophic nutrition of plants, animals must take into their bodies preformed organic molecules; they cannot construct them from inorganic chemicals. Most animals do this by **ingestion**—eating other organisms and organic material that is decomposing, called detritus.

2. Animals typically store their carbohydrate reserves as glycogen, whereas plants store theirs as starch.

3. Animal cells lack the cell walls that characterize plant cells. Also, intercellular junctions—tight junctions, desmosomes, and gap junctions (see Figure 7.34) —are found only in animals.

4. Unique among animals are two types of tissues responsible for impulse conduction and movement: nervous tissue and muscle tissue.

5. Most animals reproduce sexually, with the diploid stage usually dominating the life cycle. In most species, a small flagellated sperm fertilizes a larger, nonmotile egg to form a diploid zygote. The zygote then undergoes **cleavage,** a succession of mitotic cell divisions. During the development of most animals, cleavage leads to the formation of a multicellular stage called a **blastula,** which often takes the form of a hollow ball. (Variation in the formation and developmental fate of the blastula will be discussed in Chapter 43.) Following the blastula stage is the process of **gastrulation,** during which embryonic tissues of the adult body parts are produced. Some animals develop directly through transient stages of maturation into adults, but the life cycles of many include larval stages. The **larva** is a free-living, sexually immature form. It is morphologically distinct from the adult stage, usually eats different food, and may even have a different habitat than the adult, as in the case of a frog tadpole. Animal larvae eventually undergo **metamorphosis,** a resurgence of development that transforms the animal into an adult.

Animals inhabit nearly all environments of the biosphere. The seas, where the first animals probably arose, are still home to the greatest number of animal phyla. The freshwater fauna is extensive, but not nearly as rich in diversity as the marine fauna.

Terrestrial habitats pose special problems for animals, as they do for plants (see Chapter 27), and few animal phyla have made successful evolutionary treks onto land. Earthworms (phylum Annelida) and land snails (phylum Mollusca) are generally confined to moist soil and vegetation. Only the vertebrates and arthropods, including insects and spiders, are represented by a great diversity of animal species adapted to various terrestrial environments.

CLUES TO ANIMAL PHYLOGENY

Our understanding of the evolutionary relationships among the animal phyla is generally hazy. The origin of most animal phyla and major body plans took place in Precambrian and early Cambrian times, leaving us with a fragmented fossil record. Thus, when reconstructing the evolutionary history of animal phyla, zoologists depend largely on clues from the comparative anatomy and embryology of living forms.

Major Branches of the Animal Kingdom

Although lively debate continues, many zoologists maintain that the modern animal phyla were all derived from one group of protists. Figure 29.2 is an evolutionary tree depicting one set of ideas about how animal groups evolved. Some of the major events in the history of animal evolution are shown on the tree and are discussed below.

Sponges (phylum Porifera) have unique development and an anatomical simplicity that separates them from all other animal phyla. They lack true tissues and are called **parazoa** (meaning "beside the animals"). Other animal phyla are called **eumetazoa.**

The eumetazoa have been divided into several major branches, partly on the basis of body symmetry. Hydras, jellyfishes, and their relatives have **radial symmetry** (Figure 29.3a) and are collectively called the **radiata.** A radial animal has a top and bottom, or an oral and an aboral side, but no front and back and no left and right. The other major branch of eumetazoan evolution led to animals with **bilateral** (two-sided) **symmetry** (Figure 29.3b). A bilateral animal has not only a top (**dorsal** side) and bottom (**ventral** side) but also a head (**anterior**) end and tail (**posterior**) end and a left and right side. All such animals are collectively called the **bilateria.**

Associated with bilateral symmetry is **cephalization,** an evolutionary trend toward the concentration of sensory equipment on the anterior end, the end of a traveling animal that is usually first to encounter food, danger, and other stimuli. A head end is an adaptation for movement, such as crawling, burrowing, and swimming. The symmetry of an animal generally fits its lifestyle. Many radial animals are sessile forms (attached to a substratum) or plankton (drifting or weakly swimming aquatic forms), and their symmetry equips them to meet the environment equally well from all sides. More active animals are generally bilateral.

Symmetry alone is not a foolproof criterion for assigning an animal phylum to a particular evolutionary line. The radial symmetry of some animals has apparently evolved secondarily from a bilateral condition as an adaptation to a more sedentary lifestyle. For example, sea stars (phylum Echinodermata) are radially symmetrical, but their embryonic development and internal anatomy show clearly that they arose from a bilaterally symmetrical ancestor.

These two fundamentally different kinds of symmetry probably arose very early in the history of animal life; indeed, some zoologists think that jellyfishes and

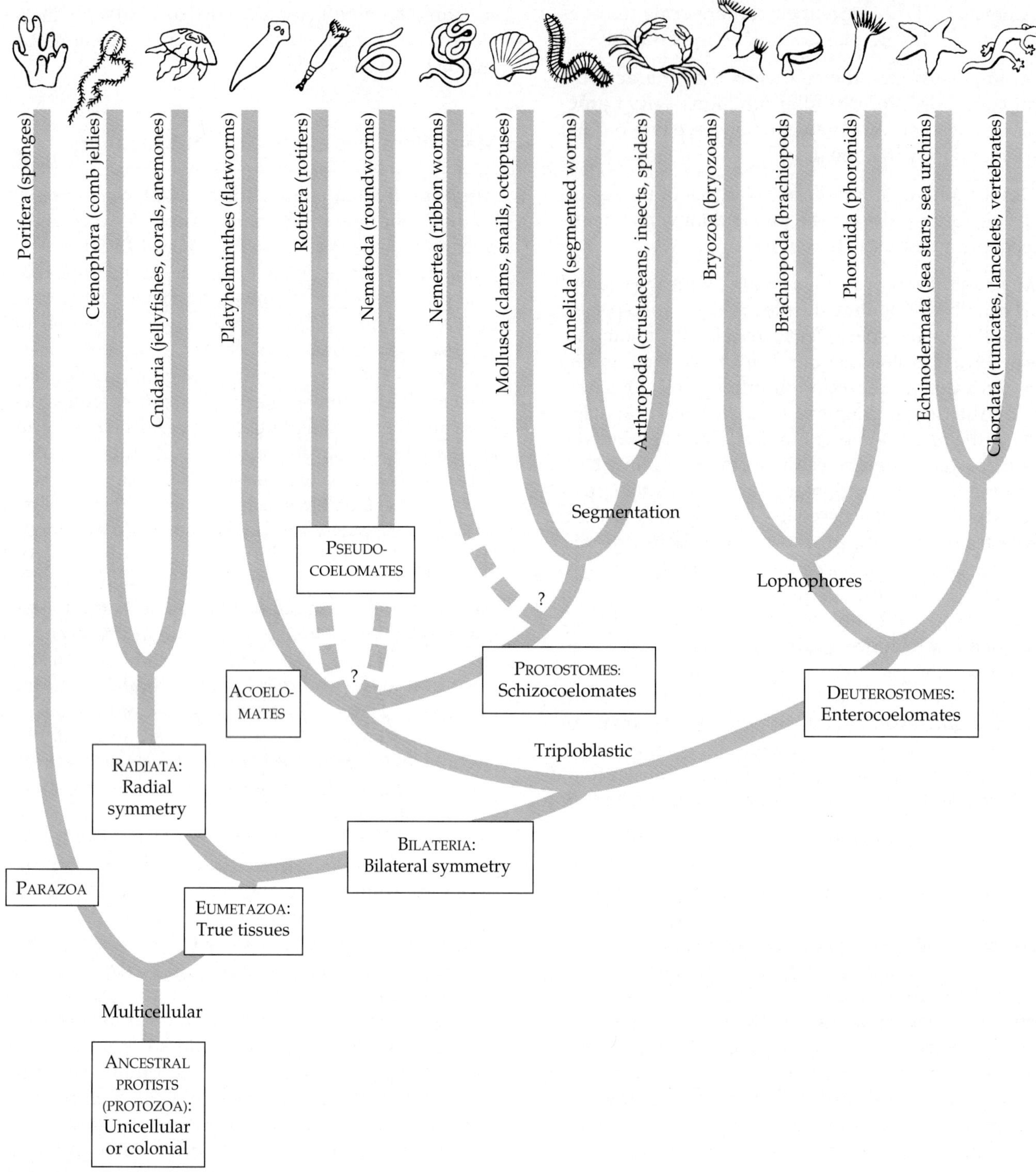

Figure 29.2
One version of animal evolution.

their relatives had different protistan ancestors than the bilateral eumetazoa.

Development and Body Plan

Early in the development of all animals except sponges, the embryo becomes layered through the process of gastrulation (Figure 29.4). As a general rule, these concentric layers, termed the **germ layers,** form the various tissues and organs of the body as development progresses. **Ectoderm,** covering the surface of the embryo, gives rise to the outer covering of the animal and, in some phyla, to the central nervous system. **Endoderm,** the innermost germ layer, lines the primitive gut, or **archenteron,** and gives rise to the lining of

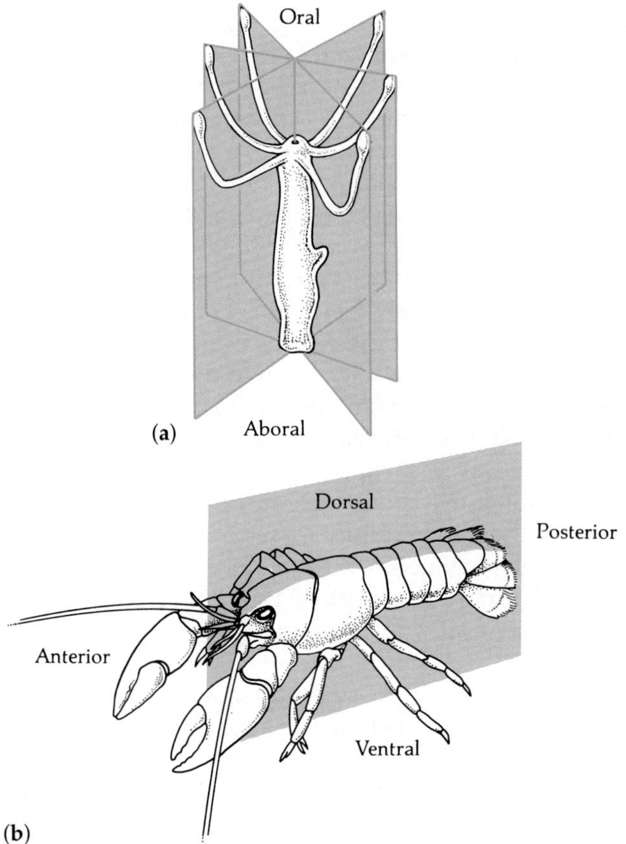

Figure 29.3
Body symmetry. (**a**) The parts of a radial animal, such as a hydra, are arranged like spokes of a wheel that radiate from the center. Any imaginary slice through the central axis would divide the animal into mirror images. (**b**) A bilateral animal has a left and right side, and only one imaginary cut would divide the animal into mirror-image halves.

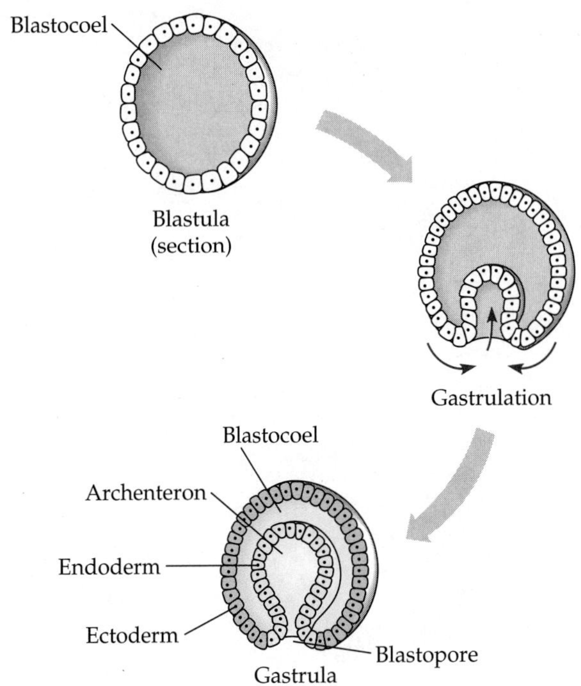

Figure 29.4
Gastrulation. All eumetazoan embryos undergo gastrulation, whereby germ layers are produced. Shown here is the simplest of several mechanisms of gastrulation. These are cross sections of three-dimensional embryos. The blastula stage, for example, is a hollow ball.

the digestive tract and its outpocketings, such as the liver and lungs of vertebrates. A few groups of animals (the cnidarians, for example) produce only these two germ layers and are thus said to be **diploblastic.** All other eumetazoa are **triploblastic** and produce a third germ layer, the **mesoderm**, between the ectoderm and endoderm. This mesoderm forms the muscles and most other organs between the gut and outer covering of the animal.

Triploblastic animals with solid bodies—that is, without a cavity between the gut and outer body wall—are referred to as the **acoelomates** (Gr. *a*, "without," and *koilos*, "a hollow"). This group includes flatworms (phylum Platyhelminthes) and animals of a few other phyla (Figure 29.5a). The other phyla of bilateral, triploblastic animals have tube-within-a-tube body plans, with a fluid-filled cavity separating the digestive tract from the outer body wall. If the cavity is not completely lined by tissue derived from mesoderm, it is termed a **pseudocoelom.** Animals with this body plan, such as rotifers (phylum Rotifera), roundworms (phylum Nematoda), and a few other phyla, are called **pseudocoelomates** (Figure 29.5b). **Coelomates** are animals with a true **coelom,** a fluid-filled body cavity completely lined by tissue derived from mesoderm. The inner and outer layers of tissue that surround the cavity connect dorsally and ventrally to form mesenteries, which suspend the internal organs (Figure 29.5c).

A body cavity has many functions. Its fluid cushions the suspended organs, helping to prevent internal injury. The cavity also enables the internal organs to grow and move independently of the outer body wall. If it were not for your coelom, every beat of your heart or ripple of your intestine could deform your body surface, and exercise would distort the shapes of the internal organs. In soft-bodied coelomates such as earthworms, the noncompressible fluid of the body cavity is under pressure and functions as a hydrostatic skeleton against which muscles can work. Though they may have first evolved as adaptations for burrowing by soft-bodied animals, coeloms evolved independently at least twice, as depicted on the tree in Figure 29.2.

The Protostome–Deuterostome Dichotomy

The coelomate phyla can be divided into two distinct evolutionary lines. Mollusks, annelids, and arthropods represent one of these lines and are collectively called

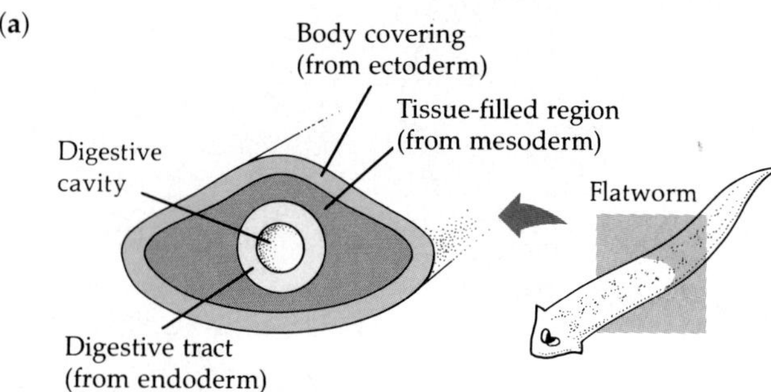

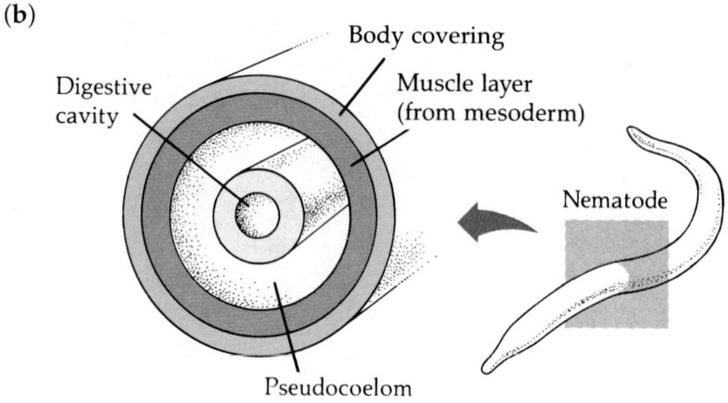

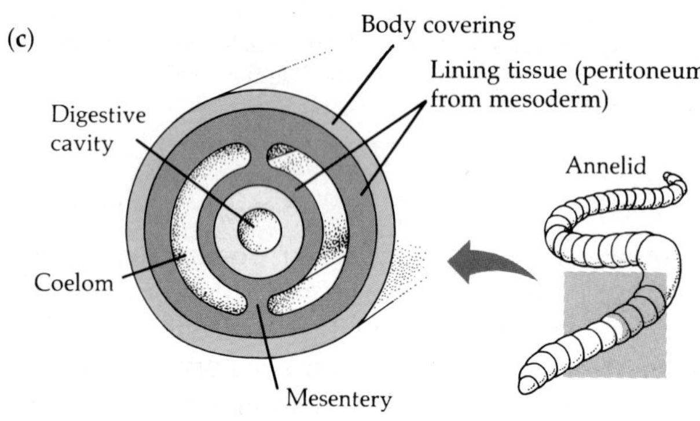

Figure 29.5
Body plans of the bilateria. The various organ systems of the animal develop from the three germ layers that form in the embryo. (**a**) Acoelomates lack a body cavity between the gut and outer body wall. (**b**) Pseudocoelomates have a body cavity only partially lined by mesodermally derived tissue. (**c**) Coelomates have a true coelom, a body cavity completely lined by mesodermally derived tissue.

protostomes. Echinoderms, chordates, and some other phyla, collectively called **deuterostomes,** represent the other line (see Figure 29.2). Protostomes and deuterostomes are distinguished by several fundamental differences in their development.

Cleavage Differences between the two coelomate lineages are evident as early as the cleavage divisions that transform the zygote into a ball of cells. Many protostomes undergo **spiral cleavage,** in which planes of cell division are diagonal to the vertical axis of the embryo. As seen in the eight-cell stage resulting from spiral cleavage, small cells lie in the grooves between larger, underlying cells (Figure 29.6a). Furthermore, the so-called **determinate cleavage** of some protostomes rigidly casts the developmental fate of each embryonic cell very early. A cell isolated at the four-cell stage from a protostome, such as a snail, forms an inviable embryo that lacks parts.

In contrast to the protostome pattern, the zygote of many deuterostomes undergoes **radial cleavage,** which also occurs in the radial animals. Here, the cleavage planes are either parallel or perpendicular to the vertical axis of the egg; as seen in the eight-cell stage, the cells are aligned, one directly above the other. Most deuterostomes are further characterized by **indeterminate cleavage,** meaning that each cell produced by early cleavage divisions retains the capacity to develop into a complete embryo. If the cells of a sea star embryo, for example, are separated at the four-cell stage, each will go on to form a normal larva. It is the indeterminate cleavage of the human zygote that makes identical twins possible.

Blastopore Fate Another difference between protostomes and deuterostomes is apparent later in development. In gastrulation, the rudimentary gut of an embryo forms as a blind pouch (the archenteron), which has a single opening to the outside known as the **blastopore** (Figure 29.6b). A second opening forms later at the opposite end of the archenteron to produce a digestive tube with a mouth and anus. The mouth of a typical protostome develops from the first opening, the blastopore, and it is for this characteristic that the protostome line is named (Gr. *protos,* "first," and stoma, "mouth"). By contrast, the mouth of a deuterostome (Gr. *deuteros,* "second") is derived from a secondary opening, and the blastopore usually forms the anus, not the mouth. This fundamental difference helps to justify splitting these two groups of animals phylogenetically.

Coelom Formation A third fundamental difference between protostomes and deuterostomes is in the development of the coelom (Figure 29.6c). As the archenteron forms in a protostome, initially solid masses of mesoderm split to form the coeloms; this is called **schizocoelous** development (Gr. *schizo,* "split"). Development of the body cavities of deuterostomes is termed **enterocoelous:** The mesoderm arises from the wall of the archenteron and hollows to become the coelomic cavities.

Keeping in mind that any phylogenetic grouping is tentative, we can now see the kingdom Animalia as arranged on the evolutionary tree in Figure 29.2. In the following pages, we will survey some of the most important phyla of invertebrate animals. Although we consider some aspects of function, such as reproduction and nutrition, as they characterize particular groups, we will leave the main discussion of animal physiology for Unit Seven.

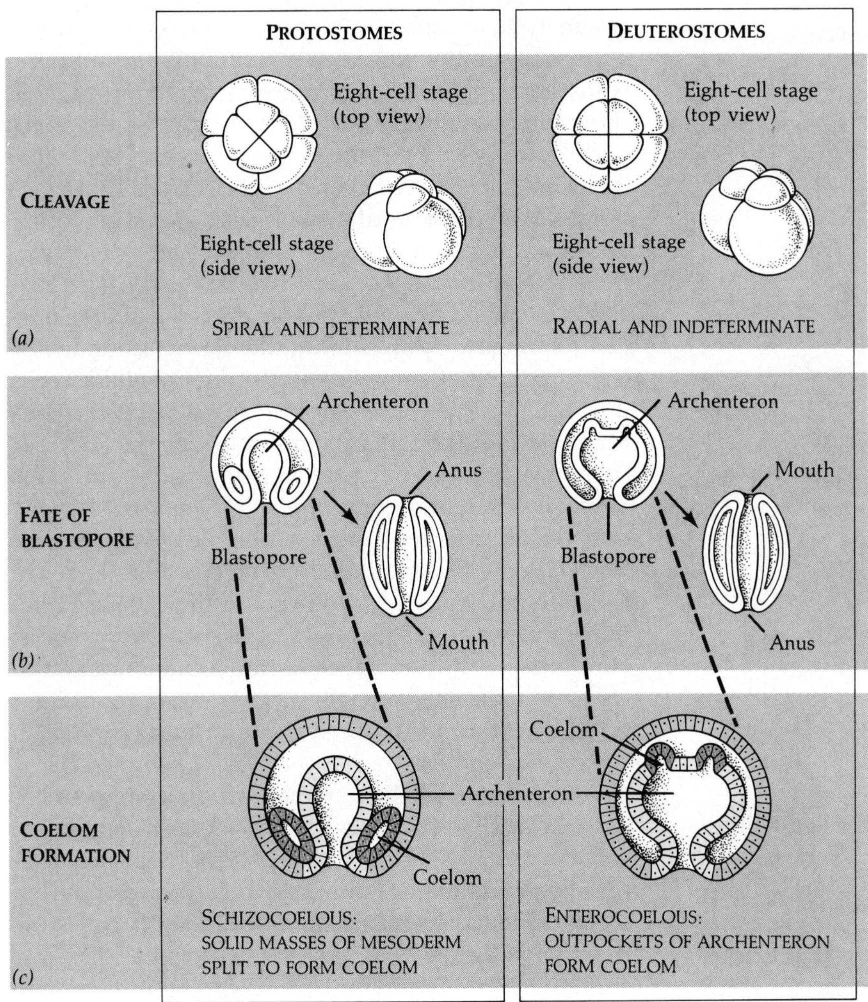

Figure 29.6
A comparison of early development in protostomes and deuterostomes. (**a**) Protostomes have spiral cleavage; deuterostomes have radial cleavage. (**b**) The blastopore of the gastrula forms the mouth in protostomes; the mouth forms from a secondary opening in deuterostomes. (**c**) Coelom formation also begins in the gastrula stage. Protostomes are called schizocoels because their coelom forms from splits in the mesoderm. Deuterostomes are called enterocoels because their coelom forms from mesodermal outpocketings of the archenteron (blue = ectoderm, yellow = endoderm, red = mesoderm).

PARAZOA (PHYLUM PORIFERA)

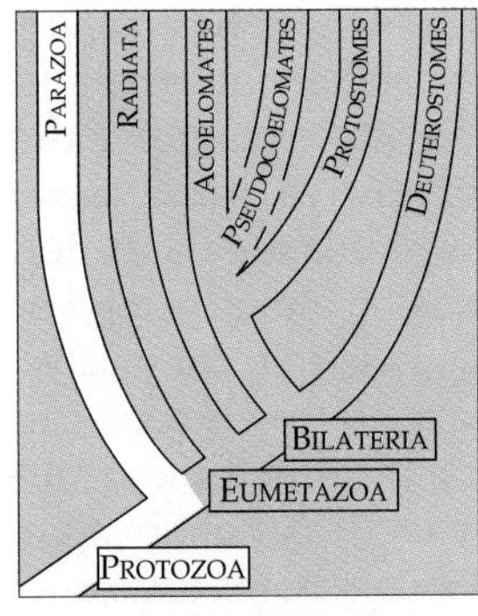

Sponges (phylum Porifera) are sessile animals that appear so sedate to the human eye that the ancient Greeks believed them to be plants (Figure 29.7). Sponges range in height from about 1 cm to 2 m. Of the 9000 or so species of sponges, only about 100 live in fresh water; the rest are marine. The body of a simple sponge resembles a sac perforated with holes. (*Porifera* means "pore bearers.") Water is drawn through the pores into a central cavity, the **spongocoel,** and then flows out of the sponge through a larger opening called the **osculum** (Figure 29.8). More complex sponges have folded body walls and branched spongocoels.

Sponges are suspension-feeders (also known as filter-feeders) that collect food particles from the water circulated through the porous body. Lining the inside

Figure 29.7
Sponges. Sessile animals without specialized organs and tissues, sponges filter food from water pumped through their porous bodies. It has been estimated that a sponge must filter about 1 ton of water to grow by 1 ounce. The diverse species of sponges vary in shape and color, some brightly pigmented by symbiotic algae. This is an azure vase sponge *(Callyspongia plicifera).*

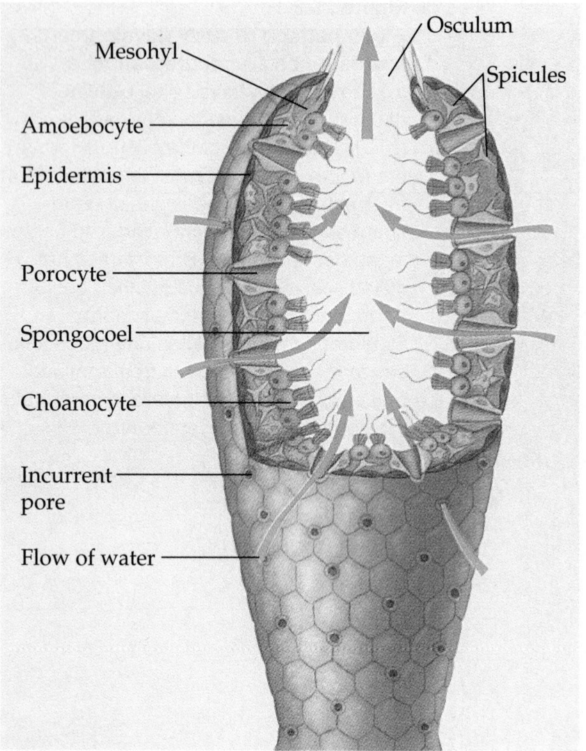

Figure 29.8
Anatomy of a sponge. The wall of this simple sponge has two layers of cells separated by a gelatinous matrix, the mesohyl (middle matter). The outer layer consists of tightly packed epidermal cells. The incurrent pores are channels through porocytes, cells shaped like elongated donuts that span the body wall. The spongocoel is lined mainly by choanocytes, each with a flagellum ringed by a collar of fingerlike projections.

of the body are flagellated **choanocytes,** or collar cells (for the membranous collar around the base of the flagellum; see Figure 29.8). The flagella generate a water current, and the collars trap food particles and ingest them by phagocytosis.

Wandering through a gelatinous layer called the **mesohyl** between the two layers of the sponge body wall are cells called **amoebocytes,** named for their use of pseudopodia. Amoebocytes have many functions. They take up food from the choanocytes, digest it, and carry nutrients to other cells. Amoebocytes also form tough skeletal fibers within the mesohyl. In some groups of sponges, these fibers are sharp spicules made from calcium carbonate or silica; other sponges produce more flexible fibers composed of a protein called spongin. Variation in the chemical composition of the skeleton is one way taxonomists group sponges into the different classes of the phylum Porifera.

Most sponges are **hermaphrodites** (Gr. *Hermes,* the male god, and *Aphrodite,* the female goddess), meaning that each individual functions as both male and female in sexual reproduction by producing sperm *and* eggs. Gametes arise from choanocytes or amoebocytes. Eggs reside in the mesohyl, but sperm are released into the spongocoel and are carried out of the sponge by the water current. Cross-fertilization results from some of the sperm being drawn into neighboring individuals. Fertilization occurs in the mesohyl, where the zygote develops into a larva covered with flagella. The swimming larvae disperse after their release into the spongocoel and their exit through the osculum. A tiny fraction of the larvae survive to settle on a suitable substratum and begin the sessile existence characteristic of sponges. Upon settling, the larva "turns inside out" during metamorphosis, and the flagellated cells then face the interior. Sponges are capable of extensive **regeneration,** the replacement of lost parts. They use this power not only for repair but also to reproduce asexually from fragments broken off a parent sponge.

Sponges are among the least complex of all animals. They lack organs, and the layers of loose federations of cells are not really tissues because the cells are relatively unspecialized. Sponges have no nerves or muscles, but the individual cells can sense and react to changes in the environment. Under certain conditions, the cells around the pores and osculum contract to close the openings.

The ancestors of the phylum Porifera were possibly choanoflagellates, collared protists that resemble the choanocytes of sponges. Some choanoflagellates form colonies, and the first sponges may have evolved from similar colonial protists (see Figure 29.43).

EUMETAZOA: RADIATA

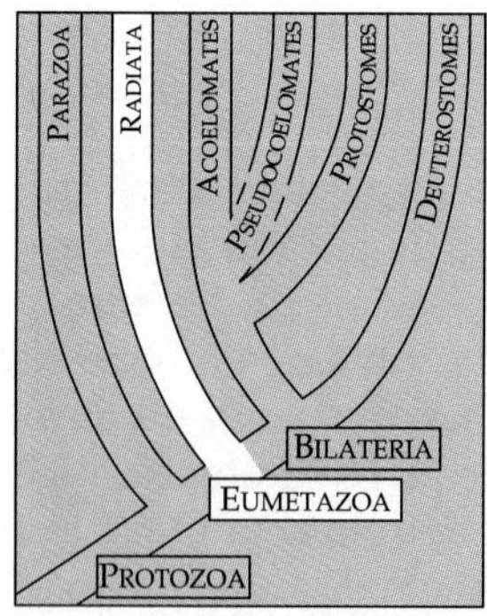

Phylum Cnidaria

Members of the phylum Cnidaria include hydras, jellyfishes, sea anemones, and corals. They are among the most primitive eumetazoa. Their diploblastic condition and radial symmetry suggest to zoologists that they arose very early off the eumetazoan line. The absence of mesoderm has limited these animals to a relatively simple body construction. In spite of these constraints, however, the cnidarians are a successful group with over 10,000 living species, most of which are marine.

The basic body plan of a cnidarian is a sac with a central digestive compartment, the **gastrovascular cavity,** and a single opening that functions as both mouth and anus. This basic body plan has two variations: the sessile polyp and the floating medusa (Figure 29.9). Examples of the polyp form are hydras and sea

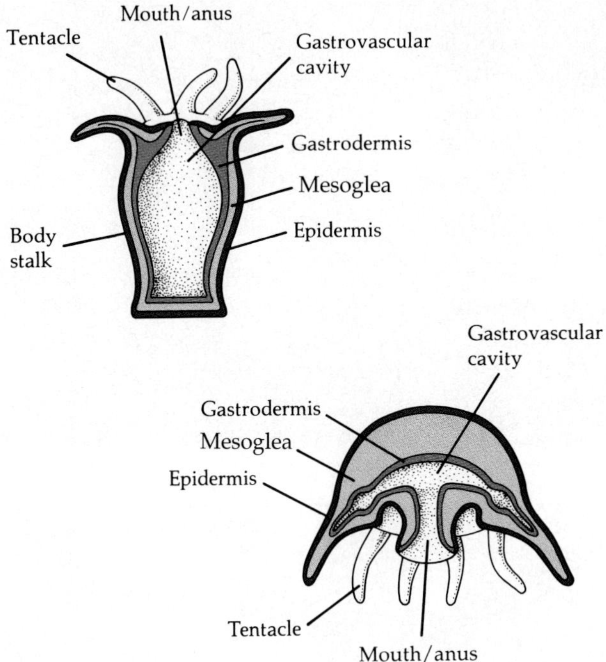

Figure 29.9
Polyp and medusa forms of cnidarians. The body wall of a polyp (top) or medusa (bottom) has two layers of cells, an outer layer of epidermis specialized for protection and an inner layer of gastrodermis for digestion. After the animal ingests food, the gastrodermis secretes digestive enzymes into the gastrovascular cavity. The gastrodermal cells engulf small pieces of the partially digested food by phagocytosis, and digestion is completed within the cells in food vacuoles. Flagella on the gastrodermal cells keep the contents of the gastrovascular cavity agitated and help distribute nutrients. Sandwiched between the epidermis and gastrodermis is a gelatinous layer of mesoglea. In some polyps, such as hydra, the mesoglea is quite thin and noncellular, but in others, such as sea anemones, it is thick and cellular. (Cellular mesoglea is called mesenchyme.) In many medusae, the mesoglea is thick and jellylike—thus their name, jellyfish.

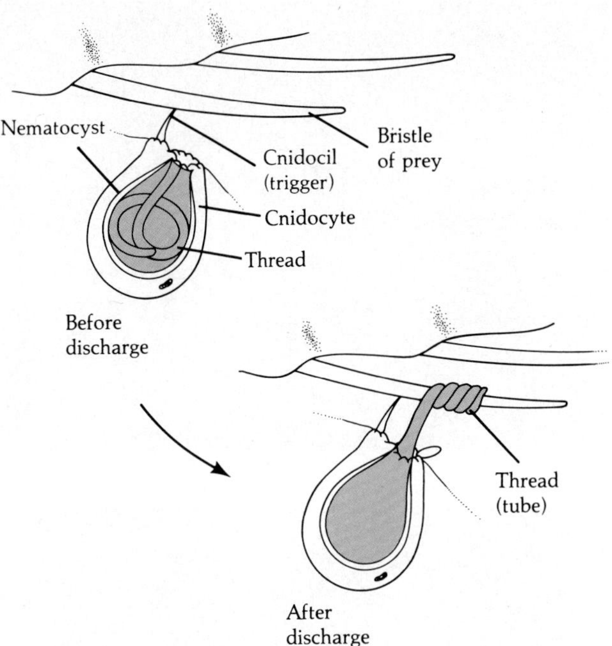

Figure 29.10
Cnidocytes of cnidaria. Each cnidocyte contains a stinging capsule, the nematocyst, which itself contains an inverted threadlike weapon. When a trigger, or cnidocil, is stimulated by touch or by certain chemicals, the thread everts and shoots out from the nematocyst. Some of the threads are long and entangle the tiny appendages of small animals that bump haplessly into the tentacles, whereas other threads puncture the prey and inject a poison that paralyzes the victim.

anemones. The animals we generally call jellyfishes are medusae. **Polyps** are cylindrical forms that adhere to the substratum by the aboral end of the body and extend their tentacles, waiting for prey. A **medusa** is a flattened, mouth-down version of the polyp. It moves freely in the water by a combination of passive drifting and weak contractions of the bell-shaped body. The tentacles of the jellyfish dangle from the oral surface, which points downward. Some cnidarians exist only as polyps, others only as medusae, and still others pass sequentially through both medusa and polyp stages in their life cycles. This dimorphic (dual body form) life history is a unique feature of the cnidarians.

Cnidarians are carnivores that use tentacles arranged in a ring around the mouth to capture prey and push the food into the gastrovascular cavity where digestion begins. The undigested remains are egested through the mouth/anus. The tentacles are armed with batteries of **cnidocytes,** unique cells that function in defense and the capture of prey (Figure 29.10). Cnidocytes, which contain stinging capsules called **nematocysts,** give the phylum Cnidaria its name (Gr. *cnide,* "nettle").

Muscles and nerves occur in their simplest forms in cnidarians. Cells of the epidermis and gastrodermis have bundles of microfilaments arranged into contractile fibers (see Chapter 7). True muscle tissue does not appear in diploblastic animals, since it is derived from mesoderm. The gastrovascular cavity acts as a hydrostatic skeleton against which the contractile cells can work. When the animal closes its mouth, the volume of the cavity is fixed, and contraction of selected cells causes the animal to change shape. The slow movements are coordinated by a nerve net. Cnidarians have no brain, and cnidarian behavior seems to be completely rigid; no one has yet trained a jellyfish. The noncentralized nerve net innervates simple sensory receptors that are distributed radially around the body. Thus, the animal can detect and respond to stimuli equally from all directions.

The phylum Cnidaria is divided into three major classes: Hydrozoa, Scyphozoa, and Anthozoa (Figure 29.11).

Figure 29.11
Representatives of the cnidarian classes. (**a**) Polyps of a colonial species belonging to the class Hydrozoa. (**b**) Jellyfish of the class Scyphozoa. The medusa is the conspicuous stage of the scyphozoan life cycle. The largest scyphozoan species have tentacles over 30 m long dangling from umbrellas 2 to 3 m in diameter. (**c**) Sea anemones and other members of the class Anthozoa exist only as polyps. (**d**) A colony of coral polyps (class Anthozoa). Many corals harbor symbiotic algae that contribute to the food supply of the polyps. Coral reefs, which provide habitats for an enormous variety of invertebrates and fishes, are restricted to warm, shallow seas. This is a star coral.

Hydra, one of the few cnidarian genera found in fresh water, is an unusual member of the class Hydrozoa in that it exists only in the polyp form. When environmental conditions are favorable, *Hydra* reproduces asexually by budding, the formation of outgrowths that pinch off from the parent to live independently (see Figure 12.2). When environmental conditions deteriorate, *Hydra* reproduces sexually, forming resistant zygotes that remain dormant until conditions improve. Most hydrozoans alternate polyp and medusa forms, as in the life cycle of *Obelia* (Figure 29.12). The colonial polyp stage is conspicuous, a characteristic of the class Hydrozoa.

In the class Scyphozoa, it is generally the medusa that prevails in the life cycle. The medusae of most species live among the plankton as the animals commonly called jellyfishes. Most coastal scyphozoans go through a small polyp stage during their life cycle, but jellyfishes that live in the open ocean have generally eliminated the sessile polyp.

Sea anemones and corals belong to the class Anthozoa ("flower animals"). They occur only as polyps. Coral animals live as solitary or colonial forms and secrete hard external skeletons of calcium carbonate. Each polyp generation builds on the skeletal remains of earlier generations to construct "rocks" with shapes characteristic of the species. These skeletons we call coral.

Phylum Ctenophora

Comb jellies, of the phylum Ctenophora, superficially resemble small cnidarian medusae, but the relationship between ctenophores and cnidarians is uncertain (Figure 29.13). There are only about 100 species of

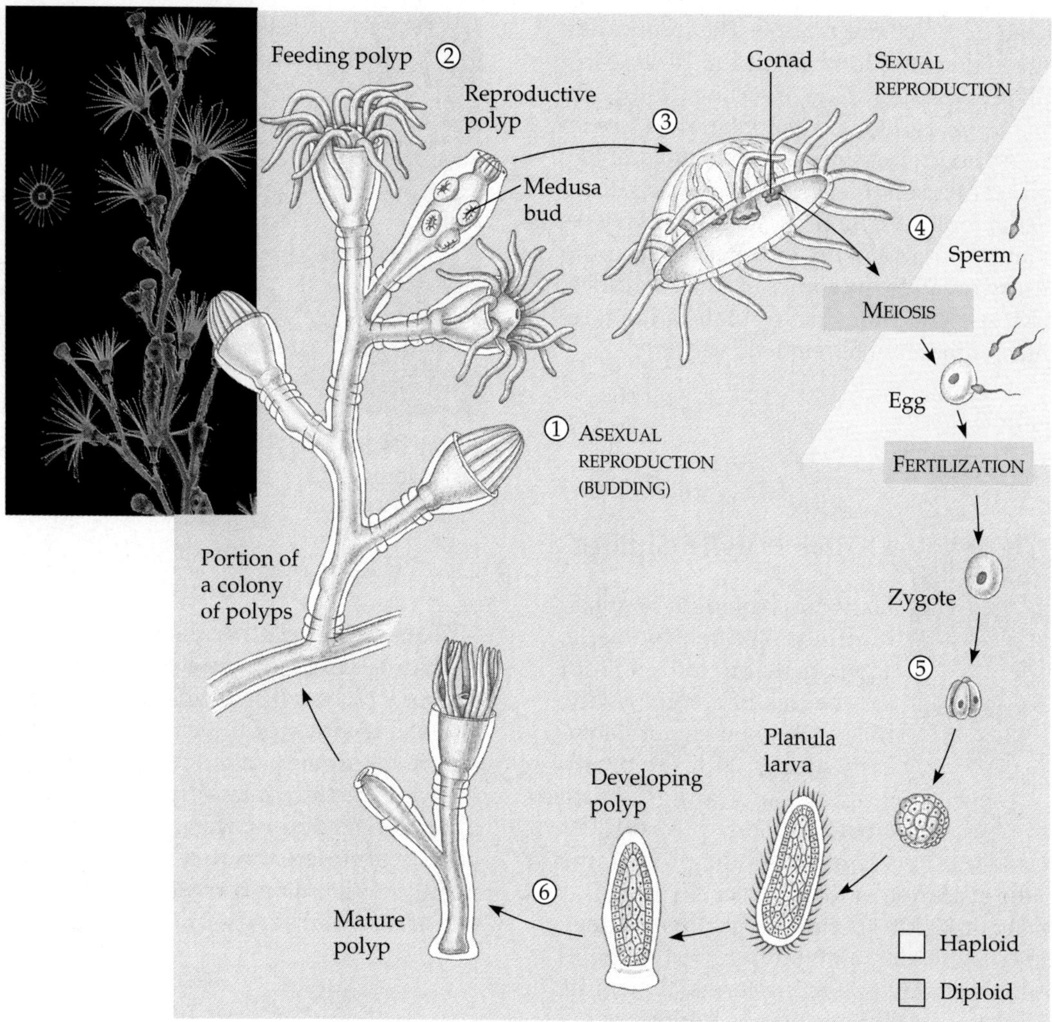

Figure 29.12
The life cycle of the hydrozoan *Obelia*. ① The polyp reproduces asexually by budding to form a colony of interconnected polyps. ② Some polyps, equipped with tentacles, are specialized for feeding. ③ Other polyps, specialized for reproduction, lack tentacles and produce tiny medusas by asexual budding. ④ The medusas swim off, grow, and reproduce sexually. ⑤ The zygote develops into a solid, ciliated larva called the planula. ⑥ The planula eventually settles and develops into a new polyp. The polyp stage is asexual and the medusa stage is sexual, and these two stages alternate, one producing the other. But do not confuse this with the alternation of generations that occurs in the plant kingdom. Both polyp and medusa are diploid organisms, whereas one of the plant generations is haploid.

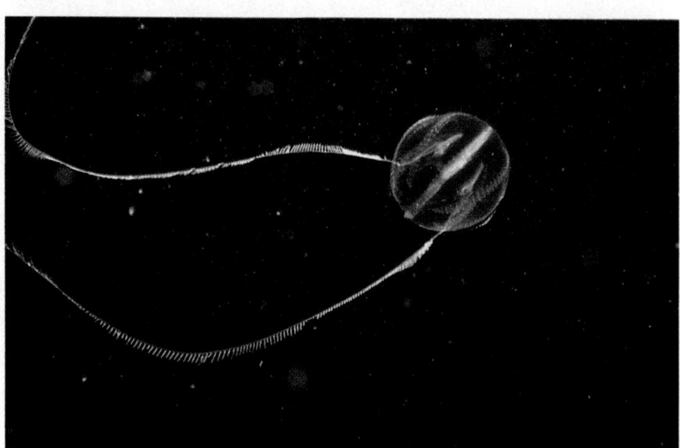

Figure 29.13
A ctenophoran, or comb jelly. This planktonic marine animal is named for its eight combs of cilia, used for locomotion. Ctenophorans and cnidarians are radiate animals, but many zoologists believe these groups have little else in common.

comb jellies, all of which are marine. The transparent animals range in diameter from about 1 to 10 cm. Most are spherical or ovoid, but there are elongate and ribbonlike forms up to 1 m long. Ctenophorans ("comb-bearers") are named for their eight rows of comblike plates composed of fused cilia. They are the largest animals to use cilia for locomotion. An aboral sensory organ acts as an orientation cue, and nerves running from the sensory organ to the combs of cilia coordinate movement. Most comb jellies have a pair of long, retractable tentacles that function in food capture.

BILATERIA: ACOELOMATES

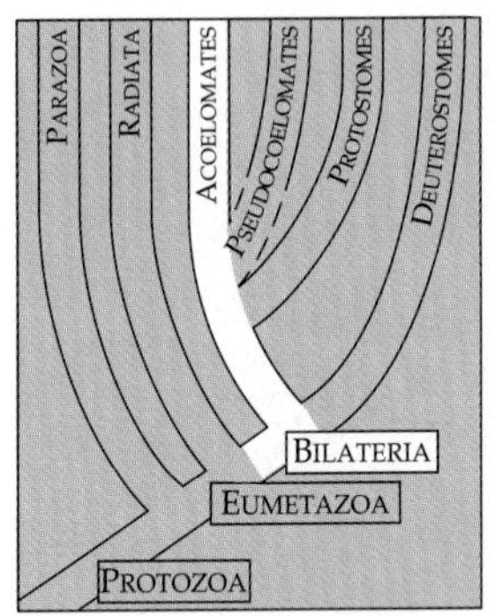

Phylum Platyhelminthes

Flatworms (phylum Platyhelminthes) are triploblastic, acoelomate bilateria. Except for the lack of a body cavity, they share many embryological features with the protostomes, including spiral cleavage. They probably diverged from the protostome line prior to the evolution of the schizocoelom.

There are about 20,000 species in this phylum; they live in marine, freshwater, and damp terrestrial habitats. In addition to many free-living forms, flatworms include the flukes and the tapeworms, which are exclusively parasitic on other animals. Their bodies are generally flattened dorsoventrally. (The word *platyhelminth* means "flat worm.") They range in size from nearly microscopic free-living species to certain tapeworms over 20 m long.

Flatworms show several important evolutionary developments compared to the cnidarians. They are bilaterally symmetrical with unidirectional movement and moderate cephalization. Furthermore, they produce embryonic mesoderm, a third germ layer that contributes to the development of more complex organs and organ systems, and true muscle tissue. However, the gut of a typical flatworm, like that of a cnidarian, is a gastrovascular cavity with only one opening. The flatworms are divided into four classes: Turbellaria (turbellarians; mostly free-living flatworms), Trematoda and Monogenea (flukes), and Cestoda (tapeworms).

Class Turbellaria Turbellarians are nearly all free-living (nonparasitic) and mostly marine (Figure 29.14), but the species called planarians abound in ponds and streams. **Planarians** are carnivores that prey on smaller animals or feed on carrion (dead animals). The anatomy of a planarian is described opposite in Figure 29.15.

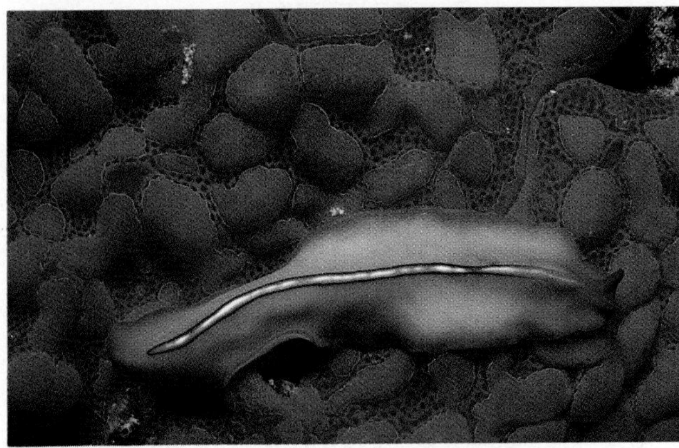

Figure 29.14
A flatworm. Class Turbellaria consists mainly of free-living marine flatworms.

Planarians and other flatworms lack organs specialized for gas exchange and circulation. The flat shape of the body places all cells close to the surrounding water, and ramification of the gastrovascular cavity distributes food throughout the animal. Nitrogenous waste in the form of ammonia diffuses directly from the cells into the surrounding water. Flatworms also have a relatively simple excretory apparatus that functions mainly to maintain osmotic balance between the animal and its surroundings. This system consists of cili-

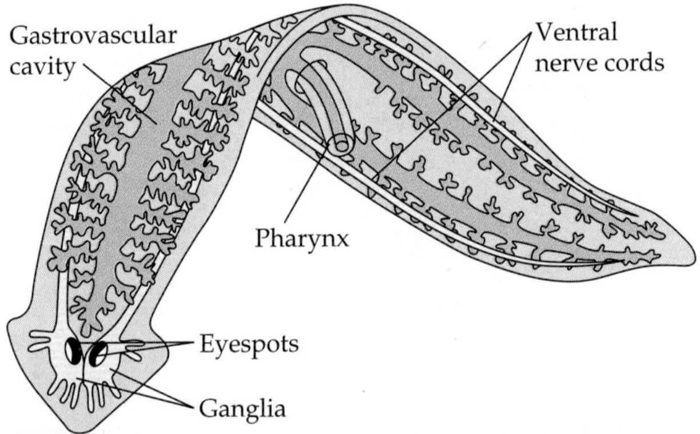

Figure 29.15
Anatomy of a planarian. The mouth is at the tip of a muscular pharynx that extends from the middle of the ventral side of the animal. Digestive juices are spilled onto prey, and the pharynx sucks small pieces of food into the gastrovascular cavity, where digestion continues. Digestion is completed within the cells lining the gastrovascular cavity, which has three branches, each ramified to provide an extensive surface area. Undigested wastes are egested through the mouth. Located at the anterior end of the worm, near the main sources of sensory input, is a pair of ganglia, dense clusters of nerve cells. From the ganglia, a pair of ventral nerve cords runs the length of the body. With cross-nerves connecting the two cords (see Figure 44.16c), the planarian nervous system is ladderlike.

ated cells known as flame cells that waft fluid through branched ducts opening to the outside. (For further discussion, see Chapter 40.) The evolution of osmoregulatory structures was a major factor in allowing some turbellarians to invade freshwater and even moist terrestrial environments.

Planarians move by using cilia on the ventral epidermis, gliding along a film of mucus they secrete. Some turbellarians occasionally also use their muscles to swim through water with an undulating motion.

A planarian has a head (is cephalized) with a pair of eyespots that detect light and lateral flaps called auricles that function mainly for smell. The planarian nervous system is more complex and centralized than the nerve nets of cnidarians. Planarians can learn to modify their responses to stimuli.

Planarians can reproduce asexually through regeneration. The parent constricts in the middle, and each half regenerates the missing end. Sexual reproduction also occurs. Although planarians are hermaphrodites, copulating mates cross-fertilize.

Classes Trematoda and Monogenea Flukes live as parasites in or on other animals. Many have suckers for attaching to internal organs of the host, and a tough covering helps protect the parasite. Reproductive organs nearly fill the interior of a mature fluke.

Flukes generally have life cycles complicated by an alternation of sexual and asexual stages, and many require an intermediate host where larvae develop before infecting the final host, where the adult fluke lives. For example, flukes that parasitize humans spend parts of their life histories in snails (Figure 29.16). The 200 million people around the world who are infected with blood flukes *(Schistosoma)* suffer body pains, anemia, and dysentery.

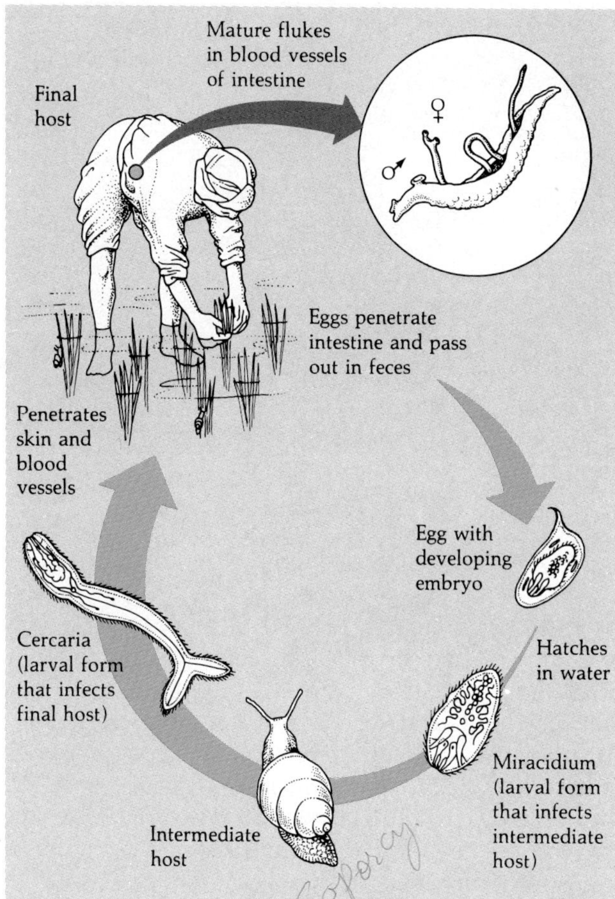

Figure 29.16
The life history of a blood fluke *(Schistosoma mansoni).* Workers in irrigated fields contaminated with human feces may be exposed to larvae of blood flukes that have escaped from their intermediate host, snails. The mature female fluke fits into a groove running the length of the larger male's body.

Class Cestoda Tapeworms, of the class Cestoda, are parasitic flatworms that as adults live mostly in vertebrates, including humans. The tapeworm head, or scolex, is armed with suckers and often menacing hooks that lock the worm to the intestinal lining of the host. Posterior to the scolex is a long ribbon of units called proglottids, which are little more than sacs of sex organs. The tapeworm lacks a digestive system; its food is predigested by the host. Mature proglottids, loaded with thousands of eggs, are released from the posterior end of the worm and leave the body with feces. In one type of cycle, human feces contaminate the food or water of intermediate hosts, such as pigs or cattle, and the tapeworm eggs develop into larvae that encyst in muscles of these animals. Humans acquire the larvae by eating undercooked meat contaminated with cysts, and the worms develop into mature adults within the human. Large tapeworms, which may be 20 m or more in length, can cause intestinal blockage and can rob enough nutrients from the human host to cause nutritional deficiencies.

Phylum Nemertea

Members of the phylum Nemertea are called ribbon worms or proboscis worms (Figure 29.17). Their evolutionary position is presently being debated, although they are clearly related to the protostome lineage. A ribbon worm's body is structurally acoelomate, like that of the flatworm, but it contains a small fluid-filled sac that some zoologists view as a true coelom. This sac is used to hydraulically operate an extensible proboscis by which the worm captures prey. The nemerteans are positioned on the evolutionary tree in Figure 29.2 to reflect their possible coelomate nature, but they are discussed here to highlight comparison with flatworms.

Proboscis worms range in length from less than 1 mm to more than 30 m. Nearly all of the 900 or so members of this phylum are marine, but a few species inhabit fresh water or damp soil. Some are active swimmers, and others burrow in the sand.

Proboscis worms probably evolved from flatworms. The two phyla have similar excretory, sensory, and

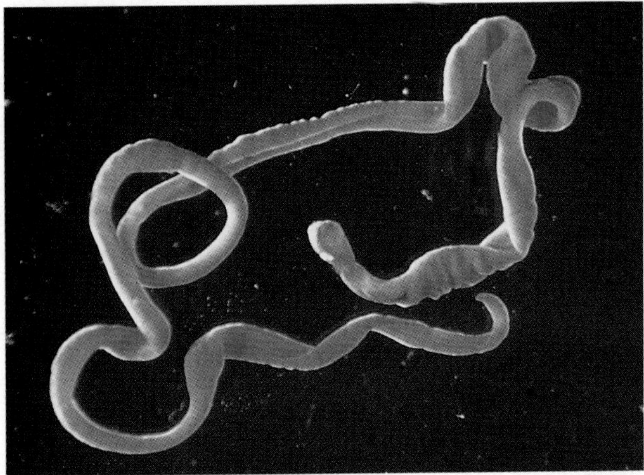

Figure 29.17
A ribbon worm, phylum Nemertea.

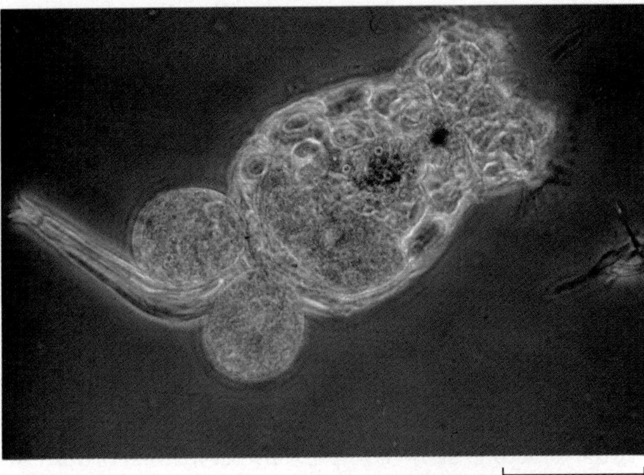

Figure 29.18
A rotifer. These pseudocoelomates, though smaller than many protozoa, are more anatomically complex than flatworms (LM).

nervous systems. But, in addition to the unique proboscis apparatus, two anatomical features not found in flatworms have evolved in the phylum Nemertea: a simple circulatory system and a **complete digestive tract**—that is, a digestive tube with a separate mouth and anus. The circulatory system consists of vessels through which blood flows, and some species have red blood cells containing a form of hemoglobin, which transports oxygen. Proboscis worms have no heart, but muscles squeeze the vessels to propel the blood. A digestive tube is a feature the nemerteans share with animals that follow in this survey of the invertebrates.

PSEUDOCOELOMATES

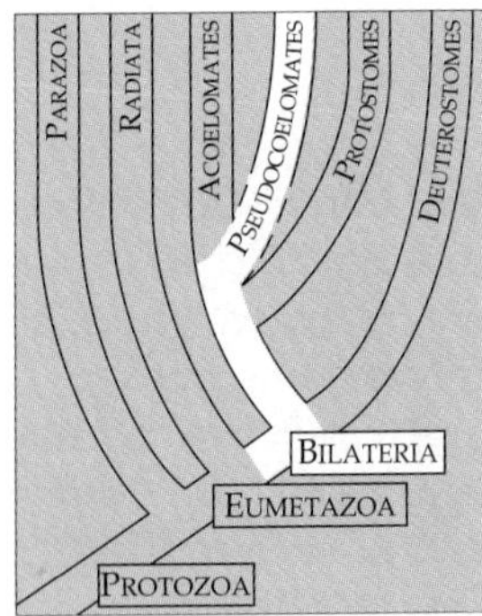

A pseudocoelomate body plan has evolved in several phyla of small animals. Their evolutionary relationships with other groups and with each other are very unclear and are the subject of much controversy. Thus, the position of these phyla on the evolutionary tree in Figure 29.2 is very tentative, suggesting only that they may be more closely related to protostomes than to deuterostomes. In fact, the pseudocoelomate condition probably arose independently several times. We discuss here only two of these phyla: Rotifera and Nematoda.

Phylum Rotifera

Rotifers (about 1800 species) are tiny animals that mainly inhabit fresh water, although some live in the sea or in damp soil. Ranging in size from only about 0.05 to 2.0 mm, smaller than many protozoa, rotifers are nevertheless truly multicellular and have a complete digestive tract and other specialized organ systems (Figure 29.18). Internal organs lie within the pseudocoelom. The pseudocoelomic fluid serves as a hydrostatic skeleton and as a medium for the diffusion of nutrients and wastes in these tiny animals.

The word *rotifer,* derived from Latin, means "wheel-bearer," a reference to the crown of cilia that draws a vortex of water into the mouth. Posterior to the mouth is a jawlike organ that grinds food, mostly microorganisms suspended in the water.

Rotifer reproduction is odd. Some species consist only of females that produce more females from unfertilized eggs, a type of reproduction called **parthenogenesis.** Other species produce two types of eggs that develop by parthenogenesis, one type forming females and the other type developing into degenerate males that cannot even feed themselves. The males produce sperm that fertilize eggs to form resistant zygotes able to survive when a pond dries up. When conditions are favorable again, the zygotes break dormancy and develop into a new female generation that then reproduces by parthenogenesis until conditions become unfavorable again.

Phylum Nematoda

Nematodes are cylindrical pseudocoelomate worms with tapered ends. They are called roundworms and are among the most numerous of all animals in both species and individuals (Figure 29.19). Roundworms are found in most aquatic habitats, in wet soil, in the moist tissues of plants, and in the body fluids and tissues of animals. About 80,000 species of nematodes are known, and perhaps ten times that number actually exist. Roundworms range from less than 1 mm to more

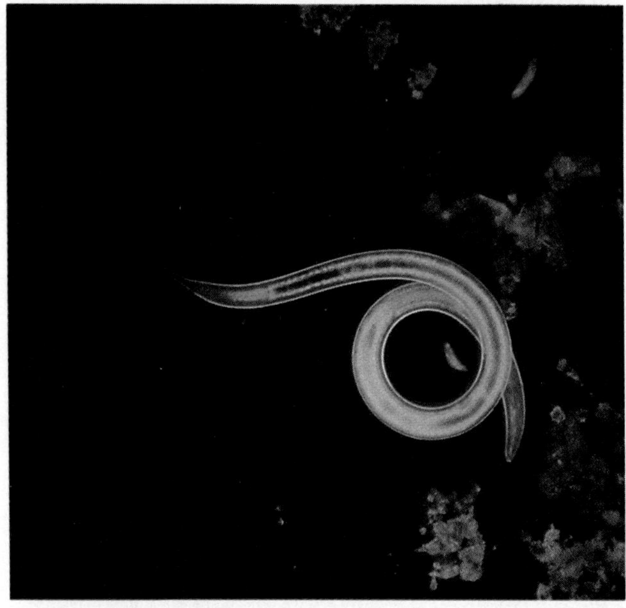

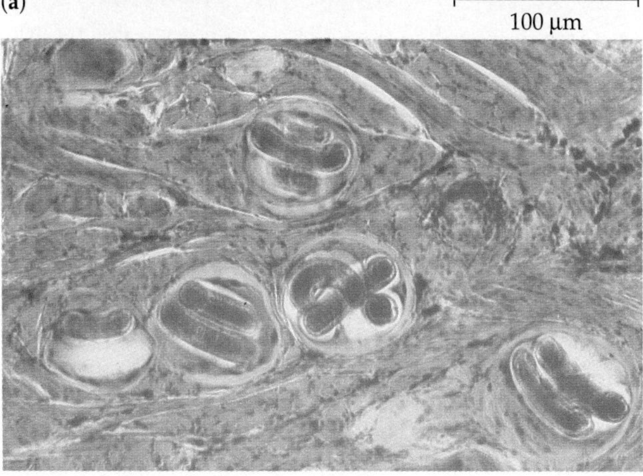

Figure 29.19
Nematodes. (**a**) A free-living nematode. Some of the internal organs of this freshwater worm are visible through the transparent skin (LM). (**b**) Juveniles of *Trichinella spiralis* encysted in human muscle tissue (LM). This parasitic nematode causes trichinosis, characterized by severe nausea and sometimes death when large numbers of the juveniles penetrate heart muscle.

than 1 m in length. They have a complete digestive tract, and the pseudocoelomic fluid helps transport nutrients throughout the body. Roundworms are covered with a tough but transparent cuticle. The muscles of nematodes are all longitudinal, and their contraction produces a thrashing motion best suited to moving in fluids with a lot of particulate matter, such as a muddy lake bottom or the stomach contents of another animal.

Reproduction of nematodes is usually sexual. The sexes are separate in most species, females generally being larger than males. Fertilization is internal, and a female may deposit 100,000 or more fertilized eggs per day. The zygotes of most species are resistant cells able to survive harsh conditions.

Great numbers of nematodes live in moist soil and in decomposing organic matter on the bottoms of lakes and oceans. These extremely numerous free-living worms play an important role in decomposition and nutrient cycling, but little is known about most species. An exception is the soil-dwelling species *Caenorhabditis elegans,* which is an important research organism and one of the best known of all animals. Researchers have been able to trace the development of all of *C. elegans'* cells, and knowledge gained in studying this nematode continues to expand our general understanding of animal development (see Chapter 43).

The phylum Nematoda also includes many important agricultural pests that attack the roots of plants. Other species of roundworms parasitize animals. Humans host at least 50 nematode species, including various pinworms and hookworms. One notorious nematode is *Trichinella spiralis,* the worm that causes trichinosis (Figure 29.19b). Humans acquire the worms by eating infected pork or other meat with juveniles encysted in the muscle tissue. Within the human intestine, the juveniles develop into sexually mature adults. Females burrow into the intestinal muscles and produce more juveniles, which bore through the body or travel in lymph vessels to encyst in other organs, including the skeletal muscles. Cooking infected meat usually kills the *Trichinella* juveniles, but microwave ovens may leave portions of the meat undercooked and the juveniles unharmed.

PROTOSTOMES (SCHIZOCOELOMATES)

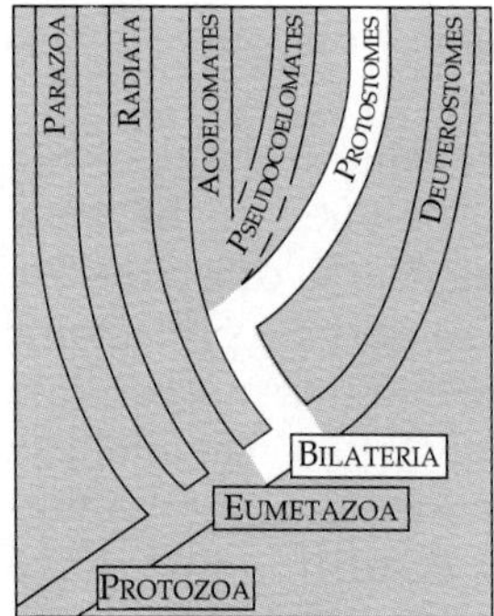

The protostome lineage of animal evolution has given rise to the mollusks, to the annelids and several other phyla of worms with a coelom that functions as a hydrostatic skeleton for burrowing, and to the arthropods, the most diverse and widespread of all animal phyla.

Phylum Mollusca

Snails and slugs, oysters and clams, and octopuses and squids are all mollusks. In all, the phylum Mollusca has more than 50,000 known species. Most mollusks are marine, though some inhabit fresh water, and there are snails and slugs that live on land. Mollusks are soft-bodied animals (L. *molluscus,* "soft"), but most are protected by a hard shell made of calcium carbonate.

Figure 29.20
The basic body plan of mollusks. Three hallmarks of the phylum are the mantle (dark gray), visceral mass (medium gray), and foot (light gray). A mantle cavity houses gills in many species. The long digestive tract is coiled in the visceral mass. Most mollusks have an open circulatory system with a dorsal heart that pumps circulatory fluid (hemolymph) through arteries into sinuses (body spaces) bathing the organs. Excretory organs called nephridia remove metabolic wastes from the hemolymph. The nervous system consists of a nerve ring around the esophagus, from which nerve cords extend. The enlargement shows the mouth region with the radula, a rasplike feeding organ present in many mollusks. The radula is a belt of backward-curved teeth that extends from the mouth and slides back and forth, scraping and scooping like a backhoe.

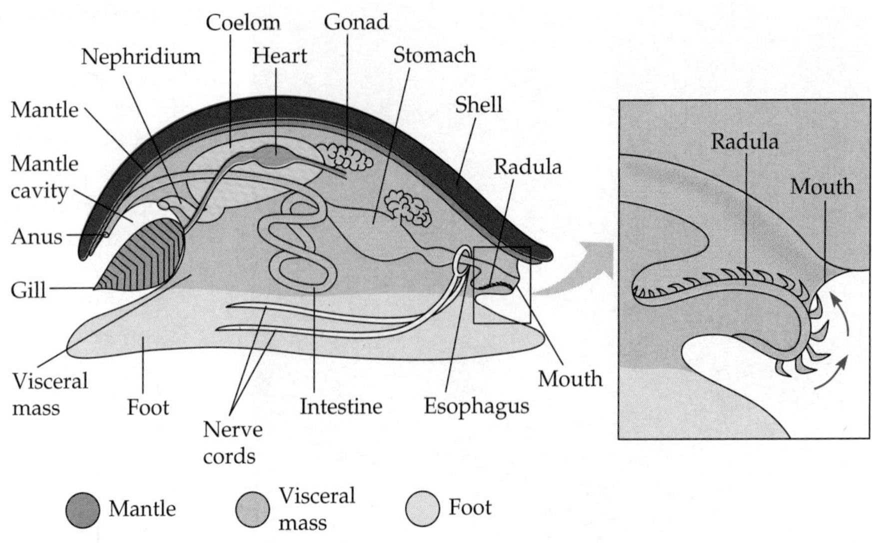

Squids and octopuses have reduced shells that have been internalized, or they have lost their shells completely during their evolution.

Despite their apparent differences, all mollusks have a similar body plan (Figure 29.20). The body has three main parts: a muscular **foot** usually used for movement, a **visceral mass** containing most of the internal organs, and a **mantle,** a heavy fold of tissue that drapes over the visceral mass and may secrete a shell. The mantle also extends away from the visceral mass, producing a water-filled chamber, the **mantle cavity,** which houses the gills, anus, and excretory pores. Many mollusks feed by using a straplike rasping organ called a **radula** to scrape up food. Most mollusks have separate sexes, with gonads (ovaries or testes) located in the visceral mass. Many snails, however, are hermaphrodites.

Zoologists debate the relationship of mollusks to other coelomate protostomes. The life cycle of many marine mollusks includes a ciliated larva called the **trochophore,** also characteristic of some other protostome phyla. But mollusks lack the one trait that most defines an annelid heritage—true segmentation. The mollusk *Neopilina* (class Monoplacophora) has some of its internal organs repeated, but this condition may have evolved secondarily from an ancestral mollusk with unrepeated organs. Many zoologists doubt that mollusks evolved from an annelidlike ancestor and favor the hypothesis that they arose earlier on the protostome line, before the evolution of segmentation.

The basic body plan of mollusks has evolved in various ways in the different classes of the phylum. Of the eight classes, we examine four here: Polyplacophora (chitons), Gastropoda (snails and slugs), Bivalvia (clams, oysters, and other bivalves), and Cephalopoda (squids, octopuses, and nautiluses).

Class Polyplacophora Chitons are marine animals with oval shapes and shells divided into eight dorsal plates (the body itself, however, is unsegmented). You can find chitons clinging to rocks along the shore during low tide (Figure 29.21). Try to dislodge a chiton by hand, and you will be surprised at how tenaciously its foot, acting as a suction cup, grips the rock. Using this muscular foot in the same fashion as a snail does, a chiton can creep slowly over the rock surface. In this way, the chiton searches out sheets of algae, which it ingests with its radula.

Class Gastropoda The largest of the molluscan classes, Gastropoda has more than 40,000 species. Most gastropods are marine, but there are also many freshwater species. Garden snails and slugs have adapted to land.

Figure 29.21
A chiton. Clinging tenaciously to rocks in the intertidal zone, this chiton (class Polyplacophora) displays the eight-plate shell characteristic of this class of mollusks.

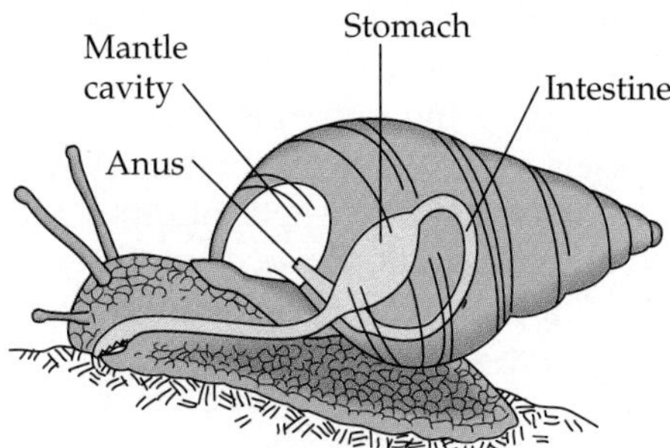

Figure 29.22
The results of torsion in a gastropod. Because of torsion (twisting of the visceral mass) during embryonic development, the digestive tract is coiled and the anus is near the mouth at the head end of the animal. After torsion, some of the organs that were bilateral subsequently atrophy on one side of the body. In the shell-less slugs and nudibranchs, the visceral mass is no longer contorted, but an asymmetric arrangement of the viscera is a clue that the ancestors of these animals carried out the torsion characteristic of gastropods. Torsion should not be confused with coiling, which is an independent process.

The most distinctive characteristic of the class Gastropoda is an embryonic process known as **torsion.** During embryonic development, one side of the visceral mass grows faster than the other. The uneven growth causes the visceral mass to rotate up to 180 degrees, so that the anus and mantle cavity are placed above the head in the adult (Figure 29.22). Some zoologists speculate that the advantage of torsion is to place the visceral mass and heavy shell more centrally over the snail's body.

Most gastropods are protected by single, spiraled shells into which the animals can retreat when threatened (Figure 29.23). The shell is often conical, but abalones and limpets have somewhat flattened shells. Slugs and nudibranchs (sea slugs) have lost their shells during their evolution. Many gastropods have distinct heads with eyes at the tips of tentacles. Gastropods inch along literally at a snail's pace by a rippling motion of the elongated foot. (Some very tiny species use cilia for locomotion.) Most gastropods use their radula to graze on plant material. Several groups, however, are predators, and the radula is modified for boring holes in the shells of other mollusks or for tearing apart tough animal tissue. In one group, the cone snails, the teeth of the radula form separate poison darts, which are fired at prey.

Gastropods are among the few invertebrate groups to have successfully populated the land. Terrestrial snails lack the gills typical of most aquatic gastropods, and instead the vascularized mantle cavity functions as a lung to exchange respiratory gases with the air.

Class Bivalvia The mollusks of the class Bivalvia, also known as Pelecypoda, include many species of clams, oysters, mussels, and scallops. Bivalves have shells divided into two halves (Figure 29.24). The two

(a)

(b)

Figure 29.23
Gastropods. (**a**) Gastropoda is one of the most diverse animal classes. (**b**) Nudibranchs, or sea slugs, have eliminated the shell during their evolution.

Figure 29.24
A bivalve. This scallop has many eyes peering out between the two halves of the hinged shell.

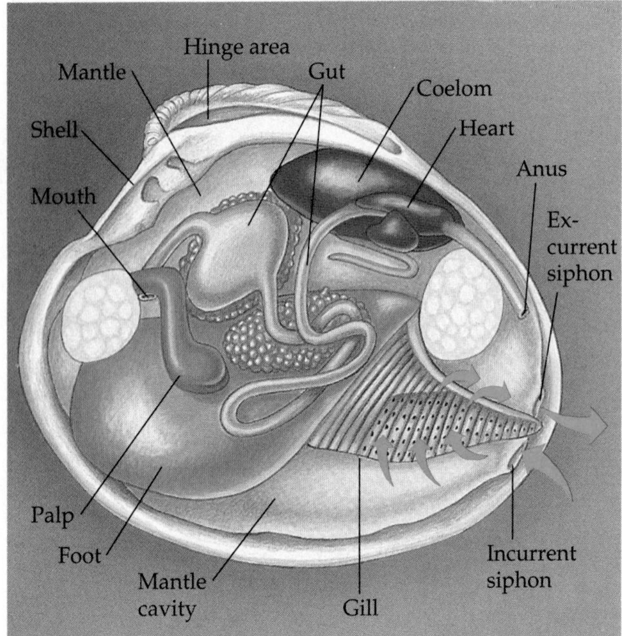

Figure 29.25
Anatomy of a clam. The left half of the bivalve shell has been removed. Food particles suspended in water that enters through the incurrent siphon are collected by the gills and passed to the mouth via elongated flaps, called palps.

parts of the shell are hinged at the mid-dorsal line, and powerful adductor muscles draw the two halves tightly together to protect the soft-bodied animal. When the shell is open, the bivalve may extend its hatchet-shaped foot for digging or anchoring.

The mantle cavity of a bivalve contains gills that are used for feeding as well as gas exchange (Figure 29.25). Bivalves have no distinct head, and the radula has been lost. Most bivalves are suspension-feeders. They trap fine food particles in mucus that coats the gills, and then use cilia to convey the particles to the mouth. Water flows into the mantle cavity through an incurrent siphon, passes over the gills, and then exits the mantle cavity through an excurrent siphon.

Being suspension-feeders, most bivalves lead rather sedentary lives. Clams can pull themselves into the sand or mud, using the muscular foot for an anchor. Sessile mussels secrete strong threads that tether to rocks, docks, boats, and the shells of other animals. In addition to digging, scallops can also skitter along the sea floor by flapping their shells, rather like the mechanical false teeth sold in novelty shops.

Class Cephalopoda Unlike the sluggish gastropods and the sedentary bivalves, cephalopods are built for speed, an adaptation that fits their carnivorous diet (Figure 29.26). Squids and octopuses use beaklike jaws to bite their prey; they then inject poison to immobilize the victim. The mouth is at the center of several long tentacles. A mantle covers the visceral mass, but the shell is usually either reduced and internal (squids) or missing altogether (octopuses). One shelled cephalopod, the chambered nautilus, survives today (Figure 29.26c).

A squid darts about, usually backward, by drawing water into its mantle cavity and then firing a jet stream of water through the excurrent siphon that points anteriorly. The animal steers by pointing the siphon in different directions. The foot of a cephalopod has become modified into this muscular siphon and probably other parts of the tentacles and head. (*Cephalopod* means "head foot.") Most species of squid are less than 75 cm long, but there are also giant squids, the largest of all invertebrates. The largest specimen on record was 17 m long (including the tentacles) and weighed about 2 tons.

Rather than swimming as squids do in the open seas, most octopuses live on the sea floor, where they creep and scurry about in search of crabs and other food.

Cephalopods are the only mollusks with **closed circulatory systems,** meaning that their blood is always contained in vessels. Cephalopods also have well-developed nervous systems with complex brains. The ability to learn and behave in a complex manner is

(a)

(b)

(c)

Figure 29.26
Cephalopods. (**a**) Squid, such as this California market squid *(Loligo opalescens)*, are speedy carnivores with beaklike jaws and well-developed eyes. (**b**) Octopuses are believed to be among the most intelligent invertebrates. (**c**) Chambered nautiluses are the only cephalopods alive today with external shells.

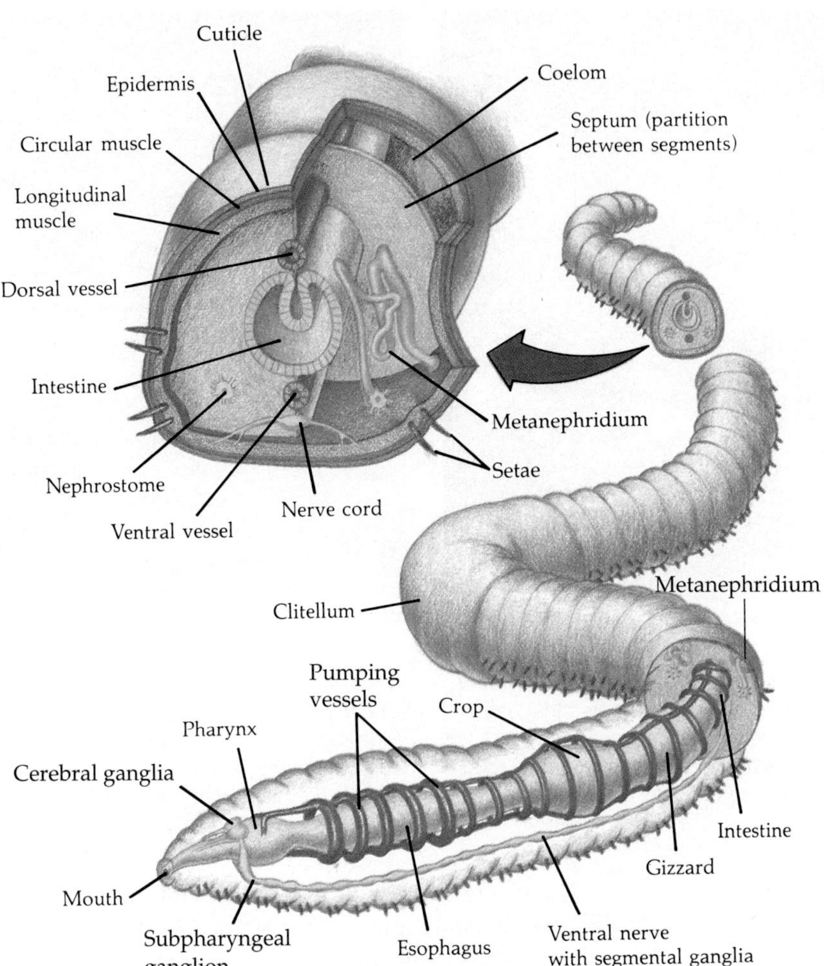

Figure 29.27
Anatomy of an earthworm. Annelids are segmented both externally and internally. Many of the internal structures are repeated, segment by segment. Externally, each segment has four pairs of setae, bristles that provide traction for burrowing.

probably more critical to fast-moving predators than to sedentary animals, such as clams. Squids and octopuses also have well-developed sense organs.

The ancestors of octopuses and squids were probably shelled mollusks that took up a predaceous lifestyle, the loss of the shell occurring in later evolution. Shelled cephalopods called **ammonites,** many of them very large, were the dominant invertebrate predators of the seas for hundreds of millions of years until their disappearance during the mass extinctions at the end of the Cretaceous period.

Phylum Annelida

Annelid worms have segmented bodies that give them a ringed appearance. (*Annelida* means "little rings.") There are about 15,000 annelid species, ranging in length from less than 1 mm to the 3-m length of a giant Australian earthworm. Annelids live in the sea, most freshwater habitats, and damp soil. We can describe the anatomy of annelids in terms of a well-known member of the phylum, the earthworm (Figure 29.27).

The coelom of the earthworm is partitioned by septa, but the digestive tract, longitudinal blood vessels, and nerve cords penetrate the septa and run the length of the animal (the major vessels have segmental branches). The digestive system has several specialized regions: the pharynx, the esophagus, the crop, the gizzard, and the intestine. The complex closed circulatory system consists of a network of vessels containing blood with oxygen-carrying hemoglobin. Dorsal and ventral vessels are connected by segmental pairs of vessels. The dorsal vessel and five pairs of vessels that circle the esophagus of an earthworm are muscular and pump blood through the circulatory system.

In each segment of the worm is a pair of excretory tubes called **metanephridia** with ciliated funnels, called nephrostomes, that remove wastes from the blood and coelomic fluid. The metanephridia lead to exterior pores, through which the metabolic wastes are discharged. Tiny blood vessels are abundant in the skin, which the earthworm uses as its respiratory organ to obtain oxygen. Because gas exchange must occur across a film of moisture, an earthworm suffocates if its skin dries.

A brainlike pair of cerebral ganglia lies above and in front of the pharynx. A ring of nerves around the pharynx connects to a subpharyngeal ganglion, from which a fused pair of nerve cords run posteriorly. All along

(a)

(b)

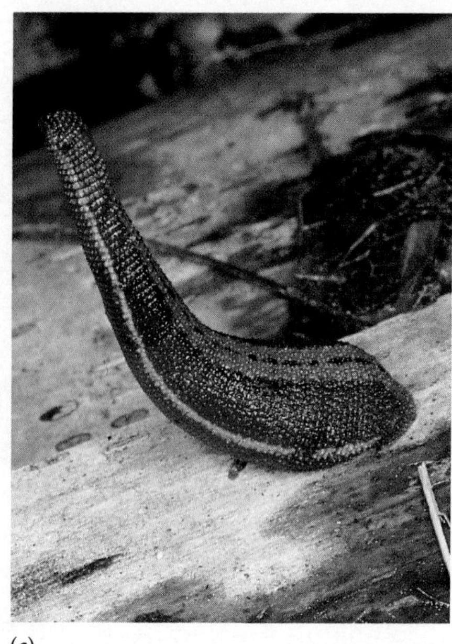

(c)

Figure 29.28
Annelids, or segmented worms. (**a**) Most annelids of the class Polychaeta are marine worms. Each segment has a pair of lateral flaps that function in movement and as gills for exchange of respiratory gases with the surrounding water. (**b**) Fanworms are tube-dwelling polychaetes that use their feathery headdresses for gas exchange and to extract suspended food particles from the seawater. This species is known as a Christmas-tree worm. (**c**) Leeches (class Hirudinea) are free-living carnivores or parasites that suck blood from other animals. This species inhabits the Malaysian rain forest.

these ventral nerve cords are segmental ganglia, also fused.

Earthworms are hermaphrodites, but they cross-fertilize. Two earthworms mate by aligning themselves in such a way that they exchange sperm, and then they separate. The received sperm are stored temporarily while a special organ, called the **clitellum,** secretes a mucous cocoon. The cocoon slides along the worm, picking up its eggs and then the stored sperm. The cocoon then slips off the worm's head and resides in the soil while the embryos develop. Some earthworms can also reproduce asexually by fragmentation followed by regeneration. (Chapter 42 covers fragmentation and other modes of asexual reproduction.)

Some aquatic annelids swim in pursuit of food, but most are bottom-dwellers that burrow in the sand and silt; earthworms, of course, are burrowers. Many annelids creep along by coordinating two sets of muscles, one longitudinal and the other circular. These muscles work against the noncompressible coelomic fluid, a hydrostatic skeleton. The muscles can alter the shape of each segment individually because the coelom is segregated into separate compartments. When the circular muscles of a segment contract, that segment becomes thinner and elongates. Contraction of the longitudinal muscles causes the segment to shorten and thicken. The worm probes forward as alternating contractions of circular and longitudinal muscles progress along the segments like waves. The coelom and segmentation may have first functioned as adaptations for this type of movement.

The phylum Annelida is divided into three classes (Figure 29.28): Oligochaeta (earthworms and their relatives), Polychaeta (polychaetes), and Hirudinea (leeches).

Class Oligochaeta This class of segmented worms includes the earthworms and a variety of aquatic species. An earthworm eats its way through the soil, extracting nutrients as the soil passes through the digestive tube. Undigested material, mixed with mucus secreted into the digestive tract, is egested as casts through the anus. Farmers value earthworms because the animals till the earth and the castings improve the texture of the soil. Charles Darwin estimated that 1 acre of British farmland had about 50,000 earthworms that produced 18 tons of castings per year.

Class Polychaeta Each segment of a polychaete ("many setae") has a pair of paddlelike or ridgelike structures called parapodia ("almost feet"). In many polychaetes these parapodia, richly vascularized, function as gills that provide an extended area of skin for gas exchange (Figure 29.28a). Parapodia are also instrumental in locomotion. Each parapodium has several setae made of the polysaccharide chitin.

Most polychaetes are marine. A few adult forms drift and swim among the plankton, many crawl on or burrow in the sea floor, and many others live in tubes, which the worms make by mixing mucus with bits of sand and broken shells. The tube-dwellers include the brightly colored fanworms, which trap microscopic

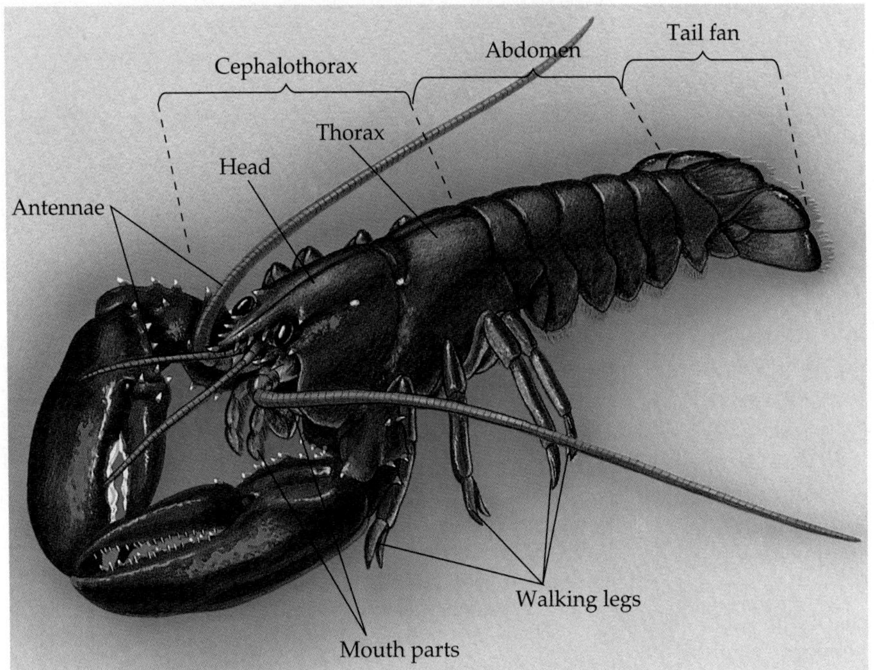

Figure 29.29
External anatomy of a lobster, an arthropod. In this dorsal view of the animal, many of the characteristic features of arthropods are apparent, including the jointed exoskeleton, sensory antennae and eyes, and multiple appendages modified for different functions. Arthropods are segmented, but this characteristic is evident only in the abdominal portion of the lobster.

food particles in feathery tentacles that extend from the opening of the tube (Figure 29.28b).

Class Hirudinea The majority of leeches inhabit fresh water, but there are also land leeches that move through moist vegetation. Many leeches feed on other small invertebrates, but others are blood-sucking parasites that feed by attaching temporarily to other animals, including humans (Figure 29.28c). Leeches range in length from about 1 to 30 cm. Some use bladelike jaws to slit the skin of the host, whereas others secrete enzymes that digest a hole through the skin. The host is usually oblivious to this attack because the leech secretes an anesthetic. After making the incision, the leech secretes another chemical, hirudin, which keeps the blood of the host from coagulating. The parasite then sucks as much blood as it can hold, often more than ten times its own weight. After this gorging, a leech can last for months without another meal. Until this century, leeches were frequently used by physicians for bloodletting. Leeches are still used for treating bruised tissues and for stimulating the circulation of blood to fingers or toes that have been sewn back to hands or feet after accidents.

Before leaving annelids, let's highlight two evolutionary innovations—the coelom and segmentation—that are well developed in this phylum. The evolutionary significance of the coelom cannot be overemphasized. In addition to providing a hydrostatic skeleton and allowing new and diverse methods of locomotion, it has other advantages. It provides body space for storage and for complex organ development; it serves as a cushion that protects internal structures; and it allows a functional separation of the action of body wall muscles from those of internal organs, such as the muscles of the gut.

Segmentation set the foundation for specialization of body regions; groups of segments became modified for different functions. This regional specialization is seen to some degree in annelids, but it reached its zenith among the arthropods.

Phylum Arthropoda

It is estimated that the arthropod population of the world, including crustaceans, spiders, and insects, numbers about 1 billion billion (10^{18}) individuals. Nearly 1 million arthropod species have been described, mostly insects. In fact, two out of every three organisms known are arthropods, and the phylum is represented in nearly all habitats of the biosphere. On the criteria of species diversity, distribution, and sheer numbers, Arthropoda must be regarded as the most successful phylum of animals ever to live.

Characteristics of Arthropods The diversity and success of arthropods are largely related to their segmentation, hard exoskeleton, and jointed appendages. (*Arthropoda* means "jointed feet.") Groups of segments and their appendages have become specialized for a great variety of functions, including sensing, feeding, locomotion, and reproduction. This evolutionary flexibility resulted not only in great diversification but also in an efficient body plan by the division of labor among regions. For example, the appendages are variously modified for walking, feeding, sensory reception, copulation, and defense (Figure 29.29).

The body of an arthropod is completely covered by the **cuticle,** an **exoskeleton** (external skeleton) constructed from layers of protein and chitin. The cuticle can be modified into thick, hard armor over some parts of the body or be paper thin and flexible in other locations, such as the joints. The exoskeleton protects the animal and provides points of attachment for the muscles that move the appendages. The skeleton of arthropods is both strong and relatively impermeable. As we will see, both of these qualities were largely responsible for the invasion of land by various arthropod groups. The rigid exoskeleton also posed some evolutionary problems. For example, in order to grow, an arthropod must occasionally shed its old exoskeleton and secrete a larger one. This process, called **molting,** is energetically expensive and leaves the animal temporarily vulnerable to predators and other dangers.

Arthropods tune in to their environment with well-developed sensory organs, including eyes, olfactory receptors for smell, and antennae for touch and smell. Cephalization is extensive, with most sensory organs concentrated on the anterior end of the animal.

Arthropods have open circulatory systems in which fluid called hemolymph (*blood* is reserved for fluid in a closed circulatory system) is propelled by a heart. Hemolymph leaves the heart via short arteries and passes into spaces called sinuses surrounding the tissues and organs. Hemolymph reenters the arthropod heart through pores that are usually equipped with valves. Collectively, the body sinuses are called the hemocoel, which is not part of the coelom. In most arthropods, the coelom that forms in the embryo becomes much reduced as development progresses, and the hemocoel becomes the main body cavity in adults. Although this condition resembles the open circulatory system of mollusks, the two probably arose independently.

A variety of organs specialized for gas exchange have evolved in arthropods. These organs must allow diffusion of respiratory gases in spite of the exoskeleton. Most aquatic species have gills with thin feathery extensions that place an extensive surface area in contact with the surrounding water. Terrestrial arthropods generally have internal surfaces specialized for gas exchange. Most insects, for instance, have tracheal systems, branched air ducts leading into the interior from pores in the cuticle.

Arthropod Phylogeny and Classification Arthropods are segmented animals that probably evolved from annelids or from a segmented protostome that was a common ancestor of annelids and arthropods. Perhaps early arthropods resembled onychophorans, a separate phylum of animals that look like walking worms (Figure 29.30). The multiple appendages of onychophorans are unjointed, but there are enough Cambrian fossils of jointed-legged animals that resemble segmented worms to support other evidence of an evolutionary link between Annelida and Arthropoda.

Figure 29.30
***Peripatus*, the walking worm (phylum Onychophora).** *Peripatus* is distinctly segmented and has excretory organs, musculature, and certain other features that are annelidlike. *Peripatus* resembles arthropods in its respiratory and circulatory systems, its cuticle made of chitin, and its jaws modified from appendages.

Arthropods evolved along four main lines, which most zoologists recognize as subphyla: **Trilobitomorpha** (the extinct trilobites); **Cheliceriformes** or **Chelicerata** (spiders, scorpions, sea spiders, and some extinct groups); **Uniramia** (insects, centipedes, millipedes); and **Crustacea** (crabs, lobsters, shrimps, barnacles, etc.).

Among the early arthropods were the **trilobites** (Figure 29.31). They were common denizens of the shallow seas throughout the Paleozoic era but disap-

Figure 29.31
A fossil arthropod. Trilobites were prevalent arthropods throughout the Paleozoic era. About 4000 trilobite species have been described from fossils.

Figure 29.32
Horseshoe crabs. These "living fossils," which have changed little in hundreds of millions of years, have survived from a rich diversity of chelicerates that once filled the seas. Horseshoe crabs are common on the Atlantic and Gulf coasts of the United States.

peared with the great Permian extinctions that closed that era, about 280 million years ago. Trilobites had pronounced segmentation, but their appendages showed little variation from segment to segment. As arthropods continued to evolve, the segments tended to fuse and become fewer in number and the appendages became specialized for a variety of functions. (Compare the trilobites with the lobster in Figure 29.29.)

The trilobites were outlasted by the **eurypterids,** or sea scorpions. These marine predators, up to 3 m long, were chelicerates. The body of a chelicerate is divided into an anterior cephalothorax and a posterior abdomen. The appendages are more specialized than those of trilobites, and the most anterior appendages are modified as either pincers or fangs. Chelicerates (Gr. *cheilos,* "lips," and *cheir,* "arm") are named for these feeding appendages, the **chelicerae.** Most of the marine chelicerates, including the eurypterids, are extinct; one survivor is the horseshoe crab (Figure 29.32). The bulk of modern chelicerates are found on land in the form of the **class Arachnida**, which includes scorpions, spiders, ticks, and mites (Figure 29.33). Arachnids have a cephalothorax with six pairs of appendages: the chelicerae, a pair of appendages called pedipalps that usually function in sensing or feeding, and four pairs of walking legs. Spiders use their fanglike chelicerae, equipped with poison glands, to attack prey. As the chelicerae masticate (chew) the prey, the spider spills digestive juices onto the torn tissues. The food softens, and the spider sucks up the liquid meal. In most spiders, gas exchange is carried out by **book lungs,** stacked plates contained in an internal chamber. Figure 29.34 shows the anatomy of a spider in more detail.

(a)

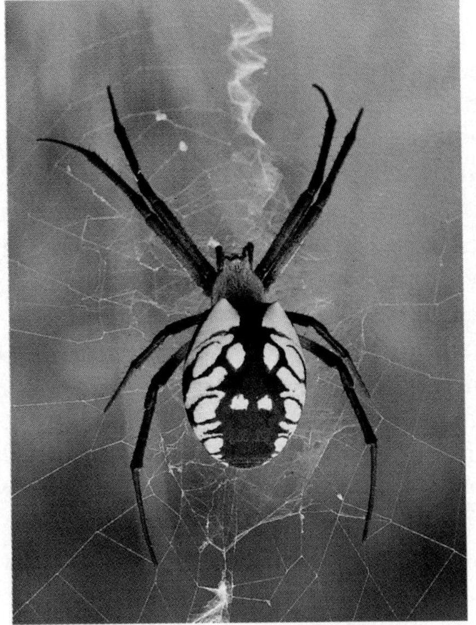

(b)

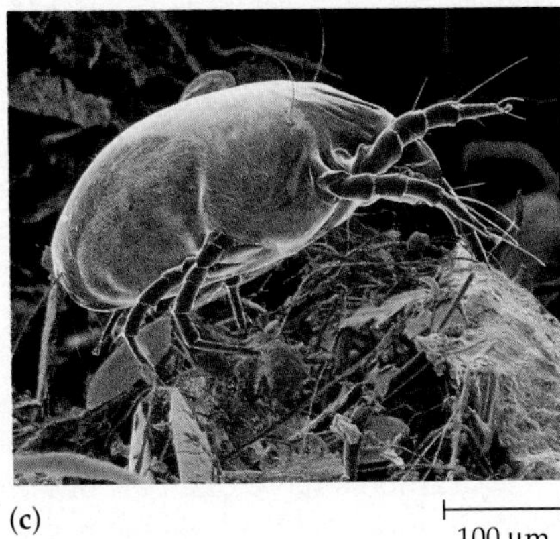

(c)

Figure 29.33
Arachnids. (**a**) Scorpions, which hunt by night, were among the first terrestrial carnivores, preying on vegetarian arthropods that fed on the early land plants. The pedipalps of scorpions are pincers specialized for defense and the capture of food. The tip of the tail bears a poisonous sting. (**b**) Spiders are generally most active during the daytime, when they hunt for prey or trap insects in webs. (**c**) This house-dust mite, magnified in an artificially colored micrograph, is a ubiquitous scavenger in human dwellings (SEM). Unlike some mites that carry disease-causing bacteria, dust mites are harmless except to people who are allergic to them.

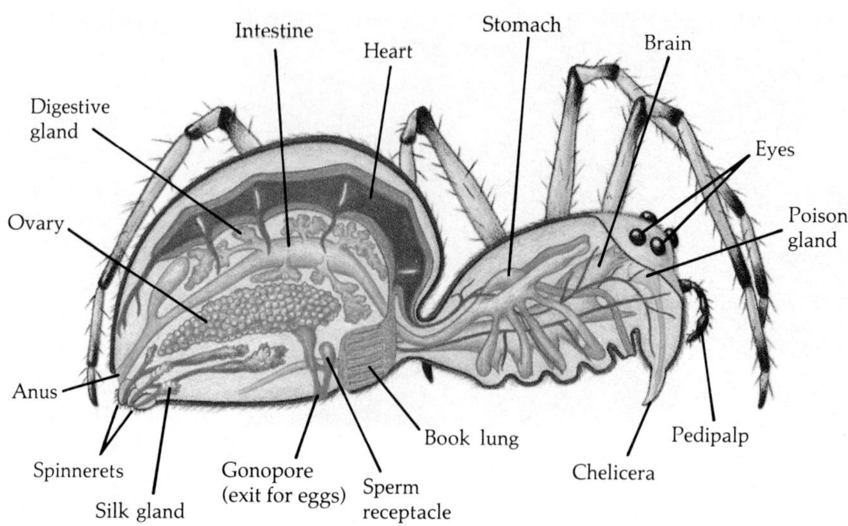

Figure 29.34
Anatomy of a spider, an arachnid.

A unique adaptation of many spiders is catching flying insects by stringing webs of silk, a protein produced as a liquid by special abdominal glands. The silk is spun by organs called spinnerets into fibers that solidify. Each spider engineers a style of web characteristic of its species and constructs the web perfectly on the first try. This complex behavior is apparently inherited. Besides building their webs from silk, various spiders use these fibers in other ways—as droplines for rapid escape, as coats that cover eggs, and even as "gift wrapping" for food that certain male spiders offer females during courtship.

Aside from the chelicerates, another major line of arthropod evolution produced the uniramians and crustaceans. Rather than having clawlike chelicerae, these arthropods have jawlike **mandibles.** They are also distinguished from chelicerates in having one or two pairs of sensory **antennae** and usually a pair of **compound eyes** (multifaceted eyes with many separate focusing elements). Chelicerates lack antennae, and most have simple eyes (eyes with a single lens).

Uniramians have one pair of antennae and uniramous (unbranched) appendages; crustaceans have two pairs of antennae and typically biramous (branched) appendages. Crustaceans are primarily aquatic and are believed to have evolved in the ocean. Uniramians, on the other hand, are believed to have evolved on land.

The move onto land by chelicerates and uniramians was made possible in part by the exoskeleton. When the exoskeleton first evolved in the seas, its main functions were probably protection and anchorage for muscles; but it eventually helped certain arthropods live on land by solving the problems of water loss and support. The arthropod cuticle is relatively impermeable to water, helping to prevent desiccation. The firm exoskeleton also solved the problem of support when arthropods left the buoyancy of water. Uniramians and chelicerates both spread onto land during the early Devonian period, following the colonization by plants. The oldest fossil evidence of terrestrial animals is burrows of millipedelike arthropods about 450 million years old. Fossilized arachnids almost as old have also been found.

Uniramians The evolutionary link between arthropods and annelids is evident in the distinct segmentation of the classes Diplopoda and Chilopoda (Figure 29.35).

Millipedes (**class Diplopoda**) are wormlike, with a large number of walking legs (two pairs per segment), though fewer than the thousand their name implies. They eat decaying leaves and other plant matter. Millipedes were probably among the earliest animals on land, living on mosses and primitive vascular plants.

Centipedes (**class Chilopoda**) are terrestrial carnivores. The head has a pair of antennae and three pairs of appendages modified as mouthparts, including the jawlike mandibles. Each segment of the trunk region has one pair of walking legs. Centipedes use poison claws on the anteriormost trunk segment to paralyze prey and to defend themselves.

In species diversity, insects **(class Insecta)** outnumber all other forms of life combined. They live in almost every terrestrial habitat and in fresh water, and flying insects fill the air. Insects are rare, though not absent, in the seas, where crustaceans are the dominant arthropods. Class Insecta is divided into about 26 orders, some of which are described in Table 29.1 (pp. 622–623). **Entomology,** the study of insects, is a vast

(a)

(b)

Figure 29.35
Diplopods (millipedes) and chilopods (centipedes). (**a**) Millipedes feed on decaying plant matter. (**b**) The house centipede *(Scutigera coleoptrata)*, a fast-moving carnivore, feeds on insects, including cockroaches, and other small invertebrates.

field with many subspecialties, including physiology, ecology, and taxonomy. Here we can only examine the general characteristics of this class of animals.

The oldest insect fossils date back to the Devonian period, which began about 400 million years ago. However, when flight evolved during the Carboniferous and Permian periods, it spurred an explosion in insect diversity. Another major radiation of insect species took place during the Cretaceous period, when insects diversified along with the flowering plants they ate and pollinated (see Chapter 27). Flight is obviously one key to the great success of insects. A flyer can escape predators, find food and mates, and disperse to new habitats much faster than animals that must crawl about on the ground.

Many insects have one or two pairs of wings that emerge from the dorsal side of the thorax (Figure 29.36). Since the wings are extensions of the cuticle and not true appendages, insects are able to fly without sacrificing any walking legs. (By contrast, the flying vertebrates—birds and bats—have one of their two pairs of walking legs modified for wings and are generally quite clumsy on the ground.) Insect wings may have first evolved as cuticular extensions that helped the insect body absorb heat, only later becoming organs for flight. Other views suggest that wings allowed the animals to glide from vegetation to the ground, or even served as gills in aquatic insects. Some insects beat their wings at speeds of several hundred cycles per second by using muscles to warp the shape of the entire cuticle covering the thorax. As the wings flap, they change angles, producing lift on both the up and down strokes.

Dragonflies, with two coordinated pairs of wings, were among the first insects to fly. Several insect orders that evolved later than dragonflies have modified flight equipment. Bees and wasps, for instance, hook their wings together and move them as a single pair. Butterflies get a similar result by overlapping their an-

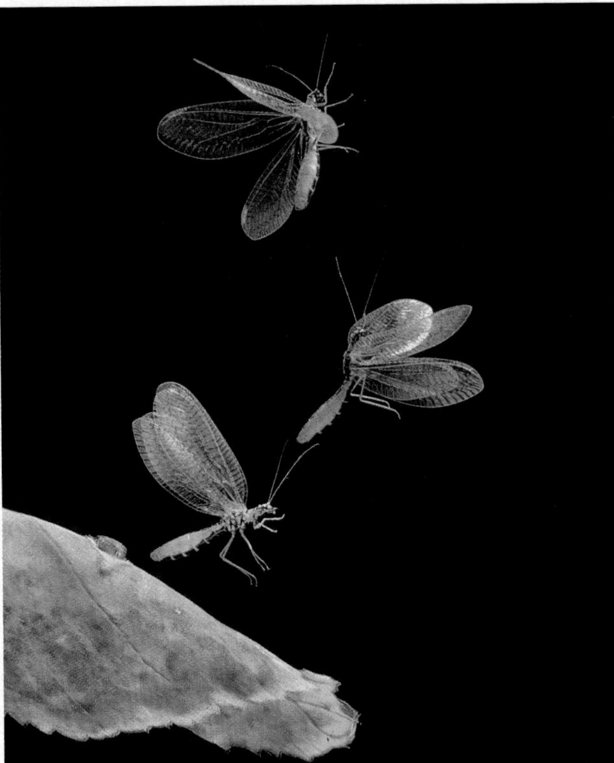

Figure 29.36
Insect flight. Insect wings are not modified appendages but are extensions of the cuticle, flapped by flight muscles that bend the cuticle of the thorax. Here, a lacewing takes off from a leaf.

Table 29.1 Some Major Orders of Insects

Order	Approximate Number of Species	Main Characteristics	Examples	
Anoplura	2,400	Wingless; sucking mouthparts; small with flattened body, reduced eyes; legs with clawlike tarsi for clinging to skin; incomplete metamorphosis; very host-specific.	Sucking lice	Human body louse
Coleoptera	500,000	Two pairs of horny, membranous wings; heavy, armored exoskeleton; biting and chewing mouthparts; complete metamorphosis.	Beetles, weevils	Japanese beetle
Dermaptera	1,000	Two pairs of leathery, membranous wings; biting mouthparts; large pincers in males; incomplete metamorphosis.	Earwigs	Earwig
Diptera	80,000	One pair of wings and halteres (balancing organs); sucking, piercing, lapping mouthparts; complete metamorphosis.	Flies, mosquitoes	Horsefly
Hemiptera	55,000	Two pairs of horny, membranous wings; piercing, sucking mouthparts; incomplete metamorphosis.	True bugs: assassin bug, bedbug, chinch bug	Leaf-footed bug
Hymenoptera	90,000	Two pairs of membranous wings; head mobile; well-developed eyes; chewing and sucking mouthparts; stinging; complete metamorphosis; many species social.	Ants, bees, wasps	Cicada-killer wasp

Table 29.1 *(Continued)*

Order	Approximate Number of Species	Main Characteristics	Examples	
Isoptera	2,000	Two pairs of wings, but some stages are wingless; chewing mouthparts; social; division of labor for reproduction, work, defense; incomplete metamorphosis.	Termites	 Termite
Lepidoptera	140,000	Two pairs of wings; hairy bodies; long coiled tongue for sucking; complete metamorphosis.	Butterflies, moths	 Swallowtail butterfly
Odonata	5,000	Two pairs of wings; biting mouthparts; incomplete metamorphosis.	Damselflies, dragonflies	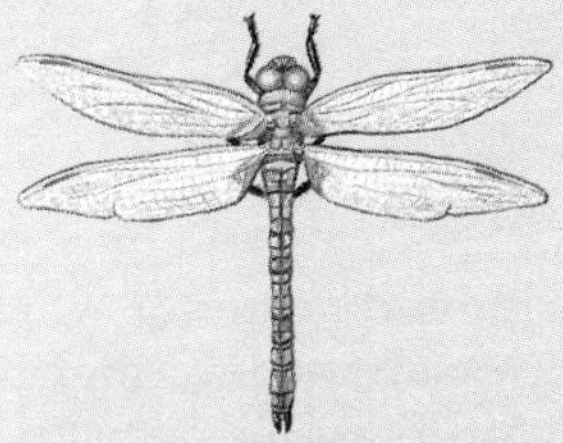 Dragonfly
Orthoptera	30,000	Two pairs of horny, membranous wings; biting and chewing mouthparts in adults; incomplete metamorphosis.	Crickets, roaches, grasshoppers, mantids	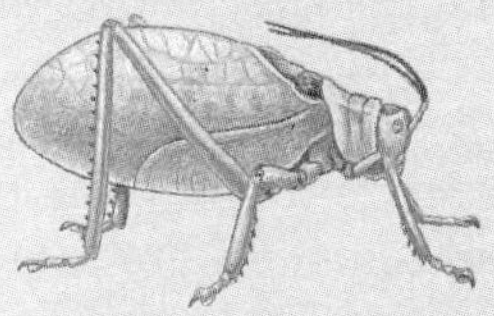 Katydid
Siphonaptera	1,200	Small, wingless, laterally compressed; piercing and sucking mouthparts; jumping legs; complete metamorphosis.	Fleas	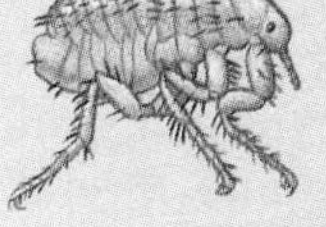 Flea
Trichoptera	7,000	Two pairs of hairy wings; lapping mouthparts; complete metamorphosis; aquatic larvae build movable cases of sand and gravel bound together by secreted silk.	Caddisflies	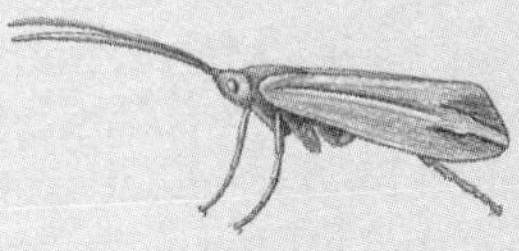 Caddisfly

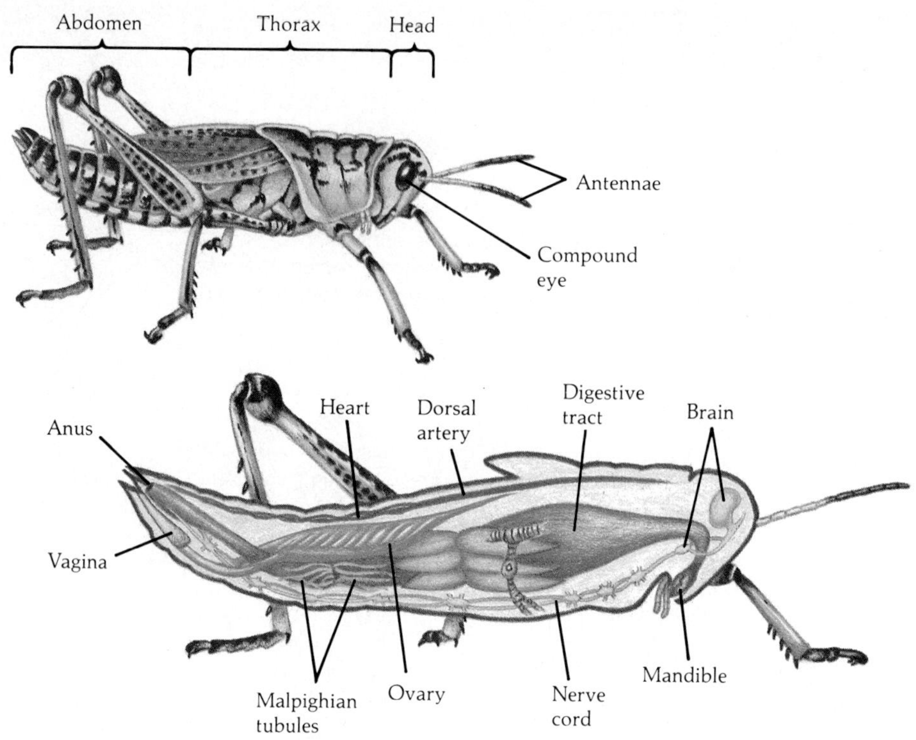

Figure 29.37
Anatomy of a grasshopper, an insect.

terior and posterior wings. In beetles, the posterior wings are used for flight, while the anterior ones are modified as covers that protect the flight wings when the beetle is on the ground or burrowing.

Figure 29.37 shows the anatomy of a representative insect, the grasshopper. The insect body has three regions: a head, a thorax, and an abdomen. Segmentation is apparent along the thorax and abdomen, but the head segments are fused. On the insect head are one pair of antennae and a pair of compound eyes. Several pairs of appendages modified for chewing (as in grasshoppers) or for lapping, piercing, and sucking (in certain other insects) form the mouthparts. The thorax of the insect bears three pairs of walking legs.

The internal anatomy of an insect includes several complex organ systems. The digestive tract is a tube pinched into several regions, each with its own function in the breakdown of food and the absorption of nutrients. Like other arthropods, an insect has an open circulatory system, with a heart pumping hemolymph through the sinuses of the hemocoel. Metabolic wastes are removed from the hemolymph by unique excretory organs called **Malpighian tubules,** which are outpocketings of the gut. Gas exchange in insects is accomplished by a **tracheal system** of branched, chitin-lined tubes that infiltrate the body and carry oxygen directly to cells. The tracheal system opens to the outside of the body through spiracles, pores that can open or close to regulate air flow and limit water loss.

The insect nervous system consists of a pair of ventral nerve cords with several segmental ganglia. The two cords meet in the head, where the ganglia of several anterior segments are fused into a dorsal brain close to the antennae, eyes, and other sense organs concentrated on the head. Insects are capable of complex behavior, though this behavior seems to be largely innate. Even the intricate social behavior of some bees and ants is apparently inherited. (We will discuss insect digestion, circulation, excretion, gas exchange, nervous control, and behavior in more detail in Unit Seven.)

Many insects undergo metamorphosis in their development. In the **incomplete metamorphosis** of grasshoppers and some other orders, the young resemble adults but are smaller and have different body proportions. The animal goes through a series of molts, each time looking more like an adult, until it reaches full size. Insects with **complete metamorphosis** have larval stages, known by such names as maggot, grub, or caterpillar, that look entirely different from the adult stage (Figure 29.38). The main job of the larva is to eat and grow. The primary function of the adult is to find a mate and reproduce. Mates come together and recognize each other as members of the same species by advertising with bright colors (butterflies), sound (crickets), or odors (moths). After mating, a female lays her eggs on an appropriate food source where the larvae can begin eating as soon as they hatch.

Reproduction in insects is usually sexual, with separate male and female animals. Fertilization is usually internal. In most species, sperm are deposited directly into the female's vagina at the time of copulation,

Figure 29.38
Metamorphosis of a butterfly. The larva (caterpillar) spends its time eating and growing, molting as it grows. After several molts, the larva encases itself in a cocoon and becomes a pupa. Within the pupa, the larval tissues are broken down and the adult is built by the division and differentiation of cells that were quiescent in the larva. Finally, the adult emerges from the cocoon. Fluid is pumped into veins of the wings, and then the fluid is withdrawn to leave the hardened veins as struts supporting the wings. Now the insect flies off and reproduces, deriving much of its nourishment from the calories stored by the feeding larva.

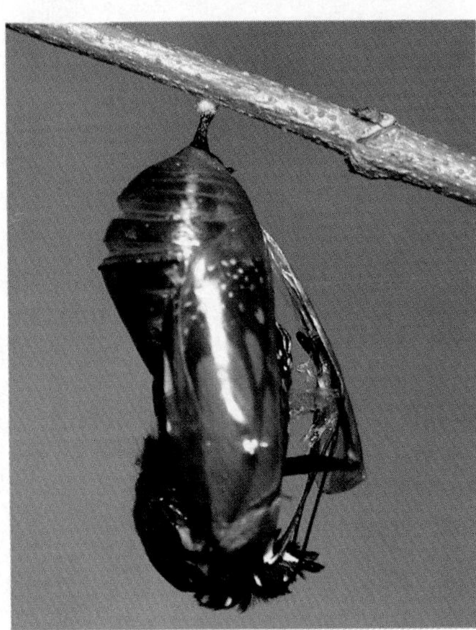

though in some species the male deposits a sperm packet outside the female, who then picks it up. Inside the female, a structure called the spermatheca stores the sperm, usually enough to fertilize more than one batch of eggs. Though most insect species produce a multitude of eggs, some flies produce live offspring, usually one at a time. Many insects mate only once in a lifetime.

Animals so numerous, diverse, and widespread as insects are bound to affect the lives of all other terrestrial organisms, including humans. On the one hand, we depend on insects to pollinate many of our crops and orchards. On the other hand, insects are carriers for many diseases, including malaria and African sleeping sickness. Furthermore, insects compete with humans for food. In parts of Africa, for instance, insects claim about 75% of the crops. Trying to minimize their losses, farmers in the United States spend billions of dollars each year on pesticides, spraying crops with massive doses of some of the deadliest poisons ever invented. Try as they may, not even humans have challenged the preeminence of insects and their arthropod kin. Or, as Cornell University's Thomas Eisner puts it: "Bugs are not going to inherit the Earth. They own it now. So we might as well make peace with the landlord."

Crustaceans While arachnids and insects thrived on land, crustaceans, for the most part, remained in the seas and ponds where they are now represented by about 40,000 species (Figure 29.39).

The multiple appendages of crustaceans are extensively specialized. Lobsters and crayfish, for instance, have a toolkit of 19 pairs of appendages (see Figure 29.29). Crustaceans are the only arthropods with two pairs of antennae. Three or more pairs of appendages are modified as mouthparts, including the hard mandibles. Walking legs are present on the thorax,

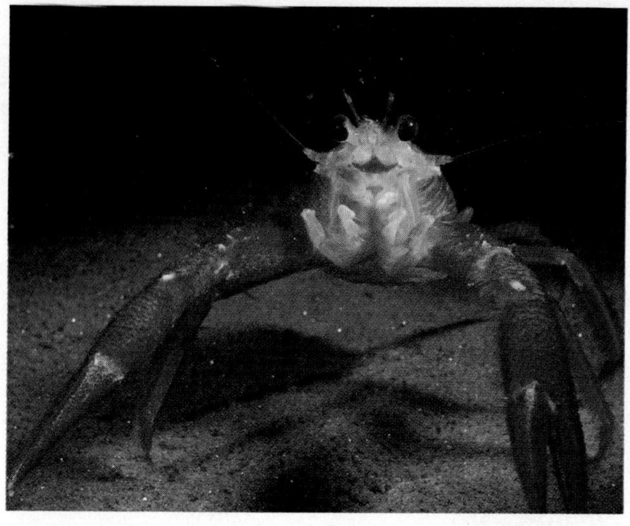
(a)

(b)

(c)

Figure 29.39
Crustaceans. (**a**) Crabs, lobsters, crayfish, and shrimp are the most familiar crustaceans. This is a red crab. (**b**) Tiny planktonic crustaceans known as krill are consumed by the ton by whales and other large suspension-feeders. The inset is a magnification of one of the many krill seen as tiny flecks in the main photograph. (**c**) Barnacles are sessile crustaceans with a shell (exoskeleton) hardened by calcium carbonate. Notice the jointed appendages projecting from the shell. These are used to capture small plankton and organic particles suspended in the water.

and, unlike insects, crustaceans have appendages on the abdomen. A lost appendage can be regenerated.

Small crustaceans exchange gases across thin areas of the cuticle, but larger forms have gills. The circulatory system is open, with a heart pumping hemolymph through arteries into sinuses that bathe the organs. Crustaceans excrete nitrogenous wastes by diffusion through thin areas of the cuticle, but a pair of glands regulates the salt balance of the hemolymph.

Sexes are separate in most crustaceans. In the case of the lobster, the male uses a specialized pair of appendages to transfer sperm to the reproductive pore of the female during copulation. Most aquatic crustaceans go through one or more swimming larval stages.

Lobsters, crayfish, crabs, and shrimp are all relatively large crustaceans called decapods. The exoskeleton, or cuticle, is hardened by calcium carbonate; the portion that covers the dorsal side of the cephalothorax forms a shield called the carapace. Most decapods are marine. Crayfish, however, live in fresh water, and some tropical crabs live on land.

The isopods are mostly small marine crustaceans, but this group also includes sow bugs and pill bugs, familiar land animals. These terrestrial crustaceans live mostly in moist soil and other damp places.

Another group of small crustaceans, the copepods, are among the most numerous of all animals. They are important members of the plankton communities that are the foundation consumers of marine and freshwater food chains. Plankton also includes larvae of many larger crustaceans. Another group of shrimplike crustaceans is the planktonic **krill,** which compose a major food source for many species of whales (Figure 29.39b).

Barnacles are sessile crustaceans with parts of their cuticles hardened into shells by calcium carbonate. They feed by using their appendages to strain food from the water (Figure 29.39c).

DEUTEROSTOMES (ENTEROCOELOMATES)

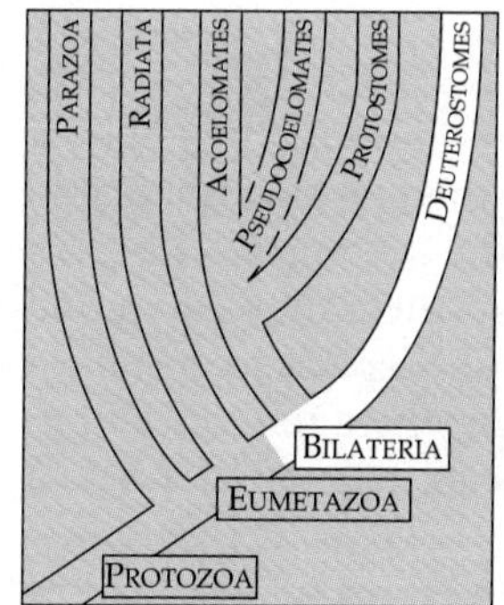

At first glance, lophophorates and echinoderms, which include the sea stars, may seem to have little in common with the phylum Chordata, which includes birds and mammals, but these animals share features characteristic of deuterostomes: radial cleavage, development of the coelom from the archenteron, and, with one exception, formation of the mouth at the end of the embryo opposite the blastopore.

The Lophophorate Animals

In subdividing coelomate animals into protostomes and deuterostomes, three phyla—Phoronida, Bryozoa, and Brachiopoda—have been the subject of a good

(a)

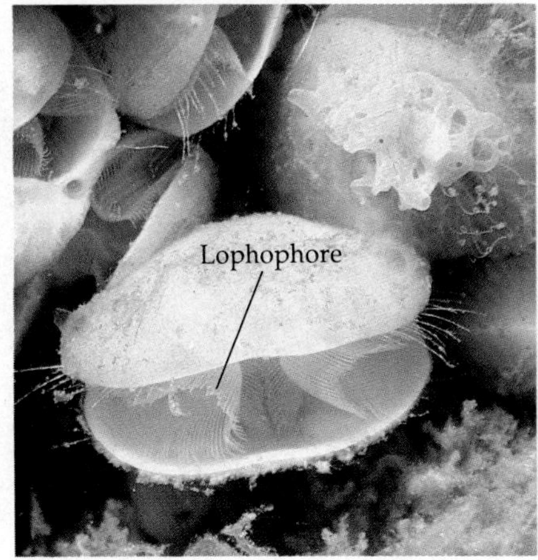

(b)

Figure 29.40
The lophophorate animals. The most distinctive characteristic of this group is the lophophore, an organ that functions in suspension-feeding. (**a**) Bryozoans, such as this common sea mat *(Membranipora membranacea)*, are colonial lophophorates, often with hard exoskeletons. (**b**) Brachiopods are lophophorates with a bivalve shell.

deal of controversy. These phyla are collectively called the **lophophorate animals,** a reference to the most distinctive structure they share—the **lophophore** (Figure 29.40). The lophophore is a horseshoe-shaped or circular fold of the body wall bearing ciliated tentacles that surround the mouth at the anterior end of the animal. The anus lies outside the whorl of tentacles. The cilia draw water toward the mouth between the tentacles, which help trap food particles for these suspension-feeders. The common occurrence of this complex apparatus in the lophophorate animals suggests that the three phyla are related. However, other similarities, such as a U-shaped digestive tract and absence of a distinct head, are adaptations to a sessile existence that may have evolved convergently.

In their embryonic development, the lophophorate animals as a group most closely resemble other deuterostomes. However, in the phoronids, the mouth arises from the embryonic blastopore, a feature associated with protostomes. Because of this and some other areas of conflict, the evolutionary position of the lophophorates remains uncertain. Based on evidence, it seems most reasonable to treat them as an early branch off the main deuterostome line.

Phoronids are tube-dwelling marine worms ranging from 1 mm to 50 cm in length. Some live buried in the sand within tubes made of chitin, extending their lophophore from the opening of the tube and withdrawing it into the tube when threatened. There are only about 15 species of phoronid worms in two genera.

Bryozoans are tiny colonial animals that superficially resemble mosses. (*Bryozoa* means "moss animals.") In most species, the colony is encased in a hard exoskeleton with pores through which the lophophores of the animals extend (Figure 29.40a). Of the 5000 species of bryozoans, most live in the sea where they are among the most widespread and numerous sessile animals. Several species are important reef builders.

Brachiopods, or lamp shells, superficially resemble clams and other bivalve mollusks, but the two halves of the brachiopod shell are dorsal and ventral to the animal rather than lateral, as in clams (Figure 29.40b). A brachiopod lives attached to its substratum by a stalk, opening its shell slightly to allow water to flow between the shells and the lophophore. All brachiopods are marine. The living brachiopods are remnants of a much richer past; only about 330 extant species are known, but there are 30,000 species of Paleozoic and Mesozoic fossils. A tie to the past is *Lingula,* a living brachiopod genus that has changed little in 400 million years.

Phylum Echinodermata

Sea stars and most other **echinoderms** (Gr. *echin,* "spiny," and *derma,* "skin") are sessile or sedentary animals with radial symmetry (Figure 29.41). The internal and external parts of the animal radiate from the center, often as five spokes. A thin skin covers an endoskeleton of hard calcareous plates. Most echinoderms are prickly, owing to skeletal bumps and spines of various functions. Unique to echinoderms is the **water vascular system,** a network of hydraulic canals branching into extensions called tube feet that function in locomotion, feeding, and gas exchange.

Sexual reproduction of echinoderms usually involves separate male and female individuals that release their gametes into the sea water. The radial adults develop by metamorphosis from bilateral larvae. The early embryology of echinoderms clearly aligns them with the deuterostomes.

The 7000 or so echinoderms, all marine, are divided into six classes: Asteroidea (sea stars), Ophiuroidea (brittle stars), Echinoidea (sea urchins and sand dollars), Crinoidea (sea lilies), Holothuroidea (sea cucumbers), and Concentricycloidea (sea daisies). The last group, the sea daisies, were discovered only recently.

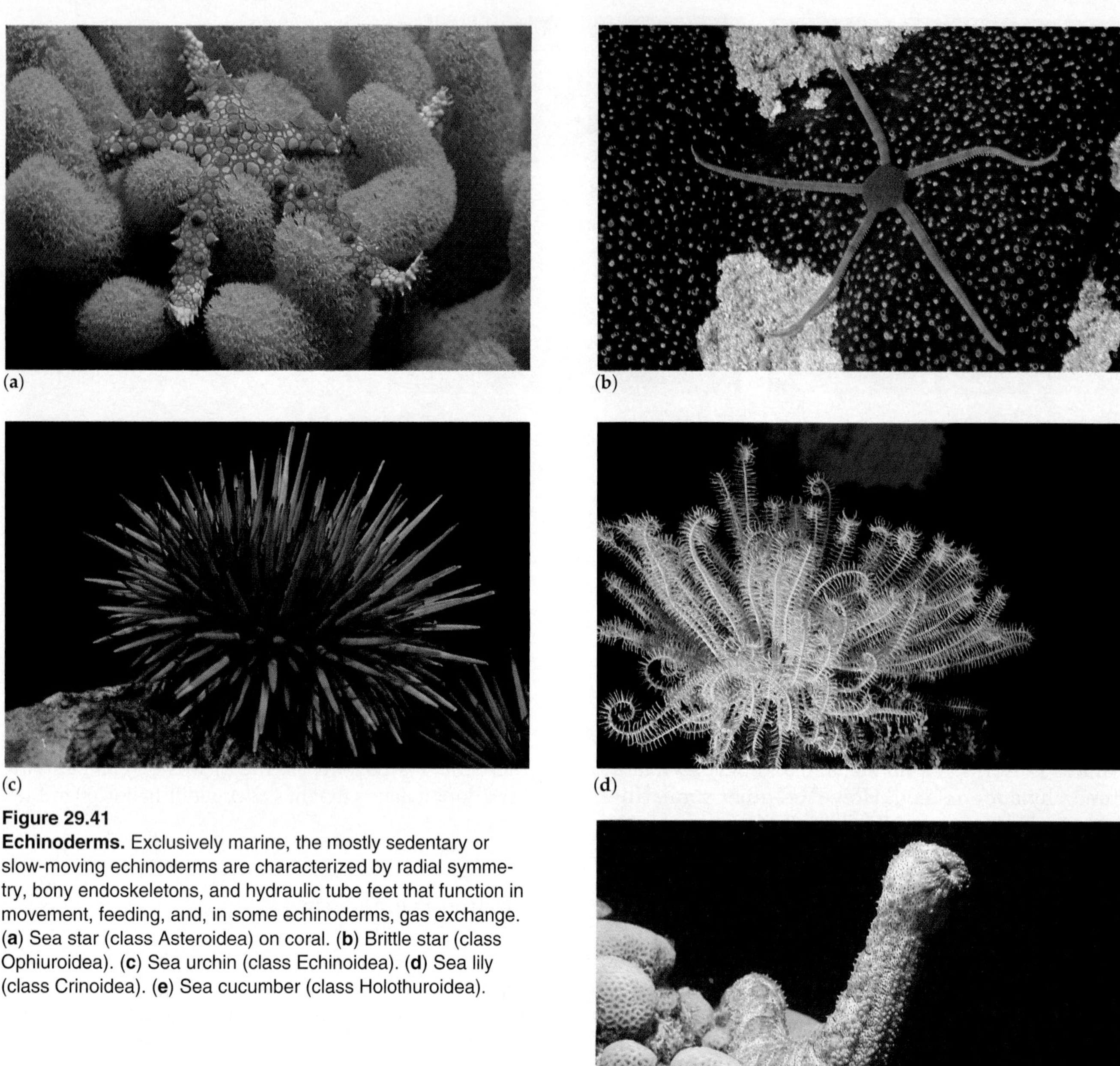

(a) (b) (c) (d) (e)

Figure 29.41
Echinoderms. Exclusively marine, the mostly sedentary or slow-moving echinoderms are characterized by radial symmetry, bony endoskeletons, and hydraulic tube feet that function in movement, feeding, and, in some echinoderms, gas exchange. (**a**) Sea star (class Asteroidea) on coral. (**b**) Brittle star (class Ophiuroidea). (**c**) Sea urchin (class Echinoidea). (**d**) Sea lily (class Crinoidea). (**e**) Sea cucumber (class Holothuroidea).

They are strange little creatures that live on waterlogged wood in the deep sea.

Sea stars have five arms (sometimes more) radiating from a central disk (Figure 29.42). The undersurfaces of the arms bear tube feet. Each tube foot can act like a suction disk. By a complex set of hydraulic and muscle actions, the suction can be created or released. The sea star coordinates its tube feet to adhere firmly to rocks or to creep along slowly as the tube feet extend, grip, contract, release, extend, and grip again. Sea stars also use their tube feet to grasp prey, such as clams and oysters. The arms of the sea star embrace the closed bivalve, hanging on tightly by the tube feet. The sea star turns its stomach inside-out, everting it through its mouth and between the shells of the bivalve. The digestive tract of the sea star secretes juices that begin digesting the soft body of the mollusk within its shell.

Sea stars and some other echinoderms have strong powers of regeneration. Sea stars can regrow lost arms, but the process is very slow. Except for one species, however, they cannot regrow an entire body from a single arm.

Brittle stars have distinct central disks, and the arms are long and flexible. Their tube feet lack suckers, and they move by serpentine lashing of the arms. Feeding mechanisms vary among the different species.

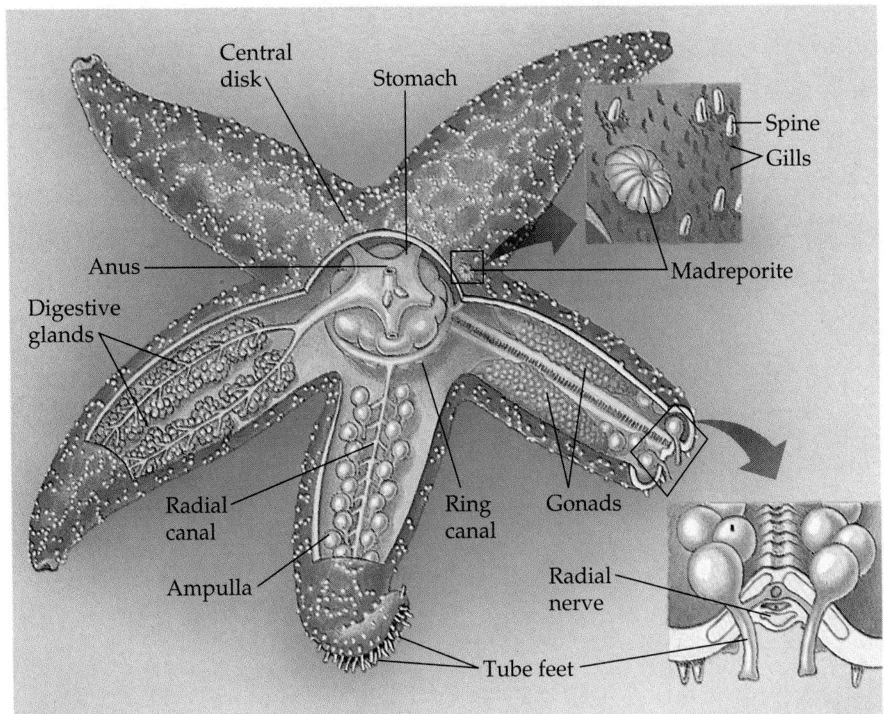

Figure 29.42
Anatomy of a sea star. The surface of a sea star is covered by spines that help defend against animals and by small gills for gas exchange. Internal organs are suspended by mesenteries in a well-developed coelom. A short digestive tract runs from the mouth on the bottom of the central disk to the anus on the top of the disk. Digestive glands secrete digestive juices and aid in the absorption and storage of nutrients. The central disk has a nerve ring and nerve cords radiating from the ring into the arms. The water vascular system consists of a ring canal in the central disk and five radial canals, each running the length of an arm in a groove. The system connects to the outside by way of the madreporite. Branching from each radial canal are hundreds of tube feet filled with fluid continuous with the rest of the water vascular system. Attached to each tube foot, and instrumental in its functioning, is a water bulb called the ampulla.

Sea urchins and sand dollars have no arms, but they do have five rows of tube feet for slow movement. These echinoderms also use muscles for pivoting their long spines, to aid in moving. The mouth of an urchin is ringed by complex jawlike structures for eating seaweeds and other food. Sea urchins are roughly spherical in shape, whereas sand dollars are flattened and disk-shaped.

Some sea lilies live attached to the substratum by stalks; others crawl about by using their long, flexible arms, which they also use in suspension-feeding. The arms circle the mouth, which is directed upward, away from the substratum. Crinoidea is an ancient class that has been very conservative in its evolution; fossilized sea lilies some 500 million years old could pass for contemporary members of the class.

On casual inspection, sea cucumbers do not look much like other echinoderms. They lack spines, and the hard endoskeleton is much reduced. Sea cucumbers are elongated in the oral-aboral axis, giving them their name and further disguising their relationship to sea stars and sea urchins. Closer examination, however, reveals five rows of tube feet, part of the water vascular system found only in echinoderms. Some of the tube feet around the mouth are developed as feeding tentacles.

Phylum Chordata

This phylum, our own, consists of two subphyla of invertebrate animals plus the subphylum Vertebrata, the animals with backbones. Grouping the chordates with echinoderms as deuterostomes on the basis of similarities in early embryonic development is not meant to imply that one phylum evolved from the other. Chordates and echinoderms have existed as distinct phyla for at least half a billion years; if the developmental similarities stem from shared ancestry, then the evolutionary paths of the two phyla must have diverged very early. We will trace the phylogeny of chordates in Chapter 30, focusing on the history of vertebrates. This chapter concludes by considering some more general questions about early diversification of the animal kingdom and summarizing the evolutionary history of the groups we have discussed.

THE ORIGIN AND DIVERSIFICATION OF ANIMALS

Ever since the Darwinian revolution, zoologists have speculated about the origins of animals from unicellular ancestors. One hypothesis, called the **syncytial hypothesis,** holds that animals arose from a multinucleate protist (perhaps a ciliate) that became subdivided by membranes into many cells. According to this hypothesis, early animals may have resembled acoel worms, a group of flatworms lacking digestive cavities (here, acoel refers to the absence of a gut). Because of problems in deriving the sponges and radiate cnidarians from a bilateral flatworm and a number of other difficulties, most zoologists adhere to a second idea,

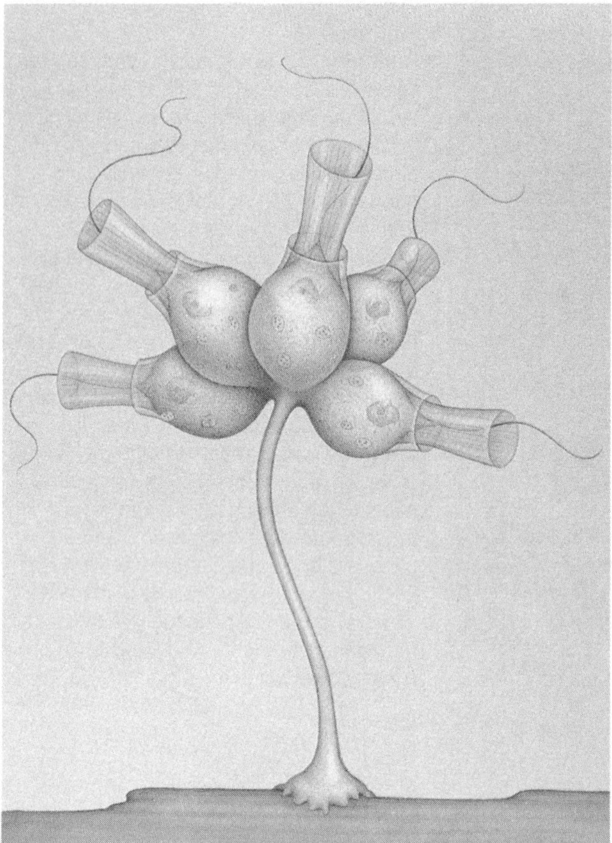

Figure 29.43
Colonial choanoflagellates. Some zoologists consider these protists, which resemble the collar cells (choanocytes) of sponges, to be similar to the ancestors of the animal kingdom. Electron microscopy has shown that mitochondria in choanoflagellates differ markedly from those of other protists but closely resemble those of animal cells.

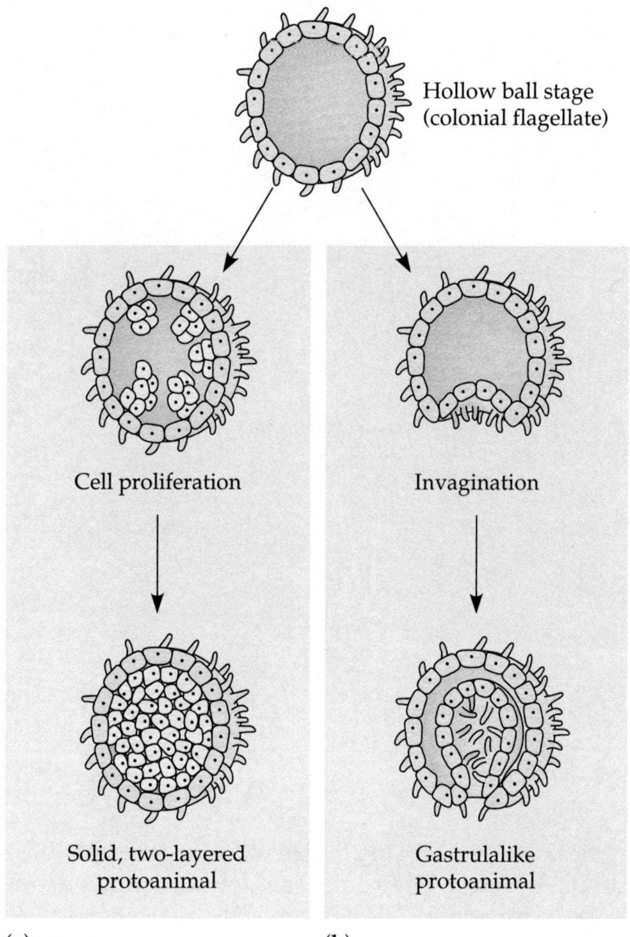

Figure 29.44
Hypothetical models for the evolution of protoanimals from hollow colonial flagellates. (**a**) Formation of a layered planuloid ancestor by internal cell proliferation. (**b**) An alternate model leading to a gastrulalike organism via invagination.

the **colonial hypothesis,** which postulates that ancestral animals were heterotrophic colonial flagellates. Such ancestral organisms would have been free-swimming, hollow spheres of heterotrophic cells with an anterior–posterior orientation and distinct reproductive and somatic cells. Many zoologists think that a group of extant protists known as choanoflagellates deserve attention (Figure 29.43). Their cells bear a collar and flagellum, like the choanocytes of sponges.

Most proponents of the colonial hypothesis believe that all animals—including sponges, the radially symmetrical cnidarians, and the bilateral phyla—evolved from hollow flagellate colonies. But several innovations must have evolved in these colonies prior to the origin of the eumetazoa. Cell layers, a fundamental feature of animals, may have developed as cells proliferated into the center of the ancestral hollow colonies. This could have produced a protoanimal composed of an outer layer of flagellated locomotor cells and an inner mass of digestive and reproductive cells. Such a hypothetical organism, called a planuloid, would have consisted of a solid mass of cells and may have resembled the planula larva of certain cnidarians (Figure 29.44a). An alternative to this model proposes that cell layers formed by invagination of the hollow colonial flagellates. This would have produced a two-layered protoanimal similar to the gastrula stage in animal development (Figure 29.44b). Yet another hypothesis holds that a contemporary animal named *Trichoplax adhaerens* resembles the form of the earliest animals (Figure 29.45).

All or none of these hypotheses may approach what actually happened in animal evolution, and the question of what protoanimals looked like remains unresolved because no transitional fossils between unicellular forms and eumetazoa have been discovered. In fact, the first animals, diverse and relatively complex, appear in the fossil record without discernible ties to their protistan predecessors. As a phylogenetic puzzle, the early diversification of animals is as tough a problem as the origin of animals from protists.

The oldest known fauna lived from about 700 mil-

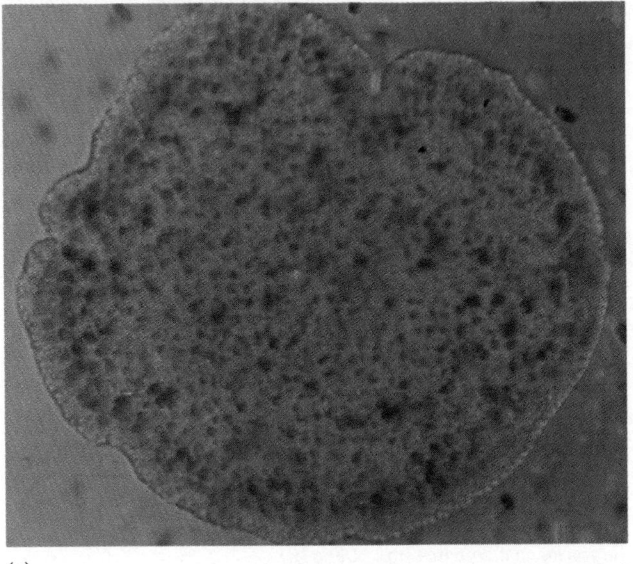

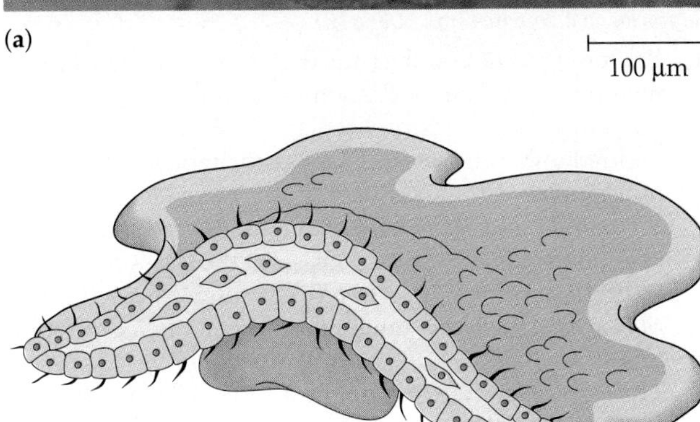

Figure 29.45
***Trichoplax adhaerens* (phylum Placozoa).** (**a**) *Trichoplax,* the simplest known animal, is a tiny, flattened marine creature with a ciliated epidermis covering a solid core of relatively unspecialized cells (LM, dorsal view). (**b**) The animal has no gut, but as it crawls over a food particle, it hunches its back to form a temporary digestive cavity around the food; a similar habit may have formed the first gut cavity.

Figure 29.46
Fossils of early animals. (**a**) This sandstone imprint of an animal from the Ediacaran Hills of southern Australia is approximately 650 million years old. Although this radially symmetrical animal resembles a jellyfish, the evolutionary relationships between the Ediacaran fauna and the Cambrian animals that followed are still uncertain. (**b**) Adaptive radiation during the Cambrian period, which began about 590 million years ago, produced nearly all known phyla of animals. This fossil of a polychaete worm was collected from the Burgess Shale of British Columbia, one of the richest Cambrian beds.

lion years ago to about 590 million years ago, a span of Earth history at the end of the Precambrian era that has been named the **Ediacaran period** (for the Ediacara Hills of Australia, where these fossils were first studied extensively). A few fossils of this same period have also been found in Africa and southern China. The Ediacaran fauna has been nicknamed the worm-jellyfish stage of animal evolution; most of the creatures were soft-bodied forms that resembled, at least superficially, cnidarian medusae, colonial cnidarians called sea pens, and annelids.

A second radiation produced a much more diverse fauna as the Paleozoic era dawned with the Cambrian period, about 590 million years ago. The increase in animal diversity may have been set off by the explosive evolution of shells and hard skeletons, which opened new adaptive zones and revolutionized predator–prey relations. Nearly all modern phyla are represented among the Cambrian fossils, along with many phyla that subsequently disappeared. Figure 29.46 compares the fossil of a Cambrian animal with a fossil representing the Ediacaran fauna.

Explosive revolution or gradual transition, the Cambrian radiation produced nearly all the basic designs we observe in modern animals. In the last half-billion years, animal evolution mainly generated new variations on old body plans. It has been the major theme of this chapter to trace the evolution of animal body plans and to describe some of the major groups of living animals that show them today. In the next chapter, we will trace the evolution of members of the phylum Chordata.

STUDY OUTLINE

1. Animal life began with the Precambrian evolution of multicellular marine forms that lived by eating other organisms.
2. Subsequent evolution has generated a diversity of animals, grouped into about 35 phyla. About 95% of animal species are invertebrates (lacking backbones).

Defining "Animals" (pp. 598–599)

1. Animals are multicellular eukaryotes distinguished by a specific type of heterotrophy called ingestion.
2. Animal cells lack walls and store carbohydrate reserves as glycogen. In most animals, cells are successively organized into tissues, organs, and organ systems.
3. Animal reproduction is primarily sexual, and the life cycle is dominated by the diploid stage, gametes generally being the only haploid cells. Asexual budding or regeneration occurs in some species.
4. In sexual reproduction, fertilization of an egg by a flagellated sperm initiates cleavage in the zygote and the formation of a hollow ball of cells called the blastula.
5. Many animals go through metamorphosis, a second stage of development that transforms a sexually immature larva into a morphologically distinct sexual adult.
6. Muscles and nerves, which control active behavior, are unique to animals.

Clues to Animal Phylogeny (pp. 599–603)

1. Taxonomists rely mainly on comparative anatomy and embryology to reconstruct animal phylogeny.
2. Eumetazoa (all animals except sponges) diverged early into two major branches, the Radiata and the Bilateria. Members of the branch Radiata are the jellyfish and their relatives, sedentary and planktonic forms with radial symmetry. Members of the branch Bilateria are characterized by bilateral symmetry and cephalization.
3. Bilateral animals develop from embryos constructed of three concentric primary germ layers: an inner endoderm, an outer ectoderm, and a middle mesoderm.
4. The body plan of Bilateria is either solid, as in the acoelomates, or has a digestive tube separated from the outer body wall by a cavity. In pseudocoelomates, the cavity is incompletely lined by embryonic mesoderm. Coelomates have a true coelom, a body cavity completely lined by mesoderm.
5. Based on features of embryonic development, coelomate phyla are divided into two main groups: the protostomes, comprising the annelids, mollusks, and arthropods; and the deuterostomes, consisting of the lophophorate animals, echinoderms, and chordates.

Parazoa (Phylum Porifera) (pp. 603–604)

1. The least complex animals are the sponges, which lack tissues and organs, such as muscles and nerves.
2. Sponges filter-feed by drawing water through pores into a central spongocoel and out an osculum.

Eumetazoa: Radiata (pp. 604–608)

1. Phylum Cnidaria consists of primarily marine carnivores possessing tentacles armed with stinging cnidocytes that aid in defense and the capture of prey. The simple, radial body exists as a sessile polyp or a floating medusa (or both stages, in some cases) organized around a central gastrovascular cavity with a single opening for both mouth and anus.
2. Phylum Cnidaria is divided into three classes. Class Hydrozoa usually alternates polyp and medusa forms, although the polyp is more conspicuous and the only stage in the genus *Hydra*. The jellyfishes belong to the class Scyphozoa, in which the medusa is the prevalent form of the life cycle. Class Anthozoa contains the sea anemones and corals, which occur only as polyps. Corals secrete external skeletons of calcium carbonate that persist as elaborate rocklike formations.
3. Phylum Ctenophora, the comb jellies, are transparent animals whose locomotion depends on eight rows of cilia.

Bilateria: Acoelomates (pp. 608–610)

1. Phylum Platyhelminthes, the flatworms, are the simplest members of Bilateria. Ribbonlike animals with a single opening to their gastrovascular cavities, the phylum is divided into four classes. Class Turbellaria is made up of mostly free-living, primarily marine species. Classes Trematoda and Monogenea, the flukes, live as parasites in animals. Class Cestoda consists of the tapeworms, parasitic flatworms with a headlike scolex connected to a ribbon of detachable units of sex organs called proglottids.
2. The proboscis worms of the phylum Nemertea are named for the retractable tube they use for defense and prey capture. This group has a simple circulatory system and a complete digestive tract with both mouth and anus.

Pseudocoelomates (pp. 610–611)

1. Phylum Rotifera is made up of tiny animals possessing a crown of cilia that draws food into the mouth.
2. The roundworms of the phylum Nematoda are among the most numerous animals in both species and individuals. Nematodes have tapered ends. The phylum includes both free-living species and parasites, including the nematode that causes trichinosis.

Protostomes (Schizocoelomates) (pp. 611–626)

1. Phylum Mollusca includes a diverse spectrum of soft-bodied species possessing various modifications of a muscular foot, a visceral mass, and an overlying mantle, which in many species secretes a shell of calcium carbonate. In addition, many mollusks have a radula, a rasping organ used in feeding.
2. Four of the eight classes of the phylum Mollusca are as follows. Class Polyplacophora is made up of the chitons, oval-shaped marine animals that cling to exposed rocks and are encased in an armor of dorsal plates. The largest molluscan class is Gastropoda, the snails and their relatives. This group is often protected by single, spiraled shells, although unshelled slugs also occur. Embryonic torsion of the body is a distinctive characteristic. The clams and their relatives of the class Bivalvia have hinged shells divided into two halves. Bivalves are sedentary, headless suspension-feeders that use gills for

both gas exchange and feeding. Class Cephalopoda includes squids and octopuses, carnivores with beaklike jaws surrounded by tentacles of the modified foot. The shell is either reduced or absent in most genera, consistent with their active lifestyles.

3. Phylum Annelida is a group characterized by body segmentation. Their progressive, wavelike locomotion results from alternating contractions of circular and longitudinal muscles against a fluid-filled, compartmentalized coelom. Annelids have well-developed regulatory and nervous systems. Many are hermaphrodites but cross-fertilize. The phylum is divided into three classes. Class Oligochaeta includes earthworms and various aquatic species. Representatives of the class Polychaeta possess vascularized, paddlelike parapodia that function as gills and aid in locomotion. Class Hirudinea consists of the leeches, some of which are blood-sucking parasites.
4. Phylum Arthropoda has more known species than all other phyla combined. The arthropods have bodies with various modified, jointed appendages and an exoskeleton that must be shed during molting to allow growth. Arthropods also possess well-developed sense organs, an open circulatory system, and a variety of specializations for gas exchange.
5. The segmentation of arthropods suggests evolution from annelids. Chelicerates are arthropods with pincer or fanglike feeding appendages. A separate line of evolution produced arthropods with jawlike mandibles. The subphylum Uniramia contains the insects, centipedes, and millipedes, which have one pair of antennae and unbranched (uniramous) appendages. The subphylum Crustacea contains primarily aquatic organisms with two pairs of antennae and branched appendages. The five principal classes of Arthropoda follow.
6. Class Arachnida includes spiders, ticks, scorpions, and mites, extant chelicerates with simple eyes, two pairs of feeding appendages, and four pairs of walking legs.
7. Classes Diplopoda and Chilopoda, the millipedes and centipedes, are animals whose segmentation attests to annelid ancestry. The vegetarian millipedes have two walking legs per segment, whereas the carnivorous centipedes have one pair per segment and are armed with poison claws.
8. In terms of species diversity, members of the class Insecta outnumber all other forms of life combined and have exploited virtually every habitat on Earth. Part of their success is due to the evolution of flight and their relationship with flowering plants. The insect head, which contains one pair of antennae, a pair of compound eyes, and variously modified mouthparts, is connected to a segmented thorax and abdomen. The thorax bears three pairs of walking legs and often one or two pairs of wings. In many species, young go through complete or incomplete metamorphosis.
9. Lobsters and crayfish are members of the class Crustacea. Among a number of multiple appendages are two pairs of antennae and appendages modified for pinching, chewing, locomotion, and copulation.

Deuterostomes (Enterocoelomates) (pp. 626–629)

1. The lophophorate animals, the phyla Phoronida, Bryozoa, and Brachiopoda, may have evolved early during deuterostome history. They are grouped together on the basis of their lophophores, horseshoe-shaped, suspension-feeding organs bearing ciliated tentacles.
2. Sea stars and their relatives comprise six classes of the marine phylum Echinodermata, radially symmetrical animals with a unique water vascular system ending in tube feet used for locomotion and feeding. A thin, bumpy or spiny skin covers a calcareous endoskeleton.
3. Phylum Chordata, discussed in detail in Chapter 30, is made up of two subphyla of invertebrates plus a vertebrate subphylum, to which humans belong.

The Origin and Diversification of Animals (pp. 629–631)

1. Sponges almost certainly evolved from colonial flagellates, but because transitional fossils are lacking, ideas on eumetazoan ancestry have been based on inferences drawn from the simple anatomy of certain modern organisms. Different hypotheses have variously envisioned protoanimals as ciliated forms that later became subdivided by membranes, colonial flagellates, or planuloids.
2. The oldest known fauna consisted of soft-bodied animals that lived during the Ediacaran period. During the ensuing Cambrian period, a much more diverse fauna evolved, which included many species with hard shells and skeletons and nearly all the modern phyla. All the basic designs observed in modern animals arose during the Cambrian radiation.

SELF-QUIZ

1. The subkingdom Parazoa consists of
   a. the radially symmetrical sponges, jellyfishes, and their relatives
   b. the only animals that do not feed by ingestion
   c. animals with protostome-type development
   d. colonial choanoflagellates
   e. animals lacking tissues and organs
2. As a group, acoelomates are characterized by
   a. gastrovascular cavities
   b. a body cavity called a hemocoel
   c. deuterostome development
   d. a coelom that is not completely lined with mesoderm
   e. a solid body without a cavity surrounding internal organs
3. Which of the following is *not* descriptive of deuterostomes?
   a. radial cleavage
   b. determinate cleavage
   c. enterocoelous formation of the body cavity
   d. development of blastopore into anus
   e. echinoderms and chordates
4. The branches Radiata and Bilateria of the eumetazoa both exhibit
   a. cephalization
   b. bilateral symmetry of larval forms
   c. dominance of diploid life stage
   d. a complete digestive tract with separate mouth and anus
   e. three germ layers in embryonic development

5. A land snail, a clam, and an octopus all share
   a. a mantle
   b. a radula
   c. gills
   d. embryonic torsion
   e. distinct cephalization
6. Which of the following is *not* a characteristic of the phylum Annelida?
   a. hydrostatic skeleton
   b. segmentation
   c. metanephridia
   d. pseudocoelom
   e. closed circulatory system
7. In arthropods, the possession of a cuticle usually dictates
   a. an aquatic environment
   b. molting
   c. metamorphosis
   d. external fertilization
   e. gills or tracheal system for gas exchange
8. Which of the following is *not* true of the chelicerates?
   a. They have antennae.
   b. Their body is divided into a cephalothorax and an abdomen.
   c. The horseshoe crab is one surviving marine member.
   d. They include ticks, scorpions, and spiders.
   e. Their anterior appendages are modified as pincers or fangs.
9. Which of the following combinations of phyla and characteristics is *incorrect?*
   a. Nemertea—proboscis worms, complete digestive tract
   b. Nematoda—roundworms, pseudocoelomate
   c. Cnidaria—radial symmetry, polyp and medusa body forms
   d. Platyhelminthes—flatworms, gastrovascular cavity, acoelomate
   e. Porifera—gastrovascular cavity, mouth from blastopore
10. Which of the following subdivisions of the animal kingdom encompasses all the others?
   a. protostomes
   b. bilateria
   c. pseudocoelomates
   d. coelomates
   e. deuterostomes

CHALLENGE QUESTIONS

1. Pick a representative acoelomate, pseudocoelomate, and coelomate and explain at least one salient embryological difference between each of your three animals.
2. In the late 1970s, J. Kingsolver and M. Koehl proposed that insect wings were evolutionary derivatives of thermoregulatory structures that projected laterally from the body. The scientists maintained that natural selection worked first on the heat-exchange benefits of the "protowings" and later on flight. Experimental data indicated that thermoregulatory effects occurred for wing lengths up to 1 cm, but wings had to be at least 1.5 cm long before aerodynamic effects became significant. What problem did this finding pose for the model, and how might it be conceptually resolved?
3. Professor Lynn Margulis of the University of Massachusetts has suggested that observing an explosion of animal diversity in Cambrian strata is like viewing Earth from a satellite over a long period of time and noticing the emergence of cities only after they are large enough to be evident at that distance. What do you think Dr. Margulis was saying about the Cambrian "explosion"?

SCIENCE, TECHNOLOGY, AND SOCIETY

1. An irrigation project has allowed farmers in an African country to grow more food. In the past, crops were planted only after spring rains. Now fields can be watered year-round. Improved crop yield has had an unexpected cost: an increase in the incidence of schistosomiasis, or blood fluke disease (see Figure 29.16). Why do you think the irrigation project increased the incidence of schistosomiasis? Suggest three methods that could be tried to prevent people from becoming infected. What ecological, economic, and social factors must you take into account in implementing your plans?
2. Under what circumstances should we regard insects as pests? Develop arguments either for or against the use of pesticides in specific situations.

FURTHER READING

Barnes, R. S. K. "Two-Layered Awakening." *Nature,* April 18, 1991. Are some animals more closely related to plants than to other animals?

Brusca, R. G., and G. J. Brusca. *Invertebrates.* Sunderland, MA: Sinauer Associates, 1990. An evolutionary approach to the animal kingdom.

Gould, S. J. "The Reversal of Hallucigenia." *Natural History,* January 1992. The possible evolutionary significance of an early Cambrian animal.

Gould, S. J. *Wonderful Life: The Burgess Shale and the Nature of History.* New York: Norton, 1989. Gould's best-seller about contingency, evolution, and the history of animals.

Levinton, J. "The Big Bang of Animal Evolution." *Scientific American,* November 1992. What evolutionary mechanisms made the Cambrian explosion possible?

Mitchell, L. G., J. A. Mutchmor, and W. D. Dolphin. *Zoology.* Menlo Park, CA: Benjamin/Cummings, 1988. A very accessible introductory text.

Wilson, E. O. "Empire of the Ants." *Discover,* March 1990. A Pulitzer Prize–winning biologist writes about his favorite animals.

Wootton, R. "The Mechanical Design of Insect Wings." *Scientific American,* November 1990. How flight evolved in the most successful animals.

30 THE VERTEBRATE GENEALOGY

PHYLUM CHORDATA
CHORDATES WITHOUT BACKBONES
THE ORIGIN OF VERTEBRATES
VERTEBRATE CHARACTERISTICS
CLASS AGNATHA
CLASS PLACODERMI
CLASS CHONDRICHTHYES
CLASS OSTEICHTHYES
CLASS AMPHIBIA
CLASS REPTILIA
CLASS AVES
CLASS MAMMALIA
THE HUMAN ANCESTRY

Figure 30.1
A vertebrate hallmark. Vertebrates are named for their backbone, which consists of a series of vertebrae. This hallmark is apparent in this skeleton of a snake. This chapter traces the evolution of vertebrates.

Most of us are curious about our genealogies. On the personal level, we wonder about our family ancestry. As biology students, we are interested in retracing human ancestry within the broader scope of the evolutionary history of the entire animal kingdom. The questions we must ask are: What were our ancestors like? How are we related to other animals? What are our closest relatives? In this chapter, we trace the phylogeny of the vertebrates, the group that includes humans and their closest relatives. Mammals, birds, reptiles, amphibians, and the various classes of fishes are all classified as **vertebrates** because they have a backbone and many other features in common (Figure 30.1). Following the plan of the other chapters in this unit, we examine some aspects of vertebrate anatomy and physiology as they pertain to our evolutionary theme. Detailed discussion of animal form and function is reserved for Unit Seven. Our first step in tracking the vertebrate genealogy is to determine where vertebrates fit in the animal kingdom.

PHYLUM CHORDATA

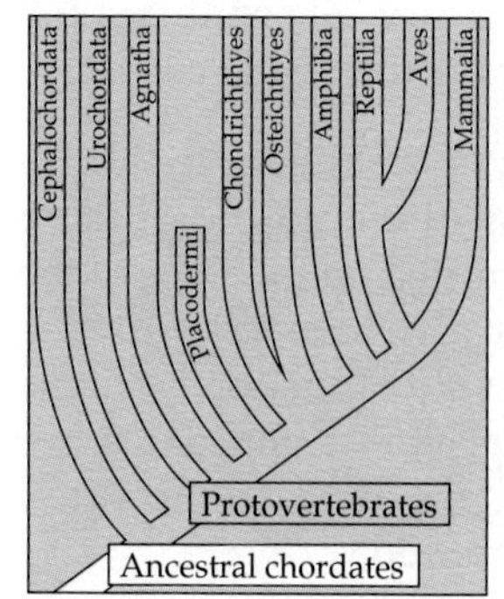

The vertebrates make up one subphylum within the phylum Chordata. **Chordates** also include two subphyla of invertebrates, the cephalochordates and the urochordates.

Chordate Characteristics

Based on certain similarities in early embryonic development, chordates are grouped as deuterostomes along with the echinoderms (see Chapter 29). Although chordates vary widely in appearance, they are distinguished as a phylum by the presence of four anatomical structures that appear at

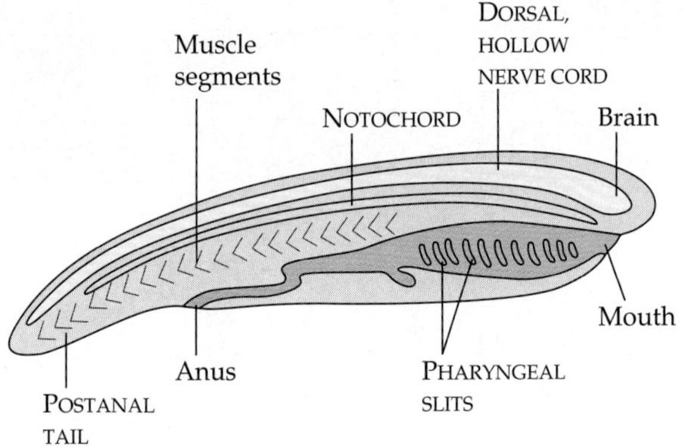

Figure 30.2
Chordate characteristics. All chordates possess the four trademarks of the phylum: a notochord; a dorsal, hollow nerve cord; pharyngeal slits; and a postanal tail.

some point during the animal's lifetime, often only during the embryo stage (Figure 30.2).

1. *Notochord.* All chordate embryos have a *notochord,* which is a longitudinal, flexible rod located between the gut and the nerve cord. The notochord is composed of large, fluid-filled cells encased in fairly stiff, fibrous tissue. It extends through most of the length of the animal as a relatively simple skeleton. Chordates are named for this structure. In some invertebrate chordates and primitive vertebrates, the notochord persists to support the adult, but in most vertebrates a more complex, jointed skeleton develops and the adult retains only remnants of the embryonic notochord—as gelatinous material of the disks between vertebrae of humans, for example.

2. *Dorsal, hollow nerve cord.* The nerve cord of a chordate embryo develops from a plate of ectoderm that rolls into a tube located dorsal to the notochord. The result is a dorsal, hollow nerve cord unique to chordates. Other animal phyla have solid nerve cords, usually ventrally located. The chordate nerve cord forms the central nervous system: the brain and spinal cord.

3. *Pharyngeal slits.* The lumen of the digestive tube of nearly all chordate embryos opens to the outside through several pairs of slits located on the sides of the pharynx, the region of the digestive tube just posterior to the mouth. The pharyngeal slits function as filter-feeding devices in many invertebrate chordates but have become modified for gas exchange and other functions during vertebrate evolution.

4. *Muscular postanal tail.* Most chordates have a tail extending beyond the anus. By contrast, most nonchordates have a digestive tract that extends nearly the whole length of the body. The chordate tail contains skeletal elements and muscles and provides much of the propulsive force in many aquatic species.

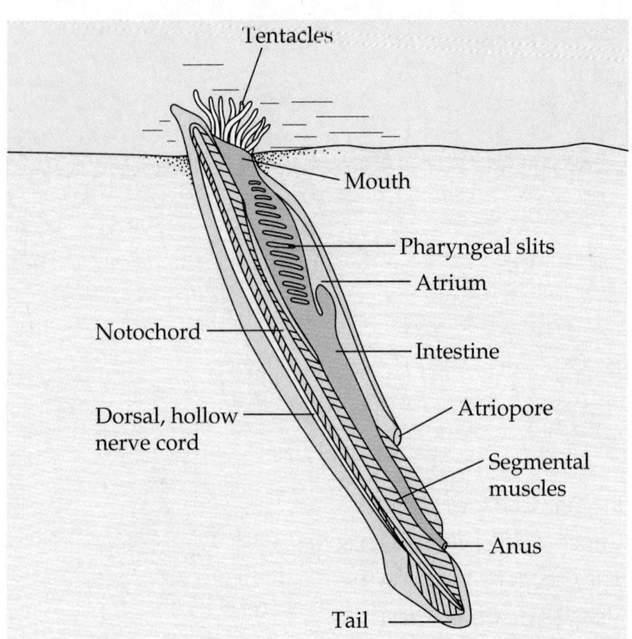

Figure 30.3
Subphylum Cephalochordata: the lancelet *Branchiostoma* (formerly called *Amphioxus*). This small invertebrate displays all four chordate characteristics. The pharyngeal slits are used for suspension-feeding. Water passes into the pharynx and through slits into a chamber called the atrium that vents to the outside via the atriopore. Food particles trapped by a mucous net are swept by cilia into the digestive tract. The muscle segments produce the sinusoidal swimming of these animals.

CHORDATES WITHOUT BACKBONES

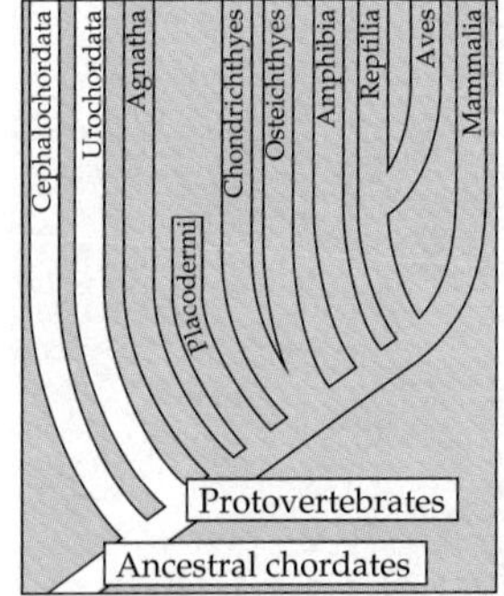

Subphylum Cephalochordata

Known as lancelets because of their bladelike shape, **cephalochordates** closely resemble the idealized chordate in Figure 30.2. The notochord, dorsal hollow nerve cord, numerous gill slits, and postanal tail all persist into the adult stage (Figure 30.3). A tiny marine animal only a few centimeters long, the lancelet wriggles backward into the sand, leaving only its anterior end exposed. The animal feeds by using a mucous net secreted across the pharyngeal slits to filter tiny food particles from seawater drawn into the mouth by ciliary pumping. The water exits through the slits, and the trapped food passes down the digestive tube.

The lancelet frequently leaves its burrow and swims to a new location. Though a feeble swimmer, the lancelet displays, in rudimentary form, the method of swimming that fishes use. Coordinated contraction of muscles serially arranged like rows of chevrons (<<<<)

(a) A tunicate

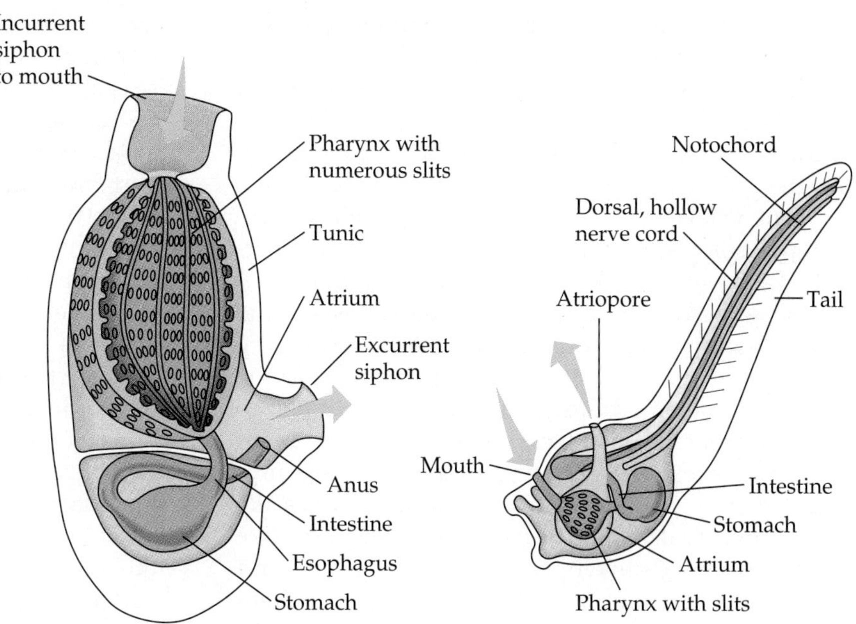

(b) Anatomy of an adult tunicate

(c) Anatomy of a larval tunicate

Figure 30.4
Subphylum Urochordata: the tunicate. (**a**) An adult tunicate, or sea squirt, is a sessile animal commonly arranged in a U shape. (**b**) In the adult, prominent pharyngeal slits function in suspension-feeding, but the other chordate characteristics are not obvious. (**c**) In the tunicate larva, which is a free-swimming suspension-feeder, the chordate characteristics are evident. Some urochordates lack the sessile adult stage, existing as free-swimming, larvalike forms throughout life.

along the sides of the notochord flexes the notochord from side to side in a sinusoidal (~) pattern. This serial musculature is evidence of the lancelet's segmentation. The muscle segments develop from blocks of mesoderm called **somites** arranged along each side of the notochord of a chordate embryo. The segmentation of chordates evolved independently of segmentation in annelids and arthropods.

Subphylum Urochordata

Urochordates are commonly called **tunicates.** Most tunicates are sessile marine animals that adhere to rocks, docks, and boats (Figure 30.4a), but others are planktonic. Some species are colonial. Seawater enters the animal through an incurrent siphon, passes through the slits of the dilated pharynx into a chamber called the atrium, and exits through an excurrent siphon (Figure 30.4b). The food filtered from this water current by a mucous net is passed by cilia into the intestine. The anus empties into the excurrent siphon. The entire animal is cloaked in a tunic made of a cellulose-like carbohydrate called tunicin. Because they shoot a jet of water through the excurrent siphon when molested, tunicates are also called sea squirts.

The adult tunicate scarcely resembles a chordate. It displays no trace of a notochord, nor is there a nerve cord or tail. Only the pharyngeal slits suggest a link to other chordates. But all four chordate trademarks are manifest in the larval form of some groups of tunicates (Figure 30.4c). The larva swims until it attaches by its head to a surface and undergoes metamorphosis, during which most of its chordate characteristics disappear.

THE ORIGIN OF VERTEBRATES

The phylum Chordata, like most animal phyla, makes its first appearance in the fossil record in Cambrian rocks. (For a review of the geological time scale, see Table 23.1.) Paleontologists have found fossilized invertebrates resembling cephalochordates in the Burgess Shale of British Columbia, Canada, which is about 550 million years old, about 50 million years older than the oldest known vertebrates. The record of the rocks is too incomplete for us to retrace the origin of the earliest vertebrates from invertebrate ancestors, but we can speculate about this evolution based on comparative anatomy and embryology.

Most zoologists think that protovertebrates were suspension-feeders (also known as filter-feeders) with all four of the fundamental chordate characteristics. They may have resembled lancelets, but cephalochordates probably represent an offshoot of the evolutionary lineage that gave rise to the vertebrates. Protovertebrates were most likely derived from a urochordatelike ancestor, perhaps similar to the free-swimming, planktonic larvae of tunicates (see Figure 30.4c).

The major milestones in protovertebrate evolution probably occurred early in the Cambrian period. During that time, the phenomenon called **paedogenesis,** the precocious attainment of sexual maturity in a larva, may have had a major impact on vertebrate evolution. Zoologists postulate that some early urochordatelike larval forms became sexually mature and reproduced before undergoing metamorphosis to the sessile tunicate stage. If reproducing larvae were very successful, natural selection may have reinforced the absence of metamorphosis, and a life cycle similar to that of vertebrates may have evolved. (Perhaps reminiscent of ancestral events, some living species of urochordates exist only as free-swimming larvalike forms.) Eventually some of the reproducing, larvalike forms may have developed segmental muscles and stronger skeletal support in their tails, enabling them to forage more actively. Actively foraging animals must deal with navigation in currents and avoidance of predators. In this context, natural selection may have favored the evolution of a protovertebrate with a distinct head equipped with a brain and acute sensory organs.

VERTEBRATE CHARACTERISTICS

Vertebrates retain chordate features while adding other specializations. They tend to be highly cephalized. Most possess well-developed sense organs associated with a distinct brain at the anterior end of the dorsal nerve cord. In addition, most vertebrates possess a column of serially arranged skeletal units, the **vertebrae** (singular, **vertebra**), that enclose the nerve cord (see Figure 30.1). In most adult vertebrates, the vertebral column is the most obvious sign of segmentation. The brain is also encased in a skeletal structure, the skull. Together, the skull and vertebral column make up much of the axial (central) skeleton. Other axial elements, present in many vertebrates, are the ribs and breastbone. Most vertebrates also have an appendicular skeleton supporting two pairs of appendages (fins or limbs) of the body (see Chapter 45). The vertebrate endoskeleton may be made of either hard bone or flexible cartilage, or some combination of these two materials. Although the skeleton consists mostly of a nonliving matrix, living cells that secrete and maintain the matrix are present. The living endoskeleton of a vertebrate can grow with the animal, unlike the nonliving exoskeleton of arthropods.

Vertebrates have closed circulatory systems, with blood pumped by a ventral chambered heart through arteries to microscopic capillaries that reach nearly every body cell. After flowing through the capillaries, blood returns to the heart in veins. Oxygen is transported by red blood cells containing hemoglobin. The blood is oxygenated as it flows through the skin or highly vascularized membranes lining gills or lungs. Waste products are removed from the blood as it passes through excretory structures concentrated in compact organs, the kidneys.

Male and female sexes are separate in most vertebrates. Reproduction is sexual, although in some

Table 30.1 The Vertebrate Classes

Class	Main Characteristics	Examples
Agnatha	Jawless vertebrates: cartilaginous skeleton; notochord persists throughout life; marine and fresh water; living species lack paired appendages.	Lampreys, hagfishes
Chondrichthyes	Cartilaginous fishes: cartilaginous skeleton; jaws; notochord replaced by vertebrae in adult; respiration through gills; internal fertilization, may lay eggs or bear live young; acute senses, including lateral line system.	Sharks, skates, rays
Osteichthyes	Bony fishes: bony skeletons and jaws; most species have external fertilization and lay large numbers of eggs; respiration mainly through gills; many have a swim bladder; marine and fresh water.	Bass, trout, perch, tuna
Amphibia	Aquatic larval stage metamorphosing into terrestrial adult (many species); may lay eggs or bear live young; respiration through lungs and/or skin.	Salamanders, newts, frogs, toads
Reptilia	Terrestrial tetrapods with scaly skin: respiration via lungs; lay amniotic shelled eggs or bear live young.	Snakes, lizards, turtles, crocodiles
Aves	Tetrapods with feathers: forelimbs modified to form wings; respiration through lungs; endothermic; internal fertilization; shelled amniotic eggs; acute vision.	Owls, sparrows, penguins, eagles
Mammalia	Tetrapods with young nourished from mammary glands of females; diaphragm to ventilate lungs; endothermic; most bear live young; possess hair.	Monotremes (platypuses); marsupials (kangaroos); placentals (rodents)

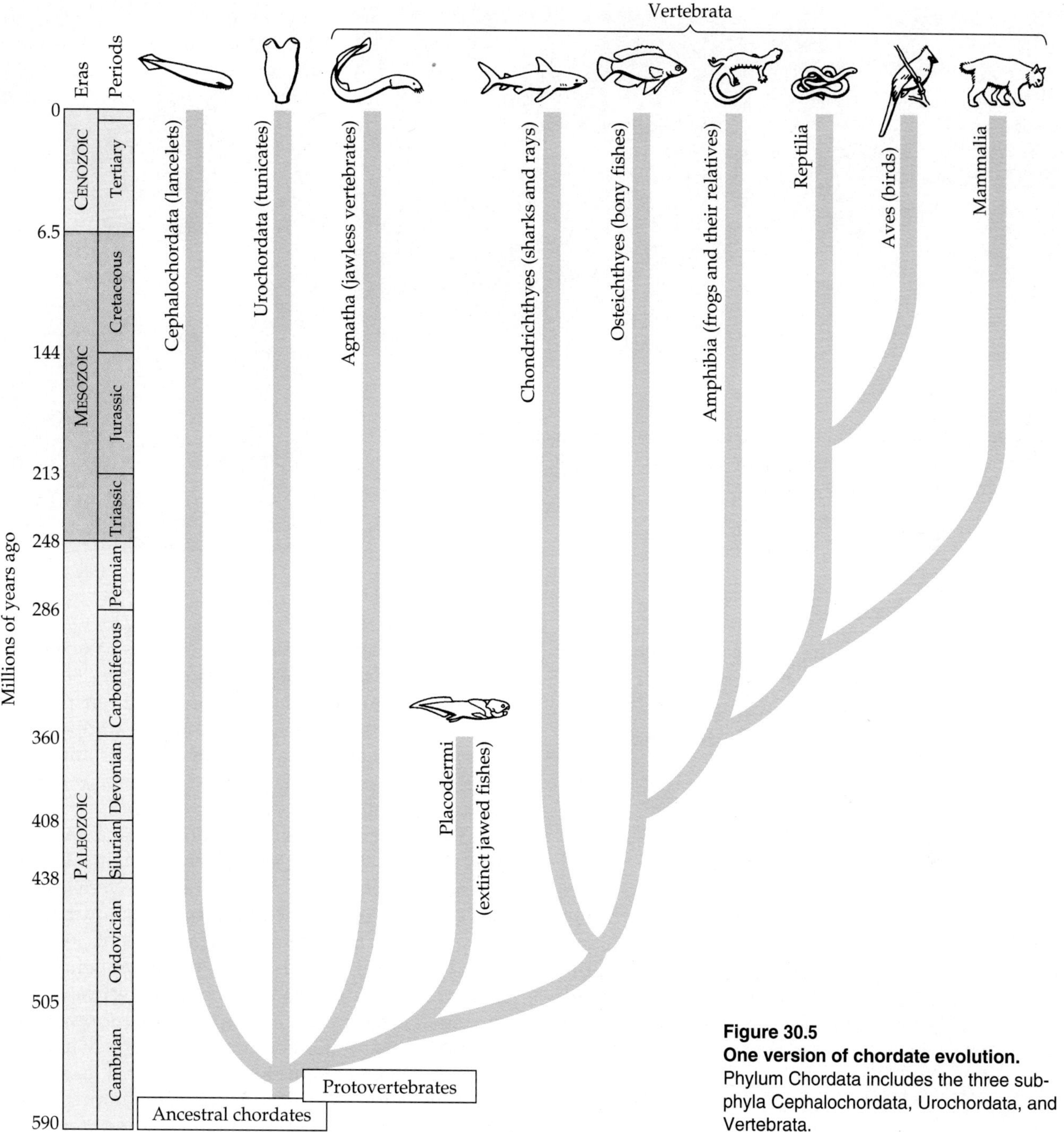

Figure 30.5
One version of chordate evolution.
Phylum Chordata includes the three subphyla Cephalochordata, Urochordata, and Vertebrata.

species in most vertebrate classes, eggs develop without fertilization, by parthenogenesis. Fertilization may be either external or internal, depending on species.

There are seven extant classes of the subphylum Vertebrata (Table 30.1). Three of these are commonly called fishes: Agnatha (jawless vertebrates), Chondrichthyes (sharks and rays), and Osteichthyes (bony fishes). The other four classes—Amphibia (frogs and salamanders), Reptilia (reptiles), Aves (birds), and Mammalia (mammals)—are collectively called **tetrapods** (Gr. *tetra*, "four," and *pod*, "foot") because most have two pairs of limbs that support them on land. In the following pages, we will survey the diversity of organisms in each of these classes and highlight some of the major evolutionary trends among the vertebrates. The survey also includes a brief description of an extinct class of early jawed vertebrates called placoderms. As a prelude to learning about the various groups of vertebrates, examine the evolutionary tree in Figure 30.5. This tree represents one set of ideas about

(a)

Figure 30.6
An agnathan. (**a**) A sea lamprey, a jawless vertebrate or agnathan. Acting as both a predator and a parasite, a lamprey uses this rasping mouth (**b**) to bore a hole in the side of a fish, living on the blood of its host.

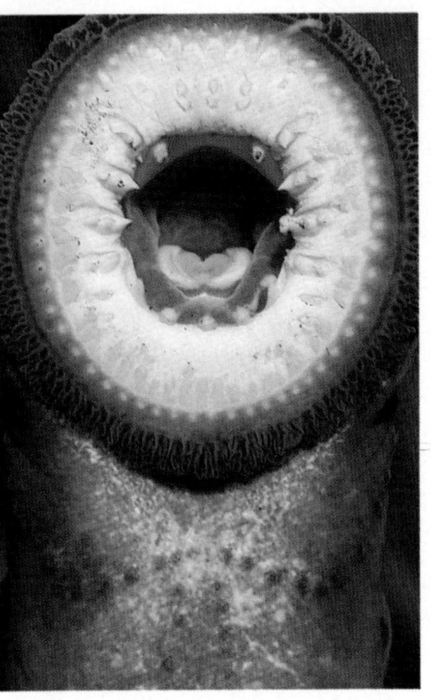

(**b**)

the relationships among the classes of the subphylum Vertebrata. Throughout the chapter you will find selected portions of this tree to help you keep track of where each class fits in the vertebrate genealogy.

CLASS AGNATHA

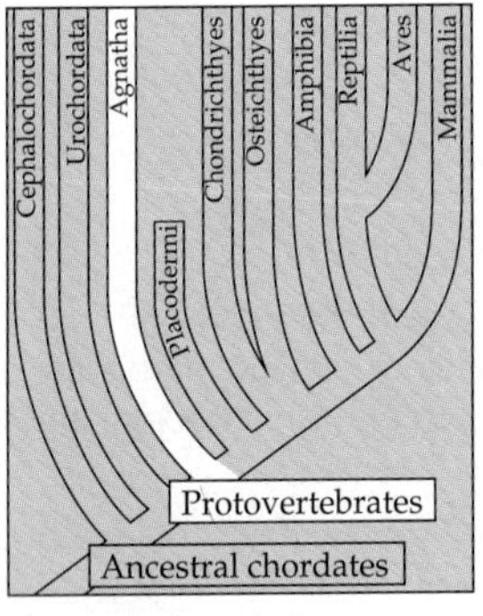

The oldest vertebrate fossils are diverse jawless creatures called **agnathans** ("without jaws"), including fishlike animals called **ostracoderms** that were encased in an armor of bony plates. Traces of agnathans are found in Cambrian strata, but most date back to the Ordovician and Silurian periods, about 400 to 500 million years ago. The fossil record provides few clues directly linking agnathans to invertebrate ancestors, and the record is not complete enough to explain the tremendous evolutionary radiation of agnathan forms that had occurred by the time of the late Silurian.

Early agnathans were generally small, less than 50 cm in length. Many lacked paired fins and apparently were bottom-dwellers that wiggled along streambeds or the sea floor, but there were also some active, mid-water forms with paired fins. All had mouths with circular or slitlike openings that lacked jaws. Most agnathans were probably mud-suckers or suspension-feeders that took in sediments or suspended organic debris through their mouths and then passed it through the gill slits where food was trapped. Thus, the pharyngeal apparatus retained the primitive feeding function, although gills in agnathans were probably also the major sites of gas exchange.

The ostracoderms and most other agnathan groups declined and finally disappeared during the Devonian period, but about 60 species of jawless vertebrates are alive today in the form of lampreys and hagfishes (Figure 30.6). In common with many extinct agnathans, lampreys and hagfishes lack paired appendages. In contrast to many ostracoderms, however, they have no external armor. The eel-shaped sea lamprey feeds by clamping its round mouth onto the flank of a live fish, using a rasping tongue to penetrate the skin of its prey, and ingesting the prey's blood. Sea lampreys live as larvae for years in freshwater streams and then migrate to the sea or lakes as they mature into adults. The larva is a suspension-feeder that looks very much like the lancelet, a cephalochordate. Some species of lampreys feed only as larvae. Following several years in streams, they attain sexual maturity, reproduce, and die within a few days.

Hagfishes superficially resemble lampreys, but they are mainly scavengers rather than blood-suckers or suspension-feeders and their mouthparts are not adapted for rasping. Some species feed on sick or dead fish, whereas other hagfishes eat marine worms. Hagfishes lack a larval stage and live entirely in salt water.

CLASS PLACODERMI

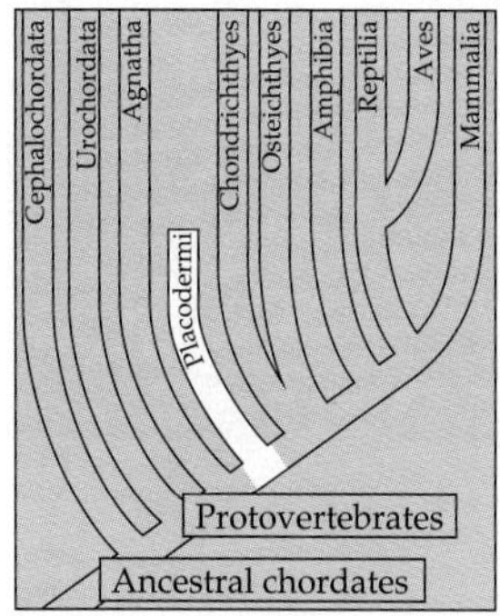

During the late Silurian and early Devonian periods, the agnathans were largely replaced by armored fishes called **placoderms.** The largest were more than 10 m long, but most were less than 1 m long. In contrast to many ostracoderms, placoderms had paired fins, which greatly enhanced their swimming ability. Placoderms also had jaws, making their mouth more than just a fixed orifice for rasping or scooping sediments. With paired fins and hinged jaws, many species were active predators, capable of chasing prey and biting off chunks of food. Thus, these two modifications of the early vertebrate body plan allowed the diversification of both lifestyles and nutrient sources.

The hinged jaws of vertebrates evolved by modification of the skeletal rods that had previously supported the anterior pharyngeal (gill) slits (Figure 30.7). The remaining gill slits, no longer required for suspension-feeding, remained as the major sites of respiratory gas exchange with the external environment.

The origin of vertebrate jaws from these skeletal parts was a major adaptive event, illustrating a general feature of evolutionary change: New adaptations usually evolve by the modification of existing structures. Hinged jaws also evolved in arthropods, but these had a totally different origin than vertebrate jaws (see Chapter 29). Arthropod jaws are modified appendages that work from side to side rather than up and down like vertebrate jaws. As a mechanism of adaptation, evolution is limited by the raw material with which it must work; evolution is generally more of a remodeling process than a creative one.

The Devonian period (about 350 to 400 million years ago) is known as the age of fishes, when the placoderms and another group of jawed fishes called acanthodians (placed in a separate class) radiated and many new forms evolved in both fresh and salt water. Placoderms and acanthodians dwindled and disappeared almost completely by the beginning of the Carboniferous period, 350 million years ago. During the Devonian or earlier, however, ancestors of placoderms and acanthodians may have given rise to sharks (class Chondrichthyes). Another group of large, jawed predators—the bony fishes (class Osteichthyes)—diverged from the common ancestor some 425 to 450 million years ago. Sharks and bony fishes are the reigning vertebrates over the watery two-thirds of the surface of Earth.

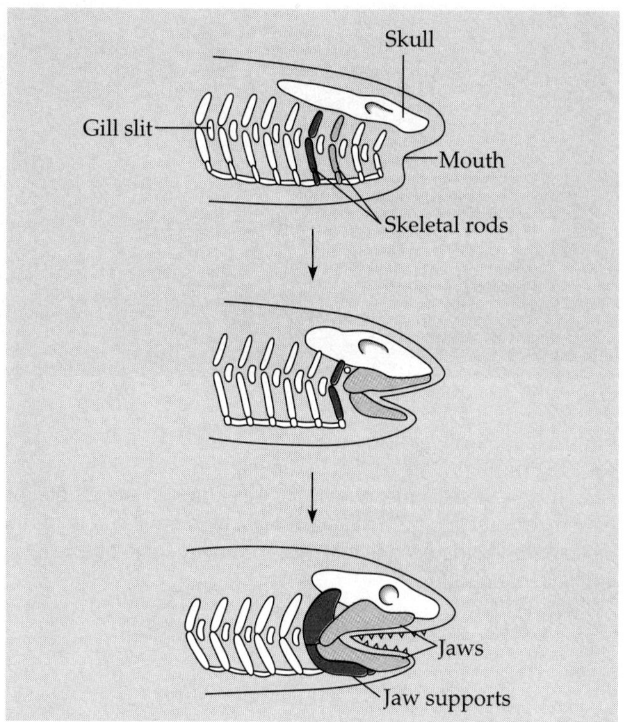

Figure 30.7
The evolution of vertebrate jaws. The skeleton of the jaws and their supports evolved from two pairs of skeletal rods located between gill slits that were near the mouth. Pairs of rods anterior to those that formed the jaws were either lost or incorporated into the jaws.

CLASS CHONDRICHTHYES

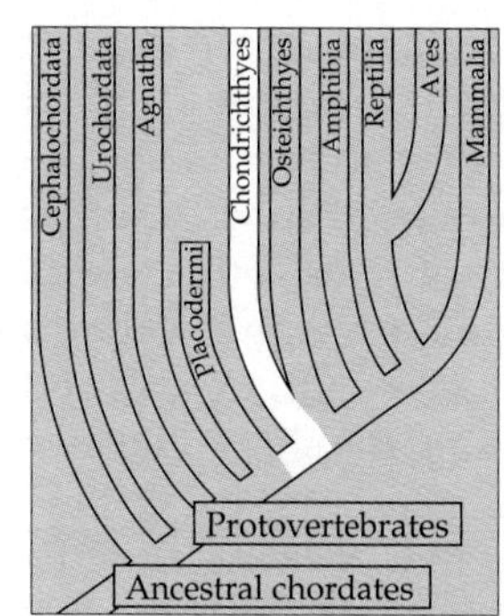

Sharks and their relatives are called cartilaginous fishes because they have somewhat flexible skeletons made of cartilage rather than bone. There are about 750 extant species in this class. Jaws and paired fins, which first appeared in the placoderms, are well developed in the cartilaginous fishes. The most widespread and diverse members of the class Chondrichthyes are the sharks and rays (Figure 30.8).

Most sharks have streamlined bodies and are swift swimmers, but they do not maneuver very well. Their stiff caudal (tail) fin helps propel the animal. The dorsal fins are used largely as stabilizers, and the paired pectoral (fore) and pelvic (hind) fins provide lift in the water. Although it gains some buoyancy by storing a large amount of oil in its huge liver, a shark is still more

(a)

(b)

Figure 30.8
Cartilaginous fishes. (**a**) Fast swimmers with acute senses and powerful jaws, sharks, such as this blacktip reef shark, are well adapted to their predatory way of life. Note the paired pectoral and pelvic fins. (**b**) Most rays are flat bottom-dwellers that crush mollusks and crustaceans for food. However, some species, such as this manta ray, cruise in open water.

dense than water, and it sinks if it stops swimming. Continual swimming also ensures that water will flow into the mouth and out through the gills, where gas exchange occurs. However, some sharks and many skates and rays spend a good deal of time resting on the sea bottom. When doing so, these fishes actively pump water over the gills.

The largest sharks and rays are suspension-feeders, depending on small planktonic animals for food. The whale shark, for example, strains 1 million L of water per hour to obtain sufficient food to support its massive body. Most sharks, however, are carnivores that swallow their prey whole or use their powerful jaws and sharp teeth to tear flesh from animals too large to swallow in one piece. Shark teeth probably evolved from the jagged scales that cover the abrasive skin. The digestive tract of many sharks is proportionately shorter than the digestive tube of many other vertebrates. Within the shark intestine is a spiral valve, a corkscrew-shaped ridge that increases surface area and prolongs the passage of food along the short digestive tract.

Acute senses are adaptations that go along with the active, carnivorous lifestyle of sharks. Sharks have sharp vision but cannot distinguish colors. The nostrils of sharks, like those of most fishes, open into dead-end cups, and they can be used only for olfaction, not for breathing. Along with eyes and nostrils, the shark head also has a pair of regions in the skin that can detect the electric fields generated by muscle contractions of nearby fish and other animals. Running the length of each flank of the shark is the **lateral line system,** a row of microscopic organs sensitive to changes in the surrounding water pressure, enabling the shark to detect minor vibrations (see Chapter 45). Sharks can also hear by sensing percussions with a pair of auditory organs. Sharks and other fishes have no eardrums, structures that terrestrial vertebrates use to transmit sound waves traveling through air into the ear toward the auditory organs. Sound reaches the shark through water, and the animal's entire body transmits the sound to the hearing organs of the inner ear.

Shark eggs are fertilized internally. The male has a pair of claspers on its pelvic fins that transfer sperm into the reproductive tract of the female. Some species of sharks are **oviparous,** laying eggs that hatch outside the mother's body. These sharks release their eggs after encasing them in protective coats. Other species are **ovoviviparous;** that is, they retain the fertilized eggs in the oviduct. Nourished by the egg yolk, the embryos develop into young that are born after hatching within the uterus. A few species are **viviparous**—the young develop within the uterus, nourished until they are born by nutrients received from the mother's blood through a placenta. The reproductive tract of the shark empties along with the excretory system and digestive tract into a common chamber, the **cloaca,** which expels through a single vent.

Although rays are closely related to sharks, they have adopted a very different lifestyle. Most rays are flattened bottom-dwellers that feed by using their jaws to crush mollusks and crustaceans. A ray's pectoral fins are greatly enlarged and used to propel the animal through the water. The tail of many rays is whiplike and, in some species, bears venomous barbs that function in defense.

(a)

(b)

Figure 30.9
Bony fishes. The species depicted here are ray-finned fishes, one of three groups of bony fishes that had evolved by the end of the Devonian period. (**a**) Longnose gar. (**b**) Perch.

CLASS OSTEICHTHYES

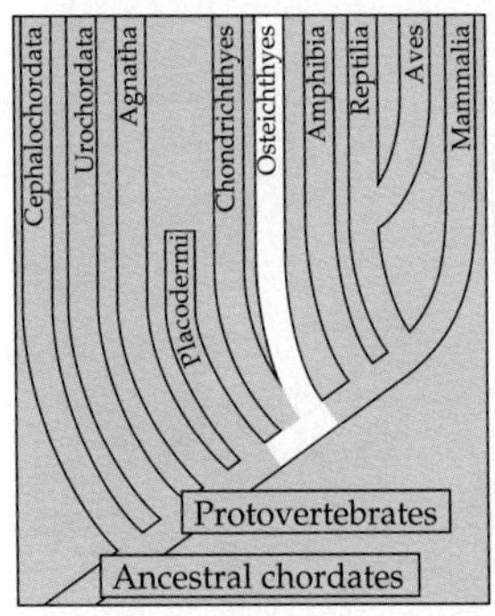

Of all vertebrate classes, bony fishes, of the class Osteichthyes, are the most numerous, both in individuals and in species (about 30,000). Fishes of this class are abundant in the seas and in nearly every freshwater habitat. Bony fishes range in size from about 1 cm to more than 6 m long. Most of the fishes familiar to us are osteichthyans (Figure 30.9).

In contrast to the cartilaginous fishes, the skeleton of most bony fishes is reinforced by a hard matrix of calcium phosphate. The skin is often covered by flattened bony scales that differ in structure from the toothlike scales of sharks. Typical of fishes, glands in the skin of a bony fish secrete a mucus that gives the animal its characteristic sliminess, an adaptation that reduces drag during swimming. In common with sharks, bony fishes have a lateral line system clearly evident as a row of tiny pits in the skin on either side of the body.

Bony fishes breathe by drawing water over four or five pairs of gills that are located in chambers covered by a protective flap, the **operculum.** Water is drawn into the mouth, through the pharynx, and out between the gills by movement of the operculum and contraction of muscles contained within the gill chambers. This arrangement enables a bony fish to breath while stationary.

Another adaptation of most bony fishes not found in sharks is the **swim bladder,** an air sac that helps control the buoyancy of the fish (Figure 30.10). Transfer of gases between the swim bladder and the blood varies the inflation of the bladder and adjusts the density of the fish. Many bony fishes, in contrast to sharks, can conserve energy by remaining almost motionless.

Bony fishes are generally maneuverable swimmers, their flexible fins better for steering and propulsion than the stiffer fins of sharks. The fastest bony fishes, which can swim in short bursts of up to 80 km per hour, have the same basic body shape as a shark. In fact, this body shape, termed fusiform (tapering on both ends), is common to all fast fishes and aquatic mammals such as seals and whales. Water is about a thousand times more dense than air, and thus the slightest bump that causes drag is even more impeding to a fish than to a bird. Regardless of their different origins, we should expect speedy fishes and marine mammals to have similar streamlined shapes, because the laws of hydrodynamics are universal. This is one more example of convergent evolution.

Details about the reproduction of bony fishes vary extensively. Most species are oviparous, reproducing by external fertilization after the female sheds large numbers of small eggs. However, internal fertilization and birthing characterize other species. Some bony fishes display complex mating rituals (see Chapter 50).

Both cartilaginous and bony fishes diversified extensively during the Devonian and Carboniferous, but whereas sharks arose in the sea, bony fishes probably originated in fresh water. The swim bladder was modified from lungs that had been used to augment the gills for gas exchange, perhaps in stagnant swamps with low oxygen content. By the end of the Devonian, three distinct subclasses of bony fishes had evolved: the ray-finned fishes (subclass Actinopterygii), the

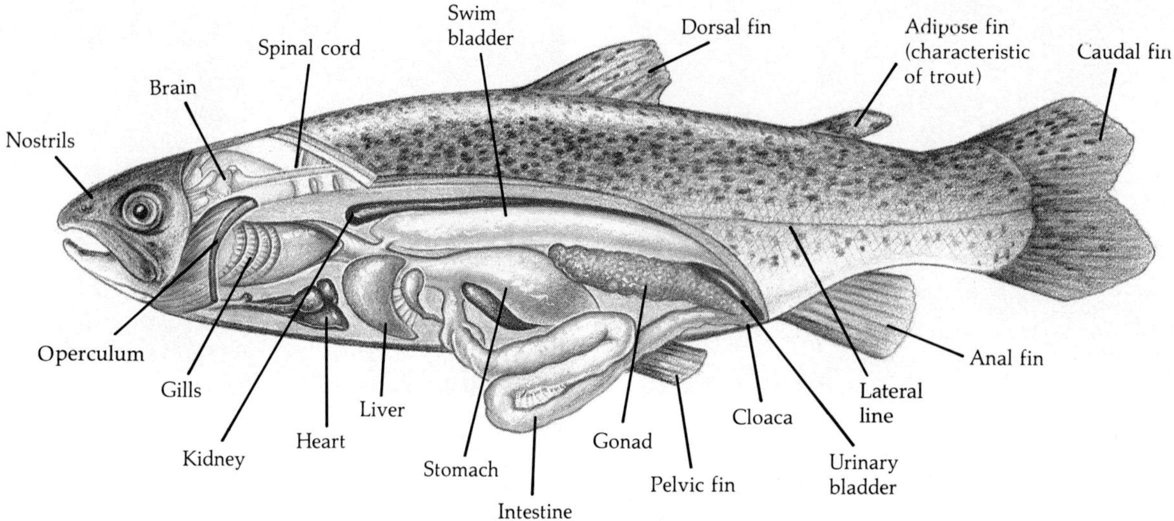

Figure 30.10
Anatomy of a trout, a representative bony fish.

lobe-finned fishes (subclass Crossopterygii), and the lungfishes (subclass Dipnoi).

Nearly all the families of fishes familiar to us are **ray-finned fishes.** The various species of bass, trout, perch, tuna, and herring are examples. The fins, supported mainly by long flexible rays, have become modified for maneuvering, defense, and other functions. Ray-finned fishes spread from fresh water to the seas during their long history. Adaptations that solve the osmotic problems of this move to salt water will be discussed in Chapter 40.

Numerous species of ray-finned fishes returned to fresh water at some point in their evolution. Some of these, including salmon and sea-run trout, replay their evolutionary round trip from fresh water to seawater back to fresh water during their life cycle.

In contrast to the ray-fins, most **lobe-finned fishes** and **lungfishes** remained in fresh water and continued to use their lungs to aid the gills in breathing. Lobe-fins and some lungfishes had fleshy, muscular pectoral and pelvic fins that were supported by extensions of the bony skeleton. Many were large, apparently bottom-dwelling forms that may have used their paired fins as aids to "walking" on the substrate under water. Some may also have been able to waddle occasionally on land.

Three genera of lungfishes live today in the Southern Hemisphere. They generally live in stagnant ponds and swamps, surfacing to gulp air into lungs connected to the pharynx of the digestive tract. When ponds shrink during the dry season, lungfishes can burrow in the mud and aestivate, which means to wait in a state of torpor.

Figure 30.11
***Latimeria*, a lobe-finned fish.** The only known living representative of the subclass Crossopterygii, *Latimeria* lives in the deep sea off Madagascar.

Lobe-finned fishes are represented today by only one known species, the coelocanth *(Latimeria).* Although most Devonian lobe-fins were probably freshwater animals with lungs, the coelocanth is a lungless species belonging to a lineage that entered the seas at some point in its evolution (Figure 30.11).

Few of us encounter lungfishes or lobe-fins, and today these animals are far less numerous than the ray-fins. However, the lobe-finned fishes of the Devonian are of great importance in the vertebrate genealogy, for they gave rise to amphibians (Figure 30.12).

Cephalochordata
Urochordata
Agnatha
Chondrichthyes
Osteichthyes
Amphibia
Reptilia
Aves
Mammalia
Placodermi
Protovertebrates
Ancestral chordates

AGNATHA
CHONDRICHTHYES
OSTEICHTHYES
AMPHIBIA
Ray-finned fishes
Lungfishes
Lobe-finned fishes
PLACODERMI (extinct)
Ostracoderms (extinct Agnathans)
Ancestral bony fishes
Ancestral jawed fishes
Protovertebrates

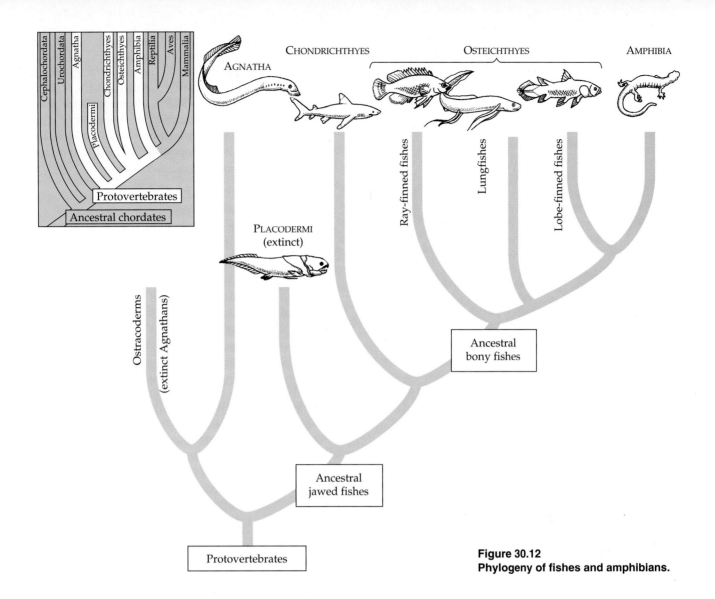

Figure 30.12
Phylogeny of fishes and amphibians.

CLASS AMPHIBIA

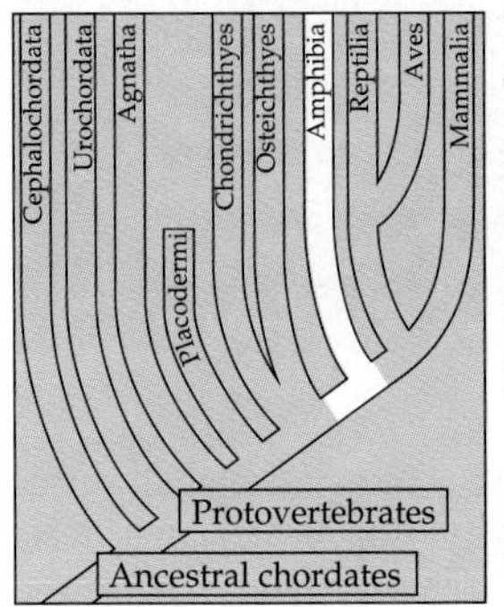

The amphibians were the first vertebrates on land. Today the class is represented by a total of about 4000 species of frogs, salamanders, and caecilians (wormlike creatures that live in tropical forests).

Early Amphibians

As is the case today, parts of the Devonian world were subject to cycles of drought, followed by heavy rainfall and then drought again. In such areas, some of the freshwater fishes apparently adapted to the unreliable conditions, and lobe-finned species with lungs prevailed. The lobe-fins are of particular interest because the skeletal structure of their fins suggests that these paired appendages could have assisted in movement on land. Some of the fossil lobe-fins, including a creature named *Eusthenopteron,* exhibited many other anatomical similarities to the earliest amphibians, making them good candidates as the ancestors of tetrapods (Figure 30.13a).

The oldest amphibian fossils date back to late Devonian times, about 350 million years ago (Figure 30.13b). New adaptive zones opened to these first vertebrates on land, with a cornucopia of food previously unexploited by backboned animals. From the start, amphibians were predators that ate insects and other invertebrates that preceded them onto land. Adaptive radiation of the earliest amphibians gave rise to a

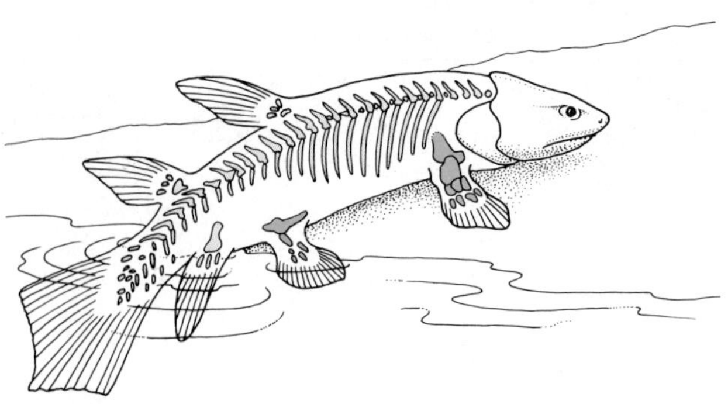

(**a**) Lobe-finned fish

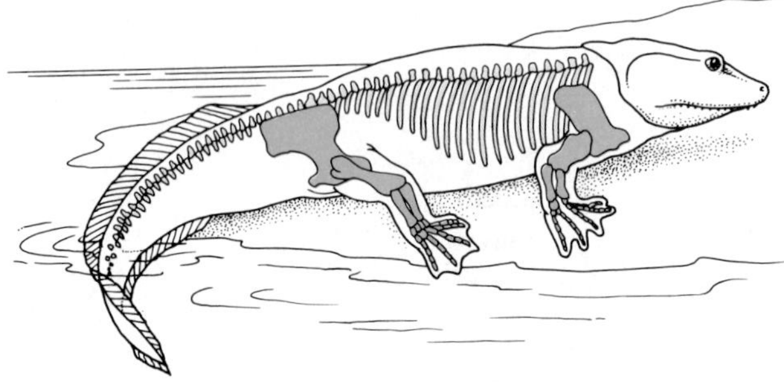

(**b**) Early amphibian

Figure 30.13
The emergence of amphibians. (**a**) The ancestors of tetrapods were lobe-finned fishes, such as *Eusthenopteron,* that had muscular fins with extensions of the skeleton that provided some support for the animal on land. (**b**) This reconstruction, based on Devonian fossils, depicts an early amphibian.

diversity of new forms. Many Carboniferous species superficially resembled reptiles. Some reached 4 m in length. Amphibians were the only vertebrates on land in late Devonian and early Carboniferous times, making the name Age of Amphibians appropriate for the Carboniferous period. Amphibians began to decline during the late Carboniferous period. As the Mesozoic era dawned with the Triassic period, about 230 million years ago, most survivors of the amphibian lineage resembled modern species.

Modern Amphibians

There are three extant orders of amphibians (Figure 30.14): urodeles ("tailed ones"—salamanders), anurans ("tail-less ones"—frogs, including toads), and apodans ("legless ones"—caecilians).

There are only about 400 species of **urodeles,** some of which are entirely aquatic, whereas others live on land as adults or throughout life. Most salamanders that live on land walk with a side-to-side bending of the body that may resemble the swagger of the early amphibians. Aquatic salamanders swim sinusoidally or walk along the bottom of streams or ponds.

Anurans, numbering nearly 3500 species, are more specialized than urodeles for moving on land. Adult frogs use their powerful hind legs to hop along the terrain. A frog nabs insects by flicking out its long sticky tongue, which is attached to the front of the mouth. Frogs display a great variety of adaptations that help

(a)

(b)

(c)

Figure 30.14
Amphibian orders. (**a**) Urodeles (salamanders) retain their tails as adults. Some are entirely aquatic, but others live on land. This northern red salamander is a member of a large family (Plethodontidae) whose members have lost their lungs during their evolution and rely entirely on their skin and moist mouth surfaces for gas exchange. (**b**) Anurans, such as this blue poison arrow frog, lack tails as adults. Poison arrow frogs inhabit tropical forests and are being depleted in numbers as the forests are being destroyed; their skin glands secrete deadly nerve toxins used by Central and South American natives to coat arrow tips. (**c**) Apodans, such as this Sri Lankan caecilian, are legless, mainly burrowing amphibians.

(a)

(b)

(c)

Figure 30.15
The life cycle of a frog *(Rana temporaria).* (**a**) The male grasps the female, stimulating her to release eggs. The eggs are laid and fertilized in water. They have a jelly coat but lack shells and desiccate in air. (**b**) The tadpole is an aquatic herbivore with a fishlike tail and internal gills. (**c**) During metamorphosis, the gills and tail are resorbed, and walking legs develop.

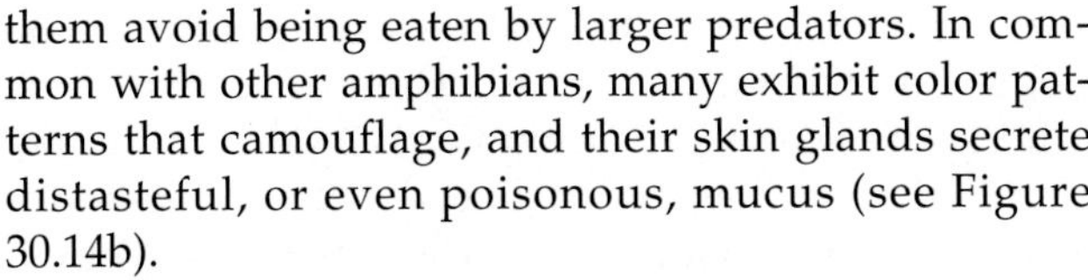

them avoid being eaten by larger predators. In common with other amphibians, many exhibit color patterns that camouflage, and their skin glands secrete distasteful, or even poisonous, mucus (see Figure 30.14b).

Apodans, the caecilians (about 150 species), are legless, nearly blind, and superficially resemble earthworms (see Figure 30.14c). Caecilians inhabit tropical areas where most species burrow in moist forest soil; a few South American apodans live in freshwater ponds and streams.

Amphibian means "two lives," a reference to the metamorphosis of many frogs (Figure 30.15). The tadpole, the larval stage of a frog, is usually an aquatic herbivore with internal gills, a lateral line system resembling that of fishes, and a long finned tail. The tadpole lacks legs and swims by undulating like its fishlike ancestors. During the metamorphosis that leads to the "second life," legs develop, and the gills and lateral line system disappear. The young tetrapod with air-breathing lungs, a pair of external eardrums, and a digestive system capable of digesting animal protein crawls onto shore and begins life as a terrestrial hunter. In spite of the name amphibian, however, many members of the class, including some frogs, do not go through the aquatic tadpole stage, and many do not live a dualistic—aquatic and terrestrial—life. There are strictly aquatic and strictly terrestrial species in all three extant orders. Moreover, salamander and caecilian larvae look much like adults, and typically both the larvae and adults are carnivorous. Paedogenesis is common among some groups of salamanders; the mudpuppy *(Necturus),* for instance, retains gills and other larval features when sexually mature.

Most amphibians maintain close ties with water and are most abundant in damp habitats such as swamps and rain forests. Even those frogs that are adapted to drier habitats spend much of their time in burrows or under moist leaves, where the humidity is high. Adult amphibians either have small, rather inefficient lungs or lack lungs altogether. Most species rely heavily on their skin to carry out gas exchange with the environment, and amphibians that live on land must cope with the problem of keeping their skin moist to allow gases to diffuse in and out. Many species also use moist surfaces of the mouth for gas exchange.

The amphibian egg has no shell, and it dehydrates quickly in dry air. Fertilization is external, with the male grasping the female and spilling his sperm over the eggs as the female sheds them (see Figure 30.15a). Oviparous amphibians generally lay their eggs in ponds or swamps or at least in moist environments. Some species lay vast numbers of eggs, and mortality is high. Desert frogs often breed explosively in temporary pools. In contrast are species that display various types of parental care and that lay relatively few eggs. Depending on the species, either males or females may incubate eggs on their back, in the mouth, or even in the stomach. Certain tropical tree frogs stir their egg masses into moist foamy nests that resist drying. There are also live-bearing amphibians (ovoviparous and even some viviparous species) that retain the eggs in the female reproductive tract, where embryos can develop without drying out.

Amphibians, particularly anurans, exhibit complex and diverse social behavior, especially during their breeding seasons. Frogs are usually quiet creatures, but many species fill the air with their mating calls during the breeding season. Males may vocalize in defense of breeding territory or to attract females. In some terrestrial species, migrations to specific breeding sites may involve vocal communication, celestial navigation, or chemical signaling.

CLASS REPTILIA

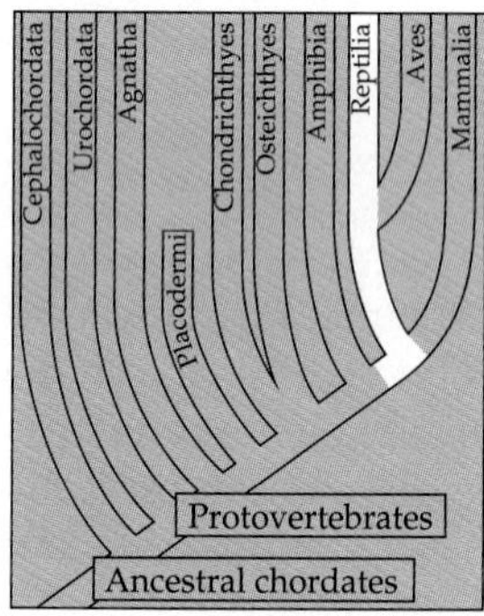

Reptiles, a diverse group with a wide array of extinct lineages, are represented today by about 7000 species of lizards, snakes, turtles, and crocodilians. Reptiles have several adaptations for terrestrial living not generally found in amphibians.

Reptilian Characteristics

Scales containing the protein keratin waterproof the skin of a reptile, helping prevent dehydration in dry air. Keratinized skin is the vertebrate analogue of the chitinized cuticle of insects and the waxy cuticle of land plants. Because they cannot breathe through their dry skin, most reptiles obtain all their oxygen with lungs. Many turtles also use the moist surfaces of their cloaca for gas exchange.

Although many viviparous reptiles exist, most species lay eggs on land; parchmentlike shells prevent desiccation. The embryo develops in the fluid of an amniotic sac within the egg. The evolution of the **amniote egg,** a shelled egg with a self-contained "pond" of amniotic fluid, enabled vertebrates to complete their life cycles on land and sever their last ties with their aquatic origins. (The seed played a similar role in the evolution of land plants; see Chapter 27.) Fertilization in reptiles must occur internally, before the shell is secreted as the egg passes through the reproductive tract of the female.

Reptiles are sometimes labeled "cold-blooded" animals because they do not use their metabolism to control body temperature. But reptiles *do* regulate body temperature by using behavioral adaptations. Many lizards can regulate their internal temperature by basking in the sun when the air is cool and seeking shade when the air is too warm. Since they absorb external heat rather than generating much of their own, reptiles are said to be **ectotherms,** a term more appropriate than cold-blooded. (Control of body temperature will be discussed in more detail in Chapter 40.) By heating directly with solar energy rather than through the metabolic breakdown of food, a reptile can survive on less than 10% of the calories required by a mammal of equivalent size. Having relatively modest food requirements and being adapted to arid conditions, many reptiles thrive in deserts.

During the Mesozoic era, reptiles were far more widespread, numerous, and diverse than they are today. Let's look briefly at this period, known as the Age of Reptiles.

The Age of Reptiles

The Origin and Early Evolutionary Radiation of Reptiles The oldest reptilian fossils are found in rocks from the upper Carboniferous period; they are about 300 million years old. Their ancestors were among the Devonian amphibians. By the end of the Carboniferous, there was a diversity of reptilian forms known as **cotylosaurs,** or **"stem reptiles."** These mostly small, lizardlike insect eaters were the ancestral stock from which the various reptilian orders arose. In two great waves of adaptive radiation, reptiles became the dominant terrestrial vertebrates in a dynasty that lasted more than 200 million years.

The first major reptilian radiation occurred during the Permian period, giving rise to terrestrial predators called **synapsids,** among other groups (Figure 30.16). One lineage of these reptiles was the progenitor of the **therapsids,** which are sometimes called mammal-like reptiles. Indeed, mammals are believed to have evolved from the therapsid line. The early therapsids were large, dog-sized predators, but a variety of other forms evolved as the Permian progressed. Some of the stem reptiles returned to the water, giving rise to the plesiosaurs and ichthyosaurs, which had seal-like or fishlike shapes. The **thecodonts** were another group that descended from the stem reptiles during the reptilian radiation of the Permian period. They are of special interest because they were the ancestors of dinosaurs, crocodilians, and birds. Both groups of ancient reptiles—synapsids and thecodonts—survived the mass extinctions of the Permian period (see

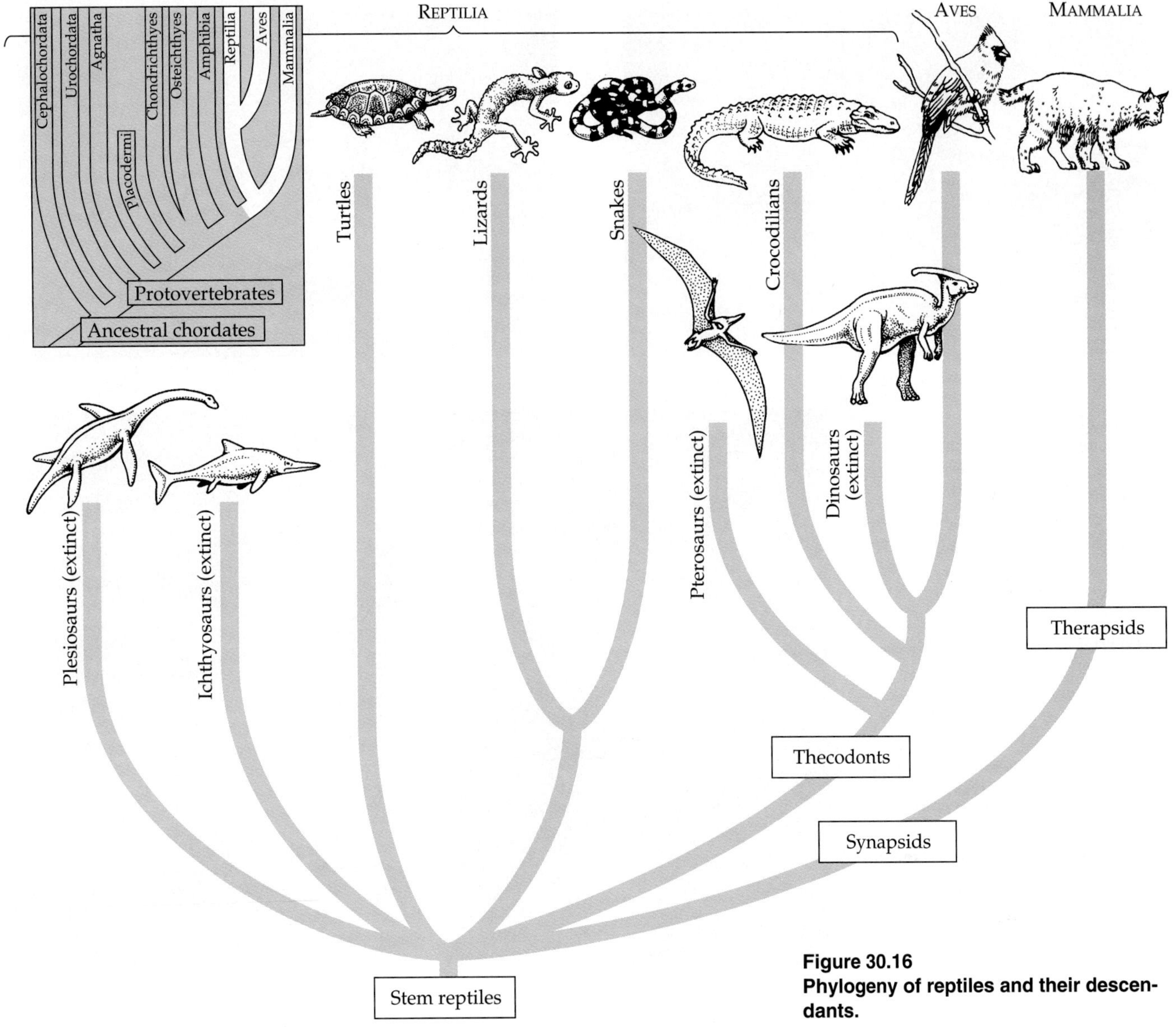

Figure 30.16
Phylogeny of reptiles and their descendants.

Chapter 23), and, in fact, carnivorous and herbivorous therapsids continued as the dominant terrestrial vertebrates until well into the Triassic period.

Dinosaurs and Their Relatives The second great reptilian radiation was under way by late Triassic times (a little more than 200 million years ago), when several lineages arose from thecodont stock (see Figure 30.16). Descendants of the thecodonts included two groups of reptiles: the dinosaurs, which lived on land, and the pterosaurs, or flying reptiles. These groups were the dominant vertebrates on Earth for millions of years. Dinosaurs, an extremely diverse group varying in body shape, size, and habitat, included the largest animals ever to inhabit land. Fossils of gigantic dinosaurs that probably weighed nearly 100 tons have recently been unearthed in New Mexico. Pterosaurs had wings formed from a membrane of skin stretched from the body wall, along the forelimb, to the tip of an elongated finger. Stiff fibers provided support for the skin of the wing.

Contradicting the long-standing view that dinosaurs were slow, sluggish creatures, there is increasing evidence that many dinosaurs were agile, fast-moving, and endothermic. (**Endothermy** is the ability to keep the body warm through metabolism.) Likewise, reevaluation of pterosaur fossils strongly indicates that many of these reptiles were active, endothermic flyers, rather than inefficient, ectothermic gliders. Despite considerable anatomical evidence supporting these hypotheses, some experts remain skeptical. The Mesozoic climate was relatively warm and

(a)

(b)

Figure 30.17
Extant reptiles. (**a**) Turtles have changed little since their evolution from stem reptiles early in the Mesozoic era. This species is a yellow-bellied turtle. (**b**) Lizards, such as this collared lizard, are the most numerous and diverse of the extant reptiles. (**c**) Snakes may have evolved from lizards that adapted to a burrowing lifestyle. This is an annulated boa. (**d**) Crocodiles and alligators are the reptiles most closely related to dinosaurs, having evolved from the same ancestral stock (thecodonts). An alligator is shown here.

(c)

(d)

consistent, and behavioral adaptations such as basking may have been sufficient for maintaining a suitable body temperature, especially for the land-dwelling dinosaurs. Also, large dinosaurs had low surface-to-volume ratios that reduced the effects of daily fluctuations in air temperature on the internal temperature of the animal.

The Cretaceous Crisis During the Cretaceous, the last period of the Mesozoic era, the climate became cooler and more variable. For reasons that are now being hotly debated, the Cretaceous was a period of mass extinctions. A quarter of the families of marine invertebrates disappeared, and the Cretaceous crisis was the curtain call for the dinosaurs. Nearly all of them were gone by about the close of the Mesozoic era, some 65 million years ago. (Fossils of some dinosaurs that survived into the early Cenozoic have recently been discovered.) There is evidence that the decline of the dinosaurs occurred over a period of 5 to 10 million years. But even this long a span is relatively sudden, on the scale of geological time, for a group so diverse and dominant to become extinct. Disappearance of the dinosaurs, one of the most engaging mysteries in the history of life, is discussed in Chapter 23. The Age of Reptiles ended when the dinosaurs disappeared, but a few reptilian groups are still successful today.

Reptiles of Today

The three largest and most diverse extant orders of reptiles are **Chelonia** (turtles), **Squamata** (lizards and snakes), and **Crocodilia** (alligators and crocodiles).

Turtles evolved from stem reptiles during the Mesozoic era and have scarcely changed since (Figure 30.17a). The usually hard shell, an adaptation that protects against predators, has certainly contributed to this long success. Those turtles that returned to water during their evolution crawl ashore to lay their eggs.

Lizards are by far the most numerous and diverse reptiles alive today (Figure 30.17b). Most are relatively small; perhaps they were able to survive the Cretaceous "crunch" by nesting in crevices and decreasing their activity during cold periods, a practice many modern lizards use.

Snakes are apparently descendants of lizards that adopted a burrowing lifestyle (Figure 30.17c). Today, most snakes live aboveground, but they have retained the limbless condition. Vestigial pelvic and limb bones in primitive snakes such as boas, however, are evidence that snakes evolved from reptiles with legs.

Snakes are carnivorous, and a number of adaptations aid them in hunting prey. They have acute chemical sensors, and though they lack eardrums, snakes are sensitive to ground vibrations, which helps them detect movements of prey. Heat-detecting organs between the eyes and nostrils of pit vipers, including rattlesnakes, are sensitive to minute temperature changes, enabling these night hunters to locate warm animals. Poisonous snakes inject their toxin through a pair of sharp hollow or grooved teeth. The flicking tongue is not poisonous but helps transmit odors toward olfactory organs on the roof of the mouth. Loosely articulated jaws enable most snakes to swallow prey larger than the diameter of the snake itself.

Crocodiles and alligators are among the largest living reptiles (some turtles are heavier) (Figure 30.17d). They spend most of their time in water, breathing air through their upturned nostrils. Crocodiles and alligators are confined to the warm regions of Africa, China, Indonesia, India, Australia, South America, and the southeastern United States, where alligators are making a strong comeback after spending years on the endangered species list. The crocodilians evolved from thecodonts and are thus the living reptiles most closely related to the dinosaurs. However, the only modern animals that seem to have descended from a group of dinosaurs are not reptiles at all, but birds.

CLASS AVES

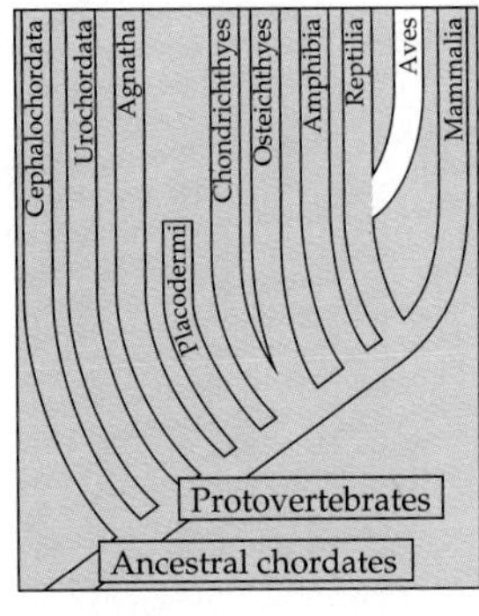

Birds (class Aves) evolved from a reptilian ancestor during the great reptilian radiation of the Mesozoic era (see Figure 30.16). Amniote eggs and scales on the legs are just two of the reptilian remnants we see in birds. But birds look quite different from lizards and other reptiles because of their distinctive flight equipment (Figure 30.18).

Figure 30.18
Flight. The special adaptations associated with flight are superimposed on ancestral reptilian traits. Birds can reach great speeds and cover enormous distances. The fastest bird is the swift, which can fly 170 km per hour. The bird that travels farthest in its annual migration is the arctic tern, which flies round trip between the North Pole and South Pole each year. The drinking bird shown here is a swallow.

Characteristics of Birds

Almost every part of a typical bird's anatomy is modified in some way that enhances flight. The bones are honeycombed—strong but light. The skeleton of a frigate bird, for instance, has a wingspan of more than 2 m but weighs only about 4 oz. Another adaptation reducing the weight of birds is the absence of some organs. Females, for instance, have only one ovary. Also, birds are toothless, an adaptation that trims the weight of the head. Food is not chewed in the mouth but ground in the gizzard, a digestive organ near the stomach. (Crocodiles also have gizzards, as did some dinosaurs.) The bird's beak, made of keratin, has proven to be very malleable during avian evolution, taking on a great variety of shapes suitable for different diets.

Flying requires a great expenditure of energy from an active metabolism. Birds are endothermic; they use their own metabolic heat to maintain a warm, constant body temperature, aided by insulation provided by feathers and a layer of fat (Figure 30.19). An efficient circulatory system with a four-chambered heart, which segregates oxygenated blood from oxygen-poor blood, supports the high metabolic rate of the bird's cells. The efficient lungs have tiny tubes leading to and from elastic air sacs that help dissipate heat and contribute to the trimming of density.

For safe flight, senses, especially vision, must be acute. Birds have excellent eyes, perhaps the best of all vertebrates. The visual areas of the brain are well developed, as are the motor areas; flight also requires fine coordination.

Figure 30.19
Emperor penguins in Antarctica. Birds are endotherms, using insulation provided by feathers and fat to slow the dissipation of body heat.

With brains proportionately larger than those of reptiles and amphibians, birds generally display very complex behavior. Avian behavior is particularly intricate during breeding season, when birds engage in elaborate rituals of courtship. Since eggs are shelled when laid, fertilization must be internal. The act of copulation is somewhat awkward because the male of most bird species has no penis. He must climb atop the female's back and then twist her tail so the mates' vents, the openings to their cloacas, can come together. After eggs are laid, the avian embryo must be kept warm through brooding by the mother, father, or both, depending on the species.

A bird's most obvious adaptation for flight is its wings. Bird wings are airfoils that illustrate the same principles of aerodynamics as the wings of an airplane. Providing power for flight, birds flap their wings by contractions of large pectoral (breast) muscles anchored to a keel on the sternum (breastbone). Some birds, such as hawks, have wings adapted for soaring on air currents and flap their wings only occasionally, whereas others, including hummingbirds, must flap continuously to stay aloft. In either case, it is the shape and arrangement of the feathers that form the wing into an airfoil.

In being both extremely light and strong, feathers are among the most remarkable of vertebrate adaptations (Figure 30.20). Feathers are made of keratin, the same protein that forms our hair and fingernails and the scales of reptiles. Feathers may have functioned first as insulation during the evolution of endothermy, only later being co-opted as flight equipment. Besides providing support and wing shape, feathers can be manipulated to control air movements around the wing.

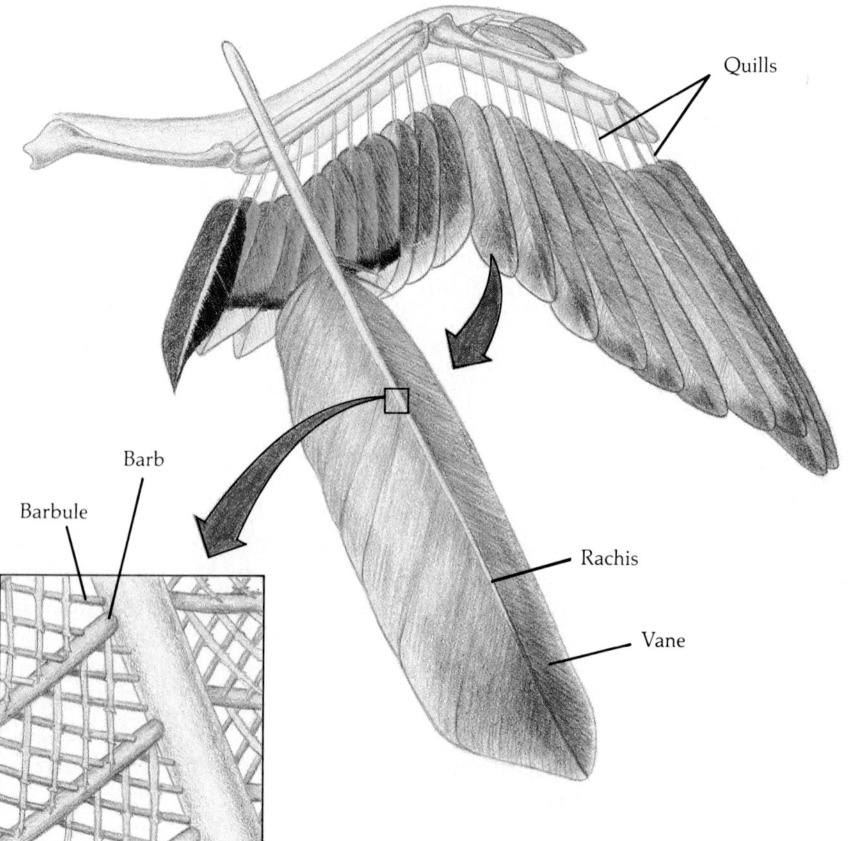

Figure 30.20
The structure of feathers. The feather consists of a central hollow shaft, the rachis, from which radiate the vanes. The vanes are made up of barbs, in turn bearing even smaller branches called barbules. Birds have contour feathers and downy feathers. Contour feathers are the stiff ones that shape the wings and body. Their barbules have hooks that cling to barbules on neighboring barbs. When a bird preens, it runs the length of a feather through the beak, engaging the hooks and uniting the barbs into a precisely shaped vane. Downy feathers lack hooks, and the free-form arrangement of barbs produces a fluffiness that provides excellent insulation because of the trapped air.

The Origin of Birds

For many zoologists, the presence of feathers is enough to classify an animal as a bird. And certainly, if we want to trace the ancestry of birds, we must search for the oldest fossils with feathers. Fossils of an ancient bird named *Archaeopteryx lithographica* have been found in Bavarian limestone in Germany, dating back some 150 million years into the Jurassic period. Unlike modern birds, but similar to reptiles, *Archaeopteryx* had clawed forelimbs, teeth, and a long tail containing vertebrae. Indeed, if it were not for the preservation of its feathers, *Archaeopteryx* would be regarded as a member of a diverse group of reptiles called the **theropods.** In common with birds, many ancient reptiles, including some theropods, were probably endothermic, built nests, and cared for their young. In fact, some zoologists argue that modern birds should be classified as reptiles.

Archaeopteryx is not considered the ancestor of modern birds, and paleontologists place it on a side branch of the avian lineage. Nonetheless, *Archaeopteryx* probably was derived from ancestral forms that also gave rise to modern birds. Its skeletal anatomy indicates that it was a weak flyer, perhaps mainly a tree-dwelling glider, and this may have been the earliest mode of flying in the bird lineage. A fossil recently discovered in Spain seems to represent a later stage of avian evolution. Dating to about 125 million years ago, this form had a tail similar to that of modern birds and forelimb bones that would have allowed stronger, more active flight than that of *Archaeopteryx.* Another fossil, discovered in Texas in 1986, has stirred some controversy about bird origin. This fossil, named *Protoavis* ("first bird"), shows a mixture of dinosaurian and avian characteristics. However, it lived 225 million years ago, about 75 million years before *Archaeopteryx.* Clearly, the final word is not yet in on the origin of birds.

The evolution of flight required radical alteration in body form, but flight provides many benefits. It allows aerial reconnaissance that enhances hunting and scavenging; it allows birds to exploit flying insects as an abundant, highly nutritious food resource; it also provides ready escape from land-bound predators and allows some birds to migrate great distances to utilize different food resources and seasonal breeding areas.

Modern Birds

There are about 8600 extant species of birds classified in about 28 orders. Flight ability is typical of birds, but there are several flightless species, including the ostrich, kiwi, and emu. Flightless birds are collectively called ratites (L., "flat-bottomed") because their breastbone lacks a keel. Ratite birds also lack the large breast muscles that attach to the sternal keel and provide flight power in flying birds.

In contrast to the ratites, other birds are said to be carinate, because they have a carina, or sternal keel supporting their large breast muscles. The demands of flight have rendered the body form of carinate birds similar to each other, yet experienced bird watchers can distinguish many species by their body profile. Carinate birds also exhibit great variety in feather colors, beak and foot shape, behavior, and flight ability (Figure 30.21). Nearly 60% of the living bird species belong in one carinate order called the passeriforms, or the perching birds. These are the familiar jays, swallows, sparrows, warblers, and many others (Figure 30.21d). Among the most unusual carinate birds are the penguins, which do not fly, but use their powerful breast muscles in swimming (see Figure 30.19).

CLASS MAMMALIA

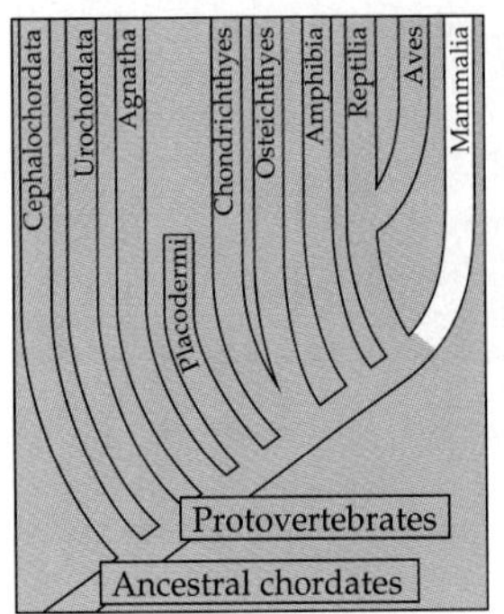

There are about 4500 species of mammals on Earth today. As mammals ourselves, we naturally have a special interest in this class of vertebrates. Let's examine some of the features we share with all other mammals.

Mammalian Characteristics

Mammals have hair, a characteristic as diagnostic as the feathers of birds. Hair, like feathers, is made of keratin, although zoologists are uncertain about its evolutionary origin. Also like feathers, hair insulates to help the animal maintain a warm and constant body temperature; mammals are endothermic. Their active metabolism is supported by an efficient respiratory system that uses a sheet of muscle called the **diaphragm** to help ventilate the lungs. The four-chambered heart of a mammal prevents the mixing of oxygen-rich blood with oxygen-poor blood.

Mammary glands that produce milk are as distinctively mammalian as hair (Figure 30.22). All mammalian mothers nourish their babies with milk, a balanced diet rich in fats, sugars, proteins, minerals, and vitamins.

Most mammals are born rather than hatched. Fertilization is internal, and the egg develops into an embryo within the uterus of the female reproductive tract. In placental mammals, the lining of the mother's uterus and membranes arising from the embryo collectively form a **placenta,** where nutrients diffuse into the embryo's blood.

(a)  (b)

(c)  (d)

Figure 30.21
Carinate birds. (**a**) In common with many species, the harlequin duck, which inhabits mountain streams and coastal areas of the Pacific Northwest, exhibits pronounced color differences between the sexes. (**b**) Most birds exhibit specific courtship and mating behavior. The western grebe virtually runs on the water surface while courting. (**c**) Hummingbirds can hover while they feed on nectar because they have short, rigid wings that can turn like propellers in virtually any direction. (**d**) The western tanager is a member of the order Passeriformes. Passeriforms are called perching birds because the toes of their feet can lock around a tree branch, allowing the bird to rest in place for long periods of time.

Figure 30.22
Mammalian trademarks. The presence of hair and mammary glands that nourish the young with milk are distinctively mammalian traits.

Figure 30.23
A therapsid, a mammal-like reptile.

Mammals have larger brains than other vertebrates of equivalent size, and they seem to be the most capable learners. The relatively long duration of parental care extends the time for parents to teach important skills to their impressionable offspring.

Differentiation of teeth is another important mammalian trait. Whereas the teeth of reptiles are generally conical and uniform in size, the teeth of mammals come in a variety of sizes and shapes adapted for chewing many kinds of foods. Our own dentition, for example, includes a mixture of knifelike teeth (incisors), modified for shearing, and grinding teeth (molars), specialized for crushing.

The Evolution of Mammals

Mammals evolved from reptilian stock even earlier than birds. The oldest fossils believed to be mammalian date back 220 million years, into the Triassic period. The ancestors of mammals were among the mammal-like reptiles known as therapsids (Figure 30.23). The therapsids disappeared during the reign of the dinosaurs, but their mammalian offshoots coexisted with the dinosaurs throughout the Mesozoic era. Mesozoic mammals were very small—about the size of shrews—and most probably ate insects. A variety of evidence, such as the size of the eye sockets, suggests that these tiny mammals were nocturnal.

The extinction of the dinosaurs at the close of the Mesozoic era reopened many adaptive zones for new animal forms, and mammals underwent an extensive adaptive radiation that filled the void. The flora was also changing. Flowering plants became abundant during the Cretaceous and replaced gymnosperms as the dominant land plants in most locations. (Some scientists speculate that this change in vegetation contributed to the decline of the dinosaurs, which may have been dependent on gymnosperms for food.) The transformations of the Cretaceous marked the end of the Mesozoic era. As the Cenozoic era dawned, mammals continued to diversify. That diversity is represented today in the three major groups: monotremes (egg-laying mammals), marsupials (mammals with pouches), and placental mammals (Figure 30.24).

(a) A monotreme

(b) A marsupial

(c) A placental mammal

Figure 30.24
The three major subclasses of mammals. (**a**) Monotremes, such as this echidna, are the only mammals that lay eggs (inset). Monotremes have hair and milk, but no nipples; young suckle through the skin of the mother. (**b**) The young of marsupials, such as this kangaroo, are born very early in their development and finish their growth while nursing from a teat in their mother's pouch. (**c**) In placental mammals, such as this zebra, young develop within the uterus of the mother. There they are nurtured by the flow of maternal blood through the dense capillary network of the placenta, the reddish portion of the afterbirth clinging to this newborn zebra.

Monotremes

The **monotremes,** the platypuses and the echidnas (spiny anteaters), are the only living mammals that lay eggs (Figure 30.24a). The egg, which is reptilian in structure and development, contains enough yolk to nourish the developing embryo. Most mammals maintain a temperature of 30°C to 39°C, but the internal temperature of a platypus is only about 30°C. Monotremes have hair and milk, two of the most important trademarks of Mammalia. On the belly of a monotreme mother are specialized glands that secrete milk. After hatching, the baby sucks the milk from the fur of the mother, who has no nipples. The mixture of ancestral reptilian traits and derived traits of mammals suggests that monotremes descended from a very early branch in the mammalian genealogy. Today, monotremes are found only in Australia and New Guinea.

Marsupials

Opossums, kangaroos, and koalas are examples of **marsupials,** mammals that complete their embryonic development in a maternal pouch called a **marsupium** (Figure 30.24b). The egg contains a moderate amount of yolk that feeds the embryo as it begins its development within the mother's reproductive tract. Marsupials are born very early in their development. A red kangaroo, for instance, is about the size of a honeybee at its birth, just 33 days after fertilization. Its hind legs are merely buds, but the forelimbs are strong enough to crawl from the exit of the reproductive tract to the mother's pouch, a journey lasting a few minutes. In the pouch, the newborn fixes its mouth to a teat and completes its development while nursing.

In Australia, marsupials have radiated and filled niches occupied by placental mammals in other parts of the world. As we might expect, convergent evolution has resulted in a diversity of marsupials that resemble their placental counterparts who have similar ecological roles (see Chapter 23). The opossums of North and South America are the only extant marsupials outside the Australian region, though South America had a diverse marsupial fauna throughout the Tertiary period. The distribution of modern and fossil marsupials begins to make sense in the context of plate tectonics and continental drift. According to recent fossil evidence, marsupials probably originated in what is now North America, spreading southward when the land masses were still joined. After the breakup of Pangaea, South America and Australia became island continents, and their marsupials diversified in isolation from the placental mammals that began an adaptive radiation on the northern continents. Australia has not been in contact with another continent since early in the Cenozoic era, about 65 million years ago. The South American fauna, meanwhile, has not remained cloistered. Placental mammals reached South America throughout the Cenozoic. The most important migrations occurred about 12 million years ago and then again about 3 million years ago, when North and South America joined at the Panamanian isthmus, and extensive two-way traffic of animals took place over the land bridge. The distribution of mammals is another example of the interplay of biological and geological evolution.

Placental Mammals

Placental mammals complete their embryonic development within the uterus, joined to the mother by the placenta (Figure 30.24c). Adaptive radiation during late Cretaceous and early Tertiary periods (about 70 to 45 million years ago) produced the orders of placental mammals we recognize today (Table 30.2 on pp. 658–659). The phylogenetic relationship between marsupials and placental mammals is somewhat obscure, but scientists believe they are more closely related to each other than either is to the monotremes. Fossil evidence indicates placentals and marsupials may have diverged from a common ancestor about 80 to 100 million years ago.

The evolutionary relationships among the many orders of placental mammals are, for the most part, also unsettled. Most mammalogists now favor a tentative genealogy that recognizes at least four main evolutionary lines of placental mammals.

One branch of placental mammals consists of those in the orders Chiroptera (bats) and Insectivora (shrews), which resemble early mammals. Bats, whose forelimbs are modified as wings, probably evolved from insectivores that fed on flying insects. In addition to the insect eaters, some bat species feed on fruit, whereas others bite mammals and lap up their blood. Most bats are nocturnal.

The second branch starts with a lineage of medium-sized herbivores that underwent a spectacular adaptive radiation during the Tertiary period, eventually giving rise to such modern orders as lagomorphs (rabbits and their relatives); perissodactyls (odd-toed ungulates, including horses and rhinoceroses; ungulates walk on the tips of toes); artiodactyls (even-toed ungulates, including deer and swine); sirenians (sea cows); proboscideans (elephants); and cetaceans (porpoises and whales). Porpoises and some whales feed on fish and squids, but the largest whales are suspension-feeders that trap huge quantities of planktonic crustaceans in grilles called baleens suspended from the roof of the whale's mouth. Some blue whales are more than 30 m long and weigh as much as 25 elephants, making them the largest animals ever to live. Cetaceans seem to be very intelligent and social, using

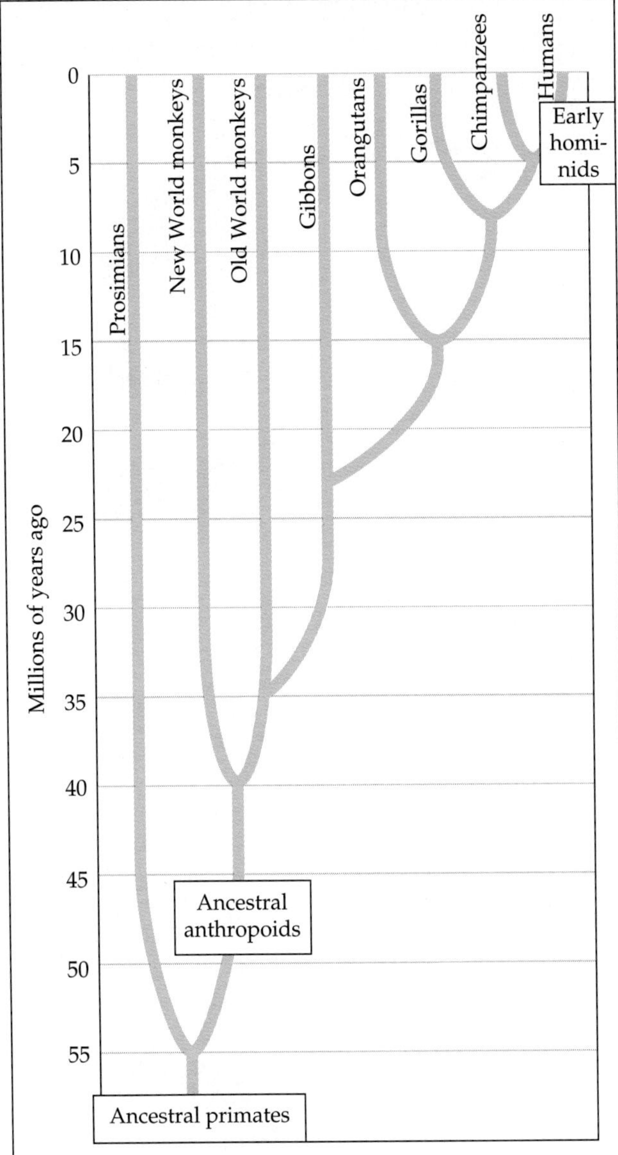

Figure 30.25
One version of primate evolution.

sound to communicate. Whales also produce high-pitched sounds that function in sonar navigation.

A third branch in the evolution of placental mammals produced the order Carnivora, which includes the cats, dogs, raccoons, skunks, and the pinnipeds (seals, sea lions, and walruses). The Carnivora probably first appeared during the early Cenozoic era. Seals and their relatives apparently evolved from middle Cenozoic carnivores that became adapted for swimming.

A fourth and very extensive adaptive radiation of placental mammals produced the primate-rodent complex. The order Rodentia includes rats, squirrels, and beavers. The order Primates includes monkeys, apes, and humans. One version of primate evolution is shown in Figure 30.25.

THE HUMAN ANCESTRY

We have now tracked the vertebrate genealogy to the mammalian order that includes *Homo sapiens* and its closest kin. We are primates, and to learn what that means, we must trace our ancestry back to the trees, where some of our most treasured traits had their beginnings as adaptations to arboreal existence.

Evolutionary Trends in Primates

The first primates were probably small arboreal mammals, and dental structure suggests that primates descended from insectivores late in the Cretaceous period. A fossil species called *Purgatorius unio,* found in Montana in beds of the Cretaceous/Tertiary boundary, is considered by many authorities to be the oldest known primate. Thus, by the end of the Mesozoic era, some 65 million years ago, our order was already defined by characteristics that had been shaped, through natural selection, by the demands of living in the trees.

For example, primates have limber shoulder joints, which make it possible to **brachiate** (swing from one hold to another). The dexterous hands of primates can hang onto branches and manipulate food. Claws have been replaced by nails in many species, and the fingers are very sensitive. The eyes of primates are close together on the front of the face. The overlapping fields of vision of the two eyes enhance depth perception, an obvious advantage when brachiating. Excellent eye-hand coordination is also important for arboreal maneuvering.

Parental care would seem to be essential for young animals in the trees. Mammals devote more energy to caring for their young than most other vertebrates, and primates are among the most devoted parents of all mammals. Most primates have single births and nurture their offspring for a long time. Though humans do not live in trees, we retain in modified form many of the traits that evolved there.

Modern Primates

Two commonly recognized suborders of the Primates are the Prosimii and Anthropoidea. The prosimians ("premonkeys") probably resemble early arboreal primates. The lemurs of Madagascar and the lorises, pottos, and tarsiers that live in tropical Africa and southern Asia are examples of prosimians (Figure 30.26). The anthropoids include monkeys, apes, and humans.

The first anthropoids to appear in the fossil record are monkeylike primates that probably evolved from prosimian stock about 40 million years ago in Africa and possibly even earlier in Asia. Africa and South America had already drifted apart, and it is not clear if

Table 30.2 Major Orders of Placental Mammals

Order	Major Characteristics	Examples	
Artiodactyla	Possess hooves with an even number of toes on each foot; herbivorous.	Sheep, pigs, cattle, deer, giraffes	Bighorn sheep
Carnivora	Carnivorous; possess sharp, pointed canine teeth and molars for shearing.	Dogs, wolves, bears, cats, weasels, otters, seals, walruses	Coyote
Cetacea	Marine forms with fish-shaped bodies, paddlelike forelimbs and no hind limbs; thick layer of insulating blubber.	Whales, dolphins, porpoises	Pacific white-sided porpoise
Chiroptera	Adapted for flying; possess a broad skinfold that extends from elongated fingers to body and legs.	Bats	Frog-eating bat
Edentata	Have reduced or no teeth.	Sloths, anteaters, armadillos	Tamandua
Insectivora	Insect-eating mammals	Moles, shrews, hedgehogs	Star-nosed mole
Lagomorpha	Posses chisel-like incisors, hind legs longer than forelegs and adapted for jumping.	Rabbits, hares, pikas	Jackrabbit

Table 30.2 *(Continued)*

Order	Major Characteristics	Examples	
Perissodactyla	Possess hooves with an odd number of toes on each foot; herbivorous.	Horses, zebras, tapirs, rhinoceroses	Indian rhinoceros
Primates	Opposable thumb; forward-facing eyes; well-developed cerebral cortex; omnivorous.	Lemurs, monkeys, apes, humans	Golden lion tamarin
Proboscidea	Have a long, muscular trunk; thick, loose skin; upper incisors elongated as tusks.	Elephants	African elephant
Sirenia	Aquatic herbivores; possess finlike forelimbs and no hind limbs.	Sea cows (manatees)	Manatee
Rodentia	Possess chisel-like, continuously growing incisor teeth.	Squirrels, beavers, rats, porcupines, mice	Red squirrel

Figure 30.26 The slender loris, a prosimian.

ancestors of New World monkeys reached South America by rafting on logs or other debris from Africa or by migration southward from North America. What is certain is that New World monkeys and Old World monkeys have been evolving along separate pathways for many millions of years (Figure 30.27). All New World monkeys are arboreal, whereas Old World monkeys include ground-dwelling as well as arboreal species. Most monkeys of both groups are diurnal (active during the day) and usually live in bands held together by social behavior.

(a)

(b)

Figure 30.27
A comparison of New World monkeys and Old World monkeys. (**a**) New World monkeys, such as spider monkeys, squirrel monkeys, and capuchins, have prehensile tails and nostrils that open to the sides. This is a Mexican spider monkey. (**b**) Old World monkeys lack prehensile tails, and their nostrils open downward. The tough seat pad is unique to the Old World group, which includes macaques, mandrills, baboons, and rhesus monkeys. This is a family of baboons.

There are four genera of apes, shown in Figure 30.28: *Hylobates* (gibbons), *Pongo* (orangutans), *Gorilla* (gorillas), and *Pan* (chimpanzees). Modern apes are confined exclusively to tropical regions of the Old World. With the exception of gibbons, modern apes are larger than monkeys, with relatively long legs and short arms and no tails. Although all the apes are capable of brachiation, only gibbons and orangutans preserve a primarily arboreal existence. Social organization varies among the genera of apes; gorillas and chimpanzees are highly social. Apes have larger brains than monkeys, and their behavior is consequently more adaptable.

The Emergence of Humankind

Paleoanthropology, the study of human origins and evolution, has a checkered history. Until about 20 years ago, researchers often gave new names to fossil forms that were undoubtedly the same species as fossils found by competing scientists. Elaborate theories have often been based on a few teeth or a fragment of jawbone. During the early part of this century, baseless speculations spawned many misconceptions about human evolution that still persist in the minds of much of the general population, long after these myths have been overthrown by fossil discoveries.

Some Popular Misconceptions Let's first dispose of the myth that our ancestors were chimpanzees or any other modern apes. Chimpanzees and humans represent two divergent branches of the anthropoid tree that evolved from a common, less-specialized ancestor.

Another misconception envisions human evolution as a ladder with a series of steps leading directly from an ancestral anthropoid to *Homo sapiens.* This is often illustrated as a parade of fossil hominids (members of the human family) becoming progressively more modern as they march across the page. If human evolution is a parade, then it is a disorderly one, with many splinter groups having traveled down dead ends. At times in hominid history, several different human species coexisted (Figure 30.29). Human phylogeny is more like a multibranched bush than a ladder, our species being the tip of the only twig that still lives. If the punctuated mode of evolution holds for humans, most change occurred as new hominid species came into existence, not by phyletic change within an unbranched hominid lineage.

One more myth we must bury is the notion that various human characteristics, such as upright posture and an enlarged brain, evolved in unison. A popular image is of early humans as half-stooped, half-witted cave-dwellers. Different features evolved at different rates—a phenomenon known as **mosaic evolution**—with erect posture, or bipedalism, leading the way. Our pedigree includes ancestors who walked upright but had brains as small as an ape's.

After dismissing some of the folklore on human evolution, however, we must admit that present understanding of our ancestry remains somewhat muddled.

Early Anthropoids The oldest known fossils of apes have been assigned to the genus *Aegyptopithecus,* the "dawn ape." This anthropoid was a cat-sized tree-dweller that lived about 35 million years ago. During

Figure 30.28
Apes. (**a**) Gibbons have long arms and are among the most acrobatic of all primates. These Asian primates are also the only monogamous apes. (**b**) The orangutan is a shy and solitary ape that lives in the rain forests of Sumatra and Borneo. Orangutans spend most of their time in trees, but they do venture onto the forest floor occasionally. (**c**) The gorilla is the largest ape, with some males almost 2 m tall and weighing about 200 kg. These gentle herbivores are confined to Africa, where they usually live in small groups of about ten to twenty individuals. (**d**) The chimpanzee lives in tropical Africa. Chimpanzees feed and sleep in trees, but also spend a great deal of time on the ground. Chimpanzees are intelligent, communicative, and social. Some biochemical evidence indicates that chimpanzees are more closely related to humans than they are to other apes.

the Miocene epoch, which began about 25 million years ago, descendants of the first apes diversified and spread into Eurasia. About 20 million years ago, the Indian plate collided with Asia and thrust up the Himalayan range. The climate became drier and the forests of what is now Africa and Asia contracted. During the late Miocene or early Pliocene, some of the apes may have made an evolutionary descent from the trees and begun living on the edge of the forest, foraging for food on the adjacent savanna (grassy plains dotted with trees). One of these African anthropoids was the ancestor of humans. Based on the fossil record and comparisons of DNA between humans and chimpanzees, most anthropologists now agree that humans and apes diverged from a common ancestor only about 5 million years ago.

Australopithecines: The First Humans In 1924, British anthropologist Raymond Dart announced that a fossilized skull discovered in a South African quarry was the remains of an early human. He named his "ape-man" *Australopithecus africanus* ("southern ape of Africa"). With the discovery of more fossils, it became clear that *Australopithecus* was a legitimate hominid that walked fully erect and had humanlike hands and

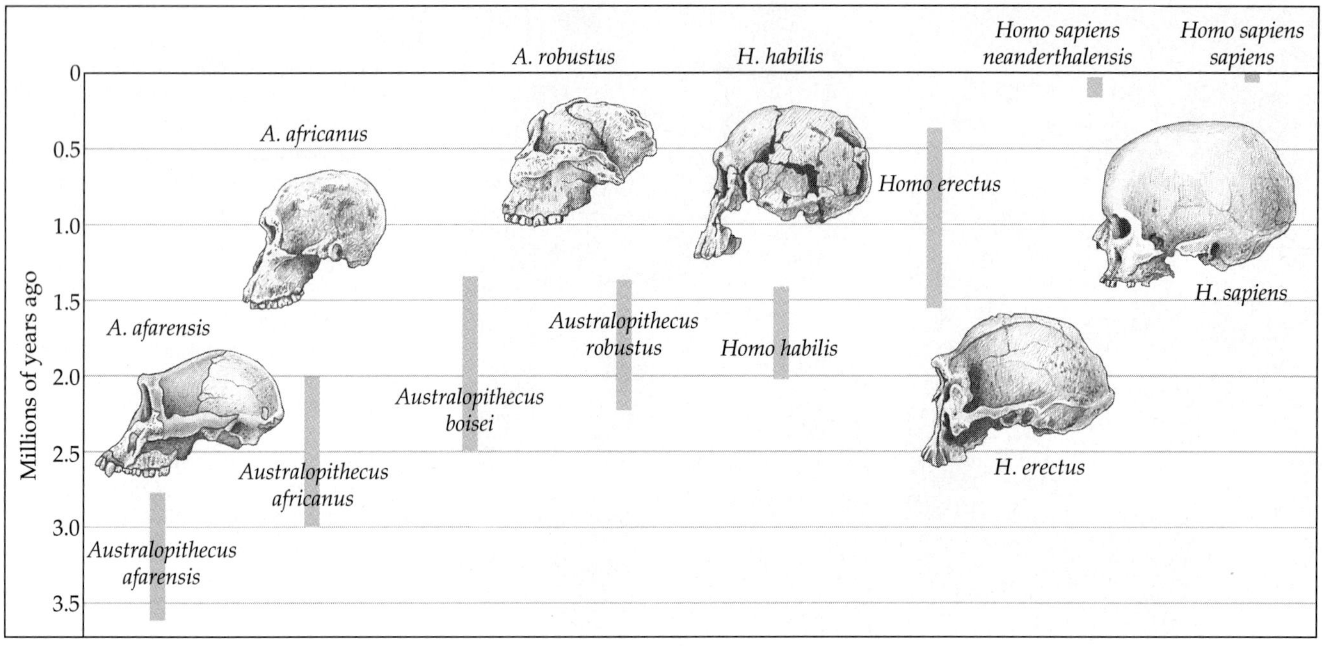

Figure 30.29
A time line of some hominid species. Notice that there have been times in the history of human evolution when two or more hominids coexisted.

teeth. However, the brain of *Australopithecus* was only about one-third the size of that of a modern human. *Australopithecus* foraged on the African savannas for nearly 2 million years, beginning at least 3 million years ago. Two forms existed, one slender and the other more robust; some paleoanthropologists assign these two forms to different species.

In 1974, in the Afar region of Ethiopia, an *Australopithecus* skeleton that was 40% complete was discovered (Figure 30.30). "Lucy," as the fossil was named, was petite—only about 1 m tall, with a head about the size of a softball. Lucy and similar fossils have been considered sufficiently different from *Australopithecus africanus* to be named a separate species, *Australopithecus afarensis* (for the Afar region). The placement of *A. afarensis* on the hominid bush has

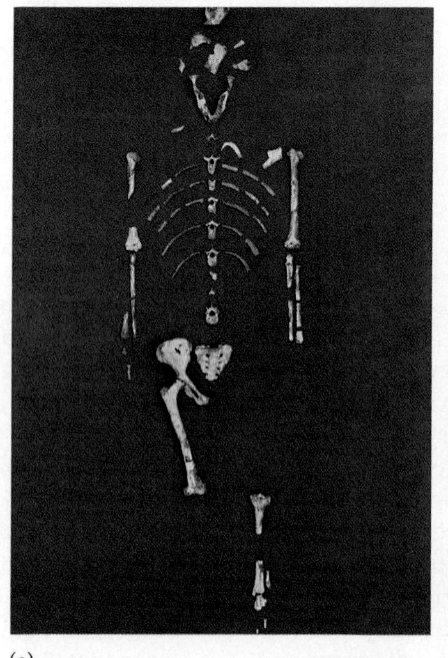
(a)

(b)

Figure 30.30
The antiquity of upright posture. (a) "Lucy" *(Australopithecus afarensis)* and her kind lived about 2.8 to 3.6 million years ago. They walked upright but retained many apelike traits, including curved toes. **(b)** The Laetoli footprints, over 3.5 million years old, confirm that upright posture evolved quite early in hominid history.

stirred considerable debate. Lucy lived about 3 million years ago, but similar fossils a million years older that may represent the same species have also been discovered. This makes *A. afarensis* the most ancient of the known australopithecines. According to one tentative genealogy, *A. afarensis* was the common ancestor of two hominid branches, one leading to *A. africanus* and the other to *Homo.* Other anthropologists, who question whether the older fossils and Lucy were one species, argue that *A. afarensis* (Lucy) was part of an australopithecine branch that had already diverged from the lineage leading to *Homo,* our genus. In any case, the discovery of Lucy confirms that upright posture evolved relatively early in hominid history. This important conclusion is supported by fossilized footprints more than 3.5 million years old that were discovered in Laetoli, Tanzania.

Australopithecus walked erect for more than a million years with no substantial enlargement of the brain. Perhaps an upright posture freed the hands for gathering food or caring for babies. The making of tools came much later. In one of his essays, Stephen Jay Gould put it this way: "Mankind stood up first and got smart later."

Enlargement of the human brain is first evident in fossils dating back to the latter part of australopithecine times, about 2 million years ago. Skulls have been found with brain capacities of about 650 cubic centimeters (cc), compared to 500 cc for *Australopithecus africanus.* Simple stone tools are sometimes found with the larger-brained fossils. Some paleoanthropologists believe the advances are great enough to place these fossils in the genus *Homo,* naming it *Homo habilis* ("handy man"). Other scientists argue that the handy man was just a variant of *Australopithecus* and that australopithecines were the first hominids to use and make tools. Regardless of the taxonomic debate, the message of the fossils is clear: After walking upright for more than a million years, hominids were finally beginning to use their brains and hands to fashion tools.

Homo habilis coexisted for nearly a million years with the smaller-brained *Australopithecus,* including *Australopithecus robustus,* named for its stocky features and heavy skull. Early *Homo* and australopithecines may not have competed directly, although all may have scavenged for food, hunted animals, and gathered fruits and vegetables. According to one theory of human origins, the late australopithecines and *Homo habilis* were distinct lines of hominids, neither evolving from the other. If this scenario is correct, then *A. africanus* was an evolutionary dead end, but *H. habilis* was on the path to modern humans, leading first to *Homo erectus,* which later gave rise to *Homo sapiens.*

***Homo erectus* and Descendants** The first hominid to migrate out of Africa into Asia and Europe was *Homo erectus* ("upright man"). The fossils known as Java Man and Beijing Man are examples of this species. *Homo erectus* was taller than *H. habilis* and had a larger brain capacity. Fossils of *H. erectus* range in age from 1.6 million years to about 300,000 years. During that period, the brain continued to expand to as large as 1200 cc, which overlaps the normal range of brain capacity for modern humans.

To survive in the colder climates of the north, humans had to live by their wits. *Homo erectus* resided in huts or caves, built fires, clothed themselves in animal skins, and designed stone tools that were more elaborate than the tools of *Homo habilis.* In anatomical and physiological adaptations, *H. erectus* was poorly equipped for life outside the tropics but made up for the deficiencies with intelligence and social cooperation. Some African, Asian, European, and Australasian (Indonesia, New Guinea, and Australia) populations of *H. erectus* gave rise to regionally diverse descendants that had even larger brains.

Among the descendants of *H. erectus* were the Neanderthals, who lived in Europe, the Middle East, and parts of Asia from about 130,000 years ago to about 35,000 years ago. (They are named Neanderthals because their fossils were first found in the Neander Valley of Germany.) Compared with us, Neanderthals had slightly heavier brow ridges and less pronounced chins, but their brains, on average, were slightly larger than ours. Neanderthals were skilled toolmakers, and they participated in burials and other rituals that required abstract thought. Much current research on Neanderthal skulls addresses an intriguing question: Did Neanderthals have the anatomical equipment necessary for speech?

The oldest known post–*H. erectus* fossils, dating back over 300,000 years, are found in Africa. Many paleoanthropologists group these African fossils along with Neanderthals and various Asian and Australasian fossils as the earliest forms of our species, *Homo sapiens.* These regionally diverse descendants of *H. erectus* are sometimes referred to as "archaic *Homo sapiens.*"

The Emergence of *Homo sapiens:* Out of Africa . . . But When? What was the fate of Neanderthals and their contemporaries who populated different parts of the world? In the view of many anthropologists, these archaic *Homo sapiens* gave rise to fully modern humans. According to this model, called the *multiregional model,* modern humans evolved in parallel in different parts of the world. If this view is correct, then the geographic diversity of humans originated relatively early, when *Homo erectus* spread from Africa into the other continents more than a million years ago. Advocates of this model see the distinctive facial features and other anatomical characteristics of each modern race as the result of this multiregional emergence of *Homo sapiens.* The model accounts for the great genetic similarity of all modern people by pointing out that occasional interbreeding among neighboring populations has always provided corridors for gene flow throughout the entire geographic range of humanity (see Chapter 21).

Based on alternative interpretations of the fossil record, some paleoanthropologists, including Christopher Stringer of University College in London, began to question the multiregional model in the past decade. The debate has focused partly on the relationship between Neanderthals and the modern humans of Europe and the Middle East. Perhaps the most famous fossils of modern *Homo sapiens*—skulls and other bones that look essentially like those of today's humans—are the remains of Cro-Magnons. Named for the French cave in which these fossils were first discovered, Cro-Magnons date back about 35,000 years. But the oldest fully modern fossils of *Homo sapiens* are found in Africa. They are about 100,000 years old, and similar fossils almost as ancient have also been discovered in caves in Israel. The Israeli fossils were found not far from other caves containing Neanderthal-like fossils ranging in vintage from about 120,000 to 60,000 years old. These two human types apparently coexisted in this region for at least 40,000 years, from the time modern forms appeared 100,000 years ago. And, according to Stringer and his supporters, the persistence of distinct Neanderthal and modern forms during their time of coexistence means that these two human types did not interbreed. If this interpretation of the Israeli fossils is correct, then the Neanderthals of that region could not have been the ancestors of the modern humans that also lived there. In fact, based on the Israeli bones and other fossil data, Stringer postulates that Neanderthals and the other so-called archaic *Homo sapiens* outside of Africa were evolutionary dead ends. Anthropologists who favor this interpretation contend that modern humans evolved from *Homo erectus* first in Africa, then migrated to other continents, replacing Neanderthals and the other regional descendants of *H. erectus*. This model, called the **monogenesis model,** contrasts sharply with the multiregional model. According to the monogenesis model, modern humanity did not emerge in many different parts of the world, but only in Africa, from which the modern humans dispersed relatively recently. If this view is correct, then the geographic diversification of modern humans into what we now recognize as races occurred within just the past 100,000 years (compared to a million years for the multiregional model).

In the late 1980s, Rebecca Cann and other geneticists in the laboratory of the late Allan Wilson at the University of California, Berkeley, made news with research results that seemed to support the monogenesis model of human origins. Instead of digging in the field for bones, Wilson's group probed in the DNA of living humans for clues about our ancestry. Specifically, these scientists compared the mitochondrial DNA (mtDNA) of a multiracial sample of more than 100 people representing four continents. The greater the difference between the mtDNA of two people, the longer ago that mtDNA diverged from a common source. Using a computer to analyze their data, the geneticists traced the source of all human mtDNA back to Africa. The surprise was the calculation that the divergence of mtDNA from this common source began just 200,000 years ago, much too late to represent the dispersal of *Homo erectus*. Advocates of the monogenesis model cheered the genetic evidence for a second, much later dispersal of Africans in the form of modern humans.

In 1992, several researchers challenged the Berkeley group's interpretation of the mtDNA data. At issue are the methods used to construct evolutionary trees from the mtDNA comparisons and the reliability of changes in mtDNA as a molecular clock for timing branch points in these trees (see Chapter 23). The criticism has encouraged proponents of the multiregional model, including Milford Wolpoff of the University of Michigan. He and many other paleoanthropologists continue to argue that multiregional evolution of modern humans fits the fossil evidence better than does African monogenesis. These scientists interpret certain fossils in various parts of the world as links between the regional versions of archaic *Homo sapiens* and the current people native to those continents. And advocates of the multiregional model do not see the Israeli fossils as evidence of long-time coexistence of Neanderthals and modern humans in that area. Instead, they interpret the Israeli collection as intergrading fossils left by interbreeding populations of archaic *Homo sapiens* that immigrated to a common region and collectively gave rise to one of the localized forms of modern humans.

And so the debate about the origin of modern humans continues (Figure 30.31). We can rephrase the problem this way: Are all contemporary people more closely related to one another than any population native to a particular continent is to the archaic *Homo sapiens* that lived earlier in that part of the world? The multiregional and monogenesis models predict opposite answers to this question. The ability to test these predictions may actually be within reach. Biochemists are now trying to recover DNA samples from fossilized skulls of Neanderthals and other archaic *H. sapiens*. The next step would be to use polymerase chain reaction (PCR) to amplify the DNA traces, producing enough for comparison to the DNA of living humans (see Chapter 19). Molecular biology and paleontology are converging on the question of human origins. *Wherever*, *whenever*, and *however* modern humanity evolved, it eventually transformed environments in all parts of the world.

Cultural Evolution: A New Force in the History of Life An erect stance was the most radical anatomical change in our evolution; it required major remodeling of the foot, pelvis, and vertebral column.

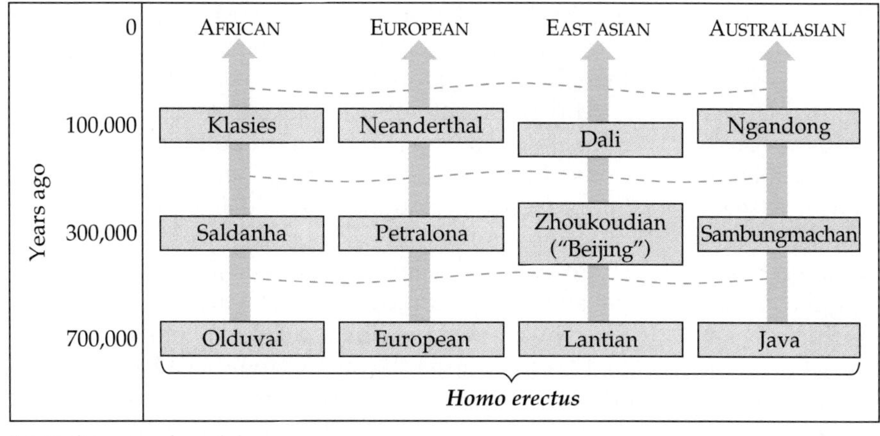

(**a**) Multiregional model

Milford Wolpoff

Years ago
0
100,000
300,000
700,000
AFRICAN
EUROPEAN
EAST ASIAN
AUSTRALASIAN
Klasies
Neanderthal
Dali
Ngandong
Saldanha
Petralona
Zhoukoudian ("Beijing")
Sambungmachan
Olduvai
European
Lantian
Java
Homo erectus

(**b**) Monogenesis model

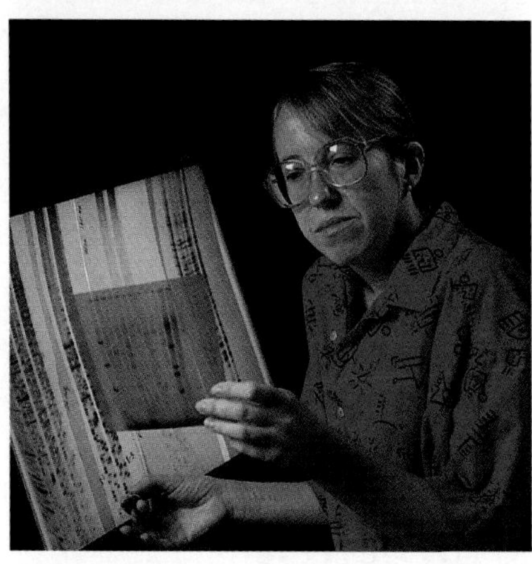

Rebecca Cann

Figure 30.31
Two models for the origin of modern humans. (**a**) According to the multiregional model, modern humans evolved in many parts of the world from regional descendants of *Homo erectus*, who dispersed from Africa about a million years ago. The boxed names indicate various fossils associated with each region. The dashed lines symbolize interbreeding and gene flow between different populations. Milford Wolpoff of the University of Michigan represents the many paleoanthropologists who believe that the multiregional model is the best interpretation of the human fossil record. (**b**) Many other paleoanthropologists interpret the fossil record in favor of a monogenesis model for the origin of modern humans. According to this model, only the African descendants of *Homo erectus* gave rise to modern humans. All other regional descendants of *H. erectus*, including Neanderthals, became extinct without contributing to the gene pool of modern humanity. Advocates of the monogenesis model argue that modern humans began spreading from Africa just 100,000 years ago, giving rise to all the diverse populations of contemporary humans. Rebecca Cann, now at the University of Hawaii, is one of the geneticists who, based on her study of mitochondrial DNA, has joined the monogenesis camp.

Enlargement of the brain was a secondary alteration made possible by prolonging the growth period of the skull and its contents. The brains of mammalian fetuses grow rapidly, but the growth usually slows down and stops not long after birth. The primate brain continues to grow after birth, and the period of growth is longer for a human than for any other primate. The extended period of human development also lengthens the time parents care for their offspring, which contributes to the child's ability to repeat the experiences of earlier generations. This is the basis of culture—the transmission of accumulated knowledge over the generations. The major means of this transmission is language, written and spoken.

The first stage in our cultural evolution began with nomads who hunted and gathered food on the African grasslands. They made tools, organized communal activities, and divided labor. The second stage came with the development of agriculture in Eurasia and the Americas about 10,000 to 15,000 years ago. Along with agriculture came permanent settlements and the first cities. The third major stage in our cultural evolution was the Industrial Revolution, which began in the eighteenth century. Since then, new technology and human population have escalated exponentially; the lives of some people spanned the flight of the Wright brothers and Neil Armstrong's walk on the moon. Through all this cultural evolution, from simple hunter-gatherers to high-tech societies and an overpopulated world, we have not changed biologically in any significant way. We are probably no more intelligent than our forebears in Africa and in Eurasian caves. The same toolmaker who chipped away at stones now fashions microchips. The know-how to build skyscrapers, computers, and spaceships is stored not in our genes but in the cumulative product of hundreds of generations of human experience, passed along by parents, teachers, and books.

Evolution of the human brain may have been anatomically simpler than acquiring an upright stance, but the consequences of cerebral expansion have been enormous. Cultural evolution made *Homo sapiens* a new force in the history of life—a species that could defy its physical limitations and shortcut biological evolution. We do not have to wait to adapt to an environment through natural selection; we simply change the environment to meet our needs. We are the most numerous and widespread of all large animals, and everywhere we go, we bring environmental change. There is nothing new about environmental change. As we have seen in this unit, the history of life is the story of biological evolution on a changing planet. But it is unlikely that change has ever been as rapid as in the age of humans. Cultural evolution outpaces biological evolution by orders of magnitude. We may be changing the world faster than many species can adapt; the rate of extinctions in this century is fifty times greater than the average for the past 100,000 years.

This rapid rate of extinction is mainly a result of habitat destruction and chemical pollution, both functions of human cultural changes and overpopulation. Simply feeding, clothing, and housing the more than 5 billion people who now exist imposes an enormous strain on Earth's capacity to sustain life. If all these people suddenly assumed the high standard of living enjoyed by many people in developed nations, it is likely that Earth's support systems would be overwhelmed. Already, for example, current rates of fossil-fuel consumption, mainly by developed nations, are so great that waste carbon dioxide may be causing the temperature of the atmosphere to increase enough to alter world climates (see Chapter 49). Today, not just individual species, but entire ecosystems, the global atmosphere, and the oceans, are seriously threatened. Tropical rain forests, which play a vital role in moderating global weather, are being cut down at a startling rate. Scientists have hardly begun to study these ecosystems, and many species in them may become extinct before they are even discovered. Of the many crises in the history of life, the impact of one species, *Homo sapiens*, is the latest and potentially the most devastating.

STUDY OUTLINE

1. Vertebrates, the backboned animals, are a subphylum of the chordates.

Phylum Chordata (pp. 635–636)

1. Chordates are a diverse group of deuterostomes, classified together by virtue of shared structures: a notochord, a dorsal hollow nerve cord, pharyngeal slits, and a postanal tail.

Chordates Without Backbones (pp. 636–637)

1. Subphylum Cephalochordata is an invertebrate group best known for the lancelet, an exemplary chordate.
2. Subphylum Urochordata, also invertebrate, includes the marine, suspension-feeding tunicates, or sea squirts.

The Origin of Vertebrates (pp. 637–638)

1. The first vertebrates may have arisen during the early Cambrian period from a suspension-feeding ancestor resembling a tunicate larva that became sexually mature before metamorphosis.

Vertebrate Characteristics (pp. 638–640)

1. One hallmark of the vertebrates is the development of highly specialized brains.
2. Most vertebrates possess a column of vertebrae that encloses the nerve cord. The skull and vertebral column are part of a bony or cartilaginous endoskeleton.
3. The vertebrate heart pumps blood through a closed circulatory system to gills, skin, or lungs, where it is oxygenated. Wastes in the blood are removed by kidneys.
4. Reproduction is almost always sexual, with separate sexes. Fertilization can be external or internal.

Class Agnatha (p. 640)

1. The oldest vertebrate fossils are jawless agnathans, a group with round or slitlike mouths, which include the armored ostracoderms. Today, class Agnatha is represented only by lampreys and hagfishes.

Class Placodermi (p. 641)

1. Placoderms, now extinct, were Devonian armored fishes with paired fins and hinged, biting jaws that evolved from the skeletal supports of gills.

Class Chondrichthyes (pp. 641–642)

1. Sharks and rays, the most widespread members of the class Chondrichthyes, have paired fins, cartilaginous skeletons, and biting jaws characteristic of the group.
2. Primarily carnivorous, the streamlined sharks have keen senses for vision, olfaction, hearing, and electroreception. A specialized lateral line system enables sharks to perceive minor vibrations.
3. Sharks have internal fertilization, and development can be oviparous, ovoviviparous, or viviparous.

Class Osteichthyes (pp. 643–644)

1. Osteichthyes is the most species-rich vertebrate class. The bony fishes have skeletons of calcium phosphate and a slimy, low-friction skin generally covered by flattened bony scales.
2. Unlike sharks, which must move to breathe, bony fishes can breathe while stationary by moving a flaplike operculum that helps draw water through the mouth and into the gill chamber.
3. Bony fishes can adjust their density and thus control their buoyancy by means of a unique swim bladder.
4. Although external fertilization and ovipary are common, bony fishes exhibit great reproductive variation.
5. The ancestor of bony fishes was likely a freshwater species, which by the end of the Devonian had spawned three distinct subclasses: ray-finned fishes, lobe-finned fishes, and lungfishes.
6. Most fish familiar to us are ray-fins. Supported by long, flexible rays, their fins are modified for various functions. During their history, various species of ray-fins have invaded the sea and returned again to fresh water.
7. Most lobe-finned fishes remained in fresh water and used lungs to aid their gills in breathing. Devonian lobe-finned fishes gave rise to amphibians.

Class Amphibia (pp. 645–648)

1. As the first terrestrial vertebrates, amphibians radiated during early Carboniferous times.
2. The lifestyles and life cycles of modern amphibians attest to their aquatic heritage. A moist skin complements the lungs in gas exchange. Oviparous forms lay their unshelled eggs in wet environments, though ovoviviparous and viviparous species also exist. Most frogs and their relatives undergo metamorphosis of an aquatic larval stage into a terrestrial adult.
3. The salamanders make up the tailed urodele order, an aquatic and terrestrial group that generally develops directly without metamorphosis.
4. Anurans (frogs) are named for their tail-less condition. Fertilization is external, and metamorphosis is common.
5. The apodans, or caecilians, are limbless, virtually blind aquatic or small burrowing species.

Class Reptilia (pp. 648–651)

1. Reptiles, a diverse group represented today by lizards, snakes, turtles, and crocodilians, have numerous terrestrial adaptations not present in amphibians.
2. Reptiles have lungs, are covered with waterproof scales, and have a shelled amniote egg that permits development in a dry environment.
3. Fertilization is internal for both viviparous and oviparous species.
4. Reptiles are ectothermic, regulating body temperature through behavioral modifications that either increase or decrease absorption of external heat.
5. The Age of Reptiles traces its origin to the Carboniferous cotylosaurs or "stem reptiles," lizardlike insectivores that evolved from Devonian amphibians.
6. The first major reptilian radiation produced the Permian synapsids, a group of terrestrial predators that gave rise to the mammal-like therapsids. Other descendants of stem reptiles, the thecodonts, survived the mass extinctions of the Permian period with the synapsids to become the stock for the evolution of dinosaurs, crocodilians, and birds.
7. The second great reptilian radiation produced flying pterosaurs and terrestrial dinosaurs from Triassic thecodonts.
8. Controversy exists about whether dinosaurs were endothermic, but there is no doubt that during the cooler, more climatically unstable Cretaceous period, something dramatic caused the abrupt demise of these magnificent reptiles.
9. The major reptilian orders that survived the Cretaceous crisis are the Chelonia, Squamata, and Crocodilia.
10. Chelonians, the turtles, generally hard-shelled reptiles that lay their eggs on land, have changed little since their evolution from stem reptiles.
11. Squamata comprises the lizards and snakes. Lizards include the most numerous and diverse reptiles today. Ancestral burrowing species gave rise to snakes.
12. Crocodilians, the crocodiles and alligators, are close relatives of the dinosaurs. Crocodilians are restricted to warm regions of the world.

Class Aves (pp. 651–653)

1. The amniote eggs and scaled legs of birds give testimony to their reptilian heritage; birds descended from dinosaurs.
2. Almost every part of avian anatomy enhances flight.
3. A body of low density is the result of honeycombed bones and lack of certain bilateral organs.
4. A heart that completely separates oxygen-poor from oxygen-rich blood and a highly efficient lung and air sac system make possible the active, endothermic metabolism required for flight.
5. Flight requires fine coordination and acute senses. Vision has reached its peak development in birds. Birds also have relatively large brains and display complex behavior, especially when breeding.
6. Fertilization is internal, and eggs are laid and kept warm by brooding until hatching.
7. Only birds have feathers, which shape wings into airfoils that provide both propulsion and lift.

Class Mammalia (pp. 653–657)

1. Hair and mammary glands are the two diagnostic characteristics of mammals. Like feathers, hair is made of the protein keratin and helps insulate the body.

2. An active endothermic metabolism is supported by a four-chambered heart and respiratory system ventilated by a muscular diaphragm.

3. Fertilization is internal, and most mammals give birth to young after development in the uterus.

4. Of all vertebrates, mammals are endowed with the largest brains and the greatest capability of learning. Relatively long periods of parental care provide the opportunity for parental teaching.

5. Mammals are characterized by highly differentiated teeth, which are adapted for chewing many kinds of foods.

6. Small, insect-eating mammals arose from the therapsids and coexisted with the dinosaurs throughout the Mesozoic era. Extinction of the dinosaurs reopened many adaptive zones and catalyzed an extensive mammalian radiation.

7. The monotremes are egg-laying mammals, today represented only by the platypuses and the echidnas found in Australia and New Guinea.

8. The marsupials include opossums, kangaroos, and the koalas, animals whose young complete their embryonic development inside a maternal pouch, the marsupium, attached to a teat from which they get their nourishment. Marsupials isolated in Australia show convergent evolution with placentals in other parts of the world.

9. The most widespread and diverse modern mammals are the placentals, a group whose young complete their embryonic development attached to a placenta inside the mother's uterus. Placental mammals and marsupials may have shared a common ancestor 80 to 100 million years ago. Subsequently, placental mammals branched out into several main evolutionary lines.

The Human Ancestry (pp. 657–666)

1. The first primates were probably small arboreal animals, which descended from insectivores in the late Cretaceous period. Life in trees demanded limber shoulder joints; dexterous, sensitive hands and fingers; close-set eyes with overlapping fields of vision for depth perception; excellent eye-hand coordination; and extended parental care—all of which extant primates retain.

2. Two subgroups of modern primates are the prosimians—lemurs and their relatives—and the anthropoids. Anthropoids probably evolved from prosimian stock and diverged early into New World and Old World monkeys that maintained separate evolutionary pathways. Modern apes—gibbons, orangutans, gorillas, and chimpanzees—are confined to the Old World.

3. The earliest known ape was *Aegyptopithecus*, a cat-sized tree-dweller that gave rise to several forms. One of these was likely the ancestor of modern apes and humans.

4. The first humanlike fossil to be discovered was *Australopithecus africanus*, a hominid with a small brain that walked erect and had human-type hands and teeth. Subsequent discovery of "Lucy," a petite fossil classified as *Australopithecus afarensis*, has confirmed the early evolution of upright posture in the hominids.

5. Enlargement of the brain and the appearance of simple stone tools occurred about 2 million years ago, as indicated by fossils of *Homo habilis*, a species that coexisted with the smaller-brained *Australopithecus* and led to the evolution of *Homo erectus*.

6. *Homo erectus* was the first hominid to venture out of the tropics and into colder climates. Intelligence and social cooperation contributed to survival.

7. *Homo erectus* gave rise to regionally diverse versions of archaic *Homo sapiens*, including Neanderthals.

8. According to the multiregional model, modern humans evolved in several locations from Neanderthals and other archaic *Homo sapiens*. In contrast, the monogenesis model views all but the African archaic *Homo sapiens* as evolutionary dead ends. According to this model, a relatively recent dispersal (100,000 years ago) of modern Africans gave rise to today's human diversity.

9. Although the erect stance was the most radical anatomical change in our evolution, enlargement of the brain and its extended period of development, which required long periods of parental care, gave rise to language and the far-reaching consequences of culture. The first stage of cultural evolution, which began in Africa with wandering hunter-gatherers, progressed to the development of agriculture about 15,000 years ago. The third stage of cultural evolution was the Industrial Revolution, which today continues as accelerating technological change. The exploding human population now threatens Earth's ecosystems.

SELF-QUIZ

1. Which of the following is *not* a general characteristic of the phylum Chordata?
   a. dorsal, hollow nerve cord
   b. vertebral column
   c. notochord
   d. pharyngeal slits
   e. postanal tail

2. Which of the following groups is entirely extinct?
   a. cephalochordates
   b. agnathans
   c. placoderms
   d. lobe-finned fishes
   e. ratite birds

3. The amniote egg first evolved in
   a. bony fishes
   b. amphibians
   c. reptiles
   d. birds
   e. mammals

4. Mammals and birds share all of the following characteristics *except*
   a. endothermy
   b. descent from reptiles
   c. a four-chambered heart
   d. teeth specialized for diverse diets
   e. the ability of some species to fly

5. If you were to observe a monkey in a zoo, which characteristic would indicate a New World origin for that monkey species?

a. distinct "seat pads"
b. eyes close together on the front of the skull
c. use of the tail to hang from a tree limb
d. occasional bipedal walking
e. downward orientation of the nostrils

6. Only an animal species with a diaphragm can be expected to have
a. hair
b. feathers
c. scales
d. lungs
e. moist skin

7. Unlike placental mammals, both monotremes and marsupials
a. lack nipples
b. have some embryonic development outside the mother's uterus
c. lay eggs
d. are found in Australia and Africa
e. include only insectivores and herbivores

8. Mosaic evolution refers to the observation that
a. reptiles had two major periods of adaptive radiation
b. preexisting structures are often co-opted for new functions
c. different anatomical features unique to a species evolve at different rates
d. some fish that evolved in fresh water and moved to the seas have returned to fresh water
e. groups that evolve in separated geographical areas often develop superficially similar animal forms that fill the same niches

9. Which of the following is *not* thought to be ancestral to humans?
a. a reptile
b. a bony fish
c. a primate
d. an amphibian
e. a bird

10. As humans diverged from other primates, which of the following is thought to have appeared first?
a. the development of culture
b. language
c. an erect stance
d. toolmaking
e. an enlarged brain

CHALLENGE QUESTIONS

1. Compare and contrast bats and birds, two flying vertebrates, in as many ways as you can.

2. Throughout chordate evolution, many existing structures have been co-opted for new functions. Describe several such cases of preadaptation.

SCIENCE, TECHNOLOGY, AND SOCIETY

1. There are 90 million acres in the U.S. National Wildlife Refuge System, representing most of the habitats within the United States and protecting thousands of animal species. Many refuges permit logging, grazing, farming, mineral exploration, and recreational activities. Each year 400,000 animals are legally taken from refuges by hunters and trappers. Pressure is mounting to permit oil and gas exploration in the Arctic National Wildlife Refuge of Alaska. What level of human activity do you think should be permitted on wildlife refuges? Should they be open to "consumptive" uses, such as logging, farming, and hunting? Or should refuges be sanctuaries where animals are touched as little as possible by human activities? Give reasons for your answers.

2. A lot of money and attention has been directed toward saving a few endangered animals, such as the black-footed ferret and the California condor. (Millions of dollars have been spent in efforts to maintain and breed these two species—more than has been spent on all endangered invertebrates, reptiles, and plants put together.) Condors and ferrets have recently been returned to the wild, but it is questionable whether they can withstand the human encroachments that endangered them in the first place. Should we work so hard to save these animals? Why or why not? Would it be better to use our resources to save more species that have a better chance of survival?

FURTHER READING

Barinaga, M. "Where Have All the Froggies Gone?" *Science,* March 2, 1990. Amphibians are in global decline. What are the causes?
Barinaga, M. "African Eve Beats a Retreat." *Science,* February 7, 1992. Conflicting interpretations of what our mitochondrial DNA tells us about the origin of modern humans.
Carroll, R. C. *Vertebrate Paleontology and Evolution.* New York: Freeman, 1987. An authoritative yet accessible text.
Cavalli-Sforza, L. L. "Genes, Peoples and Languages." *Scientific American,* November 1991. Using the evolution of language to trace human culture.
Fischman, J. "One That Got Away." *Discover,* January 1992. Using molecular systematics to trace our vertebrate roots to fish.
Gibbons, A. "Neanderthal Language Debate: Tongues Wag Anew." *Science,* April 3, 1992. Could Neanderthals speak?
Gould, S. J. "An Earful of Jaw." *Natural History,* March 1990. Extracting general evolutionary lessons from the development of the vertebrate jaw.
Ross, P. E. "Eloquent Remains." *Scientific American,* May 1992. We are on the verge of genetic analysis of prehistoric humans.
Shell, E. R. "Flesh and Bone." *Discover,* December 1991. What role did women play in early human evolution?
Tattersall, I. "Evolution Comes to Life." *Scientific American,* August 1992. The science and art of making models of organisms based on fossilized skeletons.
"The Thunder Lizards." *Natural History,* December 1991. Several articles about how dinosaurs lived.
"Where Did Modern Humans Originate?" *Scientific American,* April 1992. A pair of essays—point/counterpoint on the multiregional-monogenesis debate.
Zimmer, C. "Ruffled Feathers." *Discover,* May 1992. The controversy about the origin of birds.

UNIT SIX

Plants: Form and Function

An Interview with Virginia Walbot

When my editor made the arrangements for this interview, Virginia Walbot said we should meet in her cornfield at Stanford University. By the time the interview ended, I understood the reason for the location. Dr. Walbot doesn't just work on corn; *she feels connected to these plants in ways that make her research much more fun, insightful, and significant. A professor of biology at Stanford, Virginia Walbot represents a tradition of scientists who have found corn to be the ideal organism for answering some of the fundamental questions about genetics and development. Barbara McClintock, who passed away in 1992, discovered transposable genetic elements in corn, and eventually received a Nobel Prize when biologists finally recognized the general significance of that revelation. And George Beadle, who shared a Nobel Prize with Edward Tatum for the "one gene-one enzyme" hypothesis, lived for many summers in a small house next to a cornfield—the same Stanford field where Virginia Walbot now grows her corn—so he could be close to his research plants. During our afternoon in that field, Dr. Walbot told us what her corn plants are teaching her about the genetic basis of plant development.*

How did you acquire an interest in plants?
I spent some of my formative years on my family's truck farm in southern California. I was always interested in plants, always had a garden. When I was an undergraduate here at Stanford, I got very interested in the big picture of plants and plant evolution. Then for graduate study, I worked with Ian Sussex at Yale, and his lab always had a lot of people with different special interests—plant development, plant molecular biology, plant genetics—and because of these diverse influences, I could see the whole biological world through my work. And I think that's probably one of my strengths, that I see what I do in relation to a lot of bigger questions.

What does your current research have to do with this cornfield?
My lab is interested in the molecular biology of the expression of corn genes—specifically, in the details of how individual genes are regulated during plant development. There are two approaches to that problem. One is to isolate genes and then study them in the lab; the other is to make mutations—or let the plant make mutations for you—that alter the expression of the gene, and then see what the gene can do throughout the life of the plant. The real power of recombinant DNA techniques is that you can alter genes in a test tube readily, and then study their behavior in some defined situation. So we look for a mutation in corn, and then we find out exactly what's wrong with it, and then we might make ten or twenty related mutations in the lab.

Why is corn a good organism for this kind of work?
Corn is the plant where it's easiest to find mutants and study them. The big advantage of corn is very obvious if you've ever looked at a corn plant. The ear contains the embryo sacs with eggs, and the tassel contains the pollen where the sperm are. We cover each ear with a bag, so there's no exposure of eggs to stray pollen, and then we cover the tassels with bags (we practice safe sex in the cornfield). So every cross that's done is completely under our control, and very easy. So one corn geneticist could make, say, 300 crosses in a morning, and if each ear has a few hundred seeds on it, it's very easy to generate a population of 20, 50, 100 thousand seeds, and in a week's time you could make a million seeds. So if you're looking even for a very rare kind of event—a particular mutation, for instance,—you can do it with relatively little work in corn.

This approach to corn genetics led to Barbara McClintock's discovery of transposable genetic elements (transposons) over forty years ago, and transposons figure prominently in your current research. What are these genetic elements, and how do they affect plant development?
A transposable element is a segment of DNA that can cut itself out of the chromosome and reinsert itself elsewhere. So that's why they're called jumping genes. We study the transposable element family called *Mutator*. In corn, we can visualize the activity of the transposable element most easily when we're looking at its "cutting out" property—its ability to excise. So in these purple and beige kernels [see color photograph], the beige background occurs because the transposable element is sitting in a gene required to make the purple pigment anthocyanin. And when the transposable element hops out during development of the seed tissue, then those cells can now make the anthocyanin again. So in this case, we can think of the transposon like a genetic light switch. It's turned the gene off, and when the transposon hops out, the gene then can be turned back

on, and it is easy to see the results. Transposons can hop into any part of the gene, and they can cause a gene to turn on when it's not supposed to be on, or turn it off when it's supposed to be on, and they can alter which organs a gene is expressed in.

And this is helping you answer certain questions about plant development?
The thing that really interests me is developmental timing. What's really striking about the *Mutator* transposons we study is that they're only active after an organ such as a leaf already begins to form. So these transposons don't affect genes until after cells in the emerging organ become committed to a particular pathway of development. And this provides a way to study a basic problem in developmental biology: What's different about a cell that still has more than one potential fate in development versus a cell that's become restricted in developmental fate—it can only become an epidermal cell, for example. Transposons are helping us understand the timing and nature of these changes. And of course we have these wonderful color markers in corn, so we can just look at the color on the outside and detect when something happened in development. And with the continuous development of plants, we can observe these developmental events in an intact organism.

What do you mean by continuous development?
Plants are continuously making organs from scratch. So virtually every day a new leaf is made, and it goes through the entire leaf development pathway. It would be as if we were constantly sprouting arms, little arm primordia that grew into arms with fingers and everything. So during the life of the plant, we can watch development of the same type of organ over and over again in the same individual.

And this continuous development is a key difference between plants and animals?
Especially between vertebrates and flowering plants. During embryonic development of an animal such as a mammal, all of the organs and tissue types of the adult are formed in some way. And so after you're born, all you do, really, is grow. But a plant embryo is not a miniature adult plant. It consists of the first few organs of the plant, like the primary

root and the first leaves, but then contained within that embryo is the plan for making subsequent organs throughout the life of the plant.

How does this difference in plant and animal development relate to the different lifestyles of these organisms?
The difference has to do with how plants and animals respond to stress and changes in their environments. Mammals, for example, can use a combination of behavior, such as flight (moving out of the way), and longer-term changes in their physiology or even development (like getting more fur when it's cold) when the environment changes. Plants, because of their cell walls, are very rigid, and they're basically stuck in place; once they've germinated, that's an irreversible place decision. I guess behavior is the thing that plants really don't have. What they do have that animals lack completely is a profound developmental response to stress. Everybody knows if you buy a plant from a nursery where it's been growing in dim light, you get nice, dark green large leaves. But if you take that plant home and put it in a sunny spot in a window, the first thing it does is drop all those beautiful dark green leaves because it's getting excess light. And then what comes out? Little pale leaves, which are perfect for the new conditions. So the plant is able to mold its developmental outcome—it's making leaves all the time, always from the same genome, but what the actual leaf looks like is molded by the current conditions. And that degree of developmental flexibility is something that higher animals don't have.

What are some other important differences between plant and animal development?
Well, as with all the organ systems in a vertebrate animal, the germ line—the cells that will give rise to gametes—is laid down in the embryo. In plants, there is no fixed germ line. The meristems at the tips of shoots that continuously produce leaves can also make flowers where gametes will form. But they're waiting for environmental cues that say it's an appropriate season of the year or appropriate conditions to produce a flower. And so the plant can switch to reproductive development just like it can switch the different kinds of leaves that it would make over the course of a year.

The other incredibly important difference is the alternation of haploid and diploid generations in plants. In

animals, the haploid cells—the egg and the sperm—are quiescent from a molecular biologist's point of view, because their characteristics are preprogrammed by the genes of their diploid precursor cells. But in plants, the gametes are produced by a gametophyte, a small organism that is itself haploid. And those haploid cells of the gametophyte have to express their own genes. This provides plants with an amazing capability that animals lack entirely. If a mutation occurs in one of us, our gametes could carry that mutation and there's no way to get rid of it if it's recessive, because the diploid precursor cell that gave rise to the gametes was able to cover for the recessive mutation. But in plants, each allele is tested in a haploid organism—the gametophyte. So the gametophyte phase and this alternation of generations, which just seems to add to the complexity of diagrams in a text book, in fact eliminates most deleterious recessives before they can actually become fixed in a population.

This advantage of alternation of generations is critical if you think about where the gametes come from in a plant. In a mammal, the germ line is differentiated inside of the developing embryo, which in turn is inside the mother's body. So these gametes are really well shielded from cosmic radiation or whatever, and the mutation rate should be quite low. However, the shoot apical meristems of a plant have been undergoing continuous mitosis in the real world, and have been exposed to heat and cold and UV radiation from the sun, and all sorts of other potentially deleterious things. But alternation of generations culls out all of the detrimental recessives by selection acting on gametophyes. The plant genome that's passed on to the next diploid generation, the sporophyte, is not quite a clean slate, but many mutations have been removed.

How do you reconcile this idea with the existence of transposons that *cause* mutations in plants?

I think that because plants have this requirement to have a diverse body over time, they have a mechanism for generating diversity built into their genomes. And transposable elements provide the mechanism, particularly in corn. Plants can actually boost their mutation rate because they have a means, alternation of generations, to get rid of anything that's really bad. In fact, there are very few cases of active transposable elements in mammals, but they're just rampant in plants.

Is there any evidence that this mechanism for changing the genome is adaptive in the sense that it is a response to environmental change?

Probably the best example occurs in snapdragons, where there is a thousand-fold increase in transposon events during cold stress. Barbara McClintock devoted her Nobel Prize lecture to the plant genome's response to challenge. The general drift of her remarks was that there's some kind of tolerance zone where things are relatively homeostatic for the plant, and as you get near the edge of that, you will see the breakdown of a lot of normal processes. And some of these can actually be at the level of the chromosomes. I think it's an incipient connection between genetics and physiology. And I would say that it goes against the grain to think like this, because of Lamarck. I have given talks where people have said, you talk like someone who believes in what Lamarck said. And I don't take it as an insult, because if you look back at what Lamarck said and you think about it in modern terms, he said that the relationship between the organism and the environment matters. What he said next you may not agree with. But I think it's almost been a conviction on the part of geneticists to avoid the idea that environmental conditions could perturb the genetic process.

In ways that can actually be *adaptive?*

Yes. And it's the "ways that can be adaptive" part that we need to clarify. If you were a corn plant and you were under ideal conditions for corn, why mutate? So you should be very conservative. Global change comes, and that old genome, the genome that was just perfect under the old conditions, is now no longer optimal. You need a mechanism of rapid genetic change—you need evolution to happen. So the new thinking on this is *not* that you cause specific mutations that are adaptive, but that to adapt you need mutations. It's not that the corn plant wishes to be cold-tolerant; it's that given the genes it's got now, it never will be.

You mentioned the power of recombinant DNA methods earlier. How can cloning DNA help a developmental biologist?

A developmental biologist needs to know the progression from one stage in a cell's development to another and to divide that progression into as many small steps as possible. So you could try to understand how you get, not from Palo Alto to New York, but first from Palo Alto to San Francisco and then from San Francisco to Sacramento, and so on. One of the things that's difficult in plants is that a lot of the cell types and tissues are not very anatomically distinct over time. So there aren't that many landmarks about the developmental stage of these cells. But with molecular biology you can clone genes that produce specific proteins that may be associated with certain stages of development. You can then use the cloned gene to develop various kinds of molecular probes for detecting the RNA or protein that appears in a cell when that gene is being expressed. And this lets you follow development over a finer scale of changes than is possible if you're just looking at cell and tissue structure. And the more stages you can tease apart, the more likely you are to get the right order of which cells are doing what and when.

Why are so many researchers using molecular methods to study plant development working on the plant called *Arabidopsis?*

Well of course I love corn, but a big problem with corn is its body size. It's difficult to grow corn under highly controlled conditions throughout its life

cycle. So if you want to find a temperature-sensitive mutant of flowering, for example, that's going to be really difficult in corn. But *Arabidopsis* is very small and it tolerates crowding really well; people are able to grow tens of thousands of plants in a tray in a growth chamber and manipulate the environmental conditions precisely. And so the real advantage of *Arabidopsis* for certain problems is the experimental control over the environment. And people who aren't as lucky as I am to have a field can do sophisticated plant genetics with very little space.

Also, because *Arabidopsis* is worked on by hundreds of people and it has a small genome, it's the logical plant to complement the Human Genome Project.It's one of the organisms being used as a model for how to assemble complete maps of a genome. In other plants like corn with a fifty-times larger genome, it would be at least fifty times more work to have this same map. But I think that any investigator needs to define carefully what their question is, and then find an organism that fits. For the questions I'm asking, corn is just unrivaled as an experimental organism.

Recombinant DNA methods have also made new biotechnologies possible. How can an ability to manipulate plant genomes be applied in agriculture?
Of course traditional plant breeding has done a pretty good job of developing crops that are very well suited to current production methods and meet our idea of, for example, what's the ideal strawberry in size and smell and shape and texture. However, given problems like pathogens or changing soil conditions, people may see that another plant species offers a solution to that problem. Biotech enables people to clone a gene from one species and move that to another species. I think the kinds of commercial applications we'll see right away are gene transfers that confer resistance to certain diseases. Plant breeding could do that by starting from scratch and selecting for a more resistant individual, but with biotech you can circumvent the years of trying to find that gene within a species. I think the agricultural goals of biotech are really the same goals that plant breeding has always had—we want food that is better, cheaper, and safer. It's just a new toolbox and an accelerated pace of change.

Do you envision risks—environmental or otherwise—in tampering with plant genomes and moving genes from one species to another?
Personally, I think the risks are the same as they were with breeding technology. People have been manipulating other organisms for a long time—I mean, agriculture is a very old practice. The tendency when you find a plant variety that works well is to grow it exclusively, and this happens with or without biotech. But using only a few genetic variants out of the many that exist means that our food sources are more susceptible to a new disease or environmental conditions that are deleterious to that one type. So, just like with normal breeding, I think we need to ensure that there's a diversity of varieties that have any new trait. Maybe we'll turn the corner on this kind of responsibility with biotech, because now it's easier to move any gene into many varieties, rather than searching for just one variety that already possesses the trait.

Barbara McClintock came up earlier in the interview. Was she an important mentor for you?
Oh, yes. Of course for people interested in plant genetics she's . . . well, after Mendel, there's Barbara. I didn't really start working with corn until I was an assistant professor at Washington University in St. Louis. The University of Missouri was two hours away, and I started collaborating with people there and learning ways to determine the timing of changes during corn development. I gave a talk at Rockefeller University when my ideas were just forming, and some of Barbara's friends heard it and told her about it, and she called me. We had a series of long phone conversations, and she invited me to come to Cold Spring Harbor. This was in the 1970s, when she was still most able to share not only the message of her work but the materials. I learned a tremendous amount from her. She was a remarkable person.

What do you think drives people like you and Barbara McClintock and other scientists to work so hard?
Well, I really like what I do. I think in terms of, you know, what did Picasso do every day? He had to clean the brushes, stretch the canvas—do all those technical things to get ready to do a great painting. And I personally enjoy all of the mechanical things, from planting the corn seed to hoeing the field. And then of course, the work itself is a never-ending story. You get to a landmark—we just cloned the autonomous element for the *Mutator* system after ten years of work—and you think "Oh, that's great!", You know, we have parties, we had champagne. But then it opens up the next challenge. You don't really finish—I guess that's what I like about it.

How can our undergraduates find out if a career in science is something they would enjoy and be good at?
There's no substitute for being a summer intern in someone's lab. And I think undergraduates, often miss the opportunity to find a faculty member who will hold a discussion with them every other week for an hour on some topic of mutual interest. I've done that quite a lot. The limitation is the student's asking. It's not something that faculty members can say to a class of 500: "Please come see me." I think it is the big challenge of being in college: figuring out what it is you like to do and what you're good at, and what provides you with that "Gee, I can't wait to get to work today" feeling. Who's the baseball player who said, "I can't believe they pay me to do this"? Just that feeling that you've found a perfect fit. I think for me, the more I learn about corn and the history of it, the more connected I feel with what I do. I think that's what inspires students to work hard, when they discover something they really enjoy. It's not that you *have* to work so hard, it's that you *want* to.

31 PLANT STRUCTURE AND GROWTH

AN INTRODUCTION TO PLANT BIOLOGY: THE MAJOR THEMES

BASIC MORPHOLOGY OF FLOWERING PLANTS: AN EVOLUTIONARY PERSPECTIVE

BASIC PLANT ANATOMY: PLANT CELLS AND TISSUES

AN OVERVIEW OF PLANT GROWTH

A CLOSER LOOK AT PRIMARY GROWTH

A CLOSER LOOK AT SECONDARY GROWTH

Plants are the pillars of most terrestrial ecosystems (Figure 31.1). Their photosynthesis supports their own growth and maintenance, and it feeds, directly or indirectly, an ecosystem's various consumers, including animals. Although most humans now live far from natural ecosystems and farms, our dependence on plants is evident in our lumber, fabric, paper, medicine, and most important, our food. The study of plants began when early humans learned to distinguish edible plants from poisonous ones and began to make things from wood and other plant products. Modern plant biology still has its pragmatic side—in research aimed at improving crop productivity, for example—but the pure fun of discovery is what motivates most plant scientists. Plant biology, perhaps the oldest branch of science, is driven by a combination of curiosity and need—curiosity about how plants work and a need to apply this knowledge judiciously to feed, clothe, and house a burgeoning human population.

Figure 31.1
Plants are the producers in most ecosystems. Plants dominate most landscapes, such as this rain forest in Costa Rica. Without plants, ecosystems would quickly collapse. Photosynthesis is the metabolic portal through which energy enters an ecosystem and becomes stored in organic matter. Most other organisms, including humans, are consumers that depend on plants for their existence. To understand life on Earth, it is essential to learn how plants work. This unit surveys modern plant biology, beginning, in this chapter, with an introduction to plant structure and growth.

AN INTRODUCTION TO PLANT BIOLOGY: THE MAJOR THEMES

Plant biology is in the midst of a renaissance. New methods coupled with clever choices of experimental organisms have catalyzed a research explosion. For example, many scientists interested in the genetic control of plant development are focusing their research on *Arabidopsis thaliana,* a little weed that belongs to the mustard family (Figure 31.2). *Arabidopsis* is small enough to be grown in test tubes, and its short generation span, about six weeks, makes it an excellent model for genetic studies. But plant biologists are also attracted to *Arabidopsis* by its tiny genome; the amount of DNA per cell ranks among the smallest of all known plants. Efforts to associate specific genetic functions with certain regions of DNA are simplified, and it is likely that researchers will map the entire *Arabidopsis* genome within this decade. Already, plant biologists have pinpointed some of the genes that control the development of flowers and have learned the functions of these genes (Figure 31.2b and c). This research is just one example of what seems to be a major thrust of

(a)

(b)

(c)

Figure 31.2
***Arabidopsis*, a little weed that is helping plant biologists answer big questions about the development of plant form.** (**a**) Plant biologists have made this tiny member of the mustard family one of the organisms of choice for experimental study of how genes control plant development. To researchers, the most important characteristics of *Arabidopsis* are its small size, short life cycle, and relatively small genome. Numerous laboratories are now mapping the plant's genome and identifying the functions of many of its genes. With the help of this model system, plant biologists hope to discover how genes and environment interact to transform a zygote into a plant. (**b**) Much of this research has centered on flower development. *Arabidopsis* normally has four whorls of flower parts: sepals, petals, stamens, and carpels (see Chapter 27). (**c**) Researchers have identified several mutations that cause abnormal flowers to develop. This flower, for example, has an extra whorl of petals in place of stamens and an internal flower where normal plants have carpels.

modern plant biology: relating processes that occur at the molecular and cellular levels to what we observe at the level of the whole plant. Once again, we see the theme of a hierarchy of levels, with emergent properties arising from the ordered arrangement and interactions of component parts.

The other biological themes of this book will also guide our study of plants. The form and functions of plants are shaped by interactions with the environment on two time scales. In their long evolutionary journey from water onto land, plants became adapted by natural selection to the specific problems posed by terrestrial environments. The evolution of tissues, such as the woody tissue of trees, that can support plants against the pull of gravity is just one example (see Chapter 27). Over the short term, individual plants exhibit structural and physiological responses to environmental stimuli. As an example of how individual plants, far more than individual animals, adapt *structurally* to their environment, look at how wind has affected the growth of branches of the trees illustrated in Figure 31.3. Plants, like animals, also have many evolutionary adaptations in the form of *physiological* responses to short-term change. For example, the stomata (pores of the leaf surface) of many plants close during the hottest time of day, a response that helps the plant conserve water (see Chapter 10). As we analyze the evolutionary adaptations of plants and the responses of individual plants to their environment, our main tool will be a related theme: the correlation of structure and function. We will learn, for example, how the opening and closing of stomata are consequences of the structure of the guard cells that border these pores.

In studying the structure and function of plants, comparisons with the animal kingdom are inevitable and sometimes useful. Plants and animals confront many of the same problems, which they may solve in different ways. For example, an elephant and a redwood tree support their enormous weight on land by means of very different kinds of "skeletons"—bones in the elephant and wood in the tree. In some cases, plant and animal solutions to a common problem seem similar. For example, large organisms require internal transport systems to carry water and other substances between body parts, and networks of tubes have evolved in both plants and animals. But on closer examination, we will see that any anatomical similarities between plants and animals are superficial. These similar adaptations are, in the lexicon of biology, analogous, not homologous. The multicellular organization of plants evolved independently from that of animals, from unicellular ancestors with an entirely different mode of nutrition. The autotrophic/heterotrophic distinction has placed plants and animals on separate evolutionary paths. This is the justification for this book's

Figure 31.3
The effect of environment on plant form. The "flagging" of these firs growing on a windy ridge on Mt. Hood, Oregon, resulted from the mechanical disturbance of the prevailing winds inhibiting limb growth on the windward sides of the trees. This growth response reduces the number of limbs that are broken during strong winds. Although environment affects the growth and development of all organisms, environmental impact on plant form is particularly impressive. In contrast, animal form is much less plastic, so that wind, for example, does not alter the number, size, or placement of human limbs.

separate units on plants and animals. Setting the stage for our survey of plant biology, this chapter introduces the general body plan of flowering plants, beginning with external structure. We will see that the architecture of plants is dynamic, continuously shaped by the plants' genetically directed growth patterns and their responses to the environment.

BASIC MORPHOLOGY OF FLOWERING PLANTS: AN EVOLUTIONARY PERSPECTIVE

Plant biologists describe plant architecture on two levels: morphology and anatomy. **Plant morphology** (Gr. *morphe,* "form") is the study of the external structure of plants. The arrangement of the parts of a flower is an example of morphology. **Plant anatomy** is concerned with internal structure—the arrangement of cells and tissues within a leaf, for example. This section features the morphology of flowering plants, or angiosperms. The structure of algae, bryophytes (mosses and their relatives), ferns, and gymnosperms was covered in Unit Five, along with the evolutionary relationships of these plant groups to the angiosperms. Angiosperms are characterized by flowers and fruits, evolutionary adaptations that function in reproduction and the dispersal of seeds. With about 275,000 known species, angiosperms are by far the most diverse and widespread of all plants. Taxonomists split the angiosperms into two classes: **monocots,** named for their single cotyledon (seed leaf), and **dicots,** which have two cotyledons. Monocots and dicots have several other structural differences (Figure 31.4).

The basic morphology of plants reflects their evolutionary history as terrestrial organisms (see Chapter 27). The algal ancestors of plants were bathed in a solution of water and minerals, including bicarbonate, the source of CO_2 for photosynthesis in aquatic habitats. In contrast, the resources a terrestrial plant needs are divided between the soil and air, and the plant must inhabit these two very different environments at the same time. Soil provides water and minerals, but air is the main source of CO_2, and light does not penetrate far into the soil. The evolutionary solution to this separation of resources was differentiation of the plant body into two main systems, a subterranean **root system** and an aerial **shoot system** consisting of stems, leaves, and flowers (Figure 31.5). Neither system can live without the other. Lacking chloroplasts and living in the dark, roots would starve without sugar and other organic nutrients imported from the photosynthetic tissues of the shoot system. Conversely, the shoot system depends on water and minerals absorbed from the soil by roots. Vascular tissues, continuous through the plant, transport materials between roots and shoots. Each vein has two types of vascular tissue: **xylem,** which conveys water and dissolved minerals upward from roots into the shoots; and **phloem,** which transports food made in the leaves to the roots and to nonphotosynthetic parts of the shoot system. As we take closer looks at the morphology of roots and shoots, try to view these systems from the evolutionary perspective of adaptation to living on land.

The Root System

Roots anchor the plant, absorb minerals and water, conduct water and nutrients, and store food. The structure of roots is well adapted to these functions.

Many dicots have a **taproot** system, consisting of one large, vertical root (the taproot) that produces many smaller secondary roots (see Figures 31.4 and

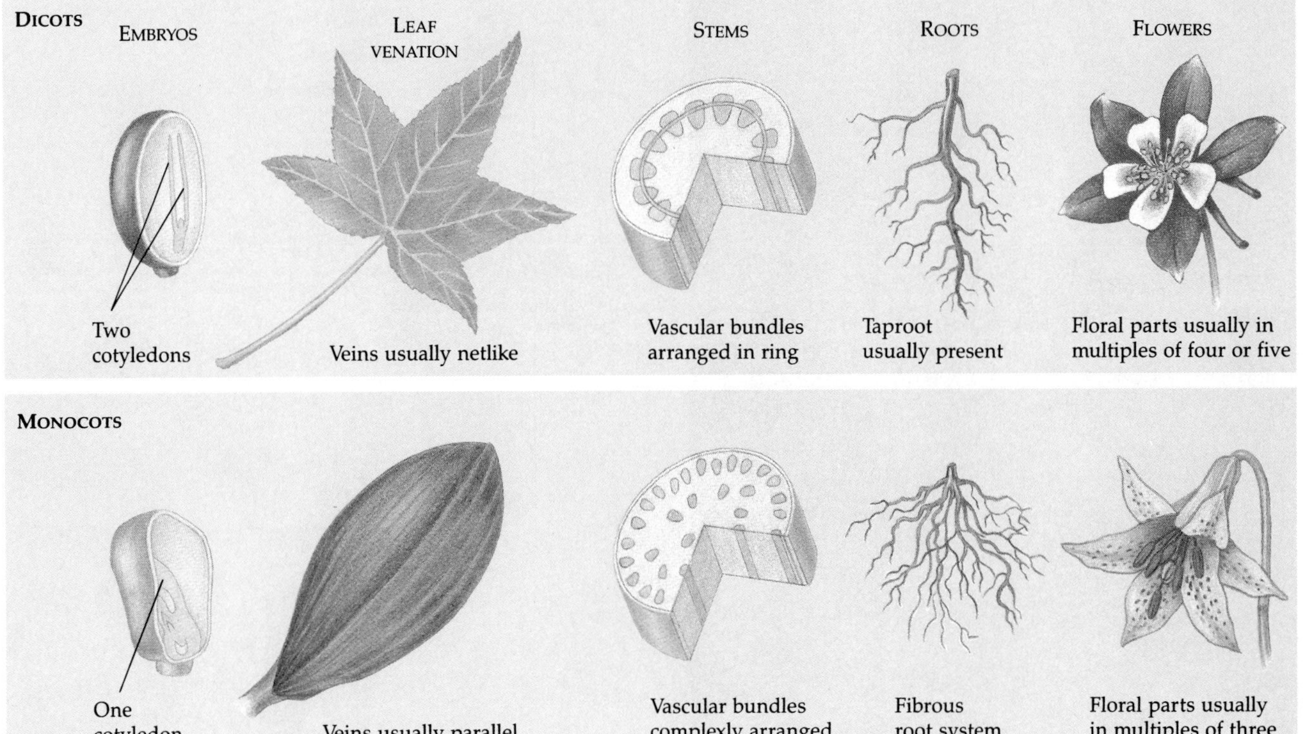

Figure 31.4
A comparison of monocots and dicots. These classes of angiosperms are named for the number of cotyledons, or seed leaves, present in the seed of the plant. Monocots include orchids, bamboos, palms, lilies, and yuccas, as well as the grasses, such as wheat, corn, and rice. A few examples of dicots are roses, beans, sunflowers, maples, and oaks. The structures in the figure are discussed over the course of this chapter.

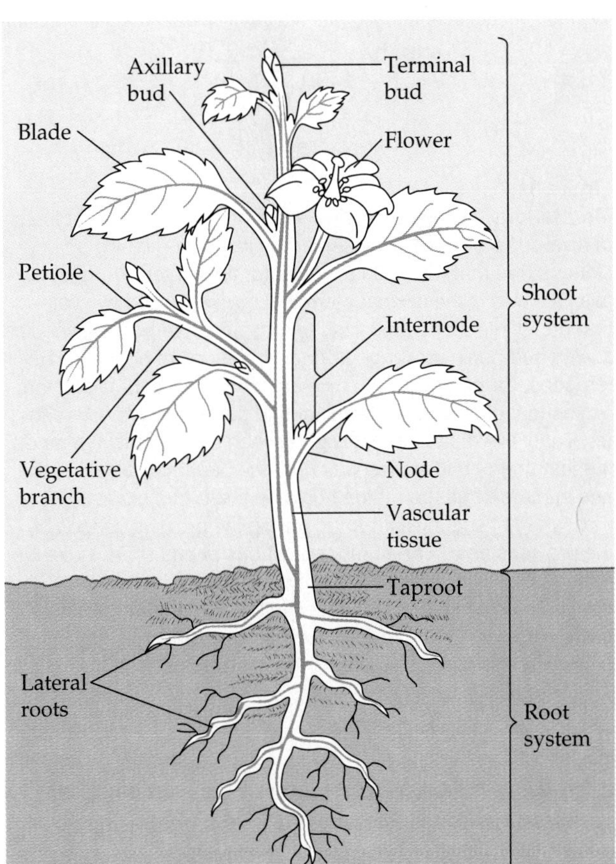

Figure 31.5
Basic morphology of a flowering plant. The plant body is divided into a root system and a shoot system, connected by vascular tissue that is continuous throughout the plant. The root system of this dicot consists of a taproot and several lateral roots. Shoots consist of stems, leaves, and flowers. The blade, the expanded portion of a leaf, is attached to a stem by a petiole. Nodes, the regions of a stem where leaves attach, are separated by internodes. At a shoot's tip is the terminal bud, the main growing point of the shoot. Axillary buds are located in the upper angles of leaves. Most of these axillary buds are dormant, but they have the potential to develop into vegetative (leaf-bearing) branches or flowers.

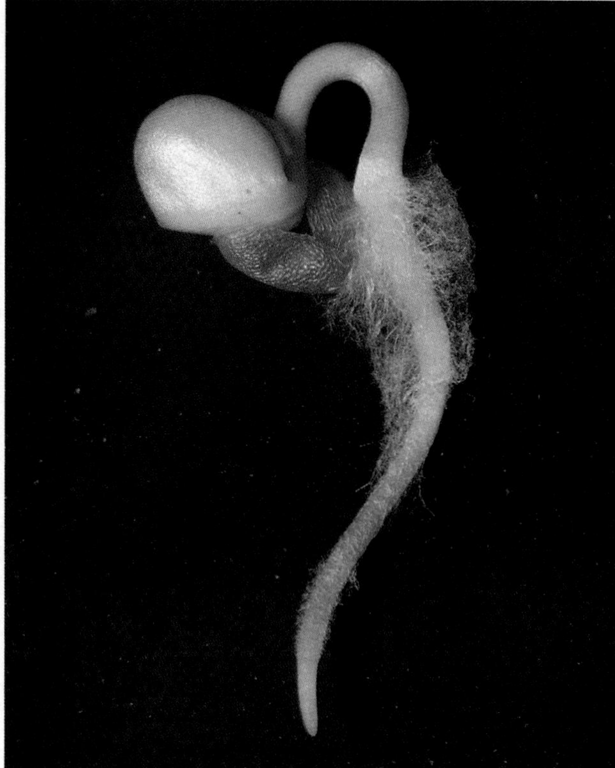

Figure 31.6
Root hairs of a radish seedling. Growing by the thousands just behind the tip of each root, the hairs cling tightly to soil particles and increase the surface area for the absorption of water and minerals by the roots.

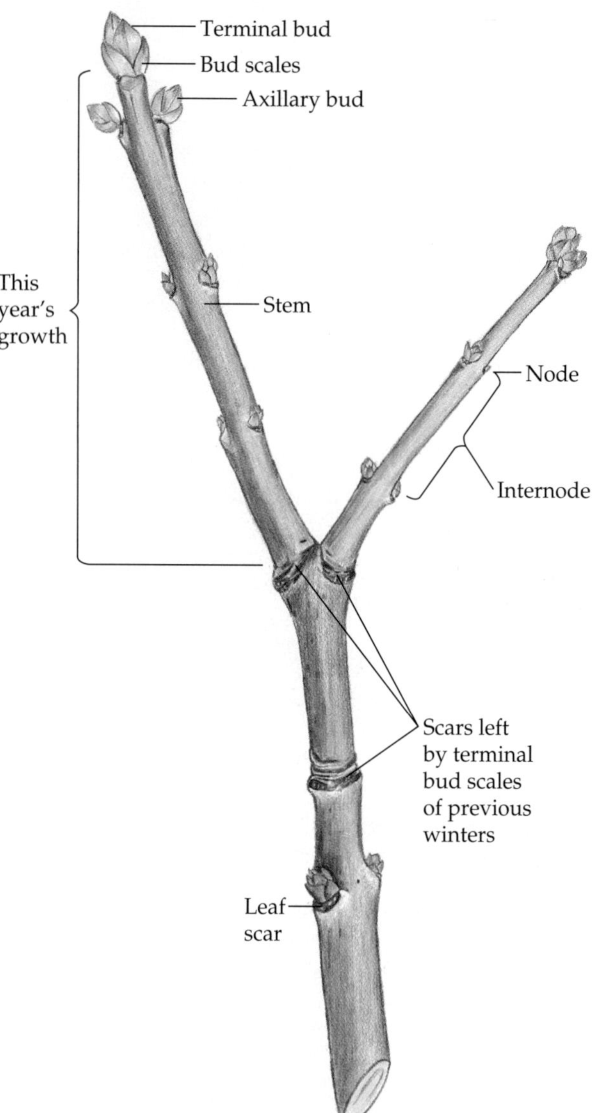

Figure 31.7
Morphology of a winter twig. Some of the external structures of a woody stem can be observed by inspecting a twig of a deciduous tree that has lost its leaves for the winter. At the tip of this lilac twig is the dormant terminal bud, enclosed by scales that protect its embryonic tissues. In spring, the bud will shed its scales and begin growing, producing a series of nodes and internodes. Farther down the twig are whorls of scars left by the scales that enclosed the terminal bud during the previous winter. Thus, the region of the twig between the terminal bud and the first ring of bud scales was produced during the past spring and summer, and the region between these bud scars and the next whorl of scars was formed during the growing season of the previous year. The number of whorls of bud scars indicates the age of a twig. Along each growth segment, nodes are marked by scars left when leaves fell during autumn. Above each leaf scar is either an axillary bud or a branch twig produced by previous growth of the axillary bud.

31.5). Penetrating deep into the soil, the taproot is a firm anchor, as you know if you have ever tried to pull up a dandelion or mustard plant. Some taproots, such as carrots, turnips, sugar beets, and sweet potatoes, are modified roots that store exceptionally large amounts of food. The plant consumes these food reserves when it flowers and produces fruit. For this reason, root crops are harvested before the plants flower.

Monocots, including grasses, generally have **fibrous root** systems consisting of a mat of threadlike roots that spread out below the soil surface. The fibrous root system gives the plant extensive exposure to soil water and minerals and anchors it tenaciously to the earth. Because their root systems are concentrated in the upper few centimeters of the soil, grasses make excellent ground cover for preventing erosion.

Although the entire root system helps anchor a plant, most absorption of water and minerals occurs near the root tips, where vast numbers of tiny **root hairs** increase the surface area of the root tremendously (Figure 31.6). A single rye-grass plant, in a growing season of only four months, produces about 600 km of roots, with root hairs accounting for most of this length.

In addition to roots that extend from the base of the shoot, some plants have roots arising aboveground from stems or even from leaves. Such roots are said to be **adventitious** (L. *adventicius*, "not belonging to"), a term that describes any plant part that grows in an unusual location. The adventitious roots of some plants,

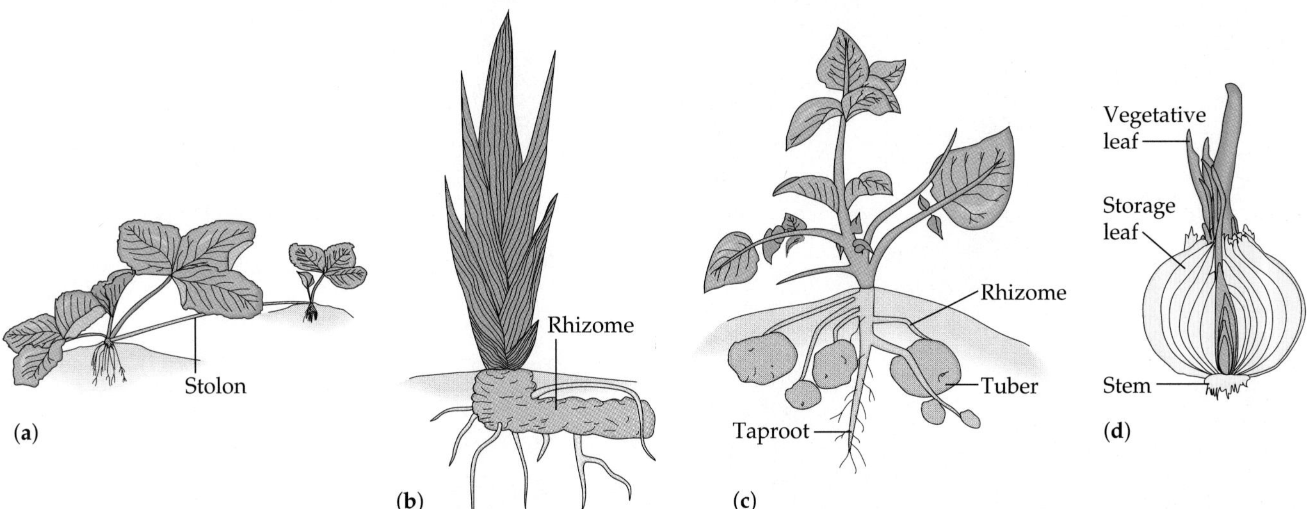

Figure 31.8
Modified stems. (**a**) Stolons, shown here on a strawberry plant, grow on the surface of the ground. (**b**) Rhizomes, like the one on this iris plant, are horizontal stems that grow underground. (**c**) Tubers are swollen ends of rhizomes specialized for storing food. The "eyes" arranged in a spiral pattern around the potato are clusters of buds that mark the nodes. (**d**) Bulbs are vertical, underground shoots consisting mostly of swollen storage leaves. You can see the many layers of modified leaves attached to the short stem by slicing an onion bulb lengthwise.

including corn (a monocot) and banyan trees (a dicot), function as props that help support stems. Prop roots are an example of how structures that evolved in one functional context can become adapted via further evolution for other functions (see Chapter 23).

The Shoot System

The shoot system consists of vegetative shoots, which bear leaves, and floral shoots, which terminate in flowers. We will postpone discussion of the structure and function of flowers until Chapter 34 and focus here on vegetative shoots. A vegetative shoot consists of a stem and the attached leaves; it may be the plant's main shoot or a side shoot, called a vegetative branch (see Figure 31.5).

Stems A stem has **nodes,** the points at which leaves are attached, and **internodes,** the stem segments between nodes (Figure 31.7; see also Figure 31.5). In the angle formed by each leaf and the stem is an **axillary bud,** which is an embryonic side shoot. Most axillary buds are dormant, however; growth is usually concentrated at the **apex** (tip) of a shoot, where there is a **terminal bud** with developing leaves and a compact series of nodes and internodes. Presence of the terminal bud is partly responsible for inhibiting the growth of axillary buds, a phenomenon called **apical dominance.** By concentrating resources on growing taller, apical dominance is an evolutionary adaptation that increases exposure of the plant to light, especially in a location with dense vegetation. However, branching is also important for increasing exposure of the shoot system to the environment, and under certain conditions, axillary buds begin growing. Some develop into shoots bearing flowers, and others become vegetative branches complete with their own terminal buds, leaves, and axillary buds. In some cases, growth of axillary buds can be stimulated by removing the terminal bud. This is the rationale for pruning trees and shrubs and "pinching back" houseplants to make them bushy.

Modified stems with diverse functions have evolved in many plants and are often mistaken for roots (Figure 31.8). Stolons are horizontal stems that grow along the surface of the ground. The "runners" of Bermuda grass and strawberry plants are examples. Rhizomes, such as those of irises, are horizontal stems that grow underground. Some rhizomes end in enlarged tubers where food is stored, as in white potatoes. Bulbs, such as those of tulips or onions, are vertical, underground shoots with leaves modified for food storage.

Leaves Leaves are the main photosynthetic organs of most plants, although green stems also perform photosynthesis. Leaves vary extensively in form, but they generally consist of a flattened blade and a stalk, the **petiole,** which joins the leaf to a node of the stem (see Figure 31.5). Grasses and most other monocots lack petioles; instead, the base of the leaf forms a sheath that envelopes the stem.

The leaves of monocots and dicots differ in how their major veins, or vascular bundles, are arranged

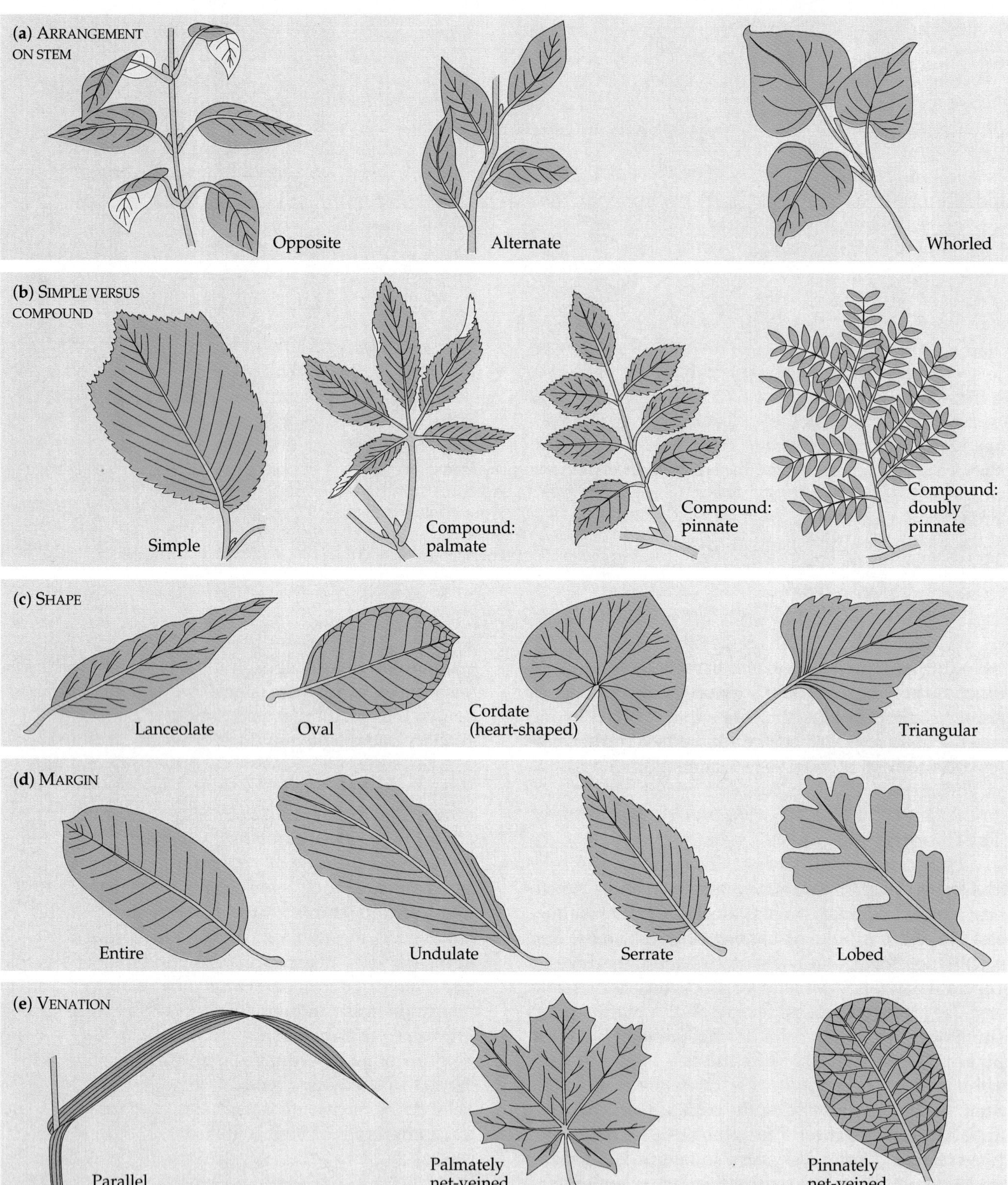

Figure 31.9

A survey of leaf morphology. (**a**) Leaves are arranged on the stem in a variety of patterns. If each node has a pair of leaves 180° apart, the leaves are said to be opposite. The leaf pattern is alternate when each node has a single leaf and the leaves of adjacent nodes point in different directions. If a node has three or more leaves attached, the arrangement is termed whorled. (**b**) Botanists refer to a leaf as being simple if it has a single, undivided blade. If the blade is divided into several leaflets, then the leaf is compound. (You can distinguish a compound leaf from a stem with several closely spaced simple leaves by examining the locations of axillary buds. There is no bud at the base of a leaflet, but there is an axillary bud where the petiole of the compound leaf joins the stem.) (**c–e**) Leaves also vary in shape, in the contour of their margins, and in their pattern of veins.

(a)

(b)

(c)

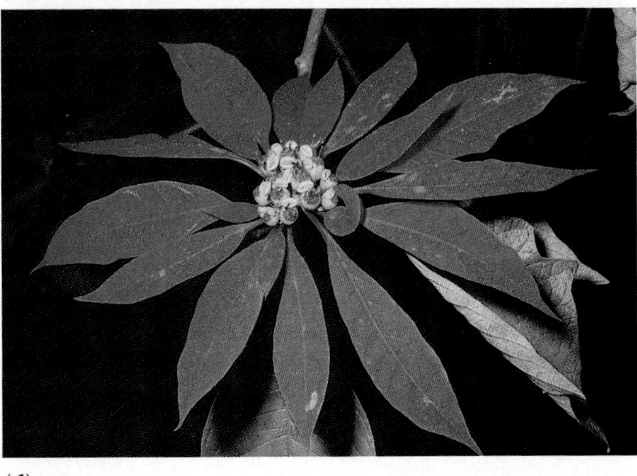

(d)

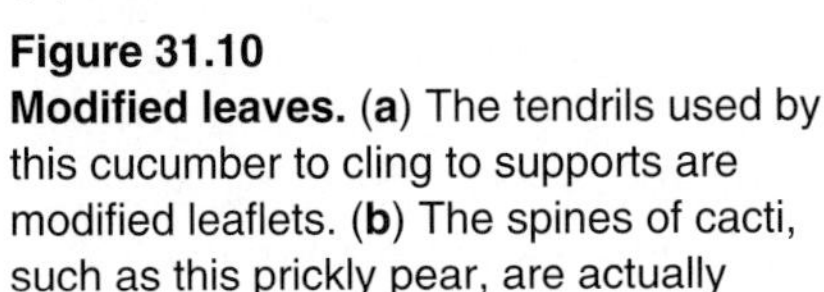

Figure 31.10
Modified leaves. (**a**) The tendrils used by this cucumber to cling to supports are modified leaflets. (**b**) The spines of cacti, such as this prickly pear, are actually leaves, and photosynthesis is carried out mainly by the fleshy green stems. (**c**) Many succulents, such as ice plant, have leaves modified for storing water. (**d**) In many plants, brightly colored leaves help attract pollinators to the flower. The red "petals" of the poinsettia are actually leaves that surround a group of flowers.

(see Figure 31.4). Most monocots have parallel major veins that run the length of the leaf blade. In contrast, dicot leaves generally have a multibranched network of major veins. All leaves have numerous minor crossveins. Vascular arrangement, leaf shape, and leaf pattern on the stem are among the characteristics used by plant taxonomists to help identify or classify plants (Figure 31.9).

Although most leaves are specialized for photosynthesis, some plants have leaves that have become adapted by evolution for other functions (Figure 31.10).

So far we have examined the structural organization of the whole plant as we see it with the unaided eye. With this overview, we can now begin to dissect the plant and explore its microscopic organization.

BASIC PLANT ANATOMY: PLANT CELLS AND TISSUES

In this section, we will learn how the structural specializations of plant cells enable them to perform cer-

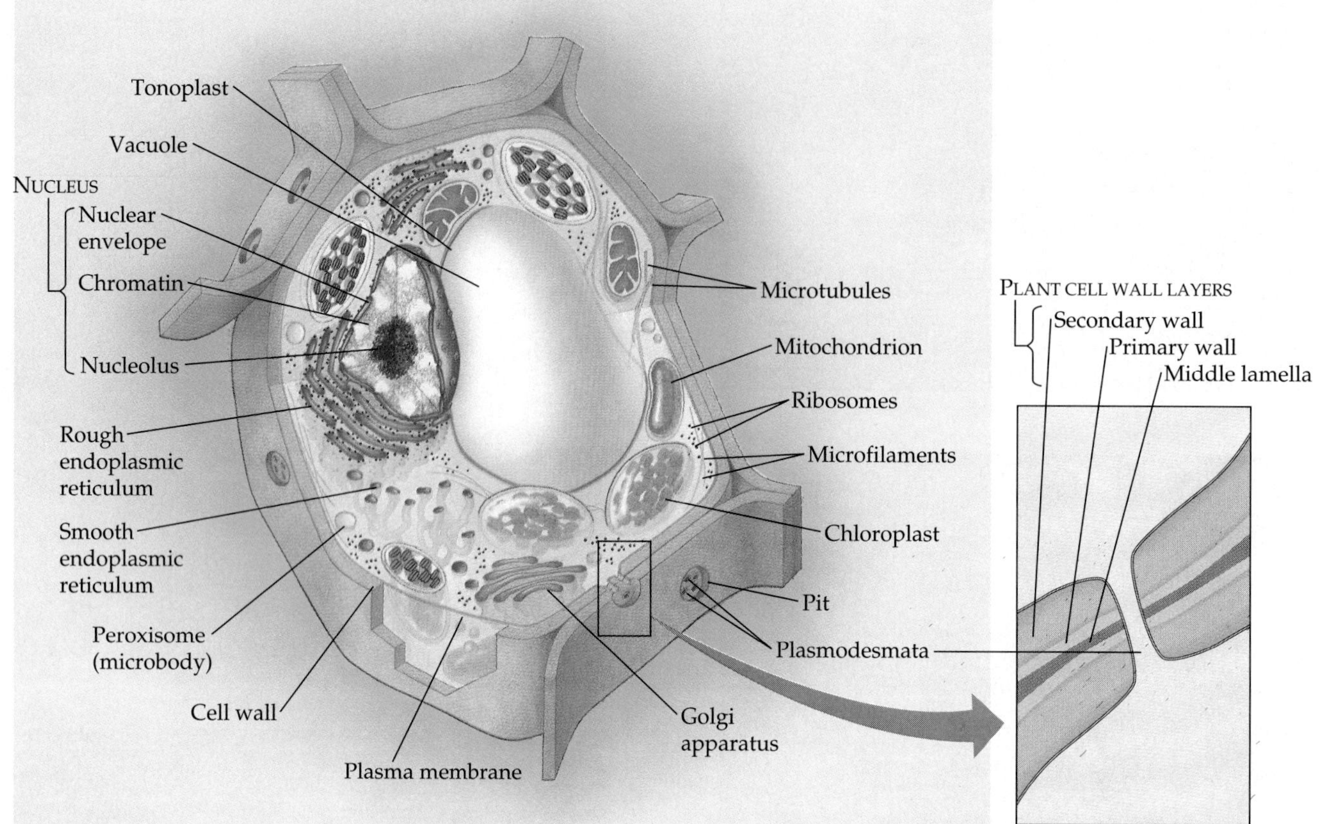

Figure 31.11
Plant cell structure: a review. A plant cell consists of a protoplast enclosed in a cell wall. The protoplast—the whole cell, excluding the cell wall—is bounded by the plasma membrane. Outside the plasma membrane is the primary cell wall and in some plants a secondary cell wall, constructed from cellulose fibers and other components. Between the primary walls of adjacent plant cells is the middle lamella, a sticky layer that cements the cells together. The protoplasts of neighboring cells are generally connected by plasmodesmata, cytoplasmic channels that pass through pores in the walls (see Figure 7.33). The plasmodesmata may be concentrated in areas called pits, where the distance between adjacent protoplasts is narrowed. When mature, most living plant cells have a large central vacuole that occupies as much as 90% of the volume of the protoplast. A membrane called the tonoplast separates the contents of the vacuole from the thin layer of cytoplasm, in which the mitochondria, plastids, and other organelles are located. Within the vacuole is the cell sap, a complex aqueous solution that helps the vacuole play an important role in maintaining the turgor, or firmness, of the cell (see Chapter 8).

tain functions. We will also learn how these cells are organized into the tissues that make up three plant organs: roots, stems, and leaves. The general structure of plant cells was described in Chapter 7. Figure 31.11 will be a helpful review of these features before we examine specific cell types.

Types of Plant Cells

What distinguishes a multicellular organism from a colony of cells is a division of labor among cells differing in structure and function (see Chapter 26). As you consider each major type of plant cell, notice the structural adaptations that make specific functions possible. In some cases, we will find distinguishing characteristics within the **protoplast,** the contents of the cell exclusive of the cell wall. For example, only the protoplasts of photosynthetic cells contain chloroplasts. But also notice that modifications of cell walls are important in how the specialized cells of a plant work.

Parenchyma Cells Because they are the least specialized of all plant cells, **parenchyma cells** are often depicted as "typical" plant cells (Figure 31.12a). Most parenchyma cells lack secondary walls, and their primary walls remain thin and flexible even when the cells are mature. The protoplast generally has a large central vacuole.

Parenchyma cells carry on most of the metabolism of the plant, synthesizing and storing various organic products. For example, photosynthesis occurs within the chloroplasts of parenchyma cells in the leaf. Some parenchyma cells in stems and roots have colorless plastids that store starch. The flesh of most fruit is composed mostly of parenchyma cells.

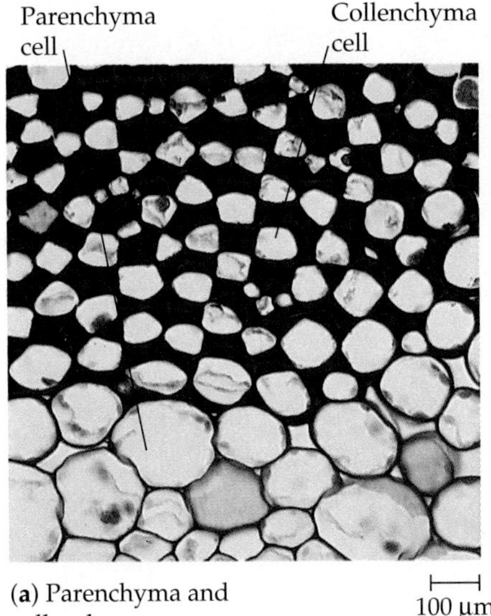

(**a**) Parenchyma and collenchyma

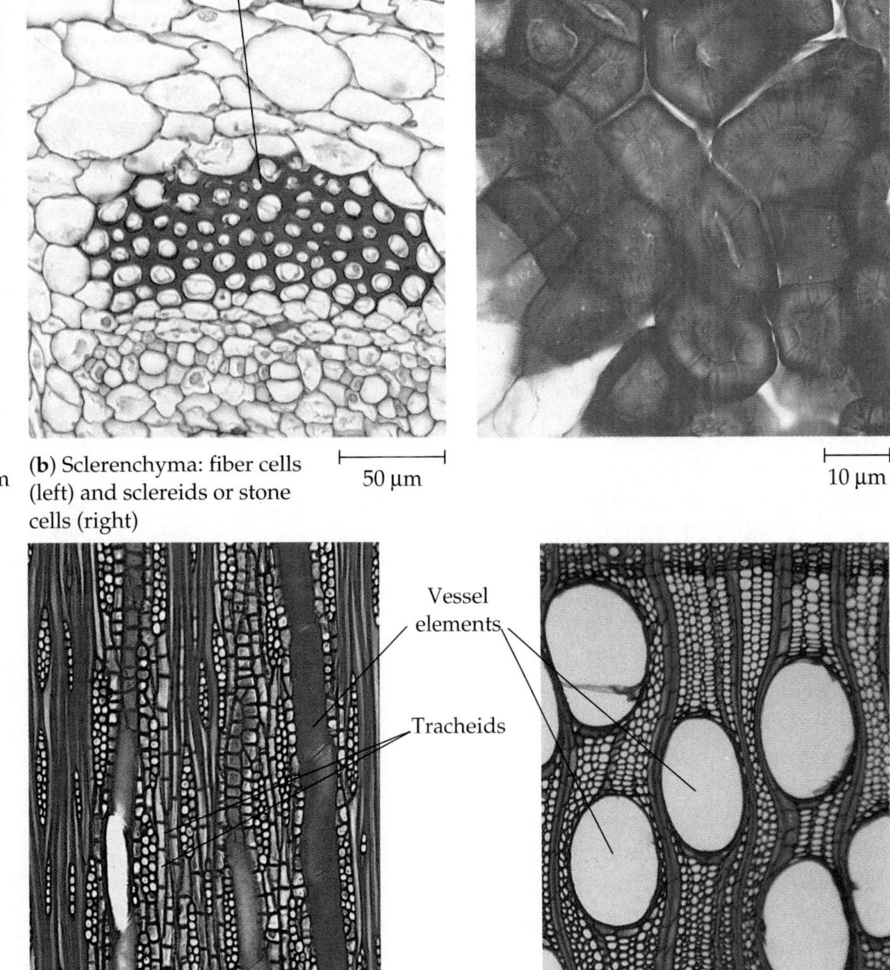

(**b**) Sclerenchyma: fiber cells (left) and sclereids or stone cells (right)

(**c**) Xylem showing vessel elements and tracheids in longitudinal (left) and transverse (right) sections

Figure 31.12
Types of plant cells. (**a**) Parenchyma cells are relatively unspecialized and usually lack secondary walls. These cells carry on most of the plant's metabolic functions. Collenchyma cells have unevenly thickened primary walls and provide support to parts of the plant that are still growing. (**b**) Sclerenchyma cells, specialized for support, have secondary walls hardened with lignin and may be dead (lacking protoplasts) at functional maturity. The fiber cells in the left micrograph are elongated sclerenchyma cells, a shape not evident when viewed in cross section. Sclereids (right) are irregularly shaped sclerenchyma cells with very thick, lignified walls. (**c**) The water-conducting cells of xylem include tapered tracheids (left) and vessel elements arranged end to end to form vessels (right). Both cell types have secondary walls and are dead at functional maturity. In gymnosperms, tracheids have the dual functions of water transport and structural support. In most angiosperms, both vessel elements and tracheids conduct water, and support is provided mainly by fiber cells. (**d**) The food-conducting cells of phloem are sieve-tube members, which are arranged end to end with perforated walls (sieve plates) between them. The cells are living at functional maturity, but lack nuclei. Alongside each sieve-tube member is a nucleated companion cell. (All LMs.)

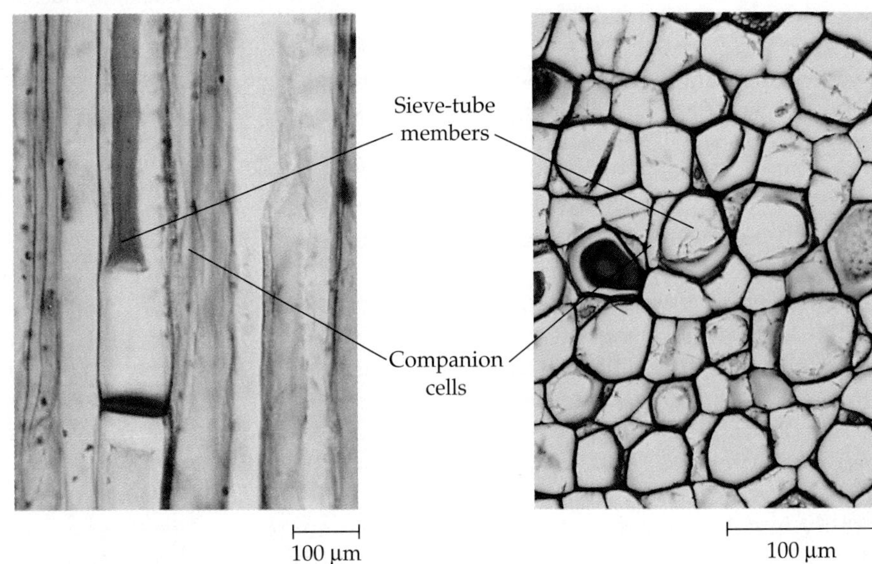

(**d**) Phloem showing sieve-tube members and companion cells in longitudinal (left) and transverse (right) sections

Developing plant cells of all types usually have the generalized structure of parenchyma cells before specializing further in structure and function. Mature parenchyma cells do not generally undergo cell division, but most of them retain the ability to divide and differentiate into other types of plant cells under special conditions—during the repair and replacement of organs after injury to the plant, for instance. It is even possible in the laboratory to regenerate an entire plant from a single parenchyma cell.

Collenchyma Cells Like parenchyma, **collenchyma cells** have protoplasts and usually lack secondary walls. Collenchyma cells have thicker primary walls than parenchyma cells, though the walls are unevenly thickened (Figure 31.12a). Usually grouped in strands or cylinders, collenchyma cells help support young parts of the plant. Young stems, for instance, often have a cylinder of collenchyma just below their surface. Because they lack secondary walls and the hardening agent lignin is absent in their primary walls, collenchyma cells provide support without restraining growth. They elongate with the stems and leaves they support.

Sclerenchyma Cells Also functioning as supporting elements in the plant, but with thick secondary walls strengthened by lignin, **sclerenchyma cells** are much more rigid than collenchyma. Mature sclerenchyma cells cannot elongate, and they occur in regions of the plant that have stopped growing in length. So specialized are sclerenchyma cells for support that many lack protoplasts at functional maturity, the stage in a cell's development when it is fully specialized for its function. Thus, at functional maturity a sclerenchyma cell may actually be dead, its rigid wall serving as scaffolding to support the plant.

The two forms of sclerenchyma cells are fibers and sclereids. Long, slender, and tapered, **fibers** usually occur in bundles (Figure 31.12b). Some plant fibers are used commercially, such as hemp fibers for making rope and flax fibers for weaving into linen. **Sclereids** are shorter than fibers and irregular in shape. Nutshells and seed coats owe their hardness to sclereids, and sclereids scattered among the soft parenchyma tissue give the pear its gritty texture.

Tracheids and Vessel Elements: Water-Conducting Cells The water-conducting elements of xylem are elongated cells of two types: **tracheids** and **vessel elements** (Figure 31.12c). Both types of cells are dead at functional maturity, but they produce secondary walls before the protoplast dies. In parts of the plant that are still elongating, the secondary walls are deposited unevenly in spiral or ring patterns that enable them to stretch like springs as the cell grows. Like the wire that reinforces the wall of a garden hose, these wall thickenings strengthen the water-conducting cells of the plant. (You will learn more about this function in Chapter 32.) Tracheids and vessel elements that form in parts of the plant that are no longer elongating usually have secondary walls with **pits,** thinner regions where only primary walls are present. A tracheid or vessel element completes its differentiation when its protoplast disintegrates, leaving behind a nonliving conduit through which water can flow (Figure 31.13).

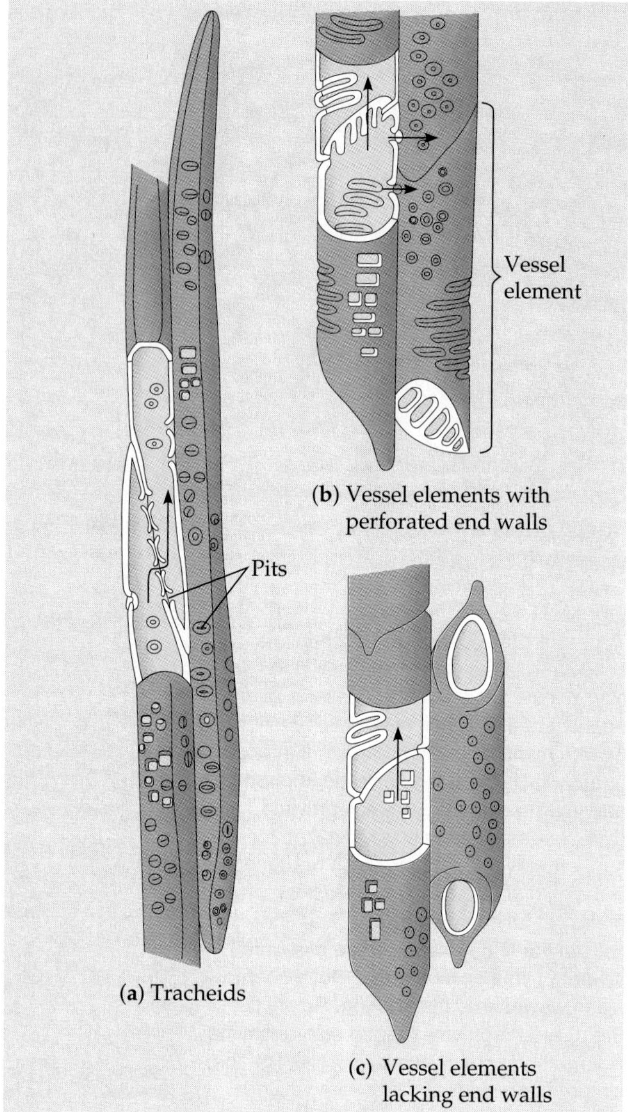

Figure 31.13
Water-conducting cells of xylem. Arrows indicate the flow of water. (**a**) Tracheids are spindle-shaped cells with pits through which water flows from cell to cell. (**b**) Vessel elements are individual cells linked together end to end to form long tubes called xylem vessels. Water streams from element to element through perforated end walls. Water can also migrate laterally between neighboring vessels through pits. (**c**) Resistance to water flow in some xylem vessels is lowered by the complete absence of walls between the vessel elements.

Tracheids are long, thin cells with tapered ends.

Water moves from cell to cell mainly through pits. Because their secondary walls are hardened with lignin, tracheids function in support as well as water transport. Vessel elements are generally wider, shorter, thinner walled, and less tapered than tracheids. The end walls of vessel elements are perforated, so water can flow freely through long xylem vessels consisting of chains of vessel elements. Most gymnosperms have only tracheids, while angiosperms generally have both tracheids and vessel elements. Vessel elements, which are generally considered to be more efficient water conductors, probably evolved from tracheids in ancient flowering plants (see Chapter 27).

Sieve-Tube Members: Food-Conducting Cells Sucrose, other organic compounds, and some mineral ions are transported within the phloem of angiosperms through tubes formed by chains of cells called **sieve-tube members** (see Figure 31.12d). In contrast to the water-conducting cells of xylem, sieve-tube members are alive at functional maturity, although their protoplasts lack such organelles as the nucleus, ribosomes, and a distinct vacuole. In angiosperms, the end walls between sieve-tube members, called **sieve plates,** have pores that presumably facilitate the flow of fluid from cell to cell along the sieve tube. The concentration of pores at the ends of the cell is a structural adaptation unique to angiosperms. In the phloem of other vascular plants, including ferns and gymnosperms, pores are distributed over the entire sieve cell.

Alongside each sieve-tube member is at least one **companion cell** (Figure 31.12d), which is connected to the sieve-tube member by numerous plasmodesmata. The nucleus and ribosomes of the companion cell may serve not only that cell but also the adjacent sieve-tube member, which has no nucleus or ribosomes of its own.

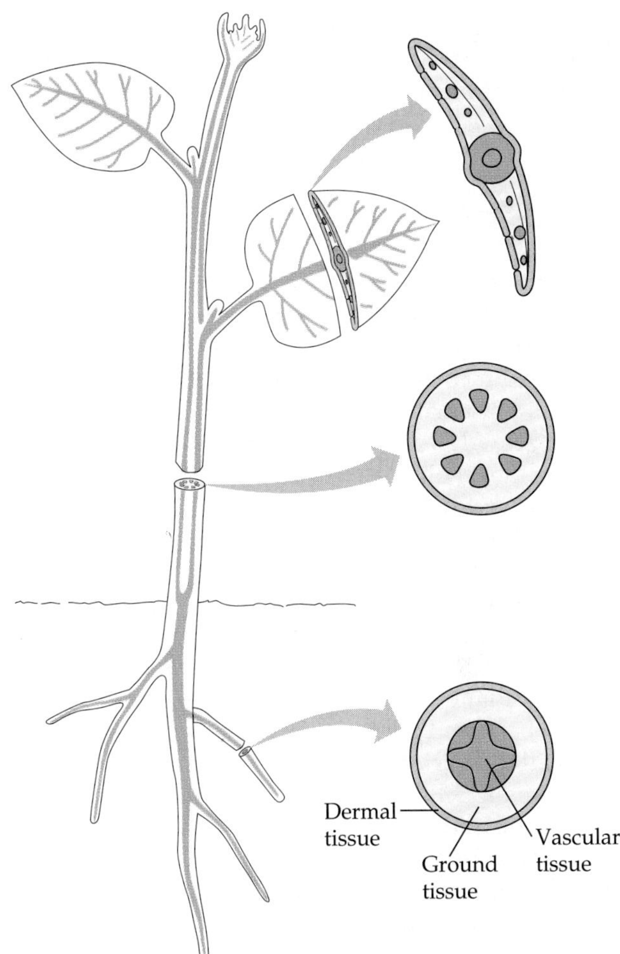

Figure 31.14
The three tissue systems. The dermal tissue system, or epidermis, is a single layer of cells that covers the entire body of a young plant. The vascular tissue system is also continuous throughout the plant, but it is arranged differently in each organ. The ground tissue system is located between the dermal tissue and vascular tissue in each organ.

The Three Tissue Systems of a Plant

The cells of a plant are organized into three tissue systems: the dermal, vascular, and ground tissue systems. Each tissue system is continuous throughout the plant body, although the specific characteristics of the tissues and their spatial relationships to one another vary in different organs of the plant (Figure 31.14). Here, we survey the three tissue systems as they occur in a young, nonwoody plant.

The **dermal tissue system,** or **epidermis,** is generally a single layer of tightly packed cells that covers and protects all young parts of the plant—the "skin" of the plant. In addition to the general function of protection, the epidermis has more specialized characteristics consistent with the function of the particular organ it covers. For example, the root hairs so important in the absorption of water and minerals are extensions of epidermal cells near the tips of roots. The epidermis of leaves and most stems secretes a waxy coating called the **cuticle** that helps the aerial parts of the plant retain water, an important adaptation to living on land. Young roots, which must absorb water from the soil, generally lack cuticles.

The continuum of xylem and phloem throughout the plant forms the **vascular tissue system,** which functions in transport and support. The specific organization of vascular tissue in stems and roots is discussed in the next section.

The **ground tissue system** makes up the bulk of a young plant, filling the space between the dermal and vascular tissue systems. Ground tissue is predominantly parenchyma, but collenchyma and sclerenchyma are also commonly present. Among the diverse functions of ground tissue are photosynthesis, storage, and support.

An examination of how a plant grows will help us understand how the tissue systems are organized in the different plant organs.

AN OVERVIEW OF PLANT GROWTH

From season to season and from year to year, the growth of plants alters our surroundings—yards, campuses, parks, vacant lots, woods, and the other landscapes in our communities. The growth of a plant from a seed is a fascinating transformation. The early stages of this growth—germination of the seed and emergence of the seedling—are among the topics of Chapter 34. Here, we will learn how plants continue to grow after their shoot and root systems are established.

Most plants continue to grow as long as they live, a condition known as **indeterminate growth.** Most animals, in contrast, are characterized by **determinate growth;** that is, they cease growing after reaching a certain size. Plants have the capacity for indeterminate growth because they have perpetually embryonic tissues called **meristems** in their regions of growth. Meristematic cells are unspecialized, and they divide to generate additional cells. Some of the products of this division remain in the meristematic region to produce still more cells, while others become specialized and are incorporated into the tissues of the growing plant. Cells that remain as wellsprings of new cells are called **initials;** those that are displaced from the meristem and are then destined to specialize within a developing tissue are called **derivatives.** While whole plants usually show indeterminate growth, certain plant organs, such as leaves and flowers, exhibit determinate growth.

The pattern of plant growth depends on the locations of the meristems (Figure 31.15). **Apical meristems,** located at the tips of roots and in the buds of shoots, supply cells for roots and shoots to grow in length. This elongation, called **primary growth,** enables roots to ramify throughout the soil and shoots to increase their exposure to light and carbon dioxide. In herbaceous (nonwoody) plants, only primary growth occurs. In woody plants, however, there is also **secondary growth,** a progressive thickening of the roots and shoots formed earlier by primary growth. Secondary growth is the product of **lateral meristems,** cylinders of dividing cells extending along the lengths of roots and shoots. These lateral meristems replace the epidermis with a secondary dermal tissue that is thicker and tougher, and they also add layers of vascular tissue. Wood is the secondary xylem that accumulates over the years.

In woody plants, primary and secondary growth occur at the same time, but in different locations. Primary growth is restricted to the youngest parts of the plant—the tips of roots and shoots, where the apical meristems

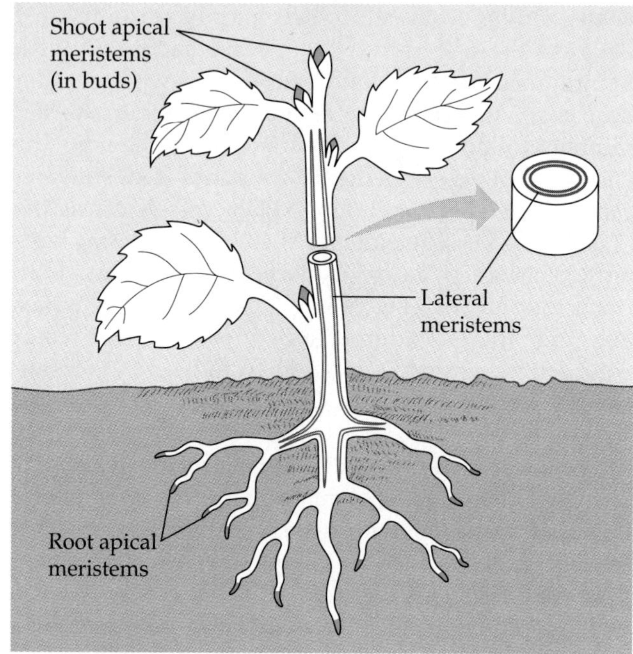

Figure 31.15
Locations of major meristems and an overview of plant growth. Meristems are self-renewing populations of cells that divide and provide cells for plant growth. Apical meristems, located near the tips of roots and shoots, are responsible for primary growth, or growth in length. Woody plants also have lateral meristems that function in secondary growth, which adds girth to roots and shoots.

are located. The lateral meristems develop in slightly older regions of the roots and shoots, some distance away from the tips. There, secondary growth adds girth to the organs. The oldest region of a root or shoot—the base of a tree branch, for example—has the greatest accumulation of secondary tissues formed by the lateral meristems. Each growing season, primary growth produces young extensions of roots and shoots, while secondary growth thickens and strengthens the older parts of the plant. Closer study of primary and secondary growth in the next two sections will help you understand the morphology and anatomy of plants.

Indeterminate growth does not imply immortality. Some plants have lifespans that are genetically programmed; such plants have a fixed longevity even when grown in constant, favorable conditions. Other plants have lifespans that are environmentally determined; that is, if the plants are grown under controlled temperature and light conditions and are protected from disease, they may live much longer than they typically do in natural environments. Plants known as **annuals** complete their life cycles—from germination through flowering and seed production to death—in a single year or less. Many wildflowers are annuals, as are the most important food crops, including the cereal grains and legumes. A plant is called a **biennial** if its life generally spans two years. Flowering usually oc-

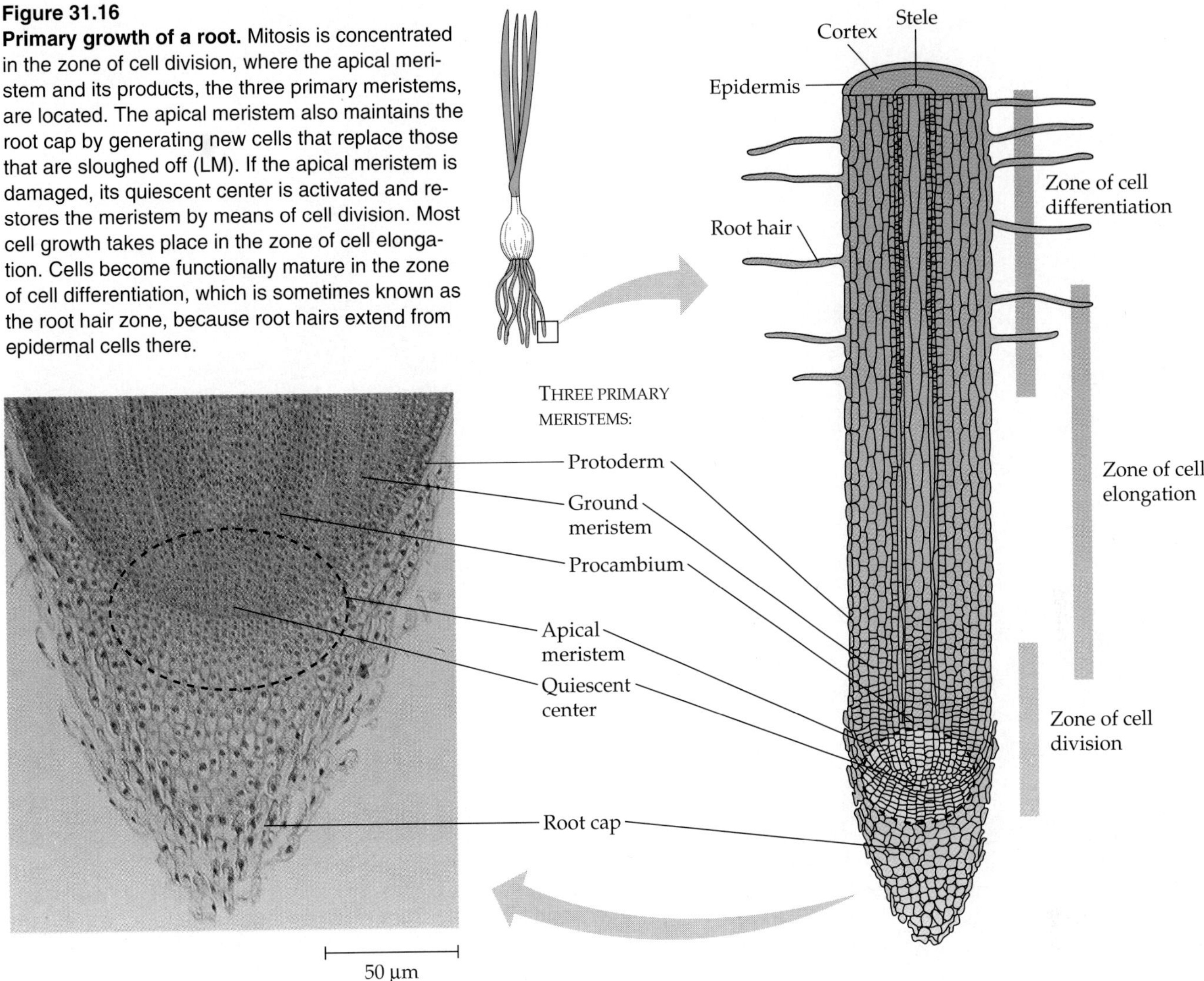

Figure 31.16
Primary growth of a root. Mitosis is concentrated in the zone of cell division, where the apical meristem and its products, the three primary meristems, are located. The apical meristem also maintains the root cap by generating new cells that replace those that are sloughed off (LM). If the apical meristem is damaged, its quiescent center is activated and restores the meristem by means of cell division. Most cell growth takes place in the zone of cell elongation. Cells become functionally mature in the zone of cell differentiation, which is sometimes known as the root hair zone, because root hairs extend from epidermal cells there.

curs during the second year, after a year of vegetative growth. Beets and carrots are biennials, but we rarely leave them in the ground long enough to see them flower. Plants that live many years, including trees, shrubs, and some grasses, are known as **perennials.** Some of the buffalo grass of the North American plains is believed to have been growing for 10,000 years from seeds that sprouted at the close of the last ice age. Perhaps these perennials have the potential to survive and continue growing for thousands more years, if they can escape disease, accidents such as floods and wildfires, climatic change, and encroachment by humans.

A CLOSER LOOK AT PRIMARY GROWTH

Primary growth produces what is called the **primary plant body,** which consists of the three tissue systems: dermal, vascular, and ground tissues (see Figure 31.14). A herbaceous plant and the youngest parts of a woody plant represent the primary plant body. Although apical meristems are responsible for the primary growth of both roots and shoots, there are important differences in the primary growth of these two kinds of organs.

Primary Growth of Roots

Primary growth pushes roots through the soil. The root tip is covered by a thimblelike **root cap,** which protects the delicate meristem as the root elongates through the abrasive soil. The cap also secretes a polysaccharide slime that lubricates the soil around the growing root tip. Growth in length is concentrated near the root's tip, where three zones of cells at successive stages of primary growth are located. From the root tip upward, they are the zone of cell division, the zone of cell elongation, and the zone of cell differentiation. These regions grade together, with no sharp boundaries (Figure 31.16).

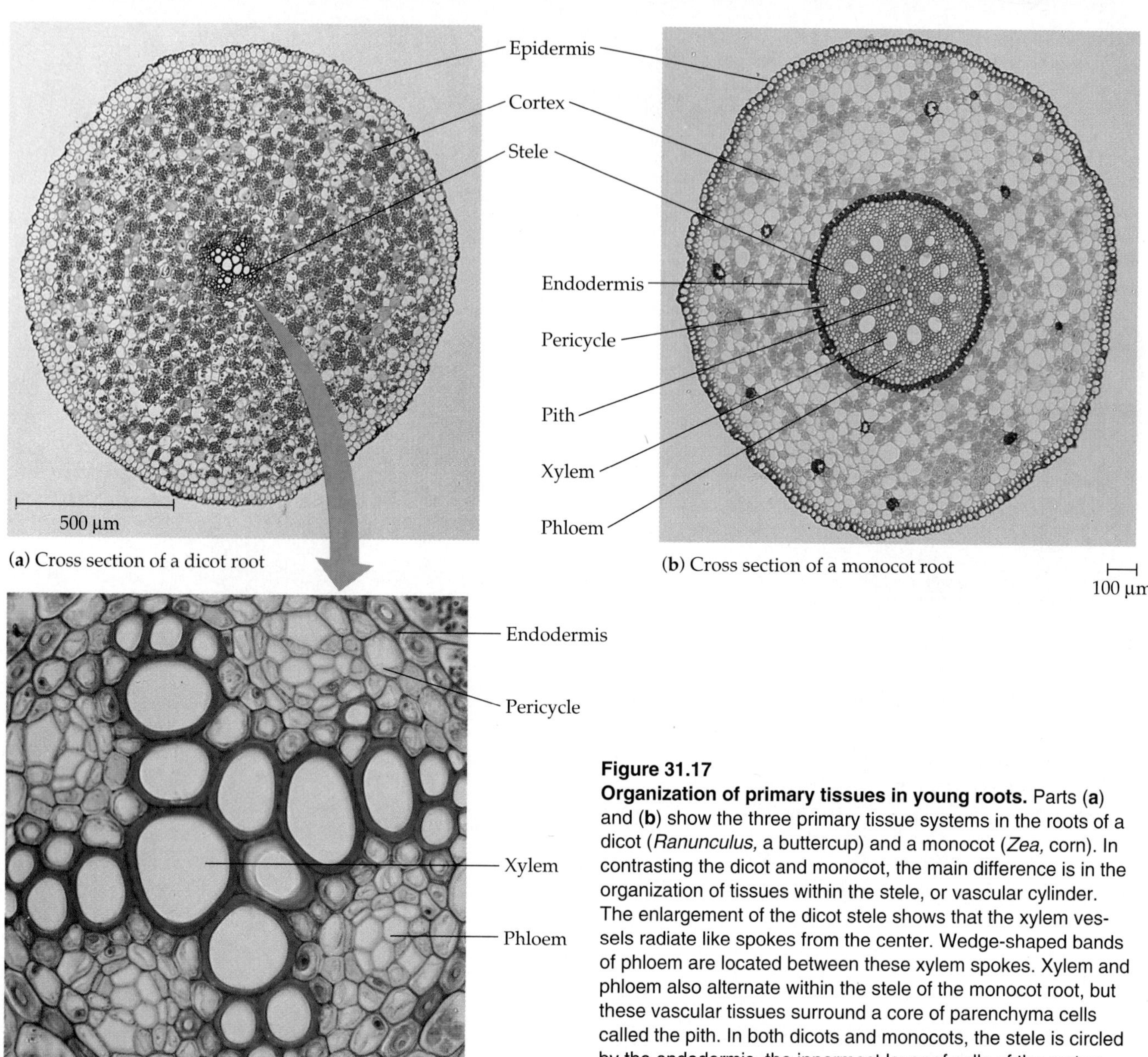

(a) Cross section of a dicot root

(b) Cross section of a monocot root

Figure 31.17
Organization of primary tissues in young roots. Parts (**a**) and (**b**) show the three primary tissue systems in the roots of a dicot (*Ranunculus*, a buttercup) and a monocot (*Zea*, corn). In contrasting the dicot and monocot, the main difference is in the organization of tissues within the stele, or vascular cylinder. The enlargement of the dicot stele shows that the xylem vessels radiate like spokes from the center. Wedge-shaped bands of phloem are located between these xylem spokes. Xylem and phloem also alternate within the stele of the monocot root, but these vascular tissues surround a core of parenchyma cells called the pith. In both dicots and monocots, the stele is circled by the endodermis, the innermost layer of cells of the cortex. Just inside the endodermis is the pericycle, a layer of cells with the potential to divide and give rise to lateral roots. (All LMs.)

The **zone of cell division** includes the apical meristem and its derivatives, called primary meristems. The apical meristem, at the heart of the zone of cell division, produces the cells of the primary meristems and also replaces cells of the root cap that are sloughed off. Near the center of the apical meristem is the **quiescent center,** a population of cells that divide much more slowly than the other meristematic cells. Cells of the quiescent center are relatively resistant to damage from radiation and toxic chemicals, and they may function as reserves that can be recruited to restore the meristem if it is somehow damaged. In experiments where part of the apical meristem is removed, cells of the quiescent center become more mitotically active and produce a new meristem. Just above the apical meristem, the products of its cell division form three concentric cylinders of cells that continue to divide for some time. These are the primary meristems—the protoderm, procambium, and ground meristem—which will produce the three primary tissue systems of the root: dermal, vascular, and ground tissues.

The zone of cell division blends into the **zone of cell elongation.** Here the cells elongate to more than ten times their original length. Although the meristem provides the new cells for growth, the elongation of cells is mainly responsible for pushing the root tip, including the meristem, ahead. The meristem sustains growth by continuously adding cells to the youngest end of the zone of elongation. Even before they finish elongating, the cells of the root begin to specialize in structure and function where the zone of elongation grades into the **zone of cell differentiation.** In this latter zone of the root, the three tissue systems produced by primary growth complete their differentiation.

Primary Tissues of Roots The three primary meristems give rise to the three primary tissues of roots, shown in Figure 31.17. The **protoderm,** the outermost primary meristem, gives rise to the epidermis, a single layer of cells covering the root. Water and minerals that enter the plant from the soil must cross the epidermis. The root hairs enhance this process by greatly increasing the surface area of epidermal cells.

The **procambium** gives rise to a central vascular cylinder, or **stele,** where xylem and phloem develop. The specific arrangement of the two vascular tissues varies. In most dicots, the xylem cells radiate from the center of the stele in two or more spokes, with phloem developing in the wedges between the spokes. The stele of a monocot generally has a central core of parenchyma cells, often called the **pith,** which is ringed by vascular tissue with the same alternating pattern of xylem and phloem as in dicots.

Between the protoderm and procambium is the **ground meristem,** which gives rise to the ground tissue system. The ground tissue, which is mostly parenchyma, fills the **cortex,** the region of the root between the stele and epidermis. Ground tissue cells of the root store food, and their plasma membranes are active in the uptake of minerals that enter the root with the soil solution. The innermost layer of the cortex is the **endodermis,** a cylinder one cell thick that forms the boundary between the cortex and the stele. The endodermis functions as a selective barrier that regulates the passage of substances from the soil solution into the vascular tissue of the stele. (The structure and function of endodermis will be discussed in detail in Chapter 32.)

An established root may sprout **lateral roots,** which arise from the outermost layer of the stele (Figure 31.18). Just inside the endodermis is the **pericycle,** a layer of cells that may become meristematic and begin dividing again. Originating as a clump of cells formed by mitosis in the pericycle, the lateral root elongates and pushes through the cortex until it emerges from the primary root. The stele of the lateral root retains its connection with the stele of the primary root, making the vascular tissue continuous throughout the root system.

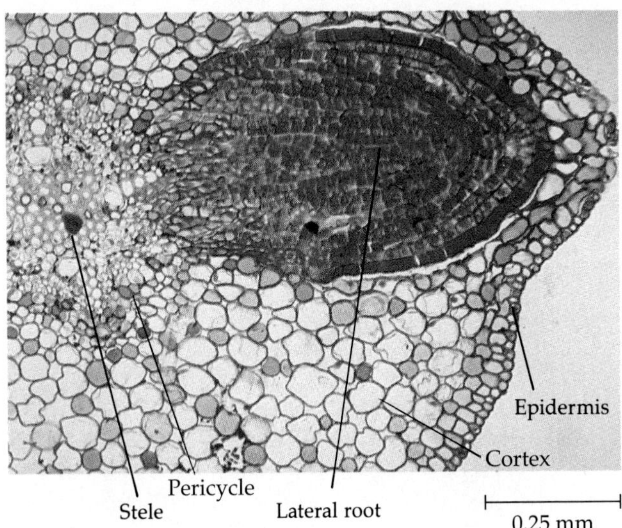

Figure 31.18
The formation of lateral roots. In this cross section of a root, a lateral root emerges from the pericycle, the outermost layer of the stele (LM).

Primary Growth of Shoots

The apical meristem of a shoot is a dome-shaped mass of dividing cells at the tip of the terminal bud (Figure 31.19a). As in the root, the apical meristem of the shoot tip gives rise to the primary meristems—protoderm, procambium, and ground meristem—which will differentiate into the three tissue systems. Leaves arise as tiny bulges, called leaf primordia, on the flanks of the apical meristem, which is dome-shaped. Axillary buds develop from islands of meristematic cells left by the apical meristem at the bases of the leaf primordia.

Within a bud, nodes, with their leaf primordia, are crowded close together, because internodes are very short (Figure 31.19b). Most of the actual elongation of the shoot occurs by the growth of slightly older internodes below the shoot apex. This growth is due to both cell division and cell elongation within the internode. In some plants, including grasses, internodes continue to elongate all along the length of the shoot over a prolonged period. This is possible because these plants have meristematic regions, called **intercalary meristems,** at the base of each internode.

Axillary buds have the potential to form branches of the shoot system at some later time (see Figure 31.5). Thus, there is an important difference in how roots and shoots form lateral extensions. Lateral roots originate from deep within a main root as outgrowths from the pericycle (see Figure 31.18). In contrast, branches of the shoot system originate from axillary buds, located at the surface of a main shoot. Only by extending from the stele can a lateral root be connected to the plant's vascular system. The vascular tissue of a stem, however, is near the surface (see Figure 31.14), and branches can develop with connections to the vascular tissue without having to originate from deep within the main shoot.

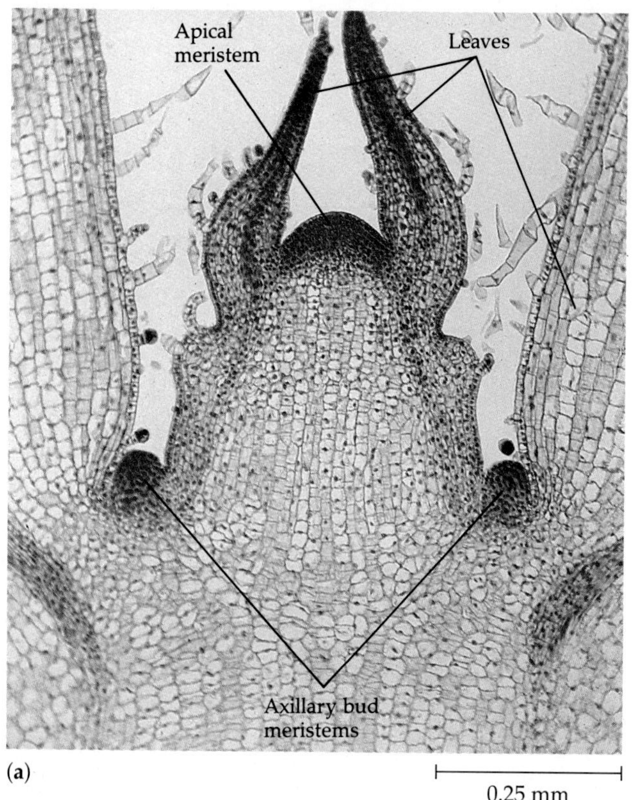

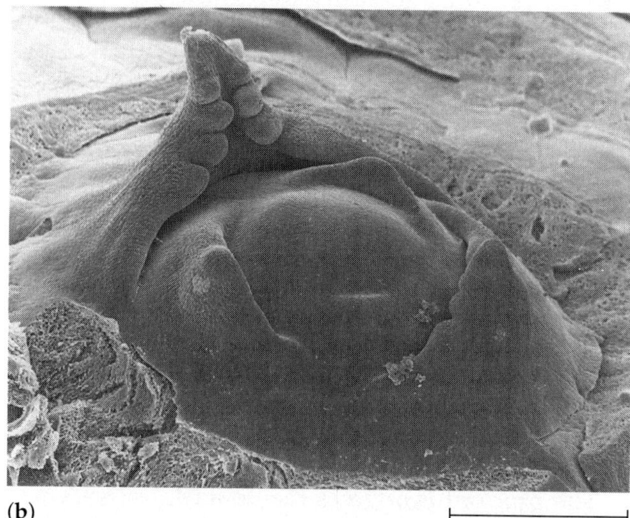

Figure 31.19
The terminal bud and primary growth of a shoot. (**a**) Leaf primordia arise from the flanks of the apical dome. The apical meristem gives rise to protoderm, procambium, and ground meristem, which in turn develop into the three tissue systems. This is a longitudinal section of the shoot tip of *Coleus* (LM). (**b**) Here we can observe the shoot apex of celery from its top side (SEM). Notice that successive leaves form very close together. Below this region, the division and growth of cells will cause internodes to elongate.

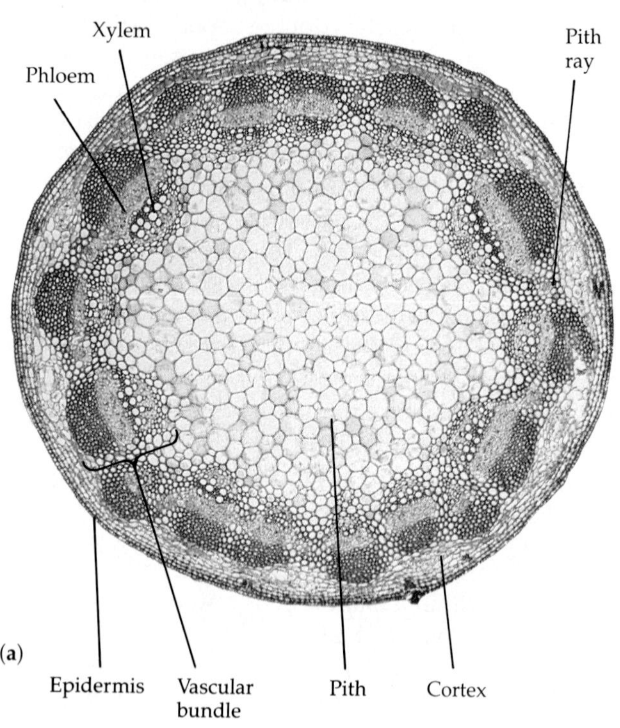

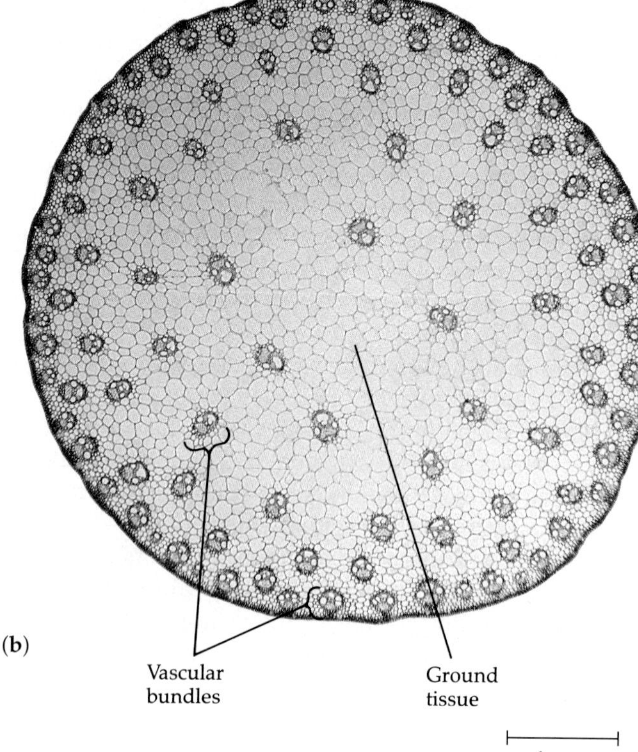

Figure 31.20
Organization of primary tissues in young stems. (**a**) A dicot stem (sunflower) with vascular bundles arranged in a ring. The ground tissue system consists of an outer cortex and an inner pith, surrounded by vascular bundles. Between the bundles are pith rays (LM). (**b**) A monocot stem (corn) with vascular bundles arranged in a complex manner throughout the ground tissue (LM).

Figure 31.21
Leaf anatomy. (**a**) This cutaway drawing of a leaf illustrates the organization of the three tissue systems: dermal tissue (epidermis), vascular tissue, and ground tissue (mesophyll, consisting of palisade parenchyma and spongy mesophyll). (**b**) This surface view of a *Tradescantia* leaf shows the cells of the epidermis and stomata, with their guard cells (LM). (**c**) Palisade and spongy regions of mesophyll are present within the leaf of a lilac, a dicot (LM).

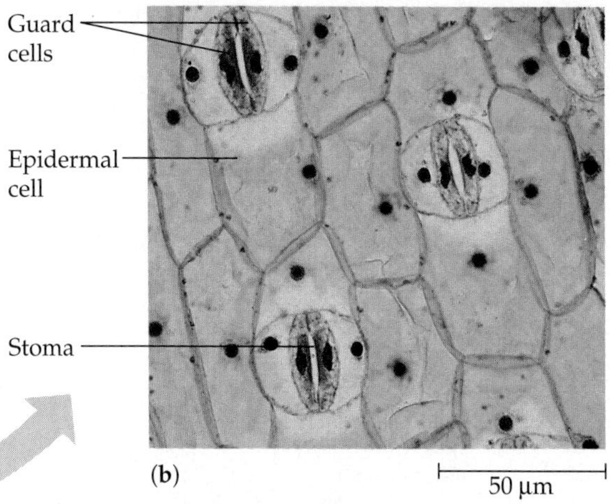

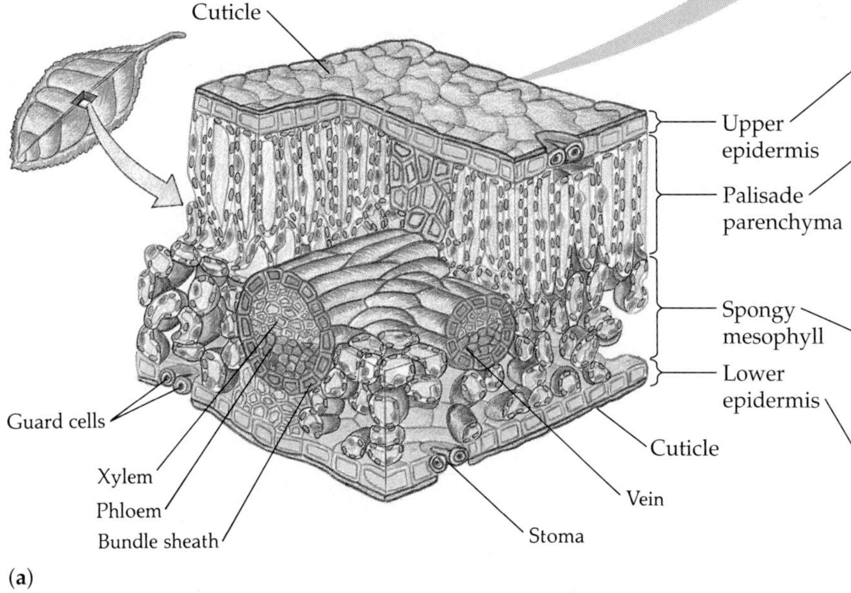

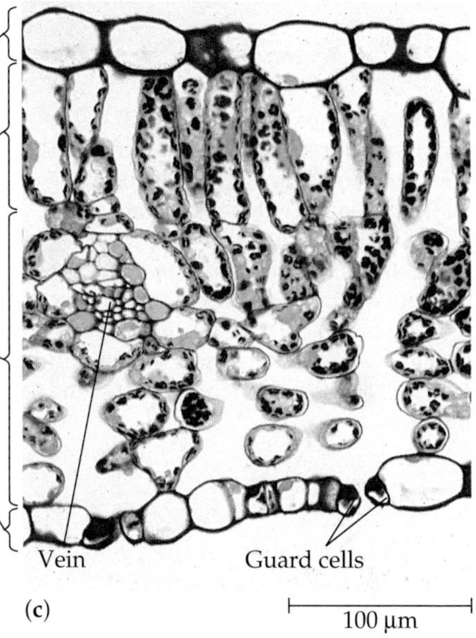

Primary Tissues of Stems Vascular tissue runs the length of a stem in several strands called **vascular bundles** (Figure 31.20), in contrast with the root, where the vascular tissue forms a single stele consisting of the entire united set of vascular bundles. At the transition zone where the shoot grades into the root, the vascular bundles converge to join the root stele.

Each vascular bundle of the stem is surrounded by ground tissue. In most dicots, the vascular bundles are arranged in a ring, with pith to the inside of the ring and cortex external to the ring. Both pith and cortex are part of the ground tissue system. The vascular bundles have their xylem facing the pith and their phloem facing the cortex side. The pith and cortex are connected by pith rays, thin layers of ground tissue between the vascular bundles. In the stems of most monocots, the vascular bundles are arranged throughout the ground tissue rather than in a ring, so the ground tissue is not divided into pith and cortex regions. The ground tissue of the stem is mostly parenchyma, but many stems are strengthened by collenchyma located just beneath the epidermis (see Figure 31.12a).

The protoderm of the terminal bud gives rise to the epidermis, which covers stems and leaves as part of the continuous dermal tissue system (see Figure 31.14).

Tissue Organization of Leaves The leaf is cloaked by its epidermis, with cells tightly interlocked like pieces of a puzzle (Figure 31.21). This epidermis, like our own skin, is a first line of defense against physical damage and pathogenic organisms. Also, the waxy cuticle of the epidermis is a barrier to the loss of water from the plant. The epidermal barrier is interrupted only by the **stomata,** tiny pores flanked by specialized epidermal cells called **guard cells.** Each stoma is actually a gap between a pair of guard cells. The stomata allow gas exchange between the surrounding air and the photosynthetic cells inside the leaf, but they are also the major avenues for the loss of water from the plant by evaporation, a process called **transpiration.** The role of the guard cells in regulating gas exchange and transpiration will be discussed in Chapter 32.

The ground tissue of a leaf, sandwiched between the upper and lower epidermis, is called **mesophyll** (Gr.

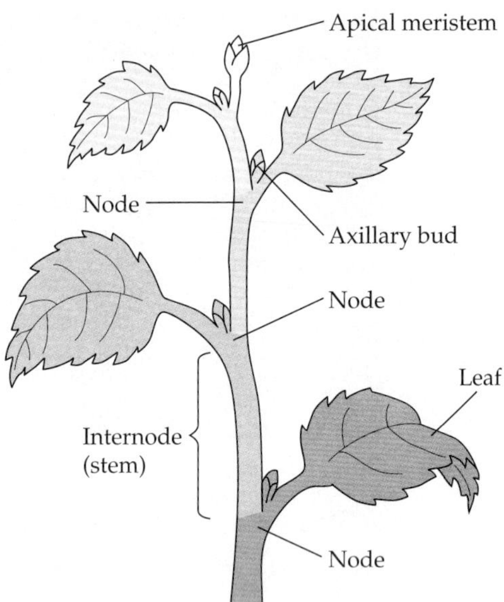

Figure 31.22
Modular construction of a shoot. Primary growth lays down a series of segments (different colors in this drawing), each consisting of a node with one or more leaves, an axillary bud in the axil of each leaf, and an internode. Miniature modules develop within the apical meristem and then grow, pushing the apex onward, where it forms the next module, and so on. This serial addition of segments at the growing end of the shoot contrasts with the development of segmentation in certain animals, in which all the segments form at about the same time in the embryo.

(a) (b)

Figure 31.23
Phase change in the shoot system of *Eucalyptus.* A silver-dollar eucalyptus *(Eucalyptus polyanthemos)* has both (**a**) juvenile leaves (the round "silver dollars") and (**b**) mature leaves (the lanceolate leaves). This dual foliage reflects a phase change in the development of the apical meristem of each shoot. In its juvenile vegetative phase, a meristem lays down modules on which round leaves develop. As the meristem changes gradually to the mature vegetative phase, the leaves become more and more lanceolate. Once a module forms, its developmental phase—juvenile or mature—is fixed; that is, round leaves do not mature into lanceolate leaves. The developmental status of the apical meristem also sets the phase of the axillary buds it forms.

mesos, "middle," and *phyll,* "leaf"). It consists mainly of parenchyma cells equipped with chloroplasts and specialized for photosynthesis. The leaves of many dicots have two distinct regions of mesophyll (Figure 31.21c). On the upper half of the leaf are one or more layers of palisade parenchyma, made up of cells that are columnar in shape. Below the palisade region is the spongy mesophyll, which gets its name from the labyrinth of air spaces through which carbon dioxide and oxygen circulate around the irregularly shaped cells and up to the palisade region. The air spaces are particularly large in the vicinity of stomata, where gas exchange with the outside air occurs. In most plants, stomata are more numerous on the bottom surface of a leaf than on top (see Figure 31.21c). This adaptation minimizes water loss, which occurs more rapidly through stomata on the sunny upper side of a leaf. Once again, it helps to view the functional structure of plants in the evolutionary context of adaptation to land.

The vascular tissue of a leaf is continuous with the xylem and phloem of the stem. Leaf traces, which are branches from vascular bundles in the stem, pass through petioles and into leaves. Within a leaf, veins subdivide repeatedly and ramify throughout the mesophyll. This brings xylem and phloem into close contact with the photosynthetic tissue, which obtains water and minerals from the xylem and loads its sugars and other organic products into the phloem for shipment to other parts of the plant. The vascular infrastructure also functions as a skeleton that supports the ground tissue (mesophyll) of the leaf (see Figure 31.21).

Modular Shoot Construction and Phase Changes During Development Serial development of nodes and internodes within the shoot apex followed by elongation of the internodes produces a shoot having a modular construction—a series of segments, each consisting of a stem, one or more leaves, and an axillary bud associated with each leaf (Figure 31.22). The development of this modular morphology should not be confused with how the segmented anatomy of certain animals such as earthworms develops. In animal development, the rudiments of all organs form in the embryo. Plants, in contrast, add organs at their tips for as long as they live. Unlike the segments of an earthworm, which are all the same age, the modules of a plant vary in age in proportion to their distance from an apical meristem.

From what we have learned so far about the shoot apex and primary growth, it would seem as if the meristem lays down a series of identical modules for as long as the shoot lives. In fact, the apical meristem can change from one developmental phase to another during its history. One of these phase changes is a gradual transition from a juvenile vegetative (leaf-producing) state to a mature vegetative state. Modification of leaf morphology is usually the most obvious sign of this phase change. In Figure 31.23, for example, we can

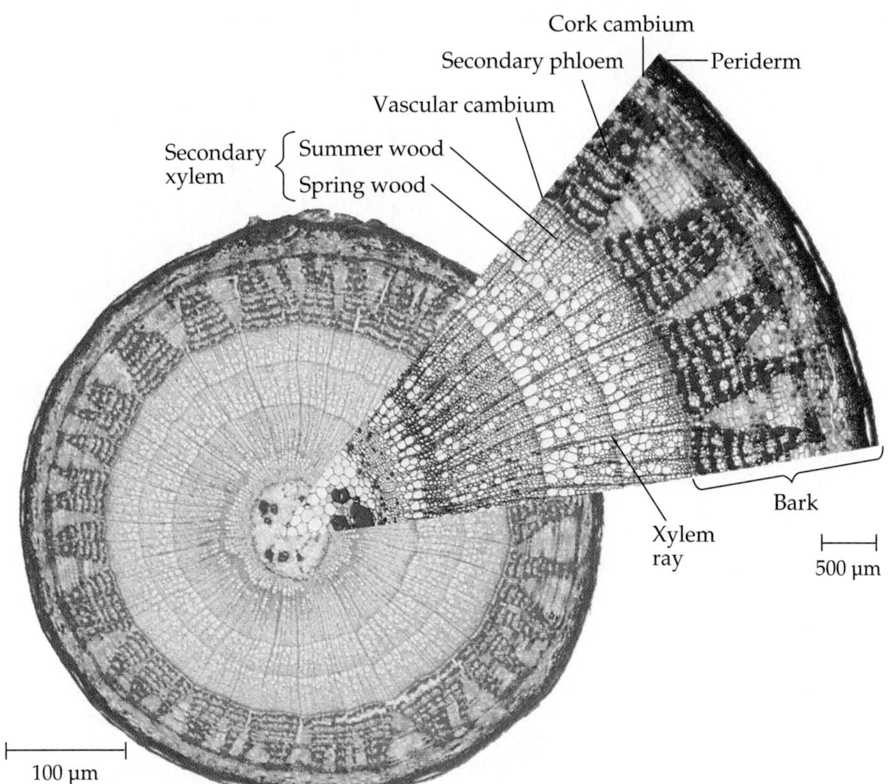

Figure 31.24
Anatomy of a woody dicot stem. Several years of secondary growth are apparent as growth rings in this cross section of a stem from *Tilia*, the American linden (basswood). At the boundary of one season's growth to the next, notice the spring wood and summer wood. Vascular cambium produces the secondary growth. Secondary xylem and secondary phloem are derived from fusiform initials, cambium cells that develop within the vascular bundles. The cambium cells that develop in the intervening pith rays are called ray initials and give rise to the xylem rays, structures that function in radial transport and storage. A second lateral meristem, the cork cambium, forms a protective covering called periderm. The bark consists of all tissues external to the vascular cambium. (Both LMs.)

compare the leaves of juvenile and mature regions of a *Eucalyptus* tree. Once the meristem has laid down juvenile nodes and internodes, they retain that status even as the shoot continues to elongate and the meristem eventually changes to the mature phase. If axillary buds give rise to branches, those shoots reflect the developmental phase of the main shoot modules from which they arise. The juvenile ⟶ mature transition is another case where it is misleading to compare plant and animal development. In an animal, this transition occurs at the level of the entire organism. In plants, phase changes during the history of apical meristems result in juvenile and mature regions coexisting along the axis of each shoot.

In some cases, a shoot apex undergoes a second phase transition from a mature vegetative state to a reproductive (flower-producing) state. Unlike vegetative growth, which is self-renewing, production of a flower by an apical meristem terminates primary growth of that shoot tip. In Chapter 34, we will study flower development in more detail, and in Chapter 35, we will examine how this phase change from vegetative growth of a shoot to flowering is controlled.

A CLOSER LOOK AT SECONDARY GROWTH

Most vascular plants grow in girth as well as length. Secondary growth produces this thickening of organs, and the **secondary plant body** consists of the secondary tissues produced during growth in diameter. Two lateral meristems function in secondary growth: the vascular cambium, which produces secondary xylem and phloem; and the cork cambium, which produces a tough covering for stems and roots that replaces the epidermis. Secondary growth occurs in all gymnosperms. Among angiosperms, secondary growth takes place in most dicot species, but is rare in monocots.

Secondary Growth of Stems

Vascular Cambium The **vascular cambium** forms from parenchyma cells that develop the capacity to divide; that is, the cells become meristematic. This transition to meristematic activity takes place in a layer between the primary xylem and primary phloem of each vascular bundle and in the pith rays between the bundles (Figure 31.24). Together, the meristematic bands in these side-by-side locations give rise to the vascular cambium as a continuous cylinder of dividing cells surrounding the primary xylem and pith of the stem. The cambium cells of the pith rays are called **ray initials.** They produce radial files of parenchyma cells known as **xylem rays,** which function as living avenues for the lateral transport of water and nutrients and in the storage of starch and other reserves. The cambium cells within the vascular bundles are the **fusiform initials,** a name that refers to the shape of

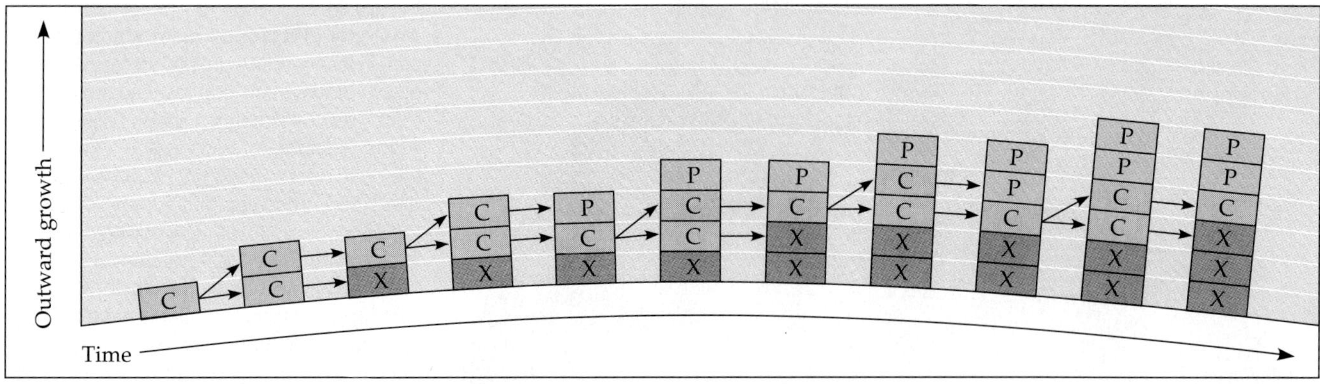

Figure 31.25
Production of secondary xylem and phloem by the vascular cambium. In this diagram, we trace the file of cells that develops from the meristematic activity of a single fusiform initial of the vascular cambium. The cambium cell (C) gives rise to xylem (X) on the inside and phloem (P) on the outside. Each time an initial divides, one daughter cell retains its status as an initial, and the other, the derivative, differentiates into a xylem or phloem cell. As layers of xylem are added, the position of the cambium becomes more distant from the center of the stem. Cell division of initials in the tangential plane of the stem enables the circumference of the vascular cambium to increase as secondary growth progresses.

these cells, which have tapered (fusiform) ends and are elongated along the axis of the stem. Fusiform initials produce new vascular tissue, forming secondary xylem to the inside of the vascular cambium and secondary phloem to the outside (Figure 31.25).

As secondary growth continues over the years, layer upon layer of secondary xylem accumulates to produce what we call wood. Wood consists mainly of tracheids, vessel elements (in angiosperms), and fibers. These cells, dead at functional maturity, have thick, lignified walls that give wood its hardness and strength. In temperate regions of the world, secondary growth in perennial plants is interrupted each year when the vascular cambium becomes dormant during winter. When secondary growth resumes in the spring, the first tracheids and vessel elements to develop usually have relatively large diameters and thin walls compared to the secondary xylem produced later in the summer. Thus, it is usually possible to distinguish spring wood from summer wood (see Figure 31.24). The annual growth rings that are evident in cross sections of most tree trunks in temperate regions result from this yearly activity of the vascular cambium: cambium dormancy, spring wood production, and summer wood production. The boundary between one year's growth and the next is usually quite conspicuous, sometimes allowing us to estimate the age of a tree by counting its annual rings.

The secondary phloem, external to the vascular cambium, does not accumulate over the years as the secondary xylem does. As a tree grows in girth, the older (outermost) secondary phloem, and all tissues external to it, develop into bark, which eventually splits and sloughs off the tree trunk.

Cork Cambium During secondary growth, the epidermis produced by primary growth splits, dries, and falls off the stem. It is replaced by new protective tissues produced by the **cork cambium,** a cylinder of meristematic tissue that first forms in the outer cortex of the stem (Figure 31.26; see also Figure 31.24). As initials in the cork cambium divide, the inner cells remain meristematic and the outer cells develop into cork cells. Thus, layers of cork cells are added outside the perimeter of the cork cambium. As these derivatives differentiate into cork cells, they deposit a waxy material called suberin in their walls and then die. It is in death that cork cells reach functional maturity, as barriers that help protect the stem from physical damage and pathogens. And because cork is waxy, it impedes water loss from stems. Together, the layers of cork and the cork cambium from which they are derived make up the **periderm,** the protective coat of the secondary plant body that replaces the epidermis of the primary body. The term **bark,** more inclusive than periderm, refers to all tissues external to the vascular cambium. Thus, in an outward direction, bark consists of phloem, cork cambium, and cork, or phloem plus periderm.

Unlike the vascular cambium, which grows in diameter, the original cork cambium is a cylinder of fixed size. After a few weeks of cork production, the cork cambium loses its meristematic activity, and its remaining cells differentiate into cork. Expansion of the stem splits the original periderm. How is it renewed to keep pace with continued secondary growth? New cork cambium forms deeper and deeper in the cortex. Eventually, no cortex is left, and the cork cambium then develops from parenchyma cells in the secondary phloem.

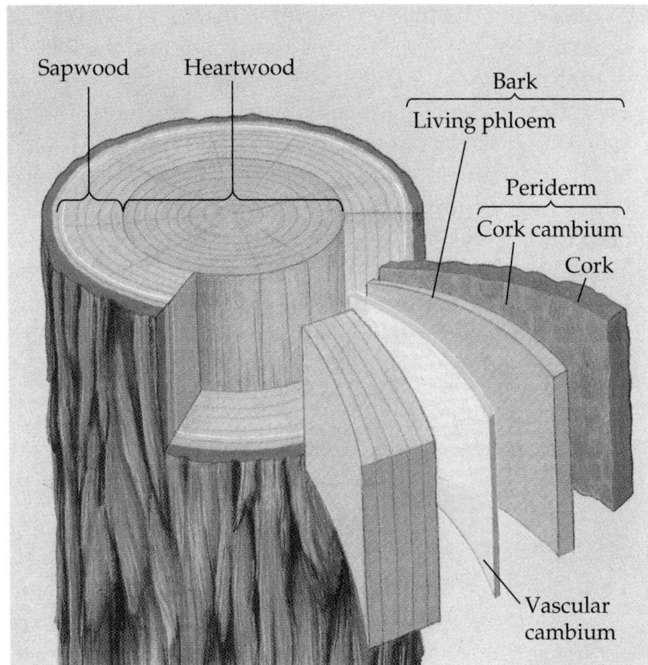

Figure 31.26
Anatomy of a tree trunk. Beginning at the center of the tree and tracing outward, we can distinguish several zones. Heartwood and sapwood both consist of secondary xylem. Heartwood is older and no longer functions in water transport; the lignified walls of its dead cells form a central column that supports the tree. This wood owes its rich color to resins and other compounds that clog the cell cavities and help protect the core of the tree from fungi and insects. Sapwood is so named because its secondary xylem cells still function in the upward transport of water and minerals (xylem sap). Since each new layer of secondary xylem has a larger circumference, secondary growth enables the xylem to transport more sap each year, providing water and minerals to an increasing number of leaves. Secondary xylem is produced by the vascular cambium, located next to the youngest xylem cells. To its outside, the vascular cambium produces secondary phloem. The outer, older layers are crushed by the expansion of the secondary xylem. The periderm, the protective coating of the secondary plant body, is made up of the cork cambium and its products, layers of dead cork cells. Bark consists of all tissues exterior to the vascular cambium—phloem and periderm (cork cambium and cork).

Only the youngest secondary phloem, which is internal to the cork cambium, functions in sugar transport. The older secondary phloem, outside the cork cambium, dies and helps protect the stem until it is sloughed off as part of the bark during later seasons of secondary growth. Spongy regions in the bark called **lenticels** make it possible for living cells within the trunk to exchange gases with the outside air for cellular respiration.

The result of many years of secondary stem growth can be seen by examining an old tree trunk in cross section (see Figure 31.26). Figure 31.27 will help you review the relationships among the primary and secondary tissues.

Secondary Growth of Roots

The two lateral meristems, vascular cambium and cork cambium, also develop and produce secondary growth in roots. The vascular cambium forms within the stele and produces secondary xylem to its inside and secondary phloem to its outside. As the stele grows in diameter, the cortex and epidermis are split and shed. A cork cambium forms from the pericycle of the stele and produces the periderm, which becomes the secondary dermal tissue. Unlike the primary epidermis of a younger root, periderm is impermeable to water. But it is only the youngest roots, those representing the primary plant body, that absorb water and

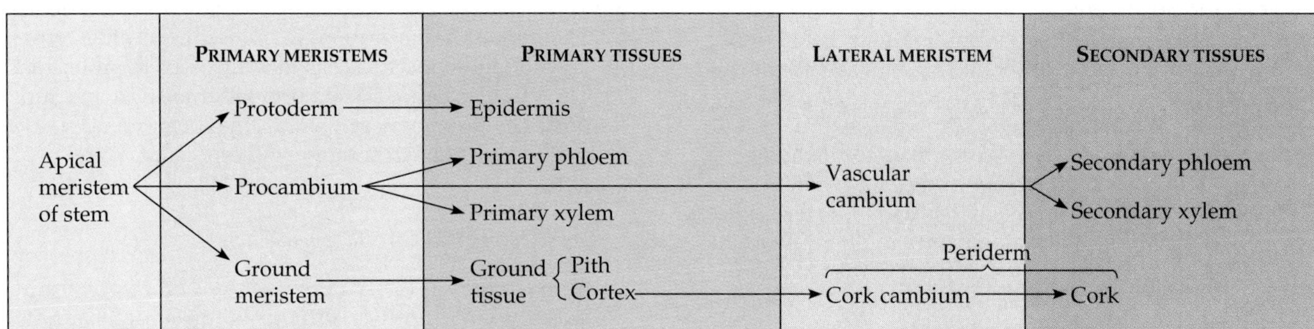

Figure 31.27
Primary and secondary growth in a woody stem: a summary.

minerals from the soil. Older roots, with secondary growth, function mainly to anchor the plant and to transport water and solutes between the younger roots and the shoot system.

Over the years, the root becomes more woody, and annual rings are usually evident in the secondary xylem. The tissues external to the vascular cambium form a thick, tough bark. After extensive secondary growth, old stems and old roots are quite similar.

* * *

In dissecting the plant to examine its parts, as we have done in this chapter, we must remember that the whole plant functions as an integrated organism, not as an aggregate of independent cells, tissues, or organs. In the following chapters, you will learn more about how materials are transported within the plant, how plants obtain nutrients, how plants reproduce and develop, and how the various functions of the plant are coordinated. Your understanding of the working plant will be enhanced by remembering that structure fits function and that interactions with the environment affect the anatomy and physiology of plants.

STUDY OUTLINE

1. Plants are the main producers in most terrestrial ecosystems.

An Introduction to Plant Biology: The Major Themes (pp. 674–676)

1. A plant represents a unit of structure and function adapted to living on land through evolution and through individual response to the environment.
2. Although plants and terrestrial animals face many of the same environmental problems, their adaptations evolved independently.

Basic Morphology of Flowering Plants: An Evolutionary Perspective (pp. 676–681)

1. Angiosperms (flowering plants) are the most diverse and widespread of plant divisions. Based on differences in anatomy, the angiosperms can be divided into two classes: monocots and dicots.
2. The differentiation of the plant body into an underground root system and an aerial shoot system is an adaptation to terrestrial life. It provides water and nutrients from the soil to a photosynthetic apparatus requiring light and carbon dioxide.
3. Vascular tissues integrate the parts of the plant body. Water and minerals move up from the roots in the xylem; sugar travels to nonphotosynthetic parts in the phloem.
4. The structure of roots is adapted to anchor the plant, absorb and conduct water and minerals, and store food. Tiny root hairs near the root tips enhance absorption.
5. The shoot system consists of stems, leaves, and flowers.
6. Leaves are attached by their petioles to the nodes of stems. The growth of axillary buds, embryonic shoots formed in the angle between the leaf and the stem, is kept in check partly by apical dominance in the presence of terminal buds. Axillary buds that are stimulated to grow may become flowers or vegetative branches. Stolons, rhizomes, and bulbs are modified stems.
7. Leaves, the primary photosynthetic organs, show extensive variation. Monocots differ from dicots in the arrangement of major veins in their leaves.

Basic Plant Anatomy: Plant Cells and Tissues (pp. 681–686)

1. Parenchyma cells are the least specialized plant cells, serving general metabolic, synthetic, and storage functions. They retain the ability to divide and differentiate into other cell types under certain conditions.
2. Collenchyma cells often occur in strands or cylinders that support young parts of the plant shoot without restraining growth. They lack secondary walls, which allows them to elongate with growing stems and leaves.
3. Sclerenchyma cells have thick, lignified secondary walls and many lack protoplasts; thus, at maturity they are unable to elongate. Both fiber and sclereid types occur, acting as scaffolding for the plant.
4. Water-conducting xylem tissue is composed of elongated tracheid and vessel element cells that are dead at functional maturity. Tracheids are long, thin, tapered cells with lignified secondary walls that function in support and permit water flow through pits. The wider, shorter, and thinner-walled vessel elements have perforated ends through which water flows freely.
5. Sieve-tube members are living cells that form phloem tubes for transport of sucrose and other organic nutrients. These cells lack nuclei and ribosomes and possess porous end walls called sieve plates. Each sieve-tube member is connected to one or more companion cells through plasmodesmata.
6. Plant tissues are arranged into three continuous systems. The dermal tissue system, or epidermis, is an external layer of tightly packed cells that functions in protection. The vascular tissue system provides transport and support. The predominantly parenchymous ground tissue system functions in organic synthesis, storage, and support.

An Overview of Plant Growth (pp. 686–687)

1. Because they possess permanently embryonic meristems, plants, unlike animals, show indeterminate growth.
2. Apical meristems at root tips and shoot buds initiate primary growth (growth in length) and the formation of the

three tissue systems. Lateral meristems are responsible for secondary growth (growth in thickness).

3. Plants that complete their life cycles in one year are called annuals, those with two-year life cycles are called biennials, and those that live many years are called perennials.

A Closer Look at Primary Growth (pp. 687-693)

1. Primary growth produces the primary plant body, which consists of the three tissue systems.
2. Root tips, protected by the root caps, grow and develop by activity of cells in the successive zones of cell division, cell elongation, and cell differentiation.
3. Just behind the apical meristem in the zone of cell division are the three primary meristems of the root. The protoderm gives rise to the epidermis, the procambium forms the central vascular stele, and the ground meristem produces the ground tissue of the cortex. Subsequent lateral roots arise from the pericycle of the stele.
4. The elongation of shoots comes from the dome-shaped apical meristem at the top of the terminal bud. Leaves arise from the sides of the apical dome from leaf primordia, and axillary buds arise from residual islands of meristematic cells at the bases of leaf primordia.
5. In contrast to the single stele of the root, the vascular tissue of stems runs in vascular bundles surrounded by ground tissue in characteristic patterns that differ between monocots and dicots.
6. Leaves are covered with a waxy epidermis. Pairs of guard cells flank openings called stomata, through which gas exchange and transpiration occur. Between the upper and lower epidermis, the ground tissue, or mesophyll, consists mainly of parenchyma cells equipped with chloroplasts for photosynthesis. A strand of vascular tissue called the leaf trace connects the veins of the leaf with the vascular tissue of the stem.
7. The shoot is a series of modules, each consisting of a node with leaves, an axillary bud, and an internode. Phase changes in the development of the shoot tip alter the morphology of modules.

A Closer Look at Secondary Growth (pp. 693-696)

1. Secondary growth produces the secondary plant body, the tissues that cause an increase in diameter.
2. The increase in the girth of stems and roots is due to secondary production of new cells by the vascular cambium and the cork cambium, two lateral meristems.
3. The vascular cambium, a continuous cylinder of meristematic cells arising between the xylem and phloem of each vascular bundle, produces secondary xylem internally and secondary phloem externally. In temperate regions, seasonal wood production cycles result in annual growth rings. The external secondary phloem eventually splits and sloughs off during growth.
4. The cork cambium, a meristematic cylinder in the outer cortex of the stem, produces waxy cork cells externally. The cork cambium and cork make up the periderm, which replaces the epidermis that sloughs off during secondary growth. Secondary phloem gives rise to new cork cambium after the original cortex is shed.
5. In roots, the vascular cambium arises between the xylem and phloem of the stele and functions similarly to that in stems. Cork cambium, produced from the pericycle of the stele, forms the periderm that replaces cortex and epidermis. Old roots resemble old stems, even in possessing annual growth rings.

SELF-QUIZ

1. Most absorption of water and dissolved minerals occurs through
   a. adventitious roots
   b. stolons
   c. taproots
   d. root caps
   e. root hairs
2. Which of the following is *not* a correctly stated difference between monocots and dicots?
   a. parallel veins in monocots; branching netlike venation in dicot leaves
   b. vascular bundles scattered in monocot stem; central vascular stele in dicot stem
   c. flower parts in threes in monocots; flower parts in multiples of four or five in dicots
   d. usually only primary growth in monocots; secondary growth in many dicots
   e. one cotyledon in monocots; two cotyledons in dicot seeds
3. The lateral roots of a young dicot originate from the
   a. pericycle of the taproot
   b. endodermis of fibrous roots
   c. meristematic cells of the protoderm
   d. vascular cambium
   e. root cortex
4. A sieve-tube member would likely lose its nucleus in which zone of growth in a root?
   a. zone of cell division
   b. zone of cell elongation
   c. zone of cell differentiation
   d. zone of cell proliferation
   e. non of the above; the functionally mature cell retains its nucleus
5. Tracheids of the primary plant body originate from the
   a. protoderm
   b. procambium
   c. ground meristem
   d. xylem rays
   e. cork cambium
6. Ivy *(Hedera helix)*, like eucalyptus, undergoes a gradual change from a juvenile vegetative state to a mature vegetative state. This results in mature leaves on upper branches having a different shape than juvenile leaves on

lower branches. If this phase transition in ivy follows the same basic pattern as the transition in eucalyptus, then the lateral buds of lower branches can develop to form __________ branches, and the lateral buds of upper branches can develop to form __________ branches.

a. only juvenile; only mature
b. only mature; only juvenile
c. juvenile or mature; only juvenile
d. only mature; juvenile or mature
e. juvenile or mature; only mature

7. Unlike primary growth, secondary growth in both roots and stems
   a. is indeterminate
   b. produces xylem and phloem
   c. involves vascular and cork cambiums
   d. results in a rapid increase in root or stem length
   e. is a function of meristematic tissue
8. Which of the following is *not* part of an older tree's bark?
   a. cork
   b. cork cambium
   c. lenticels
   d. secondary xylem
   e. secondary phloem
9. When a leaf trace extends from a vascular bundle in a dicot stem, what would be the arrangement of vascular tissues in the veins of the leaf?
   a. The xylem would be on top and the phloem on the bottom.
   b. The phloem would be on top and the xylem on the bottom.
   c. The xylem would encircle the phloem.
   d. The phloem would encircle the xylem.
   e. There is no way to determine the arrangement.
10. Which of the following cell types or structures is incorrectly paired with its meristematic origin?
   a. epidermis—protoderm
   b. stele—procambium
   c. cortex—ground meristem
   d. secondary phloem—cork cambium
   e. three primary meristems—apical meristem

CHALLENGE QUESTION

1. If you were to live for the next several years in a treehouse built on the large, lower branches of a tree, would you gain much altitude as the tree grew? Explain your answer.

SCIENCE, TECHNOLOGY, AND SOCIETY

1. Make a list of the plants and plant products you use in a single average day. Which species are represented? How many species? How do you use these various plant products? Do you think that the number of plants and plant products used in everyday life has increased or decreased in the last century? Do you think the number is likely to increase or decrease in the future? Why?

FURTHER READING

Bolz, D. M. "A World of Leaves: Familiar Forms and Surprising Twists." *Smithsonian,* April 1985. A delightful article on the adaptations of leaves, featuring exquisite photographs.

Dale, J. "How Do Leaves Grow?" *BioScience,* June 1992. Using new techniques of molecular and cell biology, researchers are answering long-standing questions about plant structure and growth.

Kaplan, D. R., and W. Hagemann. "The Relationship of Cell and Organism in Vascular Plants: Are Cells the Building Blocks of Plant Form?" *BioScience,* November 1991. Important differences in how plants and animals are organized.

Mauseth, J. D. *Botany.* Philadelphia: Saunders, 1991. Chapters 5–9 do an excellent job with primary and secondary growth.

Poethig, R. S. "Phase Change and the Regulation of Short Morphogenesis in Plants." *Science,* November 16, 1990. The molecular and cellular basis of shoot development.

"The Trees Told Him So." *Science News,* September 7, 1985. Growth rings may give clues to volcanic eruptions.

32 Transport in Plants

Basic Transport Processes in Plants: An Overview

The Absorption of Water and Minerals by Roots: A Closer Look

The Ascent of Xylem Sap: A Closer Look

The Control of Transpiration: A Closer Look

Transport in Phloem: A Closer Look

The algal ancestors of plants were completely immersed in water and dissolved minerals, and none of their cells was far from these ingredients (see Chapter 27). The evolutionary journey onto land involved the differentiation of the plant body into roots, which absorb water and minerals from soil, and shoots, which are exposed to light and atmospheric CO_2. This body plan enables plants to survive in an environment where chemical resources are divided between two media, soil and air. But the morphological solution to a dual environment posed a new problem: the need to transport materials between roots and shoots, sometimes over long distances. For example, the leaves of the aspen trees in Figure 32.1 are more than 100 m away from the roots. These remote organs are bridged by vascular tissues that transport sap throughout the plant body (see Chapter 31). Water and minerals absorbed by roots are drawn upward in the xylem to shoots. Sugar produced by photosynthesis is exported from leaves to other organs via the phloem. The whole plant depends on this inter-organ commerce to integrate the activities of its specialized parts (Figure 32.2). The mechanisms responsible for this internal transport are the subjects of this chapter.

Figure 32.1
The need for internal transport in plants. These aspen trees illustrate the spatial separation of leaves and roots and the need for internal transport in plants. A key development in the history of plants was the evolution of xylem and phloem, the vascular tissues that transport substances between the parts of plants. In this chapter, you will learn more about how internal transport in plants works.

BASIC TRANSPORT PROCESSES IN PLANTS: AN OVERVIEW

Transport in plants occurs on three levels: (1) uptake and release of water and solutes by individual cells, such as the absorption of water and minerals from the soil by cells of a root; (2) short-distance transport of substances from cell to cell at the level of tissues and organs, such as the loading of sugar from photosynthetic cells of a leaf into the sieve tubes of phloem; and (3) long-distance transport of sap within xylem and phloem at the level of the whole plant.

Transport at the Cellular Level

The transport of solutes and water across biological membranes was covered in detail in Chapter 8. Here, we re-examine a few of these transport processes in the specific context of plant cells.

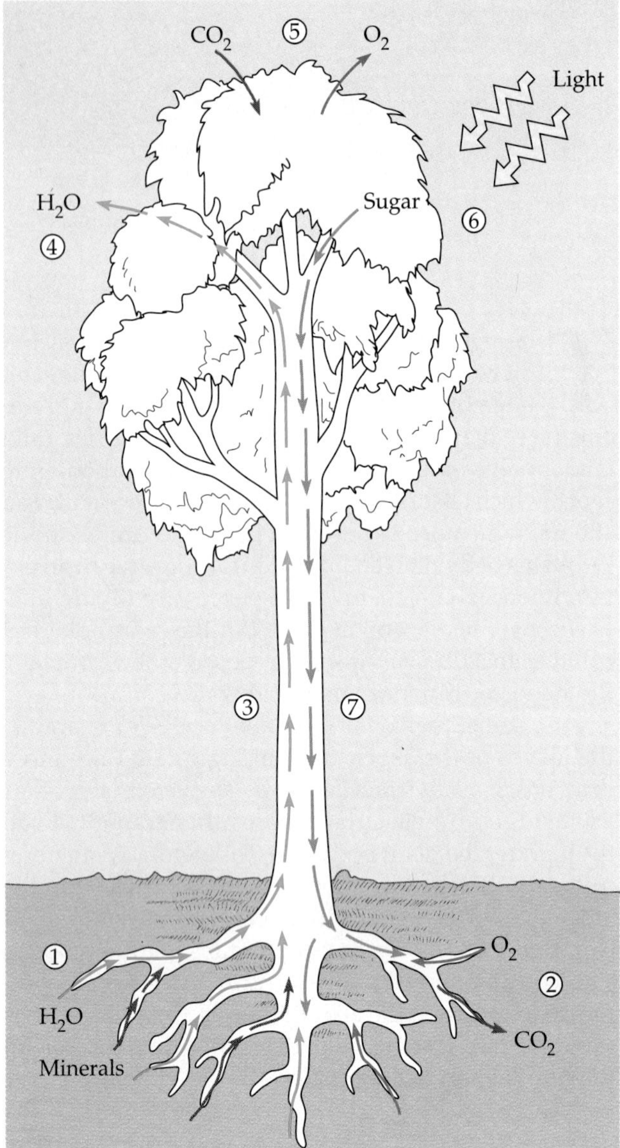

Figure 32.2
An overview of transport in plants. ① Roots absorb water and dissolved minerals from the soil. ② Roots also exchange gases with the air spaces of soil, taking in O_2 and discharging CO_2. This gas exchange supports the cellular respiration of root cells. ③ Water and minerals are transported upward within xylem, from the roots into the shoot system. ④ Transpiration, the evaporation of water from leaves (mostly through stomata), creates a force within leaves that pulls xylem sap upward. ⑤ Leaves also exchange gases through stomata, taking in the CO_2 that provides carbon for photosynthesis and expelling O_2. ⑥ Sugar is produced by photosynthesis in the leaves and ⑦ is transported within phloem to roots and other nonphotosynthetic parts of the plant.

Passive and Active Transport of Solutes The selective permeability of a plant cell's plasma membrane controls the movement of solutes between the cell and the extracellular solution. Solutes tend to diffuse down their gradients, and when this occurs across a membrane, the process is termed passive transport (*passive,* because it happens without the direct expenditure of metabolic energy by the cell). Most solutes, however, diffuse very slowly across a membrane unless they can pass through specific **transport proteins** embedded in the membrane. Some of these function as specific **carrier proteins,** which facilitate diffusion by binding selectively to a solute on one side of the membrane and releasing the solute on the opposite side. Transfer of the solute across the membrane involves a conformational (shape) change by the carrier protein. Other transport proteins function as **selective channels.** Unlike carriers, which physically transport solutes, channels are simply selective passageways across the membrane. For example, the membranes of most plant cells have potassium channels that allow potassium ions (K^+) to pass, but not similar ions, such as sodium (Na^+). Some channels are gated; environmental stimuli can cause the channels to open or close. For instance, we will see later in the chapter how the regulation of K^+ gates in the membranes of guard cells functions in the opening and closing of stomata.

Recall from Chapter 8 that active transport is the pumping of solutes across membranes against their gradients. It is termed *active* because the cell must expend metabolic energy, usually in the form of ATP, to transport a solute "uphill"—that is, counter to the direction in which the solute diffuses. One important active transporter in plant cells is the **proton pump,** which hydrolyzes ATP and uses the released energy to pump hydrogen ions (H^+) out of the cell. This results in a proton gradient, with the H^+ concentration higher outside the cell than inside the cell. The gradient is a form of stored energy, because the H^+ ions tend to diffuse "downhill," back into the cell. And because the proton pump moves positive charge, in the form of H^+, out of the cell, the pump also generates a membrane potential. Membrane potential is a voltage, a separation of opposite charges across a membrane (see Chapter 8). Proton pumping makes the inside of a plant cell negative in charge relative to the outside. This voltage is called a membrane *potential* because the charge separation is a form of potential (stored) energy that can be harnessed to perform cellular work.

Plant cells use energy stored in the proton gradient and the membrane potential to drive the transport of many different solutes (Figure 32.3). Consider, for example, one mechanism root cells use to absorb potassium from the soil solution. Because potassium ions are positively charged, and the inside of the cell is negatively charged compared to the outside, the membrane potential helps drive K^+ into the cell. Because K^+ is diffusing down its electrochemical gradient (see Chapter 8), accumulation of the ion by this mechanism represents passive transport. But it is the active transport of H^+ that maintains the membrane potential and makes it possible for the cell to accumulate K^+. In other cases, energy stored by H^+ pumping can actually be used to drive the transport of solutes *against* their elec-

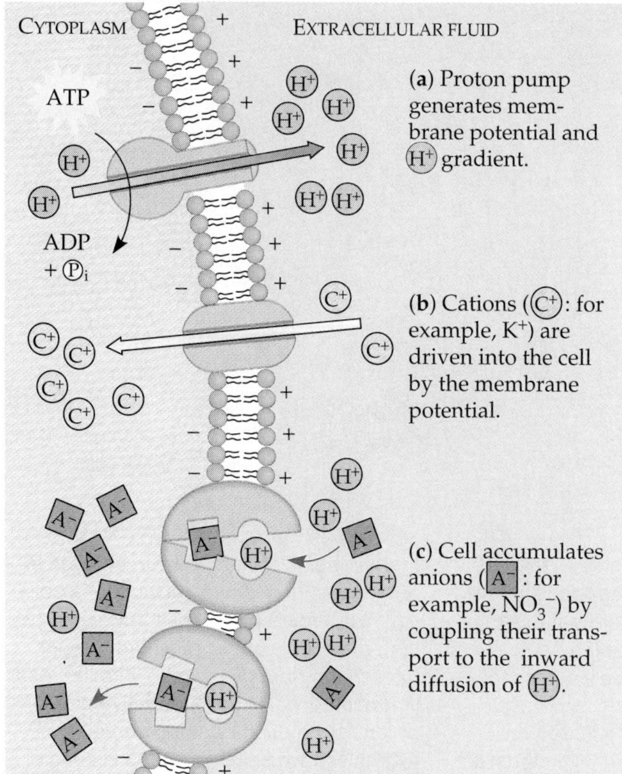

CYTOPLASM
EXTRACELLULAR FLUID
ATP
ADP
+ Ⓟi
(a) Proton pump generates membrane potential and H+ gradient.
(b) Cations (C+: for example, K+) are driven into the cell by the membrane potential.
(c) Cell accumulates anions (A−: for example, NO3−) by coupling their transport to the inward diffusion of H+.

Figure 32.3
A chemiosmotic model of solute transport in plant cells. (**a**) Much of the solute transport across the plasma membrane of a plant cell is driven indirectly by proton pumps that maintain a voltage (membrane potential) and an H^+ gradient across the membrane. (**b**) The membrane potential drives cations such as K^+ into the cell. (**c**) The cell can tap the energy stored in the H^+ gradient to accumulate anions such as NO_3^- *against* their electrochemical gradients (active transport). A transport protein functions as a symport that couples accumulation of the anion to the inward diffusion of H^+. These transport mechanisms are part of a more general chemiosmotic theory, which emphasizes the versatile role of transmembrane proton gradients in the bioenergetics of cells. In Chapters 9 and 10, for example, you learned about the central importance of chemiosmosis as an energy-coupling mechanism in cellular respiration and photosynthesis.

trochemical gradients. For example, many negatively charged minerals, such as nitrate (NO_3^-), enter root cells through carriers that also allow H^+ to re-enter the cells. This mechanism is called **cotransport** (see Chapter 8). A transport protein couples "downhill" passage of one solute (H^+) to "uphill" passage of another (NO_3^-, in this case). This coattail effect is also responsible for the uptake of the sugar sucrose by plant cells. A membrane protein cotransports sucrose with H^+, which moves down its gradient through the protein.

The role of proton pumps in the transport processes of plant cells is a specific application of the general mechanism called **chemiosmosis** (see Chapter 9). The key feature of chemiosmosis is a transmembrane proton gradient, which links energy-releasing processes to energy-consuming processes in cells. For example, you learned in Chapters 9 and 10 that mitochondria and chloroplasts use proton gradients generated by electron transport chains to drive ATP synthesis. The ATP synthases that couple H^+ diffusion to ATP synthesis during cellular respiration and photosynthesis function somewhat like the proton pumps embedded in the plasma membranes of plant cells. But compared to ATP synthases, proton pumps normally run in reverse, using ATP energy to pump H^+ against its gradient. In both cases, proton gradients are the metabolic gears that enable one process to drive another. Chemiosmosis is the unifying principle of cellular energetics. Relevant to our discussion here, chemiosmosis figures prominently in how plant cells transport solutes across their membranes.

Water Potential and Osmosis The net uptake or loss of water by a cell occurs by osmosis, the passive transport of water across a membrane (see Chapter 8). How can we predict the direction of osmosis when a cell is surrounded by a particular solution? In the case of an animal cell, it is enough to know whether the extracellular solution is hypoosmotic (lower solute concentration) or hyperosmotic (higher solute concentration) to the cell; water will move by osmosis in the hypoosmotic ⟶ hyperosmotic direction. But in the case of a plant cell, the presence of a cell wall adds a second factor affecting osmosis: pressure. The combined effects of these two factors—solute concentration and pressure—are incorporated into a single measurement called **water potential,** abbreviated by the Greek letter *psi* (ψ). The most important thing for you to learn about water potential is that water will move across a membrane from the solution with the higher water potential to the solution with the lower water potential. For example, if a plant cell is immersed in a solution having a higher water potential than the cell, osmotic uptake of water will cause the cell to swell.

Why is this relative tendency for water to move away from a location called water *potential?* By moving, water can perform work (expanding a cell, for instance). The *potential* in water potential refers to this potential energy, the capacity to perform work when water moves from a region of higher ψ to a region of lower ψ. Water potential is related to free energy (see Chapter 6), and plant biologists measure ψ in units of pressure called **megapascals** (abbreviated MPa). An MPa is equal to about 10 atmospheres of pressure. (An atmosphere is the pressure exerted at sea level by an imaginary column of air—about 1 kg of pressure per cm^2.) A couple of nonbiological examples will give some idea of the magnitude of a megapascal: A car tire is usually inflated to a pressure of about 0.2 MPa; the water pressure in home plumbing is about 0.25 MPa.

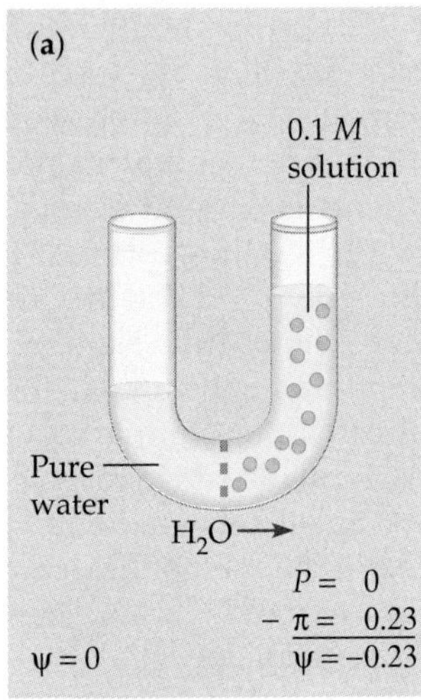

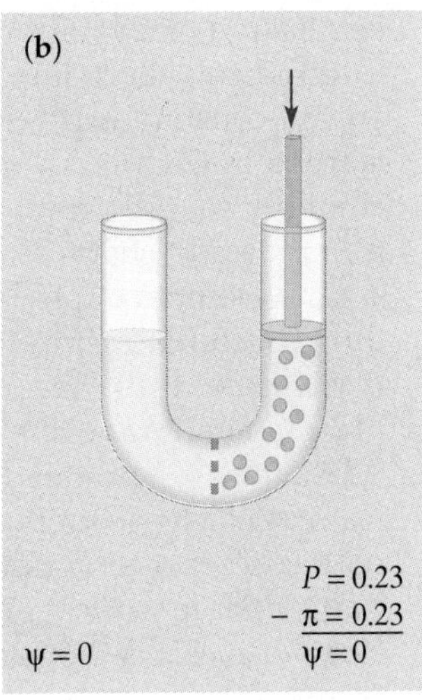

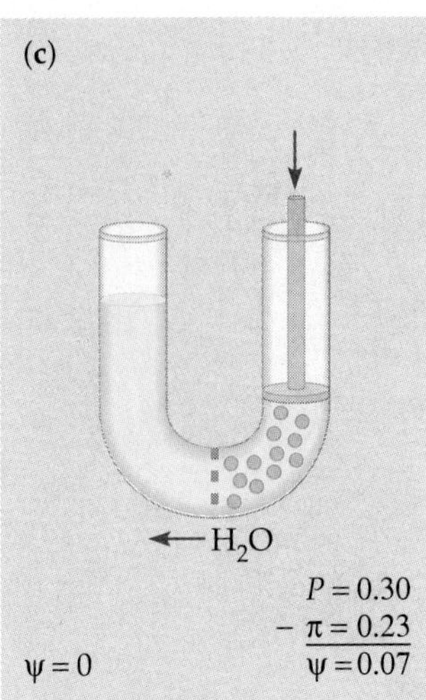

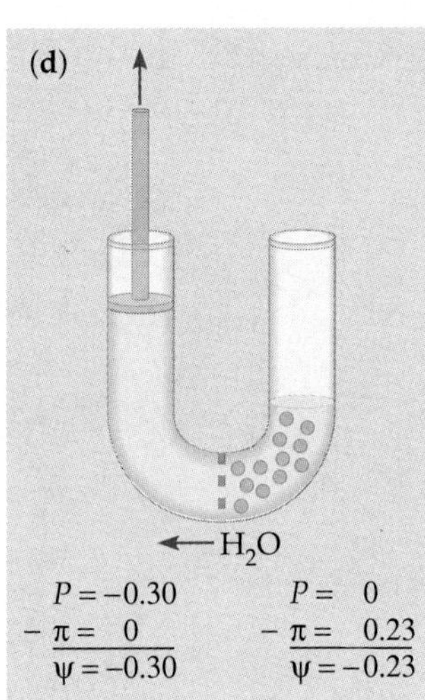

Figure 32.4
Water potential and osmosis: a mechanical model. Water moves by osmosis across a selectively permeable membrane from where water potential is higher to where it is lower. The water potential (ψ) of pure water at atmospheric pressure is 0 MPa. Addition of solutes reduces water potential (to a negative value), but application of physical pressure increases water potential. If we know the magnitude of physical pressure *(P)* and osmotic pressure (π, the tendency for water to enter a solution based on its solute concentration), we can calculate water potential: $P - \pi = \psi$. (**a**) In this U-shaped apparatus, a selectively permeable membrane separates pure water from a 0.1 molar *(M)* solution containing a particular solute that cannot pass freely across the membrane. Water will move by osmosis into the solution, increasing its volume. (The values for π and ψ are given for initial conditions, before any net movement of water.) (**b**) If we use a piston to apply just enough physical pressure to offset the osmotic pressure and increase the water potential of the solution to 0, there will be no net movement of water across the membrane. (This method can be used to determine the osmotic pressure of a solution; see Figure 8.12.) (**c**) If physical pressure exceeds osmotic pressure, we can force water from the solution into the reservoir of pure water. (**d**) A negative pressure, or tension, reduces water potential. For example, if we pull on a piston instead of pushing on it, we can lower the water potential of pure water to a negative value and cause additional water to move out of the solution.

Let's see how solute concentration and pressure affect water potential. For purposes of comparison, the water potential of pure water in a container open to the atmosphere is defined as zero megapascals (ψ = 0 MPa). Addition of solutes lowers the water potential. And since ψ is standardized as 0 MPa for pure water, any solution at atmospheric pressure has a negative water potential due to the presence of solutes. For instance, a 0.1 molar *(M)* solution of any solute has a water potential of −0.23 MPa. If this solution is separated from pure water by a selectively permeable membrane, water will move by osmosis into the solution, from the region of higher ψ (0 MPa) to the region of lower ψ (−0.23 MPa). So far, this is just another way of saying that the water is moving in the hypoosmotic ⟶ hyperosmotic direction. But we have not yet factored in the influence of pressure on ψ.

In contrast to solutes, pressure is directly proportional to water potential, and increasing pressure raises ψ. This makes sense if you remember that ψ measures the relative tendency for water to leave one location in favor of another. Physical pressure—pressing the plunger of a syringe filled with water, for example—causes water to escape via any available exits. If a solution is separated from pure water by a selectively permeable membrane, external pressure on the solution can counter its tendency to take up water due to the presence of solutes. In fact, even greater pressure will force water across the membrane from the solution to the compartment containing pure water. It is also possible to create a negative pressure, or **tension,** on water or solutions. For example, if you pull up on the plunger of a syringe, the negative pressure within the syringe can be used to draw a solution through the needle. Movement of water due to a pressure difference between two locations is called **bulk flow.** It is usually much faster than simple diffusion of water, the process responsible for water movement due to a difference of solute concentration.

The opposite effects of pressure and solute concentration on water potential are incorporated into the following equation:

$$\psi = P - \pi$$

P is physical pressure, and π (not to be confused with geometric use of the same symbol, for 3.14) is osmotic pressure (see Chapter 8). Osmotic pressure, not really a physical pressure, measures the tendency for water to enter a solution due to the presence of solutes. A 0.1 *M* solution has a π of 0.23 MPa. Thus, in the absence of a physical pressure (*P* = 0), water potential, as stated earlier, is −0.23 MPa for a 0.1 *M* solution (ψ = 0 − 0.23). If

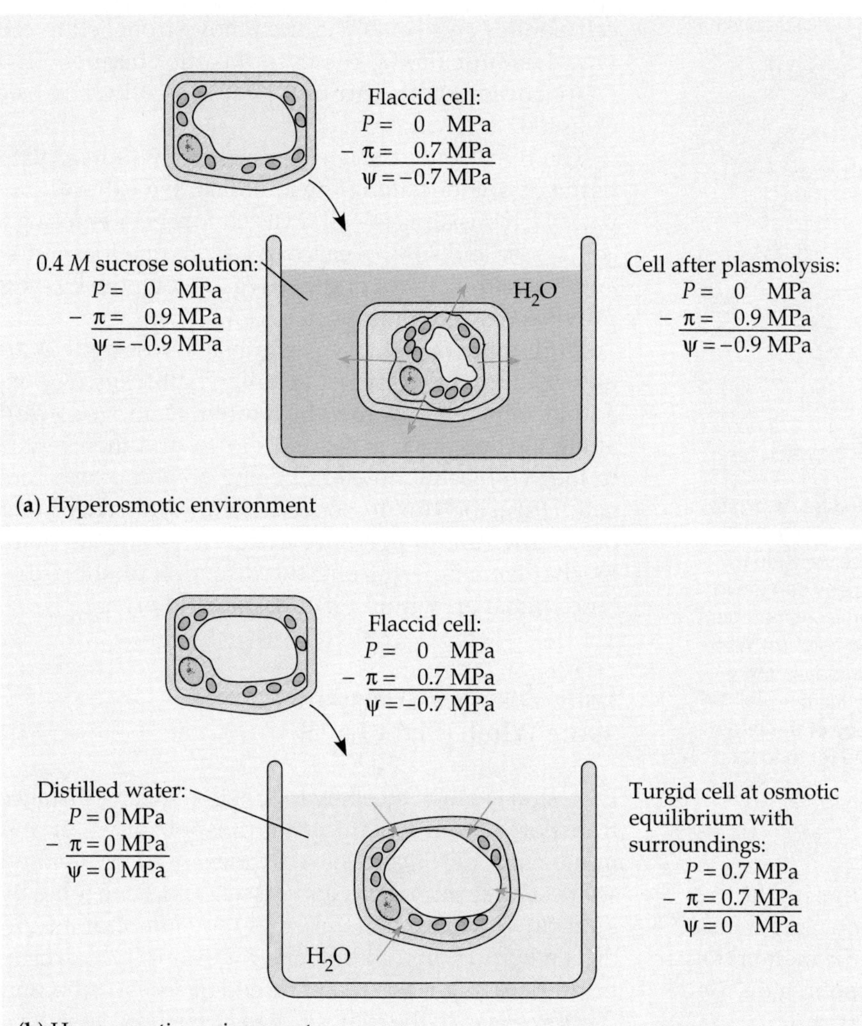

Figure 32.5
Water relations of plant cells. In these two experiments, flaccid cells are transferred from an isosmotic environment into hyperosmotic and hypoosmotic environments. (Flaccid cells are in contact with their walls, but lack turgor pressure; see Chapter 8.) (**a**) In a hyperosmotic environment, the cell initially has a greater water potential than its surroundings. The cell loses water and plasmolyzes. After plasmolysis is complete, the water potentials of the cell and its surroundings are the same. (**b**) In a hypoosmotic environment, the cell initially has a lower water potential than its surroundings. There is a net uptake of water by osmosis, causing the cell to become turgid. When this tendency for water to enter is offset by the back pressure of the elastic wall, water potentials are equal for the cell and its surroundings.

we apply a physical pressure of +0.23 MPa to this solution, we raise its water potential from a negative value to 0 ($\psi = 0.23 - 0.23$). If this pressurized solution is separated from water by a selectively permeable membrane, there will be no osmosis between the two compartments. If we increase P to +0.3 MPa, then the solution has a water potential of +0.07 MPa ($\psi = 0.3 - 0.23$), and it will lose water to a compartment containing pure water. In dissecting water potential to see these opposing effects of pressure and solutes, it is important to remember the key point: Water will move across a membrane in the direction of lower water potential. Figure 32.4 summarizes the concept of water potential.

Let's now apply what we have learned about water potential to the uptake and loss of water by plant cells (Figure 32.5). First, imagine a flaccid cell (that is, $P = 0$) bathed in a solution of higher solute concentration than the cell itself. Since the external solution has the lower water potential, water will leave the cell by osmosis, and the cell will plasmolyze, or shrink and pull away from its wall. Now let's place the same flaccid cell in pure water ($\psi = 0$). The cell has a lower water potential because of the presence of solutes, and water enters the cell by osmosis. The cell begins to swell and push against the wall to produce a **turgor pressure.** The partially elastic wall pushes back against the turgid cell. When this wall pressure is great enough to offset the tendency for water to enter because of the solutes in the cell, then $P = \pi$ and ψ (or $P - \pi$) = 0. This matches the water potential of the extracellular environment—in this example, 0 MPa. A dynamic equilibrium has been reached, and there is no further net movement of water, although a brisk, equal exchange of water across the membrane continues.

The Role of the Tonoplast Although our emphasis on transport at the cellular level has been on traffic across the plasma membrane, the **tonoplast,** the membrane of the central vacuole, is another important site of regulation. Transport proteins embedded in the tonoplast control the movement of solutes between the cytosol and the vacuole. For example, the tonoplast has proton pumps that expel H^+ from the cytoplasm into the vacuole. This augments the ability of the proton pumps of the plasma membrane to maintain a low cy-

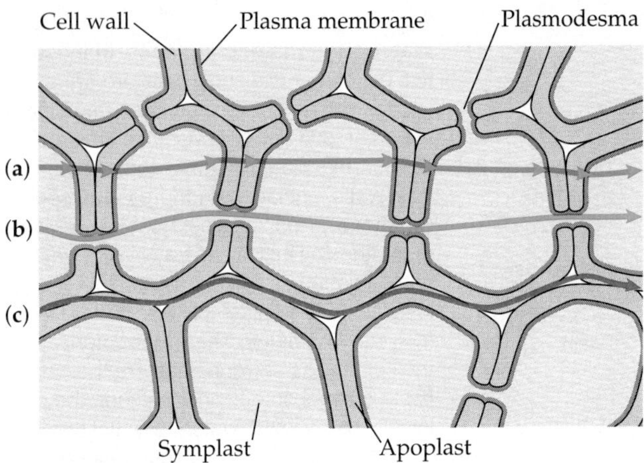

Figure 32.6
Three routes for lateral transport in a plant tissue or organ. (**a**) Solutes and water can move across an organ by the repeated crossing of the plasma membranes of the cells along the pathway. (**b**) After entering one cell, substances can move across an organ via the symplast, the cytoplasmic compartment made continuous by the presence of plasmodesmata. (**c**) Water and solutes can also travel across a tissue or organ via the apoplast, the extracellular continuum formed by the cell walls. In this diagram, substances seem confined to one of the three routes; in fact, substances may change from one route to another during their transit across an organ.

tosolic concentration of H^+. Proton pumping at the tonoplast also generates a membrane potential across that membrane, with the vacuole side of the membrane positive in charge relative to the cytosolic side. By mechanisms similar to those we saw at the plasma membrane, the proton gradient and voltage difference across the tonoplast drive transport of several solutes between the vacuole and cytosol.

Short-Distance (Lateral) Transport at the Level of Tissues and Organs

How do water and solutes move from one location to another within plant tissues and organs? For example, what mechanisms transport water and minerals absorbed by a root from the outer cells to the inner cells of the root? Such short-distance transport is sometimes called lateral transport because its usual direction is along the radial axis of plant organs, rather than up and down along the length of the plant.

Three routes are available for lateral transport (Figure 32.6). By the first route, substances move out of one cell, across the cell wall, and into the neighboring cell, which may then pass the substances along to the next cell in the pathway by the same mechanism. This route requires repeated crossings of plasma membranes, as the solutes exit one cell and enter the next.

The second route, via the **symplast,** the continuum of cytoplasm within a plant tissue, requires only one crossing of a plasma membrane. After entering one cell, solutes and water can then move from cell to cell via **plasmodesmata,** the cytoplasmic channels that connect plant cells through pores in cell walls (see Chapter 7).

The third route for lateral transport within a plant tissue or organ is along the **apoplast,** the extracellular pathway consisting of cell walls. Before ever entering a cell, water and solutes can move from one location to another within a root or other organ along the byways provided by the continuum of cell walls.

During the course of their transit across a plant organ, solutes and water can change pathways. For example, minerals absorbed by a root may move inward along the apoplast for some distance, and then switch to the symplastic route after being taken up by a root cell. Diffusion due to concentration differences and bulk flow due to pressure differences are the main mechanisms determining the direction of short-distance transport within plant tissues and organs.

Long-Distance Transport at the Whole-Plant Level

Diffusion is much too slow to function in long distance transport within a plant—the transport of water and minerals from roots to leaves, for example. Water and solutes move through xylem vessels and sieve tubes by bulk flow, which, remember, is movement of a fluid driven by pressure. In phloem, for example, hydrostatic pressure is generated at one end of a sieve tube, and this forces sap to the opposite end of the tube. In xylem, it is actually tension, a negative pressure, that drives long-distance transport. Transpiration, the evaporation of water from a leaf, reduces pressure in the leaf xylem. This creates a tension that pulls xylem sap upward from the roots.

Now that we have examined the basic mechanisms of transport at the cellular, tissue, and whole-plant levels, we are ready to take a closer look at how these mechanisms work together in the overall transport functions that enable a plant to survive on land. For example, bulk flow due to a pressure difference is the mechanism of long-distance transport of phloem sap, but it is active transport of sugar at the cellular level that maintains this pressure difference. The four transport functions we will examine in more detail are the absorption of water and minerals by roots, the ascent of xylem sap, the control of transpiration, and the transport of organic nutrients within phloem.

THE ABSORPTION OF WATER AND MINERALS BY ROOTS: A CLOSER LOOK

Water and mineral salts enter the plant through the epidermis of roots, cross the root cortex, pass into the

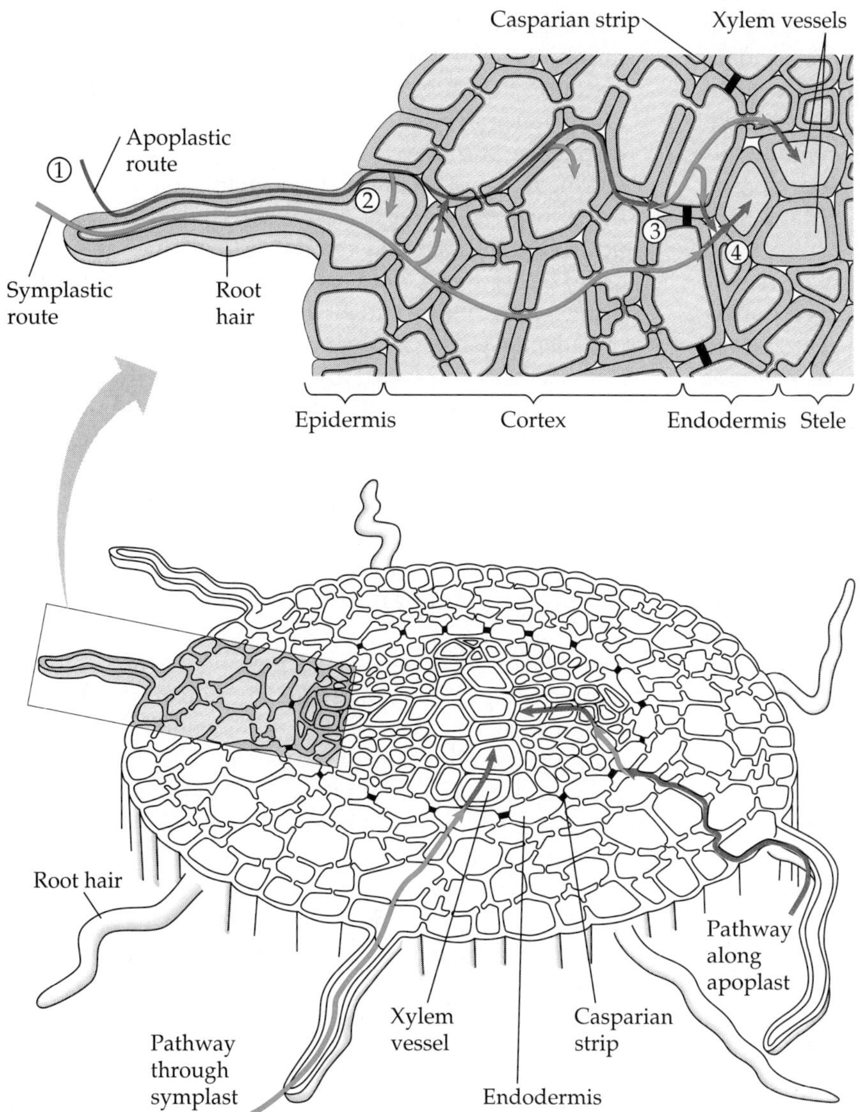

Figure 32.7
Lateral transport of minerals and water in roots. Minerals are absorbed with the soil solution by the root surface, especially by root hairs. The water and minerals then move across the root cortex to the vascular cylinder by a combination of the apoplastic (extracellular) and symplastic (the cytoplasmic continuum, with its plasmodesmata) routes (see Figure 32.6). ① Uptake of soil solution by the hydrophilic walls of the epidermis provides access to the apoplast, and the water and minerals can soak into the cortex along this matrix of cell walls. Minerals and water that cross the plasma membranes of root hairs enter the symplast. ② As soil solution moves along the apoplast, some of the water and minerals are transported into cells of the epidermis and cortex and then move inward via the symplast. ③ Water and minerals that move all the way to the endodermis through the apoplast cannot continue into the stele via the extracellular route. Within the wall of each endodermal cell is a belt of waxy material (black band) that blocks the passage of water and dissolved minerals. This barrier to apoplastic transport is called the Casparian strip. Only minerals that are already in the symplast or enter that pathway by crossing the plasma membrane of an endodermal cell can detour around the Casparian strip and pass into the stele. Thus, transport of minerals into the stele is discriminating; only those minerals that are admitted into cells by selective membranes gain access to the vascular tissue. ④ Endodermal cells, and parenchyma cells within the stele, discharge water and minerals into their walls, which, as part of the apoplast, are continuous with the xylem vessels. Water and minerals absorbed from soil are now ready for upward transport into the shoot system.

stele (vascular cylinder), and then flow up xylem vessels to the shoot system. In this section, we focus on the soil ⟶ epidermis ⟶ root cortex ⟶ stele segments of this transport pathway.

Most absorption of water and minerals occurs near root tips, where the epidermis is permeable to water and where root hairs are located. (Older regions of roots, above the tips, are not very permeable and play little role in the absorption of soil solution.) Root hairs, extensions of epidermal cells, account for most of the surface area of roots (see Chapter 31). The hairs adhere tightly to soil particles, which are usually coated with water and dissolved minerals. The soil solution flows into the hydrophilic walls of epidermal cells and passes freely along the apoplast into the root cortex. This exposes all the parenchyma cells of the cortex to soil solution, providing a much greater surface area of membrane for the uptake of water and minerals into cytoplasm than the surface area of the epidermis alone. As the soil solution moves along cell walls, some of the water and solutes are taken up by cells of the epidermis and cortex, switching pathways from the apoplast to the symplast. It is this crossing of plasma membranes that makes mineral absorption selective. The soil solution is usually very dilute, and roots can accumulate essential minerals to concentrations that are hundreds of times higher than the concentrations of these minerals in soil. For example, selective transport proteins of the plasma membrane and tonoplast enable root cells to extract K^+, an essential mineral nutrient, while the cells exclude most Na^+, which may be much more concentrated than K^+ in the soil solution.

The **endodermis,** the innermost layer of cells in the root cortex, surrounds the stele and functions as a last checkpoint for the selective passage of minerals from the cortex into the vascular tissue (Figure 32.7).

Minerals already in the symplast when they reach the endodermis continue right through the plasmodesmata of the endodermal cells and pass into the stele. These minerals were already screened by the selective membrane they had to cross to enter the symplast in the cortex. Those minerals that reach the endodermis via the apoplast find a dead end in that route, which blocks their passage into the stele. In the wall of each endodermal cell is a belt made of suberin, a waxy material that is impervious to water and dissolved minerals. This ring of wax, called the **Casparian strip**, is tangential to the stele. Thus, water and minerals cannot cross the endodermis and enter vascular tissue via the apoplast. The only way past this barrier is for the water and minerals to cross the plasma membrane of an endodermal cell and to enter the stele via the symplast. The endodermis, with its Casparian strip, ensures that no minerals can reach the vascular tissue of the root without crossing membranes. And you know that transport across membranes is selective. If minerals do not enter cells in the cortex, they must enter endodermal cells or be excluded from the vascular tissue. The structure of the endodermis and its strategic location in the root fit its function as sentry of the cortex-stele border, a function that contributes to the ability of roots to preferentially transport certain minerals from the soil into the xylem. The Casparian strip also prevents the backflow of water and minerals from the stele back into the cortex.

The last segment in the soil ⟶ xylem pathway is the passage of water and minerals that reach the stele into the tracheids and vessel elements of the xylem. These water-conducting cells lack protoplasts, and thus the lumen of the cells, as well as their walls, are part of the apoplast. But you have just learned that water and minerals enter the stele from the cytoplasm of endodermal cells. Thus, the entry of water and minerals into the xylem requires their transfer from the symplast to the apoplast. Endodermal cells and parenchyma cells within the stele discharge minerals into their walls. Both diffusion and active transport are probably involved in this transfer of solutes from symplast to apoplast, and the water and minerals are now free to enter the tracheids and xylem vessels. The water and mineral nutrients we have tracked from the soil to xylem can now be transported upward as xylem sap to the shoot system.

THE ASCENT OF XYLEM SAP: A CLOSER LOOK

Xylem sap flows upward in vessels at rates of 15 m per hour or faster. Veins branch throughout each leaf, placing xylem vessels close to each cell. Leaves depend on this efficient delivery system for their supply of water. Plants lose an astonishing amount of water by **transpiration,** the evaporation of water from leaves and other aerial parts of the plant. An average-sized maple tree, for instance, loses more than 200 L of water per hour during the summer. Put another way, an actively transpiring leaf must replace all its water each hour. Unless the transpired water is replaced by water transported up from the roots in xylem, leaves wilt and eventually die. The flow of xylem sap upward in the plant also brings mineral nutrients to the shoot system.

Xylem sap rises against gravity, without the help of any mechanical pump, to reach heights of more than 100 m in the tallest trees (see Figure 32.1). Is the sap *pushed* upward from the roots, or is it *pulled* upward by the leaves? Let's evaluate the relative contributions of these two possible mechanisms.

Pushing Xylem Sap: Root Pressure

The stele of a root behaves something like an osmometer (see Chapter 8). At night, when transpiration is very low or zero, the root cells are still expending energy to pump mineral ions into the xylem. The endodermis surrounding the stele of the root helps prevent the leakage of these ions back out of the stele. Accumulation of minerals in the stele lowers its water potential, and water flows in, generating a positive pressure that forces fluid up the xylem. This upward push of xylem sap in some plants is called **root pressure.**

Root pressure causes **guttation,** the exudation of water droplets that can be seen in the morning on tips of grass blades or the leaf margins of some small, herbaceous (nonwoody) dicots. During the night, when the rate of transpiration is low, the roots of some plants keep accumulating minerals, and root pressure pushes xylem sap into the shoot system. More water enters leaves than is transpired, and the excess is forced through specialized structures called hydathodes, which function as escape valves.

In most plants, root pressure is not the major mechanism driving the ascent of xylem sap. At most, root pressure can force water upward only a few meters, and many plants, including some of the tallest trees, generate no root pressure at all. Even in most small plants that display guttation, root pressure cannot keep pace with transpiration after sunrise. For the most part, xylem sap is not pushed from below by root pressure but pulled upward by the leaves themselves.

Pulling Xylem Sap: The Transpiration-Cohesion-Tension Mechanism

If we want to move material upward, we can either push it from below or pull it from above. However, it

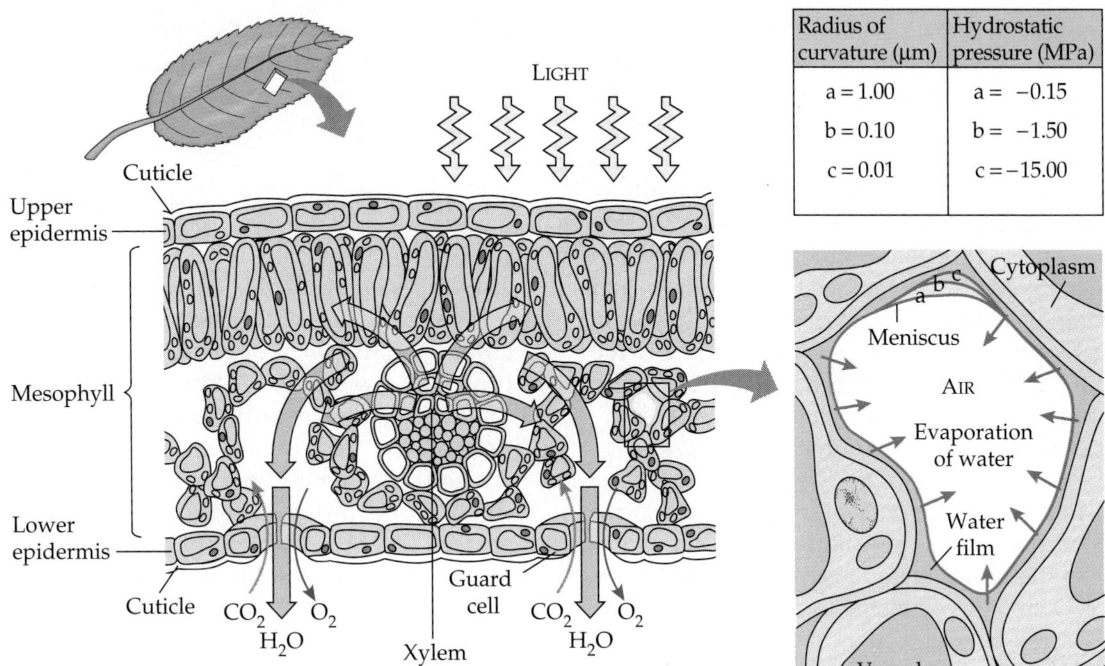

Radius of curvature (μm)	Hydrostatic pressure (MPa)
a = 1.00	a = −0.15
b = 0.10	b = −1.50
c = 0.01	c = −15.00

Figure 32.8
The generation of transpirational pull in a leaf. Water vapor diffuses from the moist air spaces of the leaf to the drier air outside via stomata, the microscopic pores bordered by guard cells. (Stomata are also the avenues for exchange of CO_2 and O_2 between the photosynthetic tissue and the atmosphere.) Evaporation from the water film coating the mesophyll cells maintains the high humidity of the air spaces. This loss of water causes the water film to form menisci, which become more and more concave as the rate of transpiration (evaporative water loss from the leaf) increases. A meniscus has a tension that is inversely proportional to the radius of the curved water surface. Thus, as the water film recedes and its menisci decrease in radii (become more concave), the tension of the water film increases. Tension is a negative pressure—a force that pulls water from locations where hydrostatic pressure is greater. The negative pressure (tension) of water lining the air spaces of the leaf is the physical basis of transpirational pull, which draws water out of xylem and through the mesophyll tissue to the surfaces near stomata. Cohesion of water due to hydrogen bonding enables transpiration to pull water up the narrow xylem vessels and tracheids without these columns of water breaking apart. In fact, transpirational pull, with the help of water's cohesion, is transmitted all the way from leaves to roots. The bulk flow of water to the top of a tree is solar-powered, because it is the absorption of sunlight by leaves that causes the evaporation responsible for transpirational pull.

is not as obvious that something as seemingly fluid as water could be pulled up a pipe. Nevertheless, that is what happens in the xylem vessels of plants. As we investigate this mechanism of transport, we will see that transpiration provides the pull, and the cohesion of water due to hydrogen bonding transmits the upward pull along the entire length of the xylem to the roots.

Transpirational Pull Stomata, the microscopic pores through the surface of a leaf, lead to a maze of internal air spaces that expose the mesophyll cells to the carbon dioxide they need for photosynthesis (Figure 32.8). The air in these spaces is saturated with water vapor because it is in contact with the moist walls of the cells. The air is drier outside the leaf on most days—that is, it has a lower water concentration than the air inside the leaf. Therefore, gaseous water, diffusing down its concentration gradient, exits the leaf via the stomata. Recall that it is this loss of water from the leaf that we call transpiration.

How is transpiration translated into a pulling force for the movement of water in a plant? The mechanism depends on generation of a negative pressure (tension) in the leaf due to the unique physical properties of water. Evaporation from the thin film of water that coats the mesophyll cells replaces the water vapor that is lost from the leaf's air spaces by transpiration. As water evaporates, the remaining film of liquid water retreats into the pores of the cell walls, attracted by adhesion to the hydrophilic walls (Figure 32.8). At the same time, cohesive forces in the water resist an increase in the surface area of the film (a surface tension effect—see Chapter 3). Combination of the two forces acting on the water—adhesion to the wall and surface tension—causes the surface of the water film to form a meniscus, or concave shape. In a sense, the water is being "pulled on" by the adhesive and cohesive forces. Thus, the water film at the surface of leaf cells has a negative pressure, a pressure less than atmospheric pressure. And the more concave the menisci (plural), the more negative the pressure of the water film. This negative pressure, or tension, is the pulling force that draws water out of the leaf xylem, through the mesophyll, and toward the cells and surface film bordering the air spaces near stomata. Transit across the mesophyll occurs via both symplastic and apoplastic routes. This water flow fits with what you learned earlier about water potential. In the equation for water potential, notice that a tension (negative pressure) *lowers* the potential. And since water moves from where its potential is higher to where it is lower, mesophyll cells will lose water to the surface film lining air spaces, which in turn loses water by transpiration. The water lost via the stomata is replaced by water that is pulled out of the leaf xylem.

Cohesion and Adhesion of Water The transpirational pull on xylem sap is transmitted all the way from the leaves to the root tips and even into the soil solution. The cohesion of water due to hydrogen bonding makes it possible to pull a column of sap from above without the water separating. Water molecules exiting the xylem in the leaf tug on adjacent molecules, and this pull is relayed, molecule by molecule, down the entire column of water in the xylem. Also helping to fight gravity is the strong adhesion of water molecules (again by hydrogen bonds) to the hydrophilic walls of the xylem cells. The very small diameter of tracheids and vessel elements contributes to the importance of this factor in overcoming the downward pull of gravity.

The upward pull on the cohesive sap creates tension within the xylem. Pressure will cause an elastic pipe to swell, but tension will pull the walls of the pipe inward. (You can actually measure a decrease in the diameter of a tree trunk on a warm day when transpirational pull puts the xylem under tension.) The rings of secondary walls (see Figure 31.26) prevent xylem vessels from collapsing, much as the wire rings maintain the shape of a vacuum hose. Transpirational pull puts the xylem under tension all the way down to the root tips, even in the tallest trees. This tension lowers water potential in the root xylem to such an extent that water flows passively from the soil, across the root cortex, and into the stele.

Transpirational pull can extend down to the roots only through an unbroken chain of water molecules. Cavitation, the formation of a water vapor pocket in a xylem vessel, such as when xylem sap freezes in winter, breaks the chain. Small plants can use root pressure to refill xylem vessels in spring, but in trees, root pressure cannot push water to the top, so a vessel with a water vapor pocket can never function as a water pipe again. However, the transpiration stream can detour around the water vapor pocket through pits between adjacent xylem vessels, and secondary growth adds a layer of new xylem vessels each year. In some angiosperm trees, including oaks and elms, only the youngest, outermost growth ring of xylem transports water, with the older xylem functioning to support the tree (see Chapter 31).

The Long-Distance Transport of Water in Plants by Solar-Powered Bulk Flow

The transpiration-cohesion-tension mechanism that transports xylem sap against gravity is an excellent example of how physical principles apply to biological problems. Long-distance transport of water from roots to leaves occurs by bulk flow, the movement of fluid driven by a pressure difference at opposite ends of a conduit. In a plant, the conduits are xylem vessels or chains of tracheids. The pressure difference is generated at the leaf end by transpirational pull, which lowers pressure (increases tension) at the "upstream" end of the xylem. On a smaller scale, gradients of water potential drive the osmotic movement of water from cell to cell within root and leaf tissue. Differences in solute concentration and pressure both contribute to this microscopic transport. In contrast, bulk flow, the mechanism for long-distance transport up xylem vessels, depends only on pressure. And the plant expends none of its own metabolic energy to lift water up to the leaves by the flow. Absorption of sunlight causes transpiration, which in turn maintains the leaf tension (negative pressure) that tugs on xylem sap. The ascent of xylem sap is solar-powered.

THE CONTROL OF TRANSPIRATION: A CLOSER LOOK

A leaf may transpire more than its weight in water each day. Leaves are kept from wilting by a transpiration stream in xylem vessels that flows as fast as 75 cm per minute, about the speed of the tip of a second hand sweeping around a wall clock. The tremendous requirement for water by a plant is part of the cost of making food by photosynthesis.

The Photosynthesis-Transpiration Compromise

To make food, a plant must spread its leaves to the sun and obtain CO_2 from the air. Carbon dioxide diffuses into the leaf, and oxygen produced as a by-product of photosynthesis diffuses out of the leaf through the stomata (see Figure 32.8). The stomata lead to the honeycomb of air spaces through which CO_2 diffuses to the photosynthetic cells of the mesophyll. The internal surface area of the leaf may be 10 to 30 times greater than the external surface area we see when we look at the leaf. This structural feature of leaves amplifies photosynthesis by increasing exposure to CO_2, but at the same time it increases the surface area for the evaporation of water, which exits the plant freely through open stomata. About 90% of the water a plant loses escapes through stomata, though these pores account for only 1% to 2% of the external leaf surface. The waxy cuticle limits water loss through the remaining surface of the leaf. The stomata of most plants are more concentrated on the lower surface of leaves, reducing transpiration because the bottom surface receives less sunlight than the top surface.

One way to evaluate how efficiently a plant uses water is to determine its transpiration-to-photosynthesis ratio, the amount of water lost per gram of CO_2 assimilated into organic material by photosynthesis. A common ratio is 600:1, meaning the plant transpires 600 g of water for each gram of CO_2 that becomes incorporated into carbohydrate. However, corn and other plants that assimilate atmospheric CO_2 by the C_4 pathway have transpiration-to-photosynthesis ratios of 300:1 or less. With the same concentration of CO_2 within the air spaces of the leaf, C_4 plants can assimilate that CO_2 at a greater rate than C_3 plants can. Since water loss is the trade-off for enabling CO_2 to diffuse into the leaf, the photosynthetic return for each gram of water sacrificed is greater for plants that can assimilate CO_2 at greater rates.

In addition to supplying water to leaves, the transpiration stream also assists in the transfer of minerals and other substances from the roots to the shoot and leaves. Transpiration also results in evaporative cooling, which can lower the temperature of a leaf by as much as 10°C to 15°C compared with the surrounding air. This prevents the leaf from reaching temperatures that could inhibit the enzymes that catalyze the photosynthetic reactions, as well as other enzymes involved in the leaf's metabolic processes. Desert succulents, however, which have low rates of transpiration, can tolerate high leaf temperatures; in this case, the loss of water due to transpiration is a greater threat than overheating.

As long as leaves can pull water from the soil fast enough to replace what they lose, transpiration is no problem. However, when transpiration exceeds the delivery of water by xylem, as when the soil begins to dry out, the leaves begin to wilt as their cells lose turgor pressure. The potential rate of transpiration will be greatest on a day that is sunny, warm, dry, and windy, because these are the climatic factors that increase the evaporation of water. Plants are not helpless against the elements, however, for they are capable of adjusting to their environment. In the photosynthesis-transpiration compromise, mechanisms that regulate the size of the stomatal openings strike a balance.

How Stomata Open and Close

Each stoma is flanked by a pair of guard cells, which are kidney-shaped in dicots and "dumbbell"-shaped in monocots. The guard cells are suspended by their epidermal neighbors over an air chamber, which leads to the leaf's maze of air spaces.

Guard cells control the diameter of the stoma by changing shape, which widens or narrows the gap between the two cells (Figure 32.9a). When guard cells take in water by osmosis, they become more turgid and swell. In most dicots, the cell walls of guard cells are not uniformly thick, and the cellulose microfibrils are oriented in a radial manner. These structural adaptations cause the guard cells to buckle outward when they are turgid, increasing the size of the gap between the cells. When the cells lose water and become flaccid, they sag together and close the space between them. This basic mechanism also applies to the stomata of monocots, although the guard cells usually have shapes somewhat different from those of dicots.

The changes in turgor pressure that open and close stomata result primarily from the reversible uptake and loss of potassium ions (K^+) by the guard cells. Stomata open when guard cells actively accumulate K^+ from neighboring epidermal cells (Figure 32.9b). This uptake of solute causes the water potential to become more negative within the guard cells, and the cells become more turgid as water enters by osmosis. Most of the K^+ and water are stored in the vacuole, and thus the tonoplast also plays a role. The cell's increase in positive charge due to the influx of K^+ is offset by the uptake of chloride (Cl^-), by the pumping out of the cell of hydrogen ions released from organic acids, and by the negative charges of these organic acids after they lose their hydrogen ions. Stomatal closing results from an exodus of K^+ from guard cells, which leads to an osmotic loss of water. The K^+ fluxes across the guard cell membrane are probably coupled to the generation of membrane potentials by proton pumps. Stomatal opening correlates with active transport of H^+ out of the guard cell. The resulting voltage (membrane potential) drives K^+ into the cell through specific membrane channels (see Figure 32.3). Plant physiologists are using a method called patch clamping to study regulation of the guard cell's proton pumps and K^+ channels (see the Methods Box, p. 711).

In general, stomata are open during daytime and closed at night. This prevents the plant from needlessly losing water when it is too dark for photosynthesis. At least three cues contribute to stomatal opening at dawn. First, light itself stimulates guard cells to accumulate potassium and become turgid. This response is triggered by illumination of a blue-light receptor in a guard cell, perhaps built into the plasma membrane. Activation of these blue-light receptors stimulates the

Figure 32.9
The control of stomatal opening and closing. (**a**) Guard cells of a dicot are illustrated in their turgid (stoma open) and flaccid (stoma closed) states. Notice the radial microfibrils in the walls, which, along with the uneven thickness of the walls, causes the pair of guard cells to buckle outward when turgid. Guard cells respond to a complex set of signals, including environmental factors and cues within the plant itself. The uptake or loss of water causes the guard cells to change their shape and widen or narrow the gap between them, a response that affects the rate of photosynthesis and controls transpiration. (**b**) Transport of K^+ (potassium ions) across the plasma membrane and tonoplast causes the turgor changes of guard cells. Stomata open when guard cells accumulate potassium (red dots), which lowers the cells' water potential and causes them to take up water by osmosis. The cells become turgid. An exodus of K^+ from guard cells causes the stomatal closure.

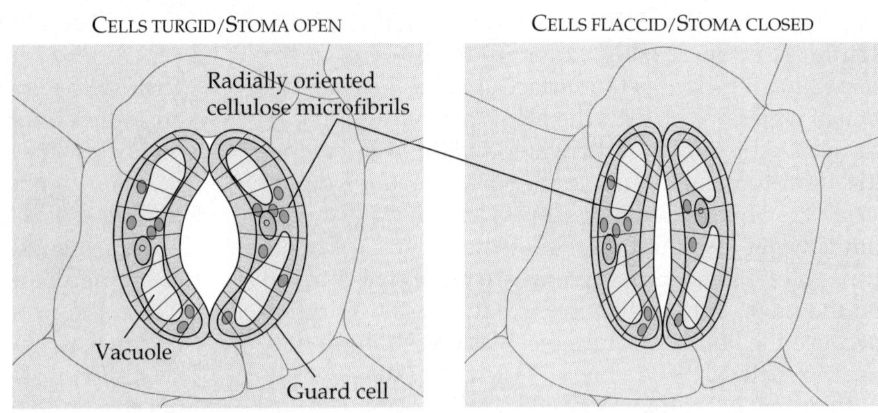

(**a**) Changes in guard cell shape and stomatal opening and closing

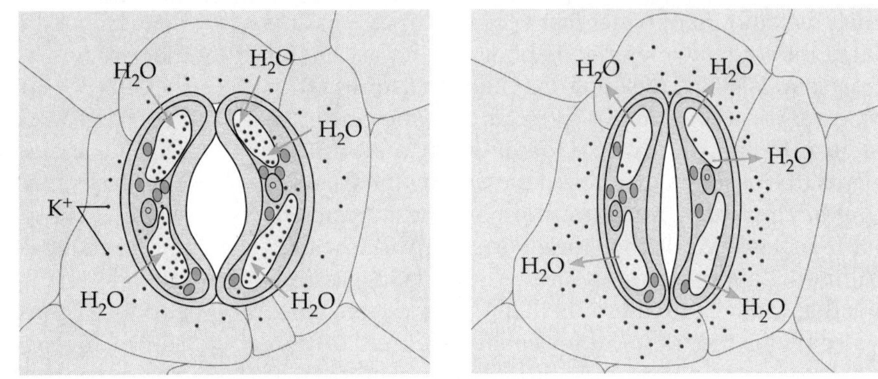

(**b**) Role of potassium in stomatal opening and closing

activity of ATP-powered proton pumps in the plasma membrane of the guard cells, which, in turn, promotes the uptake of K^+ (see the Methods Box). Light may also stimulate stomatal opening by driving photosynthesis in guard cell chloroplasts, making ATP available for the active transport of H^+ ions. A second factor causing stomata to open is depletion of CO_2 within air spaces of the leaf, which occurs when photosynthesis begins in the mesophyll. A plant can be tricked into opening its stomata at night by placing it in a chamber devoid of CO_2. A third cue in stomatal opening is an internal clock located in the guard cells. Even if you keep a plant in a dark closet, stomata will continue their daily rhythm of opening and closing (see Chapter 10, Challenge Question 1). All eukaryotic organisms have internal clocks that somehow keep track of time and regulate cyclical processes. Cycles that have intervals of approximately 24 hours are called **circadian rhythms.** You will learn more about circadian rhythms and the biological clocks that control them in Chapter 35.

Environmental stress of various kinds can cause stomata to close during the daytime. When the plant is suffering a water deficiency, guard cells may lose turgor. In addition, a hormone called abscisic acid, which is produced in the mesophyll cells in response to water deficiency, signals guard cells to close stomata. This response reduces further wilting, but it also slows down photosynthesis, which is one reason droughts reduce crop yields. High temperatures also induce stomatal closure, probably by stimulating cellular respiration and increasing CO_2 concentration within the air spaces of the leaf. High temperature and excessive transpiration may combine to cause stomata to close briefly during midday. Thus, guard cells arbitrate the photosynthesis-transpiration compromise on a moment-to-moment basis by integrating a variety of internal and external stimuli.

Evolutionary Adaptations That Reduce Transpiration

Plants adapted to arid climates are called xerophytes. They have leaves that are modified in various ways to reduce the rate of transpiration. Many xerophytes have small, thick leaves, an adaptation that limits water loss by reducing surface area relative to volume. A thick cuticle gives some of these leaves a leathery consistency. The stomata are concentrated on the lower leaf surface, and they are often located in cryptlike depressions that shelter the pores from the dry wind (Figure 32.10). During the driest months, some desert plants shed their leaves. Others, such as cacti, subsist on water the plant stores in its fleshy stems during the rainy season. (These modified stems are the photosynthetic organs of cacti; the leaves are the spines.)

Methods: Patch Clamp Recording

In 1991, Erwin Neher and Bert Sakmann of Germany's Max Planck Institute shared the Nobel Prize in physiology or medicine for their invention of a technique called patch clamping. This method enables researchers to record the traffic of ions through specific channels in membranes. Patch clamping makes it possible to isolate a very tiny "patch" of membrane and study ion movement through a single type of channel—a K^+ or H^+ channel, for instance.

To apply the patch clamp method to plant cells, it is first necessary to remove the walls. This is usually accomplished by using enzymes to digest the walls, resulting in naked (wall-less) cells, or protoplasts. Figure a is a light micrograph of guard cell protoplasts from tobacco *(Nicotiana)* leaves.

The tool for making a patch is a micropipette with a tip only about 1 μm in diameter. A larger pipette is used to hold the cell in place (Figure b). Slight suction draws a membrane "blister" into the opening of the micropipette, and the rim of the micropipette seals to the membrane. This partitions a small patch from the rest of the plasma membrane. The micropipette, hooked up to appropriate equipment, now functions as an electrode to record ion fluxes across the membrane.

Several options for studying ion transport across this patch are now available. With the whole cell still attached to the micropipette by the tight seal around the membrane patch, it is possible to record ion transport between the cytoplasm and the artificial solution that fills the micropipette, as shown in Figure c. An alternative is to pull the micropipette away from the cell, separating the membrane patch from the rest of the cell, as shown in Figure d. This enables the researcher to control the composition of solutions on *both* sides of the membrane patch. Because the patch is so small, and because the ions in the bathing solutions are known, it is possible to record the transport of a single kind of ion through selective channels or pumps in the membrane. Figure e shows such a recording for H^+ transport across a membrane patch of a guard cell. Passage of H^+ across the membrane represents an electrical current that can be measured. The graph indicates current due to H^+ flux during a recording that lasts several minutes. This particular experiment measures the effect of blue light, which increases H^+ current across the membrane.

Along with other evidence, patch clamp studies support the hypothesis that blue light and other stimuli regulate guard cells by affecting the proton (H^+) pumps of the membrane.

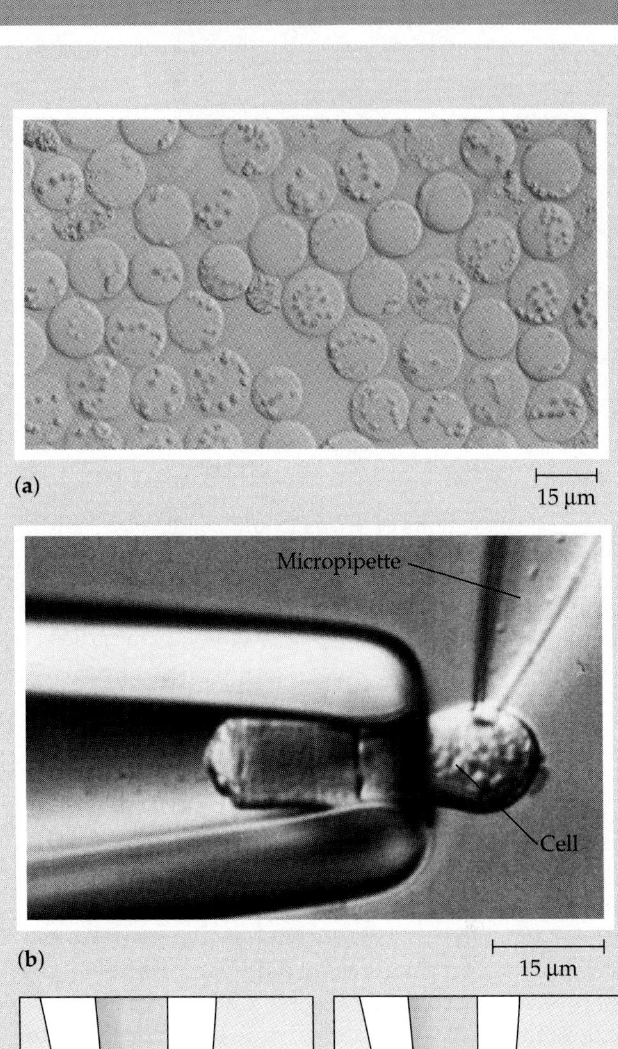

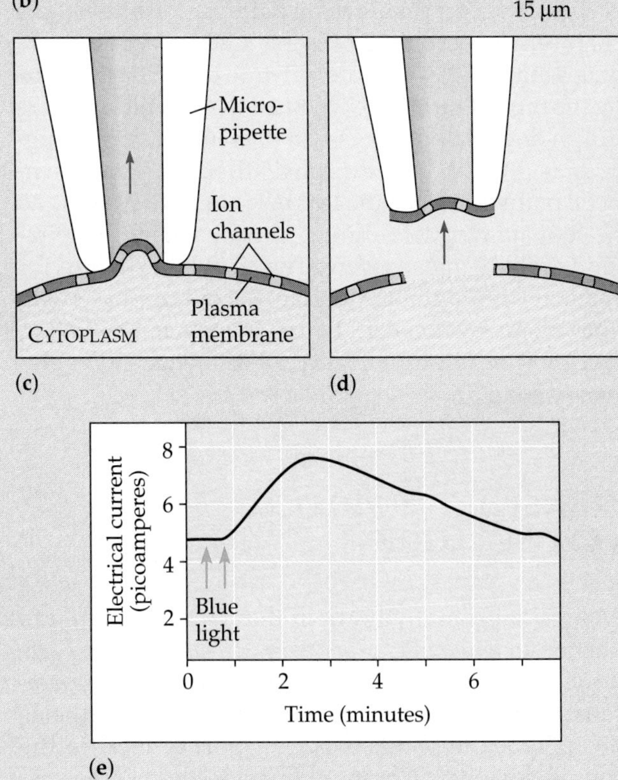

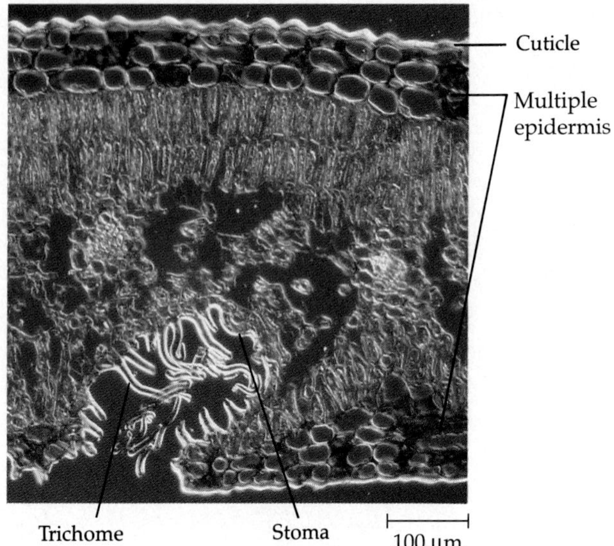

Figure 32.10
Structural adaptations of leaves to arid conditions. The stomata of this oleander, a xerophyte, are recessed in crypts equipped with hairs called trichomes (LM). This structural adaptation protects the stomata from the hot, dry wind. The upper epidermis has several layers of cells, another adaptation that reduces water loss.

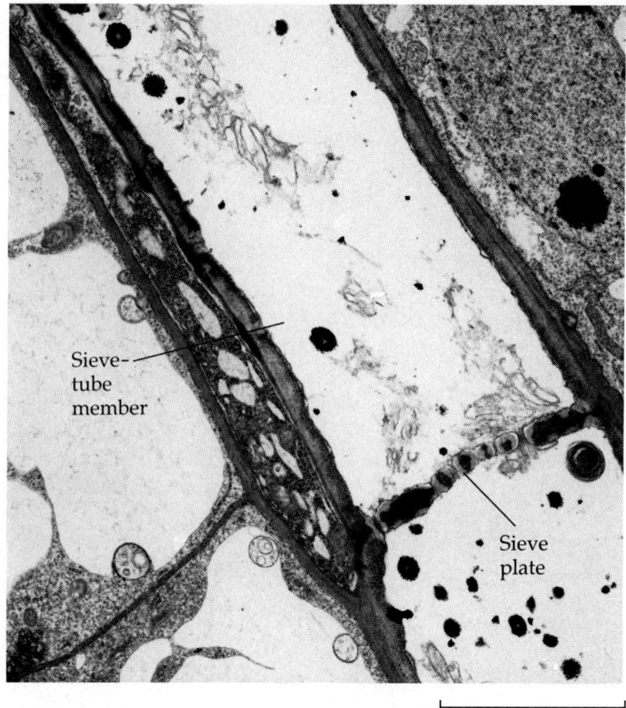

Figure 32.11
Sieve tubes and translocation. Sap flows between the sieve-tube members through porous sieve plates. Next to each sieve-tube member is a companion cell, and parenchyma cells are also present in phloem tissue (TEM).

One of the most elegant adaptations to an arid habitat is found in succulent plants of the family Crassulaceae. These plants, and a few members of other families, including pineapples, assimilate their CO_2 by an alternative photosynthetic pathway known as CAM, for crassulacean acid metabolism (see Chapter 10). Mesophyll cells in a CAM plant have enzymes that can incorporate CO_2 into organic acids during the night. During the daytime, the organic acids are broken down to release CO_2 in the same cells, and sugars are synthesized by the conventional (C_3) photosynthetic pathway. Since the leaf takes in its CO_2 at night, the stomata can close during the day when transpiration would be most severe. The circadian rhythm for stomatal opening in CAM plants is out of phase with other plants by about 12 hours. In stomatal behavior, both short-term physiological adjustments and evolutionary adaptations are manifest.

TRANSPORT IN PHLOEM: A CLOSER LOOK

Xylem sap generally flows in the wrong direction to function in exporting sugar from leaves to other parts of the plant. A second vascular tissue, the phloem, transports the organic products of photosynthesis throughout the plant. This transport of food in the plant is called **translocation.** In angiosperms, the specialized cells of phloem that function in translocation are the sieve-tube members, which are arranged end to end to form long sieve tubes (Figure 32.11). Between the members are sieve plates, porous cross walls that facilitate the flow of sap along the sieve tube.

Phloem sap is an aqueous solution that differs markedly in composition from xylem sap. By far the prevalent solute in phloem sap is sugar, primarily the disaccharide sucrose. The sucrose concentration may be as high as 30% by weight, giving the sap a syrupy thickness. Phloem sap may also contain minerals, amino acids, and hormones in transit from one part of the plant to another.

Source-to-Sink Transport

In contrast to the unidirectional transport of xylem sap, from roots to leaves, the direction that phloem sap travels is more variable. The one generalization that holds is that sieve tubes carry food from a sugar source to a sugar sink. A **source** is a plant organ in which sugar is being produced by either photosynthesis or the breakdown of starch. Leaves are usually sources. A **sink** is an organ that consumes or stores sugar. Growing roots, shoot tips, and fruits are sugar sinks supplied by phloem, as are nongreen stems and trunks. A storage organ, such as a tuber or bulb, may be either

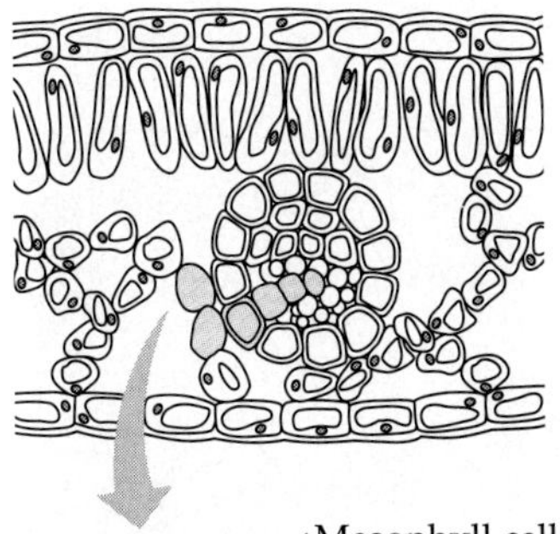

Figure 32.12
Loading of sucrose into phloem. (**a**) Sucrose manufactured in mesophyll cells can travel via the symplast (blue arrows) to sieve-tube members. In some species, sucrose exits (magenta arrow) the symplast near sieve tubes and is actively accumulated from the apoplast by sieve-tube members and their companion cells. Some companion cells, including the one shown here, are modified as transfer cells, which have wall ingrowths that increase the surface area for the transport of solutes. (**b**) A chemiosmotic mechanism is responsible for the active transport of sucrose into companion cells and sieve-tube members. Proton pumps generate an H^+ gradient, which drives sucrose accumulation with the help of a membrane protein (symport) that couples sucrose transport to the diffusion of H^+ back into the cell.

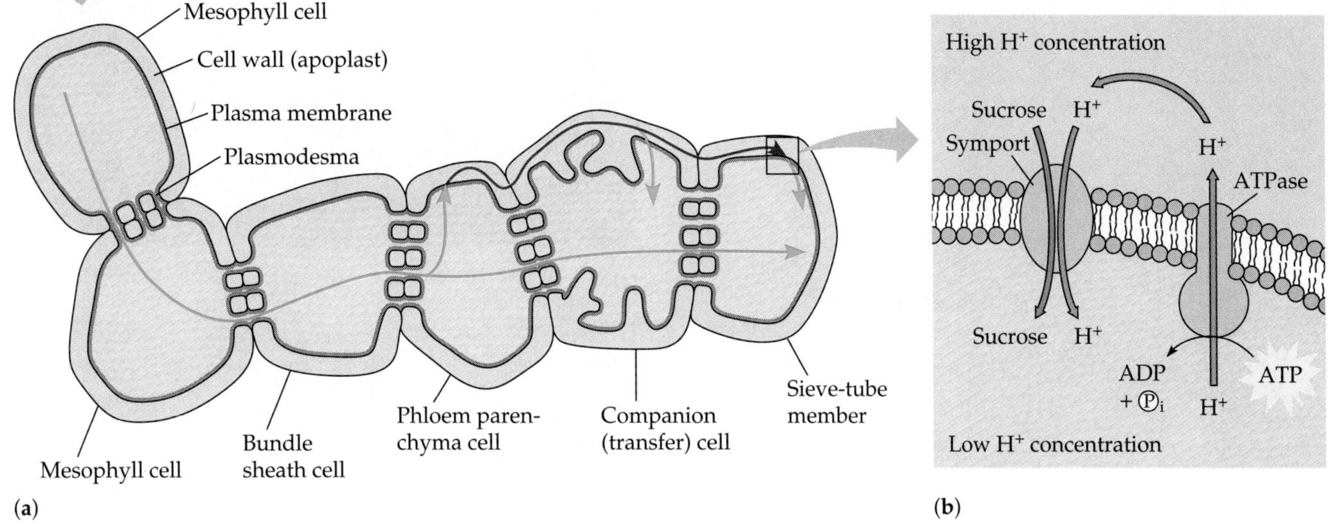

a source or a sink depending on the season. When the storage organ is stockpiling carbohydrates during the summer, it is a sugar sink. After breaking dormancy in early spring, however, the storage organ becomes a source as its starch is broken down to sugar, which is carried away in the phloem to the growing buds of the shoot system.

Other solutes may be transported to sinks along with sugar. For example, minerals that reach leaves in xylem may later be transferred in the phloem to developing fruit.

A sugar sink is usually supplied by the sources nearest to it. The upper leaves on a branch may send sugar to the growing shoot tip, whereas the lower leaves export sugar to roots. A growing fruit requires so much food that it may monopolize the sugar sources all around it. One sieve tube in a vascular bundle may carry phloem sap in one direction, while sap in a different tube in the same bundle flows in the opposite direction. For each sieve tube, the direction of transport depends only on the locations of the source and sink connected by that tube, and the direction may change with the season or developmental stage of the plant.

Phloem Loading and Unloading

Sugar produced in the mesophyll cells of a leaf must be loaded into sieve-tube members before it can be exported to sugar sinks. In some species, including certain plants of the mint and squash families, sucrose may move all the way from mesophyll to sieve members via the symplast, passing from cell to cell through plasmodesmata. In other species, sucrose reaches sieve-tube members by a combination of symplastic and apoplastic pathways (Figure 32.12a). For example, in corn leaves, sucrose diffuses through the symplast from mesophyll cells into small veins. Much of the sugar then moves out of the cells into the apoplast in the vicinity of sieve-tube members and companion cells. This sucrose is accumulated from the apoplast (walls) by the sieve-tube members and their companion cells, which pass the sugar into the sieve-tube members through plasmodesmata linking the cells. In some plants, companion cells have numerous ingrowths of their walls, an adaptation that increases the cells' surface area and enhances the transfer of solutes between apoplast and symplast. Such modified cells are called **transfer cells.**

In corn and many other plants, sieve-tube members accumulate sucrose to concentrations two to three times higher than concentrations in mesophyll, and thus phloem loading requires active transport. Proton pumps do the work that enables the cells to accumulate sucrose (Figure 32.12b). The ATP-driven pumps move H^+ out of the cell and store energy in the form of the difference in H^+ concentration across the plasma membrane. Another membrane protein uses this energy

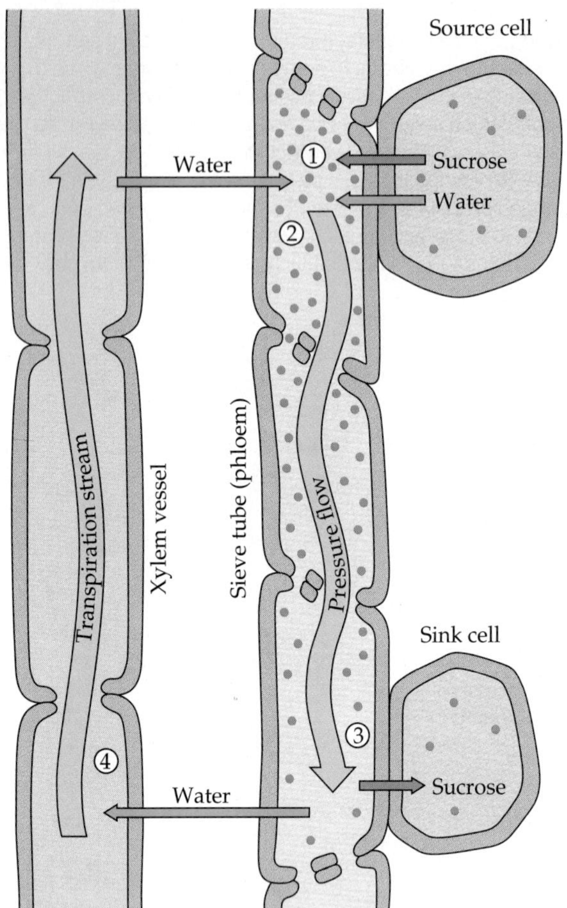

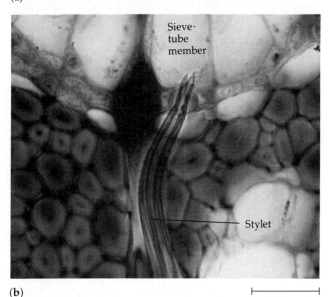

Figure 32.13
Pressure flow in a sieve tube. ① Loading of sugar into the tube at the source reduces the water potential inside sieve-tube members. This causes the sieve tubes to take up water from surrounding tissues by osmosis. ② This absorption of water generates a hydrostatic pressure that forces the sap to flow along the tube. ③ The gradient of pressure in the tube is reinforced by the unloading of sugar and the consequent loss of water at the sink. Sugar does not accumulate in sink cells because it is either consumed in metabolism or converted to storage compounds such as starch. ④ Xylem recycles water from sink to source.

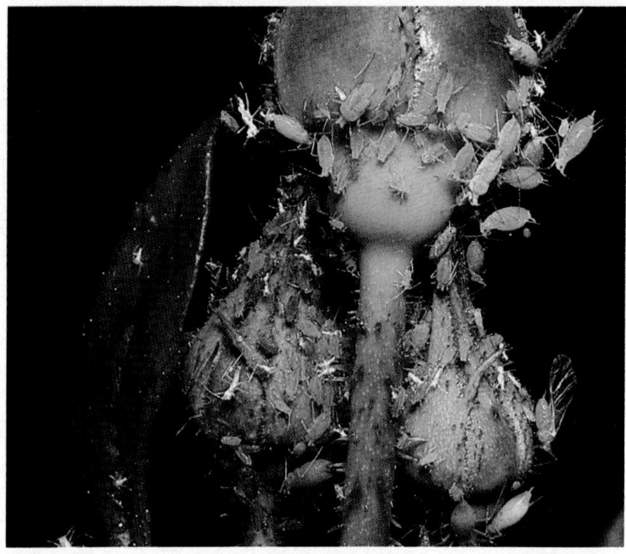

(a)

Sieve-tube member

Stylet

(b)

25 μm

Figure 32.14
Tapping phloem sap with an aphid. (**a**) Aphids feeding on rosebuds. (**b**) The insect inserts a modified mouthpart called a stylet into the plant and probes until the tip of this hypodermic-like organ penetrates a single sieve-tube member (LM). The pressure within the sieve tube force-feeds the aphid, swelling it to several times its original size. While the aphid is feeding, it can be anesthetized and severed from its stylet, which then serves the researcher as a miniature tap that exudes phloem sap for hours. The closer the stylet is to a sugar source, the faster the sap will flow out and the greater its sugar concentration. This is what we would expect if pressure is generated at the source end by an inward pumping of sugar.

source to cotransport sucrose into the cell along with returning hydrogen ions.

Downstream, at the sink end of a sieve tube, phloem unloads its sucrose. Pathways and mechanisms of sugar movement in sink organs vary, depending on species. Both symplastic and apoplastic pathways may be involved. In some plants, sucrose may be unloaded from phloem by active transport. In other species, diffusion is sufficient to move sucrose from phloem to the surrounding cells of the sink organ.

Pressure Flow (Bulk Flow) of Phloem Sap

Phloem sap flows from source to sink at rates as great as 1 m per hour, which is much too fast to be accounted for by either diffusion or cytoplasmic streaming. Phloem sap moves by bulk flow, which is driven by pressure (thus the synonym, pressure flow). Phloem loading results in a high solute concentration at the source end of a sieve tube, which lowers the water potential and causes water to flow into the tube (Figure 32.13). Hydrostatic pressure develops within the sieve tube, and the pressure is greatest at the source end of the tube. At the sink end, the pressure is relieved by the loss of water, owing to water potential being lowered outside the sieve tube by the exodus of sucrose. (In cells of some sink organs, the metabolic breakdown of sugar or the storage of sugar in starch maintains a low sucrose concentration, which favors the continuing exit of sugar from sieve tubes.) The building of pressure at one end of the tube and reduction of that pressure at the opposite end cause water to flow from source to sink, carrying the sugar along. Water is recycled back from sink to source by xylem vessels.

The pressure-flow model explains why phloem sap always flows from a sugar source to a sugar sink, regardless of their locations in the plant. Researchers have devised several kinds of experiments to test the model. Some of these innovative experiments take advantage of natural phloem probes: aphids that feed on phloem sap (Figure 32.14). The case for pressure-flow as the mechanism of translocation in angiosperms is convincing. It is not yet known, however, if this model applies to gymnosperms.

In our study of how sugar moves in plants, we have seen examples of plant transport on all three levels: at the cellular level of transport across plasma membranes (sucrose accumulation by active transport in phloem cells); at the short-distance level of lateral transport within organs (sucrose migration from mesophyll to phloem via the symplast and apoplast); and at the long-distance level of transport between organs (bulk flow within sieve tubes).

* * *

Plant physiologists still have much to learn about the mechanisms of transport in the elaborate vascular system of xylem and phloem. William Harvey, the great seventeenth-century physiologist, speculated that plants and animals have similar circulatory systems. The idea was abandoned after careful dissection failed to turn up a heart in plants. We are only now beginning to understand how the plant keeps sap flowing through its veins without the help of moving parts.

STUDY OUTLINE

1. The life of a terrestrial plant depends on an integrated system of transport between the roots and the leaf-bearing shoots.

Basic Transport Processes in Plants: An Overview (pp. 699–704)

1. Transport occurs at the level of individual cells, at the short-distance level of lateral transport within plant organs, and at the long-distance level of sap flow within xylem and phloem. Different mechanisms operate at these different levels of transport.
2. At the cellular level, solutes move across membranes by both passive transport (diffusion) and active transport. A chemiosmotic mechanism drives most active transport: Proton pumps store energy in the form of an H^+ gradient across the membrane, and specific transport proteins couple the diffusion of H^+ to the movement of other solutes.
3. Differences of water potential (ψ) drive osmotic movement of water into and out of plant cells. Solutes lower water potential and pressure increases water potential. When a plant cell is placed in a hypoosmotic environment, the lower ψ in the cell due to its higher solute concentration causes osmotic uptake of water. Eventually, pressure exerted by the elastic wall equilibrates ψ of the cell and its surroundings, and water entry and exit are now balanced for the turgid cell.
4. Lateral transport of solutes and water in plant organs can occur via the symplast (cytoplasmic continuum) or apoplast (continuum of cell walls), or by some combination of these pathways.
5. Long-distance transport of sap in xylem and phloem occurs by bulk flow, the pressure-driven movement of a fluid.

The Absorption of Water and Minerals by Roots: A Closer Look (pp. 704–706)

1. Water and dissolved solutes gain access to the roots through epidermal cells and their root hairs. The hydrophilic cell walls, in contact with those of the root cortex, provide access to the apoplastic pathway across the cortex. Minerals extracted from the apoplast by cortical cells enter the symplastic pathway.
2. The Casparian strip of the endodermis blocks apoplastic entry of water and minerals to the stele. Of minerals that reach the Casparian strip via the apoplast, only those that can cross the plasma membranes of endodermal cells gain access to the xylem. Thus, the endodermis helps regulate the mineral composition of the xylem sap.

The Ascent of Xylem Sap: A Closer Look (pp. 706–708)

1. The upward flow of xylem sap supplies minerals to the shoots and replaces water lost by transpiration.
2. The transpiration-cohesion-tension mechanism transports xylem sap. Evaporative water loss during transpiration generates surface tension of the water film coating mesophyll cells. This tension, or negative pressure, causes water to move by bulk flow out of the xylem vessels. Cohesion of water due to hydrogen bonding relays transpirational pull on xylem sap all the way down to the roots.

The Control of Transpiration: A Closer Look (pp. 708–712)

1. Plants strike a balance between gas exchange and water loss associated with the presence of stomata.

2. The transpiration-to-photosynthesis ratio is a convenient way to determine how efficiently a plant uses water. Because water loss is the trade-off for allowing carbon dioxide to diffuse into the leaf, plants with lower ratios are generally at an advantage in arid habitats.

3. Plants balance photosynthesis and transpiration by regulating the size of stomatal openings through changes in the turgor pressure of guard cells. These physical changes are due to fluxes in potassium ions that are coupled to the generation of membrane potentials by proton pumps in guard cell membranes.

4. Guard cells usually open at dawn, due to carbon dioxide depletion, an inherent circadian rhythm, and ion movements triggered by light-detecting pigments.

5. Water deficiency and high temperatures can cause stomata to close during the daytime.

6. Xerophytic plants in dry habitats have leaves with morphological or physiological adaptations that reduce transpiration.

Transport in Phloem: A Closer Look (pp. 712–715)

1. Translocation is the process of transporting photosynthetically produced food throughout the body of the plant in the phloem. In most plants, sucrose is the main solute translocated in phloem.

2. Sieve tubes of phloem carry food from a sugar source to a sugar sink. A source is an organ that produces sugar by either photosynthesis or the breakdown of starch, whereas a sink consumes or stores sugar.

3. In many plants, sucrose is loaded into phloem at a source by active transport. Proton pumps of sieve-tube members and companion cells drive sucrose loading by chemiosmotic coupling.

4. Pressure moves phloem sap along sieve tubes by bulk flow. The pressure is generated at the source end of a sieve tube by the accumulation of sucrose and the resulting osmotic uptake of water. At the sink end, pressure is relieved by the passage of sucrose and water out of the sieve tube.

SELF-QUIZ

1. Which of the following would *not* contribute to water uptake by a plant cell?
   a. an increase in the water potential of the surrounding solution
   b. a decrease in π of the surrounding solution
   c. uptake of solutes by the cell
   d. a decrease in ψ of the cytoplasm
   e. an increase in tension on the surrounding solution

2. In which of the following processes is osmosis *least* involved?
   a. long-distance transport of xylem sap
   b. the swelling of guard cells
   c. water uptake by cells immersed in a hypotonic solution
   d. root pressure
   e. water movement between neighboring cells of the root cortex

3. Stomata open when guard cells
   a. sense an increase in CO_2 in the air spaces of the leaf
   b. flop open due to a decrease in turgor pressure
   c. become more turgid due to an influx of K^+, followed by the osmotic entry of water
   d. reverse their circadian rhythm
   e. accumulate water by active transport

4. Which of the following is *not* part of the transpiration-cohesion-tension mechanism for the ascent of xylem sap?
   a. the evaporation of water from the mesophyll cells, which initiates a pull of water molecules from neighboring cells and eventually from the xylem
   b. the transfer of this pull from one water molecule to the next due to the cohesion caused by hydrogen bonds
   c. the hydrophilic walls of the narrow tracheids and xylem vessels that help maintain the column of water against the force of gravity
   d. active pumping of water into the xylem of roots
   e. the reduction of water potential in the surface film of mesophyll due to increased surface tension

5. Both ____________ and ____________ are sugar sinks.
   a. a growing root; a developing fruit
   b. a photosynthesizing leaf; a growing root
   c. a growing shoot tip; a tuber where starch is being broken down
   d. a photosynthesizing leaf; a developing fruit
   e. a tuber where starch is being broken down; a growing shoot tip

6. Which of the following does *not* appear to involve active transport?
   a. movement of mineral nutrients from the apoplast to the symplast
   b. movement of sugar from mesophyll into sieve-tube members in corn
   c. movement of sugar from one sieve-tube member to the next
   d. K^+ uptake by guard cells during stomatal opening
   e. movement of mineral nutrients into cells of the root cortex

7. The movement of sap from a sugar source to a sugar sink
   a. occurs through the apoplast of sieve-tube members
   b. may translocate sugars from the breakdown of stored starch up to developing shoots
   c. is similar to the flow of xylem sap in depending on tension, or negative pressure
   d. depends on the active pumping of water into sieve tubes at the source end
   e. results mainly from diffusion

8. The productivity of a crop begins to decline when leaves begin to wilt mainly because (explain your answer)
   a. the chlorophyll of wilting leaves decomposes
   b. flaccid mesophyll cells are incapable of photosynthesis

c. stomata close, preventing CO_2 from entering the leaf
d. photolysis, the water-splitting step of photosynthesis, cannot occur when there is a water deficiency
e. accumulation of CO_2 in the leaf inhibits the enzymes required for photosynthesis

9. Imagine cutting a live twig from a tree and examining the cut surface of the twig with a magnifying glass. You locate the vascular tissue and observe a growing droplet of fluid exuding from the cut surface. This fluid is probably (explain your answer)
   a. phloem sap
   b. xylem sap
   c. guttation fluid
   d. fluid of the transpiration stream
   e. cell sap from the broken vacuoles of cells

10. Which structure or compartment is *not* part of the plant's apoplast?
   a. the lumen of a xylem vessel
   b. the lumen of a sieve tube
   c. the cell wall of a mesophyll cell
   d. the cell walls of transfer cells
   e. the cell walls of root hairs

CHALLENGE QUESTIONS

1. A botanist used a device called a dendrometer to measure slight changes in the diameter of a tree trunk caused by water movement in the xylem. The device made simultaneous measurements at different heights, graphed below. Explain the daily pattern of variation in diameter. Do these data suggest that water is "pushed" or "pulled" up the trunk? Explain.

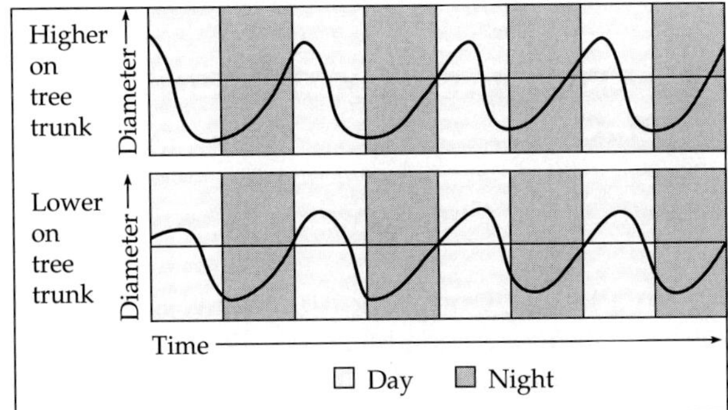

2. Wind ventilates the leaf surface, increases transpiration, and replenishes the supply of carbon dioxide for photosynthesis. Stomata close when the carbon dioxide concentration inside the leaf increases. How do you think stomatal response to carbon dioxide is related to wind speed, and what might be the adaptive value of this relationship?

3. Describe the environmental conditions that would minimize the transpiration–photosynthesis ratio for a C_3 plant.

SCIENCE, TECHNOLOGY, AND SOCIETY

1. In a reverse osmosis unit, pressure is applied to a solution, forcing water through a selectively permeable membrane but leaving solute molecules behind. Why is this called reverse osmosis? Relate reverse osmosis to the water potential concept discussed in this chapter. What might be some possible applications of reverse osmosis?

2. Many scientists believe that increased carbon dioxide in the atmosphere may result in warming of the global climate in coming decades. This could have a serious impact on agriculture. Imagine that you are in charge of a project to use recombinant DNA techniques to "engineer" crop plants that can grow under hotter, drier conditions. What traits would you try to build into the "ideal" corn or wheat plant?

FURTHER READING

Heinrich, B. "Nutcracker Sweets." *Natural History,* February 1991. How red squirrels obtain maple syrup from the sap flowing within trees.

Miller, J. A. "Plant 'Sight' from Pores and Pumps." *Science News,* November 30, 1985. Current knowledge about stomatal function may increase crop yields.

Neher, E., and B. Sakmann. "The Patch Clamp Technique." *Scientific American,* March 1992. The Nobel Prize–winning developers of this method explain how it can be used to study membrane transport.

Salisbury, F. B., and C. W. Ross. *Plant Physiology,* 4th ed. Belmont, CA: Wadsworth, 1992. A rigorous text with excellent chapters on transport.

Schulze, E.-D., R. H. Robichaux, J. Grace, P. W. Rundel, and J. R. Ehleringer. "Plant Water Balance." *BioScience,* January 1987. Evolutionary adaptations that minimize water loss.

Taiz, L., and E. Zeiger. *Plant Physiology.* Redwood City, CA: Benjamin/Cummings, 1991. Chapters 3–7 of this current text cover transport on a detailed but accessible level.

Wickelgren, I. "Plant Ion-Pump Gene Cloned, Sequenced." *Science News,* March 4, 1989. Exciting research on the proton pump that drives so much transport work in plants.

33 PLANT NUTRITION

NUTRITIONAL REQUIREMENTS OF PLANTS
SOIL
NITROGEN ASSIMILATION BY PLANTS
SOME NUTRITIONAL ADAPTATIONS OF PLANTS

Figure 33.1
A plant contacts its environment over an enormous surface. The branched body plan of a plant such as this geranium provides extensive area for the uptake of inorganic nutrients from the soil (water and minerals) and air (carbon dioxide). This chapter surveys the nutritional requirements of plants and focuses on the evolutionary adaptations that facilitate the assimilation of inorganic nutrients.

Each organism is an open system connected to its environment by continuous exchange of energy and materials (see Chapter 6). In the energy flow and chemical cycling that keep an ecosystem alive, plants and other photosynthetic autotrophs perform the key step of transforming inorganic compounds into organic ones. Autotrophic does not mean autonomous, however. Plants, of course, need sunlight as the energy source for photosynthesis. But to synthesize organic matter, plants also require raw materials in the form of inorganic substances: carbon dioxide, water, and a variety of minerals present as inorganic ions in the soil. With its ramifying root system and shoot system, a plant is extensively networked with its environment—the soil and air, which are the reservoirs of the plant's inorganic nutrients (Figure 33.1). In this chapter, you will learn more about the nutritional requirements of plants and examine some of the structural and physiological adaptations for plant nutrition that have evolved.

NUTRITIONAL REQUIREMENTS OF PLANTS

The Chemical Composition of Plants

Watch a large plant grow from a tiny seed, and you cannot help wondering where all the mass comes from. Aristotle thought soil provided the substance for plant growth because plants seemed to spring from the earth. Leaves, he believed, functioned only to shade the developing fruit. In the seventeenth century, a Belgian physician named Jean-Baptiste van Helmont performed an experiment to find out if plants grew by absorbing soil. He planted a willow seedling in a pot that contained 90.9 kg of soil. After five years, the willow had grown into a tree weighing 76.8 kg, but only 0.06 kg of soil had disappeared from the pot. Van Helmont concluded that the willow had grown mainly from the water he had added regularly. A century later, Stephen Hales, an English physiologist, postulated that plants were nourished mostly by air.

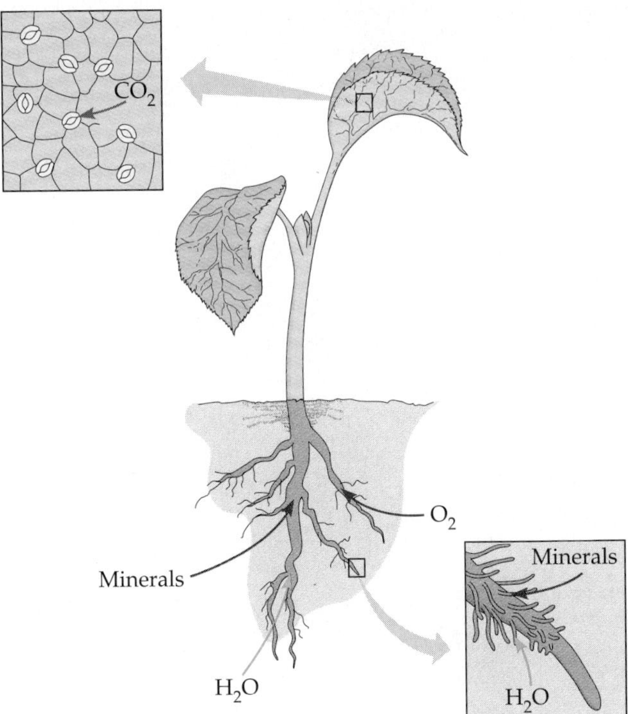

Figure 33.2
The uptake of nutrients by a plant: an overview. Roots absorb water and minerals from the soil, with root hairs greatly increasing the area of the epidermal surface that functions in this absorption. Carbon dioxide, the source of carbon for photosynthesis, diffuses into leaves from the surrounding air through stomata. (Plants also need O_2 for cellular respiration, although the plant is a net producer of O_2.) From these inorganic nutrients, the plant can produce all of its own organic material.

As it turns out, none of the early ideas about plant nutrition is entirely incorrect (Figure 33.2). Plants *do* extract minerals that are essential nutrients from the soil, but these minerals make only a small contribution to the overall mass of the plant, as van Helmont discovered. About 80% to 85% of a herbaceous (nonwoody) plant is water, and plants grow mainly by accumulating water in the central vacuoles of their cells. Furthermore, water can truly be considered a nutrient because it supplies most of the hydrogen and some of the oxygen that are incorporated into organic compounds by photosynthesis. However, only a small fraction of the water that enters a plant contributes atoms to organic molecules. Generally, more than 90% of the water absorbed by plants is lost by transpiration, and most of the water that is retained by the plant actually functions as a solvent, makes cell elongation possible, and serves to maintain the form of soft tissue by keeping cells turgid. By weight, the bulk of the organic material of a plant is derived not from water, but from the CO_2 that is assimilated from the atmosphere.

We can measure water content by comparing the weight of plant material before and after it is dried. We can then analyze the chemical composition of the dry residue. Organic substances account for about 95% of the dry weight, with inorganic minerals making up the remaining 5%. Most of the organic material is carbohydrate, including the cellulose of cell walls. Thus, carbon, oxygen, and hydrogen, the ingredients of carbohydrates, are the most abundant elements in the dry weight of a plant. Because some organic molecules contain nitrogen, sulfur, or phosphorus, these elements are also relatively abundant in plants.

More than 50 chemical elements have been identified among the inorganic substances present in plants, but it is unlikely that all these elements are essential. Roots are able to absorb minerals somewhat selectively, enabling the plant to accumulate essential elements that may be present in the soil in very minute quantities (see Chapter 32). To a certain extent, however, the minerals in a plant reflect the composition of the soil in which the plant is growing. Plants growing on mine tailings, for instance, may contain gold or silver. Studying the chemical composition of plants provides clues about their nutritional requirements, but we must distinguish elements that are essential from those that are merely present in the plant.

Essential Nutrients

A particular chemical element is considered to be an **essential nutrient** if it is required for a plant to grow from a seed and complete the life cycle, producing another generation of seeds. A method known as **hydroponic culture** can be used to determine which of the mineral elements are actually essential nutrients (see the Methods Box, p. 720). Such studies have helped identify 17 elements that are essential nutrients in all plants and a few other elements that are essential to certain groups of plants (Table 33.1). Most research has involved crop plants; little is known about the specific nutritional needs of uncultivated plants, even some of the most commercially important conifers.

Elements required by plants in relatively large amounts are called **macronutrients.** There are nine macronutrients in all, including the six major ingredients of organic compounds: carbon, oxygen, hydrogen, nitrogen, sulfur, and phosphorus. The other three macronutrients are calcium, potassium, and magnesium (Table 33.1 lists some of their functions).

Elements that plants need in very small amounts are called **micronutrients.** The eight micronutrients are iron, chlorine, copper, manganese, zinc, molybdenum, boron, and nickel. These elements function in the plant mainly as cofactors of enzymatic reactions (see Chapter 6). Iron, for example, is a metallic component of cytochromes, the proteins that function in the electron transport chains of chloroplasts and mitochondria. It is because micronutrients generally serve cat-

Methods: Hydroponic Culture to Identify Essential Nutrients

The roots of plants are bathed in solutions of various minerals dissolved in known concentrations. The water must be aerated to provide the roots with oxygen for cellular respiration. A particular mineral, such as potassium, can be omitted from the culture medium to test whether it is essential to the plants.

If the element deleted from the mineral solution is an essential nutrient, then the incomplete medium will cause plants to become abnormal in appearance compared with controls grown on a complete mineral medium. The most common symptoms of a mineral deficiency are stunted growth and discolored leaves.

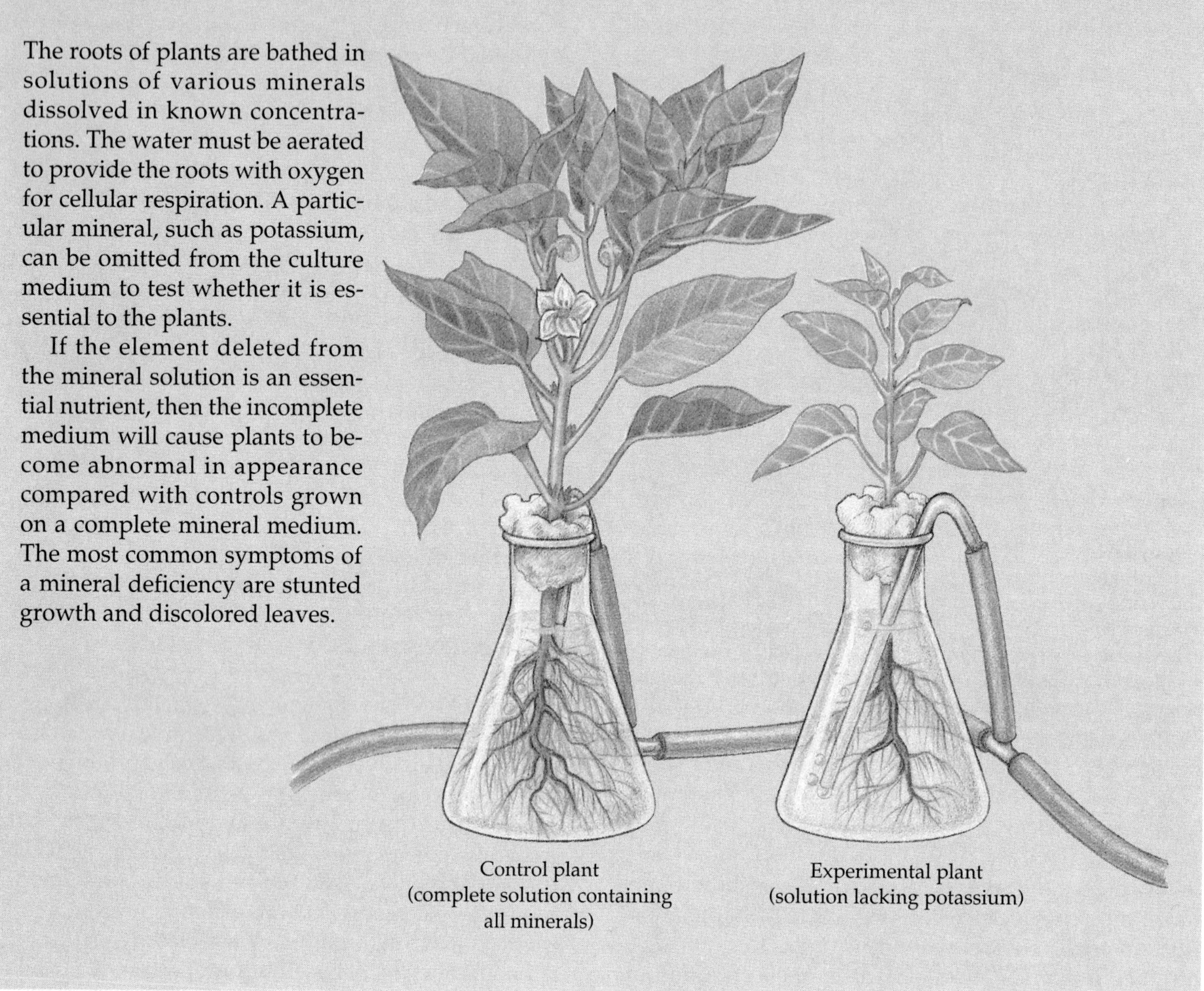

alytic roles that plants need only minute quantities of these elements. The requirement for molybdenum, for example, is so modest that there is only one atom of this rare element for every 16 million atoms of hydrogen in dried plant material. Yet a deficiency of molybdenum or any other micronutrient can weaken or kill a plant.

Mineral Deficiencies

The symptoms of a mineral deficiency depend partly on the function of that nutrient in the plant. For example, a deficiency of magnesium, an ingredient of chlorophyll, causes yellowing of the leaves, or **chlorosis** (Figure 33.3). In some cases, the relationship between a mineral deficiency and its symptoms is less direct. For instance, iron deficiency can cause chlorosis even though chlorophyll contains no iron, because this metal is required as a cofactor in one of the steps of chlorophyll synthesis.

The symptoms of a mineral deficiency depend not only on the role of that nutrient in the plant but also on its mobility within the plant. If a nutrient moves about freely from one part of the plant to another, symptoms of a deficiency will show up first in older organs, because young, growing tissues have more "drawing power" than old tissues for nutrients that are in short supply. A plant starved for magnesium, for example, will show signs of chlorosis first in its older leaves.

Table 33.1 Essential Nutrients in Plants

Element	Form Available to Plants	Major Functions
Macronutrients		
Oxygen	CO_2	Major component of plant's organic compounds.
Carbon	CO_2	Major component of plant's organic compounds.
Hydrogen	H_2O	Major component of plant's organic compounds.
Nitrogen	NO_3^-, NH_4^+	Component of nucleic acids, proteins, hormones, and coenzymes.
Potassium	K^+	Cofactor functional in protein synthesis; major solute functioning in water balance; operation of stomata.
Calcium	Ca^{2+}	Important in formation and stability of cell walls; maintenance of membrane structure and permeability; activates some enzymes; regulates many responses of cells to stimuli.
Magnesium	Mg^{2+}	Component of chlorophyll; activates many enzymes.
Phosphorus	$H_2PO_4^-$, HPO_4^{2-}	Component of nucleic acids, phospholipids, ATP, several coenzymes.
Sulfur	SO_4^{2-}	Component of proteins, coenzymes.
Micronutrients		
Chlorine	Cl^-	Activates photosynthetic elements; functions in water balance.
Iron	Fe^{3+}, Fe^{2+}	Component of cytochromes; activates some enzymes.
Boron	$H_2BO_3^-$	Cofactor in chlorophyll synthesis; may be involved in carbohydrate transport and nucleic acid synthesis.
Manganese	Mn^{2+}	Active in formation of amino acids; activates some enzymes.
Zinc	Zn^{2+}	Active in formation of chlorophyll; activates some enzymes.
Copper	Cu^+, Cu^{2+}	Component of many redox and lignin-biosynthetic enzymes.
Molybdenum	MoO_4^{2-}	Essential for nitrogen fixation; cofactor functional in nitrate reduction.
Nickel	Ni^{2+}	Cofactor for an enzyme functioning in metabolism of nitrogenous compounds.

(a)

(b)

Figure 33.3
Magnesium deficiency in tobacco. Compare the control plant shown in (**a**) to the experimental plant shown in (**b**), which shows signs of magnesium deficiency. Yellowing of the leaves (chlorosis) is the result of an inability to synthesize chlorophyll, which contains magnesium.

Figure 33.4
Hydroponic farming. In this apparatus, a nutrient solution flows over the roots of lettuce growing on a slat. Perhaps astronauts living in a space station will one day grow their vegetables hydroponically, but because of the expense, it is unlikely that this type of farming will soon relieve hunger here on Earth.

Figure 33.5
Soil horizons. This roadcut exposes a vertical profile of three soil layers, or horizons. The A horizon is the topsoil, a mixture of decomposed rock of varying textures, living organisms, and decaying organic matter. Topsoil is subject to extensive physical and chemical weathering. The B horizon contains much less organic matter than the A horizon and is less weathered. Clay particles and minerals leached from the A horizon by water may accumulate in the B horizon. The C horizon serves as the parent material for the upper layers of soil. This horizon is mainly composed of partially weathered rock.

Magnesium, which is relatively mobile in the plant, is shunted preferentially to young leaves. In contrast, a deficiency of a nutrient that is relatively immobile within a plant will affect young parts of the plant first. Older tissues may have adequate amounts of the mineral, which they are able to retain during periods of short supply. A deficiency of iron, which does not move freely in the plant, will cause yellowing of young leaves before any effect on older leaves is visible.

The symptoms of a mineral deficiency are often distinctive enough for a plant physiologist or farmer to diagnose its cause. One way to confirm the diagnosis of a specific deficiency is to analyze the mineral content of the plant and soil. Deficiencies of nitrogen, potassium, and phosphorus are the most common problems. Shortages of micronutrients are less common and tend to be geographically localized because of differences in soil composition. The amount of a micronutrient needed to correct a deficiency is usually quite small. For example, a zinc deficiency in fruit trees can usually be cured by hammering a few zinc nails into each tree trunk. Moderation is important because overdoses of some micronutrients can be toxic to plants.

One way to ensure optimal mineral nutrition is to grow plants hydroponically on nutrient solutions that can be precisely regulated (Figure 33.4). Hydroponics is currently practiced commercially, but only on a limited scale because the requirements for equipment and labor make hydroponic farming relatively expensive compared with growing crops in soil.

SOIL

The texture and chemical composition of soil are major factors determining what kinds of plants can grow well in a particular location, be it a natural ecosystem or an agricultural region. (Climate, of course, is another important factor.) Plants that grow naturally in a certain type of soil are adapted to its mineral content and texture and are able to absorb water and extract essential nutrients from that soil. In interacting with the soil that supports their growth, plants, in turn, affect the soil, as we will soon see.

Texture and Composition of Soils

Soil has its origin in the weathering of solid rock. Water that seeps into crevices and freezes in winter fractures the rock, and acids dissolved in the water also help to break down the rock. Once organisms are able to invade the rock, they accelerate the decomposition. Lichens, fungi, bacteria, mosses, and plant roots all secrete acids, and the expansion of roots growing in fissures cracks rocks and pebbles. The eventual result of all this activity is **topsoil,** a mixture of decomposed rock of varying texture, living organisms, and humus, a residue of partially decayed organic material. The topsoil and other distinct soil layers, or **horizons,** are often visible in vertical profile where there is a roadcut (Figure 33.5).

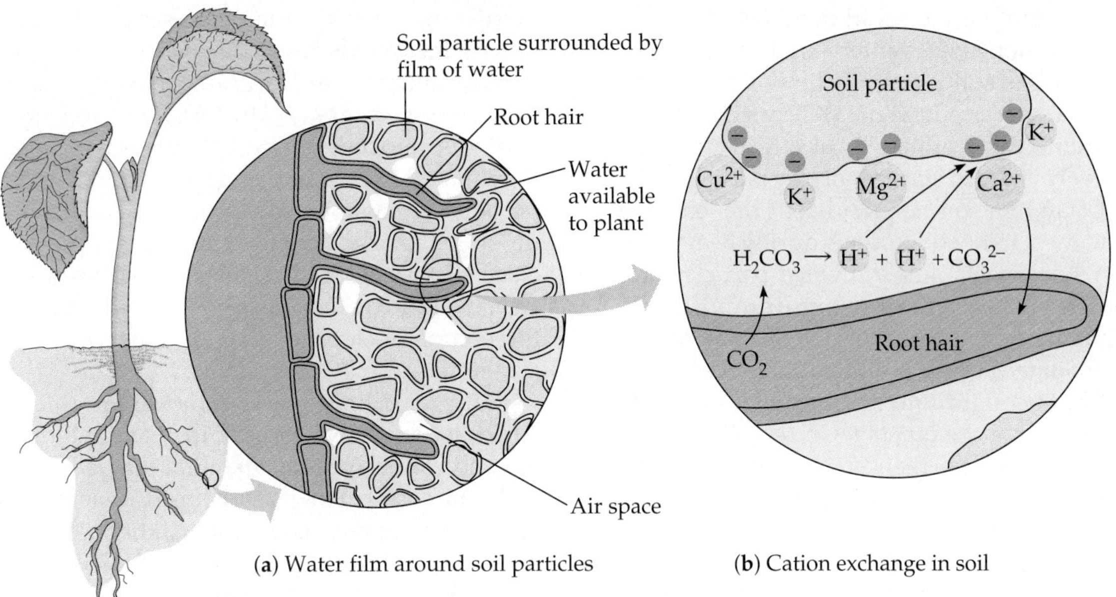

Figure 33.6
The availability of soil water and minerals. (**a**) A plant cannot extract all the water in the soil because some of it is tightly held by hydrophilic soil particles. Water retained in tiny spaces in the soil, bound less tightly to soil particles, can be imbibed by the root—especially by the epidermis, which includes the root hairs. (**b**) Cation exchange in soil. Hydrogen ions in the soil solution help make certain nutrients available to plants by displacing positively charged minerals that were bound tightly to the surface of fine soil particles. And plants contribute to the pool of H^+ in the soil. Cellular respiration in roots releases CO_2 to the soil solution, where the CO_2 reacts with water to form carbonic acid (H_2CO_3). Dissociation of this acid adds hydrogen ions to the soil.

The texture of topsoil depends on the size of its particles, which are classified in a range going from coarse sand to microscopic clay particles. The most fertile soils are usually **loams,** made up of a mixture of sand, silt, and clay. Loamy soils have enough fine particles to provide a large surface area for retaining minerals and water, which adhere to the particles. But loams also have enough coarse particles to provide air spaces containing oxygen that can be used by roots for cellular respiration. If soil does not drain adequately, roots suffocate because the air spaces are replaced by water; the roots may also be attacked by molds favored by the soaked soil. These are common hazards for houseplants that are overwatered in pots that do not drain. Some plants, however, are adapted to waterlogged soil. For example, mangroves and many other plants that inhabit swamps and marshes have some of their roots modified as hollow tubes that grow upward and function as snorkels, bringing down oxygen from the air. Variation in the degree of drainage required by plants is one of the many ways soil conditions help determine what type of vegetation prevails in a particular location.

Soil is home to an astonishing number and variety of organisms. A teaspoon of soil has about five billion bacteria that cohabit with various fungi, algae and other protists, insects, earthworms, nematodes, and the roots of plants. The activities of all these organisms affect the physical and chemical properties of the soil. Earthworms, for instance, aerate the soil by their burrowing and add mucus that holds fine soil particles together. The metabolism of bacteria alters the mineral composition of the soil. Plant roots extract water and minerals but also affect soil pH and reinforce the soil against erosion.

Humus, the decomposing organic material formed by the action of bacteria and fungi on dead organisms, feces, fallen leaves, and other organic refuse, prevents clay from packing together and builds a crumbly soil that retains water but is still porous enough for good aeration of roots. Humus is also a reservoir of mineral nutrients that are returned gradually to the soil as microorganisms decompose the organic matter.

The Availability of Soil Water and Minerals

After a heavy rainfall, water drains away from the larger spaces of the soil, but smaller spaces retain water because of its attraction for the soil particles, which have electrically charged surfaces. Some of this water adheres so tightly to the hydrophilic soil particles that it cannot be extracted by plants. In the tiniest spaces of the soil, however, is a film of water bound less tightly to the particles; this is the water generally available to plants (Figure 33.6a). It is not pure water but in fact a soil solution containing dissolved minerals. Roots absorb this soil solution (see Chapter 32).

Many minerals in soil, especially those that are positively charged, such as potassium (K^+), calcium (Ca^{2+}), and magnesium (Mg^{2+}), adhere by electrical attraction to the negatively charged surfaces of clay particles. The presence of clay in a soil helps prevent the leaching (draining away) of mineral nutrients during heavy rain or irrigation, because the finely divided particles provide so much surface area for binding minerals. Minerals that are negatively charged, such as nitrate (NO_3^-), phosphate ($H_2PO_4^-$), and sulfate (SO_4^{2-}), are usually not bound tightly to soil particles and thus tend to leach away more quickly. On the other hand, clay particles must release their bound minerals to the soil solution for roots to absorb the nutrients. Positively charged minerals are made available to the plant when hydrogen ions in the soil displace the mineral ions from the clay particles. This process, called **cation exchange,** is stimulated by the roots themselves, which release acids that add hydrogen ions to the soil solution (Figure 33.6b).

As roots absorb water and minerals, the supply of these resources is depleted in the immediate surroundings of the existing roots. However, the rapid growth of roots, with surface areas expanded by root hairs, brings the plant into contact with a larger and larger volume of soil and thus a new supply of nutrients.

Soil Management

It may take centuries for a soil to become fertile through decomposition and accumulation of organic material, but human mismanagement can destroy that fertility within a few years. Good soil management, on the other hand, can preserve fertility, resulting in sustained agricultural productivity.

To understand soil management, we must begin with the premise that agriculture is unnatural. In forests, grasslands, and other natural ecosystems, mineral nutrients are usually recycled by the decomposition of dead organic material in the soil. In contrast, when farmers harvest a crop, or a gardener uses a grass catcher on the lawn mower, essential elements are diverted from the chemical cycles going on in that location. In general, agriculture depletes the mineral content of the soil. To grow a ton of wheat grain, the soil gives up 18.2 kg of nitrogen, 3.6 kg of phosphorus, and 4.1 kg of potassium. Each year the fertility of the soil diminishes, unless fertilizers are applied to replace the lost minerals. Many crops also use far more water than the natural vegetation that once grew on that land, forcing farmers to irrigate. Prudent fertilization, thoughtful irrigation, and prevention of erosion are three of the most important goals of soil management.

Fertilizers Prehistoric farmers may have started fertilizing their fields after noticing that grass grew faster and greener where animals had defecated. The Romans used manure to fertilize their crops, and Native Americans buried fish along with seeds when they planted corn. In developed nations today, most farmers use commercially produced fertilizers containing minerals that are either mined or prepared by industrial processes. These fertilizers are usually enriched in nitrogen, phosphorus, and potassium, the three mineral elements that are most commonly deficient in farm soils.

Manure, fishmeal, and compost are referred to as "organic" fertilizers because they are of biological origin and contain organic material that is in the process of decomposing. However, before the elements in compost can be of any use to plants, the organic material must be decomposed to the inorganic nutrients that roots can absorb. In the end, the minerals a plant extracts from the soil are in the same form whether they came from organic fertilizer or from a chemical factory. Compost releases minerals gradually, however, whereas the minerals in commercial fertilizers are available immediately, but may not be retained by the soil for long. Excess minerals not taken up by the plants are wasted because they may be rapidly leached from the soil by rainwater or irrigation. To make matters worse, this mineral runoff may enter the groundwater and eventually pollute streams and lakes. Agricultural researchers are attempting to develop ways to reduce the use of fertilizers while maintaining crop yields.

To fertilize judiciously, the farmer must pay close attention to the pH of the soil. Acidity not only affects cation exchange but also influences the chemical form of all minerals. Even though an essential element may be abundant in the soil, plants may be starving for that element because it is bound too tightly to clay or it is in a chemical form the plant cannot absorb. Managing the pH of soil is touchy; a change in hydrogen ion concentration may make one mineral more available to the plant while causing another mineral to become less available. At pH 8, for instance, the plant can absorb calcium, but iron is almost completely unavailable. The pH of the soil should be matched to the specific mineral needs of the crop. If the soil is too alkaline, sulfate can be added to lower the pH. Soil that is too acidic can be adjusted by liming (adding calcium carbonate or calcium hydroxide).

Irrigation Even more than mineral deficiencies, the availability of water most often limits the growth of plants. Irrigation can transform a desert into a garden, but farming in arid regions is a huge drain on water resources. Many of the rivers in the southwestern United States have been reduced to trickles by the diversion of water for irrigation. (Quenching the thirst of growing

(a)

(b)

Figure 33.7
Irrigation. (**a**) Flood irrigation. After a field is flooded, much of the water evaporates, leaving the salts that were dissolved in the irrigation water behind to accumulate in the soil. (**b**) Drip irrigation. Instead of flooding the field or filling ditches with water, perforated pipes drip the water slowly into the soil close to the plant roots. Drip irrigation reduces the loss of water from evaporation and drainage. Both photos were taken in citrus groves in a southern California desert.

cities adds to the problem.) Another problem is that irrigation in an arid region can gradually make the soil so salty that it becomes completely infertile (Figure 33.7a). Salts dissolved in the irrigation water accumulate in the soil as the water evaporates. Eventually, the salt makes the soil solution hyperosmotic to root cells, which then *lose* water instead of absorbing it. As the world population continues to grow, more and more acres of arid land will have to be cultivated. New methods of irrigation may reduce the risks of running out of water or losing farmland to salinization (salt accumulation). For instance, drip irrigation is now used as an alternative to flooding fields for many of the crops and orchards in Israel and the western United States (see Figure 33.7b). In another approach to solving some of the problems of dryland farming, plant breeders are working to develop varieties of plants that require less water or that can tolerate more salinity.

Erosion Topsoil from thousands of acres of farmland is lost to water and wind erosion each year in the United States alone. Certain precautions can help reduce these losses. Rows of trees dividing fields make effective windbreaks, and terracing a hillside can prevent the topsoil from washing away in a heavy rain. Such crops as alfalfa and wheat provide good ground cover and protect the soil better than corn and other crops that are usually planted in rows.

If managed properly, soil is a renewable resource on which farmers can grow food for generations to come. The goal is **sustainable agriculture,** a commitment embracing a variety of farming methods that are conservation-minded, environmentally safe, and profitable (Figure 33.8).

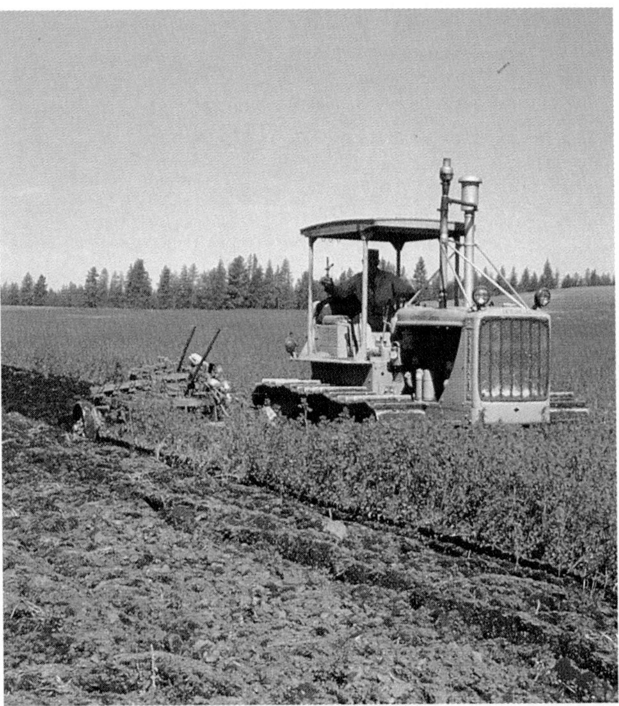

Figure 33.8
Sustainable agriculture and "green manure." The "green manure" being mulched into the soil of this Washington state farm is sweet clover. One out of every three years, clover is planted and the crop is plowed under, a practice that improves the physical structure and nutrient content of the soil for growing wheat and other crops during the other two years of the crop-rotation cycle. Substituting green manure for chemical additives, using beneficial insects that feed on crop pests as an alternative to insecticides, and choosing crops that are suited to the amount of rainfall and other climatic conditions of the locale are examples of actions aimed at developing sustainable agriculture.

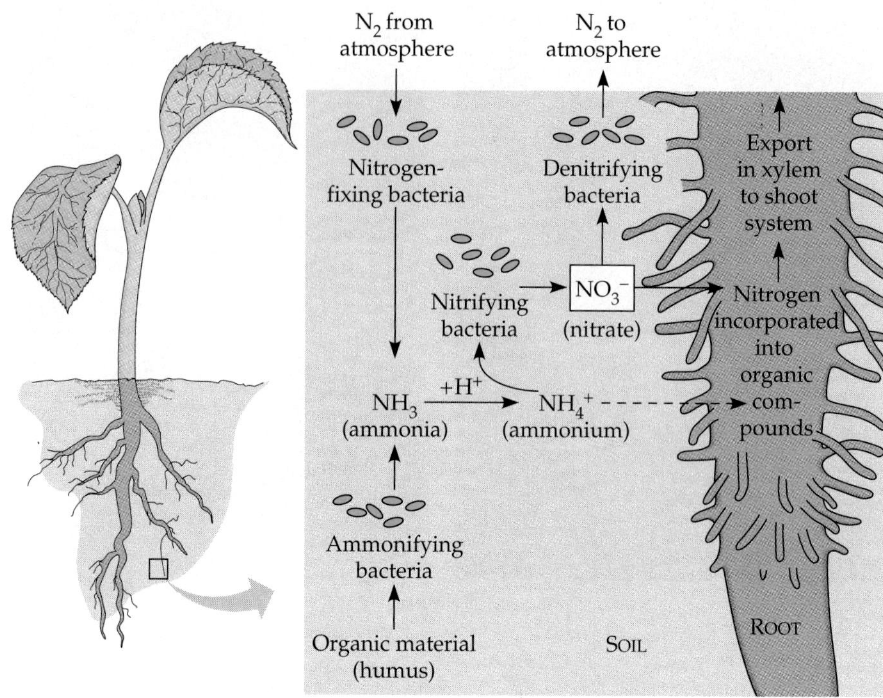

Figure 33.9
The role of soil bacteria in the nitrogen nutrition of plants. Soil bacteria form ammonium by fixing atmospheric N_2 (nitrogen-fixing bacteria) and by decomposing organic material (ammonifying bacteria). Although plants absorb some ammonium from the soil, they absorb mainly nitrate, which is produced from ammonium by nitrifying bacteria. The form of nitrogen transported up to the shoot system depends on the plant species. Most often, nitrogen from nitrate is incorporated into organic compounds such as amino acids in roots. This organic nitrogen is then exported to the shoot system via xylem.

NITROGEN ASSIMILATION BY PLANTS

Of all mineral elements, nitrogen is the one that most often limits the growth of plants and the yields of crops. Plants require nitrogen as an ingredient of proteins, nucleic acids, and other important organic molecules. It is ironic that plants sometimes suffer nitrogen deficiencies, for the atmosphere is nearly 80% nitrogen. This atmospheric nitrogen, however, is gaseous N_2, and plants cannot use nitrogen in that form. For plants to absorb nitrogen, it must first be converted to ammonium (NH_4^+) or nitrate (NO_3^-). In contrast to other minerals, the NH_4^+ and NO_3^- in soil are not derived from the breakdown of parent rock. Over the short term, the main source of nitrogenous minerals is the decomposition of humus by microbes (Figure 33.9). In this way, nitrogen present in organic compounds, such as proteins, is repackaged in inorganic compounds that can be recycled when they are absorbed as minerals by roots. However, nitrogen is lost from this local cycle when certain soil bacteria convert NO_3^- to N_2, which diffuses from the soil to the atmosphere. Still other bacteria, called **nitrogen-fixing bacteria,** restock nitrogenous minerals in the soil by converting N_2 to NH_3 (ammonia), a metabolic process called **nitrogen fixation.** The complex cycling of nitrogen in ecosystems will be traced in detail in Chapter 49. Here, we focus on nitrogen fixation and the other steps that lead directly to nitrogen assimilation by plants.

Nitrogen Fixation

All of Earth's life depends on nitrogen fixation, an exclusive function of certain prokaryotes. Soil is populated by several species of free-living bacteria that are among the nitrogen-fixing prokaryotes. Conversion of atmospheric nitrogen (N_2) to ammonia (NH_3) is a complicated, multistep process, but we can simplify nitrogen fixation by just indicating the reactants and products:

$$N_2 + 8e^- + 8H^+ + 16ATP \longrightarrow 2NH_3 + H_2 + 16ADP + 16\,Ⓟ_i$$

One enzyme complex, called **nitrogenase,** catalyzes the entire reaction sequence, which reduces N_2 to NH_3 by adding electrons along with hydrogen ions. Notice that nitrogen fixation is very expensive in terms of metabolic energy, costing the bacteria 16 ATPs for each ammonia molecule synthesized. Nitrogen-fixing bacteria are most abundant in soils rich in organic material, which provides fuel for cellular respiration.

In the soil solution, ammonia picks up another hydrogen ion to form ammonium (NH_4^+), which plants can absorb. However, plants acquire their nitrogen mainly in the form of nitrate (NO_3^-), which is produced in the soil by bacteria that oxidize ammonium (see Figure 33.9). After nitrate is absorbed by roots, the nitrogen is usually incorporated into organic compounds. Most plant species export nitrogen from roots to shoots, via xylem, in the form of amino acids and other organic compounds.

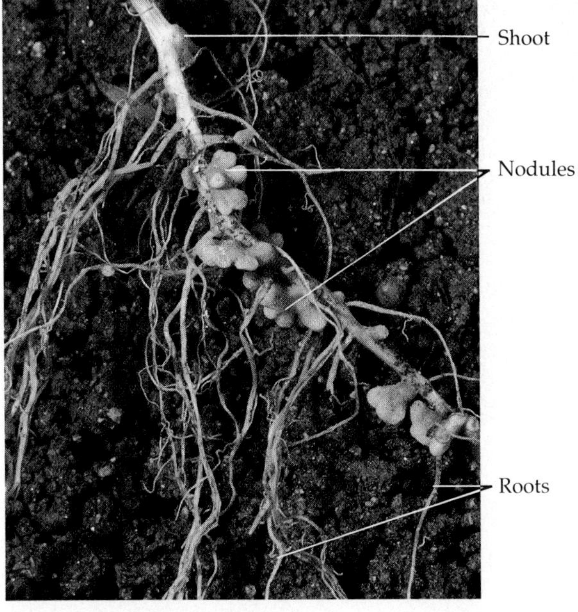

(a)

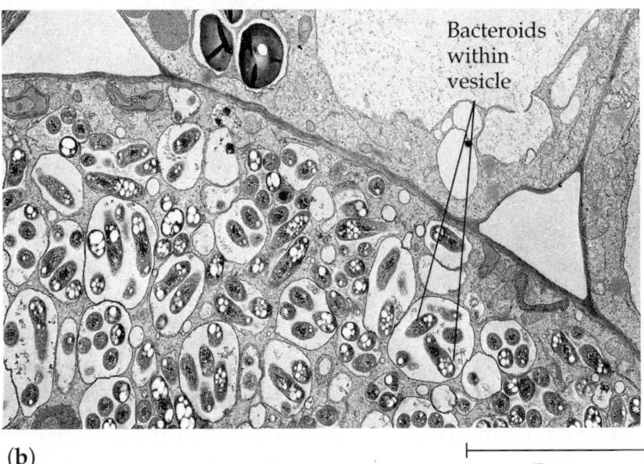

(b)

Figure 33.10
Root nodules on a legume. (**a**) The nodules of this pea root contain symbiotic bacteria that fix nitrogen and obtain photosynthetic products supplied from the plant. (**b**) In this electron micrograph, a cell from a root nodule of soybean is filled with bacteroids (TEM). The adjoining cell remains uninfected.

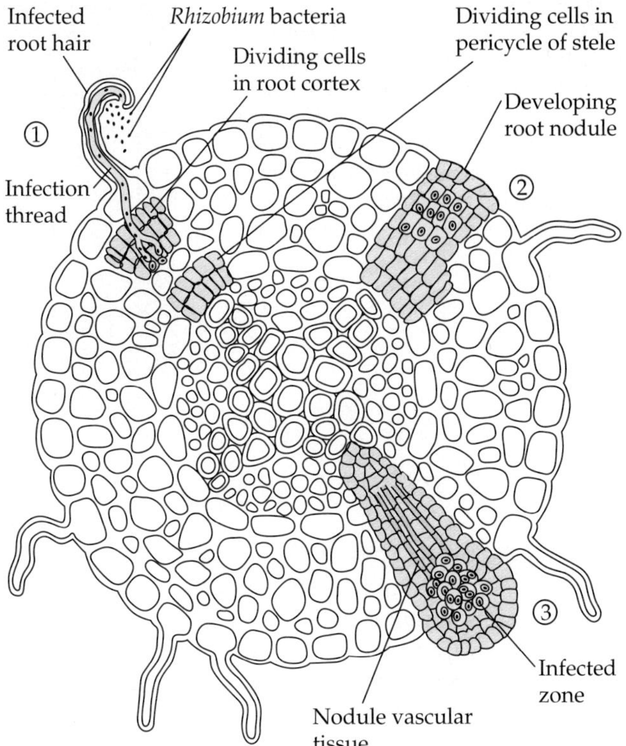

Figure 33.11
Stages in the development of a soybean root nodule. The coordinated activities of the legume and the *Rhizobium* bacteria depend on a chemical dialogue between the symbiotic partners. ① The first stage is infection. Roots secrete chemicals that attract *Rhizobium* bacteria living nearby. The bacteria, in turn, emit chemical signals that stimulate root hairs to elongate and curl around the bacterial population. The bacteria penetrate the root cortex within an infection thread. At the same time, the root begins responding to infection with a division of cells in the cortex and in the pericycle of the stele. Vesicles containing the bacteria, now known as bacteroids, bud into cortical cells from the tips of the branching infection thread. ② Growth continues in the affected regions of cortex and pericycle, and these two masses of dividing cells fuse to form the nodule. ③ The nodule continues to grow, and vascular tissue connecting the nodule to the xylem and phloem of the stele develops. This vascular tissue supplies the nodule with sugar and other organic substrates for metabolism and carries nitrogenous compounds produced in the nodule into the stele for distribution to the rest of the plant.

Symbiotic Nitrogen Fixation

Plants of the legume family, including peas, beans, soybeans, peanuts, alfalfa, and clover, have a built-in source of fixed nitrogen (Figure 33.10). Their roots have swellings called **nodules** composed of plant cells that contain nitrogen-fixing bacteria of the genus *Rhizobium* ("root living"). Inside the nodule, the *Rhizobium* assume a form called **bacteroids.** Each legume is associated with a particular species of *Rhizobium.* Figure 33.11 describes the steps in the development of root nodules.

The symbiotic relationship between a legume and its nitrogen-fixing bacteria is mutualistic, with both partners benefiting. The bacteria supply the legume with fixed nitrogen, and the plant provides the bacteria with carbohydrates and other organic compounds. The exquisite coevolution of partners is evident in their cooperative synthesis of a molecule named **leghemoglobin,** with the plant and the bacteria each making part of the molecule. Leghemoglobin is an iron-containing protein that, like the hemoglobin of human red

Figure 33.12
Agricultural application of *Rhizobium* infection. (**a**) This Rwandan farmer is mixing legume seeds with the appropriate species of *Rhizobium* in order to inoculate the seeds with the nitrogen-fixing bacteria. (**b**) A field experiment compares plants grown from the inoculated seeds (left) with uninfected plants, which show signs of nitrogen deficiency.

(a)

(b)

blood cells, binds reversibly to oxygen (*leg-* is for legume). Leghemoglobin releases oxygen for the intense respiration required to produce ATP for nitrogen fixation. More important, however, leghemoglobin keeps the concentration of free oxygen low in root nodules, an important function because the enzyme nitrogenase is inhibited by oxygen.

Most of the ammonium produced by symbiotic nitrogen fixation is used by the nodules to make amino acids, which are then transported to the shoot and leaves via xylem. When conditions are favorable, root nodules fix so much nitrogen that they actually secrete the excess ammonium, which increases the fertility of the soil for nonlegumes. This is the basis for crop rotation. One year a nonlegume such as corn is planted, and the following year alfalfa or some other legume is planted to restore the concentration of fixed nitrogen in the soil. Instead of being harvested, the legume crop is often plowed under so that it will decompose as "green manure"and add even more fixed nitrogen to the soil (see Figure 33.8). To ensure that the legume encounters its specific *Rhizobium*, the seeds are soaked in a culture of the bacteria or dusted with bacterial spores before sowing (Figure 33.12).

A few groups of nonlegumes, including alders and certain tropical grasses, host nitrogen-fixing actinomycetes (see Chapter 25). Rice, a crop of great commercial importance, benefits indirectly from symbiotic nitrogen fixation. Rice farmers culture a water fern called *Azolla* in their paddies. The fern has symbiotic cyanobacteria that fix nitrogen and increase the fertility of the rice paddy. The growing rice shades and kills the *Azolla*, and decomposition of this organic material adds more nitrogenous minerals to the paddy.

Improving the Protein Yield of Crops

The ability of plants to incorporate fixed nitrogen into proteins and other organic substances has a major impact on human welfare; the most common form of malnutrition in humans is protein deficiency. Either by choice or by economic necessity, the majority of people in the world have a predominantly vegetarian diet, and thus, particularly in developing countries, depend mainly on plants for protein. Unfortunately, many plants have a low protein content, and the proteins that are present may be deficient in one or more of the amino acids that humans need from their diet. Improving the quality and quantity of proteins in crops is a major goal of agricultural research.

Plant breeding has resulted in new varieties of corn, wheat, and rice that are enriched in protein. However, many of these "super" varieties have an extraordinary demand for nitrogen, which is usually supplied in the form of commercial fertilizer. The industrial production of ammonia and nitrate from atmospheric nitrogen is, like biological nitrogen fixation, very expensive in energy costs. A chemical factory making fertilizer consumes large quantities of fossil fuels. Generally, the countries that most need high-protein crops are the ones least able to afford to pay the fuel bill.

Improving the productivity of symbiotic nitrogen fixation is another strategy with the potential to increase protein yields of crops. Normally, the accumulation of fixed nitrogen in the nodules of legumes switches off the bacterial genes that code for nitrogenase and other enzymes involved in nitrogen fixation. Microbiologists have isolated mutant strains of *Rhizobium* that keep on making these enzymes even after fixed nitrogen accumulates. Farmers may someday grow plants infected with the mutant bacteria, a breakthrough that would increase the protein content of the legume crop and also add more fixed nitrogen to the soil. Geneticists are also working to improve the efficiency of symbiotic nitrogen fixation. The total food yield of a legume crop is often relatively low because so much of the carbohydrate produced in photosynthesis is consumed as energy for nitrogen fixation. Selecting for legume and *Rhizobium* varieties that can fix nitrogen at a lower cost in photosynthetic energy could be a boon to human nutrition.

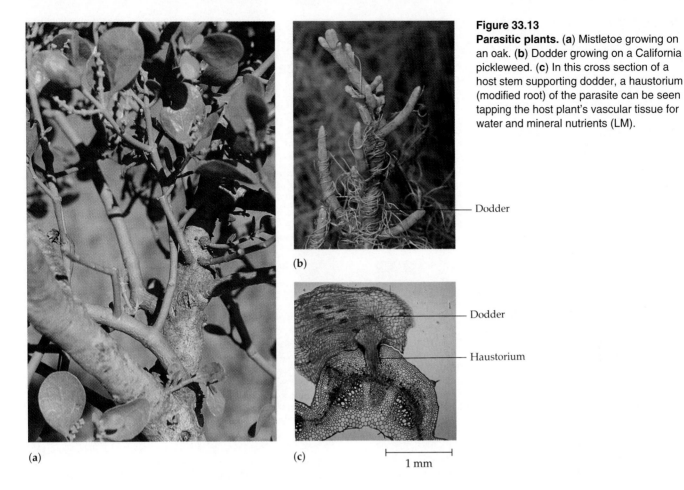

Figure 33.13
Parasitic plants. (**a**) Mistletoe growing on an oak. (**b**) Dodder growing on a California pickleweed. (**c**) In this cross section of a host stem supporting dodder, a haustorium (modified root) of the parasite can be seen tapping the host plant's vascular tissue for water and mineral nutrients (LM).

In the past few years, researchers in China and Australia have been able to induce root nodule formation in nonlegumes, including rice and wheat. The problem these workers had to solve was to eliminate the requirement for specific binding of symbiotic bacteria to the root surface, which is normally the first step in the infection of a legume by *Rhizobium.* The solution is to treat the roots of nonlegumes with chemicals that damage the cells walls of the epidermis. This enables nitrogen-fixing bacteria to infect the cortex and trigger nodule formation. In some of the experiments with wheat, the nodules fix nitrogen. More research will reveal how much nitrogen is actually fixed, and how much of this fixed nitrogen is assimilated into the proteins of the host plant.

Genetic engineering promises additional improvements in the protein yields of crops. Molecular biologists have already succeeded in transferring some of the genes required for nitrogen fixation from *Rhizobium* into other bacteria, and it may be possible through genetic engineering to create varieties of *Rhizobium* that can infect nonlegumes. Transplanting the genes for nitrogen fixation directly into the genomes of plants is also a possibility, using the plasmids of bacteria that cause a plant tumor called crown gall as vehicles for the gene transfer (see Chapter 19). This research has a long way to go from the laboratory to the field, but the prospect of wheat, potatoes, and other nonlegumes fixing their own nitrogen is a strong incentive for the work to continue.

SOME NUTRITIONAL ADAPTATIONS OF PLANTS

Symbiotic nitrogen fixation underscores the relationship between plants and their environment, which includes the other organisms that interact with plants. We complete this chapter by exploring three other types of plant adaptations that enhance nutrition through interactions with other organisms. Some plants parasitize other plants, others prey on animals, and most use fungi to absorb minerals. The correlation of structure and function is manifest in the specialized anatomical equipment of these plants.

Parasitic Plants

The mistletoe we find tacked above doorways during the holiday season lives in nature as a parasite on oaks and other trees (Figure 33.13a). Mistletoe is photosynthetic, but it supplements its nutrition by using projec-

(a)

(b)

(c)

Figure 33.14
Carnivorous plants. (**a**) The Venus flytrap is a modified leaf with two lobes that close together rapidly enough to capture an insect. An insect that enters the trap touches sensory hairs, initiating an electrical impulse that triggers closure of the trap. Movement of the trap is essentially a very rapid growth response, with cells in the outer region of each lobe accumulating water and enlarging. This changes the shape of the lobes, bringing their margins together. Glands in the trap then secrete digestive enzymes, and nutrients are later absorbed by the modified leaf. In spite of its name, the flytrap catches more ants and grasshoppers than it does flies. (**b**) Sundews capture their prey in a sticky secretion that makes the trap like flypaper. Hairs on the trap bend over the insect and restrain it, and the entire leaf cups around the prey. The hairs then secrete digestive enzymes. (**c**) Pitcher plants use a pitfall to capture insects. Insects slip into a long funnel containing water in the bottom. After the insect drowns, it is digested by enzymes secreted into the water.

tions called haustoria to siphon xylem sap from the vascular tissue of the host tree. (These structures are analogous, not homologous, to the haustoria of parasitic fungi.) Some parasitic plants, such as dodder (see Figure 33.13b and c), do not perform photosynthesis, drawing all their nutrients from other plants.

Plants called epiphytes (Gr. *epi,* "upon,"and *phyton,* "plant") are sometimes mistaken for parasites. An epiphyte is a plant that nourishes itself but grows on the surface of another plant, usually on the branches or trunks of trees. An epiphyte is anchored to its living substratum, but it absorbs water and minerals mostly from rain that falls on the leaves. Examples of epiphytes are staghorn ferns, some mosses, Spanish moss (actually an angiosperm), and many species of bromeliads and orchids.

Carnivorous Plants

Living in acid bogs and other habitats where soil conditions are poor, especially in nitrogen, are plants that fortify themselves by occasionally feeding on animals. These carnivorous plants make their own carbohydrates from photosynthesis, but they obtain some of their nitrogen and minerals by killing and digesting insects. Various kinds of insect traps have evolved by the modification of leaves (Figure 33.14). The traps are usually equipped with glands that secrete digestive juices.

Mycorrhizae

Most plants have roots modified as **mycorrhizae** (Gr. *mykos,* "fungus," and *rhiza,* "root"), which are actually mutualistic associations between the roots and fungi. The fungal partners of mycorrhizae are discussed in Chapter 28. These fungi secrete growth factors that stimulate roots to grow and branch. As the roots develop into mycorrhizae, the fungus either sheathes the root and extends hyphae among the cortex cells or invades the root cells themselves (see Figure 28.13). The other ends of the fungal mycelium function to provide the root with a greatly increased surface area through which the root gains minerals, especially phosphate, and water. The fungi are more efficient at absorbing these materials from low supply and secrete acid that increases the solubility of some minerals. Minerals taken up by the fungus are transferred to the plant, and photosynthetic products from the plant nourish the fungus. The fungi also protect the plant against certain soil pathogens.

Almost all plants are capable of forming mycorrhizae if they are exposed to the appropriate species of fungi. In most natural ecosystems, these fungi are present in soil, and seedlings develop mycorrhizae. But if seeds are collected in one environment and planted in foreign soil, the plants may show signs of malnutrition due to the absence of the plants' mycorrhizal partners. Researchers observe similar results in experiments where soil fungi are poisoned (Figure 33.15). Farmers and foresters are already applying the lessons of such research. For example, inoculation of pine seeds with spores of mycorrhizal fungi promotes the formation of mycorrhizae by the seedlings. Pine seedlings so infected grow more vigorously than those trees without the fungal association.

Figure 33.15
The benefits of mycorrhizae. The soybean plant on the left is growing in soil lacking fungi. The plant's growth is stunted, probably due to a phosphorus deficiency. The other plants have mycorrhizae, which enhance the uptake of phosphate and other minerals.

Fossil evidence indicates that the earliest plants on land had mycorrhizae (see Chapter 27). This symbiosis of plants and fungi enabled roots to extract from soil enough minerals to supply the entire plant and thus may have contributed to the colonization of land. To understand plant nutrition and everything else about life on Earth, it is essential to view organisms in their ecological contexts. Each organism represents a functional unit adapted through evolution to the physical and biological features of its environment.

STUDY OUTLINE

1. As photosynthetic autotrophs, plants produce all their own organic compounds but require inorganic nutrients in the form of carbon dioxide, water, and minerals.

Nutritional Requirements of Plants (pp. 718–722)

1. Water, the most abundant compound in most plants, is an essential nutrient in photosynthesis. Water uptake also accounts for most of the growth of a plant.
2. Atmospheric carbon dioxide is incorporated into a plant's organic material, most of which is carbohydrate.
3. Minerals (inorganic ions) are selectively absorbed from the soil by roots. Hydroponic culture has determined which minerals are essential for plant growth.
4. Plants require nine elements, the macronutrients, in fairly large amounts.
5. Eight micronutrients are required in extremely small quantities. These elements function mainly as cofactors in enzymatic reactions.
6. Mineral deficiencies reflect the composition of the soil and cause an array of symptoms that depend on the role of the nutrient and its mobility within the plant.

Soil (pp. 722–725)

1. Soil texture depends on the size of particles in topsoil. The most fertile soils are generally loams, which contain both fine particles that retain adequate water and minerals and coarse particles that provide adequate drainage.
2. An incredible number and variety of living organisms inhabit the soil. Some of these organisms are involved in producing humus, decomposing organic matter that improves the texture and mineral content of soil.

3. Fine particles of clay in soil are negatively charged and attract water and cations. By releasing acids, roots obtain these solutes in a process called cation exchange.

4. Unlike natural ecosystems, agriculture depletes the mineral content of soil, taxes water reserves, and encourages erosion. Proper soil management is essential to avoid squandering a precious resource.

Nitrogen Assimilation by Plants (pp. 726–729)

1. Plant growth and crop yields are often limited by nitrogen, an essential ingredient of proteins and nucleic acids.

2. Although the atmosphere is rich in nitrogen, plants require the assistance of bacteria living in soil to supply them with the forms of nitrogen they need. Nitrogen-fixing bacteria possess nitrogenase, an enzyme that converts atmospheric nitrogen into ammonia. The ammonia is then converted in the soil to nitrate or ammonium, which is absorbed by the plant.

3. The roots of legumes have nodular swellings that house nitrogen-fixing bacteria, which have coevolved with the plants into a mutualistic symbiotic relationship.

4. Improving the quality and quantity of protein in crops can help alleviate the problem of human malnutrition due to deficiencies in essential amino acids. Agricultural research in developing protein-rich plant breeds and enhancing nitrogen-fixing capabilities may yield practical benefits for a hungry world.

Some Nutritional Adaptations of Plants (pp. 729–731)

1. Parasitic plants either supplement their photosynthetic nutrition or give up photosynthesis entirely by tapping into the vascular tissues of host plants.

2. The chronically poor soil conditions of acid bogs are associated with the evolution of carnivorous plants. By a variety of methods, these plants obtain nitrogen and minerals by killing and digesting insects.

3. Mycorrhizae, mutualistic associations between roots and fungi, help the plant by enhancing mineral nutrition, water absorption, and pathogen resistance. The earliest land plants had mycorrhizae.

SELF-QUIZ

1. Most of the mass of organic material of a plant comes from
   a. water
   b. carbon dioxide
   c. soil minerals
   d. atmospheric oxygen
   e. nitrogen

2. Micronutrients are needed in very small amounts because
   a. most of them are mobile in the plant
   b. most function as cofactors in enzymes
   c. most are supplied in large enough quantities in seeds
   d. they play only a minor role in the health of the plant
   e. only the growing regions of the plants require them

3. Which of the following is a correct statement about nitrogen fixation?
   a. Plants convert atmospheric nitrogen to ammonia.
   b. Ammonia is converted to nitrate, which is the form of nitrogen most easily absorbed by plants.
   c. Bacteria housed in mycorrhizae are capable of producing ammonium.
   d. Mutant strains of *Rhizobium* are able to secrete excess proteins into the soil.
   e. The enzyme nitrogenase reduces N_2 to form ammonia.

4. Which of the following is *not* a true statement about organic and chemical fertilizers?
   a. Organic fertilizers supply nutrients to plants more slowly than commercial fertilizers.
   b. Commercial fertilizers are generally more expensive to produce than organic fertilizers.
   c. Chemical fertilizers are more likely to leach out of the soil than are organic fertilizers.
   d. Organic fertilizers provide already formed organic molecules to plants, whereas commercial fertilizers provide only inorganic nutrients.
   e. Commercial fertilizers are usually enriched in nitrogen, phosphorus, and sodium.

5. Mycorrhizae are
   a. nodules containing nitrogen-fixing bacteria found on the roots of legumes
   b. cellular extensions of parasitic plants that tap into the host's vascular tissue
   c. plants that use other plants as a substratum but do not take nutrients from the host
   d. symbiotic associations between roots and fungi
   e. root hairs resembling the hyphae of fungi

6. Which of the following nutrients is incorrectly paired with its function in a plant? (Consult Table 33.1.)
   a. calcium—formation of cell walls
   b. magnesium—constituent of chlorophyll
   c. iron—component of chlorophyll
   d. potassium—important in osmotic regulation
   e. phosphorus—component of nucleic acids

7. Most of the water taken up by the plant is
   a. split during photosynthesis as a source of electrons and hydrogen
   b. lost by transpiration through stomata
   c. absorbed by cells during their elongation
   d. returned to the soil by osmosis from the roots
   e. incorporated directly into organic material

8. A mineral deficiency is likely to affect older leaves more than younger leaves if
   a. the mineral is a micronutrient
   b. the mineral is very mobile within the plant
   c. the mineral is required for chlorophyll synthesis
   d. the deficiency persists for a long period of time
   e. the older leaves are in direct sunlight

9. Carnivorous adaptations of plants mainly compensate for soil that has a relatively low content of
   a. potassium
   b. nitrogen
   c. calcium
   d. water
   e. phosphate
10. Based on our retrospective view, van Helmont's famous experiment on the growth of a willow tree mainly demonstrated that
   a. the tree increased in mass mainly by absorbing water
   b. the increase in the mass of the tree could not be accounted for by the consumption of soil
   c. most of the increase in the mass of the tree was due to the uptake of O_2
   d. soil simply provides physical support for the tree without providing nutrients
   e. trees do not require water to grow

CHALLENGE QUESTIONS

1. Explain why an acre of corn actually yields more total protein than an acre of soybeans. (*Note:* Recall that nitrogen-fixation requires large amounts of metabolic energy in the form of ATP.)
2. In this photograph, you can observe artificially induced root nodules on the roots of a wheat seedling.

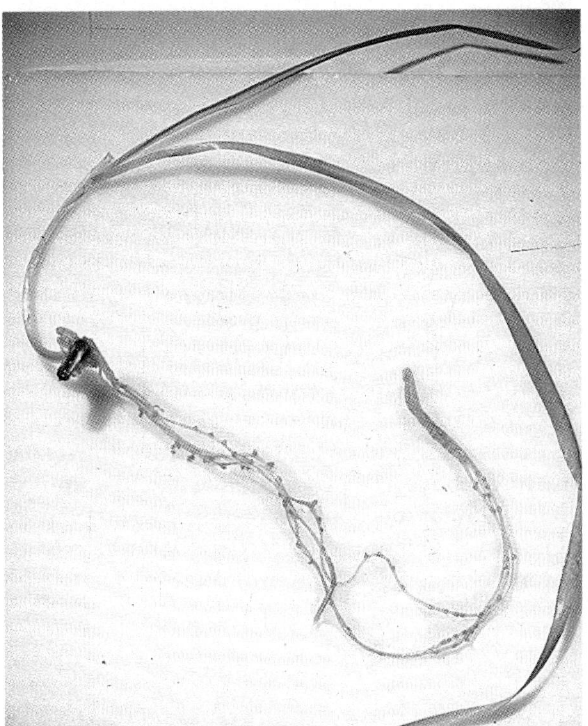

Researchers at the University of Sydney treated the roots with a chemical that softens cell walls. This enabled *Azospirillum*, a genus of nitrogen-fixing bacteria that normally live free in the soil, to infect the root cortex of seedlings. The nodules fix nitrogen. Design an experiment that uses radioactive N_2 to determine whether the nitrogen fixed in these nodules is incorporated into leaf proteins. Imagine that the results are negative (no incorporation into leaf proteins). Propose two or three hypotheses for why the fixed nitrogen fails to show up in leaf proteins. How would you test these hypotheses?

SCIENCE, TECHNOLOGY, AND SOCIETY

1. Having read this chapter, state some of the problems the world faces for future food production and outline some possible solutions. What economic and environmental costs are associated with each of these solutions?
2. For centuries, small farmers have used organic methods for growing food crops. In the United States, some large commercial farms are starting to use organic methods, and organic produce is starting to appear on supermarket shelves. What is organic farming? Does organically grown produce differ from food grown by nonorganic methods? What are some of the advantages and disadvantages of organic farming?

FURTHER READING

Albert, V., S. Williams, and M. Chase. "Carnivorous Plants: Phylogeny and Structural Evolution." *Science,* September 11, 1992. Convergent evolution has produced insect-catching species in several plant families.

Beardsley, T. "A Nitrogen Fix for Wheat." *Scientific American,* March 1991. Artificially induced root nodules on nonlegumes.

Gibbons, B. "Do We Treat Our Soil Like Dirt?" *National Geographic,* September 1984. Some thought-provoking perspectives on soil management.

Hershey, D. "Digging Deeper Into Helmont's Famous Willow Tree Experiment." *The American Biology Teacher,* November/December 1991. Dissecting a scientific classic.

Moffat, A. S. "Nitrogenase Structure Revealed." *Science,* December 14, 1990. The structure-function approach applied to one of life's most important proteins.

Moore, P. D. "Prey for the Plundering." *Nature,* March 21, 1991. How ants steal prey from Venus flytraps.

Nap, J.-P., and T. Bisseling. "Developmental Biology of a Plant-Prokaryote Symbiosis: The Legume Root Nodule." *Science,* November 16, 1990.

Pimental, D., et al. "World Agriculture and Soil Erosion." *BioScience,* April 1987. A quantitative assessment of a serious threat to world food production.

Reganold, J. P., R. I. Papendick, and J. F. Parr. "Sustainable Agriculture." *Scientific American,* June 1990.

Stone, J. "Little Bog Man." *Discover,* February 1990. Profile of a high school student who is one of the world's experts on carnivorous plants.

34 PLANT REPRODUCTION AND DEVELOPMENT

SEXUAL REPRODUCTION OF FLOWERING PLANTS

ASEXUAL REPRODUCTION

COMPARING SEXUAL AND ASEXUAL PLANT REPRODUCTION: AN EVOLUTIONARY PERSPECTIVE

SOME CELLULAR ASPECTS OF PLANT DEVELOPMENT

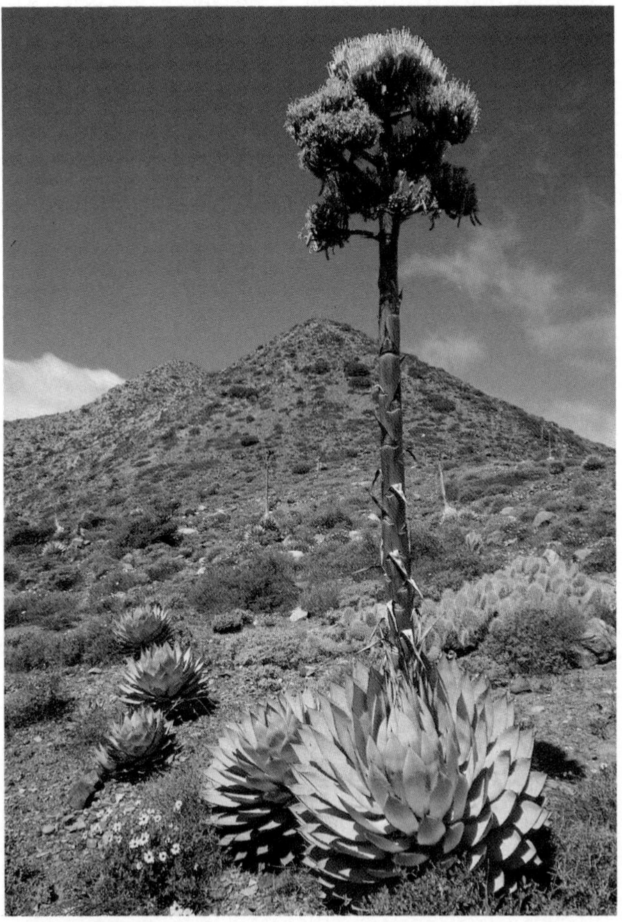

Figure 34.1
A floral stalk of the century plant. After decades of vegetative growth, the century plant *(Agave)* produces its once-in-a-lifetime bloom and consumes all its food reserves in the act of reproducing. In this chapter, you will learn how plants reproduce and how their offspring develop from seeds.

It has been said that an oak is an acorn's way of making more acorns. Indeed, in the Darwinian view of life, the fitness of an organism is measured only by its ability to replace itself with healthy, fertile offspring. Consider the century plant. It lives for decades without flowering, and then one spring, the century plant grows a floral stalk as tall as a telephone pole (Figure 34.1). That season the plant produces seeds and then withers and dies, its food reserves, minerals, and water spent in the formation of its massive bloom. Although not all flowering plants are as completely consumed as the century plant in leaving offspring, most of their other functions can be interpreted, in the broadest Darwinian sense, as mechanisms contributing to propagation.

Modifications in reproduction were key adaptations that enabled plants descended from aquatic ancestors to spread into a variety of terrestrial habitats. During the life cycles of most algae, flagellated sperm swim to eggs, and the offspring develop in the water without protection. In the conifers and angiosperms, the plants currently most widespread on land, pollen carried by wind or animals has replaced flagellated sperm as a means of bringing gametes together, and zygotes develop into embryos protected within seeds. These two developments, pollen and seeds, are among the most important adaptations of plants to life on land. Many plants can also reproduce without sex by mechanisms that promote propagation of successful individuals in specific environments. This chapter focuses on the reproduction and development of flowering plants. (The life cycles of other plant and algal groups were covered in Chapters 26 and 27.) After comparing the sexual and asexual modes of reproduction, we will examine some of the cellular mechanisms responsible for plant development. The complex interactions of hormones and environmental cues that control these events will be discussed in Chapter 35.

SEXUAL REPRODUCTION OF FLOWERING PLANTS

In this section, you will learn how flowering plants produce seeds and how these seeds develop into new plants. We begin with a review of the angiosperm life cycle, and then take a closer look at key stages.

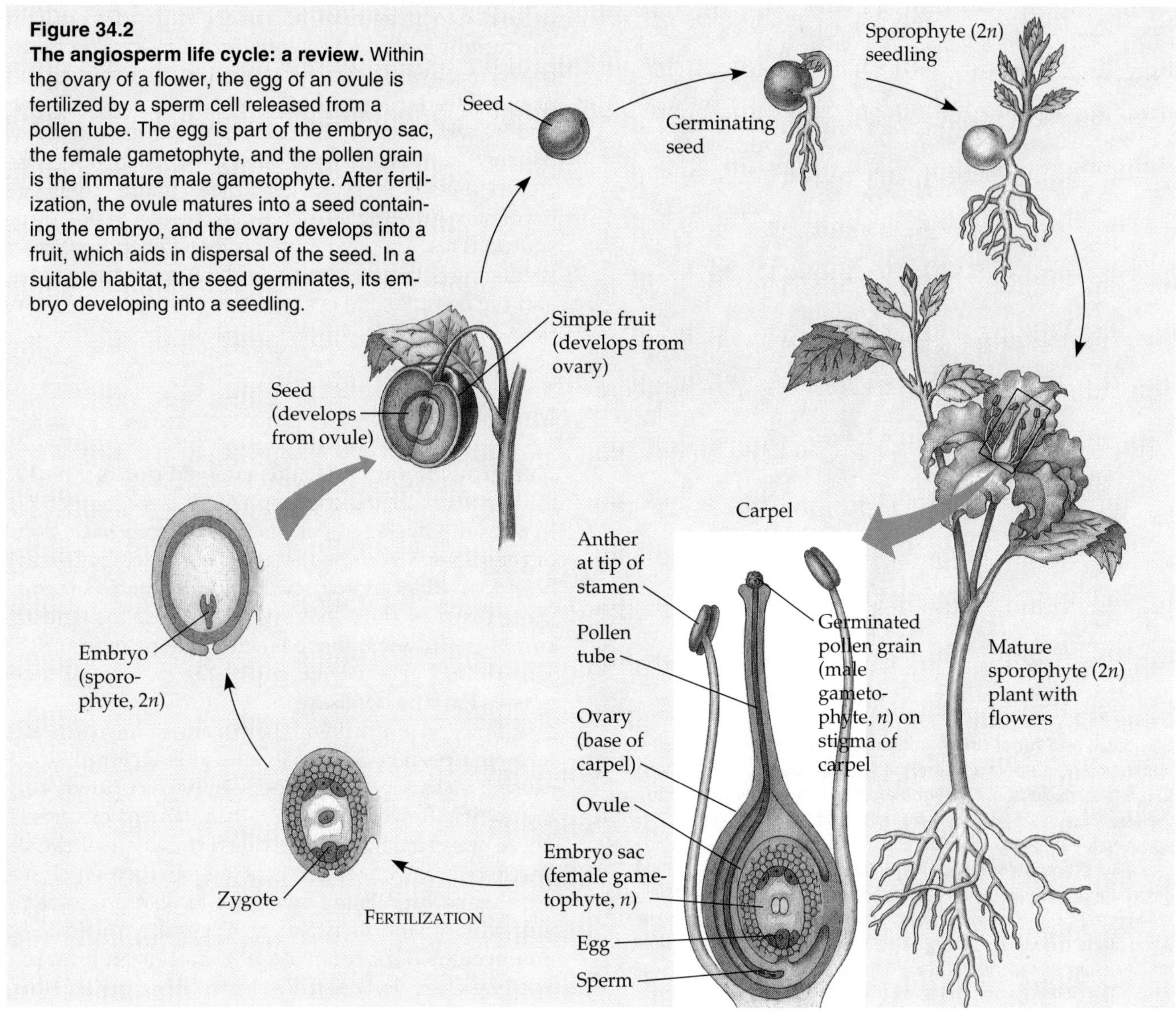

Figure 34.2
The angiosperm life cycle: a review. Within the ovary of a flower, the egg of an ovule is fertilized by a sperm cell released from a pollen tube. The egg is part of the embryo sac, the female gametophyte, and the pollen grain is the immature male gametophyte. After fertilization, the ovule matures into a seed containing the embryo, and the ovary develops into a fruit, which aids in dispersal of the seed. In a suitable habitat, the seed germinates, its embryo developing into a seedling.

The Angiosperm Life Cycle: An Overview

Figure 34.2 traces the main stages in the life cycle of a flowering plant, a topic that was covered from an evolutionary perspective in Chapter 27. The life cycles of angiosperms and other plants are characterized by an **alternation of generations,** in which haploid *(n)* and diploid (2*n*) generations take turns producing each other (see Figure 27.2). The diploid plant, called the **sporophyte,** produces haploid spores by meiosis in specialized chambers called sporangia. A spore divides by mitosis, giving rise to a multicellular male or female **gametophyte,** the haploid generation. Mitosis in the gametophytes produces gametes—sperm and eggs—and fertilization results in diploid zygotes, which divide by mitosis and form new sporophytes. In angiosperms, the sporophyte is the dominant generation in the sense that it is the conspicuous plant we see. Gametophytes became reduced during evolution to tiny structures totally contained within and dependent upon their sporophyte parents.

The reproductive structures of angiosperm sporophytes are flowers. Flowers evolved from compressed shoots with four whorls of modified leaves separated by very short internodes. These four floral organs, in their order from outside to inside of the flower, are the **sepals, petals, stamens,** and **carpels.** Their structure and function are reviewed in Figure 34.3. The stamens and carpels of flowers contain the sporangia, where male and female gametophytes, respectively, develop. The male gametophytes are sperm-containing pollen grains, which form within the chambers of anthers at the tips of stamens. The female gametophytes are egg-containing structures called embryo sacs. Embryo sacs

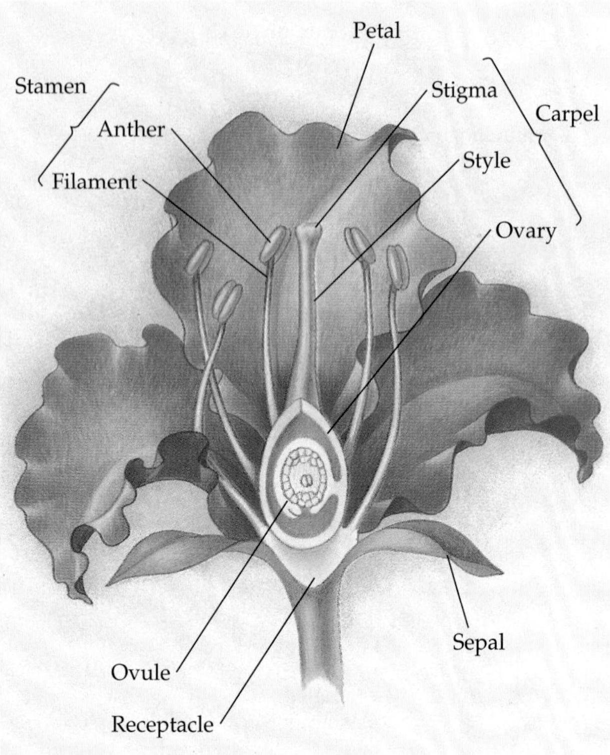

Figure 34.3
Structure and function of an idealized flower. Sepals, petals, stamens, and carpels are arranged in four whorls, all attached to the receptacle at the end of a modified stem. Usually green, the sepals of most flowers have retained a more leaflike appearance than the other floral parts. Sepals enclose and protect a floral bud before it opens. Petals, generally more brightly colored than sepals, advertise the flower to insects and other pollinators. Stamens and carpels are the fertile parts of flowers. Each stamen consists of a stalk called the filament and a terminal structure called the anther. Within the anther are chambers, where the pollen grains (male gametophytes) develop. A carpel has a slender neck, the style, leading to an ovary located at the base of the carpel. Developing within the ovary are one or more ovules, in which egg-containing embryo sacs (female gametophytes) develop. At the tip of the carpel is a sticky stigma, which serves as a landing platform for pollen brought from other flowers by wind or animals. Among the quarter of a million angiosperm species are many variations on this basic flower plan; there is really no such thing as a typical flower.

develop inside structures called ovules, which are enclosed by the ovaries (the bases of carpels). Thus, stamens and carpels are the reproductive organs, or fertile parts, of flowers, while sepals and petals are nonreproductive organs, or sterile parts.

Pollination occurs when pollen grains released from anthers and carried by wind or animals land on the sticky stigmas at the tips of carpels (though not necessarily on the same flower or plant). Pollen tubes grow down the carpels and discharge sperm into embryo sacs, resulting in the fertilization of eggs. Each zygote gives rise to an embryo, and as the embryo grows, the surrounding ovule develops into a seed. The entire ovary, meanwhile, develops into a fruit containing one or more seeds, depending on the species. Fruits, carried by wind or by animals, help disperse seeds some distance from their source plants. If deposited in sufficiently moist soil, seeds germinate; that is, their embryos start growing into a new generation of flowering sporophytes. The next gametophyte generation arises within the anthers and carpels of the new sporophytes, and the complex life cycle of flowering plants continues.

More About Flowers

Numerous floral variations evolved during the 130 million years of angiosperm history (see Chapter 27). In certain flowers, one or more of the four basic floral organs—sepals, petals, stamens, and carpels—has been lost. Plant biologists distinguish between **complete flowers,** those having all four organs, and **incomplete flowers,** those lacking one or more of the four floral parts. For example, the flowers of most grasses have no petals.

A flower equipped with both stamens and carpels is termed a **perfect flower,** even if it is incomplete because it lacks sepals or carpels. **Imperfect flowers** are incomplete flowers missing either stamens or carpels. These unisexual flowers are called staminate or carpellate, depending on which set of organs is present. If staminate and carpellate flowers are located on the same individual plant, then that plant species is said to be **monoecious** (Gr., "one house"). Corn is an example. The "ears" are derived from clusters of carpellate flowers, and the tassels consist of staminate flowers. In contrast, a **dioecious** ("two houses") species has staminate flowers and carpellate flowers on separate plants, analogous to the presence of testes and ovaries on separate male and female animals. Date palms are dioecious. Since dates develop only on the carpellate (female) palms, commercial date growers plant mostly carpellate individuals. A few males (staminate plants) provide pollen enough for hundreds of females.

In addition to these differences based on the presence of floral organs, flowers have many other variations (Figure 34.4). Much of this diversity represents adaptations of flowers to different pollinators. Indeed, the presence of animals in the environment was a key factor in angiosperm evolution (see Chapter 27).

The following sections describe in more detail the development of pollen and ovules, then trace how pollination leads to fertilization and development of seeds and fruits. As you read these sections, refer frequently to the appropriate figures, which illustrate the different processes and the relationships among them.

Figure 34.4
A few examples of floral diversity.
(**a**) This lily flower is complete, meaning that sepals, petals, stamens, and carpels are all present. In the case of the lily, the sepals and petals look the same. (**b**) Lupines are examples of plants with inflorescences, clusters of flowers. (**c**) This sunflower represents a plant family characterized by composite flowers. What appears to be a single flower is actually a collection of hundreds. The central disk consists of tiny complete flowers. What appear to be petals ringing the central disk are actually imperfect flowers called ray flowers. (**d**) The diverse shapes, colors, and odors of flowers also reflect adaptations to different modes of pollination. For example, *Hibiscus* is pollinated by hummingbirds, which are attracted to the red color. Notice that the stamens (yellow) are fused to the style of the carpel, near the stigma. As the hummingbird probes deep in the flower for nectar, its feathers become dusted with pollen, while pollen carried from other *Hibiscus* flowers is transferred to the sticky stigma. (**e**) Corn is a monoecious species with inflorescences of staminate (male) and carpellate (female) flowers on the same individual plant. The staminate inflorescences are the tassels at the tip of the plant. Several nodes below is an "ear," a collection of kernels (one-seeded fruits) that develops from an inflorescence of fertilized carpellate flowers. (**f**) *Sagittaria* is dioecious, its staminate (left) and carpellate (right) flowers on separate plants.

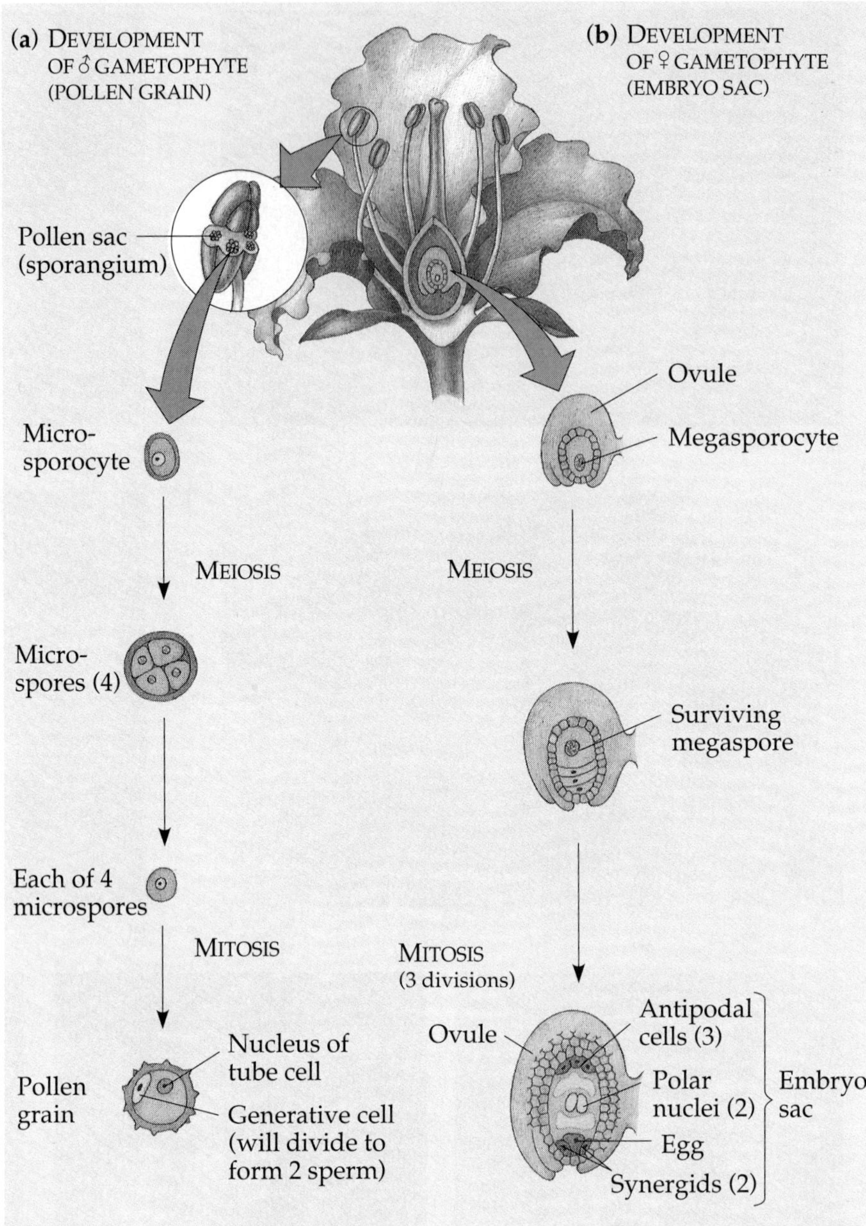

Figure 34.5
The development of angiosperm gametophytes (pollen and embryo sacs). (**a**) Pollen grains (male gametophytes) develop within the sporangia (pollen sacs) of anthers at the tips of stamens. Within each sac are numerous microsporocytes, the diploid cells that give rise to pollen. Meiosis forms four haploid microspores from each microsporocyte. Each microspore then undergoes a mitotic division to give rise to a pollen grain, an immature male gametophyte consisting of a generative cell and a tube cell. The pollen grain has a thick, tough wall. (**b**) The embryo sac (female gametophyte) develops within an ovule, itself enclosed by the ovary at the base of a carpel. A diploid cell, the megasporocyte, divides by meiosis to give rise to four cells, but only one of these survives as the megaspore. (This contrasts with pollen formation, in which all four products of meiosis go on to form gametophytes.) Three mitotic divisions of the megaspore form the embryo sac, a multicellular female gametophyte. At one end of the embryo sac are the egg and two synergid cells. Three antipodal cells are located at the opposite end. The large central cell has two nuclei called polar nuclei. The ovule now consists of the embryo sac along with the surrounding integuments (protective tissues).

(There are many variations in details in these processes, however, depending on species.)

Pollen Development: A Closer Look

Within the sporangial chambers (pollen sacs) of an anther, diploid cells called microsporocytes undergo meiosis, each forming four haploid **microspores** (Figure 34.5a). Each microspore eventually divides once by mitosis to produce two cells, a generative cell and a tube cell. The two-cell structure is encased in a thick wall that becomes sculptured into an elaborate pattern unique to the plant species. Together, the two cells and their wall constitute a pollen grain, or immature male gametophyte.

Pollen grains are extremely durable. Their tough coats are chemically different from other plant cell walls and are resistant to biodegradation. Fossilized pollen has provided many important clues to the evolutionary history of flowering plants.

The Development of Ovules: A Closer Look

Ovules, each containing a sporangium, form within the chambers of the ovary. One cell in the sporangium of each ovule, the megasporocyte, grows and then goes through meiosis to produce four haploid **megaspores** (Figure 34.5b). In most angiosperms, only one of these survives. This megaspore continues to grow, and its nucleus divides by mitosis three times, resulting in one large cell with eight haploid nuclei. Membranes then partition this mass into a multicellular structure called the **embryo sac,** which is the female gametophyte. At one end of the embryo sac are three cells: the egg cell, or female gamete, and two cells called synergids that flank the egg cell. At the opposite end are three antipodal cells. The other two nuclei, called polar nuclei, are not separated by membranes but share the cytoplasm of the large central cell of the embryo sac. The ovule, which will develop into the seed, now consists of the embryo sac (female gametophyte) and the integuments, protective layers of sporophyte tissue around the embryo sac.

Pollination and Fertilization

For the egg to be fertilized, the male and female gametophytes must meet and unite their gametes. The first step is **pollination,** the placement of pollen onto the stigma of a carpel. Some plants, including grasses and many trees, use wind as a pollinating agent. They compensate for the randomness of this dispersal by releasing enormous quantities of the tiny grains. At certain times of the year, the air is loaded with pollen, as anyone plagued with pollen allergies can attest. Many angiosperms, however, do not rely on the aimless wind to carry pollen but have relationships with animals that transfer pollen directly between flowers. The coevolution of angiosperms and their pollinators was discussed in Chapter 27.

Some flowers self-pollinate, but the majority of angiosperms have mechanisms that make it difficult or impossible for a flower to pollinate itself. In some cases, the stamens and carpels of the flower mature at different times. Many flowers that are pollinated by animals are structurally arranged in such a way that it is unlikely the pollinator could transfer pollen from anthers to the stigma of the same flower. Other flowers are **self-incompatible;** if a pollen grain from an anther happens to land on a stigma of the same flower, a biochemical block prevents the pollen from completing its development and fertilizing an egg. Self-incompatibility is based on a single gene, for which there are many alleles in a plant population. If the pollen and stigma have matching alleles (which they would if they came from the same plant), then the pollen fails to grow. Dioecious plants, of course, cannot self-pollinate because they are unisexual, being either staminate or carpellate. The various mechanisms that prevent self-pollination contribute to genetic variety by ensuring that sperm and eggs come from different parents.

After adhering to the sticky stigma, the pollen grain produces a tube that extends down between the cells of the style toward the ovary (Figure 34.6). The generative cell divides by mitosis to form two sperm, the male gametes. The pollen grain, now with tube and two enclosed sperm, constitutes the mature male gametophyte. Directed by a chemical attractant, usually calcium, the tip of the pollen tube enters the ovary, probes through the micropyle (an opening through the integuments), and discharges its two sperm within the embryo sac. One sperm fertilizes the egg to form the zygote. The other combines with the two polar nuclei to form a triploid ($3n$) nucleus in the center of the large central cell of the embryo sac. This large cell will give rise to the **endosperm,** a food-storing tissue. The union of two sperm cells with two cells of the embryo sac is termed **double fertilization.** Except for one gym-

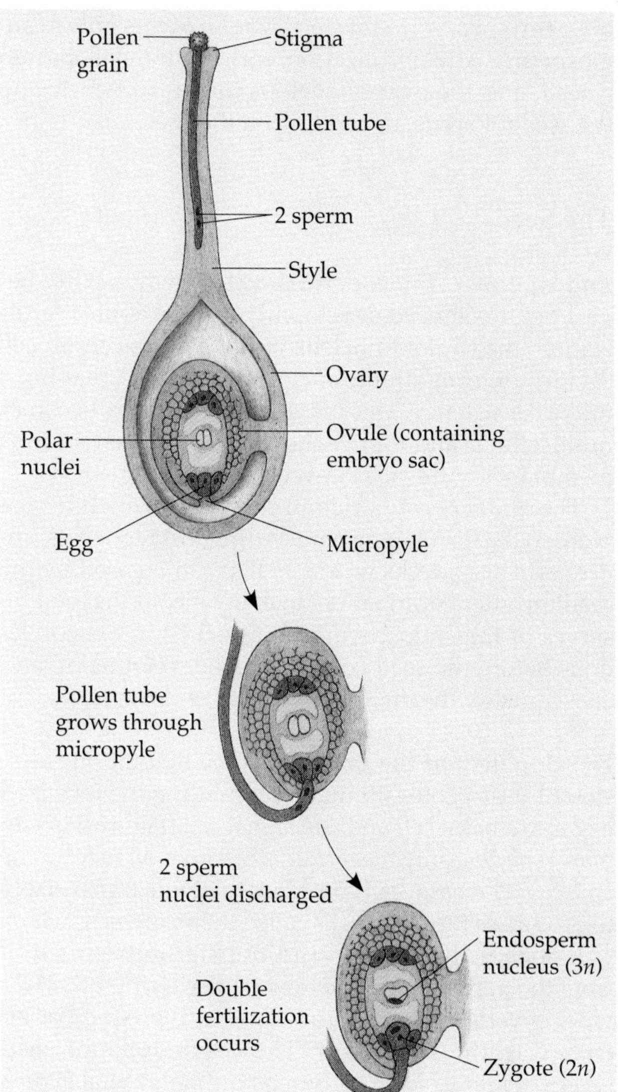

Figure 34.6
Growth of pollen tube and double fertilization in angiosperms. After a pollen grain is carried by wind or an animal to the stigma, a long pollen tube begins growing down the style toward the ovary. The tube discharges two sperm into the ovule. One sperm fertilizes the egg to form the zygote, and the other combines with two polar nuclei to form a triploid cell that will develop into a nutritive tissue called endosperm.

nosperm species, double fertilization occurs only in angiosperms. After fertilization, each ovule develops into a seed, and the ovary develops into a fruit enclosing the seed (or seeds, depending on the species).

The Seed

Endosperm Endosperm development usually begins before embryo development. After double fertilization, the triploid nucleus of the ovule's central cell divides to form a multinucleate "supercell" having a milky consistency. This mass, the endosperm, becomes multicellular and more solid when cytokinesis forms membranes and walls between the nuclei.

The endosperm is rich in nutrients, which it provides to the developing embryo. In most monocots, endosperm also stocks nutrients that can be used by the seedling after germination. In many dicots, the food reserves of the endosperm are restocked in the cotyledons before the seed completes its development, and consequently the mature seed lacks endosperm.

Development of the Embryo The first mitotic division of the zygote is transverse, splitting the fertilized egg into a basal cell and a terminal cell (Figure 34.7). In many species, only the terminal cell goes on to form the embryo. The basal cell continues to divide transversely to produce a thread of cells called the suspensor, which will anchor the embryo and transfer nutrients to it from the parent plant. Meanwhile, the terminal cell divides several times to form a spherical proembryo attached to the suspensor. The cotyledons, or seed leaves, begin to form as bumps on the proembryo. A dicot, with its two cotyledons, is heart-shaped at this stage. Only one cotyledon develops in monocots.

Soon after the rudimentary cotyledons appear, the embryo elongates. Cradled between the cotyledons is the apex of the embryonic shoot. At the opposite end of the embryo's axis, where the suspensor attaches, is the apex of the embryonic root, also with a meristem. After the seed germinates, the apical meristems at the tips of shoot and root will sustain primary growth as long as the plant lives. The three primary meristems—protoderm, procambium, and ground meristem—are also present in the embryo.

Notice that the root–shoot polarity of the embryo is determined with the very first division of the zygote. In the egg, organelles and chemicals were distributed heterogeneously throughout the cytoplasm; division of the zygote segregates different cytoplasmic components among the embryonic cells. Although these cells have equivalent nuclei, they differ in cytoplasmic composition and in their locations within the embryonic mass. The cytoplasmic environment and position of an embryonic cell help determine its developmental fate by providing cues to the genome within the nucleus. The means by which the cytoplasmic and extracellular

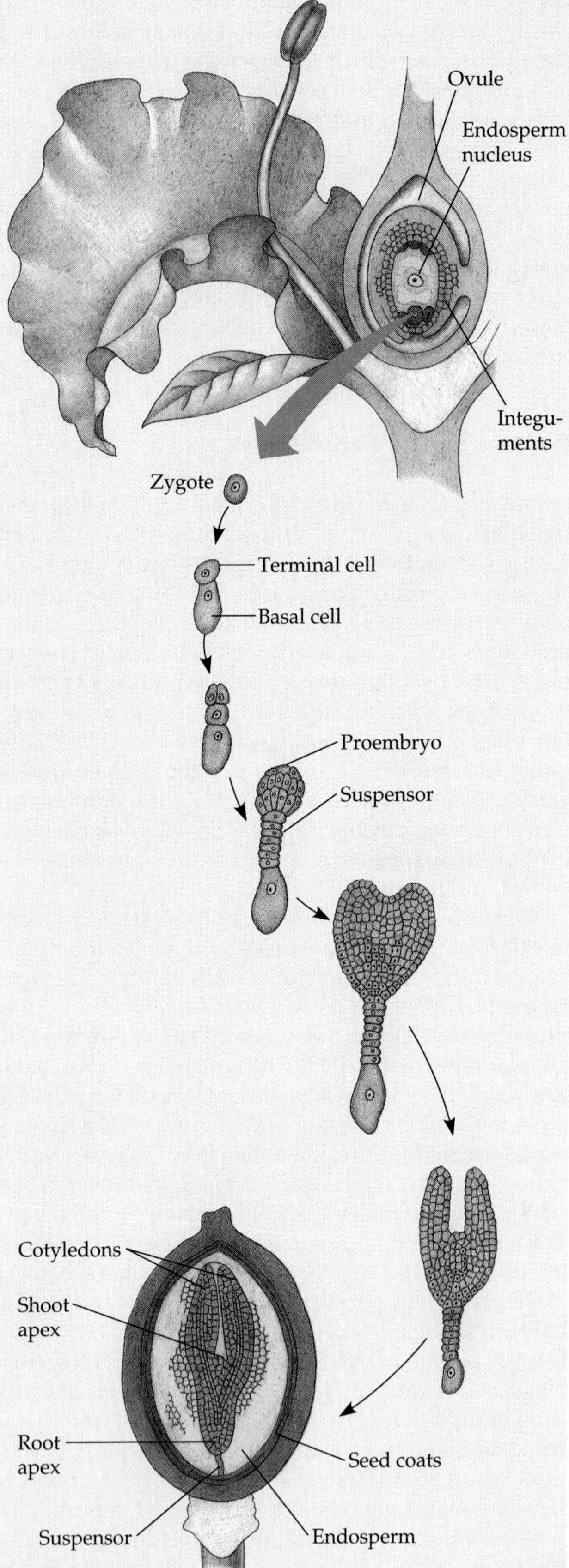

Figure 34.7
The development of a dicot plant embryo. By the time the ovule becomes a mature seed, the zygote has given rise to an embryonic plant with rudimentary organs.

environments influence the selective switching on and off of genes during cellular differentiation remains one of the most engaging problems in biology.

Structure of the Mature Seed During the last stages of its maturation, the seed dehydrates until its water content is only about 5% to 15% of its weight. By now, the embryo has ceased growing and will remain quiescent until the seed germinates. The embryo is surrounded by its enlarged cotyledons or by endosperm or by both. The embryo and its food supply are enclosed by a **seed coat** formed from the integuments of the ovule, the progenitor of the seed.

We can take a closer look at one type of dicot seed by splitting open the seed of a common bean (Figure 34.8a). At this stage, the embryo is an elongate structure, the embryonic axis, attached to fleshy cotyledons. Below the point at which the cotyledons are attached, the embryonic axis is called the **hypocotyl** (Gr. *hypo,* "under"). The hypocotyl terminates in the **radicle,** or embryonic root. The portion of the embryonic axis above the cotyledons is the **epicotyl** (Gr. *epi,* "on" or "over"). At its tip is the plumule, consisting of the shoot tip with a pair of miniature leaves.

The cotyledons of the common bean are fleshy before the seed germinates, because they absorbed food from the endosperm when the seed developed. However, the seeds of some dicots, such as castor beans, retain their food supply in the endosperm and have cotyledons that are very thin (Figure 34.8b). The cotyledons will absorb nutrients from the endosperm and transfer them to the embryo when the seed germinates.

The seed of a monocot, such as corn, has a single cotyledon (Figure 34.8c), also called the **scutellum.** This seed leaf is very thin but has a large surface area, the better to absorb nutrients from the endosperm during germination. The embryo is enclosed by a sheath consisting of a **coleorhiza,** which covers the root, and a **coleoptile,** which cloaks the embryonic shoot.

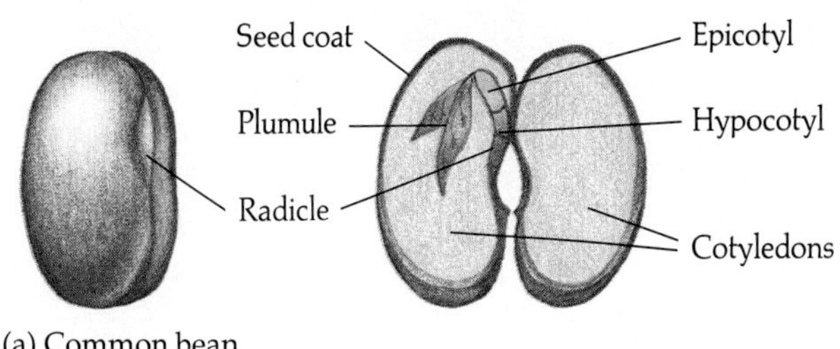

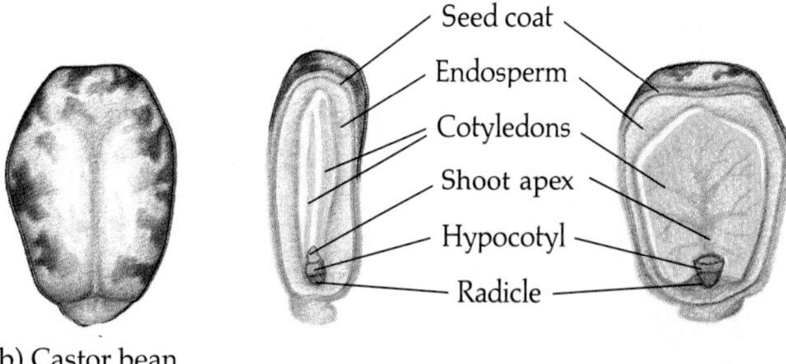

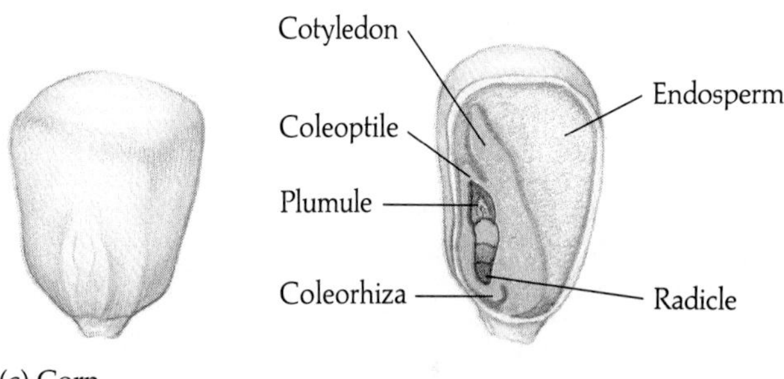

Figure 34.8
Seed structure. (**a**) The fleshy cotyledons of the common garden bean store food that was absorbed from the endosperm when the seed developed. (**b**) The castor bean has membranous cotyledons that will absorb food from the endosperm when the seed germinates. (**c**) Corn, a monocot, has only one cotyledon (scutellum). The rudimentary shoot is sheathed in a coleoptile.

Development of the Fruit

While the seeds develop from ovules, the ovary of the flower develops into a **fruit,** which protects the enclosed seeds and aids in their dispersal by wind or animals. In some angiosperms, other floral parts also contribute to what we call a fruit in grocery store vernacular. The fleshy part of an apple, for instance, is derived mainly from the fusion of flower parts located at the base of the flower, and only the core is a true fruit—a ripened ovary.

The fruit begins to develop after pollination triggers hormonal changes that cause the ovary to grow tremendously (Figure 34.9). The wall of the ovary becomes the **pericarp,** the thickened wall of the fruit. As the ovary grows, the other parts of the flower generally wither away in many plants. This transformation of the flower, called fruit set, parallels the development of the seeds. If a flower has not been pollinated, fruit usually does not set, and the entire flower withers and falls away.

Fruits are classified into several types, depending on their origin (Figure 34.10). A fruit derived from a single ovary is called a **simple fruit.** A simple fruit may be fleshy, such as a cherry, or dry, such as a soybean pod. An **aggregate fruit,** such as a strawberry, results from a single flower that has several separate carpels. A **multiple fruit,** such as a pineapple, develops from an inflorescence, a group of separate flowers tightly clustered together. When the walls of the many ovaries start to thicken, they fuse together and become incorporated into one fruit.

(a)

(b)

(c)

(d)

Figure 34.9
Fruit development. In this sequence of photographs, the flower of a pea plant is transformed into the fruit, as the wall of the ovary lengthens and thickens to become the pod.

The fruit usually ripens about the time the seeds it contains are completing their development. For a dry fruit, ripening is little more than senescence (aging) of the fruit tissues, which allows the fruit to open and release the seeds. Ripening of a fleshy fruit is more elaborate, its steps guided by complex interactions of hormones (see Chapter 35). In this case, ripening results in an edible fruit that serves as an enticement to the animals that help spread seeds. The "pulp" of the fruit becomes softer as a result of enzymes digesting components of the cell walls. There is usually a color change from green to some other color such as red, orange, or yellow. The fruit becomes sweeter as organic acids or starch molecules are converted to sugar, which may reach a concentration of as much as 20% in a ripe fruit. Unripe fruit is sour to the taste because of the high acid concentration.

By selectively breeding plants, humans have capitalized on the production of edible fruits. The apples, oranges, and other fruits we gather in grocery stores are amplified versions of much smaller natural varieties of fleshy fruits. However, the staple foods for humans are the dry, wind-dispersed fruits of grasses, which are harvested while still on the parent plant. The cereal grains of wheat, rice, corn, and other grasses are easily mistaken for seeds, but each is actually a fruit with a dry pericarp that adheres tightly to the seed coats of the single seed within.

Seed Germination

To many people, the germination of a seed symbolizes the beginning of life, but in fact the seed already contains a miniature plant, complete with an embryonic root and shoot. At germination, the plant does not begin life but rather resumes the growth and development that was temporarily suspended when the seed matured and its embryo became quiescent. Some seeds germinate as soon as they are in a suitable environment. Other seeds are dormant and will not germinate, even if sown in a proper place, until a specific environmental cue causes them to break dormancy.

Seed Dormancy Evolution of the seed was one of the most important factors in the adaptation of plants to the special problems of living and reproducing on land (see Chapter 27). Terrestrial habitats are generally less stable than lakes and seas, with fluctuating environmental conditions such as temperature and water availability. Seed dormancy increases the chances that germination will occur at a time and place most advantageous to the seedling. Seeds of desert plants, for instance, germinate only after a substantial rainfall. If they were to germinate after a modest drizzle, the soil might soon be too dry to support the seedlings. Where natural fires are common, many seeds require intense heat to break dormancy, so that seedlings are most

(a)

(b)

(c)

(d)

Figure 34.10
Types of fruit. (**a**) A cherry is a simple, fleshy fruit derived from a single ovary. (**b**) The soybean is a simple, dry fruit. (**c**) The strawberry is an aggregate fruit derived from many carpels. (**d**) A multiple fruit, the pineapple develops from many flowers.

abundant after fire has cleared away competing vegetation. Where winters are harsh, seeds may require extended exposure to cold; seeds sown during summer or fall do not germinate until the following spring. Very small seeds, such as those of some lettuce varieties, require light for germination and will break dormancy only if they are buried shallow enough for the seedlings to poke through the soil surface. Some seeds have coats that must be weakened by chemical attack as they pass through an animal's digestive tract and thus are likely to be carried some distance before germinating.

The length of time a dormant seed remains viable and capable of germinating varies from a few days to decades or even longer, depending on the species and environmental conditions. Most seeds are durable enough to last a year or two until conditions are favorable for germinating. Thus, the soil has a pool of ungerminated seeds that may have accumulated for several years. This is one reason vegetation can come back so rapidly after a fire, drought, flood, or some other disruption.

From Seed to Seedling The first step in germination of many seeds is **imbibition,** the absorption of water due to the low water potential of the dry seed (see Chapter 32). Hydration causes the seed to expand and rupture its coat and also triggers metabolic changes in the embryo that cause it to resume growth. Enzymes begin digesting the storage materials of the endosperm or cotyledons, and the nutrients are shunted to the growing regions of the embryo. This mobilization of food reserves has been studied most extensively in the grains of barley and other cereals, so we will use a cereal as an example of the process (Figure 34.11). Soon after water is imbibed, the aleurone, the thin outer layer of the endosperm, begins making α-amylase and other enzymes that digest the starch stored in the endosperm. (A similar enzyme in our saliva helps us digest bread and other foods made from the starchy endosperm of ungerminated cereal grains.) If the embryo is dissected out of the seed before water is added, no α-amylase is produced, suggesting that the embryo sends some kind of messenger to the aleurone to initiate enzyme production. This chemical signal has been identified as a gibberellin, one of the hormones discussed in more detail in Chapter 35.

The first organ to emerge from the germinating seed is the radicle, the embryonic root (Figure 34.12). Next, the shoot tip must break through the soil surface. In garden beans and many other dicots, a hook forms in the hypocotyl, and growth pushes the hook aboveground. Stimulated by light, the hypocotyl straightens, raising the cotyledons and epicotyl. Thus, the delicate shoot apex and bulky cotyledons are pulled aboveground, rather than being pushed tip-first through the abrasive soil. The epicotyl now spreads its first foliage leaves, which expand, become green, and begin mak-

Figure 34.11
Mobilization of nutrients during germination of a cereal. After water is imbibed, the embryo releases a hormone (a gibberellin) as a signal to the aleurone, which then synthesizes and secretes α-amylase and other digestive enzymes that hydrolyze stored foods in the endosperm, producing small, soluble molecules. Nutrients absorbed from the endosperm by the scutellum (cotyledon) are consumed during growth of the embryo into a seedling.

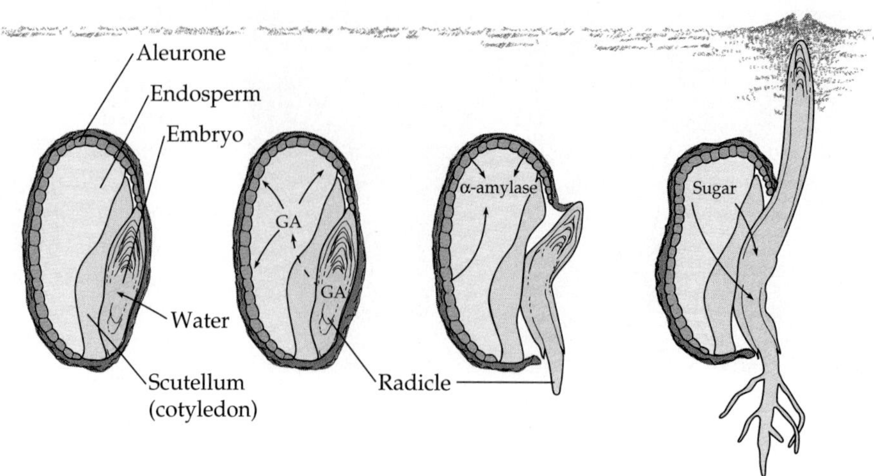

ing food by photosynthesis. The cotyledons shrivel and fall away from the seedling, their food reserves having been consumed by the germinating embryo.

Light seems to be the main cue that tells the seedling it has broken ground. The hypocotyl of a common garden bean will continue to elongate and push its hook upward until it is out of darkness. Only when the seedling senses light will the hook straighten and the epicotyl begin to elongate. We can trick a bean seedling into behaving as though it is still buried by germinating the seed in darkness, a treatment called etiolation. The etiolated seedling extends an exaggerated hypocotyl with a hook at its tip, and the foliage leaves fail to turn green. After it exhausts its food reserves, the spindly seedling stops growing and dies.

Peas, although in the same family as beans, have a different style of germinating. A hook forms in the epicotyl rather than hypocotyl, and the shoot tip is lifted gently out of the soil by elongation of the epicotyl and straightening of the hook. Pea cotyledons, unlike those of beans, remain behind in the ground.

Monocots, such as corn, use yet a different method for breaking ground when they germinate. The coleoptile, the sheath enclosing and protecting the embryonic shoot, pushes upward through the soil and into the air. The shoot tip then grows straight up through the tunnel provided by the tubular coleoptile.

Germination of a plant seed, like the birth or hatching of an animal, is a critical stage in the life cycle. The tough seed gives rise to a fragile seedling that will be confronted with predators, parasites, wind, and other hazards. In the wild, only a small fraction of seedlings endure long enough to become parents themselves. Production of enormous numbers of seeds and fruits compensates for the odds against individual survival and gives natural selection ample material to screen for the most successful genetic combinations. However, this is a very expensive means of reproduction in terms of the resources consumed in flowering and fruiting. Asexual reproduction, generally simpler and less hazardous for offspring than sexual reproduction, is an alternative means of plant propagation.

ASEXUAL REPRODUCTION

Imagine some of your fingers separating from your body, taking up life on their own, and eventually developing into entire copies of yourself. This would be an example of asexual reproduction, offspring derived from a single parent without genetic recombination. The result would be a clone, a population of asexually produced, genetically identical organisms. Some animals *can* reproduce asexually (though not humans, of course). And many plant species do clone themselves by asexual reproduction, also called **vegetative reproduction.**

Natural Mechanisms of Vegetative Reproduction

Vegetative reproduction is an extension of the capacity of plants for indeterminate growth (see Chapter 31). Plants, remember, have meristematic tissues of dividing, undifferentiated cells that can sustain or renew growth indefinitely. In addition, parenchyma cells throughout the plant can divide and differentiate into the various types of specialized cells, enabling plants to regenerate lost parts. Detached fragments of some plants can develop into whole offspring; a severed stem, for instance, may develop adventitious roots that reestablish the plant. **Fragmentation,** the separation of a parent plant into parts that re-form whole plants, is one of the most common modes of vegetative reproduction (Figure 34.13a). A variation of this process occurs in some species of dicots, in which the root system of a single parent gives rise to many adventitious

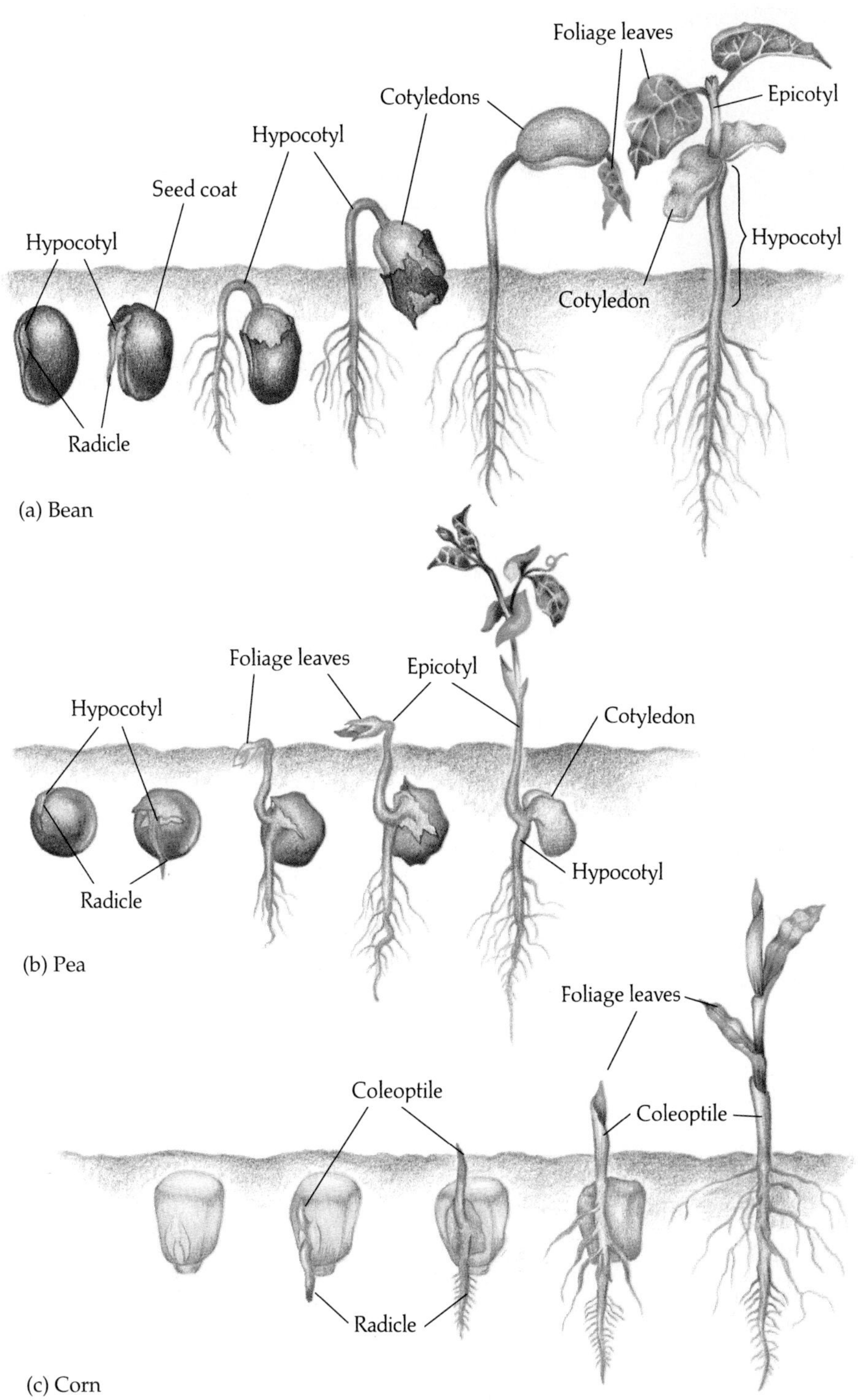

Figure 34.12
Seed germination. The radicle, the root of the embryo, emerges from the seed first. Then the shoot breaks the soil surface by one of the following mechanisms: (**a**) In beans, straightening of a hook in the hypocotyl pulls the shoot and cotyledons from the soil. (**b**) In peas, the hook is above the cotyledons on the epicotyl, and the cotyledons remain in the ground. (**c**) In corn and other monocots, the shoot grows straight up through the tube of the coleoptile.

shoots that become separate shoot systems. The result is a clone formed by asexual reproduction from one parent (Figure 34.13b). Such asexual propagation has produced the oldest of all known plant clones, a ring of creosote bushes in the Mojave Desert of California, believed to be at least 12,000 years old.

An entirely different mechanism of asexual reproduction has evolved in dandelions and some other plants, which produce seeds without their flowers being fertilized. A diploid cell in the ovule gives rise to the embryo, and the ovules mature into seeds, which in the dandelion are dispersed by windblown fruit. Thus,

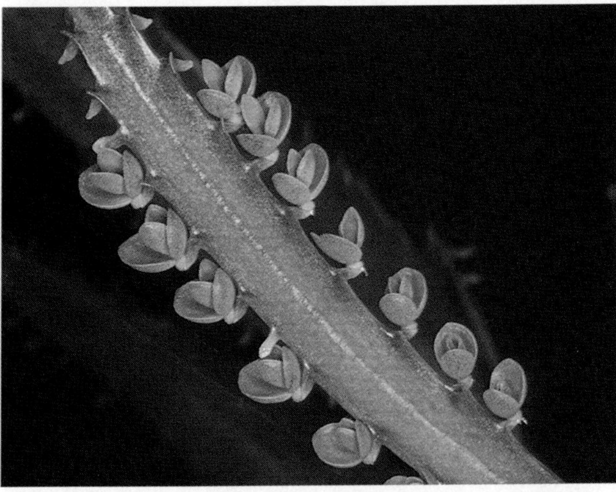

(a)

(b)

Figure 34.13
Natural mechanisms of vegetative reproduction.
(**a**) *Kalanchoe* is known as the maternity plant because of the numerous plantlets it produces along its leaf margins. The asexually produced plantlets fragment from their parent and become independent plants. (**b**) Some aspen groves, such as those shown here, are actually clones of thousands of trees descended by asexual reproduction from the root system of one parent. Notice that genetic differences among the clones result in different timing in the development of fall color and loss of leaves.

though these plants clone themselves by an asexual process, they also have the advantage of seed dispersal, an adaptation usually associated with sexual reproduction of plants. This asexual production of seeds is called **apomixis.**

Vegetative Propagation in Agriculture

With the objective of improving crops, orchards, and ornamental plants, humans have devised various methods for vegetative propagation. Most of these are based on the ability of plants to form adventitious roots or shoots.

Clones from Cuttings Most houseplants, woody ornamentals, and orchard trees are asexually propagated from plant fragments called cuttings. In some cases, shoot or stem cuttings are used. At the cut end of the shoot, a mass of dividing, undifferentiated cells called a **callus** forms, and then adventitious roots develop from the callus. Some plants, including African violets, can be propagated from single leaves rather than stems. For still other plants, the cuttings are taken from specialized storage stems. For example, a potato can be cut up into several pieces, each with a vegetative bud, or "eye," that regenerates a whole plant.

In a modification of vegetative propagation from cuttings, a twig or bud from one plant can be grafted onto a plant of a closely related species or a different variety of the same species. Grafting makes it possible to combine the best qualities of different species or varieties into a single plant. The graft is usually done when the plant is very young. The plant that provides the root system is called the **stock;** the twig grafted onto the stock is referred to as the **scion.** For example, scions from French varieties of vines that produce superior wine grapes are grafted onto root stock of American varieties, which are more resistant to certain diseases. The quality of the fruit, determined by the genes of the scion, is not diminished by the genetic makeup of the stock. In some cases of grafting, however, the stock can alter the characteristics of the shoot system that develops from the scion. For example, dwarf fruit trees are made by grafting normal twigs onto dwarf stock varieties that retard the vegetative growth of the shoot system. Since seeds are produced by the part of the plant derived from the scion, they would give rise to plants of the scion species if planted.

Test-Tube Cloning and Related Techniques In a new kind of botanical alchemy, test-tube methods are now being used to create and clone novel plant varieties. It is possible to grow whole plants by culturing small explants (pieces of tissue cut from the parent), or even single parenchyma cells, on an artificial medium containing nutrients and hormones (Figure 34.14). The cultured cells divide to form an undifferentiated callus. When the hormonal balance is manipulated in the culture medium, the callus can sprout shoots and roots with fully differentiated cells. The test-tube plantlets can then be transferred to soil, where they continue their growth. A single plant can be cloned into thousands of copies by subdividing calluses as they grow. This method is used for propagating orchids and also for cloning pine trees that deposit wood at unusually fast rates.

Cultured explants of certain plants give rise to embryolike structures that are smaller and simpler than

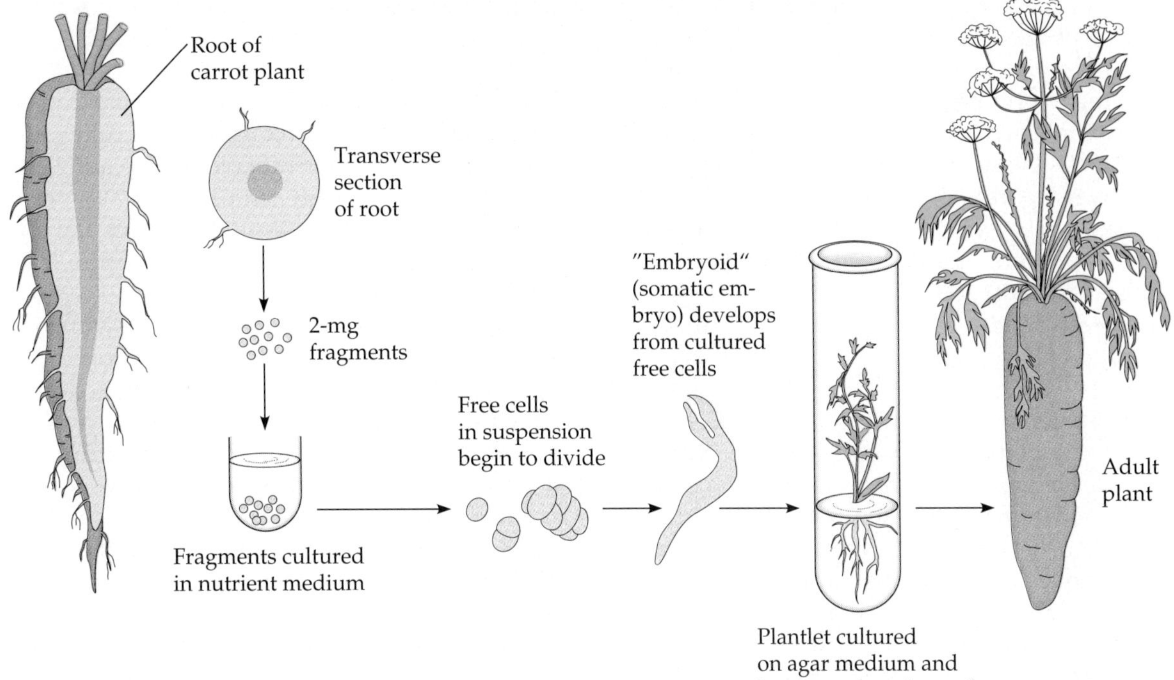

Figure 34.14
Test-tube cloning of carrots. In classic experiments conducted during the 1950s, F. C. Steward and his students at Cornell University demonstrated that whole plants could be regenerated from somatic (nonreproductive) cells dissected from a carrot. The products are genetic duplicates of the parent plant. Today, there are many agricultural applications of this basic method of vegetative propagation. For example, test-tube cloning is used to reproduce the qualities of "super" fruit trees and supply the copies to growers by the thousands.

plantlets. They are called somatic embryos because they are derived asexually from somatic (nonreproductive) cells. Plant biotechnologists have learned how to make artificial seeds by packaging somatic embryos along with nutrients within a polysaccharide gel (Figure 34.15).

Plant tissue culture also facilitates genetic engineering of plants (see Chapter 19). Most techniques for the introduction of foreign genes into plants require the use of single plant cells or small pieces of plant tissue as the starting material. Test-tube culture makes it possible to regenerate genetically altered plants from a single plant cell into which the foreign DNA has been incorporated. For example, researchers have used recombinant DNA technology to transfer a gene for bean protein into cultured cells from a sunflower plant. The experiment improved the protein quality of sunflower seeds harvested from the transgenic plants. (The potential agricultural impact of genetic engineering is discussed in more detail in Chapter 19.)

A technique known as **protoplast fusion** is being coupled with tissue culture methods to actually invent new plant varieties that can be cloned. Protoplasts are plant cells that have had their cell walls removed. Before they are cultured, the protoplasts can be screened for mutations that may improve the agricul-

Figure 34.15
Germinating artificial seeds. These capsules contained somatic embryos of alfalfa. Cultured explants from a single plant gave rise to the somatic embryos, which were then encapsulated along with nutrients and a tiny amount of fertilizer in a polysaccharide gel, the artificial seed coat. These embryos were genetically uniform because they were cloned from a single "super" parent with desirable traits. The artificial seeds in this photograph have germinated. Plants grown from artificial seeds will all mature at the same time, simplifying harvest. The main disadvantage of artificial seeds is that they are more expensive than natural ones.

tural value of the plant. It is also possible in some cases to fuse two protoplasts from different plant species that would otherwise be sexually incompatible, and then culture the hybrid protoplasts. Each of the many protoplasts can regenerate a wall and eventually form a hybrid plantlet. One success of this method has been a hybrid between potato and a wild relative called black nightshade. The nightshade is resistant to a herbicide that is commonly used to kill weeds. The hybrids are also resistant, and this makes it possible to "weed" a potato field with the herbicide without killing the potato plants.

Benefits and Risks of Monoculture By conscious effort, genetic variability in many crops has been virtually eliminated. For grains and other crops grown from seed, plant breeders have selected for varieties that self-pollinate. Whenever possible, vegetative propagation is used to clone exceptional plants. Genetic uniformity ensures that all plants in a field grow at the same rate, fruit on all the trees in an orchard ripens in unison, and yields at harvest time are dependable. There is no doubt that **monoculture,** the cultivation of large areas of land with a single plant variety, has helped farmers feed the world. But modern farms are very fragile ecosystems: Where there is little genetic variability, there is also little adaptability. A clone is a kind of superorganism; in terms of natural selection, it is one genetic individual. What is good for one is good for all, and what is bad for one threatens the entire clone. For example, monoculture was a major factor in the nineteenth-century Irish famine, which was caused by potato blight (a water mold; see Chapter 26). Spanish explorers brought potatoes to Europe from South America in the late sixteenth century. Europeans used vegetative propagation to grow the "new" food, which became the staple of the Irish diet. When the blight hit almost two centuries later, the potato plants, which were genetically quite uniform, were all susceptible to the parasite. Plant scientists fear that some plant disease could again devastate thousands of acres of an important monoculture. Plant breeders have responded by maintaining "gene banks," where they store seeds of many plant varieties that can be used to breed new hybrids.

COMPARING SEXUAL AND ASEXUAL PLANT REPRODUCTION: AN EVOLUTIONARY PERSPECTIVE

The dilemma of monoculture provides us with insight about sexual and asexual reproduction in the wild. Many plants are capable of both modes of reproduction, and each offers advantages in certain situations. Sex generates variation in a population, an asset when the environment changes (see Chapter 21). An additional benefit of sexual reproduction in plants is the seed, which can disperse to new locations and can also wait to grow until hostile environmental conditions have improved. On the other hand, a plant well suited to a stable environment can use asexual reproduction to clone many copies of itself. Moreover, the offspring of vegetative reproduction, usually mature fragments of the parent plant, are not as frail as the seedlings produced by sexual reproduction. A sprawling clone of prairie grass may cover an area so thoroughly that seedlings of other species have little chance of competing. But in the soil is a pool of seeds, waiting in the wings for some cue to germinate. After a fire, drought, or some other disturbance clears patches of the turf, seedlings can finally get a foothold when conditions improve. The seedlings are unequal in their traits, for their genomes are products of the sexual recombination of genes. A new competition ensues, in which certain plants excel and propagate themselves asexually. Both modes of reproduction, sexual and asexual, have had featured roles in the evolutionary adaptation of plant populations to their environments.

SOME CELLULAR ASPECTS OF PLANT DEVELOPMENT

Reproduction and development are closely related topics. Whether a plant arises from a sexually produced zygote or by vegetative reproduction, its transformation into a whole new individual depends on mechanisms that shape organs such as leaves and roots and generate specific patterns of specialized cells and tissues within those organs. **Development** is the sum of all of the changes that progressively elaborate an organism's body. In this section, you will learn about some of the cellular mechanisms responsible for plant development.

An Overview of Plant Development

A plant zygote is a single cell that bears no resemblance to the organism it becomes. Three overlapping developmental processes transform the fertilized egg into a plant: growth, morphogenesis, and cellular differentiation.

Growth, an irreversible increase in size, results from cell division and cell enlargement. By a series of mitotic divisions, the zygote gives rise to the multicellular embryo within a seed (see Figure 34.7). After germination, mitosis resumes, concentrated mostly in the apical meristems near the tips of roots and shoots (see

Chapter 31). But it is the enlargement of these newly made cells that accounts for most of the actual increase in the size of a plant.

If development were simply a matter of growth, then the zygote would give rise to an expanding ball of cells. In reality, growth is accompanied by **morphogenesis,** the development of form. The embryo encased in a seed has cotyledons and a rudimentary root and shoot, products of morphogenetic mechanisms that begin operating with the first division of the zygote. After the seed germinates, morphogenesis continues to shape the root and shoot systems of the growing plant. For example, morphogenesis at the shoot tip establishes the placement of leaves and other morphological features of the shoot. A fundamental difference between plants and animals is reflected in their morphogenesis: Most animals *move* through their environments; plants, in contrast, *grow* through their environments. The indeterminate growth of a plant, as you have learned, is a function of regions that remain embryonic for the life of the plant at its shoot and root tips. These regions are centers not only of continuing growth, but of continuing morphogenesis as well (Figure 34.16).

Shape alone, of course, does not enable a plant organ to perform its various functions. Each organ—a leaf, for example—has a diversity of cell types specialized for certain functions and fixed in certain locations. For example, the guard cells that border stomata differ markedly in structure and function from the surrounding cells of the epidermis. Another example is the development of xylem and phloem from the vascular cambium during secondary growth. Recall from Chapter 31 that the cambium gives rise to xylem on the inside and phloem on the outside. This acquisition of a cell's specific structural and functional features is called **cellular differentiation** (see Chapter 18).

Although we have dissected plant development into growth, morphogenesis, and differentiation, it is important to realize that these processes occur in concert as the plant develops. Their integration will become apparent as we survey some of the cellular mechanisms responsible for plant development.

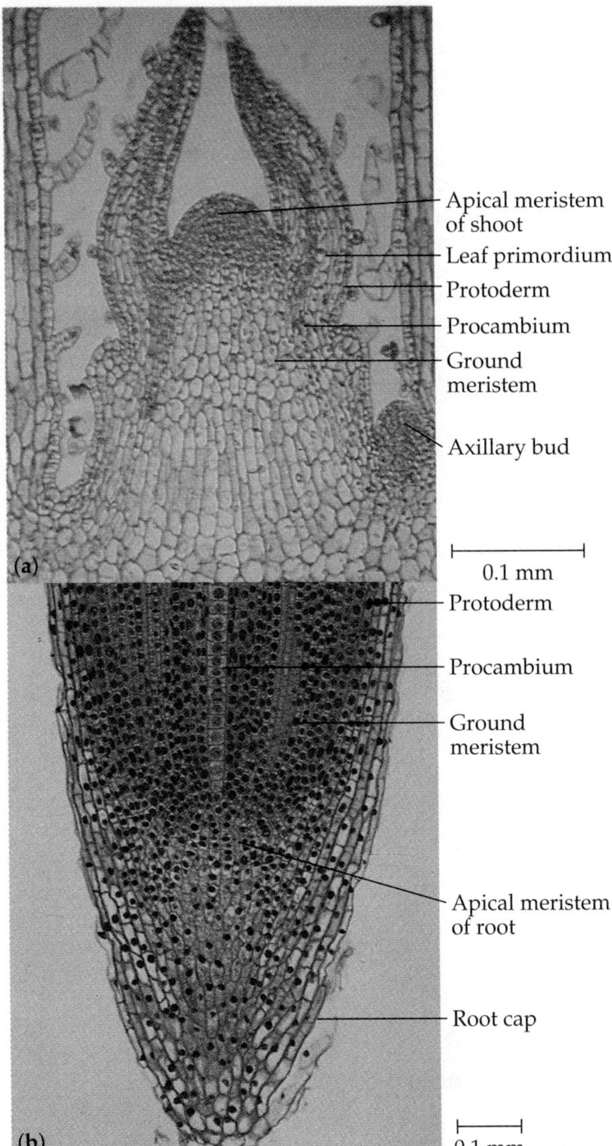

Figure 34.16
Lifelong morphogenesis in plants. In contrast to animals, which cease growing and change shape little after reaching maturity, plants exhibit indeterminate growth and persistent morphogenesis for as long as they live. For elaboration of the primary plant body, the centers of ongoing morphogenesis are at a plant's tips, in the regions of the apical meristems of (**a**) shoots and (**b**) roots. Morphogenesis is especially evident at the shoot tip, where the meristem gives rise to a succession of modules (see Chapter 31), each consisting of a leaf-bearing node, an axillary bud, and an internode (LMs).

The Roles of Cell Division and Cell Expansion in Morphogenesis Re-examine Figure 34.16, and you can see evidence of a basic principle of plant morphology: The shape of a plant organ depends mostly on the spatial orientations of cell divisions and cell expansions. In the case of the root tip, for example, notice that the young cells issue from the apical meristem in files. This is because most of the cell divisions in this region are oriented in a transverse plane, perpendicular to the long axis of the root. When the new cells begin growing, they mainly elongate; that is, most of their expansion parallels the long axis of the root.

Here we can see an important difference between plant and animal development. Animal morphogenesis also involves oriented cell division and growth, but migration of cells plays a major role as well. In contrast, plant cells cannot move about individually within a developing organ, because they are immobilized by

Figure 34.17
The preprophase band and the plane of cell division. The band is a ring of microtubules that forms just inside the plasma membrane during late interphase. Its location predicts the future plane of cell division. Although the two cells in (**a**) and (**b**) are similar in shape, they will divide in different planes. Each cell is represented by two micrographs, one unstained and the other stained with a fluorescent dye that binds specifically to microtubules, which form a "halo" (preprophase band) around the nucleus (LMs).

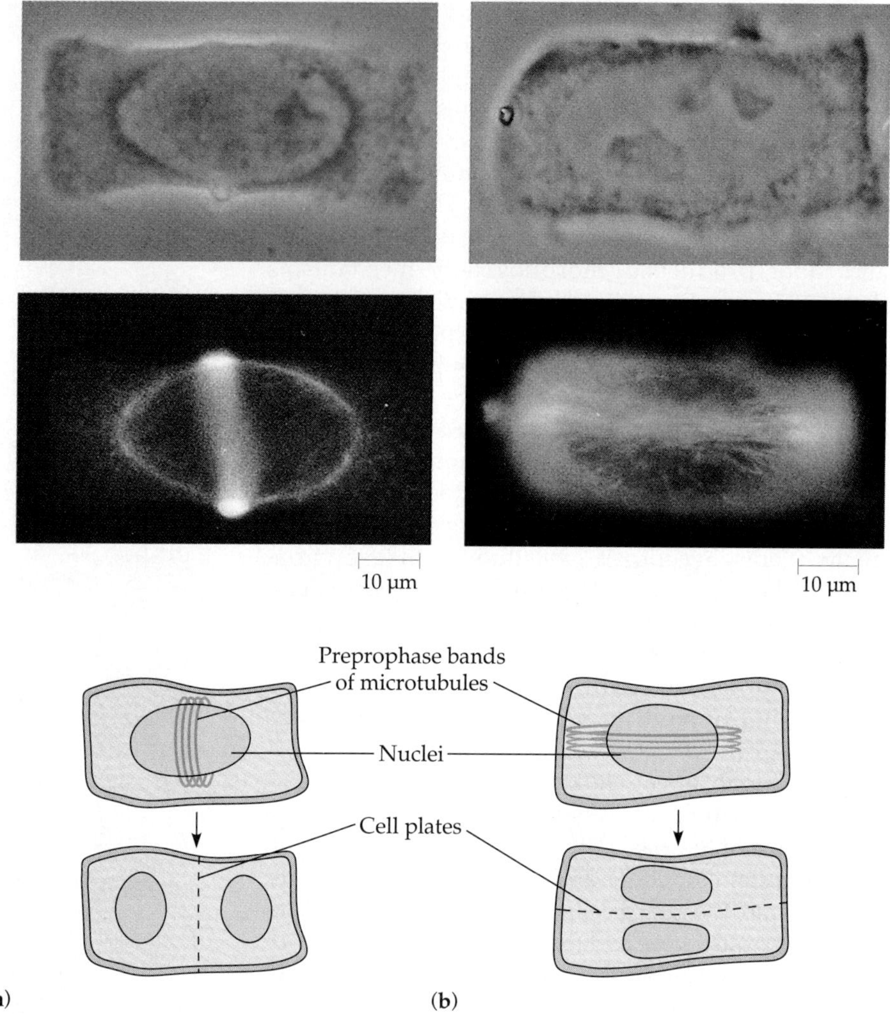

cell walls that are cemented to neighboring cells. This means that oriented cell division and expansion are the chief mechanisms of plant morphogenesis. Plant biologists have learned that the cytoskeleton controls these processes.

The Cytoskeleton: Orienting the Plane of Cell Division The plane in which a cell will divide is determined during late interphase (G_2 of the cell cycle; see Chapter 11). The first sign of this spatial orientation is a rearrangement of the cytoskeleton. Microtubules in the cortex (outer cytoplasm) of the cell become concentrated into a ring called the **preprophase band** (Figure 34.17). The band disappears before metaphase, but it has already set the future plane of cell division. The "imprint" consists of an ordered array of actin microfilaments that remain after the microtubules of the preprophase band disperse. These microfilaments hold the nucleus in a fixed orientation until the spindle forms, and later they direct movement of the vesicles that produce the cell plate. (See Chapter 11 to review the role of the cell plate in the division of plant cells.) When the cell finally divides, the walls separating the daughter cells form in the plane defined earlier by the preprophase band. Researchers are now working to learn what controls placement of the preprophase band.

The Cytoskeleton: Determining the Direction of Cell Expansion The shape of a plant organ is the outcome of the oriented growth of its cells, and we will see how the cytoskeleton controls this differential enlargement. But first we must learn a little more about *how* plant cells grow.

In a region of active growth, such as the zone of elongation in roots, cells can expand to a volume as much as fifty times their original size. This expansion occurs when the cell wall yields to the turgor pressure of the cell. Acid secreted by the cell causes chemical changes that weaken the cross-links between cellulose microfibrils in the wall. The cell is hyperosmotic to the surrounding solution, and with its wall loosened it can take up additional water by osmosis and expand. Growth continues until the wall again becomes knit tightly enough to offset the cell's turgor pressure. Once again, it is useful to highlight a difference between plants and animals. Animal cells grow mainly by synthesizing protein-rich cytoplasm, a metabolically ex-

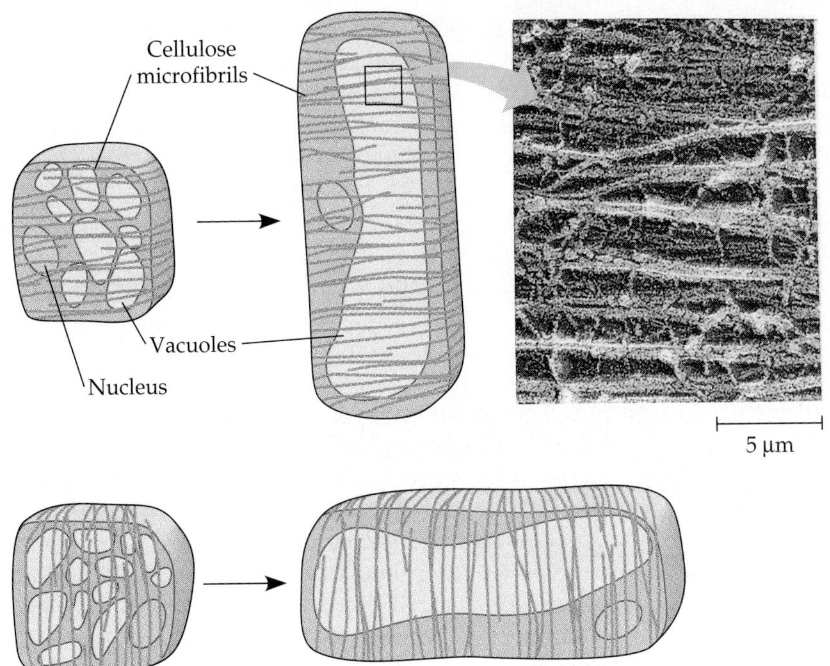

Figure 34.18
The orientation of plant cell growth. Growing plant cells expand mainly in the plane perpendicular to the orientation of cellulose microfibrils in the wall. The microfibrils are embedded in a matrix of other (noncellulose) polysaccharides, some of which form the cross-links visible in this micrograph (TEM). Weakening of these cross-links reduces restraint on the turgid cell, which can then take up more water and expand. Small vacuoles, which accumulate most of this water, coalesce to form the cell's central vacuole.

pensive process. Growing plant cells also produce additional organic material in their cytoplasm, but the uptake of water typically accounts for about 90% of a plant cell's expansion. Most of this water is packaged in the large vacuole, which forms by the coalescence of numerous smaller vacuoles as a cell grows. A plant can grow rapidly and economically because a small amount of cytoplasm can go a long way. Bamboo shoots, for instance, may elongate more than 2 m per week. Rapid extension of shoots and roots increases exposure to light and soil, an important evolutionary adaptation to the immobile lifestyle of plants.

Plant cells rarely expand equally in all directions. For example, cells near the root tip may elongate to twenty times their original length, with relatively little increase in width. The orientation of cellulose microfibrils in the innermost layers of the cell wall causes this differential growth (Figure 34.18). The microfibrils cannot stretch much, so the cell expands mainly in the direction perpendicular to the "grain" of the microfibrils. Microfibrils are synthesized by a complex of enzymes built into the plasma membrane (Figure 34.19).The pattern of microfibrils in the wall mirrors the orientation of microtubules located just across the

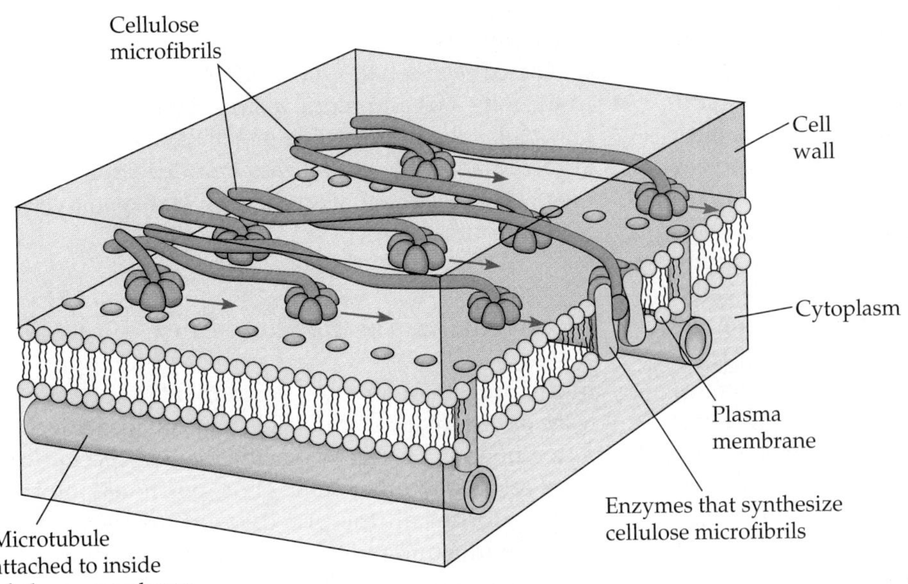

Figure 34.19
A hypothetical mechanism for how microtubules determine the orientation of cellulose microfibrils. Cellulose microfibrils are synthesized at the cell surface by complexes of enzymes that can move in the plane of the plasma membrane. According to one hypothesis, microtubules form "banks" that confine the movement of the enzymes to channels of specified direction. Each enzyme complex advances along one of these channels as the microfibril it produces becomes locked in place by cross-linking to other microfibrils.

plasma membrane in the cortex of the cell. According to one hypothesis, the microtubules control the flow of the cellulose-producing enzymes along the membrane. This specifies the alignment of microfibrils in the wall, which in turn determines the direction of cell expansion. As with the role of the cytoskeleton in the plane of cell division, an important question remains: What regulates the orientation of microtubules in the cell's cortex? Many laboratory groups are searching for an answer.

Cellular Differentiation: The Basic Problem

It is remarkable that cells as diverse as guard cells, sieve-tube members (phloem), and xylem vessel elements all descend from a common cell, the zygote. Cellular differentiation, the progressive development of a cell's specialized structure and function, is one of the most engaging research problems in modern biology.

Differentiation reflects the synthesis of different proteins in different types of cells. For example, a developing xylem cell makes the enzymes that produce lignin, which hardens the cell wall. In contrast, guard cells lack the enzymes for lignin synthesis and have flexible walls. This difference in enzyme content contributes to a structural difference correlated with the contrasting functions of these two cell types. Xylem vessels function in support (as well as transport), and their hard cell walls fit this function. Guard cells, on the other hand, control the size of stomata by changing their shape, a function that requires flexible walls (see Chapter 32). In its final stage of differentiation, a xylem cell unleashes hydrolytic enzymes that destroy the protoplast, leaving only the wall. No such metabolic episode occurs during the differentiation of a guard cell.

The problem that makes differentiation so fascinating is that the cells of a developing organism synthesize different proteins and diverge in structure and function even though they share a common genome. The cloning of whole plants from somatic cells supports the conclusion that the genome of a differentiated cell remains intact (see Figure 34.14). If a mature cell excised from a root or leaf can "dedifferentiate" in tissue culture and give rise to the diverse cell types of a plant, then it must possess all the genes necessary to make these many kinds of cells. This means that cellular differentiation depends, to a large extent, on the control of gene expression—the regulation of transcription and translation leading to specific proteins. Cells with the same genomes follow different developmental pathways because they selectively express certain genes at specific times during their differentiation. A guard cell has the genes that program self-destruction of a xylem protoplast, but it does not express those genes. A xylem vessel element does express them, but only at a specific time in its differentiation, after the cell has elongated and produced its secondary wall. Researchers are beginning to unravel the molecular mechanisms that switch specific genes on and off at critical times in a cell's development (see Chapters 18 and 43).

Pattern Formation and Positional Information

Cellular differentiation has a spatial component: Each plant organ has a characteristic pattern of tissues and of cell types within those tissues. To reinforce this feature of development, see Figure 31.14, which shows the anatomy of a dicot root versus a stem of the same species. On a larger scale, spatial organization is also apparent in the placement of a plant's organs—in the arrangement of a flower's parts, for example. The development of specific structures in specific locations is called **pattern formation.** It is a key element of morphogenesis, the overall development of an organism's form.

Pattern formation depends on **positional information,** signals of some kind that indicate each cell's location within an embryonic structure, such as a shoot tip. By affecting localized rates and planes of cell division and expansion, these signals would cause organs such as leaf primordia to emerge as "bumps" at certain places on the plant. Within a developing organ, each cell would continue to detect positional information and respond by differentiating into a particular type of cell.

Many developmental biologists speculate that gradients of chemicals called **morphogens** provide positional information. Perhaps, for example, a morphogen diffusing from a shoot's apical meristem "informs" the cells below of their distance from the shoot tip. We can imagine that the cells gauge their radial positions within the developing organ by detecting a second morphogen that emanates from the outermost cells. The gradients of these two morphogens would be sufficient for each cell to "get a fix" on its position relative to the long axis and radial axis of the rudimentary organ. This idea of diffusible morphogens is just one of several alternative hypotheses that developmental biologists are testing to learn how an embryonic cell detects its location.

Clonal Analysis of the Shoot Apex

In the process of shaping a rudimentary organ, patterns of cell division and cell expansion also affect the differentiation of cells by placing them in specific locations relative to other cells. Thus, positional information underlies all the processes of development—growth, morphogenesis, and differentiation. One approach to studying the relationships among these

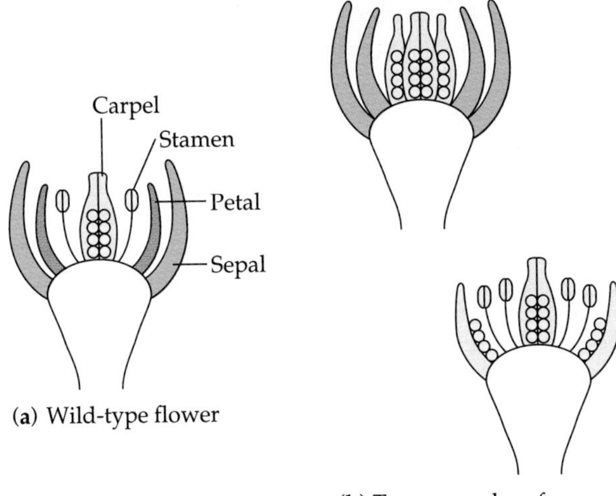

(b) Two examples of homeotic mutants

Figure 34.20
Homeotic mutations in *Arabidopsis* flower development. Use the color-coding to compare (**a**) the wild-type flower to (**b**) two of the several homeotic mutations that have been observed in *Arabidopsis.* (A homeotic mutation places one type of organ where another organ normally develops.) One of these mutants has an extra whorl of sepal-like organs where petals are normally located and extra carpel-like organs where stamens are normally found. The other mutant has "carpels" in place of sepals and "stamens" in place of petals. (The organ replacement is not actually one-for-one; for example, misplaced carpels differ in subtle ways from normal carpels.) The homeotic mutations that cause these aberrant flowers occur in genes called organ-identity genes.

processes is **clonal analysis,** in which the cell lineages (clones) derived from each cell in an apical meristem are mapped as organs develop. This is often done by using radiation or chemicals to induce somatic mutations that alter chromosome number or otherwise tag a cell in some way that distinguishes it from its neighbors in the shoot tip. The lineage of cells derived by mitosis from the mutant meristematic cell will also be "marked."

One of the important questions that clonal analysis can address is: How early is the developmental fate of a cell determined by its position in an embryonic structure? To some extent, the developmental fates of cells in the shoot apex are predictable. For example, almost all the cells derived from division of the outermost meristematic cells end up as part of the dermal tissue of leaves and stems. But it is not possible to pinpoint precisely which cells of the meristem will give rise to specific organs and tissues. Apparently random changes in rates and planes of cell division can reorganize the meristem. For example, the outermost cells *usually* divide in a plane perpendicular to the surface of the shoot tip, resulting in the addition of cells to the surface layer. But occasionally a cell at the surface divides in a plane parallel to this meristematic layer, placing one daughter cell beneath the surface among cells derived from different lineages. Thus, the cells of the meristem are not dedicated early to forming specific organs and tissues. Put another way, a cell's developmental fate is not determined by its membership in a lineage derived from a particular meristematic cell. Rather, it is the cell's *final* position in an emerging organ that determines what kind of cell it will become, presumably as a result of positional information.

Homeotic Mutations in Flower Development: In Search of the Genetic Basis of Pattern Formation

A particularly striking passage in plant development is the transition of a vegetative shoot tip into a floral meristem. While it is vegetative, the shoot tip grows indeterminately, reiterating node-internode-node-internode, module after module. But if it is converted to floral function, the shoot tip becomes determinate. Its meristem is consumed in forming the primordia for sepals, petals, stamens, and carpels.

Once a shoot meristem is induced to flower, what controls placement of the floral organs? You already know part of the answer. Positional information commits each primordium that arises on the flanks of the shoot tip to develop into an organ of specific structure and function—an anther-bearing stamen, for example. In the past few years, plant biologists have identified some of the genes that are regulated by positional information and function in this development of floral pattern. Mutations in these genes, called **homeotic mutations,** substitute one type of body part where another would normally form. For example, a particular homeotic mutation may cause an extra whorl of sepals to develop in a flower where there ought to be petals. The implication is that the wild-type alleles for these genes, called **organ-identity genes,** are responsible for the development of normal floral pattern.

Many plant biologists interested in genes that control development have focused on *Arabidopsis thaliana* as their experimental organism. Recall from Chapter 31 that this wild mustard has a relatively small genome, a characteristic that makes the search for specific genes more manageable (see Figure 31.2). Researchers have identified several homeotic mutations that affect flower development in *Arabidopsis* (Figure 34.20) and have used recombinant DNA methods to clone a few organ-identity genes. Very similar organ-identity genes have also been found in snapdragon (*Antirrhinum majus*), which is not closely related to *Arabidopsis*. This suggests that evolution has been very conservative with genes that control development of an angiosperm's basic body plan.

Organ-identity genes code for regulatory proteins called transcription factors (see Chapter 18). These proteins help control expression of *other* genes by binding to DNA at specific sites in the genome and affecting rates of RNA synthesis (transcription). Perhaps positional information determines which organ-identity gene is expressed in a particular floral-organ primordium, and the resulting transcription factor induces expression of those genes responsible for building an organ of specific structure and function. This is one of the hypotheses that is shaping continuing research on the genetic basis of pattern formation. With the help of excellent experimental systems such as *Arabidopsis* and the powerful methods of modern cell biology, plant scientists are making rapid progress toward understanding how a single cell becomes a plant.

* * *

Throughout our study of plant reproduction and development in this chapter, the role of the environment has been underemphasized. In fact, as you learned in Chapter 31, environmental factors such as light and wind have enormous impact on plant form and function. In the next chapter, you will learn how plants tune their morphology and physiology to the outside world.

STUDY OUTLINE

Sexual Reproduction of Flowering Plants (pp. 734–744)

1. The flower is an angiosperm structure of modified leaves specialized for reproduction. The floral organs are sterile sepals and petals and fertile stamens and carpels.
2. In alternation of generations in angiosperms, the dominant stage is the diploid sporophyte. The spores develop inside the flower into tiny, haploid gametophytes—the male pollen grain and the female embryo sac.
3. Pollen develops inside the sporangia of the anther.
4. Within an ovule, a haploid megaspore divides by mitosis to form the embryo sac, the female gametophyte.
5. Fertilization is preceded by pollination, the landing of pollen on the stigma of the carpel.
6. The pollen grain produces a pollen tube that extends down the style toward the embryo sac. Two sperm are released and effect a double fertilization, resulting in a diploid zygote and a triploid endosperm.
7. The zygote gives rise to an embryo with apical meristems and one or two cotyledons.
8. Mitotic division of the triploid endosperm gives rise to a multicellular, nutrient-rich mass that feeds the developing embryo and later the young seedling.
9. The integuments of the ovule form a seed coat around the embryo and its food supply.
10. A fruit is a mature ovary that protects the enclosed seeds and aids in their dispersal via wind or animals.
11. Germination begins when seeds imbibe water. This expands the seed, rupturing its coat, and triggers metabolic changes that cause the embryo to resume growth.
12. The embryonic root, or radicle, is the first structure to emerge from the germinating seed. Next, the embryonic shoot breaks through the soil surface.

Asexual Reproduction (pp. 744–748)

1. Asexual reproduction (cloning) is the production of genetically identical offspring from a single parent.
2. Fragmentation of a parent plant into parts that re-form whole plants demonstrates the versatility and latent potential of meristematic and parenchymal tissues.
3. In horticulture and agriculture, plants can be asexually propagated from isolated leaves, pieces of specialized storage stems, or cuttings (shoots).
4. Laboratory methods can clone large numbers of desired plant varieties by culturing small explants or single parenchyma cells.
5. Although monoculture of desirable clones has greatly increased agricultural productivity, the absence of genetic variability may prove disastrous in the face of uncertain future conditions.

Comparing Sexual and Asexual Plant Reproduction: An Evolutionary Perspective (p. 748)

1. Both sexual and asexual modes of reproduction have been important in the adaptation of plants to their environment by offering advantages in different situations. A stable environment favors the most successful clones of asexually produced plants. Sex generates genetic variation that makes adaptation possible.

Some Cellular Aspects of Plant Development (pp. 748–754)

1. The overlapping mechanisms of plant development are growth, morphogenesis, and cellular differentiation.
2. During growth of a plant, the planes of cell divisions and cell expansions determine the shape of each organ.
3. The cytoskeleton sets the plane of cell division by forming a preprophase band.
4. The cytoskeleton also controls the direction of cell expansion by determining the orientation of cellulose microfibrils that are deposited in the developing wall.
5. The basic problem of cellular differentiation is to explain how cells with matching genomes diverge into cells of diverse structure and function.
6. Pattern formation, the emergence of organs and tissues in specific locations, depends on the ability of developing cells to detect and respond to positional information.
7. Clonal analysis of shoot tips suggests that a cell's developmental fate is determined by its final location within a primordial organ.
8. By studying homeotic mutations that cause floral organs to develop in the wrong locations, plant biologists are investigating the genetic basis of pattern formation.

SELF-QUIZ

1. Which of the following would definitely be an imperfect flower? A flower that
   a. is also incomplete

b. lacks sepals
c. is found on a monoecious plant
d. is staminate
e. cannot self-pollinate

2. In the development of pollen, which of the following is a diploid cell that undergoes meiosis and eventually gives rise to all the others?
a. a tube cell
b. a microsporocyte
c. a sperm cell
d. a generative cell
e. a microspore

3. A seed develops from
a. an ovum
b. a generative cell
c. an ovule
d. an ovary
e. an embryo

4. A fruit is a
a. mature ovary
b. mature ovule
c. seed plus its integuments
d. fused carpel
e. enlarged embryo sac

5. Which of these conditions is needed by almost all seeds to break dormancy?
a. exposure to light
b. imbibition
c. abrasion of the seed coat
d. exposure to cold temperatures
e. covering of fertile soil

6. Which of the following is *not* an example of asexual propagation?
a. fragmentation, the separation of plant parts that develop into whole plants
b. regeneration of roots from the base of a cut stem
c. apomixis, the asexual production of seeds
d. test-tube culturing of divided calluses
e. production of fruit without fertilization in parthenocarpic plants

7. Which of the following structures is unique to the seed of a monocot?
a. coleoptile
b. radicle
c. seed coat
d. endosperm
e. cotyledon

8. Which of the following does *not* appear to have a major effect on plant morphology?
a. cell division
b. environmental factors
c. cell migration
d. cell expansion
e. selective expression of genes

9. The direction in which a plant cell will expand seems to a large extent to depend on
a. the size and position of the central vacuole
b. a ring of actin microfilaments in the cortex of the cell
c. location of the preprophase band
d. distribution of solutes in the cytoplasm
e. the orientation of microtubules just inside the plasma membrane

10. The type of mature cell that a particular embryonic plant cell will become appears to be determined by
a. selective loss of genes
b. the cell's final position in a developing organ
c. the cell's pattern of migration
d. the cell's rate of division and growth
e. the cell's particular meristematic lineage

CHALLENGE QUESTIONS

1. What are the advantages of propagating houseplants by asexual reproduction?
2. Imagine that you are a member of a team of plant biologists who have isolated and purified a suspected plant morphogen. It is secreted only by the tips of floral meristems and diffuses down the flanks of the shoot tip. Your team has been able to grow isolated bits of floral meristem in cell culture, treat them with the morphogen, and in this way induce them to form floral parts. What floral parts would you expect to see develop at relatively high and low morphogen concentrations?

SCIENCE, TECHNOLOGY, AND SOCIETY

1. Seed companies gather seeds from all over the world and use them to develop new kinds of food and medicinal plants through crossbreeding and biotechnology. The greatest diversity of potentially useful plants is in the tropics. The developing tropical countries lack the resources for conserving plants and developing new crops, but they resent large corporations exploiting their plant resources and selling seeds back to them. What would you suggest to resolve this conflict, allowing both the developing nations and the seed companies to benefit from finding, preserving, and breeding useful plants?

FURTHER READING

Coen, E. S., and E. M. Meyerowitz. "The War of the Whorls: Genetic Interactions Controlling Flower Development." *Nature*, September 5, 1991. A summary of research that is bringing us closer to understanding pattern formation.

Dale, J. "How Do Leaves Grow?" *BioScience*, June 1992. Application of cell and molecular biology in the study of plant development.

Haring, V., et al. "Self-Incompatibility: A Self-Recognition System in Plants." *Science*, November 16, 1990. How pollen can tell if it has landed on a stigma of the same plant.

Jürgens, G. "Genes to Greens: Embryonic Pattern Formation in Plants." *Science*, April 24, 1992. Genetic analysis of plant morphogenesis.

Steeves, T. A., and I. M. Sussex. *Patterns in Plant Development*, 2nd ed. Cambridge: Cambridge University Press, 1989. A short text that emphasizes experimental approaches.

Vaughan, D. A., and L. A. Sitch. "Gene Flow from the Jungle to Farmers." *BioScience*, January 1991. The problems and methods of maintaining genetic diversity in crop plants.

35 Control Systems in Plants

The Search for a Plant Hormone: A Case Study in the Scientific Process

Functions of Plant Hormones

Plant Movements

Circadian Rhythms and the Biological Clock

Photoperiodism

Signal-Transduction Pathways in Plant Cells

Figure 35.1
Plants sense and respond to their environment. The shoots of this houseplant bend toward light, a response called phototropism. In this chapter, you will learn about the control systems that enable plants to adjust their growth, development, and physiology to environmental changes.

At every stage in the life of a plant, sensitivity to the environment and coordination of responses are evident. One part of a plant can send signals to other parts. For example, the terminal bud at the apex of a shoot is able to suppress the growth of axillary buds that may be many meters away. Plants keep track of the time of day and also the time of year. They can sense gravity and the direction of light and respond to these stimuli in ways that seem to us to be completely appropriate. It is tempting to explain this response in terms of such human qualities as desire, need, wisdom, and decisiveness, or to imagine that a plant behaves a certain way to accomplish a particular result. Science, however, searches for a mechanism—a link between cause and effect. A houseplant orients its leaves toward a window, not because of conscious choice or because it is trying to find more light, but because cells on the darker side of stems and petioles grow faster than cells on the brighter side, causing the organs to curve toward the light (Figure 35.1). Natural selection has favored mechanisms of plant response that enhance reproductive success, but this implies no purposeful planning on the part of the plant.

As you learn in this chapter about the control mechanisms at work in plants, remember an important theme. Adaptation to the environment occurs on *two* time scales: Individual plants and other organisms respond adaptively to what goes on around them, but these control systems are themselves adaptations that evolved over countless generations of plants interacting with their environments.

For the most part, plants and animals respond to environmental stimuli by very different means. Animals, being mobile, respond mainly by behavioral mechanisms, moving toward positive stimuli and away from negative stimuli. Rooted to one location for life, a plant generally responds to environmental cues by adjusting its pattern of growth and development. Because the program for development of the plant remains somewhat plastic, plants of the same species vary in body form much more than animals of the same species. All lions have four limbs and approximately the same body proportions, but oak trees are less regular in their number of limbs and their shapes.

A plant's morphology and physiology are constantly tuned to the surroundings by complex interactions of environmental factors and internal signals. Some of the most important signals are chemical messengers called hormones.

THE SEARCH FOR A PLANT HORMONE: A CASE STUDY IN THE SCIENTIFIC PROCESS

The word *hormone* is derived from a Greek verb meaning "to excite." Found in all multicellular organisms, **hormones** are chemical signals that coordinate the parts of the organism. As first defined by animal physiologists, a hormone is a compound that is produced by one part of the body and is then translocated to other parts of the body, where it triggers responses in target cells and tissues. Another important characteristic of hormones is that only minute concentrations of these chemical messengers are required to induce substantial change in an organism.

The concept of chemical messengers in plants emerged from a series of classic experiments on how stems respond to light. A houseplant on a windowsill grows toward light (see Figure 35.1). If you rotate the plant, it will soon reorient its growth until its leaves again face the window. The growth of a shoot toward light is called positive **phototropism** (growth away from light is negative phototropism). In a forest or other natural ecosystem where plants may be crowded, phototropism directs growing seedlings toward the sunlight they need for photosynthesis. What is the mechanism for this adaptive response? Much of what is known about phototropism has been learned from studies of grass seedlings, particularly oats. The shoot of a grass seedling is enclosed in a sheath called the coleoptile, which grows straight upward if the seedling is kept in the dark or if it is illuminated uniformly from all sides. If the growing coleoptile is illuminated from one side, it will curve toward the light. This response results from differential growth of cells on opposite sides of the coleoptile; the cells on the darker side elongate faster than cells on the brighter side (Figure 35.2).

Some of the earliest experiments on phototropism were conducted in the late nineteenth century by Charles Darwin and his son, Francis. They observed that a grass seedling could bend toward light only if the tip of the coleoptile was present (Figure 35.3). If the tip was removed, the coleoptile would not curve. The seedling would also fail to grow toward light if the tip was covered with an opaque cap; neither a transparent cap over the tip nor an opaque shield placed farther down the coleoptile prevented the phototropic response. It was the tip of the coleoptile, the Darwins concluded, that was responsible for sensing light. However, the actual growth response, the curvature of the coleoptile, occurred some distance below the tip. Charles and Francis Darwin proposed the hypothesis that some signal was transmitted downward from the tip to the elongating region of the coleoptile. A few decades later, Peter Boysen-Jensen of Denmark tested this hypothesis and demonstrated that the signal was a mobile substance of some kind. He separated the tip from the remainder of the coleoptile by a block of gelatin, which would prevent cellular contact but through which chemicals could diffuse. These seedlings behaved normally, bending toward light. However, if the tip was segregated from the lower coleoptile by an impermeable barrier of mica, no phototropic response occurred.

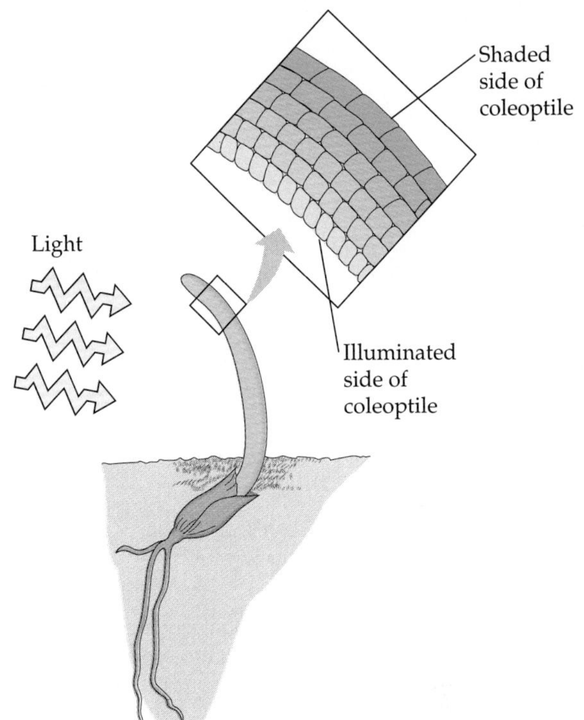

Figure 35.2
Phototropism. The coleoptile of an oat seedling grows toward light because cells on the darker side of the organ elongate faster than cells on the brighter side.

In 1926, F. W. Went, then a young plant physiologist in Holland, extracted the chemical messenger for phototropism by modifying the experiments of Boysen-Jensen (Figure 35.4). Went removed the coleoptile tip and placed it on a block of agar, a gelatinous material extracted from red algae. The chemical messenger from the tip, Went reasoned, should diffuse into the agar, and the agar block should then be able to substitute for the coleoptile tip. Went placed the agar blocks on decapitated coleoptiles that were kept in the dark. A block that was centered on top of the coleoptile caused the stem to grow straight upward. However, if the block was placed off center, then the coleoptile began to bend away from the side with the agar block, as though growing toward light. Went concluded that the agar block contained a chemical produced in the

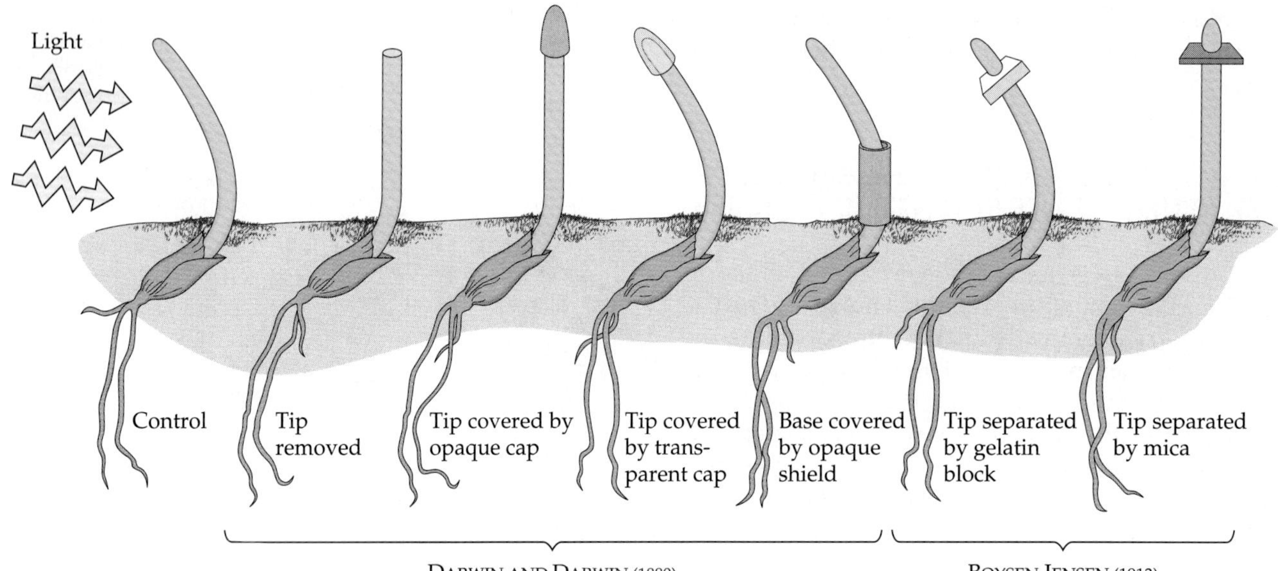

Figure 35.3
Early experiments on phototropism. Only the tip of the coleoptile can sense the direction of light, but the bending response occurs some distance below the tip. A signal of some kind must travel downward from the tip. The signal can pass through a permeable barrier (gelatin block) but not through a solid barrier (mica), suggesting that the signal for phototropism is a mobile chemical.

coleoptile tip, that this chemical stimulated growth as it passed down the coleoptile, and that a coleoptile curved toward light because of a higher concentration of the growth-promoting chemical on the darker side of the coleoptile. For this chemical messenger, or hormone, Went chose the name auxin (Gr. *auxein,* "to increase"). Auxin was later purified and its structure determined by Kenneth Thimann and his colleagues at the California Institute of Technology. Other plant hormones were subsequently discovered.

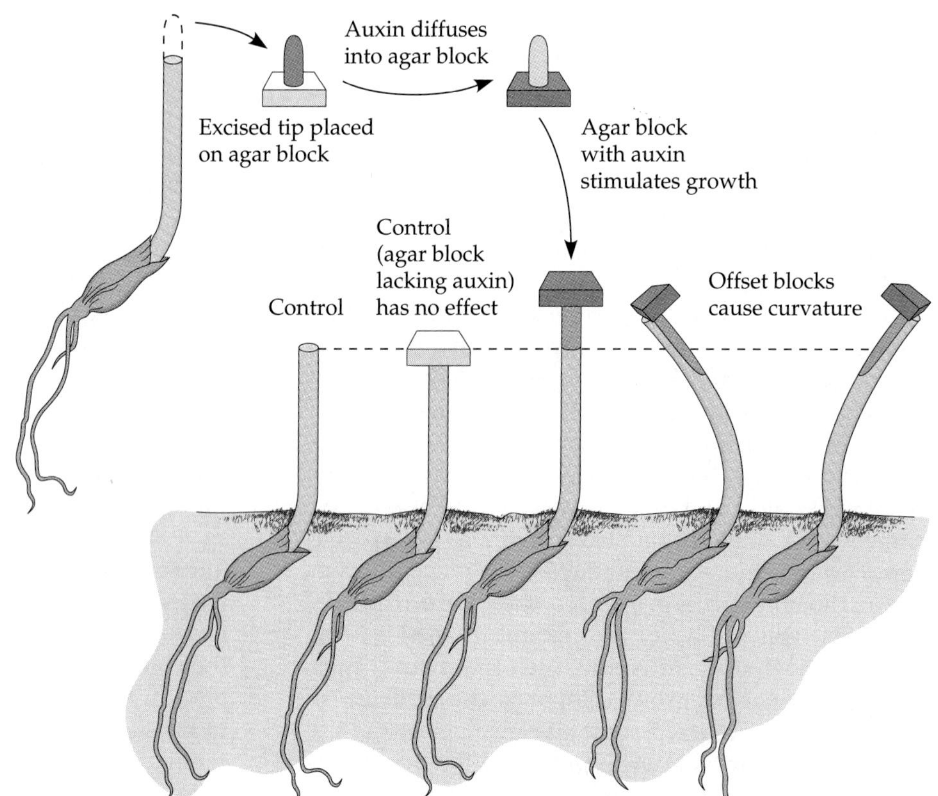

Figure 35.4
The Went experiments. Some chemical (indicated by pink) that can pass into an agar block from a coleoptile tip stimulates elongation of the coleoptile when the block is substituted for a tip. If the block is placed off center on the top of a decapitated coleoptile kept in the dark, the organ bends as if responding to illumination from the side. The chemical is the hormone auxin, which stimulates elongation of cells in the shoot.

Table 35.1 Functions of Plant Hormones

Hormone	Major Functions	Where Produced or Found in Plant
Auxin (such as IAA)	Stimulates stem elongation, root growth, differentiation and branching, development of fruit; apical dominance; phototropism and gravitropism.	Embryo of seed; meristems of apical buds and young leaves
Cytokinins (such as zeatin)	Affect root growth and differentiation; stimulate cell division and growth, germination, and flowering; delay senescence.	Synthesized in roots and transported to other organs
Gibberellins (such as GA_3)	Promote seed and bud germination, stem elongation, leaf growth; stimulate flowering and development of fruit; affect root growth and differentiation.	Meristems of apical buds, roots, and young leaves; embryo
Abscisic acid	Inhibits growth; closes stomata during water stress; counteracts breaking of dormancy.	Leaves, stems, green fruit
Ethylene	Promotes fruit ripening; opposes some auxin effects; promotes or inhibits growth and development of roots, leaves, flowers, depending on species.	Tissues of ripening fruits, nodes of stems, senescent leaves

FUNCTIONS OF PLANT HORMONES

So far, five classes of plant hormones have been positively identified. They are auxin, cytokinins (actually a class of related chemicals), gibberellins (also a class of similar chemicals), abscisic acid, and ethylene (Table 35.1). In general, these hormones control plant growth and development by affecting the division, elongation, and differentiation of cells. Each hormone has a multiplicity of effects, depending on its site of action, the developmental stage of the plant, and the concentration of the hormone. Notice in Table 35.1 that all the plant hormones are relatively small molecules. Their transport from cell to cell often involves passage across cell walls, a pathway that blocks the movement of large molecules.

Plant hormones are produced in very small concentrations, but a minute amount of hormone can have a profound effect on the growth and development of a plant organ. This implies that the hormonal signal must be amplified in some way. A hormone may act by altering the expression of genes, by affecting the activity of existing enzymes, or by changing properties of membranes. Any of these actions could redirect the metabolism and development of a cell responding to a small number of hormone molecules.

Reaction to a hormone usually depends not so much on the absolute amount of that hormone as on its relative concentration compared with other hormones. It is hormonal balance, rather than hormones acting in isolation, that may control the growth and development of the plant. These interactions will become apparent in the following survey of hormone function.

Auxin

The term **auxin** is actually used to describe any chemical substance that passes the test of promoting elongation of coleoptiles. The natural auxin that has been extracted from plants is a compound named indoleacetic acid, or IAA. In addition to this natural auxin, several other compounds, including some synthetic ones, have auxin activity. Throughout this chapter, however, the name auxin is used specifically to refer to IAA. Although auxin affects several aspects of plant development, one of its most important functions is to stimulate the elongation of cells in young developing shoots.

Auxin and Cell Elongation The apical meristem of a shoot is a major site of auxin synthesis. As auxin from the shoot apex moves down to the region of cell elongation (see Chapter 31), the hormone stimulates growth of the cells. Auxin has this effect only over a certain concentration range, from about 10^{-8} to 10^{-3} *M*. At higher concentrations, auxin may inhibit cell elongation. This is probably due to a high level of auxin inducing synthesis of another hormone, ethylene, which generally acts as an inhibitor of plant growth due to cell elongation.

The speed at which auxin is transported down the stem from the shoot apex is about 10 mm per hour—much too fast for diffusion, although slower than translocation in phloem. Auxin seems to be transported directly through parenchyma tissue, from one cell to the next. It moves only from shoot tip to base, not in the reverse direction. This unidirectional transport of auxin is called **polar transport.** Polar transport has nothing to do with gravity, for auxin travels upward in experiments where a stem or coleoptile segment is placed upside down. Polar auxin transport requires energy. Figure 35.5 illustrates how proton pumps, driven by ATP, couple metabolic energy to auxin transport. Notice another important feature of the model: Auxin exits each cell along a polar pathway via specific carrier proteins that are restricted to the basal end of the cell. The mechanism of polar auxin transport is another example of cellular work driven by chemiosmosis, the harnessing of H^+ gradients generated by proton pumps.

Proton pumps located at the plasma membrane also play a role in the growth response of cells to auxin. In a shoot's region of elongation, auxin stimulates the proton pumps, an action that lowers the pH in the wall. This acidification of the wall causes cross-links between cellulose microfibrils to break, loosening the fabric of the wall. With its wall now more plastic, the cell is free to take up additional water by osmosis and elongate (see Chapter 34). Experimental evidence supports this **acid growth hypothesis.** The mechanism works relatively fast. If auxin is applied to young stems, it induces wall loosening and cell elongation within 20 minutes. For sustained growth after this initial spurt, however, cells must make more cytoplasm and wall material. Auxin also stimulates this longer-term

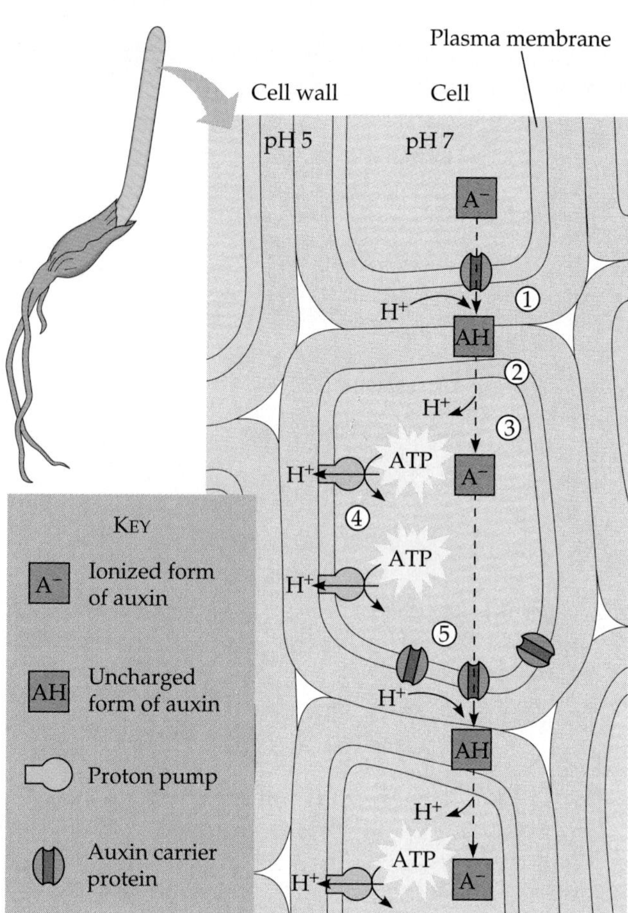

Figure 35.5
Polar auxin transport: a chemiosmotic model. In growing shoots, auxin is transported unidirectionally, from the apex down the shoot. Along this pathway, the hormone enters a cell at the apical end, exits at the basal end, diffuses across the wall, and enters the apical end of the next cell. A pH difference between the cell wall (acidic at about pH 5) and the cytoplasm (pH 7) contributes to auxin transport. In the pH 7 environment of the cell, auxin is an anion. ① When auxin encounters the acidic environment of the wall, the molecule picks up a hydrogen ion to become electrically neutral. ② As a relatively small, neutral molecule, auxin passes across the plasma membrane (see Chapter 8). ③ Once inside a cell, the pH 7 environment causes auxin to ionize. This temporarily traps the hormone within the cell, because the plasma membrane is less permeable to ions than to neutral molecules the same size. ④ ATP-driven proton pumps maintain the pH difference between the inside and outside of the cell. ⑤ Auxin can only exit the cell at the basal end, where specific carrier proteins are built into the membrane. The proton pumps contribute to this auxin efflux by generating a membrane potential (voltage) across the membrane, which favors transport of anions out of the cell. Now in the acidic environment of the wall again, auxin picks up a hydrogen ion and enters the next cell as an electrically neutral molecule. Polar auxin transport is one specific application of a basic mechanism of energy coupling in cells. This mechanism, chemiosmosis, uses proton pumps to store energy in the form of an H^+ gradient and membrane potential, and then taps this energy source to drive cellular work.

growth response. Later in the chapter, we will investigate how cells actually detect the presence of auxin and other signals and how these stimuli lead to specific responses.

Other Effects of Auxin In addition to stimulating cell elongation for primary growth, auxin affects secondary growth by inducing cell division in the vascular cambium and by influencing the differentiation of secondary xylem. Auxin also promotes the formation of adventitious roots at the cut base of a stem, an effect employed in horticulture by dipping cuttings in rooting media containing synthetic auxins. Developing seeds also synthesize auxin, which promotes the growth of fruit in many plants. Synthetic auxins sprayed on tomato vines induce fruit development without need for pollination. This makes it possible to grow seedless tomatoes by substituting for auxin that would normally be synthesized by seeds.

One of the most widely used herbicides, or weed killers, is 2,4-D, a synthetic auxin that disrupts the normal balance of plant growth. Dicots are more sensitive than monocots to this herbicide, and thus 2,4-D can be used to selectively remove dandelions and other broad-leaf weeds from a lawn or grain field.

Cytokinins

Cytokinins were discovered during trial-and-error attempts to find chemical additives that would enhance growth and development of plant cells in tissue culture. In the 1940s, Johannes van Overbeek, working at the Cold Spring Harbor Laboratory in New York, found he could stimulate the growth of plant embryos by adding coconut milk, the liquid endosperm of the giant coconut seed, to his culture medium. A decade later, Folke Skoog and Carlos O. Miller, at the University of Wisconsin, induced tobacco cells being grown in culture to divide by adding degraded samples of DNA. The active ingredients of both experimental additives turned out to be modified forms of adenine, one of the components of nucleic acids. These growth regulators were named cytokinins because they stimulate cytokinesis, or cell division. Of the variety of cytokinins that occur naturally in plants, the most common is zeatin, so named because it was first discovered in corn *(Zea mays).* Several synthetic compounds with cytokinin activity have also been produced. As you read about a few of the functions of cytokinins, notice that these hormones are complemented or countered by other hormones, especially auxin.

Control of Cell Division and Differentiation Cytokinins are produced in actively growing tissues, particularly in roots, embryos, and fruits. Cytokinins produced in the root reach their target tissues by moving up the plant in the xylem sap. Acting in concert with auxin, cytokinins stimulate cell division and influence the pathway of differentiation.

The effects of cytokinins on cells growing in tissue culture provide clues about how this class of hormones may function in an intact plant. When a piece of parenchyma tissue from a stem is cultured in the absence of cytokinins, the cells grow very large but do not divide. If cytokinins alone are added to the culture, they have no effect. If cytokinins are added along with auxin, however, the cells divide. Furthermore, the ratio of cytokinin to auxin controls the differentiation of the cells. When the concentration of the two hormones is about equal, the mass of cells continues to grow, but it remains an undifferentiated callus. If there is more cytokinin than auxin, shoot buds develop from the callus. If auxin is more concentrated than cytokinin, roots form. It is remarkable that gene expression can be controlled so extensively by manipulating the concentration of just two chemical signals.

The ability of these hormones to trigger cell division and influence differentiation could result from the fact that cytokinins stimulate RNA and protein synthesis. These newly synthesized proteins may be involved in cell division.

Control of Apical Dominance We can see another interaction of cytokinins and auxin in the control of apical dominance, the ability of the terminal bud to suppress development of axillary buds (see Chapter 31). In this case, the two hormones are antagonistic. Auxin transported down the shoot from the terminal bud restrains axillary buds from growing, causing a shoot to lengthen at the expense of lateral branching. If the terminal bud is removed, the plant may become bushier (Figure 35.6). However, cytokinins entering the shoot system from roots counter the action of auxin by signaling axillary buds to begin growing. Auxin cannot suppress growth of these buds once it has begun. Lower buds on a shoot usually break dormancy before buds closer to the apex, reflecting the relative distance away from auxin and cytokinin sources.

The check-and-balance control of lateral branching by auxin and cytokinins may be one way the plant coordinates the growth of its shoot and root systems. As roots become more extensive, the increased level of cytokinins would signal the shoot system to form more branches. The two hormones reverse their roles in the development of lateral roots; auxin stimulates and cytokinins inhibit the growth of branch roots.

Both auxin and cytokinins may regulate the growth of axillary buds indirectly by changing the concentration of still another hormone, ethylene. The levels of different nutrients in a bud may also affect its response to hormones.

Figure 35.6
Apical dominance. (**a**) Auxin from the apical bud inhibits the growth of axillary buds. This favors elongation of the shoot's main axis over lateral branching. Cytokinins, which are transported upward from roots, counter auxin, stimulating the growth of axillary buds. This explains why, in most plants, axillary buds near the shoot tip are less likely to grow than those closer to the roots. (**b**) Removal of the apical bud results in a bushier plant. Cytokinin, unchecked by auxin, stimulates the growth of axillary buds, which produce lateral branches of the shoot system.

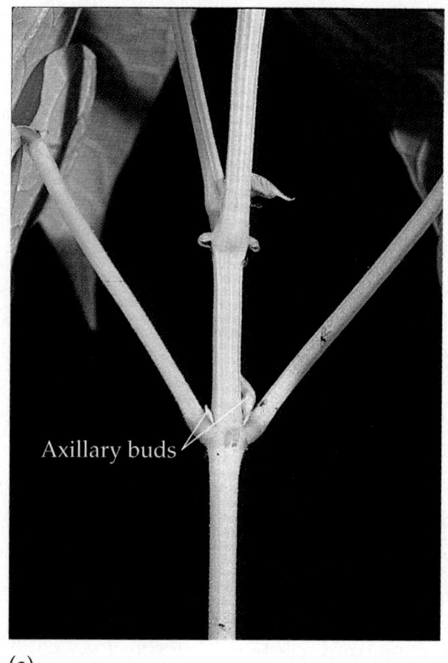

(a)

(b)

Cytokinins as Anti-Aging Hormones Cytokinins can retard the aging of some plant organs, perhaps by inhibiting protein breakdown, by stimulating RNA and protein synthesis, and by mobilizing nutrients from surrounding tissues. If leaves removed from a plant are dipped in a cytokinin solution, they stay green much longer than they otherwise would. It is probable that cytokinins also slow the deterioration of leaves on intact plants. Because of their anti-aging effect, cytokinin sprays are used to keep cut flowers fresh. They can also prolong the shelf life of fruits and vegetables after harvest, although this latter application has not yet been approved by the U.S. Food and Drug Administration (FDA).

Gibberellins

A century ago, rice farmers in Asia noticed some outrageously tall seedlings growing in their paddies. Before these rice seedlings could mature and flower, they grew so tall and spindly that they toppled over. In Japan, this aberration in growth pattern became known as *bakanae,* or "foolish seedling disease." In 1926, E. Kurosawa, a Japanese scientist, discovered that the disease was caused by a fungus of the genus *Gibberella.* By the 1930s, Japanese scientists had determined that the fungus produced hyperelongation of rice stems by secreting a chemical, which was given the name **gibberellin.** Western scientists finally learned of gibberellin after World War II. In the past 30 years, scientists have identified more than 70 different gibberellins, many of them occurring naturally in plants. All the gibberellins are subtle variations on a common molecular theme, but some forms are much more active than others in the plant. Foolish rice seedlings, it seems, suffer from an overdose of growth regulators normally found in plants in lower concentrations. Gibberellins have a variety of effects in plants.

Stem Elongation Roots and young leaves are major sites of gibberellin production. Gibberellins stimulate growth in both the leaves and the stem, but they have little effect on root growth. In a growing stem, gibberellins and auxin must be acting simultaneously in some synergistic manner we do not yet understand. It is known that gibberellins stimulate cell elongation *and* cell division in stems.

Enhancement of stem elongation by gibberellins can be seen by applying the hormones to certain dwarf varieties of plants. For instance, dwarf pea plants (including the variety Mendel studied; see Chapter 13) grow to normal height if treated with gibberellins. The extent of the growth of dwarf plants is generally correlated with the concentration of the added hormone (see the Methods Box on opposite page). If gibberellins are applied to plants of normal size, there is often no response. Apparently, these plants are already producing their own optimal dose of the hormone. Gibberellins may be at low levels or absent in dwarf varieties of some species, or the target cells may be less responsive to the hormones.

A specific case in which gibberellins cause rapid elongation of stems is **bolting,** the growth of a floral stalk. In their nonflowering stage, some plants develop a rosette form; that is, they are low to the ground with very short internodes. The plant switches to reproductive growth when a surge of gibberellins induces stems

METHODS: BIOASSAY

A technique that determines the concentration of a chemical by measuring the response of living material is called a bioassay. In this example, a plant's quantitative response to gibberellin is used to measure the concentration of the hormone. A sample of unknown concentration is applied to dwarf pea plants, and after a certain amount of growth time, their height is compared with dwarfs treated with a range of known gibberellin concentrations. Using the degree of coleoptile curvature in phototropism to determine auxin concentration is another example of a bioassay.

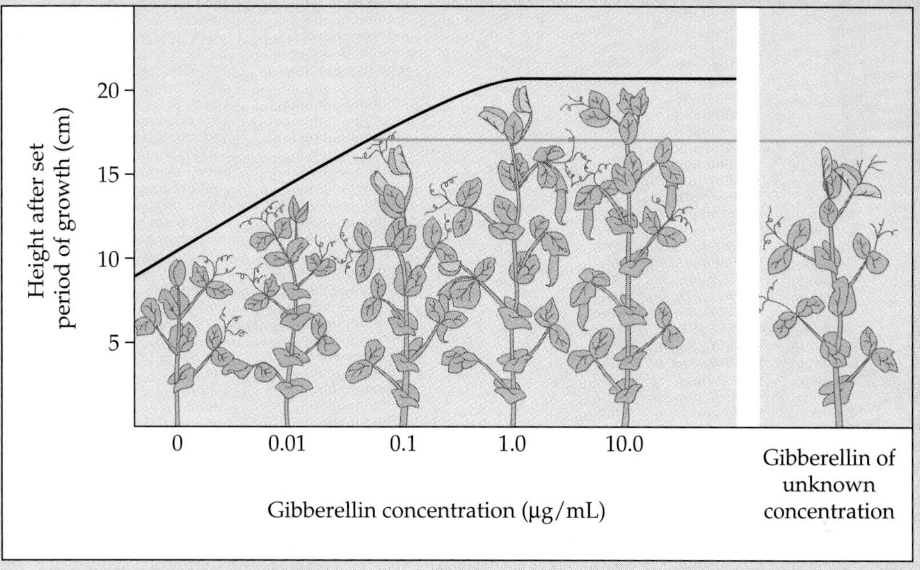

to elongate rapidly, which elevates flowers developing from buds at the tips of the stems. Cabbage plants have been selectively bred to maintain a rosette form, but a gibberellin treatment will produce bolting (Figure 35.7).

Fruit Growth Fruit development is another case in which we can observe dual control by auxin and gibberellins. In some plants, both hormones must be present for fruit to set. The most important commercial application of gibberellins is in the spraying of Thompson seedless grapes. The hormones cause the grapes to grow larger and farther apart.

Germination Many seeds have a high concentration of gibberellins, particularly in the embryo. After water is imbibed, the release of gibberellins from the embryo signals the seeds to break dormancy and germinate. Some seeds that require special environmental conditions to germinate, such as exposure to light or cold temperatures, will break dormancy if they are treated with a gibberellin solution. In nature, gibberellins in the seed are probably the link between the environmental cue and the metabolic processes that renew growth of the embryo.

Gibberellins support growth of cereal seedlings by stimulating the synthesis of digestive enzymes such as α-amylase that mobilize stored nutrients (see Chapter 34). Even before these enzymes appear, gibberellin stimulates the synthesis of messenger RNA coding for α-amylase. Here is a case in which a hormone controls development by affecting gene expression, although the hormone molecule may not be acting directly on the genome. (The last section of this chapter explores pathways that relay hormonal signals to genes.)

Gibberellins also function to break dormancy in the resumption of growth by apical buds in spring. In both seed dormancy and bud dormancy, gibberellins act antagonistically to another hormone, abscisic acid, which generally inhibits plant growth.

Figure 35.7
Bolting of cabbage induced by gibberellin treatment. As a result of selective breeding, cabbage shoots normally do not bolt unless artificially treated with large doses of gibberellin.

Abscisic Acid

The hormones we have studied so far—auxin, cytokinins, and gibberellins—usually stimulate plant growth. By contrast, there are times in the life of a plant when it is adaptive to slow down growth and assume a dormant state. The hormone **abscisic acid (ABA),** produced in the bud itself, slows growth and directs leaf primordia to develop into the scales that will protect the dormant bud during winter. The hormone also inhibits cell division in the vascular cambium. Thus, ABA helps prepare the plant for winter by suspending both primary and secondary growth. Abscisic acid was named when it was believed that the hormone caused the abscission (L. *ab,* "loss," and *caedere,* "cut") of leaves from deciduous trees during autumn, but no clear-cut role for ABA in abscission has been demonstrated.

Another stage in the life of a plant when it is advantageous to suspend growth is the onset of seed dormancy, and again it may be abscisic acid that acts as the growth inhibitor. The seed will germinate when ABA is overcome by its inactivation or removal or by increased activity of gibberellins. The seeds of some desert plants break dormancy when a heavy rain washes ABA out of the seed. Other seeds require light or some other stimulus to trigger the degradation of abscisic acid. (As noted in Chapter 34, many other environmental factors can affect germination in various plants.) In most cases, the ratio of ABA to gibberellins determines whether the seed will remain dormant or germinate. Similarly, dormancy of terminal buds is controlled more by a balance of growth regulators than by their absolute concentrations. In apple trees, for instance, the concentration of ABA is actually higher in growing buds than in dormant buds, but an excess of gibberellins overpowers the inhibitory hormone.

In addition to its role as a growth inhibitor, abscisic acid acts as a "stress" hormone, helping the plant cope with adverse conditions. For example, when a plant begins to wilt, ABA accumulates in leaves and causes stomata to close, reducing transpiration and preventing further water loss. In one variety of tomato that is deficient in ABA and suffers chronic wilting, the experimental addition of abscisic acid closes stomata by triggering potassium loss from the guard cells (see Chapter 32). In this response, we see how a small amount of hormone has its impact amplified by acting on a membrane.

Ethylene

Early in this century, citrus was ripened by "curing" the fruit in sheds equipped with kerosene stoves. Fruit growers believed it was the heat that ripened the fruit, but newer, cleaner-burning stoves did not work. Plant physiologists learned later that ripening in the sheds was actually due to **ethylene,** a gaseous by-product of kerosene combustion. It is now known that plants produce their own ethylene as a hormone, and that this hormone elicits a variety of responses in addition to fruit ripening. Unique among plant hormones because it is a gas, ethylene diffuses through the plant in the air spaces between cells. In some cases, ethylene acts to inhibit cell elongation. Many of the inhibitory effects once attributed to auxin are now believed to be the result of ethylene synthesis induced by a high concentration of auxin. It is probably ethylene, for example, that inhibits root elongation and the development of axillary buds in the presence of an excess of auxin. In addition to its role as a growth inhibitor, ethylene is also associated with a variety of aging processes in plants.

Senescence in Plants Aging, or **senescence,** is a progression of irreversible change that eventually leads to death. A normal part of plant development, senescence may occur at the level of individual cells, entire organs, or the whole plant. Tracheids and cork cells age and die before assuming their specialized functions. Autumn leaves and withering flower petals are examples of senescent organs. Plants that are annuals age and die soon after flowering. Ethylene probably has important functions in all these cases of senescence, but the aging processes where the effects of the hormone have been studied most extensively are fruit ripening and leaf abscission.

Fruit Ripening Several changes in structure and metabolism accompany the ripening of an ovary into a fruit (see Chapter 34). Some of these changes, including the degradation of cell walls, which softens the fruit, and the decrease in chlorophyll content, which causes loss of greenness, can be regarded as aging processes. Ethylene initiates or hastens these deteriorative changes and also causes some ripened fruits to drop from the plant. A chain reaction occurs during ripening, as ethylene triggers senescence, and the aging cells then release more ethylene. Because ethylene is a gas, the signal to ripen even spreads from fruit to fruit: One bad apple really does spoil the lot. If you pick or buy green fruit, you may be able to speed ripening by storing the fruit in a plastic bag, so that ethylene gas will accumulate. On a commercial scale, many kinds of fruit are ripened in huge storage containers into which ethylene gas is piped—a modern variation on the old curing shed. In other cases, measures are taken to retard ripening caused by natural ethylene. Apples, for instance, are stored in bins flushed with carbon dioxide. Circulating the air prevents ethylene from accumulating, and carbon dioxide somehow inhibits the action of whatever ethylene has not been flushed away. Stored in this way, apples picked in autumn can be shipped to grocery stores the following summer. Recently, a team of molecular biologists devised a method to manipulate the expression of one of the genes required for ethylene synthesis (Figure 35.8).

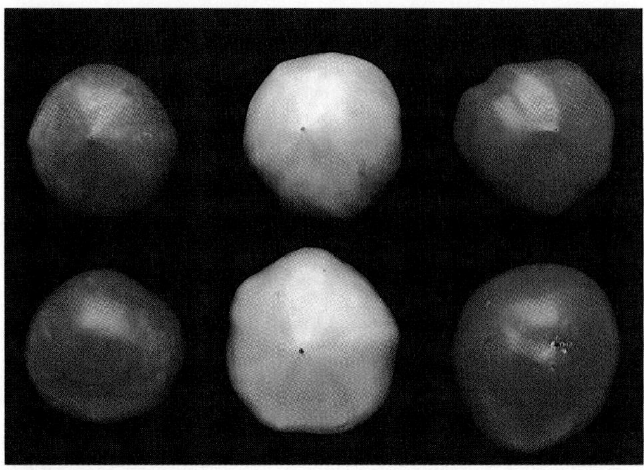

Figure 35.8
Artificial control of fruit ripening. The tomato fruits on the left have ripened naturally due to their production of the hormone ethylene. Molecular biologists suppressed ripening of the tomatoes in the center by adding an antisense RNA, which blocks transcription of one of the genes required for ethylene synthesis (see Chapter 19). The fruit can then be cued to ripen on demand by adding ethylene gas (tomatoes on the right). As such methods are refined, they may reduce spoilage of fruits and vegetables, a problem that currently ruins almost half the produce harvested in the United States.

Leaf Abscission The loss of leaves each autumn is an adaptation that keeps deciduous trees from desiccating during winter when the roots cannot absorb water from the frozen ground.

Before leaves are abscised, many of their essential elements are shunted to storage tissues in the stem. These nutrients are recycled back to developing leaves the following spring. The autumn leaf stops making new chlorophyll and loses its greenness. The fall colors are a combination of new pigments made during autumn and pigments that were already present in the leaf but concealed by the dark green chlorophyll.

When an autumn leaf falls, the breaking point is an abscission layer located near the base of the petiole (Figure 35.9). The small parenchyma cells of this layer have very thin walls, and there are no fiber cells around the vascular tissue. The abscission layer is further weakened when enzymes hydrolyze polysaccharides in the cell walls. Finally, the weight of the leaf, with the help of wind, causes a separation within the abscission layer. Even before the leaf falls, a layer of cork forms a protective scar on the twig's side of the abscission layer, preventing pathogens from invading the plant.

A change in the balance of ethylene and auxin controls abscission. An aging leaf produces less and less auxin, and this drop in concentration makes the cells of the abscission layer more sensitive to ethylene. This shift in hormonal balance is self-reinforcing, as cells in the abscission layer begin producing additional ethylene, which inhibits the leaf's synthesis of auxin. As the influence of ethylene on the abscission layer prevails, the cells produce enzymes that digest the cellulose and other components of cells walls.

The environmental stimuli for leaf abscission are the shortening days and cooler temperatures of autumn. This relationship between outside stimuli and the hormonal signals and responses within a plant brings us to an important juncture in our study of control systems. How do plants actually detect changes in their surroundings? Let's see how plant biologists are approaching this problem by studying plant movements.

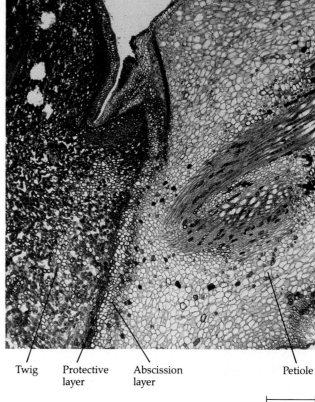

Figure 35.9
Abscission of a maple leaf. Abscission is controlled by a change in the balance of ethylene and auxin. The abscission layer can be seen here as a vertical band at the base of the petiole (LM). After the leaf falls, a protective layer becomes the leaf scar that helps prevent pathogens from invading the plant.

Twig
Protective layer
Abscission layer
Petiole
0.5 mm

PLANT MOVEMENTS

To the casual observer, most plants do not appear to be very dynamic, but time-lapse photography reveals they are capable of quite precise movements. The two types of plant movements are tropisms and turgor movements.

Tropisms

Environment has a great influence on a plant's shape. **Tropisms** (Gr. *tropos*, "turn") are growth responses that result in curvatures of whole plant organs toward or away from stimuli. The mechanism for a tropism is a differential rate of elongation of cells on opposite sides of the organ. Three of the stimuli that induce tropisms, and a consequent change of body shape, are light (phototropism), gravity (gravitropism), and touch (thigmotropism).

Phototropism We have already examined the hormonal basis for the growth of shoots toward light. Recall that cells on the darker side of a stem elongate faster than cells on the brighter side because of an asymmetric distribution of auxin moving down from the shoot tip. But what causes this unequal distribution of auxin? Experiments have demonstrated that illuminating a coleoptile from one side causes auxin to migrate laterally across the tip from the bright side to the dark side. It is not yet known how light induces this redistribution of auxin. More is known about the nature of the photoreceptor in the shoot tip, which is most sensitive to blue light and is believed to be a yellow pigment related to the vitamin riboflavin. The same receptor may be involved in stomatal opening and some other responses of plants to light.

Gravitropism Place a seedling on its side, and it will adjust its growth so that the shoot bends upward and the root curves downward. In their responses to gravity, roots display positive **gravitropism** and shoots exhibit negative gravitropism. Gravitropism functions as soon as a seed germinates, assuring that the root grows into the soil and the shoot reaches sunlight regardless of how the seed happens to be oriented when it lands.

Plants may tell up from down by the settling of **statoliths,** specialized plastids containing dense starch grains, to the low points of cells (Figure 35.10). In roots, the statoliths are located in certain cells of the root cap. According to one hypothesis, aggregation of statoliths at the low points of these cells triggers the redistribution of calcium, which in turn causes lateral transport of auxin within the root. The calcium and auxin accumulate on the lower side of the root's zone of elongation. Since these chemicals are dissolved, they do not respond to gravity but must be actively transported to

Figure 35.10
Statoliths and gravitropism. These corn roots were placed on their sides and photographed before (top) and 1.5 hours after the gravitropic response. The light micrographs show the locations of statoliths, modified plastids, within cells of the root cap. The settling of statoliths to the low points of these cells may be the gravity-sensing step that leads to the redistribution of auxin and the differential rates of elongation by cells on opposite sides of the root.

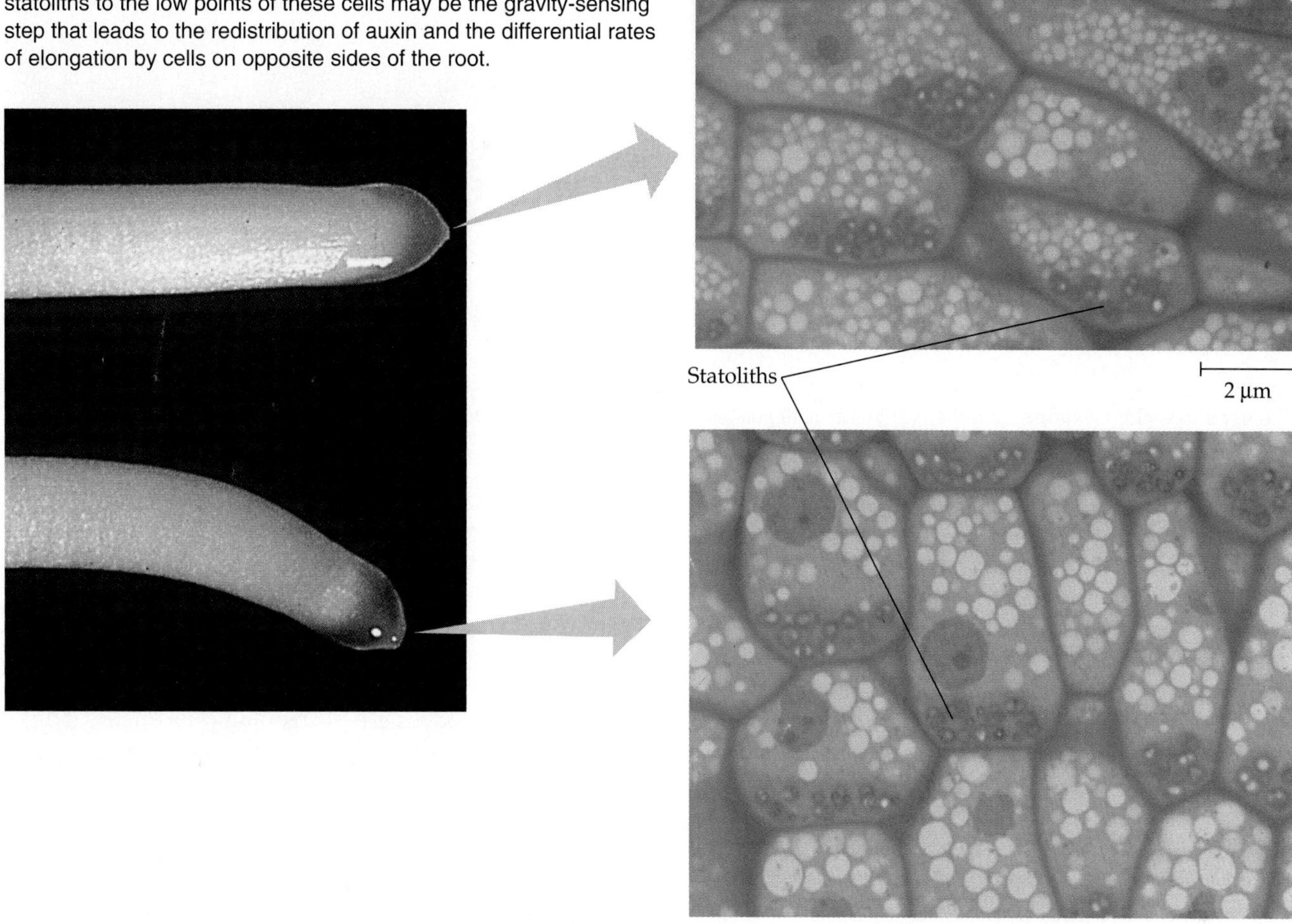

one side of the root. At high concentration, auxin inhibits cell elongation, an effect that slows growth on the lower side of the root. The more rapid elongation of cells on the upper side causes the root to curve as it grows. This tropism continues until the root is growing straight down. Less is known about the mechanism of gravitropism in shoots, and even the hypothesis for how roots grow down will be modified as research continues.

Thigmotropism Most vines and other climbing plants have tendrils that coil around supports. These grasping organs usually grow straight until they touch something; the contact stimulates a coiling response caused by differential growth of cells on opposite sides of the tendril. This directional growth in response to touch is called **thigmotropism** (Gr. *thigma,* "touch").

Mechanical stimulation can also cause a much more general response. One experiment has shown that rubbing stems with a stick a few times results in plants that are shorter and thicker than controls that are not as molested. In the wild, wind can cause a similar stunting of growth, enabling the plant to hold its ground against strong gusts. A tree growing on a windy mountain ridge, for instance, will usually have a shorter, stockier trunk than a tree of the same species growing in a more sheltered location. This decrease in stem length with concomitant thickening caused by mechanical perturbation is referred to as **thigmomorphogenesis.** It usually results from an increased production of ethylene in response to chronic mechanical stimulation.

Turgor Movements

In addition to the relatively long-lasting changes in body shape resulting from tropisms, plants are also capable of reversible movements caused by changes in the turgor pressure of specialized cells in response to stimuli.

(a)

(b)

Figure 35.11
The sensitive plant *(Mimosa pudica).* (**a**) In the unstimulated plant, leaflets are spread apart. (**b**) Within a second or two of being touched, the leaflets have folded together.

Rapid Leaf Movements When the compound leaf of the sensitive plant *Mimosa* is touched, it collapses and its leaflets fold together (Figure 35.11). This response, which takes only a second or two, results from a rapid loss of turgor by cells within pulvini, specialized motor organs located at the joints of the leaf. The motor cells suddenly become flaccid after stimulation because they lose potassium, which causes water to leave the cells by osmosis. It takes about ten minutes for the cells to regain their turgor and restore the natural form of the leaf. The function of the sensitive plant's behavior invites speculation. Perhaps by folding its leaves and reducing its surface area when jostled by strong winds, the plant conserves water. Or maybe, since collapse of the leaves exposes thorns on the stem, the rapid response of the sensitive plant discourages herbivores.

A remarkable feature of rapid leaf movements is the transmission of the stimulus through the plant. If one leaf on a sensitive plant is touched with a hot needle, first that leaf collapses, then the adjacent leaf responds, then the next leaf along the stem, and so on until all the plant's leaves are drooping. From the point of stimulation, the message that produces this response travels wavelike through the plant at a speed of about a centimeter per second. Chemical messengers probably have a role in this transmission, but an electrical impulse can also be detected by attaching electrodes to the plant. These impulses, called **action potentials,** resemble nervous messages in animals, though the action potentials of plants are thousands of times slower than those of animals. Action potentials, which have been discovered in many species of algae and plants, may be widely used as a form of internal communication. Another example is the Venus flytrap, in which action potentials are transmitted from sensory hairs in the trap to the cells that respond by closing the trap.

Sleep Movements Bean plants and many other members of the legume family lower their leaves in the evening and raise them to a horizontal position in the morning (Figure 35.12). These **sleep movements** are powered by daily changes in the turgor pressure of motor cells in pulvini similar to those of the sensitive plant. When leaves are horizontal, cells on one side of the pulvinus are turgid, whereas cells on the opposite

Figure 35.12
Sleep movements of a bean plant. Leaf position at noon (top) and leaf position at midnight (bottom). The movements are caused by reversible changes in the turgor pressure of cells on opposing sides of the pulvini, swollen regions of the petiole.

side are flaccid. This is reversed at night when the leaves close to their "sleeping" position. Paralleling the opposing changes in volume of the motor cells is a massive migration of potassium ions from one side of the pulvinus to the other. Apparently, the potassium is an osmotic agent that leads to the reversible uptake and loss of water by the motor cells. In this respect, the mechanism of sleep movements is similar to stomatal opening and closing. Sleep movements are only one example of the many responses that depend on the ability of plants to keep track of time.

CIRCADIAN RHYTHMS AND THE BIOLOGICAL CLOCK

Your pulse, blood pressure, temperature, rate of cell division, blood cell count, alertness, urine composition, metabolic rate, sex drive, and responsiveness to medications all fluctuate with the time of day. Some insects are more vulnerable to insecticides in the afternoon than in the morning. Certain fungi produce spores for several hours the same time each day. Unicellular algae that glow in the dark switch on their bioluminescence like clockwork. Plants also display rhythmic behavior—the sleep movements of legumes and the opening and closing of stomata, for example. All these rhythmic phenomena and many others are controlled by biological clocks, internal oscillators that keep accurate time. Biological clocks seem to be ubiquitous features of eukaryotic organisms, and our first evidence for biological rhythms came from studies of plants.

A physiological cycle with a frequency of about 24 hours is called a **circadian rhythm** (L. *circa,* "approximately," and *dies,* "day"). Are these rhythms truly prompted by an internal clock, or are they merely daily responses to some environmental cycle, such as the rotation of the Earth? Circadian rhythms persist even when the organism is sheltered from environmental cues. A bean plant, for example, will continue its sleep movements even if kept in constant light or constant darkness; the leaves are not simply responding to sunrise and sunset. Organisms, including humans, continue their rhythms when placed in the deepest mine shafts or orbited in satellites. All research thus far indicates that the oscillator for circadian rhythms is endogenous (internal). This clock, however, is entrained (set) to a period of precisely 24 hours by daily signals from the environment. If an organism is kept in a constant environment, circadian rhythms deviate from a 24-hour period (a "period" is the duration of one cycle). These free-running periods, as they are called, vary from about 21 to 27 hours, depending on the particular rhythmic response. The sleep movements of bean plants, for instance, have a period of 26 hours when kept under the free-running conditions of constant darkness.

Deviation of the free-running period from exactly 24 hours does not mean that biological clocks drift erratically. The clocks are still keeping perfect time, but they are not synchronized with the outside world. The light-dark cycle due to the Earth's rotation is the most common factor that entrains biological clocks. If the timing of these cues changes, it takes a few days for the clock to be reset. Thus, a plant kept for several days in the dark will be out of phase with plants growing in a normal environment where the sun rises and sets each day. The same thing happens when we cross several time zones in an airplane; when we reach our destination, the clocks on the wall are not synchronized with our internal clocks. All eukaryotes are probably prone to jet lag.

Most biologists now agree that organisms possess built-in clocks, but the nature of the internal oscillator is still unknown. Where is the clock, and how does it work? In attempting to answer these questions, we must take care to differentiate between the oscillator and the rhythmic processes it controls. The sweeping leaves of sleep movements are the "hands" of the biological clock, but these movements are not the essence of the clockwork itself. If the leaves of a bean plant are restrained for several hours so they cannot move, they will, on release, rush to the position appropriate for the time of day. We can interfere with a biological rhythm, but the clock goes right on ticking off the time. Most scientists who study circadian rhythms place the clock at the cellular level, either in membranes or in the machinery for protein synthesis.

PHOTOPERIODISM

Seasonal events are important in the life cycles of most plants. Seed germination, flowering, and the onset and breaking of bud dormancy are examples of stages in plant development that usually occur at specific times of the year. The environmental stimulus plants most often use to detect the time of year is the photoperiod, the relative lengths of night and day. A physiological response to day length, such as flowering, is called **photoperiodism.**

Photoperiodic Control of Flowering

One of the earliest clues to how plants detect the progression of seasons came from an unusual variety of tobacco studied by W. W. Garner and H. A. Allard in 1920. This variety, named Maryland Mammoth, grew exceptionally tall but failed to flower during summer when normal tobacco plants flowered. Maryland Mammoth finally bloomed in a greenhouse in December. After trying to induce earlier flowering by varying temperature, moisture, and mineral nutrition,

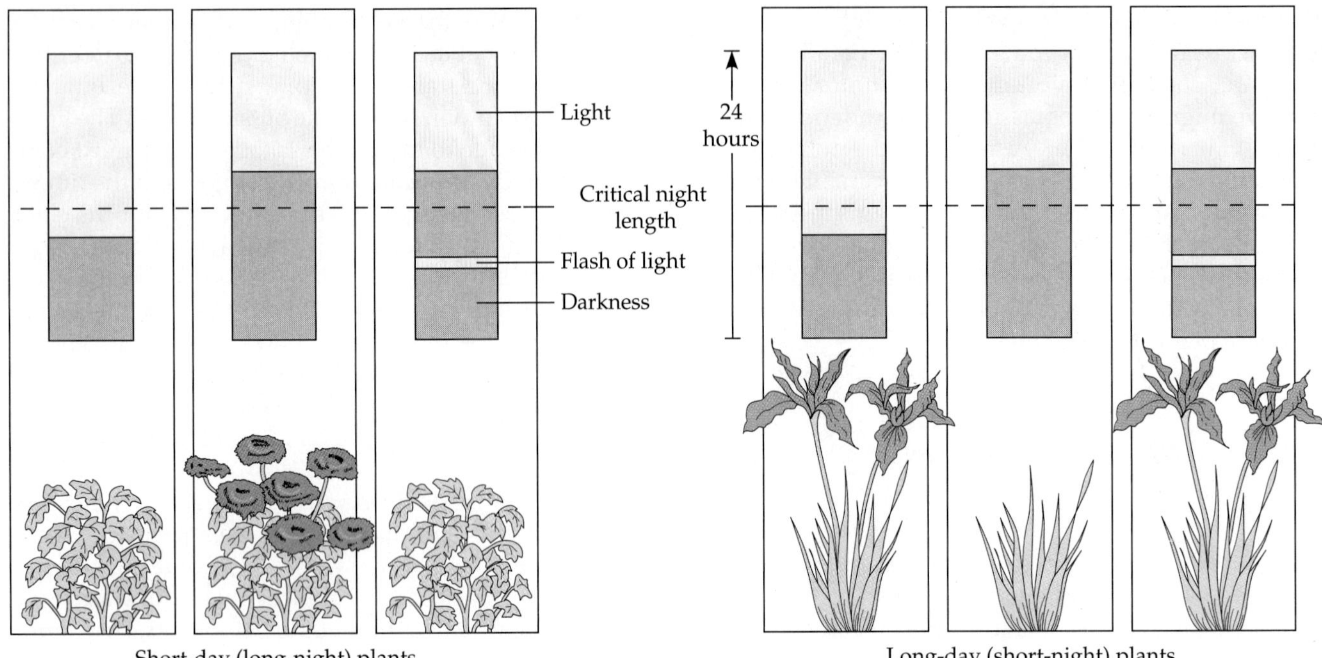

Figure 35.13
Photoperiodic control of flowering. A short-day (long-night) plant flowers when night exceeds a critical dark period. A flash of light interrupting the dark period prevents flowering. A long-day (short-night) plant flowers only if the night is shorter than a critical dark period. The night can be artificially shortened with a flash of light.

Garner and Allard learned that it was the shortening days of winter that stimulated Maryland Mammoth to flower. If the plants were kept in light-tight boxes so that lamps could be used to manipulate durations of "day" and "night," flowering would occur only if the day length were 14 hours or shorter. The Maryland Mammoths did not flower during summer because, at Maryland's latitude, the days were too long during that season.

Garner and Allard termed Maryland Mammoth a **short-day plant,** because it apparently required a light period *shorter* than a critical length to flower. Chrysanthemums, poinsettias, and some soybean varieties are a few of the other short-day plants, which generally flower in late summer, fall, or winter. Another group of plants dependent on photoperiod will flower only when the light period is *longer* than a certain number of hours. These **long-day plants** generally flower in late spring or early summer. Spinach, for example, flowers when days are 14 hours or longer. Radish, lettuce, iris, and many cereal varieties are also long-day plants. Flowering in a third group, **day-neutral plants,** is unaffected by photoperiod. Tomatoes, garden peas, rice, and dandelions are examples of day-neutral plants that flower when they reach a certain stage of maturity, regardless of day length at that time.

Critical Night Length In the 1940s, it was discovered that night length, not day length, actually controls flowering and other responses to photoperiod. Researchers worked with cocklebur, a short-day plant that flowers only when days are 16 hours or less in length (and nights are at least 8 hours long). If the daytime portion of the photoperiod is broken by a brief exposure to darkness, there is no effect on flowering. However, if the nighttime part of the photoperiod is interrupted by even a few minutes of dim light, the plants will not flower (Figure 35.13). Cocklebur requires at least 8 hours of *continuous* darkness to flower. Short-day plants are really long-night plants, but the older term is embedded firmly in the jargon of plant physiology. Long-day plants are actually short-night plants; grown on photoperiods of long nights that would not normally induce flowering, long-day plants will flower if the period of continuous darkness is shortened by a few minutes of light at any point. Thus, photoperiodic responses depend on a critical night length. Short-day plants will flower if the duration of night is *longer* than the critical length (8 hours for cocklebur), and long-day plants will flower when the night is *shorter* than the critical length. The floriculture (flower-growing) industry has applied this knowledge to produce flowers out of season. Chrysanthemums, for instance, are short-day plants that normally bloom in fall, but their blooming can be stalled until Mother's Day in May by punctuating each long night with a flash of light, thus turning one long night into two short nights.

Notice that we distinguish long-day from short-day plants *not* by an absolute night length but by whether the critical night length sets a maximum (long-day plants) or minimum (short-day plants) number of hours of darkness required for flowering.

Some plants bloom after a single exposure to the photoperiod required for flowering. Other species need several successive days of the appropriate photoperiod. Still other plants will respond to photoperiod only if they have been previously exposed to some other environmental stimulus, such as a period of cold temperatures. Winter wheat, for example, will not flower unless it has been exposed to several weeks of temperatures below 10°C. This requirement for pretreatment with cold before flowering is called **vernalization.** Several weeks after winter wheat is vernalized, a photoperiod with long days (short nights) induces flowering.

Evidence for a Flowering Hormone Buds produce flowers, but the photoperiod is detected by leaves. To induce a short-day plant or long-day plant to flower, it is enough in many species to expose a single leaf to the appropriate photoperiod. Indeed, if only one leaf is left attached to the plant, photoperiod is detected and floral buds are induced. If all leaves are removed, however, the plant is blind to photoperiod. Apparently, some message to flower is transported from leaves to buds. Most plant physiologists believe this message is a hormone, which has been named florigen (Figure 35.14). The signal to flower that travels from leaves to buds appears to be the same for short-day and long-day plants, although the two groups of plants differ in the photoperiodic conditions required for leaves to send this signal.

The evidence for a flowering hormone is compelling, but so far florigen has not been identified. It is possible that the flowering impulse is not a single chemical but a specific mixture of several hormones.

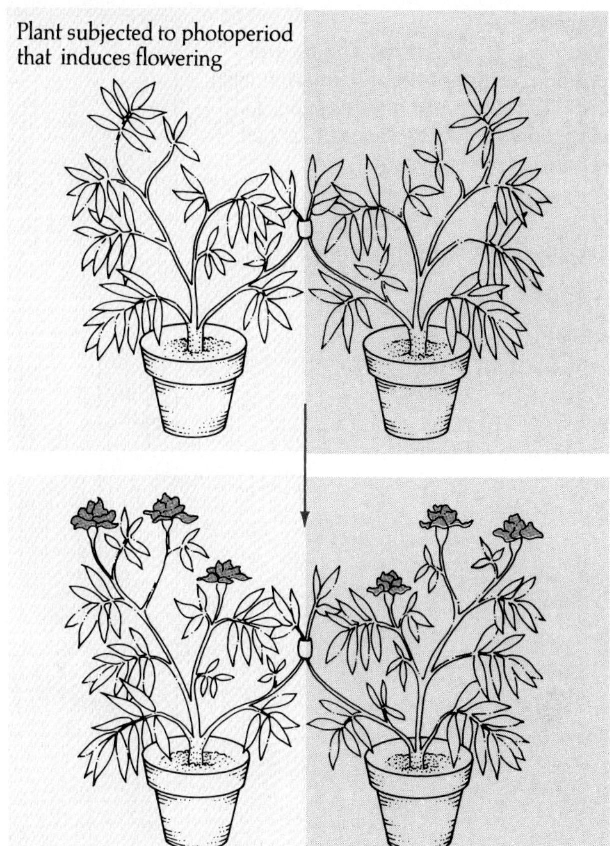

Figure 35.14
Evidence for a hormone that induces flowering. If a plant that has been induced to flower by photoperiod is grafted to a plant that has not been induced, both plants flower. This works in some cases even if one is a short-day plant and the other is a long-day plant.

Phytochrome

The discovery that night length is a critical factor controlling seasonal responses of plants poses another question: How does a plant measure the length of darkness in a photoperiod? A pigment named **phytochrome** is part of the answer. It was discovered as a result of studies on how different colors of light affect flowering, seed germination, and other responses to photoperiod.

Red light, having a wavelength of 660 nanometers (nm), is the light that is most effective in interrupting night length. A short-day plant kept under conditions of critical night length fails to flower if a brief exposure to red light breaks the dark period. Conversely, a flash of red light during the dark period will induce a long-day plant to flower even if the total night length exceeds the critical number of hours (long-day plants, remember, require nights shorter than a critical length). The effect of the red flash is to shorten the plant's perception of night length.

The shortening of night length by red light can be negated by a subsequent flash of light having a wavelength of about 730 nm. This wavelength is in the far-red part of the spectrum, just barely visible to humans. If red light (R) given during the dark period is followed by far-red (FR) light, the plant perceives no interruption in night length. A short-day plant will not flower if a night of critical length is broken by an R flash, but the plant *will* flower if it receives two flashes, first R

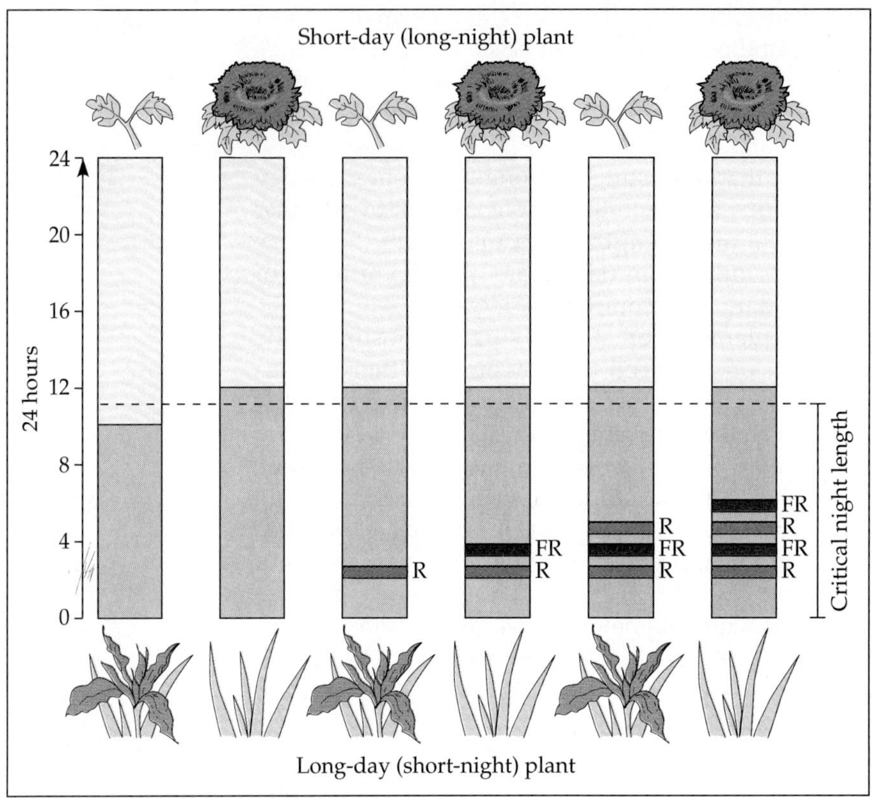

Figure 35.15
Reversible effects of red and far-red light on photoperiodic response. A flash of red light shortens the dark period. A subsequent flash of far-red light cancels the effect of the red flash.

and then FR. Reversing this sequence to FR-R prevents flowering. Each wavelength of light cancels the effect of the other. No matter how many flashes of light are given, the wavelength of only the last one affects the plant's measurement of night length. Thus, a succession of flashes with the sequence R-FR-R-FR-R prevents short-day plants from flowering, but flowering occurs if the sequence is R-FR-R-FR-R-FR. As expected, the opposite behavior occurs in long-day plants (Figure 35.15).

The photoreceptor responsible for the reversible effects of red and far-red light is phytochrome, a protein containing a chromophore (the light-absorbing portion). Phytochrome alternates between two forms that differ only slightly in structure, but one absorbs red light and the other absorbs far-red light. The two variations of phytochrome—P_r (red absorbing) and P_{fr} (far-red absorbing)—are said to be photoreversible:

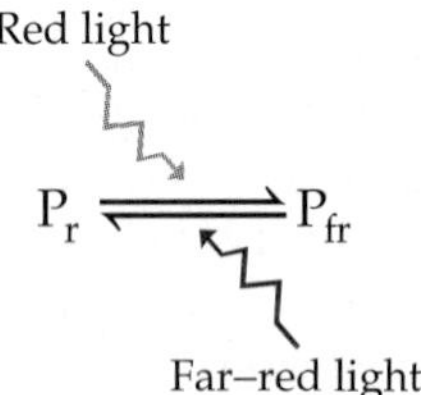

This $P_r \rightleftharpoons P_{fr}$ interconversion acts as a switching mechanism controlling various events in the life of the plant (Figure 35.16).

One function of the phytochrome system is to tell the plant that light is present. Plants synthesize phytochrome as P_r, and if they are kept in the dark, the pigment remains in that form. Then, if the phytochrome is illuminated in sunlight, some of the P_r is converted to P_{fr} because the pigment is exposed to red light (along with all the other wavelengths in sunlight) for the first time. This appearance of P_{fr} is one way plants detect sunlight. P_{fr} is the form of phytochrome that triggers many of a plant's developmental responses to light, such as the germination of seeds that require light to break dormancy. The phytochrome system also provides the plant with information about the *quality* of light. Sunlight includes both red and far-red radiation. Thus, during the day, the $P_r \rightleftharpoons P_{fr}$ photoreversion reaches a dynamic equilibrium, with the ratio of the two phytochrome forms reflecting the relative amounts of red and far-red light. This sensing mechanism enables plants to adapt to changes in light conditions. Consider, for example, a "shade-avoiding" tree that requires relatively high light intensity. If this tree becomes shaded by other trees in a forest, the phytochrome ratio shifts in favor of P_r because the forest canopy screens out more red light than far-red light. This cue induces the tree to use most of its resources to grow taller.

Other photoreceptors complement phytochrome in tuning a plant's growth and development to its environment. For example, we have already learned about

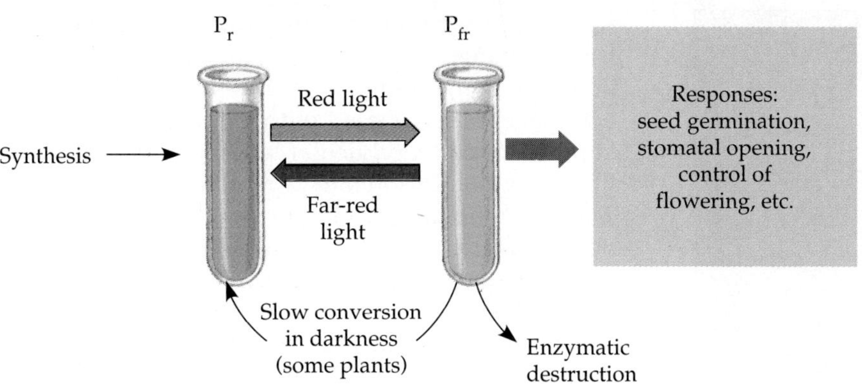

Figure 35.16
Phytochrome. The test tubes in this diagram contain solutions of the two photoreversible forms of phytochrome. Absorption of red light causes the bluish P_r to change to the blue-greenish P_{fr}. Far-red light reverses this conversion. In most cases, it is the P_{fr} form of the pigment that triggers physiological responses in the plant.

a blue-absorbing pigment in shoot tips that functions as the receptor in phototropism.

The Role of the Biological Clock in Photoperiodism

In darkness, the phytochrome ratio shifts gradually in favor of the P_r form. This is partly due to turnover in the overall phytochrome pool. The pigment is synthesized in the P_r form, and degradative enzymes destroy more P_{fr} than P_r (see Figure 35.16). In addition, in some plant species, P_{fr} present at sundown slowly converts to P_r by a biochemical mechanism. In darkness, there is no means for the accumulating P_r to be reconverted to P_{fr}. Then at sunrise, the P_{fr} level suddenly increases again by rapid photochemical conversion from P_r. Phytochrome conversion marks the beginning and end of the dark segment of a photoperiod. Does the plant use these signals to measure night length, the environmental variable that controls flowering and other responses to photoperiod? Perhaps the gradual conversion of P_{fr} to P_r in the dark is a chemical hourglass that gauges the passage of night. However, the conversion is usually completed within a few hours after sunset. If the plant used the disappearance of P_{fr} to measure the length of the dark period, it would lose track of time during the middle of the night. Night length is measured not by phytochrome but by the biological clock. The role of phytochrome in photoperiodism may be to synchronize the clock to the environment by telling it when the sun sets and rises.

If the photoperiodic requirement for flowering is met, the clock sounds some sort of alarm that causes leaves to send a flowering stimulus (perhaps florigen) to buds. Night length is measured very accurately; some short-day plants will not flower if night is even one minute shorter than the critical length. Some plant species always flower on the same day each year. According to the hypothesis described here, plants tell the season of year by using the clock, apparently entrained with the help of phytochrome, to keep track of photoperiod.

We complete our study of control systems in plants by investigating how hormonal and environmental signals are relayed within cells from the points of reception to the cellular components that actually respond.

SIGNAL-TRANSDUCTION PATHWAYS IN PLANT CELLS

In 1990, Stanford University researchers studying *Arabidopsis* growth made a serendipitous discovery. They sprayed plants with hormones to see if there was any effect on gene expression. Indeed, the treatment increased transcription of five specific genes. But when control plants were sprayed with water, activity of the same five genes increased. In fact, merely touching the plants a couple of times a day had the same effect. It was mechanical stimulation, not the hormones or water, that induced transcription of these specific genes. Touching the plants also affected morphology. Mechanically stimulated plants grew stockier and less tall than unstimulated plants (Figure 35.17). Recall that such an effect of touch on plant development is called thigmomorphogenesis. It is a reasonable hypothesis that the genes induced by touching the plants code for proteins that function in the growth response. Somehow, the touch stimulus must be transduced into an intracellular signal that is relayed to nuclei, altering gene expression. Such a mechanism linking stimulus to response in cells is called a **signal-transduction pathway.**

All the hormones and environmental stimuli you have learned about in this chapter act on plant cells via signal-transduction pathways. Plant biologists are just beginning to work out the steps of these pathways, but it is already apparent that the basic mechanisms resemble pathways that transduce signals in animal

Figure 35.17
Altering gene expression by touch in *Arabidopsis.* The stocky plant on the left was touched twice a day. Compare it to the unmolested plant on the right, which grew much taller. Researchers have discovered that this developmental response to touch is associated with the induction of five specific genes. A signal-transduction pathway is activated by mechanical stimulation of cells and relays the message to respond to the genome in the cell nuclei.

cells. We started this chapter by contrasting how plants and animals cope with environmental change. But when we trace their behavior to the cellular level, the similarities are remarkable.

We can dissect a signal-transduction pathway into three main steps: reception, transduction, and induction. Reception is the cell's detection of an environmental signal or hormone. In the case of responses to light, for example, the reception step is the absorption of light of specific wavelength by a pigment within the cell. Photoconversion of phytochrome is this type of reception. For responses to hormones, the reception step is the binding of the hormone to a specific receptor, usually a protein molecule (Figure 35.18). If the receptor is built into the plasma membrane, the hormone can trigger changes in the cell without even entering. In other cases, the hormone receptor is dissolved in the cytoplasm or associated with a particular organelle within the cell. A particular hormone can only affect those types of cells that have receptors for that signal. These are the **target cells** for that hormone. Even if other cells are exposed to the hormone, they do not respond, because there is no means to detect the stimulus. This may be one reason, for example, that cells in a shoot's zone of elongation are especially sensitive to the growth-promoting action of auxin. Thus, the reception step is one key to the specificity of responses to hormones.

The transduction step in a signal-transduction pathway amplifies the stimulus and converts it to a chemical form capable of activating the cell's responses. The critical link in transduction is a **second messenger,** some substance that increases in concentration within a cell that is stimulated by the first messenger—a hormone, for example. Binding of a hormone to its specific receptor evokes the second message. Even if the receptor is a membrane protein at the cell surface, the hormonal signal can be relayed to intracellular sites by the second messenger. For example, binding of the hormone may activate the receptor or associated membrane proteins to catalyze a chemical reaction that produces the second message on the cytoplasmic side of the membrane. Evidence is accumulating that calcium ions (Ca^{2+}) function as the second message in many plant responses. In these cases, a hormone or environmental stimulus triggers an increase in cytoplasmic Ca^{2+}. The signal may release the ion from organelles, or it may open Ca^{2+} channels in the plasma membrane, facilitating a rush of extracellular Ca^{2+} into the cell. The second messenger, Ca^{2+}, then binds to a specific protein called **calmodulin.** In a chain reaction, the calmodulin-Ca^{2+} complex activates other target molecules within the cell. This is probably the way that red light acts on phytochrome-containing cells. Absorption of red light converts P_r to P_{fr}. The P_{fr} then acts on the membrane to raise cytoplasmic Ca^{2+} concentration. The Ca^{2+} combines with calmodulin, which triggers a cascade of protein activations that are manifest as the phytochrome-mediated responses we observe.

Notice that signal transduction amplifies a stimulus. Binding of a single hormone molecule to a receptor can give rise to a large population of second messengers, which can activate an even larger number of proteins and other cellular molecules. This signal transduction contributes to the specificity of response. Two cell types that both have receptors for a particular hormone may respond differently, because the cells contain different target proteins that key on the second messengers. Although we have used the calcium burst as an example, there are other second messengers (see Chapter 41).

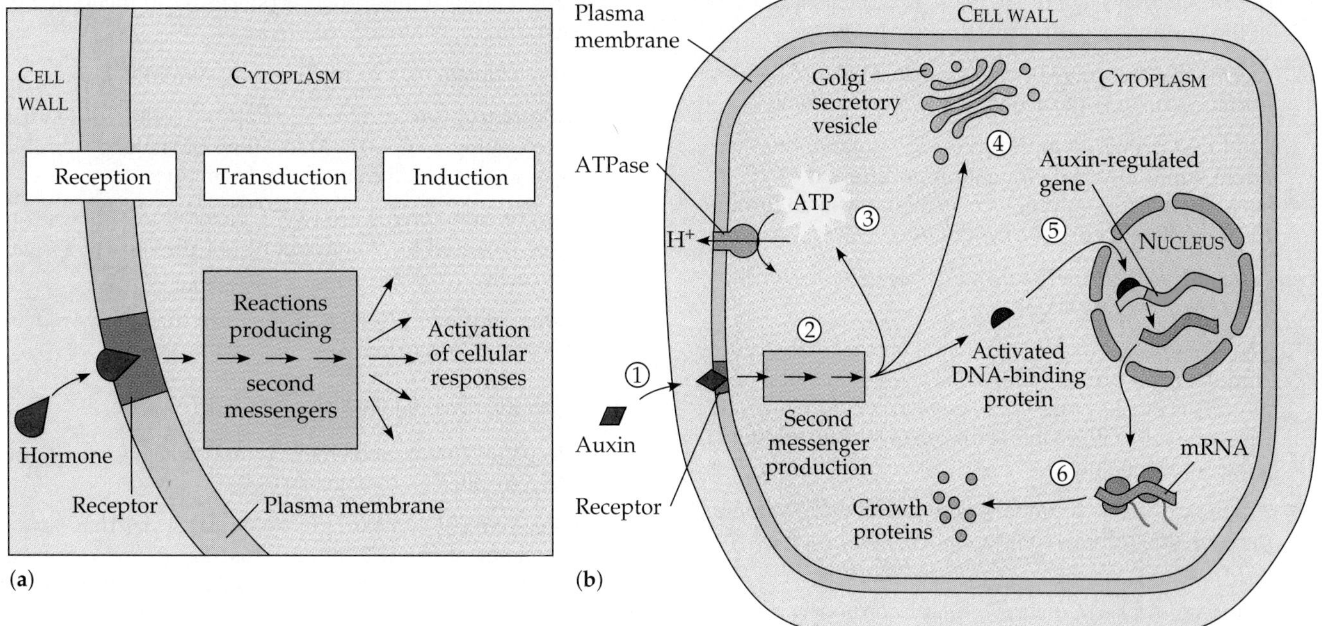

Figure 35.18
Signal-transduction pathways in plant cells. (**a**) General model. Hormone binding to a specific receptor activates chemical and transport steps that generate second messengers, which trigger the cell's various responses to the original signal. In this diagram, the receptor is on the surface of the target cell. In other cases, hormones enter cells and bind to specific receptors inside. Environmental stimuli can also initiate signal-transduction pathways. For example, phytochrome conversion is the first step in the transduction pathways that lead to a cell's responses to red light. (**b**) Specific example: a hypothetical mechanism for auxin's stimulation of cell elongation. ① The hormone binds to an auxin receptor, and ② this signal is transduced into second messengers within the cell, inducing various reponses. ③ Proton pumps are activated, and secretion of acid loosens the wall, enabling the cell to elongate. ④ The Golgi is stimulated to discharge vesicles containing materials to maintain the thickness of the cell wall. ⑤ The signal-transduction pathway also activates DNA-binding proteins that induce transcription of specific genes. ⑥ This leads to the production of proteins required for sustained growth of the cell.

In the third main step of a signal-transduction pathway, the induction step, the amplified signal induces the cell's specific responses to the stimulus. Some of these responses are relatively rapid. For example, abscissic acid can cause stomata to close within minutes by initiating a signal-transduction pathway that induces K^+ efflux from guard cells. Another example is the auxin-induced acidification of cell walls that causes cell elongation. Other responses take longer, usually because they require changes in gene expression. You learned earlier in the chapter, for instance, that auxin stimulates longer-term elongation of cells by inducing specific genes that code for proteins functioning in cell wall synthesis. And you have also learned that touch, acting on the genome through a signal-transduction pathway, alters the morphology of plants. Researchers investigating these mechanisms are getting to the heart of how a plant adapts to its environment. These scientists, along with thousands of other plant biologists working on other problems and millions of students experimenting with plants, are all extending a centuries-old tradition of curiosity about the form and function of the organisms that feed the biosphere.

STUDY OUTLINE

1. Plants adjust their growth and development in response to environmental cues.
2. Experiments on phototropism led to identification of the first plant hormone, auxin.

The Search for a Plant Hormone: A Case Study in the Scientific Process (pp. 757–758)

1. Hormones are traveling chemical messengers that coordinate functions between distant parts of an organism.

Functions of Plant Hormones (pp. 759–765)

1. Five known classes of hormones control plant growth and development by multiple effects on division, elongation, and differentiation of cells. The site of action, the

developmental stage of the plant, and the concentration of the hormone all affect reaction to a hormone.

2. Hormone effects may be due more to a balance between hormones than to the action of any one hormone.
3. Produced primarily in the apical meristem of the shoot, auxin stimulates cell elongation in different target tissues. Higher concentrations may inhibit growth through the induction of ethylene synthesis.
4. A polar transport mechanism moves auxin unidirectionally down a young shoot.
5. The acid-growth hypothesis maintains that auxin causes stimulation of proton pumps that acidify the cell wall region. This breaks cross-links between cellulose microfibrils in the cell wall, enabling the cell to take up additional water and elongate.
6. Auxin also affects secondary growth and differentiation, initiates adventitious root formation, and promotes fruit growth.
7. Cytokinins, a second class of plant hormones, are adenine derivatives that stimulate cell division.
8. Actively growing tissues, such as roots, embryos, and fruits, are rich in cytokinins, which, in concert with auxin, stimulate cell division and influence differentiation. Subtle changes in the cytokinin/auxin ratio have specific effects on plant development.
9. Cytokinins and auxin also work together to control apical dominance in a complex interaction.
10. Cytokinins retard the aging process in some plant organs by prolonging protein synthesis and mobilizing nutrients.
11. Gibberellins produced in roots and young leaves stimulate growth in leaves and stems. In conjunction with auxin, gibberellins increase elongation in stems.
12. Gibberellins and auxin also stimulate fruit development.
13. Gibberellins support germination by stimulating the synthesis of key enzymes involved in the mobilization of seed storage material.
14. Abscisic acid slows plant growth and favors the dormant state by inducing development of bud scales, inhibiting cell division in the vascular cambium, and suspending growth in buds and seeds. Whether buds and seeds break dormancy depends on the ratio of abscisic acid to gibberellins. Abscisic acid is also a "stress" hormone that helps plants cope with adverse conditions.
15. Ethylene, a gaseous hormone, diffuses through the plant in air spaces. It inhibits root growth and development of axillary buds in the presence of high auxin concentrations. Ethylene also stimulates fruit ripening and induces several aspects of senescence in plant cells and organs.
16. The mechanics of leaf abscission involve decreased auxin and increased ethylene production.

Plant Movements (pp. 766–769)

1. Tropisms are growth responses that result in curvature of entire plant organs toward or away from stimuli.
2. Light causes phototropic responses by inducing the redistribution of auxin.
3. Gravitropism may be mediated by statoliths.
4. Thigmotropism leads to an adaptive coiling of tendrils on touching a support. Thickening of stems by chronic, strong winds is an example of thigmomorphogenesis.
5. Turgor movements are rapid, reversible responses to stimuli caused by changes in turgor pressure of specialized cells.
6. Sleep movements of legumes are also powered by changes in turgor pressure in pulvini.

Circadian Rhythms and the Biological Clock (p. 769)

1. Sleep movements and other rhythmic behaviors of plants are controlled by biological clocks.
2. Circadian rhythms are physiological cycles that have a frequency of about 24 hours. Absence of environmental cues leads to free-running periods, in which the rhythms may deviate by a few hours from a 24-hour period but are still precise within their own cycles. The light-dark cycle probably entrains biological clocks.

Photoperiodism (pp. 769–773)

1. Photoperiodism, the response to relative lengths of night and day, helps regulate stages of plant development.
2. Photoperiodic control of flowering actually depends on a critical night length. This sets a minimum (short-day plants) or maximum (long-day plants) number of hours of darkness required for flowering. Flowering in day-neutral plants is unaffected by photoperiod.
3. A signal from the leaves, postulated by many to be a hormone called florigen, causes flowering of buds.
4. Phytochrome, which exists in two photoreversible states, is one factor that signals sunrise and sunset. Actual night length is measured by the biological clock.

Signal-Transduction Pathways in Plant Cells (pp. 773–775)

1. Hormones stimulate cells by binding to specific receptors (reception).
2. Reception triggers production of second messengers within the cell (transduction).
3. Second messengers activate a chain of events that mobilize the cell's specific response (induction).

SELF-QUIZ

1. Which of the following plant hormones is incorrectly paired with its function?
   a. auxin—promotes stem growth through cell elongation
   b. cytokinin—initiates senescence
   c. gibberellin—stimulates seed and bud germination
   d. abscisic acid—promotes seed and bud dormancy
   e. ethylene—inhibits cell elongation
2. Spraying some plants with a combination of auxin and gibberellins

a. promotes fruit growth
b. kills broad-leaf dicot plants
c. prevents senescence
d. promotes fruit ripening
e. is used to treat dwarfism in plants

3. Buds and sprouts often form on tree stumps. Which of the following hormones would you expect to stimulate their formation?
   a. auxin
   b. cytokinins
   c. abscisic acid
   d. ethylene
   e. gibberellins

4. Which of the following is *not* part of the acid-growth hypothesis?
   a. Auxin stimulates proton pumps in cell membranes.
   b. Lowered pH results in breakage of cross-links between cellulose microfibrils.
   c. The wall fabric becomes looser (more elastic).
   d. Auxin-activated proton pumps stimulate cell division in meristems.
   e. The turgor pressure of the cell exceeds the restraining pressure of the loosened cell wall, and the cell takes up water and elongates.

5. The hypothetical plant hormone florigen could be released prematurely in a long-day plant experimentally exposed to flashes of
   a. far-red light during the night
   b. red light during the night
   c. red light followed by far-red light during the night
   d. far-red light during the day
   e. red light during the day

6. The phytochrome system helps to entrain the biological clock by indicating to a plant that light is present when
   a. P_r is rapidly converted to P_{fr}.
   b. P_{fr} is slowly converted to P_r.
   c. P_r and P_{fr} are equal in concentration.
   d. Red light is absorbed by P_{fr}.
   e. Photosynthetic production of ATP powers phytochrome conversion.

7. Bolting of floral shoots is triggered by a high concentration of
   a. auxin
   b. gibberellins
   c. cytokinins
   d. florigen
   e. ethylene

8. If a long-day plant has a critical night length of 9 hours, which of the following 24-hour cycles would *prevent* flowering?
   a. 16 hours light/8 hours dark
   b. 14 hours light/10 hours dark
   c. 15.5 hours light/8.5 hours dark
   d. 4 hours light/8 hours dark/4 hours light/8 hours dark
   e. 8 hours light/8 hours dark/light flash/8 hours dark

9. In the control of flowering, the organs that monitor photoperiod are the
   a. floral buds
   b. lateral buds
   c. leaves
   d. roots
   e. shoot apex

10. Auxin triggers acidification of cell walls that results in rapid growth, followed by sustained, long-term cell elongation. What best explains how auxin brings about this two-stage growth response?
    a. Auxin binds to different receptors in different cells.
    b. Different concentrations of auxin have different effects.
    c. Auxin causes a second messenger to activate both proton pumps and certain genes.
    d. The dual effects are due to two different auxins.
    e. Auxin's effects are modified by other antagonistic hormones.

CHALLENGE QUESTIONS

1. Discuss how day length, phytochrome, the biological clock, gibberellins, and abscisic acid may interrelate in the germination of seeds planted just below the soil surface.
2. Explain how it is possible that a short-day plant and a long-day plant growing in the same location could flower on the same day of the year.

SCIENCE, TECHNOLOGY, AND SOCIETY

1. Imagine the following scenario: A plant scientist discovers a synthetic chemical that mimics the effects of a plant hormone. The chemical can be sprayed on apples before harvest to prevent flaking of the natural wax that is formed on the skin. This makes the apples shinier and gives them a deeper red color. What kinds of questions do you think should be answered before farmers start using this chemical on apples?
2. Certain herbicides disrupt growth by mimicking the action of plant hormones. The weed killer 2,4-D, for example, is a synthetic auxin. Recombinant DNA techniques can be used to alter crop plants to make them more resistant to herbicides, so that the herbicides can be used to "weed" crops without harming them. What are some arguments for and against this technology?

FURTHER READING

Evans, M. L., R. Moore, and K.-H. Hasenstein. "How Roots Respond to Gravity." *Scientific American*, December 1986.

Mores, P. B., and N.-H. Chua. "Light Switches and Plant Genes." *Scientific American*, April 1988. A link between environment and gene expression.

Oeller, P. W., et al. "Reversible Inhibition of Tomato Fruit Senescence by Antisense RNA." *Science*, October 18, 1991. One method that may reduce food spoilage.

Sussman, M. "Shaking *Arabidopsis thaliana*." *Science*, May 1, 1992. Cell biology of a plant's response to touch.

Taiz, L., and E. Zeiger. *Plant Physiology*. Redwood City, CA: Benjamin/Cummings, 1991. Chapters 16–21.

UNIT SEVEN Animals: Form and Function

An Interview with Karel Liem

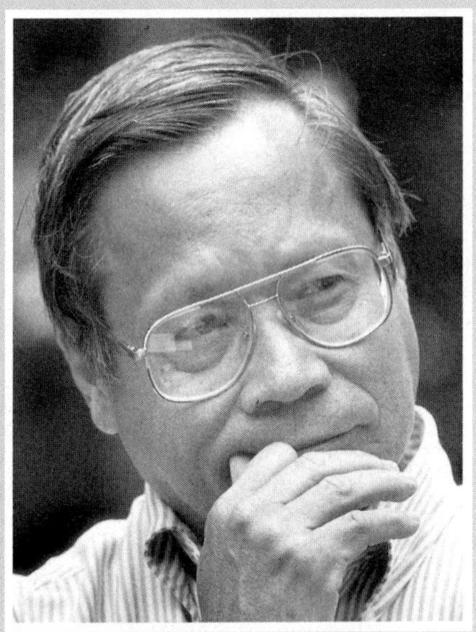

It is a common story among scientists: Some unexpected, unexplained observation captures a curious mind and focuses its general interest in nature onto a particular field of science. In the case of Karel Liem's research career, the defining moment was a nighttime encounter with a peculiar animal. Dr. Liem is now Henry Bryant Bigelow Professor of Ichthyology (the study of fishes) at Harvard University. He is recognized for both his creative research as a vertebrate biologist and his inspiring lectures as a general biology teacher. Professor Liem is past president of the American Society of Zoologists and a recipient of the Phi Beta Kappa Teaching Award, Harvard's highest commendation for classroom excellence. His unique talents as scientist and teacher are evident in this interview.

Most young people like animals, but only a small fraction of students choose zoology as a career. What experiences focused your studies on fishes and other animals?

I was fascinated with animals very early on because I grew up in Indonesia, which is a beautiful example of biodiversity in the tropics. I would look at the geckos crawling around on the ceiling, upside down, and try to figure out why they didn't fall off. Or flying lizards. I was always fascinated by the incredible adaptations that the animals had, even though I didn't know anything about adaptations. I also was interested in the behavior of animals.

I got into fish when I went to college. In my senior year I went on a trip to collect frogs for eating purposes; the class was going to have frog legs for dinner. We went out at night with lights in the rice fields around Bandung on the island of Java to collect frogs. As I walked past a grassy field I encountered a migration of animals across the field. I didn't know what they were; they looked like snakes, but they were not snakes. I picked up a dozen of these creatures and put them in a bag. We collected the frogs and had a wonderful frog-leg dinner. The next day I took the bag to the lab and emptied these organisms "X" into the sink. They were alive and well. I just could not figure out what they were. They were eel-like, about two feet long. My advisor, who came from the United States, had never seen any animals like these. He looked at them carefully and asked, "What are these?" I said, "I have no idea what they are." And he said, "Well, you find out." I found out they were fish, air-breathing fish, which were migrating over land. I wondered, what made them get out of the water to find another pond? How do they survive on land? How do they find their way? Surely a fish in tall grass can't see ahead. They have very tiny eyes. I was really interested in those questions, so I decided to do a thesis on the air-breathing fishes for my master's degree. That's the way I got into ichthyology. It was completely accidental. Originally I was going to study plant diseases. But that little accident changed my entire goal in life. Instead of being a plant pathologist, I became an ichthyologist.

Do air-breathing fishes, such as the species you studied in Indonesia, provide any clues about the vertebrates' evolutionary trek onto land?

Originally people thought, on the basis of geological evidence, that the origin of terrestrial vertebrates was triggered by drought conditions. According to this idea, bodies of water were drying out and the fish started to crawl out from the disappearing ponds and make their way to a better pond. I've done a considerable amount of research on what triggers a fish to get out of the water, and I think it's definitely associated with the drying out of ponds. But there's more to it. During the dry season, the fish would stay in water as long as possible! When the water disappeared, they burrowed into the mud. For example, lungfishes will estivate. What triggers air-breathing fishes to wander on land is competition. I have done a series of experiments in which I varied the population densities in ponds. When the population is too dense, competition for food or nest sites becomes very intense, triggering the fish to abandon the pond, but only when conditions are wet on land.

The important thing is that these fishes migrate over land only when it's very wet outside—after a rain, for instance—when the grass is wet. I think I have supported the hypothesis that vertebrates made the transition to land during hot, very wet and humid tropical conditions. Desiccation was probably the greatest problem facing organisms that were originally aquatic as they invaded terrestrial habitats. Air breathing is not a problem. Fishes can adapt easily, because most air-breathing fishes are bimodal; they engage in both gill and lung breathing. If the oxygen concentration is too low in the water, they just take up an air bubble and breathe air. So that's not the problem. I think the major problem of getting on land is desiccation.

As far as locomotion goes, I think the fishes were very well adapted to develop locomotory limbs, so to speak, because they already had very fleshy, lobed fins with bony elements in them. Even now, there are recent fishes that can walk with their fins. There is also a problem with water conservation. When fishes get on land, instead of the kidneys pumping water out all the time when they are in fresh water, they reverse that function and retain water—just like our kidneys are doing.

What about the energy it takes to move around on land? Is it harder work to travel on land than it is in water?
Very much so. Gravity and locomotion on land have probably been major constraints on amphibians, especially as they also depend so much on the external temperature to keep their metabolism up. But there are tremendous food resources on land. Vertebrates basically invaded a completely empty niche and got tremendous food resources in all the available terrestrial invertebrates. I imagine that the move onto land was kind of a compromise.

Does the descent of terrestrial vertebrates from fishes teach us any general lessons about how evolution works?
Yes. I would say it shows how existing structures may have very different functions later in evolution. Major evolutionary innovations often originate from preexisting structures.

You have also studied sex reversal in certain species of fishes. How do you explain this phenomenon?
It has something to do with fitness. There's a lot of debate about what we mean by fitness in evolution, but everyone agrees that it depends on how many surviving offspring you produce. I always tell my athletic students that in Darwinian terms they are less fit than I am because I have children and they don't. At the heart of sex reversal is fitness. So reproductive biology is definitely well-suited for "why" questions. There must be an explanation for sex reversal in terms of how it has increased the fitness in fishes that start life as a female and end life as a male and vice versa.

I worked on the same mud eel that was going from one pond to another. This species, *Monopterus albus*, turned out to be a gold mine, because it was also a sex-reversing fish. It starts as a female. Every one of these eels is born as a female; not a single individual is born as a male. Then, after being functional females for several years, they start changing sex. By the age of four to six years, every individual has become a male.

I came up with a hypothetical explanation for this. Since this fish often inhabits ponds that dry out, when the rainy season arrives they must immediately start reproducing. That's the best time in terms of food resources, water availability, and nest sites. So it is advantageous for them to reproduce right away. Now, I hypothesize that if you are born as a female first, the population is very much skewed in favor of females —there are many more females than there are males. What happens to such a population when a large pond becomes subdivided into small patches during the next dry season? Individuals trapped in very small ponds with insufficient food resources would change sex from female to male. So functional males are produced. Other populations in better ponds remain all female. When these ponds overflow and fuse together at the next rainy season, the populations of females and males are ready to reproduce, just when the environment is best. I tested this hypothesis in the laboratory. When populations are starved, sex reversal begins at an even earlier age. It is evidence that adverse conditions do trigger sex reversal.

In other species of fishes that switch sex, the causes may be entirely different. For example, some small sea basses live in schools with only one male and about 24 females. Such a reproductive school remains stable. But if you take the male out, one of the females will become a

male almost overnight. Really amazing! It's always an alpha female—a boss—that becomes the male. Apparently it is the macho behavior of the male that keeps females female. As soon as the female no longer receives the macho behavior of the male, she becomes macho herself and becomes a male.

So, how do we come up with a unified theory for all this? Biologists like to be like physicists; they would like to formulate a general theory that explains all cases of sex reversal in fishes. But in biology, there are a lot of *particularistic* explanations—explanations that apply only to a specific case. I would suggest that this is true for the phenomenon of sex reversal. There might be many different explanations. That frustrates biologists, but I think it makes biology fascinating.

You mentioned schools of sea bass. How does schooling benefit fishes?
Many fishes do school, and the general consensus is that it is basically a defense against predators. It has a confusing effect on the predator. When the predator starts zeroing in on a particular fish that somehow manages to disappear in a school, the predator has a hard time continuing that strike and misses. And schools also function as extended eyes; the individuals on the periphery see something that individuals in the center are not aware of, and the whole school moves as an ensemble. It's like little pseudopodia sticking out of the school, and these may actually determine the movement of the school. There might not be an individual leader, but there might be a leading "pseudopod" which determines whether to retreat or whether to expand and to move. That's the current theory about schooling.

Dr. Liem, much of your research is associated with an approach known as functional morphology. What is a functional morphologist?

Functional morphology really meets adaptation head on, I believe, at a level that other fields do not. Functional morphologists try to figure out how a system works, how efficiently it works, and how its performance relates to its natural habitat and in competitive interactions. We're using, basically, engineering principles and physiological principles to explain animal structure. Functional morphologists also consider the evolutionary aspects of structures.

Ultimately, functional morphologists are interested in the interactions between structures within components of an organism, for example, the leg or the head. It's the interactions between parts that may constrain the functional expression of the component. Emphasizing the nature of interactions between structures serves as a model of

how to approach organismal biology. This principle also applies to ecology, where we have different interacting factors. It is basically a network of constraints, and we would like to know how a change in the nature of the interaction may release a constraint, which results in a greater diversity. It can also be applied to the biosphere. That's an integrated system of a multitude of factors. I think functional morphology can lead the way in research strategies. Instead of searching for the optimal solution, we are searching for the nature of how components are interconnected and how these interconnections limit the functional performance. The interconnections may be of a mechanical nature. But more importantly, the interconnections may be related with developmental patterns that are genetically programmed.

Have the ideas and methods of functional morphology changed much in the past few decades?
Yes, tremendously so, I believe. There are two fronts that have revitalized morphology. One area is technology. We now have incredible tools to determine function. But it's not only technology, of course; we have also made progress conceptually. We are now dealing with the notion of adaptation in a quantitative way, in a measurable way. The combination of the technology and this conceptual progress, I think, has made functional morphology very lively.

The terms *body plan* and *design* inevitably pop up in discussions about animal form and function. What do these terms mean in their biological contexts?
They're buzzwords. I'll discuss body plan first because it's the easiest. The body plan is basically a model, an abstraction of a particular evolutionary lineage. There is the body plan of a mammal, or the body plan of an echinoderm. It's an abstract model.

A design, on the other hand, has an engineering connotation. It means the ensemble of structures that are related to a particular function. For example, the design of the limb of a digging mammal is very different from the design of a running mammal's limb. And that would include not just the bones, but also the muscles, the ligaments, and so on. The whole ensemble is called a design.

Design is a product of functional and historical factors. We often find out that the designs we see in an organism are not perfect designs; they are not like what an engineer would come up with. We ordinarily use the word *design* basically from an engineering point of view, and in that sense it is a very poor term. I explain to students that biological designs change over time, usually by changes in preexisting structures. That's not what an engineer would do. Dr. Thomas Frazzetta from the University of Illinois has a wonderful saying, that "Evolution is a change while the machine is running." We probably see a lot of that in biology, and therefore we have some rather mediocre designs, but they work well enough in nature to survive.

By viewing animal form and function in this evolutionary context, does the human body make more sense? Are there medical applications of this approach, for example?
That's a very challenging question. Here's one example. Consider the evolution of the aortic arches and the heart. It would not have been possible for placental mammals to be placental mammals if we didn't inherit the lungfish

design of the cardiovascular system. I tell my students, "Look, here's a ductus arteriosus and an opening between the left and right sides of the heart of the lungfishes. The ductus arteriosus can be opened or closed depending on whether the fish were breathing air or water. And that same ductus is now used during human fetal life to shunt blood, so we can be born as a placental. The medical doctor faces all kinds of cardiovascular problems that are really lungfish situations, stemming from a patent opening between the left and right sides of the heart and the continued existence of the ductus arteriosus." I try to make these kinds of connections whenever possible in my teaching. And I think students really enjoy it.

Teaching students about the evolution of various adaptations is also one way to emphasize biological diversity in a biology course. What other approaches do you take in class to expose your students to the diversity of life?
I try to illustrate adaptive radiations with many attractive slides during my lectures, using as big a diversity as I can get—sometimes taking examples from invertebrate diversifications, sometimes taking fish diversifications, sometimes plants, especially when you get into tropical ecology. Then they really start appreciating what diversity is about.

Are there reasons other than aesthetic and moral to value and preserve biological diversity?
Yes. I think it is very much in our personal interest to preserve the quality of life that we know. I usually argue that biologically diverse systems are often far more resilient to perturbations than a monoculture. The existence of the biosphere really depends on the diversity that we have. So I would say that conserving biological diversity is a matter of self-preservation.

How are human activities affecting the marine communities you study?
Well, consider Puget Sound, where I'm doing some research. In some parts it is still very good, but in other areas the habitat has been devastated by pollutants in the water. It is especially critical in areas like Puget Sound, which is an estuary, where there is little circulation. Chesapeake Bay is another example. The worst kind of impact is probably filling in wetlands. People decide to live on the wetlands and put buildings up. That has a tremendous impact on the inshore marine communities. Around Jamaica, the main problem is overfishing. Of course, many Jamaicans are very poor and they really need the fish for food. But the fish populations have been reduced to very small size, and that has indirect impact on the coral reefs and everything else. There are many examples of human impact on marine communities.

How can ichthyologists help us sustain fish populations into the future, even as we consume these fishes for food?
Careful management on an international basis is a key element. Ichthyologists have been playing a key role in understanding life histories of fishes. When are the eggs and larvae most susceptible to death? To preserve fisheries, we must protect the spawning grounds so that we have areas where the larvae will grow. I think a good understanding of the ecology of spawning grounds and nurseries is emerging now.

Much of the harvesting of fishes, whales, and other marine organisms goes on in international waters. How effective are national and international laws and policies at conserving these marine resources—or any resources, for that matter?
I think they're not very effective right now. It's a very tough thing. In the United States, we still struggle with the concept of who owns the water and the air. Right now, we have clearly worded laws that are supposed to help protect the water and air. But the enforcement is not very good. If you were to extend such laws internationally, they would be even less effective at this time. For instance, to convince Icelanders to preserve whales is going to take a really long time. It's in their culture. They've been whalers for centuries. It's just part of their way. They love to eat pilot whale meat. So, what can you do? And there are economic complications. When I was in Iceland, they told me that up to 90% of their total income depends on export of salmon from Iceland to the United States. So if the United States stopped importing salmon to pressure Iceland to stop whaling, Iceland would go under. When it comes to developing and enforcing environmental policies and laws, there will be a lot of inertia.

Dr. Liem, you teach a very large class. In fact, *most* students are introduced to biology in huge classes at large colleges and universities. How can these students work their way into a more personalized situation and get hands-on research experience in a professor's laboratory? And how can we encourage that interest?
I have a system, a bit like what you're doing in your book, to expose students to enthusiastic scientists. I try to get my colleagues to give optional guest lectures about what they are doing in the lab in informal settings. If you ask a top-notch researcher to give a talk to introductory biology students, he or she will see that as a challenge, and will do an incredibly good job trying to show the students that what they are doing is really fun and worthwhile. Each of those guest lecturers usually recruits students into his or her lab to do independent research or an honors thesis. Some of the guest lecturers are from Harvard Medical School. They don't have much exposure to undergraduates. And when these professors work with these students in the lab, they quickly discover that these are exciting and bright young scholars. They come to a research project without any prejudices.

36 An Introduction to Animal Structure and Function

Levels of Structural Organization
Size, Shape, and the External Environment
The Animal's Internal Environment

Many students taking a general biology course have a healthy preoccupation with human anatomy and physiology. In these chapters on animal form and function, you will learn much about how your body works. But the scope of this unit is not limited to humans or even to vertebrates. By following a comparative approach, we can see how certain common problems—the exchange of respiratory gases (O_2 and CO_2) with the surrounding environment, for instance—have been solved by animals of diverse evolutionary history and varying complexity, from protozoa to vertebrates. (Although protozoa are actually protists, we include them in this unit to illustrate unicellular analogues of certain processes that occur at higher levels of organization in animals.) In the interview preceding this chapter, Karel Liem discusses the perspective that comparative zoology provides.

When we try to understand how *any* organism works, the same two themes apply. One is the correlation between structure and function (Figure 36.1). The second theme is the capacity of life to adjust to its environment on two temporal scales: over the short term by physiological responses, and over the long term by adaptation due to natural selection (see Chapter 21). These themes will integrate our comparative study of animal form and function throughout this unit.

The objective of this chapter is to introduce you to the body plans of animals. The use of the terms *plan* and *design* in no way implies that animal forms are products of a conscious invention. The design of an animal—its shape, symmetry, internal organization, and so on—results from a pattern of development programmed by the animal's genome, itself the product of millions of years of evolution due to natural selection.

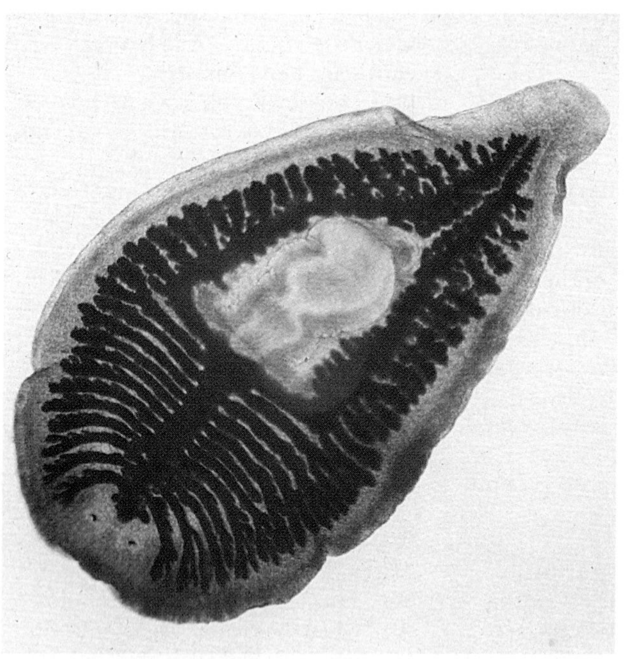

Figure 36.1
Form fits function. The gut of this flatworm *(Bdelloura candida)* is finely branched, providing a large surface area for the absorption and distribution of nutrients. We will correlate structure with function throughout our comparative study of animal anatomy and physiology. In this introductory chapter, you will learn a few basic principles of animal form and function.

LEVELS OF STRUCTURAL ORGANIZATION

Life is characterized by a hierarchy of structural order (see Chapter 1). Atoms are built into molecules, molecules are assembled into supramolecular structures such as membranes, ribosomes, and chromosomes, and these structures are organized into the living units we call cells. The cell holds a special place in the hierarchy of life because it is the lowest level of organization that can live as an organism. Protozoa, for example, have organelles that are specialized to perform

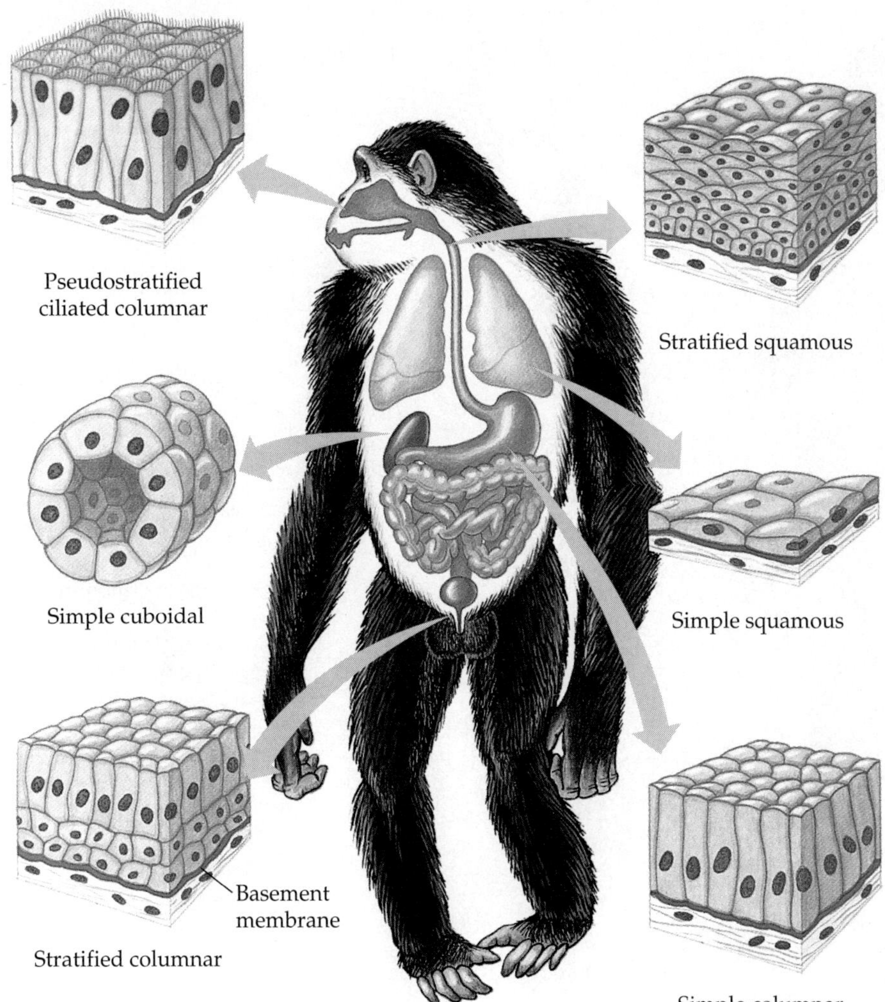

Figure 36.2
The structure and function of epithelial tissue. The structure of an epithelium fits its function. For instance, simple squamous epithelia, which are relatively leaky, are specialized for exchange of materials by diffusion. We find these epithelia in such places as the linings of blood vessels and air sacs of the lungs. A stratified squamous epithelium regenerates rapidly by cell division near the basement membrane. The new cells are pushed to the free surface as replacements for cells that are continually sloughed off. This type of epithelium is commonly found on surfaces subject to abrasion, such as the outer skin and linings of the esophagus, anus, and vagina. Columnar epithelia, having cells with relatively large cytoplasmic volumes, are often located where secretion or active absorption of substances are important functions. For example, the stomach and intestines are lined by a columnar epithelium that secretes digestive juices and absorbs nutrients. Cuboidal cells specialized for secretion make up the epithelia of kidney tubules and many glands, including the thyroid gland and salivary glands.

particular jobs, enabling them to digest food, move about, sense environmental change, excrete waste products, and reproduce—all within the confines of a single cell. Protozoa represent the cellular level of organization, the simplest level possible for an organism. Multicellular organisms, including animals, have specialized cells grouped into tissues, the next higher level of structure and function. In most animals, various tissues are combined into functional units called organs, and in the most complex animals, organs cooperate in teams called organ systems. For example, our own digestive system consists of a stomach, a small and a large intestine, a gallbladder, and several other organs, each one a composite of different types of tissues.

Animal Tissues

Tissues are groups of cells with a common structure and function. A tissue may be held together by sticky substances that coat the cells (see Chapter 7), or the cells may be woven together in a fabric of extracellular fibers. Indeed, the term *tissue* is from a Latin word meaning "weave."

Histologists, biologists who specialize in the study of animal tissues, classify tissues into four main categories: epithelial tissue, connective tissue, muscle tissue, and nervous tissue. These are present to some extent in all but the simplest animals, but the following survey emphasizes the tissues as they appear in vertebrates.

Epithelial Tissue Occurring in sheets of tightly packed cells, **epithelial tissue** covers the outside of the body and lines organs and cavities within the body. The cells of an epithelium are closely joined, with little material between them. In many epithelia, the cells are riveted together by tight junctions like those described in Chapter 7 (see Figure 7.34). This tight packing is consistent with the function of an epithelium as a barrier protecting against mechanical injury, invading microorganisms, and fluid loss. The free surface of the epithelium is exposed to air or fluid, whereas the cells at the base of the barrier are attached to a **basement membrane,** a dense layer of extracellular material.

Two criteria for classifying epithelia are the number of cell layers and the shape of the cells on the free surface (Figure 36.2). A **simple epithelium** has a single

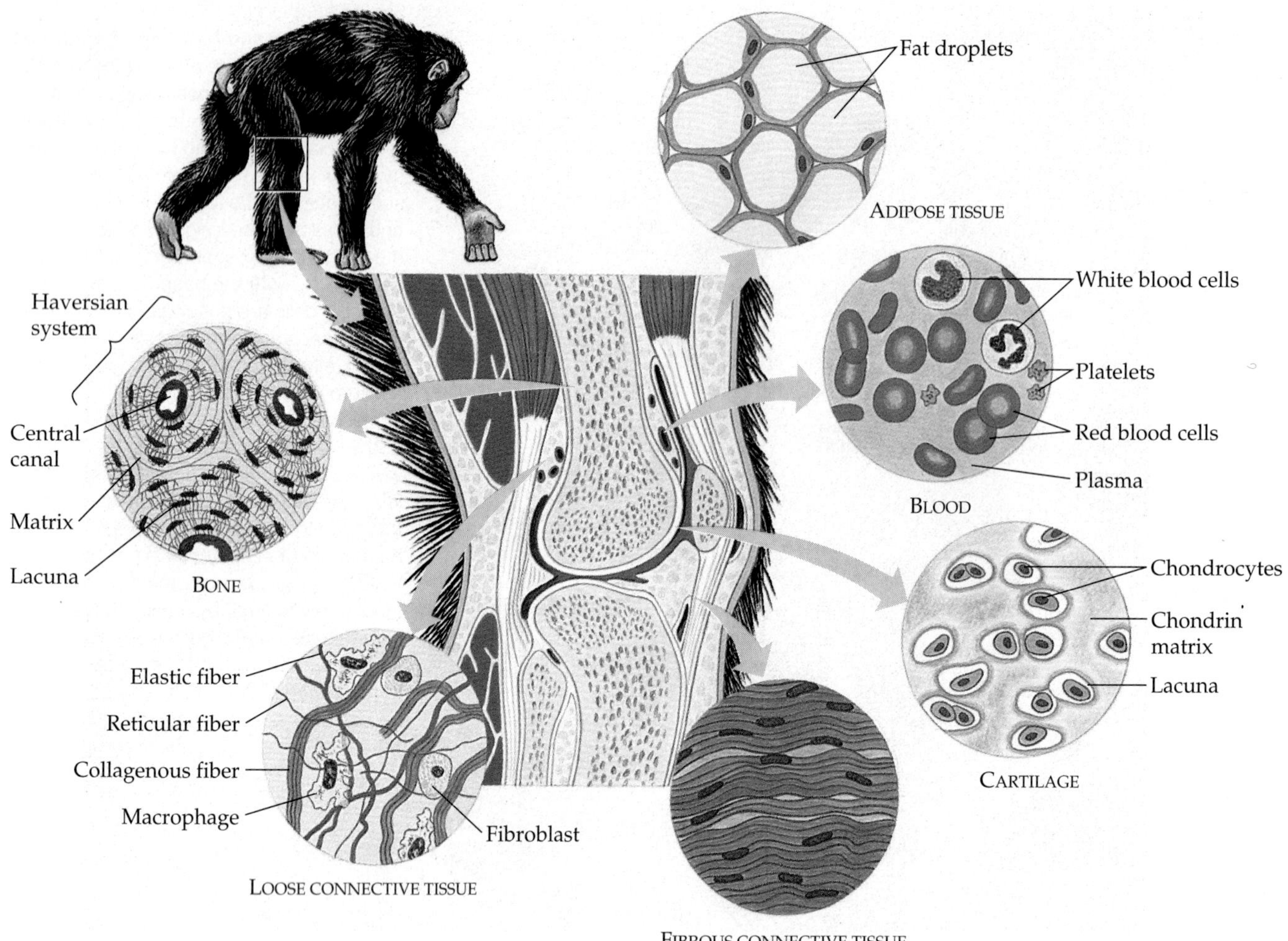

Figure 36.3
Some representative types of connective tissue. The area pictured is the region around the knee joint.

layer of cells, whereas a **stratified epithelium** has multiple tiers of cells. A pseudostratified epithelium is single-layered, but it appears stratified because the cells vary in length. The shape of the cells at the free surface of an epithelium may be **cuboidal** (like dice), **columnar** (like bricks on end), or **squamous** (like flat floor tiles). Combining the features of cell shape and number of layers, we get such terms as *simple cuboidal epithelium* and *stratified squamous epithelium.*

As well as protecting the organs they line, some epithelia are specialized for absorbing or secreting chemical solutions. For example, the epithelial cells that line the lumen (cavity) of the digestive and respiratory tracts are called **mucous membrane** because they secrete a slimy solution called mucus that lubricates the surface and keeps it moist. The mucous membrane lining the small intestine also secretes digestive enzymes and absorbs nutrients. The free epithelial surfaces of some mucous membranes have beating cilia that move the film of mucus along the surface. The ciliated epithelium of our respiratory tubes helps keep our lungs clean by trapping dust and other particles and sweeping them back up the trachea (windpipe).

Connective Tissue **Connective tissue** functions mainly to bind and support other tissues. In contrast to epithelia with their tightly packed cells, connective tissues have a sparse population of cells scattered through an extracellular **matrix.** This nonliving matrix generally consists of a web of fibers embedded in a homogeneous ground substance that may be liquid, jellylike, or solid. In most cases, the substances of the matrix are secreted by the cells of the connective tissue. The major types of connective tissue in vertebrates are loose connective tissue, adipose tissue, fibrous connective tissue, cartilage, bone, and blood. Each has a structure correlated with its specialized functions.

The most widespread connective tissue in the vertebrate body is **loose connective tissue** (Figure 36.3). It is used as binding to attach epithelia to underlying tissues and as packing material to hold organs in place. This type of connective tissue gets its name from the

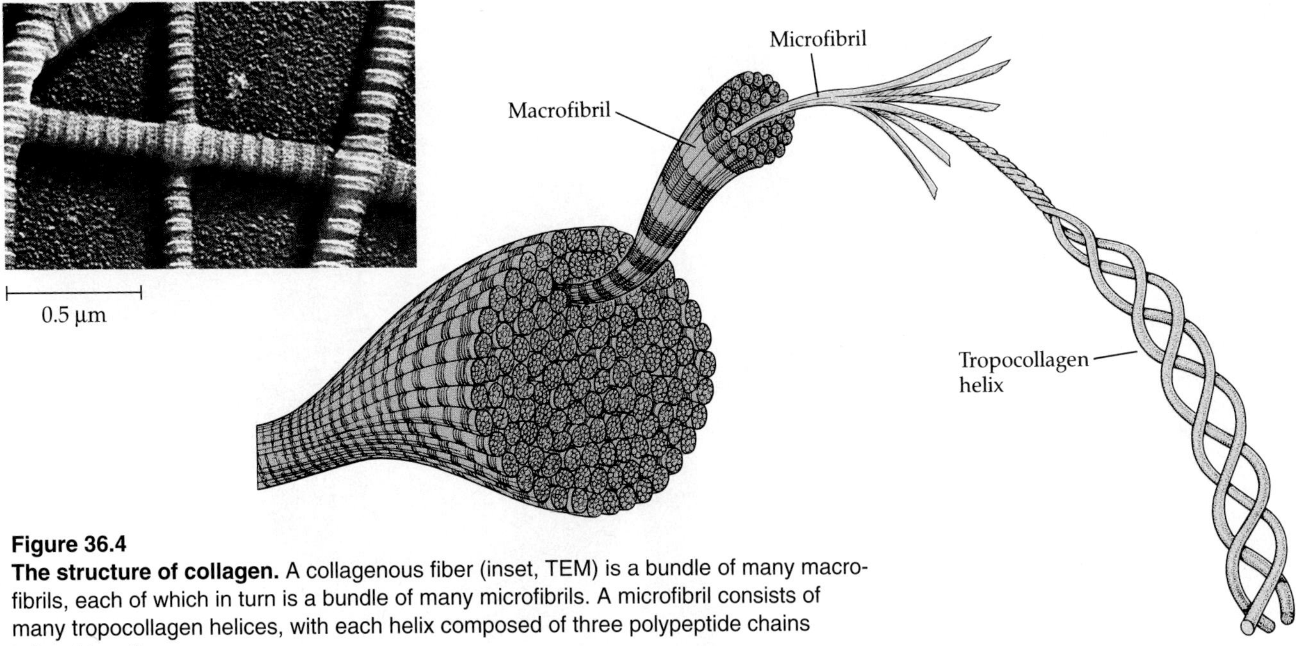

Figure 36.4
The structure of collagen. A collagenous fiber (inset, TEM) is a bundle of many macrofibrils, each of which in turn is a bundle of many microfibrils. A microfibril consists of many tropocollagen helices, with each helix composed of three polypeptide chains twisted together.

loose weave of its fibers. These fibers, which are made of protein, are of three kinds: collagenous fibers, elastic fibers, and reticular fibers. **Collagenous fibers** are made of collagen, perhaps the most abundant protein in the animal kingdom. A collagenous fiber is a bundle of several fibrils, each a ropelike coil of three collagen molecules (Figure 36.4). Collagenous fibers have great tensile strength, which means that they do not tear easily when pulled lengthwise. If you pinch and pull some skin on the back of your hand, it is mainly collagen that keeps you from tearing your flesh from the bone. **Elastic fibers** are long threads made of a protein called elastin. As their name implies, these fibers are elastic, unlike collagenous fibers, which resist stretching. Elastic fibers endow loose connective tissue with a resilience that complements the tensile strength of collagenous fibers. When you pinch the back of your hand and then let go, the rubbery elastic fibers quickly restore your skin to its original shape. **Reticular fibers** are branched and form a tightly woven fabric that joins connective tissue to adjacent tissues. Among the cells scattered in the fibrous mesh of loose connective tissue, two types predominate: fibroblasts and macrophages. **Fibroblasts** secrete the protein ingredients of the extracellular fibers. **Macrophages** are amoeboid cells that roam the maze of fibers, engulfing bacteria and the debris of dead cells by phagocytosis (see Chapter 8). They are weapons in an elaborate arsenal of defense you will learn more about in Chapter 39.

Adipose tissue is a specialized form of loose connective tissue that stores fat in adipose cells distributed throughout its matrix. Adipose tissue pads and insulates the body and stores fuel molecules. Each adipose cell contains a large fat droplet that swells when fat is stored and shrinks when fat is used by the body as fuel. Inheritance is an important factor in obesity, but there is also some evidence that the number of fat cells in our connective tissue is determined partly by the amount of fat we stored when we were babies. This might be discouraging news for dieting adults who were overweight at an early age, because it suggests that they have more fat cells than people who may have put on a few pounds later in life.

Fibrous connective tissue is dense, owing to its enrichment in collagenous fibers. The fibers are organized into parallel bundles, an arrangement that maximizes tensile strength. We find this type of connective tissue in **tendons,** which attach muscles to bones, and in **ligaments,** which join bones together at joints.

Cartilage has an abundance of collagenous fibers embedded in a rubbery ground substance called **chondrin,** which is a protein-carbohydrate complex. The chondrin and collagen are secreted by **chondrocytes,** cells confined to scattered spaces called lacunae (spaces) in the ground substance. The composite of collagenous fibers and chondrin makes cartilage a strong yet somewhat flexible support material. The skeleton of a shark is made of cartilage. Other vertebrates, including humans, have cartilaginous skeletons during the embryo stage, but the skeleton hardens into bone as the embryo matures. We nevertheless retain cartilage as flexible support in certain locations, such as the nose, the ears, the rings that reinforce the windpipe, the discs that act as cushions between our vertebrae, and the caps on the ends of some bones.

The skeletons that support the bodies of most vertebrates are made of **bone,** a mineralized connective tissue. Cells called **osteocytes** deposit a matrix of colla-

Figure 36.5
Types of vertebrate muscle. Skeletal muscle consists of bundles of long cells called fibers; each fiber, in turn, is a bundle of strands called myofibrils. The myofibrils are a linear array of sarcomeres, the basic contractile units of the muscle. Skeletal muscle is said to be striated because the alignment of sarcomere subunits in adjacent myofibrils forms light and dark bands. Cardiac muscle, also striated, has contractile properties similar to those of skeletal muscle. Unlike skeletal muscle, however, cardiac muscle fibers branch and interconnect via intercalated discs, which help synchronize the heartbeat. Visceral muscle, or smooth muscle, consists of spindle-shaped cells lacking cross-striations.

gen, but they also release calcium phosphate, which hardens within the matrix into the mineral hydroxyapatite. The combination of hard mineral and flexible collagen makes bone harder than cartilage without being brittle. The microscopic structure of hard mammalian bone consists of repeating units called **Haversian systems.** Each system has concentric layers of the mineralized matrix, which are deposited around a central canal containing blood vessels and nerves that service the bone. Osteocytes are located in lacunae, spaces surrounded by the hard matrix and connected to each other by long, thin extensions of the cells. In long bones, such as the femur (shank) of your thigh, only the hard outer region is compact bone built from Haversian systems. The interior is a spongy bone tissue also known as marrow. Blood cells are manufactured in the red marrow located near the ends of long bones. (Bones and skeletons will be examined in more detail in Chapter 45.)

Although **blood** functions differently from other connective tissues, it does meet the criterion of having an extensive extracellular matrix. In this case, the matrix is a liquid called plasma, consisting of water, salts, and a variety of dissolved proteins. Suspended in the plasma are three classes of blood cells: erythrocytes (red blood cells), leukocytes (white blood cells), and platelets. Red cells carry oxygen, white cells function in defense against viruses, bacteria, and other invaders, and platelets are involved in the clotting of blood. The composition and functions of blood will be discussed in detail in Chapters 38 and 39.

Muscle Tissue **Muscle tissue** is composed of long, excitable cells capable of considerable contraction. Arranged in parallel within the cytoplasm of muscle cells are large numbers of microfilaments made of the contractile proteins actin and myosin (see Chapter 7). Consistent with a high priority on movement, muscle is the most abundant tissue in most animals. It accounts for about two-thirds of the bulk of a well-conditioned human.

In the vertebrate body, there are three types of muscle tissue: skeletal muscle, cardiac muscle, and visceral muscle (Figure 36.5). Attached to bones by tendons, **skeletal muscle** is generally responsible for the voluntary movements of the body. Adults have a fixed number of muscle cells; weight lifting and other methods of building muscle do not increase the number of cells but simply enlarge those already present. Skeletal muscle is also called **striated muscle** because the arrangement of overlapping filaments gives the cells a striped (striated) appearance under the microscope.

Cardiac muscle forms the contractile wall of the heart. It is striated like skeletal muscle, but cardiac cells

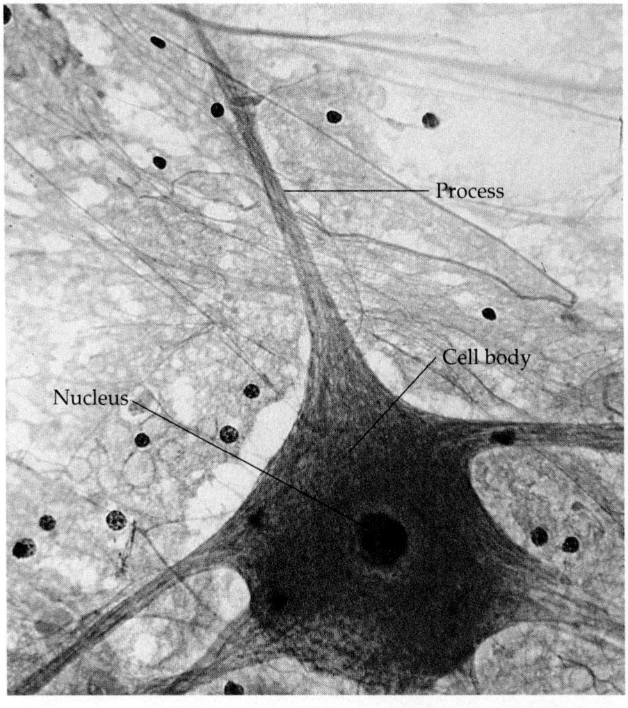

Figure 36.6
Neuron structure. This nerve cell from the spinal cord has a large cell body with multiple processes that transmit impulses (LM).

are branched. The ends of cells are joined by structures called intercalated discs, which relay the impulse to contract from cell to cell during a heartbeat.

Visceral muscle, also called **smooth muscle** because it lacks cross-striations, is found in the walls of the digestive tract, bladder, arteries, and other internal organs. The cells are spindle-shaped. They contract more slowly than skeletal muscles, but they can sustain their contracted condition for a longer period of time.

Skeletal and visceral muscles are controlled by different kinds of nerves. Skeletal muscles are called voluntary because an animal can generally contract them at will. Visceral muscles are involuntary; they are not generally subject to conscious control. You can decide to raise your hand, but you are usually unaware when visceral muscles churn your stomach or constrict your arteries. You will learn more about the control and contraction of muscles in Chapter 45.

Nervous Tissue **Nervous tissue** senses stimuli and transmits signals from one part of the animal to another. The functional unit of nervous tissue is the **neuron,** or nerve cell, which is uniquely specialized to conduct an impulse (Figure 36.6). It consists of a cell body and two or more appendages called nerve processes, which may be as long as a meter in humans. Dendrites are processes that conduct impulses toward the cell body, and axons transmit impulses away from the cell body. We postpone a detailed discussion of the structure and function of neurons until Chapter 44.

Organs and Organ Systems

In all but the simplest animals (sponges and cnidarians), different tissues are organized into the specialized centers of function called **organs.** Some organs have their tissues arranged in layers. The stomach, for example, has four major layers (Figure 36.7). The lumen is lined by a thick epithelium that secretes mucus and digestive juices. Outside this layer is a zone of connective tissue followed by a thick layer of smooth muscle. The entire stomach is encapsulated by another layer of connective tissue. A laminated organization is also apparent in the dermis, the outer skin of the body.

Many of the organs of vertebrates are suspended by sheets of connective tissue called **mesenteries** into body cavities filled with fluid. Mammals have an upper **thoracic cavity** separated from a lower **abdominal cavity** by a sheet of muscle called the diaphragm. The embryonic origin of the body cavity (coelom) and mesenteries was described in Chapter 30.

There is a level of organization still higher than organs. The major functions of vertebrates and members of most invertebrate phyla are carried out by **organ systems,** each consisting of several organs. Examples are the digestive, circulatory, respiratory, and excretory systems. Each has specific functions, but the efforts of all systems must be coordinated for the animal to survive. For instance, nutrients absorbed from the digestive tract are distributed throughout the body by the circulatory system. But the heart that pumps blood through the circulatory system depends on nutrients absorbed by the digestive tract and also on oxygen obtained from the air or water by the respiratory system. The organism, whether a protozoan or an assembly of organ systems, is a living whole greater than the sum of its parts.

SIZE, SHAPE, AND THE EXTERNAL ENVIRONMENT

An animal's size, shape, and symmetry are fundamental features of form and function that affect how the animal interacts with its environment. For instance, to take in oxygen and other materials and get rid of waste products, the animal must be arranged so every cell is bathed in an aqueous medium. Only in such an environment can dissolved substances diffuse across the plasma membrane. Because it is so small, a protozoan

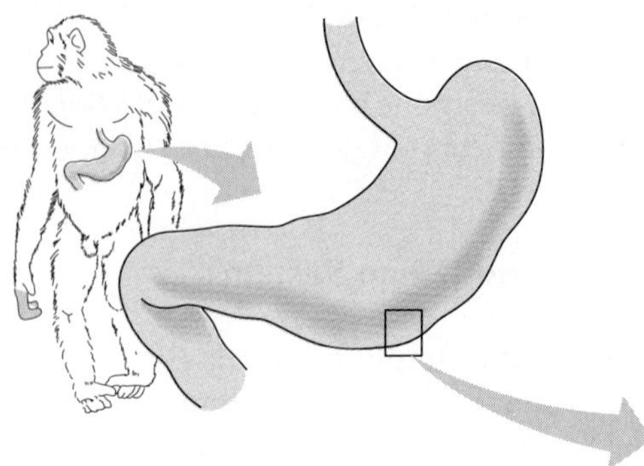

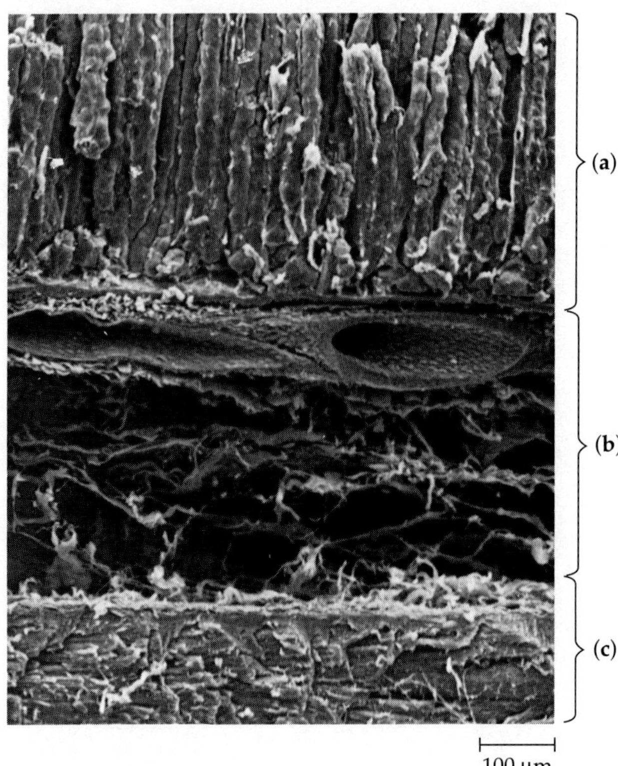

Figure 36.7
Tissue layers of the stomach, a digestive organ. Each organ of the digestive system has a wall with four main tissue layers: the mucosa, submucosa, muscularis, and serosa. The first three layers are represented in this scanning electron micrograph. (**a**) The mucosa is an epithelial layer that lines the lumen. (**b**) The submucosa is a matrix of connective tissue that contains blood vessels and nerves. (**c**) The muscularis has an inner layer of circular muscles and an outer layer of longitudinal muscles. External to the muscularis is the serosa, a thin layer of connective tissue and epithelial tissue. (From *Tissues and Organs: A Text-Atlas of Scanning Electron Microscopy* by Richard G. Kessel and Randy H. Kardon. W. H. Freeman and Company. Copyright © 1979.)

living in aqueous surroundings has a sufficient surface area of plasma membrane to service its entire volume of cytoplasm (Figure 36.8a). Geometry imposes limits on the size that is practical for a single cell. A large cell has less surface area relative to its volume than a smaller cell of the same shape (see Chapter 7). This is one reason nearly all cells are microscopic. A multicellular animal is composed of microscopic cells, each with its own plasma membrane to function as a loading and unloading platform for a modest volume of cytoplasm. But this only works if all the cells of the animal have access to a suitable aqueous environment. The freshwater invertebrate *Hydra,* built on the sac plan, has a body wall only two cell layers thick (Figure 36.8b). Because its body cavity opens to the exterior, both outer and inner layers of cells are bathed in water. A flat body shape is another way to maximize exposure to the surrounding medium. For instance, tapeworms may be several meters long, but they are very thin, and thus most cells are bathed in the intestinal fluid of the worm's vertebrate host.

Two-layered sacs and planar shapes are designs that put a large surface area in contact with the environment, but these simple forms do not allow much complexity in internal organization. Most animals are bulkier, with outer surfaces that are relatively small compared with the animal's volume. To take an extreme comparison, the surface-to-volume ratio of a whale is millions of times smaller than that of a protozoan, yet every cell in the whale must be bathed in fluid and have access to oxygen, nutrients, and other resources.

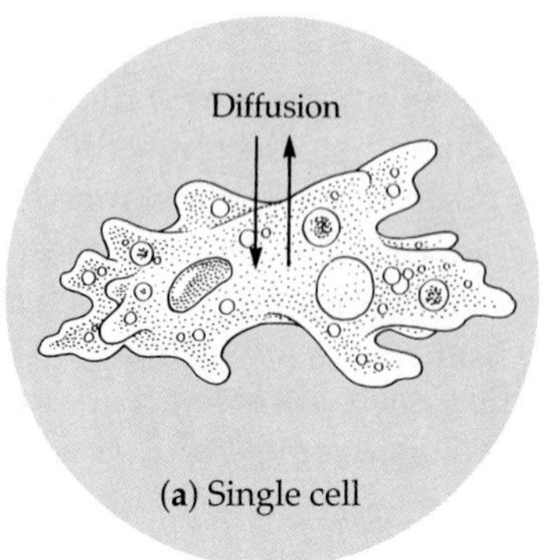

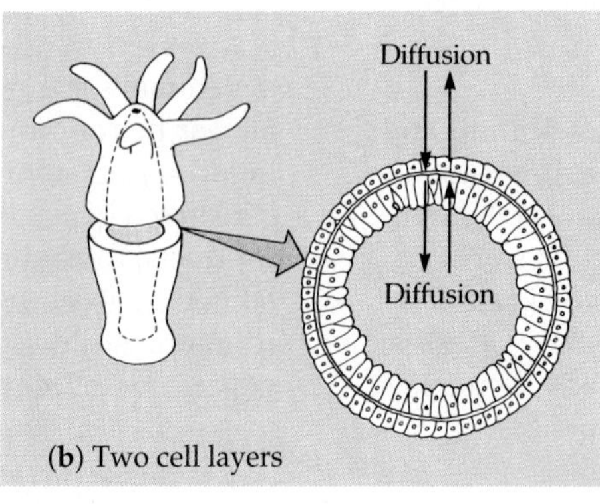

Figure 36.8
Contact with the environment. (**a**) In a unicellular organism, such as this amoeba, the entire surface area contacts the environment. Because of its small size, the cell has a large surface area relative to its volume through which to exchange materials with the external world. (**b**) Each cell in the bilayered *Hydra* directly contacts the environment and exchanges materials with it.

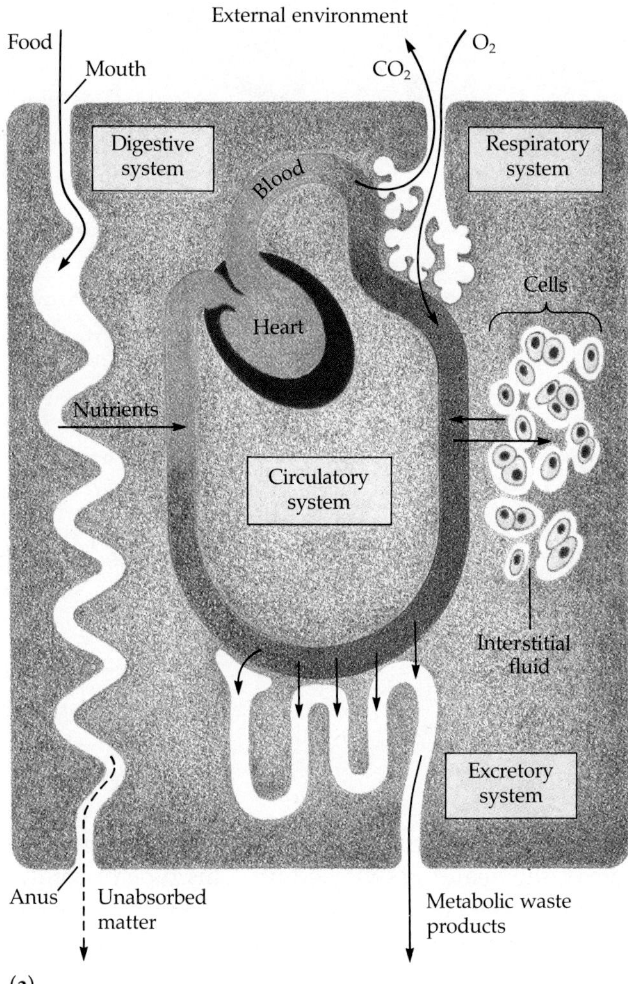

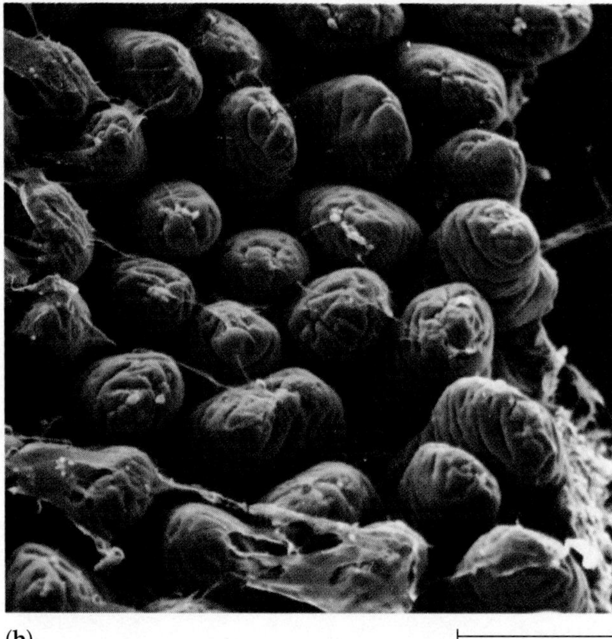

Figure 36.9
Internal surfaces of complex animals. (**a**) Rather than using the outer body surface, most complex animals use specialized surfaces for exchanging materials with the environment. Usually internal, but connected to the environment via openings on the body surface, these exchange surfaces are large in area due to a branched or folded arrangement. A circulatory system transports substances among the various internal surfaces and other parts of the body. (**b**) A large surface specialized for absorption. The lining of the mammalian small intestine has numerous projections called villi, giving the organ a very large surface area that absorbs nutrients (SEM).

Most complex animals have internal surfaces specialized for exchanging materials with the environment (Figure 36.9a). Extensive folding or ramification (branching) gives these moist internal membranes expansive surface areas. The internal membrane of our lungs, for example, is specialized for absorbing oxygen and getting rid of carbon dioxide; it lines millions of microscopic air chambers with a total surface area of about 100 m^2, the approximate area of a tennis court. The internal surface of the vertebrate kidney, consisting of about a million microscopic tubules, is specialized to filter waste products out of the blood. Still another surface, the lining of the digestive tract, absorbs nutrients. The human intestine is about 6 m long, and its wavy lining has a surface area roughly the size of a baseball diamond (Figure 36.9b). Materials are shuttled between all these exchange surfaces by the circulatory system.

Although the logistical problems of exchange with the environment are much more complicated for an animal with a compact form than for one with all its cells exposed directly to its surroundings, a compact form has some distinct benefits. Because the animal's external surface need not be bathed in water, it is possible for the animal to live on land. Also, because the immediate environment for the cells is the internal body fluid, the animal can control the quality of the solution that bathes its cells.

THE ANIMAL'S INTERNAL ENVIRONMENT

Over a century ago, French physiologist Claude Bernard made the distinction between the external environment surrounding an animal and the internal environment in which the cells of the animal actually live. The internal environment of vertebrates is called the **interstitial fluid.** This fluid, which fills the spaces between our cells, exchanges nutrients and wastes with blood contained in microscopic vessels called capillaries. Bernard also recognized the power of many animals to maintain relatively constant conditions in their internal environment, even when the external environment changes. A pond-dwelling *Hydra* is powerless to

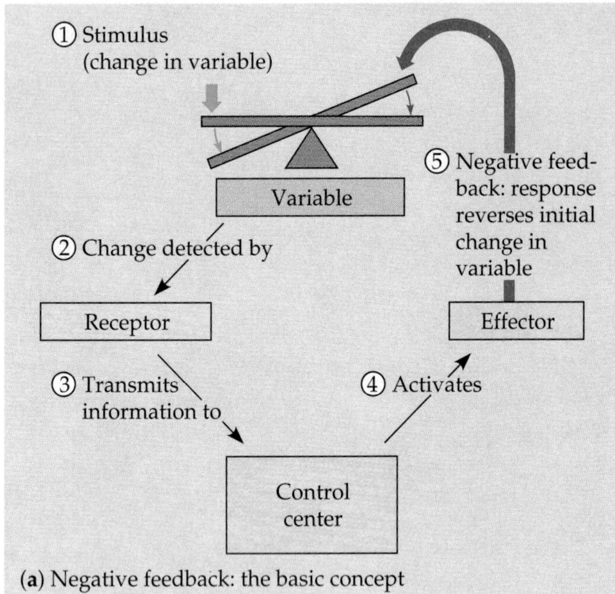

(a) Negative feedback: the basic concept

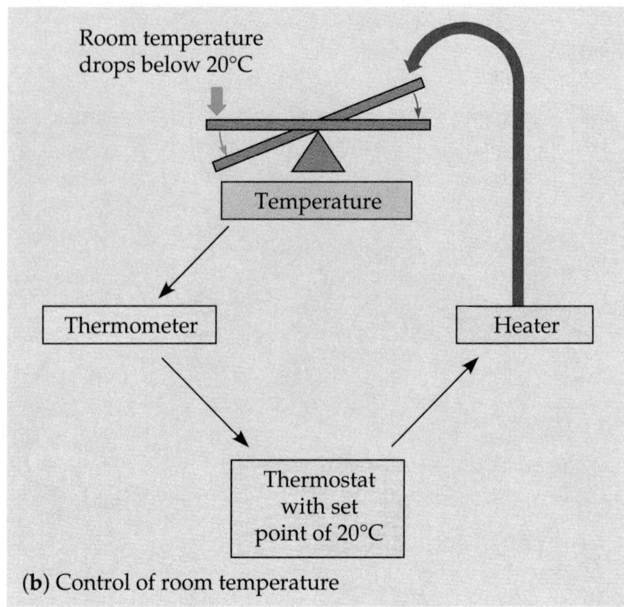

(b) Control of room temperature

Figure 36.10
Negative feedback as a mechanism of homeostasis. (**a**) The feedback circuit restores balance in some variable by countering changes. (**b**) The control of room temperature depends on negative feedback.

affect the temperature of the fluid that soaks its cells, but a human can maintain its "internal pond" at a constant temperature of 37°C. We also can control the pH of our blood and interstitial fluid to within a tenth of a pH unit of 7.4, and we regulate the amount of sugar in our blood so that it does not fluctuate for long from a concentration of 0.1%. There are times, of course, during the development of an animal when major changes in the internal environment are programmed to occur. For example, the balance of hormones in the blood of a human is altered radically during puberty. Still, the stability of the internal environment is remarkable.

Today, Bernard's "constant internal milieu" is incorporated into the concept of **homeostasis,** which means "steady state." One of the main objectives of modern physiology is to learn how animals maintain homeostasis. Any homeostatic control system has three functional components: a receptor, a control center, and an effector (Figure 36.10a). The *receptor* detects a change in some variable of the animal's internal environment, such as change in body temperature. The *control center* processes information it receives from the receptor and directs an appropriate response by the *effector.* As a nonliving analogy of how these components interact to maintain homeostasis, consider how the temperature of a room is controlled (Figure 36.10b). In this case, the receptor (thermometer) and control center are both contained in the unit called a thermostat. When room temperature falls below a **set point,** say 20°C (68°F), the thermometer activates a switch that turns on the heater (the effector). When the thermometer (receptor) informs the control center in the thermostat that the temperature has risen above the set point, the heater is switched off. This type of control circuit is called **negative feedback,** because a change in the variable being monitored triggers a response that counteracts the initial fluctuation. Owing to the lag time between sensation and response, the variable drifts slightly above and below the set point, but the fluctuations are moderate. Negative-feedback mechanisms prevent small changes from becoming too large. Most known homeostatic mechanisms in animals operate on this principle of negative feedback.

Our own body temperature is kept close to a set point of 37°C by the cooperation of several negative-feedback circuits. One of these involves sweating as a means to cool the body. A part of the brain called the hypothalamus monitors the temperature of the blood. If this thermostat detects a rise in body temperature above the set point, it sends nervous impulses directing sweat glands to increase their production of sweat, which lowers body temperature by evaporative cooling (see Chapter 3). When body temperature drops below the set point, the thermostat in the hypothalamus stops sending the signals to the glands. We will see in the chapters that follow how animals use other negative-feedback circuits to control several other characteristics of the internal environment.

We can also note physiological examples of **positive feedback,** where a change in some variable triggers mechanisms that amplify rather than reverse the change. During childbirth, for instance, the pressure of the baby's head against sensors near the opening of the uterus stimulates uterine contractions, which cause greater pressure against the uterine opening, heightening the contractions, which causes still greater pressure. Positive feedback brings childbirth to completion, and this episodic function is a very different sort of process from maintaining a physiological steady state.

It is important not to overstate the concept of a "constant internal environment." In fact, regulated change is essential to an animal's normal physiology. In some cases, the changes are cyclical, such as the changes in hormone levels responsible for the menstrual cycle in human females (see Chapter 42). In other cases, a regulated change is a reaction to some contingency. For example, the human body reacts to certain infections by raising the set point for temperature to a slightly higher level, and the resulting fever helps fight the infections. Notice that over the short term, homeostatic mechanisms are still keeping body temperature close to a set point, whatever it is at that particular time. But over the longer term, the homeostatic system allows regulated change in the body's internal environment, illustrating adaptive response as one of the themes of organismal biology.

* * *

Having examined some general principles of animal body form and physiology, we are now ready to compare how diverse animals perform such functions as digestion, circulation, gas exchange, excretion of wastes, reproduction, and coordination.

STUDY OUTLINE

1. In our study of animals, we will apply two central themes of biology: the correlation between structure and function, and the capacity to adapt to the environment by both physiological adjustments and natural selection.

Levels of Structural Organization (pp. 782–787)

1. The organization of animals emerges from the grouping of specialized cells into tissues, tissues into organs, and organs into organ systems.
2. Epithelial tissue covers the outside of the body and lines internal organs and cavities. Consistent with their barrier function, epithelial cells are tightly packed and rest on a basement membrane. Epithelia are described according to the number of cell layers (simple or stratified) and the shape of the surface cells (cuboidal, columnar, or squamous).
3. Epithelia may be specialized for absorption and secretion. The mucus secreted by the mucous membranes lining the gut and respiratory system lubricates and moistens these surfaces.
4. Connective tissues bind and support other tissues. Unlike epithelia, the cells of connective tissue are sparsely scattered through a nonliving extracellular matrix of fibers and ground substance.
5. Loose connective tissue, used as binding and packing material, consists of fibroblasts and macrophages interspersed among collagenous, elastic, and reticular fibers. Adipose (fat) tissue is a specialized type of loose connective tissue. Fibrous connective tissue, found in tendons and ligaments, is made of dense, parallel bundles of collagenous fibers. Cartilage, bone, and blood are also classified as connective tissues. Cartilage is a strong yet flexible support material consisting of collagenous fibers and a rubbery ground substance secreted by chondrocytes. The osteocytes of bone, which secrete calcium phosphate, are embedded in repeating units called Haversian systems.
6. Muscle tissue is composed of long, excitable cells containing parallel microfilaments of contractile proteins. The muscle tissue of vertebrates is subdivided into skeletal, cardiac, and visceral muscles, which differ in shape, striation, and nervous control.
7. Neurons are the functional units of nervous tissue. The sensory and transmission functions of nervous tissue are reflected in the anatomy of the neuron, which is composed of a cell body with extensions called dendrites and axons that transmit impulses.
8. In all but the simplest animals, tissues are organized into organs, functional units with complex structures. Many organs of vertebrates are suspended by mesenteries inside fluid-filled body cavities. Each organ system, or group of organs, has its specific function, but the activities of all systems are coordinated in the successful functioning of the whole organism.

Size, Shape, and the External Environment (pp. 787–789)

1. Each cell of a multicellular animal must have access to an aqueous environment, a fact that dictates body shape. Simple two-layered sacs and flat, planar shapes maximize exposure to the surrounding medium. More complex, compact body plans have highly folded, moist internal surfaces specialized for exchanging substances with the environment. A circulatory system carries these substances among the parts of an animal.

The Animal's Internal Environment (pp. 789–791)

1. The internal environment surrounding the cells is usually very different from the external environment surrounding the entire animal. In addition, the internal environment is carefully controlled and regulated, a phenomenon called homeostasis.
2. Homeostatic mechanisms usually involve negative feedback. These mechanisms also enable regulated change, as they react to occasional shifts in the body's set points for variables, such as temperature.

SELF-QUIZ

1. A histologist would describe the layered, flat, scaly epithelium of human skin as
   a. simple squamous
   b. stratified squamous
   c. stratified columnar
   d. pseudostratified cuboidal
   e. ciliated columnar
2. Which of the following structures or materials is *incorrectly* paired with a tissue?
   a. Haversian system—bone
   b. platelets—blood
   c. fibroblasts—skeletal muscle
   d. chondrin—cartilage
   e. basement membrane—epithelium
3. Which structures contribute most to the tensile strength of loose connective tissues?
   a. elastic fibers
   b. myofibrils
   c. collagenous fibers
   d. chondrin
   e. reticular fibers
4. The involuntary muscles that cause the wavelike contractions pushing food along our intestine are
   a. striated muscles
   b. cardiac muscles
   c. skeletal muscles
   d. smooth muscles
   e. intercalated muscles
5. Which of the following is *not* considered to be a tissue?
   a. cartilage
   b. mucous membrane lining the stomach
   c. blood
   d. brain
   e. cardiac muscle
6. The membranes that suspend vertebrate organs in the body cavity are called
   a. visceral muscle
   b. loose connective tissue
   c. coelomic membranes
   d. ligaments
   e. mesenteries
7. Compared to a smaller cell, a larger cell of the same shape has
   a. less surface area
   b. less surface area per unit of volume
   c. the same surface-to-volume ratio
   d. a lesser average distance between its mitochondria and the external source of oxygen
   e. a smaller cytoplasm-to-nucleus ratio
8. Which of the following vertebrate organ systems does *not* open directly to the external environment?
   a. digestive system
   b. circulatory system
   c. excretory system
   d. respiratory system
   e. reproductive system
9. Most of our cells are surrounded by
   a. blood
   b. fluid equivalent to seawater in salt composition
   c. interstitial fluid
   d. pure water
   e. air
10. Which of the following physiological responses is an example of *positive* feedback?
    a. An increase in the concentration of glucose in the blood stimulates the pancreas to secrete insulin, a hormone that lowers blood glucose concentration.
    b. A high concentration of carbon dioxide in the blood causes deeper, more rapid breathing, which expels carbon dioxide.
    c. Stimulation of a nerve cell causes sodium ions to leak into the cell, and the sodium influx triggers the inward leaking of even more sodium.
    d. The body's production of red blood cells, which transport oxygen from lungs to other organs, is stimulated by a low concentration of oxygen.
    e. The pituitary gland secretes a hormone called TSH, which stimulates the thyroid gland to secrete another hormone called thyroxine. A high concentration of thyroxine suppresses the pituitary's secretion of TSH.

CHALLENGE QUESTIONS

1. Red blood cells pick up oxygen as they travel through the capillaries (microscopic blood vessels) of the lungs, and then deposit the oxygen as they pass through capillaries in other organs. Considering this function, suggest an advantage for our blood being populated by enormous numbers of very small red blood cells rather than fewer large cells. (Assume that the *total* volume of red blood cells is the same in both cases.)
2. Choose three vertebrate tissues and describe how their structures fit their functions.

3. Identify each of the tissues below (LMs). Be as specific as possible.

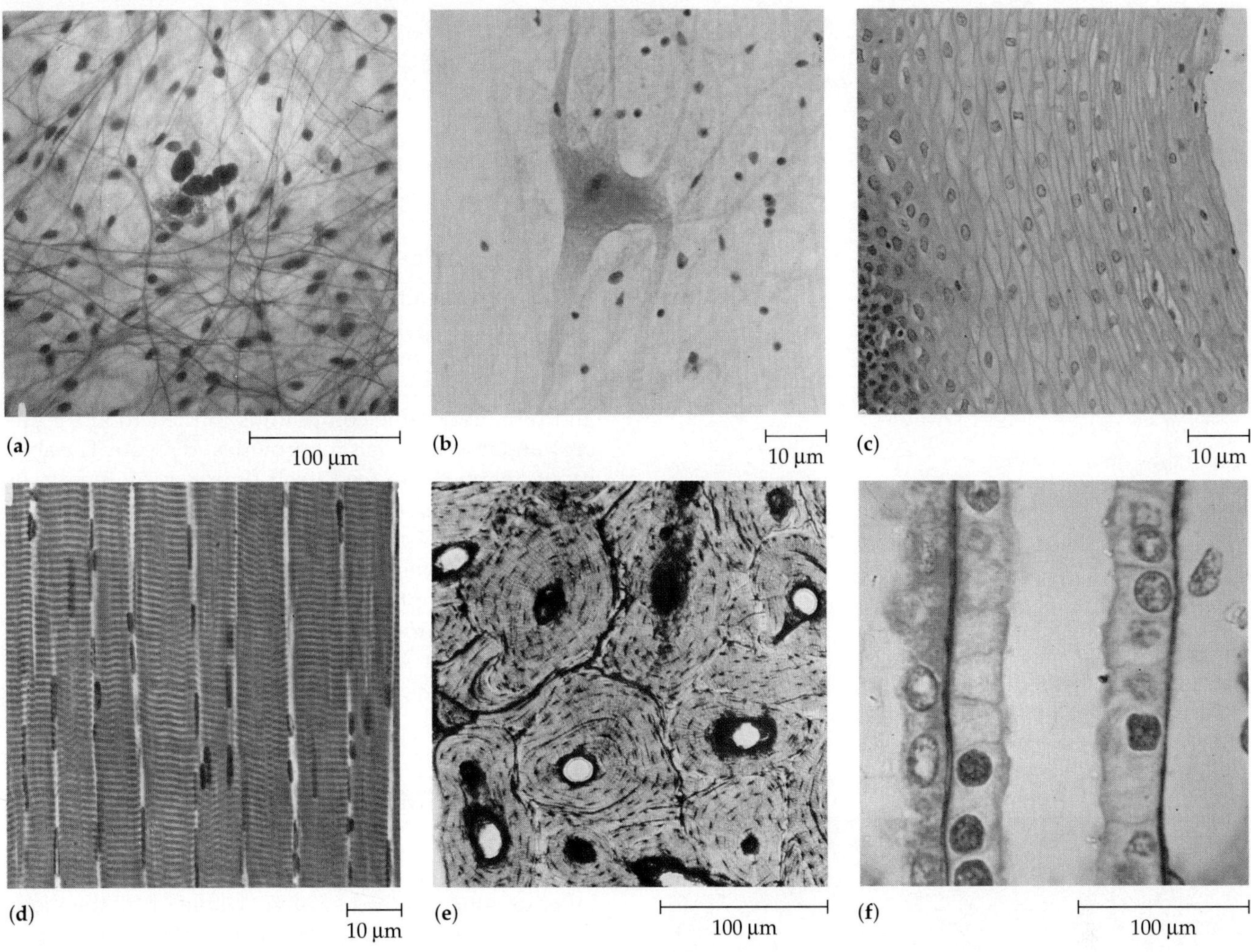

SCIENCE, TECHNOLOGY, AND SOCIETY

1. Medical researchers are investigating the possibilities of artificial substitutes for various human tissues. Examples are a liquid that could serve as "artificial blood" and a fabric that could temporarily serve as artificial skin for victims of serious burns. In what other situations might artificial blood or skin be useful? What characteristics would these substitutes need in order to function effectively in the body? Why do real tissues work better? Why not use the real things if they work better? Can you think of other artificial tissues that might be useful? Can you anticipate problems in developing and applying them?

FURTHER READING

Amato, I. "Heeding the Call of the Wild." *Science,* August 30, 1991. Materials scientists study the structure and function of skeletons and other animal products.

Caplan, A. "Cartilage." *Scientific American,* October 1984. The structure and function of an important tissue.

Houston, C. S. "Mountain Sickness." *Scientific American,* October 1992. High altitude interferes with homeostasis.

Marieb, E. *Human Anatomy and Physiology,* 2nd ed. Redwood City, CA: Benjamin/Cummings, 1992. A basic text, beautifully illustrated.

Schmidt-Nielsen, K. *Animal Physiology: Adaptation and Environment,* 4th ed. New York: Cambridge University Press, 1990. A comparative physiology text emphasizing homeostasis.

Sochurek, H., and P. Miller. "Medicine's New Vision." *National Geographic,* January 1987. Remarkable methods for photographing the human interior.

37 Animal Nutrition

Food Types and Feeding Mechanisms
Digestion: A Comparative Introduction
The Mammalian Digestive System
Evolutionary Adaptations of Vertebrate Digestive Systems
Nutritional Requirements

Figure 37.1
Ingestion, the animal mode of nutrition. Most animals nourish themselves by eating other organisms. Koalas, Australian marsupials, obtain almost all their food and water by ingesting leaves of a few eucalyptus species. It is not a very rich diet, but adaptations of the digestive tract enable koalas to specialize in a food source for which there is relatively little competition from other animals. In this chapter, you will learn about the diverse diets, feeding mechanisms, and digestive adaptations that have evolved in the animal kingdom.

Every mealtime is a reminder that we are animals, dependent on a regular supply of food derived from other organisms. As heterotrophs, animals are unable to live on inorganic nutrients alone and rely on organic compounds in their food for energy and raw materials for growth and repair. The ability of an animal to feed itself figures prominently in reproductive success, and natural selection has produced many fascinating nutritional adaptations during the long evolution of the animal kingdom (Figure 37.1). In this chapter, we will compare and contrast the diverse mechanisms by which animals obtain and process their food.

FOOD TYPES AND FEEDING MECHANISMS

Most animals are holotrophs—organisms that ingest other organisms, dead or alive, whole or by the piece (exceptions are certain parasitic animals, such as tapeworms, which absorb organic molecules directly across their outer body surface). Diets vary. **Herbivores,** including gorillas, cows, koalas, and fish that graze on algae, eat plants (the Latin *vorare* means "to eat"). **Carnivores,** such as sharks, hawks, spiders, and snakes, ingest other animals. **Omnivores** (from the Latin *omnis*, "all") consume both meat and plant material. Cockroaches, crows, raccoons, and humans, who evolved as hunters and gatherers, are examples of omnivores.

The mechanisms by which animals obtain their food also vary. Many aquatic animals are **suspension-feeders** that sift small food particles from the water. Clams and oysters, for example, use their gills to trap tiny morsels, which are then swept along with a film of mucus to the mouth by beating cilia. Baleen whales, the largest animals ever to live, are also suspension-feeders. They swim with their mouths agape, straining millions of small animals from huge volumes of water forced through screenlike plates attached to their jaws (Figure 37.2).

Substrate-feeders live in or on their food source, eating their way through the food. Examples are leaf miners, which are larvae of various insects that tunnel through the soft mesophyll in the interior of leaves

(a)

(b)

Figure 37.2
Suspension-feeding: baleen whales. (**a**) Gray whales and other baleen whales use comblike plates suspended from the upper jaw to sift small crustaceans called krill from enormous volumes of water. The whale opens its mouth and fills an expandable oral pouch with water, and then closes its mouth and contracts the pouch. This forces water out of the mouth through the baleen, leaving a mouthful of trapped food. (**b**) A behavioral adaptation enables the humpback whale to concentrate krill before filtering, as illustrated in this painting. The whale sets a "bubble net" by blowing a ring of bubbles as it swims in an upward spiral from a depth of about 15 m. Krill and other small animals near the surface tend to swim away from the bubbles, causing them to concentrate in the center of the bubble net. After herding the krill, the whale harvests its catch by swimming through the center of the ring with its mouth agape.

(Figure 37.3). Earthworms are also substrate-feeders or, more specifically, **deposit-feeders.** Eating its way through the dirt, the earthworm salvages detritus, partially decayed organic material ingested along with soil.

Fluid-feeders make their living by sucking nutrient-rich fluids from a living host (Figure 37.4). Aphids, for example, tap the phloem sap of plants, and leeches and mosquitos suck blood from animals. Because these particular fluid-feeders harm their hosts, they are considered parasites. Hummingbirds and bees, on the other hand, perform a service for their hosts, as they transfer pollen when they visit flowers to suck nectar.

Most animals, rather than filtering food from water, eating their way through the substrate, or sucking fluids, are **bulk-feeders** that ingest relatively large pieces of food (Figure 37.5). They use such diverse utensils as tentacles, pincers, claws, poisonous fangs, and jaws and teeth to kill their prey or to tear off pieces of meat or vegetation. Humans, of course, feed in this manner.

DIGESTION: A COMPARATIVE INTRODUCTION

Regardless of what it eats or how it feeds, an animal must digest its food. **Digestion** is the process of breaking food down into molecules small enough for the body to absorb. The bulk of the organic material in food consists of proteins, fats, and carbohydrates in the form of starch and other polysaccharides. Although these are suitable raw materials, animals cannot use them directly, for two reasons. First, these molecules are too large to pass through membranes and enter the

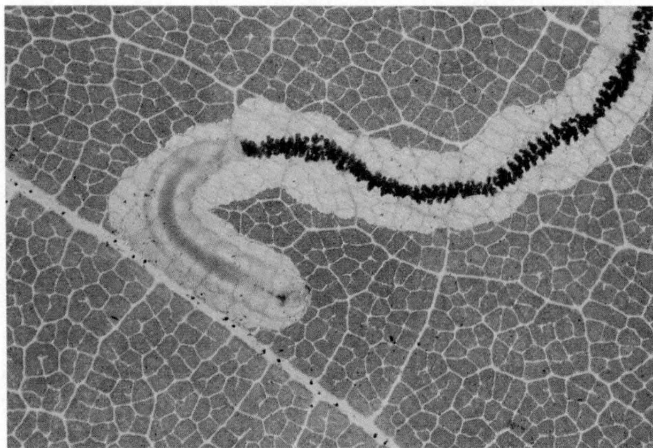

Figure 37.3
Substrate-feeding: a leaf miner. Eating its way through the soft mesophyll of an oak leaf, this caterpillar, the larva of a moth, leaves a trail of dark fecal pellets in its wake.

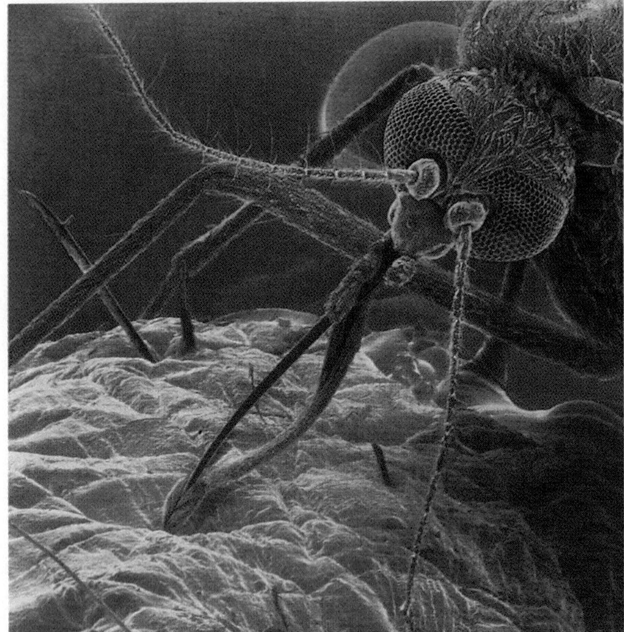

Figure 37.4
Fluid-feeding: a mosquito. This parasite has impaled its human host with a hypodermic-like mouth part and will fill its gut with a blood meal (artificially colored SEM).

cells of the animal. Second, the macromolecules that make up an animal are not identical to those of its food. In building their macromolecules, however, all organisms use common monomers. For instance, soybeans, cattle, and people all assemble their proteins from the same 20 kinds of amino acids (see Chapter 5). Digestion cleaves macromolecules into their component monomers, which the animal can then use to assemble its own brand of molecules. Polysaccharides and disaccharides are split into simple sugars, fats are digested to glycerol and fatty acids, proteins are broken down to amino acids, and nucleic acids are cleaved into nucleotides. The digestion of macromolecules is catalyzed by a special class of enzymes.

Enzymatic Hydrolysis

Recall from Chapter 5 that when a cell makes a macromolecule by linking together monomers, it does so by removing a molecule of water for each new covalent bond formed. Digestion reverses this process by breaking bonds with the enzymatic addition of water. (Actually, a hydrogen is added to one side of the bond and a hydroxyl group is added to the other side—the net effect is the addition of H_2O.) This splitting process is called **hydrolysis** (lysing with water). There are hydrolytic enzymes that digest each of the classes of macromolecules found in food. This chemical digestion is usually preceded by mechanical fragmentation of the food—by chewing, for instance. Breaking food into smaller pieces increases the surface area exposed to digestive juices containing hydrolytic enzymes.

An animal must digest its food in some type of specialized compartment where enzymes can attack the food molecules without damaging the animal's own cells. After the food is digested, amino acids, simple sugars, and other small molecules are absorbed—that is, they cross the membrane of the digestive compartment and enter the cells of the animal. Undigested material is then eliminated from the digestive compartment.

Intracellular Digestion in Food Vacuoles

The simplest digestive compartments are food vacuoles, organelles where a single cell can digest its food without the hydrolytic enzymes mixing with the cell's own cytoplasm. Protozoa digest their meals in food

Figure 37.5
Bulk-feeding: a python. Most animals ingest relatively large pieces of food, though rarely as large compared to the size of the animal as in the case illustrated here. In this amazing scene, a rock python is beginning to ingest a gazelle.

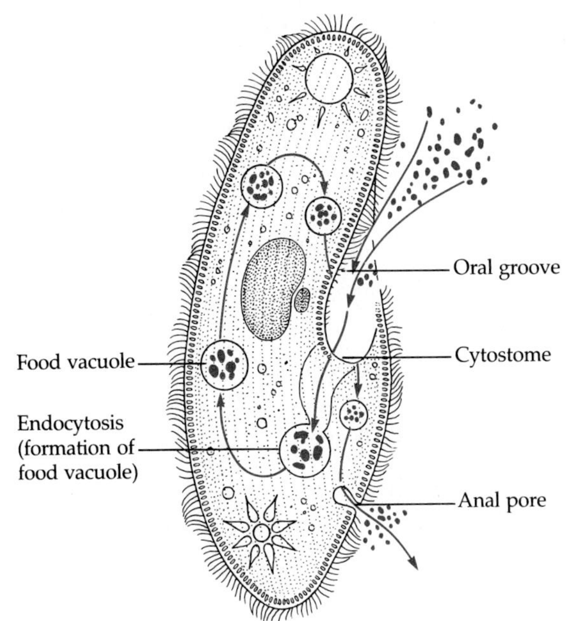

Figure 37.6
Intracellular digestion in *Paramecium*. *Paramecium* has a specialized feeding structure called the oral groove, which leads to the cell's "mouth" (cytostome). Cilia that line the groove draw water and suspended food particles toward the mouth, where the food is packaged by endocytosis into a food vacuole that functions as a miniature digestive compartment. Cytoplasmic streaming carries food vacuoles around the cell, while hydrolytic enzymes are secreted into the vacuoles. As molecules in the food are digested, the nutrients (sugars, amino acids, and other small molecules) are transported across the membrane of the vacuole into the cytoplasm. Later, the vacuole fuses with an anal pore, a specialized region of the plasma membrane where undigested material can be eliminated by exocytosis.

vacuoles, usually after engulfing the food by endocytosis (phagocytosis or pinocytosis; see Chapter 8). This digestive mechanism is termed **intracellular digestion** (Figure 37.6). Sponges are among the true animals that digest their food entirely by the intracellular mechanism (Figure 37.7). In most animals, however, at least some hydrolysis occurs by **extracellular digestion** within compartments that are continuous, via passages, with the outside of the body.

Digestion in Gastrovascular Cavities

Some of the simplest animals have digestive sacs with single openings. These pouches, called **gastrovascular cavities,** function in both digestion and distribution of nutrients throughout the body (hence, the *vascular* part of the term). The cnidarian *Hydra* provides a good example of how a gastrovascular cavity works (Figure 37.8a). *Hydra* is a carnivore that stings its prey with

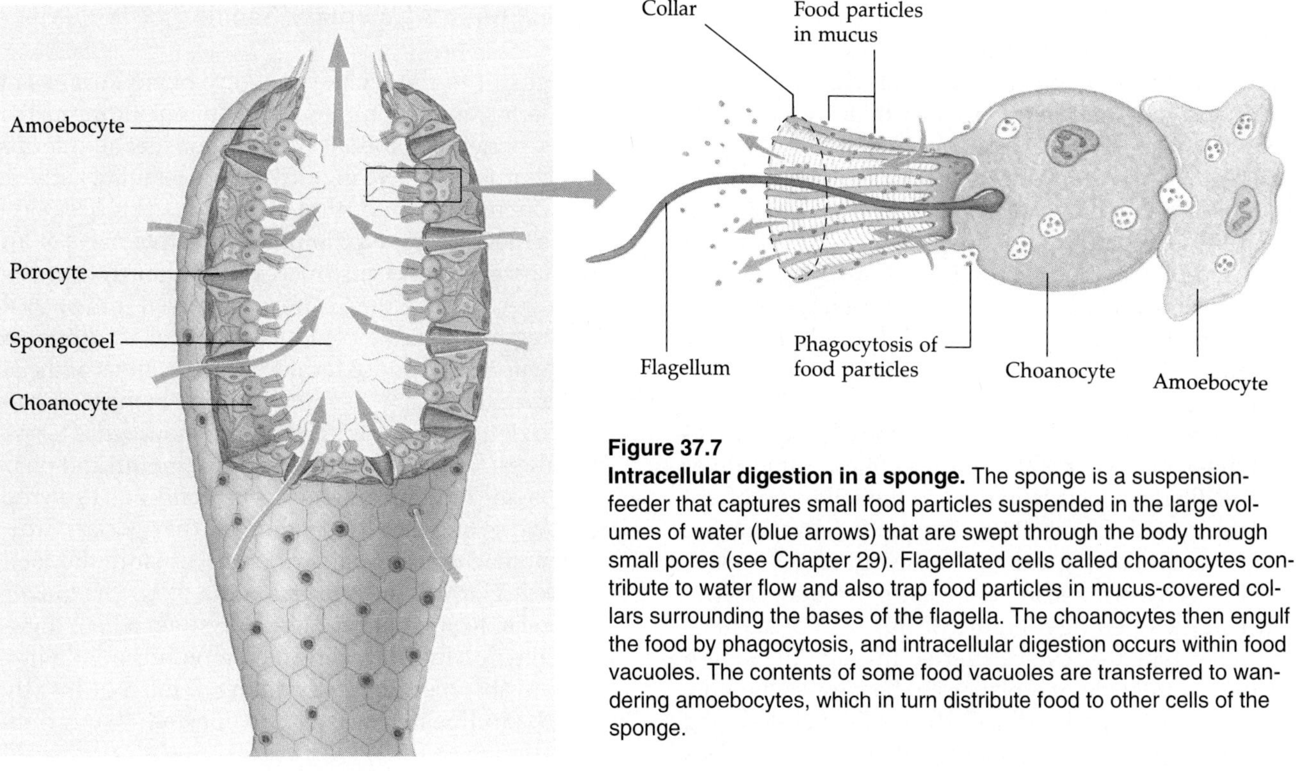

Figure 37.7
Intracellular digestion in a sponge. The sponge is a suspension-feeder that captures small food particles suspended in the large volumes of water (blue arrows) that are swept through the body through small pores (see Chapter 29). Flagellated cells called choanocytes contribute to water flow and also trap food particles in mucus-covered collars surrounding the bases of the flagella. The choanocytes then engulf the food by phagocytosis, and intracellular digestion occurs within food vacuoles. The contents of some food vacuoles are transferred to wandering amoebocytes, which in turn distribute food to other cells of the sponge.

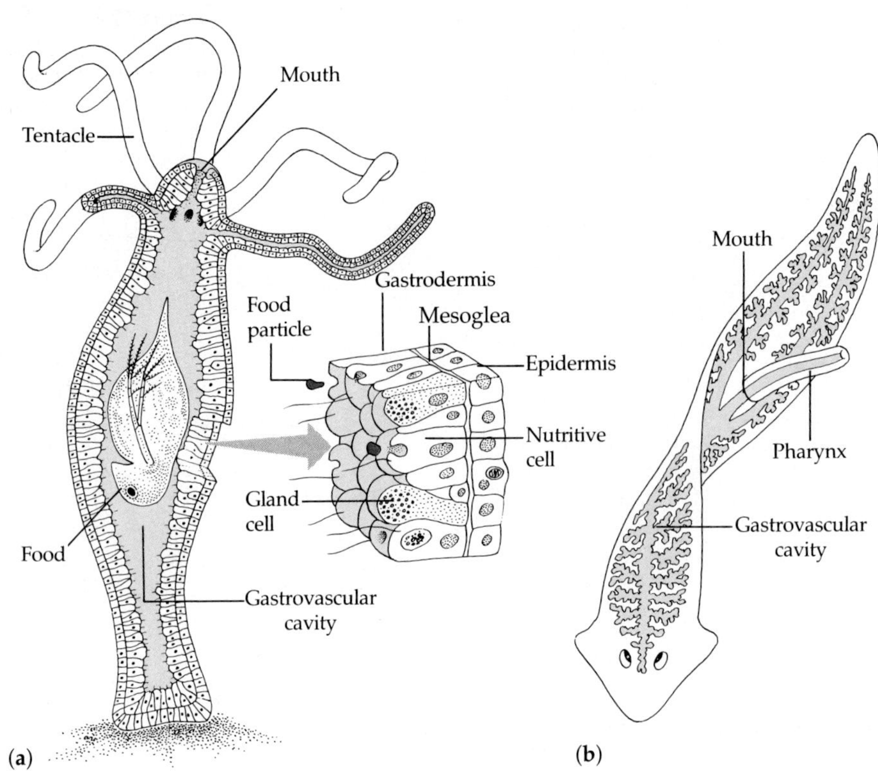

Figure 37.8
Gastrovascular cavities. (**a**) *Hydra.* The outer epidermis has protective and sensory functions, whereas the inner gastrodermis is specialized for digestion. The mesoglea, a jellylike layer, separates the two layers of cells. Enzymes released from gland cells into the gastrovascular cavity initiate digestion, which is completed intracellularly after small food particles are taken into nutritive cells by phagocytosis. Undigested waste is egested through the mouth, the single opening of the gastrovascular cavity. (**b**) *Planaria.* The gastrovascular cavity of a planarian digests food in a fashion similar to that of *Hydra.* Its extensive branching increases the surface area available for digestion and transports nutrients to all parts of the animal.

specialized organelles called nematocysts and then uses prehensile tentacles to stuff the food into a mouth that can stretch to accommodate an animal larger than the *Hydra* itself (see Chapter 29). With the food now in the gastrovascular cavity, specialized cells of the gastrodermis, the tissue layer that lines the cavity, secrete digestive enzymes that fragment the soft tissues of the prey into tiny pieces. Beating flagella on the gastrodermal cells keep the small food particles from settling and distribute them throughout the cavity. The food particles are taken into gastrodermal cells by phagocytosis, and food molecules are then hydrolyzed within food vacuoles. Extracellular digestion within the gastrovascular cavity initiates breakdown of the food, but most of the actual hydrolysis of macromolecules is accomplished by intracellular digestion. After *Hydra* has digested its meal, undigested materials remaining in the gastrovascular cavity, such as the exoskeletons of small crustaceans, are eliminated through the single opening, which functions in the dual role of mouth and anus.

If the gastrodermal cells are capable of phagocytosis and digestion, what is the function of an extracellular cavity? Phagocytosis is limited to the ingestion of microscopic food. Extracellular cavities for digestion are adaptations enabling animals to devour larger prey.

Most flatworms also have gastrovascular cavities with single openings (Figure 37.8b). The pharynx, a muscular tube that can be everted through the mouth on the ventral side of a planarian, penetrates the prey and releases a digestive juice. The partly digested food is sucked into the gastrovascular cavity, which has three main branches and many secondary ramifications that increase the surface area of the cavity extensively. As in *Hydra,* a planarian begins digesting its food within the gastrovascular cavity, but digestion is continued within cells that take up food particles by phagocytosis.

Digestion in Alimentary Canals

In contrast to the saclike anatomy of cnidarians and flatworms, more complex animals—including nematodes, annelids, mollusks, arthropods, echinoderms, and chordates—have digestive tubes running between two openings, a mouth and an anus (Figure 37.9). These tubes are called **complete digestive tracts** or **alimentary canals.** Because food moves along the canal in one direction, the tube can be organized into specialized regions that carry out digestion and absorption of nutrients in a stepwise fashion. Some regions of the alimentary canal have a similar function in most species, although the tubes are adapted in various ways to specific diets. Food ingested through the mouth and pharynx passes through an esophagus that leads to a crop, gizzard, or stomach, depending on the species. Crops and stomachs are usually organs that store the food, whereas gizzards actually grind the food. The pulverized meal next enters the long intestine, where digestive enzymes hydrolyze the food molecules, and nutrients are absorbed across the lining of the tube into the blood. Undigested wastes are egested through the **anus.**

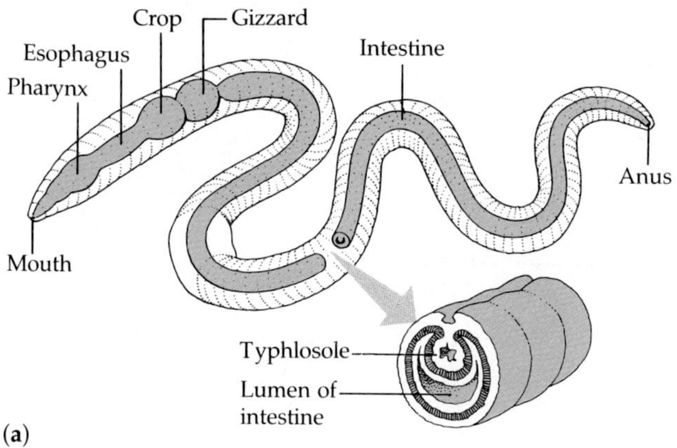

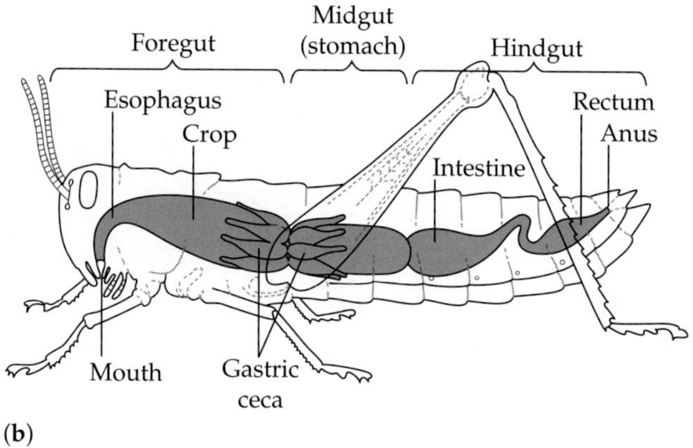

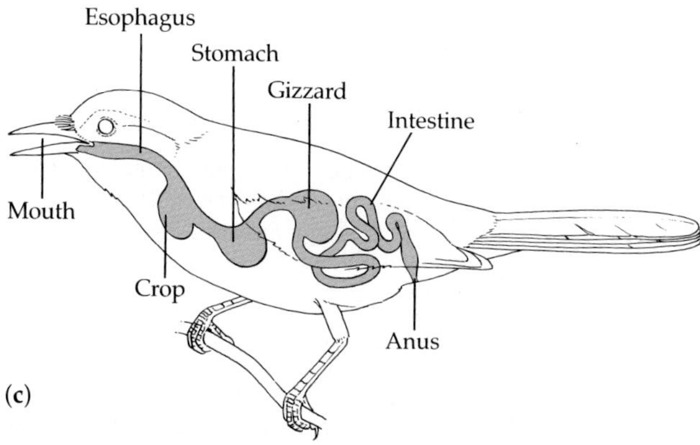

Figure 37.9
Alimentary canals. (**a**) The digestive tract of an earthworm consists of five organs. A muscular pharynx sucks food in through the mouth. Food passes through the esophagus and is stored and moistened in the crop. The muscular gizzard, which contains small bits of sand and gravel, pulverizes the food. Digestion and absorption occur in the intestine, which has a dorsal fold, the typhlosole, that increases the surface area for nutrient absorption. Finally, undigested material is expelled through the anus. (**b**) A grasshopper has several digestive chambers grouped into three main regions: a foregut, midgut, and hindgut. Food is moistened and stored in the crop, but most digestion occurs in the stomach. Gastric ceca, pouches extending from the stomach, transfer nutrients to the grasshopper's hemolymph (blood). (**c**) A bird has three separate chambers—the crop, stomach, and gizzard—where food is pulverized and churned before passing into the intestine.

THE MAMMALIAN DIGESTIVE SYSTEM

The mammalian digestive system consists of the alimentary canal and various accessory glands that secrete digestive juices into the canal through ducts. From the esophagus to the large intestine, the mammalian digestive tract has a four-layered wall (see Figure 36.7). The lumen (cavity) is lined by a mucous membrane, the mucosa. Next comes a layer made up mostly of connective tissue, followed by a layer of smooth muscle. The outermost layer is a sheath of connective tissue attached to the membrane of the body cavity. **Peristalsis,** rhythmic waves of contraction by the smooth muscles, pushes the food along the tract. At some of the junctions between specialized segments of the tube, the muscular layer is modified into ringlike valves called **sphincters,** which close off the tube like drawstrings to regulate the passage of material between chambers of the canal. In addition to the alimentary canal itself, the mammalian digestive system includes accessory glands that deliver digestive juices to the canal through ducts. These accessory organs are three pairs of **salivary glands,** the **pancreas,** and the **liver** with its storage organ, the **gallbladder.** The accessory glands originate in the embryo as outpocketings of the primordial gut; the ducts are remnants of the connections of the glands to the alimentary canal.

Using the human as an example, we can now follow a meal through the alimentary canal, seeing in more detail what happens to the food in each of the processing stations along the way (Figure 37.10).

The Oral Cavity

Both physical and chemical digestion of foods begin in the mouth. During chewing, teeth of various shapes cut, smash, and grind food, making the food easier to swallow and increasing its surface area. The presence of food in the **oral cavity** triggers a nervous reflex that causes the salivary glands to deliver saliva through ducts to the oral cavity. Even before food is actually in the mouth, salivation may occur in anticipation because of learned associations between eating and the time of day, cooking odors, or other stimuli.

In humans, more than a liter of saliva is secreted into the oral cavity each day. Dissolved in the saliva is a slippery glycoprotein (carbohydrate-protein complex) called mucin, which protects the soft lining of the mouth from abrasion and lubricates the food for easier swallowing. Saliva contains buffers that help prevent dental caries by neutralizing acid in the mouth. Also, antibacterial agents in saliva kill many of the bacteria that enter the mouth with food. Finally, **salivary amylase,** a digestive enzyme that hydrolyzes starch, is present in saliva. Salivary amylase can break only every other bond in the polysaccharide, so the smallest prod-

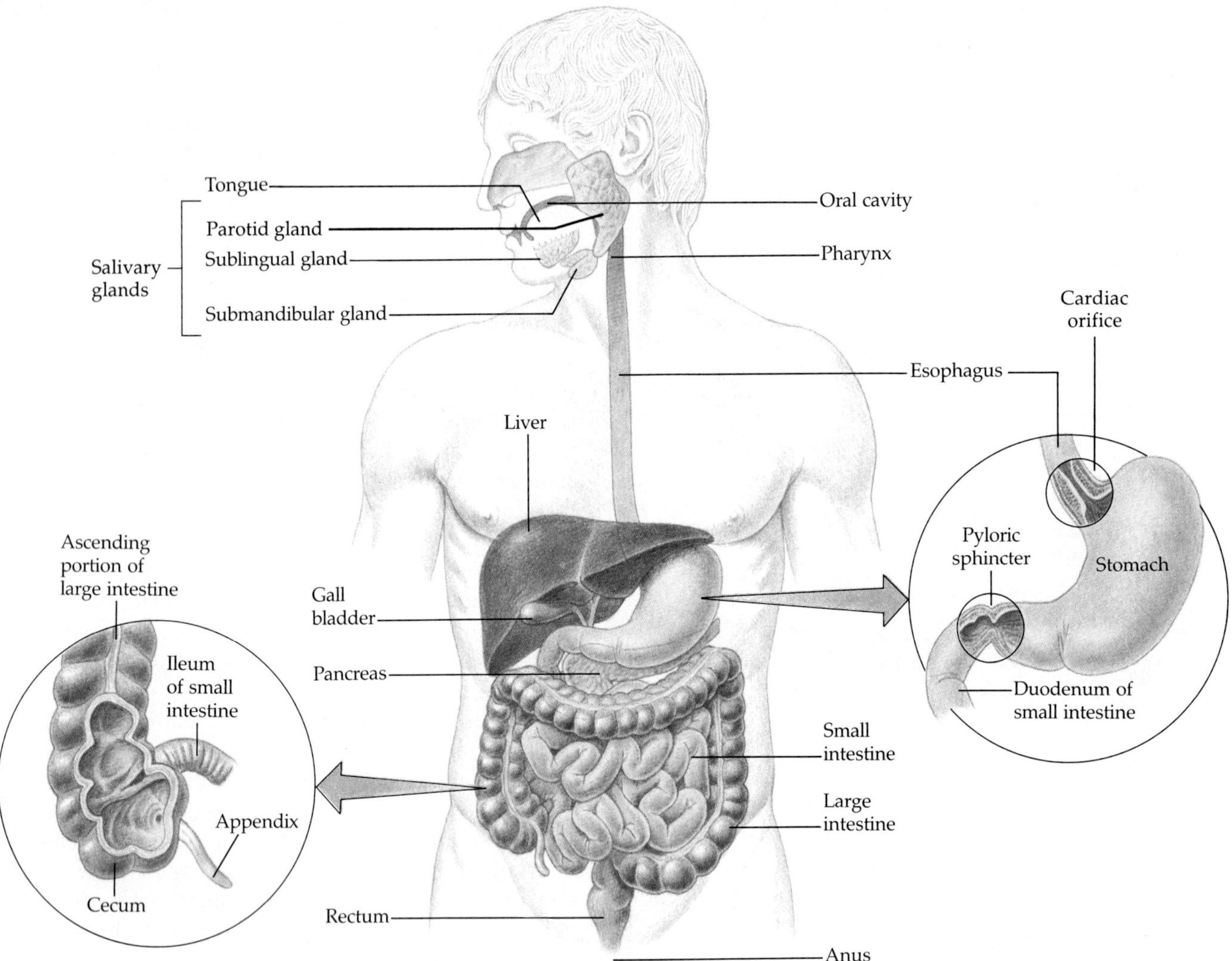

Figure 37.10
The human digestive system. Food enters the system through the mouth, where it is chewed for some seconds before being swallowed. Food takes 5 to 10 seconds to pass down the esophagus and into the stomach, where it spends 2 to 6 hours being partially digested. Final digestion and nutrient absorption occur in the small intestine over a period of 5 to 6 hours. In 12 to 24 hours, any undigested material passes through the large intestine, and feces are expelled through the anus.

uct of this digestion is the double sugar maltose. Food is not usually in the mouth long enough for amylase to do more than split starch into smaller polysaccharides; perhaps a major function of salivary amylase is to prevent an accumulation of sticky starch between teeth.

The tongue is used not only to taste but also to manipulate the food during chewing, and then to help shape the food into a ball called a **bolus.** Food is swallowed when the tongue pushes the bolus to the very back of the oral cavity and into the pharynx.

The Pharynx

The region we call our throat is the **pharynx,** an intersection that leads to both the esophagus and the windpipe (trachea). When we swallow, the top of the windpipe moves so that its opening is blocked by a cartilaginous flap called the **epiglottis** (Figure 37.11). You can see this motion in the bobbing of the "Adam's apple" during swallowing. The epiglottis and the vocal cords in the opening of the windpipe guard the respiratory system against the entry of food or fluids. The swallowing mechanism normally ensures that a bolus will be guided into the entrance of the esophagus.

The Esophagus

The **esophagus** is the segment of the alimentary canal that conducts food from the pharynx down to the stomach. Peristalsis squeezes a bolus along the narrow

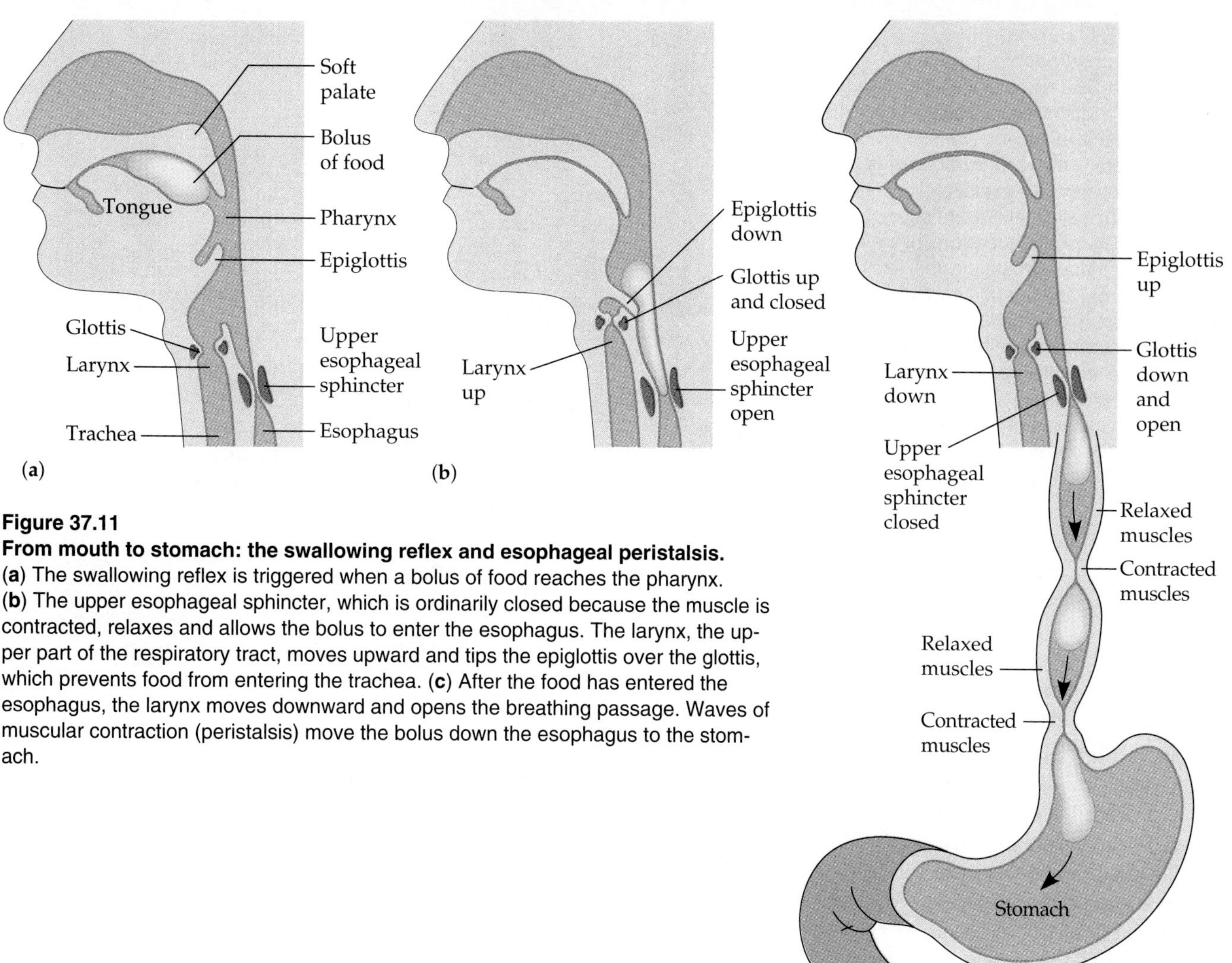

Figure 37.11
From mouth to stomach: the swallowing reflex and esophageal peristalsis.
(a) The swallowing reflex is triggered when a bolus of food reaches the pharynx. **(b)** The upper esophageal sphincter, which is ordinarily closed because the muscle is contracted, relaxes and allows the bolus to enter the esophagus. The larynx, the upper part of the respiratory tract, moves upward and tips the epiglottis over the glottis, which prevents food from entering the trachea. **(c)** After the food has entered the esophagus, the larynx moves downward and opens the breathing passage. Waves of muscular contraction (peristalsis) move the bolus down the esophagus to the stomach.

esophagus (see Figure 37.11c). Only the muscles at the very top of the esophagus are striated (voluntary). Thus, swallowing is initiated voluntarily, but then the involuntary waves of contraction by the smooth muscles take over.

The Stomach

The **stomach** is located on the left side of the abdominal cavity, just below the diaphragm. Because this large organ can store an entire meal, we do not need to eat constantly. With a very elastic wall and accordion-like folds called rugae, the stomach can stretch to accommodate about 2 L of food and fluid.

The epithelium that lines the lumen of the stomach secretes **gastric juice,** a digestive fluid that mixes with the food. With its high concentration of hydrochloric acid, the gastric juice has a pH of about 2, acidic enough to dissolve iron nails. One function of the acid is to dismantle the tissues in food by disrupting the intercellular glue that binds cells together in meat and plant material. The acid also kills most bacteria that are swallowed with the food.

Also present in gastric juice is **pepsin,** an enzyme that hydrolyzes proteins. Hydrolysis is incomplete, however, since pepsin can only break peptide bonds adjacent to specific amino acids. The result of pepsin's action, then, is to cleave proteins into smaller polypeptides. Pepsin is one of the few enzymes that works best in a strongly acidic environment. Indeed, the low pH of gastric juice denatures the proteins in food, increasing exposure of their peptide bonds to the pepsin. Even the salivary amylase that is swallowed ceases to work soon after reaching the stomach and is digested along with the food proteins.

What prevents pepsin from destroying the cells of the stomach wall that produce this hydrolytic enzyme? Pepsin is synthesized and secreted in an inactive form called **pepsinogen** (Figure 37.12). Hydrochloric acid in the gastric juice converts pepsinogen to active pepsin by removing a short segment of the protein's polypep-

Figure 37.12
The secretion of gastric juice in the stomach. (**a**) The mucosa that lines the stomach consists of simple columnar epithelium organized into tubular gastric glands. Gastric pits lead into the gastric glands, which have three types of secretory cells. The cells that line the openings of the glands secrete mucus that helps protect the stomach from digesting itself. Deeper in the gland are parietal cells, which secrete hydrochloric acid, and chief cells, which secrete pepsinogen. (**b**) Within the gastric glands, hydrochloric acid converts the inactive pepsinogen to pepsin, which, in an example of positive feedback, activates more pepsin.

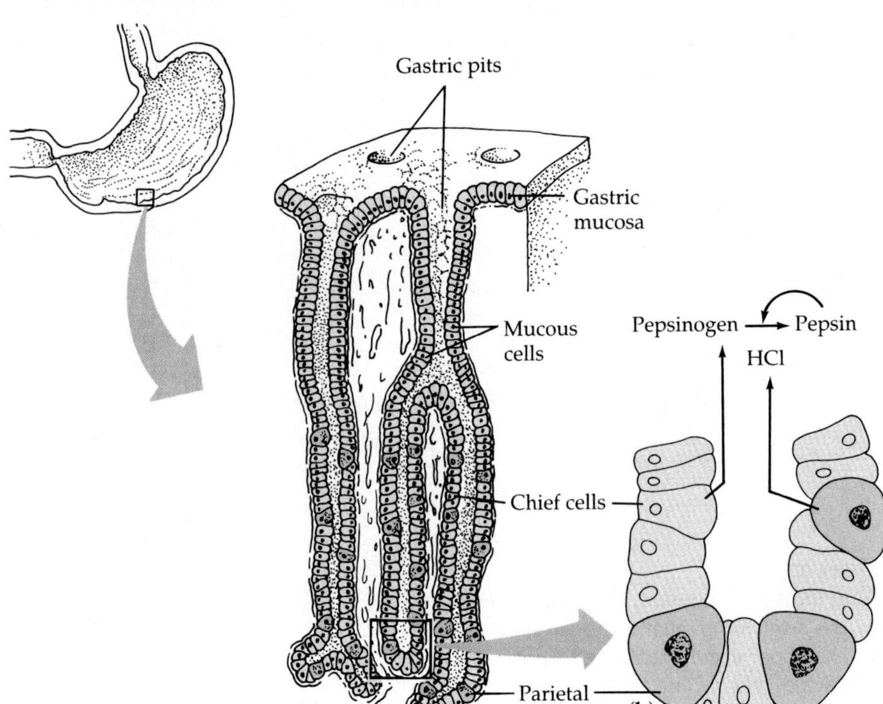

tide chain, an alteration that exposes the active site of pepsin. Because the acid and pepsinogen are secreted by different kinds of cells, the two ingredients do not mix until their release into the lumen of the stomach. Once some of the pepsinogen is activated by acid, a chain reaction occurs because pepsin itself can activate additional molecules of pepsinogen. This domino effect is an example of positive feedback. We will see other cases where protein-digesting enzymes are secreted in inactive forms, generally called **zymogens.**

A coating of mucus secreted by the epithelial cells helps protect the stomach lining from being digested by the pepsin and acid in gastric juice. Still, the epithelium is constantly eroded, and mitosis must generate enough cells to completely replace the stomach lining every three days. When pepsin and acid destroy the lining faster than it can regenerate, lesions called gastric ulcers occur.

Gastric secretion is controlled by a combination of nervous impulses and hormones. When we see, smell, or taste food, a nervous message from the brain to the stomach initiates the secretion of gastric juice. Then certain substances in the food stimulate the stomach wall to release a hormone called **gastrin** into the circulatory system. As the gastrin gradually recirculates in the bloodstream back to the stomach wall, the hormone stimulates further secretion of gastric juice. Thus, an initial burst of gastric secretion at mealtime is followed by a sustained secretion that continues to add gastric juice to the food for some time. If the pH of the stomach contents becomes too low, the acid inhibits release of gastrin, decreasing the secretion of gastric juice; this is an example of negative feedback. Each day, the stomach wall secretes about 3 L of gastric juice.

About every 20 seconds, the stomach contents are mixed by the churning action of smooth muscles. When an empty stomach churns, hunger pangs are felt. (Sensations of hunger are also associated with centers in the brain that monitor the nutritional status of the blood.) Sometimes the stomach rumbles loudly enough to be heard without the help of a stethoscope. What begins in the stomach as a meal mixed with gastric juice soon becomes a nutrient broth known as **acid chyme.**

Much of the time, the stomach is closed off at either end (see Figure 37.10). The opening from the stomach to the esophagus, the cardiac orifice, normally dilates only when a bolus driven by peristalsis arrives. Occasional backflow of acid chyme into the lower end of the esophagus causes "heartburn." (If backflow is a persistent problem, an ulcer may develop in the esophagus.) The cardiac orifice opens intermittently with each wave of peristalsis that delivers a bolus. At the bottom of the stomach is the **pyloric sphincter,** which helps regulate the passage of chyme from the stomach into the small intestine. A squirt at a time, it takes about 2 to 6 hours after a meal for the stomach to empty.

The Small Intestine

Although limited digestion of starch takes place in the oral cavity and partial digestion of proteins by pepsin in the stomach, most enzymatic hydrolysis of the macromolecules in food occurs in the **small intestine.** With a length of more than 6 m, the small intestine is the longest section of the alimentary canal (its name is based on its small diameter, compared with the diameter of the large intestine). It is not only the major organ of digestion, but also the part of the digestive system responsible for the absorption of most nutrients into the blood.

Regulation of Digestive Secretions Several other organs contribute to digestion in the small intestine by producing, storing, and secreting various digestive juices. One of these organs is the pancreas, which produces an assortment of hydrolytic enzymes as well as an alkaline solution rich in bicarbonate. The bicarbonate acts as a buffer to offset the acidity of chyme from the stomach.

Another accessory organ is the liver, which performs a wide variety of important functions in the body, including the production of **bile,** a mixture of substances that is stored in another accessory organ, the gallbladder, until needed. Bile contains no digestive enzymes, but it does contain bile salts, which aid in the digestion and absorption of fats. Bile also contains pigments that are by-products of red blood cell destruction in the liver; these bile pigments are eliminated from the body with the feces.

The first 25 cm or so of the small intestine is called the **duodenum,** and it is here that chyme seeping in from the stomach mixes with digestive juices secreted by the pancreas, liver, gallbladder, and gland cells of the intestinal wall itself.

At least four regulatory hormones help ensure that digestive secretions are present only when needed (Figure 37.13). We have already seen that gastrin is released from the stomach lining in response to the presence of food. The acidic pH of the chyme that enters the duodenum stimulates the intestinal wall to release a second hormone, **secretin.** This hormone signals the pancreas to release bicarbonate, which neutralizes the acid chyme. A third hormone, **cholecystokinin (CCK),** is produced by cells in the lining of the duodenum and causes the gallbladder to contract and release bile into the small intestine. CCK also triggers the release of pancreatic enzymes. The chyme, particularly if rich in fats, also causes the duodenum to release a fourth hormone, **enterogastrone,** which inhibits peristalsis in the stomach and thus slows down the entry of food into the small intestine.

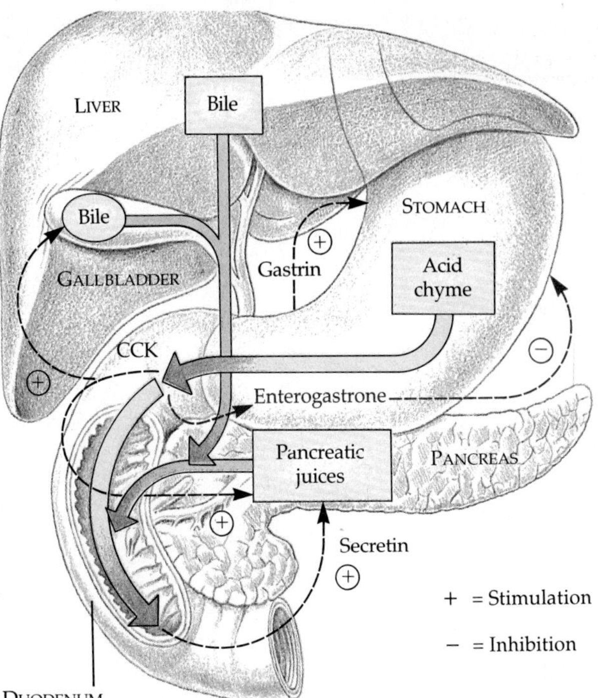

Figure 37.13
The control of digestion. Gastrin, the release of which is triggered by events associated with the introduction of food into the stomach (the presence of peptides, stomach distension, previous gastrin secretion), stimulates production of gastric juices. Most digestion of food occurs in the duodenum. The acid chyme seeps in from the stomach and is first neutralized. Secretin mediates this neutralization by stimulating the release of sodium bicarbonate by the pancreas. The presence of amino acids or fatty acids in the duodenum triggers the release of cholecystokinin (CCK), which stimulates the release of digestive enzymes by the pancreas and bile by the gallbladder. Enterogastrone slows digestion by inhibiting stomach peristalsis and acid secretion when acid chyme rich in fats (which require additional digestion time) enters the duodenum.

Digestion in the Small Intestine Let's now see how enzymes from the pancreas and intestinal wall combine to digest macromolecules. The actions of the digestive enzymes of the entire digestive system are summarized in Table 37.1.

The digestion of starch that was initiated by salivary amylase in the oral cavity is continued in the small intestine. A pancreatic amylase hydrolyzes starch into maltose, a disaccharide. Digestion is completed by the enzyme maltase, which splits the maltose and releases two molecules of the simple sugar glucose. Maltase is one of a family of **disaccharidases,** each specific for the hydrolysis of a different disaccharide. Sucrase, for instance, hydrolyzes table sugar (sucrose), and lactase digests milk sugar (lactose). (In general, adults have much less lactase than children. In some populations, such as certain tribes in Africa, lactase is missing altogether. If these people were to drink milk in quantity, they would develop cramps and diarrhea.) The disaccharidases are not dissolved within the lumen of the intestine but are actually built into the membranes and coating (glycocalyx) of the intestinal epithelium (the brush border in Table 37.1). Thus, the terminal steps in carbohydrate digestion occur at the site of sugar absorption.

Table 37.1 Enzymatic Digestion in the Human Digestive System

	Carbohydrate Digestion	Protein Digestion	Fat Digestion
Oral cavity	Starch —(SALIVARY AMYLASE)→ Maltose		
Stomach		Protein —(PEPSIN)→ Smaller polypeptides	
Lumen of small intestine	Undigested polysaccharides —(PANCREATIC AMYLASE)→ Maltose	Polypeptides —(TRYPSIN, CHYMOTRYPSIN)→ Smaller polypeptides Small polypeptides —(AMINOPEPTIDASE, CARBOXYPEPTIDASE)→ Amino acids	Fat globule —(BILE SALTS)→ Emulsified fat Fats —(LIPASE)→ Glycerol, fatty acids, glycerides
Brush border of small intestine	Disaccharides —(MALTASE, SUCRASE, LACTASE, ETC.)→ Monosaccharides	Dipeptides —(DIPEPTIDASE)→ Amino acids	

Pepsin in the stomach primes proteins for digestion by breaking them into smaller pieces, and then a team of enzymes in the small intestine completely dismantles polypeptides into their component amino acids. **Trypsin** and **chymotrypsin** are specific for peptide bonds adjacent to certain amino acids, and thus, like pepsin, break polypeptides into shorter chains. **Carboxypeptidase** splits off one amino acid at a time, beginning at the end of the polypeptide that has a free carboxyl group. **Aminopeptidase** works in the opposite direction. Notice that either aminopeptidase or carboxypeptidase alone could completely digest a protein. But teamwork between these enzymes and the trypsin and chymotrypsin that attack the interior of the protein speeds up hydrolysis tremendously. Protein digestion is further hastened by various **dipeptidases** attached to the intestinal lining that digest fragments only two or three amino acids long.

The protein-digesting enzymes, including trypsin, chymotrypsin, and carboxypeptidase, are secreted as inactive zymogens by the pancreas. An intestinal enzyme called **enterokinase** triggers activation of these enzymes within the lumen of the small intestine (Figure 37.14).

In a cooperative hydrolytic assault similar to the digestion of proteins, a team of enzymes called **nucleases** hydrolyzes DNA and RNA present in food.

Nearly all the fat in a meal reaches the small intestine completely undigested. Hydrolysis of fats is a special problem, because fat molecules are insoluble in water. Bile salts secreted into the duodenum coat tiny fat droplets and keep them from coalescing, a process called **emulsification.** Because the droplets are small, there is a large surface area of fat exposed to **lipase,** an enzyme that hydrolyzes the fat molecules.

Thus, the macromolecules from food are completely hydrolyzed to their component monomers as peristalsis moves the mixture of chyme and digestive juices along the small intestine. Most digestion is completed early in this journey, while the chyme is still in the duodenum. The remaining regions of the small intestine, the **jejunum** and **ileum,** are specialized for the absorption of nutrients.

Absorption and Distribution of Nutrients Topologically speaking, the lumen of the alimentary canal is outside the body. To enter the body, nutrients that accumulate in the lumen when food is digested must cross the lining of the digestive tract. A limited number of nutrients are absorbed in the stomach and large intestine, but most absorption occurs in the small intestine.

The lining of the small intestine has wrinkles upon wrinkles, giving it a huge surface area of 600 m^2, about the size of a baseball diamond. Large folds are decorated with fingerlike projections called **villi,** and each of the epithelial cells of a villus has many microscopic appendages, the **microvilli,** which are exposed to the

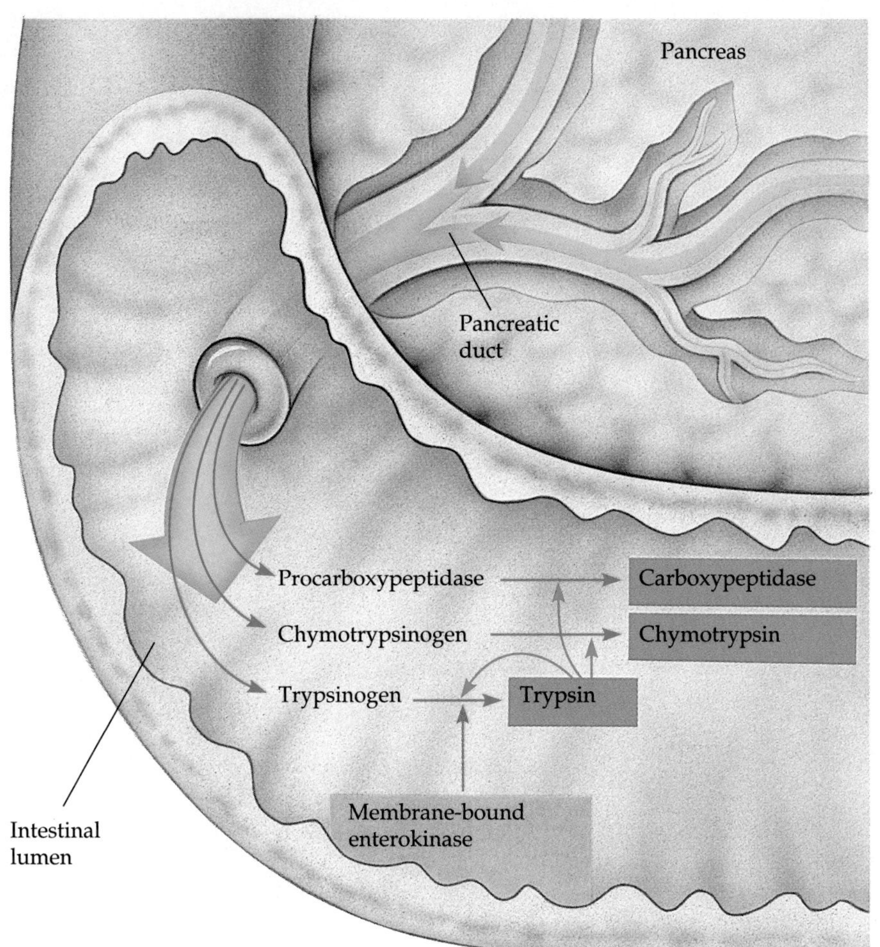

Figure 37.14
The activation of zymogens in the small intestine. The pancreas secretes protein-digesting enzymes in an inactive form, which are activated in the duodenum by enterokinase. Enterokinase, a proteolytic enzyme bound to the intestinal membrane, converts trypsinogen to trypsin by cleaving off part of the precursor peptide. The trypsin, itself a proteolytic enzyme, then activates procarboxypeptidase and chymotrypsinogen.

lumen of the intestine (hence the term *brush border*). Architecturally, the brush border is well suited for its task of absorbing nutrients (Figure 37.15).

Only two single layers of epithelial cells separate nutrients in the lumen of the intestine from the bloodstream. Penetrating the hollow core of each villus is a net of microscopic blood vessels, the capillaries. Also extending into the core of the villus is a tiny lymph vessel called a **lacteal.** (In addition to a circulatory system that carries blood, vertebrates have an auxiliary system of vessels—the lymphatic system—which carries a clear fluid called lymph. Lymph drains into the circulatory system where the two systems connect, near the left shoulder.) Nutrients are absorbed across the epithelium and then across the unicellular wall of the capillaries or lacteals. In some cases, the transport is passive. The simple sugar fructose, for example, is apparently absorbed by diffusion down its concentration gradient from the lumen of the intestine into the epithelial cells and then out of the epithelial cells into capillaries. Other nutrients, including amino acids, vitamins, glucose, and several other simple sugars, are pumped against gradients by the epithelial membranes. The absorption of some nutrients seems to be coupled to the active transport of sodium across the membranes of the epithelial cells. The membrane pumps sodium out of the cell and into the lumen, and the passive re-entry of the ions is harnessed by cotransport to drive the uptake of nutrients (see Chapter 8).

Amino acids and sugars pass through the epithelium, enter capillaries, and are carried away from the intestine by the bloodstream. After glycerol and fatty acids are absorbed by epithelial cells, they are recombined within those cells to form fats again. The fats are then coated with special proteins to make tiny globules called **chylomicrons,** which are transported by exocytosis out of the epithelial cell and into a lacteal. Some fat molecules, however, are bound to specialized proteins and are transported as **lipoproteins** into capillaries.

The capillaries that drain nutrients away from the villi all converge into a single vessel, the **hepatic portal vein,** which leads directly to the liver. The rate of flow in this large vein is about 1 L per minute. This express route ensures that the liver, which has the metabolic versatility to interconvert various organic molecules, has first access to nutrients absorbed after a meal is digested. The blood that leaves the liver may have a very

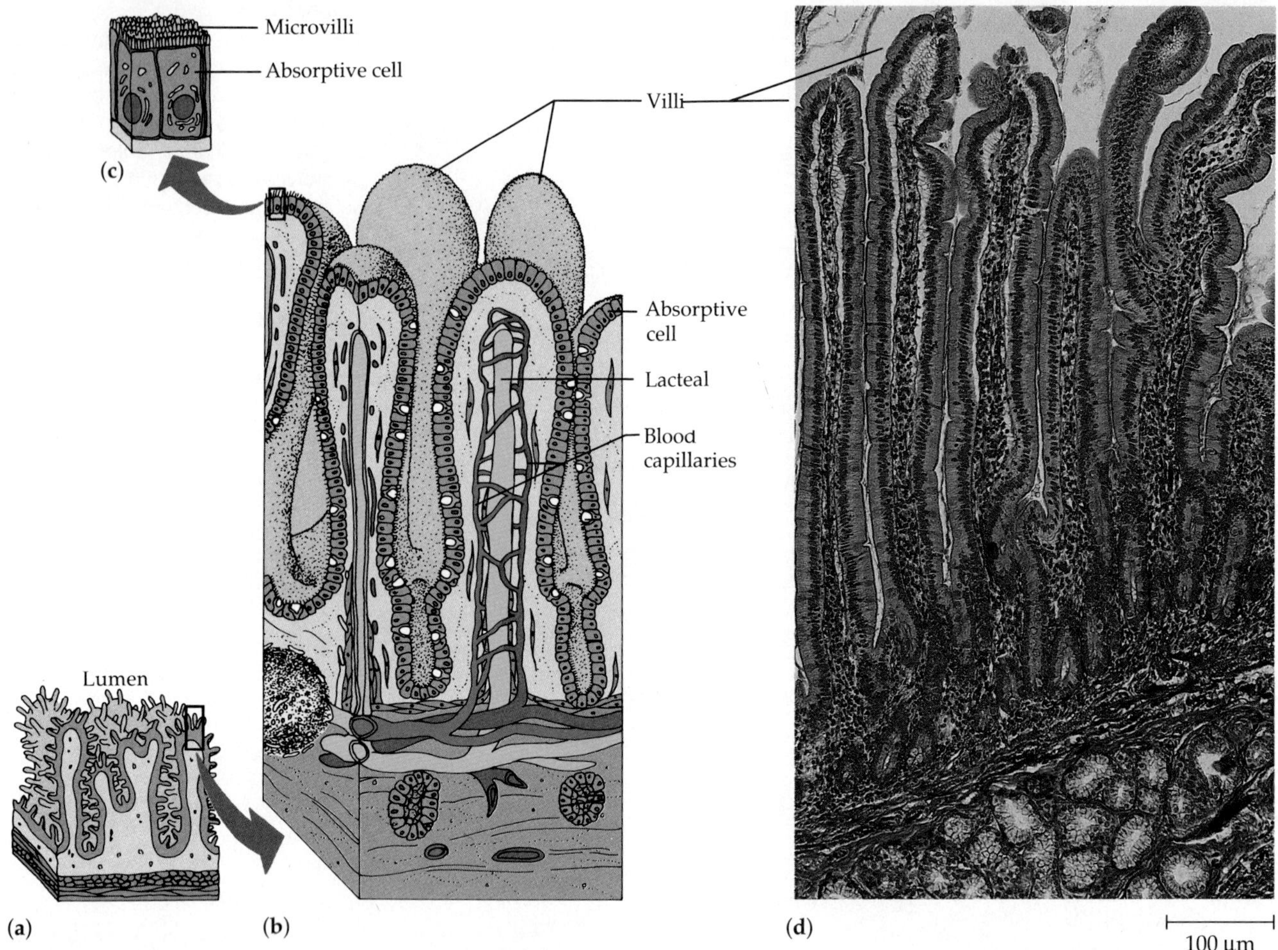

Figure 37.15
The structure of the small intestine. (**a**) Large folds of epithelium line the small intestine. (**b**) Villi project outward from the folds. Each villus has a small lymphatic vessel called a lacteal and a network of blood capillaries surrounded by a layer of epithelial cells. (**c**) The epithelial cells have microscopic projections, microvilli, that extend into the intestinal lumen as a "brush border." Microfilaments within the microvilli wiggle the tiny appendages in the nutrient broth of the intestine's lumen. (**d**) The large surface area provided by villi is evident here (LM).

different balance of nutrients from the blood that entered via the hepatic portal vein. For example, blood exiting the liver usually has a glucose concentration of very close to 0.1%, regardless of the carbohydrate content of a meal. From the liver, blood travels to the heart, which pumps the blood and the nutrients it contains to all parts of the body.

The Large Intestine

The **large intestine,** or **colon,** is connected to the small intestine at a T-shaped junction, where a sphincter acts as a valve controlling the movement of material. One arm of the T is a blind pouch called the **cecum** (see Figure 37.10). Compared to many other mammals, humans have a relatively small cecum with a fingerlike extension, the **appendix,** which is dispensable. (The appendix consists of lymphoid tissue that may make a minor contribution to defense, but it is prone to infection.) The main branch of the colon is shaped like an upside-down U about 1.5 m long.

One function of the colon is to reabsorb water that has entered the alimentary canal as the solvent of the various digestive juices. Altogether, about 7 L of fluid are secreted into the lumen of the digestive tract each day. Most reabsorption of water occurs along with the absorption of nutrients in the small intestine. The colon finishes the job by reclaiming most of the water that remains in the lumen. Together, the small intestine and colon reabsorb about 90% of the water that entered the alimentary canal. The wastes of the digestive tract, called **feces,** become more solid as they are moved along the colon by peristalsis. The movement is sluggish, and it takes about 12 to 24 hours for material to travel the length of the organ. If the lining of the colon is irritated—by a viral or bacterial infection, for instance—less water than normal may be reabsorbed, resulting in diarrhea. The opposite problem, constipation, occurs when peristalsis moves the feces along too slowly. An excess of water is reabsorbed, and the feces become compacted.

Living in the large intestine is a rich flora of mostly harmless bacteria. One of the common inhabitants of the human colon is *Escherichia coli,* a favorite research organism of molecular biologists (see Chapter 17). Intestinal bacteria live on organic material that would otherwise be eliminated with the feces. As by-products of their metabolism, the microbes generate odoriferous gases, including methane and hydrogen sulfide. Some of the bacteria produce vitamin K, which is absorbed by the host; in fact, this is probably the main source of this vitamin for humans.

The feces contain cellulose and other undigested ingredients of food. Although cellulose fibers have no caloric value to humans, their presence in the diet helps move food along the digestive tract at an orderly pace. The color of feces is due to the presence of bile pigments. An abundance of salts is also excreted by the lining of the colon. Much of the dry weight of feces consists of intestinal bacteria; the presence of *E. coli* in lakes and streams is an indication that the water is contaminated by untreated sewage.

The terminal portion of the colon is the **rectum,** where the feces are stored until they can be eliminated. Between the rectum and the anus are two sphincters, one involuntary and the other voluntary. Once or more each day, strong contractions of the colon create an urge to defecate. (As anyone who has trained a toddler knows, it is difficult to condition the child to exercise control over the voluntary sphincter.) Having followed a meal from mouth to anus, we have completed our tour of the mammalian digestive system.

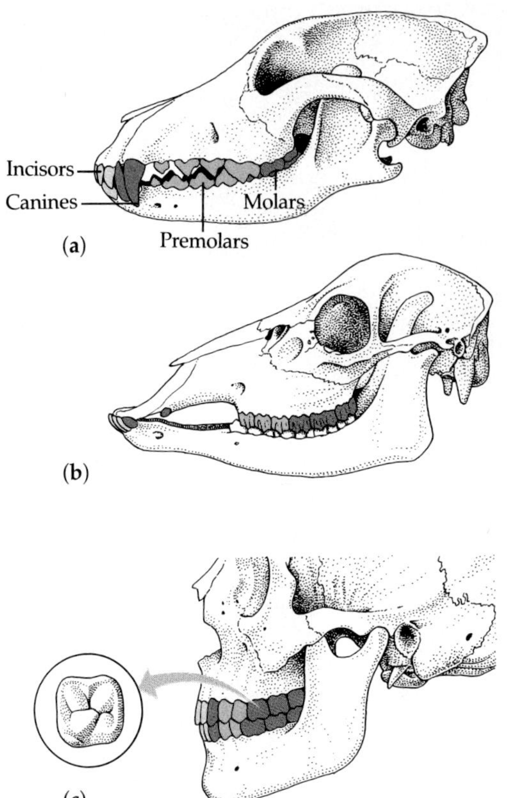

Figure 37.16
Dentition and diet. (**a**) Carnivores, such as members of the dog and cat families, generally have pointed incisors and canines that can be used to kill prey and rip away pieces of flesh. The jagged premolars and molars are modified for crushing and shredding. (**b**) In contrast, herbivorous mammals, such as horses and cows, usually have teeth with broad, ridged surfaces that work like millstones for grinding tough plant material. The incisors and canines are generally modified for chopping off pieces of vegetation. (**c**) Humans, being omnivores equipped to eat both vegetation and meat, have a relatively unspecialized dentition. The permanent (adult) set of teeth numbers 32. Beginning at the midline of the upper and lower jaw are two bladelike incisors for biting, a pointed canine for tearing, two premolars for grinding, and three molars for crushing.

EVOLUTIONARY ADAPTATIONS OF VERTEBRATE DIGESTIVE SYSTEMS

The digestive systems of mammals and other vertebrates are variations on a common plan, but there are many intriguing adaptations, often associated with the animal's diet. We will examine just a few.

Dentition, an animal's assortment of teeth, is one example of anatomical variation reflecting diet, especially in mammals. Compare the dentition of herbivores, carnivores, and omnivores in Figure 37.16. Nonmammalian vertebrates generally have less specialized dentition, but there are interesting exceptions. For example, poisonous snakes, such as rattlesnakes, are armed with fangs, modified teeth that inject venom into prey. Some fangs are hollow, like syringes, while others drip the poison along grooves on the surfaces of the teeth. Snakes in general have another important anatomical adaptation associated with feeding: The lower jaw is loosely hinged to the skull by an elastic ligament that permits the mouth and throat to open *very* wide for swallowing large prey (once again, witness the astonishing episode recorded in Figure 37.5).

The length of the vertebrate digestive system is also correlated with diet. In general, herbivores and omnivores have longer alimentary canals relative to their body size than carnivores. Vegetation is more difficult to digest than meat because it has cell walls. A longer tract furnishes more time for digestion and more surface area for absorption of nutrients, which are usually less concentrated in vegetation than in meat. A model

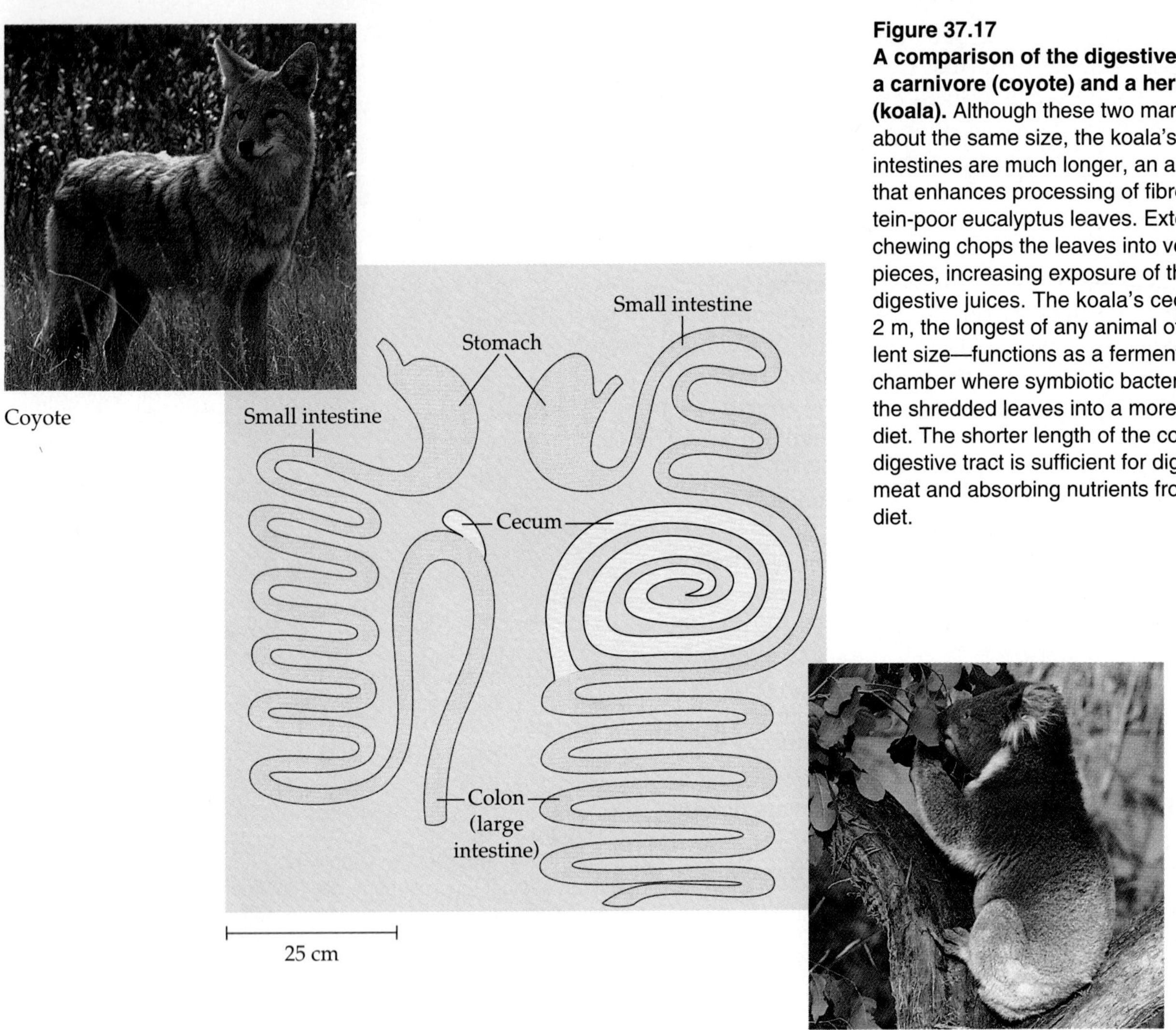

Figure 37.17
A comparison of the digestive tracts of a carnivore (coyote) and a herbivore (koala). Although these two mammals are about the same size, the koala's intestines are much longer, an adaptation that enhances processing of fibrous, protein-poor eucalyptus leaves. Extensive chewing chops the leaves into very small pieces, increasing exposure of the food to digestive juices. The koala's cecum—at 2 m, the longest of any animal of equivalent size—functions as a fermentation chamber where symbiotic bacteria convert the shredded leaves into a more nutritious diet. The shorter length of the coyote's digestive tract is sufficient for digesting meat and absorbing nutrients from this diet.

case is the frog, which changes its diet upon metamorphosis. The algae-eating tadpole has a coiled intestine that is very long relative to the size of the larva. During metamorphosis, the rest of the body grows more than the intestine, leaving the carnivorous adult with a shorter intestine relative to the frog's size.

In some vertebrates, the *functional* length of the alimentary canal is actually longer than superficial appearance reveals. For example, the shark, a carnivore, has a relatively short intestine lacking the "switchbacks" that lengthen our own intestine. However, the lining of the shark's intestine has folds that form a spiral valve, a corkscrewlike apparatus that increases surface area and provides a much longer route than if food were to move through the tube in a straight line.

In addition to being very long, the alimentary canals of many herbivorous mammals have special fermentation chambers where symbiotic bacteria and protozoa live. These microorganisms have enzymes that can digest cellulose, which the animal itself cannot digest. The microorganisms not only digest the cellulose to simple sugars but also convert the sugar to a variety of nutrients essential to the animal. Many herbivores, including horses, house their symbiotic microorganisms in a large cecum, the pouch where the small and large intestines connect. The symbiotic bacteria of rabbits and some rodents live in the large intestine as well as in the cecum. Since most nutrients are absorbed in the small intestine, however, potentially nourishing by-products of fermentation by bacteria in the large intestine are initially lost with the feces. The rabbit or rodent obtains these nutrients by eating some of its feces and passing the food through the alimentary canal a second time. (The familiar rabbit "pellets," which are not reingested, are the feces eliminated after food has passed through the digestive tract the second time.) A koala (see Figure 37.1) also has an enlarged cecum, where symbiotic bacteria ferment finely shredded eucalyptus leaves (Figure 37.17). The most elaborate adaptations for a herbivorous diet have evolved in the **ruminants,** which include cattle and sheep (Figure 37.18).

We have sampled only a few of the many evolutionary adaptations of digestive systems. We now shift our attention from the digestive equipment of animals to

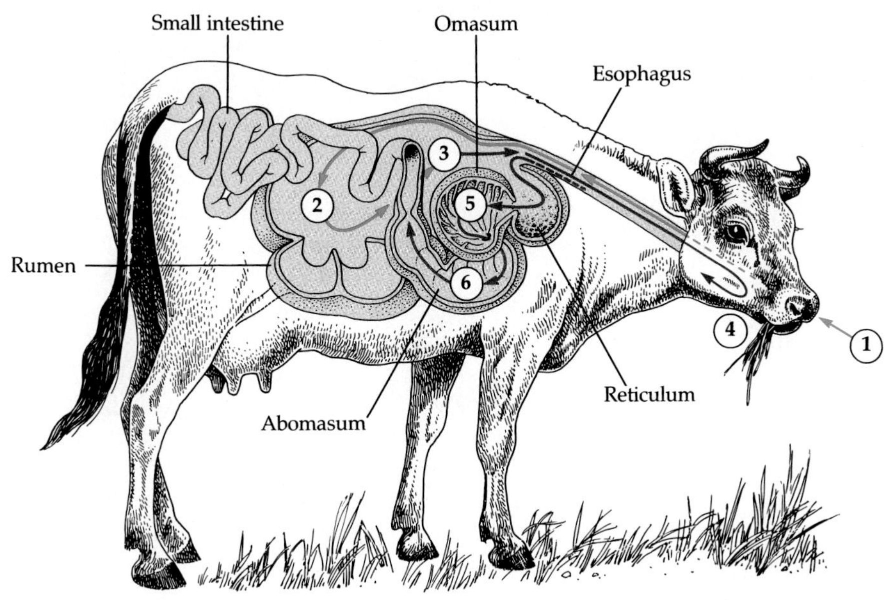

Figure 37.18
Ruminant digestion. The stomach of a ruminant has four chambers, three of which were probably modified from the lower end of the esophagus. ① When the cow first chews and swallows its food, it enters both ② the rumen and ③ the reticulum, where symbiotic bacteria go to work on the cellulose-rich meal. As by-products of their metabolism, the bacteria secrete fatty acids. ④ The cow periodically regurgitates and rechews the cud, which further breaks down the fibers, making them more accessible to further bacterial action. The cow then reswallows the cud, which ⑤ moves to the omasum, where water is removed. ⑥ The cud finally passes to the abomasum for digestion. The diet from which a ruminant actually absorbs its nutrients is much richer than the hay or grass the animal originally ate. In fact, a ruminant obtains many of its nutrients by digesting the symbiotic bacteria, which reproduce rapidly enough in the rumen to maintain a stable population.

the actual nutritional needs that must be satisfied by feeding and digestion.

NUTRITIONAL REQUIREMENTS

A healthful diet satisfies three needs: It provides fuel for cellular respiration; it supplies the organic raw materials animals use to fabricate many of their own molecules; and it supplies essential nutrients, substances the animal cannot make for itself from *any* raw material and that must be obtained in food in prefabricated form.

Food as Fuel

Animals are powered by ATP generated from the oxidation of complex organic molecules (see Chapter 9)—the carbohydrates, fats, and proteins that make up the bulk of the diet of most animals. The monomers of any of these substances can be used as fuel for cellular respiration, though priority is usually given to carbohydrates and fats. (Only when these substances are in short supply is protein used as the major fuel.)

The energy content of a food is measured in **calories** (see Chapter 3). The term *Calorie* (capital C), as used by nutritionists, is actually a **kilocalorie** (kcal). The oxidation of 1 g of fat liberates about 9.5 kcal, twice the energy liberated by 1 g of carbohydrate or protein.

Several processes must be continuously driven by metabolism for an animal to remain alive. These include breathing, the beating of the heart, and, in some animals, the maintenance of body temperature. The number of kilocalories a resting animal requires to fuel these processes for a given time is called the **basal metabolic rate (BMR).** Among vertebrate classes, birds and mammals have the highest basal metabolic rates. Birds and mammals are endothermic, which means they warm the body mainly with heat generated by metabolism (see Chapter 30). Reptiles, amphibians, and the various classes of fishes are ectothermic, which means they absorb most of their body heat from the external environment. (Thermoregulation will be discussed in detail in Chapter 40.) The caloric cost of heating (and cooling) the body is reflected in the relatively high metabolic rates of endothermic animals. Among endotherms, the caloric requirement to maintain each gram of body weight is inversely related to body size. Each gram of mouse, for instance, consumes about ten times more calories than a gram of elephant (even though the whole elephant, of course, uses far more calories than the whole mouse). Perhaps you are thinking that this relationship is a matter of geometry; the smaller the animal, the greater its surface area relative to its volume. The greater the surface-to-volume ratio, the greater the loss or absorption of heat from the surroundings and the greater the energy cost of maintaining a stable body temperature. However, the inverse relationship between body size and metabolic rate also applies to ectothermic vertebrates such as fishes, to invertebrates, and even to protozoa. Factors in addition to temperature regulation must be involved.

The BMR for humans averages about 1600 to 1800 kcal per day for adult males and about 1300 to 1500

METHODS: MEASURING METABOLIC RATE

An animal's metabolic rate is the number of calories consumed per unit of time. The apparatus shown in the left photograph helps estimate the metabolic rate of a human by measuring oxygen consumed by cellular respiration, the oxidation of food (see Chapter 9).

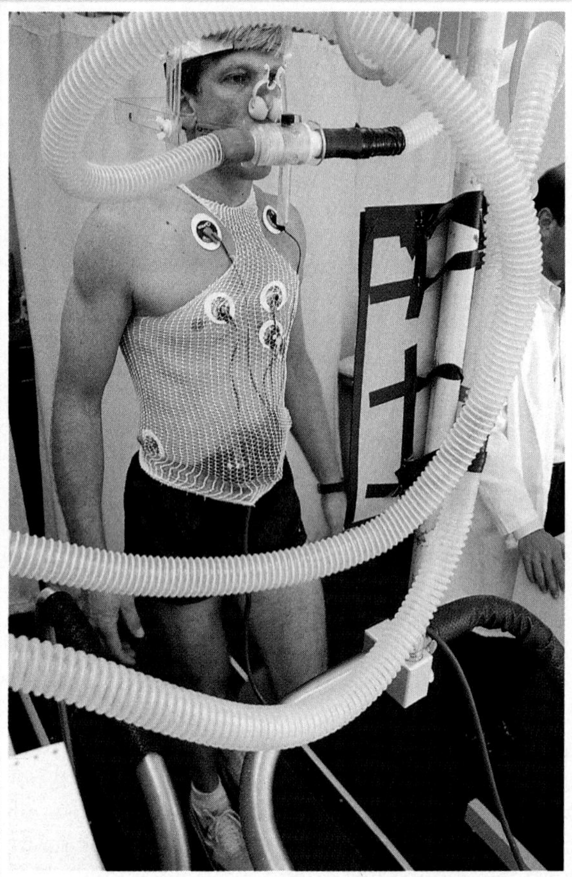

For every liter of O_2 consumed, respiration liberates about 4.83 kcal of energy from food molecules. If, for example, the subject uses 16 L of oxygen per hour, then the metabolic rate is 77.28 kcal/h (16 L/h × 4.83 kcal/L). Measured at rest on an empty stomach (digestion requires energy), this value is the subject's basal metabolic rate, or BMR. For comparisons, BMRs are standardized as kilocalories per hour per kilogram of body weight, or per square meter of body surface area. A person's age, sex, genetic makeup, and many other factors affect BMR. And, of course, actual metabolic rate usually exceeds BMR, depending on the level of physical activity. For example, the metabolic rate of this man is being monitored as he works out on a treadmill.

Similar methods can be used to monitor the metabolic rates of other animals. The ghost crab in the right photograph is on a treadmill enclosed in a plastic chamber. Air of known O_2 concentration flows through the chamber, and the metabolic rate of the crab is calculated from the difference between the amount of O_2 entering and leaving the chamber.

kcal per day for adult females (see the Methods Box). These BMRs are about equivalent to the daily energy consumption of a 100-watt light bulb. This, remember, is only a baseline—the number of calories we "burn" lying motionless. Any activity, even working quietly at your desk, consumes calories in addition to the BMR. The more strenuous the activity, the greater the caloric demand.

When an animal takes in more calories than it consumes to meet its energy requirements, the excess calories are stored. The liver and muscles store energy in the form of glycogen, a glucose polymer similar to the starch of plants (see Chapter 5). A human can store enough glycogen to supply about a day's worth of basal metabolism. If glycogen stores are full and caloric intake still exceeds caloric expenditure, then the excess food is stored primarily in adipose tissue in the form of fat. This happens even if the diet contains little fat, because the liver converts excess carbohydrates or proteins into fat, which is then distributed to adipose tissue for storage.

Between meals or when the diet is deficient in calories, fuel is taken out of storage. The glycogen stores are generally expended first, and then fat is withdrawn from adipose tissue. Most of us have enough fat to sustain us in calories for several weeks. An animal can also consume its own proteins as a source of fuel, but this rarely occurs, unless the animal is starving or stressed in some other way for a prolonged period.

An **undernourished** person or other animal is one whose diet is deficient in calories. As you have just learned, if starvation for calories persists, the body begins to break down its own proteins as a source of energy. Muscles begin to atrophy, and even the proteins of the brain are consumed. If the undernourished person survives, some of the damage is irreversible. Even a diet of a single staple such as rice or corn provides calories; thus, undernourishment is generally common only where drought, war, or some other crisis has severely disrupted the food supply. Another alarming cause of undernourishment is anorexia nervosa, a self-imposed starvation associated with a compulsive aversion to body fat.

In the United States and other developed countries, overnourishment, or obesity, is a far more common problem than undernourishment. Obesity increases the risk of heart attack, diabetes, and several other disorders. Imagine the extra strain on your heart if you were to carry a 50-lb suitcase around with you all day. How heavy we can be before we are considered unhealthy is a disputed matter. Charts of "ideal weights" for people of various heights have been liberalized in the past few years, based on statistics from insurance companies showing that "pleasantly plump" does not necessarily mean increased health risk. Nevertheless, the billions of dollars spent each year on diet books and diet aids indicate that a substantial portion of the population would like to lose weight because of a concern with health and appearance. Unfortunately, there are no magic methods for slimming, and some of the fad diets—"eat all you want as long as it's carbohydrate," for instance—can actually be dangerous. Effective weight watching is a matter of caloric bookkeeping. If we take in more calories than we need, the excess will be stored as fat. If we expend more calories than we take in, we balance the deficit by consuming body fat. Our weight is stable when caloric demand is matched by caloric supply from our diet. To lose weight, we must eat less and exercise more.

Food for Fabrication

Heterotrophs cannot make organic molecules from raw materials that are entirely inorganic. To synthesize the molecules it needs to grow and replenish itself, the animal must obtain organic precursors from its food. Given a source of organic carbon (such as sugar) and a source of organic nitrogen (such as amino acids from the digestion of protein), the animal can fabricate a great variety of organic molecules by using enzymes to rearrange the molecular skeletons of the precursors acquired from food. For instance, a single type of amino acid can supply nitrogen for the synthesis of several other types of amino acids that may not be present in the food. Also, we have seen that animals can synthesize fats from carbohydrates. In vertebrates, the conversion of nutrients from one type of organic molecule to another occurs mainly in the liver.

So we see that although animals depend on food as a source of organic carbon and nitrogen, they exhibit considerable versatility in the biosynthesis of their own organic molecules. However, there are some substances the animal cannot fabricate from any precursors.

Essential Nutrients

In addition to providing fuel and raw materials for biosynthesis, an animal's diet must also supply certain substances in preassembled form. Chemicals an animal requires but cannot make are called **essential nutrients.** They vary from species to species, depending on the biosynthetic capabilities of the animal. A particular molecule can be an essential nutrient for one animal and yet be nonessential in the diet of another animal—one that can produce the molecule for itself from some precursor.

An animal whose diet is missing one or more essential nutrients is said to be **malnourished** (recall that *undernourished* refers to caloric deficiency). For example, cattle and other herbivores may suffer mineral deficiencies if they graze on plants grown in soil that itself is deficient in key minerals. Malnutrition is much more common than undernutrition in human populations, and it is even possible for an overnourished individual to be malnourished.

There are four classes of essential nutrients: essential amino acids, essential fatty acids, vitamins, and minerals.

Essential Amino Acids Of the 20 kinds of amino acids required to make proteins, about half can be synthesized by most animals as long as their diet includes organic nitrogen. The remaining amino acids must be obtained from food in prefabricated form. Eight amino acids are essential in the adult human diet (a ninth is essential for infants). Do not let the term *essential* mislead you; an animal requires all 20 amino acids. Essential amino acids are those that must be present in the diet.

A diet that lacks one or more of the essential amino acids results in a form of malnutrition known generally as protein deficiency. The most common type of malnutrition among humans, protein deficiency is concentrated in geographical regions where there is a great gap between food supply and population size. The victims are usually children, who, if they survive infancy, are likely to be retarded in mental and physical devel-

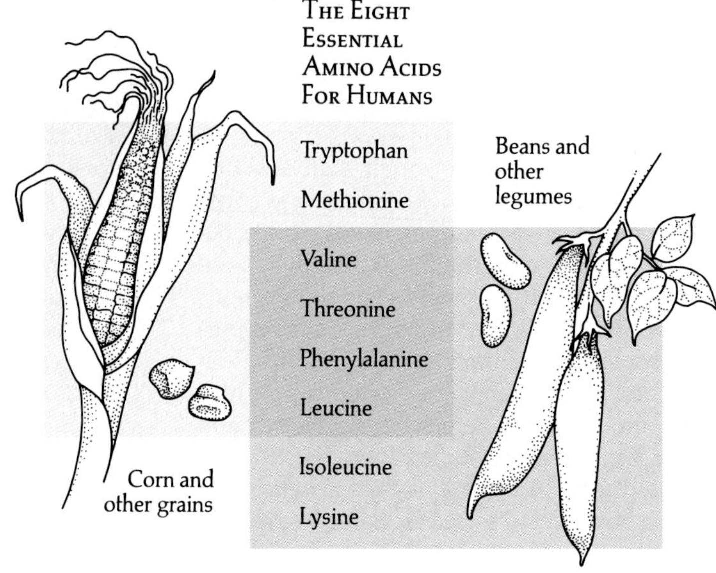

Figure 37.19
Essential amino acids from a vegetarian diet. Corn has little isoleucine and lysine. Beans have ample isoleucine and lysine, but little tryptophan and methionine. A person can obtain all eight essential amino acids by eating a meal of corn and beans.

opment. In Africa, the syndrome is named **kwashiorkor,** meaning "rejected one," a reference to the onset of the disease when a child is weaned from its mother's milk and placed on a starchy diet after a sibling is born. The problem of protein deficiency in some developing countries has been compounded by a trend away from breast-feeding altogether, an unfortunate dietary change that has been catalyzed by an aggressive marketing campaign by companies that make baby formula. Impoverished mothers often "stretch" the expensive formula by diluting it with water to such an extent that the protein content is insufficient. Also, uninformed consumers pay too little attention to hygiene when mixing and storing formula. In contrast, nursing not only provides a balanced meal, but also transfers temporary immunity to several infectious diseases from mother to child.

There is, of course, an economic component to protein deficiency. The most reliable sources of essential amino acids are meat and animal by-products (for example, eggs and cheese), which are relatively expensive. The proteins in animal products are complete, which means that they provide all the essential amino acids in their proper proportions. Most plant proteins, on the other hand, are incomplete, since they are deficient in one or more essential amino acids. Zein, for example, the principal protein in corn, is deficient in the amino acid lysine. People forced by economic necessity to obtain nearly all their calories from corn would begin showing symptoms of protein deficiency. The result would be similar for people on monotonous diets of rice, wheat, potatoes, or any other single staple. A vegetarian, however, can obtain sufficient quantities of all essential amino acids (Figure 37.19). The key is to select a combination of plant foods that complement one another; two plants with incomplete proteins may not be deficient in the same amino acids. Beans, for example, supply the lysine that is missing in corn; and whereas beans are deficient in methionine, this amino acid is present in corn. Thus, a meal of beans and corn can provide all the essential amino acids. The combination of vegetables must be consumed at the same meal. The body cannot store amino acids, so a deficiency of a single essential amino acid retards protein synthesis and limits the use of other amino acids. Most cultures have, by trial and error, developed balanced diets that prevent protein deficiency.

Essential Fatty Acids Animals are able to synthesize most of the fatty acids they need, but they are unable to make certain unsaturated fatty acids (fatty acids having double bonds; see Chapter 5). In humans, for example, linoleic acid must be present in the diet in prefabricated form. This essential fatty acid is required to make some of the phospholipids found in membranes. Most diets furnish ample quantities of essential fatty acids, and thus deficiencies are rare.

Vitamins **Vitamins** are organic molecules required in the diet in amounts that are quite small compared with the relatively large quantities of essential amino acids and fatty acids animals need. Minute doses of vitamins—from about 0.01 to 50 mg per day, depending on the vitamin—suffice because most of these molecules serve as coenzymes or parts of coenzymes and thus have catalytic functions (see Chapter 6).

Although requirements for vitamins are modest, these molecules are absolutely essential in a healthful diet. Deficiencies can cause severe syndromes. Indeed, the first vitamin to be identified, thiamine (vitamin B_1), was discovered as a result of the search for the cause of a mysterious disease called beriberi. Its symptoms include loss of appetite, fatigue, and nervous disorders. The syndrome was first described when it struck sol-

Table 37.2 Vitamin Requirements of Humans

Vitamin	Dietary Sources	Major Body Functions	Possible Outcomes of Deficiency
Water-Soluble			
Vitamin B_1 (thiamine)	Pork, organ meat, whole grains, legumes	Coenzyme in the removal of carbon dioxide	Beriberi
Vitamin B_2 (riboflavin)	Widely distributed in foods	Constituent of two coenzymes involved in energy metabolism (FAD and FMN)	Mouth sores, lesions of eyes
Niacin	Liver, lean meat, grains, legumes	Constituent of the coenzymes NAD^+ and $NADP^+$	Pellagra (skin and gastrointestinal lesions; nervous, mental disorders)
Vitamin B_6 (pyridoxine)	Meat, vegetables, whole grain cereals	Coenzyme involved in amino acid metabolism	Irritability, convulsions, muscular twitching, kidney stones
Pantothenic acid	Widely distributed in foods	Constituent of coenzyme A (see Chapter 9)	Fatigue, sleep disturbances, impaired coordination
Folacin (folic acid)	Legumes, green vegetables, whole wheat products	Coenzyme in nucleic acid and amino acid metabolism	Anemia, gastrointestinal disturbances, diarrhea
Vitamin B_{12}	Muscle meat, eggs, dairy products	Coenzyme in nucleic acid metabolism; maturation of red blood cells	Pernicious anemia, neurological disorders
Biotin	Legumes, vegetables, meat	Coenzyme in fat synthesis, amino acid metabolism, glycogen formation	Fatigue, depression, nausea, dermatitis, muscular pains
Vitamin C (ascorbic acid)	Citrus fruits, tomatoes, green peppers, salad greens	Maintains extracellular matrix of connective tissue; important in collagen synthesis	Scurvy (degeneration of skin, gums, blood vessels)
Fat-Soluble			
Vitamin A (retinol)	Provitamin A in green vegetables; retinol in dairy products	Constituent of rhodopsin (visual pigment); maintenance of epithelial tissues	Night blindness, permanent blindness
Vitamin D	Eggs, dairy products	Promotes bone growth, mineralization; increases calcium absorption	Rickets (bone deformities) in children
Vitamin E (tocopherol)	Seeds, green leafy vegetables	Antioxidant preventing cell membrane damage	Possibly anemia; never observed in humans
Vitamin K (phylloquinone)	Green leafy vegetables; small amount in cereals, fruits, meat	Blood clotting	Severe bleeding, internal hemorrhages

diers and prisoners in the Dutch East Indies in the nineteenth century. The dietary staple for these men was polished rice, which had the hulls (seed coats and other outer layers) removed to increase storage life. Not only did the men who were fed this diet develop beriberi, but so did the chickens that ate the table scraps. It was found that beriberi could be prevented in both the men and the chickens by supplementing their diets with unpolished rice. Later, the active ingredient of rice hulls was isolated. Since it belongs to the chemical family known as amines, the compound was named vitamine (a vital amine). The "e" was later dropped, and the term has persisted, even though many of the vitamins subsequently discovered are not amines.

So far, 13 vitamins essential to humans have been identified (Table 37.2). The compounds are grouped into two categories: water-soluble vitamins and fat-soluble vitamins. Water-soluble vitamins include the B

Table 37.3 Mineral Requirements of Humans

Mineral	Dietary Sources	Major Body Functions	Possible Outcomes of Deficiency
Calcium	Milk, cheese, dark-green vegetables, dried legumes	Bone and tooth formation; blood clotting; nerve transmission	Stunted growth; rickets, osteoporosis; convulsions
Phosphorus	Milk, cheese, meat, grains	Bone and tooth formation; acid-base balance; ATP formation	Weakness, demineralization of bone, loss of calcium
Sulfur	Sulfur-containing amino acids in dietary proteins	Constituent of organic compounds; cartilage and tendon development	Related to deficiency of sulfur amino acids
Potassium	Meat, milk, many fruits	Acid-base balance; body water balance; nerve function	Muscular weakness; paralysis
Chlorine	Common salt	Formation of gastric juice; acid-base balance	Muscle cramps; mental apathy; reduced appetite
Sodium	Common salt	Acid-base balance; body water balance; nerve function	Muscle cramps; mental apathy; reduced appetite
Magnesium	Whole grains, green leafy vegetables	Activates enzymes; involved in protein synthesis	Growth failure; behavioral disturbances; spasms
Iron	Eggs, meat, legumes, whole grains, green leafy vegetables	Constituent of hemoglobin and enzymes involved in energy metabolism	Iron-deficiency anemia; reduced resistance to infection
Fluorine	Drinking water, tea, seafood	May be important in maintenance of bone structure	Higher frequency of tooth decay
Zinc	Widely distributed in foods	Constituent of digestive enzymes	Growth failure
Copper	Meat, drinking water	Constituent of enzymes associated with iron metabolism	Anemia; bone changes
Manganese	Whole grain cereals, egg yolks, green vegetables	Activates several enzymes	None reported for humans
Iodine	Marine fish and shellfish, dairy products	Constituent of thyroid hormones	Goiter (enlarged thyroid)
Cobalt	Meat, milk	Constituent of vitamin B_{12}	None reported for humans

complex, which consists of several compounds that generally function as coenzymes in key metabolic processes. Vitamin C (ascorbic acid) is also water-soluble. Ascorbic acid is required for the production of connective tissue. Excesses of water-soluble vitamins are excreted with the urine, and moderate overdoses of these vitamins are probably harmless. Recently, a nervous disorder linked to an excessive intake of vitamin B_6 has been described, but only in people who have been taking at least a thousand times the daily requirement of the vitamin.

The fat-soluble vitamins are A, D, E, and K. Vitamin A is incorporated into visual pigments of the eye. Vitamin D aids in calcium absorption and bone formation. The functions of vitamin E are not yet understood, but it seems to protect the phospholipids in membranes from oxidation. Vitamin K is required for blood clotting. Excesses of fat-soluble vitamins are not excreted but instead are deposited in body fat, so overdoses may result in an accumulation of these compounds to toxic levels.

The subject of vitamin dosage has aroused heated debate. One faction believes that it is enough to fill recommended daily allowances (RDAs), which are nutrient intake recommendations for healthy people, deliberately set to be higher than most people's actual needs. Others argue that RDAs are set too low for some vitamins and that we should be thinking in terms of *op-*

timal requirements. The dust has not yet settled, particularly in the debates over optimal doses of vitamins C and E. About all that can be said with any certainty at this time is that people who eat a balanced diet are unlikely to develop symptoms of vitamin deficiency.

Research involving experimental animals is of limited value in assessing the vitamin needs of humans. A compound that is a vitamin for one species will not be essential in the diet of another species that can synthesize the compound for itself. By definition, a substance is not a vitamin if it can be made by the animal. Also, symbiotic microorganisms living within the gut of an animal may produce vitamins that supplement or substitute for vitamins in food. Rabbits, for example, require no vitamin C in their diet because their flora of intestinal bacteria produces ample quantities of the compound.

Minerals **Minerals** are inorganic nutrients, usually required in very small amounts, from less than 1 mg to about 2500 mg per day, depending on the mineral (Table 37.3). As with vitamins, mineral requirements vary with animal species. Humans and other vertebrates require relatively large quantities of calcium and phosphorus for the construction and maintenance of bone. Calcium is also necessary for the normal functioning of nerves and muscles, and phosphorus is also an ingredient of ATP and nucleic acids. Iron is a component of the cytochromes that function in cellular respiration (see Chapter 9) and of hemoglobin, the oxygen-binding protein of red blood cells. Magnesium, manganese, zinc, and cobalt are cofactors built into the structure of certain enzymes; magnesium, for example, is present in enzymes that split ATP. Vertebrates need iodine to make thyroxine, a thyroid hormone that regulates metabolic rate. Sodium, potassium, and chlorine are important in nerve function and also have a major influence on osmotic balance between cells and the interstitial fluid. Herbivores, such as deer and cattle, often seem to crave salt; vegetation generally contains a very low concentration of sodium chloride. Salt licks, either natural or placed on the range by ranchers, attract large numbers of herbivores (Figure 37.20). Most humans, on the other hand, ingest far more salt than they need. In the United States, the average person consumes enough salt to provide about twenty times the required amount of sodium.

Figure 37.20
A natural salt lick. Many herbivores cannot obtain enough sodium and chloride from plants but supplement their diet by visiting natural salt licks, as this female bighorn sheep and her lamb are doing here.

A healthful diet, then, must supply enough calories to satisfy energy needs, carbon skeletons and organic nitrogen for biosynthesis, and ample quantities of the essential nutrients.

* * *

Although the focus of this chapter has been on nutrition and digestion, nerves, blood vessels, and hormones have all been part of the discussion. No organ system is a solo act. The animal is an integrated whole, a concert of organ systems working together. In the next chapter, you will learn about the parts played by the circulatory and respiratory systems.

STUDY OUTLINE

Food Types and Feeding Mechanisms (pp. 794–795)

1. All animals are heterotrophs and must obtain their nutrients by consuming organic molecules. Most animals eat plants, animals, or both, and therefore can be classified as herbivores, carnivores, or omnivores, respectively.
2. Animals may obtain nutrients by suspension-, substrate-, deposit-, fluid-, or bulk-feeding.

Digestion: A Comparative Introduction (pp. 795–798)

1. Animal nutrition involves digestion and absorption. Digestion enzymatically breaks down the macromolecules of food into their monomers inside compartments separated from living protoplasm. The monomers are then absorbed across the membranes of the digestive compartment. Undigested material is eliminated.
2. In intracellular digestion, large food particles may be engulfed by endocytosis and digested within food vacuoles. Protozoans and sponges use this type of digestion exclusively. In the extracellular digestion of most animals, hydrolysis occurs in a separate compartment, and smaller food molecules are then absorbed into the cells.
3. The gastrovascular cavities of the simplest animals have a single opening through which food enters and undigested wastes pass. *Hydra* initiates digestion inside such

a cavity but accomplishes most hydrolysis intracellularly after the particles are small enough to enter the gastrodermal cells. A similar process occurs in flatworms.

4. More complex animals have digestive tracts, or alimentary canals, that move food through a one-way tube with specialized regions for digestion and absorption variously modified in different species.

The Mammalian Digestive System (pp. 799–807)

1. The mammalian digestive tract has a four-layered wall over most of its length. The smooth muscle layer propels food along the tract by peristalsis and regulates its passage through strategic points by means of sphincters.
2. Mammals have accessory glands that add digestive secretions to the tract through ducts. These are the salivary glands, pancreas, liver, and gallbladder.
3. Digestion begins in the oral cavity, where teeth chew food into smaller particles that are exposed to a starch-digesting enzyme in saliva, salivary amylase. Saliva also contains buffers, antibacterial agents, and mucin for lubricating the food. The tongue is adapted for tasting and manipulating food into a bolus for swallowing.
4. The pharynx is the intersection leading to the trachea and the esophagus. Food is usually prevented from entering the trachea by the epiglottis.
5. The esophagus conducts food from the pharynx to the stomach by involuntary peristaltic waves.
6. The stomach stores food and secretes gastric juice, which converts a meal to acid chyme. Gastric juice includes hydrochloric acid and the enzyme pepsin. Nervous impulses and the hormone gastrin regulate gastric motility and secretion.
7. Most of digestion and virtually all of absorption occur in the small intestine, the longest segment of the alimentary canal.
8. The pancreas, liver, and gallbladder empty by ducts into the duodenum, the first part of the small intestine. Regulatory hormones, such as secretin and cholecystokinin, modulate the activities of the accessory glands.
9. Carbohydrate digestion, begun in the mouth, continues in the duodenum in the presence of pancreatic amylase in the lumen and disaccharidases built into the plasma membranes of intestinal epithelial cells.
10. Trypsin, chymotrypsin, carboxypeptidase, and aminopeptidase, pancreatic enzymes that work together to hydrolyze polypeptides, are secreted into the small intestine as inactive zymogens that are subsequently activated by enterokinase, an intestinal enzyme. Peptidases on the intestinal cells finish protein digestion by hydrolyzing small peptide fragments.
11. Fats are broken up into smaller droplets during emulsification by bile salts in the intestinal lumen, thereby allowing maximum exposure to lipase.
12. Most digestion is completed in the duodenum, and the remaining regions of the small intestine, the jejunum and ileum, are involved in absorption.
13. The anatomy of the small intestine fits its absorptive function. The large folds of the lining have fingerlike villi, whose cells have microscopic microvilli, all greatly increasing the surface area. In the core of each villus is a network of capillaries and a lacteal that take up and distribute the products of absorption.
14. Nutrients are absorbed by passive diffusion or by active transport. Amino acids and sugars pass directly into the capillaries, but glycerol and certain fatty acids are reassembled into fats and coated with protein to form tiny chylomicrons inside intestinal cells, from which they are transported by exocytosis into lacteals that eventually empty into the bloodstream. Some lipids are attached to carrier proteins and enter the capillaries directly as lipoproteins.
15. The nutrient-laden blood from the villus capillaries travels through the large hepatic portal vein to the liver, where organic molecules are interconverted and the nutrient content of the blood is regulated.
16. A sphincter and a cecum with its attached appendix mark the junction between the small intestine and the large intestine, or colon. The colon aids the small intestine in reabsorbing water and houses bacteria, some of which synthesize vitamin K. The feces, which pass through the rectum and out the anus, contain undigested parts of food, cellulose, bile pigments, salts excreted by the colon, and intestinal bacteria.

Evolutionary Adaptations of Vertebrate Digestive Systems (pp. 807–809)

1. A mammal's dentition is generally correlated with its diet.
2. Herbivores generally have longer alimentary canals, reflecting the longer time needed to digest vegetation. Many herbivorous mammals have special fermentation chambers in the stomach, cecum, or intestines, where symbiotic microorganisms digest cellulose.

Nutritional Requirements (pp. 809–815)

1. A proper diet must supply fuel for cellular respiration, organic raw materials for synthesis of macromolecules, and essential nutrients.
2. Even at complete rest, an animal requires calories to stay alive; this is the basal metabolic rate, or BMR. Among vertebrates, the endothermic birds and mammals have a higher BMR than the ectothermic reptiles, amphibians, and fishes. In most animal groups, metabolic rate is inversely related to body size.
3. Animals store excess calories as glycogen in the liver and muscles and as fat in adipose tissue. Undernourished animals have diets deficient in calories.
4. Essential nutrients must be supplied in preassembled form because the body lacks the machinery for their biosynthesis. Malnourished animals are missing one or more of the essential nutrients.
5. Essential amino acids are those the animal cannot make from nitrogen-containing precursors.
6. Animals can synthesize most necessary fatty acids; there are a few essential unsaturated fatty acids.
7. Vitamins are organic molecules that serve as coenzymes or parts of coenzymes and thus are required in small amounts.
8. Minerals are inorganic nutrients that are required in varying amounts, depending on their role in physiology and metabolism.

SELF-QUIZ

1. In function, a paramecium's food vacuole is most analogous to the human
   a. mouth
   b. small intestine
   c. esophagus
   d. liver
   e. anus

2. Which of the following animals lack alimentary canals (complete digestive systems)?
   a. earthworms
   b. jellyfish
   c. insects
   d. fishes
   e. birds

3. Our oral cavity, with its dentition, is most functionally analogous to an earthworm's
   a. intestine
   b. pharynx
   c. gizzard
   d. stomach
   e. anus

4. Which of the following enzymes has the lowest pH optimum?
   a. salivary amylase
   b. trypsin
   c. pepsin
   d. pancreatic amylase
   e. pancreatic lipase

5. After surgical removal of an infected gallbladder, a person must be especially careful to restrict his or her dietary intake of
   a. starch
   b. protein
   c. sugar
   d. fat
   e. water

6. Trypsinogen, a pancreatic zymogen secreted into the duodenum, can be activated by
   a. chymotrypsin
   b. enterogastrone
   c. secretin
   d. trypsin
   e. pepsin

7. Carnassial teeth, which are pointed molars and premolars, would most likely be part of the dentition of a
   a. human
   b. cow
   c. lion
   d. rabbit
   e. hawk

8. What do the typhlosole of an earthworm, the spiral valve of a shark, and the villi of a mammal all have in common?
   a. All are adaptations for the efficient digestion and absorption of meat.
   b. They are all adaptations of the stomach.
   c. They are all microscopic structures.
   d. They all increase the absorptive surface area of intestinal epithelium.
   e. They are all homologous structures.

9. Basal metabolic rate, expressed on a per gram basis, would be greatest for a
   a. cheetah
   b. chipmunk
   c. shark
   d. whale
   e. human

10. If you were to sprint 100 m a few hours after lunch, which stored fuel would you probably tap?
    a. muscle proteins
    b. muscle glycogen
    c. fat stored in the liver
    d. fats stored in adipose tissue
    e. blood proteins

CHALLENGE QUESTIONS

1. Trace a bacon, lettuce, and tomato sandwich through the human alimentary canal, describing what happens to the food in each region of the tract.
2. Suggest your own hypothesis to explain the inverse relationship between body size and metabolic rate per gram of tissue. How could you test your hypothesis?
3. Some essential mineral nutrients, such as selenium and molybdenum, are required in the human diet in minuscule amounts—thousandths of a milligram. What kinds of experiments and observations would help demonstrate that a mineral is required by humans?

SCIENCE, TECHNOLOGY, AND SOCIETY

1. According to recent surveys, 75% of Americans 18 to 35 years old think they are fat. At any given time, 30 million women and 18 million men are on diets. Only 25% are *actually* medically overweight. What do you think causes so many people who are not overweight to be concerned about their weight? Other data show that more than twice as many high school girls than boys see themselves as overweight, and 45% of underweight women see themselves as fat. Why do you think there is a difference between men and women regarding their perception of their weight?
2. Famine plagues certain parts of the world today. Some people argue that distributing food more equitably to the various countries would reduce starvation, at least for a while. Others counter that it is erroneous and ultimately harmful to perceive starvation as a global problem, because the causes and long-term solutions are usually regional. Evaluate the biological, political, and ethical facets of this debate.

FURTHER READING

Christian, J. L., and J. L. Greger. *Nutrition for Living,* 3rd ed. Redwood City, CA: Benjamin/Cummings, 1991. A detailed discussion of nutrition, related to life circumstances.

Diamond, J. "The Athlete's Dilemma." *Discover,* August 1991. Have marathoners and other endurance athletes reached the limits of what metabolism can support?

Edman, J. "Biting the Hand That Feeds You." *Natural History,* July 1991. Humans as food (for mosquitoes).

Hume, I. "Reading the Entrails of Evolution." *New Scientist,* April 15, 1989. Convergent evolution of digestive systems of marsupials and placental mammals.

Lee, A., and R. Martin. "Life in the Slow Lane." *Natural History,* August 1990. Digestive adaptations of koalas.

Moog, F. "The Lining of the Small Intestine." *Scientific American,* November 1981. A look at the site of absorption.

Sanderson, S., and R. Wasserug. "Suspension Feeding Vertebrates." *Scientific American,* March 1990. This article emphasizes suspension-feeding whales.

Scrimshaw, N. "Iron Deficiency." *Scientific American,* October 1991. One of the most common nutritional deficiencies in the world.

Stiling, P. D. "Eating a Thin Line." *Natural History,* February 1988. The natural history of leaf miners, insect larvae that eat their way through the soft tissue of leaves.

38 CIRCULATION AND GAS EXCHANGE

INTERNAL TRANSPORT IN INVERTEBRATES
CIRCULATION IN VERTEBRATES
MAMMALIAN BLOOD
CARDIOVASCULAR DISEASE
GAS EXCHANGE IN ANIMALS

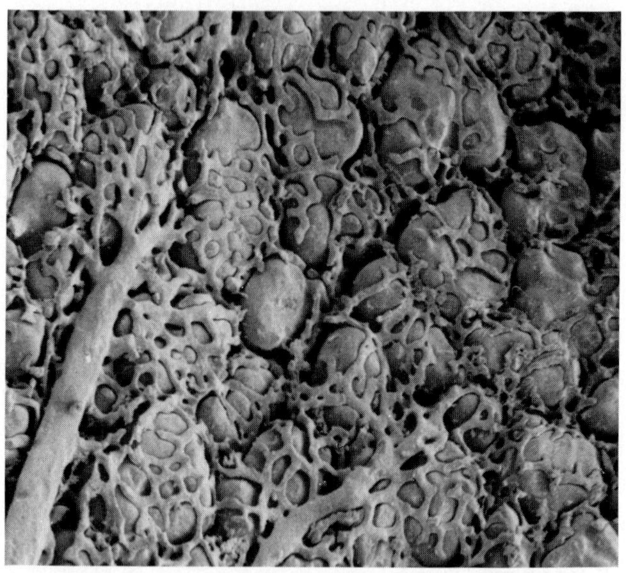

0.1 mm

Figure 38.1
Blood vessels and air sacs of a human lung. This micrograph shows the network of tiny blood vessels (capillaries) that envelop the microscopic air sacs (alveoli) within a lung (SEM). Oxygen diffuses from the sacs into the capillaries, and carbon dioxide diffuses in the opposite direction. Oxygenated blood from the lungs then flows to the heart for pumping to all other organs. In this chapter, you will learn how the diverse circulatory systems that have evolved transport materials within animals, with an emphasis on the relationship between circulation and exchange of respiratory gases.

Every organism must exchange materials with its environment. We have seen that this chemical commerce ultimately occurs at the cellular level, with substances passing across the plasma membrane between the cell and its immediate surroundings. Because substances can cross the membrane only if they are dissolved in water, every living cell must be bathed by an aqueous environment, which provides oxygen, nutrients, and other resources the cell needs and which also serves as a disposal site for carbon dioxide and other metabolic waste products that diffuse out of the cell. For a protozoan living in an aquatic habitat, the environment is the surrounding pond water or seawater, and chemical exchange is accomplished simply by diffusion or active transport across the plasma membrane. Because the protozoan is small, its external surface area is sufficient to service the entire volume of the organism. The same strategy works for the simplest multicellular animals, which have body plans that expose every cell to the surroundings (see Chapter 36).

The length of time it takes for a substance to diffuse from one place to another is proportional to the square of the distance the chemical must travel. For example, if it takes 1 second for a given quantity of glucose to diffuse 100 μm, it will take 100 seconds for the same quantity to move 1 mm, and about 3 years for that quantity to diffuse 1 m! Clearly, diffusion is inadequate for transporting chemicals over macroscopic distances in animals—for example, for moving oxygen from the lungs to the brain in humans.

All but the simplest animals have special systems for the internal transport of body fluids. Chemicals are transferred between the body fluid (blood or interstitial fluid) and the environment across the thin epithelia of organs specialized for gas exchange, nutrient absorption, or waste expulsion. In our lungs, for example, oxygen from the air we inhale diffuses across a thin epithelium and into the blood, while carbon dioxide diffuses in the opposite direction down its own concentration gradient (Figure 38.1). The circulatory system then carries the oxygen-rich blood away to all parts of the body. As the blood streams through our tissues within microscopic vessels called capillaries, chemicals are transported between the blood and the interstitial fluid that directly bathes our cells. No substance has to diffuse far to enter or leave a cell. And because our

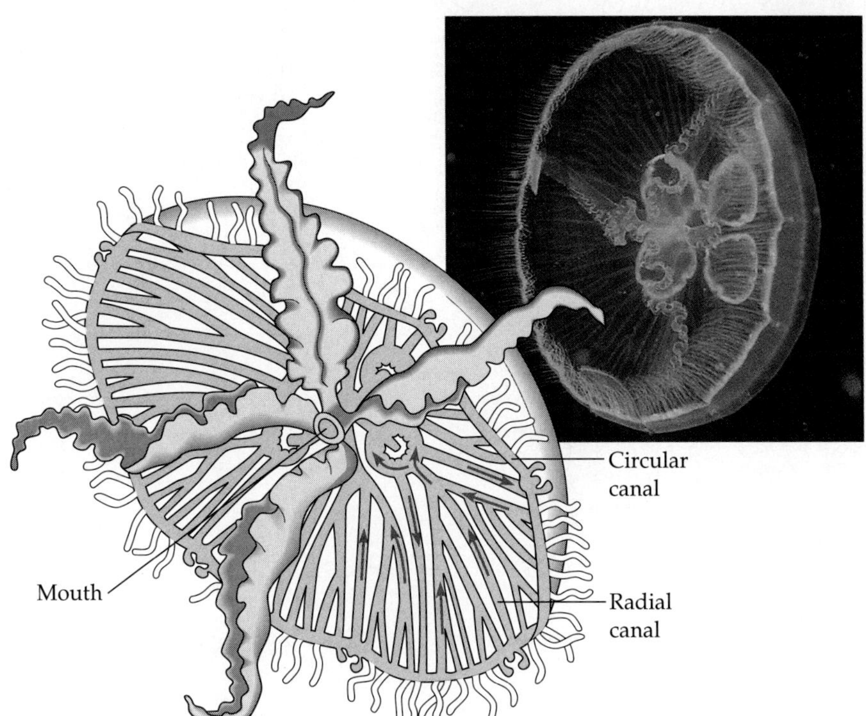

Figure 38.2
Transport in the jellyfish *Aurelia.* The mouth leads to an elaborate gastrovascular cavity (shown in gold) that has branches radiating to and from a circular canal. Ciliated cells lining the cavity circulate fluid in the directions indicated by the arrows. The animal is viewed here from its underside (oral surface).

"pond"—blood and the interstitial fluid it services—is internal, it is possible to control the chemical and physical properties of the milieu in which our cells live. By circulating the blood frequently through organs such as the liver and kidneys, where its contents of nutrients and wastes can be regulated, the circulatory system plays a central role in homeostasis (see Chapter 36).

In this chapter, you will learn about mechanisms of internal transport in animals. You will also learn about one of the most important cases of chemical transfer between animals and their environment: the exchange of the respiratory gases oxygen and carbon dioxide. We begin by surveying some of the transport systems that have evolved among the animal phyla.

INTERNAL TRANSPORT IN INVERTEBRATES

Gastrovascular Cavities

The saclike body plan of *Hydra* and other cnidarians makes any specialized system for internal transport unnecessary. A body wall only two cells thick encloses a central gastrovascular cavity, which serves the dual functions of digestion and distribution of substances throughout the body (see Figure 37.8a). The fluid inside the cavity is continuous with the water outside through a single opening; thus, both inner and outer layers of tissue are bathed by fluid. Thin strands of the gastrovascular cavity extend into the tentacles of *Hydra,* and some jellyfish have even more elaborate gastrovascular cavities (Figure 38.2). Some of the cells lining the cavity have beating flagella that stir the contents, helping distribute materials throughout the animal. Since digestion begins in the cavity, only the cells of the inner layer have direct access to nutrients, but the nutrients have only a short distance to diffuse to the cells of the outer layer.

Planarians and other flatworms also have gastrovascular cavities that exchange materials with the environment through a single opening (see Figure 37.8b). The flat shape of the body and the ramification of the gastrovascular cavity throughout the animal ensure that all cells are bathed by a suitable medium.

Open and Closed Circulatory Systems

A gastrovascular cavity is inadequate for internal transport within animals possessing many layers of cells, especially if the animals live out of water. In insects, other arthropods, and most mollusks, blood bathes the internal organs directly. This arrangement is called an **open circulatory system.** There is no distinction between blood and interstitial fluid, and the general body fluid is more correctly termed **hemolymph.** Chemical exchange between the fluid and body cells occurs as the hemolymph oozes through **sinuses,** which are spaces surrounding the organs. Hemolymph is "circulated" in a limited fashion by body movements

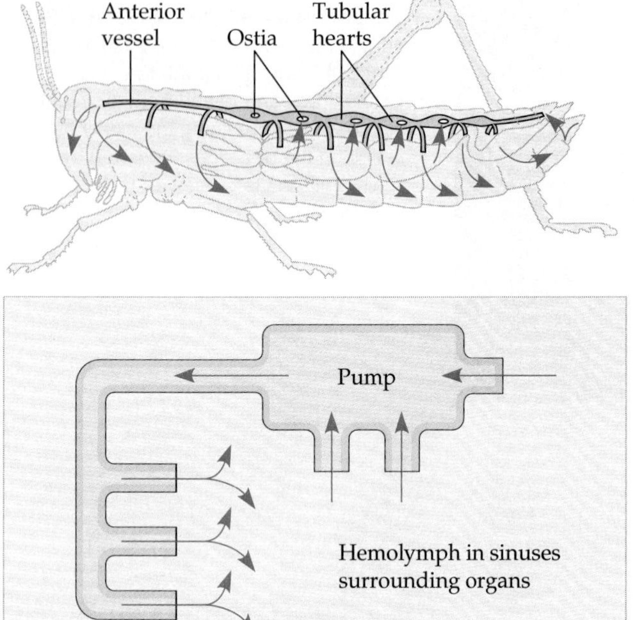

(a) **Open circulatory system.** Grasshoppers and other arthropods have an open circulatory system, in which the hemolymph has direct contact with the body tissues. Circulation is driven by the pumping of tubular hearts and by movement of the animal. Hemolymph enters the hearts through pores called ostia when the hearts relax and is pumped out when the hearts contract.

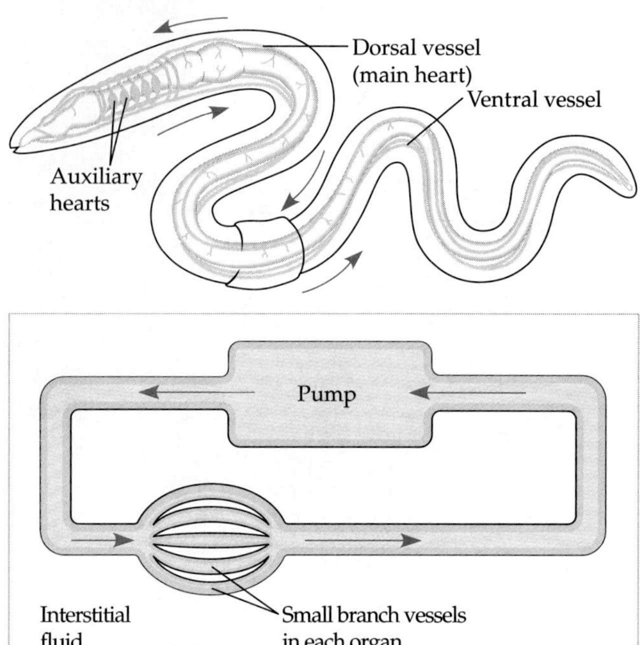

(b) **Closed circulatory system.** In the closed circulatory system of the earthworm, blood is confined to vessels. The dorsal and ventral vessels carry blood anteriorly and posteriorly, respectively, and are connected by large vessels that loop around the digestive tract. The dorsal vessel functions as the main heart. Five pairs of the connecting vessels act as auxiliary hearts.

Figure 38.3
Open and closed circulatory systems of invertebrates. The simplified diagrams contrast the basic principles of open versus closed circulation.

that squeeze the sinuses and by the contraction of hearts, usually parts of a dorsal vessel (Figure 38.3a). Contraction of the hearts pumps hemolymph through vessels, which open into the interconnected system of sinuses. When the hearts relax, they draw hemolymph in through pores called ostia, which are equipped with valves that close when the hearts contract.

An earthworm (phylum Annelida) has a **closed circulatory system,** meaning that blood is confined to vessels (Figure 38.3b). There are two major vessels, one dorsal and one ventral, from which branch smaller vessels that supply blood to the various organs. The dorsal vessel functions as the main heart, pumping blood forward by waves of peristalsis. Near the anterior end of the worm, pairs of vessels loop around the digestive tract, connecting the dorsal and ventral vessels. Five pairs of these vessels function as auxiliary hearts by pulsating. The blood exchanges materials with the interstitial fluid, which bathes the cells. Vertebrates and some mollusks (squids and octopuses) also have closed circulatory systems.

Blood percolates through an open circulatory system more slowly than it flows through a closed system. Because most classes of animals depend on their circulatory system to transport oxygen for cellular respiration, we might expect to find open systems only in animals that move sluggishly. And yet, flying insects, among the most active of all animals, have open circulatory systems. Insects, however, do not use blood to carry oxygen long distances. Oxygen infiltrates the insect body through microscopic air ducts called tracheae, which we will discuss in more detail later in the chapter.

CIRCULATION IN VERTEBRATES

Internal transport is accomplished in humans and other vertebrates by a closed circulatory system, also called the **cardiovascular system.** The components of the cardiovascular system are the heart, blood vessels, and blood. The heart consists of one **atrium** or two **atria,** the chambers that receive blood returning to the heart, and one or two **ventricles,** the chambers that pump blood out of the heart. Arteries, veins, and cap-

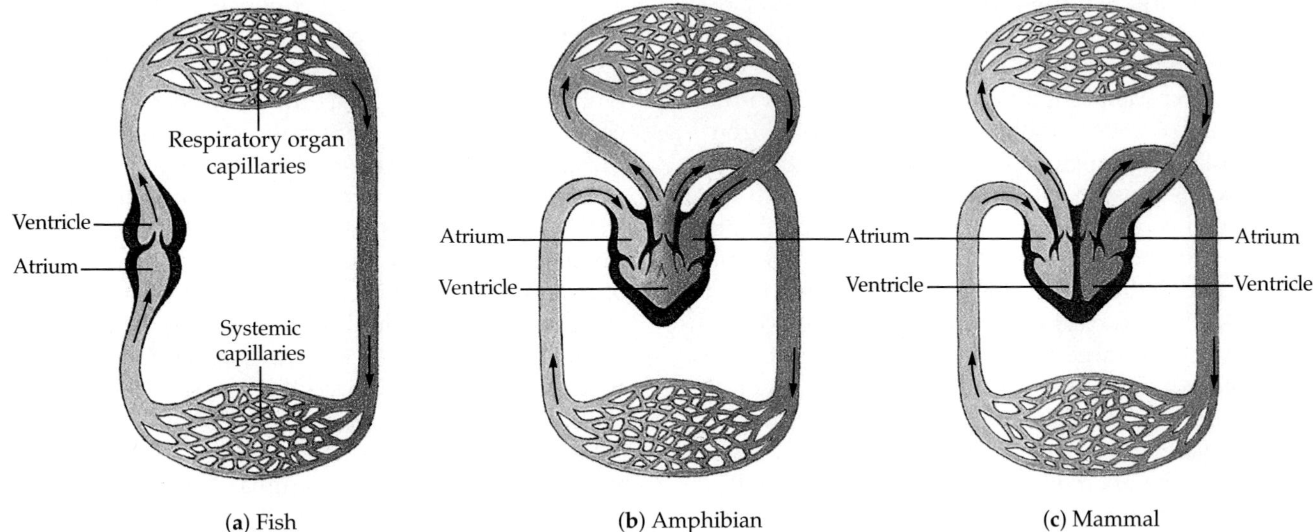

Figure 38.4
Generalized circulatory schemes of vertebrates. Red is used to symbolize oxygen-rich blood, and blue represents oxygen-poor blood. (**a**) Fishes have a two-chambered heart and a single circuit of blood flow. (**b**) Amphibians have a three-chambered heart and two circuits of blood flow: pulmonary and systemic. This double circulation delivers blood to systemic organs under high pressure. In the single ventricle, there is some mixing of oxygen-rich with oxygen-poor blood. (**c**) Mammals have four-chambered hearts and double circulation. Oxygen-rich blood is kept completely segregated from oxygen-poor blood within the heart.

illaries are the three kinds of blood vessels, which in the human body have been estimated to extend a total distance of 100,000 km. **Arteries** carry blood away from the heart to organs throughout the body. Within these organs, arteries branch into **arterioles,** tiny vessels that give rise to the capillaries. **Capillaries** form networks of microscopic vessels that infiltrate each tissue. It is across the thin walls of capillaries that chemicals are exchanged between the blood and the interstitial fluid surrounding the cells. At their "downstream" end, capillaries rejoin to form **venules,** and these small vessels converge into veins. **Veins** return blood to the heart. Notice that arteries and veins are distinguished by the *direction* in which they carry blood, not by the quality of the blood they contain. Not all arteries carry oxygenated blood, and not all veins carry blood depleted of oxygen. But all arteries do carry blood from the heart to capillaries, and only veins return blood to the heart from capillaries. We will now examine the routes of blood flow in various classes of vertebrates.

Vertebrate Circulatory Schemes: An Evolutionary Perspective

Various adaptations of the general circulatory scheme just described have evolved in the different vertebrate classes.

A fish has a two-chambered heart, with one atrium and one ventricle (Figure 38.4a). Blood pumped from the ventricle travels first to the gills, where the blood picks up oxygen and disposes of carbon dioxide across capillary walls. The gill capillaries reconvene to form a vessel that carries the oxygenated blood to capillary beds in all other parts of the body. Blood then returns in veins to the atrium of the heart. Notice that in a fish, blood must pass through *two* capillary beds during each circuit, one in the gills and a second one in some other organ. When blood flows through a capillary bed, blood pressure, the hydrostatic pressure that pushes blood through vessels, drops substantially (for reasons that will be explained shortly). Therefore, oxygenated blood leaving the gills flows to other organs in the fish quite slowly, but the process is aided by the whole-body movements during swimming.

Frogs and other amphibians have three-chambered hearts, with two atria and one ventricle (Figure 38.4b). The ventricle pumps blood into a forked artery that directs the blood through two circuits: the **pulmonary circuit** and the **systemic circuit.** The pulmonary circuit leads to the lungs and skin, where the blood picks up oxygen as it flows through capillaries. The oxygenated blood returns to the left atrium of the heart, and then most of it is pumped into the systemic circuit. The systemic circuit carries blood to all organs except the lungs and then returns the blood to the right atrium in veins. This scheme, called **double circulation,** ensures a vigorous flow of blood to the brain, muscles, and other organs because the blood is pumped a second time after it loses pressure in the capillary beds of the lungs. This is distinctly different from the single circulation in the fish, where blood flows directly from the respiratory organs (gills) to other organs under reduced pressure.

In the single ventricle of the frog, there is some mixing of oxygen-rich blood that has returned from the lungs with oxygen-poor blood that has returned from the rest of the body. However, a ridge within the ventricle diverts most of the oxygenated blood from the left atrium into the systemic circuit and most of the deoxygenated blood from the right atrium into the pulmonary circuit. In reptiles, there is even less mixing of oxygen-rich with oxygen-poor blood. Although the reptilian heart is three-chambered, the single ventricle is partially divided by a septum (wall). One order of reptiles, the crocodiles, has a complete septum that divides the ventricle into two chambers.

The four-chambered heart of a bird or mammal has two atria and two completely separated ventricles (Figure 38.4c). There is double circulation, as in amphibians and reptiles, but the heart keeps oxygen-rich blood fully segregated from oxygen-poor blood. The left side of the heart handles only oxygenated blood, and the right side receives and pumps only deoxygenated blood. With no mixing of the two kinds of blood, and with a double circulation that restores pressure after blood has passed through the lung capillaries, delivery of oxygen to all parts of the body for cellular respiration is enhanced. As endotherms, which use heat released from metabolism to warm the body, birds and mammals require more oxygen per gram of body weight than other vertebrates of equal size. Birds and mammals descended from different reptilian ancestors, and their four-chambered hearts evolved independently—an example of convergent evolution. A more detailed diagram of blood flow through the mammalian circulatory system is shown in Figure 38.5.

We will now see how the heart actually works. Although the process is described with reference to humans, all mammalian hearts work essentially the same way.

The Heart

The human heart is a cone-shaped organ about the size of a clenched fist, located just beneath the breastbone (sternum). It is enclosed in a sac having a two-layered wall. A lubricating fluid fills the space between the two membranes, enabling them to slide past each other as the heart pulsates. The wall of the heart itself consists mostly of cardiac muscle tissue (see Chapter 36). The atria have relatively thin walls and function as collection chambers for blood returning to the heart, pumping blood only the short distance to the ventricles. The ventricles have thicker walls and are much more powerful than the atria—especially the left ventricle, which must pump blood through the systemic circuit (see Figure 38.5).

The Heart Cycle The sequence of events during each heartbeat is referred to as the **heart cycle.** The cycle has two alternating phases: systole and diastole. During **systole,** the heart muscle contracts and the chambers pump blood. (Systole actually refers to only the contraction of the ventricles, but we will include atrial contraction in systole.) During **diastole,** the ventricles are filling. In an average human at rest, the entire heart cycle takes about 0.8 seconds (giving a pulse of about 65 to 75 beats per minute). Systole and diastole are generally equal in duration, lasting about 0.4 sec each. During the first 0.1 sec of systole, the atria contract, squeezing blood into the ventricles. Then, in a slow but powerful contraction, the ventricles pump blood into the arteries during the remaining 0.3 sec of systole. Notice that seven-eighths of the time—all but the first 0.1 sec of the heart cycle—the atria are relaxed and filling with blood returning in the veins. Throughout diastole, the ventricles are also filling, because blood can flow into them freely from the atria when the heart is relaxed. In fact, ventricles hold about 70% of their capacity by the end of diastole, and atrial contraction at the beginning of systole only finishes the job of filling them with blood.

Heart Valves and Heart Sounds Four valves in the heart prevent a backflow of blood when the ventricles contract (see Figure 38.5b). Between each atrium and ventricle is an **atrioventricular valve. Semilunar valves** are located at the two exits of the heart, where the aorta leaves the left ventricle and the pulmonary artery leaves the right ventricle. The valves consist of flaps of connective tissue anchored by strong fibers that prevent the valves from turning inside out. Hydrostatic pressure generated by the powerful contraction of the ventricles forces the atrioventricular valves closed, keeping blood from flowing back into the atria. The blood is pumped out into the arteries through the semilunar valves, which are forced open by ventricular contraction. The elastic walls of the arteries expand and then snap back. Along with relaxation of the ventricles, recoil of the arteries closes the semilunar valves and prevents blood from flowing back into the ventricles.

The heart sounds we can hear with a stethoscope are caused by the vigorous closing of the valves. (You can even hear them without a stethoscope by pressing your ear against the sternum of a friend.) The sound pattern is "lub-dupp, lub-dupp, lub-dupp," the first tone a lower pitch than the second. The first heart sound ("lub") is created by the forceful contraction of the ventricles and the recoil of blood against the closed atrioventricular valves. The second sound ("dupp") is the recoil of blood against the semilunar valves.

A defect in one or more of the valves causes a condition known as a heart murmur, which may be detectable as a hissing sound when a stream of blood

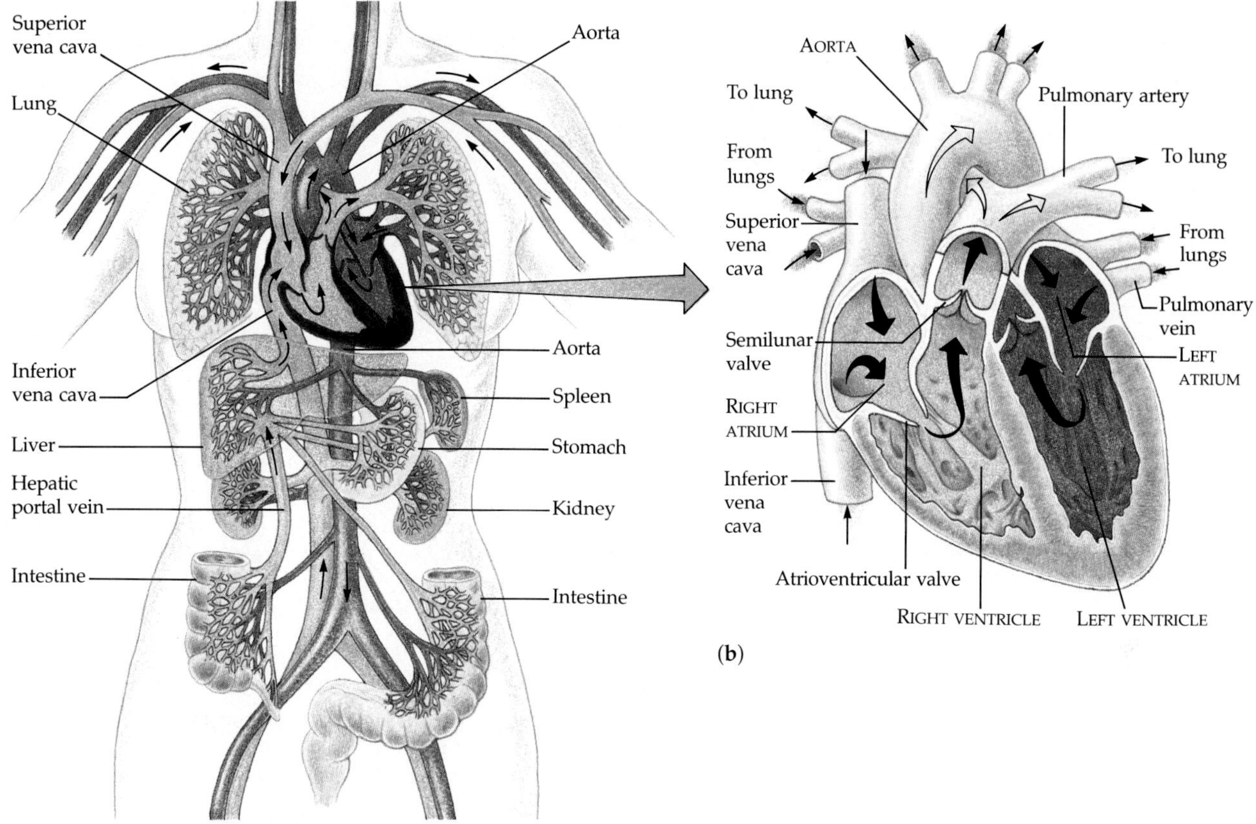

Figure 38.5
The human circulatory system.
(**a**) Beginning our tour with the systemic circuit, oxygenated blood is pumped by the left ventricle into the aorta. This artery is the largest blood vessel in the body, having a diameter about that of a quarter. The aorta arches over the heart and gives rise to arteries leading throughout the body. The first branches from the aorta are the coronary arteries, which supply blood to the heart muscle itself. Then come branches leading to the head and arms (or forelimbs). The aorta continues in a posterior direction, supplying blood to the abdominal organs and legs (or hind limbs). Within each organ, the arteries branch into arterioles, which in turn give rise to capillaries, where the blood gives up its oxygen and receives the carbon dioxide produced by cellular respiration. Capillaries rejoin to form venules, which lead to veins. The systemic veins return deoxygenated blood to the heart. Blood from the head, neck, and arms is channeled into a large vein called the superior (anterior, in other mammals) vena cava. The inferior (posterior, in other mammals) vena cava drains blood from the trunk and legs. The two venae cavae empty their blood into the right atrium, completing the systemic circuit.
The function of the pulmonary circuit is to oxygenate blood by passing it through the lungs. Blood that has returned to the heart via the systemic circuit is pumped by the right ventricle into the pulmonary artery. (Notice that an artery carries deoxygenated blood here.) The pulmonary artery forks, one branch going to each lung. As the blood flows through capillaries in the lungs, it takes up oxygen and unloads carbon dioxide. The pulmonary circuit is completed when oxygenated blood returns to the heart in pulmonary veins, which empty into the left atrium. This blood is ready to be pumped again through the systemic circuit. Part (**b**) shows the route that blood follows through the heart.

squirts backward through a valve. Some people are born with heart murmurs, while others have their valves damaged by infection (from rheumatic fever, for instance). Most heart murmurs do not reduce the efficiency of blood flow enough to warrant surgery. More serious murmurs may be corrected by replacing the damaged valves with artificial ones or with human valves taken from a cadaver.

Heart Rate and Cardiac Output Heart rate, or **pulse,** is the number of heartbeats per minute. You can easily measure your own heart rate by counting the pulsations of arteries in your wrist or neck. Each heart cycle, the contractions of the ventricles during systole pump blood with such force that the elastic arteries stretch from the pressure. For an average human at rest, the pulse is about 65 to 75 beats per minute, although individuals who exercise regularly often have slower resting pulses than those who are less fit. Your own pulse will vary, depending on your level of activity and other factors.

In comparing the heart rates of different mammals, we see an inverse relationship between size and pulse. An elephant, for instance, has a pulse of only 25 beats

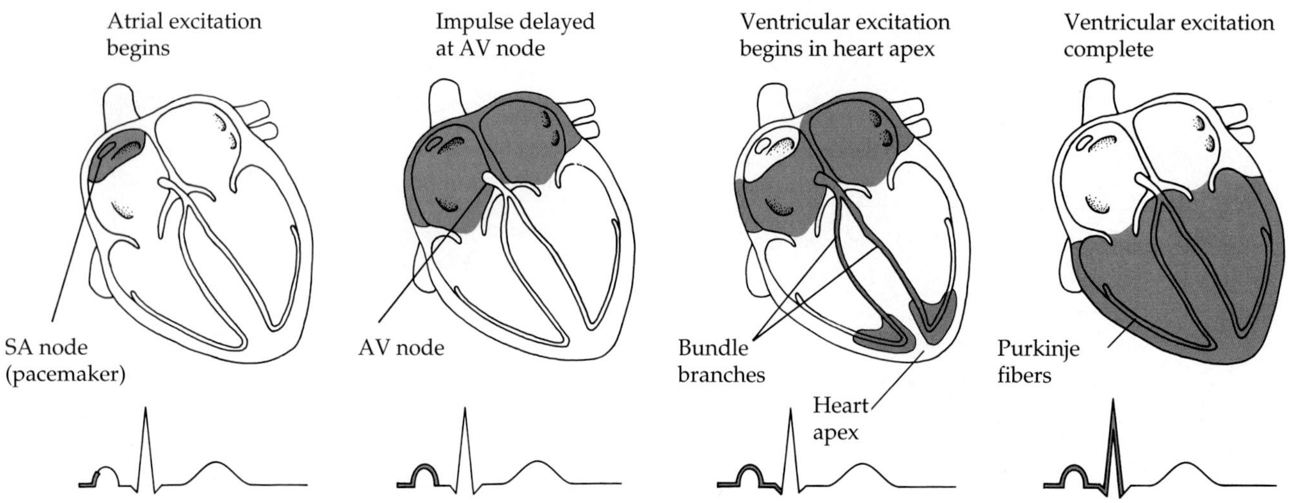

Figure 38.6
Excitation of the heart. Red is used in these diagrams to trace the sequence of excitation and the corresponding components of an electrocardiogram.

per minute, while the heart of a tiny shrew races at about 600 beats per minute. To understand the significance of this difference, recall from Chapter 37 that metabolic rate (hence, oxygen consumption) per gram of tissue is proportionately greater for smaller mammals than for larger ones. A rapid pulse is one adaptation that enhances the delivery of oxygen for cellular respiration.

The volume of blood per minute that the left ventricle pumps into the systemic circuit is called **cardiac output.** This volume depends on two factors: heart rate (pulse) and **stroke volume,** the amount of blood pumped by the left ventricle each time it contracts. The average stroke volume for a human is about 75 mL per beat. A person with this stroke volume and a resting pulse of 70 beats per minute has a cardiac output of 5.25 L/min. This is about equivalent to the total volume of blood in the human body. Cardiac output can increase about fivefold during heavy exercise.

Excitation and Control of the Heart The cells of cardiac muscle are self-excitable, or myogenic; they can contract without any signal from the nervous system. A heart removed from a frog and placed in a beaker of saline solution will continue to beat for an hour or more. Even individual cardiac muscle cells removed from the heart and viewed with a microscope can be seen to pulsate, but they do so at irregular intervals. Although the cells of cardiac muscle have an intrinsic ability to contract, they must be coordinated with one another, and the rhythm of the contractions must somehow be controlled. The rate of contraction is set by a specialized region of the heart called the **sinoatrial (SA) node,** or **pacemaker.** The SA node is located in the wall of the right atrium, near the point where the anterior vena cava enters the heart (Figure 38.6). It is composed of specialized muscle tissue that combines characteristics of both muscle and nerve. Nodal tissue contracts like muscle, but in so doing, it generates electrical impulses much like those found in nervous tissue. Each time the SA node contracts, it initiates a wave of excitation that travels through the wall of the heart. The impulse spreads rapidly, and the two atria contract in unison. (Cardiac muscle cells are electrically coupled by the intercalated discs between adjacent cells; see Chapter 36.) At the bottom of the wall separating the two atria is another patch of nodal tissue, the **atrioventricular (AV) node.** When the wave of excitation reaches the AV node, it is delayed for about 0.1 sec, which ensures that the atria will contract first and empty completely before the ventricles contract. After this delay, the signal to contract is conducted to the tips of the ventricles along bundle branches, and the wave of excitation then spreads upward through the ventricular walls via the Purkinje fibers (see Figure 38.6). The impulses that travel through cardiac muscle during the heart cycle produce electrical currents conducted through body fluids to the body surface, where the currents can be detected by electrodes placed on the skin and recorded as an **electrocardiogram** (**EKG** or **ECG**).

Contraction of the SA node sets the tempo for the entire heart, but the pacemaker itself is controlled by a variety of cues. Two sets of nerves oppose each other in adjusting heart rate; one set speeds up the pacemaker, and the other set slows it down. At any given time, heart rate is a compromise regulated by the opposing actions of these two sets of nerves. The pacemaker is also controlled by hormones secreted into the blood by glands. For example, epinephrine, the "fight-or-flight" hormone from the adrenal glands, increases heart rate (see Chapter 41). Body temperature is another factor that affects the pacemaker. A temperature increase of

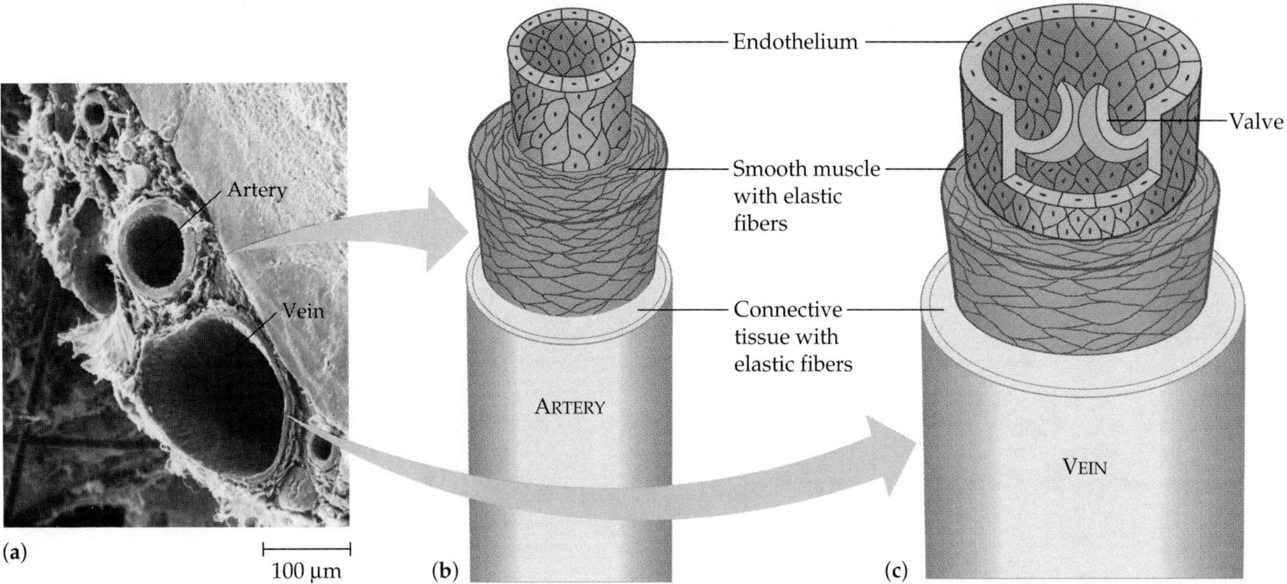

Figure 38.7
The structure of blood vessels. (**a**) In this micrograph, an artery can be seen next to a thinner-walled vein (SEM). (**b**) The wall of an artery or a vein has three layers: an inner layer of endothelium, a middle layer of smooth muscle and elastic fibers, and an outer layer of connective tissue with elastic fibers. Capillaries have much smaller diameters than arteries and veins, and their walls consist of only a single layer of endothelium. (SEM from *Tissues and Organs: A Text-Atlas of Scanning Electron Microscopy* by Richard G. Kessel and Randy H. Kardon. W. H. Freeman and Company. Copyright © 1979.)

only 1°C increases the heart rate by about 10 to 20 beats per minute. This is the reason your pulse increases substantially when you have a fever. Exercise also increases the heart rate, partly because the increased load that the returning venous blood places on the heart stimulates the SA node. This adaptation enables the circulatory system to provide the additional oxygen needed by muscles hard at work.

Blood Flow

Blood Vessel Structure The wall of an artery or vein has three layers (Figure 38.7). On the outside is a zone of connective tissue with elastic fibers that enable the vessel to stretch and recoil. The middle layer consists of smooth muscle and more elastic fibers. This layer is especially thick in the arteries, which must be stronger and more elastic than veins. (The walls of major arteries are so thick that they must themselves be supplied by blood vessels.) Blood vessels are lined by **endothelium,** a simple squamous epithelium (see Chapter 36). The endothelium, along with its basement membrane and a thin ring of connective tissue, forms the inner layer of a blood vessel. Capillaries lack the outer layers, and their very thin walls consist only of endothelium.

Blood Flow Velocity Blood does not flow through the circulatory system at a uniform speed. After it is pumped into the aorta by the left ventricle, it initially travels at a velocity of about 2 m/sec, but by the time it reaches the capillaries, it is flowing much more slowly. To understand why the blood decelerates, we need to consider the *law of continuity,* a rule that governs the flow of fluids through pipes. If a pipe changes diameter over its length, a fluid will stream through narrower segments of the pipe faster than it flows through wider segments. Since the *volume* of flow per second must be constant through the entire pipe, the fluid must flow faster as the cross-sectional area of the pipe narrows. For instance, compare the velocity of water squirted by a hose with and without a nozzle. Based on the law of continuity, it may at first seem as though blood should travel faster through capillaries than through arteries, because the diameter of capillaries is much smaller. However, it is the *total* cross-sectional area of the pipes delivering the fluid that determines flow rate. Although an individual capillary is very narrow, each artery gives rise to such an enormous number of capillaries that the *total* diameter of the conduits is actually much greater in capillary beds than in any other part of the circulatory system. Furthermore, there is more friction in capillary beds because the blood contacts a greater surface area of endothelium. Thus, blood slows down substantially as it enters capillaries from arteries, but then it speeds up again as it passes along to the veins—a result of the reduction in total cross-sectional area (Figure 38.8). Capillaries are

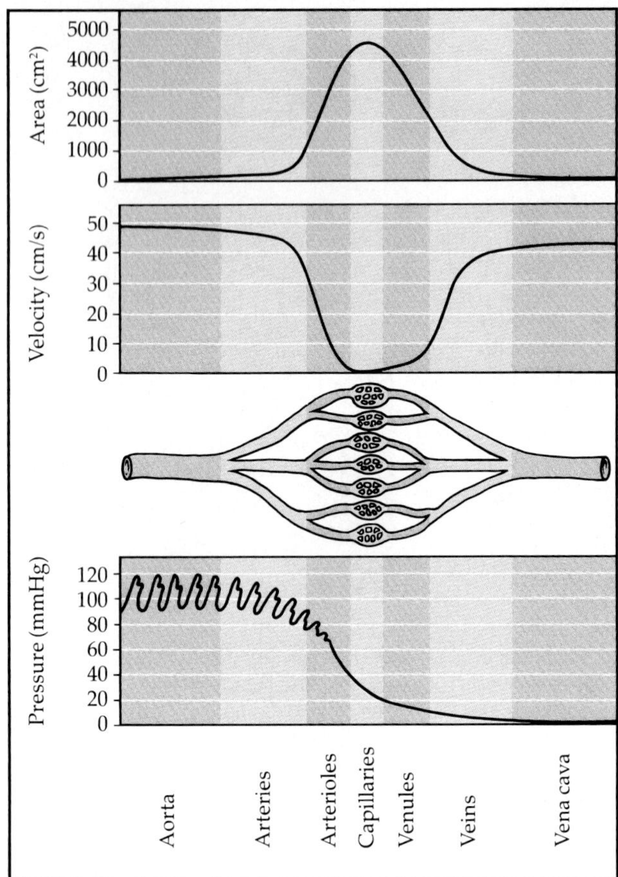

Figure 38.8
Velocity of blood flow and blood pressure. Blood slows down as it flows through a capillary bed, owing to increased friction and the large total cross-sectional area of the numerous microscopic vessels. Resistance to flow through arterioles and capillaries reduces blood pressure and eliminates the pressure peaks caused by systole.

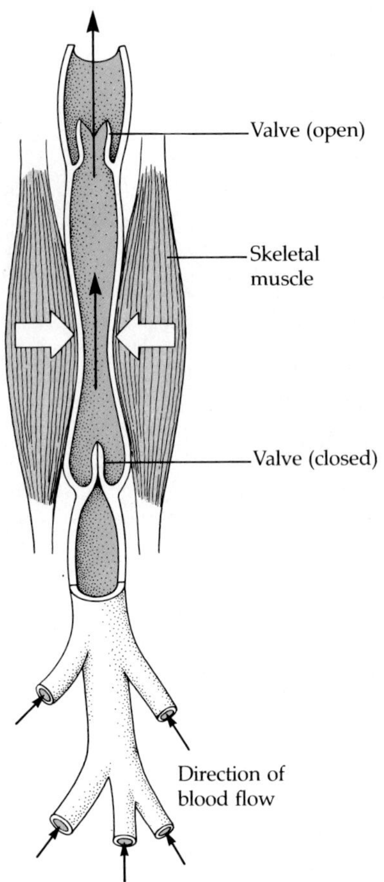

Figure 38.9
Blood flow in veins. Contracting muscles squeeze the veins, in which flaps of tissue act as one-way valves that keep blood moving toward the heart and prevent backflow. Muscular activity during exercise increases this rate of blood flow. If we sit or stand too long, the lack of muscular activity causes our feet to swell with stranded fluid unable to return to the heart. Hairdressers, assembly-line workers, and other people in occupations that require standing for long periods are prone to varicose veins in the legs, which occur when valves collapse from the unrelenting downward pull of gravity on the blood.

the only vessels with walls thin enough to permit the transfer of substances between the blood and interstitial fluid, and the leisurely flow of blood through these tiny vessels enhances this chemical exchange.

Blood Pressure Hydrostatic pressure is the force that moves fluids through pipes. (Perhaps you have experienced the annoyance of a slow-running faucet when there is inadequate water pressure.) The hydrostatic force that blood exerts against the wall of a vessel is called **blood pressure** (see the Methods Box). This pressure is much greater in arteries than in veins and is greatest in arteries during systole, when the heart contracts. When you take your pulse by placing your fingers on your wrist, you can actually feel an artery bulge with each heartbeat. The surge of pressure is partly due to the narrow openings of arterioles impeding the exit of blood from the arteries. Thus, when the heart contracts, blood enters the arteries faster than it can leave, and the vessels stretch from the pressure. The elastic walls of the arteries snap back during diastole, but the heart contracts again before enough blood has flowed into the arterioles to completely relieve pressure in the arteries. This impedance by the arterioles is called **peripheral resistance.** As a consequence of the elastic arteries working against peripheral resistance, there is a blood pressure even during diastole, driving blood into arterioles and capillaries continuously.

Blood pressure is determined partly by cardiac output and partly by the degree of peripheral resistance to blood flow due to the arterioles, the bottlenecks of the circulatory system. Contraction of smooth muscles in the walls of the arterioles constricts the tiny vessels, increases resistance, and therefore increases blood pressure in the arteries. When the muscles relax, the arterioles dilate, and pressure in the arteries falls. These muscles are controlled by nerves, hormones, and other

Methods: Measurement of Blood Pressure

Blood pressure is recorded as two numbers separated by a slash; the higher number is the systolic pressure, and the lower number is the diastolic pressure. A typical blood pressure reading for a 20-year-old is 120/70 (Figure a). The units for these numbers are millimeters of mercury (mm Hg); a blood pressure of 120 is a force that can support a column of mercury 120 mm high. A sphygmomanometer, an inflatable cuff attached to a pressure gauge, measures blood pressure in an artery (Figure b). The cuff is wrapped around the upper arm and inflated until the pressure closes the artery, so that no blood flows past the cuff. A stethoscope is used to listen for sounds of blood flow below the cuff to verify that the artery is closed. The cuff is gradually deflated until blood begins to flow into the forearm, and sounds from blood pulsing into the artery below the cuff can be heard with the stethoscope (Figure c). This occurs when the blood pressure is greater than the pressure exerted by the cuff. The pressure at this point is the systolic pressure, the high pressure exerted by the ventricles contracting. The cuff is loosened further until blood flows freely through the artery, and the sounds below the cuff disappear. The pressure at this point is the diastolic pressure, the residual pressure between heart contractions (Figure d).

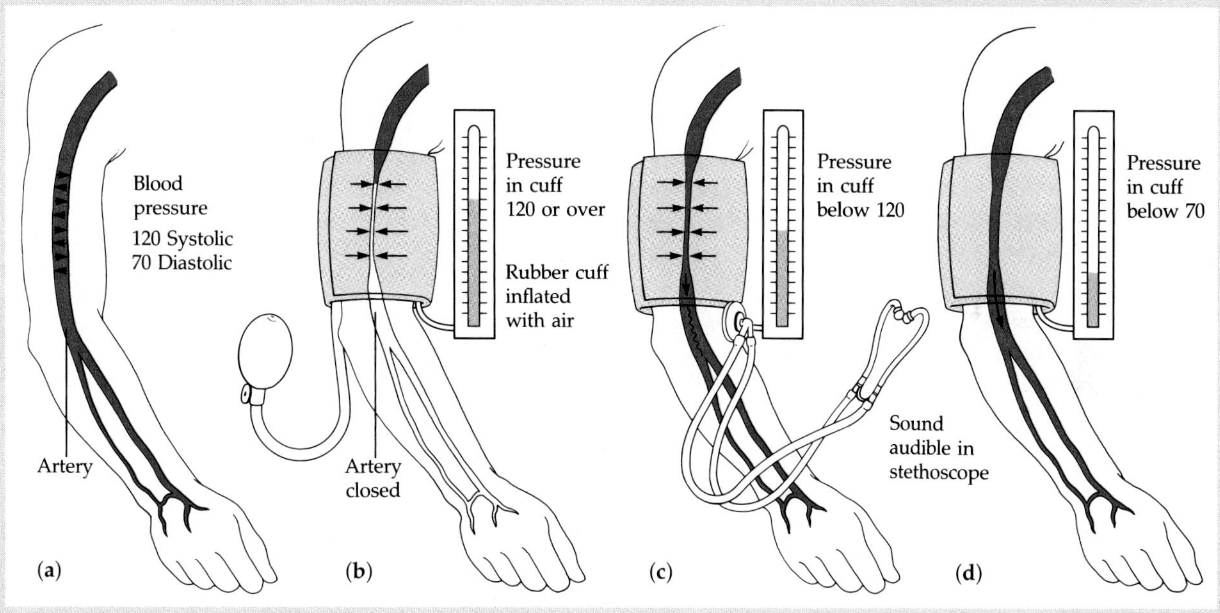

signals. Stress, both physical and emotional, can raise blood pressure by triggering neural and hormonal responses that constrict the blood vessels.

By the time blood reaches the veins, its pressure is not affected much by the heart. This is because the blood encounters so much resistance as it passes through the millions of tiny arterioles and capillaries that the force from the pumping heart can no longer propel the blood in the veins. How, then, does blood return to the heart, especially when it must travel from the lower extremities against gravity? The answer is that veins are sandwiched in between muscles, and whenever we move, our skeletal muscles pinch our veins and squeeze blood through the vessels (Figure 38.9). Within large veins are flaps of tissue that function as one-way valves, allowing the blood to flow only in the direction of the heart. (Such valves are not found in arteries, where blood pressure keeps the blood flowing in the right direction.) Muscular activity during exercise increases the rate at which blood returns to the heart, and the increased load stimulates the heart to speed up. Breathing also helps return blood to the heart; when we inhale, the change in pressure within the thoracic (chest) cavity causes the vena cava and other large veins near the heart to expand and fill.

Microcirculation and Blood Distribution Microcirculation refers to the flow of blood between arterioles

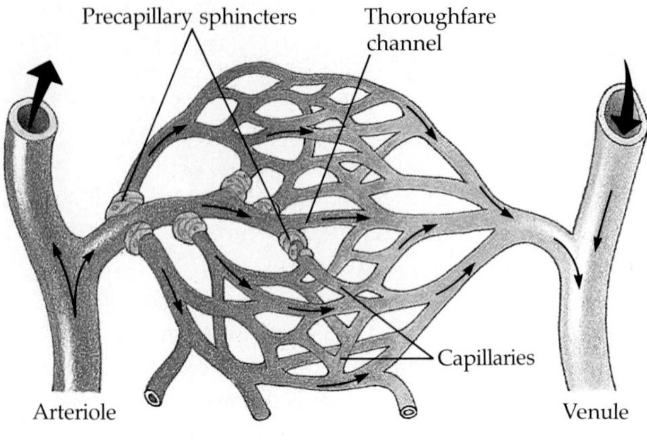

(**a**) Sphincters open

Arteriole

Venule

(**b**) Sphincters closed

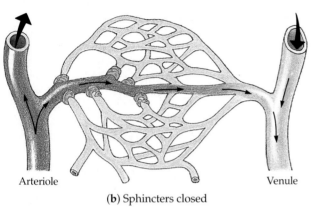

Figure 38.10
Microcirculation. Blood flows from arterioles to venules through thoroughfare channels, vessels that are always open. (**a**) When the precapillary sphincters are open, blood also flows into the capillaries. (**b**) Blood flow through capillaries is shut down when their sphincters close.

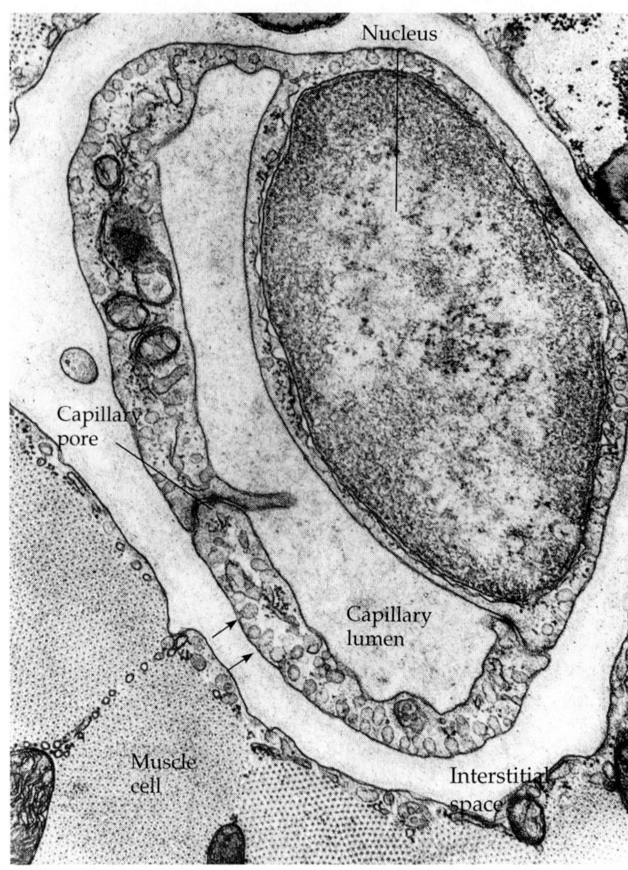

Figure 38.11
The capillary wall. Overlapping endothelial cells enclose the lumen of the capillary, as seen in this micrograph (LM). The spaces between the cells function as capillary pores. Substances cross the capillary wall by diffusion, by transport in vesicles (arrows), and by pressure-driven filtration (bulk flow) through the clefts between cells.

and venules through capillary nets (Figure 38.10). Some blood streams directly from arterioles to venules through **thoroughfare channels**, which are always open. True capillaries branch off from a thoroughfare channel. Passage of blood into the capillary is regulated by a sphincter, a ring of smooth muscle at the capillary entrance.

The distribution of blood to the various organs is controlled by the dilation and constriction of arterioles and by the precapillary sphincters. At any given time, only about 5% to 10% of the body's capillaries have blood flowing through them. Because each tissue has so many capillaries, however, every part of the body is supplied with blood at all times. The supply varies locally as blood is diverted from one destination to another. After a meal, for instance, arterioles in the wall of the digestive tract dilate, capillary sphincters open, and the digestive tract receives a larger share of blood. During strenuous exercise, blood is diverted from the digestive tract and supplied more generously to skeletal muscles. (This is one reason why heavy exercise immediately after a big meal may cause indigestion.) Most of the time, the organs most heavily perfused with blood are the brain, kidneys, liver, and the heart itself.

Capillary Exchange

Now we come to the most important business of the circulatory system: the transfer of substances between the blood and the interstitial fluid that bathes the cells. This exchange takes place across the thin walls of the capillaries. The capillary wall, remember, is the endothelium, a single layer of flattened cells that overlap at their edges (Figure 38.11).

Some materials may be carried across an endothelial cell in vesicles that form by endocytosis on one side of the cell and then release their contents by exocytosis on the opposite side; other substances simply diffuse between the blood and the interstitial fluid. Small mole-

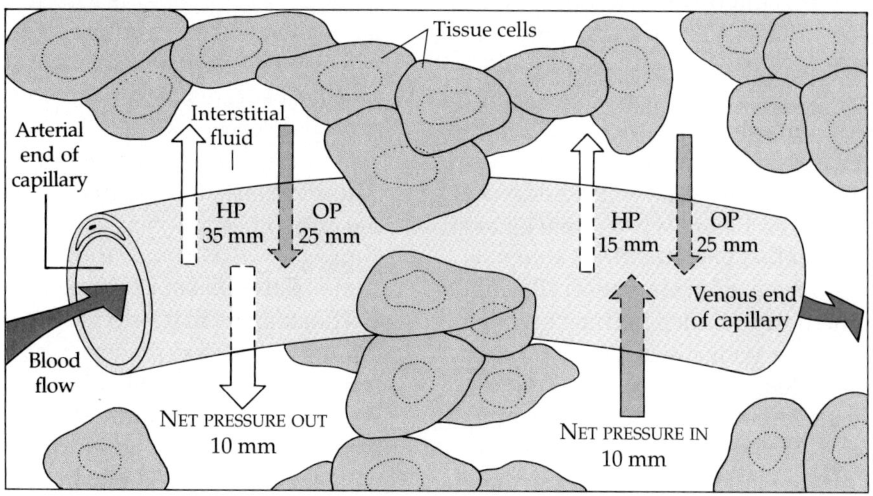

Figure 38.12
The movement of fluid in capillaries. Fluid flows out of a capillary at the upstream end near an arteriole and re-enters a capillary downstream near a venule. The direction of fluid movement at any point along the capillary depends on the difference between two opposing forces: hydrostatic pressure (HP) and osmotic pressure (OP). The hydrostatic pressure, or blood pressure, tends to force fluid out of the capillary. The osmotic pressure is a tendency for water to enter the capillary owing to the relatively high solute concentration of the blood. At the arterial end of the capillary, the hydrostatic pressure forcing fluid outward exceeds the osmotic pressure drawing water inward, resulting in a net exodus of fluid from the capillary. Because the endothelium is selectively permeable, it filters the fluid so that water and small solutes leave the capillary, while most proteins remain behind in the blood. As blood continues along the capillary, the hydrostatic pressure decreases as a result of resistance and loss of fluid volume. At the downstream end of the capillary, the tendency for fluid to exit due to hydrostatic pressure is overpowered by the tendency for fluid to enter due to osmotic pressure.

cules diffuse down their concentration gradients across the membranes of the endothelial cells. Diffusion can also occur through the clefts between adjoining cells. However, transport through these clefts occurs mainly by bulk flow, the movement of fluid due to pressure. Fluid is pushed through the leaky endothelium by hydrostatic pressure (blood pressure) within the capillary. Water and small solutes, such as sugars, salts, oxygen, and urea, move freely through these capillary clefts, but blood cells and proteins dissolved in the blood are too large to pass readily through the endothelium.

Fluid flows out of a capillary at the upstream end near an arteriole, but re-enters downstream near a venule. The mechanism behind this cycling of material between the blood and interstitial fluid is illustrated in Figure 38.12. About 99% of the fluid that leaves the blood at the arterial end of a capillary bed re-enters from the interstitial fluid at the venous end, and the remaining 1% of the fluid lost from capillaries is eventually returned to the blood by the vessels of the lymphatic system.

The Lymphatic System

Although capillaries lose only about 1% of the volume of fluid they carry, so much blood passes through the capillaries that the cumulative loss of fluid adds up to about 3 L per day. There is also some leakage of blood proteins, even though the capillary wall is not very permeable to these large molecules. The lost fluid and proteins return to the blood via the **lymphatic system** (see Figure 39.5). Fluid enters this system by diffusing into tiny lymph capillaries that are intermingled among capillaries of the true circulatory system. Once inside the lymphatic system, the fluid is called **lymph;** its composition is about the same as that of interstitial fluid. The lymphatic system drains into the circulatory system at two locations near the shoulders.

Whenever interstitial fluid accumulates rather than being returned to blood by the lymphatic system, tissues and body cavities become bloated, a condition known as edema. One type of severe, localized edema is elephantiasis, caused by parasitic worms that block the lymph vessels. Another cause of edema is severe dietary protein deficiency. When starved for amino acids, the body consumes its own blood proteins. This reduces the osmotic pressure of the blood, causing interstitial fluid to accumulate in body tissues rather than being drawn back into capillaries. A child suffering from protein deficiency may have a bloated belly because of all the fluid that collects in the body cavity.

Lymph vessels, like veins, have valves that prevent the backflow of fluid toward the capillaries. Like veins, lymph vessels depend mainly on the movement of skeletal muscles to squeeze fluid along. Rhythmic contractions of the vessel walls also help draw fluid into lymphatic capillaries.

Along a lymph vessel are specialized swellings called **lymph nodes.** By filtering the lymph and attacking viruses and bacteria, the nodes play an important role in the body's defense. Inside a lymph node is a honeycomb of connective tissue whose spaces are filled by white blood cells specialized for defense. When the body is fighting an infection, these cells multiply rapidly, and the lymph nodes become swollen and tender (which is why your physician checks for swollen nodes in your neck).

The lymphatic system, then, helps defend the body against infection and maintains the fluid level and protein concentration of the blood. In addition, as mentioned in Chapter 37, lymph capillaries penetrate the villi of the small intestine and absorb fats; thus, the lymphatic system also has the job of transporting fats from the digestive tract to the circulatory system.

MAMMALIAN BLOOD

We now shift our focus from the structure and function of blood vessels to the composition of the blood itself. Vertebrate blood is considered a connective tissue with several types of cells suspended in a liquid matrix called **plasma.** The average human body contains about 4 to 6 L of blood. If a blood sample is taken, the cells can be separated from the plasma by spinning the **whole blood** in a centrifuge. (An anticoagulant must be added to prevent the blood from clotting.) The cells, or *formed elements,* which occupy about 45% of the volume of blood, settle to the bottom of the centrifuge tube to form a dense red pellet. Above this pellet is the transparent, straw-colored plasma (Figure 38.13).

Plasma

Blood plasma consists of an extensive variety of solutes dissolved in water, which accounts for about 90% of the plasma. Among these solutes are inorganic salts, sometimes referred to as blood **electrolytes,** which are present in the plasma in the form of dissolved ions. The combined concentration of these ions is an important factor in maintaining osmotic balance of the blood and interstitial fluid. Some of the ions also help buffer the blood, which has a pH of 7.4 in humans. And the ability of muscles and nerves to function normally depends on the concentration of key ions in the interstitial fluid, which reflects their concentration in plasma. The kidney maintains plasma electrolytes at precise concentrations, an example of homeostasis.

Another important class of solutes is the plasma proteins, which have a number of functions. Collectively they act as buffers to help maintain constant pH, help determine the osmotic strength of blood, and contribute to its viscosity (thickness). The various types of plasma proteins also have specific functions. Some serve as escorts for lipids, which are insoluble in water and can travel in blood only when bound to proteins. Another class of proteins, the immunoglobulins, are the antibodies that help combat viruses and other foreign agents that invade the body (see Chapter 39). And some of the plasma proteins, called fibrinogens, are clotting factors that help plug leaks when blood vessels are injured. Blood plasma that has had these clotting factors removed is called serum.

Plasma also contains various substances in transit from one part of the body to another, including nutrients, metabolic waste products, respiratory gases, and hormones. Blood plasma and interstitial fluid are similar in composition, except that plasma has a much higher protein concentration than interstitial fluid (capillary walls, remember, are not very permeable to proteins).

Blood Cells

Dispersed throughout blood plasma are three classes of cells: red blood cells, which transport oxygen; white blood cells, which function in defense; and platelets, which are involved in blood clotting.

Red Blood Cells Red cells, or **erythrocytes,** are by far the most numerous blood cells. Each cubic millimeter of human blood contains 5 to 6 million red cells, and there are about 25 trillion of these tiny cells in the body's 5 L of blood.

The structure of the red blood cell is another excellent example of structure fitting function. A human erythrocyte is a biconcave disk, thinner in the center than at its edges. Mammalian erythrocytes lack nuclei, an unusual characteristic for living cells (the other vertebrate classes have nucleated erythrocytes). Moreover, all red blood cells lack mitochondria and generate their ATP exclusively by anaerobic metabolism. The major function of erythrocytes is to carry oxygen, and they would not be very efficient if their own metabolism were aerobic and the oxygen they carried was consumed in transit. The small size of erythrocytes also suits their function. For oxygen to be transported, it must diffuse across the plasma membranes of the red blood cells. The smaller the cells, the greater the total area of plasma membrane in a given volume of blood. The biconcave shape of the erythrocyte also adds to its surface area.

As small as a red cell is, it contains about 250 million molecules of **hemoglobin,** a protein containing iron. As red cells pass through the capillary beds of lungs, gills, or other respiratory organs, oxygen diffuses into the erythrocytes and hemoglobin binds the oxygen.

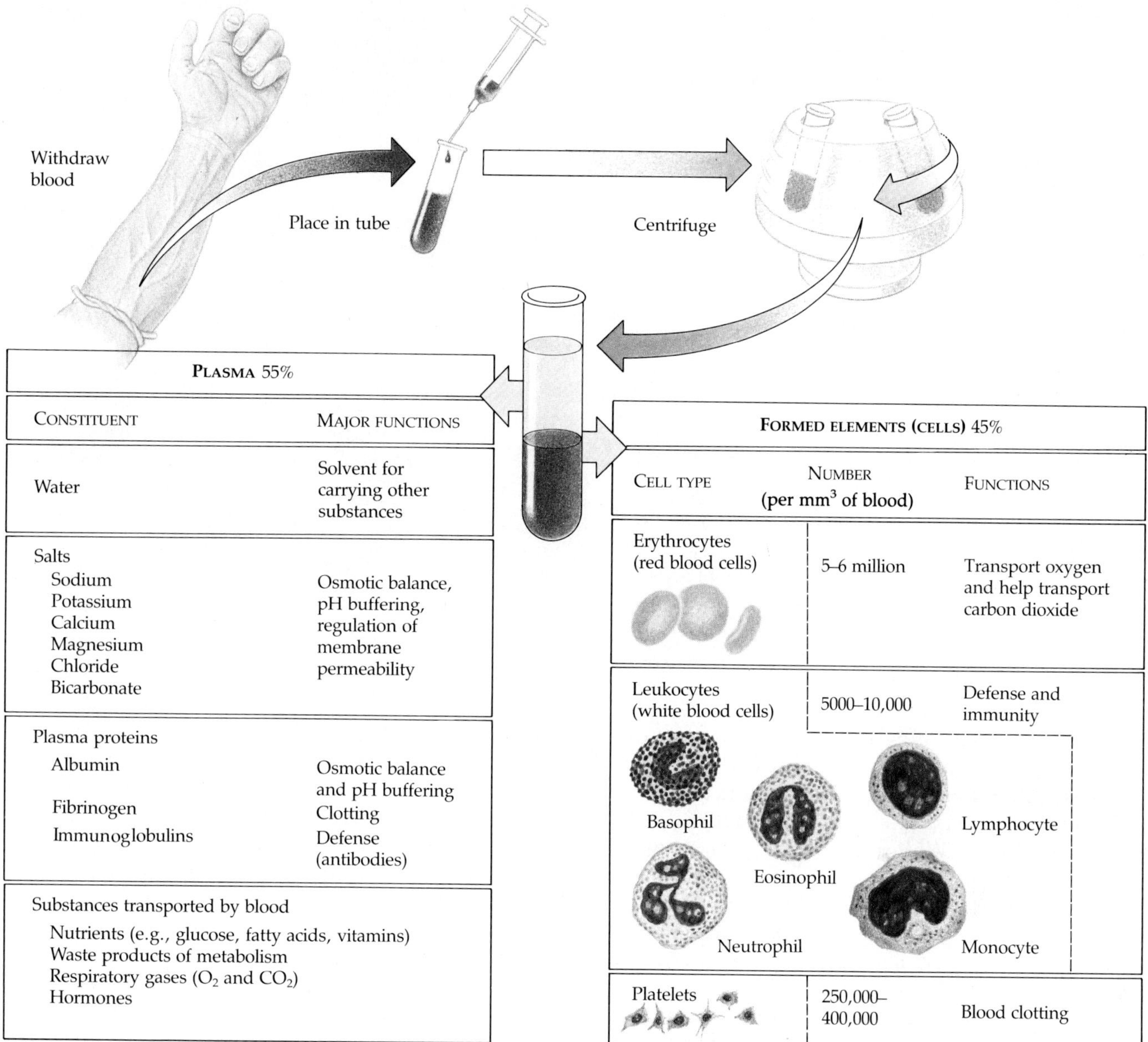

Withdraw
blood
Place in tube
Centrifuge
PLASMA 55%
CONSTITUENT
MAJOR FUNCTIONS
Water
Solvent for carrying other substances
Salts
Sodium
Potassium
Calcium
Magnesium
Chloride
Bicarbonate
Osmotic balance, pH buffering, regulation of membrane permeability
Plasma proteins
Albumin
Osmotic balance and pH buffering
Fibrinogen
Clotting
Immunoglobulins
Defense (antibodies)
Substances transported by blood
Nutrients (e.g., glucose, fatty acids, vitamins)
Waste products of metabolism
Respiratory gases (O_2 and CO_2)
Hormones
FORMED ELEMENTS (CELLS) 45%
CELL TYPE
NUMBER (per mm^3 of blood)
FUNCTIONS
Erythrocytes (red blood cells)
5–6 million
Transport oxygen and help transport carbon dioxide
Leukocytes (white blood cells)
5000–10,000
Defense and immunity
Basophil
Eosinophil
Lymphocyte
Neutrophil
Monocyte
Platelets
250,000–400,000
Blood clotting

Figure 38.13
The composition of blood.

This process is reversed in the capillaries of the systemic circuit, with the hemoglobin unloading its cargo of oxygen. (Oxygen loading and unloading will be described in more detail later in the chapter.)

Erythrocytes are formed in the red marrow of bones, particularly the ribs, vertebrae, breastbone, and pelvis. Within the marrow are **pluripotent stem cells** that can develop into any type of blood cell (Figure 38.14). Red cell production is controlled by a negative-feedback mechanism that is sensitive to the amount of oxygen reaching the tissues via the blood. If the tissues are not receiving enough oxygen, the kidney secretes a hormone called **erythropoietin,** which stimulates production of erythrocytes in the bone marrow (see Chapter 19). If blood is delivering more oxygen than the tissues can use, the level of erythropoietin is reduced, and erythrocyte production slows. On average, erythrocytes circulate for about 3 to 4 months before being destroyed by phagocytic cells located mainly in the liver. The hemoglobin is digested, and the amino acids are incorporated into other proteins made in the liver. Much of the iron of the hemoglobin is cycled back to bone marrow, where it is reused in erythrocyte production.

White Blood Cells There are five major types of white cells, or **leukocytes:** monocytes, neutrophils, basophils, eosinophils, and lymphocytes (Figure 38.13). Their collective function is to fight infections in various ways. For example, monocytes and neutrophils are

Figure 38.14
The development of blood cells. All blood cells differentiate from a common source, a population of pluripotent stem cells in red bone marrow. (*Pluripotent* refers to the potential for these cells to form any type of blood cell.) The population of pluripotent cells renews itself by mitosis. Some of these cells become lymphoid stem cells, which then develop into B cells and T cells, two classes of lymphocytes that function in the immune response (see Chapter 39). All other blood cells differentiate from myeloid stem cells, also derived from the population of pluripotent stem cells.

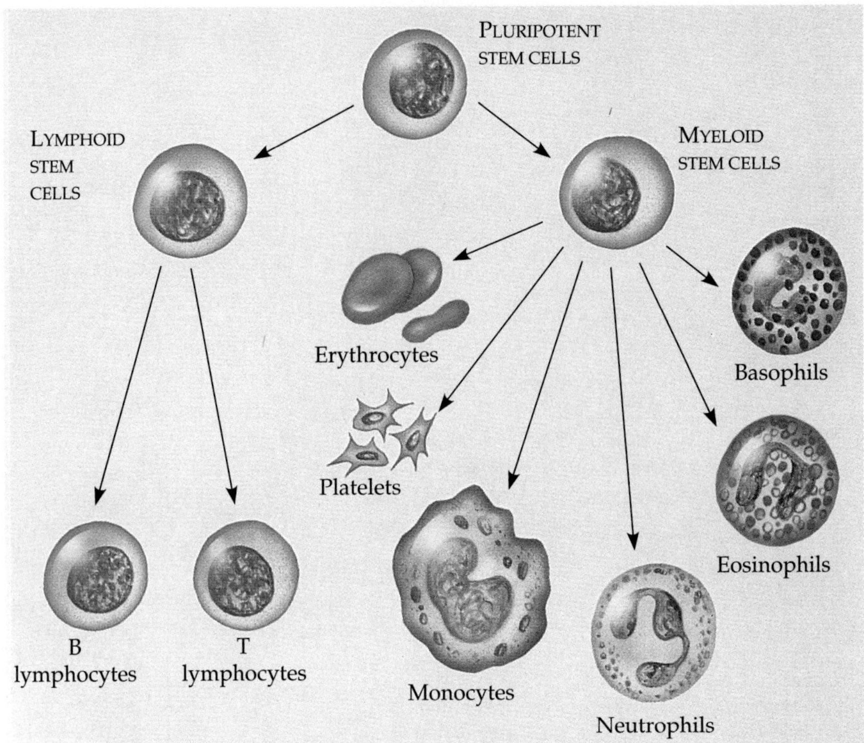

phagocytes, which eat bacteria and debris from our own dead cells. Lymphocytes give rise to the cells that produce antibodies, the plasma proteins that react against foreign substances. The leukocytes we see in blood are in transit. White cells actually spend most of their time outside the circulatory system, patrolling through interstitial fluid, where most of the battles against pathogens are waged. There are also great numbers of leukocytes in lymph nodes (see Chapter 39).

Leukocytes arise in bone marrow from the stem cells that can also differentiate into erythrocytes. Some lymphocytes mature, after leaving the marrow, in the spleen, thymus, tonsils, adenoids, and lymph nodes, all of which are called lymphoid organs. Normally, a cubic millimeter of human blood has 5,000 to 10,000 leukocytes, but the number increases whenever the body is fighting an infection.

Platelets **Platelets** are not really cells at all, but chips of cells about 2 to 3 μm in diameter. They have no nuclei and originate as pinched-off cytoplasmic fragments of large cells in the bone marrow. Platelets then enter the blood and function in the important process of blood clotting.

Blood Clotting

Each of us suffers cuts and scrapes from time to time, yet we do not bleed to death because blood contains a self-sealing material that plugs leaks in our vessels. The sealant is always present in our blood in an inactive form called **fibrinogen.** A clot forms only when this plasma protein is converted to its active form, **fibrin,** which aggregates into threads that form the fabric of the clot. The clotting mechanism usually begins with the release of clotting factors from platelets and involves a complex chain of reactions that ultimately transforms fibrinogen to fibrin (Figure 38.15). More than a dozen clotting factors have been discovered, and the mechanism is still not fully understood. An inherited defect in any step of the clotting process causes **hemophilia,** a disease characterized by excessive bleeding from even minor cuts and bruises.

Anticlotting factors in the blood normally prevent spontaneous clotting in the absence of injury. Sometimes, however, platelets clump and fibrin coagulates within a blood vessel, blocking the flow of blood. Such a clot is called a **thrombus.** These clots are more likely to form in individuals with cardiovascular disease.

CARDIOVASCULAR DISEASE

More than half of all deaths in the United States are caused by **cardiovascular disease**—diseases of the heart and blood vessels. Most often, the final blow from cardiovascular disease is either a heart attack or a stroke. These disasters are often associated with a thrombus, a blood clot that clogs a key artery. If the thrombus blocks one of the coronary arteries that supply blood to the cardiac muscle, a heart attack occurs.

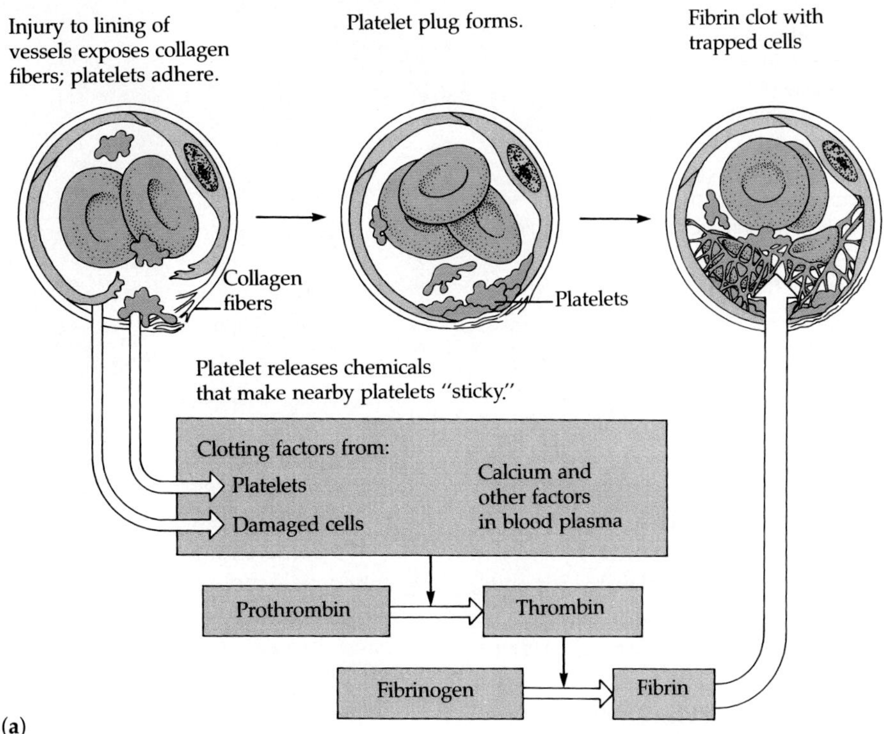

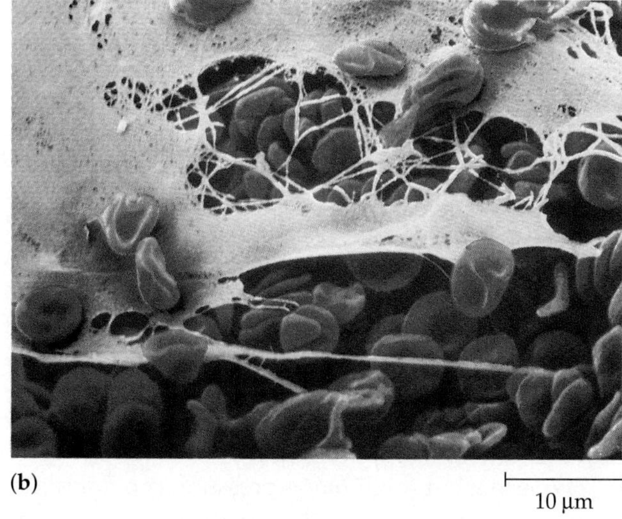

Figure 38.15
Blood clotting. (**a**) The clotting process begins when the endothelium of a vessel is damaged and connective tissue in the wall of the vessel is exposed to blood. Platelets adhere to collagen fibers in the connective tissue and release a substance that makes nearby platelets sticky. The platelets clump to form a plug that provides emergency protection against blood loss. This seal is reinforced by a clot of fibrin when damage to the vessel is more severe. Clotting factors released from the clumped platelets or damaged cells mix with clotting factors in the plasma to form an activator, which converts a plasma protein called prothrombin to its active form, thrombin. Calcium and vitamin K are among the plasma factors required for this step. Thrombin itself is an enzyme that catalyzes the final step of the clotting process, the conversion of fibrinogen to fibrin. Thus, in a cascade of reactions, injury activates prothrombin, which then activates fibrinogen. The threads of fibrin become interwoven into a patch. (**b**) Red blood cells trapped in a clot of fibrin (colorized SEM).

A thrombus that causes a heart attack may form in a coronary artery itself, or it may develop elsewhere in the circulatory system and reach a coronary artery via the bloodstream. Such a moving clot is called an embolus. The embolus is swept along until it becomes lodged in an artery too small for the clot to pass. The cardiac muscle tissue downstream from the obstruction may die. If the damage is located where it interrupts the conduction of electrical impulses through the cardiac muscle, the heart may begin beating erratically (arrhythmia) or stop altogether. (Still, the victim may survive if heartbeat is restored by cardiopulmonary resuscitation—CPR—or some other emergency procedure within a few minutes of the attack.) Similarly, many strokes are associated with a thrombus or embolus that clogs an artery in the brain. The brain tissue supplied by that artery dies. The effects of the stroke and the individual's chance of survival depend on the extent and location of the damaged tissue.

The suddenness of a heart attack or stroke belies the fact that the arteries of most victims had become gradually impaired by a chronic disease known as **atherosclerosis** (Figure 38.16). Atherosclerosis greatly increases the risk of a blood clot plugging an artery. During the course of this cardiovascular disease, growths called **plaques** develop on the inner walls of

Figure 38.16
Atherosclerosis. These light micrographs contrast a normal artery (left) with one partially occluded by an atherosclerotic plaque (right). A plaque forms when lipids such as cholesterol infiltrate a matrix of smooth muscle and connective tissue that proliferates abnormally. In some cases, the plaques become hardened by calcium deposits, resulting in a form of atherosclerosis called arteriosclerosis, or hardening of the arteries.

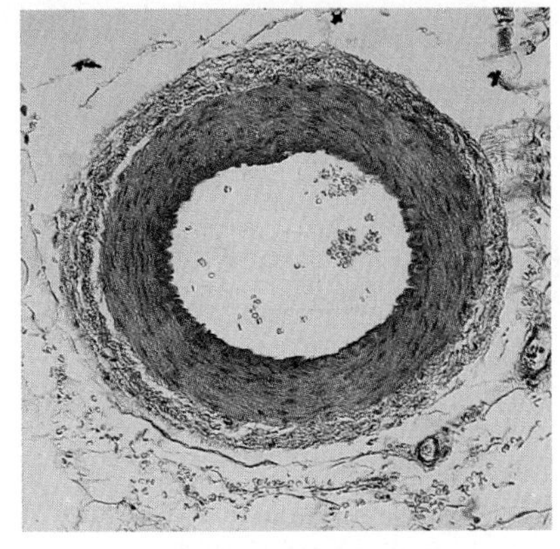

0.1 mm

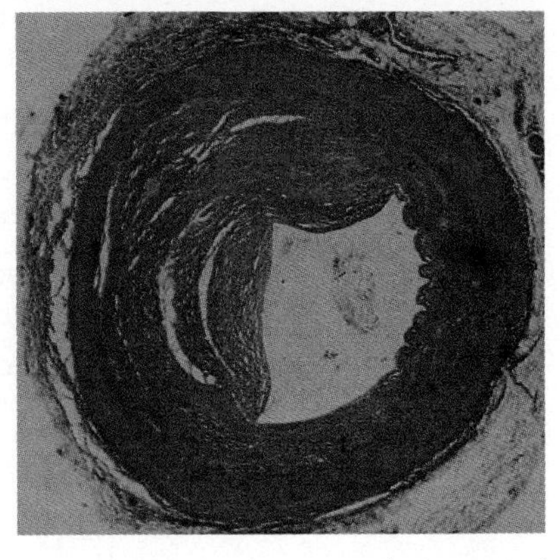

0.5 mm

the arteries and narrow the bore of the vessels. A plaque forms when lipids infiltrate an abnormally thick matrix of smooth muscle. In some cases, the plaques even become hardened by calcium deposits, resulting in a form of atherosclerosis called **arteriosclerosis,** commonly known as hardening of the arteries. An embolus is more likely to become trapped in a vessel that has been narrowed by plaques. Furthermore, plaques are common sites of thrombus formation. Healthy arteries have smooth linings. The rougher lining of an artery affected by atherosclerosis seems to encourage the adhesion of platelets, which triggers the clotting process.

As atherosclerosis progresses, arteries become more and more clogged by plaque, and the threat of heart attack or stroke becomes much greater. Sometimes, there are warnings. For example, if a coronary artery is partially blocked by atherosclerosis, a person may feel occasional chest pains, a condition known as angina pectoris. The pain is a signal that part of the heart is not receiving a sufficient supply of oxygen, and it is most likely to occur when the heart is laboring hard because of physical or emotional stress. However, many people with atherosclerosis are completely unaware of their disease until catastrophe strikes.

Hypertension (high blood pressure) promotes atherosclerosis and increases the risk of heart attacks and strokes (and, conversely, atherosclerosis tends to increase blood pressure by narrowing the bore of the vessels and reducing their elasticity). According to one hypothesis, the chronic punishment to the lining of arteries by hypertension damages the endothelium and initiates plaque formation. Acting alone or in lethal combination, hypertension and atherosclerosis, the two most common cardiovascular diseases, lead to the majority of deaths in the United States and other developed nations. Hypertension is sometimes called the "silent killer" because a person with the disease may experience no symptoms until a stroke or heart attack occurs. Fortunately, hypertension is simple to diagnose and can usually be controlled by medication, diet, exercise, or a combination of these treatments. A diastolic pressure above 90 may be cause for concern, and living with extreme hypertension—say, 200/120—is courting disaster.

To some extent, the tendency for hypertension and atherosclerosis is inherited, making certain people more predisposed than others to cardiovascular disease. We cannot do much about our genes, but the health of our cardiovascular system is not completely out of our hands. Smoking, lack of exercise, and a diet rich in animal fats and cholesterol are among the factors that have been correlated with an increased risk of cardiovascular disease.

An abnormally high concentration of cholesterol in blood plasma is one of the most important correlates of potential atherosclerosis. Cholesterol travels in the blood mainly in the form of **low-density lipoproteins (LDLs),** plasma particles consisting of thousands of cholesterol molecules and other lipids bound to a protein. In contrast to LDLs, another form of cholesterol carriers, called **high-density lipoproteins (HDLs),** actually *reduce* the depositing of cholesterol in arterial plaques. Many researchers now believe that the ratio of LDLs to HDLs is more reliable than total plasma cholesterol as an indicator of impending cardiovascular disease. Exercise tends to increase HDL concentration, while smoking has the opposite effect on the LDL to HDL ratio.

Let's end the discussion of cardiovascular disease with some good news: Over the past 20 years, the death rate from cardiovascular disease in the United States has declined by more than 25%. So far, heart transplants, bypass surgery, and other radical methods for treating heart disease have not made any statistically significant contribution to the substantial decrease in deaths from heart attacks. On the other hand, diagnosis and treatment of hypertension may be preventing a large number of heart attacks and strokes, and improved methods of intensive care for cardiovascular patients may be increasing the chances of surviving heart attacks and strokes once they occur. Also, many Americans are now more conscious of their health; as a group, we are smoking less, exercising more, and watching our diets. Education is potent medicine.

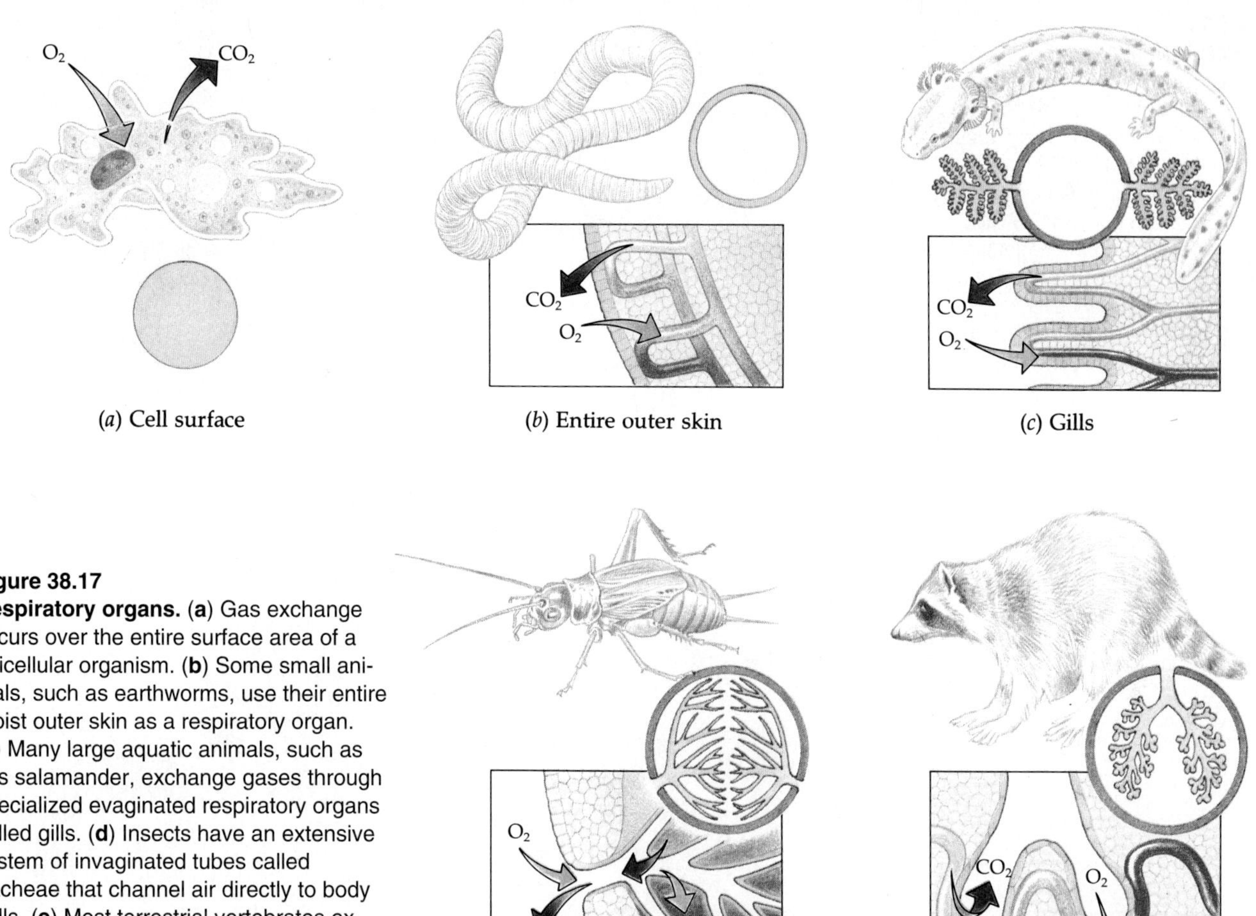

Figure 38.17
Respiratory organs. (**a**) Gas exchange occurs over the entire surface area of a unicellular organism. (**b**) Some small animals, such as earthworms, use their entire moist outer skin as a respiratory organ. (**c**) Many large aquatic animals, such as this salamander, exchange gases through specialized evaginated respiratory organs called gills. (**d**) Insects have an extensive system of invaginated tubes called tracheae that channel air directly to body cells. (**e**) Most terrestrial vertebrates exchange gases across the lining of the lungs, invaginated surfaces supplied with blood.

GAS EXCHANGE IN ANIMALS

A major function of circulatory systems is to transport oxygen and carbon dioxide between respiratory organs and other parts of the body. We now focus on the actual exchange of these gases between animals and their environments.

General Problems of Gas Exchange

Animals require a continuous supply of oxygen (O_2) for cellular respiration (see Chapter 9), and they must expel carbon dioxide (CO_2), the waste product of this process. It is important not to confuse gas exchange, the traffic of O_2 and CO_2 between the animal and its environment, with the metabolic process of cellular respiration. Gas exchange supports cellular respiration by supplying oxygen and removing carbon dioxide.

Earth's main reservoir of O_2 is the atmosphere, which is about 21% O_2. Oceans, lakes, and other bodies of water also contain oxygen in the form of dissolved O_2. The source of oxygen, called the **respiratory medium,** is air for a terrestrial animal and water for an aquatic one. The portion of an animal's surface where gas exchange with the respiratory medium occurs is called the **respiratory surface.** Oxygen and carbon dioxide cannot bubble across membranes but can only diffuse through membranes if they are first dissolved in the water that coats the respiratory surface. Thus, the respiratory surface must be moist for both aquatic and terrestrial animals, and it must also be large enough to provide O_2 and expel CO_2 for the entire body. A variety of solutions to this problem have evolved, depending mainly on the size of the animal and whether it lives in water or on land. The most common solution is specialization of a localized region of the body surface as an efficient respiratory surface to supply oxygen for the entire animal.

Respiratory Organs: General Structure and Function

The respiratory surface of a lung, gill, or other respiratory organ is a thin, moist epithelium, usually with a rich blood supply (Figure 38.17). Only this single layer

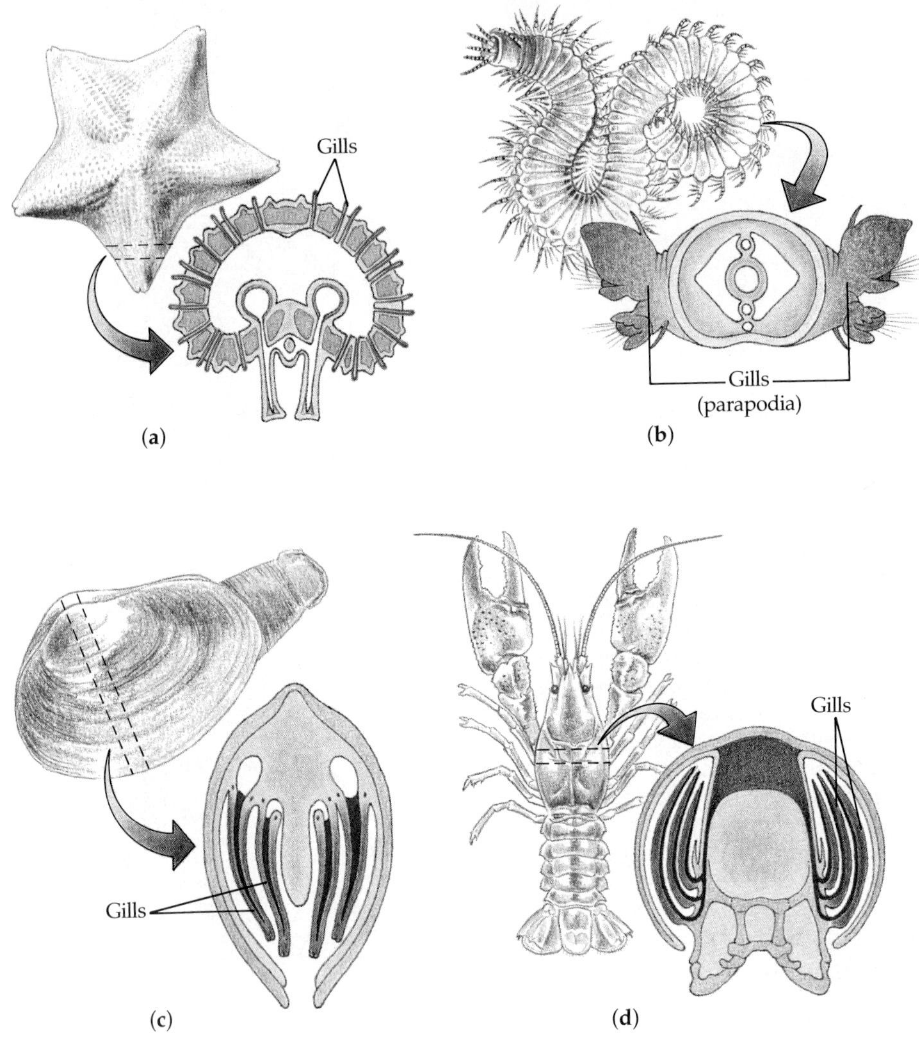

Figure 38.18
Invertebrate gills. (**a**) The gills of a sea star are evaginations of the coelom. Although they contact the external environment directly, they are protected by bony spines. (**b**) Polychaetes (phylum Annelida) have a pair of gills on each segment. (**c**) Cilia move water through the fingerlike gills of a clam. (**d**) Crayfish and other crustaceans have feathery gills beneath a thoracic exoskeleton. Modified appendages sweep water over the gills.

of cells separates the respiratory medium, whether air or water, from the blood or capillaries.

Some animals use their entire outer skin as a respiratory organ. An earthworm, for example, exchanges gases by diffusion across the general body surface. Just below the skin is a dense net of capillaries. Because the respiratory surface must be moist, earthworms and other skin breathers, including some amphibians, must live in water or damp places.

Most animals that use their general body surface as a respiratory organ are relatively small and have a long, thin (wormlike) shape or flat shape with a high ratio of surface to volume. For most other animals, the general body surface lacks sufficient area to exchange gases for the whole body. The solution is a localized region of the body surface that is extensively folded or branched, enlarging the area of the respiratory surface for gas exchange. The expanded respiratory surface of most aquatic animals is external and bathed by either fresh water or seawater. These localized extensions of the body surface are called **gills.** Gills are generally unsuitable for an animal living on land, because an expansive surface of wet membrane exposed to air would lose too much water to evaporation and also because the gills would collapse as their fine filaments, no longer supported by water, would cling together. Most terrestrial animals have their respiratory surfaces invaginated into the body, opening to the atmosphere only through narrow tubes. The lungs of terrestrial vertebrates and the tracheae of insects are two variations of this plan. We will now examine more closely the three most common respiratory organs: gills, tracheae, and lungs.

Gills: Respiratory Adaptations of Aquatic Animals

Gills are evaginations (outfoldings) of the body surface specialized for gas exchange (Figure 38.18). Some gills have simple shapes. Those of sea stars and other echinoderms, for instance, are mere bumps that dot the skin. The gills of many segmented marine worms (phylum Annelida) are flaps that extend from the sides of

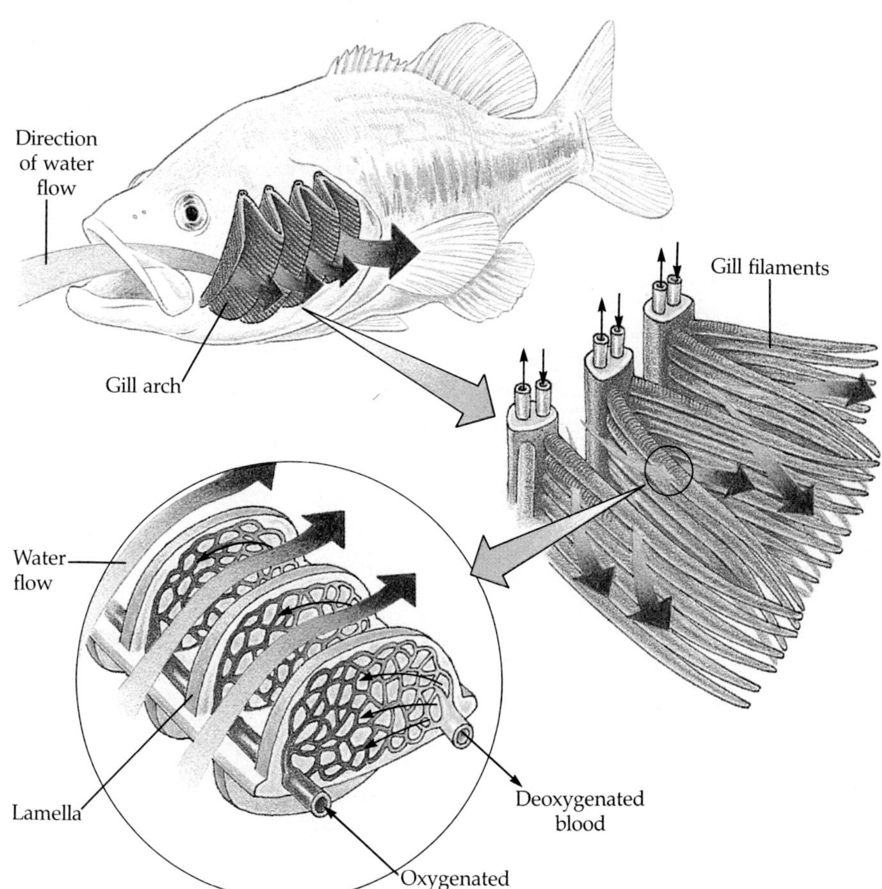

Figure 38.19
The physiology of fish gills. Fish continuously pump water through the mouth and over the gill arches, using coordinated movements of the jaws and operculum (gill cover) for this ventilation. Each arch has two rows of gill filaments, and surface area is increased even more by tiny lamellae, which protrude from the filaments (see enlargement). Blood flowing through capillaries within the lamellae picks up oxygen from the water. Notice that water flows over the lamellae in a direction opposite to the blood flow, an arrangement called a countercurrent, which enhances oxygen transfer (see Figure 38.20).

each segment of the animal. Rather than being distributed over the entire body, the complex gills of many other animals are restricted to a local region of body surface where the skin is finely dissected, forming a feathery respiratory surface having a large area. Although only a limited region of the body is devoted to gas exchange, the gills may have a total surface area that is much greater than that of the rest of the body surface. We find such finely divided gills in most mollusks, crustaceans (phylum Arthropoda), fishes, and some amphibians (some salamanders and the tadpoles of frogs). Because they are external, the delicate gills are vulnerable to physical damage and attack from other organisms; in most cases, they are sheltered by a protective cover of some kind. A flap called the operculum covers the gills of bony fishes, for instance.

As a respiratory medium, water has both advantages and disadvantages. On the positive side, there is no problem keeping the respiratory surface wet, since the gills are completely surrounded by the aqueous environment in which the animal lives. But the oxygen concentration in water is much lower than in air; and the warmer and saltier the water, the less dissolved oxygen it holds. Thus, gills must be very efficient to obtain enough oxygen from water. One process that helps is **ventilation,** a term that refers to any method of increasing flow of the respiratory medium (air or water) over the respiratory surface (lungs or gills). For example, crayfish and lobsters ventilate by using tiny appendages modified as paddles to beat a current of water over the gills. The gills of a bony fish are ventilated continuously by a current of water that enters the mouth, passes through slits in the pharynx, flows over the gills, and exits at the back of the operculum (gill cover) (Figure 38.19). If it were not for ventilation, water around the gills would soon stagnate, becoming depleted of oxygen and saturated with carbon dioxide. Ventilation brings a fresh supply of oxygen and removes carbon dioxide expelled by the gills. Because water is much denser and contains much less oxygen per unit volume than air, a fish must expend a considerable amount of energy to ventilate its gills.

The arrangement of capillaries in the gills of a fish also enhances gas exchange. Blood flows opposite to the direction in which water passes over the gills. This pattern makes it possible for oxygen to be transferred to the blood by a very efficient process called **countercurrent exchange** (Figures 38.19 and 38.20). As blood flows through the capillary, it becomes more and more loaded with oxygen, but at the same time, it is encoun-

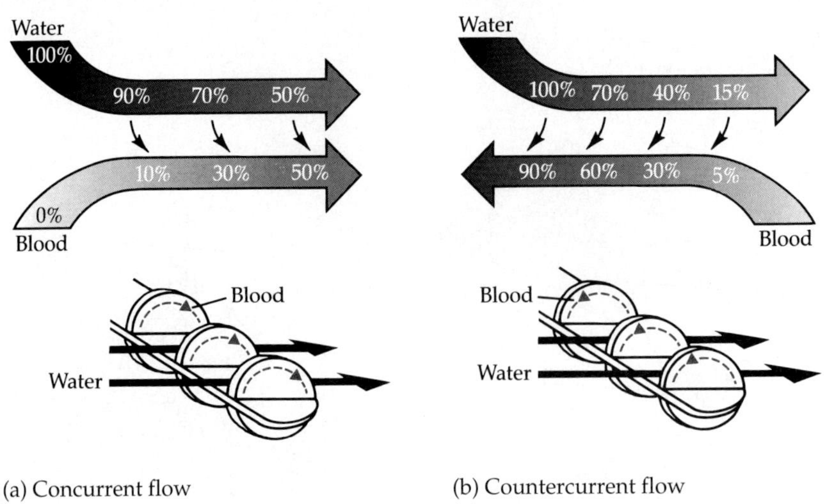

Figure 38.20
Countercurrent exchange. The vascular arrangement of fish gills maximizes oxygen transfer from the water to the blood. (**a**) If blood flowed through capillaries of the lamellae in the same direction as water flowing over the lamellae, gills could, at best, pick up only 50% of the oxygen dissolved in the water. As O_2 diffused from the water into the blood, the concentration gradient would become less and less steep until blood and water equilibrated at the same O_2 concentration. (**b**) With countercurrent flow, the blood continues to pick up oxygen from the water all along the length of the capillary. As the blood flows along the vessel and becomes more and more loaded with oxygen, it passes by water that has given up less of its O_2. Over the length of the capillary, more than 80% of the O_2 dissolved in the water is transferred by diffusion into the blood.

tering water that is more and more concentrated in oxygen because the water is just beginning its passage over the gills. This means that along the entire length of the capillary, there is a diffusion gradient favoring the transfer of oxygen from the water to the blood. So efficient is this countercurrent exchange mechanism that the gill is able to remove more than 80% of the oxygen dissolved in the water passing over the respiratory surface. The mechanism of countercurrent exchange is also important in temperature regulation and several other physiological processes, as we will see in Chapter 40.

Tracheae: Respiratory Adaptations of Insects

As a respiratory medium, air poses a different set of problems than water. Air has many advantages, not the least of which is a much higher concentration of oxygen. Also, since O_2 and CO_2 diffuse much faster in air than in water, respiratory surfaces exposed to air do not have to be ventilated as thoroughly as gills. As the respiratory surface removes oxygen from the air and expels carbon dioxide, diffusion rapidly brings more oxygen to the surface and carries the carbon dioxide away. When a terrestrial animal does ventilate, less energy is expended, because air is much easier to move than water. But offsetting these advantages of air as a respiratory medium is a problem: The respiratory surface, which must be large and moist, continuously loses water to the air by evaporation. The solution is a respiratory surface invaginated (infolded) into the body, rather than evaginated like gills. One variation of this plan is the tracheal system of insects.

Tracheae are tiny air tubes that ramify throughout the insect body (Figure 38.21). The finest branches extend to the surface of nearly every cell, where gas is exchanged by diffusion across the moist epithelium that lines the terminal ends of the tracheal system. The entrances to the tracheal system are **spiracles,** minute pores distributed over the surface of the insect's body. Small insects rely on diffusion to bring O_2 from the air into the tracheal system and carry CO_2 out of the system. Some larger insects ventilate their tracheal systems with rhythmic body movements that compress and expand the air tubes like bellows.

It is because the tracheal system exposes all cells directly to the respiratory medium and because oxygen diffuses so quickly in air that insects do not need to use their circulatory systems to transport oxygen and carbon dioxide. This is a major reason why an open circulatory system, which moves blood much more slowly than a closed system, suffices for even the most active flying insects.

Lungs: Respiratory Adaptations of Terrestrial Vertebrates

In contrast to the respiratory channels that pervade the entire insect body, **lungs** are invaginations of the body surface that are restricted to one location. Because the respiratory surface of a lung is not in direct contact with all other parts of the body, the gap must be bridged by the circulatory system, which transports oxygen from the lungs to the rest of the body. Lungs are heavily vascularized with a dense net of capillaries located just beneath the epithelium that forms the res-

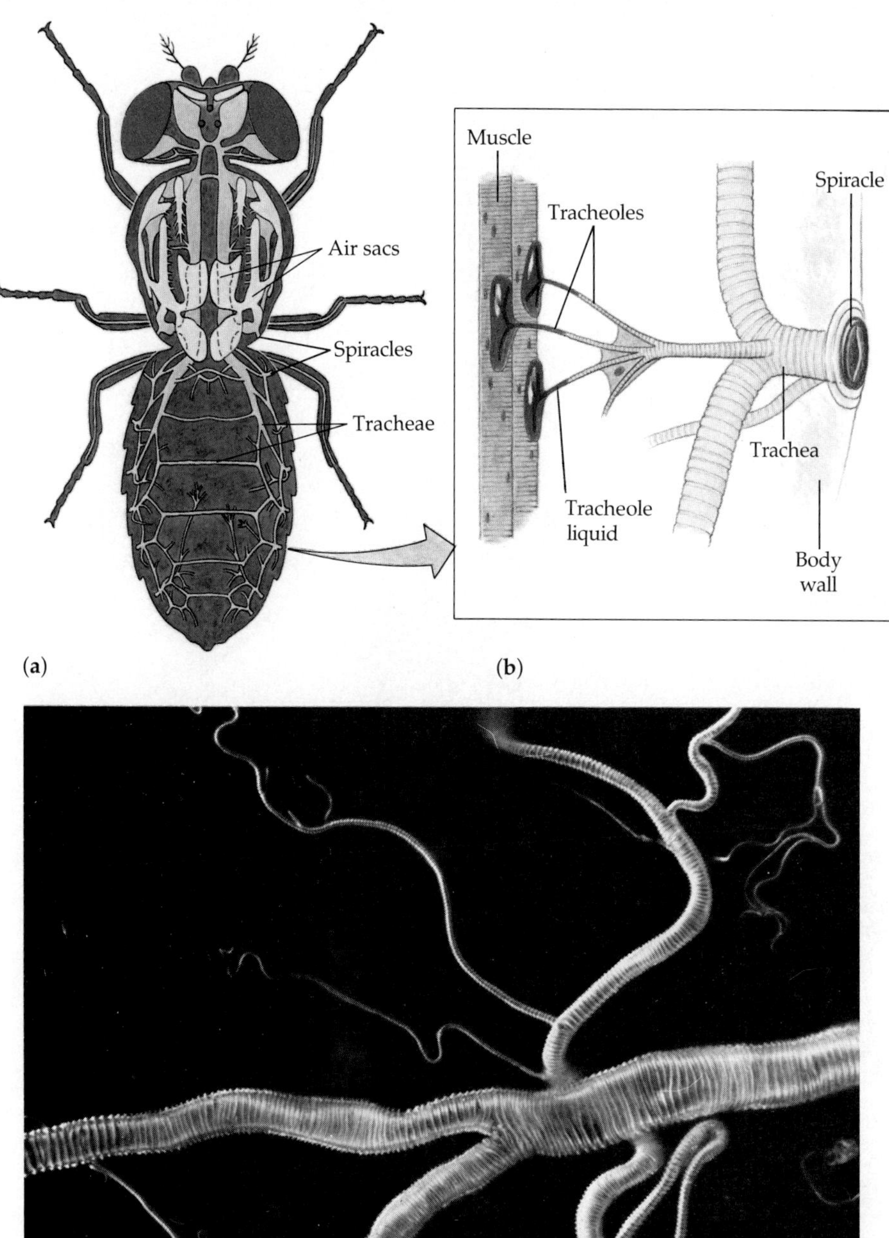

Figure 38.21
Tracheal systems. Insects and some other terrestrial arthropods (certain spiders, for instance) have a respiratory system that delivers air directly to body cells. (**a**) Tracheae branch repeatedly and extend to all parts of the body. Enlarged sections of tracheae form air sacs near organs that require a large supply of oxygen. (**b**) Air enters the respiratory system through openings called spiracles. Air passes through tracheae and the smaller tracheoles until they terminate on the plasma membrane of individual cells. The liquid in the tracheole endings regulates the amount of air that contacts the cells. When the oxygen requirement increases, some liquid is withdrawn, increasing the surface area of air in contact with cells. (**c**) In this micrograph of tracheae in a cockroach, you can see the rings of chitin that reinforce the air tubes and keep them from collapsing (LM).

piratory surface. This solution to the problem of gas exchange has evolved as a vascularized mantle in land snails, as book lungs in spiders (see Chapter 29), and as lungs in the terrestrial vertebrates—most amphibians, reptiles, birds, and mammals. The lungs of most frogs are balloonlike, with the respiratory surface limited to the outer surface area of the lungs. This is not a large area, but frogs also obtain some of their oxygen by diffusion across the moist outer skin of the body. In contrast, the lungs of mammals have a spongy texture and are honeycombed with epithelium having a total surface area much greater than the outer surface area of the lung itself. The total respiratory surface in a human is about equivalent to the surface area of a tennis court. This entire respiratory surface is coated by a thin film of moisture in which the oxygen in the lung's air spaces dissolves before diffusing across the epithelium and into the blood. Because the passageways connecting the air pockets in the lungs to the outside air are narrow, loss of water by evaporation from the respiratory surface is minimized.

Mammalian Respiratory System Anatomy The lungs of a mammal are located in the thoracic (chest) cavity, enclosed by a double-walled sac. The inner layer of the sac adheres tightly to the outside of the

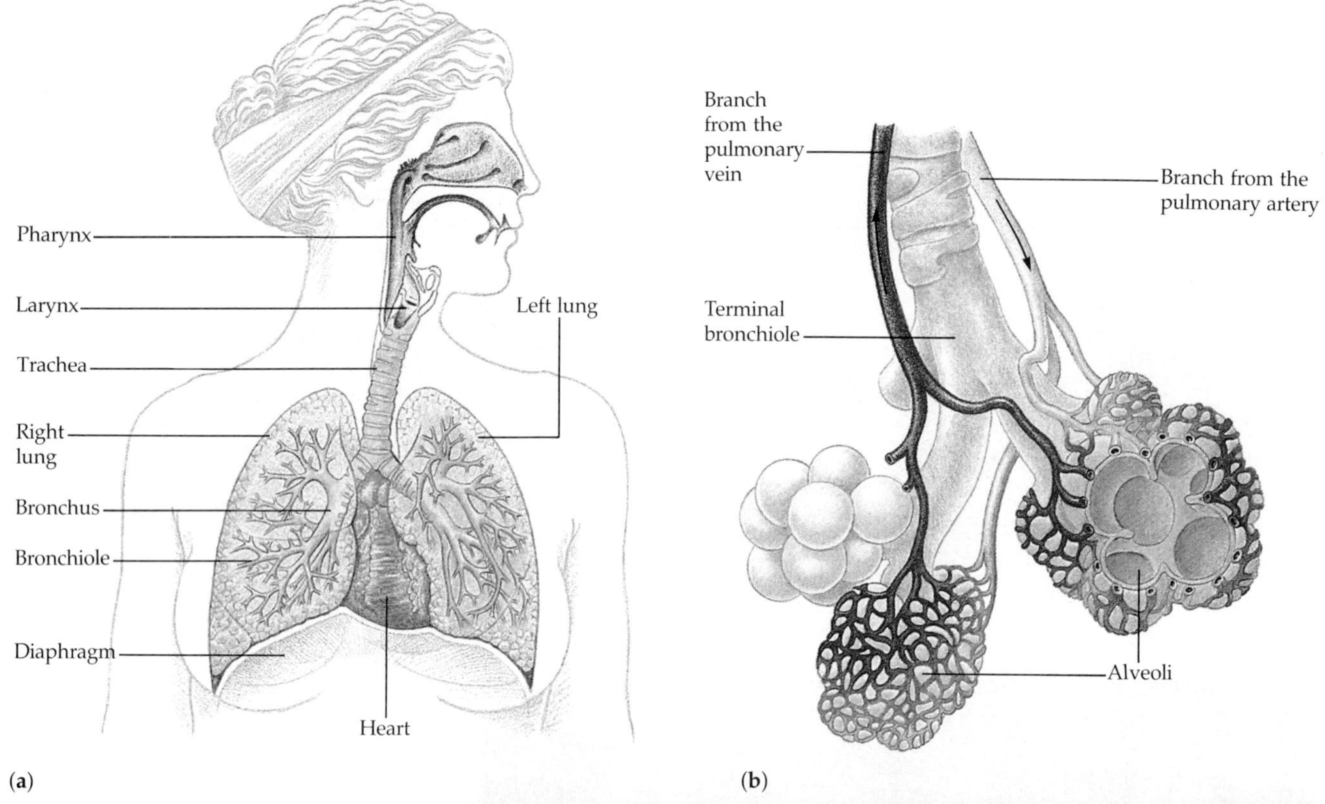

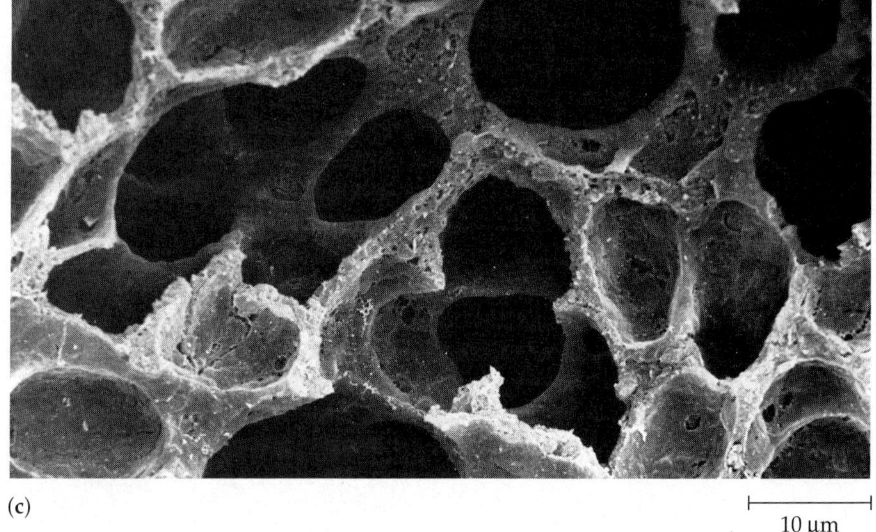

Figure 38.22
The mammalian respiratory system. (**a**) The organs of the respiratory system. Air is conveyed down the trachea and bronchi to the tiniest bronchioles, which dead-end as lobed, microscopic air sacs called alveoli. (**b**) The structure of alveoli. Oxygen within the air sac dissolves in the moist film coating the respiratory surface and diffuses across this thin epithelium into the capillaries that surround each alveolus. (**c**) A scanning electron micrograph of alveoli.

lungs, and the outer layer is attached to the wall of the chest cavity (Figure 38.22). The two layers are separated by a thin space filled with fluid. Because of surface tension, the two layers behave like two plates of glass stuck together by a film of water. The layers can slide smoothly past each other, but they cannot be pulled apart easily.

Air is conveyed to the lungs by a system of branching ducts. Air enters this system through the nostrils and is then filtered by hairs, warmed, humidified, and sampled for odors as it flows through a maze of spaces in the nasal cavity. This cavity leads to the pharynx, an intersection where the paths for air and food cross. When food is swallowed, the opening of the windpipe (glottis) is pushed against the epiglottis, and food is diverted down the esophagus to the stomach (see Figure 37.11). The rest of the time, the glottis is open, and we can breathe. The glottis leads to the **larynx** (also called the Adam's apple), which has a wall reinforced with cartilage. Humans and many other mammals use the larynx as a voice box. As air is exhaled through the chamber, a pair of **vocal cords** vibrates and produces sounds. Pitch is controlled by changing the tension of the cords.

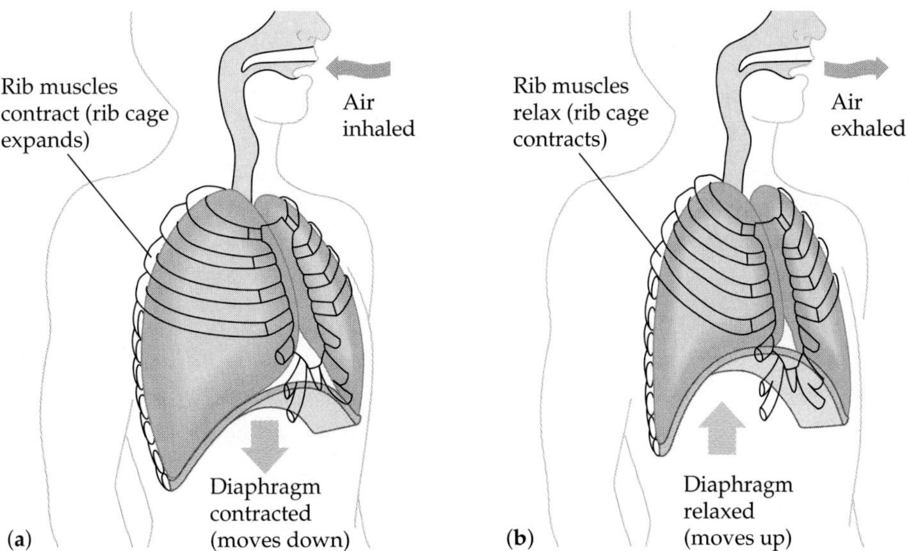

Figure 38.23
Negative pressure breathing. Mammals breathe by changing the air pressure within their lungs relative to the pressure of the outside atmosphere. They do this by changing the volume of their thoracic cavity with the rib cage and diaphragm. (**a**) To inhale, a mammal increases the volume of the thoracic cavity by contracting the muscles of the rib cage and the diaphragm. This expands the rib cage and flattens the diaphragm. Air pressure in the lungs falls below that of the atmosphere, and air rushes into the lungs. (**b**) Exhalation occurs when the rib muscles and diaphragm relax, restoring the thoracic cavity to its smaller volume.

From the larynx, air passes into the **trachea,** or windpipe. Rings of cartilage maintain the shape of the trachea, much as metal rings keep the hose of a vacuum cleaner from collapsing. The trachea forks into two **bronchi** (singular, **bronchus**), one leading to each lung. Within the lung, the bronchus branches repeatedly into finer and finer tubes called **bronchioles.** The entire system of air ducts has the appearance of an inverted tree, the trunk being the trachea. The epithelium lining the major branches of this respiratory tree is covered by cilia and a thin film of mucus. The mucus traps dust, pollen, and other contaminants, and the beating cilia move the mucus upward to the pharynx, where it can be swallowed into the esophagus. This process helps cleanse the respiratory system.

At their tips, the tiniest bronchioles dead-end as a cluster of air sacs called **alveoli** (singular, **alveolus**). The thin epithelium of the millions of alveoli in the lung serves as the respiratory surface. Oxygen in the air conveyed to the alveoli by the respiratory tree dissolves in the moist film and diffuses across the epithelium and into the web of capillaries that embraces each alveolus. Carbon dioxide diffuses from the capillaries, across the epithelium of the alveolus, and into the air space (see Figure 38.1).

Ventilating the Lungs Vertebrates ventilate their lungs by **breathing,** the alternate inhalation and exhalation of air. Ventilation maintains a maximal oxygen concentration and minimal carbon dioxide concentration within the alveoli.

A frog inhales by pushing air down its windpipe. The animal first lowers the floor of its mouth, enlarging the oral cavity and drawing air in through the open nostrils. Then, with the nostrils and mouth closed, the frog raises the floor of its mouth, which forces air down the trachea. Elastic recoil of the lungs, together with their compression by the muscular body wall, forces air back out of the lungs during exhalation. A small amount of oxygen is also absorbed across the lining of the mouth, which is ventilated by fluttering of the throat.

In contrast to frogs, mammals ventilate their lungs by **negative pressure breathing,** which works like a suction pump rather than a pressure pump: The forces are generated near the branches of the respiratory tree rather than at the trunk, so air is pulled down into the lungs rather than being pushed in from above (Figure 38.23). This is accomplished by changing the volume of the thoracic cavity, which houses the lungs. One way to do this is to use the rib muscles to pull the ribs upward from their normal position, which expands the rib cage. Movement of the lungs is coupled to movement of the rib cage by the surface tension of the fluid in the thin space between the two layers of the sac that encloses the lungs. When the rib cage expands, so do the lungs. Because this movement increases the lung volume, air pressure within the alveoli is reduced to less than atmospheric pressure. Because air always flows from a region of higher pressure to a region of lower pressure, air rushes through the nostrils and down the respiratory tree to the alveoli. When the rib

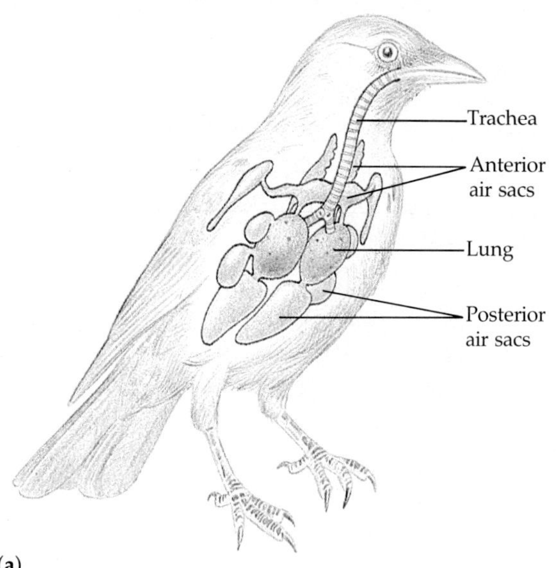

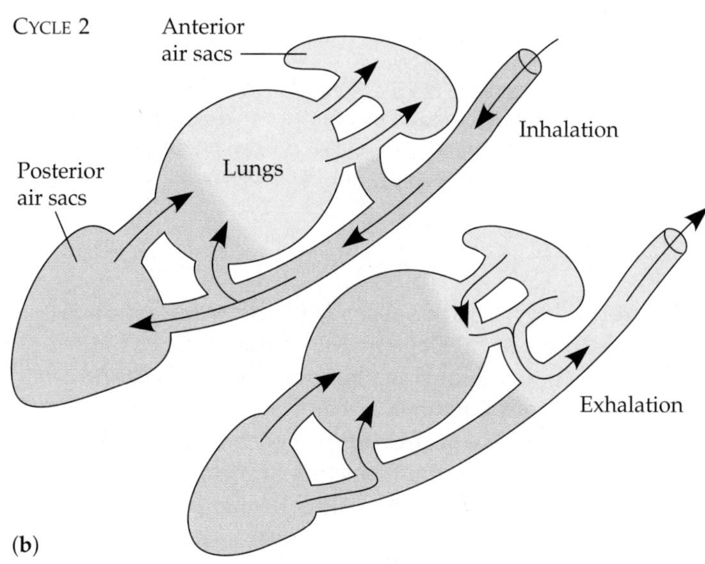

Figure 38.24
The avian respiratory system. (**a**) A bird possesses air sacs in addition to lungs. The air sacs function in ventilating the lungs, where gas exchange occurs. (**b**) In this sequence of diagrams, we follow one "breath" of fresh air (blue) through the respiratory system of a bird. Two cycles of inhalation and exhalation are required for the air to pass all the way through the system and out of the bird. During the first inhalation, most of the air bypasses the lungs and enters the posterior air sacs. That air passes through the lungs during exhalation and the next inhalation, ending up in the anterior air sacs. At the same time, the posterior sacs draw in another breath of fresh air. Together, the anterior and posterior air sacs function as bellows that keep air flowing through the air ducts of the lungs continuously and in one direction (posterior to anterior).

muscles relax, the lungs are compressed, and the increase in air pressure within the alveoli forces air up the respiratory tree and out through the nostrils. This type of breathing usually occurs only during vigorous exercise. For more shallow breathing, when we are at rest, we use the diaphragm rather than the rib muscles. The **diaphragm** is a thin sheet of muscle that forms the bottom wall of the thoracic cavity. When relaxed, the diaphragm is dome-shaped. Contraction of the diaphragm takes up the slack and causes it to flatten, which enlarges the thoracic cavity. In this way, contraction of the diaphragm lowers pressure in the lungs and causes inhalation. Air is exhaled when the diaphragm relaxes to its dome shape.

The volume of air an animal inhales and exhales with each breath is called **tidal volume.** It averages about 500 mL in humans. The maximum volume of air that can be inhaled and exhaled during forced breathing is called **vital capacity,** which is about 4000 to 5000 mL for a college-age male (somewhat less for a female). Among other factors, the vital capacity depends on the resilience (springiness) of the lungs. The lungs actually hold more air than the vital capacity, but since it is impossible to completely collapse the alveoli, a **residual volume** of air remains in the lungs even after we forcefully blow out as much air as we can. As lungs lose their resilience as a result of aging or disease (such as emphysema), residual volume increases at the expense of vital capacity.

The most complex ventilation occurs in birds (Figure 38.24). Besides their lungs, birds have eight or nine air sacs that penetrate the abdomen, neck, and even the wings. In addition to their function in breathing, these air sacs trim the density of the bird, an important adaptation for flight (see Chapter 30). The entire system, lungs and air sacs, is ventilated when the bird inhales and exhales. Air flows through the interconnected system in a circuit that passes through the lungs in one direction only, regardless of whether the bird is inhaling or exhaling. Alveoli, which are dead ends, would not be suitable in such a system. Instead, the lungs of a bird have tiny channels called **parabronchi,** through which air can flow continuously in one direction. The air sacs do not function directly in gas exchange, but act as bellows that keep air flowing through the lungs.

The Control of Breathing Although we can hold our breath voluntarily a short while or consciously breathe faster and deeper, normally our breathing is controlled by automatic mechanisms. The control center is located in the medulla, the stem of the brain that tapers into the spinal cord. We inhale when the nerves in this **breathing center** send impulses to the rib muscles or diaphragm, stimulating the muscles to contract. These nerves fire rhythmically, about 10 to 14 times per minute when we are at rest. A negative-feedback loop from the lungs to the respiratory center prevents us from overexpanding our lungs when we take a deep breath. As inhalation deepens, stretch sensors in the lung tissue send nervous impulses that inhibit the breathing center.

In addition to receiving input from the nervous system, the breathing center also monitors the pH of the blood, which begins to drop slightly as the amount of carbon dioxide in the blood increases (carbon dioxide reacts with water to form carbonic acid, which lowers pH). When the center senses a rise in the CO_2 level (a drop in pH), the tempo and depth of breathing are increased. This occurs when we exercise. Oxygen has little effect on the breathing center. Oxygen sensors in key arteries send alarm signals to the breathing center when the oxygen level is too low, but this occurs only when the deficiency is severe—at very high altitudes, for instance. Fortunately, a rise in carbon dioxide concentration is usually a good indication of a drop in oxygen concentration, because CO_2 is produced by the same process that consumes O_2—cellular respiration. It is possible, however, to trick the breathing center by hyperventilating. Excessively deep, rapid breathing purges the blood of so much CO_2 that the breathing center temporarily ceases to send impulses to the rib muscles and diaphragm. Breathing stops until the CO_2 level increases enough to switch the breathing center back on.

The breathing center, then, responds to a variety of neural and chemical signals, adjusting the rate and depth of breathing to meet the changing demands of the body. Control of breathing is only effective, however, if it is coordinated with control of the circulatory system. During exercise, for instance, cardiac output is matched to the increased breathing rate, which enhances O_2 supply and CO_2 removal as blood flows through the lungs.

Loading and Unloading of Oxygen and Carbon Dioxide To understand how gases are exchanged at various locations around the body, recall that substances diffuse down their gradients. For a gas, whether present in air or dissolved in water, diffusion depends on differences in a quantity called **partial pressure.** At sea level, the atmosphere exerts a total pressure of 760 mm Hg. This is a downward force equivalent to that exerted by a column of mercury 760 mm high. Since the atmosphere is 21% oxygen (by volume), the partial pressure of oxygen is 0.21×760, or about 160 mm Hg. (This is the portion of atmospheric pressure contributed by oxygen—hence the term *partial pressure.*) The partial pressure of carbon dioxide at sea level is only 0.23 mm Hg. These partial pressures are symbolized by P_{O_2} and P_{CO_2}. When water is exposed to air, the amount of any gas that dissolves in the water is proportional to its partial pressure in the air and its solubility in water. An equilibrium is eventually reached when the gas molecules enter and leave the solution at the same rate. At this point, the gas is said to have the same partial pressure in the solution as it does in the air. Thus, the P_{O_2} in a glass of water exposed to air is 160 mm Hg, and the P_{CO_2} is 0.23 mm Hg. A gas will always diffuse from a region of higher partial pressure to a region of lower partial pressure.

Blood arriving at a lung via the pulmonary artery has a lower P_{O_2} and a higher P_{CO_2} than the air in the alveoli (Figure 38.25). As blood enters a capillary net around an alveolus, carbon dioxide diffuses from the blood to the air within the alveolus. Oxygen in the air dissolves in the fluid that coats the epithelium and diffuses across the surface and into a capillary. By the time the blood leaves the lungs in the pulmonary veins, its P_{O_2} has been raised and its P_{CO_2} has been lowered. After returning to the heart, this blood is pumped through the systemic circuit. In the systemic capillaries, gradients of partial pressure favor the diffusion of oxygen out of the blood and carbon dioxide into the blood. This is because cellular respiration rapidly depletes the oxygen content of interstitial fluid and adds carbon dioxide to the fluid (again, by diffusion). After the blood unloads oxygen and loads carbon dioxide, it is returned to the heart by systemic veins. The blood is then pumped to the lungs again, where it exchanges gases with air in the alveoli.

Respiratory Pigments and Oxygen Transport Because oxygen is not very soluble in water, very little is transported in blood in the form of dissolved O_2. In most animals, oxygen is carried by **respiratory pigments** in the blood. These pigments are proteins that owe their color to metal atoms built into the molecules. The respiratory pigment of almost all vertebrates is hemoglobin, the iron-containing protein located in red blood cells. It is the iron of hemoglobin that actually binds to oxygen. In some animals, such as earthworms,

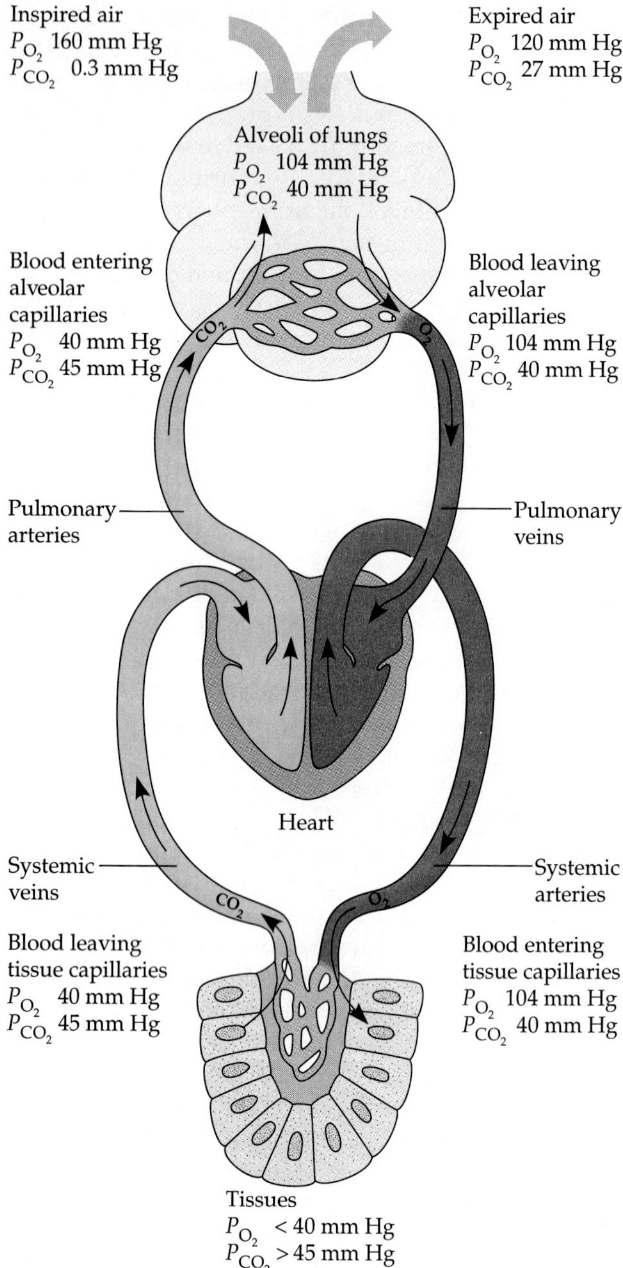

Figure 38.25
Loading and unloading of respiratory gases. Partial pressures of O_2 and CO_2 are symbolized by P_{O_2} and P_{CO_2}, respectively. Each gas diffuses from a region of higher partial pressure to a region of lower partial pressure.

the hemoglobin is dissolved directly in the blood plasma rather than being located in the cells. (Presumably, one advantage of packaging the hemoglobin in cells is that the blood can have a high concentration of this protein without increasing the osmotic pressure of the plasma.) In addition to hemoglobin, several other proteins that carry oxygen are found in the blood of various invertebrates. One of these, called **hemocyanin,** has copper rather than iron as its oxygen-binding component, coloring the blood blue. Common in arthropods and many mollusks, hemocyanin is always dissolved directly in plasma rather than being confined to cells.

Hemoglobin consists of four subunits, each with a cofactor called a heme group that has an iron atom at its center; thus, each hemoglobin molecule can carry four molecules of O_2 (see Figure 5.26b). To function as an oxygen vehicle, hemoglobin must bind the gas reversibly, loading oxygen in the lungs or gills and unloading it in other parts of the body. In this loading and unloading, there is cooperation among the four subunits of the hemoglobin molecule. (See Chapter 6 to review the concept of cooperativity in allosteric proteins.) The binding of oxygen to one subunit induces the remaining subunits to change their shape slightly so that their affinity for oxygen increases. The hesitant loading of the first O_2 molecule results in the rapid loading of three more. And when one subunit unloads its oxygen, the other three quickly follow the lead as a conformational change lowers their affinity for oxygen. This mechanism of cooperative oxygen binding and release is evident in the **dissociation curve** for hemoglobin (Figure 38.26). Over the range of oxygen concentrations where the dissociation curve has a steep slope, even a slight change in concentration will cause hemoglobin to load or unload a substantial amount of oxygen. Notice that the steep part of the curve corresponds to the range of oxygen concentrations found in body tissues. When cells in a particular location begin working harder—during exercise, for instance—oxygen concentration dips in the vicinity, as the O_2 is consumed in cellular respiration. Because of the effect of subunit cooperativity, a slight drop in O_2 concentration is enough to cause a relatively large increase in the amount of oxygen that the blood unloads.

As with all proteins, hemoglobin's conformation is sensitive to a variety of environmental factors. For example, a drop in pH lowers the affinity of hemoglobin for O_2, an effect called the *Bohr shift* (see Figure 38.26b). Because CO_2 reacts with water to form carbonic acid, an active tissue will lower the pH of its surroundings and induce hemoglobin to give up more of its oxygen, which can be used for cellular respiration. Hemoglobin is a remarkable molecule, its structure well suited for its role of transporting oxygen from regions of supply to regions of demand.

Carbon Dioxide Transport About 7% of the carbon dioxide released by respiring cells is transported as dissolved CO_2 in blood plasma. Another 23% binds to the multiple amino groups of hemoglobin. However, most carbon dioxide, about 70%, is transported in the blood in the form of bicarbonate ions. Carbon dioxide expelled by respiring cells diffuses into the blood plasma and then into the red blood cells, where the CO_2 is converted to bicarbonate. Carbon dioxide first

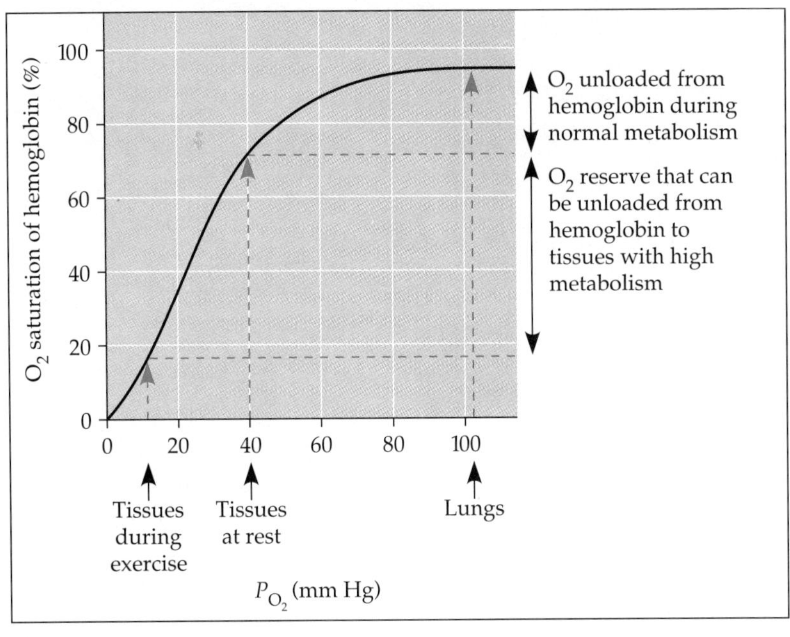

(a)

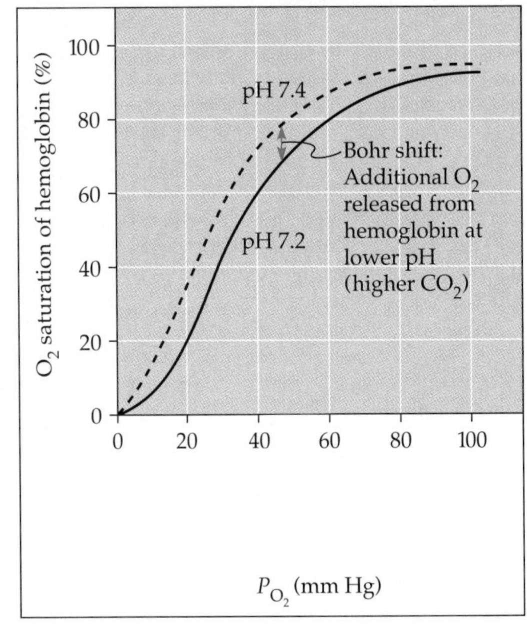

(b)

Figure 38.26
Oxygen dissociation curves for hemoglobin. (**a**) This is the dissociation curve for hemoglobin at 37°C and pH 7.4. The curve shows the relative amounts of oxygen bound to hemoglobin when the pigment is exposed to solutions varying in their partial pressure of dissolved oxygen. At a P_{O_2} of 100 mm Hg, typical in the lungs, hemoglobin is about 98% saturated with oxygen. At a P_{O_2} of 40 mm Hg, common in the vicinity of tissues, hemoglobin is only about 70% saturated; that is, it gives up about 28% of its oxygen. Notice that hemoglobin can release its reserve of O_2 to tissues that are metabolically very active—muscle tissue during exercise, for example. (**b**) Hydrogen ions affect the conformation of hemoglobin, and thus a drop in pH shifts the oxygen dissociation curve toward the right. Notice that at an equivalent P_{O_2}, say 40 mm Hg, hemoglobin gives up more oxygen at pH 7.2 than it does at pH 7.4, the normal pH of human blood. This occurs in very active tissues because the CO_2 produced by respiration reacts with water to form carbonic acid, thus lowering the pH. Hemoglobin then releases more O_2, which supports a high level of cellular respiration.

reacts with water to form carbonic acid, which then dissociates into a hydrogen ion and a bicarbonate ion. Most of the hydrogen ions produced as a by-product of this reaction attach to various sites on hemoglobin and other proteins. In this way, hemoglobin acts as a buffer, and CO_2 transport has only a small effect on the pH of blood (venous blood has a pH of 7.34, versus 7.4 for arterial blood). As blood flows through the lungs, the process is reversed. Diffusion of CO_2 out of the blood shifts the chemical equilibrium within red cells in favor of the conversion of bicarbonate to CO_2. The transport of carbon dioxide is illustrated and described in detail in Figure 38.27.

Adaptations of Diving Mammals We can further our appreciation of physiological diversity by examining some of the special adaptations that make it possible for air-breathing mammals, such as certain species of seals, whales, and dolphins, to make long underwater dives in search of food. One diving mammal that has been studied extensively is the Weddell seal (*Leptonychotes weddelli*), a large antarctic predator (adults weigh about 400 kg). Weddell seals catch large cod and other deep-sea fish by plunging to depths of 200 to 500 m during dives that typically last about 20 minutes. They occasionally submerge for more than an hour, perhaps to escape predators or explore new routes beneath the ice (Figure 38.28).

One diving adaptation of the Weddell seal is its ability to store oxygen. Compared to humans, the seal contains about twice as much oxygen per kilogram of body weight, mostly in the blood and muscles. About 36% of our total oxygen is in our lungs, and 51% is in our blood. In contrast, the Weddell seal holds only about 5% of its oxygen in its relatively small lungs, stocking 70% in the blood. This is possible partly because the seal has about twice the volume of blood per kilogram of body weight as a human. Another adaptation is the seal's huge spleen, which can store about 24 L of blood. The spleen probably contracts after a dive begins, fortifying the blood with additional erythrocytes loaded with oxygen. Diving mammals also have a higher concentration of an oxygen-storing protein called **myoglobin** in their muscles than most other mammals; the Weddell seal can stow about 25% of its oxygen in muscle, compared to only 13% in humans.

Figure 38.27
Carbon dioxide transport in the blood. Carbon dioxide produced by body tissues diffuses into the plasma and then into the erythrocytes. In the erythrocytes, the enzyme carbonic anhydrase catalyzes a reaction of CO_2 and H_2O to form carbonic acid, which dissociates into a hydrogen ion (H^+) and a bicarbonate ion. Bicarbonate diffuses out of the erythrocyte into the blood. Electrical balance of the cell is maintained by the uptake of chloride. The hydrogen ions from carbonic acid, which have the potential to acidify the blood, are mostly bound to hemoglobin (Hb) within the erythrocytes. The reversibility of the carbonic acid–bicarbonate conversion also helps buffer the blood, releasing or removing H^+, depending on pH (see Chapter 3). The processes that occur in the tissue capillaries are reversed in the lungs. CO_2 is reconstituted, diffuses out of the blood and into the lungs, and is expelled in an exhaled breath. About 70% of the carbon dioxide in blood is transported in the form of bicarbonate. The remainder travels as dissolved CO_2 or CO_2 bound to the amino groups of proteins.

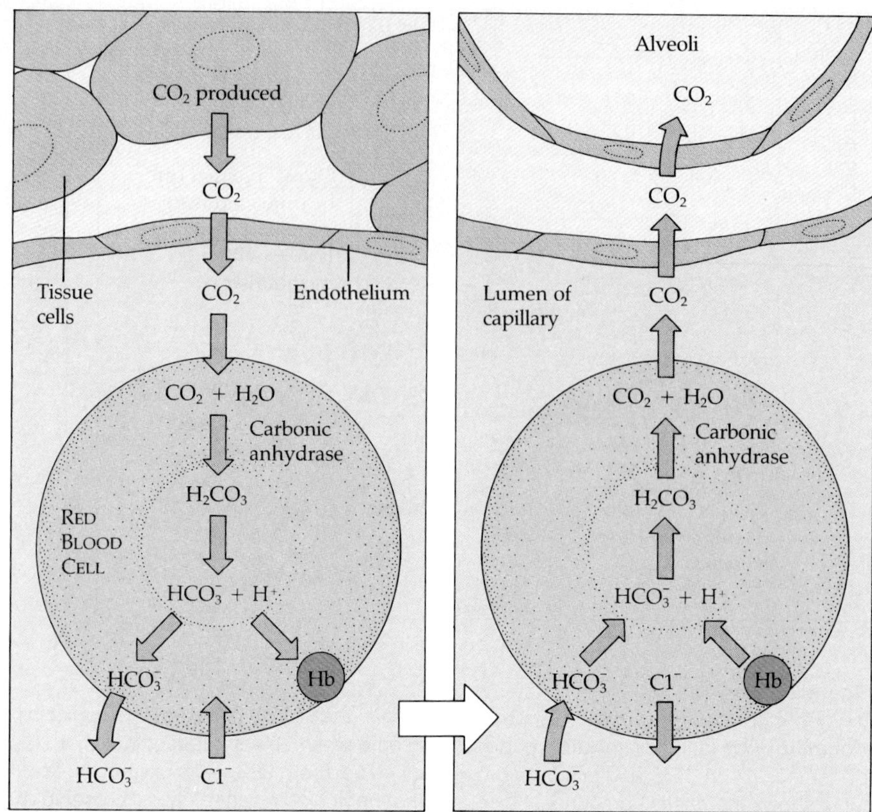

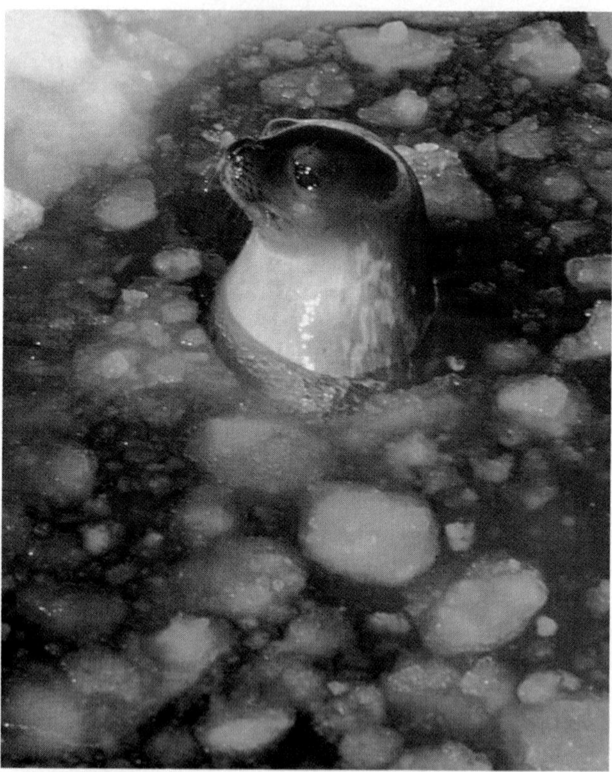

Figure 38.28
The Weddell seal, a diving mammal. Adaptations of the circulatory and respiratory systems enable these antarctic seals to travel underwater for more than an hour.

Thus, the Weddell seal owes part of its diving virtuosity to blood and muscles enriched with oxygen.

Diving mammals not only begin an underwater trip with a relatively large reservoir of oxygen, they also have adaptations that conserve that oxygen. A diving reflex slows the pulse, and an overall reduction in oxygen consumption parallels the lower cardiac output. Regulatory mechanisms affecting peripheral resistance route most blood to the brain, spinal cord, eyes, adrenal glands, and placenta (in pregnant seals). Blood supply to the muscles is restricted, and it is shut off altogether during the longest dives. Thus, during dives of more than about 20 minutes, the muscles deplete the oxygen stored in their myoglobin and then derive their ATP from fermentation rather than from respiration (see Chapter 9).

Response to environmental pressures—over the short term by physiological adjustments and over the long term by natural selection—is a central theme in our study of organisms that is showcased by the unusual adaptations of diving mammals.

* * *

We have seen throughout this chapter that the circulatory and respiratory systems cooperate extensively. Still another activity of the body that depends on the circulatory system is defense, which we will discuss in the next chapter.

STUDY OUTLINE

1. The exchange of oxygen, carbon dioxide, nutrients, and metabolic wastes between an organism's cells and the environment must occur across fluid-bathed plasma membranes.
2. In all but the simplest animals, special systems are required for transporting substances within the body.
3. Transfer of gases, nutrients, and wastes occurs across thin, specialized epithelia that separate blood or interstitial fluid from the environment.

Internal Transport in Invertebrates (pp. 819–820)

1. Cnidarians and flatworms have gastrovascular cavities that function in circulation as well as digestion.
2. Arthropods and most mollusks have open circulatory systems, in which tissues are bathed directly in hemolymph pumped by a heart into sinuses.
3. Annelids have closed circulatory systems, with blood confined to vessels, some of which pulsate and function as hearts. Vertebrates and some mollusks also have closed circulatory systems.

Circulation in Vertebrates (pp. 820–830)

1. In vertebrates, blood flows in a closed cardiovascular system consisting of blood vessels and a two- to four-chambered heart. The heart has one atrium or two atria, which receive blood from veins, and one or two ventricles, which pump blood into arteries. Arteries branch into arterioles, which give rise to capillaries—the sites of chemical exchange between blood and interstitial fluid. Capillaries rejoin into venules that converge into veins.
2. In fishes, a two-chambered heart pumps blood to gills for oxygenation, and the blood then travels to other capillary beds of the body before returning to the heart.
3. Amphibians and most reptiles have a three-chambered heart in which the single ventricle pumps blood to both lungs and body in the pulmonary and systemic circuits. These two circuits return blood to separate atria. This double circulation repumps blood returning from the capillary beds of the respiratory organ, thus ensuring a strong flow of blood to the rest of the body.
4. Birds and mammals, both endotherms, have four-chambered hearts that keep oxygen-rich and oxygen-poor blood completely separated.
5. The heart cycle consists of a period of contraction, called systole, and a period of relaxation, called diastole.
6. Heart valves dictate a one-way flow of blood through the heart.
7. Together with the stroke volume, heart rate (pulse) determines cardiac output, the volume of blood pumped into the systemic circulation per minute.
8. The intrinsic contraction of the cardiac muscle is coordinated by a conduction system originating in the sinoatrial (SA) node (pacemaker) of the right atrium. The pacemaker initiates a wave of contraction that spreads to both atria, hesitates momentarily at the atrioventricular (AV) node, and then progresses to both ventricles. The pacemaker is itself influenced by nerves, hormones, body temperature, and atrial volume changes during exercise.
9. All blood vessels, including capillaries, are lined by a single layer of endothelium. Arteries and veins have two additional outer layers composed of characteristic proportions of smooth muscle, elastic fibers, and connective tissue.
10. The velocity of blood flow varies in the circulatory system, being slowest in the capillary beds as a result of their high resistance and large total cross-sectional area. This leisurely flow enhances exchange of substances between the blood and interstitial fluid.
11. Blood pressure is determined by cardiac output and peripheral resistance due to variable constriction of the arterioles.
12. Muscular activity and pressure changes during breathing propel blood back to the heart in veins equipped with one-way valves.
13. The moment-to-moment blood supply to different organs is determined by variable constriction of arterioles and capillary sphincters.
14. Capillary exchange is the ultimate function of the circulatory system. Substances traverse the endothelium in endocytotic-exocytotic vesicles, by diffusion, or dissolved in fluids forced out by blood pressure at the arterial end of the capillary. Exuded fluid re-enters the circulation directly at the venous end of the capillary or indirectly through the lymphatic system.

Mammalian Blood (pp. 830–832)

1. Whole blood consists of cells (formed elements) suspended in a liquid matrix called plasma.
2. Plasma is a complex aqueous solution of inorganic electrolytes, proteins, nutrients, metabolic waste products, respiratory gases, and hormones.
3. Plasma proteins influence blood pH, osmotic pressure, and viscosity, and function in lipid transport, immunity (antibodies), and blood clotting (fibrinogens).
4. Red blood cells, or erythrocytes, transport oxygen, a function reflected in their small size, biconcave shape, anaerobic metabolism, and hemoglobin content. Pluripotent stem cells in red bone marrow give rise to all types of blood cells.
5. Five types of white blood cells, or leukocytes, function in defense by phagocytizing bacteria and debris or by producing antibodies.
6. Platelets are fragments of cells produced in the bone marrow that function in blood clotting, a cascade of complex reactions that converts the plasma fibrinogen into fibrin.

Cardiovascular Disease (pp. 832–834)

1. A leading cause of death in the United States is cardiovascular disease, a deterioration of the heart and blood vessels. Gradual plaque buildup during atherosclerosis or arteriosclerosis narrows the diameter of vessels and encourages the formation of thrombi or emboli that can cause heart attack or stroke by blocking strategic vessels in the heart and brain.

Gas Exchange in Animals (pp. 835–846)

1. Metabolism demands a constant supply of oxygen and removal of carbon dioxide. Animals require large, moist respiratory surfaces for adequate diffusion of respiratory

gases between their cells and the respiratory medium, either air or water.

2. A gill is any localized evagination of the body surface specialized for gas exchange. The efficiency of gas exchange in some gills, including those of fishes, is increased by ventilation and countercurrent flow of blood and water.
3. The tracheae of insects are tiny branching tubes that penetrate the body, bringing oxygen directly to cells.
4. Terrestrial vertebrates, land snails, and spiders have internal lungs restricted to one location.
5. Mammalian lungs are enclosed in a double-walled sac whose layers adhere to one another, to the lungs, and to the chest cavity. Air inhaled through the nostrils passes through the pharynx and glottis, into the trachea, bronchi, and bronchioles, and into the dead-end alveoli, where gas exchange occurs.
6. Lungs must be ventilated by breathing. Mammals ventilate with negative pressure by contracting and relaxing rib muscles and a diaphragm, which changes the volume and hence the pressure of the thoracic cavity and lungs relative to the atmosphere.
7. Tidal volume, the amount of air normally inhaled or exhaled, is less than the vital capacity of the lungs. Even after forceful exhalation, a residual volume of air remains in the alveoli.
8. Birds have one-way ventilation of the lungs, made possible by a system of air sacs, which also decrease density and dissipate heat, important adaptations for flight.
9. Breathing is regulated automatically by complex control systems involving a breathing center in the medulla of the brain stem. Rhythmic nervous impulses, blood pH changes reflecting carbon dioxide concentrations, and oxygen sensors work to adjust the rate and depth of breathing to match the metabolic demands of the body.
10. Oxygen and carbon dioxide diffuse from where their partial pressures are higher to where they are lower. In the lungs, oxygen enters the blood and carbon dioxide leaves. The blood then enters the heart through the pulmonary vein and is pumped through the systemic circuit where the diffusion process is reversed in the metabolic environment of the tissues.
11. Respiratory pigments increase the amount of oxygen that blood can carry. Arthropods and many mollusks use copper-containing hemocyanin, whereas vertebrates and some invertebrates use iron-containing hemoglobin.
12. Hemoglobin molecules have four iron-containing subunits, each one capable of binding a molecule of oxygen. The binding of the first oxygen molecule increases the affinity of the other subunits for oxygen. The dissociation curve for hemoglobin is steepest in the range of oxygen concentrations found in body tissues. A drop in pH reduces hemoglobin's affinity for oxygen, thus providing more oxygen to active tissues.
13. Most carbon dioxide generated during metabolism is transported in the form of bicarbonate ions, which result from the dissociation of carbonic acid formed in the red blood cells from the chemical union of carbon dioxide and water. Hydrogen ions from the dissociation are bound to hemoglobin and other proteins, serving to buffer the blood. The entire process is reversed when blood enters the lungs, allowing free carbon dioxide to diffuse into the environment.
14. Diving mammals have a larger volume of blood, a higher concentration of myoglobin in their muscles, and a diving reflex that slows heart rate and reroutes blood flow away from muscles and noncritical organs.

SELF-QUIZ

1. Which of the following respiratory systems is *not* closely associated with a blood supply?
   a. vertebrate lungs
   b. fish gills
   c. tracheal systems of insects
   d. the outer skin of an earthworm
   e. the parapodia of a polychaete worm
2. Blood returning to the mammalian heart in a pulmonary vein will drain *first* into the
   a. vena cava
   b. left atrium
   c. right atrium
   d. left ventricle
   e. right ventricle
3. Pulse is a direct measure of
   a. blood pressure
   b. stroke volume
   c. cardiac output
   d. heart rate
   e. breathing rate
4. The respiratory medium flows unidirectionally over the respiratory surface in the gas exchange systems of
   a. mammals
   b. frogs
   c. birds
   d. insects
   e. sea stars
5. In negative pressure breathing, inhalation results from
   a. forcing air from the throat down into the lungs
   b. contracting the diaphragm
   c. relaxing the muscles of the rib cage
   d. using muscles of the lungs to expand the alveoli
   e. contracting the abdominal muscles
6. The maximum volume of air you can forcefully exhale after taking the deepest possible breath is called
   a. tidal volume
   b. residual volume
   c. vital capacity
   d. total respiratory volume
   e. alveolar volume
7. A decrease in the pH of human blood caused by exercise would
   a. decrease breathing rate
   b. increase heart rate
   c. decrease the amount of O_2 unloaded from hemoglobin
   d. decrease cardiac output
   e. decrease CO_2 binding to hemoglobin
8. Compared to the interstitial fluid that bathes active muscle cells, blood reaching that tissue in arteries has a
   a. higher P_{O_2}
   b. higher P_{CO_2}
   c. greater bicarbonate concentration
   d. lower pH
   e. greater osmotic pressure

9. Which of the following reactions prevails in red blood cells traveling through pulmonary capillaries? (Hb = hemoglobin)
   a. $Hb + 4\ O_2 \longrightarrow Hb(O_2)_4$
   b. $Hb(O_2)_4 \longrightarrow Hb + 4\ O_2$
   c. $CO_2 + H_2O \longrightarrow H_2CO_3$
   d. $H_2CO_3 \longrightarrow H^+ + HCO_3^-$
   e. $Hb + 4\ CO_2 \longrightarrow Hb(CO_2)_4$
10. Compared to a human, a diving mammal of equal size has
   a. less blood, an adaptation that helps conserve oxygen
   b. larger lungs
   c. a larger spleen
   d. less oxygen stored in muscles
   e. less oxygen stored in blood

CHALLENGE QUESTIONS

1. Because they support their weight against gravity and repeatedly accelerate limbs from standing starts, terrestrial vertebrates consume more energy in locomotion than do fishes swimming through water. In other words, it takes more calories per gram of animal to move 1 m on land than it does to move 1 m in water (assuming, of course, that the animal is in its natural habitat, either land or water). How does this disparity fit in with the evolution of the vertebrate cardiovascular system?
2. The hemoglobin of a human fetus differs from adult hemoglobin. Compare the dissociation curves of the two hemoglobins in the following graph, and then explain the physiological significance of the difference.

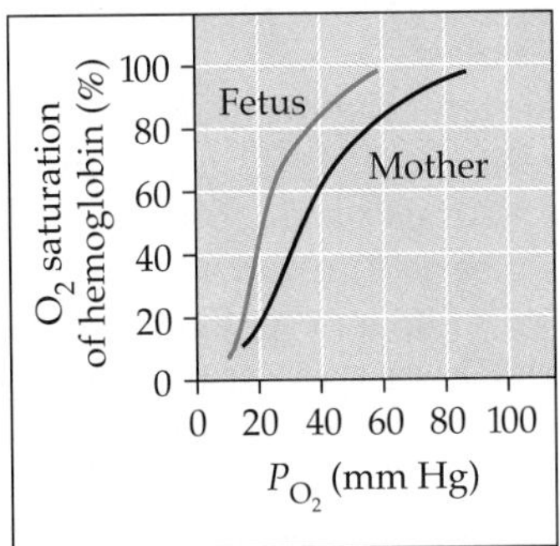

3. Describe some adaptations you think would be helpful for a mammal living at very high altitude, where the air is "thin" (relatively low P_{O_2}). Include an interpretation of the following graph in your discussion.

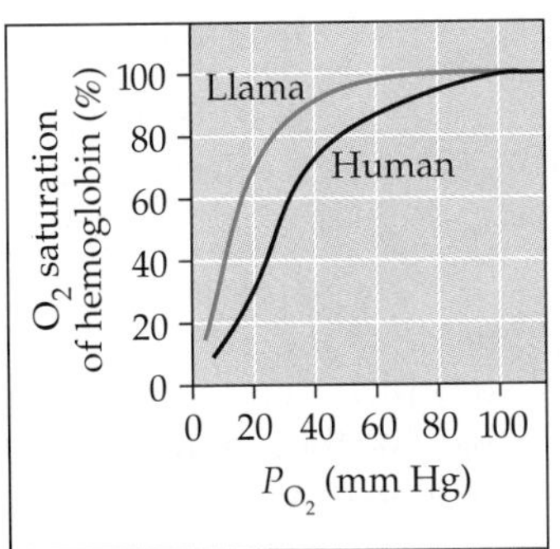

SCIENCE, TECHNOLOGY, AND SOCIETY

1. The incidence of cardiovascular disease is much lower in the Mediterranean countries (Spain, Italy, Greece, etc.) than in North America. Suggest several alternative hypotheses that might explain this difference. How could you test your hypotheses?
2. Physicians use two drugs to break up the blood clots that cause heart attacks. Drug A, a bacterial enzyme, has been used for decades. It is not protected by patents, so it is relatively cheap—$76 to $300 per dose. Drug B is a human protein produced by recombinant DNA techniques. It is patented, heavily marketed by its manufacturer, and costs $2200 per dose. Early studies suggested that Drug B is more effective than Drug A at dissolving blood clots. Many doctors switched drugs, some fearing they might be sued if they did not use the most effective treatment, regardless of cost. In later trials, the two drugs were equally effective. A third comparison of the drugs is now under way. Imagine that you are a physician treating heart attack patients. Which drug would you choose? What factors would influence your choice of treatment? If you were the patient, which drug would you prefer and why? Should physicians consider cost when they select a treatment for their patients?
3. Hundreds of studies have linked smoking with cardiovascular and lung disease. According to most health authorities, smoking is the leading cause of preventable, premature death in the United States. The government bans television advertising of cigarettes and requires the tobacco industry to place health warnings on packages and in print ads. Antismoking and health groups have proposed that cigarette advertising be banned entirely. What are some arguments in favor of a total ban on cigarette advertising? What are arguments in opposition? Do you favor or oppose such a ban? Why?

FURTHER READING

Dickerson, R. E., and I. Geis. *Hemoglobin: Structure, Function, Evolution and Pathology.* Menlo Park, CA: Benjamin/Cummings, 1983. A comprehensive treatment of the vertebrate respiratory pigment.

Golde, D. "The Stem Cell." *Scientific American,* December 1991. The master cell of blood cell production.

Lawn, R. "Lipoprotein (a) in Heart Disease." *Scientific American,* June 1992. A cholesterol-binding protein that raises the risk of cardiovascular disease.

Lillywhite, H. B. "Snakes, Blood Circulation, and Gravity." *Scientific American,* December 1988. How does a snake slithering up a tree get blood to its brain?

Pennisi, E. "Thicker Than Water." *Science News,* October 5, 1992. The hydrodynamics of blood flow.

Stroh, M. "Breathing Lessons." *Science News,* May 9, 1992. The use of computers to develop models for how breathing works.

Zapol, W. M. "Diving Adaptations of the Weddell Seal." *Scientific American,* June 1987. Lab and field studies of a living diving machine.

Zivin, J., and D. Choi. "Stroke Therapy." *Scientific American,* July 1991. Causes, effects, and experimental treatment of strokes.

39 The Body's Defenses

Nonspecific Defense Mechanisms

The Immune System and Specific Defense: Some Basic Concepts

The Humoral Response: A Closer Look

The Cell-Mediated Response: A Closer Look

The Complement System: A Component of Both Nonspecific and Specific Defense

Self Versus Nonself: Some Applications

Disorders of the Immune System

Immunity in Invertebrates

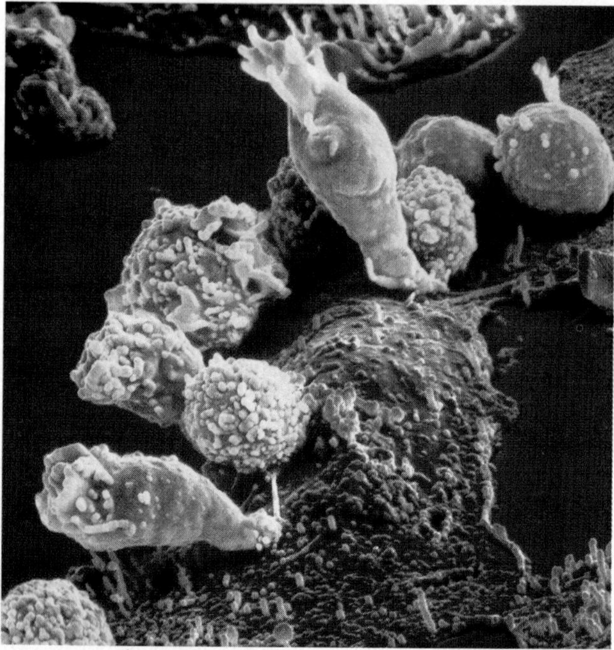

Figure 39.1
Cells of the immune system attacking a cancer cell. An animal's health is threatened not only from the outside by pathogenic microorganisms, but also from within by the body's own cells that go awry, such as cancer cells. The immune system defen against both threats. Here, a troop of cells called cyto- lls (light green) attacks a tumor cell (colorized SEM). pter, you will learn how the immune system and other chanisms protect an animal's body.

An animal must defend itself against unwelcome intruders: the many potentially dangerous viruses, bacteria, and other pathogens it encounters in the air and in food and water. It must also deal with abnormal cells that periodically appear in the body and, if not eliminated, develop into cancer (Figure 39.1). Two cooperative defense systems that counter these threats have evolved. One of these systems is nonspecific in nature—that is, it does not distinguish one infectious agent from another. This nonspecific system includes two lines of defense that an invader encounters in sequence. The first line of defense is external: It consists of the epithelial tissues that cover and line our bodies (skin and mucous membranes) and the secretions these tissues produce. The second line of defense is internal: It is triggered by chemical signals and employs antimicrobial proteins and phagocytic cells that indiscriminately attack any invader that penetrates the body's outer barriers. In some cases, inflammation is a sign that this second line of defense has been deployed.

The other defense system, the immune system, responds specifically to the particular type of invader. This *immune response* includes production of the specific defensive proteins called antibodies. It also involves the participation of several different types of cells that are descendants of the white blood cells called lymphocytes (see Chapter 38). The immune system constitutes a third line of defense, which comes into play simultaneously with the second (Figure 39.2). This chapter examines how an animal's nonspecific and specific defenses work together to protect the body. Our main focus is on the defense mechanisms of vertebrates, which include a highly developed immune system.

NONSPECIFIC DEFENSE MECHANISMS

An invading microbe must penetrate an external barrier formed by the skin and mucous membranes, which cover the surface and line the openings of an animal's body (see Chapter 36). If it succeeds in doing so, the pathogen encounters the second line of nonspecific

Nonspecific defense mechanisms		Specific defense mechanisms (immune system)
First line of defense	Second line of defense	Third line of defense
• Skin • Mucous membranes and their secretions	• Phagocytic white blood cells • Antimicrobial proteins • The inflammatory response	• Lymphocytes • Antibodies

Figure 39.2
An overview of the body's defenses.

defense: interacting mechanisms that include phagocytosis, antimicrobial proteins, and the inflammatory response.

The Skin and Mucous Membranes

The intact skin is a barrier that cannot normally be penetrated by bacteria or viruses, although minute abrasions may allow their passage. Likewise, the mucous membranes that line the digestive, respiratory, and genitourinary tracts bar the entry of potentially harmful microbes. Beyond their role as a physical barrier, the skin and mucous membranes counter pathogens with chemical defenses. In humans, for example, secretions from oil and sweat glands give the skin a pH ranging from 3 to 5, which is acidic enough to discourage many microorganisms from taking up residence there. (Bacteria that make up the normal flora of the skin are adapted to its acidic, relatively dry environment.) Saliva, tears, and mucous secretions bathe the surface of exposed epithelia, washing away many potential invaders. In addition, these secretions contain various antimicrobial proteins. One of these protective proteins is **lysozyme,** an enzyme that attacks the cell walls of many bacteria and destroys many microbes entering the upper respiratory system and the openings around the eyes.

Mucus, the thick (viscous) fluid secreted by mucous membranes, also traps particles that contact it. Microbes entering the upper respiratory system are often caught in the mucus and are then swallowed or expelled. Lining the trachea are specialized epithelial cells equipped with cilia that sweep out microbes and other particles trapped by the mucus, preventing them from entering the lungs (Figure 39.3). Microbes present in food or trapped in swallowed mucus from the upper respiratory system must pass through the highly acidic gastric juice produced by the stomach lining, which destroys most of the microbes before they can enter the intestinal tract.

Phagocytic White Cells and Natural Killer Cells

The body's internal mechanisms of nonspecific defense depend mainly on **phagocytosis,** the ingestion of invading particles by certain types of white blood cells (see Figure 8.25). The phagocytic cells called **neutrophils** comprise about 60% to 70% of all white blood cells. Attracted by chemical signals, neutrophils can leave the blood and enter infected tissue by amoeboid movement, destroying microbes there. (This migration toward the source of a chemical attractant is an example of *chemotaxis;* see Chapter 25.) However, neutrophils tend to self-destruct as they destroy foreign invaders, and their average life is only a few days.

Monocytes, although constituting only about 5% of the white blood cells, provide an even more effective phagocytic defense. After maturing, monocytes circulate in the blood for a few hours, then migrate into tis-

0.1 μm

Figure 39.3
The ciliary escalator. An important part of the body's first line of defense is found in the upper part of the respiratory system. Specialized cells (orange) produce mucus that traps microbes before they can enter the lungs. The mucous membrane is also equipped with ciliated cells, which are colored yellow here (SEM). Synchronized beating of the cilia expels mucus and the trapped microbes.

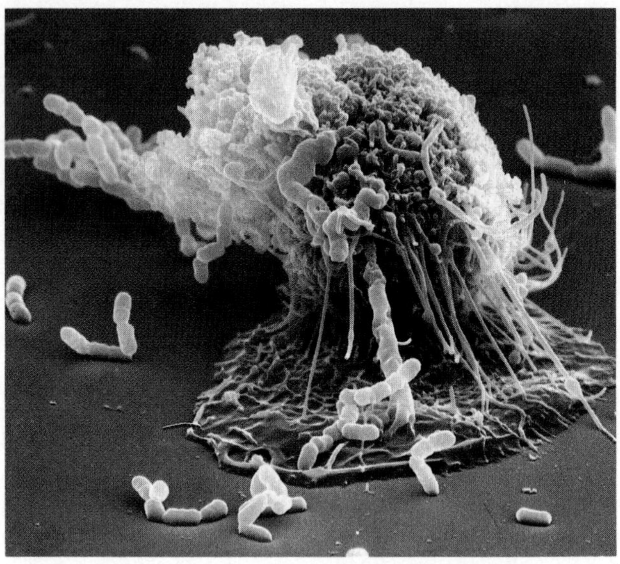

5 μm

Figure 39.4
Phagocytosis by a macrophage. This micrograph shows fibril-like pseudopodia of a macrophage ensnaring rod-shaped bacteria, which will be ingested and destroyed (colorized SEM).

sues as they enlarge and develop into **macrophages** ("big eaters"). Macrophages, the largest phagocytic cells, are especially effective, long-living phagocytes. These amoeboid cells extend pseudopodia that pull in microbes (Figure 39.4), which are then destroyed by digestive enzymes and reactive forms of oxygen within the macrophages. However, mechanisms for evading phagocytosis have evolved in some microbes. Certain bacteria, for example, have special capsules that the macrophage cannot grasp. Others have developed a resistance to the lytic enzymes of the phagocyte and can even reproduce within macrophages. The presence or absence of such an adaptation often explains why one microbe is a pathogen and a similar microbe is not.

Some macrophages reside permanently in organs and connective tissues. In the lungs, for example, they are the alveolar macrophages; in the liver, they are called Kupffer's cells. Although fixed in place, they are located where infectious agents circulating in the blood and lymph will contact them. Fixed macrophages are especially numerous in the lymph nodes and in the spleen, key organs of the lymphatic system (Figure 39.5). Other macrophages wander through all tissues of the body.

About 1.5% of the white cells are **eosinophils,** which have only limited phagocytic activity but contain destructive enzymes within cytoplasmic granules. Their main contribution to defense is against larger parasitic invad rs, such as worms. Eosinophils position them- gainst the external wall of a worm, which is ger than an eosinophil, and discharge the de- nzymes from their granules.

Nonspecific defense of the body also includes **natural killer cells.** They do not attack microorganisms directly, but rather destroy the body's own infected cells, especially cells harboring viruses, which can reproduce only within host cells. The natural killers also assault aberrant cells that could form tumors. The mode of destruction is not phagocytosis but an attack on the membrane of the target cell, which causes that cell to lyse (break open).

Antimicrobial Proteins

A variety of proteins function in nonspecific defense either by attacking microorganisms directly or by impeding their reproduction. Particularly important antimicrobial proteins are complement and the interferons. **Complement** is a group of at least 20 proteins, named for its cooperation with (complementation of) other defense mechanisms. The complement proteins act together in a cascade of activation steps that culminates with lysis of invading microbes. Some components of the complement system also function in chemotaxis as attractants for the recruitment of phagocytes to sites of infection. Since complement is an essential part of both the nonspecific and the specific defense systems, we will look at how it works after our introduction to both systems.

The **interferons** were first identified in 1957, when researchers discovered that virus-infected cells produce a substance that helps other cells resist infection by the virus—that is, interferes with the virus. There are several known types of interferons. Some are now mass-produced by recombinant DNA technology and are being tested clinically for treatment of viral infections and cancer (see Chapter 19). In the body, interferons are secreted by an infected cell as an early, nonspecific defense before specific antibodies appear. Interferons cannot save the infected cell, but they diffuse to neighboring cells, where they stimulate production of other proteins that inhibit viral replication in those cells. Thus, an infected cell can help protect uninfected cells. The defense is not virus-specific; interferon produced in response to one viral strain confers resistance to unrelated viruses. Interferons are most effective in controlling short-term infections, such as colds and influenza. In addition to its role as an antiviral agent, one type of interferon activates phagocytes, enhancing their ability to ingest and kill microorganisms.

The Inflammatory Response

Damage to tissue by a physical injury, such as a scratch, or by entry of microorganisms, triggers an **inflammatory response** (Figure 39.6). Small blood vessels in the vicinity of the injury dilate (vasodilation), in-

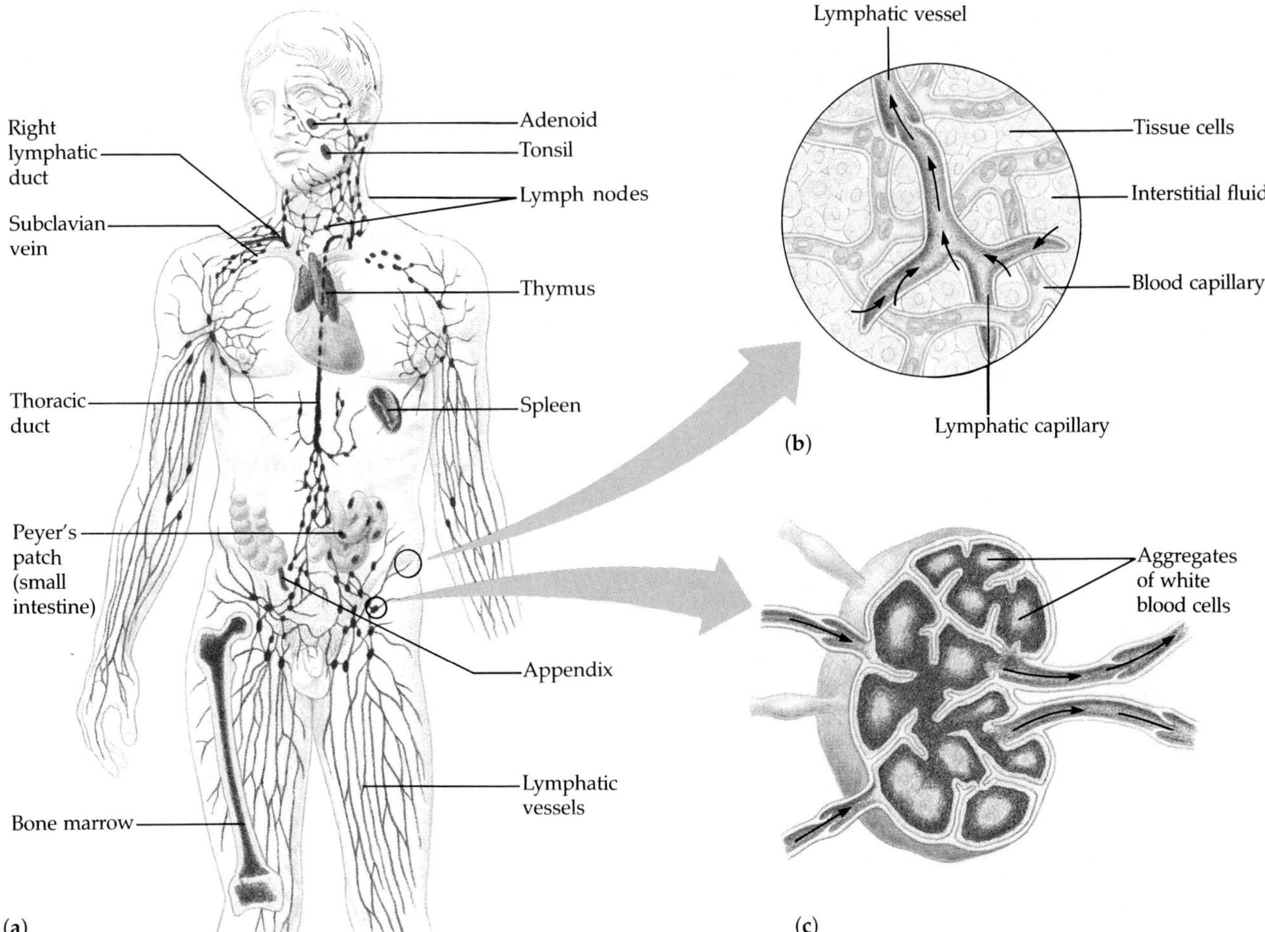

Figure 39.5
The human lymphatic system. (**a**) The lymphatic system returns fluid from the interstitial spaces to the circulatory system (see Chapter 38). In addition to the lymphatic vessels, the system includes various satellite organs that are important in the body's immune system, including the thymus, bone marrow, spleen, adenoids, tonsils, appendix, Peyer's patches, and numerous lymph nodes. (**b**) Some of the interstitial fluid that bathes tissues is taken up by the lymphatic capillaries. The fluid, now called lymph, flows through the system of vessels, eventually returning to the blood circulatory system near the shoulders where the right lymphatic duct and thoracic duct drain into veins, such as the subclavian vein. (**c**) Along the way, lymph must pass through numerous lymph nodes, where any pathogens present in lymph encounter macrophages and lymphocytes, another class of white blood cells with defensive functions.

creasing the blood supply to the injured area. This is responsible for inflammation's characteristic redness and heat (*inflammation* means "setting on fire"). The dilated blood vessels also become more permeable, and fluids move from the blood into neighboring tissues, causing the edema (swelling) associated with inflammation.

The processes that constitute the inflammatory response are initiated by chemical signals. One of these signals is **histamine,** which is contained in circulating white cells called basophils and in cells called mast cells that are found in connective tissue. Injury to these cells causes the release of histamine, which triggers local vasodilation and makes capillaries in the vicinity leakier. White blood cells and damaged tissue cells also discharge prostaglandins (see Chapter 41) and other substances that promote blood flow to the injured locale. Vasodilation and the increase in blood vessel permeability also deliver clotting elements into the injured area. Blood clotting marks the beginning of the repair process and helps block the spread of pathogenic microbes to other parts of the body.

Probably the most important element of the inflammatory response is phagocyte migration from the blood into the injured tissues, which usually begins within an hour after injury. The increased flow of blood to the site of injury and the leakage from the capillaries enhance the migration of phagocytic cells into the interstitial fluid. In this chemotaxis, several chemical mediators, including some of the proteins of the complement system, attract the phagocytes to the damaged tissue. Neutrophils arrive at the site of the injury first, followed by monocytes that develop into macrophages. The macrophages not only devour

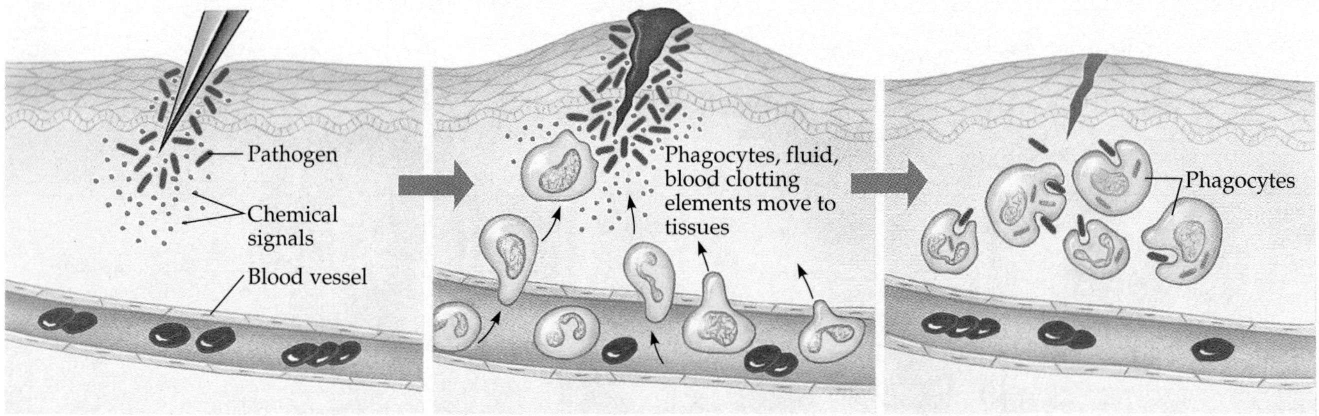

① Tissue injury; release of chemical signals (histamine, prostaglandins).

② Vasodilation (increased blood flow); increased vessel permeability; phagocyte migration.

③ Phagocytes (macrophages and neutrophils) consume pathogens and cell debris; tissue heals.

Figure 39.6
A simplified view of the inflammatory response. ① The localized response is triggered when cells of tissue injured by bacteria or physical damage release chemical signals such as histamine and prostaglandins. ② These induce increased permeability of the capillaries and blood flow to the affected area. Also released are certain chemicals that attract phagocytic cells and lymphocytes. ③ After their arrival at the site of injury, the phagocytes consume pathogens and cell debris, and the tissue heals.

pathogens, but they also clean up the remains of tissue cells and of neutrophils that have self-destructed after eliminating many microorganisms. The pus that often accumulates at the site of infection consists mostly of dead cells and the fluid that leaked from the capillaries during the inflammatory response. If pus remains in place, it is gradually absorbed by the body over a period of a few days.

The inflammatory responses described so far are localized, as in infections caused by a splinter. But the body's reaction to an infection may also be systemic—that is, more widespread. For example, injured cells emit molecules that stimulate the release of more neutrophils from bone marrow—a call for reinforcements. The number of white cells in the blood may increase severalfold within a few hours after the inflammatory reaction begins. This dramatic increase in leukocyte (white blood cell) count often indicates a severe infection, such as meningitis or appendicitis. Another systemic response to infection is fever. Toxins produced by pathogens may trigger the fever, but certain white blood cells also release molecules called **pyrogens,** which set the body's thermostat at a higher temperature. A very high fever may be dangerous, but a moderate fever contributes to defense. It inhibits the growth of some microorganisms, perhaps in part by decreasing the amount of iron available to them. Fever may also facilitate phagocytosis and, by speeding up body reactions, may speed the repair of tissues.

L... review the body's nonspecific defense sys... e first line of defense, the skin and mucous ...es, prevents most pathogens from entering the body; the second line of defense uses phagocytes, natural killer cells, antimicrobial proteins, and inflammatory responses to defend against pathogens that have managed to enter the body. These two lines of defense are termed nonspecific because they do not target specific pathogens—a certain bacterial species, for example.

THE IMMUNE SYSTEM AND SPECIFIC DEFENSE: SOME BASIC CONCEPTS

Basic Features of the Immune System

The body's third line of defense, the **immune system,** is distinguished from the nonspecific defenses by four features: specificity, diversity, self/nonself recognition, and memory.

Specificity refers to the immune system's ability to recognize and eliminate particular microorganisms and foreign molecules—a certain strain of flu virus, for example. A foreign substance that elicits this immune response is called an **antigen.** The immune system responds to an antigen by producing specialized lymphocytes and specific proteins called **antibodies.** (*Antigen* is a contraction of *anti*body-*gen*erating.) Antigens that trigger an immune response include molecules of viruses, bacteria, fungi, protozoans, and parasitic worms. Antigens also mark the surfaces of such foreign materials as pollen, insect venom, and transplanted tissue, such as skin or organs. Each antigen has a unique molecular shape and stimulates production of the very type of antibody that defends against that specific antigen. Thus, in contrast to the nonspecific defenses, each response of the immune system targets a specific invader, distinguishing it from other foreign molecules that may be very similar.

Diversity refers to the immune system's ability to respond to millions of kinds of invaders, each recognized by its antigenic markers. This diversity of response is possible because the immune system is equipped with an enormous variety of lymphocyte populations, each population keyed to a particular antigen. Among antibody-producing lymphocytes, for example, each population is stimulated by a specific antigen and responds by synthesizing and secreting the appropriate type of antibody.

Self/nonself recognition refers to the immune system's ability to distinguish the body's own molecules from foreign molecules (antigens). Failure of self/nonself recognition can lead to autoimmune disorders, in which the immune system destroys the body's own tissues. You will learn about the mechanism of self/nonself recognition and autoimmune disorders later in the chapter.

Memory refers to the immune system's ability to remember antigens it has encountered and to react against them more promptly and effectively on subsequent exposures. This feature of **acquired immunity** was recognized about 2400 years ago by Thucydides of Athens, who described how those sick and dying during an epidemic of plague were cared for by those who had recovered, "for no one was ever attacked a second time." The concept is familiar to all of us: If we had chickenpox as a child, we are unlikely to get it again.

Active Versus Passive Acquired Immunity

Immunity conferred by recovery from an infectious disease such as chickenpox is called **active immunity** because it depends on response by a person's own immune system. In this case, the active immunity is *naturally* acquired. Active immunity can also be acquired *artificially,* by vaccination. Vaccines may be inactivated bacterial toxins, killed microorganisms, or living but weakened microorganisms. These agents can no longer cause disease, but they retain the ability to act as antigens and stimulate an immune response. A vaccinated person who encounters the actual pathogen will have the same quick defensive reaction based on immunological memory as a person who has had the disease.

Antibodies can also be transferred from one individual to another, creating **passive immunity.** This occurs naturally when a pregnant woman's body passes some of her antibodies across the placenta to the fetus. A newborn's immune system is not fully operative, but if the mother is immune to chickenpox, for example, the infant will also be temporarily immune to this disease. Certain antibodies are also passed from the mother to her nursing infant in breast milk, especially in the colostrum, or first secretions. Passive immunity persists only for a few weeks or months, however, and the infant's own immune system must take over defense of the body. Passive immunity can also be transferred artificially by introducing antibodies from an animal or human who is already immune to the disease. For example, rabies is treated in humans by injecting antibodies from people who have been vaccinated against rabies. This produces an immediate immunity, which is important because rabies progresses rapidly, and the response to vaccination would take too long. The injected antibodies last only a few weeks, but by then the person's own immune system has produced antibodies against the rabies virus.

The Immune System's Dual Defense: An Overview of Humoral and Cell-Mediated Responses

The immune system can actually mount two different types of responses to antigens: a humoral response and a cell-mediated response. **Humoral immunity** results in the production of antibodies, which are secreted by certain lymphocytes and circulate as soluble proteins in blood plasma and lymph—fluids long ago called humors. Around the turn of the century, researchers transferred such fluids from one animal to another to see if they also transferred immunity. They found that immunity for some conditions could be passed along only if lymphocytes were transferred. This second type of immunity, which depends on the direct action of cells (certain types of lymphocytes) rather than antibodies, became known as **cell-mediated immunity.**

The circulating antibodies of the humoral system defend mainly against toxins, free bacteria, and viruses present in body fluids. In contrast, lymphocytes of the cell-mediated system are active against bacteria and viruses inside the host's cells and against fungi, protozoans, and worms. The cell-mediated system also attacks transplanted tissue and cancer cells, both of which it perceives as "nonself."

B Cells and T Cells The white blood cells called **lymphocytes** are responsible for humoral and cell-mediated immunity. The vertebrate body is populated by two main classes of lymphocytes: **B cells** (B lymphocytes), which carry out the humoral immune response, and **T cells** (T lymphocytes), which function in the cell-mediated immune response. (Later in the chapter, we will see that one subclass of T cells also participates in humoral immunity by stimulating B cells.)

All blood cells, including lymphocytes, originate from stem cells in the bone marrow (see Figure 38.14),

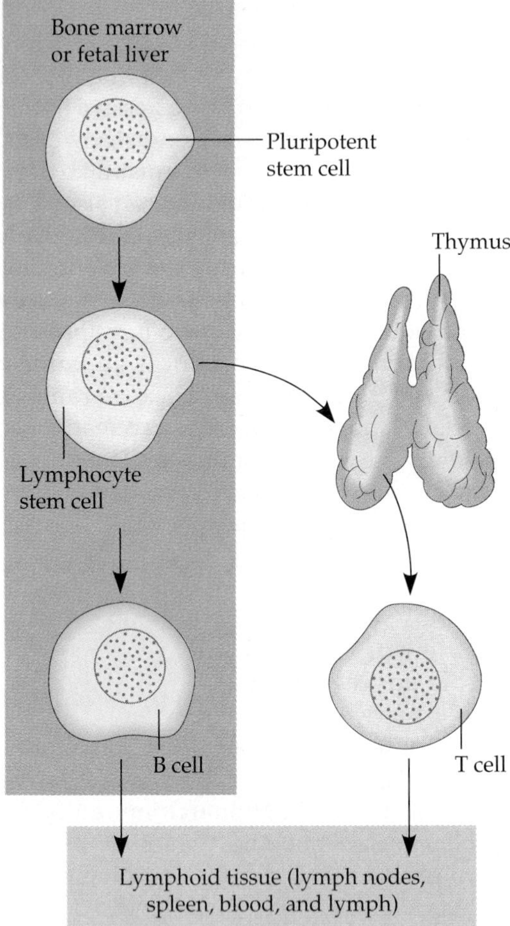

Figure 39.7
The development of lymphocytes. Like other blood cells, lymphocytes differentiate from pluripotent stem cells in bone marrow (see Figure 38.14). Lymphocytes that continue their maturation in bone marrow develop into B cells, while lymphocytes that move to the thymus and complete their maturation there differentiate into T cells. Both classes of lymphocytes populate other lymphatic organs, particularly the lymph nodes and spleen. The B and T cells are now ready for their functions in humoral immunity and cell-mediated immunity, respectively.

or in the case of the fetus, mainly in the liver. Initially, all lymphocytes are alike, but they later differentiate into T cells or B cells, depending on where they continue their maturation (Figure 39.7). Lymphocytes that migrate from the bone marrow to the thymus, a gland in the upper region of the chest, develop into T cells (T for thymus). Lymphocytes that remain in the bone marrow and continue their maturation there become B cells. (The B actually stands for the bursa of Fabricius, an organ unique to birds where B cells mature. These lymphocytes were first discovered in chickens, but it may help you remember the site of maturation for these cells if you equate B with bone.)

Mature B cells and T cells are most concentrated in lymph nodes, the spleen, and other lymphatic organs, where the lymphocytes are most likely to encounter

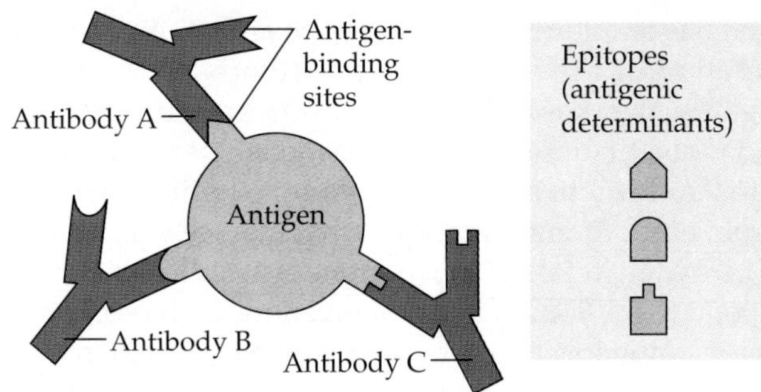

Figure 39.8
Epitopes (antigenic determinants). Antibodies bind to epitopes on the surface of an antigen. In this example, specific antibodies of three types react with different epitopes on the same large antigen molecule.

antigens (see Figure 39.5). Both B cells and T cells are equipped with specific **antigen receptors** on their plasma membranes. A B cell's antigen receptors are actually bound antibody molecules keyed to a certain antigen. The antigen receptors of T cells are not antibodies, but these **T cell receptors** recognize antigens as specifically as antibodies do. Thus, the specificity and diversity of the immune system depend on each B cell and T cell's having receptors that enable that lymphocyte to identify and respond to a particular antigen.

When antigens bind to receptors on the surface of a lymphocyte, this activates the lymphocyte to divide and give rise to a population of **effector cells,** the cells that actually defend the body in an immune response. In the case of humoral responses, B cells activated by antigen binding give rise to effector cells called **plasma cells,** which secrete antibodies that help eliminate that particular antigen. Among the effector cells derived from T cells are **cytotoxic T cells,** which destroy infected cells and cancer cells. Another subclass of effector T cells, **helper T cells,** play a pivotal role in stimulating both humoral and cell-mediated immunity. After completing our survey of basic concepts of immunology, we will examine the humoral and cell-mediated responses in more detail.

The Molecular Basis of Antigen-Antibody Specificity

Most antigens are proteins or large polysaccharides. These molecules are often outer components of the coats of viruses, the capsules and cell walls of bacteria, and the surfaces of many other types of cells. Foreign molecules associated with transplanted tissues and organs or with blood cells from other individuals or species can also incite an immune response. The surfaces of foreign substances such as pollen also include antigens. Antibodies do not generally recognize the antigen as a whole molecule (Figure 39.8). Rather, they

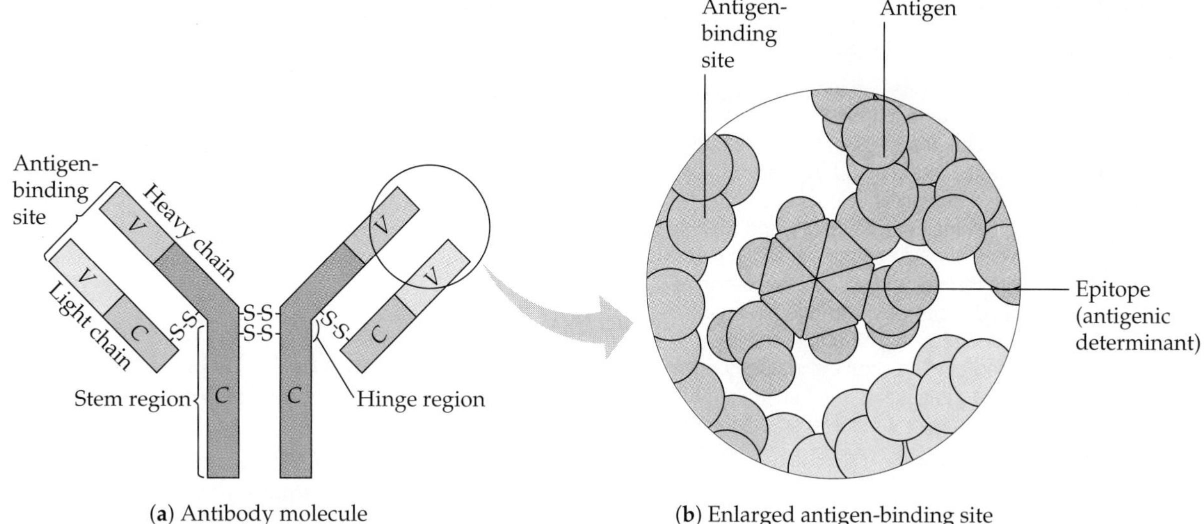

(a) Antibody molecule

(b) Enlarged antigen-binding site

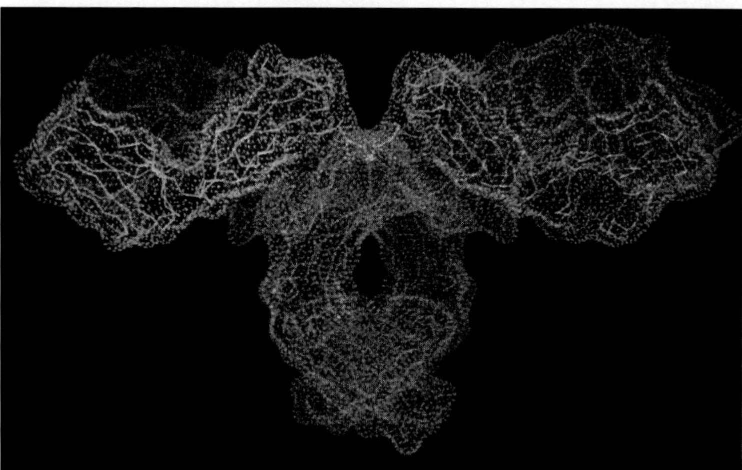

(c) Model of an antibody molecule

Figure 39.9
The structure of a typical antibody molecule.
(**a**) The Y-shaped molecule (monomer) is composed of two light and two heavy chains linked together by disulfide bridges (S-S). Most of the molecule is the same for all antibodies of the same immunoglobulin class and is termed the constant region *(C)*. The amino acid sequences of the variable regions *(V)*, which make up the two identical antigen-binding sites, are different in each specific type of antibody, giving these sites specific shapes that fit specific antigenic epitopes. The stem of the Y binds to the surface of a cell in some cases. (**b**) This enlargement shows an antigenic epitope bound to an antigen-binding site. (**c**) A computer graphic image of an antibody molecule.

identify a localized region on the surface of an antigen called an antigenic determinant, or **epitope**. A single antigen such as a bacterial protein may have several effective epitopes, stimulating the immune system to make several distinct antibodies against it. And different parts of the bacterial cell may possess different antigens. Thus, the immune system may respond to a particular species of bacterial cell by producing several types of antibodies keyed to the various epitopes marking the antigens of the bacterial cell wall, capsule, and flagella.

Antibodies constitute a class of proteins called **immunoglobulins (Igs).** Every antibody molecule has at least two identical sites that bind to the epitope that provoked its production. A typical antibody is shown in Figure 39.9a. It has four polypeptide chains joined to form a Y-shaped molecule: two identical light chains and two identical heavy chains. Both heavy and light chains have *constant* regions. These regions are called constant because their sequences of amino acids vary little among the antibodies that perform a particular type of defense, despite a wide variation in antigen specificity. At the tips of the Y-shaped molecule's two arms are the *variable* regions of the heavy and light chains, so named because their amino acid sequences vary extensively from antibody to antibody. These tips of the Y function as the *antigen-binding sites.* The dimensions and contours of the binding sites are determined by the unique amino acid sequences of the variable regions of the heavy and light chains. The association between an antigen-binding site and an epitope resembles that between an enzyme and its substrate: Several weak bonds form between contiguous chemical groups on the respective molecules (Figure 39.9b).

The antigen-binding site is responsible for an antibody's recognition function—its ability to recognize a specific epitope of an antigen. The stem of the antibody molecule, consisting of the constant regions of the polypeptide chains, is responsible for the antibody's effector function, the mechanism by which it inactivates or destroys an antigenic invader.

There are only five types of constant regions, and these determine the five major classes of mammalian immunoglobulins: IgM, IgG, IgA, IgD, and IgE. The

Table 39.1 The Five Classes of Immunoglobulins

 IgM (pentamer)	IgMs are the first circulating antibodies to appear in response to an initial exposure to an antigen. However, the concentration in the blood declines rapidly. This is diagnostically useful, because the presence of IgM usually indicates a current infection by the pathogen causing its formation. IgM consists of five Y-shaped monomers arranged in a pentamer structure. The numerous antigen-binding sites make it very effective in agglutinating antigens and in reactions involving complement. IgM is too large to cross the placenta and does not confer maternal immunity.
 IgG (monomer)	IgG is the most abundant of the circulating antibodies. It readily crosses the walls of blood vessels and enters tissue fluids. IgG also crosses the placenta and confers passive immunity from the mother to the fetus. IgG protects against bacteria, viruses, and toxins circulating in the blood and lymph, and triggers action of the complement system.
 IgA (dimer)	IgA is produced primarily in the form of two Y-shaped monomers (a dimer) by cells abundant in mucous membranes. The main function of IgA is to prevent attachment of viruses and bacteria to epithelial surfaces. IgA is also found in many body secretions, such as saliva, perspiration, tears, and colostrum (the first milk of a nursing mammal). This helps protect the infant from gastrointestinal infections.
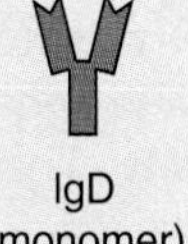 IgD (monomer)	IgD antibodies do not activate the complement system and cannot cross the placenta. They are mostly found on the external membranes of B cells, probably functioning as an antigen receptor required for initiating differentiation of B cells into forms of B cells that produce antibodies against the antigen.
 IgE (monomer)	IgE antibodies are slightly larger than IgG molecules and represent only a very small fraction of the total antibodies in the blood. The stem regions attach to receptors on mast cells and basophils and, when triggered by an antigen, cause the cells to release histamine and other chemicals that cause an allergic reaction.

shapes and functions of these basic antibody types are summarized in Table 39.1. Each class is characterized by a type of constant region that enables the antibody molecules to perform certain defense functions. For example, IgA is the only class of antibody that can be secreted across epithelia, and it is present in saliva, sweat, and tears. Within each Ig class is an enormous variety of specific antibodies with unique antigen-binding sites.

Clonal Selection: The Cellular Basis of Immunological Specificity and Diversity

We can imagine two possible mechanisms by which an animal's immune system could respond specifically to the millions of potential antigens. In one scenario, each lymphocyte is plastic in its response, adapting to a particular antigen and tailoring its action to match whatever antigen happens to be present at that time—by changing the type of antibody it secretes, for example. The evidence supports the alternate view: Each lymphocyte recognizes and responds to only one antigenic epitope, and the ability of the immune system to defend against an almost unlimited variety of antigens depends on the enormous diversity of antigen-specific lymphocytes. (The genetic basis for this immunological diversity was discussed in Chapter 18.)

Each lymphocyte's specificity for an antigenic target is rigidly predetermined during embryonic development, before an encounter with that antigen ever takes place. The mark of this specificity is the antigen receptor the lymphocyte wears on its surface. The lymphocyte may or not ever come into contact with the corresponding antigen. If that antigen *does* enter the body and binds to receptors on the specific lymphocytes, then those (and only those) lymphocytes are activated to mount an immune response. The selected cells proliferate by cell division and develop into a large number of identical effector cells—a *clone* of cells that combats the very antigen that provoked the response. For example, plasma cells that develop from an activated B cell all secrete the same type of antibody that functioned as the antigen receptor on the original B cell that first encountered the antigen. This antigen-specific selection and cloning of lymphocytes is called **clonal selection** (Figure 39.10). The concept of clonal selection is so fundamental to understanding immunity that it is worth restating. Each antigen selectively activates a tiny fraction of cells from the body's diverse pool of lymphocytes. This relatively small number of selected cells gives rise to a clone of millions of effector cells, all dedicated to eliminating the specific antigen that stimulated the humoral or cell-mediated immune response. Since most antigens, such as bacterial proteins, have many epitopes, several different clones of effector T cells or plasma cells may be generated in response to the same antigen.

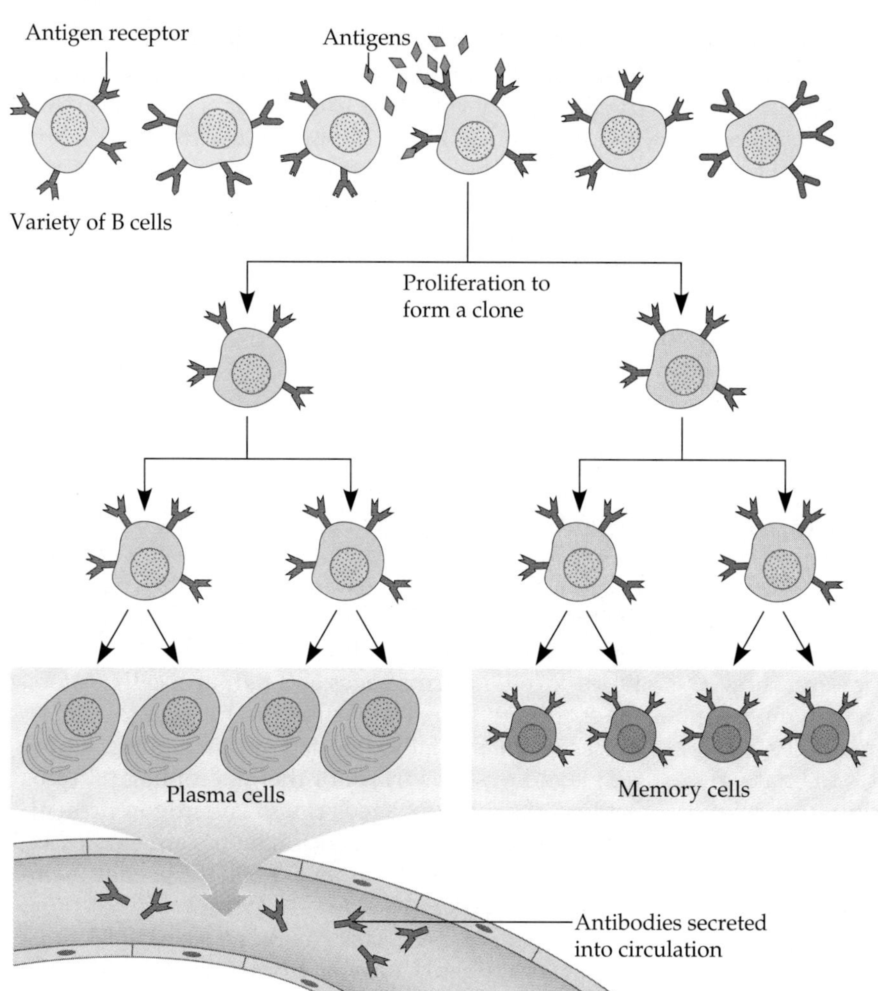

Figure 39.10
Clonal selection. The B cells and T cells of the body collectively recognize an almost infinite number of antigens, but each cell recognizes only one type of antigen. (Notice the variation in the shapes of antigen receptors on the six lymphocytes at the top of this illustration.) When an antigen binds to a B or T cell, that cell proliferates, forming a clone of effector cells with the same specificity. Here, the antigen selects a particular B cell and stimulates it to reproduce and give rise to a population of identical effector cells called plasma cells. The plasma cells secrete antibodies specific for the antigen. Notice that relatively long-lived memory cells for the antigen are also formed. These cells can respond rapidly upon subsequent exposure to the same antigen.

The Cellular Basis of Immunological Memory

The selective proliferation of lymphocytes to form clones of effector cells against an antigen constitutes the **primary immune response.** Between exposure to an antigen and maximum production of effector cells, there is a lag period of 5 to 10 days. During this lag period, the lymphocytes selected by the antigen are differentiating into effector T cells and antibody-producing plasma cells. If the body is exposed to the same antigen at some later time, the response is faster (only 3 to 5 days) and more prolonged than the primary response. This is the **secondary immune response** (Figure 39.11). The antibodies produced at this time are also more effective in binding to the antigen than those produced during the primary response.

The immune system's ability to recognize an antigen as previously encountered is called immunological memory. This ability is based on **memory cells,** which are produced along with the relatively short-lived effector cells of the primary immune response (see Figure 39.10). During the primary response, these memory cells are not active. However, they survive for long periods and proliferate rapidly when exposed

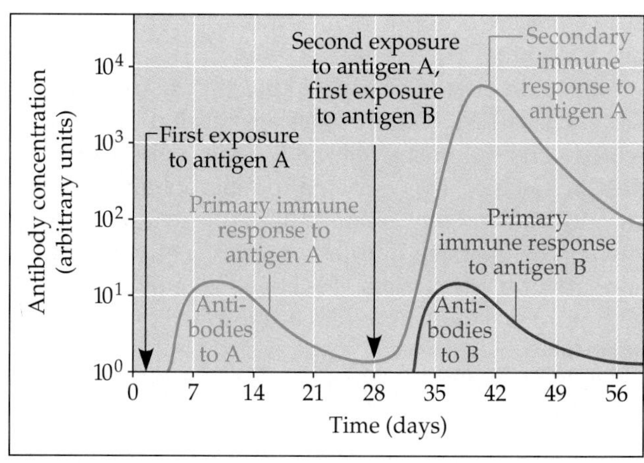

Figure 39.11
Immunological memory. An initial exposure to antigen A stimulates a primary immune response, with the eventual production of antibodies against that antigen. Notice the time lag and a relatively small response. A second exposure to antigen A at day 28 produces a faster and greater secondary immune response. If antigen B were also injected at day 28, the reaction to that antigen would be a primary response, not a secondary one. This experiment demonstrates that the secondary response, which is due to the presence of long-lived memory cells, is specific.

again to the same antigen that caused their formation. The secondary immune response gives rise to a new clone of memory cells, as well as to new effector cells. By this mechanism, childhood exposure to diseases such as chickenpox usually confers immunity for a lifetime.

The Development of Self-Tolerance

You have already learned that a key feature of the immune system is its ability to distinguish self from nonself, and that antigen receptors on the surfaces of lymphocytes are responsible for detecting foreign molecules that enter the body. There *are* no antigen receptors for an animal's *own* molecules, and this is the basis for **self-tolerance,** the lack of a destructive immune response against the body's cells. Self-tolerance develops before birth, when lymphocytes bearing antigen receptors begin to mature in the embryo. Any lymphocytes with receptors for molecules present in the body at that time are destroyed, leaving only lymphocytes with antigen receptors for foreign molecules.

Among the important "self-markers," the native molecules tolerated by an individual's immune system, are a collection of molecules called the **major histocompatibility complex (MHC).** These molecules are glycoproteins (proteins with carbohydrate attached) embedded in the plasma membranes of cells. A family of genes codes for the MHC proteins. Because there are at least 20 MHC genes and at least 50 alleles for each gene, it is virtually impossible for any two people, except identical twins, to have matching sets of MHC markers on their cells. Thus, the major histocompatibility complex is a biochemical fingerprint unique to each individual.

Two main classes of MHC molecules mark cells as "self." **Class I MHC** molecules are located on all nucleated cells—almost every cell of the body. **Class II MHC** molecules are restricted to macrophages and B lymphocytes and play an important role in interactions between cells of the immune system. You will learn more about the functions of MHC as we now examine the humoral and cell-mediated immune responses in more detail.

THE HUMORAL RESPONSE: A CLOSER LOOK

You have learned that the humoral immune response occurs when B cells with specific receptors are stimulated by an antigen to differentiate into a clone of plasma cells, which begin to secrete antibodies. Antibodies are most effective against pathogens that are circulating in the blood and lymph, such as free bacteria and viruses. You have also learned that this selective activation of B cells arms the body with long-lived memory cells that function in a secondary humoral response. In the following discussion, you will see that the humoral response involves other cells as well, including T cells. What distinguishes this branch of the immune system is its production of antibodies, which are not involved in cell-mediated defense. Notice, however, that the two branches often operate together, so that our distinctions sometimes break down.

The Activation of B Cells

In most cases, the selective activation of a B cell to form a clone of plasma cells and memory cells is a two-step process. One step, discussed earlier in the chapter, is the binding of antigen to specific receptors on the surface of the B cell. The other step involves two other types of cells, macrophages and **helper T cells.** After a macrophage engulfs pathogens by phagocytosis, pieces of the partially digested antigen molecules are displayed on the surface of the macrophage. In this context, the macrophage is called an **antigen-presenting cell (APC).** The antigen fragments are cradled by class II MHC molecules embedded in the plasma membrane of the APC. A helper T cell's specific receptors recognize this self/nonself combination of MHC and a particular antigen. Contact between the T cell and the antigen-presenting macrophage activates the T cell, which proliferates and forms a clone of helper T cells keyed to the specific antigen. The helper T cells then selectively stimulate B cells that have already encountered that particular antigen (Figure 39.12). A helper T cell uses the same mechanism to identify its complementary B cell as it used to recognize a macrophage displaying the antigen. When antigen binds to antigen receptors on a B cell, the cell takes in few of these foreign molecules and then displays antigen fragments bound to class II MHC markers on the cell surface. The helper T cell's receptor recognizes and binds to this antigen-MHC complex.

Thus, macrophages and B cells both act as antigen-presenting cells in their interactions with helper T cells, but there is an important difference: Each macrophage can display a diversity of antigens, depending on the type of pathogen it has engulfed; but each B cell, being specific, can bind to and subsequently display only one type of antigen. Helper T cells are also antigen-specific and can be activated only by those macrophages presenting the appropriate antigen in combination with a class II MHC molecule. In this way, the nonspecific defense afforded by macrophages also enhances specific defense by selectively priming T cells, which then stimulate the appropriate B cells to mount a humoral immune response against a certain antigen.

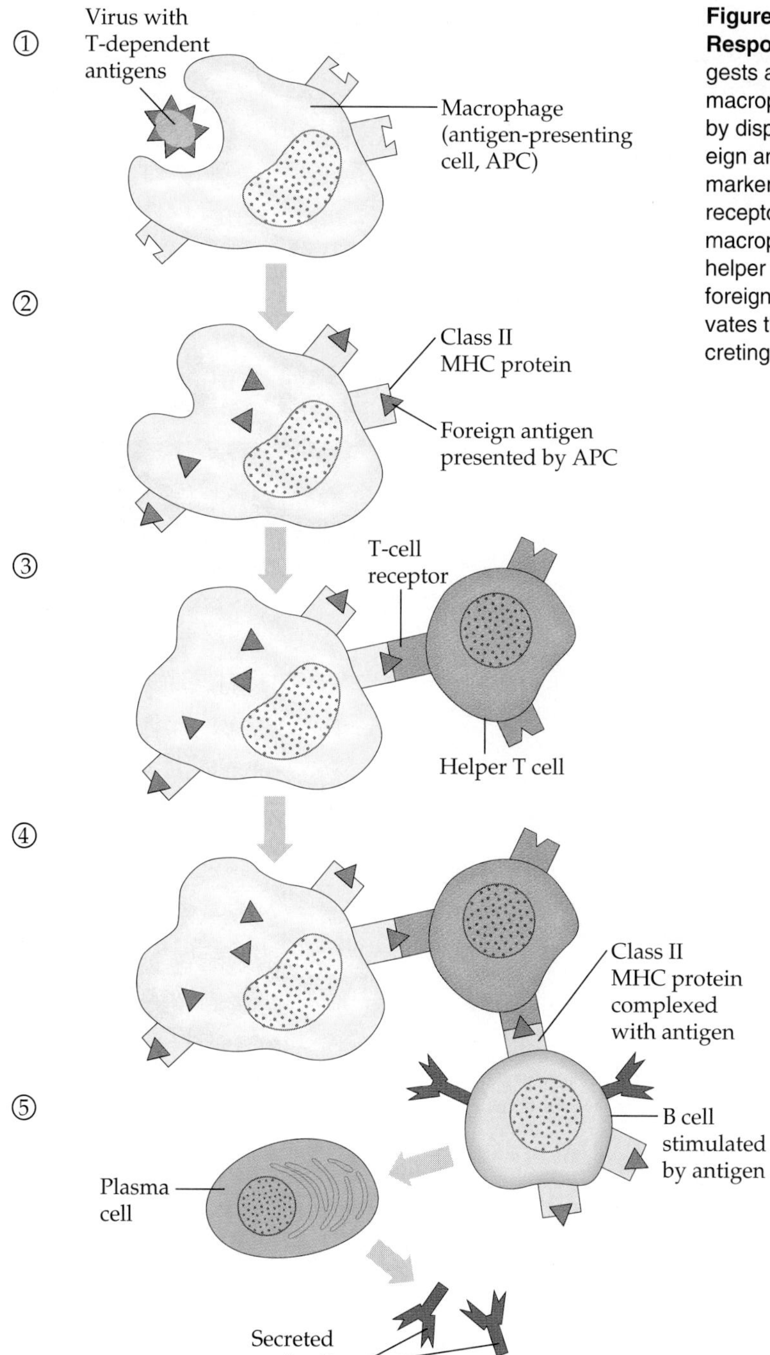

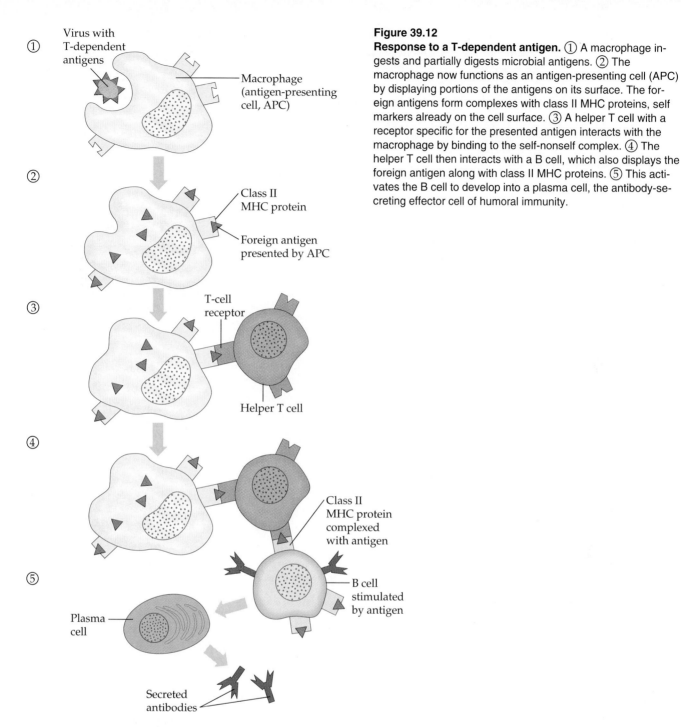

Figure 39.12
Response to a T-dependent antigen. ① A macrophage ingests and partially digests microbial antigens. ② The macrophage now functions as an antigen-presenting cell (APC) by displaying portions of the antigens on its surface. The foreign antigens form complexes with class II MHC proteins, self markers already on the cell surface. ③ A helper T cell with a receptor specific for the presented antigen interacts with the macrophage by binding to the self-nonself complex. ④ The helper T cell then interacts with a B cell, which also displays the foreign antigen along with class II MHC proteins. ⑤ This activates the B cell to develop into a plasma cell, the antibody-secreting effector cell of humoral immunity.

Antigens that evoke the cooperative response you have just learned about are known as **T-dependent antigens,** because they cannot stimulate antibody production without the involvement of T cells. Most antigens are T-dependent. However, certain types of antigens, called **T-independent antigens,** trigger humoral immune responses without involving macrophages or T cells. These antigenic molecules are usually long chains of repeating units, such as polysaccharides or protein subunits. Such antigens are found in bacterial capsules and bacterial flagella. Apparently, the numerous subunits of such antigens bind simultaneously to a number of the antigen receptors on the B-cell surface. This provides enough stimulus to the B cells without the assistance of T cells. However, the humoral response (antibody production) to T-independent antigens is generally much weaker than the response to T-dependent antigens.

Once activated by T-dependent or T-independent antigens, a B cells gives rise to a clone of plasma cells, and each of these effector cells secretes as many as 2000 antibodies per second for the 4- to 5-day lifetime of the cells. These specific immunoglobulins help eliminate the foreign invader.

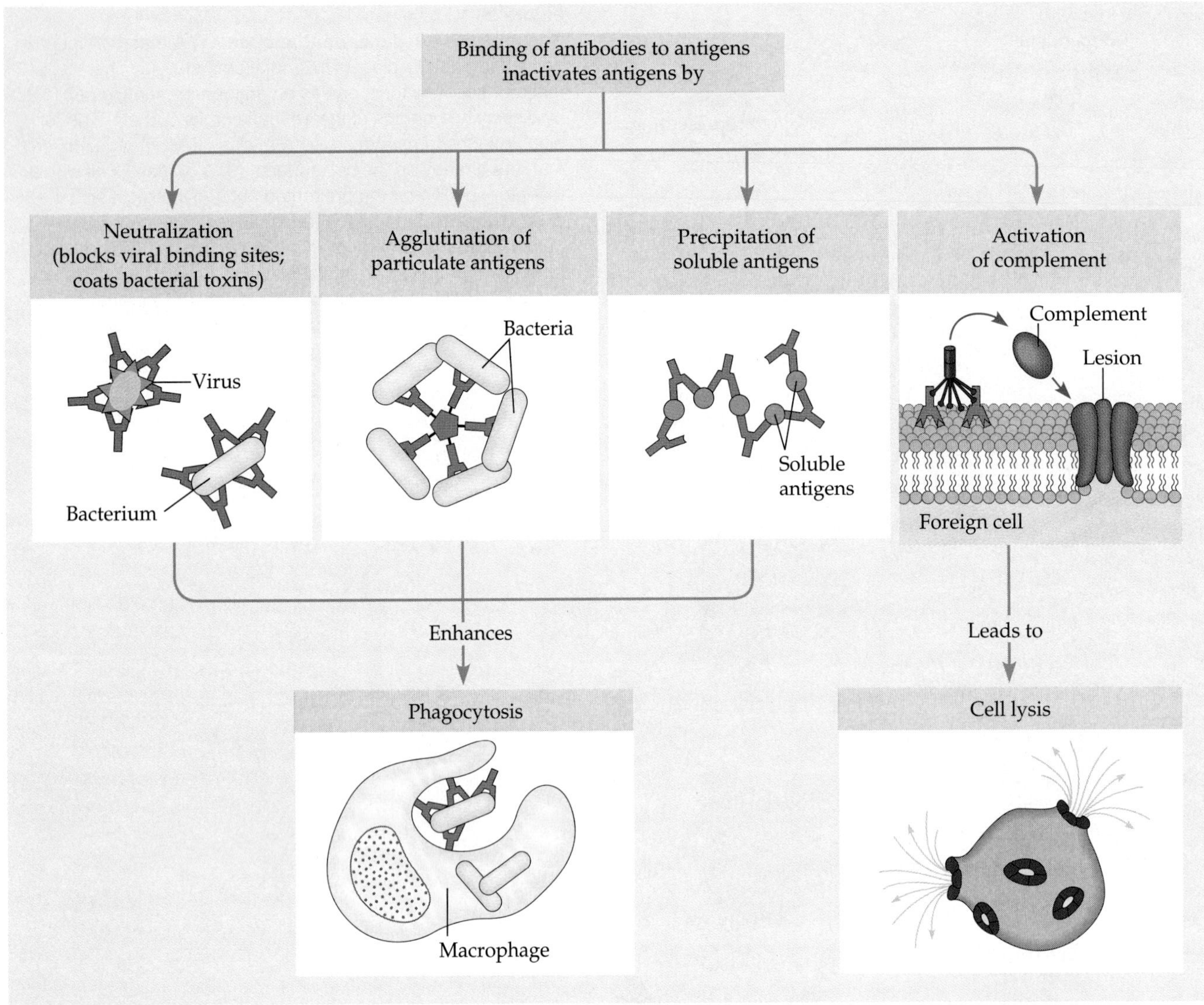

Figure 39.13
Effector mechanisms of humoral immunity. The binding of antibodies to antigens tags foreign cells and molecules for destruction by phagocytes and complement.

How Antibodies Work

An antibody does not usually destroy an antigenic invader directly. The binding of antibodies to antigens to form an antigen-antibody complex is the basis of several effector mechanisms (Figure 39.13). The simplest of these is neutralization, in which the antibody blocks certain sites on an antigen, making it ineffective. Antibodies neutralize a virus by attaching to the sites the virus must use to bind to its host cell. Similarly, coating a bacterial toxin with antibody effectively neutralizes it. Eventually, phagocytic cells dispose of the antigen-antibody complex. Another effector mechanism is the agglutination (clumping) of bacteria by antibodies. Agglutination is possible because each antibody molecule has at least two antigen-binding sites and can cross-link adjacent antigens. The bacterial clumps are easier for phagocytic cells to engulf than single bacteria. A similar mechanism is precipitation, the cross-linking of soluble antigen molecules (rather than cells) to form immobile precipitates that are captured by phagocytes.

One of the most important effector mechanisms of the humoral response is the activation of the complement system by antigen-antibody complexes. Recall from earlier in the chapter that complement is a group of proteins that acts cooperatively with elements of the nonspecific and specific defense systems. Antibodies often combine with complement proteins, activating the complement to produce lesions in a foreign cell's membrane that result in lysis (bursting) of the cell. We will examine this process in more detail later.

Monoclonal Antibodies

Until the late 1970s, the only source of antibodies for research or treatment of diseases was the blood of animals. This was expensive and inefficient. Furthermore, animals have antibodies against many different antigens circulating in their blood, making it difficult to isolate the needed ones. These problems were solved with the development of a new technique for making **monoclonal antibodies.** The term *monoclonal* means that all the cells producing such antibodies are descendants of a single cell; thus, they all produce identical antibody molecules. Monoclonal antibody technology makes it possible to produce commercial quantities of pure antibody relatively cheaply. Many new diagnostic methods for detecting pathogenic microbes in clinical samples depend on monoclonal antibodies. Monoclonal antibodies are also the basis of the over-the-counter pregnancy tests that detect a hormone excreted only in the urine of a pregnant woman.

Monoclonal antibodies will have increasing application as therapeutic agents—for example, to combat microbial toxins in the bloodstream. There is also the prospect of using monoclonal antibodies to treat cancer. The basic idea is to produce monoclonal antibodies against some patient's cancer cells and attach a toxin to the antibodies. The antibody-toxin complex, called an immunotoxin, acts as a "magic bullet" to selectively find and destroy the cancer cells.

The technology that makes monoclonal antibodies possible is the fusion of two cells to form a hybrid cell called a **hybridoma** (see the Methods Box, p. 864). The two types of cells from which a hybridoma is made have distinct abilities, neither of which can be exploited alone. One type of cell is derived from a cancerous cell known as a myeloma. These cancer cells, unlike normal cells, can be cultivated indefinitely. The other type of cell is a normal antibody-producing plasma cell, obtained from the spleen of an animal inoculated with the antigen of interest. This cell can be cultivated for only a few generations. Fusing the myeloma cell with the plasma cell to form a hybridoma combines key qualities of the two cells; the hybridoma produces a single type of antibody and can be cultivated indefinitely to manufacture that antibody on a large scale.

THE CELL-MEDIATED RESPONSE: A CLOSER LOOK

Recall that the body's specific response to antigens is a *dual* defense, consisting of humoral and cell-mediated immunity. Many pathogens, including all viruses, are obligate intracellular parasites that can reproduce only within the cells of the host. The humoral immune response helps the defense network identify and destroy free pathogens, but it is the cell-meditated arm of immunity that battles pathogens that have already entered cells. The key components of cell-mediated immunity are T lymphocytes, or T cells, the lymphocytes that complete their maturation in the thymus. And recall that mature T cells migrate from the thymus to lymphoid organs such as the lymph nodes and the spleen.

In contrast to B cells, T cells cannot be activated by free antigens present in body fluids. T cells respond only to antigenic epitopes displayed on the surfaces of the body's own cells. These bound antigens are recognized by T-cell receptors, specific proteins embedded in the plasma membranes of T cells. We have already seen one example, the binding of helper T cells to antigens displayed on the surfaces of antigen-presenting macrophages and B cells in the activation of humoral immunity (see Figure 39.12). Recall from that example that a T-cell receptor actually recognizes a combination of the antigen along with one of the body's own self (MHC) markers. It is neither the antigenic determinant alone nor the MHC protein that the T-cell receptor "sees," but rather a self-nonself complex of the two molecules together. The MHC protein that presents the antigen is shaped something like a hammock that nestles the antigen. An MHC molecule can associate with a variety of antigens. However, each combination of MHC and antigen forms a unique complex that can be recognized by specific T cells having matching receptors.

The MHC-antigen complex is like a red flag to T cells, calling them into action against cells infected by the pathogen represented by that particular antigen. The situation stimulates T cells that have the appropriate receptors to proliferate to form clones of effector cells specialized for fighting a particular pathogen. In fact, two main types of T cells—**helper T cells (T_H)** and **cytotoxic T cells (T_C)**—are called into action.

We have already examined one function of helper T cells: their role in activating B cells to secrete antibodies against T-dependent antigens during humoral responses. T_H cells also activate other types of T cells to mount cell-mediated responses to antigens.

The ability of helper T cells to stimulate other lymphocytes depends on chemical signals called cytokines. A **cytokine** is a molecule that is secreted by one cell as a regulator of neighboring cells (see Chapter 42). Binding of a helper T cell to an antigen-presenting macrophage causes that macrophage to release a cytokine called **interleukin-1**. This signals the T_H cell to release another cytokine known as **interleukin-2**. In an example of positive feedback (see Chapter 36), interleukin-2 stimulates T_H cells to grow and divide more rapidly, increasing the supply of both T_H cells and interleukin-2. The interleukin-2 and other cytokines released by this growing population of T_H cells also help activate B cells, thus stimulating the humoral response against a specific antigen. And cy-

METHODS: PRODUCTION OF MONOCLONAL ANTIBODIES WITH HYBRIDOMAS

The method described here makes it possible to produce large quantities of pure antibodies reactive against one antigenic determinant. The application illustrated in the figure is the preparation of monoclonal antibodies to detect human chorionic gonadotropin (HCG), a hormone present in the blood or urine of pregnant women. A mouse injected with an antigenic determinant characteristic of HCG will begin to produce antibodies against the antigen. The spleen can then be removed and its B lymphocytes isolated. The lymphocytes are mixed with myeloma (cancer) cells having a mutation that prevents them from surviving without a particular nutrient present in their growth medium. Some of the lymphocytes and myeloma cells fuse to become hybridomas. The hybridoma cells are identified when all cells are grown in a medium deficient in the nutrient needed by the mutant myeloma cells. Any unfused myeloma cells die, but the hybridoma cells live, because the lymphocyte DNA supplies the necessary nutrient gene and its product. Each hybridoma clone is tested for production of antibodies that react with the specific antigenic determinant. Clones that test positive are isolated and cultured for large-scale production of the antibody.

Monoclonal antibodies against HCG are now used for very sensitive pregnancy tests. A positive reaction between the antibodies and blood or urine shows the presence of the HCG antigen, indicating that the woman who supplied the sample is pregnant.

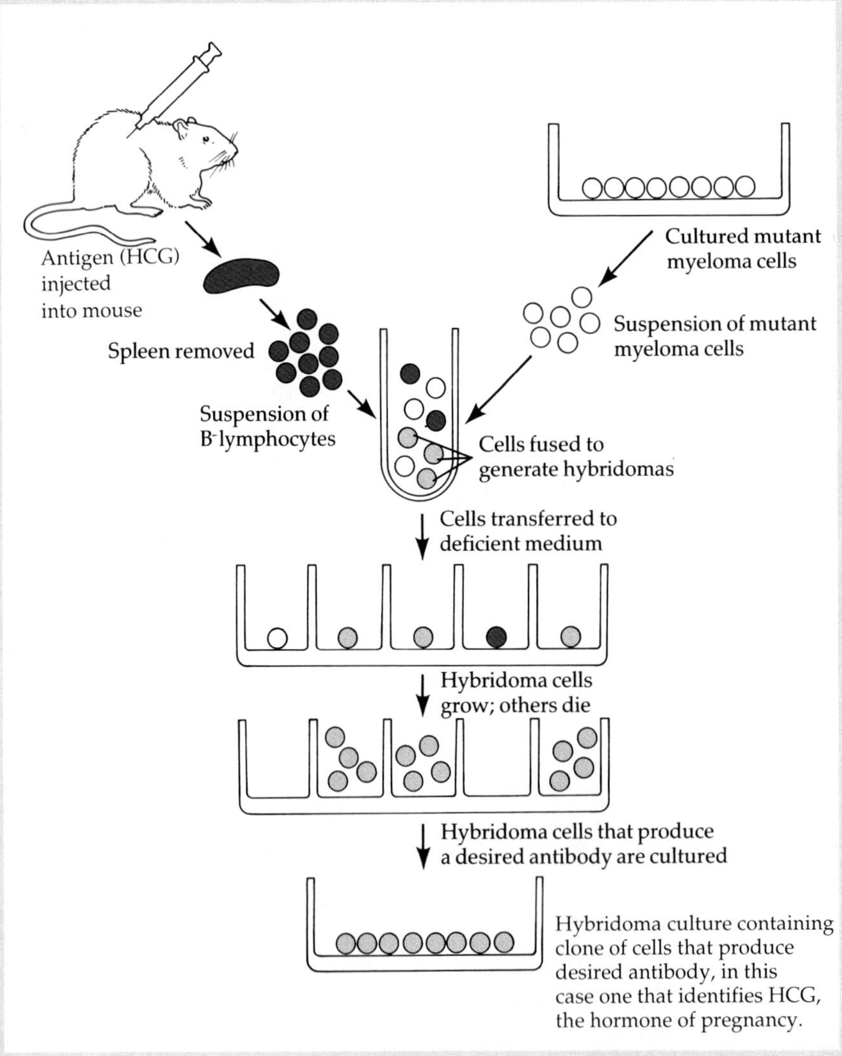

Antigen (HCG) injected into mouse
Cultured mutant myeloma cells
Suspension of mutant myeloma cells
Spleen removed
Suspension of B lymphocytes
Cells fused to generate hybridomas
Cells transferred to deficient medium
Hybridoma cells grow; others die
Hybridoma cells that produce a desired antibody are cultured
Hybridoma culture containing clone of cells that produce desired antibody, in this case one that identifies HCG, the hormone of pregnancy.

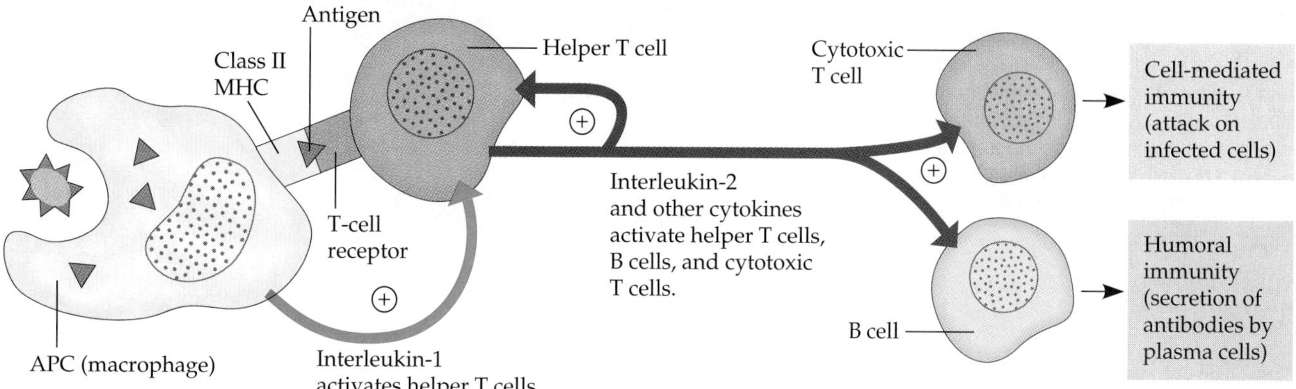

Figure 39.14
The central role of helper T cells. Helper T cells mobilize both humoral and cell-mediated arms of the immune response. The T-cell receptor of the helper recognizes the MHC-antigen complex displayed on the surface of an antigen-presenting cell, usually a macrophage. The macrophage secretes interleukin-1, a chemical signal that contributes to the activation of the helper T cells. An activated T cell grows and divides, producing a clone of helper T cells, all with receptors keyed to MHC markers in combination with the specific antigen that triggered the response. The helper cells discharge a second chemical signal, interleukin-2, which amplifies the cell-mediated response by stimulating proliferation and activity of other helper T cells, all specific for the same antigenic determinant. Interleukin-2 and other cytokines also help activate B cells, which function in humoral immunity, and cytotoxic T cells, which function in the cell-mediated immune response.

tokines from T_H cells help arm the cell-mediated response by stimulating another class of T lymphocytes to differentiate into the effector cells called cytotoxic T cells (Figure 39.14).

Cytotoxic T cells (T_C) are the only T cells that actually kill other cells (see Figure 39.1). Host cells infected by viruses or other intracellular pathogens display antigens complexed with class I MHC molecules, the self markers found on all nucleated cells of the body. Each cytotoxic T cell has a receptor that can recognize and bind to a specific antigen combined with the class I MHC marker. Notice an important difference in the functions of the receptors on T_H and T_C cells: The receptors on helper T cells enable those lymphocytes to stimulate immunity by binding specifically to antigen-presenting cells marked by *class II* MHC molecules, such as macrophages and B cells. In contrast, cytotoxic T cells, by recognizing specific antigens in association with *class I* MHC markers, can bind to any cell of the body infected by that particular antigenic invader. When a T_C cell docks on the surface of an infected cell, it releases *perforin,* a protein that forms an open lesion in the infected cell's membrane. The target cell loses cytoplasm through the lesion, leading to cell lysis (Figure 39.15). While this destroys the host cell, it also deprives the pathogen of a place to reproduce and exposes the pathogen to circulating antibodies. This cell-mediated action is an essential feature of the body's overall defense system, because antibodies cannot attack pathogens that have already invaded host cells. After destroying an infected cell, the T_C cell continues to live and can kill many other cells.

Cytotoxic T cells also function in defense against cancer. Cancer cells arise periodically in the body and, because they carry distinctive molecular markers not found on normal cells, are identified as nonself by the immune system. Cytotoxic T cells target the cancer cells and lyse them (Figure 39.15b). Cancers are most likely to become established in individuals with defective immune systems and elderly people with declining immunity.

Although T_C cells and natural killer cells attack target cells by similar mechanisms, there is an important difference between the roles of these cells in defense: Natural killer cells, part of the body's *nonspecific* defense, do not respond against specific antigens.

A third type of T lymphocyte, called a **suppressor T cell (T_S)** is not well understood. (Some immunologists believe that T_S cells are actually a type of T_H cell rather than a separate subclass of T cells.) T_S cells probably function in turning off the immune response when an antigen is no longer present (hence their name).

Figure 39.16 summarizes the humoral and cell-mediated responses, the two arms of the immune system's attack on foreign invaders. These specific responses overlap the body's nonspecific defenses. The proteins known as complement are partly responsible for this cooperation of defense mechanisms.

THE COMPLEMENT SYSTEM: A COMPONENT OF BOTH NONSPECIFIC AND SPECIFIC DEFENSE

The complement proteins interact with other specific and nonspecific defense mechanisms and are crucial to the operation of both systems. About 20 complement

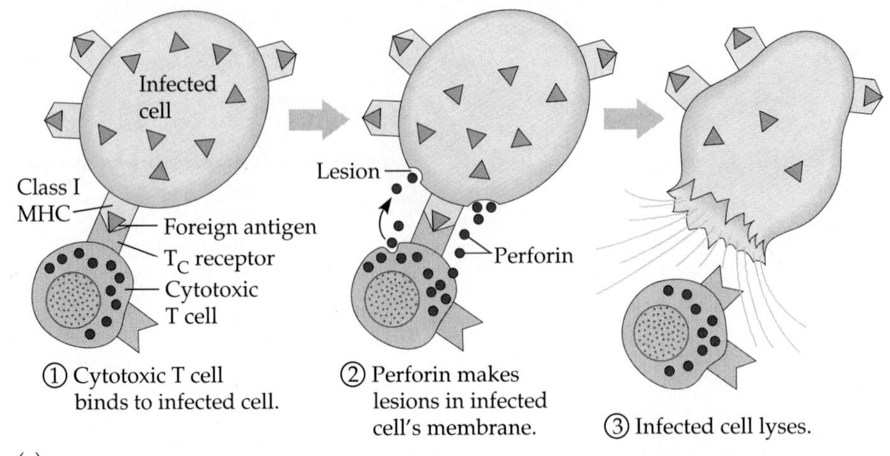

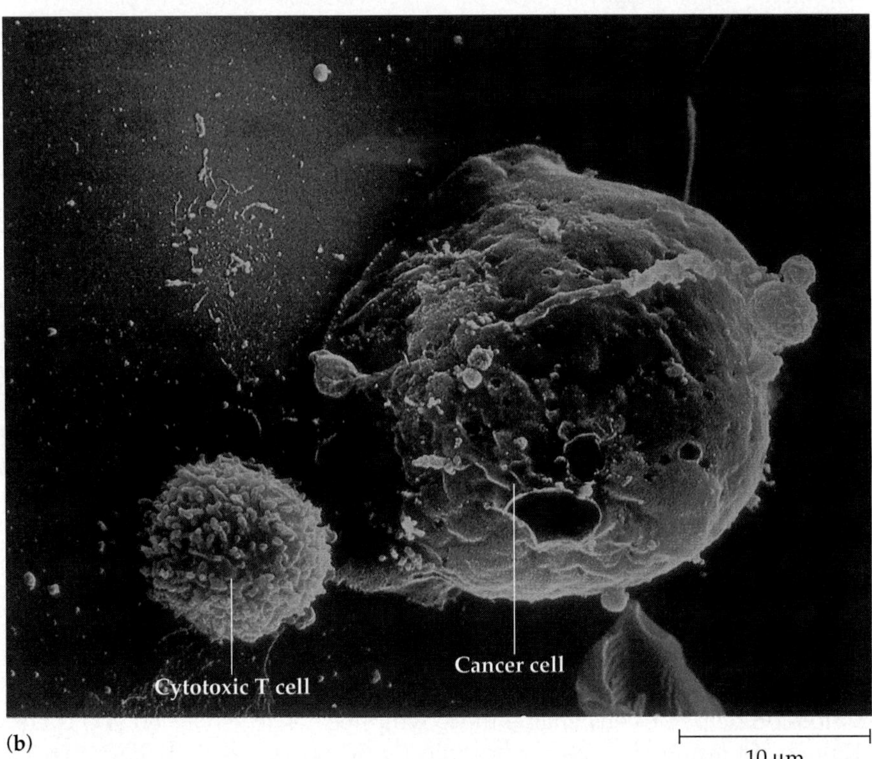

Figure 39.15
Cytotoxic T cells. These cells attack infected cells and cancer cells. (**a**) A cytotoxic T cell (T_C) docks on the surface of an infected cell and discharges a protein called perforin, which lyses the infected cell. (**b**) The T_C cell (smaller cell) has lysed a cancer cell (SEM). Also see Figure 39.1.

proteins circulate in the blood in inactive form. They are activated in cascade fashion, each one activating the next in the series.

Complement's activation pathway in the specific defense system was known first and is called the *classical pathway*. This pathway is initiated when antibodies of the immune system bind to a specific invader, such as a bacterial cell, targeting the cell for destruction. A complement protein then bridges the gap between two adjacent antibody molecules. This association of antibodies and complement activates complement proteins to form a *membrane attack complex* (Figure 39.17). The complex produces a small lesion in the pathogen's membrane, resulting in lysis of the cell. This is similar to the way T_C cells kill infected host cells, but the target cell in this case is the invader itself.

The *alternative pathway* of complement action does not require cooperation with antibodies, and the targets are therefore not specific. This alternative complement pathway is also able to generate a membrane attack complex. Substances found in many bacteria, yeasts, viruses, virus-infected cells, and protozoan parasites can activate complement to form such a complex without the help of antibodies. Another outcome of the alternative pathway is complement's contribution to inflammation. By binding to histamine-containing cells, complement proteins trigger the release of histamine, the chemical alarm to the body's defense systems that signals a local injury. Several complement proteins also attract phagocytes to the infected site.

Complement also cooperates with phagocytes in the destruction of pathogens. In a process called *opsoniza-*

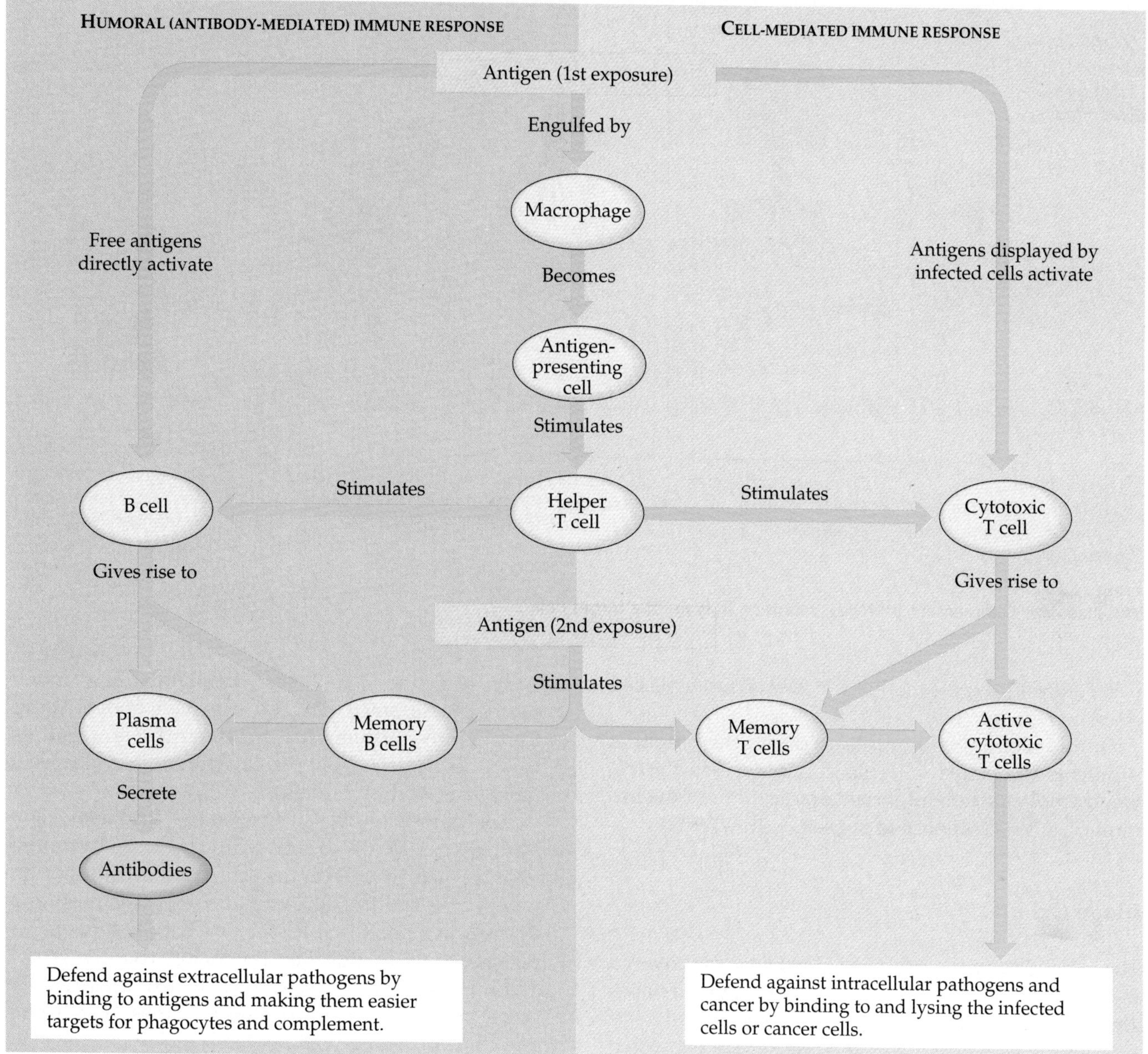

Figure 39.16
A summary of the immune responses. In this simplified flow chart, green arrows track the primary response, and blue arrows track the secondary response.

tion (L. "to relish"), complement proteins attach to foreign cells and stimulate phagocytes to ingest those cells. (Coating the pathogens with antibody molecules can have the same effect.) In another example of teamwork in the body's defense systems, complement, antibodies, and phagocytes function together in a phenomenon called *immune adherence.* A microbe coated with both antibodies and complement proteins adheres to surfaces such as the walls of blood vessels, making the pathogens easier prey for phagocytic cells circulating in the blood.

Now that we have examined how the immune system works, the next two sections will survey some important applications.

SELF VERSUS NONSELF: SOME APPLICATIONS

In addition to distinguishing between an animal's own cells and pathogens such as bacteria and viruses, the immune system wages war against cells from other individuals of the same species. For example, a skin graft from one person placed on another person (not an identical twin) will look healthy for a day or two, but after that it will be destroyed by the immune system. It is interesting that a mother does not reject a fetus as a foreign body. Apparently, the anatomical relationship between mother and fetus is the key to this acceptance

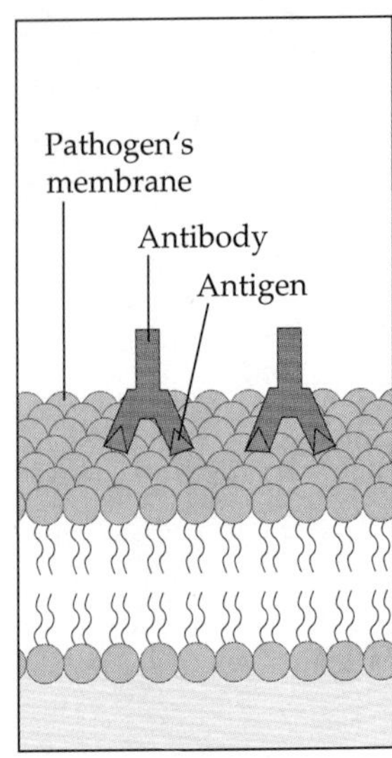

① Antibody molecules attach to antigens on pathogen's plasma membrane.

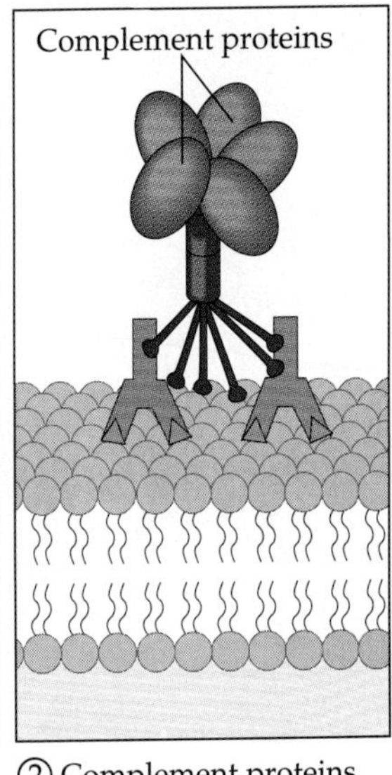

② Complement proteins attach to pair of antibodies.

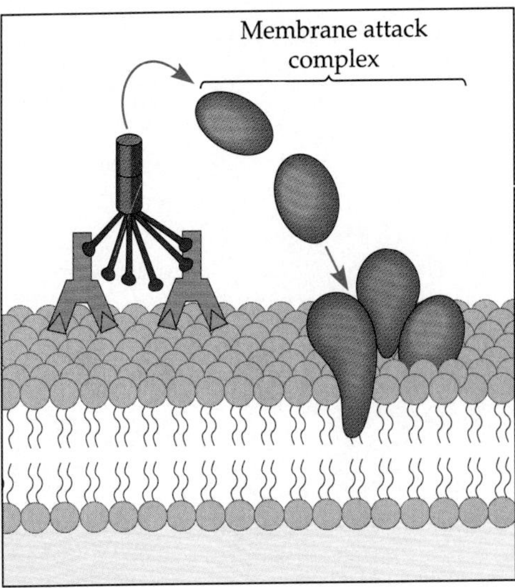

③ Activated complement proteins attach to pathogen's membrane in step-by-step sequence, forming a membrane attack complex.

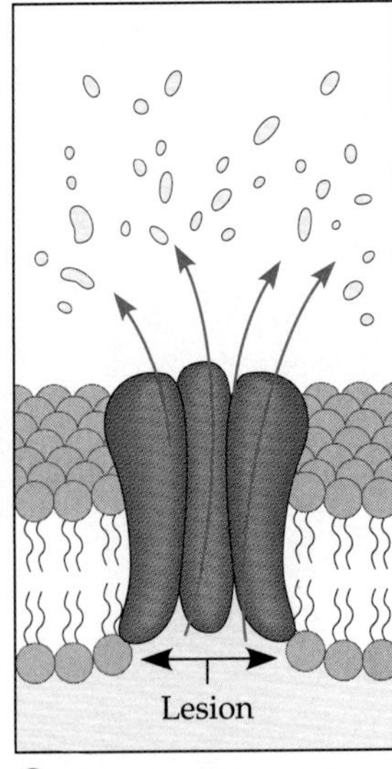

④ Complement proteins lyse target membrane, resulting in lesion and death of pathogen.

Figure 39.17
The classical complement pathway, resulting in lysis of a target cell.

(see Figure 42.18); the mother rejects fetal tissue placed elsewhere in her body.

In this section, you will learn about the potential problems associated with blood transfusions and organ transplants, two important applications of the immune system's response to nonself.

Blood Groups

The genetic basis for **ABO blood groups** was covered in Chapter 13. An individual with type A blood has A antigens on the surface of his or her red blood cells. The A molecule is referred to as an antigen here because it may be identified as foreign if placed in the body of another person; the type A antigen is *not* antigenic to its owner. And an individual belonging to blood group A will not produce antibodies against A antigens (see Figure 13.12). However, a person with type B blood will produce antibodies against A antigens (anti-A antibodies). If type A blood is introduced into the bloodstream of a type B person, the transfused red blood cells will stimulate the production of anti-A antibodies, and the transfused cells will be destroyed by complement-mediated lysis (see Figure 39.17). The reverse situation, transfusion of type B cells into a type A person, would have a similarly disastrous result. Type O individuals are known as universal donors because their blood cells carry neither A nor B antigens. Type O blood cells will not cause an individual to produce anti-A or anti-B antibodies. Type O individuals can only receive type O blood, however; since A and B antigens are foreign to them, they would make both anti-A and anti-B antibodies. Type AB individuals are universal recipients; since their red cells have both antigens, they make neither anti-A nor anti-B antibodies. However, type AB people can donate blood only to other type AB individuals.

Another red blood cell antigen, the **Rh factor,** is notorious in cases where antibodies produced by a pregnant woman react with the red blood cells of her developing fetus. The situation arises if the mother is Rh-negative (lacks the Rh factor), but the fetus is Rh-positive, having inherited the factor from the father. The mother develops antibodies against the Rh factor when small amounts of fetal blood cross the placenta and come in contact with her lymphocytes, usually late in pregnancy or during delivery of the baby. Typically, the mother's response to this first exposure is mild and without medical consequences for the baby. The real danger occurs in subsequent pregnancies, when the mother's immune response against the Rh factor has already been primed and her antibodies can cross the placenta to destroy the red blood cells of an Rh-positive fetus. To prevent this, the mother may be injected with anti-Rh antibodies after delivering her first Rh-positive baby. The antibodies destroy the Rh-positive red cells that have entered the mother's circulation before her immune system has been stimulated by the Rh antigen.

Tissue Grafts and Organ Transplants

The major histocompatibility complex (MHC), the molecular fingerprint unique to each self, is responsible for the rejection of tissue grafts and organ transplants. Foreign MHC molecules are antigenic, causing cytotoxic T cells to mount a cell-mediated response against the donated tissue or organ. To minimize this rejection, attempts are made to match the MHC of the organ donor and recipient as closely as possible. In the absence of an identical twin, siblings or parents usually provide the closest match. Various drugs are necessary to suppress the immune response when most transplants are performed. The complication with this strategy is that the organ recipient is more vulnerable to infection when drugs compromise the immune system. Such drugs as cyclosporine and FK 506 offer the advantage of suppressing only cell-mediated immunity without crippling humoral immunity.

Notice that the body's disastrous reaction to an incompatible blood transfusion or a transplanted organ is not a disorder of the immune system, but a normal action taken by a healthy immune system exposed to foreign antigens. However, as with any complex system, the immune apparatus does malfunction in some individuals, resulting in a variety of serious disorders.

DISORDERS OF THE IMMUNE SYSTEM

Autoimmune Diseases

Sometimes, for reasons immunologists are only beginning to understand, the immune system goes awry and turns against self, leading to a variety of **autoimmune diseases.** In systemic lupus erythematosus, for example, people develop immune reactions against components of their own cells, especially nucleic acids released by the normal breakdown of skin and other tissues. Rheumatoid arthritis is a crippling autoimmune disease in which inflammation damages the cartilage and bone of joints. Insulin-dependent diabetes seems to be caused by the destruction of insulin-producing cells of the pancreas by an autoimmune reaction. Rheumatic fever is an autoimmune condition that was once a major killer of young adults and is now making a reappearance in certain areas. Antibodies produced in response to streptococcal infections (such as strep throat) also react with heart muscle tissue in some people, damaging the heart valves. Repeated episodes of streptococcal infection result in more antibodies and more heart damage. In Grave's disease, antibodies attach to receptors on the thyroid gland that normally react with a thyroid-stimulating hormone produced by the pituitary. The abnormal binding of antibodies to these receptors stimulates the thyroid gland to produce thyroid hormones in excessive amounts. People suffering from Grave's disease often show bulging eyes and a greatly enlarged thyroid gland.

Allergy

Allergies are hypersensitivities of the body's defense system to certain environmental antigens called *allergens.* One hypothetical explanation for allergies is that they are evolutionary remnants of the immune system's response to parasitic worms. Infestations by such worms may have been a much more common problem in the past. The defense mechanism that combats worms is similar to the allergic response that causes such disorders as hay fever or asthma.

The most common allergies involve antibodies of the IgE class (see Table 39.1). For example, hay fever and other allergies caused by pollen occur when IgE molecules recognize these allergens. The IgE antibodies attach by their stems to mast cells, noncirculating cells found in connective tissue. In this way, a susceptible person becomes sensitized to the specific pollen antigen. Thereafter, whenever a pollen grain bridges the space between two adjacent IgE monomers, the mast cell responds with a rapid reaction called degranulation, releasing histamine and other inflammatory agents (Figure 39.18). Recall that histamine causes dilation and increased permeability of small blood vessels. In the case of an allergy, response to these histamines includes such symptoms as sneezing, a runny nose, and smooth muscle contractions that often result in breathing difficulty. Antihistamines are drugs that interfere with the action of histamine.

The most serious type of acute allergic response is anaphylactic shock, a life-threatening reaction to injected or ingested antigens. Hypersensitivity to wasp or bee stings is an example. Anaphylactic shock occurs when mast cell degranulation triggers an abrupt dilation of peripheral blood vessels and causes a precipitous drop in blood pressure. Death may occur within a few minutes. People who are extremely allergic to certain foods, like peanuts or fish, have died from eating tiny amounts of these foods. Some individuals with such severe hypersensitivities carry syringes containing the hormone epinephrine, which counteracts the allergic response.

Immunodeficiency

Certain individuals are inherently deficient in either humoral or cell-mediated immune defenses. *Both* branches of the immune system fail to function in a congenital disorder known as *severe combined immunodeficiency.* For people with this genetic disease, the only chance for long-term survival is a successful bone marrow transplant that will continue to supply functional lymphocytes. One risk of such transplants is that the transplanted cells will identify their new hosts as foreign and mount an attack against the marrow recipient (graft versus host disease).

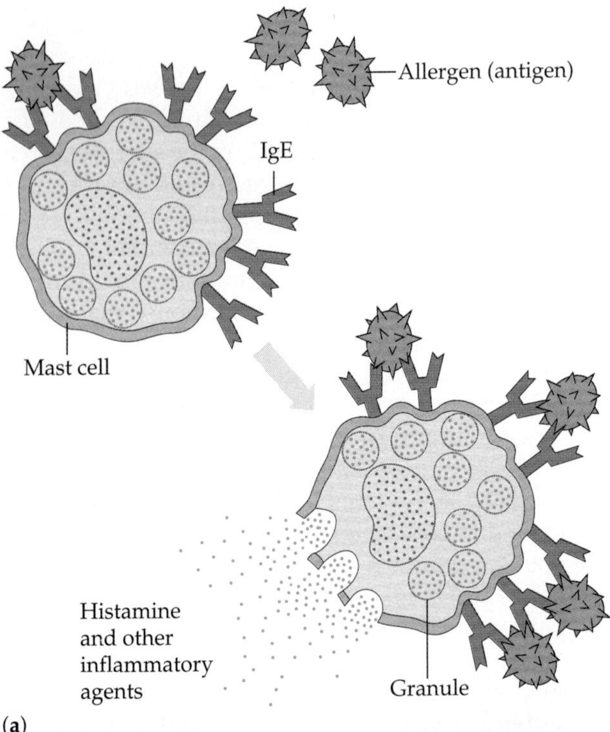

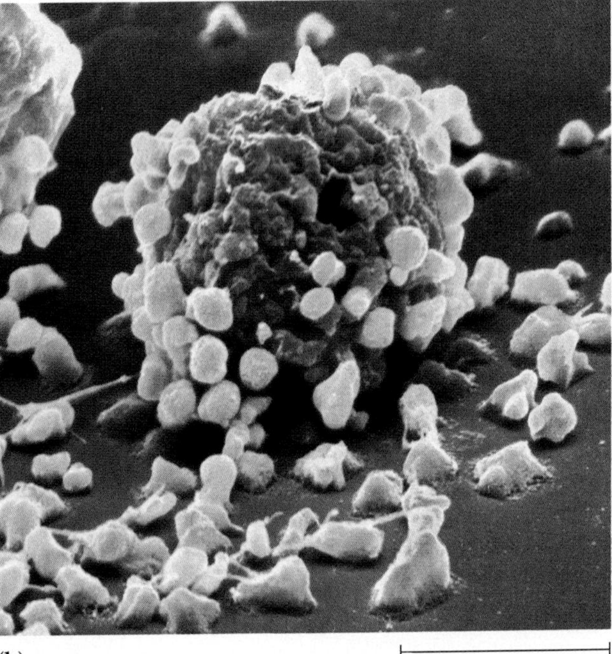

Figure 39.18
Mast cells and the allergic response. (**a**) Mast cells are coated with IgE molecules attached by their stem regions. When an antigen, such as plant pollen, bridges the gap between two of these antibodies, it triggers the degranulation of the cell. Degranulation releases histamine, leading to most of the symptoms of the allergy. (**b**) This micrograph illustrates degranulation of a mast cell (colorized SEM).

Immunodeficiency is not always an inborn condition. Certain cancers depress the immune system, especially Hodgkin's disease, which damages the lymphatic system and makes the patient susceptible to many infections. The viral disease AIDS, discussed in the next section, causes very severe depression of the immune system. Immunodeficiency also results from deliberate suppression of the immune system with drugs in order to minimize rejection of transplanted organs.

Nearly 2000 years ago, the Greek physician Galen recorded that people suffering from depression were more likely than others to develop cancer. At one time or another, most of us have heard the admonition, "Don't get run down. It'll lower your resistance and you'll get sick." Indeed, there is growing evidence that physical and emotional stress can compromise immunity. (See the interview with Candace Pert that precedes Unit One.) Hormones secreted by the adrenal glands during stress affect the numbers of white blood cells and may suppress the immune system in other ways. In one study, students were examined just after a vacation and again during final exams. Their defense systems were impaired in various ways during exam week. For example, interferon levels were lower.

Evidence suggests that there is a direct link between the nervous system and the immune system. A network of nerve fibers penetrates deep into lymphoid tissue, including the thymus. Also, receptors for chemical signals secreted by nerve cells have been discovered on the surfaces of lymphocytes. Some signals secreted by nerve cells when we are relaxed and happy may actually enhance immunity. These and other observations (and speculations) have led physiologists to take a serious look at how general health and state of mind affect immunity.

Acquired Immunodeficiency Syndrome (AIDS)

In 1981, health-care workers in the United States became aware of an increased frequency of Kaposi's sarcoma, a cancer of the skin and blood vessels, and of *Pneumocystis* pneumonia, a respiratory infection caused by a protozoan. Both of these diseases are extremely rare in the general population but occur more frequently in severely immunosuppressed individuals. This observation led to the recognition of an immune system disorder that was named **acquired immunodeficiency syndrome,** or **AIDS.** By 1983, virologists working in the United States and France had identified a causative agent, a virus now known as the **human immunodeficiency virus,** or **HIV.** With a mortality rate that apparently approaches 100%, this may be the most lethal pathogen ever encountered. HIV probably arose by evolution from another virus in central Africa and may have caused unrecognized

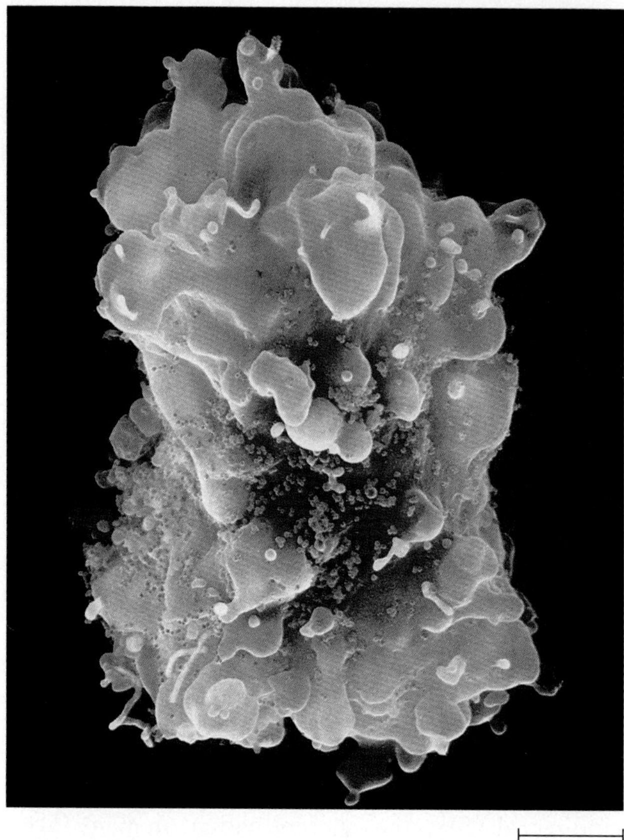

Figure 39.19
A T cell infected with HIV. The viruses (blue) bud continuously from the surface of the T cell (orange; colorized SEM). The cell will die, but only after it produces many copies of its killer.

cases of AIDS there for many years. The virus has been identified in preserved blood samples from as early as 1959 in African nations and in England.

HIV infects certain T cells, including helper T cells, which carry a receptor called CD4 on their surface. Glycoproteins on the HIV envelope bind specifically to this receptor (see Chapter 17). Other cells that carry CD4, including some macrophages and a few subclasses of B cells, may also be infected, as can a few types of tissue cells lacking this receptor. Following attachment, the virus enters the cell and begins to replicate. Newly formed viruses bud from the host cell, circulate, and infect other cells (Figure 39.19). The infected cells may produce new viruses for an extended time or may be killed quickly, either by the virus or by the response of the immune system. HIV may also remain latent for many years as a provirus assimilated into the genome of an infected cell. The provirus is invisible to the immune system.

The ability of HIV to remain latent is one reason why anti-HIV antibodies fail to eradicate the disease. Probably more important, however, are the extremely rapid mutational changes in antigens the virus undergoes during the infection. Indeed, every HIV probably differs in at least one small way from its parent. The immune system responds effectively against HIV infection at first, but it is eventually overwhelmed by the accumulation of more resistant variants. In Figure 39.20, notice that the number of viruses gradually increases as the helper T-cell population (and hence the

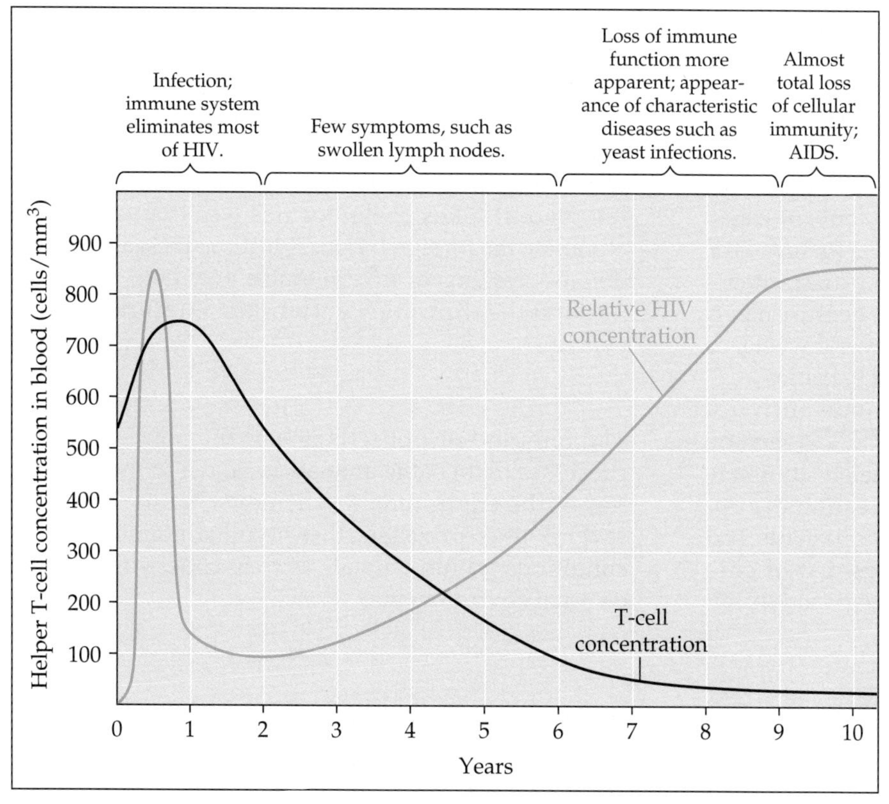

Figure 39.20
The stages of HIV infection. HIV concentration increases rapidly after the initial infection but is then almost eliminated by the immune response. However, some viruses survive to increase slowly over time, while the helper T-cell concentration decreases. AIDS is the last stage of the process.

body's protection) decreases. When the damage to the immune system reaches a certain point, cell-mediated immunity collapses, and diseases such as Kaposi's sarcoma and *Pneumocystis* pneumonia can become established. AIDS is the name given to this late stage of HIV infection. It is defined by a specified reduction of T cells and the appearance of the characteristic secondary infections.

The time required for an HIV infection to progress to AIDS varies, but it averages about ten years. During most of this time, the patient exhibits only moderate hints of illness, such as swollen lymph glands and, later, yeast infections of the mucous membranes. Infants infected by their mothers in utero often progress to AIDS much more rapidly. Those who have been exposed to HIV have circulating antibodies against the virus, and detection of these antibodies is the most common method for identifying infected individuals, who are said to be **HIV-positive.** This is the test used to screen blood donations, for example.

Transmission of HIV requires the transfer of body fluids, such as blood or semen, containing infected cells. Unprotected sex (without a condom) among male homosexuals and transmission between intravenous drug users via unsterilized needles account for most of the AIDS cases reported thus far in the United States and Europe. However, transmission of AIDS among heterosexuals is rapidly increasing as a result of unprotected sex with infected partners. In Africa and Asia transmission has been primarily by heterosexual sex, especially where there is a high incidence of other sexually transmitted diseases that result in genital lesions. These sores facilitate transmission of HIV, as the skin barrier (first line of defense) is breached and HIV-susceptible cells, such as macrophages and T cells, are attracted to the area by the inflammatory response.

HIV is not transmitted by casual contact or even kissing. Breast milk, however, has transmitted the disease from mothers to nursing infants. Blood transfusions have been almost eliminated as a route of transmission in developed countries by the test for anti-HIV antibodies. This type of test will never guarantee a completely safe blood supply, however, because a person may be infected by the virus for several weeks or months before the antibodies become detectable.

At this time, AIDS is incurable. Some antiviral drugs, such as AZT, ddC, and ddI, may extend the lives of patients, but they do not completely eliminate the virus. Many drugs are useful for treating the opportunistic infections common in AIDS patients and can prolong life, but again, these drugs do not cure AIDS. Research to develop a vaccine is very active, but the antigenic variability of the virus poses a major problem. For the present, the best approach for slowing the spread of AIDS seems to be to educate people about the practices that transmit HIV, such as unprotected sex and sharing needles. Condoms reduce the risk of transmitting the virus but do not completely eliminate it. *Anyone* who has sex—vaginal, oral, or anal—with a partner who may have had sex with another individual during the past 15 years risks exposure to HIV.

IMMUNITY IN INVERTEBRATES

This chapter has emphasized the nonspecific and specific defenses as they occur in vertebrates, because relatively little is known about how invertebrates react against pathogens that have penetrated the skin and other outer barriers. However, experiments have established that one fundamental facet of defense, the ability to distinguish self from nonself, is well developed in invertebrates. For example, if the cells of two sponges of the same species are mixed, the cells from each individual sort themselves out and aggregate, excluding cells from the other individual. In many invertebrates, amoeboid cells called coelomocytes identify and destroy foreign materials.

Studies of tissue grafting in earthworms have shown a memory response in the defense systems of these annelids. When a portion of body wall is grafted from one worm to another, the recipient's coelomocytes attack the foreign tissue. If donor and recipient are from the same population, then the grafted tissue survives for about eight months before it is completely rejected. However, if a worm receives a graft from a donor taken from a distant location (different population), the graft is rejected in just two weeks. If a second graft from that same donor to the same recipient is attempted, it takes coelomocytes less than a week to eliminate the foreign tissue. Additional research on the defense systems of invertebrates may help biologists understand how the vertebrate immune system evolved.

* * *

The immune response is one of the many adaptive processes that enable animals to adjust to the adversities of the environment. The next chapter describes several other processes that help maintain favorable conditions within animals as they cope with varying external environments.

STUDY OUTLINE

1. The immune system consists of three lines of defense that protect the vertebrate body. The first two lines are nonspecific. The third line, the immune system, responds specifically to each type of foreign invader.

Nonspecific Defense Mechanisms (pp. 850–854)

1. The first line of defense is the intact skin and mucous membranes. This defense includes the physical barrier formed by these tissues, as well as the microbe-attacking enzyme called lysozyme in body secretions; mucus that traps particles; ciliated cells lining the upper respiratory system; and gastric juices that kill microbes passing through the stomach.
2. The second line of defense is primarily dependent upon neutrophils and macrophages, phagocytic white cells in the blood and lymph. Natural killer cells also participate in nonspecific defense.
3. The most important antimicrobial proteins are interferons and complement. Interferons are proteins secreted by virus-infected cells that inhibit neighboring cells from making new viruses. Complement is a group of proteins involved in nonspecific and specific defense.
4. A local inflammatory response is triggered by tissue damage. Injured cells release histamine, a chemical signal that dilates blood vessels and increases capillary permeability, allowing large numbers of phagocytic white blood cells to enter the interstitial fluid.
5. Fever is part of the body's systemic inflammatory response.

The Immune System and Specific Defense: Some Basic Concepts (pp. 854–860)

1. The immune system of the body recognizes foreign microbes, toxins, or transplanted tissues as not belonging to itself and develops an immune response to inactivate or destroy the specific type of invader.
2. The immune system is distinguished from nonspecific defense by four features: specificity, diversity of response, self/nonself recognition, and memory of foreign invaders encountered before.
3. A foreign substance that elicits an immune response is called an antigen. The immune system responds to antigens by producing specialized lymphocytes and specific proteins called antibodies.
4. Immunity is the result of the immune system's enhanced response to a previously encountered pathogen. Active immunity can be acquired by exposure to an actual disease or to a vaccine that simulates a disease. Passive immunity can be acquired by administering antibodies formed in others, or it can be passed from mother to child via the placenta and milk.
5. The immune system has two main parts. Humoral immunity is based on circulation of antibodies in the blood and lymph, and defends against free viruses, bacteria, and other extracellular threats. Cell-mediated immunity defends against intracellular pathogens by destroying infected cells. Cell-mediated immunity also reacts against transplanted tissue and cancer cells.
6. Lymphocytes, which develop in bone marrow, are the main cells of the immune system. B lymphocytes mature in the bone marrow and function in humoral immunity. T lymphocytes mature in the thymus and function mainly in cell-mediated immunity. Mature B cells and T cells are most concentrated in the lymph nodes and other lymphatic organs.
7. Specific binding of antigens to receptors on lymphocytes triggers an immune response, in which the lymphocytes give rise to populations of infection-fighting effector cells.
8. Most antigens are proteins or large polysaccharides; they may have many epitopes (antigenic determinants) on their surfaces.
9. Antibodies constitute a class of proteins called immunoglobulins (Ig). The variable region of the Ig molecule is different for each type of antibody. Each antibody has at least two identical binding sites, formed by the variable region, that attach specifically to a single antigenic determinant.
10. The constant region of the Ig molecule is the same for all antibodies of the same class. There are five major Ig classes: IgG, IgM, IgA, IgD, and IgE.
11. Clonal selection occurs when a lymphocyte is activated by the binding of an antigen to its specific antigen receptor and proliferates to produce a clone of effector cells, all specific for that particular antigen. An activated T cell produces effector T cells, including cytotoxic T cells; an activated B cell produces plasma cells (antibody-producing cells).
12. Proliferation of a specific lymphocyte clone (the primary immune response) also produces the long-lived memory cells responsible for the enhanced response (secondary immune response) to future exposure to the antigen. The immune system's recognition of a previously encountered antigen is called immunological memory.
13. Self-tolerance develops in the embryo, when lymphocytes bearing receptors for the embryo's native molecules are destroyed. Self cells are marked by molecules of the major histocompatibility complex (MHC).

The Humoral Response: A Closer Look (pp. 860–863)

1. In humoral immunity, the production of plasma cells by a B cell often depends on the cooperation of macrophages functioning as antigen-presenting cells (APC) and specialized T cells called helper T cells. The APC displays antigenic fragments along with class II MHC markers. The helper T cell functions as a liaison between the APC and a B cell, which is activated in the process. The B cell then differentiates into plasma cells, which produce enormous numbers of specific antibodies.
2. An antibody does not usually destroy an antigen directly but targets it for elimination by complement and phagocytes.
3. Monoclonal antibody production is a method for making pure antibodies on an industrial scale for use in diagnostic tests, treatment, and research.

The Cell-Mediated Response: A Closer Look (pp. 863–865)

1. Cell-mediated immunity is mainly a function of two types of T cells, helper T cells (T_H) and cytotoxic T cells (T_C).
2. A T cell is activated when its receptor binds to a specific MHC-antigen complex.
3. Cell secretions called cytokines enable T_H cells to stimulate both humoral and cell-mediated immunity. The cytokine interleukin-1 is secreted by APCs, stimulating helper T cells to produce interleukin-2, which stimulates the proliferation of T_H cells, activation of B cells, and differentiation of cytotoxic T cells.
4. Cytotoxic T cells kill infected or cancerous cells with perforin, which opens a lesion in the target cell. The T_C cells recognize their targets by binding to a class I MHC marker complexed with a specific antigen.
5. Suppressor T cells (T_S) may turn off the immune response when antigen is no longer present.

The Complement System: A Component of Both Nonspecific and Specific Defense (pp. 865–867)

1. Complement, a group of blood proteins, can lyse a target cell by combining with antibodies (the classical pathway). In the alternative pathway, complement functions as an antimicrobial defense without the help of antibodies.

Self Versus Nonself: Some Applications (pp. 867–869)

1. The antigens on red blood cells determine whether a person has A, B, AB, or O type blood. Transfusion of incompatible blood stimulates the production of antibodies against the foreign antigens, resulting in complement-mediated lysis of the transfused cells. The Rh factor, another red blood cell antigen, may create difficulties when an Rh-negative mother carries successive Rh-positive fetuses.
2. Since the immune system rejects transplanted tissue and organs as nonself, immunosuppressive drugs are usually employed after a transplant.

Disorders of the Immune System (pp. 869–872)

1. Sometimes the immune system goes awry and turns against self, leading to autoimmune diseases such as rheumatoid arthritis and insulin-dependent diabetes.
2. Allergies, such as hay fever, cause release of histamine from mast cells. This release (degranulation) is triggered by an allergen, such as pollen, bridging the gap between two antibody molecules attached to the mast cell.
3. Some people are naturally deficient in humoral or cell-mediated immune defenses, or both. Some cancers cause immunodeficiency, as does deliberate suppression of the immune system with drugs to minimize transplant rejection.
4. Acquired immunodeficiency syndrome (AIDS) is caused by destruction of T cells and other cells by HIV, the human immunodeficiency virus, over a period of years. AIDS, the final stage of this process, is marked by a low level of T cells and certain diseases characteristic of a deficient cell-mediated immune response.

Immunity in Invertebrates (p. 872)

1. Invertebrates have the ability to distinguish between self and nonself. Amoeboid cells called coelomocytes can identify and destroy foreign substances. Experiments in earthworms show that their defense systems can both reject and form a memory response to tissue grafts.

SELF-QUIZ

1. Which of the following molecules is *incorrectly* paired with its source?
   a. lysozyme—saliva
   b. histamine—injured cells
   c. interferons—virus-infected cells
   d. immunoglobulins—neutrophils
   e. interleukin-2—helper T cell
2. Which of the following is *not* characteristic of the early stages of a localized inflammatory response?
   a. increased permeability of capillaries
   b. attack by cytotoxic T cells
   c. release of clotting proteins
   d. release of histamine
   e. dilation of blood vessels
3. The major difference between humoral immunity and cell-mediated immunity is that
   a. humoral immunity is nonspecific, whereas cell-mediated immunity is specific for particular antigens
   b. only humoral immunity is a function of lymphocytes
   c. humoral immunity cannot function independently; it is always activated by cell-mediated immunity
   d. humoral immunity acts against free-floating antigens, whereas cell-mediated immunity works against pathogens that have entered body cells
   e. only humoral immunity displays immunological memory
4. Monoclonal antibodies are
   a. produced by clones formed from memory cells
   b. used to produce large quantities of interferon
   c. produced by cultures of hybridoma cells
   d. produced by clones of T cells fused with tumor cells
   e. produced by recombinant DNA methods
5. Epitopes (antigenic determinants) bind to which portions of an antibody?
   a. variable regions
   b. constant regions
   c. only light chains
   d. only heavy chains
   e. the stem region
6. Which of the following molecules is *incorrectly* paired with its action?
   a. interleukin-1—stimulates division of helper T cells
   b. interleukin-2—increases proliferations of helper T and cytotoxic T cells
   c. interferon—helps neighboring cells resist viral infection
   d. histamine—fights allergic reactions
   e. lysozyme—attacks bacterial cell walls
7. Which of the following cells is *incorrectly* paired with its function?
   a. plasma cell—produces antibodies

b. helper T cell—lyses foreign cells
c. memory cell—rapidly proliferates into clones of effector cells when it encounters antigen
d. macrophage—engulfs bacteria and viruses
e. cytotoxic T cell—releases perforin that lyses infected cells

8. Which blood transfusion would agglutinate blood?
   a. A donor ⟶ A recipient
   b. A donor ⟶ O recipient
   c. A donor ⟶ AB recipient
   d. O donor ⟶ A recipient
   e. O donor ⟶ AB recipient

9. HIV compromises the immune system mainly by infecting
   a. cytotoxic T cells
   b. helper T cells
   c. suppressor T cells
   d. plasma cells
   e. B cells

10. The body produces antibodies complementary to foreign antigens. The process by which the body comes up with the correct antibodies to fight a given disease is most like which of the following?
    a. going to a tailor and having a suit made to fit you
    b. ordering the lunch special at a restaurant without looking at the menu
    c. going to a shoe store and trying on shoes until you find a pair that fits
    d. picking the first video that you haven't already seen
    e. selecting a lottery winner by means of a random drawing

CHALLENGE QUESTIONS

1. Assuming that immunologic memory is intact, how can you explain people getting colds or flu year after year?
2. Explain why the passive immunity conferred from mother to child is only temporary.

SCIENCE, TECHNOLOGY, AND SOCIETY

1. Concern is increasing over the rising rate of HIV infection among teenagers. In San Francisco, for example, the teenage incidence of HIV is estimated to be doubling every 16 to 18 months. Schools in some large cities have instituted programs to make condoms available to students, along with advice and counseling about safer sex. These plans have divided school boards and communities. Some citizens and church groups are opposed to giving condoms to students, because it might appear to encourage their sexual activity. Many school and public health officials view the situation differently. New York City Schools Chancellor Joseph Fernandez says, "This is not an issue of morality. It is a matter of life and death." All agree that the spread of AIDS is a serious problem. The heart of the controversy seems to be whether the schools should take such a direct role in this area of student life. What do you think the schools' role should be?
2. AIDS activists charge that the U.S. government was slow to respond to the AIDS crisis because the disease was perceived to affect mostly homosexuals and drug users, not the mainstream population. Do you think enough is being done to confront AIDS? If not, what else could be done?
3. New immunosuppressant drugs have allowed researchers to attempt transplants of tissues and organs from animals to humans. In 1992, the first baboon-to-human liver transplant was performed. The patient was a 35-year-old man whose liver had been destroyed by hepatitis B. A human transplant was not possible because the virus would have attacked the new liver. Scientists believe that the human hepatitis B virus does not infect baboons. The hospital where the operation took place was picketed by animal rights protesters opposed to such transplants, carrying signs saying "Baboons and Humans: Both Victims" and "Animals Are Not Ours to Experiment On," and chanting "Animals are not spare parts!" Do you think sacrificing the life of an animal is justified in this situation? Why or why not? How do you feel about routine use of animals as sources of organs for humans? What other scientific and medical uses of animals are justified or unjustified, in your opinion? Why?

FURTHER READING

Anderson, R., and R. May. "Understanding the AIDS Pandemic." *Scientific American,* May 1992.

Bower, B. "Questions of Mind Over Immunity." *Science News,* April 6, 1991. The link between psychology and immune function.

Johnson, H., J. Russell, and C. Pontzer. "Superantigens in Human Disease." *Scientific American,* April 1992. What causes anaphylactic shock?

Nowak, R. "Diversity Equals Death." *Discover,* May 1992. How the rapid evolution of HIV contributes to AIDS.

Oliwenstein, L. "The Bug That Can Say No." *Discover,* April 1992. Immunity in cockroaches is providing clues about the evolution of immune systems.

Robbins, A., and P. Freeman. "Obstacles to Developing Vaccines for the Third World." *Scientific American,* November 1988. Who will pay for immunizing the children of developing countries?

Tonegawa, S. "The Molecules of the Immune System." *Scientific American,* October 1985. An excellent overview of the molecules and battle plans involved in immunological defense.

Young, J. D.-E., and Z. A. Cohn. "How Killer Cells Kill." *Scientific American,* January 1988.

40 CONTROLLING THE INTERNAL ENVIRONMENT

OSMOREGULATION
EXCRETORY SYSTEMS OF INVERTEBRATES
THE VERTEBRATE KIDNEY
NITROGENOUS WASTES
THE REGULATION OF BODY TEMPERATURE
THE INTERACTION OF REGULATORY SYSTEMS

Figure 40.1
Survival in a changing environment. Brine shrimp *(Artemia salina)* are abundant in evaporation ponds and salt lakes (the habitat illustrated here is the Dead Sea in Israel). These tiny crustaceans can tolerate changes in water salinity to a level ten times saltier than seawater, surviving in what is essentially a saturated salt solution. As the salinity of the external environment changes, a brine shrimp is able to maintain its internal salt concentration at a relatively stable level. In this chapter, you will learn about some of the mechanisms of homeostasis that control the internal environment of animals.

Most animals can survive fluctuations in the external environment more extreme than any of their individual cells could tolerate. Brine shrimp, which live in salt lakes and evaporation ponds, withstand changes in external salinity that would inflict osmotic shock if such changes were experienced by cells within the body of the animal (Figure 40.1). A goldfish can endure water as acidic as pH 3 or as alkaline as pH 10 for an hour or more, but the cells within the fish will die if their internal pH drifts only slightly. During the course of a day, a human may be exposed to substantial changes in ambient temperature but will die if *internal* body temperature fluctuates more than a few degrees about a mean of 37°C. In all these examples, the animals survive changes in their external environment by maintaining their internal environment within ranges that can be tolerated by their cells, a condition known as homeostasis (see Chapter 36). Homeostasis relates directly to one of the major themes in our study of animals: the ability of organisms to cope with environmental change, over the short term by physiological compensation and over the long term by adaptation based on natural selection. (These two levels of adjustment to the environment are, of course, related. Homeostatic mechanisms can best be interpreted as adaptations that have evolved in populations facing certain environmental problems.) A second important theme in biology, the correlation between structure and function, is also evident in the form and physiology of the tissues and organs responsible for homeostasis.

In most animals, the majority of cells are not in direct contact with the external environment but are bathed by an internal body fluid (sponges and cnidarians are exceptions). This internal "pond" is typically either hemolymph, as in brine shrimp and other animals with open circulatory systems, or interstitial fluid serviced by blood, as in vertebrates and other animals with closed circulatory systems. (Note that animals with closed circulatory systems have *three* internal fluid compartments: the intracellular compartment consisting of the cytosol of cells; and two extracellular compartments, the blood plasma and interstitial fluid.) Changes in these body fluids are tempered by a variety of regulatory systems (Figure 40.2), usually involving feedback mechanisms, as discussed in Chapter 36.

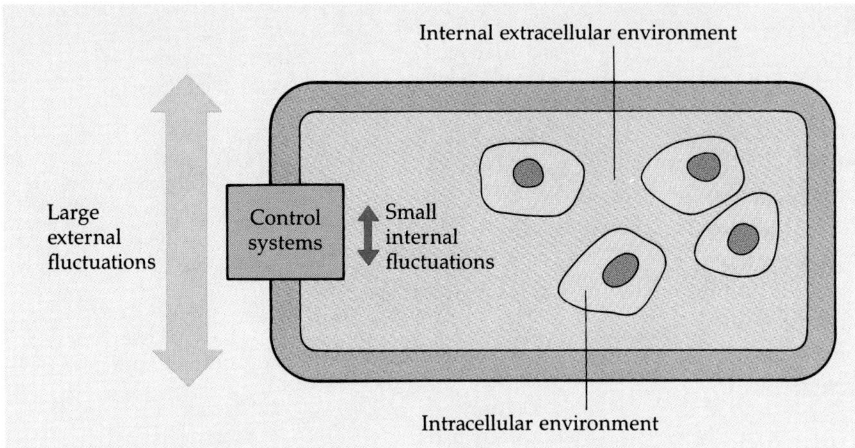

Figure 40.2
Homeostasis. Control systems cushion the internal environment from the impact of fluctuations in the external environment. For example, a brine shrimp inhabiting a very saline environment continuously swallows the salty water, but the animal's gills pump the excess salt out of the body, minimizing internal changes while the salinity of the external environment fluctuates.

The problems and solutions of controlling the internal environment vary, depending on the phylogenetic history of the animal and the environment in which the species has evolved. This chapter takes a comparative approach to homeostasis, concentrating on how animals control solute balance and the gain and loss of water (**osmoregulation**), how they get rid of the nitrogen-containing waste products of metabolism such as urea (**excretion**), and how they maintain internal temperature within a tolerable range (**thermoregulation**).

OSMOREGULATION

Think about the ways water enters and leaves your body. We acquire most of our water in our food and drink and a smaller amount by oxidative metabolism. ("Metabolic water" is produced by cellular respiration when electrons and hydrogen are added to oxygen; see Chapter 9.) We lose water by urinating and defecating, and by evaporative loss due to sweating and breathing. For aquatic animals, evaporation is unimportant, but these animals experience the uptake and loss of water across the body surface by osmosis. Even if an animal is protected by a covering that impedes water loss or gain, specialized epithelia that must be exposed to the environment in order to exchange gases (such as gills, lungs, and tracheae) cannot be waterproof.

Whether an animal inhabits land, fresh water, or saltwater, one general problem occurs: The cells of the animal cannot survive a net water gain or loss. Although water continuously enters and leaves an animal cell across the plasma membrane, uptake and loss must balance. Animal cells swell and burst if there is a net uptake of water or shrivel and die if there is a net loss of water (see Chapter 8). We will now see that there are two basic solutions to this problem of water balance.

Osmoconformers and Osmoregulators

Recall from Chapter 8 that osmosis, a special case of diffusion, is the movement of water across a selectively permeable membrane. It occurs whenever two solutions separated by the membrane differ in total solute concentration, or **osmolarity** (total solute concentration expressed as molarity, or moles of solute per liter of solution; see Chapter 3). The units of measurement for osmolarity used in this chapter are milliosmoles per liter (mosm/L). This unit is equivalent to a total solute concentration of 10^{-3} *M*. For example, the osmolarity of human blood is about 300 mosm/L, while seawater commonly has an osmolarity of about 1000 mosm/L. Two solutions are said to be isosmotic if they are equal in osmolarity. There is no *net* osmosis between isosmotic solutions. When two solutions differ in osmolarity, the one with the greater concentration of solutes is referred to as hyperosmotic and the more dilute solution as hypoosmotic. Water will flow by osmosis across a membrane from a hypoosmotic solution to a hyperosmotic one.

One way for a saltwater animal to balance water exchange with the environment is to be isosmotic with its aqueous surroundings. Such animals, which do not actively adjust their internal osmolarity, are known as **osmoconformers.** Animals that are not isosmotic with their surroundings, called **osmoregulators,** must either discharge excess water if they live in a hypoosmotic environment or continuously take in water to offset osmotic loss if they inhabit a hyperosmotic environment. All freshwater animals and many marine animals are osmoregulators, maintaining an internal osmolarity that differs from the surrounding water. Humans and other terrestrial animals, also osmoregulators, must compensate for water loss. With these general approaches to water balance in mind, we can now survey some specific examples of osmoregulatory adaptations to various environments.

Problems of Osmoregulation in Different Environments

Marine Animals Animals first evolved in the sea, which remains the most common environment for the majority of phyla. Most marine invertebrates are osmoconformers, with body fluids isosmotic to the surrounding seawater. However, these animals differ from seawater in their concentrations of specific salts. The difference is usually slight, but in some cases it is substantial. Thus, an animal that conforms to the osmolarity of its surroundings may still regulate its internal composition of ions.

The hagfish, a jawless vertebrate (class Agnatha), is isosmotic with the surrounding seawater, but most marine vertebrates osmoregulate. Sharks and most other cartilaginous fishes (class Chondrichthyes) maintain internal salt concentrations that are relatively low compared with that of seawater mainly by the use of rectal glands that pump salt out of the body through the anus. However, the shark has an osmolarity close to that of seawater, a characteristic attributed to the retention of a large amount of organic solute in the form of urea, a nitrogenous (nitrogen-containing) waste product other animals excrete rather than accumulate. (Shark meat must be soaked in fresh water before it is eaten to wash out this high concentration of urea.) Sharks produce and retain another organic compound, trimethylamine oxide (TMAO), which helps protect proteins from the damaging effects of urea. As a result of the high concentration of organic solutes in their body fluids, sharks are actually slightly hyperosmotic to seawater. They do not drink, and they balance the osmotic uptake of water by copious urination. Among other osmoregulators that maintain body fluids hyperosmotic to seawater as a result of accumulating urea are the crab-eating frogs of Southeast Asia, the only saltwater amphibians.

Bony fishes (class Osteichthyes) evolved from ancestors that entered freshwater habitats. In their subsequent evolution, many groups of bony fishes became marine but internally remained more similar to fresh water in osmolarity. Marine bony fishes constantly lose water by osmosis to their hyperosmotic surroundings. They compensate by drinking large amounts of seawater and using the epithelium of their gills to pump the excess salt out of the body. Similarly, many marine birds, including sea gulls, have nasal salt glands that rid the animals of much of the salt they obtain by drinking seawater. Marine reptiles, including the marine iguana and sea turtles, also have salt glands that function in osmoregulation.

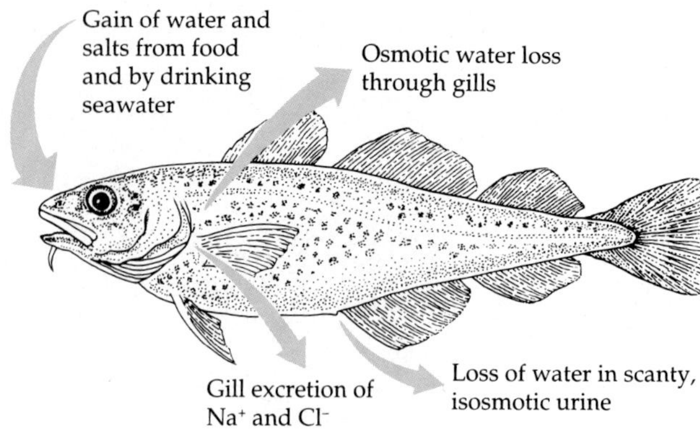

(a) Marine bony fish

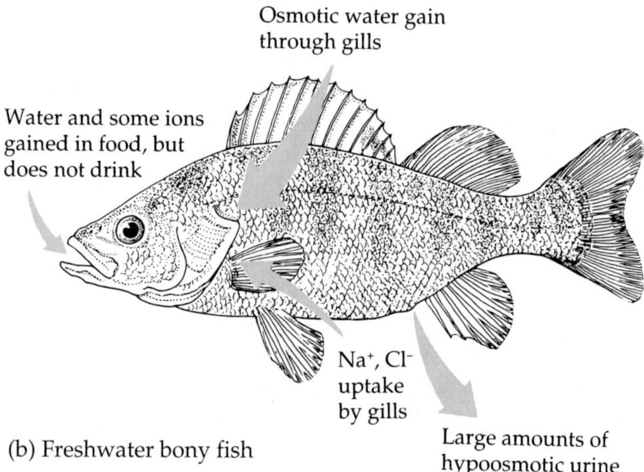

(b) Freshwater bony fish

Figure 40.3
Osmoregulation in marine and freshwater bony fishes: a comparison. (**a**) A marine fish, such as this cod, is hypoosmotic to the surrounding seawater and therefore constantly loses water by osmosis, mainly across the epithelium of the gills. The animal compensates for the fluid loss by drinking large amounts of seawater and then pumping salt out of the body across the gill epithelium. Urine is scanty and isosmotic to the body fluids of the fish. (**b**) The opposite problem is faced by freshwater fishes, such as this perch, which constantly gain water from their hypoosmotic surroundings by osmosis. This water uptake is balanced by the copious excretion of urine, which is hypoosmotic to the body fluids of the fish. Although the urine is dilute, the animal still loses important salts during excretion and compensates by the active uptake of ions across the epithelium of the gills.

Freshwater Animals The osmoregulatory problems of freshwater animals are the opposite of those of marine animals. Freshwater animals are constantly taking in water by osmosis because the osmolarity of their internal fluids is much higher than that of their immediate environment. Freshwater protozoa such as *Amoeba* and *Paramecium* have contractile vacuoles that function as bilge pumps (see Figure 26.8c). Many freshwater animals, including fishes, bail out water by excreting large amounts of very dilute urine. As they excrete water, however, they experience a small net loss of salts. Their salt content is partly replenished by eating plants and animals that have much higher salt concentrations than the water. Also, the gills of some freshwater fishes pump sodium and chloride ions from the external environment into the blood (Figure 40.3).

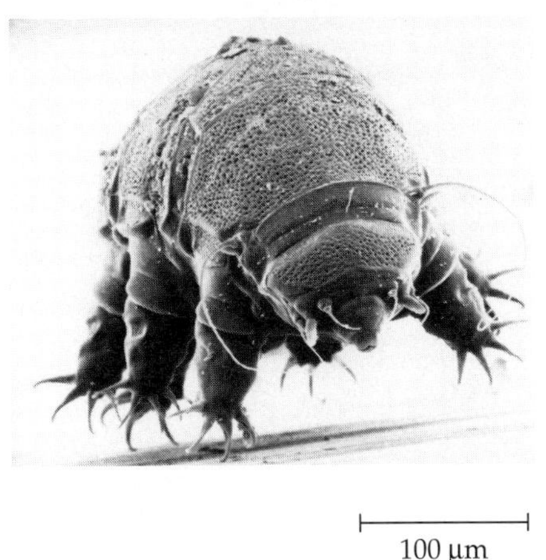

(a)

(b)

Figure 40.4
Anhydrobiosis (cryptobiosis). (**a**) Tardigrades, or water bears, inhabit ponds and films of water in soil and on mosses and lichens. (**b**) When their habitat completely evaporates, these invertebrates can lose more than 95% of their body water and survive for decades in this dehydrated, inactive state (SEMs). (The dehydrated tardigrade on the right is actually smaller than the one on the left, as you would expect; the scale of the right photograph is about twice that of the left.)

Salmon and other fishes that migrate between seawater and fresh water undergo a transition in their osmoregulation. While in the ocean, salmon drink seawater and excrete excess salt from the gills, osmoregulating like other marine fishes. With their migration to fresh water, salmon cease drinking, and the epithelia of their gills are modified for the accumulation of salt from the dilute environment.

Euryhaline Animals Most animals, whether osmoconformers or osmoregulators, cannot tolerate substantial changes in external osmolarity. Such animals are said to be **stenohaline** (Gr. *stenos,* "narrow"; haline refers to salt). However, some animals, called **euryhaline** animals (Gr. *eurys,* "broad"), do survive radical fluctuations of osmolarity in the surrounding water, either by conforming to the changes or by regulating their internal osmolarity within a narrow range even as the external osmolarity changes. Brine shrimp (see Figure 40.1) are examples of euryhaline animals that osmoregulate. Among other euryhaline animals is a diversity of invertebrates and fishes that inhabit the brackish water of estuaries, where salinity changes with each rainfall and the daily tides.

Anhydrobiosis Dehydration dooms most animals, but some aquatic invertebrates living in ponds and films of water around soil particles can lose almost all their body water and survive in a dormant state when their habitats dry up. This remarkable adaptation is called **anhydrobiosis** ("life without water"), or cryptobiosis ("hidden life"). Among the most striking examples are the tardigrades, or water bears, tiny invertebrates less than 1 mm long (Figure 40.4). In their active, hydrated state, these animals contain about 85% water by weight but can dehydrate to less than 2% water and survive in an inactive state, dry as dust, for a decade or more. Just add water, and within minutes the rehydrated tardigrades are moving about and feeding.

One mechanism that makes anhydrobiosis possible is the animal's production of a large amount of disaccharides, particularly trehalose, a double sugar consisting of two glucose units. With their multiple hydroxyl groups capable of forming hydrogen bonds, the sugars apparently replace water associated with membranes and proteins, protecting these cellular structures and molecules from extreme distortion during dehydration.

Terrestrial Animals Unfortunately, few truly terrestrial animals are capable of anhydrobiosis. Humans, for example, die if they lose about 12% of their body water. The threat of desiccation is perhaps the most important problem confronting terrestrial life, both plants and animals. The severity of this problem may be one reason why only two groups of animals, arthropods and vertebrates, have colonized the land with great success. (Although other phyla have some representatives on land, most of their species are aquatic.)

What *are* some of the evolutionary adaptations that have made it possible for animals, which consist mostly of water, to survive on land? Much as a waxy cuticle contributes to the success of plants on land (see Chapter 27), most terrestrial animals are covered by relatively impervious surfaces that help prevent dehydration. Examples are the waxy layers of the exoskeletons of insects, the shells of land snails, and the multiple layers of dead, keratinized skin cells covering most terrestrial vertebrates (see Chapter 30). Still, most terrestrial animals lose a considerable amount of water that must be replenished by drinking and eating moist foods. Behavioral adaptations, such as nervous and hormonal mechanisms that control thirst, are important osmoregulatory mechanisms in land-dwelling animals (these mechanisms are discussed later in the chapter). Many terrestrial animals, especially in deserts, are nocturnal, another important behavioral adaptation that reduces dehydration. The kidneys and other excretory organs of terrestrial animals often exhibit adaptations that help conserve water (and these adaptations will also be highlighted later in the chapter). Some mammals are so well adapted to minimizing

(a)

		Water Balance in a Kangaroo Rat	Water Balance in a Human
Water gain (mL/day)	Ingested in liquid	0	1500 (60%)
	Ingested in food	6.0 (10%)	750 (30%)
	Derived from metabolism	54.0 (90%)	250 (10%)
		60.0 (100%)	2500 (100%)
Water loss (mL/day)	Evaporation	43.9 (73%)	900 (36%)
	Urine	13.5 (23%)	1500 (60%)
	Feces	2.6 (4%)	100 (4%)
		60.0 (100%)	2500 (100%)

Source: Kangaroo rat data from Schmidt-Neilsen, *Animal Physiology: Adaptations and Environment,* 2nd ed. Cambridge: Cambridge University Press, 1979. p. 324.

(b)

Figure 40.5
Water balance in kangaroo rats. (**a**) Kangaroo rats, which live in the American southwestern desert, survive without drinking any water by regulating body fluids very efficiently. (**b**) The animal obtains 90% of its water as a product of metabolic reactions and the remaining 10% from free water present in the food it eats. A kangaroo rat excretes little water because its specialized kidneys can concentrate urine. The feces are also relatively dry. The table compares these sources of water gain and loss to water balance in humans.

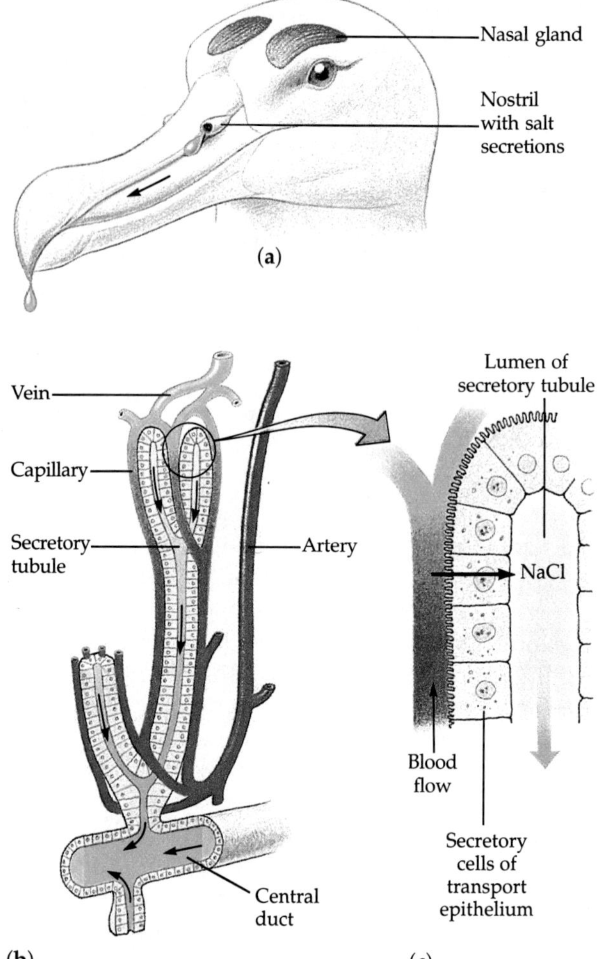

Figure 40.6
Salt glands in birds. (**a**) Many marine birds, such as this albatross, drink seawater and excrete excess salt through nasal glands. (**b**) A salt gland has many lobes, which contain several thousand tubules radiating from a central duct. Each tubule is surrounded by capillaries, where the blood flows counter to the flow of salt secretion (arrows). This countercurrent system facilitates salt transfer from the blood to the tubule (see Chapter 38). (**c**) Secretory cells of the transport epithelium lining the tubules pump salt from the blood into the tubules.

water loss that they can survive in deserts without drinking. For example, kangaroo rats conserve water so frugally that they can compensate for water loss by the production of metabolic water (mostly water produced by combining hydrogen and oxygen in cellular respiration; see Chapter 9) and the intake of very small quantities of water in their food (Figure 40.5).

Transport Epithelia and Osmoregulation

Although the problems of water balance in environments as diverse as saltwater, fresh water, and land are very different, the solutions in osmoregulators are mostly variations on a common theme: the use of specialized epithelia, called **transport epithelia,** to regulate the transport of salt (and hence water, which follows solute movement by osmosis) between the animal's internal fluids and the external environment.

A transport epithelium is usually a single sheet of cells facing the external environment or some channel that leads to the exterior through an opening on the body surface (Figure 40.6). The cells of the epithelium are joined by impermeable tight junctions (see Figure

7.34a), forming a continuous barrier at the tissue-environment boundary. This configuration, another example of how form fits function, ensures that any solute passing between the extracellular fluid and the environment must pass through the selectively permeable membranes of cells. It is the molecular composition of the epithelium's plasma membrane that determines the specific osmoregulatory functions. Transport epithelia vary in their passive permeabilities to water and salts, and in the number, type, and orientation of membrane proteins responsible for active transport. For example, differences in membrane structure and function account for the transport epithelia of the gills of marine fishes pumping salt outward, while the gills of freshwater fishes pump salt inward.

One of the most efficient transport epithelia is found in the nasal glands of marine birds, which drink seawater and excrete the excess salt via the nasal salt glands (see Figure 40.6). As mentioned earlier, sharks also use salt glands, in this case located along the epithelium of the rectum, to expel salt.

The transport epithelia of salt glands are dedicated exclusively to osmoregulation—maintaining salt and water balance. In other cases, transport epithelia function in the excretion of nitrogenous wastes as well as osmoregulation. In the excretory systems of most animals, transport epithelia are arranged into tubular networks with extensive surface areas, as in the vertebrate kidney and a diversity of invertebrate excretory systems.

EXCRETORY SYSTEMS OF INVERTEBRATES

Protonephridia: The Flame-Cell System of Flatworms

The simplest tubular excretory system is the **flame-cell system** of flatworms (phylum Platyhelminthes). These animals have neither circulatory systems nor coeloms (see Chapter 29), so the flame-cell system must regulate the contents of the interstitial fluid directly. The apparatus consists of a branched system of tubules ramifying throughout the body (Figure 40.7). Each of the smallest tubules at the tips of this excretory

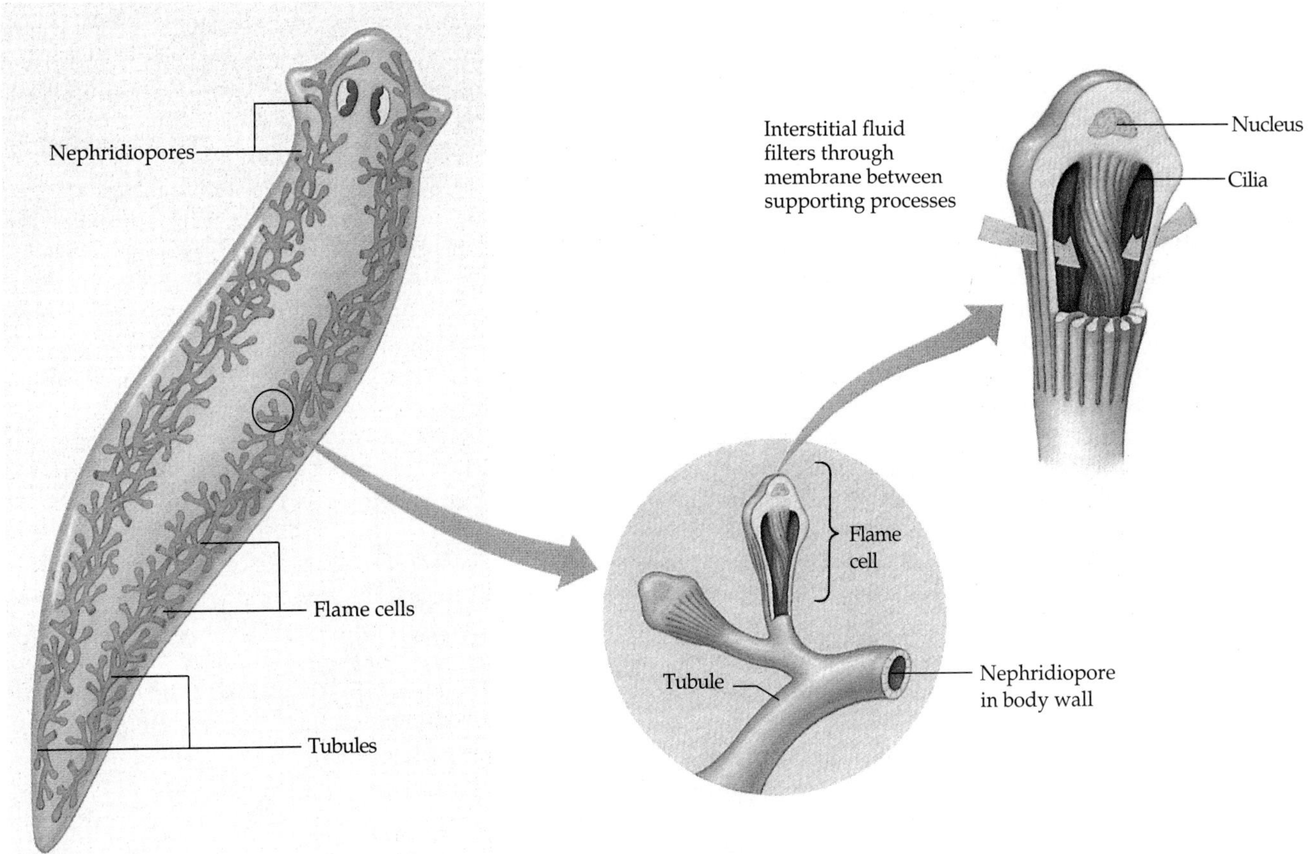

Figure 40.7
Protonephridia: the flame-cell system of planaria. Protonephridia are excretory tubules lacking internal openings. In planaria, interstitial fluid is filtered across the membranes of flame cells. Cilia projecting from the flame cells into the tubules keep the fluid moving along the tubular system, which opens to the exterior of the body through numerous pores. The transport epithelium lining the tubules functions in osmoregulation. For example, the epithelium of the flame-cell system of freshwater flatworms probably pumps salts from the tubular fluid back into the interstitial fluid, enabling the animals to excrete a dilute urine and balance the osmotic uptake of water from the hypoosmotic environment.

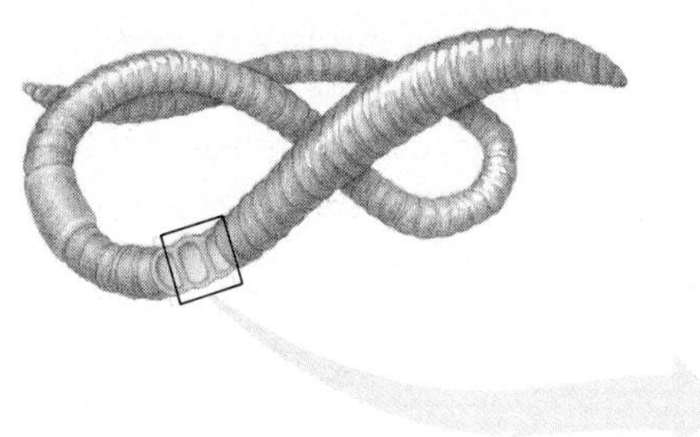

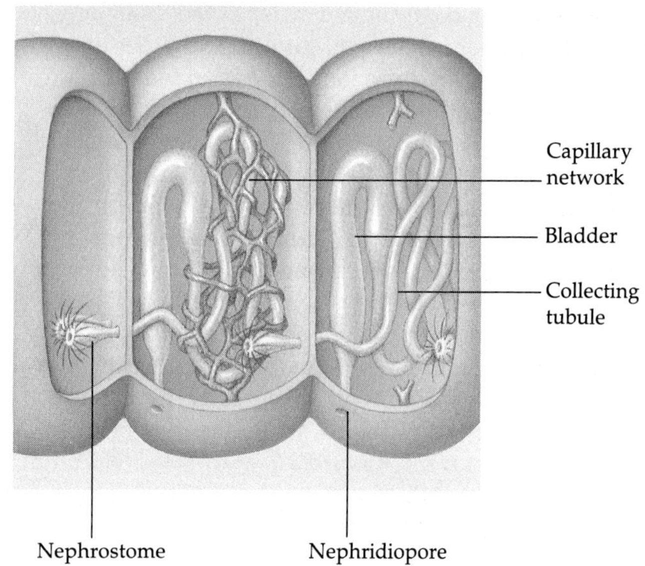

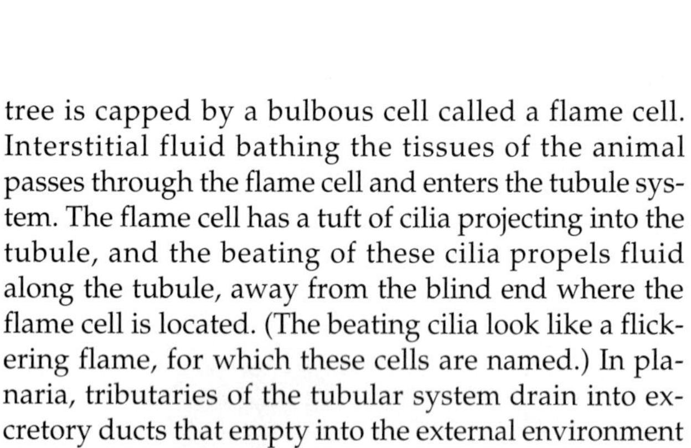

Figure 40.8
Metanephridia of annelids. Each segment of an earthworm contains a pair of metanephridia, which drain the adjacent anterior segment. Fluid enters the nephrostome, passes through the metanephridium, and empties into a storage bladder that opens to the outside through the nephridiopore. The urine becomes very dilute as the transport epithelium pumps certain salts back into the blood circulating in the capillary network enveloping the metanephridium.

tree is capped by a bulbous cell called a flame cell. Interstitial fluid bathing the tissues of the animal passes through the flame cell and enters the tubule system. The flame cell has a tuft of cilia projecting into the tubule, and the beating of these cilia propels fluid along the tubule, away from the blind end where the flame cell is located. (The beating cilia look like a flickering flame, for which these cells are named.) In planaria, tributaries of the tubular system drain into excretory ducts that empty into the external environment through numerous openings called nephridiopores.

The excreted fluid is very dilute in the case of freshwater flatworms; this helps balance the osmotic uptake of water from the hypoosmotic environment. The cellular mechanisms for this osmoregulation are unknown, however. It is likely that the lining of the tubules is a transport epithelium specialized for reabsorbing certain salts before the fluid exits the body. The flame-cell systems of freshwater flatworms function mainly in osmoregulation; most metabolic wastes are excreted into the gastrovascular cavity and eliminated through the mouth (see Chapter 36). However, some parasitic flatworms, which are isosmotic to the surrounding fluids of their host organisms, use their tubular excretory systems mainly to get rid of nitrogenous wastes. This difference in function illustrates how anatomical equipment common to a group of organisms can be adapted in diverse ways by evolution in different environments.

The flame-cell system is one example of a simple type of tubular excretory system called a **protonephridium,** a network of closed tubules lacking internal openings. In addition to the flame-cell systems of flatworms, protonephridia are also found in rotifers, some annelids, the larvae of mollusks, and lancelets, which are invertebrate chordates. (See Chapters 29 and 30 to review these animal phyla.)

Metanephridia of Earthworms

In contrast to the closed protonephridium, another type of excretory tubule, the **metanephridium,** has internal openings that collect body fluids. Metanephridia are found in most annelids, including earthworms (Figure 40.8). Each segment of the worm has its own pair of metanephridia, which are serpentine tubules immersed in the coelomic fluid of that segment. As in most animals with closed circulatory systems, blood vessels are intimately associated with the excretory tubules of the earthworm; a network of capillaries envelops each metanephridium. The tubule drains to the outside of the body through a nephridiopore. At the opposite end of a metanephridium is the nephrostome, a ciliated funnel that collects coelomic fluid from the body segment just anterior. (Notice again that the metanephridium, unlike the protonephridium of flatworms, is open at both ends.) As the fluid moves along the tubule, the transport epithelium bordering the lumen pumps essential salts out of the tubule, and the salts are reabsorbed into the blood circulating through the surrounding capillaries. The urine that exits through the nephridiopore is hypoosmotic to the body fluids of the earthworm. By excreting this dilute urine in amounts up to 60% of the body weight of the worm per day, the metanephridia offset the continuous osmosis taking place across the skin of the animal from the damp soil. (The skin must be moist and permeable to function as a respiratory organ; see Chapter 38.)

Malpighian Tubules of Insects

Insects and other terrestrial arthropods have open circulatory systems, with tissues bathed directly in hemolymph contained in sinuses (see Chapter 38). Excretory organs called **Malpighian tubules** remove nitrogenous wastes from the blood and function in osmoregulation (Figure 40.9). These organs open into the digestive tract at the juncture of the midgut and hindgut. The tubules, which dead-end at the tips away from the gut, dangle in the fluid of the body cavity. The transport epithelium that lines a tubule pumps certain solutes, including salts and nitrogenous wastes, from the blood into the lumen of the tubule. The fluid within the tubule then passes through the hindgut into the rectum. The epithelium of the rectum pumps most of the salt back into the blood, and water follows the salts by osmosis. The nitrogenous wastes are eliminated as nearly dry matter along with the feces. The insect excretory system is one adaptation that has contributed to the tremendous success of these animals on land, where conserving water is essential.

THE VERTEBRATE KIDNEY

Rather than being scattered throughout the body like the nephridia of earthworms, the excretory tubules of vertebrates, called nephrons, are collected into compact organs, the **kidneys.** Blood is cycled through the kidneys, which remove nitrogenous wastes and function in osmoregulation by adjusting the concentrations of various salts in the blood. The kidneys, the blood vessels that serve them, and the structures that carry urine formed in the kidneys out of the body are the components of the vertebrate excretory system. We will focus first on the mammalian version of the system, and then compare the excretory systems of the various vertebrate classes.

Anatomy of the Excretory System

In humans, the kidneys are a pair of bean-shaped organs about 10 cm long (Figure 40.10a). Blood enters each kidney via the **renal artery** and leaves each kidney via the **renal vein.** Although the kidneys account for less than 1% of the weight of the human body, they receive about 20% of the blood pumped with each heartbeat. **Urine,** the waste fluid formed within the kidney, exits the organ through a duct called the **ureter.** The ureters of both kidneys drain into a common **urinary bladder.** The bladder is periodically emptied by micturition (urination); this final excretion of

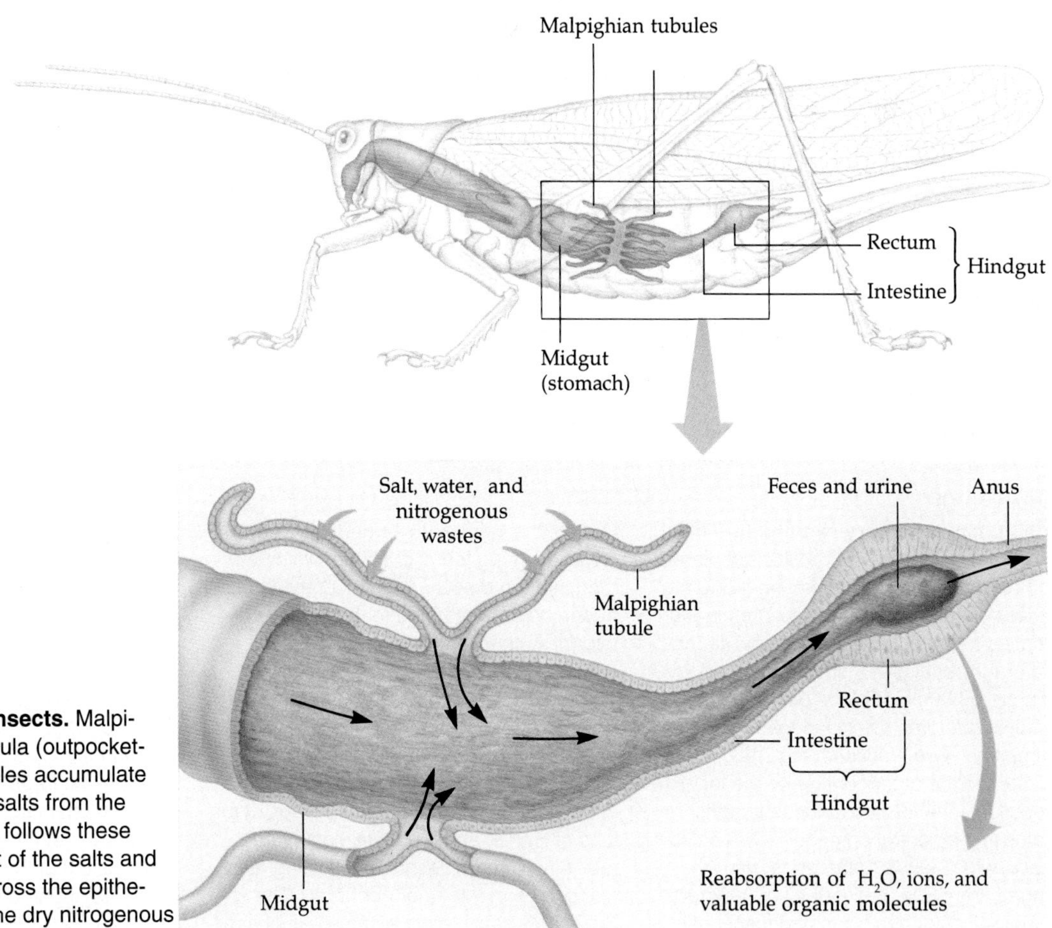

Figure 40.9
Malpighian tubules of insects. Malpighian tubules are diverticula (outpocketings) of the gut. The tubules accumulate nitrogenous wastes and salts from the coelomic fluid, and water follows these solutes by osmosis. Most of the salts and water are reabsorbed across the epithelium of the rectum, and the dry nitrogenous wastes are eliminated with the feces.

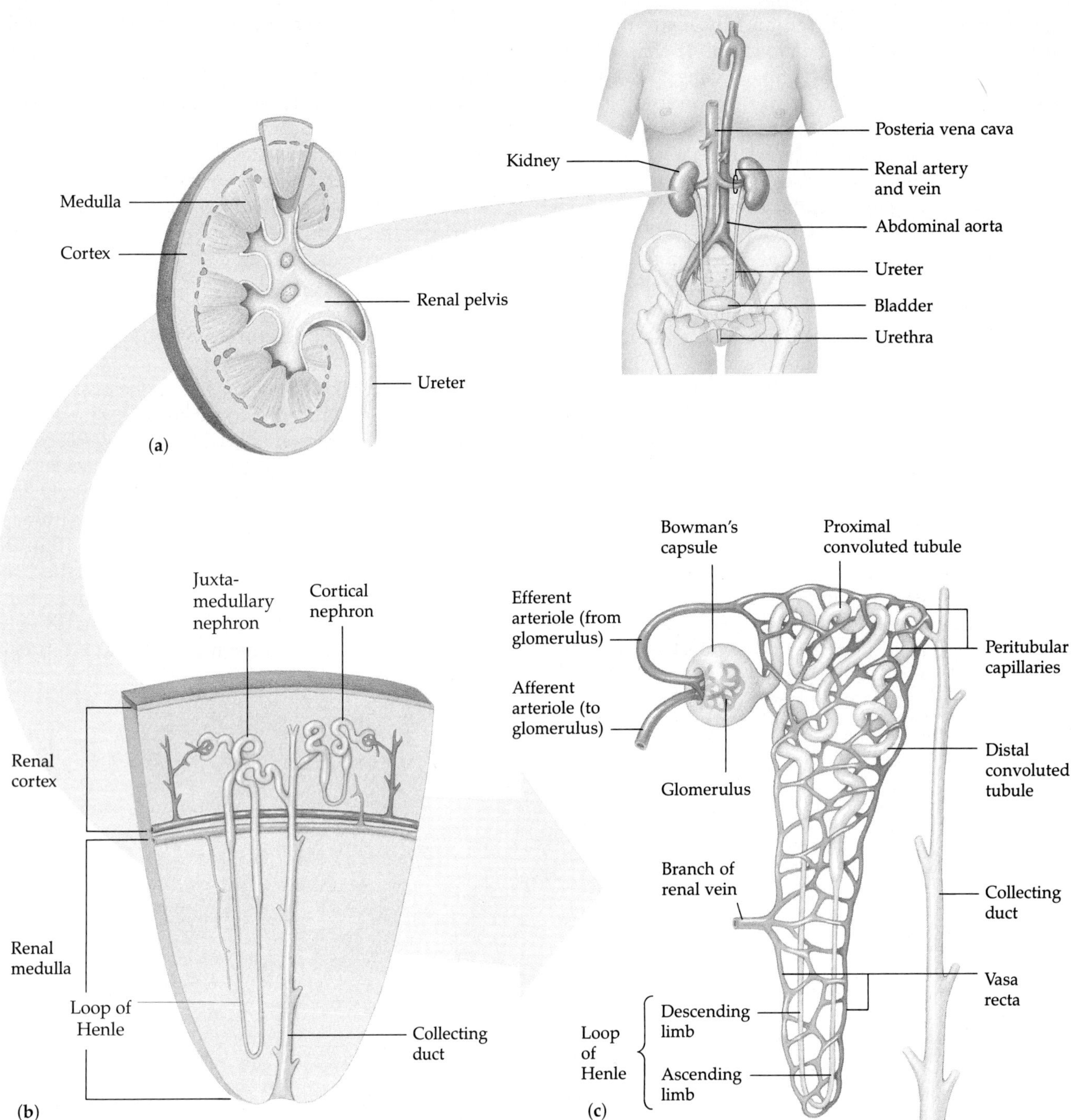

Figure 40.10

The human excretory system. (**a**) The kidneys regulate the composition of blood and form urine, which is transported to the urinary bladder in the ureters. The bladder empties to the outside of the body via the urethra. A renal artery supplies the kidney with up to 2000 L of blood per day, and a renal vein returns the blood to the general circulation. Two functional regions of the kidney are the outer cortex and the inner medulla. Urine formed in these regions drains into a central chamber, the renal pelvis, which leads to the ureter. (**b**) Arranged radially in the kidney are hundreds of thousands of nephrons, the functional units of the kidney. Juxtamedullary nephrons have long loops of Henle, which extend into the medulla of the kidney. Cortical nephrons are restricted to the kidney cortex. (**c**) A nephron consists of a renal tubule and its associated blood vessels. The receiving end of a renal tubule is the Bowman's capsule, a cup-shaped swelling of the tubule. The major regions of the tubule are the proximal convoluted tubule, the descending and ascending limbs of the loop of Henle, and the distal convoluted tubule, which drains into a collecting duct that serves many other nephrons. The Bowman's capsule envelops a ball of capillaries called the glomerulus. Blood pressure forces water and all types of small solutes from the blood plasma of these capillaries across the wall of the capsule and into the lumen of the renal tubule, forming a filtrate that is processed as it moves along the tubule. From the glomerulus, blood travels to the peritubular capillaries and the vasa recta. The interstitial fluid bathing the nephron has a continuous traffic of substances passing between the renal tubule and the nephron capillaries. Processing of the filtrate continues in the collecting duct, and the final product, now called urine, drains into the renal pelvis.

urine from the body is through a tube called the **urethra,** which empties near the vagina of females or through the penis of males. Sphincter muscles near the junction of the urethra and the bladder control micturition.

Structure of the Nephron

The functional unit of the kidney is the **nephron,** which consists of a **renal tubule** and its associated blood vessels (Figure 40.10b and c). Each kidney contains a large number of nephrons—about a million in a human kidney, representing a total of about 80 km of tubules. Water, urea, salts, and other small molecules present in blood flow from capillaries into the renal tubule, where the fluid is now called **filtrate.** The transport epithelium lining the renal tubule processes the filtrate to form the urine, which is eventually excreted from the kidney. From the 1100 to 2000 L of blood that flows through the human kidneys each day, the nephrons process about 180 L of filtrate but excrete only about 1.5 L of urine. The rest of the filtrate, including about 99% of the water, is reabsorbed into the blood.

The blind end of the renal tubule, which receives filtrate from the blood, is expanded to form a cup-shaped receptacle called the **Bowman's capsule,** which embraces a ball of capillaries, the **glomerulus.** From the Bowman's capsule, the filtrate passes successively through three main regions of the renal tubule: the **proximal convoluted tubule;** the **loop of Henle,** a long hairpin turn with a descending limb and an ascending limb; and the **distal convoluted tubule.** This last portion of the renal tubule empties its filtrate into a **collecting duct,** which receives filtrate from many other renal tubules. The many collecting ducts of the kidney then pass the filtrate, now called urine, into the renal pelvis, a chamber that in turn drains into the ureter.

The nephrons have a radial orientation in the kidney, with the loops of Henle and collecting ducts perpendicular to the kidney surface. Bowman's capsules, proximal convoluted tubules, and distal convoluted tubules are located in the outer zone of the kidney, the **cortex.** In the human kidney, about 80% of the nephrons, the **cortical nephrons,** have reduced loops of Henle and are almost entirely confined to the cortex. The other 20% of the nephrons, the **juxtamedullary nephrons,** have well-developed loops that extend to the inner zone of the kidney, the **medulla.** Only mammals and birds have juxtamedullary nephrons; loops of Henle are absent in the nephrons of the other vertebrate classes. As we will soon see, the juxtamedullary nephrons play an important role in the ability of mammals to excrete urine that is hyperosmotic to body fluids, an adaptation that conserves water.

Each nephron is supplied with blood by an **afferent arteriole,** a branch of the renal artery that subdivides to form the capillaries of the glomerulus. The capillaries converge as they leave the capsule to form an **efferent arteriole,** but then this vessel subdivides again into a second network of capillaries, the **peritubular capillaries.** These capillaries intermingle with the proximal and distal convoluted tubules of the nephron. Additional capillaries extend downward to form the **vasa recta,** the capillary system that serves the loop of Henle. The vasa recta is also a loop, with a descending vessel and an ascending vessel conveying blood in opposite directions.

Although the renal tubule and its surrounding capillaries are closely associated, they do not exchange materials directly across their walls. The tubules and capillaries are immersed in interstitial fluid, through which various substances pass back and forth between the plasma within capillaries and the filtrate within the nephron tubules.

Keeping in mind the correlation of structure and function, let's now investigate how the complex organization of the renal tubule and its blood vessels explains how the nephrons work.

General Physiology of the Nephron

Nephrons regulate the composition of blood by a combination of three processes that transfer material between the renal tubules and the capillaries that serve them:

1. *Filtration.* Blood pressure forces fluid from the capillaries of the glomerulus across the epithelium of the Bowman's capsule into the lumen of the renal tubule. The porous capillaries, along with specialized cells of the capsule called **podocytes,** function as a filter, being permeable to water and small solutes but not to blood cells or large molecules such as plasma proteins (Figure 40.11). **Filtration** is nonselective with regard to small molecules; any substance small enough to be forced through the capillary wall by blood pressure enters the renal tubule. Thus, at this point, the filtrate contains a mixture of solutes such as salts, glucose, vitamins, nitrogenous wastes such as urea, and other small molecules that mirrors the concentrations of these substances in blood plasma.

2. *Secretion.* As filtrate travels through the renal tubule, it is joined by substances that are transported across the tubule epithelium from the surrounding interstitial fluid. Because small molecules pass freely from the plasma within capillaries into the interstitial fluid, the net effect of renal **secretion** is the addition of plasma solutes to the filtrate within the tubule. The proximal and distal convoluted tubules are the most common sites of secretion. Unlike the nonselective filtration that occurs at the glomerulus-capsule interface, secretion is a very selective process involving both passive and active transport. For example, the controlled secretion of

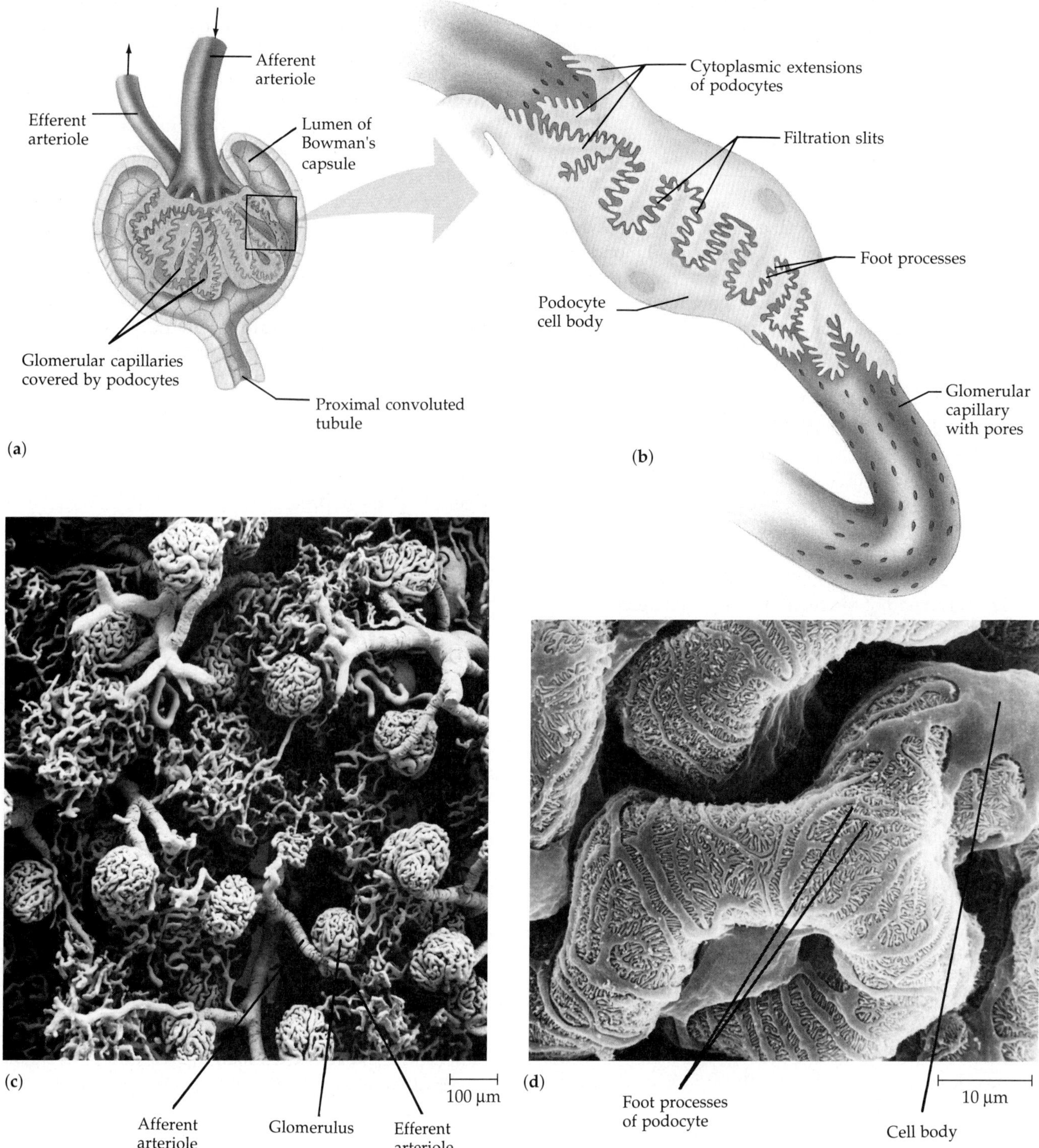

Figure 40.11
The glomerular filtration apparatus. (**a**) A Bowman's capsule has two layers of cells, a simple squamous epithelium making up the outer wall of the cup (see Chapter 36) and an inner layer of cells called podocytes that embrace the capillaries of the glomerulus. (**b**) A podocyte consists of a cell body with branched foot processes. The processes of neighboring podocytes interdigitate to form slits. These clefts, along with the numerous pores (fenestrations) of the capillaries, function in the filtering of blood, allowing blood pressure to force water and small solutes into the lumen of the capsule but restricting the passage of blood cells and macromolecules such as plasma proteins. (**c**) Glomeruli, with their afferent (incoming) and efferent (outgoing) arterioles, can be seen here (SEM). The renal tubules have been removed. (From *Tissues and Organs: A Text-Atlas of Scanning Electron Microscopy* by Richard G. Kessel and Randy H. Kardon. W. H. Freeman and Company. Copyright © 1979.) (**d**) The foot processes of podocytes envelop capillaries of the glomerulus, forming an elaborate network of slits (SEM).

hydrogen ions from the interstitial fluid into the nephron tubule is important in maintaining a constant pH for the body fluids.

3. *Reabsorption.* Because filtration is nonselective, it is important that small molecules essential to the body be returned to the interstitial fluid and blood plasma. This selective transport of substances across the epithelium of the renal tubule from the filtrate to the interstitial fluid is called **reabsorption.** The convoluted tubules and the loop of Henle all contribute to reabsorption, as does the collecting duct that receives filtrate from the tubule. Nearly all the sugar, vitamins, and other organic nutrients present in the initial filtrate are reabsorbed. Most of the water of the filtrate is also reabsorbed in the kidneys of mammals and birds, enabling these terrestrial animals to save water by decreasing urine volume. Together, reabsorption and secretion control the concentrations of various salts in body fluids, responding to imbalances by causing the kidneys to excrete more or less of a particular ion.

Thus, selective secretion and reabsorption modify the composition of the filtrate, increasing the concentrations of some substances and decreasing the concentrations of others in the urine that is finally excreted. The overall effect of filtration, secretion, and reabsorption is analogous to cleaning out a drawer (blood) by first removing all the small articles (filtration), returning useful items to the drawer (reabsorption), adding additional useless household items to the refuse pile (secretion), and then discarding all the unwanted objects (excretion).

To better understand the versatility of the nephron as an apparatus for controlling the composition of body fluids, it is necessary to examine the specialized functions of the transport epithelium that forms the wall of the various regions of the renal tubule and collecting duct.

Transport Properties of the Renal Tubule

The filtrate formed by glomerular filtration is essentially identical to blood plasma in overall osmolarity and in the concentrations of small solutes. Filtrate becomes urine as a result of the serial processing of the fluid as it flows along the tubule and collecting duct (Figure 40.12).

1. *Proximal convoluted tubule.* The transport epithelium making up the wall of the proximal convoluted tubule alters the volume and composition of filtrate substantially, both by reabsorption and secretion. For example, ammonia is secreted, passing from the peritubular capillaries into the interstitial fluid and then across the epithelium of the tubule to join the filtrate. The proximal tubule also helps maintain a constant pH in body fluids by the controlled secretion of hydrogen ions. Drugs and other poisons that have been processed in the liver are secreted into the filtrate by the epithelium of the proximal convoluted tubule. On the other hand, nutrients, including glucose and amino acids, are actively transported from the filtrate to the interstitial fluid, and then into the blood within the peritubular capillaries. Without this reabsorption, these nutrients would be lost with the urine. Potassium is also reabsorbed.

One of the most important functions of the proximal convoluted tubule is the reabsorption of NaCl and water. In fact, about 75% of the NaCl and about 70% of the water that is pushed by filtration from the blood into the renal tubule is reabsorbed across the epithelium of the proximal convoluted tubule. The epithelial cells in this region of the tubule have a structure well suited to this voluminous reabsorption. Facing the lumen of the tubule is a **brush border,** so named for the numerous microvilli projecting from the epithelial cells (these microvilli are similar to those of intestinal cells; see Chapter 37). This adaptation provides an extensive surface area for reabsorption. Salt and water present in the filtrate diffuse across the brush border and enter the epithelial cells. On the opposite side of the epithelium—the side facing the interstitial fluid outside the tubule—active transport occurs. The membranes pump Na^+ out of the cells and into the interstitial fluid. This transfer of positive charge is balanced by the passive transport of Cl^- out of the tubule. As salt moves from the filtrate to the interstitial fluid, water follows passively by osmosis. The surface of the epithelium facing the exterior of the tubule has a much smaller area than does the brush border, which minimizes the leakage of salt and water back into the tubule. Instead, the salt and water now diffuse from the interstitial fluid into the peritubular capillaries.

2. *Descending limb of the loop of Henle.* Reabsorption of water continues as the filtrate moves along the tubule to the **descending limb** of the loop of Henle. Here, the transport epithelium is freely permeable to water, but not very permeable to salt and other small solutes. For water to move out of the tubule by osmosis, the interstitial fluid bathing the tubule must be hyperosmotic to the filtrate. The osmolarity of the interstitial fluid does in fact increase gradually, becoming progressively greater along a radial axis from the outer cortex to the inner medulla of the kidney. (The mechanism that maintains this gradient will be discussed shortly.) Thus, filtrate moving downward from the cortex to the medulla within the descending limb of the loop of Henle continues to lose water to interstitial fluid of greater and greater osmolarity. At the same time, the NaCl concentration of the filtrate increases as water departs by osmosis.

3. *Ascending limb of the loop of Henle.* The filtrate reaches the tip of the loop, located deep in the kidney medulla in the case of juxtamedullary nephrons, and then heads

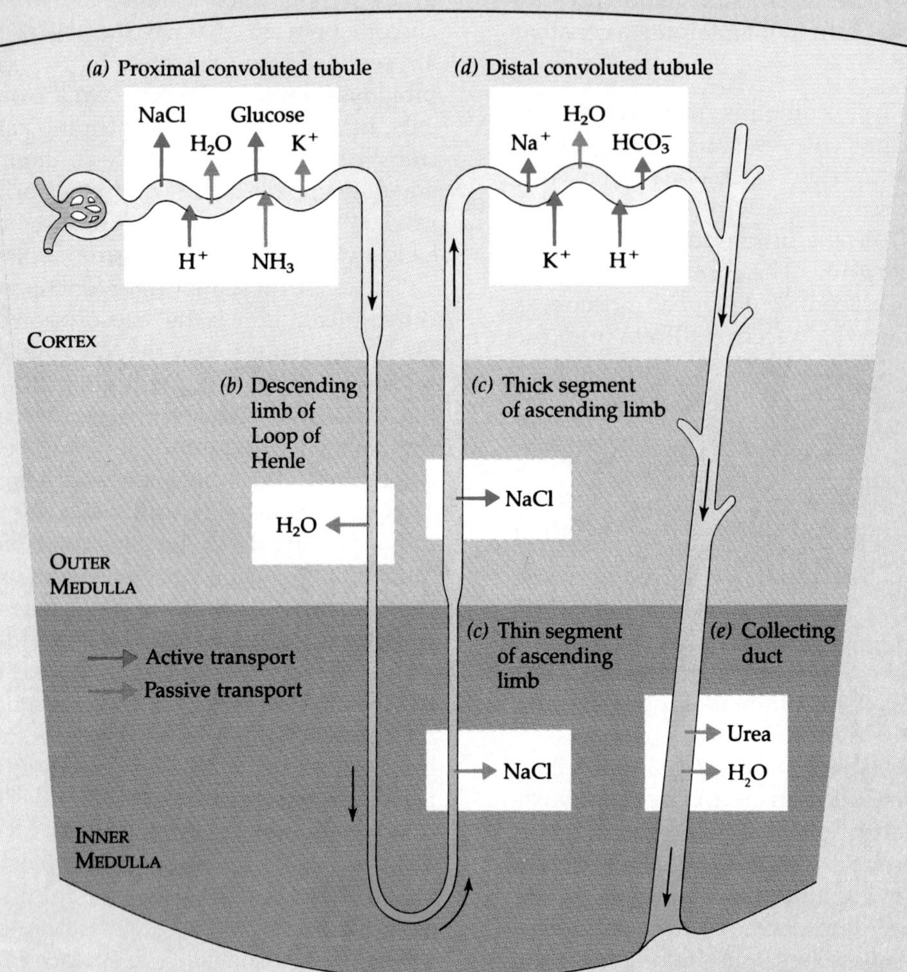

Figure 40.12
The renal tubule: regional functions of the transport epithelium. In this diagram, purple arrows indicate passive transport, and red arrows symbolize active transport. (**a**) The proximal convoluted tubule plays an important role in homeostasis by its controlled secretion and reabsorption of several substances. For example, about 75% of the NaCl and 70% of the water filtered from blood into the renal tubule are reabsorbed across the epithelium of the proximal tubule. This region also functions in the reabsorption of nutrients and in controlling pH by the secretion of H^+. (**b**) The descending limb of the loop of Henle is permeable to water but not to salt. Osmotic loss of water from the filtrate, as the descending limb penetrates the kidney medulla, concentrates NaCl in the filtrate. (**c**) The ascending limb of the loop of Henle consists of a thin segment and a thick segment. Both have epithelia that are relatively impermeable to water. The thin segment is permeable to NaCl, and the salt that was concentrated in the filtrate within the descending limb now diffuses out of the ascending limb, contributing to a high interstitial osmolarity in the inner medulla of the kidney. The thick segment continues the transfer of salt from the filtrate to the interstitial fluid, but now the transport is active. (**d**) The distal convoluted tubule is another important region of controlled secretion and reabsorption. For example, this segment of the renal tubule helps regulate blood pH by the reabsorption of bicarbonate (HCO_3^-), a buffer. The distal tubule also functions in K^+ and Na^+ homeostasis. (**e**) The specialized epithelium of the collecting duct is permeable to water but not to salt. The duct carries the filtrate toward the kidney medulla for a second time, and the filtrate becomes more and more concentrated as water is lost to the interstitial fluid. The bottom portion of the collecting duct is permeable to urea, and leakage of this solute into the interstitial fluid contributes to the high osmolarity of the kidney medulla.

up to the cortex again within the ascending limb of the loop. In contrast to the descending limb, the transport epithelium of the ascending limb is permeable to salt, but not very permeable to water. The ascending limb actually has two specialized regions, a thin segment near the loop tip and a thick segment leading to the distal convoluted tubule. As filtrate ascends in the thin segment, NaCl, which became concentrated in the descending limb, diffuses out of the tubule into the interstitial fluid. This loss of salt contributes to the high osmolarity of the interstitial fluid in the medulla. The exodus of salt from the filtrate continues in the thick segment of the ascending limb, but here the transport is active. The epithelium of the thick segment pumps Cl^- out of the tubule, and Na^+ responds to this efflux of negative charge by following the Cl^- passively. By losing salt without giving up water, the filtrate becomes progressively more dilute as it moves up to the cortex again in the ascending limb of the loop.

4. *Distal convoluted tubule.* The distal convoluted tubule is another important site of selective secretion and absorption. For example, the distal tubule plays a key role in regulating the K^+ and Na^+ concentration of body fluids by varying the amount of the K^+ that is secreted into the filtrate and the amount of Na^+ that is reabsorbed from the filtrate. The distal tubule also contributes to pH regulation, by the quantitative secretion of H^+ and by the reabsorption of bicarbonate (HCO_3^-), an important buffer in the blood and interstitial fluid.

5. *Collecting duct.* The collecting duct now carries the filtrate back in the direction of the medulla and renal pelvis. The epithelium of the collecting duct is permeable to water but not to salt. Thus, as the collecting duct traverses the gradient of osmolarity that exists in the interstitial fluid, the filtrate loses more and more water by osmosis to the hyperosmotic fluid outside the duct. Loss of water concentrates the urea in the filtrate, but not all of this urea is immediately passed along to the renal pelvis in the urine. At the bottom of the collecting duct, in the inner medulla, the epithelium of the duct is permeable to urea. Because of the high concentration of urea in the filtrate at this point, some of the urea diffuses out of the duct and into the interstitial fluid, bathing the portions of nephrons in the medulla. This interstitial urea is a major solute contributing, along with the salt, to the high osmolarity of the interstitial fluid in the kidney medulla. And it is this high osmolarity of the interstitial fluid that enables the kidney to conserve water by excreting urine that is hyperosmotic to the general body fluids.

How the Mammalian Kidney Conserves Water: A Closer Look

The osmolarity of human blood is about 300 mosm/L, but the kidney can excrete urine up to four times as concentrated—about 1200 mosm/L. The loop of Henle and the collecting duct cooperate to maintain the gradient of osmolarity in the interstitial tissue of the kidney that makes it possible to concentrate the urine. The two solutes responsible for this osmolarity gradient are NaCl, which is deposited in the kidney medulla by the loop of Henle, and urea, which leaks across the epithelium of the collecting duct in the inner medulla.

To better understand the physiology of the mammalian kidney as a water-conserving organ, let's retrace the flow of filtrate through the renal tubule, this time focusing on how the juxtamedullary nephrons maintain an osmolarity gradient in the kidney and use that gradient to excrete a hyperosmotic urine (Figure 40.13). Filtrate passing from the Bowman's capsule to the proximal convoluted tubule has an osmolarity of about 300 mosm/L, the same as blood. As the filtrate flows through the proximal convoluted tubule, located in the kidney cortex, a large amount of water *and* salt is reabsorbed; thus, the volume of filtrate decreases substantially at this stage, but the osmolarity remains about the same. Now the filtrate begins its serpentine trip, down into the medulla within the descending limb of the loop of Henle, back up to the cortex in the ascending limb, and then down to the medulla one more time within the collecting duct.

As the filtrate flows from cortex to medulla in the descending limb, water exits the renal tubule by osmosis, and the osmolarity of the filtrate increases as solutes, including NaCl, become more concentrated. Increasing gradually from cortex to medulla, the salt concentration of the filtrate peaks at the elbow of the loop of Henle. This maximizes the diffusion of salt out of the tubule as the filtrate rounds the curve and enters the ascending limb, which, remember, is leaky to salt but not to water. Thus, the two limbs of the loop of Henle cooperate in maintaining the gradient of osmolarity in the interstitial fluid of the kidney. The descending limb produces a progressively saltier filtrate, and the ascending limb exploits this concentration of NaCl to help maintain a high osmolarity in the interstitial fluid of the kidney medulla.

Notice that the loop has some of the qualities of a countercurrent system, similar in principle to the countercurrent mechanism that maximizes oxygen absorption by the gills of fishes (see Chapter 38). Although the two limbs of the loop of Henle are not in direct physical contact, they are close enough together to affect one another's chemical exchanges with a common interstitial fluid. The loop of Henle can concentrate salt in the inner medulla only because traffic in the descending limb counters the osmolarity gradient produced by the ascending limb in the interstitial fluid.

What prevents the capillaries of the kidney medulla from dissipating the osmolarity gradient by carrying away the NaCl that leaks from the ascending limb into the interstitial fluid? Notice in Figure 40.10 that the vasa recta is also a countercurrent system, with descending and ascending capillaries carrying blood in opposite directions through the kidney's osmolarity gradient. As the descending vessel conveys blood toward the inner medulla, water is lost from the blood and NaCl diffuses into the blood. These fluxes are simply reversed as blood flows back toward the cortex in the ascending vessel, with water re-entering the blood and salt diffusing out of the blood. Thus, the vasa recta can supply nutrients and other important substances carried by blood without interfering with the osmolar-

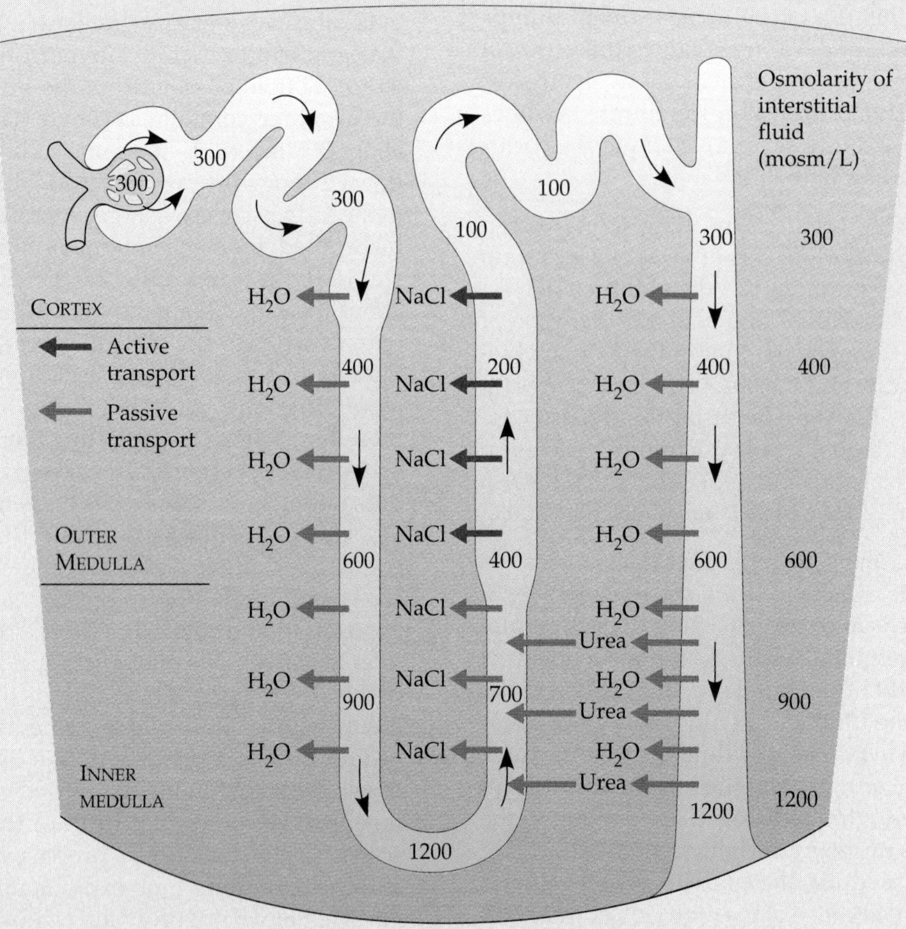

Figure 40.13
How the human kidney concentrates urine: the two-solute model. From the cortex to the inner medulla, the interstitial fluid of the kidney increases in osmolarity from about 300 to 1200 mosm/L. Two solutes contribute to this gradient of osmolarity: NaCl and urea. The loop of Henle maintains the interstitial gradient of NaCl. The filtrate concentration of this salt increases by the loss of water from the descending limb, then the ascending limb leaks the salt into the interstitial fluid. Additional salt is actively transported out of the thick segment of the ascending limb. The second solute, urea, is added to the interstitial fluid of the kidney medulla by diffusion out of the collecting duct (urea remaining in the collecting duct is excreted). Urea re-enters the tubule by diffusion into the ascending limb of the loop of Henle. The filtrate makes a total of three trips between the cortex and medulla, first down, then up, and then down one more time in the collecting duct. As the filtrate flows in the collecting duct past interstitial fluid of increasing osmolarity, more and more water moves out of the duct by osmosis, thus concentrating the solutes, including urea, that are left behind in the filtrate. Under conditions in which the kidney conserves as much water as possible, urine can reach an osmolarity of about 1200 mosm/L, considerably hyperosmotic to blood (about 300 mosm/L). This ability to excrete nitrogenous waste with a minimal loss of water is a key terrestrial adaptation of mammals.

ity gradient that makes it possible for the kidney to excrete a hyperosmotic urine.

By the time the filtrate completes its circuitous excursion in the loop of Henle, it is *not* hyperosmotic to body fluids at all, but is actually slightly hypoosmotic. This is because the thick segment of the ascending limb actively pumps NaCl out of the tubule, making the filtrate more and more dilute. Now the filtrate descends once again toward the medulla, this time in the collecting duct, which, remember, is permeable to water but not to salt. Flowing from cortex to medulla, the filtrate loses water by osmosis as it encounters interstitial fluid of increasing osmolarity. This concentrates urea in the filtrate, and some of this urea leaks out of the lower portion of the collecting duct, making a major contribution to the high interstitial osmolarity of the inner medulla. (This urea is recycled by its diffusion into the loop of Henle, but continual leaking of urea from the collecting duct maintains a high interstitial concentration of this solute.) The urea that remains in the collecting duct is excreted with a minimal loss of water from the body because osmosis causes the filtrate in the collecting duct to equal the osmolarity of the interstitial fluid, which can be as high as 1200 mosm/L in the inner medulla. Notice that urine, at its most concentrated, is actually *isosmotic* to the interstitial fluid of the inner medulla, but that makes it *hyperosmotic* to blood and interstitial fluid elsewhere in the body. The juxtamedullary nephron, with its urine-concentrating features, is a key adaptation to terrestrial life, enabling mammals to get rid of nitrogenous waste without squandering water.

Regulation of the Kidneys

While it is true that the kidneys *can* excrete hyperosmotic urine, it is not always desirable for them to do so. Nevertheless, if you are dehydrated and water is unavailable, the kidneys can excrete a small volume of hyperosmotic urine as concentrated as 1200 mosm/L, making it possible to discharge wastes with a minimal water loss. But if you have consumed an excessive amount of fluid, the kidneys can actually excrete a large volume of hypoosmotic urine as dilute as 70 mosm/L, making it possible to eliminate a lot of water without losing essential salts. The kidney is a versatile osmoregulatory organ, where water and salt reabsorption are subject to a combination of nervous and hormonal controls. (Hormones, chemical signals between various organs of the body, are discussed in detail in Chapter 41. Here, we are concerned only with the effects of a few hormones on the kidneys.)

One hormone important in osmoregulation is **antidiuretic hormone,** or **ADH** (Figure 40.14a). It is produced in a part of the brain called the hypothalamus and then stored and released from an organ called the pituitary gland, which is positioned just below the hypothalamus. **Osmoreceptor cells** located in the hypothalamus monitor the osmolarity of blood, stimulating the release of additional ADH when blood osmolarity rises above a set point of 300 mosm/L (in humans). Excessive water losses due to sweating or diarrhea are examples of crises that could cause an increase in blood osmolarity. More ADH is then discharged into the bloodstream and reaches the kidney. The main targets of ADH are the distal convoluted tubules and the collecting ducts of the kidney, where the hormone increases the permeability of the epithelium to water. This amplifies water reabsorption, which helps prevent further deviation of blood osmolarity from the set point. By negative feedback, the subsiding osmolarity of the blood reduces the activity of osmoreceptor cells in the hypothalamus, and less ADH is secreted. Only the ingestion of additional water in food and drink can bring osmolarity all the way back down to 300 mosm/L. When very little ADH is released, as would occur after a large volume of water has lowered the blood osmolarity, then the kidneys would absorb little water, resulting in copious excretion of dilute urine. (Voluminous urination is called diuresis, and it is because ADH opposes this state that it is called *anti*diuretic hormone.) Alcohol can perturb water balance by inhibiting the release of ADH, causing excessive loss of water in the urine and dehydrating the body; perhaps some of the symptoms of a hangover are due to this dehydration. Normally, however, blood osmolarity, ADH release, and water reabsorption in the kidney are all linked in a feedback loop that contributes to homeostasis.

A second mechanism that regulates kidney function involves a specialized tissue called the **juxtaglomerular apparatus (JGA),** located in the vicinity of the afferent arteriole that supplies blood to the glomerulus (Figure 40.14b). When the blood pressure in the afferent arteriole drops, or when the Na^+ concentration of the blood is too low, the JGA releases an enzyme called **renin** to the bloodstream. Within the blood, renin activates a plasma protein called **angiotensin.** The active form of this protein, called angiotensin II, functions as a hormone, with multiple effects that increase the Na^+ concentration of the blood and raise blood pressure. For example, angiotensin II causes a generalized constriction of arterioles, which raises blood pressure. Increasing pressure within the afferent arterioles of the nephrons, in turn, increases filtration rate. Angiotensin II also acts remotely on the kidney by stimulating the adrenal glands, organs located on top of the kidneys, to release another hormone called **aldosterone.** This hormone acts on the distal convoluted tubules of the nephrons, stimulating the reabsorption of Na^+. Because water follows the Na^+ out of the renal tubule by osmosis, aldosterone also increases blood volume and blood pressure. It was a drop in blood pressure or a deficiency of Na^+ that triggered renin release from the JGA in the first place, and the various responses increase blood pressure and Na^+ concentration, thus reducing the release of renin—another example of a feedback circuit functioning in homeostasis.

It may seem that the functions of ADH and aldosterone are redundant, but this is not the case. It is true that both hormones increase water reabsorption, but they are enlisted to counter different osmoregulatory problems. The release of ADH is a response to an increase in the osmolarity of the blood, as occurs when the body is dehydrated—by inadequate intake of water, for instance. But imagine a situation that causes an excessive loss of salt and body fluids—an injury, for example, or severe diarrhea. This reduces the blood's volume without increasing its osmolarity. Aldosterone would save the day by increasing water and Na^+ reabsorption in response to the drop in blood volume caused by fluid loss. Normally, ADH and aldosterone are partners in homeostasis; ADH alone would lower blood Na^+ concentration by stimulating water reabsorption in the kidney, but aldosterone helps maintain balance by stimulating Na^+ reabsorption.

Still another hormone, **atrial natriuretic protein (ANP),** opposes the renin-angiotensin-aldosterone system. The wall of the atrium of the heart releases ANP in response to an increase in blood volume and pressure, and the hormone counters by inhibiting the re-

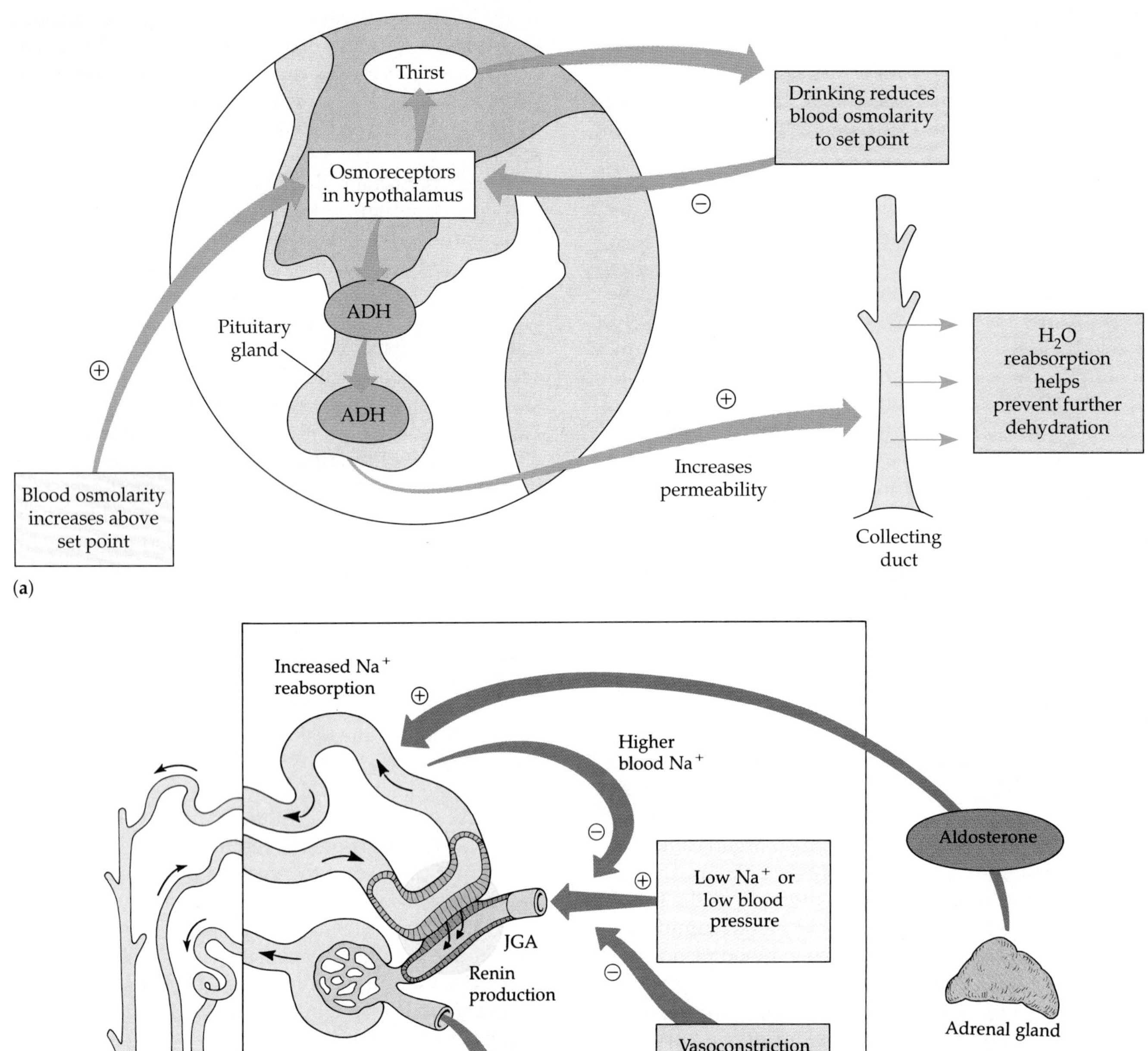

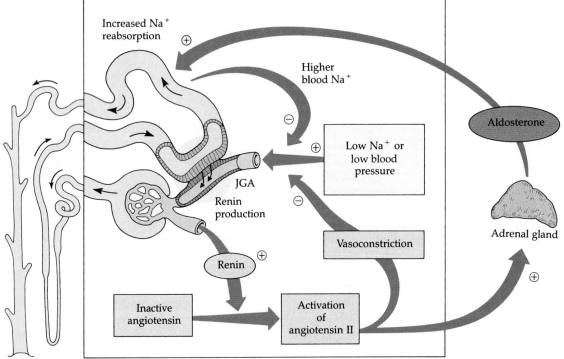

Figure 40.14
Hormonal regulation of the kidney.
(**a**) Antidiuretic hormone (ADH), produced in the hypothalamus of the brain and secreted from the pituitary gland, enhances fluid retention by increasing the permeability of the kidneys' collecting ducts to water. Release of ADH is triggered when osmoreceptor cells in the hypothalamus detect an increase in the osmolarity of the blood. In this situation, the osmoreceptor cells also promote thirst, and drinking reduces the osmolarity of the blood and inhibits the secretion of ADH, completing the feedback circuit. (**b**) The juxtaglomerular apparatus (JGA), a specialized tissue in the vicinity of the arterioles leading to glomeruli of the kidneys, responds to a decrease in blood pressure or Na^+ concentration by releasing the enzyme renin into the bloodstream. In the blood, renin initiates the conversion of angiotensin to its active form, angiotensin II. A hormone, angiotensin II acts directly to increase blood pressure by causing arterioles to constrict. Angiotensin II also signals the adrenal glands to release aldosterone, a hormone that stimulates the active reabsorption of Na^+ across the epithelium of the distal convoluted tubules. Reabsorption of Na^+, in turn, results in the reabsorption of more water. Thus, the release of renin from the JGA leads to an increase in the concentration of Na^+ in the blood and an increase in blood volume and pressure, results that complete the feedback circuit by suppressing the further release of renin.

lease of renin from the juxtaglomerular apparatus and also by directly reducing aldosterone release from the adrenal glands. These actions decrease Na^+ reabsorption and lower blood volume and pressure. Thus, ADH, the renin-angiotensin-aldosterone trio, and ANP provide an elaborate system of checks and balances that regulate the kidney's ability to control the osmolarity, salt concentration, volume, and pressure of blood.

Having considered the mammalian kidney and its regulation in detail, we can now compare the structures and functions of kidneys in other vertebrate classes.

Comparative Physiology of the Kidney

Variations in nephron structure and physiology equip the kidneys of different vertebrates for osmoregulation in their various habitats. We have seen, for instance, that nephrons of the mammalian kidney can concentrate urine and conserve water. Among mammals, those able to excrete the most hyperosmotic urine, such as kangaroo rats and other mammals adapted to the desert, have exceptionally long loops of Henle that maintain steep osmotic gradients in the kidney. This results in urine becoming very concentrated as it passes from cortex to medulla in the collecting ducts. In contrast, beavers, which rarely face problems of dehydration, have nephrons with very short loops, resulting in dilute urine.

Birds, like mammals, have kidneys with juxtamedullary nephrons that specialize in conserving water. However, the nephrons of birds have much shorter loops of Henle than those typical of mammalian nephrons, and birds are unable to concentrate urine to the osmolarities achieved by mammalian kidneys.

The kidneys of reptiles, having only cortical nephrons, produce urine that is, at best, isosmotic to body fluids. However, the epithelium of the cloaca (see Chapter 30) helps conserve fluid by reabsorbing some of the water present in urine and feces. Also, most terrestrial reptiles excrete nitrogenous wastes in an insoluble form known as uric acid, which helps conserve water because it does not contribute to the osmolarity of the urine. (This adaptation is discussed in more detail in the next section.)

In contrast to mammals and birds, freshwater fishes face the problem of excreting excess water because the animal is hyperosmotic to its surroundings. Instead of conserving water, the nephrons use cilia to sweep a large volume of very dilute urine from the body. Freshwater fishes conserve salts by efficient reabsorption of ions from the filtrate in the nephrons.

Amphibians' kidneys function much like those of freshwater fishes. When in fresh water, the skin of the frog accumulates certain salts from the water by active transport, and the kidneys excrete dilute urine. On land, where dehydration is the most pressing problem of osmoregulation, frogs conserve body fluid by reabsorbing water across the epithelium of the urinary bladder.

Bony fishes that live in seawater, being hypoosmotic to their surroundings, face the opposite problem to that of their freshwater relatives. In many species, nephrons lack glomeruli and capsules, and concentrated urine is formed by secreting ions into the renal tubules. Thus, the kidneys of marine fishes excrete very little urine and function mainly to get rid of divalent ions such as Ca^{2+}, Mg^{2+}, and SO_4^{2-}, which the fish takes in by its incessant drinking of seawater. As mentioned previously, monovalent ions such as Na^+ and Cl^- are excreted mainly by the gills, as is most nitrogenous waste in the form of NH_4^+ (ammonium).

Osmoregulation—the control of salt and water balance—was the original function of the kidney. In the course of evolutionary development, the excretion of nitrogenous wastes became a second function.

NITROGENOUS WASTES

Metabolism produces toxic by-products. Perhaps the most troublesome is the nitrogen-containing waste from the metabolism of proteins and nucleic acids. Nitrogen is removed from these nutrients when they are broken down for energy, or when they are converted to carbohydrates or fats. The nitrogenous waste product is ammonia, a small and very toxic molecule. Some animals excrete their ammonia directly; others first convert it to less toxic wastes such as urea or uric acid (Figure 40.15). We will see that the form of nitrogenous waste an animal excretes depends on both the animal's evolutionary history and its habitat.

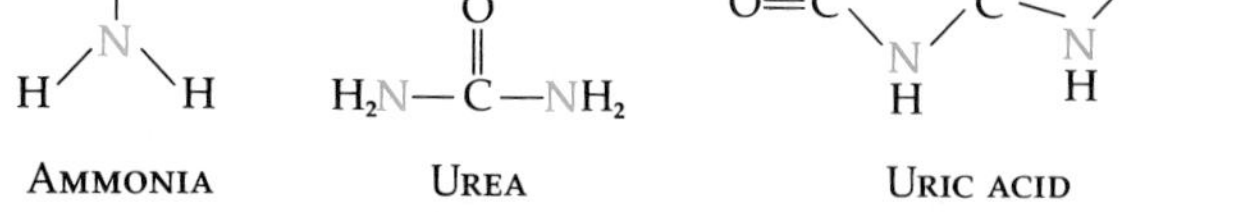

AMMONIA — UREA — URIC ACID

Figure 40.15
Nitrogenous wastes. Ammonia is a toxic by-product of the metabolic removal of nitrogen (deamination) from proteins and nucleic acids. Most aquatic animals get rid of ammonia by excreting it from body fluids. Most terrestrial animals convert the ammonia to urea or uric acid, which conserves water because these less toxic wastes can be transported in the body in more concentrated form.

Ammonia

Most aquatic animals excrete nitrogenous wastes as **ammonia.** Ammonia molecules are small and very soluble in water, so they easily permeate membranes. In soft-bodied invertebrates, ammonia diffuses across the whole body surface into the surrounding water. In fishes, most of the ammonia is lost as ammonium ions (NH_4^+) across the epithelium of the gills, with kidneys playing only a minor role in the excretion of nitrogenous waste. In freshwater fishes, the epithelium of the gills takes up Na^+ from the water in exchange for NH_4^+, which helps maintain Na^+ concentrations much higher than the Na^+ concentration in the surrounding water.

Urea

Ammonia excretion, though it works in water, is unsuitable for disposing of nitrogenous waste on land. A terrestrial animal would have to urinate copiously to get rid of ammonia, because a compound so toxic could only be transported in the animal and excreted in a very dilute solution. Instead, mammals and most adult amphibians excrete **urea.** (Many marine fishes and turtles, which have the problem of conserving water in their hyperosmotic environment, also excrete urea.) This substance can be handled in much more concentrated form because it is about 100,000 times less toxic than ammonia. Urea excretion enables the animal to sacrifice less water to discard its nitrogenous waste, an important adaptation for living on land.

Urea is produced in the liver by a metabolic cycle that combines ammonia with carbon dioxide. The circulatory system carries the urea to the kidneys. As mentioned earlier, not all urea is excreted immediately by mammalian kidneys; some of it is retained in the kidneys, where it contributes to osmoregulation by helping maintain the osmolarity gradient that functions in water reabsorption. Recall that sharks also produce urea, retained at a relatively high concentration in the blood, which helps balance the osmolarity of body fluids with the surrounding seawater.

Amphibians that undergo metamorphosis generally switch from excreting ammonia to excreting urea during the transformation from an aquatic larva, the tadpole, to the terrestrial adult. This biochemical modification, however, is not inexorably coupled with metamorphosis. Frogs that remain aquatic, such as the South African clawed toad *(Xenopus),* continue excreting ammonia after metamorphosis. But if these animals are forced to stay out of water for several weeks, they begin to produce urea. Similarly, African lungfish switch from ammonia to urea excretion if their habitat dries up and they are forced to burrow in the mud and become inactive (see Chapter 30).

Uric Acid

Land snails, insects, birds, and some reptiles excrete **uric acid** as the major nitrogenous waste. Because it is thousands of times less soluble in water than either ammonia or urea, uric acid can be excreted as a precipitate after nearly all the water has been reabsorbed from the urine. In birds and reptiles, the urine is excreted into the cloaca and eliminated in pastelike form along with feces from the intestine.

Uric acid and urea represent two different adaptations that enable terrestrial animals to excrete nitrogenous wastes with a minimal loss of water. One factor that seems to have been important in determining which of these alternatives evolved in a particular group of animals is the mode of reproduction. Soluble wastes can diffuse out of a shell-less amphibian egg or be carried away by the mother's blood in the case of a mammalian embryo. The vertebrates that excrete uric acid, however, produce shelled eggs, which are permeable to gases but not to liquids. If an embryo released ammonia or urea within a shelled egg, the soluble waste would accumulate to toxic concentrations. Uric acid precipitates out of solution and can be stored within the egg as a solid that is left behind when the animal hatches.

In grouping the various vertebrates according to the nitrogenous wastes they excrete, the boundaries are not drawn strictly along phylogenetic lines but depend also on habitat. Among reptiles, for instance, lizards, snakes, and terrestrial turtles excrete mainly uric acid; crocodiles excrete ammonia in addition to uric acid; and aquatic turtles excrete both urea and ammonia. In fact, individual turtles modify their nitrogenous wastes when their environment changes. A tortoise that usually produces urea can shift to uric acid production when the temperature increases and water becomes less available. This is another example of how response to the environment occurs on two levels: Evolution determines the limits of physiological responses for a species, but individual organisms make adjustments within these evolutionary constraints. This principle also applies to the regulation of body temperature.

THE REGULATION OF BODY TEMPERATURE

Metabolism is very sensitive to changes in the temperature of an animal's internal environment. For example, the rate of cellular respiration increases with temperature up to a certain point, and then declines when temperatures are high enough to begin denaturing enzymes (see Chapter 6). The properties of membranes also change with temperature. Although different species of animals are adapted to different temperature ranges—some animals thrive in deserts where temperatures often reach 40°C (105°F), and others flourish in frigid polar climates—each animal has an optimal temperature range. Within that range, many animals can maintain a constant internal temperature as the external temperature fluctuates. To understand the problems and mechanisms of temperature regulation, it is important to first consider the exchange of heat between organisms and their environment.

Heat Production and Transfer Between Organisms and Their Environment

An organism, like all objects, exchanges heat with its environment by four physical processes: conduction, convection, radiation, and evaporation.

Conduction is the direct transfer of thermal motion (heat) between molecules of the environment and those of the body surface, as when an animal sits in a pool of cold water or on a hot rock. Heat will always be conducted from a body of higher temperature to one of lower temperature (see Chapter 3). Water is 50 to 100 times more effective than air in conducting heat. This is one reason you can rapidly cool your body on a hot day by going for a swim.

Convection is the mass flow of air or liquid past the surface of a body, as when a breeze contributes to heat loss from the skin of an animal (or when a fan brings comfort to a human on a hot, still day). On the other hand, a wind-chill factor compounds the harshness of cold winter temperatures.

Radiation is the emission of electromagnetic waves produced by all objects warmer than absolute zero, including an animal's body and the sun. Radiation can transfer heat between objects that are not in direct contact, as when an animal absorbs heat radiating from the sun. A unique adaptation for exploiting solar radiation has recently been discovered in polar bears and arctic seals. The fur of these animals is actually clear, not white. Each hair functions somewhat like an optical fiber that transmits ultraviolet radiation to the black skin, where the energy is absorbed and converted to body heat.

Evaporation is the loss of heat from the surface of a liquid that is losing some of its molecules as gas. Evaporation of water from an animal has a significant cooling effect on the animal's surface (see Chapter 3).

If you were to sit at rest in still air at a comfortable temperature cooler than your body (for example, an air temperature of 23°C), conduction would account for only about 1% of your heat loss, convection for 40%, radiation for 50%, and evaporation for 9%. Convection and evaporation are the most variable causes of heat loss. A breeze of just 15 km/hr will increase total heat loss substantially by increasing convection fivefold.

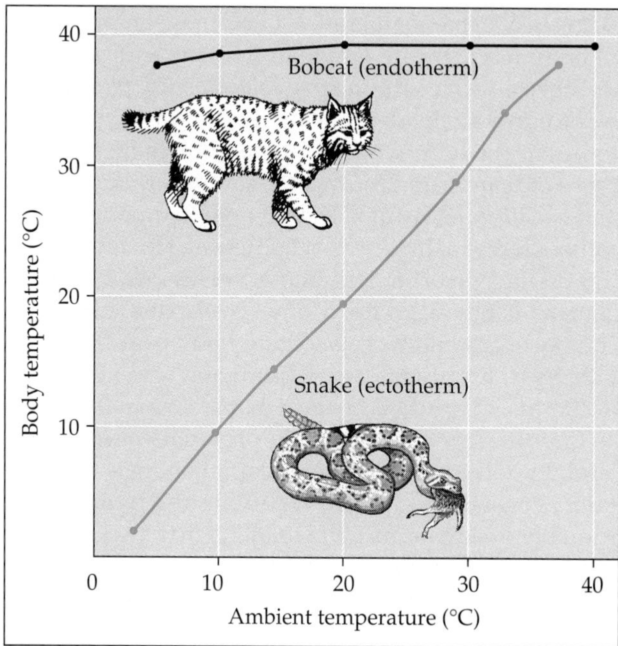

Figure 40.16
The relationship of body temperature to ambient (environmental) temperature in an ectotherm and an endotherm. An ectotherm (for example, a snake) obtains most of its body heat from the surroundings. An endotherm (for example, a bobcat) derives its body heat mainly from metabolism and uses metabolic energy for heating *and* cooling mechanisms that keep body temperature relatively constant. (Modified from P. T. Marshall and G. M. Hughes, *Physiology of Mammals and Other Vertebrates*, 2nd ed. Cambridge: Cambridge University Press, 1980.)

Evaporative cooling is increased greatly by the production of sweat. However, evaporation can only occur if the surrounding air is not saturated with water molecules (that is, the relative humidity is less than 100%). This is the biological basis for the common complaint, "The heat is not as bad as the humidity."

Ectotherms and Endotherms

One way to classify the thermal physiology of various animals is to emphasize the major source of body heat. Animals that warm their bodies mainly by absorbing heat from their surroundings are called **ectotherms.** Invertebrates, fishes, amphibians, and reptiles are generally ectotherms. In contrast, animals that derive most of their body heat from their own metabolism are called **endotherms.** Endotherms usually maintain a consistent internal temperature even as the environmental temperature fluctuates (Figure 40.16). However, a constant body temperature does not necessarily distinguish endotherms from ectotherms; for example, many marine fishes and invertebrates inhabit water with such stable temperatures that these animals vary in body temperature less than humans and other endotherms. Also, the terms *cold-blooded* and *warm-blooded* are misleading. Many lizards, which are ectotherms, have active body temperatures higher than those of mammals. Note again that the terms *ectotherm* and *endotherm* are not based on body temperature but rather on the main source of body heat. However, even this distinction of environmental versus metabolic heat sources is not absolute: Many ectotherms, including certain fishes and insects, obtain body heat from metabolism; and birds and mammals, which are endotherms, may add body heat by basking in the sun.

Endothermy solves certain problems of living on land. Endothermy enables terrestrial animals to maintain a constant body temperature in the face of environmental temperature fluctuations that are generally more severe than those an aquatic animal confronts. In general, the endothermic vertebrates—birds and mammals—are warmer than their surroundings, but these animals also have mechanisms for cooling the body in a hot environment. A consistently warm body temperature requires active metabolism, but, conversely, a warm body temperature contributes to the high levels of aerobic metabolism (cellular respiration) required for the endurance of intense physical activity. This is one reason endotherms can generally endure vigorous activity longer than ectotherms. These connections between body temperature, active aerobic metabolism, and mobility were important in the evolution of endothermy; moving on land requires considerably more effort than moving in water (see Chapter 45). The efficient circulatory and respiratory systems of birds and mammals can be thought of as adaptations accompanying the evolution of endothermy and a high metabolic rate. This is not to say that ectothermy is incompatible with terrestrial success. Among ectotherms are amphibians and reptiles, which have their own adaptations for coping with the temperature changes of terrestrial environments. In the next section, we will compare the mechanisms that determine body temperature in various endotherms and ectotherms.

Thermoregulation in Terrestrial Mammals

Heat Production Metabolism generates heat that can warm the body. Insulating layers of fat and fur help retain body heat. The rate of heat production can be increased in one of two ways: by the increased contraction of muscles (by moving around or by shivering) or by the action of certain hormones (especially epinephrine and thyroxine) that increase metabolic rate. The hormonal triggering of heat production is called **nonshivering thermogenesis.** The process takes place throughout the body, but some mammals have a tissue called **brown fat** in the neck and between the shoulders that is specialized for rapid heat production (see Figure 9.19).

Being endothermic is liberating, but it is also energetically expensive, especially in a cold environment. For example, at 20°C, an endotherm such as a human male has a basal metabolic rate of 1800 kcal per day (see Chapter 37). In contrast, an ectotherm of similar weight, such as an American alligator, has a basal metabolic rate of only 60 kcal per day. Thus, endotherms generally consume much more food (as measured in kilocalories) than ectotherms of equivalent size.

Mechanisms of Thermoregulation A land mammal maintains a relatively constant body temperature by a combination of physiological and behavioral adjustments that fall into four general categories:

1. *Changing the rate of metabolic heat production.* For example, when exposed to cold, many animals can double or triple their metabolic heat production through the increased activity of skeletal muscles and nonshivering thermogenesis.

2. *Adjusting the rate of heat exchange between an animal and its environment.* One such mechanism alters the amount of blood flowing to the skin. Increased blood flow usually results from **vasodilation.** In this process, certain nerves to the superficial blood vessels (those near the body surface) decrease their activity, relaxing the muscles of blood vessel walls and allowing blood flow through the vessels to increase. When this occurs, more heat is transferred to the environment by conduction, convection, and radiation. The reverse adjustment, **vasoconstriction** of superficial vessels, reduces blood flow and heat loss by decreasing the diameter of the vessels. In rabbits, for example, little blood flows to the large, poorly insulated ears when the rabbit sits in a cool environment. However, if the rabbit exercises, increasing its heat production and thus its body temperature, the blood vessels in its ears vasodilate, and warm blood carried to this region can lose heat to the environment. Vasodilation and vasoconstriction also contribute to regional temperature differences within the animal. For example, on a cool day, a human's temperature may be several degrees lower in the arms and legs than in the trunk, where most vital organs are located. Hair is also important in regulating heat exchange between a mammal and its environment. Most land mammals react to cold by raising their fur, which increases insulation. (Humans just get goose bumps, a vestige from our furry ancestors. Although we have relatively little hair, the muscles that raise hair in the cold still contract.)

3. *Cooling by evaporative heat loss.* Terrestrial mammals lose water from their respiratory tract surfaces and across their skin. Evaporation from the respiratory

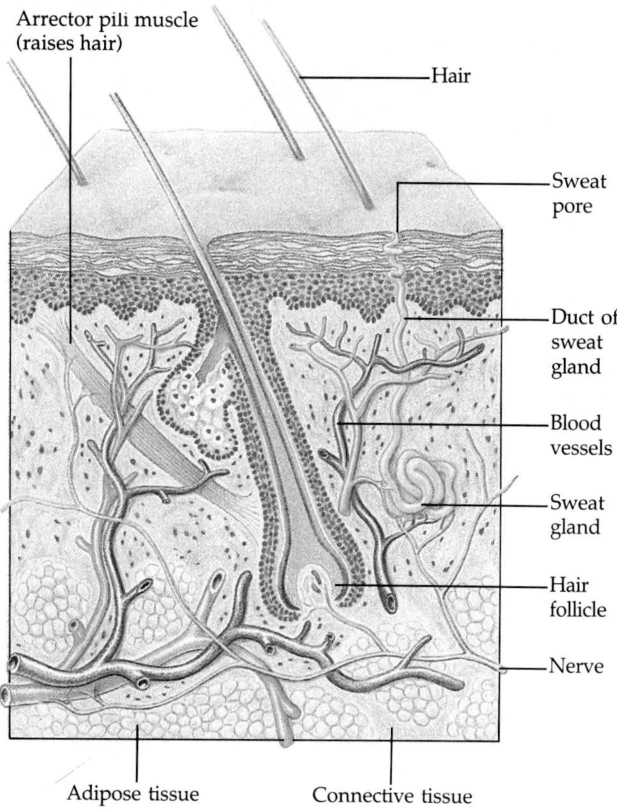

Figure 40.17
Skin as an organ of thermoregulation. Fat (adipose tissue) and hair help insulate mammals. Heat loss to the environment can be regulated by the constriction and dilation of superficial blood vessels and by the erection and compaction of fur. Sweat glands, under nervous control, function in evaporative cooling.

tract can be increased by panting. Evaporation across the skin can be increased by sweating, which is under nervous control (Figure 40.17). If the humidity of the air is low enough, the water will evaporate and cool the skin. Mammals lacking sweat glands have other mechanisms that promote evaporative cooling when heat stress is severe. For instance, some rodents spread saliva on their heads. Bats use both saliva and urine to cool themselves by evaporation.

4. *Behavioral responses.* Mammals and many other animals can increase or decrease body heat loss by relocating. They will bask in the sun in winter, find cool, damp areas or burrow in summer, or even migrate to a more suitable climate (Figure 40.18).

The Thermostat Nerve cells controlling thermoregulation, as well as those controlling other aspects of homeostasis, are concentrated in the hypothalamus of the brain (to be discussed in detail in Chapter 44). This small brain region has two thermoregulatory areas (Figure 40.19). One is called the **heating center** because it controls vasoconstriction of superficial vessels, erection of fur, shivering, and nonshivering thermogenesis. The other area is called the **cooling center** because it controls vasodilation and sweating (or panting). Nerve cells that sense temperature are located in the skin, the hypothalamus, and some other parts of the nervous system. Some of these temperature receptors, called Ruffini organs (warm receptors), increase their nervous activity when the temperature increases. Others, called bulbs of Krause (cold receptors), increase their activity when the temperature decreases. The warm receptors excite the cooling center of the hypothalamus and inhibit the heating center, whereas the cold receptors have exactly the opposite effects on the two centers. Body temperature is thus controlled by feedback mechanisms, with the hypothalamus functioning as a thermostat that triggers heating and cooling responses (see Chapter 36).

Thermoregulatory Adaptations in Other Animals

Birds The average body temperature of a bird is very high, around 40°C. Cooling mechanisms prevent their body temperature from exceeding the optimal range. Birds have no sweat glands, but instead use panting to promote evaporative heat loss. Some species have a specialized, highly vascularized pouch in the floor of the mouth that they can flutter, increasing evaporation.

Birds also have mechanisms for reducing heat loss. Feathers provide excellent insulation. All species, especially those that walk or swim in water, face the problem of losing large amounts of heat from their legs and feet. Warm blood from the core of the body must flow to the cells of these extremities. A special arrangement of arteries and veins called a **countercurrent heat**

Figure 40.18
Behavioral adaptations for thermoregulation. (**a**) Bathing in cool water brings immediate relief from the heat and continues to cool the surface for some time by evaporation. (**b**) Basking is an important mechanism for warming the body by radiation, convection, and conduction (from warm rocks, for example), especially in ectotherms such as these marine iguanas that inhabit the Galapagos Islands. (**c**) Dressing for the weather, illustrated by these Siberian children, is a thermoregulatory behavior unique to humans.

(a)

(b)

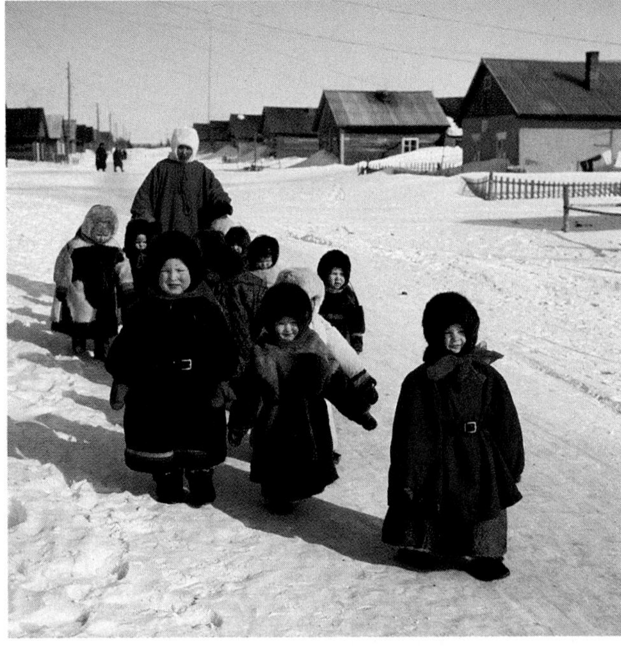

(c)

exchanger reduces heat loss (Figure 40.20a). Arteries carrying warm blood down the legs are in close contact with veins conveying blood in the opposite direction, back toward the trunk of the body. This countercurrent arrangement facilitates heat transfer from arteries to veins along the entire length of the blood vessels. Near the end of the leg, where the blood in an artery has been cooled to a temperature far below the animal's core temperature, the artery can still transfer heat to the even colder blood of a juxtaposed vein (recall that heat is conducted only from a warmer object to a cooler one). The venous blood can continue to absorb heat as it travels upward because it is passing warmer and warmer arterial blood flowing downward from the body core. By the time the venous blood reaches the top of a leg, it is almost as warm as the body core, minimizing the heat lost as a result of supplying blood to legs immersed in cold water. In some birds, blood can enter the limbs either through such an exchanger or by way of vessels that detour around the exchanger. The relative amount of blood that enters the limbs by way of the two different paths varies, controlling the amount of heat loss.

Marine Mammals Seals and whales maintain a body temperature of about 36°C to 38°C, similar to that of other mammals. All marine mammals live in water colder than their body temperature, and many live at least part of the year in nearly freezing arctic or antarctic water. Although the loss of heat to water occurs 50 to 100 times more rapidly than heat loss to air, the metabolism of marine mammals is not much higher than that of land mammals of similar size, suggesting that marine mammals conserve heat more effectively.

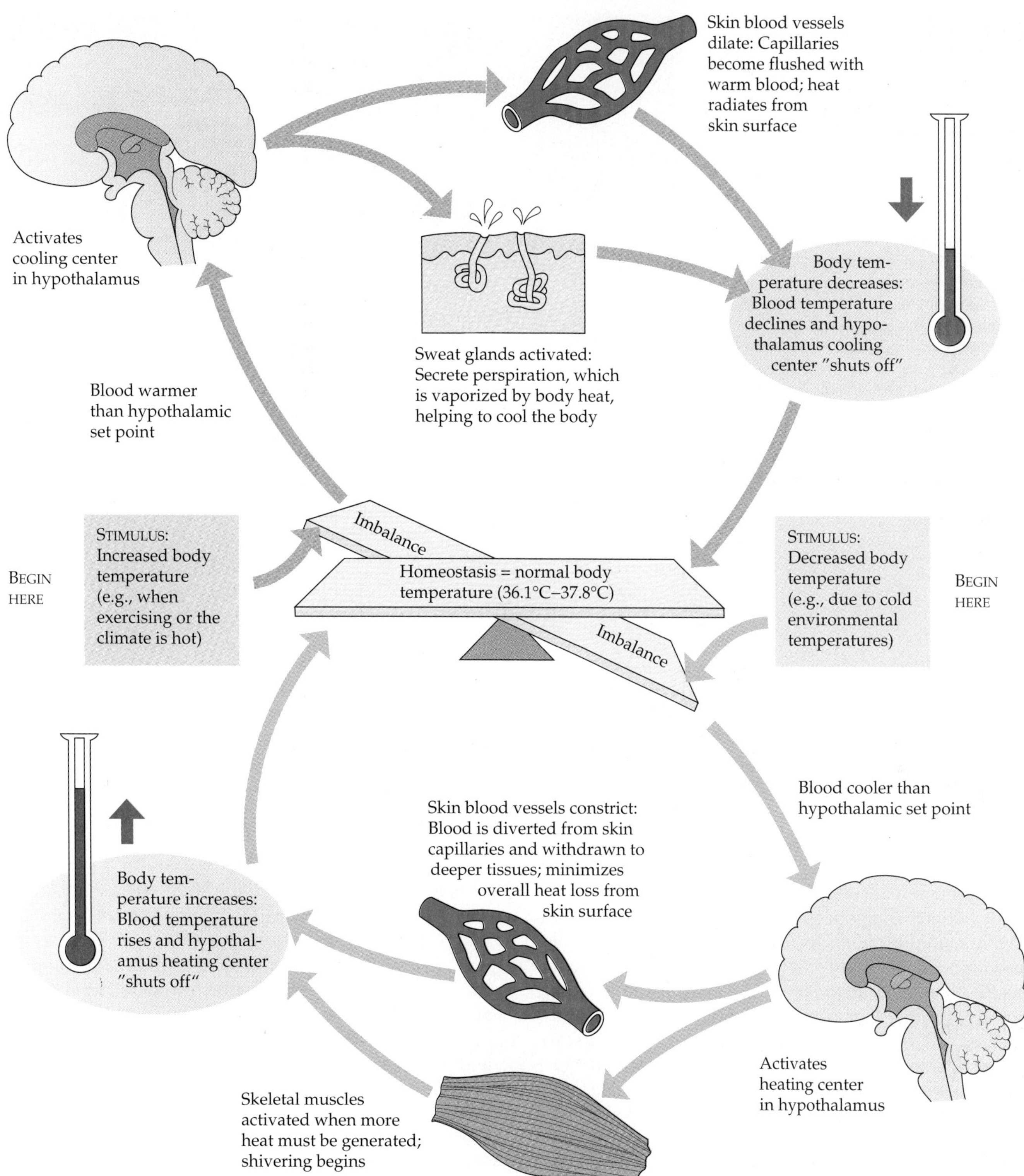
Skin blood vessels dilate: Capillaries become flushed with warm blood; heat radiates from skin surface
Activates cooling center in hypothalamus
Sweat glands activated: Secrete perspiration, which is vaporized by body heat, helping to cool the body
Body temperature decreases: Blood temperature declines and hypothalamus cooling center "shuts off"
Blood warmer than hypothalamic set point
Imbalance
STIMULUS: Increased body temperature (e.g., when exercising or the climate is hot)
BEGIN HERE
Homeostasis = normal body temperature (36.1°C–37.8°C)
STIMULUS: Decreased body temperature (e.g., due to cold environmental temperatures)
BEGIN HERE
Imbalance
Blood cooler than hypothalamic set point
Skin blood vessels constrict: Blood is diverted from skin capillaries and withdrawn to deeper tissues; minimizes overall heat loss from skin surface
Body temperature increases: Blood temperature rises and hypothalamus heating center "shuts off"
Activates heating center in hypothalamus
Skeletal muscles activated when more heat must be generated; shivering begins

Figure 40.19
The thermostat function of the hypothalamus and feedback mechanisms in human thermoregulation.

Figure 40.20
Countercurrent heat exchange.
(**a**) Some birds, such as this Canadian goose, possess countercurrent systems in their legs that reduce heat loss. The arteries carrying blood down the legs contact the veins that return blood to the body core. In a cold environment, the arterial blood transfers heat to the venous blood returning to the body (black arrows). The countercurrent flow facilitates heat exchange by setting up a thermal gradient between the arterial and venous blood over the entire length of the vessels (see Chapter 38). The body's core temperature is further protected by the constriction of surface veins. (**b**) In the flippers of marine mammals, such as this Pacific bottlenose dolphin, where there is no blubber, each artery is surrounded by several veins in a countercurrent arrangement that allows efficient heat exchange between venous and arterial blood.

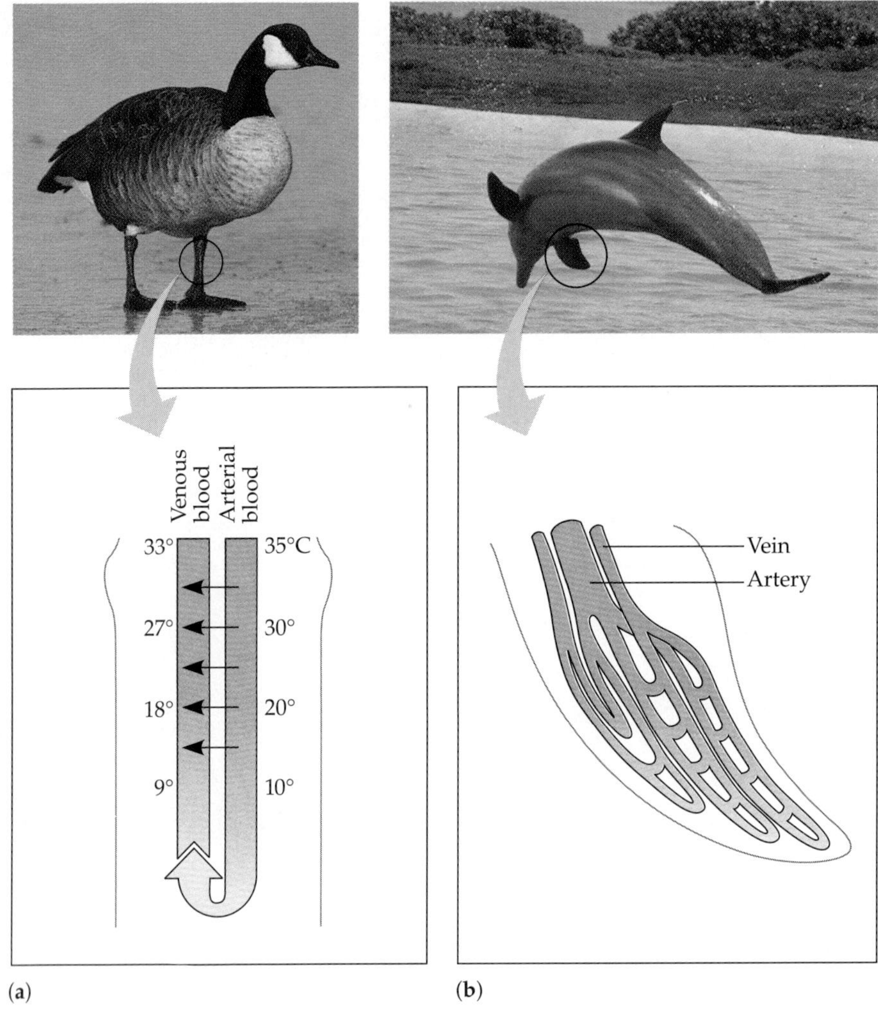

The major adaptation for retaining body heat in a cold, wet environment is a very thick layer of insulating fat called blubber, just under the skin. In the tail and flippers, where there is no blubber, countercurrent heat exchange occurs between the arterial and venous blood, as in the legs of water-dwelling birds (Figure 40.20b).

Because many marine mammals move to warmer environments in their annual migrations, they also face the challenge of dissipating metabolic heat. Blood vessels that serve the skin dilate, allowing greater amounts of blood to be cooled by the surrounding water, which even in these warmer seas is typically still well below body temperature.

Reptiles Reptiles are generally ectotherms with relatively low metabolic rates that contribute little to normal body temperature. Reptiles warm themselves mainly by behavioral adaptations. They seek warm places, orienting themselves toward heat sources to increase heat uptake and expanding the body surface exposed to a heat source. Reptiles do not simply maximize heat uptake, however; they may behave in such a way as to truly regulate their temperature within a range. If a sunny spot is too warm, for instance, a lizard will alternately sit in the sun and in the shade, or will turn in another direction, which lowers the surface area exposed to the sun. By seeking favorable microclimates within the environment, many reptiles maintain body temperatures that are quite stable.

Indications are that some of the sophisticated regulatory mechanisms found in mammals are present in reptiles, at least in a rudimentary state. (Recall from Chapter 30 that mammals and birds evolved from reptiles.) For example, in diving reptiles, body heat is conserved by routing more blood to the body core during a dive. Some reptiles can also increase thermogenesis, just as mammals do. For example, female pythons incubating their eggs generate considerable heat by shivering.

Amphibians The optimal temperature range for amphibians varies substantially with the species. For example, closely related species of salamanders have average body temperatures ranging from 7°C to 25°C.

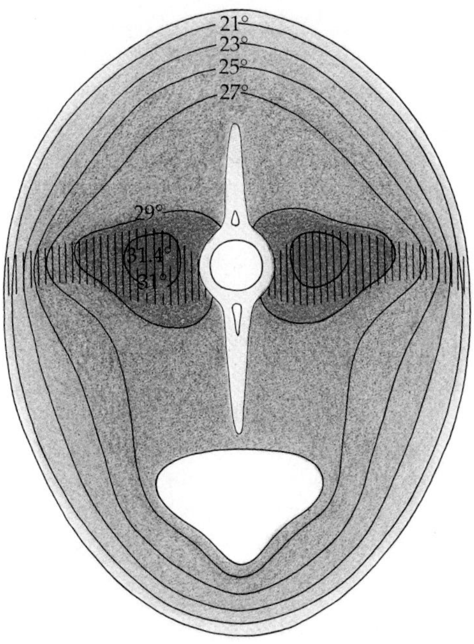

(a) Internal temperature in the bluefin tuna

Arteries
Veins

(b) Countercurrent heat exchange system in the shark

0.5 mm

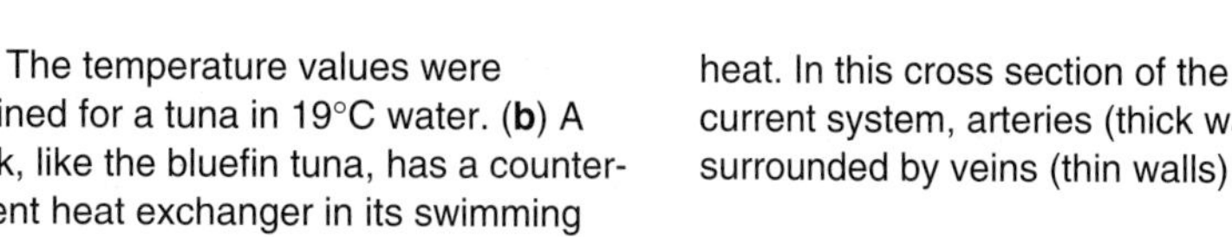

Figure 40.21
Thermoregulation in large, active fishes. (**a**) The bluefin tuna maintains an internal temperature much higher than the surrounding water. The fish is warmest around its dark swimming muscles (shading). The temperature values were obtained for a tuna in 19°C water. (**b**) A shark, like the bluefin tuna, has a countercurrent heat exchanger in its swimming muscles that reduces the loss of metabolic heat. In this cross section of the countercurrent system, arteries (thick walls) are surrounded by veins (thin walls) (LM).

Amphibians produce very little heat, and most lose heat rapidly by evaporation from their body surfaces, making it difficult to control body temperature. However, behavioral adaptations enable them to maintain body temperature within a satisfactory range most of the time, by moving to a location where solar heat is available or into water, for instance. When the surroundings are too warm, the animals seek cooler microenvironments, such as shaded areas. Some amphibians, including bullfrogs, can vary the amount of mucus they secrete from their surface, a physiological response that regulates evaporative cooling.

Fishes The body temperature of most fishes is usually within 1°C or 2°C of the surrounding water temperature. However, some large, active fishes maintain an elevated temperature at their body core. The heat generated by their swimming muscles is retained by specializations of the circulatory system. The bluefin tuna, for example, has its major blood vessels just under the skin. Branches deliver blood to the deep swimming muscles, where the small vessels are arranged into a countercurrent heat exchanger called the **rete mirabile,** which means "wonderful net." The apparatus enhances vigorous activity by keeping the swimming muscles several degrees warmer than the tissues nearer the animal's surface, which is about the same temperature as the surrounding water (Figure 40.21). Such animals, called partial endotherms, illustrate how the distinction between ectothermy and endothermy can sometimes blur.

Invertebrates Most invertebrates have very little control over their body temperature, but some do adjust temperature by behavioral or physiological mechanisms. The desert locust, for example, must reach a certain temperature to become active. It orients in a direction that maximizes the absorption of sunlight.

Some species of large flying insects, such as bees and large moths, can generate internal heat. They are able to "warm up" before taking off by contracting all the flight muscles in synchrony, so that only slight movements of the wings occur, but large amounts of heat are produced. (This is functionally analogous to shivering.) This higher temperature of the flight muscles enables the insects to sustain the intense activity required for flight (Figure 40.22a). This mechanism has been elaborated to such an extent in bumblebees that the internal temperature of the abdomen is maintained above the environmental temperature at all times, not just in preparation for flight. Equally impressive are the endothermic adaptations of winter moths, which

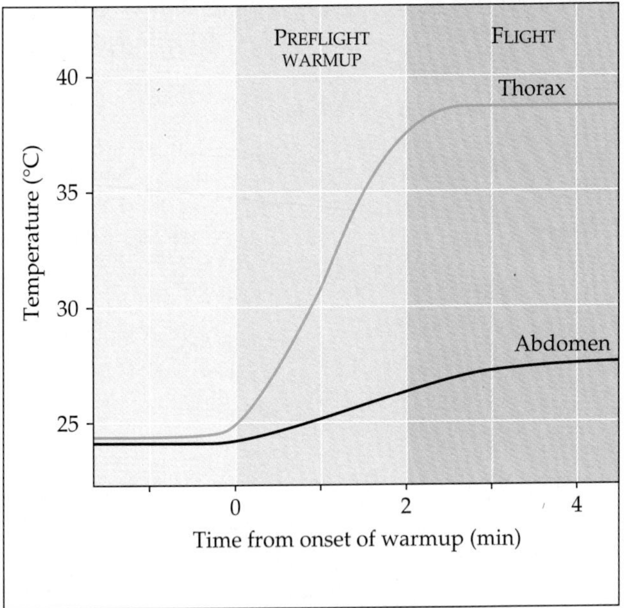

(**a**) Preflight warmup in the sphinx moth

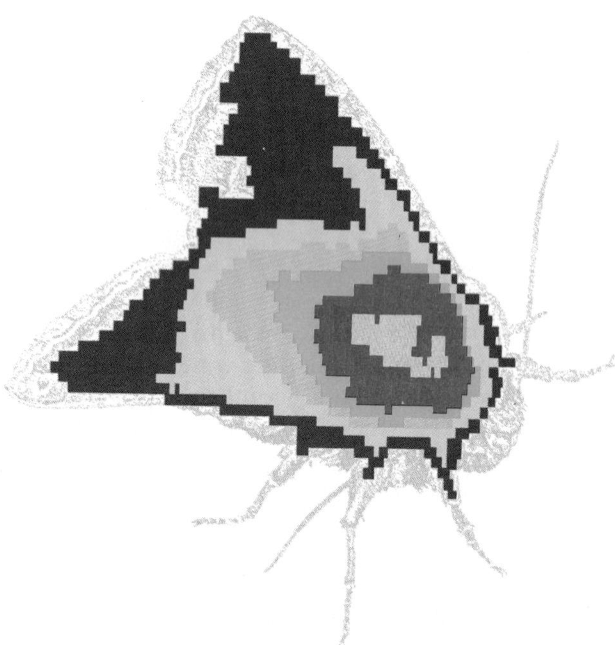

(**b**) Internal tempurature in the winter moth

Figure 40.22
Thermoregulation in moths. (**a**) The sphinx moth *(Manduca sexta)* is one of the many insect species using a shiveringlike mechanism for preflight warmup of its thoracic flight muscles. Once the moth takes off, metabolic activity of the flight muscles maintains a high thoracic temperature. (**b**) Thermoregulation is particularly impressive in some species of the Noctuidae family of moths, which are active during cold winter months. Various endothermic adaptations, including a countercurrent heat exchanger in the thorax region, help keep the flight muscles warmed to a temperature of 30°C, even though the ambient temperature may be subfreezing. An infrared map superimposed over an image of a winter moth portrays the distribution of heat immediately after a flight. Yellow, located here in the thorax region, indicates the highest temperature. Moving outward from the thorax, the variously colored zones correspond to regions of cooler and cooler body temperatures.

enable these insects to survive and fly during cold winter months. For example, the thorax of a winter moth has a countercurrent heat exchanger that helps maintain the flight muscles at a temperature of about 30°C, even on cold, snowy nights (Figure 40.22b).

Honeybees use an additional mechanism that depends on social organization to increase body temperature. In cold weather, they increase their movements and huddle together, which retains their heat. They maintain a relatively constant temperature by changing the density of the huddling. Individuals move from the cooler outer edges of the cluster to the warmer center and back again, thereby circulating and distributing the heat. Honeybees also control the temperature of their hive by transporting water to it in hot weather and fanning with their wings, which promotes evaporation and convection. The whole colony of honeybees uses many of the mechanisms of temperature control also seen in single, larger organisms.

Temperature Acclimation

Many animals can adjust to a new range of environmental temperature over a period of many days or weeks, a physiological response called **acclimation.** Seasonal change is one context in which physiological adjustments to a new temperature range are important. For example, many frogs that inhabit temperate regions can tolerate winter temperatures that are low enough to be lethal to the same frogs if encountered during summer. The upper limits of temperature tolerance can also be adjusted. The bullhead catfish, for instance, can survive water temperatures up to 36°C in summer, but a temperature above 28°C is lethal during winter.

Physiological acclimation to a new temperature range is multifaceted. Cells may increase the production of certain enzymes, helping compensate for the lowered activity of each enzyme molecule at temperatures that are not optimal. In other cases, cells produce variants of enzymes that have the same function but different temperature optima (see Chapter 6). Membranes may also change in the proportions of saturated and unsaturated lipids they contain. This response helps keep membranes fluid at different temperatures (see Chapter 8). Other physiological responses to a new range of external temperatures involve adjustments in the mechanisms that control the internal temperature of the animal.

Torpor

Torpor is an alternative physiological state in which metabolism decreases and the heart and respiratory system slow down. One type of torpor is **hibernation,** during which body temperature is maintained at a

Figure 40.23
Hibernation, one type of torpor. During hibernation, metabolic rate is slowed substantially and body temperature is lowered, adaptations that enable an animal to survive long, cold periods when food is scarce. This is a golden-mantled ground squirrel.

lower level than normal (Figure 40.23). Hibernation enables the animal to withstand long periods of cold temperatures and diminished food supplies. Another type of torpor characterized by slow metabolism and inactivity is **aestivation,** which allows an animal to survive long periods of elevated temperatures and diminished water supplies. **Diurnation** is a type torpor physiologically similar to hibernation and aestivation, but it lasts for much shorter time periods. Most bats and hummingbirds, for instance, have a daily torpor during daylight hours and also hibernate for many weeks during the winter months (see Figure 9.19).

Many ectothermic animals enter a state of torpor when their food supply decreases. In addition to the other physiological changes, their optimal temperature range becomes lower. Environmental changes that tend to increase the food supply cause these animals to come out of their torpor in a matter of hours.

Some endothermic animals also exhibit torpor that appears to be adapted to their feeding patterns. All endotherms that show a daily torpor, such as hummingbirds, bats, and shrews, are relatively small and have a very high metabolic rate when active. The daily period of torpor allows the animals to survive the hours when they do not normally feed. However, this activity-torpor cycle may not be triggered by the availability of food. Even if food is made available to a shrew all day, it still goes through its daily torpor; the cycle is a built-in rhythm controlled by the biological clock (see Chapter 35). In fact, the need for sleep in humans may be a remnant of a more pronounced daily torpor in our early mammalian ancestors.

Hibernation and aestivation are often triggered by seasonal changes in the length of daylight. As the days shorten, some animals will eat huge quantities of food before hibernating. Ground squirrels, for instance, will more than double their weight in a month of gorging.

THE INTERACTION OF REGULATORY SYSTEMS

One of the major challenges in animal physiology is unraveling how the various regulatory systems are controlled and how they interact. For example, the regulation of body temperature involves mechanisms that also have an impact on such parameters of the internal environment as osmolarity, metabolic rate, blood pressure, tissue oxygenation, and body weight. Under some conditions, usually at the physical extremes compatible with the organism's life, the demands of one system might come into conflict with those of other systems. For instance, in very warm and dry environments, the conservation of water takes precedence over evaporative heat loss. As a result, many desert animals tolerate occasional hyperthermia (abnormally high body temperature). Normally, however, the various regulatory systems act in concert to maintain homeostasis in the internal environment. The feedback circuits that integrate homeostasis involve nervous communication and hormones, which are chemical signals transmitted in the bloodstream. The next chapter addresses hormonal control in animals.

STUDY OUTLINE

1. Many animals can survive large fluctuations in their external environment by using homeostatic mechanisms to maintain a relatively constant internal environment.
2. Physiological adjustments, usually involving feedback mechanisms, regulate the hemolymph or interstitial fluid that bathes the body's cells.

Osmoregulation (pp. 877–881)

1. Water uptake must balance water loss, requiring various mechanisms of osmoregulation in different environments. Osmoconformers are isosmotic with their aqueous surroundings and do not regulate their osmolarity. Osmoregulators control water uptake and loss in a hyperosmotic or hypoosmotic environment.

2. Most marine invertebrates are osmoconformers, whereas most marine vertebrates osmoregulate. Sharks have an osmolarity slightly higher than seawater because they retain urea. Marine bony fishes lose water to their hyperosmotic environment and must compensate by drinking large quantities of seawater. Marine vertebrates excrete excess salt through rectal glands, gill epithelia, or nasal salt glands and kidneys.
3. Freshwater organisms constantly take in water from their hypoosmotic environment. Protozoa pump out excess water with contractile vacuoles, and freshwater animals excrete copious amounts of dilute urine. Salt loss is replaced by eating or ion uptake across gill epithelia.
4. Euryhaline animals are able to tolerate large osmotic changes in their environments, often by altering their osmoregulatory mechanisms.
5. Terrestrial animals combat desiccation by drinking and eating food with high water content and through the nervous and hormonal control of thirst, behavioral adaptations, and water-conserving excretory organs.
6. Osmoregulation is usually accomplished by the transport of salt across a transport epithelium, followed by the osmotic flow of water.

Excretory Systems of Invertebrates (pp. 881–883)

1. Extracellular fluid is filtered into the protonephridia of the flame-cell system in flatworms. These blind-ended tubules excrete a dilute fluid and function in osmoregulation.
2. Each segment of an earthworm has a pair of open-ended, tubular excretory organs closely associated with the capillaries. These metanephridia collect coelomic fluid, the transport epithelia pump out salts for reabsorption, and dilute urine is excreted through nephridiopores.
3. In insects, Malpighian tubules function in osmoregulation and removal of nitrogenous wastes from the hemolymph. Insects produce a relatively dry waste matter, an important adaptation to terrestrial life.

The Vertebrate Kidney (pp. 883–893)

1. Nephrons, the excretory tubules of vertebrates, are arranged into compact organs, the kidneys, which, along with associated blood vessels and excretory ducts, comprise the excretory system.
2. The major parts of the renal tubule are the Bowman's capsule, proximal convoluted tubule, loop of Henle, distal convoluted tubule, and collecting duct, which passes urine into the renal pelvis. Most nephrons are radially arranged in the kidney cortex; mammals and birds also have juxtamedullary nephrons with loops of Henle that extend into the medulla. Blood supply to the nephron is through an afferent arteriole, which divides into the capillaries of the glomerulus. An efferent arteriole carries blood away from the Bowman's capsule and subdivides into the peritubular capillaries embracing the convoluted tubules. The vasa recta is a system of capillaries that serves the loop of Henle.
3. Nephrons control the composition of the blood by filtration, secretion, and reabsorption. During filtration, blood pressure nonselectively filters water and small solutes from the glomerulus held within the Bowman's capsule into the lumen of the nephron tubule. Additional substances destined for excretion are directly secreted from the interstitial fluid into the tubule by active and passive transport. Filtered substances that must be returned to the blood, such as vital nutrients and water, are reabsorbed at various points along the nephron.
4. Most of the salt and water filtered from the blood is reabsorbed by the proximal convoluted tubule. In addition, ammonia, drugs, and hydrogen ions (for the control of body pH) are selectively secreted into the filtrate; glucose and amino acids are actively transported out of the filtrate; and potassium is reabsorbed. The descending limb of the loop of Henle is permeable to water but not to salt; water moves by osmosis into the hyperosmotic interstitial fluid. Salt diffuses out of the concentrated filtrate as it moves through the salt-permeable ascending limb of the loop of Henle. The distal convoluted tubule is specialized for selective secretion and reabsorption, playing a key role in regulating potassium concentration and blood pH. The collecting duct, which is permeable to water but not to salt, carries the filtrate through the osmolarity gradient of the medulla, and more water exits by osmosis. Urea also diffuses out of the tubule, joining salt in forming the osmotic gradient that enables the kidney to produce urine that is hyperosmotic to the blood.
5. The osmolarity of the filtrate varies as it travels through a juxtamedullary nephron. The differing permeabilities to water, salt, and urea in regions of the tubule, combined with the active transport of salt, produce the increasing osmotic gradient in the medulla that results in the production of a hyperosmotic urine.
6. The osmolarity of the urine, which can vary widely depending on the hydration needs of the body, is regulated by nervous and hormonal control of water and salt reabsorption in the kidneys. ADH, released in response to a rise in blood osmolarity signaled by osmoreceptor cells in the hypothalamus, increases water reabsorption by the tubule. The juxtaglomerular apparatus (JGA) responds to decreased blood pressure or Na^+ concentration by the release of renin, which activates angiotensin. This activated blood protein causes blood arterioles to constrict and the adrenal glands to release aldosterone, a hormone that stimulates reabsorption of Na^+ and the passive flow of water from the filtrate. Both ADH and aldosterone control mechanisms result in the reabsorption of water and more concentrated urine. ANP, released by the atrium in response to increased blood pressure, inhibits the release of renin and counters the renin-angiotensin-aldosterone system.
7. Adaptations of nephron structure and function in vertebrates are related to the requirements for osmoregulation and excretion of nitrogenous wastes in various habitats.

Nitrogenous Wastes (pp. 893–894)

1. The metabolism of proteins and nucleic acids generates ammonia, a toxic waste product that is excreted in one of three forms, depending on the habitat and evolutionary history of the animal.
2. Most aquatic animals excrete ammonia, a highly toxic but very soluble molecule that easily passes across the body surface or gill epithelia into the surrounding water.
3. The liver of mammals and most adult amphibians converts ammonia into the less toxic urea, which is carried by the circulatory system to the kidneys and excreted in concentrated form with a minimal loss of water.

4. Uric acid is an insoluble precipitate excreted in the paste-like urine of land snails, insects, birds, and some reptiles.
5. The mode of reproduction of terrestrial animals is related to their form of nitrogenous waste product. Differences in waste product within phyla are related to habitat; some organisms can actually shift their nitrogenous waste form depending on environmental conditions.

The Regulation of Body Temperature (pp. 894–903)

1. Heat transfer between the organism and the environment involves the physical processes of conduction, convection, radiation, and evaporation.
2. Ectotherms absorb their body heat from the environment. Endotherms, mainly mammals and birds, rely on the heat from their own metabolism. Endothermy allows terrestrial animals to maintain a constant body temperature and a high level of aerobic metabolism that facilitates movement on land.
3. The physiological and behavioral processes that enable land mammals to regulate their body temperature include the adjustment of the rate of metabolic heat production by shivering or nonshivering thermogenesis, vasodilation or vasoconstriction of surface blood vessels, control of evaporative heat loss through panting or sweating, and various behavioral responses.
4. The heating center and cooling center are thermoregulatory areas of the hypothalamus that receive nervous impulses from warm and cold receptors and respond by initiating either cooling or warming processes.
5. Birds may thermoregulate by panting, increasing evaporation from a vascularized pouch in the mouth, and by passing blood going to the legs through a countercurrent heat exchanger.
6. Marine mammals maintain their high body temperatures in cold water by a thick layer of insulating blubber and countercurrent heat exchange between arterial and venous blood. Vasodilation of skin vessels allows dissipation of heat in warm waters.
7. Reptiles and amphibians maintain internal temperatures within tolerable ranges mainly by various behavioral adaptations.
8. Although the body temperature of most fishes matches the environment, some large, active species maintain a higher temperature in their swimming muscles with a countercurrent heat exchanger called a rete mirabile.
9. A few invertebrates use physiological or behavioral mechanisms, such as muscle contractions, countercurrent heat exchangers, or social huddling, to adjust their temperature.
10. Many animals can physiologically acclimate to a gradual change in temperature, such as a seasonal change.
11. Torpor, including hibernation, aestivation, and diurnation, is a physiological state characterized by a decrease in metabolic, heart, and respiratory rates. This state allows the animal to temporarily withstand varying periods of unfavorable temperatures or lack of food and water.

The Interaction of Regulatory Systems (p. 903)

1. Homeostasis is a dynamic response to the environment, integrated by feedback circuits involving nervous communication and hormones.

SELF-QUIZ

1. Which of the following organisms is closest to being isosmotic to its environment?
   a. a marine bony fish
   b. a freshwater bony fish
   c. a marine jellyfish
   d. a freshwater protozoan
   e. a marine reptile
2. *Unlike* an earthworm's metanephridia, a mammalian nephron
   a. is intimately associated with a capillary network
   b. forms urine by changing the composition of fluid inside the tubule
   c. functions in both osmoregulation and excretion of nitrogenous wastes
   d. filters blood instead of coelomic fluid
   e. has a transport epithelium
3. The majority of water and salt filtered into the Bowman's capsule is reabsorbed by
   a. the brush border of the transport epithelia of the proximal convoluted tubule
   b. diffusion from the descending limb of the loop of Henle into the hyperosmotic interstitial fluid of the medulla
   c. active transport across the transport epithelium of the thick upper segment of the ascending limb of the loop of Henle
   d. selective secretion and diffusion across the distal convoluted tubule
   e. diffusion from the collecting duct into the increasing osmotic gradient of the medulla
4. The high osmotic concentration of the kidney medulla is maintained by all the following *except*
   a. diffusion of salt from the ascending limb of the loop of Henle
   b. active transport of salt from the upper region of the ascending limb
   c. the spatial arrangement of juxtamedullary nephrons
   d. diffusion of urea from the collecting duct
   e. diffusion of salt from the collecting duct
5. Antidiuretic hormone affects the kidney by
   a. stimulating the release of renin
   b. constricting arterioles and thus raising blood pressure
   c. increasing the reabsorption of Na^+ in the distal convoluted tubules
   d. countering the renin-angiotensin-aldosterone mechanism
   e. increasing the water permeability of the distal convoluted tubule and collecting duct, thus increasing the osmolarity of the urine
6. Ammonia (or ammonium) is excreted by most
   a. bony fishes
   b. organisms that produce shelled eggs
   c. adult amphibians
   d. land snails
   e. insects

7. Nonshivering thermogenesis is
   a. a behavioral adaptation for absorbing heat in ectotherms
   b. a hormone-triggered rise in metabolic rate, often associated with brown fat
   c. a countercurrent heat exchange of blood going to the limbs
   d. a heat-producing method of the rete mirabile of large, active fishes
   e. isometric muscle contractions that have been identified in winter moths
8. Physiological adjustments or acclimation by an ectotherm to cooler seasonal temperatures may include
   a. an increase in metabolic rate
   b. aestivation
   c. changes in the lipid components of cell membranes
   d. increased vasodilation and countercurrent heat exchange
   e. hibernation
9. Malpighian tubules are excretory organs found in
   a. vertebrates
   b. insects
   c. flatworms
   d. annelids
   e. jellyfish
10. Which process in the nephron is *least* selective?
    a. secretion
    b. reabsorption
    c. transport across the epithelium of a collecting duct
    d. filtration
    e. salt pumping by the loop of Henle

CHALLENGE QUESTIONS

1. A large part of the terrestrial success of arthropods and vertebrates is attributable to their osmoregulatory capabilities. Compare and contrast the Malpighian tubule with the nephron in regard to anatomy, relationship to circulation, and physiological mechanisms for conserving body water.
2. The longer a cat plays with a lizard or frog, the harder it is for the reptile or amphibian to escape. Explain this scenario in terms of the metabolic adaptations of the animals (assume that the cat has not injured its captive).

SCIENCE, TECHNOLOGY, AND SOCIETY

1. The kidneys remove many drugs from the blood, and these substances show up in the urine. Some employers require a urine drug test at the time of hiring and at intervals during the term of employment. An employee who fails a drug test can lose his or her job. What are some arguments for and against drug testing of individuals in certain kinds of occupations?
2. Kidneys were the first organs to be successfully transplanted. A donor can live a normal life with a single kidney, making it possible for individuals to donate a kidney to an ailing relative or even an unrelated individual with a similar tissue type. In some countries, poor people *sell* kidneys to transplant recipients through organ brokers. What are some of the ethical issues associated with this organ commerce?

FURTHER READING

Diamond, J. "Pearl Harbor and the Emperor's Physiologists." *Natural History,* December 1991. How our childhood locale affects our ability to tolerate hot climates.

Heinrich, B. "Thermoregulation in Winter Moths." *Scientific American,* March 1987. Endothermy in insects.

Marieb, E. N. *Human Anatomy and Physiology,* 2nd ed. Redwood City, CA: Benjamin/Cummings, 1992. Chapter 26 describes the human excretory system.

McKenzie, A. "Seeking the Mechanisms of Hibernation." *BioScience,* June 1990. The role of the brain in a fascinating adaptation.

Monastersky, R. "Dinosaurs in the Dark." *Science News,* March 19, 1988. How did dinosaurs living near the poles survive winter?

Smith, H. W. *From Fish to Philosopher.* Boston: Little, Brown, 1953. Vertebrate evolution as revealed by kidney structure and function. A classic.

Underwood, B. "Bee Cool." *Natural History,* December 1990. How bees survive Himalayan winters.

41 Chemical Signals in Animals

An Overview of Chemical Signals
Mechanisms of Hormone Action at the Cellular Level
Invertebrate Hormones
The Vertebrate Endocrine System
Endocrine Glands and the Nervous System

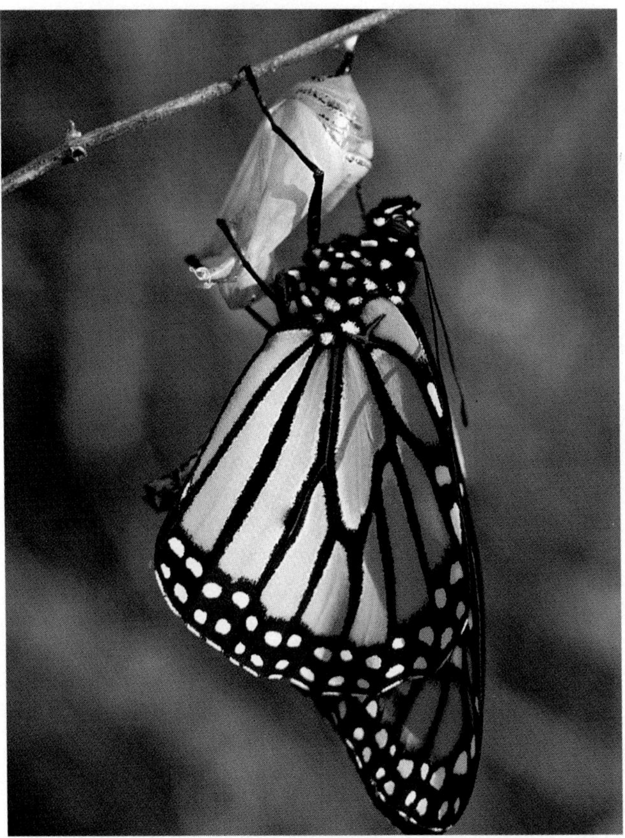

Figure 41.1
Hormones help control growth, development, reproduction, and homeostasis. For example, insect hormones are responsible for triggering the metamorphosis of a caterpillar into a butterfly. This monarch butterfly has just emerged from its cocoon. In this chapter, you will learn how hormones and other chemical signals coordinate animal bodies.

People offer hormones as an explanation for the howling of alley cats and the moodiness of teenagers. Over a million diabetics in the United States take the hormone insulin, and other hormones are used in cosmetics intended to keep the skin smooth or are added to livestock feed to fatten cows. Hormones affect metabolism, homeostasis, growth, development, emotion, and behavior. But what are they, and how do they work?

Animals coordinate the activities of their specialized parts. The two major systems of internal communication that have evolved are the **nervous system** and the **endocrine system.** The nervous system, which we will study in Chapters 44 and 45, conveys high-speed signals along specialized cells called neurons. These rapid messages function in such activities as the movement of body parts in response to sudden environmental changes—jerking one's hand away from a flame, for instance. Other biological processes use slower means of communication. Different parts of the body, for example, must be informed how fast to grow and when to develop the characteristics that distinguish male and female or the juvenile and the adult of a species (Figure 41.1). In some cases, neighboring cells communicate by chemical messages that simply diffuse through the interstitial fluid from one cell to another. In many cases, however, messages must travel over greater distances to reach their targets. The endocrine system is a major source of such dispatches in the form of the chemicals called hormones. This chapter surveys the hormonal control of animal form and function.

AN OVERVIEW OF CHEMICAL SIGNALS

All animals exhibit some form of coordination by chemical signals (Figure 41.2). The hormones of the endocrine system convey information between organs of the body, while other types of chemical messengers are used for other functions. Pheromones are like hormones in that they are chemical signals, but they are used to communicate between different individuals, as

Figure 41.2
An overview of chemical signaling.
(**a**) Endocrine cells secrete their signals, hormones, into the circulatory system. Although the bloodstream distributes a hormone throughout the body, only specific cells, the target cells, have receptors for the hormone and so respond to the chemical signal. (**b**) There are two mechanisms of local chemical signaling. In synaptic signaling, nerve cells (neurons) release neurotransmitters into a synapse, a very narrow space between the neuron and a single target cell. In paracrine signaling, a secreting cell affects nearby target cells by discharging local regulators into the interstitial fluid.

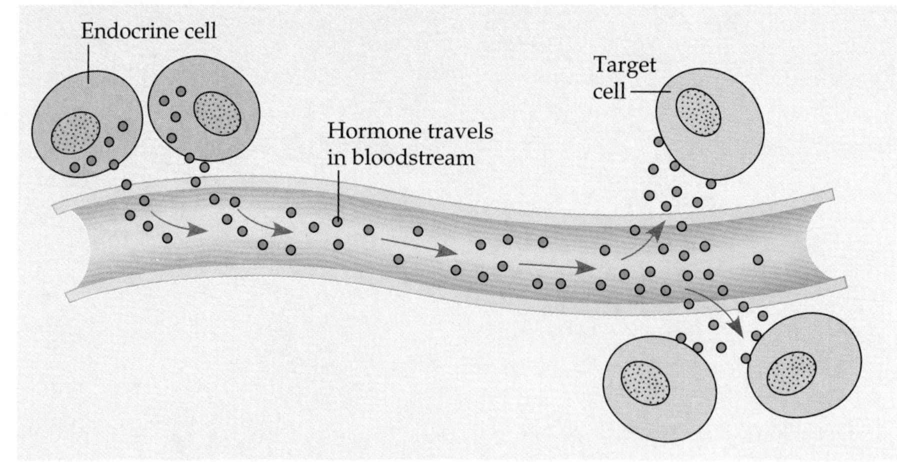

(**a**) Endocrine signaling: long-distance chemical communication

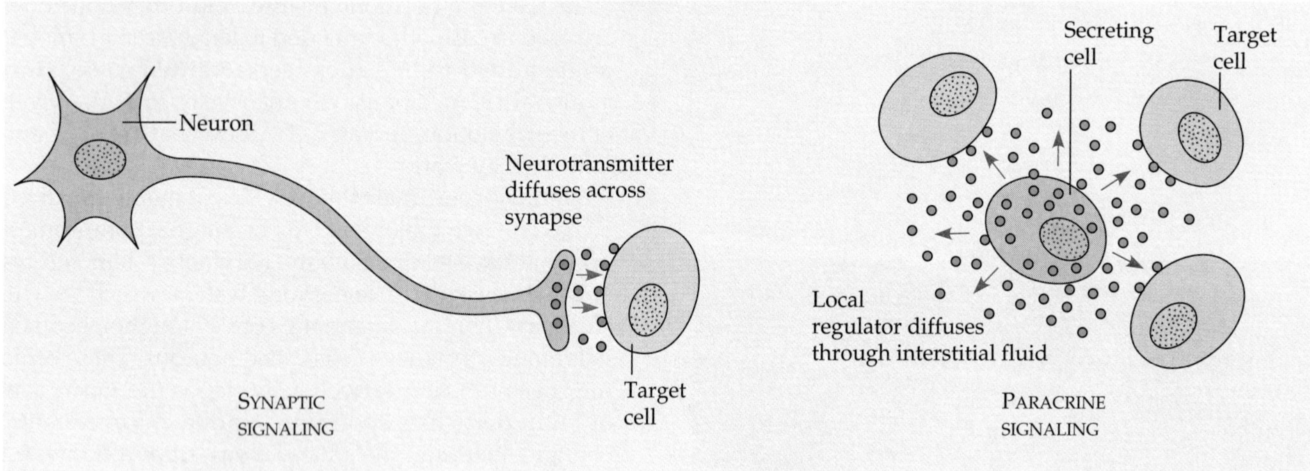

(**b**) Two forms of local signaling

in mate attraction. Still other messengers act only between cells on a localized scale. The most familiar of these local regulators are the neurotransmitters that carry information between cells of the nervous system. In this section, we compare the different types of chemical messengers.

Hormones

An animal **hormone** (from the Greek *hormon,* "excite") is a chemical signal that is transmitted throughout the body via the circulatory system. Although a hormone reaches all parts of the body, only certain types of cells, the **target cells,** are equipped to respond. Thus, a given hormone traveling in the bloodstream elicits specific responses—a change in metabolism, for example—from selected target cells, while other cell types ignore that particular hormone. Hormones are very potent regulators, being effective in minute amounts; even a slight change in a hormone's concentration can have a significant impact on the body.

Hormones are secreted into the bloodstream by specialized cells called endocrine cells, which are generally assembled into organs called endocrine glands. A gland is a secretory organ, and animal glands can be classified as exocrine or endocrine. Exocrine glands produce a variety of substances, such as sweat, mucus, and digestive enzymes, and convey their products to the appropriate locations by means of ducts. In contrast, **endocrine glands** are ductless glands. They produce hormones and secrete these chemical messengers into the bloodstream for distribution throughout the body. Many organs perform both endocrine and exocrine functions. The pancreas, for example, secretes at least two hormones directly into the circulatory system. However, endocrine cells make up only 1% to 2% of the total weight of the pancreas; the rest of the organ is exocrine tissue that produces bicarbonate ions and digestive enzymes that are carried to the small intestine through a duct (see Chapter 37).

Current research in *endocrinology,* the study of hormones, is moving so rapidly that any description of hormones and their actions must be kept flexible.

There are more than 50 known hormones in the human body, and the endocrine systems of other animals are similarly elaborate. In terms of their chemical structure, these hormones can be grouped into three general classes. **Steroid hormones,** which include the sex hormones, are fat-soluble molecules that the body fashions from cholesterol (see Chapter 5). Another group consists of hormones derived from amino acids, most frequently from tyrosine. An example is epinephrine (also known as adrenaline), the "fight-or-flight" hormone secreted by the adrenal glands. The third group of hormones, the **peptides,** is the most diverse. Peptides are chains of amino acids. The smallest is only three amino acids long, but other hormones are full-fledged proteins with as many as 200 amino acids. Insulin, a hormone secreted by the pancreas, is an example of a peptide hormone large enough to be considered a protein. Notice that our classification of hormones as steroids, amino acid derivatives, or peptides groups the hormones according to similarities in chemical structure without considering function. Two hormones belonging to the same chemical class may have completely unrelated functions.

Each hormone molecule has a specific shape that can be recognized by that hormone's target cells. The first step in a hormone's action is the specific binding of the chemical signal to a **hormone receptor,** a protein within the target cell or built into its plasma membrane. This meeting of hormone and receptor triggers the target cell's responses to the hormonal signal. Cells are unresponsive to a particular hormone if they lack the appropriate receptors.

Our overview of hormones includes another general principle: The effects of a hormone are often countered by an antagonistic (opposing) signal, often another hormone. For example, insulin acts to decrease sugar concentration in the blood, while insulin's antagonist, a hormone named glucagon, increases blood sugar level. Feedback mechanisms adjusting the balance of these two hormones maintain blood sugar at a concentration very close to a set point (Figure 41.3). We will see that antagonistic hormones function in many other examples of homeostasis.

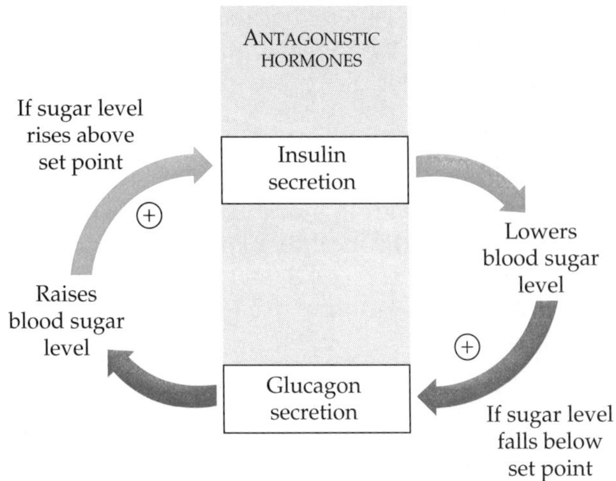

Figure 41.3
Antagonistic hormones and homeostasis: an example. Feedback mechanisms in which blood sugar concentration controls secretion of insulin and glucagon maintain sugar levels close to the set point. Later in the chapter, you will learn more about this and other cases of antagonistic hormones functioning in homeostasis.

Pheromones

Pheromones are chemical signals that function much like hormones, with one important exception: Instead of coordinating the parts of a single animal's body, pheromones are communication signals *between animals* of the same species. Pheromones are often classified according to their functions as mate attractants, territorial markers, or alarm substances, to name a few. One interesting example of a pheromone is the queen substance released by the queen of a honeybee colony. This pheromone causes the worker bees to settle down near the queen and prevents them from giving the young a special diet that makes new queens. Thus, the queen substance is essential to the social stability of the hive.

All pheromones are small, volatile molecules that can disperse easily into the environment and, like hormones, are active in minute amounts. The mate attractants of some female insects can be detected by males as much as a mile away. The pheromone of the female gypsy moth elicits behavioral responses in males at concentrations as low as 1 molecule of pheromone in 10^{17} molecules of other gases in the air.

Compared to most animals, humans do not have well-developed olfactory senses, but it is interesting to speculate whether we use pheromones to communicate. Some indirect evidence suggests we do. There have been cases of women living together for several months, as in dormitories, convents, or prisons, who begin to have synchronous menstrual cycles: They all menstruate at the same time of the month. The role of pheromones in social behavior is considered in more detail in Chapter 50.

Local Regulators

Chemical messengers that affect target cells adjacent to or near their point of secretion function in local regulation (see Figure 41.2b). The neurotransmitters are a familiar group of local regulators. These substances carry information from one neuron to another or from a neuron to a muscle, gland, or other target cell. In *synaptic signaling,* a neuron dispatches its transmitter into a synapse, a junction with a single target cell. Thus, neurotransmission is the most direct form of chemical communication. Less direct, but still very localized, is *paracrine signaling*. This occurs when a cell dispatches regulatory substances into the surrounding interstitial fluid, affecting only nearby target cells. Examples of paracrine signals are histamine and interleukins, which are among the local regulators that coordinate the immune response (see Chapter 39). Two additional groups of local regulators are growth factors and prostaglandins.

Growth Factors Attempts to culture mammalian cells on artificial media led to the discovery of various **growth factors,** proteins that must be present in the extracellular environment in order for certain types of cells to grow and develop normally. For example, embryonic neurons will develop the long processes called axons only if a protein called nerve growth factor (NGF) is present in the culture medium. Epidermal growth factor (EGF) is required to grow epithelial cells in culture, but, in spite of its name, EGF also stimulates the growth of many other types of cells. Several other growth factors have also been discovered.

A variety of experiments indicate that growth factors not only work in the artificial environment of cultured cells, but also regulate development within the animal body. For instance, injecting epidermal growth factor into fetal mice accelerates epidermal (skin) development. Also, antibodies that destroy nerve growth factor obstruct normal development of the nervous system when injected into rodent embryos.

Experiments have yet to reveal how growth factors work. It is known that the target cells of a particular growth factor have proteins on their surfaces that function as receptors for that factor. It is the binding of the growth factor to the receptor that triggers the response of the target cell to the chemical stimulus. Some oncogenes (genes that contribute to cancer; see Chapter 18) code for membrane proteins that mimic growth factor receptors. However, these aberrant receptors do not require the binding of a growth factor to stimulate the growth and division of cells, and thus the tissue grows uncontrollably and abnormally.

Prostaglandins The **prostaglandins (PGs)** are modified fatty acids, often derived from lipids of the plasma membrane. Released from most types of cells into the interstitial fluid, prostaglandins function as local regulators affecting nearby cells in various ways.

About 16 prostaglandins have been discovered, and very subtle differences in their molecular structures can result in profound differences in how these signals affect target cells. For example, two prostaglandins, called prostaglandin E and prostaglandin F, have opposite effects on the smooth muscle cells in the walls of blood vessels serving the lungs. PGE causes the muscles to relax, which dilates the blood vessels and promotes the oxygenation of blood. PGF signals the muscles to contract, which constricts the vessels and reduces blood flow through the lungs. Thus, these two chemical signals are antagonistic, and shifts in their relative concentrations contribute to moment-to-moment adjustments an animal must make to cope with changing circumstances. The use of antagonistic signals as counterbalances is a common regulatory mechanism in both chemical and nervous coordination of the body.

Some of the best-known actions of prostaglandins are on the female reproductive system. For example, prostaglandins secreted by cells of the placenta cause the nearby muscles of the uterus to contract, helping to induce labor during childbirth.

Prostaglandins also function as local regulators in the defense mechanisms of vertebrates. Various prostaglandins help induce fever and inflammation and also intensify the sensation of pain (which can be thought of as contributing to the body's defense by sounding an alarm that something harmful is going on). Aspirin may reduce these symptoms of injury or infection by inhibiting the synthesis of prostaglandins.

So far, biologists have probably recognized only a fraction of the diverse functions of prostaglandins as local regulators of animal tissues. Scientists continue to discover new hormones and other chemical signals and are learning more about their actions. In fact, a central question in endocrinology is: How do hormones actually trigger the changes they do?

MECHANISMS OF HORMONE ACTION AT THE CELLULAR LEVEL

A set of general principles seems to govern the activity of those hormones we know about to date. First, hormones can act at very low concentrations. Second, a given hormone can affect different target cells within an animal differently, or it may affect different species differently. A striking example is thyroxine, a hormone of the thyroid gland. In humans and other vertebrates, it is responsible for metabolic regulation. But thyroxin also plays diverse roles in animal development. In a specific case, thyroxine triggers the development of an adult frog from the tadpole stage, stimulating resorption of the tadpole's tail and other morphological changes during metamorphosis.

Although the diversity of responses to hormones is seemingly endless, a hormone triggers specific changes in a target cell by one of only two general mechanisms. Some hormones, especially the steroids, enter the nucleus of the target cell and influence the expression of the cell's genes. In contrast, most nonsteroid hormones, including peptides, attach to the cell surface and influence activity within the cell through

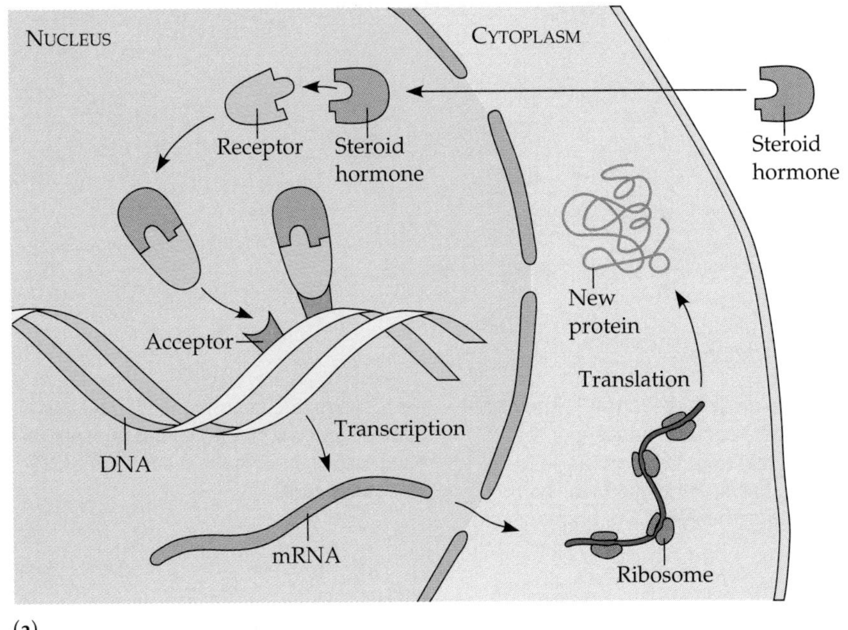

(a)

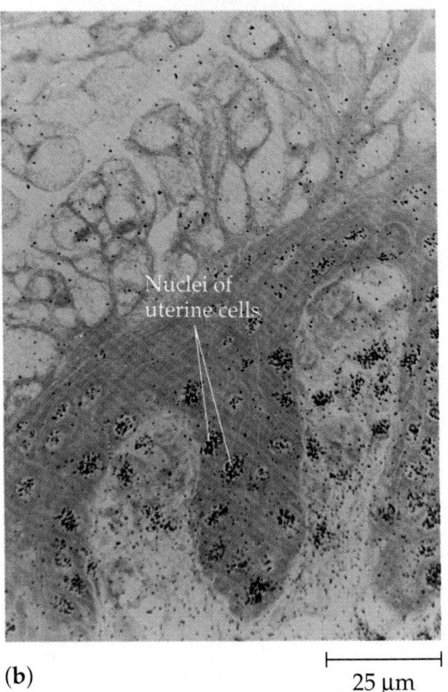

(b)

Figure 41.4
Steroid hormone action. (**a**) In this model, the lipid-soluble steroid passes through the plasma membrane and binds to a receptor protein present only in target cells. Recent evidence suggests that the receptors reside in the nucleus rather than the cytoplasm. The hormone-receptor complex then binds to specific acceptor proteins that recognize certain enhancer sites along the DNA, stimulating the expression of specific genes. (**b**) In this autoradiogram (see Chapter 2), radioactively labeled progesterone, a steroid hormone (black dots), is concentrated in the nuclei of specific target cells in the uterus of a guinea pig (LM). Other cell types in the uterus do not bind the hormone.

cytoplasmic intermediaries called second messengers. Both mechanisms involve receptor molecules to which the hormones bind. The receptors are proteins, and it is the hormone-receptor complex that actually triggers the effects of the hormone. In this section, we examine these two general mechanisms, or **signal transduction pathways,** the sequences of steps that intervene between hormone-receptor binding and the actual responses of target cells. (See Chapter 35 for an introduction to signal transduction pathways in plants.)

Steroid Hormones and Gene Expression

Steroids pass through the target cell membrane and enter the nucleus, where they affect the transcription of specific genes (see Chapter 18). (Thyroxine, though not a steroid, probably works in a similar manner.) The general signal transduction pathway for steroid hormones was discovered by studying two vertebrate hormones, estrogen and progesterone. In most mammals, including humans, these hormones are necessary for the normal development and function of the female reproductive system. In the early 1960s, researchers demonstrated that cells in the reproductive tract of female rats accumulate estrogen. The hormone was found within the nuclei of these cells, but not in the cells of such tissues as the spleen, which do not respond to estrogen. Progesterone also enters the nuclei of target cells. Such observations led to the hypothesis that cells sensitive to a steroid hormone contain receptor molecules that bind specifically to that hormone. It is now known that the steroid receptors are specialized proteins within the cells, perhaps in the nucleus.

The specificity of receptor proteins is a key to understanding steroid hormone action. According to the model illustrated in Figure 41.4a, a steroid hormone that crosses the membrane of its target cell binds to a receptor protein in the nucleus. This hormone-protein complex then has the proper conformation to bind to another specific protein, an acceptor protein that recognizes certain regions of the DNA. These acceptor proteins are probably transcription factors that stimulate transcription of specific genes by attaching to enhancer sequences along the DNA (see Chapter 18). In this way, a steroid signal can initiate the expression of selected genes. The mRNA molecules thus produced are processed and transported to the cytoplasm, where they direct the synthesis of new proteins. For example, estrogen induces certain cells in the reproductive system of a female bird to synthesize large quantities of ovalbumin, the main storage protein stockpiled in egg white. Different proteins are synthesized in the liver cells of the same animal in response to estrogen.

How can two types of cells respond differently to a common chemical signal (estrogen, in this case)? It is

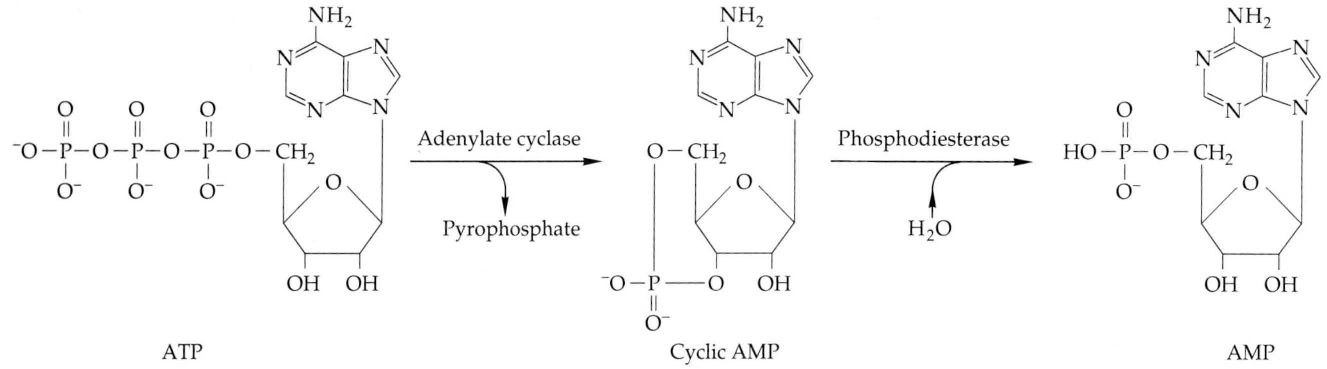

ATP → (Adenylate cyclase; Pyrophosphate) → Cyclic AMP → (Phosphodiesterase; H_2O) → AMP

Figure 41.5
Cyclic AMP, a second messenger. Binding of a peptide hormone to its specific receptor on the surface of a target cell causes adenylate cyclase, an enzyme embedded in the plasma membrane, to convert ATP to cyclic AMP (cAMP). The cAMP functions as a second messenger that relays the signal from the membrane to the metabolic machinery of the cytoplasm. When the first message, the hormonal signal, ceases, the second message is also soon silenced by the action of phosphodiesterase, an enzyme that converts cAMP to inactive AMP.

likely that the estrogen receptor is the same protein in both target cells, but the chromosomal proteins (transcription factors) that accept the hormone-receptor complex recognize enhancers controlling different genes in the two kinds of cells.

Peptide Hormones and Second Messengers

A dramatic example of hormone action is illustrated when a frog's skin turns darker or lighter, an adaptation that helps camouflage the frog as lighting changes. This color change is controlled by a peptide hormone called melanocyte-stimulating hormone (MSH), secreted by the pituitary gland at the base of the brain. Melanocytes are specialized skin cells that contain the dark brown pigment melanin in cytoplasmic organelles called melanosomes. The frog's skin appears light when the melanosomes are clustered tightly around the cell nucleus and darker when the melanosomes spread throughout the cell. When MSH is added to the interstitial fluid surrounding the pigment-containing cells, the melanosomes disperse. However, direct microinjection of MSH into individual melanocytes does not induce melanosome dispersion. Why is it that some hormones will not act directly within a target cell?

Peptide hormones and most hormones derived from amino acids are unable to pass through the plasma membrane of their target cells, but they still influence cellular activity. Unlike steroids, these hormones bind to specific receptors on the cell's outer surface. The hormone receptors are proteins embedded in the plasma membrane. The surface of a target cell typically has about 10,000 receptors keyed to a particular hormone. (Based on this large number, one might think that the membrane is densely populated by proteins devoted to receiving a single type of chemical message. However, 10,000 receptors account for only about one ten-thousandth of the total number of proteins embedded in the plasma membrane of an animal cell.) On the extracellular surface of each receptor protein is a binding site that specifically fits a particular hormone molecule. This binding of hormone to the receptor sparks a burst of biochemical activity within the cell. But how can a chemical signal received on the *outside* of a cell affect metabolism *inside* the cell? The solution is a **second messenger,** an intracellular signal produced by the cytoplasmic surface of the plasma membrane in response to extracellular binding of the first messenger, the hormone. Two important secondary messengers are cyclic AMP and inositol triphosphate.

Cyclic AMP Current understanding of how nonsteroid hormones act via second messengers has advanced from the pioneering work of Earl W. Sutherland, whose research led to a Nobel Prize in 1971. Sutherland and his colleagues at Vanderbilt University were investigating how the hormone epinephrine stimulates the hydrolysis of glycogen within liver cells and muscle cells. Glycogen is a storage polysaccharide, and its hydrolysis to release the sugar glucose-1-phosphate increases the energy supply for cells. Thus, one effect of epinephrine, secreted from the adrenal gland during times of physical or mental stress, is the mobilization of fuel reserves. Sutherland's research team discovered that epinephrine stimulates glycogen breakdown by activating a cytoplasmic enzyme called glycogen phosphorylase. However, if epinephrine was added to a test-tube mixture containing the phosphorylase and its substrate, glycogen, no hy-

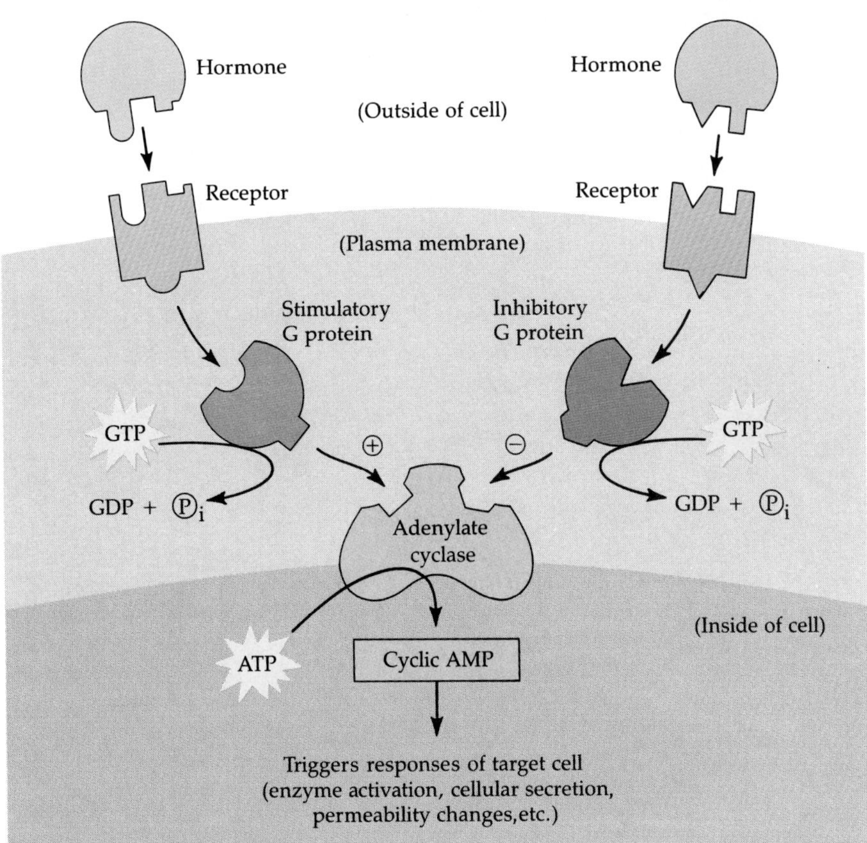

Figure 41.6
The control of adenylate cyclase. A hormone-receptor complex does not interact directly with adenylate cyclase but uses intermediaries called G proteins (because they hydrolyze GTP). The cytoplasmic concentration of cAMP, the second messenger in response to many hormones, is partly controlled by the effects of stimulatory and inhibitory G proteins on the activity of adenylate cyclase.

drolysis occurred. Epinephrine could activate glycogen phosphorylase only if the hormone was added to the extracellular solution bathing intact cells with plasma membranes. Thus, the search began for a second messenger that transmits the signal from the plasma membrane to the metabolic machinery in the cytoplasm.

Sutherland found that the binding of epinephrine to the plasma membrane of a liver cell elevates the cytoplasmic concentration of a compound called cyclic adenosine monophosphate, abbreviated **cyclic AMP** or **cAMP** (Figure 41.5). The enzyme **adenylate cyclase** converts ATP to cAMP in response to a hormonal signal—in this case, epinephrine. Adenylate cyclase is a membrane protein with its active site facing the cytoplasm. The enzyme is idle until activated by the binding of epinephrine to the hormone's receptor protein. Thus, the first messenger, the hormone, triggers the synthesis of cAMP, which functions as a second messenger that relays the signal to the cytoplasm. It is not epinephrine itself but its cytoplasmic emissary, cAMP, that activates glycogen hydrolysis within the cell. When the extracellular concentration of the hormone declines, this silences the second message. Hormones bind reversibly to their receptors by weak bonds, and less epinephrine means that there are, at any particular instant, fewer hormone-receptor complexes to activate adenylate cyclase. Also, the second messenger does not persist for long in the absence of the first message, because another enzyme converts the cAMP to an inactive product. Another surge of epinephrine will again boost the cytoplasmic concentration of the second messenger. Epinephrine is only one of many hormones derived from amino acids and peptide hormones that use cAMP as a second messenger.

Subsequent research has revealed additional players in the second-messenger scenario of hormone action. Although the hormone receptor and adenylate cyclase are both built into the plasma membrane, they do not interact directly. How, then, does the binding of hormone to the receptor activate adenylate cyclase? A third membrane protein functions as an intermediary. It is called a **G protein** because it binds to GTP (guanosine triphosphate), an energy carrier closely related to ATP (see Chapter 9). Docking of the hormone with the receptor activates the G protein, which in turn activates adenylate cyclase to produce cAMP (Figure 41.6). The energy for converting the first message to the second message is provided by GTP, which the G protein hydrolyzes to GDP (guanosine diphosphate; this exergonic process is basically the same as the ATP ⟶ ADP + Ⓟ$_i$ reaction).

A G protein provides a versatile link between a variety of hormone receptors and adenylate cyclase. The

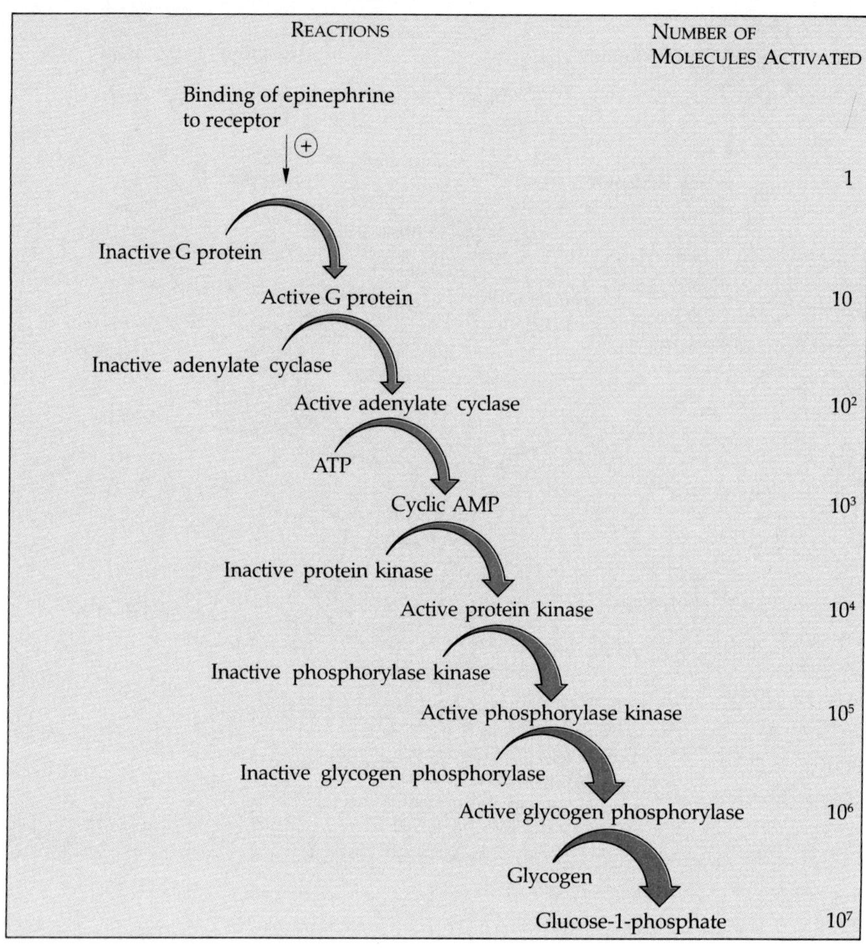

Figure 41.7
How an enzyme cascade amplifies response to a hormone. In this system, the hormone epinephrine acts through cyclic AMP to activate a succession of enzymes, leading to extensive hydrolysis of glycogen. In reality, the number of activated molecules at each step is much larger than shown here. With each step, the hormonal signal is amplified, because each enzyme molecule can activate many molecules of its substrate, the next protein in the cascade.

same G protein that stimulates cAMP production in response to epinephrine also transduces several other hormonal signals. For example, MSH, the hormone that triggers pigment dispersion in the melanocytes of a frog, works via the receptor–G protein–adenylate cyclase linkage (the receptors, however, are unique for each hormone).

A second type of G protein *down* regulates (inhibits) adenylate cyclase. Binding of an inhibitory hormone to its receptor causes this G protein to lower the activity of adenylate cyclase, thus reducing the cytoplasmic concentration of cAMP. These opposing effects of two G proteins in regulating adenylate cyclase enable the cell to fine-tune its metabolism in response to slight changes in the proportions of antagonistic hormones.

Having examined how membrane proteins convert a hormonal signal to the second message, let's next consider how cAMP is actually translated into a particular metabolic response by the cell. In the case of sugar mobilization in response to epinephrine, we can specifically ask how cAMP stimulates the activity of glycogen phosphorylase, the enzyme that catalyzes the hydrolysis of glycogen. The stimulation is indirect, using two intermediaries (Figure 41.7). First, cAMP activates an enzyme called cAMP-dependent protein kinase. A **protein kinase** is an enzyme that catalyzes the transfer of a phosphate group from ATP to proteins, the process of phosphorylation. Depending on the type of protein that receives the phosphate group, phosphorylation either increases or decreases the activity of the protein. In the case of liver cells responding to epinephrine, the cAMP-dependent protein kinase stimulates *another* kinase enzyme, phosphorylase kinase. This second kinase adds a phosphate group to glycogen phosphorylase, the enzyme that actually hydrolyzes glycogen. Thus, the metabolic response to epinephrine involves an **enzyme cascade,** where each step activates an enzyme that in turn activates the next enzyme in the series: cAMP starts the cascade within the cytoplasm by activating protein kinase, which activates phosphorylase kinase, which activates glycogen phosphorylase, which mobilizes sugar by hydrolyzing glycogen.

One advantage of this elaborate enzyme cascade is that it amplifies response to the hormone. With each catalytic step in the cascade, the number of activated products is much greater than in the preceding step. A relatively small number of adenylate cyclase molecules can produce a large number of cAMP molecules, which activate even more molecules of protein kinase, which activate still more molecules of phosphorylase kinase, and so on. As a result of this amplification, a very few molecules of epinephrine binding to receptors on the surface of a liver or muscle cell can quickly result in the release of millions of sugar molecules from glycogen.

Such an indirect mechanism of hormone action offers another advantage. An animal can use a single basic mechanism to mediate the responses of diverse cell types to many different hormones. We have already seen that two different hormones can sound the same second message via a common linkage of G proteins to adenylate cyclase. The second messenger, cAMP, alters the activity of various cytoplasmic proteins by activating a cAMP-dependent protein kinase. Given this generalized action of hormones, how can we account for specificity in the responses of various target cells to hormonal signals? First, a particular type of cell is only tuned in to certain hormones—those that fit the receptors found on the surface of that cell type. Second, cAMP-dependent protein kinases vary somewhat in structure and function from tissue to tissue. Third, the steps in the enzyme cascade that follow the activation of protein kinase differ markedly from one cell type to another. Different kinds of specialized cells contain different proteins capable of being phosphorylated by protein kinase. In a liver cell responding to epinephrine, protein kinase activates phosphorylase kinase in the enzyme cascade that leads to glycogen breakdown. In a frog melanocyte responding to MSH, protein kinase apparently activates proteins of the cytoskeleton that function in pigment dispersion. In other cases, protein kinase phosphorylates membrane proteins and alters the transport of substances across the membrane. As it differentiates, each type of cell acquires its individual responses to peptide hormones by producing a unique set of proteins regulated by cAMP-dependent protein kinases. In this way, the same second messenger, cAMP, means different things to different cells.

Inositol Triphosphate Many chemical messengers in animals, including neurotransmitters, growth factors, and some hormones, induce responses in their target cells by increasing the cytoplasmic concentration of calcium ions (Ca^{2+}). Binding of the messenger to a specific receptor on the cell surface results in the release of Ca^{2+} from a reservoir within the cell, usually the endoplasmic reticulum (see Chapter 7). The Ca^{2+} released to the cytoplasm alters the activities of specific enzymes, either directly or by first binding to a protein called **calmodulin,** which then binds to other proteins and changes their activities. Given these discoveries, calcium was once believed to be a second messenger in the responses of cells to extracellular signals. However, additional research has shown that calcium is actually a *third* messenger. The second messenger, which intermediates between the hormonal signal and the rise in cytoplasmic Ca^{2+} concentration, is a compound called **inositol triphosphate (IP_3).**

Inositol triphosphate is derived from a particular type of phospholipid present in the plasma membrane (Figure 41.8). Production of IP_3 involves a reaction sequence in the membrane analogous to the mechanism that transduces a hormonal signal into the production of cAMP. A hormone binds to its specific receptor, and the hormone-receptor complex activates a G protein that also resides in the membrane. This G protein, different from those that function in the cAMP system, stimulates a membrane enzyme called phospholipase C. This enzyme then cleaves a phospholipid into two products, IP_3 and another compound called diacylglycerol. Both products may function as second messengers. The diacylglycerol activates another membrane enzyme called protein kinase C, which triggers some of the cytoplasmic responses to the hormone by phosphorylating specific proteins (similar to the action of cAMP-dependent protein kinase, but protein kinase C activates a different set of cytoplasmic proteins). The IP_3 released from the plasma membrane probably interacts with the membrane of the endoplasmic reticulum in some way that causes Ca^{2+} to leak from the ER. The signal is finally transmitted to various enzymes and other proteins by the increased concentration of cytoplasmic Ca^{2+}, and specific responses of the target cell to the original hormonal signal occur.

Let's review the two general modes of hormone action. Steroid hormones generally work by entering the nucleus of the target cell and altering gene expression. Most other hormones, including peptides, bind to the target cell's plasma membrane and regulate cytoplasmic enzymes via second messengers, often cAMP or IP_3. Notice that steroids mainly affect the *synthesis* of proteins, while peptide hormones most often affect the *activity* of enzymes and other proteins already present in the cell. This is an important distinction. Target cells generally respond more slowly to steroids than to peptide hormones, but the changes induced by steroids usually last longer. Indeed, steroid hormones are commonly involved in the development of tissues. For example, steroid hormones secreted by a woman's ovary stimulate the gradual growth of the uterine wall over a period of a few weeks during each menstrual cycle. In contrast to this time frame, it takes only minutes for MSH, a peptide hormone, to trigger pigment dispersion in the melanocytes of frog skin.

Now that we have considered the two general modes of hormone action, let's examine the specific functions of some animal hormones.

INVERTEBRATE HORMONES

Although diverse invertebrate hormones function in homeostasis—by regulating water balance, for instance—the hormones that have been most extensively studied function in reproduction and development.

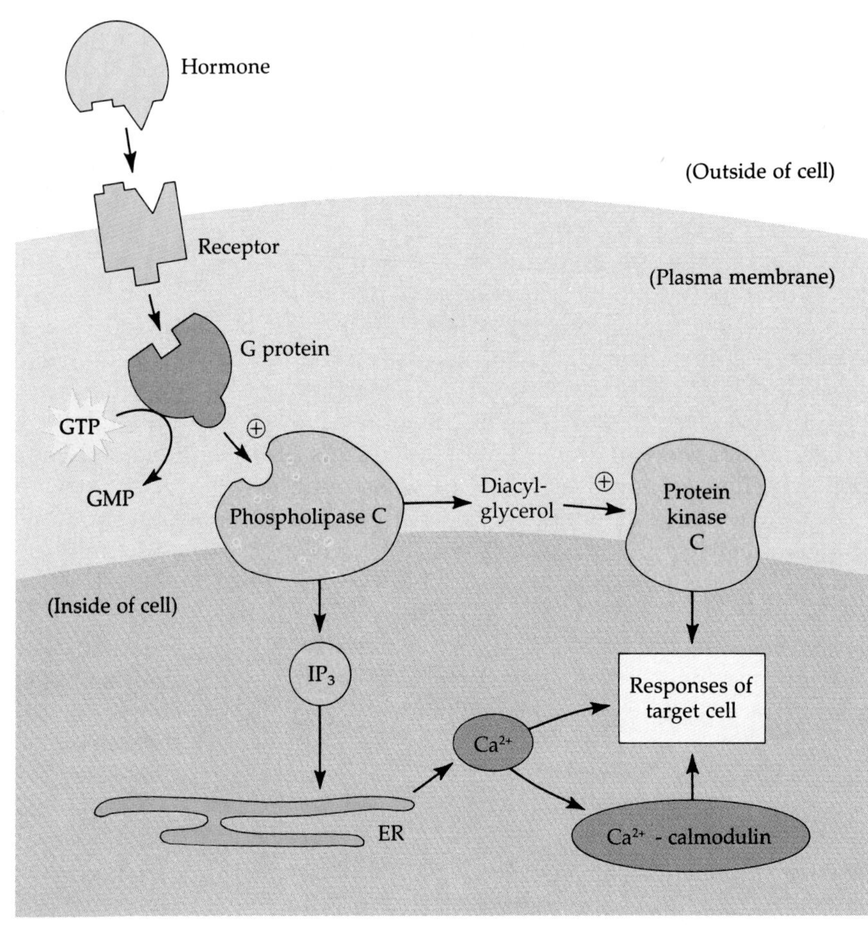

Figure 41.8
Inositol triphosphate as a second messenger. The first messenger, the hormone, binds to its specific receptor, activating a G protein that in turn stimulates phospholipase C, an enzyme that resides in the plasma membrane. Phospholipase C cleaves a membrane phospholipid into two products—inositol triphosphate (IP_3) and diacylglycerol—that function as second messengers. Diacylglycerol stimulates still another membrane enzyme, protein kinase C, which triggers various responses of the target cell by phosphorylating specific proteins. The IP_3 released from the membrane in response to the hormonal signal triggers the release of Ca^{2+} from the endoplasmic reticulum. This increases the cytoplasmic concentration of Ca^{2+}, an inorganic ion that regulates the activity of many cellular proteins, either by itself or by first binding to a protein called calmodulin.

For example, in *Hydra* one hormone stimulates growth and budding (asexual reproduction) but prevents sexual reproduction. Annelids in the genus *Nereis* produce a hormone that stimulates egg production and inhibits body growth. One well-studied example of nerve and hormone interaction in controlling both reproductive physiology and behavior in mollusks involves the egg-laying hormone of the sea slug *Aplysia.* This peptide hormone, when secreted from the specialized neurons known as bag cells, almost immediately stimulates the laying of thousands of eggs. At the same time, this egg-laying hormone inhibits feeding and locomotion, activities that interfere with reproduction.

All groups of arthropods have extensive endocrine systems. The crustaceans, for example, have hormones for growth and reproduction, water balance, movement of pigments in the integument and in the eyes, and regulation of metabolism. Because they are the most species-rich class of animals and are economically important, insects have been the most thoroughly studied arthropods. Insects have exoskeletons that cannot stretch; thus, insects appear to grow in spurts, shedding the old and secreting a new exoskeleton with each molt. Further, most insects acquire their adult characteristics in a single, terminal molt. By studying these events, scientists have identified how three important hormones interact to control insect development (Figure 41.9).

Insect molting is triggered by the steroid hormone **ecdysone,** secreted from a pair of prothoracic glands just behind the head. Besides stimulating the molt, ecdysone also favors the development of adult characteristics, as in the change from a caterpillar to a butterfly, shown in Figure 41.1. Ecdysone has the same mechanism of action as vertebrate steroid hormones, stimulating the transcription of specific genes. Ecdysone production is itself controlled by a second hormone, called **brain hormone (BH).** A peptide from the brain, this hormone promotes development by stimulating the prothoracic glands to secrete ecdysone.

Brain hormone and ecdysone are balanced by **juvenile hormone (JH),** the third hormone in this system. JH is secreted by a pair of small glands just behind the brain, the corpora allata. Juvenile hormone promotes retention of larval characteristics. In the presence of a relatively high concentration of juvenile hormone, ecdysone can still stimulate molting, but the product is a larger larva. Only when the level of juvenile hormone

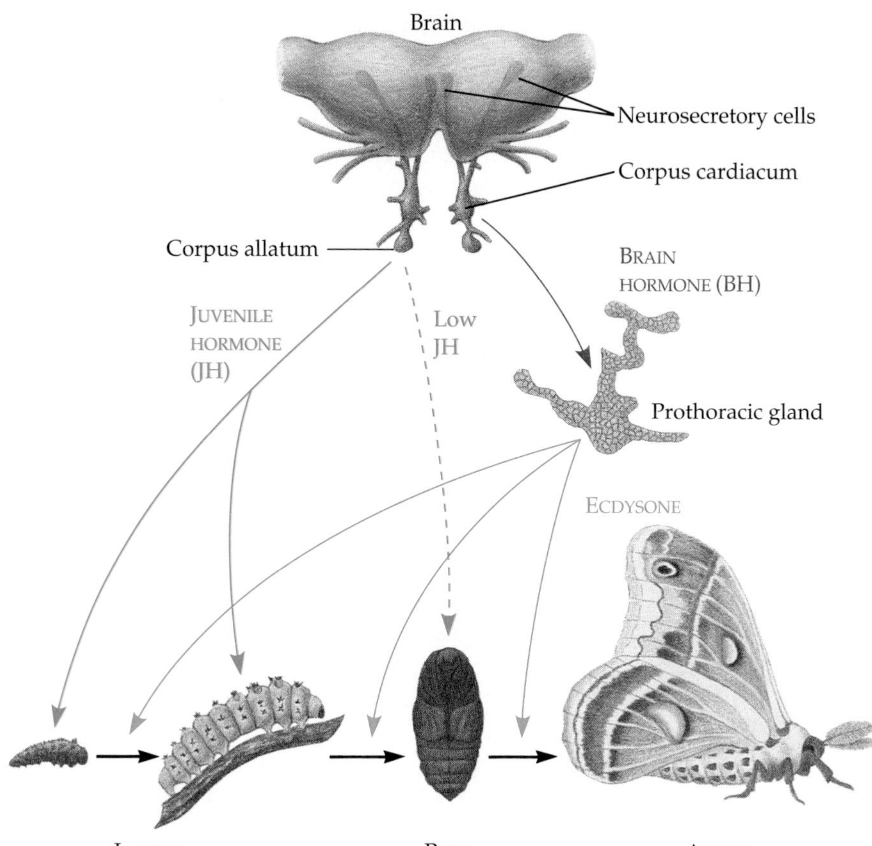

Figure 41.9
Hormonal regulation of insect development. Most insects go through a series of larval stages, with each molt (shedding of the old exoskeleton) leading to a larger larva. Molting of the last larval stage gives rise to a pupa, in which metamorphosis produces the adult form of the insect. These developmental stages are controlled by at least three hormones: brain hormone, ecdysone, and juvenile hormone. Neurosecretory cells in the brain produce brain hormone (BH), but the hormone is stored and released from an organ called the corpus cardiacum. BH signals its main target organ, the prothoracic gland, to produce the hormone ecdysone. Ecdysone secretion is episodic, with each release stimulating a molt. Juvenile hormone (JH), secreted by the corpus allatum, determines the result of the molt. At relatively high concentrations of JH, ecdysone-stimulated molting produces another larval stage. Thus, JH suppresses metamorphosis. But when levels of JH fall below some threshold concentration, a pupa forms at the next ecdysone-promoted molt. The adult insect emerges from the pupa.

wanes can ecdysone-induced molting produce a pupa. Within the pupa, metamorphosis replaces larval anatomy with the insect's adult form. (Synthetic versions of JH are now being used as insecticides to prevent insects from maturing into reproducing adults.)

In all these invertebrate examples, we see the importance of the nervous system to hormone activity. Often, the distinction between the endocrine system and the nervous system blurs. We will see these interactions repeatedly as we survey the vertebrate endocrine system.

THE VERTEBRATE ENDOCRINE SYSTEM

In vertebrates, more than a dozen tissues and organs secrete hormones. Some of these glands are endocrine specialists—secretion of one or more hormones is their major function. Figure 41.10 shows the location of the major endocrine glands in the human body, and Table 41.1 (p. 919) summarizes the functions of the major vertebrate hormones.

Hormones vary in their range of targets. Some, like the sex hormones, which promote male and female characteristics, affect most of the tissues of the body. Others affect only one or a few tissues. Some hormones have other endocrine glands as their targets; these so-called **tropic hormones** are particularly important to our understanding of chemical coordination. In fact, we will begin our discussion of the vertebrate endocrine system by looking at the major control axis of the system.

The Hypothalamus and the Pituitary Gland

The **hypothalamus** plays an important role in integrating the vertebrate endocrine and nervous systems. This region of the lower brain receives information from peripheral nerves and from other parts of the brain and then initiates endocrine signals appropriate to the environmental conditions. In many vertebrates, for example, the brain passes sensory information about seasonal changes and the availability of a mate to the hypothalamus by means of nerve signals; the hypothalamus then triggers the release of reproductive hormones required for breeding.

The hormone-releasing cells in the hypothalamus are specialized neurons that differ from both the secretory cells of most endocrine glands and other neurons.

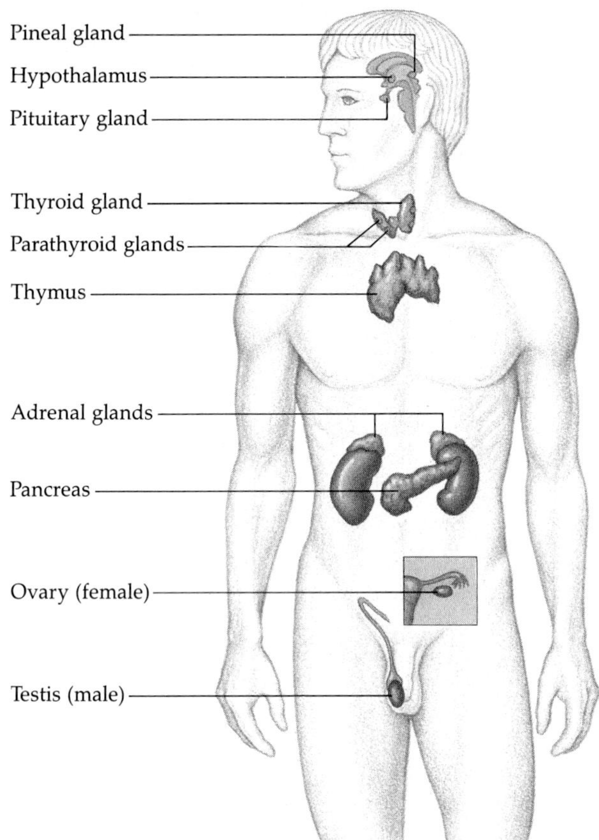

Figure 41.10
Human endocrine glands surveyed in this chapter. (The endocrine functions of the stomach, small intestine, heart, and kidneys were covered in Chapters 37, 38, and 40.)

Rather than being specialized just for hormone synthesis and release or just for the conduction of nerve impulses, they do both. Called **neurosecretory cells,** these hypothalamic cells receive signals from other nerve cells, but instead of signaling to an adjacent nerve cell or muscle, they release hormones into the bloodstream. (Notice in Figure 41.9 that neurosecretory cells are also found in insect brains.) The hypothalamus contains two sets of neurosecretory cells. One set produces the hormones of the posterior pituitary. The other produces **releasing hormones** that regulate the anterior pituitary.

The **pituitary gland,** a small appendage at the base of the hypothalamus, was formerly called the "master gland" because so many of its hormones regulate other endocrine functions. However, the pituitary itself obeys hormonal orders from the hypothalamus. The pituitary has two lobes, each with a different function (Figure 41.11, p. 920). The **posterior lobe,** or **neurohypophysis,** is an extension of the brain that stores and secretes two peptide hormones that are actually made by the hypothalamus. The two hormones released from the posterior lobe act directly on muscles and the kidneys, rather than affecting other endocrine glands. The **anterior lobe** of the pituitary, or **adenohypophysis,** produces its own hormones, several of which are tropic hormones that act on other endocrine glands. Figure 41.11 summarizes the names and targets of the pituitary hormones.

Posterior Pituitary Hormones The posterior lobe of the pituitary stores and releases the hypothalamic hormones **oxytocin** and **antidiuretic hormone (ADH).** They are small peptides of nine amino acids, only two of which differ between oxytocin and ADH. The hormones are synthesized by the cell bodies of neurosecretory cells in the hypothalamus and move within extensions of these cells into the posterior pituitary. Oxytocin induces contraction of the uterine muscles during childbirth and causes the mammary glands to eject milk during nursing. ADH acts on the kidneys to increase water retention and thus decrease urine volume.

ADH is part of an elaborate feedback scheme that helps regulate the osmolarity of the blood. This mechanism was introduced in Chapter 40, but it is worth reviewing here to illustrate how hormones contribute to homeostasis and how negative feedback controls hormone levels. Blood osmolarity is monitored by a group of nerve cells that function as osmoreceptors in the hypothalamus. When the osmolarity of the plasma increases, these cells shrink slightly (due to osmosis) and transmit nerve impulses to certain neurosecretory cells of the hypothalamus. These neurosecretory cells respond by releasing ADH from their tips (located in the posterior pituitary) into the general circulation. When ADH reaches the kidneys, it binds to receptors on the surface of the cells lining the collecting ducts. This binding increases the water permeability of the epithelium of the collecting duct by acting through a cAMP second messenger system. Because of this increased permeability, water exits the collecting ducts and enters nearby capillaries, helping to prevent a further increase in blood osmolarity above the set point. The brain's osmoreceptors also stimulate a thirst drive, and drinking water brings blood osmolarity back down to the set point. Thus, the hormonal reaction to high blood osmolarity is augmented by a behavioral response controlled by the nervous system. As the more dilute blood arrives at the brain, the hypothalamus responds to the reduction in osmolarity by slowing the release of ADH and diminishing feelings of thirst. The effects of these hormonal and behavioral responses—increased water reabsorption in the kidneys and drinking, respectively—prevent overcompensation by shutting off further secretion of the hormone and quenching thirst. We see once again how negative feedback helps maintain homeostasis (see Chapter 36). This example also highlights the central role of the hypothalamus as a member of both the endocrine system and the nervous system.

Table 41.1 Major Endocrine Glands of Vertebrates and Their Hormones

Gland	Hormone	Chemical Structure	Representative Actions (in Mammals)
Pituitary			
Posterior lobe (releases hormones produced by hypothalamus)	Oxytocin	Peptide	Stimulates contraction of uterine muscles and mammary gland cells.
	Antidiuretic hormone (ADH)	Peptide	Promotes reabsorption of water in kidneys.
Anterior lobe	Growth hormone (GH)	Protein	Stimulates general growth, particularly skeletal; affects metabolic functions.
	Prolactin (PRL)	Protein	Stimulates milk production and secretion.
	Follicle-stimulating hormone (FSH)	Glycoprotein	Stimulates ovarian follicle and spermatogenesis.
	Luteinizing hormone (LH)	Glycoprotein	Stimulates corpus luteum and ovulation in females and interstitial cells in males.
	Thyroid-stimulating hormone (TSH)	Glycoprotein	Stimulates thyroid gland to secrete hormones.
	Adrenocorticotropin (ACTH)	Peptide	Stimulates adrenal cortex to secrete glucocorticoids.
Thyroid	Triiodothyronine (T_3) and thyroxine (T_4)	Iodine-containing amino acids	Increase oxygen consumption and heat production; stimulate and maintain metabolic processes.
	Calcitonin	Peptide	Lowers blood calcium levels by inhibiting calcium release from bone.
Parathyroid	Parathyroid hormone (PTH)	Peptide	Raises plasma calcium levels by stimulating calcium release from bone.
Pancreas	Insulin	Protein	Lowers blood sugar; increases glycogen storage in liver; stimulates protein synthesis.
	Glucagon	Peptide	Stimulates glycogen breakdown in liver.
Adrenal glands			
Medulla	Epinephrine	Modified amino acid	Increases blood sugar; constricts blood vessels in skin and kidneys.
	Norepinephrine	Modified amino acid	Increases heart rate; constricts blood vessels throughout the body.
Cortex	Glucocorticoids (e.g., cortisol)	Steroids	Increase blood sugar by affecting many aspects of carbohydrate metabolism.
	Mineralocorticoids (e.g., aldosterone)	Steroids	Promote reabsorption of sodium and excretion of potassium in kidneys.
Gonads			
Testis	Androgens (e.g., testosterone)	Steroids	Support spermatogenesis; develop and maintain secondary male sex characteristics.
Ovary (follicle)	Estrogens	Steroids	Initiate building of uterine lining; develop and maintain female sex characteristics.
Ovary (corpus luteum)	Progesterone and estrogens	Steroids	Promote continued growth of uterine lining.
Pineal	Melatonin	Modified amino acid	Involved in circadian rhythms.
Thymus	Thymosin	Peptide	Stimulates T-lymphocyte development.

Figure 41.11
Hormones of the hypothalamus and pituitary glands. The pituitary gland, located at the base of the brain and surrounded by bone (see orientation diagram), consists of a posterior lobe (neurohypophysis) and an anterior lobe (adenohypophysis). The posterior lobe develops from the floor of the brain along with the hypothalamus, whereas the anterior lobe is derived from the roof of the mouth. (**a**) Posterior lobe. Neurosecretory cells in the hypothalamus synthesize antidiuretic hormone (ADH) and oxytocin, peptide hormones that are transported down the axons to the posterior pituitary, where they are stored. The pituitary gland releases the hormones into the blood, where they circulate and bind to target cells in the kidneys (ADH) and mammary glands and uterus (oxytocin). (**b**) Anterior lobe. Endocrine cells in the anterior pituitary manufacture a number of peptide hormones and secrete them into the circulation, but the release of these hormones is controlled by the hypothalamus. Neurosecretory cells in the hypothalamus secrete releasing hormones into a capillary network located in the median eminence, which is above the stalk of the pituitary. Blood containing the releasing hormones travels from the median eminence through short portal veins and into a second capillary network within the anterior pituitary. Various releasing hormones stimulate or inhibit the release of specific hormones by the pituitary cells.

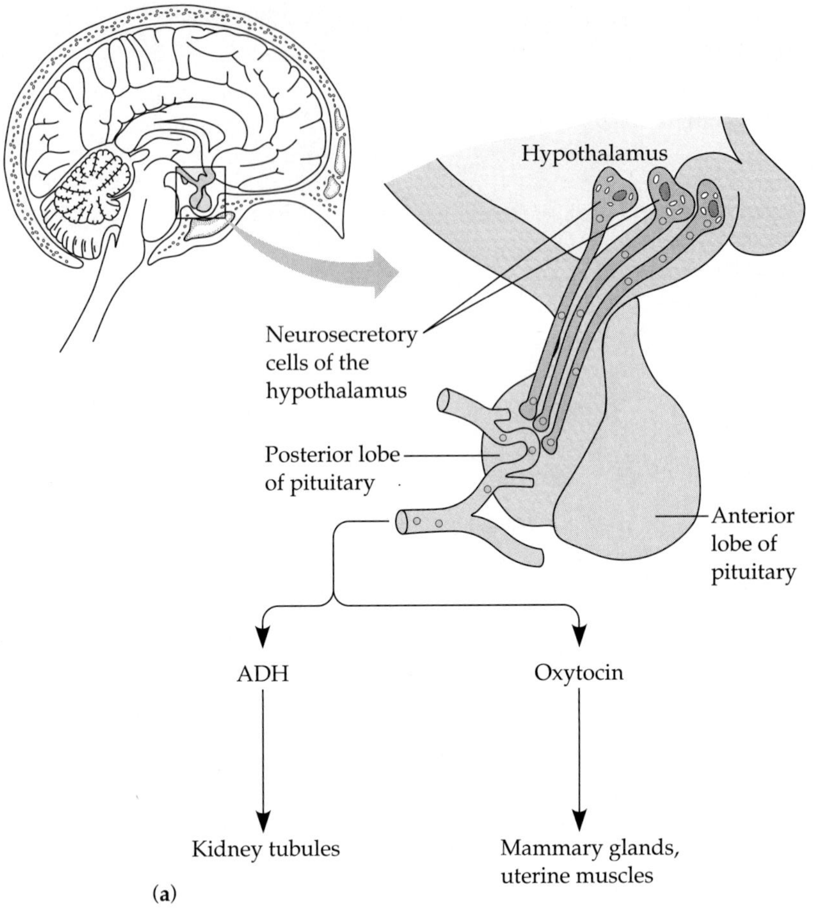

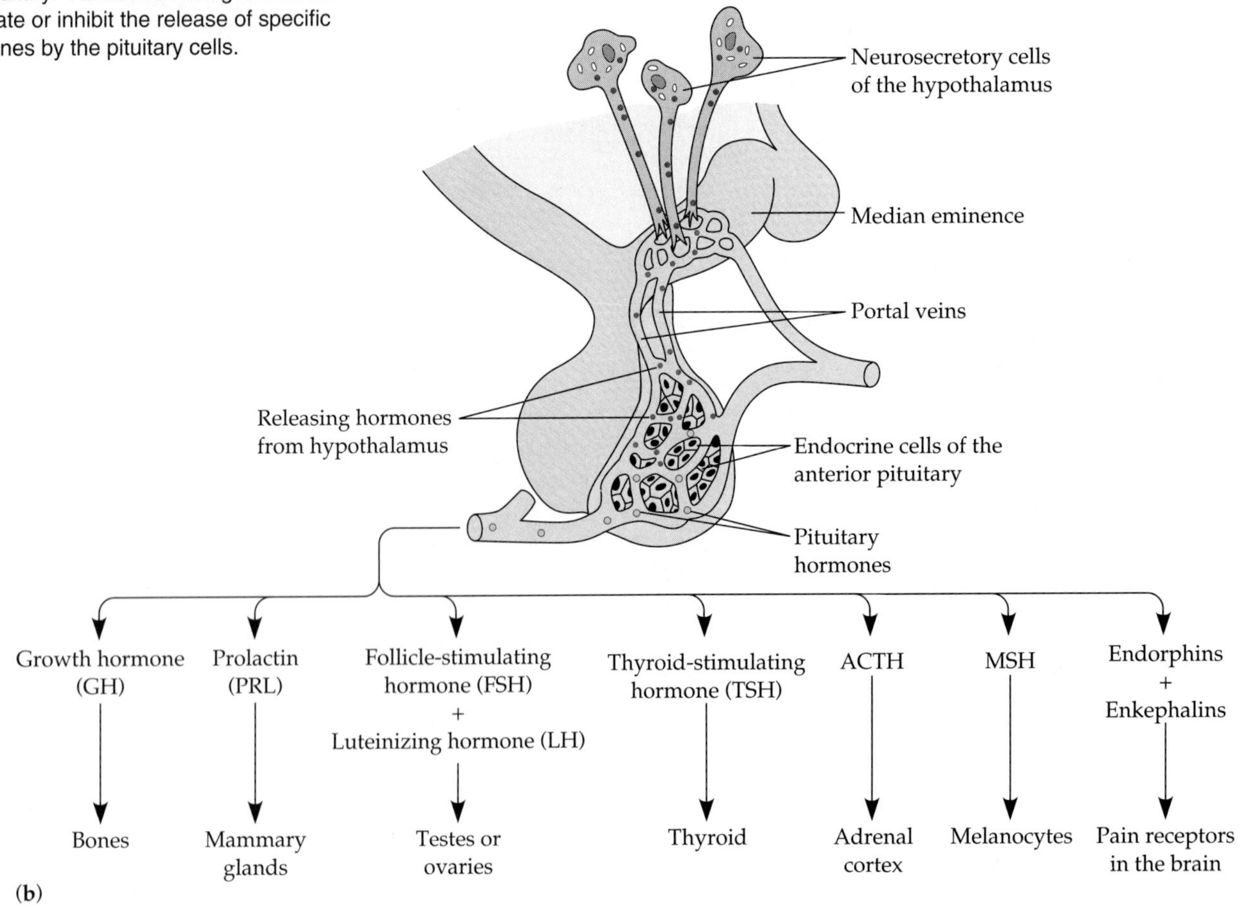

Anterior Pituitary Hormones The anterior pituitary produces many different protein and peptide hormones. Four of these are tropic hormones that stimulate the synthesis and release of hormones from other endocrine glands: Thyroid-stimulating hormone (TSH) regulates the release of thyroid hormones; adrenocorticotropin (ACTH) controls the adrenal cortex; and follicle-stimulating hormone (FSH) and luteinizing hormone (LH) govern reproduction by acting on the gonads. Other hormones produced by the anterior pituitary are growth hormone (GH), prolactin (PRL), melanocyte-stimulating hormone (MSH), and the endorphins and enkephalins.

Growth hormone (GH), a protein of almost 200 amino acids, affects a wide variety of target tissues. GH both promotes growth directly and stimulates production of other growth factors. For example, the ability of GH to stimulate growth of bones and cartilage is partly due to the hormone's signaling the liver to produce hormones called **somatomedins,** which circulate in blood plasma and directly stimulate bone and cartilage growth. (This endocrine response to growth hormone qualifies GH as a tropic hormone. And hormone secretion by the liver qualifies that organ as an endocrine gland, among its many other functions.) In the absence of GH, skeletal growth of an immature animal will stop. If GH is injected into an animal that has been deprived of its own GH, growth will be partially restored.

Several human growth disorders are related to abnormal GH production. Excessive production of GH during development can lead to gigantism, while excessive GH production during adulthood results in abnormal growth of bones in the hands, feet, and head, a condition known as acromegaly. Deficient GH production in childhood can lead to hypopituitary dwarfism. Children with GH deficiency have been treated successfully with human growth hormones isolated from cadaver pituitaries. But the supply falls short of demand, and growth hormones from most other animals are ineffective. One of the most dramatic achievements of genetic engineering has been the production of GH by bacteria with genes for human GH spliced into their genomes (see Chapter 19). The product is now being used to treat children with hypopituitary dwarfism, and some athletes are now carelessly taking GH (legally or illegally) because it promotes protein synthesis.

Prolactin (PRL) is a protein so similar to GH that it is believed they are encoded in genes that evolved from the same ancestral gene. The physiological roles of these two hormones, however, are different. Prolactin's most remarkable characteristic is the great diversity of effects it produces in different vertebrate species. For example, PRL stimulates mammary gland growth and milk synthesis in mammals; regulates fat metabolism and reproduction in birds; delays metamorphosis in amphibians, where it may also function as a larval growth hormone; and regulates salt and water balance in freshwater fish. This list suggests that prolactin is an ancient hormone whose functions have diversified during the evolution of the various vertebrate classes.

Three of the tropic hormones secreted by the anterior pituitary are closely related chemically. **Follicle-stimulating hormone (FSH), luteinizing hormone (LH),** and **thyroid-stimulating hormone (TSH)** are all similar glycoproteins, protein molecules with carbohydrate attached to them. FSH and LH are also called **gonadotropins** because they stimulate the activities of both the male and female gonads, the testes and ovaries. TSH, as the name implies, stimulates the production of hormones by the thyroid gland.

The remaining hormones from the anterior pituitary all come from a single parent molecule called **pro-opiomelanocortin.** This large protein is cleaved into several short fragments inside the pituitary cells. At least four of these fragments are active peptide hormones. **Adrenocorticotropin (ACTH)** is a tropic hormone that stimulates the production and secretion of steroid hormones by the adrenal cortex. As described earlier, **melanocyte-stimulating hormone (MSH)** regulates the activity of pigment-containing cells in the skin of some vertebrates. Very small amounts of MSH are secreted by the human pituitary, but the functions of this hormone in humans are uncertain. The other two derivatives of pro-opiomelanocortin are classes of hormones called **endorphins** (or **enkephalins**). These molecules are also produced by certain neurons in the brain. They are sometimes called the body's natural opiates because they inhibit the perception of pain. In fact, heroin and other opiate drugs mimic endorphins and bind to the same receptors in the brain. (See the interview with Candace Pert that precedes Unit One.) Some researchers speculate that the so-called runner's high results partly from the release of endorphins when stress and pain in the body reach critical levels.

Hypothalamic Releasing Hormones Let's now return to the hypothalamus to see how the anterior pituitary is controlled. The hypothalamic signals called releasing hormones are produced by neurosecretory cells and released into capillaries in a region at the base of the hypothalamus called the **median eminence** (see Figure 41.11b). Unlike most veins that drain body organs, the veins exiting the median eminence do not connect directly to the vena cava. Rather, they subdivide to form a second capillary bed within the anterior pituitary. (Recall a similar portal circulation in the liver and kidneys; see Chapters 37 and 40, respectively.) In this way, the hypothalamic releasing hormones have direct access to the gland they control.

Although they are called releasing hormones, some of these signals actually *inhibit,* rather than stimulate, the secretion of hormones from the anterior pituitary. Every anterior pituitary hormone is controlled by at least one releasing hormone, and most have both a releasing hormone and a release-inhibiting hormone.

The Thyroid Gland

In humans and other mammals, the **thyroid gland** consists of two lobes located on the ventral surface of the trachea (see Figure 41.10). In other vertebrates, the halves of the gland may be more separated on the two sides of the pharynx. The thyroid gland produces two very similar hormones derived from the amino acid tyrosine: triiodothyronine (T_3), which contains three iodine atoms, and thyroxine (T_4), which contains four iodine atoms:

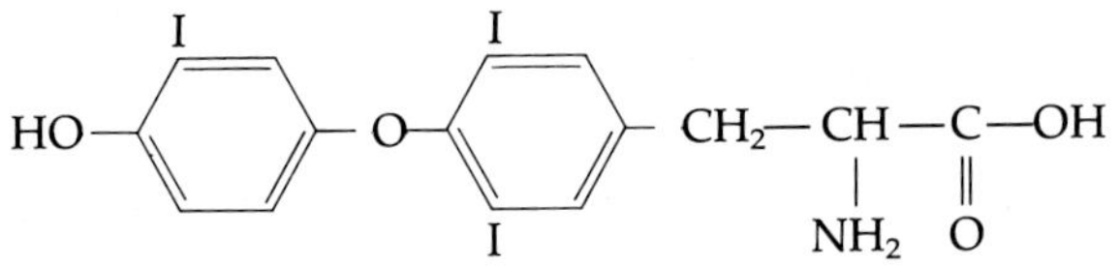

TRIIODOTHYRONINE (T_3)

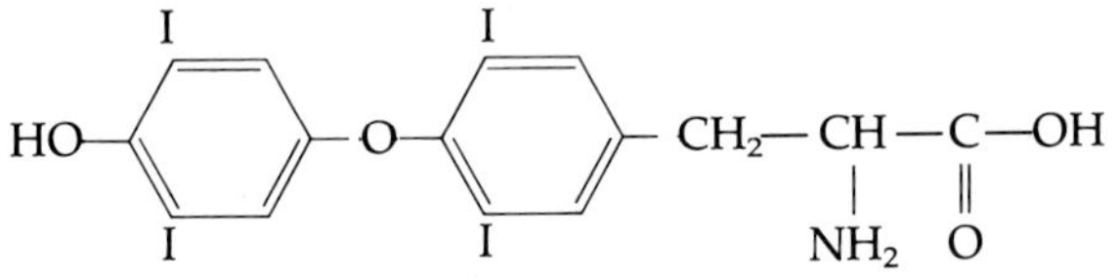

THYROXINE (T_4)

In mammals, T_3 is usually the more active of the two hormones, although both have the same effects on their target cells.

The thyroid gland plays a crucial role in the development and maturation of vertebrates. A striking example is the thyroid's control of metamorphosis of a tadpole into a frog, which involves massive reorganization of many different tissues. The thyroid is equally important in human development. An inherited condition of thyroid deficiency known as cretinism results in markedly retarded skeletal growth and poor mental development. These defects can often be overcome, at least partially, if treatment with thyroid hormones is begun early in life. Studies with animals have shown that thyroid hormones are required for the normal functioning of bone-forming cells and for branching of nerve cells during embryonic development of the brain.

The thyroid gland also regulates metabolism in mammals. In humans, an excessive secretion of thyroid hormones, known as hyperthyroidism, produces such symptoms as high body temperature, profuse sweating, weight loss, irritability, and high blood pressure. The opposite condition, hypothyroidism, can cause cretinism in infants and produce symptoms such as weight gain, lethargy, and intolerance to cold in adults. Another condition associated with a shortage of thyroid hormone is an enlargement of the thyroid called goiter, often caused by a deficiency of iodine in the diet (see Chapter 2).

The secretion of thyroid hormones is controlled by the hypothalamus and pituitary in a negative feedback loop (Figure 41.12).

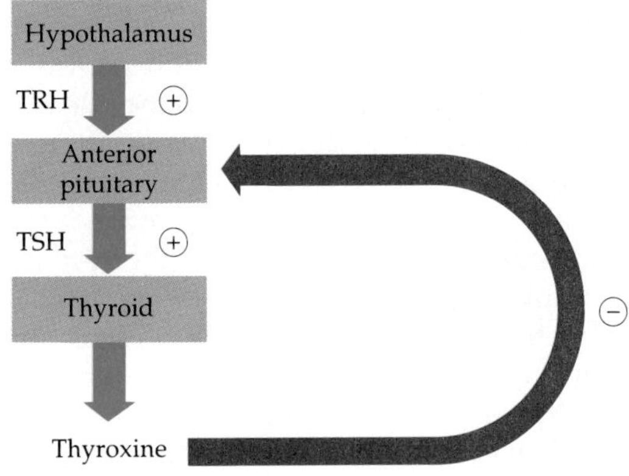

Figure 41.12
A feedback control loop regulating thyroxine secretion. The anterior pituitary gland produces thyroid-stimulating hormone, or TSH. When TSH binds to its receptors in the thyroid gland, it generates cAMP as a second messenger within the target cells, triggering the synthesis and release of the thyroid hormones T_3 and T_4 (thyroxine). Secretion of TSH itself is controlled by TSH-releasing hormone, or TRH, from the hypothalamus. The system is balanced by negative feedback, with high levels of T_3 and T_4 inhibiting TSH secretion (probably by desensitizing pituitary cells to the releasing hormone). The hypothalamus-pituitary-thyroid feedback system explains why iodine deficiencies lead to goiter. In the absence of sufficient iodine, the thyroid gland cannot synthesize adequate amounts of T_3 and T_4. Consequently, the pituitary continues to secrete TSH, leading to an enlargement of the thyroid.

The mammalian thyroid gland also contains endocrine cells that secrete **calcitonin.** This polypeptide lowers calcium levels in the blood as part of calcium homeostasis, described in the next section.

The Parathyroid Glands

The four **parathyroid glands,** embedded in the surface of the thyroid, function in homeostasis of calcium ions. They secrete **parathyroid hormone (PTH),** which raises blood levels of calcium and thus has an effect opposite to that of the thyroid hormone calcitonin. Parathyroid hormone elevates blood Ca^{2+} by stimulating Ca^{2+} reabsorption in the kidney, and by inducing specialized bone cells called osteoclasts to decompose the mineralized matrix of bone and release Ca^{2+} to the blood. Calcitonin has just the opposite effects on the kidneys and bone, thus decreasing blood Ca^{2+}. Vitamin D, synthesized in the skin and converted to its active form in many tissues, is essential to PTH function, so it is also required for complete calcium balance. A lack of PTH causes blood levels of calcium to drop dramatically, leading to convulsive contractions of the skeletal muscles. If unchecked, this condition, known as tetany, is fatal. The control of blood calcium is an example of how homeostasis is often maintained by the balancing

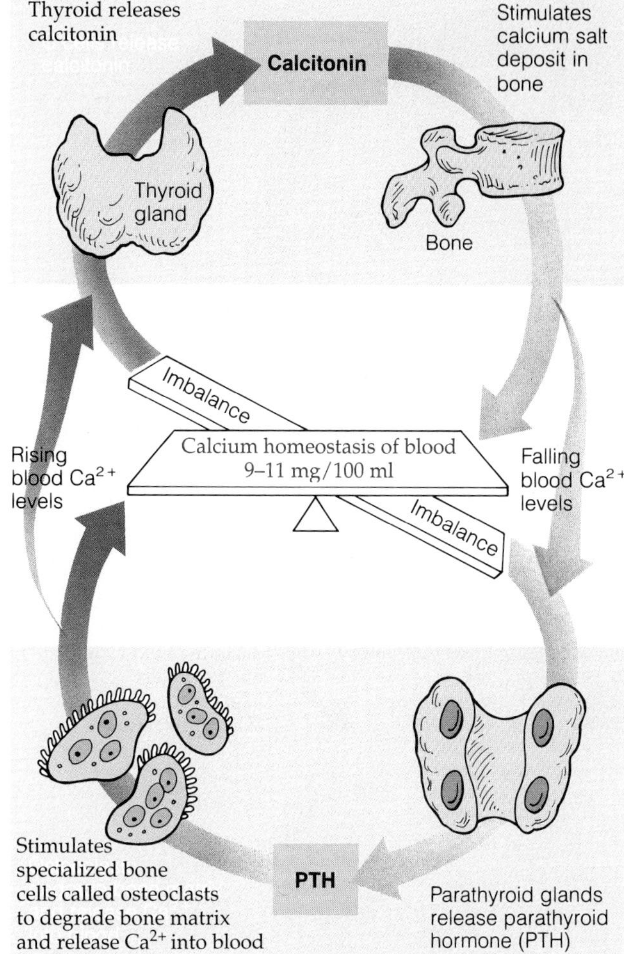

Figure 41.13
Hormonal control of calcium homeostasis in mammals. A negative feedback system involving two antagonistic hormones, calcitonin and parathyroid hormone (PTH), maintains the concentration of calcium in blood within a very narrow range of 9–11 mg/100 mL. A rise in blood Ca^{2+} induces the thyroid gland to secrete calcitonin, which lowers the Ca^{2+} concentration by increasing bone deposition and by reducing Ca^{2+} reabsorption in the kidneys (bone is the only target illustrated here). These effects on bone and the kidneys are reversed by PTH, which is secreted from the parathyroid glands when the concentration of blood Ca^{2+} falls below the set point of 10 mg/100 mL. Blood Ca^{2+} levels begin to increase, but are only allowed to rise so far before the thyroid counters by secreting more calcitonin. In classic feedback fashion, these two hormones balance one another to minimize fluctuations in the concentration of blood Ca^{2+}, an ion essential to the normal functioning of all cells.

of two antagonistic hormones—in this case, PTH and calcitonin (Figure 41.13).

The Pancreas

The **pancreas** consists mostly of exocrine tissue that produces digestive enzymes and exports them to the small intestine via the pancreatic duct (see Chapter 37). Scattered among this exocrine tissue are the **islets of Langerhans,** clusters of endocrine cells. Each islet has a population of **alpha (α) cells,** which secrete the peptide hormone **glucagon,** and a population of **beta (β) cells,** which secrete the protein hormone **insulin.**

Insulin and glucagon are antagonistic hormones that regulate the concentration of glucose in the blood (see Figure 41.3). This is a critical homeostatic function, because glucose is a major fuel for cellular respiration and a key source of carbon skeletons for the synthesis of other organic compounds (see Chapter 9). Metabolic balance depends on the maintenance of blood sugar at a concentration near a set point, which is 90 mg/mL in humans. When blood sugar exceeds this level, insulin acts to lower the glucose concentration. When blood sugar drops below the set point, glucagon increases glucose concentration. By negative feedback, blood sugar concentration determines the relative amounts of insulin and glucagon secreted by the islet cells, which are equipped with glucose receptors.

Insulin and glucagon both influence blood sugar concentration by multiple mechanisms. Insulin lowers blood sugar partly by stimulating cells of many tissues, including muscle, to take up glucose from the blood. Insulin also decreases the supply of glucose by slowing glycogen breakdown in the liver and inhibiting the conversion of amino acids and fatty acids to sugar (Figure 41.14). In contrast, glucagon increases the supply of blood sugar by stimulating glycogen hydrolysis and enhancing the conversion of amino acids and fatty acids to glucose in the liver. The liver's ability to fulfill this important role in glucose management is associated with its metabolic versatility and its access to absorbed nutrients via the hepatic portal vein, which carries blood directly from the small intestine to the liver.

When these mechanisms of glucose homeostasis go awry, there are serious consequences. Diabetes mellitus, perhaps the best known endocrine disorder, is caused by a deficiency of insulin or a loss of response to insulin in target tissues. The result is high blood sugar—so high, in fact, that the diabetic's kidneys excrete glucose, which explains why the presence of sugar in urine is one test for diabetes. As more glucose concentrates in the urine, more water is excreted with it, resulting in excessive volumes of urine and persistent thirst. (*Diabetes,* from the Greek, refers to this copious urination, and *mellitus* is from the Greek word for "honey," referring to the presence of sugar in the urine.) Because glucose is unavailable as a major fuel source for diabetics, fat must serve as the main substrate for cellular respiration. In severe cases of diabetes, acidic metabolites formed during fat breakdown accumulate in the blood, threatening life by lowering blood pH.

There are actually two major forms of diabetes with very different causes. **Type I diabetes mellitus** (insulin-dependent or juvenile-onset diabetes) is an autoimmune disorder, in which the immune system mounts an attack on the cells of the pancreas. (Chapter

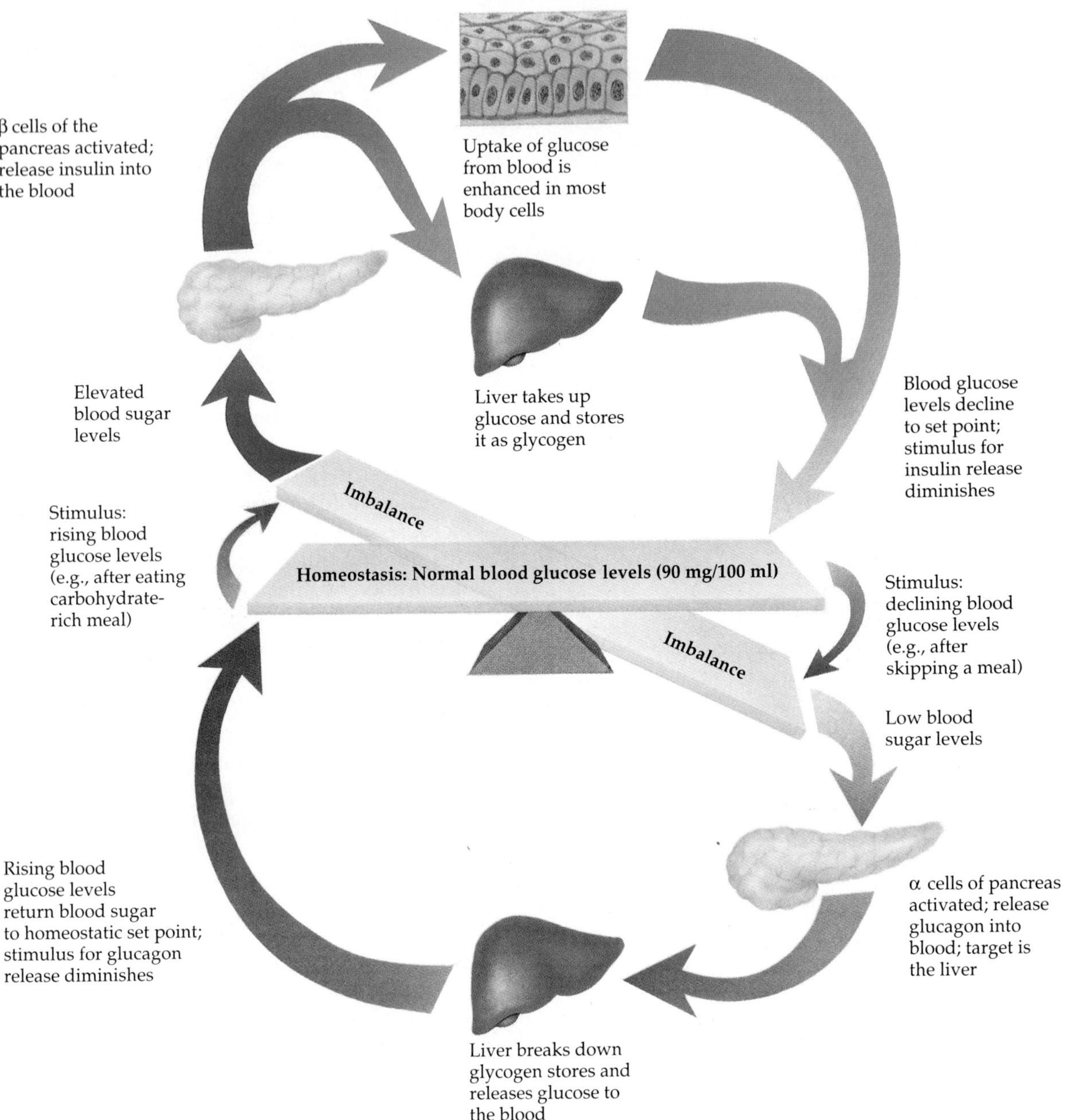
β cells of the pancreas activated; release insulin into the blood
Uptake of glucose from blood is enhanced in most body cells
Elevated blood sugar levels
Liver takes up glucose and stores it as glycogen
Blood glucose levels decline to set point; stimulus for insulin release diminishes
Imbalance
Stimulus: rising blood glucose levels (e.g., after eating carbohydrate-rich meal)
Homeostasis: Normal blood glucose levels (90 mg/100 ml)
Stimulus: declining blood glucose levels (e.g., after skipping a meal)
Imbalance
Low blood sugar levels
Rising blood glucose levels return blood sugar to homeostatic set point; stimulus for glucagon release diminishes
α cells of pancreas activated; release glucagon into blood; target is the liver
Liver breaks down glycogen stores and releases glucose to the blood

Figure 41.14
Glucose homeostasis maintained by insulin and glucagon: a closer look. A rise in blood sugar above the set point of 90 mg/100 mL in humans stimulates the pancreas to secrete insulin, which acts on its target cells to lower blood sugar. When blood sugar concentration dips below the set point, the pancreas responds by secreting glucagon, which acts on the liver to raise the blood sugar level. (See Figure 41.3 for an overview.)

39 discusses possible causes of autoimmune reactions.) This usually occurs rather suddenly during childhood, destroying the individual's ability to produce insulin. Treatment consists of insulin injections, which are usually taken several times daily. Until recently, this insulin was extracted from animal pancreases, but genetic engineering has provided a relatively inexpensive source of human insulin by inserting the gene for the hormone into bacteria (see Chapter 19). **Type II diabetes mellitus** (non–insulin-dependent or adult-onset diabetes) is characterized either by a deficiency of insulin, or, more commonly, by reduced responsiveness in target cells due to some change in insulin receptors. Type II diabetes usually occurs after about age 40, becoming more likely with increasing age. More than 90% of diabetics are type II, and many can manage their blood sugar solely by exercise and dietary control. Heredity is a major factor in type II diabetes.

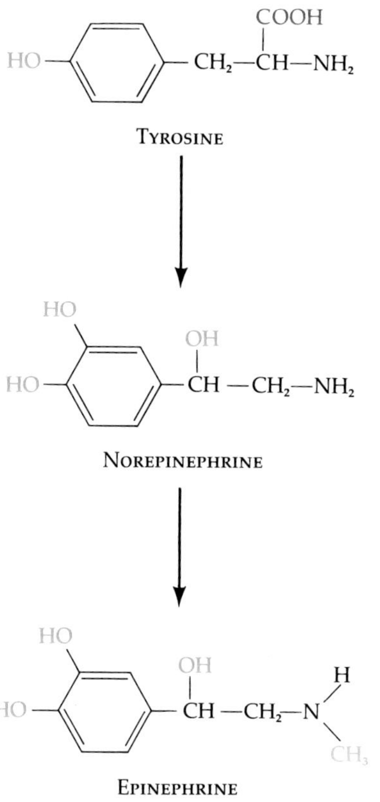

Figure 41.15
The synthesis of catecholamine hormones. Cells in the adrenal medulla called chromaffin cells synthesize the catecholamines norepinephrine and epinephrine from the amino acid tyrosine. Norepinephrine is made by removing a carboxyl group and adding hydroxyl groups. Epinephrine is made from norepinephrine by adding a methyl (—CH_3) group.

The Adrenal Glands

The **adrenal glands** are adjacent to the kidneys. In mammals, each adrenal gland is actually made up of two glands with different cell types, functions, and embryonic origins: the **cortex,** or outer portion, and the **medulla,** or central part of the gland. Nonmammalian vertebrates have quite different arrangements of the same tissues.

Adrenal Medulla What makes your heart beat faster and your skin develop goose bumps when you sense danger or approach a stressful situation, like speaking in public? These reactions are part of the "fight-or-flight" syndrome stimulated by two hormones of the adrenal medulla, **epinephrine** (also known as adrenaline) and **norepinephrine** (noradrenaline). These compounds, called **catecholamines,** are synthesized from the amino acid tyrosine (Figure 41.15).

Epinephrine is secreted in response to positive or negative stress—everything from extreme pleasure to increased cold to life-threatening danger. Release of epinephrine into the blood produces rapid and dramatic effects involving several targets. Epinephrine, using cAMP as a second messenger (see Figure 41.7), mobilizes glucose in skeletal muscle and liver cells and stimulates the release of fatty acids from fat cells. The fatty acids may be used by cells for energy. In addition to increasing the availability of energy sources, epinephrine and norepinephrine have profound effects on muscle contraction. For example, they increase both the rate and the stroke volume of the heartbeat (see Chapter 38). They also cause smooth muscles of some blood vessels to contract and muscles of other vessels to relax, with an overall effect of shunting blood away from the skin, gut, and kidneys, while increasing the blood supply to the heart, brain, and skeletal muscles.

What causes the release of epinephrine during the response to stress? The adrenal medulla is under the control of nerve cells from the sympathetic division of the autonomic nervous system (see Chapter 44). In fact, nerve endings from the sympathetic nervous system are found in the adrenal medulla in close contact with individual cells known as chromaffin cells. When the nerve cells are excited by some form of stressful stimulus, they release the neurotransmitter acetylcholine. Acetylcholine combines with receptors on the chromaffin cells, stimulating the release of epinephrine. Norepinephrine is released independently of epinephrine. Its function is primarily that of sustaining blood pressure. Both epinephrine and norepinephrine also function as neurotransmitters in the nervous system, as we will see in Chapter 44.

Adrenal Cortex The adrenal cortex, like the adrenal medulla, reacts to stress. But it responds to endocrine signals rather than to nervous input. Stressful stimuli cause the hypothalamus to secrete a releasing hormone that stimulates the anterior pituitary to release the tropic hormone ACTH. When it reaches its target via the bloodstream, ACTH stimulates cells of the adrenal cortex to synthesize and secrete a family of steroids called **corticosteroids.** In another case of negative feedback, elevated levels of corticosteroids in the blood suppress the secretion of ACTH.

Many corticosteroids have been isolated from the adrenal cortex; the two main types in humans are the **glucocorticoids,** such as cortisol, and the **mineralocorticoids,** such as aldosterone. The structures of these and several other important steroid hormones are shown in Figure 41.16.

The primary effect of glucocorticoids is on glucose metabolism. Glucocorticoids promote the synthesis of glucose from noncarbohydrate substrates, such as proteins, making more glucose available as fuel in response to a stressful situation. This effect is slower but of longer duration than the stress-induced action of epinephrine (Figure 41.17).

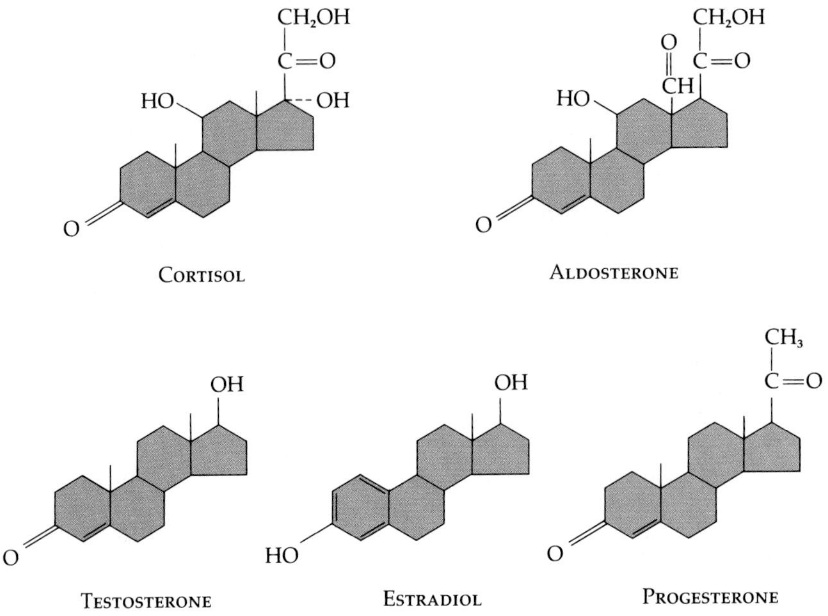

Figure 41.16
The structures of some steroid hormones. Cortisol (a glucocorticoid) and aldosterone (a mineralocorticoid) are made in the adrenal cortex. The gonads synthesize testosterone (an androgen), estradiol (an estrogen), and progesterone (a progestin). The precursor for the synthesis of steroid hormones is cholesterol (see Chapter 5).

CORTISOL

ALDOSTERONE

TESTOSTERONE

ESTRADIOL

PROGESTERONE

Abnormally high doses of glucocorticoids administered as medication suppress certain components of the body's immune system—for example, the inflammatory reaction that occurs at the site of an infection. Glucocorticoids are used to treat diseases in which excessive inflammation is a problem. Cortisone, for instance, was once thought to be a miracle drug that could cure serious inflammatory conditions such as arthritis. It has become clear, however, that long-term use of the corticoid hormones can result in increased susceptibility to infection and disease because of their immunosuppressive effects.

The second major type of adrenal steroids, the mineralocorticoids, have their major effects on salt and water balance. Aldosterone, for example, stimulates cells in the kidney to reabsorb sodium ions from the filtrate, which also causes water reabsorption and raises blood pressure. Control of aldosterone secretion is largely independent of the pituitary and the hypothalamus. Instead, it is regulated by hormones produced in the liver and kidneys in response to changes in plasma ion concentration (see Chapter 40).

The Gonads

Steroids produced in the testes of males and ovaries of females affect growth and development and also regulate reproductive cycles and behaviors. There are three major categories of gonadal steroids: androgens, estrogens, and progestins (see Figure 41.16). All three types are found in both males and females but in different proportions.

The testes primarily synthesize **androgens,** the main such hormone being **testosterone.** In general, androgens stimulate the development and maintenance of the male reproductive system. Androgens produced early in the development of an embryo determine that the fetus will develop as a male rather than a female. At puberty, high concentrations of androgens are responsible for the development of human male secondary sex characteristics, such as male patterns of hair growth and a low voice.

Estrogens, the most important of which is estradiol, have a parallel role in the maintenance of the female reproductive system and the development of female secondary sex characteristics. In mammals, **progestins,** which include progesterone, are primarily involved with preparing and maintaining the uterus, which supports the growth and development of an embryo.

Synthesis of both estrogens and androgens is controlled by gonadotropins from the anterior pituitary gland, follicle-stimulating hormone and luteinizing hormone. The secretion of FSH and LH is controlled by one hypothalamic releasing hormone, GnRH (gonadotropin-releasing hormone). The complex feedback relationships that regulate secretion of gonadal steroids will be described in detail in Chapter 42.

Other Endocrine Organs

Many organs with primarily nonendocrine functions also secrete hormones, many of which we have encountered in earlier chapters. The digestive tract, for example, is the source of at least eight hormones, including gastrin and secretin (see Chapter 37). Erythropoietin, from the kidney, stimulates red blood cell production (see Chapter 38). Atrial natriuretic hormone, secreted by the heart, helps regulate salt and water bal-

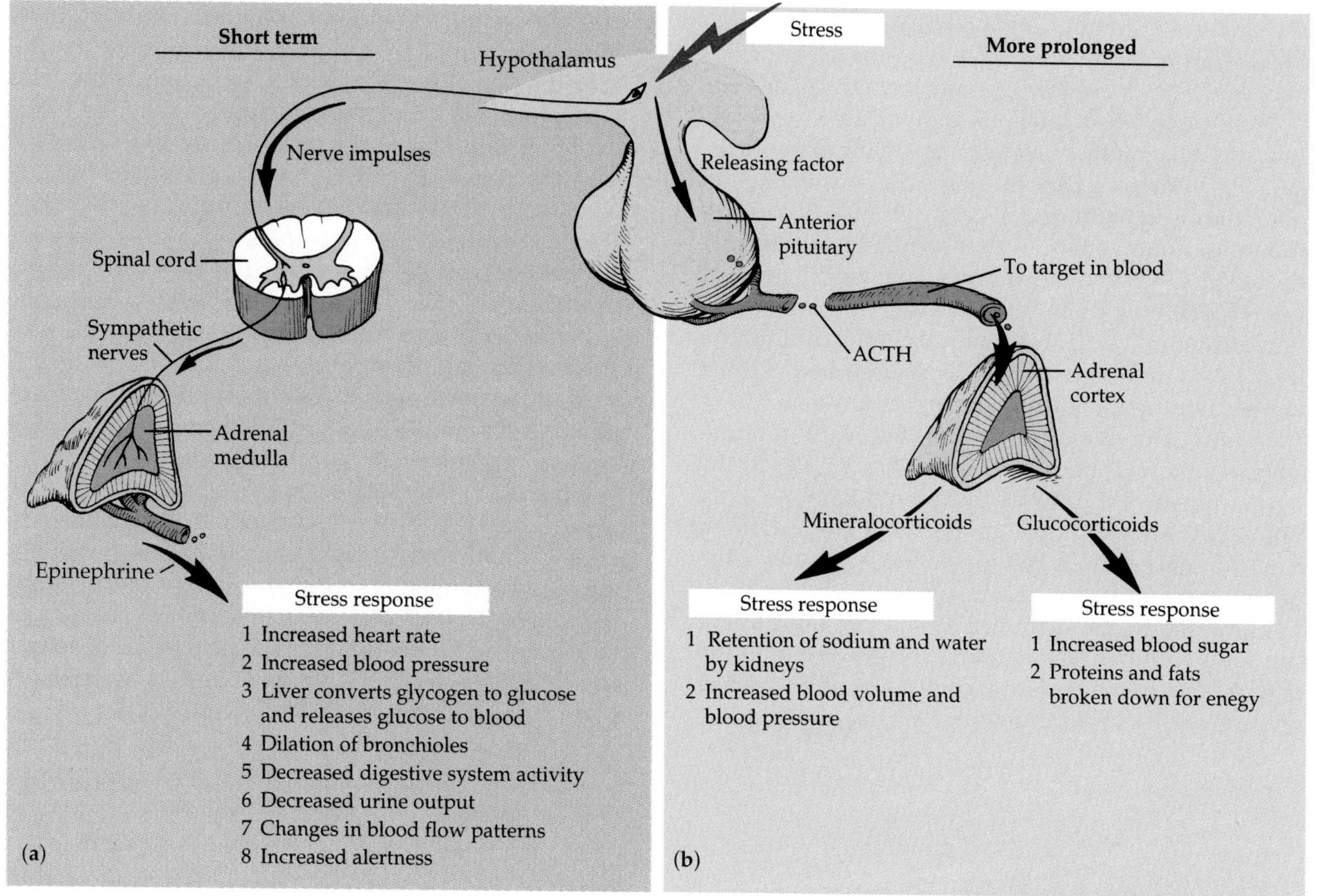

Figure 41.17
Stress and the adrenal gland. Stressful stimuli cause the hypothalamus to activate the adrenal medulla via nerve impulses and the adrenal cortex via hormonal signals. (**a**) The medulla mediates short-term responses to stress by secreting epinephrine, (**b**) while the cortex controls more prolonged responses by secreting its steroid hormones.

ance and blood pressure (see Chapter 40). Two other endocrine organs that deserve some attention are the pineal gland and the thymus.

The **pineal gland** is a small mass of tissue near the center of the mammalian brain (closer to the brain surface in some other vertebrates). Although we know considerably more about the pineal than we did when Descartes described it as the seat of the soul, it still remains largely a mystery. The pineal secretes the hormone **melatonin,** a modified amino acid. The pineal contains light-sensitive cells or has nervous connections from the eyes, and melatonin regulates functions related to light and to seasons marked by changes in day length. For example, melatonin, like MSH, affects skin pigmentation in many vertebrates. Most of the pineal's functions, though, are related to biological rhythms associated with reproduction. Since melatonin is secreted at night, the amount secreted depends on the length of the night. In winter, for example, the days are short and the nights are long, so more melatonin is secreted. Thus, melatonin production is a link between a biological clock and daily or seasonal activities, such as reproduction. However, the precise role of melatonin in mediating rhythms is not yet clear.

The **thymus** is another enigmatic gland. It was not until the 1960s that its role in the immune system was discovered. This gland lies across the front of the neck in humans and is quite large during childhood. At puberty, when the immune system is well established, the thymus begins to decline quickly and virtually disappears by adulthood. The thymus secretes several messengers, including **thymosin,** that stimulate the development and differentiation of T lymphocytes (see Chapter 39).

ENDOCRINE GLANDS AND THE NERVOUS SYSTEM

There have been many examples in this chapter of interactions between the endocrine system and the nervous system. Indeed, we have seen that the two systems are often inseparable and may function as a single unit. We can synthesize much of our understanding of the chemical communication and coordination in animal bodies by examining three types of relationships between the endocrine system and the nervous system.

First, these two regulatory systems are *structurally* related. Many endocrine glands are made of nerve tissue. The vertebrate hypothalamus and posterior pituitary, the insect brain, and the bag cells of *Aplysia* are all examples of nerve tissues that secrete hormones into the blood. Other endocrine glands that are not nervous tissue in their present form have evolved from the nervous system. The adrenal medulla is derived from a modified ganglion (cluster of nerve cell bodies) that has become separated from the nervous system.

Second, the two systems are *chemically* related. Several vertebrate hormones are used as signals by the nervous system as well as by the endocrine system. Epinephrine, for example, functions in the body both as an adrenal hormone and as a neurotransmitter in the nervous system.

Third, the two systems are *functionally* related. We can see two types of functional relationships. In the first, the coordinating system controlling physiological processes involves both nervous and hormonal components arranged in series. For example, milk letdown, the release of milk by a mother during nursing, is controlled by a neuroendocrine reflex: Suckling stimulates sensory cells in the nipples, and nervous signals to the hypothalamus trigger the release of oxytocin from the posterior pituitary. In the second type of functional relationship, each system affects the output of the other. We have seen several examples of how the nervous system controls the endocrine glands, including the stimulation of the adrenal medulla. But the endocrine system also affects both the development of the nervous system and its output—behavior.

Thus, we have seen that animals possess two coordinating systems. We have studied one, the endocrine system, in some detail, but that discussion has drawn us closer and closer to the other. In Chapters 44 and 45, we will turn to the other coordinating system, the nervous system. First, however, we will consider one of the most fundamental subjects in biology, one in which the role of the endocrine system is central not only to the survival of the individual but also to the propagation of the species. In Chapters 42 and 43, we will explore reproduction and development.

STUDY OUTLINE

1. The endocrine system utilizes chemical messengers, called hormones, to affect target organs at distant sites.

An Overview of Chemical Signals (pp. 907–910)

1. A hormone is a molecule that is secreted by an endocrine gland and that travels in blood to a target cell, where it binds with specific receptors and elicits a response. Antagonistic (opposing) hormones function in many cases of homeostasis.
2. Pheromones act as communication signals between different individuals of the same species.
3. Local regulators, such as neurotransmitters, growth factors, and prostaglandins, affect target cells in the immediate vicinity of their secretion.

Mechanisms of Hormone Action at the Cellular Level (pp. 910–915)

1. Steroid hormones penetrate the plasma membrane and bind to specific protein receptors in the nucleus. Hormone-receptor complexes then bind to an acceptor protein on a chromosome and initiate transcription.
2. Peptide hormones, which cannot pass through the cell membrane, bind to specific receptors on the plasma membrane. Through second messengers such as cyclic AMP and inositol triphosphate, these hormones trigger a cascade of metabolic reactions within the cells.
3. Binding of a hormone to a surface receptor activates G protein, which in turn activates the membrane protein adenylate cyclase to produce cAMP. The cAMP then activates cAMP-dependent protein kinase, an enzyme that phosphorylates other proteins.
4. Inositol triphosphate (IP_3) serves as a second messenger for neurotransmitters, growth factors, and some hormones. Binding of a hormone to its receptor activates a G protein that stimulates a membrane enzyme to cleave a membrane phospholipid into IP_3 and diacylglycerol. Diacylglycerol activates a protein kinase, and IP_3 causes the release of Ca^{2+} from storage in the endoplasmic reticulum. Ca^{2+}, acting alone or bound to calmodulin, alters the activities of certain enzymes.

Invertebrate Hormones (pp. 915–917)

1. Invertebrate hormones control different aspects of development, reproduction, and homeostasis.
2. Arthropods have well-developed endocrine systems. In insects, molting and development are controlled by an interplay between ecdysone and juvenile hormone.

The Vertebrate Endocrine System (pp. 917–927)

1. The hypothalamus integrates endocrine and neural function by influencing the pituitary gland. Under the direction of releasing hormones from the hypothalamus, the anterior pituitary produces several tropic hormones that act on other endocrine glands. The posterior pituitary is an extension of the brain that stores and releases two peptide hormones produced by the hypothalamus.

2. The posterior pituitary is a repository for oxytocin and antidiuretic hormone (ADH). Oxytocin induces uterine contractions and milk ejection, whereas ADH enhances water reabsorption in the kidneys.
3. The anterior pituitary produces an array of protein and peptide hormones, including thyroid-stimulating hormone (TSH), follicle-stimulating hormone (FSH), luteinizing hormone (LH), growth hormone (GH), prolactin (PRL), adrenocorticotropin (ACTH), melanocyte-stimulating hormone (MSH), and the endorphins.
4. The chemically related tropic hormones, TSH and the gonadotropins (FSH and LH), stimulate the thyroid gland and the gonads, respectively, to produce their hormones.
5. GH promotes growth, either directly or by stimulating the production of other growth factors.
6. Prolactin, named for its stimulation of lactation in mammals, has diverse effects in different vertebrates.
7. The remaining four anterior pituitary hormones, all of which come from pro-opiomelanocortin molecules, include ACTH, which has a tropic effect on the adrenal cortex, and MSH, which influences skin pigmentation in some vertebrates. Endorphins and enkephalins, the brain's natural opiates, inhibit the perception of pain.
8. Hypothalamic releasing hormones control the secretion of specific hormones from the anterior pituitary.
9. The thyroid gland produces iodine-containing hormones that stimulate metabolism and influence development and maturation in vertebrates. The thyroid also secretes calcitonin, which lowers calcium levels in the blood.
10. The parathyroid glands raise plasma calcium levels by secreting parathyroid hormone (PTH). PTH works with calcitonin to effect calcium homeostasis by actions on bone and the kidneys.
11. The endocrine portion of the pancreas consists of islet cells that secret insulin and glucagon. High plasma glucose stimulates the release of insulin, which increases cellular uptake of glucose, promotes formation and storage of glycogen in the liver, and stimulates protein synthesis and fat storage. Low plasma glucose triggers glucagon release, which increases blood glucose by stimulating the conversion of glycogen to glucose in the liver and increasing the breakdown of fat and protein. Type I diabetes mellitus is an autoimmune disorder resulting in a lack of insulin. Type II diabetes is usually caused by the loss of responsiveness of target cells to insulin.
12. The adrenal gland consists of an outer cortex and an inner medulla. The medulla releases epinephrine and norepinephrine in response to stress-activated impulses from the sympathetic nervous system. These hormones mediate a variety of "fight-or-flight" responses. The adrenal cortex releases two groups of corticosteroids, glucocorticoids and mineralocorticoids. Glucocorticoids influence glucose metabolism and the immune system; mineralocorticoids affect salt and water balance.
13. The gonads—testes and ovaries—produce varying proportions of androgens, estrogens, and progestins, steroids that affect growth, development, morphological differentiation, and reproductive cycles and behaviors.
14. The pineal gland secretes melatonin, which influences skin pigmentation, biological rhythms, and reproduction in various vertebrates.
15. The thymus gland functions during early life to stimulate the development of T lymphocytes by means of thymosin and other chemical messengers.

Endocrine Glands and the Nervous System (pp. 927–928)

1. The endocrine and nervous systems function together in chemical communication and control in animals.

SELF-QUIZ

1. Which of the following is *not* an accurate statement about hormones?
   a. Hormones are chemical messengers that travel to target cells through the circulatory system.
   b. Hormones often regulate homeostasis through antagonistic functions.
   c. Hormones of the same chemical class usually have the same general function.
   d. Hormones are secreted by specialized cells usually located in endocrine glands.
   e. Hormones are often regulated through feedback loops.
2. A major difference in the mechanism of action between steroid and peptide hormones is that
   a. steroid hormones mainly affect the synthesis of proteins, whereas peptide hormones mainly affect the activity of proteins already in the cell
   b. target cells react more rapidly to steroid hormones than they do to peptide hormones
   c. steroid hormones enter the nucleus, whereas peptide hormones stay in the cytoplasm
   d. steroid hormones bind to a receptor protein, whereas peptide hormones bind to G protein
   e. steroid hormones affect metabolism, whereas peptide hormones affect membrane permeability
3. Which of the following accurately represents the sequence of components in a cellular response to a peptide hormone?
   a. hormone binding to adenylate cyclase—G protein—protein kinase—phosphorylation of enzymes
   b. hormone binding to receptor—G protein—transcription factor—protein kinase
   c. hormone binding to cAMP—G protein—cAMP-dependent protein kinase—adenylate cyclase
   d. hormone binding to G protein—adenylate cyclase—protein kinase—phosphorylation of proteins
   e. hormone binding to receptor—G protein—adenylate cyclase—protein kinase
4. Growth factors are local regulators that
   a. are produced by the anterior pituitary
   b. are modified fatty acids that stimulate bone and cartilage growth
   c. are found on the surface of cancer cells and stimulate abnormal cell division
   d. are proteins that bind to surface receptors and stimulate target cell growth and development
   e. include histamines and interleukins and are necessary for cellular differentiation

5. Which of the following hormones is *incorrectly* paired with its action?
   a. oxytocin—stimulates uterine contraction in childbirth
   b. thyroxine—stimulates metabolic processes
   c. insulin—stimulates glycogen breakdown in liver
   d. ACTH—stimulates release of glucocorticoids by adrenal cortex
   e. melatonin—affects biological rhythms, seasonal reproduction
6. An example of antagonistic hormones controlling homeostasis is
   a. calmodulin and parathyroid hormone in calcium balance
   b. insulin and glucagon in glucose metabolism
   c. progestins and estrogens in sexual differentiation
   d. epinephrine and norepinephrine in "fight-or-flight" response
   e. oxytocin and prolactin in milk production
7. Which of the following human conditions is *incorrectly* paired with a hormone?
   a. acromegaly—growth hormone
   b. diabetes—insulin
   c. cretinism—thyroid hormones
   d. tetany—PTH
   e. hypopituitary dwarfism—ACTH
8. A portal vein carries blood from the hypothalamus directly to the
   a. thyroid
   b. pineal gland
   c. anterior pituitary
   d. posterior pituitary
   e. thymus
9. A second messenger derived from membrane lipids is
   a. cyclic AMP
   b. calmodulin
   c. inositol triphosphate
   d. protein kinase
   e. calcium
10. The main target organs for tropic hormones are
   a. muscles
   b. blood vessels
   c. endocrine glands
   d. kidneys
   e. nerves

CHALLENGE QUESTIONS

1. A woman with a hypothyroid condition is treated with thyroxine. How is this medication likely to affect levels of TSH and TSH-releasing hormone (TRH)?
2. Describe three specific cases in which the nervous and endocrine systems interact.

SCIENCE, TECHNOLOGY, AND SOCIETY

1. Growth hormone (GH) produced via DNA technology has enabled hundreds of children who suffer from hypopituitary dwarfism to grow normally and reach a stature within the normal range. Now that the hormone is readily available and relatively inexpensive, many parents who are concerned that their children are not growing fast enough want to use GH to make them grow faster and taller. There can be harmful effects, such as a reduction in body fat and an increase in muscle mass. And no one yet knows if GH injections will have serious long-term harmful effects in individuals who do not have a hypopituitary condition. Do you think GH therapy should be regulated? In your opinion, what criteria should determine the cases in which GH treatments or other hormone therapies are appropriate?

FURTHER READING

Atkinson, M., and N. MacLaren. "What Causes Diabetes?" *Scientific American,* July 1990.

Cowen, R. "Speeding Up Wound Healing the EGF Way." *Science News,* July 15, 1989. The important function of a growth factor.

Hanson, B. "Raging Hormones in the White House." *Discover,* January 1992. Stress and hyperthyroidism in a former first family (including their dog).

Linder, E., and A. Gilman. "G Proteins." *Scientific American,* July 1992. Signal transduction pathways in hormone action.

Pacchioli, D. "Potent Aromas." *Discover,* November 1991. The role of pheromones in the reproduction of mice.

Snyder, S., and D. Bredt. "Biological Roles of Nitric Oxide." *Scientific American,* May 1992. Recent research on a chemical signal with multiple functions.

Snyder, S. H. "The Molecular Basis of Communication Between Cells." *Scientific American,* October 1985. An excellent article emphasizing the relationship between the endocrine and nervous systems.

Weiss, R. "Heightened Concern Over Growth Hormone." *Science News,* December 8, 1990. The benefits, risks, and ethics of a new product of genetic engineering.

Weissmann, G. "Aspirin." *Scientific American,* January 1991. This painkiller probably suppresses the synthesis of prostaglandins.

42 Animal Reproduction

Modes of Reproduction

Mechanisms of Sexual Reproduction

Mammalian Reproduction

The many aspects of animal form and function we have studied so far can be viewed, in the broadest context, as various adaptations contributing to reproductive success. Individuals are transient. A population transcends finite lifespans only by reproduction, the creation of new individuals from existing ones (Figure 42.1). Animal reproduction is the subject of this chapter. We will first compare the diverse reproductive mechanisms that have evolved, and then examine the details of mammalian, particularly human, reproduction.

Figure 42.1
Two from one: asexual reproduction of a sea anemone *(Anthopleura elegantissima).* One individual in this photograph is undergoing longitudinal fission and will give rise to two smaller offspring. In this chapter, you will learn about the diverse mechanisms of reproduction, both asexual and sexual, that have evolved in the animal kingdom.

MODES OF REPRODUCTION

There are two principal modes of animal reproduction. **Asexual reproduction** occurs when a single individual produces offspring genetically identical to itself. All these genetically identical individuals from one lineage form a clone. In **sexual reproduction,** two individuals produce offspring having a combination of genes inherited from both parents (see Chapters 12 and 34).

Asexual Reproduction

Many invertebrates can reproduce asexually by fission, splitting off new individuals from existing ones or separating into two or more individuals of equal size (see Figure 42.1). Certain cnidarians and tunicates provide good examples of the former process, called **budding.** In this form of asexual reproduction, a new individual grows out from the body of the original (see Figure 12.2). The offspring may detach from its parent, or they may remain joined, eventually forming extensive colonies. Stony corals, which may be more than 1 m across, are colonies of several thousand individuals connected to one another. Another type of asexual reproduction is **fragmentation,** the breaking of the body into several pieces, each of which develops into a complete adult. When polychaetes, segmented worms of the phylum Annelida, prepare to fragment, the anterior ends of what will become new individuals develop heads with sensory organs before the parent actually fragments. In another form of asexual reproduction, some invertebrates release specialized groups of cells that can grow into new individuals. For example, the **gemmules** of sponges are formed when cells of several types migrate together within the sponge and become surrounded by a protective coat. Finally, **regeneration** following an injury is also a form of asexual reproduction if it results in two or more individuals where there was only one before. Echinoderms provide one of the best examples: When one arm is removed from a sea star, the animal will grow a new arm. This in itself would not qualify as reproduction. However, if the free arm has even a small piece of the central disk attached, it too will regenerate into a complete new sea star.

Asexual reproduction often allows many offspring to be produced in a short amount of time, making this reproductive mode ideal for colonizing a habitat rapidly. Theoretically, asexual reproduction is most advantageous in stable, favorable environments because it perpetuates successful genotypes precisely.

Sexual Reproduction

In animals, sexual reproduction involves the fusion of two haploid **gametes** to form a diploid **zygote.** The female gamete, the **ovum** (unfertilized egg), is usually a relatively large and nonmotile cell. The male gamete, the **spermatozoon,** is generally a small, flagellated cell (though arthropods produce nonmotile sperm that must be placed near the female reproductive tract).

Sexual reproduction increases genetic variability among offspring by generating unique combinations of genes inherited from two parents (see Chapters 12 and 13). Many biologists believe that sexual reproduction, by producing offspring having varying phenotypes, may enhance the reproductive success of parents in certain situations—in a fluctuating environment, for instance.

Reproductive Cycles and Patterns

Most animals show definite cycles in reproductive activity, often related to changing seasons. The periodic nature of reproduction allows animals to conserve resources and reproduce when more energy is available than is needed for maintenance and when environmental conditions favor the survival of offspring. Ewes (female sheep), for example, have 15-day reproductive cycles and ovulate at the midpoint of each cycle. But these cycles only occur during fall and early winter, so lambs are born in the spring. Even animals that live in apparently stable habitats, such as the tropics or the ocean, generally reproduce only at certain times of the year. Reproductive cycles are controlled by a combination of hormonal and seasonal cues, examples of the latter including such factors as temperature, rainfall, or day length.

Animals may use either asexual or sexual reproduction exclusively or alternate between the two. In aphids, rotifers, and the freshwater crustacean *Daphnia,* each female can produce eggs of two types, depending on environmental conditions such as the time of year. One type of egg is fertilized, but the other type develops by **parthenogenesis,** which means that the egg develops directly without being fertilized. The adults produced by parthenogenesis are often haploid, and their cells do not undergo meiosis in forming new eggs. In the case of *Daphnia,* the switch from sexual to asexual reproduction is often related to season. Asexual reproduction occurs under favorable conditions and sexual reproduction during times of environmental stress.

Parthenogenesis has a role in the social organization of certain species of bees, wasps, and ants. Male honeybees, or drones, are produced parthenogenetically, whereas females, both sterile workers and reproductive females (queens), develop from fertilized eggs.

Among vertebrates, several genera of fishes, amphibians, and lizards reproduce exclusively by a complex form of parthenogenesis requiring doubling of chromosomes after meiosis to create diploid "zygotes." For example, there are about 15 species of whiptail lizards (genus *Cnemidophorus*) that reproduce exclusively by parthenogenesis. There are no males in these species, but the lizards imitate courtship and mating behavior typical of sexual species of the same genus. During the breeding season, one female of each mating pair impersonates a male (Figure 42.2a). The roles change two or three times during the season, female behavior occurring when the level of the female sex hormone estrogen is rising prior to ovulation (the release of eggs) and male behavior occurring after ovulation when the level of estrogen drops (Figure 42.2b). In fact, ovulation is more likely to occur if an individual is mounted by a pseudomale during the critical time of the hormone cycle; isolated lizards lay fewer eggs than individuals that are allowed to go through the motions of sex. Apparently, these parthenogenetic lizards, which evolved from species having two sexes, still require certain sexual stimuli for maximum reproductive success.

Sexual reproduction presents a special problem for sessile or burrowing animals, such as barnacles and earthworms, or for parasites, such as tapeworms, which may have difficulty encountering a member of the opposite sex. One solution to this problem is **hermaphroditism.** Each individual has both functional male and female reproductive systems (*hermaphrodite* is a contraction of "Hermes" and "Aphrodite," names of a Greek god and goddess). Although some her-

(a)

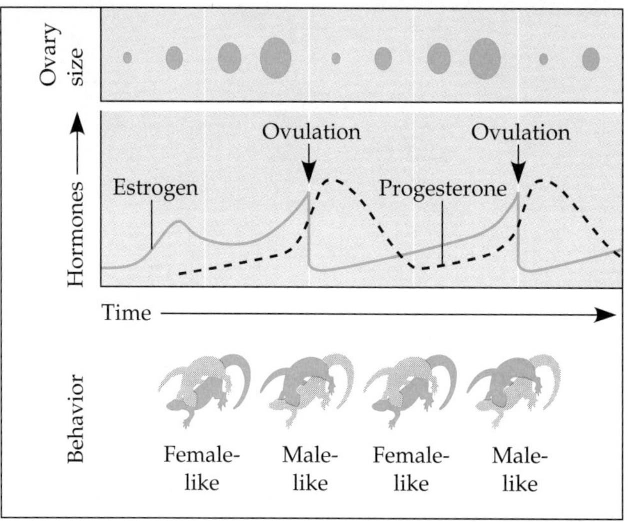

(b)

Figure 42.2
Pseudosex in parthenogenetic lizards. The desert grassland whiptail lizard *(Cnemidophorus uniparens)* is an all-female species. These reptiles reproduce by parthenogenesis; eggs undergo a chromosome doubling after meiosis and develop into lizards without being fertilized. However, ovulation is enhanced by courtship and mating rituals that imitate the behavior of closely related species that reproduce sexually. (**a**) In this photo, the lizard on top is a female playing the role of a male. Every two or three weeks during the breeding season, individuals switch sex roles. (**b**) The pseudosexual behavior of *C. uniparens* is correlated with a cycle of hormone secretion and ovulation. During the period when estrogen levels are increasing and the ovary is growing, a lizard is more likely to behave like a female. After ovulation, estrogen levels drop, the concentration of a second steroid hormone, progesterone, increases, and the individual is more likely to behave like a male. Unisex lizards lay fewer eggs if they are kept in isolation.

Figure 42.3
Sex reversal in a sequential hermaphrodite. In many species of reef fishes called wrasses, sex is not fixed but can change at some time in the animal's life. Sex reversal is often correlated with size. In this scene, a male Caribbean bluehead wrasse and two smaller females are feeding on a sea urchin. All wrasses of this species are born females, but the oldest, largest individuals change sex and complete their lives as males.

maphrodites fertilize themselves, most must mate with another member of the same species. When this occurs, each animal serves as both male and female, donating and receiving sperm. Each individual encountered is a potential mate, and twice as many offspring can be produced from such a union as could result if only one individual's eggs were fertilized.

Another remarkable reproductive pattern is **sequential hermaphroditism,** in which an individual reverses its sex during its lifetime. (In the interview that precedes this unit, Karel Liem discusses his research on this phenomenon.) In some species, the sequential hermaphrodite is **protogynous** (female first), while other species are **protandrous** (male first). In various species of reef fishes called wrasses, sex reversal is associated with age and size. For example, the Caribbean bluehead wrasse is a protogynous species in which only the largest (usually the oldest) individuals change from female to male (Figure 42.3). These fish live as harems consisting of a single male and several females. If the male dies or is removed in experiments, the largest female in the harem changes sex and becomes the new male. Within a week, the transformed individual is producing sperm instead of eggs. In this species, the male defends the harem against intruders, and thus larger size may give a greater reproductive advantage to males than it does to females. In contrast, there are protandrous animals that change from male to female when size increases. In such cases, greater size may increase the reproductive success of females more than it does males. For example, production of huge numbers of gametes is an important asset for sedentary animals, such as oysters, that release their gametes into the surrounding water. Egg cells are generally much larger than sperm cells, and thus females produce fewer ga-

Figure 42.4
Mass ovulation by a sponge. Females of this species of tube sponge *(Agelas)* release hundreds of thousands of eggs in July, in synchrony with sperm release from males. Sticky mucus prevents the eggs from being carried away from the reef by ocean currents.

metes than males. Of course, larger females produce more eggs than smaller ones, and species of oysters that are sequential hermaphrodites are generally protandrous.

The diverse reproductive cycles and patterns we observe in the animal kingdom are adaptations that have evolved by natural selection. We will see many other examples, as we survey the various mechanisms of sexual reproduction.

MECHANISMS OF SEXUAL REPRODUCTION

Patterns of Fertilization and Development

The two major mechanisms of fertilization that have evolved are external and internal fertilization, each having specific environmental and behavioral requirements.

Getting Gametes Together The mechanics of fertilization play an important part in sexual reproduction. In **external fertilization,** the eggs are shed by the female and fertilized by the male in the environment (Figure 42.4). In contrast, **internal fertilization** occurs when the sperm are deposited in (or nearby) the female reproductive tract, and egg and sperm unite within the body of the female. These patterns correlate with the habitats of species and with their phylogenetic positions.

Because external fertilization requires an environment where an egg can develop without desiccation or heat stress, it occurs almost exclusively in moist habitats. Many aquatic invertebrates simply shed their eggs and sperm into the surroundings, and fertilization occurs without the parents actually making physical contact. Still, timing is crucial to ensure that mature sperm encounter ripe eggs. Often, environmental cues such as temperature or length of daylight cause all the individuals of a population to release gametes at once, or pheromones from one individual releasing gametes trigger gamete release in others. In the South Pacific, the Palolo worm (a polychaete) lives in crevices of coral reefs but mates at the ocean surface. The entire population of worms inhabiting a Samoan reef reproduce on a single fall night during a certain phase of the moon. All the worms rise to the surface and shed their gametes simultaneously. People who live on nearby islands take advantage of this mating swarm to collect the worms, a favorite food delicacy.

Most fishes and amphibians that employ external fertilization show specific mating behaviors, resulting in one male fertilizing the eggs of one female. Courtship behavior acts as a mutual trigger for the release of gametes, with two effects: The probability of successful fertilization is increased, and the choice of mates may be somewhat selective.

Internal fertilization requires cooperative behavior that makes copulation possible. In some cases, uncharacteristic sexual behavior is eliminated by natural selection in a most direct manner; for example, female spiders will eat males if specific reproductive signals are not followed during mating. Many more examples of sexual selection and mating behavior will be discussed in Chapter 50.

Internal fertilization also requires sophisticated reproductive systems. Copulatory organs for delivery of sperm and receptacles for its storage and transport to the eggs must be present.

Protection of the Embryo External fertilization usually results in enormous numbers of zygotes, but the proportion that survive and develop further is often quite small. In contrast, internal fertilization usually produces fewer zygotes, but this may be offset by greater protection of the embryos and parental care of the young. Major types of protection of the embryo include resistant eggs, development of the embryo within the reproductive tract of the female parent, and parental protection of eggs.

Many species of terrestrial animals have eggs that can withstand harsh environments. Among the vertebrates, we can compare the eggs of fishes and amphibians, which have only a gelatinous coat that allows free exchange of gases and water, with the amniote eggs of birds and reptiles, whose calcium and protein shells are resistant to water loss and physical damage (see Chapter 30).

Rather than secreting a protective shell around the egg, many animals retain the embryo, which develops within the female reproductive tract. Among mammals, the monotremes lay eggs reminiscent of their reptilian forebears, whereas marsupials such as kangaroos and opossums retain their embryos for a short period in the uterus; the embryos then crawl out and complete fetal development attached to a mammary gland in the mother's pouch. The embryos of placental mammals develop entirely within the uterus, being nourished by the mother's blood supply through a special organ, the placenta (see Chapter 30).

When a bird hatches from an egg, or a kangaroo crawls out of its mother's pouch for the first time, or a human is born, it still is not capable of independent existence. We are familiar with adult birds feeding their young and mammals nursing their offspring, but parental care is much more widespread than we might suspect, and it takes a variety of unusual forms. In one species of tropical frog, for instance, the male carries the tadpoles in his stomach until they metamorphose and hop out as young frogs. There are also many cases of parental care among invertebrates (Figure 42.5).

Figure 42.5
Parental care in an invertebrate. Compared to most arthropods, a female scorpion produces a relatively small number of offspring, but parental protection enhances the survival of those offspring. Fertilization is internal, and the eggs are retained within the reproductive tract of the mother, where they hatch. The young (light-colored in this photograph) then cluster on the back of their mother, prolonging the parental protection.

Diversity in Reproductive Systems

To reproduce sexually, animals must have systems to produce and deliver gametes to the gametes of the opposite sex. These reproductive systems are varied. The least complex systems do not even contain distinct **gonads,** the organs that produce gametes in most animals. The most complex reproductive systems contain many sets of accessory tubes and glands to carry and protect the gametes and developing embryos. If we can make one generalization, it is that the complexity of the reproductive system is not entirely related to the phylogenetic position of the animal; the reproductive systems of parasitic flatworms, for example, are among the most complex in the animal kingdom.

Invertebrate Reproductive Systems Diverse reproductive systems have evolved among invertebrates. We will examine three examples.

Polychaete annelids are a group of mostly marine worms that have separate sexes with relatively simple reproductive systems. Most polychaetes do not have distinct gonads; rather, the eggs and sperm develop from undifferentiated cells lining the coelom. As the gametes mature, they are released from the body wall and fill the coelom. Depending on the species, mature gametes may be shed through the excretory openings, or the swelling mass of eggs may split the body open, killing the parent and spilling the eggs into the environment.

Insects have separate sexes with complex reproductive systems. In the male, sperm develop in the testes and pass through a duct to the seminal vesicles, where they are stored. The seminal vesicles empty into an ejaculatory duct, which runs through the penis. A pair of accessory glands that add fluid to the semen also may be present. Eggs in the female pass from the ovaries through the oviducts and are deposited in the

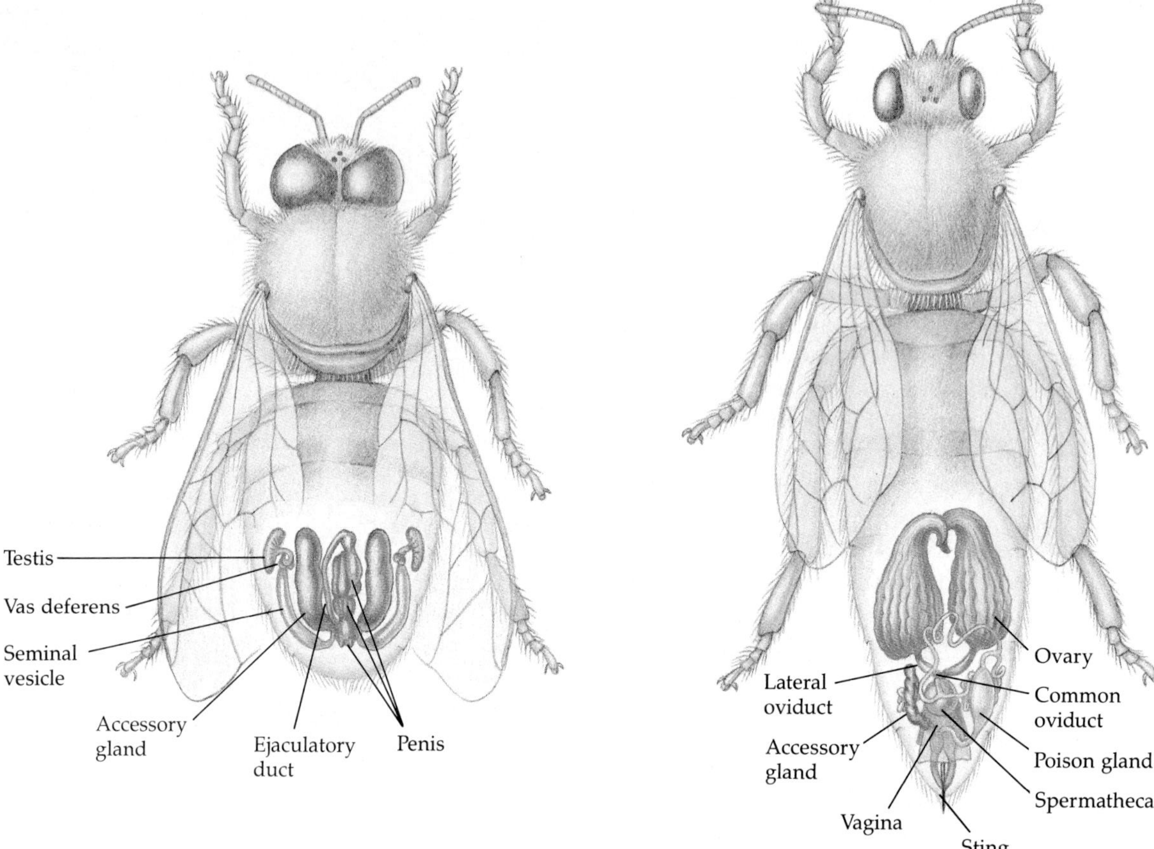

Figure 42.6
Insect reproductive anatomy. (**a**) A male honeybee. Sperm form in the testes, pass through the sperm duct (vas deferens), and are stored in the seminal vesicle. The male ejaculates sperm along with fluid from the accessory glands. Some species of insects and other arthropods possess appendages called claspers to grasp the female during copulation. (**b**) A female honeybee. Eggs develop in the ovaries, pass through the oviducts, and are deposited in the vagina. Some species have a spermatheca for the storage of sperm. Sperm fertilize eggs within the vagina.

vagina, which opens to the outside of the body (Figure 42.6). The female reproductive system also may include a **spermatheca,** a blind sac in which sperm may be stored for a year or more.

Flatworms (phylum Platyhelminthes) are hermaphrodites with extremely complex reproductive systems (Figure 42.7). In addition to those structures described for insects, the female reproductive system includes yolk and shell glands, as well as a uterus where the eggs are fertilized and development begins in some species. The male system contains a complex copulatory apparatus sometimes called a penis. Copulation in flatworms is usually mutual, with each partner inseminating the other. The mechanisms of insemination range from insertion of the penis into a vagina to hypodermic impregnation, where the penis injects sperm into the body tissues and sperm migrate to the female reproductive tract.

Vertebrate Reproductive Systems The basic plan of all vertebrate reproductive systems is quite similar, but there are some important variations on this common theme. Most mammals have separate openings for the digestive, excretory, and reproductive systems, but all other vertebrates have a common opening for all three systems, the **cloaca** (see Chapter 30). The uterus of most vertebrates is bicornate, having two separate branches. In humans and other mammals that develop only a few young in the uterus, as well as in birds and snakes, the uterus has only one branch. Differences in the male reproductive systems center around the copulatory organs. Nonmammalian vertebrates do not have well-developed penes (plural of penis) and may just evert the cloaca to ejaculate.

MAMMALIAN REPRODUCTION

In describing mammalian reproduction, we will use humans as an example, but the reproductive anatomy is similar in other mammals.

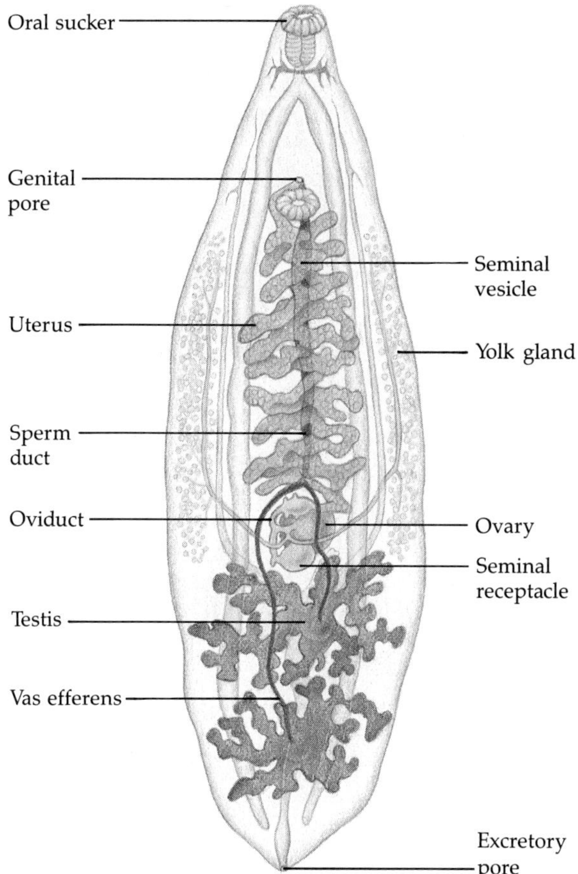

Figure 42.7
Reproductive anatomy in a parasitic flatworm. Flatworms can reproduce either asexually by fission or sexually. During the breeding season, a flatworm develops both male and female reproductive organs that open through a common genital pore. Copulation usually results in mutual insemination. The flatworm depicted here is a liver fluke (see Chapter 29).

The Anatomy of Reproduction in Humans

Male Anatomy The human reproductive system is frequently described as two sets of organs, the internal reproductive organs and the external **genitalia**. The male genitalia are the **scrotum** and **penis.** The male internal reproductive organs consist of gonads that produce gametes (sperm cells) and hormones, accessory glands that secrete products essential to sperm movement, and a set of ducts that carry the sperm and glandular secretions (Figure 42.8).

The male gonads, or **testes** (singular **testis**), are a pair of tightly coiled tubes surrounded by several layers of connective tissue. These tubes are the **seminiferous tubules,** where sperm form. The **interstitial cells** scattered between the seminiferous tubules produce testosterone and other androgens, the male sex hormones.

Sperm production cannot occur at normal body temperatures in most mammals, but the testes of humans and many other mammals hang outside the abdominal cavity in the scrotum, a fold of skin. The temperature in a scrotum is about 2°C below that in the body cavity. The testes develop high in the abdominal cavity and descend into the scrotum just before birth. (In about 1% to 2% of human males, the testes do not descend, but they can be brought down by hormone therapy or surgery.) In some mammals (not humans), the testes are drawn back into the body cavity between breeding seasons. Whales and bats are exceptional in retaining the testes within the body cavity permanently.

From the seminiferous tubules of the testes, the sperm pass into the coiled tubules of the **epididymis,** which stores sperm and is the site of their final maturation. During **ejaculation,** the sperm are propelled from the epididymis through the muscular **vas deferens.** These ducts run from the scrotum around and behind the urinary bladder where they join to form a short **ejaculatory duct.** The ejaculatory duct opens into the **urethra,** the tube draining both the excretory system and the reproductive system. Thus, the reproductive and excretory systems of the male are connected, but, as we will see, this is not the case in females. The urethra runs through the penis and opens to the outside at the tip of the penis.

In addition to the testes and ducts, the male reproductive system contains three sets of glands that add their secretions to the **semen,** the fluid that is ejaculated. The **seminal vesicles** contribute about 60% of the total volume of the semen. This pair of glands lies below and behind the bladder and empties into the ejaculatory duct. The fluid from the seminal vesicles is thick and clear, containing mucus, amino acids, and large amounts of fructose (sugar), which provides energy for the sperm. The seminal vesicles also secrete prostaglandins (see Chapter 41). These chemical signals, once in the female reproductive tract, stimulate contractions of the uterine muscles, which help move the semen up the uterus. Proteins in the seminal fluid cause the semen to coagulate after it is deposited in the female reproductive tract, making it easier for uterine contractions to move the semen.

The **prostate gland** is the largest of the accessory glands. It surrounds the initial segment of the urethra and secretes its products directly into the urethra through several small ducts. Prostatic fluid is thin, milky, and quite alkaline, which balances the acidity of any residual urine in the urethra and the natural acidity of the vagina. The prostate gland is the source of some of the most common medical problems of men over 40. Benign (noncancerous) enlargement of the prostate occurs in more than half of all men in this age group.

The **bulbourethral glands,** the final accessory structures, are a pair of small glands along the urethra be-

Figure 42.8
Reproductive anatomy of the human male. (**a**) Lateral view. (**b**) Frontal view.

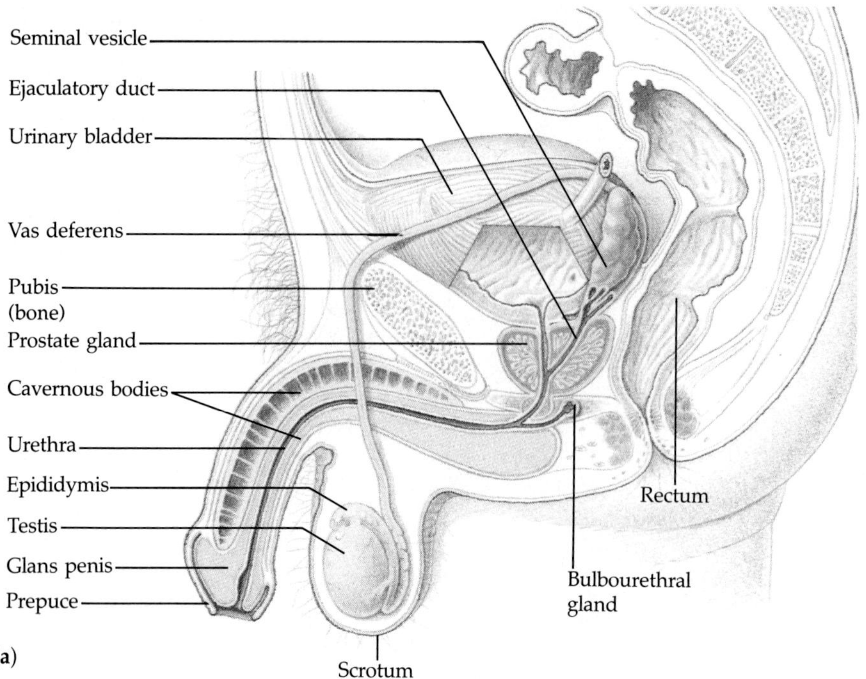

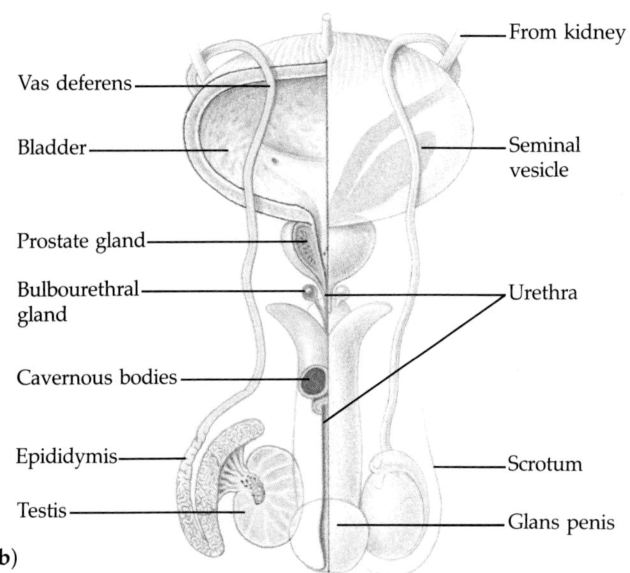

low the prostate. Their function is still in question. They secrete a viscous fluid before emission of the sperm and semen. It has been suggested that this fluid lubricates the penis and vagina, but the volume (just one or two drops) seems insufficient to be very effective for this function. Bulbourethral fluid does carry some sperm released before ejaculation, which is one factor in the high failure rate of the withdrawal method of birth control.

The human penis is composed of three cylinders of spongy tissue (cavernous bodies) derived from modified veins and capillaries. During sexual arousal, this erectile tissue fills with blood from the arteries. As it fills, the increasing pressure seals off the veins that drain the penis, causing it to engorge with blood. The resulting erection is essential to insertion of the penis into the vagina. Rodents, raccoons, walruses, and several other mammals also possess a **baculum,** a bone that stiffens the penis.

The main shaft of the penis is covered by relatively thick skin, whereas the head, or **glans penis,** has a much thinner covering and is consequently more sensitive to stimulation. The human glans is covered by a fold of skin called the foreskin, or **prepuce,** which may

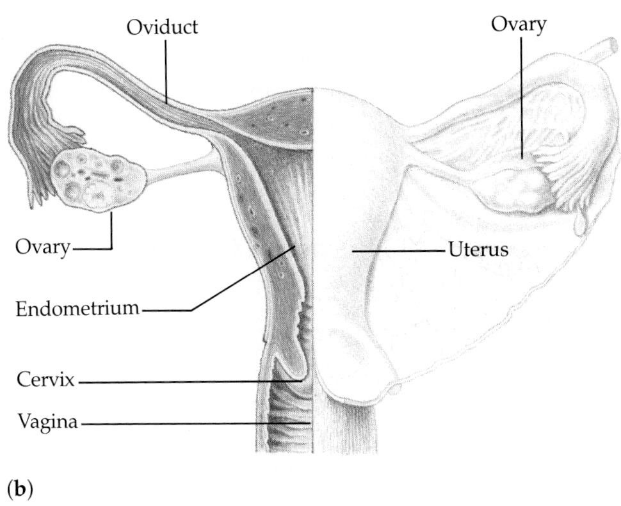

Figure 42.9
Reproductive anatomy of the human female. (**a**) Lateral view. (**b**) Frontal view.

be removed by circumcision. Circumcision, which arose from religious traditions, has no verifiable basis in health or hygiene.

Female Anatomy The female internal reproductive organs consist of a pair of gonads and a system of ducts and chambers to conduct the gametes as well as to house the embryo and fetus. Female external genitalia are the clitoris and the two sets of labia that surround the clitoris and vaginal opening (Figure 42.9). The female gonads, the **ovaries,** lie in the abdominal cavity below most of the digestive system. Each ovary is enclosed in a tough protective capsule and contains many follicles. A **follicle** consists of one egg cell surrounded by one or more layers of follicle cells, which nourish and protect the developing egg cell. All of the 400,000 follicles a woman will ever have are formed at birth. Of these, only several hundred will be released during the woman's reproductive years. After puberty, one (or rarely, two or more) follicle matures and releases its egg during each menstrual cycle. The cells of the follicle also produce the primary female sex hormones, the estrogens. When **ovulation** occurs, the egg is expelled from the follicle (much like a small volcano), and the remaining follicular tissue grows within the ovary to form a solid mass called the **corpus luteum.** The corpus luteum secretes progesterone, the hormone of pregnancy, and additional estrogen. If the egg is not fertilized, the corpus luteum disintegrates, and a new follicle matures during the next cycle.

The female reproductive system is not completely closed, and the egg cell is released into the abdominal

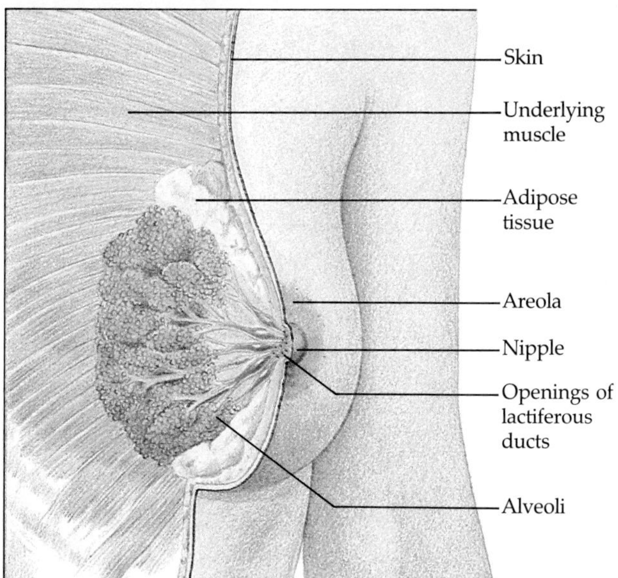

Figure 42.10
The human mammary gland.

cavity near the opening of the **oviduct,** or fallopian tube. The oviduct has a funnel-like opening, and cilia on the inner epithelium lining the duct help collect the egg cell by drawing fluid from the body cavity into the duct. The cilia also convey the egg cell down the duct to the **uterus,** commonly called the womb. The uterus is a thick, muscular organ shaped somewhat like an upside-down pear. It is remarkably small; the uterus of a woman who has never been pregnant is about 7 cm long and 4 to 5 cm wide at its widest point. The unique arrangement of muscles that make up the bulk of the uterine wall allow it to expand to accommodate a 4-kg fetus. The inner lining of the uterus, the **endometrium,** is richly supplied with blood vessels.

The narrow neck of the uterus is the **cervix,** which opens into the vagina. The **vagina** is a thin-walled chamber that forms the birth canal through which the baby is born; it is also the repository for sperm during copulation. Although much thinner than the uterine wall, the walls of the vagina are sufficiently muscular that they can contract to hold a penis tightly or expand to allow passage of a baby's head.

The vagina is the terminal portion of the female reproductive system, but it is covered externally by two pairs of skin folds that form a **vestibule** containing the vaginal orifice and the opening of the urethra. (Notice that in contrast to the male reproductive tract, the female tract opens separately from the excretory system.) At birth, and until intercourse or vigorous physical activity takes place, the human vaginal orifice is covered by a thin membrane called the **hymen,** which has no known function. The vestibule is bordered by the slender **labia minora,** which in turn are protected by a pair of thick, fatty ridges called the **labia majora.** Like the vagina, the labia minora are composed of erectile tissue and enlarge during arousal and intercourse. At the top of the vestibule is a small bulb of erectile tissue called the **clitoris,** the female equivalent of the glans of the penis. Like that organ, the clitoris is composed of erectile tissue and is one of the most sensitive points of stimulation in sexual response.

The **mammary gland,** or breast, is another structure important to mammalian reproduction, although it is not part of the reproductive tract itself (Figure 42.10). The secretory apparatus consists of a series of **alveoli,** small sacs of epithelial tissue that secrete milk. The alveoli drain into a series of lactiferous (milk) ducts that open at the nipple. Deposits of fatty tissue form the main mass of the mammary gland of a nonlactating mammal. The lack of estrogen in males prevents the development of both the secretory apparatus and the fat deposits, so the breasts remain small, and the nipple is not connected to the ducts.

The external genitalia of both sexes arise from common **primordia,** or undifferentiated embryonic structures (Figure 42.11). Sex chromosomes determine the balance of male and female sex hormones in the embryo, and in turn the hormonal balance determines whether the primordial genitalia develop into female or male structures. If androgens are present, male structures form. Female structures form in the absence of androgens.

Hormonal Control of Mammalian Reproduction

The Male Pattern The principal male sex hormones are the **androgens,** of which **testosterone** is the most important. Androgens, steroid hormones produced by the interstitial cells of the testes, are directly responsible for the primary and secondary sex characteristics of the male. The primary sex characteristics are those associated with the reproductive system: development of the vasa deferentia and other ducts, the external genitalia, and sperm production. The secondary sex characteristics are features we associate with maleness that are not directly related to the reproductive system. They include deepening of the voice; the male distributions of axillary (armpit), facial, and pubic hair; and muscle growth (androgens stimulate protein synthesis). Androgens are also potent determinants of behavior in mammals and other vertebrates. In addition to specific sexual behaviors and libido (sex drive), androgens increase general aggressiveness and are responsible for such actions as singing in birds and calling by frogs. Hormones from the anterior pituitary and hypothalamus control both androgen secretion and sperm production by the testes (Figure 42.12).

The Female Pattern The pattern of hormone secretion controlling female reproduction differs strikingly from the male pattern, reflecting the cyclic nature of fe-

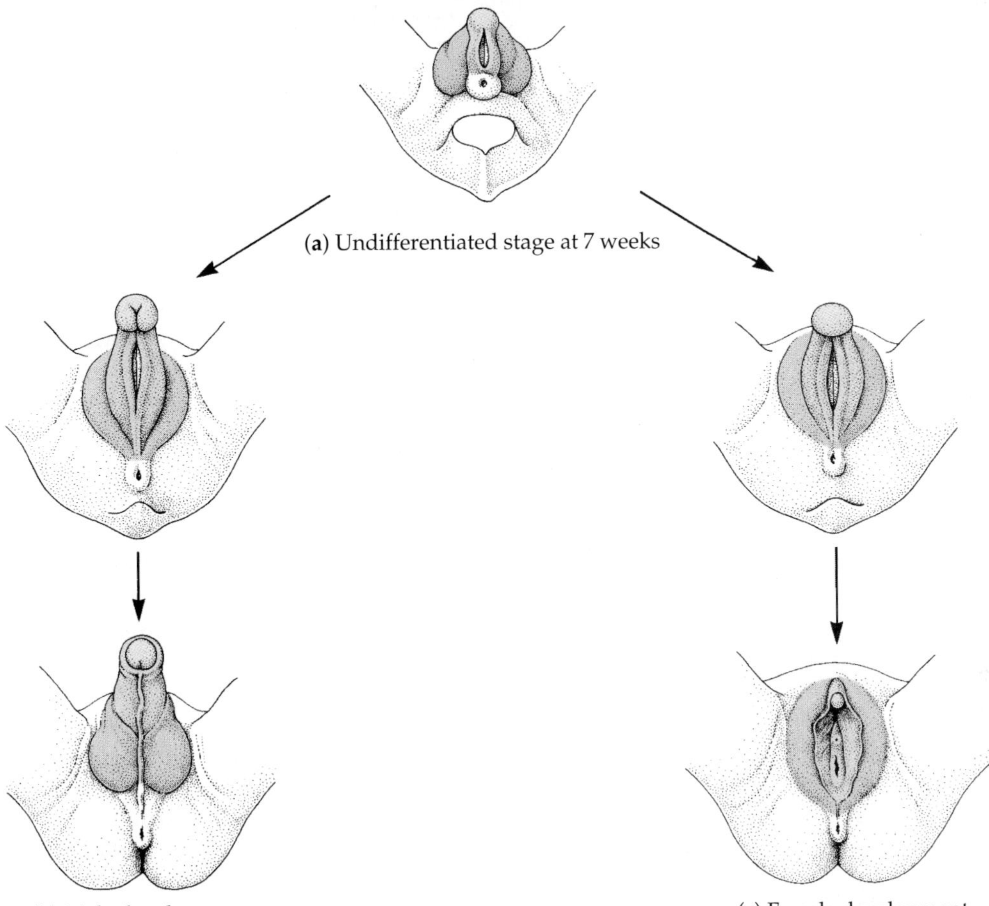

Figure 42.11
The development of external genitalia in humans. (**a**) Both external structures and internal reproductive organs are undifferentiated until about the eighth week of gestation. All embryos have a conical elevation called the genital tubercle, which has a shallow depression called the urethral groove. Urethral folds surround the groove, which in turn are bounded by labioscrotal swellings. Production of hormones, under genetic control, triggers sexual differentiation. (**b**) In the male, the urethral groove elongates and closes completely, and the urethral folds give rise to the shaft of the penis. The labioscrotal swellings develop into the scrotum. (**c**) In the female, the elongated urethral groove remains open and the urethral folds become the labia minora. The labioscrotal swellings develop into the labia majora.

male reproduction. Whereas the male produces sperm continuously, females release only one egg or a few eggs at one time during each cycle. The control of this cycle is considerably more complex than the control of male reproduction.

Two different types of cycles occur in female mammals. Humans and many other primates have **menstrual cycles,** whereas other mammals have **estrous cycles.** In both cases, ovulation occurs at a time in the cycle after the endometrium has started to thicken and has become more extensively vascularized, which prepares the uterus for the possible implantation of an embryo. One major difference between the two types of cycles involves the fate of the uterine lining if pregnancy does not occur. In menstrual cycles, the endometrium is shed from the uterus through the cervix and vagina in a bleeding called **menstruation.** In estrous cycles, the endometrium is reabsorbed by the uterus, and no bleeding occurs.

Other distinctions include more pronounced behavioral changes during estrous cycles than during menstrual cycles and stronger effects of season and climate on estrous cycles. Whereas human females may be receptive to sexual activity throughout their cycles, most mammals will copulate only during the period surrounding ovulation. This period of sexual activity is called **estrus** (L. *oestrus*, "frenzy," "passion") or heat because the body temperature increases slightly. The length and frequency of reproductive cycles vary widely among mammals. The human menstrual cycle averages 28 days, whereas the estrous cycle of the rat is only 5 days. Bears and dogs have one cycle per year, but elephants cycle several times per year.

Let's examine the menstrual cycle of the human female in more detail as a case study of how a complex function is coordinated by hormones. The cycle *averages* 28 days, but only about 30% of women have cycle lengths within a day or two of the statistical 28 days.

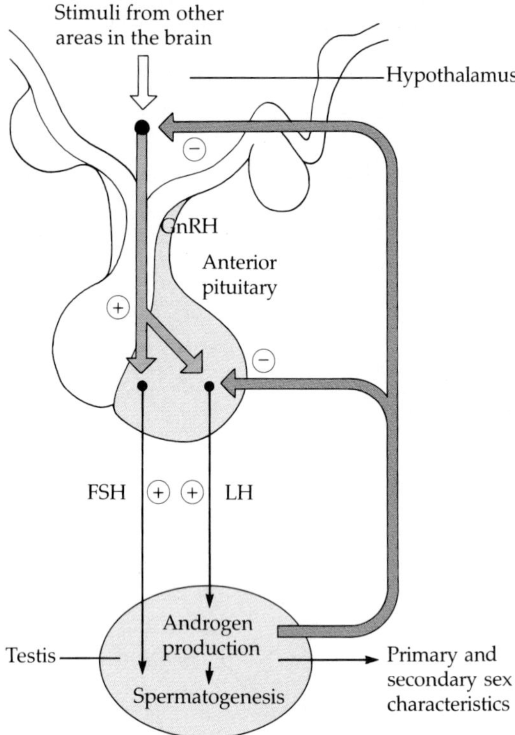

Figure 42.12
Hormonal control of the testes. The pituitary secretes two gonadotropic hormones with different effects on the testes. Luteinizing hormone (LH) stimulates androgen production by the interstitial cells. Follicle-stimulating hormone (FSH) acts on the seminiferous tubules to increase sperm production (spermatogenesis). Since androgens are also required for sperm production, LH stimulates spermatogenesis indirectly. LH and FSH are in turn regulated by a single hormone from the hypothalamus, gonadotropin-releasing hormone (GnRH). How GnRH controls the release of two different hormones at different times is still unknown. LH, FSH, and GnRH concentrations in the blood are regulated by negative feedback by androgens. GnRH is also controlled by negative feedback from the two pituitary gonadotropins. In human males, these feedback loops keep the hormones at relatively constant levels, but in many other mammalian species, there are seasonal cycles in hormone concentration associated with specific breeding seasons.

Cycles vary from one woman to another, ranging from about 20 to 40 days. In some women the cycles are usually very regular, but in other individuals the timing varies from cycle to cycle.

The term *menstrual cycle* refers specifically to the changes that occur in the uterus. By convention, the first day of a woman's "period"—that is, the first day of menstruation—is designated as day 1 of the cycle. The **menstrual flow phase** of the cycle, during which menstrual bleeding occurs, usually persists for a few days (Figure 42.13d). Then the endometrium begins to thicken for a week or two, during what is called the **proliferative phase** of the menstrual cycle. During the next phase, the **secretory phase,** usually about two weeks in duration, the endometrium continues to thicken, becomes more vascularized, and secretes a fluid rich in glycogen. If an embryo has not implanted in the uterine lining by the end of the secretory phase, a new menstrual flow commences, marking day 1 of the next cycle.

Paralleling the menstrual cycle is an **ovarian cycle** (Figure 42.13c). It begins with the **follicular phase,** during which several follicles in the ovary begin to grow. The egg cell enlarges, and the coat of follicle cells becomes multilayered. Of the several follicles that start to grow, only one usually continues to enlarge and mature, while the others disintegrate. The maturing follicle develops an internal fluid-filled cavity and grows very large, forming a bulge near the surface of the ovary. The follicular phase ends with **ovulation** when the follicle and adjacent wall of the ovary rupture, releasing the egg cell. The follicular tissue that remains in the ovary after ovulation is transformed into the corpus luteum, an endocrine tissue that secretes female hormones during what is called the **luteal phase** of the ovarian cycle. The next cycle begins with a new growth of follicles.

Hormones coordinate the menstrual and ovarian cycles in such a way that growth of the follicle and ovulation are synchronized with preparation of the uterine lining for possible implantation of an embryo. Five hormones participate in an elaborate scheme involving both positive and negative feedback. These hormones are: gonadotropin-releasing hormone (GnRH), secreted by the hypothalamus; follicle-stimulating hormone (FSH) and luteinizing hormone (LH), the two gonadotropins secreted by the anterior pituitary gland; and estrogens and progesterone, the female sex hormones secreted by the ovary. The levels of the pituitary and ovarian hormones in blood plasma are traced in Figure 42.13a and b, along with ovarian and menstrual cycles. Referring to the figure frequently as you read the following discussion will help you to understand how the female reproductive system is regulated.

During the follicular phase of the ovarian cycle, the pituitary secretes relatively small quantities of FSH and LH in response to stimulation by GnRH from the hypothalamus. At this time, the cells of immature follicles in the ovary have receptors for FSH, but not for LH. The FSH stimulates the growth of follicles, and the cells of these growing follicles secrete estrogen. Notice in Figure 42.13b that the amount of estrogen secreted during this time is small, and because low levels of estrogen inhibit secretion of the pituitary hormones, the levels of FSH and LH are also relatively low during most of the follicular phase. These hormonal relationships change radically and rather abruptly when the rate of estrogen secretion by the growing follicle begins to rise steeply. Whereas a *low* concentration of estrogen kept secretion of pituitary gonadotropins at a low

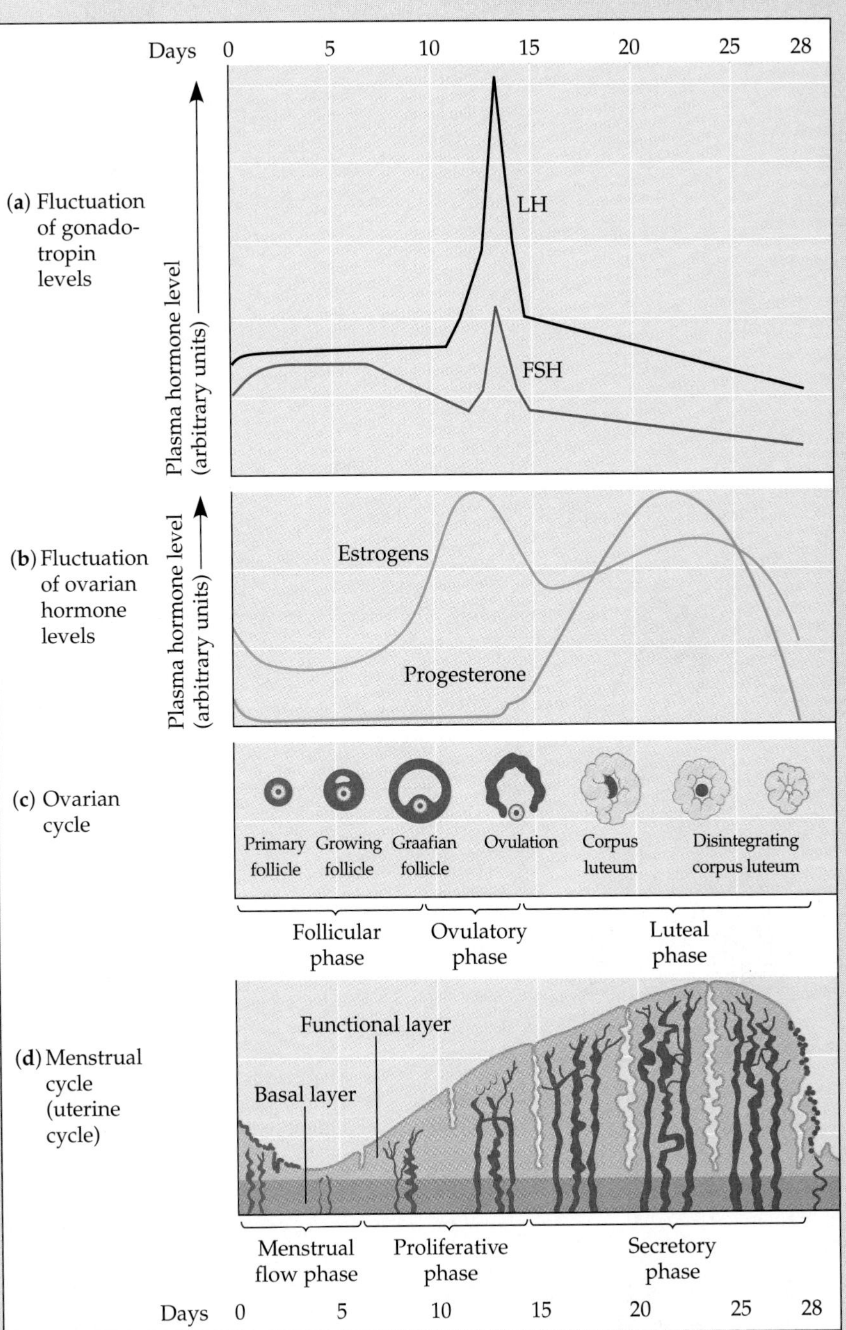

Figure 42.13
The reproductive cycle of the human female. Hormones coordinate the ovarian and menstrual cycles, preparing the uterine lining (endometrium) for implantation of an embryo even before ovulation. This figure shows an idealized 28-day cycle, but actual reproductive cycles vary in length from about 20 to 40 days. (**a**) The changes in the levels of the two pituitary gonadotropins, LH and FSH, during the reproductive cycle. (**b**) The changes in the levels of the two ovarian hormones, estrogens (a family of closely related hormones) and progesterone. Use these graphs as visual aids to support the text's discussion of the feedback circuits that regulate the levels of these hormones. (**c**) The ovarian cycle consists of a follicular phase, during which follicles grow and secrete increasing amounts of estrogens; an ovulatory phase, when ovulation occurs; and a luteal phase, during which the corpus luteum formed from the follicular tissue after ovulation secretes estrogens and progesterone. The follicular phase varies in length from woman to woman, and in some women from one cycle to the next. The luteal phase usually lasts about 13 to 15 days, regardless of the overall length of the cycle. (**d**) The menstrual cycle consists of a menstrual flow phase, a proliferative phase, and a secretory phase. Menstruation, the bleeding associated with disintegration of the endometrium, occurs during the menstrual flow phase. The first day of flow marks day 1 of the menstrual cycle. During the proliferative phase, estrogens from the growing follicle stimulate the endometrium to thicken and become increasingly vascularized. During the secretory phase, the endometrium continues to thicken, its arteries enlarge, and endometrial glands grow. These changes in the endometrium require estrogens and progesterone, which are secreted by the corpus luteum after ovulation. Thus, the secretory phase of the menstrual cycle parallels the luteal phase of the ovarian cycle. Disintegration of the corpus luteum at the end of the luteal phase sharply reduces the amount of estrogens and progesterone available to the endometrium, and so it is sloughed off. The first day of menstruation, marking the beginning of the next cycle, usually occurs about 13 to 15 days after ovulation. However, because the follicular phase of the ovary, and hence the proliferative phase of the uterus, is so variable in duration, it is not usually possible to predict how long after menstruation the next ovulation will occur. (This is one of the problems with the rhythm method of contraception.) In the event of pregnancy, additional mechanisms (discussed later in the chapter) maintain high levels of estrogens and progesterone and prevent loss of the endometrium.

level, a *high* concentration of estrogen has the opposite effect and *stimulates* secretion of gonadotropins by acting on the hypothalamus to increase its output of GnRH. You can see this response in Figure 42.13a as a steep incline of FSH and LH levels that follows closely behind the increase in estrogen concentration. The effect is greater for LH because the high concentration of estrogen, in addition to stimulating GnRH secretion, also increases the sensitivity of LH-releasing mechanisms in the pituitary to the hypothalamic signal

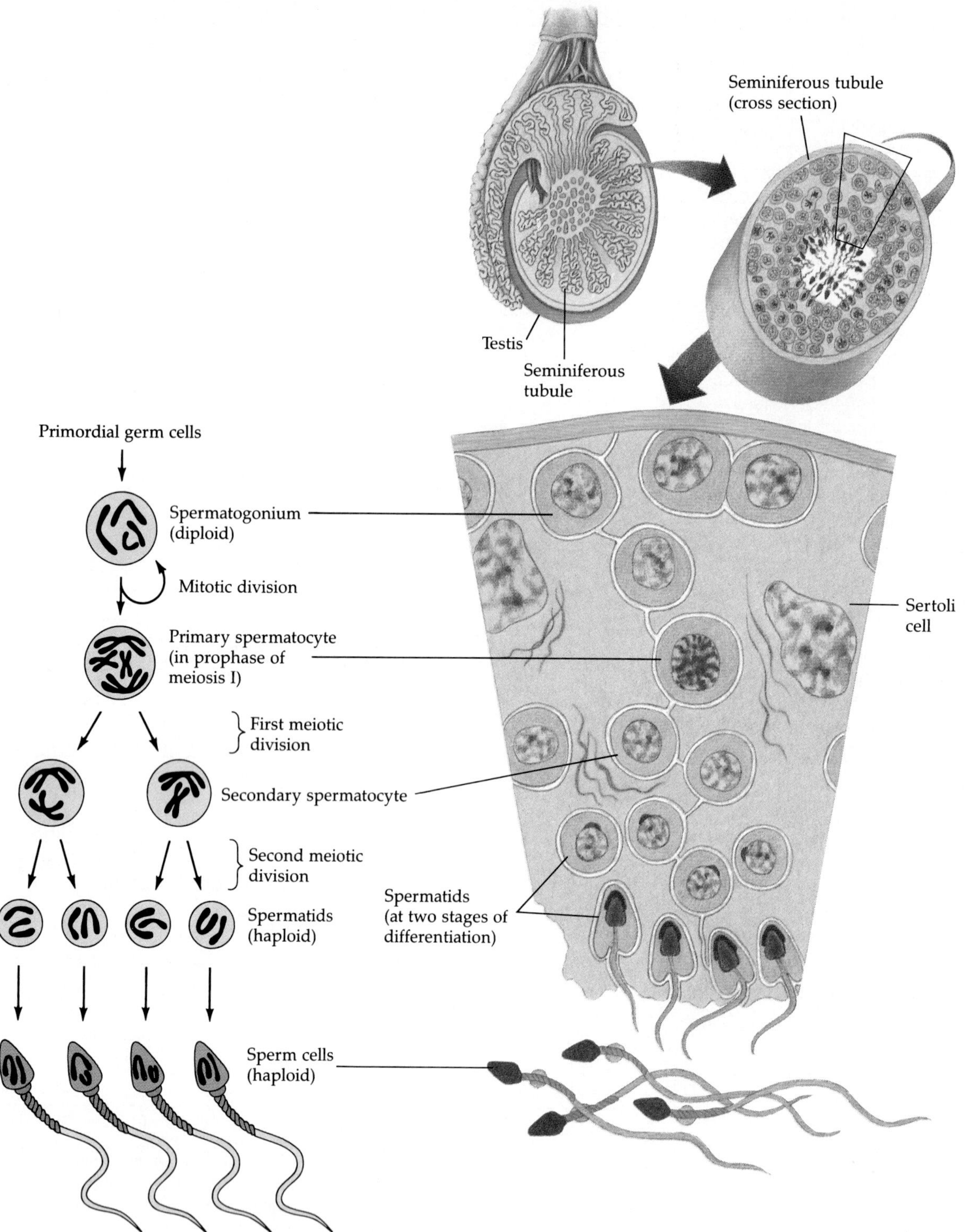

Figure 42.14
Spermatogenesis. These diagrams correlate the meiotic stages in sperm development (left) with the histology of seminiferous tubules. Primordial germ cells of the embryonic testes differentiate into spermatogonia, the diploid cells that are the precursors of sperm. Located near the outer wall of the seminiferous tubules, spermatogonia undergo repeated mitoses to produce large, renewable populations of potential sperm. In a mature male, about three million spermatogonia per day differentiate into primary spermatocytes. Meiosis, which generates haploid gametes, occurs in two steps. First, meiosis I produces two secondary spermatocytes. Second, the second meiotic division forms four spermatids. Spermatids then differentiate into mature spermatozoa, or sperm cells. This involves association of the developing sperm with large Sertoli cells, which transfer nutrients to the spermatids. During spermatogenesis, the developing sperm are gradually pushed toward the center of the seminiferous tubule and make their way to the epididymis, where they acquire motility. This process, from spermatogonia to motile sperm, takes 65 to 75 days in the human male.

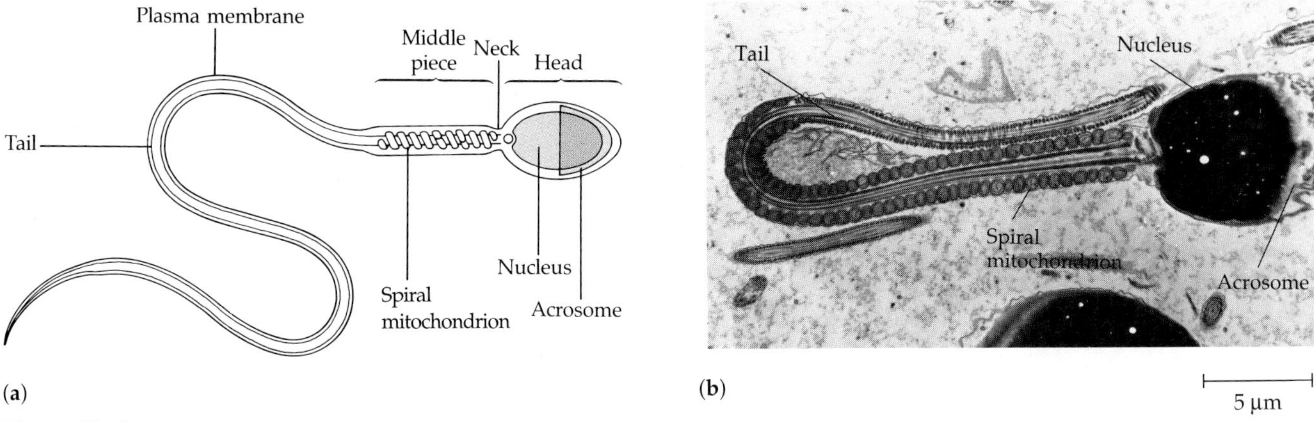

Figure 42.15
Sperm. (**a**) Human sperm structure. (**b**) Transmission electron micrograph of a sperm cell from a rhesus monkey, which is very similar to human sperm.

(GnRH). By now, the follicles have receptors for LH and can respond to this hormonal cue. In a case of positive feedback, the increase in LH concentration caused by increased estrogen secretion from the growing follicle induces final maturation of the follicle, and ovulation occurs about a day after the LH surge.

Following ovulation, LH stimulates the transformation of the follicular tissue left behind in the ovary to form the corpus luteum, a glandular structure (it is for this "luteinizing" function that luteinizing hormone [LH] is named). Under continued stimulation by LH during the luteal phase of the ovarian cycle, the corpus luteum secretes estrogen and a second steroid hormone, progesterone. The corpus luteum usually reaches its maximum development about 8 to 10 days after ovulation. As the progesterone and estrogen levels rise, the combination of these hormones exerts negative feedback on the hypothalamus and pituitary, inhibiting the secretion of the gonadotropins LH and FSH. When the LH concentration plummets, the corpus luteum, which requires LH in order to function, begins to disintegrate. Consequently, the concentrations of estrogen and progesterone decline sharply near the end of the luteal phase. The dropping levels of ovarian hormones liberate the hypothalamus and pituitary from the inhibitory effects of these hormones. The pituitary then begins to secrete enough FSH to stimulate the growth of new follicles in the ovary, initiating the follicular phase of the next ovarian cycle.

How is the ovarian cycle synchronized with the menstrual cycle? Estrogen, secreted in increasing amounts by growing follicles, is a hormonal signal to the uterus, causing the endometrium to thicken. Thus, the follicular phase of the ovarian cycle is coordinated with the proliferative phase of the menstrual cycle. *Before* ovulation, the uterus is already being prepared for a possible embryo. After ovulation, estrogen and progesterone secreted by the corpus luteum stimulate continued development and maintenance of the endometrium, including an enlargement of arteries supplying blood to the uterine lining and growth of endometrial glands that secrete a fluid with nutrients that can sustain an early embryo before it actually implants in the uterine lining. Thus, the luteal phase of the ovarian cycle is coordinated with the secretory phase of the menstrual cycle. The rapid drop in the level of ovarian hormones when the corpus luteum disintegrates causes spasms of arteries in the uterine lining that deprive the endometrium of blood. Disintegration of the endometrium results in menstruation and the beginning of a new menstrual cycle. In the meantime, ovarian follicles that will stimulate renewed thickening of the endometrium are just beginning to grow. Cycle after cycle, the maturation and release of egg cells from the ovary is integrated with changes in the uterus, the organ that must accommodate an embryo if the egg cell is fertilized. In the absence of pregnancy, a new cycle begins. We will soon see that there are "override" mechanisms that prevent disintegration of the endometrium in the event of pregnancy.

In addition to their role in coordinating reproductive cycles, estrogens are also responsible for the secondary sex characteristics of the female. The hormones induce deposition of fat in the breasts and hips, increase water retention, affect calcium metabolism, stimulate breast development, and mediate female sexual behavior.

Gamete Formation (Gametogenesis)

Spermatogenesis The production of mature sperm cells, **spermatogenesis,** is a continuous and prolific process in the adult male. Each ejaculation of a human male contains about 400 million sperm cells, and males can ejaculate daily with little loss of fertilizing capacity. Spermatogenesis occurs in the seminiferous tubules of the testes (Figure 42.14).

The structure of a spermatozoon (sperm cell) fits its function (Figure 42.15). The thick head containing the

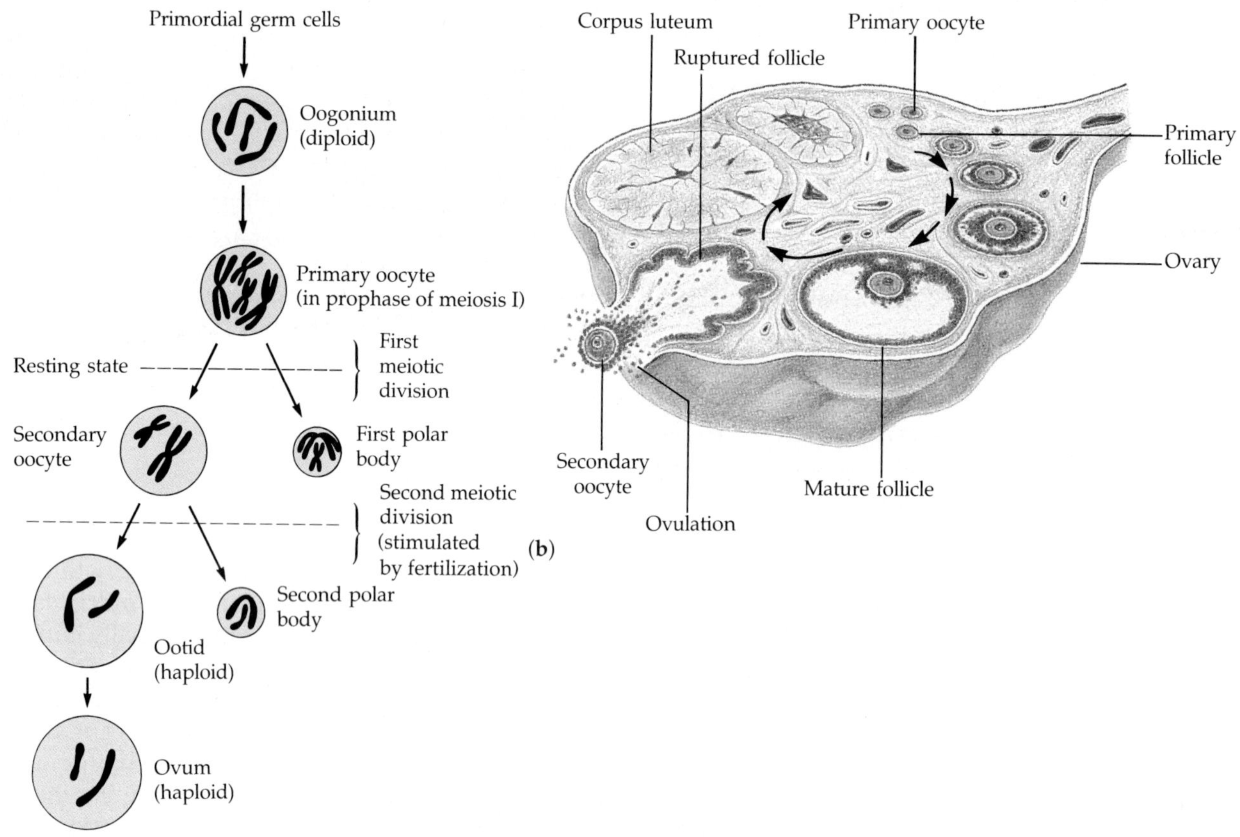

Figure 42.16
Oogenesis. (a) The correlation between oogenesis and meiosis. Oogenesis begins with mitosis of the primordial germ cells in the embryo to produce diploid oogonia, which develop into primary oocytes. During the subsequent meiotic divisions, unequal cytokinesis concentrates almost all the cytoplasm in a single large ovum. The other products of meiosis are tiny cells called polar bodies that degenerate. (This situation contrasts with spermatogenesis, which yields four sperm cells for each primary spermatocyte; see Figure 42.14.) **(b) A cutaway view of an ovary, showing the stages of oogenesis.** A human female is born with all her potential ova already present as primary oocytes. Between birth and puberty, the primary oocytes enlarge and the follicles around them grow. Under the influence of FSH in each reproductive cycle, a follicle enlarges and the first meiotic division produces a secondary oocyte and the first polar body. At this stage of oogenesis, LH triggers ovulation of the secondary oocyte. Only if a sperm cell penetrates the secondary oocyte is the second meiotic division initiated. After meiosis is completed and the second polar body divides from the ovum, the haploid nuclei of the sperm and the now-mature ovum fuse. Here, the growth of the follicle, ovulation, and formation and disintegration of the corpus luteum are schematically represented as a cycle, with the stages connected by arrows. In reality, these stages are never present simultaneously because the ovarian cycle is temporal, not spatial.

haploid nucleus is tipped with a special body, the **acrosome,** which contains enzymes that help the sperm penetrate the egg. Behind the head, the spermatozoon contains large numbers of mitochondria (or a single large one in some species) that provide ATP for movement of the tail, a flagellum. Mammalian sperm morphology is quite variable, with head shapes ranging from a slender comma through the oval of the human sperm to nearly spherical.

Oogenesis Development of ova (mature, unfertilized egg cells), or **oogenesis,** differs from spermatogenesis in three important ways (Figure 42.16). First, during the meiotic divisions of oogenesis, cytokinesis is unequal, with almost all the cytoplasm monopolized by a single daughter cell. This large cell can go on to form the ovum, and the three smaller cells, called polar bodies, soon degenerate. This contrasts with spermatogenesis, when all four products of meiosis I and II develop into mature sperm (compare Figures 42.14 and 42.16). Second, while the precursor cells of sperm continue to divide by mitosis throughout the reproductive years of a male, this is not the case for oogenesis in the female. At birth, an ovary already contains all the presumptive egg cells it will ever have, though still in an immature state; therefore, eggs are a nonrenewable resource. Third, oogenesis has long "resting" periods before the process is complete, in contrast to spermatogenesis, which produces mature sperm from precursor cells in an uninterrupted sequence. After puberty in a female, a few potential ova undergo the first meiotic division within growing follicles during each ovarian cycle. The egg cell released during ovulation is not actually a mature ovum because it has not yet completed meiosis by undergoing the second division. In humans, penetration of the egg cell by the sperm triggers the second meiotic division, and only then is oogenesis actually complete.

Sexual Maturation

Mammals cannot reproduce until they have undergone substantial growth and development after birth. For example, a human male can achieve an erection at birth but has no sperm to ejaculate. In humans, the onset of reproductive ability is called puberty. It is a gradual process that usually begins about two years earlier in females than in males. Between the ages of 8 and 14, depending on the individual, the hypothalamus begins secreting increasing amounts of GnRH, leading to higher levels of FSH followed by increased LH. These gonadotropins trigger maturation of the reproductive system and development of the secondary sex characteristics (by initiating secretion of sex hormones from the gonads). The first indication of puberty is a growth spurt, followed by first menstruation at about age 12 or 13 in girls or first ejaculation of viable sperm at age 13 or 14 in boys. The age of puberty is quite variable, and recent analysis of historical data suggests that there has been little change in the average age throughout modern times.

Human Sexual Physiology

Many vertebrates and invertebrates have elaborate and complex mating behaviors, but these are usually stereotyped interactions that involve specific sequences of reciprocal behaviors (see Chapter 50). The hallmark of human sexuality is its diversity in stimuli and response. Behind this variable sexual behavior, however, is a common physiological pattern, often called the sexual response cycle. As is also true of reproductive anatomy, endocrinology, and gametogenesis, the sexual responses of males and females have similarities as well as differences.

Two types of physiological reactions predominate in both sexes. **Vasocongestion** is the filling of a tissue with blood caused by increased blood flow through the arteries of that tissue. Erection of the penis is one example of vasocongestion, and similar responses occur in the testes, labia, clitoris, vagina, and breasts. **Myotonia,** increased muscle tension, is also a widespread phenomenon in sexual response. Both skeletal and smooth muscle may show sustained or rhythmic contractions, including those associated with orgasm.

The sexual response cycle can be divided into four phases: excitement, plateau, orgasm, and resolution. An important function of the **excitement phase** is preparation of the vagina and penis for **coitus** (sexual intercourse). During this phase, vasocongestion is particularly evident in erection of the penis and clitoris; enlargement of the testes, labia, and breasts; and vaginal lubrication. Myotonia may occur, resulting in nipple erection or tension of the arms and legs.

The **plateau phase** continues these responses. In females, the outer third of the vagina becomes vasocongested, while the inner two-thirds becomes slightly expanded. This change, coupled with the elevation of the uterus, forms a depression that receives sperm at the back of the vagina. Reactions in nonreproductive organs continue as breathing increases and heart rate rises, sometimes to 150 beats per minute—not in response to the physical effort of sexual activity but as an involuntary response to stimulation of the autonomic nervous system (see Chapter 44).

Orgasm is characterized by rhythmic, involuntary contractions of the reproductive structures in both sexes. Male orgasm has two stages. Emission is the contraction of the glands and ducts of the reproductive tract, which deposits semen in the urethra. Expulsion or ejaculation occurs when the urethra contracts and the semen is expelled. During female orgasm, the uterus and outer vagina contract, but the inner two-thirds of the vagina does not. Orgasm is the shortest phase of the sexual response cycle, usually lasting only a few seconds. In both sexes, contractions occur at about 0.8-sec intervals and may involve the anal sphincter and several abdominal muscles.

The **resolution phase** completes the cycle and reverses the responses of the earlier stages. Vasocongested organs return to their normal size and color, and muscles relax. Most of the changes during resolution are completed in 5 minutes. Loss of penile and clitoral erection, however, may take longer. An initial loss of erection, or detumescence, is rapid in both sexes, but a return of the organs to their nonaroused size may take as long as an hour.

Conception, Pregnancy, and Birth

Pregnancy, or **gestation,** is the condition of carrying one or more developing embryos in the uterus. It is preceded by **conception,** the fertilization of the egg by a spermatozoon, and continues until the birth of the baby or babies. Human pregnancy averages 266 days (38 weeks) from conception, or 40 weeks from the start of the last menstrual cycle. Duration of pregnancy in other species correlates with body size and extent of development of the young at birth. Many rodents

Figure 42.17
Postfertilization events. (**a**) After fertilization occurs, the zygote travels down the oviduct toward the uterus. (**b**) The zygote begins to divide about 24 hours after fertilization and then continues to divide rapidly (cleavage). (**c**) Three to four days after ovulation, the embryo, a ball of cells at this stage, reaches the uterus and floats freely for several days, nourished by fluid secreted by endometrial glands. (**d**) At the blastocyst stage, the embryo is implanted into the endometrium about 7 days after ovulation.

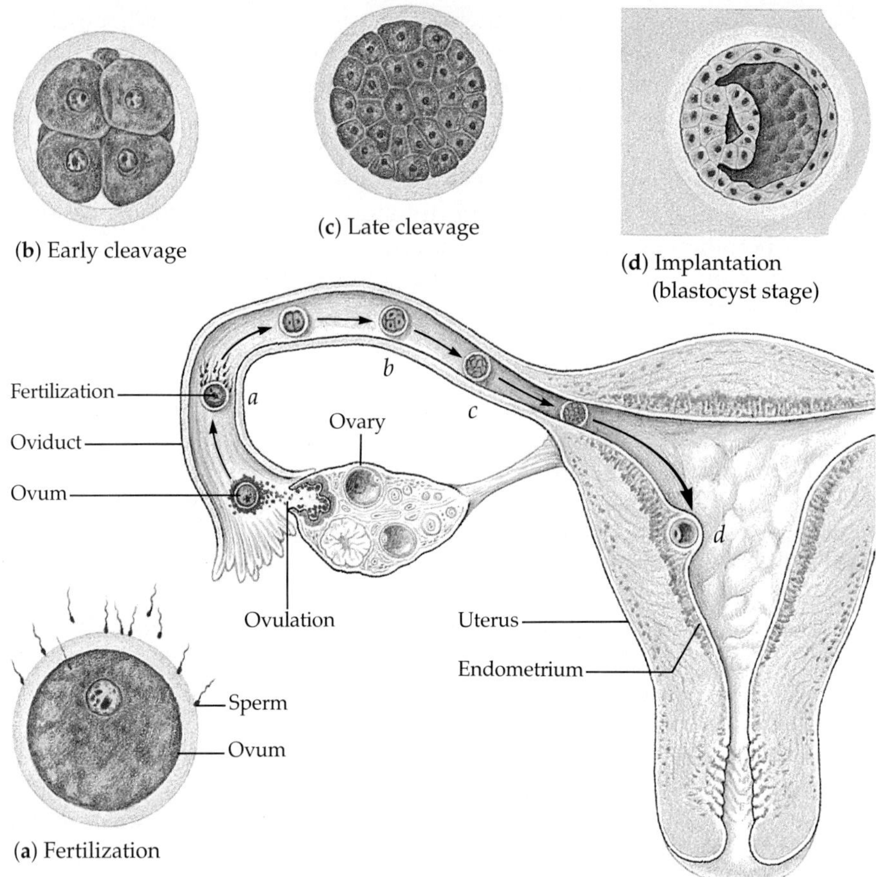

(mice and rats) have gestation periods of about 21 days, whereas those of dogs are closer to 60 days. In cows, gestation averages 270 days (almost the same as humans); in giraffes, it is about 420 days; and in elephants, gestation is more than 600 days.

Human gestation can be divided for convenience of study into three **trimesters** of about three months each. The first trimester is the time of most radical change for both the mother and the baby. The egg is fertilized by the sperm in the oviduct, and the resulting zygote (fertilized egg) travels down the oviduct into the uterus, moved by the cilia of the oviduct (Figure 42.17). **Cleavage,** or cell division, begins about 24 hours after fertilization and continues more rapidly thereafter. This occurs while the embryo is still passing down the oviduct, a trip that usually takes 3 to 5 days. A few days after reaching the uterus, or about 1 week after fertilization, the zygote has developed into a hollow ball of cells called a **blastocyst,** which implants into the endometrium. Differentiation of body structures now begins in earnest. (Embryonic development will be described in detail in Chapter 43.) During implantation, the blastocyst bores into the endometrium, which responds by growing over the blastocyst. Eventually, tissues grow out from the developing embryo and mingle with the endometrium to form the **placenta** (Figure 42.18). This disk-shaped organ, containing embryonic and maternal blood vessels, grows to about the size of a dinner plate and weighs somewhat less than 1 kg. Diffusion of material between maternal and fetal circulations furnishes a means of respiratory gas exchange and nutrient transfer as well as waste removal for the embryo. Blood from the embryo travels to the placenta through arteries of the umbilical cord and returns via the umbilical vein, passing through the liver of the embryo.

The first trimester is also the main period of **organogenesis,** the development of the body organs (Figure 42.19). The heart begins beating by the fourth week and can be detected with a stethoscope by the end of the first trimester. By the end of the eighth week, all the major structures of the adult are present in rudimentary form. At this point, the embryo is called a **fetus.** Although well differentiated, the fetus is only 5 cm long by the end of the first trimester. Because of its rapid organogenesis, the embryo is most sensitive during the first trimester to such threats as radiation and drugs that can cause birth defects.

The first trimester is also a time of rapid change for the mother. The embryo secretes hormones that signal

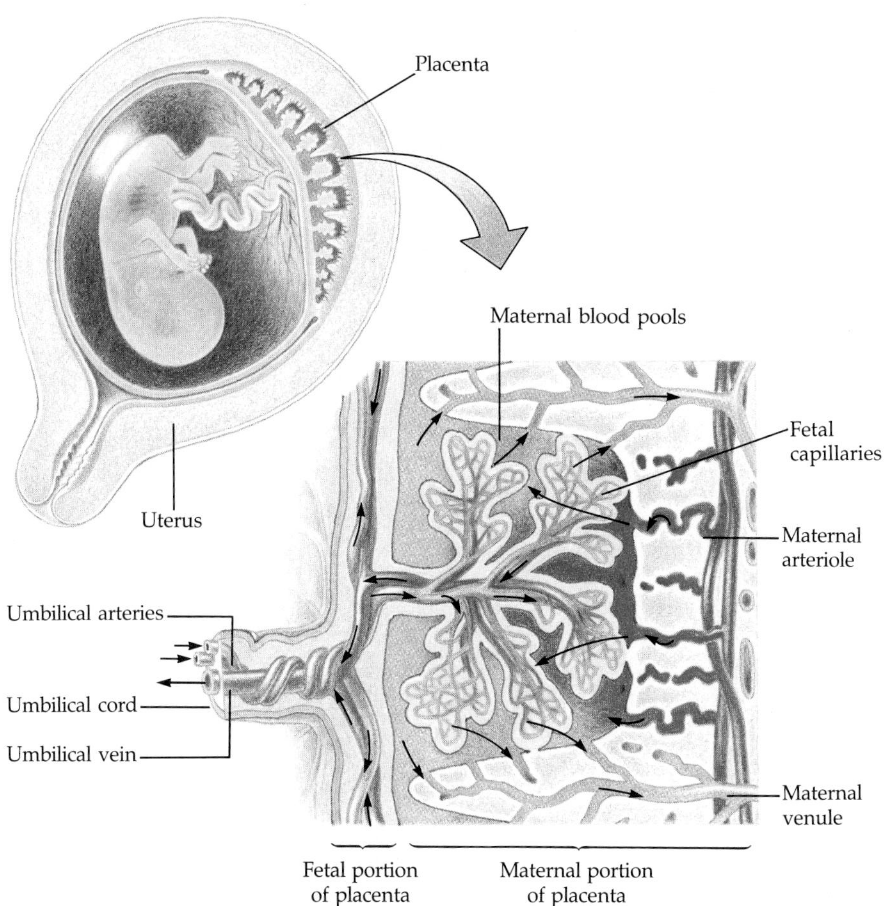

Figure 42.18
Fetal circulation. The embryo obtains nutrients directly from the endometrium during the first 2 months of development. From the third month until birth, the placenta transports nutrients and wastes to and from the embryo (now called a fetus). The placenta, a combination of maternal and fetal tissues, exchanges oxygen, carbon dioxide, glucose, and other substances between mother and fetus. The maternal blood enters the placenta in arterioles, flows through blood pools in the endometrium, and leaves via venules. The fetal blood, which remains in vessels, enters the placenta through arteries, passes through capillaries in fingerlike villi, where oxygen and nutrients are acquired, and exits through veins leading back to the fetus. Materials are exchanged by diffusion between the fetal capillary bed and the maternal blood pools.

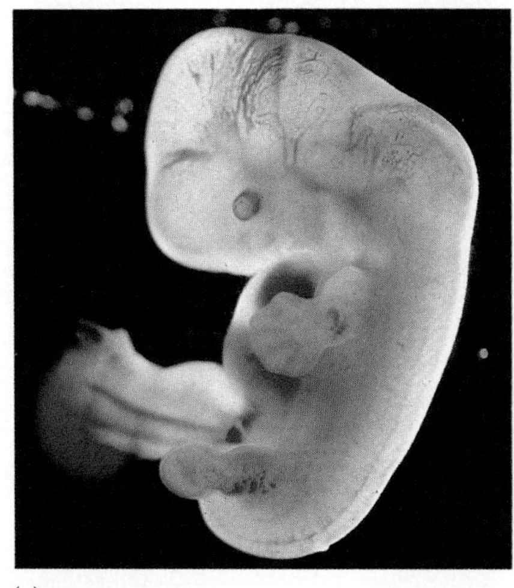
(a)

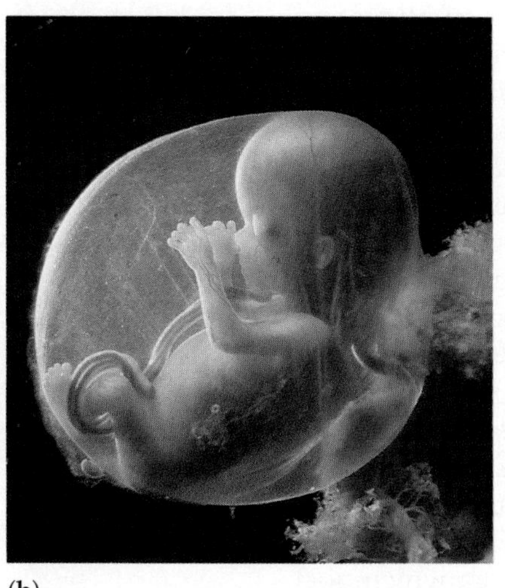
(b)

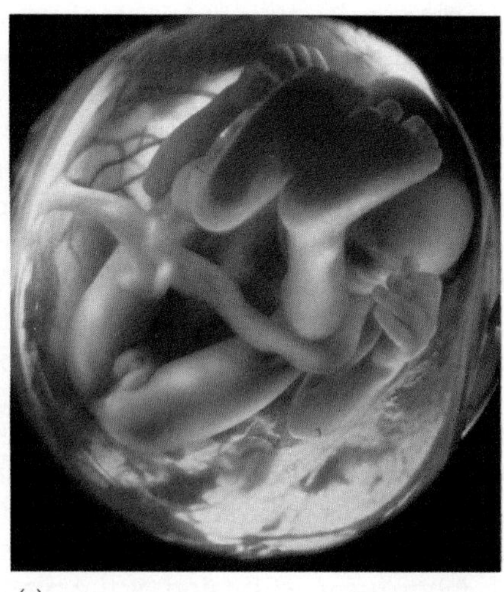
(c)

Figure 42.19
Human fetal development. (**a**) At 5 weeks, limb buds, eyes, heart, liver, and rudiments of all other organs have started to develop in the embryo, which is only 1 cm long. (**b**) Growth and development of the offspring, now called a fetus, continue during the second trimester. This fetus is 14 weeks old and about 6 cm long. (**c**) The fetus in this photograph is 20 weeks old. By the end of the second trimester (at 24 weeks), the fetus grows to about 30 cm in length.

its presence and control the mother's reproductive system. One embryonic hormone, **human chorionic gonadotropin (HCG),** acts like pituitary LH to maintain progesterone and estrogen secretion by the corpus luteum through the first trimester of pregnancy. In the absence of this hormonal override, the decline in maternal LH due to inhibition of the pituitary by progesterone would result in menstruation and spontaneous abortion of the embryo. Levels of HCG in the maternal blood are so high that some is excreted in the urine, where it can be detected in pregnancy tests. High levels of progesterone initiate changes in the pregnant woman's reproductive system that include increased mucus in the cervix to form a protective plug, growth of the maternal part of the placenta, enlargement of the uterus, and (by negative feedback on the hypothalamus and pituitary) cessation of ovulation and menstrual cycling. The breasts also enlarge rapidly and are often quite tender.

During the second trimester, the fetus grows rapidly to about 30 cm and is quite active. Movements may be felt by the mother during the early part of the second trimester and be seen through the abdominal wall by the middle of this time period. Hormone levels stabilize as HCG declines, the corpus luteum deteriorates, and the placenta secretes its own progesterone, which maintains the pregnancy. During the second trimester, the uterus will grow enough for the pregnancy to become obvious.

The third and final trimester is one of rapid growth of the fetus to about 3–3.5 kg in weight and 50 cm in length. Fetal activity may decrease as the fetus fills the available space within the embryonic membranes. As the fetus grows and the uterus expands around it, the mother's abdominal organs become compressed and displaced, leading to frequent urination, digestive blockages, and strain in the back muscles.

Birth, or **parturition,** occurs through a series of strong, rhythmic contractions of the uterus, commonly called **labor.** Prostaglandins from the uterus, oxytocin from the posterior pituitary (see Chapter 41), and nervous reflexes all play roles in regulating labor contractions. During the first stage of labor, the opening of the cervix thins out (effaces) and dilates (Figure 42.20a and b). Complete dilation of the cervix ends the first stage of labor. The second stage is the birth of the baby. The uterus is firmly attached to the floor of the abdomen, so the continued strong contractions force the fetus down and out of the uterus and vagina. The umbilical cord is commonly cut and clamped at this time. The final stage of labor is the expulsion of the placenta, which normally follows the baby.

Lactation is an aspect of postnatal care unique to mammals. After birth, decreasing levels of progesterone free the anterior pituitary from negative feedback and allow prolactin secretion. Prolactin stimulates milk production after a delay of 2 or 3 days. The

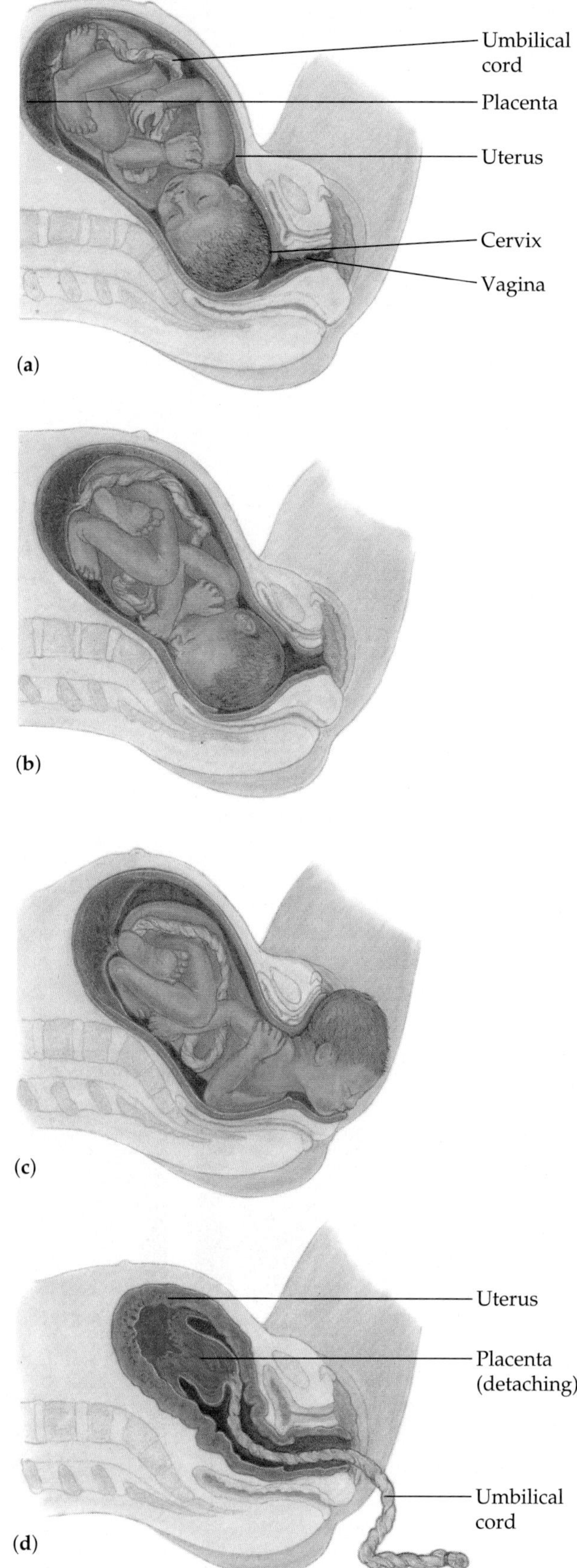

Figure 42.20
The stages of labor. (**a**) Early dilation of the cervix. (**b**) Late dilation of the cervix. (**c**) Birth of the baby. (**d**) Placental stage: detachment and delivery of the placenta.

release of milk from the mammary glands is controlled by oxytocin, described in Chapter 41.

Reproductive Immunology Pregnancy is an immunological enigma. Half of the embryo's genes are inherited from the father, and thus many of the chemical markers present on the surface of the embryo will be foreign to the mother. Why, then, does the mother not reject this foreign body, the embryo, much as she would repel a tissue or organ graft bearing antigens from another person? Reproductive immunologists are only beginning to solve this puzzle. Part of the answer is the presence of a physical barrier. A protective layer called the trophoblast prevents the embryo itself from actually contacting maternal tissue. But the trophoblast develops along with the embryo from the cells of the blastocyst, and this protective barrier, which penetrates the endometrium, may also be foreign to the mother. According to one hypothesis, the trophoblast does *not* develop paternal markers and thus does not trigger an immune response by the mother. However, several laboratories have recently discovered paternal antigens on portions of the trophoblast. There is evidence that the trophoblast produces a chemical signal that induces development of a special type of white blood cell in the uterus that prevents other white cells from mounting an attack on the foreign tissue. This suppressor cell may work by secreting a substance that blocks the action of interleukin-2, the cytokine required for a normal immune response (see Chapter 39). One hypothesis suggests that this local dampening of the immune response occurs, paradoxically, only after white blood cells in the vicinity have first identified the trophoblast as foreign tissue and have taken the first steps of the immune response. If this immunological alarm is not intense enough—that is, if the father's cellular markers are too similar to the mother's—then no suppressor cells are produced.

Some researchers speculate that if the initial immune response is too weak to trigger suppression, then the persistent immunological attack on the foreign tissue, though weak, may lead to spontaneous abortion of the embryo. According to this view, failure to suppress the immune response in the uterus may account for many of the cases of women who have multiple miscarriages for no other apparent reason. There has been some success treating frequent miscarriers by sensitizing the woman's immune system to her mate's antigens through immunization—that is, by injecting appropriate chemical markers into the mother prior to pregnancy. The idea is to intensify the subsequent response to the foreign tissue of the embryo to a level that trips the suppressor mechanism. Some critics of this interpretation argue that the psychological support that women receive during this experimental treatment counts for more than the immunotherapy itself. Only more research will resolve the interesting and important questions about how a woman's immune system tolerates a 9-month parasitic relationship with a large foreign organism.

Contraception **Contraception** literally means "against taking," in this case, the taking in of a child. The term has come to mean preventing a pregnancy through one of several methods. Some of these methods prevent the release of mature eggs and sperm from gonads, others prevent fertilization by keeping sperm and eggs apart, and still others prevent implantation of an embryo or abort the embryo (Figure 42.21). The following brief introduction to the biology of these methods makes no pretense of being a contraception manual. For more complete information, consult a respected text on human sexuality or health (see the suggested readings at the end of the chapter).

Fertilization can be prevented by abstinence from sexual intercourse or by any of several barriers that keep live sperm from contacting the egg. Temporary abstinence, often called the **rhythm method** of birth control, depends on refraining from intercourse when conception is most likely. Because the egg can survive in the oviduct for 24 to 48 hours and sperm for up to 72 hours, a couple practicing temporary abstinence should not engage in intercourse during the few days before and after ovulation, which is usually difficult to predict. The most effective methods for timing ovulation combine several indicators, including changes in cervical mucus and body temperature during the menstrual cycle. Because rhythm methods depend on regular menstrual cycles and changes that may not be apparent in all women, they are less effective than most other methods. A failure rate of 10% to 20% is typical, even if the couple follows the rules and the woman's menstrual cycle is regular. (Failure rate is the number of women who become pregnant during a year out of every 100 women using a particular contraceptive method, expressed as a percentage.)

The several **barrier methods** of contraception that block the sperm from meeting the egg, with failure rates of less than 10%, are more effective than the rhythm method. The **condom** is a thin, natural membrane or latex rubber sheath that fits over the penis to collect the semen. The diaphragm is a dome-shaped rubber cap fitted into the upper portion of the vagina before intercourse. Both of these methods are more effective when used in conjunction with a spermicidal (sperm-killing) foam or jelly. More recently introduced barriers include the cervical cap, which fits tightly around the opening of the cervix, held in place for a prolonged period by suction, and the contraceptive sponge, also inserted into the vagina.

The intrauterine device (IUD) probably prevents implantation of the blastocyst in the uterus by irritating the endometrium, but its precise mechanism of preventing pregnancy is unknown. IUDs are small,

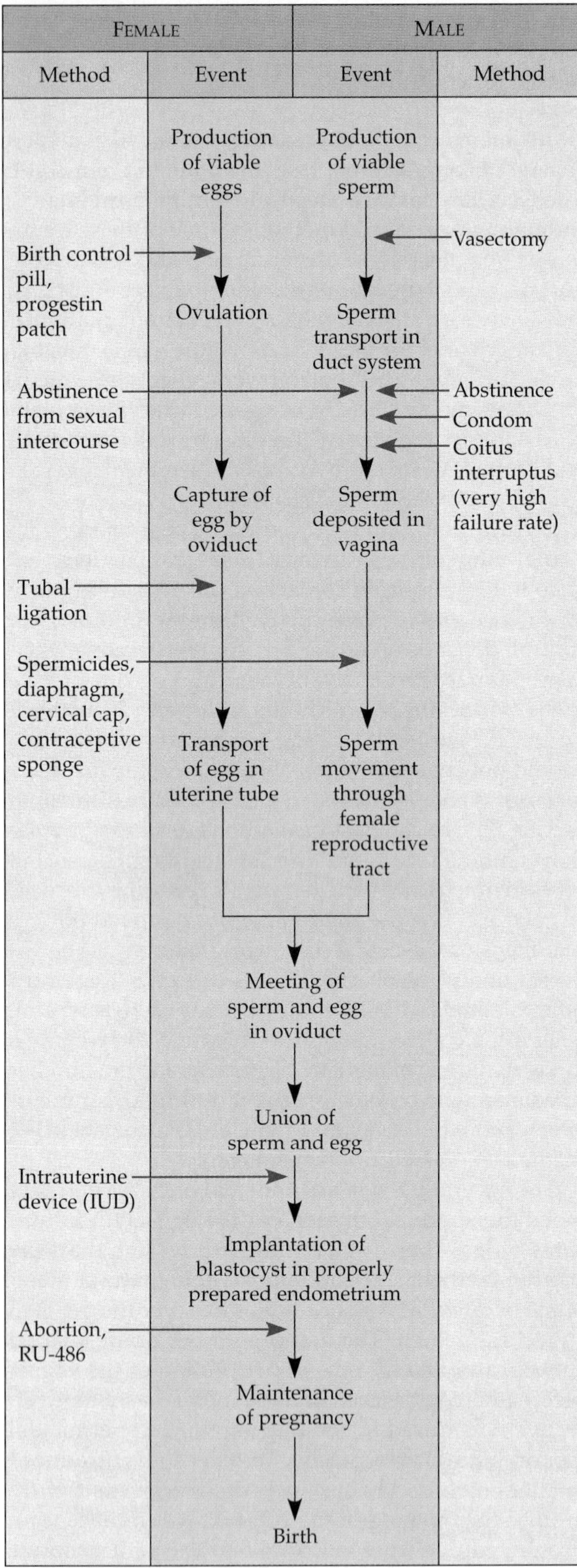

Figure 42.21
Mechanisms of some contraceptive methods. The magenta arrows indicate the steps where contraceptives work.

usually plastic devices, and they come in a variety of shapes that fit into the uterine cavity. IUDs have a low failure rate but cause harmful effects in a small percentage of women. Problems of persistent vaginal bleeding, uterine infection, perforation of the uterus, tubal pregnancy (implantation of the embryo in the oviduct), and spontaneous expulsion of the devices have been reported and have led to many lawsuits against IUD manufacturers. New IUD products are beginning to appear.

As a method of preventing fertilization, coitus interruptus, or withdrawal (removal of the penis from the vagina before ejaculation), is unreliable. Sperm may be present in secretions that precede ejaculation, and a lapse in timing or willpower can result in late withdrawal.

Besides complete abstinence from sexual intercourse, the methods that prevent the release of gametes are the most effective means of birth control. Chemical contraception—birth control pills—have failure rates of less than 1%, and sterilization is nearly 100% effective. Birth control pills are combinations of a synthetic estrogen and a synthetic progestin (progesteronelike hormone). These two hormones act by negative feedback to stop the release of GnRH by the hypothalamus and FSH (an estrogen effect) and LH (a progestin effect) by the pituitary. By blocking LH release, the progestin prevents ovulation. As a backup measure, the estrogen inhibits FSH secretion so that no follicles develop. In 1990, the FDA approved a long-lasting version of this contraceptive approach: a small patch that can be placed under the skin, where it dispenses progestin for up to five years.

Birth control pills have been the center of much debate, particularly because of the long-term harmful effects of the estrogens. No solid evidence exists for cancers caused by the pill, but cardiovascular problems are a major concern. Birth control pills have been implicated in blood clotting, atherosclerosis, and heart attacks. Smoking while using chemical contraception increases the risk of mortality tenfold or more. Although the pill places women at risk for these diseases, it eliminates the dangers of pregnancy; women on birth control pills have mortality rates about one-half those of pregnant women.

Sterilization is the permanent prevention of gamete release. **Tubal ligation** in women involves cutting a short section out of the oviduct to prevent eggs from traveling into the uterus. **Vasectomy** in men is the cutting of the vas deferens to prevent sperm from entering the urethra. Both male and female sterilization are relatively safe and free from harmful effects. Both are also difficult to reverse, so the procedures should be considered permanent.

Abortion is the termination of a pregnancy in progress. Spontaneous abortion, or miscarriage, is very common, occurring in as many as one-third of all pregnancies, often before the woman is even aware of her pregnancy. In addition, about 1.5 million women per year in the United States opt for abortions performed by physicians. In France, a drug called RU-486 is available to women who decide to terminate pregnancy within the first few weeks. An analog of progesterone, RU-486 blocks progesterone receptors in the uterus, thus preventing progesterone from maintaining pregnancy.

Of all contraceptives for sexually active individuals, latex condoms are the only ones that offer some protection against sexually transmitted diseases, including AIDS (see Chapter 39). This protection is, however, not absolute.

Reproductive Technology

Recent scientific and technological advances have made it possible to deal with problems of reproduction in striking ways. For example, it is now possible to diagnose many genetic diseases and congenital (present at birth) disorders while the fetus is in the uterus. Noninvasive procedures use high-frequency sound waves, or **ultrasound,** to detect fetal condition. Amniocentesis and chorionic villi sampling are more invasive techniques (see Figure 13.17). In **amniocentesis,** a long needle is inserted into the amnion (the sac surrounding the fetus), and a sample of fluid is withdrawn. Fetal cells in the fluid are cultured for 2 to 4 weeks, and the cultured cells can then be analyzed for genetic disorders and chromosomal problems, such as Down syndrome (see Chapter 14). **Chorionic villi sampling** is a more recent technique wherein a small sample of tissue is removed for genetic and metabolic analysis from the chorion, the fetal part of the placenta. This test carries a greater risk than amniocentesis (5% to 20% versus 1% spontaneous abortions following testing), but results may be obtained in days rather than weeks, and chorionic sampling can be performed earlier in the pregnancy.

Ultrasonography, amniocentesis, and chorionic villi sampling all pose important ethical questions. To date, essentially all detectable disorders remain untreatable in the uterus, and many cannot be corrected even after birth. Parents may be faced with difficult decisions about whether to terminate a pregnancy or cope with a child who may have profound defects and a short life expectancy. These are not easy questions; they demand careful, informed thought and competent counseling.

Another breakthrough in reproductive technology that has received considerable publicity is **in vitro fertilization.** First accomplished in 1978 in England, this procedure is now performed in major medical centers throughout the world. Women whose oviducts are blocked can have ova surgically removed from hormonally prepared follicles. The ova are then fertilized in Petri dishes in the laboratory. After about 2½ days, when the embryo has reached the eight-cell stage, it is placed in the uterus and allowed to implant. It sounds simple, but in vitro fertilization is a difficult and costly procedure. At this time, only about one out of six attempts is successful, at a cost of $4000 or more per attempt. This success rate may seem low, but in fact it is probably not different from the pregnancy rate resulting from insemination by intercourse. Multiple embryos are often placed in the uterus to increase the chances of a successful pregnancy. With several hundred children now conceived by in vitro fertilization, we have no evidence of any abnormalities associated with the procedure.

One area of reproductive research finally receiving much attention is male contraception. Male chemical contraceptives have proved quite elusive. Testosterone will block release of pituitary gonadotropins, but testosterone itself stimulates spermatogenesis. Estrogens are effective, but they inhibit libido and can be feminizing. The best prospects so far are for analogs of GnRH, which are potent inhibitors of spermatogenesis. Other treatments under study include progestins and gossypol, a compound extracted from cotton seeds.

* * *

In this chapter, we have considered the structural and physiological bases of animal reproduction, giving little attention to the mechanics of development that transform a zygote into an animal form. The next chapter focuses on embryology and other topics in animal development.

STUDY OUTLINE

Modes of Reproduction (pp. 931–934)

1. Asexual reproduction produces a clone of genetically identical offspring from a single parent. Budding, fragmentation, and regeneration are mechanisms of asexual reproduction in various invertebrates.
2. Sexual reproduction requires the fusion of male and female gametes to form a diploid zygote. The production of offspring with varying genotypes and phenotypes may enhance reproductive success in fluctuating environments.

3. Animals may be exclusively sexual or asexual; or they may alternate between the two reproductive modes, depending on environmental conditions. Variations on these two modes are made possible through parthenogenesis, hermaphroditism, and sequential hermaphroditism.
4. Reproductive cycles are controlled by hormones and seasonal changes in temperature, rainfall, or day length.

Mechanisms of Sexual Reproduction (pp. 934–936)

1. External fertilization requires critical timing, mediated by environmental cues, pheromones, and/or courtship behavior. External fertilization is most common in aquatic or moist habitats, where the zygote can develop without desiccation and heat stress.
2. Internal fertilization requires important behavioral interactions between male and female animals, as well as compatible copulatory organs.

Mammalian Reproduction (pp. 936–953)

1. Male reproductive anatomy consists of internal organs, and a scrotum and penis, the external genitalia. The gonads, or testes, reside in the cool environment of the scrotum. They possess endocrine interstitial cells surrounding sperm-forming seminiferous tubules that successively lead into the epididymis, vas deferens, ejaculatory duct, and urethra, which exits at the tip of the penis. Accessory glands add secretions to the semen.
2. Female reproductive anatomy consists of a pair of ovaries, oviducts, a uterus, a vagina, and external genitalia. The ovaries are stocked with gamete-containing follicles by the time of birth. Beginning at puberty, one or more follicles mature during each menstrual cycle. After ovulation, the remaining tissue of the follicle forms a corpus luteum that secretes progesterone and estrogen for a variable duration, depending on whether or not pregnancy occurs.
3. The oviduct draws the gamete into its open end and transports it to the uterus by ciliary action. The uterus opens through the cervix into the muscular vagina, which serves as a sperm receptacle and birth canal.
4. Although separate from the reproductive system, the mammary gland, or breast, evolved in association with parental care. Milk-secreting alveoli drain into ducts that open at the nipple.
5. In the embryo, genital primordia common to both sexes develop into male or female structures depending on the presence or absence of androgens.
6. Androgens from the testes cause the development of primary and secondary sex characteristics in the male. Androgen secretion and sperm production are both controlled by hypothalamic and pituitary hormones.
7. Female hormones are secreted in a rhythmic fashion reflected in the menstrual or estrous cycle.
8. In both types of cycles, the uterine lining thickens in preparation for possible implantation. The menstrual cycle, however, is punctuated by endometrial bleeding and lacks the clear-cut period of sexual receptivity limited to the heat period of the estrous cycle.
9. The human menstrual cycle consists of the menstrual flow phase, proliferative phase, and secretory phase. The ovarian cycle includes the follicular and luteal phases.
10. The female reproductive cycle is orchestrated by cyclic secretion of GnRH from the hypothalamus and FSH and LH from the anterior pituitary. The developing follicle produces estrogen, and the corpus luteum secretes progesterone and estrogen. Positive and negative feedback produces the changing levels of these five hormones, which coordinate the menstrual and ovarian cycles.
11. The production of sperm by spermatogenesis is a continuous and prolific process.
12. Unlike spermatogenesis, oogenesis occurs in discontinuous stages throughout life and involves unequal division of the cytoplasm during meiosis to form one large, functional gamete from each oogonium.
13. Puberty is a gradual process that involves the development of secondary sex characteristics and the onset of reproductive capability.
14. Common physiological patterns of the sexual response cycle underlie apparent differences in the rich diversity of human sexuality. Both males and females experience erection of certain body tissues due to vasocongestion and the increased muscle tension of myotonia, which culminates in orgasm.
15. Human pregnancy can be divided into three trimesters. Organogenesis is completed by eight weeks.
16. Birth, or parturition, results from strong, rhythmic uterine contractions that bring about the three stages of labor: dilation of the cervix, birth of the baby, and delivery of the placenta.
17. The ability of a pregnant woman to accept her "foreign" fetus may be due to the suppression of the immune response in her uterus.
18. Prevention of pregnancy by a variety of contraceptive measures includes prevention of the release of mature gametes from the gonads, prevention of gamete union in the female tract, or prevention of implantation of the zygote.
19. Current technological advances allow detection of fetal condition by ultrasound, amniocentesis, and chorionic villi sampling. Current technology also offers in vitro fertilization and promises to yield new developments in contraception.

SELF-QUIZ

1. Which of the following characterizes parthenogenesis?
   a. An individual may change its sex during its lifetime.
   b. Specialized groups of cells may be released and grow into new individuals.
   c. An organism is first a male and then a female.
   d. An egg develops without being fertilized.
   e. Both members of a mating pair have male and female reproductive organs.
2. Which of the following structures is *incorrectly* paired with its function?
   a. gonads—gamete-producing organs
   b. spermatheca—sperm-transferring organ found in male insects
   c. cloaca—common opening for reproductive, excretory, and digestive systems
   d. baculum—bone that stiffens penis, found in some mammals

e. endometrium—lining of the uterus, forms maternal part of placenta

3. Which of the following male and female structures are *least* alike in function?
   a. seminiferous tubules—vagina
   b. interstitial cells of testes—follicle cells
   c. testes—ovaries
   d. spermatogonia—oogonia
   e. vas deferens—oviduct
4. A difference between estrous and menstrual cycles is that
   a. nonmammalian vertebrates have estrous cycles, whereas mammals have menstrual cycles
   b. the endometrial lining is shed in menstrual cycles but reabsorbed in estrous cycles
   c. estrous cycles occur more frequently than menstrual cycles do
   d. estrous cycles are not controlled by hormones
   e. ovulation occurs before the endometrium thickens in estrous cycles
5. Peaks of LH and FSH production occur during
   a. the flow phase of the menstrual cycle
   b. the follicular phase of the ovarian cycle
   c. the period surrounding ovulation
   d. the end of the luteal phase of the ovarian cycle
   e. the secretory phase of the ovarian cycle
6. The *direct* function of GnRH is to
   a. stimulate production of estrogen and progesterone
   b. initiate ovulation
   c. inhibit secretion of pituitary hormones
   d. stimulate secretion of LH and FSH
   e. initiate the flow phase of the menstrual cycle
7. During human gestation, organogenesis occurs
   a. in the first trimester
   b. in the second trimester
   c. in the third trimester
   d. while the embryo is in the fallopian tube
   e. during the blastocyst stage
8. Except for abstinence from sexual intercourse, the contraceptive method most effective in reducing the chance of contracting AIDS and other sexually transmitted diseases is the
   a. birth control pill
   b. diaphragm
   c. latex condom
   d. IUD
   e. RU-486 pill
9. Fertilization of human eggs most often takes place in the
   a. vagina
   b. ovary
   c. uterus
   d. oviduct (fallopian tube)
   e. vas deferens
10. In mammalian males, the excretory and reproductive systems share the
    a. testes
    b. urethra
    c. ureter
    d. vas deferens
    e. prostate

CHALLENGE QUESTIONS

1. Describe how sexual and asexual reproduction differ in mechanism and result.
2. Explain why menstruation and ovulation do not occur during pregnancy.
3. Compare and contrast oogenesis and spermatogenesis.

SCIENCE, TECHNOLOGY, AND SOCIETY

1. New techniques for sorting sperm, combined with in vitro fertilization or artificial insemination, make it possible for a couple to choose their baby's sex. Would you want to do this if you had the chance? Why or why not? What potential problems can you foresee if this procedure becomes widely available?
2. Because of a uterine tumor, Julie is unable to carry a fetus to term. Eggs were surgically removed from her ovaries and fertilized in a glass dish with her husband Ron's sperm. One of the fertilized eggs was implanted in the uterus of a young woman named Michelle, whom Julie and Ron had hired as a surrogate mother. When the baby was born 9 months later, Michelle decided she wanted to keep it, and she refused to accept payment. Julie and Ron sued to gain custody of the baby. Who do you think should get the baby? Who is the "real" mother? What are your criteria for making a decision?
3. New technology has made it possible for doctors to save a small percentage of babies born 16 weeks prematurely. A baby born this early weighs just over a pound and faces months of treatment in an intensive care nursery. The cost for care may be hundreds of thousands of dollars per infant. Many of the surviving infants have mental and physical disabilities. Some people wonder whether such a huge technological and financial investment should be devoted to such a small number of babies. They feel that the resources might better be directed at providing prenatal care that could prevent many premature births. What is your opinion in this controversy? Defend your position.

FURTHER READING

Crews, D. "Courtship of Unisexual Lizards: A Model for Brain Evolution." *Scientific American*, December 1987. Research on all-female species.

Crooks, R., and K. Baur. *Our Sexuality*, 5th ed. Redwood City, CA: Benjamin/Cummings, 1993.

Diamond, J. "Turning a Man." *Discover*, June 1992. What controls whether an embryo will develop as a female or a male?

Duellman, W. "Reproductive Strategies of Frogs." *Scientific American*, July 1992. Fascinating reproductive diversity.

Freedman, D. "The Aggressive Egg." *Discover*, June 1992. Does the human egg attract sperm to it?

Moran, N. "Quantum Leapers." *Natural History*, April 1992. How aphids clone themselves.

Roberts, L. "U.S. Lags on Birth Control Developments." *Science*, February 23, 1990. An evaluation of why no essentially new contraceptive products have been developed in the United States in the past 30 years.

Ulmann, A., G. Teutsch, and D. Philibert. "RU-486." *Scientific American*, June 1990. A history of the development of this pill that induces abortion.

43 ANIMAL DEVELOPMENT

AN OVERVIEW OF DEVELOPMENTAL PROCESSES
FERTILIZATION
EARLY STAGES OF EMBRYONIC DEVELOPMENT
THE EMBRYOLOGY OF AMNIOTES
MECHANISMS OF DEVELOPMENT

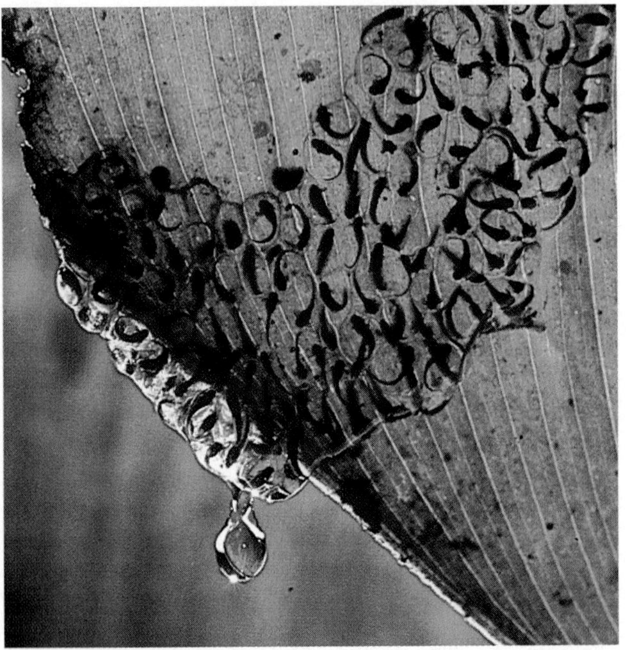

Figure 43.1
Living droplets, tadpoles of a tropical tree frog *(Hyla ebraccata).* It took only a week for these tadpoles to develop after their mother deposited her gelatinous egg mass on the underside of this leaf. When the tadpoles begin to wiggle, the droplets fall from the leaf into the swamp below, where the tadpoles undergo metamorphosis into frogs. The tiny frogs then emerge from the swamp and take to the trees. Although unique details of development have evolved in each animal species, a common set of mechanisms transforms zygotes into specific animal forms. In this chapter, you will learn about these basic mechanisms of animal development.

It is difficult to imagine that each of us began life as a single cell about the size of the period at the end of this sentence. Less than a month after conception, our brains were taking form and our rudimentary hearts had already begun to pulsate. It took a total of only about nine months—the length of a school year—to be transfigured from zygote to human. Molecular biologist Sidney Brenner has wryly stated that developmental biology is about "how to make a mouse." How does a single fertilized egg cell give rise to a specific animal form built of billions of differentiated cells organized into specialized tissues and organs? Animal development, a marvel biologists are only beginning to understand, is the subject of this chapter (Figure 43.1).

All animals develop throughout their lifetime, showing progressive changes in form and function. Among the important episodes are development of the embryo, growth and maturation of the animal after birth or hatching, metamorphosis of invertebrates and amphibians that have larval stages, regeneration (the replacement of lost body parts), wound healing, and aging. Although embryonic development is the main topic of this chapter, the basic cellular mechanisms that transform an embryo are also involved in postembryonic development.

AN OVERVIEW OF DEVELOPMENTAL PROCESSES

New research methods have made the study of development one of the frontiers of modern biology, but the question of how an egg becomes an animal has been asked for centuries. As recently as the eighteenth century, the view prevailed that the egg or sperm contains a preformed, miniature embryo that simply grows during its development. This idea of **preformation** came to include the notion that this embryo must contain all its descendants: a series of successively smaller embryos within embryos. According to this version of embryology, dissecting an egg would be analogous to opening one of those Russian dolls that comes apart to reveal a family of ever smaller dolls. Based on the preformation model, one theologian proposed that Eve, in the Garden of Eden, stored all future humanity within her.

The competing theory of embryology was an idea, originally proposed 2000 years earlier by Aristotle, that the form of an embryo emerges gradually from a relatively formless egg. This theory of the progressive development of form is called **epigenesis.** As microscopy improved during the nineteenth century, biologists could see that embryos took shape gradually, and epigenesis displaced preformation as the favored explanation among embryologists.

Modern biology, of course, has completely discarded the idea of a tiny person living in an egg or sperm cell. But when interpreted in broader terms, the concept of preformation may have some merit. Although an embryo's form emerges gradually as it develops from a fertilized egg, something *was,* in effect, preformed in the zygote. An organism's development is largely determined by the genome of the zygote and the organization of the cytoplasm of the egg cell. Messenger RNA, proteins, and other components are heterogeneously distributed in the unfertilized egg, and this has a profound impact on the development of the future embryo in most animal species. After fertilization, division of the zygote partitions the cytoplasm in such a way that nuclei of different embryonic cells are exposed to different cytoplasmic environments. This sets the stage for the expression of different genes in different cells. As the embryo develops, there will be an emergence of inherited traits that is ordered in space and time by mechanisms that selectively control gene expression (see Chapter 18).

The three component processes in embryonic development are cell division, differentiation, and morphogenesis. Through a succession of mitotic divisions, the zygote gives rise to a large number of cells—trillions, in the case of a newborn human. Cell division alone, however, would produce a great ball of identical cells, which is nothing like an animal. During embryonic development, cells not only increase in number, they also undergo **differentiation,** to become the specialized cells that are organized into the tissues and organs of the animal. How two cells with the same genes can become as different as, say, a nerve cell and a muscle cell is one of the enduring puzzles of biology. Movements of cells and tissues are also necessary to convert the cell mass of the early embryo into the characteristic three-dimensional form of the larval or juvenile stage of the species. Such movements and rearrangements are collectively called **morphogenesis,** which literally means the "building of form."

The importance of precise regulation of these three developmental processes is evident in certain human disorders that result from developmental mechanisms going awry. For example, the condition known as cleft palate, in which the upper part of the mouth cavity fails to completely close, is a defect of morphogenesis. Some malignant and early appearing forms of cancer, including a fatal cancer of the retina called retinoblastoma, seem to be primarily defects in the regulation of cell division, already discussed in Chapters 11 and 18. Some testicular cancers are examples of disorders associated with aberrations of the differentiation process. When one considers how many things *can* go wrong during cell division, differentiation, and morphogenesis, it is all the more remarkable that most animals develop normally.

These three processes—regulated cell division, differentiation, and morphogenesis—form the heart of our inquiry into how animals develop. In the first half of the chapter, we will view the role of these processes during the early stages of embryonic development, when the basic body plan of an animal takes form. In the second half of the chapter, we will take a closer look at the cellular mechanisms that control development.

FERTILIZATION

Embryonic development begins with **fertilization,** the union of a sperm and an egg. These gametes are both highly specialized cell types produced by a complex series of developmental events in the testes and ovaries of the parents. One of the functions of fertilization is to combine haploid sets of chromosomes from two individuals into a single diploid cell, the zygote. Its other function is activation of the egg: Contact of the sperm with the egg's surface initiates metabolic reactions within the egg that trigger the onset of embryonic development.

Fertilization has been studied most extensively by combining the gametes of sea urchins—an abundant source of sperm and eggs—in the laboratory. Although the details of fertilization vary with different animal groups, sea urchins (phylum Echinodermata) provide a good general model for the important events of fertilization.

The Acrosomal Reaction

The eggs of sea urchins are fertilized externally after the animals release their gametes into the surrounding seawater. When a sperm cell is exposed to molecules from the slowly dissolving jelly coat that surrounds an egg, a vesicle at the tip of the sperm called the acrosome (see Chapter 42) discharges its contents by exocytosis (Figure 43.2). This acrosomal reaction releases hydrolytic enzymes that enable an extending **acrosomal process** to penetrate the jelly coat of the egg. The tip of the process is coated with a protein called **bindin** that adheres to specific receptor molecules located on the **vitelline layer** just external to the plasma membrane of the egg. "Lock-and-key" recognition of molecules ensures that eggs will be fertilized only by sperm

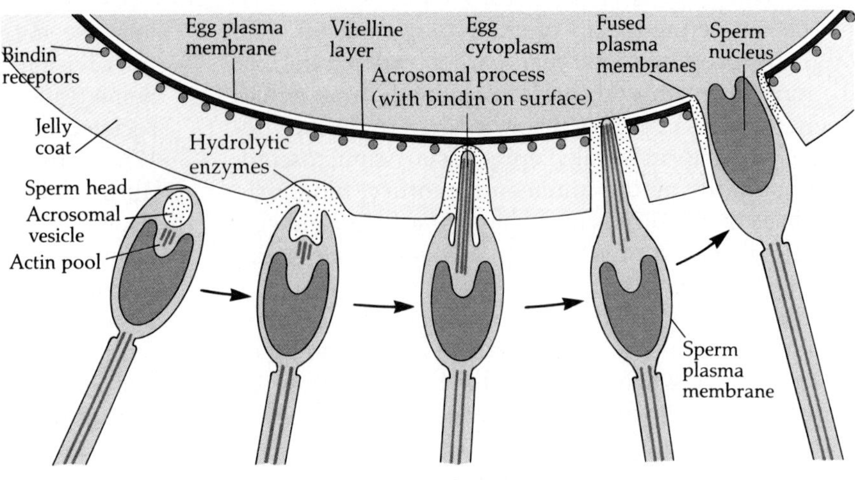

Figure 43.2
The acrosomal reaction during sea urchin fertilization. This rapid reaction is illustrated in a temporal sequence from left to right. Upon contacting the jelly coat of the egg, the acrosomal vesicle in the head of the sperm releases proteins, some of them hydrolytic enzymes that excavate a hole in the jelly. An acrosomal process extends by the polymerization of actin, a protein. Bindin protein molecules on the surface of the acrosomal process attach to receptors on the vitelline layer of the egg, providing species specificity for fertilization. The plasma membranes of the gametes fuse, and the sperm nucleus enters the egg.

of the same animal species. This is especially important when fertilization occurs externally in water, where gametes of other species are likely to be present.

Enzymes on the acrosomal process probably digest material in the vitelline layer. The tip of the process then contacts the plasma membrane of the egg. In some animal species, the area of the egg contacted by the sperm undergoes rapid reorganization of the underlying cytoskeleton and surface microvilli, resulting in a slight conical protrusion. The sperm's plasma membrane, which extends over the acrosomal process, fuses with the egg's plasma membrane, and the nucleus of the sperm enters the cytoplasm of the egg.

Even before the sperm nucleus enters the egg, fusion of the gamete membranes causes a nervelike electrical response by the plasma membrane of the egg. Union of the gamete membranes opens ion channels that allow sodium ions to flow into the egg cell and change the membrane potential, the voltage across the membrane (see Chapter 8). This membrane depolarization, as the electrical response is called, bars other sperm cells from fusing with the plasma membrane. A common phenomenon among animal species, this **fast block to polyspermy**—which occurs in much less than a second—prevents multiple fertilizations that would result in an aberrant chromosome count and abnormal mitosis.

The Cortical Reaction

The acrosomal reaction of the sperm makes it possible for the plasma membranes of the gametes to fuse, and the egg responds to this stimulus with a cortical reaction (Figure 43.3). The sperm-egg fusion somehow causes calcium to be released into the cytoplasm from a reservoir within the egg, probably the endoplasmic reticulum (Figure 43.4). The increased concentration of cytoplasmic Ca^{2+} causes vesicles located in the cortex, a gelatinous outer zone of cytoplasm found in many types of eggs, to fuse with the plasma membrane. These **cortical granules** release their contents by exocytosis into the perivitelline space between the plasma membrane and the adjacent vitelline layer. The secreted substances probably include enzymes that loosen the adhesive material between the plasma membrane and vitelline layer and a high concentration of macromolecules, which causes the resulting space to swell by the osmotic uptake of water. This swelling elevates the vitelline layer. Other enzymes polymerize molecules in the elevated vitelline layer to form a hardened **fertilization membrane,** which resists the entry of additional sperm. By this time, usually about a minute after sperm and egg fuse, the voltage across the plasma membrane has returned to normal, and the fast block to polyspermy no longer functions. But the fertilization membrane, along with other changes of the egg's surface, functions as a **slow block to polyspermy.**

Activation of the Egg

The sharp rise in the egg's cytoplasmic concentration of Ca^{2+} not only stimulates the cortical reaction but also incites metabolic changes within the egg cell. The unfertilized egg has a very slow metabolism, but within a few minutes after fertilization, the rates of cellular respiration and protein synthesis increase substantially. With these rapid changes, the egg cell is said to be **activated.** Calcium stimulates the extrusion of hydrogen ions (H^+) from the egg, changing the pH of the cytoplasm from 6.8 to 7.3. This pH change seems to be indirectly responsible for many of the metabolic responses of the egg to fertilization. Increased cytoplasmic Ca^{2+} and pH have been observed in the

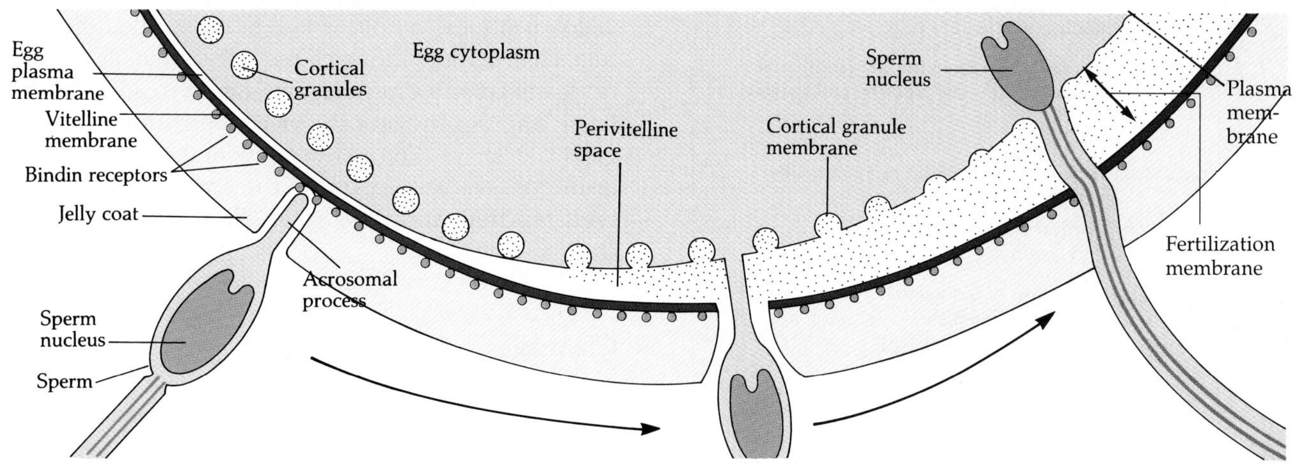

Figure 43.3
The cortical reaction during sea urchin fertilization. This reaction is illustrated in a temporal sequence from left to right. Cortical granules in the viscous cortex of the egg fuse with the plasma membrane and discharge their contents by exocytosis. Enzymes and other substances released from the cortical granules raise the vitelline layer and harden it to form a fertilization membrane that serves as a slow block to polyspermy.

fertilized eggs of sea urchins and many other species, so these may be rather general features of the activation mechanism.

Sperm entry and egg activation are distinguishable functions of fertilization. Eggs can be artificially activated by the injection of calcium or by a variety of mildly injurious treatments, such as temperature shock. This artificial activation switches on the metabolic responses of the egg and causes it to begin developing by parthenogenesis (without fertilization by a sperm). It is even possible to artificially activate an egg that has had its own nucleus removed. (Of course, embryonic development of such an egg terminates at a very early stage.) The fact that an egg lacking a nucleus can begin making new kinds of proteins upon activation means that mRNA coding for these proteins must be stockpiled in an inactive form in the cytoplasm of the unfertilized egg. This is an example of controlling gene expression at the translational rather than the transcriptional level (see Chapter 18). The increase in protein synthesis is brought about by a concentrated set of mechanisms in which stored mRNA is made available for translation, and the efficiency of the protein synthetic machinery increases severalfold.

While the activated egg gears up its metabolism, the nucleus of the sperm cell within the egg begins to swell. After about 20 minutes, the sperm nucleus merges with the egg nucleus to produce the diploid nucleus of the zygote. Replication of DNA commences, and the first cell division occurs (in the case of sea urchins) in about 90 minutes. The events of fertilization are summarized in Figure 43.5.

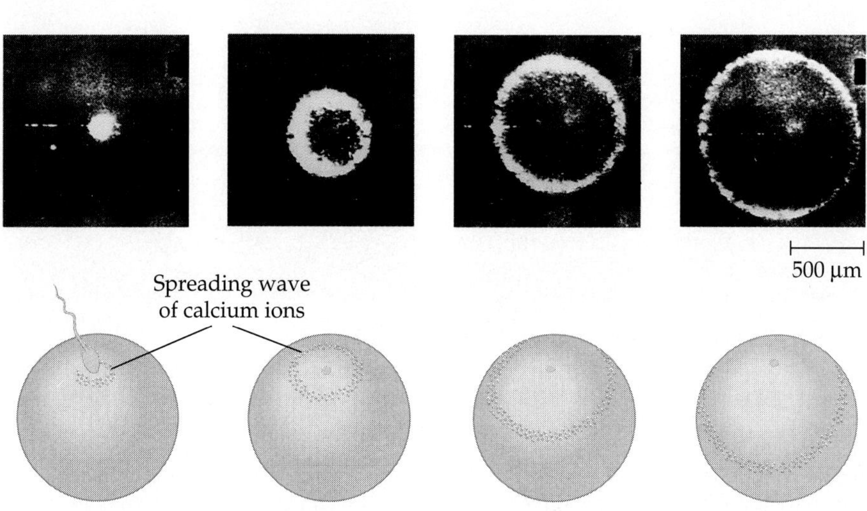

Figure 43.4
A wave of Ca^{2+} release during the cortical reaction. A fluorescent dye that glows in the presence of free Ca^{2+} was used in this experiment to track the cortical reaction from the point of sperm contact (0 sec) during the fertilization of a fish egg. The spreading wave of Ca^{2+} ions, released into the cytosol from some intracellular reservoir (probably the endoplasmic reticulum), functions as a second messenger (sperm contact being the first) that helps activate metabolic changes in the egg.

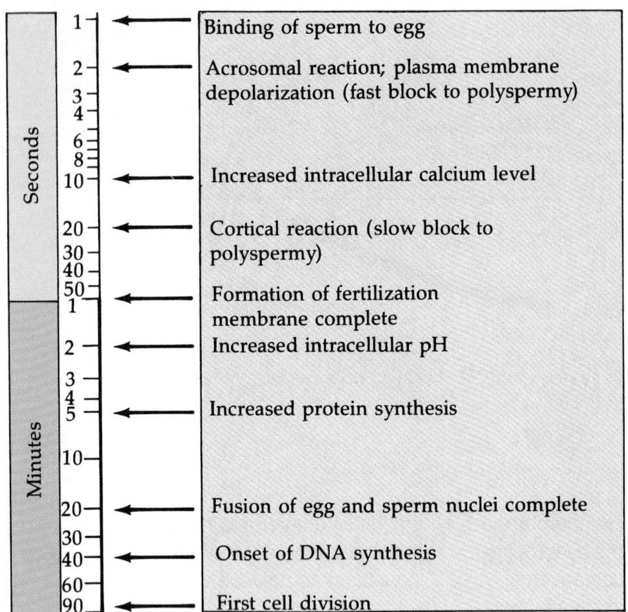

Figure 43.5
Timeline for the fertilization of sea urchin eggs. Notice that the scale is logarithmic. The process begins when a sperm cell contacts the jelly coat of an egg.

EARLY STAGES OF EMBRYONIC DEVELOPMENT

Although mechanisms of morphogenesis mold an animal throughout its embryonic development, the basic body plan is established in the early stages of this process. The key stages are cleavage, in which cell division produces a hollow ball of cells called the blastula; gastrulation, in which the cells of the blastula are reorganized into a multilayered embryo called the gastrula; and organogenesis, in which rudimentary organs take shape. For an overview of these early developments, we will examine the embryology of two well-studied animals, a sea urchin and the frog *Xenopus laevis*.

Cleavage

Fertilization is followed by **cleavage,** a succession of rapid cell divisions that produces a ball of cells from the zygote (Figure 43.6). During cleavage, the embryos of sea urchins and *Xenopus* (and most other animals) nourish themselves on **yolk,** nutrients stored in the egg. The embryo does not grow during this period of development; the cytoplasm of one large cell is simply partitioned into many smaller cells, called **blastomeres.** The cells cycle between the S (DNA synthesis) and M (mitosis) phases of cell division, essentially skipping the G_1 and G_2 phases, when most cell growth normally occurs (see Chapter 11).

By dividing the zygote into many smaller blastomeres, cleavage increases the surface-to-volume ratio of each cell, which enhances oxygen uptake and other important exchanges with the surroundings. Also, cleavage substantially reduces the volume of cytoplasm that must be controlled by each nucleus. And cleavage partitions different domains (regions) of cytoplasm present in the original undivided egg cell into separate cells. Because the domains may contain dif-

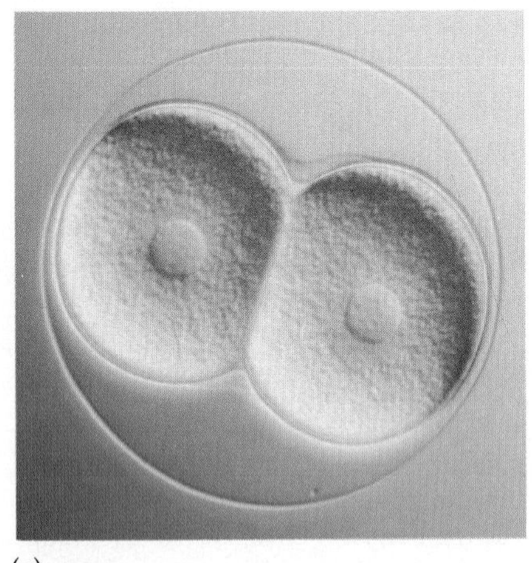
(a)

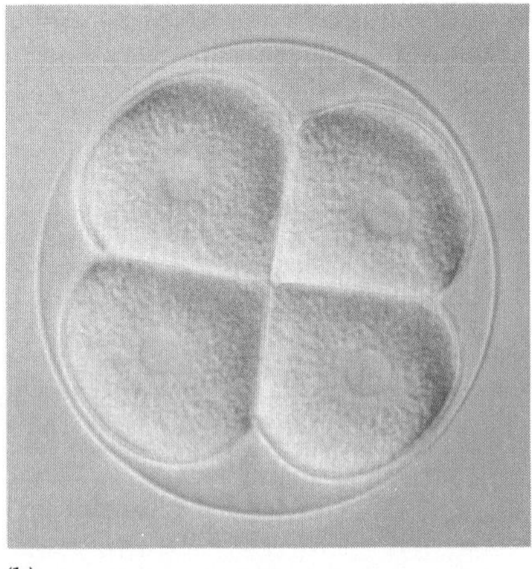
(b)

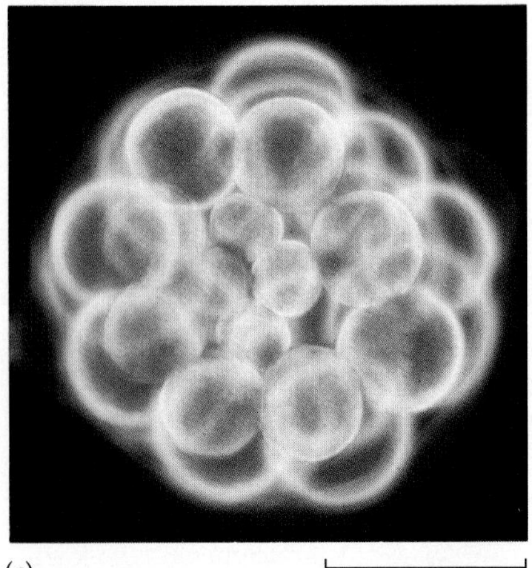
(c)

Figure 43.6
Cleavage in an echinoderm (sea urchin) zygote. Cleavage divisions consist of relatively rapid mitosis and cytokinesis without growth of the cells between successive divisions. Following fertilization, cleavage transforms a single large cell, the zygote, into a ball of much smaller cells. These are phase contrast micrographs of living sea urchin embryos. (**a**) The two-cell stage, following the first cleavage division, which occurs about 45 to 90 minutes after fertilization (notice that the fertilization membrane is still present). (**b**) The four-cell stage, following the second cleavage division. (**c**) In a few hours, repeated cleavage divisions have formed a multicellular ball. The embryo is still retained within the fertilization membrane, from which the motile larva that develops from the embryo will eventually "hatch."

ferent cytoplasmic components, this partitioning provides an early basis for cell differentiation, as discussed later in the chapter.

The eggs of most animals have a definite polarity, and cleavage planes follow a specific pattern relative to the axis of the zygote. In *Xenopus,* the axis is defined by the point, called the **animal pole,** where the polar body (superfluous maternal haploid nucleus; see Chapter 42) is budded from the cell. The opposite end of the zygote is known as the **vegetal pole.** The hemispheres of the zygote are named for their respective poles. The animal hemisphere has melanin granules embedded in the cortex, giving it a deep gray hue, while the vegetal hemisphere contains the yellow yolk. The sea urchin also has an animal-vegetal axis, but there are no colored markers (in most species) to reveal it, and we know it exists only because of certain developmental experiments covered later in the chapter.

In both sea urchins and frogs, the first two cleavage divisions are polar (vertical), resulting in four cells that each extend from animal to vegetal pole. The third division is equatorial (horizontal), producing an eight-celled embryo with two tiers of four cells each (Figure 43.7a). The general pattern up to this point is the same in both types of embryos. In fact, echinoderms, chordates, and the other animal phyla grouped as deuterostomes share many features of early embryonic development (see Figure 29.6). These similarities distinguish the deuterostomes from the protostomes, the evolutionary branch that includes the annelids, arthropods, and mollusks. For example, cleavage in deuterostomes is radial, meaning that the upper (animal) tier of four cells is aligned directly over the lower (vegetal) tier at the eight-cell stage. In contrast, most protostomes exhibit spiral cleavage, in which the cells of the upper tier sit in the grooves between the cells of the lower tier.

Continued cell division in sea urchins produces a solid ball of cells known as the **morula** (L. for "mulberry") (one is shown in Figure 43.6c). A fluid-filled cavity called the **blastocoel** forms in the center of the morula, and soon the cells are arranged in a single epithelial layer surrounding the blastocoel. This hollow-ball stage of embryonic development is the **blastula.** The larger and much yolkier egg of *Xenopus* also forms a hollow blastocoel, but it is eccentrically located in the animal hemisphere, and the blastula wall may be many cell layers thick. The yolk-laden cells of the vegetal hemisphere are much larger than those of the animal hemisphere (Figure 43.7b).

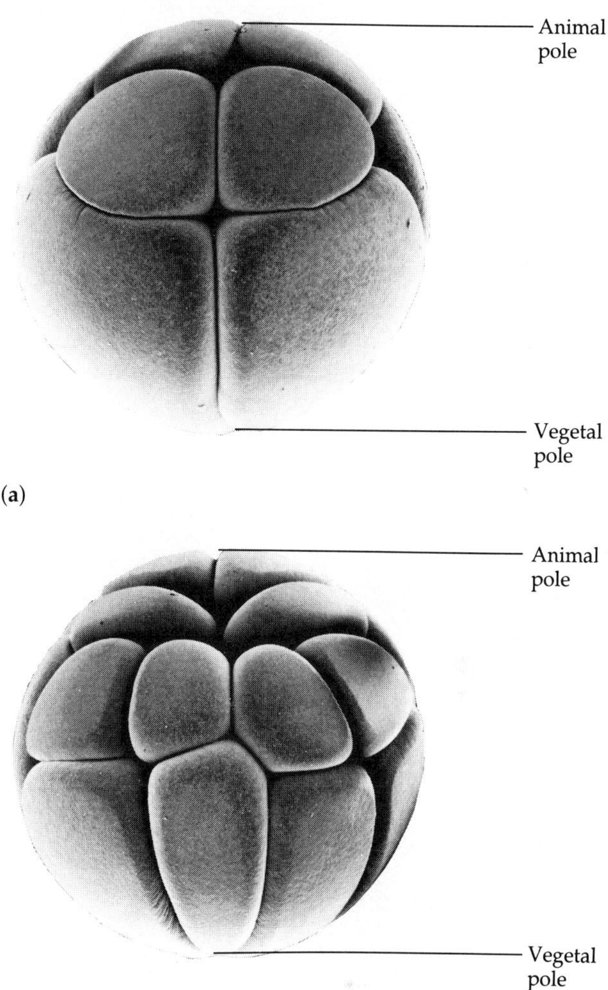

Figure 43.7
Cleavage in a frog egg. Yolk, concentrated near the vegetal pole of the egg, impedes the formation of cleavage furrows. **(a)** After two equal polar divisions, the third cleavage division is perpendicular to the polar axis but is displaced by yolk to the animal pole's side of the equator. Thus, the four blastomeres at the animal pole are smaller than the four blastomeres at the vegetal pole. **(b)** As cleavage continues, the cells at the animal pole divide more frequently than the yolk-laden cells near the vegetal pole. This accounts for the blastomeres at the animal pole being smaller and more numerous than those at the vegetal pole (SEMs).

Gastrulation

Gastrulation is a profound rearrangement of the cells of the blastula—a real morphogenetic drama. Gastrulation differs in detail from one animal group to another, but a common set of cellular changes drives this spatial rearrangement of embryos. These general cellular mechanisms are changes in cell motility, changes in cell shape, and changes in cellular adhesion to other cells and to molecules of the extracellular matrix (see Chapter 7). The essential result of gastrulation in all forms of animals is that some of the cells at or near the surface of the hollow blastula move to a new, more interior location. This transforms the blastula, a hollow ball, into a multilayered embryo called the **gastrula.**

Let's examine gastrulation in a sea urchin embryo (Figure 43.8). Recall that the wall of the sea urchin blastula consists of a single layer of cells. Gastrulation be-

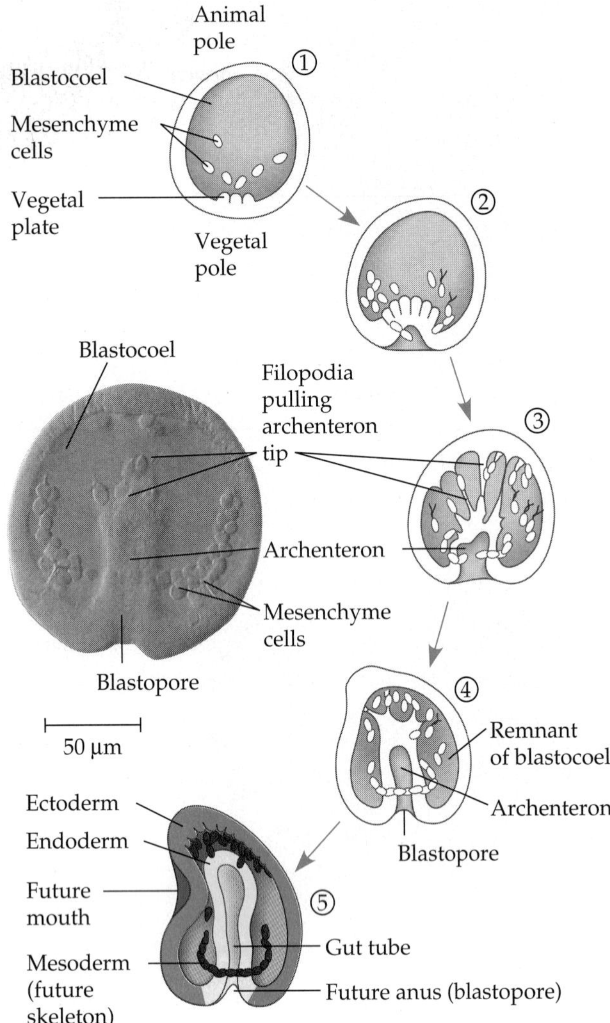

Figure 43.8
Sea urchin gastrulation. ① A vegetal plate forms, and migratory mesenchyme cells detach from the blastula wall. ② The vegetal plate invaginates (buckles inward). ③ Cells at the archenteron tip form filopodia, which search until they find the animal pole and adhere to cells there (LM). ④ Contraction of the filopodia deepens the archenteron, which becomes the gut of the larva. ⑤ The three primary germ layers—ectoderm (color-coded blue throughout this chapter), mesoderm (red), and endoderm (yellow)—are now in place.

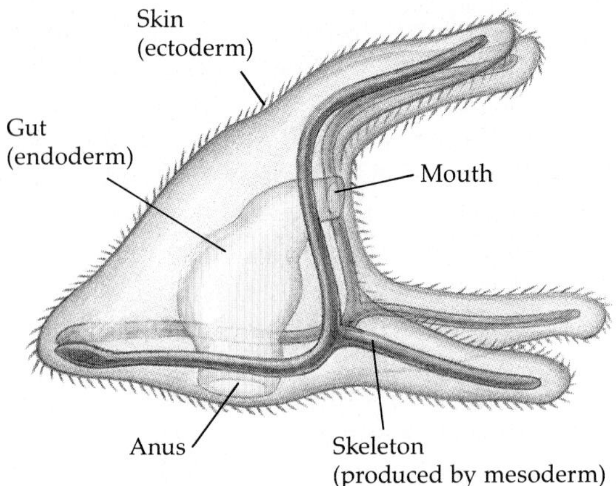

Figure 43.9
The sea urchin larva. The basic body plan organized by gastrulation is still evident in this feeding larva. The archenteron is now a functional gut with walls derived from endoderm. The gut has two openings: an anus formed from the blastopore of the gastrula and a mouth formed from a secondary opening. Some of the mesenchyme cells of the mesoderm have secreted minerals that form a simple internal skeleton. The skin that develops from ectoderm is ciliated. The larva drifts near the sea surface as plankton, feeding on bacteria and unicellular algae.

gins at the vegetal pole, where the cells flatten slightly to form a plate that buckles inward, a process called **invagination.** Other cells near the plate detach from the blastula wall and enter the blastocoel as migratory cells called mesenchyme cells. The buckled vegetal plate then undergoes extensive rearrangement of its cells, a process that transforms the shallow invagination into a deeper, narrower pouch called the **archenteron,** or primitive gut. The opening of the archenteron, which will become the anus, is called the **blastopore.** At this stage, the archenteron extends about two-thirds of the way across the blastocoel. Some additional mesenchyme cells now emigrate from the tip of the archenteron. Other cells at the archenteron tip develop long, thin cytoplasmic extensions called filopodia. When the probing filopodia contact the animal pole, they adhere to the blastocoel wall. The filopodia then contract and drag the archenteron the rest of the way across the blastocoel.

The embryo is now a gastrula with three cell layers, or **primary germ layers.** From the outside moving inward, these are the **ectoderm,** the **mesoderm** (derived from the migrating mesenchyme cells), and the **endoderm.** Thus, the triploblastic (three-layered) body plan characteristic of most animal phyla is established very early in development (see Chapter 30). The primary germ layers give rise to all the animal's organ systems. In the sea urchin, the gastrula eventually develops into the larva shown in Figure 43.9. The ectoderm has become the larva's outer skin (epidermis), and the endoderm has become the cell layer forming the gut wall. The tip of the archenteron has fused with the outer cell layer near the animal pole to form the mouth, and the blastopore has become the anus. The mesoderm has formed the muscles, pigment cells, and skeletal spicules of the sea urchin larva.

Gastrulation during frog development also produces a three-layered embryo with an archenteron, as shown in Figure 43.10. The mechanics of gastrulation

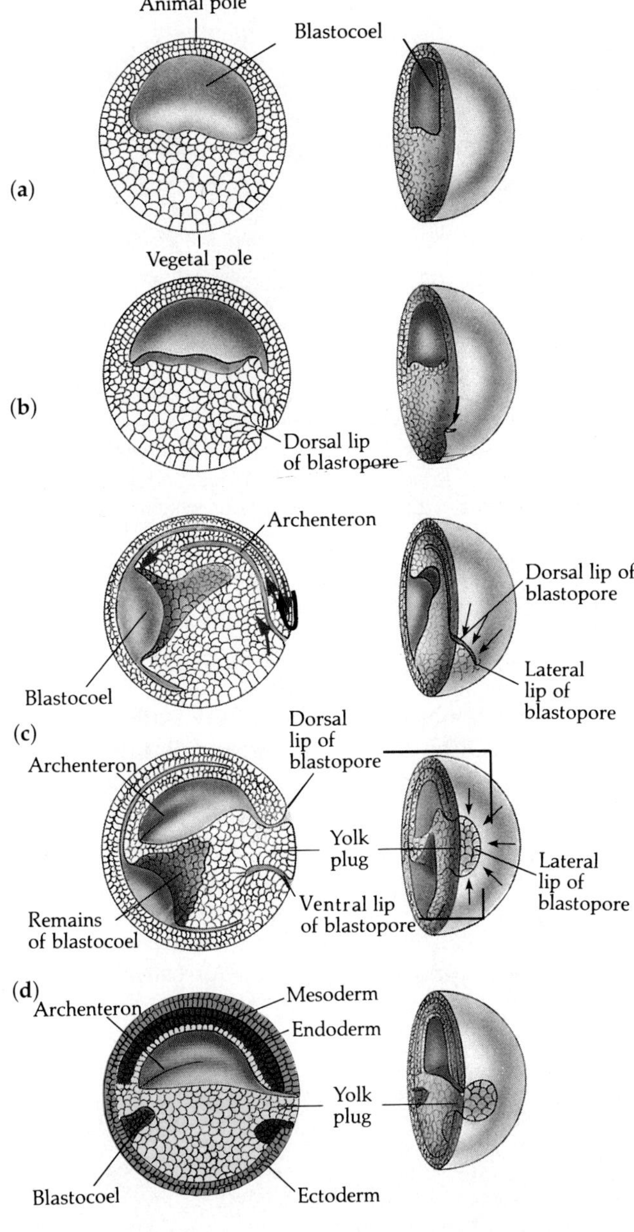

Figure 43.10
Gastrulation in a frog embryo. (**a**) The blastocoel of the frog blastula is off-center and surrounded by a wall that is more than one cell thick. (**b**) The dorsal lip of the blastopore forms by the burrowing movement (arrow) of bottle-shaped cells. (**c**) After the initial invagination, additional cells roll over the edge of the dorsal lip (involution) and move away from the blastopore within the embryo. The cells of the animal pole spread out into a thinner sheet, as cells near the blastopore disappear into the interior of the embryo. As involution occurs, the lip of the blastopore begins to extend laterally. (**d**) The blastopore continues in a circular extension until it completely rings a yolk plug of large, yolky cells derived from the vegetal pole of the embryo. By now, an archenteron lined by the involuted cells has formed. (**e**) The complex movements of gastrulation produce an embryo with three primary germ layers ready for organogenesis.

are much more complicated in the frog, because of the large, yolk-laden cells of the vegetal hemisphere and because the wall of the blastula is more than one cell thick in most species. The first sign of gastrulation is a small crease on one side of the blastula where the blastopore will eventually be located. This invagination is produced by bottle cells, named for the shape they assume when they lead the way into the interior of the embryo. As these cells burrow inward, they remain attached for a while to the surface by long necks, giving the cells their bottle shapes. The tuck produced by invagination will become the upper edge, or **dorsal lip,** of the blastopore. The location of the dorsal lip forecasts the future dorsal-ventral axis of the frog. Then, in a process called **involution,** cells on the surface of the embryo roll over the edge of the dorsal lip into the interior of the embryo, then continue migrating along the roof of the blastocoel. The blastopore wall near the animal pole is several cells thick, and these cells spread out to maintain the surface area of the embryo, as cells nearer the blastopore involute and disappear from the surface. As gastrulation progresses, increasingly lateral cells join the dorsal cells in their involution. As this occurs, the lip of the blastopore widens, first becoming frownlike in shape and finally forming a complete circle.

At this stage, the lip of the blastopore surrounds a **yolk plug** consisting of the large, food-laden cells from the vegetal pole of the embryo. The complex movements of gastrulation have nearly eliminated the blastocoel and formed the archenteron, lined by endoderm. Cells once on or near the surface of the embryo now reside in the interior as mesoderm, and cells remaining on the surface make up the ectoderm. The three primary germ layers are in place, and the embryo's organs begin to take form.

Organogenesis

Various regions of the primary germ layers develop into the rudiments of organs during the process of **organogenesis.** The organs that begin to take shape first in the embryos of frogs and other chordates are the neural tube and notochord, the skeletal rod characteristic of all chordate embryos (see Chapter 30). Figure 43.11 shows this early organogenesis as it occurs in a frog. The **notochord** is formed from dorsal mesoderm just above the archenteron, and the neural tube originates as a plate of dorsal ectoderm just above the developing notochord. The neural plate soon rolls itself into the **neural tube,** which will become the central nervous system—the brain and spinal cord. These organs are hollow in chordates because of this mechanism of development. The notochord elongates and stretches the embryo along its anterior-posterior axis.

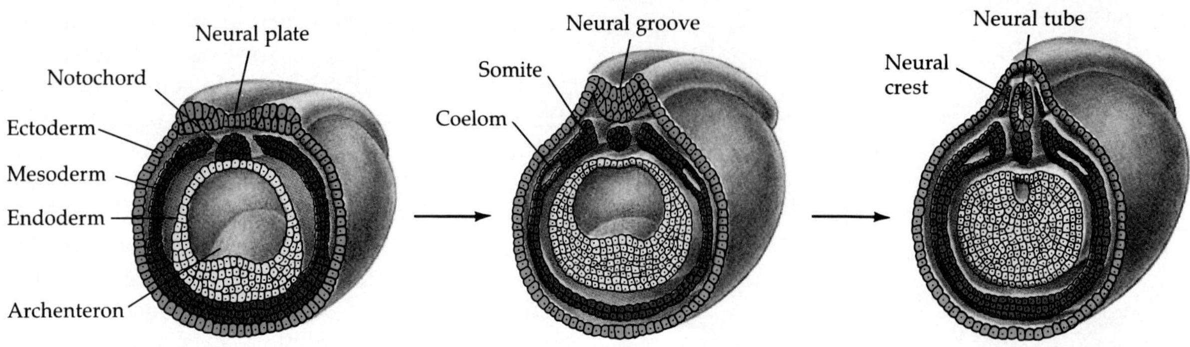

Figure 43.11
Early organogenesis in a frog embryo (cross-sectional views). Dorsal mesoderm forms the notochord. Blocks of mesoderm called somites flank the notochord and give rise to segmental structures, such as the vertebrae of the backbone and serially arranged skeletal muscles. Lateral mesoderm separates into the two tissue layers that line the coelom. Dorsal ectoderm thickens to form the neural plate, which rolls to become the neural tube. Tissue at the meeting margins of the tube separate from the tube as the neural crest, a source of migrating cells that eventually form many structures, including the bones and muscles of the skull, the pigment cells of the skin, the adrenal medulla glands, and peripheral ganglia of the nervous system.

Later, the notochord will function as a core around which mesodermal cells gather to form the vertebrae.

The strips of mesoderm lateral to the notochord separate into blocks called **somites,** which are arranged serially on both sides along the length of the notochord (Figure 43.12). Cells from the somites not only give rise to the vertebrae of the backbone, they also form the muscles associated with the axial skeleton. This serial origin of the axial skeleton and muscles reinforces the point made in Chapter 30 that chordates are basically segmented animals, although the segmentation becomes less obvious later in development. (There are signs of segmentation even in the adult, as in the series of vertebrae in a human or the segments of chevron-shaped muscles in a fish.) Lateral to the somites, the mesoderm separates into two layers that form the lining of the body cavity, or coelom.

As organogenesis progresses, many other rudimentary organs develop from the primary germ layers. In addition to forming the nervous system, ectoderm gives rise to the animal's external covering (epidermis) and associated glands, the inner ear, and the lens of the eye. Mesoderm produces the notochord and the lining of the coelom, as well as the muscles, skeleton, gonads, kidneys, and most of the circulatory system. Endoderm forms the lining of the digestive tract, and it gives rise to organs that originate as outpocketings of the archenteron, such as the liver, pancreas, and (in some vertebrates) lungs.

Along the border where the neural tube pinches off from the ectoderm, cells break loose to form the **neural crest,** a band of cells unique to vertebrate embryos. Cells of the neural crest later migrate to various parts of the embryo, forming the pigment cells in the skin, some of the bones and muscles of the skull, the teeth, the medulla of the adrenal glands, and peripheral components of the nervous system, such as sensory and sympathetic ganglia.

Embryonic development of the frog leads to a larval stage, the tadpole (see Figure 43.1), which hatches from the jelly coat that originally cloaked the egg. Later, metamorphosis will transform the frog from the aquatic, herbivorous tadpole to the terrestrial, carnivorous adult.

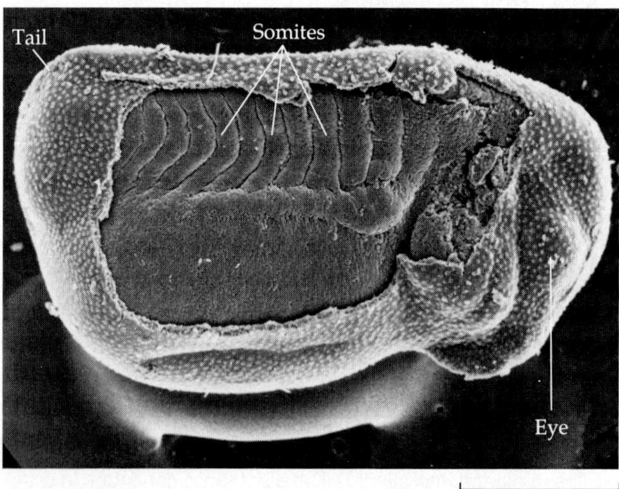

Figure 43.12
The embryonic origin of segmentation in a frog embryo. At this stage of development, called the tailbud stage, serial somites are evident along each side of the embryo. The somites will give rise to such segmental structures as the vertebrae and axial muscles. Part of the ectoderm of this embryo has been removed to reveal the somites (SEM).

Morphogenesis and cellular differentiation, which we will discuss later in the chapter, continue to refine the organs that arise from the primary germ layers. However, the fundamental plan of an animal with three concentric layers is already evident by the end of the gastrula stage.

THE EMBRYOLOGY OF AMNIOTES

All vertebrate embryos require an aqueous environment for development. In the case of fish and amphibians, the egg is laid in the surrounding sea or pond and needs no special water-filled enclosure. The vertebrate move to land required solving the problem of reproduction in dry environments, and two solutions evolved: the shelled egg of reptiles and birds and the uterus of placental mammals. Within the shell or uterus, the embryos of birds, reptiles, and mammals are surrounded by fluid-filled sac formed by a membrane called the amnion. Animals of these three classes are therefore called **amniotes.** We have already examined the embryology of a nonamniote vertebrate, the frog. For comparison, we will now study the early development of two amniotes, a bird and a mammal. Though organogenesis is very similar in all vertebrates, there are important differences during the first two stages of development, cleavage and gastrulation.

Avian Development

The part of a bird egg that we commonly call the yolk is actually the egg cell (ovum), swollen with the food reserve that is properly called yolk. This enormous cell is surrounded by a protein-rich solution (the egg white) that will provide additional nutrients for the growing embryo. Cleavage of the fertilized egg is restricted to a small disc of yolk-free cytoplasm at the animal pole of the egg cell. After fertilization, cell division partitions this yolk-free cytoplasm to form a cap of cells called the **blastodisc,** which rests on the large, yolky, undivided portion of the original egg cell (Figure 43.13a). This incomplete division of a yolk-rich egg is known as **meroblastic cleavage.** It contrasts with **holoblastic cleavage,** the term used for the complete division of eggs having little yolk (as in the sea urchin) or a moderate amount of yolk (as in the frog).

Cleavage at the animal pole of the bird egg is followed by sorting of the blastodisc cells into upper and lower layers, the epiblast and the hypoblast. The cavity between these two layers is the avian version of the blastocoel. And this embryonic stage is the avian equivalent of the blastula, although its form is different from the hollow ball of an early frog embryo.

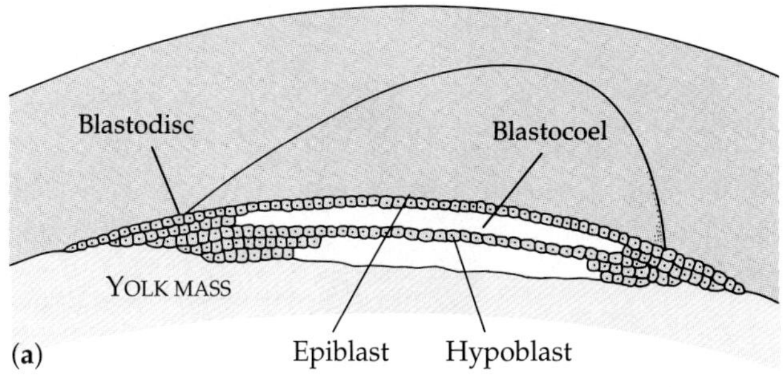

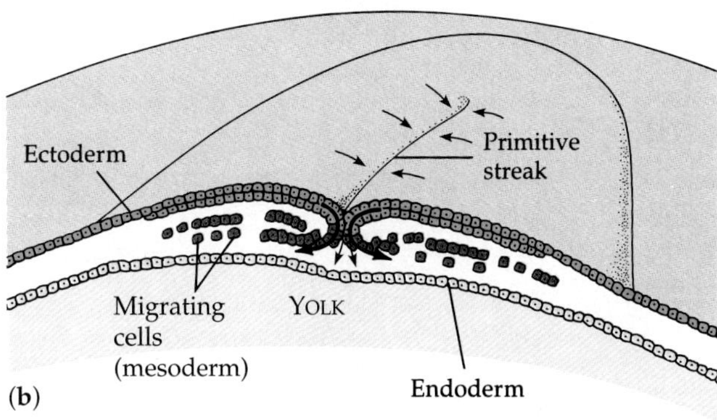

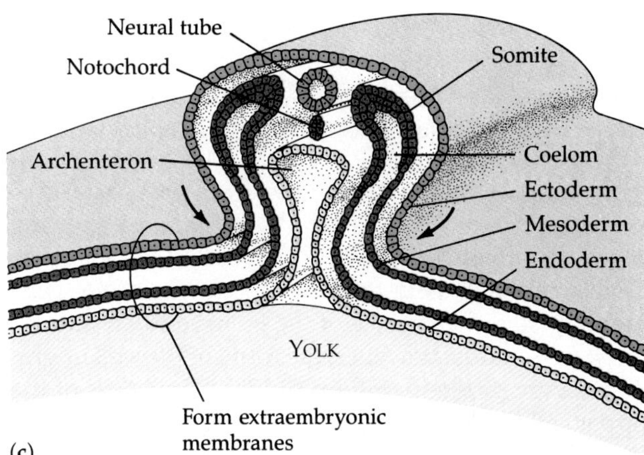

Figure 43.13
Cleavage and gastrulation in a chick embryo. (a) For the eggs of birds and reptiles, which store a very large amount of yolk, cleavage is meroblastic, or incomplete. Cell division is restricted to a small cap of cytoplasm at the animal pole. Cleavage produces a blastodisc that rests on the large, undivided mass of yolk. The blastodisc becomes arranged into two layers (epiblast and hypoblast) that bound the blastocoel, forming the avian version of a blastula. **(b)** During gastrulation, some cells of the epiblast migrate (arrows) into the interior of the embryo through the primitive streak, shown here in transverse section. Some of these cells move laterally to form mesoderm, while others invade the hypoblast to contribute to endoderm. **(c)** The rudimentary gut (archenteron) is formed when lateral folds pinch the embryo from the yolk. About midway along its length, the embryo will remain attached to the yolk by a yolk stalk. Primary germ layers excluded from the embryo proper contribute to the system of extraembryonic membranes that support further development.

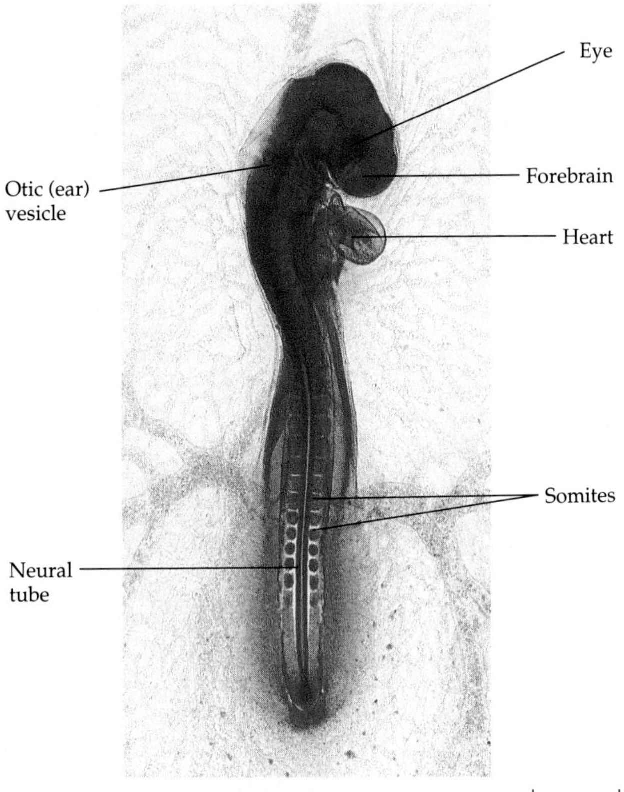

Figure 43.14
Organogenesis in a chick embryo. Rudiments of most major organs have already formed in this chick, which is about 56 hours old (LM).

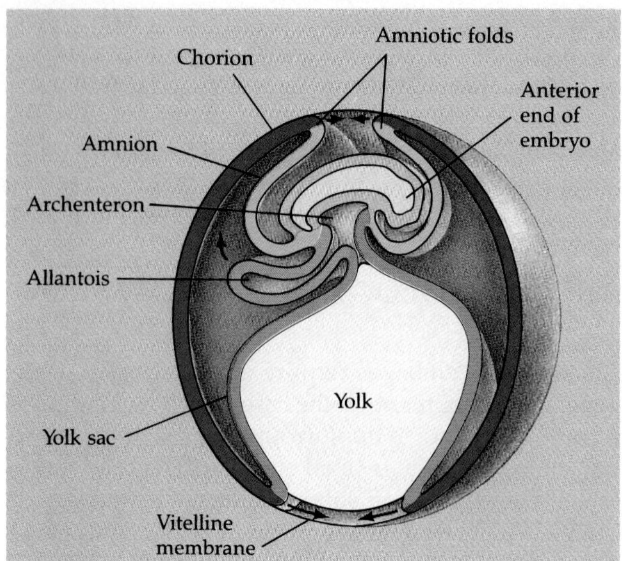

Figure 43.15
The development of extraembryonic membranes in a chick. Each of the four membranes, which provide support services for the embryo, develops from epithelial sheets that are external to the embryo proper. The yolk sac expands over the surface of the yolk mass. Cells of the yolk sac will digest yolk, and blood vessels that develop within the membrane will carry nutrients into the embryo. Lateral folds of extraembryonic tissue extend over the top of the embryo and fuse to form two more membranes, the amnion and the chorion, which are separated by an extraembryonic extension of the coelom. The amnion encloses the embryo in a fluid-filled amniotic sac, protecting the embryo from desiccation and buffering it against mechanical shocks. The fourth membrane, the allantois, originates as an outpocketing of the embryo's hindgut. The allantois is a sac that extends into the extraembryonic coelom separating the chorion and amnion. It functions as a disposal sac for uric acid, the insoluble nitrogenous waste of the embryo. As the allantois continues to expand, it presses the chorion against the vitelline membrane, the inner lining of the eggshell. Together, the allantois and chorion form a respiratory organ that serves the embryo. Blood vessels that form in the epithelium of the allantois transport oxygen to the embryonic chick. The extraembryonic membranes of reptiles and birds are adaptations associated with the special problems of development on land.

Gastrulation, as in the frog embryo, involves cells moving from the surface of the embryo to an interior location. In birds, however, the route of this cell migration is very different (Figure 43.13b). Some cells of the upper cell layer (epiblast) move toward the midline of the blastodisc, then detach and move inward toward the yolk. The medial movement on the surface and the inward movement of cells at the blastodisc's midline produce a groove called the **primitive streak.** As the streak lengthens, it marks what will become the bird's anterior-posterior axis. The primitive streak is functionally equivalent to the lip of a frog blastopore, but it is arranged like a linear tuck rather than a ring. The cells that pass through the streak move back laterally again, producing a middle mesodermal layer in the cavity between the epiblast and hypoblast. However, some of the migrating cells dive down and add to the hypoblast, which becomes the endoderm (lying on top of the yolk). Cells remaining in the epiblast form the ectoderm. The bird embryo now has its three primary germ layers, which at this stage form a stack. Eventually the borders of this embryonic disc fold downward and come together, pinching the embryo into a three-layered tube that remains attached to the yolk by a stalk at midbody (Figure 43.13c). Neural tube formation, development of the notochord and somites, and other events in organogenesis occur much as in the frog embryo (Figure 43.14).

Notice in Figure 43.13c that only part of each primary germ layer contributes to the embryo itself. The tissue layers that form external to the embryo develop into four **extraembryonic membranes** that support further embryonic development within the egg. These four membranes are the **yolk sac,** the **amnion,** the **chorion,** and the **allantois** (Figure 43.15).

Mammalian Development

Mammalian eggs are fertilized in the oviduct, and the earliest stages of development occur while the embryo completes its journey down the oviduct to the uterus (see Chapter 42). In contrast to the large, yolky eggs of birds and reptiles, the egg of a placental mammal is quite small, storing little in the way of food reserves. Cleavage of the mammalian zygote is therefore holoblastic, but gastrulation and early organogenesis follow a pattern similar to that of birds and reptiles. (Recall from Chapter 30 that mammals descended from reptilian stock during the early Mesozoic era.)

Cleavage is relatively slow in mammals. In the case of humans, the first division is complete about 36 hours after fertilization, the second division at about 60 hours, and the third division at about 72 hours. The mammalian zygote has no apparent polarity. The cleavage planes seem to be randomly oriented, and the blastomeres are equal in size.

Further development of the human embryo is shown in Figure 43.16. By 5 days after fertilization, the embryo has over 100 cells. At this stage, the cells form the tight junctions characteristic of a compact epithelium (see Figure 7.34a), which is arranged around a central cavity. This is the embryonic stage known as the **blastocyst** (see Chapter 42). Protruding into one end of the blastocyst cavity is a cluster of cells called the **inner cell mass,** which will subsequently develop into the embryo proper and some of the extraembryonic membranes. The outer epithelium surrounding the cavity is the **trophoblast,** which will form the fetal portion of the placenta.

The embryo reaches the uterus by the blastocyst stage, implanting about one week after fertilization. The trophoblast secretes enzymes that enable the blastocyst to penetrate the uterine lining. Bathed in blood spilled from eroded capillaries in the endometrium (uterine lining), the trophoblast thickens and extends fingerlike projections into the surrounding maternal tissue. The placenta eventually forms from this proliferated trophoblast and the region of endometrium it invades (see Chapter 42). At about the time the blastocyst implants in the uterus, the inner cell mass forms a flat **embryonic disc** that will develop into the embryo proper. Homologous to the blastodisc of birds and reptiles, the embryonic disc has an epiblast and hypoblast. Gastrulation occurs by the inward movement of cells from the upper layer through a primitive streak to form mesoderm and endoderm, just as it does in the chick.

Four extraembryonic membranes homologous to those of reptiles and birds form during mammalian development (see Figure 43.15). The chorion, which develops from the trophoblast, completely surrounds the embryo and the other extraembryonic membranes. The

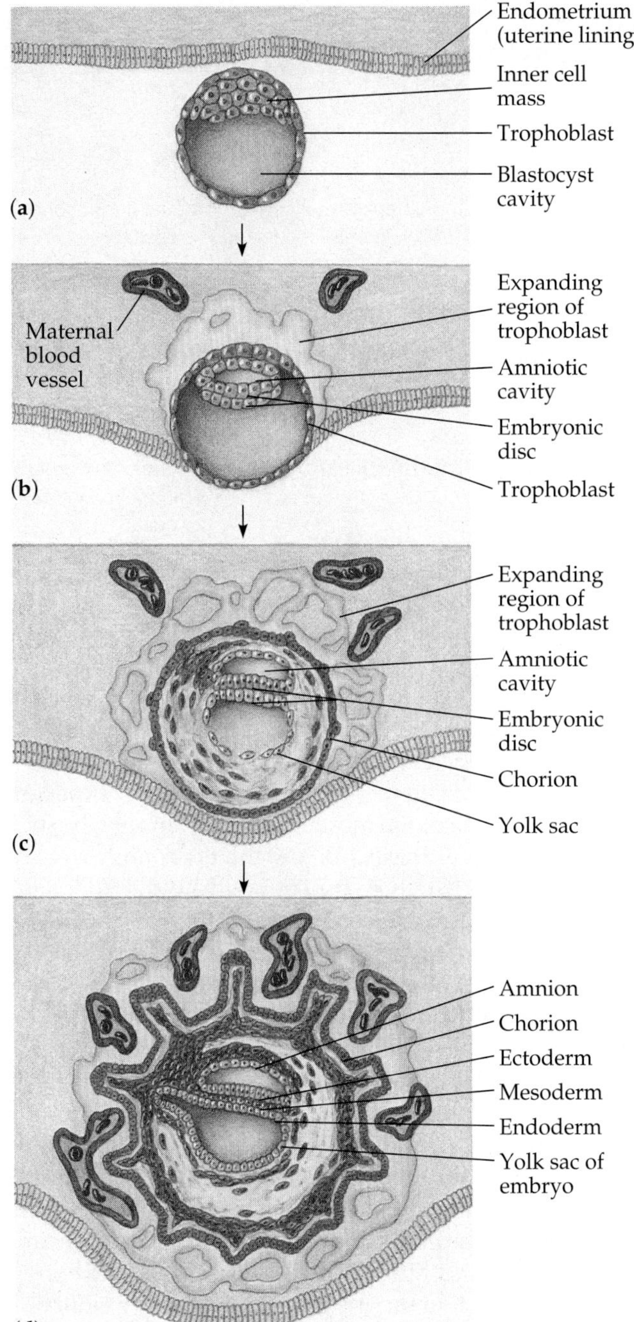

Figure 43.16
Early development of a human embryo and its extraembryonic membranes. (**a**) Cleavage produces a blastocyst, consisting of a trophoblast surrounding a cavity and an inner cell mass protruding into the cavity from one side. The blastocyst implants in the uterine lining. Although the placental mammal develops within the uterus of the mother rather than within a shelled and yolk-rich egg, the extraembryonic membranes in birds and reptiles have persisted in the evolution of mammals. (**b**) and (**c**) The mammalian embryo proper develops from a flat embryonic disc homologous to the blastodisc of the chick. (**d**) Gastrulation occurs by the inward movement of mesoderm and endoderm through a primitive streak.

Table 43.1 Derivatives of the Primary Germ Layers in Mammals

Ectoderm	Mesoderm	Endoderm
All nervous tissue	Sketetal, smooth, and cardiac muscle	Epithelium of digestive tract (except that of oral and anal cavities)
Epidermis of skin and epidermal derivatives (hairs, hair follicles, sebaceous and sweat glands, nails)	Cartilage, bone, and other connective tissues	Glandular derivatives of digestive tract (liver, pancreas)
Cornea and lens of eye	Blood, bone marrow, and lymphoid tissues	Epithelium of respiratory tract
Epithelium of oral and nasal cavities, paranasal sinuses, and anal canal	Endothelium of blood vessels and lymphatics	Thyroid, parathyroid, and thymus glands
Tooth enamel	Lining of body cavity	Epithelium of reproductive ducts and glands
Epithelium of pineal and pituitary glands and adrenal medulla	Organs of urogenital system (ureters, kidneys, gonads, and reproductive ducts)	Epithelium of urethra and bladder

Source: Adapted from Elaine N. Marieb, *Human Anatomy and Physiology,* 2nd ed. Redwood City, CA: Benjamin/Cummings, 1992.

amnion begins as a dome above the embryonic disc and encloses the embryo in a fluid-filled amniotic cavity. (The fluid from this cavity is the "water" expelled from the vagina of the mother when the amnion breaks just prior to childbirth.) Below the embryonic disc is another fluid-filled cavity enclosed by the yolk sac. Although this cavity contains no yolk, the membrane that surrounds it is given the same name as the homologous membrane in birds and reptiles. The yolk sac membrane of mammals is a site of early formation of blood cells, which later migrate into the embryo proper. The fourth extraembryonic membrane, the allantois, develops as an outpocketing of the embryo's rudimentary gut, as it does in the chick. The allantois is incorporated into the umbilical cord, where it forms blood vessels that transport oxygen and nutrients from the placenta to the embryo and rid the embryo of carbon dioxide and nitrogenous wastes. Thus, the extraembryonic membranes of shelled eggs, where embryos are nourished with yolk, were conserved as mammals diverged from reptiles in the course of evolution, but with modifications suited for development within the reproductive tract of the mother.

Organogenesis begins with the formation of the neural tube, notochord, and somites. By the end of the first trimester of human development, rudiments of all the major organs have developed from the three primary germ layers, as summarized in Table 43.1.

MECHANISMS OF DEVELOPMENT

We have seen that the basic body plan of an animal is established quite early in embryonic development. Describing a sequence of changes, however, does not explain how those changes occur. By manipulating embryos in a variety of experiments, developmental biologists have discovered some of the general mechanisms at work as diverse animals take shape. These discoveries are integrated by two basic concepts:

1. *The cytoplasmic organization of the unfertilized egg leads to regional differences in the early embryos of many animal species.* By partitioning the heterogenous cytoplasm of the egg, cleavage parcels out different substances (mRNA, proteins, etc.) to different blastomeres. These local differences in cytoplasmic composition influence gene expression and help determine the developmental fates of cells in the early embryo (Figure 43.17).

2. *Cell-cell interactions compound the influence of location on a cell's developmental fate.* One cell of an embryo can induce cytoplasmic changes that affect gene expression in a neighboring cell. This is accomplished by transmission of chemical signals or, if the cells are actually in contact, by membrane interactions (Figure 43.17b).

Keep these two related concepts in mind as we sample some experimental studies of animal development.

Polarity of the Embryo

A bilaterally symmetric animal has an anterior-posterior axis, a dorsal-ventral axis, and left and right sides (see Chapter 29). In some cases, these three polarities are already established at the time of fertilization. In the frog, for example, the embryonic axes are set before the first cleavage division.

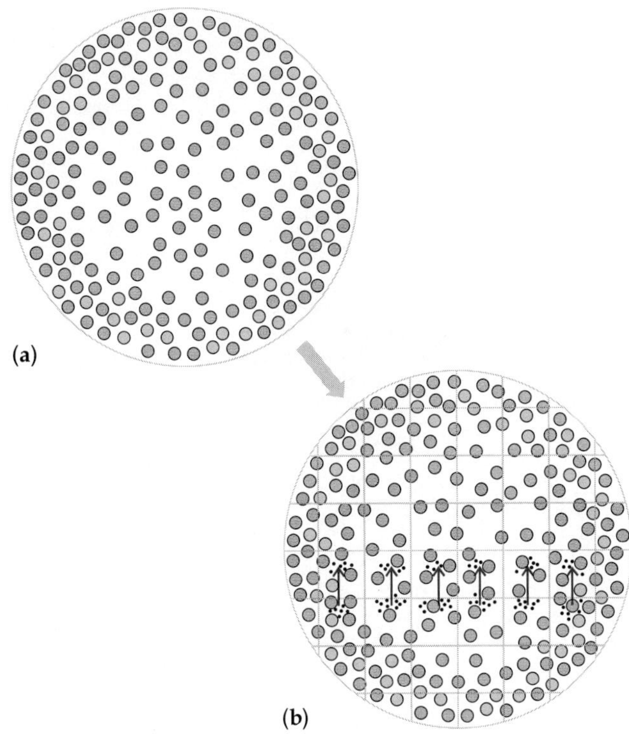

Figure 43.17
Two basic principles of early development. (**a**) The first principle is based on the fact that the unfertilized eggs of most animals have an uneven distribution of cytoplasmic substances, symbolized here by dots of different colors. (**b**) Division of this heterogeneous egg cytoplasm during cleavage results in different cells of the embryo having different cytoplasmic compositions. This sets the stage for regional variation in gene expression. A second basic principle of early development is the role of cell-cell interactions in determining the developmental fates of cells. The arrows represent the movement of chemical signals (small black dots) that enable certain cells to induce developmental events in nearby cells.

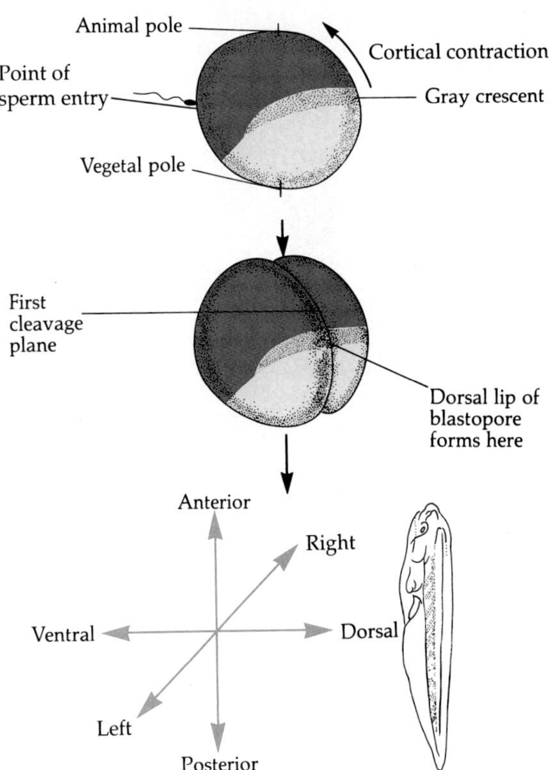

Figure 43.18
The establishment of embryonic axes in amphibians. At the time of fertilization, a gray crescent forms on the side of the zygote opposite the point where the sperm cell entered. Rotation of the pigmented cortex relative to the underlying yolk produces the crescent by exposing lighter-colored cytoplasm beneath the melanin granules. The first cleavage division bisects the gray crescent, and the dorsal lip of the blastopore will later form where the crescent was located. Thus, all three axes of the embryo are established before the zygote begins cleavage.

Recall that the eggs of most animals are already polarized along an animal-vegetal axis, owing to the heterogeneous distribution of cytoplasmic components within the egg. The amphibian egg has yolk platelets concentrated near the vegetal pole and cortical melanin granules concentrated near the animal pole. A rearrangement of the amphibian egg cytoplasm occurs at the time of fertilization (Figure 43.18). The cortex (outer cytoplasm) of the egg is rotated toward the point of sperm entry, probably because the centriole brought into the egg by the sperm cell reorganizes the cytoskeleton. Contraction of the cortex pulls the edge of the pigmented layer that is opposite the point of sperm entry toward the animal pole, exposing lighter-colored cytoplasm that had been masked by the pigment. This contraction forms the **gray crescent,** a half-moon-shaped mark near the equator of the egg, on the side opposite the point of entry. The first cleavage division will bisect the gray crescent, and the dorsal lip of the blastopore will later form where the gray crescent was located in the zygote.

We see that all three axes of the embryo are defined before the onset of cleavage. The animal-vegetal axis of the egg becomes the anterior-posterior axis of the embryo, and the location of the gray crescent determines both the dorsal-ventral axis and the left-right axis. When the egg is experimentally manipulated to prevent the rotation of the cortex that forms the gray crescent, gastrulation is abnormal, and axial structures such as the neural tube fail to develop.

Although most animals have embryonic axes that are fixed at the zygote stage or during early cleavage, there are important exceptions, including mammals. We have seen that mammalian eggs lack apparent polarity and that cell divisions are randomly oriented during early cleavage. We cannot distinguish one end of the mammalian embryo from the other until the blastocyst stage.

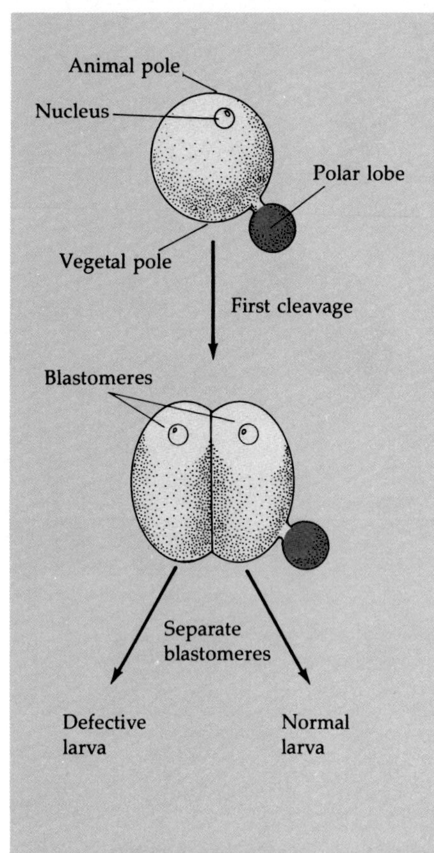

(a)

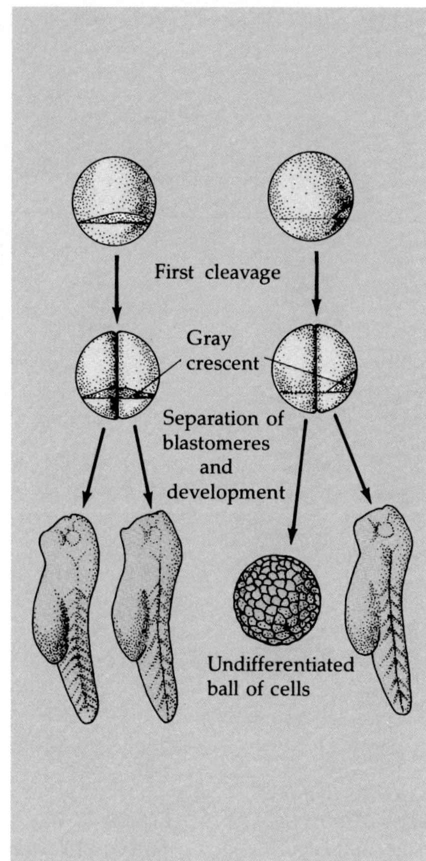

(b)

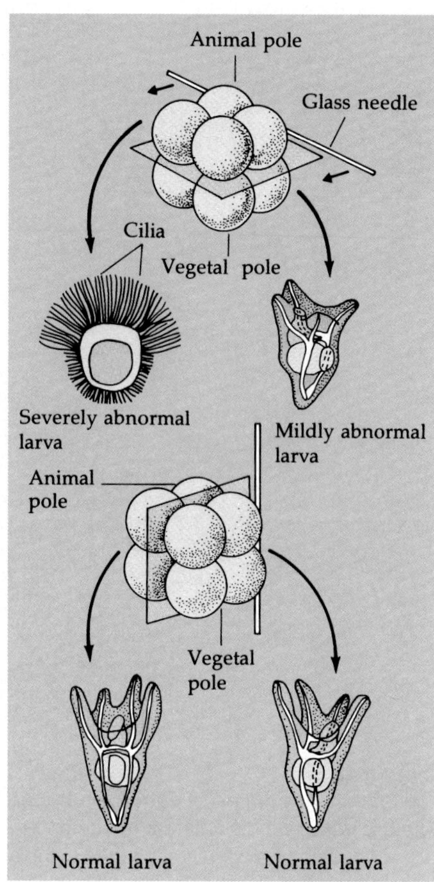

(c)

Figure 43.19
Cytoplasmic determinants. (**a**) In the development of the mollusk *Dentalium,* a polar lobe is restricted to one of the two blastomeres formed by the first cleavage. The lobe apparently contains determinants required for the development of certain structures. Thus, the cell lacking the lobe has already lost its totipotency after the very first cell division. (**b**) The first cleavage division of an amphibian zygote normally divides the gray crescent between the two blastomeres, which retain their totipotency after the division. However, in experiments where pressure is used to divert the cleavage plane from the crescent, only the blastomere obtaining material from the crescent develops normally. (**c**) Cytoplasmic determinants in the sea urchin egg apparently have a polar distribution. At the eight-cell stage, splitting the embryo vertically results in two normal larvae, but splitting the embryo horizontally produces a mildly defective larva and a severely defective one.

Localized Cytoplasmic Determinants

You have already learned that the organization of the egg cytoplasm has an important influence on early development of most animals. The key point is worth reinforcing here: If cytoplasmic components are unevenly distributed within the egg, the early cleavage divisions will partition the zygote into blastomeres that differ in cytoplasmic makeup (see Figure 43.17). The localized substances sequestered within specific blastomeres may function as **cytoplasmic determinants** that fix the developmental fates of different regions of the embryo very early, presumably by controlling gene expression (see Chapter 18). In the development of many mollusks, for example, the zygote forms a polar lobe that is restricted to one of the two blastomeres at the first cleavage division (Figure 43.19a). In experiments where the two blastomeres are pulled apart, only the cell with the polar lobe develops into a normal embryo. Moreover, even that cell will fail to develop normally if the polar lobe is removed; the larva will lack a heart, eyes, and several other organs derived from mesoderm in mollusks. Apparently, the polar lobe contains cytoplasmic determinants required for the subsequent development of these mesodermal structures.

Even if cytoplasmic determinants are asymmetrically distributed in the zygote, the first cleavage may occur along an axis that produces two blastomeres of equal developmental potential. In amphibians, for instance, blastomeres experimentally separated at the two-cell stage develop into normal tadpoles. The cells are said to be **totipotent,** meaning they retain the potential to form all parts of the animal. However, if an amphibian zygote is experimentally manipulated so that the first cleavage plane misses the gray crescent instead of bisecting it, only the blastomere that gets the gray crescent will develop into a normal tadpole (Figure 43.19b). Thus, a zygote's characteristic pattern of cleavage, along with its distribution of cytoplasmic determinants, affects the destiny of cells in the embryo.

Mosaic and Regulative Development

Development can be described as mosaic or regulative, depending on how early the blastomeres lose their totipotency. The development of a mollusk from an egg with a polar lobe is an example of **mosaic development,** because each cell is rigidly set to give rise to specific parts of the embryo, much as each piece of a mosaic has a fixed place in the overall picture. In **regulative development,** cells are totipotent longer, and their fates can be experimentally altered. The distinction is somewhat arbitrary, since the developmental fate of most cells eventually becomes narrowed in all animals. For example, sea urchin development is regulative up to the four-cell stage, but the third cleavage plane divides cytoplasmic determinants in such a way that some of the cells lose their totipotency (Figure 43.19c).

Mammalian embryos remain regulative in development much longer than most other animals. The cells do not begin to lose their totipotency until they become arranged into the trophoblast and inner cell mass of the blastocyst. This long regulative period is probably related to the egg's lack of polarity and the apparent randomness of the early cleavage planes. Up to the eight-cell stage, the blastomeres of a mouse embryo all look alike, and indeed, each can form a complete embryo if isolated. Conversely, two mouse embryos at the eight-cell stage can be fused into an enlarged morula that, when transplanted to a uterus, will develop into a normal mouse that combines the traits of two sets of parents. Embryos with mosaic development cannot adjust to similar kinds of tampering.

Fate Maps and the Analysis of Cell Lineages

In embryos where the axes are defined early in development, it should be possible to determine which parts of the embryo will be derived from each region of the zygote or blastula. In the 1920s, German embryologist W. Vogt charted a **fate map** for the amphibian embryo (Figure 43.20a). He used vital (nontoxic) dyes to label cells of different regions of the blastula surface with different colors and later sectioned the embryo to see where the colors turned up. By comparing Figures 43.20a and 43.10, notice that different regions of the embryo become rearranged between the blastula and gastrula stages. This is because of the complex movements that occur during gastrulation.

New methods and clever choices of experimental organisms have enabled developmental biologists to make even more detailed fate maps. Various techniques are used to mark an individual blastomere during cleavage and then follow the marker as it is distributed to all the mitotic descendants of that cell (Figure 43.20b). Fate mapping at this level is called cell lineage analysis. (Chapter 34 discussed lineage analysis of plants.) In the case of one organism, the nematode *Caenorhabditis elegans,* it has been possible to map the developmental fate of every cell, beginning with the first cleavage division of the zygote (Figure 43.20c).

Morphogenetic Movements

The migration of cells and changes in cell shape are fundamental processes in the development of an animal's form. Three properties of cells—extension, contraction, and adhesion—are involved in the various morphogenetic movements that shape the embryo, including gastrulation and formation of the neural tube.

Changes in the shape of a cell usually involve reorganization of the cytoskeleton. Consider, for example, how the cells of the neural plate become distorted in shape during formation of the neural tube (Figure 43.21). First, microtubules lengthen the cells of the plate in the dorsal-ventral axis of the embryo. Then, ordered arrays of microfilaments contract the apical ends of the cells, giving them a wedge shape that deforms the epithelium inward. Similar shape changes are observed for other invaginations (inpocketings) and evaginations (outpocketings) of the tissue layers throughout development.

Amoeboid movement based on extension and contraction is also important in morphogenesis. In the gastrulation of some embryos, as we have seen, invagination is initiated by a wedging of cells on the surface of the blastula, but the penetration of cells deeper into the embryo involves the extension of filopodia by cells on the leading edge of the migrating tissue. These amoeboid cells drag the sheet of epithelium into the blastocoel to form the endoderm and mesoderm of the embryo. There are also many cases in morphogenesis where amoeboid cells migrate individually, as when the cells of the neural crest disperse to various parts of the embryo.

The extracellular matrix helps guide cells in their morphogenetic movements. The matrix includes adhesive substances and fibers that may function as tracks, directing migrating cells along a particular route. One family of extracellular glycoproteins, the **fibronectins,** helps cells adhere to their substratum as they migrate (Figure 43.22). There is a close relationship between the orientation of fibronectin fibrils in the extracellular matrix and that of contractile microfilaments of the cytoskeleton within migrating cells. The extracellular fib-

Figure 43.20
Examples of fate maps and cell lineage analysis. (**a**) Cellular fates of a frog embryo were determined by marking different regions of the blastula surface with dyes of various colors, then determining the locations of dyed cells in the gastrula or at later stages of development, such as at this neural tube stage. The blastula is viewed from the left side, with the future anterior end at the upper left. (**b**) The drawings on the top depict 64-cell embryos of a tunicate, an invertebrate chordate (see Chapter 30). An individual cell can be injected with a dye used as a marker, enabling a researcher to determine which cells in a later embryo are derived from the marked cells. The two light micrographs of larvae contrast the regions that develop from two different blastomeres indicated by the two drawings. (**c**) The nematode *Caenorhabditis elegans* is transparent at all stages of its development, making it possible for researchers to trace the lineage of every cell, from the zygote to the 2000 cells of the adult worm (LM). The diagram shows a detailed lineage only for the intestine, which is derived exclusively from one of the first four cells formed during cleavage.

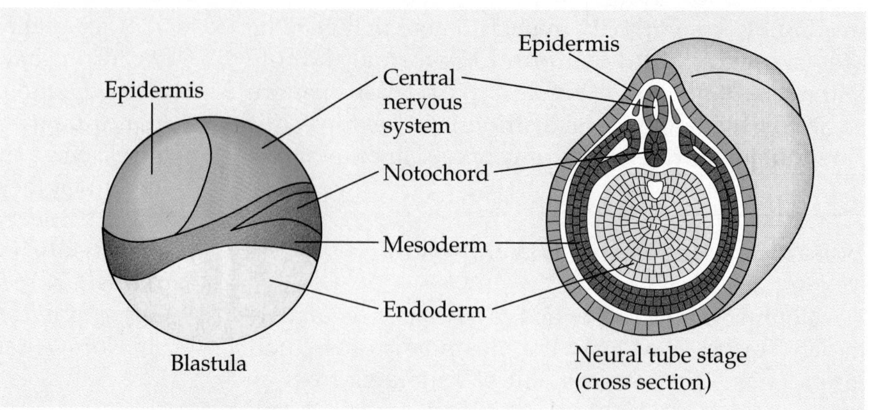

(**a**) Fate map of a frog embryo

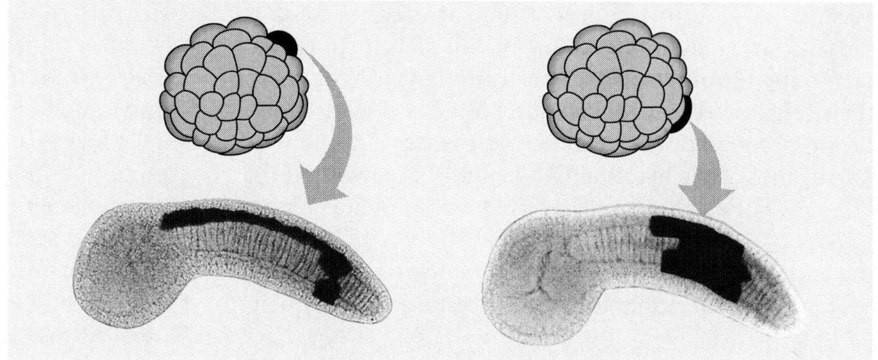

(**b**) Cell lineage analysis in a tunicate

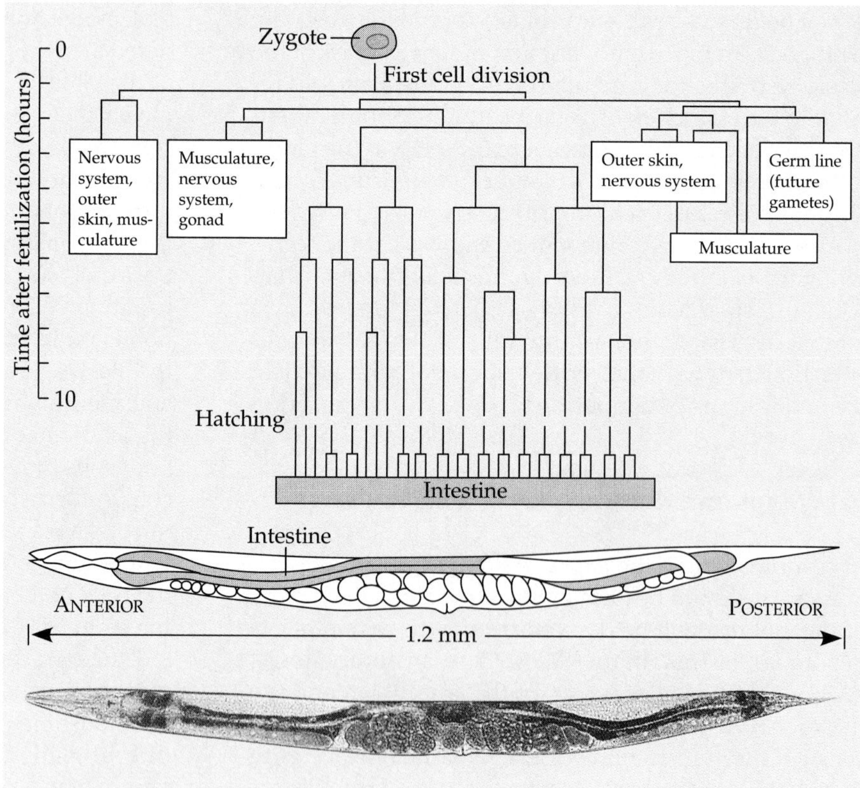

(**c**) Cell lineages of *Caenorhabditis elegans*

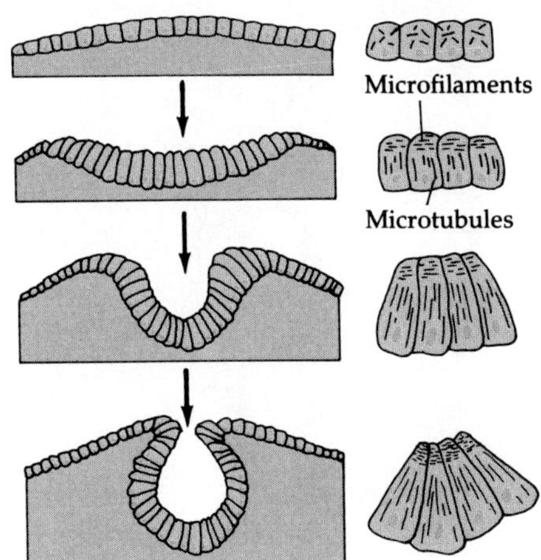

Figure 43.21
Change in cellular shape during morphogenesis. Reorganization of the cytoskeleton is believed to function in invagination and evagination of epithelial layers. In the formation of the neural tube, microtubules elongate the cells of the neural plate, then microfilaments at the apexes of the cells contract to deform the cells into wedge shapes.

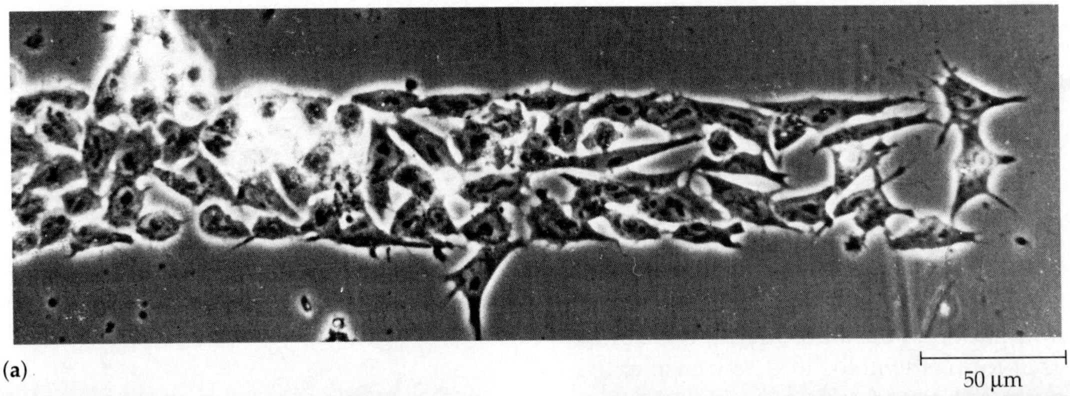

(a)

50 μm

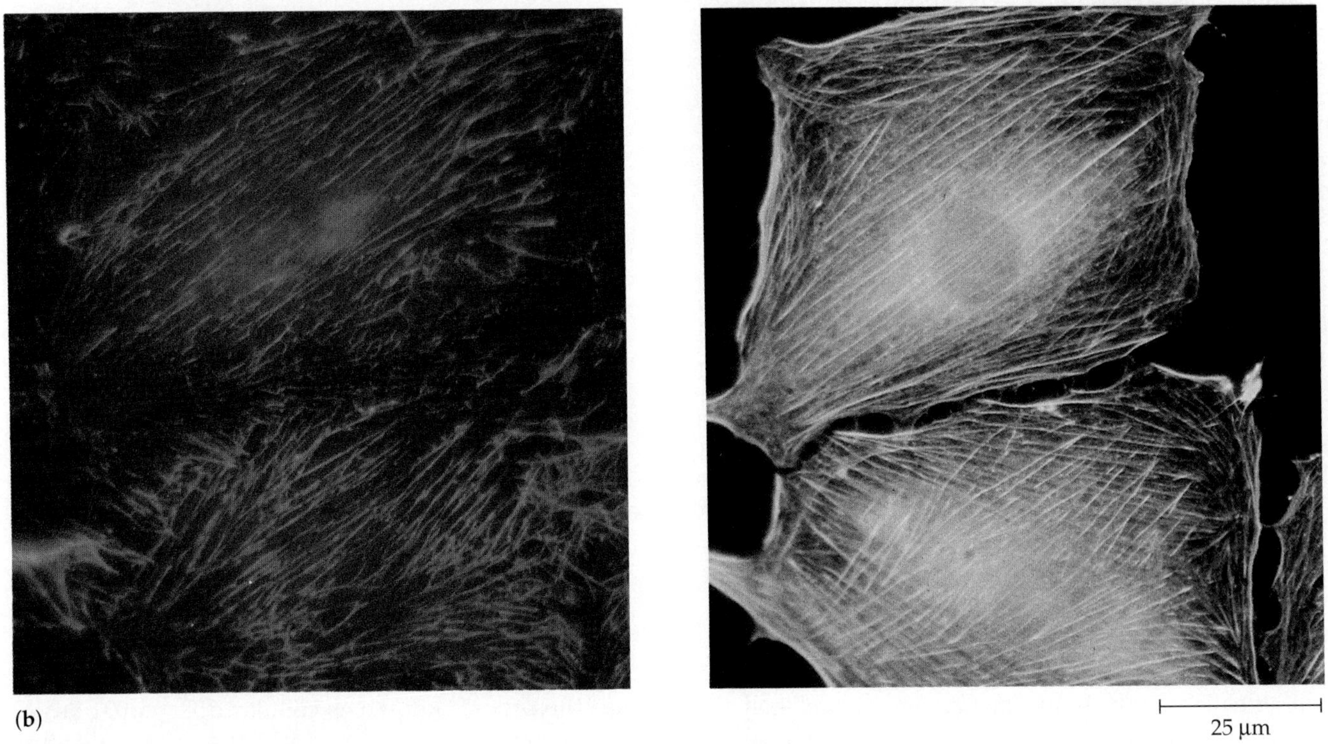

(b)

25 μm

Figure 43.22
The extracellular matrix and cell migration. (**a**) Cells from the neural crest migrate along a strip of fibronectin fibrils placed on an artificial substratum (LM). (**b**) Two different fluorescent dyes have been used to stain the fibronectin fibrils of the substratum (left) and the actin microfilaments (right) within these two cells (LMs). Notice that the orientations of the intracellular microfilaments and the extracellular fibrils correspond.

rils are themselves ordered by the orientation of the cytoskeleton in the cells that secrete the extracellular materials. In this way, one group of cells can influence the path along which another group of cells migrates during the development of tissues and organs.

Equally important is the intercellular glue that holds certain cells together as tissues and organs are shaped. For example, if the cells of a gastrula are experimentally dissociated from one another and then allowed to reaggregate, they will sort into three layers, with endoderm on the inside, ectoderm on the outside, and mesoderm in between. Contributing to this selective association are substances on the surfaces of cells called **cell adhesion molecules (CAMs),** which vary in either amount or chemical identity from one type of cell to another.

Induction

After the morphogenetic movements of gastrulation produce a three-layered embryo, interactions among the layers are important in the origin of most organ rudiments. The ability of one group of cells to influence the development of another group of cells is known as **induction.**

It is induction by the rudimentary notochord that causes the dorsal ectoderm of the gastrula to thicken into a neural plate. If chordamesoderm, the dorsal mesoderm that forms the notochord, is transplanted from its normal position to a position beneath the ectoderm in some other part of the embryo, a neural plate will form and give rise to a neural tube in an abnormal location. In a series of transplant experiments in the 1920s, German biologist Hans Spemann and his student Hilde Mangold discovered that the dorsal lip of the blastopore is responsible for setting up the interaction between chordamesoderm and the overlying ectoderm. For example, transplanting a small piece of the dorsal lip to some other part of the surface of an early amphibian gastrula results in dual notochords, which induce the formation of two neural tubes. Spemann referred to the dorsal lip of the blastopore as the primary *organizer* of the embryo because of its role in the early stages of organogenesis. Developmental biologists have recently identified a growth factor called **activin** as the chemical signal that most likely causes the cells that move inward through the dorsal lip to form chordamesoderm. The inducing substance that subsequently causes overlying ectoderm to form the neural plate is not yet known.

The induction of dorsal ectoderm for development into the neural tube is just the first of many cell-cell interactions that transform the primary germ layers into organ systems. For example, development of the vertebrate eye involves a series of reciprocal inductions between ectoderm and evaginations of the rudimentary brain (Figure 43.23).

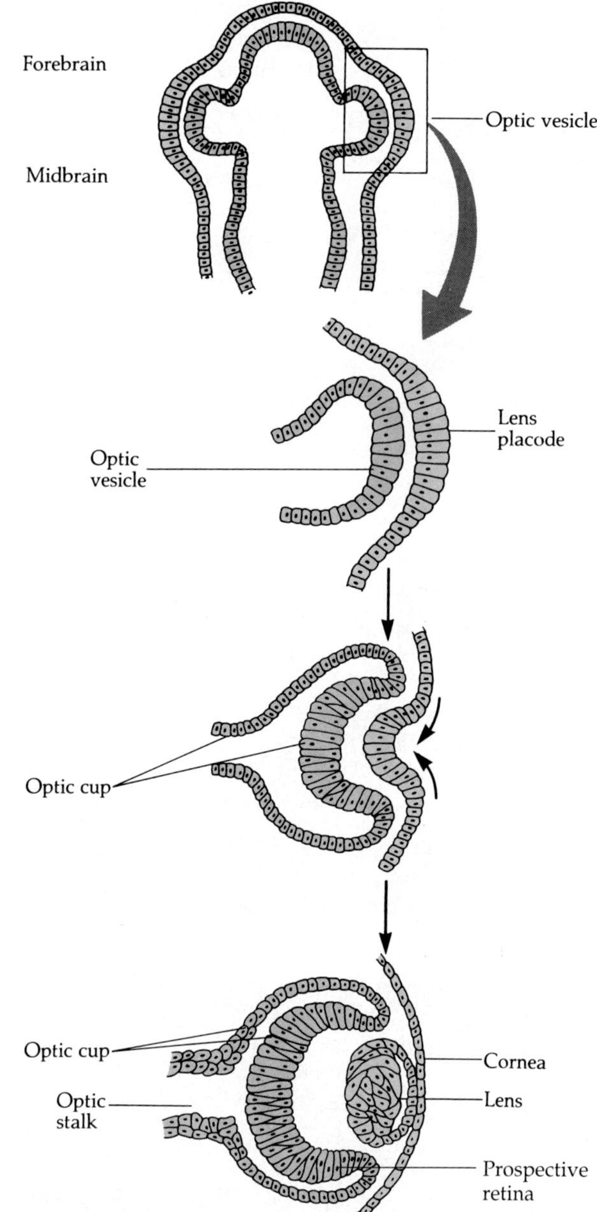

Figure 43.23
Induction during eye development. In a sequence of reciprocal inductions, the eye forms from two different layers of cells. The optic vesicle, an outgrowth of the rudimentary brain, is required for adjacent ectoderm to form a thickened region called a lens placode, which in turn induces invagination of the optic vesicle to form a cup. The optic cup then induces the lens placode to invaginate, forming the lens, which in turn induces development of the cornea.

Differentiation

As the tissues and organs of an embryo take form, their cells begin to specialize in structure and function during the process known as cellular differentiation. At the microscopic level, the first signs of specialization are alterations in cellular structure. At the molecular level, differentiation is heralded by the appearance of **tissue-specific proteins,** proteins found only in a certain type of cell.

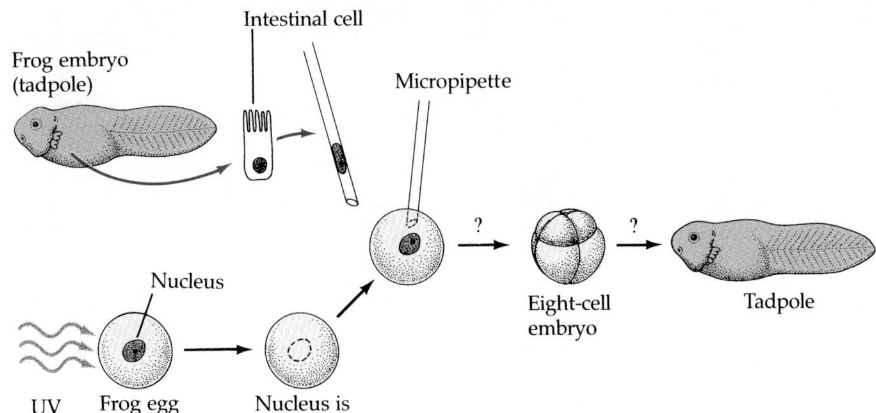

Figure 43.24
Nuclear transplantation. After the frog egg nucleus is destroyed by ultraviolet (UV) radiation, a nucleus from a more advanced developmental stage is inserted into the egg to test whether nuclei change irreversibly as cells begin to differentiate. The earlier the developmental stage from which the nucleus comes, the more likely it will support development. Nuclei from very early stages frequently prove to be totipotent, whereas nuclei from late stages (such as a tadpole) rarely are.

During differentiation, cells become specialists at making certain proteins. Developing lens cells in vertebrates, for example, synthesize large quantities of crystallins, proteins that aggregate to form transparent fibers that give the lens the ability to transmit and focus light. Because no other vertebrate cell type makes crystallins, these proteins can be used to follow the progress of differentiation by lens cells. The cue for the immature lens cell to turn on its crystallin genes is induction by the still immature retina (see Figure 43.23). In response to chemical signals from the rudimentary retinal cells, specific mRNA molecules coding for crystallin are transcribed and accumulate in the immature lens cell's cytoplasm. Then synthesis of crystallin begins, and the lens cells devote 80% of their capacity for protein synthesis to making this one type of protein. The cells elongate and flatten as they accommodate the crystallin fibers.

Genomic Equivalence If a lens cell makes crystallin but a blood cell does not, we must conclude that the lens cell expresses a gene that is not expressed in the blood cell. We could account for such differences if cells lost nonessential genes as they differentiated, but most evidence supports the conclusion that nearly all cells of an organism have **genomic equivalence**—that is, they all have the same genes. (There are a few exceptions, as discussed in Chapter 18.) What happens to these genes as a cell begins to differentiate? We can shed some light on this question by asking whether differentiation is reversible.

One way to examine the reversibility of differentiation is to replace the nucleus of a normal egg or zygote with the nucleus of a differentiated cell. If genes are irreversibly inactivated during differentiation, the transplanted nucleus will not be able to guide the development of a normal embryo. The pioneering experiments in nuclear transplantation were carried out by American embryologists Robert Briggs and Thomas King during the 1950s and were later extended by John Gurdon. These investigators removed or destroyed the nuclei of frog and toad egg cells, then transplanted nuclei from embryonic and tadpole cells of the same species into the anucleated eggs (Figure 43.24). The ability of the transplanted nuclei to support normal development turned out to be inversely related to the age of the donor embryos. If the nuclei came from the relatively undifferentiated cells of an early embryo, most of the recipient eggs developed into tadpoles. But with nuclei from the differentiated intestinal cells of a tadpole, fewer than 2% of the eggs developed into normal tadpoles, and most of the embryos failed to make it through even the earliest stages of embryonic development.

Developmental biologists still argue over the meaning of these results, but most agree on two conclusions. First, nuclei *do* change in some way as they prepare for differentiation. (Recall from Chapter 18 that genomes are rearranged in various ways.) Second, this change is not always irreversible, implying that the nucleus of a differentiated cell has all the genes required for making all other parts of the organism. The cloning of plants from somatic cells supports this conclusion (see Chapter 34). The cells of the body differ in structure and function not because they contain different genes, but because they express different portions of a common genome. Possible mechanisms for this tissue-specific gene expression were discussed in Chapter 18.

Determination Even before there is molecular or cytological evidence of differentiation, the cell may already be committed to following a particular course of development. A cell is said to be **determined** when it is possible to predict its developmental fate. One way to find out whether cells have already been determined is to transplant them to some other part of the embryo to see if their developmental fate can be altered. For example, the larvae of *Drosophila* and other insects contain structures called **imaginal disks,** islands of cells destined to form various organs when the larva undergoes metamorphosis to form an adult insect. One imaginal disk may form a leg, another may give rise to an antenna, and still another may develop into an eye. In the larva, the cells of the imaginal disks are not yet

Figure 43.25
Homeotic mutations and abnormal pattern formation in *Drosophila.* Homeotic mutations affect the development of the adult body parts radically during metamorphosis of a fly larva. These micrographs contrast the heads of two fruit flies (SEMs). Where small antennae are located in the normal fruit fly (left), one homeotic mutant has legs (right).

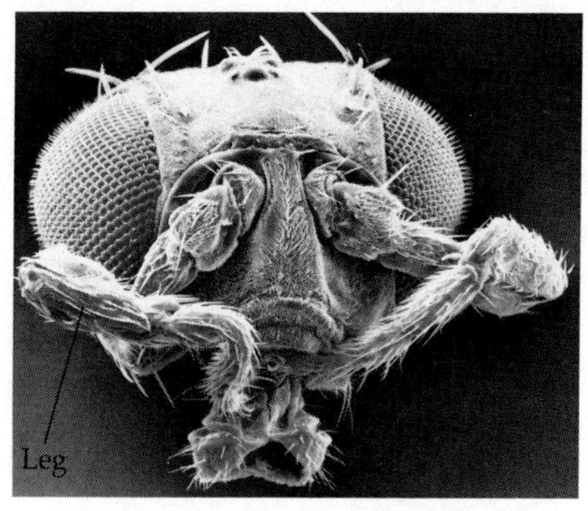

differentiated, but they *are* determined; if a disk that normally forms an antenna is experimentally removed and replaced with one destined to form a leg, the adult fly will have a leg projecting from its head where an antenna is normally found. (We will see in the next section that genetic changes called homeotic mutations can produce similarly bizarre results.) Thus, determination has a "memory" aspect. A cell that has been determined passes that developmental commitment on to daughter cells, and they "remember" what they are supposed to become, even if they are moved to a different environment.

Determination seems to be a serial process. Beginning with a totipotent zygote, the developmental options of cells become more and more narrow as development progresses. When differentiation finally occurs, it is the product of a cell's developmental history extending back to early cleavage. If dorsal ectoderm of an early amphibian gastrula is replaced with ectoderm from some other location, the transplanted tissue will form a neural plate above the notochord. We conclude from this that the transplanted ectoderm had not yet been determined to develop into epidermis and is still capable of being induced to form the neural plate. However, if ectoderm from a late-stage gastrula is transplanted to the same location, it will not respond to induction, and no neural plate will form. At some point during the gastrula stage, the ectoderm has become more restricted in its developmental potential. However, there is still some flexibility. For example, ectoderm no longer capable of forming a neural plate may still be able to differentiate into either epidermis or the lens of an eye, depending on its location. If an optic vesicle is moved to an abnormal location early enough, it will induce the overlying ectoderm to form a lens. If the manipulation is performed later, epidermis rather than a lens will form over the transplanted optic vesicle.

Determination apparently involves control of the genome by the cytoplasmic environment of the cell. Cytoplasmic cues can begin to determine cells as early as the first cleavage division, as in the mosaic development of mollusks (see Figure 43.19a). By partitioning the heterogeneous cytoplasm of an egg, cleavage exposes the nuclei of cells to different cytoplasmic determinants that may affect which genes will be expressed later, when the cell begins to differentiate. Morphogenetic movements also play a role in determination and differentiation, by placing cells in embryonic neighborhoods of differing chemical and physical environments.

Pattern Formation

There is more to building a specialized organ than the differentiation of its cells. An arm and a leg have the same mixture of tissues—muscle, connective tissue, cartilage, and skin—but with different spatial arrangements. The shaping of an animal and its individual parts involves **pattern formation,** the emergence of a body form with specialized organs and tissues all in their characteristic places. (Pattern formation during plant development was covered in Chapter 34.)

Homeotic Genes Pattern formation has its genetic basis in genes that control the overall body plan of an animal. One class of these regulatory genes, the **homeotic genes,** controls the developmental fate of groups of cells. Homeotic genes were first identified in fruit flies *(Drosophila).* A mutation in a homeotic gene might produce a fly with such bizarre characteristics as an extra set of wings, extra or missing body segments, or legs growing from the head in place of antennae (Figure 43.25).

Analysis of the DNA sequences of fruit fly homeotic genes revealed a sequence 180 nucleotides long common to all the genes. This specific DNA sequence, which was given the name **homeobox,** was soon identified in certain other fruit fly genes involved with development. When biologists looked for the homeobox sequence in other organisms, they first found it in animals that, like *Drosophila,* had body plans containing repeated segments (see Chapter 36). Some biologists speculated, therefore, that homeoboxes were involved in segmentation. However, in the past few years, homeoboxes have been found in virtually every eukaryotic organism examined—from yeast to human. These DNA sequences are the same, or nearly so, in all these organisms. Biologists now think that the homeobox sequence plays a more wide-ranging role in controlling patterns of development.

Support for a general developmental role for homeoboxes comes from several lines of evidence. Homeobox nucleotide sequences are found within a variety of protein-coding genes, most of which are associated in some way with development. They are translated into peptide sequences 60 amino acids long called *homeodomains,* which have the property of binding to specific sequences in DNA. All the homeodomain-containing proteins examined so far have turned out to be transcription factors that activate or repress transcription of other genes by binding to enhancer regions of DNA (see Chapter 18). Presumably, they regulate development by coordinating the transcription of batteries of developmental genes, switching them on or off. In early fruit fly embryos, cells in different parts of the embryo show different combinations of active and inactive homeobox genes.

Although much of the research on homeoboxes has been carried out with insects, two recent discoveries illustrate the spectrum of homeobox involvement in vertebrate development. In frogs, a homeobox-containing gene has been identified that apparently controls "posteriorness." When cells expressing this gene are moved to the anterior (front) end of an embryo, a headless frog develops. In a less sensational but equally important discovery, biologists have learned that a homeobox-containing gene is involved in turning on the genes for antibody formation in humans.

The homeodomain amino acid sequence, while varying somewhat from gene to gene and from organism to organism, has been highly conserved in evolution. For example, the homeodomain of the fruit fly protein that controls the phenotype shown in Figure 43.25 differs from one of the frog's homeodomains by only one amino acid out of 60—even though flies and frogs have evolved separately for hundreds of millions of years. Furthermore, eukaryotic homeodomains in general show some similarity to the DNA-binding domains of prokaryotic regulatory proteins. The combined evidence suggests that homeoboxes are all derived from a nucleotide sequence that arose very early in the history of life, and that they have continued to be very important in the regulation of gene expression and development throughout the living world.

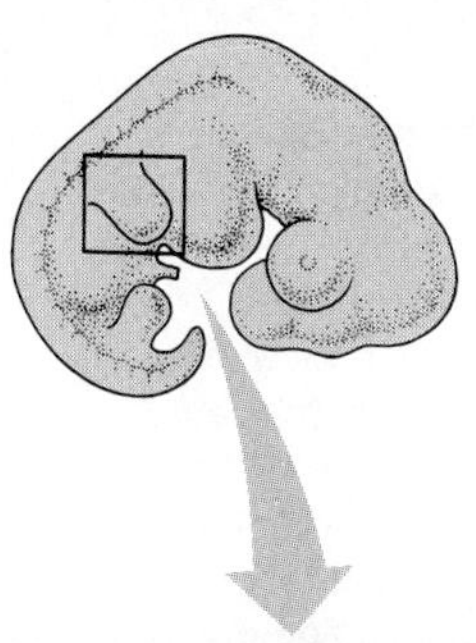

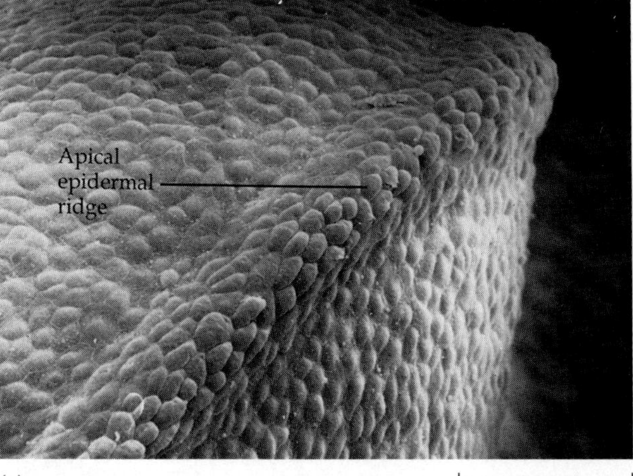

(a) 50 µm

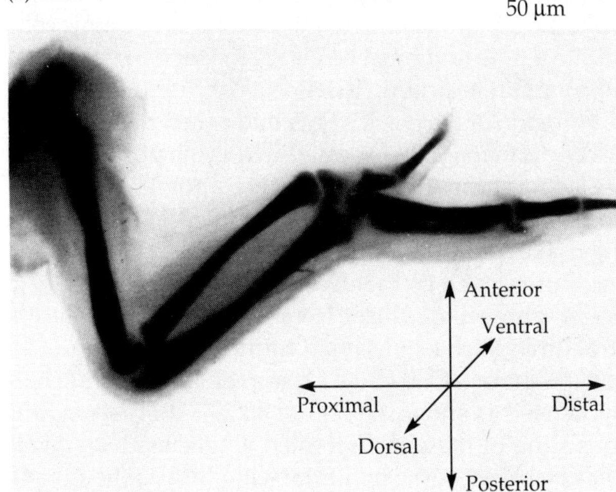

(b)

Figure 43.26
Pattern formation during vertebrate limb development.
(**a**) Vertebrate limbs develop from rudiments called limb buds (SEM). (**b**) As the bud develops into a limb, such as this wing of a chick embryo, a specific pattern of tissues emerges. This requires each embryonic cell to receive some kind of positional information indicating location along the three axes of the limb.

Positional Information There is more to the story of pattern formation than regulatory genes. Within each rudimentary organ, a specific pattern of specialized tissues emerges, even though all the cells have the same genes, including regulatory genes. Genes that regulate development must respond to some kind of **positional information**—signals that indicate a cell's location relative to other cells in an embryonic structure (see Chapter 34).

Lewis Wolpert and his colleagues have been investigating the role of positional information in the development of chick limbs. The wings and legs begin as undifferentiated limb buds that go on to develop their unique forms (Figure 43.26). Each bone and muscle

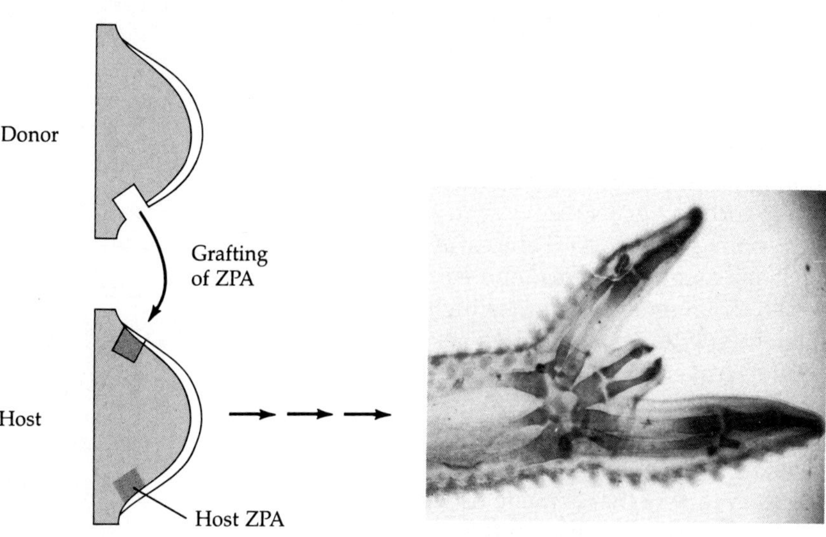

Figure 43.27
Altering positional information with grafting experiments. A region called the zone of polarizing activity (ZPA) is the reference point for indicating the position of cells along the anterior-posterior axis of a vertebrate limb bud. The ZPA is located where the posterior margin of the bud is attached to the body. In this experiment, a second ZPA is added to the anterior margin of a limb bud by transplanting the tissue from a donor. Cells near the grafted ZPA, like cells near the host bud's own ZPA, apparently receive positional information indicating "posterior." The pattern that emerges in the developing limb is a mirror image, with an arrangement of digits equivalent to two human hands joined together at the thumbs.

and every other limb component has a precise location and orientation relative to three axes: the proximal-distal axis, from the base of the limb to the tip of the digits; the anterior-posterior axis, from the front edge of the limb to the rear edge; and the dorsal-ventral axis, from the top of the limb to the bottom. Embryonic cells within a limb bud must receive some kind of positional information indicating location along these three axes.

A region of the chick's limb bud called the *zone of polarizing activity (ZPA),* located where the posterior side of the bud is attached to the body, apparently assigns position along the anterior-posterior axis to developing tissues. Cells nearest the ZPA give rise to posterior structures, such as the digit homologous to our little finger, and cells farthest from the ZPA form anterior structures, such as the avian equivalent of our forefinger and thumb. One of the experiments supporting this hypothesis is shown in Figure 43.27. Another region, consisting of mesoderm located just beneath an *apical epidermal ridge* at the tip of the wing bud, is thought to assign position to the embryonic tissue along the proximal-distal axis as the limb bud grows. Positional information along the dorsal-ventral axis probably reflects the relative distance of developing tissue from the dorsal and ventral ectoderm of the limb bud. In experiments where the ectoderm of a bud is separated from the mesoderm, then replaced with its orientation rotated by 180°, the limb develops with its dorsal-ventral orientation reversed. (This is equivalent to reversing the palm and back of your hand.)

From experiments such as these, we can conclude that pattern formation requires cells to receive and interpret environmental cues that vary from one location to another. One possibility is a chemical signal that varies in concentration along a gradient, enabling cells to resolve their position along that gradient. Such a substance is called a **morphogen** (see Chapter 34). Gradients of different morphogens along three axes would provide the positional information a cell needs to determine its location in the three-dimensional realm of a developing organ. Until recently, morphogens were only hypothetical substances, but in 1987, Christina Thaller and Gregor Eichele identified a possible morphogen, a derivative of vitamin A called retinoic acid. If this compound is dabbed onto the anterior margin of a limb bud, the result is the same as when an additional ZPA is transplanted to this location: two sets of digits forming a mirror image arrangement (see Figure 43.27). Furthermore, the researchers were able to demonstrate a gradient of retinoic acid in the normal limb buds of chick embryos, with the compound most concentrated near the ZPA. Some developmental biologists find the evidence that retinoic acid is a morphogen unconvincing, but ongoing experiments to determine the effects of retinoic acid on gene expression may help resolve the issue.

Other experiments suggest that positional information functions in the development of overall body form, not just in the patterning of individual parts, such as limbs. In one experiment, a small block of undifferentiated chick mesoderm from the part of the leg that would normally become the thigh was transplanted to a position just below the ectoderm at the tip of the wing bud. The transplanted tissue did not develop into wayward thigh tissue, nor did it form structures typical of wing tips, but instead gave rise to a toe. Apparently, the transplanted mesoderm had received its positional information in installments, first determining that it was in the posterior limb bud (leg) and subsequently sensing that it was near the tip of a developing limb.

The Hierarchy of Pattern Formation Graded cues probably cannot operate over long distances in the embryo. However, installments of positional information provided throughout development can establish the

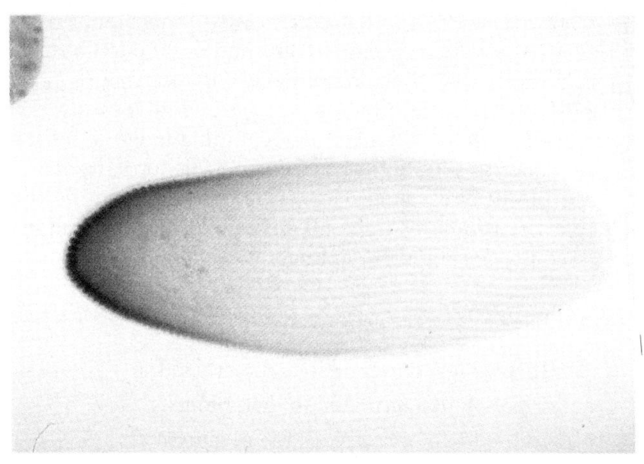

Figure 43.28
Bicoid distribution in a *Drosophila* embryo. Bicoid protein is the product of a particular egg-polarity gene that helps determine the anterior-posterior axis of the fly. Nurse cells surrounding the unfertilized egg stock mRNA for bicoid at one end of the egg, and this designates that region as the animal's future anterior end (LM). After fertilization, the bicoid gradient in the embryo functions as one of the morphogens controlling pattern formation.

gross form of the animal when the embryo is very small ("This will be the head end, this will be the tail end," and so on), and then refine its parts as development progresses. Researchers are beginning to work out the genetic basis of this hierarchy of pattern formation in *Drosophila.*

The overall polarity (anterior-posterior and dorsal-ventral axes) of a fruit fly is determined by the heterogeneous distribution of substances in the unfertilized egg. Nurse cells surrounding the egg secrete molecules of mRNA that become localized within the cytoplasm of the egg. The genes from which these mRNAs are transcribed are called egg-polarity genes. For example, nurse cells at one end of the egg secrete mRNA from a gene for a protein called bicoid, and this fixes the future anterior end of the animal. The bicoid mRNA may be held in place by the cytoskeleton. When the embryo begins to develop after fertilization, the bicoid mRNA is translated, and a gradient of bicoid protein provides positional information, with the end where the protein is most concentrated becoming the head end (Figure 43.28). Thus, bicoid is a morphogen that helps determine body plan on the coarsest level. A "posterior" morphogen and "dorsal-ventral" morphogens are also involved in this first stage of pattern formation. In the second stage, these morphogens trigger regional differences in the expression of segmentation genes, which code for proteins that control pattern formation on a finer scale. These genes determine the number of segments and basic pattern formation in the developing embryo. In the third stage of pattern formation, products of the segmentation genes determine which homeotic genes will be expressed in each segment. Recall that homeotic genes control which anatomical structures will develop in each segment. Thus, the body plan of *Drosophila* emerges from a cascade of gene activations that determine pattern on a finer and finer scale.

* * *

Through a variety of experiments, developmental biologists are beginning to learn how the one-dimensional information encoded in the nucleotide sequence of a zygote's DNA is translated into the three-dimensional form of an animal. There are enough unanswered questions, however, to entertain many future generations of scientists.

STUDY OUTLINE

1. Development encompasses changes in form and function that occur from conception to death.

An Overview of Developmental Processes (pp. 956–957)

1. An embryo is not preformed in an egg; it develops by epigenesis, the gradual, gene-directed acquisition of form. Cell division, differentiation, and morphogenesis are the three main processes that transform a unicellular zygote into a multicellular organism.

Fertilization (pp. 957–959)

1. Fertilization both reinstates diploidy and activates the egg to begin a chain of metabolic reactions that triggers the onset of embryonic development.
2. The acrosomal reaction, which occurs when the sperm meets the egg, releases hydrolytic enzymes that digest through material surrounding the egg.
3. Gamete fusion depolarizes the egg cell membrane and sets up a fast block to polyspermy.
4. Egg cell depolarization also initiates the cortical reaction, in which calcium ions stimulate cortical granules to erect a fertilization membrane that functions as a slow block to polyspermy.
5. The calcium ions also trigger metabolic changes in the fertilized cell that greatly increase cellular respiration and protein synthesis, thus activating the egg.

Early Stages of Embryonic Development (pp. 960–965)

1. The basic body plan of an animal is determined in the early stages of development.
2. Fertilization is followed by cleavage, a period of rapid cell division without growth, which results in the production of a large number of small cells called blastomeres.
3. Cleavage planes usually follow a specific pattern relative to the animal and vegetal poles of the zygote.
4. Cleavage produces a solid ball of cells, called the morula, which develops a fluid-filled blastocoel, thereby becoming the blastula.

5. Gastrulation transforms the blastula by invagination into a cup-shaped, two-layered embryo called the gastrula. At this stage, the embryo has an archenteron that opens through the blastopore.
6. A mesoderm layer develops between the outer ectoderm and the inner endoderm, completing the three primary germ layers from which all body structures are derived during organogenesis.
7. The notochord forms from dorsal mesoderm, and the neural tube develops from the ectodermal neural plate above the notochord.

The Embryology of Amniotes (pp. 965–968)

1. The shelled eggs of birds and reptiles have large yolks and undergo meroblastic cleavage restricted to a small disc of cytoplasm at the animal pole. A cap of cells called the blastodisc forms and begins gastrulation with the formation of the primitive streak. In addition to the embryo, primary germ layers give rise to the four extraembryonic membranes: yolk sac, amnion, chorion, and allantois.
2. The eggs of placental mammals are small and store little food, exhibiting a holoblastic cleavage with no apparent polarity. Gastrulation and organogenesis, however, resemble the patterns of birds and reptiles. After fertilization and early cleavage in the oviduct, the blastocyst implants in the uterus. The trophoblast initiates formation of the fetal portion of the placenta, and the embryo proper develops from an embryonic disc. Extraembryonic membranes homologous to those of birds and reptiles function in intrauterine development.

Mechanisms of Development (pp. 968–979)

1. In most animals except mammals, the polarity of the embryo is established during early cleavage.
2. The unequal distribution of cytoplasmic determinants in specific blastomeres can fix the developmental fates of different regions of the embryo very early.
3. In regulative development, cells remain totipotent longer than do the blastomeres in mosaic development.
4. Experimentally derived fate maps of polar embryos have shown that specific regions of the zygote or blastula develop into specific parts of the embryo. In *Caenorhabditis elegans,* it has been possible to trace the lineage of every cell in the body.
5. Morphogenetic movements rely on three properties of cells: extension, contraction, and adhesion. Morphogenesis by reorganization of the cytoskeleton and by amoeboid movement is aided by extracellular fibronectin fibrils and cell surface adhesion molecules.
6. Transplant experiments have demonstrated that certain groups of cells can influence the development of contiguous cells by induction. The dorsal lip of the blastopore is a primary organizer of the amphibian embryo.
7. Cell differentiation is orchestrated by the serial process of determination, which progressively narrows the developmental options of each cell.
8. Nuclear transplantation and other experiments reveal the genomic equivalence of most specialized cells. Therefore, differentiation reflects differential expression of genes selectively activated in a common genome.
9. Cells are ordered into specific three-dimensional positions to produce specific organs and body parts by pattern formation, which requires cells to receive and interpret positional information that varies with location.
10. Developmental biologists are beginning to work out the sequence of gene activations responsible for pattern formation in *Drosophila.* Homeotic genes determine the types of structures that will develop at certain locations in *Drosophila* and other animals.

SELF-QUIZ

1. The cortical reaction functions directly in the
   a. formation of a fertilization membrane
   b. production of a fast block to polyspermy
   c. release of hydrolytic enzymes from the sperm cell
   d. generation of a nervelike impulse by the egg cell
   e. the fusion of egg and sperm nuclei
2. Which of the following is common to both avian and mammalian development?
   a. holoblastic cleavage
   b. primitive streak
   c. trophoblast
   d. yolk plug
   e. gray crescent
3. The archenteron develops into
   a. the mouth in protostomes
   b. the blastocoel
   c. the endoderm
   d. the lumen of the digestive tract
   e. the placenta
4. In a frog embryo, the blastocoel is
   a. completely obliterated by yolk platelets
   b. lined with endoderm during gastrulation
   c. located primarily in the animal hemisphere
   d. restricted to the vegetal hemisphere
   e. the cavity that later forms the archenteron
5. Which embryonic membrane is closest to the shell in a bird egg?
   a. allantois
   b. amnion
   c. chorion
   d. yolk sac
   e. endoderm
6. In an amphibian embryo, a band of cells called the neural crest
   a. rolls up to form the neural tube
   b. develops into the main sections of the brain
   c. produces amoeboid cells that migrate to form teeth, skull bones, and other structures in the embryo
   d. has been shown by experiments to be the organizer region of the developing embryo
   e. induces the formation of the notochord

7. The results of transplantation of frog nuclei into anucleated eggs allow for which conclusion?
   a. Frogs cannot be cloned.
   b. All differentiated cells actually express the same genes.
   c. The later the embryonic stage, the less likely that nuclei from cells at that stage can support the development of an egg into a tadpole.
   d. The nuclei of differentiated cells actually lack some genes found in other cell types.
   e. The differentiated state is unstable.
8. The specific arrangement of tissues and organs in a body part, such as a leg, is called
   a. pattern formation
   b. induction
   c. differentiation
   d. determination
   e. organogenesis
9. In the early development of an amphibian embryo, the organizer is the
   a. neural tube
   b. notochord
   c. archenteron roof
   d. dorsal lip of the blastopore
   e. dorsal ectoderm
10. Homeotic genes are believed to play a crucial role in controlling development because
   a. they code for regulatory proteins that can bind to enhancers in DNA
   b. they appear to cause cells to differentiate
   c. they appear to cause cells to become determined
   d. they are identical in all eukaryotic organisms
   e. they have been found in all segmented animals

CHALLENGE QUESTIONS

1. Using the frog embryo as an example and being as specific as you can, describe the role of morphogenetic movements in gastrulation and the role of pattern formation in organogenesis.
2. Explain how in certain animals (such as *Drosophila*) the organization of the unfertilized egg's cytoplasm affects the form of the adult animal that develops after fertilization.
3. The "snout" of a frog tadpole bears a sucker. A salamander tadpole has a mustache-shaped structure called a balancer in the same area. If ectoderm from the side of a young salamander embryo is transplanted to the snout of a frog embryo, the frog tadpole later has a balancer. If ectoderm is transplanted from the side of a slightly older salamander embryo to the snout of a frog embryo, the frog tadpole ends up with a patch of salamander skin on its snout. Explain the results of this experiment in terms of the mechanisms of development.

SCIENCE, TECHNOLOGY, AND SOCIETY

1. Nerve cells transplanted from aborted fetuses can relieve the symptoms of Parkinson's disease, a brain disorder. Fetal tissue transplants might also be used to treat epilepsy, diabetes, Alzheimer's disease, and spinal cord injuries. Why might tissues from a fetus be particularly useful for replacing diseased or damaged cells? Since 1989, there has been controversy over whether the U.S. government should allow fetal tissues from induced abortions to be used in transplant research. Opponents would allow only tissues from miscarriages to be used. Why would most researchers prefer to use tissues from aborted fetuses? Why do you think some people oppose this? Where do you stand, and why?
2. The abortion debate in the United States has provoked thought and created controversy over the question, When does human life begin? In your opinion, at what point does human life begin? What are some other possible answers to this question? Do you think continuing study of human embryological development can give a definitive answer? Why or why not?
3. The U.S. government recommends that pregnant women abstain from drinking alcohol and has mandated that this warning label be placed on liquor bottles: "Women should not drink alcoholic beverages during pregnancy because of the risk of birth defects." Imagine the dilemma of two young servers in a Seattle restaurant in 1991. An obviously pregnant woman ordered a strawberry daiquiri. The server, unsure about what to do, approached a second server, who showed a warning label to the customer and asked her, "Ma'am, are you sure you want this drink?" The customer became angry, demanded her drink, and went to the restaurant manager, who fired the two servers. What would you have done if you had been in either server's place? Why?

FURTHER READING

Beardsley, T. "Smart Genes." *Scientific American,* August 1991. How do genes know where they are in an embryo?

Caldwell, M. "How Does a Single Cell Become a Whole Body?" *Discover,* November 1992. One of the most important questions in modern biology.

DeRobertis, E., G. Oliver, and C. Wright. "Homeobox Genes and the Vertebrate Body Plan." *Scientific American,* July 1990. The modular nature of animal development.

Ezzell, C. "Retinoic Acid Flunks as a Master Molecule." *Science News,* March 9, 1991. Debate about a possible morphogen.

Gilbert, S. F. *Developmental Biology,* 3rd ed. Sunderland, MA: Sinauer Associates, 1991. A widely used upper-division text.

Kline, D. "Activation of the Egg by the Sperm." *BioScience,* February 1991.

Malacinski, G. M., A. W. Neff, J. A. Alberts, and K. A. Souza. "Developmental Biology in Outer Space." *BioScience,* May 1989. Embryology in a weightless environment.

Marx, J. "How Embryos Tell Heads From Tails." *Science,* December 13, 1991. The molecular biology of animal polarity.

Marx, J. "Homeobox Genes Go Evolutionary." *Science,* January 24, 1992. The long history of genes that control development.

Newman, R. A. "Adaptive Plasticity in Amphibian Metamorphosis." *BioScience,* October 1992. How does environment affect an individual animal's development?

Slack, J. "Molecules of the Moment." *Nature,* January 3, 1991. Activin as an inducer of mesoderm formation.

44 Nervous Systems

Cells of the Nervous System

Transmission of Electrical Signals Along a Neuron

The Synapse: Transmission Between Cells

Invertebrate Nervous Systems

The Vertebrate Nervous System

A baseball infielder's diving catch of a line drive is a spectacular display of animal coordination. At the crack of the bat, a living computer called the brain analyzes the complex visual scene and isolates the rapidly approaching blur that is the ball, calculates its speed and trajectory, formulates the appropriate set of movements required to intercept its flight, and signals multiple muscles to thrust the hand to meet the ball—all in a fraction of a second. The coordinated behavior of which this is an example is, of course, more than a game. To survive and reproduce, an animal must respond rapidly and appropriately to its environment, and it is the task of the nervous system to carry out these interactions with the environment.

As pointed out in Chapter 41, the nervous system is one of two coordinating systems in animals, the other being the endocrine system. In many cases, these two systems cooperate and interact to control physiology and behavior. For example, hormones and nervous system signals are both involved in the fight-or-flight responses of animals to threats and stress. However, the very different structural organizations of the endocrine system and the nervous system give rise to important differences in the roles of the two coordinating systems. The endocrine system consists of individual glands scattered throughout the body, which dispatch their chemical messages (hormones) to diffuse target cells via the bloodstream. The nervous system, in contrast, is a signaling network with branches carrying information directly to and from the specific target tissues in the various parts of the organism. Because the cells of the nervous system (neurons) are specialized for the rapid electrical transmission of signals, this structural organization enables information to be transmitted much more quickly in the nervous system than in the endocrine system. Messages can travel along neurons as fast as 100 m/sec (over 200 mph), which allows signals to go from the brain of a human to the hands (or vice versa) in a few milliseconds. Compare this to the response speed in the endocrine system, which is typically seconds or minutes.

A second distinguishing feature of the nervous system is its pinpoint control. The hormones released by an endocrine gland are dispersed throughout the body by the cardiovascular system, but the nervous system

Figure 44.1
A nerve cell and a microprocessor. This micrograph shows an unusual juxtaposition of the components from which living computers and human-made computers are made (SEM). A single nerve cell (neuron) was taken out of the nervous system and allowed to grow on the surface of a Motorola 68000 microprocessor. In this chapter, you will learn how nervous systems consisting of neurons coordinate animal bodies.

can control very fine movements by stimulating only one or a few muscles to contract at a particular instant. A third difference is that the nervous system has an incomparable structural complexity that enables it to integrate more kinds of information and to elicit a broader range of responses than the endocrine system.

The nervous system is probably the most intricately organized aggregate of matter on Earth. Just 1 cm^3 of the human brain may contain several million nerve cells, each of which may communicate with thousands of other neurons in data-processing networks that make the most elaborate human-made circuit boards look primitive. These neural pathways control our every perception and movement and, somehow, enable us to learn, to remember, and to think. Human-made computers, made up of integrated circuit chips (Figure 44.1), are capable of complex tasks, but your biological computer, made up of nerve cells, is performing vastly more complex tasks as you read and comprehend these words.

In general, the nervous system has three overlapping functions: sensory input, integration, and motor output (Figure 44.2). Input is the conduction of signals from sensory receptors, such as the light-detecting cells in the eyes of the baseball infielder, to integration centers in the nervous system. Integration refers to the process by which the information from the environmental stimulation of the sensory receptors is interpreted and associated with appropriate responses of the body (for our infielder, the identification of the baseball, calculation of its predicted path, and the selection of the correct sequence of movements necessary to intercept it). For the most part, integration is carried out in the **central nervous system,** the brain and spinal cord. Motor output is the conduction of signals from the processing center, the central nervous system, to **effector cells,** the muscle cells or gland cells that actually carry out the body's responses to stimuli. The nerves communicating motor and sensory signals between the central nervous system and the rest of the body are collectively called the **peripheral nervous system.** From receptor to effector, the information is communicated from one neuron to the next by a combination of electrical and chemical signals that will be a major focus of this chapter. The chapter also compares the structural organization of nervous systems of diverse types of animals. Chapter 45 connects the nervous system to its inputs and outputs by discussing sensory receptors and the physiology of movement.

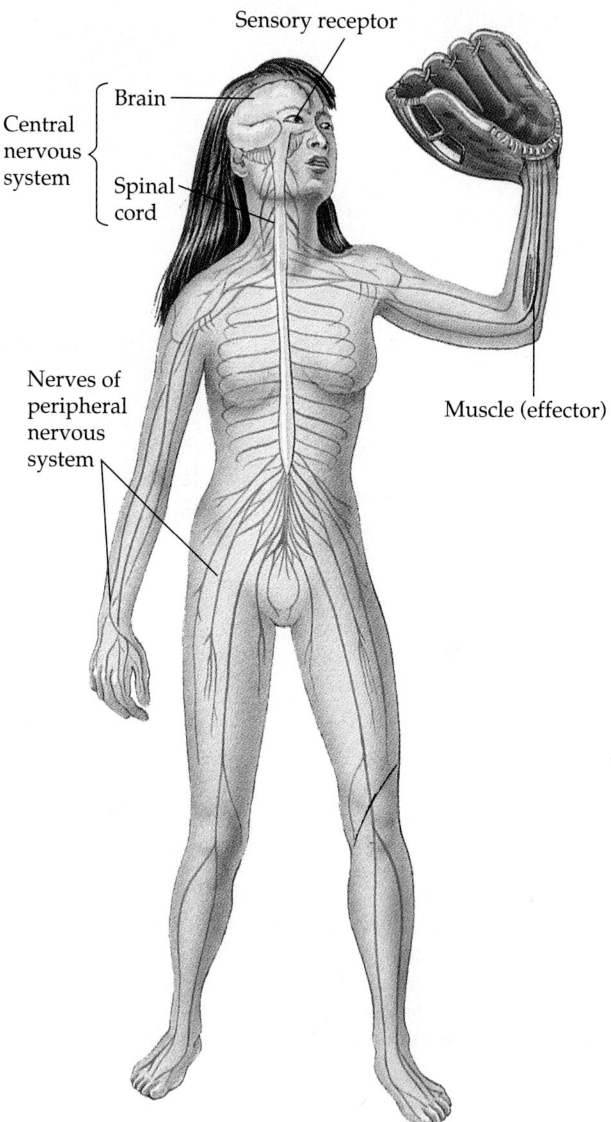

Figure 44.2
An overview of the human nervous system. The brain and spinal cord together form the central nervous system, which is responsible for the integration of information. The network of nerves that forms the peripheral nervous system carries information from sensory receptors (sensory input) to the central nervous system, and motor commands from the central nervous system (motor output) to various target organs, or effectors.

CELLS OF THE NERVOUS SYSTEM

Two main classes of cells populate the nervous system: neurons and supporting cells. Neurons are the cells that actually conduct messages along the pathways of the nervous system. More numerous are the supporting cells, which provide structural reinforcement in the nervous system and also protect, insulate, and generally assist neurons.

Neurons

Neurons are the functional units of the nervous system and are specialized for transmitting signals from one

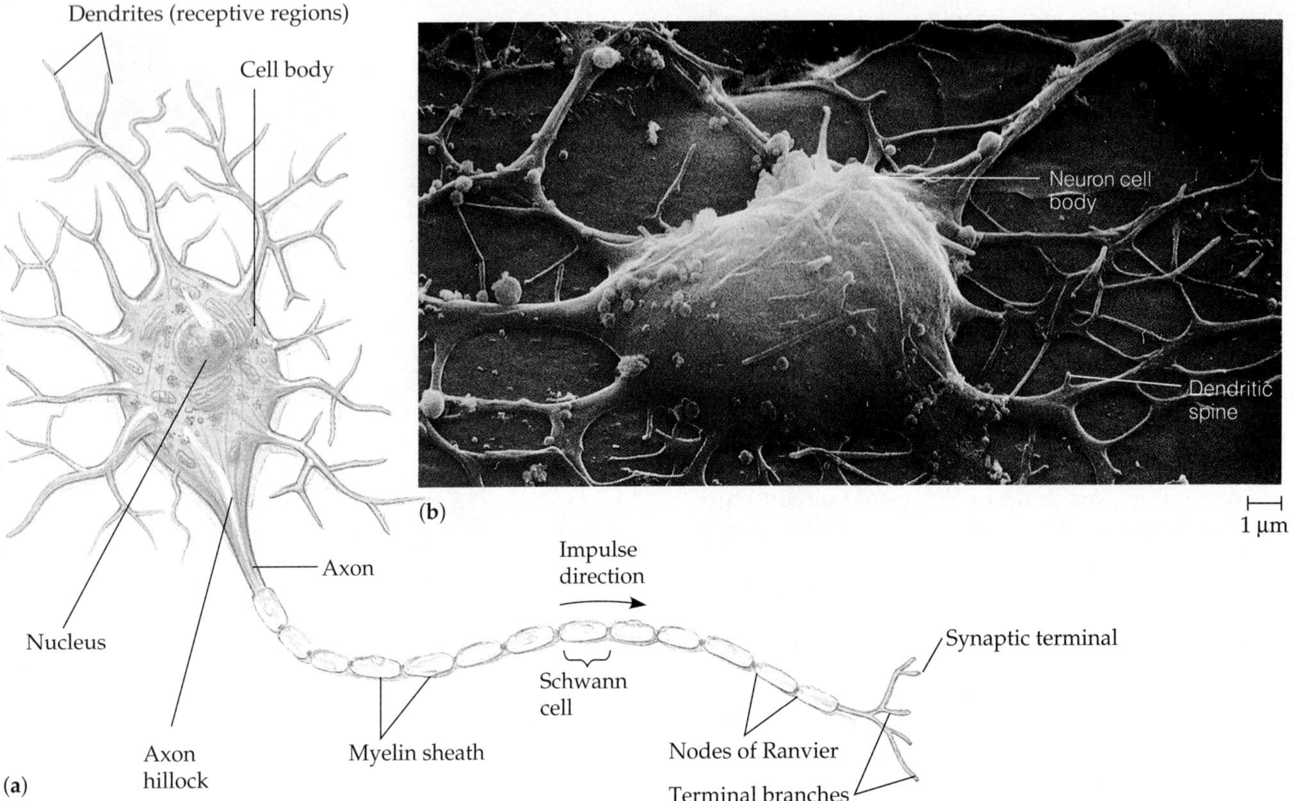

Figure 44.3
The structure of a typical vertebrate neuron. (**a**) The cell body of the neuron contains the nucleus and other organelles and gives rise to various branched projections called processes, which are of two types: dendrites and axons. The dendrites typically receive inputs and conduct signals toward the cell body, while the axon conducts signals away from the cell body. (This direction of the usual flow of information is indicated by the arrow.) At the end of the axon are numerous synaptic terminals, where the neuron makes connections with other neurons or other target cells. The axon of many neurons is wrapped by a myelin sheath of insulating Schwann cells (or oligodendrocytes, in the case of the central nervous system). The small gaps between successive Schwann cells are called the nodes of Ranvier. (**b**) Scanning electron micrograph of a neuron.

location in the body to another. Although there are many different types of neurons, varying in their structural details depending on their roles in the nervous system, most neurons share some common features (Figure 44.3). A neuron has a relatively large **cell body** containing the nucleus and a variety of other cellular organelles. The most striking features of neurons are fiberlike extensions called processes that increase the distance over which the cells can conduct messages. The processes are of two general types: **dendrites,** which convey signals toward the cell body, and **axons,** which conduct messages away from the cell body.

The dendrites of most types of neurons, such as the one in Figure 44.3a, are numerous and extensively branched (the word is derived from the Greek *dendron,* "tree"). Thus, the dendrites are structural adaptations that increase the surface area of the "receiving end" of the neuron, where the cell receives inputs from other neurons.

Many neurons have a single axon, which may be very long. In fact, the sciatic nerve in your leg has axons that extend all the way from the lower part of your spinal cord to muscles of the lower leg and foot, a distance of a meter or more. The axon hillock, or initial segment, is the portion of the cell body from which the axon originates; as we will see, this is the region where impulses that pass down the axon are usually generated. The axons of many vertebrate neurons are enclosed by a chain of supporting cells, called **Schwann cells** in the peripheral nervous system, that form an insulating layer called the **myelin sheath.** The axon may be branched, and each branch may give rise to hundreds or thousands of specialized endings called **synaptic terminals,** which relay neural signals to other cells by releasing chemical messengers called neurotransmitters. The site of contact between a synaptic terminal and a target cell (either another neuron or an effector cell, such as a muscle cell) is called a **synapse.** Thus, the synapse is where one neuron communicates with another neuron in a neural pathway.

Although Figure 44.3 shows common structural features of a vertebrate neuron, individual neurons show

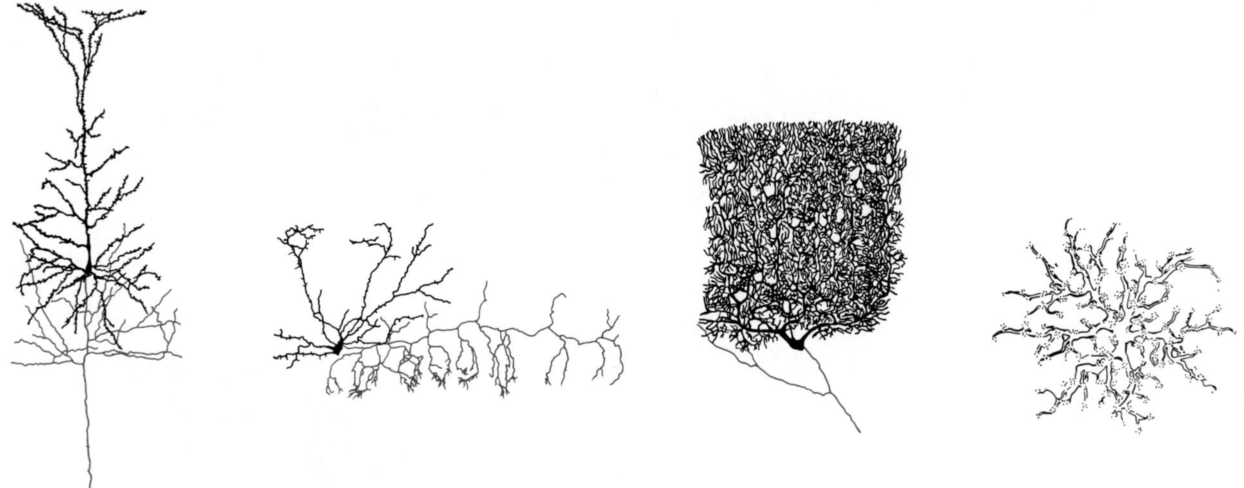

Figure 44.4
Diversity of neuron shape. There are large differences in the branching patterns of individual neurons, depending on their functions and locations in the nervous system. In each example, the cell body and dendrites are shown in black, and the axon is shown in red. One of these neuron types (far right), found in the retina of the eye, has no axon; its dendrites serve the dual function of receiving and sending signals.

a great diversity of shapes adapted for their specific functions in the nervous system. A small sampling of neurons illustrating this diversity is shown in Figure 44.4.

The cell bodies of most neurons are located in the central nervous system (the brain and spinal cord), although the cell bodies of certain types of neurons are located outside the central nervous system in clusters called ganglia. There are three major classes of neurons, based on the kinds of cells with which they make synaptic connections. **Sensory neurons** communicate information about the external and internal environments from sensory receptors to the central nervous system. **Motor neurons** convey impulses from the central nervous system to effector cells. Sensory input and motor output of the nervous system are usually integrated by **interneurons,** which are located within the central nervous system and make synaptic connections only with other neurons. In most neural pathways, sensory neurons synapse with interneurons, which in turn synapse with motor neurons (Figure 44.5).

Supporting Cells

Supporting cells outnumber neurons tenfold to fiftyfold in the nervous system. Although the supporting cells do not actually conduct nerve impulses, they are essential for the structural integrity of the nervous system and for the normal functioning of neurons.

Supporting cells in the central nervous system are called **glial cells,** meaning "glue" cells. There are several types of glial cells in the brain and spinal cord. **Astrocytes** encircle the capillaries in the brain and contribute to the **blood-brain barrier.** This barrier restricts the passage of most substances into the brain, allowing the extracellular chemical environment of the central nervous system to be tightly controlled. Glial cells called **oligodendrocytes** form insulating myelin sheaths around axons of many neurons in the central nervous system. In the peripheral nervous system, Schwann cells are the supporting cells that form the myelin sheath (see Figure 44.3).

Neurons become myelinated in a developing nervous system when Schwann cells or oligodendrocytes grow around axons in such a way that their plasma membranes form many concentric layers—somewhat like a jelly roll. The membranes are mostly lipid, which is a poor conductor of electrical currents. Thus, the myelin sheaths provide electrical insulation of the axon, analogous to the rubber insulation covering cop-

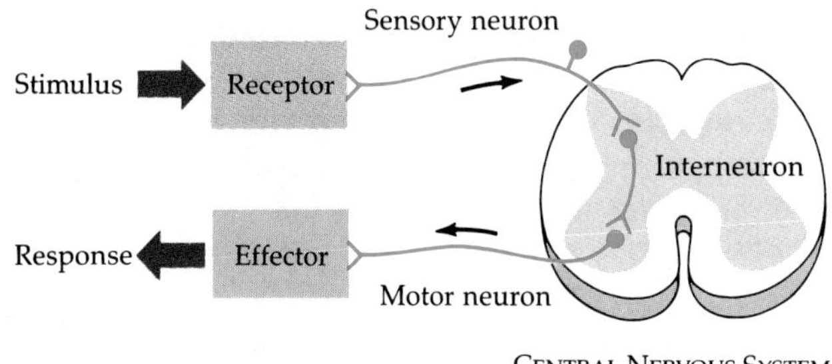

Figure 44.5
A neural pathway between a receptor and an effector.

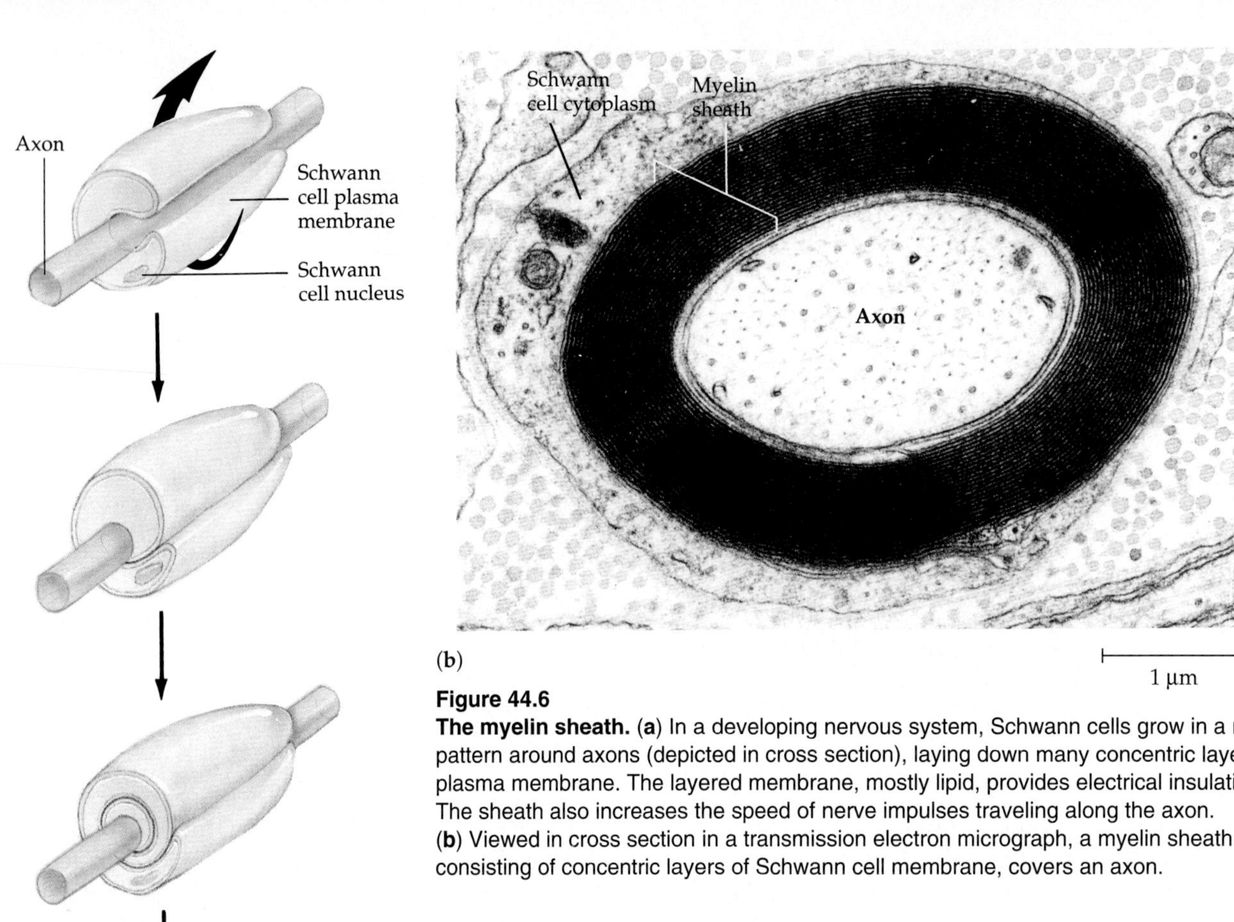

Figure 44.6
The myelin sheath. (**a**) In a developing nervous system, Schwann cells grow in a rolling pattern around axons (depicted in cross section), laying down many concentric layers of plasma membrane. The layered membrane, mostly lipid, provides electrical insulation. The sheath also increases the speed of nerve impulses traveling along the axon. (**b**) Viewed in cross section in a transmission electron micrograph, a myelin sheath, consisting of concentric layers of Schwann cell membrane, covers an axon.

per wires (Figure 44.6). We will see later in this chapter that the myelin sheath increases the speed of propagation of nerve impulses. In the degenerative disease known as multiple sclerosis, myelin sheaths gradually deteriorate, and progressive loss of coordination occurs because of the disruption in transmission of nerve impulses. Clearly, supporting cells are indispensable components of a working nervous system. That being said, let's return to the neuron and its ability to conduct an impulse.

TRANSMISSION OF ELECTRICAL SIGNALS ALONG A NEURON

The signal transmitted along the length of a neuron, from a dendrite or a cell body to the tip of an axon, is an electrical signal that depends on the flow of ions across the plasma membrane of the cell. This part of the chapter focuses on how an electrical voltage arises in a cell and on how a flow of ions *across* the membrane can be converted to a signal that travels in a perpendicular direction *along* the neuron.

The Origin of Electrical Membrane Potential

All living cells have a charge difference across their plasma membranes, the inside of the cell being more negative than the outside. This difference in charge gives rise to an electrical voltage gradient across the membrane, which can be measured with ultrafine microelectrodes (see the Methods Box). The voltage measured across the plasma membrane is called the **mem-**

Methods: Measuring Membrane Potentials

Due to a difference in the relative concentrations of cations and anions on opposite sides of the plasma membrane, the cytoplasmic side of the membrane is negative in charge compared to the extracellular side. This charge separation is called a membrane potential, or, for an unstimulated neuron, a resting potential.

Electrophysiologists can measure the membrane potential as a voltage by using microelectrodes (shown in Figure a) connected to a sensitive voltmeter or oscilloscope. Precise mechanical devices called micromanipulators (moved with the large knobs being manipulated by the scientist in Figure b) are used to position one electrode just inside the cell for comparison with a reference electrode located outside the cell. The voltmeter indicates the magnitude of the charge separation across the membrane, typically about –70 mV for an unstimulated neuron. (The minus sign indicates that the inside of the cell is negative compared to the outside.)

The giant neurons of certain invertebrates, including squids and lobsters, have been particularly useful model systems for studying the electrophysiology of nerve impulses, because the large cells are relatively easy to impale with electrodes. Some of these giant neurons have axons with diameters of about 1 mm. Once electrodes are in place, they can be used not only to measure the voltage of the resting potential, but also to record changes in voltage due to the ion currents that occur during transmission of a nerve impulse.

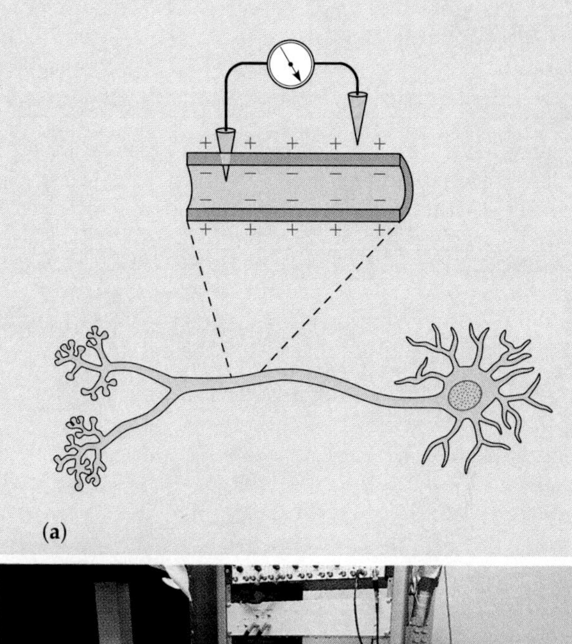
(a)

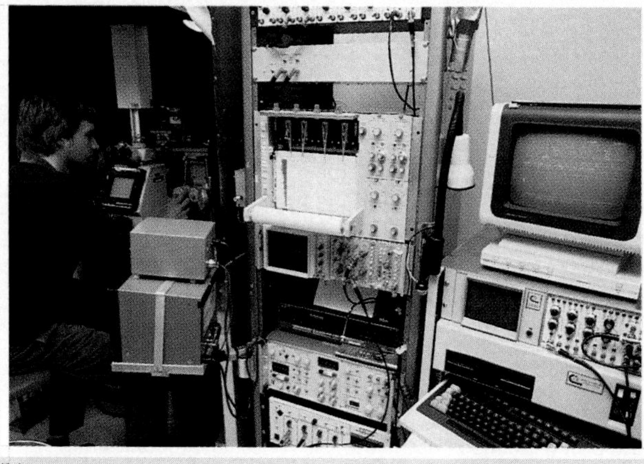
(b)

brane potential, and it is typically in the range of –50 to –100 mV in an animal cell. By convention, the voltage outside the cell is called zero, so the minus sign indicates that the inside of the cell is negative with respect to the outside. For a neuron in its resting state (that is, not transmitting an electrical signal), a membrane potential of –70 mV is typical. To indicate that this potential refers to the resting state, it is termed the **resting potential** of the neuron. Although all cells have a membrane potential, only certain kinds of cells, including neurons and muscle cells, have the ability to generate active changes in their membrane potential, which are used to conduct signals. To indicate the fact that these cells can be excited by a stimulus to produce an active electrical impulse, cells of this type are called **excitable cells.** We will consider first the origin of the resting membrane potential and then return to the special properties of excitable cells.

The membrane potential arises because of differences in the ionic composition of the intracellular and extracellular fluids and because of the selective permeability of the plasma membrane, the barrier between those two fluids. Both the intracellular and the extracellular fluids contain various kinds of dissolved substances, including a variety of electrically charged substances (ions). The ionic compositions inside and outside a typical mammalian cell are shown in Figure 44.7. The principal cation (positively charged ion) outside cells is sodium (Na^+), although there is also some potassium (K^+); inside cells, the situation is reversed, with K^+ the principal cation and Na^+ having a much lower concentration. Outside cells, the main anion

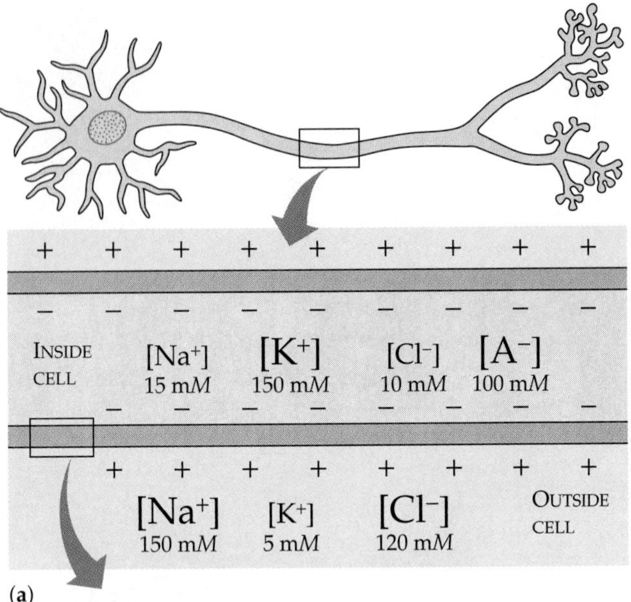

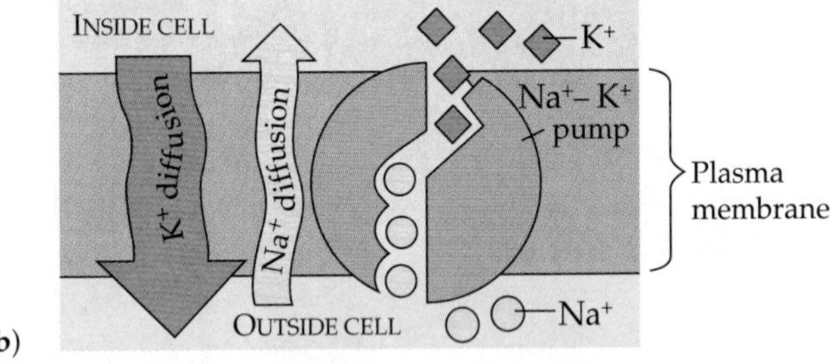

Figure 44.7
The basis of the resting membrane potential. (**a**) The intracellular and extracellular fluids have different ionic compositions. Shown here are the approximate concentrations for a mammalian cell (in millimoles per liter, abbreviated m*M*) of sodium ($[Na^+]$), potassium ($[K^+]$), chloride ($[Cl^-]$), and impermeant internal anions ($[A^-]$). (Brackets indicate concentration of the enclosed substances, e.g., $[Na^+]$ stands for sodium ion concentration.) K^+ diffuses out of the cell down its concentration gradient, but the impermeant anions cannot follow, so the interior of the cell develops a net negative charge. (**b**) At the resting membrane potential, there is a steady diffusion of K^+ out of the cell (orange arrow) and steady diffusion of Na^+ into the cell (beige arrow); the thickness of the arrows indicates the relative permeability of the membrane to K^+ and Na^+. Over time, this would cause the ionic gradients shown in (**a**) to dissipate. This is prevented by the sodium-potassium pump, which uses ATP to actively transport Na^+ out of the cell and K^+ into the cell.

(negatively charged ion) is chloride (Cl^-); there are other anions in the extracellular fluid (phosphate, bicarbonate, proteins, etc.), but they can be ignored in the present context. Inside cells, there is also some Cl^-, but the principal anion (A^-) is actually a class of negatively charged substances (proteins, negative amino acids, sulfate, phosphate, etc.) that can be lumped together and treated as a group.

Recall from Chapter 8 that the plasma membrane is a phospholipid bilayer with associated membrane proteins. Ions, being electrically charged, cannot dissolve in lipid and thus cannot directly diffuse across the lipid of the plasma membrane. In order to cross the membrane, ions must either move through ion channels, which are aqueous pores made up of specific transmembrane protein molecules, or be carried by transport proteins. There are many different kinds of selective ion channels; some channels allow only sodium ions to cross, others only K^+, and still others only Cl^-. So, depending on how many ion channels of each kind are inserted into the plasma membrane of a cell, it is possible for the membrane to have very different permeabilities to each of the different ions in the intracellular and extracellular fluids. Cells usually have much greater permeability to K^+ than to Na^+, suggesting that the membrane has many more potassium channels than sodium channels; in a resting neuron, for instance, potassium permeability is about fiftyfold higher than sodium permeability. Because the internal anions (A^-) are primarily large organic molecules (proteins and amino acids), they cannot cross the membrane and thus are an impermeant pool of internal negative charge.

How does the distribution of ions shown in Figure 44.7 give rise to a membrane potential? Consider the case of potassium ions. There is a large concentration gradient for diffusion of K^+ out of the cell, and the membrane has a high permeability to potassium. Thus, there will be a net flux of K^+ out of the cell (efflux) driven by the concentration gradient. But as K^+ exits, it transfers positive charge from the inside to the outside of the cell. Because A^- is impermeant, it is not free to follow the potassium out, and the cell interior becomes progressively more negative with respect to the exterior, as positive charge is lost while negative charge is trapped within. Notice that as this electrical gradient builds up, it provides a competing influence on the movement of K^+ across the membrane: The increasing internal negativity attracts positively charged potassium, supporting an influx of K^+ down the electrical gradient. If K^+ were the only ion that could cross the membrane, the voltage across the membrane would continue to build up until the influx of K^+ down the electrical gradient exactly balances the efflux of K^+ down the concentration gradient. At that point, there would be no further net transfer of charge across the membrane, and the membrane potential would reach a stable, resting value. For the potassium concentration gradient shown in Figure 44.7, a stable membrane potential of about –85 mV would be required to exactly counterbalance the concentration gradient in the manner just described. This value of membrane potential is called the **equilibrium potential** for potassium ions, because it is the potential at which there will be no net movement of K^+ across the membrane (that is, potassium is at equilibrium).

Potassium, however, is not the only ion to which the plasma membrane is permeable. Although the membrane is much less permeable to Na^+ than to K^+, the permeability to Na^+ is not zero. For sodium, both the concentration gradient ($[Na^+]$ greater outside cell) and the electrical gradient (negative inside cell) tend to move sodium ions into the cell; the resulting trickle of positive charge into the cell, carried by Na^+, makes the actual value of the membrane potential somewhat more positive than the –85 mV that would be expected if the membrane were permeable only to potassium ions. This explains why the resting potential of a neuron is typically about –70 mV rather than about –85 mV.

Over time, the steady influx of sodium would cause a progressive increase in internal sodium concentration. Also, because the influx of Na^+ makes the cell interior less negative than the –85 mV required to balance the potassium concentration gradient, there will be a steady efflux of potassium and a progressive decline in internal K^+ concentration. In other words, if the situation were left unchecked, the concentration gradients for Na^+ and K^+ shown in Figure 44.7 would gradually dissipate. This is prevented by a different type of plasma-membrane protein also found in abundance in neurons: the sodium-potassium pump (see Chapter 8). This protein uses energy from ATP to drive the active transport of sodium back out of the cell, against both the concentration and electrical gradients for sodium. At the same time, the pump moves potassium into the cell, thus restoring the concentration gradient for this ion as well. In essence, the cell uses metabolic energy, in the form of ATP, to maintain the ionic gradients across the membrane that give rise to the steady-state membrane potential, or resting potential of the neuron.

The Action Potential

The properties of membrane potential discussed above apply to all cells. Neurons, however, have other kinds of special ion channels, called **gated ion channels,** that allow the cell to change its membrane potential in response to stimuli that the cell receives. In the case of a sensory neuron, the stimulus may come from the organism's environment (for example, light in the case of photoreceptors in the eye, or vibrations in the air in the case of receptors in the ear). In the case of an interneuron, the stimulus will ordinarily be produced by the activation of other neurons that make inputs to the cell. The effect of the stimulus on the neuron depends on the type of gated ion channel that is opened by the stimulus. If the stimulus opens a potassium channel, efflux of potassium will increase and the membrane potential will become more negative as K^+ exits. Such an increase in the electrical gradient across the membrane is called a **hyperpolarization** (Figure 44.8a). If the channel opened by the stimulus is a sodium channel, influx of Na^+ will increase and the membrane potential will become less negative. Such a reduction in the electrical gradient is called a **depolarization** (Figure 44.8b). The voltage changes produced by stimulation of this type are called **graded potentials,** because the magnitude of the voltage change (either hyperpolarization or depolarization) depends on the strength of the stimulus: A larger stimulus will open more channels and produce a larger change in permeability.

In an excitable cell, such as a neuron, the response to a depolarizing stimulus is graded with stimulus intensity only up to a particular level of depolarization, called the **threshold.** If a depolarization reaches the threshold, a different type of response, called an **action potential,** is triggered (Figure 44.8c). The threshold potential in a neuron is typically about 15 to 20 mV more positive than the resting potential (that is, at a potential of –50 to –55 mV). Hyperpolarizing stimuli do not produce action potentials; in fact, hyperpolarization makes it less likely that an action potential will be fired by making it more difficult for a depolarizing stimulus to reach threshold.

The action potential is an **all-or-none event,** meaning that the magnitude of the action potential is independent of the strength of the depolarizing stimulus that produced it, provided the depolarization is sufficiently large to reach threshold. Once an action potential is triggered, the membrane potential goes through a stereotyped sequence of changes (Figure 44.9a). There is a large depolarizing phase, during which the membrane briefly reverses polarity, actually making the interior of the cell positive compared to the outside for a brief period. This is followed rapidly by a steep repolarizing phase, during which the membrane potential returns to its usual resting level. There may also be a period, called the undershoot, during which the membrane potential is transiently more negative than the normal resting potential. The whole event is typically over within a few milliseconds.

The action potential arises because excitable cells possess particular types of ion channels called **voltage-gated channels** or voltage-sensitive channels. These channels open and close, like gates, in response to changes in membrane potential (the voltage gradient across the membrane). Two types of voltage-gated channels contribute to the action potential: sodium channels and potassium channels. The gating of these channels during an action potential is summarized in Figure 44.9b. The voltage-gated sodium channels have two gates that are voltage-sensitive: The activation gate responds to depolarization by opening and does so rapidly; the inactivation gate responds to depolar-

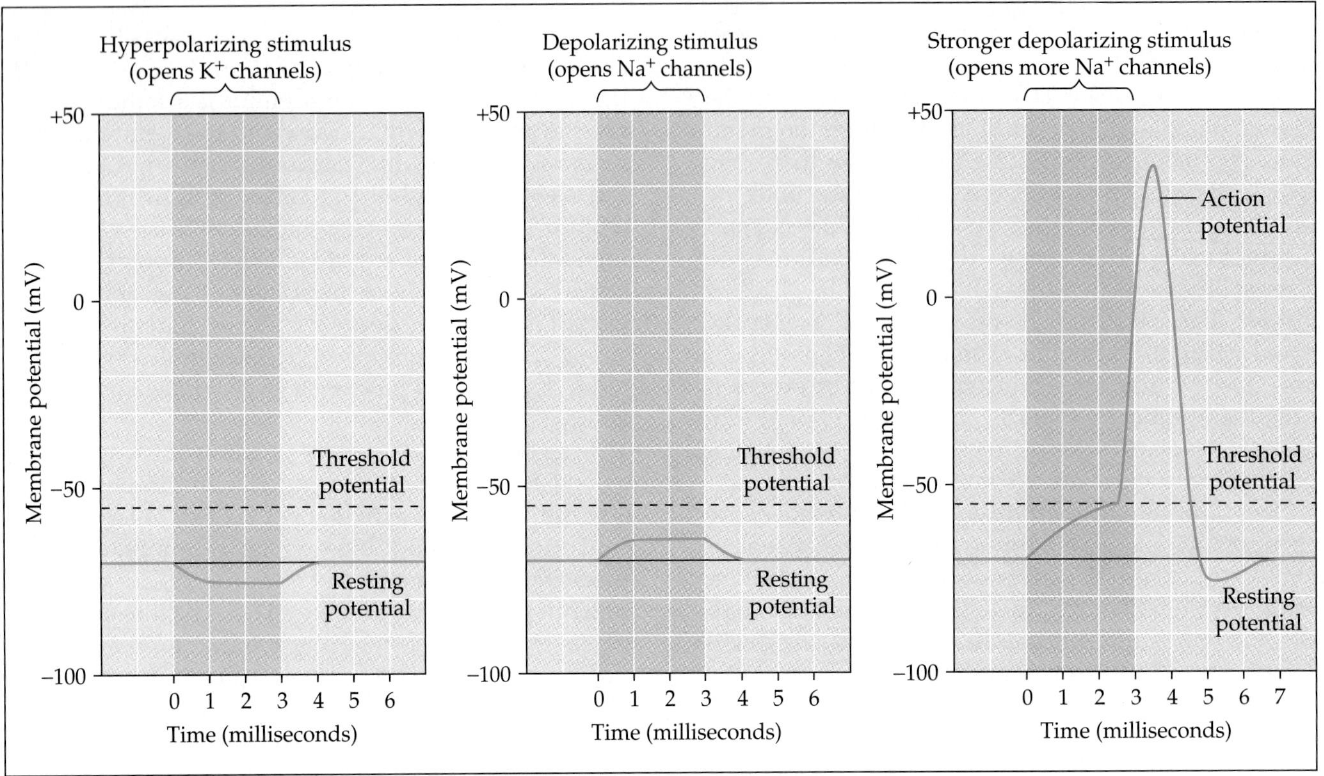

(a) Graded hyperpolarization (b) Graded depolarization (c) Action potential

Figure 44.8
Graded potentials and the action potential in a neuron. Neurons are stimulated by environmental changes that alter the membrane potential of the cell. Depending on the type of ion channel opened by the stimulus, the neuron may be caused to hyperpolarize or depolarize in response to the stimulus. (**a**) If the stimulus opens potassium channels, a hyperpolarization of the neuron would result. The size of the hyperpolarization is graded with the strength of the stimulus; therefore, the response is called a graded potential. (**b**) If the stimulus opens sodium channels, the neuron depolarizes in response to the stimulus. Once again, the size of the depolarization depends on the strength of the stimulus, so that the response is a graded depolarization. (**c**) A depolarizing stimulus of sufficient strength will change the membrane potential to a critical level called the threshold potential. This triggers a different type of response called an action potential. Unlike a graded potential, the size of the action potential is not affected by the strength of the stimulus that triggered it, provided the stimulus produces a depolarization large enough to reach the threshold. Thus, an action potential is an all-or-none event.

ization by closing, but does so slowly. At the resting potential, the inactivation gate is open but the activation gate is closed, so the channel does not allow Na^+ to enter the neuron. Upon depolarization, the activation gate opens quickly, and sodium permeability increases. This causes an influx of Na^+, which depolarizes the neuron further, opening more voltage-gated sodium channels and causing still more depolarization. This inherently explosive process, an example of positive feedback (see Chapter 36), continues until all the sodium channels are open. Depolarization to the threshold potential causes further depolarization (the action potential). At that point, the sodium permeability is about a thousandfold greater than the permeability of the resting neuron. The positive feedback that underlies the rapid depolarizing explains why the action potential, once triggered, is an all-or-none event.

Two factors underlie the rapid repolarizing phase of the action potential as membrane potential is returned to rest. First, the sodium-channel inactivation gate, which is slow to respond to changes in voltage, has time to respond to depolarization by closing, returning sodium permeability to its low resting level. Second, voltage-sensitive potassium channels open in response to the depolarization, allowing K^+ to flow rapidly out of the cell to help restore the internal negativity of the resting neuron (Figure 44.9). Like the sodium-channel activation gate, the gate controlling the potassium channel responds to depolarization by opening, but unlike the sodium-channel inactivation gate, it responds relatively slowly to changes in membrane potential. Thus, there is a period at the end of the action potential during which sodium permeability has returned to its resting level, but potassium permeability is actually higher than the resting level. The added potassium efflux during this period explains the undershoot. Notice in Figure 44.9b that at this time in the action potential, both the activation gate and the inactivation gate of the sodium channel are closed. If a second depolarizing stimulus arrives during this period,

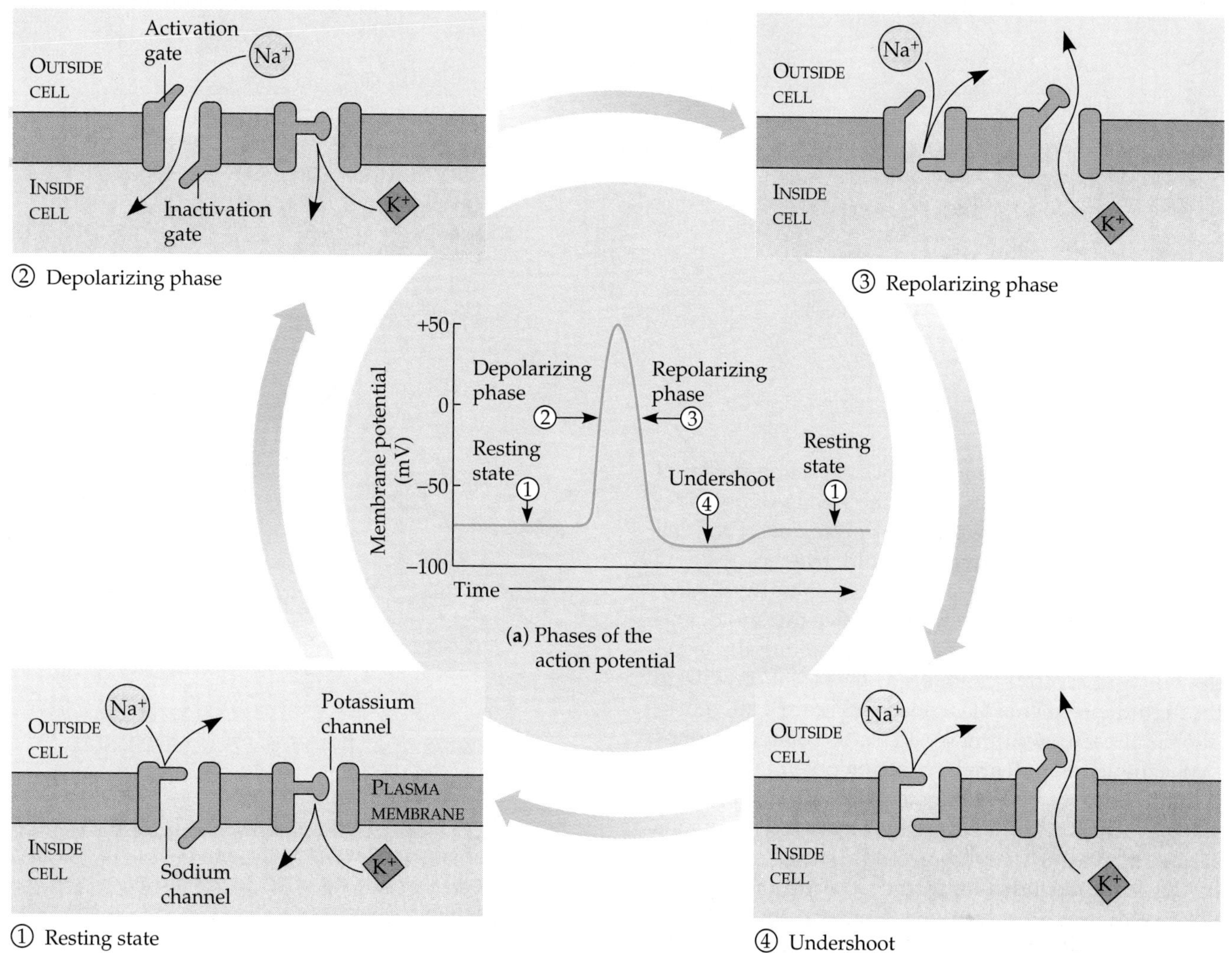

Figure 44.9
The phases of the action potential and the role of gated ion channels in those phases. (**a**) The action potential can be divided into four phases, during which the voltage-sensitive gates controlling the sodium and potassium channels are in different states. (**b**) ① In the resting state, neither channel is open. ② During the depolarizing phase of the action potential, activation gates of the sodium channels open, but potassium channels remain closed. ③ During the repolarizing phase of the action potential, inactivation gates close the sodium channels, and potassium channels open. ④ During the undershoot, sodium channels have closed, but potassium channels remain open because the relatively slow gates of the channels have not had time to respond to the repolarization of the membrane. Within another millisecond or two, the resting state is restored, and the system is ready to respond to the next stimulus.

it will be unable to trigger an action potential because the inactivation gates have not had time to reopen after the preceding action potential. This period when the neuron is insensitive to depolarization is called the **refractory period,** and it sets the limit on the maximum rate at which action potentials can be stimulated in a neuron. The properties of the gated channels involved in the action potential are summarized in Table 44.1.

If the action potential is an all-or-none event with amplitude (size) unaffected by the intensity of the stimulus, then how can the nervous system distinguish strong stimuli from weaker ones that are still sufficient to trigger action potentials? Strong stimuli result in a greater *frequency* of action potentials than weaker stimuli; if a stimulus is intense, then the neuron "fires" repeatedly, producing action potentials as rapidly as the refractory period will allow. Thus, it is the number of action potentials per second, not their amplitude, that codes for stimulus intensity in the nervous system.

Propagation of the Action Potential

A neuron is usually stimulated at its dendrites or cell body, and the resulting action potential then travels along the axon to the other end of the cell. How is the

Table 44.1 A Summary of Responses of the Voltage-Sensitive Sodium and Potassium Channels to Depolarization

Channel	Gate	Resting State of Gate	Response of Gate to Depolarization	Speed of Response
Na^+	Activation	Closed	Opens	Fast
Na^+	Inactivation	Open	Closes	Slow
K^+	Activation	Closed	Opens	Slow

action potential propagated? Once an action potential occurs at one end of a neuron, the strong depolarization guarantees that the neighboring region of the cell will be depolarized above threshold, triggering a new action potential at that position. This will, in turn, bring the next portion of the cell above threshold, and so on (Figure 44.10). At each position along the axon, the voltage-sensitive channels at that location will go through the sequence described in Figure 44.9, reproducing at each point the sequence of voltage changes that is the action potential. An action potential does not actually travel but is regenerated anew at each position along the membrane. In this way, local currents of ions across the plasma membrane give rise to a signal that travels in a perpendicular direction, along the axon.

Once the action potential is propagating along the axon, what prevents Na^+ entry from re-exciting the region *behind* the action potential, which would cause depolarization to spread back toward the cell body as well as in the direction of normal propagation? Recall that an action potential is followed by a refractory period, when inactivation gates of sodium channels are closed and an action potential cannot be triggered. A wave of depolarization passing a point along the axon cannot induce another action potential behind it, but only in the forward direction. Thus, the axon is normally a one-way avenue for the conduction of nerve impulses.

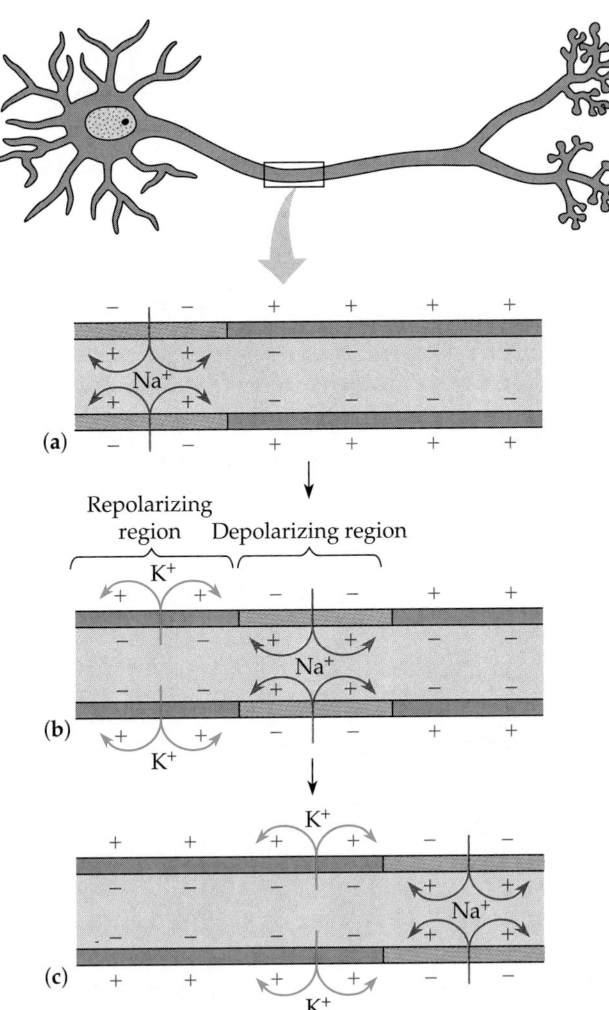

Figure 44.10
Propagation of the action potential. As sodium ions flow inward across the membrane, the resulting depolarization spreads to the portion of membrane just ahead of the impulse. Depolarization of this axon segment then initiates an action potential at that site, which in turn depolarizes the next portion of the axon. In this manner, the action potential progresses along the axon. As the action potential propagates, repolarization due to K^+ exit occurs in its wake.

Speed of Propagation of the Action Potential

One factor that affects the speed at which the action potential propagates along the axon is the diameter of the axon: The larger the diameter of the axon, the faster the speed of transmission. This is because the resistance to the flow of electrical current is inversely proportional to the cross-sectional area of the "wire" conducting the current. In a thick axon, the depolarization produced by an action potential at a particular location can effectively reach further along the interior of the axon and

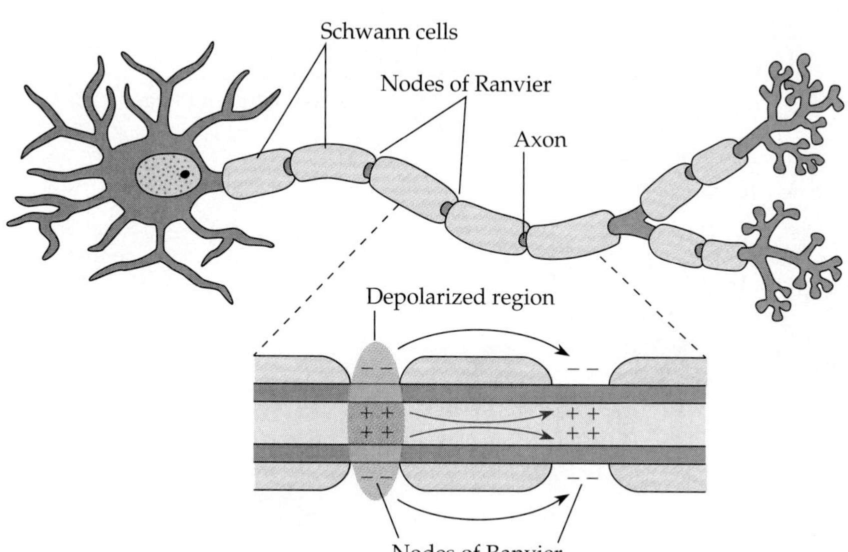

Figure 44.11
Saltatory conduction. In a myelinated axon, the depolarization resulting from an action potential at one node of Ranvier spreads along the interior of the axon to the next node (magenta arrows), triggering an action potential there. The action potential thus jumps from node to node as it propagates along the axon (black arrows).

set up a new action potential at a greater distance away than in a thin axon. Transmission speed varies from several centimeters per second for very thin axons to about 100 meters per second for the giant axons of certain invertebrates, including squids and lobsters (see the Methods Box, p. 987). These giant axons, which can be 1 mm in diameter, function in behavioral responses requiring great speed, such as the backward tail-flip that enables a threatened lobster or crayfish to escape.

A different means to speed the propagation of action potentials has evolved in vertebrates. Recall that many axons in vertebrate nervous systems are myelinated, coated with insulating layers of membranes deposited by glial cells (see Figures 44.3 and 44.6). The myelin sheath has small gaps called the **nodes of Ranvier** between successive glial cells along the axon. The voltage-sensitive ion channels that produce the action potential are concentrated in these node regions of the axon. Also, extracellular fluid is in contact with the axon membrane only at the nodes, so that the flow of ions between the inside and outside of the axon can occur only in these regions. For these reasons, the action potential does not propagate in a continuous manner over the length of the axon, but rather "jumps" from node to node, skipping the insulated regions of membrane between the nodes (Figure 44.11). This mechanism, called **saltatory conduction** (L. *saltare*, "to leap"), results in faster transmission of the nerve impulse.

We have seen how stimulating the dendrite of a neuron can produce an action potential that is propagated along the axon to the very tip of the neuron. But how is the impulse transmitted from one neuron to the next in a neural pathway? Our next step toward understanding nervous communication is to examine the synapse, the tiny space between neurons.

THE SYNAPSE: TRANSMISSION BETWEEN CELLS

The synapse is a unique junction that controls communication between neurons. Synapses are also found between sensory receptors and sensory neurons, and between motor neurons and the muscle cells they control. Here, we focus on synapses between neurons, which usually conduct signals from the axon terminals to dendrites or cell bodies of the next cells in a neural pathway. The transmitting cell is called the **presynaptic cell,** and the receiving cell is called the **postsynaptic cell.** Synapses are of two types: electrical synapses and chemical synapses.

Electrical Synapses

An **electrical synapse** allows action potentials to spread directly from the presynaptic to postsynaptic cell. The cells are connected by gap junctions (see Figure 7.34c), intercellular channels that allow the local ion currents of an action potential to flow between neurons. The giant neuron processes of lobsters and other crustaceans are actually composed of several short, thick neurons, connected end to end and coupled by electrical synapses. These make it possible for impulses to travel from neuron to neuron along the neural pathway without delay and with no loss of signal strength. Electrical synapses in the central nervous systems of vertebrates synchronize the activity of neurons responsible for some rapid, stereotyped movements. For example, electrical synapses in the brains of some fishes function in the tail-flapping reflex that helps the animal escape from a predator. However, chemical synapses

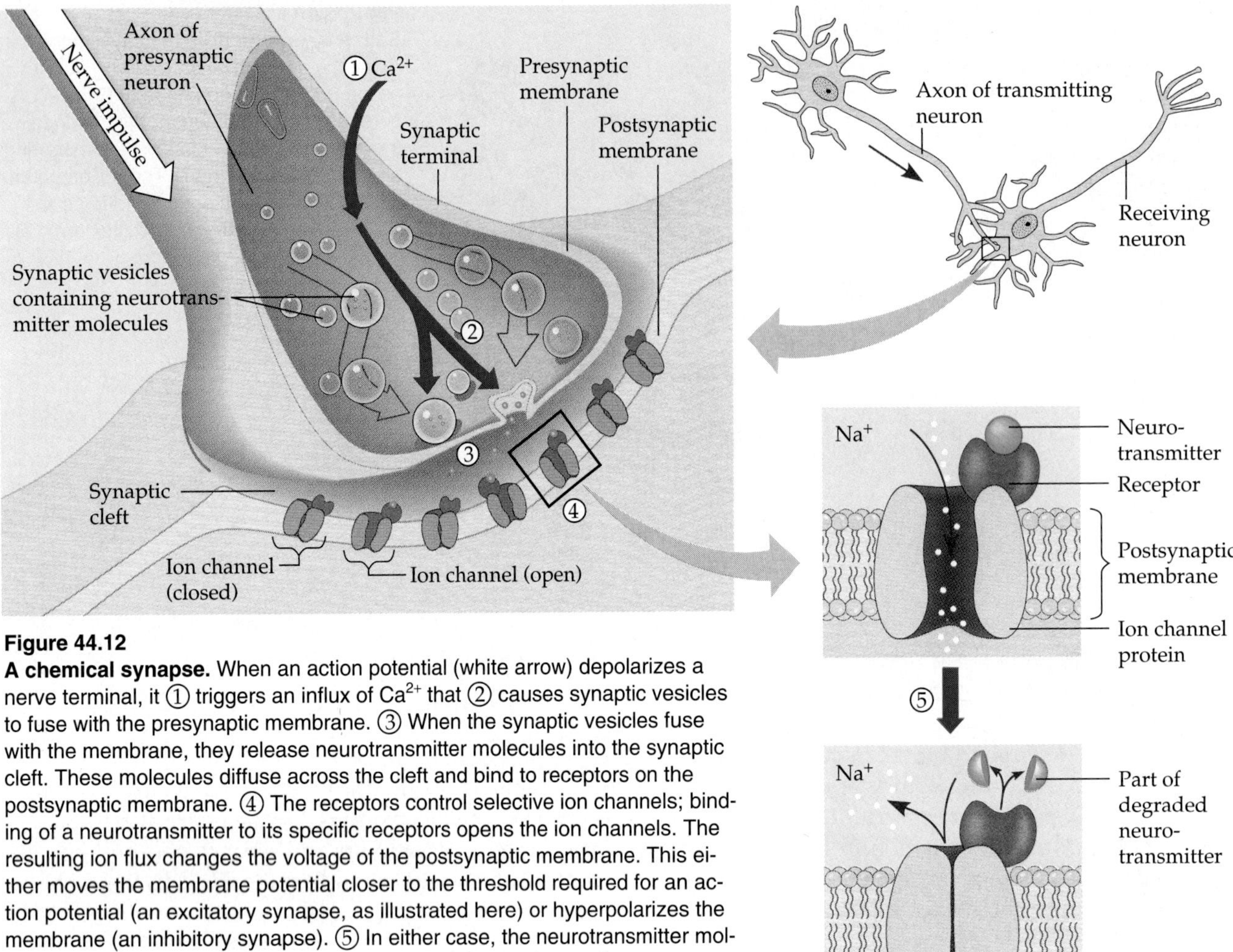

Figure 44.12
A chemical synapse. When an action potential (white arrow) depolarizes a nerve terminal, it ① triggers an influx of Ca^{2+} that ② causes synaptic vesicles to fuse with the presynaptic membrane. ③ When the synaptic vesicles fuse with the membrane, they release neurotransmitter molecules into the synaptic cleft. These molecules diffuse across the cleft and bind to receptors on the postsynaptic membrane. ④ The receptors control selective ion channels; binding of a neurotransmitter to its specific receptors opens the ion channels. The resulting ion flux changes the voltage of the postsynaptic membrane. This either moves the membrane potential closer to the threshold required for an action potential (an excitatory synapse, as illustrated here) or hyperpolarizes the membrane (an inhibitory synapse). ⑤ In either case, the neurotransmitter molecules are quickly degraded by enzymes or are taken up by another neuron, closing the ion channels and terminating the synaptic response.

are much more common than electrical synapses in vertebrates and most invertebrates.

Chemical Synapses

At a **chemical synapse,** a narrow gap, or synaptic cleft, separates the presynaptic cell from the postsynaptic cell. Because of the cleft, the cells are not electrically coupled, and an action potential occurring in the presynaptic cell cannot be transmitted directly to the membrane of the postsynaptic cell. Instead, a series of events at the synapse converts the electrical signal of the action potential arriving at the axon terminal into a chemical signal that travels across the synapse, where it is converted back into an electrical signal in the postsynaptic cell.

The key to understanding the function of a chemical synapse is to examine its structure. On the presynaptic side of the cleft is a synaptic terminal, the swelling at the end of one of the branches of the presynaptic axon (Figure 44.12). Within the cytoplasm of the synaptic terminal are numerous sacs called **synaptic vesicles.** Each of these sacs contains thousands of molecules of a **neurotransmitter,** the substance that is released as an intercellular messenger into the synaptic cleft. Many different neurotransmitters have been discovered in the nervous systems of animals. Most neurons secrete only one kind of neurotransmitter. However, a single neuron may *receive* chemical signals from a variety of neurons that secrete different neurotransmitters from their synaptic terminals.

A neuron dispatches its neurotransmitter into the synapse when the action potential being propagated along the axon arrives at the synaptic terminal and depolarizes the **presynaptic membrane,** the surface of the synaptic terminal that fronts on the cleft. Calcium ions play a central role in this conversion of the electrical impulse into a chemical signal. Depolarization of the presynaptic membrane causes Ca^{2+} to rush into the cell through voltage-sensitive channels. The sudden rise in the cytoplasmic concentration of Ca^{2+} stimulates the synaptic vesicles to fuse with the presynaptic membrane and spill the neurotransmitter into the synaptic cleft by exocytosis (see Chapter 8). Thousands of vesicles may respond in unison to a single action potential. The neurotransmitter diffuses the short distance from the presynaptic membrane to the **postsynaptic membrane,** the plasma membrane of the cell body or dendrite on the other side of the synapse.

The postsynaptic membrane is specialized to receive the chemical message. Projecting from the extracellular surface of the membrane are proteins that function as specific receptors for neurotransmitters. The receptors are associated with selective ion channels that open and close, controlling fluxes of ions across the postsynaptic membrane. A receptor is keyed to a particular type of neurotransmitter, and when it binds to this chemical, the gate of the ion channel opens, allowing specific ions, such as Na^+, K^+, or Cl^-, to cross the membrane. Thus, the ion channels of the postsynaptic membrane are chemically sensitive, in contrast to the voltage-sensitive channels responsible for the action potential. The ion movements resulting from the binding of the neurotransmitter to its receptors alter the membrane potential of the postsynaptic cell. Depending on the type of receptors and the ion channels they control, neurotransmitters binding to postsynaptic membrane may either excite the membrane by bringing its voltage closer to the threshold potential or inhibit the postsynaptic cell by hyperpolarizing its membrane. In either case, the chemical signal is quickly muted by enzymes that break the neurotransmitter down into smaller chemical components that can be recycled back to the presynaptic cell. For example, the transmitter known as acetylcholine is rapidly degraded by cholinesterase, an enzyme present in the synaptic cleft and also on the postsynaptic membrane.

Notice that one important function of the synapse is that it allows nerve impulses to be transmitted in only a single direction over a neural pathway. Synaptic vesicles are found only in the terminals at the tips of axons, and thus only the presynaptic membrane can discharge neurotransmitters. And receptors are restricted to the postsynaptic membrane, ensuring that only this membrane can receive a chemical signal from another neuron. Another important function of the synapse is to integrate information consisting of excitatory and inhibitory signals.

Summation: Neural Integration at the Cellular Level

A single neuron may receive information from numerous neighboring neurons via thousands of synapses, some of them excitatory and some of them inhibitory (Figure 44.13). Whether or not a neuron fires off an action potential at any particular instant depends on its ability to integrate these multiple positive and negative inputs.

Excitatory and inhibitory synapses have opposite effects on the membrane potential of the postsynaptic cell. At an excitatory synapse, the neurotransmitter receptors control a type of gated channel that allows Na^+ to enter the cell and K^+ to leave the cell. Because the driving force is greater for Na^+ than for K^+ (remember, the voltage and concentration gradient both drive Na^+ into the cell), the effect of opening these channels is a net flow of positive charge into the cell. This depolarizes the cell, moving the membrane potential closer to the threshold voltage and making it more likely that the postsynaptic cell will generate an action potential. Therefore, the electrical change caused by binding of the transmitter to the receptor is called an **excitatory postsynaptic potential,** or **EPSP.**

At an inhibitory synapse, binding of the neurotransmitter molecules to the postsynaptic membrane hyperpolarizes the membrane by opening ion gates that make the membrane more permeable to K^+, which rushes out of the cell; or to Cl^-, which enters the cell due to a large concentration gradient (see Figure 44.7); or to both of these ions. These ion fluxes push the membrane potential to a voltage even more negative than the resting potential, making it more difficult for an action potential to be generated. Therefore, the voltage change associated with chemical signaling at an inhibitory synapse is called an **inhibitory postsynaptic potential,** or **IPSP.** Whether a particular neurotransmitter results in an EPSP or an IPSP depends on the type of receptors and gated ion channels on the postsynaptic membrane responding to that neurotransmitter.

Both EPSPs and IPSPs are graded potentials that vary in magnitude with the number of neurotransmitter molecules binding to receptors on the postsynaptic membrane. The change in voltage, either a local depolarization or hyperpolarization, lasts only a few milliseconds because the neurotransmitters are inactivated by enzymes soon after their release into the synapse. Also, the electrical impact on the postsynaptic cell decreases with distance away from the synapse. For the postsynaptic cell to fire, the local ion currents due to EPSPs must be strong enough to depolarize the membrane in the region of the axon hillock to the threshold potential, usually about –50 mV. The axon hillock is the zone of "spike" initiation, the region where voltage-sensitive Na^+ gates open to generate an action potential when some stimulus has depolarized the membrane to the threshold.

A single EPSP at one synapse, even one close to the axon hillock, is not usually strong enough to trigger an action potential. However, several synaptic terminals acting simultaneously on the same postsynaptic cell, or a smaller number of synaptic terminals discharging neurotransmitter repeatedly in rapid-fire succession,

Figure 44.13
Integration of multiple synaptic inputs. (**a**) Each neuron, especially in the central nervous system, is on the receiving end of thousands of synapses, some of them excitatory (green) and some of them inhibitory (red). At any instant, an action potential is generated at the axon hillock if the combined effect of ion currents induced by excitatory and inhibitory synapses depolarizes the membrane in the hillock region to the threshold potential. This mechanism of integrating multiple positive and negative inputs by adding their individual effects is called summation. (**b**) This micrograph reveals numerous synaptic knobs that communicate with a single postsynaptic cell (SEM).

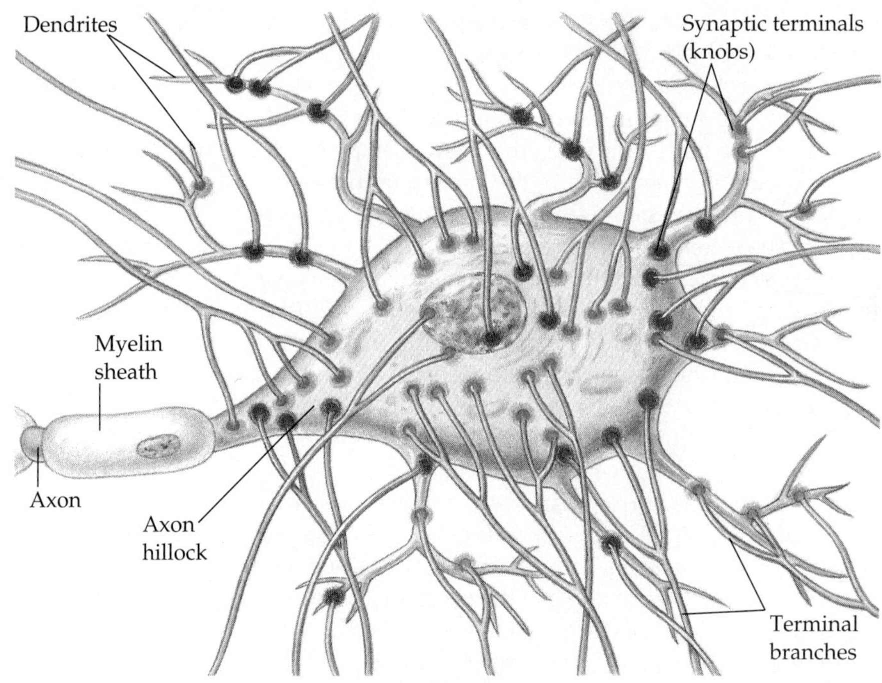

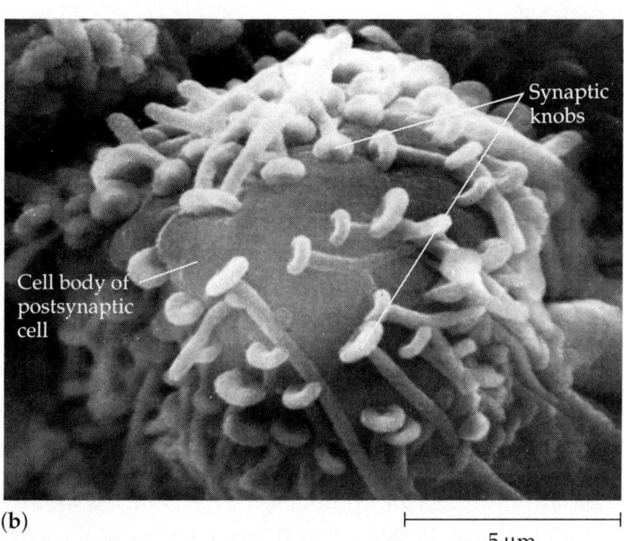

can have an accumulative impact on the membrane potential at the axon hillock. This additive effect of postsynaptic potentials is called **summation** (Figure 44.14).

There are two types of summation: temporal summation and spatial summation. In **temporal summation,** chemical transmissions from one or more synaptic terminals occur so close together in time that each postsynaptic potential affects the membrane before the voltage has returned to the resting potential after the previous stimulation (Figure 44.14b). In **spatial summation,** several different synaptic terminals, usually belonging to different presynaptic neurons, stimulate a postsynaptic cell at the same time and have an additive effect on the membrane potential (Figure 44.14c).

By reinforcing one another through temporal or spatial summation, the ion currents associated with several EPSPs can depolarize the axon hillock to the threshold voltage, causing the neuron to fire (generate an action potential). IPSPs also summate, teaming up to hyperpolarize the membrane to a voltage more negative than any single release of neurotransmitter at an inhibitory synapse can achieve. Furthermore, EPSPs summate with IPSPs, each countering the electrical effects of the other (Figure 44.14d).

The axon hillock is the neuron's integrating center; its membrane potential at any instant is an average of the summated depolarization due to all EPSPs and the summated hyperpolarization due to all IPSPs. (Of course, synapses close to the axon hillock, compared to more remote synapses, have a disproportionate impact on the membrane potential at the integrating center.) Whenever the EPSPs overpower IPSPs enough for the membrane potential at the axon hillock to reach the threshold voltage, an action potential is generated and the impulse is transmitted along the axon to the next synapse. A few milliseconds later, after the refractory

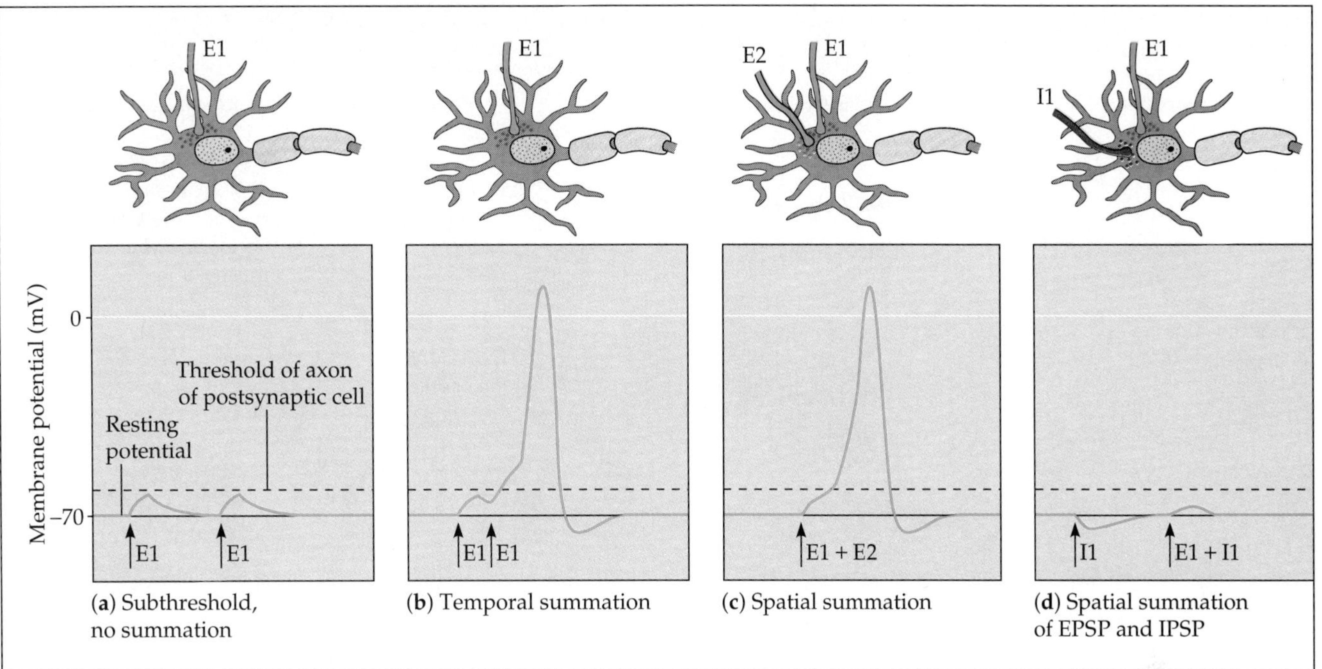

Figure 44.14
Summation of postsynaptic potentials. (**a**) Generally, a single EPSP cannot depolarize the postsynaptic membrane all the way to the threshold level, and thus no action potential is generated at the axon hillock. Repeated EPSPs do not summate if they do not overlap in time. In the graph that traces changes in the membrane potential at the axon hillock, the arrow indicates the time of the signal at one excitatory synapse, labeled E1. (**b**) Temporal summation occurs when a second release of neurotransmitter at a synapse affects the postsynaptic membrane when it is still partially depolarized from a slightly earlier stimulation. Due to temporal summation, EPSPs that are individually subthreshold can generate an action potential if they overlap in time, so that they reinforce one another. (**c**) Spatial summation occurs when two or more presynaptic cells release neurotransmitter at the same time, causing a cumulative voltage change greater than the individual EPSPs. Here, two excitatory synapses, E1 and E2, team up to depolarize the postsynaptic cell to threshold voltage. (**d**) EPSPs and IPSPs also summate, but, of course, they move the membrane potential in opposite directions. In this case, an inhibitory synapse, I1, subtracts from the depolarization due to an excitatory synapse, E1.

period, the neuron may fire again if the sum of all synaptic inputs at that moment is still sufficient to depolarize the membrane of the axon hillock to threshold level. On the other hand, by that time, the sum of all EPSPs and IPSPs may put the membrane potential at the axon hillock at a voltage more negative than the threshold, or even hyperpolarize the membrane to a potential more negative than the resting potential, desensitizing the neuron for the moment.

Action potentials, remember, are all-or-none events. But now we see that the occurrence of these nerve impulses depends on the ability of the neuron to integrate quantitative information in the form of multiple excitatory and inhibitory inputs, each involving the specific binding of a neurotransmitter to a receptor on the postsynaptic membrane.

Neurotransmitters and Receptors

Neuroscientists are only beginning to understand the chemistry of the nervous system. About ten different molecules are known to function as neurotransmitters, and dozens more are good candidates for addition to the growing list. For a particular compound to be recognized for certain as a neurotransmitter at a specific type of synapse, it must meet three criteria:

1. The presynaptic cell must contain the compound in synaptic vesicles and discharge the substance when the cell is appropriately stimulated, and the chemical must then affect the membrane potential of the postsynaptic membrane.

2. The compound should be able to cause an EPSP or IPSP when experimentally injected into the synapse with a micropipette.

3. The substance must be removed rapidly from the synapse, by either enzymatic degradation or uptake by a cell, allowing the postsynaptic membrane to return to resting potential.

As analytical methods in neurochemistry continue to improve, it is likely that many additional compounds will meet these criteria for neurotransmitters.

Table 44.2 Neurotransmitters

Neurotransmitter	Structure	Functional Class	Sites Where Secreted
Acetylcholine	$H_3C-C(=O)-O-CH_2-CH_2-N^+-(CH_3)_3$	Excitatory to vertebrate skeletal muscles; excitatory or inhibitory at other sites	Central nervous system; peripheral nervous system; vertebrate neuromuscular junction
Biogenic Amines			
Norepinephrine	$(HO)_2C_6H_3-CH(OH)-CH_2-NH_2$	Excitatory or inhibitory	Central nervous system; peripheral nervous system
Dopamine	$(HO)_2C_6H_3-CH_2-CH_2-NH_2$	Generally excitatory; may be inhibitory at some sites	Central nervous system; peripheral nervous system
Serotonin	HO-indole ring (N–H), $C-CH_2-CH_2-NH_2$	Generally inhibitory	Central nervous system
Amino Acids			
GABA (gamma aminobutyric acid)	$H_2N-CH_2-CH_2-CH_2-COOH$	Generally inhibitory	Central nervous system; inhibitory transmitter at invertebrate neuromuscular junction
Glycine	H_2N-CH_2-COOH	Generally inhibitory	Central nervous system
Glutamate	$H_2N-CH(COOH)-CH_2-CH_2-COOH$	Generally excitatory	Central nervous system; excitatory transmitter at invertebrate neuromuscular junction
Peptides			
Met-enkephalin (an endorphin)	Tyr–Gly–Gly–Phe–Met	Generally inhibitory	Central nervous system
Substance P	Arg–Pro–Lys–Pro–Gln–Gln–Phe–Phe–Gly–Leu–Met	Excitatory	Central nervous system; peripheral nervous system

Acetylcholine is one of the most common neurotransmitters in both invertebrates and vertebrates (Table 44.2). At the vertebrate neuromuscular junction, the synapse between a motor neuron and a skeletal muscle cell, acetylcholine is the transmitter released from the terminal of the motor axon to depolarize the postsynaptic muscle cell. In other cases, acetylcholine is inhibitory. For example, this transmitter slows down the hearts of vertebrates and mollusks. This versatility of a single transmitter, remember, depends on the diverse types of receptors found on different postsynaptic cells.

The **biogenic amines** are neurotransmitters derived from amino acids. One family of biogenic amines, known more specifically as catecholamines, is produced from the amino acid tyrosine. This group includes **epinephrine, norepinephrine,** and **dopamine.** Another biogenic amine, **serotonin,** is synthesized from the amino acid tryptophan. The biogenic amines most commonly function as transmitters within the central nervous system. However, norepinephrine also functions in a branch of the peripheral nervous system called the autonomic nervous system, which will be discussed shortly. Dopamine and serotonin are widespread in the brain and affect sleep, mood, attention, and learning. Imbalances of these transmitters are associated with mental illness; for example, an excess production of dopamine is one factor that has been tied to schizophrenia. Some psychoactive drugs, including LSD and mescaline, apparently induce hallucinations by binding to receptors for certain biogenic amines.

Three amino acids, **glycine, glutamate,** and **gamma aminobutyric acid (GABA)** are known to function as neurotransmitters in the central nervous system. GABA, believed to be the transmitter at most inhibitory synapses in the brain, produces IPSPs by increasing the chloride permeability of the postsynaptic membrane. The brain has hundreds of times more GABA than it does any other neurotransmitter.

Several **neuropeptides,** relatively short chains of amino acids, probably serve as neurotransmitters. The **endorphins** (or **enkephalins**) are neuropeptides that function as natural analgesics, decreasing perceptions of pain by the central nervous system. They were first discovered in the 1970s by neurochemists studying the mechanism of opium addiction. Candace Pert and Solomon Snyder of Johns Hopkins University found specific receptors for the opiates morphine and heroin on neurons in the brain (see the interview that precedes Unit One). It seemed odd that humans would have receptors keyed to chemicals from a plant (the opium poppy). Further research found that, in fact, the drugs bind to these receptors in the brain by mimicking endorphins, the natural painkillers produced in the brain during times of physical or emotional stress, such as during the labor of a pregnant woman. In addition to relieving pain, endorphins also decrease urine output (by affecting ADH secretion; see Chapter 41), depress respiration, produce euphoria, and have other emotional effects through specific pathways in the brain. An endorphin is also released from the anterior pituitary gland as a hormone that affects specific regions of the brain. Once again, we see the overlap between nervous and endocrine control.

A neurotransmitter affects the postsynaptic cell by one of two general mechanisms. Acetylcholine and the amino acid transmitters bind to receptors and alter the permeability of the postsynaptic membrane to specific ions, either depolarizing (EPSP) or hyperpolarizing (IPSP) the membrane. In contrast, the biogenic amines and the neuropeptides usually have a longer-lasting impact because they affect metabolism within the postsynaptic cell. Binding of these transmitters to their specific receptors on the postsynaptic membrane activates an enzyme that produces a second messenger within the cell, often cyclic AMP, the same molecule that functions as a second messenger in many responses of target cells to hormones (see Chapter 41).

It must be emphasized again that the action of a neurotransmitter depends on how the specific receptor on the postsynaptic membrane behaves when bound to the transmitter. Considerable research has focused on the acetylcholine receptor that functions at excitatory synapses, where acetylcholine depolarizes the postsynaptic membrane. The receptor is a protein consisting of five polypeptide subunits that span the plasma membrane. The protein is donutlike in shape, with a central pit that probably leads to a channel through which Na^+ and K^+ can pass. The mechanism by which the binding of acetylcholine to the receptor protein opens the ion channel is not yet known.

Neural Circuits and Clusters

We have now seen how neurons, the individual units of the nervous system, function. But in a nervous system, the activity of millions or more of these units must be coordinated. Some organizational principles are common to all nervous systems.

Neurons are arranged in circuits, groups of neurons that feed into and away from one another to carry information along specific pathways. The three major patterns—**convergent circuits, divergent circuits,** and **reverberating circuits**—are shown and described in Figure 44.15.

Nerve cell bodies are not uniformly distributed within the nervous system, but are arranged in functional groups called **ganglia** (singular, **ganglion**). When a ganglion occurs within the brain, it is often referred to as a **nucleus** (not to be confused with the nucleus of an individual cell). These clusters of cells allow parts of the nervous system to coordinate activities without involving the entire system. This centralization of nervous coordination is one of the key features of the evolution of nervous systems in both invertebrates and vertebrates.

INVERTEBRATE NERVOUS SYSTEMS

While there is remarkable uniformity in how nerve cells function throughout the animal kingdom, there is great diversity in how nervous systems are organized. The simplest type of nervous system occurs in *Hydra,* a cnidarian (Figure 44.16a). The cnidarian **nerve net** is a loosely organized system of nerves with no central control. Because most of the synapses in the nerve net are electrical, impulses are conducted in both directions, and stimulation at any point on the body of a hy-

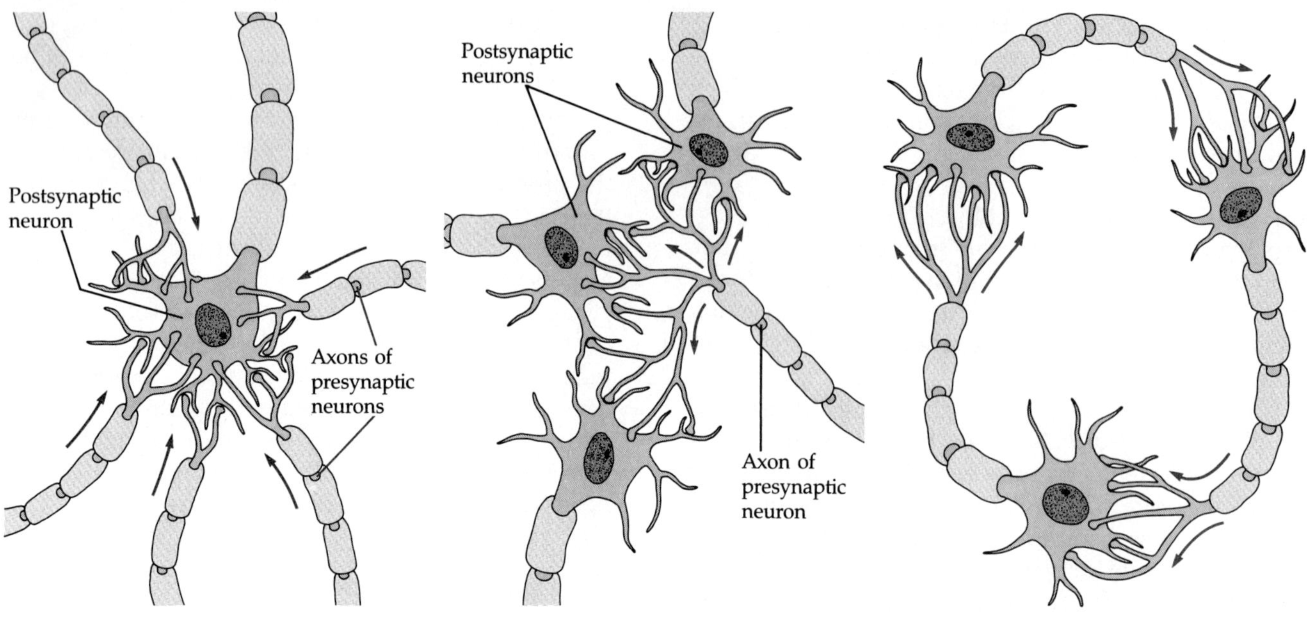

Figure 44.15
Three types of neural circuits. (**a**) In convergent circuits, information from several presynaptic neurons comes together at a single postsynaptic neuron. Convergent circuits can bring together information from several sources, such as vision, touch, and hearing, to identify an object in the environment. (**b**) In divergent circuits, information from a single neuron spreads out to several postsynaptic neurons. This type of circuit can take information from a single source, such as the eye, to several parts of the brain. (**c**) Some neurons participate in circular reverberating circuits, in which the signal returns to its source. Reverberating circuits are thought to play a part in memory storage.

dra spreads from that site to cause movements of the entire body. Other cnidarians and ctenophores show the first signs of centralization. In jellyfishes, clusters of nerve cells at the margin of the bell (associated with simple sensory structures) and pathways around the bell confer the ability to perform tasks such as swimming, which require more complex coordination of the entire body.

Such modified nerve nets are not restricted to Cnidaria and Ctenophora. The nervous systems of echinoderms are similar to those of jellyfishes (Figure 44.16b). In the sea star, for example, radial nerves travel through each arm from a central nerve ring around the oral disk. Branches of the radial nerves form an interconnected network similar to the nerve net of *Hydra.* This system coordinates movement regardless of which arm leads.

Cnidarians and echinoderms are radially symmetric and often sedentary. Bilateral animals with more active lifestyles tend to have sense organs and feeding structures localized at the anterior, or head, end. This concentration of sensory and feeding organs on the anterior end of a moving animal, the part of an animal most likely to make first contact with food or threatening stimuli, is the evolutionary trend called **cephalization** (see Chapter 29). Enlargement of anterior ganglia that receive sensory information and control feeding structures is a corresponding feature of cephalization that gave rise to the first brains.

Flatworms are the simplest animals to show two hallmarks of cephalization: a "brain" (cephalic ganglia) and one or more nerve trunks that serve as thoroughfares for information, constituting a central nervous system (Figure 44.16c). The relatively simple flatworm brain contains many large interneurons that coordinate most nervous functions, relaying signals from the sensory structures of the head to the muscles of the body. From two to several nerve trunks travel posteriorly from the brain, usually in a ladderlike system with transverse nerves connecting the main trunks. This simple nervous system controls such responses as the flatworm's ability to sense bright light and retreat to a darker place, such as the safety of a space beneath a rock.

Other invertebrates show increasing centralization of the nervous system. In contrast to the diffuse, ladderlike nervous system of flatworms, a well-defined ventral nerve cord with a prominent brain at the anterior end occurs in annelids and arthropods (Figure 44.16d and e). The nerve cords of these segmented animals often contain ganglia in each segment to coordinate the actions of that segment. The brains are much

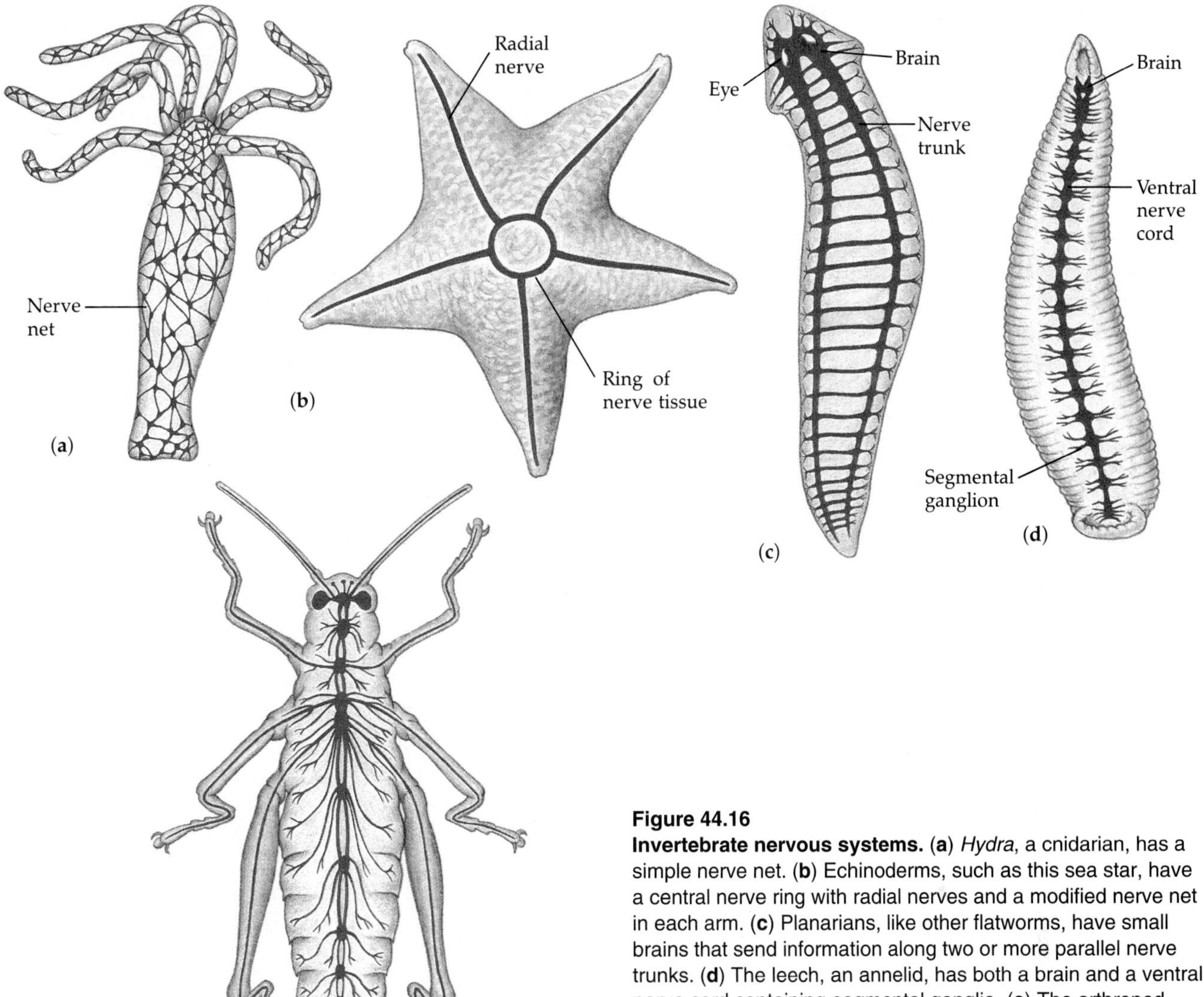

Figure 44.16
Invertebrate nervous systems. (**a**) *Hydra*, a cnidarian, has a simple nerve net. (**b**) Echinoderms, such as this sea star, have a central nerve ring with radial nerves and a modified nerve net in each arm. (**c**) Planarians, like other flatworms, have small brains that send information along two or more parallel nerve trunks. (**d**) The leech, an annelid, has both a brain and a ventral nerve cord containing segmental ganglia. (**e**) The arthropod nervous system, as shown in this insect, exhibits more extensive fusion of ganglia, and thus is less uniformly segmented than the annelid system.

larger and more complex than those of flatworms. Some of the segmental ganglia, however, have only a few large cells.

The mollusks provide good examples of how nervous system complexity correlates with habitat and natural history as well as with phylogeny. Sessile or slow-moving mollusks such as clams have little or no cephalization and only simple sense organs. Their central nervous system is a chain of ganglia circling the body. In contrast, the cephalopods have the most sophisticated nervous systems of any invertebrates, rivaling even those of some vertebrates. The large brain of the octopus, accompanied by large, image-forming eyes and rapid conduction along giant axons, correlates well with the active predatory life of the animal. The octopus is capable of learning to discriminate between visual patterns and to perform specific tasks in laboratory experiments. Indeed, learning and memory are probably everyday functions of the octopus as it interacts with its natural environment.

THE VERTEBRATE NERVOUS SYSTEM

Because the vertebrate nervous system is so complex, it is convenient to divide it into components that differ in function (Figure 44.17). The primary division is between the central nervous system, which processes in-

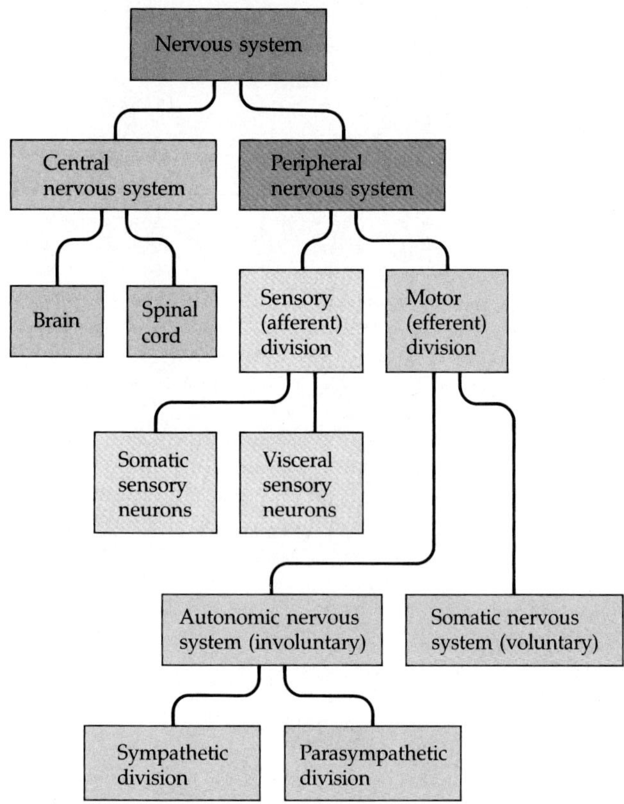

Figure 44.17
The divisions of the vertebrate nervous system.

formation, and the peripheral nervous system, which carries information to and from the central nervous system and the sensory, muscle, and gland cells.

The Peripheral Nervous System

The peripheral nervous system really consists of two separate groups of cells. The **sensory,** or **afferent, nervous system** is made up of neurons that bring information *to* the central nervous system from sensory receptors, whereas the **motor,** or **efferent, nervous system** carries signals *from* the central nervous system to effector cells.

In the human peripheral nervous system, there are 12 pairs of cranial nerves that originate in the brain and innervate organs of the head and upper body, as well as 31 pairs of spinal nerves that innervate the entire body. Most of the cranial nerves and all the spinal nerves contain both sensory and motor neurons; a few of the cranial nerves are sensory only (the olfactory and optic nerves, for example).

Nervous systems have two basic functions: to control responses to the external environment and to coordinate the functions of internal organs (that is, to maintain homeostasis). The sensory nervous system contributes to both, bringing in stimuli from the external environment and monitoring the status of the internal environment. The motor nervous system has separate divisions associated with these two functions. The motor neurons of the **somatic nervous system** carry signals to skeletal muscles mainly in response to external stimuli. The somatic nervous system is often considered voluntary because it is subject to conscious control, but a substantial proportion of skeletal muscle movement is actually determined by reflexes. A **reflex** is an automatic reaction to a stimulus, mediated by the spinal cord or lower brain. The **autonomic nervous system,** in contrast to the somatic nervous system, regulates the internal environment by controlling the smooth and cardiac muscles and the organs of the gastrointestinal, cardiovascular, excretory, and endocrine systems. This control is generally involuntary.

The autonomic nervous system consists of two divisions that are anatomically, physiologically, and chemically distinguishable. These two divisions are the **sympathetic** and **parasympathetic nervous systems** (Figure 44.18). When sympathetic and parasympathetic nerves innervate the same organ, they often (but not always) have antagonistic (opposite) effects. In general, the parasympathetic division enhances activities that gain and conserve energy, such as stimulating digestion and slowing the heart. The sympathetic division increases energy expenditures and prepares an individual for action by accelerating the heart, increasing metabolic rate, and performing related actions.

The somatic and autonomic nervous systems often cooperate. In response to a drop in temperature, for example, the hypothalamus signals the autonomic nervous system to constrict surface blood vessels to reduce heat loss, at the same time signaling the somatic nervous system to cause shivering.

The Central Nervous System

The central nervous system, or CNS, forms the bridge between the sensory and motor functions of the peripheral nervous system. The CNS consists of a bilaterally symmetric group of structures in two main organs. The **spinal cord,** which runs down the neck and back inside the vertebral column, or spine, receives information from the skin and muscles and sends out motor commands for movement. The **brain,** at the top of the spinal cord, contains centers for more complex integration of homeostasis, perception, movement, and (in humans, at least) intellect and emotions. The central nervous system is covered by a set of three protective layers of connective tissue, the **meninges.**

The axons carrying signals in the CNS are located in well-defined bundles, or tracts, whose myelin sheaths

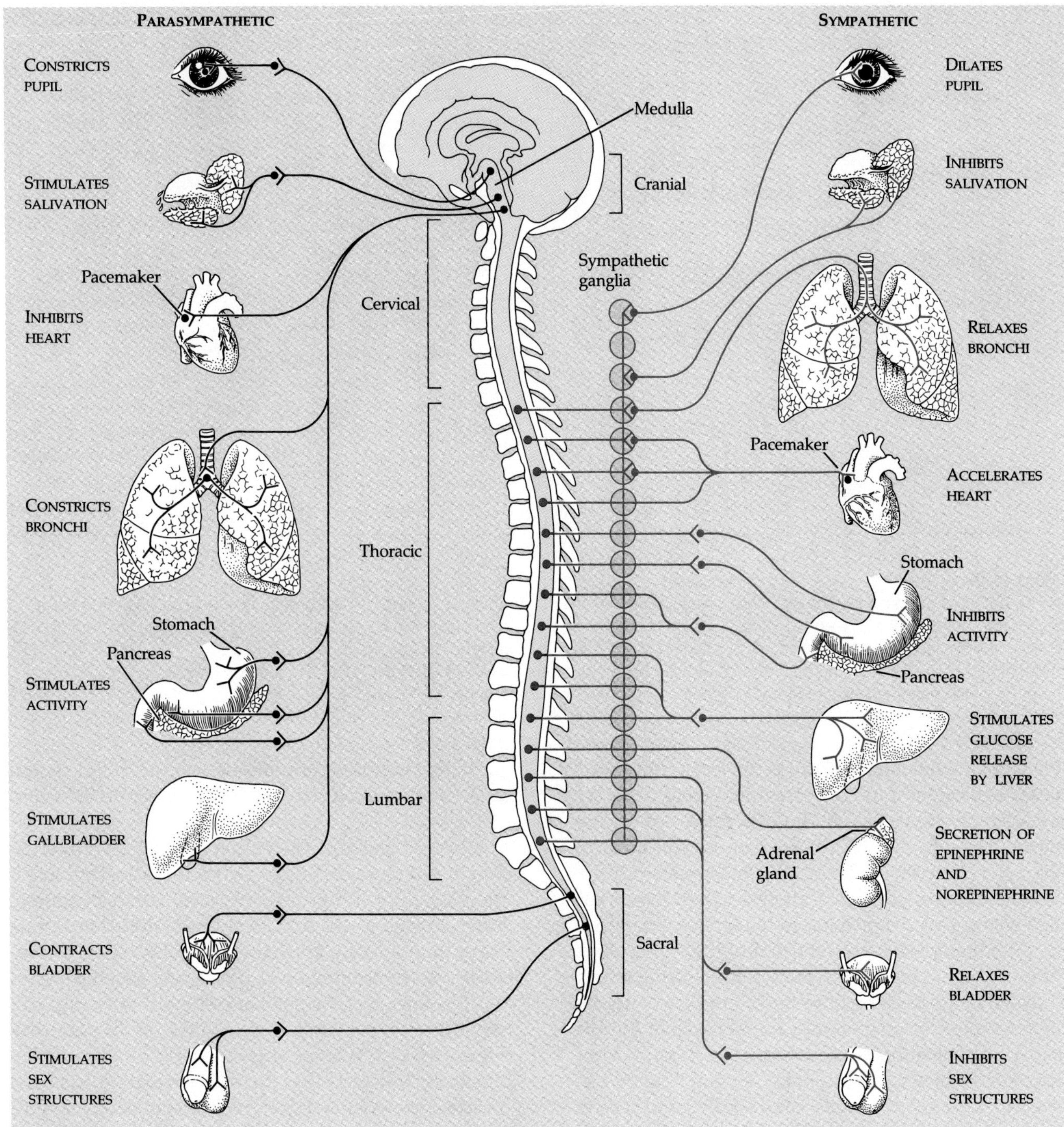

Figure 44.18
The autonomic nervous system. The sympathetic and parasympathetic systems differ anatomically and chemically. Anatomically, the divisions of the autonomic nervous system are distinguished by where their nerves originate. The nerves of the sympathetic nervous system emerge from the upper and central spinal cord (the thoracic and lumbar regions). Parasympathetic nerves originate at the top and bottom of the central nervous system, coming from some of the cranial nerves and from the sacral region of the spinal cord. Most autonomic pathways consist of a chain of two neurons in series, but the arrangements and positions of the junctions between the presynaptic and postsynaptic cells differ. Sympathetic nerves usually have short presynaptic axons that synapse with cell bodies of the second neurons in prominent sympathetic ganglia near the spinal cord; the neurotransmitter released at this synapse is acetylcholine. The long axons of the postsynaptic cells leave the sympathetic ganglia and branch to travel to the target organs. There, the neurotransmitter released from the sympathetic synaptic terminals is usually norepinephrine. The presynaptic axons of the parasympathetic nerves are typically much longer than the presynaptic axons in sympathetic nerves, and synapses with the second neurons usually occur at or near the target organ. Acetylcholine is the neurotransmitter released by parasympathetic neurons, both at the target organ and at the synapse between the two neurons in the chain.

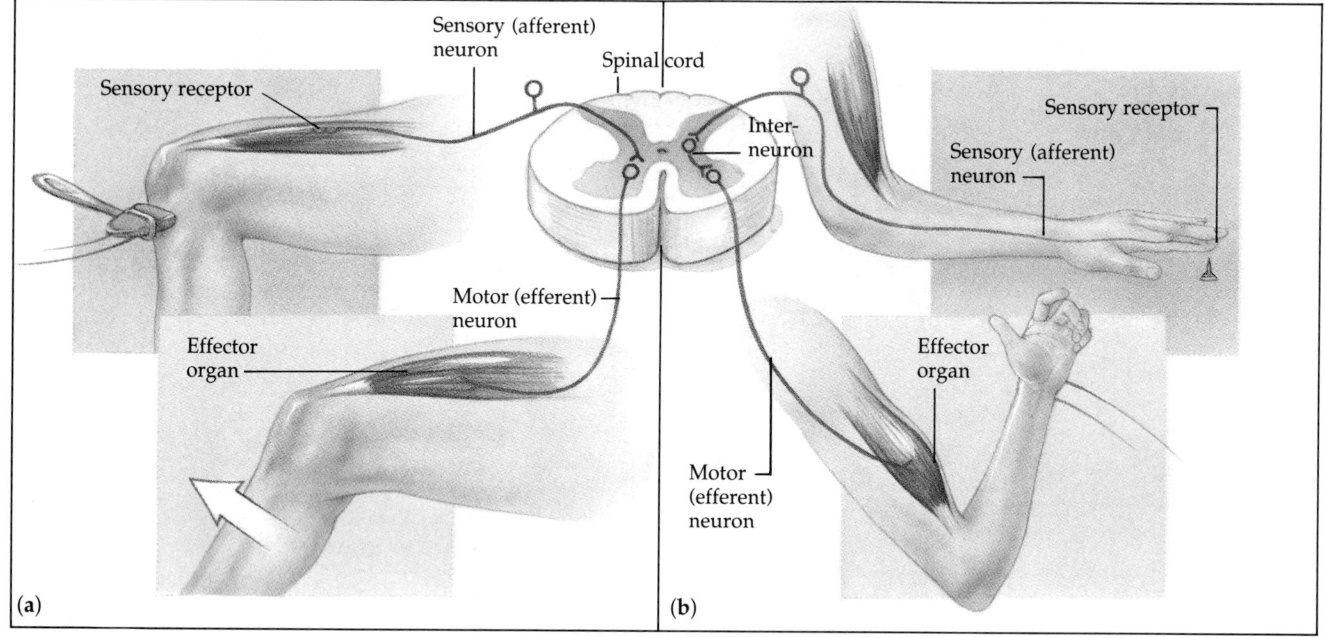

Figure 44.19
The spinal cord and spinal reflexes. The gray matter (butterfly-shaped region) in the center of the spinal cord contains the cell bodies of motor neurons and interneurons. The outer white matter consists of motor and sensory axons. The incoming sensory neurons have their cell bodies outside the spinal cord in dorsal root ganglia. (**a**) The patellar knee-jerk reflex is a simple spinal reflex involving only two neurons. (**b**) Most reflexes involve at least one interneuron coordinating sensory input with motor output.

give them a white appearance. In the brain, this **white matter** is located in the inner region, where there are pathways going to the cell bodies in the outer **gray matter.** The situation is reversed in the spinal cord, where the white matter is outside the gray matter.

All vertebrate nervous systems are hollow. Fluid-filled spaces called **ventricles** in the brain are continuous with the narrow **central canal** of the spinal cord. These spaces are filled with **cerebrospinal fluid,** which is formed in the brain by filtration of the blood. Among the most important functions of cerebrospinal fluid is absorption of shock to cushion the brain. Cerebrospinal fluid also carries out circulatory functions, bringing nutrients, hormones, and white blood cells to different parts of the brain. The cerebrospinal fluid normally circulates through the ventricles and central canal of the spinal cord and drains back into the veins.

The Spinal Cord The spinal cord, shown in cross section in Figure 44.19, has two principal functions: It integrates simple responses to certain kinds of stimuli, and it carries information to and from the brain. Spinal integration usually takes the form of a reflex, an unconscious, programmed response to a specific stimulus. The knee-jerk, or patellar, reflex is an example of the simplest type of reflex, involving only two neurons (Figure 44.19a). When a stretch receptor in the quadriceps muscle detects that the patellar tendon has been stretched (as when struck by a physician's rubber hammer), a sensory neuron carries that information up the

Figure 44.20
Regions of the generalized vertebrate brain.

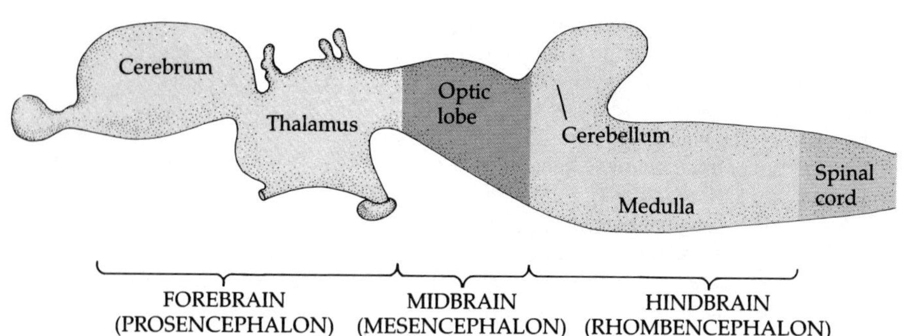

thigh to the spinal cord, where the sensory neuron synapses directly with a motor neuron. If the signal is strong enough, an action potential is generated in the motor neuron, causing contraction of the quadriceps muscle and the forward knee jerk. Most reflexes are more complex, sometimes involving one or more interneurons between the sensory and motor neurons (Figure 44.19b). Also, branches in the pathway may carry signals to other segments of the spinal cord or to the brain to generate larger-scale or more complex responses.

Evolution of the Vertebrate Brain The evolution of complex behaviors in vertebrates is largely due to increases in brain complexity. The vertebrate brain began as a set of three bulges at the anterior end of the spinal cord (Figure 44.20). These three regions, the **prosencephalon** or **forebrain,** the **mesencephalon** or **midbrain,** and the **rhombencephalon** or **hindbrain,** are present in all vertebrates. In more complex brains, they become further subdivided, providing additional capacity for the integration of complex activities.

Three trends are evident in the evolution of the vertebrate brain (Figure 44.21). First, the relative size of the brain increases in certain evolutionary lineages. Brain size is a fairly constant function of body weight among fishes, amphibians, and reptiles, but increases dramatically relative to body size in birds and mammals. A rodent weighing 100 g would have a much larger brain than a 100 g lizard. But the brain of that lizard and the brain of a 100 g fish would be approximately the same size.

A second evolutionary trend is increased compartmentalization of function. The three primitive divisions remain, but they become subdivided into areas assuming specific responsibilities. In the hindbrain, for example, the region known as the cerebellum becomes a prominent structure for coordinating movements. In the forebrain, one subdivision called the diencephalon contains the thalamus and hypothalamus and other groups of cells, while another subdivision called the telencephalon contains the cerebral cortex, the part of the brain most important in learning and memory. As these specific regions become more complex, the original divisions between the three bulges become blurred. The midbrain, hindbrain, and lower forebrain of mammals are not readily distinguishable in adults, although the three separate bulges are apparent in developing embryos.

The third trend in the evolution of the vertebrate brain is the increasing sophistication and complexity of the forebrain. As amphibians and reptiles made the transition from water to land, the vision and hearing functions of the midbrain and hindbrain became increasingly important, and natural selection favored enlargement of these regions. Beyond this, however, more complex behaviors parallel the growth of one important region of the forebrain—the **cerebrum.** Among mammals, in particular, more sophisticated behavior is associated with the relative size of the cerebrum and the presence of folds, or convolutions, that increase the surface area of the cerebrum. Because the cell bodies of the cerebrum are in the cortex, or outer layer, the surface area of the brain is more important in determining

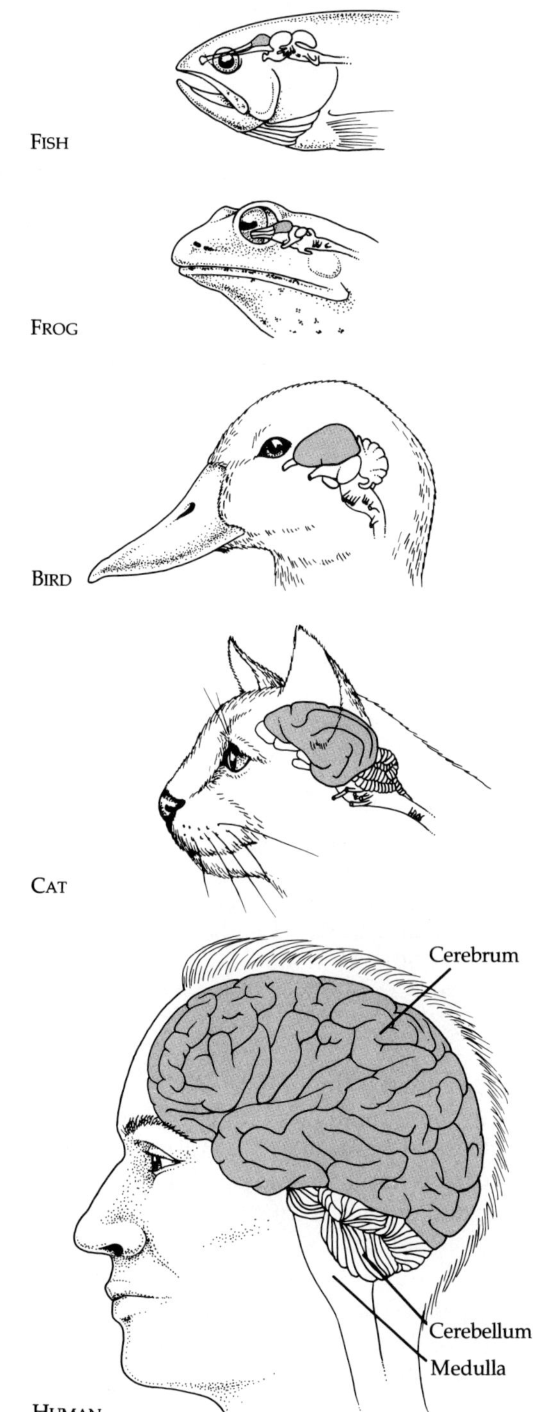

Figure 44.21
Evolution of the vertebrate brain. Three major trends are evident: an increase in overall brain size relative to body size, compartmentalization of function, and increased development of the forebrain (gold), especially in mammals.

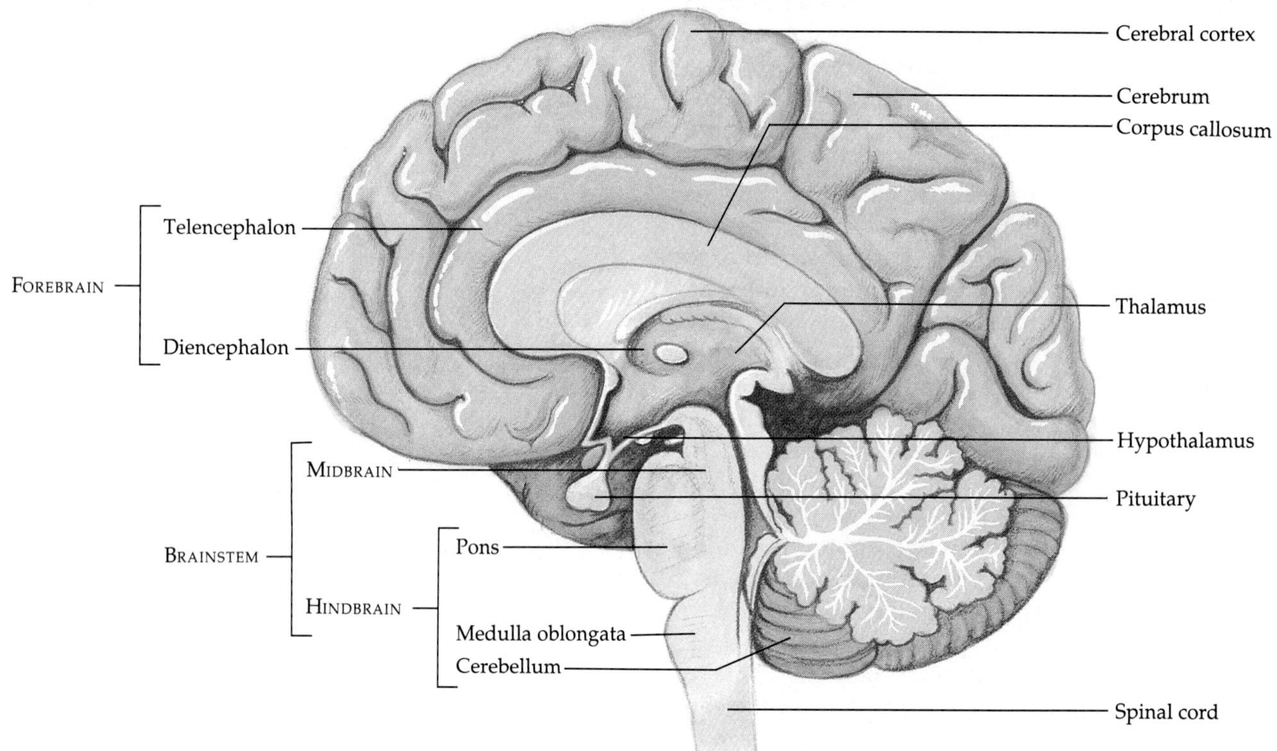

Figure 44.22
The human brain. This midsagittal section shows the major structures of the brain. The hindbrain controls homeostasis and coordinates movement, while the midbrain functions as a relay center for sensory information. The forebrain is highly developed in mammals, with the diencephalon a relay and homeostasis center and the telencephalon the locus for information processing.

performance than is the volume. Although less than 5 mm thick, the human cerebral cortex occupies over 80% of the total brain mass. Marsupials, such as the opossum, have very little folding of the cerebral cortex, while cats and other placental mammals have substantially more. Primates and cetaceans (whales and porpoises, for example) have dramatically larger and more complex cerebral cortices than any other vertebrates. In fact, the cerebral cortex of the porpoise is second in surface area (relative to body size) only to that of the human.

The Human Brain At 1.35 kg (about 3 lb), the human brain is one of the largest organs in the body. Its soft, almost squishy texture belies the density of its cells and the complexity of its structure and function. As already noted, the brain develops from three primary bulges at the anterior end of the spinal cord. These, in turn, differentiate into several distinct structures with specific functions (Figure 44.22). The hindbrain and midbrain together make up the **brainstem,** and they form a cap on the spinal cord that extends to about the middle of the brain.

The hindbrain has three parts that function in homeostasis, movement coordination, and signal conduction. The lowest parts of the brain, the **medulla oblongata** and the **pons** just above it, appear as swellings of the hindbrain at the top of the spinal cord. The medulla contains centers that control several visceral (autonomic, homeostatic) functions, including breathing, heart and blood vessel activity, swallowing, vomiting, and digestion. The pons also participates in some of these activities, having nuclei (ganglia) that regulate the breathing centers in the medulla, for example. All the bundles of axons carrying sensory information to and motor instructions from higher brain regions pass through the hindbrain, making conduction of information one of the most important functions of the medulla and pons. The hindbrain also helps coordinate large-scale body movements such as walking. Most of the descending axons carrying instructions about movement to the spinal cord from the midbrain and forebrain cross from one side of the CNS to the other as they pass through the medulla. As a result, the right side of the brain controls much of the movement of the left side of the body, and vice versa.

The primary function of the **cerebellum,** the third part of the hindbrain, is coordination of movement. This highly convoluted, semidome-shaped outgrowth on the dorsal surface of the hindbrain is tucked behind and partially beneath the cerebrum. The cerebellum receives sensory information about the position of the joints and the length of the muscles, as well as information from the auditory and visual systems. It also receives input from the motor pathways, telling it which actions are being commanded from the cerebrum. The cerebellum uses this information to provide unconscious coordination of movements and balance. If one part of the body is moved, the cerebellum will coordinate other parts to ensure smooth action and maintenance of equilibrium. Hand-eye coordination is one example of cerebellar function. If the cerebellum is damaged, the eyes can follow a moving object, but they will not stop at the same place as the object.

The upper portion of the brainstem, or midbrain, contains centers for the receipt and integration of several types of sensory information. It also serves as a projection center, sending coded sensory information along neurons to specific regions of the forebrain. The most prominent areas of the midbrain are the **superior** and **inferior colliculi,** bumps on the dorsal surface of the brainstem that are part of the visual and auditory systems. All fibers involved in hearing either terminate in or pass through the inferior colliculi. The superior colliculi are important visual centers. In nonmammalian vertebrates, they may be the only visual centers, taking the form of prominent optic lobes. In mammals, vision is integrated in the forebrain, leaving the superior colliculi to coordinate visual reflexes and limited perceptual functions. Some of the major nuclei in the midbrain are part of the **reticular formation,** which regulates states of arousal (discussed shortly).

The most sophisticated neural processing occurs in the forebrain. The intricate networks of integrating centers and sensory and motor pathways allow pattern and image formation, as well as associative functions, such as memory, learning, and emotions. Of the two major divisions of the forebrain, the lower **diencephalon** contains two integrating centers: the thalamus and the hypothalamus. The upper **telencephalon** contains the cerebrum, the most complex integrating center in the CNS.

Within the diencephalon, the most prominent integrating center is the **thalamus,** a major relay station for sensory information on its way to the cerebrum. The thalamus contains many different nuclei, each one dedicated to sensory information of a particular type. Incoming information from all the senses is sorted out in the thalamus and sent on to the appropriate higher brain centers for further interpretation and integration. The thalamus also receives input from the cerebrum and from parts of the brain controlling emotion and arousal, making it an important switching station controlling access to the cerebrum.

The **hypothalamus** has been called the "Casablanca of the nervous system," because so many mysterious messages are sorted out there. Although weighing only a few grams, this area of the brain is one of the most important sites for regulation of homeostasis. We have already seen that the hypothalamus is the source of two sets of hormones, the posterior pituitary hormones and the releasing hormones for the anterior pituitary (see Chapter 41). It also contains the body's thermostat, centers for regulating hunger and eating, thirst and drinking, and many other basic survival functions. This region also plays a role in sexual response and mating behaviors, the fight-or-flight (alarm) response, and pleasure.

The hypothalamic pleasure centers have been given that name because of the responses seen when they are stimulated in experimental animals, although we cannot really know whether a rat experiences what humans interpret as pleasurable sensations. Nevertheless, when electrodes are implanted in the pleasure centers, the animal will press a bar continually to receive electrical shocks in this region, to the exclusion of eating, drinking, and mating.

Below the cerebral cortex (the major structure of the telencephalon) lies a cluster of nuclei referred to as the basal ganglia. The basal ganglia are important centers for motor coordination, acting as switches for impulses from other motor systems. If the basal ganglia are damaged, a person may become passive and immobile because the ganglia no longer allow motor impulses to be sent to the muscles. Degeneration of cells entering the basal ganglia occurs in Parkinson's disease.

The **cerebral cortex** is the largest and most complex part of the human brain, and the part that has changed the most during vertebrate evolution. The highly folded human cortex has a surface area of about 0.5 m^2 and is divided into five lobes, four of which are visible from the surface (Figure 44.23). Like the rest of the brain, the cerebral cortex is bilaterally symmetric, and the two hemispheres are connected by a thick band of fibers known as the **corpus callosum.** The cortex contains both sensory and motor areas involved in direct processing of information, as well as association areas that integrate information from several sources. Some of the major functional areas of the cortex are indicated in Figure 44.23.

The primary sensory and motor areas are bilateral. The primary sensory area receives impulses generated by stimulation of tactile, pressure, and pain receptors scattered throughout the body. The primary motor area of the cortex initiates impulses to the various skeletal muscles. The sensory and motor cortex is a mosaic of regions corresponding to different parts of the body. The proportion of the sensory or motor cortex devoted to a particular part of the body is correlated with the importance of sensory or motor information for that part of the body. For example, more brain surface is committed to sensory and motor communication with the hands than with the entire torso of the body. Impulses transmitted from receptors to specific areas of the primary sensory region of the cerebral cortex enable us to associate pain, touch, pressure, heat, or

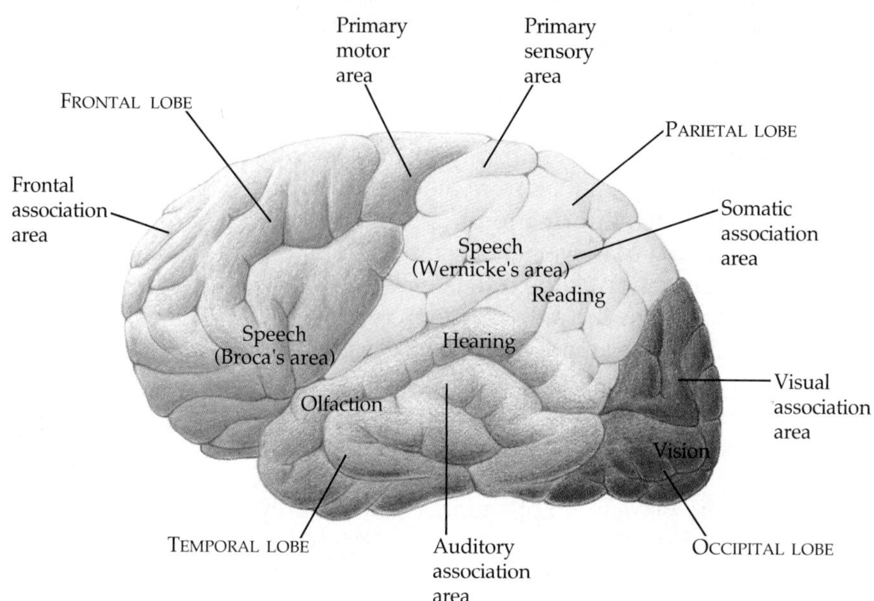

Figure 44.23
The cerebral cortex. The surface of the cerebral cortex can be divided into the four lobes visible here, with specialized functions localized in each one. The fifth lobe, the insula, is hidden within the fold separating the temporal and parietal lobes.

cold with specific parts of the body subjected to those stimuli. However, the so-called special senses—vision, hearing, smell, and taste—are integrated by other regions of the cortex (see Figure 44.23).

Integration and Higher Brain Functions

Integration of nerve impulses occurs at all levels in the human nervous system. The simplest kinds of integration are the spinal reflexes illustrated in Figure 44.19; the most complex integration enables the cerebral cortex to generate works of art or scientific discovery. Four aspects of brain activity that are particularly interesting are arousal and sleep, emotions, lateralization in the brain (differential functions of the left and right hemispheres of the brain), and memory.

Arousal and Sleep As anyone who has sat through a lecture on a warm spring day knows, attentiveness and mental alertness vary from moment to moment. Arousal is a state of awareness of the external world. The counterpart of arousal is sleep, when an individual continues to receive external stimuli but is not conscious of them. The mechanisms of arousal and sleep are fairly well worked out, but the question of why we sleep remains a compelling problem. All birds and mammals sleep and show a characteristic sleep–wakefulness cycle, which may be mediated by the hypothalamus.

Sleep and wakefulness produce different patterns in the electrical activity of the brain, which can be recorded in an **electroencephalogram,** or **EEG.** As a general rule, the less mental activity taking place, the more synchronous the brain waves of the EEG. When a normal person is lying quietly with closed eyes, slow, synchronous *alpha waves* predominate. When the eyes are opened or the person solves a complex problem, faster *beta waves* take over, indicating desynchronization of the parts of the brain. Sleep produces a third type of pattern, *delta waves,* which are quite slow and highly synchronized.

Sleep, however, is a dynamic process, and the EEG of a sleeping person is far from constant. Sleep can be divided into two patterns. One is a pattern of slow, deep delta waves. At other times, a desynchronized EEG reminiscent of wakefulness occurs; during these periods, the eyes move actively across the visual field behind the lids, and this is therefore called REM (rapid eye movement) sleep. Most dreaming occurs during REM sleep. Like sleep, dreaming has been ascribed magical or prophetic importance, but its true function remains unknown. The two patterns of sleep typically alternate during a night, with a 90-minute cycle containing a 20- to 30-minute bout of REM sleep.

Sleep and arousal are controlled by several centers in the cerebrum and the brainstem. The reticular formation is a vital link in determining states of arousal and consciousness. This group of over 90 separate nuclei (brain ganglia) extends from the medulla to the thalamus, through which almost all neuron processes reaching the cerebral cortex must pass. The reticular formation is essentially a sensory filter that selects which information reaches the cortex. The more input the cortex receives, the more alert and aware a person is. But arousal is not just a generalized phenomenon; specific sets of information can be ignored ("tuning out" stimuli) while the brain is actively processing other input. Also, specific centers regulate sleep and wakefulness. The pons and medulla contain nuclei that cause sleep when stimulated, but the midbrain has a center that causes arousal. It has been suggested that serotonin is the neurotransmitter of the sleep-producing centers. Drinking milk before bedtime might induce sleep because milk contains large amounts of tryptophan, the amino acid from which serotonin is synthesized.

Emotions What causes us to laugh, cry, love, and fight has been the subject of much biological and philosophical speculation. Some say our emotions cause facial expressions; others suggest that contracting the facial muscles stimulates the emotional centers of the brain. Some theories suggest that emotions result from feedback from the body's organs and muscles to the central nervous system. Emotions are difficult to study experimentally, because even if an experimental animal seems to *show* emotion, we cannot say conclusively that the animal *feels* the emotion in the same sense that we do.

Despite these uncertainties, we know that much of human emotion depends on interactions of the cerebral cortex and a group of nuclei in the lower part of the forebrain called the **limbic system.** Much as the reticular system selects signals affecting arousal and sleep, the limbic system selects certain emotional and behavioral responses. Surgical destruction of the amygdala, part of the limbic system, has been used as a treatment for extreme aggression, but the docility produced is not necessarily a cure and may have serious effects on other brain functions.

Right Brain/Left Brain The association areas of the cerebral cortex, unlike the primary motor and sensory areas (see Figure 44.23), are not bilaterally symmetric; each side of the brain controls different functions. Speech, language, and calculation, for example, are centered in the left hemisphere, while the right hemisphere controls artistic ability and spatial perception. Much of what we know about this **lateralization** of the brain comes from the work of Nobel Prize–winner Roger Sperry and his colleagues, who study "split-brain" patients. Some forms of epilepsy involve reverberating circuits that send massive electrical discharges between hemispheres through the corpus callosum. These patients may be treated by surgical severing of the corpus callosum. Surprisingly, this drastic procedure does not seem to affect behavior overtly, but it does have subtle effects on the patient's brain function.

A person with a severed corpus callosum may appear perfectly normal in most situations, but careful experiments reveal much about lateralization. A patient holding a key in the left hand, with both eyes open, will readily name it as a key. If blindfolded, though, the subject will recognize the key and use it to open a lock (and may be able to describe it), but will be completely unable to name it. The center for speech is in the left hemisphere, but sensory information from the left hand crosses and enters the right side of the brain. Without the corpus callosum to function as a switchboard between the two sides of the brain, knowledge of the size, texture, and function of the object cannot be transferred from the right to the left hemisphere. Thus, sensory input and spoken response are dissociated.

Language and Speech Two areas on the left hemisphere of the cerebral cortex are required for storing information related to speech (see Figure 44.23). **Wernicke's area** stores information required for speech content, arranging the words of a learned vocabulary into meaningful speech according to rules of grammar. **Broca's area** contains information required for speech production. It is told "what to say" by Wernicke's area, and then Broca's area programs the motor cortex to move the tongue, lips, and other speech muscles to articulate the words. (You can see these areas "in action" in the PET scans shown in Figure 2.6.) Damage to either area causes very different kinds of aphasia, the inability to speak coherently. Wernicke's aphasia leads to the production of long strings of words and nonsense syllables. Broca's aphasia leads to loss of fluency in speech, but at least some content remains.

Memory The manuscript for this chapter was typed on a personal computer that weighs about 11 kg and can retain about 4×10^6 bits (1200 pages) of information. Your brain, a much more powerful computer, weighs about 1.35 kg and stores literally hundreds of millions of bits of information dating back to the beginning of your life.

Memory, essential for learning, is the ability to store and retrieve information related to previous experiences. Human memory occurs in two stages. **Short-term memory** reflects immediate sensory perception of an object or idea and occurs before the image is stored. Short-term memory allows you to dial a phone number after looking it up but without looking at it directly. If you call the number frequently, it becomes stored in **long-term memory** and can be recalled several weeks after you originally looked it up. The transfer of information from short-term to long-term memory is enhanced by rehearsal ("practice makes perfect"), favorable emotional state (we learn best when we are alert and motivated), and association of new information with information previously learned and stored in long-term memory (it is much easier to learn a new card game if you already have a lot of "card sense" from playing other games).

In its ability to learn and remember, the human brain apparently distinguishes between facts and skills. When you acquire factual knowledge by memorizing dates, names, word definitions, the parts of the brain, and other information, this fact memory can be consciously and specifically retrieved from the data bank of your long-term memory. You can even recall visual images, such as the face of a friend. In contrast,

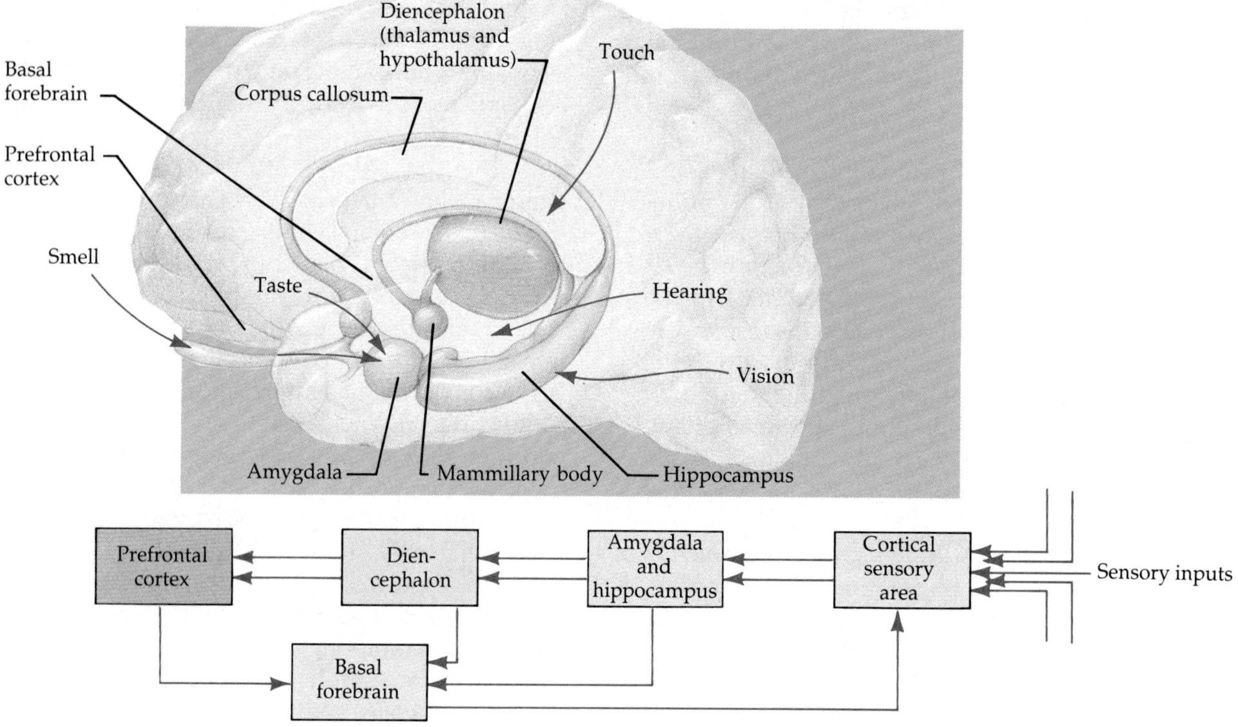

Figure 44.24
A possible memory pathway. According to this model, information received by the sensory regions of the cerebral cortex is relayed by the amygdala and hippocampus to the diencephalon (thalamus and hypothalamus), which in turn transmits impulses to forebrain regions called the prefrontal cortex and basal forebrain. Using multiple neural connections with the sensory cortex, the basal forebrain completes the memory circuit by transmitting impulses to the same sensory region that first perceived the sensory input. This may contribute to the storing of that particular experience in the cerebral cortex.

skill memory usually involves motor activities that are learned by repetition without consciously remembering specific information. You perform learned motor skills, such as walking, tying your shoes, riding a bicycle, or writing without consciously recalling the individual steps required to do these tasks correctly. Once a skill memory is learned, it is difficult to unlearn. For example, a duffer who has played golf for years with a self-taught, awkward swing has a much tougher time learning a smooth swing than does a novice just learning the game. Bad habits, as we know, are difficult to break.

By studying experimental animals and amnesia (memory loss) in humans, neuroscientists are beginning to map the major brain pathways involved in memory (Figure 44.24). In the pathway for fact memory, sensory information is transmitted from the sensory regions of the cerebral cortex to the **hippocampus** and **amygdala,** two components of the limbic system, which also function in emotions. In what is apparently a memory circuit, the hippocampus and amygdala then relay impulses to other regions of the forebrain. The circuit is completed when an integrating area called the basal forebrain, which has extensive neural connections with the sensory areas of the cortex, dispatches impulses back to the very region of the cortex where the sensory perception first occurred. Perhaps the returning impulses cause chemical or structural changes in the sensory cortex that store the event as a memory, such as a visual image, a particular sound, or a fragrance that we associate with a certain person or episode. A different neural pathway functions in skill memory.

Physiologist Karl Lashley spent several decades of this century searching for what he called the engram, the physical basis of a memory. But because partial damage to one area of the cerebral cortex does not destroy individual memories, he ultimately concluded that there is no highly localized memory trace in the nervous system. Rather, a memory seems to be stored within a certain association area of the cortex with some redundancy.

Many neuroscientists are investigating the cellular changes involved in memory or learning. According to one hypothesis, structural changes in dendrites have a function in learning. One version of this idea postulates that neural input causes a postsynaptic cell in the brain to take up calcium, which in turn activates enzymes that alter the cytoskeleton and change the shape of the dendrite in such a way that future transmission across that synapse is enhanced.

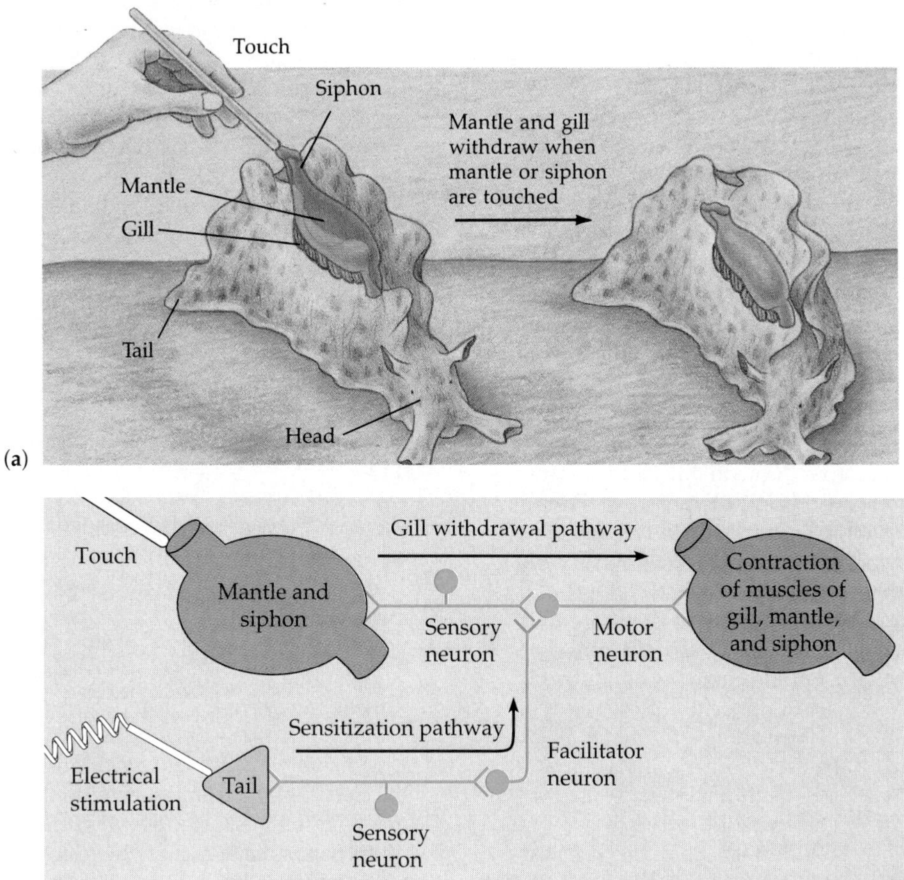

Figure 44.25
A learning pathway in the sea hare *(Aplysia).* (**a**) The mantle, with a siphon that allows water to flow over the gills, normally protrudes from the dorsal surface of this mollusk. Touching the mantle or siphon triggers a reflex that protects the mantle by withdrawing it. If the mantle is touched repeatedly, the withdrawal response becomes progressively weaker, a simple type of learning called habituation. (**b**) *Aplysia* can also be sensitized to stimuli so weak that they do not normally induce the withdrawal reflex. If the tail of the animal is stimulated prior to touching the mantle, the withdrawal reflex is stronger. If these treatments of dual stimulation are repeated several times, the enhanced withdrawal is still evident weeks later when the mantle is touched *without* stimulation of the tail. The mechanism of this simple learning involves a facilitator neuron that fires when the tail is stimulated. This neuron intervenes by secreting a neurotransmitter (serotonin) that intensifies synaptic transmission between the sensory neuron and motor neuron of the gill-withdrawal reflex by altering the transport of Ca^{2+} into the synaptic terminal. When the tail is touched and a signal reaches the terminal of the siphon's sensory neuron via the facilitator neuron, more Ca^{2+} than usual enters the sensory cell through the Ca^{2+} channels. This results in the release of more neurotransmitter from the sensory cell, strengthening the withdrawal response. By investigating learning pathways involving only a few neurons in invertebrates, neuroscientists may discover fundamental mechanisms that also work in the more complex nervous systems of humans and other vertebrates.

The human brain, with its billions of neurons, is too complex to serve as a model system to study the most basic mechanisms of memory and learning, which may, after all, be quite similar in diverse animals. For this reason, neuroscientists are probing the much simpler nervous systems of certain invertebrates. For example, the sea hare *Aplysia* (phylum Mollusca) has only about 20,000 neurons, yet it displays several forms of behavioral plasticity (learning). For instance, these animals eventually ignore mild touch stimuli that are presented repeatedly, a primitive type of learning called habituation (Figure 44.25). *Aplysia* can also be conditioned to evoke a stronger-than-normal withdrawal response to a soft touch that has been paired with another stimulus, such as a mild electrical stimulation of the tail. This learning results from changes in the properties of ion channels at the synapses between sensory and motor neurons in the central nervous system of *Aplysia.* Conditioning ("training") increases the amount of calcium that enters the synaptic terminals of the sensory neurons, resulting in more neurotransmitter being released with each action potential. Thus, a single synapse could "learn" from past experience.

Whether memory involves changes at individual synapses, or some still undiscovered mechanism, the sophistication and complexity of even relatively simple nervous systems remains one of the most exciting and fascinating aspects of modern biology.

STUDY OUTLINE

1. Although functionally related to the endocrine system, the nervous system is characterized by more rapid communication, more precise control, and broader range of responses.
2. The nervous system's three main functions are sensory input, integration, and motor output to effector cells.

Cells of the Nervous System (pp. 983–986)

1. Cells of the nervous system include neurons, which transmit the signals, and supporting cells, which support, insulate, and protect the neurons.
2. A neuron's fiberlike dendrites and axons conduct information toward and away from the cell body, respectively. Axons originate from the axon hillock and terminate in numerous branches. Synaptic terminals at the end of the axon release neurotransmitters into the synapses and thus relay neural signals to the dendrites or cell bodies of other neurons or effectors.
3. The central nervous system consists of the brain and spinal cord. The peripheral nervous system contains sensory neurons, which transmit information from internal and external environments to the central nervous system; and motor neurons, which carry information from the brain or spinal cord to effector organs. Interneurons, or association neurons, of the central nervous system integrate sensory input and motor output.
4. Glial cells—supporting cells in the central nervous system—include astrocytes, which line the capillaries in the brain and contribute to the blood-brain barrier; and oligodendrocytes, which wrap and insulate some neurons in a myelin sheath. In the peripheral nervous system, myelin sheaths are formed by supporting cells called Schwann cells.

Transmission of Electrical Signals Along a Neuron (pp. 986–993)

1. The membrane potential for a nontransmitting neuron is due to the unequal distribution of ions, particularly sodium and potassium, across the membrane; the cytoplasm is more negatively charged than the extracellular environment. Resting potential is maintained by differential ion permeabilities and the Na^+-K^+ pump.
2. A stimulus that affects the membrane's permeability to ions can either depolarize or hyperpolarize the membrane relative to the membrane's resting potential. This local voltage change is called a graded potential, and its magnitude is proportional to the strength of the stimulus.
3. An action potential is a rapid, transient depolarization of the neuron's membrane. A local depolarization to the threshold potential opens voltage-sensitive sodium channels, and the rapid influx of Na^+ brings the membrane potential to a positive value. The membrane potential is restored to its normal resting value by the delayed opening of voltage-sensitive K^+ channels and by the closing of the Na^+ channels. A refractory period follows an action potential, corresponding to the period when the voltage-sensitive Na^+ channels are inactivated.
4. The all-or-none generation of an action potential always creates the same amplitude of voltage change for a given neuron. The frequency of action potentials varies with the intensity of the stimulus.
5. Once an action potential is initiated in an axon, the large depolarization that is produced by the action potential spreads along the axon to trigger an action potential in the next portion of the axon. In this way, a wave of depolarization propagates to the end of the axon.
6. The rate of transmission of a nerve impulse is directly related to the diameter of the axon. Saltatory conduction, as action potentials jump between the nodes of Ranvier of myelinated axons, speeds nervous impulses in vertebrates.

The Synapse: Transmission Between Cells (pp. 993–999)

1. Synapses between neurons conduct impulses from the axon of a presynaptic cell to a dendrite or cell body of a postsynaptic cell.
2. Electrical synapses use gap junctions to directly pass an action potential between two neurons.
3. In a chemical synapse, a depolarization stimulates the fusion of synaptic vesicles with the presynaptic membrane and the release of their neurotransmitter into the synaptic cleft. The diffusing transmitter binds to receptor proteins on the postsynaptic membrane that are associated with particular ion channels. The selective opening of these chemically sensitive gates either brings the membrane potential closer to the threshold potential (EPSP) or hyperpolarizes the membrane (IPSP). The neurotransmitter is rapidly broken down by enzymes.
4. Whether an action potential is created in the postsynaptic cell depends on the temporal or spatial summation of EPSPs and IPSPs at the axon hillock.
5. One of the most common invertebrate and vertebrate neurotransmitters is acetylcholine. Other transmitters that have been identified include the biogenic amines (epinephrine, norepinephrine, and dopamine), several amino acids, and some neuropeptides, such as the opiumlike endorphins.
6. Groups of neurons may interact and carry information along specific pathways called circuits. Nerve cell bodies are arranged in functional groups called ganglia or nuclei.

Invertebrate Nervous Systems (pp. 999–1001)

1. Invertebrate nervous systems range from the diffuse nerve nets of the cnidarians to the highly centralized nervous systems of the cephalopods, which possess large brains capable of fairly sophisticated learning.

The Vertebrate Nervous System (pp. 1001–1011)

1. The peripheral nervous system consists of the sensory, or afferent, nervous system, which brings information from sensory receptors to the central nervous system; and the motor, or efferent, nervous system, which carries signals away from the central nervous system to effector muscles and glands.

2. The motor nervous system has a somatic portion, which carries signals to skeletal muscles, and an autonomic portion, which regulates the primarily automatic, visceral functions of smooth and cardiac muscles.
3. The autonomic system is subdivided into the parasympathetic and sympathetic nervous systems, which are anatomically, functionally, and chemically distinct and usually antagonistic in effect on target organs.
4. The central nervous system (CNS), composed of the brain and spinal cord, serves as the link between the sensory and motor subdivisions of the peripheral nervous system.
5. The spinal cord mediates many reflexes that integrate sensory input with motor output. It also has tracts of neurons that carry information to and from the brain.
6. All vertebrate brains develop and diversify from three regions: the forebrain, the midbrain, and the hindbrain.
7. Three evolutionary changes in the vertebrate brain include increases in relative size, in compartmentalization of function, and in complexity of the forebrain, particularly the cerebral cortex.
8. The human brain develops from the three primary regions, which differentiate into specialized structures.
9. The medulla oblongata and pons of the hindbrain work together to control several homeostatic functions, such as breathing and digestion. The medulla and pons also conduct sensory and motor information between the spinal cord and higher brain centers.
10. The cerebellum of the hindbrain coordinates movement and balance by integrating unconscious sensory and motor signals.
11. The midbrain receives, integrates, and projects sensory information to the forebrain.
12. The forebrain is the site of the most sophisticated neural processing, with major integrative centers in the thalamus, hypothalamus, and cerebrum.
13. The thalamus routes neural input to specific areas of the cerebral cortex, the outer gray matter of the cerebrum. The functions of the hypothalamus range from hormone production to regulation of body temperature, hunger, thirst, sexual response, and the alarm response.
14. The cortex contains distinct sensory and motor areas, which directly process information; and association areas, which integrate information.
15. Sleep and arousal are controlled by several areas in the cerebrum and brainstem, the most important of which is the reticular formation, which filters the sensory input sent to the cortex.
16. Human emotions are believed to originate from interactions between the cerebral cortex and the limbic system, a group of nuclei (ganglia) in the lower forebrain.
17. The two sides of the association areas of the cerebral cortex control different functions. Speech, language, and analytical ability are centered in the left hemisphere, whereas spatial perception and artistic ability predominate in the right. Nerve tracts of the corpus callosum link the two sides and allow the brain to function as an integrated whole.
18. Different aspects of language and speech are controlled by Wernicke's area and Broca's area.
19. Human memory consists of short-term and long-term memories. The learning and memory of facts appear to differ from those of skills. The hippocampus and amygdala, two components of the limbic system, participate in circular brain pathways involved in fact memory. Chemical or structural changes in the neurons of the sensory cortex may store memories.
20. Simpler organisms, such as *Aplysia,* are often used to study the mechanisms of memory and learning. Simple learning in this mollusk has been related to changes in the ion channels at the synapses of interneurons.

SELF-QUIZ

1. Which of the following occurs when a stimulus depolarizes a neuron's membrane?
   a. Na^+ diffuses out of the cell.
   b. The action potential approaches zero.
   c. The membrane potential changes from the resting potential to a voltage closer to the threshold potential.
   d. The depolarization is all or none.
   e. The Na^+-K^+ pump is stimulated.
2. Action potentials are usually propagated in only one direction along an axon because
   a. the nodes of Ranvier only conduct in one direction
   b. the brief refractory period prevents depolarization in the direction from which the impulse came
   c. the axon hillock has a higher membrane potential than the tips of the axon
   d. ions can only flow along the axon in one direction
   e. both sodium and potassium voltage-sensitive gates open in one direction
3. The depolarization of the presynaptic membrane of an axon *directly* causes
   a. voltage-sensitive calcium channels in the membrane to open
   b. synaptic vesicles to fuse with the membrane
   c. an action potential in the postsynaptic cell
   d. the opening of chemical-sensitive gates that allow neurotransmitter to spill into the synaptic cleft
   e. an EPSP or IPSP in the postsynaptic cell
4. Anesthetics reduce pain by blocking the transmission of nerve impulses. Which of these three chemicals might work as an anesthetic?
   a. a chemical that blocks voltage-sensitive sodium channels in membranes
   b. a chemical that opens voltage-sensitive potassium channels
   c. a chemical that blocks neurotransmitter receptors
   d. a, b, and c
   e. only b and c
5. Gray matter is
   a. made up of three protective layers called meninges
   b. located on the outside of the spinal cord
   c. restricted to the brain
   d. populated by cell bodies of neurons
   e. found in the ventricles of the vertebrate brain

6. Which of the following structures or regions is *incorrectly* paired with its function?
   a. Broca's region—screening of information between spinal cord and the brain; regulates arousal and sleep
   b. medulla oblongata—homeostatic control center
   c. cerebellum—unconscious coordination of movement and balance
   d. corpus callosum—band of fibers connecting left and right cerebral hemispheres
   e. hypothalamus—production of hormones and regulation of temperature, hunger, and thirst
7. Which of the following is a *true* statement about the mantle-withdrawal reflex of the sea hare *Aplysia*?
   a. The animal learns this behavior.
   b. Repeatedly stimulating the siphon over a short period of time intensifies the withdrawal reflex.
   c. *Aplysia* can learn to associate stimulation of the tail with stimulation of the siphon.
   d. There is evidence that the reflex involves conscious thought on the part of *Aplysia.*
   e. Even on the first try, stimulating the tail substitutes for stimulating the siphon in triggering the withdrawal reflex.
8. Receptor sites for neurotransmitters are located on the
   a. tips of axons
   b. axon membranes in the regions of the nodes of Ranvier
   c. postsynaptic membrane
   d. membranes of synaptic vesicles
   e. presynaptic membrane
9. Nerve nets are most characteristic of the nervous systems of
   a. annelids
   b. vertebrates
   c. insects
   d. cnidarians
   e. flatworms
10. All the following electrical changes of neurons are graded events *except*
    a. EPSPs
    b. IPSPs
    c. action potentials
    d. depolarizations caused by stimuli
    e. hyperpolarizations caused by stimuli

CHALLENGE QUESTIONS

1. From what you know about the action potential, propose one feasible mechanism whereby anesthetics might prevent pain.
2. Describe various ways in which drugs that are stimulants could increase activity of the nervous system by acting at the synapse.
3. Describe the role of calcium in nerve impulses.

SCIENCE, TECHNOLOGY, AND SOCIETY

1. Nancy Cruzan lay in a coma for nine years after suffering irreversible brain damage in a car accident. Her family asked that she be allowed to die, but legal authorities refused to allow the hospital to withhold life support. A long legal battle went all the way to the U.S. Supreme Court, which ruled in 1990 that a person had a right to refuse medical treatment, but that the state may require clear evidence of the patient's wishes. Nancy Cruzan died twelve days after her feeding tube was removed. Only 10% to 15% of Americans have a living will or other documentation stating what they wish to be done in the event that they are brain dead, being kept alive by artificial life support. What would you want done if you or a member of your family were in this situation? Why? Who should speak for the patient? Is it a physician's job to ask a patient his or her wishes, if that is possible? What constitutes evidence of a patient's wishes? If a patient wants treatment withheld, can a hospital or doctor refuse to comply?
2. Parkinson's disease is a brain disorder that interferes with motor control, causing slow, jerky movements. It occurs when neurons reduce production of dopamine, a neurotransmitter. Experiments show that many victims can be helped by transplanted tissues obtained from aborted fetuses, a treatment that restores the missing dopamine. In 1992, the U.S. government banned federal funding of fetal implants, fearing that using tissue from aborted fetuses will encourage women to have abortions. It has been suggested that tissue from miscarriages might be used for research. However, researchers doubt there is enough usable tissue from miscarriages for their needs. Do you think the ban on fetal tissue research is justified, or do you think this research should be funded? What are your reasons?

FURTHER READING

Begley, S. "Mapping the Brain." *Newsweek,* April 20, 1992. Powerful new imaging tools reveal the working brain.

Kemp, M. "A Squid for All Seasons." *Discover,* June 1989. An important model organism in neurobiology.

Koch, C. "What Is Consciousness?" *Discover,* November 1992.

Marx, J. "Alzheimer's Debate Boils Over." *Science,* September 4, 1992. Hypotheses on a progressive disorder of the brain.

Matthews, G. *Cellular Physiology of Nerve and Muscle,* 2nd ed. Boston: Blackwell Scientific Publications, 1991. A lucid presentation of the electrical properties of cells.

Melzack, R. "Phantom Limbs." *Scientific American,* April 1992. What causes the brain to register pain for body parts that have been amputated?

Rogers, L. "The Left and Right Brains at Work." *New Scientist,* February 11, 1989. Is lateral dominance inherited, or can it be developed?

Snyder, S., and Bredt, D. "Biological Roles of Nitric Oxide." *Scientific American,* May 1992. An inorganic molecule with diverse signaling functions in the body.

45 SENSORY AND MOTOR MECHANISMS

SENSORY RECEPTORS
VISION
HEARING AND BALANCE
TASTE AND SMELL
AN INTRODUCTION TO ANIMAL MOVEMENT
SKELETONS AND THEIR ROLES IN MOVEMENT
MUSCLES

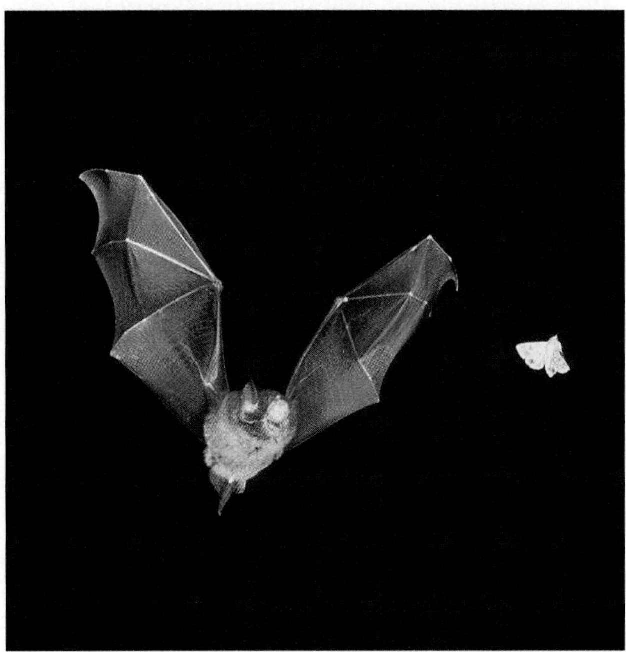

Figure 45.1
Evolution shapes animal behavior. The bat uses sonar to locate its prey; the moth has vibration sensors tuned to the bat's ultrasonic clicks. The moth responds with an escape maneuver; the bat reacts with a sharp turn that tracks the moth's movement. Natural selection has refined the behavior of both predators and prey. In this chapter, you will learn about sensory and motor adaptations that function in animal behavior.

In the gathering dusk, the male moth's antennae detect the pheromone scent of a female moth somewhere upwind. The moth takes to the air, following the scent trail toward the female. Suddenly, vibration sensors in the moth's abdomen signal the presence of ultrasonic chirps of a rapidly approaching bat. The bat is using its sonar to home in on the moth, its favorite meal. Reflexively, the moth's nervous system alters the motor output to the wing muscles, sending the moth into an evasive spiral toward the ground. The bat alters its path accordingly, attempting to intercept the flight of its intended prey (Figure 45.1). The outcome of this interaction depends on the abilities of both predator and prey to sense important environmental stimuli and to produce coordinated movement that is appropriate. Although not all of an animal's moment-by-moment interactions with the environment are as dramatic as such predator–prey struggles, detection and processing of sensory information and the generation of motor output provide the physiological basis for all animal behavior.

In Chapter 44, we saw how the nervous system transmits and integrates sensory and motor information. We will now examine the input and output of the coordinating system that controls so many aspects of animal behavior. We will first consider the sensory receptors that receive information from the environment, and then examine the structure and function of muscles, the motor effectors that bring about movement in response to that information. Skeletons will also be discussed in the context of body movement.

SENSORY RECEPTORS

Amputees often report pain or numbness in "phantom" limbs no longer present. To understand this phenomenon, we must consider the distinction between sensation and perception.

Sensation and Perception

As we saw in Chapter 44, information is transmitted in the nervous system in the form of action potentials. An action potential triggered by light striking the eye is no different from an action potential triggered by air vibrating in the ear, yet we readily distinguish sight from sound. The difference depends on the part of the brain that receives the signal. The air vibrations we call sounds, for example, are converted by the ear into nerve impulses that are received by a particular region of the cerebral cortex. These nerve impulses, conveyed as action potentials along sensory neurons to the brain, are called **sensations.** Once the brain is aware of the sensations, it interprets them, giving us the **perception** of sounds. Other kinds of input are sent to other parts of the brain and trigger different perceptions. What matters, then, is where the impulse goes, not what triggers it.

It follows that if neurons from your eyes could be crossed with those from your ears, you might perceive a camera flash as a loud "boom" and a concert as bursts of light. In a less dramatic demonstration of this principle, you have probably experienced light flashes while rubbing your eyes. The rubbing provides enough energy to generate action potentials in the sensory neurons of the optic nerves. The stimulus is pressure, but the perception is a spot of light because this is the only perception that can be generated by the part of the brain receiving the sensations. To return to the example of pain in a phantom limb, severed nerves that carried impulses from the limb of an amputee may remain alive and respond to irritation. They can still transmit sensations to the brain and trigger perception. The resulting pain is just as real as that experienced by a person who irritates the nerves in an existing arm. But how does the sensory system actually generate sensations?

The General Function of Sensory Receptors

Sensations, and the perceptions they evoke in the mind, begin with excitation of **sensory receptors,** structures that transmit information about changes in the external and internal environment of the animal. Receptors are usually modified neurons, which occur singly or in groups along with other cell types within sensory organs, such as the eyes and ears. Receptors are specialized to respond to various stimuli, including heat, light, pressure, and chemicals. All these stimuli represent forms of energy. The general function of receptor cells is to convert the energy of stimuli into the electrochemical energy of action potentials and carry those action potentials into the nervous system. This task can be broken down into five functions common to all receptor cells: reception, transduction, amplification, transmission, and integration.

Reception The ability of a cell to absorb the energy of a stimulus is called reception. Each kind of receptor has a region specifically suited to absorbing a particular type of energy. As we will see, sensory cells in the human eye, for example, have membranes that contain a light-absorbing pigment molecule.

Transduction The conversion of stimulus energy into electrochemical activity of nerve impulses is called transduction. Receipt of a stimulus changes the permeability of the receptor cell, causing a change in its membrane potential and, ultimately, changes in the number of action potentials transmitted along sensory pathways to the central nervous system. In some cases, a stimulus such as pressure can stretch the membrane and increase ion flow generally. In other cases, specific receptor molecules on the membrane of a receptor cell open or close gates to ion channels when the stimulus is present. We will discuss specific examples of sensory transduction later in this chapter.

Amplification The stimulus energy is often too weak to be carried into the nervous system and must be amplified. Amplification of the signal may occur in accessory structures of a complex sense organ, as when sound waves are amplified by a factor of more than 20 before they reach the receptors of the inner ear. Amplification also may be a part of the transduction process itself. An action potential conducted from the eye to the brain has about 100,000 times as much energy as the few photons of light that triggered it.

Transmission Once the energy in the stimulus has been transduced into changes in the membrane potential of the receptor cell, those changes must be transmitted to the nervous system. In some instances, such as in the case of "pain cells," the receptor itself is actually a sensory neuron that conducts action potentials to the central nervous system. Other receptors are separate cells that must transmit chemical signals across synapses to sensory neurons. In either case, the initial response of the receptor to the stimulus during transduction is a graded change in membrane potential called a **receptor potential.** (Recall from Chapter 44 that a graded potential is a change in the voltage across the membrane that is proportional to the strength of the stimulus.) If the receptor also functions as the sensory neuron, then the intensity of the receptor potential affects the frequency of action potentials that travel as sensations to the central nervous system. For separate receptor cells, the strength of the stimulus and receptor potential affect the amount of neurotransmitter released by the receptor at its synapse with a sensory neuron, which in turn determines the frequency of action potentials fired by the sensory neuron. Many sensory neurons fire at a low rate spontaneously, so a stimulus does not really switch the production of action potentials on or off, but rather modulates their frequency. In this way, the central nervous system is sensitive not only to the presence or absence of a stimulus but also to changes in stimulus intensity.

Integration The processing of information begins as soon as it starts to be received. Signals from receptors are integrated through summation of graded potentials, as are those within the nervous system (see Chapter 44).

One type of integration by the receptor cell is **sensory adaptation,** a decrease in sensitivity during continued stimulation (not to be confused with the term *adaptation* as used in an evolutionary context). Without sensory adaptation, you would feel every beat of your heart and every bit of clothing on your body. Receptors are selective in the information they send to the central nervous system, and adaptation reduces the likelihood that a continued stimulus will be transmitted.

Another important aspect of sensory integration is the sensitivity of the receptors. The threshold for firing in receptor cells varies with conditions. For example, the firing thresholds of glucose receptors in the human mouth and in the feet of flies can vary over several orders of magnitude of sugar concentration, as both the general state of nutrition and the amount of sugar in the diet change.

Integration of sensory information occurs at all levels within the nervous system, and the cellular actions just described are only the first steps. Complex receptors such as the eyes have higher levels of integration as signals converge on sensory nerves, and the central nervous system further processes all incoming signals.

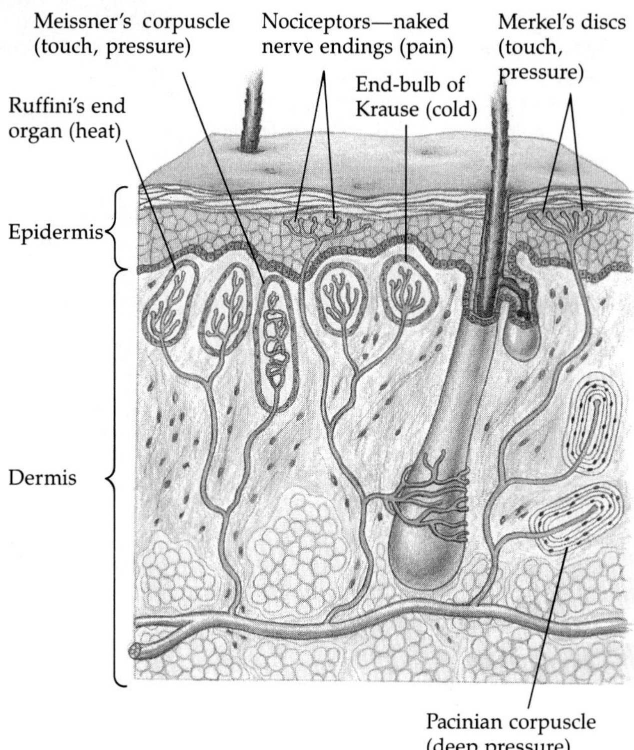

Figure 45.2
Receptors in the skin. The sensory receptors illustrated here are described throughout this chapter. All these tactile receptors are actually modified dendrites, encapsulated by connective tissue in some cases.

Types of Receptors

The various types of sensory receptors can be divided on the basis of location into two broad groups: **exteroreceptors,** which receive information from the outside world (light, sound, touch), and **interoreceptors,** which provide information about the body's internal environment. Another way of categorizing receptors is in terms of the energy stimulus to which they respond. Based on this criterion, the five types of receptors we will consider are mechanoreceptors, chemoreceptors, electromagnetic receptors, thermoreceptors, and pain receptors.

Mechanoreceptors These receptors are called **mechanoreceptors** because they are stimulated by physical deformation caused by such stimuli as pressure, touch, stretch, motion, and sound—all forms of mechanical energy. Bending or stretching of the plasma membrane of a mechanoreceptor increases its permeability to both sodium and potassium ions, resulting in a depolarization (receptor potential).

The human sense of touch relies on mechanoreceptors that are actually modified dendrites of sensory neurons. **Pacinian corpuscles** are found in deep skin layers and respond to strong pressure. Closer to the surface, detecting light touch, are **Meissner's corpuscles** and **Merkel's discs.** Figure 45.2 illustrates the various sensory receptors in the human skin.

An example of an interoreceptor stimulated by mechanical distortion is the **muscle spindle,** or stretch receptor. This mechanoreceptor is used to monitor the length of skeletal muscles. The muscle spindle contains modified muscle fibers attached to sensory neurons and runs parallel to muscle. When the muscle is stretched, the fibers of the spindle are also stretched, depolarizing the sensory neurons and triggering action potentials that are transmitted back to the spinal cord. It is the activation of the sensory neurons of the muscle spindles that provides the sensory component of the patellar stretch reflex discussed in Chapter 44.

The **hair cell** is a common type of mechanoreceptor used to detect motion. Hair cells are found in the vertebrate ear; in the lateral line organs of fishes and amphibians, where they detect movement relative to the environment (see Chapter 30); and in the balance organs of arthropods. The "hairs" are either specialized cilia or microvilli (cellular projections supported by microfilaments; see Chapter 37). They project upward from the surface of the hair cell into either an internal compartment, such as the human inner ear, or an external environment, such as a pond. When the cilia bend in one direction, they stretch the hair cell membrane and increase its permeability to sodium and potassium ions, thereby increasing the rate of impulse production in a sensory neuron. When the cilia bend in the opposite direction, ion permeability decreases, re-

Figure 45.3
Chemoreceptors in an insect. The antennae of the male silkworm moth *Bombyx mori* are covered with chemoreceptive hairs that are highly sensitive to the female sex pheromone bombykol. (Each filament of these feathery antennae actually bears hundreds of olfactory hairs, too small to be seen in this photograph.) Males show definite behavioral responses when as few as 50 of the more than 50,000 bombykol receptors on the antennae come into contact with one bombykol molecule per second.

ducing the number of action potentials in the sensory neuron. This specificity allows hair cells to respond to the direction of motion as well as to its strength and speed. The role of hair cells in hearing and balance is explored later in this chapter.

Chemoreceptors **Chemoreceptors** include both general receptors that transmit information about the total solute concentration in a solution and specific receptors that respond to individual kinds of molecules. Osmoregulators in the mammalian brain, for example, are general receptors that detect changes in the total solute concentration of the blood and stimulate drinking behavior when osmolarity increases (see Chapter 40). Osmoreceptors in the feet of house flies respond to a dilute solution of virtually any substance. Most animals also have receptors specific to important molecules, including glucose, oxygen, carbon dioxide, and amino acids. In all these examples, the stimulus molecule binds to a specific site on the membrane of the receptor cell and initiates changes in membrane permeability. Two other groups of chemoreceptors show intermediate specificity. **Gustatory** (taste) and **olfactory** (smell) **receptors** respond to *categories* of related chemicals. Humans often classify such categories as sweet, sour, salty, or bitter. (Taste and olfaction are discussed in detail later in this chapter.) One of the most sensitive and specific chemoreceptors known is found in the antennae of the male silkworm moth (Figure 45.3); it detects the female sex pheromone bombykol.

Electromagnetic Receptors Electromagnetic radiation consists of energy of different wavelengths, which takes such forms as visible light, electricity, and magnetism (see Chapter 10). **Photoreceptors,** which detect the radiation we know as visible light, are often organized into eyes. Snakes have extremely sensitive infrared receptors that detect the body heat of prey standing out against a colder background (Figure 45.4a). Some fishes discharge electric currents and use special electroreceptors to locate objects, such as prey, that disturb the electric currents. The platypus, a monotreme mammal (see Chapter 30), has electroreceptors on its bill that can probably detect electric fields generated by the muscles of prey, such as crustaceans, frogs, and small fishes. There is also evidence that some animals that home or migrate use the magnetic field lines of the Earth to help orient themselves (Figure 45.4b). Although the nature of the magnetoreceptors is unknown, the ferrous mineral magnetite has been found in the skulls of several animals. (Researchers have recently found magnetite in human skulls, but there is no evidence that these deposits are associated with a magnetic sense.)

Thermoreceptors Responding to either heat or cold, **thermoreceptors** help regulate body temperature by signaling both surface and body core temperature. There is still debate about the identity of thermoreceptors in the mammalian skin. Possible candidates are two receptors consisting of encapsulated, branched

(a)

(b)

Figure 45.4
Specialized electromagnetic receptors. (**a**) Rattlesnakes and other pit vipers, such as this Asian species, have a pair of infrared receptors, one between each eye and nostril. The organs are sensitive enough to detect the infrared radiation emitted by a warm mouse a meter away. The snake moves its head from side to side until the radiation is detected equally by the two receptors, indicating that the mouse is straight ahead. (**b**) Some migrating animals, such as these Beluga whales, can apparently sense Earth's magnetic field and use the information, along with other cues, for orientation. The mechanism of the magnetic sense is unknown.

dendrites: **Ruffini's end organs** may be heat receptors, and **end-bulbs of Krause** are possibly cold receptors (see Figure 45.2). Some researchers, however, believe that these structures are actually modified pressure receptors and propose that naked dendrites of certain sensory neurons are the actual thermoreceptors of the skin. There is more agreement that interothermoreceptors in the hypothalamus of the brain function as the major thermostat of the mammalian body.

Pain Receptors Virtually all animals experience pain, although we cannot say what perceptions they actually associate with stimulation of their pain receptors. Pain is one of the most important senses because the stimulus becomes translated into a negative reaction, such as withdrawal from danger. Rare individuals who are born without any pain sensation suffer large numbers of cuts and burns. They may even die from such conditions as a ruptured appendix because they cannot feel the associated pain and are unaware of the danger.

Pain is detected by a class of naked dendrites called **nociceptors** (see Figure 45.2). Different groups of pain receptors respond to excess heat, pressure, or specific classes of chemicals released from damaged or inflamed tissues. Some of the chemicals that trigger pain include histamine and acids. Prostaglandins increase pain by sensitizing the receptors—that is, lowering their threshold. Aspirin reduces pain by inhibiting prostaglandin synthesis. Nociceptive neurons carry impulses to the spinal cord, synapsing with neural pathways leading to the brain.

We have seen that receptors can be classified according to the type of energy they receive. In many cases, large numbers of a particular type of receptor cell are collected into complex sense organs that detect specific environmental stimuli. We will now examine the structure and function of sense organs responsible for vision, hearing, equilibrium, taste, and olfaction.

VISION

Light Receptors and Vision of Invertebrates

Most invertebrates can detect light with receptors containing light-absorbing pigments. One of the simplest light receptors is the **eye cup** of planarians. These structures provide information about light intensity and direction without actually forming an image. The receptor cells are located within a cup formed by a layer of darkly pigmented cells that block light. Light can enter the cup and stimulate the photoreceptors

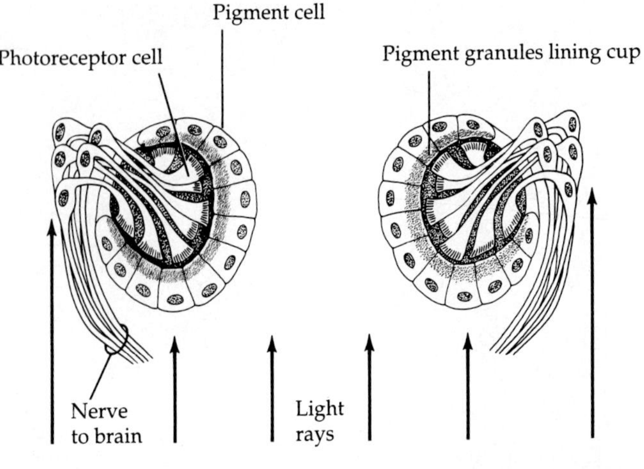

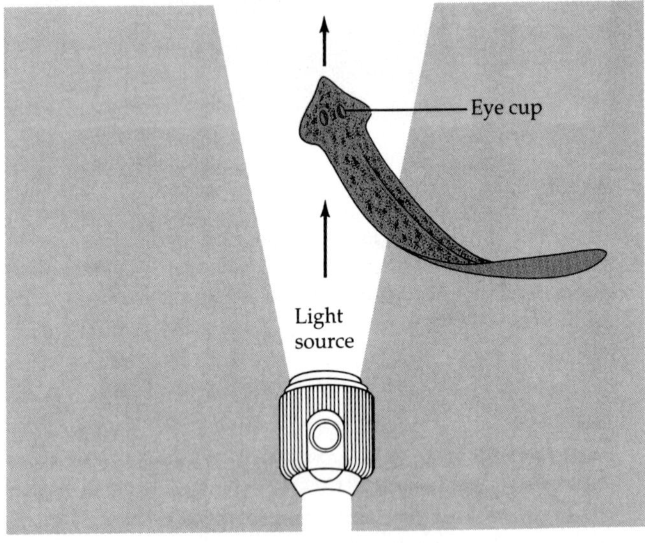

Figure 45.5
Eye cups and orientation behavior in planarians. The head of the planarian has two eye cups with photoreceptors that send nerve impulses to the brain. The wall of the cup consists of pigment cells that shade the photoreceptors. Light can reach the photoreceptors only through the mouth of the cup, which opens to the side of the head and slightly forward. The brain, comparing the frequency of nerve impulses from the two eye cups, directs the body to turn until the sensations from the two cups are equal and at a minimum. This orientation response causes the planarian to swim directly away from the light source until it reaches a dark pocket under a rock or some other shaded sanctuary where the animal is less obvious to predators.

only through an opening on one side of the cup where there are no pigment cells (Figure 45.5). The mouth of one eye cup faces left and slightly forward, and the mouth of the other cup faces right-forward. Thus, light shining from one side of the planarian can enter only the eye cup on that side of the animal. The brain compares the rate of nerve impulses coming from the two eye cups, and the animal turns until the sensations from the two cups are equal and minimal. The result is that the animal swims directly away from the light source and reaches a shaded location beneath a rock or some other object, a behavioral adaptation that helps hide the planarian from predators.

True image-forming eyes of two major types have evolved in invertebrates: the compound eye and the single-lens eye. **Compound eyes** are found in insects and crustaceans (phylum Arthropoda) and some polychaete worms (phylum Annelida). A compound eye consists of up to several thousand light detectors called **ommatidia** (the "facets" of the eye), each with its own cornea and lens. Each ommatidium registers light from a tiny portion of the field of view. Differences in the intensity of light entering the many ommatidia result in a mosaic image (Figure 45.6). Although the image is not as sharp as that produced by the human eye, the compound eye is more acute at detecting movement, an important adaptation for flying insects and small animals constantly threatened with predation. This advantage of the compound eye is partly due to the rapid recovery of the photoreceptors. The human eye can distinguish light flashes up to about 50 flashes per second; for this reason, the individual images of a movie, which flash at a faster rate, fuse together to create the perception of smooth motion. The compound eyes of some insects, however, recover from excitation rapidly enough to detect the flickering of a light flashing 330 times per second. Such an insect viewing a movie could easily resolve each frame of the film as a separate still image. Insects also have excellent color vision, and some (including bees) can see into the ultraviolet range of the spectrum, which is invisible to us. In studying animal behavior, we cannot extrapolate our sensory world to other species; different animals have different sensitivities.

Recent studies indicate that the compound eyes of certain crabs and mayflies can actually form a single unfragmented image under conditions of dim light. The ommatidia have lenses that work like prisms and parabolic mirrors, focusing the light entering several ommatidia onto a single photoreceptor, thus increasing the sensitivity of the eye to light. When lighting is brighter, these superimposition eyes, as they are called, seem to work like typical compound eyes.

The second type of invertebrate eye is the **single-lens eye,** found in some jellyfish, polychaetes, spiders, and many mollusks. It works on a principle similar to that of a camera. The single lens focuses light onto the retina, a bilayer of photosensitive receptor cells. The eyes of humans and other vertebrates are also of the camera type, but they evolved independently and differ from the single-lens eyes of invertebrates in several details.

Vertebrate Vision

The human eye, shown in Figure 45.7, can detect an almost countless variety of colors, can form images of ob-

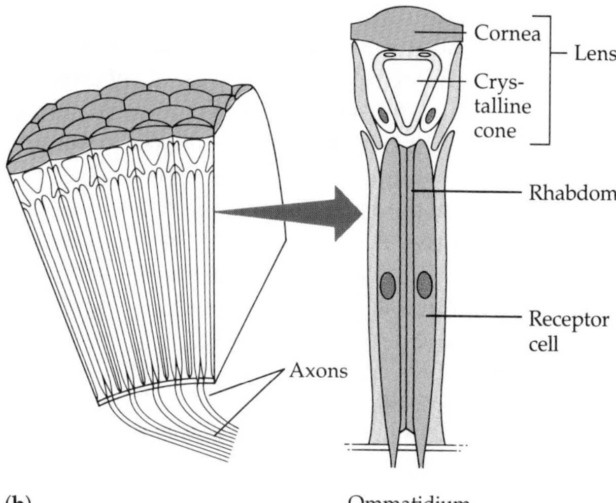

Figure 45.6
Compound eyes. (**a**) The faceted eyes of a horsefly, photographed with a stereo microscope. (**b**) The cornea and crystalline cone of each ommatidium function as lenses that focus light through a structure called a rhabdom, a stack of pigmented plates formed on the inside of a circle of receptor cells. The image is a mosaic of dots formed by the different intensities of light entering the many ommatidia.

jects miles away, and can respond to as little as one photon of light. Remember, however, that it is actually the brain that "sees." Thus, to understand vision, we must begin by learning how the vertebrate eye generates sensations (action potentials), and then follow these signals to the visual centers of the brain, where sights are perceived.

Structure and Function of the Vertebrate Eye The globe of the vertebrate eye, or eyeball, consists of a tough, white outer layer of connective tissue called the **sclera** and a thin, pigmented inner layer called the **choroid.** At the front of the eye, the sclera becomes the transparent **cornea,** which lets light into the eye and acts as a fixed lens. The anterior choroid forms the

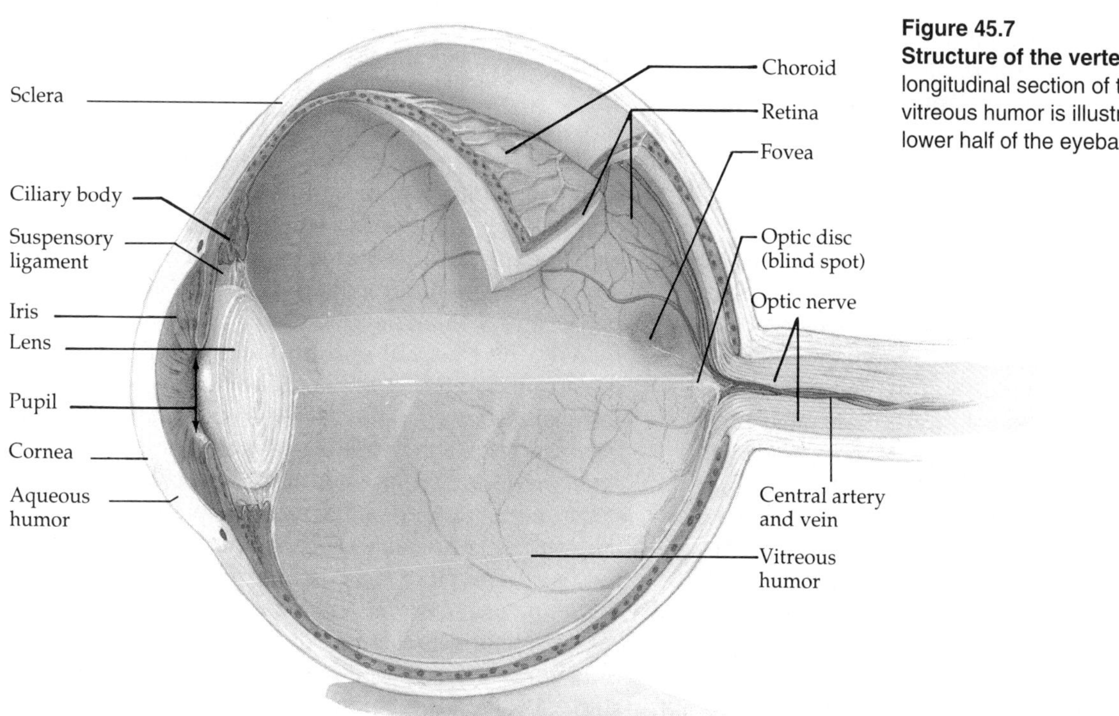

Figure 45.7
Structure of the vertebrate eye. In this longitudinal section of the eye, the jellylike vitreous humor is illustrated only in the lower half of the eyeball.

donut-shaped **iris,** which gives the eye its color. By changing size, the iris regulates the amount of light entering the **pupil,** the hole in the center of the iris. Just inside the choroid, the retina forms the innermost layer of the eyeball. The retina contains the photoreceptor cells themselves. Information from the photoreceptors leaves the eye at the optic disc, where the optic nerve attaches to the eye. Because there are no photoreceptors in the optic disc, this spot on the lower outside of the retina is a blind spot: Light focused onto that part of the retina is not detected.

The **lens** and **ciliary body** divide the eye into two cavities, one between the lens and the cornea, and a much larger cavity behind the lens within the eyeball itself. The ciliary body constantly produces the clear, watery **aqueous humor** that fills the anterior cavity of the eye. Blockage of the ducts that drain the aqueous humor can produce glaucoma, increased pressure that leads to blindness by compressing the retina. The posterior cavity, filled with the jellylike **vitreous humor,** occupies most of the volume of the eye. The aqueous and vitreous humors function as liquid lenses that help focus light onto the retina. The lens itself is a transparent protein disc that focuses an image onto the retina. Many fishes focus by moving the lens forward or backward, camera-style. Humans and other mammals, however, focus by changing the *shape* of the lens. When viewing a distant object, the lens is flat. To focus on a close object, the lens becomes almost spherical, a change termed **accommodation** (Figure 45.8).

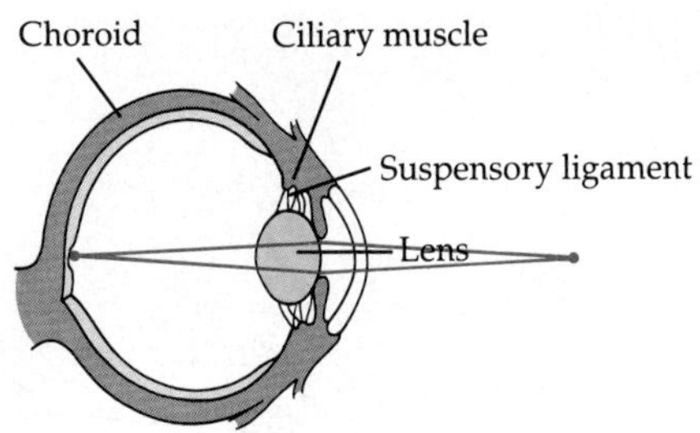

(**a**) Near vision (accommodation)

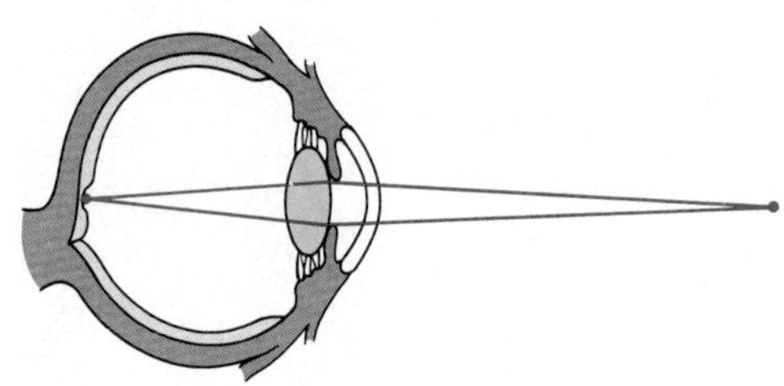

(**b**) Distance vision

Figure 45.8
Focusing in the mammalian eye. The lens bends light and focuses it onto the retina. The thicker the lens, the more sharply the light is bent. The lens is nearly spherical when focusing on near objects and much flatter when focusing at a distance. The shape of the lens is controlled by the ciliary muscle. (**a**) When viewing a close object, the ciliary muscles contract, pulling the border of the choroid layer of the eye toward the lens and causing the suspensory ligaments to slacken. With this reduced tension, the elastic lens becomes thicker and rounder, bending light in such a way that near objects can be focused on the retina. This adjustment of the lens for close viewing is called accommodation. (**b**) When viewing a distant object, the ciliary muscle relaxes, allowing the choroid to expand and put tension on the suspensory ligament. The lens is pulled into a flatter shape, and the distant object is focused onto the retina. Notice that the focusing muscle is relaxed during distance viewing and is contracted for close work.

Signal Transduction in the Eye When the lens focuses a light image onto the retina, how do the cells of the retina transduce the stimuli into sensations—action potentials that transmit this information about the environment to the brain? Built into the human retina are about 125 million **rod cells** and 6 million **cone cells,** two types of photoreceptors named for their shapes. They account for 70% of all receptors in your body, a fact that underscores the importance of the eyes and visual information in how humans perceive their environment.

Rods and cones have different functions in vision, and the relative numbers of these two photoreceptors in the retina are partly correlated with whether an animal is most active during the day or at night. Rods are more sensitive to light but do not distinguish colors; they enable us to see at night, but only in black and white. It takes more light to stimulate cones, and thus these photoreceptors do not function in night vision, but they can distinguish color during the day. Color vision is found in all vertebrate classes, though not in all species. Fishes, amphibians, reptiles, and birds have well-developed color vision, but humans and other primates are among the minority of mammals that can see color. Most mammals are nocturnal, and a maximum number of rods in the retina is an adaptation that gives these animals keen night vision. Cats, usually most active at night, have limited color vision and probably see a pastel world during the day. In the human eye, rods are found in greatest density at the lateral regions of the retina and are completely absent from the **fovea,** the center of the visual field (see Figure 45.7). If you look directly at a dim star at night, it is harder to see than if you look at it on an angle, which focuses the starlight onto the regions of the retinas most populated by rods. You achieve your sharpest day vision, however, by looking straight at the object of interest. This is because cones are most dense at the fovea, where there are about 150,000 color receptors per mm^2. Some birds have more than a million cones per mm^2, which enables such species as hawks to spot field mice and other small prey from high in the sky. In the retina of the eye, as in all biological structures, we see variations representing evolutionary adaptations.

Each rod or cone cell has an outer segment with a stack of folded membranes in which visual pigments are embedded (Figure 45.9a). The actual light-absorb-

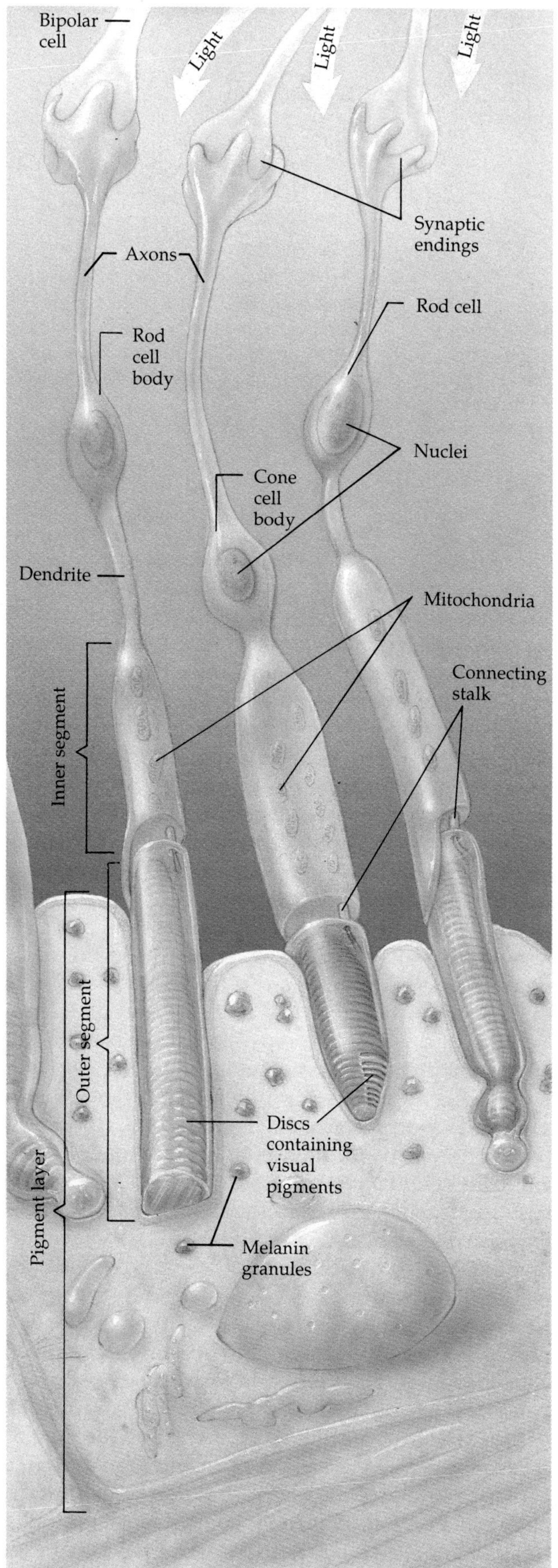

(**a**) Rods and cones, the photoreceptors in the retina

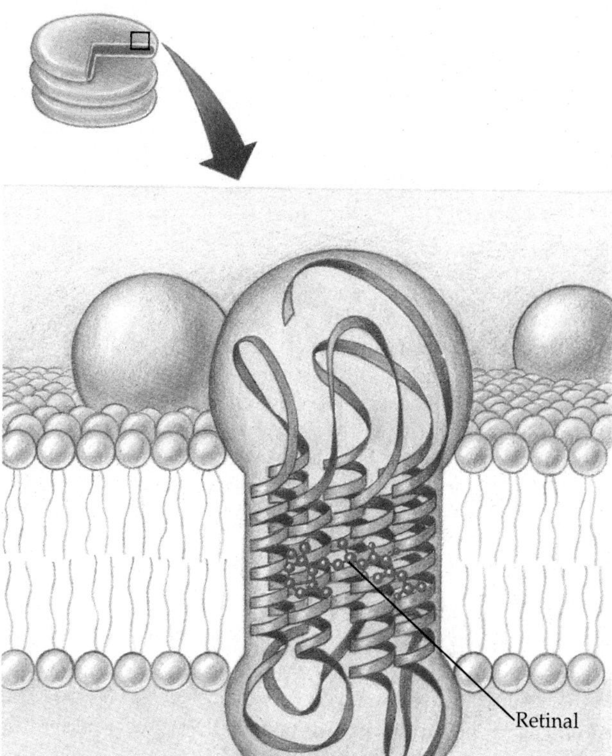

(**b**) Rhodopsin, the visual pigment in rods

Figure 45.9
Photoreceptors of the retina. (**a**) The retina is populated by two types of photoreceptors. Rods are very sensitive to light and function in black-and-white vision at night; cones are less sensitive to light and account for color vision during the day. Both types of cells are modified neurons. Each rod and cone has an outer segment partly embedded in a layer of darkly pigmented epithelial cells. The outer segment is connected by a short stalk to an inner segment, which is in turn connected to the cell body. Axons of the rods and cones synapse with other retinal neurons called bipolar cells. Within the outer segment of a rod or cone is a stack of folded membrane. The visual pigments responsible for detecting light focused onto the retina are built into these stacked membranes. (**b**) The visual pigments consist of a light-absorbing molecule called retinal (derived from vitamin A) bonded to a protein called an opsin. Each type of photoreceptor has a characteristic kind of opsin, which affects the absorption spectrum of the retinal. In the case of rods, the whole pigment complex—retinal plus the specific type of opsin—is called rhodopsin, which is the visual pigment illustrated here. Notice that the opsin has several regions of alpha helix (see Chapter 5) spanning the membrane. At the core of the opsin is the light-absorbing retinal.

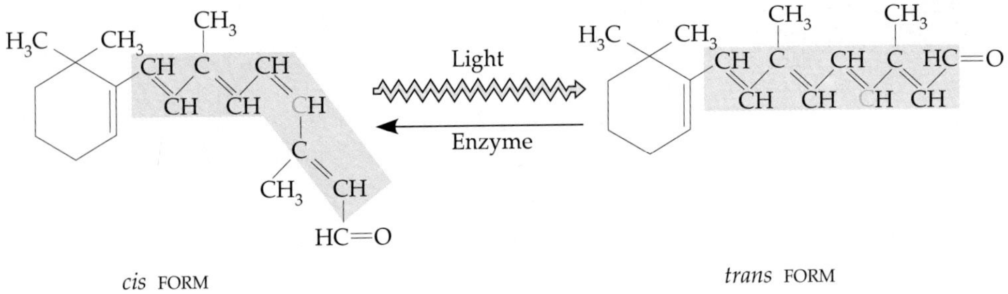

Figure 45.10
How light affects retinal. Retinal exists in two forms, isomers of one another. Absorption of light converts the pigment from the *cis* isomer to the *trans* isomer. When the photoreceptor is no longer stimulated by light, enzymes convert the retinal back to the *cis* form. The photochemical reaction changes the shape of retinal, causing a conformation change in the opsin protein to which it is linked. This event triggers a chain of metabolic responses by the photoreceptor that ultimately changes the voltage across the plasma membrane of the cell, producing a receptor potential (which, in this case, is actually a hyperpolarization, not a depolarization).

ing molecule is **retinal,** which is synthesized from vitamin A. The retinal is bonded to a membrane protein called an **opsin.** Opsins vary in structure from one type of photoreceptor to another, and the light-absorbing ability of the retinal is affected by the specific identity of its opsin partner. Rods have their own type of opsin, and, along with the retinal component, the molecule is called **rhodopsin** (Figure 45.9b). When rhodopsin absorbs light, the retinal changes shape and eventually dissociates from the opsin. This photochemical reaction is referred to as "bleaching" of the rhodopsin. In the dark, enzymes convert the retinal back to its original form, and it recombines with opsin to form rhodopsin (Figure 45.10). Bright light keeps the rhodopsin bleached and rods become unresponsive; cones take over. When you walk from a bright environment into a dark place, such as walking into a movie theater during a matinee, you are initially almost blind; there is not enough light to stimulate the cones, and it takes at least a few minutes for the bleached rods to become functional again.

Color vision results from the presence of three subclasses of cones in the retina, each with its own type of opsin associated with retinal to form visual pigments collectively called **photopsins.** These photoreceptors are known as red cones, green cones, and blue cones, referring to the colors their brands of photopsin are best at absorbing. The absorption spectra for these pigments overlap, and the perception of intermediate hues depends on the differential stimulation of two or more types of cones. For example, when both red and green cones are stimulated, we may see yellow or orange, depending on which of these two populations of cones is most strongly stimulated. Color blindness, more common in males than females because it is generally inherited as a sex-linked trait (see Chapter 14), is due to a deficiency or absence of one or more types of cones.

Transduction of light energy into action potentials is not accomplished by the photoreceptors alone, but results from the cooperation of additional types of neurons in the retina. The chemical response of retinal to light triggers a complex chain of metabolic events that ultimately alters the membrane potential of the rod or cone cell. The photoreceptors themselves do not fire action potentials. The light-induced change in voltage across the membrane is a localized receptor potential, and like all receptor potentials, it is a graded response proportional to the strength of the stimulus. The membrane does not depolarize. Light actually hyperpolarizes the membrane by decreasing the permeability to sodium ions. This results in the receptor cell releasing less neurotransmitter in the light than in the dark. Thus, it is actually a *decrease* in the chemical signal to the cells with which rods and cones synapse that serves as a message that the photoreceptors have been stimulated by light. The axons of rods and cones synapse with neurons called **bipolar cells,** which in turn synapse with **ganglion cells** (Figure 45.11). Additional types of neurons in the retina, **horizontal cells** and **amacrine cells,** help to integrate the information before it is dispatched to the brain. The axons of ganglion cells then convey the resulting sensations to the brain as action potentials along the optic nerve.

Visual Integration Processing of visual information begins in the retina itself. Signals from the rods and cones may follow either vertical or lateral pathways (see Figure 45.11). In the vertical pathway, information passes directly from the receptor cells to the bipolar cells to the ganglion cells. The horizontal and amacrine cells provide lateral integration of visual signals. Horizontal cells carry signals from one rod or cone to other receptor cells and to several bipolar cells; amacrine cells spread the information from one bipolar cell to several ganglion cells. When a rod or cone stim-

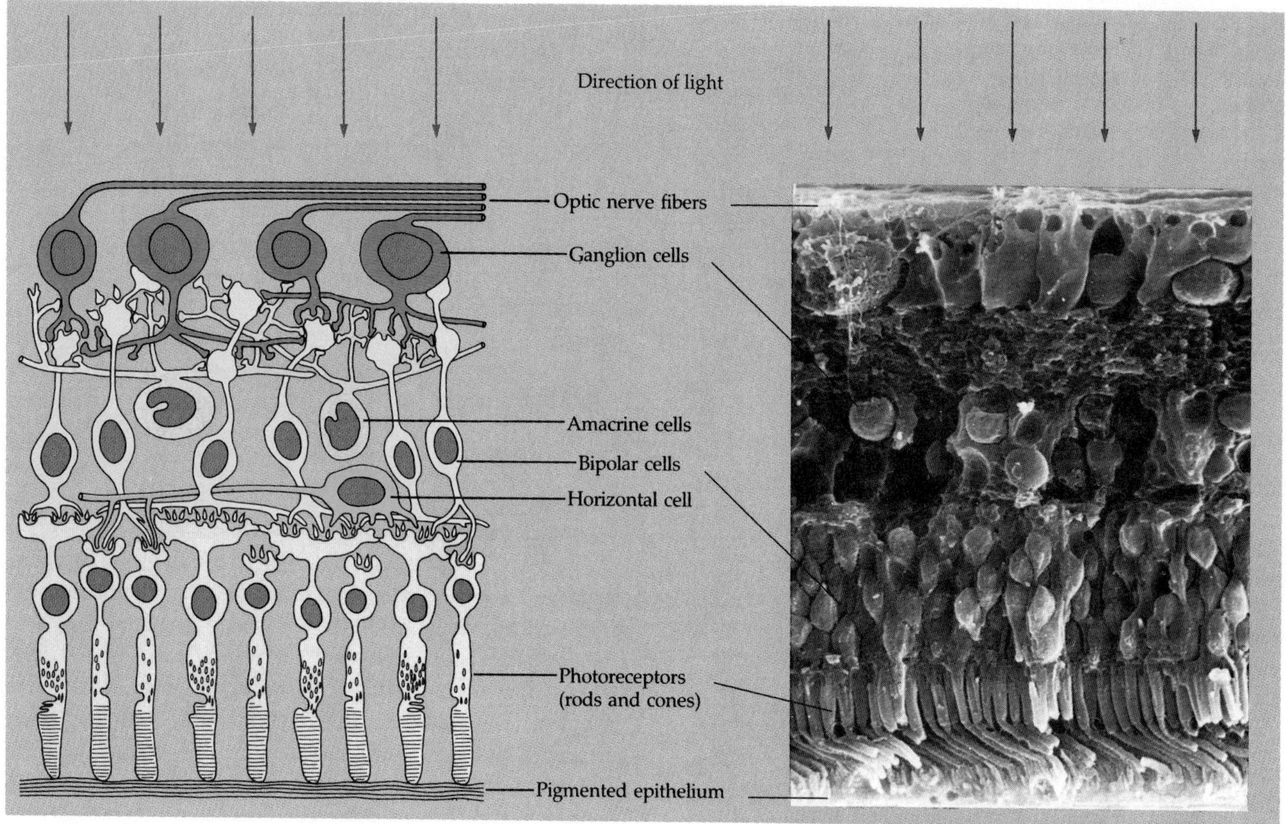

Figure 45.11
The human retina. Light must pass through several relatively transparent layers of cells before reaching the rods and cones. These photoreceptors communicate with ganglion cells via bipolar cells. The axons of the ganglion cells transmit the visual sensations (action potentials) to the brain. There is not a one-to-one relationship between the rods and cones, bipolar cells, and ganglion cells. Rather, each bipolar cell receives information from several rods or cones, and each ganglion cell from several bipolar cells. The horizontal and amacrine cells carry information across the retina to integrate the signals. All the rods or cones that feed information to one ganglion cell form the receptive field for that cell. The larger the receptive field (the more rods or cones that supply a ganglion cell), the less sharp the image, because it is less evident exactly where the light struck the retina. The ganglion cells of the fovea have very small receptive fields, so visual acuity is very sharp in this area. (SEM from *Tissues and Organs: A Text-Atlas of Scanning Electron Microscopy* by Richard G. Kessel and Randy H. Kardon. W. H. Freeman and Company. Copyright © 1979.)

ulates a horizontal cell, the horizontal cell stimulates nearby receptors, but inhibits more distant receptors and bipolar cells that are not illuminated, making the light spot appear lighter and the dark surroundings even darker. This integration, called **lateral inhibition,** sharpens edges and enhances contrast in the image. One type of retinal integration based on lateral inhibition is described in Figure 45.12. Lateral inhibition is repeated by the interactions of the amacrine cells with the ganglion cells and occurs at all levels of visual processing.

Axons of ganglion cells form the optic nerves that transmit sensations from the eyes to the brain. The optic nerves from the two eyes meet at the **optic chiasma** near the center of the base of the cerebral cortex. The optic chiasma has its nerve tracts arranged in such a way that what is sensed in the left field of view by both eyes is transmitted to the right side of the brain, and what is detected in the right field of view goes to the left side of the brain (Figure 45.13). Most of the ganglion cell axons lead to the **lateral geniculate nuclei** of the thalamus. Neurons of the lateral geniculate nuclei continue back to the **primary visual cortex** in the occipital lobe of the cerebrum. Additional interneurons carry the information to other, more sophisticated visual processing and integrating centers elsewhere in the cortex.

Point-by-point information in the visual field is projected along neurons onto the visual cortex according to its position in the retina, but the information the brain receives is highly distorted. How this coded set of spots, lines, and movement is converted into our perception and recognition of objects is still largely unknown; there is much left to learn about how we actually "see."

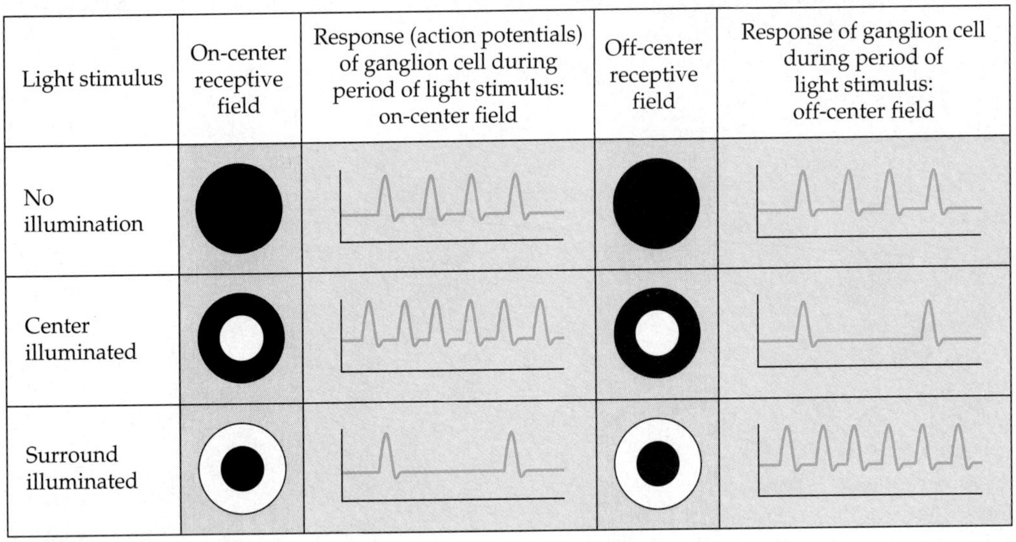

Light stimulus	On-center receptive field	Response (action potentials) of ganglion cell during period of light stimulus: on-center field	Off-center receptive field	Response of ganglion cell during period of light stimulus: off-center field
No illumination				
Center illuminated				
Surround illuminated				

Figure 45.12
Visual integration in the vertebrate retina. The retina is organized into visual receptive fields consisting of the many photoreceptors that communicate with a common ganglion cell. Action potentials are propagated along the axons of the ganglion cell to the brain with a frequency based on integration of the inputs from the receptors making up that receptive field. Bipolar cells, horizontal cells, and amacrine cells function as the integrating network of the receptive field. Even in the dark, a ganglion cell fires action potentials with a constant frequency. If the entire receptive field is uniformly illuminated, there is little effect on this basal rate of discharge by the ganglion cell. It is when only a portion of the field is illuminated that the frequency of action potentials sent from the ganglion cell to the brain changes. Two types of receptive fields are known as on-center fields and off-center fields. A ganglion cell that has an on-center receptive field is stimulated when a spot of light is focused on the center of the receptive field, and it is inhibited when only the peripheral region of the field is illuminated. A ganglion cell that has an off-center receptive field has just the opposite response: A spot of light focused onto the center of the field inhibits the ganglion cell, but illumination of the periphery of the field stimulates the cell. Thus, on-center ganglion cells respond best to bright spots against a dark background, while off-center ganglion cells prefer dark spots against a light background. Much of the visual information transmitted from the retina to the brain is based on these enhanced patterns of bright and dark spots.

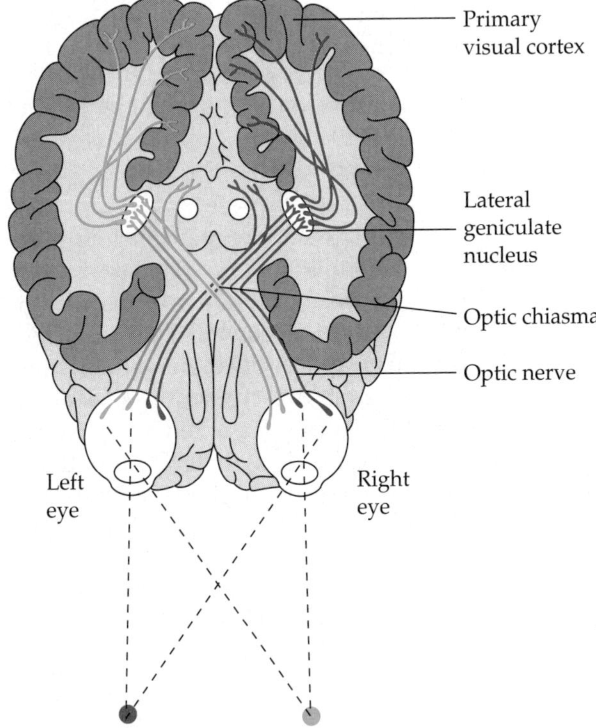

Figure 45.13
Neural pathways for vision. Objects in the left field of view are "seen" by the right side of the brain, while objects in the right field of view are projected into the left side of the brain. (We are seeing the brain from the bottom in this diagram.) The lateral geniculate nuclei, ganglia located in the thalamus, relay the sensations to the visual cortex.

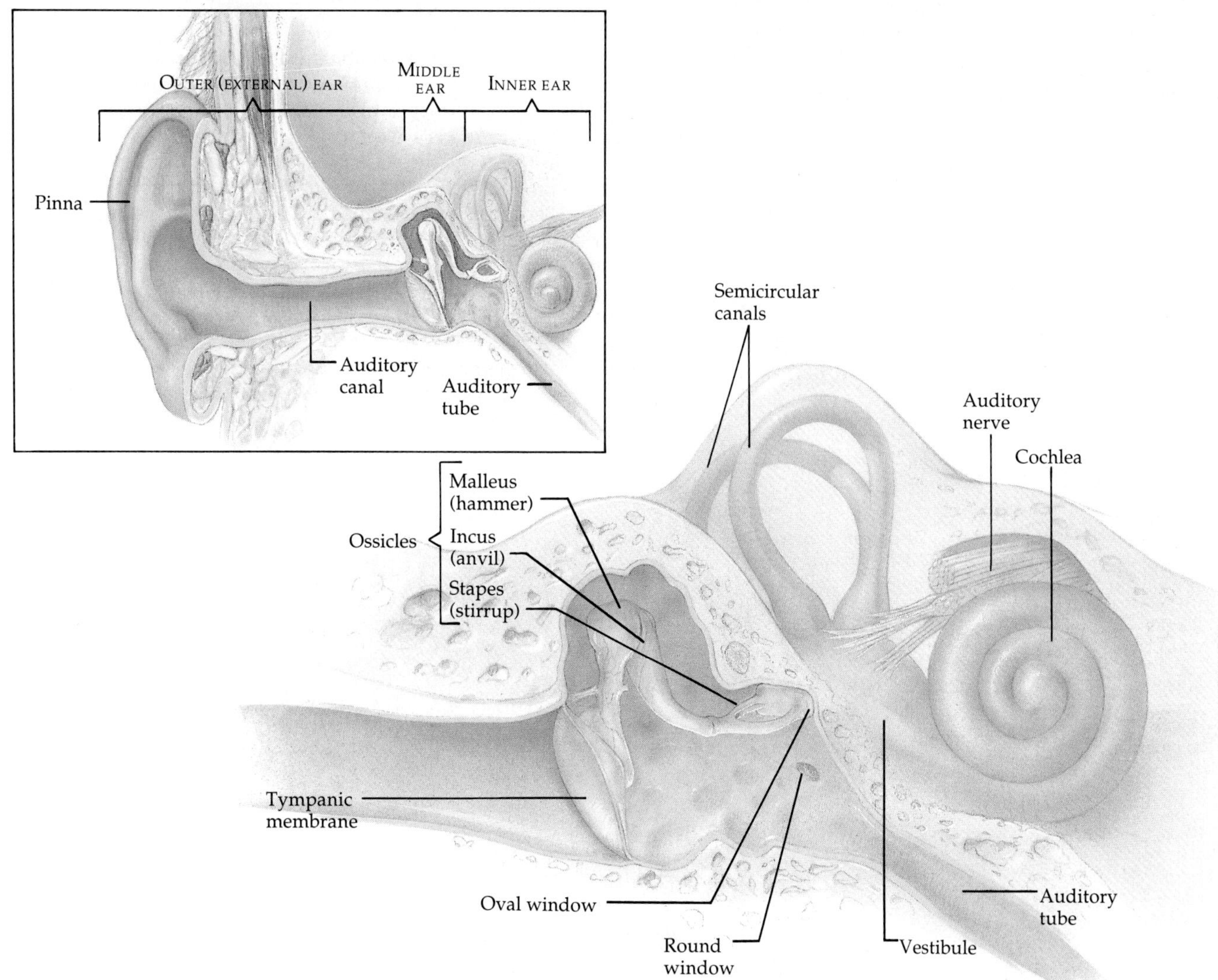

Figure 45.14
Structure of the human ear. The hair cells, the actual sensory receptors of the ear, are located within the labyrinth of canals of the inner ear. The cochlea contains hair cells that function in hearing. The vestibule and semicircular canals contain hair cells responsible for the sense of balance.

HEARING AND BALANCE

The senses of hearing and balance are related in most animals. Both senses involve mechanoreceptors containing hair cells that trigger action potentials when the hairs are bent by settling particles or moving fluid. In mammals and most other terrestrial vertebrates, the sensory organs for hearing and balance are located together within the ear.

The Mammalian Ear

The human ear is really two separate sense organs. It contains the organ of audition, or hearing, and that of equilibrium, or balance. Both of these senses are functions of hair cells in fluid-filled canals.

The ear itself can be divided into three regions (Figure 45.14). The **outer ear** consists of the external **pinna** and the **auditory canal,** which collect sound waves and channel them to the **tympanic membrane** (eardrum) of the **middle ear.** There the vibrations are conducted through three small bones—the **malleus** (hammer), **incus** (anvil), and **stapes** (stirrup)—to the **inner ear,** passing through the **oval window,** a membrane beneath the stapes. The middle ear also opens into the **auditory (Eustachian) tube,** which connects with the pharynx and equalizes pressure between the middle ear and the atmosphere—enabling you to "pop" your ears when changing altitude, for example. The inner ear consists of a labyrinth of channels within

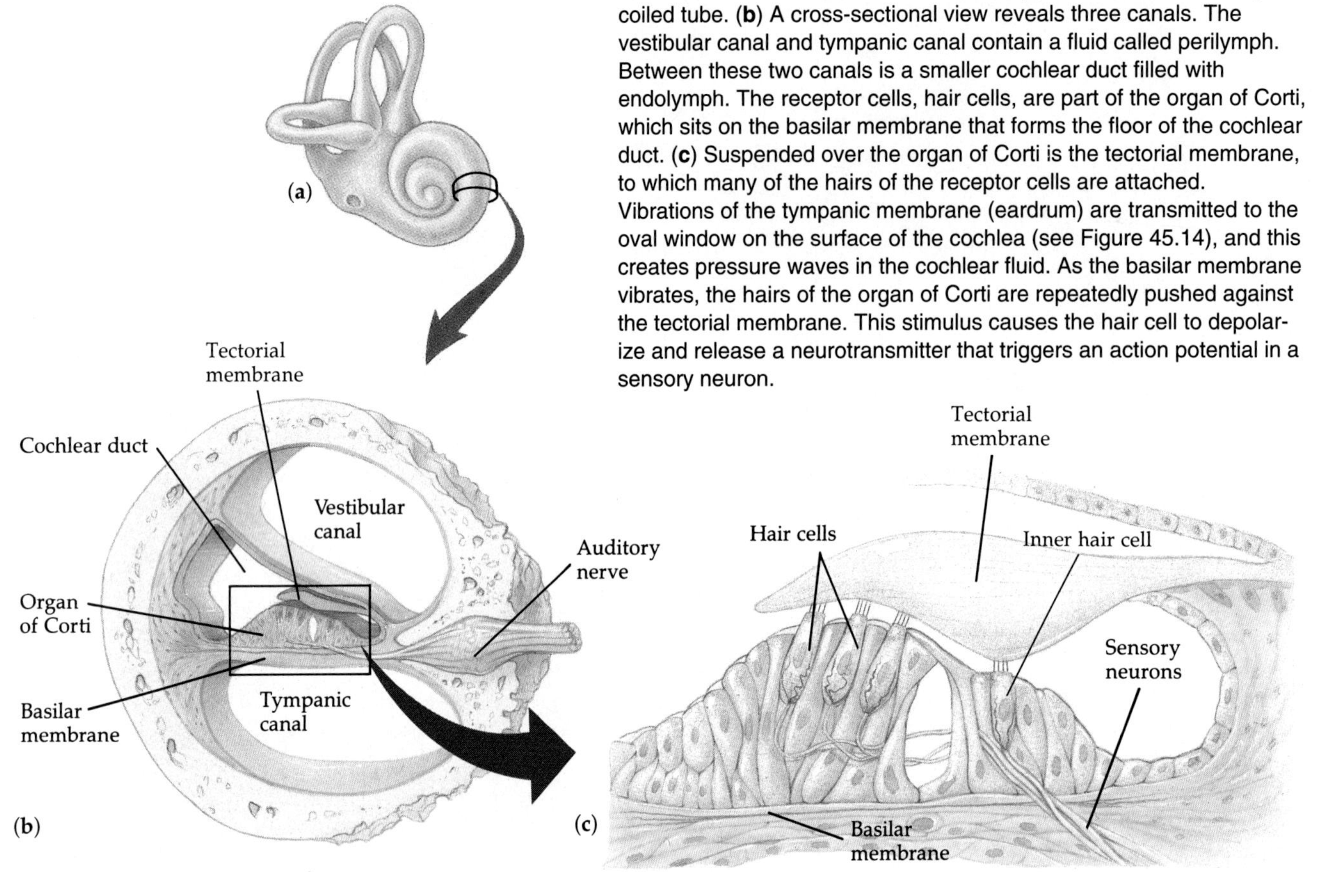

Figure 45.15
Structure and function of the cochlea. (**a**) The cochlea is a long coiled tube. (**b**) A cross-sectional view reveals three canals. The vestibular canal and tympanic canal contain a fluid called perilymph. Between these two canals is a smaller cochlear duct filled with endolymph. The receptor cells, hair cells, are part of the organ of Corti, which sits on the basilar membrane that forms the floor of the cochlear duct. (**c**) Suspended over the organ of Corti is the tectorial membrane, to which many of the hairs of the receptor cells are attached. Vibrations of the tympanic membrane (eardrum) are transmitted to the oval window on the surface of the cochlea (see Figure 45.14), and this creates pressure waves in the cochlear fluid. As the basilar membrane vibrates, the hairs of the organ of Corti are repeatedly pushed against the tectorial membrane. This stimulus causes the hair cell to depolarize and release a neurotransmitter that triggers an action potential in a sensory neuron.

a skull bone (the temporal bone). These channels are lined by a membrane and contain fluid that moves in response to sound or movement of the head.

The part of the inner ear involved in hearing is a complex coiled organ known as the **cochlea** (from the Latin for "snail"). It occupies part of the labyrinthine cavity within the temporal bone. The cochlea has two large chambers, an upper vestibular canal and a lower tympanic canal, separated by a smaller cochlear duct (Figure 45.15). The vestibular and tympanic canals contain a fluid called perilymph, and the cochlear canal is filled with a liquid named endolymph. The floor of the cochlear canal, the basilar membrane, bears the **organ of Corti.** Built into the organ of Corti are the actual receptor cells of the ear, hair cells with their hairs projecting into the cochlear canal. The tips of some of the hairs are attached to the tectorial membrane, which hangs over the organ of Corti like a shelf. Let's now see how this complex anatomy of the ear is correlated with the function of hearing.

The Process of Hearing The ear converts the energy of pressure waves traveling through air into nerve impulses that the brain perceives as sound. Vibrating objects, such as the reverberating strings of a guitar or the vocal cords of a speaking person, create percussion waves in the surrounding air. These waves cause the tympanic membrane to vibrate with the same frequency as the sound. Frequency is the number of vibrations per second, or hertz (Hz). The three bones of the middle ear amplify and transmit the mechanical movements to the oval window, a membrane on the surface of the cochlea. Vibrations of the oval window produce pressure waves in the fluid within the cochlea.

The cochlea transduces the energy of the vibrating fluid into action potentials. The stapes vibrating against the oval window creates a traveling pressure wave in the fluid of the cochlea that passes into the vestibular canal (Figure 45.16). This wave continues around the tip of the cochlea and through the tympanic canal, dissipating as it strikes the **round window.** The pressure waves in the vestibular canal push downward on the central canal and basilar membrane. As the basilar membrane vibrates in response to the pressure waves, it alternately presses the hair cells into and draws them away from the tectorial membrane. This "tweaking" of the hairs distorts the plasma membranes of the receptor cell, making it more permeable to sodium. The resulting depolarization increases neurotransmitter release from the hair cell and the frequency

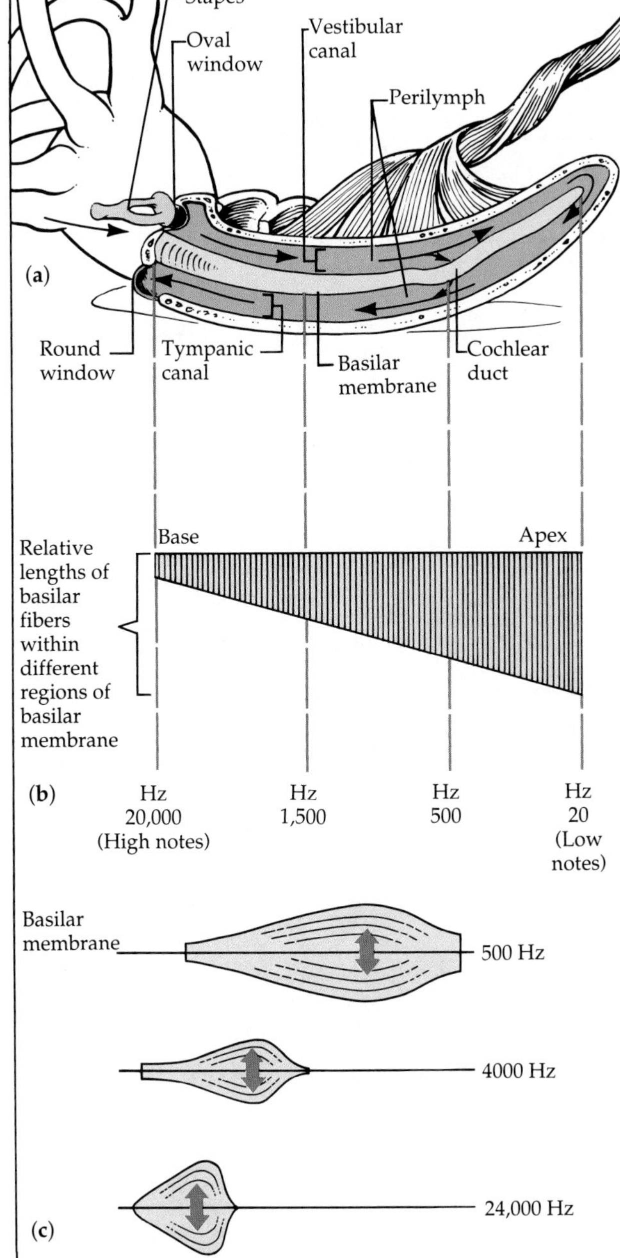

Figure 45.16
How the cochlea distinguishes pitch. (**a**) Vibrations of the stapes against the oval window agitate the fluid within the cochlea (uncoiled here), setting up pressure ripples that have a frequency equivalent to the sound waves that entered the ear. The waves pass through the vestibular canal to the apex of the cochlea, then back toward the base of the cochlea via the tympanic canal. The energy causes the cochlear duct, with its basilar membrane and organ of Corti (see Figure 45.15), to vibrate up and down. The bouncing of the basilar membrane stimulates the hair cells within the cochlear duct. (**b**) Fibers span the width of the basilar membrane. Like harp strings, these fibers vary in length, being shorter near the basal end of the membrane and longer near the apex. The length of the fibers "tunes" them to vibrate at specific frequencies. (**c**) Thus, pressure waves in the cochlea cause a specific region along the length of the basilar membrane to oscillate more vigorously than other regions not "tuned" to that frequency. The preferential stimulation of hair cells is perceived in the brain as sound of a certain pitch.

of action potentials in the sensory neuron with which the hair cell synapses. This neuron carries the sensations to the brain through the auditory nerve.

Sound is detected by increases in the frequency of impulses in the auditory neuron, but how is the quality of that sound determined? Two important aspects of sound are volume and pitch. **Volume** (loudness) is determined by the **amplitude,** or height, of the sound wave. The greater the amplitude of a sound, the more vigorous the vibrations of fluid in the cochlea, the greater the bending of the hair cells, and the more action potentials generated in the sensory neurons. **Pitch** is a function of the **frequency** of sound waves and is usually expressed in hertz. Short, high-frequency waves produce high-pitched sound, while long, low-frequency waves generate low-pitched sound. Healthy young humans can hear sounds in the range of 40 to 20,000 Hz; dogs can hear sounds as high as 40,000 Hz; and bats can emit and hear sounds of even higher frequency, using this ability to locate objects by sonar (see Figure 45.1).

Pitch can be distinguished by the cochlea because the basilar membrane is not uniform along its length. The proximal end near the oval window is relatively narrow and stiff, while the distal end near the tip is wider and more flexible (see Figure 45.16). Each region of the basilar membrane is most affected by a particular frequency of vibrations, and the region vibrating most vigorously at any instant sends the most action potentials along the auditory nerve. But the actual perception of pitch depends on the mapping of the brain. Sensory neurons from the auditory pathway project onto specific auditory regions of the cerebral cortex according to the region of the basilar membrane in which the signal originated. When a particular site of the cortex is stimulated, we perceive a sound of a particular pitch.

Balance and Equilibrium in Mammals The balance apparatus in humans and most other mammals is also centered in the inner ear. Behind the oval window is a vestibule (see Figure 45.14) that contains two chambers, the **utricle** and **saccule.** The utricle opens into three **semicircular canals** that complete the apparatus for equilibrium.

Sensations related to body position are generated much like sensations of sound in humans and most other mammals. Hair cells in the utricle and saccule respond to changes in head position with respect to gravity and movement in one direction. The hair cells are arranged in clusters, and all their hairs project into a gelatinous material containing many small calcium carbonate particles called otoliths ("ear stones"). Because this material is heavier than the endolymph within the utricle and saccule, gravity is always pulling downward on the hairs of the receptor cells, sending a constant train of action potentials along the sensory neurons of the vestibular branch of the auditory nerve.

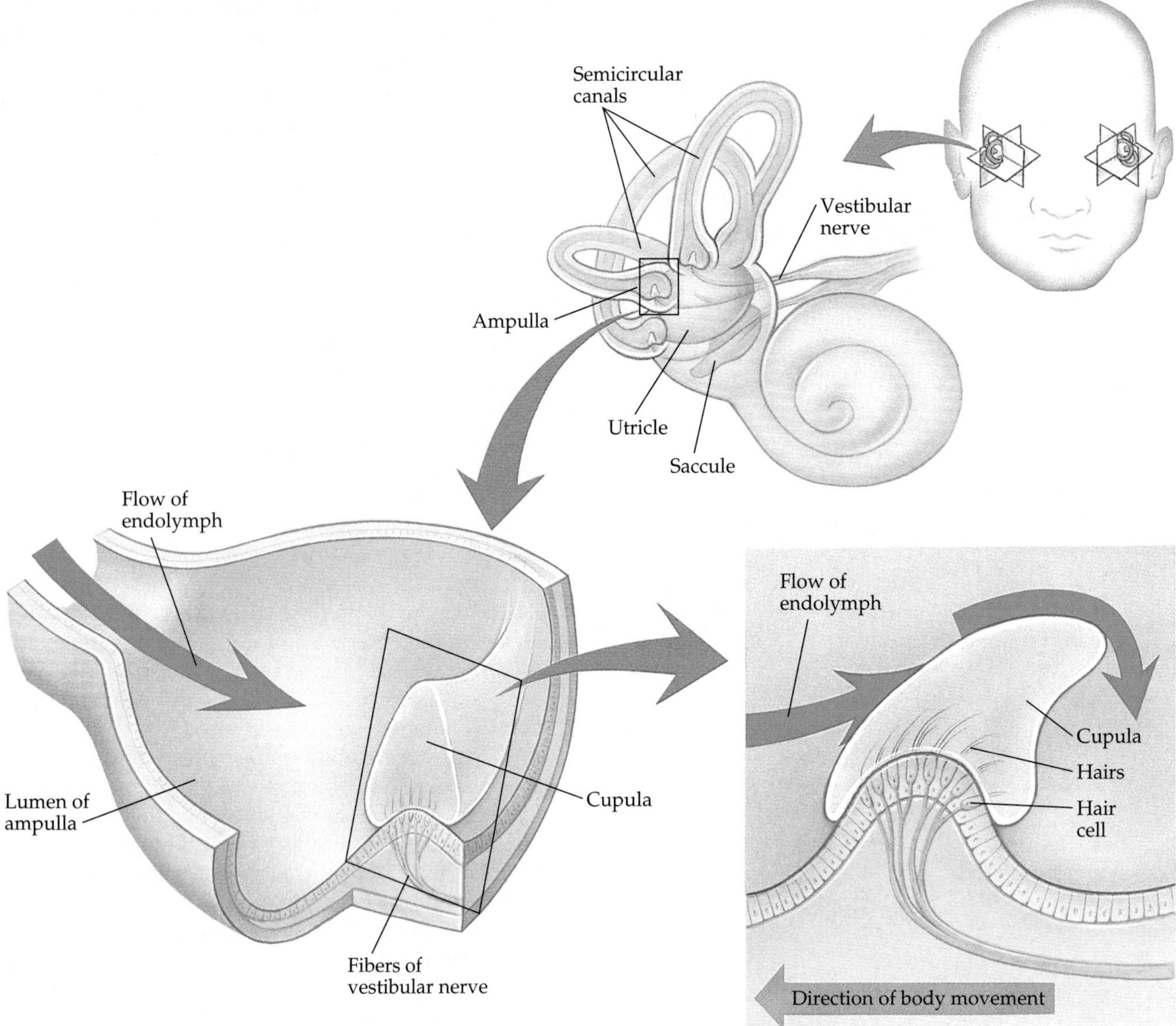

Figure 45.17
The semicircular canals and equilibrium. A swelling called an ampulla at the base of each semicircular canal contains a cluster of hair cells. The hairs from these cells project into a gelatinous mass called the cupula. When the head rotates, inertia prevents the endolymph in the canals from moving with it, so the fluid presses against the cupula, bending the hair cells. This bending increases the frequency of impulses in the sensory neurons in direct proportion to the speed of rotation. This receptor mechanism adapts quickly if rotation continues at a constant speed. Gradually, the endolymph begins to move with the head, and the pressure on the cupula is reduced. If rotation stops suddenly, however, the fluid continues to flow through the semicircular canals and again stimulates the hair cells. This new stimulus causes dizziness. The saccule and utricle contain hair cells that detect the pull of gravity on small particles called otoliths. This tells the brain which way is up and also informs the brain of any linear acceleration associated with movement.

When the position of the head changes with respect to gravity (as when the head bends forward), the force on the hair cell changes, and it increases (or decreases) its output of neurotransmitter. The brain interprets the resulting changes in impulse production by the sensory neurons to determine the position of the head. By a similar mechanism, the semicircular canals, arranged in the three planes of space, detect rotation of the head (Figure 45.17).

Hearing and Equilibrium in Other Vertebrates

Most fishes and aquatic amphibians have a **lateral line system** that runs along both sides of the body (see Chapter 30). The system contains mechanoreceptors that detect movement by a mechanism similar to the function of the inner ear. Water from the animal's surroundings enters the lateral line system through nu-

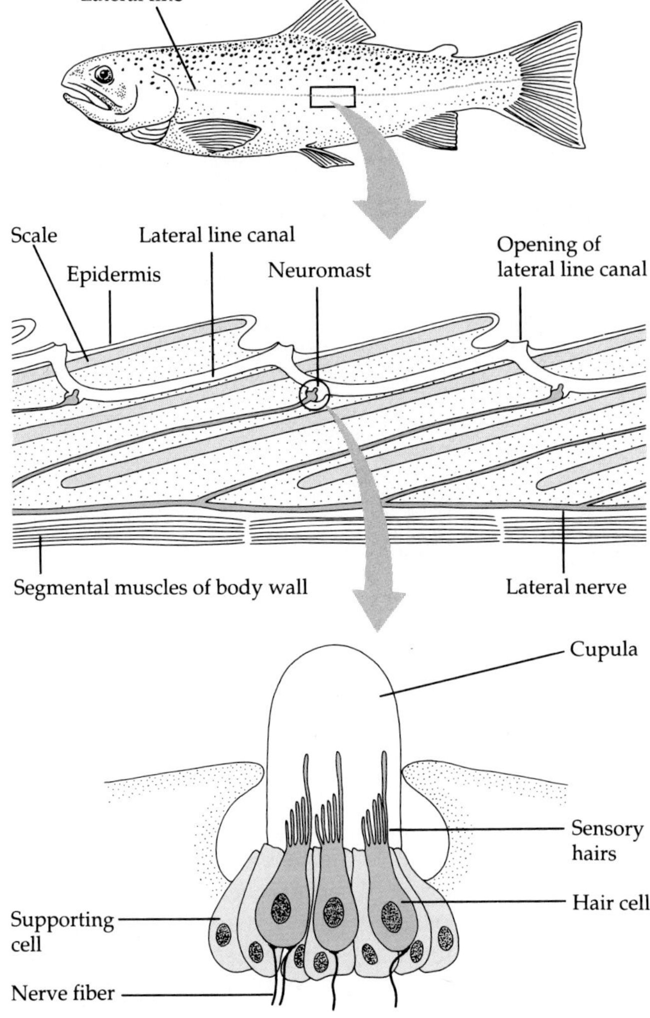

Figure 45.18
The lateral line system in fish. Water flowing through the system bends hair cells, which transduce the energy into action potentials conveyed to the brain. The lateral line system enables the fish to monitor water currents, pressure waves produced by moving objects, and low-frequency sounds conducted through water.

merous pores and flows along a tube past the mechanoreceptors (Figure 45.18). The receptor units, called **neuromasts,** resemble structures called ampullae in the semicircular canals. Each neuromast has a cluster of hair cells, with the sensory hairs embedded in a gelatinous cap called the cupula. As pressure of the moving water bends a cupula, the hair cells transduce the energy into action potentials that are transmitted along a nerve to the brain. This information helps the fish perceive its movement through water or the direction and velocity of water currents flowing over its body. The lateral line system also detects water movements or vibrations generated by other moving objects, including prey and predators.

Like other vertebrates, fishes also have inner ears located near the brain. Along with the lateral line system, the inner ears enable a fish to hear. There is no cochlea, but there are a saccule, utricle, and semicircular canals, structures homologous to the equilibrium sensors of our own ears. Within these chambers in the inner ear of a fish, sensory hairs are stimulated by the movement of otoliths, tiny granules. Unlike the mammalian hearing apparatus, the inner ear of a fish has no eardrum and does not open to the outside of the body. Vibrations of the water caused by sound waves are conducted through the skeleton of the head to the inner ears, setting the otoliths in motion and stimulating the hair cells. The air-filled swim bladder (see Chapter 30) also vibrates in response to sound and may contribute to the transfer of sound to the inner ear. Some fishes, including catfishes and minnows, have a series of bones called the **Weberian apparatus,** which conducts vibrations from the swim bladder to the inner ear. Sound waves also stimulate the hair cells of the lateral line system, but only if the sound is of relatively low frequency. The inner ears extend the hearing of fishes to higher frequencies.

The lateral line system functions only in water. In terrestrial vertebrates, the inner ear has evolved as the main organ of hearing and equilibrium. Some amphibians have a lateral line system as tadpoles, but not as adults living on land. In the ear of a terrestrial frog or toad, sound vibrations traveling in the air are conducted by a tympanic membrane and a single bone to the inner ear (the mammalian ear, remember, has three bones). There is recent evidence that the lungs of a frog also vibrate in response to sound and transmit their vibrations to the eardrum via the auditory tube. A small side pocket of the saccule functions as the main hearing organ of the frog, and it is this outgrowth of the saccule that gave rise to the more elaborate cochlea during the evolution of mammals. Birds also have a cochlea, but like amphibians and reptiles, sound is conducted from the tympanic membrane to the inner ear by a single bone, the stapes.

Sensory Organs for Hearing and Equilibrium in Invertebrates

Many invertebrates sense sounds. Hearing structures have been most extensively studied in the arthropods. For example, the body hairs of many insects vibrate in response to sound waves of specific frequencies, depending on the stiffness and length of the hairs. The hairs are commonly tuned to frequencies of sounds produced by other organisms. Male mosquitoes use fine hairs on their antennae to detect the hum produced by the beating wings of flying females, thus enabling them to locate a mate; a tuning fork that vibrates at the same frequency as a female mosquito's wings will also attract males. Some caterpillars, larvae of insects, use vibrating body hairs to hear the buzzing wings of predatory wasps, warning the caterpillars of

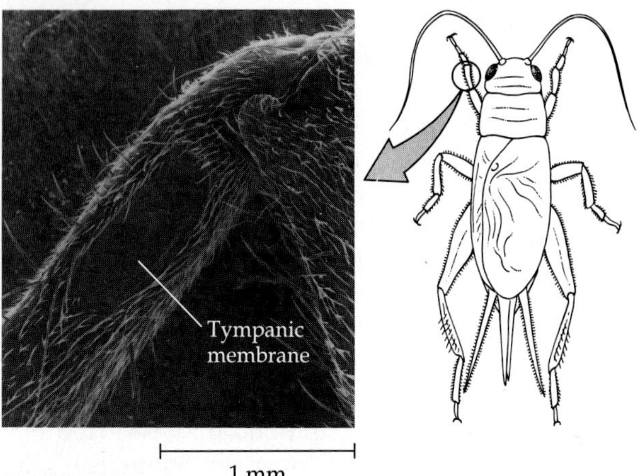

Figure 45.19
An insect ear. The tympanic membrane, this one on the front leg of a cricket, vibrates in response to sound waves (SEM). This vibration stimulates mechanoreceptor cells attached to the inside of the eardrum.

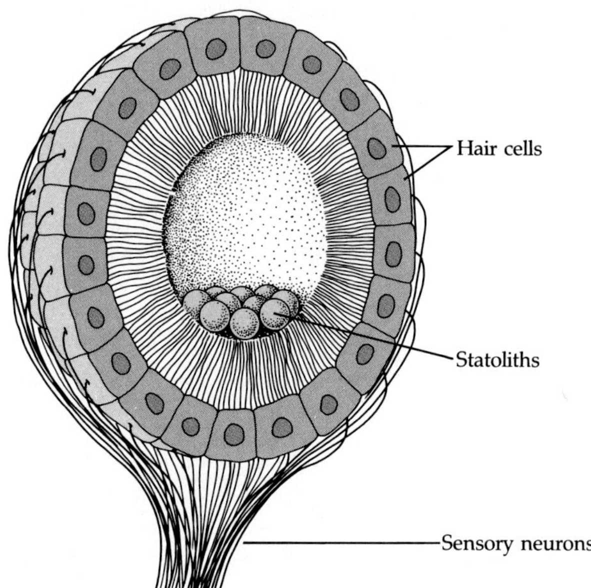

Figure 45.20
The statocyst of an invertebrate. The settling of statoliths to the low point within the chamber bends hair cells in that location, providing the brain with information about the position of the body.

danger. Many insects also have localized "ears," most commonly on their legs (Figure 45.19). A tympanic membrane (eardrum) is stretched over an internal air chamber. Sound waves vibrate the tympanic membrane, stimulating receptor cells attached to the inside of the membrane and resulting in nerve impulses that are transmitted to the brain. Some moths can hear notes of such high pitch that they detect the sounds bats produce for echolocation, and perception of these sounds triggers an escape maneuver in which the moths change directions abruptly.

Most invertebrates have mechanoreceptors called **statocysts** that function in their sense of equilibrium (Figure 45.20). A common type of statocyst consists of a layer of hair cells surrounding a chamber containing **statoliths,** which are grains of sand or other dense granules. Gravity causes the statoliths to settle to the low point within the chamber, stimulating hair cells in that location. (This is similar to how the saccule and utricle function in vertebrates, and indeed these structures in the vertebrate inner ear are considered to be specialized types of statocysts.) The statocysts of invertebrates have various locations. For example, many jellyfish have statocysts at the fringe of the "bell," giving the animals indications of body position. Lobsters and crayfish have statocysts near the bases of their antennules. Crayfish have been tricked into swimming upside down in experiments where the statoliths were replaced with metal shavings that could be pulled to the upper end of the statocysts with magnets.

TASTE AND SMELL

The senses of taste (gustation) and smell (olfaction) depend on chemoreceptors that detect specific chemicals in the environment. In the case of terrestrial animals, taste is the detection of certain chemicals that are present in a solution, and smell is the detection of airborne chemicals. However, these chemical senses are usually closely related, and there really is no distinction in aquatic environments.

Various animals use their chemical senses to find mates, to recognize territory that has been marked with some chemical substance, and to help navigate during migration (Figure 45.21). Many animals, especially social insects, also use chemicals to communicate. In Chapter 50 you will learn how the social organization of a beehive or anthill is based on chemical "conversation." In all animals, taste and smell are important in feeding behavior. For example, the cnidarian *Hydra* begins to swallow when chemoreceptors detect a compound called glutathione, which is released from prey that has been captured by the *Hydra's* tentacles. Try adding some glutathione to the water around a *Hydra* in the laboratory, and you can observe this feeding behavior for yourself. (In fact, if there are two or more of the animals on your watch glass, they may resort to cannibalism.)

The taste receptors of insects are located within sensory hairs called **setae** on the feet and mouthparts. The animals use their sense of taste to select food. A tasting hair contains several chemoreceptor cells, each especially responsive to a particular class of chemical stimuli, such as sugar or salt. By integrating sensations (nerve impulses) from these different receptor cells, the brain of the insect can apparently distinguish a very

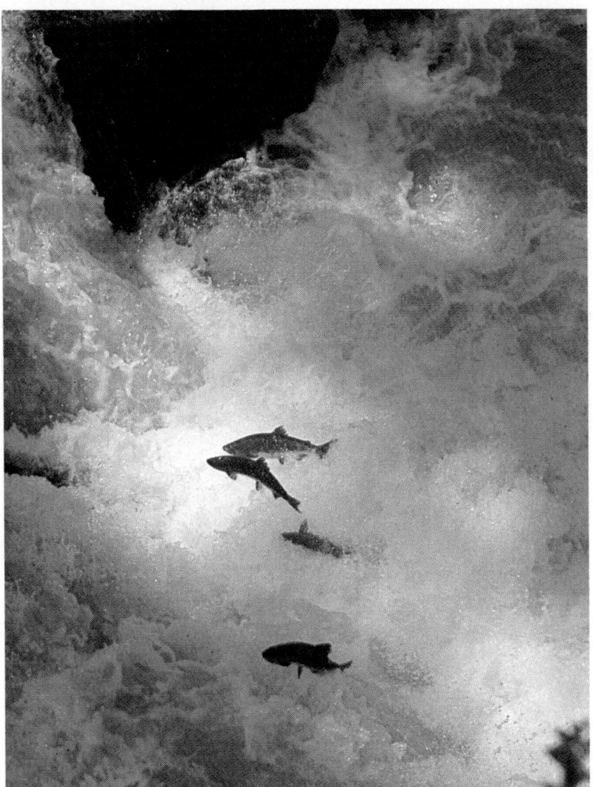

Figure 45.21
Salmon follow their noses home. Various species of Pacific salmon make a one-time round trip from small streams where the fish hatch, out to the sea via rivers such as the Columbia, then back to the stream of their origin to spawn. While in the sea, salmon from many river systems school and feed together in the Gulf of Alaska. When they are a few years old, sexually mature salmon segregate into groups of common geographic origin and migrate back toward the river from which they emerged as juveniles. During this first stage of the return, salmon may navigate by using the position of the sun. But once a salmon reaches the general region of the river leading to its home stream, a keen sense of smell takes over. The water that flows from each stream into the river carries a unique scent resulting from the types of plants, soil, and other environmental components of that stream. This scent is apparently imprinted in the memory of a young salmon before it migrates to the sea. Years later, on its return journey, a salmon follows these chemical cues at each fork in the river system. Battling white water, perhaps for hundreds of miles, the fish eventually arrives at its place of origin, spawns, and dies.

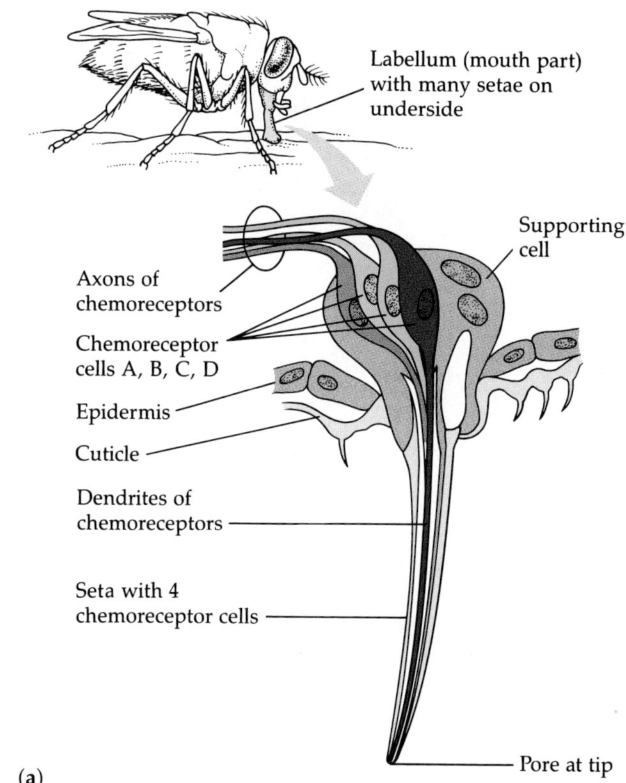

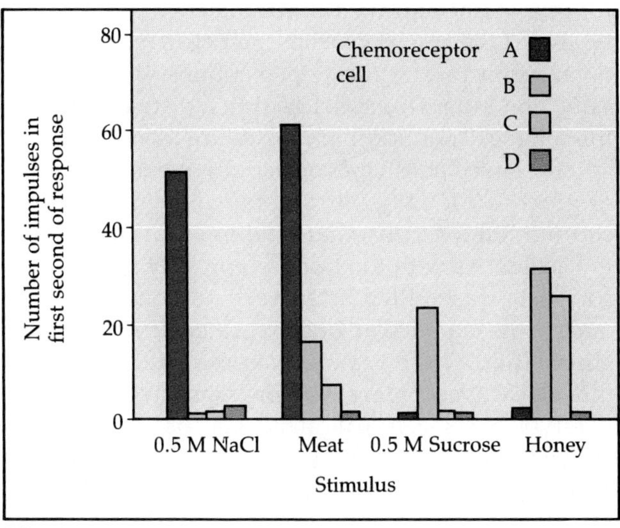

Figure 45.22
The mechanism of taste in a blowfly. (**a**) Gustatory setae (hairs) on the feet and mouthparts each contain four chemoreceptor cells with dendrites that extend to the pore at the tip of the sensory hair. These receptor cells are labeled with letters. (**b**) Each taste cell is especially sensitive to a particular class of substance; for example, the receptor labeled B is most responsive to sugars. This specificity, however, is relative; each cell can respond to some extent to a broad range of chemical stimuli. Thus, any natural food probably stimulates two or more of the receptor cells. The brain apparently integrates the frequencies of impulses arriving along the axons of the four classes of receptor cells and distinguishes a great variety of tastes.

large number of tastes (Figure 45.22). Insects can also smell airborne chemicals, using olfactory setae, usually located on the antennae (see Figure 45.3).

In humans and other mammals, the chemical senses of gustation and olfaction are functionally similar and interrelated. In both cases, a small molecule must dissolve in liquid to reach the receptor cell and stimulate the sensation. That molecule binds to a specific protein in the receptor cell membrane and triggers a depolarization of the membrane and release of neurotransmitter.

The receptor cells for taste are organized into **taste buds** scattered in several areas of the tongue and

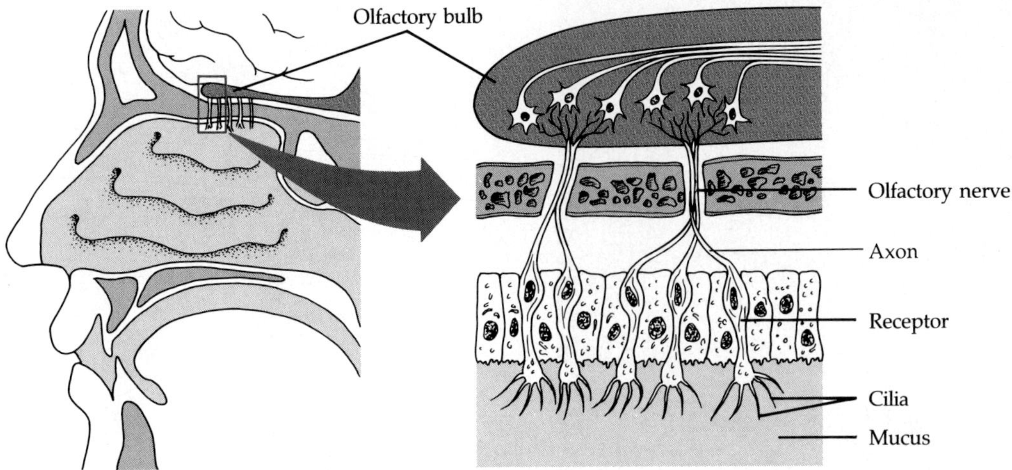

Figure 45.23
Olfaction in humans. Specific binding of molecules to the ciliary membrane of receptor cells triggers action potentials conveyed directly to the olfactory bulb of the brain by axons of the receptor cells.

mouth. Most of the taste buds are on the surface of the tongue or are associated with raised papillae on the tongue. Although we cannot distinguish different types of taste receptors from their structures, we recognize four primary taste perceptions—sweet, sour, salty, and bitter—each detected in a distinct region of the tongue. These primary tastes are associated with specific molecular shapes or charges (the ring structure of glucose for sweet, for instance, or the positive sodium ion for salty) that bind to separate receptor molecules. As with the taste receptors of insects, sensory data transmitted by sensory neurons from taste buds to the mammalian brain represent the differential stimulation of the various classes of receptors. Although each receptor cell is more reponsive to a particular type of substance, it can actually be stimulated by a broad range of chemicals. With each taste of food or sip of drink, the brain integrates the differential input from the taste buds, and a complex flavor is perceived.

The olfactory sense of mammals detects certain airborne chemicals. Olfactory receptor cells line the upper portion of the nasal cavity and send impulses along their axons directly to the olfactory bulb of the brain (Figure 45.23). The receptive ends of the cells contain cilia that extend into the layer of mucus coating the nasal cavity. When an odorous substance diffuses into this region, it binds to specific receptor molecules in the olfactory hairs and depolarizes the olfactory receptor cell. Humans can distinguish thousands of different odors, but these are probably based on a few primary odors, analogous to the primary tastes of the gustatory system.

Although the receptors and brain pathways for taste and olfaction are independent, the two senses do interact. Indeed, much of what we call taste is really smell. If the olfactory system is blocked, as by a head cold, the perception of taste is sharply reduced.

Throughout these discussions of sensory mechanisms, we have seen many examples of how these inputs to the nervous system result in the specific body movements that we observe as animal behavior. The swimming of planarians away from light, the escape behavior of a moth that hears bat sonar, the feeding movements of *Hydra* when it tastes glutathione, and the homing of a salmon that can smell its breeding stream—these are just a few cases that have been mentioned so far in this chapter. The remainder of the chapter focuses on the motor mechanisms that make these animal responses possible: how animals use their muscles and skeletons to move.

AN INTRODUCTION TO ANIMAL MOVEMENT

Movement is an animal hallmark. To catch food, an animal must either move through its environment or move the environmental milieu past itself. Sessile animals stay put, but they wave prehensile tentacles that capture prey or use beating cilia to generate water currents that draw and trap small food particles (see Chapter 37). Most animals, however, are mobile and spend a considerable portion of their time and energy actively searching for food, as well as escaping from danger and looking for mates.

The modes of animal locomotion are diverse. Several animal phyla include species that swim. On land and in the sediments on the floor of the sea and lakes, animals crawl, walk, run, or hop. Flight equipment has evolved in only a few animal classes among the insects (phylum Arthropoda) and among the reptiles, birds, and mammals (phylum Chordata).

Whatever the mode of transportation, an animal must exert enough force against its environment to overcome friction and gravity. The relative importance of these two impedances depends on the environment. Because most animals are reasonably buoyant in water, overcoming gravity is less of a problem for swimming animals than for species that move on land or through the air. On the other hand, water is a much denser medium than air, and thus the problem of resistance (friction) is a major one for aquatic animals. A sleek, fusiform (torpedolike) shape is a common adaptation of fast swimmers. On land, a walking or running animal must be able to support itself and move against gravity, but, at least at moderate speeds, air poses relatively little resistance. To move in such an environment, strong skeletal support is more important than a streamlined shape. A walking or running animal must also overcome inertia with each step by accelerating a leg from a standing start. This is one of the main reasons running animals consume more energy per meter than equivalently sized animals specialized for swimming or flying (Figure 45.24). A flying animal does not use its skeleton to support itself against the pull of gravity, but gravity poses a major problem in a different way: To be airborne, the wings must develop enough lift to enable the animal to completely overcome the downward force of gravity. In all these situations, we observe body shapes, specialized appendages, and other adaptations of morphology representing evolutionary solutions to specific problems of movement.

Underlying these diverse forms of movement are fundamental mechanisms common to all animals. At the cellular level, all animal movement is based on one of two basic contractile systems, both of which use protein strands moving against one another. These two systems of cell motility, microtubules and microfilaments, were discussed in Chapter 7. Microtubules are responsible for the beating of cilia and the undulations of flagella. Microfilaments play a major role in amoeboid movement, and they are also the contractile elements of muscle cells. It is the contraction of muscles that concerns us in this chapter, but the work of a muscle in itself cannot translate into movement of the animal. Swimming, crawling, running, and flying all result from muscles working against some type of skeleton.

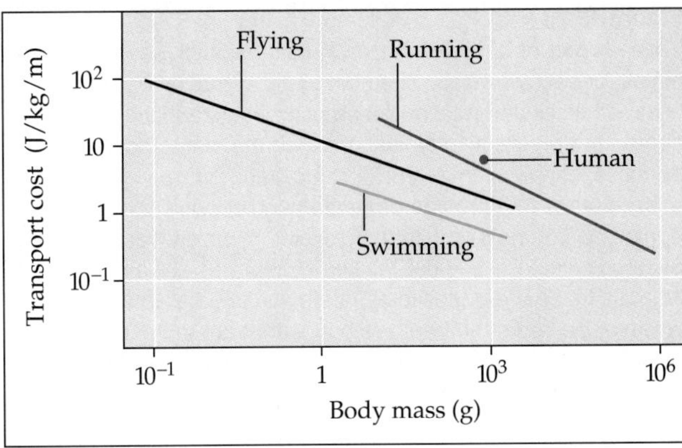

Figure 45.24
The cost of transport. This graph compares the transport cost, in joules per kilogram of body weight per meter traveled, for animals specialized for swimming, flying, and running (1 J = 0.24 calories). Notice that both axes are plotted on logarithmic scales. Swimming is the most efficient mode of transport (assuming, of course, that an animal is specialized for swimming). If we were to compare energy consumption per *minute* rather than per meter, we would find that flying animals use more energy than animals swimming or walking for the same amount of time. However, birds and flying insects are generally fast, and thus running requires more energy per *meter* than flying. To walk or run, an animal must not only work against gravity and friction, but must repeatedly accelerate limbs from standing starts, using considerable energy to overcome inertia. Notice also, however, that a larger animal travels more efficiently than a smaller species specialized for the same mode of transport. For example, a horse consumes less energy per kilogram of body weight than a cat running the same distance (of course, *total* energy consumption is greater for the larger animal). If you compare the extremes of body mass on the *x*-axis of this graph, you can see that a large, running animal actually moves at a lower energy cost than a small fish or bird.

SKELETONS AND THEIR ROLES IN MOVEMENT

Three functions of a skeleton are support, protection, and movement. Most land animals would sag from their own weight if they had no skeleton to support them. Even an animal living in water would be a formless mass with no framework to maintain its shape. Many animals have hard skeletons that protect soft tissues. For example, the vertebrate skull protects the brain, and the ribs form a cage around the heart, lungs, and other internal organs. And skeletons aid in movement by giving muscles something firm to work against. There are three main types of skeletons: hydrostatic skeletons, exoskeletons, and endoskeletons (Figure 45.25).

Hydrostatic Skeletons

A **hydrostatic skeleton** consists of fluid held under pressure in a closed body compartment. This is the main type of skeleton in most cnidarians, flatworms, nematodes, and annelids (see Chapter 29). These ani-

Figure 45.25
Three types of skeletons. (**a**) Cnidarians, such as these jellyfishes, use a hydrostatic skeleton to aid in movement. Hydrostatic skeletons, consisting of pressurized fluid in a body compartment, are also typical of flatworms, nematodes, and annelids. (**b**) Many invertebrates, including arthropods, have exoskeletons that provide protection and support. Here, an aphid is sitting next to its old exoskeleton, which was shed in a process called molting (SEM). (**c**) Vertebrates are among the animals that have endoskeletons. In this photograph of a bat embryo, the red structures are bones that are in the process of ossifying (hardening by deposit of calcium salts), and the blue structures are cartilage.

(b)

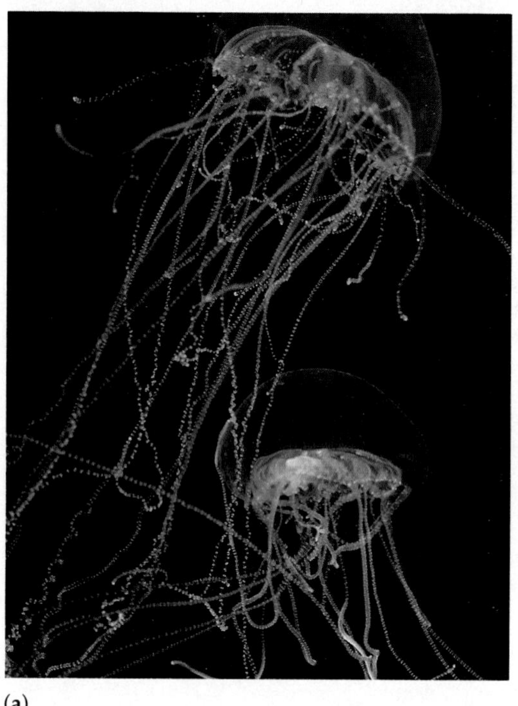

(a)

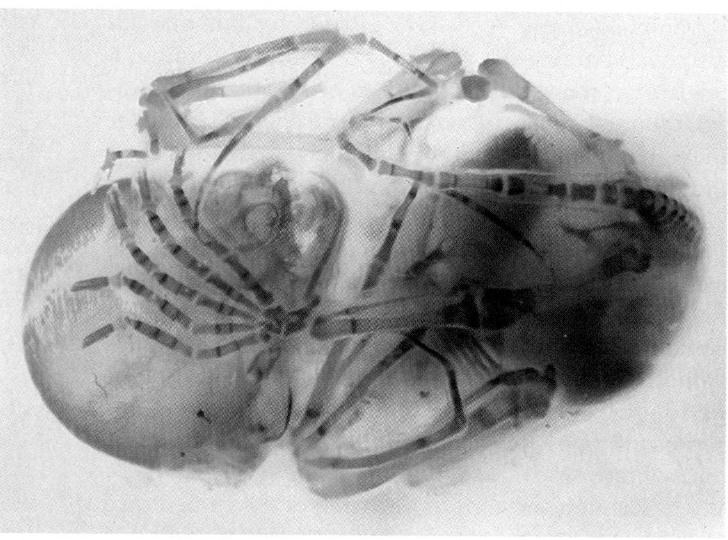

(c)

mals control their form and movement by using muscles to change the shape of the fluid-filled compartments. *Hydra,* for example, can elongate by closing its mouth and using contractile cells in the body wall to constrict the central gastrovascular cavity. Since water cannot be compressed very much, decreasing the diameter of the cavity forces it to increase in length. In planarians, the interstitial fluid is kept under pressure and functions as the main hydrostatic skeleton. Flatworms move by using muscles in the body wall to exert localized forces against this hydrostatic skeleton. Roundworms (nematodes) are able to hold the fluid in the body cavity (a pseudocoelom; see Chapter 29) at a high pressure, and contractions of longitudinal muscles result in thrashing movements. In earthworms, it is the coelomic fluid that functions as a hydrostatic skeleton. The coelomic cavity is divided by septa between the segments of the worm, and thus the animal can change the shape of each segment individually, using both circular and longitudinal muscles. Hydrostatic skeletons provide no protection, and they certainly could not support a large animal living on land.

Exoskeletons

An **exoskeleton** is a hard encasement deposited on the surface of an animal. For example, most mollusks are enclosed in calcareous (calcium carbonate) shells secreted by the mantle, a sheetlike extension of the body wall. As the animal grows, it enlarges the diameter of the shell by adding to its outer edge. Clams and other bivalves close their hinged shells using muscles attached to the inside of this exoskeleton.

The jointed exoskeleton typical of arthropods is a **cuticle,** a nonliving coat secreted by the epidermis.

Muscles are attached to knobs and plates of the cuticle that extend into the interior of the body. About 30% to 50% of the cuticle consists of **chitin,** a polysaccharide similar to cellulose (see Chapter 5). Fibrils of chitin are embedded in a matrix made of protein, forming a composite material, analogous to fiberglass, that combines strength and flexibility. Where protection is most important, the cuticle is hardened by the addition of organic compounds called quinones, which cross-link the proteins of the exoskeleton. Some crustaceans, such as crabs and lobsters, harden portions of their exoskeletons even more by adding calcium salts. In contrast, at the joints of the legs, where the cuticle must be thin and flexible, there is only a small amount of inorganic salts and little cross-linking of proteins. The exoskeleton of an arthropod cannot enlarge once it is deposited; it must periodically be shed (molted) and replaced by a larger case with each spurt of growth by the animal.

Endoskeletons

An **endoskeleton** consists of hard supporting elements, such as bones, buried within the soft tissues of an animal. Sponges are reinforced by hard spicules consisting of inorganic material or by softer fibers made of protein. Echinoderms have an endoskeleton of hard plates beneath the skin. These ossicles are composed of magnesium carbonate and calcium carbonate crystals, and the separate plates are usually bound together by protein fibers. Sea urchins have a skeleton of tightly bound ossicles, but the ossicles of sea stars are more loosely bound, allowing the animal to change the shape of its arms.

Chordates have endoskeletons consisting of cartilage, bone, or some combination of these materials (Figure 45.26). The mammalian skeleton is built from more than 200 bones, some fused together and others connected at joints by ligaments that allow freedom of movement. Anatomists divide the vertebrate frame into an axial skeleton, consisting of the skull, vertebral column (backbone), and rib cage; and an appendicular skeleton, made up of limb bones and the pectoral and pelvic girdles that anchor the appendages to the axial skeleton. In just one appendage, the vertebrate forelimb, we can identify several types of joints (Figure 45.27).

In addition to functioning as buttresses that support the body, the bones of the vertebrate skeleton act as levers when the muscles attached to them contract.

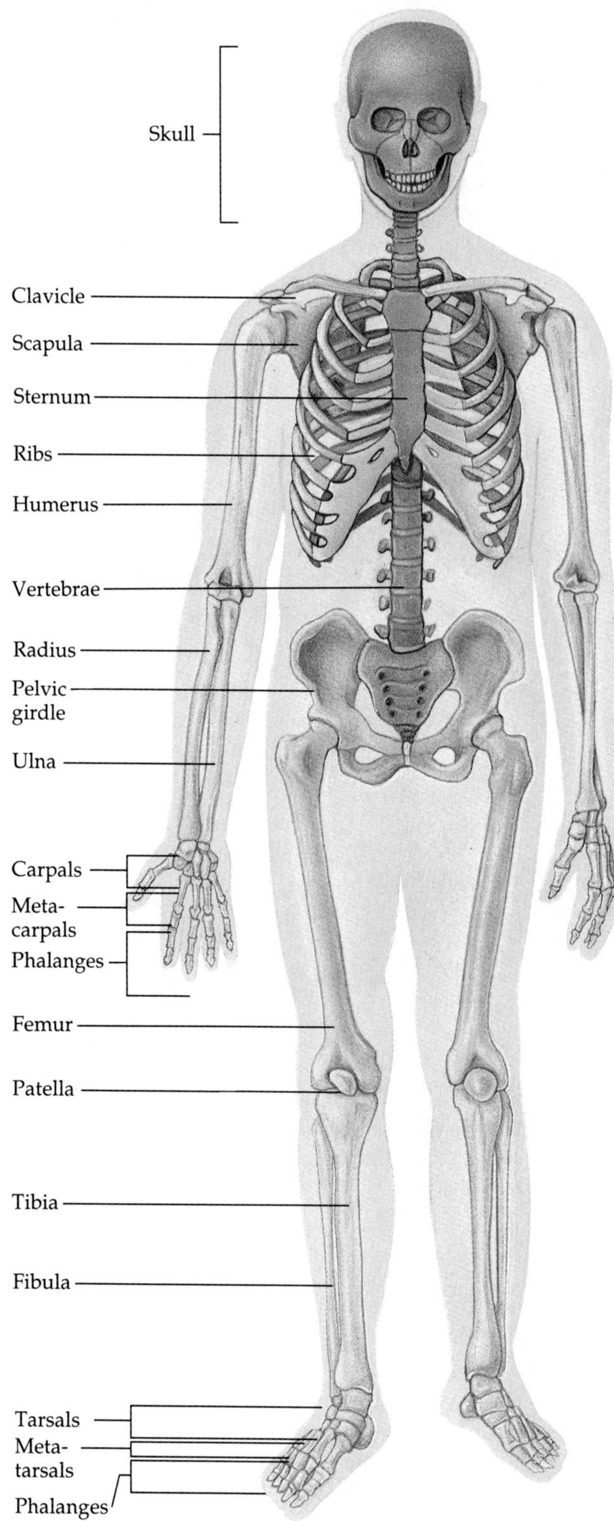

Figure 45.26
The human skeleton. The axial skeleton is colored green and the appendicular skeleton is colored gold in this illustration.

MUSCLES

Animal movement, as we have seen, is based on the contraction of muscles working against some type of skeleton. The action of a muscle is *always* to contract; muscles can extend only passively. The ability to move a part of the body in opposite directions requires that muscles be attached to the skeleton in antagonistic

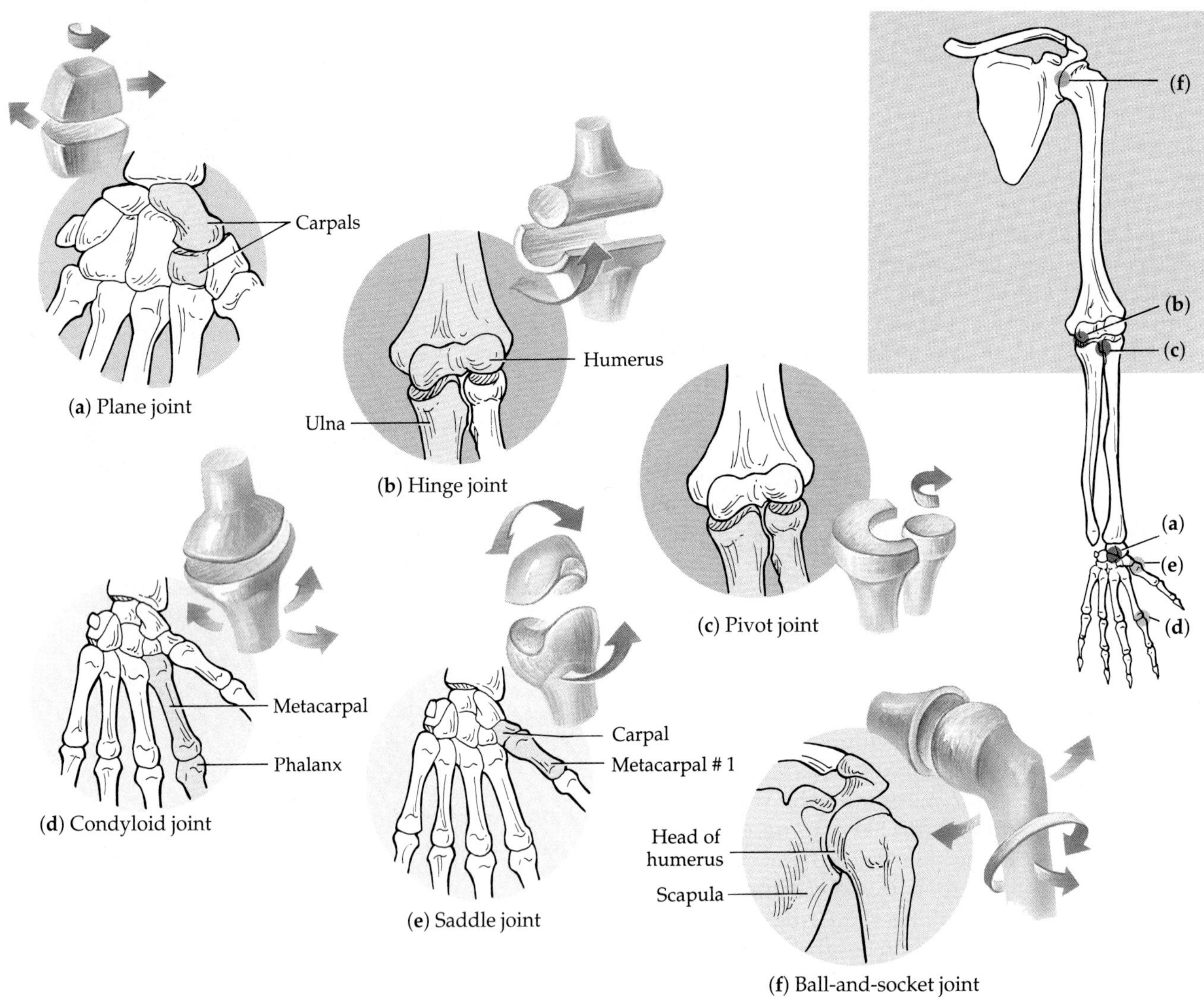

Figure 45.27
Types of joints in the vertebrate skeleton. With its many articulations, the vertebrate forelimb provides examples of six types of joints. The arrows indicate the movements facilitated by each joint.

pairs, each muscle working against the other (Figure 45.28). We flex our arm, for instance, by contracting the biceps, with the hinged joint of the elbow acting as the fulcrum of a lever. To extend the arm, we relax the biceps while the triceps on the opposite side contracts. But how does a muscle actually contract? The key to function, as usual, is structure. In this section, we will examine the structure and mechanism of contraction of vertebrate skeletal muscle and then compare this basic pattern with other types of muscle.

The Structure and Physiology of Vertebrate Skeletal Muscle

Vertebrate **skeletal muscle,** which is attached to the bones and responsible for their movement, is characterized by a hierarchy of smaller and smaller parallel units (Figure 45.29). A skeletal muscle is a bundle of long fibers running the length of the muscle. Each fiber is a single cell with many nuclei, reflecting its formation by the fusion of many embryonic cells. Each fiber is itself a bundle of smaller **myofibrils** arranged longitudinally. The myofibrils, in turn, are composed of two kinds of **myofilaments. Thin filaments** consist of two strands of actin and one strand of regulatory protein coiled around one another, while **thick filaments** are staggered arrays of myosin molecules.

Skeletal muscle is also called striated muscle because the regular arrangement of the myofilaments creates a repeating pattern of light and dark bands (Figure 45.30a). Each repeating unit is a **sarcomere,** the fundamental unit of organization of the muscle. The borders of the sarcomere, the **Z lines,** are lined up in

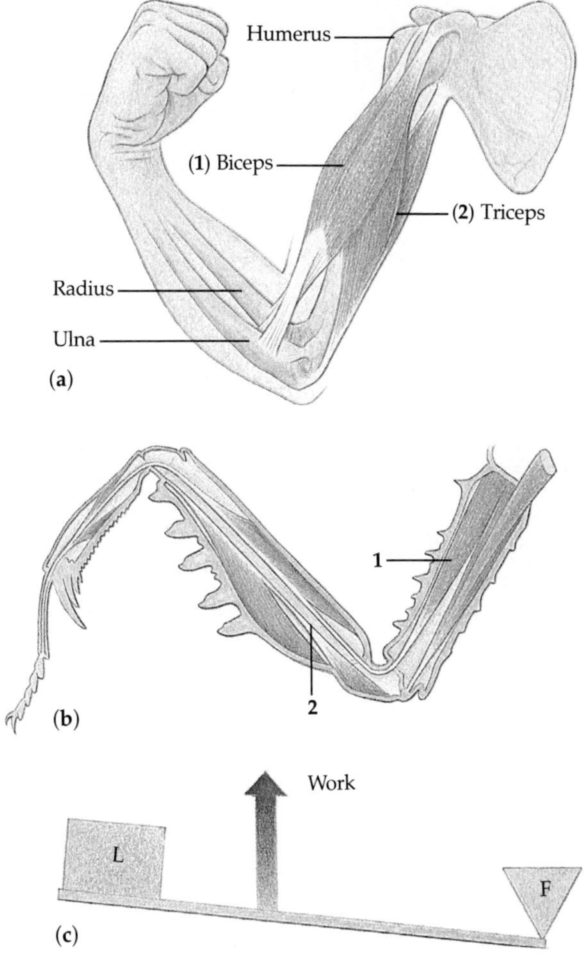

Figure 45.28
The cooperation of muscles and skeletons in movement. Muscles actively contract, but they elongate only when passively stretched. Back-and-forth movement is generally accomplished by antagonistic muscles, each working against the other. This arrangement works with either an endoskeleton or an exoskeleton. (**a**) In humans, contraction of the biceps (muscle 1) raises the forearm. Contraction of the triceps (muscle 2) lowers the forearm. (**b**) To raise the insect leg, muscle 2 contracts. When muscle 1 contracts, the leg lowers. Although contraction of muscle 1 produces an opposite effect in humans and insects, both muscle-skeletal systems work according to the same physical principle. (**c**) To raise the appendage, the skeleton acts as a lever, with the joint as a fulcrum (F) and the limb as the load (L). When the muscle applies force between the fulcrum and the load, the lever pivots upward about the fulcrum. Contraction of the muscles in the human and insect has different results because the muscles have different positions relative to the skeleton.

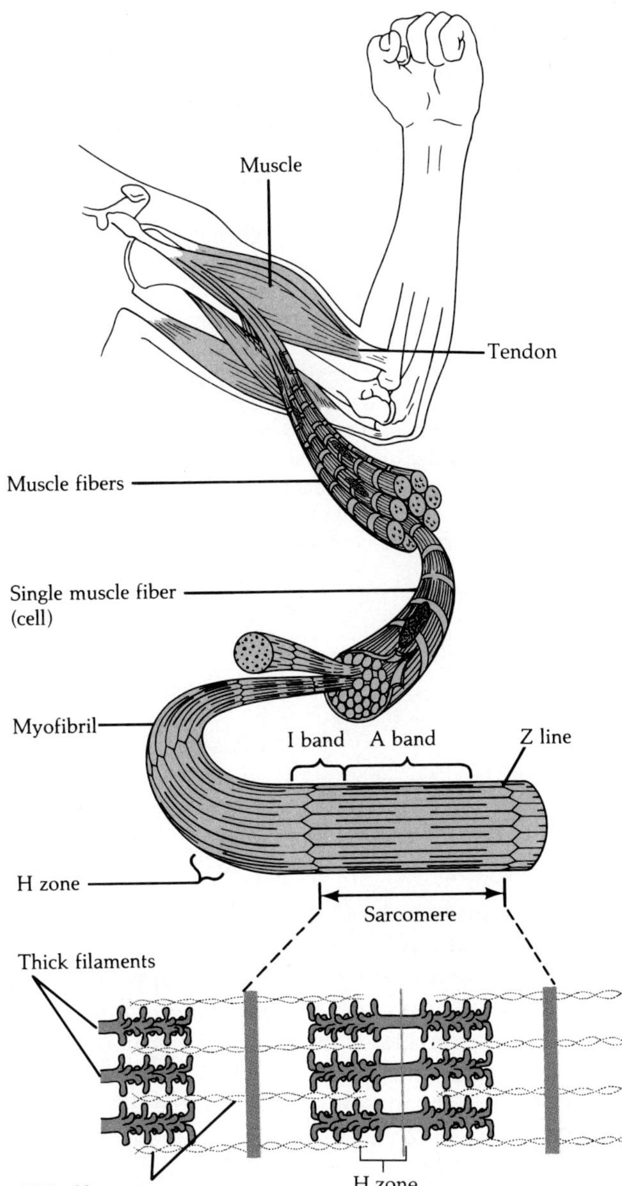

Figure 45.29
The structure of skeletal muscle. Skeletal muscle is attached by tendons to the bones that it moves. The muscle consists of bundles of multinucleated muscle fibers (cells), each consisting of a bundle of myofibrils. Each myofibril is made of thick and thin filaments aligned in a regular array of sarcomeres, units marked by Z lines at each end.

adjacent myofibrils to contribute to the striations visible with a light microscope. The thin filaments are attached to the Z lines and project toward the center of the sarcomere, while the thick filaments are centered in the sarcomere. At rest, the thick and thin filaments do not overlap completely, and the area near the edge of the sarcomere where there are only thin filaments is called the **I band.** The **A band** is the broad region that corresponds to the length of the thick filaments. The thin filaments do not extend completely across the sarcomere, so the **H zone** in the center of the A band contains only thick filaments. This arrangement of thick and thin filaments is the key to how the sarcomere, and hence the whole muscle, contracts.

Molecular Mechanism of Muscle Contraction When a muscle contracts, the length of each sarcomere is reduced; that is, the distance from one Z line to the

Figure 45.30
The sliding filament model of muscle contraction. (**a**) The alternating light and dark bands in the striated muscle from a frog show the regular organization of thick and thin filaments (TEM). As you can see in the diagram, the ends of the A bands include the regions where the thick and thin filaments overlap, while the H zone contains only the thick filaments. Only thin filaments occur in the I bands. The dark zone in the center of each I band, called the Z line, is attached to the thin filaments. The sarcomere is the entire apparatus between successive Z lines. (**b**) The contraction of one sarcomere: The muscle contracts when the thick and thin filaments slide past each other. The sarcomere becomes shorter, although the lengths of the individual filaments do not change.

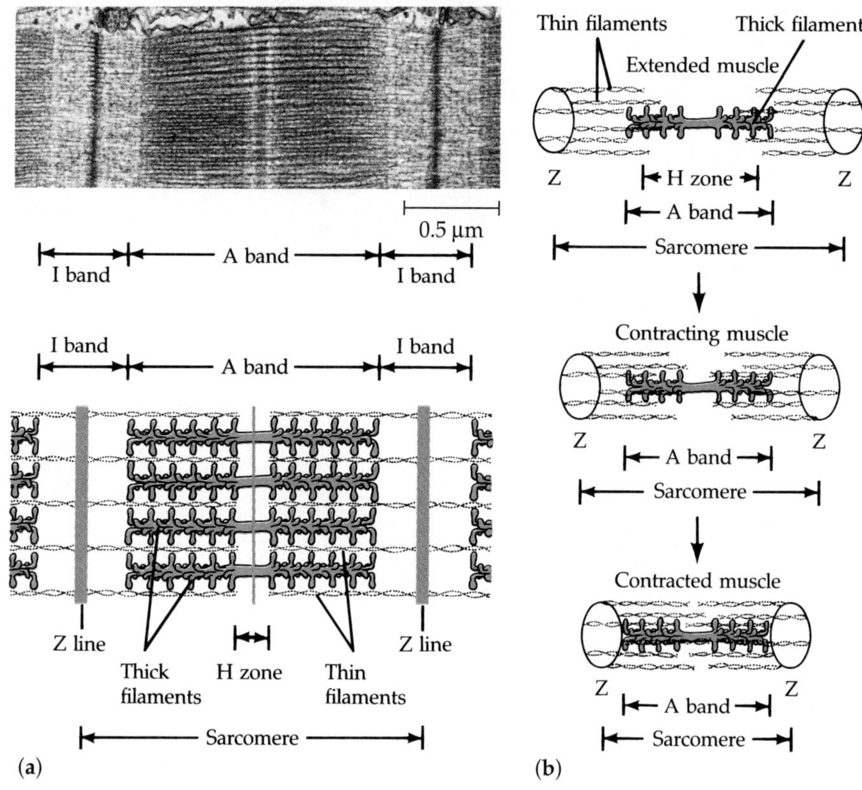

next becomes shorter (Figure 45.30b). In the contracted sarcomere, the A bands do not change in length, but the I bands shorten and the H zone disappears. This behavior can be explained by the **sliding filament model** of muscle contraction. According to this model, neither the thin filaments nor the thick filaments change in length when the muscle contracts; rather, the filaments slide past each other longitudinally, so that the degree of overlap of the thin and thick filaments increases. If the region of overlap increases, then both the length occupied only by thin filaments (the I band) and the length occupied only by thick filaments (the H zone) must decrease.

Sliding of the filaments is based on the interaction of the actin and myosin molecules that make up the thin and thick filaments. Myosin consists of a long, fibrous "tail" region, with a globular "head" region sticking off to the side. The tail is where the individual myosin molecules cohere to form the thick filament. The myosin head can bind ATP and hydrolyze it into ADP and inorganic phosphate. Some of the energy released by cleaving the ATP is transferred to the myosin, which changes shape to a high-energy configuration (Figure 45.31). This energized myosin binds to a specific site on actin, forming a **cross-bridge.** The stored energy is released, and the myosin head relaxes to its low-energy configuration. This relaxation changes the angle of attachment of the myosin head to the fibrous myosin tail; as the myosin bends inward on itself, it exerts tension on the thin filament to which it is bound, pulling the thin filament toward the center of the sarcomere. The bond between the low-energy myosin and actin is broken when a new molecule of ATP binds to the myosin head. In a repeating cycle, the free head can then cleave the new ATP to revert to the high-energy configuration and attach to a new binding site on another actin molecule farther along the thin filament. Each of the approximately 350 heads of a thick filament form and re-form about 5 cross-bridges per second, driving filaments past each other.

A muscle cell typically stores only enough ATP for a few contractions. Muscle cells also store glycogen in columns between the myofibrils, but most of the energy needed for repetitive muscle contraction is stored in substances called **phosphagens. Creatine phosphate,** the phosphagen of vertebrates, can supply a phosphate group to ADP to make ATP.

Excitation-Contraction Coupling A skeletal muscle contracts only when stimulated by a motor neuron. When the muscle is at rest, the myosin-binding sites on the actin molecules are blocked by a strand of the regulatory protein **tropomyosin.** Another set of regulatory proteins, the **troponin complex,** controls the position of tropomyosin on the thin filament (Figure 45.32). For a muscle cell to contract, the cross-bridge attach-

Figure 45.31
The cyclic interaction between myosin and actin in muscle contraction. Starting at the top, the myosin head is bound to ATP and is in its low-energy configuration. ① The myosin head hydrolyzes ATP to ADP and inorganic phosphate (Ⓟ$_i$) and is in its high-energy configuration. ② The myosin head binds to actin, forming a cross-bridge. ③ Myosin relaxes to its low-energy state, sliding the thin filament. ④ Binding of a new molecule of ATP releases the myosin head. The myosin head can then return to the high-energy configuration and begin a new cycle.

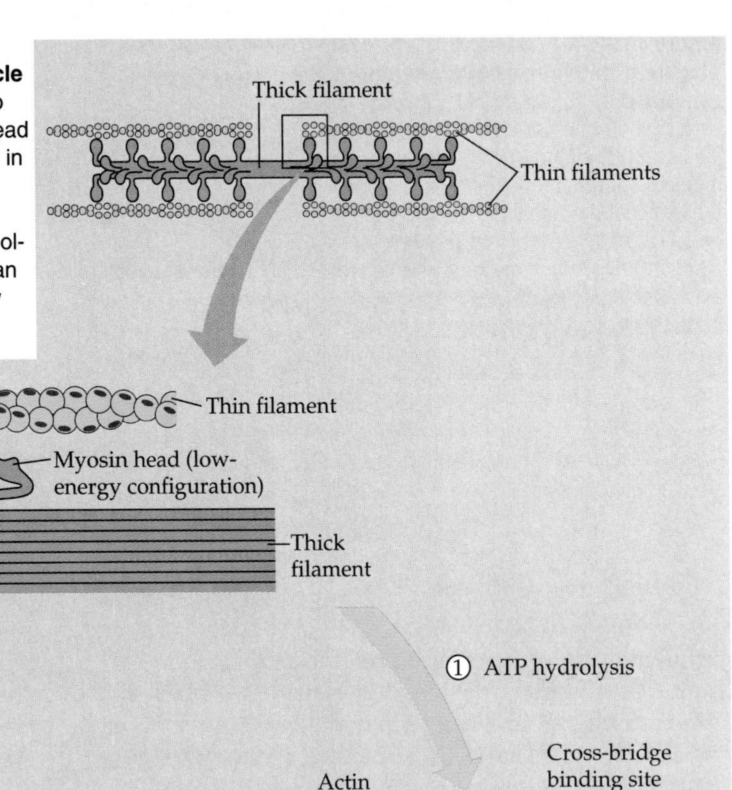

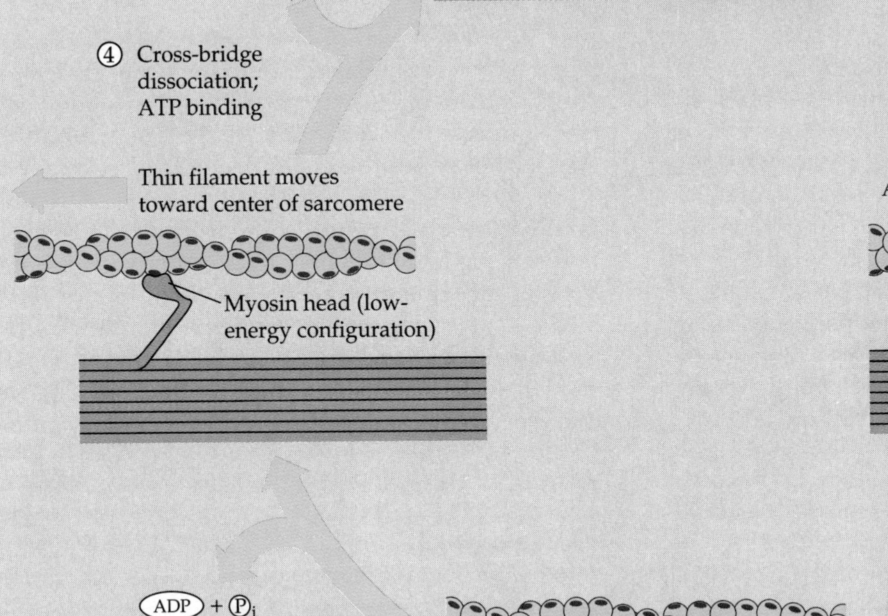

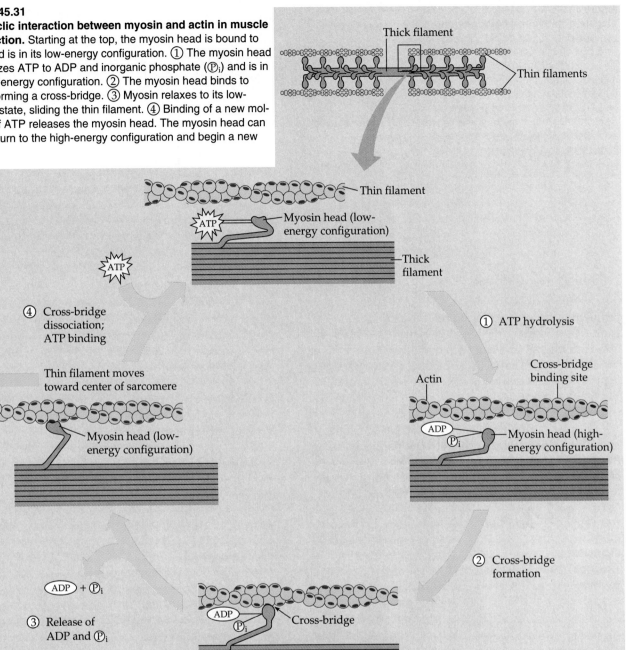

ment sites on the actin must be uncovered. This occurs when calcium ions bind to troponin, altering the interaction between troponin and tropomyosin. The Ca^{2+} binding causes the whole tropomyosin-troponin complex to change shape and expose the myosin-binding sites on actin. When calcium is present, the sliding of thin and thick filaments can occur, and the muscle contracts. When internal calcium concentration falls, the binding sites of actin are covered, and contraction stops.

Calcium concentration in the cytoplasm of the muscle cell is regulated by the **sarcoplasmic reticulum,** a specialized endoplasmic reticulum (Figure 45.33). The membrane of the sarcoplasmic reticulum actively transports calcium from the cytoplasm into the interior of the reticulum, which is thus an intracellular storehouse for calcium. The stimulus leading to contraction of a skeletal muscle cell is an action potential in the motor neuron that makes synaptic connection with the muscle cell. Recall from Chapter 44 that the synaptic

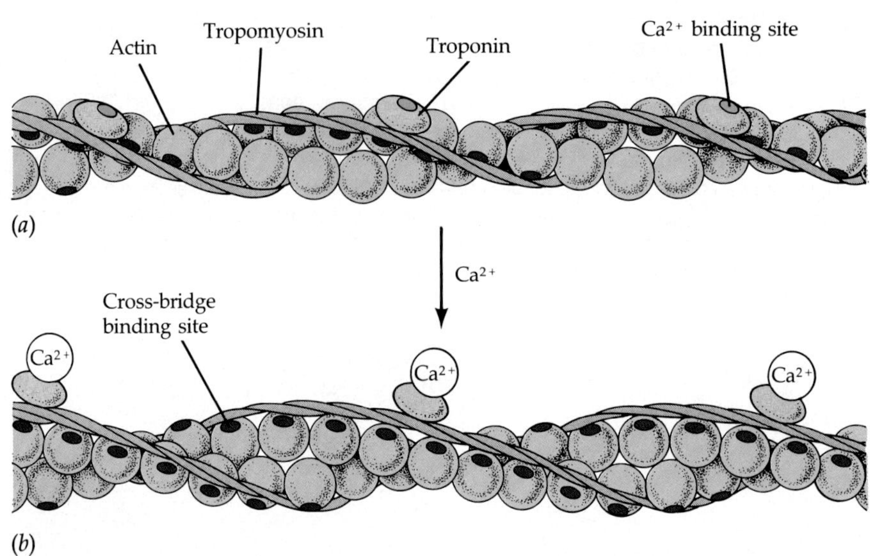

Figure 45.32
The control of muscle contraction. The thin filament has two strands of actin twisted to form a helix. (**a**) When the muscle is at rest, a long, rodlike tropomyosin molecule blocks the myosin-binding sites that are instrumental in forming cross-bridges. (**b**) When another protein complex, troponin, binds calcium ions, the actin binding sites are exposed, cross-bridges with myosin can form, and contraction occurs.

terminal of the motor neuron releases acetylcholine at the neuromuscular junction, depolarizing the postsynaptic muscle cell and triggering an action potential in the muscle cell. That action potential is the signal for contraction. The action potential spreads deep into the interior of the muscle cell along infoldings of the plasma membrane called **transverse tubules.** Where the transverse tubules contact the sarcoplasmic reticulum, the action potential changes the permeability of the sarcoplasmic reticulum, causing it to release calcium ions. These calcium ions bind to troponin, allowing the muscle to contract. The contraction is terminated as the sarcoplasmic reticulum pumps the calcium back out of the cytoplasm, and the tropomyosin-troponin complex again blocks the myosin-binding sites as the concentration of calcium falls.

Graded Contractions of Whole Muscles Stimulation of a single muscle fiber by a motor neuron results in an all-or-none contraction, a twitch (Figure 45.34). However, we know from experience that the action of a whole muscle, such as the biceps, is graded; we can vary the extent and strength of contraction. One way that the nervous system can produce a graded contrac-

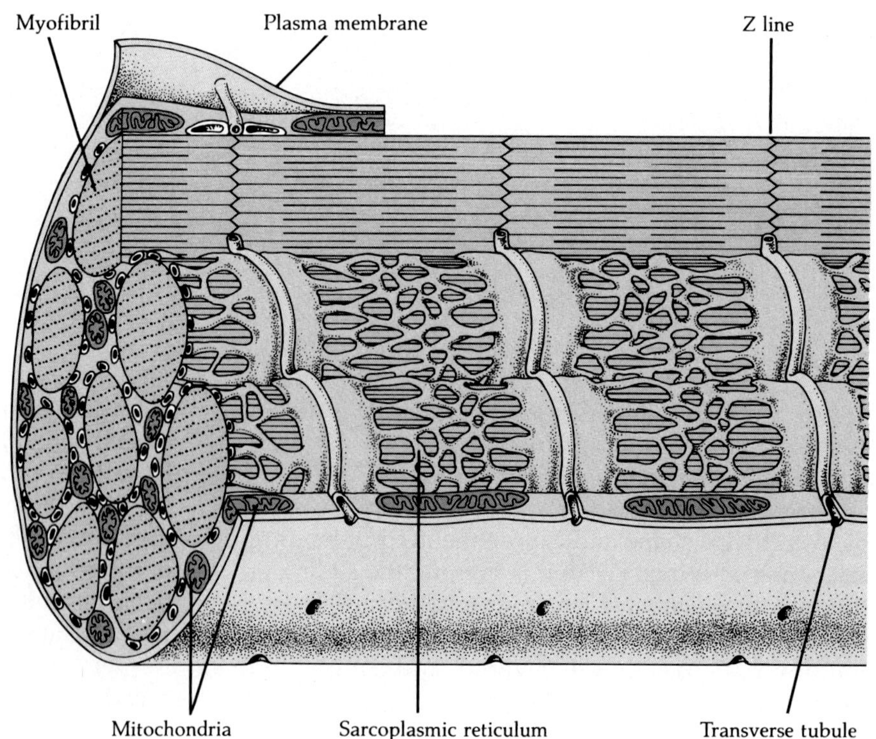

Figure 45.33
The sarcoplasmic reticulum and transverse tubules. The sarcoplasmic reticulum is a system of membranes that accumulates and releases the calcium ions that trigger muscle contraction. The transverse tubules are continuous with the plasma membrane of the muscle fiber and relay the contraction signal (action potential).

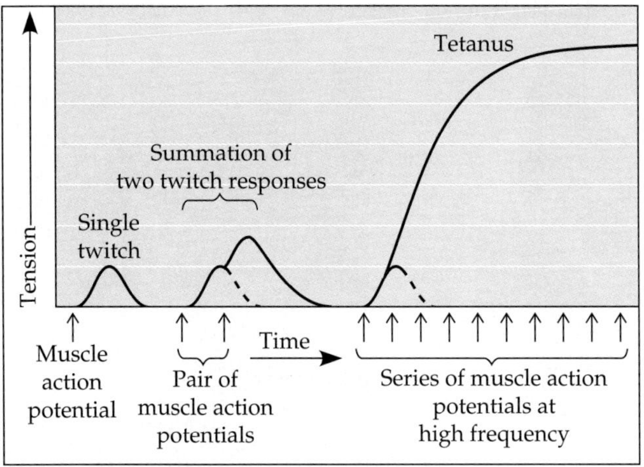

Figure 45.34
Temporal summation of muscle cell contractions. This graph compares the tension developed in a muscle in response to a single action potential, a pair of action potentials, and a train of action potentials. The dashed lines show the response that would have resulted if only the first action potential occurred.

tion of the whole muscle is by varying the frequency of action potentials in the motor neurons controlling the muscle. A single action potential will produce an increase in tension lasting about 100 milliseconds or less. If a second action potential is triggered before the response to the first is over, then the tension will sum to produce a greater response (see Figure 45.34). If a train of action potentials arrives at a muscle cell, the tension will reach a summated level that depends on the rate of stimulation. And if the rate of stimulation is fast enough, then the twitches blur into one smooth and sustained contraction called **tetanus** (not to be confused with the disease of the same name). Motor neurons usually deliver their action potentials in rapid-fire volleys, and the resulting summation of tension results in smooth contraction typical of tetanus rather than the jerky actions of individual muscle twitches.

The nervous system can also produce graded contraction of a whole muscle by taking advantage of the organization of the muscle cells into **motor units.** In a vertebrate muscle, each muscle cell is innervated by only one motor neuron, but each motor neuron may make synaptic connections with many muscle cells (Figure 45.35). There may be hundreds of motor neurons controlling an individual muscle, each with its own pool of muscle fibers scattered throughout the muscle. A motor unit consists of a single motor neuron and all the muscle fibers it controls. When the motor neuron fires, all the muscle fibers in the motor unit contract as a group. The strength of the resulting contraction will therefore depend on how many muscle fibers the motor neuron contacts. In most muscles, there is wide variation in the number of muscle fibers among motor units; some motor neurons may control only a few muscle cells, while others may control hundreds. The nervous system can thus regulate the strength of contraction in the whole muscle both by determining how many motor units are activated at a given instant and by selecting whether large or small motor units are activated. Tension in a muscle can be progressively increased by activating more and more of the motor neurons controlling the muscle, a process called **recruitment** of motor neurons. Depending on how many motor neurons your brain recruits to the task, you can lift a fork, or something much heavier, like your biology textbook.

Some muscles, especially those that hold the body up and maintain posture, are almost always partially contracted. However, prolonged contraction results in muscle fatigue caused by depletion of ATP, dissipation of the ion gradients required for normal electrical signaling (see Chapter 44), and accumulation of lactate (see Chapter 9). In a mechanism that avoids fatigue in postural muscles, the nervous system alternates the activation among the various motor units making up the muscle, so that different motor units take turns maintaining the prolonged contraction.

Fast and Slow Muscle Fibers We have seen that the action potential in a skeletal muscle fiber is only a trigger for the contraction; the actual duration of the contraction is controlled by how long the calcium concentration in the cytoplasm remains elevated. Not all skeletal muscle fibers are identical in this regard. We can identify fast and slow fibers based on the duration of their twitches. A slow fiber has less sarcoplasmic reticulum than a fast fiber, so calcium remains in the cytoplasm longer. This causes a twitch in a slow fiber to last about five times longer than in a fast fiber. Slow fibers are also specialized to make use of a steady supply of energy; they have many mitochondria, a rich blood supply, and an oxygen-storing protein called **myoglobin.** Myoglobin, the brownish-red pigment in the dark meat of poultry and fish, binds oxygen more tightly than hemoglobin, so it can effectively extract oxygen from the blood. Slow fibers are often found in muscles used to maintain posture, because they can sustain long contractions. Fast muscle fibers are used for rapid, powerful contractions. Some, such as the flight muscles of birds, may be able to sustain long periods of repeated contractions without fatiguing.

Other Types of Muscle

There are many types of muscles in the animal kingdom, but as noted before, they all share the same fundamental mechanism of contraction: the sliding of actin filaments and myosin filaments past one another.

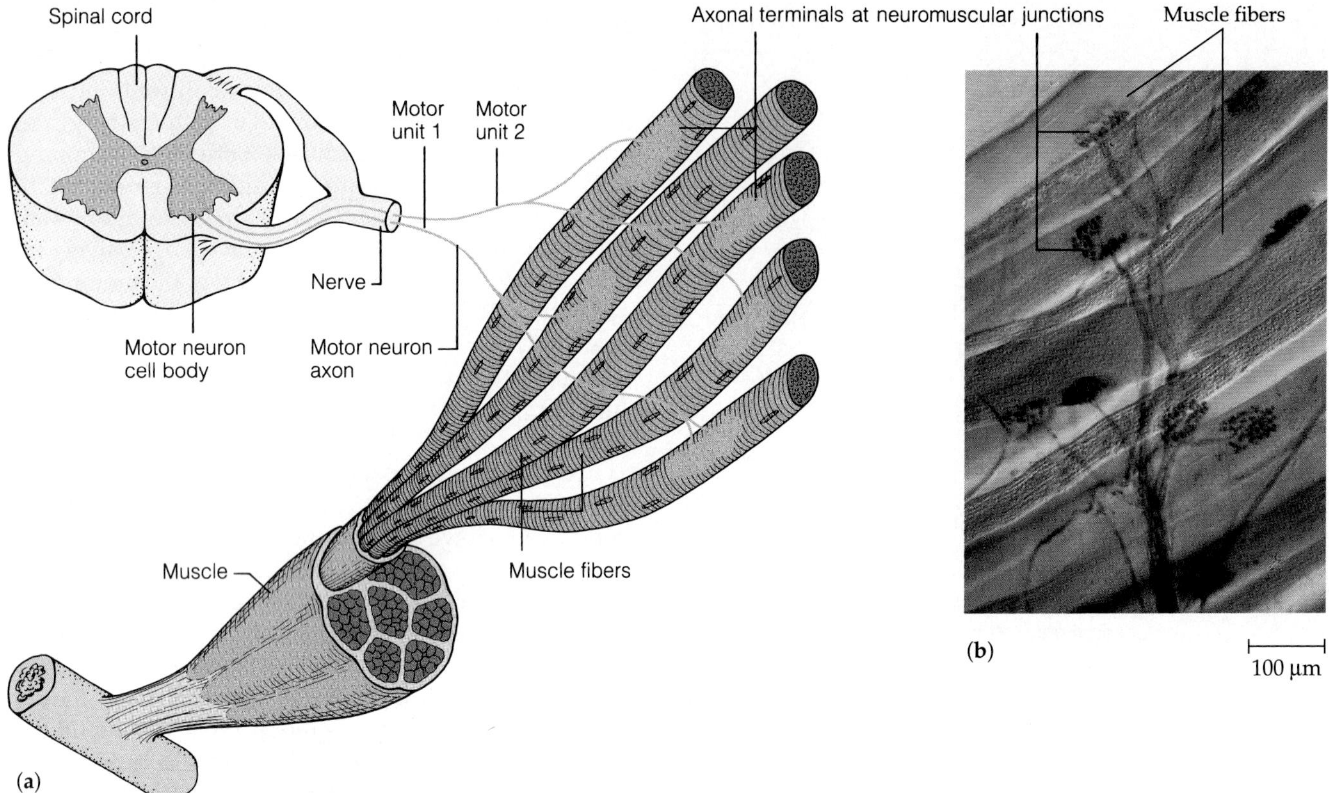

Figure 45.35
Motor units. (**a**) A branched motor neuron typically innervates several muscle fibers scattered throughout the muscle. The entire contractile apparatus—the neuron and the fibers it controls—is called a motor unit. One factor that strengthens a muscular contraction is the recruitment of additional motor units. (**b**) Here, you can see a single branched axon that synapses with several muscle fibers of a motor unit (LM).

In Chapter 36, we saw that in addition to skeletal muscle, vertebrates also have cardiac and smooth muscle.

Vertebrate **cardiac muscle** is found in only one place—the heart. Like skeletal muscle, cardiac muscle is striated. The primary differences between skeletal and cardiac muscle are in their electrical and membrane properties. The junctions between cardiac muscle cells contain specialized regions called **intercalated discs** (see Figure 36.5), where gap junctions provide direct electrical coupling among cells. Thus, an action potential generated in one part of the heart will spread to all the cardiac muscle cells, and the whole heart will contract (see Chapter 38). Skeletal muscle cells will not fire an action potential and contract unless triggered to do so by input from a controlling motor neuron; cardiac muscle cells, however, can generate action potentials on their own, without any input from the nervous system. The plasma membrane of a cardiac muscle cell has **pacemaker channels,** which cause rhythmic depolarizations that trigger action potentials and cause single cardiac muscle cells to "beat" even when isolated from the heart and placed in cell culture. The action potentials of cardiac muscle cells are also different from those of skeletal muscle cells, lasting up to twenty times longer. The action potential of a skeletal muscle cell serves only as a trigger for contraction and does not control the duration of contraction; in a cardiac cell, the duration of the action potential does play an important role in controlling the duration of contraction.

Smooth muscle lacks the striations of skeletal and cardiac muscle because the actin and myosin filaments are not all regularly arrayed along the length of the cell. Rather, the filaments may have a spiral arrangement within smooth muscle cells. Smooth muscle also contains less myosin than striated muscle, and the myosin is not associated with specific actin strands. Because of the organization, smooth muscle cannot generate nearly as much tension as striated muscle, but it can contract over a much greater range of lengths. Further, smooth muscle does not have either a transverse tubule system or a well-developed sarcoplasmic reticulum. Calcium ions must enter the cytoplasm via the plasma membrane during an action potential, and the amount reaching the filaments is rather small. Contractions are relatively slow, but there is a greater range of control than is possible in striated muscle. Smooth muscle is found mainly in the walls of hollow organs such as digestive tract organs and blood vessels. Smooth muscle propels substances through the hollow organ by alternately contracting and relaxing.

Invertebrates possess muscles similar to vertebrate skeletal and smooth muscles. Arthropod skeletal muscles are nearly identical to vertebrate skeletal muscles. However, the flight muscles of insects are capable of independent, rhythmic contraction, so the wings of some insects can actually beat faster than action potentials can arrive from the central nervous system. Another interesting evolutionary adaptation has been discovered in the muscles that hold clam shells closed. The thick filaments of these muscle fibers contain a unique protein called paramyosin that allows the muscles to remain in a fixed state of contraction for as long as a month.

*   *   *

Although we have talked about receptors and muscles separately in this chapter, they are the two ends of a single integrated system. An animal's behavior, so fundamental to how the animal interacts with its environment, is the product of a nervous system connecting certain sensations to certain responses. Behavior is discussed in Unit Eight, within the broader context of ecology: the study of interactions between organisms and their environment.

STUDY OUTLINE

1. Animal behavior depends on the ability of the nervous system to integrate the input from sensory receptors about the internal and external environments and translate the information into appropriate responses in muscles and other effectors.

Sensory Receptors (pp. 1015–1019)

1. Sensations are action potentials traveling along sensory neurons that are interpreted by different parts of the brain as perceptions.
2. Sensory receptors are usually modified neurons that collect and transmit environmental stimuli.
3. Reception is the specialized ability of a receptor cell to absorb the particular energy of a stimulus.
4. Transduction is the conversion of the stimulus energy into membrane potentials and action potentials.
5. The stimulus energy may be amplified by accessory structures of sense organs or by transduction.
6. Transmission of the receptor potential to the nervous system occurs either as action potentials (when the receptor is a sensory neuron) or as release of neurotransmitter, which then initiates action potentials in the sensory neuron with which the receptor cell synapses.
7. Integration of information begins at the receptor level by the processes of summation, adaptation, and variation in the sensitivity of the receptor. Integration and processing continue in the central nervous system.
8. Sensory receptors can be classified according to the type of stimulus energy they receive.
9. Mechanoreceptors respond to stimuli such as pressure, touch, stretch, motion, and sound.
10. Chemoreceptors respond either to generalized solute concentrations or to specific molecules.
11. Electromagnetic receptors detect energy in the form of different wavelengths of radiation.
12. Various types of thermoreceptors signal surface and core temperatures of the body.
13. Pain is detected by nociceptors, a group of diverse receptors that respond to excess temperature, pressure, or specific classes of chemicals.

Vision (pp. 1019–1026)

1. The light receptors and visual capabilities of invertebrates vary widely, ranging from the simple light-sensitive eye cup of planaria to the image-forming compound eye of insects and crustaceans and the single-lens eye found in some jellyfish, spiders, and many mollusks.
2. The vertebrate eye is composed of three concentric layers: an outer sclera with its transparent, anterior cornea; a middle pigmented choroid layer with the iris that surrounds the pupil; and the innermost retina, which contains the photoreceptor cells.
3. Transduction of the light signal occurs in specialized photoreceptors called rods and cones, which contain light-absorbing retinal bonded to a form of opsin.
4. When rods and cones absorb light, their membrane becomes hyperpolarized, and they release less neurotransmitter. This change in chemical message is transmitted to bipolar cells and then to ganglion cells, whose axons in the optic nerve convey action potentials to the brain. Horizontal cells and amacrine cells integrate information before it is sent to the brain, by contrast-enhancing lateral inhibition, for example.
5. Most axons of the optic nerves go to the lateral geniculate nuclei of the thalamus, from which neurons lead to the primary visual cortex in the occipital lobe.

Hearing and Balance (pp. 1027–1032)

1. In mammals and most terrestrial vertebrates, the hair cell mechanoreceptors for hearing and balance are located within the ear.
2. The outer ear consists of the external pinna and auditory canal. The tympanic membrane (eardrum) transmits sound waves to three small bones of the middle ear, which amplify and transmit the waves through the oval window to the fluid in the coiled cochlea of the inner ear. The pressure waves vibrate the basilar membrane and the attached organ of Corti, which contains receptor hair cells. The bending of the hairs against the tectorial membrane depolarizes the hair cells and initiates action potentials in the auditory nerve to the brain.

3. Volume is a function of the amplitude of the sound wave that results in greater bending of the hair cells. Pitch is related to frequency of sound waves. Regions of the basilar membrane vibrate more vigorously at different frequencies and transmit to specific regions of the cortex.
4. The inner ear vestibular apparatus, composed of the utricle, saccule, and three semicircular canals, functions in balance and equilibrium.
5. The detection of water movement in fishes and aquatic amphibians is accomplished by a lateral line system of clustered hair cell receptors.
6. Many arthropods use vibrating body hairs and localized "ears," consisting of a tympanic membrane and receptor cells, to sense sounds. Most invertebrates detect position in space by means of statocysts.

Taste and Smell (pp. 1032–1034)

1. Taste and smell both depend on the stimulation of receptor cells by small, dissolved molecules that bind to proteins on a chemoreceptor membrane.
2. Setae containing taste receptors are found on the feet and mouthparts of insects.
3. In mammals, taste receptors are organized into taste buds that respond to distinct shapes of molecules. Olfactory receptor cells line the upper part of the nasal cavity and send their axons to the olfactory bulb of the brain.

An Introduction to Animal Movement (pp. 1034–1035)

1. Skeletons function in support, protection, and movement.
2. Animal movement is a function of protein strands moving past each other, either in the microtubules of cilia and flagella or by microfilaments in amoeboid movement and muscle contraction.

Skeletons and Their Roles in Movement (pp. 1035–1037)

1. A hydrostatic skeleton, found in most cnidarians, flatworms, nematodes, and annelids, consists of fluid under pressure in a closed body compartment.
2. Exoskeletons, found in most mollusks and arthropods, are hard coverings deposited on the surface of an animal.
3. Endoskeletons, found in sponges, echinoderms, and chordates, are hard supporting elements embedded within the animal's body.

Muscles (pp. 1037–1045)

1. Muscles, often present in antagonistic pairs, contract and pull against the skeleton to provide movement.
2. Vertebrate skeletal muscle consists of a bundle of muscle cells, each of which contains myofibrils composed of thin filaments of actin and thick filaments of myosin.
3. Contraction begins when impulses from a motor neuron are transmitted to the muscle cell membrane through release of acetylcholine at the neuromuscular junction. Action potentials travel to the interior of the cell along the transverse tubules, stimulating the release of calcium from the sarcoplasmic reticulum. The calcium binds to the regulatory troponin-tropomyosin complex on the thin filaments, exposing the myosin-binding sites on the actin. Cross-bridges form, and bending of the myosin heads pulls the thin filaments toward the center of the sarcomere. The energy to move the myosin heads is provided by ATP, which is hydrolyzed by myosin.
4. A motor unit consists of a branched motor neuron and the muscle fibers it innervates. Multiple motor unit recruitment results in stronger contractions.
5. A muscle twitch results from a single stimulus. More rapidly delivered signals produce a graded contraction by summation. Tetanus is a state of smooth and sustained contraction, obtained when motor neurons deliver a volley of action potentials.
6. Cardiac muscle, found only in the heart, consists of striated, branching cells that are electrically connected by intercalated discs. Cardiac muscle cells can generate action potentials without neural input.
7. In smooth muscle, contractions are slow but can be sustained over long periods.

SELF-QUIZ

1. Which of the following receptors is *incorrectly* paired with its category?
   a. hair cell—mechanoreceptor
   b. muscle spindle—mechanoreceptor
   c. gustatory receptor—chemoreceptor
   d. rod—electromagnetic receptor
   e. nociceptor—deep pressure receptor
2. Some sharks close their eyes just before they bite. Although they cannot see their prey, their bites are on target. Researchers have noted that sharks often misdirect their bites at metal objects, and they can find batteries buried under the sand of an aquarium. This evidence suggests that sharks keep track of their prey during the split second before they bite in the same way that
   a. a rattlesnake finds a mouse in its burrow
   b. a male silkworm moth locates a mate
   c. a bat can find moths in the dark
   d. a platypus locates its prey in a muddy river
   e. a migrating bird finds its way on a cloudy night
3. Which of the following is an *incorrect* statement about the vertebrate eye?
   a. The vitreous humor regulates the amount of light entering the pupil.
   b. The transparent cornea is an extension of the sclera.
   c. The fovea is the center of the visual field and contains only cones.
   d. The ciliary muscle functions in accommodation.
   e. The retina lies just inside the choroid and contains the photoreceptor cells.
4. The utricle and saccule are
   a. involved in lateral integration of visual signals
   b. visual centers in the cortex
   c. part of the vestibular apparatus responsible for positional information about the head
   d. semicircular canals that respond to head rotation
   e. organs of balance found in insects and crustaceans
5. When you first walk from a brightly lit area into darkness, which of the following occurs?
   a. The photopsins in your cones become bleached.

b. Your rod cells become hyperpolarized.
c. The receptor cells release less neurotransmitter.
d. Lateral inhibition caused by your horizontal cells ceases.
e. Your rhodopsin is still dissociated into retinal and opsin, and your rods are nonfunctional.

6. The transduction of sound waves to action potentials takes place
a. within the tectorial membrane as it is stimulated by the hair cells
b. when hair cells are bent against the tectorial membrane, causing them to depolarize and release neurotransmitter molecules that stimulate sensory neurons
c. as the basilar membrane becomes more permeable to sodium and depolarizes, initiating an action potential in a sensory neuron
d. as the basilar membrane vibrates at different frequencies in response to the varying volume of sounds
e. within the middle ear as the vibrations are amplified by the malleus, incus, and stapes

7. The role of calcium in muscle contraction is
a. to break the cross-bridges as a cofactor in the hydrolysis of ATP
b. to bind with troponin, changing its shape so that the actin filament is exposed
c. to transmit the action potential across the neuromuscular junction
d. to spread the action potential through the transverse tubules
e. to reestablish the polarization of the plasma's membrane following an action potential

8. Tetanus refers to
a. the partial sustained contraction of major supporting muscles
b. the all-or-none contraction of a single muscle fiber
c. a stronger contraction resulting from multiple motor unit summation
d. the result of wave summation, which produces a smooth and sustained contraction of a muscle
e. the state of muscle fatigue caused by depletion of ATP and accumulation of lactate

9. Which of the following is a *true* statement about cardiac muscle cells?
a. They lack an orderly arrangement of actin and myosin filaments.
b. They have less extensive sarcoplasmic reticulum and thus contract more slowly than do smooth muscle cells.
c. They are connected by intercalated discs, through which action potentials spread to all cells in the heart.
d. They have a resting potential more positive than an action potential threshold.
e. They only contract when stimulated by neurons.

10. Which of the following changes occurs when a skeletal muscle contracts?
a. The A bands shorten.
b. The I bands shorten.
c. The Z lines slide farther apart.
d. The actin filaments contract.
e. The thick filaments contract.

CHALLENGE QUESTIONS

1. Explain the difference between sensation and perception. Give some examples.
2. Compare vision and hearing with respect to receptor type, transduction, amplification, transmission, and integration characteristics.
3. Although skeletal muscles generally fatigue fairly rapidly, clam shell muscles have a unique protein called paramyosin that allows them to sustain contraction for up to a month. From your knowledge of the cellular mechanism of contraction, propose a hypothesis to explain how paramyosin might work. How would you test your hypothesis experimentally?

SCIENCE, TECHNOLOGY, AND SOCIETY

1. Have you ever felt your ears ringing after listening to loud music from a sound system or at a concert? The sound often tops 90 decibels—loud enough to permanently impair hearing. Do you think people are aware of the possible danger from prolonged exposure to loud music? What, if anything, should be done to warn or protect them?
2. You may know an elderly person who has broken a bone (often a hip) as a consequence of osteoporosis. This weakening of the bones affects many women after menopause. Researchers think that prevention is the best way to deal with osteoporosis. They recommend exercise and maximum calcium intake during the teens and twenties. Is it realistic to think that young people ever view themselves as future senior citizens? How would you recommend that they be encouraged to develop health habits that might not pay off for forty or fifty years?

FURTHER READING

Alexander, R. "Muscles Fit for the Job." *New Scientist,* April 15, 1989. How are particular muscles adapted for diverse functions in movement?

Barinaga, M. "The Secret of Saltiness." *Science,* November 1, 1991. The physiology of a basic taste.

Baskin, Y. "The Rhino's Silent Call." *Discover,* April 1992. A mammal that communicates with sounds too low for us to hear.

Carafoli, E., and J. T. Penniston. "The Calcium Signal." *Scientific American,* November 1985. A close look at calcium and its regulation and function in processes like muscle contraction.

Freeman, W. "The Physiology of Perception." *Scientific American,* February 1991. How the brain processes sensory data.

Nathans, J. "The Genes for Color Vision." *Scientific American,* February 1989. The genetics and physiology of color blindness.

Nilsson, D.-E. "Vision Optics and Evolution." *BioScience,* May 1989. Comparative anatomy and physiology of eyes.

Ramachandran, V. "Blind Spots." *Scientific American,* May 1992. Numerous optical illusions based on the blind spot.

UNIT EIGHT ECOLOGY

AN INTERVIEW WITH ARIEL LUGO

In 1989, Hurricane Hugo pounded Puerto Rico, but El Yunque, the island's grandest mountain, steered the eye of the storm back out to sea, saving the city of San Juan and its million citizens. The feat reinforced El Yunque's status as Puerto Rico's sacred mountain. El Yunque is also the location of the Caribbean National Forest, the only tropical forest in the U.S. National Forest System and an important research destination for tropical ecologists. Scientists know the forest as the Luquillo Experimental Forest. We traveled to Puerto Rico to meet Ariel Lugo, director of the Institute of Tropical Forestry, a division of the U.S.D.A. Forest Service. After his undergraduate education and a master's degree at the University of Puerto Rico, Dr. Lugo earned his Ph.D. at the University of North Carolina, and served as a biology professor at the University of Florida before returning to his native island as a U.S.D.A. Forest Service scientist. To prepare for our interview with Dr. Lugo, my editor and I hiked on El Yunque for a day with a Forest Service guide. It was not enough time to learn much about the rain forest, but our field trip helped us appreciate why Puerto Ricans treasure the mountain and its life–and why tropical forests are such important ecosystems. There is global concern about deforestation in the tropics, and Dr. Lugo has emerged as a voice of reason in communicating between the diverse interest groups that view the future of tropical forests from different perspectives. In this interview, Dr. Lugo explains why land use in tropical forests is such a complex issue and expresses his optimism for rehabilitating rain forest that has been abused in the past.

Dr. Lugo, what personal experiences influenced you to become a tropical ecologist?

My father, who was a botanist, used to drag me all over the island on his field trips, but I couldn't have cared less about botany back then. Like every Puerto Rican, I was going to be a doctor. I went to the university, I went to medical school interviews, I took the exams, and just before I started medical school, I had a summer off. And in that summer I got a job with a fellow named Howard T. Odum, who was at that time the director of the Marine Sciences Institute in Texas. It was 1964, when the Cold War was really at its peak. The U.S. government was irradiating ecosystems all over the United States to find out which of them would survive if there were an atomic explosion. They decided to irradiate a tropical system, and Odum was the director of that project. He gave me a summer job. I spent the whole summer carrying bricks up the mountain to mark the trails. And that summer with Odum was it–no medical school. I decided I wanted to be an ecologist. And the fact that I got a "D" in embryology helped my decision! I remember in my interview for medical school, the doctor said, "You know, you have a "D" in embryology. Do you realize that you have to pass embryology in your first semester?" And I hated embryology. But it was Howard Odum and the rain forest project that led me into this field.

Why are tropical forests so important, and why are ecologists so concerned about destruction of these forests?

They're important first because they're half of the world's forests. They control most of the cycles of nutrients and water and the stability of the climate. Secondly, tropical forests support about a fifth of the world's population. Most of the world's poor live near tropical forests, and so lots of people depend on them for day-to-day survival. Thirdly, people are finding out that most of the world's biodiversity is in tropical forests. Once you ascribe a value to the forests--a value in terms of biodiversity, or a value in terms of climate control or in what they do for poor people–then you become horrified that they're being deforested. We're now losing 17 million hectares of tropical forest every year.

What are the reasons for deforestation in the tropics?

Some people think that greedy people are cutting the forest for the lumber. But lumber is really a minor factor in deforestation. The major fraction of the tropical deforestation is to compensate for the loss of agricultural land. People need food. To get food, they have to cut the forest down and plant their crops. And it's well known that this is a major reason why we're losing tropical forest. But what most people don't realize is that much of the agriculture in the tropics is not sustainable because it is mismanaged. In some tropical regions of Latin America and Africa, 50% to 100% of the deforestation is to make up for agricultural lands that are poorly managed and degraded. So the main reason for deforestation is to satisfy human needs for food.

Official policies are another cause of deforestation. This problem has decreased a lot recently, but still, in many countries of the tropics, people who deforest can make a profit because a parcel of land covered by tropical forest becomes more valuable if it is cleared. And these kinds of deforestations are very tough, because they come down to the land-use strategies of some tropical countries. It's a very complicated issue, this issue of deforestation.

In debates about land use in the tropics, what are some of the conflicts among different social, economic, and environmental interests?
Each sector has a different plan for the tropical forest. First of all, there are the indigenous people. Millions of people live in tropical forests, and they clearly have a plan for the forest that is their habitat, since it provides them with survival. But a conservationist who is flying in an airplane over the forest sees all that green and says, "Ah, here's where I'm going to protect the biodiversity of the world. We're not going to touch this forest, because this is going to balance what we've been doing in the rest of the world." So for them, you want to leave it alone, but not because there are people living there. You want to leave the tropical forest alone because you want to solve other problems with it. But if in that plane you have a timber baron, that person sees in this piece of forest one of the last of the Earth's untouched stores of timber. And you could say, if you were a timber baron, "Ah, I want to make money by cutting those trees and selling them." But there might be other people with different plans--the government, for example. The government sees the vast expanses of tropical forest as the future of the country. The United States made its future by cutting its forests. For example, at one time in the United States, fuel wood provided a lot of the energy. Well, *most* of the energy in poor tropical countries is from fuel wood. And that comes from the forest. So people in villages and cities who still depend on charcoal see the tropical forest as a source of fuel. But for the government, it's also an area to be developed. Many governments over history have set aside large areas of tropical forest and moved people in: "Go live there. We'll build roads; we'll provide infrastructure. Let's develop this country." So you see, that's a different view of the forest: that a forest provides space for new cities, for new economic development.

So each sector of society sees the forest in a different way, and when they talk about it, each one talks from their own perspective. The challenge is to harmonize or integrate all these legitimate claims on the forest. Now, who's going to put it all together? Who's going to come in there and organize development such that you have indigenous people, and you have conservation of the resource, and you maintain your climatic balances, and you develop some areas, and you have agriculture, and you get fuel wood--that requires management in my view, and it requires a lot of compromise, and a lot of hard thinking, and a lot of good will and political will. But these types of discussions are not yet in the forefront. In the controversy about the tropics, we're *talking* about sustainable development, but really we're still sectored.

And in the meantime, deforestation accelerates. What are some of the consequences, both on the local level and on a global scale?
Deforestation is bad locally because when you lose the forest cover, the value of the forest is lost. We have to admit that sometimes you need to deforest--we cannot be such purists as to say that we can't cut any forest. But when deforestation occurs in an unplanned, uncontrolled way, then you lose species and productivity, and degrade your soil and water resources. These regional effects fragment the landscape. And when you fragment the landscape, you disconnect nature--you lose the value of ecosystems working in synchrony.

And when you add up all the deforestation, you get into the global impact. I mentioned that tropical forests are so huge that they help regulate climate, they help regulate the cycles of nutrients and water and gases, and so on. Large-scale deforestation can tilt the balance of nutrient cycling. For example, the amount of carbon in tropical forests is sufficient to almost double the amount of CO_2 in the atmosphere if it were to be released all at once. That will never happen, but it shows you the potential. But the main thing I want to tell you about the consequences of deforestation is just the waste. If tropical forests have so much value, and we just wantonly de-

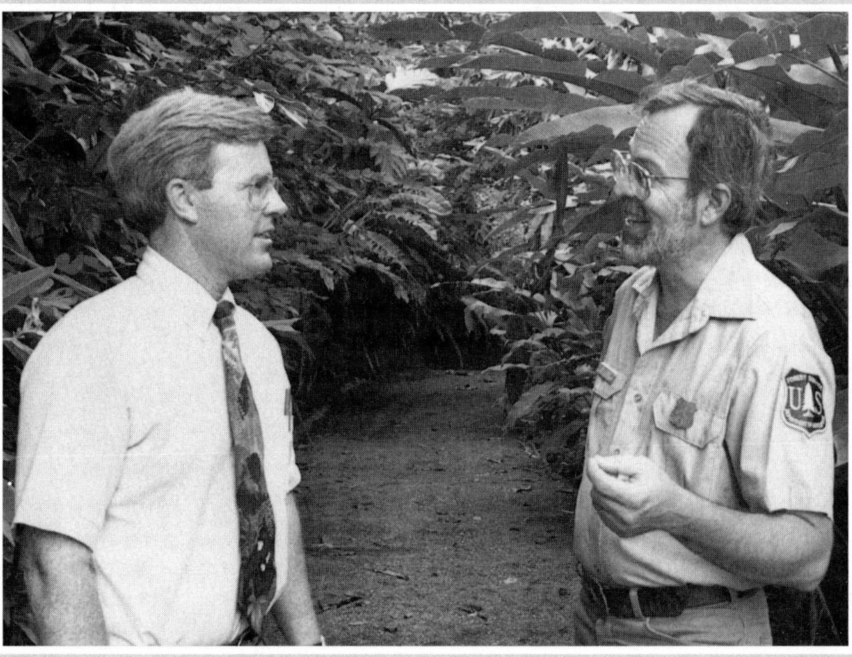

stroy them, then we are wasting a resource that could be used to improve the well-being of people.

So a greater sense of the rain forest's economic and ecological value may help slow deforestation?
In some cases, yes–what we need to understand is the capability of the land to produce value. Now value could come in many ways. It could come by leaving the forest alone, or it could come by putting pavement over the forest. But to me, it is critical that when we convert land from one use to another, we have to make sure that the use we're after is sustainable, that we can prolong the derivation of value. Many times what happens is that when land is converted, nobody gains anything except on paper. And the country as a whole might even be losing, because long term, the negatives of the new use of the land might outweigh the paper value added.

So I think that as the world gets smaller, we need to understand each hectare of land on the planet, and try to ascertain its optimal uses. And we will find that we don't want to touch some areas, because they're just absolutely critical for our overall survival on the planet. Other areas we're going to have to convert. For example, how can you not produce food where the soils are rich and where you can sustain agriculture? It would be foolish not do that, and it's more foolish to try agriculture in the worst soils or put houses on your rich soils. So I think as we develop a conservation ethic, we will learn how to use resources more wisely and more effectively.

You mentioned the importance of sustainable use of the land. What does this mean?
Sustainable productivity means that you benefit from what the land can produce without being so greedy as to wipe out this productivity in the process. How can we manage ecosystems within our own generation, get the benefits of the production, and then pass these ecosystems on with the same capacity so that the next generation is not handicapped by a reduction of productivity? That's what sustainable productivity means to me. The level to which we're going to achieve it depends on culture. I think different cultures have different expectations, different needs, and so some cultures push nature more to the limit than others.

Is there any hope of rehabilitating tropical forests in areas that have been deforested?
There's a lot of hope. And we have many examples--I always use Puerto Rico. About half of the preserve I work in, the Luquillo Experimental Forest, was used agriculturally in the recent past. When the Forest Service purchased

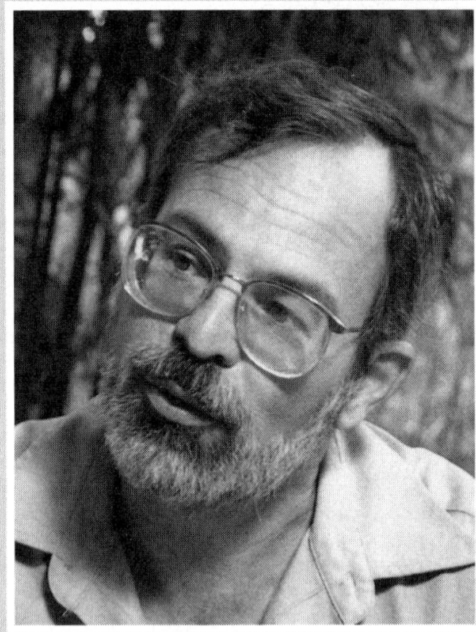

those lands in the 1930s, '40s, and '50s, the first thing we had to do was somehow get the people out without making enemies. We said, "Okay you guys, you can farm in here, but we're going to ask that between the rows of crops you put trees, and you're going to take care of the trees. And then when those trees grow up, we're going to find you an equal amount of land outside of the forest so you can farm there." That was the deal. And it was very successful.

Now, two types of abandonments took place. In some areas, people left and nature took over. In other areas, where people planted these trees that I mentioned and we took care of the trees, forest grew up nicely. And what we have today is beautiful forest, in some cases natural, with natural species, in other cases with the species that were planted up to fifty years ago. The key is taking the pressure out of the forest. It has resiliency; you can restore the diversity. In fact, when those trees matured and the Forest Service proposed to cut them, we had ten thousand people demonstrating! They didn't want us to cut this "virgin" forest. I think that's a great measure of success, when the rehabilitation is so good that people defend it as natural forest, when in fact it was a managed deal.

Even more critical is what happens when you have damaged the land severely, when you have taken bulldozers and really devastated the landscape. Can such abused land come back? Well, we have examples here in Puerto Rico where yes, we can do it. Sometimes when you bulldoze areas, native species cannot survive because the damage to the land is so traumatic that the native species just don't have the necessary adaptations. Then what you need to do is to bring in a ringer, an exotic species from the outside. You get a nitrogen fixer that can grow in the worst places in the world. The one we use here is a species of *Albizzia,* which now grows all over the island. Wherever the Highway Department has razed with bulldozers, *Albizzia* just comes in like gangbusters and makes this beautiful forest on the sides of the highways. And that's an exotic species giving us a daily lesson that you can find forest species to rehabilitate even the worst damaged land.

Now, the postscript to this is that once *Albizzia* takes over a site, builds up organic matter, changes the soil and provides shade, then birds start visiting that area, bringing seed from all over the place. We've been documenting how over a period of thirty years, you can add an average of one new native species per year into these exotic tree plantations. As these trees grow, they shade the *Albizzia,* which disappears from the area. Eventually, you have a native forest. So we're confident that there are ecological tools for managing succession and restoring the rain forest on even badly damaged lands. The problem is what I mentioned earlier--do we have the political and social will--are we ready to manage the landscape in a positive way? If you have the tools, all you need is a chance.

How do hurricanes and other natural disturbances differ from human disturbances in their impact on the rain forest?
Hurricane Hugo in 1989 was such a huge disturbance that it took out all the leaves of the Luquillo forest. But when we did the studies on this damage, we found that the actual mortality of trees was relatively low–no more than 20%. Trees die at 1% to 2% per year when there are no hurricanes, and then up to 20% of them die instantly with the hurricane.

What we're learning is that hurricanes are the main organizing force of that forest. The forest goes through a cycle that averages sixty years, starting with great impact by winds and rain of a hurricane, and then sixty years of regrowth. And in those sixty years of regrowth, you can get complete changes in the species that dominate the landscape. We see the species change, we see the growth rate change, we see the density of trees change, we see the size of the trees change, we see the density of the wood change. So as we measure these changes, this is teaching us how that forest can incorporate the energy of the hurricane. In other words, the hurricane might appear destructive, but it's actually constructive, because it makes the forest more productive; it rejuvenates the forest.

Now you ask me, how does this impact of natural disturbance compare with human impact on the rain forest? Well, there's no comparison! Nature can predict hurricanes, because they occur in cycles. And this means that life can adapt to these regular disturbances. I can tell you, for example, how the animal populations have actually exploded after a hurricane, which means that they have behavioral adaptations, they have population dynamics that allow them to respond in a positive way. But human activity is unpredictable because humans act on impulse, on the short term, or without reason. So it's much harder for nature to adapt to human activity.

Is this view of adaptation to periodic catastrophes part of a paradigm shift away from the idea that tropical rain forests are particularly fragile, unresilient ecosystems?

Absolutely. Right now, tropical ecologists are putting a lot of emphasis on gaps, such as the gaps created by tree falls, for instance. To me, that's a gigantic step forward, because of the connections between destruction and reconstruction. And then Hurricane Hugo went over a major research site, the Luquillo forest, and Hurricane Andrew went over the Everglades National Park where you had many scientists studying it, and we had Hurricane Iniki in Hawaii. All of a sudden, scientists are being made aware of these dramatic natural impacts. We're also more aware now of El Niño, which is creating "global havoc." But what we thought was havoc really has patterns--it makes fires in Borneo and does other things, and now we can tie these things

together and we realize, "Oh, this is part of the whole Earth going through one of its cycles." And I think this change in thinking is going to help in the management of ecosystems, because once we start to accept catastrophe as part of nature, the next step is to manage this other catastrophe of human intervention.

You wrote in one of your articles, "nature can no longer be studied in a social vacuum." Can you elaborate?

Well, I'll come to that with a little story about how scientists used to study nature. Scientists used to typically go and isolate themselves to study nature in some reserve, away from everybody. A friend of mine from Germany, Hans Klinge, one of the great German ecologists, used to spend a full season marking his plots of vegetation in the Amazon, and another season actually harvesting. But when he came back for the last season, one of the plots was an agricultural field. To me, that's such a telling story, because it shows: Hey scientists, the party's over where you could go and isolate yourself and do your studies, and then publish them in the Journal of Obscure Results, and that was it. Now, if you want to go work in Brazil, you'd better have a permit from the Brazilians. You can no longer take your samples to the museum of State U. like you used to—you need to have vouchers in Brazil, you need to have Brazilian colleagues, you have a responsibility to give a talk in Brazil, you have to learn Portuguese, you have to publish in their journals, and I think obviously you have a responsibility to explain what the heck you're doing and how it relates to our efforts to have a better life.

There's an enormous pressure on scientists now to show relevance. In the U.S.D.A. Forest Service, for example, we are managing an important resource, so we're getting signals everywhere that we need to be socially conscious, that we can no longer be isolated. But my statement goes beyond that. I think if you're paid by society, you have to explain what you're doing, and you have to make sure it has some relevance. And so in the tropics, where there's so much human need, where all these forces are fighting for the tropical forest, where the future climate of the world may be at stake, where the biodiversity of the world is at stake, where the day-to-day living of millions of people is at stake--I think science has to be done in that context. This does not mean I'm against basic science; on the contrary, I don't make any distinction between basic science and applied science. I think it's an artificial distinction. I think there's good science and bad science, and good science is always useful, and you should be able to explain it.

Dr. Lugo, you started by telling how Howard Odum inspired you to become an ecologist. Now that you are in the role of mentor, what advice do you have for students who are interested in ecology as a possible career, especially tropical ecology?

Regardless of one's interest, the best advice I can give students is that they better take all their science courses, and now, since we have to be socially responsible, maybe a second or third language. Odum made me take all the hard sciences–the math, the physics, the chemistry–as well as the biology. And in my view, that's what prepared me to become successful as an ecologist, because ecology is an integrative science. Now you have to be careful, because if you try to specialize too early, then you might fall into a school of thought that puts you in a tangent. I think people should just become good scientists. Once you're a good scientist, then you can adapt your career. And to succeed in science today, you must have good communication skills–the writing and speaking are so critical. They don't tell that to science students when they're going to school. You could be the best scientist in the world, but if you can't write about your work, and if you can't speak up and explain it, you go nowhere.

46 An Introduction to Ecology: Distribution and Adaptations of Organisms

The Scope and Development of Ecology
Terrestrial Biomes
Freshwater Biomes
Marine Biomes
The Environmental Diversity of the Biosphere
Responses of Organisms to Environmental Variation

Figure 46.1
A sense of home. An external view of Earth, as in this photograph of an earthrise taken by Apollo astronauts, gives a compelling sense of the planet as a finite home. This awareness helped propel the environmental movement, which has kicked into high gear during this decade. As a basic science, ecology focuses on the distribution and abundance of organisms, but ecology also helps us understand the environmental impact of humans and their technology. This chapter introduces ecology, with an emphasis on the diverse physical environments of the biosphere and the adaptations that have evolved in these varied ecological theaters. Our survey of ecology continues in the next three chapters, which feature interactions between organisms.

Ecology (from the Greek *oikos,* "home," and *logos,* "to study") is the scientific study of the interaction between organisms and their environments. This straightforward definition masks an enormously complex and exciting area of biology that is also of increasingly critical practical importance. Let's first look more closely at the key words in our definition.

As an area of *scientific* study, ecology incorporates the hypothetico-deductive method, using observations and experiments to test hypothetical explanations of ecological phenomena (see Chapter 1). As we will see shortly, however, ecologists often face extraordinary challenges in their research because of the complexity of their questions, the diversity of their subjects, and the large time and space scales over which studies must be conducted. Ecology is also challenging because of its multidisciplinary nature; ecological questions form a continuum with those from many other areas of biology, including genetics, evolution, physiology, and behavior.

The *environment,* as defined by ecologists, includes **abiotic** factors, such as temperature, light, water, and nutrients. Just as important in their effects on organisms are **biotic** factors—the other organisms that are part of any individual's environment. Other organisms may compete with an individual for food and other resources, prey upon it, or change its physical and chemical environment. As we will see, questions about the relative importance of various environmental factors are frequently at the heart of ecological studies—and the accompanying controversies.

Finally, in our dissection of the definition of ecology, we should highlight the *interaction* between organisms and their environment. Organisms are affected by their environment but, by their very presence and activities, they also change it—often dramatically. Through their metabolism, microorganisms in a lake at night reduce the oxygen and lower the pH of the lake. Trees reduce light levels on the floor of a forest as they grow, sometimes making the environment unsuitable for their own offspring. Throughout our survey of ecology, we will see many more examples of how organisms and their environments affect one another.

The basic science of ecology, the study of how organisms interact with their environment, should be distinguished from the same word that is sometimes used informally in popular writing. You have probably read accounts of how a particular pollutant might destroy the "ecology" of an area. Although this is not what is meant by our definition, there is obviously a close link between the science of ecology and the vast array of current environmental concerns, some of which extend to our ability to survive as a species. Beginning primarily in the 1960s, when many people became alarmed by the adverse effects that humans and our technology were having on the environment, "ecology" has worked its way into the public consciousness. Photographs of the planet taken by the lunar astronauts contributed to the growing awareness of Earth as a finite home in the vastness of space, rather than an unlimited frontier for human activity (Figure 46.1). Acid rain, localized famine aggravated by land misuse and population growth, the growing list of species extinct or endangered because of habitat destruction, and the poisoning of soil and streams with toxic wastes are just a few of the problems that threaten the home we share with millions of other forms of life. The science of ecology provides the necessary background for us to understand these problems and to solve them. Ariel Lugo (see the interview preceding this chapter) and most other ecologists agree that they have a responsibility to educate legislators and the general public about decisions that affect the environment. Usually, however, any response also involves the realms of ethics, economics, and politics, which dominated debate at the 1992 United Nations Earth Summit in Rio de Janeiro. These are considerations beyond our scope here, but the chapters of this unit will point out many connections between basic ecology and environmental issues.

This chapter introduces ecology by defining its scope and describing some of the large-scale ecological patterns found in various places on Earth. We will also examine the important abiotic environmental factors, especially climate, that influence organisms and the ways in which they are adapted to those factors. Biotic factors, organism–organism interactions, are emphasized in the next three chapters.

THE SCOPE AND DEVELOPMENT OF ECOLOGY

The Questions of Ecology

The questions of ecology are extremely wide-ranging. To help us keep a general focus, we can think of ecology as *the study of the distribution and abundance of organisms.* What factors determine where species are found, and what factors control their numbers in those locations? Specific aspects of these general questions apply to increasingly comprehensive levels of organization, from the interactions of individual organisms with the abiotic environment to the dynamics of ecosystems.

Organismal ecology (sometimes called autecology or physiological ecology) is concerned with the behavioral, physiological, and morphological ways in which individual organisms meet the challenges posed by physicochemical aspects of the environment. The organism's limits of tolerance for environmental stresses ultimately determine where it can live.

The next level of organization in ecology is the **population,** a group of individuals in a particular geographic area that belong to the same species. Population ecology concentrates mainly on factors that affect population size and composition, as described in Chapter 47.

A **community** consists of all the organisms that inhabit a particular area; it is an assemblage of populations of different species. Questions at this level of analysis involve the ways in which predation, competition, and other interactions among organisms affect community structure and organization.

A level of ecological study even more inclusive than the community is the **ecosystem**, which includes all the abiotic factors in addition to the community of species that exists in a certain area. Some critical questions at the ecosystem level concern energy flow and the cycling of chemicals within and among the various biotic and abiotic components.

Ecology ultimately deals with the highest levels in the hierarchy of biological organization (see Chapter 1). The web of interactions at the heart of ecological phenomena is what makes this branch of biology so engaging and challenging.

Ecology as an Experimental Science

By necessity, humans have always had an absorbing interest in other organisms and their environments. As hunters and gatherers, prehistoric people had to learn where game and edible plants could be found in greatest abundance. Observing and describing organisms in their natural habitat became an end in itself to natural historians from Aristotle to Darwin. Extraordinary insight can still be gained through this descriptive approach, and thus natural history remains fundamental to ecological science.

Although ecology has a long history as a descriptive science, most modern ecologists are also skilled experimentalists who meet the special demands of research that often involves large amounts of time and space. Consider this seemingly simple question: How does the foraging of squirrels on acorns affect the distribution and abundance of oak trees? Think about what might be required to answer experimentally: large areas of land from which squirrels could be removed,

Figure 46.2
Ecology as an experimental science. The foundations of ecology lie in natural history: the observation of organisms in their native, undisturbed environments. Over the past three decades, however, ecology has become more of an experimental science. Here, a team of researchers prepares an experiment to test hypotheses about the colonization of islands by insects and other animals. The scientists enclosed a number of small mangrove islands in the Florida Keys with scaffolding that was used to support a plastic tent over the island. The island was then fumigated to destroy its fauna of small invertebrates. When the tent was removed, the researchers could study recolonization. (For a further description of this work, see Chapter 48.)

other large areas that could serve as controls, and a long period of time in which to observe results. In spite of the difficulty of conducting such experiments, an increasing number of creative ecologists are testing hypotheses in the laboratory and are experimentally manipulating populations and communities in the field. For example, one researcher removed periwinkle snails from some tidepools and added them to others to investigate the effect of this key herbivore on algal populations. Another ecologist eliminated all the insects on small islands to study recolonization of the islands from nearby mainland populations (Figure 46.2).

Many ecologists also devise mathematical models to define and help answer ecological questions. In this approach, important variables and their hypothetical relationships are described through mathematical equations. The potential ways in which the variables interact can then be studied, usually with the help of a computer. This approach is very appealing because it allows ecologists to simulate large-scale experiments that might be impossible to conduct in the field. Of course, such simulations are only as good as the basic information on which the models are based, and obtaining that information still requires extensive fieldwork. We will study simple models of population growth in Chapter 47.

The dramatic advances ecologists have made through continuing descriptive research, laboratory and field experiments, and mathematical modeling have highlighted the complex relationships between organisms and their biotic and abiotic surroundings. Questions such as "What determines the number of species of birds in a forest?" must often be answered by "It depends on the particular case." Despite this complexity—or, rather, because of it—ecology is an extraordinarily dynamic and exciting science. As ecologists debate a wide range of issues, ideas constantly ebb and flow, some finding favor for a while until displaced by another, perhaps only to return later in a modified form. Science thrives on this type of continual questioning and testing of ideas.

Ecology and Evolution

Though the field had not been defined at the time, Charles Darwin was an ecologist. He found evidence for evolution in the geographic distribution of organisms and in their exquisite adaptations to specific environments. He also realized how the short-term interactions of organisms with their environments could have long-term effects through natural selection. Put another way, the momentary episodes enacted in the frame of what is sometimes called *ecological time* translate into effects over the longer scale of *evolutionary time.* For instance, hawks feeding on field mice have an impact on the gene pool of the population of prey by curtailing the reproductive success of certain individuals. One long-term effect of such a predator–prey interaction may be a prevalence in the mouse population of fur coloration that camouflages the animals.

Another connection between ecology and evolution is apparent in the effects of geological history on the current distribution of species. For example, the pattern of continental drift after the fragmentation of Pangaea helps explain the near absence of native placental mammals in Australia (see Chapter 23).

An evolutionary theme is inherent in these chapters on ecology: The distribution and abundance of organisms are products of long-term evolutionary changes as well as ongoing interactions with the environment. As part of our overview of ecology, let's now survey the distribution and abundance of life on a global scale by exploring Earth's major landscapes, such as forests, grasslands, and deserts.

TERRESTRIAL BIOMES

The entire portion of Earth that is inhabited by life is called the **biosphere;** it is the sum of all the planet's

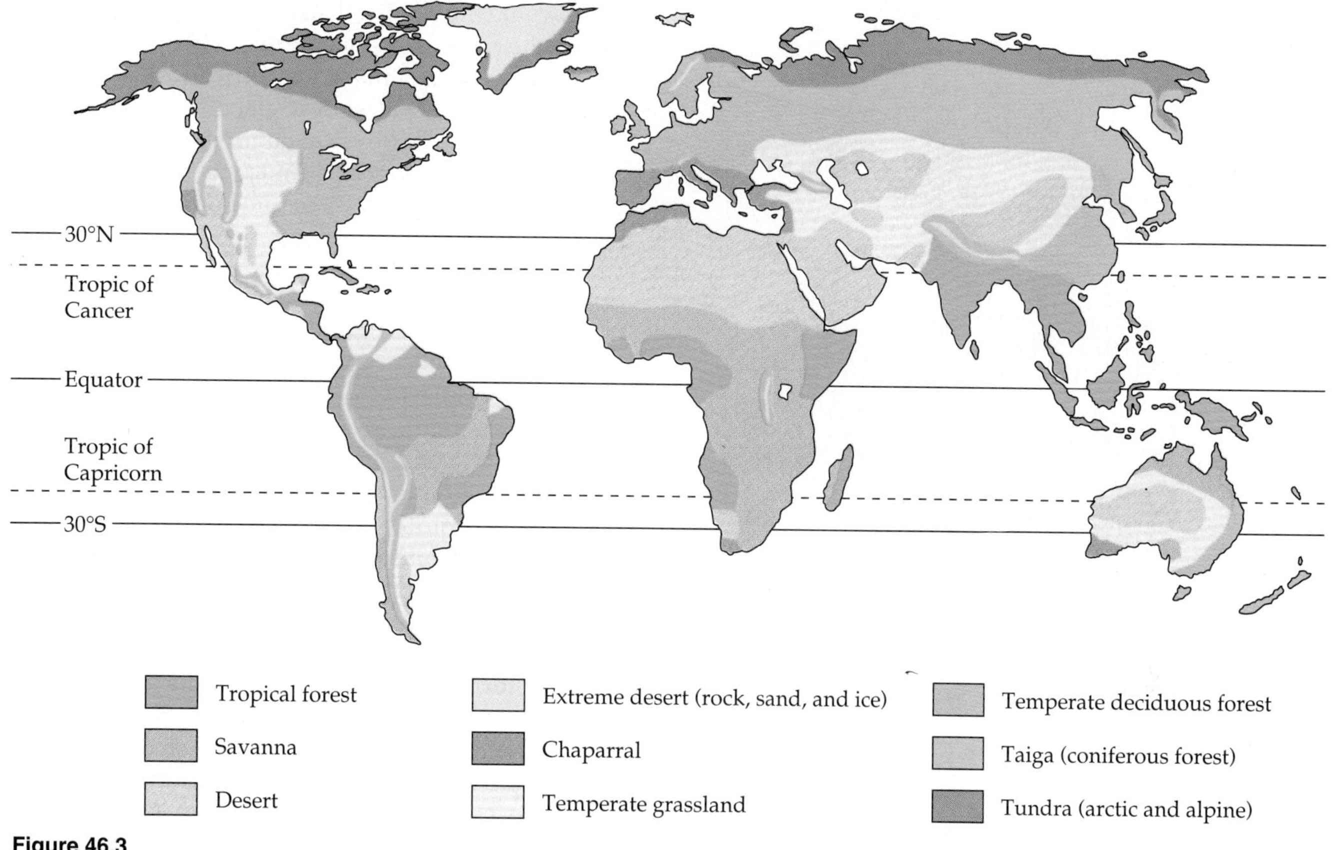

Figure 46.3
The distribution of major terrestrial biomes. Although terrestrial biomes are mapped here with sharp boundaries, biomes actually grade into one another, sometimes over relatively large areas. The "tropics" are the low latitude regions bordered by the Tropic of Cancer and the Tropic of Capricorn.

communities and ecosystems. The biosphere is a relatively thin layer consisting of seas, lakes, rivers, and streams, the land to a soil depth of a few meters, and the atmosphere to an altitude of a few kilometers. Ecologists have long recognized striking global and regional patterns in the distributions of organisms within the biosphere, and these patterns have provided the impetus for much basic ecological research.

The general appearance of communities may be relatively familiar over a large geographic area and may also be similar to communities in other, disjunct regions of the Earth. For example, coniferous forests extend in a broad band across North America, Europe, and Asia. Extensive desert areas are scattered over many parts of the Earth. The major types of communities and ecosystems that are typical of broad geographic regions are called **biomes**. Although terrestrial biomes are often named for the predominant vegetation, each biome is also characterized by microorganisms, fungi, and animals adapted to that particular environment. Grasslands, for example, are more likely than forests to be populated by large grazing mammals.

The actual species composition throughout a biome varies from one location to another. In the North American coniferous forest, red spruce is common in the east but does not occur in most other areas, where black spruce and white spruce are abundant. Although the vegetation of African deserts superficially resembles that of North American deserts, the plants are in different families. Such "ecological equivalents" can arise because of convergent evolution (see Chapter 23).

There is no strict way of defining specific biomes, and through the years different ecologists have recognized and organized biomes in different ways. Also, biomes usually grade into each other, without sharp boundaries. If the area of intergradation is itself large, it may be recognized as a separate biome (see the discussion on savanna later in this section). The major biomes are mapped in Figure 46.3. Although the photographs that accompany this section provide general portraits of the different biomes, one picture cannot adequately illustrate the variability present within each biome.

Within a biome there may, in fact, be extensive patchiness, with several communities represented.

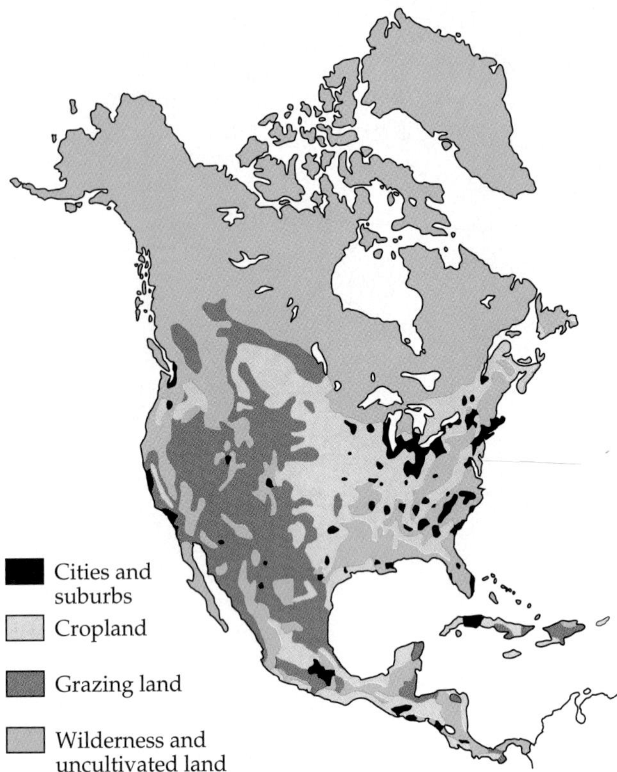

Figure 46.4
Urban and agricultural biomes. Many regions on Earth have been disturbed by intense human activity. In urban and agricultural biomes, natural communities have been replaced by housing, industry, cropland, and grazing range. Relatively few undisturbed habitats remain in most regions on the planet.

Biomes are usually recognized on the basis of the communities that develop as the result of succession (changes in community structure through time), a topic discussed in Chapter 48. Disturbances often allow representatives of earlier successional stages to occur. For example, snowfall may break branches and small trees and cause openings in the coniferous forest, allowing deciduous species, such as aspen and birch, to grow. Most of the eastern United States is classified as temperate forest, but human activity has eliminated all but a tiny percentage of the undisturbed forest that would otherwise be present. In fact, ecologists recognize an "urban biome" and an "agricultural biome" scattered over much of the Earth, where the natural development of communities has been drastically altered (Figure 46.4).

We may now ask *why* a particular biome develops in a certain area. The general answer seems to be that the prevailing climate, particularly temperature and rainfall, is the most important factor in determining the kind of biome that will develop. Other factors, such as the geology of an area, may greatly affect mineral nutrient availability and soil structure, which in turn influence the kind of vegetation that will develop. After first describing the major terrestrial and aquatic biomes on Earth, we will analyze global, regional, and local variations in the physical environment later in the chapter. Let's now survey the major terrestrial biomes, traveling generally in a direction from the equator to the poles.

Tropical Forest

A variety of **tropical forests** are found within 23.5° latitude of the equator, where the average temperature (around 23°C) and length of daylight (around 12 hours) vary little throughout the year. Rainfall, on the other hand, is quite variable in the tropics, and the amount of precipitation, rather than temperature or photoperiod, is the prime determinant of the vegetation growing in an area. In lowland areas that have a prolonged dry season and scarce rainfall at any time, **tropical thorn forests** predominate. The plants found there are a mixture of thorny shrubs and trees, and succulents. In other areas that have distinct wet and dry seasons, **tropical deciduous forests** are common. Deciduous trees and shrubs drop their leaves during the long dry season (when water lost through transpiration would exceed available supplies) and releaf only during the following heavy rains or monsoons. The luxuriant **tropical rain forest** is found in areas near the equator, where rainfall is abundant (greater than 250 cm per year) and the dry season lasts no more than a few months.

The tropical rain forest has the greatest diversity of species of all communities, perhaps harboring as many plant and animal species as all other terrestrial biomes combined (Figure 46.5). As many as 300 species of trees, some of them 50 to 60 m tall, can be found in 1 hectare (10,000 m^2, or about 2.5 acres). Because of the size and density of trees, competition for light becomes a strong selective force in the plant communities of the tropical rain forest.

Though the forest as a whole is extremely dense, individuals of many plant species are widely scattered and rely on mutualistic interactions with animals to deliver pollen. Animals are also important in dispersing fruits and seeds. The animals are typically tree-dwellers; monkeys, birds, insects, snakes, bats, and even frogs find food and shelter many meters above the ground. The warm environment allows the presence of numerous ectothermic animals, and amphibian and reptile species are often more numerous in the tropical rain forest than in any other biome.

Human impact on the tropical rain forest is currently a matter of great concern. In the interview pre-

Figure 46.5
Tropical rain forest. Tropical rain forest often has a closed canopy, with little light reaching the ground below. The rain forest illustrated here is located in Montserrat, a Caribbean Island. When an opening does occur, perhaps because of a fallen tree, other trees and large woody vines known as lianas grow rapidly, competing for light and space as they fill the gap. Many of the giant trees are covered with epiphytes (plants that grow on other plants rather than in soil), such as orchids and bromeliads, and many have wide buttress bases that assist the shallow roots in anchoring the tree in the soil.

ceding this chapter, Ariel Lugo discusses the complex issue of deforestation in the tropics and explains how it is possible to rehabilitate tropical forests. The destruction of the tropical forest is proceeding at an alarming rate. More than one-half is already gone, and projections suggest that these communities may disappear entirely by the end of this decade. The loss is more than aesthetic; destruction of tropical rain forests may cause large-scale changes in world climate, as well as large-scale extinction of species (see Chapter 49).

Savanna

Savanna is grassland with scattered individual trees (Figure 46.6). Extensive savanna covers wide tropical and subtropical areas of central South America, central and southern Africa, and parts of Australia. There are generally three distinct seasons in these regions: cool and dry, hot and dry, and warm and wet, in that sequence. Some savanna soils are quite fertile, but most are porous, resulting in the rapid drainage of water. Porous soils have only a thin layer of the rich, partially decomposed organic matter called humus.

Figure 46.6
Savanna. This Kenyan savanna (with Mount Kilimanjaro in the background) is a showcase of large herbivores and their predators. The luxuriant growth of grasses and forbs (small broad-leaved plants) during the rainy season provides a rich food source for animals. However, large grazing mammals must migrate to greener pastures and watering holes during regular periods of seasonal drought.

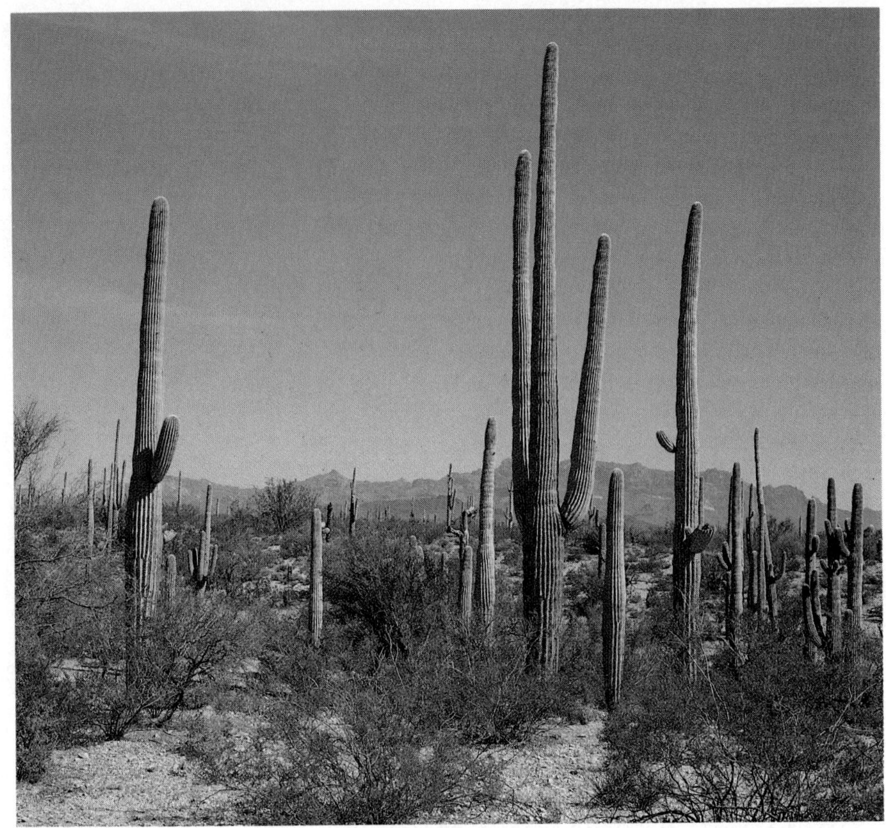

Figure 46.7
Desert. The Sonoran Desert of southern Arizona is characterized by giant saguaro cacti and deeply rooted shrubs. Evolutionary adaptations of desert plants include a remarkable array of protective devices, such as spines on cacti and poisons in the leaves of shrubs, that deter feeding by mammals and insects. Many desert plants also rely on CAM photosynthesis, a metabolic adaptation that conserves water in this arid environment (see Chapter 10).

Savanna is relatively simple in physical structure but often rich in numbers of species. Deciduous trees and shrubs are scattered at low density across the open landscape, because frequent fires and large grazing mammals kill many seedlings before they become well established. The predominant vegetation consists of low-growing grasses and **forbs** (small broad-leaved plants that grow with grasses). Grasses are wind pollinated, but the forbs produce showy flowers that attract insect pollinators, often in great numbers in the summer.

Tropical savannas on different continents are home to some of the world's large herbivores, including the giraffe, zebra, antelope, buffalo, and kangaroo. Burrowing animals, whose nest sites and shelters are primarily underground, are also common: mice, moles, gophers, ground squirrels, snakes, worms, and arthropods. Animals in the savanna are most active during the rainy season, and many species are nocturnal. In the dry season, when the aboveground vegetation is sparse, many small animals are dormant or subsist on seeds and dead plant parts.

The term *savanna* is also applied to areas where forest and grassland biomes intergrade. For example, scattered savanna occurs in North America where the temperate forest and grasslands merge, in a band running roughly from Minnesota to east Texas. Here the climatic conditions and community features are intermediate between those of forest and grassland.

Desert

Deserts (Figure 46.7) are the driest of all terrestrial biomes, characterized by low and unpredictable precipitation (less than 30 cm per year). Although some deserts can be very hot (with soil surface temperatures above 60°C during the day), cold deserts also exist; the hot deserts generally experience large daily fluctuations in temperature. Hot deserts are found in the southwestern United States, along the west coast of South America, in North Africa, and in the Middle East. Cold deserts are found west of the Rocky Mountains, in eastern Argentina, and in much of central Asia. The driest deserts, where average annual rainfall is less than 2 cm (some years having no rain at all), are the Atacama in Chile, the Sahara in Africa, and parts of central Australia.

The density of desert vegetation is largely determined by the frequency and amount of precipitation. The driest deserts receive too little rainfall to support any perennial vegetation. In less arid regions, the dominant vegetation is sparse, consisting of widely scattered drought-resistant shrubs and cacti or other succulents that store water in their tissues. For example, the "pleated" structure of saguaro cacti enables the plants to expand when they absorb water during wet periods. Periods of rainfall (for example, late winter in the Sonoran Desert of the southwestern United States) are marked by sudden and spectacular blooms of annual plants.

Seed-eating animals, such as ants, birds, and rodents, are common in deserts, feeding on the numerous small seeds produced by the plants. Reptiles, such as lizards and snakes, are important predators of these seed-eaters. Like the desert plants, most desert animals are well adapted to scarcity of water and extreme temperatures. Many animals are active only during the cooler months of the year. Others are nocturnal, spending the day in underground burrows where they are shielded from the hot, dry air and intense sunlight. Diurnal animals are often very light in color, thereby reflecting the sunlight. Most desert animals also exhibit remarkable physiological adaptations to an arid environment. Some mice, for example, never drink, deriving all their water from the metabolic breakdown of the seeds they eat. And the development of spadefoot toads from egg to tadpole to frog is completed in less than two weeks in the temporary pools where they breed.

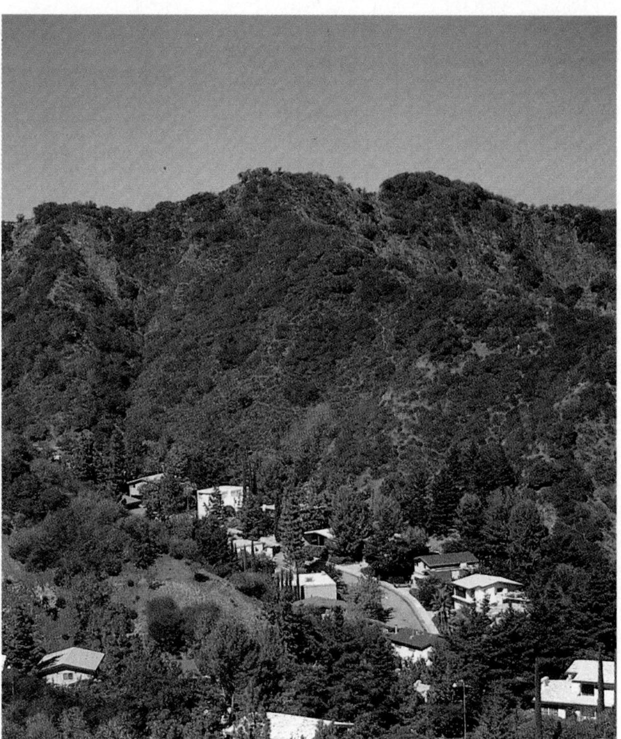

Figure 46.8
Chaparral. Plants of the chaparral, such as this California scrubland, are adapted to periodic fires. The dry, woody plants are frequently ignited by lightning and by careless human activities, creating summer and autumn brushfires in the densely populated canyons of southern California and elsewhere. The large storage roots of the plants allow them to resprout more quickly than the houses can be rebuilt.

Chaparral

Coastal areas between 30° and 40° latitude, where cool ocean currents circulate offshore, are often characterized by mild, rainy winters and long, hot, dry summers. These areas are dominated by **chaparral** (sometimes called scrubland), stands of dense, spiny shrubs with tough evergreen leaves (Figure 46.8). First described in the Mediterranean region, chaparral vegetation is also found along coastlines in California, Chile, southwestern Africa, and southwestern Australia. Plants from these various regions are unrelated but resemble each other in form and function—for instance, the low-growing eucalyptus shrubs of Australia and the scrub live oak of California. Annual plants are also common in chaparral regions during winter and early spring, when rainfall is most abundant.

Chaparral is maintained by and adapted to periodic fires. Many of the shrubs store food reserves in their fire-resistant root crown, enabling them to resprout quickly and use nutrients released by fires. In addition, many chaparral species produce seeds that will germinate only after a hot fire, whereas other species are clonal, reproducing asexually without complete reliance on seeds.

Animals characteristic of the chaparral are browsers such as deer, fruit-eating birds, and rodents that eat seeds of annual plants, as well as lizards and snakes.

Temperate Grassland

Temperate grasslands share some of the characteristics of tropical savanna, but they are found in regions of relatively cold winter temperatures. Temperate grasslands include the veldts of South Africa, the puszta of Hungary, the pampas of Argentina and Uruguay, the steppes of the former Soviet Union, and the plains and prairies of central North America (Figure 46.9).

The key to the persistence of all grasslands is seasonal drought, occasional fires, and grazing by large mammals, all of which prevent woody shrubs and trees from invading and becoming established. Grassland soils tend to be thick and nutrient-rich, and the roots of perennial grasses are often very deep. The amount of annual rainfall influences the height of grassland vegetation; in North America, "tall-grass prairie" occurs in wetter regions and "short-grass prairie" in drier areas.

Grasslands expanded in range following the retreat of glaciers, as hotter and drier climates prevailed worldwide after the last ice age. Coupled with this expansion was the proliferation of large grazing mammals. The bison of North America, the gazelles, zebras, and rhinoceri of the African veldt, and the wild horses and antelopes of the Asian steppes are some examples. Large carnivores, such as lions and wolves, feed on the grazers. Temperate grasslands are also inhabited by dense populations of burrowing rodents, such as prairie dogs, and other small mammals.

Figure 46.9
Temperate grassland. Temperate grasslands, such as this South Dakotan prairie, once covered much of central North America, supporting enormous herds of bison and other grazers. Because grassland soil is both rich in nutrients and deep, these habitats provide fertile land for agriculture. Most grassland in the United States has been converted to farmland (see Figure 46.4), and very little natural prairie exists today.

Temperate Deciduous Forest

Temperate deciduous forests occur throughout midlatitude regions where there is sufficient moisture to support the growth of large trees—most of the eastern United States, most of middle Europe, and part of eastern Asia. These temperate forests are characterized by broad-leaved, deciduous trees (Figure 46.10).

Temperatures range from very cold in the winter to hot in the summer (–30°C to 30°C), with a five- to six-month growing season. Precipitation is relatively high and fairly evenly distributed throughout the year, though groundwater may be temporarily unavailable if the soil freezes on very cold winter days. Temperate deciduous forests have a distinct annual rhythm in which trees drop leaves and become dormant in winter, then produce new leaves each spring. Although losing leaves is costly in terms of energy and nutrients, the relatively rich soils in temperate areas provide the nutrients needed for the production of new leaves in the spring. Rates of decomposition are lower in temperate forests than in the tropics, and temperate deciduous forests accumulate a thick layer of leaf litter, which conserves many of the biome's nutrients.

More open than the tropical forest and not as tall, the temperate forest has several layers of vegetation, including one or two strata of trees, an understory of shrubs, and low-growing forbs. Species composition varies widely around the world; some of the dominant trees are oak, birch, hickory, beech, and maple species.

Figure 46.10
Temperate deciduous forest. Dense stands of deciduous trees are trademarks of temperate deciduous forests, such as this one in Marquette County, northern Michigan. Deciduous forest trees drop their leaves before winter, when temperatures are too low for effective photosynthesis, and water lost through transpiration is not easily replaced from frozen soil. Many temperate deciduous forest mammals also enter a dormant winter state called hibernation, and some bird species migrate to warmer climates.

Figure 46.11
Taiga. Dense, uniform stands of coniferous trees dominate the taiga, such as this fir forest in Alberta's Banff National Park. Taiga receives heavy snowfall during winter. The conical shape of the conifers prevents much snow from accumulating on their branches. When heavy snow does break boughs or causes trees to fall, the openings in the forest allow more sunlight to reach the ground, which in turn allows deciduous trees and shrubs to grow.

Humans have dramatically altered temperate forests by logging for building materials and fuel, clearing for agriculture, and introducing exotic pests and diseases; only scattered remnants of the original worldwide forest remain today.

Because of the variety and abundance of food and habitats it offers, the temperate deciduous forest supports a rich diversity of animal life. A great variety of microorganisms, insects, and spiders, for instance, live in the soil or leaf litter or feed on the leaves and understory of trees and shrubs. The forest is also home to many species of birds and small mammals and, where human encroachment has not eliminated them, wolves, bobcats, foxes, bears, and mountain lions.

Taiga

The **taiga,** also known as coniferous or boreal forest, extends in a broad band across northern North America, Europe, and Asia to the southern border of the arctic tundra (Figure 46.11). Taiga is also found at cool high elevations in more temperate latitudes, as in much of the mountainous region of western North America. The taiga is characterized by long, cold winters and short, wet summers that are occasionally warm. There may be considerable precipitation, mostly in the form of snow. Taiga soil is usually thin, nutrient-poor, and acidic. It forms slowly, owing to the low temperatures and the waxy covering of conifer needles, which decompose slowly. Nevertheless, plants grow quickly during the long days of summer (up to 18 hours of daylight) at these high latitudes.

The conifer stands typically consist of only one or a few species of spruce, pine, fir, or hemlock, often so dense that little undergrowth is present. Deciduous species such as oak, birch, willow, alder, and aspen occur in particularly wet or disturbed habitats.

The heavy snowfall that may accumulate to several meters each winter has important ecological consequences. By insulating the soil before the coldest temperatures occur, snow prevents the soil from becoming permanently frozen. Mice and other small mammals that would quickly freeze to death above the snow remain active all winter in snow tunnels at ground level, where they continue to forage on old vegetation.

The animal populations in the taiga consist mainly of seed-eaters, such as squirrels, jays, and nutcrackers; herbivores, such as insects that eat leaves and wood; and larger browsers, such as deer, moose, elk, snowshoe hares, beavers, and porcupines. Predators of the taiga include grizzly bears, wolves, lynxes, and wolverines. Many mammals in the taiga have thick winter coats that insulate them against the cold, and some hibernate through the long winter.

Coastal coniferous forests, such as the temperate rain forests and redwood forests of the Pacific Northwest, are similar to taiga, being dominated by dense stands of only one or a few tree species. However, these unique communities are considerably warmer and moister than taiga because of their proximity to the ocean. These forests are being logged at an alarming rate, and old growth stands of these trees may soon disappear.

Tundra

The northernmost limits of plant growth occur in the **arctic tundra,** where plant forms are limited to low

Figure 46.12
Tundra. Permafrost, bitterly cold temperatures, and high winds are responsible for the absence of trees and other tall plants in this arctic tundra on the edge of Canada's Hudson Bay (photographed in autumn). Tundra covers expansive areas of the arctic, amounting to 20% of Earth's land surface. High winds and cold temperatures create similar plant communities, called alpine tundra, on very high mountaintops at all latitudes, including the tropics.

shrubby or matlike vegetation (Figure 46.12). The arctic tundra encircles the north pole and extends southward to the coniferous forests of the taiga. Similar communities, called **alpine tundras,** are found on high mountains at altitudes above those where trees can grow. The floras and faunas of the arctic and alpine tundra are generally similar, with perhaps 40% of the plant species in common. However, there are significant differences in the two environments.

In the arctic tundra, the climate is often very cold, with little light available during the long winters. Although the upper meter of topsoil may thaw during the summer, the underlying subsoil remains permanently frozen, a condition called **permafrost,** which prevents the roots of plants from growing very deep. The tundra may receive as little precipitation as some deserts, yet the combination of permafrost, low temperatures, and low evaporation leaves the soils continually saturated, further restricting the types of plants that can grow there. This environment supports dwarf perennial shrubs, sedges, grasses, mosses, and lichens. Plant growth and reproduction occur in a rapid burst during the brief summers, marked by nearly continuous daylight.

Animals of the arctic tundra withstand the cold by living in burrows or having good insulation that retains heat. Because many bird species, especially shorebirds and waterfowl, are migratory, the fauna is much richer in summer than in winter. Ectothermic animals are rare except for the abundant gnats, midges, and mosquitos in the summer. Many arthropods spend the winter at immature stages of growth, which are more resistant to cold than adult forms. The arctic tundra is also home to many herbivorous mammals, such as the large musk oxen and caribou in North America and the reindeer of Europe and Asia, as well as the smaller, cyclically abundant hares and lemmings. Common predators include arctic fox, wolves, and snowy owls, and polar bears near the coast.

Alpine tundra occurs at all latitudes, even in the tropics, if the elevation is high enough. Tropical alpine tundra is confined to the very highest mountaintops, where nightly temperatures are usually below freezing. In contrast to the arctic tundra, daylight varies little from 12 hours per day throughout the year. Also, instead of a brief, intense period of productivity, vegetation in the tropical alpine tundra exhibits slow but steady rates of photosynthesis and growth all year.

FRESHWATER BIOMES

Life first arose in water and evolved there for almost three billion years before plants and animals moved onto land and diversified in terrestrial habitats. Today, the largest part of the biosphere is still occupied by aquatic habitats. Ecologists distinguish between freshwater biomes and marine biomes on the basis of physicochemical differences that influence the communities occupying these distinctive aquatic habitats. Freshwater biomes, for example, are usually characterized by a salt concentration less than 1%, whereas marine biomes generally have higher salt concentrations. However, the general theme developed in the last section applies to all of these biomes as well: Where aquatic environments are similar in abiotic factors, similar adaptations and similar communities are usually found. Earth's major freshwater and marine biomes are mapped in Figure 46.13.

Freshwater biomes are closely linked to the terrestrial biomes through which they pass or in which they are situated. Streams and rivers are created by the runoff of water from terrestrial habitats, and ponds and lakes form where runoff accumulates in a landlocked basin. The particular characteristics of a freshwater biome are influenced by the patterns and speed of water flow, and the climate to which the biome is exposed. Although we will examine the general characteristics of two major freshwater biomes here, many other freshwater biomes, such as marshes and swamps, are also important components of the biosphere.

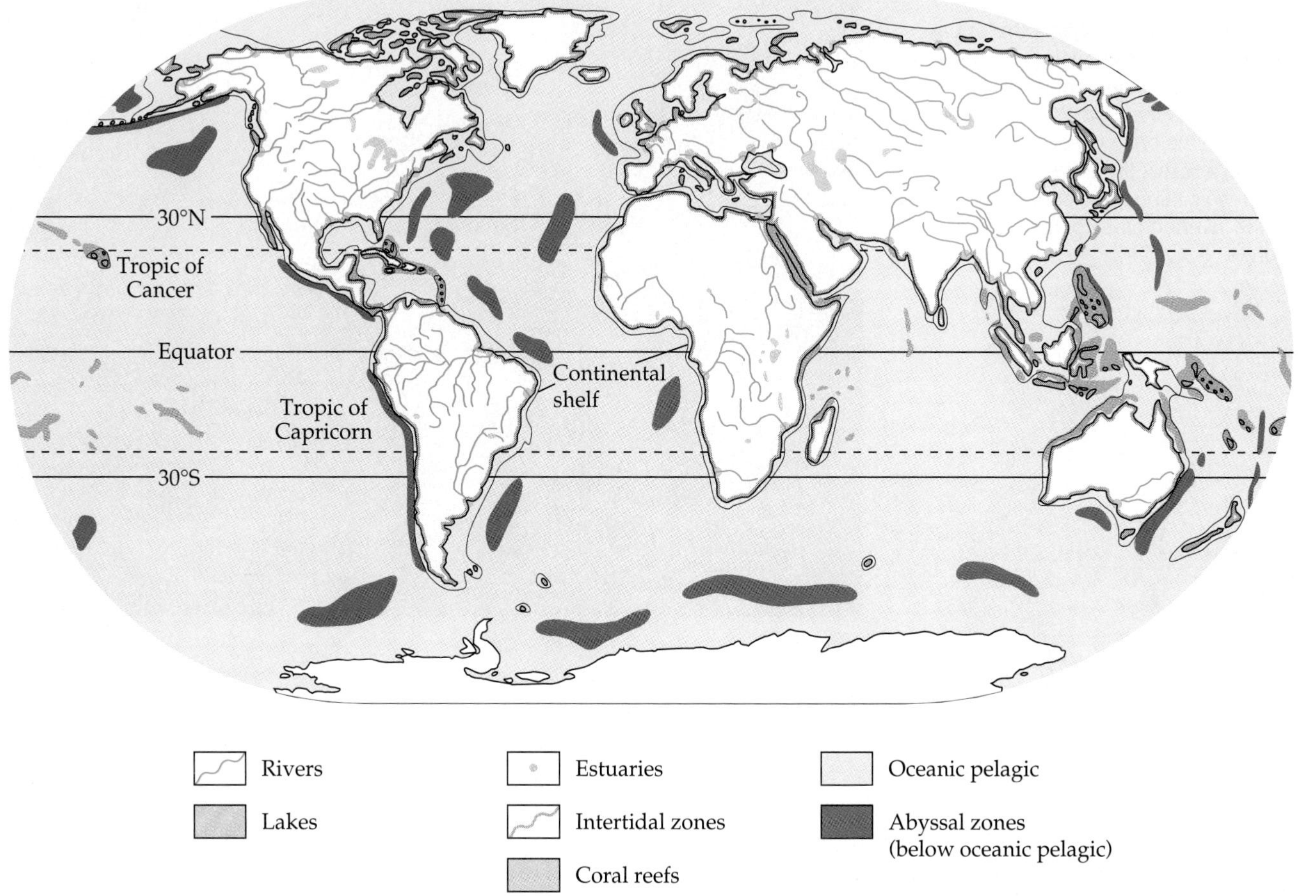

Figure 46.13
The distribution of major aquatic biomes. The characteristics of aquatic biomes are determined by the physical and chemical properties of water in different regions on Earth. Rivers and streams are created by runoff from the land that collects in channels and flows into standing bodies of water. Lakes form where water accumulates in inland basins, and estuaries are tidal habitats where rivers flow into the sea. A variety of marine biomes exist in the oceans, where salt concentration is usually high. Intertidal communities flourish along coastlines where rising tides periodically inundate the land. Coral reefs develop in the shallow tropical waters of continental shelves. Pelagic communities exist in the open ocean (over continental shelves and in deeper waters). Benthic communities are found in the substrate below all bodies of water. The map also indicates abyssal zones, the habitat of deep benthic communities.

Ponds and Lakes

Standing bodies of fresh water range from a few square meters to thousands of square kilometers in area; small bodies of fresh water are called **ponds** and larger ones **lakes** (Figure 46.14). Except in the shallowest ponds and lakes, there is usually a significant vertical stratification of important physicochemical variables. As described later in the chapter, light is rapidly absorbed both by the water itself and by the microorganisms in it, so that light intensity rapidly decreases with depth. Ecologists distinguish between the upper **photic zone,** where there is sufficient light for photosynthesis, and the lower **aphotic zone,** where little light penetrates. Water temperature also tends to be stratified, especially during summer in deeper ponds and lakes of temperate zones. Heat energy from sunlight warms the surface waters to whatever depth the sunlight penetrates, but the deeper waters remain quite cold. A narrow vertical zone of rapid temperature change called a **thermocline** separates the uniformly warm upper layer from the uniformly cold bottom water.

Communities of plants and animals are distributed within ponds and lakes according to the depth of the water and its distance from shore. Rooted and floating aquatic plants flourish in the **littoral zone,** the shallow, well-lighted, warm waters close to shore; some have stems and leaves that emerge above the water surface. The littoral community in most lakes is very diverse, including many species of attached algae, especially diatoms, and a variety of grazing snails and suspension-feeding clams, as well as herbivorous and carniv-

Figure 46.14
Ponds and lakes. Oregon's Crater Lake illustrates the pristine quality of a nutrient-poor oligotrophic lake. The shortage of nutrients limits the productivity of phytoplankton in the open water of the limnetic zone; as a result, the water is clear and oxygen-rich, supporting populations of fish and the invertebrates upon which they feed.

orous insects, crustaceans, fishes, and amphibians. For many of the insects, such as dragonflies and midges, only the egg and larval stages are strictly aquatic. The adults emerge from the water as flying insects that complete their life cycle in the air and on land, returning to the water only to lay their eggs. Aquatic and semiaquatic reptiles such as turtles and snakes, waterfowl such as ducks and swans, and some mammals often feed on the plants and animals in the littoral zone.

The well-lighted, open surface waters farther from shore, called the **limnetic zone,** are occupied by a variety of phytoplankton, consisting of algae and cyanobacteria. These organisms photosynthesize and reproduce at a high rate during spring and summer. Zooplankton, mostly rotifers and small crustaceans, graze on the phytoplankton. The zooplankton are consumed by many small fish, which in turn become food for larger fish, semiaquatic snakes and turtles, and fish-eating birds.

The small organisms of the limnetic zone are short-lived, and their remains continually sink into the **profundal zone,** the deep, aphotic regions of the pond or lake. Microbes and other organisms in the profundal zone use oxygen for cellular respiration as they decompose this **detritus,** the dead organic material that "rains" down from the limnetic zone. Because the deep waters are colder and thus denser, they do not mix with the surface waters that contact the atmosphere. Decomposers may actually deplete the oxygen supply in the profundal zone by the end of a productive summer, making the deep waters unsuitable for most organisms. The decomposition also releases large quantities of mineral nutrients from the detritus, but these nutrients are trapped in the waters at the bottom of the lake. In the temperate zone, lake water mixes twice each year, bringing oxygen to the profundal zone and nutrients to the limnetic zone. These periodic turnovers will be discussed later in the chapter.

Lakes are often classified according to their production of organic matter. **Oligotrophic** lakes are deep and nutrient-poor, and the phytoplankton in the limnetic zone are not very productive (see Figure 46.14). The water of these lakes is clear, and, because detritus from the limnetic zone is limited, the deep waters contain lots of oxygen all year. **Eutrophic** lakes, in contrast, are shallower, and the nutrient content of their water is high. As a result, the phytoplankton are very productive, the waters are murkier than those of an oligotrophic lake, and oxygen supplies may be depleted in the profundal zone in summer. Over long periods of time, oligotrophic lakes may develop into eutrophic lakes as runoff brings in large quantities of mineral nutrients and sediments. Unfortunately, human activities often speed this natural process dramatically. Runoff from fertilized lawns and agricultural fields and dumping of municipal wastes enrich the lakes with excessive amounts of nitrogen and phosphorus, nutrients that normally limit the growth of phytoplankton and plants. The result of such pollution is often a population explosion of algae, the production of much detritus, and the eventual depletion of oxygen supplies. Such "cultural eutrophication" makes the water unusable and degrades the lake's aesthetic value (see Chapter 49).

Streams and Rivers

Streams and **rivers** are bodies of waters moving continuously in one direction (Figure 46.15). At the head-

Figure 46.15
Rivers. Large rivers form when streams and smaller rivers flow into a common channel. This photograph illustrates the confluence of the Rio Negro and the Amazon, the largest river in the world, in Brazil. Where the two rivers merge, currents from the Rio Negro keep its relatively clear, dark waters separate from the muddy waters of the Amazon. The waters mix thoroughly farther downstream, creating the dark, silty waters of the lower Amazon.

waters of a stream (perhaps a spring or snowmelt), the water is cold and clear, and it carries little sediment and a few mineral nutrients. The channel is usually narrow, with a swift current passing over a rocky substrate. Farther downstream, where numerous tributaries may have joined to form a river, the water is more turbid, carrying substantially more sediment (from the erosion of soil) and nutrients. The channel near the mouth of a river is relatively wide, and the substrate is generally silty from the deposition of sediments over long periods of time.

Many factors influence the flow, the nutrient and oxygen content, and the turbidity of streams and rivers. Where shallow water flows rapidly over a rough bottom, surface **riffles** reflect the turbulent flow; where deep water flows slowly over a smooth bottom, **pools** of water are common; and where deep water flows rapidly over a flat bottom, smooth **runs** of water are apparent. Nutrient content is largely determined by the terrain and vegetation through which streams and rivers flow. Fallen leaves from dense, overhanging vegetation can add substantial amounts of organic matter, and the erosion of rocks in the underlying streambed can increase the concentration of inorganic nutrients in the flowing water. The turbulent flow of many streams constantly oxygenates the water, whereas the murky, warm waters of large rivers may contain relatively little oxygen. The temperature of the water varies with altitude and latitude, and the amount of water flowing in a stream or river varies seasonally with rainfall patterns and snowmelt.

Biological communities in streams and rivers are substantially different from those in ponds and lakes. Many fast-flowing streams and rivers do not support large stationary plankton communities because these small organisms are washed away by the flow of water. Instead, photosynthesis by attached algae and rooted plants supports the food chains. However, where dense vegetation on the banks of a narrow stream blocks the sunlight needed for photosynthesis, organic material carried into the stream by runoff provides the most important input of food for consumers. The high concentration of silt near the mouths of large rivers increases the turbidity and can also block the passage of light, making photosynthesis difficult, if not impossible.

Because of the dramatic variations in the physical environment described above, the composition of animal communities varies significantly from the headwaters of a stream to the mouth of the river. Upstream, fish such as trout may be present where their requirements for cool temperatures, high oxygen, and clear water are met. In the warmer, murkier waters further downstream, catfish and carp may be abundant. Benthic (bottom-dwelling) communities also change, and many insect species are restricted to the relatively short stretches of a stream or river that provide their specific requirements. Some large rivers are also inhabited by a variety of turtles, snakes, and, in tropical regions, crocodilians and porpoises.

Not surprisingly, stream- and river-dwelling animals exhibit evolutionary adaptations that enable them to resist being carried away by the relentless flow of water. Small animals are typically dorso-ventrally flattened and are able to attach to rocks temporarily. Many insect species live on the underside of rocks or on their downstream side, thereby exploiting a small habitat that is relatively free of turbulent flow. Other species are restricted to quiet pools where the flow of water is slower than in riffles or runs.

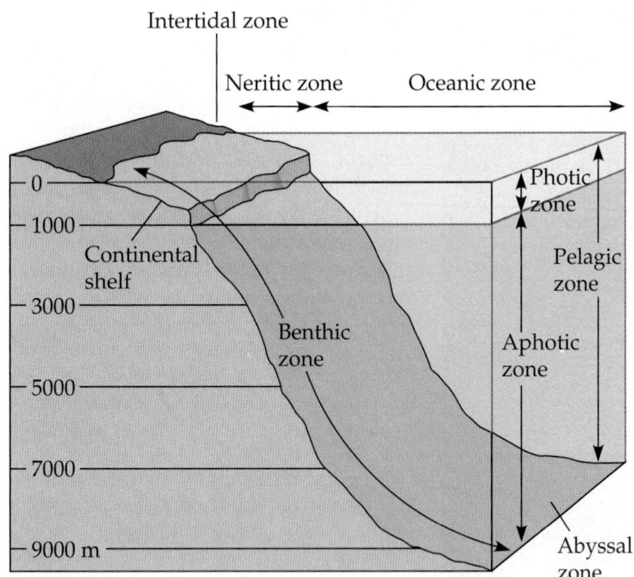

Figure 46.16
Oceanic zones. The marine environment can be classified on the basis of three physical criteria: light penetration (photic and aphotic zones), distance from shore and water depth (intertidal, neritic, and oceanic zones), and open water or bottom (pelagic and benthic zones). The abyssal zone is the benthic region in the very deepest oceans. Ecologists often use two designations, such as the oceanic pelagic zone, to identify the location of a biome.

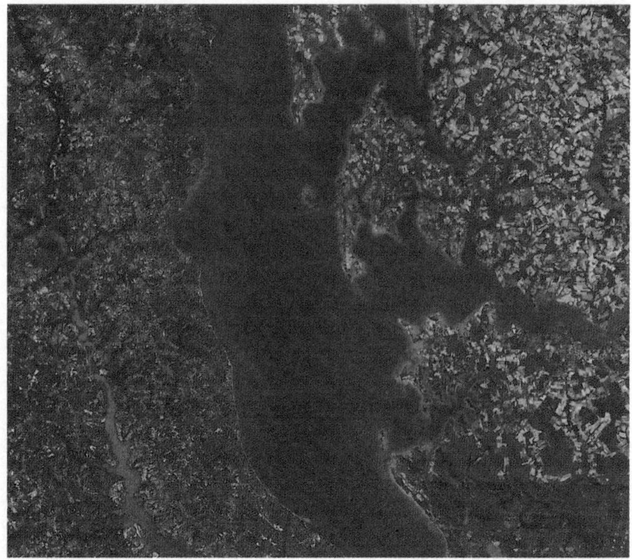

Figure 46.17
Estuaries. This aerial view of Chesapeake Bay estuaries shows the intimate association of river mouths and the marine environment into which they carry water. Unfortunately, the land surrounding Chesapeake Bay is heavily populated and industrialized, and pollution that enters the bay through four major rivers has made it unsuitable for many plant and animal species. What was once a bountiful natural source of seafood and other resources has been degraded and rendered less productive by thoughtless human activity.

MARINE BIOMES

Marine biomes are those found in the oceans, which cover nearly three-fourths of Earth's surface. Evaporation of seawater provides most of the planet's rainfall, and ocean temperatures have a major effect on world climate and wind patterns. In addition, marine algae supply a substantial portion of the world's oxygen and consume huge amounts of atmospheric carbon dioxide. Marine environments are characterized by an average salt concentration of 3%, although salinity varies greatly over space and time.

Like those in freshwater lakes, marine communities are distributed according to the depth at which they occur and their distance from shore (Figure 46.16). There is a photic zone where phytoplankton, zooplankton, and many fish species occur. Below is the aphotic zone. Because water absorbs light so well and the ocean is so deep, most of the ocean is virtually devoid of light, except for tiny amounts produced by a few luminescent fishes and invertebrates. The shallow zone where land meets water is called the **intertidal zone.** Beyond the intertidal zone is the **neritic zone,** the shallow regions over the continental shelves. Past the continental shelf is the **oceanic zone,** reaching very great depths. Open water of any depth is the **pelagic zone,** at the bottom of which is the seafloor, or **benthic zone.**

Humans harvest the ocean heavily and use it as a dump for waste, thinking that its vastness makes it essentially invulnerable. We are now seeing the effects of this short-sighted view, however, as seafood becomes scarcer; whales, dolphins, and other marine mammals are in danger of extinction; and many coastal areas are polluted. In this section, we examine five marine biomes in their relatively *undisturbed* state.

Estuaries

The area where a freshwater stream or river merges with the ocean is called an **estuary;** it is often bordered by extensive intertidal mudflats or saltmarshes (Figure 46.17). Salinity varies spatially within estuaries, from nearly that of fresh water to that of the ocean; it also varies on a daily cycle with the rise and fall of the tides. Nutrients from the river enrich estuarine waters, making estuaries one of the most biologically productive environments on Earth.

Saltmarsh grasses, algae, and phytoplankton are the major producers in estuaries. This environment also supports a variety of worms, oysters, crabs, and many of the fish species that humans consume. Many marine invertebrates and fish use estuaries as a breeding ground or migrate through them to freshwater habitats upstream. Estuaries are also crucial feeding areas for many semiaquatic vertebrates, particularly waterfowl.

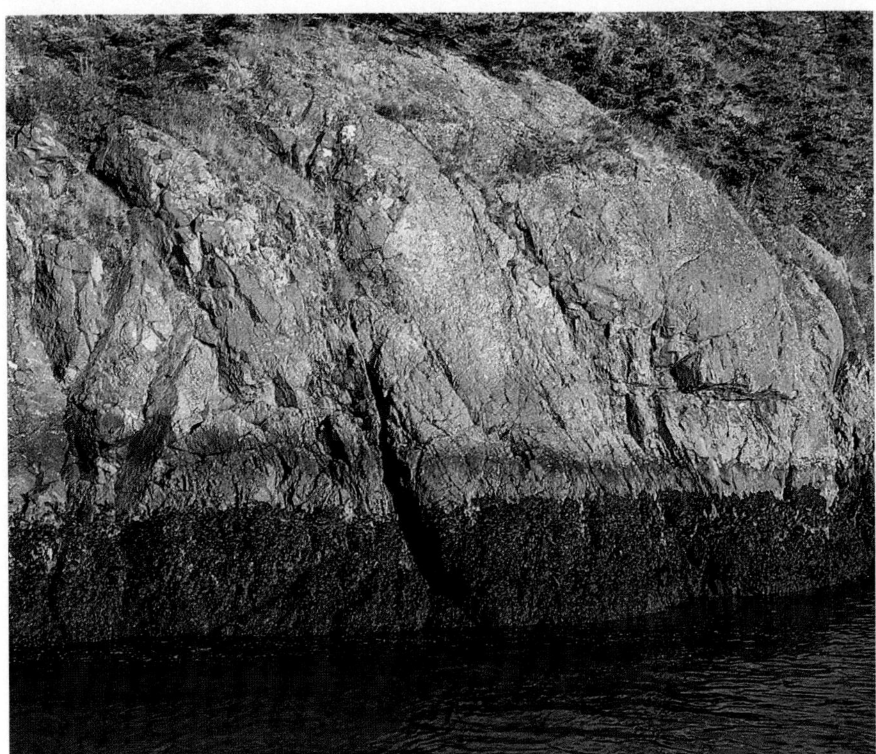

Figure 46.18
The intertidal zone. This photo of the rocky intertidal zones at Washington State's Lopez Island was taken at very low tide to illustrate the vertical zonation of algae and animals. The density of organisms in each of the three major zones is roughly proportional to the percentage of time that the zone is submerged. Organisms in the highest zone are frequently exposed to air and sun and have numerous adaptations that prevent dehydration and overheating.

Although estuaries support a wide variety of extremely valuable commercial species, areas around estuaries are also prime locations for commercial and residential developments. Unfortunately, estuaries are at the receiving end for pollutants dumped upstream. Very little undisturbed estuary habitat remains, and a large percentage has been totally eliminated by landfill and development. Many states have now, rather belatedly, taken steps to preserve remaining estuaries and other wetland habitats.

The Intertidal Zones

The intertidal zone, where the land meets the sea, is alternately submerged and exposed by the daily cycle of tides (Figure 46.18). Intertidal communities are therefore subject to huge daily variations in the availability of seawater (and the nutrients it carries) and in temperature. Perhaps most significant of all, intertidal organisms are subject to the mechanical forces of wave action, which can dislodge them from the habitat.

The rocky intertidal zone is vertically stratified. Most of the organisms have structural adaptations that allow them to attach to the hard substrate in this physically tumultuous environment. The uppermost zone, which is submerged only during the highest tides, contains relatively few species of algae, and grazing mollusks and suspension-feeding barnacles that are sometimes eaten by crabs and shorebirds. The middle zone is generally submerged at high tide and exposed at low tide. It is inhabited by a diverse array of algae, sponges, sea anemones, bryophytes, suspension-feeding barnacles and mussels, herbivorous and predatory snails, crabs, sea urchins, sea stars, and small fishes. Tidepool organisms in this zone may experience dramatic increases in salinity as evaporation decreases the volume of water in the pools during low tides. The bottom of the intertidal zone is exposed only during the lowest tides. The low intertidal zone and the neritic subtidal zone just below it house an extraordinary diversity of invertebrates and fish species that live within the dense cover of abundant and productive seaweeds.

On sandy substrates (beaches) or mudflats, the intertidal zone is not as clearly stratified. Wave action constantly moves the particles of mud and sand, and few large algae or plants occupy these habitats. Many animals, such as suspension-feeding worms and clams and predatory crustaceans, bury themselves in sand or mud, feeding when the tides bring sources of food. Other surface-dwelling organisms, such as crabs and shorebirds, are scavengers or predators on these organisms.

Coral Reefs

In warm tropical waters in the neritic zone, **coral reefs** constitute a conspicuous and distinctive biome. Currents and waves constantly renew nutrient supplies to the reefs, and sunlight penetrates to the ocean floor, allowing photosynthesis.

Figure 46.19
Coral reefs. This coral reef in Fiji illustrates the diversity of algae and animals that populate these productive tropical biomes. Free-living algae and those symbiotic with the coral animals photosynthesize during the day, but the coral animals themselves extend their polyps and feed on plankton at night. The activities of other invertebrates and fishes also follow a daily cycle, and different components of the fauna are active by day and by night.

Coral reefs are dominated by the structure of the coral, a diverse group of cnidarians that secrete a hard external skeleton made of calcium carbonate. These skeletons vary in shape, forming a substrate upon which other corals and algae grow (Figure 46.19). Although the coral animals themselves feed on microscopic organisms and particles of organic debris, they are dependent on the photosynthesis of symbiotic dinoflagellates. An immense variety of microorganisms, invertebrates, and fishes live among the coral and algae, making the reefs one of the most diverse and productive biomes on Earth. Prominent herbivores include snails, sea urchins, and fishes, and these are in turn consumed by predatory octopus, sea stars, and carnivorous fishes.

Some coral reefs cover enormous expanses of shallow ocean, but this delicate biome is easily degraded by pollution and development, as well as by souvenir hunters who gather the coral skeletons. Corals are also subject to damage from native and introduced predators such as the "crown of thorns" sea star, which has undergone a population explosion in many regions. Reef communities are very old, growing very slowly, and they may not be able to withstand additional human encroachment.

The Oceanic Pelagic Biome

Most of the ocean's water lies far from shore in the **oceanic pelagic biome,** constantly mixed by the ever-circulating ocean currents. Nutrient concentrations are typically quite low, though the waters are periodically enriched by upwellings of the ocean that carry mineral nutrients from the bottom waters to the surface. Although the temperature varies with latitude and depth, pelagic waters are generally cold.

Photosynthetic plankton grow and reproduce rapidly in the photic region of the oceanic biome, representing only the top 100 m of the open ocean. Their activity accounts for nearly half the photosynthetic activity on Earth. Zooplankton, including protozoans, worms, copepods, shrimplike krill, jellyfishes, and the small larvae of invertebrates and fishes also ride the currents, grazing on the phytoplankton. Most plankton exhibit morphological structures like bubble-trapping spines, lipid droplets, gelatinous capsules, and air bladders, which help them stay afloat within the photic zone.

The oceanic pelagic biome also includes free-swimming animals, called nekton, which can move against the currents to locate food. Large squids, fishes, sea turtles, and marine mammals feed on either plankton or each other. Although many of these animals feed in the photic region of the pelagic zone, others live at great depth where fish may have enlarged eyes, enabling them to see in the very dim light, or luminescent organs that attract mates and prey. Many pelagic birds, such as petrels, terns, albatrosses, and boobies, catch fishes in the surface waters. A number of marine animals are migratory, following seasonally available food sources or moving between summer breeding grounds and their winter feeding range.

Benthos

The ocean bottom below both the neritic and the pelagic zones is the benthic zone, occupied by communities of organisms collectively called **benthos.** Like the profundal zone of lakes, many nutrients reach the seafloor by "raining down" in the form of detritus from

Figure 46.20
Benthos: a deep-sea vent community. Benthic faunas occupy the ocean bottom from the intertidal zone to the abyssal zone. The species composition of benthos varies dramatically with water depth. Pictured here is a vent community, first discovered at a depth of 2500 m in the late 1970s. These communities are found at spreading centers on the seafloor, where hot magma superheats the water (see Chapter 23). About a dozen species of bacteria identified near the vents are chemoautotrophic producers that obtain energy by oxidizing H_2S formed by reaction of the hot water with dissolved sulfate (SO_4^{-2}). Among the animals in these communities are giant tube-dwelling worms, some more than 1 m long. They are apparently nourished by chemosynthetic bacteria that live as symbionts within the worms. Many other invertebrates, including arthropods and echinoderms, are also abundant around the vents.

Figure 46.21
The patchiness of the environment. The intersection of forest, lake, and river biomes illustrates local patchiness of the biosphere. If we were to take successively closer views of this area, we would find even more environmental variation on a smaller and smaller scale. The importance of environmental patchiness varies with the relative sizes of the patches and the organisms that inhabit them.

the waters above. Although the benthic zone in shallow, near-coastal waters may receive substantial sunlight, light and temperature decline dramatically with depth. The bottom itself is composed of sand or very fine sediments ("ooze") made up of silt and, in the deep sea, the shells of dead microscopic organisms.

Neritic benthic communities are extremely productive, consisting of bacteria, fungi, seaweeds and filamentous algae, sponges, sea anemones, worms, clams, crustaceans, sea stars, sea urchins, and fishes. Species composition of these communities varies with distance from the shore, water depth, and composition of the bottom. Many organisms live buried in soft substrates and are not apparent to the casual observer.

Deep benthic communities live in the **abyssal zone,** where continuous cold (about 3°C), extremely high water pressure, the near absence of light, and low nutrient concentrations are typical. However, oxygen is usually present in abyssal waters, and a fairly diverse community of invertebrates and fishes occupies this region. Scientists have recently discovered a unique assemblage of organisms associated with deep-sea hydrothermal vents of volcanic origin in midocean ridges (Figure 46.20). In this dark, hot, oxygen-deficient environment, the primary producers are not photosynthesizing organisms but bacteria that are chemoautotrophs (see Chapter 25). These bacteria are consumed by giant polychaete worms, arthropods, echinoderms, and fishes.

THE ENVIRONMENTAL DIVERSITY OF THE BIOSPHERE

In our survey of the biosphere, we identified differences in the physical environments that characterize various terrestrial, freshwater, and marine biomes. The biosphere is, indeed, a mosaic of habitats differing in such abiotic factors as temperature, rainfall, and light, which influence the distribution of such biotic factors as type of vegetation. The map in Figure 46.3 emphasizes the patchiness of the biosphere on a global scale, but we can also see more local patchiness, such as sunny meadows and lakes surrounded by forest (Figure 46.21). In this section, we first examine some important abiotic factors in more detail and analyze how their interactions influence the distribution of biomes worldwide. We then consider global, regional, and seasonal variations in the physical environment. Throughout this discussion it is important to remember that the physical environment varies in both space and time. Although two regions of the Earth may experience different conditions at any given moment, daily and annual fluctuations of abiotic factors sometimes blur or accentuate the distinctions between them.

Important Abiotic Factors

Temperature Environmental temperature is an important factor in the distribution of organisms because of its effect on biological processes and the inability of most organisms to regulate body temperature precisely. Cells may rupture if the water they contain freezes at temperatures below 0°C, and the proteins of most organisms denature at temperatures above 45°C. In addition, few organisms can maintain a sufficiently active metabolism at very low or very high temperatures. Within this range, however, most biochemical reactions and physiological processes occur more rapidly at higher temperature. Extraordinary adaptations enable some organisms to live outside this temperature range. The actual internal temperature of an organism is affected by heat exchange with its environment (see Chapter 40), and most organisms cannot maintain body temperatures more than a few degrees above or below the ambient temperature. As endotherms, mammals and birds are the major exceptions, but even endotherms function best within certain environmental temperature ranges, which vary with the species.

Water Water, as you know, is essential to life, but its availability varies dramatically among habitats. Freshwater and marine organisms live submerged in an aquatic environment, but they face problems of water balance if their intracellular osmolarity does not match that of the surrounding water. Organisms in terrestrial environments encounter a nearly constant threat of desiccation.

Sunlight Sunlight provides the energy that drives nearly all ecosystems, although only plants and other photosynthetic organisms use this energy source directly. Light intensity is not the most important factor limiting plant growth in many terrestrial environments, but shading by a forest canopy makes competition for light in the understory intense. In aquatic environments, the intensity and quality of light limit the distribution of photosynthetic organisms. Every meter of water selectively absorbs about 45% of the red light and about 2% of the blue light that pass through it. As a result, most photosynthesis in aquatic environments occurs relatively near the surface. However, the photosynthetic organisms themselves absorb some of the light that penetrates, further reducing light levels in the waters below.

Light is also important to the physiology, development, and behavior of the many plants and animals that are sensitive to photoperiod, the relative lengths of daytime and nighttime. Photoperiod is a more reliable indicator than temperature for cueing seasonal events, such as flowering or migration (see Chapters 35 and 50).

Wind Wind amplifies the effects of environmental temperature on organisms by increasing heat loss due to evaporation and convection (the wind-chill factor). It also contributes to water loss in organisms by increasing the rate of evaporation in animals and transpiration in plants. In addition, wind can have a substantial effect on the morphology of plants by inhibiting the growth of limbs on the windward side of trees; limbs on the leeward side grow normally, resulting in a "flagged" appearance (see Figure 31.3).

Rocks and Soil The physical structure, pH, and mineral composition of rocks and soil limit the distribution of plants and the animals that feed upon them, thus contributing to the patchiness we see in terrestrial biomes. In streams and rivers, the composition of the substrate can affect water chemistry, which in turn influences the resident plants and animals. And the structure of the substrate in the intertidal and benthic zones determines the types of organisms that can attach or burrow in those habitats.

Periodic Disturbances Catastrophic disturbances such as fires, hurricanes, tornadoes, and volcanic eruptions can devastate biological communities. After the disturbance, the area is recolonized by organisms or repopulated by survivors, but the structure of the community undergoes a succession of changes during this rebound (discussed in detail in Chapter 48). Some disturbances, such as volcanic eruptions, are so infrequent and highly unpredictable over space and time that organisms have not acquired evolutionary adaptations to them. Fire, on the other hand, although unpredictable over the short term, recurs frequently in some communities, and many plants have adapted to this periodic disturbance, as discussed earlier in the section on chaparral. And, as Ariel Lugo pointed out in the interview preceding this chapter, some tropical forests actually benefit from periodic hurricanes.

Climate and the Distribution of Biomes

The important abiotic factors just described exert a direct effect on the biology of organisms. The first four factors—temperature, water, light, and wind—are the major components of **climate,** the prevailing weather

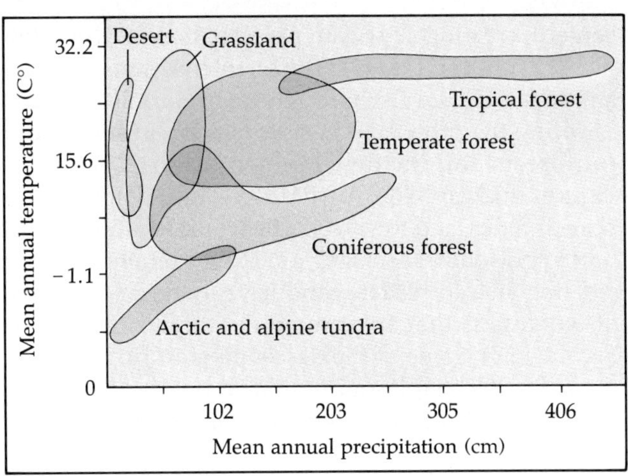

Figure 46.22
A climograph for some major North American biomes. The areas plotted here encompass the annual mean temperatures and precipitation occurring in some major North American biomes. The climograph provides only circumstantial evidence, however, that these factors are important in explaining the distribution of the biomes. The areas of overlap, for example, show that these variables alone are not sufficient to explain the observed distribution.

conditions at a locality. We can see the great impact of climate on the distributions of biomes by constructing a climograph, which is a plot of the temperatures and rainfall in a particular region, often in terms of annual means. For example, Figure 46.22 shows a climograph for some of the major North American biomes. Notice that the range of rainfall occurring in coniferous forest regions is similar to that of temperate forest areas, but that temperatures are cooler. Grasslands, however, are generally drier than either kind of forest, and deserts are drier still.

Annual means for temperature and rainfall are reasonably well correlated with the biomes we find in different regions. However, we must always be careful to distinguish a *correlation* between variables from *causation,* a cause-and-effect relationship. Although our climograph provides circumstantial evidence that temperature and rainfall are important to the distributions of biomes, it does not prove that these variables govern their geographic location. Only a detailed analysis of the water and temperature tolerances of the individual species within a biome could establish the controlling effects of these variables.

We can see in our climograph that factors other than mean temperature and precipitation must also play a role in biome patterns; notice that there are regions where biomes overlap. For example, there are areas in North America with a certain temperature and precipitation combination that support a temperate forest; but other areas with the same values for these variables support a coniferous forest; still others, a grassland. How do we explain this variation? First, remember that the climograph is based on annual *means.* Often it is not only the mean climate that is important but also the pattern of climatic variation. For example, some areas may get regular precipitation throughout the year, whereas others with the same annual amount have distinct wet and dry seasons. A similar phenomenon may occur with respect to temperature. Other factors, such as an area's geology, may greatly affect mineral nutrient availability and soil structure, which in turn affect the kind of vegetation that will develop.

With these complex considerations in mind, let's take a closer look at global climate patterns as well as local and seasonal variations in the physical environment to understand the geographic distributions of biomes.

Global Climate Patterns

The Earth's global climate patterns are largely determined by the input of solar energy and the planet's movement in space. About half the solar energy that reaches the upper layers of the atmosphere is absorbed before it reaches Earth; certain wavelengths of light (including the ultraviolet wavelengths that are damaging to biological systems) are more readily absorbed by oxygen molecules and ozone than are others. Much of the energy that strikes Earth itself is absorbed by land and water (and organisms), although some is reflected back into the atmosphere. All nonbiological components of the biosphere—atmosphere, land, and water—are heated when they absorb solar radiation, and this process establishes the temperature variations, cycles of air movement, and evaporation of water that are responsible for dramatic latitudinal variations in overall climate.

Because solar radiation is most intense when the sun is directly overhead, the spherical shape of the Earth creates latitudinal variation in the intensity of sunlight (Figure 46.23). However, the Earth is also tilted on its axis by 23.5° relative to its plane of orbit around the sun, and this tilt creates seasonal variation in the intensity of solar radiation in both the Northern and the Southern Hemispheres (Figure 46.24). The angle of incoming solar radiation changes daily everywhere as Earth rotates around the sun, but because the tilt of the Earth is permanently fixed, only the **tropics** (those regions that lie between 23.5° north latitude and 23.5° south latitude) ever receive sunlight from directly overhead. As a result, the tropics experience the greatest annual input and the least seasonal variation in solar radiation of any region on Earth. Indeed, the tropics are characterized by relatively small seasonal variations in day length and mean temperature. The seasonality of light and temperature increases steadily toward the poles; polar regions have long, cold winters with periods of continual darkness and short summers with periods of continual light.

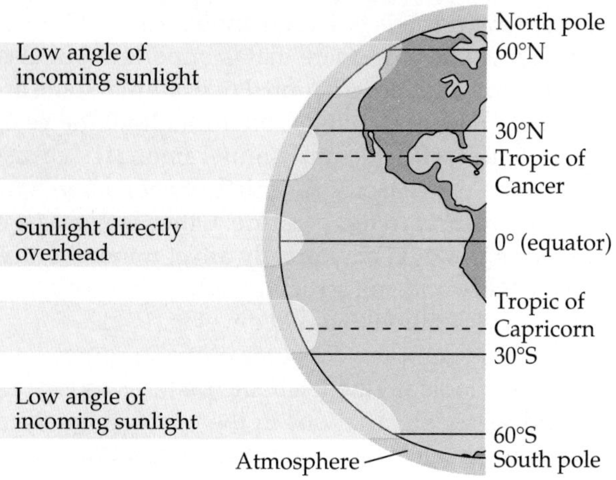

Figure 46.23
Solar radiation and latitude. Because sunlight strikes the equator perpendicularly, more heat and light are delivered there per unit of surface area than at higher northern and southern latitudes, where sunlight has a longer path through the atmosphere and strikes the curved surface of Earth at an oblique angle. This uneven distribution of solar radiation creates latitudinal differences in temperature and light intensity and establishes the vertical air currents illustrated in Figure 46.25b.

Intense solar radiation near the equator initiates a global circulation of air, creating precipitation and winds, which in turn influence the global distribution of some biomes (Figure 46.25). High temperatures in the tropics evaporate water from Earth's surface and cause warm, wet air masses to rise and flow toward the poles. The rising air masses release much of their water content, creating abundant precipitation in tropical regions. Thus, high temperatures, intense sunshine, and ample rainfall are all characteristic of a tropical climate, fostering the growth of lush vegetation in some tropical forests and the development of coral reefs (see Figures 46.3 and 46.13). The high-altitude air masses, now dry, descend toward Earth at latitudes around 30° north and south, absorbing moisture from the land and creating an arid climate conducive to the development of the deserts that are common at these latitudes (see Figure 46.3). Some of the descending air flows toward the poles at low altitude, establishing a midlatitude circulation cell that deposits abundant precipitation (though less than in the tropics), where the air masses again rise and release moisture in the vicinity of 60° latitude. Broad expanses of coniferous forest (taiga) dominate the landscape at fairly wet, but generally cool, latitudes. A third circulation cell carries some of the cold and dry rising air to the poles where it descends and flows back toward the equator, absorbing moisture and creating the comparatively rainless and bitterly cold climates of the arctic and antarctic. Although the arctic tundra receives very little annual rainfall, water cannot penetrate the underlying permafrost and accumulates in pools on the shallow topsoil during the short summer.

Local and Seasonal Effects on Climate

Proximity to bodies of water and topographic features create a climatic patchiness on a regional scale, and

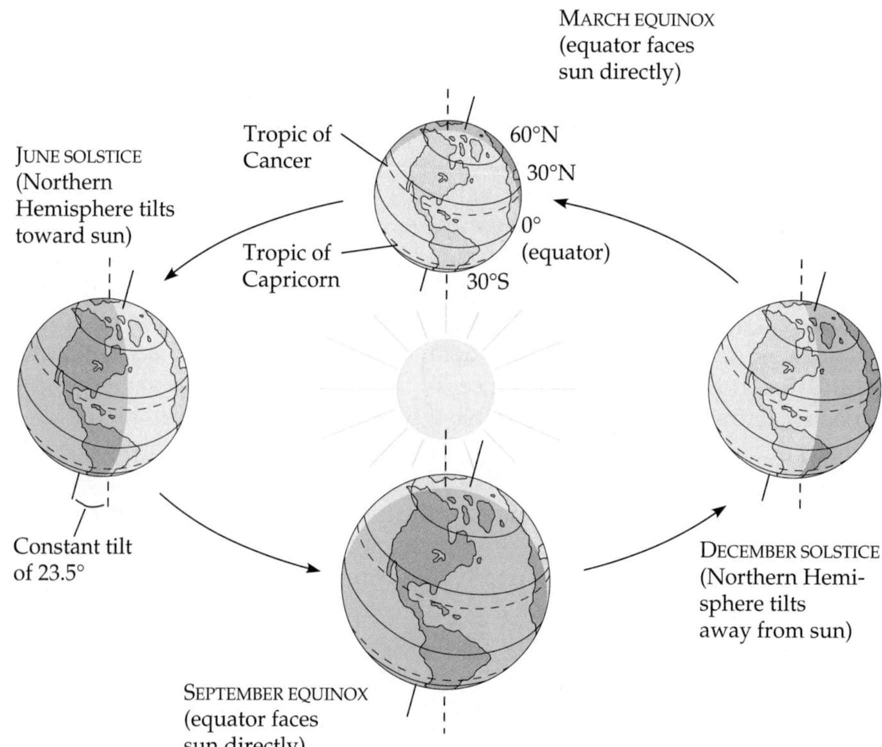

Figure 46.24
The cause of the seasons. The permanent tilt of the Earth on its axis causes seasonal variation in temperature and light intensity as the planet makes its annual pilgrimage around the sun. The December solstice marks the beginning of winter in the Northern Hemisphere, when the north pole is maximally tilted *away from* the sun. Day length is reduced, and solar radiation arrives at its most oblique angle, producing lower temperatures during short winter days. In contrast, the December solstice marks the beginning of summer in the Southern Hemisphere, when the south pole is maximally tilted *toward* the sun. Day length increases, and the sun is more nearly overhead, increasing temperatures during the long days of summer. Seasons in the two hemispheres are reversed when the north pole is tilted toward the sun and the south pole away from the sun on the June solstice. During the March and September equinoxes, neither pole is tilted toward the sun, and all regions on Earth experience 12 hours of daylight and 12 hours of darkness.

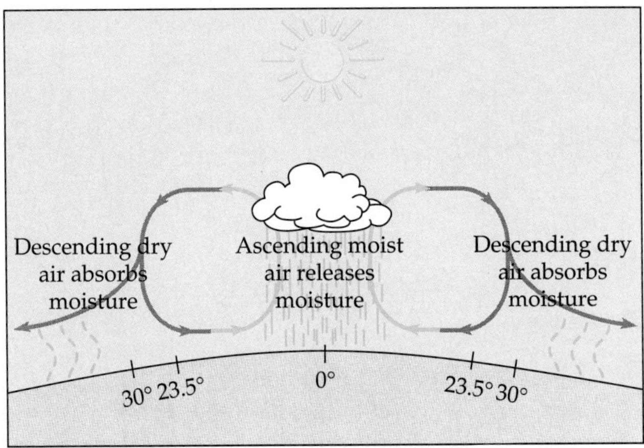

(a) Air circulation and precipitation at the equator

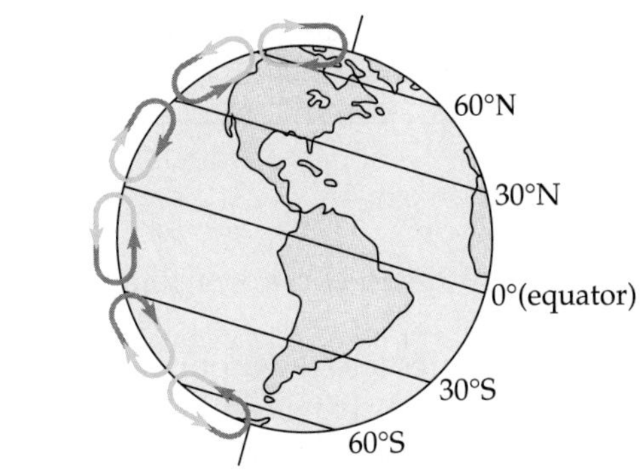

(b) Global air circulation

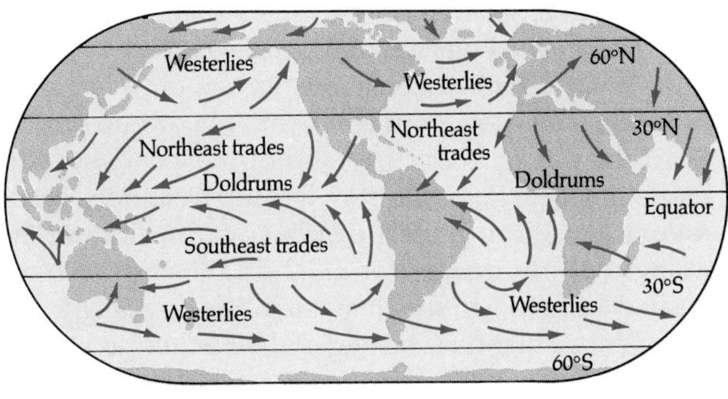

(c) Global wind patterns

Figure 46.25
Global air circulation, precipitation, and winds. (**a**) Air masses in the lower atmosphere are warmed by solar radiation and by heat radiating from Earth. Air expands, decreases in density, and rises when it is warmed. Thus, warm air at the equator rises, creating an area of light, shifting winds known as the doldrums. The warm air masses expand as they rise, and, because this expansion distributes their internal heat energy over a larger volume, they also cool as they move upward through the atmosphere. Cool air holds less water vapor than warm air, and the rising air masses drop large amounts of rain in the tropics. The now dry air masses flow toward the two poles at high altitude, cooling further as they spread away from the equator. The density of the air masses increases as they cool, and they descend, absorbing water from the land and creating bands of arid climate around 30° latitude. (**b**) The movement of heated air creates three major air circulation cells on either side of the equator. Within each circulation cell, rising air (blue) releases moisture as precipitation, and descending air (gray) absorbs moisture, creating arid conditions. The first circulation cells, described in part (**a**), are completed when some of the air that descends at 30° latitude flows back toward the equator at low altitude. The remainder of the air flows away from the equator, also at low altitude, initiating midlatitude circulation cells that acquire water and then release it as the air rises around 60° latitude. The third circulation cells carry cool, dry air to the poles where it descends and flows back toward the equator. Air from the different circulation cells merges where the cells meet, constantly mixing the gases in the lower levels of the atmosphere. (**c**) Air flowing in the lower levels of the circulation cells, close to Earth's surface, creates predictable global wind patterns. However, as Earth rotates on its axis, land near the equator moves faster than that at the poles, deflecting the winds from the vertical paths illustrated in part (**b**) and creating more easterly and westerly flows. Cooling trade winds blow from west to east in the tropics and subtropics. In contrast, the prevailing westerlies blow from west to east in the temperate zone.

smaller features of the landscape also contribute to local climatic variation. Although global climate patterns help explain the geographic distribution of some major biomes (tropical rain forest, desert, taiga, and tundra), regional and local variations in climate and soil influence the distributions of some less widely distributed communities and individual species.

Ocean currents influence climate along the coasts of continents by heating or cooling overlying air masses which may then pass across the land (Figure 46.26). Evaporation from the ocean is also greater than it is over land, and coastal regions are generally moister than inland areas at the same latitude. The rain forests of the Pacific Northwest and the large redwood groves to the south require the cool, misty climate produced by the cold California current that flows southward along the western United States. Similarly, the warm Gulf Stream tempers the climate on the west coast of the British Isles, making it warmer than the coast of New England, which is actually further south but is cooled by a branch of the Labrador Current.

As every vacationer knows, oceans and large inland bodies of water generally moderate the climate of nearby terrestrial environments on a daily cycle. During a warm summer day, when the land is hotter than a lake or the ocean, air over the land heats and rises, drawing a cool breeze from the water across the land. At night, by contrast, air over the warmer ocean

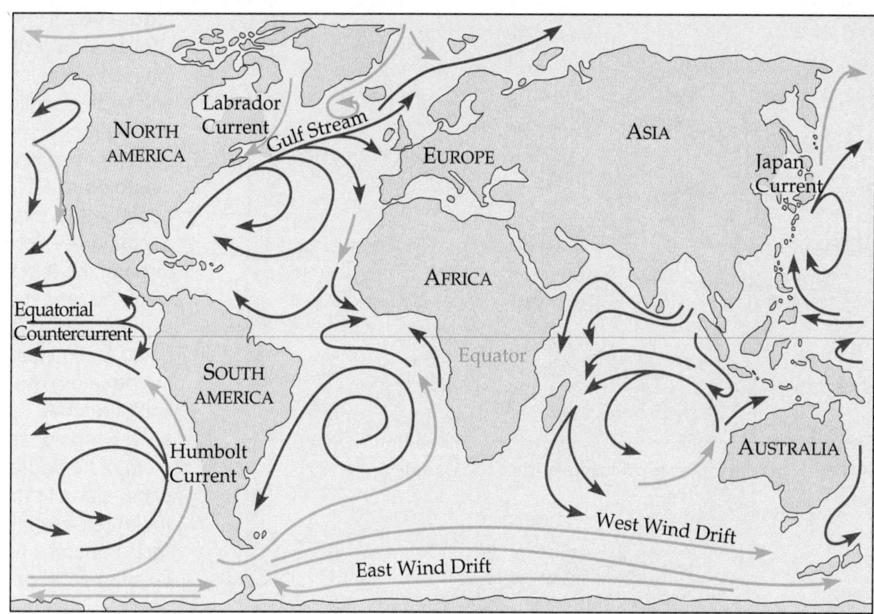

Figure 46.26
Ocean currents. Ocean currents are generated by Earth's rotation, prevailing winds, the differential heating of ocean waters, and the geographic positions of continents. In the open ocean, major currents flow clockwise in the Northern Hemisphere and counterclockwise in the Southern Hemisphere. Continents and other large land masses deflect east–west currents near the equator, diverting warm water (magenta) toward the poles along the east coasts of continents. Once cooled near the poles, cold currents (blue) flow back toward the equator along the west coasts of continents.

rises, establishing a circulation that draws cold air from the land out to sea, replacing it with warmer air from offshore. Proximity to water does not always moderate climate, however. Several regions (including the coast of central and southern California) have a "Mediterranean-type" climate; in summer, cool, dry ocean breezes are warmed when they contact the land, absorbing moisture and creating the hot, rainless summer characteristic of the chaparral biome.

Mountains also exert a significant effect on solar radiation, local temperature, and rainfall. South-facing slopes in the Northern Hemisphere receive more sunlight than nearby north-facing slopes and are therefore warmer and drier; in the mountains of western North America, spruce and other conifers occupy the north-facing slopes, whereas shrubby, drought-resistant vegetation inhabits slopes that face south. In addition, at any particular latitude, air temperature declines approximately 6°C with every 1000-m increase in elevation, paralleling the decline of temperature with latitude. In the north temperate zone, for example, a 1000-m increase in elevation produces a temperature change equivalent to that over an 880-km increase in latitude. This is one reason why mountain communities are similar to those at lower elevation further from the equator (Figure 46.27). When warm, moist air approaches a mountain, it rises and cools, releasing moisture on the windward side of the range. On the leeward side of the mountain, cooler, dry air descends, absorbing moisture and producing a rainshadow. Deserts commonly occur on the leeward sides of mountain ranges, a phenomenon evident in the Great Basin and Mojave Desert of western North America, the Gobi Desert of Asia, and in the small deserts that characterize the southwest corners of some Caribbean islands.

Seasonality generates local environmental variation in addition to the global changes in day length, solar radiation, and temperature described earlier. Because of the changing angle of the sun over the course of the year, the belts of wet and dry air on either side of the equator undergo slight seasonal shifts in latitude that produce marked wet and dry seasons around 20° latitude, where tropical deciduous forests grow. In addition, seasonal changes in wind patterns produce variations in ocean currents, sometimes causing the upwelling of cold, nutrient-rich water from deep ocean layers, thus nourishing organisms that live in the pelagic photic zone. Ponds and lakes are also extremely sensitive to seasonal temperature changes and experience a biannual mixing of their waters as a result of changing water-temperature profiles (Figure 46.28). This turnover, as it is called, brings oxygenated water from the surface to the bottom and nutrient-rich water from the bottom to the surface in both spring and autumn. These cyclic changes in the nonbiological properties of the lake are essential for the survival and growth of organisms at all levels within this biome.

Climate also varies on a very fine scale, called microclimate. For example, ecologists often refer to the microclimate on a forest floor or under a rock. Many features in the environment influence microclimates by casting shade, reducing evaporation from soil, and minimizing the effects of wind. Forests trees frequently moderate the microclimate below. Cleared areas generally experience greater temperature extremes than the forest interior, because of greater solar radiation and wind currents that are established by the rapid

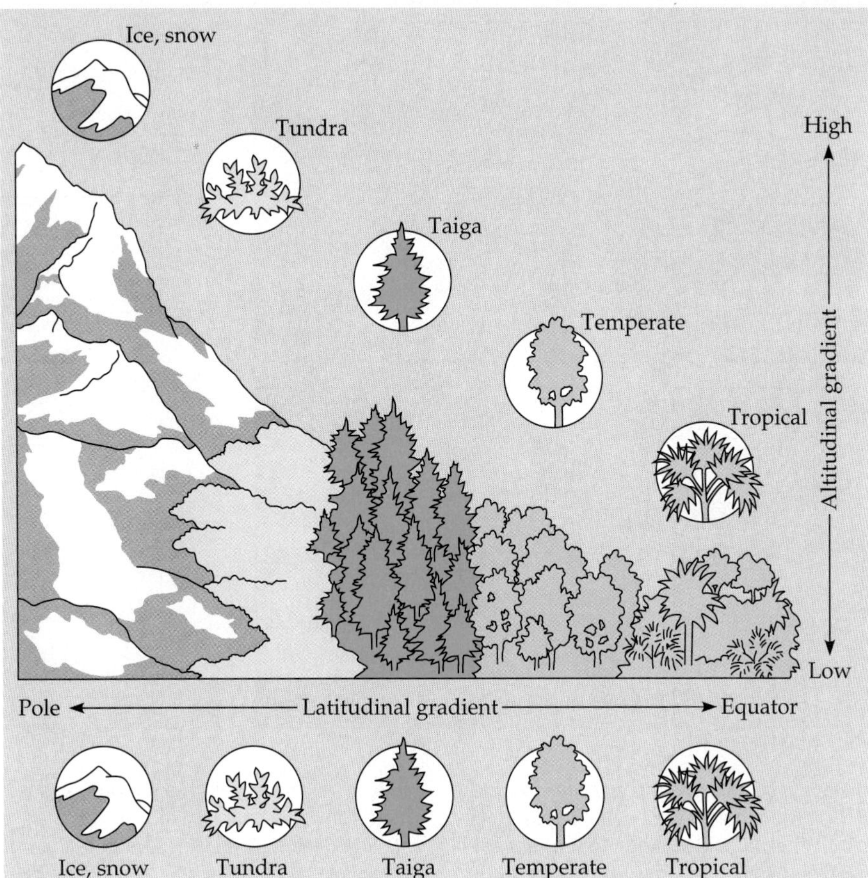

Figure 46.27
Effects of altitude and latitude. The altitudinal distribution of biomes at tropical latitudes mirrors the global latitudinal distribution of biomes. In other words, when moving along either an altitudinal (from sea level upward) or latitudinal (from the equator to a pole) gradient, you would generally find the same sequence of biomes, mainly because temperature variation is similar along both geographic gradients.

heating and cooling of open land; evaporation is generally greater in clearings as well. Low-lying ground is usually wetter than high ground and tends to be occupied by different species of trees within the same forest. If you have ever lifted a log or large stone in the woods, you are well aware that there are organisms (such as salamanders, worms, and some insects) that live in the shelter of this microenvironment, buffered from extremes of temperature and moisture. Every environment on Earth is similarly characterized by a mosaic of small-scale differences in the physical factors that influence the distributions of organisms.

RESPONSES OF ORGANISMS TO ENVIRONMENTAL VARIATION

Although we can explain the distributions of biomes with respect to global and regional climatic variation, it is important to remember that biomes and their ecological communities are composed of populations of individual organisms that are adapted to the physical environments in which they live. Natural selection has produced diverse adaptations to extremes of temperature, light, and other abiotic factors. Some organisms living in the arctic and antarctic tolerate air temperatures of –70°C, and other organisms survive desert temperatures higher than 45°C. Aquatic life can be found in water with salt concentration near zero (snowmelt) and in salt lakes several times more saline than seawater. However, no organisms can survive the full range of environmental conditions present on Earth, and the geographic distribution of a population or species is partly determined by its ability to tolerate a specific subset of environmental conditions. Various structures and physiological mechanisms, discussed in Unit Six (for plants) and Unit Seven (for animals), have evolved as adaptations to the constraints of specific environments. In this section, we will briefly survey the general types of adaptations in the context of variation in the physical environment.

Organisms actually interact with combinations of abiotic factors that are woven into integrated environments. The success of an organism at survival and reproduction reflects its overall tolerance to the entire set of environmental variables it confronts. Also, the ability to tolerate a particular factor may depend on another factor. For example, many aquatic ectothermic organisms can survive reduced oxygen at low temper-

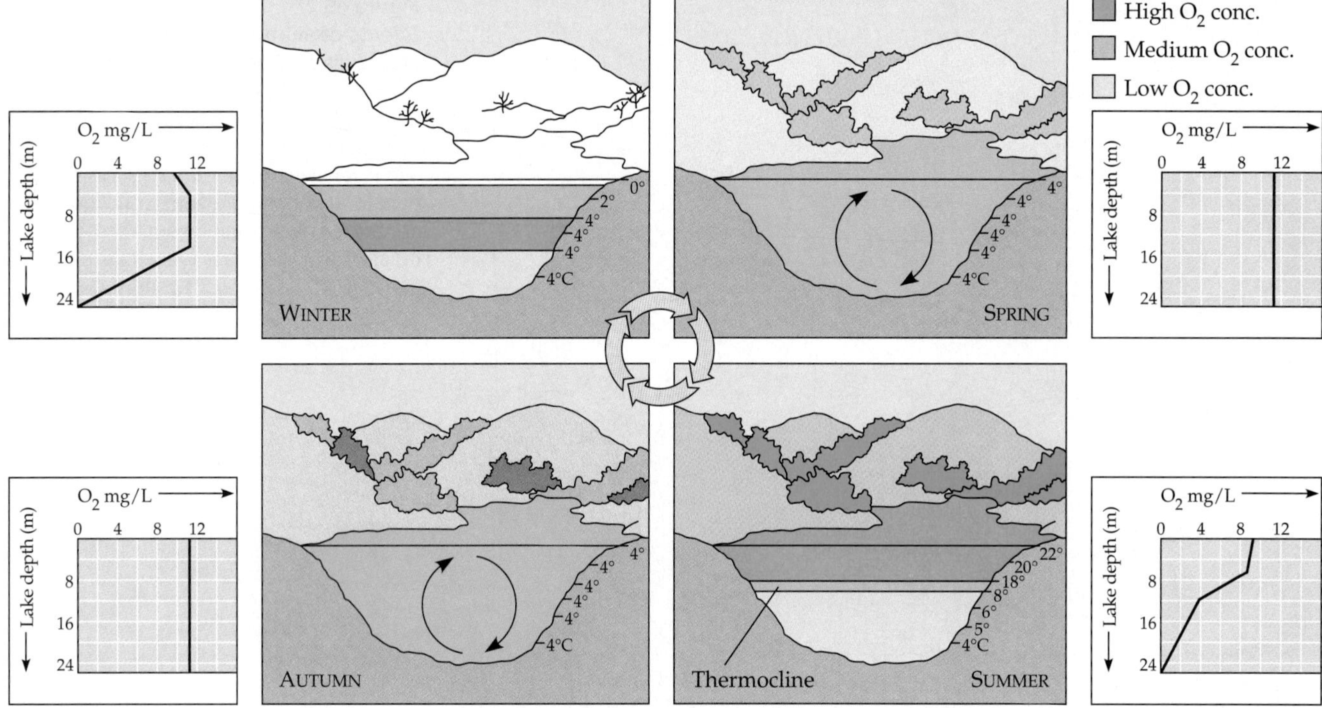

Figure 46.28
Lake turnover. The biannual mixing of lake waters occurs because water is most dense at 4°C, and water at that temperature sinks below water that is either warmer or colder. In winter, the coldest water in the lake (0°C) lies just below the surface ice; water is progressively warmer at deeper levels of the lake, typically 4° to 5°C at the bottom. In spring, as the sun melts ice and warms the surface water to 4°C, it sinks below the cooler layers immediately below, eliminating the thermal stratification established in winter. In the absence of thermal layering, spring winds mix the water to great depth, bringing oxygen to the bottom waters and nutrients to the photic zone at the surface. In summer, the lake regains a distinctive thermal profile with warm water at the surface, separated from cold bottom water by the thermocline described earlier. In autumn, as surface water cools rapidly, it sinks below the underlying layers, remixing the lake water until the surface begins to freeze and the winter temperature profile is reestablished.

atures, but not at high temperatures when their metabolic rates are also high. Coping with a set of environmental problems usually involves imperfect adaptations that represent evolutionary compromises (see Chapter 21). Panting or sweating, for instance, cools the body on a hot day but can also lead to a water deficit.

Homeostasis and the Principle of Allocation

Chapter 36 introduced the term *homeostasis* to describe the maintenance of a steady-state internal environment in the face of variations in the external environment. Many animals and plants can be described as **regulators** that use behavioral and physiological mechanisms to achieve homeostasis (Figure 46.29). Their homeostatic mechanisms buffer the effects of environmental fluctuations in temperature, moisture, light intensity, and a host of chemical factors in the environment. For example, in brine shrimp, which live at high but variable salt concentrations, the ability to maintain a stable internal salt concentration by osmoregulation has evolved (see Chapter 40). However, other organisms, particularly those that live in relatively stable environments, are often **conformers,** allowing conditions within their bodies to vary with external changes in these variables. Many marine invertebrates, for example, live in environments where the salinity is very stable. These organisms have no ability to osmoregulate, and if placed in water of varying salinity, they will lose or gain water to conform to the external environment.

Behavioral and physiological regulation requires the expenditure of energy, and in some environments, the cost of regulation may outweigh the benefits of homeostasis. For example, if temperature regulation requires a forest-dwelling lizard to travel large distances (and risk capture by a predator) to find an exposed sunny perch, it might survive longer and produce more offspring by simply allowing its body temperature to conform to that of the forest environment. Many species are conformers under certain environmental conditions but can regulate to some extent under others (Figure 46.29b). Conforming and regulating

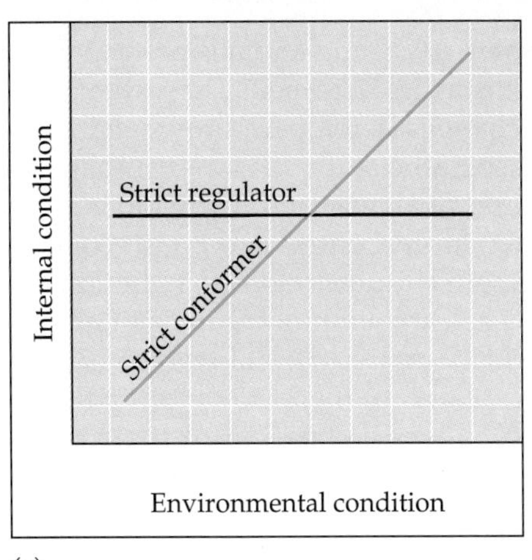

(a)

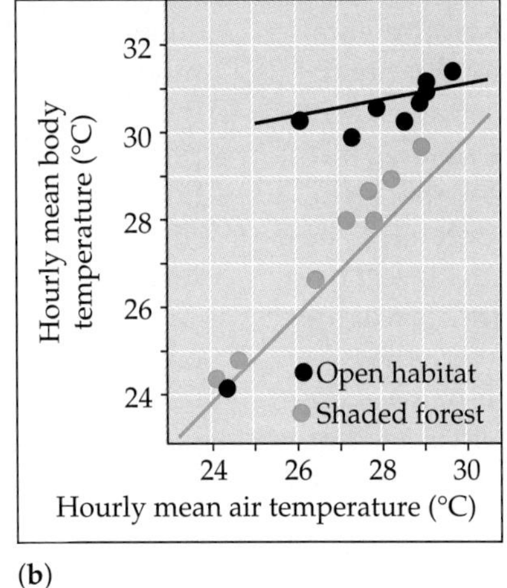

(b)

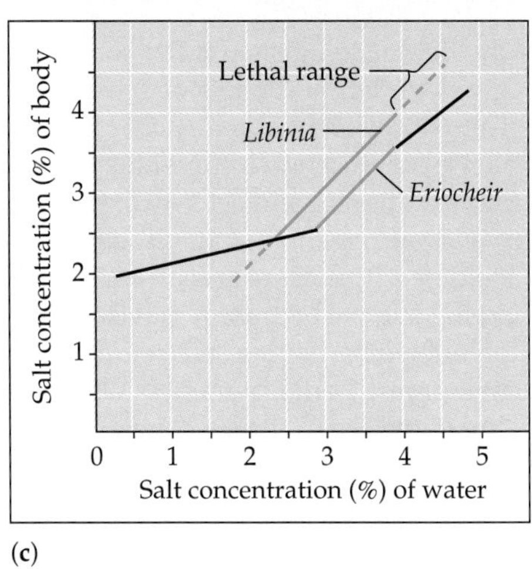

(c)

Figure 46.29
Regulators and conformers. (**a**) With respect to an environmental variable, organisms can be described as either regulators that maintain a nearly constant internal environment over a range of external conditions or conformers that allow their internal environment to vary. Although the two "strategies" are idealized in the illustration, most organisms are neither perfect regulators nor perfect conformers.
(**b**) Some species change their regulatory mode in different environments. *Anolis cristatellus*, a small lizard on Puerto Rico, behaviorally thermoregulates in open habitats where it can bask in the plentiful patches of sun (black dots), but it is a thermoconformer in shaded forests where basking sites are rarer (blue dots).
(**c**) Spider crabs *(Libinia)* are osmoconformers with little ability to physiologically regulate internal osmolarity in the narrow range of salinity where they live. If exposed to slightly higher or lower salinity in the laboratory, the crabs continue to conform and soon die. In contrast, Chinese mitten crabs *(Eriocheir)* are osmoconformers at some salinities but are able to osmoregulate at higher and lower salt concentrations. As a result, they are less restricted in their distribution and may migrate between freshwater and marine habitats.

represent extremes in a continuum of possible responses to an environmental change. Few organisms are perfect conformers or regulators, and only a detailed analysis of the relative costs and benefits of regulation in a specific environment can provide insight into the "strategies" that natural selection has produced.

One concept that organismal ecologists have found useful in assessing the responses of organisms to their complex environments is the **principle of allocation.** This principle holds that each organism has a limited amount of energy that can be allocated for obtaining nutrients, escaping from predators, coping with environmental fluctuations (maintenance), growth, and reproduction. Energy budgets are like checking accounts with a restricted balance and only a small credit line; long-term overdrafts are not allowed. And energy expended for homeostasis is therefore not available for other functions. For example, in grasshoppers, which are moderately active ectotherms, about 30% of assimilated energy remains after the animal's basic maintenance needs are met. This energy can be channeled into growth or reproduction. In contrast, for a very active endotherm, such as a weasel that uses most of its ingested energy to stay warm and active, only 2.5% of assimilated energy remains, and for a wren, only 0.5% remains. For the latter organisms, there are apparently significant evolutionary advantages to higher activity levels and endothermy to offset the disadvantage of high maintenance costs.

Different priorities in energy allocation are related to the distribution of organisms and their homeostatic mechanisms. Conformers that live in very stable environments might be able to channel more of their energy into growth and reproduction. However, the intolerance of such specialists to environmental change severely restricts their geographic distribution. In contrast, regulators that allocate a larger fraction of their energy to coping with environmental changes may grow and propagate less efficiently, but such organisms are able to survive and reproduce over a wider range of variable environments.

Environmental Grain

As already emphasized, environments vary over both time and space. The importance of environmental variation to a particular organism depends on the spatial scale of the variation in relation to the organism's size and movement patterns. For a small herbivorous insect living in a field of wildflowers, the differences among the plants may be profound because only some plants are suitable food sources. However, to a grazing mammal such as a horse, all the plants may be more or less the same, and the horse feeds indiscriminately within the field. Ecologists use the concept of **environmental grain** to define the use of spatial variation by organisms of different sizes. A coarse-grained environment is one in which environmental patches are so large (relative to the size and activity of the organism) that an individual organism can choose among patches. The field of wildflowers is coarse-grained for the insect described above because it may spend its entire larval life on one plant, having selected that plant over others. A fine-grained environment is one in which patches are small relative to the size and activities of an organism, and the organism may not even behave as though patches exist. For the horse, the field is a fine-grained environment in which there are no choices to be made. Of course, a large field in which succulent clover grows on one side and spiny thistles on the other represents a coarse-grained environment even to the horse, because it can select which side of the field it prefers.

Temporal variation in the environment can also be described as coarse-grained or fine-grained, depending on the periodicity of the variation in relation to the lifespan of the organism. A sudden cold snap might have a dramatic effect on the success of an insect but little influence on the activities of a long-lived mammal. Daily variations in environmental factors are usually fine-grained for most organisms, whereas seasonal and longer term shifts in climate are coarse-grained for even the largest organisms.

Organisms can respond to spatial and temporal variations in the environment with a variety of adaptations that differ in their activation times. Behavioral adaptations are almost instantaneous in their effects and easily reversed, whereas physiological adaptations may be implemented and changed over time scales ranging from seconds to weeks. Morphological adaptations may develop over the lifetimes of individual organisms or between generations. Adaptive genetic changes in populations are slower still, evolving over one or more generations.

Behavioral Responses

Behavioral response in the sense of muscular reaction to a stimulus is limited to animals. Behavioral mechanisms are so important to how animals interact with their environments that they will be discussed in detail in Chapter 50. Let's consider them briefly here from an environmental perspective.

The quickest response of many animals to an unfavorable change in the environment is to move to a new location. Such movement may be fairly localized. For example, lake trout avoid the heat of the upper zone of a lake during summer by moving to deeper water. Many desert animals escape intense heat by burrowing, and they maintain a reasonably constant body temperature while active by shuttling between sun and shade (see Chapter 40). Some animals are capable of migrating great distances in response to such environmental cues as changes in temperature or changes in photoperiod associated with seasonal transitions. Some ducks, geese, swallows, and many other migratory birds overwinter in Central and South America, returning to northern latitudes to breed in the summer.

Some animals are able to modify their immediate environment by cooperative social behavior. Honeybees, for instance, can cool the inside of their hive on hot days by the collective beating of their wings. During cold periods, they seal the hive, helping to retain the heat generated by their activity inside. Many small mammals huddle within burrows during cold weather, a behavioral mechanism that reduces heat loss by minimizing the total amount of surface the animals expose to the cold air.

Physiological Responses

As already noted, physiological responses to environmental change are generally slower than behavioral reactions. If you moved from Boston, which is essentially at sea level, to the mile-high city of Denver, one physiological response to the lower oxygen pressure in your new environment would be an increase in the number of your red blood cells, but this reaction would require several days to a few weeks. Some physiological responses are much faster, however. When you venture outside on a very cold day, blood vessels in your skin may constrict within seconds, a physiological response that reduces loss of body heat.

Regulation and homeostasis are the hallmarks of physiological adaptation. However, all organisms, whether regulators or conformers, function most efficiently under certain environmental conditions. We can study an organism's response to changing environmental conditions in the laboratory by varying a single abiotic factor, such as temperature, and measuring some aspect of the organism's performance. The resulting tolerance, or performance, curves are approximately bell-shaped, with peak performance at some optimal condition and the tails of the curve representing the limits of the organism's tolerance to the particular environmental variable. Figure 46.30 shows two such curves for the effect of temperature on the swimming speed of goldfish. Tolerance limits are important determinants of the geographic distribution of organisms, though biological interactions can prevent a species from occupying a habitat to which it is physiologically adapted (see Chapter 48).

Physiological responses to environmental variation can also include **acclimation,** a shift of the performance curve in the direction of the environmental change. Goldfish acclimated to cold water swim faster at lower temperatures than those acclimated to warm water

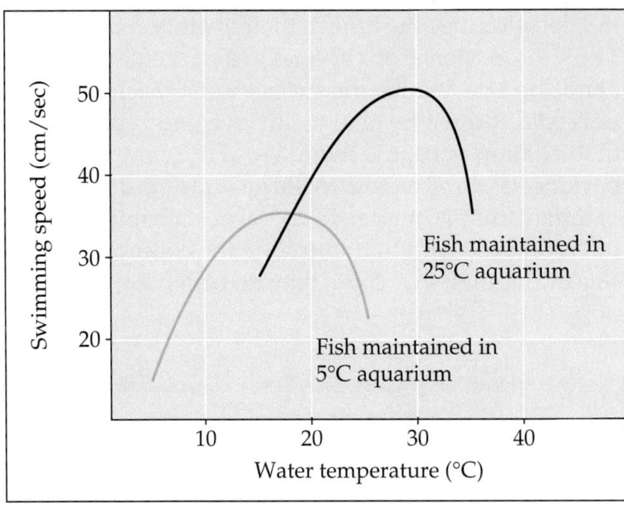

Figure 46.30
Tolerance curves and acclimation. Goldfish, like all ectothermic animals, have an optimal temperature for physiological function as well as a range of temperatures they can tolerate. In this example, swimming speed, which varies with temperature, is used as an index of the fish's general well-being. Goldfish normally live at temperatures between 25°and 30°C, but both the optimal temperature and the tolerance range can be shifted somewhat when the fish is gradually acclimated to a water temperature of 5°C. Acclimation is a slow process, and the fish would not survive a sudden, large change in water temperature. Acclimation also involves trade-offs. Notice that the cold-acclimated fish swims faster than the warm-acclimated fish at 15°C but much more slowly at 25°C.

(see Figure 46.30). Organisms can acclimate to other environmental factors as well. Acclimation is a gradual process, taking days or weeks, and the ability to acclimate is generally related to the range of environmental conditions experienced under natural conditions. Organisms that live in very hot climates, for example, usually do not acclimate to extreme cold.

Morphological Responses

Organisms may react to some change in the environment with developmental or growth responses that alter the form or internal anatomy of the body. Although these structural changes are often irreversible over the lifetime of an individual organism, some animals experience regular seasonal changes in morphology. Many mammals and birds, for example, grow a heavier coat of fur or feathers in winter; sometimes coat color changes seasonally as well, camouflaging the animal against winter snow and summer vegetation.

In general, plants are more morphologically plastic than animals; this response helps them compensate for their inability to move from one environmental patch to another (see Chapter 31). One remarkable example is the arrowleaf plant, which can grow on land, completely submerged in water, or rooted in water with its upper leaves emerging above the surface. Leaf morphology varies with the environment in which the leaves grow. Submerged leaves are flexible, bending with the currents, and, lacking a waxy cuticle, are able to absorb mineral nutrients from the surrounding water. Arrowleaf plants growing on land have more extensive root systems, and their leaves are more rigid and covered with a thick cuticle that reduces water loss.

Adaptation Over Evolutionary Time

The various behavioral, physiological, and morphological mechanisms we have examined are responses of individual organisms operating on an ecological time scale. However, it is important to remember that these responses occur within a framework of adaptations fashioned by natural selection acting over evolutionary time. For example, all plants are capable of changing the size of the stomata in their leaves (see Chapter 32), a physiological response that helps prevent desiccation under environmental conditions when transpiration would exceed delivery of water. In plants living in the desert, the ability to adjust the size of the stomatal openings in response to water stress is superimposed on many other anatomical and physiological adaptations that have accumulated over evolutionary time as these plants have evolved in their arid environments. For instance, some desert plants have their stomata in pits, protected from the hot, dry winds that accelerate transpiration. Also common among desert plants is the CAM pathway of photosynthesis (see Chapter 10), which enables the plants to keep their stomata closed during the daytime.

The distinction between short-term adjustments on the scale of ecological time and adaptation on the scale of evolutionary time begins to blur when we consider that the range of responses of an individual to changes in the environment is itself the product of evolutionary history. For example, when an endotherm such as a mammal uses physiological adjustments to maintain constant body temperature in the face of fluctuations in environmental temperature, it is employing mechanisms of homeostasis that are adaptations acquired by natural selection.

In adapting organisms to their localized environments, natural selection also places constraints on the distribution of populations. For instance, earthworms are skin breathers that obtain oxygen by diffusion across their moist body surface, a solution to the problem of gas exchange that restricts these animals to damp soils. If pine seeds are blown from the rim of the Grand Canyon to the canyon floor 2000 m below, where conditions are much hotter and drier, the seeds are unlikely to germinate and grow successfully in the new environment. Organisms locked by their adaptations into one type of environment may fail to survive if dispersed to some foreign environment, or they may face extinction if their local environment changes beyond their tolerance limits. On the other hand, the absence of a species in a particular place does not necessarily imply that the species could not survive in that location. Pines would not live even on the rim of the Grand Canyon if they had not managed to disperse to that location at some time in their evolutionary history. Thus, the existence of a species in a particular place depends on two factors: the species must reach that location, and it must be able to survive and reproduce in that location once it is there. We will evaluate the importance of these factors to the geographic distribution of organisms in Chapter 48. In the next chapter, we will focus on processes that influence the composition and size of populations where they do occur.

STUDY OUTLINE

The Scope and Development of Ecology (pp. 1053–1054)

1. Ecology is the scientific study of the distribution and abundance of organisms, as determined by interactions with the biotic and abiotic parts of their environment.
2. Ecology spans increasingly comprehensive levels of organization, from the individual organism through populations and communities to the ecosystem and biome.
3. Ecology is inextricably interwoven with evolution. The distribution and abundance of organisms depend not only on the immediate environment, but also on their long-term evolutionary history.

Terrestrial Biomes (pp. 1054–1062)

1. Biomes are major assemblages of generally similar communities.
2. Tropical forests are found near the equator, where photoperiod and temperature are nearly constant but where rainfall varies with location and season. The tropical rain forest is the most species-rich terrestrial biome.
3. Savanna is a tropical grassland with scattered trees. Precipitation varies greatly between wet and dry seasons, and fires are frequent. Large grazing herbivores and small burrowing animals are common.
4. Deserts are arid biomes, with extremes in temperature and very low precipitation.
5. Chaparral consists of fire-adapted scrublands of dense, spiny evergreen shrubs, usually found along coastlines characterized by mild, rainy winters and long, hot, dry summers.
6. Temperate grasslands occur in relatively cool climates with rich, deep soils. Periodic fires and drought inhibit the growth of woody shrubs and trees.
7. Temperate deciduous forests occur in midlatitudes where there is sufficient moisture to support the growth of large, broad-leaved deciduous trees, which show distinct seasonal rhythms of leaf drop and regrowth.
8. Taiga constitutes the dense coniferous or boreal forests, characterized by long, cold, snowy winters and short summers.
9. Tundra occurs at the northernmost limits of plant growth and at high altitudes, where plant forms are limited by cold and winds to a low shrubby or matlike morphology.

Freshwater Biomes (pp. 1062–1066)

1. Freshwater biomes are closely associated with and strongly influenced by the terrestrial biomes in which they are situated or through which they pass.
2. Lakes and ponds are usually stratified vertically in regard to light, temperature, and community structure. Phytoplankton and zooplankton are the primary food source for the rest of the community. Lakes are classified on the basis of their nutrient content and productivity.
3. Rivers and streams contain freshwater communities that change significantly from the source to the final destination in an ocean or lake. Upstream areas contain organisms associated with clear water, cool temperatures, and rocky substrates. Downstream are organisms that can tolerate murkier water and warmer temperatures.

Marine Biomes (pp. 1066–1069)

1. Marine communities are in the oceans, which comprise nearly three-fourths of Earth's surface. Oceanic zones can be classified according to degree of light penetration as photic or aphotic and according to depth as intertidal (shoreline), neritic, or oceanic (the last two being progressively deeper and more distant from the shoreline). Open water is called the pelagic zone. The bottom of the ocean in all regions is the benthic zone; in the deepest parts, this constitutes the abyssal zone, a region of constant darkness, cold temperatures, and high pressure.
2. An estuary is the transition zone between a river or stream and the ocean into which it empties. Such areas experience large fluctuations in salinity, but they are extremely productive and support an abundance of both aquatic and semiaquatic organisms.
3. The rocky intertidal is a vertically stratified biome that is periodically inundated by seawater. Rocky intertidal organisms are attached to the substrate. Those in the uppermost zone are frequently exposed to air and sun and have adaptations that prevent desiccation and overheating.

4. Coral reefs occur in nutrient-rich and warm, shallow tropical waters. Skeletons of coral animals form complex structures among which a diversity of invertebrates and fishes live.

5. The oceanic pelagic biome includes most of the open ocean where photosynthetic and predatory plankton occupy the upper layers.

6. Benthic communities subsist largely on detritus that "rains" down from the pelagic zone. Where there are hot deep sea vents, chemosynthetic bacteria are fed upon by giant worms and other invertebrates.

The Environmental Diversity of the Biosphere (pp. 1069–1075)

1. The biosphere is an environmental mosaic in which several abiotic factors have an important impact on the distribution and abundance of organisms: temperature, water availability and quality, light intensity, wind, soil characteristics, and occasional disturbances such as fire.

2. Global climates and seasonality are established by the input of solar energy and Earth's rotation around the sun. Differential heating of the atmosphere and Earth's surface produce air circulation cells and latitudinal variation in temperature and precipitation, which in turn account for the geographic distribution of major biomes.

3. Bodies of water and topographic features influence regional and local climates. Oceans and lakes moderate the climate in coastal localities, and mountains influence temperature and rainfall. Temperate zone lakes undergo a biannual mixing of the nutrients and oxygen in their waters. Small features in the landscape influence local microclimates.

Responses of Organisms to Environmental Variation (pp. 1075–1080)

1. Although natural selection has produced a wide spectrum of adaptations to the environmental diversity of the biosphere, most species tolerate only a relatively narrow range of environmental variables.

2. Organisms can be described as regulators or conformers, depending on their homeostatic capabilities.

3. According to the principle of allocation, the total amount of energy available to an organism for all of its processes, including response to physical variables, is limited. Natural selection has resulted in different "budgeting" approaches, which are usually related partly to the stability of an organism's environment.

4. Animals may respond to the environment behaviorally, by moving or migrating to more favorable locations.

5. Physiological responses often include regulation for the maintenance of homeostasis. Tolerance limits and optimal temperatures can shift through physiological adjustments involved in acclimation.

6. Some birds and mammals undergo seasonal changes in the morphology of their coats. Many plants exhibit morphological plasticity, which helps compensate for their inability to move to new locations.

7. Although behavioral, physiological, and morphological adjustments operate in the here-and-now of ecological time, they occur within the framework of adaptations resulting from natural selection over evolutionary time.

SELF-QUIZ

1. Which statement follows from the principle of allocation?
   a. The number of organisms an area can support is determined by its energy supply.
   b. Physiological adjustments to environmental changes can extend the tolerance limits of organisms.
   c. The total amount of energy available to an organism is partitioned into such processes as reproduction, obtaining nutrients, and coping with the environment.
   d. Organisms that use more energy for growth and reproduction are able to survive in a wider range of variable environments.
   e. Organisms allocate most of their energy for homeostasis.

2. Which statement about tolerance limits is *incorrect?*
   a. They can often be tested experimentally and plotted as a tolerance curve.
   b. They help determine whether organisms can live in particular environments.
   c. They can be extended by acclimation in some cases.
   d. They are likely to be greater in regulators than in conformers.
   e. They are generally greatest for organisms restricted to stable environments.

3. Which of the following biomes is *incorrectly* paired with the description of its climate?
   a. savanna—cool temperature, precipitation uniform during the year
   b. tundra—extreme cold, permafrost, brief summer
   c. chaparral—mild and wet winters, hot and dry summers
   d. temperate grasslands—relatively cool climates, periodic drought
   e. tropical forests—nearly constant photoperiod and temperature

4. A diagnostic difference between temperate deciduous forests and taiga involves
   a. the type of dominant tree species
   b. soil quality
   c. the amount of precipitation
   d. the latitude
   e. species diversity

5. Which of the following is *incorrectly* paired with its description?
   a. neritic zone—shallow area over continental shelf
   b. abyssal zone—deepest benthic region
   c. pelagic zone—area of open water
   d. aphotic zone—zone in which light penetrates
   e. intertidal zone—shallow area at the edge of the water.

6. In which area are algal blooms most likely to occur?
   a. headwaters of a stream
   b. downstream area of a river
   c. lake or pond
   d. intertidal zone of an ocean
   e. benthic zone of an ocean
7. In general, deserts are located at latitudes where
   a. dry air is descending
   b. moist air is descending
   c. dry air is rising
   d. rising air creates doldrums
   e. air masses are stationary
8. The growing season would generally be shortest in which biome?
   a. tropical rain forest
   b. savanna
   c. taiga
   d. deciduous forest
   e. temperate grassland
9. Imagine some cosmic catastrophe that jolts Earth so that its axis is perpendicular to the line between the sun and Earth. The most predictable effect of this change would be
   a. no more night and day
   b. a big change in the length of the year
   c. cooling of the equator
   d. loss of seasonal variations at northern and southern latitudes
   e. elimination of ocean currents
10. Andrea was a passenger on a plane that flew first over temperate forest, then grassland and desert, finally landing at an airport in the chaparral biome. The route of the flight was between
   a. New York and Denver
   b. New York and Los Angeles
   c. Denver and Los Angeles
   d. Washington, DC and Phoenix
   e. Seattle and Washington, DC

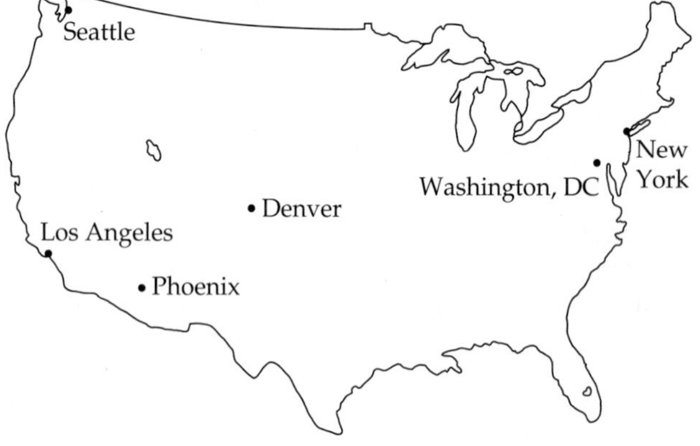

CHALLENGE QUESTIONS

1. In which terrestrial biome is your college or university located? Describe how ten local features on your campus influence local microclimates.
2. Describe five abiotic factors that might be important to the life of a tree in a temperate forest. How might the tree adjust to day-to-day changes in these factors?

SCIENCE, TECHNOLOGY, AND SOCIETY

1. Near Lawrence, Kansas, there was a rare patch of the original North American temperate grassland that had never been converted to farming. It was home to numerous native grasses, annual plants, and grassland animals. Among the species present were two endangered plants. Environmental activists thought the area should be set aside as a nature preserve, and they started to raise money to save it. In 1990, the owner of the land plowed it, stating that he did not want to be told what he could do with his property. He was within his rights, since there are no federal laws protecting endangered species on private land. What issues and values are in conflict in this situation? How do you think such conflicts should be resolved?
2. During the summer of 1988, huge forest fires burned a large portion of Yellowstone National Park. The National Park Service has a natural-burn policy: Fires that start naturally are allowed to burn unless they endanger human settlements. Lightning ignited the Yellowstone fires, so they were allowed to spread and burn themselves out, with firefighters primarily protecting people. This drew a lot of public criticism; the Park Service was accused of letting a national treasure go up in flames. Park Service scientists stuck with the natural-burn policy. Do you think this was the best decision? Support your position.
3. Although chaparral habitats burn regularly, many people in southern California and elsewhere build expensive houses in canyons and on hillsides covered with this vegetation. Should local governments regulate development in such communities? Should the government continue to provide disaster relief to people whose homes burn in these periodic fires so that these people can rebuild in the hills?

FURTHER READING

Abrahamson, W. G., T. G. Whitham, and P. W. Price. "Fads in Ecology." *BioScience,* May 1989. Changing ideas in a dynamic field.

Burman, A. "Saving Brazil's Savannas." *New Scientist,* March 2, 1991.

Gomez-Pompa, A., and A. Kaus. "Taming the Wilderness Myth." *BioScience,* April 1992. How Western beliefs affect environmental policy.

Holloway, M. "Still Negotiating." *Scientific American,* June 1992. Behind-the-scenes politics that set the stage for the United Nations Earth Summit.

"Managing Planet Earth." *Scientific American,* September 1989. A special issue devoted to the environment.

May, R. "How Many Species on Earth? " *Scientific American,* October 1992. Estimating biodiversity.

Ray, G., and J. Grassle. "Marine Biology Diversity." *BioScience,* July/August 1991. Why we need a program to conserve marine communities.

Ricklefs, R. E. *Ecology,* 3rd ed. New York: Chiron Press, 1986.

47 Population Ecology

Density and Dispersion
Demography
The Evolution of Life Histories
Models of Population Growth
The Regulation of Population Size
Human Population Growth

Every day, the morning newspaper and the evening news report local and global problems that threaten our well-being or provoke conflicts between individuals and nations. The media provide updates on global warming, acid rain, toxic waste, oil spills, accidents at nuclear power plants, and many other symptoms of an ailing biosphere. We see conflict between Western nations and those in the Middle East, aggravated in great part by the demand for oil. Videotapes and photographs of emaciated children starving to death depress everyone and remind us that the quality of life is dismal for millions of people. Closer to home, school boards argue about whether to discuss birth control with sexually active teenagers. Contributing to all these apparently unrelated problems is a common factor: the continued increase of the human population in the face of limited, and often dwindling, resources. The human population explosion is now Earth's most significant biological phenomenon. As a species of about 5.6 billion individuals, we require vast amounts of materials and space, including places to live, land to grow our food, and places to dump our waste. Endlessly expanding our presence on Earth, we have devastated the environment for many other species and now threaten to make it unfit for ourselves (Figure 47.1).

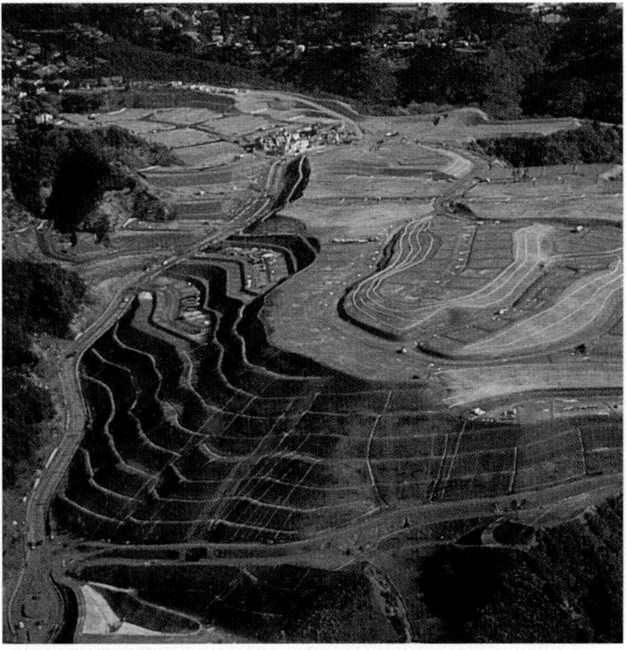

Figure 47.1
Human population growth inevitably destroys habitat for other species. Giant bulldozers have decapitated this hill in what is left of the Los Angeles chaparral, providing flat pads for construction of a housing tract. Understanding the human population explosion is a specific application of population ecology, the subject of this chapter.

To understand the problem of human population growth on more than a superficial level, we must consider the general principles of population ecology. It is obvious that no population can grow indefinitely. Species other than humans sometimes exhibit population explosions, but their populations inevitably crash. In contrast to these radical booms and busts, many populations are relatively stable over time, with only minor increases or decreases in population size. Population ecology, the subject of this chapter, is concerned with measuring changes in population size and composition, and identifying the causes of these fluctuations. The study of populations is useful to many human endeavors, such as controlling outbreaks of agricultural pests, predicting future fish harvests, maintaining populations of microorganisms used in biotechnology (see Chapter 19), and preserving what remains of natural environments.

We can think of populations in a variety of ways. In Chapter 21, our emphasis was on populations as interbreeding groups of individuals of a single species.

Table 47.1 Representative Population Densities

Organism	Density
Diatoms	5,000,000/m^3
Soil arthropods	5,000/m^2
Barnacles (adult)	2,000/m^2
Trees	50,000/km^2
Field mice	25,000/km^2
Woodland mice	1,200/km^2
Deer	4/km^2
Human beings	
Netherlands	346/km^2
Canada	2/km^2

Source: Data from C. J. Krebs, *Ecology,* 3rd ed. New York: Harper & Row, 1985.

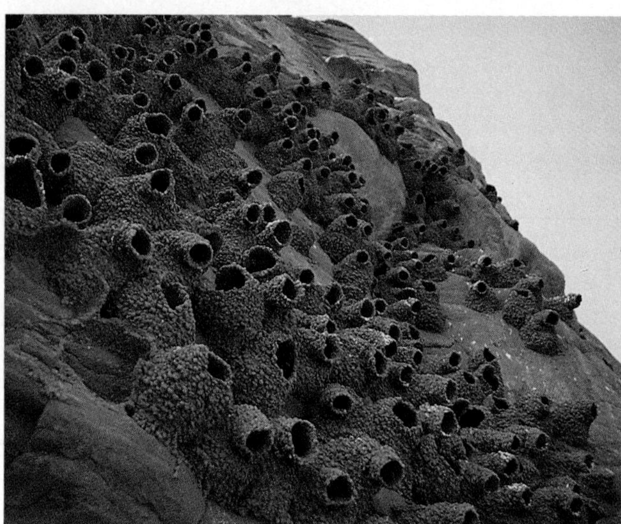

Figure 47.2
An indirect census of a bank swallow population. The number of birds nesting in this colony can be estimated by counting the number of entrance holes to the mud nests.

With a more ecological focus, we can also think of a population as individuals of one species that simultaneously occupy the same general area; they rely on the same resources and are influenced by similar environmental factors. Population ecology therefore intersects with both organismal ecology and community ecology; for example, the size of a population might be affected by the tolerance of individuals to cold temperature, as well as by predation on them by some other species. In fact, the dynamics of a population are usually determined by some complex interaction of factors. As we analyze population structure and growth in this chapter, remember this basic theme: The characteristics of a population are shaped by the interactions between individuals and their environment on both ecological and evolutionary time scales, and natural selection can modify these characteristics in a population. Later in this chapter, we will return to our discussion of the human population. For now, let's begin examining some of the ways to describe and analyze populations of any species.

DENSITY AND DISPERSION

At any given moment, every population has geographical boundaries and a population size (the number of individuals it includes). Ecologists who study the dynamics of populations begin by defining boundaries appropriate to the organisms under study and to the questions being posed. One ecologist may follow changes in the size of a barnacle population in a single New England tidepool, whereas another researcher may monitor fluctuations in the number of caribou in the entire state of Alaska. Regardless of these differences in scale, two important characteristics of any population are its density and its dispersion. Population **density** is the number of individuals per unit area or volume—the number of trees per km^2 in a forest, for example (Table 47.1). **Dispersion** refers to the pattern of spacing among individuals within the geographical boundaries of the population.

Measuring Density

In rare cases, it is possible to determine population size and density by actually counting all individuals within the boundaries of the population. We could count the number of trees of a particular species in a relatively small wooded area or tally the number of sea stars in a tidepool, for example. Herds of large mammals, such as buffalo or elephants, can sometimes be counted accurately from airplanes. In most cases, however, it is impractical or impossible to count all individuals in a population. Instead, ecologists often use a variety of sampling techniques to estimate densities and total population sizes. For example, they might estimate the number of alligators in the Florida Everglades by counting individuals in a few representative plots of an appropriate size. The estimate becomes more accurate as the sample plots increase in number or dimensions.

In some cases, population sizes are estimated not by counts of organisms but by indirect indicators, such as the numbers of nests or burrows (Figure 47.2), or signs, such as droppings or tracks. Another sampling technique commonly used to estimate wildlife populations is the **mark-recapture method,** described in the Methods Box.

Methods: A Mark-Recapture Estimate of Population Size

Traps are placed within the boundaries of the population being studied, and captured animals are marked with tags, collars, bands, or spots of dye and then released. After a few days or a few weeks, enough time for the marked animals to mix randomly with unmarked members of the population, traps are set again. The proportion of marked to unmarked animals that are captured during the second trapping gives an estimate of the size of the entire population. If there have been no births, deaths, immigration, or emigration, the following simple equation can be used to estimate the population size:

$$N = \frac{\text{Number marked} \times \text{Total catch second time}}{\text{Number of marked recaptures}}$$

For example, suppose that 50 sanderlings are captured in mist net traps, marked with leg bands (Figure a), and released. Two weeks later, 100 sanderlings are captured (Figure b). If 10 of this second catch are marked birds that have been recaptured, we would estimate that 10% of the total sanderling population is marked. Since 50 birds were originally marked, we would then estimate that the entire population consists of about 500 birds. Notice that this method assumes each marked individual has the same probability of being trapped as each unmarked individual. This is not always a safe assumption, however; an animal that has been trapped once, for instance, may be wary of the traps later.

(a)

(b)

Patterns of Dispersion

A population's **geographical range** is defined as the geographic limits within which it lives. Within the range, however, local densities may vary substantially because not all areas provide equally suitable habitat, and because individuals exhibit patterns of spacing in relation to other members of the population (Figure 47.3). The possible patterns vary in a continuum, from **clumped,** if the individuals are aggregated in patches; to **uniform,** if the spacing is even; to **random,** if spacing varies in an unpredictable way.

Clumping may result from environmental heterogeneity, with resources concentrated in patches within the range. Plants may be clumped in certain sites because soil conditions and other environmental factors vary locally; although seeds may be distributed everywhere, germination and growth are best where conditions are most suitable. For example, the eastern red cedar is often found clumped on limestone outcrops, where soil is less acidic than in nearby areas. Animals often move within the range toward a particular microenvironment that satisfies their requirements. For example, many forest insects and salamanders are clumped under logs where the humidity remains high. Herbivorous animals of a particular species are likely to be most abundant where their food plants are concentrated. Clumping of animals may also be associated with mating or other social behavior. For example, crane flies often swarm in great numbers, a behavior that increases mating chances for these short-lived insects.

An evenly spaced, or uniform, distribution often results from antagonistic interactions of individuals in

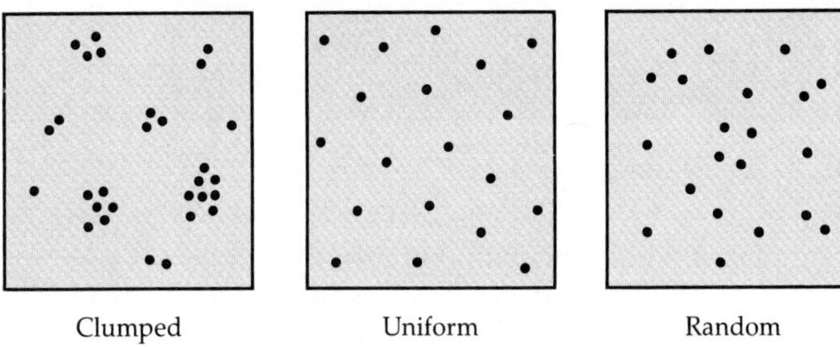

Figure 47.3
Patterns of dispersion within a population's geographical range. Individuals within a population frequently exhibit either clumped or uniform distributions within their geographical range, but random distributions are rare. In populations with a clumped distribution, the individuals *within* each clump also show a pattern of dispersion, as do the clumps themselves.

the population. For example, a tendency toward regular spacing of plants may result from shading and competition for water and minerals (Figure 47.4a); some plants also secrete chemicals that inhibit germination and growth of nearby individuals. In animal populations, regular spacing may be caused by competition for some resource or by social interactions that set up individual territories for feeding, breeding, nesting, or resting (Figure 47.4b and c). Territorial behavior is discussed later in this chapter and in Chapter 50 with other aspects of behavioral ecology.

Random spacing occurs in the absence of strong attractions or repulsions among individuals of a population. For example, forest trees are sometimes randomly distributed. Overall, however, random patterns are not very common in nature; most populations show at least a tendency toward either clumping or uniform distribution.

Figure 47.4
Examples of uniform dispersion. (**a**) The regular spacing of these creosote bushes in their desert habitat may reduce their competition for scarce water. (**b**) The uniform spacing of these king penguins on South Georgia Island reflects their maintenance of very small breeding territories. (**c**) Humans often show a uniform distribution, as illustrated by these sunbathers in Sydney, Australia.

DEMOGRAPHY

Changes in population size reflect the relative rates of processes that add individuals to the population and those that eliminate individuals from it. Additions occur through births (which we will define here to include all forms of reproduction) and immigration, the influx of new individuals from other areas. Opposing these additions are mortality (death) and emigration. In this chapter, we focus primarily on factors that influence birth rates and death rates.

The study of the vital statistics that affect population size is called **demography.** Birth and death rates usually vary among subgroups within a population, depending in particular on age and sex. It follows that future population size will be determined partly by the existing age structure and sex ratio, two of the most important demographic factors.

Age Structure and Sex Ratio

When the average lifespan of individuals in a population is greater than the time it takes to mature and reproduce, generations overlap. The coexistence of generations gives a population its **age structure,** which is the relative number of individuals of each age (see Figure 47.20).

Each age group has a characteristic birth and death rate. Often, juveniles and old individuals are more likely to die than individuals of intermediate age, who have the optimum combination of youthful vigor and the ability to find food or avoid predators that comes with maturity. Also, birth rate, the number of offspring produced during a certain amount of time, is often greatest for individuals of intermediate age (see Table 47.2). For example, as a group, a given number of 10-year-old female fur seals produce about twice as many young during the breeding season as an equal number of 5- or 18-year-old females do. In humans, the death rate is highest in the first year and, of course, in old age; the birth rate is highest among 20-year-old women. In general, a population with a large percentage of individuals of prime reproductive (or slightly younger) age will grow proportionately faster than a population that has an age structure heavily skewed toward older individuals. (The implications of this conclusion for human population growth will be discussed later in the chapter).

An important demographic feature related to age structure is **generation time,** which is the average span between the birth of individuals and the birth of their offspring. In general, generation time is strongly related to body size over a broad range of organisms (Figure 47.5). Other factors being equal, a shorter generation time will result in faster population growth (assuming, of course, that the overall birth rate is greater than the death rate). This is simply because the increases in population size attributed to births accumulate more rapidly when individuals reach sexual maturity in a shorter period of time.

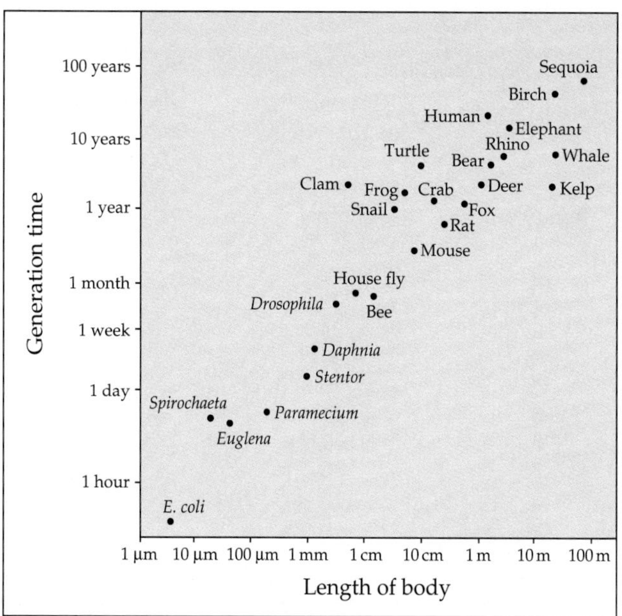

Figure 47.5
Size and generation time. Small organisms have short generation times, achieving reproductive maturity quickly. Generation time increases with body size because larger organisms take longer to reach the size at which they can reproduce.

The **sex ratio,** the proportion of individuals of each sex, is another important demographic statistic that affects population growth. The number of females is usually directly related to the number of births that can be expected, but the number of males may be less significant because in many species, a single male can mate with several females. In herds of elk, for example, there are fewer males of reproductive age than females, but this has no significant effect on the number of births in the overall population, because each male guards a "harem" of females with which he mates. In Canada geese, by contrast, individuals form monogamous pair-bonds for life; any significant reduction in males would be more likely to affect the population's birth rate. Wildlife management often reflects these demographic considerations. For example, deer-hunting regulations are usually more liberal regarding the killing of bucks than does because each buck typically mates with many does.

Life Tables and Survivorship Curves

About a century ago, when life insurance first became available, insurance companies needed to determine how much longer, on average, an individual of a given

Table 47.2 Life Table for a Population of the Barnacle *Balanus glandula**

Age (yr)	Observed Number Alive at Start of Age Interval	Proportion of Original Cohort Surviving at Start of Age Interval	Number Dying During Age Interval	Mortality Rate	Fecundity (Average No. of Eggs Produced per Female) During Age Interval
0	142	1.000	80	0.563	0
1	62	0.437	28	0.452	4,600
2	34	0.239	14	0.412	8,700
3	20	0.141	(4.5)	0.225	11,600
4	(15.5)	0.109	(4.5)	0.290	12,700
5	11	(0.078)	(4.5)	0.409	12,700
6	(6.5)	(0.046)	(4.5)	0.692	12,700
7	2	0.014	0	0.000	12,700
8	2	0.014	2	1.000	12,700
9	0	0.0	—	—	0

*Numbers in parentheses were interpolated from survivorship curves.
Source: Data from J. H. Connell, "A Predator–Prey System in the Marine Intertidal Region. I. *Balanus glandula* and Several Predatory Species of *Thais.*" *Ecological Monographs,* 40:49–78, 1970.

age could be expected to live. The result was the development of mortality summaries in what are called, ironically, **life tables.** Population ecologists have adapted this approach for nonhuman populations.

One way to construct a life table is to follow the fate of a **cohort,** a group of individuals of the same age, from birth until all are dead (Table 47.2). The essential data from which much of the table is constructed are simply the number of individuals that remain alive at successive sampling times. Obviously, this approach to constructing a life table (following the same cohort over time) can be used for only a very limited number of species. Fortunately, a life table can also be constructed by knowing the age at death of a sample of individuals and then calculating backward to their births.

The four middle columns in Table 47.2 are really just different ways of summarizing the same thing: how mortality varies with age over a time period corresponding to the maximum lifespan; each age interval in this table is one year. We can see, for example, that the mortality rate during a given age interval is simply the proportion of individuals, alive at the start of the interval, that died during the year. The last column in the life table shows the age-specific **fecundity,** the average number of eggs produced by a female of a particular age.

A graphic way of depicting some of the data in a life table is to draw a **survivorship curve,** which is simply a plot of the numbers in a cohort still alive at each age (Figure 47.6). Survivorship curves can be classified into three general types. A Type I curve is relatively flat, reflecting low death rates during early and middle life, dropping steeply as death rates increase among older age groups. Humans and many other large mammals that produce relatively few offspring but provide them with good care often exhibit this kind of curve. In contrast, a Type III curve drops sharply at the left of the graph, reflecting very high death rates for the young, but then flattens out as death rates decline for those

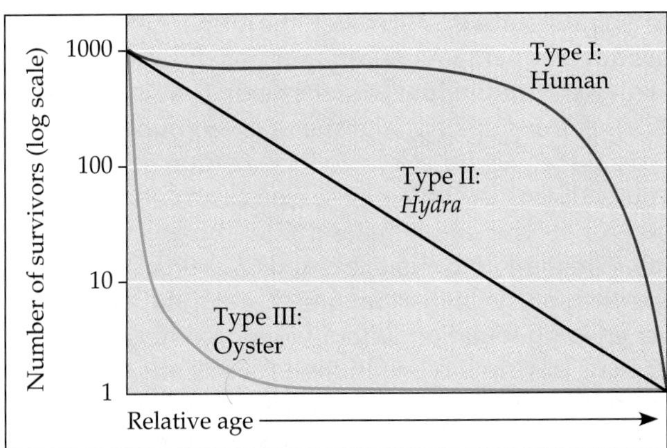

Figure 47.6
Survivorship curves. These three curves illustrate idealized patterns of survivorship in different sorts of organisms. As an example of a Type I curve, humans in developed countries experience high survival rates until old age, except during infancy, as indicated at the left of the graph. At the opposite extreme are Type III curves for organisms like oysters, which experience very high mortality as larvae but decreased mortality later in life. Type II survivorship curves are intermediate between the other two types. Notice that the *y*-axis is logarithmic and that the *x*-axis is on a relative scale, so that species with widely varying lifespans can be compared on the same graph.

few individuals that have survived to a certain critical age. This type of curve is usually associated with organisms that produce very large numbers of offspring but provide little or no care, such as many fishes and marine invertebrates. An oyster, for example, may release millions of eggs, but most offspring die as larvae from predation or other causes. Those few that manage to survive long enough to attach to a suitable substrate and begin growing a hard shell, however, will probably survive for a relatively long time. Type II curves are intermediate, with mortality more constant over the lifespan. This kind of survivorship has been observed in some annual plants, various invertebrates such as *Hydra,* some lizard species, and rodents, such as the gray squirrel. Many species, of course, fall somewhere between these basic types of survivorship or show more complex patterns. In birds, for example, mortality is often high among the youngest individuals (as in a Type III curve) but fairly constant among adults (as in a Type II curve). Some invertebrates, such as crabs, may show a "stair-stepped" curve, with brief periods of increased mortality during molts (caused by physiological problems or greater vulnerability to predation), followed by periods of lower mortality (when the exoskeleton is hard).

Survivorship is an important factor in the changes in population size through time. Let's now consider some additional characteristics that influence population dynamics.

THE EVOLUTION OF LIFE HISTORIES

Birth, reproduction, and death—the specific episodes that characterize the **life history** of any organism—influence the reproductive success of individuals and affect the growth of populations over ecological time. Of course, a particular life history pattern, like most characteristics of an organism, is the result of natural selection operating over evolutionary time. Because of varying pressures of natural selection, life histories are diverse. Pacific salmon, for example, hatch in the headwaters of a stream, then migrate to open ocean, where they require several years to mature. They eventually return to freshwater streams to spawn, producing millions of small eggs in a single reproductive opportunity, and then they die. In contrast, some lizards produce only a few large eggs during their second year, then repeat the reproductive act annually for several years. The life histories of plants are just as variable. Some species of oaks do not reproduce until the tree is 20 years old, but then produce vast numbers of large seeds each year for a century or more. Annual desert wildflowers generally germinate, grow, produce many small seeds, and then die, all in the span of a month after spring rains. Complicating matters further, important characteristics of life history may vary significantly among populations of a single species, or even among individuals in the same population.

What accounts for this variation in life histories? Every plant and animal exhibits a complex set of adaptations. Natural selection acts on this entire complex of traits—on the organism as a whole. As a result, it is usually difficult to disentangle the evolutionary advantages of each separate component of life history. However, we can find some obvious cases where features of life history are constrained by other evolutionary developments. For example, characteristics of mammalian reproduction preclude a life history in which thousands of offspring are produced each year.

We can identify three major characteristics of life history that affect the number of offspring a female will produce during her lifetime. We focus on females because, as already noted, it is the reproductive capability of females that usually determines the growth of a population. The restricted energy budgets of organisms create trade-offs among these three key characteristics, resulting in the integrated life history patterns observed in nature.

1. *Clutch size.* The number of offspring produced at each reproductive episode is called the clutch size, which varies because the energy devoted to reproduction can be packaged in different ways. Generally, organisms with a large clutch size produce small eggs or offspring; thus, each offspring is endowed with little energy to start life on its own. Large clutch size and small young are typical of organisms with a Type III survivorship pattern. In contrast, offspring from small clutches are generally larger, and each stands a better chance of surviving to adulthood, as illustrated by Type I and Type II curves. In many organisms, such as some fishes that produce large numbers of eggs, clutch size increases with body size as an animal ages. In other organisms, clutch size can vary seasonally within a single population (Figure 47.7).

2. *Number of reproductive episodes per lifetime.* Some organisms, such as the Pacific salmon and annual plants, reproduce only once in their lifetime, whereas others, including many plant and animal species, can reproduce repeatedly. The advantage of a single reproductive episode is an organism's ability to invest all its energy budget into the production of offspring; such organisms, however, do not survive to reproduce again. In contrast, organisms that reproduce repeatedly over their lifespan partition their energy into maintenance, growth, and reproduction; robust individuals may survive, perhaps attaining a larger size, to reproduce again later.

3. *Age at first reproduction.* For organisms that reproduce repeatedly during their lifespan, the timing of re-

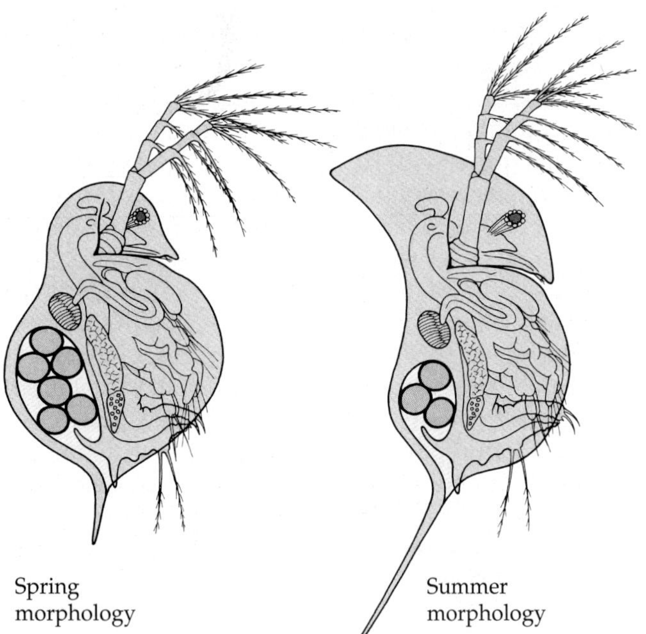

Figure 47.7
Seasonal variation in life history caused by predation. The freshwater crustacean *Daphnia retrocurva* shows marked seasonal variation in morphology and clutch size. In spring, the phytoplankton eaten by *Daphnia* are abundant, and predators that eat *Daphnia* are scarce. Under these favorable conditions, individuals develop into a rounder form with a large brood chamber containing six eggs. In summer, however, other plankton that feed on *Daphnia* are abundant. *Daphnia* developing at this time of year have large "helmets" and long tail spines, making them more difficult for predators to consume. These morphological changes, however, compress the brood chamber so that only half as many eggs can be carried. The helmeted individuals presumably survive longer and ultimately produce more offspring under these conditions than would the rounder form with more eggs. Natural selection has apparently sacrificed the larger clutch size in favor of increased survival and a better chance of producing at least some eggs.

Table 47.3 Characteristics of Opportunistic and Equilibrial Species

Characteristic	Opportunistic Species	Equilibrial Species
Homeostatic capability	Limited	Often extensive
Maturation time	Short	Long
Lifespan	Short	Long
Mortality rate	Often high	Usually low
Number of young produced per reproductive episode	Many	Few
Number of reproductions per lifetime	Usually one	Often several
Age at first reproduction	Early	Late
Size of offspring or eggs	Small	Large
Parental care	None	Often extensive

Source: Data from E. R. Pianka, *Evolutionary Ecology,* 4th ed. New York: Harper & Row, 1987.

production can have a large effect on the female's lifetime reproductive output. Consider again the trade-offs within an organism's energy budget. If a female uses some of her energy to reproduce at a younger-than-average age, she will have less energy left for maintenance and growth, reducing her potential to produce larger clutches in the future. On the other hand, a female that delays reproduction invests more energy in maintenance and growth. Although such a "strategy" might increase her potential for future reproduction, she may die before producing any offspring at all. Mathematical models of these phenomena suggest that, if older females produce much larger clutches than younger ones, and if there is a good chance of surviving to an advanced age, then female lifetime reproductive output is maximized when first reproduction is delayed. The situation for humans, however, is quite different, a point we will consider later in this chapter.

Darwinian fitness, of course, is measured not by how many offspring are produced but by how many survive to produce their own offspring: Heritable characteristics of life history that result in the most descendants over the long term will become more common within the population. If we were to construct a hypothetical life history that would yield the greatest lifetime reproductive output, we might imagine a population of individuals that begin reproducing at an early age, have large clutch sizes, and reproduce many times in a lifetime. However, natural selection cannot maximize all these variables simultaneously, because organisms have a finite energy budget that mandates trade-offs. For example, the production of many offspring with little chance of survival may result in fewer descendants than the production of a few well-cared-for offspring that can compete vigorously for limited resources in an already dense population.

In the 1960s, population ecologists began to distinguish among the vast array of life histories evident in nature, in an attempt to define the evolutionary forces that produced them (Table 47.3). Some populations of generally small species exhibit an **opportunistic life history** based on the production of large numbers of offspring in a single reproductive event (Figure 47.8a). Survivorship of offspring is generally low in these species, and the sizes of their populations fluctuate

(**a**) An opportunistic species

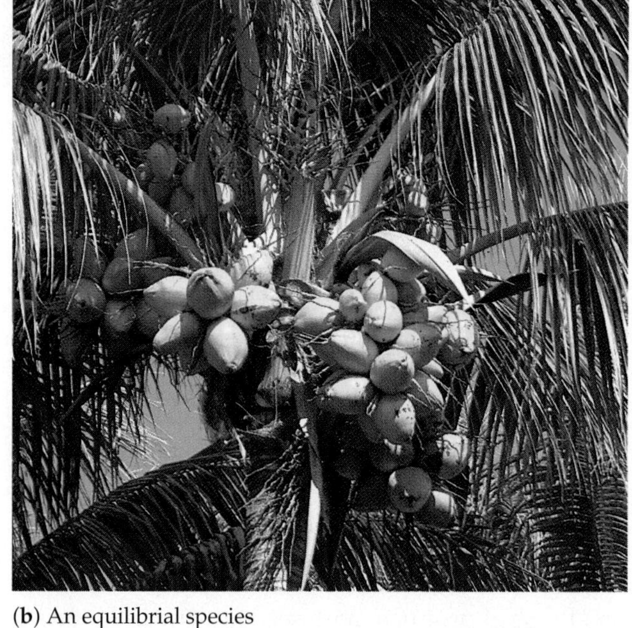

(**b**) An equilibrial species

Figure 47.8
Seed production by opportunistic and equilibrial plant species. (**a**) Most annual plants, such as this dandelion, grow quickly and produce a large number of seeds, exemplifying an opportunistic life history pattern. Although most of the seeds will not produce mature plants, their large number and ability to disperse to new habitats ensure that at least some will grow and eventually produce seeds themselves. (**b**) Some plants, such as this coconut palm, have an equilibrial life history pattern, producing a moderate number of very large seeds. The large endosperm provides nutrients for the embryo (a plant's version of parental care), an adaptation that helps ensure the success of a relatively large fraction of offspring. Animal species exhibit similar trade-offs between clutch size and the amount of nutrients provided to each offspring.

dramatically; when conditions are good in a variable environment, many individuals reproduce, but when conditions are bad, many may die without reproducing. For such species living in a relatively uncertain environment, natural selection has fostered an emphasis on the quantity of reproduction rather than individual survivorship and the generous endowment of offspring. Such organisms frequently take advantage of environmental opportunities, dispersing to open or disturbed habitats where their rapid maturation and reproduction allow the population to flourish, if only briefly. Desert annuals and garden weeds are good examples of opportunistic species.

In contrast, other populations of generally larger species exhibit an **equilibrial life history,** based on the repeated production of a smaller number of better-endowed offspring that are each more likely to survive to adulthood (Figure 47.8b). Because such species have more extensive mechanisms of homeostasis than opportunistic species, they are less affected by environmental fluctuations. Thus, population size varies less dramatically, fluctuating around some equilibrium number (this observation is discussed in more detail in the next section). Although reproductive output may vary from year to year, natural selection has resulted in the production of better-endowed offspring that can become established in the well-adapted population into which they are born. Slow maturation and parental care are typical of equilibrial species, exemplified by the life histories of many large terrestrial vertebrates.

The distinction between opportunistic and equilibrial species quickly became popular, not only because it integrated a number of important concepts but also because it generated testable hypotheses. Many field and laboratory studies generally support the view that features of life history are affected by environmental stability and the ability of populations to withstand environmental variation. However, it is important to understand that the portraits defined above represent extremes in what is actually a range of life histories; many populations have life histories and factors affecting them that fall somewhere between the two extremes. Some of the most enlightening studies compare closely related species, or different populations of the same species, that show different life history characteristics in different environments. For example, populations of the familiar dandelion may vary along the opportunistic-equilibrial continuum, depending on how severely their habitats are disturbed. Plants from populations in frequently mowed areas tend to be

Table 47.4 Percentage of Each of Four Types of Dandelions in Three Michigan Populations

Habitat	Number in Sample	A	B	C	D
		GENETIC TYPES			
		Smaller ← Size → Larger			
		Many ← Seed Number → Few			
Dry, full sun, frequently mowed	94	73	13	14	0
Dry, shade, occasional disturbance	96	53	32	14	1
Wet, semishade, undisturbed	94	17	8	11	64

Source: Data from M. Gadgil and O. T. Solbrig, *American Naturalist* 106:14–31, 1972.

smaller and produce more seeds than dandelions in undisturbed areas (Table 47.4).

As with so many important ecological ideas developed in the past 20 to 30 years, the distinction between opportunistic and equilibrial species is now seen as an oversimplification. The severity and frequency of important environmental fluctuations vary widely, and no simple classification of life histories can summarize the wealth of biological responses. Recently, some ecologists have suggested stress as another important environmental characteristic that shapes life history. In this sense, a stressful environment is not as much one that fluctuates as one in which conditions remain continuously difficult—for example, extreme cold or dim light. Here we might expect to find organisms that grow slowly for a relatively long time and produce their few offspring infrequently—characteristics that are similar to those of the equilibrial species that use homeostatic mechanisms to buffer environmental fluctuations. We have seen, too, how interactions between species, such as predation, may affect life history in variable ways even within the same population (see Figure 47.7). Increasingly, ecologists recognize that a population may show a mix of the traditional opportunistic and equilibrial characteristics; in most natural settings, some complex interplay of factors determines the most favorable life history pattern.

Now that we have analyzed some underlying patterns of survivorship, maturation, and reproduction, let's examine the effects of these phenomena on the growth of populations and the factors that regulate their sizes.

MODELS OF POPULATION GROWTH

To begin to understand the potential for population increase, consider a single bacterium that can reproduce by fission every 20 minutes under ideal laboratory conditions. At the end of this time, there would be two bacteria, four after 40 minutes, and so on. If this continued for only a day and a half—a mere 36 hours—there would be bacteria enough to form a layer a foot deep over the entire Earth. At the other life history extreme, elephants may produce only six young in a 100-year lifespan. Darwin calculated that it would take only 750 years for a single mating pair of elephants to produce a population of 19 million. Obviously, indefinite increase does not occur either in the laboratory or in nature. A population that begins at a low level in a favorable environment may increase rapidly for a while, but eventually the numbers must, as a result of limited resources and other factors, stop growing.

As we discussed in Chapter 46, it is often difficult for ecologists to apply experimental methods to their questions and to predict the consequences of changes in ecological systems. Mathematical modeling, to the extent that it is based on accurate assumptions, provides an alternative approach to some of these problems. Modeling enables an ecologist to study how variables interact or to make predictions about what would happen if some of the variables changed. Particularly in population ecology, which deals mainly with variations in numbers and rates, mathematical modeling is a common and productive approach, though it is sometimes complex. Here, we will explore only a few basic considerations and two simple models for populations with overlapping generations and age structures.

Exponential Population Growth

Imagine a hypothetical population consisting of a few individuals living in an ideal environment. Under these favorable conditions, there are no restrictions on the abilities of individuals to harvest energy, grow, and reproduce, aside from their inherent physiological limitations. The population will increase in size with every birth and with the immigration of individuals from other populations. By contrast, the population will decrease in size with every death and with the emigration of individuals out of the population. For simplicity, let's ignore the effects of immigration and emigration (a more complex formulation would certainly include these factors). We can define change in population size during a fixed time interval with this verbal equation:

$$\left[\begin{array}{c}\text{Change in population}\\ \text{size during time interval}\end{array}\right] = \left[\begin{array}{c}\text{Births during}\\ \text{time interval}\end{array}\right] - \left[\begin{array}{c}\text{Deaths during}\\ \text{time interval}\end{array}\right]$$

We use mathematical models as a simple way of generalizing the ideas that are expressed in words. If we let N = population size and t = time, we can also specify that ΔN = change in population size and Δt = the time interval (appropriate to the lifespan and generation time of the species) over which we are evaluating population growth. The verbal equation presented above can be rewritten as $\Delta N/\Delta t = B - D$, in which B = the absolute number of births in the population during the time interval and D = the absolute number of deaths.

All populations differ in size, and we want to create a general mathematical model that can be applied to any population. We therefore convert the simple model presented above into one in which births and deaths are expressed as the average number of births and deaths per individual during the specified time interval. Let b = the per capita birth rate. If, for example, a population of 1000 individuals experiences 34 births per year, the per capita birth rate is $^{34}/_{1000}$, or 0.034. If we know the per capita birth and death rates, we can calculate the expected number of births and deaths in a population of any size. For example, if you know that the annual per capita birth rate is 0.034 and the population size is 500 (instead of 1000), you could use the formula $B = bN$ to calculate the absolute number of births expected in that population: 17 per year. To make sure you are comfortable with this calculation, how many births would you expect in a population of 700 or in a population of 1700, for which $b = 0.050$ per year? Similarly, the per capita death rate, symbolized as d, allows us to calculate the expected number of deaths in a population of any size. If $d = 0.016$ per year, we would expect 16 deaths per year in a population of 1000 individuals. Using the formula $D = dN$, how many annual deaths would you expect per year if $d = 0.010$ annually in populations numbering 500, 700, and 1700? For natural populations or those in the laboratory, the per capita birth and death rates can be calculated from estimates of population size and data in a life table, such as Table 47.2.

We can revise the population growth equation again, this time using per capita birth and death rates rather than the absolute numbers of births and deaths:

$$\Delta N/\Delta t = bN - dN$$

One final simplification is in order. Because population ecologists are concerned with overall changes in population size, they use r to identify the difference in the per capita birth and death rates ($r = b - d$). This value, the per capita population growth rate, tells whether a population is actually growing (positive value of r) or declining in size (negative value of r). **Zero population growth (ZPG)** occurs when the per capita birth and death rates are equal and r equals 0. Note that births and deaths still occur in the population, but they balance each other exactly. (The relevance of ZPG for the human population and the obstacles preventing the human population from leveling off are discussed later in this chapter.

Using the population growth rate, we rewrite the equation as

$$\Delta N/\Delta t = rN$$

Finally, most ecologists use the notation of differential calculus to express population growth in terms of instantaneous growth rates:

$$dN/dt = rN$$

If you have not yet studied calculus, don't be intimidated by the form of the last equation; it is essentially the same as the previous one, except that the time intervals are very small.

We started this section by describing a population living under ideal conditions. In such a situation, the population grows at the fastest rate possible, because all members of the population have access to abundant food and are free to reproduce at their physiological capacity. This maximum population growth rate, called the **intrinsic rate of increase,** is symbolized as r_{max}. Population increase under these conditions is called **exponential population growth:**

$$dN/dt = r_{max}N$$

in which the size of the population increases rapidly, resulting in a J-shaped growth curve through time (Figure 47.9a). Notice that although the intrinsic rate of increase is constant as the population grows, the population actually accumulates more new individuals per unit of time when it is large than when it is small—the curve in Figure 47.9a gets progressively steeper through time. This occurs because population growth is dependent upon N as well as r_{max}, and larger populations experience more births (and deaths) than small ones growing at the same per capita rate (Table 47.5).

The opportunistic species described in the previous section often experience periods of exponential population growth; because their population growth rates may be close to r_{max}, they are sometimes described as ***r*-selected** populations or species. As noted above, such species often have short generation times and great reproductive potential; generation time and r_{max} are inversely related over a broad range of species (Figure 47.10).

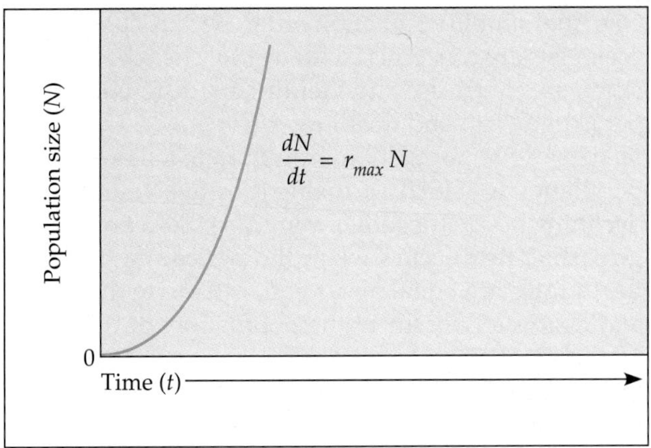

(a) **Exponential population growth.** The exponential growth model predicts unlimited population increase under ideal conditions of unlimited resources.

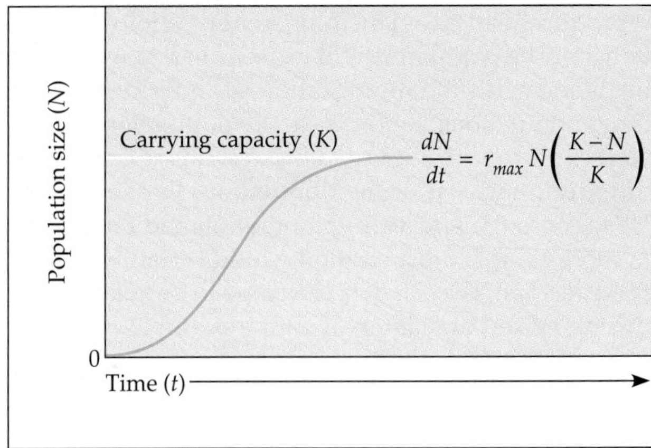

(b) **Logistic population growth.** The logistic growth model assumes that there is a maximum population size that the environment can support, termed the carrying capacity, or K. According to this model, the rate of population growth slows as the population size approaches the carrying capacity of the environment.

Figure 47.9
Population growth predicted by exponential and logistic models.

Logistic Population Growth

As our idealized population grows larger in size, its increased density may influence the ability of individuals to harvest sufficient resources for maintenance, growth, and reproduction. Many populations subsist on a finite amount of available resources, and as the population becomes more crowded, each individual has access to an increasingly smaller share. Ultimately, there is a limit to the number of individuals that can occupy a habitat. Ecologists define the **carrying capacity** as the maximum stable population size that a particular environment can support over a relatively long period of time. Carrying capacity, symbolized as K, is a property of the environment, and it varies over space and time with the abundance of limiting resources. For example, the carrying capacity for Carolina wrens or other songbirds may be high in lush habitats where insects are abundant, but lower in other habitats where food items are less numerous.

Table 47.5 Exponential Growth in a Hypothetical Population Where the Intrinsic Rate of Increase Is Constant at 0.05 per Capita per Year*

Year	N	ΔN
1	10,000	500
2	10,500	525
3	11,025	551
4	11,576	578
5	12,154	608
6	12,762	638
7	13,400	670
8	14,070	704
9	14,774	739
10	15,513	776

*The initial N (population size) is 10,000. ΔN is rounded to the nearest whole number. Notice that ΔN increases with increasing N, even though the intrinsic rate of increase (r_{max}) is constant.

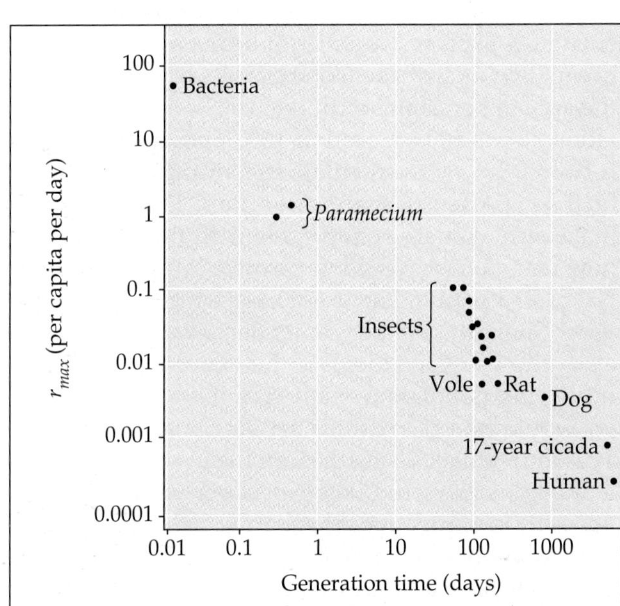

Figure 47.10
Generation time and r_{max} (maximum population growth rate). Small organisms that mature quickly tend to have short generation times and high r_{max}. Larger organisms, which take longer to reach sexual maturity, usually have long generation times and low r_{max}. Notice that here r_{max} is presented per capita per day, even for large animals.

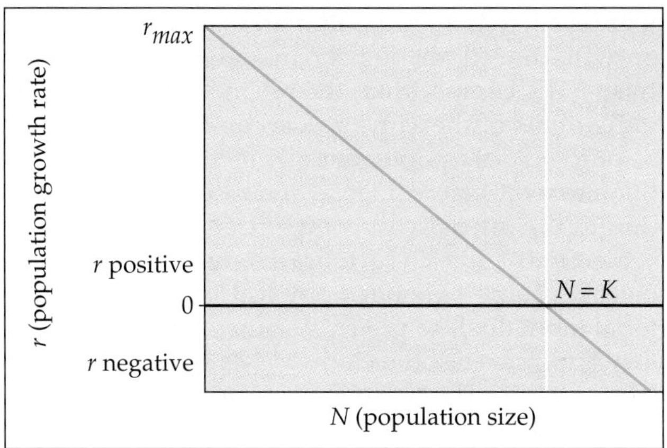

Figure 47.11
Reduction of *r* (population growth rate) with increasing *N* (population growth size). The logistic model of population growth assumes that *r* decreases as *N* increases. When *N* is close to 0, *r* equals r_{max}, and population grows rapidly. However, as *N* approaches *K*, *r* approaches 0, and population growth slows. If *N* is greater than *K*, *r* is negative, and population size decreases. (The logistic model does not allow for this condition, but temporary periods when *N* exceeds *K* are sometimes observed in actual populations.)

Table 47.6 Variations in *r* and Δ*N* in a Hypothetical Population Showing Logistic Growth Where *K* Is 1000 and the Intrinsic Rate of Increase Is Constant at 0.05 per Capita per Year*

N	(*K* – *N*)/*K*	*r*	Δ*N*
20	0.98	0.049	+ 1
100	0.90	0.045	+ 5
250	0.75	0.038	+ 9
500	0.50	0.025	+13
750	0.25	0.013	+ 9
1000	0.00	0.000	0

*Δ*N* is rounded to the nearest whole number.

Crowding and resource limitation can have a profound effect on the population growth rate. If individuals cannot harvest sufficient resources to reproduce, per capita birth rates will decline. If they cannot consume enough energy to maintain themselves, per capita death rates may increase. A decrease in *b* and/or an increase in *d* result in a smaller *r* and a lower overall rate of population growth.

We can modify our mathematical model of population growth to incorporate changes in *r* as the population size grows toward the carrying capacity (as *N* approaches *K*). A model of **logistic population growth** incorporates the effect of population density on *r*, allowing it to vary from r_{max} under ideal conditions to zero as carrying capacity is reached. To understand this model, envision the environment as a laboratory beaker. The maximum volume of the beaker is analogous to the carrying capacity of the environment, the amount of liquid it contains at any moment is analogous to the current population size, and the speed with which you fill the beaker is analogous to the population growth rate. When the beaker is empty or nearly so, you can pour liquid into it rapidly with little fear that the beaker might overflow. However, when the beaker is nearly full, you must add liquid slowly to avoid a spill. Analogously, when a population's size is below the carrying capacity, population growth is rapid according to the logistic model, but when *N* is near *K*, population growth is slow.

Mathematically, we construct the logistic model by starting with the model of exponential population growth and creating a term that reduces the value of *r* as *N* increases (Figure 47.11). If the maximum sustainable population size is *K*, the term (*K* – *N*) tells us how many new individuals the environment can accommodate, and the term (*K* – *N*)/*K* tells us what percentage of *K* is still available for population growth. By multiplying r_{max} by (*K* – *N*)/*K*, we reduce the value of *r* as *N* increases:

$$\frac{dN}{dt} = r_{max}N\left(\frac{K-N}{K}\right)$$

Table 47.6 shows hypothetical calculations of *r* and *N* at varying population sizes for a population growing according to the logistic model. Notice that when *N* is low, the term (*K* - *N*)/*K* is large, and *r* is not much discounted; but when *N* is large and resources limiting, the term (*K* - *N*)/*K* will be small, and *r* is substantially reduced below r_{max}. As in the logistic model, ZPG occurs when *b* equals *d* and *r* equals 0, in this case when *N* equals *K*. The logistic model of population growth produces a sigmoid (S-shaped) growth curve (see Figure 47.9b). New individuals are added to the population most rapidly at intermediate population sizes when there is a breeding population of substantial size but also lots of available space and other resources in the environment. The population growth rate slows dramatically as *N* approaches *K*.

The equilibrial species described in the last section may experience population growth that approximates the predictions of the logistic equation. In general, long generation times and small clutches limit reproductive potential, and their populations fluctuate much less than those of opportunistic species. Because their populations stabilize near carrying capacity, equilibrial species are sometimes described as ***K*-selected.** Carrying capacities can be determined by factors other than food resources, though energy limitation is perhaps the commonest determinant of *K*. Other limiting

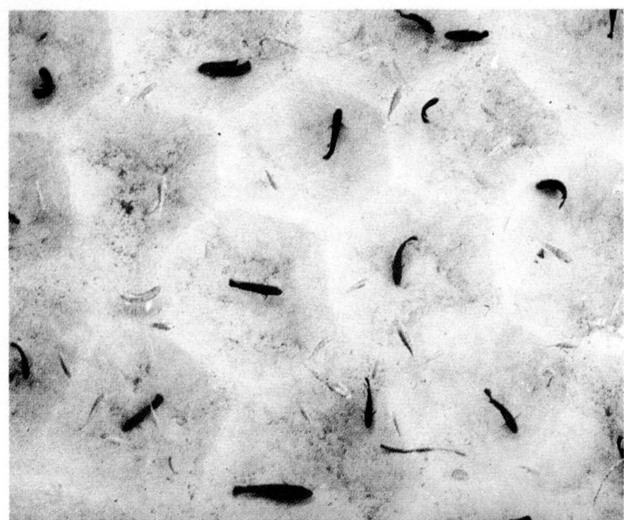

Figure 47.12
Territoriality in fish. These captive cichlids have defined territories (the hexagonal depressions in the sand) within their tank. Territoriality provides the resident with exclusive use of resources within the territory, reducing intraspecific competition for those resources. Of course, space to construct a territory is a resource itself, and these fish actively compete for space within this limited artificial environment.

factors include the availability of specialized nesting sites (as for spotted owls and other hole-nesting organisms), roosting sites (as for some bats), and shelters or refuges from potential predators.

The major biological implication of the logistic model is that increasing population density reduces the resources available for individual organisms and that resource limitation ultimately limits population growth. Indeed, the logistic model is a model of **intraspecific competition:** the reliance of two or more individuals of the same species on the same limited resources. As the population size increases, the competition becomes more intense, and *r* declines in proportion to the intensity of competition. In many vertebrates and some invertebrates, **territoriality,** the defense of a well-bounded physical space, allows territorial individuals to protect resources within the territory and prevent other individuals from using them (Figure 47.12). This behavioral mechanism may reduce intraspecific competition for food and nest sites within the territory, but the space to construct a territory becomes the resource for which individuals compete.

The accumulation of toxic metabolic wastes is also a factor in carrying capacity. In laboratory cultures of small microorganisms, for example, metabolic by-products accumulate as the population grows, poisoning the population within this limited, artificial environment. Indeed, ethanol accumulates as a waste product when yeast ferments sugar, and the alcohol content of wine is usually less than 13%, the maximum ethanol concentration that wine-producing yeast cells can tolerate.

How well does the logistic equation actually describe the growth of populations? The growth of laboratory populations of some small animals, such as beetles or crustaceans, and microorganisms, such as paramecia, yeast, and bacteria, fit S-shaped curves fairly well (Figure 47.13a, b). However, these experimental populations are grown in a constant environment lacking predators and other species that may

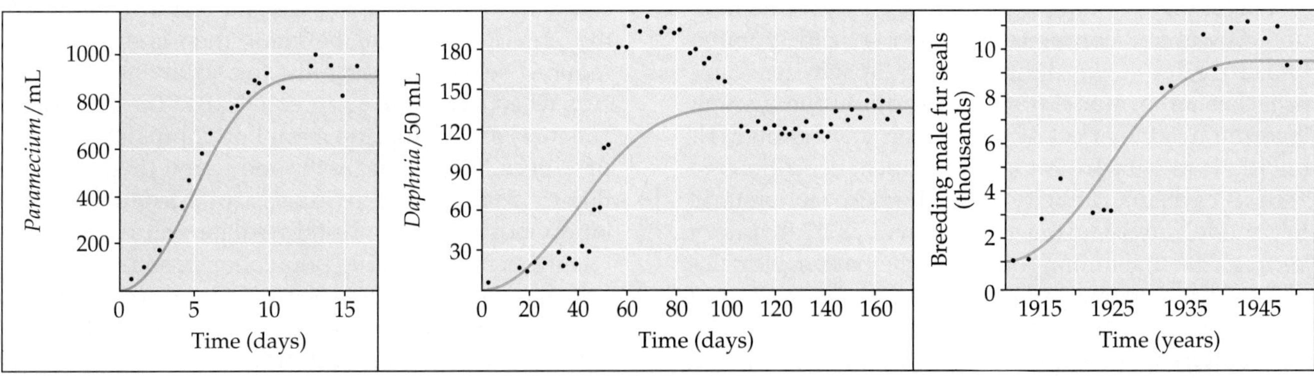

(**a**) A *Paramecium* population in laboratory culture

(**b**) A *Daphnia* population in laboratory culture

(**c**) A fur seal (*Callorhinus ursinus*) population on St. Paul Island, Alaska

Figure 47.13
Examples of logistic population growth. In each of these studies, actual data (points on the graph) are compared to an idealized curve (blue) based on the logistic model of population growth. (**a**) Growth of *Paramecium aurelia* in small laboratory cultures closely approximates logistic growth if an experimenter maintains a constant environment by regularly adding food and removing toxic wastes. (**b**) Similarly, growth of a population of *Daphnia* in a small laboratory culture also shows approximately logistic growth. Notice, however, that this population grew so quickly that it overshot the carrying capacity of its artificial environment, and then settled back to a relatively stable population size. (**c**) The numbers of male fur seals with "harems" on St. Paul Island, Alaska, were greatly depressed by hunting until 1911. After hunting was banned, the population increased dramatically and now oscillates around an equilibrium number, presumably the island's carrying capacity for this species.

compete for resources, idealized conditions that never occur in nature. Even under these laboratory conditions, not all populations stabilize at a clear carrying capacity, and most populations show some unpredictable deviations from a smooth sigmoid curve.

We rarely have the chance to observe the growth pattern of equilibrial populations in nature because we are not present to make observations when the populations begin to grow. However, studies of some organisms introduced into new habitats or populations that rebound after being decimated by disease or hunting provide general support for the concept that underlies logistic population growth (Figure 47.13c).

Some of the basic assumptions built into the logistic model clearly do not apply to all populations. For example, the model incorporates the idea that even at low population levels, each individual added to the population has the same negative effect on population growth rate; that is, *any* increase in N reduces the term $(K - N)/K$. Some populations, however, show an **Allee effect** (named after the researcher who first described it), in which individuals may have a more difficult time surviving and reproducing if the population size is too small. For example, a single plant standing alone may suffer from excessive wind but would be protected in a clump of individuals. Some oceanic birds require large numbers at their breeding grounds to provide the necessary social stimulation for reproduction. And conservationists fear that populations of rhinoceri, animals that live solitary lives, may be so small that individuals will not be able to locate mates in the breeding season. In these cases, a greater number of individuals in the population has an enhancing effect, up to a point, on population growth. In addition, when a population is small, there is a greater possibility that chance events will eliminate all individuals, or that inbreeding will lead to a general reduction in fitness (see Chapter 21).

The logistic model also incorporates the assumption that populations approach carrying capacity smoothly. In many populations, however, there is a lag time before the negative effects of an increasing population are realized. For example, as some important resource, such as food, becomes limiting for a population, reproduction will be reduced, but the birth rate may not be affected immediately because the organisms may use their energy reserves to continue producing eggs for a short time. This may cause the population to overshoot the carrying capacity. Eventually, deaths will exceed births, and the population may then drop below carrying capacity; even though reproduction begins again as numbers fall, there is a delay until new individuals actually appear. Apparently as a result of such time lags, many populations seem to oscillate about their carrying capacity or to overshoot at least once before attaining a stable size (see Figure 47.13b).

Finally, as we will discuss in the next section, populations do not necessarily remain at, or even reach, levels where population density is an important factor. In many insects and other small, quickly reproducing, opportunistic organisms that are sensitive to environmental fluctuations, physical variables such as temperature or moisture reduce the population well before resources become limiting. For describing changes in the numbers of these organisms, the idea of carrying capacity does not really apply.

Overall, the logistic model is a useful starting point for thinking about how populations grow and for constructing more complex models. Although it fits few, if any, real populations exactly, the logistic model incorporates basic ideas that, with modification, do apply to many populations. And like any good hypothesis, this model has stimulated many experiments and discussions that, whether they support the model or not, lead to a greater understanding of population ecology in general.

THE REGULATION OF POPULATION SIZE

The exponential and logistic models predict very different patterns of population growth. The exponential model sets no limit on population increase, whereas the logistic model predicts the regulation of population growth as density increases.

Ecologists have long debated the most important factors regulating population growth. At one time, population ecologists were somewhat divided into two camps on this subject: One emphasized the importance of density-dependent factors in population regulation, and the other emphasized density-independent factors. The current consensus is that the relative importance of these factors differs for opportunistic and equilibrial species and their specific circumstances, and that often both density-dependent and density-independent factors interact to affect a population.

Density-Dependent Factors

In restricting population growth, a **density-dependent factor** is one that intensifies as the population increases in size. At high population densities, such factors affect not only a larger percentage of individuals in the population but also each individual more strongly. As noted above in our description of the logistic model, density-dependent factors reduce the population growth rate by decreasing reproduction and by increasing mortality in a crowded population.

Resource limitation in crowded populations can influence the future size of a population by profoundly reducing reproduction (Figure 47.14). Available food supplies often limit the reproductive output of songbirds, for example, and, as bird population density in-

Figure 47.14
Decreased fecundity at high population densities. (**a**) The average number of seeds produced by plantain *(Plantago major)*, a small herb, decreases with increased sowing density. In this experiment, the germination rate and the proportion of plants that produced any seeds also decrease at higher densities, but mortality rate increases. (**b**) Average clutch size decreases markedly with increasing population density in a woodland population of an English songbird, the great tit *(Parus major)*. This response provides some reduction in the birth rate of large populations.

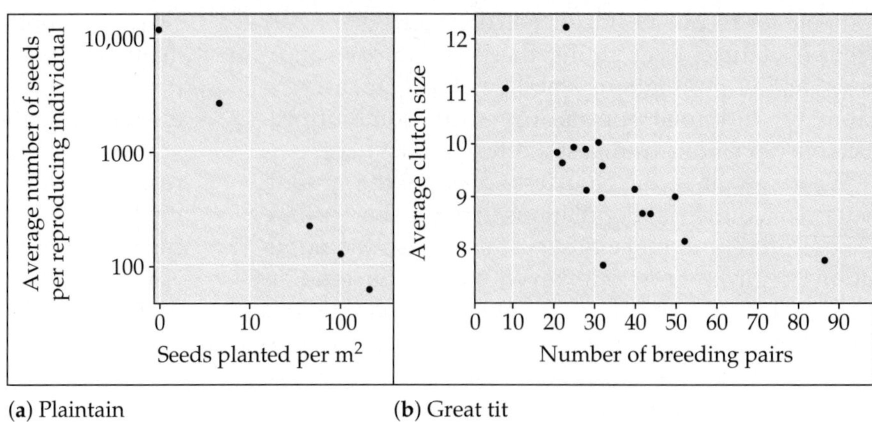

creases in a particular habitat, each female may lay fewer eggs. Seed production by plants is similarly affected by crowding. However, limiting resources may not always operate with the kind of simple, gradually increasing severity that is predicted by the logistic equation. For oceanic birds such as gannets, which nest on rocky islands where they are relatively safe from predators, the limited number of suitable nesting sites allows only a certain number of pairs to nest and reproduce. Up to a certain population size, most birds can find a suitable nest site, but few birds beyond that number breed successfully.

Population density also influences the health and survivorship probabilities of plants and animals (Figure 47.15). Plants grown under crowded conditions tend to be smaller and less robust than those grown at lower density. Small plants are less likely to survive, and those that do produce fewer flowers, fruit, and seeds, a phenomenon well recognized by backyard gardeners who thin their flowers and vegetables to produce the best possible yield. Similarly, animals often experience increased mortality at high population densities. In laboratory studies of flour beetles, for example, the percentage of eggs that hatch and survive to adulthood decreases steadily as density increases from moderate to high levels.

Predation may also be an important density-dependent regulator for some populations. A predator probably encounters and captures more prey as the population density of the prey increases. This is not a density-dependent effect, however, because the same percentage of the prey population may still be taken. Many predators, however, exhibit **switching behavior:** They begin to concentrate on a particularly com-

Figure 47.15
Decreased vigor and survivorship at high population densities. (**a**) The average weight of pigweed *(Amaranthus retroflexus)*, a large annual plant, decreases markedly with increased population density. Smaller plants are less robust than large ones and less likely to survive and reproduce. (**b**) The percentage of flour beetles *(Tribolium confusum)* surviving from egg stage to adult in a laboratory culture decreases at moderate to high population densities, reducing the numbers of adults in the next generation.

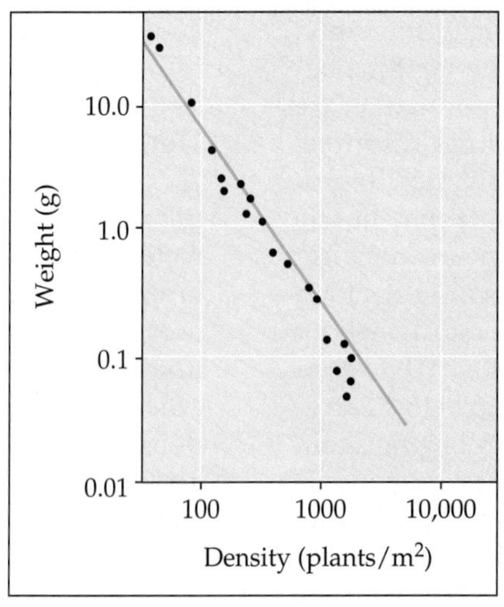

(**a**) Pigweed

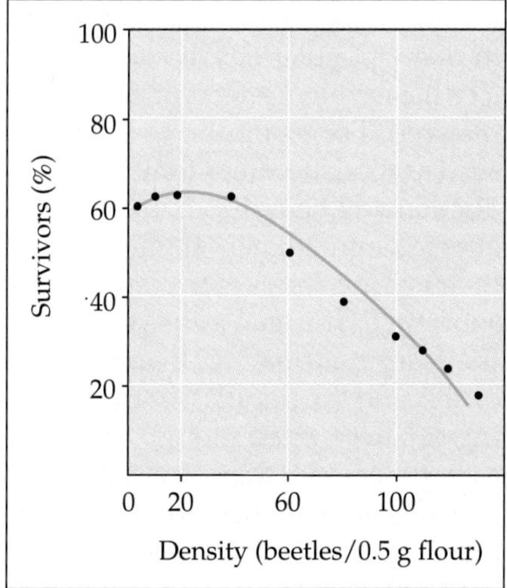

(**b**) Flour beetle

mon species of prey because it becomes energetically efficient to do so. (See the discussion on optimal foraging in Chapter 50.) For example, trout may concentrate for a few days on a particular species of insect that is emerging from its aquatic larval stage, then switch as another insect species becomes more abundant. As a prey population builds up, predators may feed preferentially on that species, consuming a higher percentage of individuals; this can cause density-dependent regulation of the prey population. In the rocky intertidal, for example, predation by sea stars limits the numbers of mussels, their favorite prey. Such interactions at the community level are important in affecting population sizes, as will be discussed in more detail in Chapter 48.

For some animal species, intrinsic factors, rather than the extrinsic factors just discussed, appear to regulate population size. White-footed mice in a small field enclosure will multiply from a few to a colony of 30 to 40 individuals, but eventually reproduction will decline until the population ceases to grow. Although this change is clearly associated with increasing density, it occurs even when food and shelter, the main resources needed by the mice, are provided in abundance. Through mechanisms that are not yet understood, high densities cause a stress syndrome in which hormonal and other physiological changes delay sexual maturation, cause reproductive organs to shrink, and generally inhibit reproduction.

Although numerous studies have demonstrated the expected effects of population density on the reproduction, growth, and survivorship of plant and animal populations, proving that density-dependent population regulation occurs in nature is, in fact, quite difficult. All other factors being equal, one must show that increases in density cause a downward adjustment of population size and that decreases in density are followed by periods of population growth. As always in ecology, other factors, including the abiotic environment and population interactions (see Chapter 48), vary widely over space and time. The observation that the size of many field and laboratory populations appears to fluctuate around the carrying capacity (see Figure 47.13) provides appealing—but only circumstantial—evidence that density-dependent factors are at work.

Density-Independent Factors

The occurrence and severity of **density-independent factors** are unrelated to population size; they affect the same percentage of individuals regardless of population size. The most common and important density-independent factors are related to weather and climate. For example, a freeze in the fall may kill a certain percentage of the insects in a population. Obviously, the timing of the first freeze and just how cold it gets are not affected by the density of the insect population.

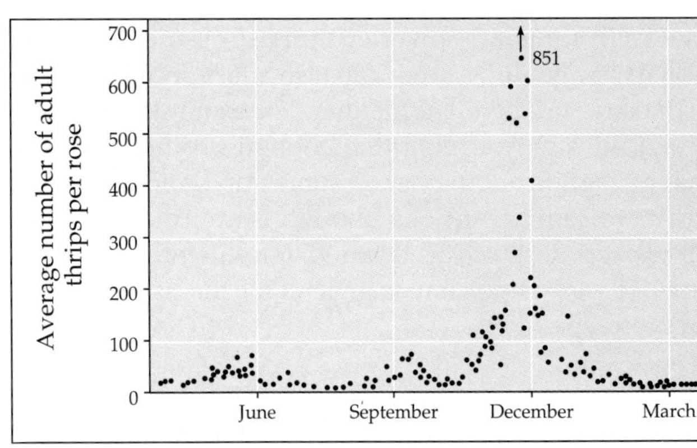

Figure 47.16
Density independence of population size. Populations of the insect genus *Thrips* grow rapidly during spring in the flowers that provide both food and shelter. Before the population reaches any carrying capacity, however, numbers are drastically reduced during the dry Australian summer, when most flowers die. (In interpreting this graph, remember that summer in the Southern Hemisphere begins in late December.)

In some natural populations, density-independent factors routinely reduce population size before resources or other density-dependent factors become important. This seems to be the case, for instance, for population cycles of small Australian insects in the genus *Thrips* (Figure 47.16). These animals feed on the pollen and flower tissues of plants belonging to the rose family, including apples, and *Thrips* can become severe pests. Population size of *Thrips* is closely linked to the number of flowers available, which in turn is correlated with seasonal weather patterns. The insect populations are relatively small during cool winters but rapidly increase during spring, as flowers bloom and warm weather increases insect rates of development and egg production. Summers, however, are extremely dry, and flowers survive only in a few sheltered locations. During the favorable spring period, the *Thrips* population increases in a manner that may approximate exponential growth; with the onset of summer, however, adult mortality causes the population to crash—well before density-dependent factors would become important and before the population shows any evidence of reaching a carrying capacity. A few individuals remain in the surviving flowers, and these allow population growth to resume again when favorable conditions return. The population size of many other species, particularly insects and other small organisms, is probably limited at some point primarily by density-independent factors.

Environmental episodes more sporadic than seasonal changes in weather can also affect populations in a density-independent manner. For example, fires and hurricanes may strike often enough in some areas to have a significant impact on some populations. In contrast, volcanic eruptions, though extraordinarily dramatic and destructive, occur so infrequently that they are not very important as general mechanisms of population regulation.

The Interaction of Regulating Factors

Over the long term, many populations remain fairly stable in size and are presumably close to a carrying capacity that is determined by density-dependent factors. Superimposed on this general stability, however, are short-term fluctuations due to density-independent factors. In one case study, researchers monitored the number of European herons, large birds, for 30 years in two different areas of England. The general pattern was the same for both areas: Populations were reasonably stable over the three-decade span, but major declines occurred following unusually cold winters.

In some cases, density-dependent and density-independent factors act together to regulate a population. For example, in very cold, snowy areas, many deer may starve to death during the winter. The severity of this factor is proportional to the season's harshness; colder temperatures increase energy requirements (and therefore the need for food), while deeper snow makes it harder to find food. But the effect is also density-dependent, because each individual in a large population gets a proportionately smaller share of what little food is available.

The relative importance of density-dependent and density-independent controls may also vary seasonally. This complex interaction of factors apparently influences populations of the bobwhite quail. The range of this bird extends northward into southern Wisconsin, where survival is difficult during winter. The number of birds alive at the end of the winter is largely determined by the depth of snow cover, a density-independent factor. If there is little snow, as much as 80% of the population will survive, but with more snow, survivors may decline to only 20%. Whatever happens in winter, however, the population size at the end of summer is quite constant from year to year. This is because surviving adults may benefit from the high mortality of others in a snowy winter by showing a high reproductive rate the following spring, reflecting the per capita abundance of resources. Even the *Thrips* populations described earlier as a classic example of density-independent limitation are probably regulated during part of the year by a density-dependent factor; that is, the number of flowers available in the summer is a limited resource that restricts the number of surviving insects. Most populations are probably regulated by some mix of density-dependent and density-independent factors.

Population Cycles

Some populations of insects, birds, and mammals fluctuate in density with remarkable regularity. For example, many species of arctic voles and lemmings have population cycles of 3 to 4 years, whereas snowshoe hares have a 10-year cycle. Although we have better data on the cycles themselves than on the underlying causes, several explanatory hypotheses have been proposed.

One idea is that crowding regulates cyclical populations, perhaps by affecting the organisms' endocrine systems. As described above for white-footed mice, stress resulting from a high population density may alter hormonal balance, which in turn reduces fertility, increases aggressiveness, and induces the mass emigrations that are often portrayed in nature films. However, it is not known if such changes are a common occurrence in the many species of animals that have population cycles.

Another hypothesis is that population cycles are caused by a time lag in the response to density-dependent factors, creating large fluctuations of population size above and below carrying capacity. Such a mechanism involving predation as the major density-dependent causative agent was once a popular explanation for a correlation in the cycles of the snowshoe hare and its predator, the Canadian lynx (Figure 47.17). However, recent research indicates that the lynx is not the major factor regulating the hare population. An alternative explanation is that high hare density causes a deterioration in the quality of their food. Studies indicate that when certain plants are damaged by herbivores, their nutrient content decreases, and they produce increased amounts of defensive chemicals. It is therefore likely that the density of the hare population may be regulated not as much by its predators as by cyclic changes in the plants they eat—or perhaps by some other resource limitation.

Perhaps the most startling population cycles known are those of periodical cicadas, grasshopperlike insects that complete their life cycles every 13 or 17 years, emerging from the ground at phenomenal densities (as high as 600 per m^2). This long life cycle may be an adaptation that avoids the type of predictable predator–prey interaction described above for hare and lynx. Cicada populations are locally controlled, however, by a fungus, whose spores can remain alive underground for the 13 or 17 years between cicada outbreaks. The

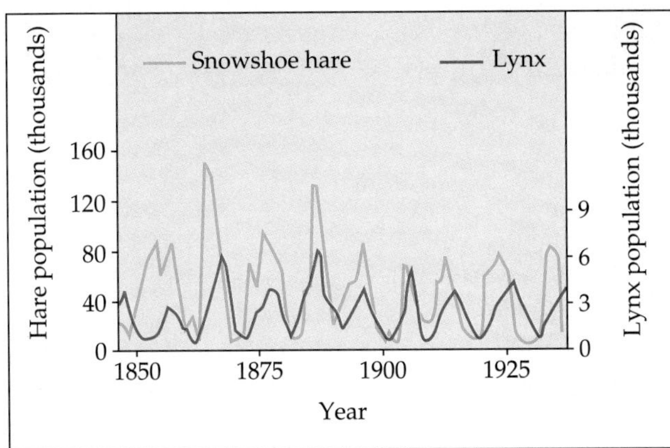

Figure 47.17
Population cycles in snowshoe hare and lynx. Oscillations in the population density of hare followed by corresponding changes in lynx density were once interpreted as strong circumstantial evidence that these populations of prey and predator regulated each other. (Population counts are based on the number of pelts sold by trappers to the Hudson Bay Company.) However, snowshoe hare populations on islands where lynx are absent show similar cycles. Periodic crashes in the hare population may be associated with changes in their food plants, induced by overfeeding. The cycles of lynx populations may, indeed, be caused by cyclic fluctuations of the hare, a major food source for lynx. Changing ideas on the hare–lynx interaction emphasize the problem of inferring process from pattern. Most patterns of population growth are likely caused by multiple interacting factors that are difficult to untangle without direct experimentation.

causes of cycles undoubtedly vary among species and perhaps among populations of the same species. Thus far, there is no satisfactory general explanation.

HUMAN POPULATION GROWTH

The human population has been growing exponentially for centuries. Indeed, it is unlikely that any other population of large animals has ever sustained exponential growth for so long. The explosive growth of our population is the primary cause of severe environmental degradation, and the environmental problems that we now confront cannot be solved without stringent regulation of our numbers.

Human population increased relatively slowly until about 1650, when approximately 500 million people inhabited the Earth (Figure 47.18). The population doubled to 1 billion within the next two centuries, doubled again to 2 billion between 1850 and 1930, and doubled still again by 1975 to more than 4 billion. The population now numbers about 5.6 billion people and increases by about 80 million each year. It takes only 3 years for world population to add the population equivalent of another United States. If the present growth rate persists, there will be 8 billion people on Earth by the year 2017.

Human population growth is based on the same general parameters that affect other animal and plant populations: birth rates and death rates. Birth rates increased and death rates decreased when agricultural societies replaced a lifestyle of hunting and gathering about 10,000 years ago. Since the Industrial Revolution, exponential growth has resulted mainly

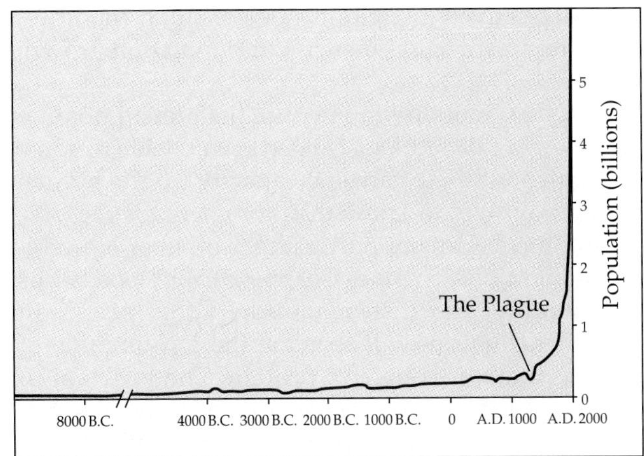

Figure 47.18
Human population growth. The human population has grown almost continuously throughout history, but it has skyrocketed since the Industrial Revolution. No other population of organisms has shown such steady population growth, and the human population must eventually either level off or decline. Whether this reduction in population growth will occur because of decreased birth rates or mass deaths is an open question, one that a careful population policy now can address.

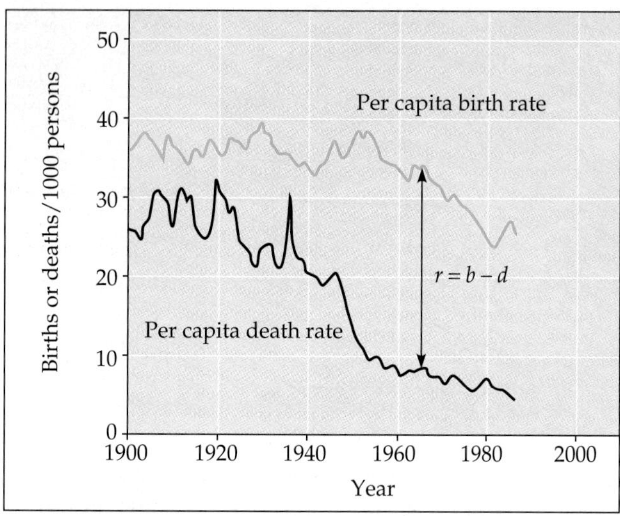

Figure 47.19
Changes in birth and death rates in Sri Lanka. Although family planning efforts and better medical care have reduced per capita birth and death rates in Sri Lanka, the population growth rate (*r*, the difference between the birth and death rates at any point in time) increased after 1940. Population control will occur only after reduction in the birth rate is greater than reduction in the death rate.

Table 47.7 Population Growth Rates, 1985–1990

Country	*r*
Africa	
Ethiopia	0.020
Kenya	0.042
Uganda	0.035
North and Central America	
Honduras	0.032
Mexico	0.022
United States	0.008
South America	
Brazil	0.021
Paraguay	0.029
Uruguay	0.008
Asia	
China	0.014
Japan	0.004
Mongolia	0.031

Source: Data from World Resources Institute, United Nations Environment Programme, and the United Nations Development Programme. *World Resources* 1990–91. 1990.

from a drop in death rates, especially infant mortality, even in the least industrialized countries (Figure 47.19). Improved nutrition, better medical care, and sanitation have all contributed to an increased percentage of newborns that survive long enough to leave offspring of their own. The effect of decreasing mortality on population growth is coupled with birth rates that are still relatively high in most developing countries, resulting in an actual increase in population growth rates.

Despite our ability to measure the human population growth rate, ecologists have been unable to agree on Earth's ultimate carrying capacity for the human population. We do know that, for energetic reasons, environments can support a larger number of herbivores than carnivores (see Chapter 49). But food habits vary widely, and it seems unlikely that people in wealthier countries will abandon the consumption of meat as a major source of protein. The problem of defining *K* for humans is confounded by the observation that carrying capacity has changed with human cultural evolution (see Chapter 30). The advent of agricultural and industrial technology has increased *K* twice during human history, and opponents of population control are counting on some new, as yet unidentified technological breakthrough that will allow our population to grow and plateau at some higher level.

Current worldwide population growth is actually a mosaic of various rates of growth in different countries (Table 47.7). Some developed countries, such as Sweden, are near zero population growth because birth and death rates balance. The human population as a whole, however, continues to grow because birth rates greatly exceed death rates in most nations, particularly in developing countries.

A major factor in the variation in *r* among countries is the variation in their age structures, an important demographic factor in present and future growth trends (Figure 47.20). The relatively uniform age distribution in Sweden, for instance, contributes to that country's stable population size; individuals of reproductive age or younger are not disproportionately represented in the population. In contrast, Mexico has an age structure that is bottom-heavy, skewed toward young individuals who will grow up and sustain the explosive growth with their own reproduction. Notice in Figure 47.20 that the age structure for the United States is relatively even except for a bulge that corresponds to the "baby boom" that lasted for about two decades after the end of World War II. Even though couples born during those years are now having an average of fewer than two children, the nation's overall birth rate still exceeds death rate because there are so many "boomers" of reproductive age. Immigration also contributes in a minor way to the growth of the U.S. population. Nevertheless, the overall growth rate of the United States is relatively low—about 1% per year compared with 1.7% for the world as a whole.

A unique feature of human population growth is our ability to control it with voluntary contraception and government-sponsored family planning. Reduced family size certainly contributes to successful population control, but the U.S. government, formerly a major funder of world family-planning efforts, has unfortunately reduced its support for such programs drastically since 1980. Social change and the rising educational and career aspirations of women in many cultures encourage them to delay marriage and postpone reproduction. Delayed reproduction dramatically decreases population growth rates; recall the inverse relationship between generation time and population growth rate illustrated in Figure 47.10. You can develop a sense of this phenomenon by imagining two human populations in which women each produce three children but begin reproduction at different ages. In one population, females first give birth at age 15 and in the other at age 30. If we start with a cohort of newborn girls, after 30 years, women in the first population will already begin to have grandchildren, whereas women in the second population will be giving birth to their first children. After 60 years, women in the first population will have a large number of great-great grandchildren (who will themselves begin to reproduce 15 years later), but women in the second population will just begin to see their grandchildren being born.

Just as it is difficult to predict future sizes of other animal and plant populations, it is difficult to foresee the future for human population growth, though for somewhat different reasons. Technology has undoubtedly increased Earth's carrying capacity for humans, but no population can continue to grow indefinitely. Exactly what the world's human carrying capacity is and under what circumstances we will approach it are topics of great concern and debate. Ideally, human populations would reach carrying capacity smoothly and then level off. This will occur when birth rates and death rates are equal, and a decrease in birth rate is more desirable than an increase in death rate. If, however, the population fluctuated about K, we would expect periods of increase followed by mass death, as has occurred during plagues, localized famines, and international military conflicts. In any case, the human population must eventually stop growing. Unlike other organisms, we can decide whether zero population growth will be attained through social changes involving individual choice or government intervention, or through increased mortality due to resource limitation and environmental degradation. For better or worse, we have the unique opportunity to decide the fate of our species and the rest of the biosphere.

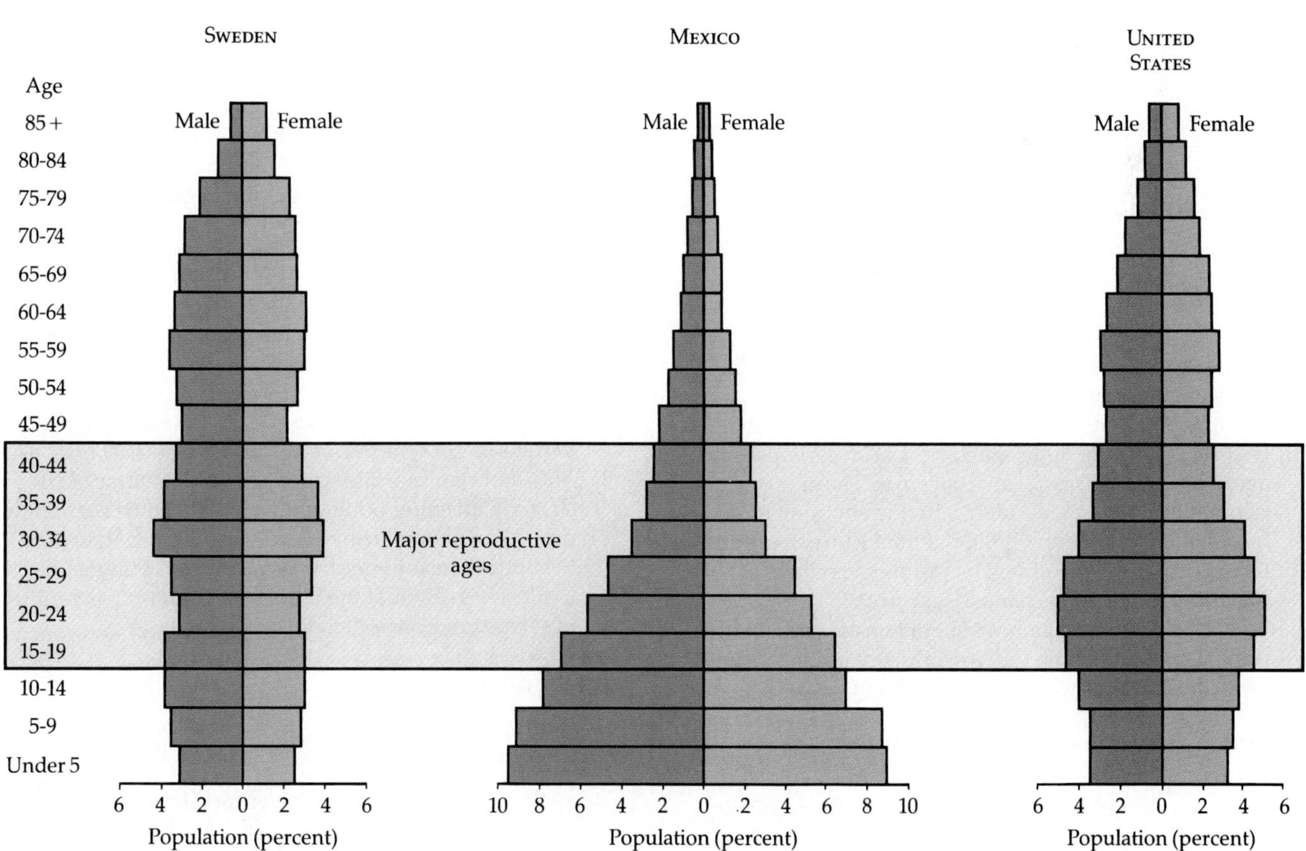

Figure 47.20
Age structures of three nations. The proportion of individuals in different age groups (1985 data) has a significant impact on the potential for future population growth. Mexico, for example, has a large fraction of individuals who are young and likely to reproduce in the near future. In contrast, Sweden's population is distributed more evenly through all age classes, with a high percentage of individuals beyond prime reproductive years. The United States has a fairly even age distribution, except for the bulge corresponding to the post–World War II baby boom.

* * *

We have seen throughout this chapter that population dynamics change during ecological time and evolutionary time as a result of interactions with the environment, which includes both physical factors and other organisms living in the same place. In the next chapter, we will examine in more detail the interaction of organisms within communities.

STUDY OUTLINE

1. Population ecology is the study of population fluctuations and factors influencing population dynamics.

Density and Dispersion (pp. 1084–1086)

1. A population is a group of individuals of the same species that inhabit a specified geographical range and exhibit a characteristic density and dispersion.
2. Density is the number of individuals per unit area or volume. Dispersion is the pattern of spacing for individuals and may range from clumped, to uniform, to random, as determined by various environmental or social factors.

Demography (pp. 1087–1089)

1. Demography is the study of vital statistics that affect population size and growth.
2. The age structure of a population influences population growth, owing to the differential reproductive capabilities and mortality probabilities of the different age groups.
3. Life tables describe the fate of a cohort of newborn organisms throughout their lives, tabulating mortality rates, the number of survivors from one age to the next, and average reproductive output.
4. Survivorship curves, which plot the number of individuals of a cohort alive at each age, can be divided into three general types depending on whether mortality is greater among the young or old or constant over all ages.

The Evolution of Life Histories (pp. 1089–1092)

1. Natural selection has led to the evolution of diverse life history strategies that maximize lifetime reproductive success.
2. The clutch size, number of reproductive episodes per lifetime, and age at first reproduction are three important life history characteristics. Natural selection balances the energetic trade-offs among these traits.
3. Some populations are said to have an opportunistic life history, producing large clutch sizes at an early parental age, usually in variable environments. The size of such populations fluctuates widely with the environment.
4. In populations with an equilibrial life history, reproductive efforts are concentrated on producing relatively few offspring with a good chance of survival.
5. Most populations have both opportunistic and equilibrial characteristics. The relative influence of the two may vary under different environmental conditions.

Models of Population Growth (pp. 1092–1097)

1. The exponential growth equation ($dN/dt = r_{max} N$) is a useful way to think about the potential of a population for explosive growth. This model predicts very simply that the larger a population becomes, the faster it grows. Such exponential growth is never sustained for extended periods of time in any population.
2. A more realistic model limits growth by incorporating the carrying capacity (K), the maximum population size that can be supported by the available resources. The logistic equation ($dN/dt = r_{max} N [K - N]/K$) fits an S-shaped curve in which population growth levels off as size approaches carrying capacity.
3. Few populations fit the logistic model exactly, but many show a generally similar pattern in which the growth rate is highest when the population size is intermediate.

The Regulation of Population Size (pp. 1097–1101)

1. Population size is regulated through density-dependent factors, though density-independent factors strongly influence the populations of some species.
2. Density-dependent factors have a greater effect as the population density increases, owing to more intense intraspecific competition for resources or increased predation and disease. The logistic equation reflects the effect of density-dependent factors, which ultimately stabilize populations around the carrying capacity.
3. Density-independent factors, such as climatic events, reduce population size by a given fraction, regardless of its density. Such factors are often superimposed on density-dependent factors.
4. Defining the importance of specific environmental factors in population regulation is complicated by the overlap of density-dependent and density-independent factors and the difficulty of determining cause-and-effect relationships.
5. Some populations have remarkably regular population cycles that may be a result of the physiological effects of crowding or time lags in response to density-dependent factors.

Human Population Growth (pp. 1101–1104)

1. Human population growth has been exponential for centuries, sustained by such factors as the Industrial Revolution and improved nutrition, medical care, and sanitation. Different countries have different rates of growth for various environmental, cultural, and historical reasons.
2. The importance of age structure as a demographic factor strongly affecting population growth can be seen in the post–World War II baby boom in the United States and in the bottom-heavy age structure of some countries.
3. Explosive growth of the human population has resulted in severe environmental degradation.

SELF-QUIZ

1. A uniform dispersion pattern for a population may indicate that
   a. the population is spreading out and increasing its range
   b. resources are heterogeneously distributed
   c. individuals of the population are competing for some resource, such as water and minerals for plants or nesting sites for animals
   d. there is an absence of strong attractions or repulsions among individuals
   e. the density of the population is low

2. In a mark-recapture study of a lake trout population, 40 fish were captured, marked, and released. In a second capture, 45 fish were captured; 9 of these were marked. What is the estimated number of individuals in the lake trout population?

a. 90 c. 360 e. 1800

b. 200 d. 800

3. The term $(K - N)/K$ influences dN/dt such that

a. the increase in actual population numbers is greatest when N is small

b. as N approaches K, r_{max} (the intrinsic rate of increase) becomes smaller

c. when N equals K, population growth is zero

d. when K is small, the influence of density-dependent factors is smaller

e. as N approaches K, birth rate approaches zero

4. The carrying capacity for a population is

a. the number of individuals in that population

b. reached when mortality exceeds natality

c. inversely related to r_{max}

d. the population size that can be supported by available resources for that species within the habitat

e. set at 8 billion for the human population

5. A Type III survivorship curve would be expected in a species in which

a. mortality occurs at a constant rate over the lifespan

b. parental care is extensive

c. a large number of offspring are produced, but parental care is minimal

d. mortality rate is quite low for the young

e. *K*-selection prevails

6. A population that has a relatively low *r* value will most likely

a. have large clutch sizes with relatively small offspring

b. be found in environments that are highly variable

c. have an early age of first reproduction and a short generation time

d. produce fewer offspring with more competitive capabilities

e. be regulated by density-independent factors

7. The example of the population cycles of the snowshoe hare and its predator, the lynx, illustrates that

a. predators are the major factor in controlling the size of prey populations, and are, in turn, regulated in their numbers by the oscillating supply of prey

b. the two species must have evolved in close contact with each other since their life histories are intertwined

c. one should not conclude a cause-and-effect relationship when viewing population patterns without careful observation and experimentation

d. both populations are controlled by density-independent factors

e. the hare population is *r*-selected whereas the lynx population is *K*-selected

8. The current size of the human population is closest to

a. 2 million d. 6 billion

b. 3 billion e. 10 billion

c. 4 billion

9. Consider five human populations that differ demographically *only* in their age structures. The population that will grow the most in the next 30 years is the one with the greatest fraction of people in which age group?

a. 10–20 c. 30–40 e. 50–60

b. 20–30 d. 40–50

10. All these terms are characteristic of the human populations in industrialized countries *except*

a. relatively small clutch size

b. several potential reproductions per lifetime

c. opportunistic life history

d. type I survivorship curve

e. bottom-heavy age structure

CHALLENGE QUESTION

1. A biologist studied a population of fish called cichlids in an African lake that is 120 hectares (1 hectare = 10,000 m^2) in area. She found that the fish lived only in scattered reed beds that accounted for one-fourth the area of the lake. She caught 185 fish, tagged them, and released them. Two days later, she netted 208 fish and found that 35 of them were tagged. About how many cichlids are in the lake? What is the density of the cichlid population (in this case, in fish per hectare of lake surface)? What is the pattern of dispersion of the fish? How might the pattern of dispersion affect your interpretation of their density?

SCIENCE, TECHNOLOGY, AND SOCIETY

1. The mountain gorilla, spotted owl, and California condor are all endangered by human activities. In general, are these *r*-selected or *K*-selected species? How might the life histories of these species leave them vulnerable to human encroachment? How would you design plans to save them that reflected their life histories?

2. Many people regard the rapid population growth of developing countries as our most serious environmental problem. Others think that the population growth in industrialized countries, though smaller, is actually a greater threat to the environment. What kinds of problems result from population growth in (a) developing countries, and (b) the industrialized world? Which do you think is the greater threat, and why?

FURTHER READING

Ackerman, L., et al. "The Successful Animal." *Science 86*, January/February 1986. Cultural and historical aspects of human population growth and control.

Daily, G. C., and P. R. Ehrlich. "Population, Sustainability, and Earth's Carrying Capacity." *BioScience*, November 1992. The relationship between current population growth and the standard of living for future generations.

Lewin, R. "Judging Paternity in the Hedge Sparrow's World." *Science*, March 31, 1989. The use of DNA fingerprinting by population biologists to study the reproductive success of individual birds.

Peters, G. L., and R. P. Larkin. *Population Geography*, 3rd ed. Dubuque, IA: Kendall/Hunt Publishing, 1989. A detailed worldwide overview of human population growth and distribution.

Pool, R. "Ecologists Flirt with Chaos." *Science*, January 20, 1989. Does chaos theory apply to population ecology?

Savonen, C. "One Salmon, Two Salmon . . . 10,000 Salmon: Counting the Fish in Alaska." *Oceans*, January/February 1985. A population density determination in action.

48 Community Ecology

Two General Views of Communities

Population Interactions

Community Structure: Patterns and Processes

Succession

Biogeographical Aspects of Diversity

Conservation Biology: Lessons from Community Ecology and Biogeography

Figure 48.1
Population interactions shape communities. Communities are characterized by the interactions among their component populations. In this Kenyan scene, vultures wait for whatever will remain of a lion's zebra kill. In this chapter, you will learn how interactions such as predation, competition, and symbiosis affect the structure of biological communities.

On your next walk through a natural field or woodland, or even across campus or through a park, try to observe some of the interactions of the species present. You may see birds using trees as nesting sites, bees pollinating flowers, shelf fungi growing on trees, spiders trapping insects in their webs, ferns growing in shade provided by trees—a sample of the many interactions that exist in any ecological theater. In addition to the physical and chemical factors discussed in Chapter 46, an organism's environment includes other individuals in its population and populations of other species living in the same area. Such an assemblage of species living close enough together for potential interaction is called a **community** (Figure 48.1).

This chapter examines the diverse kinds of biotic interactions among populations and addresses the central issue in community ecology: What factors are most significant in structuring a community—in determining its species composition and both the absolute and the relative abundance of species present? The analysis of communities is an active area of ecological research, a field replete with unanswered questions and controversies.

TWO GENERAL VIEWS OF COMMUNITIES

Why are certain combinations of species found together as members of a community? Two divergent views on this question emerged among ecologists in the 1920s and 1930s, derived primarily from observations of plant distributions. One idea, the "individualistic" concept, first enunciated by H. A. Gleason, depicted the community as a chance assemblage of species found in the same area simply because they happen to have similar physical (abiotic) requirements—for example, for temperature, rainfall, and soil type. The alternative view, the "interactive" concept, advocated by American ecologist F. E. Clements, saw the community as an assemblage of closely linked species, locked into association by mandatory biotic interactions that cause the community to function as an integrated unit. Carried to the extreme, this view interpreted communities as a kind of "superorganism." Evidence for the latter view was based on the observation that certain species of plants are consistently found together. For example, deciduous forests in the northeastern United States almost always include certain species of oak, maple, birch, and beech, along with a specific assemblage of shrubs and vines.

We can apply the hypothetico-deductive method to evaluate these two alternative hypotheses. The two views of communities make contrasting predictions about how plant species should be distributed along a gradient of environmental variables, such as altitude or temperature. The individualistic hypothesis predicts that communities should generally lack discrete geographic boundaries because each species has an independent distribution along the environmental gradient. In other words, each species will be distributed according to its tolerance ranges for abiotic factors that vary along the gradient, and plant communities change continuously along the gradient with the addition or loss of particular species (Figure 48.2a). According to the interactive hypothesis, in contrast, species should be clustered into discrete communities with distinct boundaries, because the presence or absence of a particular species is largely governed by the presence or absence of other species with which it interacts (Figure 48.2b).

In most cases, especially where there are broad regions characterized by gradients of environmental variation, the composition of plant communities does seem to change on a continuum, with each species more or less independently distributed (Figure 48.2c). Such distributions generally support the view of plant communities as relatively loose associations without discrete boundaries. However, where some key factor in the physical environment changes abruptly, adjacent communities are delineated by correspondingly sharp boundaries. In California, for example, many native plants have been replaced by introduced European grasses, but in some strictly delimited areas where serpentine rocks increase the magnesium content of the soil, native wildflowers persist (Figure 48.3). The individualistic hypothesis can account for this type of discontinuity without assuming that certain plant species occur in the same area because they are locked into mandatory community relationships.

However, these views may not apply to the animals in a community, which are sometimes linked more intimately to other organisms. For example, limpkins, long-beaked birds of Florida swamps, feed primarily on one species of snail. The birds' evolutionary adaptations make them extremely efficient in their specialized foraging, but their geographic distribution is restricted by the distribution of their prey. In contrast, gray squirrels of the eastern United States are not as strongly tied to a particular food. Although they may be found in a variety of habitats, including forests where pine is abundant, they are most common in areas of mature hardwoods, such as oak and hickory trees.

Thus, in trying to explain why certain species commonly occur together in communities, it is difficult to make simple generalizations that are both accurate and widely applicable. The distributions of all organisms are probably affected to some extent by both abiotic gradients and interactions with other species.

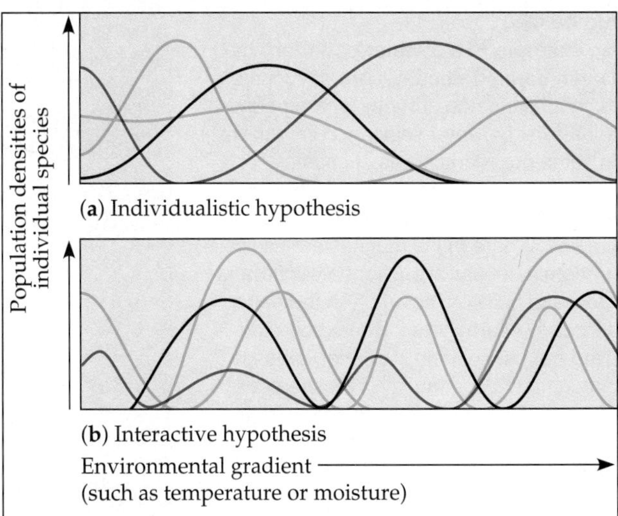

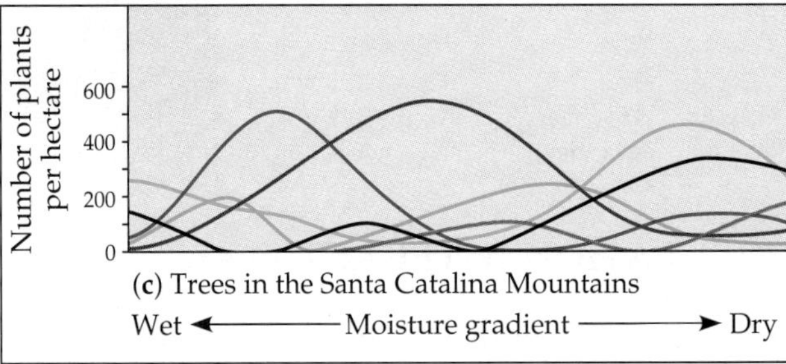

Figure 48.2
Individualistic and interactive views of communities. These two major hypotheses can be compared and tested with graphs of species distributions along abiotic environmental gradients such as temperature or moisture. Each line on the graphs represents the abundance of one species. (**a**) The individualistic hypothesis suggests that species are independently distributed along gradients and that a community is simply the assemblage of species that happen to occupy the same area. (**b**) The interactive hypothesis suggests that communities are discrete groupings of particular species that are closely interdependent and nearly always occur together. (**c**) The distribution of tree species at one elevation in the Santa Catalina Mountains of Arizona supports the individualistic view of communities. Each tree species has an independent distribution along the gradient, apparently conforming to its tolerance for moisture, and the species that live together at any point along the gradient have similar physical requirements. Because the vegetation changes continuously along the gradient, it is impossible to delimit sharp boundaries for the communities.

Figure 48.3
An example of a distinct boundary between communities. Where environmental factors change abruptly, sharp boundaries exist between adjacent communities. In some grassland areas of coastal California, weathering of certain types of rock increases the magnesium content of the soil. Where these magnesium-rich soils predominate, native California wildflowers grow and blossom colorfully in the early spring. In nearby areas where soil is derived from sandstone, however, grasses dominate the community. The sharpness of these local boundaries is apparent in this scene from the Jasper Ridge Biological Preserve near Stanford University.

POPULATION INTERACTIONS

Although the geographic distributions of many species are largely determined by their adaptations to abiotic environmental factors, all organisms are also influenced by biotic interactions with other individuals in their immediate vicinity. In Chapter 47, we examined one such interaction, intraspecific competition. In this section, we consider **interspecific interactions,** those that occur between populations of different species living together within a community. As you will see, these interactions may have positive, negative, or neutral effects on one or more of the populations involved. Before describing these interactions, however, we will briefly discuss their evolutionary contexts.

Coevolution

Just as the physical and chemical features of the environment are important factors in adaptation by natural selection, interspecific interactions are also subject to evolutionary change. In some cases, the adaptation of one species to the presence of another has a relatively obvious evolutionary basis. For example, natural selection has apparently favored peppered moths that blend in with the living lichens on which the moths sometimes rest (see Figure 20.10).

In its broadest sense, **coevolution** is the term used to describe more complex interactions involving a series of reciprocal evolutionary adaptations in two species: A change in one species acts as a new selective force on another species, and counteradaptation by the second species, in turn, affects selection on individuals in the first. Coevolution has been studied most extensively in predator–prey relationships and in symbiosis. One general example of coevolution was discussed in Chapter 27: the interactions between some species of flowering plants and their exclusive pollinators.

In many cases where two species appear to have coevolved adaptations, it is usually difficult to establish that an evolutionary change in one species creates a selective force that induces an evolutionary change in the other. As an example, let's examine a specific relationship that probably illustrates coevolution but also demonstrates the complexities that may confound our attempts to unravel evolutionary history. Passionflower vines of the genus *Passiflora* are protected against most herbivorous insects by their production of toxic compounds in young leaves and shoots. However, the larvae of butterflies of the genus *Heliconius* can tolerate these defensive chemicals, apparently by using digestive enzymes to destroy the compounds. This counteradaptation has enabled *Heliconius* larvae to become specialized feeders on plants that few other insects can eat (Figure 48.4).

Survival of the larvae is further enhanced by a behavioral adaptation of the butterflies. The eggs that female *Heliconius* butterflies lay on the leaves of passionflower vines are bright yellow, and other females generally avoid laying eggs on leaves marked by yellow dots. This behavior presumably reduces intraspecific competition among the larvae for food because only a few larvae hatch and feed on any one leaf. An infestation of *Heliconius* larvae can devastate a passionflower vine, and we may assume that these poison-resistant insects are a strong selection force favoring the evolution of additional defenses in the plants.

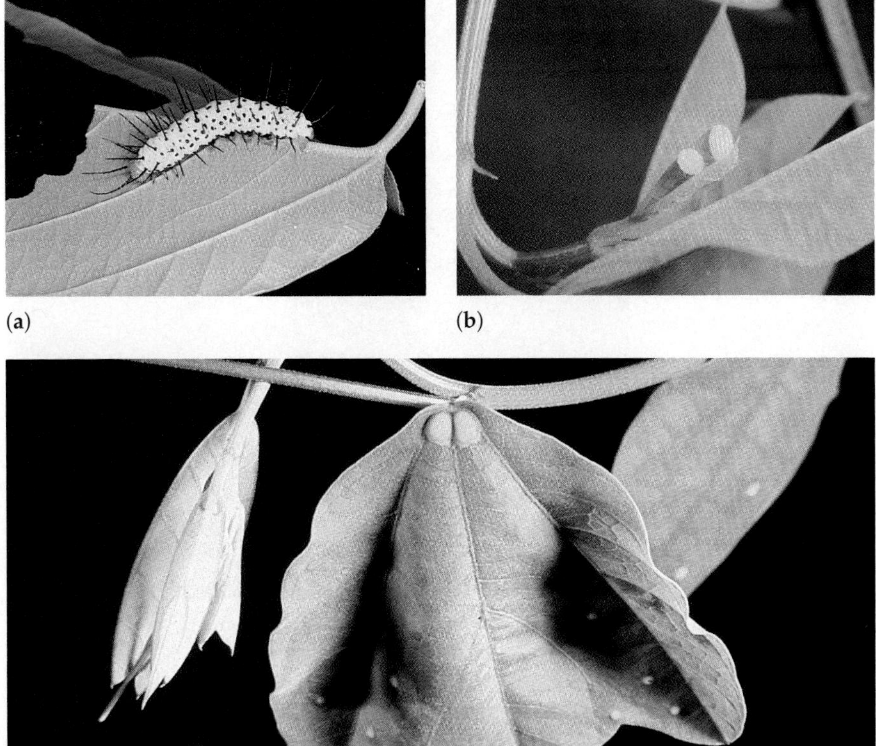

(a) (b) (c)

Figure 48.4
Evolutionary interactions between a plant and an insect. (**a**) Passionflower vines *(Passiflora)* produce toxic chemicals that help protect leaves from herbivorous insects. A counteradaptation has evolved in the butterfly *Heliconius:* The larvae can feed on the leaves because they have digestive enzymes that break down the toxic compounds. (**b**) The females of some *Heliconius* species will avoid laying eggs on passionflower leaves that already carry bright yellow egg clusters deposited by other females. This behavior reduces intraspecific competition on individual leaves and helps ensure an adequate food supply for offspring. (**c**) These egglike yellow nectaries grow on the leaves of some species of passionflowers. Experiments show that *Heliconius* females avoid laying eggs on leaves with these yellow spots. These nectaries, as well as smaller ones scattered over the leaf, also attract ants and other insects that prey on the butterfly eggs and larvae.

In some species of *Passiflora,* the leaves have conspicuous yellow spots that look like *Heliconius* eggs, an adaptation that may divert the butterflies to other plants in their search for egg-laying sites. The story is more complicated, however. The yellow "spots" on the passionflower vine are actually nectaries, which attract ants and wasps that prey on *Heliconius* eggs or larvae. There is evidence, too, that the mere presence of ants on a leaf will discourage a *Heliconius* butterfly from laying eggs there. Thus, adaptations that may seem, on superficial examination, to be coevolutionary responses between just two species may, in fact, result from interactions among many species in the community. It is difficult to sort out the importance of the various selective forces, and the simple idea of coevolution as adaptation–counteradaptation occurring exclusively between two species does not often adequately describe the interactions within communities.

Despite the problems in assessing cause and effect in the evolution of complex ecological relationships, biologists agree that adaptation of organisms to other species in a community is a fundamental characteristic of life. Put another way, interactions of species in ecological time often translate into adaptations over evolutionary time. We will encounter several examples, as we now survey the interspecific interactions of greatest importance in community ecology: predation, competition, and symbiosis.

Predation

The most conspicuous population interactions in all communities are those in which a predator eats its prey (see Figure 48.1). In most cases, the predator and prey are of different species, though cannibalism has been reported in many animals. Here, we use the terms **predator** and **prey** not only for animals that eat other animals, but also for herbivores that feed on plants. In Chapter 47, we discussed how predation might cause density-dependent regulation of both predator and prey populations. In this section, we briefly describe some adaptations of predators and focus on the varied defensive responses of their prey.

Many important feeding adaptations of predators are both obvious and familiar. Predators have acute senses that locate and identify potential prey, and many have structures such as claws, teeth, fangs, stingers, and poison to catch and subdue, or simply chew, the organisms on which they feed. Rattlesnakes and other pit vipers, for example, locate their prey with special heat-sensing organs located near their eyes, and they kill small birds and mammals by injecting them with toxins through their fangs. Similarly, many herbivorous insects locate appropriate food plants with chemical sensors on their feet, and their mouthparts are adapted to shred tough vegetation. Predators that pursue their prey are generally fast and agile, whereas those that lie in ambush are often camouflaged in their environments. Through repeated encounters with predators over evolutionary time, most prey species have evolved a variety of defensive mechanisms, which we will now survey.

Plant Defenses Against Herbivores In plants, which cannot run away from herbivores, a variety of mechanical and chemical defenses against the animals that might eat them have evolved. Thorns may discourage large herbivores such as vertebrates, and some plants have microscopic crystals in their tissues, or hooks or spines on their leaves, that make feeding difficult even for small insects.

Many plants produce chemicals that function in defense by making the vegetation distasteful or harmful to an herbivore. These chemicals, classified as **secondary compounds,** are so named because they form as metabolic by-products of a major biochemical pathway, such as glycolysis or the Krebs cycle (see Chapter 9). For many years, plant physiologists were puzzled about the widespread occurrence and high concentrations of these secondary compounds, and only recently has their importance in defense become clear.

Some well-known poisons and drugs are secondary compounds that probably function as chemical weapons against herbivores: strychnine, produced by plants of the genus *Strychnos;* morphine, from the opium poppy; nicotine, produced by tobacco; and mescaline, from the peyote cactus. Other compounds are responsible for the familiar flavors of cinnamon, cloves, and peppermint. And tannins, used by humans in the preservation of leather, are widespread, especially in oaks. Some plants even produce secondary products that are analogous to insect hormones and cause abnormal development in some insects that eat them.

The specific defenses of a plant population can act as selective agents that provoke the evolution of counteradaptations in herbivore populations. The responses of herbivores may nullify the plant defenses and allow subsequent generations of herbivores to eat the descendants of the original plant population. Additional defenses may then evolve in the plants. Ecologists usually study the products of such coevolutionary interactions after several layers of counteradaptations have already evolved, as we saw earlier in the *Passiflora-Heliconius* example. Although a plant's defenses may restrict the number of herbivore species that can feed upon it, the herbivores must eat to reproduce, and selection to overcome plant defenses is strong. Hence, no plant defense is likely to provide eternal protection.

Figure 48.5
Camouflage. Cryptic coloration of the canyon tree frog *(Hyla arenicolor)* allows it to "disappear" on a granite background.

Animal Defenses Against Predators Animals can avoid being eaten by using passive defenses, such as hiding, or active ones, such as escaping or defending themselves against predators. Fleeing is perhaps the most direct antipredator response, and many animals use a zigzag pattern of flight or a direct retreat into a shelter to avoid being caught. Active self-defense is less common, though some large grazing mammals will vigorously defend their young from predators such as lions. Many other defenses, however, rely on adaptive coloration, which has evolved repeatedly among animals in a variety of contexts.

Camouflage, called **cryptic coloration,** is the quintessential passive defense, making potential prey difficult to spot against its background (Figure 48.5). A camouflaged animal need only remain still on an appropriate substrate to avoid detection. The case of the English peppered moth (discussed in Chapter 20) is a classic example of how selection by predators can result in the evolution of cryptic coloration. The shape of an animal can also help camouflage, as the example in Figure 48.6 illustrates.

Deceptive markings are another form of adaptive coloration. Large, fake eyes or false heads can apparently deceive predators momentarily, allowing the prey to escape (Figure 48.7), or they may induce the predator to strike a nonvital area.

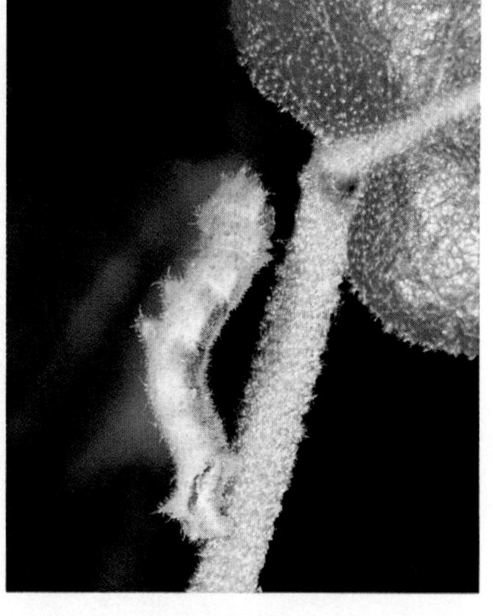

Figure 48.6
Different camouflage for different seasons: an experimental study of a remarkable adaptation. The larva (caterpillar) of *Nemoria arizonia,* a moth that lives in the southwestern United States, occurs in two seasonal forms. Caterpillars that hatch during the spring feed on catkins, the male flowers of oaks, and these caterpillars develop into forms that resemble the catkins in shape and coloration (left). A second brood of caterpillars hatches in summer. By then, catkins are gone, and the new caterpillars feed on leaves. These "summer" caterpillars develop into forms that resemble oak stems rather than catkins (right). In laboratory experiments, caterpillars fed with catkins develop in the catkinlike form, and the caterpillars fed on leaves become stemlike individuals; diet determines the morphology of the caterpillar. (In fact, the two caterpillars in these photographs are full siblings raised on different diets.) In the natural environment, each form of the caterpillar is camouflaged against the vegetation it feeds and rests on, an evolutionary adaptation that presumably makes the caterpillars less visible to predators such as birds.

Some animals that are easy to find have mechanical or chemical defenses against would-be predators (Figure 48.8). Most predators are strongly discouraged by the familiar defenses of porcupines and skunks. Some insects acquire chemical defense passively by accumulating toxins from the plants they eat. Monarch butterflies, for example, store cardiac poisons they acquire from the milkweed plants they eat as larvae; remarkably, the poisons are retained through metamorphosis and are present in adult butterflies. The monarch is resistant to these poisons, but the compounds are distasteful or harmful to most other animals. Birds that eat monarch butterflies regurgitate their prey and quickly learn to avoid feeding on that species of insect.

Figure 48.7
Deceptive coloration. The hindwing markings of the io moth *(Automerisio)* resemble the eyes of a much larger animal. Potential predators may be momentarily startled when the moth moves its forewings, allowing the moth to escape.

Animals with effective physical or chemical defenses are often brightly colored, a warning to predators known as **aposematic coloration** (Figure 48.9). This warning coloration seems to be adaptive, because predators learn more quickly to avoid harmful prey that are extremely conspicuous.

Mimicry A predator or species of prey may gain a significant advantage through **mimicry,** a phenomenon in which the **mimic** bears a superficial resemblance to another species, the **model.** Defensive mimicry in prey often involves aposematic models.

In **Batesian mimicry,** a palatable species mimics an unpalatable model. In one intriguing example, the small juveniles of a lizard species in the Kalahari Desert of southern Africa mimic the color pattern and posture of a similarly sized, aposematically colored beetle that sprays an acidic and pungent fluid when

(a)

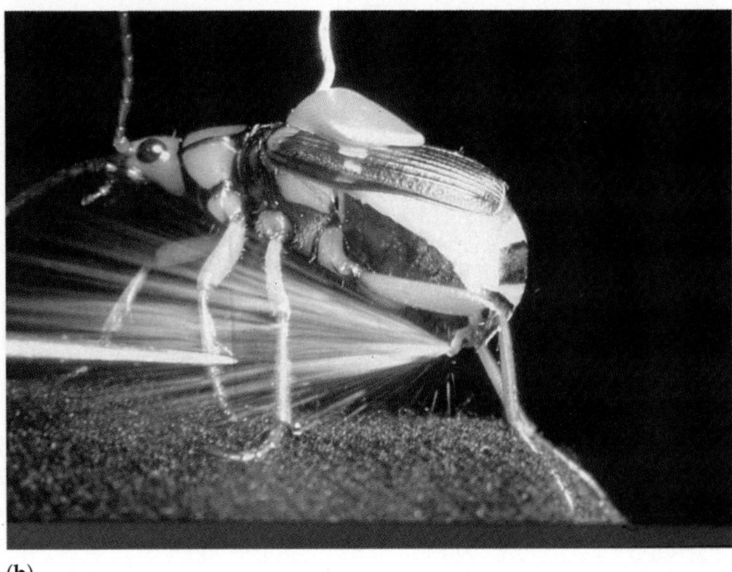

(b)

Figure 48.8
Mechanical and chemical weapons against predators. (**a**) Sharp quills are a mechanical defense for the porcupine. (**b**) The bombardier beetle sprays attackers with a toxic chemical.

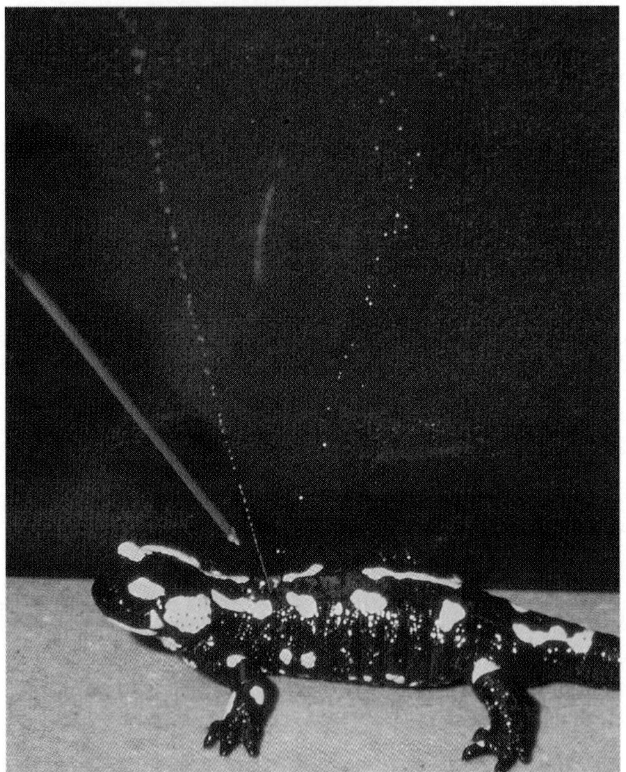

Figure 48.9
Aposematic (warning) coloration. The conspicuous markings of the fire salamander, an amphibian that can squirt a nerve poison from glands on its back, act as a warning to would-be predators. Warning coloration probably trains predators quickly to avoid such brightly colored animals.

disturbed by a predator (Figure 48.10). As the lizards grow to a size larger than that of the beetle models, they develop cryptic coloration instead. Additional examples of Batesian mimicry are the many harmless snakes that mimic the conspicuous red, white, and black markings of the poisonous coral snake. For Batesian mimicry to be effective, however, models must generally outnumber mimics; otherwise, predators would learn that animals with a particular coloration are good rather than bad to eat.

In **Müllerian mimicry,** two or more unpalatable, aposematically colored species resemble each other. Presumably each species gains an additional advantage, because predators will learn more quickly to avoid any prey with a particular appearance.

Predators also use mimicry in a variety of ways. For example, some snapping turtles have tongues that resemble a wriggling worm, thus luring small fish; any fish that tries to eat the "bait" is itself quickly consumed as the turtle's strong jaws snap closed.

Interspecific Competition

When two or more species in a community rely on similar limiting resources, they may be subject to **interspecific competition.** Competition is manifested in different ways under natural conditions: Actual fighting over resources is termed **interference competition,** whereas the consumption or use of similar resources is

(a)

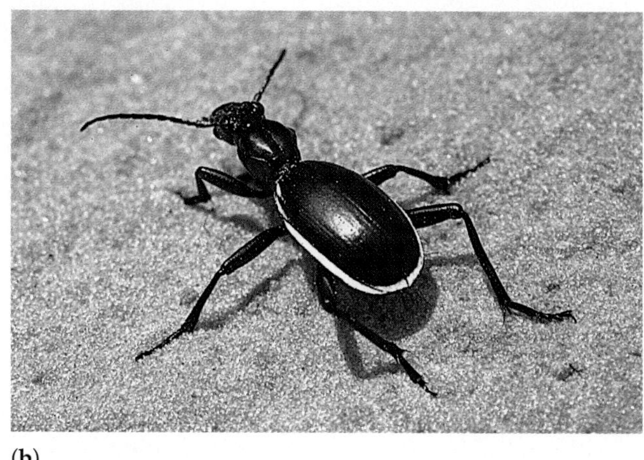

(b)

Figure 48.10
Batesian mimicry. In Batesian mimicry, a palatable species mimics the appearance of an unpalatable one that is usually aposematically colored. (**a**) Juveniles of the harmless lizard *Heliobolus lugubris* mimic the color and posture of (**b**) the oogpister beetle *(Anthia)*, a species that sprays a noxious fluid at would-be predators.

called **exploitative competition.** The density-dependent effects of interspecific competition are similar to those of intraspecific competition, discussed in Chapter 47. As population densities increase, every individual has access to a smaller share of some limiting resource; as a result, mortality rates increase, birth rates decrease, and population growth is curtailed. In interspecific competition, however, the population growth of a species may be limited by the density of competing species as well as by the density of its own population. For example, if several bird species in a forest feed on a limited population of insects, the density of each species may have a negative impact on population growth in the others. Similarly, species may compete for nesting sites, shelters, or any other resource that is in short supply.

The Competitive Exclusion Principle In the first quarter of this century, two mathematician-biologists, A. J. Lotka and V. Volterra, independently modified the logistic model of population growth (see Chapter 47) to incorporate the effects of interspecific competition. They predicted that two species with similar requirements could not co-occur in the same community; one species would inevitably harvest resources and reproduce more efficiently, driving the other to local extinction. Even a slight reproductive advantage would eventually lead to the elimination of the inferior competitor and an increase in the density of the superior one.

In 1934, Russian ecologist G. F. Gause tested this hypothesis with laboratory experiments on the effects of interspecific competition between two closely related species of the protozoan *Paramecium, P. aurelia* and *P. caudatum* (Figure 48.11). Grown in separate cultures under constant conditions and with a constant

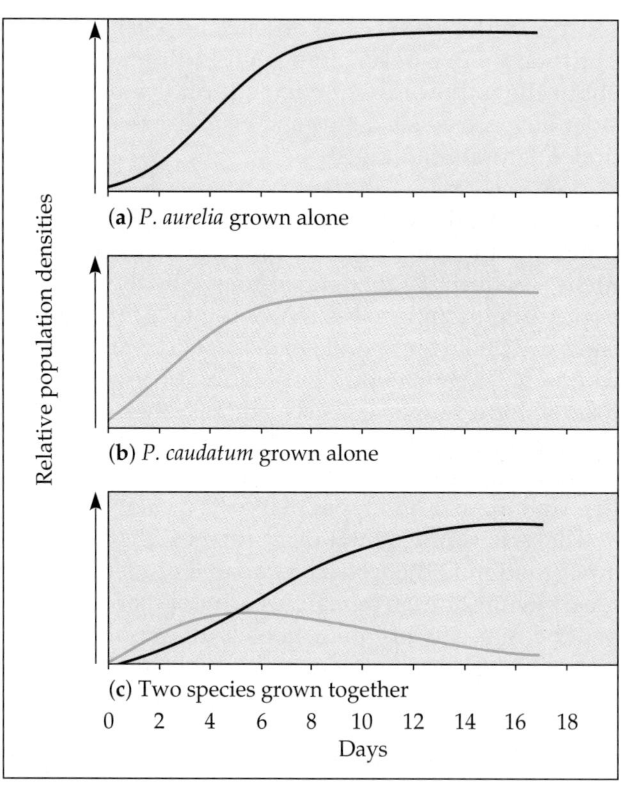

Figure 48.11
Competition in laboratory populations of *Paramecium.* In separate laboratory cultures with constant amounts of bacteria added every day for food, populations of the two species, (**a**) *P. aurelia* and (**b**) *P. caudatum,* each grow to carrying capacity. (**c**) When the two species are grown together, *P. aurelia* has a competitive edge in obtaining food, and *P. caudatum* is driven to extinction in the culture.

amount of bacteria added every day as food, each population of *Paramecium* grew until it leveled off at what was apparently the carrying capacity. When the two species were cultured together on a constant food supply, however, *P. caudatum,* apparently unable to compete with *P. aurelia,* was driven to extinction in the microcosm of the culture dish. Gause's experiments supported the hypothesis that two species with similar needs for the same limiting resources cannot coexist in the same place. This concept was later termed the **competitive exclusion principle.** Subsequent laboratory experiments on several other species of animals and plants reinforced the principle.

Although laboratory studies have generally confirmed the predictions of the competitive exclusion principle, natural communities are infinitely more complex than laboratory environments. Ecologists have argued, sometimes vociferously, about the importance of competition in nature, but only recently have they used field experiments to evaluate the issue. Most studies of competition have, instead, used indirect evidence from observations of how species use resources in the presence or absence of potential competitors. Before describing such indirect evidence about the importance of competition, we must consider how ecologists define and visualize resource use under natural conditions.

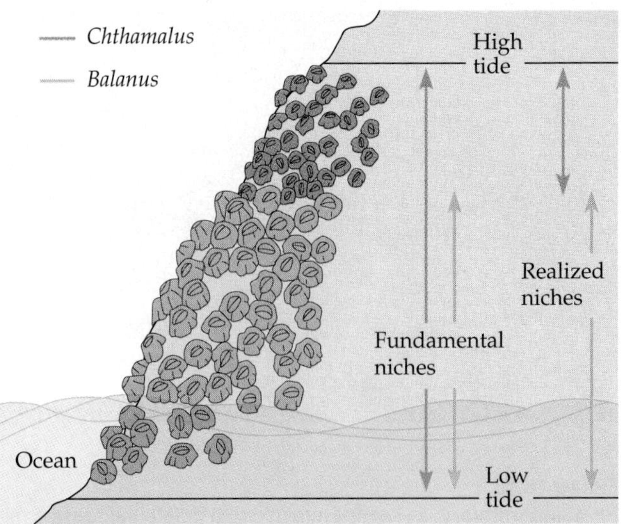

Figure 48.12
Fundamental and realized niches. *Balanus balanoides* and *Chthamalus stellatus* are two species of barnacle that grow on the same rocks along the Scottish coast. These rocks are exposed during low tide. The barnacles have a stratified distribution, with *Balanus* most concentrated in the lower portions of the shore and *Chthamalus* in the higher strata. The swimming larvae of the barnacles may settle randomly on the rocks and begin to develop into sessile adults, but *Balanus* fails to survive high on the rocks, apparently because it is unable to resist desiccation when these areas are exposed to air for several hours during low tides. Thus, its fundamental niche and its realized niche are similar. Although *Chthamalus* is concentrated primarily on the upper strata of rocks, when American ecologist Joseph H. Connell removed *Balanus* from the lower strata, the *Chthamalus* population spread into that area. Apparently *Chthamalus could* survive lower on the rocks than where it is generally found, if it were not for competition from *Balanus;* thus, its realized niche is only a fraction of its fundamental niche. Connell actually documented *Balanus* overgrowing, undercutting, and crushing *Chthamalus.*

The Ecological Niche The way in which a population "fits into" an ecosystem is called its **ecological niche.** The niche is not defined simply by the habitat a population occupies; it is the sum total of the organism's use of the biotic and abiotic resources in its environment. The niche of a population of tropical tree lizards, for example, consists, among other variables, of the temperature range it tolerates, the size of trees upon which it perches, the time of day in which it is active, and the size and types of insects it eats.

The term **fundamental niche** refers to the resources a population is theoretically capable of using under ideal circumstances. In reality, each population is embedded in a web of interactions with populations of other species, and biological constraints, such as competition, predation, or the absence of some usable resources, may force the population to use only a subset of its fundamental niche. The resources a population actually uses are collectively called its **realized niche.** Figure 48.12 illustrates the difference between fundamental and realized niches for two species of barnacle.

We can now restate the competitive exclusion principle to say that two species cannot coexist in a community if their niches are identical. However, ecologically similar species can coexist in a community if there are one or more significant differences in their niches.

Evaluating Competition in Nature If competition is as potent a force as the competitive exclusion principle suggests, we would expect it to be quite rare in natural communities. After all, weaker competitors will face extinction if the ability to avoid competition with more successful species by using a different set of resources does not evolve. This expectation poses an intellectual and practical dilemma for ecologists, because it is difficult to demonstrate the existence and importance of a force (competition) that, by its very nature, cannot operate for long periods of time. Although we cannot look directly at the evolutionary history of a community, many ecologists stress the importance of past competition, citing several lines of circumstantial evidence that suggest it has been a major factor in shaping some of the ecological relationships we see today.

One line of evidence is simply the observation that similar species always seem to exhibit some niche differences when they coexist in a community. Patterns of **resource partitioning,** in which sympatric species consume slightly different foods or use other resources in

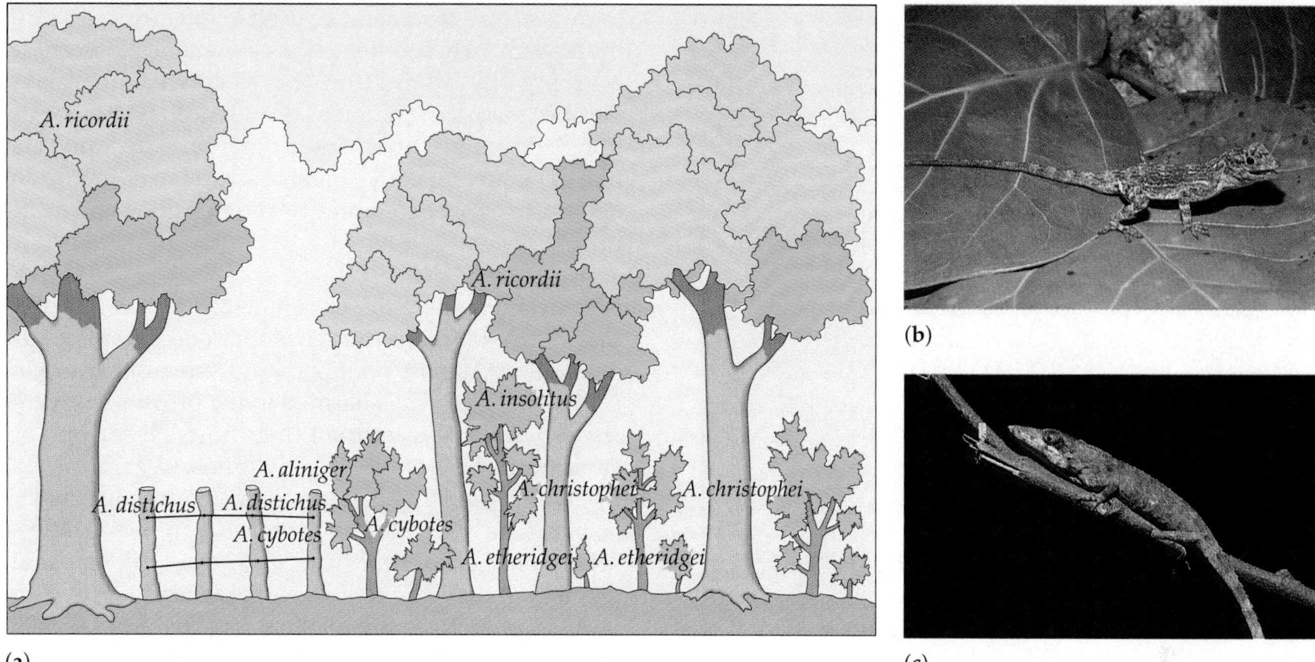

Figure 48.13
Resource partitioning in a group of coexisting lizards. (**a**) Seven species of *Anolis* lizards live in close proximity at La Palma in the Dominican Republic. Each species perches in a characteristic microhabitat, distinguished by the amount of sun it receives and the size of the vegetation. (**b**) *A. distichus*, for example, perches on fenceposts and other sunny surfaces (such as this leaf), whereas (**c**) *A. insolitus* usually perches on shady branches. Such patterns of resource partitioning probably reduce interspecific competition among the members of a community, enabling them to coexist within a small geographical area.

slightly different ways, are well documented, particularly among animals. Several species of small arboreal lizards of the genus *Anolis,* for example, often occur together in the same community. At one site in the Dominican Republic, seven *Anolis* species live in close proximity to each other, feeding on small arthropods that land within their territories. However, each species uses a characteristic perching site (Figure 48.13), and these perch differences presumably minimize competition among the lizard species. Each species also has morphological characteristics, such as body size or leg length, that adapt it to its particular microhabitat, suggesting that natural selection has favored perch site specializations among these sympatric lizards. Similar patterns of resource partitioning are evident in other *Anolis* communities throughout the American tropics.

A second line of circumstantial evidence for the importance of competition comes from comparisons of closely related species whose populations are sometimes sympatric and sometimes allopatric. Although allopatric populations of such species are similar in structure and use similar resources, sympatric populations may show a divergence in morphology and the resources they use. The Galapagos finches described in Chapter 20 provide a good example of such **character displacement** in beak sizes and, presumably, in the seeds that they can eat most efficiently (Figure 48.14). Allopatric populations of *Geospiza fuliginosa* and *G. fortis* have beaks of similar size, but on an island where both species occur, a significant difference in beak depth has evolved. This difference presumably allows the two species to avoid competition by feeding on seeds of different sizes.

Although examples of resource partitioning and character displacement are compelling, they do not really prove the importance of competition, because such studies do not demonstrate that the population density of one species influenced the population density of another when the two species were in more direct competition. However, controlled field experiments can provide such evidence. In a classic study, Joseph H. Connell manipulated the densities of the two barnacle species in Figure 48.12, which ordinarily grow in different strata of the rocky intertidal. Recall that wave action dislodges unattached organisms in this community, and the resident species presumably compete for the limited attachment space on the rocks. After Connell removed *Balanus* from the lower strata where it was most common, *Chthamalus* was able to grow there. This simple experiment, as well as direct observations of interference competition between the species, demonstrated that *Balanus* outcompetes *Chthamalus* where their fundamental niches overlap.

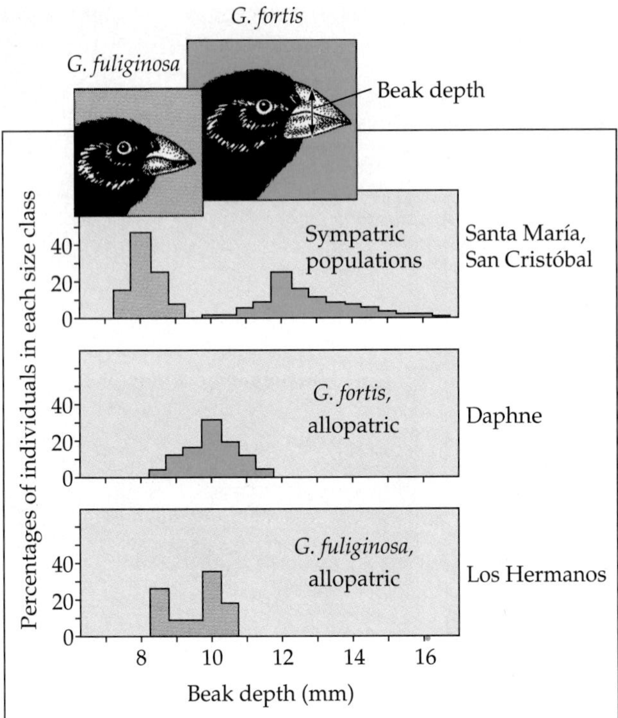

Figure 48.14
Character displacement. Although allopatric populations of potential competitors are often similar in morphology and use equivalent resources, sympatric populations may diverge in both characteristics. In this example, two species of Galapagos finches have similar beak morphologies and presumably eat similarly sized seeds, where their populations are allopatric on Daphne and Los Hermanos islands. However, where the two species are sympatric on Santa María and San Cristóbal, *Geospiza fuliginosa* has a shallower, smaller beak and *G. fortis* a deeper, larger one. Such evolutionary changes in morphology are thought to reflect resource partitioning. In this case, the two species have adapted to eating different sizes of seeds.

Similar field experiments on a variety of plant and animal species suggest that interspecific competition is strong under some circumstances, and that the effects of competition are often different for the competing species.

Symbiotic Interactions

Species commonly interact in ways other than by predation and competition. **Symbiosis** ("living together") is a term that encompasses a variety of interactions in which two species, a **host** and its **symbiont,** maintain a close association. There are three types of symbiotic interactions. In **parasitism,** one organism, the parasite, harms the host; in **commensalism,** one partner benefits without significantly affecting the other; and in **mutualism,** both partners benefit from the relationship. (The term *symbiosis* is sometimes used to refer exclusively to mutualism.) These interactions sometimes intergrade, blurring the distinctions between the categories. Because parasitic, commensal, and mutualistic interactions, like competition and predation, affect population densities, they, too, must be included as important determinants of a community's composition and structure.

Parasitism Parasitism can be treated as a special case of predation in the sense that one organism, the parasite, derives its nourishment from another organism, its host, which is harmed in the process. Parasites are generally smaller than their hosts, absorbing organic nutrients from their body fluids. Organisms that live within host tissues, such as tapeworms and malarial parasites, are called **endoparasites;** others that feed briefly on the external surface of a host, like mosquitoes and aphids, are called **ectoparasites.**

As in predator–prey relationships, natural selection favors parasites that are best able to locate hosts and feed on them. Many endoparasites acquire hosts by passive mechanisms. For example, *Ascaris,* a nematode endoparasite of the human intestine, produces an abundance of eggs that are passed from a host's digestive tract to the external environment. In places without good sanitation, other humans inadvertently ingest the eggs and acquire the parasite themselves. Ectoparasites, on the other hand, often have elaborate host-finding adaptations. Some aquatic leeches, for example, first locate a host by detecting its movement in the water and then confirm its identity on the basis of temperature and chemical cues on the host's skin.

Natural selection has also favored the evolution of defensive capabilities in potential hosts. Some of the secondary plant products that are toxic to herbivores, for instance, are also toxic to such parasites as fungi and bacteria. In vertebrates, the immune system provides a multipronged defense against internal parasites (see Chapter 39). Many parasites, particularly microorganisms, have adapted to particular hosts, often a single species. In such specific interactions, coevolution generally results in a relatively stable relationship that does not kill the host quickly—an event that would eliminate the parasite, as well.

An example will demonstrate how rapidly natural selection can temper a host–parasite relationship. In the 1940s, Australia was plagued by a population of hundreds of millions of rabbits, which were descended from just 12 pairs imported a century earlier. In 1950, the myxoma virus, which parasitizes rabbits, was introduced in an effort to control the rabbit population. The virus spread rapidly and killed 99.8% of all rabbits infected. However, a second exposure to the virus killed only 90% of the remaining rabbits, and the third infection killed only about 50%. Today, the virus has

Figure 48.15
Commensalism between a bird and a mammal. The cattle egret concentrates its feeding where grazing cattle and other large herbivores, such as this cape buffalo in Tanzania, flush insects from the vegetation. This relationship benefits the bird without apparent harm or benefit to the mammal. Of course, the relationship may help the buffalo in some way not yet discovered.

only a mild effect on the host, and the rabbit population has rebounded. Apparently, viral infection selected for host genotypes that were better able to resist the parasite. At the same time, there was selection for less virulent strains of the virus, which had greater success in being propagated and transmitted by hosts that survived. Natural selection has therefore stabilized this host–parasite relationship.

Commensalism Few absolute examples of commensalism exist because of the unlikelihood that one partner in an ecological interaction will be completely unaffected by the other. "Hitchhiking" species, such as algae that grow on the shells of aquatic turtles or barnacles that attach to whales, are sometimes considered to be commensal. However, the hitchhikers may actually decrease the reproductive success of their hosts slightly by reducing the efficiency of the hosts' movements in their search of food or escape from predators. An association that may be truly commensal is the relationship between cattle egrets (heronlike birds) and grazing cattle (Figure 48.15). The egrets concentrate their predation near grazing cattle, which flush insects and other small animals from the vegetation as they move. Because the birds increase their feeding rates when following cattle, they clearly benefit from the association with cattle. It is difficult to imagine that the cattle either benefit from or are harmed by the relationship in any way. Because commensalism benefits only one of the species involved, any evolutionary change in the relationship is likely to occur only in the beneficiary.

Figure 48.16
Mutualism between acacia trees and ants. Certain species of Central and South American acacia trees, called bull's horn acacias, have hollow thorns that house stinging ants of the genus *Pseudomyrmex.* The ants feed on sugar produced by nectaries on the tree and on protein-rich swellings called Beltian bodies that grow at the tips of leaflets. The acacia benefits from housing and feeding a population of pugnacious ants, which attack anything that touches the tree. The ants sting other insects, remove fungal spores and other debris, and clip surrounding vegetation that happens to grow close to the foliage of the acacia. A series of experiments in the 1960s demonstrated the mutualistic nature of the relationship. When the ants on experimental trees were poisoned, the trees died, probably because of damage by herbivores and failure to compete with the surrounding vegetation for light and growing space.

Mutualism In contrast to commensal relationships, mutualistic relationships require the evolution of adaptations in both participating species, because changes in either species are likely to affect the survival and reproduction of the other. Many coevolved mutualistic adaptations have been described in earlier chapters: nitrogen fixation by bacteria in the root nodules of legumes; digestion of cellulose by microorganisms in the guts of termites and ruminant mammals; photosynthesis by unicellular algae in the tissues of corals; the exchange of nutrients in mycorrhizae, the association of fungi and the roots of plants; and the specific interactions of pollinators and flowering plants. Figure 48.16 illustrates yet another interesting mutualistic interaction: the relationship between certain species of acacia and the ants that protect these trees from herbivorous insects.

Many mutualistic relationships may have evolved from predator–prey or host–parasite interactions. Certain angiosperm plants, for example, have adaptations that attract animals for pollination or seed dispersal; these probably represent counteradaptations to the herbivores' feeding on pollen and seeds. In many cases, pollen is spared when the pollinator is able to consume nectar, and seeds are dispersed by animals that eat the fruit that encloses them. Any plants that could derive some benefit by sacrificing organic materials other than pollen or seeds would increase their reproductive success, and the adaptations for mutualistic interactions would spread through the plant population.

COMMUNITY STRUCTURE: PATTERNS AND PROCESSES

Although we understand the dynamics of some ecological interactions within many communities, these relationships are inextricably tied together into a complex web that itself is much more difficult to study. Research over the past 100 years has increased our knowledge of community-level processes enormously, but ecologists have not yet discovered a set of universal principles that can generally account for the characteristics of particular communities and explain how the diverse interactions among their populations influence those characteristics. Let's begin this discussion with a description of the community characteristics that ecologists study, then evaluate how population interactions influence those attributes.

Community Characteristics

Much as populations exhibit characteristics, such as density and distribution patterns, that are not features of individual organisms, communities also exhibit several emergent properties that are unique to this level of biological organization. These properties result from the particular species composition of a community and the biological interactions among the populations.

Vegetation Structure and Associated Animals Our first impression of a terrestrial community is likely to be the overall form of the vegetation. What is the community's vertical profile? Do trees, shrubs, or grasses predominate? Is the vegetation stratified into canopy, subcanopy, and herb levels? The vegetation in a community largely determines the types of animals found there. Large grazing mammals, for example, are found only in grasslands and savanna, and tree-nesting birds occur only in habitats supporting that form of vegetation. In general, structurally complex vegetation provides a diversity of microhabitats that can be used by a variety of animal species (see Figure 48.13), whereas short vegetation does not provide a structural resource that animal populations can partition easily.

Trophic Structure The most basic observations of a natural community will yield information about what the species within the community eat. The various feeding relationships, or **trophic structure** (Gr. *trophe*, "nourishment") are determined by predator–prey, host–parasite, and plant–herbivore interactions, but studies of trophic structure also identify some interactions that might represent competition for food. As described below, these trophic relationships have a strong impact on other community characteristics. The trophic structure in a community also determines the flow of energy and the cycling of nutrients among plants, herbivores, carnivores, and the organisms that feed upon the wastes and remains of these species. Because energy flow and nutrient cycling also involve abiotic components, such as air, water, and soil, we will postpone discussion of these topics until our consideration of ecosystems in Chapter 49.

Species Richness, Relative Abundance, and Diversity Communities differ dramatically in their **species richness,** the numbers of species they contain. Ecologists also recognize, however, that some communities are comprised of a few common species and many rare ones, whereas others contain an equivalent number of species that are all equally common. The **relative abundance** of species within a community has an enormous impact on its general character. Imagine, for example, two communities, each with a total of 100 organisms distributed among four different species (A, B, C, and D). Assume, however, that the relative abundance of individuals in each species differs between the two communities as follows:

Community 1:	25A	25B	25C	25D
Community 2:	97A	1B	1C	1D

If these were, say, trees in a forest, and you were exploring Community 1, you would easily notice four different species. In Community 2, however, almost every tree you saw would be species A, and only rarely would you see an individual species of B, C, or D. Although these two communities are equally rich—they have the same number of species—Community 1 has a more even representation of the four species. Most people would intuitively think of Community 1 as more diverse. Indeed, the term **species diversity,** as used by ecologists, considers *both* components of diversity: species richness and relative abundance.

Researchers have proposed a variety of mathematical formulas, called diversity indexes, that combine data on species richness and relative abundance. As with many topics in this chapter, however, there is no general agreement on which of these indexes strikes the most appropriate balance of these two measures, or even whether variation in the evenness of species distributions provides much insight into the factors responsible for community structure. One reason for the controversy is that the relative abundance of a species is not necessarily a measure of its importance to the interactions that characterize the community. An organism's size and activity also play a major role. A large predator, for example, may strongly affect the populations of many other species, even though the predator is not particularly abundant.

Stability Community stability refers to the ability of a community to resist change and recover its original composition in the wake of some disturbance, such as a fire or a disease that kills most individuals of a dominant species. Stability depends on both the type of community and the nature of the disturbance. For example, after a fire, a grassland may return to its original species composition much faster than a forest would. As we will see, the subject of stability provides one of the best examples of the pendulumlike swings in viewpoint that characterize ecology as a vigorous field of scientific inquiry.

Determinants of Community Characteristics

Our analysis of biomes in Chapter 46 identified a number of abiotic factors that determine the nature of the vegetation that grows in various regions of the Earth. But what factors determine the other community characteristics described above? Why are some communities diverse and others not? Why are some communities relatively stable, whereas others do not readily recover from disturbances? As noted above, ecologists have no general answers to these complicated questions; but community ecology is an active area of research, and careful study has produced some insights into how population interactions influence the characteristics of particular communities. We now explore some of these insights.

The Role of Competition In the 1960s and 1970s, many ecologists proposed that competition was a major factor limiting the diversity of species that could occupy a community. This hypothesis was largely based on their frequent observations of niche differences and resource partitioning among sympatric species. A given quantity of resources, these ecologists argued, could be partitioned only so finely before the effects of competition would inevitably lead to the extinction of poorer competitors, setting a limit to the number of species that could occur together. Recall, however, that the only reliable way to demonstrate the importance of interspecific competition is to show that an increase in the density of one species leads to a decrease in the density of its presumed competitor. And the situation in nature is, of course, complicated by the fact that in diverse communities, many species may compete with each other simultaneously.

Numerous ecologists were inspired by this hypothesis to conduct field experiments in which they either removed or added potential competitors to a community and monitored the effects of these manipulations on the other species present. The barnacle study described earlier is an excellent example of this experimental approach (see Figure 48.12). Studies were undertaken in a wide variety of habitats and focused on a relatively broad taxonomic range of organisms. Competition was indeed documented in a majority of the cases examined, suggesting that it may be a potent force in structuring communities. Interspecific competition did not, however, always lead to competitive exclusion, and competing species may sometimes coexist, albeit at reduced densities.

When evaluating data collected in many studies, note that ecologists, like other scientists, investigate situations that are apt to yield interesting results. Therefore, it is reasonable to assume that the subjects of most of these studies were selected because they were expected to exhibit competitive interactions. These data may therefore overestimate the importance of competition, because few ecologists conduct competition experiments on species that are less likely to compete. Although we can conclude that competition is important in regulating the relative abundance of many species and perhaps the species diversity of many communities, nobody can say exactly how important it is.

Another general bias in studies of competition is that relatively little research has been conducted on the numerous species of herbivorous insects, organisms that are generally opportunistic and subject to density-independent population limitation (see Chapter 47). We would not expect competition to be important to the ecology of species that rarely approach their carrying capacities, and few studies have demonstrated significant competition among such species. Herbivore populations in general are rarely regulated by their food supply, and surveys of competition indicate that it is far more prevalent among populations of plants and carnivores.

Many ecologists are still reluctant to embrace competition as the major factor structuring communities because, while it is hard enough to demonstrate that two species are competing in the present, it is even more difficult to assess what has happened in their evolutionary past. Furthermore, as noted above, competition cannot be important unless population sizes are near carrying capacity and resources are limiting. Finally, there is mounting evidence that predation and other forces also play a major role in shaping ecological communities.

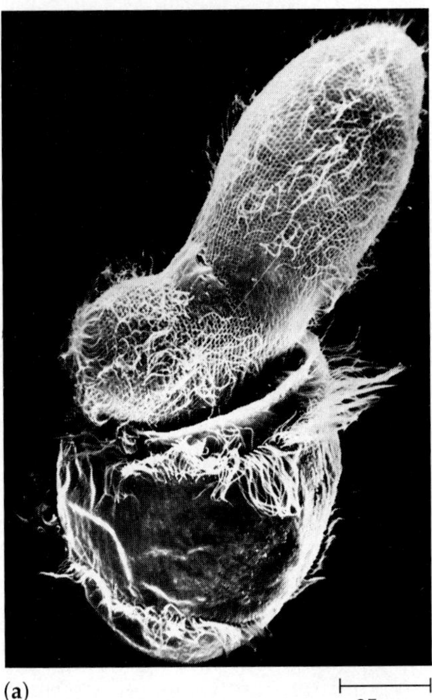

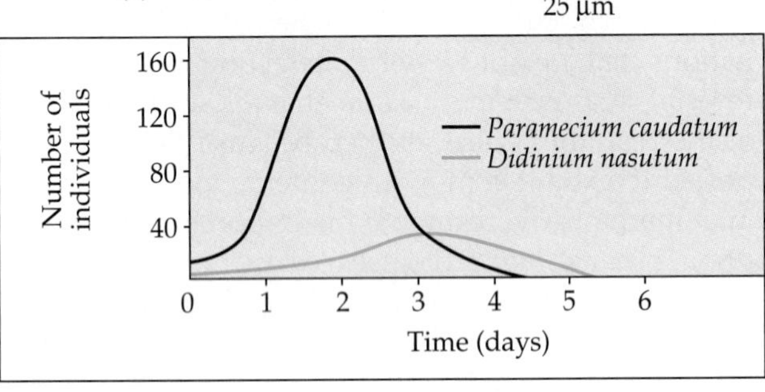

Figure 48.17
Predator–prey dynamics in a laboratory environment. (**a**) The ciliate *Didinium* (bottom) eats *Paramecium*, another ciliate (SEM). (**b**) When these two species are cultured together, *Didinium* consumes all the *Paramecium*, and then becomes extinct as a result of starvation. In natural communities, however, which are much more complex than the simple two-species system of a culture flask, predator–prey relationships can actually help stabilize species diversity rather than reduce it.

The Role of Predation In simple laboratory experiments where a single predator species is kept with a single prey species having no refuge, the predator may devour all the prey and then perish from starvation (Figure 48.17). If this scenario were commonly enacted in nature, predators would always reduce the diversity of species in communities. In fact, the role of predators in structuring communities is more complex. The defensive abilities of prey described earlier constitute one reason why predators rarely drive their prey and themselves to extinction. The switching behavior of many predators is also important. A predator may switch to an alternative food source when the population of one species of prey begins to dwindle because of intense predation or some other factor. Moderate predation may even *help* maintain the presence of some prey species, by preventing them from overshooting their carrying capacity and then experiencing a downward adjustment of population size to a level dangerously close to local extinction.

Probably the most important effect of a predator on community structure is to moderate competition among its prey species. Heavy predation can reduce the density of a very successful competitor, thereby allowing weaker competitors to persist in the community. Experiments by Robert Paine in the 1960s were among the first to provide a clear picture of this complex interaction. Paine removed the dominant predator, a sea star of the genus *Pisaster,* from experimental areas within the intertidal zone of the Washington coast. The favorite prey of *Pisaster,* the mussel *Mytilus,* then outnumbered many of the other tidepool organisms that compete for attachment space on the rocks. Because *Mytilus* was such a dominant competitor in the experimentally created predator-free environment, the species richness of the community declined from 15 to 8. This research and numerous other field experiments have given rise to the concept of a **keystone predator**, a predator that exerts an important regulating effect on other species in the community. Keystone predators maintain higher species diversity in a community by reducing the densities of strong competitors, such that competitive exclusion of other species does not occur.

The Role of Environmental Patchiness As described in Chapter 46, all environments are patchy over both space and time. For example, the mineral content of soil varies locally with the chemical composition of the rocks from which it is derived. Similarly, soil moisture varies with topography, such that low-lying areas are generally wetter than high ground. If different species are best adapted to different local conditions, environmental patchiness will increase the community's species diversity by facilitating resource partitioning among potential competitors. The distributions of some low-growing plants in the forests of Indiana, for example, are regulated by their seedlings' tolerance of calcium and organic matter. Because these factors vary locally, wild black cherry occurs in some patches and two species of violets are found in others. If the distribution of these abiotic factors were uniform throughout the forest, one of the species would probably outcompete the others, reducing species diversity in the community. Of course, the importance of patchiness varies with the size and lifespan of the organisms under study. Local soil characteristics may have a huge influence on the distributions of small plants, but they are not likely to have much impact on a large herbivorous mammal that eats the plants.

Evaluating Causative Factors As we have seen, competition, predation, symbiosis, and abiotic factors that create environmental patchiness all have significant effects on the characteristics of various communities. Sometimes these factors interact, complicating our search for general principles about the determinants of community structure and diversity. Furthermore, it is often extremely difficult to correctly identify the long-term evolutionary factors that produced the ecological relationships we study today.

Recent research on the nesting behavior of two North American bird species that occupy the same trees illustrates the difficulty of identifying causative factors. MacGillivray's warbler nests low in trees; black-headed grosbeaks nest high. At first this might be interpreted as another example of resource partitioning that reduces competition, in this case for nesting sites. But Thomas Martin has recently demonstrated that predation may be the key factor in this ecological segregation of nests; predators locate and feed on the young of these birds more successfully when nests are concentrated in the same area, rather than being more widely distributed in the trees. In other words, the partitioning we see in nest sites reduces predation, and this, rather than a reduction in competition between the bird species, may explain the observed difference.

Thus, careful studies lead to the conclusion that communities are structured by multiple interactions of organisms with their biotic environment and with abiotic factors as well. Which interactions are most important can vary from one type of community to another, and even among different components of the same community. For example, in a particular community, the diversity of herbivores may be controlled mainly by predators with prey-switching behavior, whereas the diversity of these predators may depend mainly on competition for food. Other kinds of interactions may affect numerous species to varying degrees. Because of the complexity of these community networks, there are few, if any, natural communities for which we have a good understanding of all the important relationships or how the relationships evolved.

Community ecology is made even more challenging by the fact that the structure of a community may change, sometimes over relatively short periods of time. Disturbances influence structure, and the activities of the organisms themselves may destabilize an existing structure and favor a new one. The next section examines some of these changes.

SUCCESSION

Changes in community composition and structure are most apparent after some disturbance—a flood, a fire, the advance and retreat of a glacier, volcanic eruption, or human activity—strips away the existing vegetation. The disturbed area may be colonized by a variety of species, which are gradually replaced by others. Such transitions in species composition over ecological time represent a process called **ecological succession.** In traditional views of community ecology, the community passes through a sequence of predictable transitional stages called **seres,** ultimately achieving a relatively stable state called a **climax community.**

This process is called **primary succession** if it begins in an essentially lifeless area where soil has not yet formed, such as on a new volcanic island, or on the till left behind by a retreating glacier. In the case of glaciers, which are still shrinking in places like Glacier Bay, Alaska, the barren ground is first occupied by mosses and lichens, and then by dwarf willows. After about 50 years, alders form dense stands. These eventually give way to Sitka spruce, which are later joined by hemlock to form the relatively stable spruce-hemlock forest that we recognize as taiga (see Chapter 46). The entire process takes about 200 years (Figure 48.18).

Secondary succession occurs where an existing community has been cleared by some disturbance that leaves the soil intact. Often the area begins a return to something like its original state. Old-field succession in the Piedmont region of the mid-Atlantic United States has been studied extensively. If an agricultural field is abandoned in this area, a herbaceous community develops as the first sere, followed by shrubby forms that give way to pine. The pines are eventually replaced by a climax community dominated by oaks and hickories, the vegetation that originally occupied the site before farmers cleared it for planting. As in glacial-till succession, the entire process takes about two centuries.

As noted above, succession may seem to produce a final stage called the climax community. Many ecologists once viewed the climax as a common endpoint that communities living under similar environmental conditions inevitably attain—a condition that then persists almost indefinitely. As we will see, however, the notion of a climax community is too simplistic to represent the wealth of variation in nature. Recent research into the causes of succession has moved ecologists toward a more pluralistic view of the nature of communities and their development through time.

Causes of Succession

In most cases, a variety of interrelated factors determines the course of succession. Early seres are typically characterized by *r*-selected opportunistic species that are good colonizers because of their high fecundity and excellent dispersal mechanisms. Many of these may be described as "fugitive" species that do not compete well in established communities, but

(a) (b) (c) (d) (e) (f)

Figure 48.18
Succession after the retreat of glaciers. These photographs illustrate the different stages of succession: (**a**) retreating glacier; (**b**) barren landscape after the retreat; (**c**) moss and lichen stage; (**d**) alders and cottonwoods covering the hillsides; (**e**) spruce coming into the alder and cottonwood forest; (**f**) spruce and hemlock forest. Ecologists usually deduce the process of succession by studying a variety of areas that are at different successional stages. These photographs were taken in different places, of course, because the time frame of the changes they illustrate is about 200 years, much longer than the lifespan of any ecologist or photographer.

maintain themselves by constantly colonizing new areas before better competitors can become established in the same places. Tolerance of the abiotic conditions in a barren area also affects the species composition during early successional stages. Many *K*-selected equilibrial species may colonize an area, but they will not grow abundantly if environmental conditions there are at the extremes of their tolerance limits. (Also recall that important abiotic factors differ on a local scale, as on north- and south-facing slopes of a mountain, creating a mosaic of species composition within the boundaries of a community.) Variations in the growth rates and maturation times of colonizing species are also clearly important. For example, if the seeds of two plant species, one an annual herb and the other a tree, colonize a community at the same time, the herbaceous species will have earlier prominence because of its faster growth and shorter generation time. The ecological impact of the tree species may not be realized until the trees are relatively large.

Many of the changes in community structure during succession may be induced by the organisms themselves. Such changes are said to be **autogenic.** Direct biotic interactions may be involved, including **inhibition** of some species by others through exploitative competition, interference competition, or both. However, the presence of organisms also affects the abiotic environment by modifying local conditions. This may result in **facilitation,** in which the group of organisms representing one successional stage "paves the way" for species typical of the next sere. For example, the alders that are abundant in an intermediate stage of glacial-till succession lower soil pH as their dropped leaves decompose. The change in pH facilitates the entry of spruce and hemlock, which require acidic soil. Sometimes the changes that facilitate the development of a later sere actually make the environment unsuitable for the very species responsible for the changes.

Both inhibition and facilitation may be involved throughout the successional process. For example, horseweed is one of the earliest colonizers in the old-field succession described earlier. For a year or two, these plants may inhibit other species through shading and their use of soil water. However, as the horseweed and other early species die and decompose, their presence facilitates the entry of later species by adding organic matter to the soil, which aids in holding moisture. The pine trees that dominate a later successional stage require full sunlight; their growth becomes self-inhibitory as the trees shade the ground and prevent any of their own offspring from growing. At the same time, the pines continue to add organic material to the soil, and the shade they cast keeps the forest floor moist, conditions that facilitate the germination and growth of the hardwoods that follow. At the climax stage, environmental conditions are such that the same species can continue to maintain themselves. For example, the oak-hickory forest that is the climax stage of the old-field succession maintains the moist, shaded environment that allows offspring of these species to grow, while inhibiting most of the species typical of earlier stages.

As noted earlier, some ecologists have challenged the traditional idea of a climax community. For one thing, many communities are routinely disturbed by **allogenic** ("of outside origin") factors during the course of succession. For example, as we saw in Chapter 46, prairie grasslands are maintained by fire. Without fire, some grassland areas would develop into forest, at least in moister areas. We could say that forest is the climax community for such areas, but that would make little sense if the forest community never develops. In this case, periodic fires stabilize the community at a stage that precedes the typical climax state. There is also evidence that what appear to be climax communities may not be stable over very long time frames. Studies of pollen, preserved in lake sediments, provide evidence of community composition over thousands or even millions of years. Recent research indicates that common tree species sometimes disappear from North American forests for hundreds of years, only to eventually return at a later time. The reasons for such changes are unknown.

Human Disturbance

Disturbance by human activities has had an impact on succession of communities all over the world. Logging and clearing for farmland have reduced large tracts of mature hardwood and pine forests to small patches of disconnected woodlots in many parts of the eastern and midwestern United States and throughout Europe. Similarly, agricultural development has disrupted what were once the vast grasslands of the North American prairie.

After a community is disturbed and then left alone, early stages of secondary succession, which are often dominated by weedy and shrubby vegetation, may persist for many years. This type of vegetation can be found extensively in forests that have been clear-cut, in agricultural fields no longer under cultivation, and in vacant lots and construction sites that are periodically cleared. Much of the United States is now a hodgepodge of early successional growth where mature communities once prevailed.

Human disturbance of communities is by no means limited to the United States and Europe, nor is it a recent problem. Tropical rain forests are quickly disappearing as a result of clear-cutting for lumber and pastureland. Centuries of overgrazing and agricultural disturbance have undoubtedly contributed to the current famine in parts of Africa by turning seasonal grasslands into great barren areas.

Community Equilibrium and Species Diversity

The traditional concept of ecological succession predicts that species diversity eventually reaches an equilibrium in the climax community. During the successional process, the *K*-selected species that are stronger competitors replace the *r*-selected early colonizers as population densities increase and the vegetation modifies the site. Once the longer-lived *K*-selected species have become established, the rate at which new species replace old ones slows down. Although there may be some turnover of less dominant species in the climax community, the addition of new colonists is balanced by localized extinctions. This equilibrial view of community dynamics emphasizes the importance of population interactions in structuring the community. Predation, competition, and symbiosis become more extensive and varied during succession, making increased diversity possible; and species diversity generally does increase during succession. According to this model of community development and structure, succession reaches a climax when the web of biotic interactions becomes so intricate that the community is saturated. No additional species can "fit into" the community unless resources become available through the localized extinction of species that are already present. Indeed, this pattern appears to occur on islands and in other geographically restricted locations, for reasons that we will discuss later in the chapter.

An opposing view portrays communities as being in a continual state of flux: The identities and numbers of species change during all successional stages, even in the so-called climax stage. This nonequilibrial model of community dynamics emphasizes the importance of the less predictable factors, such as dispersal and disturbance, in the development of community composition and structure. The course of succession may vary, for example, with the identity of the particular species that happened to colonize an area first. Severe disturbances such as fires, hurricanes, windstorms, and landslides may prevent a community from ever achieving a state of equilibrium, because their destruction of vegetation returns local sites to earlier stages of succession. Proponents of the nonequilibrial view of community development often describe a mature community as a **polyclimax,** a state in which the community as a whole is an unpredictable mosaic of patches at different successional stages. Local environmental heterogeneity also contributes to the polyclimax nature of many communities because different species occupy different habitat patches.

According to the nonequilibrial model, disturbance is a major determinant of community composition and species diversity. When disturbance is severe and frequent, the community may include only good colonizers typical of early stages of succession. If disturbances are mild and rare in a particular location, then the late-successional species that are most competitive will make up the community. According to the **intermediate disturbance hypothesis,** species diversity is greatest where disturbances are moderate in both frequency and severity, because organisms typical of different successional stages will be present. Studies of species diversity in tropical rain forests provide some evidence for the intermediate disturbance hypothesis. Scattered throughout these forests are small clearings where trees and the vines attached to them have fallen. In these disturbed areas, immigrations and extinctions occur in rapid succession, and species from various successional stages coexist within a relatively small area.

Diversity and Community Stability

In the 1960s, many ecologists believed that increased species diversity and the accompanying web of biotic interactions somehow caused communities to be more stable. However, there was serious disagreement about how to define "stability." Some researchers considered stability in terms of *resistance* to change. Others used the term to describe *resiliency,* the ability of a community to recover its original state after a disturbance. The work of Robert Paine, described earlier, exemplifies the difficulty of this issue. Intertidal communities are extremely diverse, and, if undisturbed, they appear to retain their characteristics over long periods of time. However, the removal of a single species, the keystone predator *Pisaster,* can radically alter the composition and diversity of this community. In the 1970s, a number of mathematical models actually predicted *decreasing* stability with increasing diversity—just the reverse of earlier views. Some ecologists rejected these models as based on unrealistic assumptions.

Today, most ecologists agree that no simple relationship exists between diversity and stability. However, debates still rage, and some discussions become entangled in a "chicken and egg" dialogue. For example, should we think of the high diversity of the tropical rain forest as a *cause* of *stability* in this community, or should we think of the high diversity as *resulting from* a certain amount of *instability?* After several decades of research, these questions are still largely unresolved.

BIOGEOGRAPHICAL ASPECTS OF DIVERSITY

Biogeography is the study of the past and present distributions of individual species as well as entire floras and faunas. This field provides a different approach to an understanding of community properties by analyzing both global and local phenomena, mostly from a historical perspective.

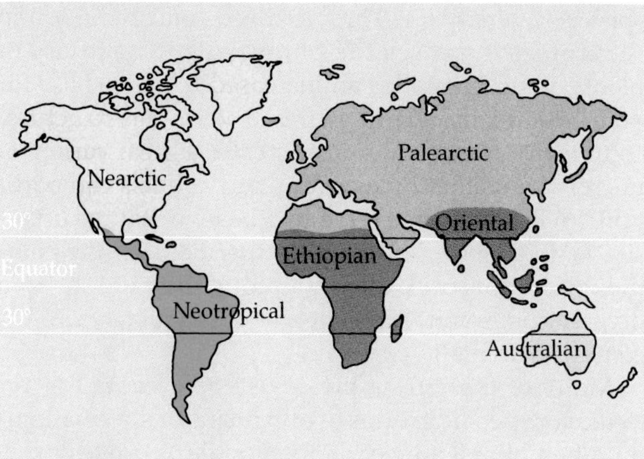

Figure 48.19
Biogeographical realms. Continental drift and barriers such as deserts and mountain ranges all contribute to the distinctive floras and faunas found in major regions of the Earth. Except for Australia, the realms are not sharply delineated but grade together in zones where taxa from adjacent realms coexist.

Traditionally, biogeographers have been concerned with the actual identities of the species that comprise particular communities, rather than the emergent properties of the communities themselves. One might, for example, use an evolutionary perspective to analyze the species compositions of South American and African rain forests, communities that share many general properties but few individual species. More than 100 years ago, historical biogeographers recognized that terrestrial life can be divided into biogeographical realms whose often diffuse boundaries are associated with the patterns of continental drift that followed the breakup of Pangaea (Figure 48.19). Thus, to a very great extent, the present distributions of species reflect their distant evolutionary history as well as more recent interactions with both biotic and abiotic components of the environment. We discussed historical aspects of biogeography in Chapters 20 and 23. Although continental drift and its effects on species distributions are still active areas of research, some biogeographers have more recently applied the principles of community ecology to the analysis of geographic distributions. The intersection of these fields is fertile ground: Biogeographical analyses have contributed as much to our understanding of community patterns and processes as the study of community ecology has contributed to our knowledge of biogeography. Here we consider four intersecting topics that have captured the attention of ecologists during the last three decades.

Limits of Species Ranges

Understanding the determinants of a species' geographical range is central to any analysis in community ecology and biogeography. Three general explanations can account for the limitation of a species to a particular range today: (1) The species may never have dispersed beyond its current boundaries; (2) pioneers that did spread beyond the observed range failed to survive; and (3) over evolutionary time, the species has retracted from a once larger range to its present boundaries. Research in paleontology and historical biogeography can identify cases for which the third explanation is valid. For example, we know from fossil evidence that close relatives of living elephants and camels once occupied North America, but local extinctions have caused retractions of the geographical ranges of both groups.

Distinguishing between the first and second explanations is more difficult, because discovering a few colonists that dispersed to a new area but then died is much harder than looking for a needle in a haystack. However, transplant experiments, in which individuals of a plant or animal species are moved to similar environments outside their range, can provide useful information. Survival of the transplants suggests that the species had not dispersed to suitable locations outside the existing range. Failure of the transplants, in contrast, suggests that the species could not expand its range because of its inability to tolerate abiotic conditions or to compete with resident species. In one simple experiment, researchers transplanted several individuals of the lizard *Anolis cristatellus* from the warm lowlands in Puerto Rico to a cool, forested site at higher elevation. These animals failed to survive for more than a few weeks, presumably because they could not attain sufficiently high body temperatures to capture and digest their food. Therefore, this species does not occupy high-elevation forests, because it is not adapted to the physical environment in such habitats.

Global Clines in Species Diversity

Ecologists have long recognized the existence of clines (gradual variation) in species diversity with major geographical gradients. The number of terrestrial bird species in North America, for example, increases steadily from the arctic to the tropics (Figure 48.20). Although similar clines are known for most other major groups of organisms, such as microbes, flowering plants, reptiles, and mammals, some more restricted taxonomic groups are notable exceptions. The shorebirds known as sandpipers are most diverse in the arctic, and coniferous trees achieve their greatest species richness in the temperate zone. In other frequently observed patterns, the diversity of terrestrial organisms at a given latitude usually decreases with altitude, and the diversity of marine benthic faunas generally increases with depth.

How do we explain such gradations in species diversity? The latitudinal gradients are the most obvious

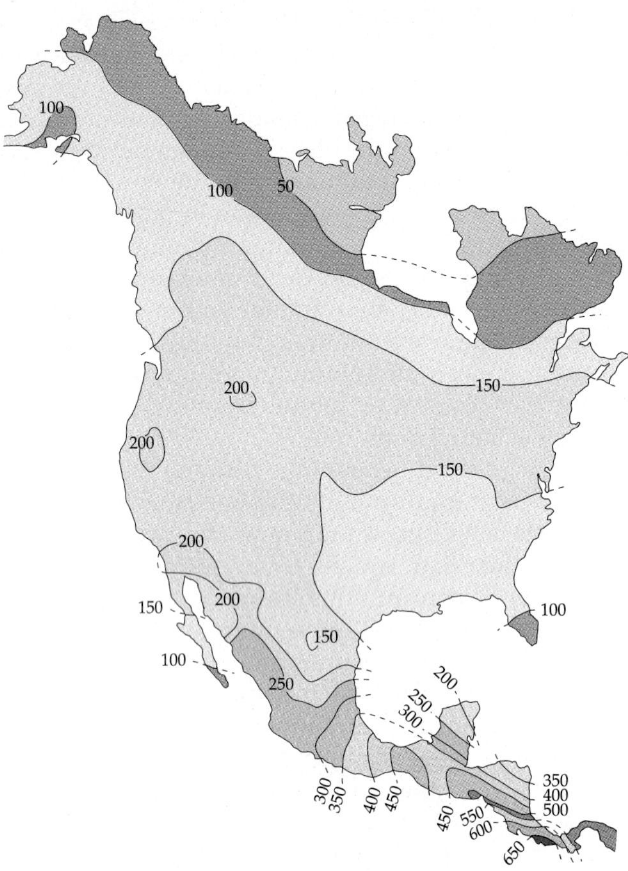

Figure 48.20
Species density of North American birds. Biogeographers often plot latitudinal trends in numbers of species in the form of a "topographic" map that illustrates how many species occupy different geographical areas. In this species density map for North American birds, we see that fewer than 100 species are found in arctic areas, whereas more than 600 occupy some tropical regions of Central America.

and dramatic of these patterns, and their generality has led ecologists to seek universal explanations for the high diversity of tropical communities. Many hypotheses have been proposed. (1) Some ecologists suggest that tropical communities are very old and rarely experience major natural disturbances; these communities may foster rapid and frequent speciation, and their age has allowed complex population interactions to coevolve and develop to a greater extent than in the temperate zone. (2) Tropical regions may generally experience intermediate levels of disturbance and have greater environmental patchiness, allowing a greater diversity of plant species to form the resource base for diverse communities of animals. (3) The stability and predictability of tropical climates might allow many organisms to specialize on a narrower range of resources; having smaller niches would reduce competition and permit a finer level of resource partitioning among species, which in turn would foster higher species diversity. (4) The increased solar radiation in the tropics increases the photosynthetic activity of plants, which provides an increased resource base for other organisms. (5) In a similar vein, the structural complexity of tropical forests creates a great variety of microhabitats that other plants and animals can partition. (6) Finally, some ecologists believe that the diversity is, in a sense, self-propagating because the complex predator–prey and symbiotic interactions in a diverse community prevent any populations from becoming dominant.

Many of these hypotheses can be accepted or rejected for specific groups of organisms or specific communities, but all are nearly impossible to prove or disprove conclusively. The question of what factors determine latitudinal gradients in species diversity is simply too large to be addressed with simple experiments either in the laboratory or in the field. Although most of the explanations listed above may be true for some organisms and some communities, complex combinations of these factors probably operate under most circumstances.

The multiple explanations of latitudinal trends of diversity in terrestrial habitats may not adequately explain the diversity gradient with depth for benthic communities in the deep-sea. Several ecologists have proposed that, despite the low level of productivity in deep sea benthic communities, the long-term stability of these habitats has fostered many coevolved relationships among the organisms present. However, no research has yet been able to establish a *causal* relationship between environmental stability and species diversity.

Island Biogeography

Because of their limited size and isolation, islands provide excellent opportunities for studying some of the factors that affect the species diversity of communities. By "islands" we mean not only oceanic islands but also habitat islands on land, such as lakes or mountain peaks separated by lowlands—in other words, any areas surrounded by an environment not suitable for the "island" species. In the 1960s, American ecologists Robert MacArthur and E. O. Wilson developed a general theory of island biogeography to identify the important determinants of species diversity on an island with a given set of physical characteristics. The study of islands may help us understand some of the interactions in more complex systems.

Imagine a newly formed oceanic island some distance from a "mainland" that will serve as a source of colonizing species. Two factors will determine the number of species that eventually inhabit the island: the rate at which new species immigrate to the island and the rate at which species become extinct on the is-

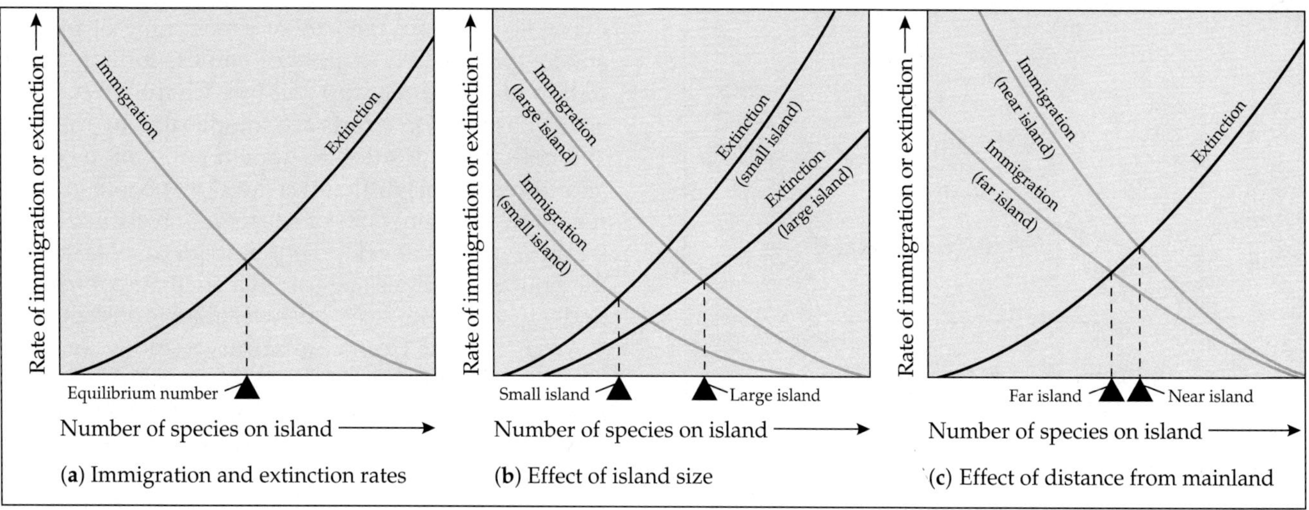

Figure 48.21
The theory of island biogeography. (**a**) The equilibrium number (black triangle) of species on an island represents a balance between the immigration of new species to the island and the extinction of species that are already there. (**b**) Large islands will ultimately have a larger equilibrium number of species than small islands because immigration rates are higher and extinction rates are lower on large islands. (**c**) Although extinction rates do not differ with an island's distance from a mainland source of species, near islands will have larger equilibrium numbers of species than far islands, because immigration rates to near islands are higher than those for more distant ones.

land. Immigration and extinction rates are, in turn, affected by two important variables: the size of the island and its distance from the mainland. Small islands will generally have lower immigration rates, because it is more difficult for potential colonizers to "find" a small island. For example, birds blown out to sea by a storm are obviously more likely to land by chance on a larger island than on a small one. Small islands will also have higher extinction rates. They generally contain fewer resources and less diverse habitats for colonizing species to partition, increasing the likelihood of competitive exclusion. Distance from the mainland is also important; for two islands of equal size, a closer island will have a higher immigration rate than one farther away.

The immigration and extinction rates are also affected at any given time by the number of species already present on the island. As the number of species on the island increases, the immigration rate of new species decreases, because any individual reaching the island is less likely to represent a species that is not yet present. At the same time, as more species inhabit an island, extinction rates on the island increase because of the greater likelihood of competitive exclusion. These relationships are summarized in Figure 48.21, where immigration and extinction rates are plotted as a function of number of species present on the island. The ultimate point of this model is that, eventually, an equilibrium will be reached where the rate of species immigration matches the rate of species extinction. The number of island species at this equilibrium point is correlated with the two variables already discussed. The theory of island biogeography predicts that species number on an island where immigration and extinction rates are equal is directly proportional to the size of the island and inversely proportional to the distance of the island from the mainland. (Any equilibrium, of course, is dynamic; immigration and extinction continue, and the exact species composition may change over time.) This theory generally applies over a relatively short period of time, where colonization is the important process determining species composition; over a longer period, evolutionary changes in island species and speciation on the island can affect the species composition and community structure.

Observations and experiments furnish evidence that new islands do, indeed, reach an equilibrium in their species richness. For example, within 35 years after a volcanic eruption killed nearly all organisms on the island of Krakatoa in 1883, the diversity of birds that repopulated the island had reached an equilibrium of about 30 species. MacArthur and Wilson's studies of the diversity of amphibians and reptiles on many island chains, including the West Indies, support the prediction that species richness increases with island size (Figure 48.22). Species counts also fit the prediction that the number of species decreases with increasing remoteness of the island.

In the late 1960s, Wilson and Daniel Simberloff tested the theory of island biogeography in experiments on small islands of mangroves off the southern tip of Florida (see Figure 46.2). Six islands, each about

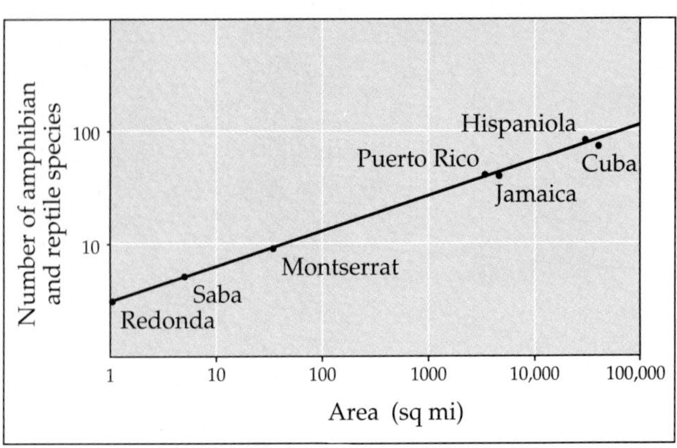

Figure 48.22
Species richness and island size. This species-area graph illustrates that the number of amphibian and reptile species found on West Indian islands is closely related to the island sizes. Large islands harbor more species because greater habitat diversity allows greater resource partitioning among the resident species, reducing the likelihood of competitive exclusion.

12 m in diameter, were enclosed in tents and fumigated with a poison to kill all resident arthropods. The pesticide used, methyl bromide, decomposes rapidly, and the islands could be recolonized by arthropods from the mainland species pool. Within about a year, species numbers equilibrated on each island, with the fewest species on the island most distant from the coast. Although the equilibrium number for each island was about the same before fumigation and after recolonization, the species composition was different. Chance events—in this case, which species of arthropods happened to disperse to which islands—affected the species composition of the communities. This is another example of how ecological theory has become less deterministic in its predictions; in general, community structure and dynamics seem much less predictable than ecologists once believed.

CONSERVATION BIOLOGY: LESSONS FROM COMMUNITY ECOLOGY AND BIOGEOGRAPHY

As human encroachment on natural communities advances at a rapid pace, concerned citizens and governments are supporting efforts to conserve as much as possible of the biological diversity that remains on Earth. At the 1992 United Nations Earth Summit in Rio de Janeiro, only the United States, out of 172 nations, failed to sign the Biodiversity Treaty, designed to protect a wide variety of species around the world. Global support for conservation stems from our esthetic appreciation for other forms of life, the recognition that products useful to humans may ultimately be derived from as yet undiscovered species (see Chapter 27), and the understanding that natural environments are necessary to maintain the proper functioning of the biosphere, particularly in light of endless human degradation of air, water, and soil (see Chapter 49). However, conservation efforts are complicated by religious and cultural opposition to human population control and the economic difficulties faced by people in every region of the planet. Strict conservationists might call for an end to all development in wilderness areas, but the political realities suggest that such an approach is unlikely to be embraced by governments anywhere.

What lessons from community ecology and biogeography can be applied to conservation so that we may maximize the preservation of biodiversity? Conservation efforts now generally focus on the creation of nature preserves, areas that governments try to protect from development and pollution for the purpose of conserving particular species and natural communities. Nature preserves exist as islands in a metaphorical sea of disturbed habitat that is unsuitable for the organisms being protected. Lessons from the study of island biogeography are being increasingly applied in the design of nature preserves; the goal is to understand *in advance* what a preserve with a particular set of characteristics might successfully protect. The species-area relationship portrayed in Figure 48.22, for example, indicates that the number of species a preserve can harbor is directly related to the size of the preserve. Species in large preserves will be less likely to suffer competitive exclusion; even rare predators that usually exist at relatively low densities might maintain their populations in very large preserves.

Ideally, we would protect vast areas of relatively undisturbed habitat, but one ecologist has calculated that even by setting aside only 10% of a community as a preserve, we might hope to protect as many as half the species that occupied it in its undisturbed state. Of course, detailed knowledge of the requirements of the species in question is essential to a successful design. Minimum sizes for successful preserves must certainly exist, particularly for large predatory animals that each require a great deal of space. A preserve for leopards in Sri Lanka, for example, would have to include 5000 km^2 of undisturbed habitat to maintain a viable population of 500 individuals, and a preserve for 500 tigers would have to be roughly twice that size. The situation is complicated by the fact that many birds and mammals will only occupy core areas at the center of a suitable habitat patch; therefore, the outer edge of many preserves will not contribute to their conservation, though other species may prefer such habitats.

Many ecologists have pondered the relative benefits of creating one large preserve versus a group of smaller preserves that collectively include the same area as a large one. If the smaller preserves all contain the same representation of species, one large preserve probably would be preferable. However, if the community being conserved includes environmental heterogeneity, such that different species occupy different areas, several small preserves would retain a larger total number of species than one large preserve. In addition, populations within the small preserves might avoid the spread of epidemic diseases that could not cross the unsuitable habitats between them. When ecologists and wildlife managers design a series of small preserves, they are also concerned with their spatial relationship to one another. Reserves that are closely spaced might allow the possibility of recolonization from one preserve to another if a species became locally extinct. Similarly, corridors of suitable habitat that connect the preserves might encourage the spread of protected species among them. Many questions about the optimal design of nature preserves still remain to be answered. This active area of research among community ecologists highlights the relationship between science and society.

STUDY OUTLINE

1. A community is an assemblage of species living close enough together for potential interaction.

Two General Views of Communities (pp. 1106–1108)

1. Many plant species seem to be independently distributed; animals, however, are frequently linked to other species. For most species, presence in a given community probably reflects some combination of factors relating to their tolerance of abiotic factors and biological interactions.

Population Interactions (pp. 1108–1118)

1. Coevolution refers to a series of reciprocal interactions between two species.
2. Predation has important implications for the evolution of both predators and prey.
3. Plants defend against herbivores by mechanical defenses and the production of compounds that are irritating or even toxic to animals.
4. Animals resist predation by cryptic coloration, deceptive markings, rapid escape, and the possession of mechanical or chemical defenses that are sometimes advertised by aposematic coloration.
5. In Batesian mimicry, a palatable prey species resembles an unpalatable one; in Müllerian mimicry, a number of unpalatable prey species resemble one another.
6. Interspecific competition may be an important force determining the densities of species in some communities.
7. The competitive exclusion principle states that two species competing for the same limiting resources cannot coexist in the same place. Although laboratory populations have demonstrated this principle, there is uncertainty about its applicability to natural communities.
8. The ecological niche is the sum total of the organism's use of the biotic and abiotic resources of its environment. Closely related species may be able to coexist if there are one or more significant differences in their niches.
9. Resource partitioning and character displacement provide indirect evidence for the importance of past competition.
10. Symbiotic relationships in the form of parasitism, commensalism, and mutualism are important biotic interactions in communities.

Community Structure: Patterns and Processes (pp. 1118–1121)

1. Communities have several important characteristics: the prevalent form of vegetation; trophic structure; species richness, relative abundance, and species diversity; and stability.
2. Competition may be important in structuring many communities, reducing the densities of competing species; however, competitive exclusion is not an inevitable result of competition.
3. Keystone predators limit the densities of good competitors, thus increasing species diversity by allowing more species to coexist.
4. Environmental patchiness can also increase species diversity, because different species are best adapted to conditions in different patches.
5. Determining the causative factors that structure communities is often difficult because of the complex ways in which populations interact.

Succession (pp. 1121–1124)

1. Succession involves changes in species composition of a community over ecological time. Primary succession occurs where no soil previously existed, whereas secondary succession begins in an area where soil remains after a disturbance.
2. Sometimes succession involves inhibition, a phenomenon in which species inhibit the growth of newcomers or, in some cases, themselves.
3. Facilitation refers to alterations in the environment by the species of one stage that enable species in the next stage to grow.
4. Disturbances have a variable effect on communities, acting either to stabilize community structure or to initiate or alter succession, depending on the severity and duration of the disturbance.
5. Some ecologists believe that communities are in continual flux, with the identity and numbers of species changing at all stages of succession, even at the climax stage. Dispersal and disturbance probably play key roles in this nonequilibrium.

Biogeographical Aspects of Diversity (pp. 1124–1128)

1. Biogeography is the study of the past and present distribution of species. The major biogeographical realms are associated with patterns of continental drift.
2. A species would be limited to a given range if it never dispersed beyond that range, if it dispersed but failed to survive in other locations, or if it retracted from a larger range. Transplant experiments have been instructive in distinguishing between the first two alternatives in specific situations.

3. A variety of interacting factors have probably contributed to biogeographic clines in species diversity.
4. A general theory of island biogeography maintains that species richness on an island levels off at some dynamic equilibrium point, where new immigrations are balanced by extinctions. Furthermore, the theory predicts that species richness is directly proportional to size and inversely proportional to distance of the island from the mainland.

Conservation Biology: Lessons from Community Ecology and Biogeography (pp. 1128–1129)

1. The design of nature preserves is frequently based on the principles established by studies of island biogeography.

SELF-QUIZ

1. The concept of trophic structure of a community emphasizes the
   a. prevalent form of vegetation
   b. keystone predator
   c. feeding relationships within a community
   d. effects of coevolution
   e. species richness of the community
2. According to the principle of competitive exclusion,
   a. two species cannot coexist in the same habitat
   b. extinction or emigration are the only possible results of competitive interactions
   c. intraspecific competition results in the success of the best adapted individuals
   d. two species cannot share the same realized niche in a habitat
   e. resource partitioning will allow a species to utilize all the resources of its fundamental niche
3. The effect of a keystone predator within a community may be to
   a. competitively exclude other predators from the community
   b. maintain species diversity by preying on the most abundant prey species
   c. increase the relative abundance of the most competitive prey species
   d. encourage the coevolution of predator and prey adaptations
   e. create nonequilibrium in species diversity
4. All the following statements are consistent with the nonequilibrial model of community structure *except*
   a. Chance events such as dispersal and disturbance play major roles in determining species diversity.
   b. Species diversity may be increased by certain types of disturbances.
   c. Even when a community represents a mature climax stage, species composition and the number of species may continue to change.
   d. Succession reaches a climax when the intricate web of interactions allows for the addition of new species only when resources are made available by extinction.
   e. The course of succession may vary depending on the chance arrival of early colonizers.
5. Transplant experiments provide evidence that
   a. transplants always fail
   b. continental drift accounts for the geographic distribution of species
   c. the theory of island biogeography is valid
   d. keystone predators maintain community structure
   e. some species can live outside their normal ranges
6. An example of cryptic coloration is
   a. the green color of a plant
   b. the bright markings of a tropical frog
   c. the stripes of a skunk
   d. the mottled coloring of moths that rest on lichens
   e. the bright colors of an insect-pollinated flower
7. An example of Müllerian mimicry is
   a. a butterfly that resembles a leaf
   b. two poisonous frogs that resemble each other in coloration
   c. a minnow with spots that look like large eyes
   d. a beetle that resembles a scorpion
   e. a carnivorous fish with a wormlike tongue that lures prey
8. Which country is *incorrectly* paired with its biogeographical realm?
   a. Canada—Palearctic
   b. United States—Nearctic
   c. Brazil—Neotropical
   d. South Africa—Ethiopian
   e. India—Oriental
9. To be certain that two species had coevolved, one would *ideally* need to establish that
   a. the two species originated about the same time
   b. local extinction of one species dooms the other species
   c. each species affects the population density of the other species
   d. one species has adaptations that *specifically* tracked evolutionary change in the other species, and vice versa
   e. the two species are adapted to a common set of environmental conditions
10. According to the theory of island biogeography, species richness would be greatest on an island that is
   a. small and remote
   b. large and remote
   c. large and close to a mainland
   d. small and close to a mainland

CHALLENGE QUESTIONS

1. An ecologist studying desert plants performed the following experiment. She staked out two identical plots that included a few sagebrush plants and numerous small annual wildflowers. She found the same five wildflower species in similar numbers in both plots. Then she enclosed one of the plots with a fence to keep out kangaroo rats, the most common herbivores in the area. After two years, four species of wildflowers were no longer

present in the fenced plot, but one wildflower species had increased drastically. The control plot had not changed significantly. Using the principles and terminology discussed in the chapter, what do you think happened?

2. In the mountains of California, ecologist Craig Heller of Stanford University found that in most locations, the least chipmunk lived in sagebrush areas and the yellow pine chipmunk lived in higher areas of mixed sagebrush and piñon pines. However, if the yellow pine chipmunk was absent from a location, the least chipmunk lived in both the sagebrush and sagebrush-piñon pine areas. If the least chipmunk was absent, the yellow pine chipmunk distribution was unchanged. How might you explain this difference in terms of the concepts of community ecology?

SCIENCE, TECHNOLOGY, AND SOCIETY

1. The Great Lakes are the world's largest fresh water ecosystem. Sometime in 1986, a ship pumped out its water ballast near Detroit, and European zebra mussels (*Dreissena polymorpha*--see photo) entered the Great Lakes. The mollusks began to multiply wildly; in some places they now cling to surfaces at densities of 20,000 per m^2. They have blocked the intake pipes of power plants and water treatment plants, fouled boat hulls, and sunk buoys. Dozens of other exotic species, including sea lampreys, have also been accidentally introduced to the Great Lakes since the St. Lawrence Seaway opened in 1959. Collectively, the introduced species have caused millions of dollars of damage and have had a harmful impact on some of the Great Lakes' native species. How might such invasions be prevented in the future? (A drastic solution would be to close the Great Lakes to transoceanic shipping, but perhaps you can suggest less extreme solutions.)

Photo and information courtesy of R. Douglas Hunter, Oakland University, Michigan

2. By 1935, hunting and trapping had eliminated wolves from the United States outside Alaska. Since wolves have been protected as an endangered species, they have moved south from Canada and have become re-established in the Rocky Mountains and northern Great Lakes. Conservationists would like to speed up this process by reintroducing wolves into Yellowstone National Park. Local ranchers are opposed to bringing back the wolves because they fear predation on their cattle and sheep. What are some reasons for reestablishing wolves in Yellowstone National Park? What effects might the reintroduction of wolves have on the ecological communities in the park? What might be done to mitigate the conflicts between ranchers and wolves?

FURTHER READING

Beardsley, T. "Desert Dynamics." *Scientific American,* November 1992. An experiment tests the effect of kangaroo rats on desert vegetation.

Begon, M., J. L. Harper, and C. R. Townsend. *Ecology,* 2nd ed. New York: Blackwell Scientific, 1990. A general ecology text with a strong treatment of community ecology.

Culotta, E. "Biological Immigrants Under Fire." *Science,* December 6, 1991. How the introduction of exotic species disrupts natural communities.

DeVries, P. J. "Singing Caterpillars, Ants, and Symbiosis." *Scientific American,* October 1992. Some butterfly larvae use ants as guardians.

MacArthur, R. H., and E. O. Wilson. *The Theory of Island Biogeography.* Princeton, NJ: Princeton University Press, 1967.

Miller, J. A. "Invasion of the Ecosystem." *Science News,* June 29, 1985. A brief highlight of the dangers involved in ecological transplant experiments.

Oliwenstein, L. "Royal Flush." *Discover,* January 1992. New debate about a classic case of mimicry (monarch and viceroy butterflies).

Rennie, J. "Living Together." *Scientific American,* January 1992. The evolution of parasitism.

49 ECOSYSTEMS

Trophic Levels and Food Webs

Energy Flow

Chemical Cycling

Human Intrusions in Ecosystem Dynamics

Biotic Effects at the Biosphere Level

An **ecosystem** consists of all the organisms living in a community as well as all the abiotic factors with which they interact. As with populations and communities, the boundaries of ecosystems are usually not discrete. This unit of study can range from a laboratory microcosm, such as a terrarium, to lakes and forests. Indeed, some ecologists regard the entire biosphere as a sort of global ecosystem, comprised of all the local ecosystems on Earth.

The most inclusive level in the hierarchy of biological organization, an ecosystem involves two processes that cannot be fully described at lower levels: energy flow and chemical cycling. Energy enters most ecosystems in the form of sunlight and is converted to chemical energy by autotrophic organisms, passed to heterotrophs in the organic compounds of food, and dissipated in the form of heat. Chemical elements such as carbon and nitrogen are cycled between abiotic and biotic components of the ecosystem. Photosynthetic organisms acquire these elements in inorganic form from the air, soil, and water and assimilate them into organic molecules, some of which are consumed by animals. The elements are returned in inorganic form to the air, soil, and water by the metabolism of plants and animals and by other organisms, such as bacteria and fungi, that break down organic wastes and dead organisms. Energy flow and chemical cycling are related because both occur by the transfer of substances through the feeding relationships in the ecosystem. However, because energy, unlike matter, cannot be recycled, an ecosystem must be powered by a continuous influx of new energy from an external source (the sun). This chapter describes the dynamics of energy flow and chemical cycling in ecosystems, and considers some of the consequences of human intrusions into these processes (Figure 49.1).

Figure 49.1
Human intrusions into ecosystems. The destruction of this tropical forest in Guatemala illustrates several major ways in which humans alter ecosystems. Clear-cutting the forest destroys vegetation and important habitats for animals; burning of wood vaporizes some of the nutrients contained in the trees and introduces large amounts of carbon dioxide into the atmosphere. In this chapter, you will learn about the dynamics of ecosystems and how they are disrupted by an exploding human population and its technology.

TROPHIC LEVELS AND FOOD WEBS

Each ecosystem has a **trophic structure** of different feeding relationships that determines the route of energy flow and the pattern of chemical cycling (see Chapter 48). Ecologists divide the species in a community or ecosystem into **trophic levels** on the basis of their main source of nutrition. The trophic level that ultimately supports all others consists of autotrophs, which are called the **primary producers** of the ecosystem. Most producers are photosynthetic organisms that use light energy to synthesize sugars and other organic compounds, which they then use as fuel for cellular respiration and as building material for growth. All other organisms in an ecosystem are consumers—heterotrophs that are directly or indirectly dependent on the photosynthetic output of producers. Herbivores, which eat plants or algae, are the **primary consumers** of photosynthetic products. The next trophic level includes all those species that are **secondary consumers,** carnivores that eat herbivores. These carnivores may in turn be eaten by other carnivores that are **tertiary consumers,** and some ecosystems have carnivores of an even higher level. Some consumers, the **detritivores,** derive their energy from organic wastes, such as feces or fallen leaves and the remains of dead organisms (detritus) from the other trophic levels.

The main producers in most terrestrial ecosystems are green plants. In streams, much of the organic material used by consumers is also supplied by terrestrial plants, entering the ecosystem as debris that falls into the water or is washed in by runoff (see Chapter 46). In the limnetic zone of lakes and in the open ocean, photosynthetic protists and cyanobacteria, which form the phytoplankton, are the most important autotrophs, whereas multicellular algae and aquatic plants are abundant producers in the shallow, near-shore areas of both freshwater and marine ecosystems. In the aphotic zone of the deep sea, however, most life depends on photosynthetic production in the photic zone; energy and nutrients rain down from above in the form of dead plankton and other detritus. One notable exception is the community of organisms that live near hot water vents on the deep-sea floor (see Figure 46.20). Chemoautotrophic bacteria that derive energy from the oxidation of hydrogen sulfide are the main producers in these ecosystems, which are therefore supported by geothermal, rather than solar, energy.

The primary consumers, or herbivores, on land are mostly insects, snails, and certain vertebrates, including grazing mammals and the numerous birds and mammals that eat seeds and fruit. In aquatic ecosystems, phytoplankton is consumed mainly by zooplankton, which includes heterotrophic protists, various small invertebrates (especially crustaceans and, in the ocean, larval stages of many species that live in the benthos as adults), and some fish.

Examples of organisms functioning as secondary consumers in terrestrial ecosystems are spiders, frogs, and insect-eating birds, and lions and other carnivorous mammals that eat grazers such as antelope. In aquatic habitats, many fish feed on zooplankton and are in turn fed upon by other fish. In the benthic zone of the seas, algae-eating invertebrates are prey to other invertebrates, such as sea stars.

The organic material that composes the living organisms in an ecosystem is eventually recycled, broken down and returned to the abiotic environment in forms that can be used by plants. Detritivores, which feed on nonliving organic material, are sometimes called decomposers, emphasizing their importance in this recycling process. The most important detritivores in most ecosystems are bacteria and fungi, which usually live on the surface of their food sources. These organisms first secrete enzymes that digest organic material and then absorb the breakdown products; some bacteria and fungi can even digest cellulose. Earthworms and such scavengers as crayfish, cockroaches, and bald eagles are also detritivores, but these animals digest organic material internally after ingesting their food. In fact, all heterotrophs, including humans, are decomposers in the sense that they break down organic material and release inorganic products, such as carbon dioxide and ammonia, to the environment. The main distinction is that detritivores usually feed on material that is already dead when they find it. Detritivores often form a major link between primary producers and the secondary and tertiary consumers in an ecosystem. A crayfish, for example, might feed on plant detritus at the bottom of a lake and then be eaten by a bass. In a forest, birds might feed on earthworms that have been feeding on leaf litter in the soil.

The pathway along which food is transferred from trophic level to trophic level, beginning with producers, is known as a **food chain** (Figure 49.2). However, few ecosystems are so simple that they are characterized by a single, unbranched food chain. Several types of primary consumers usually feed on the same plant species, and one species of primary consumer may eat several different plants. Such branching of food chains occurs at the other trophic levels as well. For example, frogs, which are secondary consumers, eat several insect species, which may also be eaten by various birds. In addition, some consumers feed at several different trophic levels. An owl, for instance, may eat primary consumers such as field mice and also feed on higher-level consumers such as snakes. Omnivores, including humans, eat producers as well as consumers of different levels. Thus, the feeding relationships in an ecosystem are usually woven into elaborate **food webs** (Figure 49.3).

ENERGY FLOW

All organisms require energy for growth, maintenance, reproduction, and, in some species, locomotion. Primary producers use light energy to synthesize energy-rich organic molecules, which can subsequently be broken down to make ATP (see Chapter 10). Consumers acquire their organic fuels secondhand (or

Figure 49.2
Examples of terrestrial and marine food chains. Energy and nutrients pass through the trophic levels of an ecosystem when organisms feed on one another. Detritivores, important components of every ecosystem, are not included here.

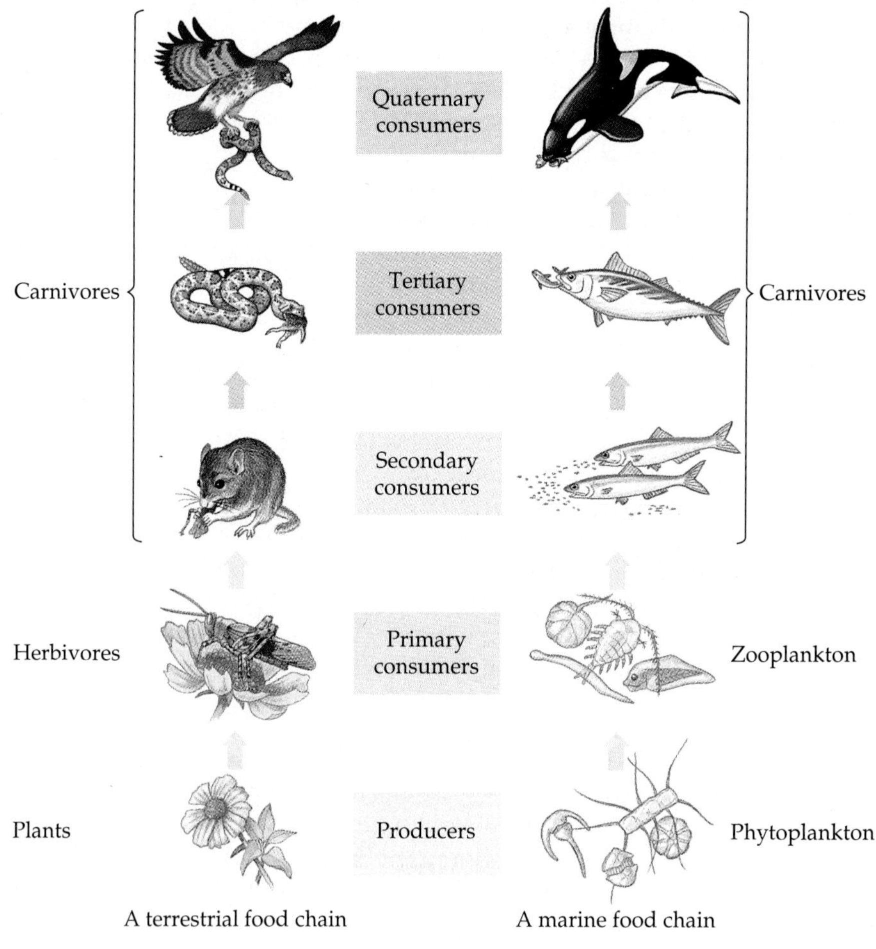

even thirdhand or fourthhand) through food webs. Therefore, the extent of photosynthetic activity sets the spending limit for the energy budget of the entire ecosystem.

The Global Energy Budget

Every day, planet Earth is bombarded by 10^{22} joules (J) of solar radiation, the energy equivalent of 100 million atomic bombs the size of the one dropped on Hiroshima. As described in Chapter 46, the intensity of the solar energy striking Earth and its atmosphere varies with latitude, such that the tropics receive the strongest input. Most solar radiation is absorbed, scattered, or reflected by the atmosphere in an asymmetrical pattern determined by variations in cloud cover and the quantity of dust in the air over different regions. As a result, the amount of solar radiation striking Earth's surface shows dramatic regional variation (Figure 49.4). These variations in the amount of solar radiation reaching the globe ultimately limit the photosynthetic output of ecosystems in different places.

Much of the solar radiation that reaches the biosphere lands on bare ground and bodies of water that either absorb or reflect the incoming energy. Only a small fraction actually strikes algae and the leaves of plants, and only some of this is of wavelengths suitable for photosynthesis (see Chapter 10). Of the visible light that does reach leaves and algae, only about 1% is converted to chemical energy by photosynthesis, and this photosynthetic efficiency varies with the type of plant, light level, and other factors. Although the fraction of the total incoming solar radiation that is ultimately trapped by photosynthesis is very small, primary producers on Earth collectively create about 170 billion tons of organic material per year—an impressive quantity.

Primary Productivity

The rate at which light energy is converted to chemical energy (organic compounds) by the autotrophs of an ecosystem is called **primary productivity.** The total of this productivity is known as **gross primary produc-**

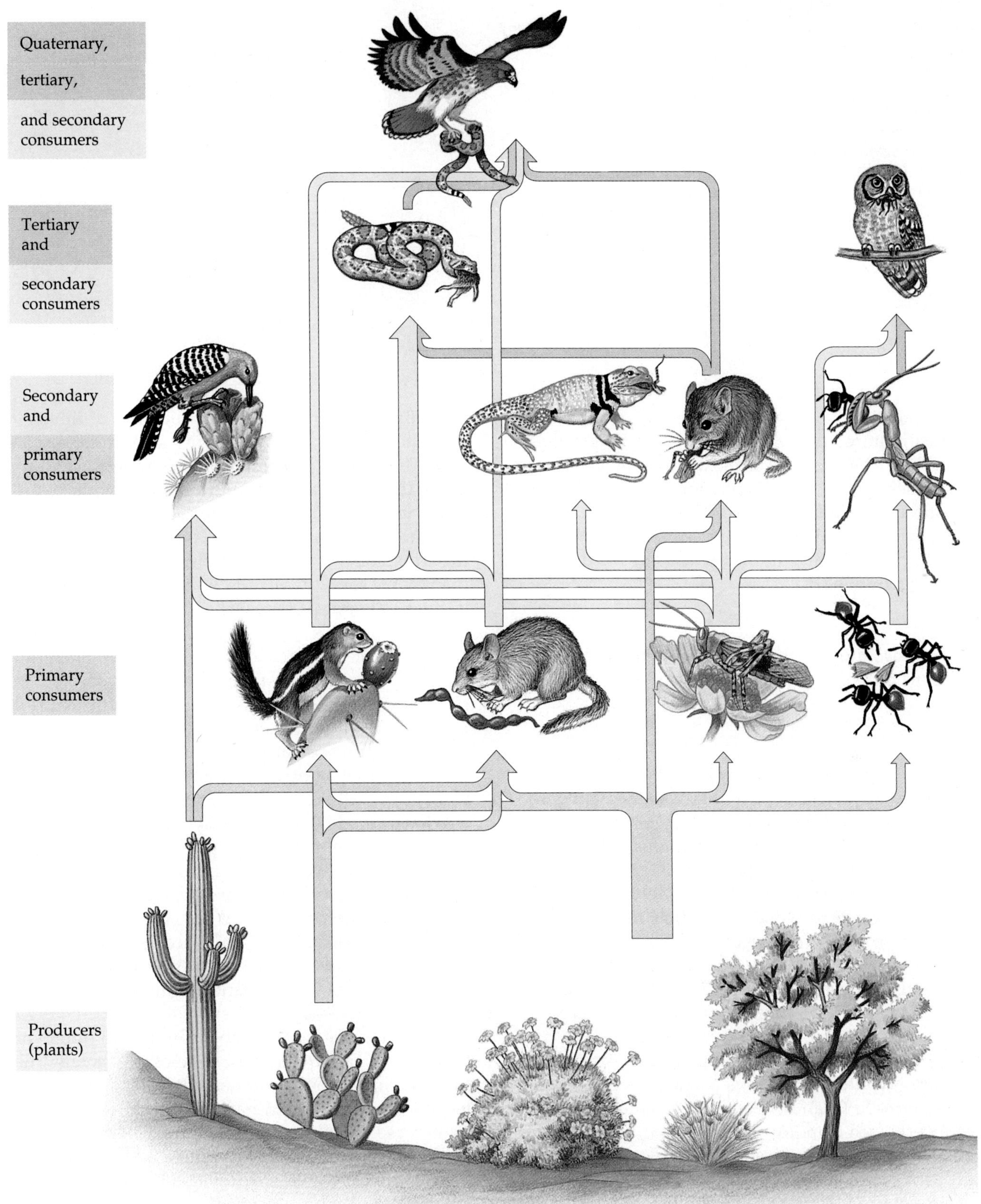

Figure 49.3
A food web. This simplified diagram of feeding relationships in the Sonoran Desert in the southwestern United States does not include all species that feed at each trophic level. Detritivores, which feed on the remains of organisms in every trophic level, are also omitted. The arrows are color-coded to indicate the trophic levels at which different species feed.

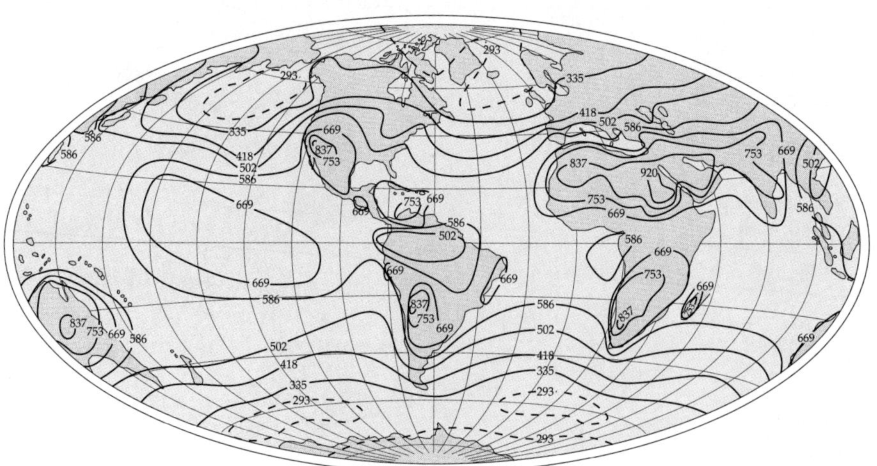

Figure 49.4
Mean annual solar radiation at the surface of the Earth. The amount of solar radiation ($J/cm^2/yr$) that strikes Earth varies with latitude as well as local atmospheric conditions and topographic features (1 J = 0.239 cal). The solar energy that reaches any ecosystem is one factor that limits its photosynthetic productivity. Although this illustration presents the mean annual solar energy on a worldwide scale, recall (from Chapter 46) that seasonal variability in solar radiation increases from the equator to the poles.

tivity (see the Methods Box, p. 1138). Not all of this product is stored as organic material in the growing plants, because the plants use some of the molecules as fuel in their own cellular respiration (the maintenance costs described in Chapter 46). The **net primary productivity (NPP)** is thus equal to the gross primary productivity (GPP) minus the energy used by the producers for respiration (Rs):

$$\text{NPP} = \text{GPP} - \text{Rs}$$

We may also think of this relationship in terms of the equations for photosynthesis and respiration:

$$6\,CO_2 + 6\,H_2O \underset{\text{Respiration}}{\overset{\text{Photosynthesis}}{\rightleftarrows}} C_6H_{12}O_6 + 6\,O_2$$

Gross primary productivity results from the rightward process (photosynthesis); net primary productivity is the difference between the yield of photosynthesis and the consumption of organic fuel symbolized by the leftward process. In less technical terms, net primary productivity accounts for the accumulation of organic material that we see as the growth of plants.

Net primary productivity is the measurement of interest to us because it represents the storage of chemical energy available to consumers in an ecosystem. For most primary producers, between 50% and 90% of the gross primary productivity remains as net primary productivity after their energetic needs are fulfilled. The NPP-to-GPP ratio is generally smaller for large producers with elaborate nonphotosynthetic structures, such as trees, which support large and metabolically active stem and root systems.

Primary productivity can be expressed in terms of energy per unit area per unit time (for example, $J/m^2/yr$), or as **biomass** (weight) of vegetation added to the ecosystem per unit area per unit time ($g/m^2/yr$). Biomass is usually expressed in terms of the dry weight of organic material because water molecules contain no usable energy and because the water content of plants varies over short periods of time. An ecosystem's primary productivity should not be confused with the total biomass of plants present at a given time, the **standing crop biomass;** primary productivity is the *rate* at which the vegetation synthesizes *new* biomass. Although a forest has a very large standing crop biomass, its productivity may actually be less than that of some grasslands, which do not accumulate vegetation because of grazing and because many of the plants are annuals.

Different ecosystems vary considerably in their productivity as well as in their contribution to the total productivity on Earth (Figure 49.5). Tropical rain forests are among the most productive terrestrial ecosystems, and because they cover a large portion of the Earth, they contribute a large proportion of the planet's overall productivity. Estuaries and coral reefs also have very high productivity, but their total contribution to global productivity is relatively small because these ecosystems are not very extensive. The open ocean contributes more primary productivity than any other ecosystem, but this is because of its very large size; productivity per unit area is relatively low. Deserts and tundra also have low productivity.

The factors most important in limiting productivity depend on the type of ecosystem and, for a single ecosystem, on seasonal changes in the environment. Productivity in terrestrial ecosystems is generally correlated with precipitation, temperature, and light intensity. Farmers often irrigate their fields, for example, to increase productivity in habitats where the availability of water limits photosynthetic activity; heat and light, as well as water, are provided to plants grown in greenhouses. Within any one type of terrestrial ecosystem, productivity increases with its proximity to the equator because water, heat, and light are more readily available in the tropics (see Chapter 46).

Inorganic nutrients may also be important in limiting the productivity of many terrestrial ecosystems. Recall from Chapter 33 that plants need a variety of nu-

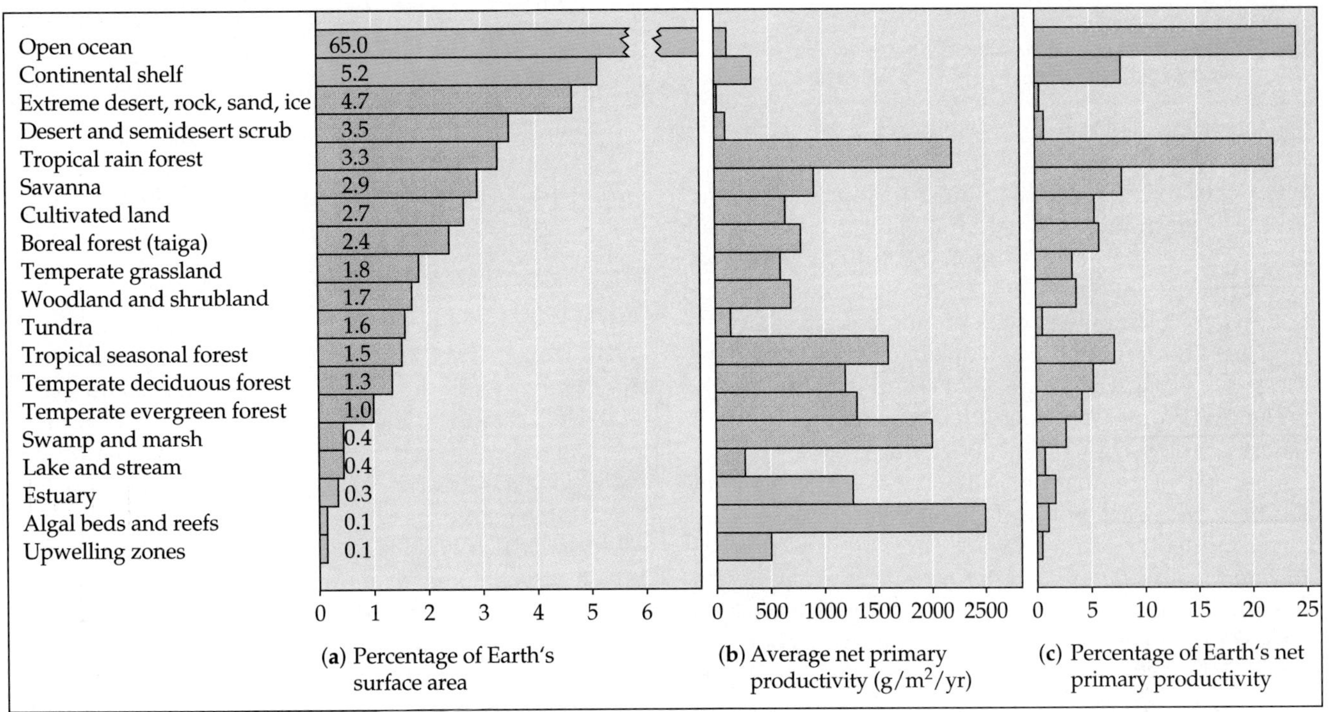

Figure 49.5
Productivity of different ecosystems. (**a**) The geographical extent and (**b**) the productivity per unit area of different ecosystems determine their contribution to (**c**) worldwide primary productivity. Open ocean, for example, contributes a lot to the planet's productivity because of its large size, whereas tropical forest contributes a lot because of its high productivity.

trients, some in relatively large quantities and others only in trace amounts—but all are crucial. The exact proportions of mineral nutrients needed by different species vary, as do the exact proportions available in the soil or water. Primary productivity removes nutrients from a system, sometimes faster than they are returned. At some point, productivity may slow or cease because a specific nutrient is no longer present in sufficient quantity. It is unlikely that all nutrients will be exhausted simultaneously, so that further productivity is limited by the single nutrient—called the **limiting nutrient**—that is no longer present in adequate supply. Adding other nutrients to the system will not stimulate renewed productivity because they are already present in sufficient quantity. However, the addition of the limiting nutrient will stimulate the system to resume growth until some other nutrient or the same one becomes limiting at a higher level of productivity. In many ecosystems, either nitrogen or phosphorus is a key limiting nutrient; both are required by plants in large amounts, but they are often present in only moderate or small quantities in natural environments. Some evidence also suggests that CO_2 sometimes limits productivity. Increasing the concentration of CO_2 around a plant may increase its productivity, but sometimes only by a small amount because another nutrient then becomes limiting. Later in the chapter we will discuss how human intrusions affecting nutrient balance have disrupted terrestrial and aquatic ecosystems.

Productivity in the seas is generally greatest in the shallow waters near continents and along coral reefs where abundant nutrients, light, and heat stimulate algal growth. In the open oceans, light intensity and temperature affect the productivity of phytoplankton communities. Productivity is generally greatest near the surface and declines sharply with depth, as light is rapidly absorbed by water and plankton. The primary productivity per unit area of the open ocean is relatively low because some inorganic nutrients, especially nitrogen and phosphorus, are in short supply near the surface; at great depth, where nutrients are abundant, there is insufficient light to support photosynthesis. Phytoplankton communities are most productive where upwelling currents bring nitrogen and phosphorus to the surface. This phenomenon is apparent, for instance, in antarctic seas, which, in spite of the cold water and low light intensity, are actually more productive than most tropical seas. The chemoautotrophic ecosystems near hot water vents are also very productive, but these communities are not widespread, and their overall contribution to marine productivity is small.

In freshwater ecosystems, as in the open ocean, light intensity and its variation with depth appears to be an important determinant of productivity. Water temper-

Methods: Measuring Gross Primary Productivity in Aquatic Habitats

How do we measure the rate of productivity of the tiny phytoplankton in aquatic habitats? Ecologists have developed an ingenious technique based on comparing changes in O_2 concentration of illuminated versus unilluminated phytoplankton samples.

A transparent bottle and an opaque bottle are each filled with a water sample taken from the same depth. The bottles are then stoppered and returned to that depth for a specified time, often one day. A basic assumption of this method is that the plankton communities inside each bottle are similar and represent the unconfined community in the area from which the water samples were taken. The bottles are removed from the pond after the predetermined time, and their O_2 concentration is measured by a chemical method. These values are compared with the initial O_2 concentration measured in a separate sample of water taken from the same depth at the beginning of the study.

Since no photosynthesis can occur in the dark bottle, its O_2 concentration will decrease from the initial value because of respiration by the phytoplankton. The magnitude of the decrease estimates the amount of respiration in the dark bottle sample. In the light bottle, by contrast, O_2 concentration should increase over the initial value because O_2 produced by photosynthesis should exceed the amount used by the phytoplankton for their own respiration. The magnitude of the increase in O_2 concentration in the light bottle measures the amount of O_2 produced by photosynthesis *in excess* of that sample's respiration. However, we have no direct measure of how much O_2 in the light bottle was used by that sample for its own respiration. If respiration rates are comparable in the two bottles, the decrease in O_2 concentration in the dark bottle is also an estimate of the amount used for respiration in the light bottle. Thus, the difference in O_2 concentration between the two bottles provides a measure of the total amount of O_2 produced by photosynthesis of phytoplankton in the light bottle. This method can be used to determine the gross primary productivity, based on the chemical equation for photosynthesis:

$$6\ CO_2 + 6\ H_2O \rightleftharpoons C_6H_{12}O_6 + 6\ O_2$$

This "light bottle/dark bottle" method is relatively insensitive to small changes in oxygen concentration, and therefore may be difficult to use in unproductive waters like the open ocean. The most common method today is a much more sensitive one that uses the radioactive tracer ^{14}C. Bicarbonate labeled with this form of carbon is introduced to a water sample in a bottle, which is then incubated at the depth from which it was taken, or under similar light/temperature conditions on board a ship. The label is assimilated as $^{14}CO_2$ by the phytoplankton and incorporated into the products of photosynthesis. After the incubation period, which may be only an hour or so, the plankton are filtered out of the sample, and their radioactivity level is measured using an instrument called a scintillation counter (see Chapter 2). Since the radioactivity represents carbon retained by the algae, it can be used to estimate net primary production.

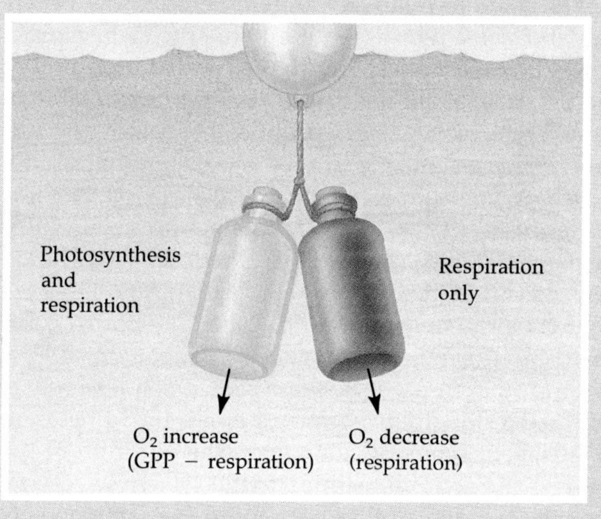

ature is also important, causing marked seasonal fluctuation of productivity in temperate regions. The availability of inorganic nutrients may also limit productivity in freshwater ecosystems as it does in the oceans, but the biannual turnover of lakes mixes the waters, carrying nutrients to the well-illuminated surface layers (see Chapter 46).

Energy Transfers and Ecological Pyramids

The rate at which an ecosystem's consumers—herbivores, carnivores, and detritivores—convert the chemical energy of the food they eat into their own new biomass is called the **secondary productivity** of the ecosystem.

Figure 49.6
An idealized pyramid of net productivity. In the case illustrated, 10% of the energy available at each trophic level is converted to new biomass in the trophic level above it. Notice that producers convert only about 1% of the energy in the sunlight available to them into primary productivity. In actual ecosystems, the decline in productivity with the transfer of energy between trophic levels varies with the particular species present; a 10% transfer of energy is a rough average.

Productivity inevitably declines with each transfer of energy through the trophic hierarchy. This decline is partly a consequence of the laws of thermodynamics: Total energy must be conserved (first law), but working organisms cannot avoid converting some of the energy they consume into heat, which is dissipated from the ecosystem (second law) (see Chapter 6). Thus, not all the chemical energy stored as biomass by net primary productivity can be converted to secondary productivity, even if consumers were to eat all the organic matter in the trophic level below them, which, of course, they do not.

Consider, for instance, the transfer of organic matter from producers to herbivores, the primary consumers. In most ecosystems, herbivores manage to eat only a small fraction of the plant material produced, and they cannot digest all the organic compounds that they do ingest. For example, caterpillars digest and absorb only about half the organic material they eat, passing the indigestible wastes as feces. If, for example, a caterpillar ate 200 J (48 cal) of leaves, it would absorb only about 100 J of energy. Typically, about two-thirds of the organic material absorbed by herbivores is used as fuel for cellular respiration, which degrades the food molecules to inorganic waste products and heat. The remainder of the organic material a primary consumer assimilates can add biomass to that trophic level. Hence, our caterpillar would use about 67 of its 100 assimilated J for its own maintenance (cellular respiration) and the remaining 33 J for growth. Only the chemical energy stored as growth (or production of offspring) by herbivores is available as food to secondary consumers. In one sense, our hypothetical example actually overestimates the conversion of primary productivity into secondary productivity because we did not account for all the primary productivity that the caterpillar did not even consume. In a real grassland, for example, grasshoppers and other insects may convert only about 4% of the net primary productivity to secondary productivity.

Carnivores are slightly more efficient at converting food into biomass, mainly because meat is more digestible than vegetation. But in many cases, secondary consumers use more of the energy they assimilate for cellular respiration, which dramatically decreases the amount of chemical energy available to the next trophic level. Endotherms, in particular, devote an enormous proportion of their assimilated energy to maintaining a high and constant body temperature.

Ecological efficiency is the ratio of net productivity at one trophic level to net productivity at the level below. Ecological efficiencies vary greatly among organisms; however, a value of 10% is commonly used as a rough approximation. In other words, about 90% of the energy available at one trophic level never transfers to the next. This multiplicative loss of energy from a food chain can be represented diagrammatically by a **pyramid of productivity,** in which the trophic levels are stacked in blocks, with primary producers forming the foundation of the pyramid. The size of each block is proportional to the productivity of each trophic level (per unit time). Pyramids of productivity are typically quite bottom-heavy because ecological efficiencies are low: Only about 10% of the energy at one level transfers as productivity to the level above (Figure 49.6).

One important ecological consequence of decreasing energy transfers through a food web can be represented in a **biomass pyramid,** in which each tier is proportional to the standing crop biomass (the total dry weight of all organisms) in a trophic level at any given moment. Biomass represents the chemical energy stored in the organic matter of a trophic level. Biomass pyramids generally narrow sharply from producers at the base to top-level carnivores at the apex (Figure 49.7a) because energy transfers between trophic levels are so inefficient. Some aquatic ecosystems, however, have inverted biomass pyramids, with primary consumers outweighing producers. In the waters of the English Channel, for example, the weight of zooplankton is five times the weight of phytoplankton (Figure

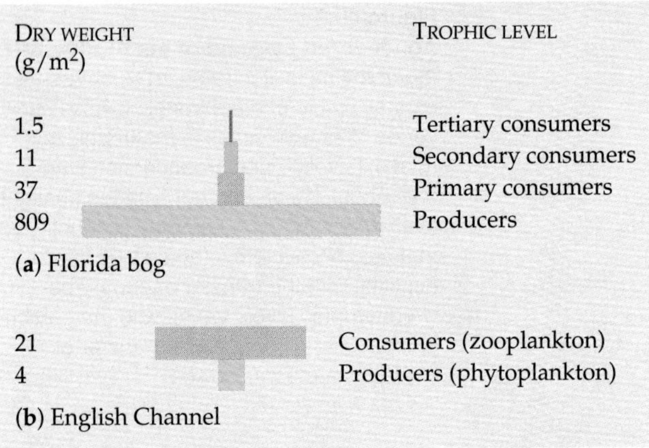

Figure 49.7
Pyramids of standing crop biomass. Numbers denote the dry weight (g/m^2) for all organisms at a trophic level. (**a**) Most biomass pyramids show a sharp decrease in biomass at successively higher trophic levels, as illustrated by data from a bog at Silver Springs, Florida. (**b**) However, in some aquatic ecosystems, such as the English Channel, a small standing crop of producers (phytoplankton) supports a larger standing crop of primary consumers (zooplankton) because the phytoplankton reproduce rapidly and are consumed at a high rate.

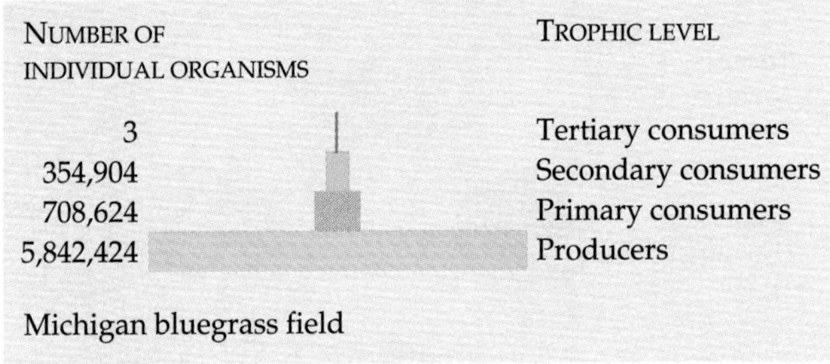

Figure 49.8
A pyramid of numbers. Because top-level predators are generally large animals, the small biomass at the top of a food chain is contained in a relatively small number of individuals. In this pyramid of numbers for a bluegrass field in Michigan, only three top carnivores are supported in an ecosystem based on the productivity of nearly 6 million plants.

49.7b). Such inverted biomass pyramids occur because the zooplankton consume the phytoplankton so quickly that the producers never develop a large population size or standing crop. Instead, the phytoplankton have a high *turnover rate:* They grow, reproduce, and are consumed rapidly. Nevertheless, productivity by the phytoplankton is fairly high, and the energy pyramid for this ecosystem is upright because phytoplankton have a higher productivity than zooplankton.

The multiplicative loss of energy from food chains severely limits the overall biomass of top-level carnivores that any ecosystem can support. Only about one-thousandth of the chemical energy fixed by photosynthesis can flow all the way through a food web to a tertiary consumer, such as a hawk or a shark. This partly explains why food webs usually include only three to five trophic levels. There are, for example, no nonhuman predators of lions, eagles, and killer whales because their biomass is insufficient to support yet another trophic level. Biomass, remember, refers to the total dry weight of all the animals in a trophic level, and, because individual predators tend to be fairly large animals, the limited biomass at the top of an ecological pyramid is concentrated in a relatively small number of individuals. This phenomenon is reflected in a **pyramid of numbers,** in which the size of each block is proportional to the number of individual organisms present in each trophic level (Figure 49.8). Populations of top predators are typically very small, and the animals may be widely spaced within their habitats. As a result, predators are highly susceptible to extinction (as well as to the evolutionary consequences of small population size discussed in Chapter 21) when their ecosystems are disturbed.

The concept of an energy or biomass pyramid also has implications for the human population. Eating meat is a relatively inefficient way of tapping photosynthetic productivity. A human obtains far more calories by eating grain directly as a primary consumer than by processing that same amount of grain through another trophic level and eating grain-fed beef. Worldwide agriculture could, in fact, successfully feed many more people than it does today if we all consumed only plant material, feeding more efficiently as primary consumers. Of course, calories are not the only factor in healthful nutrition (see Chapter 37).

CHEMICAL CYCLING

Although ecosystems receive an essentially inexhaustible influx of solar energy, chemical elements are available only in limited supply because the meteorites that occasionally strike Earth are the only extraterrestrial source of matter. Life on Earth therefore depends on the recycling of essential chemical elements. Organisms borrow their atoms from other components of the biosphere. Even while an individual organism is alive, much of its chemical stock is rotated continuously, as nutrients are absorbed and waste products released. Atoms present in the complex molecules of an organism at its time of death are returned as simpler compounds to the atmosphere, water, or soil by the action of detritivores. This decomposition replenishes the pools of inorganic nutrients that plants and other autotrophs use to build new organic matter. Because nutrient circuits involve both biotic and abiotic components of ecosystems, they are also called **biogeochemical cycles.** Water is an important vehicle for the transfer of chemicals in some biogeochemical cycles, and

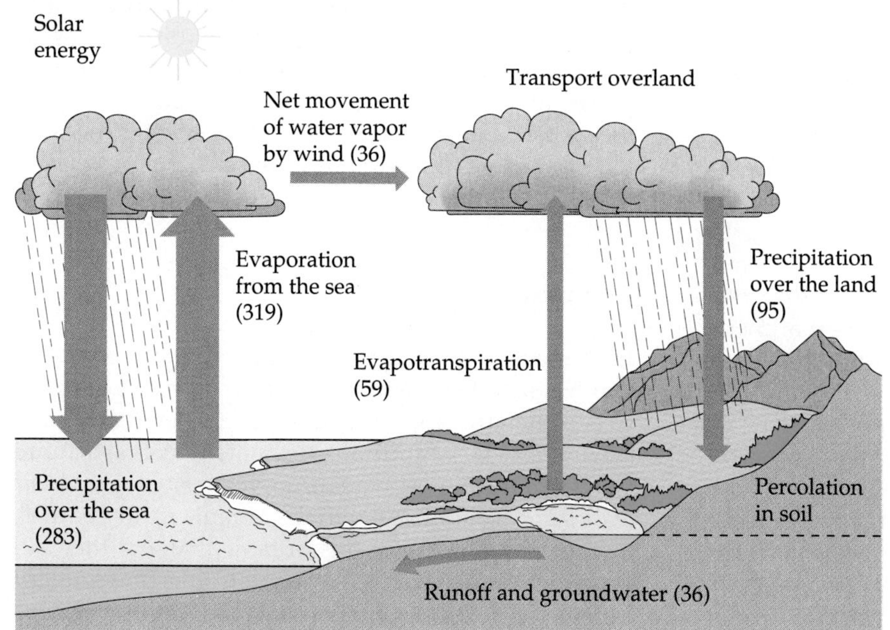

Figure 49.9
The hydrologic cycle. On a global scale, evaporation exceeds precipitation over the oceans. The result is a net movement of water vapor, carried by winds, over the land. The excess of precipitation over evaporation on land results in the formation of surface and groundwater systems that flow back to the sea, completing the major part of the cycle. Over the sea, evaporation forms most water vapor. On land, however, 90% or more of the vaporization is due to plant transpiration. The numbers in this diagram indicate water flow in billion billion (10^{18}) grams per year.

water itself moves in a global hydrologic cycle (Figure 49.9).

A chemical's specific route through an ecosystem varies with the particular element and the trophic structure of the ecosystem. We can, however, recognize two general categories of biogeochemical cycles. Gaseous forms of carbon, oxygen, sulfur, and nitrogen occur in the atmosphere, and cycles of these elements are essentially global. For example, some of the carbon and oxygen atoms a plant acquires from the air as CO_2 may have been released into the atmosphere by the respiration of an animal in some distant locale. Other elements that are less mobile in the environment, including phosphorus, potassium, calcium, and the trace elements, generally cycle on a more localized scale, at least over the short term. Soil is the main abiotic reservoir of these elements, which are absorbed by plant roots and eventually returned to the soil by decomposers, usually in the same general vicinity.

Before examining the details of some individual cycles, let's look at a general model of nutrient cycling that shows the main reservoirs of elements and the processes that transfer elements among reservoirs (Figure 49.10). Most nutrients accumulate in four reser-

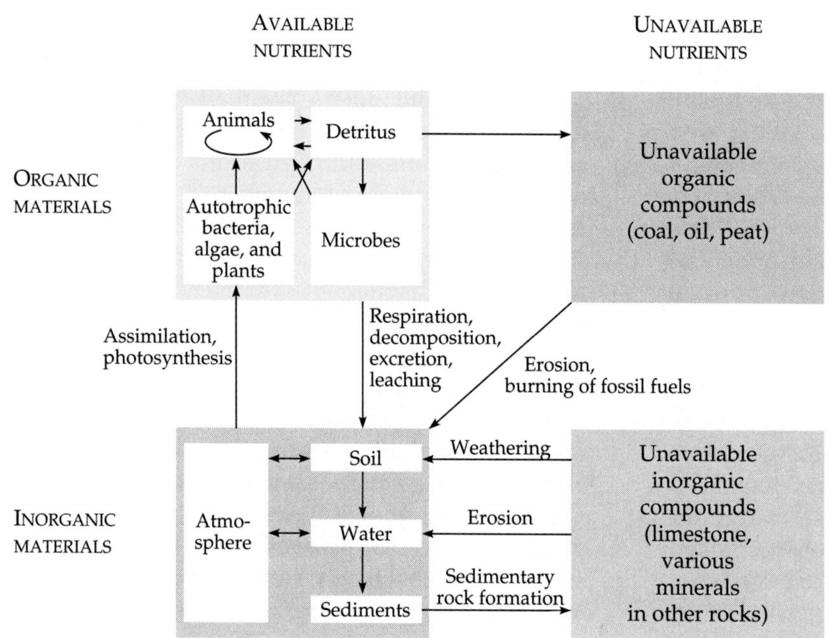

Figure 49.10
A general model of nutrient cycling. Most nutrients cycle within the biosphere among four major compartments, or reservoirs. The mechanisms that move nutrients from one compartment to another are indicated in this diagram.

voirs, or compartments. Each compartment is defined by two characteristics: whether it contains organic or inorganic materials, and whether or not the materials are directly available for use by organisms. One compartment of organic materials is composed of the living organisms themselves; these nutrients are available to other living organisms when they feed on one another. The other organic compartment includes "fossilized" deposits of once-living organisms (coal, oil, and peat), from which nutrients cannot be assimilated directly. Material moved from the living organic compartment to the fossilized organic compartment long ago, when organisms died and were buried by sedimentation over millions of years to become coal or oil.

Nutrients also occur in two inorganic compartments, one in which they are available for use by organisms and one in which they are not. The available inorganic compartment includes all the elements, ions, and molecules that are dissolved in water or present in soil or air. Organisms assimilate these materials from this compartment directly and return nutrients to it through the fairly rapid processes of respiration, excretion, and decomposition. Elements in the unavailable inorganic compartment are tied up in limestone and the minerals of other rocks. Although organisms cannot tap into this compartment directly, nutrients slowly become available for use through the action of weathering and erosion. Similarly, the unavailable organic materials described above also move into the available inorganic nutrients compartment through erosion or when fossil fuels are burned and their elements are vaporized.

Describing biogeochemical cycles in general theory is much simpler than actually tracing elements through these cycles in the field. Not only are ecosystems exceedingly complex, they usually exchange at least some of their materials with other regions. A pond, for example, has discrete boundaries, but several processes add and remove key nutrients. Minerals dissolved in rainwater or runoff from the neighboring land are added to the pond. Pollen, fallen leaves, and other airborne material rich in nutrients may also land there. And, of course, carbon, oxygen, and nitrogen cycle between the pond and the atmosphere. Birds may feed on fish or the aquatic larvae of insects, which derived their store of nutrients from the pond, and some of those nutrients may then be excreted or eliminated on land far from the pond's drainage area. Keeping track of the inflow and outflow is even more challenging in less clearly delineated terrestrial ecosystems. Nevertheless, ecologists have worked out the general schemes for chemical cycling in several ecosystems, often by adding tiny amounts of radioactive tracers that allow them to follow chemical elements through the various biotic and abiotic constituents of the ecosystems.

To illustrate some of the variations and complexities involved, we will now trace in more detail the cycling of three important elements: carbon, nitrogen, and phosphorus. As you study the particulars of each cycle, try to understand them in terms of the general compartment model in Figure 49.10. Also keep in mind that these three examples only illustrate general processes. Every nutrient has a unique cycle, and each one is potentially limiting in some ecosystem.

The Carbon Cycle

In the carbon cycle, the reciprocal processes of photosynthesis and cellular respiration provide a link between the atmosphere and terrestrial environments (Figure 49.11). Autotrophs (primarily plants) acquire CO_2 from the atmosphere through the stomata of their leaves, incorporating it into the organic matter of their own biomass; some of this organic material then becomes the carbon source for heterotrophs (the primary and higher-level consumers). Respiration by both autotrophs and heterotrophs returns CO_2 to the atmosphere.

Although CO_2 is present in the atmosphere at a relatively low concentration (about 0.03%), carbon recycles at a relatively fast rate because plants have a high demand for this gas. Each year, plants remove about one-seventh of the CO_2 in the atmosphere; this is approximately (but not exactly) balanced by respiration. Some carbon may be diverted from the cycle for longer periods. This happens, for example, when it is accumulated in wood and other durable organic material. Decomposition eventually recycles even this carbon to the atmosphere as CO_2, although fires can oxidize such organic material to CO_2 much faster. Some processes, however, can remove carbon from short-term cycling for millions of years. In some environments, organic litter accumulates much more quickly than decomposers can break it down. Under certain conditions, these deposits eventually formed coal and petroleum (the fossil fuels) that became locked away in the compartment that includes unavailable organic nutrients.

The amount of CO_2 in the atmosphere varies slightly with the seasons. Concentrations of CO_2 are lowest during the Northern Hemisphere's summer and highest during winter. This seasonal pulse of CO_2 concentration occurs because there is more land in the Northern Hemisphere than in the Southern, and therefore more vegetation. The vegetation has maximal photosynthetic activity during summer, reducing the global amount of atmospheric CO_2. During winter, the vegetation respires more CO_2 than it uses for photosynthesis, causing a global increase in the gas (see Figure 49.19).

Superimposed on this seasonal fluctuation is a continuing increase in the overall concentration of atmospheric CO_2 caused by the combustion of fossil fuels by humans. From a long-term perspective, this can be viewed as a return to the atmosphere of CO_2 that was

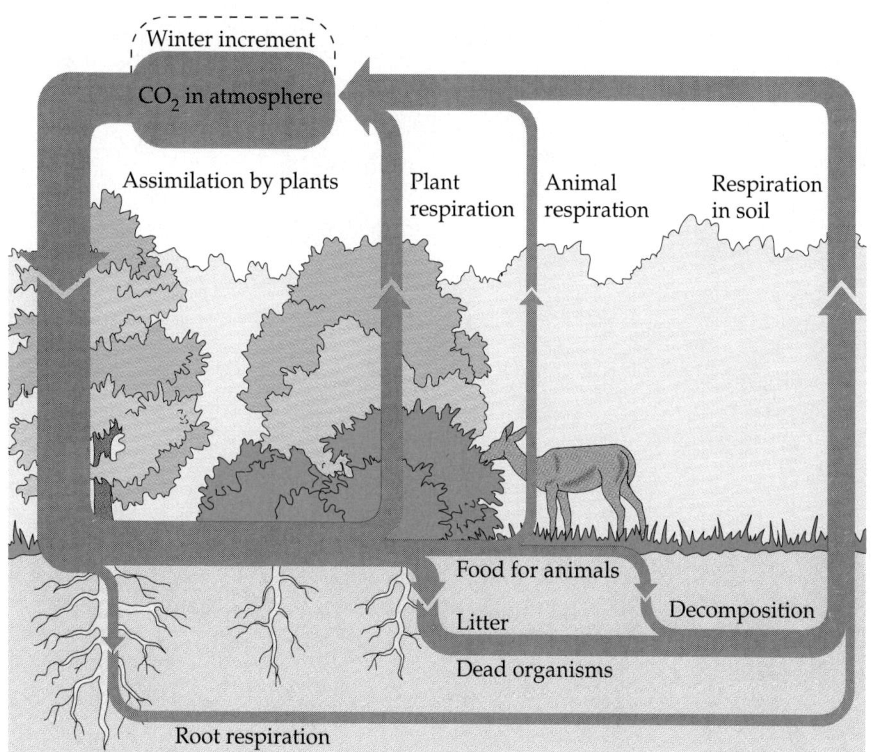

Figure 49.11
The carbon cycle. The reciprocal processes of photosynthesis and cellular respiration are responsible for the major transformations and movements of carbon. A seasonal pulse in atmospheric CO_2 is caused by variation in the distribution of vegetation on Earth (see the text). On a global scale, the return of CO_2 to the atmosphere by respiration closely balances its removal by photosynthesis. However, the burning of wood and fossil fuels adds more CO_2 to the atmosphere; as a result, the amount of atmospheric CO_2 is steadily increasing. Atmospheric CO_2 also moves into or out of aquatic systems, where it is involved in a dynamic equilibrium with other inorganic forms, including bicarbonates.

removed by photosynthesis long ago. But in the millions of years while this material was effectively out of circulation, a new equilibrium developed in the global carbon cycle. Now that balance is being disrupted, with uncertain consequences that we will consider later.

The same basic processes of respiration and photosynthesis are important in aquatic environments, but the cycling of carbon is more complicated because of the interaction of CO_2 with water and limestone. Dissolved carbon dioxide reacts with water to form carbonic acid (H_2CO_3). Carbonic acid in turn reacts with the limestone ($CaCO_3$) that is abundant in many waters, including the ocean, to form bicarbonates and carbonate ions:

$$H_2O + CO_2 \rightleftharpoons H_2CO_3$$
$$H_2CO_3 + CaCO_3 \rightleftharpoons Ca(HCO_3)_2 \rightleftharpoons Ca^{2+} + 2\,HCO_3^-$$
$$\underset{\text{Bicarbonate}}{2\,HCO_3^-} \rightleftharpoons 2\,H^+ + \underset{\text{Carbonate}}{2\,CO_3^{2-}}$$

As CO_2 is used in photosynthesis in aquatic and marine environments, the equilibrium of this reaction series shifts toward the left, converting bicarbonates back to CO_2. Thus, bicarbonates serve as a CO_2 reservoir. Aquatic autotrophs may also use dissolved bicarbonate directly as their source of carbon. Overall, the amount of carbon present in various inorganic forms in the ocean, not including sediments, is about fifty times that available in the atmosphere. Because of these inorganic reactions of CO_2 in water, as well as its uptake by marine phytoplankton, the ocean may serve as an important "buffer" that will absorb some of the CO_2 being added to the atmosphere by the burning of fossil fuels.

The Nitrogen Cycle

Earth's atmosphere is almost 80% N_2, but plants cannot assimilate this form of nitrogen. Only certain prokaryotes can fix nitrogen—that is, reduce atmospheric nitrogen to ammonia (NH_3), which can be used to synthesize nitrogenous organic compounds such as amino acids. Indeed, prokaryotes are vital links at several points in the nitrogen cycle (Figure 49.12).

Nitrogen is fixed in terrestrial ecosystems by free-living (nonsymbiotic) soil bacteria as well as by symbiotic bacteria *(Rhizobium)* in the root nodules of legumes and certain other plants (see Chapter 33). Some cyanobacteria fix nitrogen in aquatic ecosystems. Organisms that fix nitrogen, of course, are fulfilling their own metabolic requirements, but they also release excess ammonia that becomes available to other organisms. Small amounts of nitrogen are also fixed by lightning in the atmosphere. Far more important, however, is industrial fixation for the production of fertilizer, which now makes a significant contribution to the pool of nitrogenous materials in the soil and waters of agricultural regions (and also contributes to environmental problems discussed later in the chapter).

Figure 49.12
The nitrogen cycle. Most of the nitrogen cycling through food webs is taken up by plants in the form of nitrate. Most of this, in turn, comes from the nitrification of ammonia that results from the decomposition of organic material. The addition of nitrogen from the atmosphere through fixation and its return via denitrification involve relatively small amounts compared to the local recycling that occurs in the soil or water.

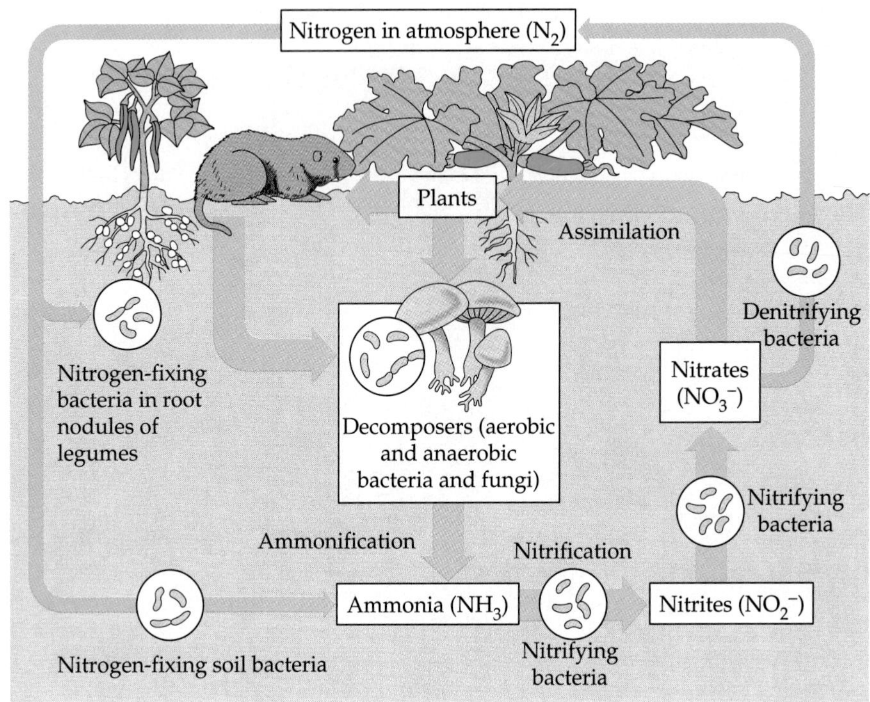

Although plants can use ammonia directly, most of the ammonia in soil is used by certain aerobic bacteria as an energy source; their activity oxidizes ammonia to nitrite (NO_2^-) and then to nitrate (NO_3^-), a process called **nitrification.** Nitrate released from these bacteria can then be assimilated by plants and converted to organic forms, such as amino acids and proteins. Animals can assimilate only organic nitrogen, by eating plants or other animals. Some bacteria can obtain the oxygen they need for metabolism from nitrate rather than from O_2. As a result of this **denitrification** process, some nitrate is converted back to N_2, returning to the atmosphere.

Decomposition of organic nitrogen back to ammonia, a process called **ammonification,** is carried out by many bacteria and eukaryotes. Detritivores are especially important in recycling large amounts of nitrogen to the soil through this process.

Overall, most of the nitrogen cycling in natural systems involves the nitrogenous compounds in soil and water, not atmospheric N_2. Although nitrogen fixation has been important in the buildup of a pool of available nitrogen, it contributes only a tiny fraction of the nitrogen assimilated annually by total vegetation. Nevertheless, many common species of plants depend on their association with nitrogen-fixing bacteria to provide this essential nutrient in a form they can assimilate. The amount of N_2 returned to the atmosphere by denitrification is also relatively small. The great bulk of assimilated nitrogen comes from nitrate, which is efficiently recycled from organic forms by ammonification and nitrification.

The Phosphorus Cycle

In some respects the phosphorus cycle is simpler than either the carbon or the nitrogen cycle. Cycling does not include movement through the atmosphere because there are no significant phosphorus-containing gases. In addition, phosphorus occurs in only one important inorganic form, phosphate (PO_4^{3-}), which plants absorb and use for organic synthesis. Weathering of rocks gradually adds phosphate to soil (Figure 49.13). After producers incorporate phosphorus into biological molecules, it is transferred to consumers in organic form, and added back to the soil by the excretion of phosphate by animals and by the action of decomposers on detritus. Humus and soil particles bind phosphate (see Chapter 33), so that the recycling of phosphorus tends to be quite localized in ecosystems. However, phosphorus does leach into the water table, gradually draining from terrestrial ecosystems to the sea. Severe erosion can hasten this drain, but in most natural ecosystems, weathering of rock can keep pace with the loss of phosphate. Phosphate that reaches the ocean gradually accumulates in sediments and becomes incorporated into rocks that may later be included in terrestrial ecosystems as a result of geological processes that raise the seafloor or lower sea level at a particular location. Thus, two phosphate cycles occur according to very different time scales. Most phosphate recycles locally among soil, plants, and consumers on the scale of ecological time. A parallel sedimentary cycle removes and restores terrestrial phosphorus over geological time. The same general

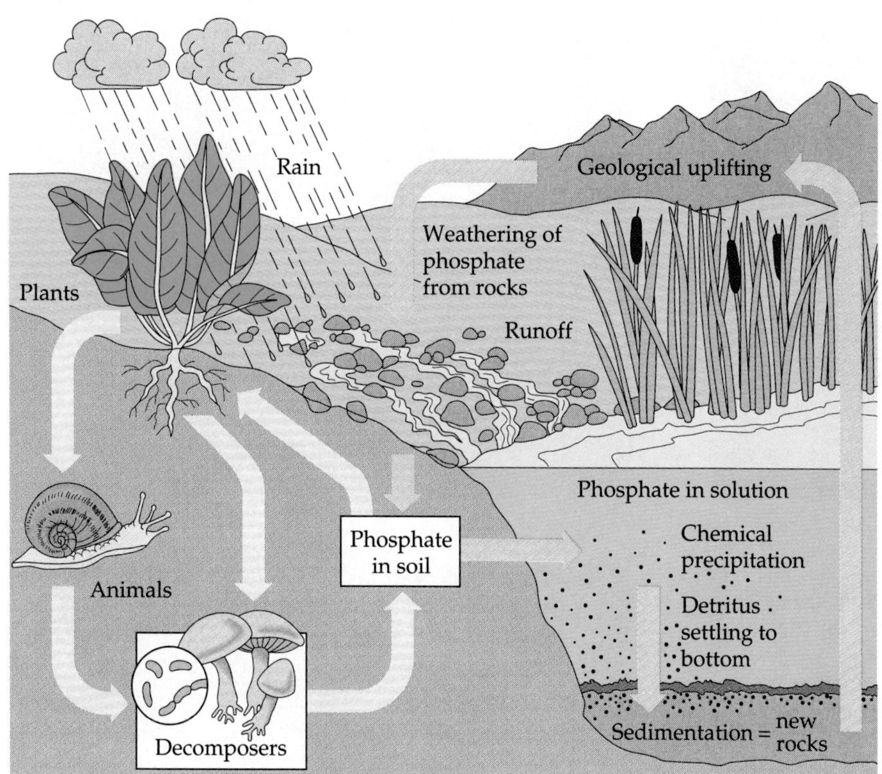

Figure 49.13
The phosphorus cycle. Phosphorus, which does not have an atmospheric component, tends to cycle locally (yellow arrows). Exact rates vary in different systems. Generally, small losses from terrestrial systems caused by leaching are balanced by gains from the weathering of rocks. In aquatic systems, as in terrestrial systems, phosphorus is cycled through food webs. Typically, however, some phosphorus is lost from the ecosystem because of chemical processes that cause precipitation or through settling of detritus to the bottom, where sedimentation may lock away some of the nutrient before biotic processes can reclaim it. On a much longer time scale, this phosphorus may become available to ecosystems again through geological processes such as uplifting (gold arrows). This general pattern applies to many other nutrients, including trace elements.

pattern applies to other nutrients that lack atmospheric forms.

Variations in Nutrient Cycling Time

In some parts of a tropical rain forest, key nutrients such as phosphorus occur in the soil at levels far below those typical of a temperate forest. At first this might seem paradoxical because tropical forests generally have very high productivity. The key to this apparent riddle is that organic material decomposes rapidly in tropical areas because of the warm temperatures and abundant precipitation. In addition, the immense biomass of these forests creates a high demand for nutrients, which are absorbed almost as soon as they become available through the action of detritivores. As a result of the rapid decomposition, relatively little organic material (about 1% to 2% of the total) accumulates as leaf litter on the floor of tropical forests, and virtually all the nutrients in the ecosystem are present in the living vegetation. In temperate forests, by contrast, decomposition is much slower, and leaf litter may account for 5% to 20% of all the organic material in the ecosystem. Many of the nutrients present in temperate forest ecosystems are therefore included in the detritus and soil, where they may reside for fairly long periods of time before being assimilated by plants. Thus, the relatively low concentrations of some nutrients in the soil of tropical rain forests results from a fast cycling time, not an overall scantiness of these elements in the ecosystem. Among the factors that influence nutrient cycling times in ecosystems are rate of decomposition, size of the standing crop, local soil chemistry, and the frequency of fires.

HUMAN INTRUSIONS IN ECOSYSTEM DYNAMICS

As the human population has grown to an immense size, our activities and technological capabilities have intruded in one way or another into the functioning of many communities and ecosystems. Even where we have not completely destroyed a natural system, our actions have disrupted the trophic structure, energy flow, and chemical cycling of ecosystems in most areas of the world. The effects are sometimes local or regional, but the ecological impact of humans can be widespread or even global. Acid rain, for example, which was discussed in detail in Chapter 3, may be carried by prevailing winds and fall hundreds or thousands of miles from the smokestacks emitting the chemicals that produce it. Here we address a sampling of the many other concerns ecologists and environmentalists have about human intrusions into the dynamics of ecosystems.

(a)

(b)

Figure 49.14
Ecological effects of warfare. Military conflicts devastate entire landscapes as well as particular ecosystems. (**a**) The use of a defoliant (agent orange) during the Vietnam War quickly destroyed vast areas of tropical forest. The photograph on the top left shows a Vietnamese tropical forest before defoliation, and the photograph on the top right shows the same area afterward. One of the chemicals in agent orange is a carcinogen. (**b**) In this satellite view of Kuwait at the end of the 1991 Persian Gulf War, hundreds of burning oil wells are visible as orange dots. The fires blackened the skies, deposited a thick layer of oily soot on surrounding areas, and squandered precious fossil fuel.

Habitat Destruction and the Biodiversity Crisis

Human encroachment on natural ecosystems has reached epidemic proportions, as indicated in the map of agricultural and urban biomes in Figure 46.4. The clearing of natural ecosystems, which is usually necessary for agricultural, industrial, and residential development, undoubtedly causes the greatest local disruption of natural environments. Clear-cut harvesting of timber also destroys vast tracts of forest. Relatively little undisturbed habitat still exists in many countries, and ecologists estimate that more than 75% of the world's original forests have been cleared or severely disrupted. In the United States, for example, only 15% of the original primary forest (most of it in Alaska) and less than 1% of the original tallgrass prairie remain; the statistics on forests are even worse in some other regions, such as Europe, China, and Australia. In recent years, environmentalists have focused attention on the destruction of tropical forests, among the most productive ecosystems on Earth. Some estimates suggest that if tropical forests continue to be cut at the present rate (about 500,000 km^2 worldwide per year), nearly all large tracts of this ecosystem will be eliminated within a decade or two.

Development and logging are certainly not the only human actions that disrupt entire ecosystems. The ecological consequences of war are both dramatic and devastating (Figure 49.14). During the Vietnam War, for example, the United States used large quantities of chemical defoliants to kill the vegetation in which enemy soldiers concealed themselves. More recently, as the Persian Gulf War ended, Iraqi soldiers created other disasters that ravaged the landscape of the disputed territory. Huge oil spills polluted marine ecosystems, and the burning of oil wells blackened the skies and left sooty deposits on everything nearby. Even in peacetime, large military drills can devastate an ecosystem through the activity of vehicles and personnel, pollution of the air by military machinery, and the production of toxins and radioactive materials in anticipation of future conflicts.

One major concern about the wholesale destruction of any natural habitat, particularly tropical forest, is that many species may become extinct as the ecosystems disappear. The **biodiversity crisis** received enormous attention at the 1992 United Nations Earth

(a)

(b)

Figure 49.15
Nutrient recycling in the Hubbard Brook Experimental Forest: an example of LTER (long-term ecological research). (**a**) Concrete weirs built across streams at the bottom of watersheds enabled researchers to monitor the outflow of water and mineral nutrients from the ecosystem. Because these small dams were anchored to the impervious bedrock, all water draining from the valleys had to pass through the weirs. (**b**) Some watersheds were completely logged to study the effects of clear-cutting on drainage and nutrient cycling. All the logs and rubble from the clear-cutting operation were left in place, and care was taken not to disturb the soil any more than necessary.

Summit in Rio de Janeiro, and a global treaty to preserve biodiversity was signed by the leaders of virtually all nations (the United States was a notable exception). The preservation of endangered species is a difficult task under any circumstances, because ecologists are uncertain about minimum viable population sizes and the minimum expanses of suitable habitat that need to be protected. The problem is particularly complicated for migratory species that winter in one country and breed in another; successful conservation efforts for such species require closely coordinated international cooperation and the careful preservation of habitat in both parts of the species' range. Some North American songbird species, for example, experienced population declines of 1% to 6% during a single recent decade (1978 to 1987) because of habitat destruction in both their northern breeding grounds and their tropical wintering grounds; preservation of forests in North America or South America alone will not save these species from extinction. Nobody can accurately estimate the magnitude of the biodiversity crisis, largely because taxonomists have identified and described only a fraction of the species present on Earth. Millions of plant and animal species are probably still unknown to scientists, and it is a sad irony when human activities cause the extinction of species, including ones that we don't even know exist, which could be potential sources of valuable products.

Effects of Deforestation on Chemical Cycling: The Hubbard Brook Forest Study

Deforestation not only eliminates the trees and associated animals from a region, it also severely disrupts water and nutrient cycles. For the past 30 years, a team of scientists has conducted a long-term study of nutrient cycling in a forest ecosystem under both natural conditions and after severe human intrusion. The study site is the Hubbard Brook Experimental Forest in the White Mountains of New Hampshire. It is a nearly mature deciduous forest with several valleys, each drained by a small creek that is a tributary of Hubbard Brook. Bedrock impenetrable to water is close to the surface of the soil, and each valley constitutes a watershed that can drain only through its creek.

The research team first determined the mineral budget for each of six valleys by measuring the inflow and outflow of several key nutrients. They collected rainfall at several sites to measure the amount of water and dissolved minerals added to the ecosystem. To monitor the loss of water and minerals, they constructed V-shaped, concrete weirs (dams) across the creek at the bottom of each valley (Figure 49.15a). About 60% of the water added to the ecosystem as rainfall and snow exits through the stream, and the remaining 40% is lost by transpiration from plants and evaporation from the soil.

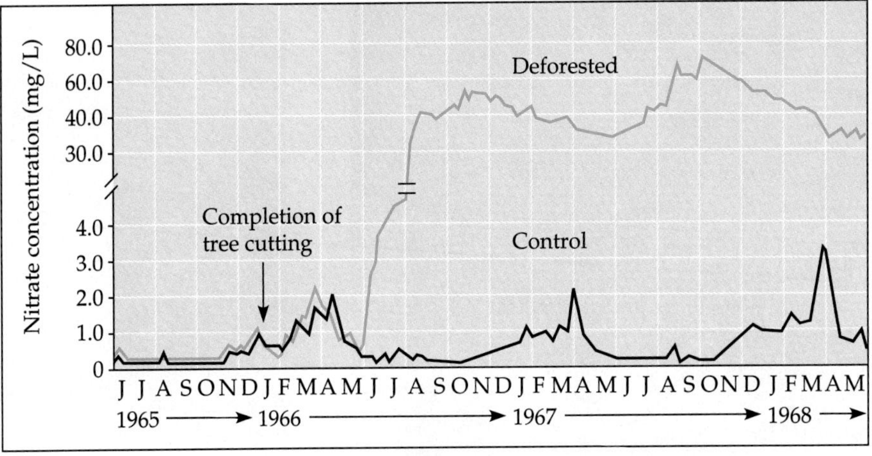

Figure 49.16
Loss of nitrate from a deforested watershed in the Hubbard Brook Experimental Forest. The concentration of nitrate in runoff from the deforested watershed was sixty times greater than in a control unlogged watershed.

Preliminary studies confirmed that internal cycling within a mature terrestrial ecosystem conserves most of the mineral nutrients. Mineral inflow and outflow balanced and were relatively small compared with the quantity of minerals being recycled within the forest ecosystem. For example, only about 0.3% more Ca^{2+} left a valley via its creek than was added by rainwater, and this small net loss was probably replaced by chemical decomposition of the bedrock. During most years, the forest actually registered small net gains of a few mineral nutrients, including nitrogenous ones.

In 1966, one of the valleys, with an area of 15.6 hectares, was completely logged in an experiment that tested the effect of deforestation on nutrient cycling (Figure 49.15b). The inflow and outflow of water and minerals in the experimentally altered watershed were compared with those in a control watershed for three years. Water runoff from the valley increased by 30% to 40% after deforestation, apparently because there were no trees to absorb and transpire water from the soil. Net losses of minerals from the deforested watershed were huge. The concentration of Ca^{2+} in the creek increased fourfold, for example, and the concentration of K^+ increased by a factor of 15. Most remarkable was the loss of nitrate, which increased in concentration in the creek sixtyfold (Figure 49.16). Not only was this vital mineral nutrient drained from the ecosystem, but nitrate in the creek also reached a level considered unsafe for drinking water.

Similar studies in other experimental forests have corroborated the evidence from Hubbard Brook that deforestation severely alters both water drainage and nutrient cycling. These studies are examples of **long-term ecological research (LTER),** an ambitious strategy for tracking the dynamics of natural ecosystems and the effects of human intrusions.

Agricultural Effects on Nutrient Cycling

Human activity often intrudes in nutrient cycles by removing nutrients from one part of the biosphere and adding them to another. This may result in the depletion of key nutrients in one area as well as the disruption of the natural equilibrium in both areas. Consider, for example, the ecological impact of agriculture on a natural ecosystem. After natural vegetation is cleared from an area, crops may be grown for some time without an additional supplement of nutrients because of the existing reserve of both organic and inorganic nutrients in the soil. However, in agricultural ecosystems, a substantial fraction of these nutrients is not recycled but is, instead, exported from the area, often to distant places, in the form of crop biomass. The "free" period for crop production—when there is no need to add nutrients to the soil—varies greatly, depending on the system. When some of the early North American prairie lands were first tilled, for example, good crops could be produced for many years because the large store of organic materials in the soil continued to provide nutrients through decomposition. By contrast, some cleared land in the tropics can be farmed for only one or two years. Eventually, in any area under intensive agriculture, the natural store of nutrients becomes exhausted, and fertilizer must be added. The industrially synthesized fertilizers used extensively today are produced at considerable expense in terms of both money and energy (see Chapter 33).

Many of the nutrients added to agricultural systems are quickly "short-circuited" into aquatic ecosystems. As we have seen, some nutrients, such as phosphate, are essentially lost from terrestrial systems once they exit to aquatic ones. Normally, the rate of loss is low because leaching is generally a slow process. But nutrients in crops soon appear in the wastes of humans and livestock and then appear in streams and lakes through runoff from fields and discharge as sewage. Someone eating a piece of broccoli in Washington, DC, is consuming nutrients that only days before might have been in the soil in California; and a short time later, some of these nutrients will be in the Potomac River on their way to the sea, having passed through an individual's digestive system and the local sewage facili-

Figure 49.17
The experimental eutrophication of a lake. The far basin of this lake was separated from the near basin by a plastic curtain and fertilized with inorganic sources of carbon, nitrogen, and phosphorus. Within two months, the far basin was covered with an algal bloom, as can be seen here. The near basin, which was treated with only carbon and nitrogen, remained unchanged. In this case, phosphorus was the key limiting nutrient, and its addition stimulated the explosive growth of algal populations.

ties. Once in aquatic systems, the nutrients may stimulate excessive growth of algae, a related problem to be discussed shortly. Researchers are attempting to find ways to reclaim nutrients at sewage treatment facilities and return them to the land to reduce the need for constant fertilizer supplements. There has been some public resistance to this idea because of an understandable concern about adding the products of human waste to the land that produces the food we eat. Technological advances in sewage treatment, however, make it possible to recycle these nutrients without any hazard to human health. Clearly, with nutrient cycling along with many other ecological considerations, it is in our long-term interest to take new and realistic approaches to the way we interact with the rest of the biosphere.

Accelerated Eutrophication of Lakes

The chemical composition and character of lakes change naturally (see Chapter 46). In a young oligotrophic lake, primary productivity is relatively low because the nutrients required by phytoplankton are scarce. Gradually, runoff from the land adds additional nutrients that are captured by the primary producers and then continuously recycled through the lake's food webs. Thus, the overall productivity increases, and a mature lake is said to be more **eutrophic** (Gr. for "well nourished"). Under natural conditions, lakes eventually reach an equilibrium where the inflow of new nutrients is balanced by losses from the system due to outflow and sedimentation. Natural eutrophication is seldom a problem. A "rich" natural lake, one in which there is moderate growth of algae and plants, is usually aesthetically pleasing and valuable for human recreation.

Unfortunately, human intrusion has disrupted almost all freshwater ecosystems by what is termed *cultural eutrophication.* Sewage and factory wastes, runoff of animal waste from pastures and stockyards, and leaching of fertilizer from agricultural, recreational, and urban areas has overloaded many streams, rivers, and lakes with inorganic nutrients. This enrichment often results in an explosive increase in the density of photosynthetic organisms (Figure 49.17). Shallower areas become weed-choked, making boating and fishing impossible. Large algal blooms become common, possibly resulting in increased oxygen production during the day but greatly reduced oxygen levels at night, because of respiration by the excessive biomass. As the algae die and organic material accumulates at the lake bottom, the metabolism of decomposers consumes all the oxygen in the deeper waters. All of these effects may make it impossible for some organisms to survive. For example, eutrophication of Lake Erie resulting from pollution wiped out commercially important fishes such as blue pike, whitefish, and lake trout by the 1960s. Since then, tighter regulations on the dumping of wastes into the lake have enabled some fish populations to rebound, but many of the native species of fishes and invertebrates have not recovered.

Poisons in Food Chains

Humans produce an immense variety of toxic chemicals, including thousands of synthetics previously unknown in nature, that have been dumped into ecosystems with little regard for the ecological consequences. Many of these poisons cannot be degraded by microorganisms and consequently persist in the environment for years or even decades. In other cases, the

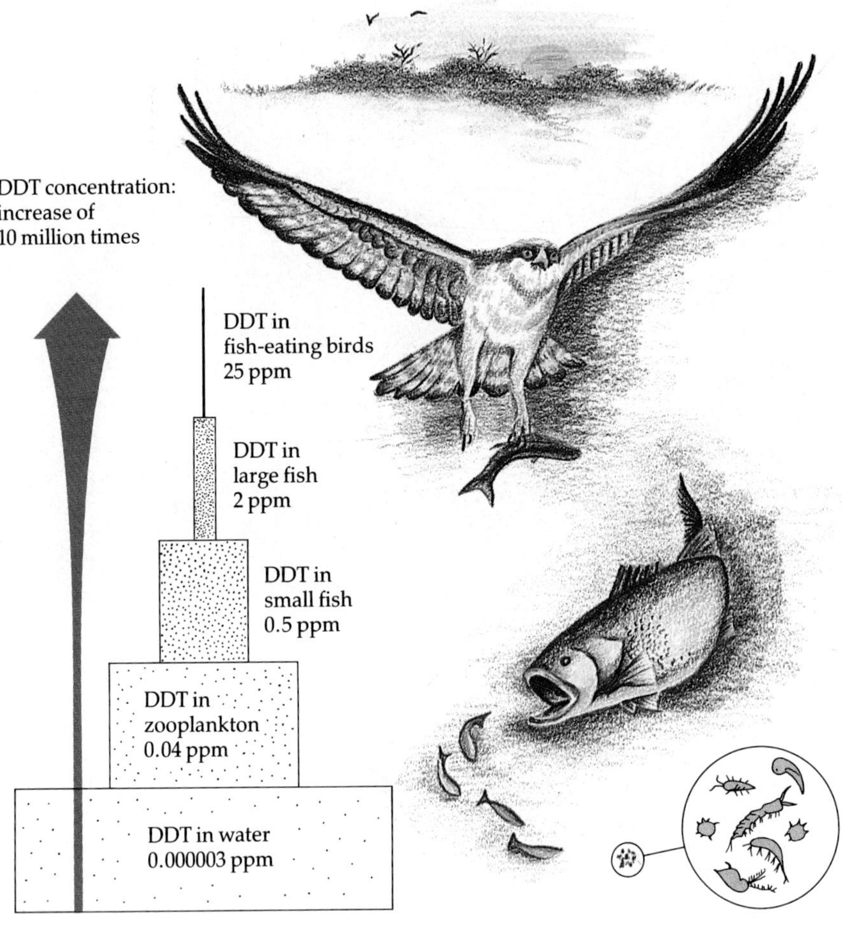

Figure 49.18
Biological magnification of DDT in a food chain. DDT concentration in a Long Island Sound food chain was magnified by a factor of about 10 million, from just 0.000003 parts per million (ppm) as a pollutant in the seawater to a concentration of 25 ppm in a fish-eating bird, the osprey.

chemicals released into the environment may be relatively harmless, but they are converted to more toxic products by reaction with other substances or by the metabolism of microorganisms. For example, mercury, a by-product of plastic production, was once expelled into rivers and the sea in an insoluble form. Bacteria in the bottom mud, however, converted the waste to methyl mercury, an extremely toxic soluble compound that then accumulated in the tissues of organisms.

Organisms acquire toxic substances from the environment along with nutrients and water. Some of the poisons are metabolized and excreted, but others accumulate in specific tissues. Some toxins become more concentrated in successive levels of a food web, a process called **biological magnification.** Magnification occurs because the biomass at any given trophic level is produced from a much larger biomass ingested from the level below. Thus, top-level carnivores tend to be the organisms most severely affected by toxic compounds that have been released into the environment.

A well-known example of biological magnification involves the pesticide DDT, which has been used to control insects such as mosquitoes and agricultural pests. This pesticide is now banned in the United States. DDT persists in the environment and is transported by water to areas far from where it is applied, and it rapidly became a global problem. Because the compound is soluble in lipids, it collects in the fatty tissues of animals, and its concentration is magnified in higher trophic levels (Figure 49.18). Traces of DDT have been found in nearly every organism tested; it has even been found in human breast milk in many countries. One of the first signs that DDT was a serious environmental problem was a decline in the populations of pelicans and eagles, birds that feed at the top of their food chains. The accumulation of DDT (and DDE, a product of its partial breakdown) in the tissues of these birds interfered with the deposition of calcium in their eggshells; when these birds tried to incubate their eggs, the weight of the parents broke the affected eggs, resulting in catastrophic declines in their reproduction rates.

One of the most serious environmental threats that we now face stems from the release of radioisotopes in nuclear accidents and the unsafe storage of radioactive waste. The 1986 power plant disaster at Chernobyl, Ukraine (in the former Soviet Union), belched great clouds of radioactive materials into the air, contaminating crops and natural ecosystems downwind from the power plant. In the region around Chernobyl, the rate of birth defects among farm animals and the rates of thyroid cancer, leukemia, and cataracts in humans have increased dramatically since the accident. In the United States, minor mishaps at nuclear plants are a common occurrence; a potential disaster that might have been much larger than the one at Chernobyl was narrowly averted at the Three Mile Island plant in Harrisburg, Pennsylvania, in 1979. The danger from these contaminants can last for many years, because radioactive isotopes with long half-lives persist in ecosystems and are subject to the same pattern of biological magnification as other toxic materials.

Intrusions in the Atmosphere

Many human activities release a variety of gaseous waste products, and we once thought the vastness of the atmosphere could absorb these materials without significant consequences. We now know, of course, that, like the oceans, the atmosphere is finite in size and that human intrusions can cause fundamental changes in its composition and its interactions with the rest of the biosphere. Two pressing problems created by human intrusions into the atmosphere are rising levels of carbon dioxide and degradation of atmospheric ozone.

Carbon Dioxide Emissions Since the Industrial Revolution, the concentration of CO_2 in the atmosphere has been increasing as a result of the combustion of fossil fuels and burning of enormous quantities of wood removed by deforestation (see Figure 49.1). Various methods have estimated that the average carbon dioxide concentration in the atmosphere before 1850 was about 274 parts per million (ppm). When a monitoring station on Hawaii's Mauna Loa began making very accurate measurements in 1958, the CO_2 concentration was 316 ppm (Figure 49.19). Today, the concentration of CO_2 in the atmosphere exceeds 351 ppm, an increase of more than 11% in just the past 30 years. If CO_2 emissions continue to increase at the present rate, in the year 2075 atmospheric concentration of this gas will be double what it was at the start of the Industrial Revolution.

Increased productivity by vegetation is one predictable consequence of increasing CO_2 levels. In fact, when CO_2 concentrations are raised in experimental chambers, such as greenhouses, most plants respond with increased growth. However, because C_3 plants are more limited than C_4 plants by CO_2 availability (see Chapter 10), one effect of increasing CO_2 concentrations on a global scale may be the spread of C_3 species into terrestrial habitats previously favoring C_4 plants. This may have important agricultural implications. For example, corn, a C_4 plant and the most important grain crop in the United States, may be replaced on farms by wheat and soybeans, C_3 crops that will outproduce corn in a CO_2-enriched environment. However, no one can predict the gradual and complex effects that rising

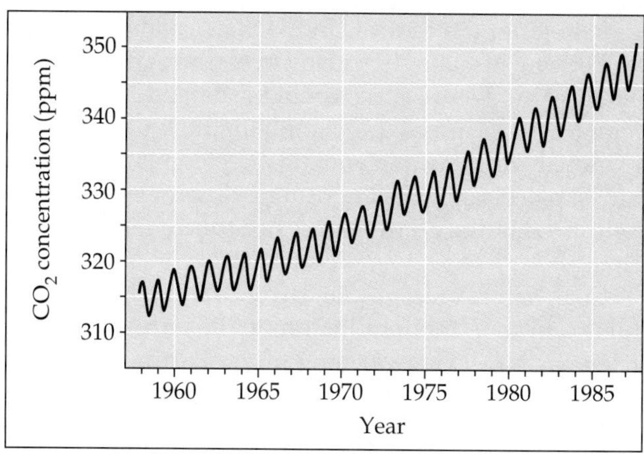

Figure 49.19
The increase in atmospheric carbon dioxide from 1958 to 1988. In addition to normal seasonal fluctuations in CO_2 levels, this graph shows a steady increase in the total amount of CO_2 in the atmosphere. These measurements are being taken at a relatively remote site in Hawaii, where the air is "clean" and free from the variable short-term effects that occur near large urban areas. The vertical scale on this graph exaggerates the apparent increase, which is actually about 11% over 30 years.

CO_2 levels will have on species composition in natural communities.

One factor that complicates predictions about the long-term effects of rising atmospheric CO_2 concentration is its possible influence on Earth's heat budget. As noted earlier, much of the solar radiation that strikes the planet is reflected back toward space. Although CO_2 and water vapor in the atmosphere are transparent to visible light, they intercept and absorb much of the reflected infrared radiation, rereflecting it back toward Earth. This process retains some of the solar heat. Were it not for this **greenhouse effect,** the average air temperature at the Earth's surface would be only –18°C. The marked increase in atmospheric CO_2 concentrations during the last 150 years has recently caused concern among ecologists and environmentalists because of its potential effect on global temperature.

Scientists continue to debate how increasing levels of atmospheric CO_2 will affect the temperature. Several mathematical models predict that a doubling of CO_2 concentration, which could occur by the end of the next century, might produce an average temperature increase of 3°C to 4°C. An increase of only 1.3°C would make the world warmer than at any time in the past 100,000 years. A worst-case scenario suggests that the warming would be greatest near the poles; the resultant melting of polar ice might raise sea levels by an estimated 100 m, gradually flooding coastal areas 150 or more km inland from the current coastline. New York, Miami, Los Angeles, and many other currently populated areas would then be under water. Additionally, a warming trend would probably alter the geographic distribution of precipitation, making major agricultural areas of the central United States and the former Soviet Union much drier. However, the various mathematical models disagree about the details of how each region will be affected. By studying how past periods of global warming and cooling affected plant communities, paleoecologists are using a different strategy to help predict the consequences of future temperature changes.

Coal, natural gas, gasoline, wood, and other organic fuels central to modern life cannot be burned without releasing CO_2. The warming of the Earth now under way as a result of the addition of CO_2 to the atmosphere is a problem of uncertain consequences and no simple solutions. At the 1992 Earth Summit, leaders of the world's industrialized countries signed a treaty that provides a framework for stabilizing CO_2 emissions by the end of this century. However, given the importance of combustion to our increasingly industrialized societies, implementation of these guidelines will be difficult without a concerted international effort and the acceptance of dramatic changes in both personal lifestyles and industrial processes.

Depletion of Atmospheric Ozone Life on Earth is protected from the damaging effects of ultraviolet radiation by a protective layer of ozone molecules (O_3) that is present in the lower stratosphere between 17 and 25 km above the surface of the Earth. Ozone absorbs ultraviolet radiation, preventing much of it from contacting organisms in the biosphere. Satellite studies of the atmosphere suggest that the ozone layer has been gradually depleted, or "thinned," since 1975, and that the depletion continues at an increasing rate.

Research has shown that the destruction of atmospheric ozone largely results from the accumulation of chlorofluorocarbons, chemicals used for refrigeration, as propellants in aerosol cans, and in certain manufacturing processes. When the breakdown products from these chemicals rise to the stratosphere, the chlorine they contained reacts with ozone, reducing it to molecular O_2. Subsequent chemical reactions liberate the chlorine, allowing it to react with other ozone molecules in a never-ending catalytic cycle. The effect is most apparent over Antarctica, where cold winter temperatures facilitate these atmospheric reactions. Scientists first described the "ozone hole" over Antarctica in 1985 and have since documented that it is a seasonal phenomenon that grows and shrinks on an annual cycle. However, the magnitude of ozone depletion and the size of the ozone hole have increased steadily in recent years, and the hole sometimes extends as far as the southernmost portions of Australia, New Zealand, and South America. At the more heavily populated middle latitudes, ozone levels have decreased 2% to 10% during the past 20 years.

The consequences of ozone depletion for life on Earth may be quite severe. Some scientists expect increases in both lethal and nonlethal forms of skin cancer and cataracts among humans as well as unpredictable effects on crops and natural communities, especially the phytoplankton that are responsible for a large proportion of Earth's primary productivity (see Figure 49.5). The danger posed by ozone depletion is so great that many nations have agreed to end the production of chlorofluorocarbons within a decade. Unfortunately, even if all chlorofluorocarbons were banned today, the chlorine molecules that are already in the atmosphere will continue to influence stratospheric ozone levels for at least a century.

The Introduction of Exotic Species In Chapter 48, we discussed the utility of transplant experiments to identify why species are restricted to particular geographical ranges. Such an approach is favored by many ecologists, but most transplant experiments have, in fact, been undertaken by travelers who inadvertently carry hitchhiking seeds or insects, or by people who intentionally introduce foreign plants or animals for agricultural or ornamental purposes. Many (perhaps most) transplanted species fail to survive outside their normal range. However, there are numerous examples of successful transplants. In the 1800s, starlings and English sparrows were imported to the northeastern United States from Europe; both species are now abundant in cities and the countryside throughout much of North America. Carp, freshwater fish from China, were introduced about the same time and have become common members of lake and stream communities.

Although new species will probably have some negative effects on native species, some transplants have been especially noteworthy in this regard. The case of rabbits in Australia was mentioned in Chapter 48. Gypsy moths were imported to the United States from Europe by a scientist who was trying to crossbreed several silk-producing moths. In 1869, some of the moths escaped from the scientist's home near Medford, Massachusetts, and gypsy moths have been spreading southward ever since, defoliating large areas of forest. Another fiasco was the importation of Africanized honeybees by Brazilian honey producers in 1956. These bees, hybrids of the common European honeybee and an African variety, are very aggressive and have occasionally swarmed and attacked humans and livestock. In the past 30 years, the range of Africanized bees has spread across South America and through Central America and Mexico, and sightings have already been reported in the United States. The normal range of these bees was limited by barriers to dispersal, including the Atlantic Ocean. A similar problem is the northward march of fire ants, which were accidentally introduced to the southern United States from Brazil in the 1940s (Figure 49.20).

Recently, some ecologists and environmentalists have become concerned about the ecological consequences of releasing genetically engineered organisms

(a)

Figure 49.20
March of the fire ants. (**a**) The fire ant *(Solenopsis invicta)* can inflict a beelike sting and tends to attack in large numbers. In the southern United States, more than 70,000 sting victims seek medical help each year. Shown here is a winged female fire ant, along with a much smaller worker. (**b**) Fire ants were accidentally imported from Brazil about 45 years ago, probably in the hold of a ship that docked in Mobile, Alabama. The range of the ant has been spreading ever since, as traced by this series of maps.

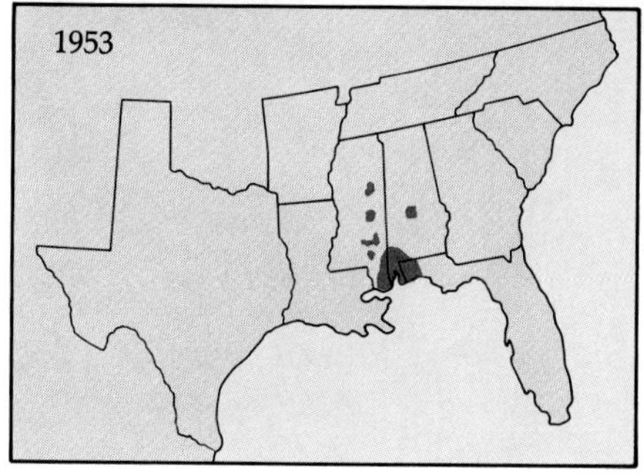

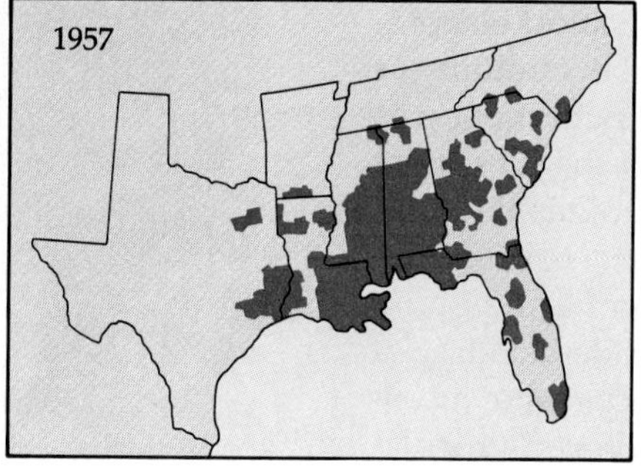

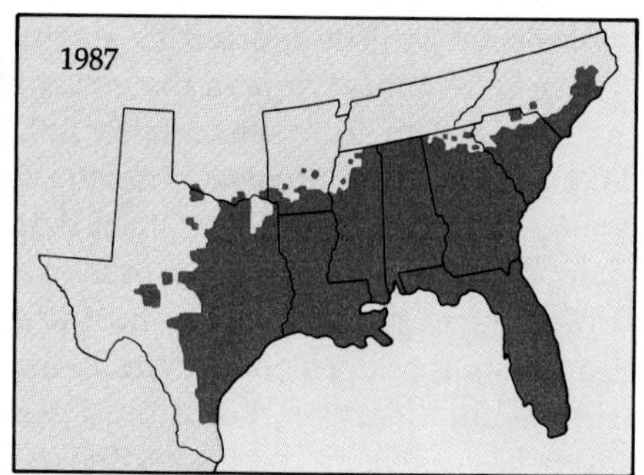

(b)

into natural and agricultural environments. Advances in gene-transfer technology now enable scientists to move genetic material from one species into another. Such transfers are useful for developing herbicide resistance or pest resistance in agricultural crops, and field tests of engineered plants in the tomato family have been under way for several years.

Potential ecological problems associated with genetic engineering of crop plants were discussed in detail in Chapter 19, but these concerns are worth reviewing here. Some ecologists worry that the introduced traits might allow the crop plants themselves to become weeds or pests if they escape from the fields in which they are grown. In addition, the crop plants might interbreed with naturally occurring plants, increasing their ability to invade areas where they might not occur naturally. For example, if a weed species accidentally acquired an introduced gene for herbicide resistance, new mechanisms to control its spread would have to be developed. Similarly, genes for insect resistance in crop plants could increase the rate at which insect pests coevolve mechanisms that overcome plant defenses. Although it is now hard to predict the long-term consequences of accidental gene transfers on community and ecosystem structure, it is possible to imagine the evolution of a "superweed" that might outcompete its competitors in natural environments. The introduction of insect resistance into crop plants and its accidental transfer to noncrop species could also influence the population dynamics of insects that feed on these plants and on the higher-level consumers that in turn feed on them. It is too early to evaluate the potential dangers of introducing genetically engineered organisms into natural environments, but this question is certain to provide fertile ground for much ecological research and continued debate.

BIOTIC EFFECTS AT THE BIOSPHERE LEVEL

Some scientists postulate that life plays a major role in *regulating* the overall environment of the Earth and maintaining it within certain limits suitable to life. This idea is known as the Gaia hypothesis, named for a mythical goddess of Earth.

According to the Gaia hypothesis, first enunciated in 1979 by British scientist James Lovelock, life actually shapes the climate and atmosphere of Earth. For example, transpiration by tropical forests produces clouds that affect global weather patterns. Photosynthesis and chemical cycling in ecosystems have generated an atmosphere with a far higher concentration of O_2 than would be found on any lifeless planet, and that concentration has been maintained within a range of 15% to 25% of the atmosphere for the past 200 million years.

Besides adjustments in the amount of photosynthesis and respiration occurring in the biosphere, another factor in controlling O_2 concentration is the release of methane (CH_4) by certain bacteria and by huge numbers of termites. In fact, termites probably account for about half of the methane in the atmosphere. Methane reacts with O_2 to form CO_2 and H_2O, and thus methane-producing organisms may help regulate atmospheric O_2 concentrations. It is also possible that organisms moderate fluctuations in atmospheric CO_2 concentrations to some extent. For example, occasional blooms (population explosions) of marine photoplankton may remove enormous amounts of CO_2 from the atmosphere.

To many ecologists, the original form of the Gaia hypothesis was little more than a semimystical elaboration of Clements' view of the ecological community as a "superorganism" (see Chapter 48). Strict proponents of the Gaia hypothesis suggested that the biosphere had a self-regulated metabolism that countered fluctuations in the physical environment. This concept implied coordinated evolution of life on Earth, a phenomenon that does not fit within our understanding of the mechanisms of evolution. Recently, some advocates have reframed the hypothesis to portray regulation as a natural consequence of the checks and balances inherent in the dynamics of ecosystems. Most scientists would probably agree that organisms have significant effects on the environment, with some feedback processes also occurring.

In this more moderate view, Gaia is merely a metaphor for the interconnectedness of nature. It is clear from the study of ecosystems that processes occurring in one location can affect life far away. For this [illegible] own behavior, [illegible] ople through- [illegible] f any, ecosys- [illegible] we know that [illegible] tly fifty times [illegible] 00,000 years, [illegible] natural habi- [illegible]

[illegible] ntal crisis re- [illegible] biosphere's [illegible] ety of Amer- [illegible] professional [illegible] w research [illegible] **tiative.** The [illegible] ire the basic [illegible] telligent de- [illegible] n of Earth's [illegible] s studies of global change, including interactions between climate and ecological processes; biological diversity and its role in maintaining ecological processes; and the ways in which the productivity of natural and artificial ecosystems can be sustained. This initiative will require the reorientation of scientific research to meet the goals, the development of new interdisciplinary approaches to study these complex phenomena, and the commitment of substantial financial resources to undertake the research. Ultimately, the effort must be far bigger than either the Human Genome Project or the Superconducting Supercollider, two of the largest scientific projects that this country has ever undertaken. However, the importance of this research agenda cannot be overstated because life itself may ultimately be at stake.

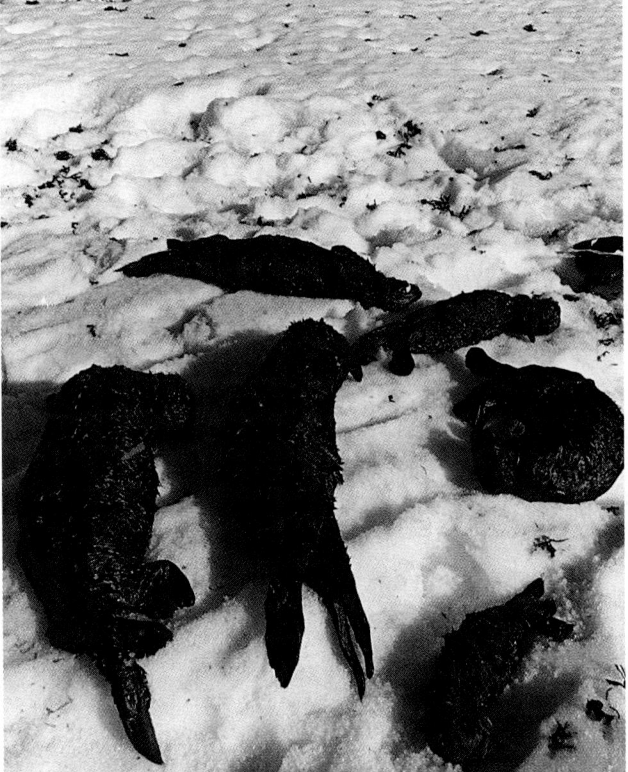

Figure 49.21
A remote ecosystem damaged by human activity. On March 24, 1989, a supertanker ran aground and spilled more than ten million gallons of crude oil into the pristine waters of Prince William Sound in Alaska. These sea otters were among the victims. This ecological disaster was a shocking reminder that our technological tentacles reach far: We may be driving automobiles in Los Angeles, Chicago, or New York, but the impact of our demand for oil is felt thousands of miles away.

The current state of the Earth demonstrates clearly that we are treading dangerously on uncharted ecological ground. Perhaps our guide for future activity should be the words of the great American naturalist Aldo Leopold: "If the biota, in the course of aeons, has built something we like but do not understand, then who but a fool would discard seemingly useless parts? To keep every cog and wheel is the first precaution of intelligent tinkering."

STUDY OUTLINE

1. An ecosystem is the most inclusive level in the hierarchy of biological organization, consisting of the community of organisms in a defined area and the abiotic factors making up the physical environment.
2. Energy flow and chemical cycling are two interrelated processes that occur by the transfer of substances through the feeding levels of ecosystems.

Trophic Levels and Food Webs (pp. 1132-1133)

1. The specific route of energy flow and the pattern of chemical cycling are dictated by the trophic structure of a community.
2. Primary producers are autotrophic organisms that support all other organisms of the community, from the lowest trophic level.
3. Heterotrophic organisms are consumers. Herbivores are primary consumers that eat the autotrophs; other species, the secondary and tertiary consumers, are carnivores that eat herbivores and those that eat other carnivores, respectively. Detritivores feed on organic wastes and dead organisms from all trophic levels.
4. The main primary producers in most ecosystems are photosynthetic autotrophs, although a few marine communities are powered by geothermal energy harnessed by chemoautotrophic bacteria.
5. The pathway along which food from producers is transferred to the various levels of consumers is known as a food chain. However, natural feeding relationships are usually more like webs, because consumers feed at more than one trophic level or on a variety of items at the same level.

Energy Flow (pp. 1133–1140)

1. Only a small proportion of the solar energy striking Earth is used by autotrophs for photosynthesis.
2. Primary productivity is the rate at which solar energy is converted to the chemical energy of organic compounds by autotrophs. It sets the spending limit for the energy budget of ecosystems.
3. Only net primary productivity, defined as gross primary productivity minus the energy used by the primary producers for respiration, is available to consumers.
4. The various levels of heterotrophs convert net primary productivity into new forms of biomass. Such secondary productivity declines with each trophic level. Generally, only about 10% of the chemical energy available at one trophic level appears at the next level.

Chemical Cycling (pp. 1140–1145)

1. Nutrient cycles involve both biotic and abiotic components of ecosystems; they are also called biogeochemical cycles.
2. Biogeochemical cycles can be global or localized, depending on the mobility of the element in the environment, which in turn depends on its chemical nature and its route through the trophic structure of the ecosystem.
3. Nutrients cycle through four compartments of organic and inorganic materials that are either directly available or unavailable for use by organisms.
4. The carbon cycle primarily reflects the reciprocal processes of photosynthesis and cellular respiration. Producers convert inorganic CO_2 into organic molecules that are eventually degraded by consumers into CO_2, which is reused by the autotrophs. In aquatic environments, CO_2 is also involved in a complex equilibrium with water and $CaCO_3$.
5. In the nitrogen cycle, certain prokaryotes fix abundant nitrogen in the atmosphere into ammonia, which other bacteria convert into nitrites and nitrates. Plants absorb ammonia and nitrates and convert them into protein that can be passed into the food chain. Detritivores decompose nitrogenous waste into ammonia, which can be reused by plants or reconverted into nitrites or nitrates. Nitrogen in the soil is returned to the atmosphere in the form of free nitrogen by denitrifying bacteria.
6. The phosphorus cycle does not involve any atmospheric forms. Over ecological time and on a localized scale, phosphorus in the form of phosphate is absorbed by plants, transferred to consumers, and returned to the soil by animal excretion and the action of decomposers. A sedimentary cycle operates over geological time and involves the weathering of rock, eventual leaching of the phosphorus into the sea, and incorporation into oceanic sediments that are later exposed to the air and weathered by geological processes.
7. The proportion of a nutrient in a particular form and its cycling time in that form vary among ecosystems. In the tropics, for example, nutrients pass quickly through the decomposition stage back into living biomass.

Human Intrusions in Ecosystem Dynamics (pp. 1145–1153)

1. Local and regional human impact on trophic structure, energy flow, and chemical cycling can have widespread or global ecological consequences.
2. Human encroachment on natural systems has destroyed many habitats, creating a biodiversity crisis.
3. Long-term ecological research (LTER) at Hubbard Brook in New Hampshire showed that complete logging of a mature deciduous forest increased water runoff and caused huge losses of minerals, effectively destroying the nutrient equilibrium that previously existed.
4. Agricultural practices result in the constant removal of nutrients from ecosystems, so that large supplements are continually required. Distribution of the nutrients in crops to humans and animals often results in "short-circuiting" to aquatic environments.
5. Dumping of nutrient-rich wastes into aquatic habitats accelerates eutrophication, and the increased biomass depletes O_2.
6. The release of toxic wastes has polluted the environment with harmful substances that often persist for long periods of time. Some of these substances become concentrated along the food chain by biological magnification, causing far-reaching deleterious effects.
7. Because of the burning of wood and fossil fuels, CO_2 in the atmosphere has been steadily increasing. The ultimate effects are uncertain but may include significant warming and other influences on climate. The release of

chlorofluorocarbons has resulted in a depletion of the protective ozone layer in the stratosphere.

8. The introduction of exotic species has disrupted many communities and ecosystems. The effects of releasing genetically engineered species are not yet understood.

Biotic Effects at the Biosphere Level (pp. 1153–1154)

1. The controversial Gaia hypothesis maintains that climate and the environment are regulated by the diverse actions of living organisms.
2. Whether or not the Gaia idea truly reflects the nature of the biosphere, the study of ecosystems has shown the complex interconnectedness of both biotic and abiotic components.
3. Ecologists are embarking on a gigantic new research agenda, the Sustainable Biosphere Initiative, to gather basic information about how ecosystems function.

SELF-QUIZ

1. Which of the following organisms is *incorrectly* paired with its trophic level?
   a. cyanobacteria—primary producer
   b. grasshopper—primary consumer
   c. zooplankton—secondary consumer
   d. eagle—tertiary consumer
   e. fungi—detritivore

2. One of the lessons from a pyramid of productivity is that
   a. only one-half of the energy in one trophic level is passed on to the next level
   b. most of the energy from one trophic level is incorporated into the biomass of the next level
   c. the energy lost as heat or in cellular respiration is 10% of the available energy of each trophic level
   d. ecological efficiency is highest for primary consumers
   e. eating grain-fed beef is an inefficient means of obtaining the energy trapped by photosynthesis

3. The role of detritivores in the nitrogen cycle is to
   a. fix N_2 into ammonia
   b. release ammonia from organic compounds, thus returning it to the soil
   c. denitrify ammonia, to return N_2 to the atmosphere
   d. convert ammonia to nitrate, which can then be absorbed by plants
   e. incorporate nitrogen into amino acids and organic compounds

4. The Hubbard Brook Forest Study demonstrated all of the following *except* that
   a. most minerals were recycled within a forest ecosystem
   b. mineral inflow and outflow within a natural watershed were nearly balanced
   c. deforestation resulted in an increase in water runoff
   d. the nitrate concentration in waters draining the deforested area became dangerously high
   e. the regrowth of seedlings quickly restored normal nutrient cycling following deforestation

5. The recent increase in atmospheric CO_2 concentration is mainly a result of an increase in
   a. primary productivity
   b. methane production by termites and some bacteria
   c. the absorption of infrared radiation escaping from Earth
   d. the burning of fossil fuels and wood
   e. cellular respiration by the exploding human population

6. Which of the following is a result of biological magnification?
   a. Top-level predators may be most harmed by toxic environmental chemicals.
   b. DDT has spread throughout every ecosystem and is found in almost every organism.
   c. The greenhouse effect will be most significant at the poles.
   d. The biosphere is able to regulate the environment and adjust to disruptions.
   e. Many nutrients are being removed from agricultural lands and shunted into aquatic ecosystems.

7. Which of these ecosystems has the lowest primary productivity per square meter?
   a. a salt marsh
   b. an open ocean
   c. a tidepool
   d. a grassland
   e. a tropical forest

8. Quantities of mineral nutrients in soils of tropical rain forests are relatively low because
   a. the standing crop is small
   b. microorganisms that recycle chemicals are not very abundant in tropical soils
   c. the decomposition of organic refuse and reassimilation of chemicals by plants occur rapidly
   d. nutrient cycles occur at a relatively slow rate in tropical soils
   e. the high temperatures destroy the nutrients

9. In the nitrogen cycle, the bacteria that restock the atmosphere with N_2 are
   a. nitrogen-fixing bacteria
   b. denitrifying bacteria
   c. nitrifying bacteria
   d. ammonifying bacteria
   e. *Rhizobium* bacteria of root nodules

10. Which of the following statements concerning the hydrologic cycle is *incorrect?*
    a. There is a net movement of water vapor from oceans to terrestrial environments.
    b. Precipitation exceeds evaporation on land.
    c. Most of the water that evaporates from oceans is returned by runoff from land.
    d. Transpiration makes a significant contribution to evaporative loss of water from terrestrial ecosystems.
    e. Evaporation exceeds precipitation over the seas.

CHALLENGE QUESTIONS

1. Herbivores have generally been considered to have a negative impact on their plant prey. However, a controlled study of a natural community in which the crustacean herbivore *Daphnia pulex* feeds on algae showed that *Daphnia* also had a stimulatory effect on the algal populations that approximately balanced its impact on algal mortality. From your understanding of the nitrogen cycle, what do you think was the mechanism for the *Daphnia*-induced stimulation of algal growth?

2. Imagine that you have been chosen as the biologist for the design team for a self-contained habitat, Space Station ECOS, to be assembled in orbit. It will be stocked with organisms you choose to form a self-sustaining ecosystem that will support you and five other people for two years. Describe the main functions you expect the organism to perform. List the types of organisms you would select, and explain why you chose them.

SCIENCE, TECHNOLOGY, AND SOCIETY

1. The amount of CO_2 in the atmosphere is increasing, and global temperature has increased over the last century; however, scientists do not agree about the extent to which the two phenomena are related. Most say that greenhouse warming is under way, and we need to take action now to avoid drastic environmental change. Some say it is too soon to tell, and we should gather more data before we act. The latter group dominated the U.S. delegation to the 1992 U.N. Earth Summit. Because of U.S. opposition, specific CO_2 targets were eliminated from the global warming agreement. What are the advantages and disadvantages of doing something now to slow global warming? What are the advantages and disadvantages of waiting until more data are available?

2. Until recently, response to environmental problems has been fragmented—an antipollution law here, incentives for recycling there. Meanwhile, the problems of the contrasts between rich and poor nations, diminishing resources, and pollution continue to grow. Now people and governments are starting to envision a sustainable society—one in which each generation inherits sufficient natural and economic resources and a relatively stable environment. The Worldwatch Institute, a respected environmental policy organization, estimates that we must reach sustainability by the year 2030 to avoid economic and environmental disaster. To get there, we must begin shaping a sustainable society during this decade. In what ways is our present system not sustainable? What might we do to work toward sustainability, and what are the major roadblocks to achieving it? How would your life be different in a sustainable society?

FURTHER READING

Aber, J. D., K. J. Nadelhoffer, P. Steudler, and J. M. Melillo. "Nitrogen Saturation in Northern Forest Ecosystems." *BioScience,* June 1989. Nitrogenous exhaust from fossil fuel consumption may stress the biosphere.

Barlow, C., and T. Volk. "Gaia and Evolutionary Biology." *BioScience,* October 1992. Analysis of a debate about biosphere dynamics.

Barton, J. H. "Biodiversity at Rio." *BioScience,* November 1992. A summary of the U.N. Earth Summit.

Broeker, W. "Global Warming on Trial." *Natural History,* April 1992. How solid is the evidence that Earth is heating up?

Brown, L. R. *State of the World.* New York: Norton, 1992. A new edition of this book, prepared by the staff of the Worldwatch Institute, appears each year to provide an up-to-the-minute analysis of human intrusions in ecosystems.

Gillis, A. "Why Can't We Balance the Globes's Carbon Budget?" *BioScience,* July/August 1991. The atmospheric impact of oceanic chemistry.

Hammond, A. L. *World Resources, 1990–1991.* New York: Oxford University Press, 1990. A revised version of this book, prepared by the staff of the World Resources Institute, appears every other year as an updated guide to the global environment. The book is prepared in collaboration with the United Nations Environment Programme and the United Nations Development Programme.

National Wildlife. The February issue of this magazine provides an annual update on the state of the Earth.

Odum, H. T. *Systems Ecology: An Introduction.* New York: Wiley, 1984. A college-level text that emphasizes important aspects of ecosystems.

Repetto, R. "Accounting for Environmental Assets." *Scientific American,* June 1992. Why we should revise our economic indicators to reflect the true cost of damaging ecosystems.

Risser, P. G., J. Lubchenco, and S. A. Levin. "Biological Research Priorities—A Sustainable Biosphere." *BioScience,* October 1991. A brief description of the Sustainable Biosphere Initiative.

Wilson, E. O. "The Diversity of Life." *Discover,* September 1992. Why we should be concerned about the extinction of other speicies.

50 BEHAVIOR

THE EVOLUTIONARY LOGIC OF BEHAVIORAL ECOLOGY
PROXIMATE VERSUS ULTIMATE CAUSATION
INNATE COMPONENTS OF BEHAVIOR
LEARNING AND BEHAVIOR
BEHAVIORAL RHYTHMS
MOVEMENT
FORAGING BEHAVIOR
SOCIAL INTERACTIONS
MATING BEHAVIOR
COMMUNICATION
SELFISH VERSUS ALTRUISTIC BEHAVIOR
ANIMAL COGNITION
HUMAN SOCIOBIOLOGY

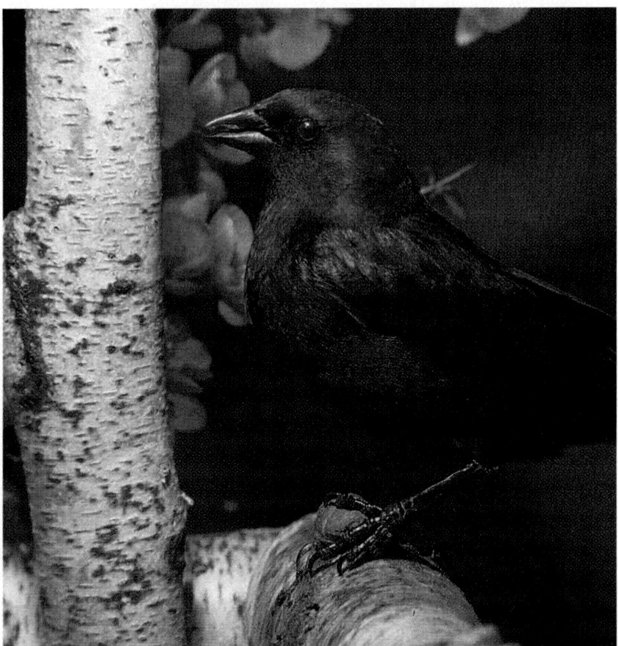

Figure 50.1
Why is this bird singing? The vocal virtuosity of this brown-headed cowbird *(Molothrus ater)* developed from an interplay of instinct and learning—from genes and experience. A male cowbird has a repertoire of many tunes. Why has natural selection favored this behavior over the expert vocalization of a single song? In this chapter, you will learn how biologists studying animal behavior are approaching such questions. Our emphasis will be on behavioral ecology, a research field that seeks evolutionary explanations for animal behavior.

Almost everyone has at one time or another listened to an infant lying in its crib going through an array of coos, babbles, and other baby sounds. Few people realize that this talking to oneself also occurs in thousands of species of other animals, most notably in young songbirds. Unlike the vocalizations of older humans and adult birds, the disjointed and poorly formed sounds of infants and young birds do not seem to be directed to anyone in particular. Why, then, do the youngsters go through this process? Even when no one else is nearby, there is a listener: the youngster itself. Apparently, baby sounds provide a mechanism for the young animal to learn to match its vocalizations to inborn or innate sound templates. The process of matching is akin to the human process of learning a new song from a musical score. The score contains the directions for playing, but a musician needs to listen to his or her own practicing in order to get the piece right. Somewhere in the brain there is a similar score: a neural network that dictates the nature of baby sounds. These first sounds of people and songbirds eventually give way to an array of learned sounds. Youngsters hear sounds from conspecifics (members of the same species) and then learn to match these other individuals' songs just as they learned to match their innately programmed baby sounds.

Bird song provides an excellent introduction to the subject of behavior, because biologists are beginning to understand a great deal about its development, functions, and consequences at virtually all levels of biological organization, from molecules to whole organisms to entire populations. Unfortunately, this cannot be said for many other complex behaviors. In addition to being amenable to modern experimental methods, bird song is attractive to researchers because of several striking parallels between it and human speech, including the one already noted. Bird song has also become a model system for animal behavior research because it demonstrates a very important generalization: Behavior is influenced by both innate and learned factors (Figure 50.1). Early baby sounds may be coded by innate neural templates, but young animals must go through a learning process to produce the sounds.

Before we discuss bird song and many other types of behavior in more detail, we need to consider how behavior is defined and how biologists go about studying it. A dictionary definition of **behavior** is likely to read something like "to act, react, or function in a particular way to some stimulus." Much of behavior consists of externally observable muscular activity, the "act" and "react" components of this definition. But a young bird that hears an adult song may show no obviously related muscular activity. Instead, the memory of the song may be stored in the bird's brain. Although this memory may change the brain's functioning, any observable muscular response will come later, perhaps even months later, when the bird begins to learn to match this stored memory of an adult song. Thus, if we think of behavior as what an animal does and *how* it does it, this definition will encompass the nonmotor components of behavior as well as an animal's observable actions.

The study of animal behavior is undoubtedly one of the oldest aspects of biology. Tens of thousands of years ago, such knowledge was essential to human survival. By learning the habits of the animals around them, early humans increased their chances of securing a meal and decreased their chances of becoming a meal. Thus, our ancestors' study of animal behavior ultimately enhanced their Darwinian fitness.

THE EVOLUTIONARY LOGIC OF BEHAVIORAL ECOLOGY

Darwinian fitness is still central to our study of animal behavior, but in a very different way. While early people increased their *own* fitness by studying animal behavior, today's biologists use the fitness of their animal subjects as a guiding principle for their research. The principle is a simple one: Because natural selection works on the enormous amount of genetic variation generated by mutation and recombination, we expect organisms to possess features that maximize their genetic representation in the next generation (see Chapter 20). Applied to behavior, this principle means that we expect animals to behave in ways that maximize their fitness. For example, feeding behavior is likely to optimize efficiency, or net energy gain, and the choice of a mate tends to maximize the number of healthy offspring produced.

This expectation of optimal behavior is valid only if genes influence behavior, because if there were no genetic influence, behavior could not be subjected to natural selection and could not evolve. Genes do exert a strong influence on many behaviors (Figures 50.1 and 50.2). Even learned behaviors typically depend on genes that create a neural system receptive to learning. The research approach based on the expectation that animals increase their Darwinian fitness by optimal behavior is called **behavioral ecology.** This approach has dominated the study of animal behavior for the last 15 years. Behavioral ecology is appealing because it is heuristic, meaning that it leads to the formulation of questions or predictions. This, of course, is what science is all about.

Suppose you became interested in the observation that many songbirds have a repertoire of song types. Some of these songs sound identical to us but can be easily distinguished when analyzed with a sound spectrograph (Figure 50.3). If you followed the practices of behavioral ecology, you could formulate several testable hypotheses, all starting with "A repertoire increases fitness because . . ." (Recall from Chapter 1 that hypotheses are phrased as possible explanations.) You might hypothesize that a repertoire increases fitness because it makes an older, more experienced male more attractive to females. For this to be true, (1) males learn more song types as they get older, so that repertoire size is a reliable indicator of age; and (2) females prefer to mate with males with large repertoires. Thus, your hypothesis makes two clearly testable predictions. To test the first prediction, you can determine whether there is a correlation between a male's age and the size of his song repertoire. If there is not, your hypothesis will be invalidated, which would be informative although perhaps disappointing. As it turns out, some songbird species show this correlation, while others do not. Next, you can determine whether females are more sexually stimulated by a large song repertoire as opposed to a small one. This can be done by playing tape-recorded male songs to females that have been made especially receptive to male song by the temporary administration of a female hormone. Such females indicate their song preferences by assuming a copulatory posture, even though there is no male around. Figure 50.4 shows another way to assess female responses. All this work may lead to an evolutionary explanation: Bird-song repertoires result in females mating more often with experienced males, who will give their offspring a better chance to survive.

Now suppose you did not employ behavioral ecology to guide your research on repertoires. Without a heuristic approach that generates testable hypotheses and predictions, you would probably record observations of numerous aspects of singing behavior. Although these efforts may produce interesting data, they would not *explain* the behavior. Alternatively, you might hypothesize that repertoires have nothing to do with fitness, but that male birds simply find variety more pleasurable than singing the same boring song over and over again. The method for testing such a hypothesis is not clear, emphasizing again how much more productive it is to use evolutionary principles as a guide to behavioral research.

Evolution is the core theme of biology (see Chapter 1), and behavioral ecology, with its emphasis on evolutionary explanations, is the focal point of this chapter.

Figure 50.2
The genetic basis of behavior: a case study. Several species of brightly colored African parrots, commonly known as lovebirds, build cup-shaped nests inside tree cavities. Females typically make nests with thin strips of vegetation (or, in the laboratory, paper) that they cut with their beaks. (**a**) In one species, Fischer's lovebird *(Agapornis fischeri)*, the bird cuts relatively long strips and carries them back to the nest one at a time in her beak. (**b**) In contrast, the peach-faced lovebird *(A. roseicollis)* cuts shorter strips and usually carries several at a time by tucking them into the feathers of the lower back. Tucking is a fairly complex behavior because the strips must be held just right and pushed in firmly, and the feathers then smoothed over. (**c**) These two species are closely related and have been experimentally interbred. The resultant hybrid females exhibited an intermediate kind of nest-building behavior. The strips cut by the hybrid birds were intermediate in length; even more interesting was the birds' hybrid manner of handling the strips. They usually made some attempt to tuck them into their rear feathers, but in some cases they did not let go after turning and pushing them a short distance. In other cases, the strips were manipulated or inserted improperly or simply dropped. The result was almost total failure to transport strips by this method. Eventually, the birds learned to transport the strips in their beaks. Even so, they always made at least token tucking attempts. (**d**) After several years, the birds still turned their heads to the rear before flying off with a strip. These observations demonstrate that the phenotypic differences in the behavior of the two species are based on different genotypes. We also see that innate behavior can be modified by experience; the hybrid birds eventually learned to transport the strips.

(**a**) Nests made with long strips—no tucking behavior

(**b**) Nests made with short strips—tucking behavior

(**c**) Hybrid nests made with intermediate-length strips—in first mating season, unsuccessful tucking behavior

(**d**) In later seasons, only head-turning behavior

PROXIMATE VERSUS ULTIMATE CAUSATION

When we observe a certain behavior, we are apt to ask two types of questions. When our questions are about *why* the animal has the behavior, we are asking about **ultimate causation**—that is, the reason something exists. In the study of animal behavior, questions about ultimate causation are evolutionary questions: We want to know why natural selection favored this behavior and not a different one. This evolutionary approach is the central feature of behavioral ecology. Hypotheses that address a "why" question always take the form of ultimate explanations; in behavioral ecology, such hypotheses propose that the behavior maximized fitness in some particular way. In contrast, we often find ourselves asking questions about *how* an animal carries out a particular behavior. When we ask "how" questions, we are asking about the immediate cause, or **proximate causation,** of something; in the study of animal behavior, this means that we want to understand the mechanisms underlying the particular behavior. Proximate causes can include the environmental stimulus, if any, responsible for triggering a behavior and internal processes of the animal's nervous, muscular, and endocrine systems.

To a large extent, the chapters in Unit Seven emphasized proximate causes of animal physiology and behavior, and this chapter will also deal with proximate causation to some extent. Although behavioral ecologists are primarily concerned with the evolutionary

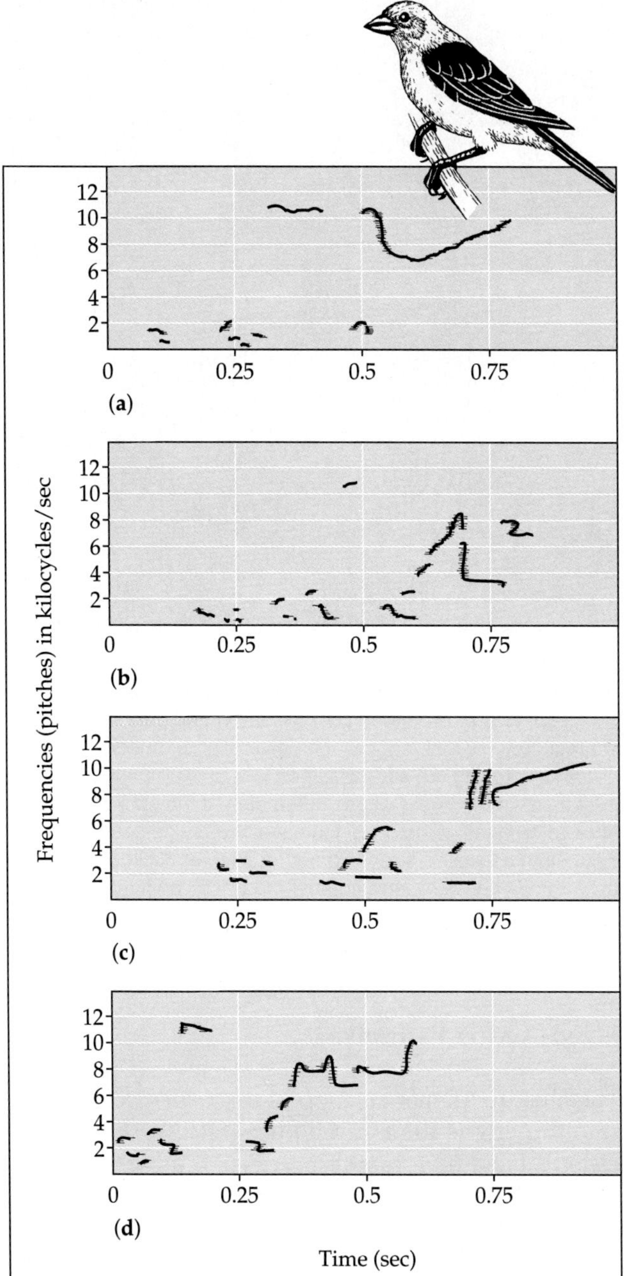

Figure 50.3
The repertoire of a songbird. These sonograms, or "voiceprints," show a graph of a sound's frequency (perceived as pitch) versus time. These are four distinct song types of one male brown-headed cowbird. The song on top is quite distinct and easily distinguishable from the other three, which would sound similar to us but probably not to a bird. Individual cowbirds generally have three to six song types, but other species have dozens or even hundreds of types.

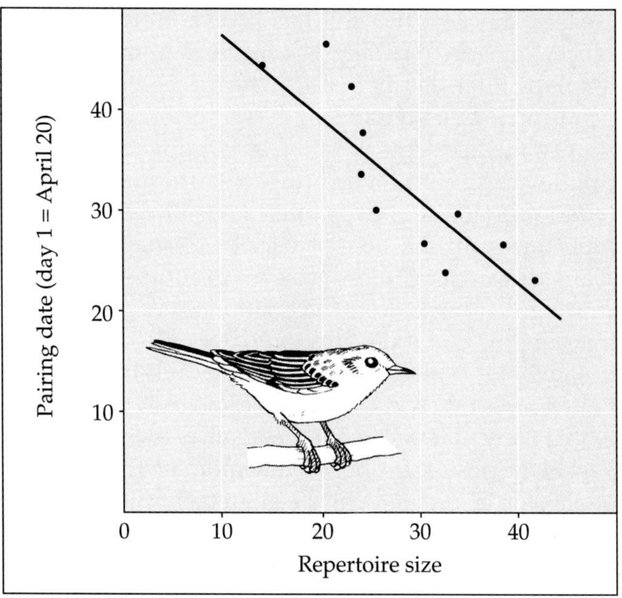

Figure 50.4
Female warblers prefer males with large song repertoires. Male sedge warblers in Europe with large repertoires attract females to pair with them earlier in the breeding season than males with small repertoires. Since the first females who choose mates pick males with large repertoires, it is likely that females prefer large repertoires. Males with large repertoires benefit from pairing early because breeding early tends to be more successful than breeding late in the season.

significance of a behavior rather than exactly how the behavior is carried out, they need to be aware of proximate causation. By limiting the range of behaviors upon which natural selection can act, proximate causes limit the behavior we expect to see. For example, it would be unreasonable for a behavioral ecologist studying a songbird to propose, at least as the initial hypothesis, that odor is an important cue in mate choice. For this to be true, a bird would have to be able to smell its prospective mate from some distance, but most birds have little sense of smell. Besides its relevance to behavioral ecology, proximate causation is important in its own right. For example, understanding the brain processes that enable you to read this page and to comprehend and store the information in your memory is one of the most engaging quests in all of science.

To drive home the distinction between ultimate and proximate causation, consider the observation that bluegill sunfish (and many other animals) breed in spring and early summer. In terms of ultimate causation, the answer to why they breed during this period is that this is when breeding is most productive or adaptive. The warmth of the water allows for fast growth of young, as does an especially abundant food supply. Fish that breed at other times would be at a selective disadvantage. In terms of proximate causation, the answer is that breeding is triggered by the effect of increased day length on a fish's pineal gland (see Chapter 41). Bluegills and many other animals can be stimulated to begin breeding by experimentally lengthening their daily exposure to light. This stimulus results in neural and hormonal changes that induce nest building and other reproductive behavior. The increased day length itself has little adaptive significance, but since day length is the most reliable indicator of the time of year, a proximate mechanism that depends on increased day length has been selected for.

As a further demonstration of proximate versus ultimate causation, let's return to our study of song repertoires and to the idea that some birds have large repertoires for the sake of variety. Perhaps an ingenious researcher could devise an experiment demonstrating that birds do find variation more pleasurable than singing one song. However, it would be incorrect to say that this pleasurable response caused the evolution of repertoires. A behavioral ecologist would argue that if there is in fact such a response, the pleasure is part of the proximate, but not the ultimate causation. In other words, the male bird's enjoyment increases the likelihood that he will have a repertoire, and this is adaptive for some reason totally unrelated to the enjoyment. In a similar way, the enjoyment nearly all people derive from sweet foods is a proximate mechanism that increases the likelihood of eating sweet, high-energy foods. Such food was in short supply until the rise of modern agriculture, and the fitness associated with its consumption is the ultimate reason for the natural selection of the proximate mechanism we call a sweet tooth. Thus, the "how" and "why" questions about animal behavior are related in their evolutionary basis: Proximate mechanisms produce behaviors that ultimately evolved because they increase fitness in some way.

Figure 50.5
A cuckoo hatchling evicting eggs: an innate behavior. Some species of European cuckoos are brood parasites, meaning that they lay their eggs in the nests of other species. Within hours of hatching, an otherwise helpless cuckoo methodically pushes all of the host's eggs out the nest. If the cuckoo hatches after some or all of the nestlings have already hatched, it will evict them as well. There is no opportunity for cuckoo nestlings to learn this behavior from another individual, so it must be innately programmed.

INNATE COMPONENTS OF BEHAVIOR

Before the advent of behavioral ecology, an older discipline called *ethology* dominated the study of animal behavior. Ethology originated in the 1930s with naturalists who tried to understand how a variety of animals behave in their natural habitats. One of the major findings of ethology was that animals can carry out many behaviors without ever having seen them performed. In other words, many behaviors are innately programmed (Figure 50.5). And while such behaviors seem purposeful because they are clearly beneficial, they are carried out in ways that show the animals are unaware of the significance of their actions. The work that led to these and other findings was initiated primarily by ethologists Konrad Lorenz, Niko Tinbergen, and Karl von Frisch, who received a Nobel Prize in 1973 for their discoveries. Their research provided a valuable description of the ways animals behave, but little of their work involved the formulation and testing of explanatory hypotheses. However, behavioral ecology was able to build on the firm descriptive basis provided by ethology to transform the study of behavior into a modern hypothesis-oriented part of biology.

Fixed-Action Patterns

The most fundamental concept associated with classical ethology is the **fixed-action pattern (FAP),** also called a fixed-motor pattern. A FAP is a highly stereotyped behavior that is innate. Once an animal initiates a FAP, it usually carries the action pattern to completion even if other stimuli impinge on the animal or the activity becomes inappropriate. A FAP is triggered, or released, by an external sensory stimulus known as a **sign stimulus** or **releaser.** Thus, we can think of a FAP as the innate ability of an animal to detect a certain stimulus, combined with an innate behavioral program that is activated by the stimulus to direct some kind of motor activity. In many cases the sign stimulus is some feature of another species. For example, some moths instantly fold their wings and drop to the ground in response to the ultrasonic signals sent out by predatory bats (see Chapter 45).

A classic case of releasers in social behavior can be observed in the male three-spined stickleback fish, which attacks other males that invade his territory. The releaser for the attack behavior is the red belly of the intruder. The stickleback will not attack an invading fish lacking a red underside but will readily attack nonfish-like models as long as some red is present (Figure 50.6).

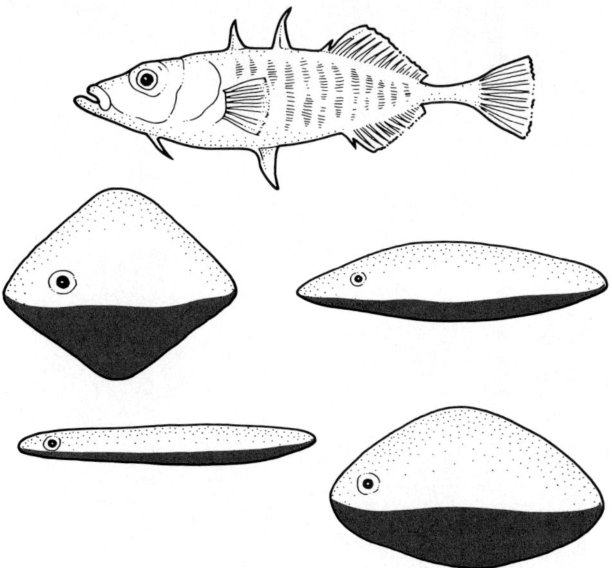

Figure 50.6
Sign stimuli used to demonstrate FAPs. Aggression in male three-spined stickleback fish is triggered by a simple cue. The realistic model at the top without a red underside produces no response. All the others produce strong responses because they have the required red underside.

Figure 50.7
A graylag goose retrieving an egg. The goose stands up, extends her neck, and retrieves the egg by pulling it slowly with the underside of her bill as she settles back into the nest. The egg is lopsided and tends to roll away, but the goose uses a side-to-side motion of her head to keep the egg on course.

Tinbergen, who first reported these findings, was inspired to look into the matter by his casual observation that his fish responded aggressively when a red truck passed their tank.

Interactions of parents and young often involve FAPs and have been studied extensively in birds. In many bird species, when a parent returns to the nest with food, the newly hatched, blind young respond immediately with begging behavior, raising their heads, opening their mouths (gaping), and cheeping loudly. The releaser for this begging is the impact of the parent landing on the nest. Later, when the birds' eyes are more developed, the sight of a parent releases gaping, but the specific stimuli required still are not complex; simple models of the parents work very well. In turn, the releaser for the parent's feeding behavior is remarkably simple, merely a gaping mouth.

A ground-nesting bird such as the graylag goose of Europe sometimes accidentally bumps one of its eggs out of the nest. The sight of an approximately spherical object near the nest is a releaser for the goose's egg-retrieval behavior, in which it rolls the egg back into its nest with side-to-side head motions (Figure 50.7). The adaptiveness of this behavior is obvious, but it has two features that are very strange from a human perspective. First, if the egg slips away (or is experimentally pulled away) as it is being retrieved, the goose stops her side-to-side head motion but continues the normal retrieval movement with her head extended, as though she were still pulling the egg. Only after she sits down does she "notice" that an egg is still outside the nest, and then she begins another retrieval. If the egg is again removed, the goose goes through the retrieval maneuver again. Second, if an unusual object such as a doorknob or a toy dog is placed near a nest, the goose will retrieve it as well. The toy dog may be discarded after the goose attempts to sit on it, but a smoothly rounded object like a doorknob may be incubated for a long time.

The releaser for parental protective behavior in turkeys is the cheeping sound produced by the young chicks. A mother will ignore her offspring if it is placed under a thick glass dome that deadens all sound but leaves the young bird in full view. Even more remarkably, a deaf mother will kill her offspring because she can no longer detect the appropriate releaser for mothering behavior. It is likely that a turkey hatchling whose cheeps cannot be heard releases a predatory response in its mother.

By now, you can see the automatic way animals behave in many situations. If a stickleback processed information like a human, it would realize that the models in Figure 50.6 are not real fish, despite their red bellies. Humans respond much more to an entire situation than do most other animals, and we generally base our actions on more diverse information. The classic experiments just described, and many others, suggest that the animals are not taking thoughtful or intelligent action, at least in these cases. By intelligent we mean integrating various kinds of information and then making a decision; in the case of a FAP, the animals are behaving more like robots. FAPs do occur in humans, however. Infants grasp strongly with their hands in response to a tactile stimulus. An infant's smile is also a FAP; it is readily induced by simple stimuli such as a sound or a figure consisting of two dark spots on a white circle, a kind of rudimentary representation of a face.

In all these cases, adaptive responses to specific stimuli have been selected for. The variations on experimental releasers that ethologists introduce often present animals with situations that almost never arise naturally. If egglike objects were commonly found near goose nests, natural selection probably would have resulted in a discrimination mechanism that allows geese to identify their own eggs. We know that such mechanisms are possible, because they *have* evolved in many birds. For example, some species can differentiate between their own eggs and those of brood parasites such as cuckoos. These parasites lay their eggs in the nests of another species (host species), and their young often cause the death of the host's own young (see Figure 50.5). In terms of the reproductive success (Darwinian fitness) of the host bird, there is a great advantage in recognizing the cuckoo egg as foreign and removing it from the nest (Figure 50.8). We see here that natural selection can lead to complex and sensitive behavioral mechanisms when there is a strong selection pressure, as with host species parasitized by cuckoos. Yet animals will show what seems to us to be gross incompetence when confronted with novel situations unrelated to past selection pressures, as with geese that retrieve nonegg objects—a situation unlikely to occur in the natural habitat.

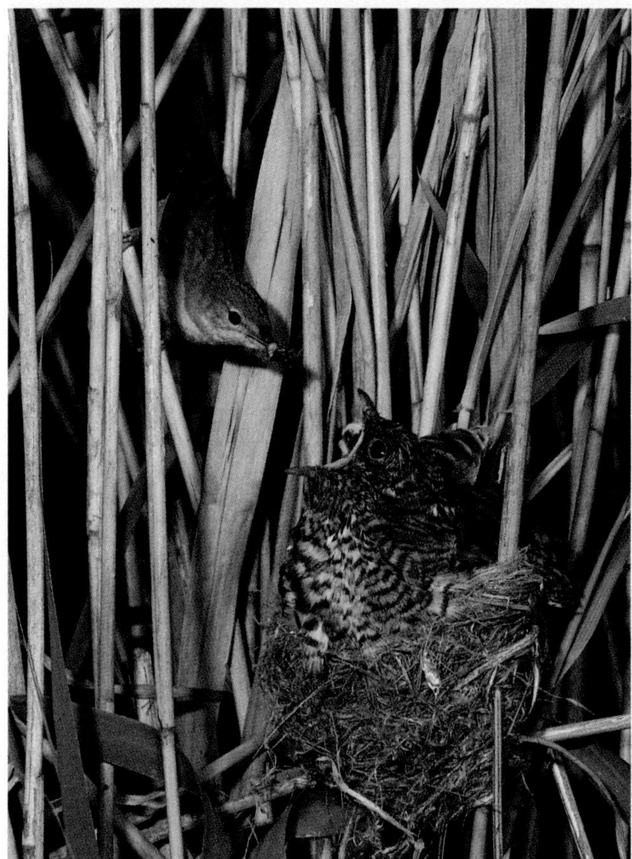

Figure 50.8
A reed warbler feeding a baby cuckoo. This cuckoo hatched from an egg laid by its mother directly into the warbler's nest. After it hatched, the cuckoo pushed all the warbler's young out of the nest (see Figure 50.5). Even though it looks nothing like a nestling warbler, the cuckoo is fed by its foster parents, because the releaser for parental feeding is simply a gaping mouth. Although they do not differentiate among nestlings, reed warblers and many other birds recognize their own eggs and remove cuckoo eggs that are dissimilar. This has resulted in the evolution of cuckoo eggs that are nearly identical to those of many host species.

The mechanistic component of behavior based on FAPs is demonstrated most clearly by some animals that seem to have an endless capacity for repeating stereotyped behavior. One example, first described by the nineteenth-century naturalist Jean-Henri Fabre, involves digger wasps. The female wasp builds a nest in the ground, places a paralyzed cricket in it, and then lays an egg that will eventually hatch into a larva that consumes the cricket. The wasp accomplishes all this with a highly stereotyped series of FAPs. She places the cricket a short but rather precise distance (about 2.5 cm) from the nest, enters the nest briefly, apparently for a final inspection, then comes out to get the cricket and carry it into the nest. If an observer moves the cricket a short distance while the wasp is inside, she searches for it after emerging and, after retrieving it, puts it back in its original spot near the entrance. She then enters the nest for another inspection, even though she has just made one, and emerges again a short time later. If the cricket has been moved again, the wasp repeats the cycle—and will continue to repeat it at least forty times, showing no sign of either tiring of the repetition or of circumventing the scientist's game by changing behavior. The wasp can only get past the "inspect-the-nest"step by immediately finding the cricket near the nest.

The Nature of Sign Stimuli

Ethologists have found from experiments that sign stimuli for a FAP are generally based on one or two simple characteristics. In many cases, the stimulus is the most obvious (or the only) characteristic of a particular situation; for example, the ultrasonic signals of bats are the obvious cue to trigger avoidance behavior in moths. In other cases, it appears that animals have settled on certain characteristics out of an array of possible choices. When an adult herring gull brings food to its chick, it bends its head down and moves its beak, which has a red spot. The chick pecks the beak, stimulating the adult to regurgitate the food. The chick might be cued to peck by a variety of stimuli, including such obvious things as a lump at the end of a rectangle (simulating food at the end of a bill). However, careful studies have demonstrated that the releaser is a red spot swung horizontally at the end of a long, vertically oriented object. We expect natural selection to have favored cues that have a high probability of association with the relevant object or activity, but when there are many possible cues, there is probably some randomness in which one becomes fixed upon as the sign stimulus for a FAP.

Close correlations exist between an animal's sensitivity to general stimuli and the specific sign stimuli to which it responds. For example, frogs have retinal cells that are especially good at detecting movement, and it is the movement of an object that releases the frog's tongue-shooting behavior during feeding. A frog will starve to death if surrounded by dead or motionless flies but will attack one readily if it moves. Sometimes, exaggeration of the relevant stimulus produces a stronger response. Wider gaping by young birds more readily releases the feeding behavior of the adult. Because a young bird gapes more widely in response to increased hunger, the hungriest young are most likely to be fed. This aspect of sign stimuli can be seen in simple experiments in which animals are presented with a *supernormal stimulus,* an artificial stimulus that elicits a stronger response than any naturally occurring stimuli. A graylag goose will attempt to roll a volleyball into her nest instead of her own egg, and an oystercatcher (a bird of seashore areas) will attempt to incubate a giant model of her egg instead of the real thing.

Some ethologists have suggested that the use of simple cues to release programmed behavior prevents an animal from wasting time processing or integrating a wide variety of input. Perhaps a better way of interpreting the situation has to do with the limitations of innate programming. A frog's sensory-neural network for detecting movement is probably much less complex than the apparatus that would be necessary to rapidly distinguish a fly from another object of similar size. In any case, simple cues usually work quite well in an animal's normal sensory world, though not in the experimental world ethologists often create.

Note that computer terms, like "programmed," used to describe behavior are only metaphors for what *seems* to be occurring. These familiar terms suggest a model consistent with what is observed, but they are not intended to imply that the mechanisms of animal behavior are the same as those of a computer.

LEARNING AND BEHAVIOR

Learning, which is formally defined as the modification of behavior by experience, often affects even innately programmed behaviors such as FAPs. Before we discuss the various ways in which learning occurs, we will review what has been referred to as the nature-versus-nurture controversy.

Nature Versus Nurture

While the early ethologists, who were mostly Europeans, were developing their approach, psychologists in America were studying animal behavior from a very different perspective. Focusing on learning, they typically studied a few species of captive reared animals, such as laboratory rats. Psychologists were impressed with how effectively they could shape animal behavior with learning, while ethologists were equally impressed with the degree to which animal behavior seemed innate. A clash was perhaps inevitable, and it came to be known as the *nature-versus-nurture controversy.* Few scientists ever believed that behavior is either all genetic or all learned, but this is the way outsiders often viewed the debate between ethologists and psychologists. Instead, the debate was mostly over which input—instinct or learning—is of primary importance.

Nearly all biologists today agree that all behavior is a consequence of genetic and environmental influences. Even a behavior as simple and automatic as a knee-jerk reflex is not determined totally by genes. Although you do not have to learn how to perform a knee-jerk response when a doctor taps your knee, this behavior is still shaped by the environment. If you grew up with insufficient nutrients, you would likely have an abnormal knee-jerk response. Similarly, even though an animal may not have to witness a FAP because the basic behavior is innate, learning is still involved. Most FAPs improve with performance, as animals learn to carry them out more efficiently. Is instinct also important to all behaviors? You might say that some things, such as the different human languages, are completely learned. It is true that the ability to speak either English or Spanish has no genetic basis, but the ability to learn *any* language is a function of a complex brain that develops in a particular environmental context under the guidance of a human genome.

Despite the dictum that all behavior is some mix of learning and instinct, or of environmental and genetic inputs, it is still helpful to identify aspects of behavior that seem to be due primarily to one or the other of these influences. The objective is not to tally up the results and see which influence is more important overall. But sorting out the relative importance of genes and environment for a particular behavior helps us understand variations in that behavior among individuals of a species. For example, in the fruit fly *Drosophila,* individuals typically exhibit an activity cycle of about 24 hours. Occasionally, though, one finds an individual with a 19-hour cycle. Breeding experiments show that this difference is due to different alleles at a single gene locus. Although the difference between 19- and 24-hour *Drosophila* is genetically determined, we cannot conclude that everything else about activity cycles is genetically determined. We have already identified language differences among people as a completely learned behavioral variation of an innate ability. There are many parallel situations among other animals, including one involving vocalizations. Many songbirds have different songs in different populations. These "dialects" arise because male birds learn to sing like other males, generally early in life, and this sort of learning is prone to occasional variations that will cause populations to differ. Birds reared in one dialect but moved to another population early in life will learn the new dialect, just as a person born in the deep south who moves north early in life will speak like a northerner. In both cases, genetic differences have little or no significance. As we discuss the various forms of learning, we will see additional examples in which it is productive to analyze a behavior and identify those aspects for which genetic or environmental influences are of primary importance.

Learning Versus Maturation

Some behaviors that are clearly innate, in the sense that one individual does not have to learn them from another, may be performed more quickly or effectively as time goes on. However, this does not necessarily mean that learning has occurred. Behavior may improve because of ongoing developmental changes in neuromuscular systems, a process called **maturation.** We commonly speak of birds "learning" to fly, and you may have seen a hapless fledgling awkwardly fluttering about as if it were practicing. However, young birds have been experimentally reared in restrictive devices so that they could never flap their wings until an age when their normal kin were already flying. Such birds flew immediately and normally when released. Thus, the improvement must result from neuromuscular maturation, not from learning.

Earlier we described the pecking behavior that young gull chicks direct toward their parents' bills. Recently hatched chicks are quite indiscriminate and will peck at a wide variety of objects, but chicks a week or two old show their strongest response to accurate models of their parents' bills. Is this a maturation or a learning process? Cross-fostering experiments involving herring gulls and laughing gulls show that maturation is not involved. A laughing gull chick that has been reared by a herring gull will respond more strongly to its foster parent's bill type than to the bill of its own species. The reverse is true for a cross-fostered herring gull chick. This is an example of how learning can modify a behavior that is basically instinctive.

Habituation

Habituation is a very simple type of learning that involves a loss of responsiveness to unimportant stimuli or to stimuli that do not provide appropriate feedback. Examples are widespread. *Hydra* stop contracting if disturbed too often by water currents. Gray squirrels and many other animals recognize alarm calls of conspecifics threatened by a predator, but they eventually stop responding if these calls are not followed by an actual attack (the "cry-wolf" effect). Many species of orchids have flowers that resemble female bees or wasps. Male insects that attempt to mate with these flowers bring about pollination (which explains how plants benefit from this mimicry), but they do not experience all the stimuli that would occur in an actual mating. The insects eventually learn that these flowers are not really mates and stop responding to them. Habituation makes good adaptive sense to a behavioral ecologist. Trying to mate all day with flowers will not do much to increase a male insect's fitness.

Imprinting

Some of the most interesting cases where learning interacts closely with innate behavior involve the phenomenon known as **imprinting,** the only form of learning that attracted much interest from ethologists. You have probably seen young ducks or geese following

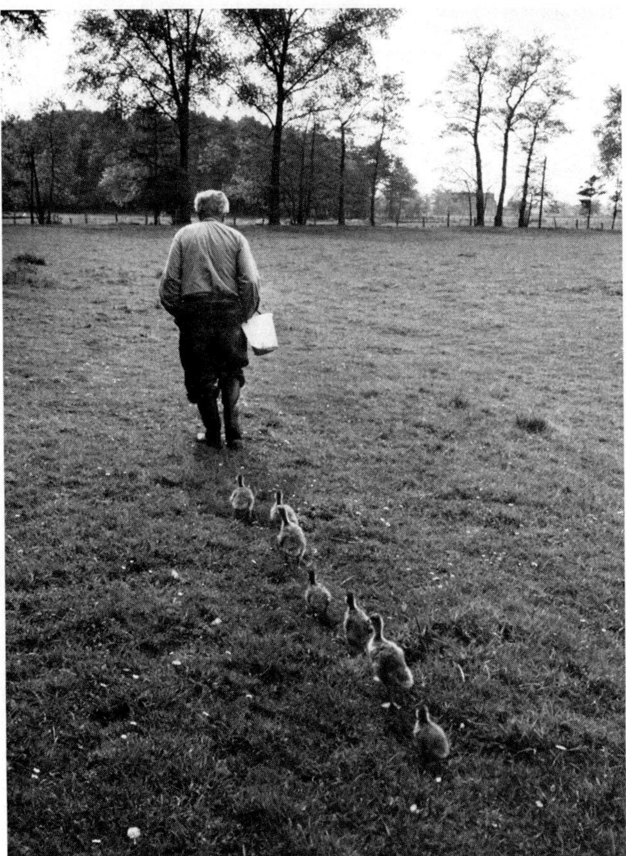

Figure 50.9
Imprinting. Konrad Lorenz was "mother" to these imprinted geese.

their mother. This behavior makes good adaptive sense to us, since the adult bird is likely to have more experience than the young in finding food, avoiding predators, and otherwise getting along in the world. But how do the young know whom—or what—to follow? In his most famous study, Konrad Lorenz divided a clutch of graylag goose eggs, leaving some with the mother and putting the rest in an incubator. The young reared by the mother showed normal behavior, following her about as goslings and eventually growing up to interact and mate with other geese. When the artificially incubated eggs hatched, the geese spent their first few hours with the researcher instead of with their mother. From that day on, they steadfastly followed Lorenz and showed no recognition of their own mother or other adults of their own species (Figure 50.9). As adults, the birds continued to prefer the company of Lorenz and other humans over that of their own species, and they sometimes even tried to mate with humans.

Apparently, graylag geese have no innate sense of "mother" or "I am a goose, you are a goose." Instead, they simply respond to and identify with the first object they encounter that has certain simple characteristics. What is innate in these birds is the ability or tendency to respond; the outside world provides the *imprinting stimulus,* something to which the response will be directed. The most important imprinting stimulus in Lorenz's graylag geese was movement of an object away from the young, although the effect was greater if the object emitted some sound. The sound did not have to be that of a goose, however; Lorenz found that a box with a ticking clock in it was readily and permanently accepted as a "mother."

Many other examples of imprinting are now known. Salmon are noteworthy for their long migrations in the open ocean, where they feed, grow, and mature after hatching in freshwater streams (see Figure 45.21). Some species remain at sea for several years. Nonetheless, as demonstrated by the tagging of thousands of young fishes, each returning individual goes back to its own home stream to spawn. This involves not only finding the proper major stream to enter at the coast, but also making a number of correct choices among smaller and smaller tributaries as the fish winds its way upstream. Extensive research has demonstrated conclusively that this ability is based on olfactory imprinting. Young fishes artificially reared and exposed to a chemical called morpholine can be directed into a stream upon their return if morpholine is introduced into it. Under natural conditions, the fishes apparently imprint on the complex bouquet of odors unique to their specific stream, and can recognize those odors and swim toward their source even after being long distances from the stream for long periods.

A **critical period,** which is a limited time during which learning can occur, is one of the two features that distinguish imprinting from other types of learning. The second feature is that what is learned is irreversible. Lorenz found, for example, that geese totally isolated from any moving objects during the first two days after hatching, which is the critical period for imprinting on parents, failed to imprint afterward on anything. Imprinting has commonly been thought of as involving very young animals and rather short critical periods. But it is now clear that a similar learning process occurs in older animals, and that the critical period may be of various durations. For example, just as a young bird requires imprinting to "know" its parents, the adults must also imprint to recognize their young. For a day or two after their young hatch, adult herring gulls will accept and even defend a strange chick introduced to their nesting territory. After imprinting, which is probably based largely on individually variable cues such as the call notes of chicks, the adults will kill and eat any strange chick.

By imprinting on their parents, young birds not only learn who will care for them, they also learn species identity and the kind of bird they should mate with later in life. Sexual imprinting generally occurs later than parental imprinting, and the critical period lasts longer. For example, in one study involving two closely related species of finches, young males of one species were reared first with members of their own species, then with members of the other species during their several-week-long critical period for sexual identity. Later, when exposed to females of their own species, they mated quite reluctantly. They readily mated with females of the other species, however, even when they had not seen any members of that species for as long as eight years. Identification with the second species had been permanently imprinted. While a critical period and irreversibility characterize imprinting, it is now recognized that these phenomena are not always rigidly fixed. For example, the cross-fostered finches did eventually mate with females of their own species.

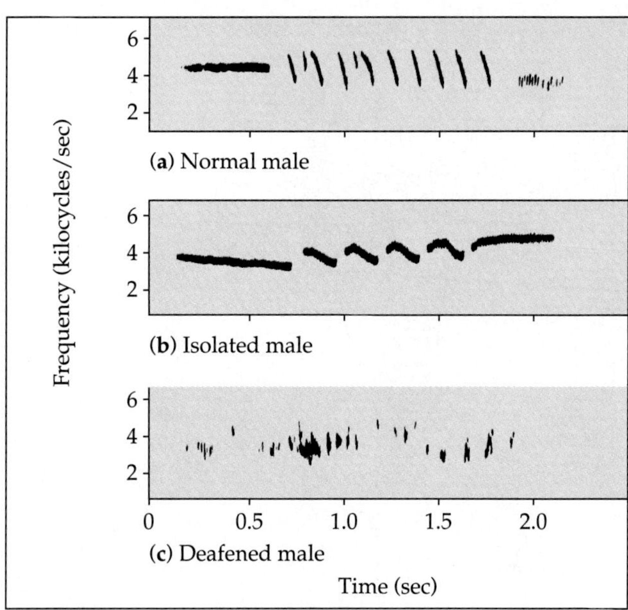

Figure 50.10
Songs of white-crowned sparrows reared under three conditions. (**a**) Males that heard tape recordings of their species' song before reaching 50 days of age had experiences simulating those in nature and learned to sing a normal song several months later. (**b**) Isolated males were reared in soundproof chambers, which ensured that they never heard their species' song. (**c**) Males in the last category were deafened *after* they had heard recordings of their species' song, but *before* they ever started to practice singing.

Song Development in Birds: A Study of Imprinting

This chapter began with a brief discussion of vocal learning in young birds, a behavior that has been the subject of extensive research. Both genes and learning influence song development, which in many species conforms closely to an imprinting process. A closer look at how birds learn to sing will reinforce some of the concepts of animal behavior that you have learned so far.

Many studies of song development have focused on common North American species. Thirty years ago, experiments showed that male white-crowned sparrows raised in isolation in soundproof chambers developed abnormal songs. However, if tape recordings of normal white-crowned sparrow songs were played to the birds when they were 10 to 50 days old, they developed a normal song several months later. The normal song developed as a result of a bird's comparing its own singing with its memory of the taped song. This was shown by experimentally deafening some birds after exposing them to the taped playbacks: The deaf birds developed songs even simpler and more abnormal than isolated birds that could hear but were never exposed to songs (Figure 50.10). We see that there are two learning processes in the sparrow's song development. First, a bird must acquire a song type by hearing an adult; and second, it must learn to match this song by listening to itself. Young males raised in soundproof chambers and exposed to taped songs of white-crowned sparrows after 50 days of age sang like birds that never heard these songs. In other words, the sparrows were receptive to learning the taped songs only during a critical period, which is why song learning can be considered a type of imprinting. Innate influences were also demonstrated: Young white-crowned sparrows that heard taped songs of other sparrow species during the critical period did not adopt these songs. Apparently, they were genetically predisposed to learn their own species' song.

In recent years, the story has become even more complex. When isolated white-crowned sparrows more than 50 days old were exposed to live singing adults of another species, they learned the song of the other species. Because a bird can interact socially with it, a live singing adult is a much stronger stimulus than a taped song. So this strong stimulus can overcome the innately programmed tendency to acquire only a white-crowned sparrow song. Again we see that an innate tendency can be modified by experience and is not necessarily inflexible. Furthermore, the critical period proved to be longer when the stimulus was a live bird than when it was a taped playback. People also have a critical period for learning vocalizations: It is well known that foreign languages are learned most easily up until the teen years. Of course, this critical period is not rigidly fixed. Adults can learn a new language, but they usually require much more effort and time to become fluent than a child does.

Bird species other than the white-crowned sparrow have different programs for song development. The closely related song sparrow develops a nearly normal song even if reared in total isolation. But males have larger repertoires if they can learn song variants from other birds. Mockingbirds have huge repertoires of at least 150 song types, many of which are accurate mimics of sounds made by other kinds of birds and animals and even telephone rings. The adaptive value of a large repertoire is apparently so great for mockingbirds that genetic programming places few or no restrictions on the range of songs they can acquire. Until recently, it was assumed that all birds had to listen to themselves in order to develop song. But eastern phoebes experimentally deafened early in life, well before they began to sing, developed normal songs. Just why birds show such an array of developmental programs for their songs is unclear, but the diversity suggests that this is a fairly flexible behavioral system in many groups of birds, and one that is highly responsive to different selection pressures.

Classical Conditioning

Many animals can learn to associate one stimulus with another, a process known as **associative learning.** A type of associative learning called **classical conditioning** is well known from the work of Ivan Pavlov, a Russian physiologist, around 1900. Pavlov sprayed powdered meat into dogs' mouths, causing them to salivate (primarily a physiological rather than a behavioral response). Just before the spraying, however, he exposed the dogs to a sound such as a ringing bell or clicking metronome. Eventually, the dogs salivated readily in response to the sound alone, which they had learned to associate with the normal stimulus. Similar types of conditioning experiments have been carried out with other animals. Many kinds of perceptual sharpening probably involve this same kind of conditioning. For example, the tendency of a young gull to respond to a red spot becomes more generalized to include the entire adult.

Operant Conditioning

Another type of associative learning that directly affects behavior is **operant conditioning,** also called trial-and-error learning. Here, an animal learns to associate one of its own behaviors with a reward or punishment and then tends to repeat or avoid that behavior. The best known laboratory work involving operant conditioning is that of American psychologist B. F. Skinner. A rat or other animal placed in a "Skinner box" has a choice of various levers it can push, some of which reward the animal by releasing food. The animal's choices may be random at first, but it quickly learns to choose those levers that provide food. Such learning is the basis for most of the animal training done by humans, in which the trainer typically encourages a behavior by rewarding the animal. Eventually the animal performs the behavior on command, without always receiving a reward.

Operant conditioning is undoubtedly very common in nature. For example, animals quickly learn to associate eating particular food items with good or bad tastes and modify their behavior accordingly. Remember that what tastes good or bad is likely to be programmed by natural selection on the basis of a food's value. Thus, as with imprinting, genes influence the outcome of operant conditioning. An interesting example of operant conditioning occurred in England during the early 1950s, when tits (chickadeelike birds) learned to peck through the paper tops of milk bottles left on doorsteps and drink the cream from the top. Apparently, one or more of the birds discovered that its general pecking, probing behavior was rewarded if directed at the bottles.

Observational Learning

It is obvious that many vertebrates monitor the behavior of other individuals and thereby learn important information. The tit's habit of attacking milk bottles spread so rapidly to other parts of England, and even to other species of tits, that it is clear the birds learned from one another. **Observational learning** allows for new traditions to become established and to be passed on to succeeding generations. As we have discussed, song development in many birds involves this sort of learning, as individuals listen to and eventually adopt the songs of older birds.

Play

Many mammals and some birds engage in behavior that can best be described as **play.** Such behavior has no apparent external goal but involves movements closely associated with goal-directed behaviors. For example, playful stalking and attacking of conspecifics occurs in many predator species, such as those in the dog and cat families. Although this behavior does not usually involve painful bites, the animals grab and mouth one another, using movements similar to those used to capture and kill prey. Play is most common when there are no apparent external stimuli to attract an animal's attention. A common feature of play is that it is potentially dangerous or costly. Baboons sometimes kill and eat juvenile vervet monkeys, and they are most successful in doing so when the monkeys are playing in groups away from adults. In a study of caged ibex, a type of wild goat, play resulted in at least five out of fourteen kids sustaining injuries that resulted in limps.

Play obviously consumes energy, and the risks to life and limb result in significant additional costs. What could be the ultimate adaptive basis for such seemingly pointless behavior? The "practice hypothesis" suggests that play is a type of learning that allows ani-

Figure 50.11
Play behavior. In the view of behavioral ecologists, the rough-housing behavior of these lion cubs evolved in spite of the energy it consumes and the risks it poses, because play improves individual fitness in some way. Practicing survival behavior, such as the capture of prey, and building a healthy body through exercise are two possible benefits of play.

mals to perfect behaviors needed in functional circumstances. This hypothesis is supported by the observation that play is most common in young animals. However, movements used in play show little improvement after their first few practices. An equally likely ultimate explanation for play is the "exercise hypothesis," which suggests that play is adaptive because it keeps the muscular and cardiovascular systems in top condition. In other words, according to this hypothesis, play evolved for the same reason that people engage in aerobic exercise. The exercise hypothesis also predicts that play should be especially common in young animals, because they typically do not have to exert themselves in useful activities while under the protection and care of their parents (Figure 50.11).

Insight

Insight learning is the ability to perform a correct or appropriate behavior on the first attempt in a situation with which the animal has no prior direct experience. (Some scientists prefer to call this reasoning, rather than learning.) If a chimpanzee is placed in an area with a banana hung too high above its head to be reached and several boxes on the floor, the chimp can "size up" the situation and stack the boxes, enabling it to reach the food. In general, insight is best developed in mammals, especially primates, but even in these groups the amount of insight often varies substantially from one situation or species to another. Although the great majority of animals seem to have little or no ability to use insight (Figure 50.12), some recent work on bees suggests that an insightlike capability may be found even among invertebrates, as we will discuss in a later section on animal cognition.

The Ultimate and Proximate Bases of Learning

Although animals frequently appear to do complex things, most of their behavior can be understood as relatively fixed patterns that are controlled by simple stimuli and often modified in their details by simple kinds of learning. Nonetheless, their limited abilities work very well in normal circumstances. The capacity to learn is a product of evolution and an adaptation that enhances survival and reproductive success. It is clear that learning is subject to natural selection, because much of it is genetically programmed to occur at specific times, as in imprinting, and to be based on a limited set of stimuli. Unfortunately, the internal proximate mechanisms of learning are very poorly known, and only recently has progress begun on linking some simple kinds of learning to internal biochemical or physiological changes (see Chapter 44).

Now that we have examined the interplay of innate and learned components of behavior, we can apply the basic concepts to a sampling of behavioral adaptations that have evolved in the animal kingdom.

BEHAVIORAL RHYTHMS

Animals exhibit all kinds of regularly repeated behaviors: feeding in the day and sleeping at night (or vice versa), reproducing every spring, migrating in spring and fall, and so on. What determines when a squirrel sleeps and when it awakens? This sort of question has long been the subject of sophisticated experimentation designed to explain the proximate mechanisms that control behavioral rhythms. The ultimate basis for rhythms is usually quite obvious, as animals typically carry out behaviors when their particular ecological niches can be exploited most safely and profitably. However, as with most questions about behavior, the proximate bases of rhythms are neither as obvious nor as simple as they might seem. We have already discussed circadian (daily) rhythms in some detail in relation to plants (see Chapter 35). Animals, too, are subject to such rhythms, which are regulated by various environmental cues. But what would happen to rhythmic behavior if an animal were placed in an environ-

Figure 50.12
Lack of insight. The dog cannot apply insight to its problem—that is, "size up" the situation and proceed correctly to the food on the first try. Through trial and error, however, the dog will eventually find its way to the food. From then on, it will "solve" a similar problem quickly.

ment with no external cues about time? Or to put it another way, is rhythmic behavior based on *exogenous* (external) timers, *endogenous* (internal) components, or both?

Numerous studies have been performed to assess the relative importance of endogenous and exogenous proximate mechanisms in the rhythmic behavior of many species. These studies show that circadian rhythms usually have a strong endogenous component, referred to as a *biological clock,* but because the endogenous rhythm does not exactly match that of the environment, an exogenous cue is necessary to keep cycles timed to the outside world. The exogenous cue is sometimes called a *zeitgeber* (German for "time-giver"). Light is undoubtedly the most common zeitgeber for circadian rhythms. For example, activity of the North American flying squirrel normally begins with the onset of darkness and ends at dawn, which suggests that light is an important exogenous regulator. However, if a squirrel is placed in constant light or constant darkness, the rhythmic activity is not abolished; in fact, it holds up quite well for at least a month or so. The duration of each cycle (one period of activity plus one period of inactivity) deviates slightly from 24 hours, so that under constant (free-running) conditions, an animal gradually drifts out of phase with the outside world. Eventually, it will be active or inactive at the reverse times of animals in the forest (Figure 50.13). For some animals, on the other hand, light and dark are not important regulators of the activity cycle. Fiddler crabs living in the intertidal zone forage actively when the tide is out and retreat to their burrows when it comes in. If crabs are experimentally given a simulated continuous low-tide condition, they will show an activity rhythm based on the 12.4-hour tidal cycle.

Human circadian rhythms have been studied by placing individuals in comfortable living quarters deep underground, where they could make their own schedules with no external cues of any kind. Under these free-running conditions, the biological clock of humans seems to have a period of about 25 hours, but with much individual variation; as with other animals, humans use external cues (zeitgebers) to adjust their rhythms to 24 hours in the real world.

Whether endogenous timekeepers are important in rhythmic behavior that involves longer cycles is much less clear. In many species, **circannual behaviors,** such as breeding and hibernation, are based at least in part on physiological and hormonal changes directly linked to exogenous factors such as day length. This was the case, for example, with the bluegill fish discussed earlier. Little is known about possible endogenous factors, in large part because of practical difficulties in conducting experiments: Animals would have to be maintained for years (instead of days) under constant conditions. Some long-term studies on ground squirrels indicate that fat deposition, associated with hibernation, occurs regularly in a constant environment, but almost nothing is known about possible endogenous control of other kinds of circannual phenomena, particularly behaviors.

Even with short-term rhythms, in which endogenous components are clearly present, the internal, proximate timing mechanism is unknown. Current hypotheses posit some kind of biochemical clock, perhaps derived from molecular interactions that occur regularly. As yet, however, no satisfactory model has been developed, much less experimentally verified.

MOVEMENT

Animal movements are another aspect of behavior with a long tradition of experimental study. As with rhythms, the emphasis has been on proximate causes, especially the mechanisms animals use to detect and respond to the external cues that guide movements. The ultimate bases for oriented movements are often

(a)

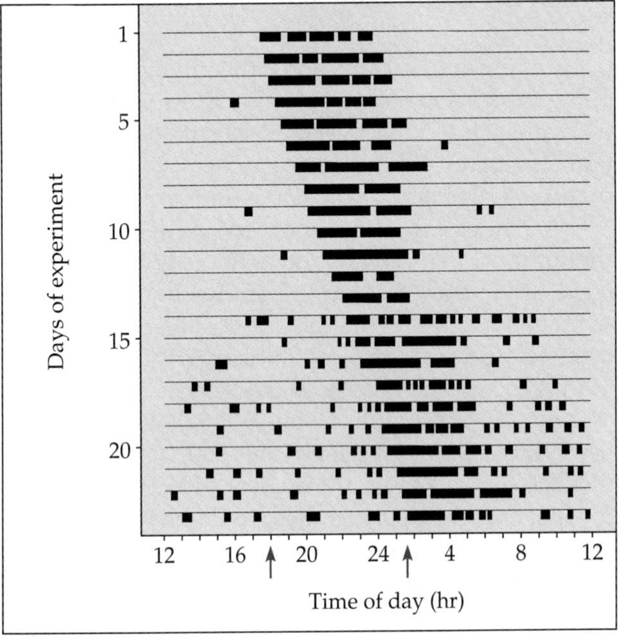

(b)

Figure 50.13
Rhythmic activity of a flying squirrel. (**a**) The nocturnal flying squirrel is most active during the first few hours after sunset. In order to trace the activity level of a squirrel kept in constant darkness for 23 days, the animal was placed in a cage with an exercise wheel connected electronically to a chart recorder (which moves graph paper past a pen at a fixed speed). Whenever the animal ran, the wheel caused the pen to mark the graph paper, producing the graph shown in (**b**). The thicker horizontal lines indicate the squirrel's periods of activity. The pattern displays a definite rhythm, but the period of greatest activity is shifted slightly each day. The free-running cycle for this squirrel was 24 hours, 21 minutes. Thus, after 23 days, its activity cycle was about 8 hours out of synchronization with the actual timing of sunrise and sunset (that is, 23 times 21 min equals 483 min, or 8 hr). The magenta arrows indicate the times when sustained activity began on days 1 and 23 of the experiment. In a normal environment, the daily cues of alternating light and dark periods would adjust the biological clock to a 24-hour cycle. Biological clocks are intrinsic to organisms, but the timing of rhythmic behavior must be adjusted to important events in the environment.

obvious. Different animals are adapted to different environments and typically have behaviors that bring them to those environments, often at a particular time of year.

Animals can use a variety of environmental cues to guide them as they move from one place to another. A **kinesis,** the simplest sort of movement, involves a change in activity rate in response to a stimulus. Sowbugs or woodlice become more active in dry areas and less active in humid ones, a simple behavior that tends to keep these animals in moist environments. Notice that a kinesis is actually randomly directed: The animals do not move toward or away from specific conditions, but since they slow down in a favorable environment, they tend to stay there. In contrast, a **taxis** is an oriented movement, more or less automatic, toward or away from some stimulus. For example, housefly larvae are negatively phototactic after feeding, automatically moving away from light; this simple response presumably ensures that the flies remain in an area where they are harder for predators to detect. Trout are positively rheotactic (Gr. *rheos,* "current"); they automatically swim or orient themselves in an upstream direction, which keeps them from being swept away.

The seasonal movement of animals over relatively long distances is called **migration.** Migrating animals generally make one round trip between two regions each year. The most notable examples are the migrations of birds, whales, a few butterfly species, and certain oceangoing fish. How is it, for example, that golden plovers find their way over 8000 miles from their arctic breeding grounds to southeastern South America? Even more remarkably, some populations of these birds return to the Hawaiian Islands, a small piece of land in a vast expanse of ocean (Figure 50.14).

Migrating animals use three mechanisms to find their way. In *piloting,* an animal moves from one familiar landmark to another until it reaches its destination. Piloting is used mostly for short-distance movements and has little value at night or over oceans. In *orientation,* an animal can detect compass directions and travels in a particular straight-line path for a certain distance or until it reaches its destination. The most complex process is *navigation,* which involves determining one's present location relative to other locations, in addition to detecting compass direction (orientation). If you were dropped off at an unfamiliar spot and told that your home was directly to the north, you could use a compass and straight-line travel to get home; that is, you could use orientation. But a compass alone would not be adequate if you were not told which way to go—to choose the right direction you would need to determine where you were in relation to home. Figure 50.15 reinforces this distinction between orientation and navigation.

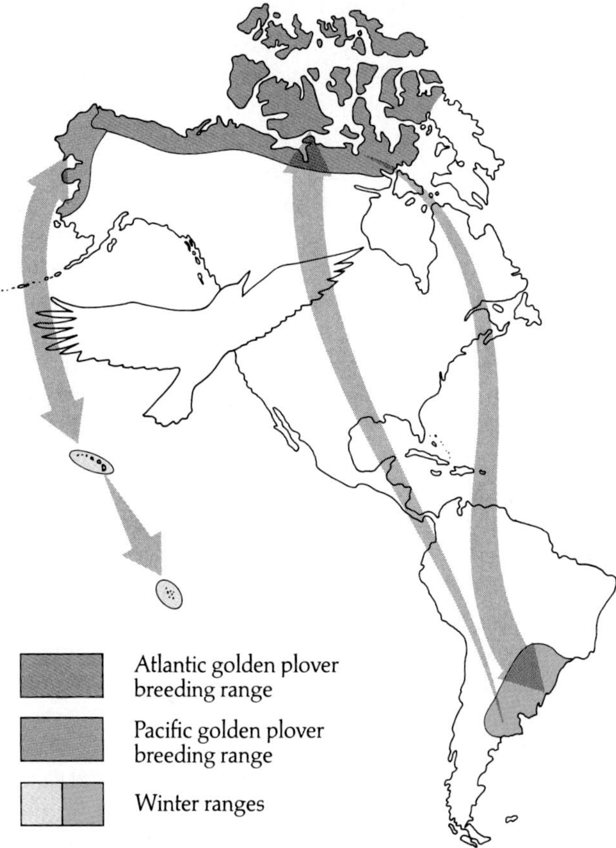

Figure 50.14
Migration routes of the golden plover. These birds are able to navigate across vast expanses of ocean to the relatively small Hawaiian and Marquesas islands.

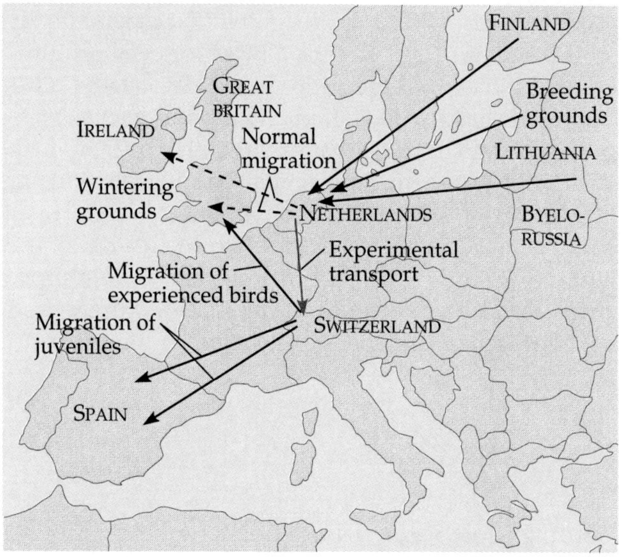

Figure 50.15
Orientation versus navigation in juvenile and adult starlings. About 11,000 starlings were captured in the Netherlands on their migration from their breeding grounds in northeastern Europe to their wintering grounds in Great Britain, Ireland, and northern France. After being transported to Switzerland and released, juvenile starlings, which had never made the journey before, continued to fly west and southwest, which brought them to Spain. Adults, all of whom had made the trip at least once before, flew northwest, an atypical direction but one that took them to their usual wintering grounds. Both age groups were able to detect direction, but only the adults showed true navigation because only they determined where their original goal was relative to the site to which they were transported.

What sorts of cues do animals use for orientation and navigation? The answer, surprising at first but now firmly established, is that some species of birds and other animals commonly use the same means of navigation as early sailors: the heavens. The sun (for day-migrating species) and stars (for night migrants) provide excellent, albeit complex, cues to direction.

Navigating by the sun or stellar constellations requires an internal timing device to compensate for the continuous daily movement of celestial objects. Consider what would happen if you started walking one day, orienting yourself by keeping the sun on your left. In the morning you would be heading south, but by evening you would be heading back north, having made a circle and gotten nowhere. At night, the stars also shift their apparent position as the Earth rotates. One night-migrating bird called the indigo bunting avoids the need for a timing mechanism by the same means as ancient navigators: fixing on the North Star, which moves little. Many migrants, however, use some type of internal clock. For example, if an experimental sun is held in a constant position, starlings change their orientation steadily at a rate of about 15° per hour. How they do it is unknown. The problem is very complex because the apparent location of the sun shifts at a variable rate, being fastest at mid-day. Furthermore, the apparent position of celestial objects changes as the animal moves over its migration route. As with the various internal rhythms discussed earlier, almost nothing is known of the mechanisms that could account for this extraordinary ability.

With some bird species, experimentally obscuring the sky to simulate cloudy weather causes them to flutter about in all directions or to cease migratory behavior. Many species, however, are known to continue migrating quite accurately under clouds or even through fog. There is now good evidence that some birds have the ability to detect and orient themselves to Earth's magnetic field, which varies geographically. The migratory orientation of some species can be experimentally manipulated by changing the magnetic field around them. Very little is known about how birds detect magnetism, but it is intriguing that magnetite, the iron-containing ore once used by sailors as a primitive compass, has been found in the heads of some birds. Magnetite has also been found in the abdomens of bees and in certain bacteria that orient with respect to magnetic field. Magnetic sensing may be a widespread, important orienting mechanism among animals, overlooked until recently because we seem to lack this sense ourselves. Although researchers have recently discovered magnetite in human skulls, there is no other evidence for a magnetic sense in humans.

FORAGING BEHAVIOR

Feeding is obviously a behavior essential to survival and reproductive success, but what determines exactly what an animal eats? This is one of the most frequent questions behavioral ecologists consider. Animals feed in many ways, using various feeding behaviors that are closely linked to morphological traits. Suspension-feeding, for example, requires behavior different from that required for active predation. Ecological and evolutionary considerations are also extremely important; food habits are a fundamental part of an animal's niche and may be shaped in part by competition with other species. In this section, we will examine some aspects of feeding from a behavioral perspective.

Animals in many species can theoretically choose from a large array of potential foods. Some animals tend to be generalists, feeding on a wide variety of items. Gulls feed on material that may be living or dead, aquatic or terrestrial, plant or animal. In contrast, limpkins (wading birds of swamps in the southeastern United States) are specialists that feed only on one kind of snail. Specialists usually have morphological and behavioral adaptations that are highly specific to their food, and as a result they are extremely efficient at foraging. Generalists cannot be as efficient at securing any one type of food, but they have the advantage of having other options if a food becomes unavailable.

Most generalists do not choose their food randomly. Often, an animal will concentrate on a particular item when it is abundant, sometimes to the exclusion of other foods. The animal is said to have a **search image** for the favored item. We can understand search images from our own experience. If you were looking for a particular package on a kitchen shelf, you would probably scan rapidly, looking for a package of particular size and color rather than reading labels. Eventually, if the favored item becomes scarce in relation to others, the animal will develop a new search image. Search images therefore allow an animal to combine efficient short-term specialization with the flexibility of generalization.

The switching described here, and other aspects of the choices animals make while foraging, have generated enormous interest among behavioral ecologists; in fact, a research emphasis on *optimal foraging* has developed in recent years. We would expect natural selection to favor animals that choose foraging strategies that maximize the differential between benefits and costs. Benefits are usually considered in terms of energy (calories) gained. However, other optimization criteria, such as specific nutrients, are sometimes more important than energy. Costs or tradeoffs associated with foraging consist of the energy needed to locate, catch, and eat food; the risk of being caught by a predator during feeding; and time taken away from other vital activities, such as searching for a mate.

Many trade-offs must be considered in an optimal foraging study. A given food item may be large and therefore contain considerable energy, but if it is farther away than a small item, moving to it will require more energy. In addition, the larger item may require more time to subdue or manipulate before swallowing, time in which other prey could be pursued. The smallmouth bass, for example, readily consumes both minnows and crayfish. The fact that it does not show an overall preference suggests that the trade-offs balance, with minnows being the optimal prey in some situations and crayfish in others. Minnows contain more usable energy per unit weight (the crayfish has a lot of hard-to-digest exoskeleton), but they may require more energy expenditure to pursue. However, even though crayfish may be easier to catch, their large claws and aggressive resistance make them harder to subdue. Trade-offs also include the relative abundance and size of each type of prey.

Behavioral ecologists use these and many other factors to predict how an animal will forage given a particular set of conditions. Their goal is not to test whether animals in fact forage optimally, but to use the expectation of optimality as a guide to organizing research and generating testable predictions. When their predictions are borne out, behavioral ecologists come closer to understanding the major factors that shape an animal's foraging behavior. When their predictions are not borne out, they have still made progress, because they know they must consider additional factors. Predictions in optimal-foraging studies are often highly quantitative; they are frequently based on direct measurements of the calories an animal must expend to secure particular food items and the calories it gains by doing so. Numerous studies of many species show that animals tend to modify their behavior to keep their overall ratio of energy intake to energy expenditure high. Their ability to do this is sometimes quite surprising. The smallmouth bass is somehow able to factor in all the relevant variables and forage in a highly efficient manner, switching between minnows

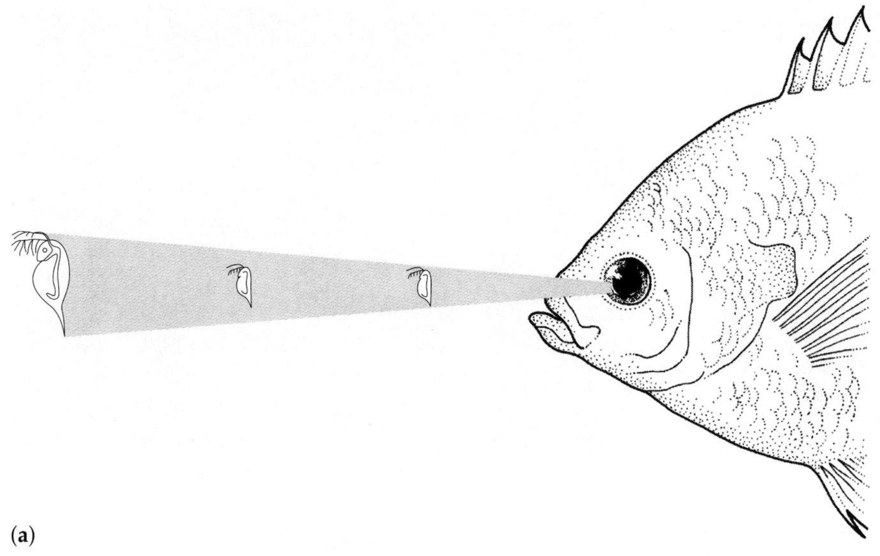

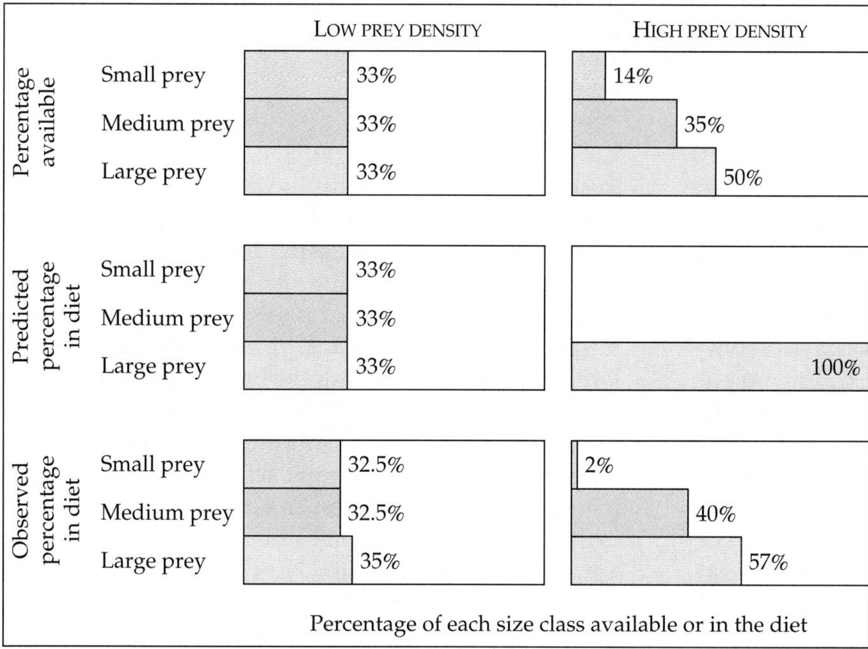

Figure 50.16
Feeding by young bluegill sunfish. (a) The fish do not feed randomly but tend to select larger *Daphnia.* This seems to be accomplished by use of "apparent size." Confronted with various potential prey, the animal will pursue the one that looks largest. Small prey (low energy yield) at the middle distance in this example may be ignored. But small prey at the closest distance may be taken because the necessary energy expenditure is smaller. More distant but larger prey will require more energy expenditure but will provide a high yield and thus may be chosen over small prey at middle and even close distances. (**b**) When prey are at low density, calculations based on optimal foraging theory predict that bluegill sunfish will not be selective, but will eat any size prey as it is encountered. At higher prey densities, the ratio of energy intake to energy expended can be maximized by concentrating only on larger prey. In the experiments described here, we see that bluegills did forage nonselectively at lower prey density. At higher prey density, they favored larger prey, though not to the extent predicted.

and crayfish as conditions change. We know little about how animals do this. Presumably much of the ability is innate, although experience undoubtedly also plays a role.

The bluegill sunfish provides an interesting example of how animals maximize the energy intake-to-expenditure ratio. These animals feed on small crustaceans called *Daphnia,* generally selecting larger prey, which supply the most energy. However, smaller prey will be selected if a larger one is too far away (Figure 50.16a). The optimal foraging approach predicts that the proportion of small to large prey eaten will also vary with the overall density of prey. At very low prey densities, bluegill sunfish should exhibit little size selectivity, because all the prey encountered are needed to meet energy requirements. At higher prey densities, it is more efficient to concentrate on larger crustaceans. In actual experiments, bluegill sunfish did become more selective at higher prey densities, though not to the extent that would theoretically maximize efficiency (Figure 50.16b). Young bluegill sunfish forage fairly efficiently, but not as close to the optimum as older individuals, who are apparently able to make more complex distinctions. It may be that younger fish judge size and distance less accurately because their vision is not yet completely developed. It is unclear, however, whether the improvement in the older fish is due solely to maturation or whether it also involves learning.

(a)

(b)

Figure 50.17
Cooperative prey capture. (**a**) This pack of African wild dogs is killing a wildebeest much larger than themselves. (**b**) This line of white pelicans is moving toward a school of fish. The cooperative behavior makes it difficult for fish to escape by swimming around the birds. Although all participants benefit from these sorts of cooperation, each individual is behaving in a manner that maximizes its own benefits.

SOCIAL INTERACTIONS

Social behavior, broadly defined, is any kind of interaction of two or more animals, usually of the same species. Since members of a population have a common niche, there is a strong potential for conflict, especially among members of species that normally maintain densities near carrying capacity. Sometimes social behavior seems to involve cooperation, as when a group carries out behavior more efficiently than is possible for a single individual (Figure 50.17). Keep in mind that even when behavior requires some cooperation and seems to be mutually beneficial to interacting individuals, as in mating behavior, each participant usually acts in a way that will maximize its benefits, even if this is at a cost to the other participant. In the remainder of this section, we will explore competitive social interactions, where this "selfish" aspect of behavior is most obvious. In a later section, we will examine altruistic ("unselfish") behavior.

Agonistic Behavior

In **agonistic behavior,** a contest involving both threatening and submissive behavior determines which competitor gains access to some resource, such as food or a mate. Sometimes the encounter involves tests of strength. More commonly, the contestants engage in threat displays that make them look large or fierce, often with exaggerated posturing and vocalizations. Eventually one individual stops threatening and ends with a submissive or appeasement display, in effect surrendering. Much of this behavior includes **ritual,** the use of symbolic activity, so that usually no serious harm occurs to either combatant (Figure 50.18). Dogs and wolves show aggression by baring their teeth; erecting their ears, tail, and fur; standing upright; and looking directly at their opponent—all of which make the animal appear large and threatening. The eventual loser, on the other hand, sleeks its fur, tucks its tail, and looks away. This appeasement display inhibits any further aggressive activity. Although in some cases animals do inflict injury, natural selection favors a strong tendency to end the contest as soon as the winner is established, because violent combat could injure the victor as well as the defeated. Any future interaction of

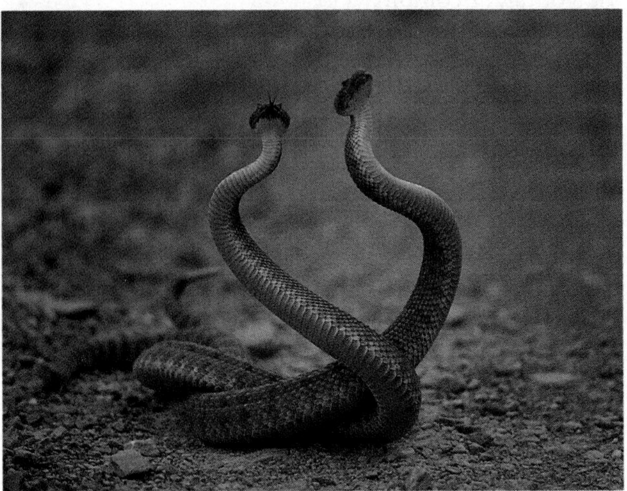

Figure 50.18
Ritual wrestling by rattlesnakes. Rattlesnakes attempt to pin each other to the ground, but they never use their deadly fangs in such combat.

the same two animals is usually settled much more quickly in favor of the original victor. Examples of ritualized agonistic behavior are widespread.

Dominance Hierarchies

If several hens who are unfamiliar with one another are put together, they respond by pecking each other and skirmishing about. Eventually, the group establishes a clear "pecking order"—a more or less linear **dominance hierarchy.** Within a group, the alpha (top-ranked) hen controls the behavior of all others, often by mere threats rather than actual pecking. The beta (second-ranked) hen similarly subdues all others except the alpha, and so on down the line to the omega, or lowest, animal. The advantage to the top-ranked bird is obvious, since it is assured of access to resources such as food. Even for lower-ranked animals, the system ensures that they do not waste energy or risk harm in futile combat.

Wolves typically function in packs (cooperation being necessary for killing large prey) in which there is a dominance hierarchy among the females. The top female controls the mating of the others. When food is generally abundant, the top female mates and allows others to do so also; when food is scarce, she allows less mating by other females, thereby making more food available for her own young. Although these examples illustrate dominance hierarchies among females, male hierarchies are also common.

Territoriality

A **territory** is an area that an individual defends, usually excluding other members of its own species. Territories are typically used for feeding, mating, rearing young, or combinations of these activities. Usually a territory is fixed in location, its size varying with the species, the territory's function, and the amount of resources available. Song sparrow pairs, for example, may have territories of about 3000 m^2, in which they carry out all activities during the several months of their breeding season. Gannets and other seabirds, in contrast, mate and nest in territories of only a few square meters or less and feed away from their territories (Figure 50.19). Bull sea lions defend small territories used only for mating, whereas red squirrels have rather large territories apparently based on feeding patterns. In many species that defend territories only during the breeding season, individuals may form social groups at other times.

Note the distinction between a territory and a home range, which is simply the area in which an animal roams about and which is often not defended. In some species, as with breeding song sparrows, territory and home range are the same; but for other species such as

Figure 50.19
Territories. Gannets nest virtually only a peck apart and defend their territories by calling and pecking at each other. This population is on Bonaventure Island, Quebec.

gannets, a territory is considerably smaller than the home range. The distinction cannot always be made clearly. Gray squirrels, for example, typically have home ranges that overlap extensively, but one individual may defend part of the range from competitors.

Territories are established and defended through agonistic behavior, and an individual that has gained a territory is often difficult to dislodge. Why do owners usually win? One cogent explanation in behavioral ecology is that a territory is worth more to an owner than to an intruder because the owner is already familiar with it. Thus, because it has more to gain from a territory, an owner is more likely to escalate a battle than is an intruder. Ownership is usually continually proclaimed; this is a primary function of most familiar bird songs, as well as the noisy bellowing of sea lions and the chatter of the red squirrel. Other animals may use scent marks or frequent patrols that warn potential invaders (Figure 50.20). Defense is usually directed only at conspecifics; a white-crowned sparrow may live within a song sparrow's territory because, as we saw in Chapter 48, a different species usually has a different niche and is less likely to be a direct competitor. Another adaptive reason for concentrating defense on conspecifics is that they are likely to mate with a territory holder's mate.

Although dominance hierarchies and territoriality evolved as a result of their advantages to individuals, such systems have important effects at the population level, because they tend to stabilize density. If resources were allocated evenly among all members of a population, the "fair share" that each individual received might not be enough to sustain anyone, leading to occasional population crashes. Dominance and terri-

(a)

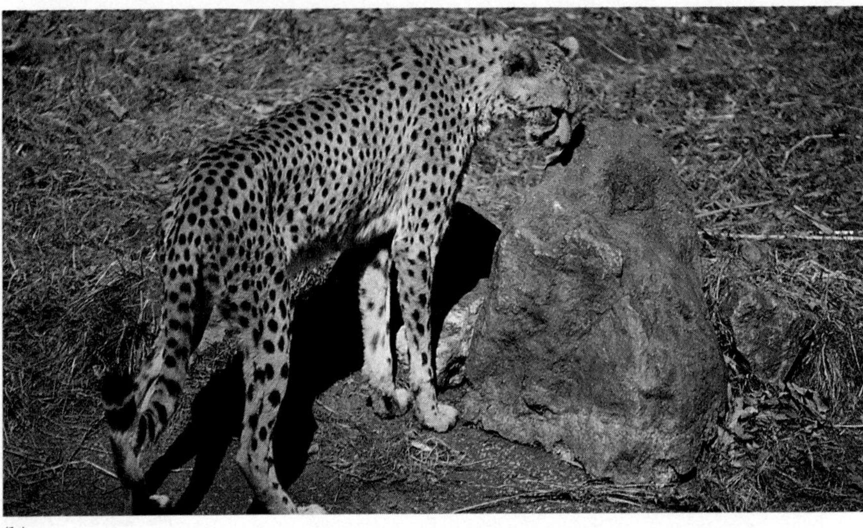

(b)

Figure 50.20
Staking out territory with chemical markers. (**a**) This male cheetah, a resident of Africa's Serengeti National Park, is spray-urinating on rocks. The odor will serve as a chemical "No Trespassing" sign to other males. (**b**) Another male cheetah sniffs a marked rock. With their keen olfactory sense, cheetahs can distinguish their own marks from those left by others.

toriality usually mean that at least some individuals receive an adequate amount of a resource. Often, in fact, if a resource such as food becomes scarce, territories expand somewhat. In addition, there are usually individuals low in the hierarchy or lacking territories who are ready to move up or step in if one of the successful individuals dies. The result is relatively stable populations from year to year.

MATING BEHAVIOR

All aspects of reproductive behavior receive extensive attention from behavioral ecologists. The reason is that researchers can often determine the number of young an individual produces, which comes very close to determining an animal's fitness. By contrast, the correlation between measurable behaviors and fitness may not be as strong for other aspects of behavioral ecology, such as optimal foraging.

Courtship

Most animals probably do not have any conscious sense of reproduction as an important function in their lives, nor do they have the kind of continuous attraction to the opposite sex found in most humans. Often there is a strong tendency for an animal to view any organism of the same species as a threatening competitor to be driven off, if possible. Even within many social species, individuals usually avoid contact with one another. How, then, is mating accomplished? In many animals, potential partners must go through a complex courtship interaction, unique to the species, before mating. This complex behavior often consists of a series of fixed-action patterns, each triggered by some action of the other partner and initiating, in turn, the other partner's next required behavior. This sequence of events assures each animal not only that the other is not a threat, but also that the other animal's species, sex, and physiological condition are all correct.

In some species, courtship also plays an important role in allowing one or both sexes to choose a mate from a number of candidates. When individuals choose mates, females nearly always show greater discrimination than males, because in most species they have more parental investment in each offspring. **Parental investment** is defined as the time and resources that an individual must expend to produce an offspring. Eggs are usually much larger and much more costly to produce than sperm. Placental mammals have less disparity in gamete size than most other animals, but female placental mammals invest considerable time and energy in carrying the young before birth. In most species, females are very choosy, since picking a poor-quality male can be a costly error. By contrast, males of most animal species mate with as many females as possible. Males usually compete with one another for mates, sometimes by trying to impress females. Thus, males of most species perform more intense courtship displays than females; in many species, in fact, it is only males who court. Secondary sex characteristics are usually much more pronounced in males; bright plumage in birds and antlers in deer are examples. Differential reproductive success that results from advantages in attracting or competing for mates is called *sexual selection,* which was discussed in more detail in Chapter 21.

In some animals, competition among individuals of the same sex (usually males) almost entirely determines which animals of that sex will mate. But in other species, individuals, usually females, actively choose among potential mates. There are two ultimate bases for such choices. First, if the other sex provides parental care, it is advantageous to choose as competent a mate as possible. For example, male common terns (a bird species related to gulls) carry fish and display them to potential mates as part of their mating ritual. Eventually, a male may begin to feed fish to a female. This behavior may be a good proximate indicator of a male's ability to provide food for tern chicks. In some animal species, females prefer males who are capable of the most extreme and energetic courtship displays or who have the most extreme secondary sex characteristics, such as long tails. Perhaps these characteristics are proximate indicators that the males are vigorous and in good health.

The second ultimate basis for mate choice is genetic quality. This is most likely to be important when males provide no parental care and sperm are their only contribution to offspring. In a number of bird and insect species, males display communally in a small area called a lek. Females visit the lek and choose among the displaying males. After mating occurs, there is no further contact between females and males. In terms of a female's fitness, it is beneficial for her to mate with a male that has favorable characteristics, since the success of her offspring will depend on both her genes and those of her mate. The proximate basis for this sort of choice is also likely to be a preference for males that court the most vigorously and are adorned with the most extreme secondary sex characteristics. An especially important factor in male genetic quality is the ability to counter pathogens and parasites (see Figure 21.12). Even when the quality of a male's parental care is a factor in female choice, genetic quality may also be important.

It is often very difficult to determine which of these bases for female choice is more important, and in some cases it is even difficult to determine if differential mating success among males is due to male-male competition, female choice, or both. One of the best known examples of ritualized courtship is that of the three-spined stickleback fish mentioned earlier (Figure 50.21). Although the courtship sequence is based on releasers and FAPs, it does not necessarily proceed

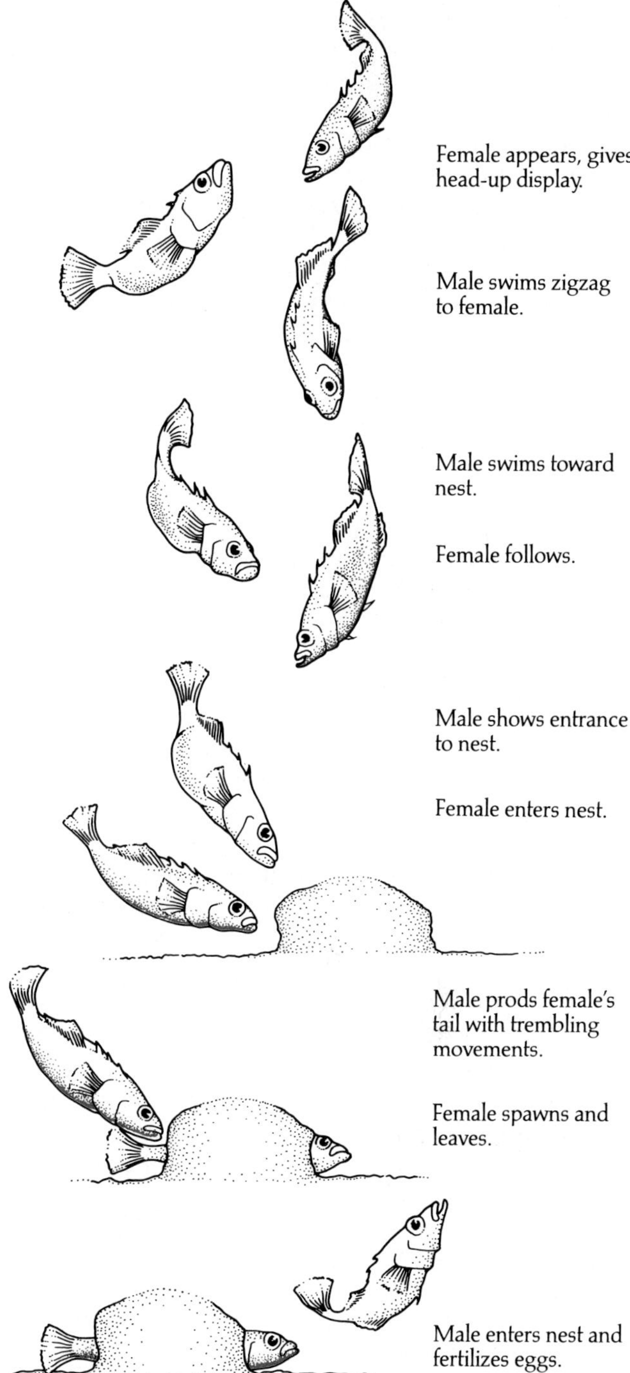
Female appears, gives head-up display.
Male swims zigzag to female.
Male swims toward nest.
Female follows.
Male shows entrance to nest.
Female enters nest.
Male prods female's tail with trembling movements.
Female spawns and leaves.
Male enters nest and fertilizes eggs.

Figure 50.21
Courtship behavior in the three-spined stickleback. Males are strongly territorial, defending an area in which they have built a nest. If a gravid (egg-carrying) female approaches, her swollen belly releases zigzag swimming in the male. This entices her to swim closer, which in turn stimulates him to swim to the nest and stick his snout inside. This action stimulates the female to wriggle into the tunnel. The male then nuzzles her tail, which stimulates her to spawn, after which she swims out the other end of the nest. The male then enters and deposits sperm on the eggs, after which he immediately and quite aggressively drives the female out of the area.

smoothly or quickly. Often a female will begin to follow a male and then hesitate. She may cut off the sequence at any point, sometimes because she is simultaneously being courted by other nearby males. The final result is successful mating by some but not necessarily all males. This is a case in which female choice clearly occurs. In an ultimate sense, the choice is probably based completely or largely on the quality of a male's parental care, because only males provide care in sticklebacks. Notice that "choice" in this context does not imply conscious decision making on the part of these animals. If a certain behavior enhances an individual's reproductive success, the process of natural selection can account for the prevalence of that behavior in a population.

Ritualized acts, whether of courtship or agonistic behavior, probably evolved from actions whose meaning was at one time more direct. We can see an interesting example of this process in the balloon flies of the family Empididae. In a few species, males spin oval balloons of silk and carry them while flying in a swarm, which is approached by females seeking mates. After a female chooses a male, she accepts his balloon, and the two fly off to copulate. A clue about how this ritual originated comes from examining the behavior of other species in the family. In some predatory species, the male brings a dead insect for the female to eat while he mates with her; perhaps this keeps her from attacking him. In other species, the dead insect is carried inside a silk balloon like the one just described, a variation that may have evolved because silk was helpful in subduing the insect or because it made the male's gift look larger. In balloon flies, which are not predators but instead eat nectar, the ritual has apparently evolved into merely bringing something formerly associated with food. As one entomologist remarked, it is as though over the course of evolutionary time, a suitor wooed a lady first with diamonds, then with a box containing diamonds, and finally with nothing but the box itself.

Mating Systems

The mating relationship between males and females varies a great deal among species. In many species, mating is **promiscuous,** with no strong pair-bonds or lasting relationships. In those species where the mates remain together for a longer period, the relationship may be **monogamous** (one male mating with one female) or **polygamous** (an individual of one sex mating with several of the other). Polygamous relationships most often involve a single male and many females (**polygyny**), which can be explained in terms of a difference in parental investment, as discussed earlier. However, there are also species in which a single female mates with several males (**polyandry**).

The needs of the young are an important ultimate factor in the evolution of mating systems. Most newly hatched birds are unable to care for themselves, requiring a large, continuous food supply that one parent may not be able to provide. In such a system, a male may ultimately leave more viable offspring by helping a single mate than by going off to seek more mates. This may explain why most birds are monogamous. In birds with young that feed and care for themselves almost immediately after hatching, there is less need for parents to stay together. Males of these species can maximize their reproductive success by seeking other mates, and polygyny is relatively common in such birds. In the case of mammals, the lactating female is often the only food source for the young. Males usually play no role, or, if they protect the females and young, they typically take care of many at once in a harem.

Another factor influencing mating systems and parental care is *certainty of paternity.* Young born or eggs laid by a female definitely contain the female's genes. But even within a normally monogamous relationship, these young may have been fathered by a male other than the female's usual mate. For example, this was the case in a population of red-winged blackbirds studied by researchers who used DNA fingerprinting to determine the paternity of chicks (see the Methods Box). The certainty of paternity is relatively low in most species with internal fertilization because the acts of mating and birth (or mating and egg laying) are separated over time. This could explain why parental care is exclusively by males in very few bird or mammal species. However, certainty of paternity is much higher when egg laying and mating occur together, as in external fertilization. This may explain why parental care in fishes and amphibians, when it occurs at all, is at least as likely to be by males as by females. The trend occurs even within these classes: Male parental care occurs in only 2 (7%) out of 28 fish and amphibian families with internal fertilization but in 61 (69%) out of 89 families with external fertilization. In fish, even when parental care is given exclusively by males, the mating system is often polygynous, with several females laying eggs in a nest tended by one male. It is important to point out again that when behavioral ecologists use terms such as "certainty of paternity," they do not mean that animals are aware of those factors when they behave a certain way. Parental behavior correlated with certainty of paternity exists because it has been reinforced over generations by natural selection.

COMMUNICATION

Much of what we have discussed under social interactions and mating behaviors involves instances in

Methods: Using Molecular Biology to Answer Behavioral Questions

Molecular biology has had an impact on every field of biology, including behavioral ecology. Here, we examine the specific case of a research project that applied DNA fingerprinting to determine the paternity of chicks born in a population of red-winged blackbirds (Figure a). The DNA fingerprinting was based on restriction fragments (RFLPs), as described in Chapter 19.

In most bird species, mating relationships are fixed, typically involving one male and one to several females that are joint inhabitants of the male's territory. By counting the number of young in the nest or nests within a male's territory, it should be possible to compare the relative reproductive success (fitness) of different males. Behavioral ecologists have sometimes used this approach to compare the effects of behavioral variations on the fitness of different birds in a population. But there is a problem with this method of measuring how many offspring each male sires. Females may occasionally mate with males from neighboring territories. Although a male generally guards his mate (or mates) during her fertile period, there is evidence that this vigilance is imperfect at preventing the females from mating with other males. In the 1970s, biologists at the U. S. Fish and Wildlife Service tried to control populations of red-winged blackbirds, which sometimes threaten agricultural crops. Rather than kill the birds, the biologists performed vasectomies on the males. Surprisingly, the researchers found that 50% or more of the eggs in each vasectomized male's territory had been fertilized, presumably by neighboring males who had not had the procedure. But perhaps vasectomies altered male behavior, increasing the frequency of extrapair copulations (EPCs). Thus, there was still the possibility that EPCs were rare for females mated to normal, untreated males.

The problem was not resolved until 1990, when a team of researchers published a DNA fingerprinting analysis of paternity in a population of redwings. Analysis of DNA from blood samples of nestlings and adults revealed that 45% of the nests had at least one offspring fathered by a male other than the one claiming that territory. The DNA fingerprinting even allowed the researchers to identify which males had ventured out of their territories and fathered offspring in the territories of neighbors. In Figure b, the green areas represent the territories of individual males. The fractions indicate the number of offspring sired by a male over the total number of young in his territory. The arrows show the direction of EPCs and the number of "extra" young sired by each male. Thus, male X fathered all three of the young in his territory, as well as three offspring in the neighboring territories of males Y and Z. A relatively large number of EPCs occur in the territories of males that have a larger-than-average number of females. Perhaps this is because a male with multiple mates cannot guard them all effectively.

In the past few years, DNA fingerprinting studies have shown that EPCs are common in many bird species that were previously assumed to be faithful in their pair-bonding.

(a)

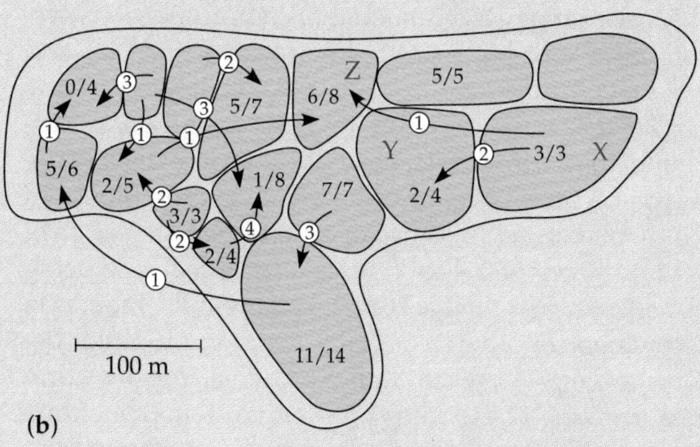

(b)

which animals intentionally, although not necessarily consciously, transmit information by special behaviors called *displays*. The intentional transmission of information between individuals is the usual definition of communication in behavioral ecology. Singing by male birds is, in effect, a display that transmits the information, "This is my territory. Keep out!" This is almost certainly an important message of singing; if we play the tape-recorded songs of another male in a male bird's territory, he becomes highly agitated, approaching and sometimes even attacking the speaker. Someone else has not only ignored his warning but has claimed the territory for himself. This simple experiment is so infallible that some bird watchers routinely use it to find and see secretive birds that would otherwise stay hidden. The playback procedure makes another important point. We cannot get into an animal's brain to determine whether it has received a message sent by another individual. How, then, do we know when communication has occurred? We usually say that communication has occurred when an act by a "sender" produces a detectable change in the behavior of another individual, the "receiver." Bird song is communication because it produces a response.

Most ethologists assumed that communications systems evolve in ways that maximize the quantity and accuracy of information. Behavioral ecologists have a very different view. Following basic evolutionary reasoning, they argue that a behavior such as a display develops because of its positive effects on the fitness of the displayer, not because of its effects on other individuals (receivers). Individuals often benefit by being dishonest rather than accurate, and animals frequently deceive one another, especially in communication between different species. Male and female *Photinus* fireflies find each other and mate when females give a characteristic pattern of light flashes in response to flashes given by males. However, females of another firefly genus, *Photurus*, respond to these male *Photinus* flashes with the same flashing pattern. A male *Photinus* receives the incorrect message that a female *Photurus* is a suitable mate, and when he finds her she does not mate with him, but kills and eats him. Such cases of mimicry, in which communication is adaptive for the sender but maladaptive for the receiver, are widespread.

Deceit also occurs within species. In some mammals, after a dominant male assumes control of a social group, he kills young that are born too soon to be his offspring. Without dependent young, the females ovulate sooner, allowing the new dominant male to father their offspring. In an Asian monkey known as the hanuman langur, females in the early stages of pregnancy actively solicit copulations from new males. These females then give birth only shortly before young fathered by these males would appear. This may deceive a male into treating the female's young as if they were his own, when in fact they are not. Again, none of this implies that langurs have consciously shaped their behavior to maximize fitness. Instead, it simply implies that females with a tendency to solicit copulations from new males have had a selective advantage over other females without such a tendency.

Another important evolutionary consideration in communication is the sensory mode used to transmit information. Animals transmit information with visual, auditory, chemical (olfactory), tactile, and electrical signals. Which mode is used to transmit information is closely related to an animal's basic lifestyle. Most mammals are nocturnal, which makes visual displays relatively ineffective. But olfactory and auditory signals work as well in the dark as in the light, and most mammalian species stress these signals (see Figure 50.20). Birds, by contrast, are mostly diurnal and use mostly visual and auditory signals. They almost never use olfactory signals, probably because they can fly faster than chemical signals can travel. (It is hard to imagine a system in which it would be adaptive for a messenger to arrive before its message.) Unlike most mammals, humans are diurnal and stress the same visual and auditory communication used by birds. Therefore, we can detect the songs and bright colors that birds use to communicate with each other. This may explain why bird watching is so popular. If humans had the well developed olfactory abilities of most mammals and could detect their rich world of chemical cues, mammal sniffing might be as popular as bird watching.

Animals that communicate by odors emit chemical signals called **pheromones.** These are especially common among mammals and insects and often relate to reproductive behavior (see Chapter 41). Female silkworm moths, for example, emit a pheromone that can attract males from several kilometers away. Once the moths are together, pheromones are also important as releasers for specific courtship behaviors. Another example is the familiar trailing behavior of ants, in which scouts release scents that guide other ants to the food.

One of the most complex communication systems—certainly among invertebrates—is that of honeybees. For maximum foraging efficiency, workers must convey to one another the location of good food sources, which may change frequently as various flowers bloom or new patches are located. How do honeybees communicate? The problem was studied in the 1940s by Austrian zoologist Karl von Frisch, who shared the Nobel Prize with Lorenz and Tinbergen. By carefully watching individual bees when they returned to the vertical face of an open hive, von Frisch discovered that other bees gather around the returned one. The returned bee then goes through a "dance." Based on his experiments, von Frisch proposed that the dance indicates the location of food. If the source is relatively close—less than about 50 m—the bee moves rapidly sideways in tight circles (the "round dance"), causing the others to become excited (Figure 50.22a). Often the

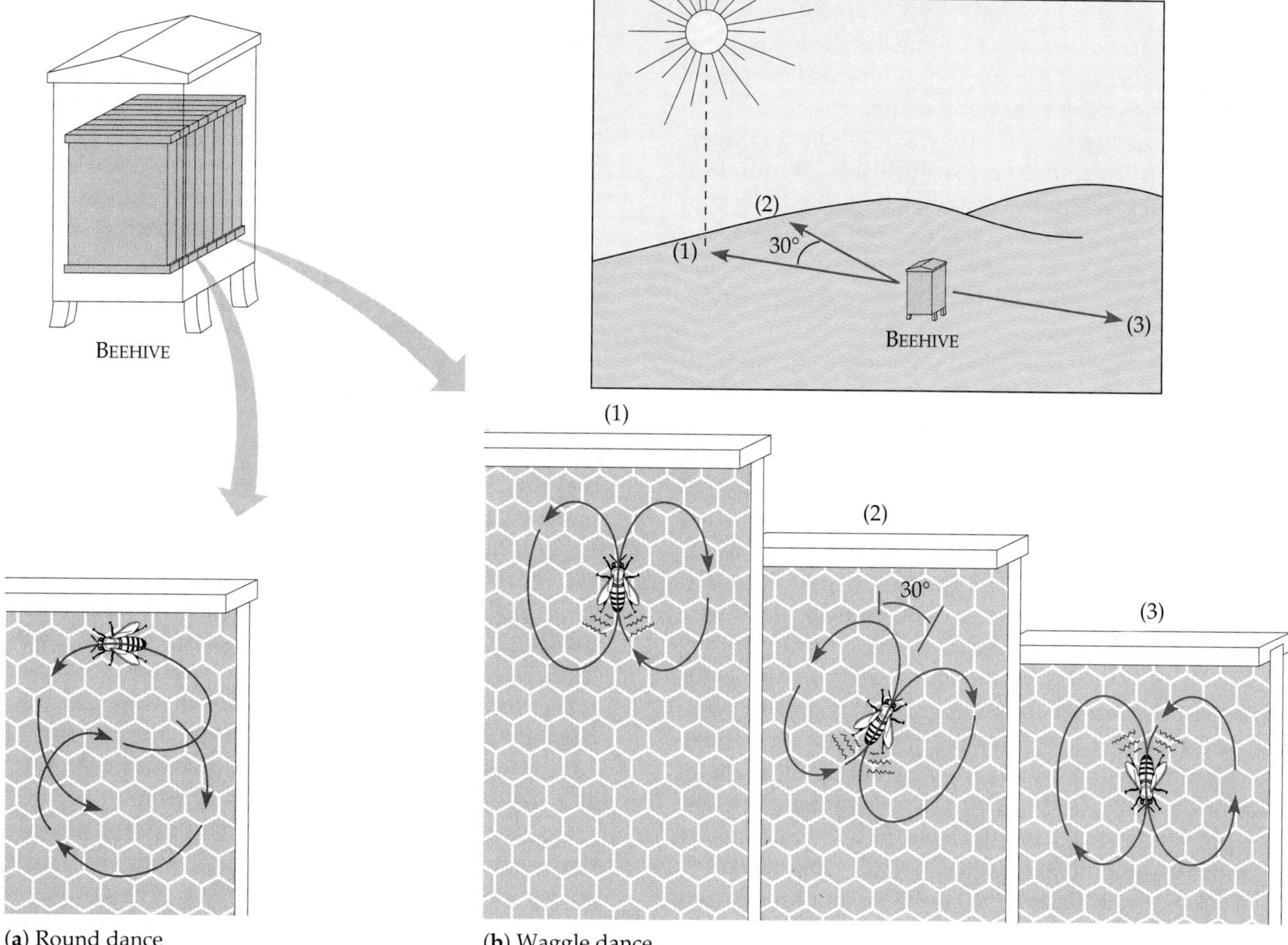

(a) Round dance

(b) Waggle dance

Figure 50.22
Communication in bees: one hypothesis. (**a**) The round dance indicates that food is near. (**b**) The waggle dance is performed when food is distant. According to von Frisch's hypothesis, the waggle dance indicates both distance and direction. Distance is indicated by the speed at which a bee wags its abdomen during the dance. Direction is indicated by the angle (in relation to the vertical surface of the hive) of the straight run that forms part of the dance itself. (1) For instance, if the straight run is directly upward, this signals that food is in the same direction as the sun. (2) If the angle is 30° to the right of vertical, the food is 30° to the right of the sun. (3) If the straight run is directly downward, the food is in the direction opposite the sun.

dancer regurgitates some nectar that the others taste. The workers then leave the hive and begin foraging nearby. Although the round dance does not indicate direction, tasting the nectar likely helps the bees identify a scent to fly toward.

If the food is farther away, more information is needed. A worker returning from a longer distance does a "waggle dance": a half-circle swing in one direction, followed by a straight run and then a half-circle swing in the other direction (see Figure 50.22b). According to von Frisch, this dance indicates both direction and distance. The angle of the straight run in relation to the vertical surface of the open hive is the same as the horizontal angle of the food in relation to the sun. If the bee runs at a 30° angle to the left of vertical, the other workers will fly 30° to the left of the horizontal direction of the sun. If the dancer runs directly upward, the others will fly directly toward the sun, and so forth. Distance to the food is indicated by variation in the speed at which a bee wags its abdomen during the straight run. A high rate (40 wags/sec) translates as "about 100 m away," whereas a slow rate (18 wags/sec) means "about 1000 m away," with intermediate rates corresponding to intermediate distances. Again, food is typically regurgitated, so that when a bee leaves to forage, it already "knows" the type of food to seek, its distance, and its direction. Some researchers question the evidence for von Frisch's model, and are exploring alternative hypotheses.

SELFISH VERSUS ALTRUISTIC BEHAVIOR

Most behaviors are selfish, meaning that they benefit the individual at the expense of others. A bird that establishes a territory deprives other individuals of one, and if there is not enough habitat, these other individuals may be unable to breed. Even in species in which individuals do not engage in agonistic behavior, most adaptations that benefit one individual will indirectly harm others. Superior foraging ability by one individual may leave less food for others. It is easy to understand the pervasive nature of selfishness if natural selection shapes behavior. Behavior that maximizes an individual's reproductive success will be favored by selection, regardless of how much damage such behavior does to another individual, a local population, or even an entire species.

How, then, can we explain occasional examples of animals behaving altruistically—that is, unselfishly? Animals do sometimes take action that reduces their own personal welfare but benefits others. Consider the example of Belding's ground squirrels, which live in some mountainous regions of the western United States and are vulnerable to predators such as coyotes and hawks. If a predator approaches, one of the squirrels often gives a high-pitched alarm call. This alerts unaware individuals, who then retreat to their burrows. Careful observations have confirmed that the conspicuous alarm behavior increases the risk of being killed, because it identifies the caller's location. How can a squirrel enhance its fitness by aiding other members of the population, which are apt to be its closest competitors? Fitness would seemingly be maximized by quietly allowing a predator to take a competitor.

A classic example of altruistic behavior occurs in bee societies, in which the workers are sterile. The workers themselves will never reproduce, but they labor on behalf of a single fertile queen. Furthermore, the workers sting intruders, a behavior that helps defend the hive but results in the death of the individual. Another interesting case of altruism involves red-cockaded woodpeckers, an endangered species of the southeastern United States. Within a territory, a single pair nests and produces young in a cavity high in an old pine tree. With them are two to four younger birds that do not breed but that assist in practically all other aspects of reproduction. They help defend the territory, dig out the cavity, and even incubate the eggs and feed the young.

How can altruistic behavior evolve if it does not enhance—and, in fact, may even reduce—the reproductive success of the self-sacrificing individuals? Natural selection favors anatomical, physiological, and *behavioral* traits that increase reproductive success, which in turn propagates the genes responsible for those traits. When parents sacrifice their own personal well-being to produce and aid offspring, this actually increases the fitness of the parents, because it maximizes their genetic representation in the population. But what about helping other close relatives? Like parents and offspring, siblings have half of their genes in common. So selection might also favor helping one's parents produce more siblings or even helping siblings directly. In the 1960s, W. D. Hamilton, then a graduate student in England, first realized that selection could result in animals increasing their genetic representation in the next generation by "altruistically" helping close relatives other than their own offspring. This realization led to the concept of **inclusive fitness,** which describes the total effect an individual has on proliferating its genes by producing its own offspring *and* by providing aid that allows other close relatives to produce their offspring.

An important quantitative measure of inclusive fitness is the *coefficient of relatedness,* which is the proportion of genes that are identical in two individuals because of common ancestors. Figure 50.23 explains how to calculate the coefficient, but intuitively you know that it must be higher for siblings than for cousins (0.5 versus 0.125) and higher for parents and offspring than for grandparents and grandchildren (0.5 versus 0.25). We expect that the higher the coefficient of relatedness, the more likely an individual is to aid a relative: The individual's altruism can result in more genes identical to its own in the next generation if it aids a sibling rather than a cousin. This mechanism of increasing inclusive fitness is called **kin selection.** The contribution of kin selection to inclusive fitness varies among species. Kin selection may be rare to nonexistent in species that are never social or that disperse so widely that individuals never live near close relatives. In these species, an animal's inclusive fitness is essentially the same as its individual fitness—its fitness based on production of only its own offspring.

British geneticist J. B. S. Haldane anticipated the concepts of inclusive fitness and kin selection by jokingly saying that he would lay down his life for two brothers or eight cousins. In today's terms, we would say that he would do this because either two brothers or eight cousins would result in as much representation of Haldane's genes as would two of his own offspring. But this assumes equal reproductive potential for all individuals. We assume that Haldane was as likely to have as many children as each of his individual brothers and cousins. Often, however, reproductive potentials differ. A sterile individual should theoretically sacrifice its life for one fertile brother or even one cousin, otherwise its inclusive fitness will be zero. Even differences in reproductive potentials that are not as great as between sterile and fertile individuals can still contribute to kin selection. For example, territories might be in limited supply, and one animal might have one while his brother does not. The homeless animal might have a greater increase in its inclusive fitness by helping its brother than by trying to win a territory of its own, if the second option has only a small chance of success. In trying to predict whether an individual animal will aid relatives, behavioral ecologists have derived a formula that combines coefficients of relatedness, costs to the altruist, and benefits to the recipient

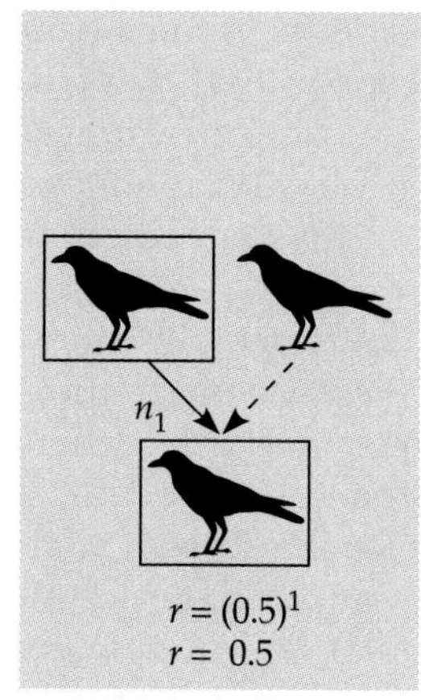

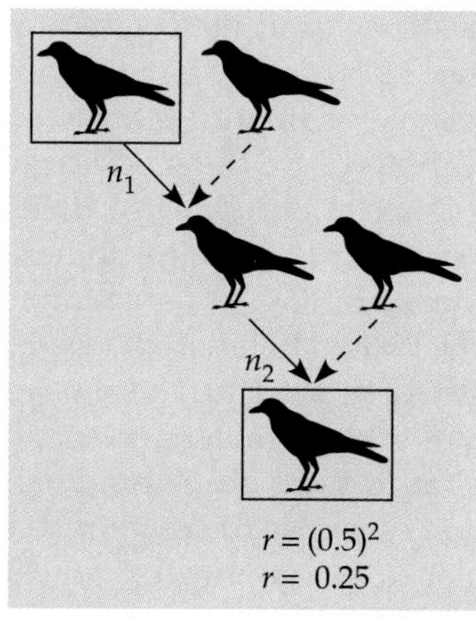

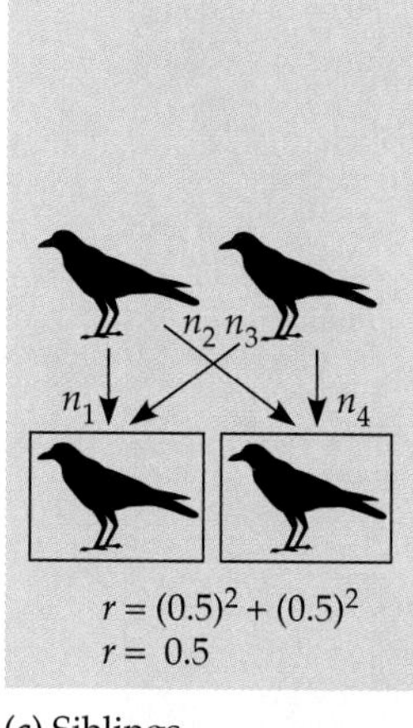

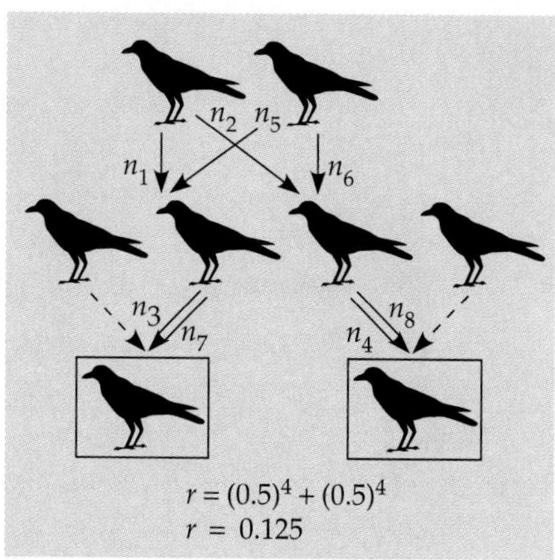

(**a**) Parent and offspring (**b**) Grandparent and grandchild (**c**) Siblings (**d**) Full cousins

Figure 50.23
Calculating the coefficient of relatedness. In each of these four examples, we determine the coefficient of relatedness between the two crows identified in each case by boxes. The coefficient of relatedness, *r*, is the proportion of one individual's entire genome that is identical by descent to another individual's genome. Stated in more practical terms, it is the probability that any particular allele is identical in two individuals because it was inherited from common ancestors. The solid arrows indicate the links *(n)* in family history that are relevant for calculating *r* between the boxed birds. The dashed arrows indicate the links we can ignore in our calculations. The key point to remember is that at each generational link, meiosis results in a 0.5 probability that an allele present in a parent will be passed on to a particular offspring. The coefficient of relatedness is calculated as 0.5^n, where *n* is the number of generational links. Thus, in (**a**), the coefficient of relatedness between the boxed individuals (parent and offspring) is 0.5 (0.5^n, where $n = 1$ because there is only one generational link). If we had decided instead to calculate *r* between this offspring and the other parent, the answer would also be 0.5. In (**b**), where there are two generational links, we calculate an *r* of 0.25 between grandparent and grandchild. In (**c**) and (**d**), there is more than one pathway possible for the indicated individuals to inherit alleles from common ancestors. In these cases, we must add all 0.5^n values to calculate the overall coefficient of relatedness.

of aid. (Animals, of course, do not work out the kin selection formula for themselves or calculate behavior that increases their inclusive fitness.)

If kin selection explains altruism, then the examples of unselfish behavior we observe should involve close relatives. This expectation is met, but often in complex ways. As in most mammals, female Belding's squirrels settle close to their site of birth, while males settle at distant sites. Thus, only females are likely to live near close relatives, and nearly all alarm calls are given by females (Figure 50.24). However, if all of a female's close relatives are dead, she rarely gives alarm calls. In the case of worker bees, the individuals are sterile, and anything they do to help the entire hive benefits the only permanent member who is reproductively active, the queen, who is their mother. In the red-cockaded woodpecker and many other bird species that sometimes have nest helpers (including the familiar American crow), the helpers are almost always older offspring of the pair being helped or siblings of one pair member. Also, helpers are usually individuals who have not been able to establish a territory because the entire suitable habitat is occupied. For these nest assistants, helping a close relative may not be much of a sacrifice. And sometimes these helpers inherit the territory when their relatives die, which may be the surest way to secure a territory in a crowded landscape. This could mean that helping in some cases evolved primarily from a direct advantage to the individual, rather than because of kin selection.

Some animals occasionally behave altruistically toward others who are not relatives. A baboon may help an unrelated companion in a fight, or a wolf may offer food to another wolf even though they share no kinship. Such behavior can be adaptive if the aided individual returns the favor in the future. This sort of exchange of aid is called **reciprocal altruism** and is commonly invoked to explain altruism in humans. Reciprocal altruism is rare in other animals; it is limited largely to species with social groups stable enough that individuals have many chances to exchange aid. It is likely that all behavior that seems altruistic actually increases fitness in some way. Thus, some behavioral ecologists argue that *true* altruism never really occurs, except, perhaps, in humans.

ANIMAL COGNITION

One dictionary defines *cognition* as "knowing, including both awareness and judgment." A simple but pro-

(a)

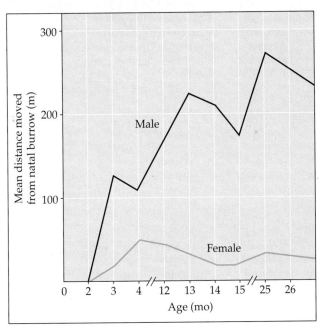

(b)

Figure 50.24
Altruistic behavior and a male-female difference in its expression. (a) By sounding an alarm call, this ground squirrel warns others of danger, such as an approaching predator. Nearly all alarm calls are given by females. **(b)** This graph helps explain the male-female difference in altruistic behavior of ground squirrels. After they are weaned, males disperse much farther from their birthplaces than do females. Therefore, females are more likely to live near close relatives, and warning these relatives increases the inclusive fitness of the altruist.

found question is whether nonhuman animals are cognitive beings. Are other animals consciously *aware* of themselves and of the world around them? Do they "feel" pain or pleasure or sadness in the same way as we do? As yet, we have no way of answering these questions or of really getting inside the mind of another creature. We have seen, of course, that other animals may sometimes behave more like programmed computers, and they certainly do not have the ability to integrate information (to "think") to the same extent as humans do. But is this a matter of degree—a continuum of abilities—or are humans fundamentally different in some behavioral respect? Because of the difficulty of answering these questions, most researchers have taken a mechanistic approach referred to as **behaviorism,** describing behavior in terms of stimulus/response or, more recently, in computer terminology ("programmed," for example). However, many people who have spent a lot of time with pets or wild animals find it difficult to think of these organisms as merely sophisticated robots.

Donald Griffin of Princeton University has been the foremost recent proponent of **cognitive ethology,** a view that sees conscious thinking as an inherent and routine part of the behavior of other animals. He argues that if other animals behave in ways we associate with conscious processing in ourselves, perhaps it makes sense to assume that there is the same underlying awareness. In her famous field studies, Jane Goodall has documented cognitive decision making in chimpanzees. Griffin suggests that such abilities may extend to many nonprimate branches of the phylogenetic tree. This view sees cognitive ability arising through the normal process of natural selection, and, like many other major animal functions, forming a phylogenetic continuum that extends far back in evolutionary history. James Gould, also at Princeton, has reported that bees can form and use "mental maps" of their foraging areas. As he notes, this kind of cognitive function is usually considered to be a form of thinking. Ultimately, answers to questions about animal cognition may profoundly affect how we interact with other animals, and how we view ourselves as well.

HUMAN SOCIOBIOLOGY

In 1975, E. O. Wilson of Harvard University published a text entitled *Sociobiology,* which describes social systems in various animal species. The book's thesis was that social behavior has an evolutionary basis; that is, that behavioral characteristics, like specific anatomical and physiological traits, exist because they are expressions of genes that have been perpetuated by natural selection. This is, of course, the basic premise of behavioral ecology, and the origin of this modern approach is due in large part to Wilson's book. In the last chapter of *Sociobiology,* Wilson speculates on the evolutionary basis of certain kinds of social behavior in humans. The book rekindled the nature-versus-nurture controversy, and the debate over sociobiology remains heated almost two decades later.

Let's examine the debate in the context of a specific example: avoidance of incest. Such avoidance is adaptive because inbreeding may increase the frequency of certain kinds of genetic disorders. Many birds and mammals clearly avoid incest. Similarly, nearly all human cultures have laws or taboos forbidding sexual relations or marriage between brother and sister. Is there an innate aversion to incest that we share with other species, or do we acquire this behavior as part of our socialization? Someone favoring the "nurture" side of the debate might argue that if this behavior were innate, cultural taboos would be superfluous. According to this argument, avoidance of incest is a learned behavior, and the social stigma attached to incest may be based on experience; people who break the taboo are more likely to have defective children. Some sociobiologists, on the other hand, would argue that the occurrence of a specific behavior in diverse cultures is evidence that the behavior has an innate component. According to this argument, incest taboos are simply proximate mechanisms that reinforce a behavior that ultimately evolved because of its effect on fitness.

Some sociobiologists have cited the experience of the Israeli communal societies known as kibbutzim as evidence that innate repulsion of humans to incest is strong enough to counter cultural factors. From the time of birth, kibbutz children used to spend most of their time in large child-care centers, which gave them the equivalent of a sibling relationship with all other children in the kibbutz. Early in the kibbutz movement, parents encouraged their children to marry within their own kibbutz. However, according to a study of more than 5000 people, such marriages were extremely rare, in spite of the wishes of parents and kibbutz leaders. Apparently, people who spend nearly all their time together as children have little sexual attraction for one another as adults. (In most situations, of course, individuals who cohabit throughout childhood are true siblings or other close relatives.) A sociobiologist might argue from this example that the human resistance to incest—in this case, with "artificial" siblings—overrides cultural pressure encouraging it.

Some sociobiologists, including Wilson, envision cultural and genetic components of social behavior as linked in a cycle of reinforcement. If most members of a society share an innate avoidance to incest, the aversion is likely to become formalized in laws and taboos. The cultural stigma, in turn, acts as an environmental factor in natural selection, amplifying the evolutionary component of the behavior. In the case of incest taboos, society may shun or imprison offenders, thus lowering their reproductive fitness. According to this view, genes and culture are integrated in human nature.

The spectrum of possible social behaviors may be circumscribed by our genetic potential, but this is very different from saying that genes are rigid determinants of behavior. Sociobiology does not reduce us to robots stamped out of a common, rigid genetic mold. Individuals vary extensively in anatomical features, and we should expect inherent variations in behavior as well. Furthermore, though we are locked into our genotypes, our nervous systems are not "hard wired." Environment intervenes in the pathway from genotype to phenotype for physical traits, and even more, in general, for behavioral traits. And human behavior is probably more plastic than that of any other animal. Over our recent evolutionary history, we humans have built up structured societies with governments, laws, cultural values, and religions that define what is acceptable behavior and what is not, even when unacceptable behavior might enhance an individual's Darwinian fitness. Perhaps it is our social and cultural institutions that make us truly unique and that provide the only feature in which there is no continuum between humans and other animals (Figure 50.25).

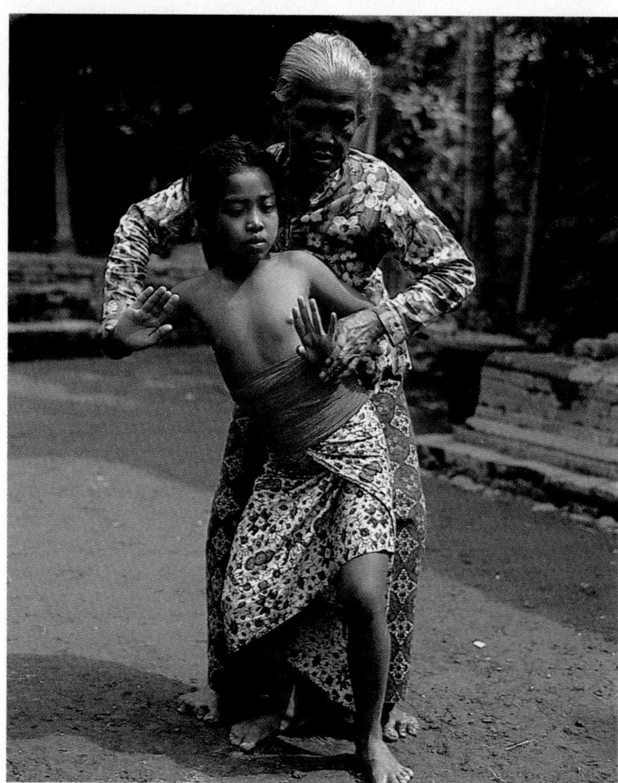

Figure 50.25
Both genes and culture build human nature. Teaching of the younger generation by the older is one of the basic ways in which all cultures are transmitted. Sociobiologists see tutoring as an innate tendency with adaptive value that has evolved in the human species.

* * *

The many facets of biology are reflected in the study of behavior; it is an intersection of biochemistry, genetics, physiology, evolutionary theory, and ecology. The study of behavior, especially social behavior, also connects biology to the social sciences and humanities. Through the study of life in general, it is inevitable that we will learn more about ourselves.

STUDY OUTLINE

1. Most animal behavior reflects some combinations of innate and learned components.

The Evolutionary Logic of Behavioral Ecology (p. 1159)

1. Behavioral ecology is based on the expectation that animals behave in ways that increase their Darwinian fitness (reproductive success). This approach lends itself to the hypothetico-deductive method.

Proximate Versus Ultimate Causation (pp. 1160–1162)

1. Questions about the stimuli that provoke behaviors and about the actual mechanisms of response address the proximate causes of animal behavior.
2. Evolutionary explanations in terms of why natural selection may have favored a particular behavior address the ultimate causes of animal behavior.

Innate Components of Behavior (pp. 1162–1165)

1. A fixed-action pattern (FAP) is a highly stereotyped, innate behavior that continues to completion after initiation by an external sign stimulus. Releasers are specific sign stimuli that function as communication signals between individuals of the same species.
2. Sign stimuli tend to be based on simple cues associated with the relevant object or activity.

Learning and Behavior (pp. 1165–1170)

1. Behavior can be modified by learning in various ways. However, some apparent learning is due mostly to inherent maturation.
2. Habituation is a simple kind of learning involving loss of sensitivity to unimportant stimuli.
3. Imprinting, first described by Konrad Lorenz with newly hatched geese, is a type of learned behavior with a significant innate component, acquired during a limited critical period.
4. Studies of song development in birds have revealed an interplay between the innate and learned components of imprinting.
5. Associative learning involves linking one stimulus with another, as demonstrated by classical conditioning.
6. Operant conditioning, or trial-and-error learning, is a type of associative learning exhibited by many animals.
7. In observational learning, animals copy the behavior of other individuals in the population.
8. Insight learning involves the ability to reason by correctly performing a task on the first attempt in a situation in which the animal has had no earlier experience.

Behavioral Rhythms (pp. 1170–1171)

1. Various daily behaviors are governed by endogenous clocks, which in turn require exogenous cues to keep the behavior properly timed with the real world.
2. Circannual behaviors, such as breeding and hibernation, seem to be controlled primarily by physiological and hormonal changes influenced by such exogenous factors as day length.

Movement (pp. 1171–1174)

1. Kinesis is a random movement displaying a stimulus-specific change in activity rate, whereas taxis is an automatic movement toward or away from a specific stimulus.
2. The most sophisticated mechanisms of orientation and navigation are used by migrating animals. Some species of migratory birds are guided by the sun or stars. Others may be able to use the Earth's magnetic field.

Foraging Behavior (pp. 1174–1175)

1. The large array of animal feeding patterns has generated a varied repertoire of foraging behaviors.
2. Species may be specialists or generalists as foragers. Even generalists do not forage randomly, but form search images that can be changed if the food item becomes scarce.
3. According to the premise of optimal foraging, animals forage efficiently in nature by modifying their behavior in complex ways that favor a high energy intake-to-expenditure ratio.

Social Interactions (pp. 1176–1178)

1. Social behavior encompasses the spectrum of interactions between two or more animals, usually of the same species.
2. Agonistic behavior involves a contest in which one competitor gains an advantage in obtaining access to a limited resource, such as food or a mate. Natural selection has generally favored the evolution of symbolic, exaggerated ritual to resolve the conflict involving both aggressive and submissive behavior.
3. Some animals show dominance hierarchies, with differentially ranked individuals permitted options appropriate to their status in the "pecking order."
4. Territoriality is a behavior in which an animal defends a specific, fixed portion of its home range against intrusion by other animals of the same species through agonistic interactions.

Mating Behavior (pp. 1178–1180)

1. Courtship interactions are complex, species-specific behaviors that typically include a series of fixed-action patterns and releasers, ensuring that the participating individuals are nonthreatening and of the proper species, sex, and physiological condition for mating.
2. Courtship may also involve selection of a specific mate from an array of potential candidates.
3. Some segments of courtship behavior likely evolved from agonistic interactions, and the ritualistic aspects of both courtship and agonistic behavior probably evolved from actions whose meaning was more direct.
4. Evolution affects mating relationships, depending partly on parental investment in the next generation, which varies according to the degree of prenatal and postnatal input the parents are required to make.
5. Mating relationships, which vary widely among different species, include promiscuity, monogamy, and

polygamy, the latter involving polygyny more often than polyandry.

Communication (pp. 1180–1183)

1. Animals communicate with one another through their various senses. Odors are particularly effective signals in many species, as demonstrated by the functions of pheromones.
2. Honeybees have an impressive intraspecific communication system, which has been studied extensively by Karl von Frisch. Complex "round" and "waggle" dances executed by worker bees at different speeds signal the presence of food and its distance and direction from the hive.

Selfish Versus Altruistic Behavior (pp. 1183–1185)

1. Altruistic behavior, which benefits a conspecific at the potential expense of the helpful individual, seems to be best explained by kin selection, a theory of inclusive fitness maintaining that genes enhance survival of copies of themselves by directing organisms to care for others who share those genes.
2. Altruism commonly involves close relatives, although not always. Reciprocal altruism between unrelated individuals may be explained as ultimately advantageous to the altruistic individual in the future when another individual "returns the favor."

Animal Cognition (pp. 1185–1186)

1. Because of the impossibility of knowing what goes on inside the minds of other animals, most researchers have taken a mechanistic approach, called behaviorism, to the study of animal behavior.
2. Recently, some workers have argued in favor of cognitive ethology, which sees cognitive ability as a normal function of other animals that has arisen, like other traits, through natural selection.

Human Sociobiology (pp. 1186–1187)

1. The final chapter of E. O. Wilson's *Sociobiology*, published in 1975, suggests that human social behavior could be understood in evolutionary terms, igniting a heated and emotional debate in the scientific community.
2. Sociobiologists acknowledge human behavioral plasticity, which is manifested in an impressive spectrum of social behaviors that is circumscribed, but not rigidly determined, by our genetic potential.

SELF-QUIZ

1. Bees can see colors we cannot see and detect minute amounts of chemicals we cannot smell. But unlike many insects, bees cannot hear very well. Which of the following statements fits best into the perspective of behavioral ecology?
   a. Bees are too small to have functional ears.
   b. Hearing must not contribute much to a bee's fitness.
   c. If a bee could hear, its tiny brain would be swamped with information.
   d. This is an example of a fixed-action pattern.
   e. If bees could hear, the noise of the hive would distract the bees from their work.
2. The nature-versus-nurture controversy centers on
   a. the distinction between proximate and ultimate causes of behavior
   b. the role of genes in learning
   c. whether animals have conscious feelings or thoughts
   d. the extent to which an animal's behavior is innate or learned
   e. the importance of good parental care
3. Which of the following statements is *not* true of fixed-action patterns?
   a. They are highly stereotyped, instinctive behaviors.
   b. They seem to lack adaptive significance.
   c. They are triggered by sign stimuli in the environment and, once begun, are continued to completion.
   d. They are often released by one or two simple cues associated with the relevant object or organism.
   e. A supernormal stimulus often produces a stronger response.
4. The return of salmon to their home streams to spawn is an example of
   a. olfactory imprinting
   b. insight
   c. associative learning
   d. operant conditioning
   e. habituation
5. Every morning Patricia turns on the light in the room and then feeds the fish in her aquarium. After a couple of weeks of this routine, she noticed that the fish came to the surface whenever she turned the light on, whether or not any food was present. This most likely illustrates
   a. habituation
   b. positive phototaxis
   c. imprinting
   d. associative learning
   e. observational learning
6. Which of the following is *not* true of agonistic behavior?
   a. It is most common among members of the same species.
   b. It may be used to establish and defend territories.
   c. It often involves symbolic conflict and does not cause serious harm to either the winner or the loser in the encounter.
   d. It is a uniquely male behavior.
   e. It may be used to establish dominance hierarchies.
7. Which of the following suggests that some animals have internal clocks?
   a. Some birds can sense changes in Earth's magnetic field.
   b. A crab that has moved away from the shore can still sense the tides.
   c. A squirrel kept in darkness drifts away from a 24-hour cycle.
   d. Many animals become active at dawn and settle down at sunset.
   e. Rats kept in constant light show a daily rhythm of activity.

8. The core idea of sociobiology is that
   a. human behavior is rigidly predetermined by inheritance
   b. humans cannot learn to alter their social behavior
   c. many aspects of social behavior have an evolutionary basis
   d. the social behavior of humans is comparable to that of honeybees
   e. environment outweighs genes in human behavior
9. Learning to ignore insignificant stimuli is specifically called
   a. conditioning
   b. imprinting
   c. habituation
   d. insight
   e. critical learning
10. A honeybee returning to the hive from a food source performs a waggle dance with the run oriented straight to the left on the vertical surface. This means that the food is located
   a. 90° left of the hive
   b. 90° left of the line from the hive to the sun
   c. in the opposite direction, straight to the right of the hive
   d. above the hive and slightly to the left
   e. very close to the hive

CHALLENGE QUESTIONS

1. European birds called wagtails eat insects. During winter when food is scarce, a wagtail may defend a territory where it captures an average of 20 insects per hour. Or it may join a flock that ranges widely over the countryside. A bird in a flock averages 25 insects per hour. Discuss the two strategies in terms of optimal foraging. Why do you think the wagtails do not always travel in flocks?
2. Starting with the very first time a bee leaves the hive, it always flies in a circle around the hive before heading out on a foraging trip. If it is prevented from seeing the hive when it leaves or if the hive is moved while the bee is gone, the bee is not able to locate the hive when it returns. For this reason, beekeepers know that a hive should only be moved at night, when all the bees are inside. What component of this "orientation flight" behavior appears to be innate? What component shows learning?
3. Many species of gulls nest very close together on flat-topped islands, but herring gulls space their nests much farther apart. Kittiwakes nest in tiny pockets on the faces of cliffs. Suggest experiments to determine whether these birds are programmed to imprint on their own eggs or young. Which birds do you think will be found to imprint on their own eggs? Their own young? Why?

SCIENCE, TECHNOLOGY, AND SOCIETY

1. Researchers are very interested in studying identical twins who were separated at birth and raised apart. So far, data suggest that twins are much more alike than researchers would have predicted. They have similar personalities, mannerisms, habits, and interests. What kind of general question do you think researchers hope to answer by studying twins that have been raised apart? Why do twins make good subjects for this kind of research? What do the results suggest to you? What are the potential pitfalls of this research? What potential abuses might occur if the studies are not evaluated critically and the results are carelessly cited in support of a particular social agenda?
2. Chimpanzees, ants, lions, and wolves sometimes attack other groups of their own species and kill individuals, but human beings are the only animals that carry out warfare in an organized way on a mass scale. What does our knowledge of animal behavior tell us about human aggression and violence? Are humans more aggressive than other animals? Do other animals control their aggression in ways that we do not? What role does our technology play in the scale and ferocity of our wars? Is knowledge of animal behavior irrelevant when it comes to the subject of war? Why or why not?

FURTHER READING

Alcock, J. *Animal Behavior: An Evolutionary Approach,* 4th ed. Sunderland, MA: Sinauer, 1989.

Blaustein, A. R., and R. K. O'Hara. "Kin Recognition in Tadpoles." *Scientific American,* January 1986. The genetic basis of kin selection.

Davies, N., and M. Brooke. "Coevolution of the Cuckoo and Its Hosts." *Scientific American,* January 1991. The ultimate basis of a fascinating behavior.

Diamond, J. "Sexual Deception." *Discover,* August 1989. Speculations on the evolutionary basis of seduction.

Funk, D. H. "The Mating of Tree Crickets." *Scientific American,* August 1989. The role of a familiar insect song.

Jolly, A. "A New Science That Sees Animals as Conscious Beings." *Smithsonian,* March 1985. Can animals think?

Lohman, K. "How Sea Turtles Navigate." *Scientific American,* January 1992. One of the most remarkable homing phenomena in nature.

Lorenz, K. *On Aggression.* New York: Harcourt Brace Jovanovich, 1966. A classic book on competitive social interactions.

Nelson, B. "The Gannets of Cape Kidnappers." *Natural History,* January 1992. Different populations of the same species exhibit regional dialects in their ritualized displays.

O'Brien, W., H. Browman, and B. Evans. "Search Strategies of Foraging Animals." *American Scientist,* March/April 1990. The relative merits of "cruising" versus "ambushing" as foraging behaviors.

Palmer, J. "The Rhythmic Lives of Crabs." *BioScience,* May 1990. A case study in circadian behavior.

Ridley, M. "Swallows and Scorpionflies Find Symmetry Is Beautiful." *Science,* July 17, 1992. By preferring mates with certain physical traits, female animals have a profound impact on evolution.

Rosenthal, E. "The Forgotten Female." *Discover,* December 1991. When females choose mates, they have a powerful influence on evolution.

Wilson, E. O. *Sociobiology: The New Synthesis.* Cambridge, MA: Harvard University Press, 1975.

APPENDIX ONE SELF-QUIZ ANSWERS

Chapter 2

1. b
2. a
3. b
4. c
5. c
6. c
7. c
8. a
9. b
10. b

Chapter 3

1. d
2. c
3. b
4. c
5. b
6. d
7. d
8. c
9. c
10. c

Chapter 4

1. b
2. c
3. d
4. d
5. a
6. b
7. b
8. d
9. d
10. b

Chapter 5

1. d
2. c
3. d
4. a
5. b
6. b
7. b
8. b
9. c
10. b

Chapter 6

1. b
2. c
3. b
4. b
5. d
6. a
7. c
8. c
9. d
10. c

Chapter 7

1. a
2. c
3. c
4. d
5. (a) centrioles—may help organize microtubules in animal cells. (b) mitochondrion—cellular respiration. (c) nucleus—genetic control cell. (d) Golgi apparatus (dictyosome)—storage, refinement, sorting, and secretion of cellular products. (e) chloroplast—photosynthesis. (f) rough ER—membrane synthesis; synthesis of secretory proteins.
6. d
7. b
8. e
9. c
10. e

Chapter 8

1. b
2. c
3. a
4. a
5. c
6. a
7. fructose
8. glucose
9. into cell
10. b

Chapter 9

1. d
2. b
3. c
4. d
5. a
6. a
7. b
8. b
9. b
10. b

Chapter 10

1. d
2. b
3. d
4. b
5. b
6. c
7. c
8. a
9. a
10. d

Chapter 11

1. c
2. b
3. b
4. e
5. c
6. a
7. a
8. c
9. c

Chapter 12

1. d
2. b
3. d
4. d
5. c
6. a
7. d
8. a
9. c
10. b

Chapter 13 Genetics Problems

1. a. $1/64$ b. $1/64$ c. $1/8$ d. $1/32$

2. Albino is a recessive trait; black is dominant.

Parents	Gametes	Offspring
$BB \times bb$	B and b	All Bb
$bb \times Bb$	b and $1/2$ B, $1/2$ b	$1/2$ Bb, $1/2$ bb

3. Incomplete dominance, with heterozygotes being blue in color. Mating a blue rooster with a black hen should yield approximately equal numbers of blue and black offspring.

4. F_1 cross is $AARR \times aarr$. Genotype of progeny is $AaRr$, phenotype is all axial-pink. F_2 cross is $AaRr \times AaRr$. Genotypes of progeny are 4 $AaRr$: 2 $AaRR$: 2 $AARr$: 2 $aaRr$: 2 $Aarr$: 1 $AARR$: 1 $aaRR$: 1 $AArr$: 1 aarr. Phenotypes of progeny are 6 axial-pink : 3 axial-red : 3 axial-white : 2 terminal-pink : 1 terminal-white : 1 terminal-red.

5. a. $PPLl \times PPLl$, $PpLl$, or $ppLl$
   b. $ppLl \times ppLl$
   c. $PPLL \times$ any of the 9 possible gentypes
   d. $PpLl \times Ppll$
   e. $PpLl \times PpLl$

6. Man I^Ai; woman I^Bi; child ii. Other genotypes for children are $1/4$ I^AI^B, $1/4$ I^Ai, $1/4$ I^BI.

7. Four

8. a. $3/4 \times 3/4 \times 3/4 = 27/64$
   b. $1 - 27/64 = 37/64$
   c. $1/4 \times 1/4 \times 1/4 = 1/64$
   d. $1 - 1/64 = 63/64$

9. a. $1/256$ b. $1/16$ c. $1/256$ d. $1/64$ e. $1/128$

10. If the "curl" allele is dominant, then the original mutant crossed with noncurl cats will produce both curl and noncurl offspring. If the mutation is recessive, then curl offspring will only result from curl × curl matings. You know that cats are true-breeding when curl × curl matings produce only curl offspring. A pure-bred curl cat is homozygous (as it turns out, for the dominant allele, which causes the curled ears).

11 a. 1 b. $1/32$ c. $1/8$ d. $1/2$

12. $1/9$

13. $1/16$

14. Recessive; George = Aa, Arlene = aa, Sandra = AA or Aa, Tom = aa, Sam = Aa, Wilma = aa, Ann = Aa, Michael = Aa, Daniel = Aa, Alan = Aa, Tina = AA or Aa, Carla = aa, Christopher = AA or Aa

15. $1/4$

Chapter 14

1. b
2. b
3. a
4. b
5. d
6. b
7. a

Genetics Problems

1. 0; ½; 1/16

2. Recessive. If the disorder were dominant, it would affect at least one parent of a child born with the disorder. For a girl to have the disorder, she would have to inherit recessive alleles from *both* parents. This would be very rare, especially since males with the allele die in their early teens.

3. ¼ for each daughter (½ chance that child will be female × ½ chance of a homozygous recessive genotype); ½ for first son.

4. 17%

5. 6%. Wild type (heterozygous for normal wings and red eyes) × recessive homozygote with vestigial wings and purple eyes.

6. Between *T* and *A*, 12%; between *A* and S, 5%.

7. Between *T* and *S*, 18%. Sequence of genes is *T-A-S*.

8. Both children would be blind; the son's children would have numbness, and the daughter's children would be blind.

9. The disorder would always be inherited from the mother.

Chapter 15

1. c
2. e
3. d
4. d
5. b
6. c
7. b
8. d
9. a
10. d

Chapter 16

1. a
2. d
3. b
4. d
5. c
6. c
7. a
8. d
9. e
10. b

Chapter 17

1. a
2. e
3. d
4. b
5. d
6. c
7. e
8. d
9. a
10. c

Chapter 18

1. c
2. a
3. d
4. a
5. e
6. a
7. e
8. c
9. b
10. b

Chapter 19

1. b
2. b
3. c
4. b
5. e
6. c
7. e
8. c
9. d
10. b

Chapter 20

1. a
2. b
3. a
4. c
5. d
6. d
7. c
8. b
9. b

Chapter 21

1. b
2. d
3. c
4. a
5. a
6. a
7. c
8. b
9. e
10. b

Chapter 22

1. b
2. b
3. b
4. a
5. d
6. c
7. e
8. d
9. d
10. b

Chapter 23

1. d
2. b
3. a
4. e
5. c
6. b
7. a
8. d
9. c
10. b

Chapter 24

1. b
2. e
3. c
4. d
5. c
6. b
7. a
8. c
9. b
10. d

Chapter 25

1. c
2. a
3. d
4. b
5. b
6. a
7. e
8. c
9. a
10. b

Chapter 26

1. b
2. c
3. e
4. c
5. d
6. c
7. b
8. d
9. d
10. b

Chapter 27

1. c
2. a
3. a
4. b
5. a
6. d
7. a
8. b
9. a
10. c

Chapter 28

1. c
2. b
3. c
4. d
5. d
6. a
7. e
8. a
9. b
10. b

Chapter 29

1. e
2. e
3. b
4. c
5. a
6. d
7. b
8. a
9. e
10. b

Chapter 30

1. b
2. c
3. c
4. d
5. c
6. a
7. b
8. c
9. e
10. c

Chapter 31

1. e
2. b
3. a
4. c
5. b
6. a
7. c
8. d

9. a
10. d

Chapter 32
1. e
2. a
3. c
4. d
5. a
6. c
7. b
8. c
9. a
10. b

Chapter 33
1. b
2. b
3. e
4. d
5. d
6. c
7. b
8. b
9. b
10. b

Chapter 34
1. d
2. b
3. c
4. a
5. b
6. e
7. a
8. c
9. e
10. b

Chapter 35
1. b
2. a
3. b
4. d
5. b
6. a
7. b
8. b
9. c
10. c

Chapter 36
1. b
2. c
3. c
4. d
5. d
6. e
7. b
8. b
9. c
10. c

Chapter 37
1. b
2. b
3. c
4. c
5. d
6. d
7. c
8. d
9. b
10. b

Chapter 38
1. c
2. b
3. d
4. c
5. b
6. c
7. b
8. a
9. a
10. c

Chapter 39
1. d
2. b
3. d
4. c
5. a
6. d
7. b
8. b
9. b
10. c

Chapter 40
1. c
2. d
3. a
4. e
5. e
6. a
7. b
8. c
9. b
10. d

Chapter 41
1. c
2. a
3. e
4. d
5. c
6. b
7. e
8. c
9. c
10. c

Chapter 42
1. d
2. b
3. a
4. b
5. c
6. d
7. a
8. c
9. d
10. b

Chapter 43
1. a
2. b
3. d
4. c
5. c
6. c
7. c
8. a
9. d
10. a

Chapter 44
1. c
2. b
3. a
4. d
5. d
6. a
7. c
8. c
9. d
10. c

Chapter 45
1. e
2. d
3. a
4. c
5. e
6. b
7. b
8. d
9. c
10. b

Chapter 46
1. c
2. e
3. a
4. a
5. d
6. c
7. a
8. c
9. d
10. b

Chapter 47
1. c
2. b
3. c
4. d
5. c
6. e
7. c
8. d
9. a
10. c

Chapter 48
1. c
2. d
3. b
4. d
5. e
6. d
7. b
8. a
9. d
10. c

Chapter 49
1. c
2. e
3. b
4. e
5. d
6. a
7. b
8. c
9. b
10. c

Chapter 50
1. b
2. d
3. b
4. a
5. d
6. d
7. e
8. c
9. c
10. b

APPENDIX TWO CLASSIFICATION OF LIFE

This appendix presents the taxonomic classification used for the major groups of organisms discussed in this text; not all phyla are included. Plant and fungal divisions are the taxonomic equivalents of phyla.

Kingdom Monera*

Archaebacteria
- Methanogens
- Extreme halophiles
- Thermoacidophiles

Eubacteria
- Cyanobacteria
- Phototrophic bacteria
- Chemoautotrophic bacteria
- Pseudomonads
- Nitrogen-fixing aerobic bacteria
- Spirochetes
- Endospore-forming bacteria
- Enteric bacteria
- Rickettsias and chlamydias
- Mycoplasmas
- Actinomycetes
- Myxobacteria

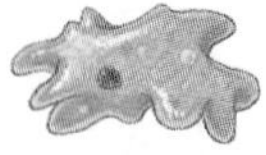

Kingdom Protista

Phylum Rhizopoda (amoebas)
Phylum Actinopoda (heliozoans, radiolarians)
Phylum Foraminifera (forams)
Phylum Apicomplexa (apicomplexans)
Phylum Zoomastigophora (zooflagellates)
Phylum Ciliophora (ciliates)
Phylum Dinoflagellata (dinoflagellates)
Phylum Chrysophyta (golden algae)
Phylum Bacillariophyta (diatoms)
Phylum Euglenophyta (euglenoids)
Phylum Chlorophyta (green algae)
Phylum Phaeophyta (brown algae)
Phylum Rhodophyta (red algae)
Phylum Myxomycota (plasmodial slime molds)
Phylum Acrasiomycota (cellular slime molds)
Phylum Oomycota (water molds)
Phylum Chytridiomycota (chytrids)

Kingdom Plantae

Division Bryophyta (mosses)
Division Hepatophyta (liverworts)
Division Anthocerophyta (hornworts)
Division Psilophyta (whisk ferns)
Division Lycophyta (club mosses)
Division Sphenophyta (horsetails)
Division Pterophyta (ferns)
Division Coniferophyta (conifers)

*Informal names are used here for the bacterial groups because there is not yet a consensus on how to divide the kingdom Monera into phyla.

Division Cycadophyta (cycads)
Division Ginkgophyta (ginkgos)
Division Gnetophyta (gnetae)
Division Anthophyta (flowering plants)
  Class Monocotyledones (monocots)
  Class Dicotyledones (dicots)

Kingdom Fungi

Division Zygomycota (zygomycetes)
Division Ascomycota (sac fungi)
Division Basidiomycota (club fungi)
Division Deuteromycota (imperfect fungi)
Lichens (symbiotic associations of algae and fungi)

Kingdom Animalia

Phylum Porifera (sponges)
Phylum Cnidaria
  Class Hydrozoa (hydrozoans)
  Class Scyphozoa (jellyfishes)
  Class Anthozoa (sea anemones and coral animals)
Phylum Ctenophora (comb jellies)
Phylum Platyhelminthes (flatworms)
  Class Turbellaria (free-living flatworms)
  Class Trematoda (flukes)
  Class Monogenea (flukes)
  Class Cestoda (tapeworms)
Phylum Nemertea (proboscis worms)
Phylum Rotifera (rotifers)
Phylum Nematoda (roundworms)
Phylum Mollusca (mollusks)
  Class Polyplacophora (chitons)
  Class Gastropoda (gastropods: snails and their relatives)
  Class Bivalvia (bivalves)
  Class Cephalopoda (cephalopods: squids and octopuses)
Phylum Onychophora
Phylum Annelida (segmented worms)
  Class Oligochaeta (oligochaetes)
  Class Polychaeta (polychaetes)
  Class Hirudinea (leeches)
Phylum Arthropoda (arthropods)
  Subphylum Trilobitomorpha (trilobites)
  Subphylum Cheliceriformes (or Chelicerata, chelicerates)
    Class Arachnida (arachnids: spiders, ticks, scorpions)
  Subphylum Uniramia (uniramians)
    Class Diplopoda (millipedes)
    Class Chilopoda (centipedes)
    Class Insecta (insects)
  Subphylum Crustacea (crustaceans)
Phylum Phoronida (phoronids)
Phylum Bryozoa (bryozoans)
Phylum Brachiopoda (brachiopods: lamp shells)
Phylum Echinodermata (echinoderms)
  Class Asteroidea (sea stars)
  Class Ophiuroidea (brittle stars)
  Class Echinoidea (sea urchins and sand dollars)
  Class Crinoidea (sea lilies)
  Class Concentricycloidea (sea daisies)
  Class Holothuroidea (sea cucumbers)
Phylum Chordata (chordates)
  Subphylum Cephalochordata (cephalochordates: lancelets)
  Subphylum Urochordata (urochordates: tunicates)
  Subphylum Vertebrata (vertebrates)
    Class Agnatha (jawless vertebrates)
    Class Placodermi (extinct jawed fishes)
    Class Chondrichthyes (cartilaginous fishes)
    Class Osteichthyes (bony fishes)
    Class Amphibia (amphibians)
    Class Reptilia (reptiles)
    Class Aves (birds)

APPENDIX THREE THE METRIC SYSTEM

Measurement	Unit and Abbreviation	Metric Equivalent	Metric-to-English Conversion Factor	English-to-Metric Conversion Factor
Length	1 kilometer (km)	= 1000 (10^3) meters	1 km = 0.62 mile	1 mile = 1.61 km
	1 meter (m)	= 100 (10^2) centimeters = 1000 millimeters	1 m = 1.09 yards 1 m = 3.28 feet 1 m = 39.37 inches	1 yard = 0.914 m 1 foot = 0.305 m
	1 centimeter (cm)	= 0.01 (10^{-2}) meter	1 cm = 0.394 inch	1 foot = 30.5 cm 1 inch = 2.54 cm
	1 millimeter (mm)	= 0.001 (10^{-3}) meter	1 mm = 0.039 inch	
	1 micrometer (μm) (formerly micron, μ)	= 10^{-6} meter (10^{-3} mm)		
	1 nanometer (nm) (formerly millimicron, mμ)	= 10^{-9} meter (10^{-3} μm)		
	1 angstrom (Å)	= 10^{-10}meter (10^{-4}μm)		
Area	1 hectare (ha)	= 10,000 square meters	1 ha = 2.47 acres	1 acre = 0.0405 ha
	1 square meter (m^2)	= 10,000 square centimeters	1 m^2 = 1.196 square yards 1 m^2 = 10.764 square feet	1 square yard = 0.8361 m^2 1 square foot = 0.0929 m^2
	1 square centimeter (cm^2)	= 100 square millimeters	1 cm^2 = 0.155 square inch	1 square inch = 6.4516 cm^2
Mass	1 metric ton (t)	= 1000 kilograms	1 t = 1.103 tons	1 ton = 0.907 t
	1 kilogram (kg)	= 1000 grams	1 kg = 2.205 pounds	1 pound = 0.4536 kg
	1 gram (g)	= 1000 milligrams	1 g = 0.0353 ounce 1 g = 15.432 grains	1 ounce = 28.35 g
	1 milligram (mg)	= 10^{-3} gram	1 mg = approx. 0.015 grain	
	1 microgram (μg)	= 10^{-6} gram		
Volume (solids)	1 cubic meter (m^3)	= 1,000,000 cubic centimeters	1m^3 = 1.308 cubic yards 1m^3 = 35.315 cubic feet	1 cubic yard = 0.7646 m^3 1 cubic foot = 0.0283 m^3
	1 cubic centimeter (cm^3 or cc)	= 10^{-6} cubic meter	1 cm^3 = 0.061 cubic inch	1 cubic inch = 16.387 cm^3
	1 cubic millimeter (mm^3)	= 10^{-9} cubic meter (10^{-3} cubic centimeter)		

Measurement	Unit and Abbreviation	Metric Equivalent	Metric-to-English Conversion Factor	English-to-Metric Conversion Factor
Volume (Liquids and Gases)	1 kiloliter (kl or kL)	= 1000 liters	1 kL = 264.17 gallons	1 gallon = 3.785 L
	1 liter (L)	= 1000 milliliters	1 L = 0.264 gallons 1 L = 1.057 quarts	1 quart = 0.946 L
	1 milliliter (mL)	= 10^{-3} liter = 1 cubic centimeter	1 mL = 0.034 fluid ounce 1 mL = approx. ¼ tea-spoon 1 ml = approx. 15–16 drops (gtt.)	1 quart = 946 mL 1 pint = 473 mL 1 fluid ounce = 29.57 mL 1 teaspoon = approx. 5 mL
	1 microliter (ml or mL)	= 10^{-6} liter (10^{-3} milliliters)		
Time	1 second (s)	= 1/60 minute		
	1 millisecond (ms)	= 10^{-3} second		
Temperature	Degrees Celsius (°C) (Absolute zero, when all molecular motion ceases, is –273 °C. The Kelvin (K) scale, which has the same size degrees as Celsius, has its zero point at absolute zero. Thus, 0° K = –273°C.)		°F = 9/5 °C + 32	°C = 5/9 (°F – 32)

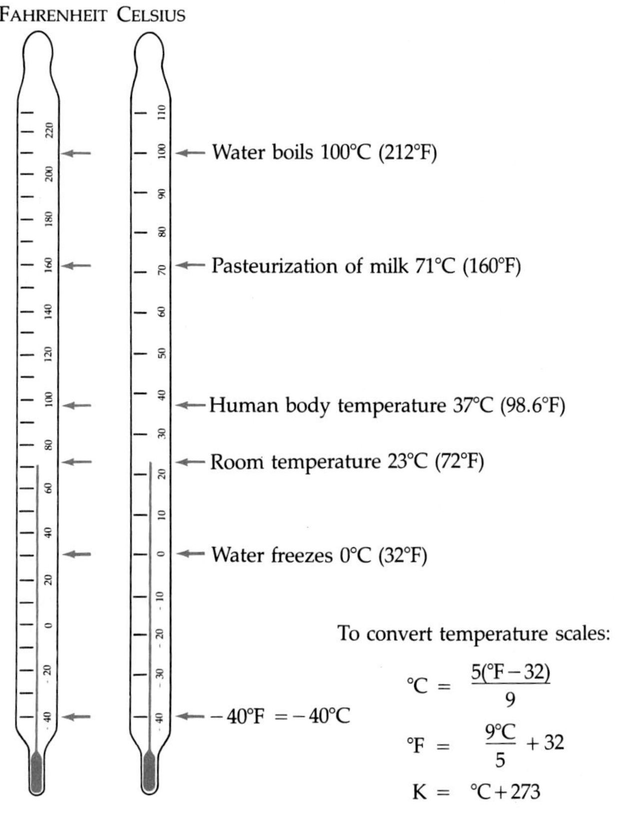

Fahrenheit and celsius scales compared.

APPENDIX FOUR CONCEPT MAPPING: A TECHNIQUE FOR LEARNING

Concept mapping is a process—a process of organizing your knowledge to increase your understanding and to help you learn. A concept map is a diagram that shows the organization of ideas and the relationship among concepts in a particular subject area. The *structure* of a concept map is a hierarchically organized cluster of concepts, enclosed in boxes and connected with lines that explicitly state the relationships among the concepts. The *function* of a concept map is to help you structure your understanding of a topic and create personal meaning. The *value* of a concept map arises from the process of thinking and evaluating what you must do in order to create the map.

Biology is a rich and diverse subject, filled with information and terminology that are necessary to understand the concepts and principles of the field. Beginning students often focus their attention on memorizing bits of information, while missing how the pieces fit into the "big picture." A student must assimilate these details into broader concepts that are then related to other concepts, groups of concepts, and organizing principles. Neil Campbell's *BIOLOGY* helps the student in this process by regularly pointing out the themes that run through the study of biology: hierarchy of organization, emergent properties, correlation between structure and function, and evolution. Every chapter of this book has a set of key ideas or principles that are supported by examples and explanations. A student must recognize how these ideas are organized and develop a personal conceptual framework that can house the details of that chapter. Sketching a concept map will help create such a structure.

To develop a concept map for a particular area, you must first identify the most important ideas or concepts. That process alone will help you sort out details from the organizing principles. Then you evaluate the relative importance of these key concepts (those that are most inclusive, those that are subordinate to other concepts); arrange the concepts in a meaningful cluster; and label the connections or relationships among the concepts.

A concept map is an individual picture of the understanding you had developed at the time you made the map. Meanings change and grow as you gain more knowledge and experience in an area. You develop a richer picture; you can make more connections among concepts and relate ideas in a more meaningful way. Knowledge is not static; it grows. As your understanding of an area develops, your concept map will evolve—sometimes becoming more simplified and streamlined, sometimes becoming more complex and interrelated.

Concept maps are context-dependent. The same group of concepts can be organized in several ways, depending on the focus of the map. The following examples show similar clusters of concepts from Neil Campbell's first chapter, on themes in the study of life. Notice how the hierarchy of concepts changes depending on whether you are looking at an overview of biology or at one of the specific themes. Read through these maps; read up and down and across the connections that are drawn. Recognize that these maps provide one way of structuring the ideas of this chapter. There is no one "right" or "wrong" concept map; they are individual representations of understanding. Maps may, however, be more or less accurate or valid, and sharing and talking through concept maps is a good way to assess your understanding.

Remember, the real value in concept mapping is in the process, in the weighing and relating and organizing of concepts that each individual must do to develop a personal and meaningful understanding of this fascinating subject of biology.

Many examples and concept mapping exercises are included in the *Student Study Guide* that accompanies Campbell's *BIOLOGY*. Additional information about concept maps may be found in *Learning How to Learn* by J. D. Novak and D. Gowin (New York: Cambridge University Press, 1984).

Martha Taylor
Cornell University

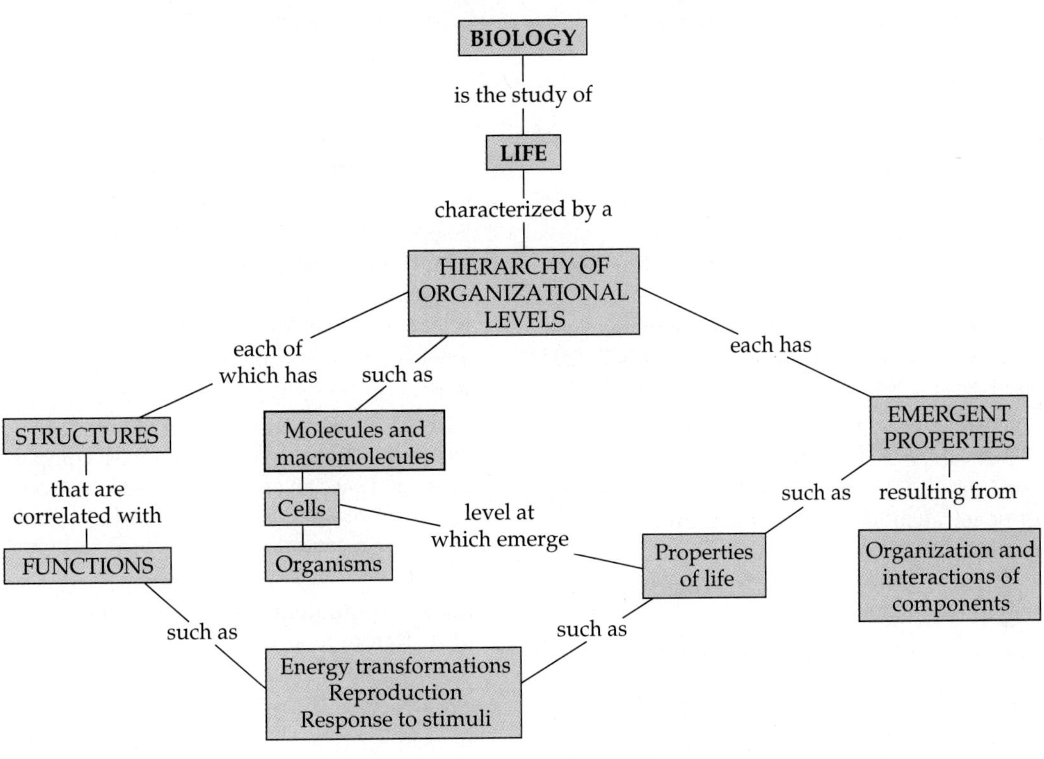
BIOLOGY
is the study of
LIFE
characterized by a
HIERARCHY OF ORGANIZATIONAL LEVELS
each of which has
such as
each has
STRUCTURES
Molecules and macromolecules
EMERGENT PROPERTIES
that are correlated with
Cells
level at which emerge
such as
resulting from
FUNCTIONS
Organisms
Properties of life
Organization and interactions of components
such as
such as
Energy transformations
Reproduction
Response to stimuli

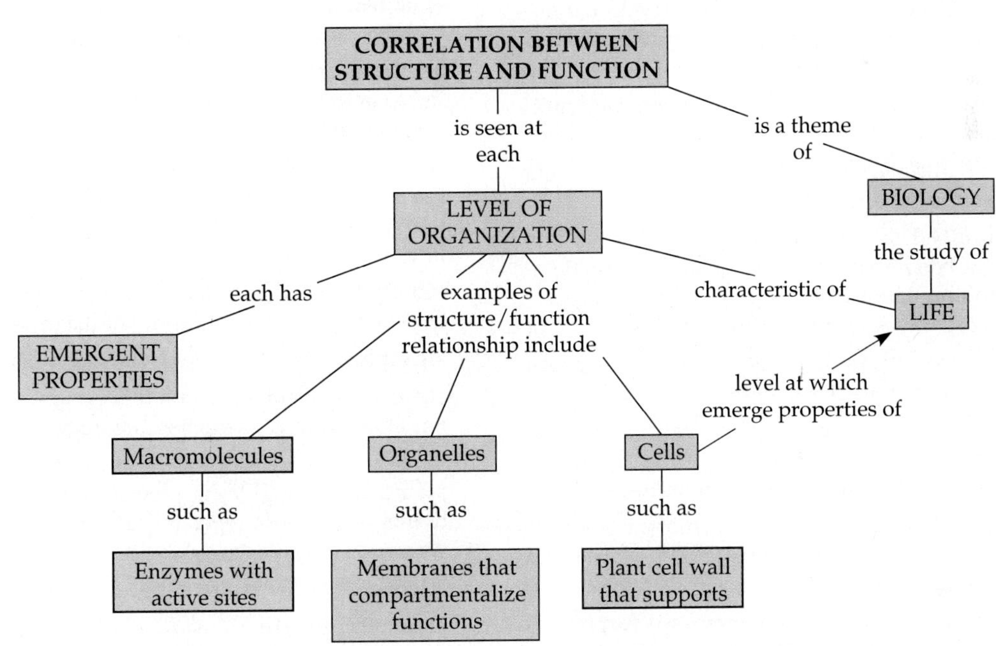
CORRELATION BETWEEN STRUCTURE AND FUNCTION
is seen at each
is a theme of
BIOLOGY
LEVEL OF ORGANIZATION
the study of
each has
examples of structure/function relationship include
characteristic of
LIFE
EMERGENT PROPERTIES
level at which emerge properties of
Macromolecules
Organelles
Cells
such as
such as
such as
Enzymes with active sites
Membranes that compartmentalize functions
Plant cell wall that supports

Glossary

A site Aminoacyl-tRNA site; the binding site on a ribosome that holds the tRNA carrying the next amino acid to be added to a growing polypeptide chain.

abiotic Pertaining to nonliving environmental factors, such as temperature, light, water, and nutrients.

abscisic acid (ABA) *(ab-SIS-ik)* A plant hormone that generally acts to inhibit growth, promote dormancy, and help the plant withstand stressful conditions.

absorption spectrum The range of a pigment's ability to absorb various wavelengths of light.

abyssal zone *(uh-BIS-ul)* The portion of the ocean floor where light does not penetrate and where temperatures are cold and pressures intense.

acclimation *(AK-lim-AY-shun)* Physiological adjustment to a change in an environmental factor.

accommodation Automatic adjustment of an eye to focus on near objects.

acetylcholine One of the most common neurotransmitters; functions by binding to receptors and altering the permeability of the postsynaptic membrane to specific ions, either depolarizing or hyperpolarizing the membrane.

acetyl CoA The entry compound for the Krebs cycle in cellular respiration; formed from a fragment of pyruvate attached to a coenzyme.

acid A substance that increases the hydrogen ion concentration in a solution.

acoelomate *(a-SEEL-oh-mate)* A solid-bodied animal lacking a cavity between the gut and outer body wall.

acquired immunity The type of immunity achieved when antigens enter the body naturally or artificially. Acquired immunity is due to stimulation of antibody production and production of memory cells keyed to the antigen.

acrosome *(AK-ruh-sowm)* An organelle at the tip of a sperm cell that helps the sperm penetrate the egg.

actin *(AK-tin)* A globular protein that links into chains, two of which twist helically about each other, forming microfilaments in muscle and other contractile elements in cells.

action potential The rapid change in the membrane potential of an excitable cell, caused by stimulus-triggered, selective opening and closing of voltage-sensitive ion channels.

activation Induction by the sperm of increased respiration and protein synthesis in the egg; the earliest trigger of embryonic development.

active site The specific portion of an enzyme that attaches to the substrate by means of weak chemical bonds.

active transport The movement of a substance across a biological membrane against its concentration or electrochemical gradient, with the help of energy input and specific transport proteins.

adaptive peak An equilibrium state in a population when the gene pool has allele frequencies that maximize the average fitness of a population's members.

adaptive radiation The emergence of numerous species from a common ancestor introduced into an environment, presenting a diversity of new opportunities and problems.

adhesion *(ad-HEE-zhun)* The clinging of one substance to another.

adrenal gland *(uh-DREEN-ul)* An endocrine gland located adjacent to the kidney in mammals; composed of two glandular portions: an outer cortex, which responds to endocrine signals in reacting to stress and effecting salt and water balance, and a central medulla, which responds to nervous inputs in reacting to stress.

aerobic respiration Harvesting chemical energy in the form of ATP from food molecules, with oxygen as the final electron acceptor.

age structure The relative number of individuals of each age in a population.

aggregate fruit A ripened ovary, or fruit, resulting from a single flower with several separate carpels.

agnathan *(AG-naa-thun)* A member of a jawless class of vertebrates represented today by the lampreys and hagfishes.

agonistic behavior *(ag-on-IS-tik)* A type of behavior involving a contest of some kind that determines which competitor gains access to some resource, such as food or mates.

AIDS (acquired immunodeficiency syndrome) The name of the late stages of HIV infection; defined by a specified reduction of T cells and the appearance of characteristic secondary infections.

aldehyde *(AL-duh-hyde)* An organic molecule with a carbonyl group located at the end of the carbon skeleton.

alga (plural, **algae**) A photosynthetic, plantlike protist.

all-or-none event An action that occurs either completely or not at all, such as the generation of an action potential by a neuron.

allantois *(AL-an-TOE-iss)* One of four extraembryonic membranes; serves as a repository for the embryo's nitrogenous waste.

allele *(uh-LEEL)* An alternative form of a gene.

allometric growth *(AL-oh-MET-rik)* The variation in the relative rates of growth of various parts of the body, which helps shape the organism.

allopatric speciation *(AL-oh-PAT-rik)* A mode of speciation induced when the ancestral population becomes segregated by a geographical barrier.

allopolyploid *(AL-oh-POL-ee-ploid)* A common type of polyploid species resulting from two different species interbreeding and combining their chromosomes.

allosteric site *(AL-oh-STEER-ik)* A specific receptor site on an enzyme molecule remote from the active site. Molecules bind to the allosteric site and change the shape of the active site, making it either more or less receptive to the substrate.

alpha (α) helix A spiral shape constituting one form of the secondary structure of proteins, arising from a specific hydrogen-bonding structure.

alternation of generations A life cycle in which there is both a multicellular diploid form, the sporophyte, and a multicellular haploid form, the gametophyte; characteristic of plants.

altruistic behavior *(AL-troo-ISS-tik)* The aiding of another individual at one's own risk or expense.

alveolus *(al-VEE-oh-lus)* (plural, **alveoli**) (1) One of the dead-end, multilobed air sacs that constitute the gas exchange surface of the lungs. (2) One of the milk-secreting sacs of epithelial tissue in the mammary glands.

amino acid *(uh-ME-no)* An organic molecule possessing both carboxyl and amino groups. Amino acids serve as the monomers of proteins.

amino group A functional group that consists of a nitrogen atom bonded to two hydrogen atoms; can act as a base in solution, accepting a hydrogen ion and acquiring a charge of +1.

aminoacyl-tRNA synthetases A family of enzymes, at least one for each amino acid, that catalyzes the attachment of an amino acid to its specific tRNA molecule; also called amino acid-activating enzymes.

amniocentesis *(AM-nee-oh-sen-TEE-sis)* A technique for determining genetic abnormalities in a fetus by the presence of certain chemicals or defective fetal cells in the amniotic fluid, obtained by aspiration from a needle inserted into the uterus.

amnion *(AM-nee-on)* The innermost of four extraembryonic membranes; encloses a fluid-filled sac in which the embryo is suspended.

amniote A vertebrate possessing an amnion surrounding the embryo; reptiles, birds, and mammals are amniotes.

amniote egg A shelled egg with a self-contained reservoir of amniotic fluid, which enables some vertebrates to complete their life cycle on dry land.

anabolic pathway *(AN-uh-BOLL-ik)* A metabolic pathway that consumes energy to build complicated molecules from simpler ones.

anaerobic respiration A form of respiration that occurs in a few groups of bacteria living in anaerobic environments such as soil; the final electron acceptors are sulfate and nitrate.

anagenesis *(AN-uh-JEN-eh-sis)* A pattern of evolutionary change involving the transformation of an entire population, sometimes to a state different enough from the ancestral population to justify renaming it as a separate species; also called phyletic evolution.

analogy *(uh-NAL-oh-jee)* The similarity of structure between two species that are not closely related; attributable to convergent evolution.

androgens *(AN-droh-jens)* The principal male steroid hormones, such as testosterone, which stimulate the development and maintenance of the male reproductive system and secondary sex characteristics.

aneuploidy *(AN-yoo-ploy-dee)* A chromosomal aberration in which certain chromosomes are present in extra copies or are deficient in number.

angiosperm *(AN-jee-oh-spurm)* A flowering plant, which forms seeds inside a protective chamber called an ovary.

animal pole In amphibian, reptile, and bird blastulas, the pole of the egg with the least amount of yolk.

anion *(AN-eye-on)* A negatively charged ion.

annual A plant that completes its entire life cycle in a single year or growing season.

anterior Referring to the head end of a bilaterally symmetric animal.

anther *(AN-thur)* The terminal pollen sac of a stamen, inside which pollen grains with male gametes form in the flower of an angiosperm.

antibiotic A chemical that kills or inhibits the growth of bacteria, often via transcriptional or translational regulation.

antibody An antigen-binding immunoglobulin, produced by B cells, that functions as the effector in an immune response.

anticodon *(AN-tee-CO-don)* A specialized base triplet on one end of a tRNA molecule that recognizes a particular complementary codon on an mRNA molecule.

antidiuretic hormone (ADH) A hormone important in osmoregulation.

antigen *(AN-teh-jen)* A foreign macromolecule that does not belong to the host organism and that elicits an immune response.

aphotic zone *(ay-FOE-tik)* The part of the ocean beneath the photic zone, where light does not penetrate sufficiently for photosynthesis to occur.

apical dominance *(AY-pik-ul)* Concentration of growth at the tip of a plant shoot, where a terminal bud partially inhibits axillary bud growth.

apical meristem *(AY-pik-ul MARE-eh-stem)* Embryonic plant tissue in the tips of roots and in the buds of shoots that supplies cells for the plant to grow in length.

apomorphic character *(AP-oh-MORE-fik)* A derived phenotypic character, or homology, that evolved after a branch diverged from a phylogenetic tree.

apoplast *(AP-oh-plast)* In plants, the nonliving continuum formed by the extracellular pathway provided by the continuous matrix of cell walls.

aposematic coloration *(AP-oh-so-MAT-ik)* The bright coloration of animals with effective physical or chemical defenses that acts as a warning to predators.

aqueous solution *(AY-kwee-us)* A solution in which water is the solvent.

archaebacteria *(AR-kuh-bak-TEER-ee-uh)* An ancient lineage of prokaryotes, represented today by a few groups of bacteria inhabiting extreme environments. Some taxonomists place archaebacteria in their own kingdom, separate from the other bacteria.

archenteron *(ark-EN-ter-on)* The endoderm-lined cavity formed during the gastrulation process that develops into the digestive tract of an animal.

arteries Vessels that carry blood away from the heart to organs throughout the body.

arteriole *(ar-TEER-ee-ole)* A branch of an artery that gives rise to capillaries.

arteriosclerosis A cardiovascular disease caused by the formation of hard plaques within the arteries.

artificial selection The selective breeding of domesticated plants and animals to encourage the occurrence of desirable traits.

ascus (plural, **asci**) A saclike spore capsule located at the tip of the ascocarp in dikaryotic hyphae; defining feature of the Ascomycota division of fungi.

asexual reproduction A type of reproduction involving only one parent that produces genetically identical offspring by budding or by the division of a single cell or the entire organism into two or more parts.

associative learning The acquired ability to associate one stimulus with another; also called classical conditioning.

assortative mating A type of nonrandom mating in which mating partners resemble each other in certain phenotypic characters.

asymmetric carbon A carbon atom covalently bonded to four different atoms or groups of atoms.

atomic number The number of protons in the nucleus of an atom, unique for each element and designated by a subscript to the left of the elemental symbol.

atomic weight The total atomic mass, which is the mass in grams of one mole of the atom.

ATP (adenosine triphosphate) *(uh-DEN-oh-sin try-FOS-fate)* An adenine-containing nucleoside triphosphate that releases free energy when its phosphate bonds are hydrolyzed. This energy is used to drive endergonic reactions in cells.

ATP synthase A protein complex that produces ATP.

atrioventricular valve A valve in the heart between each atrium and ventricle that prevents a backflow of blood when the ventricles contract.

atrium *(AY-tree-um)* (plural, **atria**) A chamber that receives blood returning to the vertebrate heart.

autogenous model According to this model, eukaryotic cells evolved by the specialization of internal membranes originally derived from prokaryotic plasma membranes.

autoimmune disease An immunological disorder in which the immune system turns against itself.

autonomic nervous system *(AWT-uh-NAHM-ik)* A subdivision of the motor nervous system of vertebrates that regulates the internal environment; consists of the sympathetic and parasympathetic divisions.

autopolyploid *(AW-toe-POL-ee-ploid)* A type of polyploid species resulting from one species doubling its chromosome number to become tetraploids, which may self-fertilize or mate with other tetraploids.

autosome *(AW-tuh-some)* A chromosome that is not directly involved in determining sex, as opposed to the sex chromosomes.

autotrophic nutrition *(AW-tuh-TROE-fik)* A mode of obtaining organic food molecules without eating other organisms. Autotrophs use energy from the sun or from the oxidation of inorganic substances to make organic molecules from inorganic ones.

auxins *(AWK-sins)* A class of plant hormones, including indoleacetic acid (IAA), that has a variety of effects, such as phototropic response through stimulation of cell elongation, stimulation of secondary growth, and development of leaf traces and fruit.

auxotroph *(AWK-soh-trofe)* A nutritional mutant that is unable to synthesize and that cannot grow on media lacking certain essential molecules normally synthesized by wild-type strains of the same species.

axillary bud *(AKS-ill-air-ee)* An embryonic shoot present in the angle formed by a leaf and stem.

axon *(AKS-on)* A typically long outgrowth, or process, from a neuron that carries nerve impulses away from the cell body toward target cells.

B cell A type of lymphocyte that develops in the bone marrow and later produces antibodies, which mediate humoral immunity.

bacterium (plural, **bacteria**) A unicellular microorganism, also called a prokaryote, which has no true nucleus. Bacteria are classified into two groups based on a difference in cell walls, as determined by Gram staining.

balanced polymorphism A type of polymorphism in which the frequencies of the coexisting forms do not change noticeably over many generations.

Barr body The dense object that lies along the inside of the nuclear envelope in cells of female mammals, representing the one inactivated X chromosome.

basal body A cell structure identical to a centriole that organizes and anchors the microtubule assembly of a cilium or flagellum.

basal metabolic rate (BMR) The minimal number of kilocalories a resting animal requires to fuel itself for a given time.

base A substance that reduces the hydrogen ion concentration in a solution.

basement membrane The floor of an epithelial membrane on which the basal cells rest.

basidium (plural, **basidia**) A reproductive appendage that produces sexual spores on the gills of mushrooms. The fungal division Basidiomycota is named for this structure.

Batesian mimicry *(BAYTZ-ee-un MIM-ih-kree)* A type of mimicry in which a harmless species looks like a different species that is poisonous or otherwise harmful to predators.

behavioral ecology A heuristic approach based on the expectation that Darwinian fitness (reproductive success) is improved by optimal behavior.

benign tumor A noncancerous abnormal growth composed of cells that multiply excessively but remain at their place of origin in the body.

benthic zone The bottom surfaces of aquatic environments.

beta (β) pleated sheet A zigzag shape constituting one form of the secondary structure of proteins, formed of hydrogen bonds between polypeptide segments running in opposite directions.

biennial *(by-EN-ee-ul)* A plant that requires two years to complete its life cycle.

bilateral symmetry Characterizing a body form with a central longitudinal plane dividing the body into two equal but opposite halves.

bilateria *(BY-leh-TEER-ee-uh)* Members of the branch of eumetazoans possessing bilateral symmetry.

bile A mixture of substances containing bile salts, which emulsify fats and aid in their digestion and absorption.

binary fission The type of cell division by which prokaryotes reproduce; each dividing daughter cell receives a copy of the single parental chromosome.

biogeochemical cycles The various nutrient circuits, which involve both biotic and abiotic components of ecosystems.

biogeography The study of the past and present distribution of species.

biological magnification A trophic process in which retained substances become more concentrated with each link in the food chain.

biological species A population or group of populations whose members have the potential to interbreed.

biomass The dry weight of organic matter comprising a group of organisms in a particular habitat.

biome *(BY-ome)* One of the world's major communities, classified according to the predominant vegetation and characterized by adaptations of organisms to that particular environment.

biosphere *(BY-oh-sfeer)* The entire portion of Earth that is inhabited by life; the sum of all the planet's communities and ecosystems.

biotechnology The industrial use of living organisms or their components to improve human health and food production.

biotic *(by-OT-ik)* Pertaining to the living organisms in the environment.

blastocoel *(BLAS-toe-seel)* The fluid-filled cavity that forms in the center of the blastula embryo.

blastopore *(BLAS-toe-por)* The opening of the archenteron in the gastrula that develops into the mouth in protostomes and the anus in deuterostomes.

blastula *(BLAS-tyoo-la)* The hollow ball of cells marking the end stage of cleavage during early embryonic development.

blood-brain barrier A specialized capillary arrangement in the brain that restricts the passage of most substances into the brain, thereby preventing dramatic fluctuations in the brain's environment.

blood pressure The hydrostatic force that blood exerts against the wall of a vessel.

bond energy The quantity of energy that must be absorbed to break a particular kind of chemical bond; equal to the quantity of energy the bond releases when it forms.

book lungs Organs of gas exchange in spiders, consisting of stacked plates contained in an internal chamber.

bottleneck effect Genetic drift resulting from reduction of a population, typically by a natural disaster, such that the surviving population is no longer genetically representative of the original population.

Bowman's capsule *(BOH-munz)* A cup-shaped receptacle in the vertebrate kidney that is the initial, expanded segment of the nephron where filtrate enters from the blood.

brainstem The hindbrain and midbrain of the vertebrate central nervous system. In the human, it forms a cap on the anterior end of the spinal cord, extending to about the middle of the brain.

bryophytes *(BRY-oh-fites)* The mosses, liverworts, and hornworts; a group of nonvascular plants that inhabit the land but lack many of the terrestrial adaptations of vascular plants.

budding An asexual means of propagation in which outgrowths from the parent form and pinch off to live independently or else remain attached to eventually form extensive colonies.

buffer A substance that consists of acid and base forms in solution and that minimizes changes in pH when extraneous acids or bases are added to the solution.

bulk flow Movement of water due to a difference in pressure between two locations.

C_3 plant A plant that uses the Calvin cycle for the initial steps that incorporate CO_2 into organic material, forming a three-carbon compound as the first stable intermediate.

C_4 plant A plant that prefaces the Calvin cycle with reactions that incorporate CO_2 into four-carbon compounds, the end-product of which supplies CO_2 for the Calvin cycle.

calmodulin *(kal-MOD-yoo-lin)* An intracellular protein to which calcium binds in its function as a second messenger in hormone action.

calorie The amount of heat energy required to raise the temperature of 1 g of water 1°C; the amount of heat energy that 1 g of water releases when it cools by 1°C. The Calorie (with a capital C), usually used to indicate the energy content of food, is a kilocalorie.

Calvin cycle The second of two major stages in photosynthesis (following the light reactions), involving atmospheric CO_2 fixation and reduction of the fixed carbon into carbohydrate.

CAM plant A plant that uses crassulacean acid metabolism, an adaptation for photosynthesis in arid conditions, first discovered in the family Crassulaceae. Carbon dioxide entering open stomata during the night is converted into organic acids, which release CO_2 for the Calvin cycle during the day, when stomata are closed.

capillaries *(KAP-ill-air-ees)* Microscopic blood vessels penetrating the tissues and consisting of a single layer of endothelial cells that allows exchange between the blood and interstitial fluid.

capsid The protein shell that encloses the viral genome; rod-shaped, polyhedral, or more completely shaped.

carbohydrate *(KAR-bo-HI-drate)* A sugar (monosaccharide) or one of its dimers (disaccharides) or polymers (polysaccharides).

carbon fixation The initial incorporation of carbon dioxide into organic compounds.

carbonyl group *(KAR-buh-nil)* A functional group present in aldehydes and ketones, consisting of a carbon atom double-bonded to an oxygen atom.

carboxyl group *(kar-BOX-ul)* A functional group present in organic acids, consisting of a single carbon atom double-bonded to an oxygen atom and also bonded to a hydroxyl group.

carcinogen *(kar-SIN-oh-jen)* A chemical agent that causes cancer.

cardiac muscle *(KAR-dee-ak)* A type of muscle that forms the contractile wall of the heart; its cells are joined by intercalated discs that relay each heartbeat.

cardiac output The volume of blood per minute pumped by the left ventricle of the heart.

carnivore An animal, such as a shark, hawk, or spider, that eats other animals.

carpel *(KAR-pel)* The female reproductive organ of a flower, consisting of the stigma, style, and ovary.

carrier protein A transport protein involved in facilitated diffusion, possessing a specific binding site for a specific substance.

carrying capacity The maximum population size that can be supported by the available resources, symbolized as *K*.

cartilage *(KAR-til-ij)* A type of flexible connective tissue with an abundance of collagenous fibers embedded in chondrin.

Casparian strip *(kas-PAR-ee-un)* A water-impermeable ring of wax around endodermal cells in plants that blocks the passive flow of water and solutes into the stele by way of cell walls.

catabolic pathway *(KAT-uh-BOL-ik)* A metabolic pathway that releases energy by breaking down complex molecules into simpler compounds.

catabolite activator protein (CAP) *(ka-TAB-ul-LITE)* In *E. coli,* a helper protein that stimulates gene expression by binding within the promoter region of an operon and enhancing the promoter's ability to associate with RNA polymerase.

cation *(KAT-eye-on)* An ion with a positive charge, produced by the loss of one or more electrons.

cation exchange A process in which positively charged minerals are made available to a plant when hydrogen ions in the soil displace mineral ions from the clay particles.

cell adhesion molecules (CAMs) A diverse group of molecules on the surface of cells that contribute to selective cell association during embryonic development.

cell center A region in the cytoplasm near the nucleus from which microtubules originate and radiate.

cell cycle An ordered sequence of events in the life of a dividing cell, composed of the M, G_1, S, and G_2 phases.

cell fractionation The disruption of a cell and separation of its organelles by centrifugation.

cell-mediated immunity The type of immunity that functions in defense against fungi, protists, bacteria, and viruses inside host cells and against tissue transplants with highly specialized cells that circulate in the blood and lymphoid tissue.

cell plate A double membrane across the midline of a dividing plant cell, between which the new cell wall forms during cytokinesis.

cell wall Unique to plant cells, a wall formed of cellulose fibers embedded in a polysaccharide-protein matrix. The primary cell wall is thin and flexible, whereas the secondary cell wall is stronger and more rigid and the primary constituent of wood.

cellular differentiation Divergence in structure and function of cells as they become specialized during a multicellular organism's development; depends on the control of gene expression.

cellular respiration The most prevalent and efficient catabolic pathway for the production of ATP, in which oxygen is consumed as a reactant along with the organic fuel.

cellulose *(SELL-yoo-lose)* A structural polysaccharide of cell walls, consisting of glucose monomers joined by β-1, 4-glycosidic linkages.

Celsius scale *(SELL-see-us)* A temperature scale (°C) equal to $\frac{5}{9}$(°F − 32) that measures the freezing point of water at 0°C and the boiling point of water at 100°C.

central nervous system (CNS) In vertebrate animals, the brain and spinal cord.

centriole *(SEN-tree-ole)* One of two structures in the center of animal cells, composed of cylinders of nine triplet microtubules in a ring. Centrioles help organize microtubule assembly during cell division.

centromere *(SEN-troh-mere)* The centralized region joining two sister chromatids.

centrosome Material present in the cytoplasm of all eukaryotic cells and important during cell division; also called microtubule-organizing center.

cephalization *(SEF-uh-le-ZA-shun)* The evolution of a head (anterior) end with sensory structures and a highly specialized brain to process sensory input; a feature of bilaterally symmetric animals, especially the vertebrates.

cephalochordate A chordate without a backbone, represented by lancelets, tiny marine animals.

cerebellum *(SEH-reh-BELL-um)* Part of the vertebrate hindbrain (rhombencephalon) located dorsally; functions in unconscious coordination of movement and balance.

cerebral cortex *(seh-REEB-brul)* The surface of the cerebrum; the largest and most complex part of the mammalian brain, containing sensory and motor nerve cell bodies of the cerebrum; the part of the vertebrate brain most changed through evolution.

cerebrum *(seh-REEB-brum)* The dorsal portion, composed of right and left hemispheres, of the vertebrate forebrain; the integrating center for memory, learning, emotions, and other highly complex functions of the central nervous system.

chaparral *(SHAP-uh-RAL)* A scrubland biome of dense, spiny evergreen shrubs found at midlatitudes along coasts where cold ocean currents circulate offshore; characterized by mild, rainy winters and long, hot, dry summers.

character displacement The divergence of overlapping characteristics in two species living in the same environment as a result of resource partitioning.

chemical equilibrium In a reversible chemical reaction, the point at which the rate of the forward reaction equals the rate of the reverse reaction.

chemical synapse *(SIN-aps)* A junction between two neurons or between a neuron and a muscle, receptor, or gland that uses chemicals to transmit information between the cells.

chemiosmosis *(KEE-mee-os-MOH-sis)* The ability of certain membranes to use chemical energy to pump hydrogen ions and then harness the energy stored in the H^+ gradient to drive cellular work, including ATP synthesis.

chemoautotroph *(KEE-moh-AW-toe-trohf)* An organism that needs only carbon dioxide as a carbon source but that obtains energy by oxidizing inorganic substances.

chemoheterotroph *(KEE-moh-HET-er-oh-trohf)* An organism that must consume organic molecules both for energy and carbon.

chemoreceptor *(KEE-moh-ree-SEP-tur)* A receptor that transmits information about the total solute concentration in a solution or about individual kinds of molecules.

chiasma *(KYE-as-muh)* (plural, **chiasmata**) The X-shaped, microscopically visible region representing homologous chromatids that have exchanged genetic material through crossing over during meiosis.

chitin *(KY-tin)* A structural polysaccharide of an amino sugar found in many fungi and in the exoskeletons of all arthropods.

chlorophyll A green pigment located within the chloroplasts of plants; chlorophyll *a* can participate directly in the light reactions, which convert solar energy to chemical energy.

chloroplast *(KLOR-oh-plast)* An organelle found only in plants and photosynthetic protists that absorbs sunlight and uses it to drive the synthesis of organic compounds from carbon dioxide and water.

cholesterol *(kol-ESS-teh-rol)* A steroid forming an essential component of animal cell membranes and acting as a precursor molecule for the synthesis of other biologically important steroids.

chondrin A protein-carbohydrate complex secreted by chondrocytes; chondrin and collagen fibers form cartilage.

chordate *(KOR-date)* A member of a diverse phylum of animals that possess a notochord; a dorsal, hollow nerve cord; pharyngeal gill slits; and a postanal tail as embryos.

chorion *(KOR-ee-on)* The outermost of four extraembryonic membranes; contributes to formation of the mammalian placenta.

chorionic villi sampling *(KOR-ee-on-ik VILL-eye)* A technique for diagnosing genetic and congenital defects while the fetus is in the uterus. A small sample of the fetal portion of the placenta is removed and analyzed.

chromatin *(KRO-muh-tin)* The aggregate mass of dispersed genetic material formed of DNA and protein and observed between periods of cell division in eukaryotic cells.

chromosome *(KRO-muh-some)* A long, threadlike association of genes in the nucleus of all eukaryotic cells and most visible during mitosis and meiosis. Chromosomes consist of DNA and protein.

cilium *(SILL-ee-um)* (plural, **cilia**) A short cellular appendage specialized for locomotion, formed from a core of nine outer doublet microtubules and two inner single microtubules ensheathed in an extension of plasma membrane.

circadian rhythm *(sur-KAY-dee-un)* A physiological cycle of about 24 hours, present in all eukaryotic organisms, that persists even in the absence of external cues.

cladistics *(kluh-DIS-tiks)* A taxonomic approach that classifies organisms according to the order in time at which branches arise along a phylogenetic tree, without considering the degree of morphological divergence.

cladogenesis *(KLAY-doh-GEN-eh-sis)* A pattern of evolutionary change that produces biological diversity by budding one or more new species from a parent species that continues to exist; also called branching evolution.

classical conditioning A type of associative learning; association of a normally irrelevant stimulus with a fixed behavioral response.

classical evolutionary taxonomy An approach to biological classification and phylogeny that considers both overall homology and evolutionary branching sequences.

cleavage The process of cytokinesis in animal cells, characterized by pinching of the plasma membrane; also, the succession of rapid cell divisions without growth during early embryonic development that converts the zygote into a ball of cells.

cleavage furrow The first sign of cleavage in an animal cell; a shallow groove in the cell surface near the old metaphase plate.

cline *(KLYNE)* Variation in features of individuals in a population that parallels a gradient in the environment.

cloaca *(kloh-AY-kuh)* A common opening for the digestive, urinary, and reproductive tracts in all vertebrates except most mammals.

clonal analysis *(KLON-ul)* Analysis of cell lineages in a developing organism.

clonal selection The mechanism that determines specificity and accounts for memory of antigens in the immune system; occurs because an antigen introduced into the body selectively activates only a tiny fraction of inactive lymphocytes, which proliferate to form a clone of effector cells specific for the stimulating antigen.

clone *(KLONE)* A lineage of genetically identical individuals.

cloning vector An agent used to transfer DNA in genetic engineering, such as a plasmid that moves recombinant DNA from a test tube back into a cell, or a virus that transfers recombinant DNA by infection.

closed circulatory system A type of internal transport in which blood is confined to vessels.

cochlea *(KOH-klee-uh)* The complex, coiled organ of hearing that contains the organ of Corti.

codominance A phenotypic situation in which both alleles are expressed in the heterozygote.

codon *(KOH-don)* A three-nucleotide sequence of DNA or mRNA that specifies a particular amino acid or termination signal and that functions as the basic unit of the genetic code.

coelom *(SEE-loam)* A body cavity completely lined with mesoderm.

coelomate *(SEE-loh-mate)* An animal whose body cavity is completely lined by mesoderm, the layers of which connect dorsally and ventrally to form mesenteries.

coenocytic *(SEN-no-SIT-ik)* Referring to a multinucleated condition resulting from repeated division of nuclei without cytoplasmic division.

coenzyme *(ko-EN-zyme)* An organic molecule serving as a cofactor. Most vitamins function as coenzymes in important metabolic reactions.

coevolution The mutual influence on the evolution of two different species interacting with each other and reciprocally influencing each other's adaptations.

cofactor Any nonprotein molecule or ion that is required for the proper functioning of an enzyme. Cofactors can be permanently bound to the active site or may bind loosely with the substrate during catalysis.

cohesion *(ko-HEE-zhun)* The binding together of like molecules, often by hydrogen bonds.

coleoptile *(KOAL-ee-OP-tyle)* The structure that covers the embryonic shoot in germinating monocot seeds.

collecting duct The location in the kidney where filtrate from renal tubules is collected; the filtrate is now called urine.

collenchyma cell *(koal-EN-keh-muh)* A flexible plant cell type that occurs in strands or cylinders that support young parts of the plant without restraining growth.

commensalism *(kuh-MEN-sul-iz-um)* A symbiotic relationship in which the symbiont benefits but the host is neither helped nor harmed.

community All the organisms that inhabit a particular area; an assemblage of populations of different species living close enough together for potential interaction.

companion cell A type of plant cell that is connected to a sieve-tube member by many plasmodesmata and whose nucleus and ribosomes may serve one or more adjacent sieve-tube members.

competitive exclusion principle The concept that when the populations of two species compete for the same limited resources, one population will use the resources more efficiently and have a reproductive advantage that will eventually lead to the elimination of the other population.

competitive inhibitor A substance that reduces the activity of an enzyme by entering the active site in place of the substrate whose structure it mimics.

complement A group of at least 20 blood proteins that cooperate with other defense mechanisms; may amplify the inflammatory response, enhance phagocytosis, or directly lyse pathogens; activated by the onset of the immune response or by surface chemicals on microorganisms.

complementary DNA (cDNA) DNA that is identical to a native DNA containing a gene of interest except that the cDNA lacks noncoding regions (introns) because it is synthesized in the laboratory using mRNA templates.

complete digestive tract A digestive tube that runs between a mouth and an anus; also called alimentary canal. An incomplete digestive tract has only one opening.

complete flower A flower that has sepals, petals, stamens, and carpels.

compound A chemical combination, in a fixed ratio, of two or more elements.

compound eye A type of multifaceted eye in insects and crustaceans consisting of up to several thousand light-detecting, focusing ommatidia; especially good at detecting movement.

condensation synthesis A process of polymer formation in which monomers are linked together by removing a molecule of water.

cone cell One of two types of photoreceptors in the vertebrate eye; detects color during the day.

conformation The three-dimensional structure of a protein, determined by a specific linear sequence of amino acids that make up a polypeptide chain.

conjugation *(KON-joo-GAY-shun)* A recombination mechanism that results in the transfer of genetic material between two bacterial cells that are temporarily joined.

contraception The prevention of pregnancy.

contractile vacuole An organelle that pumps excess water out of many freshwater protist cells.

convection *(kon-VEK-shun)* The mass movement of warmed air or liquid to or from the surface of a body or object.

convergent evolution The independent development of similarity between species as a result of their having similar ecological roles and selection pressures.

cooperativity *(koh-OP-ur-uh-TIV-eh-tee)* The interaction of the constituent subunits of a protein that causes a conformational change in one subunit to be transmitted to all the others.

corepressor *(KOH-re-PRESS-ur)* A metabolite that cooperates with a repressor protein to block operon activity.

cork cambium *(KAM-bee-um)* A cylinder of meristematic tissue in plants that produces cork cells to replace the epidermis during secondary growth.

corpus callosum *(KOR-pus KAHL-lose-um)* A thick band of nerve fibers connecting the right and left cerebral hemispheres of the vertebrate brain.

corpus luteum *(KOR-pus LOO-tee-um)* Secreting tissue in the ovary that forms from the collapsed follicle after ovulation and produces progesterone.

cortex The region of the root between the stele and epidermis filled with ground tissue.

cotransport The coupling of the "downhill" diffusion of one substance to the "uphill" transport of another against its own concentration gradient.

cotyledons *(KOT-eh-LEE-dons)* The one (monocot) or two (dicot) seed leaves of an angiosperm embryo.

countercurrent exchange The opposite flow of adjacent fluids that maximizes transfer rates; for example, blood in the gills flows in the opposite direction in which water passes over the gills, maximizing oxygen uptake and carbon dioxide loss.

covalent bond *(koh-VALE-ent)* A type of strong chemical bond in which two atoms share one pair of electrons in a mutual valence shell.

crista *(KRIS-tuh)* (plural, **cristae**) An infolding of the inner membrane of a mitochondrion that houses the electron transport chain and the enzyme catalyzing the synthesis of ATP.

crossing over The reciprocal exchange of genetic material between nonsister chromatids during synapsis of meiosis I.

cryptic coloration *(KRIP-tik)* A type of camouflage that makes potential prey difficult to spot against its background.

cuticle *(KYOO-teh-kul)* (1) A waxy covering on the surface of stems and leaves that acts as an adaptation to prevent desiccation in terrestrial plants. (2) The exoskeleton of an arthropod, consisting of layers of protein and chitin that are variously modified for different functions.

cyanobacteria *(sigh-AN-oh-bak-TEER-ee-uh)* Photosynthetic, oxygen-producing bacteria (formerly known as blue-green algae).

cyclic AMP (cAMP) (cyclic adenosine monophosphate) A small, ring-shaped molecule that acts as a chemical signal in slime molds, as an intracellular second messenger in vertebrate endocrine systems, and as a regulator of the *lac* operon.

cyclic electron flow A route of electron flow during the light reactions of photosynthesis that involves only photosystem I and produces ATP but not NADPH or oxygen.

cyclic photophosphorylation *(FO-to-foss-FOR-i-lay-shun)* The production of ATP during the cyclic pathway of electron flow.

cytochrome *(SIGH-tuh-krome)* An iron-containing protein, a component of electron transport chains in mitochondria and chloroplasts.

cytokinesis *(SIGH-toh-kin-EE-sis)* The division of the cytoplasm to form two separate daughter cells immediately after mitosis.

cytokinins *(SIGH-toh-KY-nins)* A class of related plant hormones that retard aging and act in concert with auxins to stimulate cell division, influence the pathway of differentiation, and control apical dominance.

cytoplasm *(SIGH-toh-plaz-um)* The entire contents of the cell, exclusive of the nucleus, and bounded by the plasma membrane.

cytoplasmic determinants Localized cellular substances that become partitioned into specific blastomeres of an embryo, thereby affecting the developmental fates of different regions very early.

cytoskeleton *(SIGH-toh-SKEL-eh-ton)* A network of microtubules, microfilaments, and intermediate filaments that ramify throughout the cytoplasm that serve a variety of mechanical and transport functions.

cytosol *(SIGH-toh-sol)* The semifluid portion of the cytoplasm.

dalton *(DAWL-ton)* The atomic mass unit. A measure of mass for atoms and subatomic particles.

Darwinian fitness A measure of the relative contribution of an individual to the gene pool of the next generation.

day-neutral plant A plant whose flowering is not affected by photoperiod.

decomposers Saprotrophic fungi and bacteria that absorb nutrients from nonliving organic material such as corpses, fallen plant material, and the wastes of live organisms, and convert them into inorganic forms.

deletion (1) A deficiency in a chromosome resulting from loss of a fragment through breakage. (2) A mutational loss of a nucleotide from a gene.

demography *(de-MOG-ruf-ee)* The study of statistics relating to births and deaths in populations.

denaturation A process in which a protein unravels and loses its native conformation, thereby becoming biologically inactive. Denaturation occurs under extreme conditions of pH, salt concentration, and temperature.

dendrite *(DEN-dryt)* One of usually numerous, short, highly branched processes of a neuron that conveys nerve impulses toward the cell body.

density The number of individuals per unit area or volume.

density-dependent factor Any factor influencing population regulation that has a greater impact as population density increases.

density-dependent inhibition The phenomenon observed in normal animal cells that causes them to stop dividing when they come into contact with one another.

density-independent factor Any factor influencing population regulation that acts to reduce population by the same percentage, regardless of size.

deoxyribonucleic acid (DNA) *(DEE-ox-ee-rye-boh-noo-KLAY-ik)* A double-stranded, helical nucleic acid molecule capable of replicating and determining the inherited structure of a cell's proteins.

depolarization *(dee-POL-ur-iz-AY-shun)* An electrical state in an excitable cell whereby the inside of the cell is made less negative relative to the outside than was the case at resting potential. A neuron membrane is depolarized if a stimulus decreases its voltage from the resting potential of –70 mV in the direction of zero voltage.

deposit-feeder A heterotroph, such as an earthworm, that eats its way through detritus, salvaging bits and pieces of decaying organic matter.

dermal tissue system The protective covering of plants; generally a single layer of tightly packed epidermal cells covering young plant organs formed by primary growth.

desmosome *(DEZ-muh-soam)* A type of intercellular junction in animal cells that functions as an anchor.

determinate cleavage A type of embryonic development in protostomes that rigidly casts the developmental fate of each embryonic cell very early.

determinate growth A type of growth characteristic of animals, in which the organism stops growing after it reaches a certain size.

determination The progressive restriction of developmental potential, causing the possible fate of each cell to become more limited as the embryo develops.

detritivore *(deh-TRYT-eh-vore)* A member of a class of consumers that derives its energy from organic wastes and dead organisms representing all trophic levels.

detritus *(deh-TRY-tis)* Dead organic matter.

deuterostomes *(DOO-ter-oh-stomes)* One of two distinct evolutionary lines of coelomates, consisting of the echinoderms and chordates and characterized by radial, indeterminate cleavage, enterocoelous formation of the coelom, and development of the anus from the blastopore.

diaphragm A sheet of muscle that forms the bottom wall of the thoracic cavity in mammals; active in ventilating the lungs.

diastole *(die-ASS-tuh-lee)* The stage of the heart cycle in which the heart muscle is relaxed, allowing the chambers to fill with blood.

dicot *(DIE-kot)* A subdivision of flowering plants whose members possess two embryonic seed leaves, or cotyledons.

differentiation *See* cellular differentiation.

diffusion *(deh-FYU-shun)* The spontaneous tendency of a substance to move down its concentration gradient from a more concentrated to a less concentrated area.

digestion The process of breaking down food into molecules small enough for the body to absorb.

dihybrid cross *(DIE-HIGH-brid)* A breeding experiment in which parental varieties differing in two traits are mated.

dikaryon *(die-KAH-ree-on)* A mycelium of certain septate fungi that possesses two separate haploid nuclei per cell.

dioecious *(die-EE-shus)* Referring to a plant species that has staminate and carpellate flowers on separate plants.

diploid cell *(DIP-loyd)* A cell containing two sets of chromosomes ($2n$), one set inherited from each parent.

directional selection Natural selection that favors individuals on one end of the phenotypic range.

disaccharide *(die-SAK-ur-ide)* A double sugar, consisting of two monosaccharides joined by dehydration synthesis.

dispersion The distribution of individuals within geographical population boundaries.

diversifying selection Natural selection that favors extreme over intermediate phenotypes.

DNA-DNA hybridization The comparison of whole genomes of two species by estimating the extent of hydrogen bonding that occurs between single-stranded DNA obtained from the two species.

DNA ligase *(LYE-gaze)* A linking enzyme essential for DNA replication; catalyzes the covalent bonding of the $3'$ end of a new DNA fragment to the $5'$ end of a growing chain.

DNA methylation The addition of methyl groups ($—CH_3$) to bases of DNA after DNA synthesis; may serve as a long-term control of gene expression.

DNA polymerase An enzyme that catalyzes the elongation of new DNA at a replication fork in the $5' \longrightarrow 3'$ direction by the addition of nucleotides to the existing chain.

DNA probe A chemically synthesized, radioactively labeled segment of nucleic acid used to find a gene of interest by hydrogen-bonding to a complementary sequence.

domain (1) A structural and functional portion of a polypeptide that may be coded for by a specific exon; a globular region of a protein with tertiary structure. (2) A taxonomic category above the kingdom level; the three domains are archaebacteria, eubacteria, and eukaryotes.

dominance hierarchy A linear "pecking order" of animals, where position dictates characteristic social behaviors.

dominant allele In a heterozygote, the allele that is fully expressed in the phenotype.

dorsal Referring to the top (or back) half of a bilaterally symmetric animal.

dorsal lip The upper edge of the blastopore produced by invagination during gastrula formation in amphibian embryos; the site toward which surface cells of the gastrula converge and migrate inward along the roof of the blastocoel in the process of involution.

double circulation A circulation scheme with separate pulmonary and systemic circuits, which ensures vigorous blood flow to all organs.

double fertilization A mechanism of fertilization in angiosperms, in which two sperm cells unite with two cells in the embryo sac to form the zygote and endosperm.

double helix The form of native DNA, referring to its two adjacent polynucleotide strands wound into a spiral shape.

Down syndrome A human genetic disease resulting from having an extra chromosome 21, characterized by mental retardation and heart and respiratory defects.

duplication An aberration in chromosome structure resulting from an error in meiosis or mutagens; duplication of a portion of a chromosome resulting from fusion with a fragment from a homologous chromosome.

dynein *(DY-nin)* A large contractile protein forming the sidearms of microtubule doublets in cilia and flagella.

ecdysone *(EK-deh-sone)* A steroid hormone that triggers molting in arthropods.

ecological efficiency The ratio of net productivity at one trophic level to net productivity at the next lower level.

ecological niche *(NICH)* The sum total of an organism's utilization of the biotic and abiotic resources of its environment.

ecological succession Transition in the species composition of a biological community, often following ecological disturbance of the community; the establishment of a biological community in an area virtually barren of life.

ecology *(ee-KOL-uh-jee)* The study of how organisms interact with their environments.

ecosystem *(EE-koh-sis-tum)* A level of ecological study that includes all the organisms in a given area as well as the abiotic factors with which they interact; a community and its physical environment.

ectoderm *(EK-tuh-durm)* The outermost of the three primary germ layers in animal embryos; gives rise to the outer covering and, in some phyla, to the nervous system, inner ear, and lens of the eye.

ectotherm *(EK-toh-thurm)* An animal, such as a reptile, fish, or amphibian, that must use environmental energy and behavioral adaptations to regulate its body temperature.

effector cell A muscle cell or gland cell that performs the body's responses to stimuli; responds to signals from the brain or other processing center of the nervous system.

electrical synapse *(SIN-aps)* A junction between two neurons separated only by a gap junction, in which the local currents sparking the action potential pass directly between the cells.

electrocardiogram (EKG or ECG) A plot of electrical activity of the heart over the cardiac cycle; measured via multiple skin electrodes.

electrochemical gradient The diffusion gradient of an ion, representing a type of potential energy that accounts for both the concentration difference of the ion across a membrane and its tendency to move relative to the membrane potential.

electrogenic pump An ion transport protein generating voltage across the membrane.

electromagnetic spectrum The entire spectrum of radiation; ranges in wavelength from less than a nanometer to more than a kilometer.

electron microscope (EM) A microscope that focuses an electron beam through a specimen, resulting in resolving power a thousandfold greater than that of a light microscope. A transmission electron microscope (TEM) is used to study the internal structure of thin sections of cells. A scanning electron microscope (SEM) is used to study the fine details of cell surfaces.

electron transport chain A group of molecules in the inner membrane of a mitochondrion that synthesize ATP by means of an exergonic slide of electrons. Thylakoid membranes of chloroplasts are also equipped with electron transport chains.

electronegativity *(eh-LEK-troh-neg-uh-TIV-eh-tee)* The tendency for an atom to pull electrons toward itself.

element Any substance that cannot be broken down to any other substance.

embryo sac The female gametophyte of angiosperms, formed from the growth and division of the megaspore into a multicellular structure with eight haploid nuclei.

endergonic reaction *(EN-dur-GON-ik)* A nonspontaneous chemical reaction in which free energy is absorbed from the surroundings.

endocrine glands *(EN-doh-krin)* Ductless glands that secrete hormones directly into the bloodstream.

endocrine system The internal system of chemical communication involving hormones, the ductless glands that secrete hormones, and the molecular receptors on or in target cells that respond to hormones; functions in concert with the nervous system to effect internal regulation and maintain homeostasis.

endocytosis *(EN-do-sigh-TOE-sis)* The cellular uptake of macromolecules and particulate substances by localized regions of the plasma membrane that surround the substance and pinch off to form an intracellular vesicle.

endoderm *(EN-doh-durm)* The innermost of the three primary germ layers in animal embryos; lines the archenteron and gives rise to the liver, pancreas, lungs, and the lining of the digestive tract.

endodermis *(EN-doh-DER-mis)* The innermost layer of the cortex in plant roots; a cylinder one cell thick that forms the boundary between the cortex and the stele.

endomembrane system The collection of membranes inside and around a eukaryotic cell, related either through direct physical contact or by transfer of membranous vesicles.

endometrium *(EN-doh-MEE-tree-um)* The inner lining of the uterus, which is richly supplied with blood vessels that provide the maternal part of the placenta and nourish the developing embryo.

endoplasmic reticulum (ER) *(EN-doh-plaz-mik reh-TIK-yoo-lum)* An extensive membranous network in eukaryotic cells, continuous with the outer nuclear membrane and composed of ribosome-studded (rough) and ribosome-free (smooth) regions.

endorphin *(en-DOR-fin)* A hormone produced in the brain and anterior pituitary that inhibits pain perception.

endoskeleton *(EN-doh-SKEL-eh-ton)* A hard skeleton buried within the soft tissues of an animal, such as the spicules of sponges, the plates of echinoderms, and the bony skeletons of vertebrates.

endosperm *(EN-doh-spurm)* A nutrient-rich tissue formed by union of a sperm cell with two polar nuclei during double fertilization, which provides nourishment to the developing embryo in angiosperm seeds.

endosymbiotic model *(EN-doh-SIM-by-OT-ik)* A hypothesis about the origin of the eukaryotic cell maintaining that the forerunners of eukaryotic cells were symbiotic associations of prokaryotic cells living inside larger prokaryotes.

endothelium *(EN-doh-THEEL-ee-um)* The innermost, simple squamous layer of cells lining the blood vessels; the only constituent structure of capillaries.

endotherm *(EN-doh-thurm)* An animal that uses metabolic energy to maintain a constant body temperature, such as a bird or mammal.

endotoxin *(EN-doh-TOKS-in)* A component of the outer membranes of certain gram-negative bacteria responsible for generalized symptoms of fever and ache.

energy The capacity to do work by moving matter against an opposing force.

enhancer A DNA sequence that recognizes certain transcription factors that can stimulate transcription of nearby genes.

enkephalin *(en-KEF-uh-lin)* A hormone produced in the brain and anterior pituitary that inhibits pain perception.

entropy *(EN-truh-pee)* A quantitative measure of disorder or randomness, symbolized by *S*.

enzymes A class of proteins serving as catalysts, chemical agents that change the rate of a reaction without being consumed by the reaction.

epicotyl *(EP-eh-KOT-ul)* In angiosperm seeds, the portion of the embryonic axis above the cotyledons.

epidermis *(EP-eh-DER-mis)* (1) The dermal tissue system in plants. (2) The outer covering of animals.

epigenesis *(EP-eh-JEN-eh-sis)* The progressive development of form in an embryo.

epinephrine A hormone produced as a response to stress; also called adrenaline.

epiphyte *(EP-eh-fite)* A plant that nourishes itself but grows on the surface of another plant for support, usually on the branches or trunks of tropical trees.

episome *(EP-eh-soam)* A plasmid capable of integrating into the bacterial chromosome.

epistasis A phenomenon in which one gene alters the expression of another gene that is independently inherited.

epithelial tissue *(EP-eh-THEEL-ee-ul)* Sheets of tightly packed cells line organs and body cavities.

epitope A localized region on the surface of an antigen that is chemically recognized by antibodies; also called antigenic determinant.

equilibrium potential The membrane potential for a given ion at which the voltage exactly balances the chemical diffusion gradient for that ion.

erythrocyte *(er-RITH-roh-site)* A red blood cell; contains hemoglobin, which functions in transporting oxygen in the circulatory system.

essential nutrient (1) A chemical element required for a plant to grow from a seed and complete the life cycle. (2) A nutrient substance that an animal cannot make itself from raw materials but that must be obtained in food in prefabricated form.

estivation *(ES-teh-VAY-shun)* A physiological state characterized by slow metabolism and inactivity, which permits survival during long periods of elevated temperature and diminished water supplies.

estrogens *(ES-troh-jens)* The primary female steroid sex hormones, which are produced in the ovary by the developing follicle during the first half of the cycle and in smaller quantities by the corpus luteum during the second half. Estrogens maintain the female reproductive system and develop secondary female sex characteristics.

estrous cycle *(ES-trus)* A type of reproductive cycle in all female mammals except higher primates, in which the nonpregnant endometrium is reabsorbed rather than shed, and sexual response occurs only during midcycle at estrus.

estrus *(ES-trus)* The limited period of heat or sexual receptivity that occurs around ovulation in female mammals having estrous cycles.

ethylene *(ETH-ul-een)* The only gaseous plant hormone, responsible for fruit ripening, growth inhibition, leaf abscission, and aging.

eubacteria *(YOO-bak-TEER-ee-uh)* The lineage of prokaryotes that includes the cyanobacteria and all other contemporary bacteria except archaebacteria.

euchromatin *(yoo-KROW-muh-tin)* The more open, unraveled form of eukaryotic chromatin, which is available for transcription.

eukaryotic cell *(YOO-kar-ee-OT-ik)* A type of cell with a membrane-enclosed nucleus and membrane-enclosed organelles, present in protists, plants, fungi, and animals. Also called eukaryote.

eumetazoa *(YOO-met-uh-ZOH-uh)* Members of the subkingdom that includes all animals except sponges.

eutrophic lake A highly productive lake, having a high rate of biological productivity supported by a high rate of nutrient cycling.

evaporative cooling The property of a liquid whereby the surface becomes cooler during evaporation, owing to the loss of highly kinetic molecules to the gaseous state.

evolution All the changes that have transformed life on Earth from its earliest beginnings to the diversity that characterizes it today.

excitable cells A cell, such as a neuron or a muscle cell, that can use changes in its membrane potential to conduct signals.

excitatory postsynaptic potential (EPSP) *(POST-sin-AP-tik)* An electrical change (depolarization) in the membrane of a postsynaptic neuron caused by the binding of an excitatory neurotransmitter from a presynaptic cell to a postsynaptic receptor; makes it more likely for a postsynaptic neuron to generate an action potential.

excretion *(ex-KREE-shun)* The disposal of nitrogen-containing waste products of metabolism.

exergonic reaction *(EX-ur-GON-ik)* A spontaneous chemical reaction in which there is a net release of free energy.

exocytosis *(EX-oh-sigh-TOE-sis)* The cellular secretion of macromolecules by the fusion of vesicles with the plasma membrane.

exon The coding region of a eukaryotic gene that is expressed. Exons are separated from each other by introns.

exoskeleton A hard encasement on the surface of an animal, such as the shells of mollusks or the cuticles of arthropods, that provides protection and points of attachment for muscles.

exotoxin *(EX-oh-TOX-in)* A toxic protein secreted by a bacterial cell that produces specific symptoms even in the absence of the bacterium.

exponential population growth The geometric increase of a population as it grows in an ideal, unlimited environment.

extraembryonic membranes *(EX-truh-EM-bree-AHN-ik)* Four membranes (yolk sac, amnion, chorion, allantois) that support the developing embryo in reptiles, birds, and mammals.

F_1 generation The first filial or hybrid offspring in a genetic cross-fertilization.

F_2 generation Offspring resulting from interbreeding of the hybrid F_1 generation.

F plasmid The fertility factor in bacteria, a plasmid that confers the ability to form pili for conjugation and associated functions required for transfer of DNA from donor to recipient.

facilitated diffusion The spontaneous passage of molecules and ions, bound to specific carrier proteins, across a biological membrane down their concentration gradients.

facultative anaerobe *(FAK-ul-tay-tiv AN-uh-robe)* An organism that makes ATP by aerobic respiration if oxygen is present but that switches to fermentation under anaerobic conditions.

fat (triacylglycerol) *(tri-A-sil-GLI-ser-all)* A biological compound consisting of three fatty acids linked to one glycerol molecule.

fate map A means of tracing the fates of cells during embryonic development.

fatty acid A long carbon chain carboxylic acid. Fatty acids vary in length and in the number and location of double bonds; three fatty acids linked to a glycerol molecule form fat.

feedback inhibition A method of metabolic control in which the end-product of a metabolic pathway acts as an inhibitor of an enzyme within that pathway.

fermentation A catabolic process that makes a limited amount of ATP from glucose without an electron transport chain and that produces a characteristic end-product, such as ethyl alcohol or lactic acid.

fertilization The union of haploid gametes to produce a diploid zygote.

fetus The human embryo with all the major structures of the adult present in rudimentary form after about eight weeks.

fiber A lignified cell type that reinforces the xylem of angiosperms and functions in mechanical support; a slender, tapered sclerenchyma cell that usually occurs in bundles.

fibrin *(FY-brin)* The activated form of the blood-clotting protein fibrinogen, which aggregates into threads that form the fabric of the clot.

fibroblast *(FY-broh-blast)* A type of cell in loose connective tissue that secretes the protein ingredients of the extracellular fibers.

fibronectins *(FY-broh-NEK-tins)* A family of extracellular glycoproteins that helps embryonic cells adhere to their substrate as they migrate.

first law of thermodynamics *(THUR-moe-die-NAM-iks)* The principle of conservation of energy. Energy can be transferred and transformed, but it cannot be created or destroyed.

fixed-action pattern (FAP) A highly stereotyped behavior that is innate and must be carried to completion once initiated.

flaccid *(FLAS-ed)* Limp; walled cells are flaccid in isotonic surroundings, where there is no tendency for water to enter.

flagellum *(fluh-JEL-um)* (plural, **flagella**) A long cellular appendage specialized for locomotion, formed from a core of nine outer doublet microtubules and two inner single microtubules, ensheathed in an extension of plasma membrane.

flame-cell system The simplest tubular excretory system, present in flatworms and acting to directly regulate the contents of the extracellular fluid.

fluid-feeder An animal that lives by sucking nutrient-rich fluids from another living organism.

fluid mosaic model The currently accepted model of cell membrane structure, which envisions the membrane as a mosaic of individually inserted protein molecules drifting laterally in a fluid bilayer of phospholipids.

follicles *(FOL-eh-kuls)* Microscopic structures in the ovary that contain developing ova and secrete estrogens.

food chain The pathway along which food is transferred from trophic level to trophic level, beginning with producers.

food web The elaborate, interconnected feeding relationships in an ecosystem.

fossil A relic or impression of an organism from the past, preserved in rock.

founder effect A cause of genetic drift attributable to colonization by a limited number of individuals from a parent population.

fragmentation A mechanism of asexual reproduction in which the parent plant or animal separates into parts that reform whole organisms.

frameshift mutation A mutation occurring when the number of nucleotides inserted or deleted is not a multiple of 3, thus resulting in improper grouping into codons.

free energy A quantity of energy that interrelates entropy (S) and the system's total energy (H); symbolized by G. The change in free energy of a system is calculated by the equation $G = \Delta H - T\Delta S$, where T is absolute temperature.

free energy of activation The initial investment of energy necessary to start a chemical reaction; also called activation energy.

frequency-dependent selection A decline in the reproductive success of a morph resulting from the morph's phenotype becoming too common in a population; a cause of balanced polymorphism in populations.

fruit A mature ovary of a flower that protects dormant seeds and aids in their dispersal.

functional group A specific configuration of atoms commonly attached to the carbon skeletons of organic molecules and usually involved in chemical reactions.

fundamental niche The total resources an organism is theoretically capable of utilizing.

G protein A membrane protein that functions as an intermediary between hormone receptors in cell membranes and the enzyme adenylate cyclase, which converts ATP to cAMP in the second messenger (cAMP) system in nonsteroid hormone action. Depending on the type of hormone, G proteins may increase or decrease the activity of adenylate cyclase in producing cAMP.

G_1 phase The first growth phase of the cell cycle, consisting of the portion of interphase before DNA synthesis begins.

G_2 phase The second growth phase of the cell cycle, consisting of the portion of interphase after DNA synthesis occurs.

gametangium *(GAM-eh-TANJ-ee-um)* (plural, **gametangia**) The reproductive organ of bryophytes, consisting of the male antheridium and female archegonium; a multichambered jacket of sterile cells in which gametes are formed.

gametes *(GAM-eets)* Haploid egg or sperm cells that unite during sexual reproduction to produce a diploid zygote.

gametophyte *(guh-ME-toh-fite)* The multicellular haploid form in organisms undergoing alternation of generations, which mitotically produces haploid gametes that unite and grow into the sporophyte generation.

ganglion *(GANG-lee-un)* (plural, **ganglia**) A cluster (functional group) of nerve cell bodies in a centralized nervous system.

gap junction A type of intercellular junction in animal cells that allows the passage of material or current between cells.

gastrovascular cavity The central digestive compartment, usually with a single opening that functions as both mouth and anus.

gastrula *(GAS-troo-la)* The two-layered, cup-shaped embryonic stage.

gastrulation *(GAS-truh-LAY-shun)* The formation of a gastrula from a blastula.

gated ion channel A specfic ion channel that opens and closes to allow the cell to alter its membrane potential.

gel electrophoresis *(JELL eh-LEK-troh-for-EE-sis)* Separation of nucleic acids or proteins, on the basis of their size and electric charge, by measuring their rate of movement through an electric field in a gel.

gene One of many discrete units of hereditary information located on the chromosomes and consisting of DNA.

gene amplification The selective synthesis of DNA, which results in multiple copies of a single gene, thereby enhancing expression.

gene cloning Formation by a bacterium, carrying foreign genes in a recombinant plasmid, of a clone of identical cells containing the replicated foreign genes.

gene flow The loss or gain of alleles from a population due to the emigration or immigration of fertile individuals, or the transfer of gametes, between populations.

gene pool The total aggregate of genes in a population at any one time.

generation time The average age when females of a population begin reproducing, which greatly influences the intrinsic rate of increase.

genetic drift Changes in the gene pool of a small population due to chance.

genetic recombination The general term for the production of offspring that combine traits of the two parents.

genetics The science of heredity; the study of heritable information.

genome *(JEE-nome)* The complete complement of an organism's genes; an organism's genetic material.

genomic equivalence *(jen-OME-ik)* The presence of all of an organism's genes in all of its cells.

genomic imprinting The parental effect on gene expression. Identical alleles may have different effects on offspring depending on whether they arrive in the zygote via the ovum or via the sperm.

genomic library A set of thousands of DNA segments from a genome, each carried by a plasmid or phage.

genotype *(JEE-noh-type)* The genetic makeup of an organism.

genus *(JEE-nus)* (plural, **genera**) A taxonomic category above the species level, designated by the first word of a species' binomial Latin name.

geographical range The geographic area in which a population lives.

gibberellins *(JIB-ur-EL-ins)* A class of related plant hormones that stimulate growth in the stem and leaves, trigger germination of seeds and breaking of bud dormancy, and stimulate fruit development with auxin.

gills Localized extensions of the body surfaces of many aquatic animals specialized for gas exchange.

glial cell *(GLEE-ul)* A nonconducting cell of the nervous system that provides support, insulation, and protection for the neurons.

glomerulus *(glum-AIR-yoo-lus)* A ball of capillaries surrounded by the Bowman's capsule in the nephron and serving as the site of filtration in the vertebrate kidney.

glycocalyx *(GLY-koh-KAY-liks)* A fuzzy coat on the outside of animal cells, made of sticky oligosaccharides.

glycogen *(GLY-koh-jen)* An extensively branched glucose storage polysaccharide found in the liver and muscle of animals; the animal equivalent of starch.

glycolysis *(gly-KOL-eh-sis)* The splitting of glucose into pyruvate. Glycolysis is the one metabolic pathway that occurs in all living cells, serving as the starting point for fermentation or aerobic respiration.

Golgi apparatus *(GOAL-jee)* An organelle in eukaryotic cells consisting of stacks of membranes that modify, store, and route products of the endoplasmic reticulum.

gonadotropins *(go-NAD-oh-TROH-pins)* Hormones that stimulate the activities of the testes and ovaries; a term for follicle-stimulating and luteinizing hormones.

gonads *(GONE-adz)* The male and female sex organs; the gamete-producing organs in most animals.

graded potential A local voltage change in a neuron membrane induced by stimulation of a neuron, with strength proportional to the strength of the stimulus and lasting about a millisecond.

gradualism A view of Earth's history that attributes profound change to the cumulative product of slow but continuous processes.

granum *(GRAN-um)* (plural, **grana**) A stacked portion of the thylakoid membrane in the chloroplast. Grana function in the light reactions of photosynthesis.

gravitropism *(GRAV-eh-TROH-piz-um)* A response of a plant or animal in relationship to gravity.

greenhouse effect The warming of the Earth due to atmospheric accumulation of carbon dioxide, which absorbs infrared radiation and slows its escape from the irradiated Earth.

gross primary productivity The total primary productivity of an ecosystem.

ground tissue system A tissue of mostly parenchyma cells that makes up the bulk of a young plant and fills the space between the dermal and vascular tissue systems.

growth factor A protein that must be present in the extracellular environment (culture medium or animal body) for the growth and normal development of certain types of cells.

guard cell A specialized epidermal plant cell that forms the boundaries of the stomata.

gymnosperm *(JIM-noh-spurm)* A vascular plant that bears naked seeds not enclosed in any specialized chambers.

habituation *(huh-BIT-choo-AY-shun)* A simple kind of learning involving loss of sensitivity to unimportant stimuli, allowing an animal to conserve time and energy.

hair cell A common type of mechanoreceptor in the vertebrate ear, in the lateral line system of fishes and amphibians, and in arthropod statocysts, that detects motion of various kinds.

half-life The average amount of time it takes for one-half of a specified quantity of a substance to decay or disappear.

haploid cell *(HAP-loid)* A cell containing only one set of chromosomes *(n)*.

Hardy-Weinberg equilibrium Stability in the frequency of alleles and genotypes in a population generation after generation; a state of equilibrium in a population's gene pool.

Hardy-Weinberg theorem An axiom maintaining that the sexual shuffling of genes alone cannot alter the overall genetic makeup of a population.

haustorium (plural, **haustoria**) In parasitic fungi, a nutrient-absorbing hyphal tip that penetrates the tissues of the host but remains outside the host cell membranes.

Haversian system *(ha-VER-shun)* One of many structural units of vertebrate bone, consisting of concentric layers of mineralized bone matrix surrounding lacunae, which contain osteocytes, and a central canal, which contains blood vessels and nerves.

heat The total amount of kinetic energy due to molecular motion in a body of matter. Heat is energy in its most random form.

helper T cells (T_H) A type of T cell that is required by some B cells to help them make antibodies or that helps other T cells respond to antigens or secrete lymphokines or interleukins.

hemoglobin *(HEE-moh-gloh-bin)* An iron-containing protein in red blood cells that reversibly binds oxygen.

hemolymph In invertebrates with open criculatory systems, the body fluid that bathes tissues.

hemophilia *(HEEM-uh-FIL-ee-uh)* A genetic disease resulting from an abnormal sex-linked recessive gene, characterized by excessive bleeding following injury.

herbivore A heterotrophic animal that eats plants.

hermaphrodite *(her-MAF-roh-dite)* An individual that functions as both male and female in sexual reproduction by producing both sperm and eggs.

heterochromatin *(HET-ur-oh-KROH-muh-tin)* Nontranscribed eukaryotic chromatin that is so highly compacted that it is visible with a light microscope during interphase.

heterochrony Evolutionary changes in the timing or rate of development.

heterocyst *(HET-ur-oh-sist)* A specialized cell that engages in nitrogen fixation on some filamentous cyanobacteria.

heteromorphic *(HET-ur-oh-MOR-fik)* A condition in the life cycle of all modern plants in which the sporophyte and gametophyte generations differ in morphology.

heterosporous *(HET-ur-OS-pur-us)* Referring to plants in which the sporophyte produces two kinds of spores that develop into unisexual gametophytes, either female or male.

heterotrophic nutrition *(HET-ur-oh-TROH-fik)* A mode of obtaining organic food molecules by eating other organisms or their by-products.

heterozygote advantage *(HET-ur-oh-ZY-gote)* A mechanism that preserves variation in eukaryotic gene pools by conferring greater reproductive success on heterozygotes over individuals homozygous for any one of the associated alleles.

heterozygous *(HET-ur-oh-ZY-gus)* Having two different alleles for a given trait.

hibernation A physiological state that allows survival during long periods of cold and diminished food, in which metabolism decreases, the heart and respiratory system slow down, and body temperature is maintained at a lower level than normal.

histamine *(HISS-tuh-meen)* A substance released by injured cells that causes blood vessels to dilate during an inflammatory response.

histone *(HISS-tone)* A small protein with a high proportion of positively charged amino acids that binds to the negatively charged DNA and plays a key role in its folding into chromatin.

HIV (human immunodeficiency virus) The infectious agent that causes AIDS; HIV is an RNA retrovirus.

holoblastic cleavage *(HO-loh-BLAS-tik)* A type of cleavage in which there is complete division of the egg, as in eggs having little yolk (sea urchin) or a moderate amount of yolk (frog).

homeobox Specific sequences of DNA that regulate patterns of differentiation during development of an organism.

homeostasis *(HOME-ee-oh-STAY-sis)* The steady-state physiological condition of the body.

homeotic genes *(HOME-ee-OT-ik jeens)* Genes that control the overall body plan of animals by controlling the developmental fate of groups of cells.

homeotic mutation A mutation in genes regulated by positional information that results in the abnormal substitution of one type of body part in place of another.

homologous chromosomes *(home-OL-uh-gus)* Chromosome pairs of the same length, centromere position, and staining pattern that possess genes for the same traits at corresponding loci. One homologous chromosome is inherited from the organism's father, the other from the mother.

homologous structures Structures in different species that are similar because of common ancestry.

homology *(home-OL-uh-gee)* Similarity in characteristics resulting from a shared ancestry.

homosporous *(home-OS-pur-us)* Referring to plants in which a single type of spore develops into a bisexual gametophyte having both male and female sex organs.

homozygous *(HOME-oh-ZY-gus)* Having two identical alleles for a given trait.

hormone *(HOR-mone)* One of many types of circulating chemical signals in all multicellular organisms, that are formed in specialized cells, travel in body fluids, and coordinate the various parts of the organism by interacting with target cells.

host range The limited number of host species, tissues, or cells that a parasite (including viruses and bacteria) can infect.

humoral immunity *(HYOO-mur-al)* The type of immunity that fights bacteria and viruses in body fluids with antibodies that circulate in blood plasma and lymph, fluids formerly called humors.

hybrid vigor Increased vitality (compared to that of either parent stock) in the hybrid offspring of two different, inbred parents.

hybridoma *(HY-brid-OH-muh)* A hybrid cell that produces monoclonal antibodies in culture, formed by the fusion of a myeloma cell with a normal antibody-producing lymphocyte.

hydrocarbon *(HY-droh-kar-bon)* An organic molecule consisting only of carbon and hydrogen.

hydrogen bond A type of weak chemical bond formed when the slightly positive hydrogen atom of a polar covalent bond in one molecule is attracted to the slightly negative atom of a polar covalent bond in another molecule.

hydrogen ion A single proton with a charge of +1. The dissociation of a water molecule (H_2O) leads to the generation of a hydroxide ion (OH^-) and a hydrogen ion (H^+).

hydrolysis *(hy-DROL-eh-sis)* A chemical process that lyses or splits molecules by the addition of water; an essential process in digestion.

hydrophilic *(HY-droh-FIL-ik)* Having an affinity for water.

hydrophobic *(HY-droh-FOBE-ik)* Having an aversion to water; tending to coalesce and form droplets in water.

hydrophobic interaction *(HY-droh-FOH-bik)* A type of weak chemical bond formed when molecules that do not mix with water coalesce to exclude the water.

hydrostatic skeleton *(HY-droh-STAT-ik)* A skeletal system composed of fluid held under pressure in a closed body compartment; the main skeleton of most cnidarians, flatworms, nematodes, and annelids.

hydroxyl group *(hy-DROKS-ul)* A functional group consisting of a hydrogen atom joined to an oxygen atom by a polar covalent bond. Molecules possessing this group are soluble in water and are called alcohols.

hyperosmotic solution *(HY-pur-os-MAH-tik)* A solution with a greater solute concentration than another, a hypoosmotic solution.

hyperpolarization *(HY-pur-POLE-ur-i-ZAY-shun)* An electrical state whereby the inside of the cell is made more negative relative to the outside than was the case at resting potential. A neuron membrane is hyperpolarized if a stimulus increases its voltage from the resting potential of –70 mV; reducing the chance that the neuron will transmit a nerve impulse.

hypha *(HY-fa)* (plural, **hyphae**) A filament that collectively makes up the body of a fungus.

hypocotyl *(HY-poh-kot-ul)* The embryonic axis of a plant below the point at which the cotyledons are attached, terminating in the radicle.

hypoosmotic solution *(HY-poh-oz-MAH-tik)* A solution with a lesser solute concentration than another, a hyperosmotic solution.

hypothalamus *(HY-poh-THAL-uh-mus)* The ventral part of the vertebrate forebrain; functions in maintaining homeostasis, especially in coordinating the endocrine and nervous systems; secretes hormones of the posterior pituitary and releasing factors, which regulate the anterior pituitary.

imaginal disk *(i-MAJ-in-ul)* An island of undifferentiated cells in an insect larva, which are committed (determined) to form a particular organ during metamorphosis to the adult.

imbibition *(IM-bih-BISH-un)* The soaking of water into a porous material that is hydrophilic.

immunoglobulins (Ig) *(IM-myoo-noh-GLOB-yoo-lins)* The class of proteins comprising the antibodies.

imprinting A type of learned behavior with a significant innate component, acquired during a limited critical period.

incomplete dominance A type of inheritance in which F_1 hybrids have an appearance that is intermediate between the phenotypes of the parental varieties.

incomplete flower A flower lacking sepals, petals, stamens, or carpels.

incomplete metamorphosis *(MET-uh-MOR-foh-sis)* A type of development in certain insects, such as grasshoppers, in which the larvae resemble adults but are smaller and have different body proportions. The animal goes through a series of molts, each time looking more like an adult, until it reaches full size.

indeterminate cleavage A type of embryonic development in deuterostomes, in which each cell produced by early cleavage divisions retains the capacity to develop into a complete embryo.

indeterminate growth A type of growth characteristic of plants, in which the organism continues to grow as long as it lives.

induced fit The change in shape of the active site of an enzyme so that it binds more snugly to the substrate, induced by entry of the substrate.

induction *(in-DUK-shun)* The ability of one group of embryonic cells to influence the development of another.

inflammatory response A line of defense triggered by penetration of the skin or mucous membranes, in which small blood vessels in the vicinity of an injury dilate and become leakier, enhancing infiltration of leukocytes; may also be widespread in the body.

ingestion *(in-JEST-shun)* A heterotrophic mode of nutrition in which other organisms or detritus are eaten whole or in pieces.

inhibitory postsynaptic potential (IPSP) *(POST-sin-AP-tik)* An electrical charge (hyperpolarization) in the membrane of a postsynaptic neuron caused by the binding of an inhibitory neurotransmitter from a presynaptic cell to a postsynaptic receptor; makes it more difficult for a postsynaptic neuron to generate an action potential.

inner cell mass A cluster of cells in a mammalian blastocyst that protrudes into one end of the cavity and subsequently develops into the embryo proper and some of the extraembryonic membranes.

inositol triphosphate (IP_3) *(in-NOS-i-tahl)* The second messenger, which functions as an intermediate between certain nonsteroid hormones and the third messenger, a rise in cytoplasmic Ca^{2+} concentration.

insertion A mutation involving the addition of one or more nucleotide pairs to a gene.

insight learning The ability of an animal to perform a correct or appropriate behavior on the first attempt in a situation with which it has had no prior experience.

insulin *(IN-sul-in)* The vertebrate hormone that lowers blood sugar levels by promoting the uptake of glucose by most body cells and promoting the synthesis and storage of glycogen in the liver; also stimulates protein and fat synthesis; secreted by endocrine cells of the pancreas called islets of Langerhans.

integral protein A protein of biological membranes that penetrates into or spans the membrane.

interferon *(IN-tur-FEER-on)* A chemical messenger of the immune system, produced by virus-infected cells and capable of helping other cells resist the virus.

interleukin-1 *(IN-tur-luke-in)* A chemical regulator (cytokine) secreted by macrophages that have ingested a pathogen or foreign molecule and have bound with a helper T cell; stimulates T cells to grow and divide and elevates body temperature. Interleukin-2, secreted by activated T cells, stimulates helper T cells to proliferate more rapidly.

intermediate filament A component of the cytoskeleton that includes all filaments intermediate in size between microtubules and microfilaments.

interneuron *(IN-tur-NOOR-ahn)* An association neuron; a nerve cell within the central nervous system that forms synapses with sensory and motor neurons and integrates sensory input and motor output.

internode *(IN-tur-node)* The segment of a plant stem between the points where leaves are attached.

interstitial cells *(IN-tur-STISH-ul)* Cells scattered among the seminiferous tubules of the vertebrate testis that secrete testosterone and other androgens, the male sex hormones.

interstitial fluid The internal environment of vertebrates, consisting of the fluid filling the spaces between cells.

intertidal zone The shallow zone of the ocean where land meets water.

intrinsic rate of increase The difference between number of births and number of deaths, symbolized as r_{max}; maximum population growth rate.

introgression *(IN-troh-GRES-shun)* Transplantation of genes between species resulting from fertile hybrids mating successfully with one of the parent species.

intron *(IN-tron)* The noncoding, intervening sequence of coding region (exon) in eukaryotic genes.

invagination *(in-VAJ-eh-NAY-shun)* The buckling inward of a cell layer, caused by rearrangements of microfilaments and microtubules; an important phenomenon in embryonic development.

inversion An aberration in chromosome structure resulting from an error in meiosis or from mutagens; reattachment in a reverse orientation of a chromosomal fragment to the chromosome from which the fragment originated.

invertebrate An animal without a backbone; invertebrates make up 95% of animal species.

in vitro fertilization *(VEE-troh)* Fertilization of ova in laboratory containers followed by artificial implantation of the early embryo in the mother's uterus.

ion *(EYE-on)* An atom that has gained or lost electrons, thus acquiring a charge.

ionic bond *(eye-ON-ik)* A chemical bond due to the attraction between oppositely charged ions.

isogamy *(eye-SOG-uh-mee)* A condition in which male and female gametes are morphologically indistinguishable.

isomer *(EYE-sum-ur)* One of several organic compounds with the same molecular formula but different structures and therefore different properties. The three types are structural isomers, geometric isomers, and optical isomers.

isosmotic solutions *(EYE-soz-MAH-tik)* Solutions of equal solute concentration.

isotope *(EYE-so-tope)* One of several atomic forms of an element, each containing a different number of neutrons and thus differing in atomic mass.

juvenile hormone (JH) A hormone in arthropods secreted by the corpora allata glands, which promotes the retention of larval characteristics.

juxtaglomerular apparatus (JGA) Specialized tissue located near the afferent arteriole that supplies blood to the kidney glomerulus; JGA raises blood pressure by producing renin, which activates angiotensin.

***K*-selection** The concept that in some (*K*-selected) populations, life history is centered around producing relatively few offspring that have a good chance of survival.

karyogamy The fusion of nuclei of two cells, as part of syngamy.

karyotype *(KAR-ee-oh-type)* A method of organizing the chromosomes of a cell in relation to number, size, and type.

keystone predator A predatory species that helps maintain species richness in a community by reducing the density of populations of the best competitors so that populations of less competitive species are maintained.

kilocalorie A thousand calories; the amount of heat energy required to raise the temperature of 1 kilogram of water by 1°C.

kin selection A phenomenon of inclusive fitness, used to explain altruistic behavior between related individuals.

kinesis *(kih-NEE-sis)* A change in activity rate in response to a stimulus.

kinetic energy *(kih-NET-ik)* The energy of motion, which is directly related to the speed of that motion. Moving matter does work by transferring some of its kinetic energy to other matter.

kinetochore *(kih-NET-oh-kore)* A specialized region on the centromere that links each sister chromatid to the mitotic spindle.

kingdom A taxonomic category, the second broadest after domain.

Krebs cycle A chemical cycle involving eight steps that completes the metabolic breakdown of glucose molecules to carbon dioxide; occurs within the mitochondrion; the second major stage in cellular respiration. Also called citric acid cycle or tricarboxylic acid (TCA) cycle.

lacteal *(lak-TEEL)* A tiny lymph vessel extending into the core of the intestinal villus and serving as the destination for absorbed chylomicrons.

lagging strand A discontinuously synthesized DNA strand that elongates in a direction away from the replication fork.

larva *(LAR-vuh)* (plural, **larvae**) A free-living, sexually immature form in some animal life cycles that may differ from the adult in morphology, nutrition, and habitat.

lateral line system A mechanoreceptor system consisting of a series of pores and receptor units (neuromasts) along the sides of the body of fishes and aquatic amphibians; detects water movements made by an animal itself and by other moving objects.

lateral meristems *(MARE-eh-stems)* The vascular and cork cambiums, cylinders of dividing cells that run most of the length of stems and roots and are responsible for secondary growth.

law of independent assortment Mendel's second law, stating that each allele pair segregates independently during gamete formation; applies when genes for two traits are located on different pairs of homologous chromosomes.

law of segregration Mendel's first law, stating that allele pairs separate during gamete formation, and then randomly re-form pairs during the fusion of gametes at fertilization.

leading strand The new continuous complementary DNA strand synthesized along the template strand in the mandatory $5' \longrightarrow 3'$ direction.

leukocyte *(LUKE-oh-site)* A white blood cell; typically functions in immunity, such as phagocytosis or antibody production.

lichen An organism formed by the symbiotic association between a fungus and a photosynthetic alga.

life table A table of data summarizing mortality in a population.

ligament *(LIG-uh-ment)* A type of fibrous connective tissue that joins bones together at joints.

ligand *(LIG-und)* A molecule that binds specifically to a receptor site of another molecule.

light microscope (LM) An optical instrument with lenses that refract (bend) visible light to magnify images of specimens.

light reactions The steps in photosynthesis that occur on the thylakoid membranes of the chloroplast and convert solar energy into the chemical energy of ATP and NADPH, evolving oxygen in the process.

lignin *(LIG-nin)* A hard material embedded in the cellulose matrix of vascular plant cell walls that functions as an important adaptation for support in terrestrial species.

limbic system *(LIM-bik)* A group of nuclei (clusters of nerve cell bodies) in the lower part of the mammalian forebrain that interact with the cerebral cortex in determining emotions; includes the hippocampus and the amygdala.

linked genes Genes that are located on the same chromosome.

lipid *(LIH-pid)* One of a family of compounds, including fats, phospholipids, and steroids, that are insoluble in water.

lipoprotein A protein bonded to a lipid; includes the low-density lipoproteins (LDLs) and high-density lipoproteins (HDLs) that transport fats and cholesterol in blood.

locus *(LOH-kus)* A particular place along the length of a certain chromosome where a given gene is located.

logistic population growth A model describing population growth that levels off as population size approaches carrying capacity.

long-day plant A plant that flowers, usually in late spring or early summer, only when the light period is longer than a critical length.

loop of Henle The long hairpin turn, with a descending and ascending limb, of the renal tubule in the vertebrate kidney; functions in water and salt reabsorption.

lungs The invaginated respiratory surfaces of terrestrial vertebrates, land snails, and spiders that connect to the atmosphere by narrow tubes.

lymph *(LIMF)* The colorless fluid, derived from interstitial fluid, in the lymphatic system of vertebrate animals.

lymphatic system *(lim-FAT-ik)* A system of vessels and lymph nodes separate from the circulatory system that returns fluid and protein to the blood.

lymphocyte A white blood cell. The lymphocytes that complete their development in the bone marrow are called B cells, whereas the lymphocytes that mature in the thymus are called T cells.

lysogenic cycle A type of viral replication cycle in which the viral genome becomes incorporated into the bacterial host chromosome as a prophage.

lysosome *(LYE-so-soam)* A membrane-enclosed bag of hydrolytic enzymes found in the cytoplasm of eukaryotic cells.

lysozyme *(LYE-so-zime)* An enzyme in perspiration, tears, and saliva that attacks bacterial cell walls.

lytic cycle *(LIT-ik)* A type of viral replication cycle resulting in the release of new phages by death or lysis of the host cell.

M phase The mitotic phase of the cell cycle, which includes mitosis and cytokinesis.

macroevolution Evolutionary change on a grand scale, encompassing the origin of novel designs, evolutionary trends, adaptive radiation, and mass extinction.

macromolecule A giant molecule of living matter formed by the joining of smaller molecules, usually by condensation synthesis. Polysaccharides, proteins, and nucleic acids are macromolecules.

macrophage *(MAK-roh-fage)* An amoeboid cell that moves through tissue fibers, engulfing bacteria and dead cells by phagocytosis.

major histocompatibility complex (MHC) *(HIS-toe-kum-pat-uh-BIL-eh-tee)* A large set of cell surface antigens encoded by a family of genes. Foreign MHC markers trigger T-cell responses that may lead to rejection of transplanted tissues and organs.

malignant tumor A cancerous growth; an abnormal growth whose cells multiply excessively, have altered surfaces, and may have unusual numbers of chromosomes and/or aberrant metabolic processes.

Malpighian tubule *(mal-PIG-ee-un)* A unique excretory organ of insects that empties into the digestive tract, removes nitrogenous wastes from the blood, and functions in osmoregulation.

mantle A heavy fold of tissue in mollusks that drapes over the visceral mass and may secrete a shell.

marsupial *(mar-SOOP-ee-ul)* A mammal, such as a koala, kangaroo, or opossum, whose young complete their embryonic development inside a maternal pouch called the marsupium.

mass number The sum of the number of protons plus the number of neutrons in the nucleus of an atom; unique for each element and designated by a superscript to the left of the elemental symbol.

matrix The nonliving component of connective tissue, consisting of a web of fibers embedded in homogeneous ground substance that may be liquid, jellylike, or solid.

matter Anything that takes up space and has mass.

mechanoreceptor A sensory receptor that detects physical deformations in the body environment associated with pressure, touch, stretch, motion, and sound.

medulla oblongata *(meh-DULL-uh OBB-long-GAH-tuh)* The lowest part of the vertebrate brain; a swelling of the hindbrain dorsal to the anterior spinal cord that controls autonomic, homeostatic functions, including breathing, heart and blood vessel activity, swallowing, digestion, and vomiting.

medusa *(meh-DOO-suh)* The floating, flattened, mouth-down version of the cnidarian body plan. The alternate form is the polyp.

megapascal (MPa) *(MEG-uh-pass-KAL)* A unit of pressure equivalent to 10 atmospheres of pressure.

meiosis *(my-OH-sis)* A two-stage type of cell division in sexually reproducing organisms that results in gametes with half the chromosome number of the original cell.

membrane potential The charge difference between the cytoplasm and extracellular fluid in all cells, due to the differential distribution of ions. Membrane potential affects the activity of excitable cells and the transmembrane movement of all charged substances.

menstrual cycle *(MEN-stroo-ul)* A type of reproductive cycle in higher female primates, in which the nonpregnant endometrium is shed as a bloody discharge through the cervix into the vagina.

menstruation *(MEN-stroo-AY-shun)* Vaginal bleeding resulting from shedding of the uterine lining during a menstrual cycle.

meristem *(MARE-eh-stem)* Plant tissue that remains embryonic as long as the plant lives, allowing for indeterminate growth.

meroblastic cleavage *(MARE-oh-BLAS-tik KLEE-vij)* A type of cleavage in which there is incomplete division of yolk-rich egg, characteristic of avian development.

mesenteries *(MEZ-en-ter-eez)* Membranes that suspend many of the organs of vertebrates inside fluid-filled body cavities.

mesoderm *(MEZ-oh-durm)* The middle primary germ layer of an early embryo that develops into the notochord, the lining of the coelom, muscles, skeleton, gonads, kidneys, and most of the circulatory system.

mesophyll *(MEZ-oh-fil)* The ground tissue of a leaf, sandwiched between the upper and lower epidermis and specialized for photosynthesis.

mesophyll cell A cell in a leaf's mesophyll, active in carbon fixation and the transport of fixed carbon to the leaf's bundle sheath.

messenger RNA (mRNA) A type of RNA synthesized from DNA in the genetic material that attaches to ribosomes in the cytoplasm and specifies the primary structure of a protein.

metabolism *(meh-TAB-oh-liz-um)* The totality of an organism's chemical processes, consisting of catabolic and anabolic pathways.

metamorphosis *(MET-uh-MOR-fuh-sis)* The resurgence of development in an animal larva that transforms it into a sexually mature adult.

metanephridium *(MET-uh-NEH-frid-ee-um)* (plural, **metanephridia**) A type of excretory tubule in annelid worms that has internal openings called nephrostomes that collect body fluids and external openings called nephridiopores.

metastasis *(meh-TAS-teh-sis)* The spread of cancer cells beyond their original site.

microevolution A change in the gene pool of a population over a succession of generations.

microfilament A solid rod of actin protein in the cytoplasm of almost all eukaryotic cells, making up part of the cytoskeleton and acting alone or with myosin to cause cell contraction.

microtubule A hollow rod of tubulin protein in the cytoplasm of all eukaryotic cells and in cilia, flagella, and the cytoskeleton.

microtubule-organizing center *See* centrosome.

microvillus (plural, **microvilli**) Collectively, fine, fingerlike projections of the epithelial cells in the lumen of the small intestine that increase its surface area.

middle lamella *(luh-MEL-uh)* A thin layer of adhesive extracellular material, primarily pectins, found between the primary walls of adjacent young plant cells.

mimicry A phenomenon in which one species benefits by a superficial resemblance to an unrelated species. A predator or species of prey may gain a significant advantage through mimicry.

missense mutation The most common type of mutation involving a base-pair substitution within a gene that changes a codon, but the new codon makes sense, in that it still codes for an amino acid.

mitochondrial matrix The compartment of the mitochondrion enclosed by the inner membrane and containing enzymes and substrates for the Krebs cycle.

mitochondrion *(MY-toe-KON-dree-un)* (plural, **mitochondria**) An organelle in eukaryotic cells that serves as the site of cellular respiration.

mitosis *(my-TOE-sis)* A process of cell division in eukaryotic cells conventionally divided into the growth period (interphase) and four stages: prophase, metaphase, anaphase, and telophase. The stages conserve chromosome number by equally allocating replicated chromosomes to each of the daughter cells.

modern synthesis A comprehensive theory of evolution emphasizing natural selection, gradualism, and populations as the fundamental units of evolutionary change; also called Neo-Darwinism.

molarity *(moh-LARE-eh-tee)* A common measure of solute concentration, referring to the number of moles of solute in 1 L of solution.

mold A rapidly growing, asexually reproducing fungus.

mole The number of grams of a substance that equals its molecular weight in daltons and contains Avogadro's number of molecules.

molecular formula A type of molecular notation indicating only the quantity of the constituent atoms.

molecule Two or more atoms held together by chemical bonds.

molting A process in arthropods in which the exoskeleton is shed at intervals to allow growth by secretion of a larger exoskeleton.

monoclonal antibody *(MON-oh-KLONE-ul)* A defensive protein produced by cells descended from a single cell; an antibody that is secreted by a clone of cells and, consequently, is specific for a single antigenic determinant.

monocot *(MON-oh-kot)* A subdivision of flowering plants whose members possess one embryonic seed leaf, or cotyledon.

monoculture Cultivation of large land areas with a single plant variety.

monoecious *(mon-EE-shus)* Referring to a plant species that has both staminate and carpellate flowers on the same individual.

monohybrid cross *(MON-oh-HY-brid)* A breeding experiment that employs parental varieties differing in a single character.

monomer *(MON-uh-mer)* The subunit that serves as the building block of a polymer.

monophyletic *(MON-oh-fye-LEH-tik)* Pertaining to a taxon derived from a single ancestral species that gave rise to no species in any other taxa.

monosaccharide *(MON-oh-SAK-ur-ide)* The simplest carbohydrate, active alone or serving as a monomer for disaccharides and polysaccharides. Also known as simple sugars, the molecular formulas of monosaccharides are generally some multiple of CH_2O.

monotreme *(MON-uh-treem)* An egg-laying mammal, represented by the platypus and echidna.

morphogenesis *(MOR-foe-JEN-eh-sis)* The development of body shape and organization during ontogeny.

morphospecies Species defined by their anatomical features.

mosaic development A pattern of development, such as that of a mollusk, in which the early blastomeres each give rise to a specific part of the embryo. In some animals, the fate of the blastomeres is established in the zygote.

mosaic evolution The evolution of different features of an organism at different rates.

motor neuron A nerve cell that transmits signals from the brain or spinal cord to muscles or glands.

motor, or efferent, nervous system In vertebrates, the component of the peripheral nervous system that transmits signals from the central nervous system to effector cells.

motor unit A single motor neuron and all the muscle fibers it controls.

MPF M-phase promoting factor: A protein complex required for a cell to progress from late interphase to mitosis; the active form consists of cyclin and *cdc2,* a protein kinase.

mucous membrane A type of epithelium that secretes a lubricant, lining the digestive tract and air tubes.

Müllerian mimicry *(myoo-LER-ee-un)* A mutual mimicry by two unpalatable species.

multigene family A collection of genes with similar or identical sequences, presumably of common origin.

multiple fruit A fruit, such as a pineapple, that develops from an inflorescence, a group of flowers clustered tightly together.

mutagen *(MYOOT-uh-jen)* A chemical or physical agent that interacts with DNA and causes a mutation.

mutagenesis *(MYOOT-uh-JEN-uh-sis)* The creation of mutations.

mutation *(myoo-TAY-shun)* A rare change in the DNA of genes that ultimately creates genetic diversity.

mutualism *(MYOO-chew-ul-iz-um)* A symbiotic relationship in which both the host and the symbiont benefit.

mycelium *(my-SEEL-ee-um)* The densely branched network of hyphae in a fungus.

mycorrhizae *(MY-koh-RYE-zee)* Mutualistic associations of plant roots and fungi.

myelin sheath *(MY-eh-lin)* An insulating coat of cell membrane from Schwann cells that is interrupted by nodes of Ranvier where saltatory conduction occurs.

myofibrils *(MY-oh-FIBE-rills)* Fibrils arranged in longitudinal bundles in muscle cells (fibers); composed of thin filaments of actin and a regulatory protein and thick filaments of myosin.

myoglobin *(MY-uh-glow-bin)* An oxygen-storing, pigmented protein in muscle cells.

myosin *(MY-uh-sin)* A type of protein filament that interacts with actin filaments to cause cell contraction.

NAD^+ Nicotinamide adenine dinucleotide (oxidized); a coenzyme present in all cells that assists enzymes in transferring electrons during the redox reactions of metabolism.

natural selection Differential success in the reproduction of different phenotypes resulting from the interaction of organisms with their environment. Evolution occurs when natural selection causes changes in relative frequencies of alleles in the gene pool.

negative feedback A primary mechanism of homeostasis, whereby a change in a physiological variable that is being monitored triggers a response that counteracts the initial fluctuation.

nephron *(NEF-ron)* The tubular excretory unit of the vertebrate kidney.

neritic zone *(neh-RIT-ik)* The shallow regions of the ocean overlying the continental shelves.

net primary productivity The gross primary productivity minus the energy used by the producers for cellular respiration; represents the storage of chemical energy in an ecosystem available to consumers.

neural crest *(NOOR-ul)* A band of cells along the border where the neural tube pinches off from the ectoderm; the cells migrate to various parts of the embryo and form the pigment cells in the skin, bones of the skull, the teeth, the adrenal glands, and parts of the peripheral nervous system.

neuron *(NOOR-on)* A nerve cell; the fundamental unit of the nervous system, having structure and properties that allow it to conduct signals by taking advantage of the electrical charge across its cell membrane.

neurosecretory cells Hypothalamus cells that receive signals from other nerve cells, but instead of signaling to an adjacent nerve cell or muscle, release hormones into the bloodstream.

neurotransmitter The chemical messenger released from the synaptic terminals of a neuron at a chemical synapse that diffuses across the synaptic cleft and binds to and stimulates the postsynaptic cell.

neutral variation Genetic diversity that confers no apparent selective advantage.

niche *See* ecological niche.

nitrogen fixation The assimilation of atmospheric nitrogen by certain prokaryotes into nitrogenous compounds that can be directly used by plants.

nodes A point along the stem of a plant at which leaves are attached.

nodes of Ranvier *(ran-VEER)* The small gaps in the myelin sheath between successive glial cells along the axon of a neuron; also, the site of high concentration of voltage-sensitive ion channels.

nodule An enlarged root section where nitrogen fixation takes place in legumes such as in peas or beans; composed of plant cells containing bacteroids with symbiotic bacteria.

noncompetitive inhibitor A substance that reduces the activity of an enzyme by binding to a location remote from the active site, changing its conformation so that it no longer binds to the substrate.

noncyclic electron flow A route of electron flow during the light reactions of photosynthesis that involves both photosystems and produces ATP, NADPH, and oxygen; the net electron flow is from water to $NADP^+$.

noncyclic photophosphorylation *(FO-toe-foss-FOR-i-lay-shun)* The production of ATP by noncyclic electron flow.

nondisjunction An accident of meiosis or mitosis, in which both members of a pair of homologous chromosomes or both sister chromatids fail to move apart properly.

nonpolar covalent bond A type of covalent bond in which electrons are shared equally between two atoms of similar electronegativity.

nonsense mutation A mutation that changes an amino acid codon to one of the three stop codons, resulting in a shorter and usually nonfunctional protein.

norm of reaction The range of phenotypic possibilities for a single genotype, as influenced by the environment.

notochord *(NO-toe-kord)* A longitudinal, flexible rod formed from dorsal mesoderm and located between the gut and the nerve cord in all chordate embryos.

nuclear envelope The membrane in eukaryotes that encloses the nucleus, separating it from the cytoplasm.

nucleic acid (polynucleotide) *(PAHL-ee-NOO-klee-o-tide)* A biological molecule (such as RNA or DNA) that allows organisms to reproduce; polymers composed of monomers called nucleotides joined by covalent bonds (phosphodiester linkages) between the phosphate of one nucleotide and the sugar of the next nucleotide.

nucleoid region *(NOO-klee-oid)* The region in a prokaryotic cell consisting of a concentrated mass of DNA.

nucleolus *(noo-KLEE-oh-lus)* (plural, **nucleoli**) A specialized structure in the nucleus, formed from various chromosomes and active in the synthesis of ribosomes.

nucleoside *(NOO-klee-oh-side)* An organic molecule consisting of a nitrogenous base joined to a five-carbon sugar.

nucleosome *(NOO-klee-oh-soam)* The basic, beadlike unit of DNA packaging in eukaryotes, consisting of a segment of DNA wound around a protein core composed of two copies of each of four types of histone.

nucleotide *(NOO-klee-oh-tide)* The building block of a nucleic acid, consisting of a five-carbon sugar covalently bonded to a nitrogenous base and a phosphate group.

obligate aerobe *(OB-lig-it AIR-obe)* An organism that requires oxygen for cellular respiration and cannot live without it.

obligate anaerobe *(AN-ur-obe)* An organism that cannot use oxygen and is poisoned by it.

oceanic zone The region of ocean water lying over deep areas beyond the continental shelf.

Okazaki fragment A short segment of the lagging ($3' \longrightarrow 5'$) strand of DNA, synthesized in the $5' \longrightarrow 3'$ direction. A series of Okazaki fragments unite to produce the lagging strand at a DNA replication fork.

oligotrophic lake A nutrient-poor, clear, deep lake with minimum phytoplankton.

ommatidium *(OM-uh-TID-ee-um)* (plural, **ommatidia**) A light-detecting unit in a compound eye that has its own cornea and lens.

omnivore *(OM-neh-vore)* A heterotrophic animal that consumes both meat and plant material.

oncogene *(ON-koh-jeen)* A gene found in viruses or as part of the normal genome that is involved in triggering cancerous characteristics.

ontogeny *(on-TOJ-en-ee)* The embryonic development of an organism.

oogamy *(oh-OG-um-ee)* A condition in which male and female gametes differ, such that a small, flagellated sperm fertilizes a large, nonmotile egg.

oogenesis *(OO-oh-JEN-eh-sis)* The process in the ovary that results in the production of female gametes.

open circulatory system An arrangement of internal transport in which blood bathes the organs directly and there is no distinction between blood and interstitial fluid.

operant conditioning *(OP-ur-ent)* A type of associative learning that directly affects behavior in a natural context; also called trial-and-error learning.

operator In *E. coli,* a segment of DNA between a promoter and structural genes that functions as an on-off switch, controlling an operon.

operon *(OP-ur-on)* A unit of genetic function common in bacteria and phages and consisting of regulated clusters of genes with related functions.

organ A specialized center of body function composed of several different types of tissues.

organ of Corti The actual hearing organ of the vertebrate ear, located in the floor of the cochlear canal in the inner ear; contains the receptor cells (hair cells) of the ear.

organelle *(OR-gan-EL)* One of several formed bodies with a specialized function, suspended in the cytoplasm and found in eukaryotic cells.

organic chemistry The study of carbon compounds (organic compounds).

organogenesis *(or-GAN-oh-JEN-eh-sis)* An early period of rapid embryonic development in which the organs take form from the primary germ layers.

orgasm Rythmic, involuntary contractions of certain reproductive structures in both sexes during the sexual response cylce.

osmoconformer *(OZ-moh-kun-FOR-mur)* An animal that does not actively adjust its internal osmolarity because it is isosmotic with its environment.

osmolarity *(OZ-moh-LAR-eh-tee)* Solute concentration expressed as molarity.

osmoregulation *(OZ-moh-reg-yoo-LAY-shun)* Adaptations to control the water balance in organisms living in hyperosmotic, hypoosmotic, or terrestrial environments.

osmoregulator An animal whose body fluids have a different osmolarity than the environment, and that must either discharge excess water if it lives in a hypoosmotic environment or take in water if it inhabits a hyperosmotic environment.

osmosis *(oz-MOH-sis)* The diffusion of water across a selectively permeable membrane.

osmotic pressure *(oz-MOT-ik)* A measure of the tendency of a solution to take up water when separated from pure water by a selectively permeable membrane.

osteocyte A cell located in lacunae of vertebrate bone, that deposits a matrix of collagen and calcium phosphate that forms a mineralized connective tissue.

ostracoderm *(os-TRAK-uh-durm)* An extinct agnathan; a fish-like creature encased in an armor of bony plates.

ovarian cycle *(oh-VAIR-ee-un)* The cyclic recurrence of the follicular phase, ovulation, and the luteal phase in the mammalian ovary, regulated by hormones.

ovary *(OH-vur-ee)* (1) In flowers, the portion of a carpel in which the egg-containing ovules develop. (2) In animals, the structure that produces female gametes and reproductive hormones.

oviduct *(OH-veh-dukt)* A tube passing from the ovary to the vagina in invertebrates or to the uterus in vertebrates.

oviparous *(oh-VIP-ur-us)* Referring to a type of development in which young hatch from eggs laid outside the mother's body.

ovoviviparous *(OH-vo-vy-VIP-ur-us)* Referring to a type of development in which young hatch from eggs that are retained in the mother's uterus.

ovulation The release of an egg from ovaries. In humans, an ovarian follicle releases an egg during each menstrual cycle.

ovule *(OV-yool)* A structure that develops in the plant ovary and contains the female gametophyte.

ovum *(OH-vum)* The female gamete; the haploid, unfertilized egg, which is usually a relatively large, nonmotile cell.

oxidation The loss of electrons from a substance involved in a redox reaction.

oxidative phosphorylation *(FOS-for-eh-LAY-shun)* The production of ATP using energy derived from the redox reactions of the electron transport chain.

oxidizing agent The electron acceptor in a redox reaction.

P site Peptidyl-tRNA site; the binding site on a ribosome that holds the tRNA carrying a growing polypeptide chain.

pacemaker A specialized region of the right atrium of the mammalian heart that sets the rate of contraction; also called sinoatrial (SA) node.

paedomorphosis *(PEE-doh-mor-FOE-sis)* The retention in an adult organism of the juvenile features of its evolutionary ancestors.

paleontology *(PAY-lee-un-TOL-uh-gee)* The scientific study of fossils.

Pangaea *(pan-JEE-uh)* The supercontinent formed near the end of the Paleozoic era when plate movements brought all the land masses of Earth together.

paraphyletic *(PAR-uh-FYE-leh-tik)* Pertaining to a taxon that excludes some members that share a common ancestor with members included in the taxon.

parasite *(PAR-uh-site)* An organism that absorbs nutrients from the body fluids of living hosts.

parasitism A symbiotic relationship in which the symbiont (parasite) benefits at the expense of the host by either living within the host (endoparasite) or outside of the host (ectoparasite).

parasympathetic nervous system *(PAR-uh-SIM-peh-THET-ik)* One of two divisions of the autonomic nervous system; generally enhances body activities that gain and conserve energy, such as digestion and reduced heart rate.

parazoa *(PAR-uh-ZOH-uh)* Members of the subkingdom of animals consisting of the sponges.

parenchyma cell *(pur-EN-kim-uh)* A relatively unspecialized plant cell type that carries on most of the metabolism, synthesizes and stores organic products, and develops into more differentiated cell types.

parental types Offspring whose phenotype is the same as one of their parents.

parthenogenesis *(PAR-then-oh-JEN-eh-sis)* A type of reproduction in which females produce offspring from unfertilized eggs.

partial pressure The concentration of gases; a fraction of total pressure.

parturition The act of giving birth to offspring.

passive transport The diffusion of a substance across a biological membrane.

pattern formation The ordering of cells into specific three-dimensional structures, an essential part of shaping an organism and its individual parts during development.

pedigree A family tree describing the occurrence of heritable characters in parents and offspring across as many generations as possible.

pelagic zone *(pel-AY-jik)* The area of the ocean past the continental shelf, with areas of open water often reaching to very great depths.

PEP carboxylase *(KAR-box-i-lase)* The enzyme that adds CO_2 to phosphoenolpyruvate in the Calvin cycle of photosynthesis.

peptide bond The covalent bond between two amino acid units, formed by condensation synthesis.

peptidoglycan *(PEP-tid-oh-GLY-kan)* A type of polymer in bacterial cell walls consisting of modified sugars cross-linked by short polypeptides.

perception The interpretation of sensations by the brain.

perennial *(per-EN-ee-ul)* A plant that lives for many years.

pericarp *(PER-eh-karp)* The thickened wall of a ripened plant ovary, or fruit.

pericycle *(PER-eh-sigh-kul)* A layer of cells just inside the endodermis of a root that may become meristematic and begin dividing again.

periderm *(PER-eh-durm)* The protective coat that replaces the epidermis in plants during secondary growth, formed of the cork and cork cambium.

peripheral nervous system The sensory and motor neurons that connect to the central nervous system.

peripheral protein A protein on the surface of biological membranes.

peristalsis *(PER-is-TAL-sis)* Rhythmic waves of contraction of digestive smooth muscle that push food along the tract.

peroxisome *(per-OX-eh-soam)* A microbody containing enzymes that transfer hydrogen from various substrates to oxygen, producing and then degrading hydrogen peroxide.

petiole *(PET-ee-ole)* The stalk of a leaf, which joins the leaf to a node of the stem.

phage *(fage)* A virus that infects bacteria; also called a bacteriophage.

phagocytosis *(FAY-go-sigh-TOE-sis)* A type of endocytosis involving large, particulate substances.

pharynx *(FAH-rinks)* An area in the vertebrate throat where air and food passages cross; in flatworms, the muscular tube that protrudes from the ventral side of the worm and ends in the mouth.

phenetics *(feh-NEH-tiks)* An approach to taxonomy based entirely on measurable similarities and differences in phenotypic characters, without consideration of homology, analogy, or phylogeny.

phenotype *(FEE-nuh-type)* The physical and physiological traits of an organism.

pheromone *(FAIR-uh-mone)* A small, volatile chemical signal that functions in communication between animals and acts much like a hormone in influencing physiology and behavior.

phloem *(FLOAM)* The portion of the vascular system in plants consisting of living cells arranged into elongated tubes that transport sugar and other organic nutrients throughout the plant.

phosphate group *(FOS-fate)* A functional group important in energy transfer.

phospholipids *(FOS-foe-LIP-ids)* Molecules that constitute the inner bilayer of biological membranes, having a polar, hydrophilic head and a nonpolar, hydrophobic tail.

photic zone *(FOE-tik)* The narrow top slice of the ocean, where light penetrates sufficiently for photosynthesis to occur.

photoautotroph *(FOE-toe-AW-to-trohf)* An organism that harnesses light energy to drive the synthesis of organic compounds from carbon dioxide.

photoheterotroph *(FOE-toe-HET-ur-oh-trohf)* An organism that uses light to generate ATP but that must obtain carbon in organic form.

photon *(FOE-tahn)* A quantum, or discrete amount, of light energy.

photoperiodism *(FOE-toe-PEER-ee-od-iz-um)* A physiological response to day length, such as flowering in plants.

photophosphorylation *(FOE-toe-fos-for-ul-AY-shun)* The process of generating ATP from ADP and phosphate by means of a proton-motive force generated by the thylakoid membrane of the chloroplast during the light reactions of photosynthesis.

photorespiration A metabolic pathway that consumes oxygen, evolves carbon dioxide, generates no ATP, and decreases photosynthetic output; generally occurs on hot, dry, bright days, when stomata close and the oxygen concentration in the leaf exceeds that of carbon dioxide.

photosynthesis The conversion of light energy into chemical energy that is stored in glucose or other organic compounds; occurs in plants, algae, and certain prokaryotes.

photosystem The light-harvesting unit in photosynthesis, located on the thylakoid membrane of the chloroplast and consisting of the antenna complex, the reaction-center chlorophyll *a*, and the primary electron acceptor. There are two type of photosystems, I and II; they absorb light best at different wavelengths.

phototropism *(foh-TOH-treh-PIZ-um)* Growth of a plant shoot toward or away from light.

pH scale A measure of hydrogen ion concentration equal to $-\log [H^+]$ and ranging in value from 0 to 14.

phyletic evolution Transformation of an unbranched lineage of organisms sometimes to a state different enough from the ancestral population to justify renaming it as a new species; also called anagenesis.

phylogeny *(fye-LOJ-en-ee)* The evolutionary history of a species or group of related species.

phylum A taxonomic category; phyla are divided into classes.

phytochrome *(FYE-tuh-krome)* A pigment involved in many responses of plants to light.

pilus *(PILL-us)* (plural, **pili**) A surface appendage in certain bacteria that functions in adherence and the transfer of DNA during conjugation.

pineal gland *(PIN-ee-ul)* A small endocrine gland on the dorsal surface of the vertebrate forebrain; secretes the hormone melatonin, which regulates body functions related to seasonal day length.

pinocytosis *(PY-noh-sigh-TOE-sis)* A type of endocytosis in which the cell ingests extracellular fluid and its dissolved solutes.

pith The core of the central vascular cylinder of monocot roots, consisting of parenchyma cells, which are ringed by vascular tissue; ground tissue interior to vascular bundles in dicot stems.

pituitary gland *(pi-TOO-i-tair-ee)* An endocrine gland at the base of the hypothalamus; consists of a posterior lobe (neurohypophysis), which stores and releases two hormones produced by the hypothalamus, and an anterior lobe (adenohypophysis), which produces and secretes many hormones that regulate diverse body functions.

placenta *(pluh-SEN-tuh)* A structure in the pregnant uterus for nourishing a viviparous fetus with the mother's blood supply; formed from the uterine lining and embryonic membranes.

placental mammal A member of a group of mammals, including humans, whose young complete their embryonic development in the uterus, joined to the mother by a placenta.

placoderm *(PLAK-oh-durm)* A member of an extinct class of fishlike vertebrates that had jaws and were enclosed in a tough, outer armor.

plankton Mostly microscopic organisms that drift passively or swim weakly near the surface of oceans, ponds, and lakes.

plasma *(PLAZ-muh)* The liquid matrix of blood in which the cells are suspended.

plasma cells Derivatives of B cells that secrete antibodies.

plasma membrane The membrane at the boundary of every cell that acts as a selective barrier that regulates the cell's chemical composition.

plasmid *(PLAZ-mid)* A small ring of DNA that carries accessory genes separate from those of a bacterial chromosome.

plasmodesma *(PLAZ-moh-DEZ-muh)* (plural, **plasmodesmata**) An open channel in the cell wall of plants through which strands of cytoplasm connect from adjacent cells.

plasmogamy The fusion of the cytoplasm of cells from two individuals; occurs as one stage of syngamy.

plasmolysis *(plaz-MOL-eh-sis)* A phenomenon in walled cells in which the cytoplasm shrivels and the plasma membrane pulls away from the cell wall when the cell loses water to a hyperosmotic environment.

plastid One of a family of closely related plant organelles, including chloroplasts, chromoplasts, and amyloplasts (leucoplasts).

platelet A small anucleate blood cell important in blood clotting; derived from large cells in the bone marrow.

pleiotropy *(PLY-eh-troh-pee)* The ability of a single gene to have multiple effects.

plesiomorphic character *(PLEEZ-ee-oh-MORE-fik)* A primitive phenotypic character possessed by a remote ancestor.

pluripotent stem cell A cell within bone marrow that is a progenitor for any kind of blood cell.

point mutation A change in the chromosome at a single nucleotide within a gene.

polar covalent bond A type of covalent bond between atoms that differ in electronegativity. The shared electrons are pulled closer to the more electronegative atom, making it slightly negative and the other atom slightly positive.

polar molecule A molecule (such as water) with opposite charges on opposite sides.

pollen grain An immature male gametophyte that develops within the anthers of stamens in a flower.

pollination *(POL-eh-NAY-shun)* The placement of pollen onto the stigma of a carpel by wind or animal carriers, a prerequisite to fertilization.

polyandry *(POL-ee-AN-dree)* A polygamous mating system involving one female and many males.

polygenic inheritance *(POL-ee-JEN-ik)* An additive effect of two or more gene loci on a single phenotypic character.

polygyny *(pol-IJ-en-ee)* A polygamous mating system involving one male and many females.

polymer *(POL-eh-mur)* A large molecule consisting of many identical or similar monomers linked together.

polymerase chain reaction (PCR) A technique for amplifying DNA in vitro by incubating with special primers, DNA polymerase molecules and nucleotides.

polymorphic *(POL-ee-MOR-fik)* Referring to a population in which two or more morphs are present in readily noticeable frequencies.

polymorphism *(POL-ee-MOR-fiz-um)* The coexistence of two or more distinct forms of individuals (polymorphic characters) in the same population.

polyp *(POL-ip)* The sessile variant of the cnidarian body plan. The alternate form is the medusa.

polypeptide chain *(POL-ee-PEP-tide)* A polymer of many amino acids linked by peptide bonds.

polyphyletic *(POL-ee-fye-LET-ik)* Pertaining to a taxon whose members were derived from two or more ancestral forms not common to all members.

polyploidy *(POL-uh-ploid-ee)* A chromosomal alteration in which the organism possesses more than two complete chromosome sets.

polyribosome *(POL-ee-RI-bo-some)* An aggregation of several ribosomes attached to one messenger RNA molecule.

polysaccharide *(POL-ee-SAK-ur-ide)* A polymer of up to over a thousand monosaccharides, formed by condensation synthesis.

population A group of individuals of one species that live in a particular geographic area.

population genetics The scientific study of gene pools and genetic variation in biological populations.

positional information Signals, to which genes regulating development respond, indicating a cell's location relative to other cells in an embryonic structure.

positive feedback A physiological control mechanism in which a change in some variable triggers mechanisms that amplify the change.

postsynaptic membrane *(post-sin-AP-tik)* The surface of the cell on the opposite side of the synapse from the synaptic terminal of the stimulating neuron that contains receptor proteins and degradative enzymes for the neurotransmitter.

postzygotic barrier *(POST-zye-OGT-ik)* Any of several species-isolating mechanisms that prevent hybrids produced by two different species from developing into viable, fertile adults.

potential energy The energy stored by matter as a result of its location or spatial arrangement.

preadaptation A structure that evolves and functions in one environmental context, but can perform additional functions when placed in some new environment.

prezygotic barrier *(PREE-zye-GOT-ik)* A reproductive barrier that impedes mating between species or hinders fertilization of ova if interspecific mating is attempted.

primary consumers Herbivores; organisms in the trophic level of an ecosystem that eat plants or algae.

primary germ layers The three layers (ectoderm, mesoderm, endoderm) of the late gastrula, which develop into all parts of an animal.

primary growth Growth initiated by the apical meristems of a plant root or shoot.

primary immune response The initial immune response to an antigen, which appears after a lag of several days.

primary producers Autotrophs, which make up the trophic level of an ecosystem that ultimately supports all other levels; mostly photosynthetic organisms.

primary productivity The rate at which light energy or inorganic chemical energy is converted to the chemical energy of organic compounds by autotrophs in an ecosystem.

primary structure The level of protein structure referring to the specific sequence of amino acids.

primary succession A type of ecological succession that occurs in an area where there were originally no organisms.

primer An already existing DNA chain bound to the template DNA to which nucleotides must be added during DNA synthesis.

principle of allocation The concept that each organism has an energy budget, or a limited amount of total energy apportionable to all of its needs for maintenance and reproduction.

procambium *(pro-KAM-bee-um)* A primary meristem of roots and shoots that forms the vascular tissue.

prokaryotic cell *(pro-KAR-ee-OT-ik)* A type of cell lacking a membrane-enclosed nucleus and membrane-enclosed organelles; found only in the kingdom Monera. Also called prokaryote.

promoter A specific nucleotide sequence in DNA, flanking the start of a gene; instructs RNA polymerase where to start transcribing RNA.

prophage *(PRO-fage)* A phage genome that has been inserted into a specific site on the bacterial chromosome.

prostaglandins (PGs) *(PROS-tuh-GLAN-dins)* A group of modified fatty acids secreted by virtually all tissues and serving a wide variety of functions as messengers.

protein *(PRO-teen)* A three-dimensional biological polymer constructed from a set of 20 different monomers called amino acids.

protoderm *(PRO-toe-durm)* The outermost primary meristem, which gives rise to the epidermis of roots and shoots.

proton-motive force The potential energy stored in the form of an electrochemical gradient, generated by the pumping of hydrogen ions across biological membranes in chemiosmosis.

proton pump *(PRO-tahn)* An active transport mechanism in cell membranes that consumes ATP to force hydrogen ions out of a cell and, in the process, generates a membrane potential.

protonephridium *(PRO-toe-NEF-rid-ee-um)* An excretory system, such as the flame-cell system of flatworms, consisting of a network of closed tubules having external openings called nephridiopores and lacking internal openings.

proto-oncogene *(PRO-toe-ONK-oh-jeen)* A normal cellular gene corresponding to an oncogene; a gene with a potential to cause cancer, but that requires some alteration to become an oncogene.

protostome *(PRO-toh-stoam)* A member of one of two distinct evolutionary lines of coelomates, consisting of the annelids, mollusks, and arthropods, and characterized by spiral, determinate cleavage, schizocoelous formation of the coelom, and development of the mouth from the blastopore.

protozoan (plural, **protozoa**) A protist that lives primarily by ingesting food, an animal-like mode of nutrition.

provirus Viral DNA that inserts into a host genome.

proximate causation The hypothesis about why natural selection favored a particular animal behavior.

pseudocoelomate *(SOO-doe-SEEL-oh-mate)* An animal, such as a rotifer or roundworm, whose body cavity is not completely lined by mesoderm.

pseudogene *(SOO-doe-JEEN)* A noncoding DNA segment with sequences similar to functional genes but lacking the sites necessary for gene expression.

pseudopodium *(SOO-doe-POE-dee-um)* (plural, **pseudopodia**) A cellular extension of amoeboid cells used in moving and feeding.

punctuated equilibrium A theory of evolution advocating spurts of relatively rapid change followed by long periods of stasis.

purine *(PYOOR-een)* One of two families of nitrogenous bases found in nucleotides consisting of two members: adenine (A) and guanine (G).

pyrimidine *(pir-IM-eh-deen)* One of two families of nitrogenous bases found in nucleotides consisting of three members: cytosine (C), thymine (T), and uracil (U).

quantitative character A heritable feature in a population that varies continuously as a result of environmental influences and the additive effect of two or more genes (polygenic inheritance).

quaternary structure *(KWAT-ur-nair-ee)* The particular shape of a complex, aggregate protein, defined by the characteristic three-dimensional arrangement of its constituent subunits, each a polypeptide.

quiescent center A region located within the zone of cell division in plant roots, containing meristematic cells that divide very slowly.

***r*-selection** The concept that in some (*r*-selected) populations a high reproductive rate is the chief determinant of life history.

radial cleavage A type of embryonic development in deuterostomes in which the planes of cell division that transform the zygote into a ball of cells are either parallel or perpendicular to the polar axis, thereby aligning tiers of cells one above the other.

radial symmetry Characterizing a body shaped like a pie or barrel, with many equal parts radiating outward like the spokes of a wheel; present in cnidarians and echinoderms.

radiata Members of the radially symmetric animal phyla, including cnidarians.

radicle An embryonic root of a plant.

radioactive Having a nucleus that gives off particles and energy; characteristic of unstable isotopes of a chemical element.

radioactive dating A method of determining the age of fossils and rocks using half-lives of radioactive isotopes.

reaction center The location of one or a pair of specialized chlorophyll *a* molecules in the pigment assembly system of the light reactions of photosynthesis.

realized niche The environmental resources that an organism actually uses; the niche that an organism actually occupies.

receptor-mediated endocytosis *(EN-do-si-TO-sis)* The movement of specific molecules into a cell by the inward budding of membranous vesicles containing proteins with receptor sites specific to the molecules being taken in; enables a cell to acquire bulk quantities of specific substances.

receptor potential An initial response of a receptor cell to a stimulus, consisting of a change in voltage across the receptor membrane proportional to the stimulus strength. The intensity of the receptor potential determines the frequency of action potentials traveling to the nervous system.

recessive allele In a heterozygote, the allele that is completely masked in the phenotype.

reciprocal altruism *(AL-troo-IZ-um)* Altruistic behavior between unrelated individuals; believed to produce some benefit to the altruistic individual in the future when the current beneficiary reciprocates.

recognition concept of species A definition of species based on mate-recognition mechanisms; assumes that reproductive adaptations of a species consist of a set of features that maximize successful mating with members of the same population; an alternative to the biological species concept.

recombinant DNA A technique in which gene segments from different sources are recombined in vitro and transferred into cells, where the DNA may be expressed.

recombinants Offspring whose phenotype differs from that of the parents.

redox reaction *(REE-doks)* A chemical reaction involving the transfer of one or more electrons from one reactant to another; also called oxidation-reduction reaction.

reducing agent The electron donor in a redox reaction.

reduction The gaining of electrons by a substance involved in a redox reaction.

reflex An automatic reaction to a stimulus, mediated by the spinal cord or lower brain.

refractory period *(ree-FRAK-tor-ee)* The short time immediately after an action potential in which the neuron cannot respond to another stimulus, owing to an increase in potassium permeability.

regeneration The replacement of parts of an organism that are lost due to injury.

regulative development A pattern of development, such as that of a mammal, in which the early blastomeres retain the potential to form the entire animal.

relative abundance The frequency of individuals of a species in a biological community; also called species equitability.

relative fitness The contribution of one genotype to the next generation compared to that of alternative genotypes for the same locus.

releaser A signal stimulus that functions as a communication signal between individuals of the same species.

releasing hormone A hormone produced by neurosecretory cells in the hypothalamus of the vertebrate brain that stimulates or inhibits the secretion of hormones by the anterior pituitary.

replication fork A Y-shaped point on a replicating DNA molecule where new strands are growing.

repressible enzyme An enzyme whose synthesis is inhibited by a specific metabolite.

repressor A protein that suppresses the expression of prophage or operon genes.

resolving power A measure of the clarity of an image; the minimum distance that two points can be separated and still be distinguished as two separate points.

resource partitioning The division of environmental resources by coexisting species populations such that the niche of each species differs by one or more significant factors from the niches of all coexisting species populations.

resting potential The membrane potential characteristic of a nonconducting, excitable cell, with the inside of the cell more negative than the outside.

restriction enzyme A degradative enzyme that recognizes and cuts up DNA (including that of certain phages) that is foreign to a cell.

restriction fragment length polymorphisms (RFLPs) Differences in DNA sequence on homologous chromosomes that result in different patterns of restriction fragment lengths (DNA segments resulting from treatment with restriction enzymes); useful as genetic markers for making linkage maps.

restriction mapping A method of determining the similarity between the genomes or genes of two species by the electrophoretic comparison of restriction fragments of DNA of the two species.

retina *(REH-tin-uh)* The innermost layer of the vertebrate eye, containing photoreceptor cells (rods and cones) and neurons; transmits images formed by the lens to the brain via the optic nerve.

retinal The light-absorbing pigment in rods and cones of the vertebrate eye.

retrovirus *(REH-troh-VIRE-us)* An RNA virus that reproduces by transcribing its RNA into DNA and then inserting the DNA into a cellular chromosome; an important class of cancer-causing viruses.

reverse transcriptase *(trans-KRIP-tase)* An enzyme encoded by some RNA viruses that uses RNA as a template for DNA synthesis.

rhodopsin A visual pigment consisting of retinal and opsin. When rhodopsin absorbs light, the retinal changes shape and dissociates from the opsin, after which it is converted back to its original form.

ribonucleic acid (RNA) *(RYE-bow-noo-KLAY-ik)* A single-stranded nucleic acid molecule involved in protein synthesis, the structure of which is specified by DNA.

ribosomal RNA (rRNA) The most abundant type of RNA. Together with proteins, it forms the structure of ribosomes that coordinate the sequential coupling of tRNA molecules to the series of mRNA codons.

ribosome A cell organelle constructed in the nucleolus, consisting of two subunits and functioning as the site of protein synthesis in the cytoplasm.

ribozyme An enzymatic RNA molecule; catalyzes reactions during RNA splicing.

RNA polymerase *(pul-IM-ur-ase)* An enzyme that links together the growing chain of ribonucleotides during transcription.

RNA processing Modification of RNA before it leaves the nucleus, a process unique to eukaryotes.

RNA splicing The removal of noncoding portions of the RNA molecule after initial synthesis.

rod cell One of two kinds of photoreceptors in the vertebrate retina; sensitive to black and white and enables night vision.

root hair A tiny projection growing just behind the root tips of plants, increasing surface area for absorption of water and minerals.

root pressure The upward push of water within the stele of vascular plants, caused by active pumping of minerals into the xylem by root cells.

rough ER That portion of the endoplasmic reticulum studded with ribosomes.

rubisco Ribulose carboxylase, the enzyme that catalyzes the first step (the addition of CO_2 to RuBP, or ribulose bisphosphate) of the Calvin cycle.

ruminant An animal, such as a cow or a sheep, with an elaborate, multicompartmentalized stomach specialized for a herbivorous diet.

S phase The synthesis phase of the cell cycle, constituting the portion of interphase during which DNA is replicated.

SA (sinoatrial) node The pacemaker of the heart, located in the wall of the right atrium. At the base of the wall separating the two atria is another patch of nodal tissue called the atrioventricular node (AV).

saltatory conduction *(SAHL-tuh-TORR-ee)* Rapid transmission of a nerve impulse along an axon resulting from the action potential jumping from one node of Ranvier to another, skipping the myelin-sheathed regions of membrane.

saprobe An organism that acts as a decomposer by absorbing nutrients from dead organic matter.

sarcomere *(SAR-koh-meer)* The fundamental, repeating unit of striated muscle, delimited by the Z lines.

sarcoplasmic reticulum *(SAR-koh-PLAZ-mik reh-TIK-yoo-lum)* A modified form of endoplasmic reticulum in striated muscle cells that stores calcium used to trigger contraction during stimulation.

saturated fatty acid A fatty acid in which all carbons in the hydrocarbon tail are connected by single bonds, thus maximizing the number of hydrogen atoms that can attach to the carbon skeleton.

savanna *(suh-VAN-uh)* A tropical grassland biome with scattered individual trees, large herbivores, and three distinct seasons based primarily on rainfall, maintained by occasional fires and drought.

Schwann cells A chain of supporting cells enclosing the axons of many neurons and forming an insulating layer called the myelin sheath.

sclereid *(SKLER-ee-id)* A short, irregular sclerenchyma cell in nutshells and seed coats and scattered through the parenchyma of some plants.

sclerenchyma cell *(skler-EN-kim-uh)* A rigid, supportive plant cell type usually lacking protoplasts and possessing thick secondary walls strengthened by lignin at maturity.

scutellum Also called cotyledon; thin seed leaf with large surface area to absorb nutrients during germination; found in monocots such as corn.

second law of thermodynamics The principle whereby every energy transfer or transformation increases the entropy of the universe. Ordered forms of energy are at least partly converted to heat, and in spontaneous reactions, the free energy of the system also decreases.

second messenger A chemical signal, such as calcium ions or cyclic AMP, that relays a hormonal message from a cell's surface to its interior.

secondary compounds Chemical compounds synthesized through the diversion of products of major metabolic pathways for use in defense by prey species.

secondary consumers The trophic level of an ecosystem consisting of carnivores that eat herbivores.

secondary growth The increase in girth of the stems and roots of many plants, especially woody, perennial dicots.

secondary immune response The immune response elicited when an animal encounters the same antigen at some later time. The secondary immune response is more rapid, of greater magnitude, and of longer duration than the primary immune response.

secondary productivity The rate at which all the heterotrophs in an ecosystem incorporate organic material into new biomass, which can be equated to chemical energy.

secondary structure The localized, repetitive folding of the polypeptide backbone of a protein due to hydrogen-bond formation between peptide linkages.

secondary succession A type of succession that occurs where an existing community has been severely cleared by some disturbance.

sedimentary rock *(SED-eh-MEN-tar-ee)* Rock formed from sand and mud that once settled in layers on the bottom of seas, lakes, and marshes. Sedimentary rocks are often rich in fossils.

seed An adaptation for terrestrial plants consisting of an embryo packaged along with a store of food within a resistant coat.

selection coefficient The difference between two fitness values, representing a relative measure of selection against an inferior genotype.

selective permeability A property of biological membranes that allows some substances to cross more easily than others.

semen *(SEE-men)* The fluid that is ejaculated by the male during orgasm; contains sperm and secretions from several glands of the male reproductive tract.

semicircular canals A three-part chamber of the inner ear that functions in maintaining equilibrium.

semilunar valve A valve located at the two exits of the heart, where the aorta leaves the left ventricle and the pulmonary artery leaves the right ventricle.

seminal vesicle *(SEM-in-ul VES-eh-kul)* A part of the male reproductive tract that stores sperm in invertebrates and produces semen in vertebrates.

seminiferous tubules *(SEM-in-IF-er-us)* Highly coiled tubes in the testes in which sperm are produced.

senescence *(seh-NESS-ents)* Aging; a progression of irreversible change in a living organism, eventually leading to death.

sensation An impulse sent to the brain from activated receptors and sensory neurons.

sensory neuron A nerve cell that receives information from the internal and external environments and transmits the signals to the central nervous system.

sensory, or afferent, nervous system In vertebrates, the component of the peripheral nervous system that transmits signals to the central nervous system from sensory receptors.

sensory receptor A specialized structure that responds to specific stimuli from an animal's external or internal environment; transmits the information of an environmental stimulus to the animal's nervous system by converting stimulus energy to the electrochemical energy of action potentials.

sepal *(SEE-pul)* A whorl of modified leaves in angiosperms that encloses and protects the flower bud before it opens.

set point Set point describes the threshold of the regulatory machinery of an organism to maintain homeostasis.

sex chromosomes The pair of chromosomes that differentiate between and are responsible for determining the sex of an individual.

sex-linked genes Genes located on one sex chromosome but not the other.

sexual dimorphism *(dye-MOR-fiz-um)* A special case of polymorphism based on the distinction between secondary sex characteristics of males and females.

sexual reproduction A type of reproduction in which two parents give rise to offspring that have unique combinations of genes inherited from the gametes of the two parents.

sexual selection Selection based on variation in secondary sex characteristics, leading to the enhancement of sexual dimorphism.

shoot system The aerial portion of a plant body, consisting of stems, leaves, and flowers.

short-day plant A plant that flowers, usually in late summer, fall, or winter, only when the light period is shorter than a critical length.

sieve-tube member A chain of living cells that form sieve tubes in phloem.

sign stimulus An external sensory stimulus that triggers a fixed-action pattern.

signal-transduction pathway A mechanism linking a mechanical or chemical stimulus to a cellular response.

simple fruit A fruit derived from a single ovary.

sister chromatids *(KROH-muh-tids)* Replicated forms of a chromosome joined together by the centromere and eventually separated during mitosis or meiosis II.

skeletal muscle Striated muscle generally responsible for the voluntary movements of the body.

sliding filament model The theory explaining how muscle contracts, based on change within a sarcomere, the basic unit of muscle organization, stating that thin (actin) filaments slide across thick filaments, shortening the sarcomere; shortening of all sarcomeres in a myofibril shortens the entire myofibril.

small nuclear ribonucleoproteins (snRNPs) *(RIBE-oh-NUKE-lee-oh-pro-teens)* A variety of small particles in the cell nucleus, composed of RNA and protein molecules; functions are not fully understood, but some form parts of spliceosomes, active in RNA splicing.

smooth ER That portion of the endoplasmic reticulum that is free of ribosomes.

smooth muscle A type of muscle lacking the striations of skeletal and cardiac muscle because of the uniform distribution of myosin filaments in the cell.

sodium-potassium pump A special transport protein in the plasma membrane of animal cells that transports sodium out of and potassium into the cell against their concentration gradients.

solute *(SOL-yoot)* A substance that is dissolved in a solution.

solution A homogeneous, liquid mixture of two or more substances.

solvent The dissolving agent of a solution. Water is the most versatile solvent known.

somatic cell *(SOWM-at-ik)* Any cell in a multicellular organism except a sperm or egg cell.

somatic nervous system The division of the motor nervous system of vertebrates composed of motor neurons that carry signals to skeletal muscles in response to external stimuli.

somite *(SO-mite)* A block of mesoderm along each side of a chordate embryo.

spatial summation An accumulative effect on the membrane potential of a postsynaptic cell, caused by several different synaptic terminals stimulating a postsynaptic cell simultaneously.

speciation *(SPEE-see-AY-shun)* The origin of new species in evolution.

species A particular kind of organism; members possess similar anatomical characteristics and have the ability to interbreed.

species diversity The number and relative abundance of species in a biological community.

species richness The number of species in a biological community.

species selection A theory maintaining that species living the longest and generating the greatest number of species determine the direction of major evolutionary trends.

specific heat The amount of heat that must be absorbed or lost for 1 g of a substance to change its temperature by 1°C.

spermatogenesis The continuous and prolific production of mature sperm cells in the testis.

spermatozoon (plural, **spermatozoa**) The male gamete; a haploid, usually small, flagellated cell.

sphincter *(SFINK-ter)* A ringlike valve, consisting of modified muscles in a muscular tube, such as a digestive tract; closes off the tube like a drawstring.

spindle An assemblage of microtubules that orchestrates chromosome movement during eukaryotic cell division.

spiracles Minute pores distributed over the surface of an insect's body that serve as entrances to the tracheal system.

spiral cleavage A type of embryonic development in protostomes, in which the planes of cell division that transform the zygote into a ball of cells occur obliquely to the polar axis, resulting in cells of each tier sitting in the grooves between cells of adjacent tiers.

spliceosome *(SPLISE-ee-oh-SOME)* A complex assembly that interacts with the ends of an RNA intron in splicing RNA; releases an intron and joins two adjacent exons.

sporangium (plural, **sporangia**) A capsule in fungi and plants in which meiosis occurs and haploid spores develop.

spore In the life cycle of a plant or alga undergoing alternation of generations, a meiotically produced haploid cell that divides mitotically, generating a multicellular individual, the gametophyte, without fusing with another cell.

sporophyll A spore-producing leaf specialized for reproduction.

sporophyte The multicellular diploid form in organisms undergoing alternation of generations that results from a union of gametes and that meiotically produces haploid spores that grow into the gametophyte generation.

stabilizing selection Natural selection that favors intermediate variants by acting against extreme phenotypes.

stamen The pollen-producing male reproductive organ of a flower, consisting of an anther and filament.

starch A storage polysaccharide in plants consisting entirely of glucose.

statocyst *(STAT-eh-SIST)* A type of mechanoreceptor that functions in equilibrium in invertebrates through the use of statoliths, which stimulate hair cells in relation to gravity.

statolith (1) A starch grain in a plant cell that settles to the low point of the cells in relation to gravity. (2) In animals, a granule of sand or limestone that, in responding to gravity, stimulates hair cells of a statocyst.

stele The central vascular cylinder in roots where xylem and phloem are located.

stereoisomer A molecule that is a mirror image of another molecule with the same molecular formula.

steroids A class of lipids characterized by a carbon skeleton consisting of four rings with various functional groups attached.

stigma The sticky tip of the carpel that receives the pollen in a flower.

stoma (plural, **stomata**) A microscopic pore surrounded by guard cells in the epidermis of leaves and stems that allows gas exchange between the environment and the interior of the plant.

strict aerobe *(AIR-obe)* An organism that can survive only in an atmosphere of oxygen, which is used in aerobic respiration.

strict anaerobe An organism that cannot survive in an atmosphere of oxygen. Other substances, such as sulfate or nitrate, are the terminal electron acceptors in the electron transport chains that generate their ATP.

stroma The fluid of the chloroplast surrounding the thylakoid membrane; involved in the synthesis of organic molecules from carbon dioxide and water.

stromatolite Rock made of banded domes of sediment in which are found the most ancient forms of life: prokaryotes dating back as far as 3.5 billion years.

structural formula A type of molecular notation in which the constituent atoms are joined by lines representing covalent bonds.

structural gene A gene that codes for a polypeptide.

style The slender, necklike portion of the carpel that leads to the ovary.

substrate The substance on which an enzyme works.

substrate-level phosphorylation The formation of ATP by directly transferring a phosphate group to ADP from an intermediate substrate in catabolism.

summation A phenomenon of neural integration in which the membrane potential of the postsynaptic cell in a chemical synapse is determined by the total activity of all excitatory and inhibitory presynaptic impulses acting on it at any one time.

suppressor T cells (T_S) A type of T cell that causes B cells as well as other cells to ignore antigens.

surface tension A measure of how difficult it is to stretch or break the surface of a liquid. Water has a high surface tension because of the hydrogen bonding of surface molecules.

survivorship curve A plot of the number of members of a cohort that are still alive at each age; one way to represent age-specific mortality.

suspension-feeder An aquatic animal, such as a clam or a baleen whale, that sifts small food particles from the water; also called filter-feeder.

sustainable agriculture Long-term productive farming methods that are environmentally safe.

symbiont *(SIM-bye-ont)* The smaller participant in a symbiotic relationship, living in or on the host.

symbiosis An ecological relationship between organisms of two different species that live together in direct contact.

sympathetic nervous system One of two divisions of the autonomic nervous system of vertebrates; generally increases energy expenditure and prepares the body for action.

sympatric speciation A mode of speciation occurring as a result of a radical change in the genome that produces a reproductively isolated subpopulation in the midst of its parent population.

symplast In plants, the continuum of cytoplasm connected by plasmodesmata between cells.

synapse *(SIN-aps)* The locus where one neuron communicates with another neuron in a neural pathway; a narrow gap between a synaptic terminal of an axon and a signal-receiving portion (dendrite or cell body) of another neuron or effector cell. Neurotransmitters released by synaptic terminals diffuse across the synapse, relaying messages to the dendrite or effector.

synapsid *(SIN-ap-sid)* A predatory reptile that evolved during the Permian period and gave rise to several lineages, including the mammal-like therapsids.

synapsis The pairing of replicated homologous chromosomes during prophase I of meiosis.

synaptic terminal One of many swollen bulbs at the end of an axon, in which neurotransmitter is stored and released.

synaptic vesicle One of many membrane-enclosed sacs in a synaptic terminal that contains neurotransmitter.

synaptonemal complex *(sin-APP-toe-NEEM-ul)* A protein structure, visible in electron micrographs of prophase I of meiosis; function is unknown but correlated with, and essential for, chiasmata formation and crossing over; often zipper-like in appearance.

syngamy *(SIN-gam-ee)* The process of cellular union during fertilization.

systematics The branch of biology that studies the diversity of life. Systematics encompasses taxonomy and is involved in reconstructing phylogenetic history.

systole The stage of the heart cycle in which the heart muscle contracts and the chambers pump blood.

T cell A type of lymphocyte responsible for cell-mediated immunity that differentiates under the influence of the thymus.

taiga *(TYE-guh)* The coniferous or boreal forest biome, characterized by considerable snow, harsh winters, short summers, and evergreen trees.

target cell A cell bearing specific receptors for a particular signal, such as a hormone.

taxis *(TAKS-iss)* A movement toward or away from a stimulus.

taxon (plural, **taxa**) The named taxonomic unit at any given level.

taxonomy The branch of biology concerned with naming and classifying the diverse forms of life.

temperate deciduous forest A biome located throughout midlatitude regions where there is sufficient moisture to support the growth of large, broad-leaved deciduous trees.

temperate virus A virus that can reproduce without killing the host.

temperature A measure of the intensity of heat in degrees, reflecting the average kinetic energy of the molecules.

temporal summation An accumulative effect on the membrane potential of a postsynaptic cell; caused by chemical transmissions from one or more synaptic terminals occurring so close together in time that each postsynaptic potential affects the membrane before the voltage has returned to the resting potential after the previous stimulation.

tendon A type of fibrous connective tissue that attaches muscle to bone.

terminal bud Embryonic tissue at the tip of a shoot, consisting of developing leaves and a compact series of nodes and internodes.

tertiary structure *(TUR-she-air-ee)* Irregular contortions of a protein molecule due to interactions of side chains involved in hydrophobic interactions, ionic bonds, hydrogen bonds, and disulfide bridges.

testcross Breeding of an organism of unknown genotype with a homozygous recessive individual to determine the unknown genotype. The ratio of phenotypes in the offspring determines the unknown genotype.

testis (plural, **testes**) The male reproductive organ, or gonad, in which sperm and reproductive hormones are produced.

testosterone The most abundant androgen in the male body.

tetanus *(TET-un-us)* The maximal, sustained contraction of a skeletal muscle, caused by a very fast frequency of action potentials elicited by continual stimulation.

tetrapod A vertebrate possessing two pairs of limbs, such as amphibians, reptiles, birds, and mammals.

thalamus *(THAAL-uh-mus)* One of two integrating centers of the vertebrate forebrain. Neurons with cell bodies in the thalamus relay neural input to specific areas in the cerebral cortex and regulate what information goes to the cerebral cortex.

thecodont A reptile that descended from the stem reptiles during the Permian period and that served as the ancestors of dinosaurs, crocodiles, and birds.

therapsid *(thur-AP-sid)* A member of a reptile lineage that descended from synapsids during the Permian period and eventually gave rise to mammals.

thermocline A narrow vertical zone of rapid temperature change that differentiates the warm upper layer from the cold bottom water of a lake.

thermodynamics The study of energy transformations that occur in a collection of matter.

thermoregulation The maintenance of internal temperature within a tolerable range.

thick filament A filament composed of staggered arrays of myosin molecules; a component of myofibrils in muscle fibers.

thigmotropism *(THIG-moh-TROH-piz-um)* The directional growth of a plant in relation to touch.

thin filament A filament consisting of two strands of actin and one strand of regulatory protein coiled around one another; one component of myofibrils in muscle fibers.

threshold The potential an excitable cell membrane must reach for an action potential to be initiated.

thylakoid *(THIGH-luh-koid)* A flattened membrane sac inside the chloroplast, used to convert light energy to chemical energy.

thymus *(THYE-mus)* An endocrine gland in the neck region of mammals that is active in establishing the immune system; secretes several messengers, including thymosin, that stimulate T cells.

thyroid-stimulating hormone (TSH) A hormone produced by the anterior pituitary gland that regulates the release of thyroid hormones.

Ti plasmid A plasmid of a tumor-inducing bacterium that integrates a segment of its DNA into the host chromosome of a plant; frequently used as a vector for genetic engineering in plants.

tight junction A type of intercellular junction in animal cells that prevents the leakage of material between cells.

tissue An integrated group of cells with a common structure and function.

totipotent Referring to embryonic cells that retain the potential to form all parts of the animal.

trace element An element indispensable for life but required in extremely minute amounts.

trachea *(TRAY-kee-uh)* The windpipe; that portion of the respiratory tube that has C-shaped cartilagenous rings and passes from the larynx to two bronchi.

tracheae *(TRAY-kee-ee)* Tiny air tubes that ramify throughout the insect body for gas exchange.

tracheal system A gas exchange system of branched, chitin-lined tubes that infiltrate the body and carry oxygen directly to cells in insects.

tracheid *(TRAY-kee-id)* A water-conducting and supportive element of xylem composed of long, thin cells with tapered ends and walls hardened with lignin.

transcription The transfer of information from a DNA molecule into an RNA molecule.

transcription unit A unit comprising the entire stretch of DNA that is transcribed into a single RNA molecule marked by the intiation and termination sites of a gene.

transduction Viral transfer of DNA from one bacterial cell to another; in generalized transduction, recombination occurs between genetic material from two cells without selective transfer, whereas in specialized transduction, selective transfer occurs from DNA located near the integration site of the prophage.

transfer RNA (tRNA) An RNA molecule that functions as an interpreter between nucleic acid and protein language by picking up specific amino acids and recognizing the appropriate codons in the mRNA.

transformation (1) The conversion of a normal animal cell to a cancerous cell. (2) A phenomenon in which external genetic material is assimilated by a cell.

transgenic organism An organism containing certain genes from another species; produced, for example, by injecting foreign DNA into the nuclei of egg cells or early embryos.

translation The transfer of information from an RNA molecule into a polypeptide, involving a change of language from nucleic acids to amino acids.

translocation (1) An aberration in chromosome structure resulting from an error in meiosis or from mutagens; attachment of a chromosomal fragment to a nonhomologous chromosome. (2) During protein synthesis, the third stage in the elongation cycle when the RNA carrying the growing polypeptide moves from the A site to the P site on the ribosome. (3) The transport via phloem of food in a plant.

transpiration The evaporative loss of water from a plant.

transport vesicle A vesicle that moves material from the ER to the Golgi apparatus.

transposon *(trans-POE-son)* A transposable genetic element; a mobile segment of DNA that serves as an agent of genetic change.

triplet code A set of three-nucleotide-long words that specify the amino acids for polypeptide chains.

triploblastic Possessing three germ layers: the endoderm, mesoderm, and ectoderm. Most eumetazoa are triploblastic.

trisomic A type of aneuploidy in which one chromosome is present in triplicate in the fertilized egg.

trophic level The division of species in an ecosystem on the basis of their main nutritional source. The trophic level that ultimately supports all others consists of autotrophs, or primary producers.

trophic structure The different feeding relationships in an ecosystem that determine the route of energy flow and the pattern of chemical cycling.

trophoblast The outer epithelium of the blastocyst, which forms the fetal part of the placenta.

tropic hormone A hormone that has another endocrine gland as a target.

tropical rain forest The most complex of all communities, located near the equator where rainfall is abundant; harbors more species of plants and animals than all other terrestrial biomes combined.

tropism A growth response that results in the curvature of whole plant organs toward or away from stimuli due to differential rates of cell elongation.

tumor A mass that forms within otherwise normal tissue, caused by the uncontrolled growth of a transformed cell.

tundra A biome at the extreme limits of plant growth; at the northernmost limits, it is called arctic tundra, and at high altitudes, where plant forms are limited to low shrubby or matlike vegetation, it is called alpine tundra.

turgid *(TUR-jid)* Firm; walled cells become turgid as a result of the entry of water from a hypoosmotic environment.

turgor pressure Force directed against a cell wall after the influx of water and the swelling of a walled cell due to osmosis.

ultimate causation The hypothetical evolutionary explanation for the existence of a certain pattern of animal behavior.

unsaturated fatty acid A fatty acid possessing one or more double bonds between the carbons in the hydrocarbon tail. Such bonding reduces the number of hydrogen atoms attached to the carbon skeleton.

urea A soluble form of nitrogenous waste excreted by mammals and most adult amphibians.

ureter A duct leading from the kidney to the urinary bladder.

urethra A tube that releases urine from the body near the vagina in females or through the penis in males; also serves in males as the exit tube for the reproductive system.

uric acid An insoluble precipitate of nitrogenous waste excreted by land snails, insects, birds, and some reptiles.

urochordate A chordate without a backbone, commonly called a tunicate, a sessile marine animal.

uterus A female reproductive organ where eggs are fertilized and/or development of the young occurs.

vaccine A harmless variant or derivative of a pathogen that stimulates a host's immune system to mount defenses against the pathogen.

vacuole A membrane-enclosed sac taking up most of the interior of a mature plant cell and containing a variety of substances important in plant reproduction, growth, and development.

vagina A thin-walled chamber that forms the birth canal and is the repository for sperm during copulation.

valence shell The outermost energy shell of an atom, containing the valence electrons involved in the chemical reactions of that atom.

vascular cambium A continuous cylinder of meristematic cells surrounding the xylem and pith that produces secondary xylem and phloem.

vascular plants Plants with vascular tissue, consisting of all modern species except the mosses and their relatives.

vascular tissue Plant tissue consisting of cells joined into tubes that transport water and nutrients throughout the plant body.

vascular tissue system A system formed by xylem and phloem throughout the plant, serving as a transport system for water and nutrients, respectively.

vas deferens The tube in the male reproductive system in which sperm travel from the epididymis to the urethra.

vegetal pole The end of a zygote opposite the animal pole; the yolky end of a frog, bird, or reptile embryo.

vegetative reproduction Cloning of plants by asexual means.

veins Vessels that return blood to the heart.

ventilation Any method of increasing contact between the respiratory medium and the respiratory surface.

ventral The bottom (or belly) half of a bilaterally symmetric animal.

ventricle *(VEN-treh-kul)* (1) A chamber that pumps blood out of the vertebrate heart. (2) A space in the vertebrate brain filled with cerebrospinal fluid.

venules Vessels into which capillaries converge and that empty into veins.

vertebrates Chordate animals with backbones, consisting of the mammals, birds, reptiles, amphibians, and various classes of fishes.

vessel elements Specialized short, wide cells in angiosperms, arranged end to end to form continuous tubes for water transport.

vestigial organ A type of homologous structure that is rudimentary and of marginal or no use to the organism.

viroid *(VY-roid)* A plant pathogen composed of molecules of naked RNA only several hundred nucleotides long.

virulent virus A virus capable of causing a disease state.

visceral muscle Smooth muscle found in the walls of the digestive tract, bladder, arteries, and other internal organs.

visible light That portion of the electromagnetic spectrum detected as various colors by the human eye, ranging in wavelength from about 400 nm to about 700 nm.

vitalism The belief that natural phenomena are governed by a life force outside the realm of physical and chemical laws.

vitamin An organic molecule required in the diet in very small amounts; vitamins serve primarily as coenzymes or parts of coenzymes.

viviparous *(vy-VIP-er-us)* Referring to a type of development in which the young are born alive after having been nourished in the uterus by blood from the placenta.

voltage-sensitive channel An ion channel in the membrane of excitable cells that is constructed of protein molecules and allows only certain ions to pass under specific changes of electric potential; also called voltage-gated channel.

water potential The physical property predicting the direction in which water will flow, governed by solute concentration and applied pressure.

water vascular system A network of hydraulic canals unique to echinoderms that branches into extensions called tube feet, which function in locomotion, feeding, and gas exchange.

wavelength The distance between crests of waves, such as those of the electromagnetic spectrum.

wild type An individual with the normal phenotype.

wobble A violation of the base-pairing rules in that the third nucleotide (5′ end) of a tRNA anticodon can form hydrogen bonds with more than one kind of base in the third position (3′ end) of a codon.

xylem *(ZY-lum)* The tube-shaped, nonliving portion of the vascular system in plants that carries water and minerals from the roots to the rest of the plant.

yeast A unicellular fungus that lives in liquid or moist habitats, primarily reproducing asexually by simple cell division or by budding of a parent cell.

yolk sac One of four extraembryonic membranes that supports embryonic development; the first site of blood cells and circulatory system function.

zygote The diploid product of the union of haploid gametes in conception; a fertilized egg.

PHOTOGRAPH CREDITS

Frontispiece © Joseph Holmes.

Front Matter *The Campbell Interviews:* Unit One: Annalisa Kraft. Unit Two: Robin Heyden. Unit Three: Pedro Perez. Unit Four: Annalisa Kraft. Unit Five: Annalisa Kraft. Unit Six: Alain McGlaughlin. Unit Seven: Ron Wurzer. Unit Eight: Alain McGlaughlin. *Table of Contents* . Unit One: Oxford Molecular Biophysics Laboratory/SPL/Photo Researchers, Inc. Unit Two: © Dr. Jeremy Burgess/Science Photo Library/Photo Researchers, Inc. Unit Three: © Dr. Ram Verma/PhotoTake, Inc. Unit Four: © A. Kerstitch/Visuals Unlimited. Unit Five: © Stephen J. Krasemann/Photo Researchers, Inc. Unit Six: © John D. Cunningham/Visuals Unlimited. Unit Seven: © Frans Lanting/Minden Pictures. Unit Eight: © L.L.T. Rhoades/Animals Animals.

Chapter 1 1.1a: © 1984 Rick Friedman/Black Star. 1.1b: © Alain McGlaughlin. 1.1c: University of California, San Francisco. 1.1d: Courtesy of Dr. Ariel Lugo. 1.2a: © Robert Fletterick. 1.2b, d: © Dr. Jeremy Burgess/SPL/Photo Researchers, Inc. 1.2c: © Manfred Kage/Peter Arnold, Inc. 1.2e: © John Shaw/Tom Stack and Associates. 1.2f: *Aspens, Northern New Mexico.* Ansel Adams Publishing Rights Trust © 1992. 1.3a: © Jane Burton/Bruce Coleman, Inc. 1.3b: © Esao Hashimoto/Animals Animals. 1.3c: © Michael Fogden/Bruce Coleman, Inc. 1.3d: © David C. Fritts/Animals Animals. 1.3e: © Gunter Ziesler/Peter Arnold, Inc. 1.3f: © G. C. Kelley, 1986/Photo Researchers, Inc. 1.3g: © Breck P. Kent/Animals Animals. 1.4a: M. W. Steer and E. H. Newcomb, University of Wisconsin, Madison/BPS. 1.4b: © J. J. Cardamone, Jr., University of Pittsburgh/BPS. 1.5: N. L. Max, University of California/BPS. 1.6a: © Frans Lanting/Minden Pictures. 1.6b: © Janice Sheldon. 1.6c: © Don Fawcett/Science Source/Photo Researchers, Inc. 1.6d: Courtesy of N. Simeonescu. 1.9a: © Manfred Kage/Peter Arnold, Inc. 1.9b: © D. P. Wilson/Photo Researchers, Inc. 1.9c: Patti Murray/Animals Animals. 1.9d: © Michael Fogden/Earth Scenes. 1.9e: © Gunter Ziesler/Peter Arnold, Inc. 1.10a: © Biophoto Associates/Photo Researchers, Inc. 1.10b: © John Walsh/Photo Researchers, Inc. 1.10c.2: © Photo Researchers, Inc. 1.11: © Kevin Schafer. 1.12: Artist: George Richmond; Photographer: Larry Burrows/Life Magazine © Time Warner, Inc. 1.13a: © E. S. Ross, California Academy of Sciences. 1.13b: © K. G. Preston Mafham/Animals Animals. 1.13c: © P. and W. Ward/Animals Animals. 1.14 © Helen Rodd. 1.15a © Hayden Mattingly.

Unit One Interview Annalisa Kraft.

Chapter 2 2.1: © Herbert Schwind/Okapia 1989/Photo Researchers, Inc. 2.3a–c: Stephen Frisch. 2.4: © Grant Heilman. 2.6: Courtesy of M. E. Raichle. 2.7: © Jack Bostrack 1988. 2.16: Stephen Frisch. 2.18: © Anglo Australian Telescope Board 1987. Methods Box: © Kevin Schafer. From M. C. Rattazzi et al., *Am. J. Human Genet.,* 28:143–154, 1976.

Chapter 3 3.1: W. Reid Thompson, Cornell University. 3.3 left: © Linda Hopson/Visuals Unlimited. 3.3 right: R. G. Kessel and C. Y. Shih, *Scanning Electron Microscopy in Biology,* New York: Springer-Verlag, 1974, p. 147. 3.4a: © Hans Pfetschinger/Peter Arnold, Inc. 3.4b: © G. I. Bernard/Animals Animals. 3.5: John Biever © *Sports Illustrated,* Time, Inc. 3.7: © Flip Nicklin 1990. 3.11: © Silvestris/Natural History Photographic Agency. 3.12: Robert F. Sisson, © National Geographic Society.

Chapter 4 4.1: © Graphics Systems Research, IBM UK Scientific Centre. 4.2: © The Bettman Archive. 4.3: © Roger Ressmeyer/Starlight. 4.7: Courtesy of Howard Green, Harvard Medical School; from H. Green and M. Meuth, *Cell* 3(1974):127–133. 4.9 right: © W. J. Weber/Visuals Unlimited. 4.9: © Stephen J. Krasemann/Photo Researchers, Inc. 4.10: © J. H. Robinson/Photo Researchers, Inc.

Chapter 5 5.1: © Syd Greenberg/Photo Researchers, Inc. 5.2a: © Dr. Linus Pauling. 5.2b: Courtesy University of California, Riverside. 5.8a: © E. H. Newcomb, University of Wisconsin, Madison/BPS. 5.8b: © R. Rodewald, University of Virginia/BPS. 5.10: Courtesy of Eva Frei and R. D. Preston, Lees, England. 5.11: © F. Collett/Photo Researchers, Inc. 5.12: © Frans Lanting. 5.13a: Courtesy American Dairy Association. 5.13b: © Lara Hartley. 5.15: © Ed Reschke. 5.19: © CNRI/SPL/Photo Researchers, Inc. 5.22a, b: M. Murayama, Murayama Research Laboratory/BPS. 5.24: F. Vollrath, *Nature* 340(1989):306. 5.29a–d: Courtesy of Frederic M. Richards, Yale University. 5.29d: Courtesy University of California, Riverside.

Chapter 6 6.1: © Ken Lucas/BPS. 6.3: © Eunice Harris 1987/Photo Researchers, Inc. 6.4a: Local Color Painting, San Francisco, CA/Bruce Nelson, contractor. 6.4b, c: Christine Case, Skyline College. 6.5a: © John D. Cunningham/Visuals Unlimited. 6.5b: Barbara J. Miller/BPS. 6.14a, b: Courtesy of Thomas Steitz. 6.17: © Paul Chesley, Photographers/Aspen. 6.20: © R. Rodewald, University of Virginia/BPS. Challenge Question 6: © J. M. Rowland/Robert and Linda Mitchell.

Unit Two Interview Robin Heyden.

Chapter 7 7.1: Mark S. Ladinsky and J. Richard McIntosh, University of Colorado. 7.4a, b: W. L. Dentler. 7.6b: S. C. Holt, University of Texas Health Center/BPS. 7.8a: Courtesy J. David Robertson. 7.11a: L. Orci and A. Perrelet, *Freeze-Etch Histology,* Heidelberg: Springer-Verlag, 1975. 7.11d: A. C. Faberge, *Cell Tiss. Res.* 151(1974):403. Springer-Verlag, 1974. 7.12: © D. W. Fawcett/Photo Researchers, Inc. 7.13: Courtesy of Lan Bo Chen, Harvard Medical School. 7.14: J. F. King, University of California School of Medicine/BPS. 7.15: G. T. Cole, University of Texas, Austin/BPS. 7.16a: R. Rodewald, University of Virginia/BPS. 7.16b: Courtesy Daniel Friend. 7.18: Courtesy of E. H. Newcomb. 7.20: © S. E. Frederick and E. H. Newcomb, *J. Cell Biol.* 43 *(1969):343.* Reproduced by copyright permission of The Rockefeller University Press. Provided by E. H. Newcomb. 7.21: Courtesy of Daniel Friend. 7.22: Courtesy of W. P. Wergin and E. H. Newcomb, University of Wisconsin, Madison/BPS. 7.23 Micrograph produced by Dr. John E. Heuser, Washington University School of Medicine, St. Louis, MO. 7.24: Courtesy of Mark S. Ladinsky and J. Richard McIntosh, University of Colorado. 7.26a: Courtesy Leah Haimo and Cathy Thaler, University of California, Riverside. 7.26b: Courtesy of Katherine Luby-Phelps, Center for Fluorescence Research in Biomedical Science, Carnegie Mellon University. 7.27: Courtesy of Kent McDonald. 7.28: Courtesy D. D. Kunkel, University of

Washington/BPS. 7.30a: © Photo Researchers, Inc. 7.30b: Courtesy W. L. Dentler, University of Kansas/BPS. 7.30c: Courtesy of William L. Dentler, University of Kansas. 7.32a: Micrograph produced by Dr. John E. Heuser, Washington University School of Medicine, St. Louis, MO. 7.32b: *J. Cell Biol.* 94(1982):425, Hirokawa Nobutaka. Reproduced by copyright permission of The Rockefeller University Press. 7.33 right: © G. F. Leedale/Photo Researchers, Inc. 7.34a: © Douglas J. Kelly, *J. Cell Biol.* 28(1966):51. Reproduced by copyright permission of The Rockefeller University Press. 7.34b: L. Orci and A. Perrelet, *Freeze-Etch Histology*, Heidelberg: Springer-Verlag, 1975. 7.34c: C. Peracchia and A. F. Dulhunty, *J. Cell Biol.* 70(1976):419. Reproduced by copyright permission of The Rockefeller University Press. 7.35: Boehringer Ingelheim International GmbH, photo © Lennart Nilsson, *The Body Victorious*, Delacourte Press, Dell Publishing Co., Inc. Self-Quiz 7a: © Biophoto Associate/Photo Researchers, Inc. Self-Quiz 7b: © K. R. Porter/Photo Researchers, Inc. Self-Quiz 7c: © D. W. Fawcett/Photo Researchers, Inc. Self-Quiz 7d: © Biophoto Associates/Photo Researchers, Inc. Self Quiz 7e: Kathryn Platt-Aloia and William W. Thomson. Self-Quiz 7f: © D. W. Fawcett/Photo Researchers, Inc. T7.1 top left: © Biophoto Associates/Photo Researchers, Inc. T7.1 middle left: © Ed Reschke. T7.1 lower left, upper and middle right: © David M. Phillips/Visuals Unlimited. T7.1 lower right: Courtesy of Noran Instruments. T7.2 left: Mark S. Ladinsky and J. Richard McIntosh, University of Colorado. T7.2 middle: Courtesy of Drs. Frank Solomon and J. Dinsmore, Massachusetts Institute of Technology. T7.2 right: Mark S. Ladinsky and J. Richard McIntosh, University of Colorado.

Chapter 8 8.1: D. W. Deamer and P. B. Armstrong, University of California, Davis. 8.4: Courtesy J. David Robertson. 8.16a, b: © Cabisco/Visuals Unlimited. 8.17a, b: © Jim Strawser/Grant Heilman. 8.24a, b: © Dr. Volker Herzog, *J. Cell Biol.* 70(1976):698. Reproduced by copyright permission of The Rockefeller University Press. 8.25: H. S. Pankratz, Michigan State University/BPS. 8.25: © D. W. Fawcett/Photo Researchers, Inc. 8.25 (2 photos): M. M. Perry and A. B. Gilbert, *J. Cell Sci.* 39(1979): 257. Copyright 1979 by The Company of Biologists Ltd. 8.26: J. V. Small and Gottfried Rinnerthaler, *Scientific American*, October 1985, p. 111. Methods Box: Courtesy of Dr. Philippa Claude, University of Wisconsin Primate Center.

Chapter 9 9.1: © Stephen Dalton/Photo Reseachers, Inc. 9.14: © Keith R. Porter/Photo Researchers, Inc. 9.17b: A. Tzagoloff, *Mitochondria*, New York: Plenum, 1982. 9.19a: © Michael Fogden/Bruce Coleman, Inc. 9.19b: Hayward and Lyman, *Mammalian Hibernation III*, Kenneth C. Fisher, ed. New York: Elsevier, 1967. 9.19c: Courtesy of Dr. Meeuse, University of Washington, Seattle. 9.22a: Egyptian Expedition of The Metropolitan Museum of Art, Rogers Fund, 1915. 9.22b: Obester Winery, Half Moon Bay, CA. Photo by Kevin Schafer. Self-Quiz 1: Courtesy of John A. Richardson, Southern Illinois University, Carbondale.

Chapter 10 10.1: Courtesy of National Optical Astronomy Observatories. 10.2a: © Gary Braasch/AllStock, Inc. 10.2b: © Harold Dorr. 10.2c, d: © Tom Adams/Peter Arnold, Inc. 10.2e: P. W. Johnson and J. McN. Sieburth, University of Rhode Island/BPS. 10.4a: E. H. Newcomb, University of Wisconsin, Madison/BPS. 10.4b: H. S. Pankratz, Michigan State University/BPS. 10.4c: N. J. Lang, Universtiy of California, Davis/BPS. 10.19 left: © John D. Cunningham/Visuals Unlimited. 10.19 right: © Helen Marcus 1981/Photo Researchers, Inc.

Chapter 11 11.2a: © Biophoto Associates/Photo Researchers, Inc. 11.2b: C. R. Wyttenbach, University of Kansas/BPS. 11.2c: © J&L Weber/Peter Arnold, Inc. 11.4b: G. F. Bahr, Armed Forces Institute of Pathology. 11.6a–g: Ed Reschke. 11.7b: Courtesy of Dr. Matthew Schibler, from *Protoplasma* 137(1987):29–44. 11.7a: Courtesy of Richard Macintosh. 11.9a, b: Courtesy of Jeremy Pickett-Heaps, University of Melbourne. 11.10: © David M. Phillips/Visuals Unlimited. 11.11b: E. H. Newcomb, University of Wisconsin, Madison/BPS. 11.12 (7 photos): J. R. Waaland, University of Washington. Self-Quiz: Carolina Biological Supply. Methods Box: Courtesy of Dr. Gunter Albrecht-Buehler, Northwestern University.

Unit Three Interview Pedro Perez.

Chapter 12 12.1: From Henri Lhote's expedition. 12.2: R. D. Campbell, University of California, Irvine/BPS. 12.6a: © Runk/Schoenberger/Grant Heilman, Inc. 12.6b: © Stan Elems/Visuals Unlimited. Methods Box: © SIU/Visuals Unlimited and © Dept. of Clinical Cytogenetics, Adden-Brookes Hospital, Cambridge/Science Photo Library/Photo Researchers, Inc.

Chapter 13 13.1: Courtesy Bettmann Archives. 13.9 (4 photos): Science Education Resources Pty. Ltd., Victoria, Australia. 13.14.left: © Stan Goldblatt/Photo Researchers, Inc. 13.14 right: © Gilbert Grant/Photo Researchers, Inc. 13.16: © Roger Ressmeyer/Starlight. End of chapter: Breeder/owner: Patricia Speciale; photographer: Norma JubinVille.

Chapter 14 14.1: From Peter Lichter and David Ward. *Science* 247 (5 January 1990), cover. 14.3a.b: © Darwin Dale. 14.12: © Ron Kimball. 14.15a: © Science Photo Library/Photo Researchers, Inc. 14.15b: © Bob Daemmrich/Stock, Boston. 14.18: © Barry L. Runk/Grant Heilman, Inc.

Chapter 15 15.1: From J. D. Watson, *The Double Helix*, New York: Atheneum, 1968, p. 215. © 1968 by J. D. Watson. 15.3: © Lee D. Simon/Photo Researchers, Inc. 15.5a: Cold Spring Harbor Archives. 15.5b: From J. D. Watson, *The Double Helix*, New York: Atheneum, 1968, p. 215. © 1968 by J. D. Watson. 15.10b: From D. J. Burks and P. J. Stambrook. *J. Cell Biol.* 77(1978):762. Reprinted by copyright permission of The Rockefeller University Press. 15.18: Courtesy of J. M. Murray, Department of Anatomy, University of Pennsylvania.

Chapter 16 16.1: Courtesy of Y. Menadue and James B. Reid, University of Tasmania. 16.6: Keith Wood. 16.10c: Institut de Biologie Moleclaire et Cellulaire, Strasbourg. 16.16b: B. Hamkalo and O. L. Miller, Jr. 16.18: O. L. Miller Jr., B. A. Hamkalo, and C. A. Thomas, Jr., *Science* 169(24 July 1970):392. Copyright 1970 by the American Association for the Advancement of Science.

Chapter 17 17.2: Visuals Unlimited. 17.3a, b: Robley C. Williams, University of California, Berkeley/BPS. 17.3c: John J. Cardamone, Jr., University of Pittsburgh/Biological Photo Service. 17.3d: R. C. Williams, University of California, Berkeley/BPS. 17.12: Courtesy of Charles C. Brinton, Jr., and Judith Carnehan. 17.17: AP/Wide World Photos.

Chapter 18 18.1: Photo courtesy of M. B. Roth and J. Gall. 18.2a top: S. C. Holt, University of Texas Health Science Center, San Antonio/BPS. 18.2a bottom: A. L. Olins, University of Tennessee/BPS. 18.2b: Courtesy of Barbara Hamkalo. 18.2c: Courtesy of J. R. Paulsen and U. K. Laemmli, *Cell*. 18.2d: G. F. Bahr, Armed Forces Institute of Pathology. 18.3: Courtesy of O. L. Miller Jr., Dept. of Biology, University of Virginia. 18.11: Courtesy of J. M. Amabis.

Chapter 19 19.1: © Nina Winter. 19.11: Cellmark Diognostics. Methods Box: Damon Biotech, Inc.

Unit Four Interview Annalisa Kraft.

Chapter 20 20.2: Courtesy of Charles H. Phillips. 20.4: © John Cancalosi/Tom Stack & Associates. 20.5: © Tom Till. 20.6b: © Erwin & Peggy Bauer. 20.7a: © Frans Lanting/Minden Pictures. 20.7b: © Alan Root/Bruce Coleman, Inc. 20.8a: © C. A. Morgan/Peter Arnold, Inc. 20.8b: Rudi Kuiter © *Discover Magazine*. 20.9: © John Colwell/Grant Heilman, Inc. 20.10a, b: © Breck P. Kent/Animals Animals. 20.11a: Larry Brown, University of Wyoming, *J. of Mammalogy* 46(1965):464. 20.13: Philip Gingerich © 1991 *Discover Magazine*. 20.15a: ©

Lennart Nilsson, *A Child Is Born*, Dell Publishing Company. 20.15b: Willis Mathews, *Ency. of Descriptive Embryology*, 4th ed. New York: MacMillan Publishing Co.

Chapter 21 21.1: © J. Antonovics/Visuals Unlimited. 21.2a: © David Cavagnaro. 21.2b: Produced from USAF DMSP (Defense Meteorological Satellite Program) film transparencies archived for NOAA/NESDIS at the University of Colorado, CIRES/National Snow and Ice Data Center. 21.5: © Anup Shah/Animals Animals. 21.6: Courtesy of E. D. Brodie III, University of California, Berkeley. 21.9: Courtesy of Thomas B. Smith, San Francisco State University. 21.10 © Animals Animals/OSF/J. A. L. Cooke. 21.12: Richard Kolar, San Diego Zoo.

Chapter 22 22.1: © Erwin & Peggy Bauer. 22.3: Courtesy of Ernst Mayr, Museum of Comparative Zoology, Harvard University. 22.4a: © L. L. T. Rhodes/Tony Stone Worldwide. 22.4b top: © Don and Pat Valenti/Tom Stack & Associates. 22.4b bottom: © John Shaw/Tom Stack & Associates. 22.6: © Joe McDonald/Visuals Unlimited. 22.7: © Robert Lee/Photo Researchers, Inc. 22.8a: © Bob McKeever 1981/Tom Stack & Associates. 22.8b: © Mike Andrews/Animals Animals. 22.11: © David Cavagnaro. 22.13: Courtesy of J. D. Thompson. 22.14: Courtesy of Kenneth Kaneshiro. Self-Quiz: © Russell C. Hansen.

Chapter 23 23.1: © Doug Lee. 23.2a: © Barbara J. Miller/BPS. 23.2b: © Margo Crabtree. 23.2c: © Tom Till. 23.2d, f: © Manfred Kage/Peter Arnold, Inc. 23.2e: © W. H. Hodge/Peter Arnold, Inc. 23.4a, b: © John Shaw/NHPA. 23.5: © Jane Burton/Photo Researchers, Inc. 23.10b: © Kevin Schafer. 23.10c: © Sygma. 23.16 (2 photos): © Tom McHugh/Photo Researchers, Inc. 23.17 (3 photos): © Erwin & Peggy Bauer. 23.18a: David Giannasi, The University of Georgia. 23.18b: © Tom McHugh 1975, NMNH/Photo Researchers, Inc.

Unit Five Interview Annalisa Kraft.

Chapter 24 24.1: © Jagga Gudlaugnd. 24.2: S. M. Awramik, University of California/BPS. 24.4a: Courtesy John Stoltz. 24.4b, c: S. M. Awramik, University of California/ BPS. 24.6: Sidney Fox, University of Miami/ BPS. 24.9a, b: Courtesy of David Deamer, from *Scientific American* 264(February 1991):124.

Chapter 25 25.1: © Science Photo Library/Photo Researchers, Inc. 25.2a–c: © David M. Phillips/Visuals Unlimited. 25.3: S. Abraham and E. H. Beachey, VA Medical Center, Memphis/BPS. 25.4: Courtesy J. Adler. 25.5a, b: R. M. Macnab and M. K. Ornston, *J. Mol. Biol.* 112(1977): 1-20. 25.6a: Courtesy of S. W. Watson © *Journal of Bacteriology*, American Society of Microbiology. 25.6b: N. J. Lang, University of California, Davis/BPS. 25.7: © Janice Sheldon/Photo 20-20. 25.8: Helen E. Carr/BPS. 25.11: © V. Paulson/BPS. T25.1.1: Courtesy of Tom Cross. T25.1.2: J. R. Waaland, University of Washington/BPS. T25.1.3: P. W. Johnson and J. McN. Sieburth, University of Rhode Island/BPS. T25.1.4: H. S. Pankratz, Michigan State University/BPS. T25.1.5: Michael Gabridge, University of Illinois/BPS. T25.1.6: K. Stephens, Stanford University/BPS. T25.1.7: H. S. Pankratz and R. L. Uffen, Michigan State University/BPS. T25.1.8: CNRI/SPL/Photo Researchers, Inc. Methods Box: Christine L. Case, Skyline College.

Chapter 26 26.1: J. R. Waaland, University of Washington/BPS. 26.3: © Peter Parks-OSF/Animals Animals. 26.4a, b: © Eric Grave/Photo Researchers, Inc. 26.5a: © Manfred Kage/Peter Arnold, Inc. 26.5b: © Paolo Koch/Photo Researchers, Inc. 26.6 inset: Masamichi Aikawa, *Science* 137(January 1990):403. 26.7 and 26.8a: © Eric Grave/Photo Researchers, Inc. 26.7b: © Roland Birke/Peter Arnold, Inc. 26.8b: © Manfred Kage/Peter Arnold Inc. 26.8c: © M. Abbey/Visuals Unlimited. 26.10a: © Biophoto Associates/Photo Researchers, Inc. 26.10b: © Eric Grave/Photo Researchers, Inc. 26.11: J. R. Waaland, University of Washington/BPS. 26.12a: © Biophoto Associates/Photo Researchers, Inc. 26.12b: Courtesy of Fred Rhoades. 26.14a: © Manfred Kage/Peter Arnold, Inc. 26.14b: Estate of Dr. J. Metzner/Peter Arnold Inc. 26.14c: © M. I. Walker/Science Source. 26.15.1: Courtesy of W. L. Dentler, University of Kansas. 26.16: BPS. 26.17: © Maurice Dube. 26.18: © Lawrence Naylor/Photo Researchers, Inc. 26.19 and 26.20a: © J. Robert Waaland, University of Washington/BPS. 26.20b: © D. P. Wilson and Eric and David Hosking/Photo Researchers, Inc. 26.20c: © Gary R. Robinson/Visuals Unlimited. 26.21.1: Ray Simons/Photo Researchers, Inc. 26.21.2: © R. Calentine/Visuals Unlimited. 26.22: © David Scharf/Peter Arnold, Inc. 26.23: © Fred Rhoades/Western Washington University. 26.24: J. R. Waaland, University of Washington/BPS.

Chapter 27 27.1: © Grant Heilman. 27.4: © Stephen Kraseman/DRK Photo. 27.5.1: © James W. Richardson/Visuals Unlimited. 27.6: © John Shaw/Tom Stack & Associates. 27.8: © Rod Planck/Tom Stack & Associates. 27.10: © Dale and Marian Zimmerman/Bruce Coleman, Inc. 27.11: © Robert and Linda Mitchell. 27.12: © E. S. Ross. 27.13: © Glenn Oliver/Visuals Unlimited. 27.13 inset: © Walter H. Hodge/Peter Arnold Inc. 27.14: © Milton Rand/Tom Stack & Associates. 27.15a: © James Castner. 27.15b: © Runk/Schoenberger/Grant Heilman Photography. 27.15c: © Michael Fogden/Bruce Coleman, Inc. 27.15d: © C. P. Hickman/Visuals Unlimited. 27.16: James W. Behnke. 27.17: © Dr. William M. Harlow/Photo Researchers, Inc. 27.18a: © John Shaw/Tom Stack & Associates. 27.18b: © W. H. Hodge/Peter Arnold, Inc. 27.21: © John Colwell/Grant Heilman. 27.23a: © D. Wilder. 27.23b: © Thomas C. Boyden. 27.23c: Merlin D. Tuttle, Bat Conservation International. 27.24a: © Lara Hartley. 27.24b: © G. Prance/Visuals Unlimited. 27.25: © John Bennett.

Chapter 28 28.1: © John D. Cunningham/Visuals Unlimited. 28.2.1 and 28.2.2: © Fred Rhoades/Western Washington University. 28.4.1 and 2: © Ed Reschke/Peter Arnold, Inc. 28.5a: © Fred Rhoades/Western Washington University. 28.5b: © Jacana/Photo Researchers, Inc. 28.5c: © J. L. Lepore/Photo Researchers, Inc. 28.6.1: © E. R. Degginger/Animals Animals. 28.7a: © Kerry T. Givens/Tom Stack & Associates. 28.7b: © Frans Lanting. 28.7c: © A. Davies/Bruce Coleman, Inc. 28.8.1: © Biophoto Associates/Photo Researchers, Inc. 28.9: © Jack Bostrack/Visuals Unlimited. 28.9 inset: © M. F. Brown/Visuals Unlimited. 28.10: N. Allin and G. L. Barron, University of Guelph/BPS. 28.11: © Fred Rhoades/Western Washington University. 28.13a, b: © Bruce Iverson 28.14: © Runk/Schoenberger/Grant Heilman. 28.15: © Mark Wicklaus/Western Washington University.

Chapter 29 29.1: © Anne Wertheim/Animals Animals. 29.7: © Andrew J. Martinez/Photo Researchers, Inc. 29.11a: © Fred McConnaughey/Photo Researchers, Inc. 29.11b: © Robert Brons/BPS. 29.11c: © Chuck Davis. 29.11d: © Chris Huss/Ellis Wildlife collection. 29.12.1: © Robert Brons/BPS. 29.13: © Fred Bavendam/Peter Arnold, Inc. 29.14: © Bill Wood/Bruce Coleman, Inc. 29.17: © William C. Jorgensen/Visuals Unlimited. 29.18: © William H. Amos/Bruce Coleman, Inc. 29.19a, b: © L. S. Stepanowicz/Photo Researchers, Inc. 29.21: © Jeff Foott/Tom Stack & Associates. 29.22: © James M. Cribb/Bruce Coleman, Inc. 29.23a: © Kevin Schafer. 29.23b: © Chris Huss. 29.24: © James M. Cribb/Bruce Coleman, Inc. 29.26a: © Tom McHugh/Steinhart Aquarium/Photo Researchers, Inc. 29.26b: © Fred Bavendam/Peter Arnold, Inc. 29.26c: © Douglas Faulkner/Photo Researchers, Inc. 29.28a: © Sea Studios. 29.28b: © Kjell Sandved/Sandved and Coleman. 29.28c: © Robert and Linda Mitchell. 29.30: © Cliff B. Frith/Bruce Coleman, Inc. 29.31: © Ronald F. Thomas/Bruce Coleman. 29.32: © Milton H. Tierney, Jr./Visuals Unlimited. 29.33a: © Robert and Linda Mitchell. 29.33b: © Paul Skelcher/Rainbow. 29.33c: © David Scharf. 29.35a, b: © Robert and Linda Mitchell. 29.36: © Stephen Dalton/Animals Animals. 29.38 upper (2) and lower middle and right: © John Shaw/Tom Stack & Associates. 29.38 lower left: © Frans Lanting/Minden Pictures. 29.39a: © Marty Snyderman. 29.39b: © Flip Nicklin. 29.39c: © C. R. Wyttenbach,

University of Kansas/BPS. 29.40a: © Colin Milkins/Oxford Scientific Films/Animals Animals. 29.40b: © Fred Bavendam/Peter Arnold, Inc. 29.41a: © Jeff Rotman/Peter Arnold, Inc. 29.41b: © Dave Woodward/Tom Stack & Associates. 29.41c: © Marty Snyderman. 29.41d: © Carl Roessler/Animals Animals. 29.41e: © Jeff Rotman/Peter Arnold, Inc. 29.45a: Courtesy of Prof. K. G. Grell, Tubingen. 29.46a: © Carole Hickman/South Australian Museum. 29.46b: S. Conway Morris.

Chapter 30 30.1: © Biophoto Associates/Photo Researchers, Inc. 30.4a: © Scott Johnson/Animals Animals. 30.6a: © Hervé Berthoule/Jacana/Photo Researchers, Inc. 30.6b: © Tom McHugh. 30.8a: © Tom McHugh/Photo Researchers, Inc. 30.8b: © Dave B. Fleetham/Visuals Unlimited. 30.9a: © E. R. Degginger/Bruce Coleman, Inc. 30.9b: © J. M. Labat/Jacana/Photo Researchers, Inc. 30.11: © Steve Martin/Tom Stack & Associates. 30.14a: © Zig Leszczynski/Animals Animals. 30.14b: © M. P. L. Fogden/Bruce Coleman, Inc. 30.14c: © R. Andrew Odum/Peter Arnold, Inc. 30.15a–c: © Hans Pfletschinger/Peter Arnold, Inc. 30.17a: © John Gerlack/Tom Stack & Associates. 30.17b: © John R. MacGregor/Peter Arnold, Inc. 30.17c: © R. Andrew Odum/Peter Arnold, Inc. 30.17d: © Patricia Caulfield/Photo Researchers, Inc. 30.18: © Stephen Dalton/Animals Animals. 30.19: © Kevin Schafer. 30.21a: © John Henry Dick, Academy of Natural Sciences, Philadelphia/Vireo. 30.21b: © G. Neuchterlein, Academy of Natural Sciences, Philadelphia/Vireo. 30.21c: © Bob and Clara Calhoun/Bruce Coleman, Inc. 30.21d: © Art Wolfe. 30.22: © Gunter Ziesler/Peter Arnold, Inc. 30.24a inset: Photo courtesy of Mervyn Griffiths, from *Australian Natural History* 17(September 1972):222–226; © Ederic Slater. 30.24a: © D. Parer and E. Parer-Cook/Auscape. 30.24b: © Tom McHugh/Photo Researchers, Inc. 30.24c: © Mitch Reardon/Photo Researchers, Inc. 30.26: © E. H. Rao/Photo Researchers, Inc. 30.27a: © Tom McHugh/Photo Researchers, Inc. 30.27b: © Norman Owen Tomalin/Bruce Coleman, Inc. 30.28a: © E. R. Degginger/Animals Animals. 30.28b: © Ulrich Nebelsiek/Peter Arnold, Inc. 30.28c: © Nancy Adams E.P.I./Tom Stack & Associates. 30.28d: © Tom McHugh/Photo Researchers, Inc. 30.30a: © Cleveland Museum of Natural History. 30.30b: © John Reader 1982. 30.31 upper: © Andrew Sacks. 30.31 lower: © James D. Wilson/Woodfin Camp.

Unit Six Interview Alain McGlaughlin. Inset of corn: Courtesy of Virginia Walbot.

Chapter 31 31.1: © Doug Wechsler/Animals Animals/Earth Scenes. 31.2a: Elliot Meyerowitz, *Science* 254(11 October 1991):206. 31.2b, c: Elliot Meyerowitz and John Bowman, *Development* 112(1991):1-20. 31.3: © David Cavagnaro. 31.6: © Doug Wechsler/Animals Animals/Earth Scenes. 31.10a: © Larry Mellichamp/Visuals Unlimited. 31.10b, c: © Kevin Schafer. 31.10d: © E. S. Ross. 31.12a: © Nels Lersten, University of Iowa. 31.12b, c: © Bruce Iverson. 31.12d, e, 1 and 2: © George J. Wilder/Visuals Unlimited. 31.16: © Bruce Iverson. 31.17a: © Ed Reschke. 31.17b: © Carolina Biological Supply. 31.17c: © Ed Reschke. 31.18: J. R. Waaland, University of Washington/BPS. 31.19a: © Ed Reschke. 31.19b: R. D. Meicenheimer, *Scientific American* 242(1980):133-151. 31.20a, b and 31.21b, c: © Ed Rescke. 31.23a, b: Courtesy of Howard F. Towner, Loyola Marymount University. 31.24a, b: © Bruce Iverson.

Chapter 32 32.1: © Branson Reynolds. 32.10: © Alfred Pasieka/Taurus Photos. 32.11: © G. F. Leedale/Biophoto Associates/Photo Researchers, Inc. 32.14a: © E. S. Ross. 32.14b: M. H. Zimmerman, *Science* 133(13 January 1961): 76, Fig. 4. Copyright 1961 by The American Association for the Advancement of Science. Methods Box a: Courtesy of Gary Tallman, Pepperdine University. Methods Box b: From T. D. Lamb, H. R. Matthews, and V. Torre, *J. Phys.* 372(1986):315–349.

Chapter 33 33.1: © Runk/Schoenberger/Grant Heilman. 33.3a, b: Courtesy of James C. Pushnik, California State University, Chico. 33.4: © John H. Hoffman/Bruce Coleman, Inc. 33.5: © Grant Heilman. 33.7a: © Runk/Schoenberger/Grant Heilman, Inc. 33.7b: © Alan Pitcairn/Grant Heilman, Inc. 33.8: John Reganhold, *Scientific American* 262(June 1991):115. 33.10a: © Breck P. Kent/Earth Scenes/Animals Animals. 33.10b: © E. H. Newcomb, University of Wisconsin/BPS. 33.11: Breck P. Kent/Earth Scenes/Animals Animals. 33.12a, b: UN Food and Agriculture Organization, *Science* 250(9 November 1990):748. 33.13a: © Runk/Schoenberger/Grant Heilman. 33.13b: © Kevin Schafer. 33.13c: © Biophoto Associates/Photo Researchers, Inc. 33.14a: © Jeff Lepore/Photo Researchers, Inc. 33.14b: © John Shaw/Tom Stack & Associates. 33.14c: © Robert and Linda Mitchell. 33.15: © R. Ronacordi/Visuals Unlimited. Self-Quiz: Courtesy of Y. T. Tchan, A. M. Zeman, and Ivan R. Kennedy, The University of Sydney.

Chapter 34 34.1: © Frans Lanting 34.4a: © Tom Branch 1978/Photo Researchers, Inc. 34.4b: © Jeanne White 1993/Photo Researchers, Inc. 34.4c: © Link,1985/Visuals Unlimited. 34.4e: © Runk/Schoenberger/Grant Heilman. 34.4f: © Ray Coleman 1985/Photo Researchers, Inc. 34.4d: © Stephen Dalton/Photo Researchers, Inc. 34.4g: © Gregory K. Scott/Photo Researchers, Inc. 34.9a–c: © W. H. Hodge/Peter Arnold, Inc. 34.9d: © Robert P. Carr/Bruce Coleman, Inc. 34.10a: © Phil Farnes/Photo Researchers, Inc. 34.10b: © John Colwell/Grant Heilman Photography. 34.10c: © Zig Leszczynski/Animals Animals. 34.10d: © Paolo Koch/Photo Researchers, Inc. 34.13a: © James L. Castner. 34.13b: © David Cavagnaro/DRK Photo. 34.15: © Plantek 1985/Photo Researchers, Inc. 34.16a: © Randy Moore/Visuals Unlimited. 34.16b: © John D. Cunningham/Visuals Unlimited. 34.17a.1: Courtesy of Susan Wick, University of Minnesota. 34.18: B. Wells and K. Roberts.

Chapter 35 35.1: © John Kaprielian/Photo Researchers, Inc. 35.6a, b: © Malcolm Wilkins. 35.7: Courtesy Sylvan Wittwer. 35.8: S. Gladfelter, Stanford University, Visual Art Services. 35.9b: © Ed Reschke. 35.10 left: Photo courtesy of Michael Evans, Ohio State University. 35.10 right: M. L. Evans, R. Moore, and K. H. Hasenstein, *Scientific American* 254(1986):112–119. 35.11a, b: © E. R. Degginger/Bruce Coleman, Inc. 35.12a, b: Courtesy Frank Salisbury. 35.17: Courtesy of Janet Braam, from *Cell* 60(9 February 1990): cover.

Unit Seven Interview Ron Wurzer.

Chapter 36 36.1: © Walter Dawn/Photo Researchers, Inc. 36.4: Courtesy of Dr. Jerome Gross, Developmental Biology Laboratory, Massachusetts General Hospital, Boston. 36.6: © Ed Reschke. 36.7: From *Tissues and Organs: A Text Atlas of Scanning Electron Microscopy,* by Richard G. Kessel and Randy H. Kardon. W. H. Freeman and Company. Copyright 1979. 36.9b: © Biophoto Associates/Photo Researchers, Inc. Self-Quiz 1a: Marion Rice. Self-Quiz 1b, c: © Bruce Iverson. Self-Quiz 1d: © Eric Grave/Photo Researchers, Inc. Self-Quiz 1e: © Bruce Iverson. Self-Quiz 1f: © Biophoto Associates/Photo Researchers, Inc.

Chapter 37 37.1: © Hans and Judy Beste/Animals Animals. 37.2a: © C. Allen Morgan/Peter Arnold, Inc. 37.2b: Painting by Richard Schlecht © National Geographic Society. 37.3: © Tom Eisner, Cornell University. 37.4: © Lennart Nilsson, Boehringer Ingelheim International GmbH. 37.5: © Gunter Zeisler/Peter Arnold, Inc. 37.15d: © Manfred Kage/Peter Arnold. 37.17, top: © Brian Milne/Animals Animals. 37.17, bottom: © Mickey Gibson/Animals Animals. 37.20: © Fred Bavendam/Peter Arnold, Inc. Methods Box, left: © William Thompson. Methods Box, right: Courtesy of Robert Full.

Chapter 38 38.1: From *Tissues and Organs: A Text-Atlas of Scanning Electron Microscopy,* by Richard G. Kessel and Randy H. Kardon. W. H. Freeman and Company. © 1979. 38.2: © Norbert Wu. 38.11: © D. W. Fawcett/Photo Researchers, Inc. 38.15b: © Lennart Nilsson, Boehringer Ingelheim/International GmbH. 38.16a: © Ed Reschke. 38.16b © W. Ober/Visuals Unlimited. 38.21c: © Tom Eisner,

Cornell University. 38.22c: © David M. Phillips/Visuals Unlimited. 38.28: © Kevin Schafer.

Chapter 39 39.1, 39.4, 39.15b, 39.18b, and 39.19: © Lennart Nilsson, Boehringer Ingelheim International GmbH. 39.3: © Science Photo Library/Photo Researchers, Inc. 39.9c: © A. J. Olson, Scripps Clinic and Research Foundation, 1986.

Chapter 40 40.1: © Eric and David Hosking, Photo Researchers, Inc. Inset: © E. S. Ross, California Academy of Sciences. 40.4a, b: John Crowe. 40.5a: © Tom McHugh/Photo Researchers, Inc. 40.11d: Peter Andrews, reprinted with permission from *Biomed. Res (Suppl.)*, 2:293–305. 40.18a: © E. Lyons/Bruce Coleman, Inc. 40.18b: © Frans Lanting/Minden Pictures. 40.18c: Sovfoto. 40.20a: © Jeff Lepore/Photo Researchers, Inc. 40.20b: © Dave B. Fleetham/Visuals Unlimited. 40.21b: Courtesy of Francis Carey. 40.22b: B. Heinrich, *Scientific American*, March 1987. 40.23: © Warren Garst/Tom Stack & Associates.

Chapter 41 41.1: © Frans Lanting/Minden Pictures. 41.4b: Drs. Walter Stumpf and Madhabananda Sar, University of North Carolina.

Chapter 42 42.1: © David Wrobel 1992/Monterey Bay Aquarium. 42.2a: Courtesy of David Crews, Photo by P. DeVries. 42.3: © Stephan Myers. 42.4: © Franklin Viola. 42.5: © K. H. Switak/Photo Researchers, Inc. 42.15b: © David M. Phillips. 42.19a–c: © Lennart Nilsson, Boehringer Ingelheim International GmbH.

Chapter 43 43.1: © G. I. Bernard/NHPA. 43.4 (4 photos): J. C. Gilkey, L. F. Jaffe, E. B. Ridgeway, and G. T. Reynolds, *J. Cell Biol.* 76 (1978):448–466. © The Rockefeller University Press. 43.6: George Watchmaker. 43.7a, b: Robert Waterman. 43.8: Courtesy of Charles A. Ettensohn, Carnegie Mellon University. 43.12: Courtesy of Thomas Poole. 43.14: Carolina Biological Supply. 43.2b: Hiroki Nishida, *Dev. Biol.* 121(1987):526. 43.20c: J. E. Sulston and H. R. Horvitz, *Dev. Biol.* 56(1977):110–156. 43.22a: Dr. Jean-Paul Thiery. Reproduced from the *Journal of Cell Biology* 1983, Vol. 96, pp. 462-473 by copyright permission of The Rockefeller University Press. 43.22b left and right: Richard Hynes, *Scientific American*, June 1986, p. 44. 43.25 left: © David Scharf. 43.25 right: F. R. Turner, University of Indiana. 43.26a: Katherine Tosney, University of Michigan. 43.26b and 43.27: Dennis Summerbell. 43.28: Dr. Wolfgang Driever and Christiane Nüsslein-Volhard, Institute for Dev. Biology, Tubingen, Germany.

Chapter 44 44.1: © John Stevens 1992/FPG. 44.3b: © Manfred Kage/Peter Arnold, Inc. 44.6b: Courtesy of G. L. Scott, J. A. Feilbach, and T. A. Duff. 44.13b: Dr. E. R. Lewis, ERL-EECS, University of California, Berkeley. Methods Box: © William Thompson.

Chapter 45 45.1: © Russell C. Hansen. 45.3: OSF/Animals Animals. 45.4a: © R. Andrew Odum/Peter Arnold, Inc. 45.4b: © Russ Kinne/Photo Researchers, Inc. 45.6a: Janice Sheldon. 45.19: From *Fundamentals of Entomology*, 3rd ed., by Richard J. Elzinga (1987). Reprinted by permission of Prentice-Hall Inc. 45.21: © Joe Monroe/Photo Researchers, Inc. 45.25a: P. Parks/OSF/Animals Animals. 45.25b: © Biophoto Associates/Photo Researchers, Inc. 45.25c: © James Hanken/Photo Researchers, Inc. 45.30 top left: Courtesy of Clara Franzini-Armstrong. 45.35b: © Eric Grave/Photo Researchers, Inc.

Unit Eight Interview Alain McGlaughlin.

Chapter 46 46.1: Courtesy of NASA. 46.2: Courtesy of Dan Simberloff. 46.5: © H. A. Miller/Visuals Unlimited. 46.6: © Jonathan Scott/Planet Earth Pictures. 46.7: © Charlie Ott/Photo Researchers, Inc. 46.8: © John D. Cunningham/Visuals Unlimited. 46.9: © Tom McHugh/Photo Researchers, Inc. 46.10: © John Shaw. 46.11: © Lars Egede-Nissen/BPS. 46.12: © B. S. Cushing/Visuals Unlimited. 46.14: Bruce F. Molnia/Terraphotographics. 46.15: J. Allan Cash Ltd. 46.17: Landsat/U.S. Geological Survey. 46.18: © Chris Huss. 46.19: © David Hall, 1984/Photo Researchers, Inc. 46.20: © Dudley Foster, Woods Hole Oceanographic Institution. 46.21: © Stephen Kraseman/Photo Researchers, Inc.

Chapter 47 47.1: © Frank M. Hanna/Visuals Unlimited. 47.2: © Victor H. Hutchinson/Visuals Unlimited. 47.4a: © Robert P. Carr/Bruce Coleman, Inc. 47.4b: Frans Lanting/Minden Pictures. 47.4c: © Bill Bachman/Photo Researchers, Inc. 47.8a: © R. Calentine/Visuals Unlimited. 47.8b: © Max and Bea Hunn/Visuals Unlimited. 47.12: Courtesy of George W. Barlow, University of Caifornia, Berkeley. 47.17 left: © Alan Carey/Photo Researchers, Inc. Methods Box: © Frans Lanting.

Chapter 48 48.1: © Sven-Olof Lindblad/Photo Researchers, Inc. 48.3: Courtesy of Paul Hertz. 48.4a–c: © L. E. Gilbert, University of Texas, Austin/BPS. 48.5: © C. Allan Morgan/Peter Arnold, Inc. 48.6a, b: Erick Greene, *Science* 243(February 3, 1989), 644, Fig. 1. Copyright 1989 by the American Association for the Advancement of Science. 48.7: © John L. Tveten. 48.8a: © Stephen Krasemen/Peter Arnold, Inc. 48.8b: © Tom Eisner and Daniel Aneshansley. 48.9: E. D. Brodie, Jr., *Scientific American* 263(July 1990):28. 48.10a, b: R. Huey, *Science* 195(1977):201–203. 48.13b: © 1985 Joseph T. Collins/Photo Researchers, Inc. 48.13c: Kevin deQueiroz, National Museum of Natural History. 48.15: © Stephen Krasemen/Photo Researchers, Inc. 48.16: © Robert and Linda Mitchell. 48.17a: © Dr. Gregory A. Antipa. 48.18a–f: Tom Bean.

Chapter 49 49.1: © David Hiser/Photographers Aspen. 49.14a: Matthew Meselson, Harvard University. 49.14: © Earth Observation Satellite Co, Lanham, Maryland/Science Photo Library/Photo Researchers, Inc. 49.15: © John D. Cunningham/Visuals Unlimited. 49.15: Furnished by Dr. Robert Pierce/U.S. Department of Agriculture, Northeastern Experimental Forest Station, Durham, NH. 49.17: D. W. Schindler, *Science* 184(24 May 1974):897, Figure 1. 49.20: © Walter R. Tschinkel, © *Discover Magazine*. 49.21: UPI/Bettmann News Photos.

Chapter 50 50.1: © Animals Animals. 50.5: © Ian Wyllie/Survival Anglia. 50.8: © John Burton/Bruce Coleman. 50.9: Lincoln P. Brower. 50.11: © Ted Kerasote/Photo Researchers, Inc. 50.11: © E. R. Degginger/Animals Animals. 50.13a: Courtesy of Pat De Coursey, University of South Carolina. 50.17a: © Bruce Davidson/Animals Animals. 50.17b: © Breck P. Kent/Animals Animals. 50.19: © Jean-Michel Labat/Jacana/Photo Researchers, Inc. Methods Box: E. R. Degginger/Animals Animals.

BPS = Biological Photo Service.

Illustration Credits

Front matter Aspen Logo: Bob Ting

Interview with Candace Pert: Figure 1. From *Drugs and the Brain,* by Solomon H. Snyder. © 1986 by Scientific American Books, Inc. Reprinted with permission of W. H. Freeman and Company.

Figure 4.7. Matthews and van Holde, *Biochemistry* , © 1990 by Benjamin/ Cummings Publishing Co. (Redwood City: Benjamin/Cummings, 1990) Figure 9.2a, p. 301. **Figure 4.8.** Adapted from Becker and Deamer, *The World of the Cell,* 2e, © 1991 by Benjamin/Cummings Publishing Co. (Redwood City: Benjamin/Cummings, 1991) Figure 2.6, p. 20.

Figure 5.10. Adapted from K.D. Johnson, D.L. Rayle, and H.L. Wedberg, *Biology: An Introduction,* © 1984 by Benjamin/Cummings Publishing Co. (Redwood City: Benjamin/ Cummings, 1984), p.21. **Figure 5.14.** From *Biology: The Science of Life,* 3e by Robert A. Wallace, Gerald P. Sanders, and Robert J. Ferl. Copyright © 1991 by HarperCollins Publishers. Reprinted by permission. **Figures 5.26a, and 5.27c, d.** © Irving Geis.

Figure 6.2. B. Alberts, D. Bray, J. Lewis, M. Raff, K. Roberts, J.D. Watson, *Molecular Biology of the Cell,* 2nd Edition. (New York: Garland Publishing, 1989).

Figure 7.7. Adapted from Matthews and van Holde, *Biochemistry,* © 1990 by Benjamin/ Cummings Publishing Co. (Redwood City: Benjamin/Cummings, 1990). **Figure 7.14.** Adapted from Marieb, *Human Anatomy and Physiology,* 2e, ©, 1992 by Benjamin/ Cummings Publishing Co. (Redwood City: Benjamin/Cummings, 1992). **Figure 7.25.** Reproduced from the *Journal of Cell Biology,* 1988, 107:1785, T. Schroer, B. Schnapp, T.S. Reese, M. Sheetz by copyright permission of The Rockefeller University Press; and from *CELL,* 1985, 40:559, R.D. Vale, B.J. Schnapp, T.S. Reese, M.P. Sheetz. © Cell Press. **Figure 7.31.** Adapted from B. Alberts, D. Bray, J. Lewis, M. Raff, K. Roberts, J. D. Watson, *Molecular Biology of the Cell,* 2nd Edition. (New York: Garland Publishing, 1989). Figure 11.55. **Figure 7.27.** Adapted from Marieb, *Human Anatomy and Physiology,* © 1989 by Benjamin/Cummings Publishing Co. (Redwood City: Benjamin/Cummings) Figure 3.18, p. 81.

Figure 9.6. Adapted from B. Alberts, O. Bray, J. Lewis, M. Raff, K. Roberts, J. D. Watson, *Molecular Biology of the Cell,* 2nd Edition. (New York: Garland Publishing, 1989). **Figure 9.9.** Adapted from Matthews and van Holde, *Biochemistry,* © 1990 by Benjamin/ Cummings Publishing Co. (Redwood City: Benjamin/ Cummings, 1989) Figure 13.2, p. 435. **Figure 9.18.** Adapted from Matthews and van Holde, *Biochemistry,* © 1990 by Benjamin/Cummings Publishing Co. (Redwood City: Benjamin/Cummings, 1989) Figure 15.19, p. 526.

Figure 10.17. Adapted from B. Alberts, O. Bray, J. Lewis, M. Raff, K. Roberts, J. D. Watson, *Molecular Biology of the Cell,* 2nd Edition. (New York: Garland Publishing, 1989) Figure 7.43.

Methods Box 11.1. Adapted from R.I. Freshney, *Culture of Animal Cells: A Manual of Basic Technique,* 2nd Edition. © 1989 John Wiley & Sons, Inc. Reprinted by permission of Wiley-Liss, a division of John Wiley and Sons, Inc. **Figure 11.8.** Adapted from the *Journal of Cell Biology* 1987:104:9 and 1988:106:1185, G.J. Gorbsky, P.J. Sammak, G. Borisy by copyright permission of The Rockefeller University Press and the authors. **Figure 11.9c.** © Scientific American Inc., George V. Kelvin. **Figure 11.13.** Adapted from B. Alberts, O. Bray, J. Lewis, M. Raff, K. Roberts, J. D. Watson, *Molecular Biology of the Cell,* 2nd Edition. (New York: Garland Publishing, 1989). **Figure 11.16a.** Adapted from "What Controls the Cell Cycle," Andrew W. Murray and Marc W. Kirschner, *Scientific American,* March 1991, p. 62. **Figure 11.16b.** Adapted from B. Alberts, O. Bray, J. Lewis, M. Raff, K. Roberts, J. D. Watson, *Molecular Biology of the Cell,* 2nd Edition. (New York: Garland Publishing, 1989) Figure 13.15.

Figure 15.1. From *An Introduction of Genetic Analysis,* 4e, David T. Suzuki et al., © 1976, 1981, 1986, 1989 by W. H. Freeman and Co. Reprinted with permission. **Figure 15.6.** Becker and Deamer, *World of the Cell,* 2e, © 1991 by Benjamin/Cummings Publishing Co. (Redwood City: Benjamin/Cummings, 1991). **Figure 15.15.** B. Alberts, O. Bray, J. Lewis, M. Raff, K. Roberts, J. D. Watson, *Molecular Biology of the Cell,* 2nd Edition. (New York: Garland Publishing, 1989).

Figure 16.1. B. Alberts, O. Bray, J. Lewis, M. Raff, K. Roberts, J. D. Watson, *Molecular Biology of the Cell,* 2nd Edition. (New York: Garland Publishing, 1989).

Figure 17.1. Watson, Hopkins, et al., *Molecular Biology of the Gene,* 4e, © 1987 by Benjamin/ Cummings Publishing Co. (Redwood City: Benjamin/Cummings, 1987). **Figure 17.3.** Tortora, Funke, Case, *Microbiology: An Introduction,* 4e, © 1992 by Benjamin/ Cummings Publishing Co. (Redwood City: Benjamin/Cummings, 1992). **Figure 17.4.** B. Alberts, O. Bray, J. Lewis, M. Raff, K. Roberts, J. D. Watson, *Molecular Biology of the Cell,* 2nd Edition. (New York: Garland Publishing, 1989). **Figure 17.8.** Darnell, Lodish, Baltimore, *Molecular Cell Biology,* 2e, Scientific American Books, W. H. Freeman and Company, 1990. **Figure 17.14.** B. Alberts, O. Bray, J. Lewis, M. Raff, K. Roberts, J. D. Watson, *Molecular Biology of the Cell,* 2nd Edition. (New York: Garland Publishing, 1989).

Methods Box 19.1. Becker and Deamer, *World of the Cell,* 2e, © 1991 by Benjamin/ Cummings Publishing Co. (Redwood City: Benjamin/Cummings, 1991). **Figure 19.2.** Tortora, Funke, Case, *Microbiology: An Introduction,* 4e, © 1991 by Benjamin/ Cummings Publishing Co. (Redwood City: Benjamin/Cummings, 1991).

Figure 20.12. Adapted from illustration by Jason Kuffer, "Natural Selection and Darwin's Finches," by Peter R. Grant, p. 86, © October 1991 by Scientific American, Inc. All rights reserved. **Figure 20.16.** Biruta Hansen, © 1986 *Discover Magazine.*

Methods Box 23.1. Richard Cowen, *History of Life,* Blackwell Scientific Publications, 1990, p. 37, Figure 2.4. Reprinted by permission of Blackwell Scientific Publications, Inc. **Figure 23.13.** Raup, D., "Extinction: Bad Genes or Bad Luck?" *New Scientist,* September 14, 1991.

Figure 24.9. B. Alberts, O. Bray, J. Lewis, M. Raff, K. Roberts, J. D. Watson, *Molecular Biology of the Cell*, 2nd Edition. (New York: Garland Publishing, 1989) p. 10, 1989. **Figure 24.10.** © John Ryan, from "Closing the Gap Between Proteins and DNA," *Science* 29 June 1990, p. 1609.

Figure 25.9. Adapted from Woese et al. *Proceedings of the National Academy of Sciences,* Vol. 87 p. 4578, June 1990.

Figure 30.5. Harland, W.B., Cox, A.V., Llewellyn, P.C., Pickton, C.A.G., Smith, A.G., Walters, R., *A Geologic Time Scale* (Cambridge: Cambridge University Press, 1982). **Figure 30.31.** From an illustration by Laurie Grace, "The Recent African Genesis of Humans," A. C. Wilson, R. L. Cann, © April 1992 by *Scientific American*; illustration by Joe Lertola, "The Emergence of Modern Humans," C.B. Stringer, © December 1990 by *Scientific American, Inc.* All rights reserved.

Figures 31.15 and 31.22. B. Alberts, O. Bray, J. Lewis, M. Raff, K. Roberts, J. D. Watson, *Molecular Biology of the Cell*, 2nd Edition. (New York: Garland Publishing, 1989) p. 1174.

Figure 33.11. Reproduced with permission from the *Annual Review of Plant Physiology and Plant Molecular Biology*, Vol. 39 © 1988 by Annual Reviews Inc.

Figure 34.19. B. Alberts, O. Bray, J. Lewis, M. Raff, K. Roberts, J. D. Watson, *Molecular Biology of the Cell*, 2nd Edition. (New York: Garland Publishing, 1989). **Figure 34.20.** Bowman, J.L., Smyth, D.R., Meyerowitz, E.M. *Development*, v. 112, 1991, p. 15.

Figure 39.9. Tortora, Funke, Case, *Microbiology: An Introduction*, 4e, © 1992 by Benjamin/Cummings Publishing Co. (Redwood City: Benjamin/Cummings,1992). **Figure 39.17.** Adapted from Lennart Nilsson, Jan Lindberg, *The Body Victorious,* DeLacorte Press, © 1987, p. 27.

Figure 40.16. Adapted from Marshall and Hughes, *Physiology of Mammals and Other Vertebrates*, 2e. (Cambridge: Cambridge University Press, 1980). **Figure 40.22.** Eckert and Randall, *Animal Physiology*, 3e, Figure 16.21, p.575, W. H. Freeman and Company, 1988).

Figure 43.4. B. Alberts, O. Bray, J. Lewis, M. Raff, K. Roberts, J. D. Watson, *Molecular Biology of the Cell*, 2nd Edition. (New York: Garland Publishing, 1989) Figure 4.35. **Figure 43.20b.** B. Alberts, O. Bray, J. Lewis, M. Raff, K. Roberts, J. D. Watson, *Molecular Biology of the Cell*, 2nd Edition. (New York: Garland Publishing, 1989) p. 902, Figure 16.29A.

Figure 44.3. Marieb, *Human Anatomy and Physiology*, 2e, © 1992 by Benjamin/ Cummings Publishing Co. (Redwood City: Benjamin/Cummings, 1992) p. 344, Figure 11.3. **Figure 44.4.** Adapted from B.B. Boycott, L. Peichl, H. Wassle, *Proceeding of the Royal Society of London* B. 203(1978):229. **Figure 44.9.** From Matthews, G.,*Cellular Physiology of Nerve & Muscle*, Blackwell Scientific Publications, 1986. **Figure 44.22.** Based on A. S. Romer, P. S. Parsons, *The Vertebrate Body,* 6e. (Philadelphia: W.B. Saunders, 1986) p. 569.

Figure 45.34. From Matthews, G., *Cellular Physiology of Nerve & Muscle*, Blackwell Scientific Publications, 1986.

Figure 46.29b. *Science* 184:1001-1003, Raymond B. Huey. © 1974 by the American Association for the Advancement of Science.

Figure 47.3. From *Evolutionary Ecology,* 3e, by E.R. Pianka. © 1983 by Harper & Row Publishers. Reprinted by permission of HarperCollins. **Figure 47.5.** John Tyler Bonner, *Size and Cycle,* Princeton University Press, 1965, p. 17. **Figure 47.14b.** *Journal of Animal Ecology*, v. 34, p. 622, 1965. Blackwell Scientific Publications. **Figure 47.15a.** Adapted from J. L. Harper, *Population Biology of Plants,* Academic Press, 1977, p. 180. **Figure 47.15b.** *Journal of Animal Ecology*, v. 50, p.141 upper left from graph 3, 1981, T. S. Bellows, Jr. Blackwell Scientific Publications. **Figure 47.16.** Andrewartha and Birch, *The Distribution and Abundance of Animals*, p. 572, University of Chicago Press © 1954. **Figure 47.20.** Illustration by James Egleson, "The Human Population," E. S. Deevy, Jr., © September 1960 by Scientific American, All rights reserved.

Figure 48.13. A.S. Rand and E.E. Williams, "The anoles of La Palma: Aspects of their ecological relationships," *Breviora* no. 327, 1969. **Figure 48.14.** Adapted from David Lambert Lack, *Darwin's Finches.* (Cambridge: Cambridge University Press, 1947) p. 57, Figure 9 and p. 82, figure 17. **Figure 48.22.** McArthur and Wilson, *The Theory of Island Biogeography*, Princeton University Press, 1967, p.8.

Figure 49.4. From *Ecology and Field Biology,* 3e, by Robert Leo Smith. © 1990 by Robert Leo Smith. Reprinted by permission of HarperCollins Publishers. **Figure 49.4.** Reprinted from "Solar Radiation at the Earth's Surface," *Solar Energy* 5:95-98. © 1961, H. E. Landsberg, with permission from Pergamon Press Ltd., Headington Hill Hall, Oxford OX3 OBW, UK. **Figure 49.7.** From *Ecology*, 2e, by R.E. Ricklefs. © 1988 by Chiron Press. Reprinted by permission of W. H. Freeman. **Figure 49.8.** P. Colinvaux, *Ecology*, Copyright © 1986 John Wiley & Sons. Reprinted by permission of John Wiley & Sons, Inc. **Figure 49.8.** E.P. Odum, *Fundamentals of Ecology.* (Philadelpia: W.B. Saunders, 1959) p.62. **Figure 49.10.** Adapted from R.E. Ricklefs, *Ecology*, 3e, W.H. Freeman and Company, 1990, p. 213, Figure 12.3. **Figure 49.12.** From "Effects of Forest Cutting and Herbicide Treatment on Nutrient Budgets in the Hubbard Brook Watershed Ecosystem," by G.E. Likens et al., *Ecological Monographs,* 1970, 40. Copyright © 1966 by Ecological Society of America. Reprinted by permission. **Figure 49.18.** From *Living in the Environment* by G. Tyler Miller, 2nd ed, Copyright 1979, Wadsworth Publishing Co., page 87. All rights reserved. **Figure 49.20b.** Tom Moore, © 1986 *Discover Magazine.*

Methods Box 50.1. Gibbs, Lisle, et al., "Realized Reproductive Successes of polygonous Red-Winged Blackbirds Revealed by Markers," *Science* Vol. 250, Nov/Dec 1990, pp. 1394–1397. **Figure 50.4.** N. Tinbergen, *The Study of Instinct*, Oxford University Press, 1951. **Figure 50.4.** Adapted from C.K. Catchpool, *Behaviour* 74:148–155, E.J. Brill, Leiden p. 153, Figure 1. **Figure 50.10.** Courtesy of Masakazu Konishi. **Figure 50.23.** Adapted from Grier, J., *Biology of Animal Behavior.* (St. Louis: Times-Mirror Publishing/C.V. Mosby, 1984) Figure 8.15. **Figure 50.24.** K.E. Holekamp, *Behavioral Ecology and Sociobiology*, 1984, 16:21–30.

INDEX

Note: A *t* following a page number indicates tabular material and an *f* following a page number indicates an illustration.

ABA. *See* Abscisic acid
A bands, 1039
Abdominal cavity, 787
Abiotic factors in environment, 1052
 diversity and, 1070
 organisms' interaction with, 1075–1080
Abiotic synthesis of organic compounds
 in origin of genetic information, 509–511
  alternative views and, 511
 in origin of life, 54, 507*f*
  monomers, 507–508
  polymers, 508
ABO blood groups
 multiple alleles and, 269
 self vs. nonself and, 868
Abortion, 952–953
Abscisic acid, 759*t*, 764
Abscission, leaf, ethylene in, 765, 766*f*
Absolute dating, 477–478
 radioactive, 477, 479
Absorption
 of minerals, by roots, 704–706
 of nutrients
  in animals, 804–806
  in fungi, 584
  in plants, 719
 of water, in large intestine, 806–807
Absorption spectrum, 205–206
 for photosynthesis, 205*f*
 spectrophotometer for determining, 205, 206
Abyssal zone, 1069
Acclimation, 1078–1079
 to temperature, 902
Accommodation, 1022
Acetic acid, 59, 60*t*
Acetone, 59, 60*t*
Acetylcholine, 998
Acetyl CoA
 formation of, 182
 in Krebs cycle, 182*f*, 183*f*
Achondroplasia, inheritance pattern of, 274
Acid chyme, 802
Acid growth hypothesis, 760
Acid rain, 49–51, 1145
Acids, 47–48
Acoelomates, 600*f*, 601, 608–610
 body plan of, 602*f*
Acquired immunodeficiency syndrome (AIDS), 352, 870–872. *See also* Human immunodeficiency virus
Acrasiomycota (cellular slime molds), 553, 554*t*
Acritarchs, eukaryote fossils among, 535
Acrosomal process, 957
Acrosomal reaction, 957–958
Acrosome, 946
ACTH. *See* Adrenocorticotropin
Actin, in muscle contraction, 141, 1040, 1041*f*
Actinomycetes, 526*t*
 nitrogen-fixing, 728
Actinopoda, 537*t*, 538
Actinopterygii (ray-finned fishes), 643, 644
Action potentials
 neuronal, 989–991
  in hearing, 1028–1029
  propagation of, 991–992
   speed of, 992–993
  in vision, 1024, 1025*f*
 in plants, rapid leaf movements and, 768
Action spectrum, 206
 for photosynthesis, 205*f*, 206
Activation energy
 definition of, 101
 lowering of, enzymes and, 100–102
Active immunity, 855
Active site, of enzyme, 102, 103*f*
Active transport, 164–167
 in plant cells, 700–701
Adam's apple. *See* Larynx
Adaptation
 to environmental change, 1075–1080
 evolutionary, 1079–1080
  as property of life, 6*f*
 sensory, 1017
Adaptive coloration, as animal defense, 1110–1112
Adaptive evolution, 450–453. *See also* Natural selection
Adaptive landscape, 468
Adaptive peaks, 469
 shifts in speciation, 468–469
Adaptive radiation
 on island chains, in allopatric speciation, 463–465
 macroevolution and, 486–487
Adaptive zone, 486
Addition rule, in inheritance, 264
Adenine, 87*f*
Adenine arabinoside, 352–353
Adenohypophysis (anterior pituitary), hormones produced by, 919*t*, 920*f*, 921
Adenosine diphosphate. *See* ADP
Adenosine triphosphate. *See* ATP
Adenoviruses, 345, 346*f*, 350*t*
Adenylate cyclase, control of, 913
ADH. *See* Antidiuretic hormone
Adhesion of water, 41, 46*t*
 in transpirational pull on xylem sap, 708
Adipose tissue, 785
 excess food stored in, 810
ADP (adenosine diphosphate)
 from ATP hydrolysis, 98–99, 174, 175*f*
 in Calvin cycle, 204, 214*f*
Adrenal glands, 919*t*, 925–926
 cortex of, 919*t*, 925–926
 hormones produced by, 919*t*, 925–926
 medulla of, 919*t*, 925
Adrenocorticotropin, 919*t*, 920*f*, 921, 925
Adult-onset diabetes, 924
Adventitious roots, 678
Aegyptopithecus, 660–661
Aerobes
 nitrogen-fixing, 527*t*
 obligate (strict), 192, 522
Aerobic respiration, 192
 anaerobic respiration and fermentation compared with, 192–193
Aestivation, 903
Afferent arteriole, 884*f*, 885
Afferent neurons (sensory neurons), 985, 1002
Aflatoxins, 594
African sleeping sickness, *Trypanosoma* spp causing, 539, 540*f*
Agaricus mushrooms, 595
Age, maternal, risk of Down syndrome and, 293–294
Age structure, population size and, 1087
 in humans, 1102, 1103*f*
Agglutination, as effector mechanism of humoral immunity, 862
Aggregate fruit, 741, 743*f*
Aging, in plants
 cytokinins affecting, 762
 ethylene affecting, 764–765
Agnatha (agnathans, jawless vertebrates), 638*t*, 639*f*, 640
Agonistic behavior, 1176–1177
Agricultural biome, 1056
Agriculture. *See also* Crop plants
 angiosperms and, 579
 DNA technology used in, 410–412, 729
 nematodes in, 611
 nutrient cycling affected by, 1148–1149
 sustainable, 725
 vegetative propagation in, 746–748
AIDS (acquired immunodeficiency syndrome), 352, 870–872. *See also* HIV
Al. *See* Aluminum
Ala. *See* Alanine
Alanine, 75*f*
Alcohol fermentation, 191
Alcohols, 58, 60*t*
Aldehydes, 58, 60*t*
Aldoses (aldehyde sugars), 66, 67*f*
 glucose as, 66, 67*f*
Aldosterone, 926*f*
 blood pressure affected by, 891
Aleurone, 743
Algae, 534, 542–551. *See also specific phyla*
 brown (Phaeophyta), 543*t*, 549–550
 conjugating, 547
 ecologic importance of, 542
 evolutionary relationships among, 551
 golden (Chrysophyta), 543*t*, 544
 green (Chlorophyta), 543*t*, 545–548
  as plant ancestors, 562
 life cycle of, 249
 red (Rhodophyta), 543*t*, 550–551
Algal blooms
 dinoflagellate, 543–544
 in lakes, 1064
Alimentary canal. *See also* Digestive tract
 digestion in, 798, 799*f*
Allantois
 in birds, 966
 in mammals, 968
Allard, H.A., 759, 770
Allee effect, 1097
Alleles, 260–261
 dominant, 261, 268–269
 genetic symbols for, 282
 multiple, 269
 recessive, 261
Allergens, 869
Allergies, 869, 870*f*
Alligators (Crocodilia), 650, 651
Allocation, 1077
Allogenic factors, communities affected by, 1123
Allometric growth, 480, 481*f*
Allopatric populations, 462
Allopatric speciation, 462, 462–464
 and adaptive radiation on island chains, 463–465
 conditions favoring, 463
Allopolyploid (allopolyploidy), sympatric speciation by, 465, 466*f*
All-or-none event, action potential as, 989
Allosteric regulation
 of cellular respiration, 195
 of enzyme activity, 105–106
Allosteric sites, 105, 106*f*
Alpha carbon, in amino acids, 74
Alpha cells, in islets of Langerhans, 923
Alpha-globin, multigene families encoding for, 375–376, 377*f*
Alpha glucose rings, 69, 70*f*
Alpha helix, 80
Alpha-tubulin, 137
Alpha waves, of EEG, 1008
Alpine tundras, 1062

Alternation of generations, 249, 560
in angiosperms, 735
in Chlorophyta (green algae), 547–548
in mosses, 563, 564*f*
in Phaeophyta (brown algae), 550
Alternative complement pathway, 866
Altitude, climate affected by, 1074, 1075*f*
Altruistic behavior, 1183–1185, 1186*f*
Aluminum, electron configuration of, 32*f*
Alvarez, L., 488
Alvarez, W., 488
Alveolar macrophages, 852
Alveoli (alveolus)
mammary gland, 940
respiratory, 841
Amacrine cells, in retina, 1024, 1025*f*
Ames, B., 339
Ames test, 338, 339
Amines, 60
Amino acids, 60–61, 74–77
catabolism of, 193, 194*f*
for dating fossils, 477
essential, 811–812
genetic codes for, 319–321
as neurotransmitters, 998*t*, 999
substitution of, protein conformation and function affected by, 79
Aminoacyl-tRNA site (A site), 327, 328*f*
Aminoacyl-tRNA synthetases, 326–327
Amino group, 60–61, 60*t*
Aminopeptidase, 804
Ammonia
excretion of, 893, 894
in nitrogen cycle, 1144
Ammonification, 1144
Amniocentesis, 276, 953
complications of, 277
Amnion
in birds, 966
in mammals, 967*f*, 968
Amniotes, embryology of, 965–968
Amoebas, 537
Amoebocytes, 604
Amphibia (amphibians), 638*t*, 639*f*, 645–648
circulatory system of, 821–822
early, 645–646
modern, 646–648
phylogeny of, 645*f*
thermoregulation in, 900–901
Amphipathic molecule, 152
Amplification
gene, 378
by sensory receptors, 1016
Amygdala
in emotions, 1009
in memory pathway, 1010
Amylase
pancreatic, 803
salivary, 799
Amylopectin, 68
Amyloplasts, 135
Amylose, 68
Anabaena spp, 526*t*
Anabolic pathways, 92, 194
Anaerobes
facultative, 192, 522
obligate (strict), 192, 522
Anaerobic respiration, 192. *See also* Fermentation
aerobic respiration and fermentation compared with, 192–193
Anagenesis (phyletic evolution), 456, 457*f*
Analogy, homology differentiated from, 491
Anaphase, in mitosis, 225, 227*f*
cell elongation during, 228, 229*f*
chromosome separation during, 225, 229*f*
in plant cell, 231*f*
Anaphase I, in meiosis I, 251*f*, 253
Anaphase II, in meiosis II, 251*f*
Anaphylactic shock, 869
Anatomy, comparative, as evidence of evolution, 433–434
Androgens, 919*t*, 926, 940, 942*f*
Anemia, sickle-cell, amino acid substitution causing, 79
Aneuploidy, 291–292
human diseases caused by, 293–294
Angelman syndrome, genomic imprinting and, 295, 296
Angiosperms (flowering plants), 562*t*, 574–579
agriculture and, 579
emergence of, 561
evolution of, 577
life cycle of, 572*t*, 576–577, 735–736
morphology of, 677*f*
relationship between animals and, 577–579
sexual reproduction in, 734–744
Angiotensin, 891, 892*f*
Angiotensin II, 891, 892*f*
Anhydrobiosis, 879
Animal behavior. *See* Behavior
Animal cells, 124*f*
in culture, for gene cloning, 394
cytokinesis in, 228, 230*f*
glycocalyx of, 144
water balance of, 162
Animal fats, as saturated fats, 71–73
Animal husbandry, DNA technology used in, 410
Animalia (animal kingdom), 512. *See also* Animals
major branches of, 599–600
Animal nutrition, 794–817
digestion and, 795–799
evolutionary adaptations in, 807–809
in mammals, 799–807
feeding mechanisms and, 794–795
food types and, 794–795
requirements for, 809–815
Animal pole, 961, 969
Animals. *See also specific type*
body plan for, 600–601
chemical signals in, 907–930
cognition and, 1185–1186
defenses of against predators, 1110–1112
definition of, 598–599
development of, 600–601, 956–981
protostome-deuterostome dichotomy and, 601–602*f*
diversification of, 629–631
evolution of, 599–600, 600*f*, 629–631
hypothetical models for, 630*f*
movement of, 1034–1035
behavior and, 1171–1174
muscles in, 1037–1045
skeleton in, 1035–1037
nutrition in. *See* Animal nutrition
organs and organ systems in, 787
phylogeny of, 599–602, 603*f*
relationship between angiosperms and, 577–579
reproduction in
asexual, 931–932
sexual, 932, 934–936
structure and function of, 782–793
interaction among size, shape, and environment in, 787–789
internal environment and, 789–791
levels of structural organization and, 782–787
tissue types in, 783–787. *See also specific type*
Animal viruses, 350–354
classes of, 352*f*
Anion, 35
Anisogamy, 547
Annelida (segmented worms), 600*f*, 615–617
metanephridia of, 882
nervous systems in, 1000–1001
reproductive systems of, 935–936
Annuals, 686
Anopheles mosquitoes, in *Plasmodium* life cycle, 539, 540*f*
Anoplura, 622*t*
ANP. *See* Atrial natriuretic protein
Anteaters, spiny (Echidnas), 655*f*, 656
Antennae, arthropod, 620
Anterior pituitary, 919*t*, 920*f*, 921
Anther, 575, 735, 736*f*
Antheridium, 562
Anthocerophyta (hornworts), 562*t*, 563
Anthophyta (flowering plants), 562*t*, 574. *See also* Angiosperms
Anthopleura elegantissima, asexual reproduction in, 931*f*
Anthozoa, 606
Anthropoids, early, 660–661
Antibiotics
mechanism of action of, and peptidoglycan in bacterial cell walls, 518
production of by microorganisms, 521
resistance to, R plasmids and, 361
Antibodies, 854
effector mechanisms of, 862
genes for, DNA rearrangement in maturation of, 379, 380*f*
monoclonal, 863, 864
structure of, 857
Anticodon, 325, 326*f*
Antidiuretic hormone, 918, 919*t*, 920*f*
in kidney regulation, 891, 892*f*, 918
Antigen-antibody complex, as effector mechanism of humoral immunity, 862
Antigen-antibody specificity, molecular basis of, 856–857
Antigen-binding sites, 857
Antigenic determinants (epitopes), 856*f*, 857
Antigen-presenting cell, 860
Antigens, 854
receptors for, on B and T cells, 856
T-dependent, 861
T-independent, 861
Antimicrobial proteins, in nonspecific body defense, 852
Antiparallel DNA strands, replication and, 309–311
Antiport, 158, 159*f*
Antiquity of life, 505–506
Antisense nucleic acid, in disease treatment, 409
Antiviral drugs, 352–353
Anurans (frogs/toads), 636–637
early embryonic development of, 960–965
fate map of, 972*f*
metamorphosis of, 647
organogenesis in, 963–965
APC. *See* Antigen-presenting cell
Aphasia, 1009
Aphotic zone
freshwater, 1063
marine, 1066
Apical dominance, 679
cytokinins in control of, 761, 762*f*
Apical epidermal ridge, 978
Apical meristems, 686
in primary growth of roots, 687–689
in primary growth of shoots, 689
Apicomplexa, 537*t*, 539
Aplysia, learning in, 1011
Apodans (caecilians), 646, 647
Apomixis, 746
Apomorphic character, in cladistics, 495–496
Apoplast, in lateral transport in plants, 704
Aposematic coloration, as animal defense, 1111, 1112f
Appendicular skeleton, vertebrate, 638
Appendix, 806
Aquatic animals. *See also specific type*
gills in, 836–838
osmoregulation in, 878, 878–879
Aqueous humor, 1022
Aqueous solutions, 46–49, 46*t*
definition of, 45
Ar. *See* Argon
Ara-A, 352–353
Arabidopsis
mapping genome of, 405
plant genetics studied in, 674–675
Arachnida (arachnids), 619
Archaebacteria, 523–524
Archaeopteryx lithographica, 653
Archegonium, 562
Archenteron, 600, 601*f*, 962
Arctic tundra, 1061–1062
Arg. *See* Arginine
Arginine, 75*f*
Argon, electron configuration of, 32*f*
Aristotle, 422
Arousal, 1008–1009
Arteries, 821
hardening of (arteriosclerosis), 834
structure of, 825
Arteriosclerosis, 834
Arthropoda (arthropods), 600*f*, 617–626. *See also specific type*
characteristics of, 617–618
classification of, 618–626
endocrine systems of, 916
nervous systems of, 1000–1001
phylogeny of, 618–626
Artificial membranes, 152

Artificial selection, 428
Artiodactyla, 658*t*
Asci, 588
Ascocarps, 588, 595
Ascomycota (ascomycetes), 587–589
 life cycle of, 588–589
Ascorbic acid (vitamin C), 813*t*, 814
Ascospores, 588*f*, 589
Asexual reproduction
 in animals, 931–932
 in plants, 744–748
 sexual reproduction compared with, 245, 748
A site, 327, 328*f*
Asn. *See* Asparagine
Asp. *See* Aspartic acid
Asparagine, 75*f*
Aspartic acid, 75*f*
Aspergillus, food spoiled by, 594
Aspilla, 580
Associative learning, 1169
Assortative mating, microevolution and, 445
Assortment, independent, Mendel's law of, 265–267
 genetic variation and, 253–254
 recombination of unlinked genes and, 284–285
Asteroidea (sea stars), 627, 628
 nervous systems of, 1000, 1001*f*
Asters, 226*f*
Astrocytes, 985
Asymmetric carbon, 57, 58*f*
Atherosclerosis, 833–834
 saturated fats and, 73
Atmosphere, human activities affecting, 1151–1153
Atomic number, 28
Atomic weight, 28
Atoms. *See also specific type*
 behavior of, 27–32
 charged (ions), 35
 definition of, 27
 structure of, 27–32
 valence of, 33
ATP (adenosine triphosphate)
 in active transport, 164
 in Calvin cycle, 203–204, 212–214
 cellular work and, 97–100, 174, 175*f*
 electron transport and, 184–189
 in glycolysis, 179, 180–181*f*
 hydrolysis of, 98–99
 in Krebs cycle, 182*f*, 183*f*
 maximum yield of, 189*f*
 regeneration of, 99
 structure of, 98–99
 synthesis of, 178*f*, 179. *See also* Oxidative phosphorylation
  by cellular respiration, 178–190
  in chloroplasts vs. mitochondria, 211–212
  electron flow and, 185–189
  by fermentation, 190–192
  origins of mechanisms for, 524–525
  by photosynthesis, 203
 work performed by, 99
ATPase, Na^+–K^+ (sodium-potassium pump), 164–166. *See also* Proton pump
ATP cycle, 99
ATP synthases, in chemiosmosis, 185–187
Atria (atrium), of heart, 820
Atrial natriuretic protein, blood pressure affected by, 891
Atrioventricular node, 824
Atrioventricular valves, 822
Auditory canal, 1027
Auditory tube (Eustachian tube), 1027
AUG codon, 321
Australopithecus, 660–661
Autecology (organismal ecology), 1053
Autodigestion, 131
Autogenic changes in community structure, 1123
Autogenous model, of eukaryotic origins, 535
Autoimmune diseases, 869
Autonomic nervous system, 1002, 1003*f*
Autophagy, 132
Autopolyploid (autopolyploidy), sympatric speciation by, 465, 466*f*
Autoradiographs, RFLP, for forensics, 409
Autosomes, 248
Autotrophic nutrition, 199–200
Auxin, 758, 759–761, 759*t*
Auxotrophs, 317
Avery, O., 302
Aves (birds), 638*t*, 639*f*, 651–653, 654*f*
 characteristics of, 651–652
 development of, 965–966
 flightless, 653
 modern, 653, 654*f*
 origin of, 653
 thermoregulation in, 897–898
 ventilation in, 842–843
AV node. *See* Atrioventricular node
Avogadro's number, 47
Axial skeleton, vertebrate, 638
Axillary bud, 677*f*, 678*f*, 679
 auxin and cytokinin interactions in development of, 761, 762*f*
Axon hillock, 984
Axons, 787, 984
Axopodia, 538
Azotobacter spp, 527*t*

B. *See* Boron
Bacillariophyta (diatoms), 543*t*, 545
Bacilli, 516
Bacillus spp, 526*t*
Bacteria, 515. *See also* Prokaryotes
 capsule of, 518
 cell walls in, 516–518
 chemical cycles and, 528
 chromosomes of, 355, 519–520
  in cell division, 222, 223*f*, 520–521
  mapping, 359–360
  replication of, 355, 356*f*
 commercial importance of, 530
 conjugation in, 357–361, 521. *See also* Plasmids
 disease caused by, 529–530
 diversity of, 521–522, 522–524
 gene expression in, control of, 363–368
 genetic recombination in, 356–363, 521
 gene transfer in, 356–363, 521
 genomes of, 355–356, 519–520
 gram-negative, 517–518
 gram-positive, 517–518
 importance of, 528–530
 internal membranes of, 519, 520*f*
 in large intestine, 807
 morphology of, 516
 motile, 518–519
 mutation of, 355–356, 521
 nitrogen-fixing, 527*t*, 726
  DNA technology affecting capability of, 411–412
  in nitrogen cycle, 1143, 1144*f*
  in root nodules, 727
 protein synthesis in, 332–333
 reproduction in, 222, 223*f*, 520–521
 RNA polymerase of, 322
 symbiotic, 528–529
 taxis in, 519
 transduction in, 357, 358*f*, 521
 transformation of, 301–302, 357, 521
 transposons in, 361–363
 use of in genetic engineering, 398–399
Bacterial colonies, 516, 520, 521*f*
Bacterial mats, history of life studied with, 505–506
Bacteriophages. *See* Phages
Bacterium. *See* Bacteria
Bacteroids, 727
Baculum, 938
Baker's yeast, 592
Balance (equilibrium), 1027–1032
 in invertebrates, 1031–1032
 in mammals, 1029–1030
 in non-mammalian vertebrates, 1030–1031
Balanced polymorphism, genetic variation and, 448–449
Bark, 694
Barnacles, 618, 626
Barr body, 290, 291*f*, 375
Barrier methods of contraception, 951
Basal body, 140
Basal ganglia, 1007
Basal metabolic rate, 809–810
 measurement of, 810
Base analogues, as chemical mutagens, 338
Basement membrane, of epithelial cells, 783
Base-pairing rules
 for codon-anticodon bonding, 326
 in DNA, 305–306
  replication and, 306, 307*f*
Base-pair substitution, 336, 337*f*
Bases, 47–48
 amino group as, 61
Basic research, DNA technology used in, 404
Basidiocarps, 589, 590*f*
Basidiomycota (basidiomycetes), 589–590
 life cycle of, 589, 590*f*
Basidium, 589
Basophils, 831*f*
 development of, 832*f*
Batesian mimicry, 1111–1112, 1113*f*
B cells (B lymphocytes), 855–856
 activation of, 860–861
 immunoglobulins made by, DNA rearrangement and, 379, 380*f*
Be. *See* Beryllium
Beadle, G., 317, 318*f*
Beagle (ship), Darwin's voyage on, 424–425
Bees
 communication systems in, 1182–1183
 flight in, 621
Beetles, flight in, 624
Behavior, 1158–1190
 agonistic, 1176–1177
 altruistic, 1183–1185, 1186*f*
 animal cognition and, 1185–1186
 circannual, 1171
 communication and, 1180–1183
 definition of, 1159
 evolution and, 1015
 foraging, 1174–1175
 genetic basis of, 1158*f*, 1159, 1160*f*
 innate components of, 1162–1165
 learning and, 1165–1170
 mating, 1178–1180
 migratory, 1172–1174
 molecular biology in study of, 1180, 1181
 movements and, 1171–1174
 play, 1169–1170
 and proximate vs. ultimate causation, 1160–1162
 in response to environment, 1078
 rhythms of, 1170–1171, 1172*f*
 selfish vs. altruistic, 1183–1185, 1186*f*
 social, 1176–1178
  in amphibians, 648
Behavioral ecology, 1159, 1160*f*. *See also* Behavior
Behavioral isolation, as reproductive barrier, 460
Behaviorism, 1186
Beijerinck, M., 345
Beijing Man, 663
Benthos (benthic communities)
 freshwater, 1065
 oceanic, 1066, 1068–1069
Benzene, carbon skeleton diversity and, 57*f*
Beriberi, vitamin B_1 deficiency causing, 812–813
Bernard, C., 789
Beryllium, electron configuration of, 32*f*
Berzelius, J.J., 54
Beta cells, in islets of Langerhans, 923
Beta-globin, multigene families encoding for, 375–376, 377*f*
Beta glucose rings, 69, 70*f*
Beta oxidation, 193–194
Beta pleated sheet, 80, 81*f*
Beta-tubulin, 137
Beta waves, of EEG, 1008
BH. *See* Brain hormone
Bicarbonate secretion, in digestion, 803
Bicoid, 979
Biennials, 686
Bilateral symmetry, 599, 600*f*, 601*f*
Bilateria, 599, 600*f*, 608–610
 body plans of, 602*f*
Bilayer, phospholipid, 74, 152
 permeability of, 168
Bile, 803
Binary fission
 in bacterial reproduction, 222, 223*f*, 355, 520
 in ciliate reproduction, 541
Bindin, 957
Binomial, 489
Bioassay, 763
Biodiversity crisis, 1146–1147
Bioenergetics, 92

Biogenic amines, 998–999, 998*t*
Biogeochemical cycles, 1140–1145
Biogeography, 1124–1128
conservation biology and, 1128–1129
of islands, 1126–1128
as sign of evolution, 431–432
of speciation, 462–467
Biological clock
in animals, 1171
in plants, 769
photoperiodism and, 773
Biological diversity. *See also* Diversity
Darwinism and, 420, 421*f*
punctuations in history of, 486–489
unity in, 10–11, 13*f*
Biological magnification, 1150
Biological organization (biological order)
emergent properties and, 4–7
hierarchy of, 3–4, 5*f*, 25*f*
Biological species, 458. *See also* Species
limitations of concept of, 458–459
Biomass, 1136
Biomass pyramid, 1139–1140
Biomes. *See also specific type*
climate affecting distribution of, 1070–1071
definition of, 1055
freshwater, 1062–1065
distribution of, 1063*f*
marine, 1066–1069
distribution of, 1063*f*
terrestrial, 1054–1062
distribution of, 1055*f*
Biophilia, 2
Biosphere
biotic effects at level of, 1153–1154
definition of, 1054–1055
environmental diversity of, 1069–1075
Biosynthesis, 194
Biotechnology, 391. *See also* DNA technology
contributions of to medicine, 405–409
Biotic factors in environment, 1052
biosphere affected by, 1153–1154
Biotin, 813*t*
Bipolar cells, in retina, 1024, 1025*f*
Birds (Aves), 638*t*, 639*f*, 651–653, 654*f*
characteristics of, 651–652
development of, 965–966
flightless, 653
modern, 653, 654*f*
origin of, 653
thermoregulation in, 897–898
ventilation in, 842–843
Bird song
development of, imprinting and, 1168–1169
as model system for animal behavior research, 1158
Birth, 950
Birth control (contraception), 951–953
Birth control pills, 952
Bishop, M., 112–115, 237
Biston betularia, as example of natural selection, 429–430
Bivalvia (bivalves, Pelecypoda), 613–614
Black bread mold, 586–587, 591
Bladder (swim), 643, 1031
Bladder (urinary), 883
Blades, of seaweed, 549
Blastocoel, 961
Blastocyst, 948, 967
Blastodisc, 965
Blastomeres, 960
Blastopore, 962
in protostomes vs. deuterostomes, 602, 603*f*
Blastula, 599, 961
Blood, 786, 830–832
composition of, 830, 831*f*
nephron in regulation of, 885–887
Blood-brain barrier, 985
Blood cells, 830–832. *See also specific type*
development of, 832
Blood clotting, 832, 833*f*
Blood flow, 825–828
velocity of, 825–826
Blood flukes, life history of, 609
Blood groups, 868
Blood pressure, 826–827
high (hypertension), 834
measurement of, 827
Blood vessels, structure of, 825
Blubber, in thermoregulation, 900
B lymphocytes (B cells), 855–856
activation of, 860–861
immunoglobulins made by, DNA rearrangement and, 379, 380*f*
BMR. *See* Basal metabolic rate
Body temperature
homeostasis in maintenance of, 790
regulation of, 894–903
Bohr shift, 844, 845*f*
Bolting, gibberellins and, 762–763, 764*f*
Bonding capacity (valence), 33
for major elements of organic molecules, 55*f*
Bone, 785
Bony fishes (Osteichythyes), 638*t*, 639*f*, 643–644
Book lungs, in arachnids, 619, 620*f*
Boreal forest (taiga), 1061
Boron
electron configuration of, 32*f*
plant requirements for, 721*t*
Borrelia spp, 527*t*
Bottleneck effect, in genetic drift, 443, 444*f*
Bound ribosomes, 331
Boveri, T., 280
Bowman's capsule, 884*f*, 885
Boysen-Jensen, P., 757
Brachiate, 657
Brachiopoda (brachiopods), 600*f*, 627
Brain
in birds, 651
in cephalopods, 614
higher functions of, 1008–1011
human, 1006–1008
in mammals, 655
right vs. left, 1009
vertebrate, 1002
evolution of, 1005–1006
regions of, 1004*f*
Brain hormone, 916
Brainstem, 1006–1007
Branching evolution (cladogenesis), 456, 457*f*, 482, 483*f*
Bread mold, 586–587, 589
Breast. *See* Mammary glands
Breathing, 841–843
control of, 843
negative pressure, 841
Breathing center, 843
Brenner, S., 956
Brewer's yeast, 592
Briggs, R., 975
Brightfield microscopy, 118*t*
Brittle stars (Ophiuroidea), 627, 628
Broca's area, 1008*f*, 1009
Bronchi (bronchus), 841
Bronchioles, 841
Brown algae (Phaeophyta), 543*t*, 549–550
Brush border
of proximal renal tubule, 887
of small intestine, 805, 806*f*
Bryophyta (mosses), 562*t*, 563
life cycle of, 563, 564*f*, 572*t*
Bryophytes (nonvascular plants), 562–563, 562*t*
Bryopsis spp, 546*f*, 547
Bryozoa (bryozoans), 600*f*, 627
"Bubble," replication, 309
Budding, 931
Buds
axillary, 677*f*, 678*f*, 679
terminal, 677*f*, 678*f*, 679
Buffers, 48–49
Bulbourethral glands, 937–938
Bulbs of Krause, 897, 1017*f*, 1019
Bulk-feeders, 795, 796*f*. *See also specific species*
Bulk flow
and long-distance transport of water in plants, 708
of phloem sap, 714–715
Bundle-sheath cells, in C_4 plants, 215
Bursa of Fabricius, 856
Butane, carbon skeleton diversity and, 57*f*
1-Butene, carbon skeleton diversity and, 57*f*
2-Butene, carbon skeleton diversity and, 57*f*
Butterflies, flight in, 621–624

C. *See* Carbon
C_3 plants, 215
C_4 plants, 215, 216*f*
photosynthesis in, vs. CAM photosynthesis, 217*f*
Ca. *See* Calcium
Cacti, photosynthesis in, 215–216, 712
Caecilians (apodans), 646, 647
Caenorhabditis elegans, 611
cell lineages of, 971, 972*f*
cal. *See* Calorie
Calcitonin, 919*t*
Calcium, 26
hormonal control of homeostasis of, 922–923
human requirements for, 814*t*, 815
plant requirements for, 721*t*
Calcium channels, in signal transduction in plant cells, 774
Calcium ions
release of during cortical reaction, 958, 959*f*
and response of egg cell to fertilization, 958–959
as second messengers, 915
in plant cells, 774
Callus, at cut end of shoot, 746
Calmodulin, 915
Calorie, 43, 809
Calvin, M., 203
Calvin cycle, in photosynthesis, 203–204, 212–214
stroma as site of, 203*f*, 204
Cambium
cork, 693*f*, 694–695
vascular, 693–694
Camouflage (cryptic coloration), as animal defense, 1110, 1111*f*
cAMP. *See* Cyclic AMP
cAMP-dependent protein kinase, 914
CAM plants, 215–216, 712
photosynthesis in, vs. C_4 photosynthesis, 217*f*, 712
CAMs. *See* Cell adhesion molecules
Cancer
abnormal cell growth in, 235–237
cytotoxic T cells in defense against, 865
gene expression and, 384–386
viral causation of, 353–354
Cancer cells, abnormal cell division in, 235–237
Candida, 592
CAP-cAMP complex, 368
CAP (catabolite activator protein), in positive gene regulation, 367–368
Capillaries, 821
in fish gills, gas exchange and, 837–838
fluid exchange and, 828–829
renal, 884*f*, 885
small intestinal, 805, 806*f*
structure of, 825
wall of, 828*f*
Capsids, viral, 345–346
Capsule, bacterial, 518
Carbohydrate membrane, in cell-cell recognition, 157–158
Carbohydrates, 66–71. *See also specific type*
catabolism of, 193, 194*f*
disaccharides, 66–67
monosaccharides, 66
photosynthesis in formation of, 201–202, 212–214
fate of, 216–218
polysaccharides, 67–71
storage, 68–69
structural, 69–71
Carbon, 26
asymmetric, 57, 58*f*
electron configuration of, 32*f*, 55
isotopes of, 28
for radioactive dating, 477, 479
molecular diversity and, 53–63
plant requirements for, 721*t*
valence for, 55*f*
versatility of in molecular architecture, 55–56
Carbon-14, for radioactive dating, 477, 479
Carbon cycle, 1142–1143
Carbon dioxide
atmospheric, emissions affecting levels of, 1151
breathing center response to, 843
covalent bonding in, 55
loading and unloading of, 843, 844*f*
partial pressure of, 843, 844*f*
transport of, 844–845, 846*f*
Carbon fixation, in Calvin cycle, 203
"Carbon fungus," 587*f*

Carbonic acid, in carbon cycle in aquatic environment, 1143
Carboniferous period
gymnosperms in, 570, 572
lycopods in, 567
seedless vascular plants formed during, 569–570
Carbon skeletons, variation in, 56–58
Carbonyl group, 58–59, 60*t*
Carboxyl group, 59, 60*t*
Carboxylic acids (organic acids), 59, 60*t*
Carboxypeptidase, 804
Carcinogens, 354, 384
Cardiac muscle, 786–787, 1044
contraction of, 824–825
Cardiac output, 824
Cardiovascular system, 820–830. *See also specific structure or organ*
diseases of, 832–834
saturated fats and, 73
Carinate birds, 653, 654*f*
Carnivora (carnivores), 656, 658*t*, 794
Carnivorous plants, 730–731
Carotenoids, 207, 559
Carpels, 575, 735, 736*f*
Carrier proteins, in plant cell transport, 700
Carrier recognition, 275–276
DNA technology in, 405
Carrying capacity, 1094–1096
for humans, 1102
Cartilage, 785
Casparian strip, in water and mineral absorption by roots, 706
Cassette mechanism, for mating-type switches in yeast, 379
Catabolic pathways, 91
and regeneration of ATP, 99
Catabolism. *See also* Cellular respiration; Fermentation
of glucose, 173–193.
of miscellaneous molecules, 193–194
Catabolite activator protein (CAP), in positive gene regulation, 367–368
Catalytic cycle, of enzymes, 102–104
Catastrophism (catastrophic disturbances)
biological communities affected by, 1070
evolution and, 422–423
species diversity and, 1124
Catecholamines, 925, 998
Cation, 35
Cavitation, and transpirational pull on xylem sap, 708
CCK. *See* Cholecystokinin
CD4 cells, in AIDS, 871
cdc2, in MPF, 234–235, 236*f*
cDNA
making, 397
as source for genes for cloning, 394, 397–398
cDNA library, 398
Cecum, 806, 808
Cell adhesion, membrane proteins in, 157*f*
Cell adhesion molecules, 974
Cell body, neuronal, 984
location of, 985
Cell-cell recognition, membrane carbohydrates in, 157–158
Cell culture, 234
Cell cycle, 224. *See also* Cell division
Cell division, 221–239
abnormal (cancer cells), 235–237
in bacterial cells, 222, 223*f*, 520–521
cell adhesion affecting, 232–233
cell cycle and, 224
control of, 230–235
cytokinesis in, 228–230, 761
density-dependent inhibition of, 231–232, 233*f*
cancer cells and, 235
eukaryotic chromosomes and, 223–224
functions of, 222*f*
mechanics of, 225–230
meiotic, 248, 250–253. *See also* Meiosis
mitotic, 224, 225, 226–227*f*, 231*f*. *See also* Mitosis
mitotic spindle in, 225–228
plane of, cytoskeleton in orienting, 750
in plant cell, 231*f*
morphogenesis and, 749–750
stages of, 225, 226–227*f*
zone of, in primary root growth, 688
Cell expansion
direction of, cytoskeleton in determining, 750–752
in plant cell morphogenesis, 749–750
Cell fractionation, 120–121
Cell lineages, analysis of, 971, 972*f*
Cell-mediated immunity, 855, 863–865, 867*f*
Cell membrane (plasma membrane), 122, 123*f*
phospholipids in, 74
sidedness of, 158*f*
structure of, 156*f*
Cell movement, endocytosis-exocytosis cycle in, 170
Cell plate, 228–229, 230*f*
Cells, 116–150. *See also specific type and structural element*
adhesion of, cell division regulated by, 232–233
bacterial, reproduction of, 222, 223*f*, 520–521
as basis of life, 7–8
chloroplasts in, 134–136
compartmental organization of, 122–123
cytoskeleton of, 136–143
density of, cell division regulated by, 231–232, 233*f*
cancer cells and, 235
differentiation of
cytokinins in control of, 761
gene expression and, 372
in plants, 749, 752
zone of, in primary root growth, 689
elongation of
during anaphase, 228, 229*f*
zone of
auxin and, 760–761
during primary root growth, 688–689
endomembrane system of, 128–133
endoplasmic reticulum of, 128–129
geography of, 121–127
glycocalyx of, 144
Golgi apparatus in, 130–131
integrated function of, 144–146
intercellular junctions of, 144
intermediate filaments in, 142–143
lysosomes in, 131–132
methods for study of, 117–121
microfilaments in, 141–142
microscopy in study of, 117–120
microtubules in, 137–141
mitochondria in, 134–136
nucleus of, 123–127
peroxisomes (microbodies) in, 133–134
reproduction of, 221–239. *See also* Cell division
ribosomes in, 127–128
size of, 121–122
surface of, 143–144
vacuoles in, 132–133
Cellular differentiation. *See* Differentiation
Cellular respiration, 173–198
aerobic, anaerobic catabolism compared with, 192–193
anaerobic, 190–192
aerobic catabolism compared with, 192–193
ATP in, 99, 174, 175*f*
biosynthesis and, 194
control of, 194–195
electron transport chains in, 177–178, 184–189
glycolysis in, 179, 180*f*
evolutionary significance of, 193
Krebs cycle in, 179, 182–184
of miscellaneous molecules, 193–194
origins of, 525–528
overview of, 178–179
as oxidation-reduction process, 174–178
oxidative phosphorylation in, 179, 184–189
summary of, 189–190
Cellular slime molds (Acrasiomycota), 553, 554*t*
Cellular work, ATP and, 97–100, 174, 175*f*
Cellulose, 69–70
bacterial digestion of to glucose, 70
beta glucose rings in, 69, 70*f*
enzymes digesting, 69–70
in plant cell wall, 69, 70*f*
structure of, starch structure compared with, 69, 70*f*
Cellulose fibril, 69, 70*f*
Cell wall, 143, 144*f*, 559
cellulose in, 69, 70*f*, 560
Celsius scale, 43
Cenozoic era, 477, 478*f*
Centimorgan, 286
Centipedes, 618, 620, 621*f*
Central canal, of spinal cord, 1004
Central nervous system, 1002–1008
neuron cell bodies in, 985
Central vacuole, 132, 133*f*
Centrioles, 137, 139*f*
Centromere, 223*f*, 224
Centrosome, 226*f*. *See also* Microtubule-organizing center
spindle microtubule assembly initiated in, 225
Cephalization, 599, 1000
in arthropods, 618
Cephalochordata (cephalochordates, lancelets), 636–637, 639*f*
Cephalopoda (cephalopods), 614–615
Cerebellum, 1006–1007
Cerebral cortex, 1005–1006, 1007, 1008*f*
Cerebrospinal fluid, 1004
Cerebrum, 1005–1006, 1007
Cervical cap, for contraception, 951
Cervix, 940
Cestoda (cestodes), 609
Chambered nautilus, 614
Channels. *See* Ion channels
Chaparral, 1059
Character, 259
Character displacement, 1115, 1116*f*
Chargaff, E., 302
Chargaff's rules, 303
Watson-Crick model and, 306
Chase, M., 302, 303*f*
Chelicerae, 619
Cheliceriformes (Chelicerata, chelicerates), 618, 619
Chelonia (turtles), 650, 650–651
Chemical bonds. *See also specific type*
definition of, 32
making and breaking of (chemical reactions), 36–37
molecule formation and, 32–36
Chemical cycles, 1140–1145
agricultural effects on, 1148–1149
in Hubbard Brook Experimental Forest, 1147–1148
prokaryotes and, 528
Chemical elements. *See* Elements
Chemical energy, 96–97. *See also* Energy
conversion of to heat, 94
transformation of, 93
ecological pyramids and, 1138–1140
laws of thermodynamics governing, 93–95
Chemical equilibrium, 37
Chemical evolution
and abiotic synthesis of organic compounds, 54, 507*f*
monomers, 507–508
polymers, 508
alternative views and, 511
DNA and proteins in, 87–88
origin of life and, 37–38
Chemical mutagens, 338
Chemical reactions, 36–37
endergonic, 97
energy profile for, 101*f*
exergonic, 96–97
free-energy changes in, 96–97
Chemical signals, 907–910
in local regulation, 908*f*, 909–910
Chemical synapses, 994–995
Chemiosmosis, 185–189
in chloroplasts vs. mitochondria, 211–212
origin of, 524–525
in phloem loading and unloading, 713–714
in plant cells, 701
Chemoautotrophs, 199, 522
characteristics of, 526*t*
as producers, 1133
Chemoheterotrophs, 522
Chemoreceptors, 1018
Chemotaxis, 519
Chernobyl, radioactive contamination and, 1150–1151
Chiasmata (chiasma), 252*f*, 253
Chilopoda (chilopods), 620, 621*f*
Chiroptera, 658*t*

Chitin, 71
in cuticle, 1037
Chitons (polyplacorphora), 612
Chlamydia spp, 527*t*
Chlamydomonas, 546
life cycle of, 547
Chloride ions, membrane potential and, 988–989
Chlorine
electron configuration of, 32*f*
human requirements for, 814*t*
plant requirements for, 721*t*
Chlorophyll, 135, 200
excitation of, 207, 208*f*
photooxidation of, 207–208
structure of, 207*f*
Chlorophyll *a*, 559
absorption spectrum for, 205, 206
in algae, 542
color of, 207
structure of, 207*f*
Chlorophyll *b*, 206–207, 559
absorption spectrum for, 207
color of, 207
structure of, 206–207
Chlorophyta (green algae), 543*t*, 545–548
as plant ancestors, 562
Chloroplasts, 134–136, 200–201, 559
chemiosmosis in, 188–189
vs. mitochondria, 211–212
as eubacteria descendants, 536
in photosynthesis, 200–201. *See also* Photosynthesis
starch stored in, 68, 69*f*
Chlorosis, 720, 721*f*
Choanocytes, 604
Choanoflagellates, colonial, and hypotheses of origin of animals, 630
Cholecystokinin, in digestion, 803
Cholesterol, 74
membrane fluidity affected by, 155*f*, 156
serum levels of, atherosclerosis and, 834
structure of, 74*f*
Chondrichthyes, 638*t*, 639*f*, 641–642
Chordata (chordates), 600f, 626, 629, 635–636. *See also specific type*
without backbones, 636–637
characteristics of, 635–636
vertebrate
characteristics of, 638–640
origin of, 637–638
Chorion
in birds, 966
in mammals, 967–968
Chorionic gonadotropin, human, 950
Chorionic villi sampling, 276, 953
Choroid, 1021
Chromatids, sister, 223*f*, 224
Chromatin, 123, 224
interphase, 375
packing of, 373–375
Chromatium spp, 527*t*
Chromoplasts, 135
Chromosome puffs, 383
Chromosomes, 123, 223–224, 244–245
alterations in, 290–295
in chromosome number, 291–292
in chromosome structure, 292–293
human disease and, 293–295
bacterial, 355, 519–520
in cell division, 222, 223*f*, 520–521
mapping, 359–360
replication of, 355, 356*f*
eukaryotic, 223–224
independent assortment of
genetic variation and, 253–254
Mendel's law and, 265–267
recombination of unlinked genes and, 284–285
inheritance based on, 280–299
theory underlying, 280–284
mutations in, genetic variation and, 447
separation of during anaphase, 225, 229*f*
sex, 248. *See also* X chromosome; Y chromosome
Chromosome theory of inheritance, 280–284
Chromosome walking, 405, 406
Chronic myelogenous leukemia, chromosomal translocations and, 294–295
Chroococcus spp, 526*t*
Chrysophyta (golden algae), 543*t*, 544
Chylomicrons, 805
Chyme, acid, 802
Chymotrypsin, 804
Chytridiomycota, 554, 554*t*, 555*f*
Cilia, 139–141, 534, 541
in body defense, 851
Ciliary body, 1022
Ciliary escalator, in body defense, 851*f*
Ciliophora (ciliates), 537*t*, 541–542
Circadian rhythms
animals affected by, 1170–1171, 1172*f*
plants affected by, 769
Circannual behaviors, 1171
Circuits, neural, 999, 1000*t*
Circulatory system
closed, 820
in birds, 651
in cephalopods, 614–615
in humans, 823*f*
in vertebrates, 638, 820–830
evolutionary perspective of, 821–822
open, 819–820
in arthropods, 618
in insects, 624, 820
cis face, of Golgi apparatus, 130
Cl. *See* Chlorine
Cladistics, 495–496
Cladogenesis (branching evolution), 456, 457*f*, 482, 483*f*
Cladogram, 495*f*
Clams, 611–612, 613–614
nervous system in, 1001
Classes, taxonomic, 489
Classical complement pathway, 866, 868*f*
Classical conditioning, 1169
Clathrin, 168
Cleavage, 228, 599, 602, 948, 960–961
in birds, 965
holoblastic, 965
mammalian, 967
meroblastic, 965
in protostomes vs. deuterostomes, 602, 603*f*
Cleavage furrow, 228
Clements, F.E., 1106
Climate
biome distribution and, 1070–1071
global patterns of, 1071–1072
local and seasonal effects on, 1072–1075
Climax community, 1121
Clines, 446
global, in species diversity, 1125–1126
Clitoris, 939, 940
Cloaca, 642, 936
Clonal analysis, 753
Clonal selection, in immunologic specificity and diversity, 858, 859*f*
Cloning
gene. *See also* DNA technology
in eukaryotic cells, 394
in plasmids, 391, 392*f*, 394, 395*f*
sources of genes for, 394–398
by vegetative reproduction, 744
from cuttings, 746
test-tube, 746–748
Cloning vectors, 394
Closed circulatory system. *See* Circulatory system, closed
Clostridium spp, 526*t*
Club fungi, 589
Club mosses (Lycophyta), 562*t*, 567
Clumped population dispersion, 1085, 1086*f*
Clusters, neural, 999
Clutch size, life history affected by, 1089, 1090*f*
CML. *See* Chronic myelogenous leukemia
Cnemidophorus uniparens, parthenogenesis in, 932, 933*f*
Cnidaria, 600*f*, 604–606
dinoflagellates and, 544
nervous system in, 605, 999–1000
Cnidocytes, 604
Coal forests, 569-570
Coated pits, 168
Coated vesicle, 168
Cobalt, human requirements for, 814*t*
Cocci, 516
Cochlea, 1028
Cochlear duct, 1028
Coding strands, 320
Codon-anticodon binding, 326
Codons, 320
recognition of, in elongation stage of translation, 328–330
Coefficient of relatedness, 1184, 1185*f*
Coelom, 601, 602*f*
in annelids, 615, 617
in protostomes vs. deuterostomes, 602, 603*f*
Coelomates, 601
body plan of, 602*f*
evolutionary lines of, 600*f*, 601–602, 603*f*
Coevolution, population interactions and, 1108–1109
Cofactors, enzyme activity affected by, 105
Cognition, animal, 1185–1186
Cognitive ethology, 1186
Cohesion of water, 41, 42*f*, 46*t*
in transpirational pull on xylem sap, 708
Cohorts, life tables constructed from, 1088
Coitus, 947
Coitus interruptus (withdrawal), for contraception, 952
Cold, receptors for (end-bulbs of Krause), 897, 1017*f*, 1019
Coleoptera, 622*t*
Collagen, 785
quaternary structure of, 81, 82*f*
Collagenous fibers, 785
Collecting duct, 884*f*, 885
transport properties of, 889
Collenchyma cells, 683*f*, 684
Colliculi, superior and inferior, 1007
Colonial hypothesis, 630
Colonies, bacterial, 516, 520, 521*f*
Colon (large intestine), 806–807
Coloration, adaptive, as animal defense, 1110–1112
Color blindness, 1024
sex-linked inheritance of, 289
Colorectal cancer, development of, 385–386
Color vision, cones in, 1022, 1024
Columnar epithelial cells, 784
Comb jellies (Ctenophora), 600*f*, 606–608
nerve nets in, 1000
Commensalism, 528, 1116, 1117
Common descent, in Darwinism, 427
Communication, behavior in, 1180–1183
Communities
biomes and, 1056
boundaries between, 1107, 1108*f*
characteristics of, 1118–1119
determinants of, 1119–1121
climax, 1121
definition of, 1106
ecology of, 1053, 1106–1131
conservation biology and, 1128–1129
equilibrium of, species diversity and, 1124
individualistic vs. interactive view of, 1106–1107
periodic disturbances affecting, 1070
and population interactions, 1108–1118
stability of, 1119
species diversity and, 1124
structure of, 1118–1121
succession affecting, 1121–1124
Companion cells, 683*f*, 685
Comparative anatomy, as evidence of evolution, 433–434
Comparative embryology, as evidence of evolution, 434
Competition
community characteristics determined by, 1119
competitive exclusion principle and, 1113–1114
evaluation of in nature, 1114–1116
interspecific, 1112–1116
intraspecific, 1096
Competitive exclusion principle, 1113–1114
Competitive inhibitors, enzyme, 105
Complementary DNA
making, 397
as source of genes for cloning, 394, 397–398
Complement system, 852, 865–867
activation of
as effector mechanism of humoral immunity, 862
pathways for, 866, 868*f*
Complete flowers, 736
Complete growth medium, 317
Complete metamorphosis, in insects, 624

Complex transposons, 362–363
Compound eyes, 620, 1020, 1021f
Compounds
  definition of, 26
  emergent properties of, 26f
Computer modeling, of macromolecule structure, 65f, 85
Concentration, solute, 46–47
  osmosis and, 160–162
Concentration gradient, in diffusion, 160
Concentricycloidea (sea daisies), 627–628
Conception, 947. *See also* Fertilization
Condensation synthesis, 65, 66f
Conditioning
  classical, 1169
  operant, 1169
Condom, 951
Conduction, for heat transfer, 895
Cones, 1022, 1023f
  opsins of, 1024
Confocal microscopy, 118t
Conformation, protein, 77–78
  factors determining, 81–83
  native, 78
Conformers, 1076, 1077f
Conidia, 588
Coniferophyta (conifers), 561, 562t, 570f, 571–572
  life cycle of, 572, 573f
Coniferous forest (taiga), 1061
Conjugating algae, 547
Conjugation, 357–361, 521, 542
  interrupting, bacterial chromosome mapping and, 359–360
  by *Paramecium caudatum*, 542f
Connective tissue, 784–786
Connell, J.H., 1115
Consanguinity (consanguineous matings), recessively inherited disorders and, 274
Conservation biology, 1128–1129
Conservation of energy (first law of thermodynamics), 93
  ecological pyramids and, 1139
Conservative model, of DNA replication, 306, 307f, 308f
Constant regions, immunoglobulin, 857–858
Constitutive genes, 367
Consumers, primary, secondary, and tertiary, 1133
Continental drift, and biogeography of macroevolution, 484–486, 487f
Continuity, law of, 825
Contraception, 951–953
  male, 953
  and population control, 1102–1103
Contraceptive sponge, 951
Contractile proteins, functions of, 76t
Contractile ring, 228
Contractile vacuoles, 132
Contraction
  muscle cell, 1039–1042
    actin and myosin in, 141, 1040, 1041f
    temporal summation of, 1043f
  in whole muscle
    graded, 1042–1043
    molecular mechanism of, 1039–1040
Control center, in homeostatic system, 790
"Controlling elements," 363. *See also* Transposons
Convection, for heat transfer, 895
Convergent circuits, neural, 999, 1000f
Convergent evolution, 491
Cooksonia, 566
Cooling, evaporative, 43–44
Cooling center, 897
Cooperativity, enzyme activity regulated by, 106
Coordinately controlled genes, arrangement of, 376–378
Copepods, 626
Copper
  in hemocyanin, 844
  human requirements for, 814t
  plant requirements for, 721t
Coral reefs, 1067–1068
Corals, 604–606, 1068
Corepressor, in control of gene expression, 366
Corey, R., 80
Cork cambium, 693f, 694–695
Cornea, 1021
Corpus callosum, 1007
Corpus luteum, 939
  hormones produced by, 919t, 939
Correns, K., 296
Cortex
  adrenal, 919t, 925–926
  cerebral, 1005–1006, 1007, 1008f
  motor, 1007, 1008f
  primary visual, 1025, 1026f
  of roots, 689
  sensory, 1007, 1008f
Cortical granules, 958
Cortical reaction, 958, 959f
Corticosteroids, 925–926
Cortisol, 926f
Corti, organ of, 1028
Cotransport, 166–167
  in plant cells, 701
Cotyledons, 577, 743, 744
Cotylosaurs ("stem reptiles"), 648, 649f
Countercurrent exchange, in gills, 837–838
Countercurrent heat exchanger, in birds, 898, 900f
Coupled transcription and translation in bacteria, 332
Courtship, 1178–1180
Covalent bonds, 33–34
  double, 33
    in carbon dioxide, 55
  nonpolar, 34
  in organic molecules, 55–56
  polar, 34
  in urea, 55
Cowpox, discovery of vaccines and, 352
Crabs, 618, 626
Cranial nerves, 1002
Crassulacean acid metabolism (CAM), 215–216, 712
Crayfish, 618, 625, 626
Creatine phosphate, 1040
Creationism, evolution and, 422
Cretaceous period
  angiosperms in, 577
  dinosaur extinction during, 60
Crick, F., 4, 304
Cri du chat syndrome, 294
Crinoidea (sea lilies), 627, 628f, 629
Cristae, 135
Critical period, in imprinting, 1167
Crocodilia (alligators/crocodiles), 650, 651
Cro-Magnons, 664
Crop plants. *See also* Agriculture
  DNA technology affecting, 410–411, 729
  improving protein yield of, 728–729
Crop rotation, 728
Crossing over
  error during, deletions and duplications caused by, 292
  genetic maps constructed with data from, 285–288
  genetic variation and, 254–255
  recombination of linked genes and, 285, 286f
Crossopterygii (lobe-finned fishes), 644
Cross-pollination, 577
  in Mendel's experiments, 259
Crowding. *See* Density
Crustacea (crustaceans), 618, 620, 625–626
Crustal plates, and continental drift in macroevolution, 485
Cryptic coloration, as animal defense, 1110, 1111f
Cryptobiosis, 879
Crystallins, in lens cells, 975
Crystallography, X-ray
  in discovery of double helix, 304, 305f
  in molecular computer modeling, 85
Ctenophora (comb jellies), 600f, 606–608
  nerve nets in, 1000
C-terminus, of polypeptide chain, 76f, 77
Cuboidal epithelial cells, 784
Cultural eutrophication, 1064, 1149
Cultural evolution, 664–666
Culture, cell, 234
Cuticle
  of arthropods, 618, 1036–1037
  plant, 559, 685
  as terrestrial adaptation, 560
Cuttings, clones from, 746
Cuvier, G., 423
CVS. *See* Chorionic villi sampling
Cyanide, electron transport chain affected by, 187
Cyanobacteria
  characteristics of, 526t
  chemical cycles and, 528
  evolution of, 525–528
  reproduction of, 222, 223f
Cycadophyta (cycads), 562t, 570f, 571
Cyclic AMP
  and CAP in positive gene regulation, 368
  as second messenger, 912–915
Cyclic electron flow, in light reactions, 209–210
Cyclin, in MPF, 235, 236f
Cyclohexane, carbon skeleton diversity and, 57f
Cyclosporin, cell-mediated immunity suppressed by, 869
Cys. *See* Cystine
Cystic fibrosis
  cloning of gene for, 405
  inheritance pattern of, 273
Cystine, 75f
Cysts
  in diatom reproduction, 545
  in protists' life cycle, 534
Cytochromes, 185f
  in electron transport chain, 184, 185f
Cytokines, 863
Cytokinesis, 224, 227f, 228–230
  in animal cell, 228, 230f
  in meiosis I, 251f
  in meiosis II, 251f
  in plant cell, 228–229, 230f, 231f
Cytokinins, 759t, 761–762
Cytological maps, 287–288
Cytology, 119
Cytoplasm, division of. *See* Cytokinesis
Cytoplasmic determinants, in development, 970–971
Cytoplasmic genes, 296–297
Cytoplasmic inheritance, 296–297
Cytoplasmic streaming, 142
Cytosine, 87f
Cytoskeleton, 135–143. *See also specific component*
  attachment of, membrane proteins in, 157f
  and direction of cell expansion, 750–752
  motor molecules and, 138f
  and plane of cell division, 750
Cytotoxic T cells, 856, 863, 865, 866f

Dalton, J., 27
Dalton, as unit of measurement, 27
Danielli, J., 152
Darkfield microscopy, 118t
Darkness (night length)
  circadian rhythms and, 1171, 1172f
  photoperiodism and, 770–771
    phytochrome affecting, 771–773
Darwin, C., 11, 13f, 14, 420, 424, 456, 757
Darwin, F., 757
Darwinism
  dual meaning of, 427–431
  in historical context, 421f
  modern acceptance of, 438–439
  natural selection and, 420, 426, 427–431, 438–439
  origin of, 424–426
  as theory, 435–436
Davson, H., 152
Davson-Danielli model, 152–153
Day-neutral plants, 770
DDT, biological magnification and, 1150
Decapods, 626
Deceptive coloration, as animal defense, 1110, 1111f
Deciduous forests
  temperate, 1060–1061
  tropical, 1056
Decomposers, 199, 528
  basidiomycetes as, 589
  fungi as, 593–594
Deduction, in scientific process, 15–19
Defense mechanisms, 850–875
  antimicrobial proteins as, 852
  complement system in, 865–867
  inflammatory response as, 852–854
  natural killer cells as, 851–852
  nonspecific, 850–854
  phagocytic white cells as, 851–852
  and self vs. nonself, 867–869
  skin as, 851

specific (immune response), 854–864
cell-mediated, 863–865
disorders of, 869–872
humoral, 860–863
in invertebrates, 872
Defensive proteins, functions of, 76*t*
Deforestation, 1057
chemical cycling affected by, 1147–1148
Deletion, 292, 336–338
human diseases caused by, 294
Delta waves, of EEG, 1008
Demography, 1087–1089
Denaturation, protein, 82–83
Dendrites, 787, 984
Density, population, 1084, 1085
size regulation and, 1097–1099
Density-dependent inhibition of cell division, 231–232, 233*f*
cancer cells and, 235
Dentition. *See* Teeth
Deoxyribonucleic acid. *See* DNA
Deoxyribose, 86
Depolarization, 989, 990*f*
Deposit-feeders, 795. *See also specific species*
Derivatives, 686
Dermal tissue system, plant, 685
Dermaptera, 622*t*
Descent with modification, 427
Deserts, 1058–1059
Desmosomes, 144, 145*f*
Determinate cleavage, 602
Determinate growth, 686
Determination, 975–976
Detritivores, 1133
Detritus, 1064
in food web, 1133
Detumescence, 947
Deuteromycota, 591
Deuterostomes, 600*f*, 602, 626–629
development of, vs. protostomes, 601–602, 603*f*
Development. *See also* Embryonic development; Growth
of amniotes, 965–968
animal, 600–601, 956–981
mechanisms of, 968–979
of birds, 965–966
localized cytoplasmic determinants of, 970–971
and macroevolution, 480–482
mammalian, 967–968
morphogenetic movements in, 971–974
mosaic, 970
plant, 748–754
definition of, 748
as property of life, 6*f*
regulative, 970
Devonian period
lycopods in, 567
plants in, 566
de Vries, H., 465
Diabetes mellitus, 923–924
Diagnosis of diseases, DNA technology in, 405–407
Diaphragm, in mammals, 653, 842
Diaphragm (contraceptive), 951
Diastole, 822
Diatomaceous earth, 545
Diatoms (Bacillariophyta), 543*t*, 544
Dicots (dicotyledons), 574, 577, 676, 677*f*
Diencephalon, 1007
Diet. *See also* Nutrition
and evolutionary adaptations in vertebrate digestive systems, 807
vegetarian, essential amino acids from, 812
Differential-interference-contrast microscopy (Nomarski microscopy), 118*t*
Differentiation, cellular, 957, 974–976
cytokinins in control of, 761
gene expression and, 372
in plants, 749, 752
in primary root growth, 689
reversibility of, 975
Diffusion, 159–160
facilitated, 163–164
Digestion, 795–798, 800*f*
in alimentary canals, 798, 799*f*
enzymes in, 803, 804*t*
evolutionary adaptations and, 807–809
extracellular, 797
in gastrovascular cavities, 797–798
as hydrolysis example, 66
intracellular, in food vacuoles, 796–797
regulation of, 803
in small intestine, 803–804
digestive secretions in, 803
Digestive tract. *See also specific organ or structure*
complete, 798, 799*f*
in insects, 624
in mammals, 799–807
in Nemertea, 610
in vertebrates, evolutionary adaptations of, 807–809
Dihybrid cross, 265
Dihydroxyacetone, structure and classification of, 67*f*
Dikaryon, in fungal reproduction, 585
Dimorphism, sexual, 452
Dinitrophenol, proton current affected by, 188
Dinoflagellata (dinoflagellates), 543–544, 543*t*
Dinosaurs, 649–650
Dioecious plants, 736
Dipeptidases, in protein digestion, 804
Diploblastic animals, 601
Diploid cells (diploidy), 248
genetic variation and, 448
Diploid nucleus, 959
Diplopoda (diplopods), 620, 621*f*
Dipnoi (lungfishes), 644
Diptera, 622*t*
Directional selection, 451–452
Disaccharidases, in digestion, 803
Disaccharides, 66–67, 68*f*
catabolism of, 193
Disease diagnosis, DNA technology in, 405–407
Disequilibrium, metabolic, 99–100
Disomy, uniparental, genomic imprinting and, 296
Dispersion, population, 1084, 1085–1086
Dispersive model, of DNA replication, 306, 307*f*, 308*f*
Displays, in behavioral communication, 1182
Dissociation, in carboxyl group, 59
Dissociation curve for hemoglobin, 844, 845*f*
Distal convoluted tubule, 884*f*, 885
transport properties of, 889
Disturbance, communities affected by, 1070, 1124
humans causing, 1123
Disulfide bridges, 81
Diurnation, 903
Divergence, speciation by, 467–468
Divergent circuits, neural, 999, 1000*f*
Diversity
of animals, 629–631
biogeographical aspects of, 1124–1128
environmental, 1069–1075
community characteristics determined by, 1120
floral, 736, 737*f*
of fungi, 586–590
global clines in, 1125–1126
immunologic, 855
clonal selection as cellular basis of, 858, 859*f*
metabolic, 521–522
origins of, 524–528
molecular
of carbon, 53–63
of polymers, 64–65
nutritional, in protists, 534
plant, value of, 579–580
of prokaryotes, 521–522, 522–524
in reproductive systems, 935–936
species, within community, 1118–1119
community equilibrium and, 1124
community stability and, 1124
unity in, 10–11, 13*f*
Diversifying selection, 452
Diving mammals, respiratory adaptations in, 845–846
Division (plant group), 561
DNA, 8, 83, 224, 244–245, 300–315. *See also* Genome
amplification of, by polymerase chain reaction, 399–402
chemical structure of, 302–303, 304*f*
complementary
making, 397
as source for genes for cloning, 394, 397–398
in conjugation, 357–358
discovery of, 300–303
double helix of, 86–87, 88*f*
discovery of, 303–306
eukaryotic, organization of
at microscopic level, 373–375
at molecular level, 375–378
evolution and, 87–88, 434–435
functions of, 83–84
as genetic material of cells, evidence for, 300–303
inserting into cells, 394
isolation of, as source of genes for cloning, 394–397
in lytic cycle, 348
methylation of, 378
mitochondrial (mtDNA), 297
restriction mapping for comparison of, 493
packing of, 373–375
prokaryotic, 519–520
organization of
at microscopic level, 373
at molecular level, 375
rearrangement of, 378–379
differentiation and, 975
recombinant, 390, 391–399. *See also* DNA technology
repair of, 313
repetitive, multigene families and, 376
replication of, 306–308, 308–312
and antiparallel structure of DNA strands, 309–311
conservative model of, 306, 307*f*, 308*f*
dispersive model of, 306, 307*f*, 308*f*
DNA ligase in, 311
DNA polymerases in, 309–310, 311*f*
and elongating a new DNA strand, 309–312
helicase in, 311–312
origins of, 309
primase in, 311
priming and, 311, 312*f*
proofreading and, 312
semiconservative model of, 306, 307*f*, 308*f*
single strand binding proteins in, 312
restriction mapping of, in molecular systematics, 493
satellite, 375
"selfish," 362. *See also* Insertion sequences
sequencing of, 399
human genome, 405
in molecular systematics, 493–494
by Sanger method, 400
synthesis of, 399
priming, 311, 312*f*
taxonomic classification based on, 492–494
in transduction, 357, 358*f*, 521
transformation of bacteria by, 301–302, 357, 521
viral, programming cells by, 302
in viral replication, 347
DNA-binding domains, gene expression affected by, 380–381, 382*f*
DNA clocks, in systematics, 494–495
DNA comparison, in molecular systematics, 492–494
DNA-DNA hybridization, in molecular systematics, 492–493
DNA fingerprint, 409–410
certainty of paternity determined by, 1180, 1181
DNA ligase
in DNA replication, 311
recombinant DNA production and, 393
DNA polymerases, 309–310, 311*f*
DNA probes. *See* Probes
DNA technology, 390–415
agricultural uses of, 410–412, 729
in animal husbandry, 410
applications of, 404–412
bacteria used in, 398–399
in basic research, 404
basic strategies of, 391–404
DNA amplification in, 399–402
DNA synthesis and sequencing in, 399
ethical issues in, 412–413, 1153
forensic uses of, 409–410
gene cloning and, 391–399
gene products production and, 398–399
human gene therapy and, 407–408
in human genome project, 404–405
medical uses of, 405–409

DNA technology (*cont.*)
nitrogen-fixation and, 411–412, 729
pharmaceutical product production and, 408–409
plant gene manipulation and, 410–411
plasmids used in, 391, 392*f*, 394, 395*f*
polymerase chain reaction in, 399–402
recombinant DNA and, 391–399
RFLP analysis in, 402–404
safety of, 412–413, 1153
sequencing in, 399
vaccine development and, 408
DNA viruses, 345, 350*t*
classes of, 352*f*
structure of, 346*f*
DNP. *See* Dinitrophenol
Domains, 336
gene expression affected by, 380–381, 382*f*
looped, 373–375
Domains of life, 524
Dominance (genetic)
complete, 268
incomplete, 267, 268–269
Dominance hierarchies, 1177
Dominant allele/dominant trait, 260, 261
inherited disorders and, 274–275
lethal, 274–275
Dominant disorders, 274–275
Dopamine, 998–999, 998*t*
Dormancy, seed, 742–743
Dorsal lip, 963
Double circulation, 821
Double covalent bond, 33
in carbon dioxide, 55
Double fertilization, in angiosperms, 577, 739–740
Double helix, 86–87, 88*f*
discovery of, 303–306
Double-stranded DNA viruses, 350*t*
Double-stranded RNA viruses, 350*t*
Down syndrome, 293–294
Downy mildews, 553–554
Dragonflies, flight in, 621
Drosophila melanogaster
genetic research on, 280–283
linked genes in, 283–284
Duchenne's muscular dystrophy
cloning of gene for, 405
sex-linked inheritance of, 289
Duodenum, 803
Duplication, 292
Dynein side-arm, 140, 142*f*
Dynein "walking," 140–141, 142*f*

Ear
insect, 1032
mammalian, 1027–1030
Eardrum (tympanic membrane), 1027
Earthworms, 615–617
circulatory system of, 820
metanephridia in, 882
Ecdysone
gene expression in insects affected by, 383
insect molting triggered by, 916
ECG. *See* Electrocardiogram
Echidnas (spiny anteaters), 655*f*, 656
Echinodermata (echinoderms), 600*f*, 626, 627–629
nervous systems of, 1000, 1001*f*
Echinoidea (sea urchins/sand dollars), 627, 628*f*, 629
early embryonic development of, 960–965
fertilization in, 957–959, 960*f*
Ecological efficiency, 1139
Ecological niche, 1114
Ecological pyramids, energy transfer and, 1138–1140
Ecological succession, 1121–1124
Ecology, 1052–1082. *See also* Environment
behavioral, 1159, 1160*f*. *See also* Behavior
community, 1053, 1106–1131
conservation biology and, 1128–1129
definition of, 1052
evolution and, 1054
as experimental science, 1053–1054
long-term research in, 1148
organismal (physiological), 1053
population, 1053
questions of, 1053
scope and development of, 1053–1054
Ecosystems, 1053, 1132–1157
and biotic effects at biosphere level, 1153–1154
chemical cycling and, 1140–1145
definition of, 1132
energy flow and, 1133–1140
human intrusion into, 1132*f*, 1145–1153
productivity of
primary, 1134–1138
secondary, 1138–1140
trophic structure of, 1132–1133
Ectoderm, 600, 601*f*, 962
derivatives of in mammals, 968*t*
Ectomycorrhizae, 594*f*
Ectoparasites, 1116
Ectotherms, 895
reptiles as, 648
Edema, 829
Edentata, 658*t*
Ediacaran period, animals in, 631
Edwards syndrome, 294
EEG. *See* Electroencephalogram
Effector, in homeostatic control system, 790
Effector cells, 856
Efferent arteriole, 884*f*, 885
Efferent neurons (motor neurons), 985, 1002
EGF. *See* Epidermal growth factor
Eggs. *See* Ova
Eichele, G., 978
Ejaculation, 937, 947
Ejaculatory duct, 937
EKG. *See* Electrocardiogram
Elastic fibers, 785
Eldredge, N., 469
Electrical synapses, 993–994
Electrocardiogram, 824
Electrochemical gradient, 166
Electroencephalogram, in arousal and sleep, 1008
Electrogenic pump, 166, 167*f*
Electromagnetic energy, light as, 204
Electromagnetic receptors, 1018, 1019*f*
Electromagnetic spectrum, 204
Electron acceptor, primary, in photooxidation of chlorophyll, 207, 208*f*
Electron configuration, 31–32
of carbon, 55
Electron density maps, in molecular computer modeling, 85
Electronegativity, 34
Electron flow, in light reactions
cyclic, 209–210
noncyclic, 210–211
Electron micrographs, 119*f*
Electron microscope, 118–119
light microscope compared with, 119*f*
Electron orbitals, 31
Electrons, 27
chemical properties of, 31–32
distribution of in atom's electron shells, 31–32
potential energy states for, 30–31
transfer of, and ionic bonding, 35*f*
valence, 32
Electron shells (energy levels), 30–31
Electron transport chains, 177–178, 184–189
origin of, 524–525, 525*f*
Electrophoresis
gel, 396
in DNA sequencing, 399
restriction fragments separated by, 393
genetic variations studied with, 446
Electroporation, 394
Elements, 26, 27*f*, 61
definition of, 26
electron configuration of, 31–32
in human body, 26*t*
inert, 32
trace, 26
Elongation
in transcription, 324
in translation, 328–330
Elongation factors, 328, 329*f*
EM. *See* Electron microscope
Embryo
development of. *See* Embryonic development
plant, development of, 740–741
polarity of, 968–969
protection of, 935
Embryology, 956–981. *See also* Development; Embryonic development
comparative, as evidence of evolution, 434
Embryonic axes, 969
Embryonic development, 948–950
of amniotes, 965–968
early stages of, 960–965
localized cytoplasmic determinants of, 970–971
mammalian, 967–968
morphogenetic movements in, 971–974
mosaic, 970
regulative, 970
Embryonic disc, 967
Embryo sac, in angiosperms, 577, 735–736, 739
Emergent properties, 4–7
Emotions, 1009
Emulsification of fats, 804
End-bulbs of Krause, 897, 1017*f*, 1019
Endergonic reactions, 97
Endler, J., 16–18
Endocrine signaling, 908*f*
Endocrine system (endocrine glands), 907, 908. *See also specific glands and hormones*
nervous system and, 927–928
in vertebrates, 917–927
Endocytosis, 167–170
receptor-mediated, 168, 169*f*
Endoderm, 600, 601*f*, 962
derivatives of in mammals, 968*t*
Endodermis, root, 689
in water and mineral absorption, 705–706
Endolymph, 1028
Endomembrane system, 128–133. *See also specific component*
relationships of, 133, 134*f*
Endometrium, uterine, 940
Endomycorrhizae, 594*f*
Endoparasites, 1116
Endoplasmic reticulum, 128–129
rough, 128
membrane production and, 129
protein synthesis and, 129
smooth, 128
functions of, 128–129
transitional, 129
Endorphins, 22, 920*f*, 921
as neurotransmitters, 998*t*, 999
Endoskeleton, 638, 1036*f*, 1037
Endosperm, 577, 739, 740
Endospores, 520
characteristics of bacteria forming, 526*t*
Endosymbiotic model of eukaryotic origins, 535–536, 551
Endothelium, 825
Endothermy (endotherms), 895–896
birds, 651
dinosaurs, 649
mammals, 653
Endotoxins, 529
Energy
activation
definition of, 101
lowering of, enzymes and, 100–102
basic principles of, 92–96
chemical
conversion of to heat, 94
transformation of, 93
ecological pyramids and, 1138–1140
laws of thermodynamics governing, 93–95
conservation of (first law of thermodynamics), 93
ecological pyramids and, 1139
definition of, 30, 92
forms of, 92
free, 95–96
of activation
definition of, 101
lowering of, enzymes and, 100–102
and changes in chemical reactions, 96–97
definition of, 96
kinetic, 92, 93*f*
definition of, 92
potential, 92, 93*f*
definition of, 30, 92
transformations of, 92–93
ecological pyramids and, 1138–1140
laws of thermodynamics governing, 93–95
utilization of
in ecosystem, 1133–1140
as property of life, 6*f*

Energy budget, 1134
Energy coupling, 92
by phosphate transfer, 99f
Energy flow, 1133–1140
Energy levels (electron shells), 30–31
Energy storage, by fats, 73
Energy transducers, mitochondria and chloroplasts as, 134–136
Engelmann, T., 205f
English peppered moth, as example of natural selection, 429–430
Enhancers, 376, 377f
gene expression affected by, 380, 381f
Enkephalins, 22, 920f, 921
as neurotransmitters, 998t, 999
Entamoeba histolytica, 538
Enteric bacteria, 526t
Enterocoelomates (deuterostomes), 600f, 602, 626–629
development of, vs. protostomes, 601–602, 603f
Enterocoelous development, 602, 603f
Enterogastrone, 803
Enterokinase, 804
Entomology, 620–621. *See also* Insects
Entropy, 93, 94f
Envelopes, viral, 346
reproduction of viruses with, 350–351
Environment
animal size and shape affected by, 787–789
definition of, 1052
diversity of, 1069–1075
community characteristics determined by, 1120
enzyme activity affected by, 104–105
fitness of, and ability of ice to float, 45, 46t
internal, of animals, 789–791
controlling, 876–906. *See also specific mechanism*
morphological, 1079
organisms' interaction with, 10, 1052–1082. *See also* Ecology
patchiness of, 1069–1075
community characteristics determined by, 1120
plant form and function affected by, 675, 676f
response to, 1075–1080
behavioral, 1078. *See also* Behavior
physiological, 1078–1079
as property of life, 6f
Environmental grain, 1077–1078
Enzymatic proteins, functions of, 76t, 157f
Enzyme inhibitors, 105
Enzymes, 100–106
activation energy lowering and, 100–102
allosteric regulation of, 105–106
catalytic cycle of, 102–104
cofactors affecting, 105
cooperativity and, 106
digestive, 803, 804t
environmental conditions affecting, 104–105
factors affecting activity of, 104–106
hydrolytic, in digestion, 796
inhibitors of, 105
in metabolism, 100–106
repressible vs. inducible, 366–367
restriction, 349, 391–393
specificity of, 102
Eosinophils, 831f
development of, 832f
in nonspecific body defense, 852
Epicotyl, 741, 743
Epidermal growth factor, 910
Epidermis, plant, 685
of leaves, 691
Epididymis, 937
Epiglottis, 800, 801f
Epinephrine, 919t, 925, 998
Epiphytes, 567, 730
Episome, 358
F, 359, 360f
Epistasis, 269–270
Epithelial tissue, 783–784
structure and function of, 783f
transport, osmoregulation and, 880–881
Epithet, specific, 489
Epitopes (antigenic determinants), 856f, 857
EPO. *See* Erythropoietin
EPSP. *See* Excitatory postsynaptic potential
Epstein-Barr virus, cancer and, 353
Equilibrial life history, 1091–1092
Equilibrial model of community development, 1124
Equilibrium. *See also* Balance
chemical, 37
community, species diversity and, 1124
Hardy-Weinberg, 441
keeping metabolism away from, 99–100
Equilibrium potential, 988
Equisetum, 567, 568f
ER. *See* Endoplasmic reticulum
Ergotism, 594
Ergots, on rye, 594, 595f
Erosion, control of in soil management, 725
Erythrocytes (red blood cells), 830–831, 831f
development of, 832f
Erythropoietin, 831
DNA technology in production of, 409
Escherichia coli, 526t
in large intestine, 807
Esophagus, 800–801
Essential amino acids, 811–812
Essential fatty acids, 812
Essentialism, evolution and, 422
Essential nutrients
for animals, 811–815
for plants, 719–720, 721t
Estradiol, 926f
functional group of, 59f
Estrogens, 919t, 926
gene expression and, 911–912
in menstrual and ovarian cycle control, 942–945
Estrous cycle, 941
Estrus, 941
Estuaries, 1066–1067
Ethane
carbon skeleton diversity and, 57f
shape of, 55, 56f
Ethanol, 60t
Ethene, shape of, 56f
Ethics, in DNA technology, 412–413, 1153
Ethology, 1162
cognitive, 1186
Ethylene, as plant hormone, 759t, 764–765, 766f
Eubacteria, 523, 524, 526–527t
Euchromatin, 375
Euglena, 545, 546f
Euglenophyta (euglenophytes), 543t, 545, 546f
Eukaryotes (eukaryotic cells), 7, 121. *See also* Cells
antiquity of, 535
chromosomes of, 223–224
DNA as genetic material in, evidence of, 302–303
gene expression in
cancer and, 384–386
chromosome puffs and, 383
control of, 379–383
small molecules in, 383–384
steroid hormones affecting, 383–384
genome of, 372–379
organization of
at microscopic level, 373–375
at molecular level, 375–378
plasticity of, 378–379
rearrangements in, 378–379
differentiation and, 975
internal complexity of, 122–123
origins of, 534–536
models of, 535–538
and origins of multicellularity, 554–556
processing of, 333–336
protein synthesis in, 319f, 332–333, 340f
protein synthesis in prokaryotes compared with, 332–333
RNA polymerases in, 322–323
structure and function of, 146t
transcription, 319f, 332–333
transcription unit in, 322
translation in, 319f, 332–333, 340f
Eumetazoa, 599, 600f, 604–608
Euryhaline animals, osmoregulation in, 879
Eurypterids, 619
Eustachian tube (auditory tube), 1027
Eutrophication of lakes, 1064
accelerated, 1149
Evaporation
cooling by, 43–44, 896–897
in heat transfer, 895
Evolution, 420–437. *See also* Macroevolution; Natural selection; Speciation
abiotic synthesis of organic compounds and, 54, 507f
alternative views and, 511
monomers, 507–508
polymers, 508
adaptive, 450–453
animal behavior and, 1015
biogeography as evidence for, 431–432
branching (cladogenesis), 456, 457f, 482, 483f
chemical reactions creating conditions for, 37–38. *See also* Chemical evolution
alternative views and, 511
comparative anatomy as evidence for, 433–434
comparative embryology as evidence for, 434
convergent, 491
as core theme of biology, 11–14
cultural, 664–666
and Cuvier's discovery of fossils, 422–423
Darwinian view of, 424–427. *See also* Darwinism
dual meaning of, 427–431
as theory only, 435–436
definition of, 420
DNA and proteins in, 87–88, 434–435
ecology and, 1054
experimental evidence for, 17–18
fossil record as evidence for, 13f, 422–423, 432, 474f, 475–478
genetic code and, 321–322
genetic variation and, 255–256
glycolysis in cellular respiration and, 193
introns and, 335–336
Lamarck's theory of, 424
of mitosis, 229–230, 232f
modern acceptance of, 438–439
molecular, neutral theory of, 450
molecular biologic evidence for, 87–88, 434–435
mutations in, 435, 439
natural theology and, 422
and necessity of new synthesis, 496–497
novelties in, origin of, 480–482
perfection and, 453
phyletic (anagenesis), 456, 457f
population interactions and, 1108–1109
of populations, 438–455. *See also* Populations
pre-Darwinian views of, 420–424
as property of life, 6f
scale of life and, 422
second law of thermodynamics and, 95
signs of, 431–435
taxonomy as evidence for, 422, 432–433
trends in, difficulty in interpretation of, 482–484
of viruses, 354–355
Evolutionary adaptation, as property of life, 6f
Excision repair, of DNA, 313
Excitable cells, 987
Excitation-contraction coupling, 1040–1042
Excitatory postsynaptic potential, 995
Excited state, of pigment molecule, 207
Excitement phase, of human sexual response, 947
Excretory systems
human, 884f
invertebrate, 881–883
vertebrate, anatomy of, 883–885
"Exercise hypothesis," for play behavior, 1170
Exergonic reactions, 96–97
and regeneration of ATP, 99
Exocrine glands, 908
Exocytosis, 167–170
Exons, 333, 376, 377f
Exoskeletons, 618, 1036–1037, 1036f
chitin in, 71
in decapods, 626
Exotic species, introduction of by humans, 1152–1153
Exotoxins, 529
Exploitative competition, 1113
Exponential population growth, 1092–1093, 1094f, 1094t
External fertilization, 934
Exteroreceptors, 1017
Extinctions
mass, macroevolution and, 487–489
of plant species, rain forest destruction and, 579–580
Extracellular digestion, 797
Extracellular matrix
cell migration and, 971–974
in connective tissue, 784

Extraembryonic membranes
in birds, 966
in mammals, 967–968
Extranuclear inheritance, 296–297
Extreme halophiles, 523
Eyeball, 1021
Eye cup, in planarians, 1019–1020
Eyes. *See also* Vision
compound, 620, 1020, 1021*f*
human, 1020–1021
induction during development of, 974
single-lens, 1020
vertebrate
signal transduction in, 1022–1024
structure and function of, 1021–1022

F. *See* Fluorine
F_1 generation, 260
F_2 generation, 260
F^+ cell, 359, 360*f*
F^- cell, 359, 360*f*
Facilitated diffusion, 163–164
Facilitation, succession and, 1123
Facultative anaerobes, 192, 522
Fairy ring, 583*f*, 589–590
Fallopian tube (oviduct), 940
Family, taxonomic, 489
FAP. *See* Fixed-action patterns
Far-red light, photoperiodic response affected by, 771–772
Fast block to polyspermy, 958
Fast fibers, 1043
Fat. *See* Adipose tissue
Fate maps, 971, 972*f*
Fats, 71–73
catabolism of, 193, 194*f*
digestion of, 804
excess food stored as, 810
function of, 73
hydrocarbons affecting characteristics of, 57*f*
hydrolysis of, 804
saturated, 71–73
structure of, 72*f*
unsaturated, 71–73
Fat-soluble vitamins, 813*t*, 814
Fatty acids, 71
essential, 812
saturated, 71, 72*f*
structure of, 72*f*
unsaturated, 71, 72*f*
Feathers, 652
Feces, 806–807
Fecundity
density affecting, 1097–1098
in life tables, 1088
Feedback
negative (feedback inhibition), 790
in control of cellular respiration, 194
of metabolism, 106–107
positive, 790–791
Feeding behavior, 1174–1175
Feeding relationships, 1118, 1132–1133
A Feeling for the Organism (book), 8
Female reproductive system, human, 939–940
Female sex hormones, 940–945
functional groups of, 59*f*
F episome, 359, 360*f*
Fermentation, 173, 190–192
aerobic and anaerobic respiration compared with, 192–193
Ferns (Pterophyta), 562*t*, 568
life cycle of, 568, 569*f*, 572*t*
Fertilization, 248, 934–935, 957–959, 960*f*. *See also* Reproduction
in angiosperms, 739–740
as chance event, 264*f*
double, in angiosperms, 577, 739–740
events after, 948*f*. *See also* Embryonic development
external, 934
internal, 934–935
in mammals, 967
random, genetic variation and, 255
species variations in timing of, 248–249
in vitro, 953
Fertilization membrane, 958
Fertilizers, in soil management, 724
Fetal circulation, 949*f*
Fetal development, 948–950
Fetal testing, for genetic disorders, 276–277
Fetoscopy, 276
complications of, 277
Fetus, 948
Fiber
dietary, 70. *See also* Cellulose
plant, 683*f*, 684
xylem, 574
Fibrin, 832
Fibrinogen, 832
Fibroblasts, 785
Fibroin, beta pleated sheets in, 80, 81*f*
Fibronectins, morphogenetic movements and, 971–974
Fibrous connective tissue, 785
Fibrous root system, 678
Filtrate, 885
concentration of by kidneys, 889–891
Filtration, glomerular, 885, 886*f*
Fingerprint, DNA, 409–410
certainty of paternity determined by, 1180, 1181
First law of thermodynamics, 93
and ecological pyramids, 1139
Fishes, 639
bony, 638*t*, 639*f*, 643–644
cartilaginous, 641–642
circulatory system of, 821
lateral line system in, 642, 1030–1031
phylogeny of, 645*f*
thermoregulation in, 901
Fission, binary, in bacterial reproduction, 222, 223*f*, 355, 520–521
Fitness. *See also* Natural selection
and adaptive evolution, 450–451
altruistic behavior and, 1184–1185
animal behavior and, 1159, 1160*f*, 1182
inclusive, 1184
Five-kingdom system, 512
modifications of, Protista and, 537
Fixed-action patterns (fixed-motor patterns), 1162–1165
FK 506, cell-mediated immunity suppressed by, 869
Flaccid cell, water balance and, 163
Flagella
eukaryotic, 139–141
of protists, 534
ultrastructure of, 141*f*
prokaryotic, 518–519
of sperm cells, 946
Flame-cell system, 881–882
Flatworms (Platyhelminthes), 600*f*, 608–609
flame-cell system of, 881–882
nervous system of, 1000, 1001*f*
reproductive systems of, 936, 937*f*
Flavin mononucleotide, in electron transport chain, 184, 185*f*
Flight
in birds, 651–652
in insects, 621
Flightless birds, 653
Florigen, 771
Flowering plants. *See* Angiosperms
Flowers, 575, 735, 736–737
development of, homeotic mutations in, 753–754
function of, 736
photoperiodic control of formation of, 769–771
structure of, 575*f*, 736*f*
variations in, 736, 737*f*
Fluid-feeders, 795, 796*f*. *See also specific species*
Fluidity, of membranes, 155–156
Fluid mosaic model of membranes, 151, 153, 155–158
Flukes, 609
Fluorescence, 207
Fluorine
electron configuration of, 32*f*
human requirements for, 814*t*
Focus, in vision, 1022
Folacin (folic acid), 813*t*
Folding, protein, 83, 84*f*
Folic acid (folacin), 813*t*
Follicle, ovarian, 939
Follicle-stimulating hormone, 919*t*, 920*f*, 921
in menstrual and ovarian cycle control, 942–945
in testicular control, 942*f*
Follicular phase, of ovarian cycle, 942, 943*f*
Food. *See also* Nutrients; Nutrition
energy content of, 809
for fabrication, 811
for fuel, 809–811
fungi as, 594–595
Food chain, 1133, 1134*f*
poisons in, 1149–1151
Food vacuoles, 131–132, 132
intracellular digestion in, 796–797
Food webs, 1133, 1135*f*. *See also* Food chain
Foot
of cephalopod, 614
of mollusks, 612
Foraging behavior, 1174–1175
Foraminifera (forams), 537*t*, 539
Forbs, 1058
Forebrain (prosencephalon), 1005, 1007
Forensics, DNA technology in, 409–410
Foreskin, 938–939
Fork, replication, 309
activities at, 310, 311*f*, 312*f*
Form and function, 9–10
Formic acid, 59
ants manufacturing, 59
Fossil fuels, burning, carbon dioxide release and, 1152
Fossils (fossil record)
as evidence of evolution, 13*f*, 422–423, 432, 474*f*, 475–478
formation of, 475
and geological time scale, 477–478
limitations of, 475–477
in macroevolution, 474*f*, 475–478
in sedimentary rock, 422–423, 475, 476*f*
speciation patterns shown by, 456
Founder effect, in genetic drift, 443–444
Fovea, 1021*f*, 1022
Fox, S., 508
F plasmid, in conjugation, 358, 359, 360*f*
Fractionation, cell, 120–121
Fragile-X syndrome, 296
Fragmentation, 931–932
in vegetative reproduction, 744
Frameshift mutation, 337*f*, 338
Franklin, R., 303, 304, 305*f*
Free energy, 95–96
of activation (activation energy), lowering of, ezymes and, 100–102
and changes in chemical reactions, 96–97
definition of, 101
Free ribosomes, 331
Freeze-etch, 154
Freeze-fracture, 154
Frequency-dependent selection, genetic variation and, 449
Freshwater animals, osmoregulation in, 878–879
Freshwater biomes, 1062–1065. *See also specific type*
distribution of, 1063*f*
von Frisch, K., 1162, 1182
Frogs (Anurans), 636
early embryonic development of, 960–965
fate map of, 972*f*
metamorphosis of, 647
organogenesis in, 963–965
Fructose, 66, 67*f*
Fruit, 575–576
development of, 741–742
effect of gibberellins on, 763
ripening of, effect of ethylene on, 765
types of, 743*f*
Fruit fly (*Drosophila melanogaster*)
genetic research on, 280–283
linked genes in, 283–284
FSH. *See* Follicle-stimulating hormone
Function, relationship of to structure, 9–10
Functional groups, 58–61. *See also specific type*
amino, 60–61
carbonyl, 58–59
carboxyl, 59
definition of, 58
hydroxyl, 58
for male vs. female hormones, 59*f*
phosphate, 61
sulfhydryl, 61
Fundamental niche, 1114
Fungi, 512, 583–597
characteristics of, 583–585, 586*f*
as decomposers, 593–594

diversity of, 586–590
ecological importance of, 593–595
edible, 594–595
evolution of, 595
growth of, 585, 586*f*
habitats of, 583–584
imperfect, 591
lifestyles of, 590–593
nutrition of, 583–584
pathogenic, 594
reproduction of, 585, 586*f*
sac (ascomycetes), 587–589
as spoilers, 594
structure of, 584–585
zygote (zygomycetes), 586–587
Funguslike protists, 551–554
Fusiform initials, 693–694

G_0 phase, 233
G_1 phase, 224
and control of cell division, 233
G_2 phase, 224, 226*f*
GABA. *See* Gamma aminobutyric acid
Galactose, 66, 67*f*
Galapagos Islands
Darwin's observations about species on, 425, 426
as model of speciation, 456*f*
Gallbladder, 799
Gametangia, 560
as terrestrial adaptation, 560
Gametes, 932
angiosperm, 735
fertilization and, 934–935
formation of, 944*f*, 945–947
meiosis in production of, 248
and origins of multicellularity, 556
recognition of, reproductive barriers and, 460–461
Gametic isolation, as reproductive barrier, 460–461
Gametogenesis, 944*f*, 945–947
Gametophytes, 249, 548, 560
angiosperm, 735
development of, 738*f*
Gamma aminobutyric acid, 998*t*, 999
Ganglia (ganglion), 999
Ganglion cells, in retina, 1024, 1025*f*
Gap junctions, 144, 145*f*
pre- and postsynaptic cells connected by, 993
Garner, W.W., 759, 770
Garrod, A., 316
Gas exchange. *See also* Respiratory organs
in animals, 835–846
in amphibians, 647
in arthropods, 618
in bony fishes, 643
in crustaceans, 626
general problems of, 835
gills for, 836–838
in insects, 624
in reptiles, 648
in spiders, 619
in vertebrates, 638
Gastric juice, secretion of, 801, 802*f*
Gastrin, 802, 803
Gastrodermal cells, 798
Gastrodermis, 798
Gastrointestinal system. *See* Digestive tract
Gastropoda (gastropods), 612–613
Gastrovascular cavities, 604
digestion in, 797–798
internal transport and, 819
Gastrula, 961
Gastrulation, 599, 600, 601*f*, 961–963
in birds, 965*f*, 966
morphogenetic movements of, 971–974
Gated ion channels, 989
in action potential, 989–991, 992*t*
voltage-sensitive, 989–990
Gause, G.F., 1113
Gel electrophoresis, 396
in DNA sequencing, 399
restriction fragments separated by, 393
Gemmae, in liverwort reproduction, 563, 565*f*
Gemmules, 932
Gender, chromosomal basis of, in humans, 288–289. *See also* Sex

Gene expression
cancer and, 384–386
control of, 363–368, 379–383
chromosome puffs and, 383
posttranscriptional, 381–383
posttranslational, 382–383
small molecules in, 383–384
steroid hormones in, 383–384
transcriptional, 380–381
translational, 382–383
in plants, touch affecting, 773, 774*f*
rearrangements in genome affecting, 378–379
differentiation and, 975
steroid hormones and, 911–912
Gene flow, microevolution caused by, 442*t*, 444
Gene manipulation. *See* DNA technology
Gene pool, 439
Hardy-Weinberg theorem and, 440–442
microevolution and, 439
Genera. *See* Genus
Generalized transduction, 357, 358*f*
Generations
alternation of, 249, 560
in angiosperms, 735
in Chlorophyta (green algae), 547–548
in mosses, 563, 564*f*
in Phaeophyta (brown algae), 550
heteromorphic, 550, 560
isomorphic generations, 550
Generation time, population size and, 1087, 1093, 1094*f*
Genes, 8, 83, 87, 244–245. *See also* Genetics; DNA; Gene pool
alternative versions of (alleles), 260–261
amplification of, 378
cloning. *See also* DNA technology
in eukaryotic cells, 394
in plasmids, 391, 392*f*, 394, 395*f*
sources of genes for, 394–398
yeast cells for, 394, 399
coding strands of, 320
constitutive, 367
coordinately controlled, arrangement of, 376–378
cytoplasmic, 296–297
definition of, 338–339
expression of. *See* Gene expression
immunoglobulin, DNA rearrangement and, 379, 380*f*
"jumping," 361. *See also* Transposons
linked, 283–284
recombination of, crossing over and, 285
location of on chromosomes, 280*f*, 282
identification of, 282–283
loss of, selective, 378
mating-type, 379
metabolism controlled by, 317, 318*f*
mitochondrial, 297
negative regulation of, 366–367
parental imprinting of, 295–296
positive regulation of, 367–368
probes in identification of, 398
products of, genetic engineering in production of, 398–399
proteins specified by, 316–343. *See also specific type*
regulatory, 365*f*, 366
selective loss of, 378
sex-linked, 283
structural, in operon, 364, 365
tandem duplication of, multigene families and, 376
unlinked, recombination of, independent assortment and, 284–285
Gene therapy, RNA technology in, 407–408
Genetic code, 319–321
dictionary of, 321*f*
evolutionary significance of, 321–322
exceptions to universality of, 322
Genetic counseling, 275–277
Genetic cross, 259*f*, 260, 261*f*
Genetic disorders
consanguinity and, 274
DNA technology in diagnosis of, 405–407
dominantly inherited, 274–275
genetic counseling and, 275–277
multifactorial, 277
mutations and, 336
preventive approach to, 275–277

recessively inherited, 273–274
screening for, 275–277
Genetic drift
in allopatric speciation, 463
bottleneck effect in, 443
founder effect in, 443–444
microevolution caused by, 442–444, 442*t*
Genetic engineering, 390. *See also* DNA technology
in animal husbandry, 410
bacteria used in, 398–399
gene products made using, 398–399
in plants, 410–411
plasmids used in, 391, 392*f*, 394, 395*f*
safety and ethical issues in, 412–413, 1153
Genetic information
DNA transmitting, 83–84
origin of, 509–511
RNA transmitting, 84
Genetic mapping
crossover data for, 285–288
of human genome, DNA technology in, 404
Genetic recombination
in bacteria, 356–363, 521
chromosomal basis of, 284–285
of linked genes, crossing over and, 285, 286*f*
of unlinked genes, independent assortment and, 284–285
variation and, 447–448
Genetics, 244
bacterial, 355–356
Genetic screening, 275–277
Genetic symbols, 282
Genetic transfer, in bacteria, 222, 223*f*, 356–363, 521
Genetic variation, 244, 445–450
as adaptive event, 450
basis of, 445–450
crossing over and, 254–255
evolution and, 255–256
geographical, 446–447
independent assortment of chromosomes and, 253–254
mutation as source of, 447–448
nature and extent of, within and between populations, 445–447
neutral, 450
preservation of, 448–450
random fertilization and, 255
recombination as source of, 448
sexual sources of, 253–255
sources of, 447–448
Geniculate nuclei, lateral, 1025, 1026*f*
Genitalia
development of from primordia, 940, 941*f*
female, 939
male, 937
Genomes, 221. *See also* DNA
bacterial, 355–356, 519–520
eukaryotic, 372–379
human, mapping, DNA technology in, 404–405
organization of
at microscopic level, 373–375
at molecular level, 375–378
plasticity of, 378–379
prokaryotic, 355–356, 519–520
rearrangements in, 378–379
differentiation and, 975
viral, 345
Genomic equivalence, 975
Genomic imprinting, 295–296
Genomic library, 397
Genophore, 519–520
Genotype
definition of, 263
phenotype differentiated from, 263*f*
Genus (genera), 489
Geographic isolation, in allopatric speciation, 462–463
Geographic range
of population, 1085–1086
of species, limits of, 1125
Geographic variation, 446
Geological time scale, 477–478
Geology, gradualism in, evolution and, 423
Geometric isomers, carbon skeleton diversity and, 57, 58*f*

Germination, 743–744
effect of gibberellins on, 763
imbibition in, 743
mobilization of nutrients during, 744*f*
seed, 742–744
Germ layers, 600, 601*f*
primary, 962
derivatives of in mammals, 968*t*
Germs. *See* Bacteria
Gestation, 947–953
GH. *See* Growth hormone
Gibberellins, 759*t*, 762–763, 743
Gills, 836–838
physiology of, 837*f*
Ginkgophyta (ginkgo), 562*t*, 570*f*, 571
Glans penis, 938
Gleason, H.A., 1106
Glial cells, 985
Global clines, in species diversity, 1125–1126
Global energy budget, 1134
Global hydrologic cycle, 1141
Globe, of vertebrate eye, 1021
Globins
multigene families and, 375–376, 377*f*
pseudogenes for, 376, 377*f*
Glomerular filtration, 885, 886*f*
urine concentration and, 889–891
Glu. *See* Glutamic acid; Glutamine
Glucagon, 919*t*, 923
Glucocorticoids, 919*t*, 925–926
Glucose, 66, 67*f*
alpha and beta rings of, 69, 70*f*
catabolism of, 173–193. *See also* Cellular respiration
cellulose-digesting bacteria and, 70
classification of, 67*f*
hydrolysis of from starch, 68–69
insulin and glucagon in maintaining homeostasis of, 924
linear and ring forms of, 66, 67*f*
photosynthesis in formation of, 201–202
Glutamate, as neurotransmitter, 998*t*, 999
Glutamic acid, 75*f*
Glutamine, 75*f*
Gly. *See* Glycine
Glyceraldehyde, structure and classification of, 67*f*
Glyceraldehyde 3-phosphate, Calvin cycle producing, 212–214
Glycerol phosphate, 60*t*
Glycine, 60–61, 60*t*, 75*f*
as neurotransmitter, 998*t*, 999
Glycocalyx, 144
Glycogen, 69
catabolism of, 193
energy stored as, 810
shape of, 68*f*
Glycolysis, 179, 180–181*f*
evolutionary significance of, 193
origin of, 524
Glycoproteins, 129
Glycosidic linkage, in disaccharides, 66
Gnetophyta (gnetae), 562*t*, 570*f*, 571
GnRH. *See* Gonadotropin-releasing hormone
Golden algae (Chrysophyta), 543*t*, 544
Golgi apparatus, 130–131
sorting functions of, 131
Gonadotropin-releasing hormone
in menstrual and ovarian cycle control, 942
in testicular control, 942*f*
Gonads, hormones produced by, 919*t*, 926
Goodall, J., 1186
Gorter, E., 152
Gould, S.J., 14, 469, 500–503, 504
GPP. *See* Gross primary productivity
G proteins, 913–914
Graded potentials, 989, 990*f*
EPSPs and IPSPs in, 995
Gradualism, evolution and, 423
Grafting, for plant reproduction, 746
Graft versus host disease, 869
Grain, environmental, 1077–1078
Gram-negative bacteria, 517–518
Gram-positive bacteria, 517–518
Gram stain, 517
Grana, 135, 200
Grasslands, temperate, 1059, 1060*f*
Gravitropism, 766–767
Gray crescent, 969
Gray matter, 1004
Green algae (Chlorophyta), 543*t*, 545–548
as plant ancestors, 562
Greenhouse effect, 1151
"Green manure," 725*f*, 728
Grendel, F., 152
Griffin, D., 1186
Griffith, F., 301
Gross primary productivity, 1134–1136, 1138
Ground meristem, root, 689
Ground pines (*Selaginella*), 567
Ground state, of pigment molecule, 207
Ground tissue system, plant, 685
Growth
allometric, 480, 481*f*
bacterial, 520–521
plant, 686–687, 748–749
hormones affecting, 757–765
primary, 686, 687–693
secondary, 686, 693–696
as property of life, 6*f*
Growth factors, 231, 910
Growth hormone, 919*t*, 920*f*, 921
human, DNA technology in production of, 409
Guanine, 87*f*
Guard cells, in leaf epidermis, 691
and stomatal opening and closing, 709–710
Gurdon, J., 975
Gustation, 1032–1034
Gustatory receptors (taste receptors), 1018, 1032–1034
Guttation, 706
Gymnosperms, 561, 562*t*, 571–572, 573*f*
four divisions of, 570*f*, 571
history of, 572

H. *See* Hydrogen
H1 histone, 373, 374*f*
Habitat destruction, biodiversity threatened by, 1146–1147
Habitat isolation, as reproductive barrier, 460
Habituation, 1166
Hagfishes, 638*t*, 640
Hair, as mammalian trademark, 653, 654*f*
Hair cells, 1017–1018, 1027, 1028, 1029
Haldane, J.B.S., 507, 1184
Hales, S., 718
Half-life, and radioactive dating, 477, 479
Halophiles, extreme, 523
Hamilton, W.D., 1184
Haploid cell, 248
Hardening of the arteries (arteriosclerosis), 834
Hardy-Weinberg equilibrium, 441
Hardy-Weinberg theorem, 440–442
Haustoria, 585, 730
Haversian system, 786
Hawaiian archipelago, adaptive radiation on, in allopatric speciation, 464–465
HCG. *See* Human chorionic gonadotropin
HCl. See Hydrochloric acid
HDLs. *See* High-density lipoproteins
He. *See* Helium
Hearing, 1027–1032
in invertebrates, 1031–1032
process of, 1028–1029
in vertebrates, 1030–1031
Heart, 820–821, 822–825
control of, 824
evolutionary adaptations in, 821–822
excitation of, 824–825
Heart attack, 832–833
Heart cycle, 822
Heart murmur, 822–823
Heart rate, 823–824
Heart sounds, 822–823
Heart valves, 822
Heat, 42–43
in energy transformations, 93–95
production of, for thermoregulation in mammals, 896
receptors for (Ruffini's end organs), 897, 1017*f*, 1019
specific
definition of, 43
of water, 42–43, 46*t*
transfer of between organisms and environment, 895
adjustment of in thermoregulation, 896
Heat exchanger, countercurrent, in birds, 898, 900*f*
Heating center, 897
Heat of vaporization
definition of, 43
of water, 43–44, 46*t*
Heavy chains, immunoglobulin, 857
HeLa cells, 236
Helicase, in DNA replication, 311–312
Heliozoans, 538
Helium, electron configuration of, 32*f*
Helium atom
electron orbital of, 31
model of, 27*f*
Helix, alpha, 80
Helix-turn-helix domain, gene expression affected by, 380–381, 382*f*
Helper T cells, 856, 863–865
in AIDS, 871
in B cell activation, 860, 861*t*, 863
Heme group, in electron transport chain, 184, 185*f*
Hemiptera, 622*t*
Hemizygous, 289
Hemocyanin, 844
Hemoglobin, 843–844
dissociation curve for, 844, 845*f*
multigene families and, 375–376, 377*f*
normal, primary structure of, 79*f*
polypeptide chain of, evolutionary relationships and, 88*t*
quaternary structure of, 81, 82*f*
in red blood cells, 830–831
sickle-cell, primary structure of, 79*f*
Hemolymph, 618, 819, 820*f*
Hemophilia, 832
cloning of gene for, 405
sex-linked inheritance of, 289–290
Henle's loop. *See* Loop of Henle
Hepatic portal vein, 805
Hepatitis B virus, cancer and, 353
Hepatophyta (liverworts), 562*t*, 563, 565*f*
Herbivores, 794
plant defenses against, 1110
as primary consumers, 1133
Hereditary diseases. *See* Genetic disorders
"Hereditary factors" (Mendel's). *See* Genes
Heritable disorders. *See* Genetic disorders
Heritable information, 8
Heritable variations. *See* Genetic variation
Hermaphrodites, sponges as, 604
Hermaphroditism, 932–934
Herpesviruses, 350*t*
cancer and, 353
reproduction of, 351
Hershey, A., 302, 303*f*
Hershey-Chase experiment, 302, 303*f*
Heterochromatin, 375
Heterochrony, 481, 482
Heterogeneous nuclear RNA (hnRNA), 333
Heteromorphic generations, 550, 560
Heterosporous plants, 567
angiosperms as, 576
Heterotrophic nutrition, 199–200. *See also* Animal nutrition
and food for fabrication, 811
in fungi, 583–584
Heterozygote advantage, genetic variation and, 448–449
Heterozygous, 262
Hexose sugars, 67*f*
Hfr cell, 359, 360*f*
HGH. *See* Human growth hormone
Hibernation, 902–903
High-density lipoproteins, 834
Hindbrain (rhombencephalon), 1005, 1006
Hippocampus, in memory pathway, 1010
Hirudinea, 617
His. *See* Histidine
Histamine, in inflammatory response, 853
Histidine, 75*f*
Histones, 373, 374*f*
multigene families and, 375
HIV (human immunodeficiency virus), 351–352, 870–872. *See also* AIDS
DNA technology in identification of, 405
life cycle of, 353*f*
structure of, 353*f*
HIV-positive condition, 872

hnRNA. *See* Heterogeneous nuclear RNA
Holdfast, of seaweed, 549
Holoblastic cleavage, 965
Holothuroidea (sea cucumbers), 627, 628*f*, 629
Holotrophs, animals as, 794
Homeoboxes, 976–977
Homeodomains, 977
Homeostasis, 790, 876, 877*f*, 1076–1077
antagonistic hormones in, 909
as property of life, 6*f*
Homeotic changes, in evolution, 481, 482
Homeotic genes, in pattern formation, 976–977
Homo erectus, 663
Homo habilis, 663
Homologous chromosomes (homologues), 246
Homologous structures, as evidence of evolution, 433
Homology
analogy differentiated from, 491–492
as evidence of evolution, 433
Homo sapiens. *See also* Humans
emergence of, 663–664
Homosporous plants, 567
Homozygous, 262
Honeybees
communication systems in, 1182–1183
flight in, 621
Hooke, R., 7, 117
Horizons, soil, 722
Horizontal cells, in retina, 1024, 1025*f*
Horizontal transmission, of plant viruses, 354
Hormonal proteins, functions of, 76*t*
Hormone receptor, 909. *See also* Receptor proteins
Hormones, 908–909. *See also specific type*
adrenal, 919*t*, 925–926
antagonistic, in homeostasis, 909
hypothalamic, 917–921
hypothalamic releasing, 921
invertebrate, 915–917
mammalian reproduction controlled by, 940–945
mechanisms of action of, at cellular level, 910–915
pancreatic, 919*t*, 923–924
parathyroid, 919*t*, 922–923
pituitary, 917–921, 919*t*
plant, 757–765. *See also specific type*
flowering and, 771
production of
by gonads, 919*t*, 926
by pineal gland, 919*t*
by thymus gland, 919*t*
thyroid, 919*t*, 922
tropic, 917
Hornwarts (Anthocerophyta), 562*t*, 563
Horses, as example of branched evolution, 482, 483*f*
Horseshoe crab, 619
Horsetails (Sphenophyta), 562*t*, 567, 568*f*
Host, in symbiosis, 1116
Host range of virus, 347
HTLV-1, cancer and, 353
Hubbard Brook Experimental Forest, 1147–1148
Human chorionic gonadotropin, 950
Human gene therapy, RNA technology in, 407–408
Human Genome Project, DNA technology in, 404–405
Human growth hormone, DNA technology in production of, 409
Human immunodeficiency virus (HIV), 351–352, 870–872. *See also* Acquired immunodeficiency syndrome (AIDS)
DNA technology in identification of, 405
life cycle of, 353*f*
structure of, 353*f*
Human pedigrees, 272–273
Humans. *See also Homo sapiens*
biodiversity threatened by, 1146–1147
brain in, 1006–1008
communities disturbed by, 1123, 1146–1147
diseases caused by nondisjunction and translocation in, 294–295
evolution of, 657–666
emergence and, 660–666
misconceptions about, 660
hearing and equilibrium in, 1027–1030
intrusions of into ecosystems, 1132*f*, 1145–1153
Mendelian inheritance in, 272–277
nervous system of, 983*f*
olfaction in, 1034*f*
population growth of, 1083*f*, 1101–1104
reproduction in
anatomy of, 937–940
hormonal control of, 940–945
physiology of, 947
Human sexual physiology, 947
Human sociobiology, 1186–1187
Humoral immunity, 855, 860–863, 867*f*
effector mechanisms of, 862
Humus, 723
in savanna, 1057
Huntington's disease
inheritance pattern of, 275
RFLP analysis in diagnosis of, 406–407
Hutton, J., 423
Hybrid breakdown, as reproductive barrier, 460*t*, 461
Hybrid inviability, as reproductive barrier, 460*t*, 461
Hybridization, 260
Hybridomas, 863, 864
Hybrid sterility, as reproductive barrier, 460*t*, 461
Hydra, 604–606
nervous system of, 999–1000, 1001*f*
Hydrocarbons
carbon skeleton diversity and, 56, 57*f*
fat characteristics and, 57*f*
Hydrochloric acid, in gastric juice, 801–802
Hydrogen, 26
covalent bonds of, 33
electron configuration of, 32*f*
formation of, 33*f*
plant requirements for, 721*t*
valence for, 55*f*
Hydrogen bonds, 36
Hydrologic cycle, 1141
Hydrolysis
of ATP, 98–99
enzymatic, in digestion, 796
glucose drawn from starch by, 68–69
polymers disassembled by, 65–66, 66*f*
Hydronium ion, 47
formation of, 47*f*
Hydrophilic substance, 46
phospholipid head as, 74, 152
Hydrophobic interaction, tertiary structure of proteins and, 80–81
Hydrophobic substance, 46
lipids as, 71
phospholipid tails as, 73–74, 152
Hydroponic culture, 719, 720
and regulation of nutrients, 722
Hydrostatic skeletons, 1035–1036, 1036*f*
Hydroxide ion, 47
Hydroxyl group, 58, 60*t*
Hydrozoa, 606
life cycle of, 606, 607*f*
Hymen, 940
Hymenoptera, 622*t*
Hyperosmotic solution, 160
Hyperpolarization, 989, 990*f*
Hypersensitivity, in allergies, 869
Hypertension, 834
Hyphae
in Chytridiomycota, 555*f*
fungal, 584, 585*f*
Hypocotyl, 741, 743, 744
Hypoosmotic solution, 160
Hypothalamic releasing hormones, 921
Hypothalamus, 917–918, 1007
hormones produced by, 919*t*, 920*f*, 921
Hypotheses, in scientific process, 15–19
Hypothetico-deductive method, 15–19
H zone, 1039

I bands, 1039
Ice, 44–45
floating, in fitness of environment, 45, 46*t*
structure of, 44*f*
Idealism, evolution and, 422
IgA, 858*f*
IgD, 858*f*
IgE, 858*f*
in allergies, 869
IgG, 858*f*
IgM, 858*f*
Igs. *See* Immunoglobulins
Ile. *See* Isoleucine
Imaginal disks, 975–976
Imbibition, 743
Immune adherence, 866–867
Immune response. *See also* Immune system
primary, 859
secondary, 859
Immune system, 854–865. *See also* Immunity
basic features of, 854–855
disorders of, 869–872
in invertebrates, 872
nervous system interconnection and, 870
Immunity, 867*f*
acquired, 855
active vs. passive, 855
cell-mediated, 855, 863–865, 867*f*
humoral (antibody-mediated), 855, 860–863, 867*f*
effector mechanisms of, 862
in invertebrates, 872
Immunization, 855
Immunodeficiency, 869–870
acquired, 870, 870–872
Immunoglobulins, 857, 858*f*. *See also specific type under Ig*
genes for, DNA rearrangement and, 379, 380*f*
Immunology, reproductive, 951
Immunotoxins, 863
Imperfect flowers, 736
Imperfect fungi, 591
Implantation, 948
Imprinting, parental (genomic), 295–296
Imprinting behavior, 1166–1168
and song development in birds, 1168–1169
Imprinting stimulus, 1167
Inborn errors of metabolism, as evidence that genes specify proteins, 316
Inbreeding, microevolution and, 445
Incest, human avoidance of, 1187
Inclusive fitness, 1184
Incomplete dominance, 267
Incomplete flowers, 736
Incomplete metamorphosis, in insects, 624
Incus (anvil), 1027
Independent assortment, Mendel's law of, 265–267
genetic variation and, 253–254
and recombination of unlinked genes, 284–285
Indeterminate cleavage, 602
Indeterminate growth, 686
Index fossils, 477
Induced fit, between enzyme and substrate, 102, 103*f*
Inducible enzymes, 366–367
Induction, 974
in plant cells, 775
Inert elements, 32
Infection, viral, 346–347
in animals, 352–353
Inferior colliculi, 1007
Inflammatory response, in body defense, 852–854
Influenza viruses, 346*f*
Information, heritable, 8
Ingestion, animal nutrition and, 598
Inheritance, 244–245, 258–279, 280–299. *See also* Genes
chromosomal basis of, 280–299
alterations and, 290–295
recombination and, 284–285
sex chromosomes and, 288–290
theory underlying, 280–284
cytoplasmic, 296–297
extranuclear, 296–297
and genetic maps based on crossover data, 285–288
Mendelian, 258–279
chromosomal basis of, 281*f*
and Darwin's theory of natural selection, 438–439
extension of concepts of, 267–272
in humans, 272–277
and integration of heredity and variation, 272
molecular biology and, 316
molecular basis of, 300–315. *See also* DNA
and parental imprinting of genes, 295–296
polygenic, 270, 271*f*
probability rules and, 263–264
rule of addition in, 264

Inheritance *(cont.)*
rule of multiplication in, 264
sex-linked, 283*f*, 288–290
statistical nature of, 264–265
Inherited disorders. *See* Genetic disorders
Inhibition
feedback (negative feedback), 790
in control of cellular respiration, 194
of metabolism, 106–107
succession and, 1123
Inhibitors, enzyme, 105
competitive, 105
noncompetitive, 105
Inhibitory postsynaptic potential, 995
Initials, 686
fusiform, 693–694
ray, 693
Initiation
of transcription, 323–324
of translation, 327–328
Initiation factors, 327
Initiator tRNA, 327, 328*f*
Innate behaviors, 1162–1165
Inner cell mass, 967
Inner ear, 1027, 1027–1028
balance apparatus in, 1029
Inosine, in tRNAs, 326
Inositol triphosphate, as second messenger, 915, 916*f*
Insecta, 620. *See also* Insects
Insectivora, 658*t*
Insects, 618, 620–625
anatomy of, 624
ears of, 1032
endocrine systems in, 916
flight in, 621
hormonal regulation of development of, 916–917
malpighian tubules in, 624, 882–883
and regulatory role of steroids in gene expression, 383
reproductive systems of, 935–936
respiratory adaptations in, 838
Insertion mutations, 336–338
Insertion sequences, 361–362
Insight learning, 1170, 1171*f*
Instinct, vs. learning, in behavior, 1165–1166
Insulin, 919*t*, 923
amino acid sequence of, discovery of, 79
carbon atoms in, 53*f*
DNA technology in production of, 399, 408–409
synthesis of genes for, 399
Insulin-dependent diabetes, 923–924
Integral proteins, 156–157
Integration
of nerve impulses, 1008–1011
sensory, 1016–1017
visual, 1024–1025, 1026*f*
Intercalary meristems, 689
Intercalated discs, 787, 1044
Intercellular junctions, 144, 145*f*
Interference competition, 1112
Interferons, in body defense, 852
Interleukin-1, 863
Interleukin-2, 863
Intermediate disturbance hypothesis, 1124
Intermediate filaments, 142–143
structure and function of, 138*f*
Internal environment, animal, 789–791
control of, 876–906. *See also specific mechanism*
interactions of systems in, 903
Internal fertilization, 934–935
Interneurons, 985
Internodes, of plant stem, 677*f*, 678*f*, 679
Interoreceptors, 1017
Interphase, between mitotic divisions, 224, 225
in plant cell, 231*f*
Interphase I, before meiosis I, 250*f*
Interspecific competition, 1112–1116
Interspecific interactions, 1108–1118. *See also specific type*
Interstitial cells, testicular, 937
Interstitial fluid, 789
Intertidal zone, 1066, 1067
evolutionary adaptation of seaweed to life in, 548–549
Intracellular digestion, in food vacuoles, 796–797
Intraspecific competition, 1096
Intrauterine device, for contraception, 951–952
Intrinsic rate of increase, 1093, 1094*f*
Introgression, 461–462
Introns, 333, 376, 377*f*
functional and evolutionary importance of, 335–336
Invagination, as part of gastrulation process, 962
Inversion, 292
Invertebrates, 598–634. *See also specific type*
equilibrium in, 1031–1032
internal transport in, 819–820
nervous systems of, 999–1001
reproductive systems of, 935–936
thermoregulation in, 901–902
vision in, 1019–1020
In vitro fertilization, 953
Involuntary muscles, 787
Involution, 963
Iodine, human requirements for, 814*t*, 815
Ion channels
gated, 989
in action potential, 989–991, 992*t*
membrane permeability affected by, 988
selective, in plant cell transport, 700
voltage-gated, 989–990
Ionic bonds, 34
electron transfer and, 35*f*
Ionic compounds (salts), 35
Ions, 35
Ion transport, 166
IP_3. *See* Inositol triphosphate
IPSP. *See* Inhibitory postsynaptic potential
Iris, of eye, 1022
Iron
in hemoglobin, 843–844
human requirements for, 814*t*
plant requirements for, 721*t*
Iron-sulfur protein, in electron transport chain, 184, 185*f*
Irrigation, in soil management, 724–725
Island biogeography, 1126–1128
Island chains, adaptive radiation on, in allopatric speciation, 463–465
Islets of Langerhans, 923
Isobutane, carbon skeleton diversity and, 57*f*
Isogamy, 547
Isoleucine, 75*f*
Isomers
carbon skeleton diversity and, 56–58
geometric, 57, 58*f*
structural, 56–57, 58*f*
types of, 58*f*
Isomorphic generations, 550
Isomotic solution, 160
Isopods, 626
Isoptera, 623*t*
Isotopes, 28–30
radioactive, 28
in absolute dating, 477, 479
harmful effects of, 30
use of in biology, 28, 29
IUD. *See* Intrauterine device
Ivanowsky, D., 345

Jacob, F., 364
Java Man, 663
Jawless vertebrates (Agnatha), 638*t*, 639*f*, 640
Jaws, vertebrate, evolution of, 641
Jellyfishes, 604–606
Jenner, E., 352
JGA. *See* Juxtaglomerular apparatus
JH. *See* Juvenile hormone
Joints, in vertebrate skeleton, 1037, 1038*f*
Judeo-Christian culture, evolution and, 422
"Jumping genes," 361. *See also* Transposons
Juvenile hormone, 916–917
Juvenile-onset diabetes, 923–924
Juxtaglomerular apparatus, in kidney regulation, 891, 892*f*
Juxtamedullary nephrons, 884*f*, 885

K. *See* Carrying capacity
K. See Potassium
Kangaroos, 655*f*, 656
Karyogamy
in ascomycetes, 588–589
in fungi, 585, 586*f*
Karyotype, 246
preparation of, 247
kcal. *See* Kilocalorie
Kelps, 549–550
Keratin, in scales of reptiles, 648
Ketones, 58, 60*t*
Ketoses (ketone sugars), 66, 67*f*
hexose as, 66, 67*f*
Keystone predator, 1120
Kidneys, 883–893
comparative physiology of, 893
regulation of, 891–893
water conserved by, 889–891
Killer cells, natural, in nonspecific body defense, 852
Kilocalorie, 43, 809
Kinesis, 1172
Kinetic energy, 92, 93*f*
definition of, 92
heat and, 42
temperature and, 42
Kinetochore, 225, 226*f*, 228*f*
Kinetochore microtubules, 225, 226*f*
chromosome separation during anaphase and, 225, 229*f*
King, T., 975
Kingdoms of life, 11, 12*f*, 489, 512–513
Kin selection, 1184
Klinefelter syndrome, 294
Koalas, 656
Koch, R., 529
Koch's postulates, 529
Kolbe, H., 54
Krause's end-bulbs, 897, 1017*f*, 1019
Krebs, H., 182
Krebs cycle, 179, 182–184
Krill, 626
K-selected species, 1095
succession and, 1123
Kupffer's cells, 852
Kurosawa, E., 762
Kwashiorkor, 812

Labia, 939
majora, 940
minora, 940
Labor, of birth, 950
lac operon, 366, 367*f*
Lactase, in digestion, 803
deficiency of, 803
Lactation, 950–951
Lacteal, 805, 806*f*
Lactic acid fermentation, 191
Lactiferous ducts, 940
Lactose, 67
control of synthesis of, 366–367
Lagging strand, synthesis of, 310, 311*f*
Lagomorpha, 658*t*
Lakes, 1063–1064
eutrophication of, 1064
accelerated, 1149
seasonal temperatures in, 1074, 1075*f*
Lamarck, J.P., 424
Lamarckian theory of evolution, 424
Lambda phages
as cloning vectors, 394
as episomes, 358
lysogenic and lytic cycles of, 349–350
Lamella, middle, 143
Laminaria
alternation of generations in life cycle of, 550
human use of for food, 549
Lampreys, 640
Lamp shells, 627
Lancelets (cephalochordates), 636–637, 639*f*
Langerhans' islets, 923
Langmuir, I., 152
Language, 1008*f*, 1009
Large intestine (colon), 806–807
Larva, 599
of sea urchin, 962
Larynx, 840
Lashley, K., 1010
Latency, viral, 351
Lateral geniculate nuclei, 1025, 1026*f*
Lateral inhibition, in retina, 1025
Lateralization, 1009
Lateral line system, 642, 1030–1031
Lateral meristems, 686
Lateral roots, 689

Lateral transport
in plants, 704
in roots, 704–706
Latimeria, 644
Latitude
climate affected by, 1074, 1075*f*
solar radiation and, 1071, 1072*f*
Law of continuity, 825
Law of independent assortment, Mendel's, 265–267
genetic variation and, 253–254
recombination of unlinked genes and, 284–285
Law of segregation, Mendel's, 260–263
Laws of thermodynamics, 93–95
ecological pyramids and, 1139
LDLs. *See* Low-density lipoproteins
Leading strand, synthesis of, 310, 311*f*
Leaf, plant. *See* Leaves, plant
Leaf traces, 692
Learning
associative, 1169
behavior and, 1165–1170
definition of, 1165
insight, 1170, 1171*f*
maturation differentiated from, 1166
memory and, 1009–1011
nature vs. nurture controversy in, 1165–1166
observational, 1169
by play, 1169–1170
in sea hare, 1011
trial-and-error, 1169, 1171*f*
Leaves, plant, 677*f*, 679–681, 691–692
abscission of, ethylene in, 765, 766*f*
anatomy of, 691*f*
evolutionary adaptations in for reduction of transpiration, 710–712
morphology of, 680*f*, 681
rapid movements of, 768
tissue organization of, 691–692
Leeches, 617
Leeuwenhoek, A., 7
Left brain vs. right brain, 1009
Leghemoglobin, 727–728
Lens, of eye, 1022
Lepidoptera, 623*t*
Leptospira spp, 527*t*
Leu. *See* Leucine
Leucine, 75*f*
Leucine zipper domain, gene expression affected by, 380–381, 382*f*
Leukemia, chromosomal translocations and, 294–295
Leukocytes (white blood cells), 831–832, 831*f*
development of, 832*f*
in nonspecific body defense, 851–852
LH. *See* Luteinizing hormone
Li. *See* Lithium
Library, genomic, 397
Lichens, 546, 592–593
ascomycetes in, 588, 592
chlorophytes in, 546
Liem, K., 480, 778–781, 933
Life
antiquity of, 505–506
cellular basis of, 7–8
diversity as property of, 10–11
early Earth and, 504–513
heritable information and, 8
and interaction of organisms with environment, 10
kingdoms of, 11, 12*f*, 489, 512–513
major episodes in history of, 505*f*
origin of, 506–512. *See also* Evolution
abiotic synthesis of organic compounds and, 54, 507*f*
monomers, 507–508
polymers, 508
alternative views of, 511
chemical evolution and, 37–38
formation of protobionts and, 508–509
genetic information and, 509–511
Platonic view of, 422
properties of, 4, 6*f*
Life cycles. *See also* Reproduction
human, 246–248
species variations and, 248–250
viral, 347
Life history. *See also* Reproduction
equilibrial, 1091–1092
evolution of, population dynamics and, 1089–1092
opportunistic, 1090–1091, 1091–1092
seasonal variation in, 1089, 1090*f*
variation in characteristics of, 16–17
Life tables, 1087–1089
Ligaments, 785
Ligands, 168
Ligase, DNA
in DNA replication, 311
recombinant DNA production and, 393
Light
circadian rhythms and, 1171, 1172*f*
reaction of with matter, 205
Light chains, immunoglobulin, 857
Light microscopes, 117
transmission electron microscopes compared with, 119*f*
types of, 118*t*
Light reactions in photosynthesis, 203, 204–212
and nature of sunlight, 204–205
thylakoids as site of, 203*f*, 204
Light receptors, vision of invertebrates and, 1019–1020
Lignin, 565
Limbic system, 1009
Limnetic zone, 1064
Linkage mapping
crossover data for, 285–288
of human genome, use of DNA technology in, 404
Linked genes, 283–284
recombination of, crossing over and, 285
Linnaeus, C., 422, 432, 489
Lipase, in hydrolysis of fats, 804
Lipids, 71–74. *See also specific type*
fats, 71–73
phospholipids, 73–74
steroids, 74
Lipoproteins, 805
high- and low-density, 834
Lithium, electron configuration of, 32*f*
Littoral zone, 1063–1064
Liver, in digestion, 803
Liverworts (Hepatophyta), 562*t*, 563, 565*f*
Lizards, parthenogenetic, 932, 933*f*
Lizards (Squamata), 650, 651
Loams, 723
Lobe-finned fishes (Crossopterygii), 644
Lobsters, 618, 625, 626
Local regulators, 908*f*, 909–910
Logistic population growth, 1094–1097, 1094*f*
Long-day plants, 770
Long-distance transport in plants, 704
and solar-powered bulk flow, 708
Long-term ecological research, 1148
Long-term memory, 1009
Looped domains, 373–375
Loop of Henle, 884*f*, 885
transport properties of
ascending limb, 887–889
descending limb, 887
Loose connective tissue, 784–785
Lophophorates, 626–627
Lophophore, 627
Lorenz, K., 1162, 1167
Lotka, A.J., 1113
Low-density lipoproteins, 834
LTER. *See* Long-term ecological research
"Lucy," 662–663
Lugo, A., 1048–1051, 1053
Lungfishes (Dipnoi), 644
Lungs, 838–846
ventilation of, 841–843
Luteal phase of ovarian cycle, 942, 943*f*
Luteinizing hormone, 919*t*, 920*f*, 921
in menstrual and ovarian cycle control, 942–945
in testicular control, 942*f*
Lycophyta (club mosses), 562*t*, 567
Lycopodium, 567
Lycopods, 567
Lyell, C., 423, 425
Lymph, 829
Lymphatic system, 829–830
B and T cells located in, 856
macrophages in, 852, 853*f*
Lymph nodes, 830
Lymphocytes, 831*f*, 832
development of, 832*f*, 855–856
in immune response, 855–856. *See also* B cells; T cells
Lyon, M., 290
Lys. *See* Lysine
Lysine, 75*f*
Lysogenic cycle, 349–350
Lysosomes, 131–132
formation and functions of, 132*f*
human disease and, 132
Lysozyme, 78
beta pleated sheets in, 80
in lytic cycle, 348
native conformation of, 78*f*
primary structure of, 78*f*
secondary structure of, 80*f*
Lytic cycle, 348–349

MacArthur, R., 1126, 1127
McCarty, M., 302
McClintock, B., 8, 363
MacLeod, C., 302
Macroevolution, 474–499
adaptive radiations in, 486–487
biogeography of, continental drift and, 484–486, 487*f*
definition of, 474
development and, 480–482
fossil record in, 474*f*, 475–478
mass extinctions and, 487–489
mechanisms of, 478–489
and origins of evolutionary novelties, 480–482
preadaptation in, 480
species selection hypothesis of, 482–484, 484*f*
trends in, difficulty in interpreting, 482–484
Macromolecules, 64–90. *See also specific type*
carbohydrates, 66–71
definition of, 64
DNA and products, 87–88
lipids, 71–74
nucleic acids, 83–87
polymers, 64–66
proteins, 74–83
structure and function of, 64–90
building models to study, 65*f*, 85
Macronucleus, in ciliates, 541
Macronutrients, plant requirements for, 719, 721*t*
Macrophages, 785
in B cell activation, 860, 861*f*
phagocytosis by, 852
Magnesium
deficiency of, in plants, 720, 721*f*
electron configuration of, 32*f*
human requirements for, 814*t*
plant requirements for, 721*t*
Magnetite, and orienting to magnetic field, 1174
Magnification, in microscopy, 117
Maidenhair tree (ginkgo), 562*t*, 570*f*, 571
Major histocompatibility complex, 860, 863
and tissue graft and organ transplant rejection, 869
Malaria, *Plasmodium* causing, 539, 540*f*
Male contraception, 953
Male reproductive system, human, 937–939
Male sex hormones, 940
functional groups of, 59*f*
Malignant tumor, 237. *See also* Cancer
Malleus (hammer), 1027
Malnourishment, 811
Malpighian tubules, in insects, 624, 882–883
Maltase, in digestion, 803
Malthus, R.T., 428
Maltose (malt sugar), 66–67
condensation synthesis of, 68*f*
Mammalia (mammals), 638*t*, 639*f*, 653–657. *See also specific type*
blood of, 830–832
characteristics of, 653–655
circulatory systems of, 821, 822
development of, 967–968
diving, respiratory adaptations in, 845–846
equilibrium in, 1029–1030
evolution of, 655
hearing in, 1027–1029
marine, thermoregulation in, 898–900
placental, 655*f*, 656–657, 658–659*t*

Mammalia (cont.)
reproduction in, 653, 936–953
hormonal control of, 940–945
respiratory anatomy of, 839–841
terrestrial, thermoregulation in, 896–897
Mammary glands, 653, 654f
in humans, 940
Mandibles, arthropod, 620
Manganese
human requirements for, 814t
plant requirements for, 721t
Mangold, H., 974
Mantle, 612
Mantle cavity, 612
Map units, 286
Margulis, L., 506f
Marine animals
gills in, 836–838
osmoregulation in, 878
thermoregulation in, 898–900
Marine biomes, 1066–1069. *See also specific type*
distribution of, 1063f
Mark-recapture method, population size measured by, 1084, 1085
Marsupials, 655f, 656
Marsupium, 655f, 656
Mass, 24
Mass number, 28
Mast cell degranulation, in allergic response, 869, 870f
MAT. *See* Mating-type locus
Maternal age, risk of Down syndrome and, 293–294
Mating behavior, 1178–1180
assortative, microevolution and, 445
nonrandom, microevolution and, 442t, 444–445
Mating systems, 1180
Mating-type genes, 379
Mating-type locus, 379
Matrix
extracellular, in connective tissue, 784
mitochondrial, 135
Matter, 24–26
chemical reactions changing, 36–37
definition of, 24
reactions of with light, 205
Matteuccia, 568f
Maturation
behavior changes associated with, 1166
sexual, in humans, 947
Maturation promoting factor (MPF), in cell division, 233–235, 236f
Mayer, A., 345
Mayr, E., 416–419, 427, 456, 457f
Mechanical isolation, as reproductive barrier, 460
Mechanism, theory of, 55
Mechanoreceptors, 1017–1018
Medulla, renal, 884f, 885
Medulla oblongata, 1006
Medusa, cnidarian, 604
Megapascals, water potential measured in, 701
Megaphylls, 568
conifer needles as, 571
Megaspores, 567, 738–739
Meiosis, 224, 248, 250–253. *See also* Cell division
mitosis compared with, 252f, 253
species variations in timing of, 248–249
stages of, 250f
Meiosis I, 250f, 251, 253
Meiosis II, 251, 252f, 253
Meiotic nondisjunction, 291–292
Meissner's corpuscles, 1017
Melanocyte-stimulating hormone, 920f, 921
Melatonin, 919t, 927
Membrane attack complex, 866, 868f
Membrane potential
in ion transport, 166
measuring, 987
origin of, 986–989
in plant cell transport, 700, 701f
Membrane production, rough endoplasmic reticulum and, 129
Membrane proteins, 152. *See also specific type*
drifting of, 156f
functions of, 156–157
Membranes, 151–172. *See also specific type*
carbohydrates in, in cell-cell recognition, 157–158
cell, 122, 123f. *See also* Plasma membrane
Davson-Danielli model of, 152–153
fluidity of, 155–156
fluid mosaic model of, 151, 153, 155–158
function of, 151–172
mitochondrial, 135
as mosaics, 156–158
proteins in. *See* Membrane proteins
selective permeability of, 158–159
structure of, 151–172
models of, 151–158
transport across
of large molecules, 167–170
of small molecles, 158–167
Memory, 1009–1011
immunologic, 855
cellular basis of, 859–860
model for pathway for, 1010
Memory cells, 859
Mendel, G., 258f, 259, 438
Mendelian inheritance, 258–279
chromosomal basis of, 281f
and Darwin's theory of natural selection, 438–439
extension of concepts of, 267–272
in humans, 272–277
and integration of heredity and variation, 272
molecular biology and, 316
Mendel's law of independent assortment, 265–267
genetic variation and, 253–254
recombination of unlinked genes and, 284–285
Mendel's law of segregation, 260–263
Mendel's model, 258–267
experimental approach in, 259–260
Meninges, 1002
Menstrual cycle, 941–942
in humans, 941–942, 943f
Menstrual flow phase of menstrual cycle, 942, 943f
Menstruation, 941
Mercaptoethanol, 60t
Meristems, 686
in primary growth of roots, 687–689
in primary growth of shoots, 689
Merkel's discs, 1017
Meroblastic cleavage, 965
Meselson, M., 306, 308f
Meselson-Stahl experiment, 306, 308f
Mesencephalon (midbrain), 1005, 1007
Mesenteries, 787
Mesoderm, 600, 601, 962
derivatives of in mammals, 968t
Mesohyl, 604
Mesophyll, 200, 691–692
in C_4 plants, 215
in CAM plants, 215–216, 712
unloading sugar in, 713–714
Mesozoic era, 477, 478f
Messenger RNA (mRNA), 84, 125–127, 334t
alteration of ends of, in RNA processing, 333
complementary DNA for cloning and, 394, 397–398
genetic code and, 320
in protein synthesis, 84, 86f
transcription and, 318, 322–324
regulation of degradation of, in posttranscriptional control of gene expression, 382
Met. *See* Methionine
Metabolic disequilibrium, 99–100
Metabolic diversity
origins of, 524–528
of prokaryotes, 521–522, 522–524
Metabolism, 91–111
ATP and cellular work in, 97–100
control of, 106–107
definition of, 91
disequilibrium and, 99–100
emergent properties and, 108
energy and
basic principles, 92–96
chemical, 96–97
enzymes in, 100–106
feedback inhibition and, 106–107
genetic control of, 317, 318f
heat generated by, 896
inborn errors of, as evidence that genes specify proteins, 316
overview of, 2
structural order and, 107
Metafemales (trisomy X), 294
Metamorphosis
of animal larvae, 599
in insects, 624
Metanephridia, 615, 882
Metaphase, of mitosis, 225, 227f, 252f
in plant cell, 231f
Metaphase I, of meiosis I, 250f, 252f, 253
Metaphase II, of meiosis II, 251f
Metaphase plate, 225, 227f
Metastasis, of malignant tumors, 237
Met-enkephalin, as neurotransmitter, 998t, 999
Methane, shape of, 34f, 55, 56f
Methanogens, 523
in sewage treatment, 530
Methionine, 75f
Methylation, of DNA, 378
Mg. *See* Magnesium
MHC. *See* Major histocompatibility complex
MHC-antigen complex, 863
Microspores, 567
Microbodies (peroxisomes), 133–134
Microcirculation, 827–828
Microclimate, 1074–1075
Microevolution
causes of, 442–445
gene flow and, 442t, 444
gene pool and, 439
genetic drift and, 442–444, 442t
Hardy-Weinberg theorem and, 440–442
mutation and, 442t, 444
natural selection and, 442t, 445
nonrandom mating and, 442t, 444–445
Microfibrils, 69, 70f
and direction of cell expansion, 751–752
Microfilaments, 141–142
functions of, 138f, 143f
structure and function of, 138f
Micronuclei, in ciliates, 541
Micronutrients, plant requirements for, 719, 721t
Microphylls, 568
Microscopy, 117–120
types of, 118t
Microspores, in angiosperm reproduction, 738
Microtubule-organizing center, 137, 225. *See also* Centrosome
Microtubules, 137–141
kinetochore, 225, 226f
chromosome separation during anaphase and, 225, 229f
nonkinetochore, 225, 226f
function of, 228
structure and function of, 138f
Microvilli
of proximal renal tubule, 887
small intestinal, 804–805, 806f
Micturition (urination), 883–885
Midbrain (mesencephalon), 1005, 1007
Middle ear, 1027
Middle lamella, 143, 144f
Migration, 1172–1174
Mildews, downy, 553–554
Milk ducts, 940
Miller, C.O., 761
Miller, S., 54, 507
Miller-Urey model, 507
Millipedes, 618, 620, 621f
Mimicry, as animal defense, 1111–1112, 1113f
Mineralocorticoids, 919t, 925, 926
Minerals
absorption of by plant roots, 704–706
deficiencies of, in plants, 720–722
human requirements for, 814t, 815
in soil, availability of to plants, 723–724
Minimal medium, 317, 318f
Missense mutations, 336, 337f
Mitchell, P., 188
Mites, 619
Mitochondria, 134–136
chemiosmosis in, 186
vs. chemiosmosis in chloroplasts, 211–212
as eubacteria descendants, 536
structure of, 184f
Mitochondrial DNA (mtDNA), 297
restriction mapping for comparison of, 493
Mitochondrial genes, 297
Mitochondrial matrix, 135

Mitosis, 224. *See also* Cell division
evolution of, 229–230, 232*f*
meiosis compared with, 252*f*, 253
in plant cell, 231*f*
stages of, 225, 226–227*f*
Mitotic clock, 236*f*
Mitotic spindle, structure and function of, 225–228
Mixotrophs, 534
Model, in mimicry, 1111
Modern synthesis, 439
Molarity, 47
Molds, 583, 591. *See also* Fungi
black bread, 586–587, 591
slime, 551–554
cellular (Acrasiomycota), 553, 554*t*
plasmodial (Myxomycota), 552–553, 554*t*
water, 553–554
life cycle of, 555*f*
Mole, as unit of measurement, 46
Molecular biology
certainty of paternity determined by, 1180, 1181
Mendelian genetics and, 316
signs of evolution in, 434–435
Molecular clocks, in systematics, 494–495
Molecular diversity
of carbon, 53–63
of polymers, 64–65
Molecular formula, 33
Molecular shape, biological importance of, 34
Molecular systematics, 492–495
Molecular weight, 47
Molecules
chemical bonds in formation of, 32–36
large, transport of across membranes, 167–170
small
in regulation of eukaryotic gene expression, 383–384
transport of across membranes, 158–167
Mollusca (mollusks), 600*f*, 611–615
body plan of, 612
nervous systems of, 1001
Molting, 618
hormones controlling, 916
Molybdenum, plant requirements for, 721*t*
Monera, 512*f*, 513. *See also* Bacteria; Prokaryotes
Monoclonal antibodies, 863, 864
Monocots (monocotyledons), 574, 577, 676, 677*f*
Monoculture, benefits and risks of, 748
Monocytes, 831–832, 831*f*
development of, 832*f*
phagocytosis by, 851–852
Monod, J., 364
Monoecious plants, 736
Monogamous mating, 1180
Monogenea, 609
Monogenesis model, and evolution of *Homo sapiens*, 664, 665
Monohybrid cross, 260
Monomers, 64
abiotic synthesis of, in origin of life, 507–508
joining of by condensation synthesis, 65, 66*f*
polymers disassembled to, by hydrolysis, 65–66, 66*f*
Monophyletic taxon, 491
Monosaccharides, 66
structure and classification of, 67*f*
Monosomy, 292
Monosomy X (Turner syndrome), 294
Monotremes, 655*f*, 656
Morels, 587*f*, 588
Morgan, T.H., 280
Morphogenesis, 957
cell shape changes in, 971, 973*f*
in plants
cell division and cell expansion in, 749–750
growth accompanied by, 749
Morphogenetic movements, 971–974
Morphogens, 752, 978
Morphospecies, 457
Morphs, 446
Morula, 960*f*, 961
Mosaic development, 971
Mosaics, membranes as, 156–158
Moss bog, 563
Mosses (Bryophyta), 562*t*, 563
life cycle of, 563, 564*f*, 572*t*
Motor cortex, 1007, 1008*f*
Motor molecules, 137, 138*f*
Motor neurons, 985, 1002
Motor units, 1043, 1044*f*
Mountains, climate affected by, 1074, 1075*f*
Mouth (oral cavity), 799–800
Movement
animal, 1034–1035
behavior and, 1171–1174
cost of, 1035*f*
muscles in, 1037–1045
seasonal (migration), 1172–1174
skeleton in, 1035–1037
plant, 766–769
MPF. *See* M-phase promoting factor
M phase, 224
M-phase promoting factor (MPF), in cell division, 233–235, 236*f*
mRNA. *See* Messenger RNA
MSH. *See* Melanocyte-stimulating hormone
mtDNA. *See* Mitochondrial DNA
Mucin, in saliva, 799
Mucous membranes, 784
as defense mechanism, 851
Muller, H., 338
Müllerian mimicry, 1112
Multicellularity, origins of, 554–556
Multifactorial characters, 271
Multifactorial inherited disorders, 277
Multigene families, 375–376
Multiple alleles, 269
Multiple fruit, 741, 743*f*
Multiplication rule, in inheritance, 264
Multiregional model, and evolution of *Homo sapiens*, 663–664, 665*f*
Mu phage, 362
Murmur, heart, 822–823
Muscle cell contraction, 1039–1042
actin and myosin in, 141, 1040, 1041*f*
temporal summation of, 1043*f*
Muscle fibers, fast and slow, 1043
Muscles. *See also specific type*
in animal movement, 1037–1045
skeletons and, 1039*f*
in cnidaria, 604
contraction of
graded, 1042–1043
molecular mechanism of, 1039–1040
vertebrate, 786–787
Muscle spindle (stretch receptor), 1017
Muscle tissue, 786–787
Muscular dystrophy, Duchenne's
cloning of gene for, 405
sex-linked inheritance of, 289
Mushrooms
as basidiocarps, 589
edible, 595
ring of (fairy ring), 583*f*, 589–590
Mussels, 613–614
Mutagenesis, 338
Mutagens, 338
Mutant phenotypes, 282
Mutations, 336–338
bacterial, 355–356, 521
consequences of, 337*f*
creation of (mutagenesis), 338
definition of, 336
deletion, 336–338
in evolution, 435, 439
frameshift, 337*f*, 338
genetic variation and, 447–448
insertion, 336–338
microevolution caused by, 442*t*, 444
missense, 336, 337*f*
nonsense, 336, 337*f*
proteins affected by, 336–338
in speciation, 469
spontaneous, 338
substitution, 336
types of, 336–338
Mutualism, 528, 1116, 1117–1118
in fungi, 584
in protists, 534
Mycelium, fungal, 584
Mycobacterium spp, 526*t*
Mycoplasma, 527*t*
size of, 121
Mycoplasma spp, 527*t*
Mycorrhizae, 586, 593, 594*f*, 731
sheathing, 594*f*
Myelin sheath, 984, 985–986
Myeloma cells, in hybridomas, 863
Myofibrils, 1038
Myofilaments, 1038
Myoglobin, 1043
Myosin, 141
in muscle contraction, 1040, 1041*f*
Myotonia, in sexual response, 947
Myxobacteria, 527*t*
Myxococcus spp, 527*t*
Myxomycota (plasmodial slime molds), 552–553, 554*t*

N. *See* Nitrogen
Na. *See* Sodium
NaCl. *See* Sodium chloride
NAD^+
in fermentation, 190–191
in glycolysis, 179, 181*f*
in Krebs cycle, 182*f*, 183*f*
reduction of in cellular respiration, 176, 177*f*
NADH
and electron transfer to oxygen, 177–178
in fermentation, 190–191
formation of, 176, 177*f*
in glycolysis, 179, 181*f*
in Krebs cycle, 182*f*, 183*f*
$NADP^+$, in photosynthesis, 203
NADPH, in Calvin cycle, 203–204, 212–214
Na^+–K^+ ATPase (Na^+–K^+ pump, sodium-potassium pump), 164–166. *See also* Proton pump
Native conformation, protein, 78
Natural killer cells, in nonspecific body defense, 852
Natural selection, 14, 420, 426, 427–431, 450–453. *See also* Evolution; Fitness
adaptation and, 427–431
behavior and, 1158*f*, 1159, 1160*f*
altruistic, 1184–1185
directional, 451
diversifying, 451
examples of, 429–431
experimental evidence for, 17–18
heritable variations and, 255–256
host-parasite relationship affected by, 1116–1117
Mendelian inheritance as explanation for, 438–439
microevolution caused by, 442*t*, 445
modern acceptance of, 438–439
modes of, 451–452
perfection and, 453
sexual selection and, 452–453, 1179
stabilizing, 451, 452*f*
subtleties of, 428–429
Natural theology, evolution and, 422
Nature vs. nurture controversy, 1165–1166
Nautilus, chambered, 614
Navigation, in migration, 1172–1174
Ne. *See* Neon
Nealson, K., 506*f*
Neanderthals, 663
Negative feedback (feedback inhibition), 790
in control of cellular respiration, 194
of metabolism, 106–107
Negative gene regulation, 366–367
Negative pressure breathing, 841
Neher, E., 711
Nematocysts, 605
Nematoda (nematodes, roundworms), 600*f*, 610–611
Nemertea (ribbon worms, proboscis worms), 600*f*, 609–610
Neo-Darwinism, 439
Neon, electron configuration of, 32*f*
Neopillina, 612
Nephron
physiology of, 885–887
structure of, 884*f*, 885
Neritic zone, 1066, 1069
Nerve cells. *See* Neurons
Nerve growth factor, 910
Nerve net, cnidarian, 999
Nerve processes, 787
Nervous system (nerves), 907, 982–1014. *See also* Neurons
cells of, 983–986

Nervous system *(cont.)*
central, 1002–1008
neuron cell bodies in, 985
in cephalopods, 614
in cnidarians, 605, 999–1000
endocrine glands and, 927–928
human, 983*f*
immune system interconnection with, 879
in insects, 624
in invertebrates, 999–1001
peripheral, 1002
in vertebrates, 1001–1011
divisions of, 1002*f*. *See also specific division*
Nervous tissue, 787
Net primary productivity, 1136, 1139
Neural crest, 964
Neural integration, at cellular level, 995–997
Neural pathway, 985
Neural tube, 963
Neurohypophysis (posterior pituitary), 918
hormones produced by, 918, 919*t*, 920*f*
Neuromasts, 1031
Neurons (nerve cells), 787, 983–985
action potential in, 989–991
propagation of, 989–991
speed of, 992–993
diversity of shape of, 985*f*
membrane potential of, 986–989
motor, 985
sensory, 985
structure of, 787, 984*f*
transmission of electrical signals along, 986–993
Neuropeptides. *See also* Peptide hormones
as neurotransmitters, 998*t*, 999
Neurosecretory cells, 918
Neurotransmitters, 994, 997–999. *See also specific type*
as local regulators, 908*f*, 909–910
Neutralization, as effector mechanism in humoral immunity, 862
Neutral theory of molecular evolution, 450
Neutral variation, 450
Neutrons, 27
Neutrophils, 831–832, 831*f*
development of, 832*f*
phagocytosis by, 851
Newborn screening, for diagnosis of genetic disorders, 277
NGF. *See* Nerve growth factor
Niacin, 813*t*
Niche, ecological, 1114
Nickel, plant requirements for, 721*t*
Nicolson, G., 153
Nicotinamide adenine dinucleotide. *See* NAD$^+$
Nicotinamide adenine dinucleotide phosphate. *See* NADP$^+$
Night length (darkness)
circadian rhythms and, 1171, 1172*f*
photoperiodism and, 770–771
phytochrome affecting, 771–773
Night vision, rods in, 1022
Nirenberg, M., 320
Nitrification, 1144
Nitrobacter spp, 526*t*
Nitrogen, 26
electron configuration of, 32*f*
plant assimilation of, 726–729
plant requirements for, 721*t*
valence of, 55*f*
Nitrogenase, 726
Nitrogen cycle, 1143–1144
Nitrogen fixation, 522, 527*t*, 726, 1143
DNA technology affecting capability for, 411–412
in root nodules, 727
symbiotic, 727–728
Nitrogenous wastes, 893–894
Nitrosomonas spp, 526*t*
Nociceptors, 1017*f*, 1019
Nodes, of plant stem, 677*f*, 678*f*, 679
Nodes of Ranvier, 993
Nodules, root, nitrogen-fixing bacteria in, 727, 1143
Nomarski microscopy (differential-interference-contrast microscopy), 118*t*
Noncompetitive inhibitors, enzyme, 105
Noncyclic electron flow, in light reactions, 210–211
Noncyclic photophosphorylation, 211
Nondisjunction, 291–292
human diseases caused by, 294
Nonequilibrial model of community development, 1124
Non-insulin dependent diabetes, 923–924
Nonkinetochore microtubules, 225, 226*f*
function of, 228
Nonpolar covalent bonds, 34
Nonrandom mating, microevolution caused by, 442*t*, 444–445
Nonsense mutations, 336, 337*f*
Nonshivering thermogenesis, 896
Nonvascular plants, 562–563, 562*t*
Norepinephrine, 919*t*, 925, 998–999, 998*t*
Norm of reaction, 271
Nostoc spp, 526*t*
Notochord, 646, 963–964
induction by, 974
NPP. *See* Net primary productivity
N-terminus, of polypeptide chain, 76*f*, 77
Nuclear envelope, 123, 126*f*
Nuclear RNA
heterogeneous (hnRNA), 333
small (snRNA), 334, 334*t*, 335*f*
Nuclear transplantation, 975
Nucleases, 804
Nucleic acid probes
in basic research, 404
for gene identification, 398
Nucleic acids, 83–87. *See also* DNA; RNA
double helix and, 86–87, 88*f*
functions of, 83–84
nucleotides in, 84–86, 87*f*
polynucleotides and, 86
Nucleoid (nucleoid region), 355, 519
Nucleolar organizers, 125
Nucleolus, 123–125
Nucleosides, 86
Nucleoside triphosphates, 309, 310*f*
Nucleosome, 373, 374*f*
Nucleotide bases, pairing of in DNA double helix structure, 305–306
replication and, 306, 307*f*
Nucleotides, 84–86, 87*f*, 245
Nucleus, cell, 123–127
in nerve cell, 999
transplantation of, 975
Nutrient cycling, 1141–1142
agricultural effects on, 1148–1149
in Hubbard Brook Experimental Forest, 1147–1148
variations in duration of, 1145
Nutrients
absorption and distribution of
in animals, 804–806
in plants, 719
essential
for animals, 811–815
for plants, 719–720, 721*t*
Nutrition. *See also* Diet
animal, 794–817. *See also* Animal nutrition
autotrophic, 199–200
heterotrophic, 199–200
plant, 718–733. *See also* Plant nutrition
Nutritional diversity, in protists, 534

O. *See* Oxygen
Obelia, life cycle of, 606, 607*f*
Obesity, 811
Obligate aerobes (strict aerobes), 192, 522
Obligate anaerobes, 192, 522
Observational learning, 1169
Ocean currents, climate affected by, 1073–1074
Oceanic pelagic biome, 1066, 1068
Oceanic zone, 1066
Octopuses, 611–612, 614–615
nervous systems in, 1001
Odonata, 623*t*
Okazaki fragments, 310–311
Oleic acid, as unsaturated fatty acid, 72*f*
Olfaction, 1032–1034
Olfactory imprinting, 1167
Olfactory receptors (smell receptors), 1018, 1034
Oligochaeta, 616
Oligodendrocytes, 985
Oligomycin, ATP synthase affected by, 187
Oligotrophic lakes, 1064
Ommatidia, 1020
Omnivores, 794
Oncogenes, 354, 384
in virus-induced cancer, 386
One-gene–one-enzyme hypothesis, 317, 318*f*
One-gene–one-polypeptide hypothesis, 317, 338
Ontogeny, as recapitulation of phylogeny, 434
Oogamy, 547
Oogenesis, 946–947
Oomycota (water molds), 553–554, 554*t*
life cycle of, 555*f*
Oparin, A.I., 507, 509
Oparin-Haldane hypothesis, 507
Open circulatory system. *See* Circulatory system, open
Operant conditioning, 1169
Operator, in operon, 364–365
Operculum, 643
gills protected by, 837
Operons, 364–366, 378
Ophiuroidea (brittle stars), 627, 628
Opossums, 656
Opportunistic bacteria, 529
Opportunistic life history, 1090–1091, 1091–1092
Opsins, retinal bonded to, 1024
Opsonization, 866–867
Optic chiasm, 1025, 1026*f*
Optimal foraging, 1174
Oral cavity, 799–800
Orbitals, electron, 31
Order, as property of life, 6*f*
Order, taxonomic, 489
Organ of Corti, 1028
Organelles, 118. *See also specific type*
Organic acids (carboxylic acids), 59, 60*t*
Organic chemistry
applications of, 54
definition of, 53, 55
foundations of, 53–55
Organic compounds, abiotic synthesis of, 54, 507*f*
monomers, 507–508
polymers, 508
Organic phosphates, 60*t*
Organismal ecology, 1053
Organogenesis, 948, 949*f*, 963–965
in mammals, 968
Organs, 787
vestigial, as evidence of evolution, 433–434
Organ systems, 787
Organ transplants, self vs. nonself and, 868–869
Orgasm, 947
Origin of life, 506–512. *See also* Evolution
abiotic synthesis of organic compounds and, 54, 507*f*
monomers, 507–508
polymers, 508
alternative views and, 511
chemical evolution and, 37–38
formation of protobionts and, 508–509
genetic information and, 509–511
On the Origin of Species by Means of Natural Selection (book), 11, 420, 421
Origins of replication, 309
Orthogenesis, 439
Orthomyxoviruses, 350*t*
Orthoptera, 623*t*
Oscillatoria spp, 526*t*
Osculum, 603, 604*f*
Osmoconformers, 877
Osmolarity, 877
Osmoreceptors, 1018
in kidney regulation, 891, 892*f*
Osmoregulation, 162, 163*f*, 877–881
effect of environment on, 878–880
transport epithelia and, 880–881
Osmoregulators, 877, 1018
Osmosis, 160–163
water potential and, in plant cells, 701–703
Osmotic pressure, 161–162
Osteichthyes (bony fishes), 638*t*, 639*f*
Osteocytes, 785–786
Ostracoderms, 640
Ostrich fern, 568*f*
Otoliths, 1029
Outer ear, 1027
Oval window, of middle ear, 1027
Ova (ovum, eggs), 932
activation of, 958–959, 960*f*

avian, 965
chromosome count of, 248
development of, 946–947
mammalian, 967
polarity of, 961
Ovarian cycle, 942, 943*f*
Ovaries
of flower, 575, 736
hormones produced by, 919*t*, 939
in humans, 939
Overnourishment, 811
Overton, C., 152
Oviduct, 940
Oviparous sharks, 642
Ovoviviparous sharks, 642
Ovulation, 939, 941, 942
Ovules, 575, 577, 736
development of, 738–739
Ovum. *See* Ova
Oxidation, beta, 193–194
Oxidation-reduction reactions (redox reactions)
introduction to, 174–175
methane combustion as, 175, 176*f*
photosynthesis as, 203
respiration as, 174–178
stepwise, 176–178
Oxidative phosphorylation, 179
and chemiosmotic coupling, 186*f*
Oxidizing agent, in oxidation-reduction reaction, 175
Oxygen, 26
covalent bond of, 33
electron configuration of, 32*f*
binding of hemoglobin to, 830–831
loading of, 830–831, 843, 844*f*
partial pressure of, 843, 844*f*
plant requirements for, 721*t*
respiratory supply of, 835
unloading of, 831, 843, 844*f*
valence for, 55*f*
Oxygen dissociation curve for hemoglobin, 844, 845*f*
Oxygen revolution, 525–528
and origins of eukaryotic life, 535
Oxytocin, 918, 919*t*, 920*f*
Oysters, 611–612, 613–614
Ozone, atmospheric, depletion of, 1152

P. *See* Phosphorus
P680, 209
P700, 209
Pacemaker, 824
contraction of, 824–825
Pacemaker channels, 1044
Pacinian corpuscles, 1017
Paedogenesis, in vertebrate evolution, 638
Paedomorphosis, 480, 481*f*
Paine, R., 1120
Pain receptors, 1019
Paleoanthropology, 660
Paleontology (paleontologists), 423, 475. *See also* Fossils
Paleozoic era, 477, 478*f*
Palmitic acid, 72*f*
Palolo worm, reproduction in, 934
Pancreas, 923–924
in digestion, 803, 804
hormones produced by, 799, 919*t*, 923–924
Pancreatic amylase, 803
Pangaea, 485, 487*f*, 504
Panspermia, 511
Pantothenic acid, 813*t*
Papilloma viruses, cancer and, 353
Papovaviruses, 350*t*
cancer and, 353
Parabronchi, 843
Paracrine signaling, 908*f*, 910
Paralytic shellfish poisoning, 544
Paramecium spp
competition between, 1113–1114
conjugation by, 542*f*
osmoregulation in, 162, 163*f*
Paramyxoviruses, 350*t*
Paraphyletic taxon, 491
Parasites, 522, 528
amoebic, 538
apicomplexan, 539
fungal, 584
plants as, 729–730
zoomastigotes, 539
Parasitism, 528, 1116–1117
Parasympathetic nervous system, 1002, 1003*f*
Parathyroid glands, 922–923
hormones produced by, 919*t*, 922–923
Parathyroid hormone, 919*t*, 922
Parazoa, 599, 600*f*, 603–604
Parenchyma cells, 682–684
Parental imprinting, 295–296
Parental investment, 1178
Parental types, 285
Parthenogenesis, 610, 932, 933*f*
Partial pressure, 843, 844*f*
Parturition (birth), 950
Parvoviruses, 350*t*
Passive immunity, 855
Passive transport, 159–160
facilitated diffusion as, 163–164
osmosis as, 160–163
in plant cells, 700–701
Patau syndrome, 294
Patch clamping, 709, 711
Patchiness of environment, 1069–1075
community characteristics determined by, 1120
Paternity, certainty of determining, 1180, 1181
Pathogenic bacteria, 529
Pathogenic fungi, 594
Pattern formation, 752, 976–979
genetic basis of, and homeotic mutations in flower development, 753–754
hierarchy of, 978–979
Pauling, L., 80, 303
Pavlov, I., 1169
PCO2 (partial pressure of carbon dioxide), 843, 844*f*
PCR. *See* Polymerase chain reaction
PDGF. *See* Platelet-derived growth factor
Peak shifts, speciation by, 468–469
"Pecking order," 1177
Pectins, in middle lamella, 143
Pedigree, 272–273
Pedipalps, 619
Pelagic zone, 1066, 1068
Penguins, 652*f*, 653
Penicillium mold, 591
Penis, 936, 937, 938–939
Pentose sugars, 67*f*
PEP. *See* Phosphoenolpyruvate
PEP carboxylase, in C_4 photosynthesis, 215
Peppered moth, as example of natural selection, 429–430
Pepsin, 801, 804
Pepsinogen, activation of, 801–802
Peptide bonds, 76*f*, 77
formation of in elongation stage of translation, 330
Peptide hormones, 909. *See also* Neuropeptides
second messengers and, 912–915
Peptidoglycan, in bacterial cell walls, 517
and antibiotic mechanism of action, 518
Peptidyl tRNA site (P site), in translation, 327, 328*f*
Perception, sensation differentiated from, 1015–1016
Perennials, 686
Perfect flowers, 736
Perforin, 865, 866*f*
Pericycle, 689
Periderm, 694
Perilymph, 1028
Periodic table, abbreviated, 31, 32*f*
Peripheral isolate, in allopatric speciation, 463
Peripheral nervous system, 1002
Peripheral proteins, 156–157
Peripheral resistance, 826*f*
Perissodactyla, 659*t*
Peritubular capillaries, 884*f*, 885
Permafrost, 1062
Permeability
of phospholipid bilayer, 158
selective, 158–159
Permian period, reptilian radiation during, 648
Peroxisomes (microbodies), 133–134
Pert, C., 20–23, 78, 870, 921
Petals, 575, 735, 736*f*
Petiole, 677*f*, 679
P generation, 260
PGs. *See* Prostaglandins
pH, 48
of acid rain, 49–50
of aqueous solutions, 49*f*
enzyme activity affected by, 104
Phaeophyta (brown algae), 543*t*, 549–550
Phage library, 397*f*
Phages (bacteriophages), 302, 346, 348–350
as cloning vectors, 394
lambda
as cloning vectors, 394
as episomes, 358
lysogenic and lytic cycles of, 349–350
mu, 362
reproduction of, 348–350
T4, lytic cycle of, 348–349
T-even
replication of, 347
structure of, 344*f*, 346
Phagocytosis, 131–132, 168, 169*f*, 851–852
in inflammatory response, 853–854
Phagotrophs, 522
Pharynx, 800
Phase-contrast microscopy, 118*t*
Phe. *See* Phenylalanine
Phenetics, 495
Phenotype
definition of, 263
environmental impact on, 270–271
genotype differentiated from, 263*f*
mutant, 282
wild-type, 282
Phenylalanine, 75*f*
Phenylketonuria
alleles for, use of Hardy-Weinberg equation in calculating, 442
cloning of gene for, 405
newborn screening for, 277
Pheromones, 909, 1182
pH gradient, across thylakoid membrane, 212
Phloem, 565, 683*f*, 685
loading and unloading of, 713–714
secondary growth in production of, 694
transport in, 704, 712–715
Phloem sap, 712
pressure flow of, 714–715
Phoronids, 627
Phosphagens, 1040
Phosphate group, 60*t*, 61
Phosphates
formation of, dissociation of phosphoric acid in, 61
organic, 60*t*
Phosphate transfer, energy coupling by, 99*f*
Phosphatidylcholine, 73*f*
Phosphodiester linkages, in polynucleotides, 86, 87*f*
Phosphoenolpyruvate, in C_4 photosynthesis, 215
Phosphofructokinase, in control of cellular respiration, 195
Phospholipid bilayer, 74, 152
permeability of, 168
Phospholipids, 73–74
function of, 74
in membranes, 152
structure of, 73*f*
Phosphoric acid, dissociation of, phosphate formed by, 61
Phosphorus, 26
electron configuration of, 32*f*
human requirements for, 814*t*
plant requirements for, 721*t*
Phosphorus cycle, 1144–1145
Phosphorylated intermediate, 99
Phosphorylation
oxidative, 179
chemiosmotic coupling and, 186*f*
substrate level, 178*f*, 179
Photic zone
freshwater, 1063
marine, 1066
Photoautotrophs, 199, 200*f*, 522, 527*t*
Photoheterotrophs, 522
Photons, 204
Photooxidation of chlorophyll, 207–208
Photoperiodism, 769–773
Photophosphorylation, 203
noncyclic, 211

Photopsins, 1024
Photoreceptors, 1018, 1022, 1023*f*
Photorespiration, 214–215
Photosynthesis, 36–37, 199–220
 C_4, 215, 216*f*, 217*f*
 Calvin cycle in, 203, 212–214
 CAM, 215–216, 217*f*, 712
 chlorophyll photooxidation in, 207–208
 chloroplasts as site of, 200–201
 electron flow in
  cyclic, 209–210
  noncyclic, 210–211
 energy budget set by, 1134
 fate of products of, 216–218
 light reactions in, 203, 204–212
 and nature of sunlight, 204–205
 origin of, 525
 overview of, 201–204
 photorespiration and, 214–215
 photosystems and, 208–209
 pigments in, 205–207. *See also* Photosynthetic pigments
 as redox process, 203
 splitting of water for, 202–203
 stages of, 203–204
 summary of, 216–218
 transpiration and, 708–709
Photosynthetic pigments, 205–207. *See also* Chlorophyll
 in algae, 542–543
 excited state of, 207
 ground state of, 207
Photosystem I, 209, 210*f*
Photosystem II, 209, 210–211, 210*f*
Photosystems, 208–209
Phototaxis, 519
Phototropism, 757–758, 766
pH scale, 48. *See also* pH
Phyla (phylum), 489
Phyletic evolution (anagenesis), 456, 457*f*
Phylloquinone (vitamin K), 813*t*, 814
Phylogenetic tree, 489, 490*f*
Phylogeny
 animal, clues to, 599–602, 603*f*
 arthropod, 618–626
 definition of, 489
 ontogeny as recapitulation of, 434
 systematics and, 489–496
Physical mutagens, 338
Physiological ecology, 1053
Phytochrome, 771–773
Phytoplankton, dinoflagellates in, 543
Picornaviruses, 350*t*
Pigments, photosynthetic, 205–207. *See also* Chlorophyll
 in algae, 542–543
 excited state of, 207
 ground state of, 207
Pili
 bacterial, 518
 sex, in conjugation, 358, 359*f*
Pilobolus, 587
Piloting, in migration, 1172
Pineal gland, 919*t*, 927
Pineapples, photosynthesis in, 215–216, 712
Pines, 570*f*, 571–572
 life cycle of, 572, 573*f*
Pinna, 1027
Pinocytosis, 168, 169*f*
Pitch, 1029
Pith, 689
Pits
 coated, 168
 in tracheids and vessel elements, 684
Pituitary gland, 918–921
 anterior lobe of, hormones produced by, 919*t*, 920*f*, 921
 hormones produced by, 920*f*
 posterior lobe of, 918
  hormones produced by, 918, 919*t*, 920*f*
PKU. *See* Phenylketonuria
Placenta, 948, 949*f*, 967
Placental mammals, 655*f*, 656–657, 658–659*t*
Placodermi (placoderms), 639, 641
Planarians, 608
 eye cup in, 1019–1020
 flame-cell system of, 881–882
Plankton
 actinopoda in, 538
 crustaceans in, 626
 protists in, 534
Plantae, 512. *See also* Plants
 divisions in, 561, 562*t*
 introduction to, 559–560
Plant biology, major themes of, 674–676
Plant body
 primary, 687
 secondary, 693
Plant cells, 125*f*, 682–685
 cytokinesis in, 228–229, 230*f*, 231*f*
 mitosis in, 231*f*
 photosynthesis in, 199–220. *See also* Photosynthesis
 signal-transduction pathways in, 773–775
 structure of, 682*f*
 types of, 682–685
 vacuoles in, 131–132
 water balance of, 162–163
Plant cell wall. *See* Cell wall
Plant fats, as unsaturated fats, 71–73
Plant hormones, 757–765. *See also specific type*
 flowering and, 771
Plant nutrition, 718–733
 adaptations for enhancement of, 729–731
 nitrogen assimilation and, 726–729
 requirement for, 718–722
 soil and, 722–725
Plants, 559–582. *See also specific type*
 anatomy of, 681–686
  definition of, 676
 carnivorous, 730–731
 cells of. *See* Plant cells
 chemical composition of, 718–719
 classification of, 561, 562*t*
 control systems in, 756–777
  effect of circadian rhythms and biological clock on, 769, 773
  hormones and, 757–765
  movements and, 766–769
  photoperiodism and, 769–773
  signal-transduction pathways and, 773–775
 day-neutral, 770
 defenses of against herbivores, 1110
 definition of, 559
 development of, 748–754
  genetic control of, 674–675
 diversity of, value of, 579–580
 in ecosystems, role of, 674
 effect of environment on form and function of, 675, 676*f*
 evolution of
  highlights of, 560–561
  and move onto land, 562–566
 flowering. *See* Angiosperms
 green algae as ancestors of, 562
 growth of, 686–687, 748–749
  effect of hormones on, 757–765
  primary, 686, 687–693
  secondary, 686, 693–696
 leaves of, 677*f*, 679–681, 691–692
  abscission of, ethylene in, 765, 766*f*
  anatomy of, 691*f*
  and evolutionary adaptations for reduction of transpiration, 710–712
  morphology of, 680*f*, 681
  rapid movements of, 768
  tissue organization of, 691–692
 life cycle of, 249, 560
 long-day, 770
 manipulating genes of, DNA technology for, 410–411
 morphology of, 676–681
 movements of, 766–769
 nitrogen assimilated by, 726–729
 nonvascular (bryophytes), 562–563
 nutrition in. *See* Plant nutrition
 parasitic, 729–730
 as producers, 1133
 reproduction in, 560, 734–755
  asexual, 744–748
  evolutionary perspective of, 748
  sexual, 734–744
 root system of, 676–679
  primary growth of, 687–689
  secondary growth of, 695–696
 seed, 562*t*, 570–579
  terrestrial adaptations of, 570–571
 seedless, 566–570
 shoot system of, 676, 679–681
  and clonal analysis of shoot apex, 752–753
  modular construction of, phase changes and, 692–693
  primary growth of, 689–693
 short-day, 770
 stems of, 679
  anatomy of, 693*f*
  primary growth of, 690*f*, 691, 695*f*
  secondary growth of, 693–695, 695*f*
 tissue systems of, 685–686. *See also specific type*
 transport in, 699–717
  at cellular level, 699–704
  long-distance
   and solar-powered bulk flow, 708
   at whole-plant level, 704
  in phloem, 704, 712–715
  short-distance (lateral), at tissue and organ level, 704
  water, 704–706
  xylem sap, 706–708
 tropisms in, 766–767
 vascular, 566
  early, 566
  terrestrial adaptations of, 563–566
 vascular system of, 565*f*
Plant viruses, 354
Planuloid, 630
Plaques, atherosclerotic, 833–834
Plasma, 830, 831*f*
Plasma cells, 856
 in hybridomas, 863
Plasma membrane (cell membrane), 122, 123*f*
 phospholipids in, 74
 sidedness of, 158*f*
 structure of, 156*f*
Plasmid library, 397*f*
Plasmids, 520
 F, in conjugation, 358, 359, 360*f*
 gene cloning in, 391, 392*f*, 394, 395*f*
 general characteristics of, 358–359
 in genetic engineering, 391, 392*f*
 R, antibiotic resistance and, 361
 Ti, in genetic engineering in plants, 410–411
 in viral evolution, 355
Plasmodesma (plasmodesmata), 144
 in lateral transport in plants, 704
Plasmodial slime molds (Myxomycota), 552–553, 554*t*
Plasmodium (feeding stage of plasmodial slime mold), 552–553
Plasmodium spp, life cycle of, 539, 540*f*
Plasmogamy
 in ascomycetes, 588
 in fungi, 585, 586*f*
Plasticity, genome, 378–379
Plastids, 135
 starch stored in, 68, 69*f*
Plateau phase, in sexual response, 947
Platelet-derived growth factor, 231
Platelets, 831*f*, 832
 development of, 832*f*
Plate movements, and continental drift in macroevolution, 485, 486*f*
Plate tectonics, 485, 486*f*
Plato, 422
Platyhelminthes (flatworms), 600*f*, 608–609
 flame-cell system of, 881–882
 nervous system of, 1000, 1001*f*
 reproductive systems of, 936, 937*f*
Platypuses, 656
Play behavior, 1169–1170
Pleated sheet, beta, 80, 81*f*
Pleiotropy, 269
Plesiomorphic character, in cladistics, 495
Pluripotent stem cells, 831, 832*f*
PO_2 (partial pressure of oxygen), 843, 844*f*
Point mutations, 336–338
 genetic variations and, 447
Poisons
 chemiosmosis and, 187–188
 in food chain, 1149–1151
Polar auxin transport, 760–761
Polar covalent bonds, 34

Polarity
of eggs, 961
of embryo, 968–969
pattern formation and, 979
Polarizing activity, zone of, 978
Polar molecule, water as, 41
Pollen, development of, 738
Pollen grains, 576–577, 735, 736, 738
Pollen sac, 738
Pollination, in angiosperms, 574, 577, 736, 739–740
animals and, 578
Polyandry, 1180
Poly-A tail, 333
Polychaeta (polychaete annelids), 616–617
reproductive systems of, 935
Polyclimax, 1124
Polygamous mating, 1180
Polygenic inheritance, 270, 271*f*
Polygyny, 1180
Polymerase chain reaction, 399–402
in genomic mapping, 405
procedure for, 401
Polymerases
DNA, 309–310, 311*f*
RNA, 322–324
binding of, in initiation of transcription, 323–324
Polymers, 64–66
abiotic synthesis of, 508
making and breaking of, 65–66
molecular diversity and, 64–65
Polymorphic population, 446
Polymorphism, 446
balanced, genetic variation and, 448–449
Polynucleotides, 86, 87*f*
Polypeptide, synthesis of. *See* Translation
Polypeptide chains, 76*f*, 77
in quaternary structure of proteins, 81
Polypeptides, functional proteins derived from, 331
Polyphyletic taxon, 491
Polyplacophora (chitons), 612
Polyploidy, 292
Polyps, cnidarian, 604
Polyribosomes, 331
Polysaccharides, 67–71
storage, 68–69
structural, 69–71
Polyspermy
fast block to, 958
slow block to, 958
Pompe's disease, 132
Ponds, 1063–1064
seasonal temperatures in, 1074
Pons, 1006
Pools of water, 1065
Population cycles, 1100–1101
Population ecology, 1053, 1083–1105
demography and, 1087–1089
density and dispersion in, 1084–1086
human population growth and, 1083*f*, 1101–1104
life history evolution and, 1089–1092
models of population growth and, 1092–1097
regulation of population size and, 1097–1101
Population genetics, 439, 439–442
Hardy-Weinberg theorem and, 440–442
microevolution and, 439
Population growth, 1092–1097
controlling, 1097–1101
density-dependent factors in, 1097–1099
density-independent factors in, 1099–1100
exponential, 1092–1093, 1094*f*, 1094*t*
human, 1083*f*, 1101–1104
interations of regulating factors in control of, 1100
logistic, 1094–1097, 1094*f*
maximum, 1093, 1094*f*
models of, 1092–1097
zero, 1093
Populations, 438–455
allopatric, 462
definition of, 439
density of, 1084, 1085
and control of population size, 1097–1099, 1100
dispersion of, 1084, 1085–1086
distribution of, 439, 440*f*
ecology of. *See* Population ecology
evolution of, 438–455
gene pool of, 439
genetics of. *See* Population genetics
genetic variation within and between, nature and extent of, 445–447
growth of. *See* Population growth
interactions among, 1108–1118. *See also specific type*
nonevolving, Hardy-Weinberg theorem and, 440–442
polymorphic, 446
size of, 1084–1086
regulation of, 1097–1101
study of (demography), 1087–1089
sympatric, 462
"Population thinking," 445
Porifera (sponges), 599, 600*f*, 603–604
mass ovulation by, 934*f*
Porphyra, human use of for food, 549
Positional information, pattern formation and, 752, 977–978
Positive feedback, 790–791
Postelsia, 549
Posterior pituitary, 918
hormones produced by, 918, 919*t*, 920*f*
Postsynaptic cell, 993
Postsynaptic membrane, 995
Postsynaptic potentials, summation of, 995–997, 997*f*
Posttranscriptional control of gene expression, 381–383
Posttranslational control of gene expression, 382–383
Potassium, 26
human requirements for, 814*t*
plant requirements for, 721*t*
in stomatal opening and closing, 709–710
Potassium channels
in action potentials, 989–990, 991*f*, 992*t*
in plant cell transport, 700
Potassium ions, membrane potential and, 987–989
Potential energy, 92, 93*f*
definition of, 30, 92
Poxviruses, 350*t*
"Practice hypothesis," for play behavior, 1169–1170
Prader-Willi syndrome, genomic imprinting and, 295, 296
Preadaptation, 480
Precambrian era, 477, 478*f*
Predation, 1106*f*, 1109–1110
community characteristics determined by, 1120
parasitism as, 1116
Predators, 1109, 1110
animal defenses against, 1110–1112
keystone, 1120
Preformation, 956
Pregnancy, 947–953
immunology of, 951
Preprophase band, 750
Prepuce, 938–939
Pressure flow, of phloem sap, 714–715
Presynaptic cell, 993
Presynaptic membrane, 994–995
Prey, 1109, 1110
Primary cell wall, 143, 144*f*
Primary consumers, 1133
Primary electron acceptor, in photooxidation of chlorophyll, 207, 208f
Primary germ layers, 962
derivatives of in mammals, 968*t*
Primary growth, 686, 687–693
Primary immune response, 859
Primary motor area, 1007, 1008*f*
Primary plant body, 687
Primary producers, 1133
Primary productivity, 1134–1138
Primary sensory area, 1007, 1008*f*
Primary structure, of protein, 78–79, 82*f*
Primary succession, 1121
Primary visual cortex, 1025, 1026*f*
Primase, in DNA replication, 311, 312*f*
Primate-rodent complex, 656
Primates, 659*t*
evolution of, 657
modern, 657–660
Primer, RNA, in DNA replication, 311, 312*f*
Primitive streak, 966
Primordia, genitalia derived from, 940, 941*f*
Principle of allocation, 1077
PRL. *See* Prolactin
Pro. *See* Proline
Probability, inheritance and, 263–264
Probes, nucleic acid
in basic research, 404
for gene identification, 398
Proboscidea, 659*t*
Proboscis worms (Nemertea, ribbon worms), 600*f*, 609–610
Procambium, root, 689
Producers, primary, 1133
Productivity
primary, 1134–1138
pyramids of, 1139
Products, in chemical reactions, 36
rate of reaction affected by concentration of, 37
Profundal zone, 1064
Progesterone, 919*t*, 926*f*
gene expression and, 911
in menstrual and ovarian cycle control, 942–945
Progestins, 926
Prokaryotes (prokaryotic cells), 7, 121, 515–532. *See also* Bacteria
cell surface of, 516–518
chemical cycles and, 528
commercial uses for, 530
disease and, 529–530
diversity of, 516, 521–522, 522–524
origins of, 524–528
form and function of, 516–522
fossils of, 505
gene exchange in, 222, 223*f*, 521
gene expression in, control of, 363–368
genesis of, and origin of life, 506–507
genome of, 519–520
organization of
at microscopic level, 373
at molecular level, 375
growth of, 520–521
importance of, 528–530
internal membranous organization in, 519, 520*f*
morphology of, 516
motility of, 518–519
in nitrogen cycle, 1143, 1144*f*
photosynthesis in, 201
protein synthesis in, 319*f*
protein synthesis in eukaryotes compared with, 332–333
reproduction of, 222, 223*f*, 520–521
RNA polymerase in, 322
symbiosis among, 528–529
transcription and translation in, 319*f*, 332–333
transcription unit in, 322
Prolactin, 919*t*, 920*f*, 921
milk production stimulated by, 950–951
Proliferative phase of menstrual cycle, 942, 943*f*
Proline, 75*f*
Prometaphase, 225, 226*f*
Promiscuous mating, 1180
Promoters
in initiation of transcription, 323–324
in operon, 365, 366
Proofreading, in DNA replication, 312–313
Pro-opiomelanocortin, 921
Propagation, vegetative. *See* Vegetative reproduction
Propanal, 59, 60*t*
Propane, carbon skeleton diversity and, 57*f*
Prophage, 350
Prophase of mitosis, 225, 226*f*, 252*f*
in plant cell, 231*f*
Prophase I, of meiosis I, 250*f*, 252*f*, 253
Prophase II, of meiosis II, 251*f*
Proplastids, 135
Prosencephalon (forebrain), 1005, 1007
Prostaglandins, 910
Prostate gland, 937
Protandrous hermaphrodite, 933
Protein clocks, in systematics, 494–495
Protein comparison, in molecular systematics, 492
Protein deficiency, 811–812
Protein folding, 83, 84*f*
Protein kinase, 914–915
MPF as, 233
Proteinoids, 508
Proteins, 74–83. *See also specific type*
amino acids of, 74–77

Proteins *(cont.)*
antimicrobial, in nonspecific body defense, 852
catabolism of, 193, 194*f*
conformation of, 77–78
factors determining, 81–83
native, 78
contractile, functions of, 76*t*
defensive, functions of, 76*t*
denaturation of, 82–83
digestion of, 804
enzymatic, functions of, 76*t*, 157*f*
evolution and, 87–88, 434–435
folding and, 83, 84*f*
functions of, 76*t*
genes specifying, 316–343
hormonal, functions of, 76*t*
improving crop yield of, 728–729
integral, 156–157
membrane, 152. *See also specific type*
drifting of, 156*f*
functions of, 156–157
mutations affecting, 336–338
native conformation of, 78
peripheral, 156–157
polypeptide chains and, 77
primary structure of, 78–79, 82*f*
quaternary structure of, 81, 82*f*
receptor
in brain, 77–78
functions of, 76*t*, 157*f*
renaturation of, 83
secondary structure of, 79–80, 82*f*
single-strand binding, in DNA replication, 312
storage, functions of, 76*t*
structural
functions of, 76*t*
silk as, 80, 81*f*
structure of, levels of, 78–81, 82*f*
synthesis of, 317–319
DNA in, 84, 86*f*
in eukaryotes, 319*f*, 332–333, 340*f*
in prokaryotes, 319*f*, 332–333
RNA in, 317–319
rough endoplasmic reticulum and, 129
taxonomic classification based on, 492
tertiary structure of, 80–81, 82*f*
transport, 158–159
in active transport, 164
in facilitated diffusion, 163–164
functions of, 76*t*, 157*f*
Protein targeting, 331–332
signal sequence for, 331, 332*f*
Protista (protists), 512f, 513, 533–558. *See also specific phyla*
characteristics of, 534
funguslike, 551–554
organisms included in, 536–537
taxonomy of, controversy about, 537
Protobionts, formation of, 508–509
Protoderm, 689
Protogynous hermaphrodite, 933
Protonephridium, 882
Proton gradient
across thylakoid membrane, 212
in plant cell transport, 700, 701*f*
Proton pump, 166, 167*f*
in phloem loading and unloading, 713–714
in plant cell transport, 700, 703–704
in polar auxin transport, 760–761
in stomatal opening and closing, 709–710
Protons, 27
Proto-oncogenes, 384
Protoplast, 682
Protoplast fusion, in test-tube cloning of plants, 747–748
Protostomes, 600f, 602, 611–626. *See also specific type*
development of, vs. deuterostomes, 601–602, 603*f*
Protozoa, 534, 537–542. *See also specific phyla*
Provirus, 351
Proximal convoluted tubule, 884*f*, 885
transport properties of, 887
Proximate causation, behavioral ecology and, 1160–1162
Pseudocoelom, 601
Pseudocoelomates, 600*f*, 601, 610–611
body plan of, 602*f*
Pseudogenes, 376
Pseudomonads, 527*t*
commercial importance of, 530
Pseudomonas spp, 527*t*
Pseudopodia, 142
of actinopoda, 538
of amoeba, 537–538
and endocytosis-exocytosis cycle in cell movement, 170
Pseudosex, in parthenogenetic reproduction, 932, 933f
Psilophyta (whiskferns), 562*t*, 566–567
Psilotum, 566
P site, in translation, 327, 328*f*
Pterophyta (ferns), 562*t*, 568
life cycle of, 568, 569*f*, 572*t*
PTH. *See* Parathyroid hormone
Puberty, 947
Puffball, 589*f*
Pulmonary circuit, in amphibians, 821
Pulse, 823–824
Punctuated equilibrium, 469, 470*f*
Punnett square, 262
Pupil of eye, 1022
Purines, 86, 87*f*
Pyramid of numbers, 1140
Pyramid of productivity, 1139
Pyridoxine (vitamin B_6), 813*t*
overdosage of, 814
Pyrimidines, 84–86, 87*f*
Pyruvate
in catabolism, 193*f*
in fermentation, 191–192
in glycolysis, 179, 181*f*
in Krebs cycle, 182
in substrate-level phosphorylation, 178*f*

Quanta, 204
Quantitative characters, 270
Quaternary structure, of proteins, 81, 82*f*
Quiescent center, in primary root growth, 688

r_{max}. *See* Intrinsic rate of increase
Radial cleavage, 602, 603*f*
Radial symmetry, 599, 600*f*, 601*f*
Radiata, 599, 600*f*, 604–608
Radiation
adaptive
on island chains, in allopatric speciation, 463–465
macroevolution and, 486–487
for heat transfer, 895
light as, 204
solar
global climate patterns and, 1071–1072
regional variation of, 1134, 1136*f*
Radicle, 743
Radioactive dating, 477, 479
Radioactive isotopes (radioisotopes), 28
in absolute dating, 477, 479
environmental threat posed by, 1150–1151
harmful effects of, 30
use of in biology, 28, 29
Radiolarians, 538
Radula, 612
Rain forests, tropical, 1056–1057
destruction of, 1057
species destruction and, 579–580
Random fertilization, genetic variation and, 255
Random population dispersion, 1085, 1086
Range, geographical
of population, 1085–1086
of species, limits of, 1125
Ranvier, nodes of, 993
Rapid eye movement (REM) sleep, 1008
Rapid leaf movements, 768
Ratites, 653
Ray-finned fishes (Actinopterygii), 643, 644
Ray initials, 693
Rays (Chondrichthyes), 638*t*, 639*f*, 641–642
Reabsorption, renal tubular, 887, 887–889
Reactants, in chemical reactions, 36
rate of reaction affected by concentration of, 37
Reaction center, for light reactions, 209
Reading frame, 321
Realized niche, 1114
Rebek, J., 511
Recapitulation theory ("ontogeny recapitulates phylogeny"), 434
Reception, sensory, 1016
Receptor-mediated endocytosis, 168, 169*f*
Receptor potential, 1016
in vision, 1024
Receptor proteins (hormone receptors), 909
in brain, 77–78
functions of, 76*t*, 157*f*
in steroid hormone action, 911–912
Receptors. *See also specific type*
in homeostatic control system, 790
in signal transduction in plant cells, 774, 775*f*
Recessive allele/recessive trait, 260, 261
inherited disorders and, 273–274
Recessive disorders, 273–274
Reciprocal altruism, 1185
Reciprocal translocation, 292–293
Recognition concept of species, 468
Recombinant DNA, 390, 391–399. *See also* DNA technology
Recombination
in bacteria, 356–363, 521
chromosomal basis of, 284–285
genetic variation and, 448
of linked genes, crossing over and, 285, 286*f*
of unlinked genes, independent assortment and, 284–285
Rectum, 807
Red algae (Rhodophyta), 543*t*, 550–551
Red blood cells (erythrocytes), 830–831, 831*f*
development of, 832*f*
Red light, photoperiodic response affected by, 771–772
Redox reactions. *See* Oxidation-reduction reactions
Red tides, dinoflagellate blooms causing, 543–544
Reducing agent, in oxidation-reduction reactions, 175
Reduction, 175
Reflex, 1002
spinal, 1004–1005, 1004*f*
Refractory period, 991
Regeneration, 932
in sponges, 604
Regulative development, 971
Regulators, 1076, 1077*f*
Regulatory genes, 365*f*, 366
Relatedness, coefficient of, 1184, 1185*f*
Relative abundance, of species within community, 1118–1119
Relative dating, 477
Relative fitness, and adaptive evolution, 450
Releaser, for fixed-action pattern, 1162
Releasing hormones, 918
REM sleep, 1008
Renal artery, 883
Renal secretion, 885–887
Renal tubules, 884*f*, 885
reabsorption by, 887
transport properties of, 887–889
Renal vein, 883
Renaturation, protein, 83
Renin, effects of on blood pressure, 891, 892*f*
Reoviruses, 350*t*
Repair, DNA, 313
Repair enzymes, 313
Repetitive sequences, 375
Replication
bacterial, 355–356, 520–521
DNA. *See* DNA, replication of
RNA, self, 509
viral, 345–346, 346–347
Replication "bubble," 309
Replication fork, 309
activities at, 310, 311*f*, 312*f*
Repressible enzymes, 366–367
Repressor, in control of gene expression, 366
Reproduction. *See also* Life cycles
in amoebas, 538
in amphibians, 647–648
in angiosperms (flowering plants), 572*t*, 576–577
in animals, 599, 931–955. *See also specific species*
asexual, 931–932
cycles and patterns of, 932–934
modes of, 931–934
sexual, 932, 934–936
in apicomplexans, 539

  in ascomycetes, 588–589
  asexual. *See* Asexual reproduction
  in basidiomycetes, 589, 590*f*
  in birds, 652
  in bony fishes, 643
  in brown algae, 550
  in cellular slime molds, 553
  in ciliates, 541–542
  in conifers, 572, 573*f*
  in diatoms, 545
  in earthworms, 616
  in ferns, 568, 569*f*, 572*t*
  in fungi, 585, 586*f*
  in green algae, 547–548
  in humans
    anatomy of, 937–940
    hormonal control of, 940–945
    physiology of, 947
  in hydrozoans, 606, 607*f*
  in insects, 624–625
  in life histories, 1089–1090
  in mammals, 653, 936–953
    hormonal control of, 940–945
  in mosses, 562–563, 564*f*, 565*f*, 572*t*
  in nematodes, 611
  in plants, 560, 734–755
    asexual, 744–748
    evolutionary perspective of, 748
    sexual, 734–744
  in plasmodial slime molds, 552–553
  population density, effect on, 1097–1098
  in prokaryotes, 520–521
  as property of life, 6*f*
  in red algae, 551
  in reptiles, 648
  in rotifers, 610
  sexual. *See* Sexual reproduction
  in sharks, 642
  in sponges, 604
  technological advances and, 953
  vegetative, 744
    in agriculture, 746–748
    natural mechanisms of, 744–746
  in vertebrates, 638–639
  in water molds, 553–554, 555*f*
Reproductive barriers, 459–462
  introgression and, 461–462
  postzygotic, 460*t*, 461
  prezygotic, 460*t*, 460–461
Reproductive immunology, 951
Reproductive systems
  diversity in, 935–936
  in humans
    female, 939–940
    male, 937–939
  in invertebrates, 935–936
  in vertebrates, 936
Reproductive technology, 953
Reptilia (reptiles), 638*t*, 639*f*, 648–651
  characteristics of, 648
  in Cretaceous period, 650
  dinosaurs, extinct ancestors of, 649–650
  extant, 650–651
  origin and early evolutionary radiation of, 648–650
  phylogeny of, 649*f*
  "stem" (Cotylosaurs), 648, 649*f*
  thermoregulation in, 900
RER. *See* Rough endoplasmic reticulum
Residual volume, 842
Resolution phase, of sexual response, 947
Resolving power, of microscope, 117
Resource partitioning, 1114–1115
Respiration, cellular. *See* Cellular respiration
Respiratory medium, 835
Respiratory organs. *See also specific type*
  adaptations of
    in aquatic animals, 836–838
    in insects, 838
    in terrestrial vertebrates, 838–846
  avian, 842–843
  in mammals
    adaptations of in diving mammals, 845–846
    anatomy of, 839–841
  structure and function of, 835–836
Respiratory pigments, 843–844
Respiratory poisons, chemiosmosis and, 187–188
Respiratory surface, 835
Resting membrane potential, 987
  basis of, 987–988
Restriction enzymes, 349, 391–393
Restriction fragment length polymorphisms. *See* RFLPs
Restriction fragments, 393
Restriction mapping, in molecular systematics, 493
Restriction point ("start"), in cell division, 233
  cancer cells and, 236
Reticular fibers, 785
Reticular formation, 1007
  arousal and, 1008
Retina, 1022, 1023*f*
  signal transduction and, 1022–1024
  visual integration in, 1024–1025, 1026*f*
Retinal, 1023*f*, 1024
Retinoic acid, as morphogen, 978
Retinol (vitamin A), 813*t*
Retroviruses, 350*t*, 351–352
  cancer and, 353
  life cycle of, 353*f*
Reverberating circuits, neural, 999, 1000*f*
Reverse transcriptase, 351
Reversibility, of chemical reactions, 37
Reznick, D., 16–18
RFLP analysis, 402–404
  in diagnosis of genetic disorders, 406–407
  in genetic mapping of human genome, 404
RFLP autoradiographs, for forensics, 409
RFLPs, 402
  DNA from different alleles distinguished by, 403*f*
R groups (side chains), of amino acids, 74
  classification according to, 75*f*, 77
  in tertiary structure of proteins, 80, 81*f*
Rhabdoviruses, 350*t*
Rh factor, and self vs. nonself, 868
Rhizobium spp
  characteristics of, 527*t*
  in root nodules, 727, 1143
  in soil enrichment, 727–728
Rhizomes, of *Psilotum*, 567
Rhizopoda, 537–538, 537*t*, 538*f*
Rhizopus stolonifer (black bread mold), 586–587, 589
Rhodophyta (red algae), 543*t*, 550–551
Rhodopsin, 1023*f*, 1024
Rhodospirillum spp, 527*t*
Rhodotorula, 592
Rhombencephalon (hindbrain), 1005, 1006
Rhynia, 566
Rhythm method, for contraception, 951
Ribbon worms (Nemertea, proboscis worms), 600*f*, 609–610
Riboflavin (vitamin B_2), 813*t*
Ribonuclease, structure of, computer model for, 85
Ribonucleic acid. *See* RNA
Ribonucleoproteins, small nuclear, 334
Ribose, 86
  structure and classification of, 67*f*
Ribosomal RNA (rRNA), 327, 334*t*
  multigene families and, 375, 376*f*
  sequencing of, in molecular systematics, 494
Ribosomes, 127–128
  anatomy of, 328*f*
  bound, 331
  free, 331
  in translation, 327, 328*f*
Ribozymes, 335
Ribulose, structure and classification of, 67*f*
Ribulose bisphosphate, in Calvin cycle, 212, 214*f*
Rickettsias, 527*t*
Riffles, shallow water and, 1065
Right brain vs. left brain, 1009
Ritual, in agonistic behavior, 1176
Rivers, 1064–1065
RNA, 83
  abiotic replication of, 509
  elongation of strand of in transcription, 324
  export of, and posttranscriptional control of gene expression, 381–382
  functions of, 84
  heterogeneous nuclear (hnRNA), 333
  messenger. *See* Messenger RNA
  processing of
    in eukaryotes, 319, 333–336
    and posttranscriptional control of gene expression, 381–382
  in protein synthesis, 317–319
  ribosomal (rRNA), 327, 334*t*
    multigene families and, 375, 376*f*
    sequencing of, in molecular systematics, 494
  self-replication of, 509
  small nuclear (snRNA), 334, 334*t*, 335*f*
  splicing of, 333–336
  synthesis of. *See* Transcription
  transfer (tRNA), 325–326, 334*t*
    structure of, 325, 326*f*
  types of, 334*t*
  in viral replication, 347
RNA polymerase binding, initiation of transcription and, 323–324
RNA polymerase II, 324
RNA polymerases, 322–324
RNA primer, in DNA replication, 311, 312*f*
RNA processing, 319, 333–336
RNA splicing, 333–336
  mechanisms of, 334–335
RNA viruses, 345, 350*t*, 351–352
  classes of, 352*f*
  structure of, 346*f*
Rocks, environmental diversity and, 1070
Rodentia, 656, 659*t*
Rods, 1022, 1023*f*
Root cap, 687
Root hairs, 678
Root nodules, nitrogen-fixing bacteria in, 727, 1143
Root pressure, in xylem sap transport, 706
Root system, 676–679
  primary growth of, 687–689
  primary tissues of, 688, 689*f*
  secondary growth of, 695–696
  water absorption of, 704–706
Rotifera (rotifers), 600*f*, 610
Rough endoplasmic reticulum, 128
  membrane production and, 129
  protein synthesis and, 129
Round window of middle ear, 1028
Roundworms (nematodes), 600*f*, 610–611
R plasmids
  antibiotic resistance and, 361
  complex transposons and, 362–363
rRNA. *See* Ribosomal RNA
RU-486, 952*f*, 953
Rubisco
  in Calvin cycle, 212, 214*f*
  in photorespiration, 214
RuBP. *See* Ribulose bisphosphate
Ruffini's end organs, 897, 1017*f*, 1019
Rugae, 801
Rule of addition, in inheritance, 264
Rule of multiplication, in inheritance, 264
Ruminants, digestive tract adaptations in, 808, 809*f*
Runs of water, 1065

S. *See* Sulfur
Saccharomyces cerevisiae, 592
  mapping genome of, 405
Saccule, 1029
Sac fungi (ascomycetes), 587–589
Sakmann, B., 711
Salamanders (Urodeles), 636
Salinity, enzyme activity affected by, 104–105
Saliva, secretion of, 799
Salivary amylase, 799
Salivary glands, 799
Salmonella spp, 526*t*
Salt. *See* Sodium; Sodium chloride
Saltatory conduction, 993
Salt concentration, enzyme activity affected by, 104–105
Salt glands, in birds, 880*f*
Salts (ionic compounds), 35
Sand dollars (Echinoidea), 627, 629
Sandwich model of membranes (Davson-Danielli model), 152–153
Sanger, F., 79, 400
Sanger method, for sequencing DNA, 400
SA node. *See* Sinoatrial node
Sap
  phloem, 712
  xylem, transport of, 706–708

Saprobes, 522
basidiomycetes as, 589
fungi as, 584, 587
water molds as, 554
Sarcomere, 1038
Sarcoplasmic reticulum, 1041, 1042*f*
Satellite DNA, 375
Saturated fats, 71–73
Saturated fatty acids, 71, 72*f*
Savannas, 1057–1058
Scale of life, evolution and, 422
Scallops, 613–614
Scanning electron microscope, 118–119
Schistosoma spp, life history of, 609
Schizocoelomates (protostomes), 600f, 602, 611–626. *See also specific type*
development of, vs. deuterostomes, 601–602, 603*f*
Schizocoelous development, 602, 603*f*
Schleiden, M., 7
Schwann, T., 7
Schwann cells, 984, 985
Science
hypotheses and deduction in, 15–19
society and, 19
technology and, 19
Scientific method, 15–19
Scion, in plant grafting, 746
Sclera, 1021
Sclereids, 683*f*, 684
Sclerenchyma cells, 683*f*, 684
Scorpions, 618, 619
Scrotum, 937
Scrubland (chaparral), 1059
Scutellum, 741
Scyphozoa, 606
Sea anemones, 604–606
asexual reproduction in, 931*f*
Sea cucumbers (Holothuroidea), 627, 628*f*, 629
Sea daisies (Concentricycloidea), 627–628
Sea hare, learning in, 1011
Sea lilies (Crinoidea), 627, 628*f*, 629
Sea palm, 549
Search image, 1174
Sea scorpions (sea spiders), 619
Seasons
carbon cycle affected by, 1142
cause of, 1072*f*
climate affected by, 1072–1075
ecosystem productivity affected by, 1136
Sea stars (Asteroidea), 627, 628
nervous systems of, 1000, 1001*f*
Sea urchins (Echinoidea), 627, 628*f*, 629
early embryonic development of, 960–965
fertilization in, 957–959, 960*f*
larva of, 962
Seaweed, evolutionary adaptations of, 548–549
Secondary cell wall, 143, 144*f*
Secondary compounds, in plant defenses against herbivores, 1110
Secondary consumers, 1133
Secondary growth, 686, 693–696
Secondary immune response, 859
Secondary plant body, 693
Secondary structure, of proteins, 79–80, 82*f*
Secondary succession, 1121
Second law of thermodynamics, 93
ecological pyramids and, 1139
Second messengers, 912–915
in signal transduction in plant cells, 774
Secretin, 803
Secretion, renal, 885–887
Secretory phase of menstrual cycle, 942, 943*f*
Sedimentary rocks, fossils in, 422–423, 475, 476*f*
Seedless plants, 562*t*, 566–570. *See also specific type*
Seed plants, 562*t*, 570–579. *See also specific type*
terrestrial adaptations of, 570–571
Seeds, 561, 577, 736, 740–741
asexual production of (apomixis), 746
dormancy of, 742–743
evolution of, 571
germination of, 742–744, 745*f*
mature, 741
Segmentation, in annelids, 617
Segmented worms (Annelida), 600*f*, 615–617
nervous systems in, 1000–1001
Segregation
as chance event, 264*f*
hypotheses for, 265*f*
Mendel's law of, 260–263
Selaginella, 567
Selective channels, in plant cell transport, 700
Selective permeability, 158–159
in plant cell transport, 700
Self-assembly, in viral replication, 347
Self-fertilization, microevolution and, 445
Self-incompatibility, 739
Selfish behavior, vs. altruistic behavior, 1183–1185
"Selfish DNA," 362. *See also* Insertion sequences
Self/nonself recognition, 855, 860, 867–869
Self-pollination, 739
Self-replication
of RNA, 509
of simple organic molecule, 511, 512*f*
Self-tolerance, 860
Semen, 937
Semicircular canals, 1029, 1030*f*
Semiconservative model, of DNA replication, 306, 307*f*, 308*f*
Semilunar valves, 822
Seminal vesicles, 937
Seminiferous tubules, 937
Senescence, in plants
effect of cytokinins on, 762
effect of ethylene on, 764–765
Sensation, perception differentiated from, 1015–1016
Sensory cortex, 1007, 1008*f*
Sensory neurons, 985, 1002
Sensory receptors, 1015–1019
general function of, 1016–1017
in skin, 1017*f*
types of, 1017–1019
Sepals, 575, 735, 736*f*
Septa, of fungal hyphae, 584, 585*f*
Sequencing, DNA, 399
of human genome, 405
by Sanger method, 400
Sequential hermaphroditism, 933–934
Ser. *See* Serine
SER. *See* Smooth endoplasmic reticulum
Seres, 1121
Serine, 75*f*
Serotonin, 998–999, 998*t*
Setae, 1032, 1033*f*
Set point, in homeostasis, 790
Severe combined immunodeficiency, 869
Sewage treatment, methanogens in, 530
Sex (gender), chromosomal basis of, 288–289
Sex chromosomes, 248, 288–290
abnormalities of number of in humans, 294*t*
nondisjunction of, 294
Sex hormones
functional groups of, male vs. female, 59*f*
gene expression in vertebrates affected by, 383–384
Sex-linked genes/sex-linked traits, 283
transmission of, 289–290
Sex-linked inheritance, 283*f*
disorders in humans transmitted by, 289–290
Sex pili, in bacterial conjugation, 358, 359*f*
Sex ratio, population size and, 1087
Sex reversal, in sequential hermaphrodites, 933–934
Sexual dimorphism, 452
Sexual imprinting, 1167–1168
Sexual life cycles. *See also* Sexual reproduction
human, 246–248
species variations and, 248–250
Sexual maturation, in humans, 947
Sexual physiology, human, 947
Sexual reproduction, 245, 246*f*
in animals, 932, 934–936
asexual reproduction compared with, 245, 748
genetic variation and, 253–255
in humans
anatomy of, 937–940
hormonal control of, 940–945
physiology of, 947
in plants, 560, 734–744, 748
Sexual selection, 452–453, 1179
and speciation by divergence, 467–468
Sharks (Chondrichthyes), 638*t*, 639*f*, 641–642
Sheldon, P., 470
Shelf fungi, 589
Shellfish poisoning, paralytic, 544
Shoot system, 676, 679–681. *See also* Leaves, plant; Stems, plant
and clonal analysis of shoot apex, 752–753
modular construction of, phase changes and, 692–693
primary growth of, 689–693
Short-day plants, 770
Short-term memory, 1009
Shrimps, 618, 626
Sickle-cell anemia
amino acid substitution causing, 79
inheritance pattern of, 274
Sickle-cell hemoglobin, primary structure of, 79*f*
Sickle-cell trait, 274
Side chains (R groups), of amino acids, 74
classification according to, 75*f*, 77
in tertiary structure of proteins, 80, 81*f*
Sieve plates, 685, 712
Sieve-tube members, 683*f*, 685, 712
loading, 713–714
Sieve tubes, 712
pressure flow in, 714–715
Signal sequence, for protein targeting, 331, 332*f*
Signal transduction
in eye, 1022–1024
in plant cells, 773–775
Sign stimulus
for fixed-action pattern, 1162
nature of, 1165
Silicon, electron configuration of, 32*f*
Silk protein, beta pleated sheets in, 80, 81*f*
Silurian period
fungi in, 595
plants in, 566
Simberloff, D., 1127
Simple epithelium, 783–784
Simple fruit, 741, 743*f*
Singer, S.J., 153
Single-lens eye, 1020
Single-strand binding proteins, in DNA replication, 312
Single-stranded DNA viruses, 350*t*
Sink, phloem sap transport to, 712–713
Sinoatrial node, 824
contraction of, 824–825
Sinuses, in open circulatory system, 819, 820*f*
Siphonaptera, 623*t*
Sirenia, 659*t*
Sister chromatids, 223*f*, 224
Skeletal muscle, 786, 787
contraction of, molecular mechanism of, 1039–1040
structure and physiology of, 1038–1043
Skeletons
in animal movement, 1035–1037
muscles and, 1039*f*
in bony fishes, 643
external
in arthropods, 618
chitin in, 71
human, 1037*f*
hydrostatic, 1035–1036, 1036*f*
types of, 1036*f*
vertebrate, 638, 785–786, 1036*f*
joints in, 1037, 1038*f*
Skin
color of in humans, polygenic inheritance of, 270, 271*f*
as defense mechanism, 851
keratinized, in reptiles, 648
sensory receptors in, 1017*f*
Skinner, B.F., 1169
"Skinner box," 1169
Skoog, F., 761
Sl. *See* Silicon
Sleep, 1008–1009
Sleep movements, of plants, 768–769
Sliding filament model, of muscle contraction, 1040
Slime molds, 551–554
cellular (Acrasiomycota), 553, 554*t*
plasmodial (Myxomycota), 552–553, 554*t*
Slow block to polyspermy, 958
Slow fibers, 1043
Slugs, 611–612, 612–613
Small intestine, 802–806
absorption and distribution of nutrients from, 804–806

digestion in, 803–804
regulation of secretion and, 803
structure of, 806*f*
Small nuclear ribonucleoproteins (snRNPs), 334
Small nuclear RNA (snRNA), 334, 334*t*, 335*f*
Smallpox vaccine, 352
Smell, 1032–1034
Smell receptors (olfactory receptors), 1018, 1034
Smooth endoplasmic reticulum, 128
functions of, 128–129
Smooth muscle, 787, 1044–1045
Snails, 611–612, 612–613
Snakes (Squamata), 650, 651
snRNPs. *See* Small nuclear ribonucleoproteins
Social behavior (social interactions), 1176–1178
in amphibians, 648
Sociobiology, human, 1186–1187
Sodium
electron configuration of, 32*f*
human requirements for, 814*t*
Sodium channels, in action potential, 989–990, 991*f*, 992*t*
Sodium chloride, 35
Sodium ions, membrane potential and, 987–989
Sodium-potassium pump, 164–166. *See also* Proton pump
Soil, 722–725
environmental diversity and, 1070
horizons (layers) of, 722
management of, 724–725
mineral content of, 723–724
texture and composition of, 722–723
water content of, 723–724
Solar radiation
global climate patterns and, 1071–1072
regional variation of, 1134, 1136*f*
Solute
concentration of, 46–47
osmosis and, 160–162
definition of, 45
transport of in plant cells, 700–701
Solution, definition of, 45
Solvent
definition of, 45
water as, 45–46, 46*t*
Somatic cells, 246
Somatic nervous system, 1002
Somites, 964
Song development in birds
imprinting and, 1168–1169
as model system for animal behavior research, 1158
Soredia, in lichens, 592, 593*f*
Source-to-sink transport, 712–713
Spatial summation, 996, 997*f*
Specialized transduction, 357, 358*f*
Speciation, 456–473. *See also* Evolution
allopatric, 462, 462–464
and adaptive radiation on island chains, 463–464
conditions favoring, 463–465
behavioral isolation and, 460
biogeography of, 462–467
by divergence, 467–468
gametic isolation and, 460–461
genetic change required for, 469
genetic mechanisms of, 467–469
gradual interpretation of, 469–471
habitat isolation and, 460
mechanical isolation and, 460
patterns of, 456, 457*f*
by peak shifts, 468–469
punctuated interpretation of, 469–471
reproductive barriers and, 459–462
sympatric, 462, 465–467
temporal isolation and, 460
Species
in binomial nomenclature, 489
biological concept of, 457
limitations of, 458–459
competition among, 1112–1116
competition within, 1096
diversity of. *See* Species diversity
exotic, introduction of by humans, 1152–1153
geographical ranges of, limits of, 1125
interactions among, 1108–1118. *See also specific type*
morphospecies concept of, 457
natural selection, effect of on, 420, 426, 427–431
origin of, 456–473. *See also* Darwinism; Evolution; Speciation
problems with concepts of, 456–459
recognition concept of, 468
relative abundance of in community, 1118–1119
richness of in community, 1118–1119
Species diversity
in community, 1118–1119
community equilibrium and, 1124
community stability and, 1124
global clines in, 1125–1126
Species selection, 482–484, 484*f*
Specific epithet, 489
Specific heat
definition of, 43
of water, 42–43, 46*t*
Specificity
of enzymes, 102
immunologic, 854
clonal selection as cellular basis of, 858, 859*f*
Spectrophotometer, for determining an absorption spectrum, 205, 206
Speech, 1008*f*, 1009
Spemann, H., 974
Spermatheca, 936
Spermatogenesis, 944–945, 944*f*
Spermatozoon (sperm cell), 932
chromosome count of, 248
structure of, 945–946
Spermicidals, 951, 952*f*
S phase, 224
Sphenophyta (horsetails), 562*t*, 567, 568*f*
Sphincters, anal, 807
Spiders, 618, 619–620
Spinal cord, 1002, 1004–1005, 1004*f*
Spinal reflexes, 1004–1005, 1004*f*
Spindle, mitotic, 225–228
Spindle fibers, 226*f*
Spiny anteaters (Echidnas), 655*f*, 656
Spiracles, 838
Spiral cleavage, 602, 603*f*
Spirilla, 516
Spirochetes, 527*t*
motility mechanism in, 519
Spirogyra spp, 546*f*, 547
Spirulina spp, 526*t*
Spliceosome, 334
Splicing, RNA, 333–336
Sponges (Porifera), 599, 600*f*, 603–604
mass ovulation by, 934*f*
Spongocoel, 603, 604*f*
Spontaneous mutations, 338
Sporangium (sporangia), 563
angiosperm, 735, 738
Spores, 249
in fungal reproduction, 585, 586*f*
Sporophylls, 567
Sporophytes, 249, 548, 560
angiosperm, 735
of *Psilotum*, 566*f*
Sporozoans. *See* Apicomplexa
Sporozoites, 539
Squamata (lizards/snakes), 650, 651
Squamous epithelial cells, 784
Squids, 611–612, 614–615
Stability, community, 1119
species diversity and, 1124
Stabilizing selection, 451, 452*f*
Stahl, F., 306, 308*f*
Stamens, 575, 735, 736*f*
Standing crop biomass, 1136
pyramids of, 1139, 1140*f*
Stanley, S., 482
Stanley, W., 345
Stapes (stirrup), 1027
Starch, 68–69
alpha glucose rings in, 69, 70*f*
catabolism of, 193
forms of, 68*f*
structure of, cellulose structure compared with, 69, 70*f*
sugar hydrolyzed from, 68–69
Statocysts, 1032
Statoliths, 1032
and gravitropism, 766, 767*f*
Stearic acid, as saturated fatty acid, 72*f*
Stele
root, 689
in xylem sap transport, 706
Stem cells
in blood cell development, 831, 832
in lymphocyte development, 855–856
"Stem reptiles" (Cotylosaurs), 648, 649*f*
Stems, plant, 679
anatomy of, 693*f*
elongation of, effect of gibberellins on, 762–763
primary growth of, 690*f*, 691, 695*f*
secondary growth of, 693–695, 695*f*
Stenohaline animals, 879
Stentor, 541
Stepwise redox reaction, 176–178
Stereoisomers, carbon skeleton diversity and, 57–58
Sterilization, 952
Steroid hormones, 74, 909, 916, 925–926
gene expression affected by, 911–912
in insects, 383
in vertebrates, 383–384
structure of, 74*f*, 926*f*
Stigma, 575
Stipe, of seaweed, 549
Stock, for plant grafting, 746
Stolons, 679
Stomach, 801–802
layers of, 787, 788*f*
Stoma (stomata), leaf, 200, 559, 691
opening and closing of, 709–710
in CAM plants, 712
in xerophytes, 710
Stone cells, 683*f*, 684
Storage diseases, 132
Storage polysaccharides, 68–69
stockpiles of, 69*f*
Storage proteins, functions of, 76*t*
Strata, in sedimentary rock, fossils in, evolution and, 422–423
Stratified epithelium, 783–784
Streams, 1064–1065
Streptomyces spp, 526*t*
Stress
adrenal hormones and, 925, 927*f*
high population density, effect of on, 1099
immunity affected by, 870
life history affected by, 1092
Stretch receptor (muscle spindle), 1017
Striated muscle, 786, 1038. *See also* Skeletal muscle
Strict aerobes (obligate aerobes), 192, 522
Stroke, 832, 833
Stroke volume, 824
Stroma, chloroplast, 135, 200
Calvin cycle in, 203*f*, 204
Stromatolites, history of life studied with, 505–506
Structural formula, 33
Structural genes, in operon, 364, 365
Structural isomers, carbon skeleton diversity and, 56–57, 58*f*
Structural order, metabolism and, 107
Structural polysaccharides, 69–71
Structural proteins
functions of, 76*t*
silk as, 80, 81*f*
Structure, relationship of to function, 9–10
Sturtevant, A.H., 286
Style, 575
Subatomic particles, 27
Substance P, as neurotransmitter, 998*t*
Substitution mutations, 336
Substrate, 102
enzyme binding to, 102, 103*f*
Substrate-feeders, 794–795. *See also specific species*
Substrate-level phosphorylation, 178*f*, 179
Subtidal zone, evolutionary adaptation of seaweed to life in, 548–549
Subunits, protein, 81
Succession, 1121–1124
Succulent plants, photosynthesis in, 215–216, 712
Sucrase, in digestion, 803
Sucrose, 67
catabolism of, 193
condensation synthesis of, 68*f*
Sugars. *See also specific type*
classification of, 66, 67*f*
disaccharide, 66–67, 68*f*
hydrolysis of from starch, 68–69

Sugars *(cont.)*
monosaccharide, 66, 67*f*
polysaccharide, 67–71
structure of, 67*f*
synthesis of, by Calvin cycle, 203–204, 212–214
fate of, 216–218
table. *See* Sucrose
Sugar sink, 713
Sulfhydryl group, 60*t*, 61
Sulfur, 26
electron configuration of, 32*f*
human requirements for, 814*t*
plant requirements for, 721*t*
Summation, 995–997, 997*f*
Sunlight
environmental diversity and, 1070
global climate patterns and, 1071–1072
nature of, 204–205
regional variation of, 1134, 1136*f*
Superior colliculi, 1007
Supernovae, 37
Supporting cells, of nervous system, 984, 985–986
Suppressor T cells, 865
Surface-receptor drugs, in disease treatment, 409
Surface tension, of water, 41, 42*f*, 46*t*
Survivorship curves, 1088–1089
Suspension-feeders, 794, 795f. *See also specific species*
Sustainable agriculture, 725
Sutherland, E.W., 912
Sutton, W.S., 280
Suzuki, D., 240–243, 408
Swallowing, 800, 801*f*
Swim bladder, 643, 1031
Switching behavior, 1098–1099
community characteristics determined by, 1120
Symbionts, 528, 1116
Symbiosis, 528–529, 1116–1118
in fungi, 584
Symmetry
bilateral, 599, 600*f*, 601*f*
radial, 599, 600*f*, 601*f*
Sympathetic nervous system, 1002, 1003*f*
Sympatric populations, 462
Sympatric speciation, 462, 465–467
Symplast, lateral transport in plants via, 704
Symport, 158, 159*f*
Synapses, 984, 993–999
chemical, 994–995
electrical, 993–994
Synapsids, 648, 649*f*
Synapsis, 252*f*, 253
Synaptic cleft, at chemical synapse, 994
Synaptic signaling, 908*f*, 909–910
Synaptic terminals, 984
Synaptic vesicles, 994
Synaptonemal complex, 254
Syncytial hypothesis, 629
Syngamy, 248, 534
in fungi, 585
Systematics, 489–496
definition of, 489
and homology vs. analogy, 491–492
molecular, 492–495
taxonomy and, 489–491, 495–496
Systemic circuit, in amphibians, 821
Systole, 822

T_C. *See* Cytotoxic T cells
T_H. *See* Helper T cells
T_S. *See* Suppressor T cells
T_3. *See* Triiodothyronine
T_4. *See* Thyroxine
T4 phage, lytic cycle of, 348–349
Table sugar. *See* Sucrose
Tadpole, 956*f*, 964
Taiga (coniferous/boreal forest), 1061
Tandem gene duplication, multigene families and, 376
Tapeworms, 609
Taproot system, 676–678
Target cells, 774, 908
Taste, 1032–1034
Taste buds, 1033–1034
Taste receptors (gustatory receptors), 1018, 1032–1034
TATA box, 324
Tatum, E., 317, 318*f*
Taxis, 519
Taxonomy, 10–11, 489–491
classical evolutionary, 496
and development of theory of evolution, 422, 432–433
schools of, 495–496
Taxon (taxa), 490, 491*f*
monophyletic, 491
paraphyletic, 491
polyphyletic, 491
Tay-Sachs disease, 132
inheritance pattern of, 273–274
T-cell receptors, 856, 863
T cells (T lymphocytes), 855–856. *See also specific type*
activation of, 863
in AIDS, 871
T-dependent antigens, 861
Teeth
evolutionary adaptations in, 807
in mammals, 655
Telencephalon, 1007
Telophase, in mitosis, 225, 227*f*
in plant cell, 231*f*
Telophase I, in meiosis I, 251*f*
Telophase II, in meiosis II, 251*f*
Temperate deciduous forests, 1060–1061
Temperate grasslands, 1059, 1060*f*
Temperate viruses, 349
as episomes, 358
Temperature, 42–43
ambient, relationship of body temperature to, 895*f*
atmospheric carbon dioxide, effect of on global, 1151
body
homeostasis in maintenance of, 790
regulation of, 894–903
environmental diversity and, 1070
enzyme activity affected by, 104
water in stabilization of, 443
Temperature acclimation, 902
Templeton, A., 467
Temporal isolation, as reproductive barrier, 460
Temporal summation, 996, 997*f*
Tendons, 785
Terminal bud, 677*f*, 678*f*, 679
in primary growth of shoots, 689, 690*f*
Termination
of transcription, 324
of translation, 330
Termination codon, 330
Terrestrial biomes, 1054–1062. *See also specific type*
distribution of, 1055*f*
Terrestrial life
animal
osmoregulation in, 879–880
respiratory adaptations for, in vertebrates, 838–846
and thermoregulation in mammals, 896–897
plant, adaptations for, 563–566, 570–571
Territoriality, 1177–1178
population growth and, 1096
Territory, 1177
Tertiary consumers, 1133
Tertiary structure, of proteins, 80–81, 82*f*
Testcross, 263
Testes (testis), 919*t*, 937
hormonal control of, 940, 942*f*
hormones produced by, 919*t*
Testosterone, 919*t*, 926, 940
functional group of, 59*f*
structure of, 926*f*
Test-tube cloning, of plants, 746–748
Tetanus, 1043
Tetraploidy, 292
Tetrapods, 639
Tetravalence, of carbon, versatility and, 55
T-even phages
replication of, 347
structure of, 344*f*, 346
Thalamus, 1007
Thalidomide, stereoisomers in teratogenic effect of, 57–58
Thaller, C., 978
Thallus, of seaweed, 549
Thecodonts, 648, 649*f*
Theology, natural, evolution and, 422
Therapsids, 648, 649*f*, 655
Thermoacidophiles, 523
Thermocline, 1063
Thermodynamics
definition of, 93
laws of, 93–95
ecological pyramids and, 1139
Thermoreceptors, 1018–1019
Thermoregulation, 894–903
behavior adaptations for, 897, 898*f*
mechanisms of, 896–897
in terrestrial mammals, 896–897
Thermostat, in thermoregulation
in humans, 899*f*
in mammals, 897
Theropods, 653
Thiamine (vitamin B_1), 813*t*
deficiency of, 812–813
Thiarubrine-A, in *Aspilla*, 580*f*
Thick(myosin) filaments, 1038, 1039
Thigmotropism, 767
Thigomomorphogenesis, 767, 773
Thimann, K., 758
Thin (actin) filaments, 1038, 1039
Thiols, 60*t*, 61
Thoracic cavity, 787
Thorn forests, tropical, 1056
Thoroughfare channels, 828
Thr. *See* Threonine
Threonine, 75*f*
Threshold, 989
Throat (pharynx), 800
Thrombus, 832
cardiovascular disease caused by, 833
Thylakoids, 135, 200–201
light reactions in, 203*f*, 204
organization of, 212, 213*f*
Thymine, 87*f*
Thymosin, 919*t*, 927
Thymus, 919*t*, 920*t*, 927
Thyroid gland, 922
hormones produced by, 919*t*, 922
Thyroid-stimulating hormone, 919*t*, 920*f*, 921
Thyroxine, 919*t*, 922
Ticks, 619
Tidal volume, 842
Tight junctions, 144, 145*f*
Time scale, geological, 477–478
Tinbergen, N., 1162, 1163
T-independent antigens, 861
Ti plasmid, in genetic engineering of plants, 410–411
Tissue grafts, self vs. nonself and, 868–869
Tissues
animal, 783–787. *See also specific type*
plant
primary
of leaves, 691–692
of roots, 688*f*, 689
of stems, 690*f*, 691
short-distance (lateral) transport at level of, 704
Tissue-specific proteins, 974–975
T lymphocytes (T cells), 855–856. *See also specific type*
activation of, 863
in AIDS, 871
TMV. *See* Tobacco mosaic virus
Toads (Anurans), 636
Tobacco mosaic disease, discovery of viruses and, 345
Tobacco mosaic virus, 345
structure of, 345, 346*f*
Tocopherol (vitamin E), 813*t*, 814
Togaviruses, 350*t*
Tolerance curves, 1078, 1079*f*
Tongue, 800
Tonoplast, 132, 133*t*
in plant cell transport, 703–704
in stomatal opening and closing, 709
Topsoil, 722, 723. *See also* Soil
Torpor, 902–903
Torsion, in gastropods, 613
Totipotent cells, 970
Touch, plant growth affected by (thigomomorphogenesis), 758
Toxins, in food chain, 1149–1151
Trace elements, 26
Trace fossils, 475

Trachea, 841
Tracheae (tracheal systems), in insects, 624, 838, 839*f*
Tracheids, 574, 683*f*, 684–685
Trait, 259
dominant, 260, 261
recessive, 260, 261
inherited disorders and, 273–274
sickle cell, 274
Transcriptase, reverse, 351
Transcription, 318–319, 322–324
coupled, 332
elongation of RNA strand in, 324
in eukaryotes, 319*f*, 332–333, 340*f*, 376, 377*f*
initiation of, RNA polymerase binding and, 323–324
in prokaryotes, 319*f*, 332–333
termination of, 324
Transcriptional control of gene expression, 380–381
Transcription factors, 324, 380
gene expression affected by, 380–381
Transcription unit, 322, 323*f*
Transduction, 357, 358*f*, 521
in plant cells, 774, 775*f*
sensory, 1016
trans face, of Golgi apparatus, 130
Transfer cells, 713
Transfer RNA (tRNA), 325–326, 334*t*
structure of, 325, 326*f*
Transformation
cancerous, 237
genetic, of bacteria, 301–302, 357, 521
Transgenic organisms, in animal husbandry, 410
Transitional endoplasmic reticulum, 129
Transition state, of chemical reactants, 101
Translation, 318–319, 325–331
aminoacyl-tRNA synthetases in, 326–327
coupled, 332
elongation stage of, 328–330
in eukaryotes, 319*f*, 332–333, 340*f*
initiation stage of, 327–328
polyribosomes in, 331
in prokaryotes, 319*f*, 332–333
ribosomes in, 327
termination of, 330
transfer RNA in, 325–326
Translational control of gene expression, 382–383
Translocation
chromosomal, 292
human diseases caused by, 294–295
reciprocal, 292–293
of food, in plants, 712–715
of tRNA on ribosome, in elongation stage of translation, 330
Transmission, by sensory receptors, 1016
Transmission electron microscope, 118–119
light microscope compared with, 119*f*
Transpiration, 691, 708–712
evolutionary adaptations in reduction of, 710–712
in long-distance transport, 704, 708
photosynthesis and, 708–709
Transpirational pull, in xylem sap transport, 707–708
Transpiration-cohesion-tension mechanism, for xylem sap transport, 706–708
Transplants, self vs. nonself and, 868–869
Transport
active, 164–167
in plant cells, 700–701
in animals, 818–819. *See also* Circulatory system
internal, in vertebrates, 819–820
ion, 166
passive, 159–160
facilitated diffusion as, 163–164
osmosis as, 160–163
in plant cells, 700–701
in plant cells, 699–704
polar auxin, 760–761
Transportation, animal. *See* Movement
Transport epithelia
osmoregulation and, 880–881
in renal tubule, 887–889
Transport proteins, 158–159
in active transport, 164
in facilitated diffusion, 163–164
functions of, 76*t*, 157*f*
in plant cell transport, 700, 703
Transposons, 361–363, 379
complex, 362–363
insertion sequences as, 361–362
in viral evolution, 355
Transverse tubules, 1042
Tree trunk, anatomy of, 695
Trematoda, 609
Treponema spp, 527*t*
Triacylglycerol, 71, 72*f*
Trial-and-error learning, 1169, 1170*f*
Trichinella spiralis, 611
Trichinosis, 611
Trichonympha, 540*f*
Trichoplax adhaerens, and hypotheses of origin of animals, 630, 631*f*
Trichoptera, 623*t*
Trihybrid cross, 267
Triiodothyronine, 919*t*, 922
Trilobites, 618–619
gradual vs. punctuated speciation in, 470–471
Trilobitomorpha, 618
Trimesters, of human gestation, 948–950
Triose sugars, 67*f*
Triplet code, 320
Triploblastic animals, 601
Triploidy, 292
Trisomy, 292
Trisomy 13 (Patau syndrome), 294
Trisomy 18 (Edwards syndrome), 294
Trisomy 21 (Down syndrome), 293
Trisomy X (metafemales), 294
Trochophore, in mollusk life cycle, 612
Trophic levels, 1132–1133
Trophic structure, 1118, 1132
Trophoblast, 967
Tropical forests, 1056–1057
deciduous, 1056
rain, 1056–1057
destruction of, 1057
species destruction and, 579–580
thorn, 1056
Tropical savannas, 1057–1058
Tropic hormones, 917
Tropisms, 766–767
Tropomyosin, 1040, 1042*f*
Troponin complex, 1040, 1042*f*
Trp. *See* Tryptophan
trp operon, 365
True-breeding, 260
Truffles, 587*f*, 588, 595
Trypanosoma spp, 539, 540*f*
Trypsin, 804
Tryptophan, 75*f*
control of synthesis of, 364–366
Tsetse fly, in *Trypanosoma* life cycle, 539, 540*f*
TSH. *See* Thyroid-stimulating hormone
Tubal ligation, 952
Tubulins, 137
Tumor, 237
Tumor-suppressor genes, 384
Tumor viruses, 353–354
Tundra, 1061–1062
Tunicates (urochordates), 637, 639*f*
cell lineage analysis in, 972*f*
Turbellaria, 608–609
Turgid cell, water balance and, 163
Turgor movements of plants, 767–769
Turner syndrome (monosomy X), 294
Turtles (Chelonia), 650, 650–651
Tympanic membrane (eardrum), 1027
"Typological thinking," 422
Tyr. *See* Tyrosine
Tyrosine, 75*f*

Ubiquinone, in electron transport chain, 184, 185*f*
Ulcers, gastric, 802
Ultimate causation, behavioral ecology and, 1160–1162
Ultrasound, 276, 953
Ulva, alternation of generations in life cycle of, 547–548
Uncouplers, proton current affected by, 187–188
Undernourishment, 811
Uniformitarianism, evolution and, 423
Uniform population dispersion, 1085–1086, 1086*f*
Uniparental disomy, genomic imprinting and, 296
Uniport, 158, 159*f*
Unique sequences, 375
Uniramia (Uniramians), 618, 620–625
Unlinked genes, recombination of, independent assortment and, 284–285
Unsaturated fats, 71–73
Unsaturated fatty acids, 71, 72*f*
Uracil, 87*f*
Urban biome, 1056
Urea
covalent bonding in, 55
excretion of, 894
Ureter, 883
Urethra, 885
male, 937
Urey, H., 507
Uric acid, excretion of, 894
Urinary bladder, 883
Urination (micturition), 883–885
Urine, 883
concentration of by kidneys, 889–891
Urochordata (urochordates/tunicates), 637, 639*f*
cell lineage analysis in, 972*f*
Urodeles (salamanders), 636
Uterus, 936, 940
Utricle, 1029

Vaccination, 855
Vaccines, 352
DNA technology in development of, 408
Vacuoles, 132–133
Vagina, 940
Val. *See* Valine
Valence, 33
for major elements of organic molecules, 55
Valence electrons, 32
Valence shell, 32
Valine, 75*f*
Valves
heart, 822
lymph vessel, 829
venous, 827
van Halmont, J-B, 718
van Leeuwenhoek, A., 7
van Niel, C.B., 202
van Overbeek, J., 761
Vaporization, heat of
definition of, 43
of water, 43–44, 46*t*
Variable regions, immunoglobulin, 857
Variation, 244, 445–450
crossing over and, 254–255
evolution and, 255–256
genetic basis of, 445–450
geographical, 446–447
independent assortment of chromosomes and, 253–254
mutation as source of, 447–448
nature and extent of, within and between populations, 445–447
neutral, 450
preservation of, 448–450
random fertilization and, 255
recombination as source of, 448
sexual sources of, 253–255
sources of, 447–448
Varmus, H., 112–115, 237
Vasa recta, 884*f*, 885
Vascular cambium, 693–694
Vascular plants, 562*t*
diversification of, 560
early, 566
seedless, 562*t*, 566–570. *See also specific type*
with seeds, 562*t*. *See also specific type*
terrestrial adaptations of, 563–566
Vascular system, water, in echinoderms, 627
Vascular tissue, plant, 685
angiosperm evolution and, 574–575
as terrestrial adaptation, 560
Vas deferens, 937
Vasectomy, 952
Vasocongestion, in sexual response, 947
Vegetal pole, 961, 969
Vegetarian diet, essential amino acids from, 812
Vegetation, animals in community determined by, 1118
Vegetative branch, 677*f*, 679

Vegetative reproduction, 744
in agriculture, 746–748
natural mechanisms of, 744–746
Vegetative shoots, 679–681
Veins, 821
blood flow in, 826*f*
structure of, 825
Ventilation
in gills, 837
in lungs, 841–843
Ventricles
in brain, 1004
in heart, 820
Venules, 821
Vernalization, 771
Vertebrae (vertebral column), 635*f*, 638
Vertebrates, 635–669. *See also specific type*
characteristics of, 638–640
circulation in, 638, 820–830
evolutionary perspective on, 821–822
classes of, 638*t*
equilibrium in, 1030–1031
jawless (Agnatha), 638*t*, 639*f*, 640
nervous system of, 1001–1011
divisions of, 1002*f*. *See also specific division*
origin of, 637–638, 639*f*
and regulatory role of steroids in gene expression, 383–384
reproductive systems of, 936
skeletons of, 638, 785–786, 1036*f*
joints in, 1037, 1038*f*
terrestrial, respiratory adaptations in, 838–846
vision in, 1020–1025, 1026*f*
Vertical transmission, of plant viruses, 354
Vesicles, 132
coated, 168
Vessel elements, 574, 683*f*, 684, 685
Vestigial organs, as evidence of evolution, 433–434
Vibrio spp, 526*t*
Vidarabine, 352–353
Villi, small intestinal, 804–805, 806*f*
Viral DNA, programming of cells by, 302
Viroids, 354
Viruses
animal, 350–354
classes of, 352*f*
reproductive cycles of, 350–352
bacterial, 348–350. *See also* Phages
cancer associated with, 353–354
oncogenes and, 386
capsids of, 345–346
classes of, 350*t*
discovery of, 344–345
DNA, 345, 350*t*
classes of, 352*f*
structure of, 346*f*
with envelopes, 346
reproduction of, 350–351
evolutionary origin of, 354–355
genomes of, 345
infections caused by, 346–347
in animals, 352–353
drugs for treatment of, 352–353
lysogenic cycle of, 349–350
lytic cycle of, 348–349
plant, 354
replication of, 345–346, 346–347
RNA, 345, 350*t*, 351–352
classes of, 352*f*
structure of, 346*f*
structure of, 345–346
temperate, 349
as episomes, 358
tumor, 353–354
Visceral mass, 612
Visceral muscle, 787
Vision, 1019–1025, 1026*f*. *See also* Eyes
in invertebrates, 1019–1020
neural pathways for, 1025, 1026*f*
in vertebrates, 1020–1025, 1026*f*
Visual cortex, primary, 1025, 1026*f*
Visual integration, 1024–1025, 1026*f*
Vital capacity, 842
Vitalism, theory of 4, 54
Vitamin A (retinol), 813*t*, 814
Vitamin B_1 (thiamine), 813*t*
deficiency of, 812–813
Vitamin B_2 (riboflavin), 813*t*
Vitamin B_6 (pyridoxine), 813*t*
overdosage of, 814
Vitamin B_{12}, 813*t*
Vitamin C (ascorbic acid), 804, 813*t*
Vitamin D, 813*t*, 814
Vitamin E (tocopherol), 813*t*, 814
Vitamin K (phylloquinone), 813*t*, 814
Vitamins, 812–815
optimal doses of, 814–815
Vitelline layer, 957, 958
Vitreous humor, 1022
Viviparous sharks, 642
Vocal cords, 840
Voltage-gated channels, 989–990
Volterra, V., 1113
Volume (loudness), hearing and, 1029
Voluntary muscles, 787
Volvox spp, 546*f*, 547
von Frisch, K., 1162, 1182

Walbot, V., 670–673
Wallace, A., 426
Warfare, environmental effects of, 1146
Warning (aposematic) coloration, as animal defense, 1111, 1112*f*
Wasps, flight in, 621
Water, 40–52. *See also* Water molecules
availability of, environmental diversity and, 1070
biomes in. *See* Freshwater biomes; Marine biomes
cohesiveness of, 41, 42*f*, 46*t*
expansion of when frozen, 44–45, 46*t*
heat of vaporization of, 43–44, 46*t*
molecular shape of, 34*f*
as plant nutrient, 719
properties of, 41–46, 46*t*
reabsorption of in large intestine, 806–807
as respiratory medium, and gills for gas exchange, 837
in soil, availability of to plants, 723–724
as solvent, 45–46, 46*t*. *See also* Aqueous solutions
specific heat of, 42–43, 46*t*
splitting, in photosynthesis, 202–203
temperature of, stratification of, 1063
temperature stabilized by, 43
transport of in plants, 699–717
and absorption by roots, 704–706
cohesion in, 41, 42*f*
Water balance
of animal cell, 162
maintenance of (osmoregulation), 877–881
of plant cell, 162–163
Water molds (Oomycota), 553–554, 554*t*
life cycle of, 555*f*
Water molecules, 34*f*. *See also* Water
dissociation of, 47
hydrogen bonds in, 40–41
polar covalent bonds in, 34
Water potential, osmosis and, in plant cells, 701–703
Water-soluble vitamins, 813–814, 813*t*
Water vascular system, in echinoderms, 627
Watson, J., 4, 304
Wavelength, 204
Weak bonds, biological importance of, 36
Weather. *See* Climate
Weberian apparatus, 1031
Weight, 24
Welwitschia, 570*f*
Went, F.W., 757, 758
Wernicke's area, 1008*f*, 1009
Whiptail lizards, parthenogenetic, 932, 933*f*
Whiskferns (Psilophyta), 562*t*, 566–567
White blood cells (leukocytes), 831–832, 831*f*
development of, 832*f*
in nonspecific body defense, 851–852
White matter, 1004
White rusts, 553–554
Whittaker, R.H., 536
Wild-type phenotype, 282
Wilkins, M., 303, 304
Wilson, E.O., 2, 1126, 1127, 1186
Wind, environmental diversity and, 1070
Wings, 652
Withdrawal (coitus interruptus), for contraception, 952
Wobble, 326
Woese, C., 523
Wogt, W., 971
Wöhler, F., 54
Wolpert, L., 977
Womb (uterus), 940
Wood, secondary growth in production of, 694
Work. *See also* Cellular work
ATP performing, 99
Worms, segmented (Annelida), 600*f*, 615–617
Wrasses, sequential hermaphriditism in, 933–934
Wright, S., 468

X chromosome, 246–248, 288–289
in fragile-X syndrome, 296
genes carried on, 282–283, 289
inactivation of (Barr bodies), 290, 291*f*
Xenopus laevis, early embryonic development of, 960–965
Xerophytes, 710
X inactivation, 290, 291*f*
X-linked characters, transmission of, 289–290
X-ray crystallography
in discovery of double helix, 304, 305*f*
in molecular computer modeling, 85
X-rays, mutagenic effects of, 338
Xylem, 565, 574
secondary growth in production of, 694
transport in, 704
water-conducting elements of, 683*f*, 684
Xylem cells, evolution of, 574
Xylem fibers, 574
Xylem rays, 693
Xylem sap, transport of, 706–708
XYY syndrome, 294

Y chromosome, 246–248, 288–289
genes carried on, 283
Yeast
mapping genome of, 405
mating-type genes in, 379
Yeast cells, for gene cloning, 394, 399
Yeasts, 591–592
Yolk, 960
avian, 965
Yolk plug, 963
Yolk sac
in birds, 966
in mammals, 968

Zeatin, 761
Zeitgebers, behavioral rhythms and, 1171
Zero population growth, 1093
Zinc
human requirements for, 814*t*
plant requirements for, 721*t*
Zinc-finger domain, gene expression affected by, 380–381, 382*f*
Z lines, 1038–1039
Zone of cell differentiation, in primary root growth, 689
Zone of cell division, in primary root growth, 688
Zone of cell elongation, in primary root growth, 688–689
Zone of polarizing activity, 978
Zoomastigophora (zooflagellates), 537*t*, 539, 540*f*
Zosterophyllum, 566
ZPA. *See* Zone of polarizing activity
Zygomycota (zygomycetes, zygote fungi), 586–587
Zygosporangia, 586
Zygote, 248, 932
Zymogens, 802, 804